Large-Scale Motions in the Universe: A Vatican Study Week

Large-Scale Motions in the Universe: A Vatican Study Week

Edited by

Vera C. Rubin and George V. Coyne, S.J.

Princeton University Press

Princeton, New Jersey

Distributed by: in Italy and Vatican City State: Libreria Editrice Vaticana
V-00120 Vatican City State
Elsewhere: Princeton University Press
Princeton, New Jersey 08540
USA

CONTENTS

III. MOTION OF THE LOCAL GROUP

IV. SMALL-SCALE MICROWAVE FLUCTUATIONS

MESSAGE

Leiden, February 25, 1988.

If it is true that putting the right questions is already half the solution of a problem, the Organizing Committee of the Vatican Study Week on *Large-Scale Motions in the Universe* may have gone a long way towards the realization of the aim of the symposium prior to the meeting itself. Those who have read the questions but, like myself, were not part of the small group attending will be anxious to learn the answers reached which are described in the present book. Judging from previous experience with these Study Weeks I am confident that the discussions will have have been extremely fertile. The principal question I am left with myself is: Have we perhaps been unable to formulate the prime illuminating question?

J. H. OORT

LIST OF PARTICIPANTS

N.A. Bahcall, Space Telescope Science Institute, 3700 San Martin Drive, Baltimore, MD 21218, USA.

J.R. Bond, Institute for Theoretical Physics, 60 St. George St., Toronto, Ontario M5S 1A1, Canada.

D. Burstein, Department of Physics, Arizona State University, Tempe, AZ 85287, USA.

G.V. Coyne, S.J., Specola Vaticana, V-00120 Vatican City State.

M. Davis, Department of Astronomy, Campbell Hall, The University of California, Berkeley, CA 94720, USA.

A. Dekel, Racah Institute of Physics, The Hebrew University, Jerusalem, 91904 Israel.

G. Efstathiou, Institutc of Astronomy, Madingley Road, Cambridge CB3 OHA, United Kingdom.

S.M. Faber, Lick Observatory, The University of California, Santa Cruz, CA 95064, USA.

M.J. Geller, Harvard-Smithsonian Center for Astrophysics, 60 Garden St., Cambridge, MA 02138, USA.

M.P. Haynes, Department of Astronomy, Space Sciences Building, Cornell University, Ithaca, NY 14853, USA.

J.P. Huchra, Harvard-Smithsonian Center for Astrophysics, 60 Garden St., Cambridge, MA 02138, USA.

N. Kaiser, Institute of Astronomy, Madingley Road, Cambridge CB3 OHA, United Kingdom.

D.C. Koo, Lick Observatory, The University of California, Santa Cruz, CA 95064, USA.

A.N. Lasenby, Cavendish Laboratory, University of Cambridge, Madingley Road, Cambridge CB3 OHE, United Kingdom.

D. Lynden-Bell, Institute of Astronomy, Madingley Road, Cambridge CB3 OHA, United Kingdom.

J. MOULD, Division of Physics, Mathematics and Astronomy, California Institute of Technology, Pasadena, CA 91125, USA.

P.J.E. PEEBLES, Joseph Henry Laboratories, Princeton University, P.O. Box 708, Princeton, NJ 08544, USA.

V.C. RUBIN, Department of Terrestrial Magnetism, Carnegie Institution of Washington, 5241 Broad Branch Road, N.W., Washington, D.C. 20015, USA.

W.R. STOEGER, S.J., Specola Vaticana, V-00120 Vatican City State.

A. SZALAY, Department of Atomic Physics, Eötvös University, H-1088 Budapest, Hungary.

R.B. TULLY, Institute for Astronomy, University of Hawaii at Manoa, 2680 Woodlawn Drive, Honolulu, HI 96822, USA.

N. VITTORIO, Istituto Astronomico, Università degli Studi di Roma, La Sapienza, Via Lancisi, 29, I-00161 Rome, Italy.

A. YAHIL, Department of Earth and Space Sciences, State University of New York at Stony Brook, Stony Brook, NY 11794, USA.

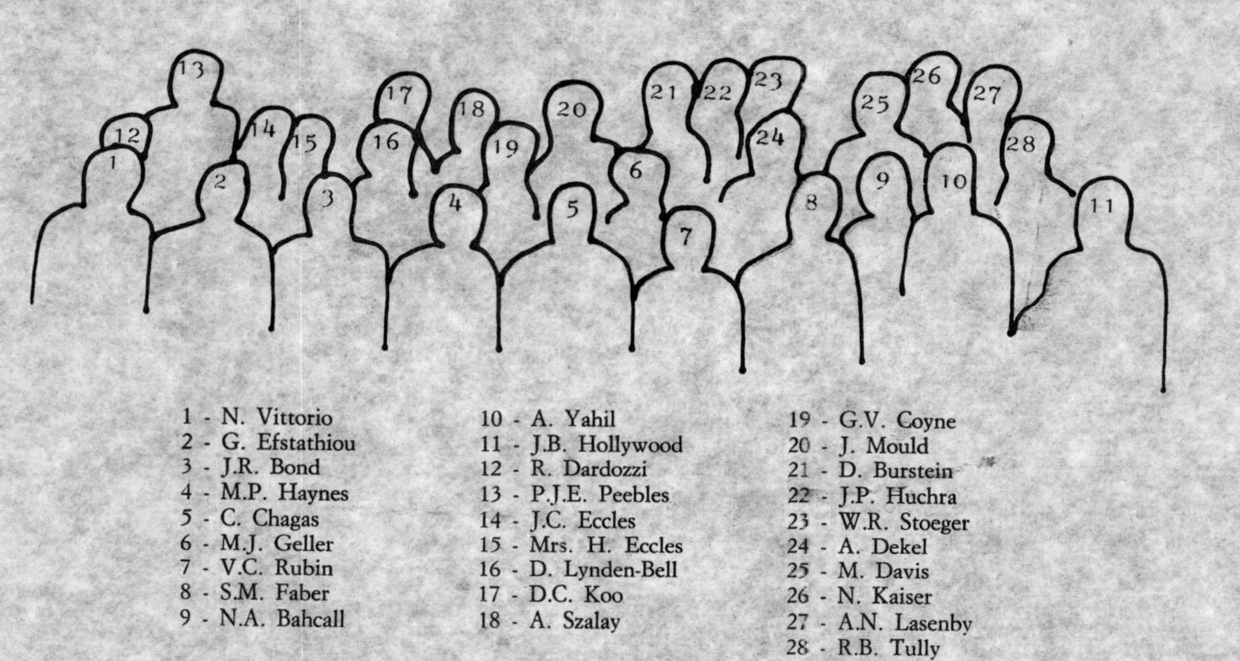

1 - N. Vittorio
2 - G. Efstathiou
3 - J.R. Bond
4 - M.P. Haynes
5 - C. Chagas
6 - M.J. Geller
7 - V.C. Rubin
8 - S.M. Faber
9 - N.A. Bahcall
10 - A. Yahil
11 - J.B. Hollywood
12 - R. Dardozzi
13 - P.J.E. Peebles
14 - J.C. Eccles
15 - Mrs. H. Eccles
16 - D. Lynden-Bell
17 - D.C. Koo
18 - A. Szalay
19 - G.V. Coyne
20 - J. Mould
21 - D. Burstein
22 - J.P. Huchra
23 - W.R. Stoeger
24 - A. Dekel
25 - M. Davis
26 - N. Kaiser
27 - A.N. Lasenby
28 - R.B. Tully

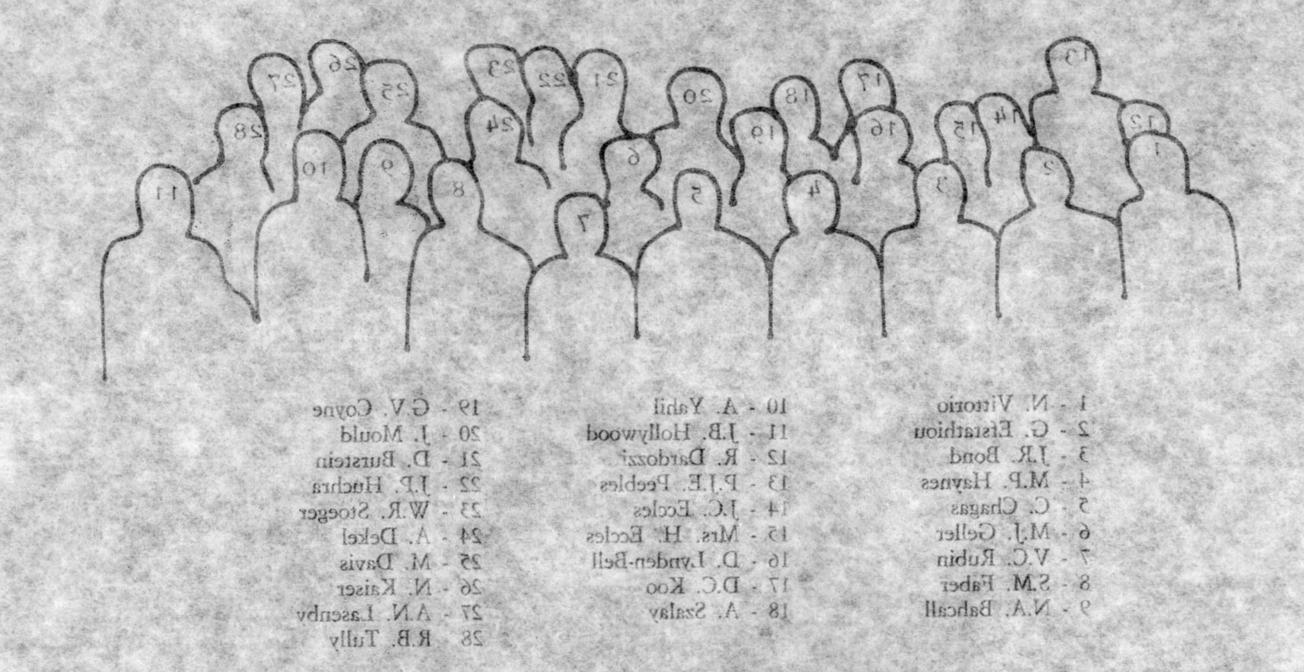

1 - N. Vittorio
2 - G. Efstathiou
3 - J.R. Bond
4 - M.P. Haynes
5 - C. Chagas
6 - M.J. Geller
7 - V.C. Rubin
8 - S.M. Faber
9 - N.A. Bahcall

10 - A. Yahil
11 - J.B. Hollywood
12 - R. Dardozzi
13 - P.J.E. Peebles
14 - J.C. Eccles
15 - Mrs. H. Eccles
16 - D. Lynden-Bell
17 - D.C. Koo
18 - A. Szalay

19 - G.V. Coyne
20 - J. Mould
21 - D. Burstein
22 - J.P. Huchra
23 - W.R. Stoeger
24 - A. Dekel
25 - M. Davis
26 - N. Kaiser
27 - A.N. Lasenby
28 - R.B. Tully

PREFACE

It is only 200 years since Sir William Herschel used the proper motions of 13 stars to discover the motion of the sun. During the week of 9 to 14 November, 1987, twenty-three astronomers met for a Study Week in Vatican City at the invitation of the Pontifical Academy of Sciences to discuss a contemporary equivalent of that study, *Large-Scale Motions in the Universe*. For the participants, the week was a gift; a gift of time during which we could describe our recent work to our colleagues. Unlike most scientific meetings in which only positive forward progress is reported, we hoped to include in our discussions also the uncertainties inherent in these initial steps in untangling the structure and the motions of our region of the universe. What's unclear, what's hidden, what's misinterpreted, what's misunderstood? All of these questions were to be raised.

To guide our deliberations many of the participants had used the occasion of the IAU Symposium 130 at Lake Balaton, Hungary in June, 1987 to identify some thought-provoking questions which would be relevant to the Vatican Study Week. These questions were distributed to the participants well in advance of the Study Week, along with a preliminary program devised by the Organizing Committee. For each of the nine sessions, a topic was identified, as well as a chairperson to introduce the subject, and a summarizer to act as a good listener and to remind us where difficulties still lay.

Within this framework and the limits of time, all participants were free to offer contributions. The preliminary questions are reprinted here; the program of the sessions can be recovered approximately by examining the Table of Contents of this volume. Written texts were not required, or even expected, at the time of the meeting. It was hoped that texts prepared immediately following the Study Week would better reproduce the flavor of discovery and uncertainty characteristic of the sessions and of the discussions. For the final session, each participant presented a less-than-five-minute statement (view-graphs not permitted) of anything on his/her mind related to the subject matter of the meeting. These statements and Avishai Dekel's summary contribution were recorded and are reproduced here after minor editing. Among this group of independent astronomers, those who violated the time and viewgraph restriction can now be identified.

We missed the participation of Marc Aaronson, who had accepted an invitation to be with us, but whose tragic death intervened. We also missed Yakov Zeldovich who was unable to attend. We note with sorrow that Prof. Zeldovich died a few weeks after our meeting. We acknowledge our scientific debt to both of them; they were missed.

All of the participants wish to thank President Carlos Chagas of the Pontifical Academy of Sciences for offering us the opportunity to exchange ideas in such a pleasant setting, and to the Director of the Chancery of the Academy, Ing. Renato Dardozzi, for administering the meeting arrangements. We also thank George V. Coyne, S.J., Director of the Vatican Observatory, for taking care of all the details which made the meeting go so well, and for his hospitality at Castel Gandalfo where we spent a delightful afternoon and evening. Jim Peebles, in his after dinner toast, correctly identified George as the "prime mover" of the Study Week. We thank also Rita Callegari, Secretary of the Vatican Observatory, and the staff of the Academy for their roles in making problems seem small scale. And I thank the other members of the Organizing Committee; Sandra Faber, Donald Lynden-Bell, and Alex Szalay, for their help with all stages in the planning for the Study Week. I also note with thanks that Jim Peebles offered valued advice as an ex-officio member of the Organizing Committee. To Janice Dunlap who helped in many ways including preparing the index I am grateful.

Often in planning this Study Week I was reminded of an old Peanuts cartoon. Peanuts is a popular American cartoon, apparently equally popular all over the world as well. In this cartoon Lucy is instructing her fall guy, Charlie Brown, as follows. On the oceans of the world are many ships, and some of these ships carry passengers. One of the activities the passengers like most is to sit on the deck and watch the ocean. Some of these passengers arrange their deck chairs to face where they are going; some of the passengers arrange their chairs to see where they have been. Lucy asks, "On the great ship of life, Charlie Brown, which way are you going to place your chair, to see where you are going, or to see where you have been?" And Charlie Brown answers, "I can't even seem to get my chair unfolded."

Some of us are trying to find out where we are going, some are looking at where we have been. For many of us, the Vatican Study Week offered an incomparable opportunity to attempt to unfold our chairs.

VERA C. RUBIN, CHAIRPERSON

SOME QUESTIONS CONCERNING THE STRUCTURE AND MOTIONS IN THE UNIVERSE

1. *What is a good description of the 3-dimensional structure of the Universe, $V < 10{,}000\ km\ s^{-1}$? at largest distances?*

   — What is the convincing evidence that some structures are strings, chains, filaments, sponges, bubbles, sheets, voids?
   — Do structures exist which exceed 100 Mpc?
   — What is the amplitude of structures on the largest scale?
   — Did these structures originate in the very early universe?

2. *What is the spectrum of peculiar velocities as a function of scale?*

   — Can details of the velocity field of the Local Supercluster be mapped and understood in the context of the distribution of galaxies? Can gravitation account for the magnitude and direction of the motion?
   — How reliable is our current knowledge of the distribution and motions of clusters and superclusters, and how seriously should we take the present conclusions based on the data?
   — How large are the internal motions in the 3-dimensional structures?

3. *What do the observations tell us about small-scale microwave background fluctuations?*
   - What are the current constraints on $\Delta T/T$?
   - What is needed to pin $\Delta T/T$ down?
   - How do background fluctuations relate to the velocity fields in various models?
   - Could background fluctuations have been wiped out by re-ionization of the intergalactic medium?

4. *What theoretical frameworks are there for understanding the structures and motions in the universe?*
   - Which approaches appear most profitable?
   - What remains unexplained?
   - What initial density perturbations appear possible?
   - Are primordial scale invariant fluctuation spectra ruled out by the observations? What fluctuation spectra for the early universe fit the data?
   - Do the combined observations of the peculiar acceleration field place constraints on primordial spectra and/or scenarios?

5. *What are realistic prospects for N-body simulations?*
   - What do we learn from a comparison of simulations and observations, both in density and in velocity?
   - How can we quantify the spatial distribution of galaxies in theories and observations?
   - Are there alternatives to ξ, the correlation function?
   - What are the observational and theoretical constraints on biasing?

6. *What caused the formation of superclusters, clusters, and galaxies?*
   - Did this process not generally occur much before $z = 3$, and can we understand why?
   - Why did quasar formation stop essentially after $z = 3$ or so?
   - What can we learn from the clustering of high redshift objects: galaxies, clusters, quasars, Ly-α clouds?
   - What do galaxies look like at $z = 3$? 2?
   - What limits can be placed on interactions of galaxies since $z = 1$?
   - How big are galaxy halos?
   - What empirical handles do we have on the mass distribution?

7. *What do we have to observe to understand the universe?*
   - What are the most definite testable predictions of the competing cosmological scenarios: cold dark matter, hot dark matter, strings, explosions?
   - How compelling is the evidence that the density of the universe equals the critical density?

8. *Ten years hence, what will we wish we had discussed in November, 1987?*

I

TWO-DIMENSIONAL AND THREE-DIMENSIONAL STRUCTURE

GALAXY AND CLUSTER REDSHIFT SURVEYS

MARGARET J. GELLER and JOHN P. HUCHRA
Harvard-Smithsonian Center for Astrophysics, Cambridge

ABSTRACT

We discuss the status of galaxy and cluster redshift surveys. We concentrate on the CfA redshift survey and a deep Abell cluster redshift survey.

Five strips of the CfA redshift survey are now complete. The data continue to support a picture in which galaxies are on thin sheets which nearly surround vast low density voids. In this and similar surveys the largest structures are comparable with the extent of the survey. Voids like the one in Boötes are a common feature of the galaxy distribution and present a serious challenge for the models. We suggest some statistics which may be useful for comparing the data with the models.

The deep cluster survey of Huchra *et al.* (1988) is nearly independent of the sample analyzed earlier by Bahcall and Soneira (1983). For this new sample the amplitude of the correlation function is a factor of ~ 2 less than for the earlier sample. However, the difference may not be significant because the cluster samples are sufficiently small that they may be dominated by single systems. The deeper sample also fails to support the claim by Bahcall, Soneira, and Burgett (1986) of ~ 2000 km s^{-1} peculiar motions; the limit for the deep sample is $\lesssim$ 1000 km s^{-1}.

1. INTRODUCTION

In 1978 Jôeveer, Einasto, and Tago suggested that the large-scale distribution of galaxies has a "cellular" pattern in which rich clusters are connected by "filamentary" structures. The data at that time were incomplete and could only hint at such structure. The discovery of the void in Boötes (Kirshner *et al.* 1981, KOSS hereafter) and the 21-cm survey of the Pisces-Perseus chain (Haynes and Giovanelli 1986, Giovanelli *et al.* 1986) soon lent support to this picture. These first surveys gave no clear message about the frequency of the structures.

It has become increasingly clear that large-scale features in the galaxy distribution are ubiquitous. The deep surveys of Koo, Kron, and Szalay (1986) show that voids are common even at high redshift. Because surveys are one-dimensional, the constraints on the sizes of the voids are poor. The AAT surveys also reveal voids along with thin structure perpendicular to the line-of-sight (Peterson *et al.* 1986). The continuing Arecibo survey delineates nearby voids and appears to support the interpretation of the Pisces-Perseus chain as a filamentary structure.

The extension of the Center for Astrophysics (CfA) redshift survey (discussed in Section 2) indicates that bright galaxies are distributed on thin sheets — two-dimensional structure— which surround (or nearly surround) vast voids. Recent completion of a southern hemisphere survey (da Costa *et al.* 1988) gives some support to this picture although the survey is not sufficiently deep or dense to be directly comparable with the recent CfA results. The 21-cm data (Haynes and Giovanelli 1986) also reveal sheet-like structures in the Perseus-Pisces region. The message of all surveys is now clear: large structures are a common feature of all surveys big enough to contain them.

Because redshift surveys of individual galaxies and redshift surveys of clusters of galaxies do not overlap substantially, the relationship between the large-scale distribution of rich clusters and the large-scale distribution of galaxies remains unclear. Analyses of existing surveys indicate that clusters of galaxies and individual galaxies are not equivalent tracers of the large-scale matter distribution. Section 3 is a discussion of a recently completed deep survey (Huchra *et al.* 1988) which is largely independent of the nearby sample analyzed by Bahcall and Soneira (1983).

We take $H_o = 100h$ km s^{-1} Mpc^{-1}, with $h = 1$ unless otherwise specified.

2. The Large-Scale Galaxy Distribution

Over the last few years each new approach to mapping out the distribution of individual galaxies has uncovered unexpectedly large structures. Surveys like the one of the Boötes (KOSS 1987) region are an efficient method of finding large voids. Surveys like the CfA redshift survey extension (Huchra and Geller 1988) which are complete over a region of large angular scale are less efficient for identifying individual large structures, but they are necessary for quantitative characterization of the distribution of galaxies over a range of scales. Table 1 is a list of existing surveys.

The goal of the Center for Astrophysics redshift survey extension is to measure redshifts for all galaxies in a merge of the Zwicky *et al.* (1961-1968)

TABLE 1
GALAXY AND CLUSTER REDSHIFT SURVEYS

FORMAT Name	Authors	N_{obj}	Date
A. LARGE AREA COMPLETE			
GALAXIES			
HMS	Humason *et al.*	500	1956
RSA	Sandage and Tammann	1300	1981
RC1 + RC2	de Vaucouleurs *et al.*	1200	1964, 1976
CFA1	Huchra *et al.*	2400	1982
UGC Gals	Bothun *et al.*	4000+	1985
CFA Slice	de Lapparent *et al.*	1100	1986
SSRS	da Costa *et al.*	1800	1987
Nearby Gals	Tully and Fisher	3000	1987
CFA2	Huchra and Geller	15000	1990
ABELL CLUSTERS			
D4	Hoessel *et al.*	116	1980
SAO	Karachentsev *et al.*	100+	1983, 1986
D5 ($m_{10} \leq 16.5$)	Postman *et al.*	561	1985, 1988
Abell-Corwin	Corwin, Olowin *et al.*	500+	1989
B. LARGE AREA INCOMPLETE			
GALAXIES			
Markarian	Arakelian *et al.*	1300	1970's
Böotes	KOSS	400	1983, 1987
Arecibo	Giovanelli and Haynes	4000+	1985+
IRAS Gals	Strauss *et al.*	2600	1988
C. SMALL AREA (Usually Deep)			
GALAXIES			
AAT	Ellis, Shanks, Fong *et al.*	260	1985, 1988
KPNO	Koo and Kron	250+	1987
Cor Bor	Postman *et al.*	250	1987
Century	Geller *et al.*	2500	1990
Southern strip	KOSS	2000+	1990
Southern AAT	Efstathiou *et al.*	5000+	
ABELL CLUSTERS			
Superclusters	Ciardullo *et al.*	5	1985
Deep Abell	Huchra *et al.*	145	1988

and Nilson (1973) catalogs which have $m_{B(0)} \leq 15.5$ and $|b| \geq 40°$. There will be about 15,000 galaxies in the complete survey; more than half of these already have measured redshifts. About 2,500 of the galaxies with measured redshifts lie in three "slices" where the survey is now complete: (1) a slice with $8^h \leq \alpha \leq 17^h$ and $26.5° \leq \delta < 32.5°$ (de Lapparent, Geller, and Huchra 1986; dLGH hereafter); (2) a slice with $8^h \leq \alpha \leq 17^h$ and $32.5° \leq \delta < 38.5°$; and (3) a slice with $8^h \leq \alpha \leq 17^h$ and $38.5° \leq \delta < 44.5°$. In the southern Galactic hemisphere 2 slices covering $6° \leq \delta < 12°$ and $18° \leq \delta < 24°$ with $0^h \leq \alpha \leq 4^h$ and $20^h \leq \alpha \leq 24^h$ are also complete. More than 60% of the redshifts were measured with the 1.5-meter telescope and the MMT at Mt. Hopkins. The mean external error in the redshift measurements is ~ 35 km s^{-1}.

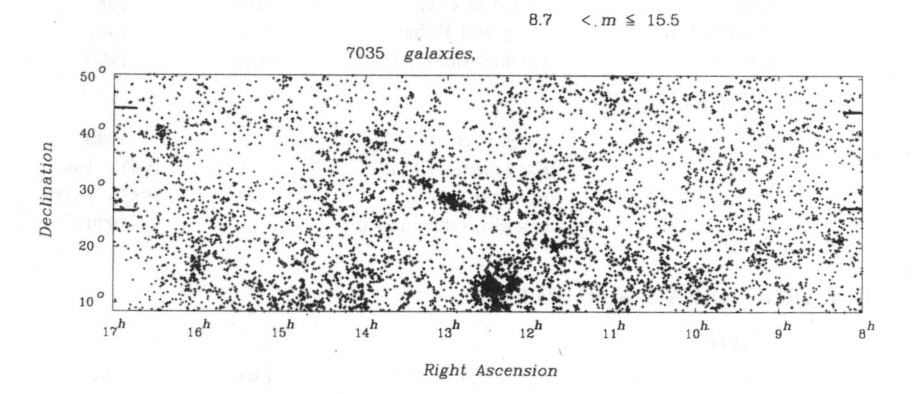

FIG. 1. Positions of galaxies in the merged Zwicky-Nilson catalogs with $m_{B(0)} \leq 15.5$ in the northern galactic cap. The bold ticks indicate the limit of the complete strips.

Figure 1 shows the positions of 7035 galaxies in the Zwicky-Nilson merge which have $m_{B(0)} \leq 15.5$ and $8^h \leq \alpha \leq 17^h$ and $8.5° \leq \delta < 50.5°$. The grid is Cartesian in α and δ. The deficiency of galaxies west of 9^h and east of 16^h is caused by galactic obscuration. The bold ticks indicate the limits of the three complete strips of the survey. The Coma cluster is the dense knot at $\alpha = 13^h$, $\delta = 30°$. Figure 2 is a similar map for the region of the southern galactic hemisphere covered by the CfA survey. Again the bold ticks indicate the limits of declination strips which are complete.

Figure 3 is a plot of the observed velocity versus right ascension for the strip centered at 29.5° (dLGH): the strip is 6° wide in declination. The plot includes only the 1065 galaxies with redshifts $\leq 15{,}000$ km s^{-1}. A galaxy of

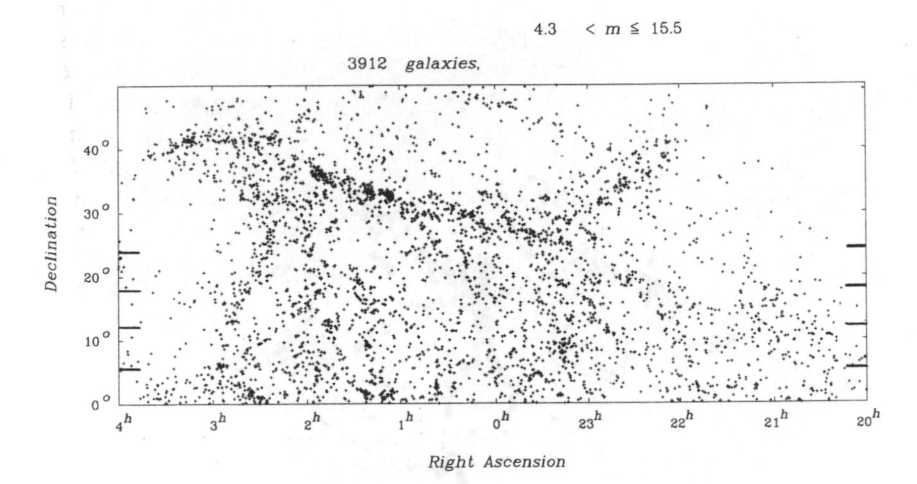

FIG. 2. Position of galaxies in the merged Zwicky-Nilson catalogs with $m_{B(0)} \leq 15.5$ and in the southern galactic cap. The bold ticks indicate the limits of the complete strips.

characteristic luminosity, M^*, is at roughly 10,000 km s^{-1} in this survey. Nearly every galaxy in this slice lies in an extended thin structure. The boundaries of the empty regions are remarkably sharp. Several of the voids are surrounded by thin structures in which the inter-galaxy separation is small compared with the radius of the empty region. The edges of some of the largest structures may be outside the right ascension limits of the survey. The only pronounced velocity finger in this slice is the Coma cluster at ~ 13^h.

The first slice alone demonstrates that the thin structures in the distribution of galaxies are cuts through two-dimensional sheets, not one-dimensional filaments. If the ~ 150 Mpc long structure which extends across the entire survey (from 9^h to 16^h between 7,000 km s^{-1} and 10,000 km s^{-1}) is a filament, a thin linear structure should be visible on the sky. This statement is particularly strong because the structure lies near the survey limit. The required filamentary structure is absent from Figure 1. Because structure on the sky can be caused by patchy obscuration and/or by inhomogeneities in the galaxy catalog, structure on the sky cannot provide complete proof (or disproof) of the filamentary nature of a structure in redshift space. A second argument against the filamentary nature of the structures in Figure 3 is that several thin, elongated structures lie in this single survey slice: the intersection of a slice with a three-dimensional network of filaments is *a priori* unlikely to be a two-dimensional network of filaments.

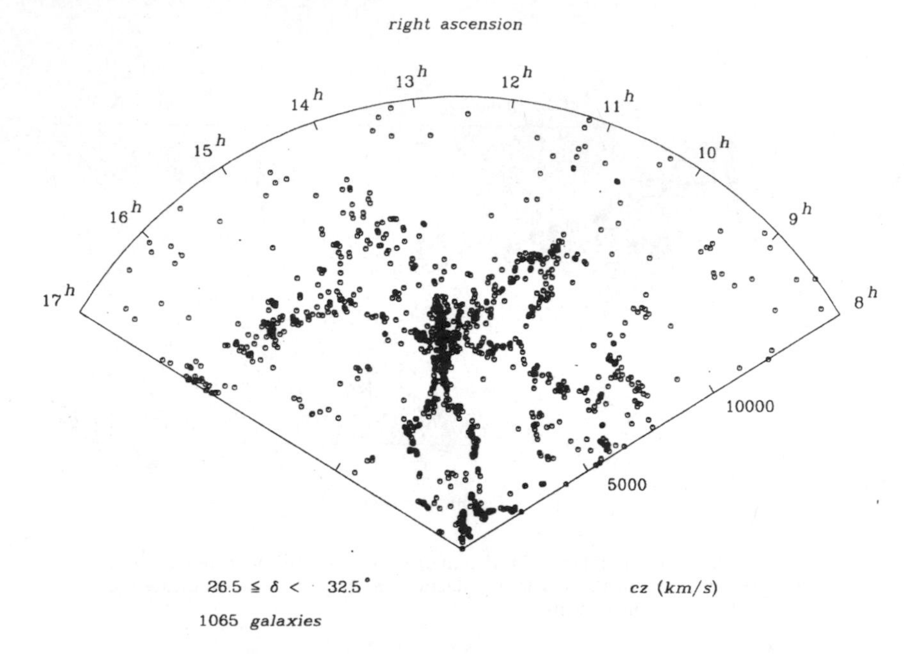

FIG. 3. Observed velocity versus right ascension for the survey strip centered at $\delta = 29.5°$. The strip is 6° in declination. Only the galaxies with velocities $\leq$ 15,000 km s^{-1} are shown.

A geometric structure in which thin sheets surround or nearly surround voids accounts for the data. Examples include "bubble-like" and "sponge-like" geometries. We use the word "bubble" to convey the image of a structure dominated by thin sheets and holes. Note that the "bubbles" are not necessarily round. In this picture the 150 Mpc "filament" is made up of portions of adjacent "bubbles" and the richest clusters like Coma lie in the interstitial regions (where several "bubbles" come together).

In this interpretation of the data, we assume that the maps are similar in redshift space and in real physical space. Simulations of both the cold dark matter (White *et al.* 1987) and the adiabatic models (Centrella *et al.* 1988) suggest that this assumption is reasonable, i.e. distortions caused by large scale flows are minimal on these scales (but see Kaiser 1987).

Maps of the adjacent slices support the qualitative, homogeneous picture suggested by the first slice. Figure 4 shows the slice centered at 35.5°, just to the north of the slice in Figure 3. Once again the galaxies are in thin structures. Furthermore these structures are natural extensions of the structures in Figure 3; the structures are highly correlated in the two slices. The two closed structures at ~ 11^h (9000 km s^{-1} $\leq$ cz $\leq$ 11,000 km s^{-1}) and at 14^h (7000

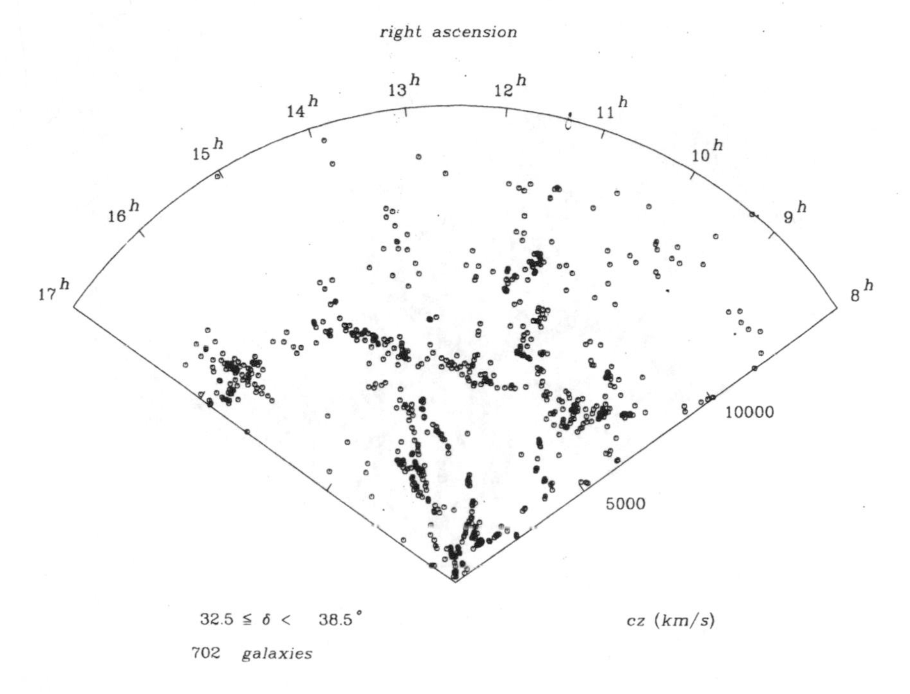

FIG. 4. Same as Figure 3 for a slice centered at $\delta = 35.5°$.

km s^{-1} $\lesssim$ cz $\lesssim$ 11,000 km s^{-1}) are not so clearly delineated in Figure 4 as in Figure 3.

Figure 5 shows the cone diagram for the two slices taken together. The distribution remains remarkably inhomogeneous with empty voids outlined by thin structures. Because the surfaces are curved or inclined relative to the plane of the slice, the structures are thicker here than in Figures 3 and 4. The largest low density region in the survey is located between 13^h20^m and 17^h with 4000 $\lesssim$ cz $\lesssim$ 9000 km s^{-1}. The diameter of this void is ~ 5000 km s^{-1} or 50 Mpc in the absence of large-scale flows. Recent infra-red Tully-Fisher measurements indicate that the diameter of the void is the same in physical space as in redshift space to within ~ 300 km s^{-1} (Geller *et al.* 1988a). The underdensity in the region ($\lesssim$ 20% of the mean) and the scale of the structure are comparable to the corresponding parameters for the void in Boötes.

Note that voids are not empty: they are regions which are underdense relative to the global mean. The galaxies inside the large void have normal properties and their infra-red Tully-Fisher distances put them at the relative distances indicated by their redshifts (Geller *et al.* 1988a). These galaxies may form a tenuous structure which would not be detected in a sparse survey like

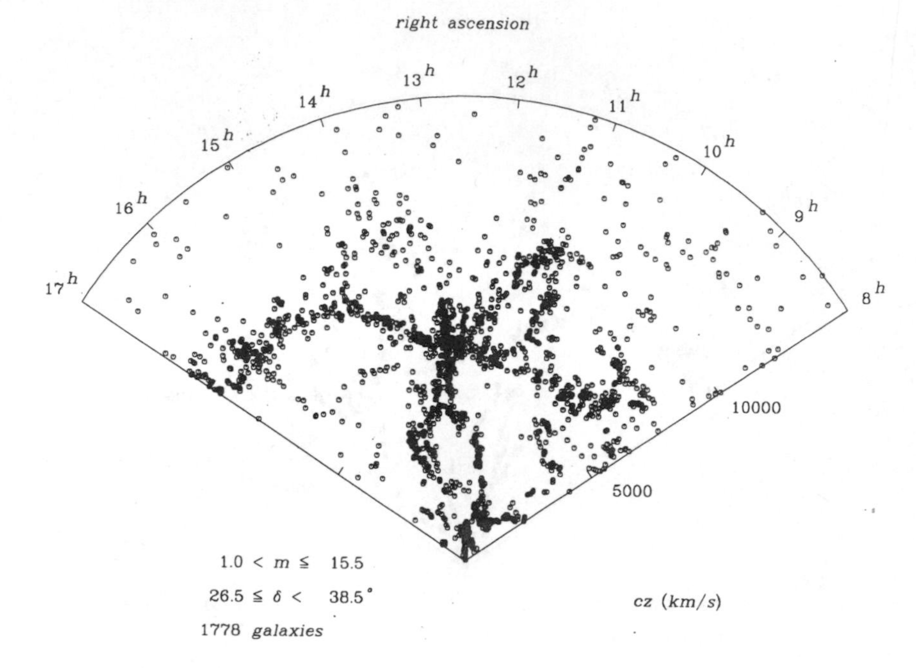

FIG. 5. Cone diagram for the slices in Figures 3 and 4 taken together. This is 12° wide in declination and is centered at 32.5°. The sample contains ~ 1700 galaxies.

the KOSS survey of Boötes. Their spectroscopic properties are similar to those in other portions of the survey.

Figure 6 shows the cone diagram for the third slice centered at $\delta = 41.5°$. At first glange this slice gives a somewhat different visual impression from the first two. The reason for the difference is that some of the surfaces lie in this slice; in particular, a portion of the structure surrounding the largest void is nearly in the plane of this slice and appears diffuse. Comparison of this slice with Figure 5 continues to support the large-scale coherence of the structures.

It is instructive to compare these new surveys with the original CfA survey (Davis *et al.* 1982, Huchra *et al.* 1983) to a limiting $m_{B(0)} = 14.5$. Figure 7 shows the 29.5° slice to this limit. Here an M^* galaxy has a velocity of 6000 km s^{-1}. Because the "effective depth" of this survey is comparable with the scale of the largest structures in Figures 3-5, these structures could not be detected.

The nearby small void centered at 13^h20^m and 3500 km s^{-1} is visible in both Figures 3 and 7. Note that the fainter galaxies fill in the gaps along the perimeter of the void. This comparison and other deep probes through the

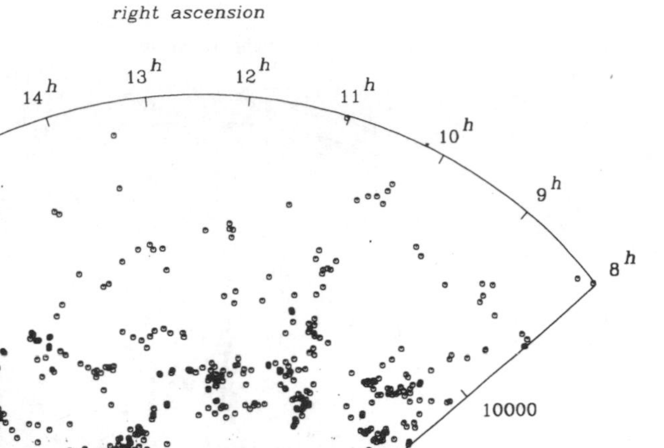

FIG. 6. Same as Figures 3 and 4 for slice centered at $\delta = 41.5°$.

29.5° slice (Postman, Huchra, and Geller 1986) indicate that the distribution is insensitive to luminosity for $M_{B(0)} \lesssim -17.4$. More precisely, the density contrast between the high and low density regions (about a factor of 20) is insensitive to absolute magnitude over this range. Surveys to fainter limiting magnitudes *do* turn up galaxies in the voids, but they turn up proportionately more in the dense structures.

The above conclusion is at odds with the analysis by Giovanelli and Haynes (1988) of structures in the Perseus-Pisces region. The definition of large-scale structure could be a function of properties of individual galaxies other than absolute luminosity. For example, there is a well-known relation between the morphology of a galaxy and local density (Hubble and Humason 1931, Dressler 1980, Dressler 1984, Postman and Geller 1984). It seems very likely that the Giovanelli and Haynes (1988) results are heavily affected by the morphology-density relation.

The change in the fractional coverage of the perimeter of a void as a function of luminosity complicates topological studies like those by Gott, Melott, and Dickinson (1986), and Hamilton, Gott, and Weinberg (1987). The data so far are inadequate to discriminate clearly between "bubble-like" (connected high density regions; isolated low density ones) and "sponge-like" (both high

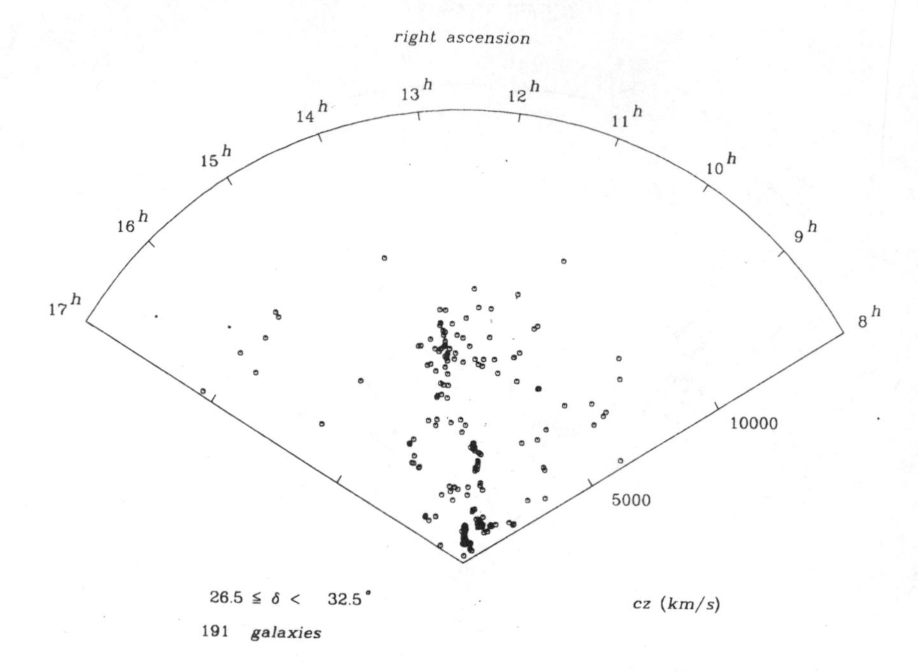

FIG. 7. Observed velocity versus right ascension for the survey strip centered at $\delta = 29.5°$ (Figures 3) but with a magnitude limit $m_{B(0)} = 14.5$.

and low density regions form a connected network) topologies. This distinction is a clue to the initial conditions for large-scale structure formation; a "bubble-like" topology points to non-random phases in the initial perturbation spectrum.

The search for a relation between surface brightness and local density has been prompted by cold dark matter models for the formation of large-scale structure (see e.g. Dekel and Silk 1986). Davis and Djorgovski (1985) published the first analysis of existing data and concluded that a correlation exists in the sense that lower surface brightness objects inhabit regions of lower average density. Later Bothun *et al.* (1986) and Thuan, Gott and Schneider (1987) obtained new data and reached the opposite conclusion: the low and high surface brightness objects trace the same structures as demonstrated by the velocity histograms for galaxies in the direction of the Virgo-Coma void (Figure 8). The Davis and Djorgovski (1985) analysis is incorrect because it was based on a sample with poorly understood selection biases.

The dependence of structure on the properties of individual galaxies can also be examined by comparing surveys based on the IRAS catalog with those based on optical catalogs. At least two complete studies (Smith *et al.* 1987,

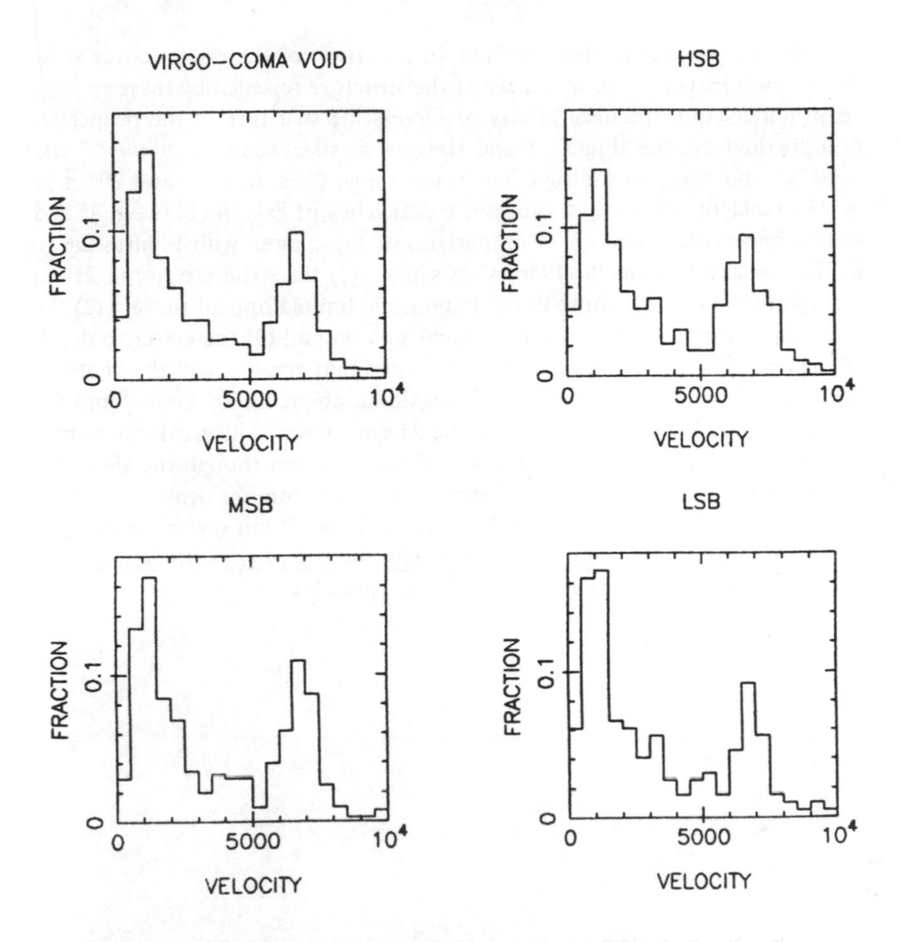

FIG. 8. The velocity distribution of all the Nilson galaxies in the direction of the Virgo-Coma void compared with the velocity distributions of the high, middle and low thirds in surface brightness (from Bothum *et al.* 1986). The underdensity between Virgo and Coma persists at all surface brightness levels.

Strauss and Huchra 1988) indicate that the galaxies in the IRAS catalog have a distribution similar to those in optical catalogs except that the IRAS galaxies are absent from the dense cores of rich clusters like Coma. Smith *et al.* (1987) cover the CfA redshift survey slice in Figure 2 and Strauss and Huchra (1988) survey the void in Boötes. The IRAS survey is not deep enough to see the far edge of the Boötes void, but the underdensity in the velocity range of the void is consistent with the claims based on optical data (~ 20% of the mean). In addition, both the photometric and spectroscopic properties of the IRAS galaxies found in the void are similar to the properties of those galaxies in the boundaries.

The CfA survey is also complete in a portion of the region covered by the Arecibo surveys. The similarity of the structure revealed by these surveys again underscores the insensitivity of large-scale structure to the properties of individual galaxies. Figures 9 and 10 show 6° slices centered at $\delta = 9°$ and $\delta = 21°$ and covering the right ascension range $0^h \leq \alpha \leq 4^h$ and $20^h \leq \alpha \leq 24^h$. Galactic obscuration causes the deficiency of galaxies between 3^h and 4^h and between 20^h and 21^h. Comparison of these plots with Figures 2a—c in Haynes and Giovanelli (1986) shows that: (1) the structure in the 21-cm survey are the same as those in the magnitude limited optical survey; (2) the sampling density is greater in the optical survey; and (3) the effective depth of the optical survey to $m_{B(0)} \leq 15.5$ is somewehat greater than the depth of the 21-cm surveys. For example, the galaxies at about 12,000 km s^{-1} and between 1^h and 23^h are not detected in the 21-cm survey. The greater apparent coherence of the structures in the optical survey (even though the slices are thinner than those in Haynes and Giovanelli) results from the denser sampling.

An important message of both the optical and 21-cm surveys is that the largest inhomogeneities are comparable with the size of the sample. None of the existing redshift surveys are thus large enough to be "fair".

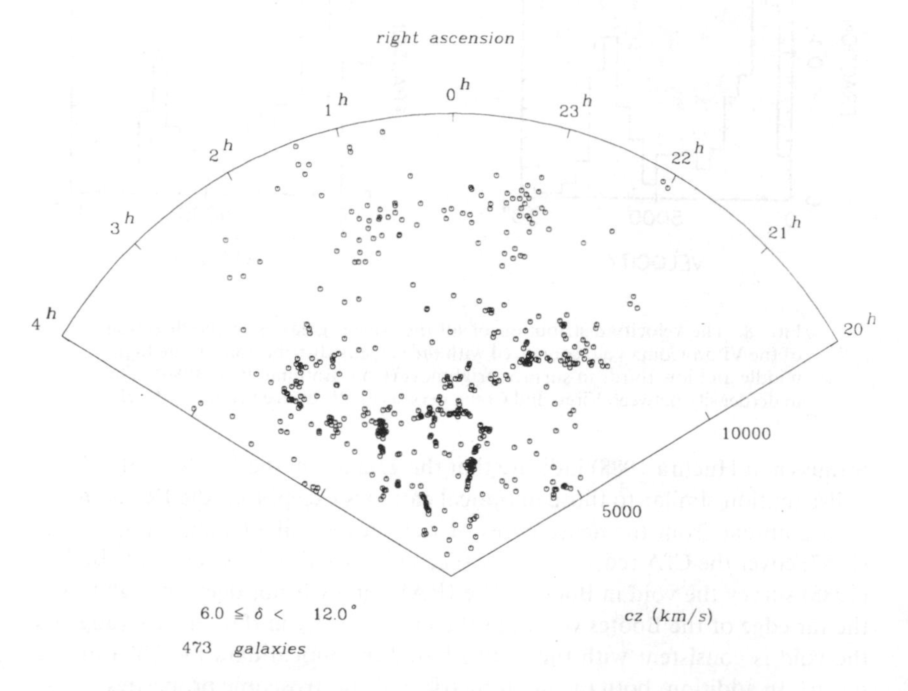

FIG. 9. Observed velocity versus right ascension for the survey strip in the southern galactic cap centered at $\delta = 9°$. The strip is 6° in declination. Only the galaxies with velocities $\leq$ 15,000 km s^{-1} are shown.

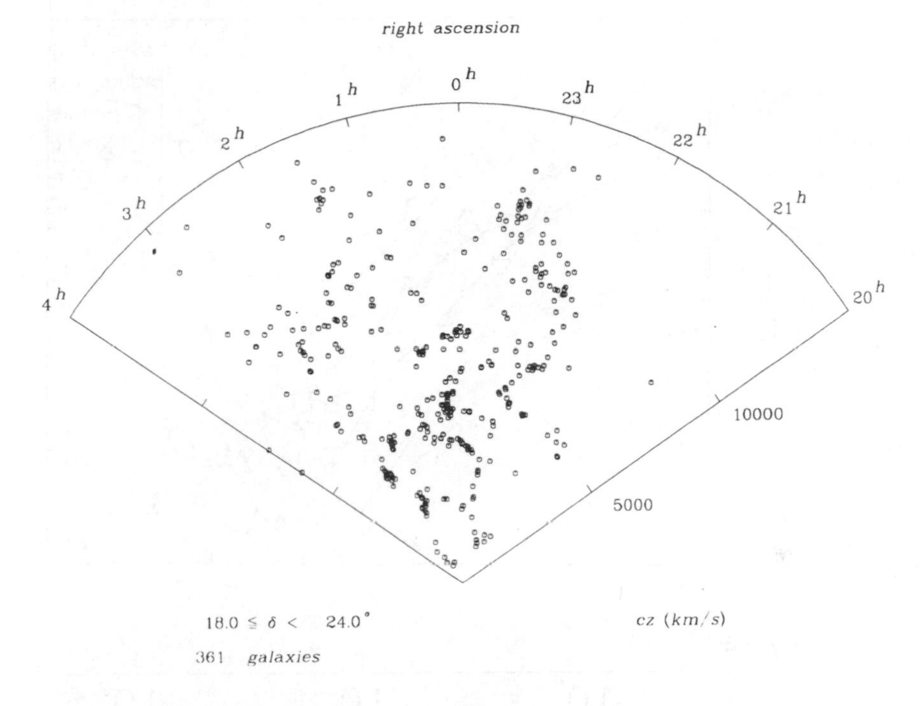

FIG. 10. Same as Figure 9 for a slice centered at $\delta = 21°$.

The largest inhomogeneities we detect are the largest we *could* detect within the limits set by the extent of the survey — we have few, if any, reliable direct limits on larger structures in the distribution of light-emitting matter. The size of the inhomogeneities relative to the volume of the surveys may underlie unexplained variations in traditional statistics of the galaxy distribution like the luminosity function (Schechter 1976, KOSS 1983, Bean *et al.* 1983, Davis and Huchra 1982) and the two-point correlation function at large scale (Groth and Peebles 1977, Davis and Peebles 1983, Kirshner, Oemler, and Schechter 1979, Shanks *et al.* 1983). When the inhomogeneities are large compared with the sample volume, mean quantities are not well-defined.

The domination of the sample by large-scale coherent structures and the related ~ 25% uncertainty in the mean density (de Lapparent, Geller, and Huchra 1988) imply that the two-point correlation function is more poorly constrained than previously thought. Figure 11 shows the two-point correlation function $\xi(s)$ where

$$s = \frac{(V_i^2 + V_j^2 - 2V_iV_j\cos\theta_{ij})^{1/2}}{H_o} \qquad (1)$$

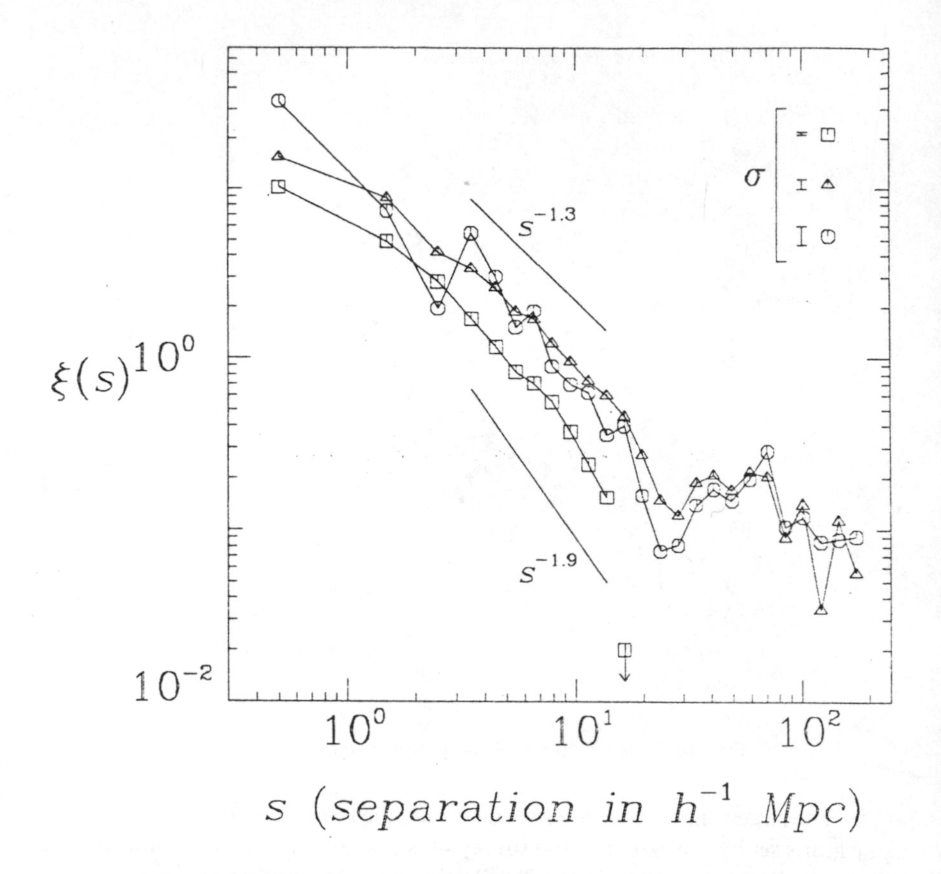

FIG. 11. The two-point correlation function $\xi(s)$ for the sample in Figure 5. The symbols denote ξ_{11} (□), $\xi_{\phi\phi}$ (Δ), and $\xi_{1\phi}$ (○). Note that the amplitude varies by a factor of two among these estimators.

for the 12° slice with declination between 26.5° and 38.5° (Figure 5). Here V_i and V_j are the velocities of two galaxies separated by θ_{ij} on the sky and H_o is the Hubble constant. We make no correction for the r.m.s. pairwise peculiar velocities of $\lesssim$ 350 km s^{-1} (Davis and Peebles 1983, de Lapparent, Geller, and Huchra 1988). The calculation of this correlation function is not seriously affected by the presence of the Coma cluster or of other more poorly sampled clusters in the sample.

Because of the large-scale coherent structures in the sample, the weighting scheme for the calculation of the correlation function *does* affect the result substantially. The calculation of the correlation function and the weighting schemes follow prescriptions in Davis and Peebles (1983). It is probable that none of these schemes are unbiased.

One estimator gives equal weight to both galaxies in the pair. Another is to give each galaxy a weight $\phi(V)^{-1}$, where $\phi(V)$ is the selection function for a magnitude limited sample. The selection function $\phi(V)$ is the probability that a galaxy with velocity V will be included in the sample. This weighting corrects for the variations in sampling as a function of redshift; it weights all volumes of space equally. This weighting increases the noise in the determination of $\xi(s)$. An intermediate procedure weights only one galaxy in the pair by $\phi(V)^{-1}$. The resulting correlation functions are ξ_{11}, $\xi_{\phi\phi}$, and $\xi_{1\phi}$, respectively. For these calculations the parameters of the luminosity function are $M^* = -19.15$, $\alpha = -1.2$, and $\phi^* \simeq 0.025\ h^3\ \mathrm{Mpc}^{-3}\ \mathrm{mag}^{-1}$ (de Lapparent, Geller, and Huchra 1988).

The differences among the estimators in Figure 11 are symptoms of the lack of a fair sample. In the absence of biases introduced by large-scale inhomogeneities in the sample, all the estimators should yield the same result to within the statistical noise shown by the error bars at the left of the Figure. The elongated structure at $\sim$ 10,000 km s^{-1} increases the amplitude of the correlation function for estimates weighted inversely with $\phi(V)$ relative to the unweighted estimate. This coherent structure contains about a thousand galaxies, nearly half of sample. The selection function $\phi(10{,}000\ \mathrm{km\ s}^{-1}) = 0.01\ \phi(1{,}000\ \mathrm{km\ s}^{-1})$. The distribution appears more highly correlated when this structure is more heavily weighted. (Note that in the CfA survey to $m_{B(0)} = 14.5$, the opposite effect occurs because the most dense structure, the local Supercluster is nearby in that sample (see de Lapparent, Geller, and Huchra 1988).

For least squares fits of the data to the standard power law form,

$$\xi(s) = \left(\frac{s_o}{s}\right)^{\gamma}, \tag{2}$$

in the range 3.5-9.5h^{-1} Mpc we find that the ranges in the slope and amplitude are, respectively, 1.3-1.9 and 5-12h^{-1} Mpc. On scales larger than 20h^{-1} Mpc the correlation function is indeterminate because the amplitude of the correlation function is comparable with the uncertainty in the mean density. For an average of the estimators in Figure 11, we obtain

$$\gamma = 1.5 \tag{3}$$

and

$$s_o = 7.5h^{-1}\ \mathrm{Mpc}. \tag{4}$$

These values are in agreement with the ones which can be measured from Figure 1 of Davis and Peebles (1983) which shows $\xi(s)$ for the 14.5 CfA sample. The frequently quoted smaller scale and steeper slope are derived from $\omega(r_p)$, the correlation function as a function of projected separation. N-body models have generally been normalized to this smaller $r_{go} = 5.4 \pm 0.3h^{-1}$ Mpc (Davis and Peebles 1983, Davis *et al.* 1985). A direct match of $\xi(s)$ to the models is probably a better procedure. This function is free of assumptions about the peculiar velocities which should be the same in the data and in the models.

Calculation of the correlation function $\xi(r_\varrho, \pi)$ (Davis and Peebles 1983), where

$$r_\varrho = \frac{V_i + V_j}{H_o} \tan \frac{\theta_{ij}}{2} \qquad (5)$$

and

$$\pi = V_i - V_j, \qquad (6)$$

is a method of examining the distortions in redshift space caused by peculiar velocities. The r.m.s. pairwise dispersion for the CfA survey slices is consistent with previous determinations, 400 ± 100 km s^{-1} on a scale of 2h^{-1} Mpc. Because the amplitude of the correlation function is small at large scale, this measure is insensitive to large-scale peculiar velocities.

Because of the limited amount of data and the difficulty of the measurements, the relationship between large-scale coherent flows and the structure in redshift surveys is a wide open question. However, coherent flows should, for example, be associated with large low density regions. If the matter density inside a void is low compared with the average surroundings (i.e. if the galaxy density contrast is a measure of the matter density contrast), the voids expand relative to the average cosmological flow and the structures should appear elongated in redshift space. The exact amplitude of the flow depends upon the underlying physics for the formation of the structure (Schwarz, Ostriker, and Yahil 1975, Ikeuchi, Tomisaka and Ostriker 1983, Peebles 1982, Fillmore and Goldreich 1984, Hoffman, Salpeter, and Wasserman 1983, Bertschinger 1985). For an isolated self-similar void the outward peculiar velocity $v_{pec} \simeq 0.3\, v_H$ where v_H is the radius of the void in redshift space. The effect of interaction between adjacent shells on peculiar velocities has not been calculated. In a large enough sample containing many voids the intrinsic spatial geometry of the voids averages out and any net elongation could be interpreted as a residual expansion.

The measurement of distances to galaxies in the structures offers a direct probe for large-scale flows associated with voids. These flows are a possible discriminant among theoretical models. In the biased cold dark matter models where the matter density contrast is much smaller than the galaxy density contrast, the outflow velocities should be small. The galaxies on the edges of the voids did not move across the low density regions to their current positions; they merely lit up there. Because many spirals lie in the extended sheets, the infra-red Tully-Fisher technique (Aaronson, Huchra, and Mould 1979) can be used to obtain limits at the few hundred kilometers per second level on scales of fifty megaparsecs, within the theoretically predicted range (Geller *et al.* 1988a). The ~ 300 km s^{-1} limit on peculiar velocities on the large shell in Figure 3 implies $v_{pec}/v_H \lesssim 0.2$, in the murky range for interpretation.

The net elongation of voids is one of several statistics which may be useful for comparing the data with models. In the absence of "fair" samples, one can still examine the properties of individual structures. In discussing these structures a "void" is a region where the density is less than the global average and the contrast is below some well-defined threshold; analogously, the "sheets" are regions above a threshold. Both the voids and the sheets can be characterized quantitatively.

The frequent mention of the "size" of the Boötes void is a demonstration of the power of a measure of the scale of the "largest" observed structure. The spectrum of void sizes is an important test of models; the small-scale end is a constraint on hot dark matter models (Zeldovich 1987, Doroshkevich *et al.* 1980, Centrella and Melott 1983, Centrella *et al.* 1988) and the large-scale end is most demanding for cold dark matter models (Davis *et al.* 1985, White *et al.* 1987) and for the explosive models (Ostriker and Cowie 1981, Ikeuchi 1981, Saarinen, Dekel, and Carr 1986). Determination of the distribution of sizes of voids requires samples much larger than those currently available.

Both cold (White *et al.* 1987) and hot (Centrella *et al.* 1988) dark matter models produce large voids and some thin coherent structures. White *et al.* (1987) argue that a standard cold dark matter model with biased galaxy formation produces a distribution of galaxies which is hard to distinguish from the data in Figures 3 - 6 and Figures 9 and 10 (see their Figure 10). Sampling according to the procedure followed by KOSS, White *et al.* (1987) find that 3 of 25 simulations contain a void as large or larger than Boötes (~ 5000 km s^{-1}).

It is not clear whether the simulations can meet the challenge posed by the increasing number of surveys with more dense sampling than the Boötes survey. There are now five surveys large enough to contain ~ 5000 km s^{-1} voids — all of them do. The sharpness of the structures in Figures 3 - 6 and 9 and 10, and the possibility that they imply non-random phase initial conditions motivate consideration of alternatives to the standard gravitational models for large-scale structure formation (Ostriker and Cowie 1981, Ikeuchi 1981, Ostriker, Thompson, and Witten 1986).

The thickness, coherence, and filling factor of the sheets provide further constraints. The FWHM of the sheets is $\lesssim$ 500 km s^{-1} (de Lapparent, Geller, and Huchra 1988). The thickness as a function of orientation with respect to the line-of-sight restricts physical models. If the sheets were collapsing pancakes, we would expect them to be thinner when they are perpendicular to the line-of-sight than when they are parallel to it. On the other hand, any internal velocity dispersion will make the sheets appear thicker when they are perpendicular to the line-of-sight.

The fraction of the survey volume filled by the coherent structures in the distribution of individual galaxies can be calculated by appropriately binning and smoothing the data. The galaxies fill $\lesssim$ 20% of the volume and the typical separation of galaxies in the sheets is $3h^{-1}$ Mpc at the survey depth D to which M^* galaxies are included. Remarkably this density is comparable with the surface density of the structures in deep probes (Koo, Kron, and Szalay 1986).

The "uniformity" of the sheets may provide a constraint on Ω (see Peebles 1986). If $\Omega = 1$ and the distribution of galaxies marks the distribution of matter, it is unlikely that a smooth shell can persist for a Hubble time; gravity causes the galaxies to clump up and "fingers" in redshift space should be apparent. If the actual matter density contrast in the sheets is small and the voids are full of nearly uniformly distributed dark matter (with Ω close to 1), the structures could still be in the linear regime. If, on the other hand, $\Omega = 0.2$ or less (as indicated by the dynamical estimates and by analysis of the abundance of the light elements), the structure could set in early on and then just stretch with the universal expansion.

Small groups of galaxies are embedded in the sheets apparent in the slice centered at 29.5° (Ramella, Geller, and Huchra 1988). The properties of these groups are similar to the properties of groups in the 14.5 survey; the median line-of-sight velocity dispersion is $\sigma \simeq 220$ km s^{-1}, the typical scale of the systems is 500 kpc, and the median mass-to-light ratio is 420 $M_\odot/L_\odot$. The centers of these groups trace out the coherent structures in the survey.

In contrast, the richest clusters of galaxies appear to occupy interstitial regions between adjacent shells. The Coma cluster (Figure 3) is an obvious example. The survey provides a warning that the objects in cluster catalogs may not always have the properties expected for real physical systems; in other words fingers in redshift space are not always present at the appropriate location. In Figure 4, for example, the finger at 16.5^h is a cluster identified by Zwicky but not by Abell. Abell includes a cluster between 9^h and 10^h at ~ 7500 km s^{-1}, but its location does not correspond to the true finger in redshift space. When the CfA survey is complete, it will include tens of Abell clusters and should provide a first *direct* measure of the relationship between individual galaxies and clusters of galaxies as tracers of the large-scale matter distribution.

3. The Distribution of Rich Clusters

Once it is known that galaxies cluster, it is natural to ask whether there is higher order structure. Do clusters of galaxies cluster? If so, are clusters of galaxies and individual galaxies equivalent tracers of the large-scale matter distribution in the universe? One important approach to this issue is the extraction of general statistics from cluster catalogs.

The language of correlation functions provides one way of expressing the difference between galaxies and clusters of galaxies as tracers. Bahcall and Soneira (1983) calculated the two-point correlation function for the Hoessel, Gunn and Thuan (1980; HGT hereafter) survey of 104 nearby Abell clusters (distance class $D \leq 4$, richness class $R \geq 1$). Postman, Geller, and Huchra (1986) calculated the cluster correlation function for a variety of other samples drawn from both the Abell (1958) and Zwicky catalogs and Shectman (1985) calculated the cluster correlation function for a sample derived from the Shane-Wirtanen (1967) counts. For all of these samples, the cluster correlation function is consistent with a power law

$$\xi_c(r) \simeq (r_{co}/r)^{1.8}. \tag{7}$$

Both the power law and the amplitude r_{co} are uncertain. The exponents generally appear to be consistent with the value for the galaxy correlation function $\gamma = 1.8$. With the slope constrained to 1.8, the uncertainty in the amplitude r_{co} for a sample of $\sim$ 150 clusters is $\sim$ 50%. For the samples analyzed so far

$$14h^{-1}\ \text{Mpc} \lesssim r_{co} \lesssim 24h^{-1}\ \text{Mpc}. \tag{8}$$

The largest value of r_{co} was obtained in the Bahcall-Soneira (1983) analysis. Values at the low end of the range were obtained by Sutherland (1988), by Shectman (1985), and by Huchra *et al.* (1988).

In general the lower values of r_{co} are for samples which contain less rich clusters, but the value is also affected by the details of the analysis (see Postman, Geller, and Huchra 1986 and Sutherland 1988). Bahcall and Soneira (1983) claim a dependence of the amplitude of the correlation function on the "richness" of systems in the catalog. This dependence seems to be confirmed by Shectman (1985) and by Postman, Geller, and Huchra (1986).

Huchra *et al.* (1988) analyzed a deep sample of 145 clusters with $R \geq 0$ (103 with $R \geq 1$). The survey covers a 561 square degree region at high galactic latitude,

$$58° \leq \delta \leq 78°,$$

and

$$10^h \leq \alpha \leq 15^h.$$

Figure 12 shows the distribution of the clusters on the sky.

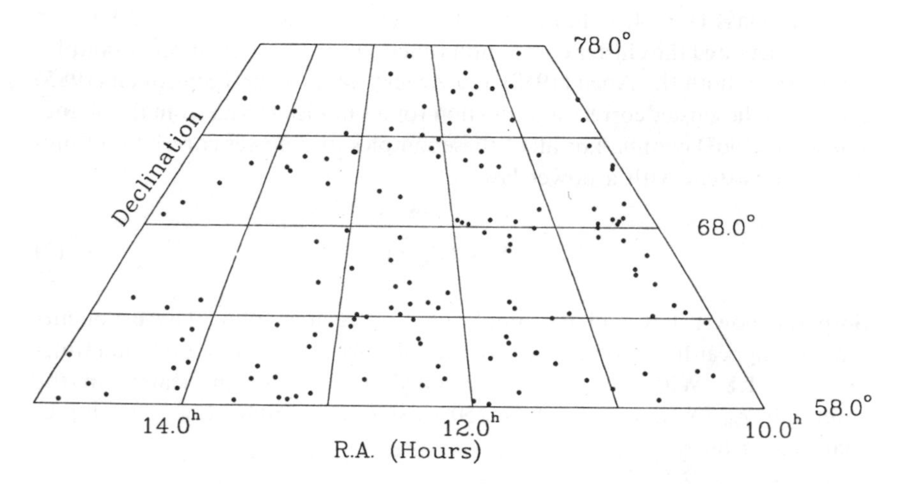

FIG. 12. The distribution on the sky of the 145 Abell clusters in the deep sample of Huchra *et al.* (1988).

The deep sample has little overlap with the HGT sample analyzed by Bahcall and Soneira; 137 of the clusters in the sample have distance class $D \geq 5$. The effective volumes of the two surveys are comparable. The area of the HGT sample is 25 times larger, but the deep survey extends ~ 3 times farther in redshift. The median redshift for the deep survey is 0.17; for the Bahcall-Soneira sample, it is 0.07.

Figure 13 is the cone diagram for the survey. Because clusters are sparse tracers of large-scale structure the apparent voids in the distribution have low significance (see Otto *et al.* 1986).

The amplitude of the correlation function for the $R > 0$ subset of the deep sample is $r_{co} = 17.8h^{-1}$ Mpc (Figure 14). In other words the amplitude of the cluster correlation function for this sample is a factor of ~ 2 less than for the HGT sample; the uncertainty in the amplitude is ~ 40%. The apparent variations in the amplitude of $\xi_c(r)$ are probably not significant. For samples

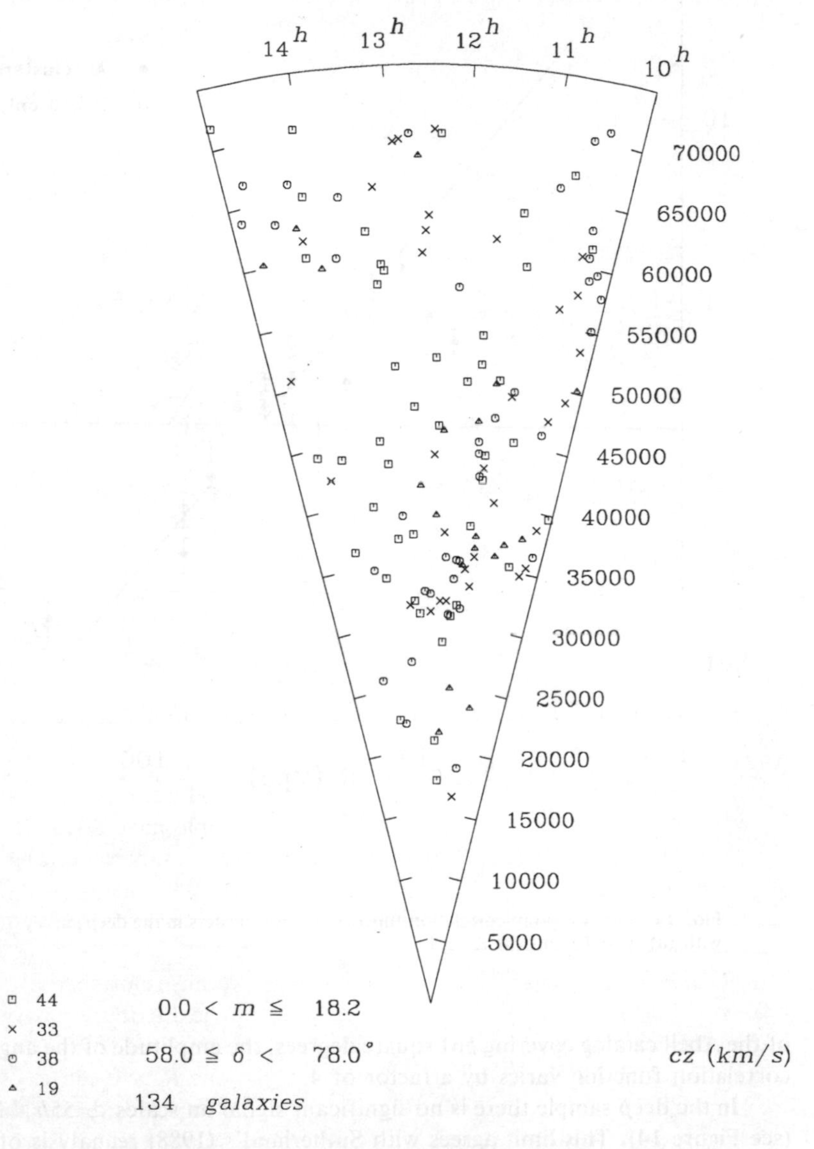

FIG. 13. Observed velocity versus right ascension for the cluster survey strip. Clusters in different declination ranges are plotted as different symbols — box = $58° \leq \delta \leq 63°$, cross = $63° \leq \delta \leq 68°$, circle = $68° \leq \delta \leq 73°$, and triangle = $73° \leq \delta \leq 78°$.

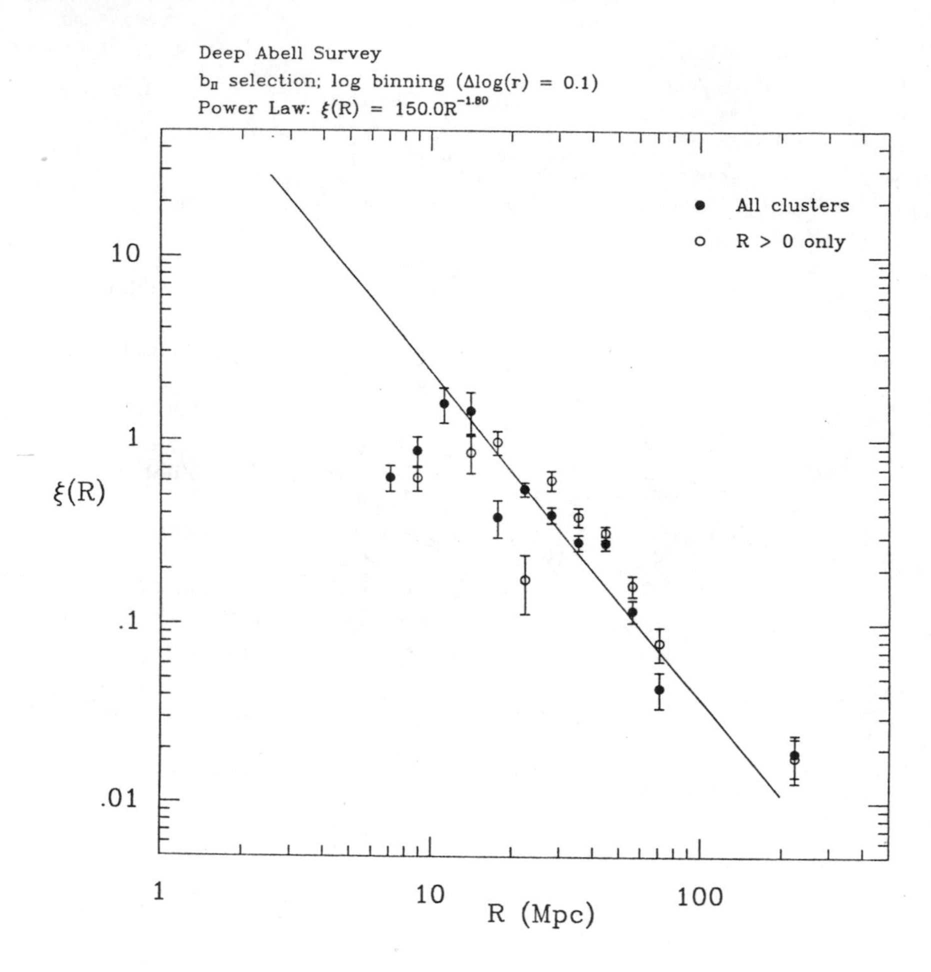

FIG. 14. The two-point correlation function for the clusters in the deep survey with (a) R $\geq$ 0, and (b) R $\geq$ 1.

of the Abell catalog covering 561 square degrees, the amplitude of the angular correlation function varies by a factor of 4.

In the deep sample there is no significant signal on scales $\gtrsim 55h^{-1}$ Mpc (see Figure 14). This limit agrees with Sutherland's (1988) reanalysis of the Bahcall-Soneira samples and with the analyses by Postman, Geller, and Huchra (1986). It is inconsistent with claims of significant signal on scales as large as 150 h^{-1} Mpc (Bahcall and Soneira 1983, Batuski *et al.* 1988).

Large-scale peculiar velocities are a further issue posed by the surveys of rich clusters. Bahcall and Soneira (1983) examined the distribution of redshift separations for cluster pairs in various ranges of angular separation. At small angular separations, they claim a broadening of the redshift distribution caused by relative pairwise peculiar motions of $\sim$ 2000 km s^{-1} (Bahcall, Soneira, and Burgett 1986). This broadening corresponds to a net elongation of structures along the redshift direction. In the sample of 104 clusters about 30% of the power in the correlation function is contributed by the Corona Borealis supercluster which is elongated along the line-of-sight. It is possible that the elongation reflects the intrinsic geometry of this particular system rather than large-scale peculiar motions. The correlation function approach is only valid when the sample is large enough for the statistics to represent an average over many systems.

Large-scale peculiar motions are not supported by the analysis of the Huchra *et al.* (1988) deep sample. For this sample, relative pairwise peculiar motions are $\lesssim$ 1000 km s^{-1}. There are several possible explanations for the different results. One is the effect of a single system like Corona Borealis on the average statistics. Another is the greater difficulty of separating close pairs of clusters in the deeper sample; in other words, if Corona Borealis were present in the deeper sample, a few of the members might not be counted as separate clusters. Undercounting of close pairs could lead to underestimation of r_{co}.

Several caveats are important in consideration of this and other cluster surveys. The redshift measurements for the Huchra *et al.* (1988) sample demonstrate that a large fraction of the clusters are complex systems; many consist of several lesser agglomerations projected along the line-of-sight. Of the 90 clusters in the sample with multiple redshifts, 40% (36 clusters) show multiple redshift systems or significant ($\geq$ 5000 km s^{-1}) extent in redshift space. In these cases, the redshifts in Figure 13 are a best guess of the Abell "cluster" redshift. The problems of foreground and background contamination are significant for all richness and distance classes in the sample. Indeed, in a few cases, it is impossible to identify any cluster at all.

4. Conclusions

Analyses of galaxy and cluster redshift surveys raise a number of important issues. In both cases, the samples are probably not large enough to be representative. In the case of galaxy surveys, the largest well-defined structures are comparable with the extent of the survey. In the case of cluster surveys single systems could dominate the average statistics.

Some have argued (see e.g. da Costa *et al.* 1988) that the large-scale galaxy distribution is heterogeneous. In other words, the distribution could contain filaments, sheets, voids and diffuse structures. Until the selection criteria for different surveys can be carefully compared, it will not be clear whether this conclusion is a function of sampling or of genuinely variable structure.

A general description of large-scale structure also depends upon having surveys which are large enough to be representative. Deeper surveys extending over large angular scales are important for enabling the identification of structures larger than those which have already been observed; the largest structures provide one of the tightest constraints on models. At least two surveys are underway to meet this goal: one in the North (Geller *et al.* 1988b) and one in the South (KOSS 1988). Both of these surveys reach to a limiting apparent magnitude $m_{B(0)} = 17.5$ and span $\sim 100°$ across the sky.

The largest cluster redshift surveys are limited to $\sim$ 150 objects. The sampling fluctuations alone lead to variations of a factor of $\sim$ 4 in the amplitude of the correlation function. Perhaps even more serious, cluster catalogs are subject to a large number of poorly understood selection biases. Not the least of these is the problem of superpositions along the line-of-sight.

Further progress in mapping the large-scale structure of the universe requires reliable photometric catalogs. Even at the depth of the Zwicky catalog, there is no uniform survey of the sky. Systematic variations in the magnitudes from one region to another in a single catalog (not to mention variations from one catalog to another) almost surely compromise detailed analyses of the properties of the structures. The advent of large format CCD's has made fundamentally important digital surveys possible. An intermediate depth (m $\leq 20^{th}$ magnitude), multicolor survey should be the next major step in the study of large-scale structures.

Acknowledgements

We would like to thank Pat Henry for data in advance of publication. We also thank Valérie de Lapparent and Marc Postman for several of the plots displayed here. This work is supported by NASA Grant NAGW-201 and by the Smithsonian Institution. Results reported here are based partly on observations obtained with the Multiple Mirror Telescope, a joint facility of the Smithsonian Institution and the University of Arizona.

REFERENCES

Aaronson, M., Huchra, J.P., and Mould, J. 1979. *Ap J.* **229,** 1.

Abell, G.O. 1958. *Ap J Suppl.* **3,** 211.

Bahcall, N.A. and Soneira, R.M. 1983. *Ap J.* **270,** 20.

Bahcall, N.A., Soneira, R.M., and Burgett, W.S. 1986. *Ap J.* **311,** 15.

Batuski, D.J. *et al.* 1988. preprint.

Bean, A.J., Efstathiou, G., Ellis, R.S., Peterson, B.A., and Shanks, T. 1983. *MN.* **205,** 605.

Bertschinger, E. 1985. *Ap J.* **295,** 1.

Bothun, G.D., Beers, T.C., Mould, J., and Huchra, J.P. 1986. *Ap J.* **308,** 510.

Centrella, J.M. and Melott, A.S. 1983. *Nature.* **305,** 196.

Centrella, J.M., Gallagher, J.S., Melott, A.S., and Bushouse, H.A. 1988. preprint.

da Costa, L.N., Pellegrini, P.S., Sargent, W.L.W., Tonry, J., Davis, M., Meiksin, A., and Latham, D.W. 1988. *Ap J.* **327,** 544.

Davis, M. and Djorgovski, S. 1985. *Ap J Letters.* **299,** L15.

Davis, M. Efstathiou, G., Frenk, C. and White, S.D.M. 1985. *Ap J.* **292,** 371.

Davis, M. and Huchra, J.P. 1982. *Ap J.* **254,** 437.

Davis, M. and Peebles, P.J.E. 1983. *Ap J.* **267,** 465.

Davis, M., Huchra, J.P., Latham, D.W., and Tonry, J. 1982. *Ap J.* **253,** 423.

Dekel, A. and Silk, J. 1986. *Ap J.* **303,** 39.

de Lapparent, V., Geller, M.J., and Huchra, J.P. 1986. *Ap J Letters.* **202,** L1. (dLGH)

_____. 1988. *Ap J.* **332,** 44.

Doroshkevich, A.G., Kotok, E.V., Novikov, I.D., Polyudiv, A.N., Shandarin, S.F., and Sigov, Yu.S. 1980. *MN.* **192,** 321.

Dressler, A. 1980. *Ap J.* **236,** 351.

_____. 1984. *Ann Rev Astron Astrophys.* **22,** 185.

Einasto, J., Jôeveer, M., and Saar, E. 1980. *MN.* **193,** 353.

Fillmore, J.A. and Goldreich, P. 1984. *Ap J.* **281,** 9.

Geller, M.J. *et al.* 1988a. in preparation.

Geller, M.J. *et al.* 1988b. in preparation.

Giovanelli, R. and Haynes, M.P. 1988. in *Large Scale Structures of the Universe.* eds. J. Audouze, M.-C. Pelletan, and A. Szalay, p. 113. Dordrecht: Kluwer Academic Publishers.

Giovanelli, R., Haynes, M.P., and Chincarini, G. 1986. *Ap J.* **300,** 77.

Gott, J.R., Melott, A. and Dickinson, M. 1986. *Ap J.* **306,** 341.

Groth, E.J. and Peebles, P.J.E. 1977. *Ap J.* **217,** 385.

Hamilton, A.J.S., Gott, R., and Weinberg, D. 1987. *Ap J.* **309,** 1.

Haynes, M.P. and Giovanelli, R. 1986. *Ap J Letters.* **306,** L55.
Hoessel, J.G., Gunn, J.E., and Thuan, T.X. 1980. *Ap J.* **241,** 486. (HGT)
Hoffman, G.L., Salpeter, E.E., and Wasserman, I. 1983. *Ap J.* **268,** 527.
Hubble, E. and Humason, M.L. 1931. *Ap J.* **74,** 43.
Huchra, J.P. and Geller, M.J. 1988. in preparation.
Huchra, J.P., Davis, M., Latham, D.W., and Tonry, J. 1983. *Ap J Suppl.* **52,** 89.
Huchra, J.P., Henry, P., Postman, M., and Geller, M.J. 1988. in preparation.
Ikeuchi, S. 1981. *PASJ.* **33,** 211.
Ikeuchi, S., Tomisaka, K., and Ostriker, J.P. 1983. *Ap J.* **265,** 538.
Jôeveer, M., Einasto, J., and Tago, E. 1978. *MN.* **185,** 357.
Kaiser, N. 1987. *MN.* **227,** 1.
Kirshner, R.P., Oemler, A., and Schechter, P. 1979. *AJ.* **84,** 951.
Kirshner, R.P., Oemler, A., Schechter, P.L., and Shectman, S.A. 1981. *Ap J Letters.* **248,** L57. (KOSS 1981)
Kirshner, R.P., Oemler, A. Schechter, P.L., and Shectman, S.A. 1983. *AJ.* **88,** 1285. (KOSS 1983)
Kirshner, R.P., Oemler, A., Schechter, P.L., and Shectman, S.A. 1987. *Ap J.* **314,** 493. (KOSS 1987)
Kirshner, R.P., Oemler, A., Schechter, P.L., and Shectman, S.A. 1988, private communication. (KOSS 1988)
Koo, D. Kron, R. and Szalay, A. 1986. in *13th Texas Symposium on Relativistic Astrophysics.* ed. M. Ulmer. Singapore: World Scientific.
Nilson, P. 1973. *Uppsala General Catalogue of Galaxies, Uppsala Astr. Obs. Ann.* **6.**
Ostriker, J.P. and Cowie, L.L. 1981. *Ap J Letters.* **243,** L127.
Ostriker, J.P., Thompson, C., and Witten, E. 1986. *Phys Lett B.* **180,** 231.
Otto, S., Politzer, D., Preskill, J., and Wise, M. 1986. *Ap J.* **304,** 62.
Peebles, P.J.E. 1982. *Ap J.* **257,** 438.
_____. 1980. *Large-Scale Structure in the Universe.* Princeton: Princeton University Press.
Peterson, B.A., Ellis, R.S., Efstathiou, G., Shanks, T., Bean, A.J., Fong, R. and Zen-Long, Z. 1986. *MN.* **221,** 233.
Postman, M. and Geller, M.J. 1984. *Ap J.* **291,** 85.
Postman, M., Geller, M.J., and Huchra, J.P. 1986. *AJ.* **91,** 1267.
Postman, M., Huchra, J.P., and Geller, M.J. 1986. *AJ.* **92,** 1238.
Ramella, M., Geller, M.J., and Huchra, J.P. 1988, in preparation.
Saarinen, S. Dekel, A., and Carr, B.J. 1986. preprint.
Schechter, P.L. 1976. *Ap J.* **203,** 297.
Schwarz, J., Ostriker, J.P., and Yahil, A. 1985. *Società Italiana di Fisica.* **1,** 157.
Shane, C.D. and Wirtanen, C.A. 1967. *Publ Lick Obs.* **Vol. XXII,** Part 1.
Shanks, T., Bean, A.J., Efstathiou, G., Ellis, R.S., Fong, R., and Peterson, B.A. 1983. *Ap J.* **274,** 529.

Shectman, S.A. 1985. *Ap J Suppl.* **57**, 77.

Smith, B., Kleinmann, S., Huchra, J. and Low, F. 1987. *Ap J.* **318**, 161.

Strauss, M. and Huchra, J. 1988. *AJ.* **95**, 1602.

Sutherland, W. 1988. preprint.

Thuan, T.X., Gott, J.R., and Schneider, S.E. 1987. *Ap J Letters.* **315**, L93.

White, S.D.M., Frenk, C.S., Davis, M., and Efstathiou, G. 1987. *Ap J.* **313**, 505.

Zeldovich, Ya. B. 1970. *AA.* **5**, 84.

Zwicky, F., Herzog, W., Wild, P., Karpowicz, M. and Kowal, C. 1961-1968. *Catalog of Galaxies and of Clusters of Galaxies.* Pasadena: California Institute of Technology.

LARGE-SCALE STRUCTURE IN THE LOCAL UNIVERSE: THE PISCES-PERSEUS SUPERCLUSTER

MARTHA P. HAYNES

National Astronomy and Ionosphere Center and Department of Astronomy, Cornell University

and

RICCARDO GIOVANELLI

National Astronomy and Ionosphere Center

1. INTRODUCTION

At the time of the Study Week on *Astrophysical Cosmology* in 1981 the inhomogeneity of the large-scale structure and the lack of some obvious preferred scale were already well known. Oort (1982) presented the results providing evidence at that time that coherent features with dimensions on the order of $70h^{-1}$ Mpc exist and that a rich variety of structures characterized the galaxy distribution: cells, filaments, voids. In the intervening years, further terminology has come into common usage in the taxonomy of the galaxy distribution: fractals, sheets, strings, bubbles, swiss cheese, meatballs and sponges, in addition to those already mentioned. Some topologies are the direct result of model predictions while others cannot be uniquely represented.

From the observational perspective of the question of large-scale structure, the 1980's has been the decade of an explosion in the number of extragalactic redshift measurements. Technological advances made primarily in detectors and spectrometers available both to optical and radio astronomers have increased dramatically the efficiency of measuring accurate redshifts for galaxies of moderate brightness. The acquisition of redshift information for a large number of galaxies now allows us to investigate the three dimensional nature of the local universe. In some regions of the sky, redshift surveys are sufficiently complete that we can begin to study coherent structures as individual units. Questions of characteristic scales and density contrasts, of environmental alteration and initial conditions and of Hubble flow and peculiar motions can be addressed to specific locations within the galaxy distribution.

In this review, we shall discuss what is currently known about one of the major large-scale features seen in the northern sky: the Pisces-Perseus (P-P) supercluster. Located at a characteristic distance corresponding to cz = +5000 km s^{-1}, the main ridge of P-P appears as a linear structure at least $45h^{-1}$ Mpc in length and includes the rich clusters A 262, A 347, and A 426, as well as lower density enhancements in Pegasus and around NGC 383 and NGC 507. At the present time, redshifts have been measured for some 5000 galaxies in the region occupied by the supercluster. A great deal of study remains to be made of the supercluster, but we have begun to use it as a laboratory for understanding the dynamics and structure of the galaxy distribution on large scales. Because of its relative proximity, observational data can be obtained for large numbers of supercluster members at a variety of wavelengths. Surely, P-P may not be a representative environment within the local universe. Rather it allows us to probe in depth the range of densities characteristic of local volumes.

2. The Pisces-Perseus Region

The P-P supercluster is evident clearly in maps of the brightest galaxies such as those catalogued by Shapley and Ames (1932). Even though most galaxies of similar apparent brightness are significantly closer, the main ridge of the supercluster stands out among the distribution of NGC galaxies and thus Bernheimer (1932), among others, was able to recognize its "metagalactic" nature. Figure la shows the distribution of the 235 galaxies brighter than B_T = 13.2 mag included in the *Revised Shapley-Ames Catalog* (Sandage and Tammann 1981; RSA) on a equal area projection of the portion of sky that we shall refer to as the "P-P region", that is, the area bounded by $22^h <$ R.A. $< 04^h$, $0° <$ Dec. $< +50°$. In the lower panel Figure 1b, only those 88 with heliocentric recessional velocities V_0 in the range $+4500 < V_0 < +6000$ km s^{-1}, characteristic of the P-P supercluster, are included. The main ridge of the supercluster includes the diagonal string of galaxies betweem 0^h 30^m and 2^h 30^m R.A. and +20° to +40° Dec.

With the acquisition of small numbers of redshifts, the reality of the connectivity of the structure became apparent. Gregory *et al.* (1981) noticed that most of the clusters in the region were located at the same distance and hence belonged to a single supercluster of approximately the same depth as width. Einasto *et al.* (1980) published maps suggesting connective structures in the region deduced from an examination of the distribution of galaxies and clusters in the *Catalog of Galaxies and Clusters of Galaxies* (Zwicky *et al.* 1961-8; CGCG). Although relatively few redshifts were available for galaxies in the

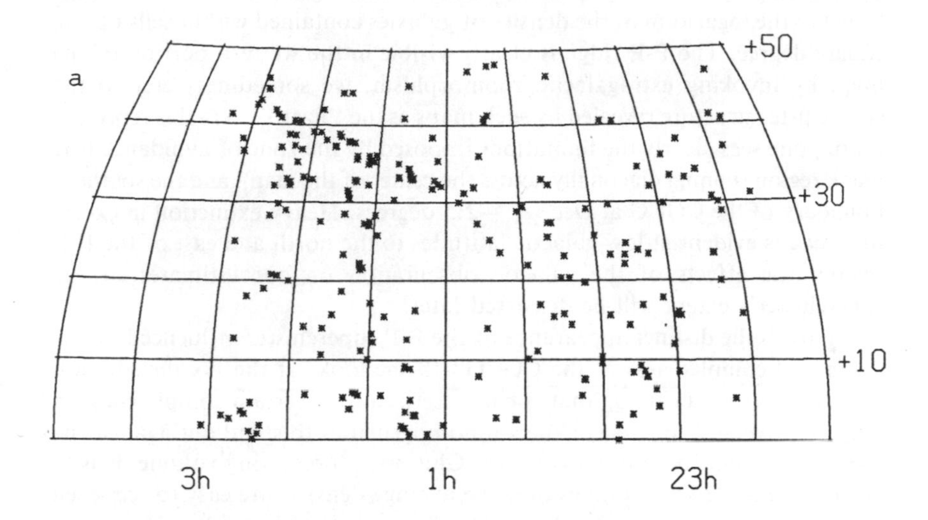

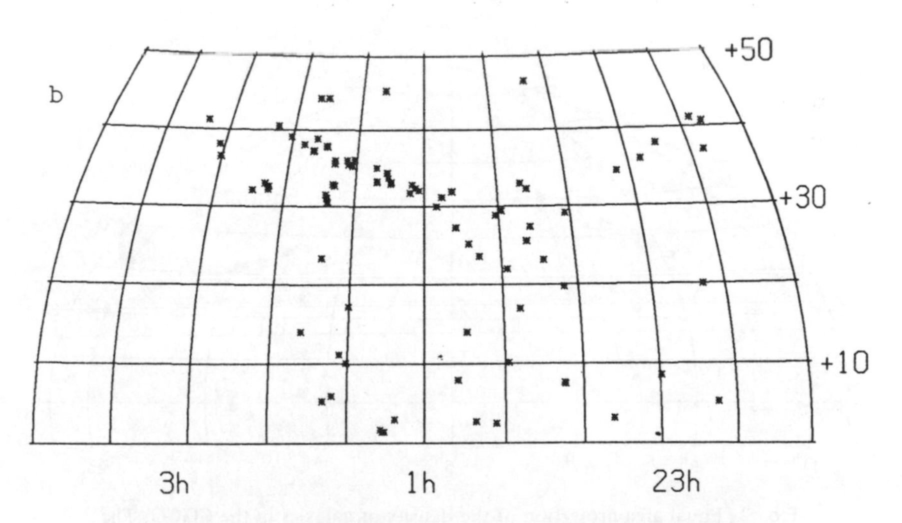

FIG. 1. (a) Equal area projection of the distribution of 273 galaxies brighter than $B_T = 13.2$ included in the RSA in the region of the P-P supercluster; (b) Distribution of the 88 galaxies shown in (a) with redshifts in the range $+4500 < V_o < +6000$ km s^{-1}.

supercluster at that time, the latter authors recognized much of the complexity of the large-scale structure.

Figure 2, reproduced from Giovanelli and Haynes (1982), illustrates the surface density distribution of galaxies in one quarter of the sky on an Aitoff equal area projection. The shade intensity (on an eight step scale) is proportional to the logarithm of the density of galaxies contained within cells of one square degree. The P-P ridge is clearly visible in the western portion of the map; by invoking extragalactic zoomorphism, we sometimes refer to the supercluster structure revealed by such maps as the "lizard". Via this representation, one sees clearly the limitations imposed by the zone of avoidance (the blank region running diagonally across the center of the map), and the southern boundary of the CGCG at Dec. = —2.5 degrees. Heavy extinction in excess of 1 mag is evident at low galactic latitudes to the north and east of the P-P region; the effects of the galactic obscuration on our estimates of the supercluster's extent will be discussed later.

How is the distinct appearance of the P-P supercluster influenced by the depth and completeness of the CGCG? If one looks at the sky distribution of galaxies from a catalog that is characterized by a certain completeness in magnitude or size, one may expect inhomogeneities to stand out against the background and foreground galaxies. Clusters, representing volume density enhancements of several times over the average density, are easy to see, even in maps of deep catalogs. Shallower density enhancements will appear more

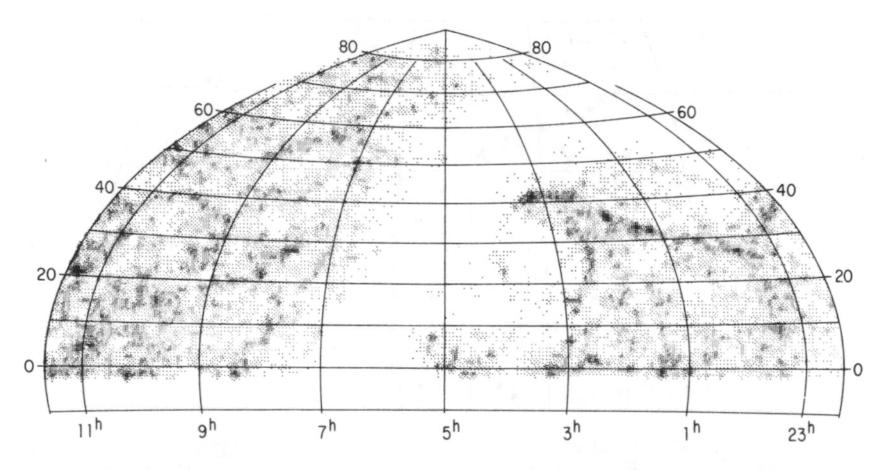

FIG. 2. Equal area projection of the density of galaxies in the CGCG. The shade intensity is proportional to the number of galaxies in each 1° by 1° cell. The blank region cutting across the map is the zone of avoidance. The catalog has a southern boundary at about Dec. = – 3°. (Giovanelli, Haynes, and Chincarini 1986).

or less conspicuous in a surface density map, depending not only on the volume density contrast of the enhancement and its extent along the line of sight, but also on the relative size of its angular extent vis-a-vis that of the sampled region and, especially, on its distance from the observer and the depth of the catalog in question. In a volume — or absolute magnitude — limited catalog, i.e. one that contains all galaxies brighter than a given luminosity limit, out to a given distance r_{lim}, a density enhancement of $1 + \Delta\varrho/\langle\varrho\rangle$ will yield the same surface density contrast $1 + (\Delta\varrho/\langle\varrho\rangle)(\Delta r/r_{lim})$ where Δr is the line of sight extent of the enhancement, independent of its distance from the observer. In an absolute magnitude-limited sample, on the other hand, a homogeneously distributed population of galaxies at that redshift yields a peaked redshift distribution, i.e., a subsample of galaxies at that redshift yelds the highest surface density counts. In this case, the peak of the distribution depends on the catalog limit and on the characteristics of the luminosity function. A density enhancement will therefore be more conspicuous in a sky map if it is located at a redshift near the peak. For example, a Gaussian density enhancement of a full width at half maximum of 400 km s^{-1} and a volume overdensity of 90, centered at $cz = +5000$ km s^{-1} will produce a surface density enhancement of about 10 in the counts of CGCG galaxies. On the other hand, a similar overdensity would be practically washed out in the Lick galaxy counts. In order to produce a surface density enhancement comparable to that produced in the CGCG, a given volume density enhancement should be three times deeper and three times further away (Giovanelli *et al.* 1986).

For any magnitude-limited sample we can estimate the redshift distribution one would have expected from a homogeneously distributed, well-behaved population of galaxies (i.e. one that adheres to a universal luminosity function). In general, we can write the number of galaxies with apparent magnitude between m and m + dm, located in the distance interval (r, r + dr), within a solid angle $\Delta\Omega$ as

$$n(m,r)\, dr\, dm\Delta\Omega = D(r)\, \Phi\, [m - 5\log(r/10)]\, \Delta\Omega r^2\, dr\, dm \quad (1)$$

where D(r) is the space density of galaxies and we adopt a Schechter (1976) luminosity function with the parameters $M^* = -20.6$ and $\alpha = -1.25$ (appropriate for blue magnitudes and $h = 1/2$). The number counts integrated for all r for any complete magnitude limited sample are given by

$$n(m)\, dm = n_0 \int_0^\infty \Phi\, [m - 5\log(r/10)]\, D(r)\, r^2\, dr\, dm. \quad (2)$$

Substituting $r = v/H$ to obtain a radial velocity distribution, where v is the radial velocity, the expected distribution for each magnitude interval is

$$N(m,v)\, dv\, dm\, \Delta\Omega = D(v/H)\, \Phi\, [(m - 5\log(v/10H)]\, \Delta\Omega\, (v/H)^2 d\,(v/H)\, dm. \quad (3)$$

The expected velocity distribution is then obtained simply by integrating

$$N(v)\,dv = dv \int N(v,m)\,dm. \qquad (4)$$

One can similarly show that the distribution of a sample complete to a magnitude limit m_{lim} peaks at a distance, corresponding to a velocity v*, that is obtained from the relation

$$v^* = 5.248 \text{ dex } (0.2\, m_{lim}). \qquad (5)$$

Thus, for a limiting magnitude of $m_{lim} = +14.9$, the redshift distribution is expected to peak at $v^* = +5000$ km s^{-1}. The CGCG is supposed to be roughly complete to a magnitude limit of +15.7. At the faintest magnitudes, errors in the magnitude measurements increase, and some systematic errors in the magnitude scale become difficult to correct for galaxies fainter than about m = +15.4 (Giovanelli and Haynes 1984). However, if a universal luminosity function applies to all galaxies, then the CGCG best traces volume density enhancements that have characteristic sizes on the order of 5 to 20 Mpc if they are located at a redshift of about 5000 to 7000 km s^{-1}. The distinctiveness of such features is not an artifact of the CGCG, but rather the CGCG will best illustrate such overdensities if they exist.

Other features are also apparent in similar displays of the other quarter of the sky also contained in the CGCG (Fontanelli 1984; de Lapparent *et al.* 1986; Bicay 1987). The Coma-A 1367 and Hercules superclusters are also apparent, although with differing clarity because they both lie at larger distances. In some locations, the observed increase in the surface density is an overstimate because of the chance projection of overlapping structures at different distances. For many purposes of identifying structures, the CGCG has been a useful indicator of obvious large-scale structure in the local universe, and the P-P supercluster is perhaps the most obvious large-scale structure outside the Local Supercluster visible in the CGCG maps.

3. The HI Line Survey

Redshift surveys have been approached from a number of perspectives, dependent partly on the scientific objectives and partly on the availability of telescopes and telescope time. Over the last decade, we and our co-workers (Giovanelli and Haynes 1984; Giovanelli, Haynes and Chincarini 1986; Giovanelli *et al.* 1986; Haynes *et al.* 1988a) have been conducting a survey of spiral galaxies in the P-P region primarily using the 21 cm line of HI as the carrier of redshift information. Twenty years ago 21 cm line emission had been measured in about 140 galaxies (Roberts 1969). Today, HI spectra exist for some 6000 objects. From the initial observations conducted as part of our P-P survey a decade ago to today, our efficiency of measuring redshifts for galaxies has increased by about a factor of five, so that although we are today observing galaxies that are smaller and fainter than the targets of the 1970's, our detection rate remains essentially unchanged for the same observing time. Of historical interest, one should keep in mind that prior to 1975, Arecibo did not operate at 21 cm. In the intervening years, the resurfacing of the antenna surface, reduction of the noise temperature of the receiver, remote tuning of the feed and construction of a wide-bandwidth 1.6 bit correlator have produced a system whereby redshifts are measured for some 1500-2000 galaxies per year even though the telescope is used only 15% of the time for such observations.

The survey we are currently undertaking aims to measure redshifts for a sample of galaxies throughout the P-P region that is complete in magnitude to $m = +15.7$, the CGCG limit, and in angular size to $a = 1.0$ arcmin, the limit of the *Uppsala General Catalog* (Nilson 1973; UGC). Three fourths of the catalogued galaxies in the region are spirals and as such are good targets for 21 cm emission searches. Objects at declinations south of Dec. = +38° have been observed with the Arecibo 305 m telescope, while those to the north have been surveyed with the 91 m antenna of the National Radio Astronomy Observatory at Green Bank. The total redshift sample includes almost 5000 measurements of which about 3300 were obtained from 21 cm spectra. The completion percentages vary somewhat over the P-P region and are lower in the northern — and southern — most portions as a result of lowered sensitivity at high zenith angles with the current feed system available at Arecibo and the lower sensitivity and limited tracking at Green Bank. At Arecibo the 21 cm observations are normally conducted over a frequency range corresponding to $v < + 15000$ kms^{-1}; searches at higher velocities are presently marred by man-made interference, a circumstance that planned telescope upgrading should alleviate. At Green Bank, the typical search is not quite so deep: $v < +13000$ km s^{-1}. It is likely that most 21 cm non-detections in fact lie at redshifts beyond the search range. In addition, the 21 cm line observations are restricted to HI rich objects, galaxies with morphologies of S0a and later, so that the extension of the survey to earlier types depends on optical redshift measurements. It should be noted that we have obtained morphological classifications and have measured angular sizes of all CGCG galaxies in the P-P region that are not included in the UGC. Such measurements were performed on glass reproductions of the Palomar Sky Survey blue plates. Fortunately, the morphological segregation that is responsible for the more pronounced clustering of early-type objects and their characteristically higher optical surface brightness make E's and S0's favorable targets of optical observers. The combination of optical and HI samples is thus complementary and yields an evenly-representated sample.

To illustrate the current limitations that are characteristic of the survey, Figure 3 illustrates the current completeness statistics for the survey. Figure 3a shows the completeness percentages for galaxies contained in the CGCG across the P-P region with $+21°30' < \text{Dec.} < +33°30'$. Of the 1460 redshift measurements available in this strip, 70% were obtained from 21 cm emission spectra. Completion here is defined for each bin of 0.1 mag as the ratio of the number of galaxies in the strip that are listed in the CGCG, and is plotted separately for all types (triangles) and for spirals (open circles). When all types are included, completion is better than 50% for all bins except the faintest $m = +15.7$. The current survey covering the entire P-P region achieves that level of completion for magnitudes of $m = +15.5$ and brighter. Figure 3b shows a similar diagram with declination being the tracked variable; separate indications are given for samples complete to the indicated apparent magnitudes. Note that the completeness is high for declinations between about $+3° < \text{Dec.} < +33°$, the region of high sensitivity for the Arecibo telescope.

The population ratios E:S0:Sp of the redshift sample included in Figure 3 are 6:11:83, marginally more spiral-rich than those found for the RSA sample of bright field galaxies. Complementary optical observations and efforts by other groups are helping to reach completion of the early-type populations.

For any well-designed redshift program, one should be able to estimate a "completeness function" and then employ that correction function to estimate the redshift distribution that one would have expected from a random population of similar objects. In the case of magnitude limited samples, we can follow the derivation of the expected redshift distribution given for a homogeneous distribution of galaxies characterized by a Schechter luminosity function $\Phi(M)$. The completeness function $c(m)$ then is simply the ratio of the observed apparent magnitude distribution for the sample $n_s(m)\,dm$ and that of a complete sample $n(m)\,dm$. Under the assumption that, within each bin of apparent magnitude, the galaxies for which redshifts have not been measured are not distinguished by any other characteristic, then we can estimate the expected redshift distribution for the incomplete observation sample by computing.

$$N(v)\,dv = dv \int_0^\infty N(v,m)\,c(m)\,dm. \tag{6}$$

For the simplest models, the counts of galaxies are expected to grow with apparent magnitude as dex (0.6 mag). In fact, simulations using a clustered distribution over a large area indicate that the irregularities are smoothed out, and the dex (0.6 mag) law is preserved (Giovanelli *et al.* 1986). A serious problem in the northern and eastern portions of the P-P region, however, arises because of the encroaching galactic extinction. The effect of Δm magnitudes of galactic obscuration on the observed number counts for a catalog complete

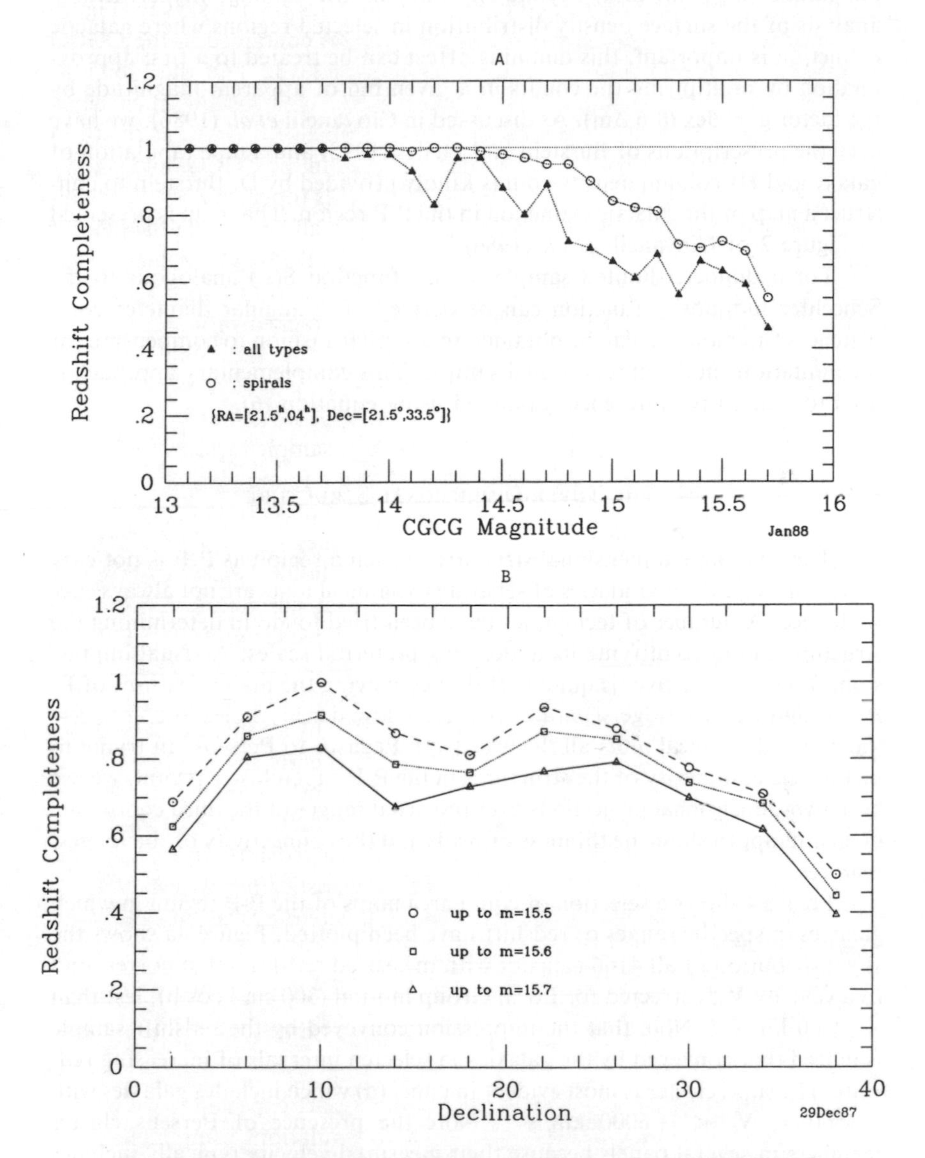

FIG. 3. Panel (a) shows the completeness fraction in 0.1 magnitude bins of the redshift sample within a declination strip across the P-P region with $+21^{\circ}30' <$ Dec. $< +33^{\circ}30'$. Shown separately is the completeness for all types (triangles) and spirals only (open hexagons). Panel (b) shows the completeness as a function of declination, throughout the P-P region, with separate symbols for galaxy samples complete to the indicated limiting magnitude.

to a limiting magnitude m_{lim} is to dim the galaxies in each interval of apparent magnitude (m_{lim}- m, m_{lim}) beyond the limit of the catalog. In performing analysis of the surface density distribution in selected regions where galactic extinction is important, this dimming effect can be treated to a first approximation by multiplying the counts in a given bin of apparent magnitude by the factor g = dex (0.6 Δm). As discussed in Giovanelli *et al.* (1986), we have used the prescriptions of Burstein and Heiles (1978) and a tape tabulation of galaxy and HI column density counts kindly provided by D. Burstein to construct a map of the galactic extinction in the P-P region. That map is presented as Figure 2 of Giovanelli *et al.* (1986).

For a diameter-limited sample, a size function S(L) analogous to the Schechter luminosity function can be derived. The angular diameter completeness function c(a) can be obtained in a similar fashion to compensate for the limitations in the observational sample. This complementary approach is used to tighten the inferences obtained using equation (6).

4. The Three-Dimensional Structure

The true three-dimensional structure in such a region as P-P is not easy to describe and the boundaries of separate dynamical units are not always easy to deduce. A number of techniques have been tried to aid in determining the structure and in identifying its underlying preferred scales. A visual impression, though subjective, is quite useful in conveying the major features of P-P: its high density ridge separates relatively low density holes and connects the major dynamical units all the way from Pegasus to Perseus. In trying to picture the complexity of the structures in the P-P supercluster region, we use both two-dimensional projections over restricted ranges of the third coordinate in an attempt to show the thinness of walls and the connectivity of the longest features.

Figure 4 shows a selection of equal area maps of the P-P region in which galaxies in specific ranges of redshift have been plotted. Figure 4a shows the sky distribution of all 4166 galaxies with measured redshifts that correspond to a velocity V_0, corrected for Local Group motion (300 sin l cos b), less than + 12000 km s^{-1}. Note that the impression conveyed by the redshift sample is indeed that conveyed by the galaxies in selected intervals of increasing redshift. The supercluster is most evident in panel (d) which includes galaxies with $+4500 < V_0 < +6000$ km s^{-1}. Note the presence of Perseus cluster members in several panels because their measured velocity typically includes a large component of dynamical distortion of a smooth Hubble flow.

Figure 5 and 6 show a selection of cone diagrams for galaxies located in separate slices in declination (Figure 5) and right ascension (Figure 6). The size of each wedge is 10° in Figure 5 and 30^m in Figure 6. While stretching

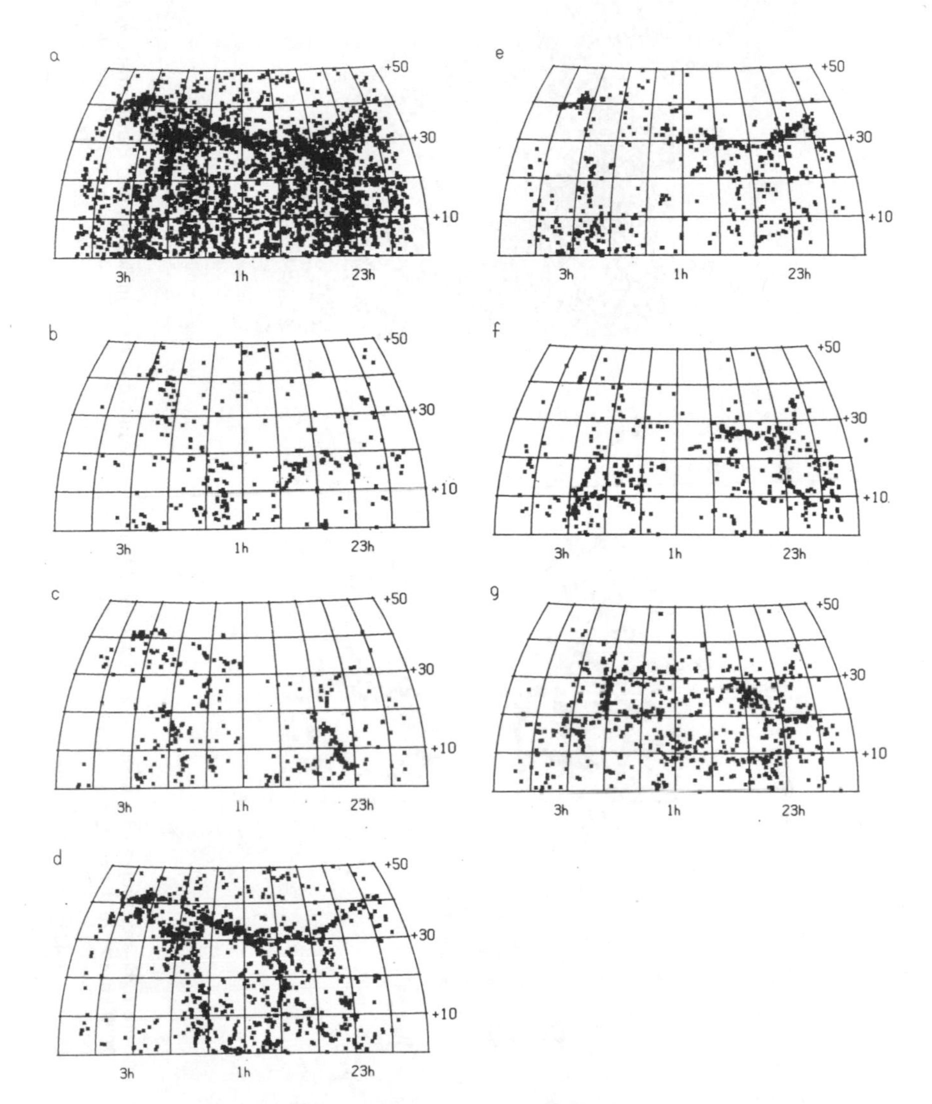

FIG. 4. Sky distribution of galaxies of known redshift on an equal area projection. Panel (a) shows all 4166 galaxies with $V_o < +12000$ km s^{-1}; (b) 325 galaxies with $V_o < +3000$ km s^{-1}; (c) 412 galaxies with $+3000 < V_o < +4500$ km s^{-1}; (d) 1305 galaxies with $+4500 < V_o < +6000$ km s^{-1}; (e) 732 galaxies with $+6000 < V_o < +7500$ km s^{-1}; (f) 584 galaxies with $+7500 < V_o < +9000$ km s^{-1}; and (g) 798 galaxies with $+9000 < V_o < +12000$ km s^{-1}.

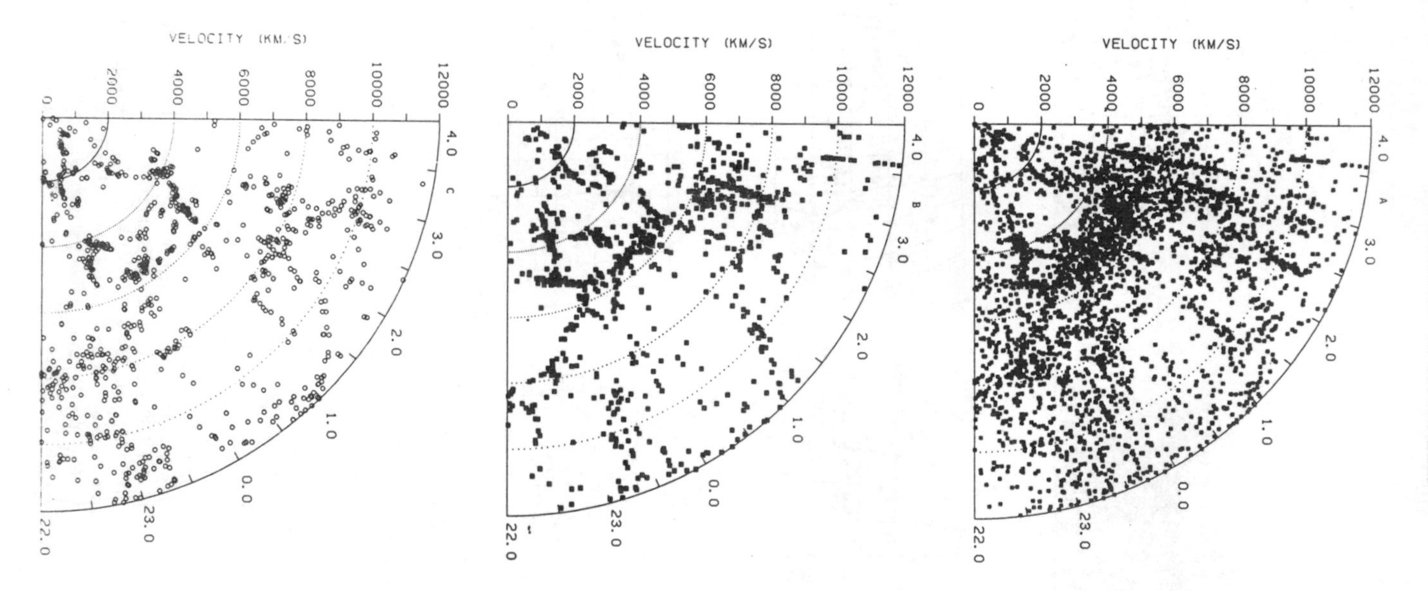

FIG. 5. Cone diagrams with right ascension as the angular coordinate for different zones of ten degrees in declination. Panel (a) shows all 4176 galaxies with $V_o < +12000$ km s^{-1} in the entire region; these are the same objects that are plotted in Fig. 4a. Panel (b) includes 1044 galaxies in the declination strip $0^o <$ Dec. $< +10^o$; (c) 937 galaxies with $+10^o <$ Dec. $< +20^o$; (d) 963 galaxies with $+20^o <$ Dec. $< +30^o$; (e) 847 galaxies with $+30^o <$ Dec. $< +40^o$; (f) $+40^o <$ Dec. $< +50^o$.

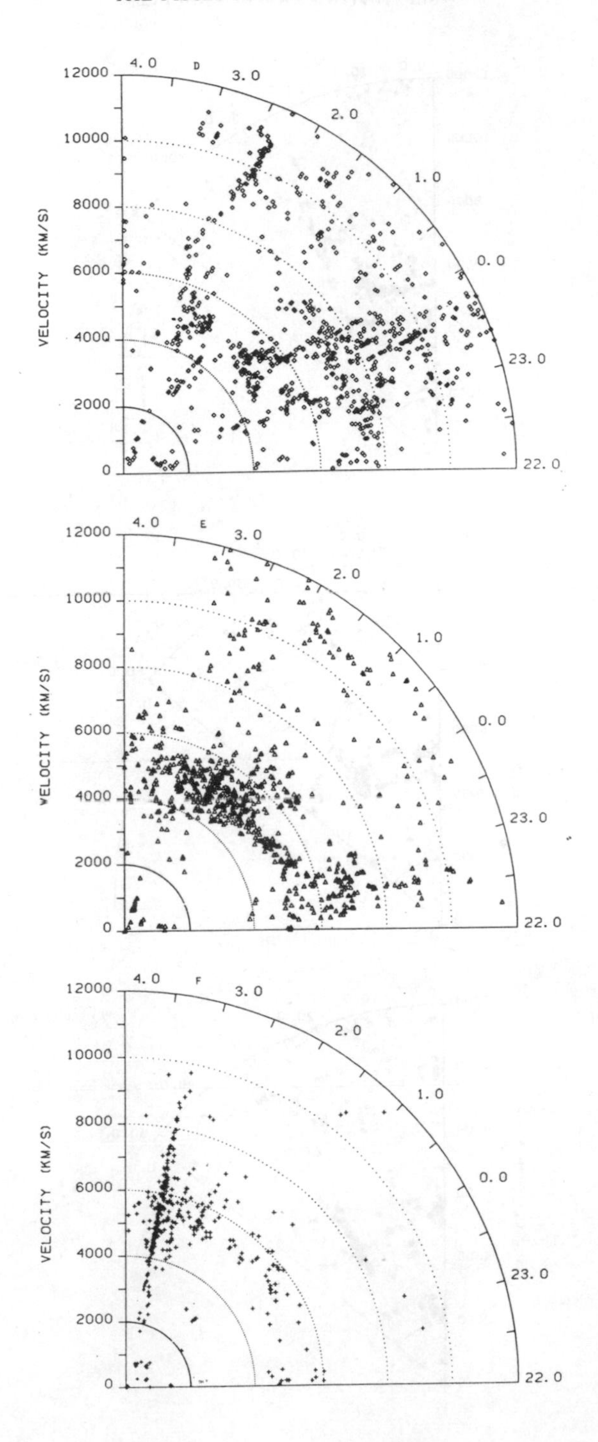
D
E
F
VELOCITY (KM/S)
12000
10000
8000
6000
4000
2000
0
4.0
3.0
2.0
1.0
0.0
23.0
22.0

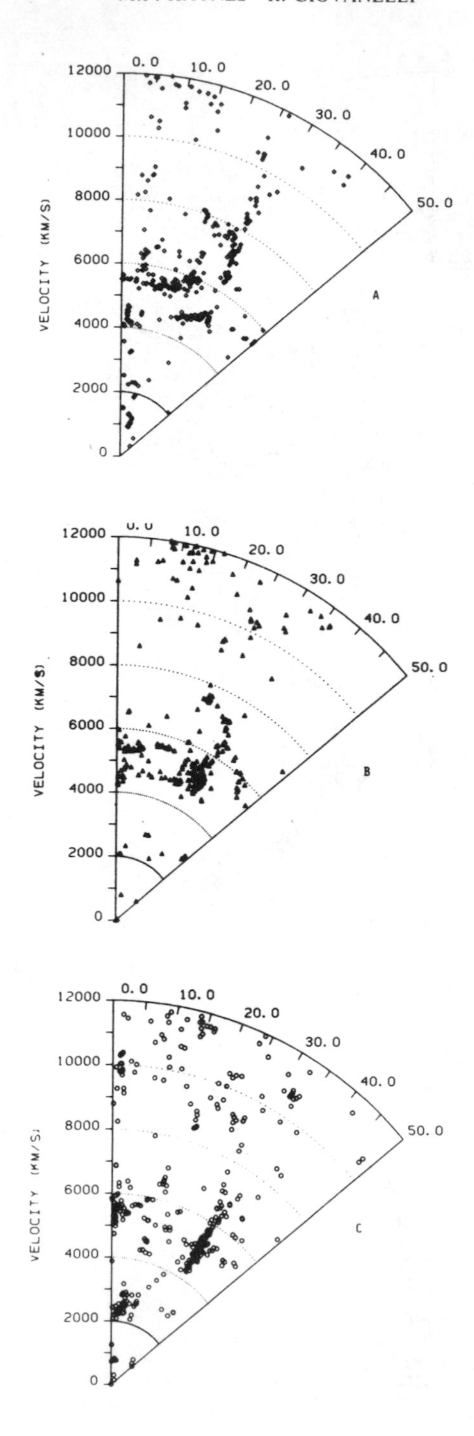

12000
10000
8000
6000
4000
2000
0
VELOCITY (KM/S)
0. 0
10. 0
20. 0
30. 0
40. 0
50. 0
A
B
C

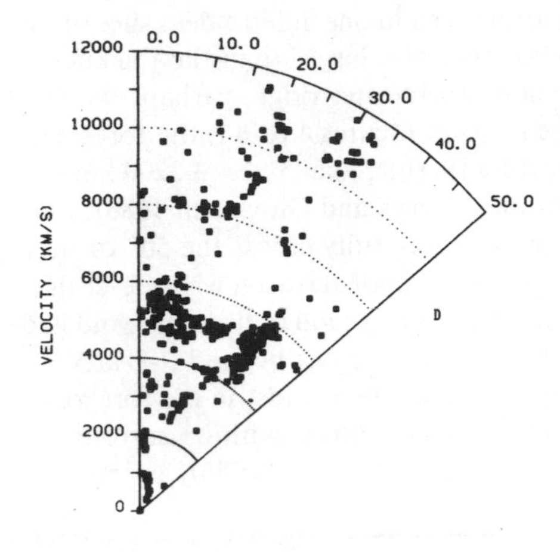

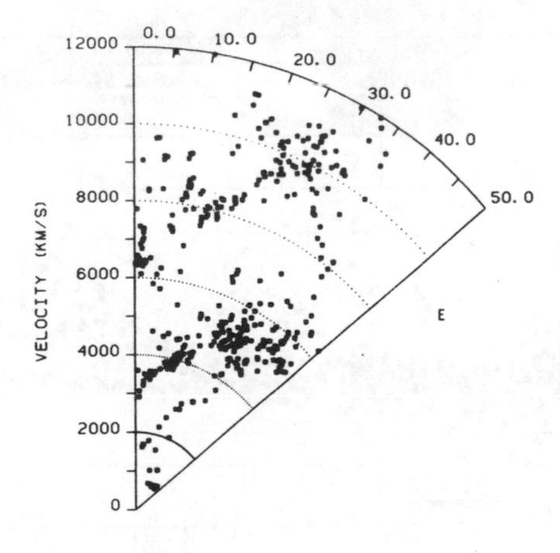

FIG. 6. Cone diagrams with declination as the angular coordinate for different 30 minute slices in right ascension. Panel (a) includes 351 galaxies with $0^h < \text{R.A.} < 0^h30^m$; (b) 301 galaxies with $0^h30^m < \text{R.A.} < 1^h$; (c) 384 galaxies with $1^h < \text{R.A.} < 1^h30^m$; (d) 402 galaxies with $1^h30^m < \text{R.A.} < 2^h$; (e) 412 galaxies with $2^h < \text{R.A.} < 2.5^h$.

along the line of sight is seen in the direction of clusters, such as the Perseus cluster in Figure 5f, in many of the other plots "strings" of galaxies that trace what appear to be narrow structures do not seem to arise from cluster dynamics. Moreover, the structure seen in one independent slice often is seen in the next, indicating coherence on scales larger than the thickness of a single wedge.

Besides the main supercluster ridge, perhaps the most prominent other feature seen in the maps in Figures 4 to 6 is the foreground void between us and P-P. This structure is centered at $V_0 = +3200$ km s^{-1} and covers at least 15° in right ascension (Haynes and Giovanelli 1986). Furthermore, it is best described as a tube extending fully across the 50° of declination (Haynes *et al.* 1988a). There is no observational reason whatsoever that galaxies of average luminosity would not have been found to lie in this void had such existed there. Note also the absence of galaxies just beyond P-P at $V_0 = +6000$ to $+7000$ km s^{-1}. The presence of these two voids, in the foreground and background of P-P, further enhances its contrast with its surroundings.

Figure 7, from Giovanelli *et al.* (1986), examines the distribution of

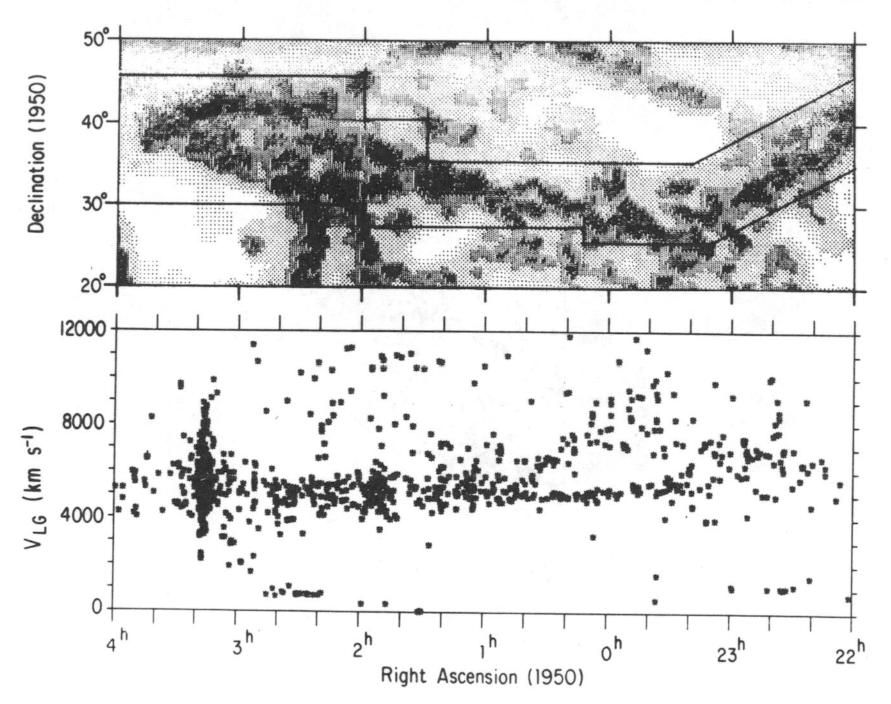

FIG. 7. Redshift distribution along main supercluster ridge. The upper panel presents a shade diagram where the shade intensity indicates the local density distribution along the ridge. In the bottom panel, individual redshift measurements are shown as a function of right ascension for galaxies within the highest density enhancement outlined in the upper panel (from Giovanelli *et al.* 1986).

measured velocities for galaxies contained within the highest density ridge of the P-P supercluster, the lizard's tail, back and head. The upper panel is a shade display in which the shade intensity is proportional to a quantity that measures the local density of galaxies. The sampled high density regime is outlined. In the lower panel the measured velocity is shown at the location in R.A. of each galaxy in the outlined region in the upper panel. It is clear that the main ridge is characterized both by a small width on the sky and depth in the radial direction. The velocity confinement into narrow lanes connecting the larger velocity dispersion clusters is striking. Because systematic velocity gradients across the ridge have not been removed in constructing Figure 7, the observed histogram at any narrow interval of right ascension would appear markedly sharper. Velocity broadening is also contributed by the large velocity dispersions found in clusters. Figure 8 attempts to show the contrast between the two portions of the supercluster, the northern, high density ridge, and the southern, more dispersed component. On a single cone diagram, with right ascension as the angular coordinate, galaxies in two adjacent declination strips are indicated by separate symbols. Galaxies found in the higher den-

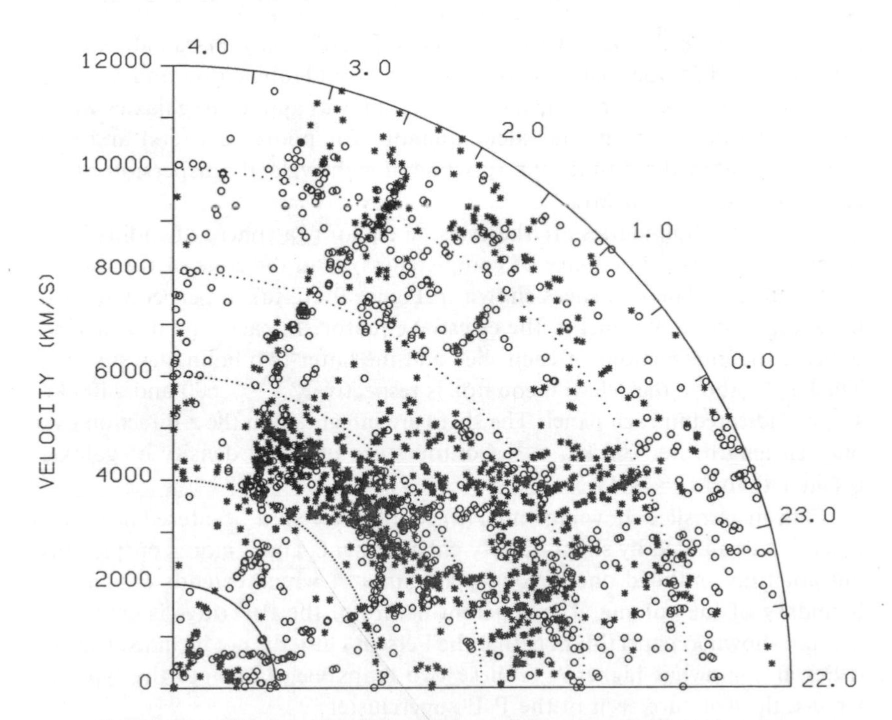

FIG. 8. Cone diagram with right ascension as the angular coordinate for galaxies in the P-P region with +25° < Dec. < +35° (asterisks) and +15° < Dec. < +25° (open circles).

sity portions to the north with $+25° < \text{Dec.} < +35°$ are indicated by asterisks, while open circles show the locations of the more southerly objects in the range $+15° < \text{Dec.} < +25°$ Notice the degree of correlation among objects in the independent samples. To varying degrees, the foreground is underdense.

In trying to describe the kinds of structures seen in the P-P supercluster and the distribution of galaxies within it, we have used the redshift sample to obtain an absolute magnitude limited subsample very nearly complete to a distance of $V_0 = +7500$ km s^{-1} and M = —19.5 over the P-P region. This subsample includes only about 20% of the total number of galaxies, but with minor corrections for incompleteness, it should well reflect the true three-dimensional distribution and hence allow us to obtain a volume density map of the supercluster as outlined by the bright galaxies. The question of whether this same picture would also be displayed by the lower luminosity objects will be deferred to Section 5.

Slices of this volume density map are presented in Figure 9, from Giovanelli and Haynes (1988). The gridding of the volume is cartesian, centered on the Earth and with cells of 200 km s^{-1} on a side. The x-axis points toward R.A. = 22^h, Dec. = 0° the y-axis toward R.A. = 4^h, Dec. = 0° and the z-axis, toward Dec. = 90°. Before the calculation of n (x, y, z), radial velocities were corrected for local large-scale deviations from Hubble flow and for virial distortions within clusters. The latter correction was applied to galaxies within one Abell radius (or an equivalent quantity for poorer clusters) and with velocities within three times the rms line-of-sight velocity dispersion of the cluster's systemic redshift.

The outer boundaries are the celestial equator, a sphere of radius V_0 = +7500 km s^{-1} (the boundary of completeness), and the cone of the Dec. = +45° surface. The two slices shown in Figure 9 are taken perpendicular to the z-axis, that is, parallel to the celestial equator. In each, the dashed lines indicate the intersections of each slice and the latter two boundary surfaces. The height above the celestial equator is respectively V_z = 600 and 2400 km s^{-1}, as indicated in each panel. The slices are integrated in the z-direction over one cell length, i.e. 200 km s^{-1}. Contours are of equal density of galaxies brighter than M = —19.5.

The nearer slice shown in panel (a) contains a "void" centered near 3000 km s^{-1}, surrounded by slight density enhancements and a more conspicuous, but not fully outlined one beyond 6000 km s^{-1}, which extends outside the boundary of the volume. The coherent nature of the P-P ridge is evident in the slice shown in panel (b); note that the Perseus cluster is not included because it lies at somewhat higher V_z. These two maps well illustrate the types of large-scale structure seen in the P-P supercluster.

Such representations allow one to sample the types of structures that can be identified in this region after removing the "polar bias" that affects Earth-

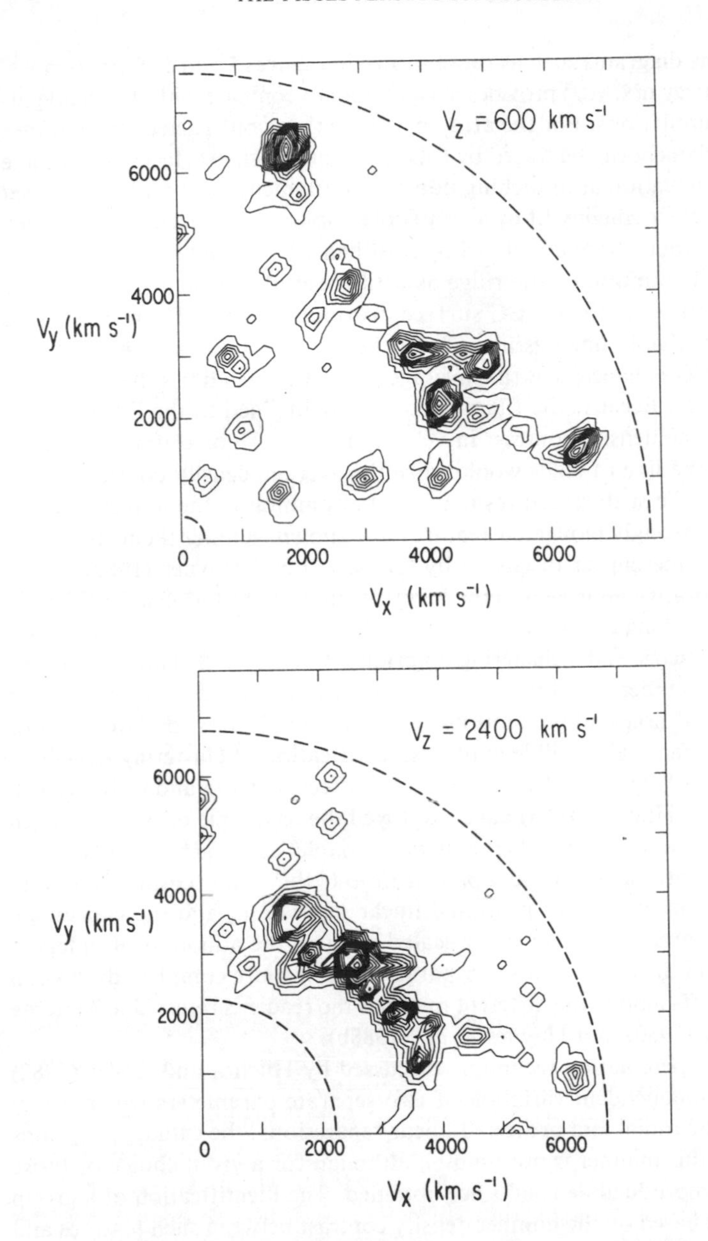

FIG. 9. Volume density contours in two slices across the density array constructed for the absolute magnitude limited subsample. Both are parallel to the the (V_x, V_y) plane (the celestial equator). The outer dashed curve is the intersection of the slice with the sphere of radial velocity V_o = +7500 km s^{-1}; the inner dashed curve is the intersection with the cone of the Dec. = +45° surface.

centered cone diagrams such as those shown in Figures 5 and 6. Furthermore, the density array n(x, y, z) provides a useful tool to compare with the analogous numerical simulations. For the array n(x, y, z) thus constructed, the outlines of the P-P filament can be traced by following equidensity surfaces. With some subjective truncation at branching points, the filament can best be described as a single surface roughly fit by a very prolate spheroid with a major to minor axis ratio of about 10 and inclined by less than 12° to the line of sight. This topological description of the ridge as a filament also explains further why it is so easy to see in the CGCG surface distribution maps. Even so, its true contrast is probably underestimated. In these maps, the main ridge of P-P shows a density enhancement that averages more than ten times the mean density of galaxies. Because the filament is slightly inclined to the line of sight, smearing of the density contrast in velocity results. A correction for the inclination to the line of sight would raise the average density contrast closer to twenty. While such structures may not be common in the universe, other examples of strongly elongated features are known, among them the Lynx-Ursa Major supercluster discussed by Giovanelli and Haynes (1982).

A quantitative approach to the identification of clustering can make use of numerical techniques to distinguish units that are chosen based on the definition of some metric and a clustering criterion (Materne 1978, Tully 1980). The simplest three-dimensional metric is the redshift distance $d = V_o/H_o$, but in the presence of non-Hubble motions, especially in clusters, the adoption of this metric at face value will lead to misidentification of clustering elongated along the line of sight, the "finger of God" effect. Huchra and Geller (1982) and Geller and Huchra (1983) have employed the technique of percolation in which the separation of objects is a binary variable, either greater than or less than some cutoff value. For any pair of objects the mean redshift is used in converting from angular to projected linear separation, and the coordinate in the radial direction is separately scaled through the adoption of independent cutoffs in the radial and sky separations. For more complete discussion of the trade-offs among the different methods the reader is referred to Materne (1978), Tully (1980) and Haynes *et al.* (1988b).

Since the percolation technique discussed by Huchra and Geller (1982) involves the independent variation of two separate parameters which affect the metric, the radial and projected linear separations, the catalog of groups produced in this manner is not unique, although for a given choice of those parameters, reproducibile results are obtained. The identification of a group is established based on the number density contrast between such features and the mean density of the sample under study. As defined by equation (4) of Huchra and Geller (1982), the contrast in number density is a quantitative estimate of the density enhancement relative to the mean and is governed by the choice of cutoff in the projected separation and the assumed luminosity function, since the cutoff is scaled to compensate for the fact that the sampling of the luminosity function varies with increasing redshift (the Malmquist bias). Our choice of parameters for the luminosity function, reference redshifts, and cutoffs are based on fits to the present sample for the luminosity function and trials of several sets of choices (e.g. Haynes *et al.* 1988b).

As a preliminary test of the percolation technique, we have applied the technique of Huchra and Geller (1982) to a section of the P-P region where the survey coverage is more complete and less affected by galactic extinction: 0^h < R.A. < 03^h, $+10° <$ Dec. $< +35°$. In that subregion, the redshift sample contains 1162 galaxies with recessional velocity less than +12000 km s^{-1}, the value chosen as the cutoff redshift. At present, we impose a magnitude limit for the complete sample of +14.9 over this range in order to correct properly for the Malmquist bias. This restriction reduces the number of galaxies used in the percolation to 425.

Even with the limitations of this sample, application of the percolation technique easily identifies 51 overdensities with a contrast of 20 or greater and containing at least two members; only 77 of the galaxies are found in regions less dense. At the present stage of our analysis, we are most interested in the properties of groups within the P-P supercluster, that is restricted to the range of recessional velocity $+4000$ km $s^{-1} < V_0 < +6500$ km s^{-1}. The ten groups that are likely supercluster members and that contain more than five members as identified by this process are listed in Table 1. Included in the table are the number of galaxies, the coordinates ⟨R.A.⟩ and ⟨Dec.⟩ and the mean velocity $\langle V_0 \rangle$ of the group centroid, the line-of-sight velocity dispersion σ, and hl, the maximum projected separation of any pair in Mpc. These ten aggregates contain masses implied by the virial theorem from 10^{13} to 10^{15} $M_\odot$ and have mass-to-light ratios from 90 to 1500.

TABLE 1

GROUP PROPERTIES FOUND BY PERCOLATION

Group	N	<R.A.> hhmm	<Dec.> sddmm	$\langle V_0 \rangle$ km s^{-1}	σ km s^{-1}	hl Mpc
1	6	0006	+3232	4557	108	2.9
2	5	0013	+1730	4737	528	4.4
3	11	0017	+2944	5146	1076	6.5
4	12	0020	+2313	5792	1418	6.4
5	31	0044	+3045	5027	736	11.
6	66	0117	+3250	4767	634	12.
7	11	0146	+1131	4928	275	4.1
8	8	0157	+2447	4721	109	3.8
9	41	0216	+3202	4649	437	13.
10	12	0229	+2109	4279	623	6.8

Figure 10 shows the distribution of mean velocities $\langle V_0 \rangle$ for all groups ($N > 2$) at the P-P redshift, in the northern portion of this restricted region $+20° < \text{Dec.} < +35°$. The number of galaxies included in each group is indicated by the size of the circle; the groups with larger numbers of galaxies are included in Table 1. Note again, the narrow range of redshift characteristic of the group found in this region. A relaxation of the contrast level from 20 to 10 will permit almost the entire sample to percolate into a single structure.

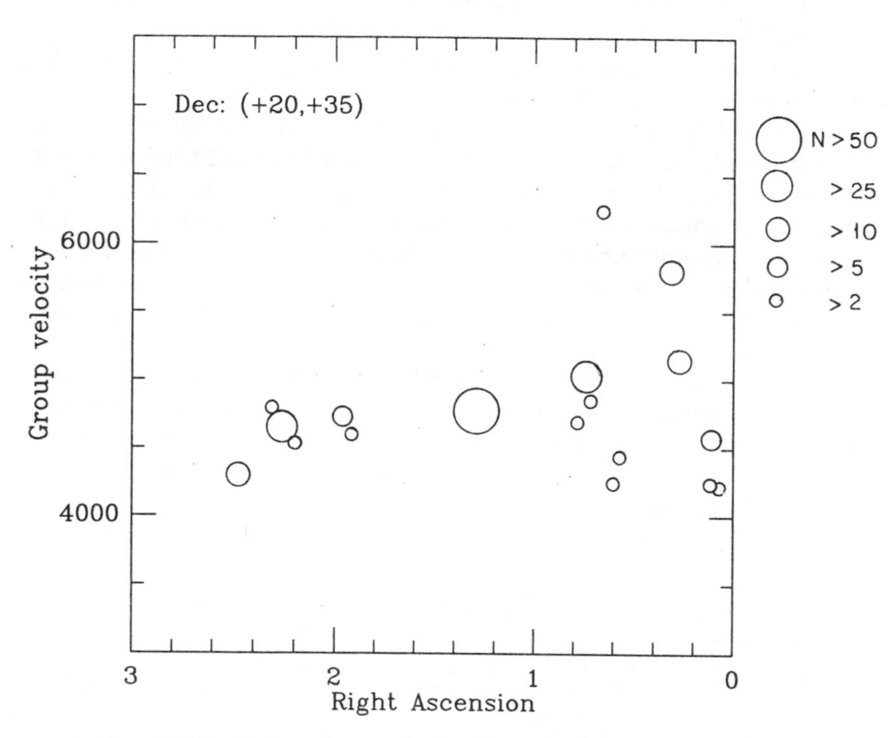

FIG. 10. Distribution of mean velocities for the 18 groups with $+4000 < V_0 < +6500$ km s^{-1} found to lie between $0^h < \text{R.A.} < 3^h$, $+20° < \text{Dec.} < +35°$. The number of galaxies included in each group is indicated by the size of the circle.

Much additional work in the application of numerical algorithms for the recognition of three-dimensional structure needs to be done. Completion of the sample to a faint limiting magnitude will aid the search for larger-scale structures because identification of the largest connective structures by such numerical techniques requires sampling in the low density regions. Just as the brightest galaxies outline the main ridge in Figure 1b, it takes a sample complete to a fainter magnitude to delineate the true extent of the supercluster from Pegasus to Perseus, because most of the links are of lower contrast than the groups and clusters they connect.

5. Segregation of Galaxy Properties

Segregation of morphological types in clusters of galaxies was first noticed over a half century ago (Hubble and Humason 1931). More recently, numerous authors (Oemler 1974; Melnick and Sargent 1977; Dressler 1980) have found that the relative population of ellipticals, lenticulars, and spirals in clusters and their peripheries is mainly a function of the local galaxian density. While some 80% of field galaxies are spirals, as few as 15% of the galaxies in condensed clusters like Coma show spiral structure. In his quantitative study of 55 rich clusters, Dressler found that the morphology-density relation appeared to be independent of cluster morphology and degree of central concentration.

The monotonic variation in the population fraction with local density found by Dressler extends in a uniform manner from the high-density cluster cores to the low density regimes of typical groups. Nearby loose groups are almost always dominated by spirals. Excluding the Virgo region, only two groups in the list of de Vaucouleurs (1975) contain predominantly bright early-type galaxies, and in both, a bright X-ray galaxy sits at the center of a higher density core surrounded by a more diffuse cloud of spirals. The more quantitative treatment of galaxy groups undertaken by Postman and Geller (1984) has shown that Dressler's morphology-density relation extends smoothly over six orders of magnitude in space density. Only at very low densities where the dynamical timescale is comparable to or greater than the Hubble time does the population fraction not reflect variations in the local galaxy density.

As established by Dressler the variation in population fraction with local density is monotonic but slow. In the highest density regions mechanisms that lead to alteration of a galaxy's morphology, such as galaxy-galaxy and galaxy-intracluster gas interactions, are believed to be relatively efficient. The known segregation of ellipticals to high density regions raises the question of whether the observed differentiation in morphology is inbred during the era of galaxy formation or shortly thereafter, or whether it is a continuing process. The scale over which segregation occurs and the degree of continuity of the variation in population fraction throughout all regimes of density will constrain models predicting the relative formation times of galaxies and their surrounding large-scale structure. Because its members are found in a wide variety of local densities, P-P will serve as a useful laboratory for studying the segregation of morphologies and other galaxian properties.

The most comonly-used statistic employed to quantify the degree and scale of clustering is the two-point spatial correlation function $\xi(r)$ and its angular form w (θ). Both are normally fit by power law forms so that the angular correlation function can be written in terms of an amplitude A and slope β, $w(\theta) = A\,\theta^{\beta}$. Using the UGC which contains the largest homogeneous compilation of galaxies including their morphological classification, Davis and Geller (1976) have compared $w(\theta)$ computed for galaxies brighter than $m = +14.5$ separately for different morphologies, and have found that the clustering properties for each type subsample are, in fact, different. Elliptical-elliptical clustering is characterized by a power law with a slope steeper that that appropriate for spiral-spiral clustering. The case for S0-S0 clustering is intermediate. For their sample, the number of galaxies found within 1 Mpc of a random elliptical galaxy is about twice that found within the same distance of a random spiral.

Since we know from Section 3 that the majority of galaxies contained in the P-P region are in fact members of the supercluster, we can gain a visual impression of the degree of morphological segregation simply by plotting separately the sky distribution of UGC galaxies of different morphological classes as was done in Figure 6 of Giovanelli *et al.* (1986). Here, we make use of the supercluster's distance. Figure 11 shows three panels which illustrate the sky distribution of different morphologies in the current redshift sample restricted to the velocity range $=3500$ km s^{-1} $< V_0 <$ $+7000$ km s^{-1}. The upper panel shows the distribution of all galaxies with available morphological types. The lower two panels display respectively only the elliptical and S0 galaxies (middle) and only the Sbc and Sc galaxies (bottom). Notice that the number of objects in the bottom two panels is nearly identical. Galaxies of progressively earlier morphology are segregated to the higher density regimes; the supercluster ridge is most evident in the distribution of elliptical and S0 galaxies. The visual impression of large-scale structure conveyed by the later-type spirals is quite different from that suggested by the early-type objects.

The angular correlation function $w(\theta)$ can also be used on the P-P sample. In dealing with an observational sample such as that compiled for Pisces-Perseus, the calculation of $w(\theta)$ must take into account the serious sample biases. Not only does the sample itself have observational constraints and selection effects, but the variability of the galactic extinction over such a large region adds to unevenness of the sampling. In practice, the correlation function must be calculated as the excess probability of finding pairs of galaxies of a given separation in the sample catalog as compared to that found in an identically-selected catalog containing a random distribution of objects. The operational definition of the angular correlation function becomes.

$$w(\theta) = \frac{N_g(\theta)}{N_r(\theta)} - 1 \qquad (7)$$

where $N_g(\theta)$ is the number of pairs in the observed sample with separations in the range $(\theta - d\theta, \theta + d\theta)$ and $N_r(\theta)$ is the number of pairs with angular

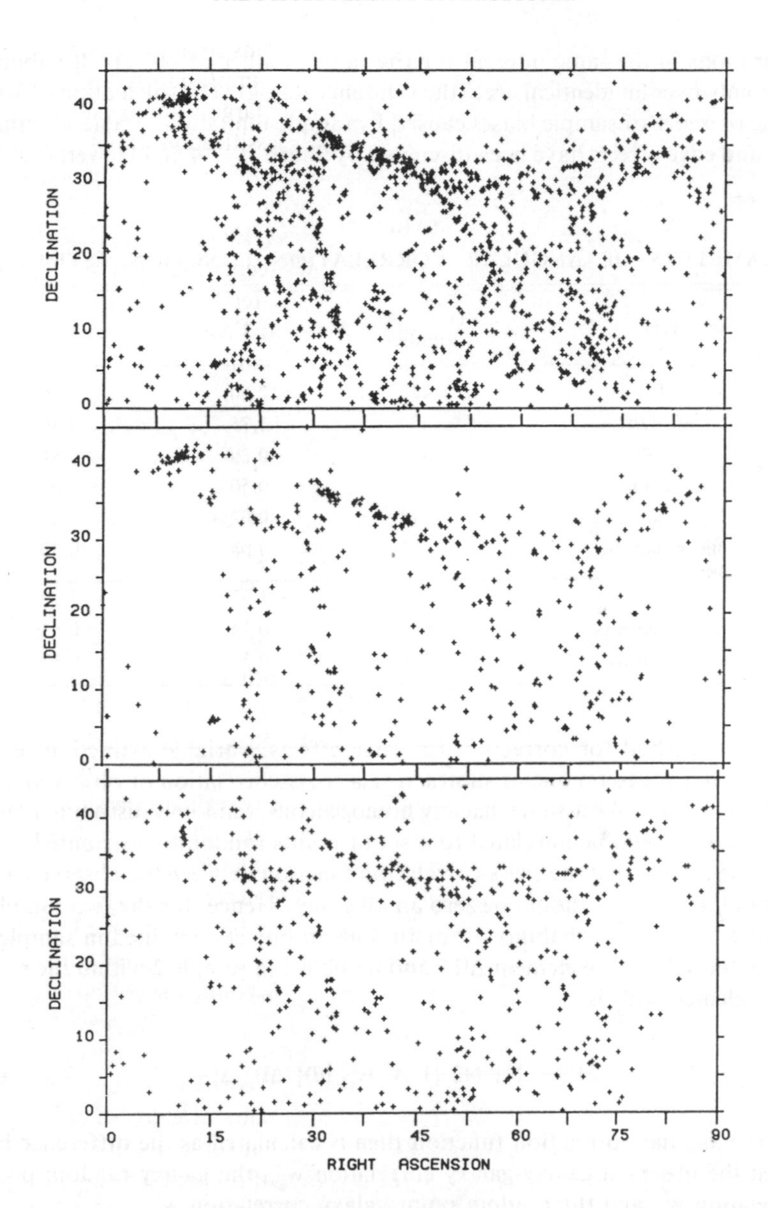

FIG. 11. Sky distribution of galaxies of separate morphological classes in the current redshift catalog restricted to the velocity range $+3500 < V_0 < +7000$ km s^{-1}. Classification has been performed for CGCG objects on the Palomar Sky Survey prints following the UGC system. The upper panel shows the distribution of all objects within those redshift limits. Elliptical and S0 galaxies are included in the middle panel while the bottom panel contains Sbc and Sc galaxies.

separations in the same interval for the same number of objects distributed randomly over an identical area (the random catalog). The calculation of w(θ) in the presence of sample biases caused by sample limitations, galactic extinction and edge effects have been discussed by Sharp (1979) and Hewett (1982).

TABLE 2
PARAMETERS OF ANGULAR CORRELATION FUNCTION ESTIMATES

type	$w(\Theta) = A\Theta^{\beta}$ number	A	β
E	227	2.60	−1.06
S0, S0a	423	1.75	−0.84
Sa, Sab	312	0.29	−0.81
Sb, Sbc	566	0.50	−0.63
Sc	678	0.62	−0.47
later than Sc	681	0.58	−0.30
early	725	1.92	−0.90
early spirals	689	0.74	−0.65
late spirals	1548	0.54	−0.37

Hewett's method for correction for edge effects, variable extinction, etc., follows Sharp's technique of subtracting a cross-correlation of observed and random samples. We assume that any homogeneous, randomly distributed sample of data should be unrelated to a set of points randomly distributed over the same area so that the cross-correlation function between the observed and the random samples should be zero an all scales. Hence, for the two samples 1 and 2, the joint probability δP of finding an object contained in sample 1 within the solid angle element $\Delta\Omega_1$, and an object in sample 2 within the solid angle element $\Delta\Omega_2$ is

$$\delta P = N1\ N2\ [1 + w_{12}(\theta)]\ \Delta\Omega_1\ \Delta\Omega_2 \quad (8)$$

and the angular correlation function then is calculated as the difference between the observed galaxy-galaxy correlation w_{gg}, the galaxy-random point correlation w_{gr} and the random point-galaxy correlation w_{rg}

$$w(\theta) = w_{gg}(\theta) - w_{gr}(\theta) - w_{rg}(\theta). \quad (9)$$

Since random points should always be randomly distributed around any galaxy when summed, the galaxy-random point correlation should always equal zero.

On the other hand, since galaxies are not necessarily randomly distributed with respect to random points, $w_{rg}(\theta)$ is not always zero, and is partially determined by the location of the galaxies relative to the sample boundaries.

The application of Hewett's method to the P-P region was made by Giovanelli *et al.* (1986) whose results for A and β are summarized in Table 2 and Figure 12. Figure 12 shows the calculated correlation function for three subsets of galaxy morphology. In addition to the global variation of clustering characteristics among ellipticals, S0's and spirals noted by Davis and Geller (1976), the analysis of galaxies in the P-P region shows the spiral classes themselves show a smooth variation: spirals of type Sa and Sab tend to cluster on smaller angular scales than do later spirals. The correlation analysis thus quantifies the visual impression of segregation evident in Figure 11.

Similar to the morphology-density relation of Dressler (1980), the population fraction can be calculated over bins of projected local density. As illustrated in Figure 12 of Giovanelli *et al.* (1986) a progressive change is seen in the slope of the variation of population fraction with density, not only from early to late morphologies as reported previously. The trend toward segregation observed in the cores of rich clusters and their peripheries appears to continue monotonically to the regions of lowest population fraction. In fact, depending on the choice of morphological class, a gradient in the population fraction can be observed at nearly every value of local density. Of perhaps equal importance, the progressive variation in the population fraction, as functions of both density and type, is striking not just within the major subgroupings but within the spiral sequence itself. The smooth change in the observed population fraction among the spiral types implies that the conditions that lead to the currently observed morphological segregation arise to a large extent from the matter density at the time of galaxy formation, or at least shortly thereafter. Diffusion of galaxies formed in different density environments is not viable over these scales within a Hubble time.

Morphology may not be the only parameter that shows segregation according to density. The study of the distribution of dwarf galaxies may be an important link to the understanding of morphological segregation and the likelihood of biased galaxy formation. Sharp, Jones and Jones (1978) have examined the distribution of DDO dwarf galaxies relative to all galaxies found in the CGCG. Those authors concluded that dwarfs do in fact tend to avoid clusters, but otherwise are found preferentially in the vicinity of bright galaxies and are intrinsically different from ordinary galaxies of the same apparent magnitude.

An additional approach examines the clustering properties in terms of the galaxies' surface brightness characteristics, after allowing for the morphological dependence on clustering. Both Davis and Djorgovski (1985) and Bothun *et al.* (1986) have considered the clustering of galaxies as a function of surface brightness. On the one hand, Davis and Djorgovski concluded that

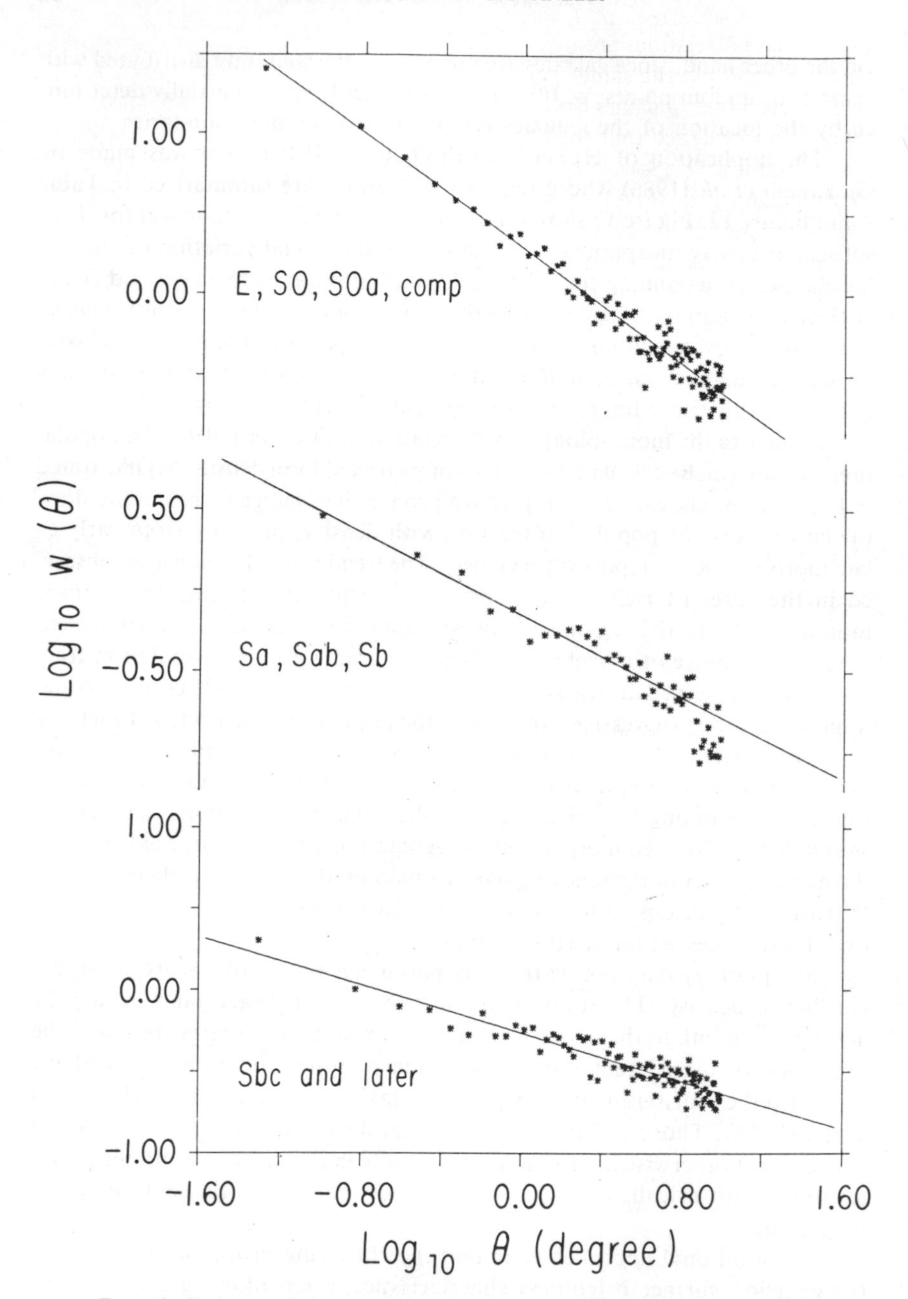

FIG. 12. Estimate of the two-point angular correlation function $w(\theta)$ versus θ, on logarithmic scales, for the subgroupings of morphological types indicated. Parameters of the best-fit power laws are listed in Table 1 (from Giovanelli *et al.* 1986).

the angular correlations are weaker and shallower for the low surface brightness (LSB) objects. At the same time, Bothun *et al.* contend that while the higher surface brightness galaxies have a smaller spatial clustering scale than do LSB's, there is no evidence that the redshift voids are filled by LSB objects.

If the distribution of light in the universe is a good tracer of the mass distribution, then the spatial correlation function $\xi(r)$ should be the same for both giant and dwarf galaxies. Environmental influences may result in a reinforcement of the clustering of ellipticals and S0's with respect to spirals as discussed above. But another critical key to the understanding of galaxy formation is held by the distribution of galaxy luminosities - the luminosity function. The recent study of the luminosity function of galaxies in the Virgo cluster by Binggeli *et al.* (1985) has proved the difference in the shapes of the luminosity function derived for different morphologies. Those authors have found that the luminosity distribution is different for ellipticals, S0's, and spirals and among the Sa/Sb, Sc, and Sd/Sm galaxies as well. The universality of the luminosity function is commonly assumed but is not well established. In fact, there is growing evidence that the galaxy luminosity function varies both with morphological type and local density. Future studies will address the question of whether this variation in the luminosity function with local environment results from different mixing of morphological types or from true luminosity segregation.

From the density array constructed for our volume - and absolute magnitude limited sample, we can estimate the dependence of galaxian properties on the value of the local density n. The value of n associated with each galaxy is that obtained for the occupied cell dxdydz. In order to compare the distribution of objects of different luminosities or in different volumes, we must construct density arrays as before but now complete to somewhat different depths in absolute magnitude. It should be noted that the densities obtained from subsamples are not directly commensurable because they have been constructed over different luminosity depths.

Figure 13 shows the contrast between the actual number of galaxies found within a given redshift shell and the number expected from a uniform density distribution for three subsamples. The upper curve, marked "filament", contains galaxies within the solid angle subtended by the density enhancement represented by the main P-P ridge. As noted previously, the observed density enhancement is a factor of ten over the mean, but the true contrast is probably even greater. A slight underdensity is present in the foreground, related to the obvious foreground void.

Similary, from the density array, we can outline the solid angle subtended by the foreground void and can compare galaxy counts as observed with those expected for a homogeneous population. The lower two curves in Figure 13 illustrate the results for the void region, plotting separately the "bright" ($M < 19.5$) galaxies and "faint" ($M > 19.5$) galaxies. The latter galaxies

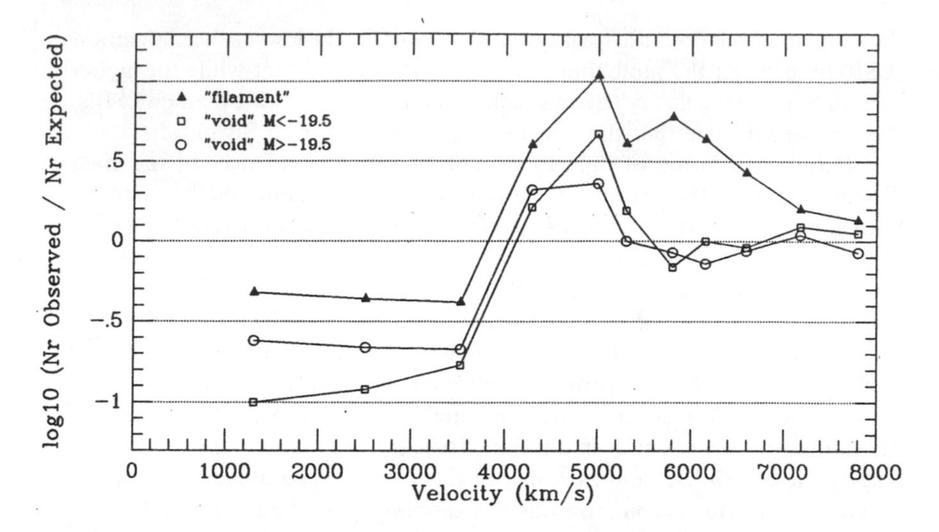

FIG. 13. Density contrast profiles showing the distribution relative to the mean of galaxies in several subunits of the P-P region. Each point corresponds to the ratio of the observed number of objects to that expected for a random distribution of galaxies within a shell in redshift centered on the point's abscissa. The "filament" line refers to the solid angle occupied by the P-P ridge. The two "void" curves show the results for two different bins of absolute magnitude in the region surrounding the void (from Giovanelli and Haynes 1988).

include all objects to the catalog limit within a given redshift shell. The partial overlap of solid angles subtended by the void and the filament is responsible for the overdensity near $+5000$ km s^{-1}. Figure 13 shows that the mean underdensity of the void is at least a factor of five. In agreement with Bothun *et al.* (1986) we also see that bright and faint galaxies outline the same general structures, but the density contrast is better outlined by the bright galaxies than by the faint ones.

Based on this analysis Giovanelli and Haynes (1988) conclude that on scales larger than cluster size, density contrasts within the galaxy distribution between high and low density regions are observed to be as large as a factor of 100, and that delineation of large-scale structure and the degree of density enhancement appear to be dependent on the luminosity of the galaxies counted.

The determination of a luminosity dependence on any parameter that may be related to distance is threatened by a Malmquist bias. Because the linear scale of the density inhomogeneities is comparable to that of the volume sampled, high and low density regions are not well-mixed, so that the median distances of cells of low and high density are not the same. As a result, a plot of luminosity versus density that included all of the galaxies in the sample would exhibit a Malmquist bias in the sense that higher luminosities would be obtained for the densities with the higher median distance. In our sampled volume, the higher density regions tend to be found in the P-P supercluster and lie somewhat farther from us than do the low density regions such as the foreground void. In order to avoid the bias introduced by this difference, we have analyzed separately different intervals of redshift, requiring that within each interval the high and low density cells must have similar median redshifts. Figure 14, from Giovanelli and Haynes (1988), shows the results for two redshifts intervals. The

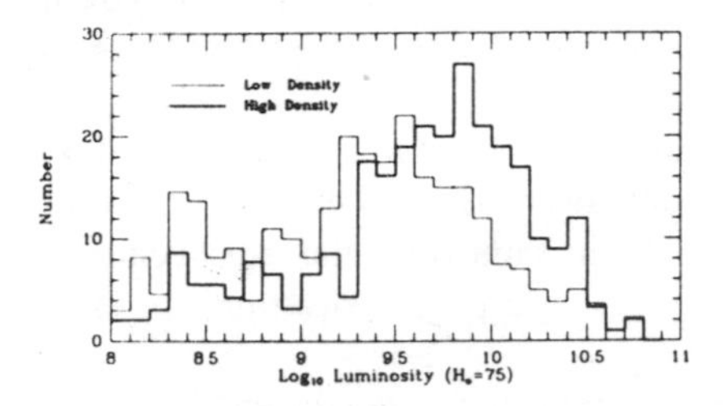

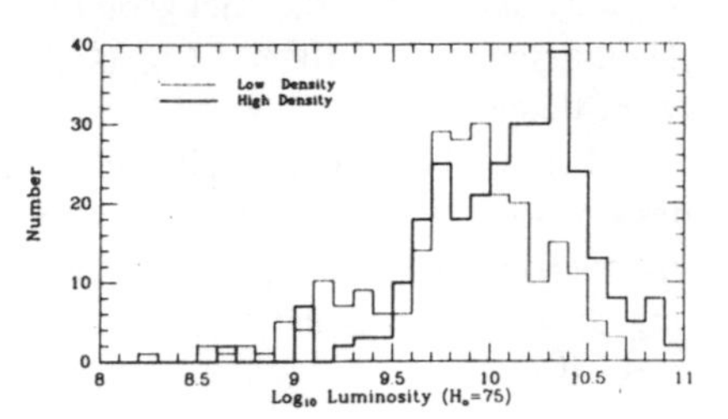

FIG. 14. Luminosity distribution, in solar units, of galaxies in high and low density regions for galaxies in separate regimes of redshift. The left panel shows galaxies with velocities $0 < V_0 < +4000$ km s^{-1}. Space densities were derived from a subsample nearly complete to M = — 18.4. The right panel includes galaxies with $+4000 < V_0 < +6200$ km s^{-1}, and the subsample is nearly complete to M = — 19.0 (from Giovanelli and Haynes 1988).

left panel (a) shows histograms of luminosity for galaxies in the redshift range $0 < V_0 < +4000$ km s^{-1}, separately for galaxies in high and low density regions; the right panel (b) shows similar histograms for galaxies with $+4000 < V_0 < +6200$ km s^{-1}. In order to optimize statistics, we have adopted different cutoffs in the absolute magnitude in defining density in the two redshift regimes. In the nearer, density is defined by the subsample of galaxies brighter than —18.4, while in the more distant sample, all galaxies brighter than —19.0 are included. In both instances, there are relatively fewer high luminosity galaxies in the low density regions. Although the present incompleteness demands some caution, the opposite appears to be true for the low luminosity galaxies. A similar analysis using surface brightness leads to the conclusion that galaxies in high density regions, which as shown in Figure 14 tend to be brighter, also have a higher surface magnitude than galaxies in low density volumes.

Some currently popular theories of galaxy formation require that the overall mass distribution must be significantly smoother than that traced by the luminous matter. One way to meet these conditions is to postulate the formation of luminous galaxies in a biased process (e.g. Dekel and Rees 1987). This bias would translate precisely to a spatial segregation of the low luminosity, low surface brightness galaxies from the bright, massive ones so that the low luminosity galaxies would fill the voids which themselves were outlined only by the bright objects. As such, the hole in the foreground of P-P mentioned in Section 3 provides a suitable location to search for the "missing" dwarfs. However, Oemler (1987) reported that a search by Eder, Schombert, Dekel, and Oemler for dwarf galaxies found no galaxies in the void. Furthermore, the most commonly-used total-power (beam switching) mode of extragalactic 21 cm line observing samples a range of redshift in each observation and in each position (ON- and OFF-source). For all of the directions in which galaxies were detected in the background of the void, no void members with typical HI masses of $4 \times 10^7 \, h^{-2} \, M_\odot$ (for an assumed velocity width of 30 km s^{-1}) were detected.

From all of these studies, we conclude that neither normal bright galaxies nor low surface brightness galaxies, nor dwarf-HI-rich galaxies, as commonly found elsewhere, inhabit the voids. At the same time, the typical clustering scales are smaller for high luminosity earlier type galaxies than for later type objects.

There are still many questions that remain unanswered (and probably more, unasked). The relationship of the ridge to the gravitational potential and conditions during the formation epoch remain to be explored, and the possible coherence of structure in Perseus and the surrounding supercluster should be checked. Strom and Strom (1978) showed that Perseus cluster ellipticals are aligned so that their major axes preferentially follow the main cluster chain at a position of 70° - 80°. On larger scales, Gregory *et al.* (1981) found evidence for alignment and perpendicularity within the supercluster ridge. The range of local densities represented by the P-P supercluster implies that the environmental influences on an individual galaxy, both past and present, have been dramatically different from one location to another. P-P will continue to be a prime site for the study of the mechanisms that might produce the observed segregation.

6. Connections to Other Structure

One of the most striking aspects of the sky distribution of galaxies in the P-P region is the location of the most massive dynamical unit, the Perseus cluster, at the northeasternmost end of the main density enhancement, as seen, for example, in Figure 2. To the north and east of Perseus, the galactic obscuration seriously limits tracing the large-scale structure and few galaxies are included in the CGCG or UGC. Kent and Sargent (1983) found an average visual extinction of 0.6 magnitudes in the direction of A426 itself, and the extinction quickly exceeds one magnitude in the region east of R.A. = 3^h. After allowing for the obscuration, Hauschildt (1987) has traced the supercluster as far east as the 3C129 cluster at R.A. = 4^h45^m, Dec. = + 45°.

After an analysis of the distribution of redshifts of radio galaxies in the region of the zone of avoidance, Burns and Owen (1979) suggested that the supercluster might extend across the obscured region toward A569, located at R.A. = $07^h\ 05^m$, Dec. = $+48°\ 43'$, and a redshift $V_o = +5800$ km s^{-1} (Struble and Rood 1987). Giovanelli and Haynes (1982) identified a low-density enhancement in Lynx-Ursa Major at about the same redshift as P-P but containing no high density clusters; they suggested that this low contrast filament could be the eastern extension. Focardi *et al.* (1984) noted the presence of an excess of galaxies behind the galactic plane detected in the red survey of galaxies within two degrees of the galactic equator compiled by Weinberger (1980). Focardi *et al.* (1984) concluded that P-P does indeed extend across the zone of avoidance and connects to the cloud around A569 rather than the Lynx-Ursa Major supercluster. However, the survey of the latter authors actually misses much of the Lynx-Ursa Major supercluster and hence their arguments are not applicable. With the data currently available outside the P-P region the redshift structure cannot be well-enough understood. The Perseus cluster itself is characterized by a systemic velocity somewhat inbetween the Lynx-Ursa Major supercluster and A569, which is itself complex, and it is unknown whether P-P continues across the zone of avoidance as a single feature in redshift or branches out somewhere in the region of obscuration (Hauschildt 1987; Haynes *et al.* 1988a).

The effects of galactic obscuration are a hindrance both to optical and to radio observers; for practical reasons, the latter must know where to point the radio telescope. The existence of galaxies invisible to optical instruments but detectable in the 21 cm line even within a few degrees of the galactic plane has been demonstrated by Kerr and Henning (1987). However, undertaking such a blind search at 21 cm over a wide area suspected to be part of the Perseus supercluster would be extremely time-consuming. The total power mode of observation, typically employed in extragalactic 21 cm line studies, always samples two points on the sky (one "on" source, and another reference "off" source position); typical redshift searches cover frequency ranges corresponding to at least 6000 km s^{-1}. In the survey of Haynes *et al.* (1988a), fewer than one percent of observations reveal another galaxy at any redshift in the off-source spectrum.

Spiral galaxies in the Perseus supercluster whose optical images are hidden in the zone of avoidance may be among the point sources detected by the Infrared Astronomical Satellite IRAS. Dow *et al.* (1988) have discussed the feasibility of using far infrared color criteria to find galaxies close to the galactic plane. Haynes *et al.* (1988a) have obtained 21 cm line spectra for a small sample of galaxies detected by IRAS close to the zone of avoidance, in order to demonstrate the potential for using this method to trace the P-P supercluster to the east beyond Perseus. Future 21 cm observations are planned to search for supercluster members among the point sources detected by IRAS.

The current limitations on the number of available redshifts preclude the compilation of statistically complete samples, and we must confess that it has been frustrating to try to analyze the redshift distribution in this region. However, it seems so unlikely that a supercluster structure as rich as P-P would end so abruptly at the Perseus cluster, that we feel confident that there must be many galaxies and perhaps clusters behind the zone of avoidance. We hope so, because we have already begun the hunt to find them.

Haynes and Giovanelli (1986) have underscored the existence of a well-defined connection between P-P and the Local Supercluster, apparent in Figure 5b in the form of two filamentary-like features which also outline the foreground void discussed earlier. Other authors have employed various clustering techniques to attempt to trace structure on even larger scales, not by tracing the galaxies but by treating the clusters as individual points. Batuski and Burns (1985) have used a percolation analysis to examine the distribution of Abell clusters with $m_{10} < 16.5$ and medium-distant CGCG clusters in a large region of the south galactic cap that overlaps the P-P region. Those authors find evidence that the P-P supercluster is part of a possible filament of galaxies and galaxy clusters that extends over $300h^{-1}$ Mpc. Tully (1986, 1987) has suggested that the relatively nearby P-P supercluster links with the more distant Pisces-Cetus complex. In fact, Tully proposes that P-P is a filamentary enhancement within a planar structure that has coherence on a scale of 0.1c.

7. Search for Non-Hubble Streaming

Deviations from the Hubble flow in the local universe have been detected as the motions of the Milky Way with respect to the microwave background, as infall motion into the Virgo cluster (Tully and Shaya 1984), and as the familiar elongated shapes in redshift space seen toward rich clusters. On scales larger than those of clusters, the magnitude of the peculiar velocities is largely unknown, and fast becoming more intriguing as evidenced by the lively discussions during this Study Week. The questions of possible large-scale streaming motions have attracted new attention with the availability of deeper redshift surveys of larger areas of sky. The "CfA slice" (de Lapparent *et al.* 1986) and the P-P survey have shown both similar and dissimilar structures: the apparent sheets or bubbles seen in both surveys and the high density main ridge unique to the P-P supercluster. Especially interesting is the apparently common appearance of large "voids", regions of low number density of galaxies surrounded by regions of higher number density. The streaming motion around such voids has been modelled as galaxies evacuating the region of the void (Davis *et al.* 1982) and alternatively as evolving negative density perturbations (Hoffman *et al.* 1983). The former approach, which is equivalent in result to that of explosive galaxy formation scenarios, predicts streaming motions around a void on the order of 50% of the Hubble expansion; in contrast, the latter favors velocities of only about 10% of the Hubble velocity. Biased galaxy formation pictures may not require any streaming motions at all.

In order to investigate the dynamics in a supercluster environment and to measure possible deviations from the Hubble flow, it is necessary to obtain redshift-independent relative distances to galaxies, i.e. distances up to a scale factor h in the Hubble constant. In the past, there have been essentially two opposite approaches to the task of predicting luminosities for galaxies, and thereby deriving their distances: the first, deriving a luminosity distance from an indicator that is distance independent and has been determined for a large number of galaxies; and the second, the so-called method of "sosies" that is based on the comparison of individually-matched galaxies for which many properties have been measured and are similar. The first approach has been pioneered by Tully and Fisher (1977). An example of the application of the "sosies" method is the measurement of the distance to the Hercules supercluster presented by de Vaucouleurs and Corwin (1986).

The acquisition of large numbers of HI line data in the P-P region makes attractive the future use of the Tully-Fisher relation to predict luminosity distances. At present, the detection of possible streaming motion is hampered by the limited success of deriving redshift-independent distances of galaxies via application of the Tully-Fisher relation. Current limits on the detection of large-scale streaming motions in surveys such as this one are imposed by: (1) the uncertainties in the available blue magnitudes and the necessary corrections; (2) the large intrinsic scatter in the Tully-Fisher relation for blue magnitudes; and (3) the small number of galaxies suitable for application of the relation within regions that can reasonably be assumed to be part of the dynamical structure under study.

The original application of the Tully-Fisher relation using blue or photographic magnitudes relies on an understanding of the numerous corrections for galactic and internal extinction, inclination, and the varying contribution of the disk and bulge (e.g. Tully and Fouqué 1985). Magnitudes have often been measured on different systems and with different apertures. In few instances have data on a complete sample been assembled from a homogeneous survey.

The movement of the luminosity measurement toward longer wavelengths has been driven primarily by the desire to minimize internal extinction effects so critical in the blue and at the same time, so uncertain. A secondary reason for the use of redder magnitudes has been the desire to lessen the influence of recent, and possibly variable, star formation which also would contribute more in the blue. The improvement in the application of the Tully-Fisher relation as a secondary distance indicator using infrared magnitudes has been demonstrated (Aaronson and Mould 1983; Aaronson *et al.* 1986). Because of the need for CCD images for large numbers of galaxies, use of the R or I bands is more practical. While the sky background is still noisy and the internal extinction corrections are not negligible at I relative to H, the sensitivity of current detectors is sufficient to provide quality I-band images in typically ten minutes or less (Bothun and Mould 1987). We have begun a program to obtain I-band images for galaxies in the P-P supercluster. The imagery will be used to derive magnitudes, diameters, and disk scale lengths in an attempt to reduce the scatter in the Tully-Fisher relation by means of multivariate analysis.

The general goal is to predict the distance d of a galaxy from $i = 1 \ldots n$ observed quantities a_i, given that at least one of them is distance-independent. One needs to determine how many independent parameters are needed to describe the scatter of the sample in parameter space. The use of Principal Component Analysis (PCA) has been discussed by such authors as Brosche (1973) and Whitmore (1984), and offers the promise of reducing the scatter in the Tully-Fisher relation for spirals in much the same way as a three-parameter fit has helped the analogous method for ellipticals. Application of such an analysis to a similar sample in the Hercules supercluster has been presented by Freudling *et al.* (1988).

At the present time, the large dispersion in the distance estimates makes it impossible to derive the non-Hubble velocity for any single object. However, although the luminosity distances derived from the Tully-Fisher relation are currently less accurate than the Hubble distances, they should be independent of peculiar velocity and hence unbiased. By investigating the difference between the Hubble distance and the luminosity distance as a function of the luminosity distance, we can test that the average of those residuals is zero for a sample of galaxies that is assumed to have a random peculiar distribution and is used as a comparison with other samples that might themselves show a streaming motion. An all-sky sample of Sc galaxies is being constructed for this purpose. Such an approach avoids the apparent distance dependent increase of the Hubble constant (Giraud 1985) in a residual versus redshift diagram, the result of the unavoidable Malmquist bias of the sample (Teerikorpi 1984). However, the large dispersion in the luminosity distance introduces the effect that the residuals will, in the end, be averaged over distance bins. Galaxies will frequently be counted in a "wrong" distance bin, that is the distance bin assigned to it by the luminosity distance rather than the true distance. A galaxy with a distance too high for its bin will appear to have a positive peculiar velocity. This usually does not matter, because for each galaxy too distant for its bin, there will be another one too close. However, if the number of galaxies is much higher at a certain distance (for a cluster or because the completeness of the survey depends on distance), then the distance bin before this concentration will contain more galaxies from higher distances than from lower ones. The opposite happens behind a cluster. As a result, an apparent streaming motion will be seen towards any concentration of observed galaxies. This spurious result can be avoided for all bins but the zero velocity bin and the bin with the highest distance (if the bin size is larger than the errors for individual galaxies) by taking a sample of galaxies which is equally distributed in redshift or by weighting the average according to the local density of observed galaxies. The former approach is usually more straightforward, but unfortunately results in a significant reduction in the number of galaxies actually used in the analysis. By drawing samples of similar objects in each redshift bin from an all-sky sample and from the sample within a supercluster or around a void, this method should be able to detect streaming motions. Freudling *et al.* (1988) have set a preliminary upper limit of 400 kms^{-1} on infall motions within the Hercules supercluster. Similar results are obtained for P-P. The future acquisition of homogeneous data samples and good magnitudes will hopefully permit measurement of departures from the Hubble flow within P-P and in the vicinity of the foreground void.

8. Summary and Conclusions

Of all the nearby superclusters Pisces-Perseus is perhaps the easiest to recognize in maps of the surface density distribution of CGCG galaxies, partly because of its geometry and richness and partly because of its distance from us relative to the depth of the catalog. The P-P region is not a randomly selected volume of space, but rather has been deliberately chosen for study because of its obvious structure. The main ridge is at least $45h^{-1}$ Mpc in length and has an axial ratio close to ten. The supercluster also contains a more widely dispersed population that is separated from foreground and background structures by relatively empty volumes. Connective density enhancements surround the foreground void and appear to link the Local Supercluster to P-P. The wealth of structure displayed by P-P reminds us that the large-scale topology is complex and not easily described. The most conspicuous high density enhancements do however exhibit filamentary morphology.

The range of local densities characteristic of the supercluster environment extends from the highest density cluster cores found in A426 and A262 to the outer supercluster periphery. In the latter locale, the interaction timescale between galaxies is much longer than the Hubble time, and objects there must have existed for virtually their entire lifetimes in such low density regions. Hence, they are objects unaffected by neighbors and occupy volumes of both low gas and galaxy density. Gas removal and disruption mechanisms believed to be responsible for the alteration of morphologies in high density environments cannot be expected to be efficient in the supercluster periphery. Comparison of the intrinsic properties of galaxies in different density regimes will allow us to determine the relative properties of galaxies in the varying environments.

Because of its location and extent, P-P is also difficult to trace at optical wavelengths and hence is especially suited to an HI line redshift survey. Its maximum extent and possible connection with Lynx-Ursa Major or A 569 in the east or the Pisces-Cetus complex in the west must still be confirmed. The boundaries of the foregound void need further delineation via a large sample of low luminosity, low surface brightness objects. If biasing mechanisms do govern the morphological segregation at early epochs, the void might still be expected to contain a population of HI-poor, low luminosity galaxies or gas clouds.

While time will provide more observations so that completion of the sample to some fainter apparent magnitude is possible, problems with understanding incompleteness and the definition of a "fair" sample will remain. If the luminosity function is dependent not only on the the morphological type of a galaxy, but separately also on the local density, then samples of unequal morphological mix may not carry the same indication of luminosity distribution. As long as galaxies are not standard candles, we will have to worry about whether the galaxies in the observational sample are properly representative. Spirals and ellipticals do not occupy the same volumes; some allowance for this segretation must be included. The interpretation of large-scale flows from the derivation of luminosity distances currently assumes that the intrinsic properties of galaxies are everywhere constant. If the luminosity function varies, even for a single type, the use of the Tully-Fisher relation to predict distances will have to be carefully applied to galaxies located in different density regimes.

In many respects, fortune provides us with a special view of the Pisces-Perseus supercluster. Is its complex and filamentary structure unique or is such structure just easy to recognize? We can ask whether we would still identify it if its main ridge did not lie nearly in the plane of the sky, if its axial ratio were not so great or if its typical volume density enhancement were not so large. Attempts to trace the extent of Perseus deliver the warning that the size of the largest apparent structure is comparable to the depth of the volume surveyed; will even greater structures be found in deeper surveys? Pisces-Perseus provides a laboratory for the exploration of intergalactic environments and their effects on galaxies and will continue to serve as a testing ground for our ideas of the origin of large-scale structure.

We wish to acknowledge that the National Astronomy and Ionosphere Center is operated by Cornell University under a cooperative agreement with the National Science Foundation.

REFERENCES

Aaronson, M. and Mould, J. 1983. *Ap J.* **265,** 1.
Aaronson, M., Bothun, G., Mould, J., Huchra, J., Schommer, R., and Cornell, M. 1986. *Ap J.* **302,** 536.
Batuski, D.J. and Burns, J.O. 1985. *Ap J.* **299,** 5.
Bernheimer, W.E. 1932. *Nature.* **130,** 132.
Bicay, M.D. 1987. *Ph. D. thesis.* Stanford University.
Binggeli, B., Sandage, A., and Tammann, G.A. 1985. *AJ.* **90,** 1759.
Bothun, G.D. and Mould, J.R. 1987. *Ap J.* **313,** 629.
Bothun, G.D., Beers, T.C., Mould, J.R., and Huchra, J.P. 1986. *Ap J.* **308,** 510.
Brosche, A. 1973. *AA.* **23,** 259.
Burns, J.O. and Owen, F.N. 1979. *AJ.* **84,** 1478.
Burstein, D. and Heiles, C.1978. *Ap J.* **225,** 40.
Davis, M. and Djorgovski, S. 1985. *Ap J.* **299,** 15.
Davis, M. and Geller, M.J. 1976. *Ap J.* **208,** 13.
Davis, M., Huchra, J., Latham, D.W., and Tonry, J. 1982. *Ap J.* **253,** 423.
Dekel, A. and Rees, M.J. 1987. *Nature.* **326,** 455.
de Lapparent, V., Geller, M.J., and Huchra, J.P. 1986. *Ap J Letters.* **302,** L1.
de Vaucouleurs, G. 1975. in *Galaxies and the Universe.* ed. by A. Sandage, M. Sandage, and J. Kristian. p. 557. Chicago: University of Chicago Press.
de Vaucouleurs, G. and Corwin, H.G. 1986. *Ap J.* **308,** 487.
Dow, M.W., Lu., N.Y., Houck, J.R., Salpeter, E.E., and Lewis, B.M. 1988. *Ap J Letters.* **324,** L51.
Dressler, A. 1980. *Ap J.* **236,** 351.
Einasto, J., Jôeveer, M., and Saar, E. 1980. *MN.* **193,** 353.
Focardi, P., Marano, B., and Vettolani, G. 1984. *AA.* **136,** 178.
Fontanelli, P. 1984. *AA.* **138,** 85.
Freudling, W., Haynes, M.P., and Giovanelli, R. 1988. *AJ.* submitted.
Geller, M.J. and Huchra, J.P. 1983. *Ap J Suppl.* **52,** 61.
Giovanelli, R. and Haynes, M.P. 1982. *AJ.* **87,** 1355.
_____. 1984. *AJ.* **89,** 1.
_____. 1988. in *Large-Scale Structures of the Universe.* ed. J. Audouze, M.-C. Pelletan, and A. Szalay, p. 113, Dordrecht: Kluwer Academic Publishers.
Giovanelli, R., Haynes, M.P., and Chincarini, G.L. 1986. *Ap J.* **300,** 77.
Giovanelli, R., Haynes, M.P., Myers, S.T., and Roth, J. 1986. *AJ.* **92,** 250.
Giraud, E. 1985. *AA.* **153,** 125.
Gregory, S.A., Thompson, K.A., and Tifft, W.G. 1981. *Ap J.* **243,** 411.
Hauschilt, M. 1987. *AA.* **184,** 43.

Haynes, M.P. and Giovanelli, R. 1986. *Ap J Letters.* **303,** L127.
Haynes, M.P., Giovanelli, R., Starosta, B.M., and Magri, C. 1988a *AJ.* **95**, 607.
Haynes, M.P., Giovanelli, R., and Magri, C. 1988b (in preparation).
Hewett, P.C. 1982. *MN.* **201,** 867.
Hoffman, G.L., Salpeter, E.E., and Wasserman, I. 1983. *Ap J.* **268,** 527.
Hubble, E. and Humason, M.L. 1931. *Ap J.* **74,** 43.
Huchra, J.P. and Geller, M.J. 1982. *Ap J.* **257,** 423.
Kent, S.M. and Sargent, W.L.W. 1983. *AJ.* **88,** 697.
Kerr, F.J. and Henning, P.A. 1987. *Ap J Letters.* **320,** L99.
Materne, J. 1978. *AA.* **63,** 401.
Melnick, J. and Sargent, W.L.W. 1977. *Ap J.* **215,** 401.
Nilson, P. 1973. *Uppsala General Catalog, Uppsala Astr. Obs. Ann.* **6** (UGC).
Oemler, G. 1974. *Ap J.* **194,** 1.
_____. 1987. in *Nearly Normal Galaxies.* ed. S.M. Faber. p. 213. New York: Springer-Verlag.
Oort, J.H. 1982. in *Astrophysical Cosmology.* eds. H.A. Bruck, G.V. Coyne, and M.S. Longair. p. 127. Città del Vaticano: Pont. Acad. Scient.
Postman, M. and Geller, M.J. 1984. *Ap J.* **281,** 95.
Roberts, M. 1969. *AJ.* **74,** 859.
Sandage, A. and Tamman, G. 1981. *Revised Shapley-Ames Catalog.* Washington: Carnegie Institution.
Schechter, P. 1976. *Ap J.* **203,** 297.
Shapley, H. and Ames, A. 1932. *Harvard Obs Ann.* **88,** No. 2.
Sharp, N.A. 1979. *AA.* **74,** 308.
Sharp, N.A., Jones, B.J.T., and Jones, J.E. 1978. *MN.* **185,** 457.
Strom, S. and Strom, K. 1978. *AJ.* **83,** 732.
Struble, M.P. and Rood, H.J. 1987. *Ap J Suppl.* **63,** 543.
Teerikorpi, P. 1984. *AA.* **141,** 407.
Tully, R.B. 1980. *Ap J.* **237,** 390.
_____. 1986. *Ap J.* **303,** 25.
_____. 1987. *Ap J.* **323,** 1.
Tully, R.B. and Fisher, J.R. 1977. *AA.* **54,** 661.
Tully, R.B. and Fouqué, P. 1985. *Ap J Suppl.* **58,** 67.
Tully, R.B. and Shaya, E.J. 1984. *Ap J.* **281,** 31.
Weinberger, R. 1980. *AA. Suppl.* **40,** 123.
Whitmore, B.C. 1984. *Ap J.* **278,** 61.
Zwicky, F., Herzog, E., Karpowicz, M., Kowal, C.T., and Wild, P. 1961-68. *Catalog of Galaxies and Clusters of Galaxies.* Pasadena: California Institute of Technology, six volumes.

MORPHOLOGY OF LARGE-SCALE STRUCTURE

R. BRENT TULLY
Institute for Astronomy, University of Hawaii

Abstract

The opportunity will be taken to provide a brief summary of some characteristics of large-scale structure. In the first section, properties of the distribution of galaxies within 3000 km s^{-1} will be described. Then, in the second, there will be a review of why one should believe in structure on a scale of 30,000 km s^{-1}.

1. Structure Delineated by Galaxies Within 3000 km s^{-1}

The *Nearby Galaxies Atlas* (Tully and Fisher 1987) provides a graphic description of the distribution of galaxies with velocities less than 3000 km s^{-1}. Some general characteristics of the galaxies within this volume are discussed by Tully (1988) and, specifically, the delineation of a clustering hierarchy is presented by Tully (1987a). Some of the important points made in those publications and in Figs. 1 and 2 are as follows:

a. Essentially all nearby galaxies can be assigned to entities called 'clouds'. Roughly 70% of these can be assigned to 'groups' that are probably bound and mostly collapsed, a further 20% lie in 'associations' that are probably *not* bound, and 10% lie at-large in the clouds. There is no convincing example of even a single isolated galaxy within the surveyed volume.

b. The Local Group sits right at the edge of what will be called the 'Local Void'. This void is apparently empty of galaxies across a diameter of 1600 km s^{-1} and only contains 14 known galaxies across a diameter of 2200 km s^{-1} (12 of these lie in two small clouds). Unfortunately, the region is split by the zone of obscuration, which diminishes the significance of the above assertions. However, if voids delineated by luminous galaxies were actually filled by gas-rich dwarfs, such entities would have been seen easily in the Local Void.

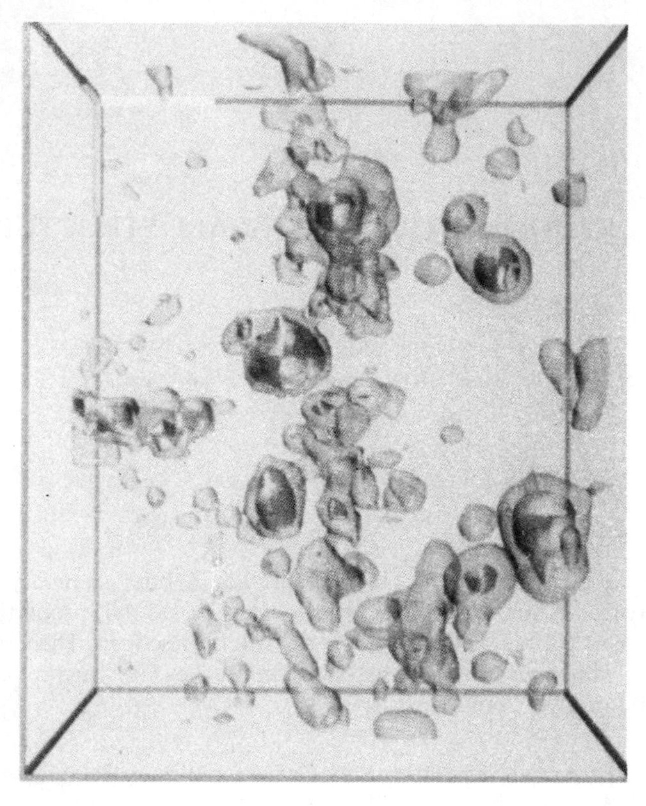

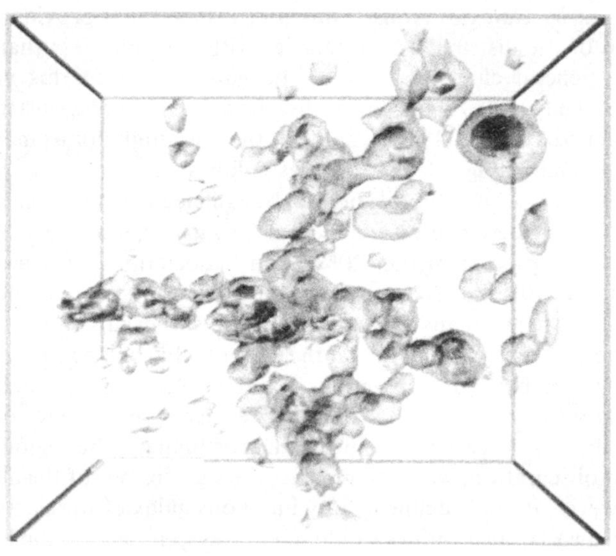

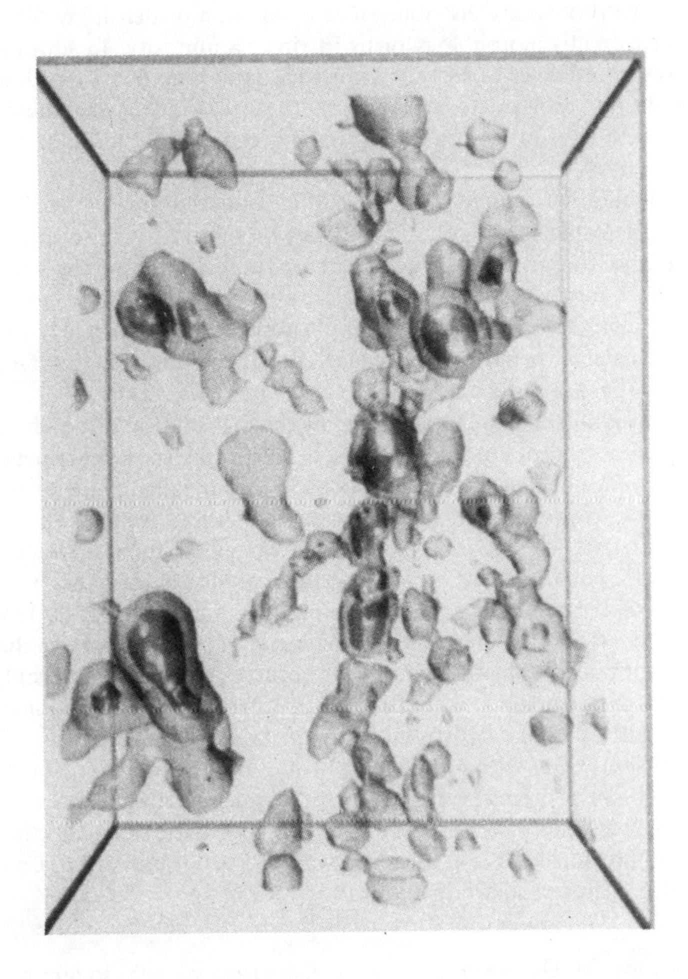

FIG. 1. Three orthogonal views of the structure in the distribution of galaxies in the Local Supercluster. Dimensions of the cube are 3750 km s^{-1} by 3000 km s^{-1} by 2625 km s^{-1}. The outer contour represents a surface of density 0.5 galaxies Mpc^{-3} (H_0 = 75 km s^{-1} Mpc^{-1}). The Virgo Cluster is located at the region of highest density near the center of the volume. The plane of the Local Supercluster is horizontal in the lower-left projection and vertical in the upper-right projection. The 'louvered wall' is vertical in both the lower-left and upper-left projections.

c. A more quantitative assessment of the dependence on environment is possible. The local density in the vicinity of each nearby galaxy was determined, and then the probability distribution as a function of density was found for galaxies classed by morphology or by intrinsic luminosity. In addition to the well-known preference of early systems to inhabit high-density regions, there is a subtle but progressive variation in the preferred environment of late systems, with later or less luminous systems more likely to lie in regions of lower density.

d. Qualitatively, there is a remarkable connectedness to the clouds that creates a filamentary network. There is a problem in trying to quantify this claim because the structures are large compared with the survey volume and are usually bounded by the zone of obscuration or by serious incompletion due to distance. However, the general phenomenon is particularly striking in the south galactic hemisphere because, outside of the few interconnected filaments, space is quite empty.

e. An extraordinary characteristic of the distribution of nearby galaxies is the apparent tendency of galaxies to lie in layers. The most prominent feature is the adherence of many nearby galaxies to the so-called plane of the Local Supercluster. However, almost all clouds, even those well off the principal plane, are stretched out in structures elongated at modest angles to the supergalactic equator. The principal plane extends across and beyond the boundary of the presently surveyed volume, a diameter of 6000 km s^{-1}.

f. There is another apparent characteristic of nearby structure that again defies quantitative evaluation because it occurs on a scale comparable to the dimensions of the survey. It would appear that there is a concentration of galaxies in a 'wall' *perpendicular* to the plane of the Local Supercluster. The wall has dimensions of at least 6000 km s^{-1} (the sample dimension) by 3000 km s^{-1} by 1500 km s^{-1}. However, viewed face-on to the major dimensions, the wall is fragmented into the apparent layers parallel to the supergalactic equator mentioned previously (it is a 'louvered wall'). The Virgo, Antlia, Centaurus, and Hydra I clusters all lie in the wall.

2. Structure Delineated by Rich Clusters within 30,000 km s^{-1}

The claim was made by Tully (1986, 1987b) that rich clusters tend to congregate in 'supercluster complexes'. There are parts of five such complexes in the well-observed region of the sky within 0.1c, each one apparently containing roughly 50 rich clusters and extending across 30,000 km s^{-1}. A supercluster complex would contain roughly $10^{18}\ M_{\odot}$ and a million big galaxies. Roughly two-thirds of nearby rich clusters are associated with a supercluster complex.

It is claimed that we reside in the 'Pisces-Cetus Supercluster Complex'. In addition to the general properties described above, this entity has the special

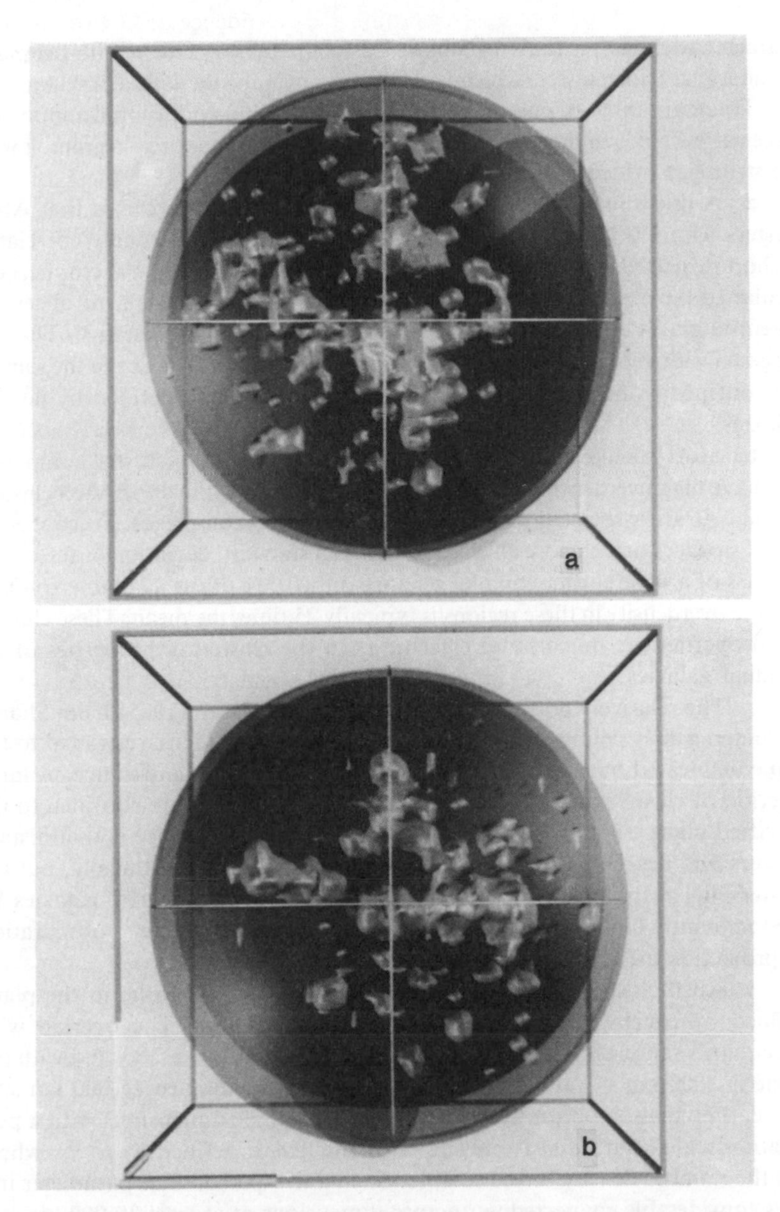

FIG. 2. Two views of the structure in the distribution of rich clusters. The radius of the sphere is 30,000 km s^{-1}. The outer contour represents a surface density of 1.7×10^{-5} clusters Mpc^{-3} (H_0 = 75 km s^{-1} Mpc^{-1}). (a) North galactic pole, which contains the Hercules-Corona Borealis (top), Leo (lower right), and Ursa Major (middle left) supercluster complexes. (b) South galactic pole, which contains the Pisces-Cetus (middle right) and Aquarius (top) supercluster complexes.

characteristic that it is flattened about 7:1 to a plane coincident with the plane of the Local Supercluster, which is delineated by galaxies within 3000 km s^{-1}.

There are three strong pieces of evidence and two additional supportive pieces of evidence to back up these claims. These are presented beginning with the strongest evidence and progressing to the weakest:

a. A one-dimensional two-point correlation analysis reveals that Abell clusters within 0.1c in the south galactic hemisphere are strongly correlated at short projected spacings when projections are taken against a vector perpendicular to the plane defined by nearby galaxies. This analysis neutralizes the potential biases due to obscuration and incompletion with distance. The test suggests, with 99% confidence, that a large fraction of the clusters in the sample are confined to one or more strata parallel to the plane defined by nearby galaxies.

b. Most Abell clusters lie in a small number of supercluster complexes that have high overdensities and low filling factors.Two-thirds of Abell clusters within 0.1c are associated with only five supercluster complexes. Triaxial ellipsoidal surfaces that enclose all clusters within a standard deviation of the center of mass of a supercluster complex enclose only 0.5% of the available volume, and the overdensity in these regions is typically 25 times the mean. These clustering properties are much more extreme than the clustering properties of individual galaxies.

c. The relatively poor clusters defined by Shectman (1985) from Shane-Wirtanen galaxy counts in the south galactic hemisphere are restricted to the plane delineated by nearly galaxies to a high degree of significance. A large fraction of clusters at a typical velocity of 15,000 km s^{-1} are confined to the specified plane with FWHM of 2000 km s^{-1}. The coincidence is that distant clusters and nearby galaxies lie on the same great circle on the sky, but the clusters lie in the south galactic hemisphere while the nearby galaxies lie predominantly in the north. Hence, the coincidence is not due to obscuration or projection effects.

d. Abell clusters within 0.1c in the south also tend to lie in the plane defined by nearby galaxies—the supergalactic equator. The overlap with Shectman's sample is small. The FWHM of Abell clusters associated with the plane is 3000 km s^{-1} at a typical distance corresponding to 15,000 km s^{-1}.

e. Percolation occurs across of order the sample dimensions with a percolation scale length that is only 70% of the length defined by $n^{-1/3}$, where n is the number density of the sample. Such a low percolation parameter implies considerable connectedness across dimensions of at least 30,000 km s^{-1}.

In summary, the evidence for structure on a scale approaching 0.1c is surprisingly strong. If true, the clustering of clusters is much more extreme than the clustering of galaxies. For example, it would not have been suspected that galaxies cluster strongly with an essentially all-sky sample of only 300-400 systems. There have been fears that the largest-scale structure is an artifact of sample biases. However, it is unlikely that biases would lead to pronounced structure in three dimensions that is not obvious in two dimensions, very unlikely to produce a feature as thin in three-dimensional projection as the Pisces-Cetus plane, and most unlikely to produce the coincidence in position angle of the Pisces-Cetus and Local Supercluster planes demonstrated by the one-dimensional two-point correlation analysis.

REFERENCES

Shectman, S. A. 1985. *Ap J Suppl.* **57**, 77.

Tully, R.B. 1986. *Ap J.* **303**, 25.

_____ . 1987a. *Ap J.* **321**, 280.

_____ . 1987b. *Ap J.* **323**, 1.

_____ . 1988. *AJ.* **96**, 73.

Tully, R. B. and Fisher, J. R. 1987. *Nearby Galaxies Atlas.* Cambridge: Cambridge University Press.

LARGE-SCALE STRUCTURE AND MOTION TRACED BY GALAXY CLUSTERS

NETA A. BAHCALL
Space Telescope Science Institute, Baltimore

1. Introduction

The study of the large-scale structure and motion in the universe is critical to our understanding of the formation and evolution of galaxies and structure. The large-scale structures observed today provide the cosmic fossils of conditions that existed in the early universe. The classic method of investigating structure in the universe is the observation of the spatial distribution of galaxies. Extensive surveys of thousands of galaxy redshifts are needed in order to cover large enough volumes and scales. Such has been carried out by several groups (see review papers by Oort 1983, Chincarini and Vettolani 1987, and Rood 1988; also Gregory and Thompson 1978, Gregory *et al.* 1981, Davis *et al.* 1982, Giovanelli *et al.* 1986, de Lapparent *et al.* 1986, da Costa *et al.* 1988).

A different approach is emphasized in this paper: using the high density peaks of the galaxy distribution, i.e., the rich clusters of galaxies, as tracers of the large-scale structure (see Bahcall 1988a for a more comprehensive review). As the mountain peaks trace mountain-chains on earth, rich clusters, with their low space density and large mean separation, serve as an efficient tracer of the largest-scale structures. While $\sim 10^6$ galaxies cover the high-latitude northern hemisphere to $\sim 18^m$, only $\sim$ 500 rich clusters highlight the structure in the same volume of space. Recent results, summarized in this paper, show that clusters do indeed provide an efficient and effective tracer of the large-scale structure of the universe.

A Hubble constant of $H_o = 100h$ km s^{-1} Mpc^{-1} is used throughout this paper.

2. The Cluster Correlation Function

2.1 - Abell Clusters

The Abell (1958) catalog of rich clusters has been analyzed by many investigators using different techniques in an attempt to determine the spatial distribution of rich clusters. Abell (1958, 1961) found that the surface distribution of the clusters in his statistical sample was highly nonrandom and reported evidence suggesting the existence of superclusters; Bogart and Wagoner (1973), Hauser and Peebles (1973), and Rood (1976) (see also references therein) also found, using nearest-neighbor distributions and/or angular correlation functions, strong evidence for superclustering among the Abell clusters. The studies dealt primarily with the surface distribution of clusters and, in some cases, used approximate estimates for cluster redshifts. More recently, Bahcall and Soneira (1983, 1984) and, independently Klypin and Kopylov (1983), used redshift measurements of complete samples of clusters to determine directly the spatial distribution of rich clusters. I discuss below these results, as well as more recent results obtained by other investigators.

The two point spatial correlation function of clusters, $\xi_{cc}(r)$, was determined by Bahcall and Soneira (1983) (hereafter BS83) using Abell's (1958) statistical sample of rich clusters of galaxies of distance class $D \leq 4$ ($z \lesssim 0.1$), with redshifts for all clusters reported by Hoessel *et al.* (1980). This sample includes all 104 Abell clusters at $D \leq 4$ that are of richness class $R \geq 1$ and are located at high galactic latitude ($|b| \geq 30°$). A summary of the sample properties and its division into distance and richness classes, as well as into hemispheres, is presented in BS83. Also listed in the above reference are properties of the much larger and deeper $D = 5+6$ statistical sample ($z \lesssim 0.2$) that includes 1547 clusters. While only a small fraction of the redshifts are measured for this sample, it was used, because of its much larger number of clusters, in various comparison tests to strenghten and confirm the results obtained from the $D \leq 4$ sample.

The frequency distribution, $F(r)$, for all pairs of clusters with separation r in the sample was determined. In order to minimize the influence of selection effects on the determination of $\xi(r)$, a set of 1000 random catalogs was constructed, each containing 104 clusters randomly distributed within the angular boundaries of the survey region, but with the same selection functions in both redshift, $n(z)$, and latitude, $P(b)$, as the Abell redshift sample. The frequency distribution of cluster pairs was determined in both the real and random catalogs and the results were compared. This procedure ensures that the selection effects and boundary conditions will affect the data and random catalogs in the same manner.

The spatial correlation function was determined from the relation

$$\xi_{cc}(r) = F(r) \,/\, F^R(r) - 1, \tag{1}$$

where $F(r)$ is the observed frequency of pairs in the Abell sample and $F^R(r)$ is the corresponding frequency of random pairs (as determined by the ensemble average frequency of the 1000 random catalogs).

The resulting correlation function is presented in Figure 1. Strong spatial correlations are observed at separations $\lesssim 25h^{-1}$ Mpc. Weaker correlations are observed to larger separations of at least $\sim 50h^{-1}$ Mpc, and possibly $\sim 100h^{-1}$ Mpc, where $\xi_{cc} \sim 0.1$; beyond $150h^{-1}$ Mpc no statistically significant correlations are observed in the present sample.

The correlation function of Figure 1 can be well approximated by a single power law relation of the form $\xi_{cc}(r) = 300r^{-1.8}$, for $5 \lesssim r \lesssim 150h^{-1}$ Mpc. The function is smooth, with little scatter at $r \lesssim 50h^{-1}$ Mpc. At $r > 50h^{-1}$ Mpc, the scatter and uncertainties increase, but weak correlations of the order of 0.2 are still detected at these very large separations. When corrected for velocity broadening among clusters, the intrinsic rich ($R \geq 1$) cluster correlation function was determined by Bahcall and Soneira to be:

$$\xi_{cc}(r) = 360r^{-1.8} = (r/26)^{-1.8} \qquad r \lesssim 100h^{-1} \text{ Mpc}. \tag{2}$$

In comparison, the correlation function of galaxies is given by (Groth and Peebles 1977, Davis and Peebles 1983):

$$\xi_{cc}(r) = 20r^{-1.8} = (r/5)^{-1.8} \qquad r \lesssim 20h^{-1} \text{ Mpc}. \tag{3}$$

The rich cluster correlation function has the same shape and slope as that of the galaxy correlation function, but is considerably stronger at any given scale, by a factor of ~ 18, than the correlation function of galaxies. The cluster correlations also extend to greater separations than the scales observed in the galaxy correlations. The cluster correlation scale-length, i.e., the scale at which the correlation function is unity, is $r_0 \simeq 26h^{-1}$ Mpc (eq. 2), as compared with $r_0 \simeq 5h^{-1}$ Mpc for galaxies. The extent of the rich cluster correlation function beyond the reported $\simeq 15h^{-1}$ Mpc break in the galaxy correlation function (Groth and Peebles 1977) suggests the existence of large-scale structure in the universe ($\gtrsim 15h^{-1}$ Mpc). While the reason for the strong increase of correlation strength and scale from galaxies to clusters is still a theoretical challenge, some possible explanations are discussed in Section 6 and in Bahcall (1988a). The cluster correlation function determined above places constraints on models for the formation of galaxies and structure.

In order to ensure that the spatial correlation function is not due to some special peculiarities in the nearby $D \leq 4$ sample, Bahcall and Soneira carried out several tests that are discussed below.

First, the angular correlation function of the much larger and deeper $D = 5+6$ sample (1547 $R \geq 1$ clusters to $z \lesssim 0.2$) was determined and compared

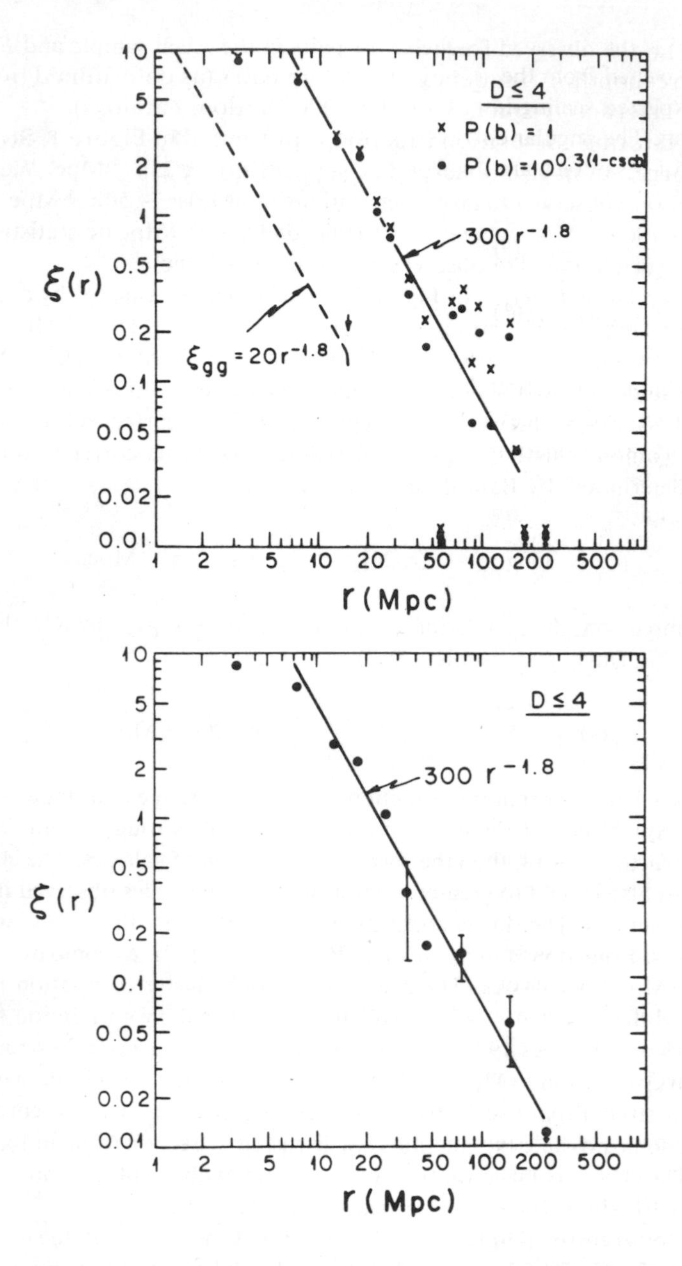

FIG. 1. Top: The spatial correlation function of the $D \leq 4$ Abell cluster sample (BS83). Crosses refer to no correction for latitude selection function; dots refer to the full correction of $P(b)$. The solid line is the best fit 1.8 power law to the data. The dashed line is the galaxy-galaxy correlation function of Peebles and co-workers. Bottom: same, but plotted in larger bins at large separations.

with that expected from the spatial correlation function above (2), as well as from the expected scaling-law (Peebles 1980a) of the $D \leq 4$ angular correlation function. The angular correlation functions of the nearby $D \leq 4$ and distant $D = 5+6$ samples are determined to be (BS83)

$$w_{D \leq 4}(\theta) \simeq 3\theta^{-1} \qquad 0\overset{\circ}{.}5 \lesssim \theta \lesssim 25^\circ \tag{4}$$

$$w_{D=5+6}(\theta) \simeq 0.8\theta^{-1} \qquad 0\overset{\circ}{.}2 \lesssim \theta \lesssim 14^\circ . \tag{5}$$

The angular correlation scale as expected from the scaling law applied to their respective distances. A comparison of the scaled functions is shown in Figure 2. If the correlations were mainly due to patchy obscuration or other omissions by Abell, the (observed) scaling would not be expected. The scaling agreement indicates that any possible projection biases in the catalog (e.g., Sutherland 1988) are rather small and do not significantly affect the correlation results (see also Dekel 1988). The reduced correlation scale suggested by Sutherland may result from overcorrecting the actual correlation power on large scales. A comparison of the $D = 5+6$ angular function with that expected from the spatial correlation function of equation (2), when integrated over the relevant redshift distribution, is given by BS83. The agreement between the $D \leq 4$ and $D = 5+6$ functions is excellent. This agreement indicates that the $D \leq 4$ redshift sample is a fair sample of the much larger sample, and that the observed correlations represent real correlations of clusters in space. The scaling-law of the angular functions was also studied by Hauser and Peebles (1973) who reached similar conclusions with regard to the reality of the intrinsic correlations.

Second, the angular correlation function was compared with the pure redshift (i.e, line-of-sight) correlations of the clusters. If the correlations were mostly due to patchy obscuration on the sky or other similar biases, no extensive redshift correlations would be expected. It is observed (BS83) that the redshift correlations are indeed positive and extend to large scales similar to those in the projected correlations, further strengthening the reality of the correlations.

Third, the angular correlation function of the $D = 5+6$ sample was determined in different regions of the sky, yielding consistent results within the uncertainties (Figure 3).

These tests, and those listed in Section 2.5 below, suggest that the observed cluster correlation function is mostly due to physical clustering of rich clusters of galaxies that extends to large scales.

Since the correlation strength appears to increase from galaxies to clusters, Bahcall and Soneira also investigated whether a similar trend is observed between the correlation function of poor and rich clusters. The angular correla-

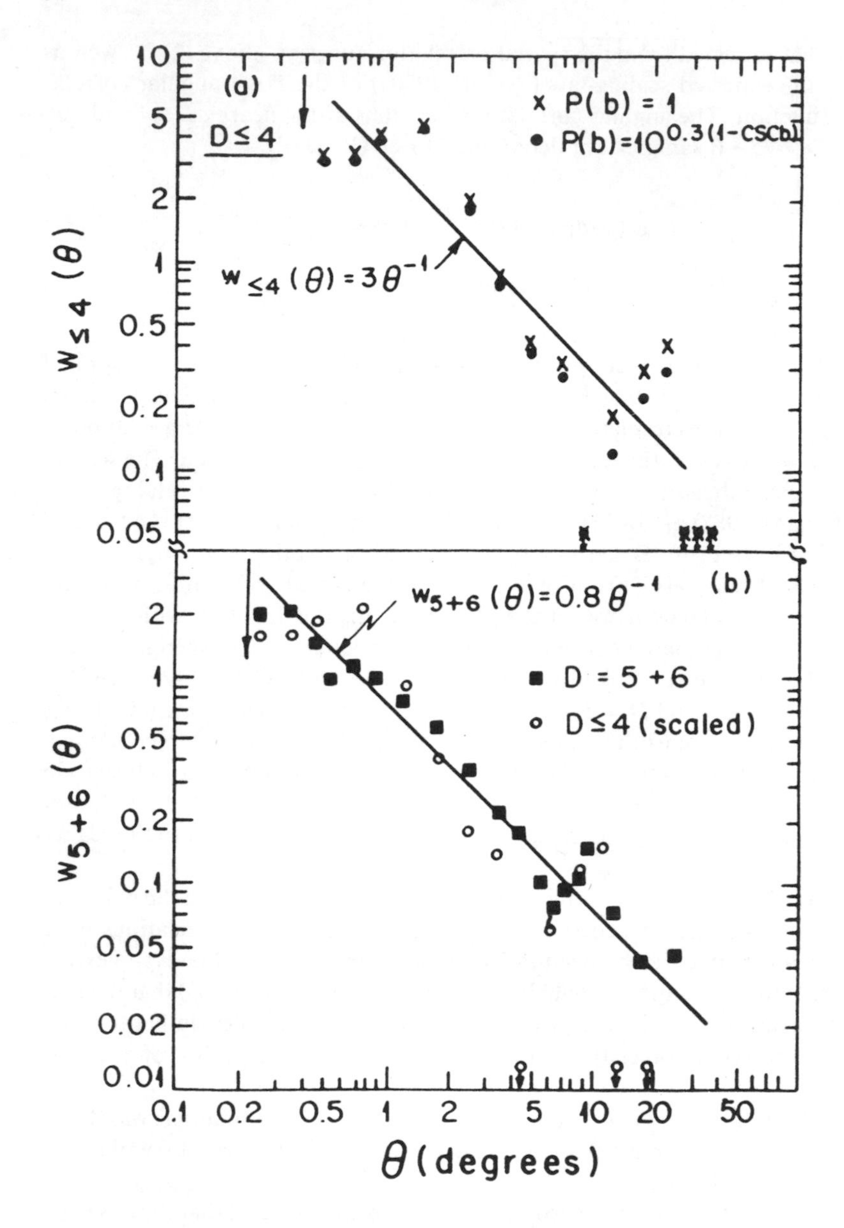

FIG. 2. Top: The angular correlation function of the $D \leq 4$ Abell cluster sample (BS83); Bottom: the angular correlation function of the deep $D = 5 + 6$ sample (squares). Open circles are the angular correlation function of the $D \leq 4$ sample (Figure 2) scaled by the standard scaling law of an intrinsic spatial correlation using the distance ratio of the two samples (Sec. 2). Correlations of $\lesssim 0.05$ are rather uncertain. The position of the mean Abell radius is indicated by the arrow.

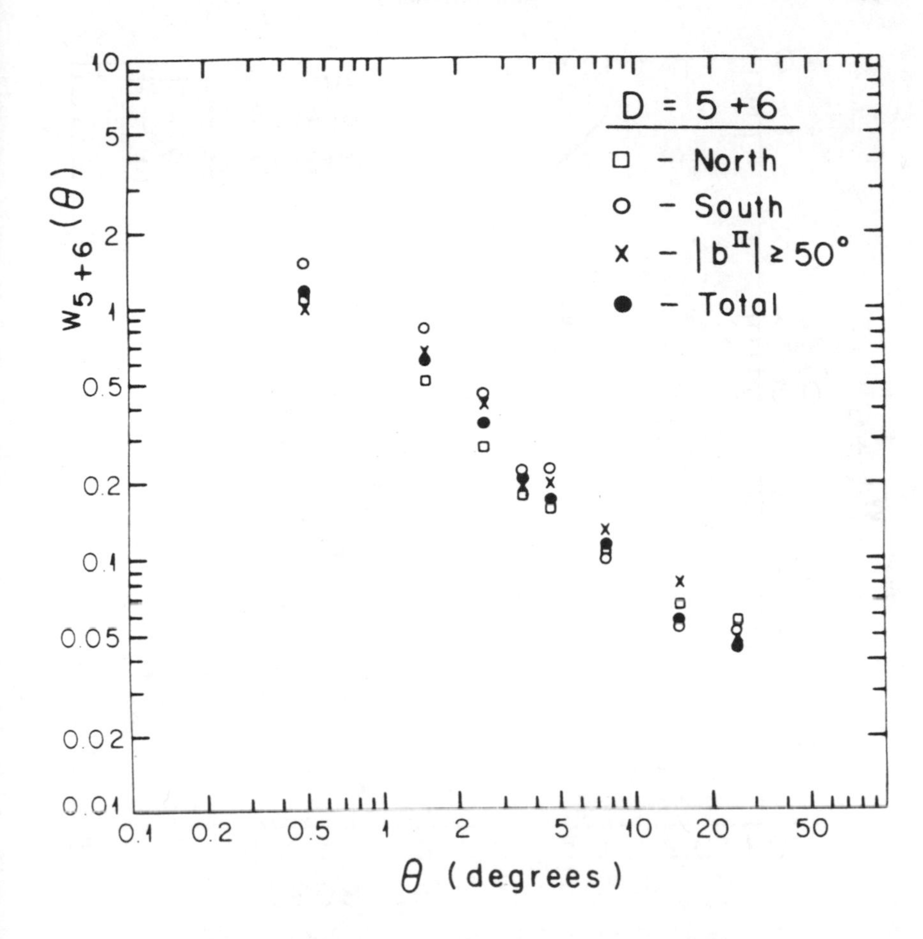

FIG. 3. The angular correlation function of the $D = 5 + 6$ Abell cluster sample for different zones: northern hemisphere (*squares*); southern hemisphere (*circles*); high latitude, $|b| \geq 50°$ (*crosses*); and the total sample (*dots*). A very strong south polar apparent supercluster increases somewhat the southern correlations at small separations. Other zones of low latitude ($|b| = 30° - 50°$) and different longitudes yield similar results (BS83).

tion functions of different richness classes ($R = 1$ and $R \geq 2$) were determined for the large $D = 5+6$ sample (1125 $R = 1$ clusters, 422 $R \geq 2$ clusters). The amplitude of the correlation function was found to be strongly dependent on cluster richness, with richer clusters ($R \geq 2$) showing stronger correlations by a factor of ~ 3 as compared with the poorer ($R = 1$) clusters. The results are shown in Figure 4. Both richness classes exhibit the same power-law shape correlation function as observed in the total sample; they satisfy

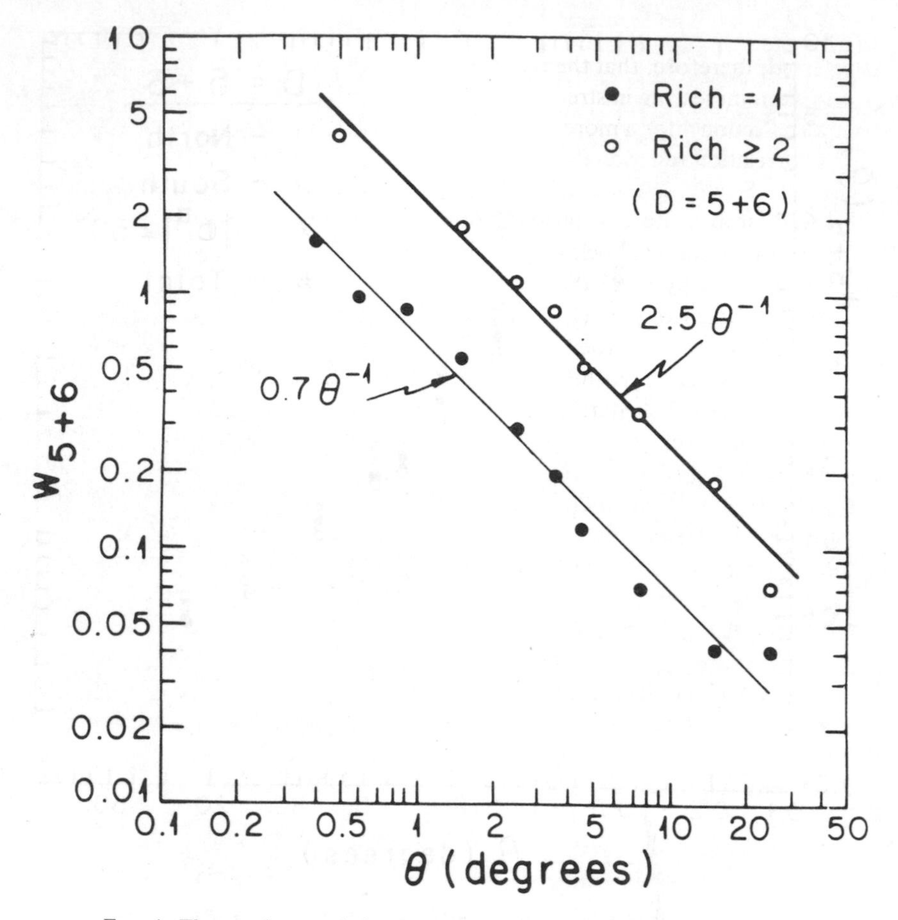

FIG. 4. The angular correlation function of richness 1 and richness ≥ 2 Abell clusters in the $D = 5 + 6$ sample (BS83).

$$w_{5+6}(\theta) \simeq 0.7\theta^{-1} \qquad R = 1 \tag{6}$$

$$w_{5+6}(\theta) \simeq 2.5\theta^{-1} \qquad R \geq 2. \tag{7}$$

The implied spatial correlation can then be represented by:

$$\xi_{cc}(r) \simeq (r/24)^{-1.8} \qquad R = 1 \tag{8}$$

$$\xi_{cc}(r) \simeq (r/48)^{-1.8} \qquad R \geq 2. \tag{9}$$

The amplitude of the total ($R \geq 1$) correlation function is dominated by the lower amplitude of the poorer, but more numerous, $R = 1$ clusters. BS83 suggested, therefore, that the correlation function depends on the richness of the system, increasing in strength from single galaxies to poor and rich clusters (see also Section 4 for a more updated richness dependence). The galaxy-cluster cross-correlation function (Seldner and Peebles 1977; see, however, Efstathiou 1988; also Section 2.4) is consistent with the cluster correlations and the trend observed above. Recent observations of clusters of different types and richnesses (see summary below) yield results that are consistent with the richness trend suggested by BS83 (Section 4).

Klypin and Kopylov (1983) investigated the spatial correlation function of a nearby sample of Abell clusters similar to the one described above, supplementing available redshift data with their own observations. Their sample includes 158 Abell clusters of all richness classes ($R \geq 0$; i.e., including the somewhat incomplete class of $R = 0$ clusters) in distance group $D \leq 4$, and located at $|b| \geq 30°$. Their results are consistent with those of BS83. They find $\xi_{cc}(r) = (r/25)^{-1.6}$ for their observed range of $r \lesssim 50\text{h}^{-1}$ Mpc. The approximately 10% difference in slope is within the 1σ uncertainty of the slope determination estimated by BS83.

The earlier work of Hauser and Peebles (1973) used power-spectrum analysis and angular correlations to investigate the distribution of clusters in the Abell catalog. They also find evidence for strong superclustering of clusters, and show that the degree and angular scale of the apparent superclustering varies with distance in the manner expected if the clustering is intrinsic to the spatial distribution rather than a consequence of patchy local obscuration.

Additional recent investigations of the spatial distribution of rich clusters of galaxies in the Abell catalog include Kalinkov *et al.* (1985), Batuski and Burns (1985), Postman *et al.* (1986), Shvartsman (1988), Szalay *et al.* (1988), and Huchra (1988). These studies include investigations of different subsamples of the catalog, to different distances, regions, and/or richnesses, as well as apply different techniques and/or corrections. All the investigations yield consistent results with those described above, as summarized below.

Kalinkov *et al.* (1985) find a spatial correlation function for rich ($R \geq 1$) Abell clusters, using new redshift estimator calibrations and richness corrections, of $\xi_{cc}(r) = (r/22.4)^{-1.9}$ for $r \lesssim 80\text{h}^{-1}$ Mpc.

Batuski and Burns (1985) determined the spatial correlation function for Abell clusters of all richness groups ($R \geq 0$) to $z \simeq 0.085$. The sample includes 226 clusters. (The higher spatial density of this sample as compared with the $R \geq 1$ sample is due to the inclusion of the $R = 0$ clusters). For this sample they find $\xi_{cc}^{R \geq 0}(r) = 65r^{-1.5}$ for $r \lesssim 150\text{h}^{-1}$ Mpc. The somewhat shallower slope, while within 2σ of the 1.8 slope, may be partially due to the use of some estimated rather than measured redshifts, which reduces the correlations on small scales and flattens the slope (see BS83). When approximated as a 1.8 power-law slope, the function is $\xi_{cc}^{R \geq 0}(r) \simeq 200(r)^{-1.8} \simeq (r/19)^{-1.8}$. This correlation function is one order of magnitude stronger than the galaxy correlations, and about 50% lower than BS83 correlation function for $R \geq 1$ clusters. The somewhat reduced correlation strength is consistent with the richness dependence suggested by BS83 and Bahcall and Burgett (1986).

Postman *et al.* (1986) re-analyzed the $D \leq 4$ sample used by Bahcall and Soneira, as well as a sample of 152 Abell clusters to $z \leq 0.1$. Their results are consistent with the BS83 correlation functions.

Shvartsman (1988) and Kopylov *et al.* (1987) used the 6-meter USSR telescope to measure redshifts of all very rich ($R \geq 2$) Abell clusters to $z \lesssim 0.23$, located at $b > 60°$. They calculated the spatial correlation function of this deep sample of very rich clusters, that includes 50 clusters in the redshift range $0.10 \leq z \leq 0.23$. They find $\xi_{cc}^{R \geq 2}(r) = (r/40)^{-1.5 \pm 0.5}$ for the range $5 \lesssim r \lesssim 50h^{-1}$ Mpc, consistent with the BS83 correlations of very rich ($R \geq 2$) clusters, and with the suggested increase of correalation strength (and length) with richness. The correlation scale for the $R \geq 2$ clusters is $\sim 40h^{-1}$ Mpc, while the correlation scale for the $R \geq 1$ clusters is $\sim 25h^{-1}$ Mpc. The above authors also report weak but positive correlations at much larger separations: $\xi(100\text{-}150h^{-1}\ \text{Mpc}) = 0.47 \pm 0.14$. A similar result is suggested by Batuski *et al.* (1988). This is comparable to the supercluster correlation results of Bahcall and Burgett (1986) (Sec. 3) where similar marginal (3σ) correlations are detected. Systematic effects, however, which may be important on these scales, are difficult to assess.

Geller and Huchra (1988) used a deep redshift sample ($z \lesssim 0.2$) of Abell clusters complete over a small region of the northern sky. They find $\xi_{cc}^{R \geq 0}(r) \sim (r/20)^{-1.8}$, consistent with the results discussed above.

The new southern hemisphere catalog of rich clusters (Abell, Corwin, and Olowin 1988) can also be analyzed for structure. Bahcall *et al.* (1988b) analyzed the distribution of clusters in this catalog. Preliminary results suggest that the correlation function of clusters in the southern sky is consistent with the results presented above for northern clusters.

2.2 - *Shectman Clusters*

Shectman (1985) used the Shane-Wirtanen (1967) counts to identify clusters of galaxies by finding local density maxima above a threshold value, after slightly smoothing the data to reduce the effect of the sampling grid. A total of 646 clusters of galaxies was identified using the specified selection algorithm.

The radial velocity distribution of these clusters is similar to the radial velocity distribution of Abell clusters of distance class $D \leq 4$ as determined by Shectman from comparisons of velocity data for a complete sample of 112 clusters. The space density of the Shectman clusters is therefore ~ 6 times greater than the space density of the 104 $R \geq 1$, $D \leq 4$ Abell cluster sample.

The angular two-point correlation function of the Shectman clusters at $|b| \geq 50°$ (a sample of 488 clusters in total) was determined by Shectman

(1985). The implied spatial correlation function is $\xi_{cc}(r) \simeq 180r^{-1.8} \simeq (r/18)^{-1.8}$. This correlation function is about ten times larger than the galaxy correlation (eq. 3), and is about a factor of two lower than the rich ($R \geq 1$) cluster correlations (eq. 2). Since the space density of the Shectman clusters is ~ 6 times higher than the density of the $R \geq 1$ clusters, and the identifications of the former are therefore with poorer clusters, the results of the Shectman cluster correlations are consistent with those of the Abell clusters, and consistent with the trend suggested by Bahcall and Soneira of increased correlation strength with cluster richness (Sec. 4).

2.3 - Zwicky Clusters

The angular distribution of clusters in the Zwicky catalog (1968) was analyzed by Postman *et al.* (1986). The cluster selection algorithm in the Zwicky catalog differs markedly from the cluster selection definition of Abell (e.g., Bahcall 1977, 1988a). Abell's definition of a cluster relates to the cluster intrinsic properties (i.e., number of galaxies within a given linear scale, and a given absolute magnitude range) and thus is independent of redshift (except for standard selection biases). Zwicky's clusters are defined relative to the mean density of the field, with varying cluster sizes and contours, and consider all galaxies down to the plate limit. Therefore, the cluster selection is by definition strongly dependent on redshift. A direct comparison between the correlation function of Zwicky and Abell clusters is, therefore, not straightforward. However, an uncorrected comparison will test to some extent the universality of the cluster correlation function, with its suggested dependence on richness, as well as further test the sensitivity of the correlation function to the cluster identification procedure.

It is found that in the distance range where Abell and Zwicky identify clusters of comparable overdensity (1173 distant Zwicky clusters with $z \simeq 0.1 — 0.14$), the correlation functions of the Abell and Zwicky clusters are indeed the same in the scale range studied ($r \lesssim 60h^{-1}$ Mpc). The angular correlation functions of the two nearer samples of the Zwicky clusters (377 Near clusters and 680 Medium-Distant clusters) are observed to be weaker (when scaled to the same depth as the $D \leq 4$ Abell sample) than the rich ($R \geq 1$) Abell clusters. Since these nearer Zwicky clusters are by definition much poorer clusters, with a considerably higher space density than the $R \geq 1$ Abell clusters, they are expected to have a weaker correlation strength (BS83 and Sec. 4).

A comparison of the cluster correlation functions determined by the various investigators discussed above using different catalogs and samples is summarized in Figures 5a-5b. A general agreement is observed among all the results. The consistency of the correlation functions determined from different catalogs, cluster selection criteria, redshift and richness ranges, and by different investigators, strongly supports the reality and universality of the cluster correlations described in this section.

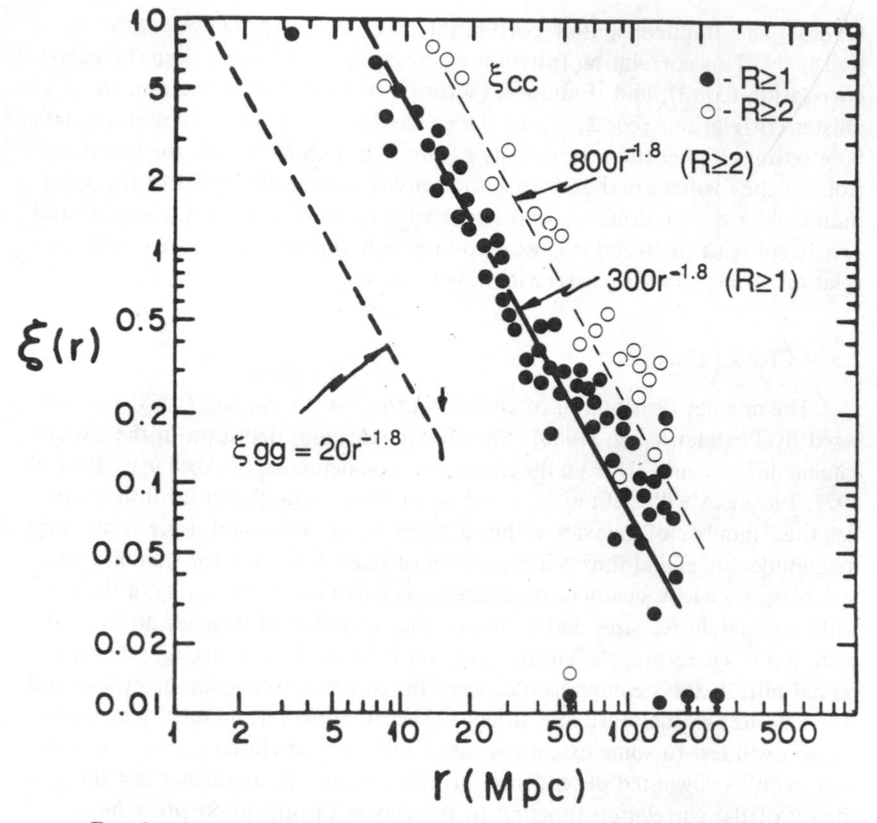

FIG. 5a. A composite of the spatial cluster correlation function determined by different investigators from different cluster samples. Abell clusters are to different depths ($z \simeq 0.08$ through $z \simeq 0.24$), different richnesses, and different regions; (Sec. 2). The BS83 correlation function, $300r^{-1.8}$, is shown (see Figure 1); the results of the different samples are all consistent with this function. The richer $R \geq 2$ clusters exhibit stronger correlations as suggested by BS83 (Sec. 4).

2.4 - Galaxy-Cluster Cross-Correlations

The angular cross-correlation between the galaxy distribution in the Shane-Wirtanen galaxy counts and the positions of rich Abell clusters was studied by Seldner and Peebles (1977), and more recently by Efstathiou (1988). This cross-correlation function, $w_{gc}(\theta)$, measures the excess probability, over random, of finding a galaxy within a given separation from a cluster; i.e., it describes the enhanced density of galaxies around a cluster.

Seldner and Peebles (1977) find that the angular function $w_{gc}(\theta)$ scales with cluster distance-class D as expected from the galaxy luminosity function.

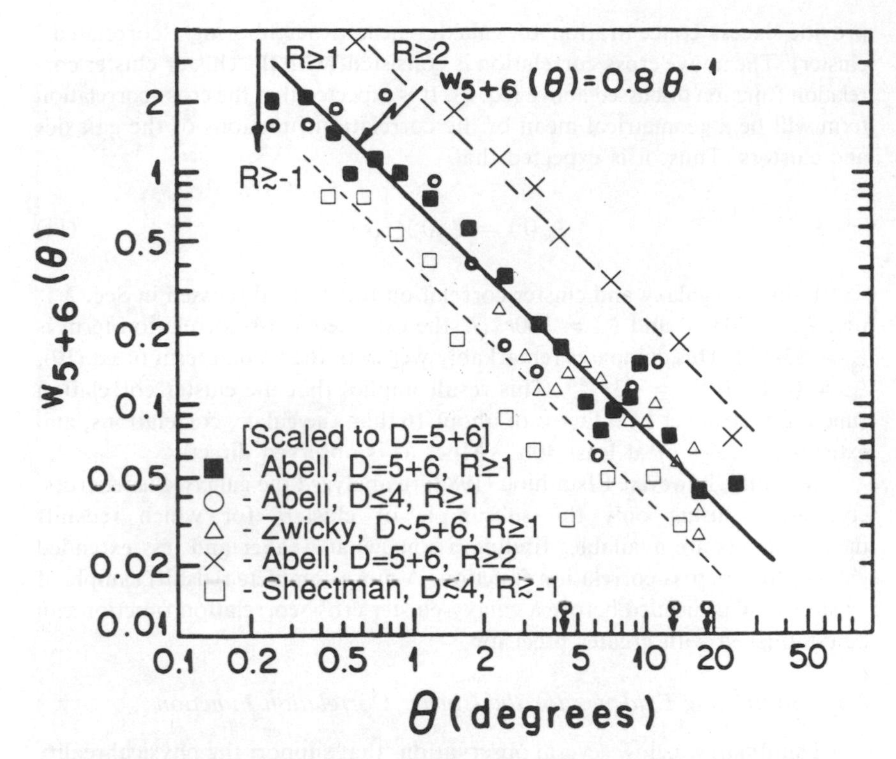

FIG. 5b. A composite of the angular cluster correlation function determined from different catalogs and samples, indicated by the different symbols (Sec. 2). All results are scaled to the $D = 5 + 6$ distance. The BS83 correlation function for the $D = 5 + 6$ clusters (Figure 2) is indicated by the solid line. The consistency among the different samples, as well as the dependence of the correlation strength on richness (Sec. 4), are apparent.

The $w_{gc}(\theta)$ estimates are reasonably well fitted by a two-power-law model for the function, $\xi_{gc}(r)$ (Peebles 1980b):

$$\xi_{gc}(r) = (r/7)^{-2.5} + (r/12.5)^{-1.7} \qquad r \lesssim 40h^{-1}\text{Mpc}. \qquad (10)$$

The enhancement of Lick counts around cluster centers is traced to $r \sim 40h^{-1}$ Mpc before it is lost in the noise.

The first term of the galaxy-cluster cross-correlation (eq. 10) represents the "standard" internal density profile of galaxies in a cluster (which generally has the shape of a bounded isothermal sphere; e.g., Bahcall 1977). The more slowly varying part of the cross-correlation function found at larger scales and represented by the $r^{-1.7}$ part of eq.(10) is produced by the clustering of clusters, as discussed in the previous subsections (i.e., galaxies from one cluster provide excess concentration of galaxies near a neighboring "correlated" cluster). The above cross-correlation is consistent with the cluster-cluster correlation function discussed above (eq. 2). It is expected that the cross-correlation term will be a geometrical mean of the correlation functions of the galaxies and clusters. Thus, it is expected that

$$\xi_{gc}(r) \simeq \xi_{gg}^{\frac{1}{2}}(r)\xi_{cc}^{\frac{1}{2}}(r). \tag{11}$$

Using the galaxy and cluster correlation functions discussed in Sec. 2.1, i.e., $\xi_{gg} \simeq 20r^{-1.8}$ and $\xi_{cc} \simeq 360r^{-1.8}$, the expected cross-correlation term is $\xi_{gc} \simeq 85r^{-1.8}$. This compares remarkably well with the second term of eq.(10), $\xi_{gc} \simeq (r/12.5)^{-1.7} \simeq 73r^{-1.7}$. This result implies that the cluster correlation function is stronger by a factor of about 16 than the galaxy correlations, and extends to scales of at least $40h^{-1}$ Mpc, as is observed directly.

Recently, however, Efstathiou (1988) re-analyzed the galaxy-cluster cross-correlations using only the subsample of clusters for which redshift measurements are available, finding a somewhat weaker and less extended galaxy-cluster cross-correlation function. A more complete redshift sample of clusters may be needed before a galaxy-cluster cross-correlation function can be established with greater precision.

2.5 - Supporting Evidence for the Cluster Correlation Function

I summarize below several observations that support the physical reality of the cluster correlation function discussed above.

a. The angular cluster correlation function scales with depth as expected from spatial correlations, rather than from patchy obscuration or systematic omission (Hauser and Peebles 1973, BS83).

b. The projected and redshift cluster correlation functions yield consistent results, thus indicating the physical reality of the correlations.

c. The cluster correlation function yields consistent results in different large regions of the sky (e.g., north versus south, high versus low latitudes, different longitude ranges; BS83). The estimated scatter, or uncertainty, in the correlation function is approximately $\pm 15\%$ in the correlation scale (i.e., approximately $\pm 25\%$ in the amplitude).

d. The cluster correlation determined from the Abell sample is consistent with more recent results using other samples and catalogs (e.g., Shectman clusters - Shectman 1985; Zwicky clusters - Postman *et al.* 1986; and subsamples of different regions, redshifts, and richnesses in the Abell catalog - BS83; Batuski and Burns 1985; Postman *et al.* 1986; Shvartsman 1988; Szalay *et al.* 1988; see Figure 5.). These comparisons provide a strong test of the sensitivity of the correlation function to the cluster identification procedure. The scatter in the correlation scale-length is $\lesssim \pm 15\%$, as discussed above.

e. The cluster correlation function is consistent with the galaxy-cluster cross-correlation function determined by Seldner and Peebles (1977). It is less consistent, however, with the cross-correlation results of Efstathiou (1988).

f. A preliminary estimate of the completeness limit of the *nearby* Abell sample obtained by comparisons with X-ray data of galaxy clusters yields a reasonably high completeness level for the sample (work in progress by Bahcall, Maccacaro, Gioia, *et al.*). Of the 25 nearest clusters in the sample, all but one are detected as extended X-ray sources with luminosities appropriate to rich clusters (the 25th cluster has an upper limit consistent with the expected luminosity). In addition, preliminary results of the Einstein Medium Deep X-ray Survey show that no extended X-ray cluster is found that should have been a rich cluster in Abell's nearby sample, but was missed (out of approximately four real clusters expected within the survey area; i.e., completeness of better than ~75%).

The suggestion of a possible "projection" effect in the Abell catalog, in which the existence of one cluster close to a neighboring cluster enhances the richness of the latter (Sutherland 1988) has only a minimal effect on the resulting correlation function (e.g., Dekel 1988). At a separation of ~ $10h^{-1}$ Mpc from the cluster center, typically less than ~1-3 galaxies from the tail of the foreground cluster will contribute to the neighboring cluster. A negligible effect is expected on scales much larger than the above. A significant projection effect would also be inconsistent with the observed scaling-law between the nearer and distant samples. The consistent results obtained for the correlation functions of clusters from different catalogs and samples also indicate that any such selection effects are likely to have only a minimal impact on the resulting correlations.

The evidence listed above supports the reality of the cluster correlation function and suggests that it is unlikely that the correlations are mainly a result of catalog biases or omissions. A determination of the cluster correlation function from catalogs with automated selection procedures will improve the accuracy of the intrinsic cluster correlations, especially at large separations where the correlations are rather weak.

3. Supercluster Correlations

Bahcall and Burgett (1986) carried the study of rich galaxy clusters one step further by studying the spatial distribution of superclusters. The sample used was the Bahcall-Soneira (1984) complete catalog of superclusters to $z \leq 0.08$, where superclusters are defined as groups of rich clusters and identified by a spatial density enhancement of clusters. All volumes of space with a spatial density of clusters f times larger than the mean cluster density are identified in the above catalog as superclusters for a specified value of f. The supercluster selection process was repeated for various overdensity values f, from

$f = 10$ to $f = 400$, yielding specific supercluster catalogs for each f value. A total of 16 superclusters are cataloged for $R \geq 1$ and $f = 20$, and 26 superclusters for $R \geq 0$ and $f = 20$.

The spatial correlation among the superclusters was determined by Bahcall and Burgett (1986) for samples of different richness and overdensity. Because of the large size of the superclusters themselves, no meaningful correlations are expected at small separations ($\lesssim 50\ h^{-1}$ Mpc). In addition, no detectable correlations are expected at very large separations ($> 200\ h^{-1}$ Mpc), since this scale is comparable to the limits of the sample. Any observable correlations are therefore expected only in a separation "window" around $\sim 100\ h^{-1}$ Mpc.

The results, presented in Figure 6, reveal correlations among superclusters on a very large scale: $\sim$ 100-150 h^{-1} Mpc. Because of the small size of the supercluster sample, the statistical uncertainty is appreciable; the observed effect is at the 3σ level (as determined by comparisons with numerical simulations of random catalogs). In addition, all the samples with different overdensities and cluster richnesses show a similar effect at a similar scale length. The results imply the existence of very large-scale structures with scales of $\sim$ 100-150 h^{-1} Mpc.

Similar results have been recently obtained by Kopylov *et al.* (1987) by studying correlations of very rich clusters to $z \lesssim 0.2$ (Sec. 2.1). They report ξ_{cc}(100-150h^{-1} Mpc) $= 0.47 \pm 0.14$. Tully's (1988, 1987) observations of very large-scale structures in the cluster distribution, up to $\sim 300h^{-1}$ Mpc, may also reflect the above observed tendency of superclusters to cluster.

Figure 6 shows that the supercluster correlation strength is stronger than that of the rich cluster correlations by a factor of approximately 4. It is approximately two orders of magnitude stronger than the galaxy correlation amplitude. While this enhancement is observed in the $\sim$ 100-150h^{-1} Mpc range, it is possible that the supercluster correlation function also follows an $r^{-1.8}$ law. If the correlations follow an $r^{-1.8}$ law, then the function would satisfy the relation

$$\xi_{sc,sc}(r) \simeq 1500 r^{-1.8} \simeq (r/60)^{-1.8}. \quad (12)$$

The implied correlation scale of superclusters would be $60h^{-1}$ Mpc, as compared with $5h^{-1}$ Mpc for the correlation scale of galaxies (Groth and Peebles 1977) and $25h^{-1}$ Mpc for rich ($R \geq 1$) clusters (BS83). This apparent increase in correlation strength is consistent with the earlier prediction of BS83 of increased correlations with richness (luminosity) of the system. The supercluster correlation amplitude fits well the predicted trend (Sec. 4).

4. Richness Dependence of the Correlations

As discussed above, the cluster correlation function appears to depend strongly on cluster richness (BS83), with richer clusters showing stronger cor-

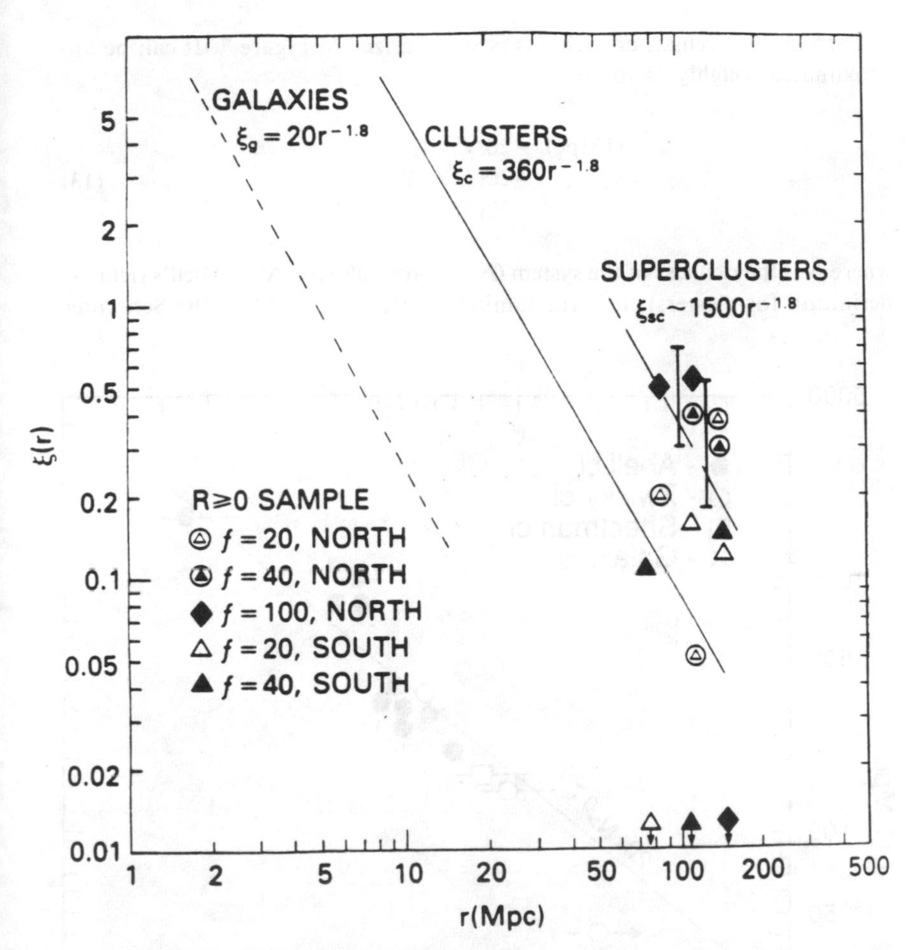

FIG. 6. The spatial correlation of superclusters for the $R \geq 0$ sample (Bahcall and Burgett 1986). Different sub-samples are indicated by different symbols. No meaningful correlations are expected below $\sim 50h^{-1}$ Mpc.

relations than poorer clusters. This result, combined with the lower correlation amplitude of individual galaxies, led Bahcall and Soneira to the conclusion that progressively stronger correlations exist, at a given separation, for richer or more luminous galaxy systems (Sec. 2.1). Several recent studies of the correlations of other types and richnesses of clusters, reviewed above (Batuski and Burns 1985; Shectman 1985, and Postman *et al.* 1986 for poorer clusters; Kopylov *et al.* 1987 for richer clusters; Bahcall and Burgett 1986 for superclusters) appear to be consistent with the trend suggested by Bahcall and Soneira and later expanded by Bahcall and Burgett (1986). This dependence of correlation strength on richness is summarized in Figure 7. It can be approximated roughly as follows:

$$\begin{aligned}\xi(1\text{Mpc}) &\sim 20N^{0.7} \\ &\sim 20(L/L^*)^{0.7} \\ &\sim 20(M/10^{12}M_\odot)^{0.5}\end{aligned} \tag{13}$$

where N is the richness of the system ($N = 1$ for galaxies; $N =$ Abell's richness definition for clusters), L is the luminosity (relative to L^* in the Schechter

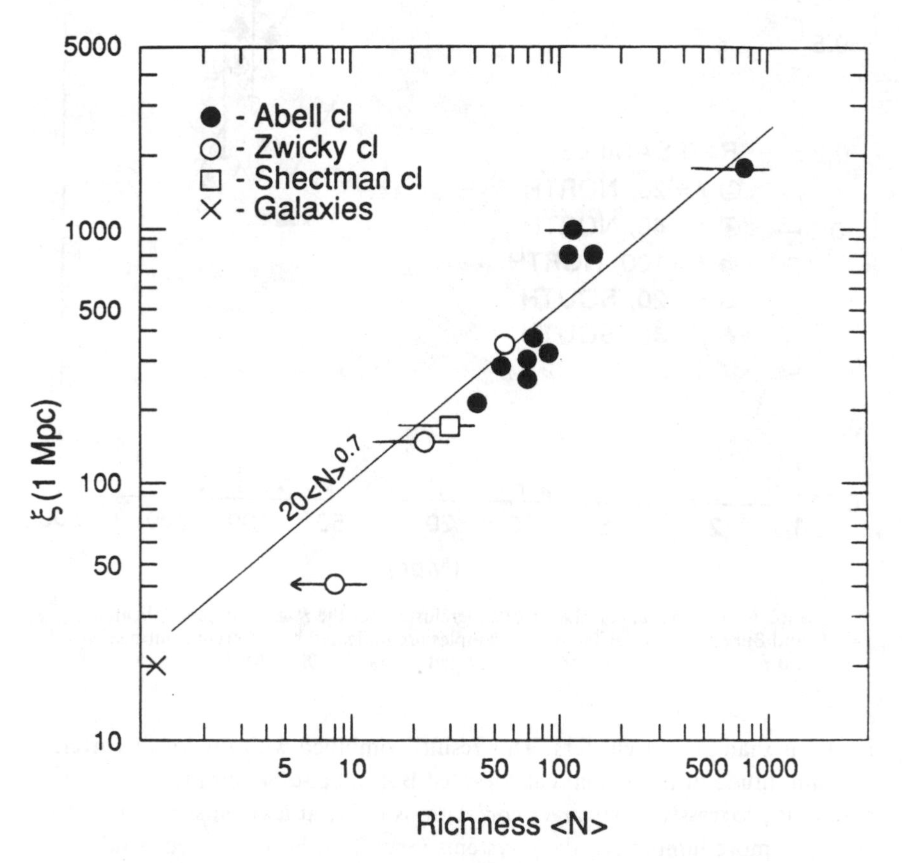

FIG. 7. The dependence of the correlation function strength on the mean richness (∝ luminosity) of the system (Bahcall and Burgett 1986). The results are for clusters from different catalogs (Abell, Zwicky, and Shectman, as indicated by the symbols), determined by different investigators for samples of different richnesses, depths, and regions (Sec. 2). The correlation strength for galaxies and superclusters are also included. The solid line indicates the approximate dependence on richness.

luminosity function), and M is the mass of the system. This relation suggests an average trend in the data and should not be regarded as an exact formula. (Obviously, the relation between N, L, and M is not unique; for a given N, different L's and M's may apply and vice versa. The difference between the M versus the L slope is due to the higher observed M/L ratios for clusters than for galaxies).

The correlation-richness dependence suggests that rich clusters populate the large-scale structures, or superclusters, more abundantly than galaxies do relative to their mean space densities. It also implies that rich clusters are indeed an efficient tracer of large-scale structure in the universe.

Several galaxy formation models such as biased cold dark matter, hybrid scenarios, and cosmic strings can reproduce a trend of increasing correlation strength from galaxies to clusters.

5. A Universal Correlation Function

The increase of correlation strength with richness implies that rich, luminous systems are more strongly clustered, at a given separation, than poorer systems. The power-law of the correlation functions is also observed to be identical in the various systems studied. Either initial conditions, or subsequent evolution, may be responsible for the observed phenomena. Since the observed correlation functions follow the same power law ($r^{-1.8}$), the effect of increased correlation strength with richness (at a given separation) can also be expressed as a scale shift in the correlation functions (Szalay and Schramm 1985). In Figure 8 I plot the amplitude of the correlation functions of the various systems (galaxies, poor and rich clusters, superclusters) as a function of the mean separation of objects in the sample, d (see Bahcall and Burgett 1986; Bahcall 1987). The mean separation is related to the mean spatial density of objects in the sample, n, through $d = n^{-1/3}$. For example the mean separation of galaxies is about 5 Mpc, while the mean separations of $R \geq 1$ and $R \geq 2$ clusters are, respectively, about 50 Mpc and 70 Mpc.

It is apparent from Figure 8 that the correlation strength increases with the sample's mean separation. Moreover, a dimensionless correlation function normalized to the sample's mean separation, d, appears to yield a constant, universal function for nearly all the systems studied (some enhancement is required for galaxies, as described below). This universal dimensionless correlation function has the form

$$\xi_i(r) \simeq 0.3(r/d_i)^{-1.8} \simeq (r/0.5d_i)^{-1.8}, \tag{14}$$

where the index i refers to the system being considered, and d_i is its mean separation. Relation (14) implies a universal dimensionless correlation

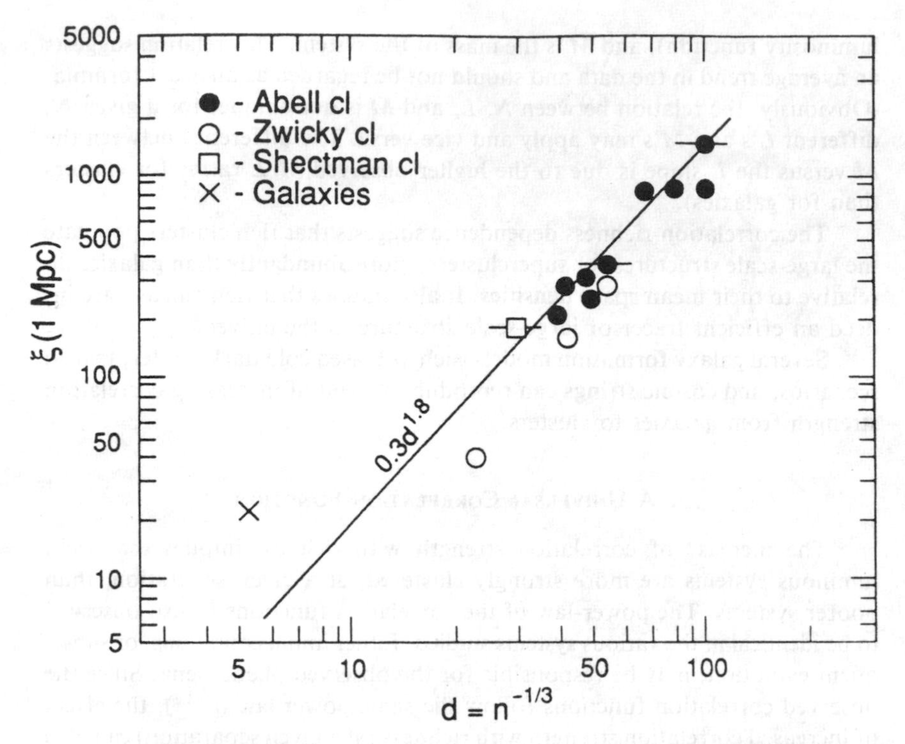

FIG. 8. The dependence of the correlation function on the mean separation of objects in the system. The results are for clusters from different catalogs (Abell, Zwicky, and Shectman, as indicated by the symbols), determined by different investigators for samples of different mean densities (i.e., mean separations). The correlation strength for galaxies and superclusters are also included. The solid line represents a $d^{1.8}$ dependence (e.g., Szalay and Schramm 1985, Bahcall and Burgett 1986).

amplitude of ~ 0.3, and, equivalently, a universal correlation scale of $r_o \simeq 0.5d_i$. The correlation function of galaxies is stronger than expressed by relation (14) by a factor of about four (Figure 8). The universality of the correlation function suggests a scale-invariant clustering process (Szalay and Schramm 1985). The stronger dimensionless galaxy correlations may imply gravitational enhancement on smaller scales. If a non-linear process, other than gravity, participates in galaxy formation, and this process is scale-invariant, the created structure will have a single power law correlation function, the slope of which (α) is related to the geometry of the structure, i.e., its fractal dimension (β). The latter is related to the correlation function slope via $\alpha = \beta - 3$ (see, e.g., Mandelbrot 1982). The fractal dimension of the universal structure implied by the above data is therefore $\beta \simeq 1.2$. Small-scale gravitational clustering may break the scale invariance and increase the dimensionless correlation amplitude for galaxies.

We do not know yet what physical process can create a scale invariant structure with $\beta \simeq 1.2$. An innovative suggestion involves cosmic strings as the primary agent in the formation of galaxies and clusters; this model appears to create such a scale-invariant infrastructure (Turok 1985). The model yields a scale-invariant correlation function similar to that observed, with a power-law of -2 (as implied by one-dimensional "string" structures with fractal dimension of unity). More detailed calculations with cosmic strings models are currently being carried out by several investigators (Bouchet and Bennet 1988, Turok 1988).

6. Phenomenological Clustering Models

6.1 - Long Tails to Galaxy Clusters

The galaxy correlations depend, at least partially, on the rich cluster correlations since clusters contain galaxies. If all galaxies were members of rich clusters, the two correlation functions should be approximately the same on large scales. The fraction of galaxies in clusters is clearly less than unity. The fraction of galaxies, f, that are associated with rich clusters, represents the probability that a randomly chosen galaxy is correlated with a rich cluster. These associations may include large structures (tens of Mpc), comparable to the separations observed in the cluster correlation function (and well above the standard Abell radius of $1.5h^{-1}$ Mpc).

The galaxy correlation function contains contributions from three terms (Bahcall 1986): galaxy pairs from the fraction f of galaxies that are cluster members; pairs from the fraction 1-f of galaxies that are non-cluster members ("field"); and cross-term pairs. Inserting the analytic expressions for each of these terms into the expression for the overall galaxy correlation function yields:

$$\left(\frac{\xi_{cc}}{\xi_{gg}}\right)^{\frac{1}{2}} = \frac{1 - (1 - f)\,(\xi^f_{gg}/\xi_{gg})^{\frac{1}{2}}}{f}. \qquad (15)$$

The above ratio of the cluster to galaxy correlation strength depends on two parameters: the fraction of galaxies in clusters, f, and the ratio of the "field" galaxy correlation strength, ξ^f_{gg} (i.e., the correlation of the 1-f fraction of galaxies outside the clusters) to the overall galaxy correlation ξ_{gg}. If all galaxies were associated with rich clusters, *i.e.*, $f = 1$, then the galaxy and cluster correlations are identical on large scales, as expected. However, for any fraction $f < 1$, the galaxy correlations will be weaker than parent cluster correlations due to the reducing effect of the less clustered "field" galaxies. Figure 9 represents graphically relation (15). The curves are the expected $\xi_{cc}/\xi_{gg}(f)$ relations for selected values of the parameter $\chi \equiv \xi^f_{gg}/\xi_{gg}$. The observed cor-

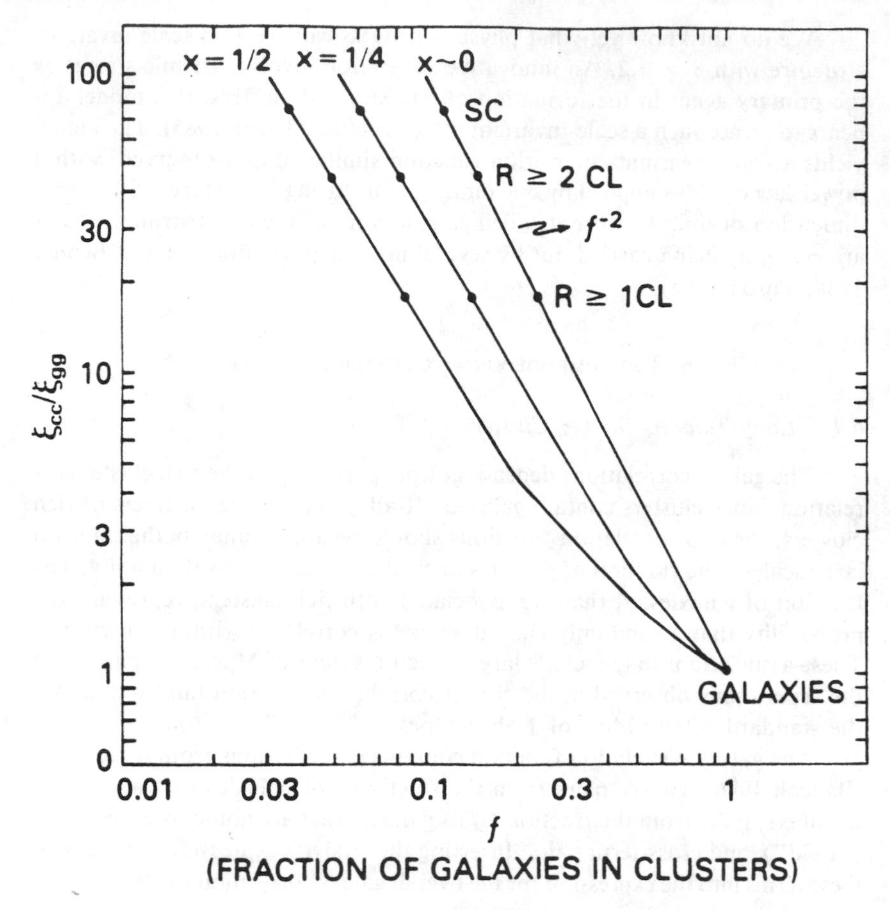

FIG. 9. The ratio of the cluster to galaxy correlation functions predicted from Bahcall's (1986) model (eq. 15) is plotted as a function of the fraction f of galaxies associated with the clusters. The different curves represent different values of the "field" (non-cluster) correlation strength, ξ^f_{gg}, in terms of the ratio parameter $x = \xi^f_{gg}/\xi_{gg}$. The observed correlation strengths of $R \geq 1$ clusters, $R \geq 2$ clusters, and superclusters are indicated by the points.

relation strengths are represented by the data points. The observed ratio $\xi_{cc}/\xi_{gg} \simeq 18$ for $R \geq 1$ clusters yields a fraction of galaxies in clusters that ranges from $f \simeq 25\%$ for $\xi^f_{gg}/\xi_{gg} \simeq 0$ to $f \simeq 15\%$ for $\xi^f_{gg}/\xi_{gg} \simeq \frac{1}{4}$. Therefore, if approximately 20% of all galaxies are associated with rich ($R \geq 1$) clusters, the galaxy correlation function will be, as observed, ~ 18 times weaker than the cluster correlations.

The model suggests (Bahcall 1986) that the fraction of galaxies associated with rich clusters is considerably larger than previously believed; most of these galaxies would be distributed in the outer tails of the clusters, which may extend to at least $\sim 30h^{-1}$ Mpc. Most clusters are therefore predicted to be embedded within much larger structures.

The model makes testable predictions that can be studied with complete redshift surveys. Redshift surveys of galaxies (e.g., Gregory and Thompson 1978; Gregory *et al.* 1981; Chincarini *et al.* 1981; Haynes and Giovanelli 1986; de Lapparent *et al.* 1986) indeed suggest that clusters are generally embedded in large elongated structures that contain a considerable fraction of galaxies. This picture is qualitatively consistent with the phenomenological model described above.

The long-tail model may also explain the negligible correlations observed among *Ly-α* clouds in QSO spectra (Sargent *et al.* 1980). If the clouds can only exist in the field (non-cluster environment) because clusters have too large an ambient pressure, they would not be expected to have significant correlations.

6.2 - The Shell Model

Galaxies may be distributed on surfaces of shells (or cells), with rich clusters located at shell intersections. This picture is suggested by redshift surveys of galaxies (Gregory *et al.* 1981, Giovanelli *et al.* 1986, de Lapparent *et al.* 1986). In order to test this "shell" model and its agreement with the observed galaxy and cluster correlations, Bahcall, Henriksen, and Smith (1988) (see also Bahcall 1988b) made simulations in which they placed galaxies on surfaces of randomly distributed shells, and formed clusters at the shell intersections. They found that the model cluster correlations are consistent with the observed cluster correlations, including the large increase in correlation strength from galaxies to clusters. The results are not very sensitive to the exact parameters used. The model galaxy correlations appear to be consistent with observations on small scales, but exhibit a tail of weak positive correlations at larger separations that are not seen in the data.

An example of results from a typical model is shown in Figure 10; more details are given in Bahcall *et al.* (1988a). The results suggest that the observed cluster correlations may be simply due to the geometry of clusters positioned on randomly placed shells or similar structures; the typical structure size is best fit with a radius of approximately $20h^{-1}$ Mpc.

7. Peculiar Motion of Clusters

The discussion in the previous sections summarizes evidence for the existence of structures on the scale of $\sim 10\text{-}150h^{-1}$ Mpc. A question of critical importance is what are the velocity fields in these structures. Peculiar velocities of clusters on these scales may indicate the existence of large amounts of (dark)

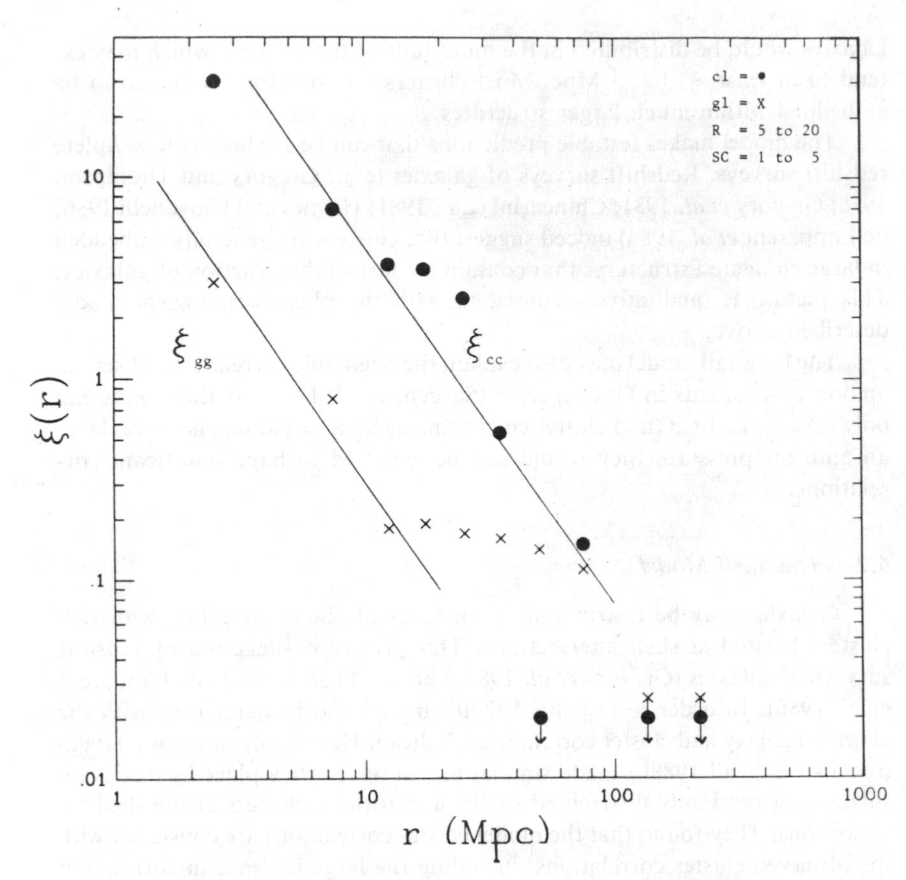

FIG. 10. Shell-model correlation functions (Bahcall *et al.* 1988) for clusters (dots) and galaxies (crosses), and their comparisons with observations (ξ_{gg}, and ξ_{cc} lines). The plotted model represents shell radii distribution in the range 5 to 20 h^{-1} Mpc, and, depending on size, 1 to 5 clusters forming at shell intersections (Sec. 6.2).

matter and are of fundamental importance for models of galaxy and structure formation. Early discussions of possible peculiar velocities among clusters in superclusters are presented by Abell (1961) and Noonan (1977). Noonan observed a tendency of clusters with neighboring Abell clusters to have a greater scatter on the Hubble diagram, which was interpreted as a gravitational perturbation on the cluster redshifts due to the neighboring clusters. More recently, Bahcall *et al.* (1986) used the complete redshift sample of $D \leq 4$ rich Abell clusters to study the possible existence of peculiar motion and/or structural anisotropy on large scales. They find strong broadening in the redshift distribution that corresponds to a cluster velocity of $\sim 10^3$ km s^{-1}. These findings are summarized below.

7.1 - Redshift Elongation: The "Finger-of-God" Effect

The distribution in space of the $D \leq 4$ redshift sample of Abell clusters was studied by Bahcall *et al.* (1986) by separating the three-dimensional distribution into its components along the line-of-sight (redshift) axis and the perpendicular axes projected on the sky. All clusters were assumed to be located at their Hubble distances as indicated by their redshifts, and their pair separations in Mpc were determined in the three components. A scatter-diagram of the cluster pair separations in the redshift (z) direction (R_z) versus their separations in α or δ (R_α or R_δ) was then determined.

If all clusters were located at their Hubble distances with negligible peculiar motion, and if the sample was not dominated by elongated structures in a given direction, a symmetric scatter-diagram should be observed. If a large peculiar velocity exists among clusters, it would manifest itself as an elongated distribution along the z-direction in the R_z-R_α and R_z-R_δ diagrams. This elongation, i.e., the so-called "Finger-of-God" effect, is normally interpreted as peculiar motion. However, the effect may also be caused by geometrically elongated structures, if they dominate the sample (with elongation in the z-direction; see below).

The results are presented in Figures 11 to 13. The scatter diagrams are plotted in Figure 11 for both the $R \geq 0$ and $R \geq 1$ samples. Frequency distributions representing these diagrams are presented in Figure 13. A strong and systematic elongation in the z-direction exists in all the real samples studied. Scatter-diagrams for sets of random catalogs do not exhibit any conspicuous elongation (Figure 12), as expected; a symmetric distribution in all directions is observed. As an additional test, Bahcall *et al.* (1986) also determined the scatter-diagrams in the projected plane, R_α - R_δ, of the cluster sample (Figure 12). Again, as expected, a symmetric distribution is observed in this plane. These tests strengthen the conclusion that the observed elongation is real. The effect of elongation is strong; statistically it corresponds to approximately 8σ in a single sample (assuming, for illustrative simplicity, Gaussian statistics). It is therefore unlikely that the observed redshift elongation is a chance fluctuation. The effect becomes more apparent in the larger $R \geq 0$ sample; this increase is expected if the effect is real.

A similar effect was observed by BS83 in their comparison between the cluster correlation function in the redshift and spatial directions. A broadening in the redshift direction was observed in that study, similar to the present finding. The elongation is unlikely to be caused by background/foreground contamination of galaxies and clusters (e.g. Sutherland 1988), since this would yield an excess of pairs at any R_z separation, as well as any R_α or R_δ, rather than the excess (i.e., broadening) observed specifically at small separations ($\Delta z \lesssim 0.015$). The effect is also much larger than either the uncertainties in the redshift measurement or the uncertainties caused by the internal velocity dispersion within the clusters (see below).

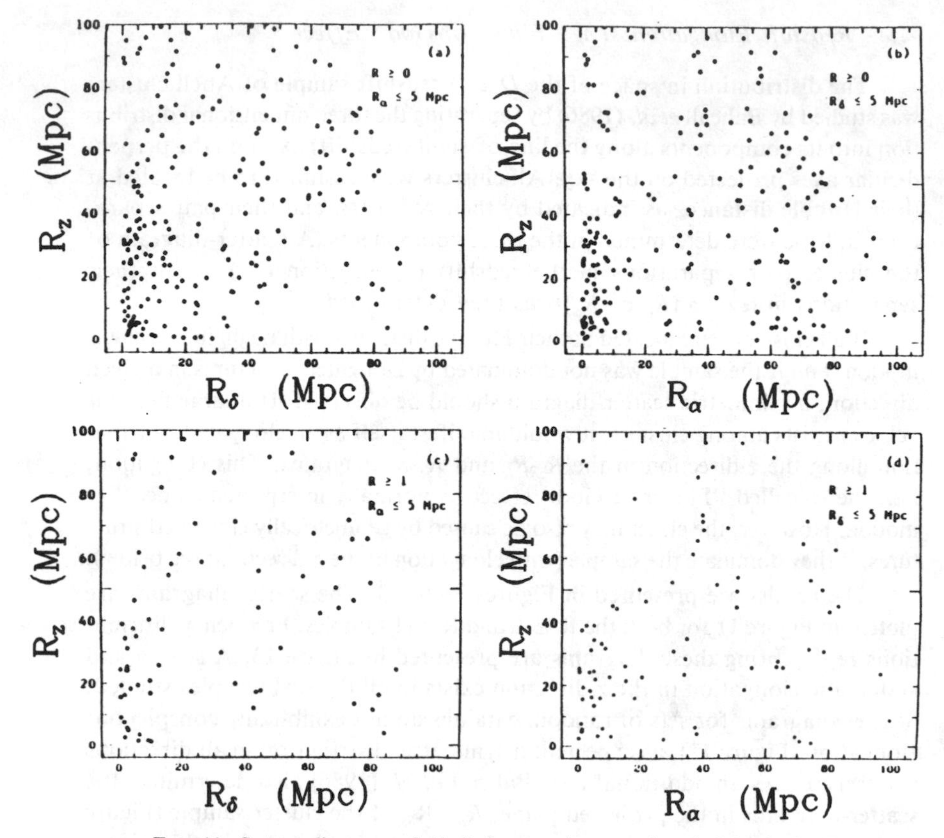

FIG. 11. Scatter-diagrams of Abell cluster pair separations in Mpc in the R_z-R_α and R_z-R_δ planes (Bahcall *et al.* 1986). (The pair separations along the third axis, perpendicular to each plane, are limited to $\leq$ 5 Mpc). All cluster pairs with a total spatial separation $\leq$ 100 Mpc are included. Figures a, b and c, d represent, respectively, the $R \geq 0$ and $R \geq 1$ richness samples. The elongation in the redshift direction is apparent in all cases.

To determine what velocity could cause the observed effect, the authors convolved the frequency distribution observed along the projected axis, which is unperturbed by peculiar motion, with a Gaussian velocity distribution. A Gaussian form is assumed for convenience in estimating the velocity broadening. This convolved distribution should match the broadened distribution observed in the redshift direction. The best fit is obtained for a velocity width of $\sqrt{2}\sigma \simeq 2000$ km s^{-1}. The estimated uncertainty on this mean velocity is approximately $+1000/-500$ km s^{-1}. The above result is consistent with the result of BS83, using the redshift broadening observed in the cluster correlation function.

The above value for the velocity width includes all contributions to the

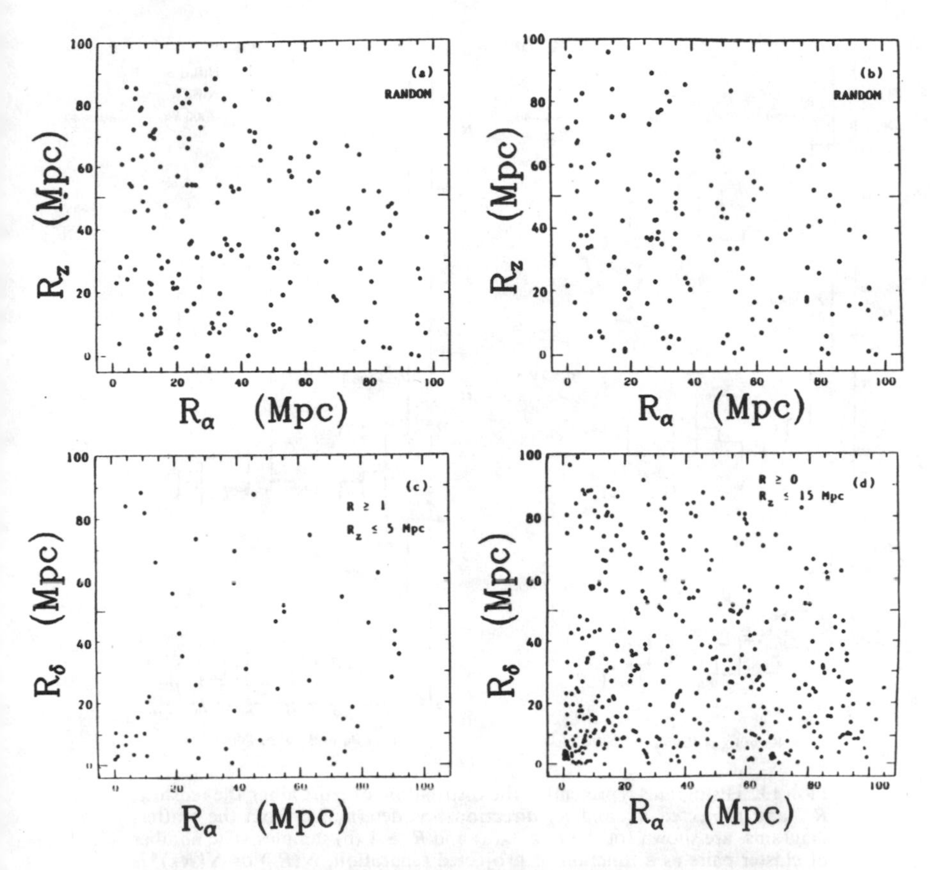

FIG. 12. Same as Figure 11 but for typical random distribution of clusters (Fig. a, b), and for the *projected* distribution (i.e. R_δ-R_α plane) of the actual cluster samples (Fig. c, d). No elongation is expected in either case and none is observed. The clustering of clusters is apparent in the data sample of Fig. c, d.

broadening effect, such as redshift measuring uncertainty and possible deviations from the true cluster redshift due to individual galaxy velocites in the clusters. Redshift measuring uncertainties are negligible compared to the 2000 km s^{-1} observed. The effect of peculiar motion within the clusters (for those clusters that have only a small number of measured galaxy redshifts) was estimated by comparing cluster redshifts from the current sample with those obtained using a larger number of measured galaxy redshifts, when available. For the latter study, the redshift catalogs of Sarazin, Rood, and Struble (1982), and Fetisova (1981) were used. A root mean square deviation for these cluster redshifts of approximately 300 km s^{-1} is observed due to the above effect. This value is reasonable considering that the full velocity dispersion in clusters

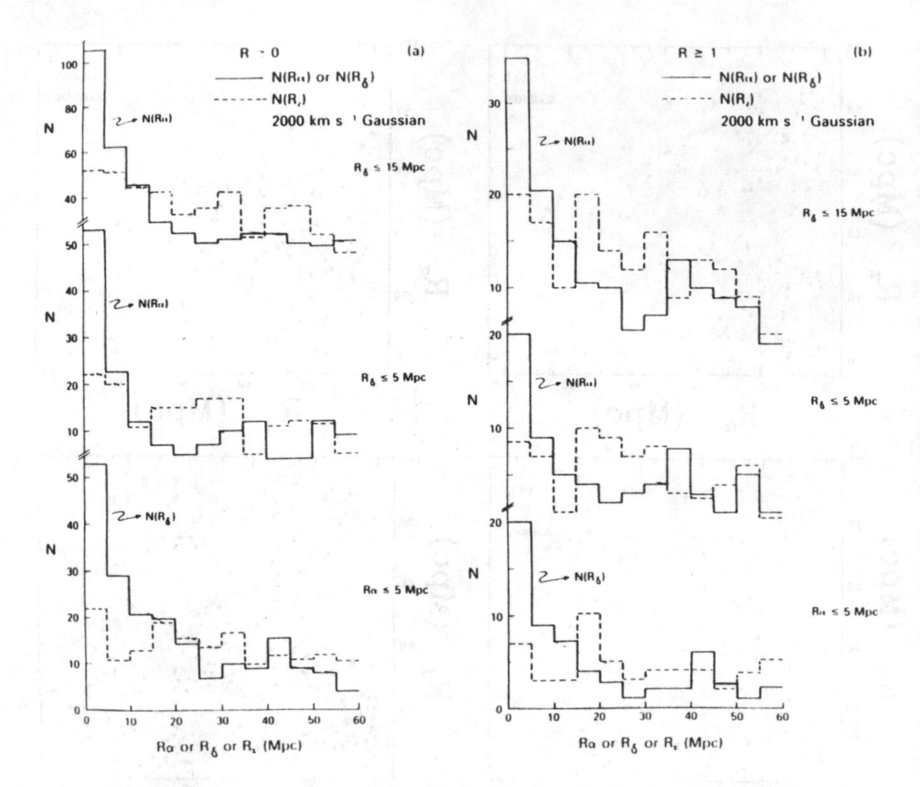

FIG. 13. Histograms representing the distribution of pairs along the redshift, R_z, and projected R_α and R_δ directions, as determined from the scatter-diagrams, are shown for the $R \geq 0$ (a) and $R \geq 1$ (b) samples. The number of cluster pairs as a function of projected separation, $N(R_\alpha)$ or $N(R_\delta)$, is represented by the solid histogram. The number of cluster pairs in the redshift direction, $N(R_z)$, is represented by the dashed histogram. The dotted curve represents a convolution of the projected distribution, $N(R_\alpha)$ or $N(R_\delta)$, with a Gaussian of 2000 km s^{-1} width. This convolved profile is in general agreement with the broadened distribution observed in the redshift direction, $N(R_z)$.

is typically ~ 1000 km s^{-1}, and that the redshifts measured are for the brightest centrally located galaxies; these galaxies are generally close to the central velocity of the cluster. Subtracting quadratically a possible deviation of $\sqrt{2}$ 300 km s^{-1} from the observed 2000 km s^{-1} yields 1950 km s^{-1}, i.e., a negligible change. Even if we assume, conservatively, ~ 700 km s^{-1} for the internal broadening, the net cluster pair velocity is still 1740 km s^{-1}. Thus, a considerable elongation effect of approximately 10^3 km s^{-1} per cluster remains after correction for internal motion.

The observed elongation may be caused by either peculiar motion of clusters or a true geometrical elongation of superclusters. These are briefly discussed below.

7.2 - Explanations of the Redshift Elongation

If the observed elongation is caused primarily by peculiar motion of clusters in superclusters, the net cluster pair motion in the line-of-sight is approximately 1700 km s^{-1}, or, equivalently, about 1200 km s^{-1} for single cluster motion. Most of this effect arises in the central parts of the rich superclusters. A large peculiar velocity could be caused by the gravitational potential of the superclusters or by non-gravitational effects such as explosions.

To estimate a supercluster mass which may support this velocity, a typical supercluster size of ~ $25h^{-1}$ Mpc (=cluster correlation scale-length) is used and the virial relation $M \propto v^2 r$ is assumed. This yields a typical supercluster mass of

$$M_{sc} \simeq 2 \times 10^{16} M_{\odot}, \qquad r \lesssim 25h^{-1}\text{Mpc}. \tag{16}$$

This mass is comparable to the mass of ~ 20 rich clusters, while typically only ~ 3-5 rich clusters are members of a supercluster. Even when the luminous tails of clusters are accounted for, the results may still suggest an excess of dark matter in superclusters as compared with clusters. Using an observed luminosity and/or density profile of r^{-3} or $r^{-2.5}$ around a rich cluster, we estimate an M/L for superclusters that is typically twice that of rich clusters, i.e., M/L ~ 500.

Redshift observations of two individual higher-redshift superclusters (Ciardullo *et al.* 1983) appear to indicate a much lower velocity for the superclusters than suggested even by a free expansion. This suggests, for these two systems, either a flat face-on geometry of the superclusters or a slow-down of the initial expansion due to the supercluster mass. In either case, it is likely that individual superclusters are at different stages of their evolution as well as at different observed orientations. The Corona Borealis supercluster (Bahcall *et al.* 1986) appears to show a redshift elongation in the distribution of both its clusters and the galaxies.

The elongation observed in the scatter diagrams may also be caused, at least partially, by a geometrical elongation of superclusters. If the most prominent superclusters are elongated in the line-of-sight direction, an apparent elongation in the distribution of pair separation along this axis may result. I discuss below an observational test to distinguish between peculiar velocity and geometrical elongation of large-scale structures.

If the observed redshift elongation is caused by geometrical elongation, cluster redshifts should be correlated with the magnitude of their standard galaxies, following Hubbles's law. No such magnitude-redshift correlation should be present if the effect is entirely due to peculiar velocity. More generally, an independent distance-indicator (such as the magnitude of the brightest cluster galaxy or Tully-Fisher type relations) could be used to determine the actual distances to the clusters, and thus to interpret the origin of the observed

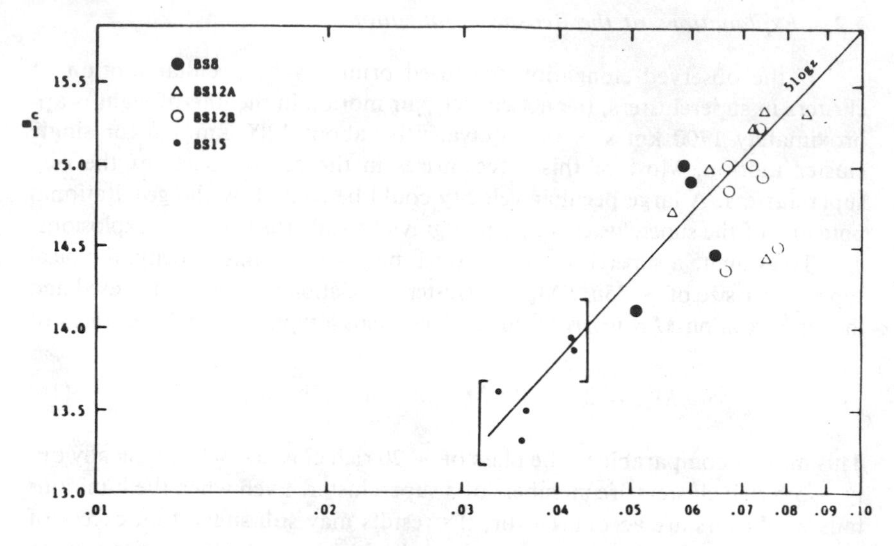

FIG. 14. The magnitude-redshift relation for the brightest cluster galaxies for clusters in the richest Bahcall-Soneira superclusters (Sec. 7.2).

redshift broadening (by comparing the actual distances with the observed redshifts).

The dependence of galaxy magnitudes on redshifts in the close cluster pairs was studied by Bahcall *et al.* (1986). The magnitude of the brightest galaxy in each cluster, m_1^c, corrected for the cluster morphological type and richness as given by Hoessel *et al.* (1980), was used as a distance indicator. If the observed redshift elongation is caused by geometrical anisotropy, a proper (Hubble) correlation of m_1^c with z is expected within individual superclusters. This correlation should not exist if peculiar velocity is the cause of the observed elongation. The expected magnitude difference for a cluster pair with a redshift separation of about 0.01-0.015, assuming Hubble distances, is ~ 0.3 to 0.5 mag (depending on z). This difference is large enough to be measured with accurate observations of standard galaxy magnitudes. A marginal m_1^c dependence was found (Bahcall *et al.* 1986) for some individual superclusters (Figure 14), suggesting that at least some of the redshift broadening observed may be due to geometrical elongation of the large structures. Increased accuracy and greater statistics for galaxy magnitudes may clarify the significance of the results. It is possible that both geometrical elongation and peculiar velocity of clusters contribute to the observed redshift broadening. Other distance indicators, such as Tully-Fisher or Faber-Jackson relations, should also be applied to the problem in order to help distinguish between peculiar motion and geometry. Recently, comparable velocities of ~ 10^3 km s^{-1} between some cluster pairs were also suggested by Mould (1988) and Burstein (1988) using actual distance indicators of galaxies.

REFERENCES

Abell, G. O. 1958. *Ap J Suppl.* **3**, 211.

_____ . *AJ.* **66**, 607.

Abell, G. O., Corwin, H. G., and Olowin, R. 1988. *Ap J Suppl.*, to be published.

Bahcall, N. A. 1977. *Ann Rev Astron Astrophys.* **15**, 505.

_____ . 1986. *Ap J Letters.* **302**, L41.

_____ . 1987. *Comm Astrophys.* **11**, 283.

_____ . 1988a. *Ann Rev Astron Astrophys.* 26.

_____ . 1988b. in *Large-Scale Structures of the Universe*, eds. J. Audouze, M.-C. Pelletan, and A. Szalay. Dordrecht: Kluwer Academic Publishers, p. 229.

Bahcall, N. A. and Burgett W. S. 1986. *Ap J Letters.* **300**, L35.

Bahcall, N. A. and Soneira, R. M. 1983. *Ap J.* **270**, 20 (BS83).

_____ . 1984. *AP J.* **277**, 27 (BS84).

Bahcall, N. A., Soneria, R. M. and Burgett, W. S. 1986. *Ap J.* **311**, 15.

Bahcall, N. A., Herinksen, M. J. and Smith, T. E. 1988. *Ap. J.*, to be submitted.

Bahcall, N. A., Batuski, D. J., and Olowin, R. 1988b. *Ap J. Letters.* **333**, L13.

Batuski, D. J. and Burns, J. O. 1985. *Ap J.* **299**, 5.

Batuski,D. J., Burns, J. O., Laubscher, B. E., and Elston, R. J. 1988. *Ap J.*, to be published.

Bogart, R. S. and Wagoner, R. V. 1973. *Ap J.* **181**, 609.

Bouchet, F. R. and Bennet, D. P. 1988, in *Large-Scale Structures of the Universe*, eds. J. Audouze, M.-C. Pelletan, and A. Szalay. Dordrecht: Kluwer Academic Publishers, p. 289.

Burstein, D. 1988. this volume.

Chincarini, G. and Vettolani, G. 1987. in *Observational Cosmology,* eds. A. Hewitt *et al.* Reidel, p. 275.

Chincarini, G, Rood, H. J. and Thompson, L. A. 1981. *Ap J Letters.* **249**, L47.

Ciardullo, R., Ford, H., Bastko, F., and Harms, R. 1983. *Ap J.* **273**, 24.

da Costa, L.N., Pellegrini, P.S., Sargent, W.L.W., Tonry, J., Davis, M., Meiksin, A., and Latham, D. 1988. *Ap J.* **327**, 544.

Davis, M. and Peebles, P.J.E. 1983. *Ap J.* **267**, 465.

Davis, M., Huchra, J., Latham, D.W., and Tonry, J. 1982. *Ap J.* **253**, 423.

Dekel, A. 1988, this volume.

de Lapparent, V., Geller, M., and Huchra, J. 1986. *Ap J Letters.* **302**, L1.

Efstathiou, G. 1988, this volume.

Fetisova, T.S. 1981. *Astr Zh.* **58**, 1137.

Geller, M.J. and Huchra, J. 1988, this volume.

Giovanelli, R., Haynes, M.P., and Chincarini, G. 1986. *Ap J.* **300**, 77.

Gregory, S.A. and Thompson, L.A. 1978. *Ap J.* **222,** 784.
Gregory, S.A., Thompson, L.A., and Tifft, W.G. 1981. *Ap J.* **243,** 411.
Groth, E. and Peebles, P.J.E. 1977. *Ap J.* **217,** 385.
Hauser, M.G. and Peebles, P.J.E. 1973. *Ap J.* **185,** 757.
Haynes, M.P. and Giovanelli, R. 1986. *Ap J Letters.* **306,** L55.
Hoessel, J.G., Gunn, J.E., and Thuan, T.X. 1980. *Ap J.* **241,** 486.
Kalinkov, M., Stavrev, K., and Kuneva, I. 1985. *AN.* **306,** 283.
Klypin, A.A. and Kopylov, A.I. 1983. *Sov Astron Letters.* **9,** 41.
Kopylov, A.I., Kuznetsov, D.Yu., Fetisova, T.S., Shvartsman, V.F. 1987. *The Large Scale Structure of the Universe, Seminar Proceedings.* Special Astrophysical Observatory, September 1986.
Mandelbrot, B.B. 1982. *The Fractal Geometry of Nature.* San Francisco: Freeman.
Mould, J. 1988, this volume.
Noonan, T. 1977. *AA.* **54,** 57.
Oort, J. 1983. *Ann Rev Astron Astrophys.* **21,** 373.
Peebles, P.J.E. 1980a. *The Large Scale Structure of the Universe.* Princeton: Princeton University Press.
_____ . 1980b. *Physical Cosmology, Les Houches, Session XXXII.* eds. R. Balian *et al.*
Postman, M., Geller, M., and Huchra, J. 1986. *AJ.* **91,** 1267.
Rood, H.J. 1976. *Ap J.* **207,** , 16.
_____ . 1988. *Ann Rev Astron Astrophys.* **26.**
Sarazin, C.L., Rood, H.J., and Struble, M.F. 1982. *AA Letters.* **108,** L7.
Sargent, W.L.W., Young, P.J., Boksenberg, A., and Tytler, D. 1980. *Ap J Suppl.* **42,** 41.
Seldner, M. and Peebles, P.J.E. 1977. *Ap J.* **215,** 703.
Shane, C.D. and Wirtanen, C.A. 1967. *Pub Lick Obs.* **22,** 1.
Shectman, S. 1985. *Ap J Suppl.* **57,** 77.
Shvartsman, V.F. 1988. in *Large-Scale Structures of the Universe*, eds. J. Audouze, M.-C. Pelletan, and A. Szalay. Dordrecht: Kluwer Academic Publishers, p. 129.
Sutherland, W. 1988. *MN.* to be published.
Szalay, A.S. and Schramm, D.N. 1985. *Nature.* **314,** 718.
Szalay, A.S., Hollosi, J., and Toth, G. 1988. preprint.
Tully, R.B. 1986. *Ap J.* **303,** 25.
_____ . R.B. 1987. *Ap J.* **323**, 1.
Turok, N. 1985. *Phys Rev Letters.* **55,** 1801.
_____ . 1988. in *Large-Scale Structures of the Universe*, eds. J. Audouze, M.-C. Pelletan, and A. Szalay. Dordrecht: Kluwer Academic Publishers, p. 281.
Zwicky, F., Herzog, E., Wild, P., Karpowicz, M., and Kowal, C.T. 1961-1968. *Catalog of Galaxies and Clusters of Galaxies, 6 Volumes.* Pasadena: Calif. Inst. of Technology.

SUMMARY OF SESSION ONE

G. EFSTATHIOU
Institute of Astronomy, Cambridge

We have heard and seen a lot of interesting things in this session. I have been sitting here wondering how it would appear to an extraterrestrial being. For the sake of argument, I will call him Pisces-Cetus, or PC for short. Let me tell you about PC. He has no eyes! We might consider this to be a great disability, but PC isn't bothered because he is blessed with an extraordinary capacity for assimilating and understanding numerical data. Correlation functions, likelihood functions, and all the rest, pose no problems for PC; he digests them with lightning speed. What would PC have made of this session?

Brent Tully has shown us pictures of structures that extend almost right across the observable universe. PC can't assess this, so he looks at the statistic. But wait a minute, wasn't the statistic motivated by the visual appearance of the map? PC is confused.

Neta Bahcall and John Huchra presented important statistical results on the clustering of Abell clusters. They are clustered much more strongly than galaxies. PC is impressed, but is a bit unsure about the Abell catalogue. Wasn't it constructed by eye? Are these humans really as objective as they claim? Nick Kaiser discussed work by Will Sutherland which suggests that spurious clustering in two-dimensions enhances the amplitude of the three-dimensional function measured in redshift space. The implication is that Abell found an enhanced number of clusters nearby other clusters on the sky. PC is worried and asks if the amplitudes of the angular correlation functions scale with distance class in the expected way; but there is no quantitative answer. Avishai Dekel discussed one mechanism that could lead to a bias, and concluded that it probably couldn't produce enough in the way of spurious clustering. Are there really problems with the Abell catalogue?

Marc Davis discussed results from the new Southern Sky redshift survey. The amplitude of the galaxy correlation function and the peculiar velocities between pairs of galaxies agrees well with the results from the CfA survey. PC thinks this is very good. The results have stabilised with sample size, so our applications of the cosmic virial theorem and our determinations of Ω look secure. But Margaret Geller argued that the 15th magnitude CfA redshift survey indicates that we had not yet achieved a fair sample of the universe. For example, she found that the amplitude of the galaxy correlation function was sensitive to her choice of weighting scheme. Furthermore, she pointed out inhomogeneities on the scale of the new survey. Jim Peebles remarked that the definition of a fair sample depends on the statistic that you want to measure. PC nodded in agreement.

Martha Haynes showed beautiful results on the distribution of galaxies in the Pisces-Perseus Supercluster. In particular, the luminous galaxies seem to be more clustered than the faint ones. Is this evidence in support of the idea that galaxies are biased tracers of the mass distribution? Does this, as Marc Davis contended, confirm the predictions of the "biased" cold dark matter model?

Well, PC has gone home to report on all this to his friends. I hope he is understanding! Our lack of precision in discussing these issues stems from two sources. Firstly, surveys of large-scale structure are difficult and very demanding of telescope time. So we don't have enough data. We must continue the effort to get more. Secondly, we have eyes! So we tend to get excited about interesting shapes and patterns instead of focussing our efforts on well designed statistical tests. Wouldn't it be better to first decide on a sensible statistic and then to design an optimal observing program to measure it? Before PC left, he whispered in my ear that if you smoothed the galaxy distribution on scales of 2000 km/s, you would be left with Gaussian random noise. Perhaps he was just being malicious.

II

LARGE-SCALE VELOCITY FIELDS

MOTIONS OF GALAXIES IN THE NEIGHBORHOOD OF THE LOCAL GROUP

S. M. FABER
Lick Observatory, University of California, Santa Cruz

and

DAVID BURSTEIN
Department of Physics, Arizona State University

Abstract

The velocity field of galaxies relative to the cosmic microwave background is investigated to a distance of 3000 km s^{-1} using two samples of spiral galaxies as well as elliptical galaxies. Velocity-field models are optimized that include motions due to a spherically symmetric Great Attractor, a Virgocentric flow, and a Local Anomaly, of which the Local Group is a part. The predictions of these models are compared graphically in a number of ways. We find that the spiral samples agree well in a formal sense with the Great-Attractor-Virgo model recently proposed to explain the motions of elliptical galaxies. However, new observations indicate that the Great Attractor is not spherically symmetric in its inner regions, which could require future modifications to this model. The Local Group shows an anomaly of 360 km s^{-1} with respect to the above model, which is shared by the cloud of galaxies around it out to 700 km s^{-1}. The amplitude and dimensions of this Local Anomaly seem to be typical of other deviant patches in the velocity field. It is likely that the Local Anomaly is the result of the irregular gravitational attraction of nearby, visible galaxies. Virgocentric infall models are heavily modified by the inclusion of the Great Attractor flow. The local Virgocentric infall velocity at the Local Group is poorly determined, but is quite a bit smaller (85-133 km s^{-1}) than conventional values. The gross properties of the velocity field that are defined by the present data are sketched in terms of a minimal, cylindrical, geometric model.

1. Introduction

We are part of a group currently investigating non-uniformities in the Hubble expansion based on the motions of nearby elliptical galaxies (Dressler *et al.* 1987, Lynden-Bell *et al.* 1988 [hereafter LFBDDTW]). We have discovered evidence for a large-scale flow of galaxies towards the constellation Centaurus in the southern hemisphere. This flow reaches a velocity of up to 1000 km s^{-1} in places and encompasses the entire Local Supercluster, including Virgo, Fornax, Eridanus, Ursa Major, and Leo (see Tully and Fisher 1987, hereafter TF87), as well as the region that has come to be known as the Hydra-Centaurus Supercluster (e.g., Chincarini and Rood 1979). Virgocentric infall exists as one sub-flow within it; in this paper we will present evidence for the existence of other (but smaller) such flows.

Guided by the familiar Virgo flow models, we have modeled the larger flow using a spherically symmetric infall centered on a point 4350 km s^{-1} distant (LFBDDTW). This infall center is located in the same direction as the Centaurus clusters but lies well beyond them. Because the model implies a major mass concentration quite distant from us, it has garnered the nickname "Great Attractor" (attributed to Alan Dressler).

A major difference between Virgo infall and the Great Attractor (hereafter GA) is the fact that the present data penetrate barely to the center of mass, far enough to show us infall towards the GA, but not infall from the back side. While this is a weakness, we will show that the data do strongly prefer a convergent flow model over simple bulk motion. The signature for this is a quadrupole term in the velocity field, which is equivalent to a tide in the spherical infall model. This quadrupole shows itself as a compression of the Hubble flow perpendicular to the direction of the GA and a stretching of the flow towards it. The interpretation of a quadrupole as convergent infall was first suggested by Lilje, Yahil, and Jones (1986, hereafter LYJ). In addition to noting the quadrupole in the elliptical galaxy data, LFBDDTW also reviewed evidence for a peak in the density of galaxies at the right distance and direction to be the GA. This evidence consisted of a map of galaxies on the sky in Centaurus, compiled by Ofer Lahav, plus published radial velocity surveys in the region.

Since that time, we have been testing the GA model against other data sets. There are four data sets of interest: the catalog of Tully-Fisher distances to nearby spiral galaxies compiled by Aaronson *et al.* (1982a) and supplemented by Bothun *et al.* (1984, the combination of both data sets will be referred to as Aaronson *et al.*); a similar catalog compiled by de Vaucouleurs and Peters (1984, hereafter DVP), a catalog of Tully-Fisher distances to nearby rich clusters by Aaronson *et al.* (1986, hereafter ABM86), and Sc I spiral distances by Rubin *et al.* (1976a, b). The present paper is limited to motions of nearby galaxies and concentrates on the first two samples, which have their major weight within 2500 km s^{-1}. The ABM86 cluster sample is used briefly to set the far-field zero-point for the Hubble expansion. This is equivalent to determining the Hubble constant in most studies, but since our distances are always given in units of km s^{-1}, the analogous scale factor for us is a dimensionless constant (see LFBDDTW).

In the present study, we investigate the following questions:

a. How well do both spiral and elliptical galaxies with distances less than 3000 km s^{-1} fit the adopted velocity field model, the dominant component of which is motion induced by the GA?

b. How well do the galaxies (almost all spirals) closest to the Local Group fit the model? LFBDDTW noted that the Local Group itself fits poorly, with a residual velocity of about 360 km s^{-1}. If there is a Local Anomaly, how far does it extend, and how does it merge with the flow pattern at larger distances?

c. Existing studies of Virgo infall (e.g., Aaronson *et al.* 1982b, hereafter AHMST) have usually modeled Virgo infall in the rest frame of the sample galaxies. Virgocentric infall and bulk motion have thus been removed, but the tidal field of the GA is not modeled. Since the local velocity of the GA flow ($\sim$ 550 km s^{-1}) is greater than previous estimates of Virgo infall (250 km s^{-1}), the tidal distortion of the former has potentially influenced previous derivations of the amplitude of Virgo infall. We therefore examine what is left of Virgo infall after the GA-induced motion has been removed.

d. What is the typical coherence length for peculiar motions relative to the GA-induced flow? Is there evidence for localized infalling regions other than the Virgo cluster? How typical is the residual motion of the Local Anomaly?

e. For completeness, we also review the most recent data on the size and structure of the GA.

Most of the quantitative results of the present paper are based on a series of three-component, maximum-likelihood fits to the spiral galaxy velocity data alone. As such, these complement the models of LFBDDTW, which use only the elliptical galaxy data. Since it is difficult to represent three-dimensional motions on a two-dimensional page, we show graphical representations of the data before and after fitting to various models. By studying these graphs in conjunction with maps of the galaxy distribution provided here (Figs. 1 and 2) and the more informative maps of TF87, it should be possible for the reader to construct a mental image of the local velocity field.

An interesting by-product of this work is the discovery that the spiral data sets are of varying accuracy and are also internally inhomogeneous, including even the Aaronson *et al.* field data. The DVP data are so heterogeneous, in fact, that we have used only those galaxies with Tully-Fisher distances. The best Aaronson *et al.* data seem to have an error that is appreciably better than that usually quoted for the Tully-Fisher method.

We also discuss briefly the pros and cons of various methods of maximum-likelihood fits, but details will appear in a later version of this work.

2. The Local Topography

Tully and Fisher have made a major contribution to our understanding of the nearby space distribution of galaxies with their recent *Nearby Galaxies Atlas* (TF87). It is unfortunate that we cannot reproduce their beautiful maps here, but a combination of the cartoon in Fig. 1 and the maps of Fig. 2 will have to suffice. Fig. 1 shows a cube of the local volume 10,000 km s^{-1} on a side in supergalactic coordinates. The major structures — Virgo, Ursa Major, Centaurus, Perseus-Pisces, and Pavo-Indus-Telescopium — lie either

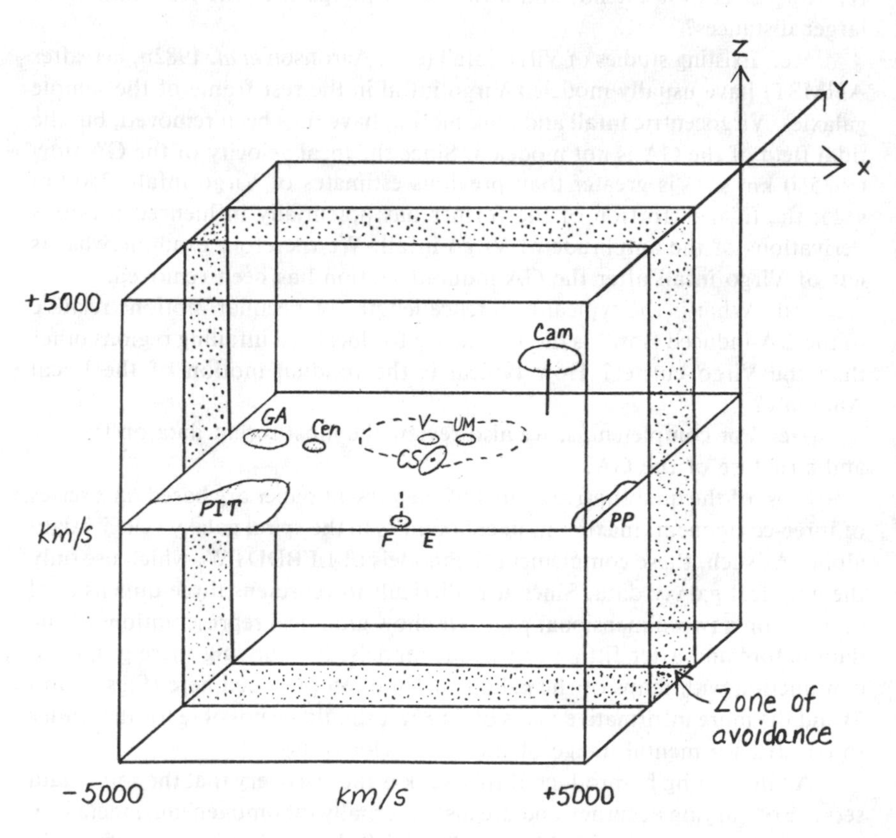

Fig. 1. A cartoon of the local volume of space that is sampled by the data sets used in this paper. The positions of the main structures are noted: (GA = Great Attractor; C = Centaurus; V = Virgo cluster; CS = Coma-Sculptor Cloud; UM = Ursa Major cluster; FE = Fornax-Eridanus; Cam = Camelopardalis; PP = Perseus-Pisces; PIT = Pavo-Indus-Telescopium.

directly in the supergalactic plane or close to it. Fornax and its neighbor, Eridanus, lie somewhat below the plane in the south, and the Leo clouds (not shown) lie below the plane under Virgo-Ursa Major. TF87 identify an important new grouping, the highly flattened Coma-Sculptor cloud located at the center of the cube, of which more is said below. The Local Group is a resident of this cloud, which in turn appears to be an appendage to the Virgo-Ursa Major complex and also part of the supergalactic plane.

The position of the GA as inferred from the elliptical galaxy data is shown by the large dot in the upper left-hand corner of the SG plane. The GA lies close to the plane, but the mass distribution in its vicinity is not well understood. The two Centaurus clusters, plus several smaller neighbors, lie in the foreground of the GA and are falling into it (away from us) with velocities approaching 1000 km s^{-1}.

Figs. 2a, b, and c show three orthogonal supergalactic projections of all galaxies with measured distances, both spirals and ellipticals, with X, Y and Z components defined in the usual manner. The data from the four samples are plotted with different symbols, as given in the caption. Several groupings from Fig. 1 are labelled, as well as the direction of the Perseus-Pisces complex. The SG plane is most visible in the Y-Z projection, which also shows the highly flattened Coma-Sculptor Cloud (center). Fig. 2 agrees with the analogous diagrams in TF87 very well, despite the fact that the diagrams in TF87 are based on radial velocity distances (a point which is discussed further below). Qualitatively the galaxies with measured distances here appear to be a fair tracer of the general galaxy population, with about 20% of the galaxies in TF87 having measured distances.

A major feature of Fig. 2 is the galactic zone of avoidance. Fortuitously, this is almost exactly perpendicular to the SG plane and also to the Y axis. The wedges empty of galaxies are especially visible in the X-Y and Y-Z projections.

3. Assumed Flow Models

The flow model we have been investigating is a three-component flow that is an embellishment of the model in LFBDDTW. It has the following components:

a. GA flow: This flow is assumed to be spherically symmetric about a point located at galactic coordinates $l_A = 309°$, $b_A = +18°$ at a distance 4200 km s^{-1} from the Local Group (slightly revised from LFBDDTW). From recent work, we have realized that the velocities of ellipticals in that direction peak at 3000 km s^{-1} and fall beyond (see Fig. 10k), indicating that the flow model requires a core radius. The velocity model adopted is

$$u_A = v_A \, [r_A/d_A] \, [(d_A^2 + c_A^2)/(r_A^2 + c_A^2)]^{(n_A+1)/2} \qquad (1)$$

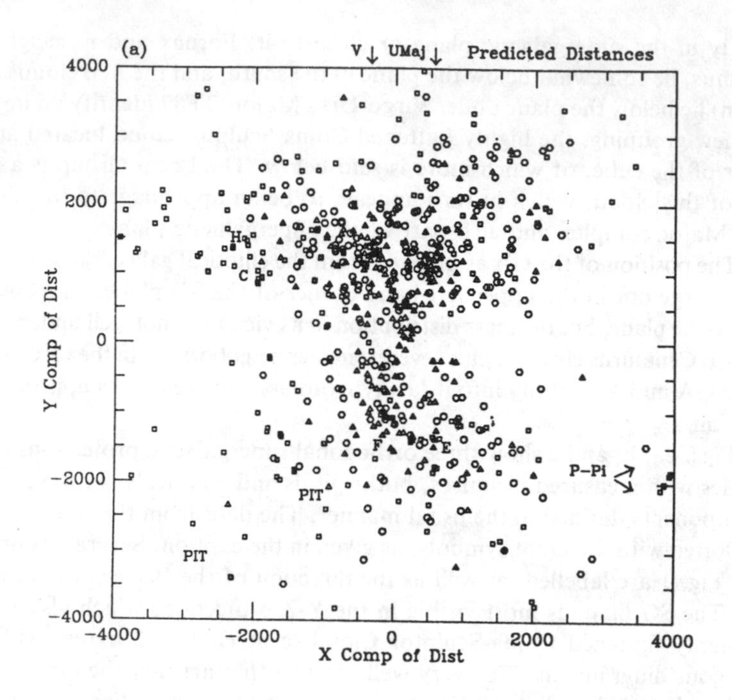
(a)
V
UMaj
Predicted Distances
Y Comp of Dist
X Comp of Dist
4000
2000
0
−2000
−4000
H
C
F
E
PIT
PIT
P-Pi

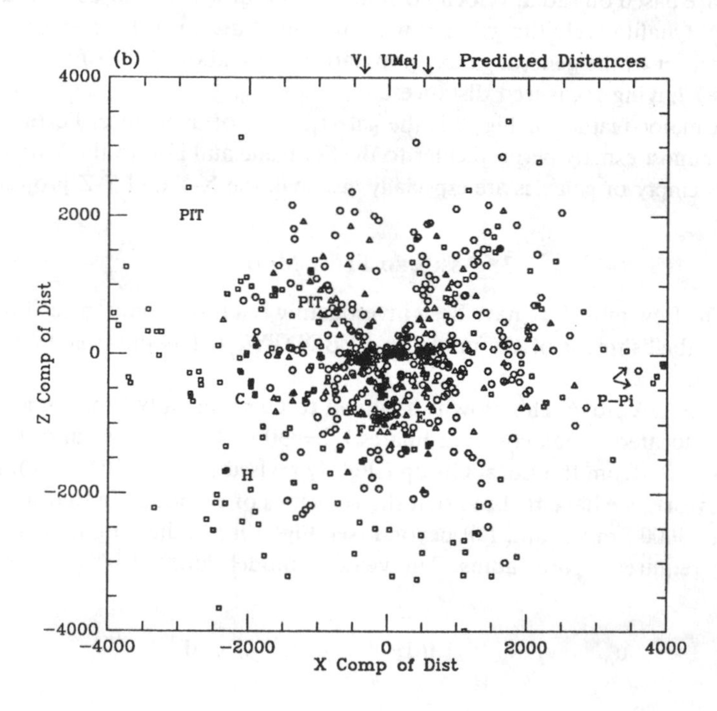
(b)
V
UMaj
Predicted Distances
Z Comp of Dist
X Comp of Dist
4000
2000
0
−2000
−4000
PIT
PIT
C
H
F
E
P-Pi

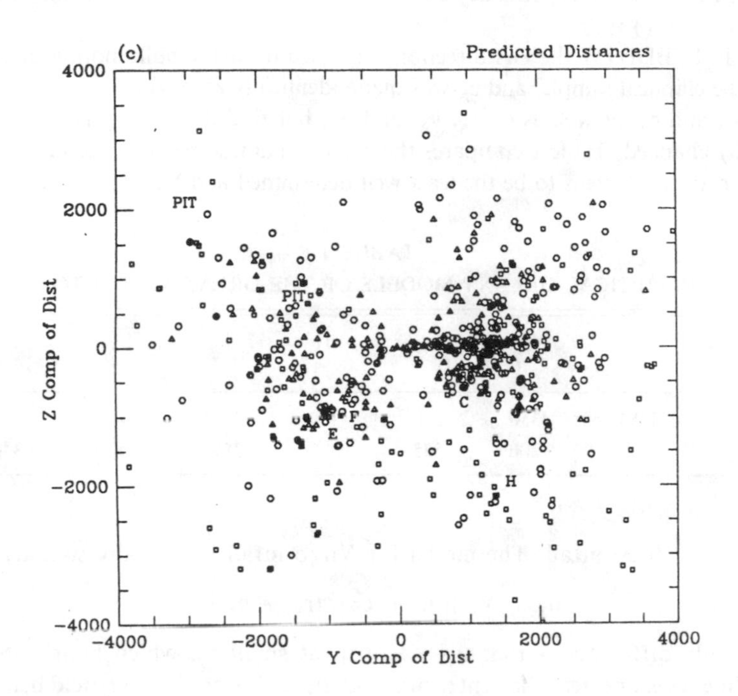

FIG. 2. Projected positions of galaxies in supergalactic coordinates, according to the predicted distance of each galaxy (Malmquist-bias corrected as discussed in the text). The four data samples are coded by different symbols: triangles denote Aaronson *et al.* "Good" data; squares denote elliptical galaxy data from LFBDDTW; hexagons denote Aaronson *et al.* "Fair" data; and circles denote de Vaucouleurs and Peters Tully-Fisher data. The general positions of clusters are either marked or pointed to by arrows: V = Virgo cluster; UMaj = Ursa Major; F = Fornax; E = Eridanus; P-Pi = Perseus-Pisces region (mostly off the maps); PIT = the Pavo-Indus-Telescopium region, which is marked in two places. The X, Y and Z coordinates are as defined in the RC2. Fig. 2a, the X-Y plane; Fig. 2b, the X-Z plane; Fig. 2c, the Y-Z plane. These maps compare well with those of TF87 (who used primarily radial velocities) with respect to both the correspondence of samples and to the relative positions of groups and clusters.

where u_A is total space velocity radially directed toward the infall center. The quantity r_A is the distance of a galaxy from the infall center, d_A is the distance of the infall center from the Local Group, v_A is the flow velocity at the radius of the Local Group, and c_A is the core radius of the flow. The quantity n_A controls the radial dependence of the flow: at small r_A, $u_A \sim 0$, while at large r_A, $u_A \sim (r_A)^{-n_A}$.

In LFBDDTW the GA direction was fixed to be the bulk motion direction for the elliptical sample, and c_A was made identically zero. Here we optimize on l_A, b_A, and c_A, as well as on r_A, v_A, and n_A, but find that l_A, b_A, r_A and v_A are hardly changed. Table 1 compares the old GA parameters with the new. Of all parameters, n_A seems to be the least well determined and has changed the most.

TABLE 1
ELLIPTICAL GALAXY MODELS OF THE GREAT ATTRACTOR

Model	d_A	v_A	n_A	l_A	b_A	c_A
	(km s^{-1})					
Old (LFBDDTW)	4350	570	1.0	307°	9°	—
New	4200	535	1.7	309°	18°	$0.34d_A$

b. Virgo infall: The model for Virgo infall is basically similar:

$$u_V = v_V \, [(d_V^2 + c_V^2)/(r_V^2 + c_V^2)]^{n_V/2} \tag{2}$$

The only difference is that $u_V \sim$ const. at small r_V, which better fits the nonlinear regime near the center of a collapsed cluster. The far-field behavior is again $u_V \sim (r_V)^{-n_V}$.

c. Local Anomaly: We show below that the velocity anomaly associated with the local Group extends to a region that is at least 700 km s^{-1} in radius around it. We therefore have incorporated a "Local Anomaly Switch", which turns on a bulk-velocity correction vector of 360 km s^{-1} for this region, making it blend smoothly with the surroundings.

d. Errors and Hubble constant: Maximum likelihood requires an error estimate for each point. As in LFBDDTW, this is represented by a measurement error Δ per point (given in magnitudes), plus a "random noise" term, σ_f, for Hubble flow noise (in km s^{-1}). The two errors are adjusted separately to maximize the likelihood. There is also a scale factor analogous to the Hubble constant that relates distance to velocity. How this is determined is explained below.

In treating the spirals we do not group them into clusters as we did with the ellipticals. Each is a separate point of equal weight within each data category.

4. Methods of Fit

We have broadened our methods of maximum-likelihood fitting in the same spirit as AHMST, who also tried multiple methods. The goal is to get from the data to the "best" velocity-field model. So far we have tried two approaches:

Method 1: Get the distance of each galaxy using the distance estimator (e.g., Tully-Fisher for spirals). Use these distances to calculate the velocity residual of every object relative to a smooth Hubble flow, and thus make a "picture" of the velocity-field residuals. Devise a mathematical model that incorporates the basic features of this velocity field, and find the free parameters by minimizing the scatter (maximizing the likelihood) in the velocity-field residuals. This was the method employed in LFBDDTW.

Method 2: Get the distance of each galaxy using its observed velocity, corrected by an *a priori* velocity-field model. With these distances, make a "picture" of the basic distance-indicator relation (for Tully-Fisher, this would be a graph of rotation velocity vs. absolute magnitude). Vary the parameters in the velocity-field model to minimize the scatter (maximize the likelihood) in the distance-indicator relation.

Each fitting method has its pros and cons, and each one is at some stage indispensable. Method 1 must be used initially to the point of making a picture of the observed velocity field. This is required to decide what terms to include in the velocity-field model. Method 1 also has residuals in km s^{-1}, and hence Hubble-flow noise yields Gaussian errors for all galaxies, near and far. This is important for nearby galaxies, for which Hubble-flow noise dominates. Without Gaussian errors, maximum-likelihood becomes much more complicated.

However, the raw distance estimates in Method 1 are biased too small due to Malmquist-bias effects and must be corrected (see LFBDDTW). For a uniform space distribution, the distance correction is a multiplicative constant given by:

$$r = r_{raw} \ (1 + 0.74 \ \Delta^2) \quad (3)$$

where Δ is the observational error in magnitudes.

The least accurate spiral samples have distance errors of $\pm$ 0.5 mag and Malmquist corrections of 19%, which are large. Worse, in clumpy regions, the correction depends on the density gradient along the line-of-sight and can even be negative. Using Virgo as an example, we show below how this effect can introduce spurious features into the velocity field in clumpy regions.

Perhaps the worst feature of Method 1 is that the galaxy distance estimator, being inherently logarithmic, does not yield symmetric, Gaussian errors in km s^{-1}. Distance errors of even $\pm$ 0.4 mag produce a seriously skewed error distribution in velocity space. This in turn introduces a bias into certain parameters, notably the scale factor (Hubble constant), and the distance error, Δ. Errors in the latter quantity are serious because they fold back into the Malmquist-bias correction.

Method 2 is the opposite of Method 1 in almost every way. It yields a picture of the distance-indicator relation rather than the velocity field. This picture can be used to fit the mathematical form of the relation from *field* galaxies. However, this approach can introduce undesirable cross-talk between the shape of the relation and features in the velocity field (LFBDDTW). It is preferable if possible to obtain the shape of the relation from independent data, e.g., from calibrating clusters. Unfortunately the AHM86 cluster galaxies do not span quite enough range in magnitude for this, and some appeal must be made to field data after all. (Yes, Jim Peebles, we have sinned, but, like all sinners we claim we can't help it.).

A great advantage of Method 2 is the fact that scatter about the Tully-Fisher relation is naturally Gaussian and constant with distance. The Gaussian maximum-likelihood method is therefore well suited to all but the nearest galaxies, and most of the flow parameters are therefore unbiased. This is crucial for galaxies beyond 1000 km s^{-1}, for which distance errors generally dominate over Hubble-flow noise. Method 2 minimizes naturally in log v rather than v.

Although the Malmquist-bias problem is smaller for Method 2, Malmquist bias still afflicts the distances — caused in this case by distance errors due to Hubble-flow noise. However, the typical distance error is usually smaller — say 10% rather than 20-25% with Tully-Fisher — and the resulting Malmquist bias is only a few percent if velocity-distance is near linear.

On the debit side, Method 2 gives no picture of the velocity-field: for that we must resort to Method 1. A second drawback is that Hubble-flow noise yields non-Gaussian residuals for nearby galaxies, for which the method thus gives biased results. (This situation is just the opposite of Method 1, which had trouble with distant galaxies, for a similar reason). The derived Hubble-flow noise parameter and the Hubble constant for nearby galaxies are most affected. LFBDDTW also show that the likelihood in Method 2 is not as "pure" as in Method 1 and maximizes in a slightly different location in parameter space. The effect on the derived velocity-field parameters is not yet clear.

As long as one is committed to Gaussian maximum-likelihood, there is clearly no single method that rigorously treats both near and far galaxies and deals well with Malmquist bias. Instead one must use an interplay of both methods, depending on need. AHMST also explored both methods.

A final drawback to Method 2 is the fact it cannot yield unique distances to galaxies in the so-called triple-valued regions around infall centers. In the local volume, there are two at least two such regions, the familiar Virgo region and a larger one around the GA. As a temporary way of dealing with this we are using a hybrid version of Method 2 that employs radial velocity as the basic distance indicator but Tully-Fisher distance to calculate the velocity flow correction to the model. In effect, this method has the Malmquist-bias properties of Method 1 but the other properties of Method 2. We will examine the "pure" Method 2 in a later paper.

In all models we have limited the sample to a certain subset of the Aaronson *et al.* spirals with exceptionally small errors (see below). In view of the systematic biases that can creep in due to errors, we feel that it is better to derive quantitative results from the best data only and then to compare the other samples graphically. Such comparisons are shown in Section 6 below.

The last issue is the scale factor, or Hubble constant. This is equivalent to fixing the absolute magnitude zero-point of the Tully-Fisher relation. From the discussion above, it should be clear that Method 2 yields a good zero-point when applied to sufficiently distant galaxies. Accordingly, we have used a standard GA-Virgocentric flow model to derive velocity-field distances to the ABM86 cluster spirals. Since these clusters are 3500 to 11,000 km s^{-1} distant and are in the opposite hemisphere from the GA, their corrected velocity-field distances should be fairly accurate. With these clusters determining the zero-point, the peculiar velocity of Coma with respect to cosmic rest turns out to be -200 ± 300 km s^{-1}, which agrees well with the value of -220 km s^{-1} from the ellipticals in LFBDDTW.

5. The Local Anomaly

The first step in model fitting is to discover how far the Local Anomaly extends. To find this, we carried out bulk-motion solutions on the Aaronson *et al.* field spirals inside successively larger shells around the Local Group (Table 2). (Since these involved nearby galaxies, Method 1 was used.) The resultant velocity vector agrees well with the motion of the Local Group within the errors (50-100 km s^{-1}) out to a radius of 700-800 km s^{-1}, at which point it begins to swing toward the GA at (309, 18). This shift in the motion of nearby galaxies was first pointed out by de Vaucouleurs and Peters (1968) and later explored by de Vaucouleurs and Peters (1984) using a variety of distance-indicator methods. The shared motion of the Local Group and nearby galaxies was also emphasized recently by Peebles (1987) using the same Aaronson *et al.* spirals as here. This locally coherent motion is the reason why Hubble-flow noise has been estimated as low as 100 km s^{-1} (e.g., Sandage 1972).

The homogeneous motion of the local patch is illustrated in Fig. 3, which plots radial velocity residuals relative to cosmic rest versus the cosine of the angle to the direction of the Local Group microwave vector (269, 28). Galax-

TABLE 2
BULK MOTIONS OF NEARBY GALAXIES

Shell Limits (km s^{-1})	v_{BLK} (km s^{-1})	l_{BLK}	b_{BLK}	No. of Galaxies	Errors[a] (km s^{-1})
0-500	623 ± 89	275° ± 12°	27° ± 4°	18	138, 43, 31
500-700	491 ± 94	275° ± 15°	29° ± 7°	14	132, 76, 46
700-900	645 ± 78	295° ± 9°	22° ± 5°	14	94, 78, 57
900-1100	497 ± 79	309° ± 10°	28° ± 4°	22	97, 76, 48
1100-1500	659 ± 85	309° ± 8°	1° ± 3°	53	94, 77, 30
Loc. Gr.	614	269°	28° with respect to the CMB		
Local GA flow vector	535 ± 70	309° ± 10°	18° ± 10°		

[a] The three axes of the error ellipse. Because of the flattening of the local cloud, the largest error tends to point roughly perpendicular to the SG plane (47,6 in galactic coordinates). The smallest error is perpendicular to the galactic plane, and the middle error is perpendicular to the other two.

ies within a sphere of radius 850 km s^{-1} are plotted, and galaxies in specific regions (i.e., towards the Virgo cluster, away from the Virgo cluster, or in the Leo Spur) are plotted with different symbols. Galaxies streaming along with the Local Group should lie on the straight line, which has total amplitude ± 614 km s^{-1}, the CMB velocity of the Local Group. A maximum-likelihood fit for the velocity dispersion inside 700 km s^{-1} is only 80-90 km s^{-1}, indeed confirming that the local Hubble flow inside the patch is quite quiet (Sandage 1972).

The local patch is the only one for which we can unambiguously determine the full three-dimensional space motion and compare it to the geometry of the local galaxy distribution. Figs. 4a, b, and c show supergalactic projections inside the local volume analogous to Fig. 2. The highly flattened cloud at the center is the Coma-Sculptor Cloud of TF87. If this cloud has the same motion as the Local Group, its total bulk velocity with respect to the CMB is X, Y, Z = (−417, 249, −376) in SG coordinates. A more interesting quantity is the difference between this motion and that of the surrounding galaxies, which agree well with the GA-Virgo model. This difference is about 360

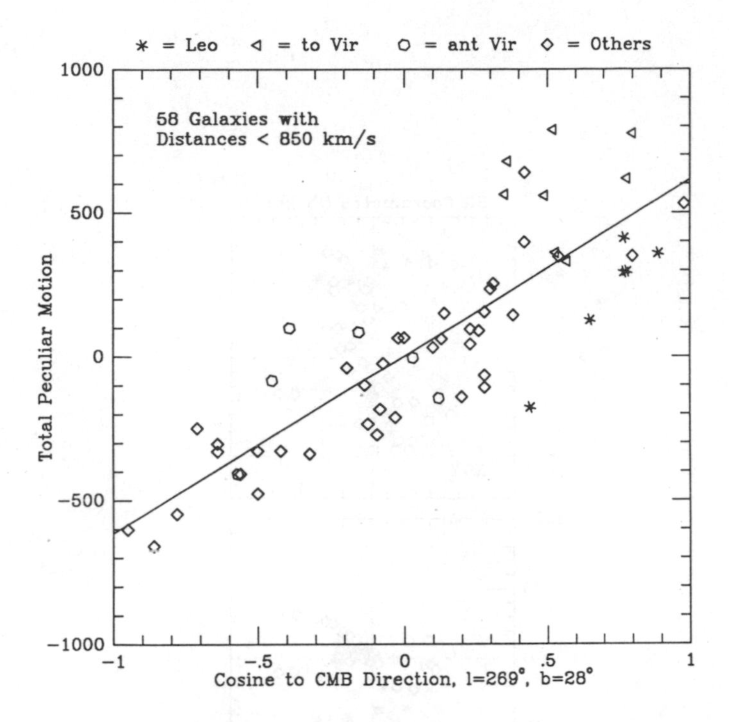

FIG. 3. Peculiar velocities relative to the CMB rest frame of all galaxies with predicted distances less than 850 km s^{-1}, plotted versus the cosine of the angle to each galaxy with respect to the direction of the Local Group's CMB motion (l = 269°, b = 28°). Galaxies moving with the Local Group should lie on the line. Galaxies located near the Virgo cluster are denoted by triangles; galaxies located the furthest from Virgo are denoted by hexagons; galaxies located in the Leo 'Spur' (TF87) are denoted by stars, and all remaining galaxies are denoted by diamonds. The most deviant galaxies (non-diamonds) tend to lie at the edges of the volume (see Fig. 4).

km s^{-1} towards (199,0), or (169, −30, −317) in SG coordinates. Since this component is locally generated in the gravitational instability picture (see Lynden-Bell and Lahav 1988), there should be a correlated feature in the nearby galaxy distribution. There is no conspicuous peak towards (199,0), but there is a striking dearth of galaxies in the opposite direction, which TF87 call the Local Void. This is visible in Fig. 2 as the empty volume towards minus X and positive Y (note that the Local Void is partially obscured by the Galactic plane). Lynden-Bell and Lahav (1988) call attention to the impact of the Local Void in swinging the motion of the Local Group away from the GA. The strong negative Z velocity of −317 km s^{-1} is probably due to the strong excess of galaxies below the SG Plane (see TF87 and also Fig. 2).

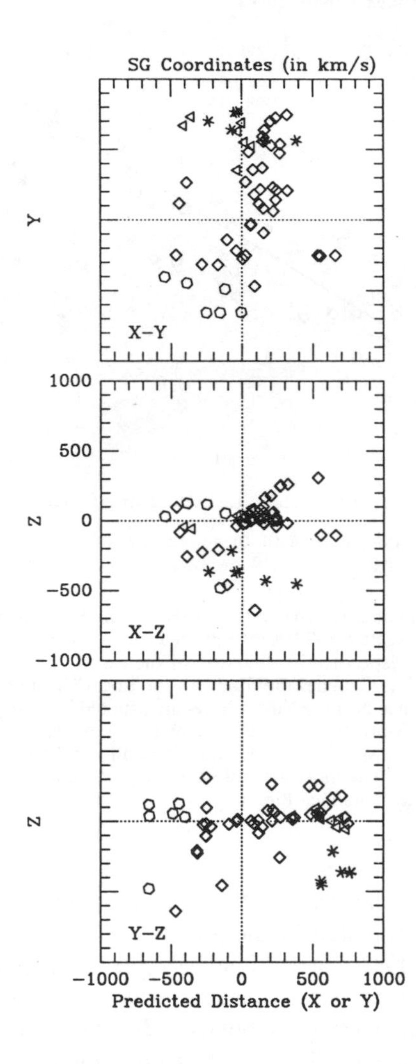

FIG. 4. The distribution in supergalactic coordinates of the galaxies in Fig. 3 (same symbols). The highly flattened structure defined by the diamonds is the Coma-Sculptor cloud of TF87. Non-diamonds are galaxies that deviate the most in Fig. 3. The peculiar motions of the six galaxies at high Z agree well with the overall cloud motion.

In short, there is a strong probability that the local peculiar motion of the Coma-Sculptor cloud (and the Local Group) can be accounted for via the gravitational attraction of visible nearby matter. The flattening of the Cloud is therefore also probably due to gravity-induced motions. This fact sets important upper limits on the role of non-gravitational forces, such as cosmic explosions, in determining the distribution and kinematics of local matter (Peebles 1987).

Peebles has also searched for internal collapse motions within the Cloud and has set stringent upper limits on flow velocities perpendicular to the plane of flattening. Once the surrounding galaxy density is mapped (e.g., using IRAS-selected or optically-selected galaxies) such motions can be used to estimate Ω in a manner analogous to Virgo infall. The only points that deviate significantly in Fig. 3 are six galaxies with large Y distance near Virgo at the edge of the volume (see Fig. 4), which are perturbed by Virgo infall and the GA tidal field, and four galaxies in the spur connected to the Leo Cloud. In particular, the six Coma-Sculptor galaxies with significant Z distances (and not in the Leo Spur) have infall motions toward the Supergalactic plane that appear to be small.

To summarize, the local Coma-Sculptor cloud is a highly flattened structure with low internal velocity dispersion but large bulk motion both perpendicular to, and parallel to, the flattening plane. It is striking that the boundary of coherent motion coincides closely with the boundary of the Cloud. The peculiar motion of the Coma-Sculptor cloud probably originates in the asymmetric distribution of nearby matter, notably in the paucity of galaxies in the Local Void and the excess of galaxies on one side of the SG plane compared to the other. The peculiar velocity relative to a large-scale flow model is about 360 km s^{-1}, which converts to 200 km s^{-1} line-of-sight. This amplitude and the dimensions of the Cloud — some 1500 km s^{-1} in diameter — are qualitatively typical of the sizes and motions of patches of galaxies elsewhere in the local volume (see Fig. 10).

6. Picturing the Velocity Field

6.1 - Distance Indicator Relationships

The hybrid form of Method 2 is used in Fig. 5 to construct four versions of the distance indicator relationship (Tully-Fisher for spirals, D_n - central velocity dispersion for ellipticals) for galaxies with distances less than 4600 km s^{-1}. Four different velocity-field models are used: (Panel 1) observed radial velocity with respct to the CMB; (Panel 2) radial velocity in CMB coordinates corrected for a Virgocentric infall of 250 km s^{-1}; (Panel 3) velocity in the rest-frame of the Aaronson *et al.* sample (i.e., with the bulk motion of the galaxies removed), corrected for a Virgocentric infall of 200 km s^{-1};

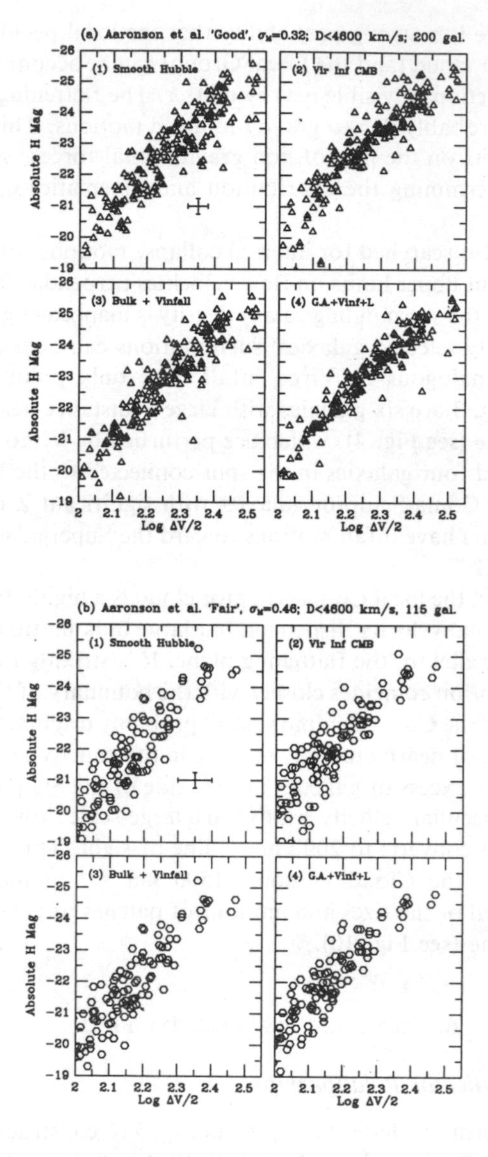

FIG. 5. These figures show the basic distance-indicator relations for the four data samples under four different assumption about the velocity field: 1) a smooth Hubble flow relative to the CMB rest frame, with no other motions; 2) a smooth Hubble flow combined with only a Virgocentric infall model, with $v_V = 250$ km s^{-1}, $n_V = 1.0$, $c_V = 0.0$; 3) a bulk motion for the whole volume, combined with the Virgocentric flow of panel 2, but with $v_V = 200$ km s^{-1}; 4) the Great Attractor + Virgocentric infall + Local Anomaly model having $v_A = 530$ km s^{-1}, $r_A = 4350$ km s^{-1}, $n_A = 1.62$, $c_A = 0.34$, $l_A = 309°$, $b_A = 18°$, $v_V = 150$ km s^{-1}, $n_V = 1.0$ and $c_V = 0.0$. Only galaxies with

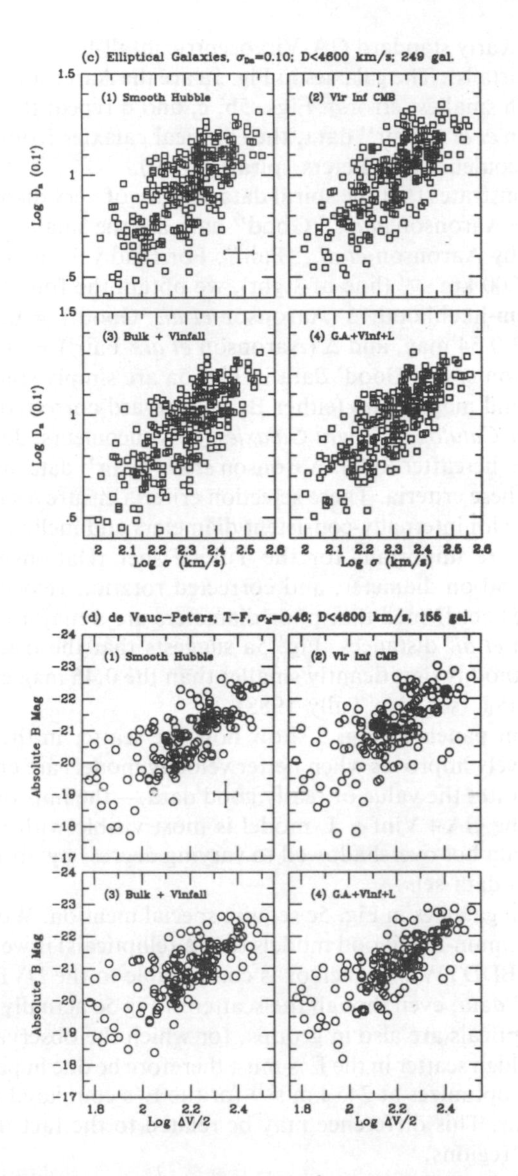

predicted distances less than 4600 km s^{-1} are plotted. Fig. 5a gives Aaronson et al. 'Good' sample, 200 galaxies; Fig. 5b gives Aaronson *et al.* "Fair" sample, 115 galaxies; Fig. 5c gives elliptical galaxies, 249 galaxies; and Fig. 5d gives de Vaucouleurs and Peters galaxies that do not overlap with the Aaronson *et al.* data, 155 galaxies. Note the improvement of the relationship for each data sample with each successive velocity field model; the improvement is most marked for that sample with the smallest observational error, the Aaronson *et al.* "Good" sample in Fig. 5a.

and (Panel 4) a nearly standard GA-Virgocentric infall-Local Anomaly model (see caption for details). The galaxies in Fig. 5a are the Aaronson *et al.* "Good" spirals (data with smallest errors). Figs. 5b, c, and d repeat the same process for the Aaronson *et al.* "Fair" data, the elliptical galaxies from LFBDDTW, and the de Vaucouleurs and Peters spirals (DVP).

Fig. 5 demonstrates that the spiral data sets are of very disparate quality. The errors in the Aaronson *et al.* "Good" data are the smallest, followed by DVP, and then by Aaronson *et al.* "Fair". For field velocity dispersion, σ_f, in the range 80-200 km s^{-1} (line-of-sight), we obtain the following values of Δ from maximum-likelihood: Δ (Aaronson *et al.* 'Good') = 0.34-0.40 mag, Δ (DVP) = 0.42-0.54 mag, and Δ (Aaronson *et al.* 'Fair') = 0.49-0.52 mag.

The Aaronson *et al.* 'Good' data in Fig. 5a are simply spirals that have both diameters and magnitudes (either B_T or Harvard-corrected) given in the *Second Reference Catalog of Bright Galaxies* (de Vaucouleurs, de Vaucouleurs, and Corwin 1976; hereafter RC2). Aaronson *et al.* "Fair" data are those galaxies not meeting these criteria. These selection criteria ensure a sample of well-studied galaxies with internally-consistent diameters and inclinations. Both of these quantities are important for the Tully-Fisher relationship, as the H magnitudes depend on diameter, and corrected rotation velocity on inclination. Tully (1988) confirmed that poor diameters are a major cause of error in the Aaronson *et al.* distances. Fig. 5a suggests that the true error of the Tully-Fisher method is significantly smaller than the 0.45 mag conservatively quoted by AHMST (see also Tully 1988).

The different panels in Fig. 5 show how the scatter in the Tully-Fisher relation successively improves when better velocity models are employed. The figure also illustrates the value of really good data — the improvement in the last step using the GA + Vinf + L model is most visible with the Aaronson *et al.* "Good" data but overshadowed to varying degress by observational errors in the other data sets.

The elliptical galaxies in Fig. 5c require special mention. We believe from clusters and maximum-likelihood models that Δ (ellipticals) is well-determined at 0.45 mag (LFBDDTW). This error is comparable to the DVP and Aaronson *et al.* "Fair" data, even though the scatter in Fig 5c actually looks worse. Many of the ellipticals are also in groups, for which the observational errors are smaller. The high scatter in the E's must therefore be due in part to a higher σ_f, which indeed optimizes at 245 km s^{-1} for the E's compared to 80-200 km s^{-1} for the spirals. This difference may be related to the fact that ellipticals populate denser regions.

6.2 - Velocity Field Maps

Method 1 can be used to construct pictures of the observed velocity-field residuals relative to the cosmic rest frame and to various models. The following diagrams compare velocity field maps for the four data sets used in Fig. 5.

Figs. 6a, b, c and d present plots similar those given in LFBDDTW. These figures show observed peculiar velocities relative to the CMB for galaxies within $\pm$ 22.5° of four planes with poles given in each panel. Together, these four planes cover the sky. The second panel in each figure coincides with the supergalactic plane, and the axes correspond closely to SG coordinates. The fourth panel is perpendicular to the SG plane but nearly parallel to the galactic plane. The axis about which the planes are rotated is coincident with the X-axis of Figs. 2 and 4.

In these and the following pictures of the velocity field, we have applied the Malmquist-bias correction of Eq. 3, which moves the galaxies away by 8-18% depending on the sample. This correction is not valid for clumpy regions such as Virgo, but is required for more uniform regions elsewhere. Thus, without correcting for the space density of the sample precisely, there is no way of making velocity-field plots that are rigorously correct everywhere. Model velocity fields that are fit to these residuals suffer from a similar bias. This is demonstrated below by quantitative model-fitting to Virgocentric infall. The effect of the Malmquist-bias correction on the velocity field around Virgo is illustrated explicitly in Fig. 11.

Comparing the four data sets in Figs. 6a, b, c and d one sees the same features in all of them: Virgocentric infall, a strong flow toward negative X, which is the main GA flow, and a general compression inwards along the Y-axis, which we interpret as the tidal signature of the GA. It is this compression that indicates radial convergence towards a point several thousand km s^{-1} distant in Centaurus. In the spiral samples, which are basically local, this tidal field is well modeled by a quadrupole with long axis towards the convergent point and short axes perpendicular to it. Pioneering this method on the Aaronson *et al.* data, LYJ found a convergent point some 3000 km s^{-1} distant in the direction of the GA. We find the same distance using a full infall model, but we argue below that this distance to the GA is not as accurate as found by other methods.

Figs. 6e, f, g and h show the residual velocities of the same galaxies after fitting to the standard GA + Vinf + L model used in Fig. 5 (fourth panel there). The improvement in every data set is dramatic. Galaxies from different samples clearly tend to have similar residual motions in the same regions of space. However, some of this is probably due to correlated Malmquist-bias errors in clumpy regions, which show up the same way in all samples.

Finally, Figs. 7a and b summarize our current picture of the overall velocity field by plotting all four data sets together. The agreement among the samples is striking indeed.

Figs. 6 and 7 also illustrate important differences in the spatial coverage of the various samples. The spirals blanket the local volume more densely than the ellipticals, many of which lie outside the volume shown here. However, the spiral samples lack coverage in the southern hemisphere and have relatively

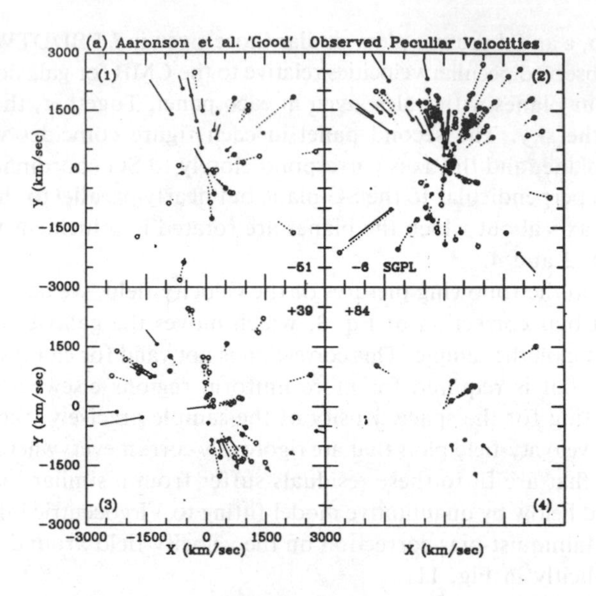

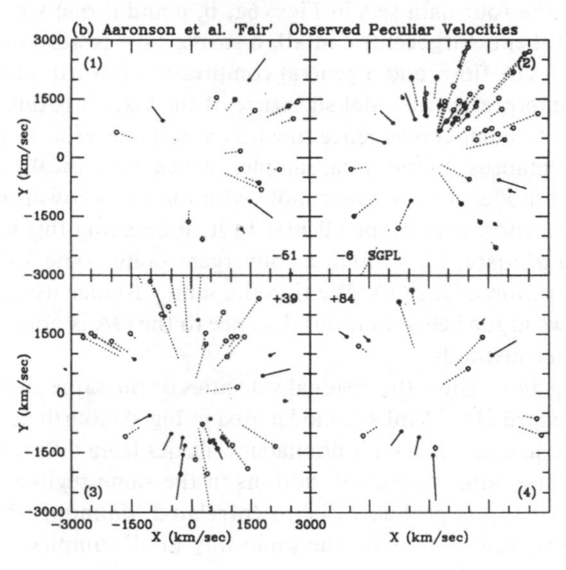

FIG. 6. Velocity field maps like those in LFBDDTW. All galaxies within 22.5° of the plane perpendicular to l = 227°, b = t°, where t is indicated in each panel. The four planes so defined cover the whole sky; the second panel (t = -6°) is the supergalactic plane and the fourth panel (t = + 84°) corresponds closely to the galactic plane. (Note that the X-axes on these diagrams are the mirror-image of those published by LFBDDTW.) Motions away from the Local Group are denoted by solid circles and solid lines; motions towards the Local Group are denoted by open circles and dotted lines. The observed peculiar velocities

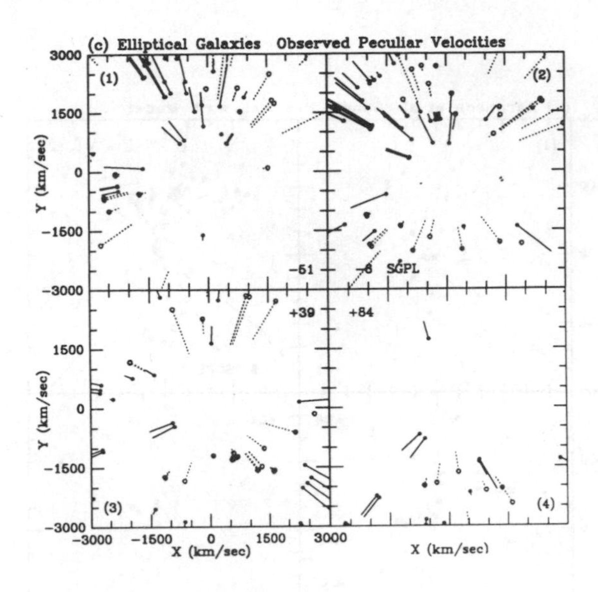

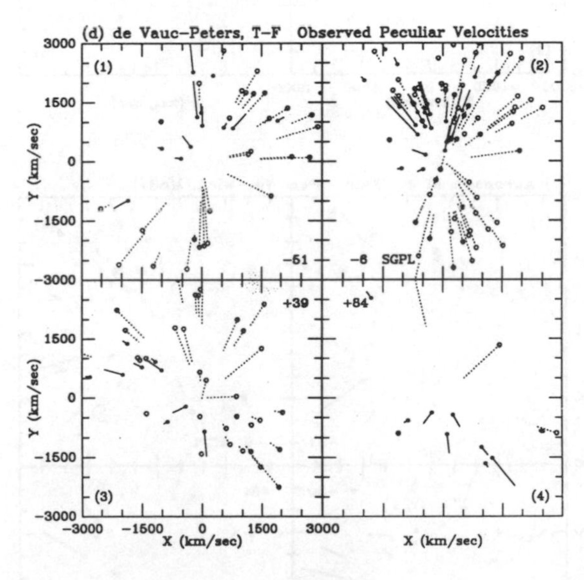

for galaxies in the CMB rest frame relative to a smooth Hubble flow are given in Figs. (a) for the Aaronson *et al.* "Good" sample; (b) for the Aaronson *et al.* "Fair" sample; (c) for the elliptical galaxies analyzed by LFBDDTW; and (d) for the de Vaucouleurs and Peters Tully-Fisher galaxies that do not overlap the Aaronson *et al.* sample. Analogous plots for peculiar velocities with respect to the standard GA + Vinf + L model used in Fig. 5 are given for each data set in Figures 6(e) — (h).

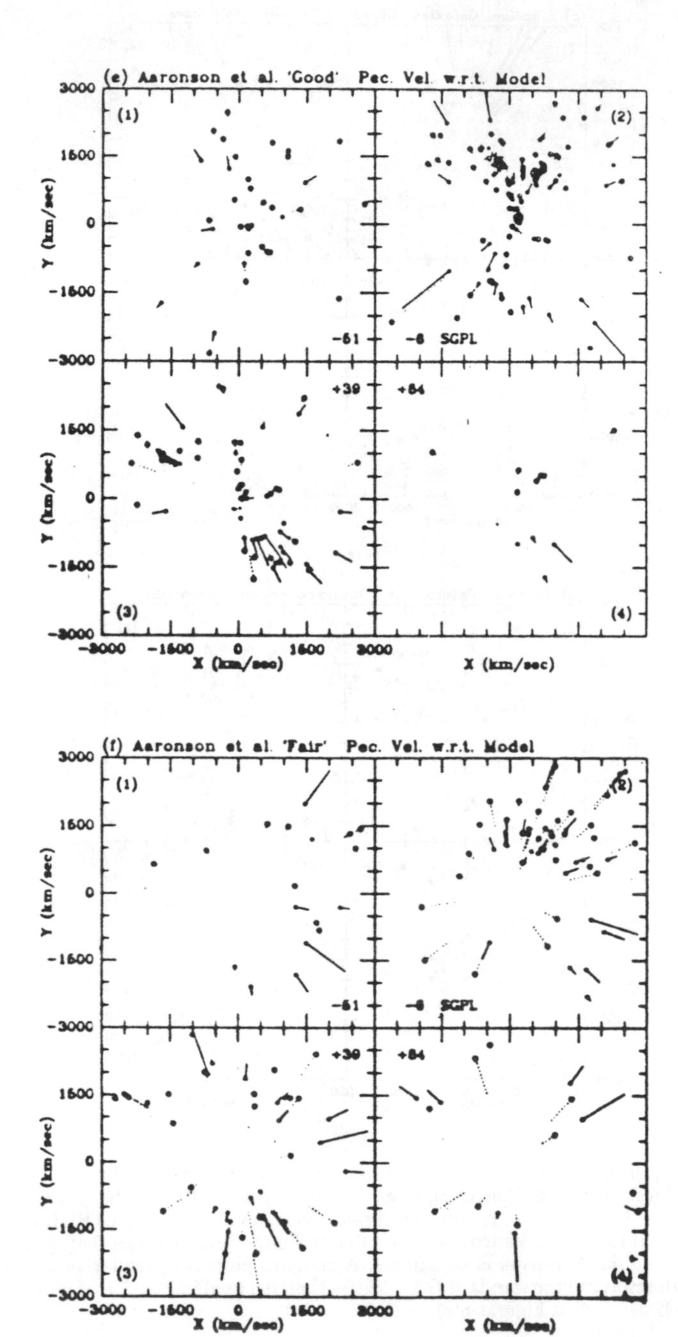
(e) Aaronson et al. 'Good' Pec. Vel. w.r.t. Model
3000
1500
0
-1500
-3000
Y (km/sec)
(1)
(2)
(3)
(4)
-51
-8 SGPL
+39
+84
-3000 -1500 0 1500 3000
X (km/sec)
X (km/sec)
(f) Aaronson et al. 'Fair' Pec. Vel. w.r.t. Model
3000
1500
0
-1500
-3000
Y (km/sec)
(1)
(2)
(3)
(4)
-51
-8 SGPL
+39
+84
-3000 -1500 0 1500 3000
X (km/sec)
X (km/sec)

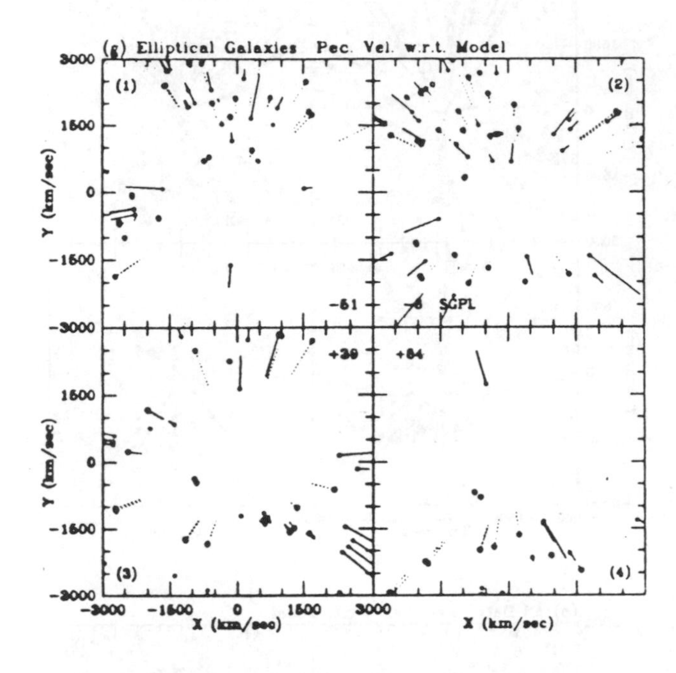
(g) Elliptical Galaxies Pec. Vel. w.r.t. Model
(1)
(2)
(3)
(4)
-51
-6 SGPL
+39
+84
Y (km/sec)
X (km/sec)
3000
1500
0
-1500
-3000

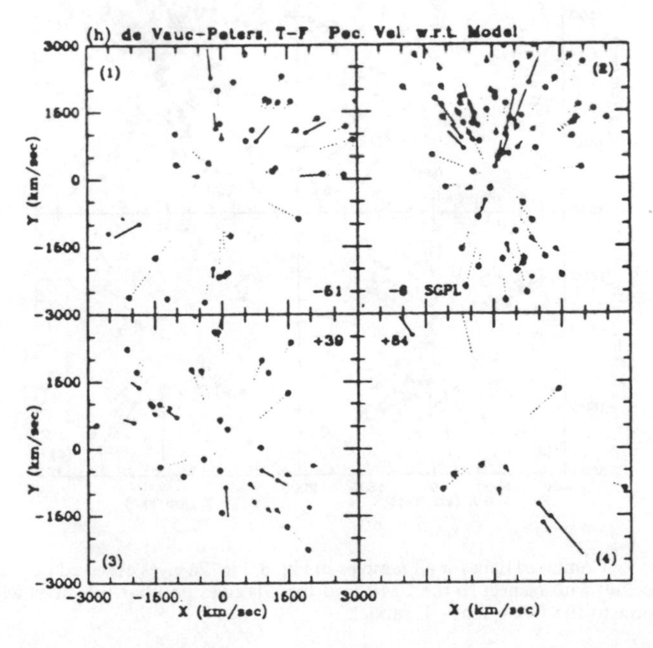
(h) de Vauc-Peters, T-F Pec. Vel. w.r.t. Model
(1)
(2)
(3)
(4)
-51
-6 SGPL
+39
+84
Y (km/sec)
X (km/sec)
3000
1500
0
-1500
-3000

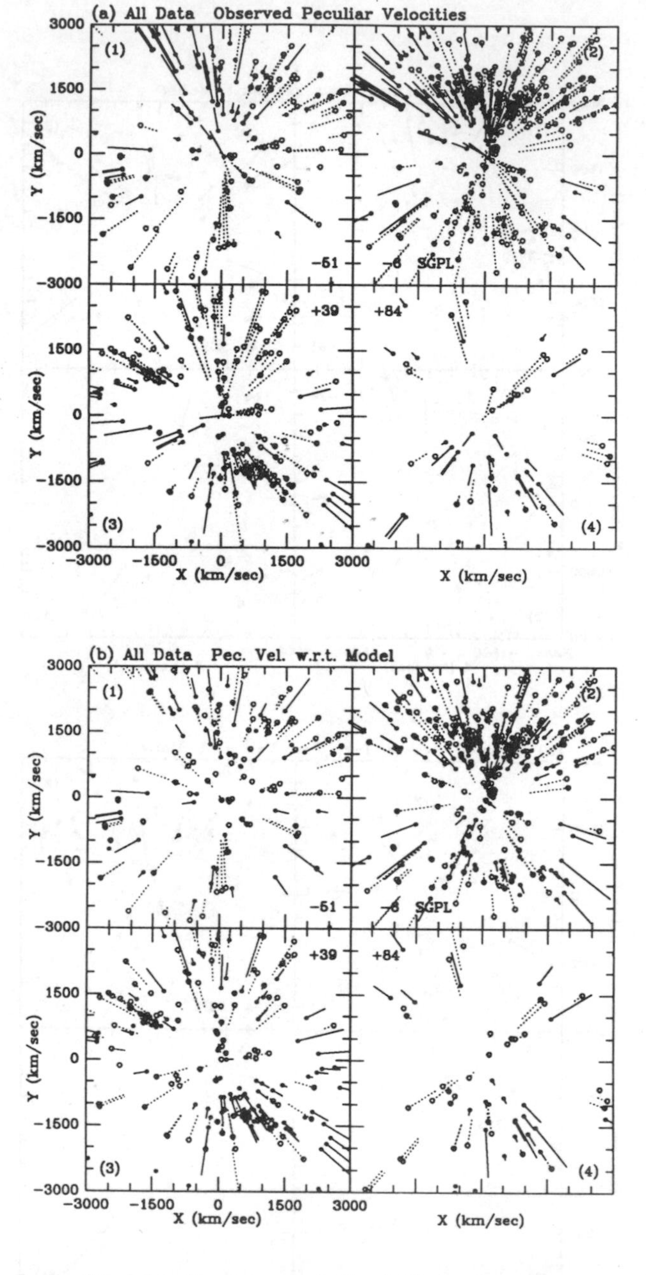

FIG. 7. Combined data for all samples in Fig. 6. Fig. 7a gives observed peculiar velocities with respect to the CMB and Fig. 7b gives peculiar velocities with respect to the GA + Vinf + L model.

few objects close to the GA (negative X axis). The information they provide about the GA is largely local, through the quadrupole term, and they can say relatively little about how the GA flow increases as one approaches its center of mass.

A major result of this Study Week was a better correspondence of the velocity-field maps with the density maps from the IRAS survey (see Yahil 1988 and Strauss and Davis 1988). Fig. 8 presents an enlarged version of Fig. 7a (second panel) giving the observed peculiar motions for galaxies within ±22.5° of the SG plane. This map is designed to match a similar one by Yahil (1988) of the velocity field predicted from IRAS in the SG plane. As presented at the Study Week, the two showed encouraging agreement, and Strauss and Davis (1988) present additional comparisons that suggest the two methods are beginning to converge.

Our impression from comparing the two maps at the conference is that the local flow due to Virgo is well matched but that the IRAS map relative

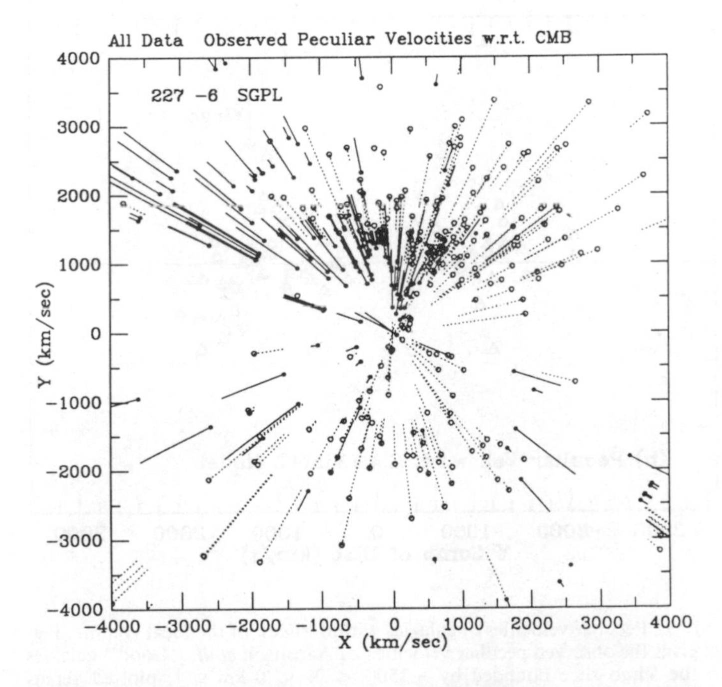

FIG. 8. An enlarged version of Fig. 7a, panel 2, showing the observed peculiar velocities of all galaxies within ±22.5° of the supergalactic plane and over a larger region. This figure should be compared to a similar figure published by Yahil (1988), which predicts peculiar velocities based on density distributions derived from IRAS galaxies.

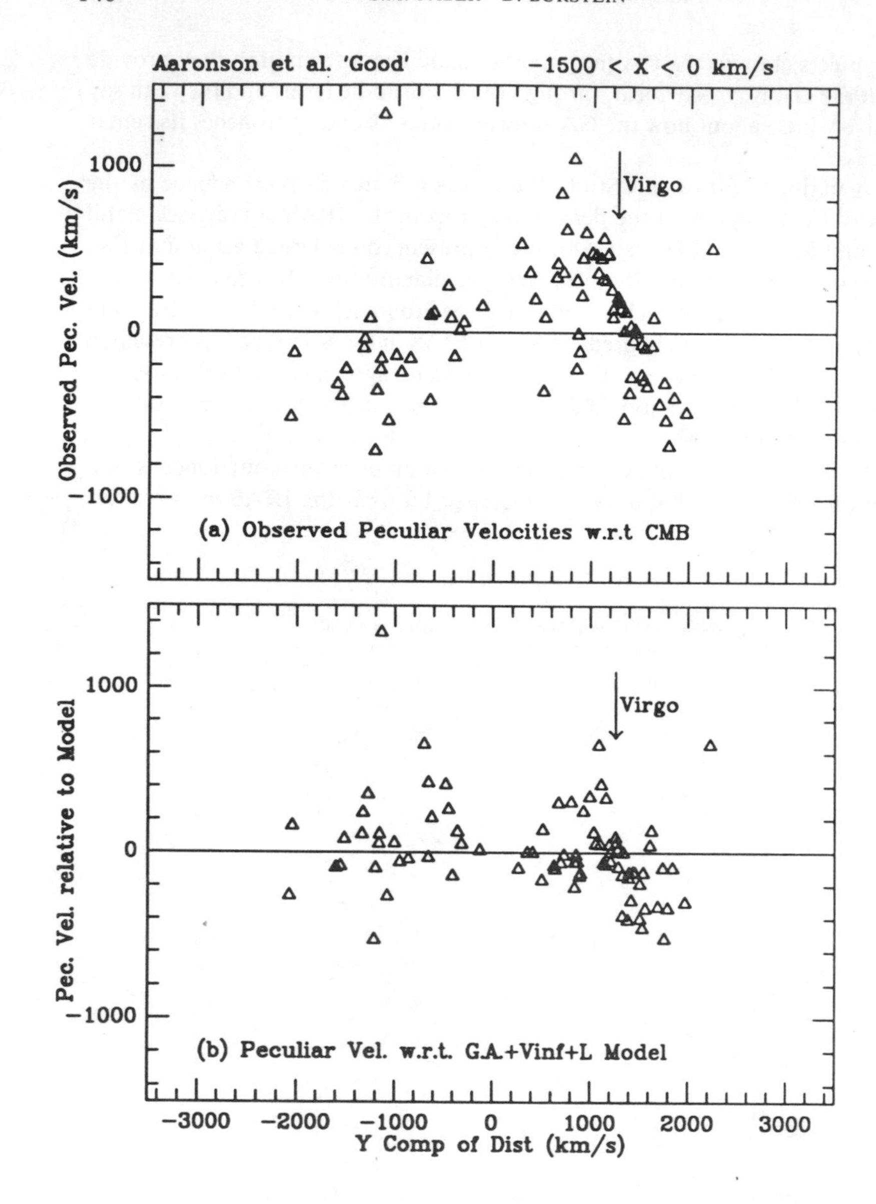

FIG. 9. Peculiar velocities of galaxies within 'slices' of the local volume. Fig. 9a gives the observed peculiar velocities of Aaronson *et al.* "Good" galaxies in the Virgo slice bounded by $-1500 < X < 0$ km s^{-1}, plotted versus Y distance. The Y position of the Virgo cluster is marked. Fig. 9b gives the peculiar velocities of the same galaxies as in Fig. 9a, but this time velocities are with respect to the standard GA + Vinf + L model of Fig. 5. Figs. 9c and d give the same Virgo 'slice' as Figs. 9a and b, but show the Aaronson *et al.* "Fair" sample (hexagons), elliptical galaxy sample (squares) and

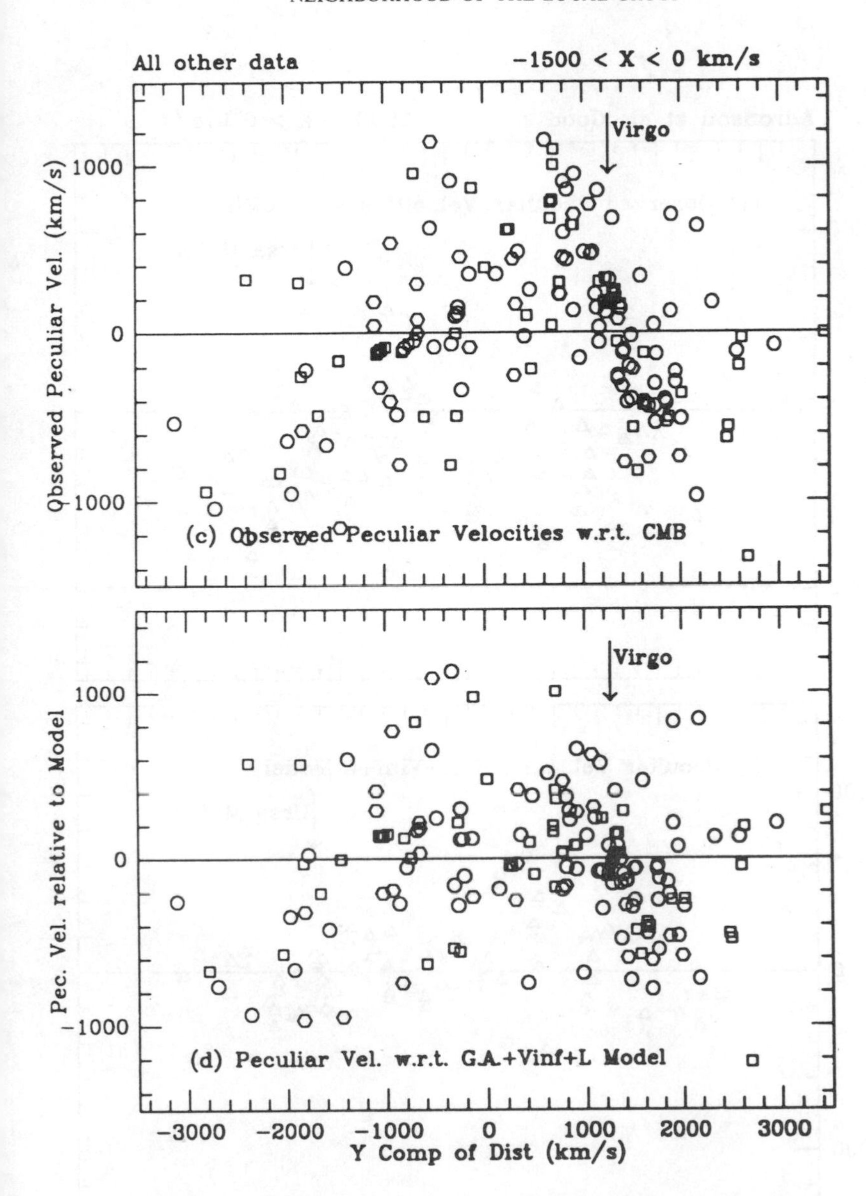

de Vaucouleurs-Peters sample (circles). Figs. 9e and f give peculiar velocities of Aaronson *et al.* "Good" galaxies in the Ursa Major slice bounded by 1500 $< X < 0$ km s^{-1}, plotted versus Y distance. The Y position of the Ursa Major cluster is marked. Fig. 9e gives observed peculiar motion with respect to the CMB, Fig. 9f shows the residual motion after fitting to the standard model. Figs. 9g and h are the same as Figs. 9e and f but show the other three samples. Details are given in the text.

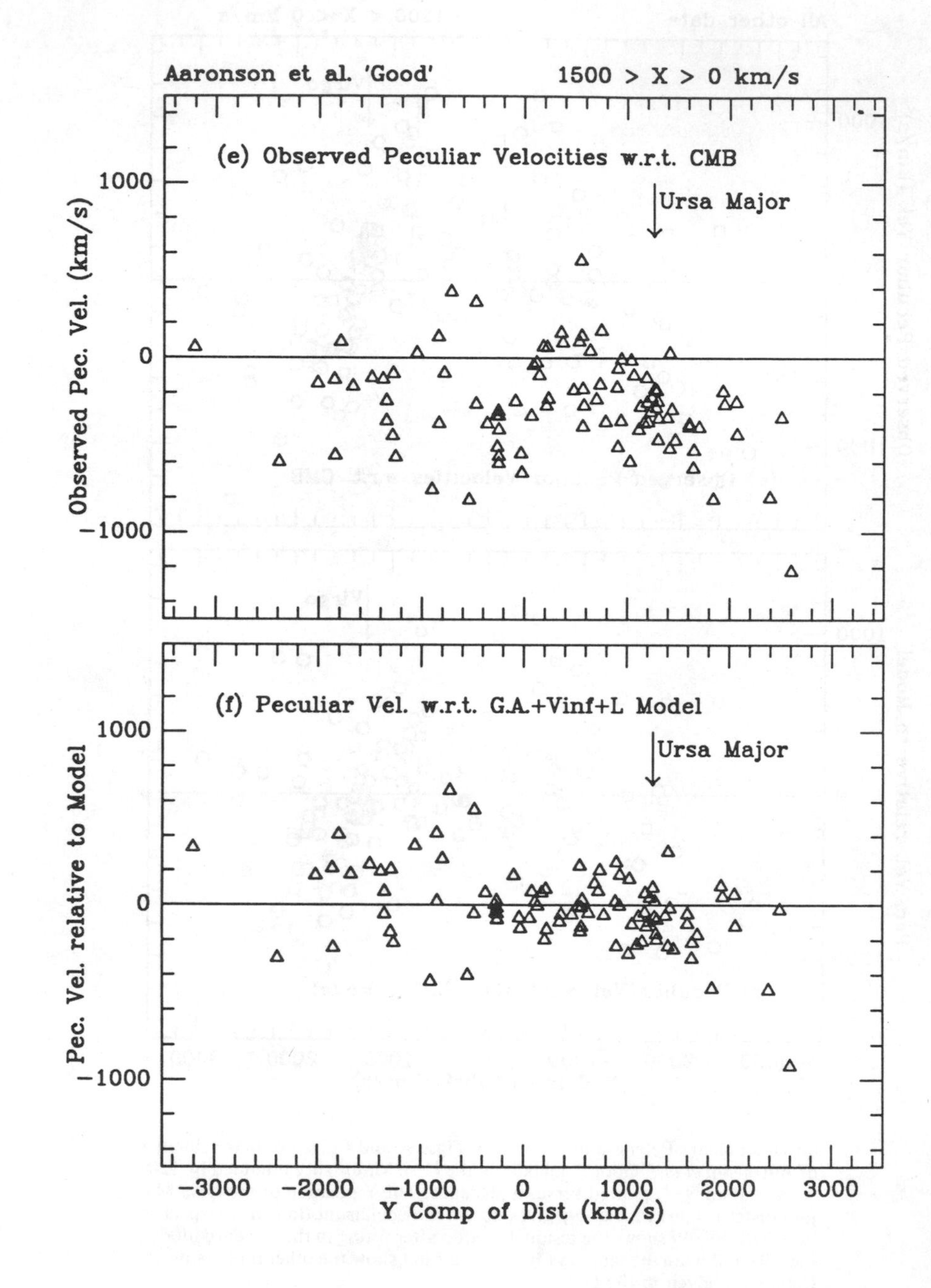
Aaronson et al. 'Good'
1500 > X > 0 km/s
(e) Observed Peculiar Velocities w.r.t. CMB
Ursa Major
Observed Pec. Vel. (km/s)
1000
0
−1000
(f) Peculiar Vel. w.r.t. G.A.+Vinf+L Model
Ursa Major
Pec. Vel. relative to Model
1000
0
−1000
−3000
−2000
−1000
0
1000
2000
3000
Y Comp of Dist (km/s)

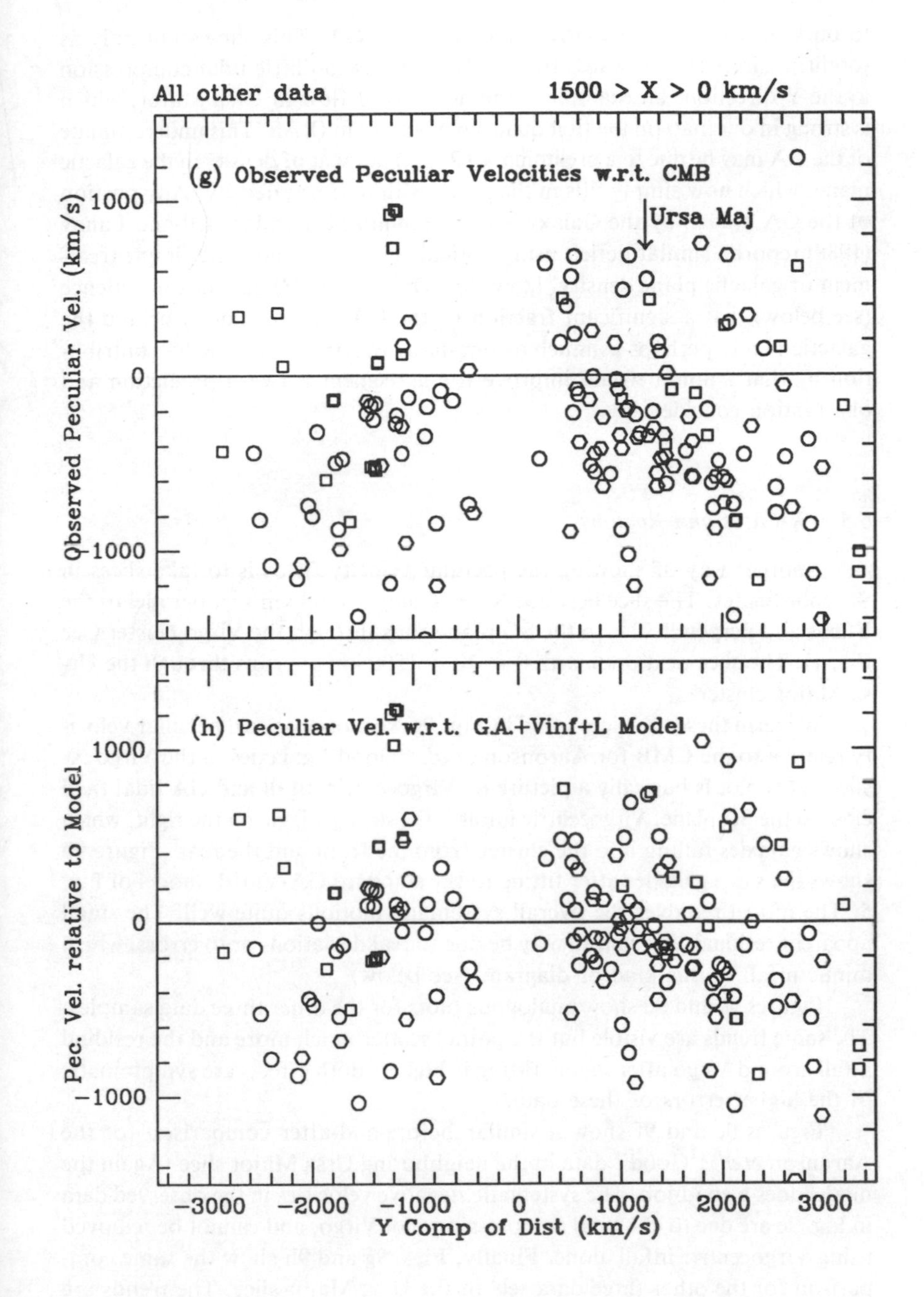
All other data
1500 > X > 0 km/s
(g) Observed Peculiar Velocities w.r.t. CMB
Ursa Maj
Observed Peculiar Vel. (km/s)
1000
0
-1000
(h) Peculiar Vel. w.r.t. G.A.+Vinf+L Model
Pec. Vel. relative to Model
1000
0
-1000
-3000
-2000
-1000
0
1000
2000
3000
Y Comp of Dist (km/s)

to ours shows too little contribution from the GA. This shows not only as too little direct flow towards the GA but also as too little tidal compression in the Y direction; an example is the net inward flow in Ursa Major, which is strong in our map (in the first quadrant) but not in IRAS. This underestimate of the GA may be due to a preliminary IRAS treatment of density in the galactic plane, which now simply fills in the plane with uniform density. Any portion of the GA hidden by the Galaxy is thus minimized. Lynden-Bell and Lahav (1988) report a similar deficit using optical counts and the same simple treatment of galactic plane density. However, Dressler (1988) has shown evidence (see below) that a significant fraction of the GA may be hidden behind the galactic plane, perhaps as much as one-half. Beefing up the IRAS contribution by that amount would improve the agreement between prediction and observation considerably.

6.3 - X-Slices and Regions

Another way of showing the peculiar velocity fields is to take slices in SG coordinates. The slice between X = 0 and -1500 km s^{-1}, parallel to the Y axis and perpendicular to the SG plane runs through the Virgo cluster (see Fig. 1). The slice next door with $0 < X < 1500$ km s^{-1} runs through the Ursa Major cluster.

We term these 'X-slices', and in Fig. 9a we plot observed peculiar velocity relative to the CMB for Aaronson *et al.* "Good" galaxies in the Virgo X-slice. This plot is basically a picture of Virgocentric infall and GA tidal field close to the SG plane. Virgocentric infall is the strong wiggle on the right, which shows galaxies falling into the cluster from the front and the rear. Figure 9b shows the same galaxies after fitting to the standard GA-Vinf-L model of Fig. 5. The model removes the overall systematic motions quite well. The small apparent residual Virgo infall may be due to real deviations or to errors, which mimic infall in this kind of diagram (see below).

Figures 9c and 9d show analogous plots for the other three data samples. The same trends are visible but the points scatter much more and the residual infall around Virgo after model fitting is higher. Both effects are symptomatic of the higher errors of these data.

Figures 9e and 9f show a similar before-and-after comparison for the Aaronson *et al.* "Good" data in the neighboring Ursa Major slice. Again the model does a good job. The systematic negative velocities in the observed data in Fig. 9e are due to GA tidal compression, not Virgo, and cannot be removed using Virgocentric infall alone. Finally, Figs. 9g and 9h show the same comparison for the other three data sets in the Ursa Major slice. The trends are similar but the scatter again is larger. The elliptical galaxies with strong positive residuals belong to the N1600-N1700 group, which is very badly fit by the model (see LFBDDTW).

Complementary to the slices are figures that plot peculiar velocity versus distance for different regions of the sky. In choosing such regions we were guided by the spatial limits of groups and clusters identified by TF87. In Fig. 10 we present "region" plots for 16 selected regions around the sky. Regions are defined for convenience either in galactic coordinates (Figs. 10a-1) or supergalactic coordinates (Figs. 10m-p). These plots cover all of the principal groups and clusters within the sampled volume. Two plots are given for each figure: the left-hand (LH) side shows observed peculiar velocity relative to the CMB versus predicted distance; the right-hand (RH) side shows residual velocity relative to the standard GA + Vinf + L model of Fig. 5, and also versus predicted distance. The line drawn in the RH plot corresponds to a distance error of ± 250 km s^{-1}.

The negative slope of the distance error line presents a fundamental ambiguity in interpreting this kind of diagram. The positive-negative pattern of distance errors looks like true infall when centered on a cluster. Indeed, if infall velocity varies as r from the center, infall precisely mimics observational errors. With *a priori* knowledge of the observational errors it is possible to assess the reality of infall in each case. With this as prologue, we give short comments about the motions in each region, against which the reader can compare his/her own interpretations:

N5846 (Fig. 10a): This region includes the well-known cluster whose name it bears. Nearly all of the galaxies along this line-of-sight seem to be associated with the cluster, which has a net motion of about 200 km s^{-1} relative to the CMB (LH side) and a motion of ~ 100 km s^{-1} relative to the model (RH side). A distance error of 17% for the Aaronson *et al.* "Good" data implies a scatter of about ± 300 km s^{-1}, which is slightly less than actually observed. The evidence for infall is suggestive but marginal.

Leo Cloud (Fig. 10b): Upon correction for the model, what looks like one large collapsing cloud, with a range in peculiar velocity of over 1000 km s^{-1}, appears to break up into two separate structures with the present data. The further cloud has a residual of ~ -200 km s^{-1} with respect to the model, and each cloud extends ~ 1000 km s^{-1} in line-of-sight distance. The apparent infall is consistent with errors.

N1549 (Fig. 10c): Nearly all the galaxies along this line-of-sight are associated with the N1549 group. As with N5846, the evidence for infall is suggestive but marginal. The center of mass has no significant left-over motion with respect to the model.

Fornax and Eridanus (Figs. 10d, e, and f): The galaxies in Fornax are localized to the cluster center and have velocities that look virialized (i.e., no correlation with predicted distance). The galaxies in Eridanus are more spread out along the line-of-sight and are less virialized. Both clusters are somewhat

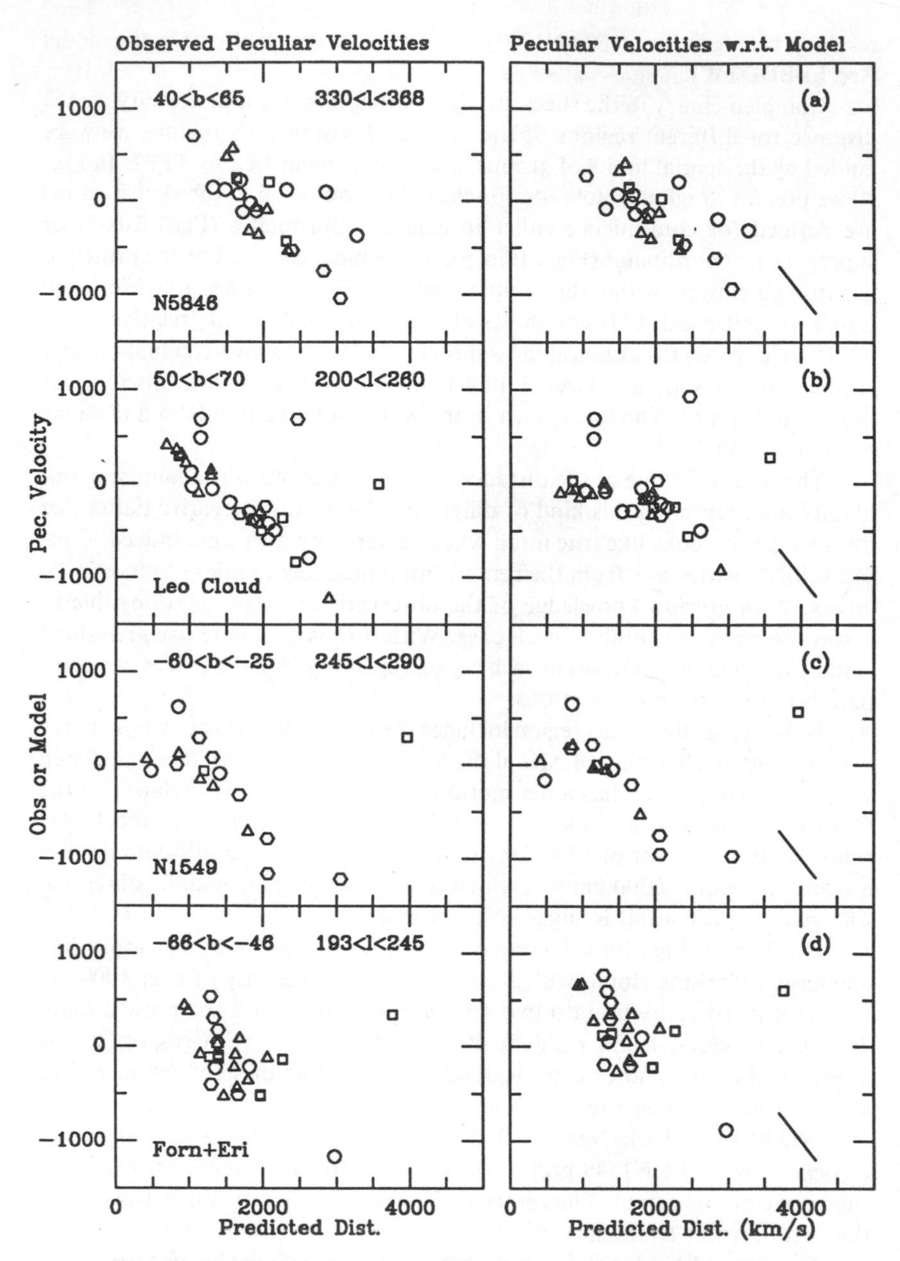

FIG. 10. Peculiar velocity versus predicted distance for galaxies in different 'regions' on the sky. The name and coordinate limits for each region are given in the left-hand figure; Figs. 10a to 10l in galactic coordinates; Figs. 10m to 10p are in supergalactic coordinates. The data for galaxies from all four samples are used; symbols have the same meaning as in Fig. 2. The left-hand side of

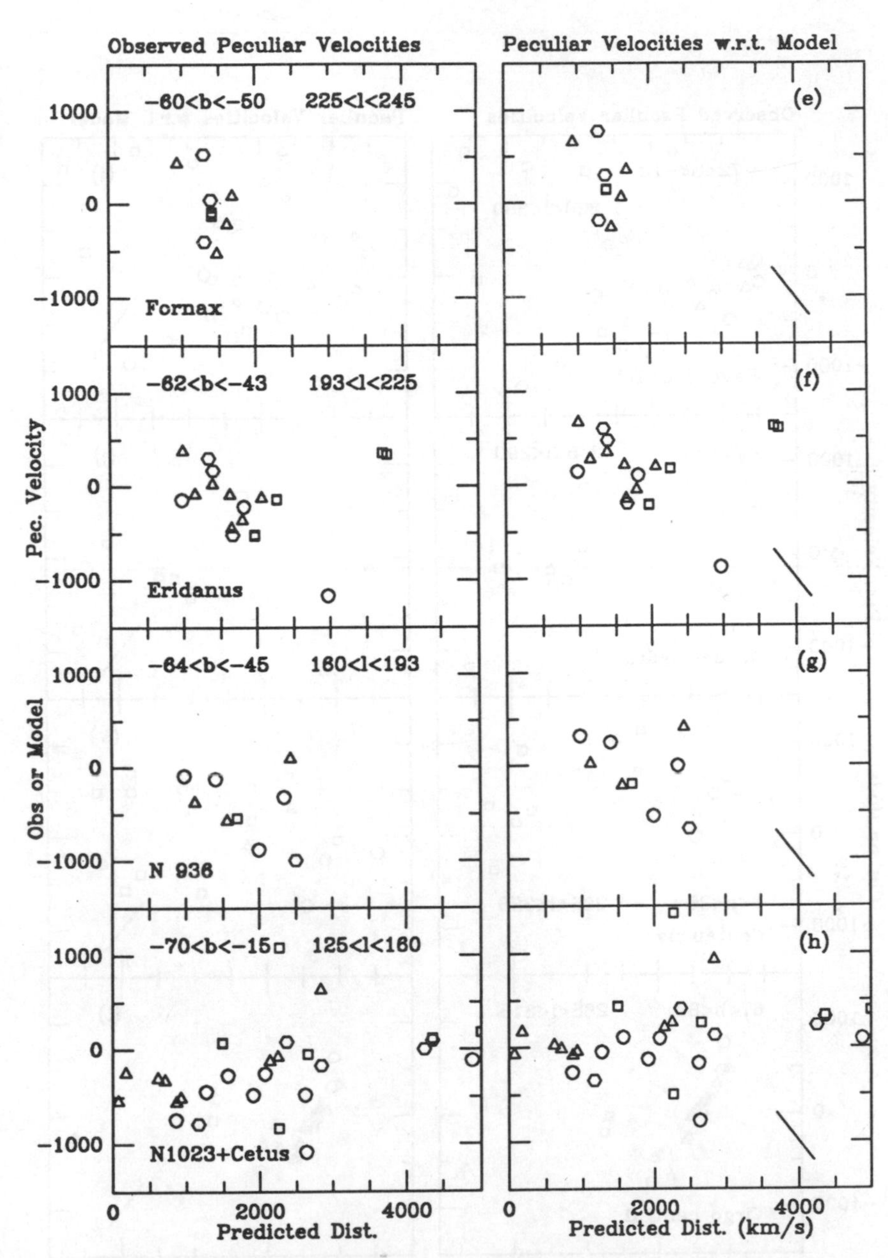

each figure plots observed peculiar velocity in the CMB rest frame versus predicted distance; the right-hand side plots the peculiar velocity with respect to the standard model. The line shown on the right-hand side of each figure represents a "distance-error" slope of ± 250 km s^{-1}. Details are discussed in the text.

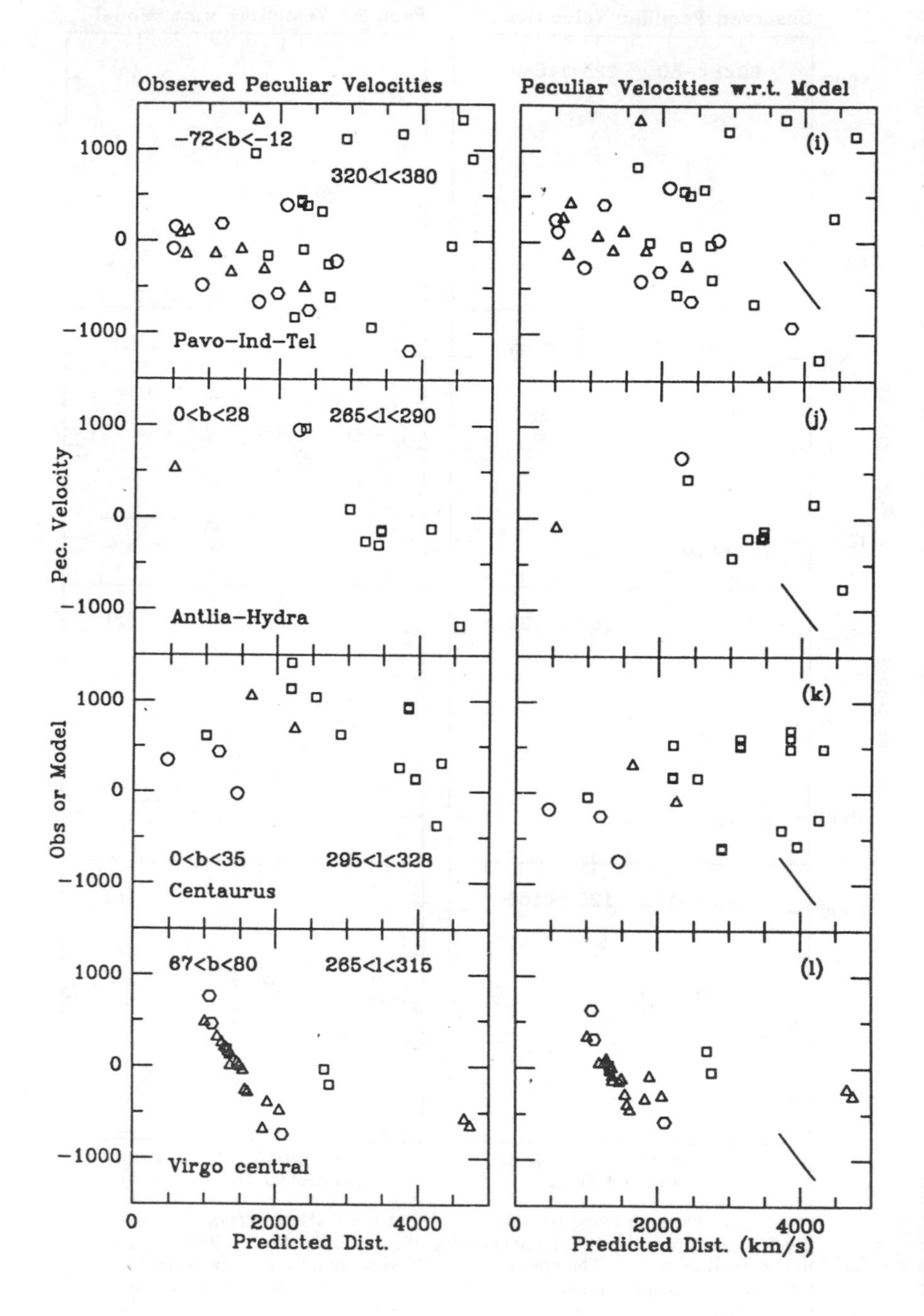
Observed Peculiar Velocities
Peculiar Velocities w.r.t. Model
-72<b<-12
320<l<380
Pavo-Ind-Tel
(i)
0<b<28
265<l<290
Antlia-Hydra
(j)
0<b<35
295<l<328
Centaurus
(k)
67<b<80
265<l<315
Virgo central
(l)
Pec. Velocity
Obs or Model
1000
0
-1000
0
2000
4000
Predicted Dist.
Predicted Dist. (km/s)

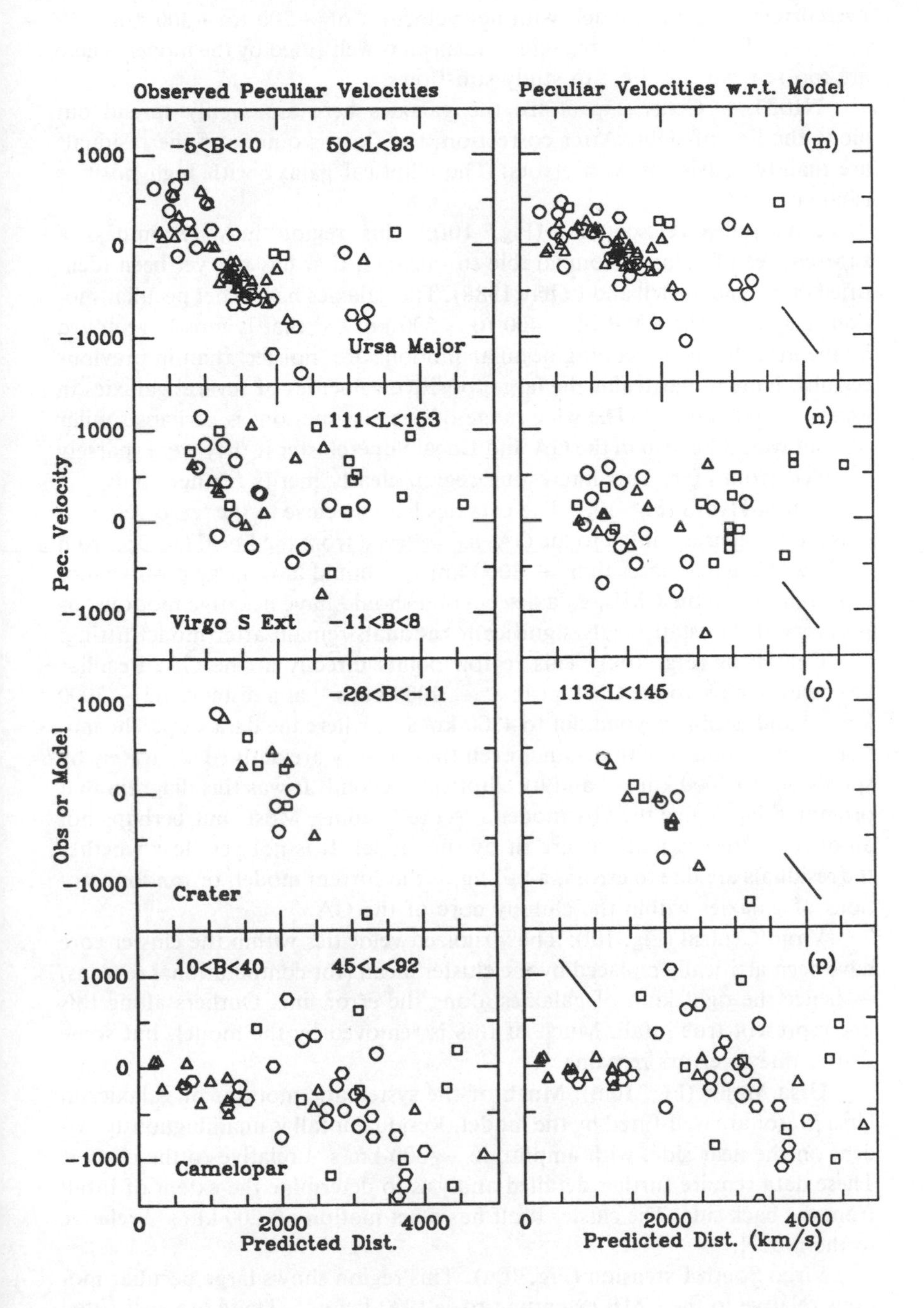
Observed Peculiar Velocities
Peculiar Velocities w.r.t. Model
-5<B<10
50<L<93
(m)
Ursa Major
111<L<153
(n)
Virgo S Ext
-11<B<8
-26<B<-11
113<L<145
(o)
Crater
10<B<40
45<L<92
(p)
Camelopar
Pec. Velocity
Obs or Model
1000
0
-1000
0
2000
4000
Predicted Dist.
Predicted Dist. (km/s)

overcorrected by the model, with net velocities of $+200$ to $+300$ km s^{-1}.

N936 (Fig. 10g): This region in the mean is well-fitted by the model. There are too few galaxies here to study sub-flows.

N1023 + Cetus (Fig. 10h): the galaxies here are evenly spread out along the line-of-sight. After correction, the flow is quiet and the residuals are mainly consistent with errors. The elliptical galaxy with high positive residual is N661.

Pavo-Indus-Telescopium (Fig. 10i): This region may encompass a supercluster of galaxies comparable to our own that has not yet been identified (see Lynden-Bell and Lahav 1988). The galaxies have a net peculiar motion relative to the CMB of -400 to -500 km s^{-1} that is mostly removed by the model. The remaining peculiar motions are 'noisier' than in previous regions. Note in particular the large, positive velocities of several galaxies in excess of 1000 km s^{-1}. The wide range of peculiar motions is perhaps similar to what would be seen in the GA and Local Supercluster if they were sparsely sampled from afar. This interesting region clearly merits further study.

Antlia-Hydra (Fig. 10j): The galaxies here lie close to the 'zero-velocity-circle' that separates infall to the GA, as observed from the Local Group, from outflow. Galaxies closer than ~ 3000 km s^{-1} should have large positive motions relative to the CMB; galaxies beyond should have negative motions, as is observed. No statistically significant residuals remain after model fitting.

Centaurus (Fig. 10k): This region points directly at the GA. Peculiar velocities relative to the CMB peak at ~ 1000 km s^{-1} at a distance of ~ 3000 km s^{-1} and decline beyond out to 4300 km s^{-1}, where the data stop. The motions vary smoothly with distance even though they are defined separately by spirals out to 2000 km s^{-1} and by ellipticals beyond. It was this diagram that prompted us to give the GA model a "core" radius. Most, but perhaps not all of the observed motions are fit by the model. It is not yet clear whether the residuals are due to errors, a failing of the current model, or random motions of galaxies within the clumpy core of the GA.

Virgo Central (Fig. 10l): The virialized velocities within the cluster core have been artifically replaced by the cluster mean (for computational reasons) — hence the tight knot of galaxies along the error line. Outliers along this line represent true infall. Much of this is removed by the model, but some scatter due to errors remains.

Ursa Major (Fig. 10m): Much of the systematic motions of galaxies in Ursa Major are well-fitted by the model. Residual infall is unambiguously evident on the near side, with amplitude ~ 300 km s^{-1} relative to the cluster. These data require further detailed analysis to determine the extent of infall from the back side. The cluster itself has a net motion of 200 km s^{-1} relative to the model.

Virgo South Extension (Fig. 10n): This region shows large peculiar motions relative to the CMB extending to $+1000$ km s^{-1}. These are well-fitted by the model on average, but large scatter remains. The 'noisy' residuals resemble those in Pavo-Indus-Telescopium.

Crater (Fig. 10o): As with the N1549 and N5846, the galaxies here appear to be associated with one group. The group as a whole has a peculiar motion of $\sim$ 300 km s^{-1} relative to the CMB that is moderately well-fitted by the model, leaving a smaller net velocity of $\sim$ 100 km s^{-1}. The evidence for true infall into this cluster is somewhat stronger than for the other two clusters (the Aaronson *et al.* "Good" data show a wider scatter than expected), but is still marginal.

Camelopardalis (Fig. 10p): Within 1700 km s^{-1}, the motion is quiet and coherent. More distant galaxies show 'noisy' residuals, but much of this may be due to errors.

Three general features of the peculiar velocity field are evident from the 'X-slice' and 'region' diagrams. First, the dominant motion in this volume is due to the GA, as shown both by direct acceleration and by tidal compression. This large-scale coherence is why the distribution of galaxies in distance-space (Fig. 2) appears very similar to the distribution of galaxies in redshift-space (TF87).

Second, the motions of galaxies that remain *relative* to the GA flow are of three kinds: a) "Bulk" motions that have a low internal velocity dispersion. Examples include the Local Anomaly, the Leo Cloud, N1023/Cetus and the near region of Camelopardalis; b) Localized clusters with possible infall (Virgo, Ursa Major, N5846, N1549, Crater, Fornax and Eridanus). The evidence for infall is convincing for Virgo and Ursa Major but is less so for the other five. Virgo is certainly the most massive of these nearby clusters; c) Regions with internally 'noisy' motions. These include Pavo-Indus-Telescopium, Virgo South Extension, Centaurus and possibly the more distant galaxies in Camelopardalis.

Finally, systematic deviations from the standard GA model occur with amplitudes of up to 300 km s^{-1}. Deviant regions include groups (Crater, Fornax, Ursa Major) and also larger regions 1000 km s^{-1} or more across (near-side Ursa Major, Leo, Eridanus and perhaps the near region in Camelopardalis). Compared to these, the local Coma-Sculptor cloud seems fairly typical in terms of its scale size (1400 km s^{-1}) and residual motion (360 km s^{-1} total space velocity, 200 km s^{-1} line-of-sight).

7. Quantitative Models

This section reports on quantitative model fits using Eqs. 1 and 2. The results are preliminary since we are still in the stage of revising both the data and the fitting methods (hence the models here sometimes differ slightly from those in the graphs). The Aaronson *et al.* "Good" data are used exclusively, and the "Local switch" is always turned on so that the Local Anomaly out to 700 km s^{-1} (see Table 2) is smoothed out. The errors Δ and σ_f have been optimized separately in each case to give the best likelihood, but we do not take their relative values too seriously (see below).

7.1 - Malmquist Effects in Virgo: Method 1 vs. Method 2

The first two models (Table 3) show the sensitivity of Virgo infall to the Malmquist-bias correction in Method 1. In Model 1 we have applied a bias correction from Eq. 3 that corresponds to uniform space density and a Δ of 0.37 mag. Model 2 is the same without the bias correction. Both models assume the "new" GA flow from Table 1.

The resultant Virgo infall parameters are given in the last columns. The Malmquist-bias correction in Model 1 clearly pumps up the Virgo flow. The quantity v_V is larger, and the fall-off away from Virgo is more gradual (smaller n_V). An intuitive appreciation can be had from Fig. 11, which shows the observed peculiar velocities relative to the CMB around Virgo in both fits. Figure 11a shows the residuals with Malmquist-bias included, Fig. 11b without. Since the correction multiplies all distances by a constant factor, the distances of all galaxies in Fig. 11a are shifted outward relative to Fig. 11b. This has a systematic effect on the velocity vectors behind Virgo and Ursa Major, which are lengthened. The model responds by choosing a slower fall-off of Virgocentric infall velocity away from the cluster center.

This Malmquist-bias correction is clearly invalid in a clumpy region like Virgo-Ursa Major. The example shows how an invalid Malmquist-bias correction can create an illusory velocity field. However, many of the Aaronson *et al.* galaxies are more uniformly distributed in space, and for them the uniform-density correction may not be a bad approximation. The situation serves to show how hard it is to to get an unbiased picture of the velocity field. It also illustrates how important it is, when making pictures of the velocity field, to have Δ as small as possible. Since the Malmquist correction goes as Δ^2, the differences between Figs. 11a and 11b would be significantly larger if the Aaronson *et al.* "Fair" or DVP data sets were used.

7.2 - Virgo Infall and the Great Attractor

The rest of the models use the hybrid version of Method 2, which effectively minimizes in log v but otherwise has the same Malmquist-bias properties as Method 1. Because we are primarily interested in Virgo we have turned off the uniform-density Malmquist correction based on the previous section, which showed that the correction biases the Virgo infall solution unphysically. Model 3 is a repeat of Model 2 with this new method. The differences, though slight, illustrate the effects of minimizing in log v rather than in v.

TABLE 3

VIRGO INFALL MODELS USING AARONSON *ET AL.* "GOOD DATA" ONLY

Model	GA Parameters						σ_f^a (km s^{-1})	Δ^a mag	Virgo Parameters		
	d_A (km s^{-1})	v_A	n_A	c_A	l_A	b_A (°)			v_V (km s^{-1})	n_V	c_V
1. Method 1 Malm. on GA solution from ellipticals	4200	535	1.7	0.34	308	18	145	0.37	260	0.5	0.00
2. Method 1 Malm. off	4200	535	1.7	0.34	308	18	77	0.45	133	1.0	0.10
3. Method 2	4200	535	1.7	0.34	308	18	90	0.41	100	1.5	0.25
4. Method 2 GA solution from spirals	3000	555	1.1	0.20	305	18	79	0.41	85	1.9	0.30
5. Method 2 AHMST clone	None, use bulk flow only						80	0.41	300	0.6	0.40

All models listed here use Aaronson *et al.* "Good" data only. GA parameters are taken either from Table 1 (Nos. 1, 2 or 3) or optimized here (No. 4).

[a] The relative values of σ_f and Δ are highly uncertain — see text.

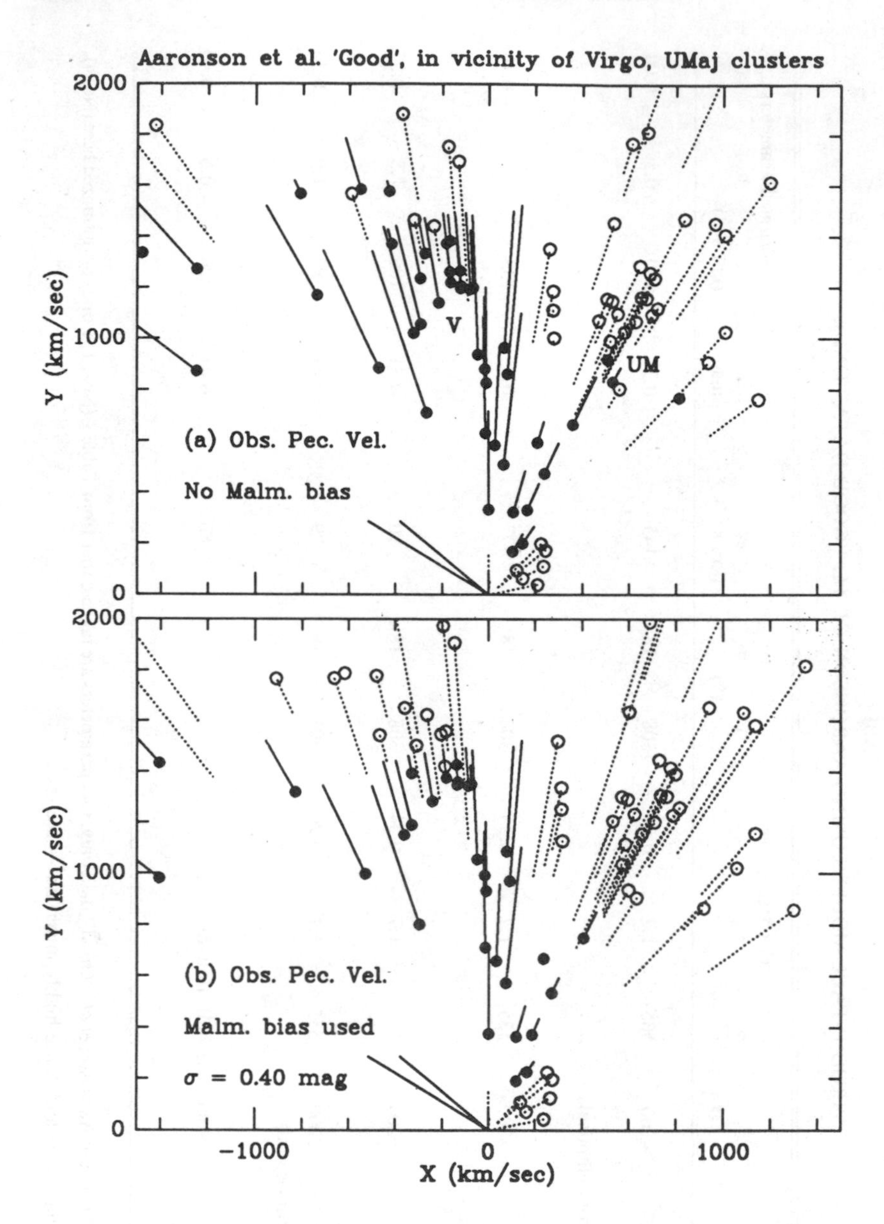

FIG. 11. The observed peculiar velocity field in the CMB rest frame, in the vicinity of the Virgo cluster, as defined by the galaxies in the Aaronson *et al.* "Good" sample, under two different assumptions: (a) Peculiar velocity predicted with no Malmquist-bias correction. (b) Peculiar velocity with a Malmquist-bias correction as in Eq. 3 and an observational error of 0.40 mag. Note the apparent increase of infall velocity on the back side of the Virgo cluster in (b) relative to (a).

Both models have a steep fall-off (high n_V) and low v_V. With its steep profile and low infall velocity at the Local Group, this picture of Virgo infall differs considerably from the conventional one, but agrees with the visual picture presented in the last section. More on this presently.

Whereas Model 3 takes its GA parameters from the elliptical galaxies, Model 4 optimizes them from the Aaronson *et al.* "Good" data themselves. The two solutions agree well with regard to flow velocity, v_A, at the Local Group and the GA direction on the sky. The fall-off away from the GA is shallower in Model 4, but this is not strongly constrained. On the other hand, Model 4 puts the GA at only 3000 km s^{-1}, compared to 4200 km s^{-1} from the ellipticals, a difference which is several times the formal error. However, the spirals determine the distance to the GA mostly through the quadrupole term, which is easily perturbed by the small-scale systematic flows that we know exist, but are not contained in the model (see above). If the recently-measured spirals in Centaurus were included (see Mould 1988) their outward velocities would push the GA further away, as do the ellipticals (see LFBDDTW). With these differences taken into account, the agreement between the local spirals and the GA model is excellent.

In computing Model 4, we were struck by its small value of local Virgo infall (85 km s^{-1}). Aside from Model 1, of dubious validity, our models never return infall velocities in the 250-300 km s^{-1} range, despite numerous attempts to perturb them. When we finally tried to recreate a model exactly equivalent to AHMST, the situation became clear. In this model (Model 5), there is no GA flow, and all radial velocities are taken with respect to the rest frame of the Aaronson *et al.* "Good" sample. The fit prefers a high v_V of 300 km s^{-1} and a very shallow fall-off with $n_V = 0.6$.

The crucial difference in this model is the lack of a GA, specifically the GA tidal compression along the Y axis. Because it produces a similar compression, Virgo infall increases to take up the slack. This is illustrated in Figs. 12a, b and c, which repeat the X-slice through Virgo shown in Figs. 9a-d. Fig. 12a shows observed peculiar velocities relative to the CMB, as in Fig. 9a. Fig. 12b shows the Virgocentric infall signal that is left after a standard GA model is removed. This is comparable to the Virgocentric infall signal in Models 2-4, where the GA is fit simultaneously with Virgo. One sees how well the GA removes all the large-scale trends, leaving only a small-scale deviation in the immediate neighborhood of Virgo. Fig. 12c shows the radial velocities after transformation to the sample rest frame, which corresponds to the Virgo signal in Model 5 and in AHMST. Here the signal continues through the position of the Local Group to the left, where negative velocities look like Virgocentric inflow. This is the primary source of cross-talk between Virgocentric flow models and the GA flow.

Figs. 12d-f show the same three cases for the neighboring Ursa Major X-slice. Removing the GA flow again has a major effect. The GA accounts

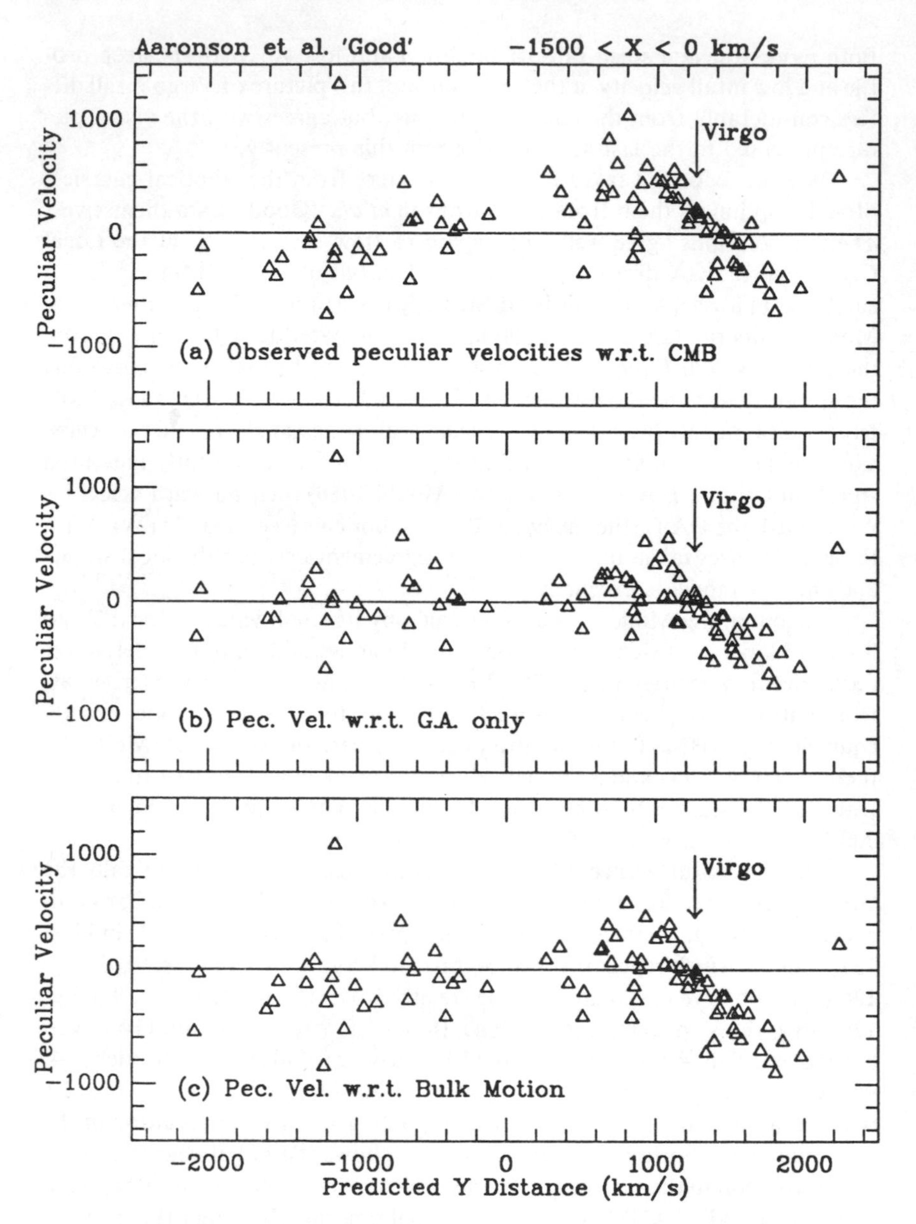

FIG. 12. 'X-Slices' illustrating the cross-talk between the peculiar velocity fields of the GA and Virgo. Only data from the Aaronson *et al.* "Good" sample are used. Figs. 12a-c plot the peculiar velocities of galaxies versus Y distance in a slice bounded by $-1500 < X < 0$ km s^{-1} that highlight motions relative to the Virgo cluster. Figs. 12d-f highlight motions of galaxies relative to the Ursa Major cluster in a slice bounded by $1500 < X < 0$ km s^{-1}. These plots differ from those in Fig. 8 in that only galaxies within a distance of 2500 km

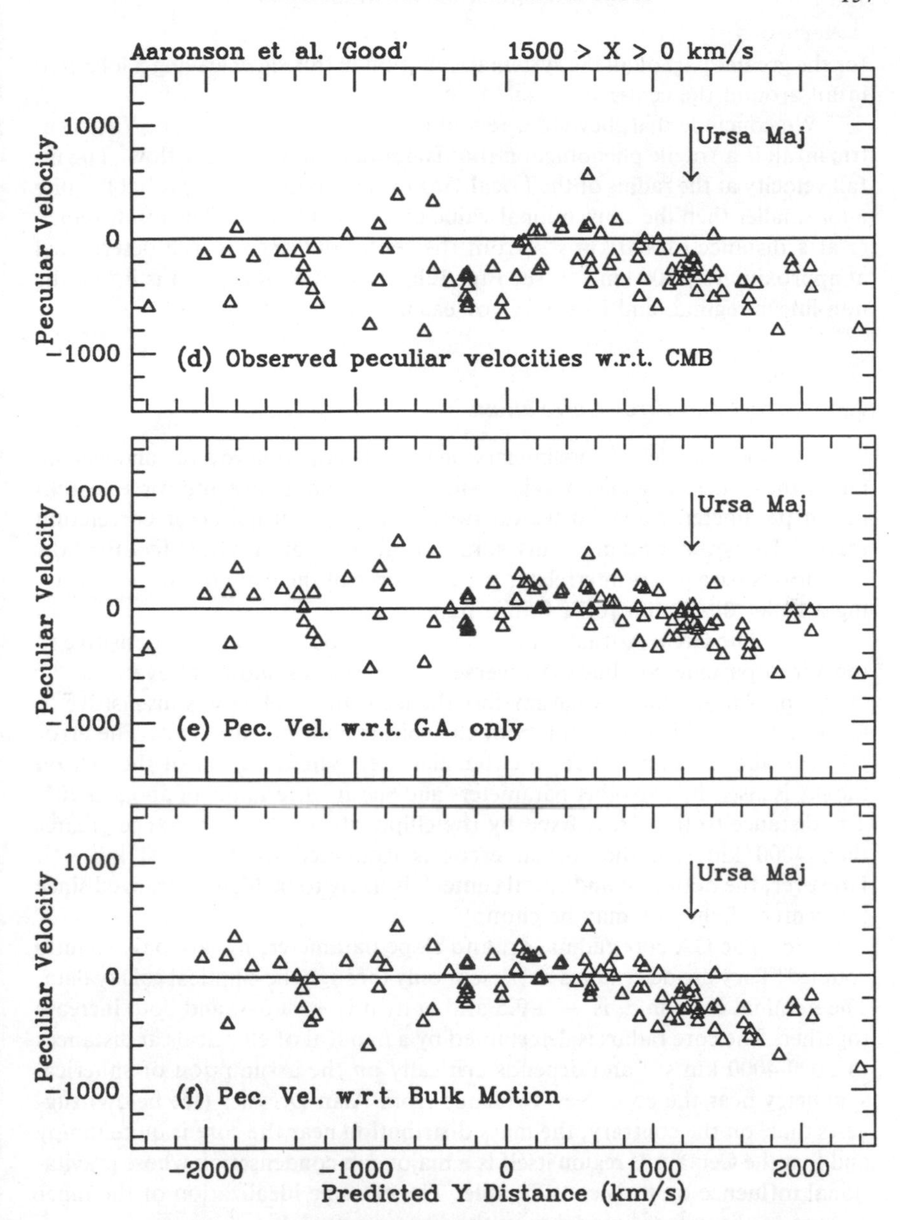

s^{-1} are used. Figs. 12a and d show observed peculiar velocities relative to the CMB. Figs. 12b and e plot peculiar velocities relative to a model that subtracts out only the velocity field due to the GA. Figs. 12c and f plot peculiar velocities relative to the bulk motion of all galaxies (in all four data samples) with distances less than 2500 km s^{-1}. With the GA subtracted, the remaining Virgo infall is highly localized to the cluster. Subtracting just the bulk motion leaves behind negative velocities at negative Y, which mimic Virgo infall.

for the greater part of all the systematic motion in this slice, leaving a localized infall around the center of Ursa Major.

We conclude that, beyond a few hundred km s^{-1} from Virgo, Virgocentric infall is a fragile phenomenon that is dominated by the GA flow. The infall velocity at the radius of the Local Group is uncertain but is probably quite a lot smaller than the conventional value of 250-300 km s^{-1}. The infall velocity at a distance of 400 km s^{-1} from the center of Virgo is well-determined at approximately 500 km s^{-1} (see Fig. 12b). However, this point is inside the non-linear regime, and its utility for calculating Ω is doubtful.

7.3 - Errors and Error Correlations

The above models are preliminary, and we do not yet have a full understanding of the errors. The likelihood maximization is done by brute-force searching in parameter space, so we do not produce the usual error correlation matrix. However, running many solutions gives us an intuitive feel for how the various parameters interrelate. We give some of the main results here, leaving a fuller discussion for a future paper.

a. The Great Attractor parameters as a group are highly insensitive to the Virgo parameters, but the reverse is not true, as shown above.

b. Among the GA parameters the local flow velocity is insensitive to all others, and this is true for both the spirals and the ellipticals. The error on v_A is about 70 km s^{-1} from either data set. The direction of the GA on the sky is insensitive to other parameters and has an uncertainty of about $\pm 10°$. The distance to the GA is fixed by the ellipticals to be at a distance greater than 4000 km s^{-1}; the formal error is estimated to be ± 300 km s^{-1}. However, the notion of an "infall center" is likely to be highly idealized since the center of the GA may be clumpy.

c. The GA core radius, c_A, and slope parameter, n_A, are only slightly coupled. They are determined at present only through the elliptical galaxy data. The nominal error in c_A is $\sim \pm 0.5$ and in n_A it is $\sim \pm 0.3$, and both increase together. The core radius is determined by a handful of ellipticals at distances of 3000-4000 km s^{-1} and depends critically on the assumption of spherical symmetry near the core. New evidence from Alan Dressler (see below) suggests that, on the contrary, the mass distribution near the core is quite lumpy and that the Centaurus region itself is a major sub-condensation whose gravitational influence on us is considerable. The extreme idealization of the inner regions in the present model may thus prove untenable.

d. If a fixed GA flow is assumed, then the two most closely coupled parameters for Virgo are v_V and n_V. Because the infall amplitude is firmly anchored at a few hundred km s^{-1} from the cluster (cf. Fig. 12b), a steeper slope must be compensated by lower infall farther out. With a fixed GA, the nominal uncertainty in v_V is ± 30 km s^{-1}, but the weakness of the signal far out and the high sensitivity to the large-scale velocity field of the GA imply larger systematic errors.

e. The error estimates for Δ and σ_f are strongly anti-correlated. Maximum-likelihood distinguishes between them by the way they fall off with distance: in magnitude units, measurement error is constant with distance, whereas field-dispersion error declines. We have seen that σ_f is well-determined locally to be 80-90 km s^{-1} out to 700 km s^{-1}. The best-fitting model over the whole volume adjusts by choosing a relatively low σ_f of 90-120 km s^{-1} and a high compensating Δ of 0.39-0.41 mag (Aaronson *et al.* "Good").

Maximum-likelihood assumes that σ_f is constant everywhere, which is risky. Patches like the Local Anomaly may be quiet internally (80-90 km s^{-1}) but deviate from the flow model by significantly larger amounts (see Sec. 6). The local value of σ_f may thus be too low to typify the whole volume. As a guide, we can use the Local Anomaly, whose magnitude of 360 km s^{-1} converts to 200 km s^{-1} line-of-sight. Taking this as an upper limit to σ_f gives = 0.34 mag for the Aaronson *et al.* "Good" data. This probably brackets the range. A better value could come from good diameters and inclinations for the spirals in clusters (see discussion by Tully 1988), which can give Δ independent of σ_f.

7.4 - *Parameter Significance*

The significance of each parameter is given by the change in likelihood that results from its addition to the model. Adding an extra parameter should decrease the likelihood by 0.5 units just by random chance. Likelihoods for Method 2 are given in Table 4 for four models. In each case, Δ and σ_f have been optimized to give the best likelihood. The first model has no motions whatsoever and provides a zeroth-order comparison. The second model is the best bulk model with no internal flows (i.e., local switch "off"). The third model turns on the local switch, which adds three parameters, and reoptimizes the bulk motion. The fourth model adds Virgo infall, which adds another three parameters, and again reoptimizes the bulk motion. Finally the fifth model adds the Great Attractor but takes away the bulk flow, which is no longer needed. This adds six parameters but takes away three, for a net gain of three. The number of free parameters in each model is given in the table.

The table shows that the addition of extra parameters is strongly merited at every stage, giving a likelihood change that at minimum is at least four times larger than expected by chance. The improvement due to the GA comes in two stages: from Model 1 to Model 2, which adds the bulk flow, and from Model 3 to Model 4, which adds the tidal term. Thus, the GA is somewhat more important than Virgo in maximizing the total likelihood.

TABLE 4
LIKELIHOOD CHANGES

No.	Model	No. Param.	Likelihood (Method 2)	Δ (Likelihood)
1	No motions	1[a]	−201.41	——
2	Bulk motion, no internal flows	4[b]	−164.82	+36.6
3	Turn on local anomaly	7[c]	−158.62	+42.8
4	Turn on Virgo[e] infall	10[d]	−121.86	+79.6
5	Turn on GA[g] turn off bulk	13[f]	−112.96	+88.5

a No motions permitted. Free parameter is the Hubble constant.
b Hubble constant plus three bulk-motion components.
c Add three Local Anomaly components.
d Add three Virgo parameters: v_V, n_V, c_V.
e Virgo parameters: $v_V = 225$ km s^{-1}, $n_V = 0.8$, $c_V = 0.3\ d_V$. These differ from Model 5, Table 3 owing to slightly different σ_f and Δ.
f Add six GA parameters: d_A, v_A, n_A, c_A, l_A, b_A. Take away three bulk-motion components.
g GA parameters are "new" from Table 1. Virgo parameters: $v_V = 100$ km s^{-1}, $n_V = 1.7$, $c_V = 0.25\ d_V$.

8. The Nature of the Great Attractor

In LFBDDTW we reviewed the scant information available at that time on the structure of the Great Attractor. With the help of Ofer Lahav we constructed a map of optical galaxies that showed a marked concentration of galaxies in that direction covering a large solid angle on the sky. A crude comparison to Virgo indicated 15 to 30 times as many galaxies within the region of strong overdensity, with perhaps as many more obscured by the galactic plane. This ratio was consistent with the mass excess inferred from scaling the GA flow to Virgo infall. The mass excess in absolute units was estimated to be about $5 \times 10^{16}\ h^{-2}\ M_\odot$.

We also referenced a radial velocity survey by da Costa *et al.* (1986) that showed a prominent peak at 4500 km s^{-1} in the same direction. If this were the center of the GA, we reasoned, it would not be expected to move very fast relative to the CMB, and hence its observed radial velocity would be in good agrement with the distance of 4350 km s^{-1} predicted by the elliptical galaxies.

A major unanswered question, though, was our failure to detect any ellipticals falling into the GA from the backside. The sample in the north penetrated far enough: why were the comparable galaxies missing in the GA? This lack was especially worrisome since the the claimed over-density in the Great Attractor ought to enhance the population of early-type galaxies.

The answer to this question has since been at least partially clarified. The GA is indeed rich in early-type galaxies, as shown by Hubble types in the ESO catalog. However, there is also a tendency in the same catalog to classify small E′s as S0′s, so they were omitted from our original target sample. Furthermore, our target sample coverage is slightly less complete in the south than in the north. Togther these two effects mean that we penetrate about 30% less deep in the south, and this seems to be enough to make the difference. Dressler and Faber are currently beginning a deeper survey of the region around the Great Attractor in hopes of detecting ellipticals on the far side.

The major advance since LFBDDTW is a large radial velocity survey of the GA region by Dressler (1988), who has targeted 1400 galaxies in the large region shown in Fig. 13. Galaxies were selected in the same way, i.e., by apparent diameter, as the Southern Sky Redshift Survey (SSRS) of da Costa *et al.* (1987), which provides a comparison sample. The present results are based on a first sample of 900 galaxies chosen randomly from the whole. Fig. 14 shows a velocity histogram of these galaxies compared to a properly scaled histogram from the SSRS and one also from the northern Harvard Redshift Survey (Geller and Huchra 1988), which is also comparable. Fig. 14 shows two strong peaks centered at 3000 km s^{-1} and 4500 km s^{-1}. The nearer peak is spatially confined on the sky to a region around the Centaurus clusters, which are visible in Fig. 13 near (298, 22).

The velocity of this peak is identical to the velocity of the closer of the two Centaurus clusters. Both the ellipticals and the new spiral survey by Aaronson *et al.* (Mould 1988) show that the nearer Centaurus cluster has a distance of around 2400 km s^{-1}, and that it has a large positive peculiar motion away from us. If this conclusion applies to all the galaxies in the 3000 km s^{-1} peak, then as a group they lie closer to us than their radial velocities would suggest.

The situation with the 4500 km s^{-1} peak is even more complex. The velocity of this peak coincides with that of the more distant Centaurus cluster, but the galaxies that comprise it are spread over virtually the entire area of Fig. 13 (see also Lucey *et al.* 1986b). Furthermore, both LFBDDTW and Aaronson *et al.* show that the more distant Centaurus cluster, like the nearer cluster, is moving rapidly away from us towards the GA. A possible interpretation is that a part of the 4500 km s^{-1} peak is closer to us but that most of it — the diffuse background — is the GA and is at rest with respect to the CMB. This interpretation is consistent with the GA model, but it is clearly not unique. Independent distance estimates to galaxies in the region are badly needed.

If the above interpretation is correct, it is possible to estimate the local overdensity in the Centaurus-GA region by comparing to the other two surveys,

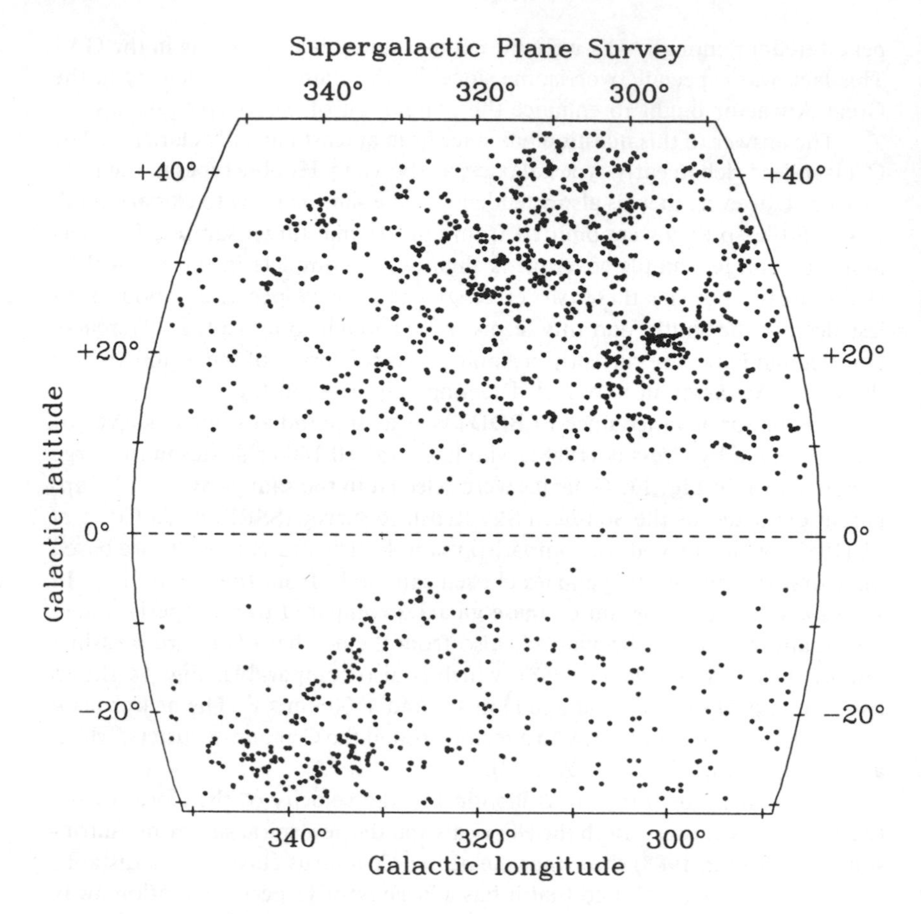

FIG. 13. The locations of galaxies targeted in Dressler's radial velocity survey of the GA region (Dressler 1988).

and then to integrate this average value to find the overdensity within a sphere centered on the GA inside the Local Group. Dressler finds enough galaxies to infer a peak $\Delta\varrho / \varrho$ of about 3 and an integrated $\Delta\varrho / \varrho$ of 1.5. This implies $\Omega = 0.1$-0.2 using the usual infall formula, so the visible mass is adequate to induce the local flow velocity of $\sim$ 550 km s^{-1} attributed to the GA. More interesting are the relative masses of the 3000 km s^{-1} and 4500 km s^{-1} peaks. Allowing for the shallower sampling and larger volume of the 4500 km s^{-1} peak, Dressler concludes that its mass is actually four times larger than the nearer peak, unlike the impression given by Fig. 14. This bolsters the interpretation of the far peak as the GA.

However, this mass ratio also predicts that the two regions are comparable in their gravitational effects on the Local Group. This says that sub-clumping

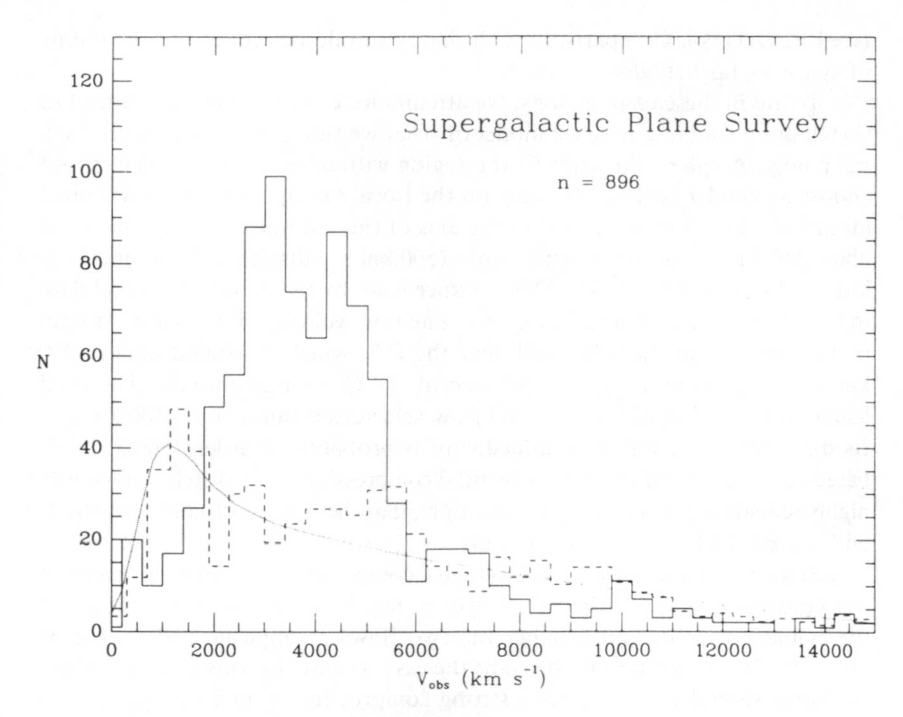

FIG. 14. Velocity histrogram for 896 galaxies measured by Dressler (solid line). These are compared to the histogram of 1657 galaxies measured by SSRS (observed data = dashed line, smoothed data = dotted line), normalized to Dressler's area of 0.85 steradian.

(including, perhaps, the Centaurus clusters) within the GA is gravitationally important. At present our homogenized, spherically symmetric model ignores such effects. Sub-clumping has two important implications: 1) We may expect to detect strong departures from radial symmetry when motions of other galaxies in the GA region are mapped. The local spirals may already be showing this in their quadrupole term; 2) It may be highly misleading to construct a spherically symmetric infall model of the GA based on just the volume currently sampled. The mass in the Centaurus region that is foreground to the GA could be as near as 2400 km s^{-1}, and hence one of the major contributors to the local acceleration field. If a mass concentration like Centaurus is not present on the far side of the GA, then the hypothetical infall velocity there at our distance from the GA could be as low as 300 km s^{-1}, all other things being equal. The spherically-averaged infall velocity at a radius of $\sim$ 4000 km s^{-1} from the center of infall thus probably lies somewhere between 400 km s^{-1} and 550 km s^{-1}. This is important in certain kinds of cosmological estimates, such as the spherical approximation for infall velocity versus $\Delta M/M$

(see LFBDDTW). Comparisons with theory should be limited to the volume of space so far actually surveyed.

To aid in these comparisons, we attempt here (with some trepidation) to sketch out a simple geometric model of what we think the present data show minimally. A sphere does not fit the region with good data very well, so we choose a cylinder centered roughly on the Local Group with long axis aimed at the GA. The flow is parallel to the axis of this cylinder, with a velocity of about 500 km s^{-1} at the far end, some 1500 km s^{-1} distant from us and 5700 km s^{-1} distant from the GA. This distance is set by the limit of the good data in that direction, as shown in Fig. 15. The flow velocity rises to a maximum of 1000 km s^{-1} at the other end near the GA, which is located about 3000 km s^{-1} away from us, at the distance of the Centaurus clusters. The total length of the region with substantial flow velocities is thus about 4500 km s^{-1}. Its diameter is not well determined, but is probably of order 3500 km s^{-1}, based on the extent of the cross-wise tidal compression. This sketch is obviously highly schematic and neglects, for example, that the flow is actually convergent and not parallel to the cylinder walls.

Since we measure the radial velocity component only, Peebles asked at this conference whether the flow might in fact break down in the middle of the cylinder near us and whether the two flows at opposite ends might be unrelated. We do not believe this to be the case because, looking sideways where the break should occur, we see a strong compression from both sides that is most simply ascribed to tidal compression by a coherent, large-scale infall. A multi-flow picture offers no explanation for this. It will be possible to answer the question definitively soon when the IRAS density maps (or comparable data) allow us to trace the three-dimensional flow pattern through the region of the Local Group. The preliminary IRAS maps at this conference already suggest that the flow pattern is indeed continuous and coherent through our location, as the cylinder model implies.

As a last point we stress that a large fraction of the GA may be hidden by the galactic plane. Two studies at this conference actually suggest this. Lynden-Bell and Lahav (1988) report that the optical peculiar acceleration converges too rapidly on nearby galaxies, whereas they expected a larger contribution from the GA — perhaps as much as a factor of two more. Strauss and Davis (1988) and Yahil (1988) likewise show diagrams that suggest that the velocity field of the GA is about a factor of two too weak. Extra mass of this order could be hidden by the galactic plane (see Fig. 13). Not knowing how much mass is hidden could be a real handicap to future studies of the region.

9. Summary

Our current major conclusions on the velocity field in the neighborhood of the Local Group are as follows:

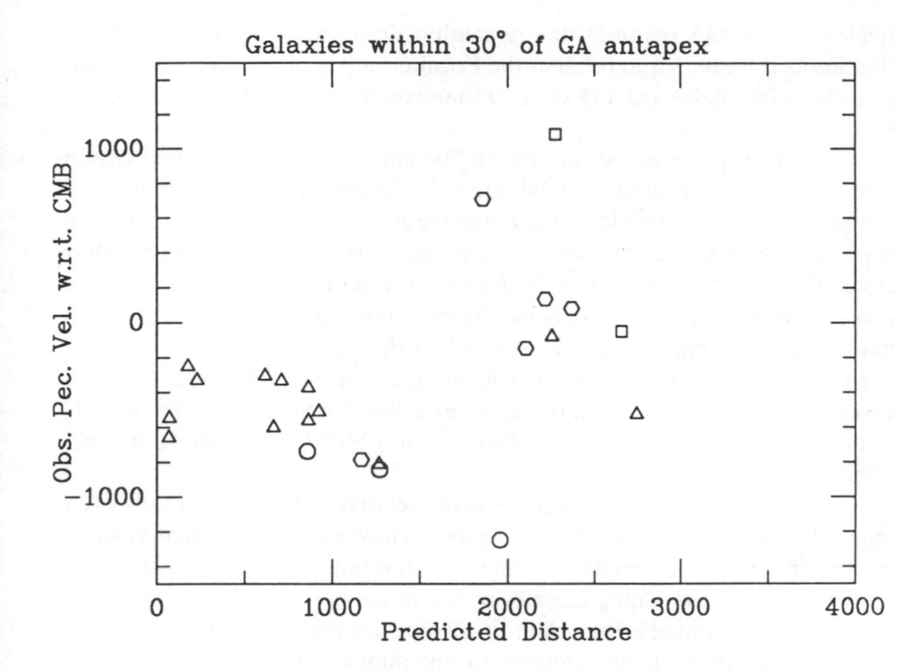

FIG. 15. The observed peculiar velocities with respect to the CMB for galaxies within 30° of the antapex of the GA (l = 129°, b = − 18°), plotted versus predicted distance. Galaxies from all four data sets are plotted as in Fig. 2. The flow to the GA appears to persist to at least 1500 km s^{-1}.

1. The local spirals within 2500 km s^{-1} generally corroborate the velocity flow pattern seen in ellipticals. When fit to a Virgo infall— GA model, they yield the same direction for the GA and v_A, but put the GA closer to us, at 3000 km s^{-1} rather than 4200 km s^{-1}. However, if new spiral measurements in Centaurus were added, the distance would be pushed farther away, closer to the elliptical value.

2. From a radial velocity survey by Dressler, evidence is mounting for a substantial overdensity of galaxies in the GA direction at about the right distance and amplitude to cause the observed flow pattern. The exact spatial distribution of galaxies there is unknown but appears to be highly clumped. The Centaurus association, with the same velocity as the closer Centaurus cluster, is a major foreground subcondensation that contributes a substantial share of the gravitational acceleration at the Local Group, in addition to the GA.

3. Virgocentric infall models are heavily modified by inclusion of the GA flow. Much of the apparent infall behind Virgo is due to the GA tidal

field. With the GA removed, the remaining Virgo infall signature far from the cluster is weak. Virgo infall at the Local Group is poorly determined but is likely to be smaller (85-133 km s^{-1}) than conventional values (250-300 km s^{-1}).

4. The sphere of galaxies out to 700 km s^{-1} around the Local Group shows an anomalous motion of 360 km s^{-1} with respect to the Virgo infall—GA model. This motion is identical within the errors to that of the Local Group itself. The patch of coherent motion is coextensive with the local Coma-Sculptor cloud identified by TF87, which is also the local portion of the Supergalactic plane. The anomalous motion is mostly perpendicular to this plane but also has a component parallel to it. It seems likely that the motion can be ascribed to irregularities in the local gravity field and is caused partly by the Local Void and partly by an excess of nearby galaxies below the SG plane. The internal velocity dispersion within the local Cloud is only 80-90 km s^{-1} along the line-of-sight.

5. The peculiar velocities of galaxies relative to the Virgo infall— GA model fall into three categories: quiet bulk flow with low internal velocity dispersion; group membership combined with infall to the group center; and "noisy" regions containing galaxies with a range of peculiar motions larger than the observational errors. Almost all galaxies are found in one of these kinds of regions. Residual velocities in one dimension can range up to 200 km s^{-1} on scales 1000 km s^{-1} and larger. The size and motion of the Local Anomaly are typical in comparison to other nearby regions.

6. Although a spherically symmetric GA model fits the present data well for both spirals and ellipticals, it is dangerous to use it for cosmology. On account of significant sub-clumping, the flow pattern around the GA is likely to be quite anisotropic. We have sketched a minimal, interim model that might be useful for cosmological calculations. The model involves a coherent flow pattern along a cylinder aimed at the GA that is roughly 4500 km s^{-1} long and 3500 km s^{-1} wide. The flow velocity varies from 500 km s^{-1} at the far end of the cylinder up to 1000 km s^{-1} near the Great Attractor, and then declines closer to the center of mass. The quadrupole term in the local galaxies suggests that the flows at opposite ends of this cylinder are, in fact, smoothly connected into one coherent flow through the location of the Local Group.

ACKNOWLEDGEMENTS

We would like to thank our colleagues Roger Davies, Alan Dressler, Donald Lynden-Bell, Roberto Terlevich, and Gary Wegner, who could not be co-authors on this preliminary report for lack of time. They have contributed much to our understanding of this subject in recent years, and we expect that they will be able to join us on the final version. We also thank Ofer Lahav for his efforts on our behalf.

REFERENCES

Aaronson, M., Huchra, J., Mould, J., Schechter, P.L., and Tully, R.B. 1982b. *Ap J.* **258**, 64. (AHMST)

Aaronson, M., Bothun, G.D., Mould, J.R., Huchra, J., Schommer, R., and Cornell, M. 1986. *Ap J.* **302**, 536. (ABM86)

Aaronson, M., Huchra, J., Mould, J.R., Tully, R.B., Fisher, J.R., van Woerden, H., Goss, W.M., Chamaraux, P., Mebold, U., Siegman, B., Berriman, G., and Persson, S.E. 1982a. *Ap J Suppl.* **50**, 241.

Bothun, G.D., Aaronson, M., Schommer, R., Huchra, J., and Mould, J. 1984. *Ap J.* **278**, 475.

Chincarini, G. and Rood, H.J. 1979. *Ap J.* **230**, 648.

da Costa, L.N., Nunes, M.A., Pellegrini, P.S., and Willmer, C. 1986. *AJ.* **91**, 6.

da Costa, L.N., Pellegrini, P.S., Sargent, W.L.W., Tonry, J., Davis, M., and Latham, D.W. 1987. preprint.

de Vaucouleurs, G. and Peters, W.L. 1968. *Nature.* **220**, 868.

______. 1984. *Ap J.* **287**, 1. (DVP)

de Vaucouleurs, G., de Vaucouleurs, A., and Corwin, H.G. 1976. *Second Reference Catalog of Bright Galaxies.* Austin: University of Texas Press. (RC2)

Dressler, A. 1988. *Ap J.* in press.

Dressler, A., Faber, S.M., Burstein, D., Davies, R.L., Lynden-Bell, D., Terlevich, R.J., and Wegner, G.W. 1987. *Ap J Letters.* **313**, L37.

Geller, M.J. and Huchra, J.P. 1988. this volume.

Lilje, P., Yahil, A., and Jones, B.J.T. 1986. *Ap J.* **307**, 91. (LYJ)

Lucey, J.R., Currie, M.J., and Dickens, R.J. 1986a. *MN.* **221**, 453.

______. 1986b. *MN.* **222**, 417.

Lynden-Bell, D. and Lahav, O. 1988. this volume.

Lynden-Bell, D., Faber, S.M., Burstein, D., Davies, R.L., Dressler, A., Terlevich, R.J., and Wegner, G.W. 1988. *Ap J.* **326**, 19. (LFBDDTW).

Mould, J. 1988. this volume.

Peebles, P.J.E. 1987, preprint.

Rubin, V.C., Thonnard, N., Ford, W.K., and Roberts, M.S. 1976a. *AJ.* **81**, 719.

Rubin, V.C., Thonnard, N., Ford, W.K., Roberts, M.S. and Graham J.A. 1976b. *AJ.* **81**, 687.

Sandage, A. 1972. *Ap. J.* **178**, 1.

Strauss, M.A. and Davis, M. 1988. this volume.

Tully, R.B. 1988. this volume.

Tully, R.B. and Fisher, J.R. 1987. *Nearby Galaxies Atlas.* New York: Cambridge University Press. (TF87)

Yahil, A. 1988. this volume.

DISTANCES TO GALAXIES IN THE FIELD

R. BRENT TULLY
Institute for Astronomy, University of Hawaii

ABSTRACT

Three topics will be discussed that are related to the determination of distances to individual galaxies and the Hubble Constant. First, there will be a brief description of recent work on the calibration of luminosity—line width relations now that photometric information based on CCD observations is available. Second, the problem of Malmquist bias will be reviewed, and it will be argued that there is a straightforward way to avoid bias in our situation. Third, the preliminary results from a program to determine the distances to individual galaxies will be summarized. It will be argued that there is a 'local velocity anomaly' and that it is this happenstance that is a principal reason for the perpetuation of a controversy over the value of H_0.

1. CALIBRATION OF LUMINOSITY—LINE WIDTH RELATIONS

New data are leading to improvements in the correlations between luminosity and H I profile line width (Tully and Fisher 1977), and there are positive implications regarding the use of these relations for the estimation of distances. Pierce and Tully (1988) discuss the calibration of B, R, and I-band relationships based on CCD imaging data and the recalibration of the H-band relationship for samples drawn from the Virgo and Ursa Major clusters. Figure 1 is taken from this reference.

The results of this new work are summarized as follows: (1) CCD photometry of galaxies in the Virgo and Ursa Major clusters was accomplished with fields-of-view that are large compared with the dimensions of the galaxies. Images were acquired in B, R, and I passbands. Resultant measured fluxes should be accurate to ± 0.03 mag. (2) In the Ursa Major Cluster, an almost complete sample of spiral galaxies brighter than $B_T = 13.3$ mag was observed.

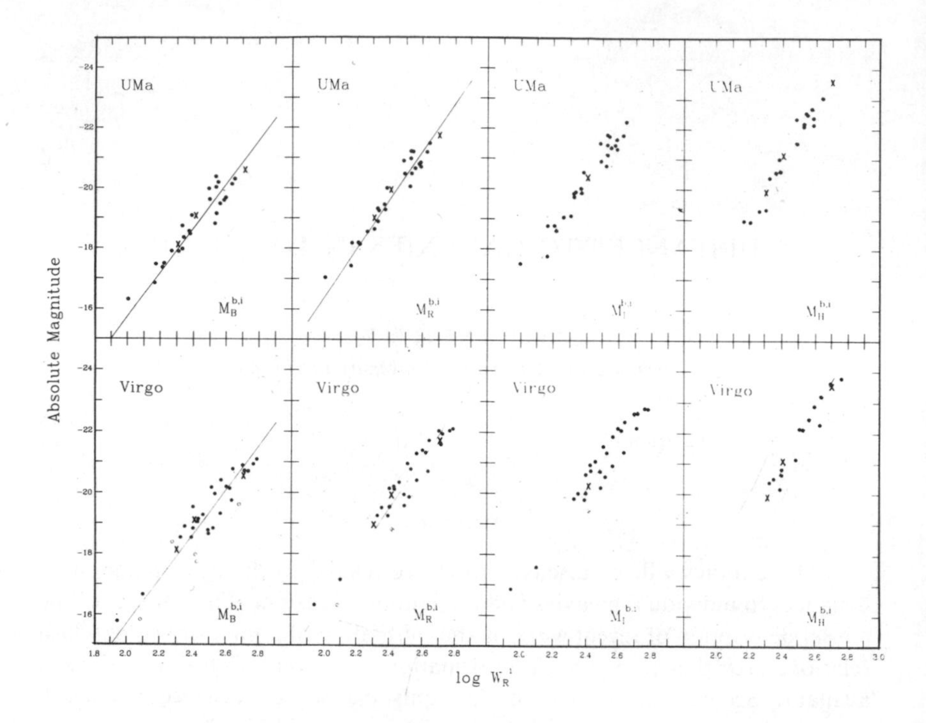

FIG. 1. Luminosity-line width relations. Top panels: Ursa Major Cluster; bottom panels: Virgo Cluster. Panels from left to right: B, R, I, and H passbands. Data have been shifted to fit to the three local calibrators (M31, M33, NGC 2403) indicated by crosses. The Virgo Cluster fits do not include the open circles, which correspond to galaxies thought to be viewed in projection falling into the cluster.

In the Virgo Cluster completion is only to 12.0 mag. Galaxies as faint as 15.2 mag were observed in these clusters. (3) There is now relatively little problem with the quality of the H I profile information for systems in these two clusters. In particular, galaxies in the Virgo Cluster can be observed with the Arecibo telescope (e.g., Helou, Hoffman, and Salpeter 1984). (4) A significant improvement has come about because the CCD images can be used to give relatively reliable inclinations, so line width deprojections are more accurate. Uncertainties are now ± 3° rms, down from ± 7° in earlier work. This information has been used to tighten the $H_{-0.5}$ band relationship (Aaronson, Huchra, and Mould 1979). (5) Scatter is lower in the Ursa Major relationships than in the Virgo relationships and lowest in the R, I, and H passbands. The observed rms scatter in Ursa Major is 0.30 mag and, given the inferred depth of the cluster, the scatter for a single galaxy must be about 0.25 mag. (6) It is argued that the greater scatter in the case of the Virgo Cluster is probably due to inclusion of systems slightly to the foreground and background of the cluster that are falling in. (7) There is only weak evidence of separation by morphological type. A 2σ effect is seen in the B-band samples, and only a 1σ effect is seen at R, I, and H bands. (8) The slope of the relationships steepen toward longer wavelengths. However, slopes are shallower with total luminosities than with luminosities within the restricted aperture used with the H-band work. The new calibration suggests that in the infrared $L_T \sim V^{3.2}$. (9) The three nearby systems M31, M33, and NGC 2403 are used to calibrate the zero point. The two Local Group galaxies were observed in B and R with a "1-inch" telescope that provided a field of view of 5°. All three calibrators have H-band measurements so this band can be given a consistent calibration. In the I band NGC 2403 provides the sole calibration. (10) The two clusters are subsequently found to be at essentially the same distances (Virgo: 15.6 ± 1.5 Mpc; U Ma: 15.5 ± 1.2 Mpc). (11) Given observed velocities and a model of Virgocentric retardation that influences our motion by 300 km s^{-1}, a value of 85 km s^{-1} Mpc^{-1} is found for the Hubble Constant.

2. Malmquist Bias

A lot has been said recently about Malmquist bias in the specific context of the luminosity—line width relations. Sandage (1988) says it is a serious problem. Bottinelli *et al.* (1986) discuss biased and unbiased domains. Giraud (1986) agrees there is a bias but feels these authors overestimate the effect. Lynden-Bell *et al.* (1988) describe a way of compensating for the problem in the case of the related $D_n - \sigma$ relationship. Schechter (1980) and Aaronson *et al.* (1982) prescribe a recipe that should largely negate the bias.

It will be argued here that the latter set of authors were on the right track but that a simpler procedure can be used that is still bias-free (see also Kraan-Korteweg *et al.* 1986). It is essential, however, that line widths be measured in an unbiased fashion. Also, one must have access to a complete sample that is representative of the sample to be studied, both in intrinsic properties and in the quality of the data.

The standard procedure in the past has been to fit a straight line to data in a plot of magnitude versus log W^i (W^i is the deprojected line width), where either the line is the regression that involves minimization of residuals in magnitudes or it represents the combination of two orthogonal regressions. In either case, if one assumes that a field galaxy with observed W^i has the absolute magnitude of the mean relationship, then the distance modulus $(m - M)$ that is determined will tend to be too low because an apparent magnitude cutoff to the sample will select in favor of brighter galaxies at a given distance. In other words, galaxies fainter than the mean will be rejected in greater numbers than galaxies brighter than the mean. The result is Malmquist bias.

However, if the line formed from the regression that involves minimization of residuals in line width is taken, and the rest of the operation is performed as before, there is no bias. As long as a straight line adequately describes the relationship and there is a normal distribution about the mean, there should be as many objects of given absolute magnitude drawn from the sample with W^i larger than the mean as smaller than the mean. This is the key point. To rephrase it, the regression on line width bisects the sample in the unbiased dimension. Every galaxy of given absolute magnitude and distance in the sample under investigation is drawn from a distribution in which there is equal *a priori* probability of a complementary galaxy of identical absolute magnitude and distance but with a line width that differs from the mean for that magnitude by the opposite sign. Neither of these galaxies has preferred entry into the sample over the other. Hence, equal numbers of objects in the sample will be drawn from the left of the regression as from the right and the assumption that the object has the mean luminosity for a given observed W^i is as likely to be too high as too low.

This intuitive concept was tested with simulated data that had the general characteristics of the actual data available. The details of the simulation will be described elsewhere (Tully 1988b), but the result was confirmation of the proposal made here. In practice, it is the combined Virgo and Ursa Major Cluster samples that provide the necessary calibration of the regression on W^i. The assumption is required that the cluster galaxies are representative of systems in other environments. At present, the Ursa Major sample is essentially complete to $M_B = -17.7$, providing a volume-limited sample complete to that limit.

3. The Local Velocity Anomaly and the Hubble Constant

In response to the challenge of this Study Week, two of us seem to have independently discovered manifestations of the same phenomenon and used the same terminology to describe it. Burstein, Faber, and Lynden-Bell have reported on work with their collaborators that leads them as well to deduce the existence of the 'local velocity anomaly'. It will be argued that this anomaly conspired to generate the controversy that has arisen over the value of the Hubble Constant.

The H-band sample discussed by Aaronson *et al.* (1982) will be used for the ensuing discussion. The calibration is in the system described in Section 1, and distances are determined in a way that should be statistically free of bias, as described in Section 2. The other information that is used is the specification of cloud membership for each object in the sample provided by the *Nearby Galaxies Catalog* (Tully 1988a).

Now, the ratio of the estimated distance to a predicted distance from a kinematic model can be determined for each galaxy. Two specific kinematic models will be considered: (I) pure Hubble expansion and (II) the Virgocentric retardation model discussed by Tully and Shaya (1984). Means and dispersions in the ratio of estimated to predicted distance can now be found for all members of a common cloud. If the kinematic model has been based on the proper value of the Hubble Constant, then on the average, the mean ratio for all clouds should be unity. Deviations from unity would arise if some clouds have motions that are poorly described by the specific kinematic model.

The results for all clouds with at least five independent measurements are shown in Figure 2a (for model I) and in Figure 2b (for model II). In the case of model I, a value of $H_0 = 80$ km s^{-1} Mpc^{-1} was required to force to unity the mean ratio: measured distance/kinematic distance. However, two of the three nearest clouds are seen to lie high in Figure 2a. We reside in the nearest one: the Coma-Sculptor Cloud. The other, very high point is associated with the Leo Spur, which is an appendage to our Coma-Sculptor Cloud. These two nearby entities contain many of the galaxies that historically have played a major role in the determination of H_0. If only galaxies within these two clouds are considered, then H_0 would have a value in the range 60-65.

In the case of model II, the two nearby clouds stand out much more prominently as special cases. After 3σ rejection of three anomalous points, a value of $H_0 = 95$ km s^{-1} Mpc^{-1} drives the mean ratio of measured to kinematic distances to unity. Still, if only data from the two nearby anomalous clouds are considered, one would conclude $H_0 \simeq 63$ km s^{-1} Mpc^{-1}.

These results suggest very strongly the resolution to the decade-old controversy over the value of the Hubble Constant. One hint is the interconnectedness of H_0 and the Virgocentric mass model as reflected in the mean ratio of the observed to predicted distances in Figures 2a and 2b. Another hint is the way the two nearest clouds particularly stand out in this ratio in the Virgocentric mass model case. Combine these hints with a statement that must be true in general: an observer *within* a region of retarded expansion due to a mass concentration and looking only at systems within the same region will tend to measure a value of H_0 that is *too low*.

This latter point is surely the explanation for why $H_0 = 80$ gives the best fit for model I (no mass concentration) and $H_0 = 95$ gives the best fit for model II (roughly $10^{15}\ M_\odot$ in Virgo). This is the same problem that must be confronted if the Virgo Cluster is used alone to estimate H_0, where observed velocity must be corrected for 'infall'. If one accepts that the Virgo Cluster has a significant gravitational effect on us, then it follows that direct measurements of H_0 from field galaxies in the local vicinity will probably give results that are *biased low*.

Provisionally accept that model II provides a better description of the Local Supercluster than the pure Hubble expansion description of model I. However, this model manifestly fails to explain the motions observed in our cloud and the Leo Spur, since the ratios of observed to expected distances are

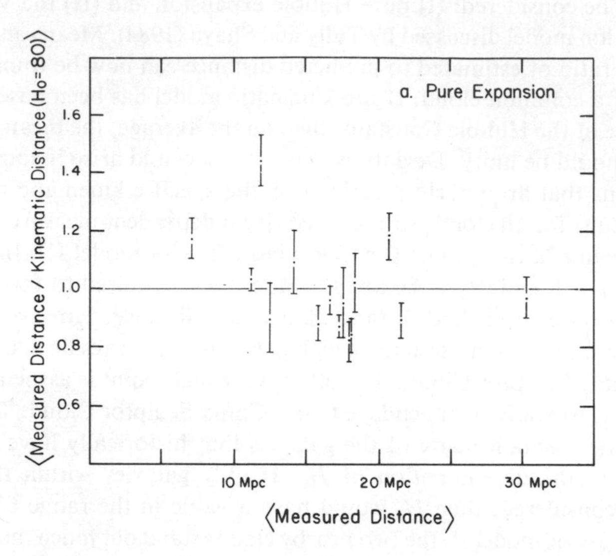

FIG. 2a. Ratio of measured to expected distances: pure expansion model. Each data point corresponds to the mean ratio associated with all galaxies with measured distances within a single cloud. Only clouds with at least five measurements are included. Model assumes $H_0 = 80$ km s^{-1} Mpc^{-1}.

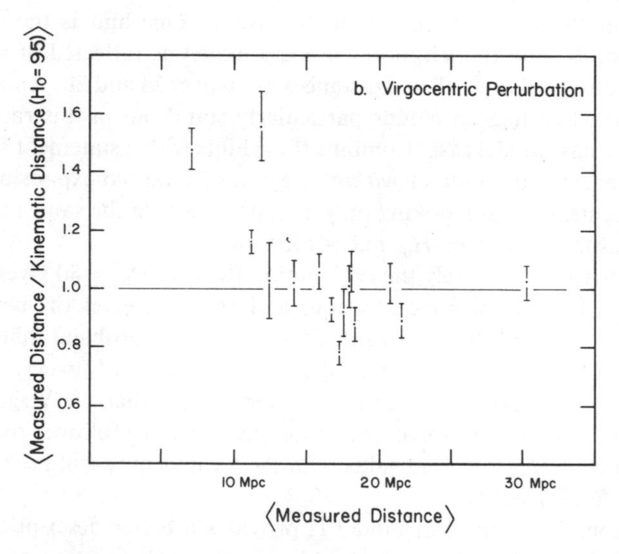

FIG. 2b. Ratio of measured to expected distances: Virgocentric retardation model. Data points correspond to the same clouds as in Fig. 2a. Model assumes $H_0 = 95$ km s^{-1} Mpc^{-1}.

much greater than unity in these two cases. It is submitted that a second effect, to be referred to as the 'local velocity anomaly', is being brought into focus.

An evident possibility is that there is sufficient mass within the Coma-Sculptor Cloud and associated Leo Spur to cause a local retardation to expansion. Within the Coma-Sculptor Cloud, typical deviant velocities are -200 km s^{-1} at observed velocities of 450 km s^{-1} and distances of 7 Mpc. An effect of this amplitude could be caused by $1 \times 10^{14}\ M_{\odot}$. The cloud has an integrated blue luminosity of $5 \times 10^{11}\ L_{\odot}$. A mass-to-light ratio of 200 – 300 $M_{\odot}/L_{\odot}$ is implied.

In actual fact, model II may well *overestimate* the influence of the Virgo Cluster. That model was developed to describe the pattern of motions in close proximity to the cluster and only to a lesser extent was influenced by motions across the full domain of the Local Supercluster. With model II, points associated with clouds in the south galactic hemisphere (the opposite hemisphere from Virgo) tend to lie *higher than the mean* in Figure 2b, whereas with model I, the equivalent points lie *lower than the mean*. This situation suggests that a correct description is *intermediate* between models I and II.

In addition, it is now becoming rather convincing that we are under the influence of a mass concentration on a scale larger than Virgocentric: that of the Great Attractor (Lynden-Bell *et al.* 1988). There is evidence in support of such a proposition in Figure 2b; points associated with clouds nearest the Great Attractor lie low, suggesting there are motions away from us relative to the motions anticipated by the Virgocentric mass model.

4. Summary

We are at a very early stage in our efforts to model the motions of nearby galaxies. Evidence is accumulating for the Great Attractor mass concentration of $5 \times 10^{16}\ M_{\odot}$ at 4300 km s^{-1}, there is a general consensus for 10^{15} $M_{\odot}$ centered on the Virgo Cluster at 1000 km s^{-1}, and now it is suggested there is a local velocity anomaly caused by $10^{14}\ M_{\odot}$ at 450 km s^{-1}. Judging by the clumpiness in the distribution of galaxies, an accurate description would have to include even more components.

This situation has a severe consequence because of the systematic underestimation of H_0 that occurs when the observer and the sample are both within an overdense region. In the specific case that confronts us, the nearest mass concentrations probably have the largest effects (local anomaly: 200 km s^{-1} at 450 km s^{-1}; Virgo: 300 km s^{-1} at 1000 km s^{-1}; Great Attractor: 600 km s^{-1} at 4300 km s^{-1}). The result is an increase in the apparent value of the Hubble Constant with distance.

Such a situation could also be a consequence of Malmquist bias. That has been the point-of-view of Bottinelli *et al.* (1986), Tammann (1987), and

Sandage (1988). It was argued in Section 2, though, that the present analysis avoids Malmquist bias. This claim is sure to be met with scepticism, but note that Figures 2a and 2b do not display the signature of Malmquist bias, which would be a steady or accelerating decrease in the ratio of measured to expected distance with increasing distance. Instead, this ratio is apparently constant with distance if one disregards the local cloud and its spur. (Kraan-Korteweg, Cameron, and Tammann (1986) also discovered a manifestation of the same phenomenon, but they dismissed it as a probable observational artifact.)

Detailed modeling of the velocity field in the Local Supercluster is in progress. The case is substantial, though, for the existence of a 'local velocity anomaly' that, abetted by the Virgo Cluster velocity anomaly, gives rise to deceptively low values of the Hubble parameter in samples dominated by nearby galaxies and creates the artifact that H_0 increases with distance, a phenomenon misinterpreted as Malmquist bias. It is argued that this perversity of nature is at the origin of the controversy over the value of H_0 and, incidentally, may have provided some sustenance to the claim that the Hubble law has a quadratic form (Segal 1976). However, this perversity would recur commonly throughout the universe since most cosmic observers probably live near mass concentrations.

With the zero-point calibrations used in this analysis, a value of $H_o \simeq 63$ km s^{-1} Mpc^{-1} would have been determined from galaxies within the local cloud and its spur, if the effect being discussed is ignored. On this zero-point, the true value is almost surely greater than the 80 km s^{-1} Mpc^{-1} suggested by the pure expansion model I, but probably not as high as the value of 95 km s^{-1} Mpc^{-1} required by model II. The most recent Aaronson *et al.* (1986) result based on clusters within 10,000 km s^{-1}, with a minor adjustment to the zero-point used in this paper, is $H_0 = 88$ km s^{-1} Mpc^{-1}. This obviously compatible value is derived over a domain that is larger than the presently documented mass concentrations and may represent a measurement of H_0 unaffected by peculiar motions.

REFERENCES

Aaronson, M., Huchra, J. P., and Mould, J. R. 1979. *Ap J.* **229,** 1.
Aaronson, M., Huchra, J. P., Mould, J. R., Schechter, P. L., and Tully, R. B. 1982. *Ap J.* **258,** 64.
Aaronson, M., Bothun, G., Mould, J. R., Huchra, J. P., Schommer, R. A., and Cornell, M. E. 1986. *Ap J.* **302,** 536.
Bottinelli, L., Gouguenheim, L., Paturel, G., and Teerikorpi, P. 1986. *AA.* **156,** 157.
Giraud, E. 1986. *AA.* **174,** 23.
Helou, G., Hoffman, G. L., and Salpeter, E. E. 1984. *Ap J Suppl.* **55,** 433.
Kraan-Korteweg, R. C., Cameron, L., and Tammann, G. A. 1986. in *Galaxy Distances and Deviations from Universal Expansion.* ed. B. F. Madore and R. B. Tully. p. 65. Dordrecht: Reidel.
Lynden-Bell, D., Burstein, D., Davies, R. L., Dressler, A., Faber, S. M., Terlevich, R., and Wegner, G. 1988. *Ap J.* **326,** 19.
Pierce, M. J. and Tully, R. B. 1988. *Ap J.* **330,** 579.
Sandage, A. 1988. *Ap J.* **331,** 583.
Schechter, P. L. 1980. *AJ.* **85,** 801.
Segal, I. E. 1976. *Mathematical Cosmology and Extragalactic Astronomy.* New York: Academic Press.
Tammann, G. A. 1987. in *IAU Symp. 124. Observational Cosmology.* ed. A. Hewitt, G. Burbidge, and L. Fang. p. 151. Dordrecht: Reidel.
Tully, R. B. 1988a. *Nearby Galaxies Catalog.* Cambridge: Cambridge University Press.
_____. 1988b. *Nature,* **334,** 209.
Tully, R. B. and Fisher, J. R. 1977. *AA.* **54,** 661.
Tully, R. B. and Shaya, E. J. 1984. *Ap J.* **281,** 31.

REFERENCES

Aaronson, M., Huchra, J., and Mould, J. 1979, Ap. J., 229, 1.
Aaronson, M., Huchra, J. P., Mould, J. R., Schechter, P. L., and Tully, R. B. 1982, Ap. J., 258, 64.
Aaronson, M., Bothun, G., Mould, J. R., Huchra, J. P., Schommer, R. A., and Cornell, M. E. 1986, Ap. J., 302, 536.
[illegible], G., [illegible], L., [illegible], and Teerikorpi, [illegible] 1990, [illegible] 177.
[illegible], C. 1986, Ap. J., 174, [illegible]
[illegible], G., Hellaby, [illegible] 1987, Ap. J. Suppl., [illegible]
[illegible] 1986, in Galaxy Distances and Deviations from Universal Expansion, ed. B. F. Madore and R. B. Tully (Dordrecht: Reidel).
Lynden-Bell, D., Burstein, D., Davies, R. L., Dressler, A., Faber, S. M., Terlevich, R., and Wegner, G. 1988, Ap. J., 326, 19.
Pierce, M. J., and Tully, R. B. 1988, Ap. J., 330, 579.
Sandage, A. 1988, Ap. J., 331, 583.
Schechter, P. L. 1980, A. J., 85, 801.
[illegible], L. 19[illegible], Mathematical [illegible] and [illegible] (New York: Academic Press).
Tammann, G. A. 1987, in IAU Symp. [illegible], Observational Cosmology, ed. A. Hewitt, G. Burbidge, and L. Z. Fang (Dordrecht: Reidel).
Tully, R. B. 1988, Nearby Galaxies Catalog (Cambridge: Cambridge University Press).
———. 1988, Nature, 334, 209.
Tully, R. B., and Fisher, J. R. 1977, A. & A., 54, 661.
Tully, R. B., and Shaya, E. J. 1984, Ap. J., 281, 31.

LARGE PECULIAR VELOCITIES IN THE HYDRA-CENTAURUS SUPERCLUSTER

JEREMY MOULD

Palomar Observatory, California Institute of Technology

ABSTRACT

Six clusters forming part of the Hydra-Cen Supercluster and its extension on the opposite side of the galactic plane have been reported to show large peculiar velocities relative to the Hubble flow. Additional observations required to verify the infrared Tully-Fisher distances on which these peculiar velocities are based include improved measurements of the isophotal diameters of the galaxy sample, and investigation of the Tully-Fisher and Faber-Jackson relations with environment. Examination of the local volume bounded by 8000 km/sec redshift suggests clusters in the Hydra-Cen complex which may provide the gravitational acceleration of the measured clusters. These are the Hydra and IC 4329 clusters on the north side of the galactic plane, and the Indus cluster and NGC 6769 group on the south side. The Cen 45 subcluster does not appear to be one of the attractors, however; it shows a very large positive peculiar velocity. It is still curious, moreover, that large peculiar velocities have not been observed in other similar mass concentrations, namely the Coma supercluster and the Perseus-Pisces supercluster. A detailed program of mapping these velocity fields with Tully-Fisher distances remains ahead.

1. INTRODUCTION

At the recent Balatonfured meeting Aaronson *et al.* (1988) presented preliminary evidence that five clusters forming part of the Hydra-Centaurus Supercluster have positive peculiar velocities ranging from zero to 1000 km/sec. A sixth cluster in Pavo showed a similar large outflow velocity. These velocities are measured with respect to a frame comoving with the expansion and in which an observer would see no microwave dipole anisotropy. The peculiar veocities are inferred from IR Tully-Fisher distances to the six clusters, based on 21 cm data from A.N.R.A.O. at Parkes and infrared photometry from Las Campanas Observatory. In the present discussion I want to slightly update the observational data and suggest some reasons for keeping an open mind about these peculiar velocities, until a number of additional critical observations are available.

2. Isophotal Diameters

CCD photometry of Hydra-Centaurus galaxies has been obtained at Cerro Tololo Interamerican Observatory by Mark Cornell of the University of Arizona early in 1987. $H_{-0.5}$ magnitudes, from which infrared Tully-Fisher distances are inferred, are defined in terms of the light contained in one third (-0.5 dex) the radius at which the galaxy's surface brightness (corrected for inclination) has fallen to 25 B mag arcsec^{-2}. Accurate isophotal diameter measurements (Burstein 1988) can reduce the random scatter in the IR Tully-Fisher relation relative to eye-estimated diameters (Nilson 1973), which we are sometimes forced to rely on, and remove (Cornell *et al.* 1987) systematic errors which would otherwise compromise our results.

Aaronson *et al.* (1987) expressed concern about the quality of the optical diameters for the Hydra-Cen and Pavo galaxies. Only 10 of the 64 galaxies used to estimate the 6 cluster distances have diameters from the Second Reference Catalog (RC2: de Vaucouleurs, de Vaucouleurs, and Corwin 1976); the remaining galaxy diameters are eye-estimates (Lauberts 1982) transformed to the RC2 system (Mould and Ziebell 1982). Recently, Mark Cornell has sent me accurate isophotal diameters for 6 of the sample galaxies (only one of which was in the RC2), showing an average (systematic) difference from the transformed diameters of 0.03 ± 0.01 in log D_{25}. In addition, a new study of the transformation equations between ESO and the RC2 has appeared (Paturel *et al.* 1987). The similarity between these equations limits systematic errors in the adopted diameters to 0.03 to -0.01 in log D_{25} (the range resulting from a difference in scale factor in the transformations). From the average slope of the infrared growth curve for spiral galaxies ($\Delta H/\Delta logA \approx 2.3$) one can calculate a limit on the implied error in the deduced peculiar velocities of 100 km/sec. Although this is reassuring, it in no way decreases the importance of measuring accurate isophotal diameters for *all* the sample galaxies.

3. Anomalous Surface Brightnesses

Aaronson *et al.* (1986) pointed out that, even after accurate isophotal diameters had been substituted for the eye-estimates in the Arecibo cluster sample, different clusters populated systematically different relations between mean surface brightness and 21 cm velocity width. These are distance independent quantities and (whatever the physical basis for such a relation) would be expected to define a consistent relation, if cluster galaxies are all drawn from the same population. If cluster galaxies are not all drawn from the same population, it is dangerous to deduce distances from the same observables deployed in a distance dependent manner (Kraan-Korteweg 1983).

Now, surface brightness is much more dependent on accurate isophotal diameters than $H_{-0.5}$ is. The 0.03 dex systematic error I mentioned earlier leads to a 0.15 mag error in Σ_H. However, the inter-cluster anomalies in surface brightness found by Aaronson *et al.* (1986) range up to 0.5 mag in Σ_H (see their figure 1), and even larger anomalies are seen in some of the Hydra-Cen and Pavo clusters with their current, mainly eye-estimated, galaxy diameters. This problem will be examined carefully when we have measured accurate isophotal diameters. Also under examination is the question of using surface brightness as a second parameter in the Tully-Fisher relation.

In the meantime, a further, related cause for concern is the indication of similar surface brightness anomalies, possibly affecting the Faber-Jackson relation for elliptical galaxies. The data plotted in Figure 1 show all the galaxies assigned the redshifts of the Virgo, Coma and Centaurus clusters by Burstein *et al.* (1987) and Davies *et al.* (1987). The effective surface brightness defined in the first paper is plotted against the velocity dispersion adopted in the second. As with the spiral galaxies, the question is, are galaxies in different clusters drawn from a common surface brightness population?

That this matter is not a red herring for either the Tully-Fisher relation or the Faber-Jackson relation can be seen from the following thought experiment. Imagine two galaxies of identical distance and mass distribution. Now take the first galaxy and turn up the luminous output of all the stars in the galaxy by, say, a factor of two. The surface brightness of the galaxy increases everywhere by a factor of two. The dynamical parameters, σ or ΔV, of the two galaxies remain the same. The D_n parameter (the diameter within which the integrated surface brightness is $\Sigma = 20.75$ mag arcsec^{-2}) becomes larger by at least a factor of $\sqrt{2}$, and the $H_{-0.5}$ magnitude becomes brighter by at least 0.75 mag. Application of the (D_n, σ) relation or the IR Tully-Fisher relation then places the first galaxy at a larger distance than the second. The counter-argument is that the first galaxy would not now fit the plane populated by elliptical galaxies in (R_e, σ, I_e) space (Faber *et al.* 1987), or, as I was told at the meeting, "galaxies like that just don't exist." The surface brightness dependence of the Tully-Fisher relation, on the other hand, remains to be fully investigated (but see Bothun and Mould 1986 for a start in that direction.)

Furthermore, we know that in the outer parts of galaxies there are gradients in M/L. On larger scales the notion of biasing, used to reconcile the observed density of the universe with $\Omega = 1$, suggests that clusters contain galaxies with atypically large dark halos. Can we still be confident that on galactic scales luminous material is linked in a fixed proportion to cold dark

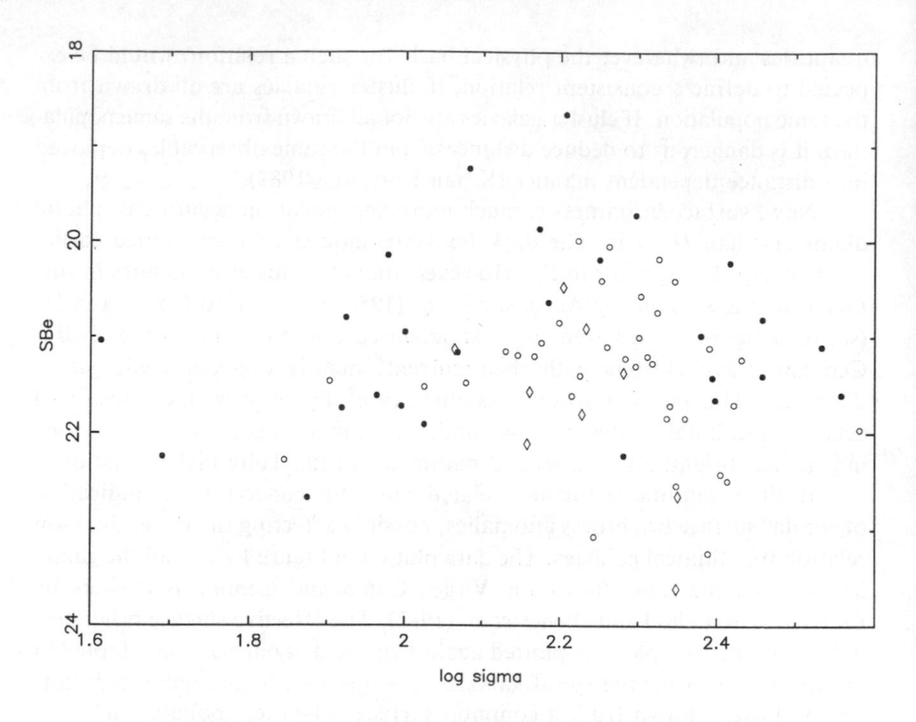

FIG. 1. Effective surface brightness versus the logarithm of the velocity dispersion for elliptical galaxies in Virgo (solid symbols), Coma (open circles), and Centaurus (diamonds). Data from Burstein *et al.* (1987) and Davies *et al.* (1987).

matter, a condition which we have known for a long time to be a prerequisite for invariant Tully-Fisher and Faber-Jackson relations (Freeman 1979, Faber 1982)? A vital observational test is to discover whether these relations are dependent on local galaxy density.

4. THE LOCAL REGION OUT TO 8000 KM/SEC

Aaronson *et al.* (1986) presented distances to 10 clusters generally with statistical errors of 5% or less. At 8000 km/sec this accuracy permits a 2σ detection of an 800 km/sec peculiar velocity. So, if we are going to consider the frequency or uniqueness of large peculiar velocities in Hydra-Cen, we should probably restrict ourselves to such a volume. Figure 2 shows the distribution of clusters within that volume, projected on to the supergalactic plane. This projection is the optimum one; the apparent relative positions of the clusters closely approximate the true one. With the exclusion of the Cancer cluster at supergalactic latitude -48°, the dispersion of the clusters about the supergalactic plane is 18°. Filled symbols indicate clusters from the combined Arecibo

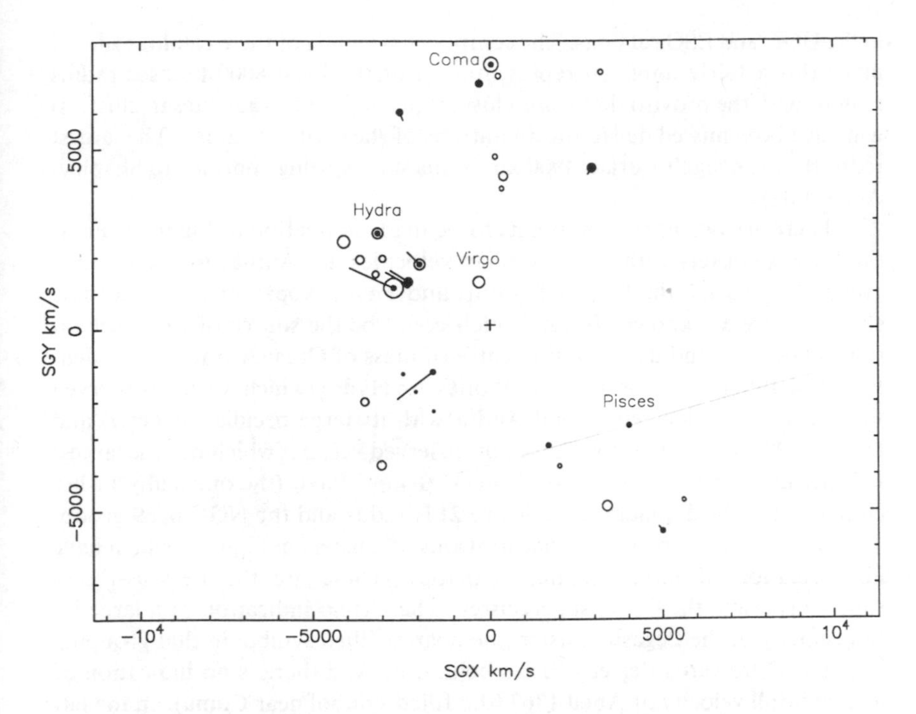

FIG. 2. Peculiar velocities for clusters and groups (denoted as solid symbols) are plotted in a spatial projection on the supergalactic plane. The Local Group is at the origin, and the arrows point radially, since that is the only measureable component. Other clusters without measured peculiar velocities and within the local 8000 km/sec are denoted by open symbols. The circle size indicates the richness of the cluster. The four major superclusters in this volume are labelled.

and Parkes samples to which IR Tully-Fisher distances have been measured, and they are plotted at their inferred velocity in the Hubble flow. A line segment joins this location to the observed velocity of the cluster, indicating the size and sense of the radial component of the peculiar velocity. The resultant peculiar velocities do not depend on H_0 or on the calibration of the Tully-Fisher relation. A different calibration would yield its own value of H_0 but not change the location of the clusters in Figure 2. Open symbols indicate the location of clusters in the samples of Huchra and Geller (1982) from the (northern) CfA survey and of Sandage (1975) from his southern redshift survey. These clusters are plotted at their observed velocities after correction for the motion of the Local Group towards the apparent microwave background apex. The size of the cluster is indicated by the radius of the circle via the following scheme. Galaxies in Virgo larger than 6′ in the UGC were counted in a 15° box. Galaxies were counted in other clusters with scaled angular dimensions in the UGC and ESO catalogs. The completeness limits of these catalogs should make this a fairly unbiased representation of the local 8000 km/sec radius region, with the proviso that some clusters (as opposed to galaxies in clusters) will have been missed due to the limitations of the cluster catalogs. The largest redshift in Sandage's work is 4900 km/s, his survey being confined to Shapley-Ames galaxies.

There are two important points to be made in relation to Figure 2. First, the Parkes clusters with large peculiar velocities are Antlia, the NGC 3557 group, Centaurus, the ESO 508 group and Pavo. Apparently behind these clusters there are known clusters which could be the source of the acceleration. These are candidates for the centre of mass of Great Attractor (Lynden-Bell *et al.* 1988). The most prominent ones are Hydra (which we have observed to be at rest and located behind Antlia with its large peculiar velocity) and the IC 4329 cluster (which we have not observed yet, but which has the largest redshift in the Hydra-Cen supercluster). Behind Pavo (the one Tully-Fisher cluster in the third quadrant of Figure 2) is Indus and the NGC 6769 group.

Second, there are other concentrations of clusters in Figure 2 which have no indication of large peculiar velocities. These are the Perseus-Pisces supercluster and the Coma supercluster. There is no indication of a large infall velocity of the Pegasus cluster (the nearest filled symbol in that grouping in Figure 2) towards Perseus-Pisces behind it. And there is no indication of a large infall velocity of Abell 1367 (the filled symbol near Coma). In the latter case the motion could be real but transverse.

Acknowledgements

In closing, I want to pay tribute to my late colleague and friend, Marc Aaronson, whose leadership and uncompromising zeal for accuracy in measurement has furnished us with so much of the data that make this subject of the large-scale motions in the universe a viable one. The efforts of our many collaborators referenced immediately below are gratefully acknowledged, as is funding by the National Science Foundation of the United States for this project.

REFERENCES

Aaronson, M., Bothun, G., Mould, J., Huchra, J., Schommer, R., and Cornell, M. 1986. *Ap J.* **302,** 536.

Aaronson, M., Bothun, G., Budge, K., Dawe, J., Dickens, R., Hall, P., Lacey, J., Mould, J., Murray, J., Schommer, R., and Wright, A. 1987. in *Large-Scale Structures of the Universe,* eds. J. Audouze, M.-C. Pelletan, and A. Szalay. Dordrecht: Kluwer Academic Publishers, p. 185.

Bothun, G. and Mould, J. 1986. *Ap J.* **313,** 629.

Burstein, D. 1988. this volume.

Burstein, D., Davies, R., Dressler, A., Faber, S. Stone, R., Lynden-Bell, D., Terlevich, R., and Wegner, G. 1987. *Ap J Suppl.* **64,** 601.

Cornell, M., Aaronson, M., Bothun, G., Mould, J., Huchra, J., and Schommer, R. 1987. *Ap J Suppl.* **64,** 507.

da Costa, L., Nunes, M., Pellegrini, P., and Willmer, C. 1986. *Ap J.* **91,** 6.

Davies, R., Burstein, D., Dressler, A., Faber, S., Lynden-Bell, D., Terlevich, R. and Wegner, G. 1987. *Ap J Suppl.* **64,** 581.

de Vaucouleurs, G., de Vaucouleurs, A., and Corwin, H. 1976. *Second Reference Catalogue of Bright Galaxies.* Austin: University of Texas Press. (RC2)

Dressler, A. 1987. *preprint.*

Faber, S.M. 1982. in *Astrophysical Cosmology.* eds. H.A. Bruck, G.V. Coyne, and M.S. Longair, p. 191. Vatican City State: Pontifical Academy.

Faber, S., Dressler, A., Davies, R., Burstein, D., Lynden-Bell, D., Terlevich, R., and Wegner, G. 1987. in *Nearly Normal Galaxies.* ed. S.M. Faber. p. 175. New York: Springer Verlag.

Freeman, K. 1979. in *Photometry, Kinematics and Dynamics of Galaxies.* ed. D. Evans. p. 85. Austin: University of Texas Press.

Huchra, J., and Geller, M. 1982. *Ap J Suppl.* **52,** 61.

Kraan-Korteweg, R. 1983. *AA.* **125,** 109.

Lauberts, A. 1982. *The ESO / Uppsala Survey of the ESO (B) Atlas.* Munich: European Southern Observatory.

Lucey, J., Currie, M., and Dickens, R. 1986. *MN.* **221,** 453.

Lynden-Bell, D., Faber, S., Burstein, D., Davies, R., Dressler, A., Terlevich, R., and Wegner, G. 1988. *Ap J.* **326,** 19.

Mould, J. and Ziebell, D. 1982. *PASP.* **94,** 221.

Nilson, P. 1973. *Uppsala General Catalogue of Galaxies.* Upsala Astr. Obs. Ann. Vol. 6. (UGC)

Paturel, G., Fouqué, P., Lauberts, A., Valentijn, E., Corwin, H., and de Vaucouleurs, G. 1987. *AA.* **184,** 86.

Sandage, A., 1975. *Ap J.* **202,** 563.

LARGE-SCALE MOTIONS FROM A NEW SAMPLE OF SPIRAL GALAXIES: FIELD AND CLUSTER

VERA C. RUBIN
Department of Terrestrial Magnetism,
Carnegie Institution of Washington

1. Introduction

At present, extensive observational programs are required in order to obtain sufficient data to permit examination of large-scale deviations from a smooth Hubble flow. Unfortunately, significant telescope time is necessary to enlarge the body of suitable samples. Meaningful answers can only come from high quality data; such data and answers accrue only slowly. Recently I decided to investigate the properties of one set of spirals with observations already at hand: those spirals which my colleagues and I had studied spectroscopically for individual rotation properties. My hope was that these galaxies could produce results concerning large-scale bulk motions without requiring additional observations. Note that I am not here discussing the Sc I spiral sample which Kent Ford and I and our colleagues studied 12 years ago (Rubin *et al.* 1976). For that early study we observed only central velocities of spirals (many of them nearly face-on), and determined distances by assigning to each galaxy the absolute magnitude appropriate to an Sc I galaxy. For the present analysis I use the rotational properties of a sample of field Sa, Sb, and Sc spirals of proper inclination.

2. Rotation Curves as a Diagnostic of Galaxy Luminosity

Rotation curves are a good diagnostic of galaxy luminosity, as is well known. Rotation velocities increase rapidly with radius for high luminosity galaxies; rotational velocities rise to higher values for high luminosity galaxies than for low luminosity galaxies. These properties lead to the Tully-Fisher relation. These properties also make it possible to construct synthetic rotation curves (Rubin *et al.* 1985), template rotation curves which can be employed to estimate the absolute magnitude for any spiral which has a known rotation curve and Hubble type.

2.1 - Field Galaxy Sample

We have recently published rotation velocities for 53 field spiral galaxies (Rubin *et al.* 1985; Rubin and Graham 1987) with systematic velocities smaller than 5100 km/s. Although a few of these galaxies lie in environments of moderate density (Whitmore 1984) and one, NGC 4321, is a member of the Virgo cluster, the results which I describe below are not altered when these few galaxies are omitted. The galaxies are well distributed in velocity from nearby to about V = 3800 km/s, with only 6 galaxies between 3800 and 5100 km/s. The median velocity is 1985 km/s. A lower velocity cut-off of 3800 km/s (used in Rubin 1988) does not alter the results presented below. I retain the higher velocity limit here, in order to compare the results for field galaxies with those of clusters at distances near 4000 km/s.

The distribution of the sample on the sky is shown in Figure 1; galaxies are identified in Table 1. The spirals are reasonably well distributed on the sky for declinations above $-10°$, but there are almost no galaxies below $\delta = -30°$. Sky coverage is just acceptable. No consideration as to sky coverage

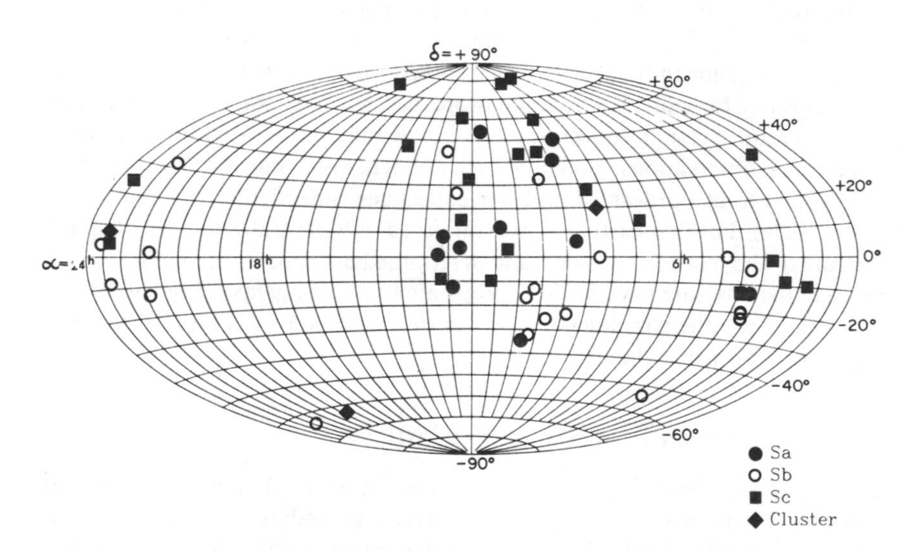

FIG.1. An all-sky plot (right ascension, declination) showing the distribution of spirals used in the analysis. The Milky Way crosses the Equator at 6^h and 18^h, with a corresponding lack of observed galaxies in those regions.

TABLE 1
RESIDUAL VELOCITIES FOR FIELD AND CLUSTER GALAXIES

Galaxy	V_{LG} km/s	m (i,b)	M(dy)	cos (Apex)	V_{LG}-$V_{(Hubble)}$ km/s	V_{MWB}-$V_{(Hubble)}$ km/s
Field Galaxies						
N701	1825	12.34	−20.7	−0.3826	−203.	−520.
N753	5077	12.55	−22.6	−0.7949	−281.	−727.
N1035	1277	12.27	−20.0	−0.2816	−145.	−378.
N1087	1523	11.14	−20.4	−0.3567	507.	250.
N1325	1499	11.52	−20.5	0.0150	231.	168.
N1353	1422	11.59	−21.6	0.0273	−751.	−802.
N1357	1967	12.24	−21.2	−0.0587	−471.	−557.
N1417	4070	12.28	−22.7	−0.1426	−884.	−996.
N1421	1985	10.86	−21.7	−0.0334	360.	293.
N1515	955	10.69	−20.4	0.4746	129.	299.
N1620	3418	12.70	−21.8	−0.0253	−554.	−557.
U3691	2076	12.15	−20.1	0.2351	667.	920.
I467	2213	11.90	−20.6	−0.5846	632.	433.
N2590	4787	13.00	−22.1	0.6075	−449.	15.
N2608	2059	12.35	−19.9	0.2194	650.	938.
N2639	3238	12.22	−22.2	−0.1179	−590.	−487.
N2715	1487	11.18	−20.1	−0.5277	585.	427.
N2742	1363	11.66	−21.3	−0.2629	−591.	−574.
N2775	1185	10.92	−21.7	0.5829	−486.	−11.
N2815	2270	11.25	−22.5	0.8946	−542.	40.
N2844	1491	13.10	−19.1	0.0850	114.	339.
N2998	4781	12.11	−22.4	0.0374	791.	991.
N3054	2152	11.54	−21.8	0.9395	−176.	429.
N3067	1411	11.93	−19.8	0.2421	302.	617.
N3145	3435	11.60	−22.4	0.8540	280.	873.
N3198	691	10.27	−20.2	0.0243	70.	263.
N3200	3266	11.20	−22.5	0.9025	518.	1124.
N3223	2617	10.96	−22.3	0.9856	373.	980.
N3281	3115	11.83	−20.5	0.9893	1653.	2259.
N3495	925	10.63	−21.1	0.6866	−184.	347.
N3593	520	11.00	−19.1	0.5579	−4.	468.
N3672	1655	11.08	−21.8	0.8270	−229.	350.
N3898	1264	11.30	−21.0	−0.1714	−178.	−108.
N4062	742	11.39	−20.9	0.2263	−693.	−399.

Galaxy	V_{LG} km/s	m(i,b)	M(dy)	cos(Apex)	V_{LG}-$V_{(Hubble)}$ km/s	V_{MWB}-$V_{(Hubble)}$ km/s
N4321	1478	9.98	−22.2	0.4564	114.	517.
N4378	2431	12.22	−21.8	0.6031	−753.	−285.
N4448	650	11.00	−21.1	0.2533	−665.	−364.
N4594	927	8.51	−22.6	0.7666	93.	614.
N4605	288	10.26	−18.4	−0.2930	18.	8.
N4682	2152	12.34	−21.5	0.7403	−779.	−272.
N4698	932	10.93	−20.5	0.5224	−34.	387.
N4800	874	11.91	−20.5	−0.0679	−643.	−524.
N4845	998	11.20	−19.8	0.5940	206.	651.
N5676	2268	11.04	−23.0	−0.2569	−945.	−963.
N6643	1735	11.02	−22.1	−0.7341	−369.	−691.
N7083	2979	11.51	−21.6	0.3031	885.	893.
U11810	4913	13.80	−20.8	−0.7295	754.	208.
N7171	2865	12.40	−21.5	−0.5442	−148.	−616.
N7217	1236	10.56	−22.5	−0.9717	−810.	−1417.
N7537	2863	12.68	−19.6	−0.7780	1434.	870.
N7541	2873	11.50	−22.9	−0.7783	−920.	−1484.
N7606	2371	10.71	−22.3	−0.6166	371.	−126.
N7664	3709	12.66	−22.2	−0.9413	−979.	−1582.
Cluster Galaxies						
U4329	4556	13.61	−20.0	0.3039	1920.	2245.
N2558	4556	12.97	−21.8	0.3140	59.	389.
U4386	4556	12.49	−21.6	0.3145	1268.	1600.
DC18-42	4353	13.76	−20.4	0.4543	957.	1074.
DC18-2	4353	13.97	−20.1	0.4474	1095.	1208.
DC18-66	4353	13.93	−20.1	0.4329	1154.	1258.
DC18-8	4353	13.71	−20.1	0.4425	1463.	1572.
DC18-10	4353	15.12	−20.2	0.4401	−1441.	−1333.
DC18-24	4353	14.54	−19.8	0.4384	663.	770.
U12417	4062	14.05	−21.1	−0.7925	−1296.	−1866.
N7591	4062	12.89	−21.2	−0.7985	774.	203.
U12498	4062	14.04	−19.2	−0.8135	1839.	1263.
N7608	4062	13.90	−19.4	−0.8159	1777.	1200.
N7631	4062	12.93	−20.9	−0.8135	1145.	570.

entered the original observation program, so it is remarkable that the sky distribution is even this good.

For each program spiral an absolute magnitude is estimated from its rotation curve, as shown in Figure 2. This procedure, while little different from the conventional use of the Tully-Fisher relation to estimate an absolute magnitude from a value of V_{max}, permits the use of a rotation curve which does not extend as far as V_{max}, and a rotation curve which exhibits velocity peculiarities. From the absolute magnitude, coupled with a corrected apparent magnitude, a distance is obtained which is, of course, independent of observed central velocity. For a value of $H_0 = 50$ km sec^{-1} Mpc^{-1} (the value used to calculate the absolute magnitude scale for the template rotation curves), I then calculate ΔVel, the difference between the observed velocity and the velocity predicted at its distance for a smooth Hubble flow. To search for a bulk motion of the set of galaxies, I look for a dipole on the sky which will minimize the residuals of ΔVel. For each galaxy, the velocity in the rest frame of the Local Group V_{LG}, the corrected apparent magnitude m(i, b), and the absolute magnitude estimated from the rotation curve M(dy), are listed in col-

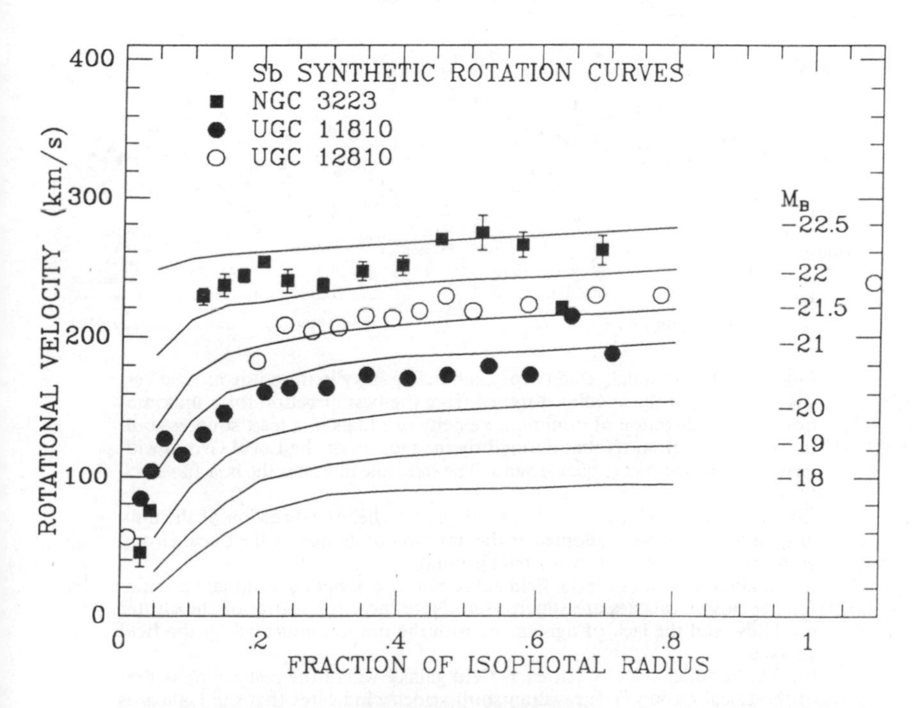

FIG.2. Optical rotation curves for three Sb galaxies, superposed upon synthetic rotation curves for Sb galaxies. For each galaxy, a value of M_B is estimated from the fit.

umns 2,3, and 4 of Table 1. For one solution, shown below, the rest frame is defined by the Local Group, and the value $\Delta\text{Vel} = V_{LG} - V_{Hubble}$ is listed in Column 6. For all remaining solutions the rest frame is that of the cosmic microwave background (MWB) radiation; $\Delta\text{Vel} = V_{MWB} - V_{Hubble}$ is in Column 7. The inclusion of a Virgo infall component does not alter the solution.

Results from several least squares solutions are shown in Figure 3, where I plot ΔVel for each galaxy versus the cosine of its angular distance from the

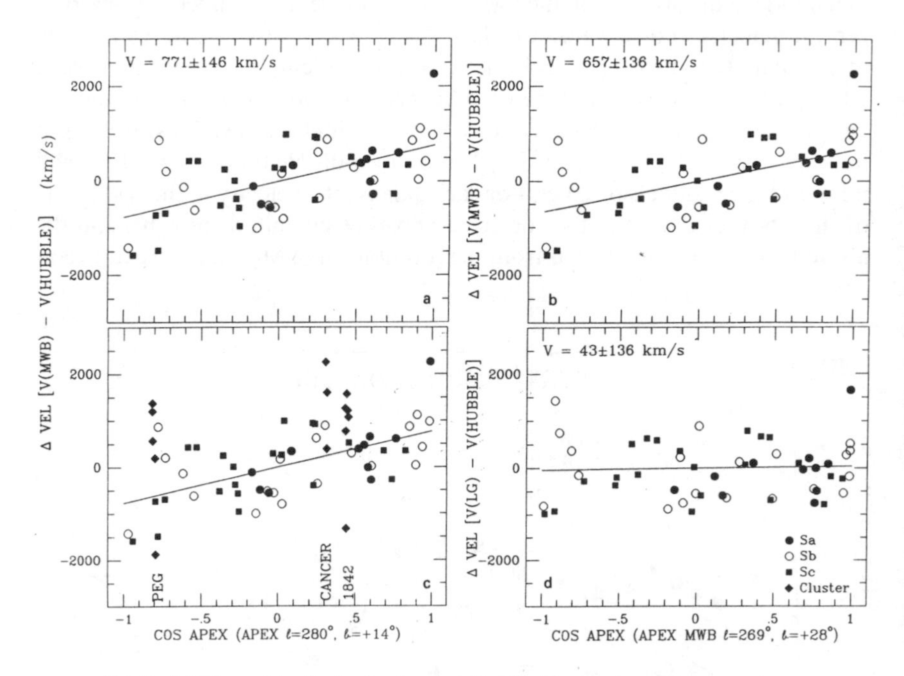

FIG.3. (a) The residual velocity for each field galaxy in the study plotted versus the cosine of the angular distance from the best direction for a bulk motion (i.e. the direction of minimum velocity residuals in a least squares solution). The rest frame is that defined by the motion of the Local Group with respect to the microwave background. The solid line indicates the best fit velocity of 771 km/s.
(b) The residual velocity for each field galaxy when the direction of the bulk motion of the sample is adopted as the direction of motion of the Local Group with respect to the microwave background.
(c) Residual velocities for the field galaxies as in (a), but the residual velocities for the cluster galaxies are superposed. Note the large scatter of the cluster residuals, and the lack of agreement with the motion indicated by the field galaxies.
(d) The residual velocity for each field galaxy where the rest frame is that of the Local Group. The resultant null velocity indicates that the Galaxy is at rest with respect to this sample, and that the entire sample is moving in concert. Note that this figure is essentially just that in (b), rotated so that V is approximately zero.

direction of the bulk motion determined by the least squares procedure. A bulk motion is indicated by a diagonal line on the plot, with the amplitude of the motion equal to the value of ΔVel at cos(Apex) = 1.

The best solution (i.e., the minimum scatter in ΔVel, Figure 3a, and column 5, Table 1) indicates a bulk motion of V = 771 ± 146 km/s toward l = 280°, b = + 14°. This is close in direction and in amplitude to the motion attributed to the Great Attractor (Faber and Burstein 1988, Burstein *et al.* 1986), and is close to the motion of the Local Group with respect to the MWB, for which V = 614 km/s toward l = 269°, b = + 28° (Lubin and Villela 1986 and references therein). In fact, a solution for a bulk motion, in which the apex of the MWB dipole is adopted as the apex, (Figure 3b) differs little from the unconstrained solution. This can be illustrated another way. If we chose as the rest frame not the MWB but the Local Group, then a solution for a bulk motion in the direction of the MWB gives a null result (Figure 3d), as does a solution which is totally unconstrained. Hence, the Local Group is virtually at rest with respect to the spirals which constitute this sample. Because we attribute to the Local Group that motion which produces the MWB dipole, this motion thus becomes a bulk motion for the entire sample as well.

The large-scale motion indicated by this spiral sample agrees well with recent determination from other samples (Lynden-Bell 1988); its concordance with the individual velocity vectors predicted by the IRAS observations is described by Davis (1988). The direction of the motion is also close to the 1976 Rubin and Ford result, after that solution is transformed to the MWB rest frame. Part of the agreement of all of the apices arises because each is the vector sum of a motion of the LG with respect to the sample, plus the larger motion of the LG with respect to the rest frame defined by the MWB.

However, even this good agreement does not put to rest some concerns about the procedures and the results. Are the properties of galaxies, as a function of direction, of distance, and of environment, so systematic that zero-points and slopes of correlations among parameters in one region are reproduced exactly in all other directions? Or are there differences among galaxies in different regions of the sky, which we erroneously interpret as a velocity? At a distance corresponding to V = 2500 km/s, a difference in absolute magnitude by about 0.4 magnitude will mimic a velocity of 500 km/s. Because it seems unlikely that a variation across the sky of twice this amount will have gone unnoticed, a coherent motion for the nearby galaxies seems well established.

2.2 - *Cluster Galaxy Sample*

More recently, we have initiated observational programs to investigate the properties of spiral galaxies in various environments. We have obtained observations (Rubin *et al.* 1988) of rotation properties for a small sample of spirals in the clusters of Cancer, Pegasus I, and DC 1842-63 (Dressler 1980). I can analyze the dynamics of these cluster galaxies as I have analyzed the field spirals, and ask if their dynamics indicates the same bulk motion as do the field spirals. Because each cluster galaxy is placed at the mean cluster distance, considerable smoothing and reduced scatter is expected on a cosine plot. Surprisingly, this is not the case, as is apparent from Figure 3c, where I superpose the residual cluster velocities on the solution based on field spirals, Figure 3a. Residual velocities for the cluster galaxies show a scatter within a single cluster that is larger than the scatter due to the field galaxies, and do not indicate the bulk motion found for the spirals.

One of the clusters, DC 1842-63, is located only 39° from the direction to the Great Attractor (but 90° from the apex defined by the MWB), and its mean velocity, V = 4353 km/s, is just that accepted for the (distance of the) Great Attractor. Residual velocities for galaxies in this cluster are not too discrepant (Figure 3c). However, residual velocities for galaxies in the Peg I cluster are predominantly positive, in a direction in which all bulk motion models predict negative residuals.

One explanation for the field/cluster discrepancy is that with small number statistics we have managed to observe deviant cluster spirals. However, our selection procedure favored galaxies of relatively normal morphology. More likely, a single bulk motion is too simple an explanation for the large motions observed. Such a conclusion would certainly follow from the flow diagrams produced by Yahil. Still an alternative possibility is that rotation properties of spirals in clusters do not match those of spirals in the field, so that the synthetic rotation curves formed from field spirals are not the proper templates for cluster galaxies. Although we discuss such possibilities elsewhere in this volume, only more observations of extended rotation curves of spirals in clusters will settle this question. No such difference for field/cluster dynamical properties is observed for elliptical galaxies according to Faber, and the cluster spirals observed at HI by Aaronson *et al.* (1986) show no deviant characteristics.

3. Unanswered Questions

Major questions remain. Are the slightly different motions returned by different samples significant, or do the differences merely reflect the different mix of objects investigated? How well do different samples agree in regions of overlap in direction and distance? Initial answers to these questions are given in the contributions here of Faber and Burstein (1988), and of Strauss and Davis (1988). Is the dipole observed in the MWB gravitationally induced, and how do the deviations from a smooth Hubble flow which we have discussed this week relate to this MWB dipole?

An important question arises for studies which mix observations of field galaxies, of cluster galaxies, observations of neutral atomic gas, and ionized gas: are the characteristics of spirals in regions of high galactic density so similar to those isolated in the field that the Tully-Fisher and Faber-Jackson relations have the same slopes and zero points independent of environment? Are these relations so general that they can be correctly determined using 21-cm observations of galaxies whose neutral hydrogen gas disks are known to be asymmetrical and tidally truncated near the cores of rich clusters? And what of the ionized gas disks? Do the asymmetries and truncations observed in the neutral hydrogen disks exist also in the ionized gas disks, or are optical and 21-cm observations looking at features with such different evolutionary histories that their correlations with other galaxy parameters are not likely to be similar? Have bulges acquired disks, have disks acquired bulges, have early type galaxies merged, all-the-while retaining similar correlations among the physical properties we measure? If, as it appears at present, the evolutionary effects on galaxies in clusters have been greatest for the early type galaxies, then perhaps we are not erring in combining 21-cm observations of principally late-type spirals in clusters and in the field with optical observations of principally field spirals. A few of these questions are addressed further in my contribution later in this volume. These are some of the questions which brought us together, and they are questions which will concern many astronomers in the coming years.

ACKNOWLEDGEMENTS

Spectra of the field and cluster galaxies were obtained at Kitt Peak, Cerro Tololo Inter American, and Las Campanas Observatories; I thank the Directors for observing time. I also thank my colleagues for continued support, and Dr. John Graham for comments on an early draft of this paper.

REFERENCES

Aaronson, M., Bothun, G., Mould, J., Huchra, J., Schommer, R., and Cornell, M. E. 1986. *Ap J.* **302,** 536.

Burstein, D., Davies, R. L., Dressler, A., Faber, S. M., Lynden-Bell, D., Terlevich, R. J., and Wegner, G. A. 1986. in *Galaxy Distances and Deviations from Universal Expansion.* eds. B. Madore and R. B. Tully. p. 255. Boston: Reidel.

Dressler, A. 1980. *Ap J Suppl.* **42,** 569.

Faber, S. M., and Burstein, D. 1988, this volume.

Lubin, P. M., and Villela, T. 1986. in *Galaxy Distances and Deviations from Universal Expansion.* eds. B. Madore and R. B. Tully. p. 169. Boston: Reidel.

Lynden-Bell, D. 1988, this volume.

Rubin, V. C. 1988. in *Large-Scale Structures of the Universe.* eds. J. Audouze, M.-C. Pelletan, and A. Szalay. Dordrecht: Kluwer Academic Publishers, p. 181.

Rubin, V. C., and Graham, J. A. 1987. *Ap J Letters.* **316,** 67.

Rubin, V. C., Thonnard, N., Ford, W. K., and Roberts, M. S. 1976. *AJ.* **81,** 719.

Rubin, V. C., Burstein, D., Ford, W. K., and Thonnard, N. 1985. *Ap J.* **289,** 81.

Rubin, V. C., Whitmore, B. C. and Ford, W. K. 1988. *Ap J.* **333,** 522.

Strauss, M. A., and Davis, M. 1988, this volume.

Whitmore, B. C. 1984. *Ap J.* **278,** 61.

III

MOTION OF THE LOCAL GROUP

WHENCE ARISES THE LOCAL FLOW OF GALAXIES?

D. LYNDEN-BELL and O. LAHAV
Mount Stromlo Observatory, Australia
and Institute of Astronomy, Cambridge

SUMMARY

If the apparent large-scale coherence of the velocity field of galaxies out to 3500 km/s is not a chance superposition, the gravity that causes it must arise from distant sources such as the Great Attractor. If light traces mass, the origin of the local gravity field can be found by tracing the origin of the locally observed net flux of extragalactic light. If possible problems with the $-2.5° > \delta > -17.5°$ strip, the zone of avoidance and convergence of the dipole at faint fluxes are ignored, then half of the local extragalactic gravity field arises from galaxies with diameters greater than 4′. The Great Attractor model would predict only 2′ corresponding to galaxies at twice the distance. This optical result agrees with a similar IRAS result. Thus the gravity field of the Great Attractor model is not confirmed, although a combination of obscuration in the Milky Way and insuffcent depth in the galaxy catalogues could be denying us the true picture.

A mapping of the supergalactic plane in depth provides further evidence that Virgo is at one side of that structure whose centre lies in Centaurus or even further south in the zone of avoidance. Even the presently catalogued distribution of extragalactic light shows that Virgo is relatively weak causing 108 km/s infall and that even here the infall towards Centaurus is twice as strong.

The Local Void gives a significant push across the supergalactic plane which slews the Local Group's motion from $l = 290°$ $b = 28°$ to the observed $l = 261°$ $b = 29°$.

1. MOTION OF THE LOCAL GROUP

We first review the determination of the Local Group's motion relative to the Cosmic Microwave Background (CMB), concentrating on uncertainties in the deduction of the Sun's motion within the Local Group and the peculiar motion of the Local Group relative to the mean flow of galaxies in its neighbourhood.

We then turn to the study of the origin of that motion from the gravity field of external galaxies. Since both gravity and light fall off as distance-squared, the assumption that light traces mass allows us to determine the gravity field from the locally measured flux of extragalactic light. For galaxies greater than 1 arcminute in diameter, existing galaxy catalogues (Nilson 1973, Lauberts 1982) are sufficient to do this provided they are properly calibrated. Exceptions occur in the zone of avoidance and in the band $-2.5° > \delta > -17.5°$ where the existing galaxy catalogue (Vorontsov Velyaminov and Arkipova, 1963-68) is too incomplete and inhomogeneous. To get a first result these zones have been filled in with the mean distibution of galaxies observed at high galactic latitudes $|b| > 40°$. Somewhat similar problems occur with the distribution of IRAS galaxies which has small missing strips at high latitudes and a large uncertain region near the galactic plane caused by galactic cirrus and confusion with galactic sources.

In both cases the assumption that light traces mass can not be precisely true since most of the mass presumably resides in dark halos around galaxies or clusters rather than in light sources. For visible light the assumption may be sufficient to give a good account of the gravity field since the M/L ratios of galaxies are not very widely distributed. Faber *et al.* (1987) deduce a factor of ~ 2 in ellipticals. For IRAS fluxes there is a far wider spread in mass to infra-red ratio with ellipticals undetected and some active galaxies found with luminosities similar to quasars. Thus IRAS flux at best only follows the mass density statistically and underestimates the high density regions where early type galaxies predominate. With this broad luminosity function it is no surprise that the distribution of IRAS sources on the sky looks like a washed out version of the distribution of optical sources (see e.g., Meurs and Harmon 1988).

The estimation of the gravity field due to the sources of greatest apparent brightness can not be done from IRAS fluxes since they are dominated by 'shot noise' of individual nearby sources which happen to be strong infra-red emitters. In practice this problem has been reduced either by binning in flux and number weighting within each flux bin (Yahil *et al.* 1986) or by ignoring flux altogether and calculating the direction of the number weighted dipole (Meiksin and Davis 1986). While the latter is useful to get a more stable direction, it loses much of the original motivation since the inverse-distance-weighting is removed. Indeed, in a Poisson universe in which galaxies are placed at random the direction of the number weighted dipole would not converge, but would perform a random walk across the sky as the limiting flux of the sources counted was reduced. Even with the inverse-distance-squared weighting the strength of the dipole in such a universe only converges as distance to the minus one half power.

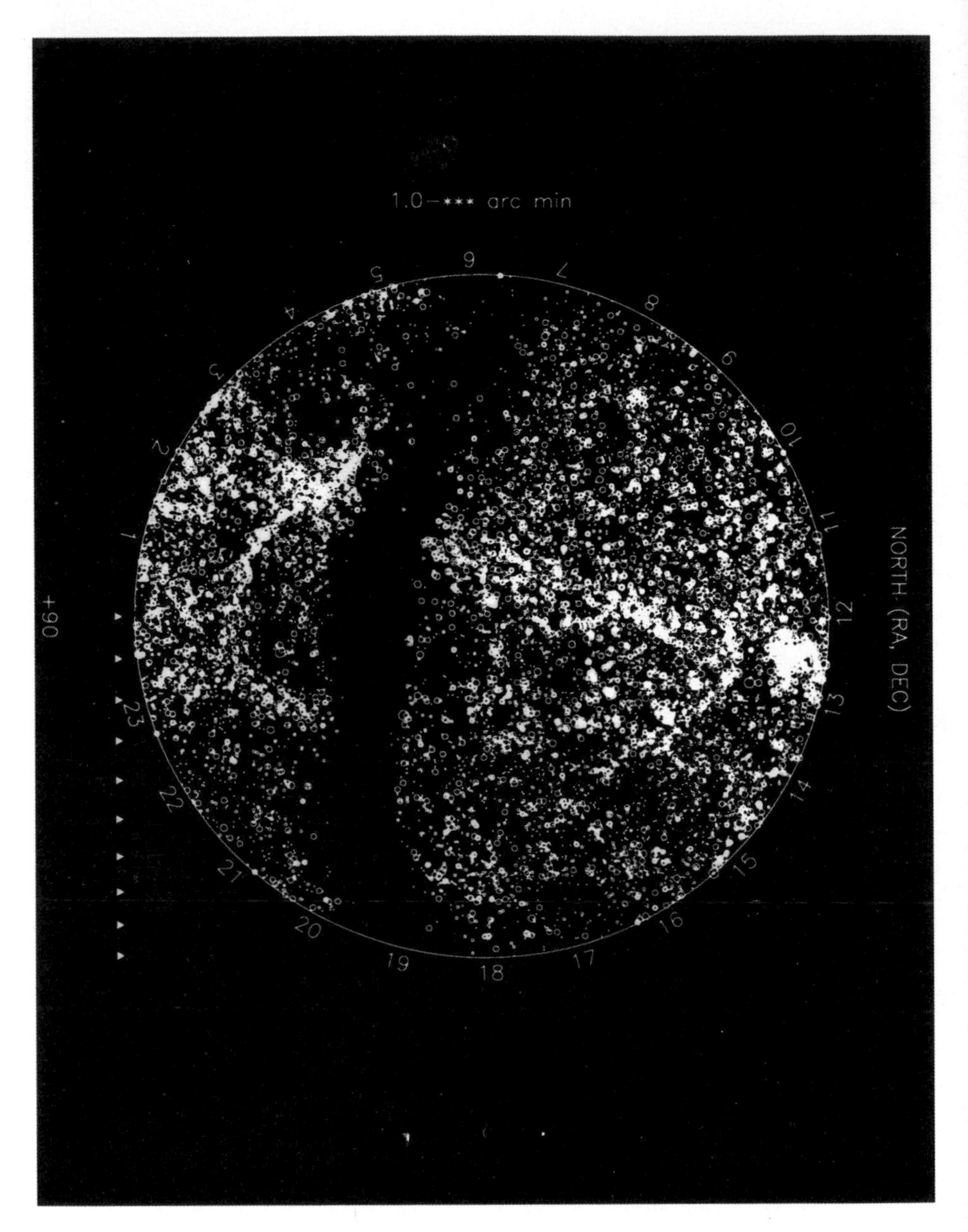

FIG. 1. Equal Area Projection of the North Celestial Hemisphere. Galaxies in Nilson's Uppsala General Catalogue coded by diameter.

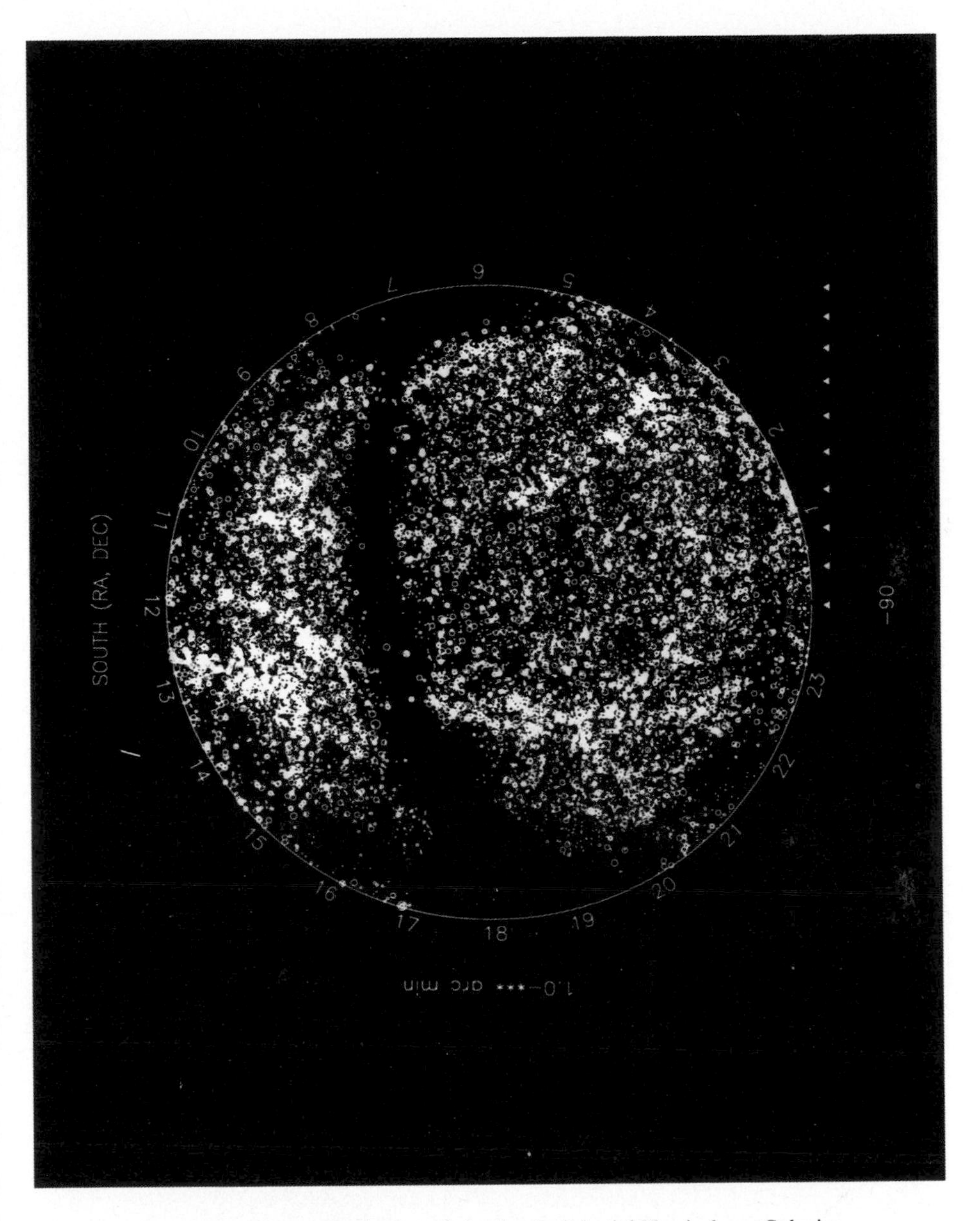

FIG. 2. Equal Area Projection of the South Celestial Hemisphere. Galaxies in Lauberts's ESO/Uppsala Catalogue coded by diameter for $\delta < 17.5°$. For $0° > \delta > -2.5°$ from UGC as Figure 1. For $-2.5° > \delta > -17.5°$ the galaxies from the Morphological Catalogue are plotted but this is readily seen to be less complete.

In the optical it is more credible to argue that with suitable modifications for galaxy type, light might trace mass; in which case the gravity field's 'shot noise' due to the apparently brightest half dozen galaxies may really cause an extra peculiar motion of the Local Group with respect to the mean flow of galaxies near here. It would be of interest to see if there is some correlation between the directions of these two vectors.

The distribution of extra-galactic light across the sky is pictured in equal area projection in figures 1-5. Figures 1 and 2 show the northern and southern celestial hemispheres as depicted from the UGC and ESO catalogues. In Figure 2 the missing strip $-2.5° > \delta > -17.5°$ has been filled in from the Morphological Catalogue, but the obvious inhomogeneity of this catalogue not only in declination but apparently also in R.A. is seen by the variable step in galaxy density where it joints the ESO catalogue at $\delta = -17.5°$. The remaining figures show the distribution of galaxies in different diameter ranges in galactic coordinates. Study of these allows one to get a good idea of the structure of the local galaxy distribution not only over the sky but also in depth. As we shall see presently a galaxy with characteristic diameter D_* has a diameter of $1'$ when placed at a distance corresponding to 7000 km/s.

Section 2 discusses the deduction of the optical dipole from these maps and performs a number of experiments to discover where it comes from. The two major components are a pull towards $l = 290°$, $b = 28°$ (between Centaurus and Hydra) and a push from roughly $l = 74°$, $b = -35°$ which arises from a void in the local galaxy distribution at distances < 2000 km/s.

Section 3 gives the directions of observed galaxy streaming and discusses their relationship to the CMB, optical and IRAS dipoles. Section 4 shows how to deduce the selection present in Huchra's radial velocity catalogue by comparing it with diameter limited catalogues. This enables us to get a crude map of galaxy density as a function of velocity within 15° of the supergalactic plane.

2. The Velocity of the Local Group

The microwave observations (Lubin and Villela 1986) show that the variation in $\Delta T/T$ around the sky is well fitted by a dipole whose direction and magnitude are independent of observing frequency. This dipole is well explained by a motion of the Sun relative to the CMB of 360 ± 27 km/s towards $l = 265° \pm 2°$, $b = 50° \pm 2°$.

Although the line joining us to Andromeda lies in the plane defined by non-local-group bright galaxies, the fact that Andromeda is both much nearer and approaching us has led to the belief that we are both gravitationally bound into the Local Group. Once this concept is adopted the Local Group's motion becomes of greater interest than the Sun's. Unfortunately the Sun's velocity relative to the barycentre of the Local Group is imperfectly known. Table 1 lists determinations starting with the first one by Humason and Wahlquist

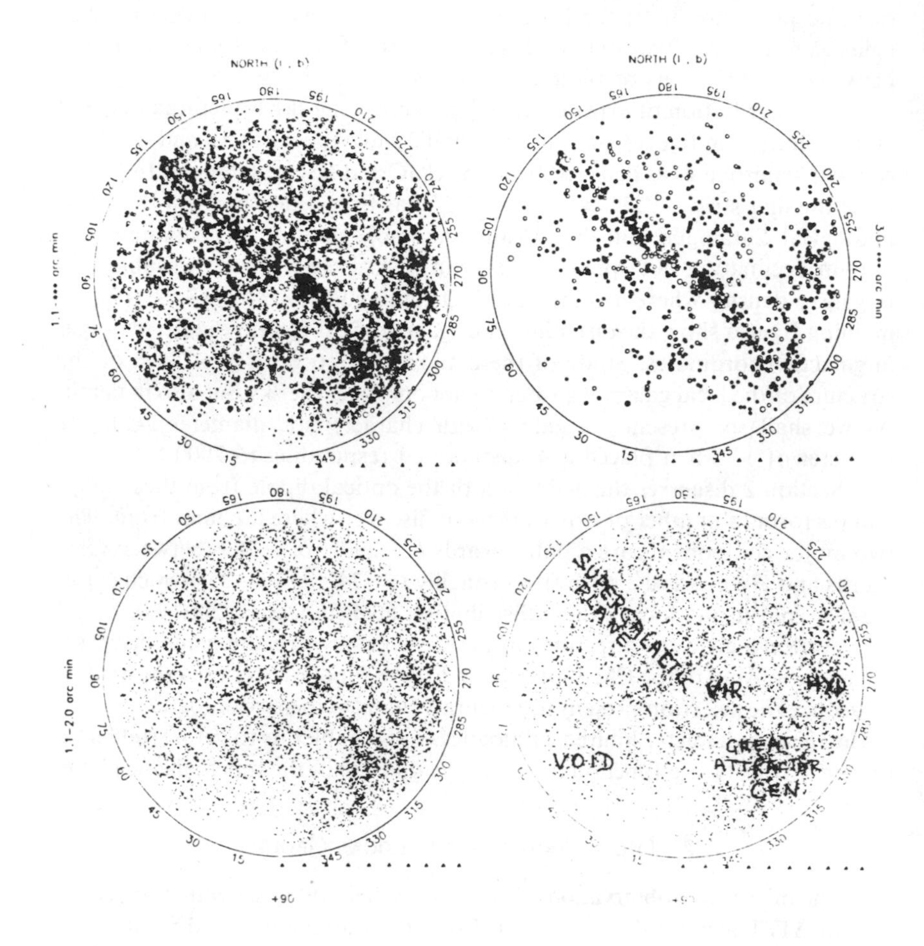

FIG. 3abcd. Equal Area Projection of the North Galactic Hemisphere from the galaxies contained in Figures 1 and 2: (a) all galaxies; (b) bright galaxies $\Theta > 3'$; (c) faint galaxies $2.0' > \Theta > 1.1'$; (d) naming chart.

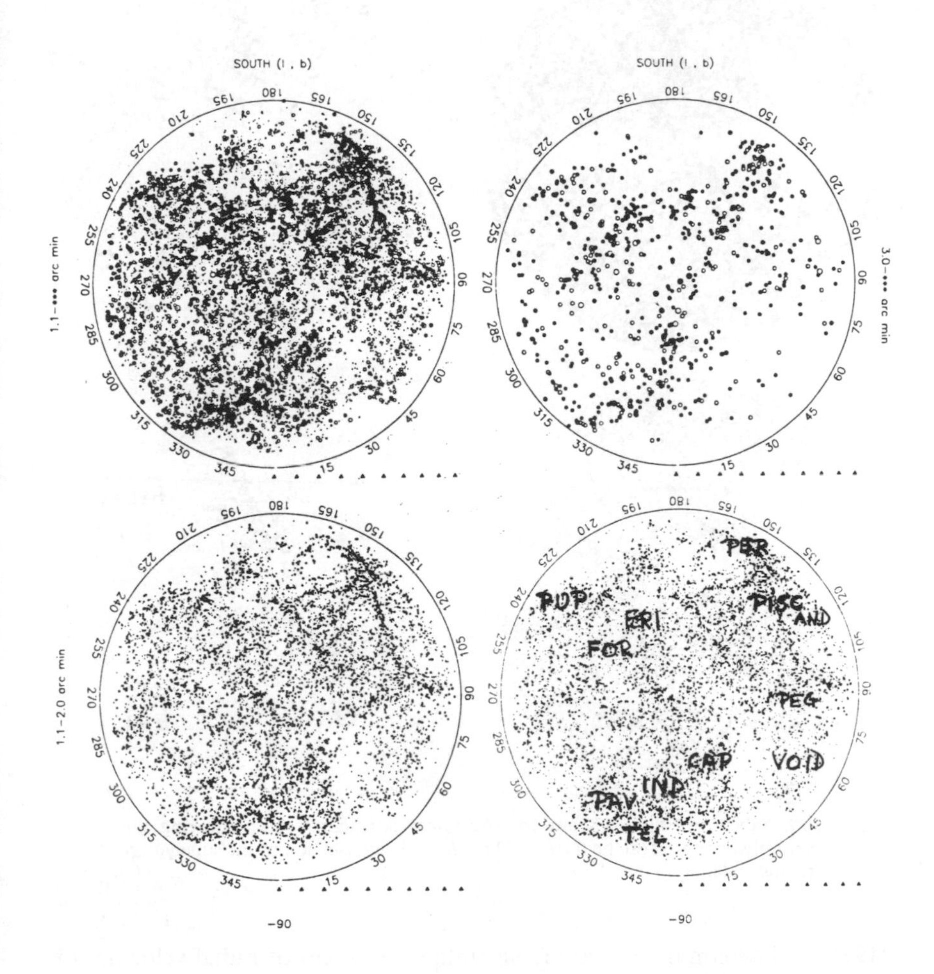

FIG. 4abcd. As Figure 3 but for the South Galactic Hemisphere.

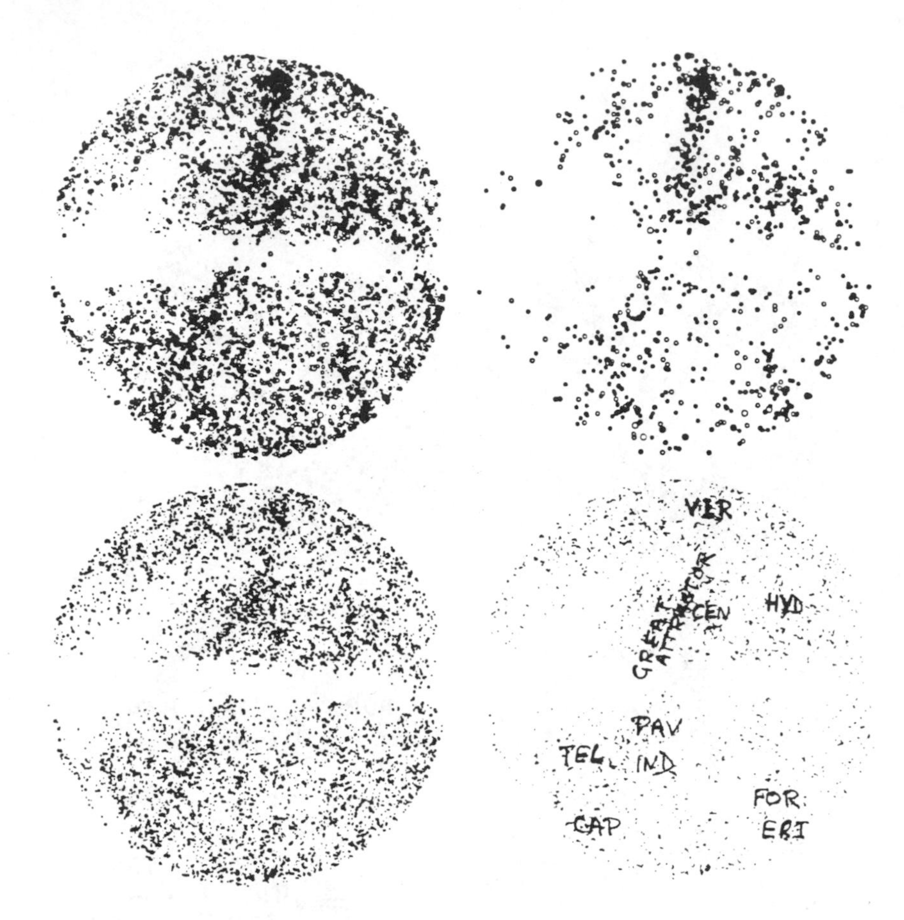

FIG. 5abcd. As Figure 3 but for the Great Attractor region showing the hemisphere centred on the the $l = 317°$, $b = 0°$ intersection of the supergalactic and galactic planes.

(1955). All determinations are from suitable averages of radial velocities of Local Group members. They differ in their treatment of whether members of the M31 subgroup and the Galaxy's subgroup are averaged together or treated as though they were independently moving. However, the main differences stem from whether the solution is an unconstrained kinematic one or is subjected to a dynamical constraint such as no net linear momentum in the barycentre's frame. The most constrained solutions assume that the two major subgroups move on a rectilinear orbit. Since none of the solutions differ from these by more than their probable errors the most constrained solutions should probably be preferred. While the differences are never large, they can be important when considering small motions such as the Local Group's

TABLE 1
THE SUN's MOTION RELATIVE TO THE LOCAL GROUP

Authors	v cos l cos b	v sin l cos b	v sin b	Number
Humason and Wahlquist (1955)	−85	277	−35	1
Yahil, Sandage, and Tammann (1977)	−79	295	−38	2
Richter, Tammann, and Huchtmeier (1987)	−50	307	−19	3
Lynden-Bell and Lin (1977) (constrained momentum	−25	313	− 9	4
Sandage (1986) (constrained to line)	−37	291	−28	5
This article (constrained to line)	−30	297	−27	6
de Vaucouleurs *et al.* (1976) standard	0	300	0	7

Sun's motion relative to CMB
$V_\odot - V_{CB} = 360 \pm 27\ (265° \pm 2°, 50° \pm 2°) = (-20, -231, 276)$
(Lubin and Villela 1986)
Local Group's motion relative to CMB by subtraction of 7.
$V_{LG} - V_{CB} = (-20, -531, 276) = 600 \pm 27\ (l = 268°, b = 27°)$

peculiar velocity relative to nearby galaxies. In Table 1 the solutions constrained to a line are compounded of the motion of the Sun relative to the LSR, the motion of the LSR around the Galaxy and the motion of the Galaxy relative to the centre of mass of it and Andromeda. Thus writing $\mu = M_A/(M_A + M_G)$ for the fraction of the Local Group's mass which resides in the M31 subgroup, we have

$$V_\odot - V_{LG} = (V_\odot - V_{LSR}) + (V_{LSR} - V_G) + \mu\,(V_G - V_A)\,\hat{r}_A$$

where $\hat{r}_A$ is the unit vector to Andromeda. Evaluating each term with IAU recommendations gives

$$V_\odot - V_{LG} = (9, 12, 7) + (0, 220, 0) + 2/3\,(123)\hat{r}_A = (-30, 297, -27)\ \text{km/s}.$$

This differs little from de Vaucouleurs' standard (0, 300, 0). We therefore use de Vaucouleurs' standard (de Vaucouleurs *et al.* 1976), since for most applications standardisation is important and it is within the errors of our preferred solution.

The peculiar motion of the Local Group relative to the mean flow of nearby galaxies is not easily detectable. Richter, Tammann, and Huchtmeier (1987) first find an apparently significant result and then reject it on the ground that the transformation to Local Group axes is equally uncertain. With de

Vaucouleurs' standard reduction their result is a motion (−28, −9, −94) ± 17 km/s, while with our preferred solution (Table 1, no. 6) it becomes (2, −6, −66). The new results of Faber and Burstein (1988) from the best Fisher-Tully determinations is smaller than its probable error, 66 ± 105 towards (170° ± 97°, 6° ± 31°) = (−64, 11, 7) km/s in de Vaucouleurs' frame, our preferred frame (−34, 14, 34). These may be contrasted with the larger values found by Aaronson *et al.* (1982) of (−172, −39, 24) in de Vaucouleurs' frame. The low values are probably to be preferred and the Local Group's peculiar velocity has not yet been detected.

3. The Optical Dipole

Lahav (Lahav *et al.* 1988) has determined the diameter function of both UGC and ESO galaxies using the northern and southern complete redshift surveys. The cumulative diameter function fits a Yahil form well. Let D_* be the characteristic diameter and $t = (D/D_*)^2$. Then Yahil's form for the cumulative function is

$$\Phi\,(>t) = \Phi_*\, t^{-\mu}\,(1 + t/\nu)^{-\nu}.$$

The UGC and ESO diameter distributions can both be fitted with one functional form with $\mu = 0.2$ $\nu = 3.4$ but with different D_*. For UGC, $D_* = 6186 \pm 269$ arcminutes km/s, with the strange diameter unit following because we use km/s as a measurement of distance (6875 arcminutes km/s = 20 h^{-1} kpc, for there are 3437.7 arcminutes in a radian). For ESO, $D_* = 6973 \pm 303$ arcminutes km/s. Assuming that in reality the true diameter functions are the same then ESO measurements of diameters are 1.13 ± .05 times larger than the UGC measurements. This is no surprise as the ESO plates are deeper and they give greater galaxy counts. This measurement difference is confirmed by the numbers of objects in the faintest bins in which the numbers counted in both UGC and ESO approximate those expected of a uniform universe. To get agreement in numbers a UGC galaxy of 1.00 arcminute diameter has to correspond to an ESO galaxy of 1.10 arcminutes diameter. This is in good agreement with the 1.13 ± .05 factor deduced above from the diameter functions. After putting all diameters onto the ESO system the common value of Φ_* is 0.011 galaxies per $(100\ \text{km/s})^3$. This joint calibration now replaces that used by Lahav (1987) previously which was the preliminary photoelectric calibration of Fouqué and Paturel (1985, which they have since revised, Paturel *et al.* 1987).

Because the ESO catalogue has no magnitude calibration we use the square of the angular diameter in its place. Notice that this automatically introduces the desired inverse square distance weighting. To get the mass weighting we have to assume that the square of the diameter of a galaxy is proportional to its mass. This would true if the product of the M/L ratio and the surface brightness were independent of mass. Since M/L ratios increase like $L^{1/4}$ (Dressler *et al.* 1987, Faber *et al.* 1987) while for bright ellipticals the surface brightnesses decrease as luminosities increase, the product can not be far from constant. It could be that using diameter squared is as good as, if not better than, using luminous flux.

Our calculation of the observed optical dipole now proceeds as follows: (1) for $|b| < 15°$ so few galaxies are in the catalogues that we have replaced this whole band with mean sky. Notice that this gives no contribution to the net dipole. We do the same in the missing band, $-2.5° > \delta > -17.5°$, but due to its offset from a great circle the mean sky replacement does give a contribution to the dipole; (2) above $|b| = 15°$ we allow for exinction as follows: (i) we increase the measured angular diameters by the factor $10^{A_B/5}$. Diameter squared then still behaves like luminosity; (ii) we increase the limiting diameter down to which the catalogue is deemed complete by the same factor. Thus Θ_c now changes over the sky; (iii) whenever the catalogue's absorption-corrected completion limit, Θ_c, is larger than the minimum true diameter, Θ_{min} down to which we are calculating a dipole, we increase the contribution of each galaxy by a factor that accounts for this incompletion. This factor is

$$1 + \frac{M(\Theta_{min}) - M(\Theta_c)}{N(\Theta_c)\ \Theta^2} ,$$

where $M(>\Theta)$ is the cumulative $\Sigma\Theta^2$ of all galaxies at the pole with angular diameter greater than Θ and $N(>\Theta)$ is their cumulative number.

M and N have been calculated from the $b < -40°$ region of the ESO catalogue to avoid any distortions caused by the proximity of Virgo and Coma in the North. Their normalization is irrelevant to the above formula, which extrapolates the part of the cumulative dipole contribution that can still be seen despite absorption, with a continuous curve of the shape seen in unobscurred parts of the sky. The formula used for absorption was that of Fisher and Tully (for details see Lahav *et al.* 1988). Thus the dipole actually calculated is

$$P(>\Theta_{min}) = 3/(4\pi)\ \Sigma \left[\Theta^2 + \frac{M(\Theta_{min}) - M(\Theta_c)}{N(\Theta_c)} \right] \hat{r} \qquad (3\text{-}1)$$

where $\hat{r}$ is the unit vector to the galaxy counted, Θ is its absorption-corrected angular diameter and Θ_c is the similarly corrected limit to the catalogue in that direction. The sum is over all galaxies with $\Theta > \Theta_{min}$ outside the excluded zones described above, and to that sum the contribution from mean sky in the uncounted zones is added. All diameters are put on the ESO scale at the beginning of the process using the factor 1.13 for the UGC diameters. To our delight the direction of the optical dipole so determined lies at $l = 261°$, $b = 29°$, very close to the direction of the Local Group's motion relative to the CMB (268,27). The monopole is defined as the true $\Sigma\ \Theta^2$ per unit solid angle, so it is given by the expression on the rhs of formula 3-1 omitting both the 3 and the $\hat{r}$. The magnitude of the dipole is 4.4×10^3 (arcmin)^2 which is 24% of the monopole of 18.3×10^3 (arcmin)^2. [Meiksin and Davis 1986 define their percentage dipoles without a 3 in the dipole term. To put their % dipoles on the scale used here, multiply their values by 3. Our definition has the property that with a truly dipolar sky with a p% dipole the flux from a small region of sky in the dipole's direction is 1 + p times the mean.] However, much of this dipole arises from relatively large galaxies. 3.8×10^3 (arcmin)^2 comes from galaxies of diameter $> 2'$, 3.0×10^3 (arcmin)^2 from galaxies of diameter $> 4'$ and 1.7×10^3 (arcmin)^2 from galaxies of diameter $> 8'$. Thus over a third arises from galaxies of distance $\lesssim$ 1000 km/sec and almost two thirds from galaxies of distance $\lesssim$ 2000 km/s. Only 1/7 of the dipole arises from galaxies $\geq$ 4000 km/sec. These numbers are in stark contrast to those predicted by the Great Attractor model for which half the gravity is generated from distances beyond the centre of the attractor at ~ 4300 km/sec distance.

While much of the optical dipole arises nearby, the contribution of the Virgo Cluster is relatively small. The excess contribution from the circle of 12° diameter enclosing the cluster, over that expected from mean sky is only 17% of the whole dipole. Increase of the circle's diameter to 20° only raises that to 18%. The Great Attractor area of sky in Centaurus is more significant. A cone of 30° total angle gives an excess which is 41% of the whole dipole. If we presume that our total dipole of 4437 (arcmin)^2 should not be increased by further contributions from fainter galaxies, nor from the excluded zones, then the gravity corresponding to the whole dipole generates the Local Group's infall of 600 km/s. On such a scale the galaxies in the 20° circle around Virgo would generate an infall of only 108 km/s here and the Centaurus region 246 km/s. Although that number sounds more hopeful for the Great Attractor interpretation a number of the galaxies included are closer and a significant fraction of the signal comes from them. We now recalibrate the dipole in terms of the net optical flux passing through unit area at the Local Group. The total optical dipole of 4437 (arcmin)^2 corresponds to a net flux from the direction of the dipole $4\pi/3$ times larger. To convert this to an approximate optical flux we plotted total B magnitudes against log Θ(max) for numerous galaxies. We find the correlation between Θ and B mag, with Θ in arc minutes, is given by $B = 15.00 - 2.5 \log \Theta^2$. Inserting 4437 $(4\pi/3)$ for Θ^2 gives the dipole's flux as $B_D = 4^m.33$. Assuming solar colours this corresponds to $m_{BOL} = 3.6$ and the optical flux, F, past the Local Group is thus approximately $F = 9 \times 10^{-7}$ ergs cm^{-2} s^{-1}. This generates an infall of 600 km/s at the Local Group. On this basis each extra $\Theta = 10'$ or $B = 10^m$ galaxy would contribute an infall velocity of 3.2 km/s.

To discover more precisely which areas of sky are responsible for the optical dipole we oriented a cone towards the total dipole and calculated the contribution from within the cone after subtracting the contribution that would have arisen within the cone from a sky that was totally uniform. In Figure 6 we show how this excess dipole within the cone varies as a function of the fraction of the whole sky within the cone. There is a strong rise up to 70% of the total as the cone's semi-angle rises to 45°, but there is also a notable

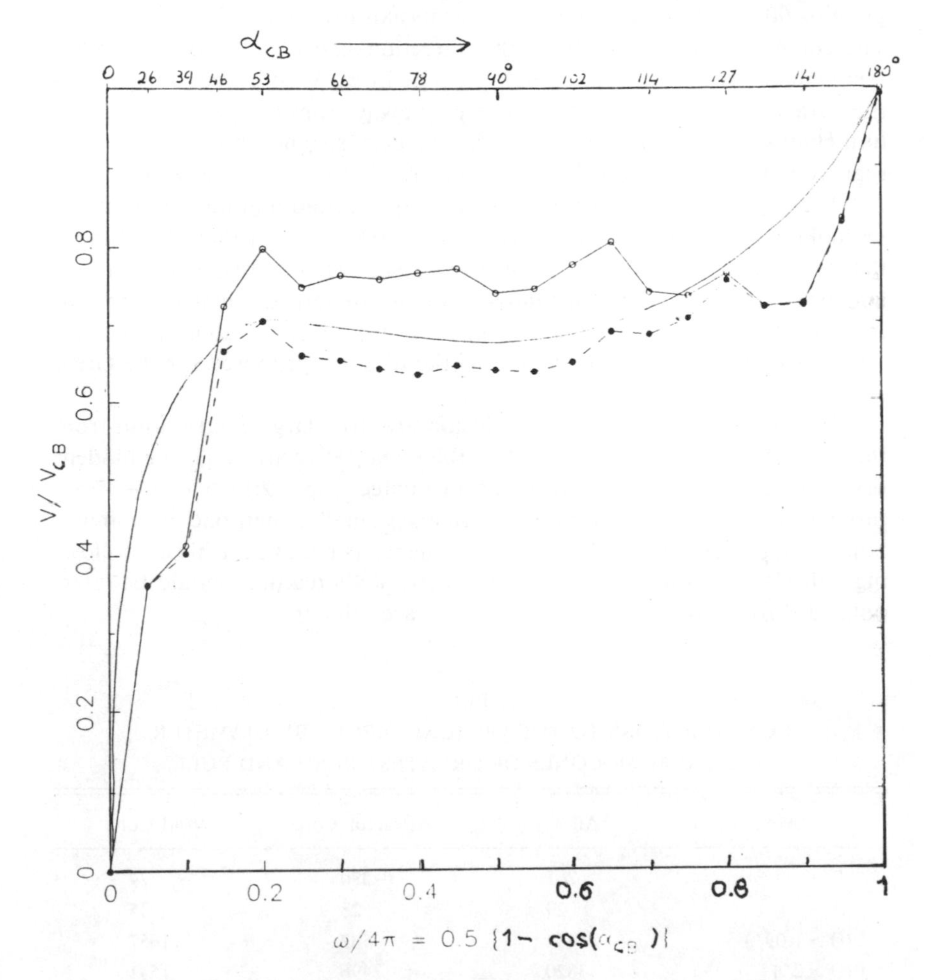

FIG. 6. Fraction of the total optical dipole as a function of the fraction of the sky within the cone centred on $l = 268°$, $b = 27°$ with different opening angles. Full line total dipole in cone; dashed line projection of total dipole onto $l = 268°$, $b = 27°$; light line prediction of $\varrho \propto |r - r_m|^{-2}$ Great Attractor model, $1/2\ [(\pi - \alpha) \sin \alpha + 1 - \cos \alpha - (\pi^2/8) \sin^2 \alpha)]$.

push from the decrement at the back of the cone. At first we thought this push from behind would be missing in the Great Attractor model but the actual prediction is not so bad. To make a better fit the Attractor itself must be more spread out and the void more concentrated but the relative magnitudes of the contributions are about right. To see if the directions of the push and the pull are really colinear we asked for the direction of that 90° (45° semi angle) cone that enclosed the greatest excess dipole. This lies not in the direction of the final dipole, but is directed toward $l = 290°$, $b = 28°$. Similarly the backward pointing 90° cone with the most push has its axis toward $l = 74°$, $b = -35°$. The compound of a push from this direction and a pull from the other represents most of the total dipole. Thus the Local Void is not colinear with the attraction and its push slews the dipole away from the Centaurus region into Hydra. These results may in reality be even stronger because the zones that we have painted with the mean sky are involved in these regions.

We may again get an idea how far away the void and the attractor are by looking at the excess dipoles from these 90° cones. From Table 2 more than half the void's contribution comes from galaxies of more than 8′ diameter and the contribution is almost complete for galaxies greater than 4′. The contribution from the attractive cone comes from galaxies of about half those sizes but that is still at only about half the distance proposed for the Great Attractor.

Possible escapes from this conclusion are: (i) a large contribution from the zone of avoidance, where most of the Great Attractor might lie hidden; (ii) larger dipole contributions from the uncounted strip $-2.5° > \delta > -17.5°$; (iii) contributions to the dipole from galaxies smaller than one arc minute. Some idea of the importance of these refinements could be achieved following Faber's suggestion that the extragalactic sky's features should be interpolated through the zones where the data are missing.

TABLE 2

CONTRIBUTIONS TO THE OPTICAL DIPOLE BY DIAMETER AND FROM 90° CONES OF GREATEST PUSH AND PULL

Area	All sky	Attractor Cone	Void Cone
l	261	290	74
b	29	28	−35
$P(\Theta > 1.03')$	4437	3562	1657
$P(\Theta > 2')$	3820	2496	1541
$P(\Theta > 4')$	3058	1794	1338
$P(\Theta > 8')$	1708	1118	906

Note: All values of P are in $(\text{arcmin})^2$ on the ESO diameter measures.

4. Comparison of Dipoles and Directions of Interest

Figure 7 shows the direction of the motion of the Local Group relative to the CMB and the motions of different galaxy samples also in the CMB frame. These are not strictly comparable since Virgo infall has been modelled and removed from the galaxy sample but not from the Local Group's motion. If Virgo infalls of 100, 200 or 300 km/s were removed the remaining motion of the Local group would be in the directions labelled 100, 200 or 300, respec-

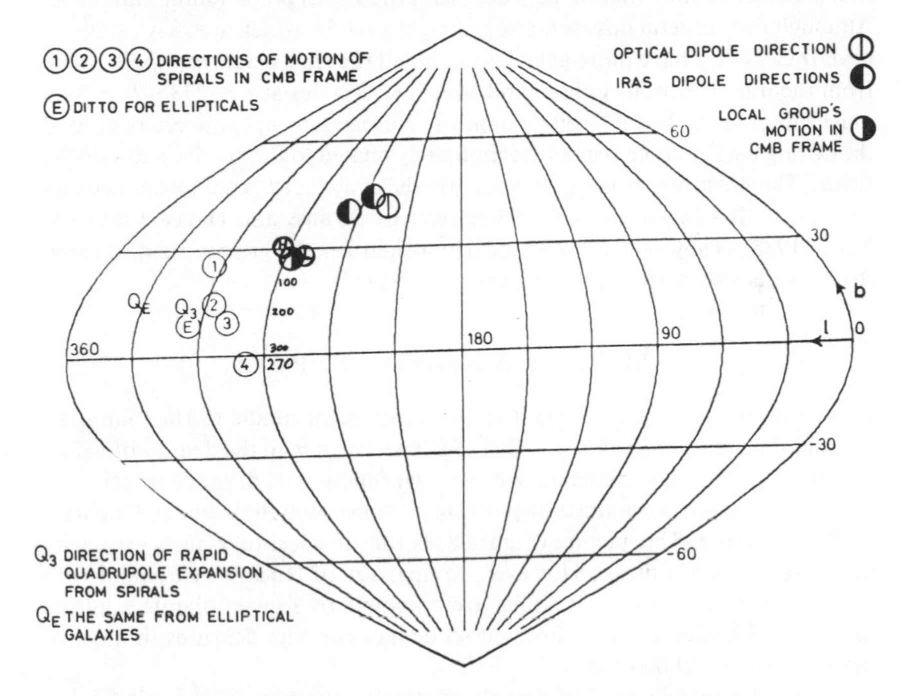

Fig. 7. Interesting directions in galactic coordinates. All mean motions are in the CMB frame.

1	Mean motion of Rubin *et al.* (1976) spirals: 862 km/s
2	Mean motion of de Vaucouleurs and Peters (1985) spirals: 460 km/s
3	Mean motion of AHM spirals according to Lilje, Yahil, and Jones (1986): 502 km/s
E	Mean motion of ellipticals with Re < 8000 km/s: 521 ± 89 km/s (Lynden-Bell *et al.* 1988, Lynden-Bell 1987)
◑	Motion of Local Group: 600 ± 27 km/s
L	Revised optical dipole of Lahav, Rowan-Robinson and Lynden-Bell (1988)
H	Revised IRAS dipole of Harmon, Lahav and Meurs (1987)
100, 200, 300	Directions of Local Group's motion relative to CMB after removal of a 100, 200, or 300 km/s Virgo infall component.

tively. The new optical dipole described here (Lahav *et al.* 1988) has shifted some 35° from the old direction (Lahav 1987). While some 10° of this shift is due to better allowance for absorption, most of the shift is caused by the new diameter calibration, through the redshift survey in place of the Fouqué and Paturel calibration. The directions of the IRAS dipoles of Yahil *et al.* (1986) and the number weighted IRAS dipole of of Meiksin and Davis (1986) are also plotted. Meurs and Harmon (1988) have made a new study of selection procedures for isolating galaxies from the IRAS point source catalogue. Although their criteria miss some large bright galaxies which are IRAS sources, nevertheless they have more galaxies in all and the flux weighted dipole found from them by Harmon, Lahav, and Meurs (1987) lies at $l = 273°$, $b = 31°$, much closer to the Local Group's motion. This agreement again confirms that the extra-galactic dipole vectors are intimately related to the gravity and velocity fields. The most recent work on these fields as deduced from radial velocity surveys of IRAS galaxies is described here by Strauss and Davis (1988) and Yahil (1988). They also deduce a nearby origin for the gravity field, a result that has also been found by Clowes *et al.* (1987).

5. Mapping the Supergalactic Plane

To investigate the reality of the Great Attractor model of the 'Samurai' (Lynden-Bell *et al.* 1988, Lynden-Bell 1987) we try to map the density of galaxies within 15° of the supergalactic plane as a function of distance or velocity. A first orientation is obtained by plotting all galaxies in that zone in Huchra's (1981) catalogue. This picture (Figure 8) is full of selection effects especially in its celestial N-S contrast. However, comparison of Huchra's catalogue with the ESO or UGC catalogues which are complete to a given diameter allows us to find the selection function and so deduce the true densities in most of the region $v < 10{,}000$ km/s.

Consider galaxies in a solid angle of sky Ω centred on the direction $\hat{r}$. Let $N_H(v, \Theta, \hat{r})$ be the number of galaxies in Huchra's catalogue per unit range of radial velocity v, angular diameter Θ and within the solid angle $\Omega(\hat{r})$. We correct all the velocities to the Local Group frame using de Vaucouleurs' standard prescription. To map the supergalactic plane in the space of radial velocities and angles in the sky we use v as though it were the distance. The true diameter of the galaxy whose angular diameter is Θ is then $D = v\Theta$. Let $\Phi(D)\,dD$ be the differential diameter function giving the fraction of all galaxies whose diameters lie in the range D to D + dD. Further we suppose that Huchra's catalogue has severe selections which are functions of both direction in the sky, $\hat{r}$, and angular diameter of the galaxy Θ. Then

$$N_H(v, \Theta, \hat{r}) = \Omega(\hat{r})\, n(v)\, v^3\, \Phi(v\Theta)\, S(\Theta, \hat{r}) \tag{5-1}$$

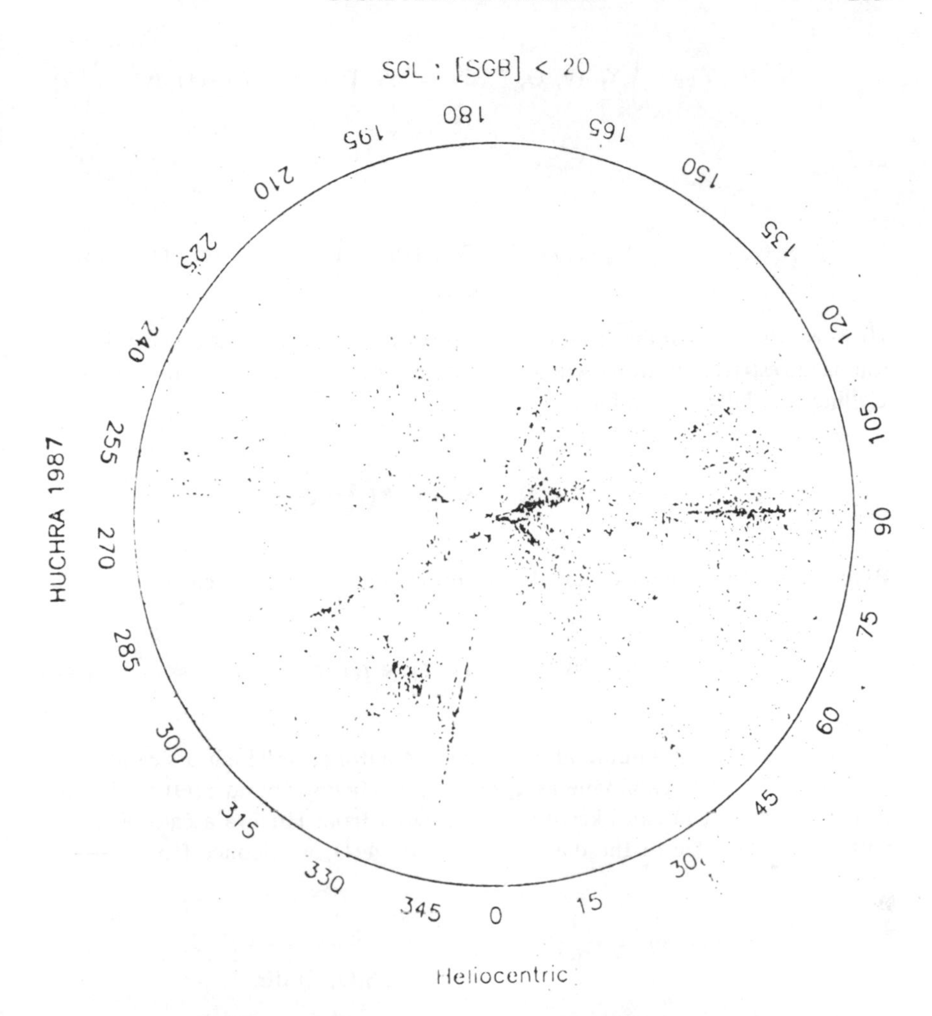

FIG. 8. Chart of all galaxies in Huchra's (1987) catalogue lying within 20° of the supergalactic plane as a function of supergalactic longitude and corrected radial velocity. The region out to 10,000 km/s is shown.

where S is the selection function giving the fraction of the objects that have measured velocities and diameters in Huchra's catalogue and n(v, $\hat{r}$) is the true number density of galaxies at radial velocity, v.

There are too few galaxies in the catalogue to determine N_H as a function of so many arguments. However integrals over N_H can be obtained from the data, in particular the integrals over radial velocity or over angular diameter. We therefore define

$$N_\Theta(\Theta, \hat{r}) = \int \mathcal{N}_H(v, \Theta, \hat{r})\, dv = \Omega S \int n(v)\, v^3\, \Phi(v\Theta)\, dv \qquad (5\text{-}2)$$

and

$$N_v(v, \hat{r}) = \int \mathcal{N}_H(v, \Theta, \hat{r})\, d\Theta = \Omega n v^3 \int \Phi(v\Theta)\, S(\Theta)\, d\Theta. \qquad (5\text{-}3)$$

However the true number count of all galaxies in a diameter limited catalogue that is unrestricted to those whose velocity has been measured, is also known. Calling this $N(\Theta, \hat{r})$ we have

$$N(\Theta, \hat{r}) = \Omega \int n(v)\, v^3\, \Phi(v\Theta)\, dv. \qquad (5\text{-}4)$$

Evidently we may deduce that the selection function S is given by

$$S(\Theta, \hat{r}) = N_\Theta(\Theta, \hat{r}) / N(\Theta, \hat{r}) \qquad (5\text{-}5)$$

We may evaluate both numerator and denominator from Huchra's catalogue and the UGC/ESO catalogue as appropriate. Having found $S(\Theta)$ we insert it into 5-3. $N_v(v, \hat{r})$ can likewise be calculated from Huchra'a catalogue so with our knowledge of the diameter function $\Phi(D)$ we deduce from 5-3

$$n(v, \hat{r}) = \frac{1}{\Omega v^3} \cdot \frac{N_v(v, \hat{r})}{\int \Phi(v\Theta)\, S(\Theta, \hat{r})\, d\Theta}.$$

Figure 9 shows a plot of $n(v, \hat{r})$ with radial velocity plotted radially and supergalactic longitude plotted as azimuth. To get sufficient numbers to calculate S the solid angles chosen are cones of 15° semi-angle centred close to the supergalactic plane tracing the maxima of the bright galaxy distribution.

It is already clear from this crude map that Virgo is at one side of the density enhancement in which it lies, the centre of the distribution being displaced towards Centaurus or Pavo.

The need for more reliable cartography of the galaxy distribution is clear. A great start on the nearby galaxies has been made by Tully and Fishers' Atlas (1987). The redshift surveys are rapidly extending this but difficulties will remain in the zone of avoidance.

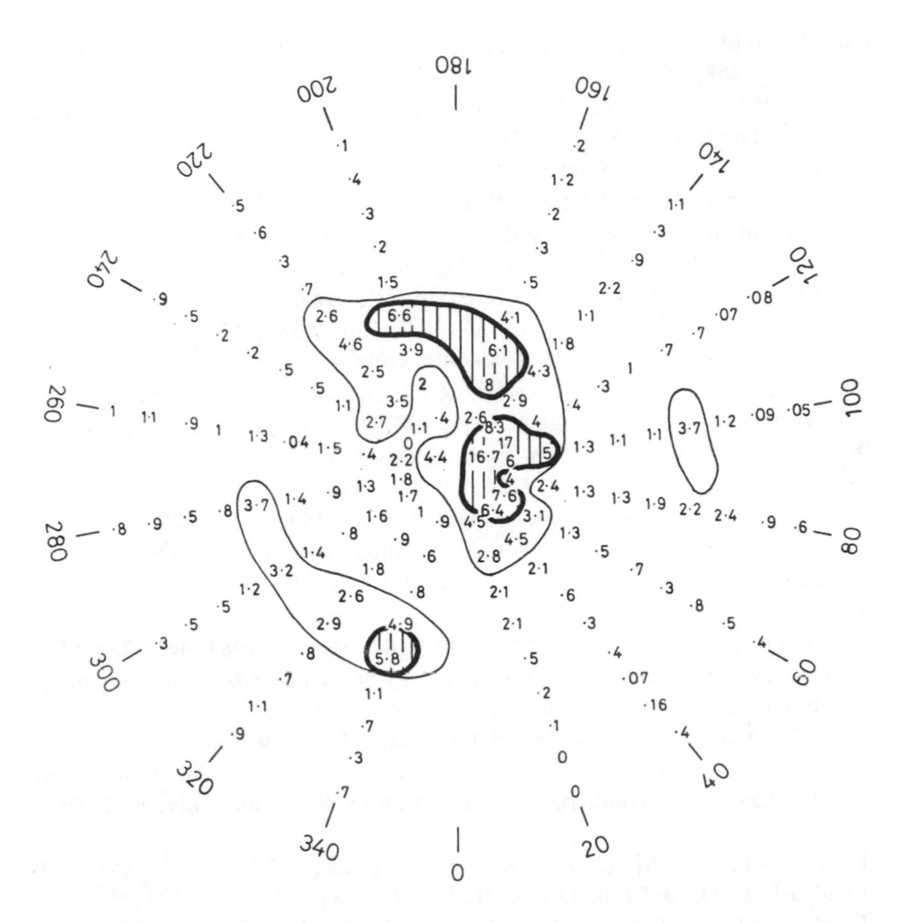

FIG. 9. Map of the number density of galaxies near the supergalactic plane as a function of supergalactic longitude and radial velocity corrected to the LG. The region out to 10,000 km/s is shown with contours at 2.5 and 5 times mean density of the universe. The Perseus-Pisces chain, the Virgo-Centaurus-Pavo complex, and the Coma region all stand out. Densities are averaged over 30° cones.

REFERENCES

Aaronson, M., Huchra, J., Mould, J., Schechter, P.L., and Tully, R.B. 1982. *Ap J.* **258,** 64.

Clowes, R.G., Savage, A., Wang, G., Leggett, S.K., MacGillivray, H.T., and Wolstencroft, R.D. 1987. *MN.* **229,** 27p.

Davis, M. 1988. this volume.

de Vaucouleurs, G. and Peters, W.L. 1985. *Ap J.* **297,** 27.

de Vaucouleurs, G., de Vaucouleurs, A., and Corwin, H.G. 1976. *Second Reference Catalogue of Bright Galaxies.* Austin: University of Texas.

Dressler, A., Lynden-Bell, D., Burstein, D., Davies, R.L., Faber, S.M., Terlevich, R.J., and Wegner, G. 1987. *Ap J.* **313,** 42.

Faber, S.M. and Burstein, D. 1988. this volume.

Faber, S.M., Dressler, A., Davies, R.L., Burstein, D., Lynden-Bell, D., Terlevich, R.J. and Wegner, G.A. 1987. in *Nearly Normal Galaxies.* p. 175. ed. Faber, S.M. New York: Springer Verlag.

Fouqué, P. and Paturel, G. 1985. *AA.* **150,** 192.

Harmon, R.L., Lahav, O., and Meurs, E.J.A., 1987. *MN.* **228,** 5p.

Huchra, J. 1987, ZCAT. tape from Smithsonian Astrophysical Observatory.

Humason, M.L. and Wahlquist, H.D. 1955. *AJ.* **60,** 254.

Lahav, O. 1987. *MN.* **225,** 213.

Lahav, O., Rowan-Robinson, M., and Lynden-Bell, D. 1988. *MN.* **234,** 677.

Lauberts, A. 1982. *The ESO/Uppsala Survey of the ESO(B).* Atlas. Garching: European Southern Observatory.

Lilje, P., Yahil, A., and Jones, B.J.T. 1986. *Ap J.* **307,** 91.

Lubin, P. and Villela, T. 1986. in *Galaxy Distances and Deviations from Universal Expansion.* p. 169. eds. Madore, B.F. and Tully, R.B. Dordrecht: Reidel.

Lynden-Bell, D. 1987. *Quart J Roy Astr Soc.* **28,** 187.

Lynden-Bell, D. and Lin, D.N.C. 1977. *MN.* **181,** 37.

Lynden-Bell, D., Faber, S.M., Burstein, D. Davies, R.L., Dressler, A., Terlevich, R.J., and Wegner, G. 1988. *Ap J.* 326, 19.

Meiksin, A. and Davis, M. 1986. *AJ.* **91,** 191.

Meurs, E.J.A. and Harmon, R.L. 1988. in preparation.

Nilson, P. 1973. *Uppsala General Catalogue of Galaxies.* Upsala Astr. Obs. Ann 6.

Paturel, G., Fouqué, P., Lauberts, A., Valentijn, E.A., Corwin, H.G., and de Vaucouleurs, G. 1987. *AA.* **184,** 86.

Richter, O.G., Tammann, G.A., and Huchtmeier, W.K. 1987. *AA.* **171,** 33.

Rubin, V.C., Ford, W.K., Thonnard, N., Roberts, M.S., and Graham, J.A. 1976. *AJ.* **81,** 687.

Rubin, V.C., Ford, W.K., Thonnard, N., and Roberts, M.S. 1976. *AJ.* **81**, 719.
Sandage, A. 1986. *Ap J.* **307,** 1.
_____ . 1987. *Ap J.* **317,** 557.
Strauss, M.A. and Davis, M. 1988. this volume.
Tully, R.B. and Fisher, J.R. 1987. *Nearby Galaxies Atlas.* Cambridge University Press.
Vorontsov Velyaminov , B.A. and Arkipova, A.A. 1963-68. *Morphological Catalogue of Galaxies.* Moscow: Moscow State University.
Yahil, A. 1988. this volume.
Yahil, A., Sandage, A.R., and Tammann, G.A. 1977. *Ap J.* **217,** 903.
Yahil, A., Walker, D., and Rowan-Robinson, M. 1986. *Ap J.* **301,** L1.

THE STRUCTURE OF THE UNIVERSE TO 10,000 KM S^{-1} AS DETERMINED BY IRAS GALAXIES

AMOS YAHIL
Astronomy Program, State University of New York at Stony Brook

ABSTRACT

A report is presented of the current status of a redshift survey of IRAS galaxies, which maps their density structure over 76% of the sky, to a distance of $100h^{-1}$ Mpc. The peculiar gravitational field is then calculated, assuming that the IRAS galaxies trace the mass. Special attention is paid to correct the Hubble positions of the galaxies self-consistently for their peculiar velocities.

The picture of peculiar gravity which emerges is more complex than had been imagined in earlier parametric models. It is dominated by two large mass concentrations, with modest over-density, one in the direction of Hydra-Centaurus (the "Great Attractor"), and the other around Perseus-Pisces. As a result, the gravitational field bifurcates not far from the position of the Local Group.

A comparison between the predicted peculiar velocities and the observed ones shows good overall agreement, confirming the gravitational origin of the peculiar velocities. It is planned to tackle the remaining differences mainly by extending the redshift survey down to $|b| \geq 5$, and by calculating peculiar velocities nonlinearly.

1. INTRODUCTION

This paper reports on the current status of a program to map the density structure of the universe to $\sim$ 10,000 km s^{-1}, through a redshift survey of tracer galaxies detected by the Infrared Astronomical Satellite (IRAS). The initial investigations of the distribution of these galaxies (Yahil, Walker, and Rowan-Robinson 1986; Meiksin and Davis 1986), used only their positions and fluxes. These studies showed a dipole anisotropy in the surface brightness of the IRAS galaxies that was coincident, within the errors, with that of the cosmic microwave background (CMB). This suggested both that the IRAS galaxies traced the mass, and that the peculiar gravitational field was due to density perturbations within the volume surveyed by the IRAS galaxies.

Unfortunately, owing to the broad luminosity function, fluxes are poor distance indicators, and offer only a limited description of the density structure. In order to improve this situation, a redshift survey was launched to obtain individual distances for the IRAS galaxies. Collaborators in this effort have included M. Davis, J. Huchra, M. Strauss, J. Tonry, and A. Yahil. While the results from this ongoing project are not yet final, the redshifts available to date have already significantly modified our picture of the universe around us, superseding the previous studies based on fluxes. This paper, and the complementary one by Strauss and Davis (1988), are reports on the current status of the project.

The IRAS redshift survey is not different in principle from the initial RSA (Sandage and Tammann 1981) or CfA (Huchra *et al.* 1983) optical surveys. The big difference is in the volume sampled. The current IRAS survey covers 76% of the sky, and densities can be determined reasonably accurately to a distance of $100h^{-1}$ Mpc ($H_0 = 100h$ km s^{-1} Mpc^{-1}). Observations are now underway, which would increase the sky coverage to 87%. This should be compared with the RSA sample, which covered 50% of the sky to $40h^{-1}$ Mpc, and the CfA catalogue, which covered 20% of the sky to $80h^{-1}$ Mpc.

The structure of this paper is as follows. The selection criteria of the IRAS sample are delineated in Sec. 2. The calculation of the density structure, presented in Sec. 4, is preceded by a discussion of the luminosity function in Sec. 3. Special attention is paid to the underlying assumption of a universal luminosity function, which is shown to be consistent with the data. The calculation of the gravitational field is taken up in Sec. 5, assuming that the IRAS galaxies indeed trace the matter. In performing this calculation, care is taken not to assume a smooth Hubble flow, with galaxies at distances given by their redshifts. Instead, a new method is devised, in which distances are obtained by correcting the redshifts self-consistently for the peculiar velocities induced by the peculiar gravity, which itself depends on the corrected distances. The essence of the paper are the results of the calculations of Sec. 4 and 5, which are presented in Sec. 6 in the form of maps of both density and gravity. The detailed predictions of the calculation can then be confronted with observations, by comparing the inferred peculiar velocities with the observed ones. This is done in Sec. 7, using the compiled data on peculiar velocities presented at this conference by Faber and Burstein (1988).

2. Sample

From the observational point of view, the IRAS galaxies are ideal tracers, since they are homogeneously detected over most of the sky, and their fluxes are unaffected by galactic extinction. The present sample covers 76% of the sky. Excluded are a band $|b| < 10°$ around the galactic plane, a strip in which the IRAS coverage was incomplete, and a few patches of known "cirrus" infrared emission.

Candidates for the redshift survey included all the sources in the Second IRAS Point Source Catalogue (PSC), with high quality 60μ flux $S_{60} > 1.936$ Jy, and satisfying the color criterion $S_{12} < S_{60}/3$ (Meiksin and Davis 1986). This condition, corresponding to a spectral inedx $\alpha < -0.68$, cannot be used for $S_{60} < 0.75$ Jy, since 12μ fluxes are not available below 0.25 Jy, but the problem does not arise with our higher flux limit. Yahil *et al.* (1986) used a somewhat less restrictive condition, $S_{25} < 3S_{60}$, corresponding to $\alpha < +1.25$, which can be used down to the 60μ limit of the PSC at 0.5 Jy. In the flux range of our redshift survey, however, this more liberal color criterion yields only a few percent more candidates, and was not used.

Prior to the the redshift survey, a major problem was contamination of the sample by infrared sources in our own interstellar medium, the so-called "cirrus". These sources have spectra that are similar to those of external galaxies, and are, therefore, included in the list of candidates generated by the above algorithm. While they are mainly confined to the plane, they do extend to higher galactic latitude in several directions of molecular cloud complexes, such as Ophiuchus and Orion, or of other galaxies of the Local Group. The existence of this cirrus contamination has prompted several attempts to use more sophisticated and restrictive color criteria to exclude them (Harmon, Lahav, and Meurs 1987; Rowan-Robinson 1988). In our redshift survey the cirrus problem was overcome by individual inspection of all the candidates on sky survey prints. This method is by far the most reliable one, since a nearby galaxy can easily be identified, even through a large amount of extinction. In any case of doubt, a spectrum of the object was obtained.

We have been so encouraged by the success of the identification program, that we decided to extend the survey to the previously excluded patches of cirrus, and down to $|b| \geq 5°$. When this survey is completed, we will have covered 87% of the sky. I have myself performed all the identifications of the candidates in these regions. My confidence in the identification is based on three factors: (1) I personally found it rather easy to make the identifications; (2) sources classified as cirrus were invariably clustered together in areas of larger obscuration; and (3) sources classified as galaxies, or questionable, were more smoothly distributed, with approximately the same sky surface density as the well identified galaxies at high galactic latitude. A quantitative test of my identifications will be available shortly for a part of the sample, where a "blind" Arecibo survey is taking spectra of all the IRAS galaxy candidates (Dow *et al.* 1988).

The above procedure for selecting galaxy candidates, as well as the subsequent determination of the luminosity function (see Sec. 3), presuppose accurate IRAS fluxes. There are several causes of concern in this regard. First, the PSC is known to underestimate the fluxes of extended sources. This would create a bias against large nearby extended galaxies, and may seriously affect our estimate of density in the Virgo supercluster. In order to overcome this difficulty, we asked Tom Soifer and Elizabeth Smith of the Infrared Processing and Analysis Center (IPAC) to addscan all the sources flagged as extended in the PSC, and used these addscanned fluxes. We also used the co-added fluxes of Rice *et al.* (1988) for galaxies with angular diameters greater than 10′. Secondly, we have noticed that the correlation coefficient flag in the PSC, which measures how well a point source template fits the scans, is anticorrelated with the ratio of addscanned to PSC fluxes. There are over a thousand sources which are not flagged as extended, but have poor correlation coefficients, and their addscanning is now in progress. Initial indications are that the problematic sources are virtually all cirrus objects. Finally, errors result from intensity enhanced detector responsivity (''hysteresis''), when crossing the galactic plane. This effect, which is expected to be important only at very low galactic latitudes, is now under investigation.

Our current redshift survey, which does not yet include those new areas of higher optical extintion, consists of 2244 galaxies, whose sky distribution is shown in Fig. 1, in both galactic and supergalactic coordinates. The excluded areas are also shown. The first impression is that the sky distribution of the galaxies is fairly smooth. Well known nearby clusters can be identified, but the contrast with their surroundings is far less marked than in similar plots of optically selected galaxies. Fig. 2 shows the distribution of the IRAS galaxies with distance, and compares it with that predicted for a homogeneous universe, with the luminosity function determined in Sec. 3. The biggest deviation is an over-density in the Virgo supercluster, but otherwise no obvious clustering is seen in this distribution, which is an average over all the 76% of the sky which we cover.

The smooth distributions in Figs. 1 and 2 might be construed to indicate that the survey extends beyond the ''local'' irregularities, and the volume it encompasses begins to approximate a ''fair sample'' of the universe. Based on our calculation of the gravitational field (see Sec. 5), we believe that this is indeed the case. The visual impression, however, is somewhat misleading, because of the absence of early-type galaxies in the IRAS sample. We are therefore seriously undersampling the high density regions, which are known to have a higher fraction of early-type galaxies (Dressler 1980, Postman and Geller 1984). This point is taken up again in Sec. 4.

3. Luminosity Function

The mapping of the density structure is performed by counting galaxies as a function of position. In order to convert these counts into densities, it

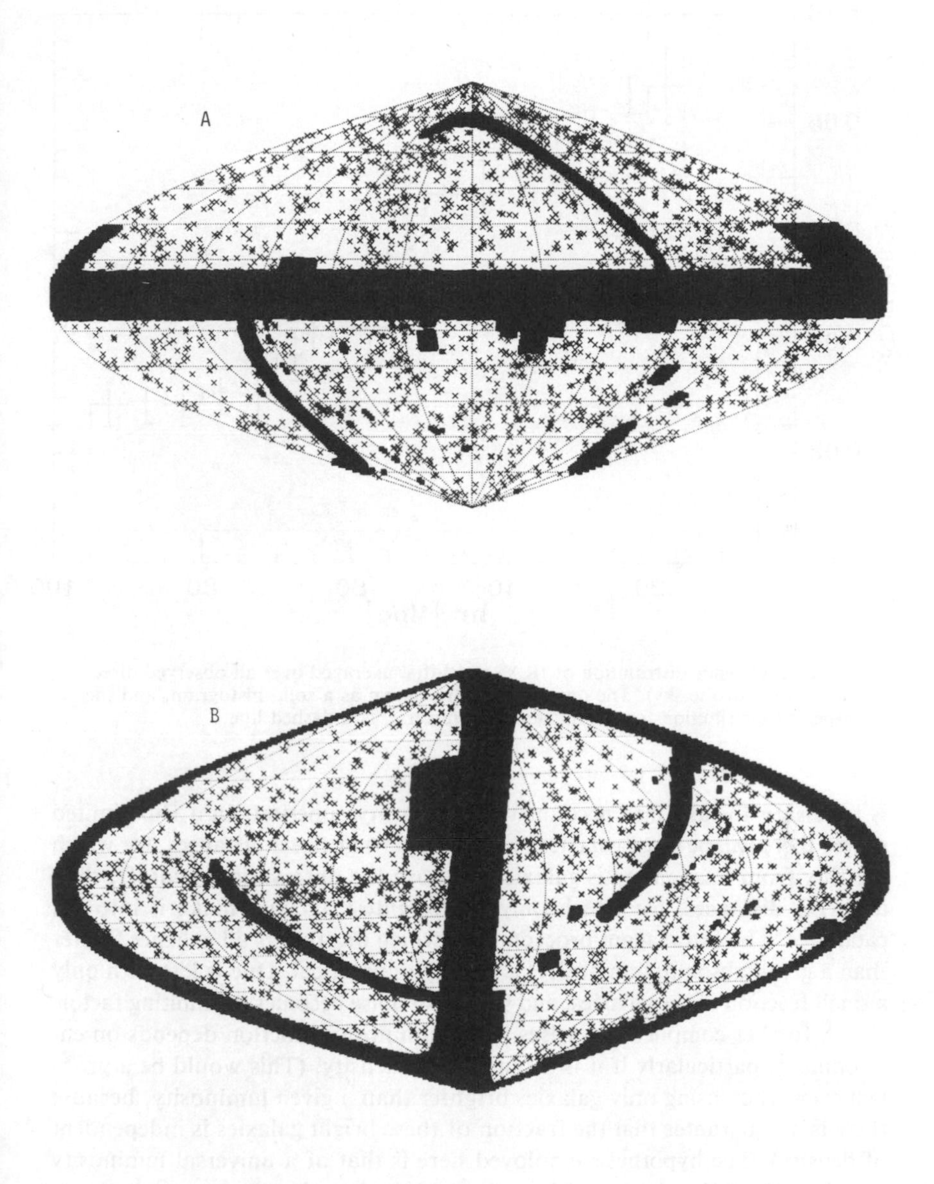

FIG.1. Sky map of the current IRAS redshift survey, showing both the area covered and the distribution of observed galaxies: (a) in galactic coordinates, (b) in supergalactic coordinates. Meridians start at 0° on the left of the plot, and end at 360° on the right.

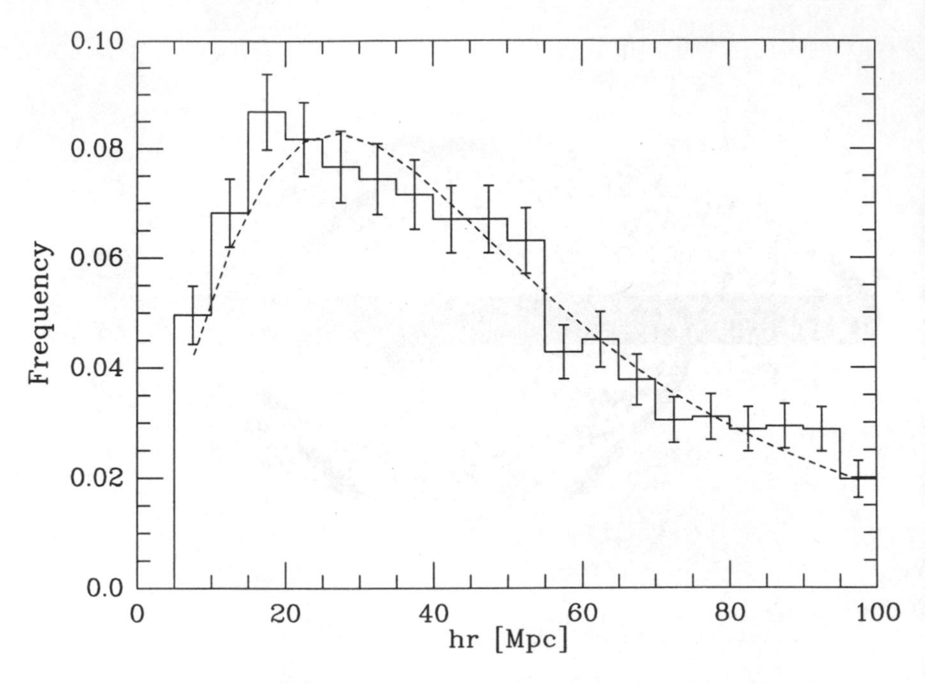

FIG.2. Frequency distribution of IRAS redshifts, averaged over all observed directions (76% of the sky). The observations are shown as a solid histogram, and the expected distribution for a homogeneous universe as a dashed line.

is necessary to take into account the luminosity function. In a flux-limited survey there will be more galaxies per unit volume at closer distances, for which the flux from less luminous galaxies is above the observational cutoff, than at larger distances, for which only the brightest galaxies make it into the catalogue. The extravagant procedure, in which one rejects all galaxies fainter than a given absolute luminosity, is totally impractical. One is left with only a small fraction of the galaxies, and statistical noise becomes the limiting factor.

A further complication arises if the luminosity function depends on environment, particularly if it is a function of density. (This would be a problem even when using only galaxies brighter than a given luminosity, because there is no guarantee that the fraction of these bright galaxies is independent of density). The hypothesis employed here is that of a universal luminosity function (Yahil, Sandage, and Tammann 1980), i.e., that the luminosity function is everywhere identical in *shape,* and differs from one location to another only in *normalization.* Indeed, it is this normalization which determines the local density (see Sec. 4). This hypothesis, therefore, needs to be confirmed, at least *a posteriori*.

The determination of the luminosity function in a density-independent fashion follows the method of Sandage, Tammann and Yahil (1979), except that the functional form of the luminosity function due to Schechter (1976), is replaced by the *cumulative* luminosity function

$$\Phi(L) = C\left(\frac{L}{L_*}\right)^{-\alpha}\left(1+\frac{L}{\beta L_*}\right)^{-\beta}, \tag{1}$$

with its corresponding *differential* luminosity function

$$\phi(L) = -d\Phi(L)/dL = \left(\frac{\alpha}{L}+\frac{\beta}{\beta L_*+L}\right)\Phi(L). \tag{2}$$

The *conditional* luminosity function, for a galaxy at distance r, is therefore

$$f(L/r) = \begin{cases} \phi(L)/\Phi(L_m), & \text{if } L \geq L_m; \\ 0, & \text{otherwise,} \end{cases} \tag{3}$$

where $L_m(r) = 4\pi r^2 \nu_{60} S_m$ is the minimum luminosity that can be seen at distance r above the flux limit S_m. (The frequency $\nu_{60} = 5\times10^{12}$ Hz is used to convert flux into "luminosity".)

This probability function is density-independent, i.e., it is not a function of the normalization constant C in eq. (1), which cancels in the ratio $\phi(L)/\Phi(L_m)$. It also has a unit integral, and is therefore the correct probability function to use in a maximum-likelihood fit (Sandage *et al.* 1979), which is equivalent to minimizing

$$\Lambda = -2\sum_i \ln f(L_i/r_i), \tag{4}$$

with respect to the three parameters which define the shape of the luminosity function: α, β, and L_*. The minimum value of Λ is arbitrary, but the deviations from the minimum follow the usual χ^2 rules, and provide error estimates for the parameters.

Our best fitted values are listed in Table 1. Fits are presented both for "redshift space", in which distances are not corrected for peculiar velocities, and for the case in which a self-consistent correction to distance was applied (see Sec. 5). Virgo galaxies—defined to be all galaxies within 10° of M87 with velocities smaller than 2500 km s^{-1}—were not included in any of the fits to the luminosity function. Neither were galaxies with distances smaller than $5h^{-1}$ Mpc or larger than $100h^{-1}$ Mpc.

It is useful to generalize the conditional luminosity function of a single galaxy to that of for a *set* of galaxies with distances $r_1, \ldots, r_N$:

$$f(L/r_1, \ldots, r_N) = \sum_{i=1}^{N} f(L/r_i), \tag{5}$$

where the distribution function is here normalized to an integral of N. Any combination of galaxies can be studied using this probability function, as long as the galaxies are not selected by luminosity. It is possible to test for the universality of the luminosity function by considering galaxies at different distances, densities, morphologies, or even colors (provided the colors are available without limits on luminosity). For example, Fig. 3 is a check of the dependence of the derived luminosity function on distance. The top panel shows the entire sample, while the lower ones break up the counts into distance bins. The fits of eq. (5) to the data (dashed lines), with the χ^2 values given in the figure, show that there is no Malmquist-like bias with distance. Note that the same fit is used in all the panels.

TABLE 1

FITTED PARAMETERS OF THE LUMINOSITY FUNCTION

	α	β	L_* $[h^{-2} L_\odot]$	C $[h^3 \text{ Mpc}^{-3}]$
Redshifts Only	0.60 ± 0.08	1.83 ± 0.15	$(3.5 \pm 1.0) \times 10^9$	5.96×10^{-3}
Corrected Distances	0.55 ± 0.08	1.92 ± 0.16	$(3.6 \pm 0.9) \times 10^9$	5.59×10^{-3}

4. Density

The calculation of local density follows the method of weights developed by Yahil *et al.* (1980) and Davis and Huchra (1982). For a flux-limited sample of galaxies with a universal luminosity function, the conversion from number counts to density is obtained by assigning each galaxy a distance-dependent weight

$$w_i = C/\Phi(L_m). \tag{6}$$

Beyond a certain distance, the density derived in this manner becomes subject to large statistical noise, and is no longer useful. A cutoff distance of $100h^{-1}$ Mpc was, therefore, imposed. The normalization constant C in eq. (6) is determined by requiring the sum of the weights to equal the surveyed volume. The mean weighted density per unit volume is then unity by construction.

In order to calculate the gravitational field, it is necessary to make some assumption about the distribution of galaxies in the masked 24% of the sky which was not surveyed. This is most easily accomplished by adding to the

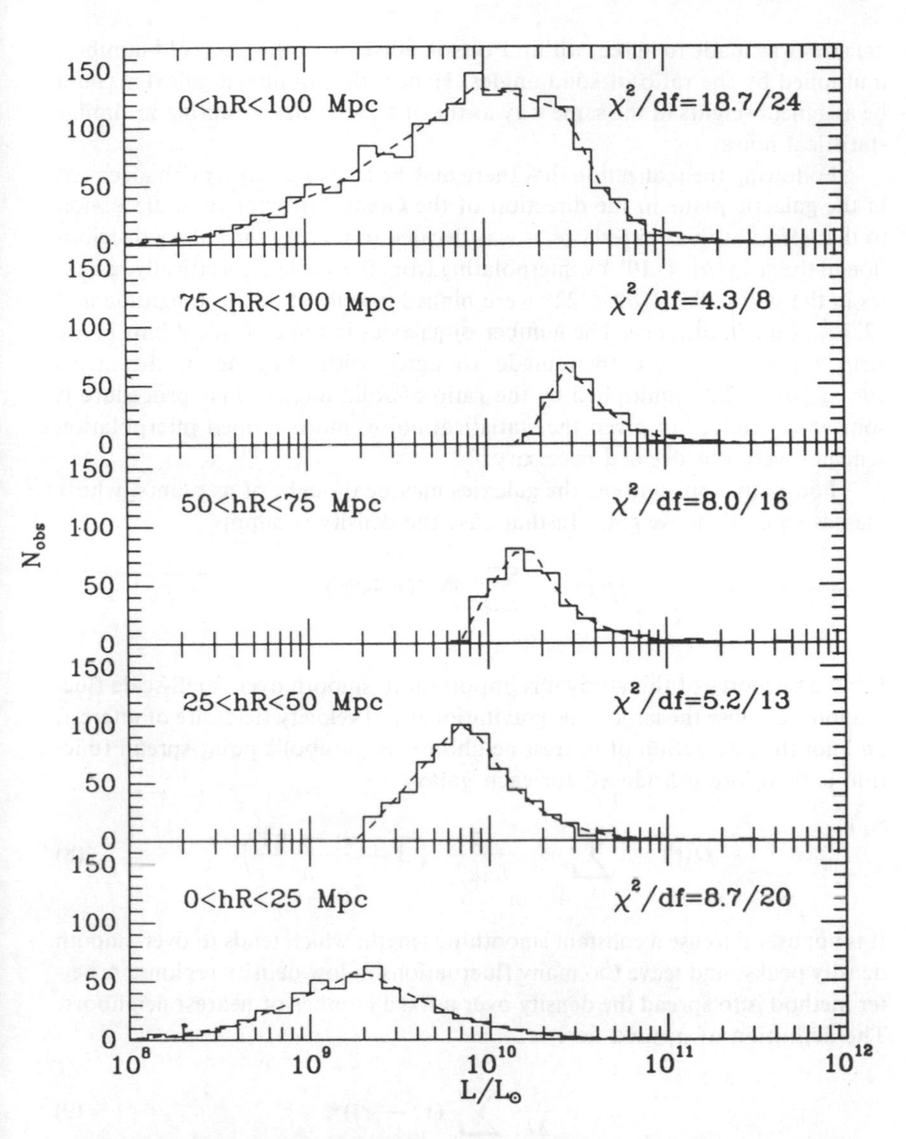

FIG.3. Distributions of infrared luminosity at different distances, compared with the expected function. All the data are fit to the same luminosity function given in Table 1. The quality of the fit is indicated by the χ^2 (measured for bins containing 5 or more galaxies). There is no Malmquist-like bias with distance.

observational data a simulated set of galaxies in the masked region. Initially, a spatially homogeneous distribution was assumed, i.e., random positions, and distances drawn from the distribution shown in Fig. 2. The number of simulated galaxies was made to agree, within Poisson noise, with the observed number, multiplied by the ratio of solid angles. Hence, the simulated galaxies could be assigned weights in the same way as the observed ones, resulting in similar statistical noise.

Following the realization that there may be a strong density enhancement in the galactic plane in the direction of the Great Attractor, and discussion to that effect at this conference, it was decided to modify the galaxy distribution in the strip $|b| < 10°$ by interpolating from both sides. Specifically, galaxies in the strips $10° < |b| < 22°$ were binned in cells of 30° in longitude and $12.5h^{-1}$ Mpc in distance. The number of galaxies in the equivalent bins in the strip $|b| < 10°$ were then made to agree with the one in the strips $10° < |b| < 22°$, multiplied by the ratio of solid angles. This procedure is somewhat crude, but given the statistical noise, more refined interpolation schemes were not deemed necessary.

For many applications, the galaxies may be thought of as points, whose masses equal their weights. In that case the density is simply

$$D(\mathbf{r}) = \sum_i w_i \delta^3(\mathbf{r} - \mathbf{r}_i). \qquad (7)$$

For the purposes of this study it is important to smooth over small-scale fluctuations, because the large-scale gravitational and velocity fields are of interest, and not the interaction of nearest neighbors. A parabolic point-spread function is therefore introduced for each galaxy

$$D(\mathbf{r}) = \sum_{\mathrm{i}} w_i \frac{15}{8\pi a_i^3} \left(1 - \frac{(\mathbf{r} - \mathbf{r}_i)^2}{a_i^2}\right). \qquad (8)$$

It is not useful to use a constant smoothing length, which tends to over-smooth density peaks, and leave too many fluctuations in low density regions. A better method is to spread the density over a fixed number of nearest neighbors. The definition of a_i used here is

$$a^2{}_i = \frac{5}{3k} \sum_{j=1}^{k} (\mathbf{r}_i - \mathbf{r}_j)^2, \qquad (9)$$

with k = 10. (For Virgo galaxies, which were all placed at the center of the cluster, k was set to the number of Virgo galaxies plus 10). This procedure treats all densities in the same way, and should therefore introduce a minimal bias in density.

An underlying assumption of the entire procedure is the existence of a universal luminosity function. The density calculation would need to be completely re-evaluated if the luminosity function proved, for example, to b density-dependent. It is possible to test for a density bias as for a distance bias Fig. 4 shows a breakup of the observed counts into density bins with the same

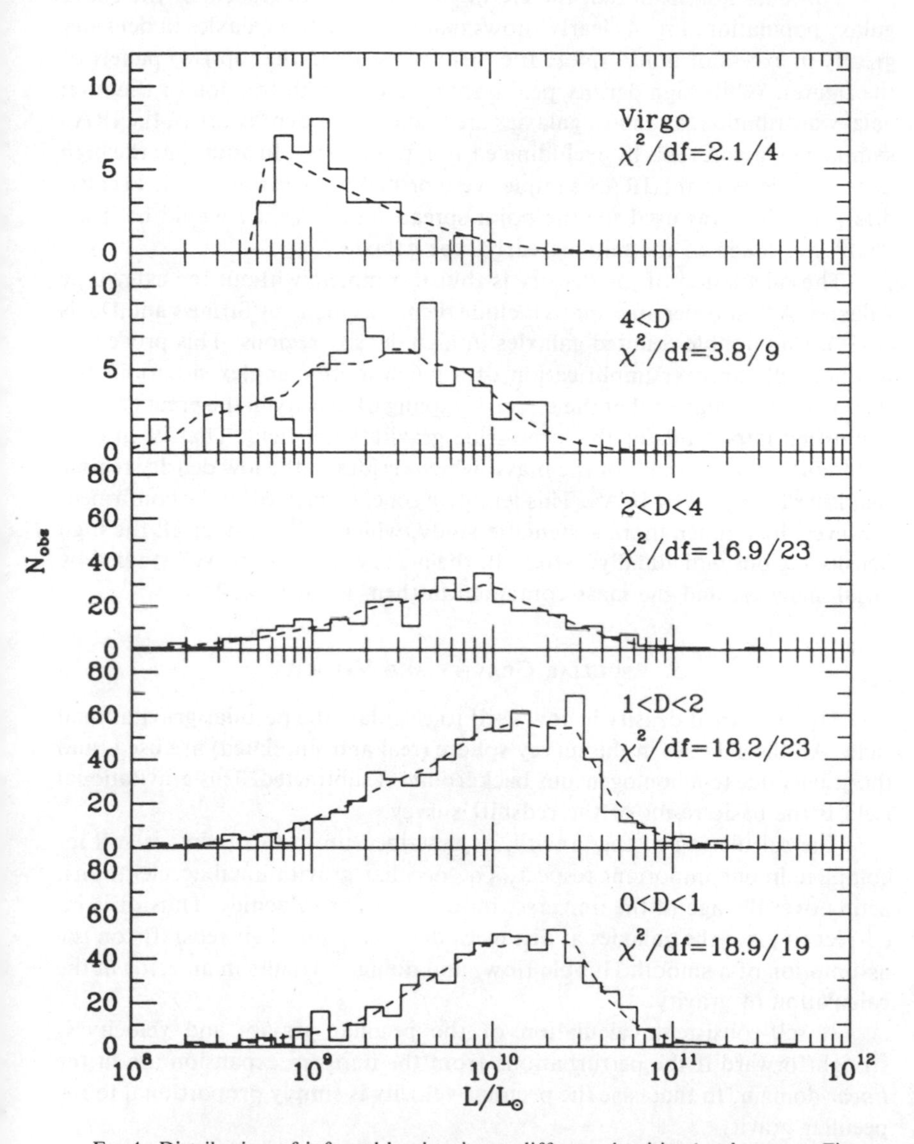

FIG.4. Distribution of infrared luminosity at different densities (analogous to Fig. 3). The absence of a density-dependent bias supports the assumption of a universal *infrared* luminosity function, but there remains the problem of undersampling of early-type galaxies in rich clusters of galaxies.

fit shown in Fig. 3. Again, the χ^2 show no significant deviations. Thus, the assumption of a universal *infrared* luminosity function appears to be consistent with the data.

This does not mean that the IRAS galaxies are fair tracers of the entire galaxy population. Fig. 4 clearly shows that there are few galaxies in densities greatly in excess of unity. (Note the different scale in the top two panels of the figure). While high density peaks are indeed a small fraction of the total galaxy distribution, and most galaxies are found in low density areas, the IRAS sample exaggerates this by excluding early-type galaxies. In addition, the high density regions in the IRAS sample were probably over-smoothed, because a larger radius was used for the point-spread function than would be if account was taken of the missing early-type galaxies.

The calculation of the density is thus incomplete without the early-type galaxies. A first order attempt to include them was made by Strauss and Davis (1988), who double counted galaxies in high density regions. This procedure is admittedly an oversimplification of a much more complex situation, but their results do suggest that the under-sampling of density in the great clusters is not that important for the large-scale gravitational field. That is, gravity is dominated by the bulk of the mass, which resides in the low density regime that is well sampled by IRAS. This tentative conclusion needs to be confirmed, however, by a much more systematic study, which will consider all the high density regions individually. After all, the nearby clusters are well studied by virial analyses, and the mass contained in them is fairly well known.

5. Peculiar Gravity and Velocity

The smoothed density is now used to calculate the peculiar gravitational field. All the galaxies in the survey sphere (real and simulated) are used, and the gravity due to a homogeneous background is subtracted. This gravitational field is the basic result of the redshift survey.

The calculation of the gravity, as described up to this point, is still incomplete in one important respect. The peculiar gravitational accelerations, acting over the age of the universe, induce peculiar velocities. Thus, it is incorrect to place the galaxies at distances deduced from their redshifts on the assumption of a smooth Hubble flow, and doing so results in an error in the calculation of gravity.

A self-consistent calculation of the peculiar gravity and velocity is straightforward if the perturbations from the uniform expansion are in the *linear* domain. In that case the peculiar velocity is simply proportional to the peculiar gravity

$$\mathbf{u}_i = \frac{2}{3} H_0^{-1} \Omega_0^{-0.4} \mathbf{g}_i(\mathbf{r}_i). \tag{10}$$

This can be combined with the usual relation between distance and the observe radial velocity in the Local Group frame (Yahil *et al.* 1977)

$$v_i = H_0 r_i + (\mathbf{u}_i - \mathbf{u}_{LG}) \cdot \hat{\mathbf{r}}_i, \tag{11}$$

to yield a closed implicit set of equations for the true distances r_i. These equations have only one free parameter, the cosmological density parameter, Ω_0. They are independent of the Hubble constant, except as a scaling factor to convert km s^{-1} to Mpc.

Given the observed velocities v_i, and a value of Ω_0, eqs. (10) and (11) can therefore be solved for the distances r_i of the galaxies. This was done iteratively, starting by assuming the distances to be Hubble ones, and using eq. (10) to obtain a first estimate of the peculiar velocities. These were then used to obtain new distances from eq. (11). (The angular coordinates of the galaxies on the sky were, of course, unchanged). The process was repeated, with an updated luminosity function and density distribution determining a corrected **g** at each iteration. Convergence to an accuracy $\sim$ 20 km s^{-1} was obtained after a few iterations. Numerical oscillations were damped by setting the new distance in each iteration to the average of the previous one, and the one suggested by the current iteration.

The choice of Ω_0 needs some clarification. The models presented at the conference were for $\Omega_0 = 1$. This value was chosen because the *magnitude* of the derived peculiar velocity of the Local Group was then of the order of its velocity with respect to the CMB, 610 km s^{-1} (Lubin and Villela 1986). With the new interpolation scheme for the galactic plane strip, this was no longer so. The procedure was, therefore, modified by changing Ω_0 in each iteration, with the resultant rescaling of all peculiar velocities, so that the magnitude of the peculiar velocity of the Local Group became exactly 610 km s^{-1}. The final value of the cosmological density parameter was $\Omega_0 = 0.5$, significantly lower than earlier determinations from infrared galaxies (Yahil *et al.* 1986; Villumsen and Strauss 1987; Lahav, Rowan-Robinson, and Lynden-Bell 1988), and closer to the value deduced from the optical dipole (Lahav 1987; Dressler 1987; Lahav *et al.* 1988).

However, this estimate of Ω_0 should be viewed as very preliminary, because the predicted peculiar velocity of the Local Group is toward $l = 231°$ and $b = 48°$, which is 35° away from the *direction* of the CMB, $l = 272°$ and $b = 30°$ (Lubin and Villela 1986). It cannot be argued that Ω_0 has been properly determined, until this difference in directions is understood.

A major fault could be the assumption of linear perturbations. In the spherical case, the correct peculiar velocity is smaller than the linear approximation by a factor $\simeq \langle D \rangle^{-1/4}$, where $\langle D \rangle$ is the mean density in the sphere interior to the galaxy in question (Yahil 1985). Linear theory therefore *overestimates* the effect of high density regimes on the peculiar velocities. The Virgo supercluster is one region where the linear calculation might thus be in error (although this is offset by the undersampling of early-type galaxies in the central cluster). Lowering the contribution of the Virgo supercluster will bring the peculiar velocity of the Local Group into better agreement with the direction of the CMB.

In fact, N-body calculations show that, in the *nonlinear* regime, peculiar velocities and accelerations are typically misaligned by $\sim 25°$ (Villumsen and Davis 1986), comparable to the discrepancy found here. A careful comparison of the predicted peculiar velocity field with the observed one will therefore probably require a self-consistent nonlinear calculation of the gravitational field. Such a calculation cannot use only the positions of the galaxies at the present epoch, as is done in the linear approximation. Instead, a guess has to be made of their positions at an earlier epoch, say $z = 10$, when linear perturbations might be a better approximation, and integrated nonlinearly to the present epoch, using an N-body code. We are now in the process of implementing such a scheme, but the results presented here still use the linear approximation.

Even in the nonlinear calculation, it will be impossible to determine the original positions of galaxies that are now in virialized clusters, where phase-mixing has completely obliterated any memory of initial conditions. In fact, the suspicion is that any shell crossing will make it very difficult to determine which galaxy came from where. Fortunately, the IRAS galaxies avoid the virialized regions of high density. Fig. 5 shows a breakup of the observed pairwise separation in redshift space, $\Delta\mathbf{v}_{ij}$, into a radial (line-of-sight) component π, and a tangential component σ. (Only one component of σ is shown, by projecting it along a random azimuthal axis.) In such a scatter diagram, virialized velocities are seen as the familiar "finger of God" concentration of pairs for which $\pi \gg \sigma$. No such effect is seen. The same conclusion can be drawn from Fig. 6, which shows a random distribution of the angle α which $\Delta\mathbf{v}_{ij}$ makes with the line-of sight (Turner and Sargent 1977).

Another question of major interest is how far one needs to integrate over density perturbations for the gravitational acceleration to converge. Put differently, where are the perturbations responsible for the gravity? Fig. 7 addresses this question by plotting the *cumulative* peculiar velocities due to concentric shells around the Local Group, both their absolute values and their three Cartesian coordinates in supergalactic coordinates. (See Sec. 6 for the coordinate convention used here.) It is seen that the bulk of the peculiar velocity of the Local Group is generated within $40h^{-1}$ Mpc, although there is a slow growth beyond that distance (Vittorio and Juszkiewicz 1987). The same picture emerges from Fig. 8, which plots the *differential* contributions of the concentric shells to the peculiar velocity of the Local Group. Poisson error bars are shown here as well, so the reader may obtain a sense of the ratio of signal to shot noise in the calculation of gravity.

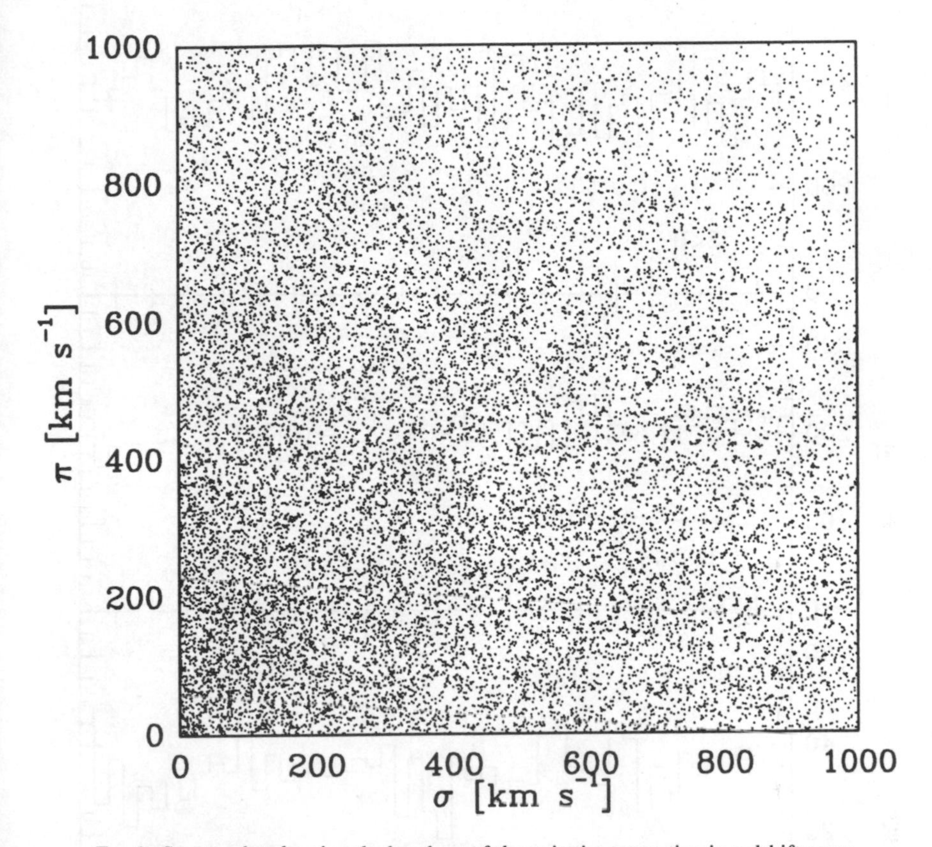

FIG.5. Scatter plot showing the breakup of the pairwise separation in redshift space, $\Delta \mathbf{v}_{ij}$, into a radial (line-of-sight) component π and a tangential component σ. (Only one component of σ is shown, by projecting it along a random azimuthal axis.) The lack of a "finger of God" concentration of pairs, for which $\pi \gg \sigma$, shows that the IRAS galaxies avoid virialized clusters.

If the gravitational acceleration acting on the Local Group is largely due to perturbations within the local supercluster ($r < 40h^{-1}$ Mpc), then it cannot be dominated by the more distant Great Attractor, whose contribution to gravity should not converge until well beyond that distance. The IRAS redshift survey does see the Great Attractor as a large complex in the direction of Hydra-Centaurus, but it is largely counteracted by the Perseus-Pisces complex in the opposite direction. The detailed maps presented in Sec. 6 show that, in fact, the gravitational field bifurcates close to the position of the Local Group.

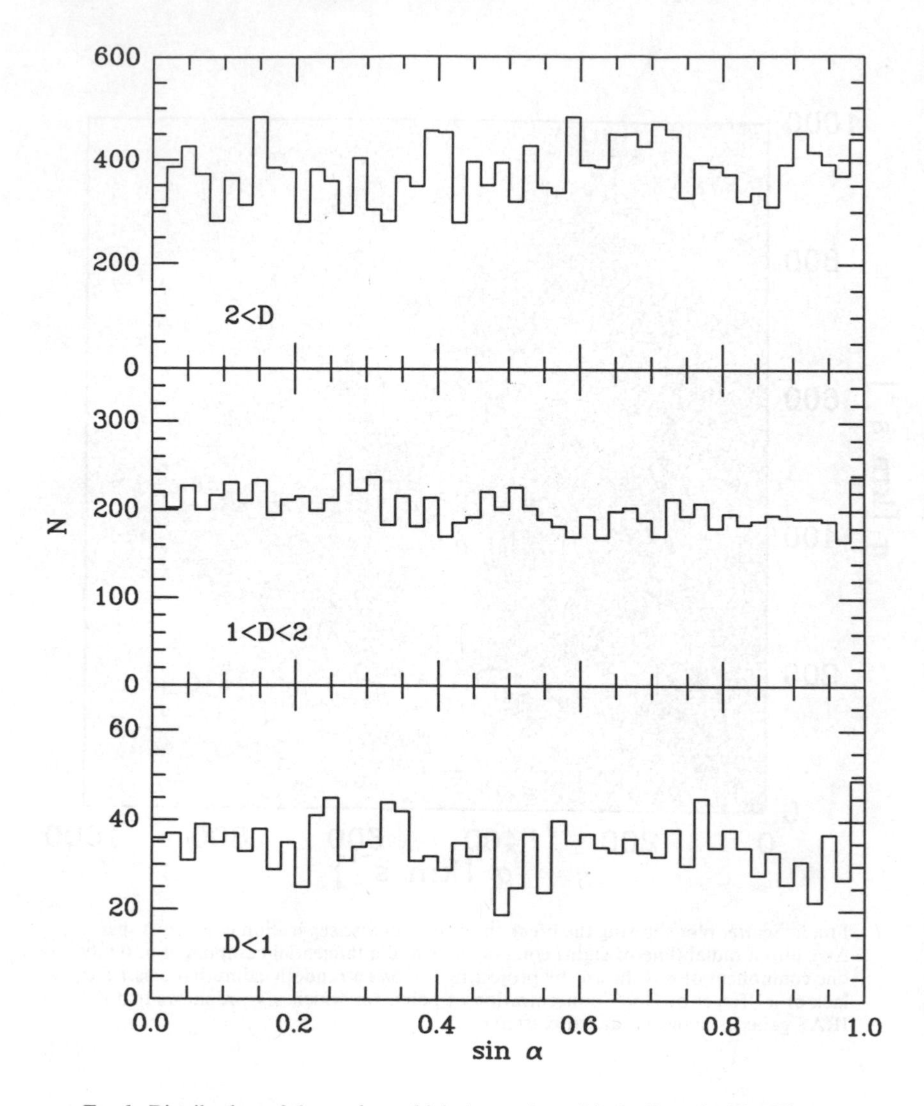

FIG.6. Distribution of the angle α which $\Delta\mathbf{v}_{vj}$ makes with the line-of- sight. The randomness of this distribution, and the lack of a concentration around $\alpha = 90°$, is another demonstration of the paucity of virialized IRAS galaxies.

6. MAPS

A major difficulty in the study of any 3-d vector field is visualization. In an effort to overcome the problem, this section present a series of maps of the density structure and gravitational field from different vantage points.

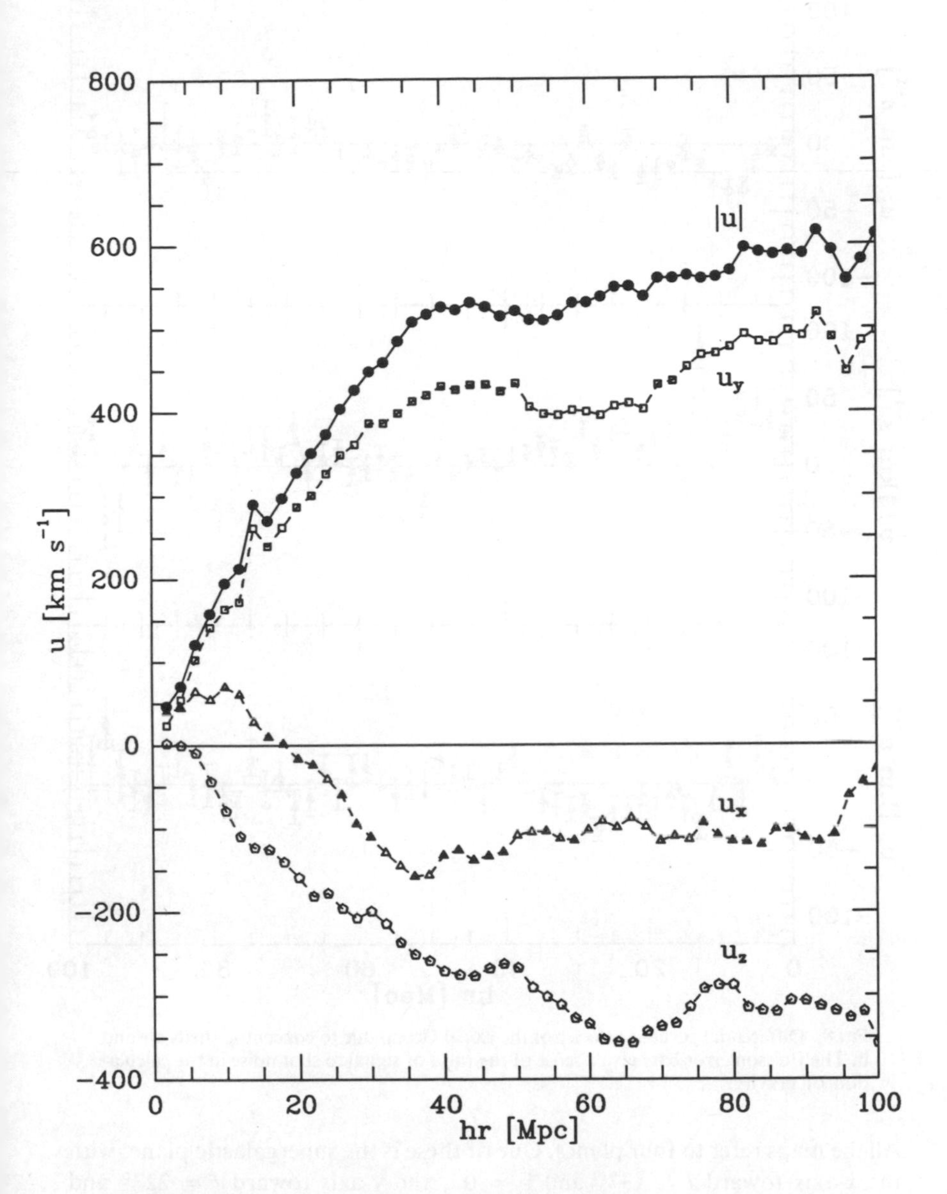

FIG.7. Cumulative peculiar velocity acting on the Local Group due to concentric shells around it. The cartesian components are in supergalactic coordinates. The bulk of the peculiar velocity of the Local Group is generated within $40h^{-1}$ Mpc. It is therefore *not* dominated by the Great Attractor, whose center is at $42h^{-1}$ Mpc, and whose contribution should not converge until well beyond that distance.

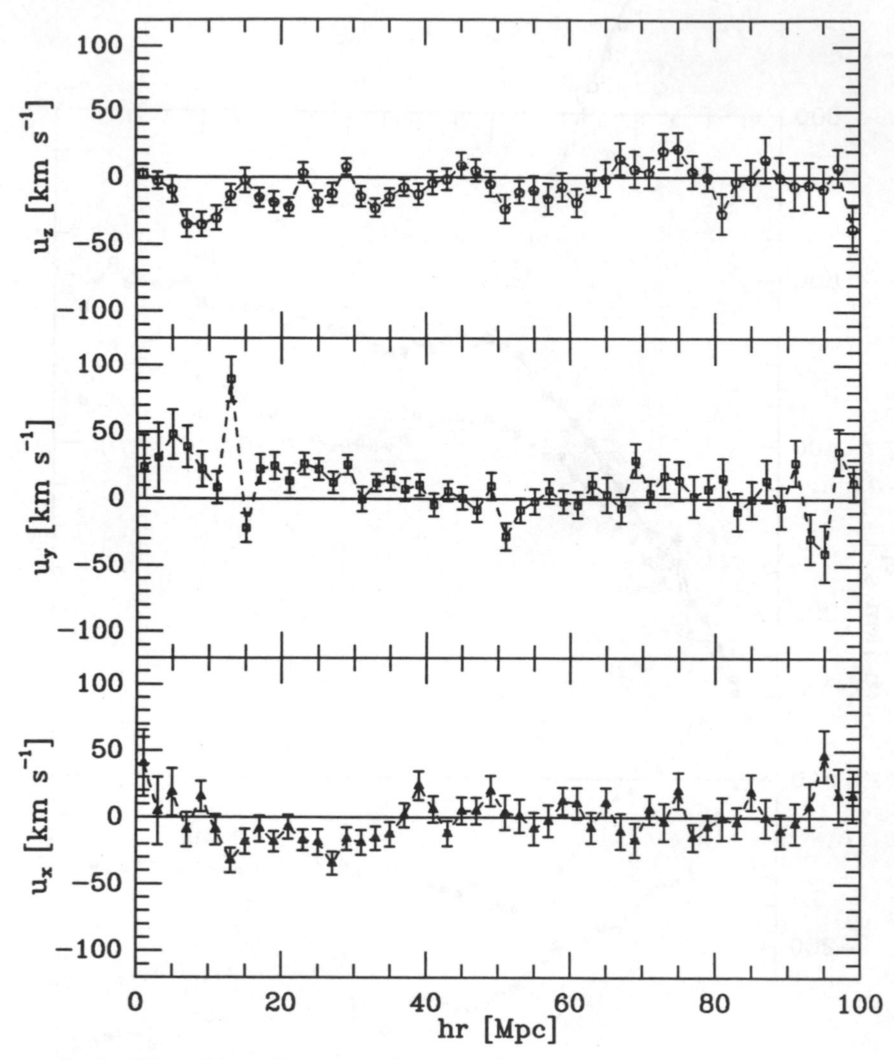

FIG.8. Differential peculiar velocity of the Local Group due to concentric shells around it. The Poisson error bars give a sense of the ratio of signal to shot noise in the calculation of gravity.

All the maps refer to four planes. One of these is the supergalactic plane, with the x-axis toward $l = 137°$ and b = 0°, the y-axis toward $l = 227°$ and $b = 84°$, and the z-axis, normal to the plane, toward $l = 47°$ and $b = 6°$. The Local Group is at the origin of the plot. The planes of the other plots are obtained by rotating by 45° around the x-axis, which is thus identical in all plots. The galactic coordinates of the z-axis are marked on each plot.

Fig. 9 shows the *smoothed* density structure in the four planes. The contours are spaced logarithmically at intervals of 2 decibels. Contours for which $D \geq 1$ are shown as solid lines, and the ones for which $D < 1$ as dashed lines. The contour plots are of thin slices, because the point-spread functions of the galaxies allow the calculation of density at any point in space. In reality, however, the density represents some mean density above and below the plane with scale height depending on location, as explained in Sec. 4. Careful inspection of the figures shows all the familiar nearby clusters of galaxies. Most of the galaxies, however, are in large-scale agglomerates with relatively low density, and these dominate gravity. One such large concentration is in the direction of the Great Attractor, along the (common) negative x-axes of all the plots. There is, however, also a significant over-density in the opposite direction, toward Perseus-Pisces. Of additional note is a big void beyond Fornax, in a direction roughly opposite to that of Virgo.

The gravitational field, calculated in Sec. 5, is presented in Fig. 10. All galaxies (both real and simulated) that are within $\pm 22.5°$ of the plane of the plot are shown as vectors, whose lengths are proportional to the components of their peculiar gravities in the plane. While the overall scale of the vectors is arbitrary, their relative lengths and directions correspond to the gravitational acceleration. Fig. 10 clearly shows that the two large-scale centers of attraction are indeed the Great Attractor and Perseus-Pisces. The gravitational field shows a distinct bifurcation between these two large complexes, not far from the position of the Local Group. In addition, there is strong gravity toward Virgo, but this local field looks far from spherical symmetry.

The major conclusion from Figs. 9 and 10 is that the local gravitational field is fairly complex, and is not adequately described by parametric models of the sort used to date. First, the Virgocentric field is not spherical. (Incidentally, the peculiar velocity of the Local Group relative to Virgo in this calculation was 410 km s^{-1}, of which the component in the direction of Virgo was 330 km s^{-1}). Secondly, the quadrupolar approximation of the shear field (Lilje, Yahil, and Jones 1986) is valid over a limited range in distance. Finally, the Great Attractor is not the only large-scale source of pull; Perseus-Pisces and the void beyond Fornax also play their role. In fact, this was already anticipated by Lilje *et al.* (1986), who showed that the eigenvalues of the quadrupole shear tensor were not in the ratio $-1 : -1 : +2$, that would be expected for a single Great Attractor.

The inadequacy of the parametric models is understandable in view of the limitations of the data from which they were derived. Previous determinations were limited in sky coverage, and in the case of the RSA catalogue also in depth. Purely kinematic models (Lynden-Bell *et al.* 1988, Faber and Burstein 1988), on the other hand, are based on the inherently incomplete information contained in the one observable (radial) component of peculiar velocities. They may also suffer from some incompletion in sky or depth

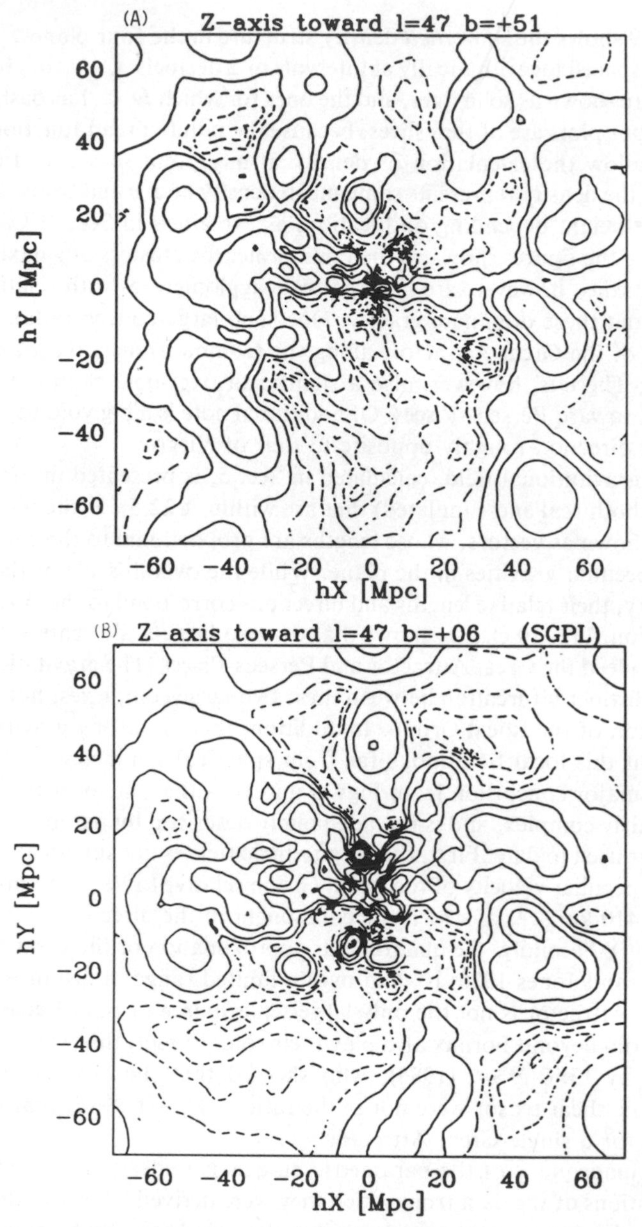

FIG.9. Smoothed density structure in four planes. The contours are spaced logarithmically at intervals of 2 decibels. Contours for which $D \geq 1$ are shown as solid lines, and ones for which $D < 1$ as dashed lines. The orientation of the planes is such that the x-axis always points in the supergalactic x direction, $l = 137°$ and $b = 0°$. The direction of the z—axis, normal to the plane, is shown at the top of each frame.

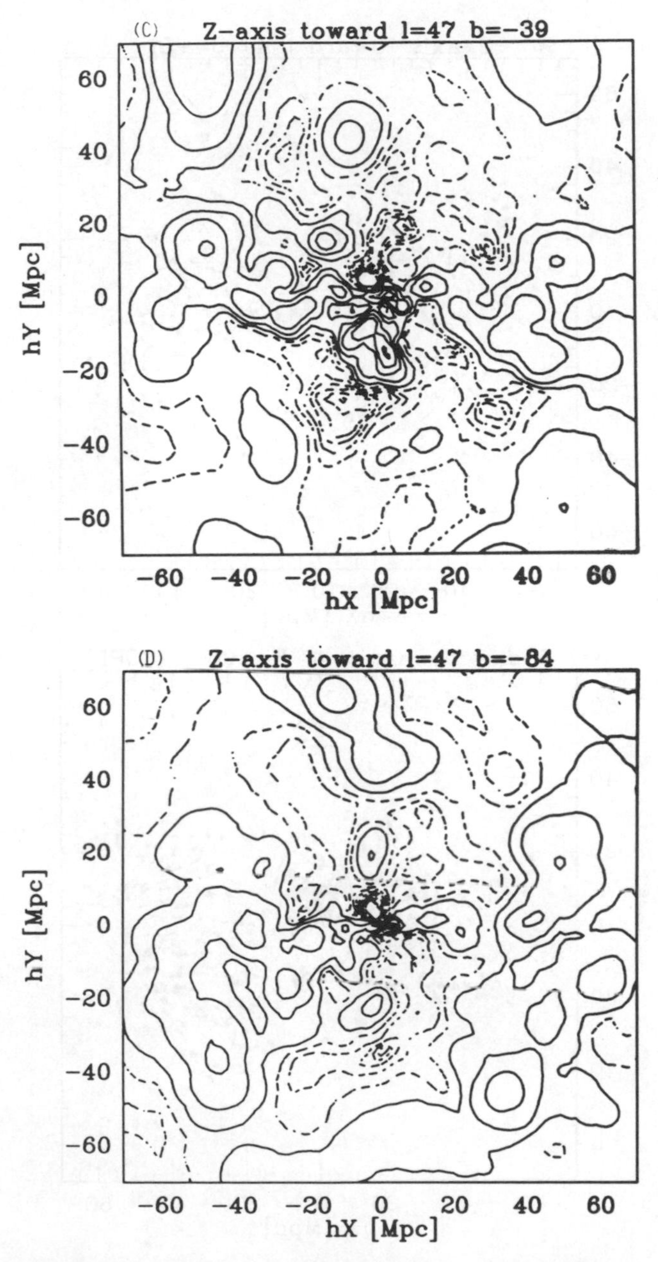

Frame (b) is the supergalactic plane (SGPL). The largest mass agglomerates are seen to be two gigantic perturbations of relatively small over-density. The Great Attractor is along the negative x-axis, and the Perseus-Pisces complex is along the positive x-axis. There is also a void beyond Fornax, roughly opposite the direction of Virgo (the highest density peak).

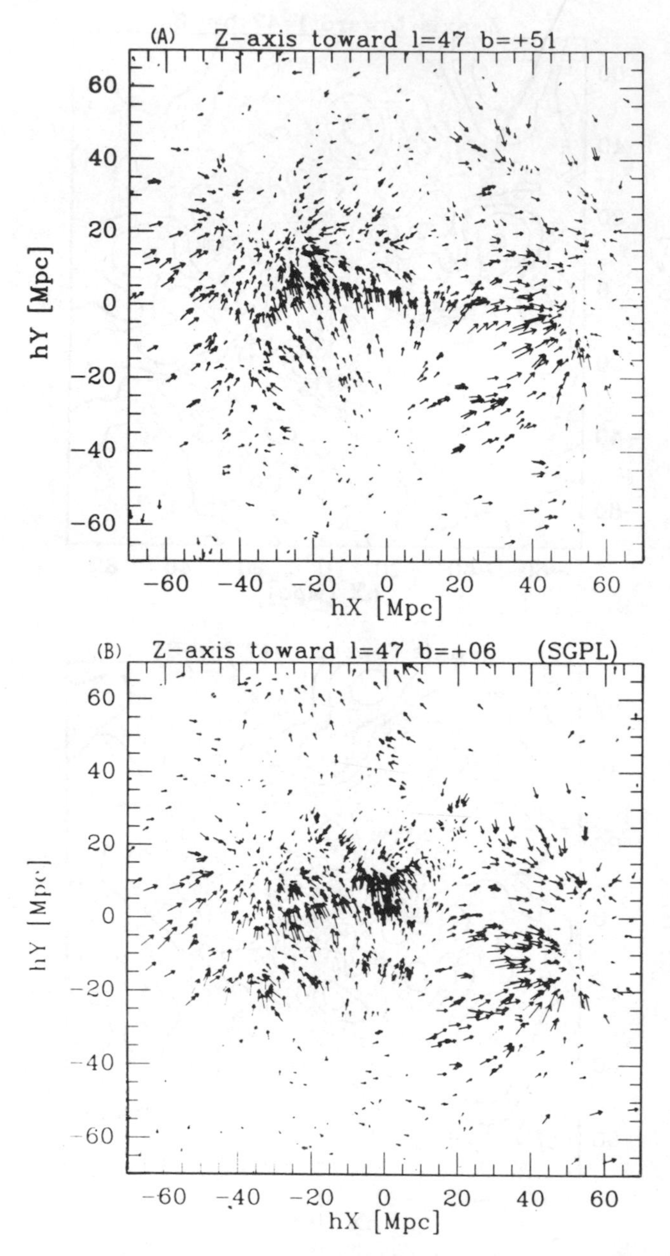

FIG. 10. The gravitational field predicted by the IRAS galaxy distribution. All galaxies (real and simulated) that are within ± 22.5° of the planes of Fig. 9 are shown as vectors, whose lengths are proportional to the components of their peculiar gravities in the plane. (The positions of the galaxies are at the tails of the vectors.) The overall scale of the vectors is arbitrary, but their relative lengths and directions correspond

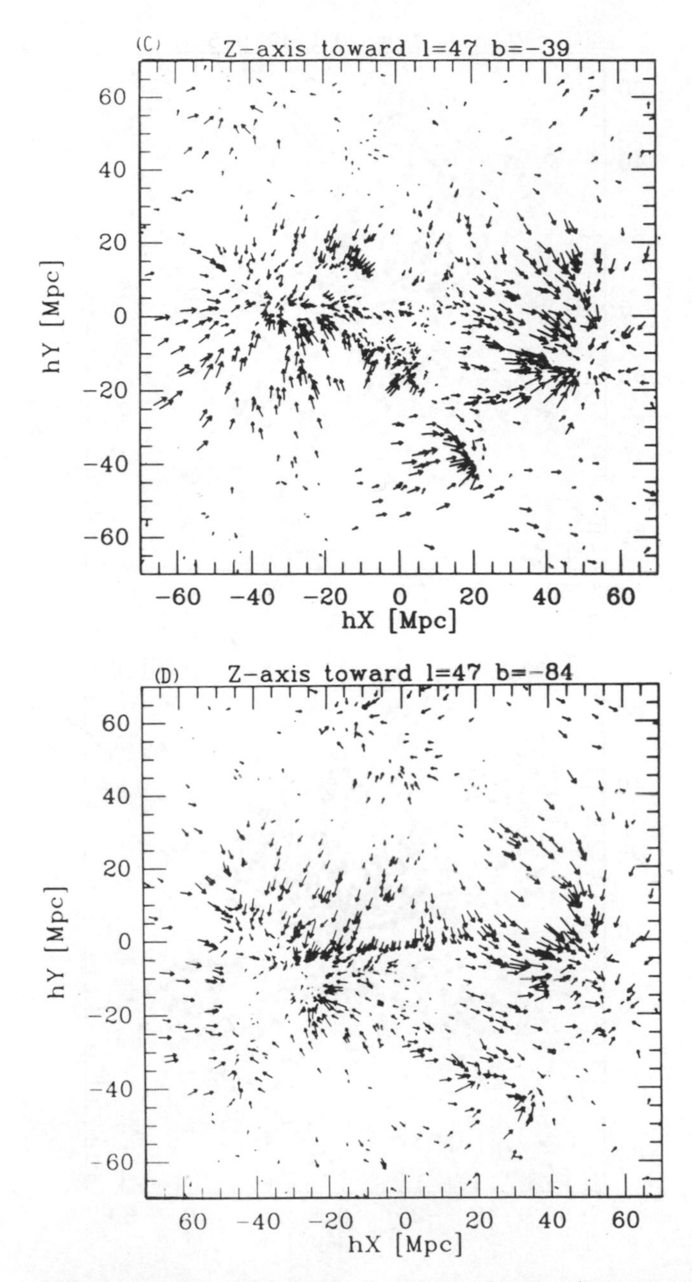

to the gravitational acceleration. The dominance of the two large mass concentrations is clearly seen, resulting in a bifurcation of the gravitational field close to the Local Group. The Virgo supercluster also exerts a significant pull, but its gravity is far from the spherical model frequently adopted in the past.

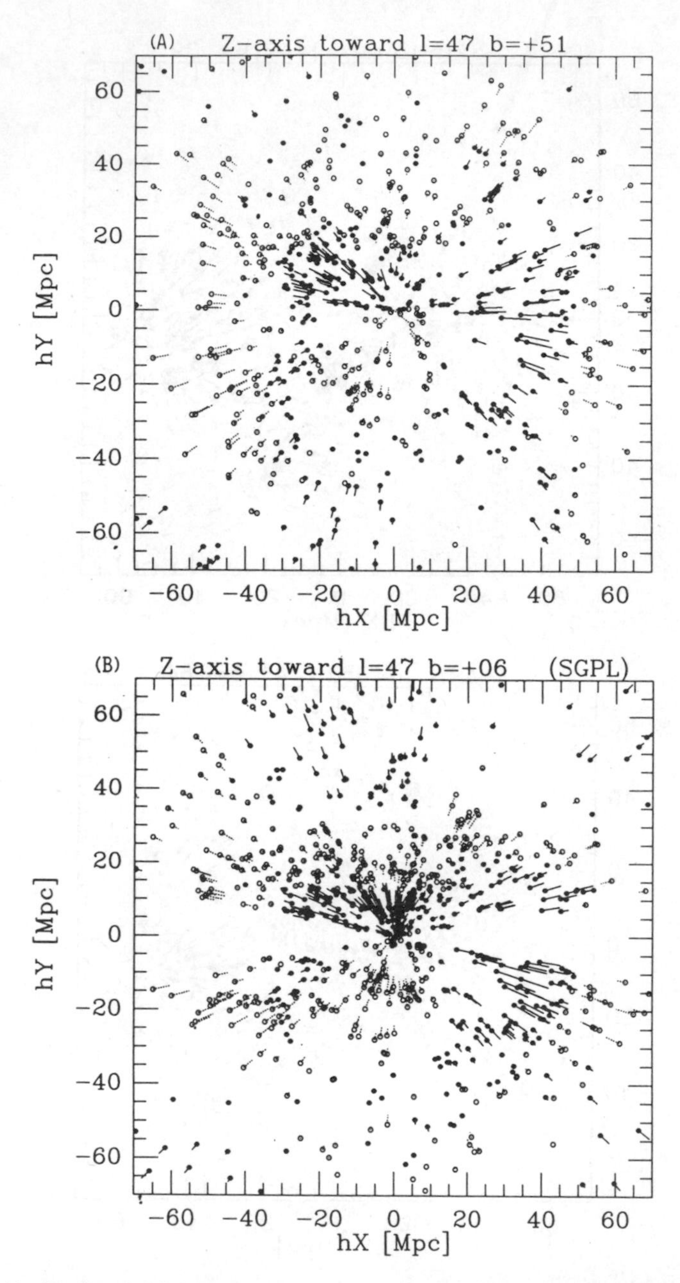

FIG. 11. Predicted radial components of the peculiar velocities of the observed IRAS galaxies, which can be compared directly with the observations presented in Figs. 6-8 of FB in this volume. Galaxies with receding peculiar velocities (in the frame of the CMB) are marked by filled circles and solid lines, and the ones with approaching

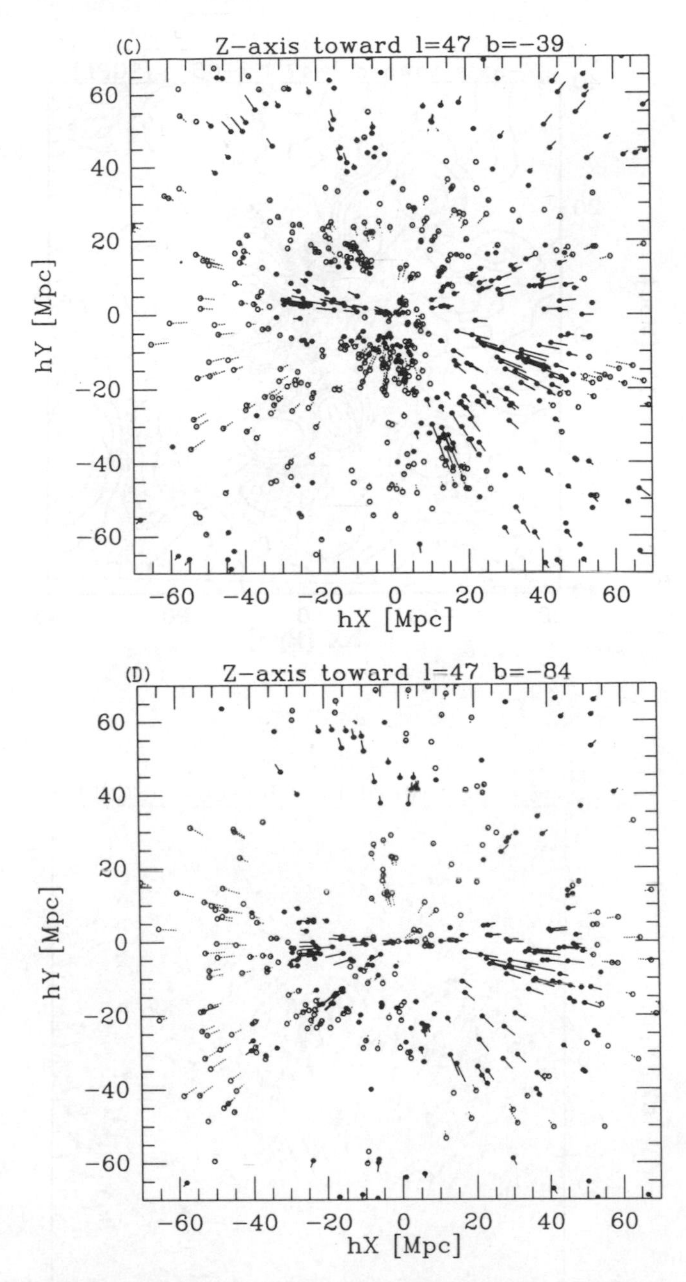

peculiar velocities are denoted by open circles and dashed lines. The location of each galaxy, after correction for peculiar velocity, is at the center of the circle, and the length of the line is the magnitude of the radial component of the peculiar velocity.

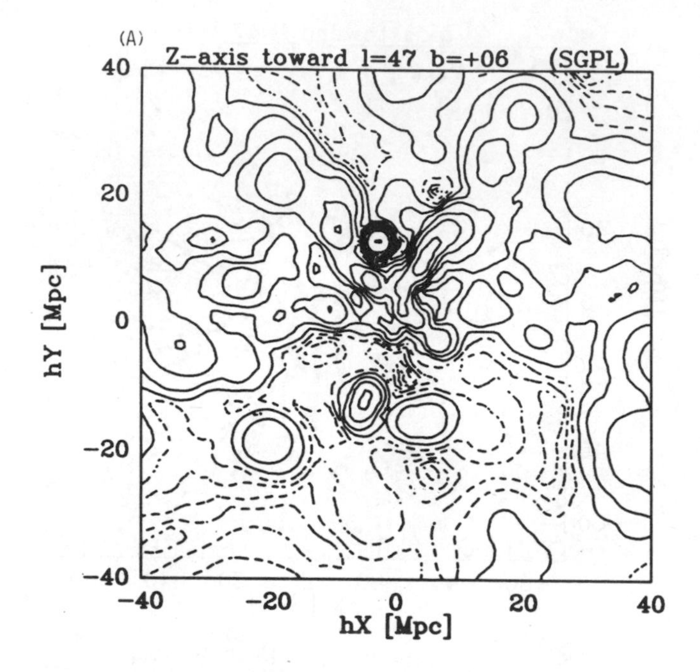
(A)
Z-axis toward l=47 b=+06 (SGPL)
hY [Mpc]
hX [Mpc]
40
20
0
-20
-40
-40
-20
0
20
40

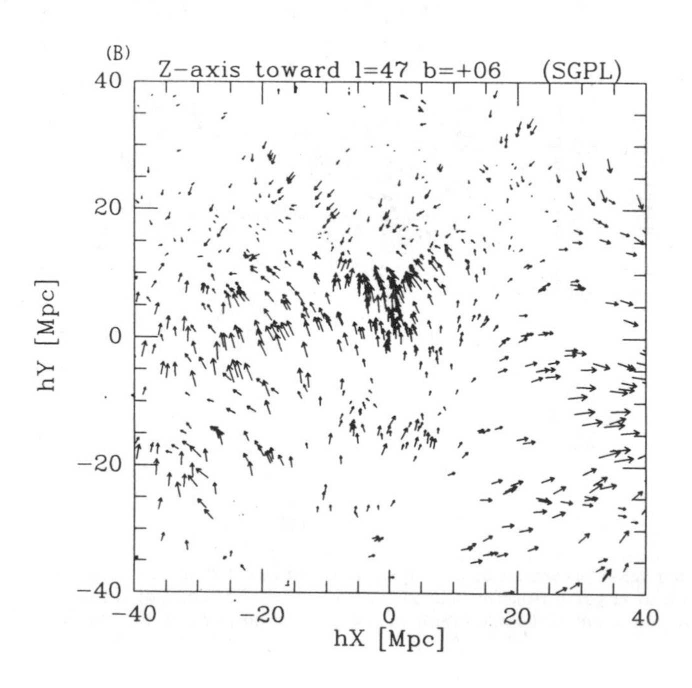
(B)
Z-axis toward l=47 b=+06 (SGPL)
hY [Mpc]
hX [Mpc]
40
20
0
-20
-40
-40
-20
0
20
40

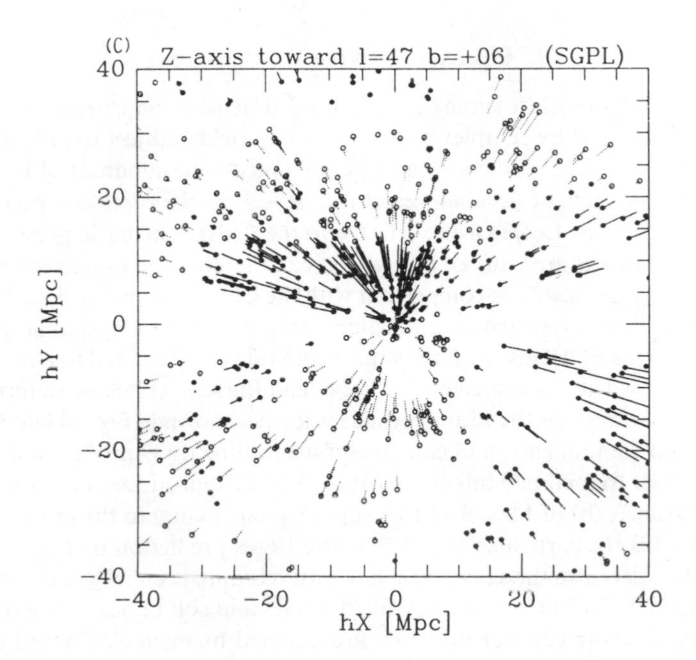

FIG.12. Blow-up of frames (b) of Figs. 9-11 (the supergalactic plane). The scale of Figs. 12a, b, c is close to that of Fig. 8 of FB. Comparison of Fig. 12c with Fig. 8 of FB shows good overall agreement, demonstrating the gravitational origin of the peculiar velocities. There is lack of agreement in the first quadrant, in the region of Ursa Major.

coverage; paucity of data in the direction of Perseus-Pisces, and at larger distances in the direction of the Great Attractor are particular problems.

The time has therefore perhaps come to replace the parametric models by detailed 3-d tables of density, such as provided by the IRAS redshift survey, and future surveys of equivalent sky coverage and depth. These surveys make detailed predictions of the peculiar velocity field, which can then be confronted with observations. An initial attempt in this direction is made in the next section.

7. Comparison with Observations

The IRAS redshift survey provides a 3-d density map, from which one can *predict* the entire complex peculiar velocity field, subject to only one free parameter, Ω_0. This prediction can be confronted with hundreds of measured radial components of peculiar velocities, as well as the three components of the velocity of the Local Group relative to the CMB. The basic premise, that peculiar velocities are induced by gravity, can therefore be subjected to a very over-constrained test. The comparison with the CMB was already made in Sec. 6. This section is devoted to a detailed comparison of the peculiar velocity field predicted by IRAS, with the superb summary and re-analysis of existing data, presented at this conference by Faber and Burstein (1988, henceforth FB).

Fig. 11 translates the vector peculiar velocities shown in Fig. 10 into observable radial components, and can be compared directly with Figs. 6 and 8 of FB. In order to facilitate this comparison, Fig. 12 reproduces a blown up version of frames (b) of Figs. 9-11 (the supergalactic plane) to the exact scale of Fig. 8 of FB. In particular, Fig. 12c is the IRAS prediction of Fig. 8 of FB. The notation is also the same. Galaxies with receding peculiar velocities in the frame of the CMB are marked by filled circles and solid lines, while the ones with approaching peculiar velocities are denoted by open circles and dashed lines. The location of each galaxy, after correction for peculiar velocity, is at the center of the circle, and the length of the line is the magnitude of the radial component of the peculiar velocity. Further comparisons are provided in Figs. 13-15, which are the analogues of Figs. 9-11 of FB.

A careful examination of the predicted versus observed peculiar velocities shows good overall agreement. This confirms both that the IRAS galaxies trace the mass, and that the peculiar velocities are induced by gravity. There is just too much correlation between peculiar gravity and velocity for alternative mechanisms to dominate.

There is, however, some disagreement in detail. The most striking difference is in the first quadrant of the supergalactic plane, where the gravity maps predict peculiar velocities moving away from the Local Group, while the data call for velocities toward it. (Recall that all velocities are in the frame of the CMB.) This difference, together with the (possibly related) misalign-

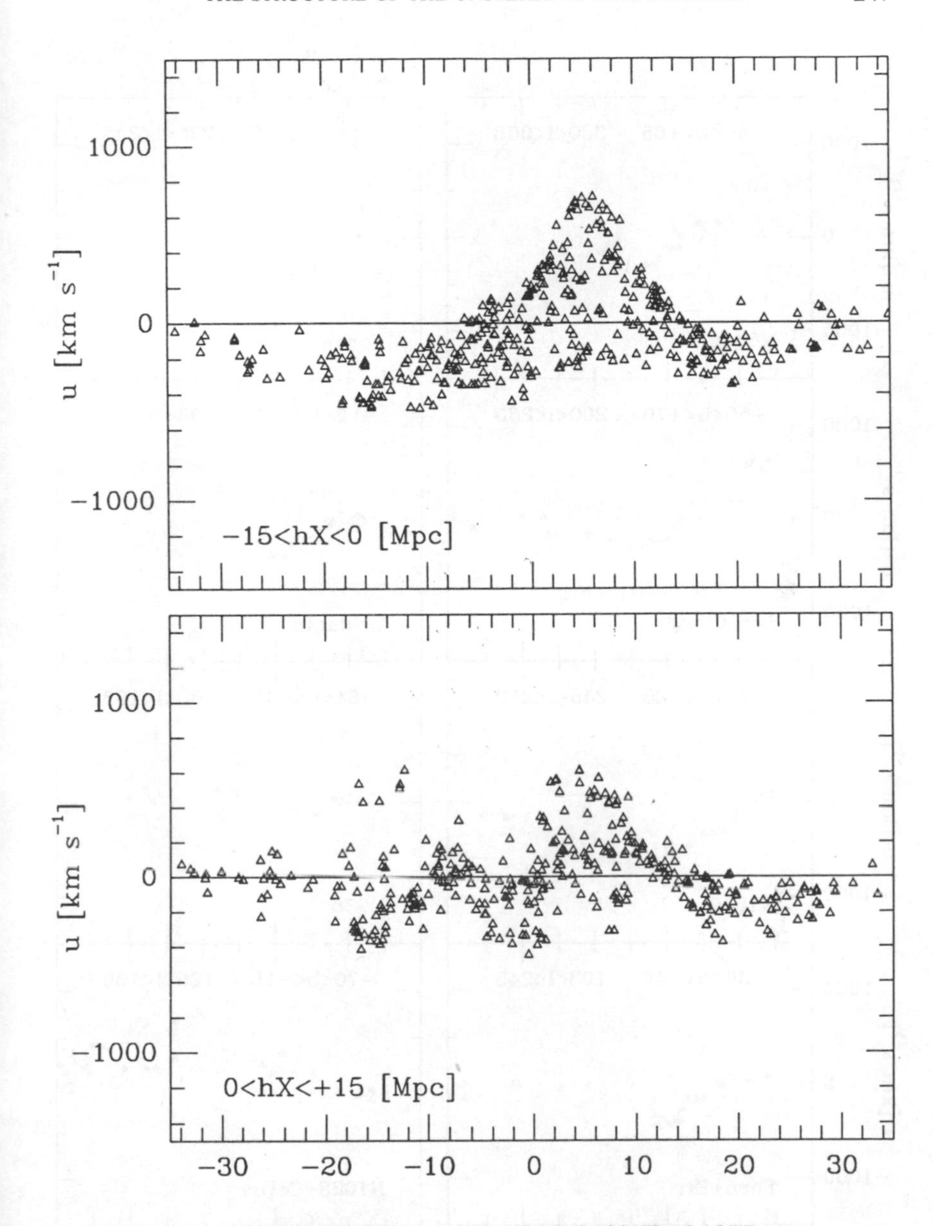

FIG.13. IRAS predictions to be compared with Fig. 9 of FB.

ment of the peculiar gravity acting on the Local Group and its velocity with respect to the CMB, are the major difficulties left to overcome. Four possible corrections to the calculation of gravity come to mind:

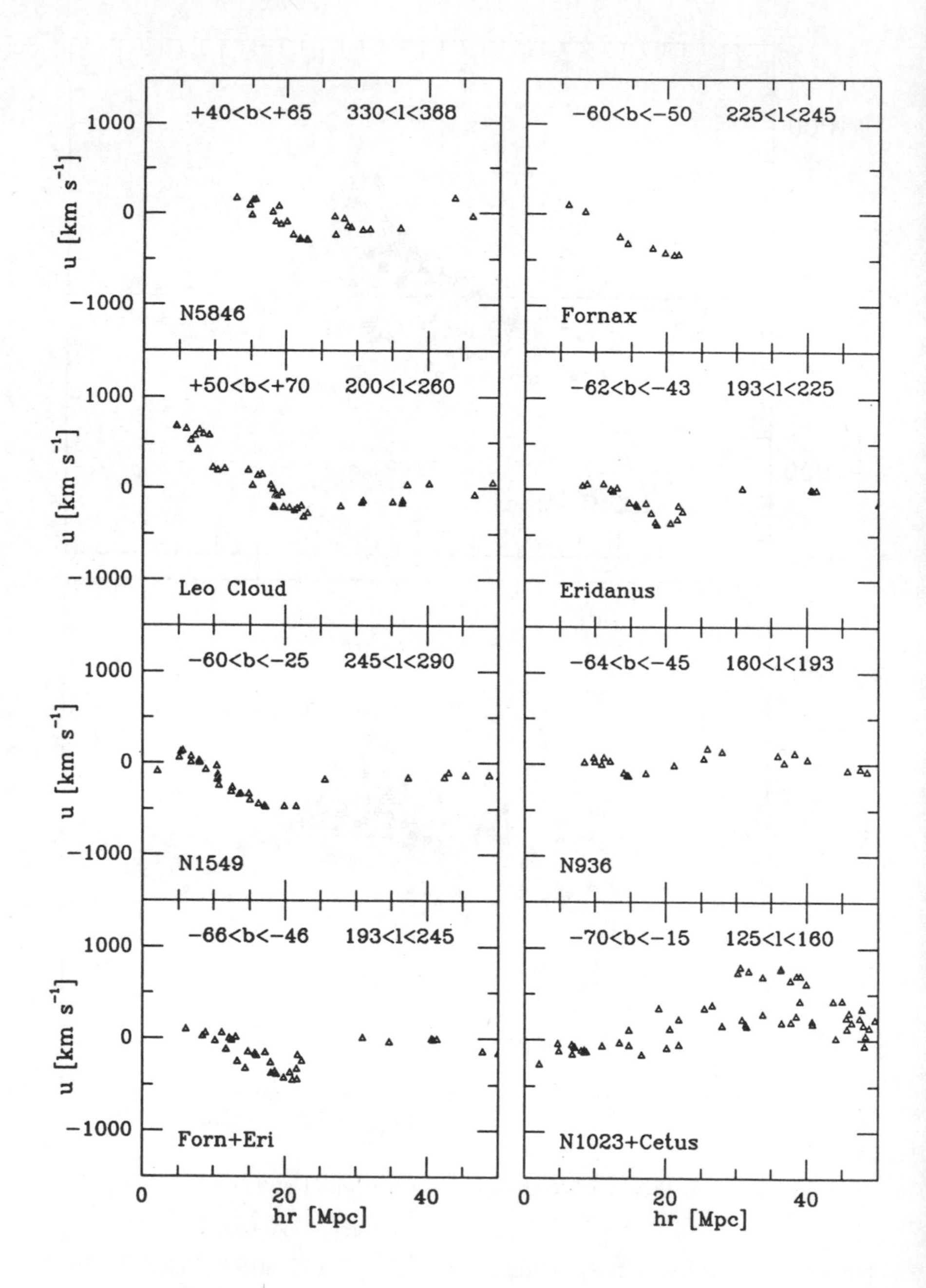

FIG.14. IRAS predictions to be compared with Fig. 10 of FB.

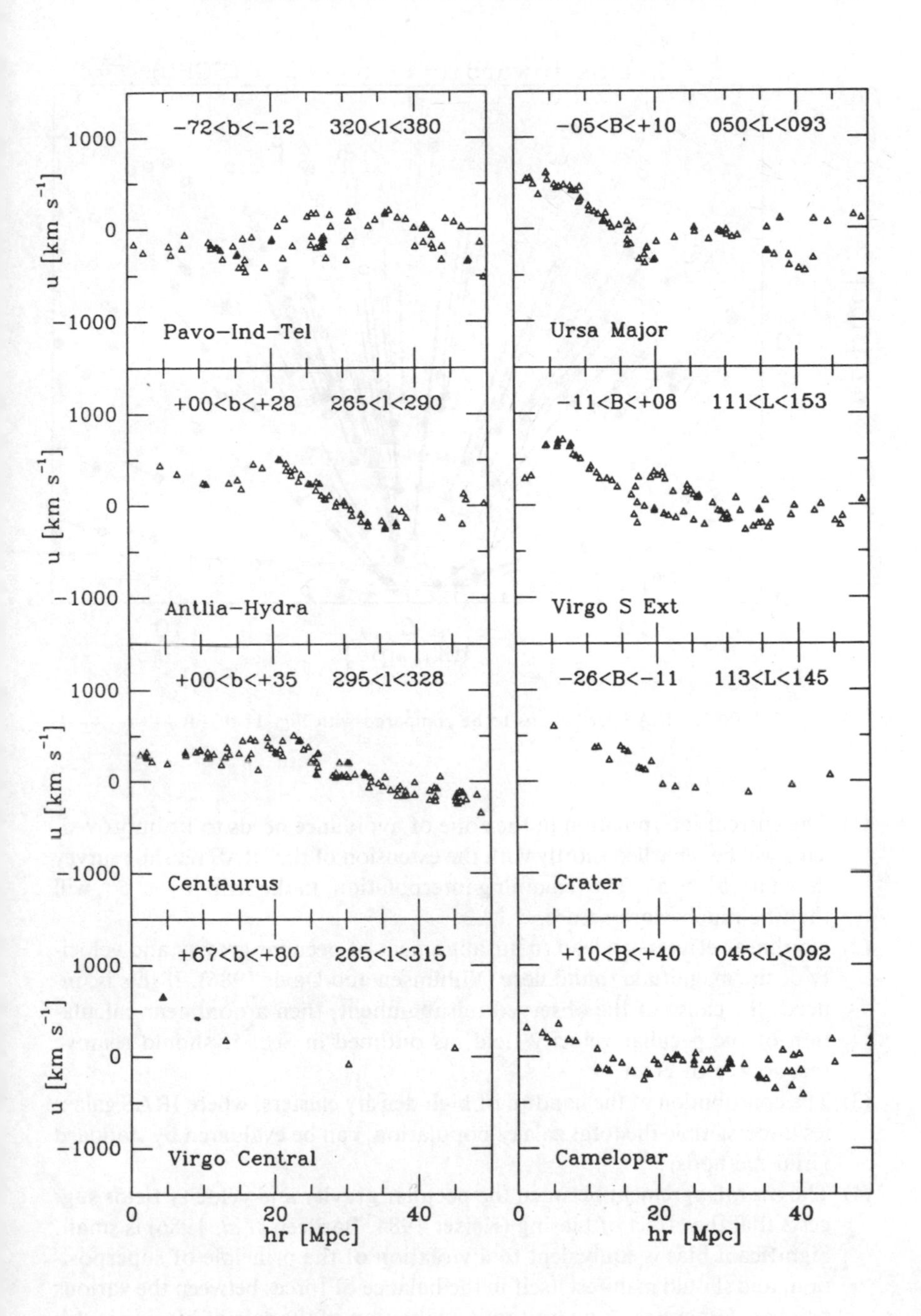
-72<b<-12 320<l<380
Pavo-Ind-Tel
-05<B<+10 050<L<093
Ursa Major
+00<b<+28 265<l<290
Antlia-Hydra
-11<B<+08 111<L<153
Virgo S Ext
+00<b<+35 295<l<328
Centaurus
-26<B<-11 113<L<145
Crater
+67<b<+80 265<l<315
Virgo Central
+10<B<+40 045<L<092
Camelopar
u [km s-1]
1000
0
-1000
hr [Mpc]
0
20
40

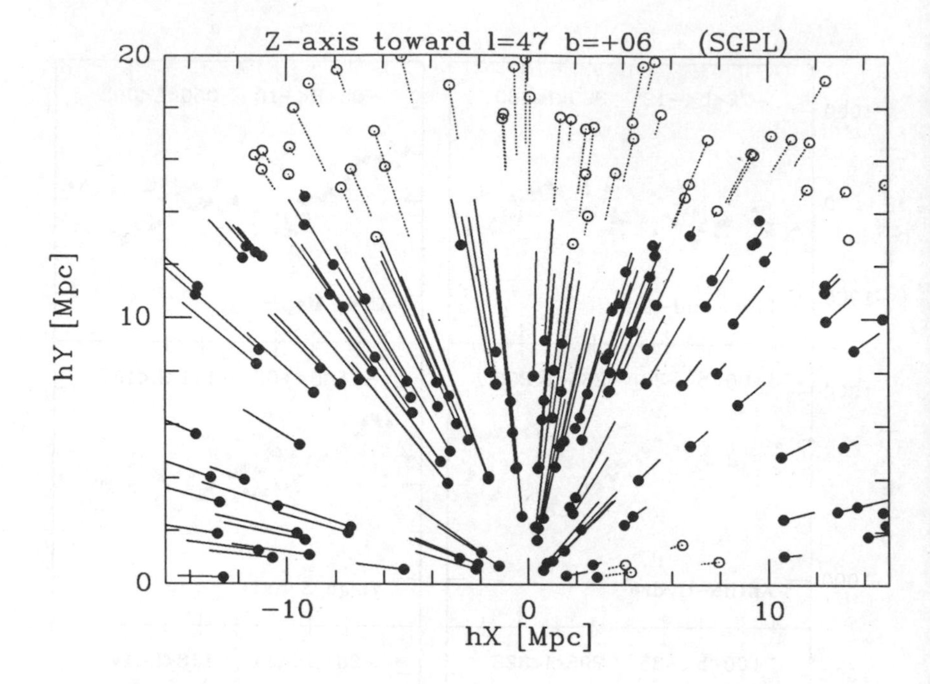

FIG. 15. IRAS predictions to be compared with Fig. 11 of FB.

(1) The current interpolation in the zone of avoidance needs to be improved. This will be remedied shortly with the extension of the IRAS redshift survey down to $|b| \geq 5°$. The remaining interpolation, in the strip $|b| < 5°$, will then be much more secure.

(2) Nonlinear effects can lead to misalignment of peculiar gravity and velocity of the magnitude found here (Villumsen and Davis 1986). If this is, indeed, the cause of the observed misalignment, then a nonlinear calculation of the peculiar velocity field, as outlined in Sec. 5, should remove this source of error.

(3) The contribution of the handful of high density clusters, where IRAS galaxies undersample the total galaxy population, can be evaluated by standard virial methods.

(4) The overall agrement between the peculiar gravity and velocity fields suggests that the effect of biasing (Kaiser 1984, Bardeen *et al.* 1986) is small. Significant bias is equivalent to a violation of the principle of superposition, and should manifest itself in the balance of forces between the various mass agglomerates. A quantitative evaluation of the role of biasing is still lacking, however, and some biasing may be called for.

8. Conclusions

The IRAS redshift survey has been the first to provide density tracers with enough sky coverage (76%) and depth ($100h^{-1}$ Mpc) to allow a reliable determination of the local peculiar gravitational field. The infrared luminosity function of the IRAS galaxies appears to be universal, with no bias as a function of density. It is therefore possible to determine density, and hence gravity, using the method of weights of Yahil *et al.* (1980) and Davis and Huchra (1982). This calculation of density and gravity requires a correction of the Hubble distances of the galaxies for peculiar velocities. This has been accomplished by a self-consistent calculation of density, gravity, and peculiar velocity in a linear approximation for gravitational instabilities.

The density structure of the IRAS galaxies, and the resultant peculiar gravity, have been presented as a series of maps. These show that the gravitational field is dominated by two large concentrations ranging over a large volume, but having a modest overdensity. One is the Great Attractor in the direction of Hydra-Centaurus, along the negative supergalactic *x*-axis; the other is the Perseus-Pisces complex, in roughly the opposite direction in the sky. As a result, the gravitational field bifurcates not far from the position of the Local Group. In addition, a non-spherical Virgo supercluster exerts a significant pull, and there is a void beyond Fornax, roughly in the opposite direction.

The gravitational field is thus far more complex than had been imagined in earlier studies, and it is questionable whether it can be represented by simple parametric models. It is, however, possible to confront the detailed predictions of the IRAS redshift survey with a growing body of data on peculiar velocities of galaxies, summarized at this conference by Faber and Burstein (1988). This comparison shows good overall agreement, and thus confirms the gravitational origin of the peculiar velocities, a major astrophysical result. It also demonstrates that the IRAS galaxies trace the mass, thus providing a limit on biasing, although that is yet to be quantified.

There are, however, also disagreements. The predicted peculiar gravitational acceleration of the Local Group is misaligned by 35° relative to its peculiar velocity with respect to the CMB. Although this is comparable to nonlinear misalignments found in N-body calculations (Villumsen and Davis 1986), it is not yet known whether the linear approximation used here is the cause of the discrepancy in this case. There is also a difference between the predicted and observed peculiar velocities in the first quadrant of the supergalactic plane, in the direction of Ursa Major.

These difficulties may be resolved by a number of improvements planned for the near future, which include: (1) an extension of the survey closer to the galactic plane, down to $|b| \geq 5°$; (2) a nonlinear calculation of the predicted peculiar velocities; (3) a careful evaluation of the effect of high density clusters, where the fraction of IRAS galaxies is smaller than in the general field; and (4) the possible inclusion of some biasing.

Further observations should provide additional constraints. Most needed are peculiar velocities in the direction of Perseus-Pisces, and on the back side of both the Great Attractor and Perseus-Pisces. Ultimately, if and when the observations of density and peculiar velocity are brought into acceptable agreement, there will emerge a reliable determination of the cosmological density parameter, Ω_0.

Acknowledgements

It is a great pleasure to thank my collaborators—M. Davis, J. Huchra, M. Strauss, and J. Tonry—for many fruitful exchanges, as well as permission to use the IRAS redshift data prior to our joint publications. Special thanks go to D. Burstein and S. Faber for their useful comments, and for making their preprint available in a timely fashion, allowing a one-to-one comparison of the IRAS predictions and the observed peculiar velocities. T. Soifer and E. Smith were very helpful in providing addscans for hundreds of IRAS sources.

REFERENCES

Bardeen, J. M., Bond, J. R., Kaiser, N., and Szalay, A. S. 1986. *Ap J.* **304,** 15.

Davis, M. and Huchra, J. 1982. *Ap J.* **254,** 437.

Dow, M. W., Lu, N. Y., Houck, J. R., Salpeter, E. E., and Lewis, B. M. 1988. *Ap J Letters.* **324,** L51.

Dressler, A. 1980, *Ap J.* **236,** 351.

_____. 1987, preprint.

Faber, S. M. and Burstein, D. 1988. this volume. (FB)

Harmon, R. T., Lahav, O., and Meurs, E. J. A. 1987. *MN.* **228,** 5p.

Huchra, J., Davis, M., Latham, D., and Tonry, J. 1983. *Ap J Suppl.* **52,** 89.

Kaiser, N. 1984. *Ap. J Letters.* **284,** L9.

Lahav, O. 1987. MN. **225,** 213.

Lahav, O., Rowan-Robinson, M., and Lynden-Bell, D. 1988. *MN.* **234,** 677.

Lilje, P. B., Yahil, A., and Jones, B. T. J. 1986. *Ap J.* **307,** 91.

Lubin, P. and Villela, T. 1986. in *Galaxy Distances and Deviations from Universal Expansion.* p. 169. eds. B. F. Madore and R. B. Tully. Dordrecht: Reidel.

Lynden-Bell, D., Faber, S. M., Burstein, D., Davies, R. L., Dressler, A., Terlevich, R. J., and Wegner, G. W. 1988. *Ap J.* **326,** 19.

Meiksin, A. and Davis, M. 1986. *AJ.* **91,** 191.

Postman, M. and Geller, M. J. 1984. *Ap J.* **281,** 95.

Rice, W. *et al.* 1988. *Ap J Suppl.* **68,** 91.

Rowan-Robinson, M. 1988. in *Comets to Cosmology: 3rd IRAS Conference.* ed. A. Lawrence. Berlin: Springer. in press.

Sandage, A. and Tammann, G. A. 1981. *Revised Shapley Ames Catalogue.* Washington: Carnegie Institution.

Sandage, A., Tammann, G. A., and Yahil, A. 1979. *Ap J.* **232,** 352.

Schechter, P. L. 1976. *Ap J.* **203,** 297.

Strauss, M. and Davis, M. 1988. this volume.

Turner, E. E. and Sargent, W. L. W. 1977. *Ap J Letters.* **22,** L3.

Villumsen, J. V. and Davis, M. 1986. *Ap J.* **308,** 499.

Villumsen, J. V. and Strauss, M. A. 1987. *Ap J.* **322,** 37.

Vittorio, N., and Juszkiewicz, R. 1987. in *Nearly Normal Galaxies.* p. 451. ed. S. M. Faber, Berlin: Springer.

Yahil, A. 1985. in *The Virgo Cluster of Galaxies.* p. 359. eds. O. G. Richter and B. Binggeli, Garching: European Southern Observatory.

Yahil, A., Sandage, A., and Tammann, G. A. 1980. *Ap J.* **242,** 448.

Yahil, A., Tammann, G.A., and Sandage, A. 1977. *Ap J.* **217,** 903.

Yahil, A., Walker, D., and Rowan-Robinson, M. 1986. *Ap J Letters.* **301,** L1.

THE PECULIAR VELOCITY FIELD PREDICTED BY THE DISTRIBUTION OF IRAS GALAXIES

MICHAEL A. STRAUSS and MARC DAVIS
Departments of Astronomy and Physics, University of California, Berkeley

ABSTRACT

We have recently completed a redshift survey of ~ 2300 *IRAS* galaxies selected uniformly over 76% of the sky, and with a characteristic depth of ~ 4000 km s^{-1}. The sky coverage is unprecedented, allowing us to map the mass distribution in the local universe in great detail, under the assumption that *IRAS* galaxies trace the mass. In particular, as the distribution of matter around any nearby galaxy is known, we can predict its peculiar velocity from linear perturbation theory. Our own peculiar acceleration points within 22° of our microwave velocity vector, and the majority of the material inducing our acceleration is located within a redshift of 3000 km s^{-1}. The existence of a Great Attractor in the galactic plane with a power-law radial density distribution that dominates our motion is inconsistent with the convergence of the *IRAS* peculiar acceleration. However, we measure an overdensity in the sphere reaching to us centered on the Great Attractor of 40%, consistent with the prediction of Lynden-Bell *et al.* (1988). The center of mass of this overdensity is displaced ≈ 500 km s^{-1} towards us.

The *IRAS* galaxy redshift distribution is completely consistent with that of the observed optical galaxies in the direction of the Great Attractor. *IRAS* predicts large bulk flows with coherence length ~ $20h^{-1}$ Mpc. The *IRAS* velocity field qualitatively reproduces the recent observations of peculiar velocities of spiral and elliptical galaxies. In particular, the substantial outflows reported for the Hydra-Centaurus-Pavo-Indus region are also qualitatively in agreement with the velocity field expected from the distribution of *IRAS* galaxies. Although we cannot rule out the existence of excess mass in the galactic plane in the direction of the Great Attractor, it is not needed to explain the observed peculiar velocities.

1. Introduction

The *IRAS* database has in many ways answered the dreams of those who wish to study the large-scale structure of the distribution of galaxies in the local universe. Peebles (1980) shows in linear perturbation theory that our peculiar velocity is directly proportional to the dipole moment of the matter distribution around us, the constant of proportionality depending only on the value of the density parameter Ω. In order to measure the dipole moment directly, one needs a redshift survey of a sample of galaxies covering the entire sky, free from systematic biases between the northern and southern hemispheres. The *IRAS* Point Source Catalog contains some 25,000 galaxies (Soifer, Houck, and Neugebauer 1987) selected in a uniform way over the sky. Furthermore, galactic extinction is negligible in the infrared, so a galaxy catalogue may be compiled from the Point Source Catalog (PSC) with a well-defined flux limit over all parts of the sky that are not limited by confusion.

Meiksin and Davis (1986), and Yahil, Walker, and Rowan-Robinson (1986) were the first to extract galaxy catalogs from the PSC, and showed that the angular dipole moment of the galaxy distribution points within 30° of the peculiar velocity vector of the Local Group, as inferred from the dipole anisotropy of the Cosmic Microwave Background. The implication is that the *IRAS* galaxies trace the mass that gives rise to our peculiar velocity. We have measured redshifts for a sample of objects extracted from the PSC based on the criteria of Meiksin and Davis (1986): high flux quality at 60μm, $F_{60}/F_{12} > 3$, and $F_{60} > 1.936$ Jy. This last condition is imposed to keep the sample to a manageable size. Objects within ten degrees of the galactic plane, as well as a few regions of high-latitude star formation, are excluded to avoid excessive cirrus contamination. A preliminary discussion of our results has appeared in the proceedings of IAU Symposium 130 (Strauss and Davis 1988, hereafter SD88); this paper is an update of that report, and is complementary to the paper of Amos Yahil (1988) in these proceedings. In Sec. 2 we discuss recent changes in our sample. In Sec. 3, we detail the calculation of peculiar velocities and show the effect they have on our peculiar acceleration. A detailed comparison of the *IRAS* density field with the Great Attractor model of Lynden-Bell *et al.* (1988) is presented in Sec. 4. We show full sky maps of the peculiar velocities and compare them to peculiar velocities measured by Aaronson *et al.* (1982), Burstein *et al.* (1987), and Rubin (1988). Sec. 5 contains our conclusions.

2. Refining the Galaxy Catalogue

There have been several important developments since the writing of SD88. A major worry at that time was the problem of extended sources, objects of sufficiently large angular diameter to be resolved by the *IRAS* beam. The PSC fluxes were determined by fitting the scan across a source to a template of the expected response to a point source, resulting in systematically underestimated fluxes for resolved objects. Most extended sources are flagged as such in the PSC; we sent a list of *all* such objects satisfying our color criterion, regardless of flux, to Tom Soifer and Elizabeth Smith of IPAC, who had them ADDSCANed. This involves creating the median of all scans crossing a given source, and integrating the result between zero-crossings. The results are encouraging; 86 objects were added to our list and the fluxes of 400 others were corrected. As we indicated in SD88, increasing the fluxes of the galaxies flagged as extended makes a big difference in the calculation of our peculiar gravity.

We have also received from Walter Rice co-added *IRAS* fluxes for galaxies with angular diameters greater than 10′ (Rice *et al.* 1988). We have obtained a machine-readable copy of the Point Source Catalog, Version 2.0, and have updated our galaxy sample accordingly. With these changes, our sample consists of 2285 galaxies with fluxes greater than 1.936 Jy at 60μm, of which 29 still require observation.

We discovered a strong anti-correlation between the correlation coefficient (CC), a measure of the goodness-of-fit of the point source template to a scan, and the ratio of ADDSCAN to PSC flux; that is, sources with poor CC have PSC fluxes that are systematically underestimated. There are almost 1400 objects in the PSC satisfying our color criterion with poor CC, but which are not flagged as extended. The ADDSCANing of these sources is in progress; preliminary indications are that virtually all of these sources are associated with foreground cirrus, and thus it is unlikely that these sources present any systematic problem with our catalog.

Finally, the *IRAS* detectors suffered hysteresis after passing over regions of large flux density, in particular after passing over the galactic plane. We are in the process of assessing how large an effect this is.

3. Calculation of and Correction for Peculiar Velocities

Given the distribution of *IRAS* galaxies around us, and armed with the assumption that they trace the mass on the large scale, we can calculate our peculiar acceleration due to the inhomogeneity of that distribution. As explained above, linear theory then relates that directly to our peculiar velocity, allowing us to obtain an estimate for Ω. Because of the large sky coverage of the present sample, we can calculate the gravitational acceleration of other points in space, as we know the distribution of matter around them. In practice, we do the following. From the observations we calculate the number density and selection function of galaxies in the survey, using the methods of Davis and Huchra (1982). The selection function is simply the fraction of the luminosity function sampled at any given redshift in a flux-limited sample. With this we fill the 24% of the sky not covered by the survey (principally the region within 10° of the galactic plane) with random galaxies with the same number density and selection function, yielding a galaxy catalogue with true 100% sky coverage. Nearby dwarf galaxies not observable to at least 500 km s^{-1} are deleted from the analysis, but otherwise the catalogue is flux limited. Each galaxy in the sample is labelled with the value of the selection function at that distance. The peculiar velocity of a point *P* in the sample is then estimated as:

$$V_P = \frac{H_0 \Omega^{0.6}}{4\pi n_I} \sum_i \frac{1}{\phi(r_i)} \frac{\mathbf{r}_i - \mathbf{r}_P}{|\mathbf{r}_i - \mathbf{r}_P|^3}, \quad (1)$$

where r_i is the vector to galaxy *i*, $\mathbf{r}_P$ is the vector to galaxy *P*, n_1 is the galaxy density calculated as in Davis and Huchra (1982), and $\phi(r_i)$ is the selection function at the distance of the galaxy *i*. The observer is at the origin. Note that the right-hand side of equation (1) is independent of H_0. Thus given a value of Ω, we can estimate the velocity flow field in the local universe. In all of the following, we shall set Ω equal to 1 for simplicity.

Strictly speaking, equation (1) is correct only when the sum extends over all of space, while our survey has a finite depth. We have found (SD88) that our own peculiar acceleration converges within 4000 km s^{-1}, so we carry out the sum in Equation (1) for all galaxies for which 400 km s^{-1} $< |\mathbf{r}_i - \mathbf{r}_P|$ $<$ 5000 km s^{-1}. As our sample becomes very sparse for redshifts greater than 10,000 km s^{-1} we do not compute peculiar velocities for test particles more distant than r_P = 5000 km s^{-1}. The small scale cutoff in the sum is a smoothing intended to eliminate small scale nonlinear behavior, where Equation (1) will not apply. Note that our sample is flux limited, not volume limited, and the shot noise errors in the computed velocity increase with distance, becoming quite substantial by a redshift of 5000 km s^{-1}.

The first thing we may do with this technique is to correct our measured redshifts for the peculiar velocities of the galaxies in our sample. We initially assume pure Hubble flow, and thus put each galaxy at the distance indicated by its velocity. The one exception to this is that galaxies within 6° of the center of the Virgo cluster, with redshifts less than 2500 km s^{-1}, are all positioned at the distance of Virgo; no further Virgocentric correction is applied. The peculiar velocity of each galaxy in our sample within 5000 km s^{-1} is calculated using Equation (1), and the redshift is corrected accordingly. This will change the density field, of course, and we recalculate the peculiar velocity field, continuing until the process converges. We find that using the average of the peculiar velocity of a given galaxy found in the previous two iterations as the initial guess, the algorithm converges to an accuracy of typically 20 km s^{-1} per galaxy within four iterations. During this procedure one must update the estimate of the selection function, which also changes slightly. In

Figure 1, we show how this process affects the calculation of our peculiar acceleration. The quantity plotted is $V\Omega^{-0.6}$ as a function of the distance to which the sum in equation (1) is carried out. This is the analogue to Figure 2 of SD88. The open symbols show our peculiar acceleration due to our sample when each galaxy is placed at its redshift distance, while the solid symbols show the effect of correcting for peculiar motions of each galaxy, and our own motion. The net effect is really quite minor. The amplitude of the acceleration is $V\Omega^{-0.6} \approx 500$ km s^{-1}, and is approximately constant from 3400 km s^{-1} outward after correcting for random shot noise. This amplitude would imply $\Omega = 1.2$ from Equation (1), although we should point out that we had to assume a value for Ω to calculate the peculiar velocities; there is some circular reasoning involved.

The cumulative direction with respect to the microwave velocity averages 22° between 3400 km s^{-1} and 7000 km s^{-1} radius, but shot noise from the

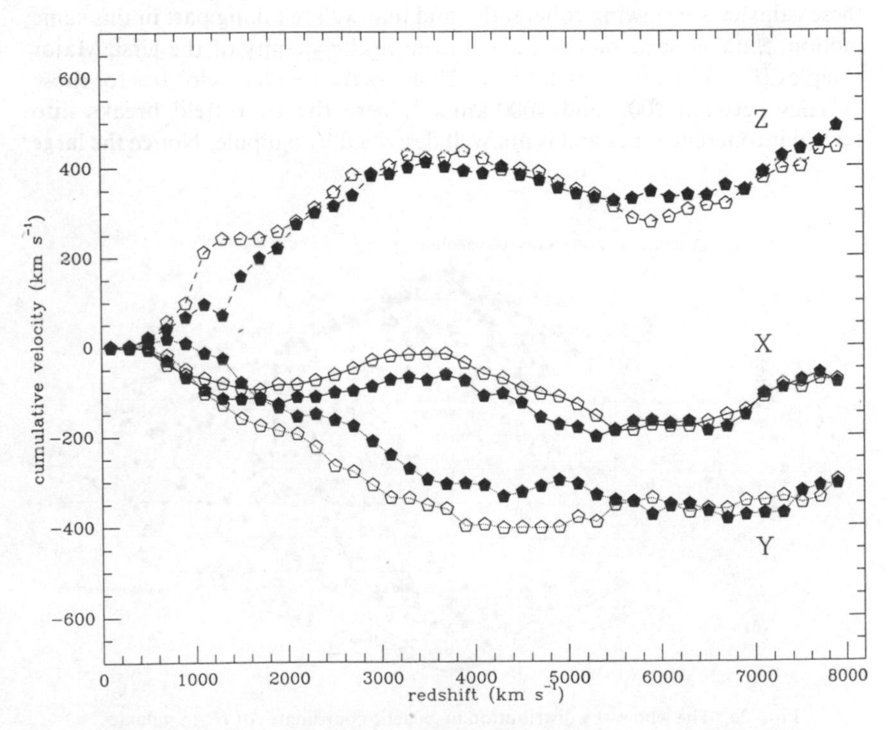

FIG. 1. The three components of the cumulative peculiar gravity. X points toward the galactic center, Y is in the galactic plane and Z is toward the North Galactic Pole. The open symbols result assuming pure Hubble flow. The solid symbols result after correcting each galaxy's distance for its inferred peculiar gravity and for our own motion.

dilute sampling causes the cumulative direction to begin to random walk for radii beyond 7000 km/s. We have not yet determined how to perform the minimum variance extraction of the "asymptotic" direction and amplitude of our acceleration in the presence of this shot noise, particularly since the expected convergence behavior depends on the power spectrum of large-scale perturbations.

We show the radial peculiar velocity field directly in Figures 2*a* and 2*b*, in galactic coordinates. In Figure 2*a*, the radial peculiar velocities (motion relative to the comoving frame) of all galaxies in our sample within 2000 km s^{-1} are indicated; those objects undergoing outflow relative to the comoving frame are indicated with solid symbols, while those pointing towards us are indicated with open symbols. The number of sides on the symbols indicates the magnitude of the peculiar velocity; triangles are for objects with V_r <200 km s^{-1}, squares for 200 < V_r < 400 km s^{-1}, and so on. The peculiar velocities of the local galaxies exhibit a strong dipole field which points approximately in the direction of our own peculiar motion, indicating that these galaxies are flowing coherently, and that we are taking part in this same motion. Smaller-scale motions are evident in the vicinity of the Ursa Major complex (l = 150°; b = 60°). Figure 2*b* shows the peculiar velocities for those galaxies between 2000 and 4000 km s^{-1}; here the flow field breaks into several incoherent zones and is not well described by a dipole. Notice the large

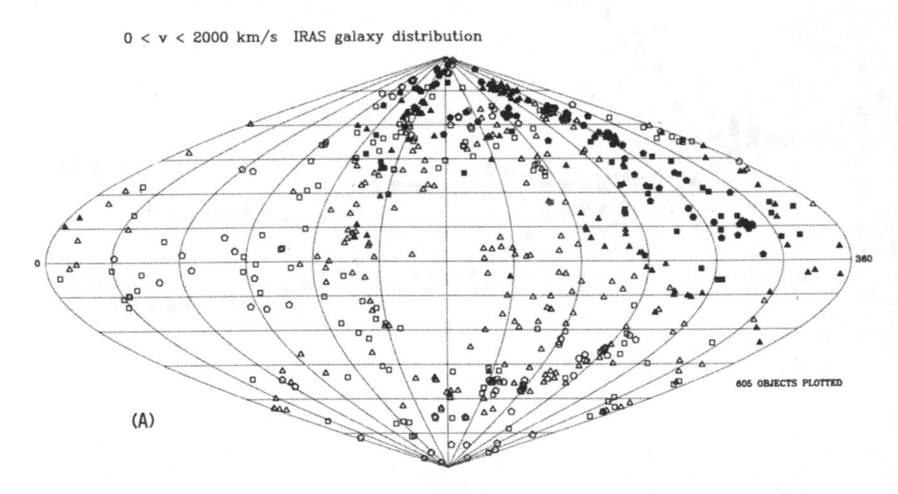

FIG. 2*a*. The whole-sky distribution in galactic coordinates of *IRAS* galaxies with redshift v < 2000 km s^{-1}. Open symbols represent galaxies with negative inferred radial peculiar velocity v_r relative to the comoving frame; solid symbols represent galaxies with v_r > 0. Triangles plot galaxies with $|v_r|$ < 200 km s^{-1}; squares are for galaxies with 200 <$|v_r|$ < 400 km s^{-1}; pentagons, hexagons, etc. are defined in an obvious sequence.

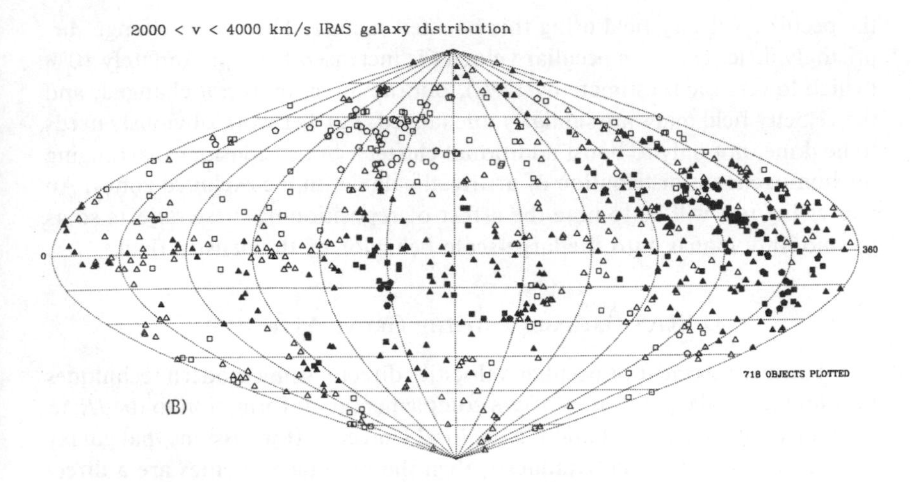

FIG. 2*b*. The whole-sky distribution in Galactic coordinates of *IRAS* galaxies with redshifts $2000 < v < 4000$ km s^{-1}. The symbols are the same as in Figure 2*a*.

outward velocities in the Hydra-Centaurus-Pavo-Indus region ($270° < l < 360°$, $-30° < b < 30°$). This is the region in which large positive peculiar velocities have been reported for the elliptical and spiral galaxies (Lynden-Bell *et al.* 1988; Aaronson *et al.* 1988), and at least qualitatively the *IRAS* maps are consistent with this expectation. A second region of outflow is in the foreground of the massive Perseus supercluster ($90° < l < 200°$; $-45° < b < 15°$). The band of galaxies with $|b| < 10°$ is part of the homogeneous random sample added to bring the survey to full sky coverage.

All of this discussion assumes that *IRAS* galaxies trace the mass distribution. However, we know that this is not true in at least one important respect. *IRAS* galaxies are mostly dusty, late-type systems, and thus our sample is deficient in elliptical and S0 galaxies. The cores of rich clusters are overdense in early-type galaxies, so our sample systematically under-represents the matter distribution in the rich clusters. The well-known nearby cluster cores; Virgo, Hydra, Centaurus, Perseus, Pavo, show lower contrast and are less conspicuous in *IRAS* than in optically selected catalogs. For example, within the 6° radius centered on the Virgo cluster to a limiting redshift of 2400 km s^{-1}, the number contrast relative to background in the CfA optical survey is twice that seen in the *IRAS* survey. We are currently experimenting with schemes to add the missing early galaxies to the *IRAS* sample. One simple scheme is to add galaxies to cluster cores until the density contrast matches that of the CfA survey. Thus we doubled the 33 galaxies within 6° of the core of Virgo, and the 8 galaxies within 3° of the center of Centaurus. We then further iterated the peculiar velocity field using this boosted sample. The results change surprisingly little. Our own peculiar velocity is increased by approximately 10% (which lowers the Ω estimate by 16%), but the direction is not changed, and the velocity field maps are similarly unchanged. Further work obviously needs to be done, not only to boost additional clusters, but to consider rearranging the homogeneous distribution of artificial galaxies in the excluded zones. An interesting test will be to note the effect of repositioning these objects so as to smoothly match onto the large-scale behavior of the density field.

4. IRAS Galaxies and the Great Attractor

The measurement of peculiar velocities directly using modern techniques for obtaining galaxy distances offers complementary information to the *IRAS* estimate of the mass distribution in the local Universe. If we assume that galaxy motions are gravitationally induced, then the peculiar velocities are a direct measure of the inhomogeneities of the density distribution, regardless of the distribution of galaxies. Thus in principle, the measurement of peculiar velocities may lead to new insights into the relative distribution of dark and luminous matter and the nature of the bias (see Davis and Efstathiou 1988). It then becomes particularly important to compare the predictions of the *IRAS* galaxies with direct measurements of peculiar velocities.

Certainly a prime motivation for the present conference, and a central topic of much of the discussion, is the model of Lynden-Bell *et al.* (1988, hereafter collectively called the Seven Samurai; see also Faber and Burstein 1988) of a so-called Great Attractor (hereafter GA). The large outward peculiar velocities seen in the clusters of Centaurus, Pavo, and Telescopium suggest that they lie directly in front of a huge mass concentration. The model is quite specific; this GA is centered at a redshift of 4,350 km s^{-1}, at galactic coordinates $l = 307°$, $b = +9°$, and has a density distribution that drops as the square of the distance from the center. There are several questions that *IRAS* can hope to answer *vis-à-vis* the GA model. Do we see any evidence for such a structure in the distribution of galaxies in our sample? Is it possible that there is a very large amount of material hidden by the galactic plane? In particular, we wish to address the Seven Samurai's claim that the GA is responsible for essentially the entire component of our peculiar velocity in that direction. Do we see any evidence for this in the distribution of *IRAS* galaxies? On the face of things, the answer is no; Figure 1 shows that *our* peculiar acceleration converges rather nicely by 4000 km s^{-1}, pointing within 22° of our peculiar velocity vector. In Figure 3 we quantify this statement. The points plotted are the amplitude of the induced peculiar acceleration, after correcting for shot noise. The convergence at 4000 km s^{-1} is particularly clear here. The smooth curves are the expected induced acceleration from a GA at $b = 0°$, v = 4200 km s^{-1}, with a power-law density distribution $\varrho \propto r^{-\gamma}$, for various

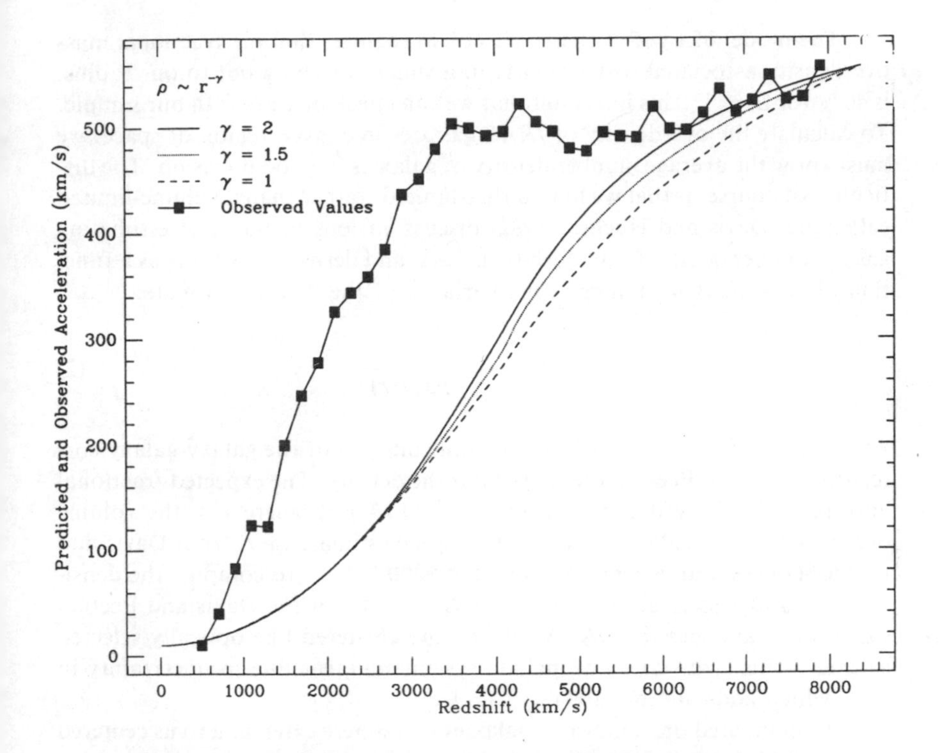

FIG. 3. The plotted squares show the amplitude of our peculiar acceleration as a function of redshift, as calculated from the fully iterated catalogue. A correction for shot noise has been made, so this is not simply the quadratic sum of the components in Figure 1. The smooth curves are the expected peculiar acceleration due to a GA placed at a redshift of 4200 km s^{-1}. The density distribution around the GA was taken to be a power law. The GA was centered squarely on the galactic plane ($b = 0°$), and the material within 10° of the plane was ignored, so as to duplicate the observational situation. All curves are normalized to the same peak value.

values of γ (see Faber and Burstein 1988; they find $\gamma = 1.7$; they also give the GA a core radius, which has not been modeled here.) An excluded zone around the galactic plane twenty degrees thick has been included in the calculation, to mimic our observational situation. All the curves have been normalized to the same peak value. Although putting the GA in the center of the plane minimizes its effect, the smooth curves all rise steeply from 3000 to 6000 km s^{-1}, in sharp contrast to what is observed. It is true that adding excess mass in the vicinity of the GA to our sample will pull the *IRAS* acceleration vector closer to the peculiar velocity vector. However, Figure 3 shows that a model of the GA *which dominates our motion* is not consistent with our results.

The model of Lynden-Bell *et al.* (1988) requires that the fractional mass overdensity associated with the GA, in a sphere reaching out to our radius, is 40% for $\Omega = 1$. This is a result that we can check directly with our sample. To calculate the overdensity of *IRAS* galaxies in a given region of space, we must know the average number density of galaxies to good precision. The difficulty, of course, is that we have a flux-limited, rather than a volume-limited catalogue. Davis and Huchra (1982) discuss various methods of estimating galaxy number density from redshift surveys, and derive a weighting as a function of redshift that minimizes the variance of the density estimate:

$$w(r) = \frac{1}{1 + \bar{n} J_3 \phi(r)}, \tag{2}$$

where $J_3 \equiv 4\pi \int_0^\infty \xi(r) r^2 \, dr$ is the volume integral of the galaxy-galaxy correlation function (Peebles 1980), and $\bar{n}$ is the density. The expected fractional uncertainty in derived density is given by $(J_3/V)^{1/2}$, where V is the volume over which the calculation is done. Using the estimate of J_3 from Davis and Peebles (1983), and using galaxies within 8000 km s^{-1} to compute the density, the density uncertainty should be 8%. The use of the Davis and Peebles value for J_3 assumes that *IRAS* galaxies are clustered like optically selected galaxies. This error does not include systematic error due to uncertainty in $\phi(r)$, which adds another 10%.

We computed the number of galaxies in a sphere extending to us centered on the position of the GA. The fully iterated distribution was used, in which we filled the galactic plane uniformly, and the counts were weighted according to Equation (2). We found $(\delta\varrho/\varrho)_{GA} = 0.4 \pm 0.2$, a large value indeed. The error quoted is that due to the uncertainty in the background density alone. This overdensity is very insensitive to the value of J_3 used in Equation (2), and does not depend on any absolute luminosity limit placed on the sample. Much work remains to characterize this overdensity and compare it with the Seven Samurai model. The center of mass of the overdensity within this sphere is centered approximately 500 km s^{-1} from the GA in our direction, which can be thought of as the mass-weighted mean position of the rich clusters in the vicinity: Hydra, Centaurus, Pavo, Indus, etc.

Dressler (1988) has done a redshift survey of ESO galaxies in the vicinity of the GA, and shows in his Figure 2 (reproduced in Faber and Burstein 1988) the redshift distribution of 890 galaxies in the region $290° < l < 350°$, $-35° < b < 45°$, $|b| > 10°$. Two large peaks are seen in the distribution; one at 3000 km s^{-1} is due to the Centaurus cluster, the other is centered at 4500 km s^{-1}, which Dressler cites as direct evidence for the GA.

Figure 4 shows the redshift distribution of the 236 *IRAS* galaxies in the same region of sky as the Dressler survey. *IRAS* is a much more dilute sample than that of the ESO galaxies, but the agreement between the histograms is

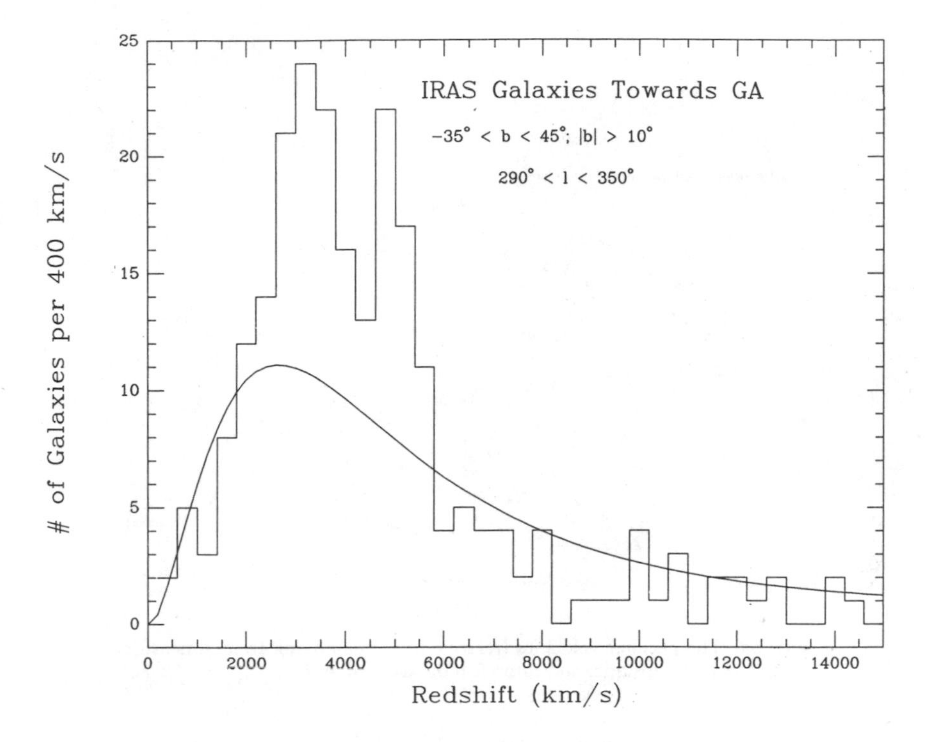

FIG. 4. The radial distribution of *IRAS* galaxies in the region $290° < l < 350°$; $-35° < b < 45°$; $|b| > 10°$. Compare this plot to Figure 2 of Dressler (1988). The smooth curve plots the expected distribution if the galaxies were homogeneously distributed with mean density as determined over the entire *IRAS* sample.

remarkable, implying that the *IRAS* galaxies trace the distribution of the ESO galaxies very well in this region. The smooth curve is the expected distribution for a homogeneous universe with a mean density as given by the average over the full *IRAS* sample. Dressler plots the redshift distribution in four separate zones of galactic latitude within his survey, and again the distribution of *IRAS* galaxies within each region matches his plots within the statistical uncertainty. From Figure 4 it is apparent that the density contrast in the GA region is approximately 2 ± 1, similar to the value inferred by Dressler, whose control sample was less well defined. However, there may be a large amount of excess hidden in the plane; *IRAS* has the potential to probe the galaxy distribution even closer to the galactic plane. We are in the process of identifying and obtaining redshifts for *IRAS* galaxies in the range $5° < |b| < 10°$, which will increase our sky coverage to 85%.

Can the distribution of *IRAS* galaxies in the present survey explain the large peculiar velocities reported in the literature which motivated the GA

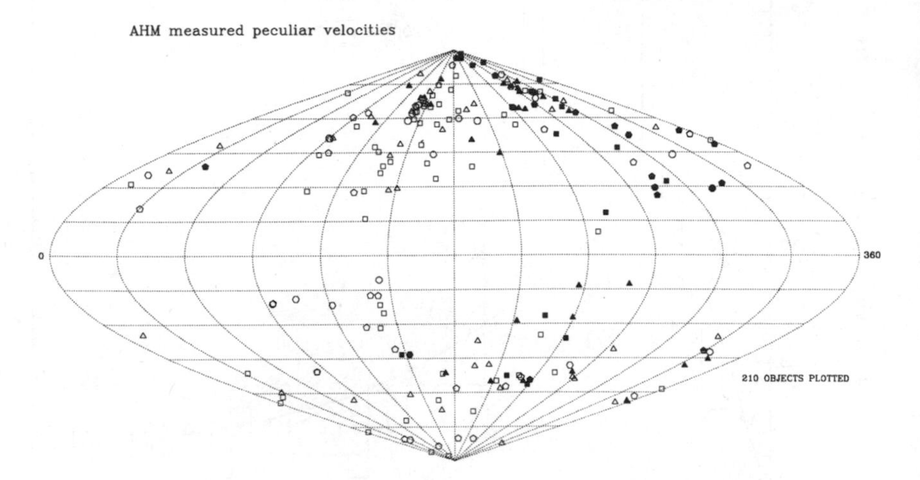

FIG. 5*a*. The peculiar velocities (relative to the comoving frame) measured using the IRTF distance indicator for the data of AHM.

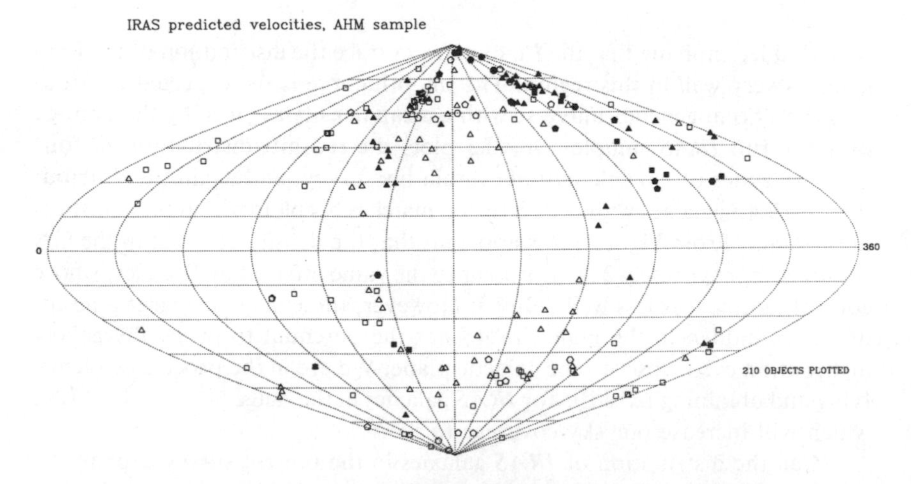

FIG. 5*b*. The *IRAS* inferred peculiar velocities for the AHM galaxies.

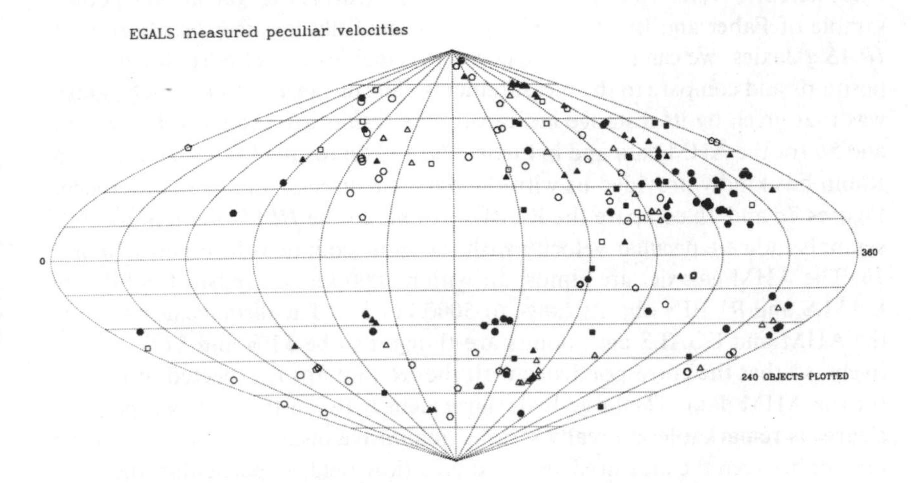

FIG. 6*a*. The peculiar velocities measured by Burstein *et al.* for the EGALS sample.

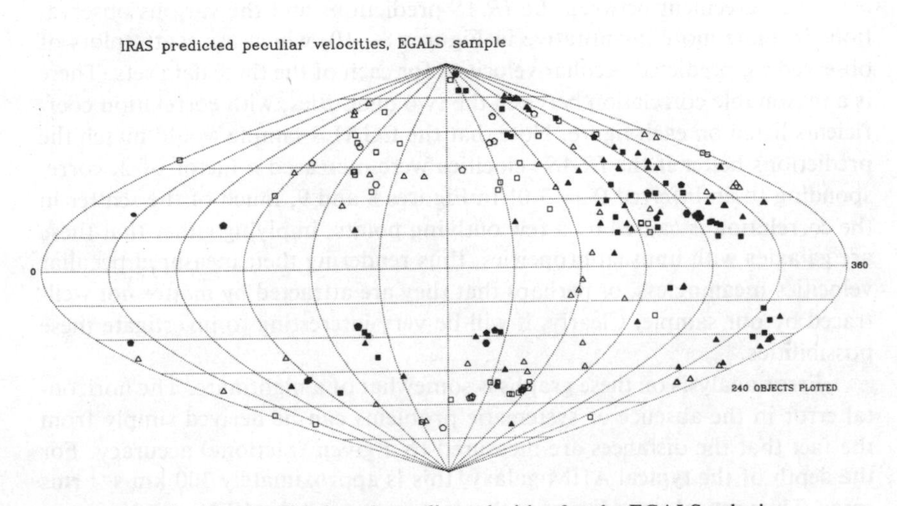

FIG. 6*b*. The *IRAS* inferred peculiar velocities for the EGALS galaxies.

model? David Burstein has kindly sent us the data for the best quality subset of the Seven Samurai program galaxies (EGALS) as well as the subset of the Aaronson *et al.* (1982) spirals having the best optical magnitude and diameter data, hereafter referred to as the AHM sample (this is the Aaronson "good" sample of Faber and Burstein 1988.) Using the fully iterated distribution of *IRAS* galaxies, we can predict the peculiar velocities of test particles at those positions and compare to the measurements. The distance used for each galaxy was that given by its distance indicator. The results are shown in Figures 5*a* and 5*b* for the AHM data and in Figures 6*a* and 6*b* for the EGALS data. Vera Rubin has kindly provided us with the data she discussed in this conference; Figures 7*a* and 7*b* compare the RUBIN sample to the *IRAS* predictions. The symbols indicate peculiar velocity with the same coding as in Figures 2*a* and 2*b*. The AHM galaxies are almost all within 3000 km s^{-1} redshift, while the EGALS and RUBIN objects range to 5000 km s^{-1}. The distance accuracy of the AHM and EGALS data points are thought to be 13% and 21% respectively, so that the correspondence with the *IRAS* maps is expected to be best for the AHM data. The large-scale agreement between the first two pairs of figures is remarkable; no really serious qualitative disagreements seem to be present between the measured and predicted flow field. In particular, the large motions seen in the vicinity of the nominal GA are reproduced very well. The RUBIN sample also gives qualitative agreement with the *IRAS* expectations, but the scatter is somewhat larger. These plots all use the unboosted *IRAS* sample to compute the velocity field; using the boosted sample increases the *IRAS* predicted velocity of many galaxies but changes the sign of relatively few. We are currently studying alternative schemes of boosting the influence of the cluster centers.

The agreement between the *IRAS* predictions and the various observations is made more quantitative in Figures 8—10, which are scatterplots of observed *vs.* predicted peculiar velocities for each of the three data sets. There is a reasonable correlation between the two quantities, with correlation coefficients listed on each figure. Note that the EGALS sample would match the predictions better if the *IRAS* velocities were increased a factor of 2, corresponding to an inferred $\Omega = 3.0$! In Figures 8 and 9, much of the scatter in the correlation is caused by a few outlying points, implying either that these are galaxies with unusual properties, thus rendering their measured peculiar velocities meaningless, or perhaps that they are attracted by matter not well-traced by our sample. Clearly, it will be very interesting to investigate these possibilities.

Error analysis on these graphs is somewhat of a nightmare. The horizontal error in the absence of systematic problems can be derived simply from the fact that the distances are measured to a given fractional accuracy. For the depth of the typical AHM galaxy, this is approximately 300 km s^{-1} rms error. The vertical error bar of each point, assuming the *IRAS* galaxies trace

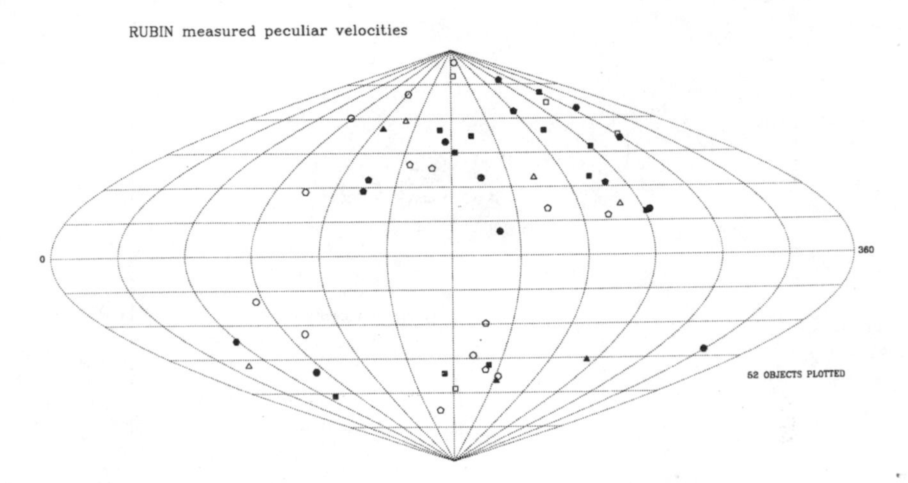

FIG. 7*a*. The peculiar velocities measured by Rubin and collaborators for the RUBIN sample.

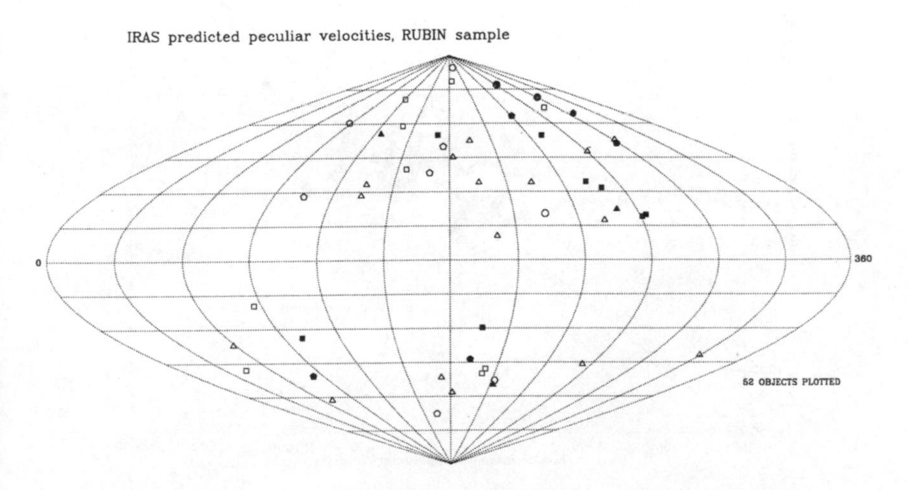

FIG. 7*b*. The *IRAS* inferred peculiar velocities for the RUBIN galaxies.

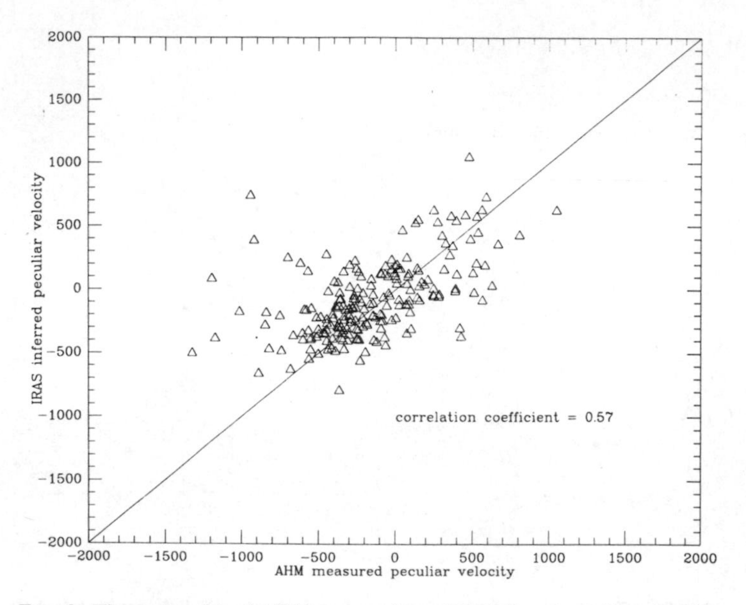

FIG. 8. The scatter plot of AHM measured *vs.* *IRAS* inferred peculiar velocities for the AHM galaxies. No attempt at compensating for the underweighted cluster centers has been made.

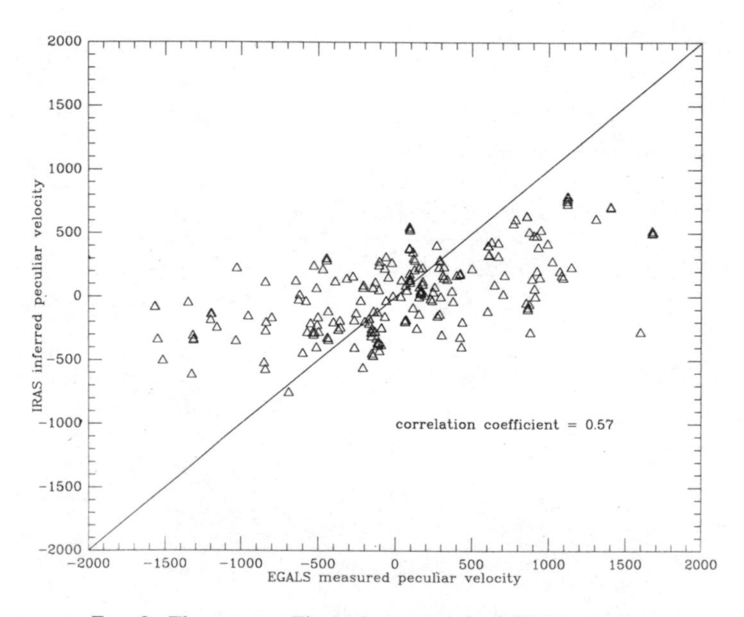

FIG. 9. The same as Figure 8, but for the EGALS sample.

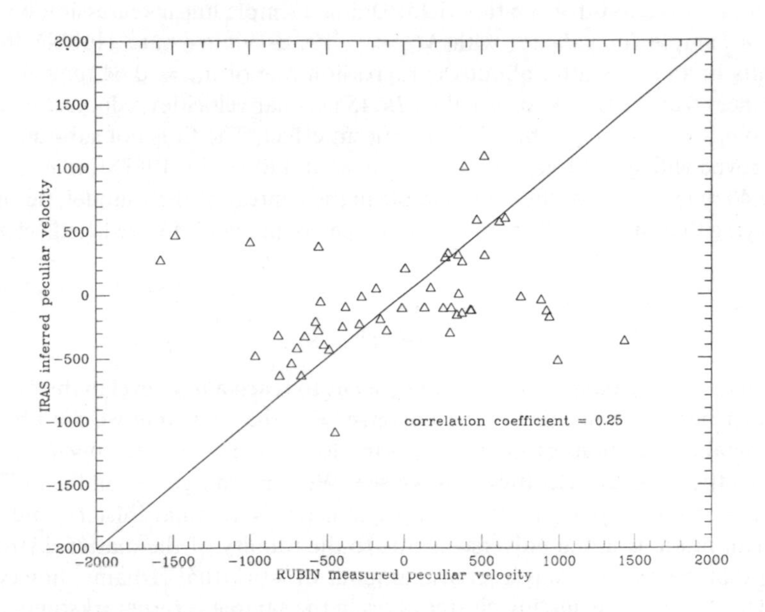

FIG. 10. The same as Figure 8, but for the RUBIN sample.

the mass and that deviations from linear theory are negligible (!?), is comprised of three terms: the shot noise in the computation of the *IRAS* acceleration which is a strong function of the distance of the point, the neglect of the acceleration due to galaxies within 400 km s^{-1} of the galaxies, and the shear in the velocity field given that we have computed the acceleration exactly at the distance given by the secondary estimator which is uncertain at the 10-20% level. Given all that, the 57% covariance in Figure 8 is probably as good as could be hoped for, although it may be that much of the scatter is due to certain regions in which *IRAS* galaxies are poor tracers of the mass distribution. The plots of the EGALS data and the RUBIN data (Figure 9 and 10) show a similar trend but larger scatter because the distance estimators are not quite as precise and the galaxies are more distant. The correlation coefficients for the latter two cases are 0.57 and 0.25 respectively. A preliminary experiment of adding more material in the galactic plane toward the GA direction did *not* improve this scatter, implying that the absence of such a mass concentration in our data is not the source of the scatter. Further work will attempt to quantify this statement.

An alternative analysis is simply to measure the scatter in the diagram of velocity width or dispersion versus absolute luminosity, both with and without corrections for the flowfield. Doing a simple linear regression on the AHM sample for galaxies with $M_H < -19$, assuming pure Hubble flow, results in an rms scatter about the regression line of $\sigma_M = 0.64$ mag. Using distances from redshifts corrected by *IRAS* peculiar velocities reduces this scatter to $\sigma_M = 0.50$ mag, which is a significant effect. The fit is not substantially improved with a quadratic IRTF law. Faber and Burstein (1988) find a scatter of 0.40 mag or less for the same sample in the context of their model, perhaps implying that there is still room for improvement in the *IRAS* predicted velocity field.

5. Conclusions

The *IRAS* galaxy sample has allowed us to reach a new level in the discussion of peculiar velocities in the universe. For the first time we are able to ask detailed questions of how the distribution of matter in the universe gives rise to the peculiar velocities that we see. We now can go beyond simplistic models of spherical infall to perfectly symmetric isothermal clusters, and can explain motions in the universe as due to the totality of the matter distribution that we see. There is a terrific amount of work that remains, however. The problem of the missing cluster cores in the sample is rather irksome, and our simple attempt at compensating can certainly be improved. We need to quantify our errors in peculiar velocity determinations; the *IRAS* sample is a very dilute sample of galaxies, and we suspect that we are largely dominated by shot noise in the calculations of peculiar gravity. Quantifying this and other effects will take large Monte-Carlo and *N*-body simulations. Detailed quantitative comparisons of observed and predicted peculiar velocities in various regions of the sky remain to be done; points of disagreement between the two may teach us interesting things about the relative distribution of *IRAS* galaxies and dark matter.

At present, we can state several disparate conclusions with regard to the GA model; (*a*) The convergence of our peculiar acceleration by 4000 km s^{-1} is inconsistent with the GA model; (*b*) The addition of a density enhancement in the direction of the GA center would bring the *IRAS* dipole vector closer to the microwave dipole vector; (*c*) There is a density enhancement in *IRAS* galaxies of 40 ± 20% in a sphere centered on the GA position and reaching out to our radius; the center of mass of the overdensity is 500 km s^{-1} from the GA position, displaced in our direction; (*d*) The redshift distribution of *IRAS* galaxies in the region agrees very well with that of optically selected galaxies; a large enhancement is seen at ~ 4000 km s^{-1}; (*e*) We find good agreement between our predicted peculiar velocities and those measured by other workers, *without* the addition of extra material in the plane. In particular, we see coherent flow within 2000 km s^{-1}, and large outflow in Centaurus and Pavo-Indus.

Thus we find ourselves halfway towards an understanding of the GA model. We suspect that the origin of the confusion is the Seven Samurai's assumption that the GA is smooth and spherically symmetric; we look forward to studying this in detail, and we predict that a picture that we all can agree upon will emerge over the next year or two.

Finally, it is time to ask detailed questions of how our results compare with predictions of various cosmological scenarios; the exciting possibility exists that the comparison of predicted and measured flow fields will be able to discriminate between competing theories.

Perhaps the most gratifying aspect of this work to come out of this workshop is the sense that it has inspired those hard-working observers who measure galaxy distances to continue their work with renewed effort. The call for more and better distances is stronger than ever; we look forward to the comparison with new data sets.

Acknowledgements

We thank our collaborators John Huchra, Amos Yahil, and John Tonry for permission to use our data before publication. Avery Meiksin is thanked as always for his many helpful suggestions and comments on this project. We thank Dave Burstein for supplying us with the Seven Samurai and AHM data in machine-readable form, and Vera Rubin for her data. MAS gratefully acknowledges the support of a Berkeley Graduate Fellowship. This research was supported by the NSF and NASA.

REFERENCES

Aaronson, M., Bothun, G., Budge, K., Dawe, J., Dickens, R., Hall, P., Mould, J., Murray, J., Persson, E., Schommer, R., and Wright, A. 1988. in *Large-Scale Structures of the Universe.* eds. J. Audouze, M.-C. Pelletan, and A. Szalay. Dordrecht: Kluwer Academic Publishers, p. 185.

Aaronson, M., Huchra, J., Mould, J. R., Tully, R. B., Fisher, J. R., van Woerden, H., Goss, W. M., Chamaraux, P., Mebold, U., Siegman, B., Berriman, G., and Persson, S. E., 1982. *Ap J Suppl.* **50,** 241.

Burstein, D., Davies, R. L., Dressler, A., Faber, S. M., Stone, R. P. S., Lynden-Bell, D., Terlevich, R. J., and Wegner, G. A. 1987. *Ap J Suppl.* **64,** 601.

Davis, M. and Huchra, J. 1982. *Ap J.* **254,** 437.

Davis, M. and Peebles, P. J. E. 1983. *Ap J.* **267,** 465.

Davis, M. and Efstathiou, G. 1988. this volume.

Dressler, A. 1988, *Ap J.* **329,** 519.

Faber, S. and Burstein, D. 1988. this volume.

Lynden-Bell, D., Faber, S. M., Burstein, D., Davies, R. L., Dressler, A., Terlevich, R. J., and Wegner, G. 1988. *Ap J.* **326,** 19. (Seven Samurai).

Meiksin, A., and Davis, M., 1986. *AJ.* **91,** 191.

Peebles, P. J. E. 1980. *The Large-Scale Structure of the Universe.* Princeton: Princeton University Press, Sec. 14.

Rice, W. L., Persson, C. J., Soifer, B. T., Neugebauer, G., and Kopan, E. L. 1988. *AP J Suppl.* **68,** 91.

Rubin, V. 1988. this volume.

Soifer, B. T., Houck, J. R., and Neugebauer, G. 1987. *Ann Rev Astron Astrophys.* **25,** 187.

Strauss, M. A., and Davis, M. 1988. in *Large-Scale Structures of the Universe.* eds. J. Audouze, M.-C. Pelletan, and A. Szalay. Dordrecht: Kluwer Academic Publishers, p. 191 (SD 88).

Yahil, A. 1988. this volume.

Yahil, A., Walker, D., and Rowan-Robinson, M. 1986. *Ap J Letters,* **301,** L1.

IV

SMALL-SCALE MICROWAVE FLUCTUATIONS

SMALL AND INTERMEDIATE SCALE ANISOTROPIES OF THE MICROWAVE BACKGROUND RADIATION

A. N. LASENBY
Mullard Radio Astronomy Observatory, Cavendish Laboratory

and

R. D. DAVIES
Nuffield Radio Astronomy Laboratories, Jodrell Bank

ABSTRACT

Anisotropies of the Cosmic Microwave Background Radiation (MWB) have, for some years now, been one of the key observational constraints against which all theories of galaxy formation and evolution have had to be tested. In this contribution, we should like to review some of the latest results in this field and perhaps raise some questions arising from them, pertinent to the main theme of this Study Week, namely the organization of structure and velocities on large scales. This is a particularly exciting period in MWB anisotropy studies, with advances in equipment, and the work of many groups around the world, bringing us closer to discovery level across a wide range of angular scales. In particular, recent results from a twin-horn experiment in Tenerife (Davies *et al.* 1987), provide evidence for a possible detection of anisotropy on a scale of 8°, although, as will be indicated below, much work still needs to be done in order to eliminate other possible non-cosmological origins for this signal. It is these new results from several different groups we want to concentrate on, and the discussion will be ordered in terms of increasing angular scale, as follows: Sec. 1. Latest small-scale VLA and Jodrell Bank results ($\theta \lesssim 1'$); Sec. 2. Owens Valley results versus the Uson and Wilkinson experiment ($2' \lesssim \theta \lesssim 7'$); Sec. 3. Tenerife results on intermediate scales ($5° \lesssim \theta \lesssim 8°$).

Then in Section 4 we consider future experimental prospects — what *is* needed to determine $\Delta T/T$ on small and intermediate angular scales? In Section 5 we offer some thoughts on the current relation between theory and experiment.

1. Latest Small-Scale VLA and Jodrell Bank Results

To set the scene for subarcminute scale MWB anisotropies, we recall that the angular scale subtended by a spherical density perturbation of mass M at the epoch of recombination is given by

$$\theta \approx 9' (\Omega^2 h)^{1/3} (M/10^{15} M_\odot)^{1/3}$$

where $h = H_0/50$ km s^{-1} Mpc^{-1}.

Anisotropies on subarcmin scales, therefore, correspond to *galactic* sized masses — $10^{12} M_\odot$ and below. However, on these scales, a variety of processes operate during the epochs up to and including recombination to render such primordial perturbations more or less invisible, at least on current theories. Thus if anisotropies were discovered on subarcmin scales, a more natural interpretation for them would be in terms of secondary effects associated with the later, presumably non-linear, stages of galaxy formation. For example, an integrated Sunyaev-Zeldovich effect from hot gas in the cores of young galaxy clusters (Cole and Kaiser 1988) or (coupled with a Doppler scattering effect) from the early stages of non-linear growth occuring in hydrodynamical theories of galaxy formation (Ostriker and Vishniac 1986) could provide quite sizeable (10^{-5} to 10^{-4} in $\Delta T/T$) fluctuations in the MWB, on scales up to a typical core size, $\sim 1'$, and with a substantial covering fraction with respect to the sky. Note that such fluctuations would not be expected to have the Gaussian signature associated with primordial perturbations.

Two instruments have been used recently in the search for these fine-scale anisotropies; the Very Large Array (VLA) and the Jodrell Bank MkIA-MkII interferometer. At the VLA an exciting position was reached about 18 months ago, with the two independent groups involved both claiming an excess of "noise" near the centre of their maps, correlated with the primary beamshape. With interferometer observations (which are necessary at radio freqencies in order to achieve the fine angular resolution), receiver and atmospheric noise should be spread uniformly over the entire sky plane, and one can look for microwave background fluctuations by looking for a component of the noise that instead peaks towards the centre, where the primary beam envelope means one has greatest sensitivity to real structures on the sky. Unfortunately, imperfections in the interferometer electronics also tend to lead to excess noise near the centre of the map, but by a combination of improvements to the VLA hardware, and much hard work by the two groups aimed at identifying and removing such sources of systematic error, the situation was reached (see e.g. Kellermann *et al.* 1986) where it was clear that there were excess signals near the map centre, due to astronomy rather than electronics. These had a surprisingly large amplitude, $\Delta T/T \sim 10^{-4}$, not easily accounted for by extrapolation of the radio source counts known at higher levels down to the flux levels probed with the new measurements (Martin and Partridge 1987).

However, as revealed by Fomalont and Kellermann at the "ΔT/Tea" meeting at Toronto in May 1987 and Wilkinson at IAU 130 in June 1987, it is clear that in the field used by Fomalont *et al.*, some of the excess is due to radio source emission from a cluster of faint galaxies (detected via a CCD) near the field centre, and that the rest can be accounted for as due to the interaction of the CLEAN algorithm, used in making the final maps, with effects due to weak sources near the confusion limit. Briefly, the missing short spacings occuring in any interferometer system (the minimum baseline on the VLA is 45 m corresponding to a scale of ~ 4.5 arcmin at the 5 GHz frequency of the observations) mean that the synthesized beam is surrounded by an extended negative bowl, so that confusing sources can give negative responses, which look like negative-going microwave background fluctuations, as well as positive ones. This would be accounted for analytically once the source counts (and source size distribution) were known at the appropriate flux levels, except that interaction of the CLEAN algorithm with an initial dirty map containing extended features (the negative bowls and possibly the sources) is nonlinear and unpredictable. Instead, the Fomalont group found that full scale numerical simulations had to be carried out, with simulated data produced and then CLEANed using the VLA package. This revealed that the excess fluctuations *could* be accounted for with reasonable extrapolations of the source counts from higher levels. Thus their results have now become upper limits to $\Delta T/T$, as follows (Fomalont, "ΔT/Tea" meeting, Toronto, 1987):

Scale	2σ *Upper limit*
10″	$< 4.4 \times 10^{-4}$
17″	$< 1.6 \times 10^{-4}$
30″	$< 1.1 \times 10^{-4}$
50″	$< 5.0 \times 10^{-5}$

The limit on the 50″ scale is in fact very competitive with the best single-dish results so far, and was achieved in far shorter integration time (100 hours versus 174 out of 445 scheduled hours for Uson and Wilkinson [1984a,b] for example), showing that once the new class of problems they introduce has been understood and dealt with, interferometers may well be the best way forward for the future in pushing down $\Delta T/T$ limits (more on this in Section 4).

The final results from the other group using the VLA for these measurements (Partridge *et al.* 1988) should be available shortly. However, they have recently analyzed their data for polarization anisotropies, and have published interesting upper limits on these, at around the 10^{-4} level on a scale of 18″ to 160″ (Partridge *et al.* 1988).

Although care is taken to reduce the effects of systematic errors on the VLA, it would obviously be of interest to have measurements available from an independent system, not so prone to the combination of correlator offsets and dish-dish crosstalk which has made VLA observations difficult. Such a system is the MkIA-MkII broadband interferometer at Jodrell Bank (Padin *et al.* 1987). This operates at 5 GHz with a bandwidth of 380 MHz and telescope spacing of 425 m. Observations totalling 315 hours have been made in 1985 and 1987, resulting in limits to $\Delta T/T$ of about $(8 - 10) \times 10^{-4}$ on scales from 6″ to 50″ (see Figs. 1 and 2, taken from Waymont *et al.* 1988). The novel feature of these observations lies in the improved method of analysis used throughout, which for the first time applies likelihood techniques to interferometer observations. This means that we can give a precise definition of the 'angular scale' on which results are given — it is the coherence angle $\theta_c = (-C''(0)/C(0))^{-1/2}$ for the assumed sky covariance function $C(\theta)$ (see Section 2) and our limits are derived for the rms $[(C(0))^{1/2}]$ of an assumed Gaussian covariance function:

$$C(\theta) = C(0) \exp \left(- \frac{\theta^2}{2\theta_c^2} \right).$$

This gives a slightly pessimistic view of the sensitivity of the experiment, since previous results (e.g. from the VLA) have only been given in terms of limits on the sky rms when already convolved with the interferometer synthesized beam. We have attempted a deconvolution to set limits to the *intrinsic* rms and hence our results will appear to be higher (by a factor $\sim \sqrt{2}$) from this cause. Note that problems of interaction of CLEAN with weak confusing sources do not arise. The fields were chosen, from VLA observations, to be blank at the levels required for the MWB experiment, and no CLEANing was carried out, except for a single source far from the main beam, which was removed via direct manipulations with the uv data.

A further novel feature, is that for the first time we have set limits directly to the power spectrum of microwave background fluctuations (Figs. 1a and

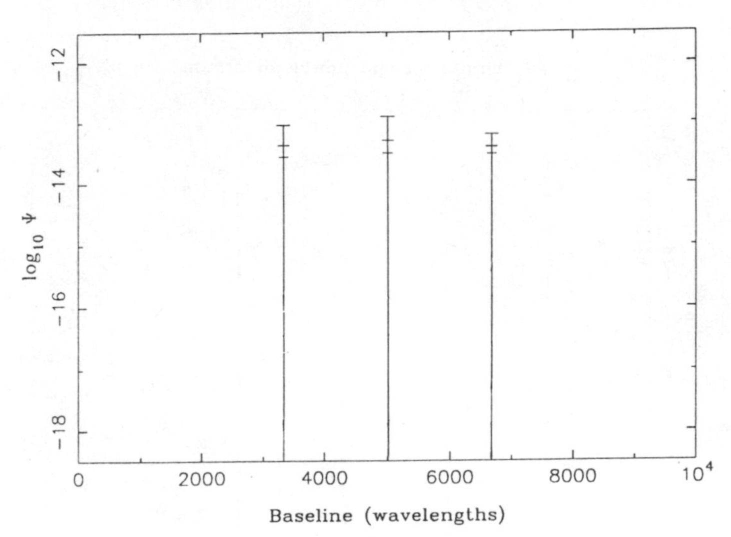

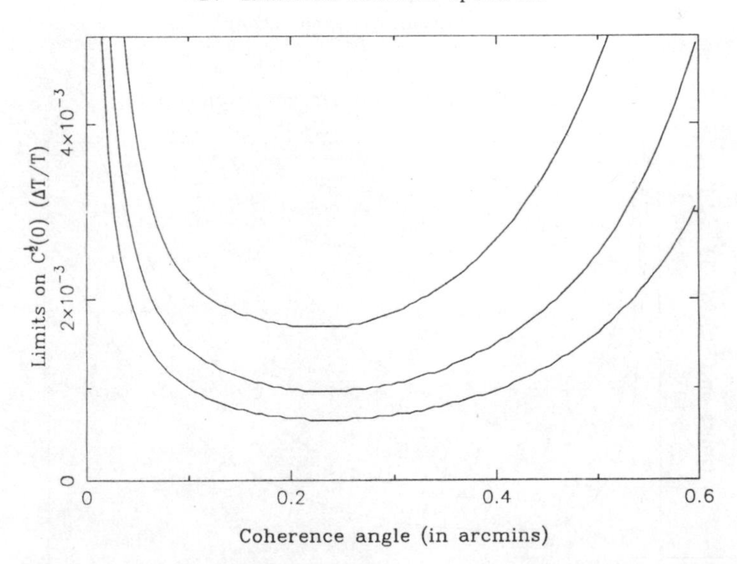

FIG. 1. Limits to small scale MWB anisotropies from Jodrell Bank MkIA-MKII observations made in 1985:
(a) Limits to the power spectrum Ψ as a function of baseline measured in wavelengths. The horizontal bars correspond to 90, 95 and 99% confidence limits respectively (working up from the bottom).
(b) Limits to $\sqrt{C_0}$ as a function of coherence angle θ_c for an assumed Gaussian form of the spectrum. The curves correspond to 90, 95 and 99% confidence limits.

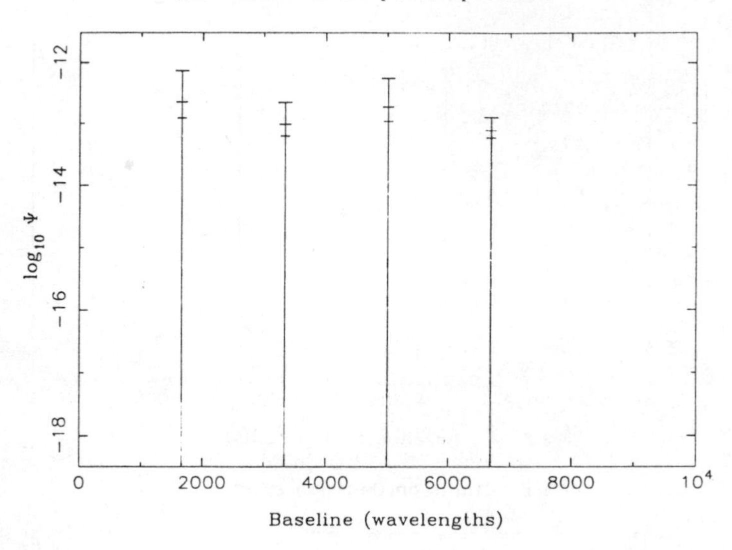

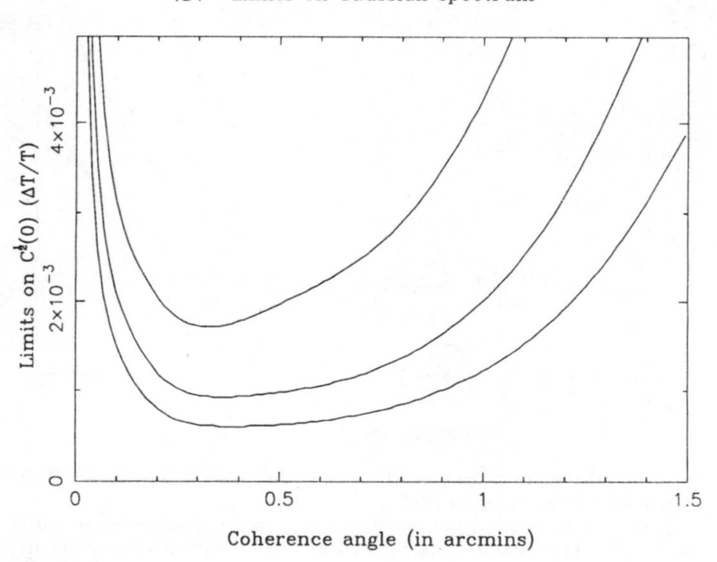

FIG. 2. Same as for Fig. 1a and b except the limits are for the 1987 data. A lower declination field was used so that smaller baselines would be included. Note the increased angular scale coverage in (b).

2a). The ultimate goal of observations of MWB anisotropy is to discover the true power spectrum (fluctuation power versus angular scale) of the microwave sky, and compare it with the theoretical predictions of, for example, cold dark matter (CDM) models. An interferometer provides an opportunity of doing this directly, since the response on a given baseline is directly proportional to the Fourier amplitude at the angular scale corresponding to the baseline. We employ a likelihood technique to provide confidence intervals for the value of the power-spectrum at 3 or 4 coarsely spaced bins of baseline length. For these small angular scales, the relation between 'baseline', measured in wavelengths — call this k — and the spherical harmonic component l more usually used to represent the power spectrum, is given by $l = 2\pi k$. With this change to the labeling of the x-axis, Figs. 1(a) and 2(a) can now be compared directly with predictions for the power spectrum such as those in e.g. Bond and Efstathiou (1987, their Figures 7a and b). If this comparison is carried out, one sees that, as expected on these scales, the theoretical predictions for primordial anisotropy are well below the upper limits, here by orders of magnitude, reflecting the inappropriateness of trying to look for CDM type fluctuations on arcsecond scales. However, we see the importance of the present results as consisting in the way they demonstrate a completely new method of getting from the observations to what we really want to know — the power spectrum. It should be stressed again that this really is model independent — the 'parameters' varied in the likelihood function are the trial values of the power spectrum at the centres of the defined bins, so apart from the unavoidable discretization of the spectrum, it is completely general in its form. We hope to apply such methods to the interferometer experiments with the 5-km telescope, and planned Very Small Array, currently being worked upon at Cambridge (see Section 4).

2. Recent Owens Valley Results and the Uson and Wilkinson Experiment

Readhead and collaborators at Caltech (see for example Readhead *et al.* (1988) and Lawrence *et al.* (1988)), have recently announced first results from a programme of observations using the Owens Valley 40 m telescope at 20 GHz. They have measured the temperature of a central spot relative to the mean of two outside positions at each of 8 centres on a circle 1° from the North Celestial Pole (NCP). The observing configuration involves rapid switching between two beams, which at a slower rate are alternately positioned over the central spot. Each beam has a FWHM of 110″ and the beamthrow is 7.′15. The temperatures they observe are:

RA of Field Centre	ΔT microKelvin
1^h	-57 ± 31
3^h	18 ± 30
5^h	-26 ± 24
7^h	144 ± 25
9^h	20 ± 23
11^h	-21 ± 23
13^h	-18 ± 29
15^h	-32 ± 35

The field at 7^h appears to include a known radio source (and the value is also highly variable), hence this field is discarded before subsequent analysis. A matter of great interest is to be able to compare these results with those from Uson and Wilkinson using the NRAO 140 foot telescope, again at 20 GHz (Uson and Wilkinson 1984a and b). They also used a combination of beam-switching and telescope motion to create a triple beam pattern and had a beam-width of 90″ and a beamthrow of 4.′5. The numbers for the twelve Uson and Wilkinson fields are (Wilkinson, private communication):

Field Number	ΔT microKelvin
1	68 ± 128
2	-23 ± 125
3	73 ± 105
4	-68 ± 131
5	-106 ± 138
6	68 ± 169
7	290 ± 181
8	-130 ± 151
9	167 ± 182
10	189 ± 176
11	-54 ± 101
12	102 ± 135

A useful method of comparing these results, and in particular of seeing what angular scales they set strictest limits to (e.g. will these be in the ratio of the beamthrow or the beamsize?), is by means of the likelihood function.

Again assuming the Gaussian form given in Section 1 for the true sky autocovariance function (acf), and that each field is independent from the others (this comes closest to being untrue in the Owens Valley case, where the minimum interfield distance is $\approx 30'$), we have for the likelihood function

$$L(\mathbf{X}|C_0, \theta_c) \propto \prod_{i=1}^{n} \frac{1}{(\sigma_i^2 + \sigma_s^2)^{1/2}} \exp\left\{\frac{-X_i^2}{2(\sigma_i^2 + \sigma_s^2)}\right\}.$$

Here $X_i \pm \sigma_i$ $(i = 1, \ldots, n; n = 7$ or $12)$ corresponds to the numbers given in the respective tables and σ_s — that part of the rms scatter in the observations due to a real signal from the sky — is derived as follows. Consider the effect of smearing in a single Gaussian beam, with dispersion σ $(\equiv \frac{\text{FWHM}}{2\sqrt{2\ln 2}})$. We have $C(\theta) \mapsto C_M(\theta)$ where in the Gaussian sky case

$$C_M(\theta) = \frac{C_0\theta_c^2}{2\sigma^2 + \theta_c^2} \exp\left\{\frac{-\theta^2}{2(2\sigma^2 + \theta_c^2)}\right\}.$$

Now consider the combination of beamswitching and telescope wagging which gives the triple beamshape. If θ_b is the beamthrow and T_C, T_E, T_W the "central, east, and west" temperatures respectively, we have

$$\Delta T = \frac{1}{2}\{(T_C - T_E) - (T_W - T_C)\} = T_C - \frac{1}{2}(T_E + T_W).$$

and thus

$$\sigma^2 = \text{Var}(\Delta T) = \frac{3}{2} C_M(0) - 2C_M(\theta_b) + \frac{1}{2} C_M(2\theta_b).$$

One can plot contours of likelihood in the (C_0, θ_c) plane and look either for a peak (corresponding to a detection of anisotropy) or, failing this, locate the angular scale of maximum sensitivity, where the contours fall (from a maximum at $C_0 = 0$) most rapidly with C_0. Fig. 3 shows such a contour plot for the data of Uson and Wilkinson. There is no peak away from the origin and the scale of maximum sensitivity is $\theta_c = 1'.8$, A plot of a slice through the likelihood contours at $\theta_c = 1'.8$ is shown in Fig. 4, where the point where the likelihood ratio falls to 1/20th of its peak value (at 0) is shown. This is 156 μK. Without wanting to call this value of $\sqrt{C_0}$ a 'confidence limit', which begs several questions, we do want to take such a value as representative of the sensitivity of an experiment on a given angular scale. As an aside, it may be worth pointing out that having constructed the likelihood function one then has several choices as to how to use it to construct limits at a given level of

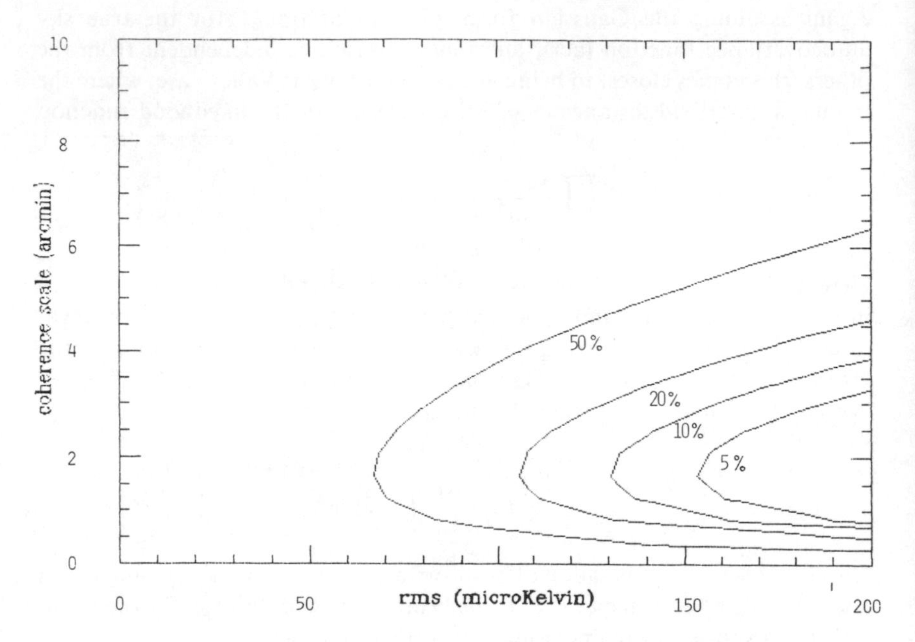

FIG. 3. Contours of likelihood in the $(\sqrt{C_0},\ \theta_c)$ plane for the Uson and Wilkinson data. The percentages are expressed relative to the peak value, which occurs along the axes.

confidence. Currently all such methods — Bayesian, frequentist and fiducial — seem to contain serious problems and inadequacies, but it is arguable that the method with least difficulties, providing a suitable prior can be found, is the Bayesian method. In this spirit, the limits given for the Jodrell Bank results in Section 1 were constructed via a Bayesian integration of the area under the likelihood function multiplied by a prior we believe most suitable for the purpose. In the present context, however, simple comparison of the values of the likelihood functions themselves is sufficient — see Kaiser and Lasenby (1988).

The corresponding contour plot for the Readhead *et al.* (1988) data is shown in Fig. 5. We see immediately that the angular scale of greatest sensitivity is $\theta_c \approx 2\rlap{.}'5$, larger than Uson and Wilkinson, (note that the corresponding FWHM is $5\rlap{.}'9$), with a corresponding 1/20th likelihood point of 56 μK, about three times lower than in the Uson and Wilkinson case. Looking at the 1/20th point of the likelihood as a function of angular scale and thus (essentially by turning the contours on their sides) obtaining a plot of $\sqrt{C_0}$ versus θ_c, a comparison of Fig. 3 with Fig. 5 yields a comparison of the Owens Valley results versus Uson and Wilkinson over a full range of angular scales,

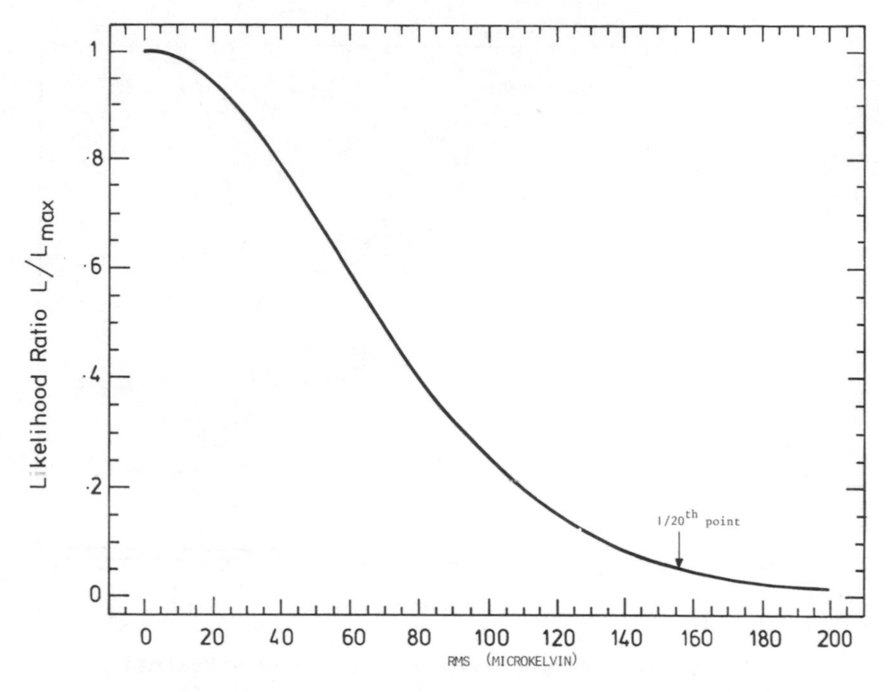

FIG. 4. Slice through the Uson and Wilkinson likelihood contours at the coherence scale of maximum sensitivity, $\theta_c \approx 1'.8$.

and shows that the Readhead *et al.* results *do* represent a very significant advance on the already powerful Uson and Wilkinson upper limits.

Readhead *et al.* (1988) themselves employ a Bayesian method of setting upper limits, with a uniform prior, and find $\Delta T/T \lesssim 1.5 \times 10^{-5}$ at 95% confidence, on a coherence scale of $\approx 2'.5$ [Lawrence has recently (private communication) revised this to 1.9×10^{-5}.] We shall return to the significance of these results in Section 5.

3. Tenerife Results on Intermediate Angular Scales

We should start this section by emphasizing straightaway the collaborative nature of this work, which is carried out jointly with Bob Watson, Ted Daintree and John Hopkins of Jodrell Bank, and John Beckman and Raphael Rebolo of the Instituto de Astrofisica de Canarias. We have been looking at intermediate to large angular scales in the MWB, with resolutions so far of 8° and 5°. It is worth saying that one expects the MWB fluctuations on these scales to be essentially primordial and that the fluctuations in matter density they are associated with will not yet have gone non-linear. This means

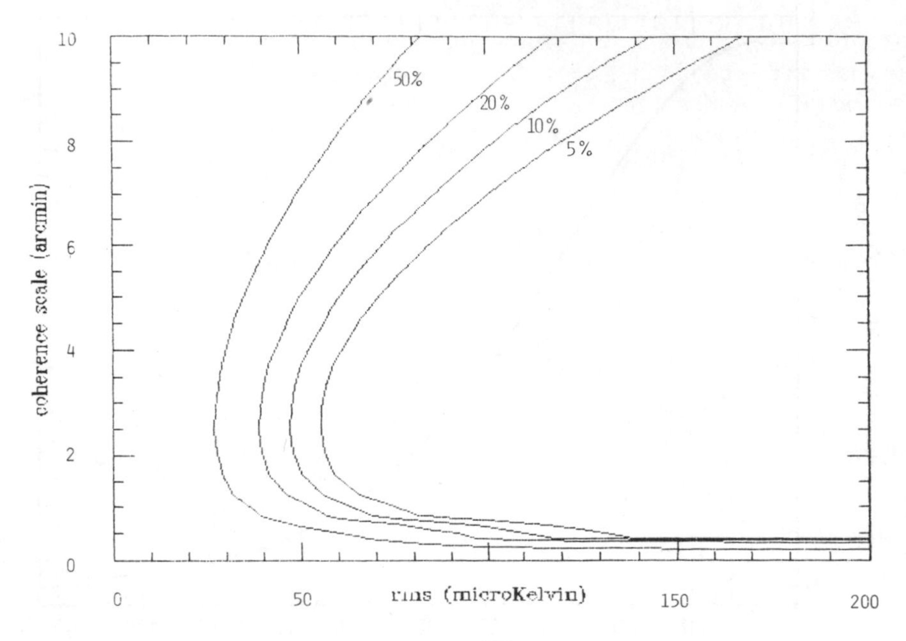

FIG. 5. Contours of likelihood in the ($\sqrt{C_0}$, θ_c) plane for the Readhead *et al.* data.

that on these scales one gets a direct picture of the long wavelength tail of the primordial perturbation spectrum with no complicated physics intervening (including no erasure by possible later reionization, which is not expected to affect scales $\gtrsim 3° - 5°$ on *any* model). Referring to the equation for angular scale in Section 1, we see that the corresponding mass scales are of order 10^{20} $M_{\odot}$, and therefore there is no immediate evidence (from the structures we see around us) that there should be MWB fluctuations on these scales. However, current theories for galaxy formation, e.g. CDM and baryon isocurvature models (Peebles 1987), *do* predict a long wavelength tail to the perturbation spectrum with $\Delta T/T$'s perhaps 5×10^{-6} in an 8° beam for a CDM model, and somewhat higher (because of the different spectral index needed) in baryon isocurvature models. All such values are strongly dependent on Ω and the power law slope n of the primordial perturbation spectrum, and the *detection* of anisotropy on intermediate scales, and knowledge of how it varies as a function of angular scale, would enable one in principle to 'read off' the values of Ω and n, and check the details of galaxy formation theory in a quite unprecedented way. The only comparable information we have comes from the peculiar velocity field on large scales, since this is also expected to correspond to perturbations still in the linear regime.

The experimental set-up used for the Tenerife observations consists of a 10. 4 GHz twin horn drift scan system with a wagging mirror. One can adjust the inclination of the mirror to control the declination being surveyed, and then as the earth rotates the triple beam difference $\Delta T = T_C - \frac{1}{2}(T_E + T_W)$, formed as usual via the combination of beam-switching and azimuthal wagging, is recorded automatically at 80 second intervals, building up a 24 hour RA coverage. Data taken when the Sun is above horizon are discarded, which means that observations of a particular declination strip have to be spaced over a year in order to give complete coverage. The analysis of a section of data at Dec 40° has already been published (Davies *et al.* 1987) and we want here to concentrate on the analysis of the *total* data set, which consists of coverages as shown in the following table:

Declination	*Number of Scans*
45°	22
40°	74
35°	22
25°	21
15°	16
5°	26
0°	57
−5°	12
−15°	25

(Note that there are two independent receivers, and therefore the number of days' observations is the number of scans divided by 2.)

Stacking these implies an rms error on the two regions with greatest coverage (Decs 40° and 0°) of ~ 220 μK per 1° bin of RA. Preliminary versions of the stacked scans for all declinations are shown in Fig. 6. The two galactic plane crossings (one strong, one weak) are clearly visible, illustrating vividly why it is necessary to restrict the analysis for MWB fluctuations to well away from the plane.

The analysis reported previously was of the lowest noise section of the Dec 40° data, from 12^h to 16.6^h in RA, with $|b| > 35°$ throughout. The same likelihood technique as described in Section 2 was used, except that here each "field" overlaps with its neighbours to a very considerable extent, since the beams are ~ 8° wide, but the data are represented by 1° bins. This means one has to compute the likelihood function via a multinormal distribution having a broad diagonal band down the covariance matrix. This turns out still to be feasable however and likelihood contours having a strong peak away

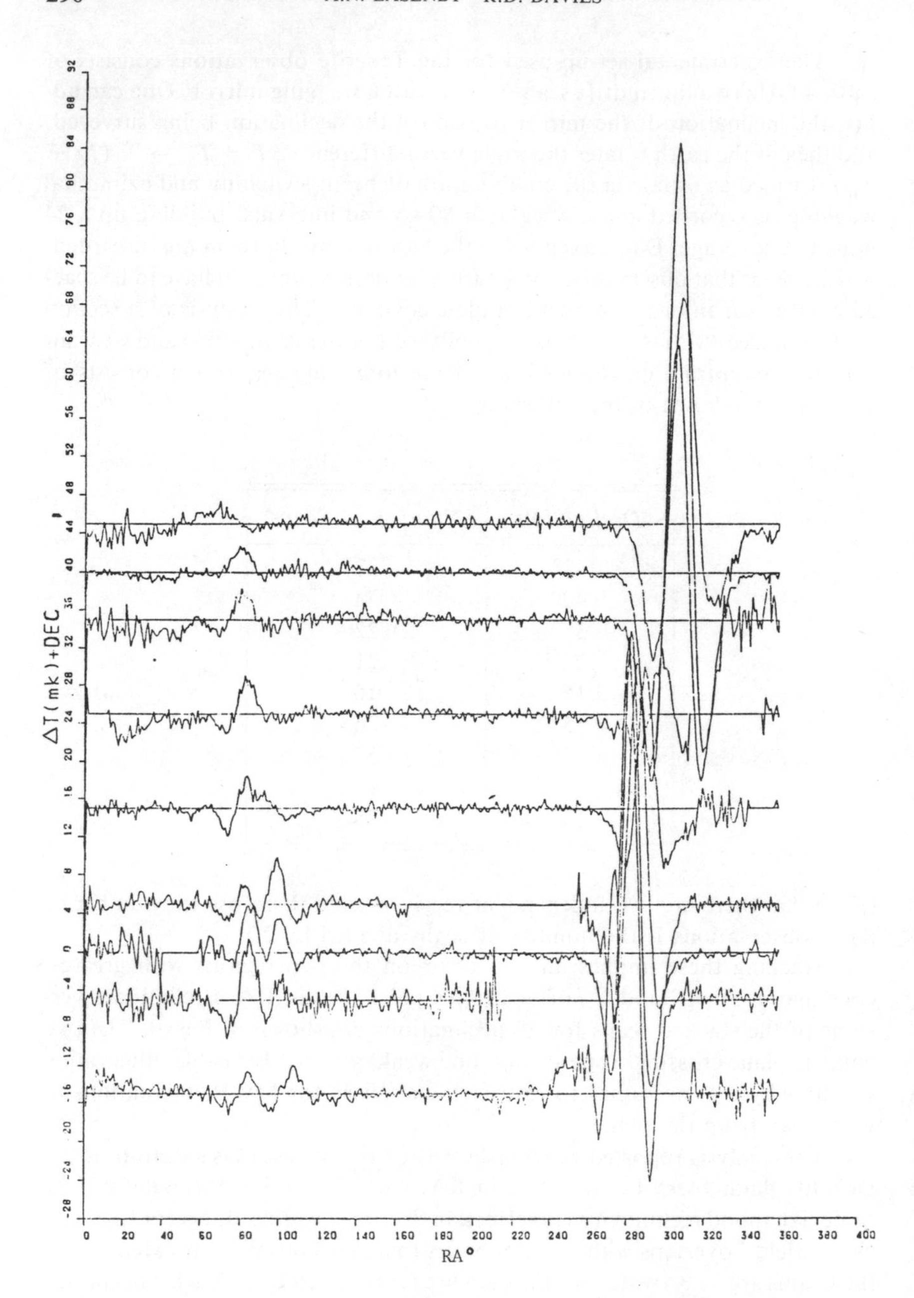

FIG. 6. Preliminary versions of the stacked scans for all the 8° data. The second difference ΔT is plotted against Right Ascension, with the vertical spacing indicating the relative Declinations.

from the origin in the (C_0, θ_c) plane were found (Davies *et al.* 1987) indicating the *detection* of anisotropy, of whatever origin, for this region. The level given was $\sim 3 \times 10^{-5}$ in $\Delta T/T$ in an 8° beam.

It must be recorded here now that there is a slight worry about whether the significance (i.e. height of the likelihood peak) was overestimated in the original analysis due to non-independence of the successive 80 sec data points which go into a 1° RA bin for a given day, perhaps due to large-scale atmospheric irregularities causing offsets with coherence time > 80 secs as they pass across the instrument. Taking such points as statistically independent results in an *under*estimation of the true scatter in the data, and hence an *over*estimation of the significance of any peaks away from zero in the likelihood function. This is currently being investigated, but it appears so far that with proper account taken of possible correlations between these points (by estimating scatter exclusively from variations day to day), the location of the peak in the likelihood contours for the Dec 40° data is not affected, although its significance is reduced.

Another problem with the data is slowly varying baselines, different from day to day but varying on timescales of several hours. These are apparent on visual inspection of some of the data sets and typically have amplitudes of 1 — 2 mK. The instrument is not sensitive to structures on the sky having these scales, so this is not a problem for the 'detection', but the worry is that they make a properly weighted stacking of the data difficult (the results given in Davies *et al.* (1987) were for this reason *un*weighted averages). In order to combine *all* the data from the different declinations properly and deal with the baseline problem, a program has been written which carries out a simultaneous χ^2 minimization with respect to (a) a baseline for each day (represented via Fourier components, with a minimum period of 8 hours) and (b) a two-dimensional model of the microwave sky, from $-30°$ to $+60°$ in declination. Currently it uses a maximum entropy regularizing function as a constraint on the sky model. This is to overcome non-uniqueness problems in the deconvolution of the triple beam pattern and the fact that in some places there is a 10° gap between neighbouring declinations. The results so far look very interesting (Fig. 7a), and suggest that at least as far as the maximum entropy reconstruction goes, there is definite evidence for real structure in the data (as already found statistically via the likelihood technique). Fig. 7a is the result of using the method with the 8° data from declinations 45°, 40° and 35° simultaneously. To help in understanding this grey-scale image (shown with logarithmic levels), also shown, in Fig. 7b, is the equivalent plot for the 408 MHz all-sky survey (Haslam *et al.* 1982), kindly provided in digital form by Dr. J. Osbourne of Durham University. It is clear that the structures near the galactic plane repeat well, but that the variations in the 10.4 GHz data

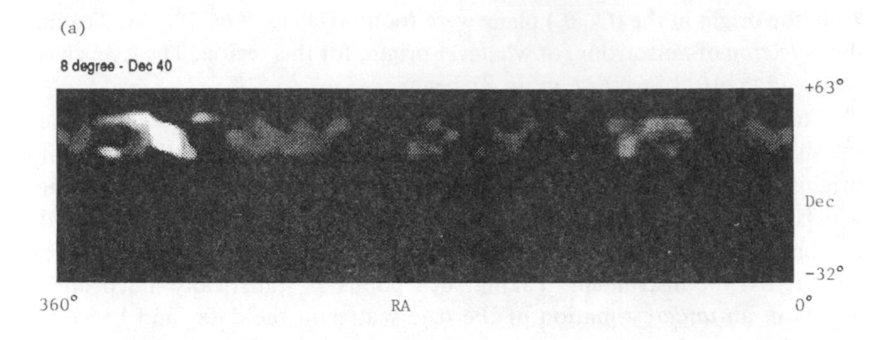

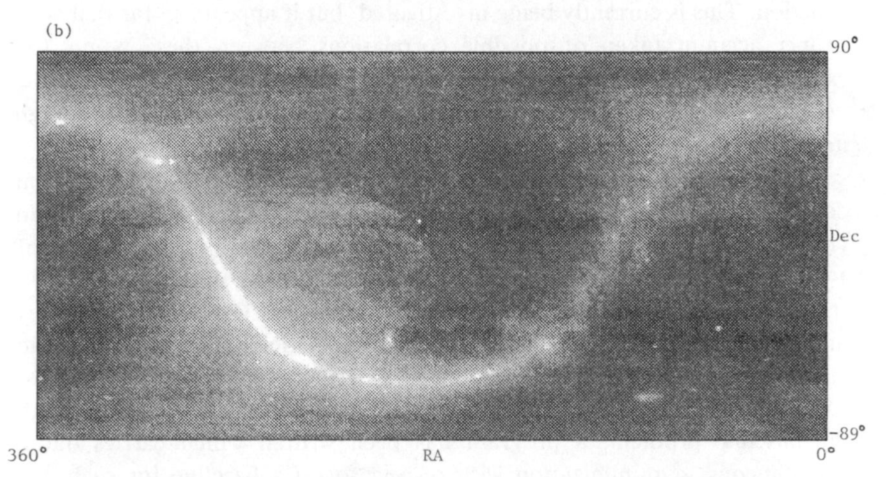

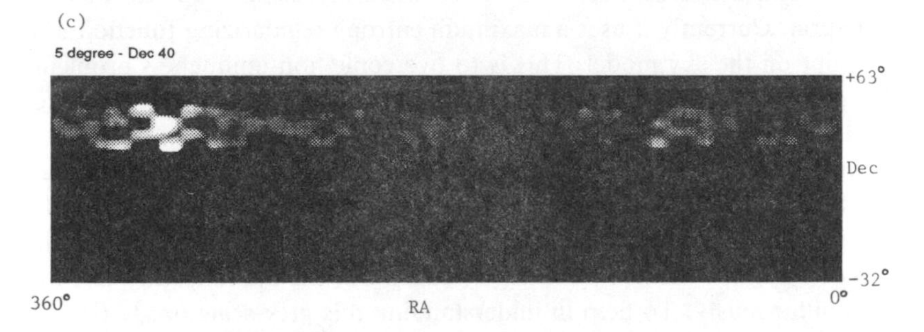

FIG. 7. (a) 2-dimensional reconstruction of the 10.4 GHz microwave sky at Dec 40°, using the 8° data.
(b) 408 MHz all sky continuum survey in the same rectangular projection.
(c) 2-dimensional reconstruction of the 10.4 GHz microwave sky at Dec 40°, using the new 5° data.

at high galactic latitudes do not correspond to what one would predict from the lower frequency data.

Two possibilities immediately present themselves aside from the interpretation that one is looking at a picture of primordial perturbations in the MWB: (a) the maximum entropy reconstruction is producing spurious features, corresponding to an overinterpretation of the 'noise', or (b) the galactic spectral index is *not* uniform, and varies sufficiently to produce the structure seen. Clearly (b) is partially true, in that the heights of galactic plane crossings indicate a .408 — 10.4 GHz index of 2.3-2.4, while in the Dec 0° case, if the high latitude areas had an index this low, we would see much larger fluctuations in the Tenerife data than are observed. These considerations put a lower limit of 2.8-2.9 to the high latitude spectral index. Such a range — from ~ 2.3 at the plane to ~ 2.8 at high latitudes — is consistent with the spectral indices found by Lawson *et al.* (1987), working from lower frequency surveys. It is difficult to repeat the exercise at Dec 40° since the high latitude region here is remarkably flat in the 408 MHz map and the same lower limits to the high latitude index do not arise. Put differently, we believe the possible galactic component is fairly small here (Davies *et al.* 1987). However, to pursue this further, it is obviously necessary to have some data at intermediate frequencies, and to this end observations are being made at Jodrell Bank using a 5 GHz twin horn interferometer system operating in the same drift scan mode as the switched beam system at Tenerife, and with a resolution of ~ 2°. Preliminary results from this are already becoming available, and while they are encouraging for a non-galactic origin of the fluctuations seen at Dec 40°, it is still too early to give a definitive answer. Ultimately, the best way to prove the cosmological origin of the 10.4 GHz fluctuations will be to work at higher frequencies, where the galaxy will unquestionably have sunk to low enough levels, in both thermal and non-thermal emission, to be negligible (see Section 4).

As regards (a), the only satisfactory way is to repeat the experiment to obtain an independent data set, and this has now been done, at a resolution of 5° rather than 8° (achieved by extending the horns, to make them longer and broader. Note the latter has the disadvantage of increasing the sensitivity to discrete radio sources, and these start to be someting that must be taken into account in assessing the 5° results. This will be carried out in future work to be published, but the results presented below have not been corrected for this). The 5° data set has comparable coverage and temperature sensitivity to the 8° data. The results of the χ^2 fitting program for four declinations near 40° (Decs 45°, 40°, 37°.5 and 35°) are shown in Fig 7c. Clearly the resolution is different, but there is an exciting similarity to Fig 7a in the position of the main features in the reconstructuion, suggesting that these features are not mere artifacts of the maximum entropy process. Again, these results are preliminary only, but they point forward to perhaps a new direction in MWB anisotropy studies, where we can start discussing individual features on maps, rather than statistical analyses only.

Before leaving this area, one should mention the interesting agreement between the levels of microwave anisotropy indicated by these studies (upper limits to the primordial anisotropy if the galactic contribution is significant) and the 1981 results of Melchiorri *et al.* who worked with a mm/sub-bolometer system, with 6° beams. Both imply $\Delta T/T$ levels of $\sim 3 \times 10^{-5}$, but in the Melchiorri case also, significant contamination from galactic emission (at their wavelengths due to dust) is probably present, although they attempted to subtract this. Direct comparison of the two data sets is difficult due to the completely different beam switching geometries and scanning techniques, but if a 2-dimensional equivalent single beam reconstruction could be formed from the Italian data, perhaps by methods similar to those used to produce Fig. 7, then the comparison could be very significant.

4. Future Experiments

Obviously many groups around the world are currently working on new experiments and here only a limited list will be attempted, comprising experiments related to work we are involved in at our respective institutions. However, two broad lessons for the future can perhaps already be drawn. The first of these is that on small (arcminute to 1°) scales, single dishes are beginning to approach the limits of their usefulness and it is likely that interferometers will be the key to future breakthroughs to sensitivity levels of a few $\times 10^{-6}$ (rather than 10^{-5}) in $\Delta T/T$. One pressing reason for this is that studies as sensitive as those at Owens Valley are now beginning to approach the 'confusion limit', where signals from discrete, weak radio sources, at a level of $\sim$ 1 per beam, begin to limit further progress. An interferometer can avoid this, by providing a tuneable range of angular scales, in which discrete radio sources can be filtered out (on the basis of their small angular size in relation to the MWB fluctuations sought) and sensitivity to fluctuations retained, provided the interferometer has enough short baselines. At Cambridge, the 5-km telescope is being enhanced in sensitivity by a factor 20 (via a combination of cooled front ends and broadband electronics) and the dishes moved closer together to turn it into a premier instrument for observing MWB fluctuations on sub-arcmin and arcmin scales. Exciting results can be expected from this, particularly perhaps in the area of Sunyaev-Zeldovich astronomy (Saunders 1988), in the near future. For work on slightly larger scales, a few arcmin up to maybe 2°, Cambridge is proposing a completely new instrument, the Very Small Array (VSA), which would consist of 10-20 elements in a dense pack configuration, with maximum baseline 12 m, operating at 15 GHz. The angular scales involved here come within the province of the second "lesson", namely that we will see an increasing trend towards experiments on intermediate scales where (a) the currently popular theories indicate the $\Delta T/T$ fluctuation power should be largest (e.g. the coherence scale for CDM fluctuations is ~ 15 arcmin) and (b) as already sketched in Section 3, the physics involved becomes simpler as one goes up in angular scale, with fewer complicated non-linear processes separating us from the genuine primordial perturbations.

On these larger scales, several Jodrell Bank experiments are planned as follow-ups to the 8° and 5° Tenerife experiments. These are: (1) A 10.4 GHz ~ 2° resolution experiment, based on small paraboloids rather than horns, and already in place in Tenerife. [This is in fact an adaptation of the equipment originally used by Mandolesi *et al.* (1986)]; (2) A 15 GHz ~ 4° resolution switched beam twin horn experiment expected to come into operation in Tenerife late 1988; (3) A 30 Ghz ~ 3° resolution switched beam twin horn experiment sited initially in Tenerife, but with a possible future move to the drier observing conditions of Antartica.

It is via instruments (2) and (3) in particular, that the possible fluctuations found at 10.4 GHz can be confirmed as cosmological in origin or not.

5. Relation to Theory

Several aspects of the consequences for theory of anisotropies at a level $\sim 3 \times 10^{-5}$ in $\Delta T/T$ in an 8° beam have already been discussed in Davies *et al.* (1987). Here we just want to highlight what seem to be some pertinent additional current issues in MWB anisotropy studies.

1. The recent Berkeley/Nagoya result (Matsumoto *et al.* 1988), indicating a radiation excess in the Wien region of the MWB spectrum may be about to change completely our picture of conditions at intermediate redshift (z = 10 — 100). This will have important consequences for FIR anisotropies and, more relevant to the radio studies discussed here, it may mean we have to consider seriously scenarios in which the universe is reionized completely at early epochs. This would then point attention strongly towards those theories (e.g. the recent baryon isocurvature models) in which reionization is a natural ingredient.

2. Given that reionization occured, how does it affect e.g. the Readhead *et al.* and Uson and Wilkinson results? The recently reported 'Vishniac effect' is exciting in showing us that on small enough scales, fluctuations are actually boosted by the reionization (see contribution by Efstathiou, 1988).

3. In the context of bayron-dominated models with reionization, the joint constraints on small scales (e.g. Readhead *et al.* and Uson and Wilkinson) and intermediate scales (Davies *et al.* and Melchiorri *et al.*) seem close to ruling out $\Omega < 1$ universes completely (again see Efstathiou contribution). In this case, if we want to retain a baryon-dominated cosmology presumably we would have to accept "non-standard" nucleosynthesis, in which there is a segrega-

tion of protons and neutrons following the quark-hadron phase transition (see Applegate and Hogan 1985) in order to avoid previous nucleosynthesis constraints on light element production, which implied $\Omega_{baryon} \lesssim 0.1$.

4. If the Davies *et al.* and Melchiorri *et al.* results are taken as detections of primordial anisotropy, then what are the consequences for predictions of large-scale streaming motions? Obviously this is a key question for this Study Week. Our previous impression has been that both the reported large-scale motions and the MWB anisotropy (if confirmed) would be too large for the CDM picture, and in fact a severe problem for it. This still seems to be true for the MWB anisotropies. However, if what Nick Kaiser says is correct (Kaiser and Lahav 1988), namely that the Great Attractor picture and its associated velocity field are compatible with CDM, then that presumably means that the MWB anisotropy 'detections' at the 3×10^{-5} level on scales of $6° — 8°$ are *too large for the velocity fields we observe*. What would be very helpful is a relatively model independent route from a measurement of MWB anisotropy (on intermediate scales) to a prediction for rms velocities on a given scale, and vice versa. Since this contribution was given, a paper by Juszkiewicz *et al.* (1987) has appeared, providing a means of setting at least lower bounds to each quantity in terms of the other. This apparently confirms that the Davies *et al.* result is bigger than the minimum required by a factor $\sim$ 10, if the coherence scale and amplitude for the velocity streaming as found from the elliptical galaxy studies (Dressler *et al.* 1987) are correct. A coherence scale approximately 3 times larger however, $\sim 180\ h^{-1}$ Mpc (h here in units of 100 km s^{-1} Mpc^{-1}), would dramatically alter the situation, leading to minimum fluctuations at about the level seen.

5. Finally, will structures such as the Great Attractor or, on the other hand, large voids, about which we have heard much this week, themselves be directly evident in the MWB at detectable levels (rather than via their effect, in surrogate forms, in terms of their partners at earlier epoch)? This is an old subject (see e.g. Rees and Sciama 1968) but also a promising line of enquiry for the future.

Acknowledgements

We would again like to thank the many colleagues and collaborators whose work has made the Tenerife observations possible, among them Bob Watson, John Hopkins, Ted Daintree, John Beckman, Jorge Sanchez-Almeida, and Raphael Rebolo.

REFERENCES

Applegate, J.H. and Hogan, C.J. 1985. *Phys. Rev.* **D31,** 3037.

Bond, J.R. and Efstathiou, G. 1987. *MN.* **226,** 655.

Cole, S. and Kaiser, N. 1988. Submitted to *MN.*

Davies, R.D., Lasenby, A.N., Watson, R.A., Daintree, E.J., Hopkins, J., Beckman, J., Sanchez-Almeida, J., and Rebolo, R. 1987. *Nature.* **326,** 462.

Dressler, A., Faber, S.M., Burstein, D., Davies, R.L., Lynden-Bell, D., Terlevich, R.J., and Wegner, G. 1987. *Ap J.* **313,** L37.

Efstathiou, G. 1988, this volume.

Haslam, C.G.T., Salter, C.J., Stoffel, H., and Wilson, W.E. 1982. AA *Suppl.* **47,** 1.

Juszkiewicz, R., Górski, K., and Silk, J. 1987. *Ap J.* **323,** L1.

Kaiser N. and Lahav, 0. 1988, this volume.

Kaiser, N. and Lasenby, A.N. 1988. Preprint.

Kellermann, K.I., Fomalont, E.B., Weistrop, D., and Wall, J. 1986. in J.-P. Swings, editor, *Highlights of Astronomy — Volume 7,* p. 367.

Lawrence, C.R., Readhead, A.C.S., and Meyers, S.T. 1988. *The Post Recombination Universe.* eds. N. Kaiser and A.N. Lasenby (NATO ASI), in press.

Lawson, K.D., Mayer, C.J., Osbourne, J.L., and Parkinson, M.L. 1987. MN **225**, 307.

Mandolesi, N., Calzolari, P., Cortiglioni, S., Delpino, F., Sironi, G., Inzani, P., De Amici, G., Solheim, J.-E., Berger, L., Partridge, R.B., Martenis, P.L., Sangree, C.H., and Harvey, R.C. 1986. *Nature.* **319,** 751.

Martin, H. and Partridge, R.B. 1987. Preprint.

Matsumoto, T., Hayakawa, S., Matsuo, H., Murakami, H., Sato, S., Lange, A.E., and Richards, P.L. 1988. *Ap J.* In press.

Melchiorri, F., Melchiorri, B.O., Ceccarelli, C., and Pietranera, L. 1981. *Ap J.* **250,** L1.

Ostriker, J.P. and Vishniac, E.T. 1986. *Ap J.* **306,** L51.

Padin, S., Davis, R.J., and Lasenby, A.N. 1987. *MN.* **224,** 685.

Peebles, P.J.E. 1987. *Ap J.* **315,** L73.

Readhead, A.C.S., Lawrence, C.R., Meyers, S.T., and Sargent, W.L.W. 1988. *The Post Recombination Universe* eds. N. Kaiser and A.N. Lasenby (NATO ASI). In press.

Rees, M.J. and Sciama, D.W. 1968. *Nature.* **217,** 511.

Saunders, R.D.E. 1988. *The Post Recombination Universe* eds. N. Kaiser and A.N. Lasenby (NATO ASI). In press.

Uson, J. and Wilkinson, D.T. 1984a. *Ap J.* **283,** 471.

_____. 1984b. *Nature.* **312,** 427.

Waymont, D.K., Lasenby, A.N., Davies, R.D., Davis, R.J., and Padin, S. 1988. In preparation.

EFFECTS OF REIONIZATION ON MICROWAVE BACKGROUND ANISOTROPIES

G. EFSTATHIOU
Institute of Astronomy, Cambridge

SUMMARY

Star formation, or other sources of energy, could have reionized the intergalactic medium at high redshifts. Primary anisotropies in the microwave background would be erased on small angular scales, but new fluctuations would be generated as a direct result of the peculiar motions of the scatterers. We summarize the key physical effects responsible for these secondary fluctuations and show that the new anisotropy limits from Owens Valley set stringent constraints on purely baryonic models with Gaussian isocurvature initial conditions.

1. INTRODUCTION

Limits on the temperature fluctuations in the microwave background radiation have proved to be powerful constraints on theories of galaxy formation. On large angular scales ($\theta \gtrsim 5°$), anisotropies are generated by spatial fluctuations in the curvature (the Sachs-Wolfe effect, Sachs and Wolfe 1967), and by fluctuations in the entropy (the "isocurvature" effect, Efstathiou and Bond 1986). These are insensitive to the recombination history, so we can set tight limits on the shape of the primordial fluctuation spectrum (see *e.g.* the article by Vittorio 1988) without worrying too much about energy injection into the intergalactic medium (IGM), except insofar as this might modify the matter distribution on small scales thereby affecting the normalization of the initial fluctuations.

On angular scales $\lesssim 1°$, the main source of anisotropy arises from the scattering of photons off moving electrons. Inferences from small-scale anisotropy limits are, therefore, extremely sensitive to the assumed recombination history. Most authors have assumed that recombination occurs at the usual time of $z \sim 1000$ (Peebles 1968) and that the IGM remains neutral thereafter. Yet the lack of a Lyα absorption trough in quasar spectra implies that the IGM must have been reionized at redshifts $z \gtrsim 4$. If reionization of the IGM happened early enough, the small-scale anisotropy pattern generated at $z \gtrsim 1000$ (here called the "primary" anisotropies) would have been severely modified.

The possibility that secondary reheating has affected fluctuations in the microwave background radiation deserves to be taken seriously for at least the following reasons:

1. Suppose that the IGM remained fully ionized; the optical depth from Thomson scattering between redshift z and the present epoch would be

$$\zeta = \frac{c\sigma_T H_0}{4\pi G m_p} (1 - Y_{He}/2)\ (\Omega_B/\Omega^2)\ (2 - 3\Omega + (1 + \Omega z)^{1/2}\ (\Omega z + 3\Omega - 2)), \quad (1)$$

where Y_{He} is the abundance of Helium (by mass), Ω is the cosmological density parameter at the present epoch, and Ω_B is the contribution of baryons to Ω. In dark matter dominated universes, the redshift z_c at which $\zeta = 1$ can be quite high, *e.g.*, if $\Omega \approx 1$, $\Omega_B \approx 0.1$, $h \approx 0.75$, then $z_c \approx 50$; but for baryon dominated models z_c can be substantially lower. Clearly there is no requirement that energy be injected into the IGM at high redshits ($z \gg z_c$) for reionization to have substantially modified the primary anisotropy pattern.

2. The IGM need not have been heated to a high temperature ($T_e \gg 10^4$K). Correspondingly, there are no general contraints on reionization (at interesting redshifts) based on either the energy outlay or on spectral distortions of the cosmic microwave background.

3. Non-linear structures must have formed before z_c to reheat the IGM (*e.g.* by photoionization due to massive stars). This would be quite natural in some models of galaxy formation (*e.g.* isocurvature baryon models).

4. In some specific models (*e.g.* pancake, or scale-invariant cold dark matter models), the first non-linear structures are expected to form at recent epochs, so reionization at $z \gtrsim z_c$ might seem implausible. However, this depends critically on the assumption of a pure power law initial spectrum extending to small scales. Reionization of the IGM could have occurred if "primordial seeds" (*e.g.* cosmic strings) caused a small fraction of the baryons to collapse into stars at $z \gtrsim z_c$.

One might imagine that the main effect of reionization at $z > z_c$ would be to erase anisotropies on all scales smaller than the horizon length at z_c (subtending an angle $\theta_c \sim (\Omega/z_c)^{1/2}$). However, just as peculiar motions at $z \sim 1000$ generate small-angle anisotropies, peculiar motions at $z \sim z_c$ can generate significant secondary anisotropies. These secondary temperature fluctuations can be analysed in considerable detail without recourse to complicated numerical calculation. In Section 2 we review computations of velocity-induced temperature fluctuations using first-order perturbation theory. These are generally far below current upper limits for universes with $\Omega_B \gtrsim 0.1$. However, Vishniac (1987) has recently shown that second-order contributions can dominate over the first-order effects on arcminute scales; this effect is reviewed in Section 3.

Throughout, we will apply the results to baryon isocurvature models. Peebles (1988) has mentioned some of the attractive features of this type of model (see also Peebles 1987a). Firstly, baryons are assumed to dominate the mass density of the universe; thus $\Omega \approx \Omega_B$ and there is no need to appeal to exotic forms of dark matter such as axions or gravitinos. Secondly, the value of Ω_B deduced from dynamical studies (*e.g.* Davis and Peebles 1983, Bean *et al.* 1983) is in reasonable agreement with the low value implied by the standard theory of primordial nucleosynthesis (Yang *et al.* 1984). This may indicate that all of the "missing mass" is baryonic and that the universe is open. Thirdly, for certain choices of initial conditions, isocurvature baryon models produce peculiar velocity fields with a large coherence length (Peebles 1987b), consistent with observations of coherent flows reported by, for example, Rubin *et al.* (1976) and Dressler *et al.* (1987).

Unfortunately, there is no known physical mechanism that could give rise to entropy fluctuations with an acceptable spectrum. The most likely outcome of an early inflationary phase would be a flat universe with scale invariant perturbations. In a baryon isocurvature model, these initial conditions would lead to large angle temperature fluctuations well in excess of the observational limits (see equation 31 below). Of course, the inflationary picture could be wrong since it is based on a bold extrapolation of particle physics to extraordinarily high energies. If we abandon current ideas on inflation, we are allowed considerable freedom in the choice of initial conditions. For example, to define a baryon isocurvature model, we need to specify the following free parameters and functions: the background cosmology is unconstrained, so we can vary Ω_B, the Hubble constant H_0 and the cosmological constant Λ; the initial spectrum of entropy fluctuations need not be a power-law, but even if it were, the spectral index n and the amplitude are free parameters; we know nothing about the statistics of the initial fluctuations, we have no reason to assume that they are Gaussian; the shape of the post-recombination fluctuation spectrum depends on the history of the IGM. Peebles (1987a) has applied the term "minimal" to the baryon isocurvature model, presumably because hypothetical dark matter particles are not needed, but it should be recognised that a great deal of new physics is required to fix the initial conditions.

There is such a large family of baryon isocurvature models that it would be difficult to test each one against observations using time consuming techniques such as N-body simulations. Nevertheless, these models are worth investigating because it would be extremely important if we could demonstrate that gravitational instability in a purely baryonic universe could not be made to work. Calculations of the microwave background anisotropies provide an inexpensive method of narrowing the options. The calculations of Efstathiou and Bond (1987) show that low density baryon isocurvature models with a standard recombination history are convincingly excluded by anisotropy limits on arcminute scales. This is a weak constraint since, as argued above, reionization is quite likely in these models. We will show below that the new anisotropy limits obtained by Readhead and collaborators at Owens Valley set strong constraints even if the IGM were reionized at $z \gtrsim z_c$.

2. First-Order Velocity-Induced Anisotropies in Reionized Models

The perturbation to the photon distribution function may be expressed as $\delta f_T(\mathbf{q}, \mathbf{x}, t) = T(\partial \bar{f}/\partial T)\Delta_T/4$, where $\bar{f}$ is the Planck function corresponding to temperature T and $\mathbf{q}$ is the comoving photon momentum. The Fourier transform of the equation describing radiative transfer is

$$\dot{\Delta}_T + ik\mu\Delta_T = \sigma_T\tilde{n}_e a(\Delta_{T0} - \Delta_T + 4\mu v_B), \ \mu = \hat{\mathbf{k}}.\hat{\mathbf{q}}, \tag{2}$$

where a is the cosmological scale factor, dots denote differentiation with respect to conformal time $\tau = \int dt/a$ and Δ_{T0} is the isotropic component of $\Delta_T(k, \mu, \tau)$. In writing equation (2) we have assumed that Thomson scattering is isotropic and we have ignored gravitational source terms on the *rhs* which are unimportant for the short wavelength perturbations of interest here. Furthermore, in this Section we ignore spatial variations in the electron density n_e, *i.e.* we have used the spatial average $\tilde{n}_e$ in (2); we shall see in the next Section that fluctuations in n_e can lead to important second order contributions to the small angle anisotropies (Ostriker and Vishniac 1986, Vishniac 1987). The baryon peculiar velocity v_B is related to the fractional perturbation in the baryon density by the equation of continuity

$$v_B = -\frac{\dot{\delta}_B}{ik}. \tag{3}$$

Since z_c is generally $\ll 150(\Omega h^2)^{1/5}$ (*i.e.* we can ignore Compton drag) but $>$ $(1/\Omega - 1)$ (*i.e.* $a \propto \tau^2$), we may assume that δ_B follows the usual growth rate for linear perturbations in an $\Omega = 1$ universe, $\delta_B \propto \tau^2$.

The solution to (2) is given by

$$\Delta_T(\tau, \mu, k) = \int_0^\tau [\Delta_{T0} + 4\mu v_B]_{\tau'}\, g(\tau, \tau')e^{ik\mu(\tau' - \tau)}\, d\tau', \tag{4}$$

where

$$g(\tau, \tau') = [\sigma_T\bar{n}_e a]_{\tau'}\, exp\, [-\textstyle\int_{\tau'}^{\tau}\sigma_T\bar{n}_e a\, d\tau],$$

and the terms in square brackets are evaluated at τ'. Solutions to (2) in the short-wavelength limit $k\tau \gg 1$ have been discussed by a number of authors (*e.g.* Sunyaev 1978, Davis and Boynton 1980, Silk 1982) who ignored the isotropic term Δ_{T0} on the *rhs,* and by Kaiser (1984) who showed that the isotropic term in (4) contributes a $\Delta T/T$ of similar magnitude to the baryon velocity term.

The zeroth moment of (2) gives

$$\Delta_{T0} = \int_0^\tau \Delta_{T0} g(\tau, \tau') j_0(k(\tau - \tau'))\, d\tau' - \int_0^\tau 4v_B i g(\tau, \tau') j_1(k(\tau - \tau'))\, d\tau'. \tag{5}$$

The integrands in (5) give significant contributions if $k(\tau - \tau') \lesssim 1$, while for $k(\tau - \tau') \gg 1$ the spherical bessel functions oscillate between positive and negative values. For short wavelengths ($k\tau \gg 1$), the slowly varying terms g, Δ_{T0} and v_B may be taken out of the integrals and replaced with their values at τ. It is this "localized" nature which allows a simple analytic solution for short wavelength perturbations (Kaiser 1984). The first integral in (5) is negligible, since it is order $\Delta_{T0}\zeta_\lambda$, where ζ_λ is the optical depth across one wavelength ($\lambda = 2\pi/k$). The isotropic component is thus

$$\Delta_{T0} \approx -\left[\frac{4v_B i}{k}\right]_\tau g(\tau, \tau), \tag{6}$$

and the solution (4) may be written

$$\Delta_T(k, \mu, \tau) \approx \left[\frac{8\delta_B}{(k\tau_c)^2}\right]_{\tau_c} e^{-ik\mu\tau} \int_0^\tau \tau'[g(\tau', \tau') + ik\mu] g(\tau, \tau') e^{ik\mu\tau'}\, d\tau', \tag{7}$$

where we have used equation (3) and assumed $\delta_B \propto \tau^2$.

Our goal is to compute the observed temperature pattern on the sky. This can be described statistically by the temperature autocorrelation function,

$$C_T(\theta) = \langle \Delta T/T(\mathbf{q}, \mathbf{x}, \tau_o)\, \Delta T/T(\mathbf{q}', \mathbf{x}, \tau_o)\rangle, \ \cos\theta = \hat{\mathbf{q}}.\hat{\mathbf{q}}'.$$

Generally, the background radiation is observed with a finite beam-width which we will approximate as a Gaussian of width σ. A convenient expression for the smoothed autocorrelation function $C_T(\theta, \sigma)$, valid for small angles $\theta \ll 1$, has been given by Doroshkevich, Sunyaev, and Zeldovich (1978),

$$C_T(\theta, \sigma) = \frac{V_x}{32\pi^2} \int_0^\infty \frac{1}{2} \int_{-1}^{+1} |\Delta(\tau, k, \mu)|^2 J_o(kR_c\theta(1 - \mu^2)^{1/2}) \exp(-(kR_c\sigma)^2(1 - \mu^2))\, d\mu k^2\, dk, \tag{8}$$

where

$$R_c = L_c \sinh(\tau_o/L_c),\ L_c = H_o^{-1}(1 - \Omega)^{-1/2} a_o^{-1},$$

$\tau \gg \tau_c$, and the perturbations are assumed to be periodic in a large box of volume V_x. Equation (8) can be simplified by noting the following points: (i) For free-streaming photons (neglecting gravity), the quantity $|\Delta_T(k, \mu, \tau)|^2$ is conserved. Provided scattering is unimportant at $\tau \ll \tau_0$, which is generally true (see equation (1)), we may choose any arbitrary epoch τ in (8). (Another consequence of $\tau_c \ll \tau_0$ is that the angle-length relation $(\theta \sim 1/(kR_c))$ is independent of τ_c). (ii) The quantity $|\Delta_T(k, \mu, \tau)|^2$ is a highly peaked function with width $\mu \sim 1/(k\tau_c)$. This can be seen immediately from (7), since the integral is approximately the Fourier transform of a broad function of width $\sim \tau_c$. This simply expresses the fact that for short wavelength plane waves, $k\tau_c \gg 1$, many wavelengths will lie across the last-scattering shell (and so their contributions to $\Delta T/T$ will cancel) unless the wavenumber $\mathbf{k}$ is oriented almost at right angles to the direction of the observer. We can therefore approximate (8) as

$$C_T(\theta, \sigma) \approx \frac{V_\chi}{32\pi^2} \int_0^\infty W_T^2(k) J_o(kR_c\theta) \exp(-(kR_c\sigma)^2) k^2\, dk, \quad (9a)$$

where

$$W_T^2(k) = \frac{1}{2} \int_{-1}^{+1} |\Delta_T(k, \mu, \tau_0)|^2\, d\mu. \quad (9b)$$

Applying Parseval's theorem to (7) we find

$$W_T^2(k) = \left[\frac{64\delta_B^2(\tau_c)}{(k\tau_c)^5}\right] \pi I_T, \quad (10a)$$

$$I_T \approx \int_0^\infty (G_I - dG_v/dx)^2\, dx,\ x \equiv \tau'/\tau_c, \quad (10b)$$

and

$$G_I = (\tau/\tau_c) g(\tau, \tau) g(\tau_o, \tau) \tau_c^2,$$
$$G_V = (\tau/\tau_c) g(\tau_0, \tau) \tau_c.$$

The integral (10b) can be evaluated analytically for many interesting ionization histories. For example, if all of the baryonic material remains ionized $\tilde{n}_e \propto a^{-3}$, then

$$g(\tau_o, x) \approx 3/\tau_c x^{-4} e^{(-1/x^3)}, \quad (11a)$$

and so

$$I_T \approx 81 \int_0^\infty x^{-8} e^{(-2/x^3)} = \frac{27}{2^{7/3}} \Gamma(7/3) = 6.38. \qquad (11b)$$

It is also easy to verify that in a fully ionized universe, retaining only the velocity term G_v reduces I_T by a factor of about 2 while retaining only the isotropic term increases I_T by a factor of about 2. As noted by Kaiser (1984), the velocity and isotropic terms tend to cancel in the full integral (10b).

We will now apply these results to purely baryonic universes with isocurvature initial conditions (Peebles 1987a). The initial perturbation in the entropy per baryon is assumed to be a power law, $|S_\gamma|^2 \propto k^n$. The power spectrum for linear baryon density fluctuations at the present epoch is related to $S_\gamma(k)$ by a transfer function $T(k)$,

$$P(k, \tau_o) = \langle|\delta_B(k, \tau_o)|^2\rangle = T^2(k)|S_\gamma(k)|^2.$$

On scales smaller than about the horizon length at the epoch when radiation and matter have equal densities ($k_{equ}^{-1} \approx 5\ (\Omega_B h^2)^{-1}$ Mpc), $P(k)$ retains the initial power law form $\propto k^n$, but on scales much larger than k_{equ}^{-1}, $T(k) \propto k^2$. Transfer functions for various baryon isocurvature models are plotted by Peebles (1987b) and Efstathiou and Bond (1987). We therefore assume

$$|\delta_B(k, \tau_c)|^2 = Ak^n, \ k \gg k_{equ} \gg 1/\tau_c, \qquad (12)$$

and so from (10a) $W_T^2 \propto k^{(n-5)}$. From equation (8), we see that the correlation function at zero lag diverges if $n < 2$, *i.e.* the contribution to $C_T(0)$ is dominated by long wavelength perturbations and so is sensitive to the detailed shape of the transfer function $T(k)$ and to the behaviour of long wavelength perturbations with $k\tau_c \sim 1$. However, the sky variance measured by "triple-beam" arrangements, such as that used in the Uson and Wilkinson (1984) experiment, is related to $C_T(\theta, \sigma)$ by

$$\left(\frac{\Delta T}{T}\right)_\theta^2 = 2[C_T(0) - C_T(\theta)] - \frac{1}{2}[C_T(0) - C_T(2\theta)], \qquad (13)$$

(Bond and Efstathiou 1984) which is, of course, insensitive to long wavelengths and converges if $n > -2$. Evaluating (13), we find

$$\left(\frac{\Delta T}{T}\right)_\theta^2 = \frac{2I_T}{\pi\tau_c^5} AV_x(R_c\sigma)^{(2-n)}\varphi^2(n, \theta/(2\sigma)), \qquad (14a)$$

where the function φ^2 is

$$\varphi^2(n, x) = \sum_{m=2}^{\infty} \frac{(-1)^m}{(m!)^2} \Gamma(n/2 - 1 + m) x^{2m} [2^{(2m-2)} - 1], \quad (14b)$$

and is plotted in Figure 1 for three values of the spectral index n. For $\chi \gg 1$, the assymptotic behaviour of φ^2 is

$$\begin{aligned} \varphi^2(n, x) &\rightarrow \frac{\Gamma(n/2 + 1)}{\Gamma(2 - n/2)} \frac{4}{n(2 - n)} [1 - 2^{-n}] x^{(2-n)}, \quad x \gg 1, \; n \neq 0, \\ &\rightarrow 2x^2 \ln 2 \quad x \gg 1, \quad n = 0. \end{aligned} \quad (14c)$$

To normalize the amplitude of $\Delta T/T$, we match the second moment of the matter autocorrelation function

$$J_3(x_o) = \int_0^{x_o} \xi x^2 \, dx, \quad (15a)$$

at $x_o = 10h^{-1}$ Mpc, to the value $J_3(10h^{-1}\text{ Mpc}) \approx 270\ h^{-3}\text{ Mpc}^3$, deduced using the galaxy correlation function of the CfA redshift survey (Davis and Peebles 1983). Thus

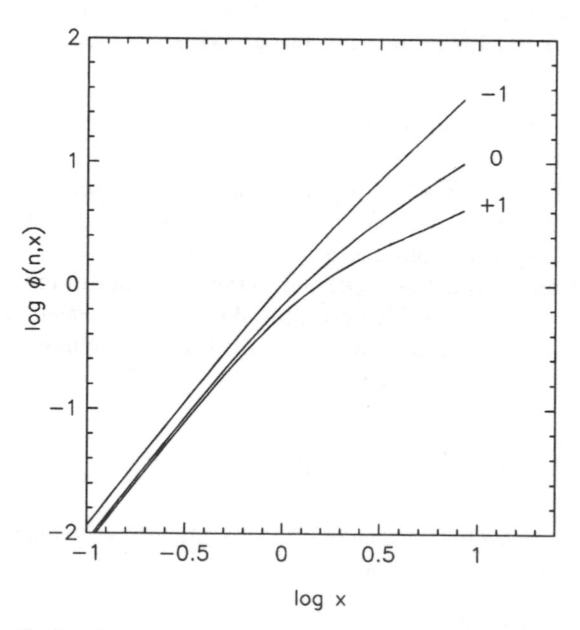

FIG. 1. The function $\varphi(n, x)$ defined in equation (14b) plotted for three values of the spectral index n, $n = -1, 0, +1$.

$$AV_x = \frac{2\pi^2}{p_n^2} \frac{J_3(x_o)x_o^n n}{\Gamma(n+2)\sin(n\pi/2)} \left[\frac{D(\tau_c)}{D(\tau_o)}\right]^2, \tag{15b}$$

where $D(\tau)$ is the growth factor for linear perturbations (Peebles 1980, equation 11.16) and the factor p_n is unity if the power spectrum is a pure power law. Values of p_n are tabulated by Efstathiou and Bond (1987) for various baryon isocurvature models, but even if $\Omega = 1$ and $n \sim -2$, $p_n \lesssim 2$. Notice that equations (14a) and (15b) imply the approximate scaling

$$(\Delta T/T)_\theta \propto h^{-1/6}\, \Omega^{(3n-5)/6}.$$

Thomson scattering can induce a net linear polarization in the microwave background fluctuations (Kaiser 1983). A linear polarization amplitude Δ_P may be defined in an analogous way to the total amplitude Δ_T and obeys the equation

$$\dot{\Delta}_P + ik\mu\Delta_P = \sigma_T \bar{n}_e a[1/2(1 - P_2(\mu))\,(\Delta_{P0} - \Delta_{T2} - \Delta_{P2}) - \Delta_P]. \tag{16}$$

If $k\tau_c \gg 1$, the quadrupole component Δ_{T2} is the dominant source term in (16), the polarization amplitudes being smaller by a factor of $\sim \zeta_\lambda$. The second moment of (7) gives

$$\Delta_{T2} \approx -\left[\frac{8\delta_B}{(k\tau)^3}\right]_{\tau_c} \frac{\pi}{4}\, \tau_c\, \frac{d[\tau g(\tau, \tau)]}{d\tau}. \tag{17}$$

Thus,

$$W_P^2 = \frac{9\pi^3}{4} \frac{\delta_B^2(\tau_c)}{(k\tau_c)^7} I_P \tag{18a}$$

where

$$I_P \approx \int_0^\infty G_P^2 dx,\; G_P \equiv \tau_c^2 \frac{d[xg(\tau, \tau)]}{dx} g(\tau_o, \tau). \tag{18b}$$

For constant ionization fraction,

$$I_P \approx (27)^2 \int_0^\infty x^{-16} e^{(-2/x^3)}\, dx = (27)^2/4.$$

Equations (18) show that the polarization fluctuations induced by optically thin perturbations are negligible, in agreement with the arguments given by Kaiser (1984).

In Table 1, we list values for $(\Delta T/T)_\theta$ for several baryon isocurvature models for the experimental set-up used by Uson and Wilkinson ($\theta = 4.5'$, $\sigma \approx 0.64'$). Typically, the predicted anisotropies are of order 10^{-6} and are well below the level of present experiments on arcminute scales. The first-order anisotropies would only approach the current upper limits if Ω were much less than ~ 0.1 and $n \lesssim -1$. These results agree qualitatively with the conclusions of other workers (*e.g.* Sunyaev 1978, Kaiser 1984).

TABLE 1

FIRST ORDER VELOCITY INDUCED ANISOTROPIES*

n	$\Delta T/T$ ($\times 10^{-6}$) $\Omega=0.1$	$\Omega=0.2$	$\Omega=1.0$
+1	0.7	0.7	0.5
0	1.3	0.9	0.3
−1	3.8	1.7	0.4

* Computed for the Uson and Wilkinson (1984) experiment for baryonic models with h = 0.5.

An analysis of the regime $k\tau_c \sim 1$ requires a full numerical solution of the Boltzman equation. In Figure (2a), we show the behaviour of $W_T^2(k)$ for two baryon isocurvature models with $\Omega = 0.2$ and $h = 0.5$. The solid line shows $W_T^2(k)$ for a model with standard recombination history while the dotted line shows what happens if the matter stays fully ionized (Efstathiou and Bond 1987). The k^{-5} tail predicted by equation (10) for the latter case is shown clearly in Figure (2b). The prominent peaks in Figure (2a) arise from Doppler contributions at wavenumbers $k \sim 2\pi/\tau_{rec} \sim 10^{-2}\ \mathrm{Mpc}^{-1}$ for standard recombination and $k \sim 2\pi/\tau_c \sim 10^{-3}\ \mathrm{Mpc}^{-1}$ in the fully ionized case. A prominent peak at $k\tau_c \sim 1$ is a characteristic feature of the numerical solutions.

Figure 3 shows a similar comparison between two adiabatic, scale-invariant cold dark matter models with $\Omega = 1$, $\Omega_B = 0.1$, $h = 0.5$. The solid lines show the case of standard recombination (see also Bond and Efstathiou 1987) and the dotted lines correspond to a universe in which the baryons never recombine. As in Figure 2, the fully ionized model displays a large peak in both W_T^2 and W_P^2 at wavenumbers $k \sim 2\pi/\tau_c \sim 4 \times 10^{-3}\ \mathrm{Mpc}^{-1}$. In fact, the values of the correlation functions at zero lag for these models (normalized according to the "mass traces light" assumption described above) are $C_T^{1/2}(0) = 3.5 \times 10^{-5}$, $C_P^{1/2}(0) = 2.0 \times 10^{-6}$ (standard recombination) and $C_T^{1/2}(0) = 2.3 \times 10^{-5}$, $C_P^{1/2}(0) = 1.7 \times 10^{-6}$ (fully ionized). Reionization has indeed erased the temperature fluctuations from short wavelengths, but the velocity

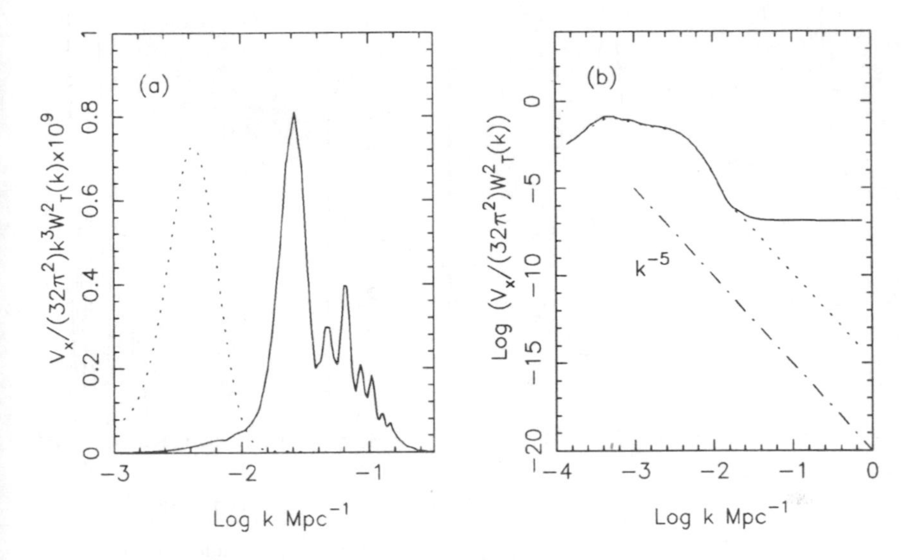

FIG. 2. Figure 2a shows $k^3 W_T^2(k)$ (including only first-order velocity induced anisotropies) for two baryon isocurvature models with $n = 0$, $\Omega = 0.2$, $h = 0.5$. Results for standard recombination are shown by the solid line while the dotted line shows results for a fully ionized universe. For clarity, the solid line has been divided by 25. Figure 2b shows $W_T^2(k)$ on a log-log plot for the fully ionized model shown in Figure 2a. The dotted line shows the k^{-5} tail predicted by equation (10a) describing first-order velocity induced anisotropies. The second-order velocity induced anisotropies described in Section 3 give rise to the constant offset (for $n = 0$) in $W_T^2(k)$ at large **k**.

induced secondary fluctuations at $k \sim 1/\tau_c$ lead to *rms* temperature and polarization fluctuations which are almost as large as in the case with standard recombination. From equation (18) we could have anticipated relatively large polarization fluctuations at $k \sim 1/\tau_c$. Polarization amplitudes at the level of a few percent, coherent over angles $\sim (\Omega/z_c)^{1/2}$, are characteristic of reionized models. Since polarization is a direct result of Thomson scattering, the coherence angle for linear polarization provides a measure of the width of the last scattering shell. A polarization pattern coherent on scales $\gtrsim 10'\Omega^{1/2}$ would provide very strong evidence that reionization had occurred (Hogan, Kaiser, and Rees 1982, Efstathiou and Bond 1987).

The results of this Section may seem a little depressing; the anisotropies predicted for experiments with beam throws of $\sim$ arcminutes are extremely small and would in any case be difficult to observe given the expected confusion at cm wavelengths from faint radio sources (Danese, de Zotti, and

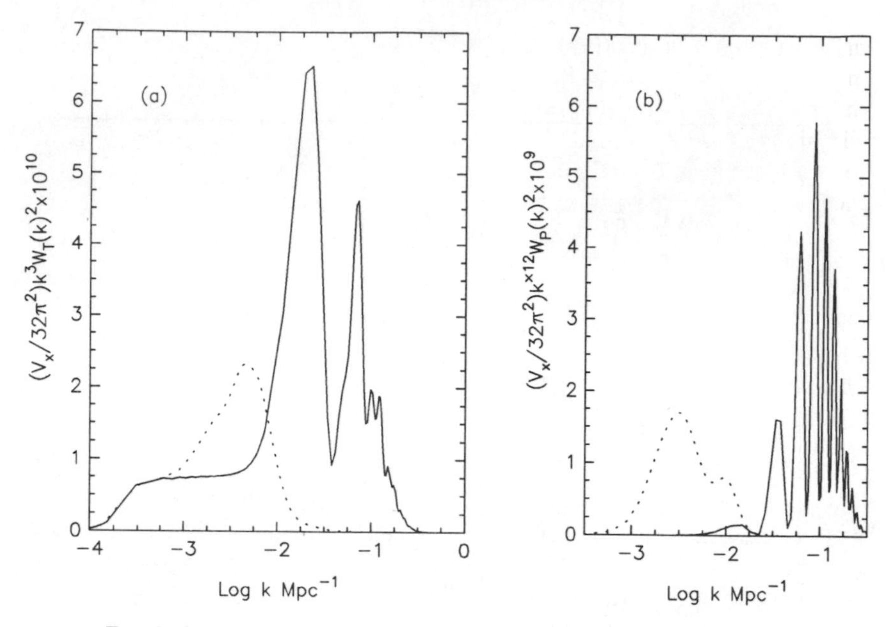

FIG. 3. The area under each curve gives the amplitude of the radiation correlation at zero lag ($C(0)$) for a scale-invariant adiabatic cold-dark matter universe with $\Omega = 1$, $\Omega_b = 0.1$, $h = 0.5$. The solid lines show results for the case of standard recombination and dotted lines show what happens if the IGM stays fully ionized. Figure 3a shows the integrand for $C_T(0)$ while Figure 3b shows the integrand of the polarization autocorrelation function $C_P(0)$.

Manolesi 1983). Substantial secondary anisotropies are generated for relatively large wavenumbers, of size comparable to the horizon at z_c, and these could be observed by experiments with beam widths of $\sigma \lesssim 1°$ and beam throws $\gg \sigma$. However, such experiments have not been performed. Fortunately, there is yet another physical effect, related to the velocity induced anisotropies analysed here, which can produce larger fluctuations on arcminute scales and may well allow us to exclude a wide class of purely baryonic models. This is described in the next Section.

3. Second-Order Velocity-Induced Anisotropies

The first-order effects described above give a small contribution to $\Delta T/T$ on small angular scales; as we have shown in the previous section, the temperature perturbation in linear theory from a single plane wave is small because the contributions from peaks and troughs cancel apart from the slow evolution of the peculiar velocity and optical depth in the light travel time across a wavelength. Ostriker and Vishniac (1986) and Vishniac (1987) have pointed out that second-order effects can, under certain circumstances, dominate over the first-order anisotropies. The physical origin of Vishniac's effect can be understood as follows. If the electron density n_e is spatially variable, then we may see a net $\Delta T/T$ on a patch of the sky which depends on the motions of the scatterers along the line-of-sight and on the fluctuation in electron density within the beam. To compute the effect on $\Delta(k, \mu, \tau)$ we are therefore required to perform a "mode-coupling" sum over products of the baryon velocity and the fractional perturbation in n_e.

The equation of radiative transfer, before Fourier transforming, is

$$\dot{\Delta}_T + \gamma_\alpha \partial\Delta_T/\partial x^\alpha = \sigma_T n_e a[\Delta_{T0} + 4\gamma_\alpha v^\alpha - \Delta_T], \tag{19}$$

where the γ_α are the direction cosines defined by the photon momentum $\hat{q}$. If we define the fractional perturbation in the electron density δ_e by

$$n_e = \bar{n}_e(1 + \delta_e), \tag{20}$$

and Fourier transform (19) with $\mathbf{k}$ along the z-axis, we recover equation (2) together with additional second-order terms arising from the fluctuating part of n_e. The most important of these second-order terms is retained in the following equation

$$\begin{aligned} \dot{\Delta}_T(k) + i\mu k\Delta_T(k) &= \sigma_T\bar{n}_e a[\Delta_{T0}(k) + 4\mu v_B(k) - \Delta_T(k)] \\ &+ \sigma_T\bar{n}_e a[4\hat{\gamma}.\sum_{\mathbf{k}'} \mathbf{v}_B(\mathbf{k}')\delta_e(\mathbf{k} - \mathbf{k}')]. \end{aligned} \tag{21}$$

If we now split $\Delta_T(k)$ into first-and second-order pieces,

$$\Delta_T(k) = \Delta_T^1(k) + \Delta_T^2(k),$$

the solution for $\Delta_T^2(k)$ is,

$$\Delta_T^2(k, \mu, \tau) \approx 4\int_0^\tau [\hat{\gamma}.\sum_{\mathbf{k}'} \mathbf{v}_B(\mathbf{k}')\delta_e(\mathbf{k} - \mathbf{k}')]g(\tau, \tau')e^{ik\mu(\tau' - \tau)}\, d\tau'. \tag{22}$$

For simplicity, we will assume that $\delta_e(\mathrm{k}) = \delta_B(k)$; this is reasonable on scales where the matter fluctuations are linear if most of the baryonic material remains in a diffuse ionized form. Symmetrizing the sum in (22) we obtain

$$\begin{aligned} \Delta_T^2(k, \mu, \tau) \approx -\frac{4}{i}e^{-i\mu k\tau}\Big[\frac{1}{\tau_c}\sum_{\mathbf{k}'} \delta_B(\tau_c, \mathbf{k}')\delta_B(\tau_c, \mathbf{k} - \mathbf{k}')\hat{\gamma}.\mathbf{D}(\mathbf{k}, \mathbf{k}') \\ \int_0^\tau (\frac{\tau}{\tau_c})^3 g(\tau, \tau')e^{ik\mu(\tau')}\, d\tau'\Big], \end{aligned} \tag{23}$$

where

$$\mathbf{D}(\mathbf{k}, \mathbf{k}') = \left(\frac{\mathbf{k}'}{|\mathbf{k}'|^2} + \frac{\mathbf{k} - \mathbf{k}'}{|\mathbf{k} - \mathbf{k}'|^2} \right).$$

As for equation (7), the integral in (23) may be approximated as a Fourier transform $\hat{G}^2(k\mu\tau_c)$. Assuming random phases,

$$|\Delta_T^2(k)|^2 = 32 \sum_{\mathbf{k}'} |\delta_B(\tau_c, \mathbf{k}')|^2 |\delta_B(\tau_c, \mathbf{k} - \mathbf{k}')|^2 (\hat{\gamma}.\mathbf{D}(\mathbf{k}, \mathbf{k}'))^2 \frac{|\hat{G}^2(k\mu\tau_c)|^2}{\tau_c^2}.$$

Thus, the contribution to W_T^2 (equation 9b) may be written

$$W_T^2 = \frac{32\pi}{k\tau_c^3} I_1 \frac{V_\chi}{(2\pi)^3} \int |\delta_B(\tau_c, \mathbf{k}')|^2 |\delta_B(\tau_c, \mathbf{k} - \mathbf{k}')|^2 (\hat{\gamma}.\mathbf{D}(\mathbf{k}, \mathbf{k}'))^2 \, d^3\mathbf{k}', \tag{24a}$$

where

$$I_1 = \frac{1}{2\pi} \int_{-1}^{+1} |\hat{G}^2(k\mu\tau_c)|^2 \, d(k\mu\tau_c). \tag{24b}$$

The integral in (24a) may be expressed in terms of the power-spectrum of the baryon perturbations at τ_c, $P(k, \tau_c)$, and one of the angular integrals may be performed analytically giving,

$$W_T^2 = \frac{4V_\chi P^2(k, \tau_c)}{\pi\tau_c^3} I_1 I_2(k) \tag{25a}$$

where

$$I_2(k) = \int_0^\infty \int_{-1}^{+1} dy d\mu \, \frac{(1-\mu^2)(1-2\mu y)^2}{(1+y^2-2\mu y)^2} \frac{P[k(1+y^2-2\mu y)^{1/2}, \tau_c]}{P(k, \tau_c)} \frac{P[ky, \tau_c]}{P(k, \tau_c)}. \tag{25b}$$

The integral I_1 may be evaluated using Parseval's theorem,

$$I_1 \approx \int_0^{\tau_o} (\tau/\tau_c)^6 (g(\tau_o, \tau)\tau_c)^2 \, d(\tau/\tau_c). \tag{25c}$$

If all of the matter remains ionized, (25c) gives

$$I_1 \approx 9\Gamma(4/3)/2^{1/3} \approx 6.38. \tag{26}$$

The second-order contributions to the polarization may easily be computed following the procedures outlined in Section (2) and above. The result is

$$W_P^2 = \frac{9\pi V_x P^2(k, \tau_c)}{64\tau_c^3(k\tau_c)^3} I_3 I_2(k), \tag{27a}$$

where

$$I_3 \approx \int_0^{\tau_o} (\tau/\tau_c)^6 (g(\tau, \tau) g(\tau_o, \tau)\tau_c^2)^2 \, d(\tau/\tau_c) \tag{27b}$$

and is approximately equal to

$$I_3 \approx 27/4 \tag{28}$$

for constant ionization fraction. As with the first-order contribution to the polarization described in Section 2, the second-order anisotropies do not have a significant polarization signature because $k\tau_c \gg 1$ on the scales of interest.

For the baryon isocurvature models, the integral I_2 does not diverge at long wavelengths for $n < -1$ because the transfer function falls steeply for $k \lesssim k_{equ}$. For short wavelengths I_2 is approximately constant, so W_T^2 in (25a) will vary as a power-law $\propto k^{2n}$. The radiation correlation function may be evaluated from (9a)

$$C_T(\theta, \sigma) = (AV_x)^2 \frac{I_1 I_2}{16\pi^3\tau_c^3} \frac{\Gamma(n + 3/2)}{(R_c\sigma)^{(2n+3)}} \, {}_1F_1(n + 3/2; 1; -\theta^2/4\sigma^2). \tag{29}$$

If $n > -1.5$ the contribution to $C_T(0, \sigma)$ is dominated by short wavelength perturbations and converges for a finite beam width σ. For $\theta \gg \sigma$, the degenerate hypergeometric function in (19) varies as $\theta^{-(2n+3)}$ for $n \neq -0.5$ and as $\exp(-\theta^2/4\sigma^2)$ for $n = -0.5$ (there is a sign change in the assymptotic behaviour at $n = -0.5$). Thus if $n > -1.5$ a double, or triple beam experiment will essentially measure uncorrelated noise and the amplitude of the temperature fluctuations will be determined by the size of the beam width. Note that equation (29) implies the approximate scaling

$$C_T^{1/2}(0, \sigma) \propto h^{5/6} \Omega^{(2+3n)/3}.$$

However, the actual scaling with Ω in the results described below is somewhat weaker than implied by this relation (this is because $C_T^{1/2}$ depends on the square of the growth factor D).

Readhead and collaborators have recently made sensitive measurements at Owens Valley which are potentially capable of detecting the fluctuations predicted by (29) (Hardebeck, Lawrence, Moffel, Myers, Readhead and Sargent, in preparation; see also Sargent 1987). A full discussion of this experiment should properly be postponed until their final analysis is published, but we can get a good idea of what their results imply as follows. Their experimental arrangement resembles the triple beam arrangement of Uson and Wilkinson; the temperature difference measured for each of eight fields is the difference between the temperature at a field point and the mean temperature in two reference beams covering arcs of 30° on the sky at 7.15′ on either side of each field point. Their beam width is $\sigma \sim 0.78'$ (1.8′ FWHM), and their experiment suggests an upper limit of 1.5×10^{-5} at the 95% confidence level (Readhead, private communication). If $n > -1.5$, we can neglect correlations on scales of order of the beam throw; furthermore, since the signal in the reference beams is averaged over a relatively large area it should be a good approximation to assume that their sky variance estimate is given by $\langle(\Delta T/T)^2\rangle \sim C_T(0, \sigma)$. In Table 2 we list values of $\sqrt{C_T(0, \sigma)}$ computed from (29) for their beam width. The results are typically an order of magnitude larger than those listed in Table 1 and in many cases exceed the Owens Valley limit.

TABLE 2

SECOND ORDER VELOCITY INDUCED ANISOTROPIES*

n	$C^{1/2}(0, \sigma)$ ($\times$ 10^{-5})			
	$\Omega = 0.1$	$\Omega = 0.2$	$\Omega = 0.4$	$\Omega = 1.0$
0	0.1	0.3	0.9	3.0
− 0.5	0.2	0.4	1.1	1.7
− 1	0.9	1.1	1.7	1.4

* Computed for a beam width $\sigma = 0.78'$ for baryonic models with h = 0.5.

4. Discussion

Figure 4 shows a schematic plot of the constraints on n and Ω implied by the Owens Valley results. The contours in the figure correspond approximately to the positions where $\Delta T/T = 1.5 \times 10^{-5}$ from the first-and second-order anisotropies discussed above. As Tables 1 and 2 show, the predictions are fairly slowly varying functions of the parameters n and Ω, so the region excluded by the contours should not be taken too literally. For example if we had plotted contours at twice the Owens Valley limit, we would hardly exclude any of the area shown in Figure 4. A more detailed statistical comparison with the observations (*e.g.* using maximum likelihood) is required to establish accurate limits on the models. Figure 4 has been plotted for $h = 0.5$; the scal-

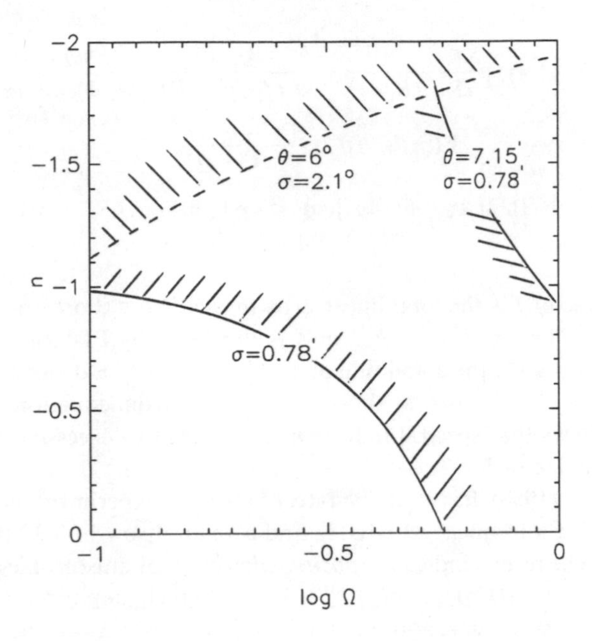

FIG. 4. Diagram showing approximate constraints on baryon-dominated isocurvature models (with $h = 0.5$). The shaded regions show areas excluded by limits on microwave backround anisotropies. The dotted line shows the region excluded by limits on angular scales of 6° ($\Delta T/T \lesssim 5 \times 10^{-5}$, Melchiorri *et al.* 1981), while the solid lines show regions excluded by the recent Owens Valley experiment ($\Delta T/T \lesssim 1.5 \times 10^{-5}$).

ing derived in Section 3 shows that that the anisotropies would have nearly twice the amplitude if $h = 1$.

The dashed line in Figure 4 shows the constraints imposed by the experiment at 6° (interpreted as an upper limit of $\Delta T/T < 5\times 10^{-5}$) performed by Melchiorri and collaborators (Melchiorri *et al.* 1981). For wavenumbers $k < k_{equ}$, the temperature perturbations can be written in terms of the initial entropy fluctuation spectrum,

$$|\Delta_T(k, \mu, \tau_o)|^2 = (8/5)^2 |S_\gamma(k)|^2, \tag{30}$$

where 4/3 of the factor 8/5 comes from the initial entropy perturbation and the remaining 4/15 comes from the Sachs-Wolfe effect arising from the gravitational potential fluctuations induced by the motion of the matter relative to the radiation. From equation (8), we derive the following expression for the temperature autocorrelation function at large angular scales,

$$C(\theta, \sigma) = \frac{\sqrt{\pi}}{25} \frac{1}{(R_C\sigma)^{(n+3)}} \frac{1}{(fp_n)^2} \frac{J_3(x_o)x_o^n n}{\Gamma(n + 2)\sin(n\pi/2)}$$
$$\Gamma((n + 3)/2) \frac{\Gamma(-(n + 1)/2)}{\Gamma(-n/2)}$$
$${}_1F_1[(n + 3)/2; 1; -\theta^2/4\sigma^2], \; n < -1, \; \sigma > \tau_c/R_c \sim (\Omega/z_c)^{1/2}, \tag{31}$$

where the factor f is the total linear growth factor for short wavelength perturbations ($T(k) = f, k \gg k_{equ}$ and is tabulated by Efstathiou and Bond (1987). (If $n > -1$, the amplitude of $C_T(\theta, \sigma)$ for $\theta \sim \sigma$ depends on the way that $|\Delta_T(k, \mu\, \tau_o)|^2$ departs, at short wavelengths, from equation (30)). Equation (31) shows that spectral indices $n \lesssim -2$ lead to excessive anisotropies at large angular scales.

Lasenby (1988) has described results of an experiment performed at Tenerife using a frequency 10.4GHz and a beam throw of 8.3° (beam width $\sigma = 3.5°$). The results indicate a *positive* detection of anisotropies at the level $\Delta T/T \sim 4 \times 10^{-5}$ (Davies *et al.* 1987). It is extremely important to determine whether this is a detection of intrinsic fluctuations in the microwave background, rather than radio continuum emission from the Galaxy. This is currently being checked by Lasenby and collaborators with further experiments at angular scales of 5° and at frequencies of 5GHz. In fact the Melchiorri *et al.* (1981) experiment also detected fluctuations at the level of $\Delta T/T = 4.1 \pm 0.7 \times 10^{-5}$. Since they used a bolometer with a wide spectral coverage in the infrared (500 — 3000μm; see Fabbri *et al.* 1982 for experimental details), we have conservatively interpreted this measurement as an upper limit of 5×10^{-5} because it is quite likely that the detection is caused by dust in the Galaxy. However, it is interesting that the detection is in qualitative agreement with the Tenerife results.

The Tenerife experiments lead to very similar constraints as those for the Melchiorri *et al.* (1981) experiment shown in Figure 4. If these two experiments have detected primordial anisotropies in the background radiation, then the results apparently conflict with the Owens Valley upper limits unless $\Omega_B \approx 1$ and $n \approx -1.5$. If we really are forced to high Ω_B, then the baryon isocurvature model loses many of its attractive features; a high value of Ω_B is incompatible with standard primordial nucleosynthesis and conflicts with local determinations of Ω unless galaxies are more clustered than the mass distribution. There are ways of avoiding this conclusion, for example, we have assumed throughout that the IGM stays fully ionized from a redshift $z_* > z_c$ to the present. But if a large fraction of the the ionized IGM condensed into dark matter, or into high density neutral clumps, at a redshift $z' > z_c$, then the second order anisotropies would be reduced by about $(z_c/z')^{5/4}$.

The physical effects described in this article illustrate some of the interesting consequences of velocity fields at high redshift. In a reionized universe Thomson scattering off moving electrons produces several potentially observable effects. The clearest indicator of reionization would be a polarization pattern that is coherent over scales $\gtrsim 30'$. The second order anisotropies described in Section 3 have a characteristic "spiky" pattern and so should be readily distinguishable from the much smoother pattern expected in models with a standard recombination history. As we have shown, the secondary fluctuations can be quite large and could be detected by modern experiments.

Acknowledgements

I thank my $\Delta T/T$ collaborator, Dick Bond, for many interesting discussions and contributions. I am grateful to Anthony Readhead for supplying me with details of the Owens Valley experiment.

REFERENCES

Bean, A.J., Efstathiou, G., Ellis, R.S., Peterson, B.A., and Shanks, T. 1983. *MN.* **205,** 605.

Bond, J.R. and Efstathiou, G. 1984. *Ap J.* **285,** L45.

_____. 1987. *MN.* **226,** 655.

Danese, L., de Zotti, G., and Mandolesi, N. 1983. *AA.* **121,** 114.

Davies, R.D., Lasenby, A.N., Watson, R.A., Daintree, E.J., Hopkins, J., Beckman, J., Sanchez-Almeida, J., and Rebelo, R. 1987. *Nature.* **326,** 462.

Davis, M. and Boynton, P.E. 1980. *Ap J.* **237,** 365.

Davis, M. and Peebles, P.J.E. 1983. *Ap J.* **267,** 465.

Doroshkevich, A.G., Zeldovich, Ya. B., and Sunyaev, R.A. 1978. *Soviet Astron.* **22,** 523.

Dressler, A., Faber, S.M., Burstein, D., Davies, R.L., Lynden-Bell, D., Terlevich, R.S., and Wegner, G. 1987. *Ap J.* **313,** L37.

Efstathiou, G. and Bond, J.R. 1986. *MN.* **218,** 103.

_____. 1987. *MN.* **227,** 33p.

Fabbri, R., Guidi, I., Melchiorri, F. and Natale, V. 1982. in *Proceedings of the Second Marcel Grossman Meeting on General Relativity.* p. 889. ed. R. Ruffini. North-Holland.

Hogan, C.J., Kaiser, N., and Rees, M.J. 1982. *Phil Trans R Soc Lond A.* **307,** 97.

Kaiser, N. 1983. *MN.* **202,** 1169.

_____. 1984. *Ap J.* **282,** 374.

Lasenby, A. 1988. this volume.

Melchiorri, F., Melchiorri, B.O., Ceccarelli, C. and Pietranera, L. 1981. *Ap J.* **250,** L1.

Ostriker, J.P. and Vishniac, E.T. 1986. *Ap J.* **306,** L51.

Peebles, P.J.E. 1968. *Ap J.* **153,** 1.

_____. 1980. *The Large-Scale Structure of the Universe.* Princeton: Princeton University Press.

_____. 1987a. *Ap J.* **315,** L73.

_____. 1987b. *Nature.* **327,** 210.

_____. 1988. this volume.

Rubin, V.C., Thonnard, N., Ford, W.K., and Roberts, M.S. 1976. *AJ.* **81,** 719.

Sachs, R.K. and Wolfe, A.M. 1967. *Ap J.* **147,** 73.

Sargent, W.L.W. 1987. *Observatory.* **107,** 235.

Silk, J. 1982. *Acta Cosmologica.* **11,** 75.

Sunyaev, R.A. 1978. in *IAU Symp. 79, The Large-Scale Structure of the Universe.* p. 393. eds. M.S. Longair and J. Einasto. Dordrecht: Reidel.
Uson, J.M. and Wilkinson, D.T. 1984. *Ap J.* **277**, L1.
Vishniac, E.T. 1987. *Ap J.* **322**, 597.
Vittorio, N. 1988. this volume.
Yang, J., Turner, M.S., Steigman, G., Schramm, D.N., and Olive, K.A. 1984. *Ap J.* **281**, 493.

V

THEORY: *AB INITIO*

COHERENCE OF LARGE-SCALE VELOCITIES

ALEXANDER S. SZALAY
Department of Physics and Astronomy, The Johns Hopkins University and Department of Atomic Physics, Eötvös University, Budapest

ABSTRACT

The radial components of the galaxy peculiar velocities are related to primordial density fluctuations via a tensor window function. For a given shell of galaxies the radial velocity as a function of direction can be expanded in multipoles, the $l = 1$ term corresponding to the bulk motion, the $l = 2$ to a quadrupole anisotropy of the velocity field. The $l = 0$ 'breathing mode', even if present, is unobservable; it would be absorbed into the local value of the Hubble constant.

We derive the distribution of the bulk velocity of such a shell and compare it to the results obtained with the usual scalar window function. The estimated velocity and the statistical scatters are significantly affected by sampling anisotropies in the galaxy selection function. The observed bulk velocity is most likely to be close to the galactic plane, aligned with the dipole anisotropy of the galaxy distribution. This effect is nongravitational, purely a result of nonuniform sampling.

We estimate the rms values of the various multipole moments of the velocity field and generate some Monte-Carlo realizations assuming the cold dark matter (CDM) spectrum. Generally the CDM velocities are too low to be compatible with the current observations.

1. INTRODUCTION

In the last few years several groups attempted to determine the bulk motion of a relatively large region (about 10 - 50 Mpc in radius) of the universe centered on us (Rubin *et al.* 1976a, b, Hart and Davies 1982, de Vaucouleurs and Peters 1984, Aaronson *et al.* 1986, Burstein *et al.* 1986, Collins *et al.* 1986).

These regions are large enough to be well described by linear theory, since nonlinear motions within the region cancel out, only the collective part remains (Clutton-Brock and Peebles 1981, Kaiser 1983). Using an independent distance determination to subtract the Hubble-flow from the observed redshift, the radial component of the peculiar velocity can be determined for each object. Distances are calculated from secondary distance indicators, different for elliptical galaxies (Burstein *et al.* 1986) and spirals (Collins *et al.* 1986). These methods usually carry systematic errors of the order of 10-20 percent.

The net center-of-mass (CM) velocity of the observed region can be determined from the raw data by using variants of the least squares fit or maximum likelihood techniques. This can be compared to the primordial fluctuation spectrum directly or indirectly (Kaiser 1988). The results have large systematic errors partly due to uncertainties in the distance estimators, partly due to the poor statistics.

Motion on such a large scale is linear; therefore, it can be related to the primordial linear fluctuation spectrum in a straightforward way as follows. The fluctuations are described by a dimensionless density fluctuation field and the peculiar velocity field relative to the Hubble flow. Both are functions of the comoving spatial coordinate $\mathbf{x}$ and time t. They can be expressed in terms of their Fourier transforms:

$$\delta(\mathbf{x}, t) = \int d^3\mathbf{x}\; e^{i\mathbf{kx}}\, \delta_{\mathbf{k}}\;;\; \mathbf{v}(\mathbf{x}, t) = \int d^3\mathbf{k}\; e^{i\mathbf{kx}}\, \mathbf{v}_{\mathbf{k}}, \tag{1}$$

where $\delta_{\mathbf{k}}$ has random phases and $P(k) = |\delta_{\mathbf{k}}|^2$ is the power spectrum. From the continuity equation we obtain

$$\mathbf{v}_{\mathbf{k}} = -i(H_0 a f)\,\frac{\mathbf{k}}{k^2}\,\delta_{\mathbf{k}} = \hat{\mathbf{k}} v_{\mathbf{k}}, \tag{2}$$

with $H_0 f = (\dot{D}/D)$ and $\hat{\mathbf{k}} = \mathbf{k}/k$. $H_0 = (\dot{a}/a)$ is the Hubble-constant, a is the expansion factor, and D is the growing solution of the linear equation on $\delta_{\mathbf{k}}$ (Peebles 1980). Given a selection function describing the observed region one can predict the distribution of the CM velocity, based upon the fluctuation spectrum. The quantity used for this prediction is the rms (root mean square) value, which is compared to the observed velocity.

The expected value of the CM velocity for a given region is the integral of $\mathbf{v}(\mathbf{x})$ over the selected area, described by a selection function $\Psi(\mathbf{s})$, normalized to $\int d^3\mathbf{s}\; \Psi(\mathbf{s}) = 1$,

$$\tilde{\mathbf{U}}(\mathbf{x}_0) = \int d^3\mathbf{s}\; \Psi(\mathbf{s})\mathbf{v}(\mathbf{s}) = \int d^3\mathbf{k}\; v_{\mathbf{k}}\, e^{i\mathbf{k}\mathbf{x}_0}\, \tilde{W}(k), \tag{3}$$

using $\mathbf{x} = \mathbf{x}_0 + \mathbf{s}$, where $\mathbf{x}_0$ is the position of the observer, and s is the relative distance. This derivation assumes that we can measure the individual peculiar velocities $\mathbf{v}(\mathbf{s})$ in full, which is not the case. All theoretical calculations of this kind have also assumed an isotropic $\Psi(s)$, whose Fourier transform is the scalar window function $\tilde{W}(k)$:

$$\tilde{W}(k) = \int d^3\mathbf{s}\ \Psi(\mathbf{s})e^{i\mathbf{k}\mathbf{s}} = 4\pi \int_0^\infty ds\ s^2\Psi(s)\ j_0(ks). \tag{4}$$

Here $j_0(ks)$ is the spherical Bessel function of 0th order. The dispersion $\langle\tilde{\mathbf{U}}^2\rangle$ can be expressed with these quantities.

$$\langle\tilde{\mathbf{U}}^2\rangle = \int d^3\mathbf{k}\ |v_\mathbf{k}|^2\,\tilde{W}(k)^2. \tag{5}$$

This isotropic rms value has been the one compared to the observed velocity so far. We show below that significant deviations occur from this simple prediction from experiment to experiment due to the way the observations and the data analysis were carried out, and we suggest that theoretical predictions should match the particular observations individually.

2. The Tensor Window Function

The observed peculiar velocities are only the radial components of the full velocity of the ith galaxy in the sample : $v^i = \hat{\mathbf{s}}^i\mathbf{v}^i$, with errors σ_i. First let us consider the vector sum of the projected velocities

$$\mathbf{V}(\mathbf{x}_0) = \frac{\Sigma_i v^i\hat{\mathbf{s}}^i/\sigma_i^2}{\Sigma_i\ 1/\sigma_i^2} = \langle v_r(\mathbf{s})\hat{\mathbf{s}}\rangle = \int d^3\mathbf{s}\ \Phi(\mathbf{s})\hat{\mathbf{s}}v_r(\mathbf{s}). \tag{6}$$

$\hat{\mathbf{s}} = \mathbf{s}/s$ is a vector of unit length parallel to $\mathbf{s}$, and $\Phi(\mathbf{s})$ is the *generalized selection function,* containing the errors as well. The errors are non-negligible, they mostly come from the distance estimation, so they are proportional to distance, s^i. They can be well approximated as $\sigma_i^2 = \sigma_f^2 + \Delta^2 s_i^2$ (Lynden-Bell *et al.* 1988). This will modify the radial selection function, more heavily weighting smaller scales. When we determine $\mathbf{V}$ in a Cartesian coordinate system, we need a tensor window-function, similar to the one discussed by Grinstein *et al.* (1987) arising from the projector $\hat{s}_\alpha\hat{s}_\beta$:

$$V_\alpha = \int d^3\mathbf{s}\ \Phi(\mathbf{s})\hat{s}_\alpha\hat{s}_\beta v_\beta(\mathbf{x}) = \int d^3\mathbf{k}\ e^{i\mathbf{k}\mathbf{x}_0}\ W_{\alpha\beta}(\mathbf{k})v_{\mathbf{k}\beta} \tag{7}$$

$$W_{\alpha\beta}(\mathbf{k}) = \int d^3\mathbf{s}\ \Phi(\mathbf{s})e^{i\mathbf{k}\mathbf{s}}\ \hat{s}_\alpha\hat{s}_\beta. \tag{8}$$

Let us consider first a completely isotropic selection function $\Phi(s)$. We can evaluate $W_{\alpha\beta}$ in this case:

$$W_{\alpha\beta}(\mathbf{k}) = 4\pi \int_0^\infty ds\, s^2\, \Phi(s) \left\{ \frac{1}{3}\, \delta_{\alpha\beta}[j_0(ks) + j_2(ks)] - \hat{k}_\alpha \hat{k}_\beta\, j_2(ks) \right\}, \tag{9}$$

significantly different from the window function in Eq. (4).

The center of mass velocity **U** can be determined from the observed radial velocities using a least squares fit:

$$\chi^2 = \sum_i \frac{(v^i - \mathbf{U}\hat{\mathbf{s}}^i)^2}{\sigma_i^2} \quad \text{leading to} \quad \mathbf{U} = \mathbf{M}^{-1}\mathbf{V}. \tag{10}$$

M is a weighted projection matrix:

$$M_{\alpha\beta} = \sum_i \frac{\hat{s}^i_\alpha \hat{s}^i_\beta}{\sigma_i^2} \Big/ \sum_i 1/\sigma_i^2 = \int d^3\mathbf{s}\; \Phi(\mathbf{s})\hat{s}_\alpha \hat{s}_\beta = \mathrm{W}_{\alpha\beta}(0). \tag{11}$$

For an isotropic selection function $M_{\alpha\beta}$ is diagonal, and the elements are $\frac{1}{3}$, so only the diagonal part of $W_{\alpha\beta}$ contributes to the dispersion of U:

$$\langle U^2 \rangle = \int d^3\mathbf{k}\; |v_\mathbf{k}|^2\, W(k)^2 \tag{12}$$

$$W(k) = 4\pi \int_0^\infty ds\, s^2\, \Phi(s)\, (j_0(ks) - 2j_2(ks)). \tag{13}$$

This window function will differ from the scalar one by the presence of $-2j_2(ks)$. In the case of an isotropic sample this is the correct window function for calculating the dispersion of the observed bulk velocity. Its deviation from the istotropic scalar $\tilde{W}(k)$ can be seen in Fig. 1a. We estimate the ratio of the dispersions due to the radial projection to be $\langle U^2 \rangle / \langle \tilde{U}^2 \rangle \approx 0.75$, with a mild dependence on the radius of the Gaussian selection function shown on Fig. 1b., calculated using the CDM spectrum.

3. Effect of Anisotropies

The assumption of isotropy is generally not correct due to clumpy distribution of galaxies, galactic extinction, and fluctuations on the scale of the survey. The anisotropies will influence the measured bulk velocity, since the radial velocities only measure a projected part of the bulk flow. Here we take the sampling anisotropies explicitly into account, and show that they have a strong effect, summarizing a more extended discussion (Regös and Szalay 1988). We expand the galaxy selection function in multipoles:

$$\Phi(\mathbf{s}) = \frac{1}{\sqrt{4\pi}} \sum_{lm} \phi_{lm}(s)\, Y_{lm}(\Omega_s). \tag{14}$$

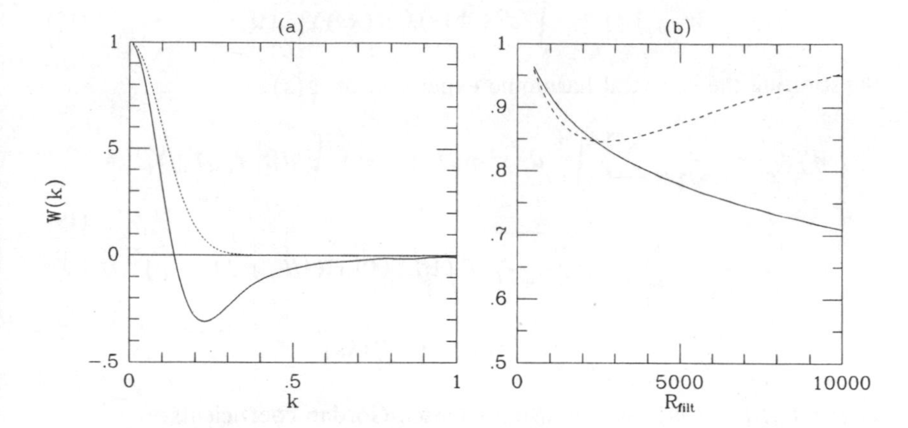

FIG. 1. (a) Deviation of the tensor window function (solid line) from the scalar (broken line) for an isotropic Gaussian radial selection (width R). (b) The solid line shows the correction to be applied to the scalar velocity prediction to obtain the tensor one. The broken line shows the same, but the window function also includes the distance error weights.

The normalization is $\int d^3\mathbf{s}\, \Phi(\mathbf{s}) = \int ds\, s^2 \phi_{00}(s) = 1$. The hermiticity of the expansion requires that $\phi_{lm} = (-1)^m \phi^*_{l,-m}$. From now on we will use spherical coordinates. In this system the unit vectors are

$$\hat{s}_\mu = \sqrt{\frac{4\pi}{3}}\, Y_{l\mu}(\Omega_s)\ ;\ \hat{k}_\mu = \sqrt{\frac{4\pi}{3}}\, Y_{1\mu}(\Omega_\mathbf{k}). \tag{15}$$

The component of the 'vector-sum' velocity $\mathbf{V}$ corresponding to $\mu = 0, \pm 1$ can be expressed with the help of the tensor window function. Let us expand $V_\mu(\mathbf{x}_0)$ in multipoles and determine the coefficients.

$$V_\mu(\mathbf{x}_0) = i(H_0 af) \int d^3\mathbf{k}\, \frac{\delta_\mathbf{k}}{k}\, e^{i\mathbf{k}\mathbf{x}_0} \int d^3\mathbf{s}\, \Phi(\mathbf{s}) \hat{s}_\mu (\hat{s}^*_\nu \hat{k}_\nu) e^{i\mathbf{k}\mathbf{s}}. \tag{16}$$

Using the Rayleigh expansion one can write

$$V_\mu(\mathbf{x}_0) = 4\pi(H_0 af) \int d^3\mathbf{k}\, \frac{\delta_\mathbf{k}}{k}\, e^{i\mathbf{k}\mathbf{x}_0} \sum_{L,M} i^L Y_{LM}(\Omega_\mathbf{k}) \int d^3\mathbf{s}\, \Phi(\mathbf{s}) \hat{s}_\mu j'_L(ks)\, Y^*_{LM}(\Omega \mathbf{s}). \tag{17}$$

Here $j_L(ks)$ is the spherical Bessel function of order L, and $j'_L(ks)$ is the derivative with respect to ks. One can define the tensor window function in spherical coordinates as

$$W_{LM,\mu}(k) = \int d^3\mathbf{s}\ \Phi(\mathbf{s})\hat{s}_\mu j_L'(ks)Y^*_{LM}(\Omega_s). \tag{18}$$

Substituting the spherical harmonic expansion of $\Phi(\mathbf{s})$

$$\begin{aligned} W_{LM,\mu} &= \frac{1}{\sqrt{3}} \sum_{l,m} \int_0^\infty ds\ s^2\ \phi_{lm}(s) j_L'(ks) \int d\Omega_s\ Y^*_{LM} Y_{1\mu} Y_{lm} = \\ &= \frac{1}{\sqrt{4\pi}} \sum_l \sqrt{\frac{2l+1}{2L+1}}\ C(1lL;\ 00)\ C(1lL;\ \mu\ M-\mu) \\ &\qquad \int_0^\infty ds\ s^2 \phi l, M_{-\mu}(s) j_L'(ks). \end{aligned} \tag{19}$$

Here $C(l_1 l_2 l_3;\ m_1 m_2)$ are the usual Clebsch-Gordan coefficients.

For a given selection function one can easily express the expectation value

$$\langle V^*_\mu V_\nu \rangle = 4\pi(H_0 af)^2 \int d^3\mathbf{k}\ \frac{|\delta_\mathbf{k}|^2}{k^2} \sum_{L,M} W^*_{LM,\mu} W_{LM,\nu}. \tag{20}$$

The bulk motion components U_μ are correlated as

$$\langle U^*_\mu U_\nu \rangle = (M^{-1})^*_{\mu\varrho}(M^{-1})_{\nu\eta} \langle V^*_\varrho V_\eta \rangle. \tag{21}$$

Now let us calculate $\mathbf{M}$ in this system. We find that only the $l = 0$ and $l = 2$ terms are non-vanishing, thus $\mathbf{M}$ depends only on the quadruple anisotropy of the generalized selection function. Here $a_{lm} = \int ds\ s^2 \phi_{lm}(s)$.

$$\begin{aligned} M_{\mu\nu} &= \frac{1}{\sqrt{4\pi}} \sum_{lm} \int_0^\infty ds\ s^2 \phi_{lm}(s) \int d\Omega_s\ Y_{lm} \frac{4\pi}{3}\ Y^*_{1\nu} Y_{1\mu} \\ &= \frac{1}{3}\left[\delta_{\mu\nu} - \sqrt{2} a_{2,\nu-\mu} C(211;\ \nu-\mu,\ \mu)\right]. \end{aligned} \tag{22}$$

3.1 - Statistical (sampling) Effect

The anisotropies have an important effect on the statistical scatter in the value of $\langle U^*_\mu U_\nu \rangle$ due to the discrete sampling and the error in the radial velocities. This dispersion can most easily be visualized in the extreme, when there are galaxies only at the north and south galactic poles. In this case we can determine the component of the bulk motion along the axis, but there is no information on the components in the galactic plane. Any quadrupole anisotropy due to galactic extinction will have such an effect, so the component in the plane has always a larger scatter than the axial one. Here we calculate its magnitude. First we define the mean statistical scatter of the velocity errors as

$$\langle \sigma^2 \rangle = \left(\frac{1}{N} \sum_{i=1}^{N} \frac{1}{\sigma_i^2} \right)^{-1} . \tag{23}$$

The variance matrix of the bulk velocity components becomes

$$\mathrm{Var}\,(U_\mu^* U_\nu) = \frac{\langle \sigma^2 \rangle}{N_e} (M^{-1})^*_{\mu\varrho} (M^{-1})_{\nu\eta} M_{\eta\varrho} = \frac{\langle \sigma^2 \rangle}{N_e} (M^{-1})^*_{\mu\nu}, \tag{24}$$

where N_e is the effective degrees of freedom. For an isotropic selection function we obtain the trivial expression

$$\mathrm{Var}(U_\mu^* U_\nu) = 3\delta_{\mu\nu} \frac{\langle \sigma^2 \rangle}{N_e} . \tag{25}$$

If we try to simulate the galactic extinction with a rotationally symmetric quadrupole a_{20}, **M** is still diagonal, but

$$\mathrm{Var}\,(|U_0|^2) = 3 \frac{\langle \sigma^2 \rangle}{N_e} \frac{1}{1 + 2a_{20} / \sqrt{5}}$$

$$\mathrm{Var}(|U \pm 1|^2) = 3 \frac{\langle \sigma^2 \rangle}{N_e} \frac{1}{1 - a_{20} / \sqrt{5}} . \tag{26}$$

3.2 - Systematic effects

Even if there were no statistical errors ($\langle \sigma^2 \rangle = 0$), i.e. we measured each radial velocity totally accurately or we have a very large number of galaxies, the anisotropy in the selection function would still affect the expectation value $\langle U_\mu^* U_\nu \rangle$.

Below we calculate the contribution of various multipole moments ϕ_{lm} to the tensor window function. It turns out that, for reasonable selection functions, the single most important contribution arises (not surprisingly) from the dipole anisotropy of galaxy distribution. We should stress, however, that this is not a gravitational but a sampling effect.

If we consider a single dipole, we can choose our coordinate system by $\phi_{1\mu} = \delta_{\mu 0} \phi_{10}$, the dipole is along the z-axis. One can simplify notation below by using

$$w_{lm;L}(k) = \int_0^\infty ds\, s^2 \phi_{lm}(s) j_L(ks) \tag{27}$$

for the Lth Bessel-moment of the lm term in the expansion. The tensor window function becomes

$$W_{LM,\mu} = \frac{\delta_{M\mu}}{\sqrt{12\pi}} \left[\delta_{L1} w_{00;1} + 3\delta_{L0}\delta_{M0} w_{10;0} + \delta_{L2} \sqrt{\frac{4}{5}}\, w_{10;2} \right]. \quad (28)$$

Calculating the expectation value

$$\langle V_\mu^* V_\nu \rangle = \frac{1}{3} \left[\delta_{\mu\nu} (H_0 a f)^2 \int d^3\mathbf{k} \frac{|\delta_\mathbf{k}|^2}{k^2} \right.$$
$$\left. \left(|w_{00;1}|^2 + 3\delta_{\mu 0} |w_{10;0}|^2 + \frac{4}{45} |w_{10;2}|^2 \right) \right]. \quad (29)$$

Normalizing the various moments to the integral of $|w_{00;1}|^2$, the term corresponding to the isotropic tensor window function is

$$A_{lm;L} = \frac{\int d^3\mathbf{k}\, (|\delta_\mathbf{k}|^2/k^2) |w_{lm;L}|^2}{\int d^3\mathbf{k}\, (|\delta_\mathbf{k}|^2/k^2) |w_{00;1}|^2}. \quad (30)$$

Assuming that there is no quadrupole, we obtain the coefficients K_ν, the correction to be applied to the results of calculations based upon the assumption of isotropy:

$$K_\nu = \frac{\langle |U_\nu|^2 \rangle}{\langle |U_\nu|^2 \rangle_{isotr}} = 1 + \frac{4}{15} A_{10;2} + \delta_{\nu 0}\, 9A_{10;0}. \quad (31)$$

For a numerical evaluation of the coefficients, we assume that the radial dependence of the selection function ϕ_{10} is the same as the overal ϕ_{00}, the dipole amplitude is D, the case of nonuniform sampling,

$$\phi_{00}(r) \propto e^{-(r/R)^2/2} \,/\, (\sigma_f^2 + \Delta^2 r^2) \,;\, \phi_{10}(r) = D\phi_{00}(r). \quad (32)$$

We use the adiabatic cold dark matter spectrum to get the coefficients. Only the $L = 0$ term has a significant contribution, the $L = 2$ becomes negligible: $A_{10;0} = 2.44D^2$, $A_{10;2} = 0.25D^2$. Thus the dipole moment will change the correction considerably:

$$K_\nu = 1 + 0.067\, D^2 + \delta_{\nu 0}\, 21.96\, D^2. \quad (33)$$

Even a small $D = 0.2$ dipole anisotropy will increase the dispersion of the observed velocity component along the direction of the dipole by a factor of 2. This effect is also coming from the nonuniform sampling. A real dipole anisotropy of the galaxy distribution would cause additional effects. Besides, another complication will arise from the use of magnitude limited samples, as discussed by Fall and Jones (1976).

3.3 - Applications to the Elliptical Galaxy Sample

We used the Dressler *et al.* (1987) elliptical galaxy sample to estimate the magnitude of these effects. We take a Gaussian of width $R = 3000$ km/s as the radial selection for both multipoles and use an error model of

$$\sigma_i^2 = \sigma_f^2 + \Delta^2 r_i^2/\sqrt{N_i}\,, \tag{34}$$

where $\sigma_f = 300$ km/s is the field-dispersion, $\Delta = 0.23$ is the error of the distance determination, and N_i is the multiplicity assigned to groups as in Lynden-Bell *et al.* (1988). The effective degrees of freedom in this case will not be integer, $N_e = 252.7$. Calculating the mean scatter for the sample we obtain

$$\langle\sigma^2\rangle = 1.8\,\Delta^2 R^2 = 1.8 \times 690^2(\mathrm{km/s})^2. \tag{35}$$

We calculated the various multipole moments of the selection function numerically. The quadrupole moment became $a_{20} = 0.44$, the various dipole components are $a_{10} = 0.156$, $a_{11} = (-0.027) + i(-0.213)$. This dipole has an amplitude $D = 0.342$, points to $l = 262.7^o$, $b = 27.2^o$, very close to the direction of the microwave anisotropy. The distribution of galaxies and the dipole direction is shown on Fig. 2. In order to simplify we transform the results to rectangular coordinates (U_x, U_y, U_z) in the galactic system.

The sampling variance in the velocities becomes in units $(\mathrm{km/s})^2$

$$\mathrm{Var}(U_\alpha U_\beta) = \begin{pmatrix} 12720 & 0 & 0 \\ 0 & 12720 & 0 \\ 0 & 0 & 7280 \end{pmatrix}. \tag{36}$$

The variance matrix of the bulk velocity, using the CDM spectrum, no biasing, normalizing to the isotropic tensor velocity dispersion $\langle U^2\rangle$ becomes

$$\mathrm{Var}(U_\alpha U_\beta) = \frac{1}{3}\,\langle U^2\rangle \begin{pmatrix} 1.04 & -0.26 & 0.13 \\ -0.26 & 3.25 & -1.03 \\ 0.13 & -1.03 & 1.44 \end{pmatrix}. \tag{37}$$

The trace of the matrix is 1.91 $\langle U^2\rangle$, corresponding to a total rms velocity of $1.38U$, equal to 430 km/s for $\langle U^2\rangle = (310\ \mathrm{km/s})^2$.

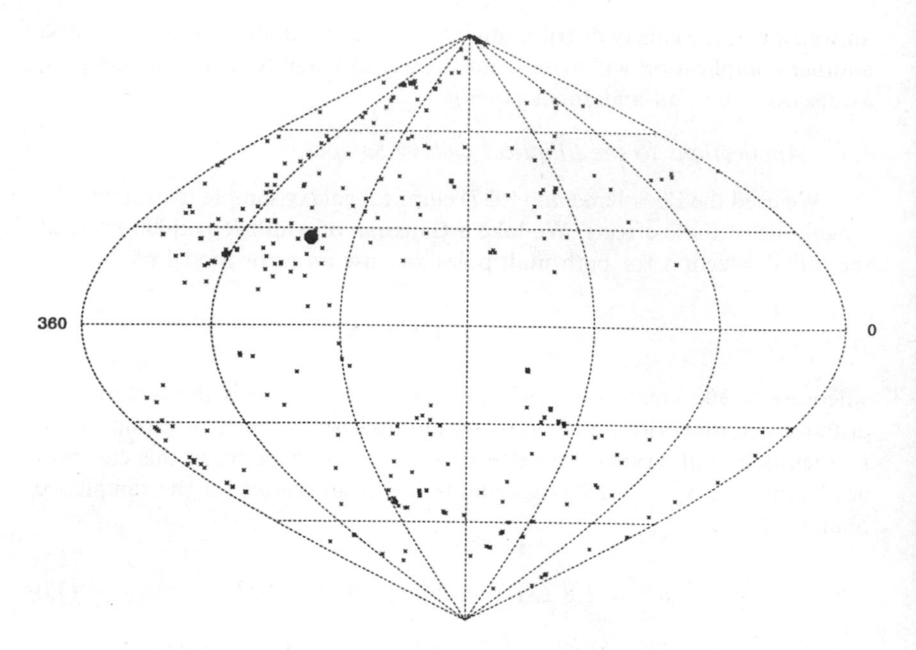

FIG. 2. The distribution of the elliptical galaxy sample of Burstein *et al.* (1986). The large black mark shows the direction of the dipole moment, $D = 0.342$, $l = 267.7^o$, $b = 27.2^o$.

4. Multipole Expansion of the Radial Velocity

If we select a relatively thin shell of galaxies, then the radial velocity on the surface of the sphere is a scalar function of the angles α, δ. It can be expanded in terms of spherical harmonics:

$$V_r(\Omega_s) = \sum_{l,m} V_{lm} Y_{lm}(\Omega_s), \tag{38}$$

with $V_{lm} = (-1)^m V^*_{l,-m}$. For that particular shell we can calculate the ensemble average of the multipole moment correlation matrix

$$\langle V^*_{lm} V_{l'm'}\rangle \delta_{ll'} \delta_{mm'} 4\pi (H_0 af)^2 \int d^3\mathbf{k} \frac{|\delta_\mathbf{k}|^2}{k^2} |W_l(k)|^2 = \delta_{ll'} \delta_{mm'} 4\pi u_l^2, \tag{39}$$

$$\text{where } W_l(k) = \int d^3\mathbf{s}\, \Phi(\mathbf{s}) j_l'(ks). \tag{40}$$

In the first part of this paper we already considered the $l = 1$ term corresponding to the dipole 'bulk' motion. There is a monopole $l = 0$, corresponding to an overall uniform expansion or inward flow, caused by the local over- or under-density, which is absorbed into the local value of the Hubble constant. The presence of this term can possibly be observed by taking larger and larger shells.

If we have two separate shells, described by their respective window functions $W_l^{(a)}(k)$ and $W_{l'}^{(b)}(k)$, the correlation of the velocity coefficients is given by

$$\langle V_{lm}^{(a)*} V_{l'm'}^{(b)} \rangle = \delta_{ll'}\, \delta_{mm'}\, 4\pi (H_0 a f)^2 \int d^3\mathbf{k}\ \frac{|\delta\mathbf{k}|^2}{k^2}\ W_l^{(a)*}(k)\ W_l^{(b)}(k). \quad (41)$$

The relative magnitude of the u_l cofficients will determine the coherence, or patchiness of the large-scale velocity field. This coherence can also be described by using the multipole coefficients: a smooth velocity field has a cutoff for the higher harmonics. Here we suggest that this patchiness carries significant information about the shape of the fluctuation spectrum, independent of the normalization. In Fig. 3 we plot the coefficients u_l^2 as a function of l for the CDM spectrum, using R = 3000 km/s, weighted with errors, normalized to a mass-variance of 1 for a top-hat radius of 800 km/s. It is obvious that the l = 0 monopole term is the largest, and the others are falling off quite rapidly. For this normalization,

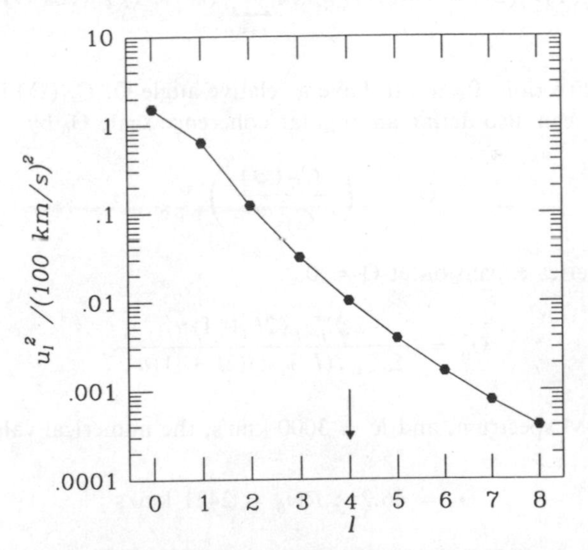

FIG. 3. The dependence of the velocity multipole moments u_l^2 on the harmonic number l, using the CDM spectrum, normalized to unit top-hat mass variance at 800 km/s, no biasing. The sample is approximated by a Gaussian selection of R = 3000 km/s radius, and error weighting. σ_f = 300 km/s, and Δ = 0.23. The arrow indicates the angular coherence scale of the CDM picture.

$$\begin{aligned} u_0^2 &= 2.147 \\ u_1^2 &= 1.032 \\ u_2^2 &= 0.188 \end{aligned} \tag{42}$$

in units of $(100 \text{ km/s})^2$. One should note that these are the dispersions for each respective $|V_{lm}|^2$. The rms 'monopole' velocity is about 150 km/s, about 5% distortion on the Hubble flow. The dispersion of the isotropic bulk motion is $\langle U^2 \rangle = 9u_1^2 = (310 \text{ km/s})^2$. For the quadrupole, there are 5 components of the irreducible tensor, and dispersion of the anisotropy in the quadrupole components can be estimated to be $(337 \text{ km/s})^2$, comparable to the bulk motion prediction.

Currently, in Lynden-Bell *et al.* (1988) there is a discussion of a quadrupole anisotropy of 0.1 — 0.2 in the Hubble flow for the various samples (300 - 600 km/s). Besides, the velocity field shows remarkable coherence, on linear scales of 500 — 2000 km/s.

5. Angular Velocity Correlations

A good way to quantify the coherence of the velocity fields is to calculate the angular velocity correlation function. From the multipole expansion

$$\langle V_r(1)V_r(2)\rangle = C_{12}(\Theta) = 4\pi \sum_{l=0}^{\infty} (2l + 1)\, P_l(\cos \Theta)\, u_l^2 \tag{43}$$

where the directions Ω_1 and Ω_2 have a relative angle Θ. $C_{12}(\Theta)$ is shown on Fig. 4. One can also define an angular coherence scale Θ_0 by

$$\Theta_0^2 = \left(\frac{C_{12}(\Theta)}{C_{12}''(\Theta)} \right)_{\Theta = 0}. \tag{44}$$

Using the series expansion at $\Theta = 0$

$$\Theta_0^2 = \frac{2\, \Sigma_{l=0}^{\infty}(2l + 1)u_l^2}{\Sigma_{l=0}^{\infty}\, l(l + 1)(2l + 1)u_l^2}. \tag{45}$$

For the CDM spectrum, and $R = 3000$ km/s, the numerical value becomes

$$\Theta_0 = 46.2^o \; ; \; R\Theta_0 = 2421 \text{ km/s} \tag{46}$$

corresponding to $l = 4$. This is indeed very close to the observed patchiness of the velocity field. Note, that this quantity does not depend on the absolute normalization of the density fluctuation spectrum, nor on the amount of biasing.

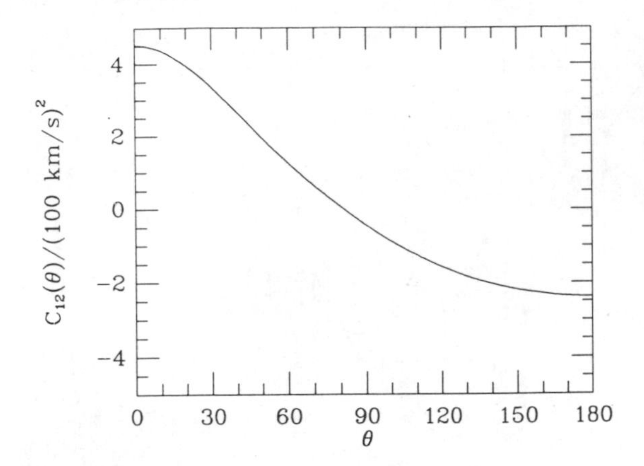

FIG. 4. The angular velocity correlation function, in units of $(100\ km/s)^2$.

The angular correlation function for the elliptical galaxy data has been calculated numerically (Burstein and Szalay 1988). This function, smoothed in 10° bins is plotted in Fig. 5.

6. MONTE-CARLO REALIZATIONS

Using the multipole expansion, we generate some images of the angular distribution of the radial velocities. For a given selection function we know the dispersion $|V_{lm}|^2$ of the expansion coefficients, and the velocity field is described by Gaussian statistics. Each V_{lm} is a random complex number,

$$V_{lm} = \sqrt{2\pi}\, u_l(x_{lm} + iy_{lm}) \tag{47}$$

where x_{lm} and y_{lm} are normal Gaussian random numbers, with unit dispersions. The hermiticity requires that

$$\begin{aligned} V_{l,-m} &= (-1)^m V^*_{lm} = (-1)^m \sqrt{2\pi}\, u_l(x_{lm} - iy_{lm}) \\ V_{lo} &= \sqrt{4\pi}\, u'_l x_{lm} \end{aligned} \tag{48}$$

Then we just add up the series, to a harmonic number $l = 10$, and generate equal area plots of the resulting velocity field. These are shown on Fig. 6. Also for comparison we smoothed the original elliptical galaxy data with various Gaussian filter lengths and these results are shown in Figs. 7 and 8.

7. CONCLUSION

We have made several improvements over the usual ways of calculating large-scale motions in the universe, taking into account that observers measure

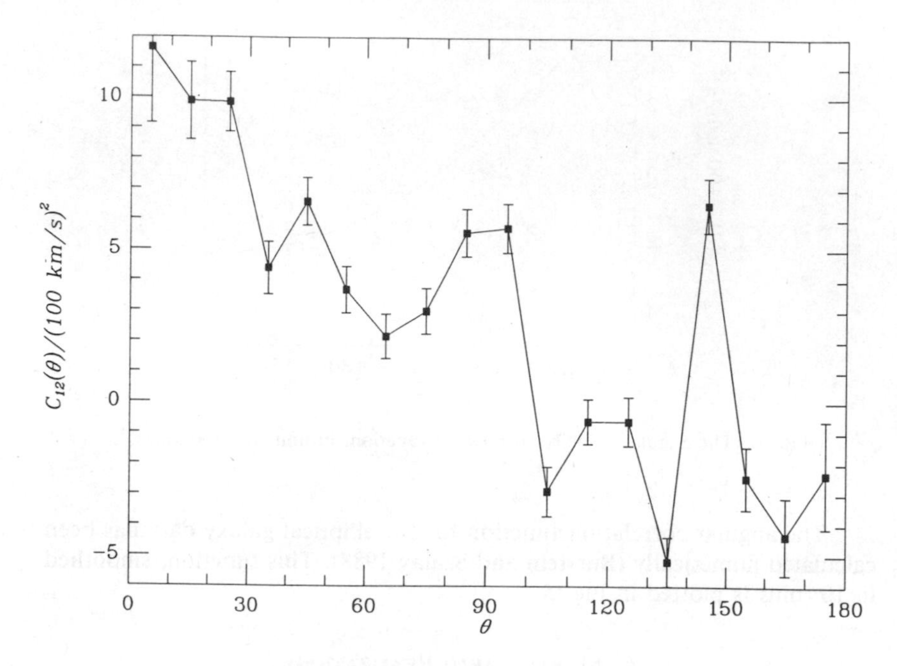

FIG. 5. The angular velocity correlation function in 10° bins, estimated from the elliptical galaxy sample (Burstein and Szalay 1988).

radial velocities, resulting in a tensor window function. Furthermore, we have calculated the twofold effect of selection anisotropies in the galaxy distribution previously not considered and we have shown that they have important consequences on the bulk motions.

The fluctuation spectrum causes a large-scale coherence in the velocity field, quantified by the multipole components of the projected velocities. The use of radial projection mixes these modes corresponding to different l, and a nonuniform sampling results in nonseparable errors. With an anisotropic sample the spherical harmonics become nonorthogonal, and the usual methods of least squares or maximum likelihood yield systematically biased estimations of the bulk motion. The major effect comes from the dipole and the quadrupole anisotropy: the direction of the velocity will be correlated with the dipole moment of the galaxy distribution and statistical uncertainties align perpendicular to the quadrupole. This will cause the measured bulk velocity to be preferentially close to the galactic plane and aligned with the dipole anisotropy, very close to what the actual observations show, although there is a real density enhancement in that direction.

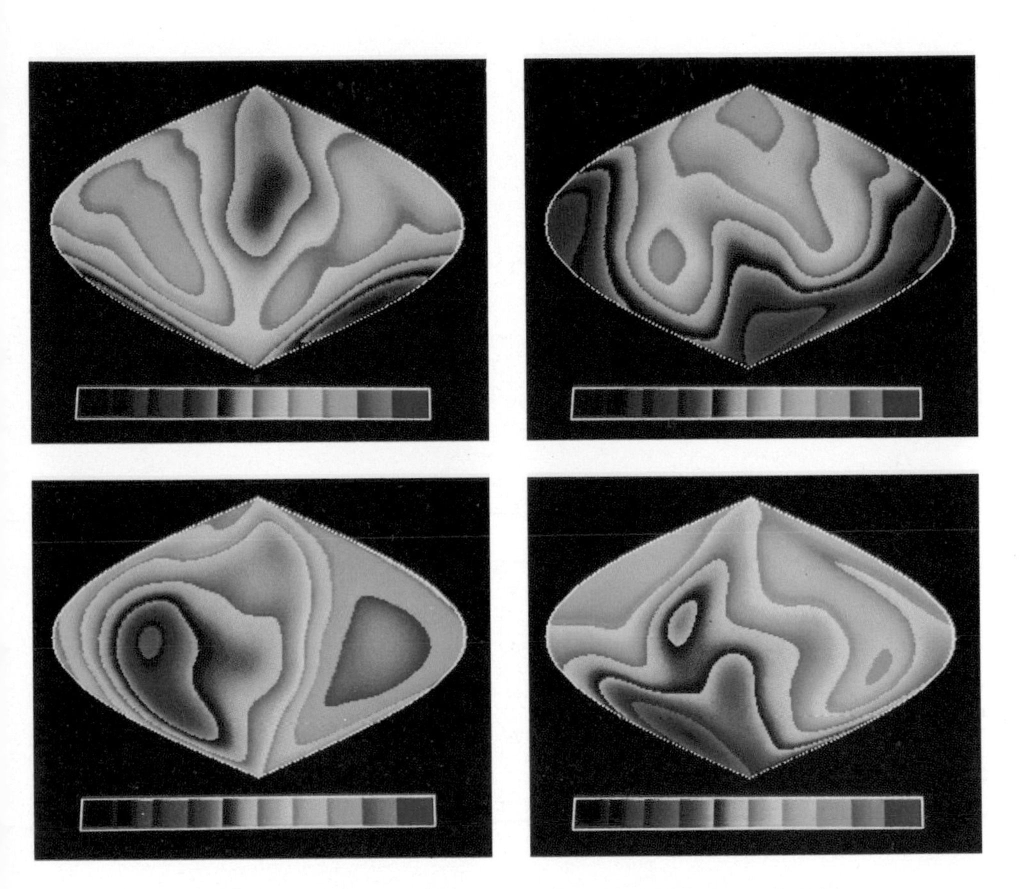

FIG. 6. Realizations of the velocity field in the unbiased CDM scenario, where the survey size is limited to 3200 km/s, and the selection function is error weighted, normalization is as above (see Fig. 3). A total of 16 different random cases have been created, the false colors span the range of -1000 km/s through 1000 km/s, and the small bins with intensity modulation are 200 km/s wide. Red colors correspond to redshift, blue to blueshift. Out of the 16 models created about 2 have a large dipole of about 600 km/s in magnitude, like the case of the Great Attractor, and several others have much smaller velocities. With a biasing factor of a few, such a large dipole would be extremely unlikely.

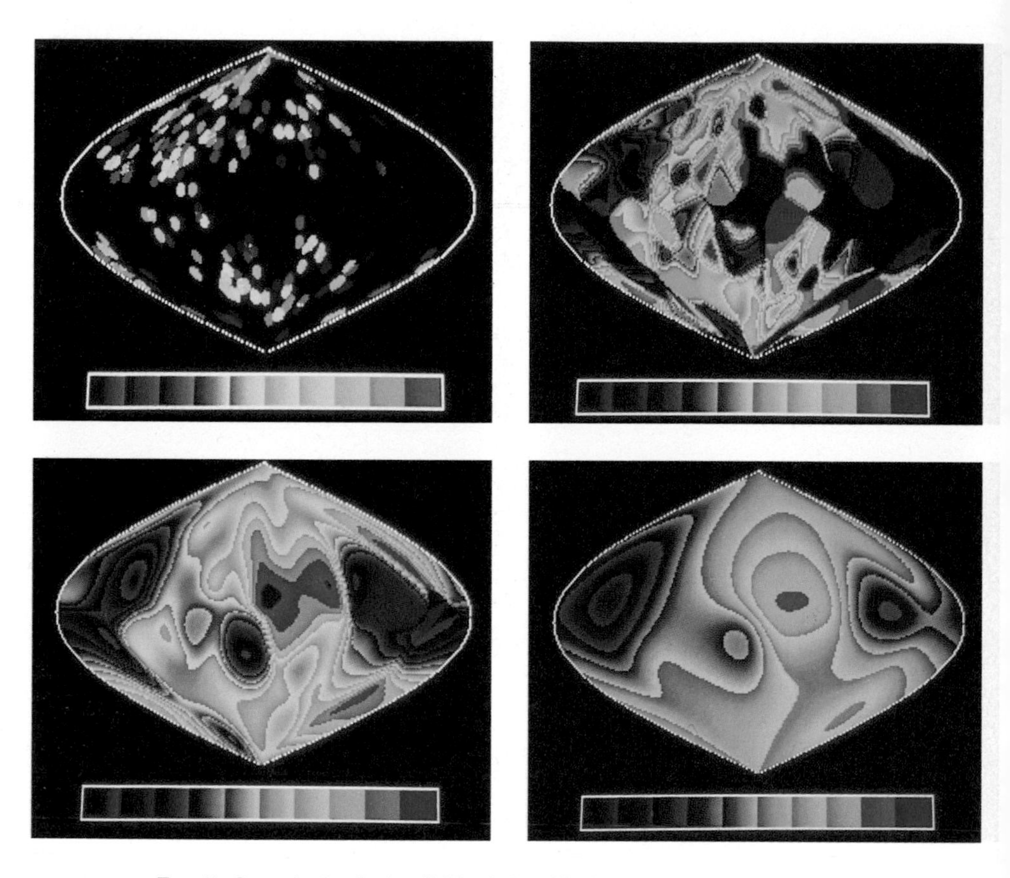

FIG. 7. Smoothed velocity field of the elliptical galaxy data of Burstein *et al.* (1986) with distances out to 8000 km sec^{-1}. Each galaxy has been smoothed with a Gaussian filter of a radius 1,5, 10, and 20 degrees respectively. The Gaussian was cut off at four filter radii. The regions with no galaxies within that distance were left blank. Velocities are coded as in Fig. 6.

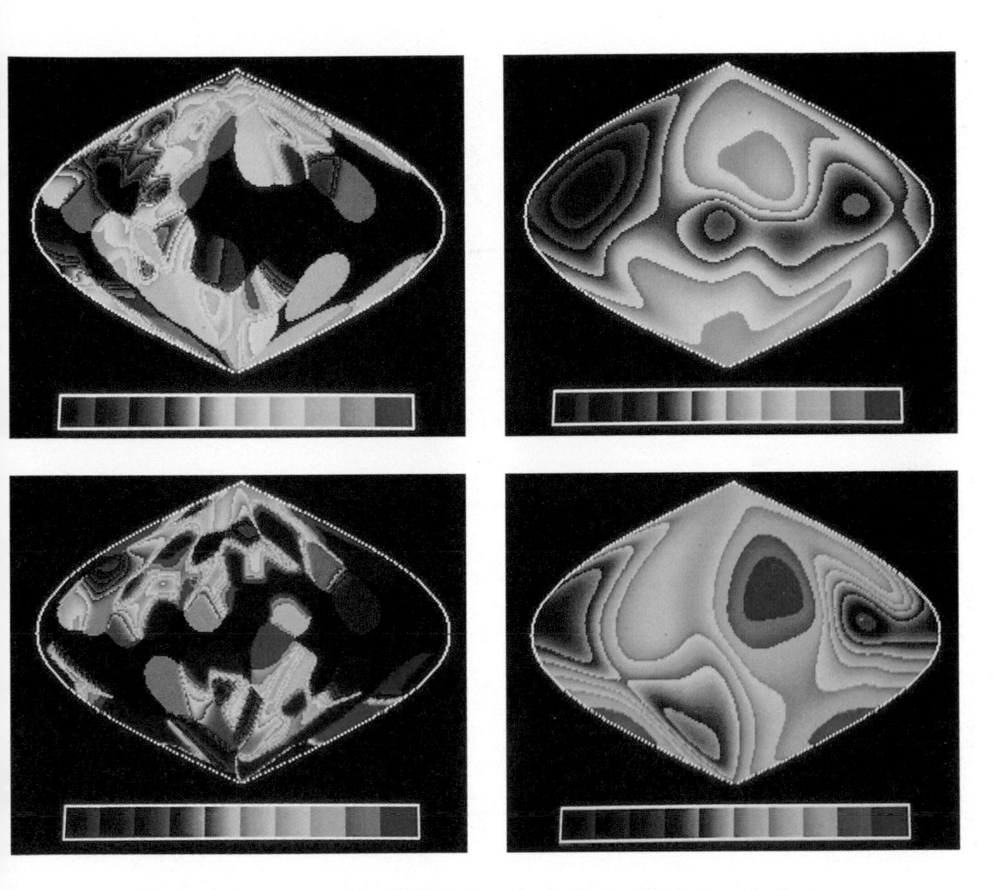

FIG. 8. Smoothed velocity field of the galaxies in the elliptical set: in the top two panels are the near galaxies (V < 3200 km sec^{-1}); in the bottom two panels are the far galaxies (3200 < V < 8000 km sec^{-1}). Smoothing lengths are 5 and 20 degrees; other details are as in Fig. 7. Note that the dipole in both the near and far fields is accompanied by a monopole term of comparable magnitude.

The radial velocity field observed over the whole sky can also be expanded in terms of spherical harmonics, and each coefficient has a different physical meaning. The monopole term describes a uniform distortion in the expansion, the dipole is the bulk motion, the quadrupole appears as an anisotropy of the Hubble flow. The general coherent appeareance of the velocity field can be well described by an angular coherence scale. Using the CDM spectrum to estimate this angle for the elliptical galaxy sample we obtain an excellent agreement, $\Theta_0 = 46.2^o$, independent of normalizations. The shape of the correlation function (the higher multipole moments) carries interesting information on the shape of the power spectrum around the 10 Mpc scales. On the whole, the standard normalization of the CDM picture gives somewhat low velocities, but the relative ratios of the bulk motion to quadrupole and the correlation length are quite acceptable.

ACKNOWLEDGEMENTS

The author would like to acknowledge useful conversations with Enikö Regös, Dave Burstein, Vera Rubin, and Nick Kaiser. This work was supported by a grant of the Hungarian Academy of Sciences (OTKA) in Hungary, and by the National Science Foundation of the United States.

REFERENCES

Aaronson, M., Bothun, G., Mould, J., Huchra, J., Schommer, R.A. and Cornell, M.E. 1986. *Ap J.* **302,** 536.

Burstein, D., Davies, R.L., Dressler, A., Faber, S.M., Lynden-Bell, D., Terlevich, R.J., and Wegner, G. 1986. in *Galaxy Distances and Deviations from Universal Expansion* (NATO ASI Series) ed. B. Madore and B. Tully. p. 123. Dordrecht: Reidel.

Burstein, D. and Szalay, A.S. 1988. in preparation.

Clutton-Brock, M. and Peebles, P.J.E. 1981. *AJ.* **86,** 1115.

Collins, C.A., Joseph, R.D., and Roberts, N.A. 1986. *Nature.* **320,** 506.

de Vaucouleurs, G. and Peters, W.L. 1984. *Ap J.* **287,** 1.

Dressler, A., Lynden-Bell, D., Burstein, D., Davies, R.L., Faber, S.M., Terlevich, R.J., and Wegner, G. 1987. *Ap J.* **313,** 42.

Fall, S.M. and Jones, B.J.T. 1976. *Nature.* **262,** 457.

Grinstein, B., Politzer, H.D., Rey, S.J., and Wise, M.B. 1987. *Ap J.* **314,** 431.

Hart, L. and Davies, R.D., 1982. *Nature.* **298,** 191.

Kaiser, N. 1983. *Ap J Letters.* **273,** L17.

_____. 1988. *MN.* **231,** 149.

Lynden-Bell, D., Faber, S.M., Burstein, D., Davies, R.L., Dressler, A., Terlevich, R.J., and Wegner, G. 1988. *Ap J.* **326,** 19.

Peebles, P.J.E. 1980. *The Large Scale Structure of the Universe.* Princeton: Princeton University Press.

Regös, E. and Szalay, A.S. 1988. *Ap J.* submitted.

Rubin, V.C., Thonnard, N., Ford, W.K. Jr., Roberts, M.S., and Graham, J.A. 1976a. *AJ.* **81,** 687.

Rubin, V.C., Thonnard, N., Ford, W.K. Jr., and Roberts, M.S. 1976b. *AJ.* **81,** 719.

THEORETICAL IMPLICATIONS OF COSMOLOGICAL DIPOLES

N. KAISER and O. LAHAV
Institute of Astronomy, Cambridge

ABSTRACT

We have compared the dipole anisotropy of the IRAS sky and the peculiar velocity bulk-flow solutions with the cold dark matter (CDM) predictions using linear theory. The model predictions fit very nicely with the observations; the χ^2 statistics are perfectly acceptable, and the likelihood analysis using these probes of large-scale structure gave essentially identical normalisation parameters as those we favour from considerations of clustering on much smaller scales.

We have developed and applied an improved version of Gott's test using the correlations between a number of angular dipoles and bulk flow solutions, rather than the traditional implementation using only the flux weighted dipole and the Local Group. We obtain a value for the bias parameter for IRAS galaxies of $b \simeq 1.5$, corresponding to an Ω_0 of around 0.4.

Following the modeling of the peculiar velocity field, we have also made simple models with one or more spherical perturbations. These models, in which roughly 300 km/s of our peculiar motion derives from inhomogeneity at distances between about 40 and 80 Mpc/h, are compatible with the observed dipoles, and also reproduce the apparent *lack* of acceleration seen at these distances in redshift space. The acceleration from distant perturbations in these models is similar to that predicted in CDM.

1. INTRODUCTION

In the cold dark matter model (hereafter CDM), it is assumed that at some very early time there were Gaussian isentropic density fluctuations $\Delta(\mathbf{r}) \equiv (\varrho(\mathbf{r}) - \bar{\varrho})/\bar{\varrho}$ with the 'Harrison-Zeldovich' constant curvature spectrum.

With the additional assumptions that the universe has critical density and is dominated by a cold collisionless particle, the power spectrum $P(k) \equiv \langle\Delta(k)\Delta^*(k)\rangle$ emerging in the matter dominated era can be calculated (e.g. Bond and Efstathiou 1984), and this gives a complete statistical description of the density and associated velocity field.

We wish to confront this model with observations of large-scale structure. One approach is to use estimates of the power spectrum or autocorrelation function, which effectively measure the rms fluctuations. On small scales these can be determined to high precision, but as we approach the depth of the survey we no longer sample enough volume to determine the properties of the ensemble. There are, however, certainly statistically significant detections of density fluctuations at very large scales from the dipole moments of galaxy counts for instance, and there are analogous detections of large-scale peculiar velocities. Our goal here is to extract the theroretical predictions for these observations. We would like to make clear at the outset that, while we will consider only *dipole* moments of the various fields, we do not assume that the dipole moments alone provide a good description of the inhomogeniety around us. The dipoles are simply a convenient set of statistics for which it seems to us to be reasonable to calculate the probability distribution function and compare with the theory as though the statistics had been chosen *a priori*. We could in principle perform a similar analysis to calculate the probability distribution function for the coefficients of a Great Attractor model for instance, but since the form of this model was chosen *a posteori* we would find it more difficult to interpret the results.

Before we can make these predictions it is necessary to specify two parameters which determine the amplitude of the fluctuations. The first parameter controls the amplitude of the *mass* fluctuations, for which an arbitrary but convenient measure is the rms fluctuation in spheres of radius 8 Mpc/h, and which we denote by σ_ϱ. Another parameter is the bias parameter b which describes the amount by which the fluctuations in galaxy number density are amplified relative to the underlying matter concentrations. These parameters are constrained, albeit rather roughly, from considerations of mass and galaxy clustering on smaller scales. Assuming that clusters of galaxies correspond to rare high peaks of the initial mass fluctuations when filtered on a suitable scale, we find (see e.g. Bardeen *et al.* 1986) that $\sigma_\varrho \simeq 0.6 - 0.7$. The greatest uncertainty here is in the determination of the masses of clusters from velocity dispersions. Since the rms fluctuations in galaxies in 8 Mpc/h spheres, $b\sigma_\varrho$, is roughly unity, this would indicate $b \simeq 1.4 - 1.8$.

With these normalisation parameters set, the model has great predictive power and is vulnerable to a multitude of observational tests, particularly those which probe structure on scales much larger than the scale where the normalisation is performed. An added bonus is that fluctuations on such scales should be very well described by linear theory, and the primordial Gaussian statistics should be preserved.

If we imagine observers throughout the universe each of whom makes a dipole measurement with identical selection procedures, then, in this model, the results form a Gaussian random vector field. For a single dipole, the statistical distribution is very simple. We can, however, generate a number of dipoles by splitting the data into flux and distance intervals, and hopefully improve the power of the test. These dipoles are necessarily quite strongly correlated with each other, so they must be interpreted with care. We shall adopt two approaches. First we shall attempt an *absolute* assessment of the CDM model with the normalisation as given above. To this end we calculate the rms values for the individual dipoles. This simple test enables one to see if there are any obvious discrepancies. Then, in order to properly incorporate the correlations and changes in direction we calculate a 'chi-squared' statistic. Secondly, as a rather crude first attempt to make a *relative* assessment of a range of theoretical models, we perform likelihood analysis with the normalisation parameters as free variables. The hope here is that if we live in a world in which there is more power at large scales than in CDM-world, then this analysis should give larger values for the normalisation parameters than those determined above. We appreciate that there are widely differing views on how to do statistics, particularly for problems like this. We hope that the combination of statistics used here will have some appeal to both 'frequentists' and 'Bayesians'.

As well as asking whether the amplitudes of the observed dipoles are compatible with the very specific CDM model, one can also use these data to test the more fundamental underlying assumption that the fluctuations in galaxy number density trace the mass fluctuations (albeit in a linearly biased manner). Traditionally, this test has been applied using the flux-weighted dipole of the galaxy counts, which gives an estimate of the net acceleration acting on the Local Group, and this can be compared with the Local Group motion relative to the microwave frame. There are two issues here. One can ask whether the velocity is parallel to the apparent acceleration — as predicted (in linear theory) in all models where structure grows by gravitational instability, though not in the case of the explosive scenario for instance. If these vectors are at least approximately parallel then it is also interesting to look at their ratio. Before 'biasing' became popular, this ratio was interpreted as a determination of Ω. In the present context we *assume* $\Omega \equiv 1$ and interpret the ratio as a measure of the bias parameter b. An appealing feature of this test is that it appears to give a determination of b which is independent of any assumption about the spatial distribution of sources of acceleration. Realistically, considerations of the finite number density of galaxies and flux limits in the catalogues tend to decouple the estimates of acceleration and velocity somewhat, and to estimate the magnitude of the error requires a model to tell us which wavelengths contribute significantly to the acceleration. We find that, in CDM-world at least, the scatter is perhaps disappointingly large using the standard choice of dipoles, but that a considerable improvment can be made making use of other dipoles which have now become available. Crudely speaking, just as an idealised flux-weighted dipole correlates very well with the motion of the Local Group, there are other dipoles which correlate well with the motion of the Local Supercluster or of even larger regions. These can be measured quite accurately now, and give an estimate of b (or $b\Omega^{-0.6}$ according to ones taste) with considerably smaller statistical uncertainty. An additional advantage, though one which is harder to quantify, is that in using the motion of large aggregates rather than the Local Group one can be much more confident in applying linear theory.

A largely independent view of large-scale structure is provided by redshift surveys. Initial estimates (Strauss and Davis 1987; Yahil 1988) of the acceleration vector from the 2 Jy IRAS redshift survey seemed to indicate that the source of the acceleration is very localised. This conflicts strongly with the view of a rather deep source for the bulk of the Local Group motion coming from studies of departures from Hubble flow. A loophole which may remove this apparent discrepancy is that no allowance was made for the effect of peculiar velocities, and these are expected to have a profound effect on the clustering pattern in redshift space (Kaiser 1987). We use a variety of models to explore this possibility and show that this effect does indeed seem to largely resolve the apparent inconsistency. We also calculate the statistical uncertainty in the acceleration vector in the CDM model, and compare with the precision of the acceleration estimate obtained from the angular dipoles.

The implications of all the observations we analyse below have been considered before in one way or another. An important feature of our analysis is that we attempt to model as realistically as possible the actual observations. We show that for each observation there corresponds a unique 'window function' defined unambiguously by the weighting scheme adopted by the observer. The observation can be written as a linear convolution of the density or velocity field and the window, plus an independent random 'noise' component deriving from the discreteness of galaxies, statistical errors in peculiar velocities or whatever. We have also attempted to model this statistical error realistically, and include this in our comparison with the CDM predictions. We find that for the velocity observations in particular, previous theoretical studies have tended to overestimate the effective depth of the surveys, and this led to the misleading impression that our universe is more inhomogeneous on large scales than CDM-world. Similarly, the anisotropy of the IRAS sky was initially thought to be indicative of density fluctuations on scales of order 100 Mpc/h (Rowan-Robinson *et al.* 1986), again with exciting implications for cosmogonical theories. Our analysis will show that the simple dipole statistics at least, are in as good agreement with the CDM predictions as could be expected. Whether the same will prove true for more detailed representations of the data we cannot say. Also, the perhaps more interesting question of which alternative models are *incompatible* with observations is only touched on slightly here.

Perhaps the most unrealistic feature of our analysis is that we assume that in all of the observations we have full and and uniform sky coverage; this approximation is not strictly necessary, and zones of obscuration etc. can be allowed for within the framework set out below. The assumption of spherical symmetry does, however, greatly simplify the formulae, and is not a bad approximation for most of the data we consider.

The layout is as follows. In Sec. 2 we consider the dipole moments of the IRAS galaxy counts. In this section we set out the details of the theoretical model and illustrate the technique used for constructing the probability distribution function for a real observation. In Sec. 3 we analyse in a completely analogous manner the dipole moments of the peculiar velocity field. In Sec. 4 we combine these data, as described above, to determine the bias factor. These sections are an extract from the more complete exposition of Kaiser and Lahav (1988) where fuller detail is given and the dipoles obtained from optical catalogues are also considered. In Sec. 5 we turn to the interpretation of the apparent acceleration vector as inferred from the 2 Jy IRAS redshift survey. Finally we summarise our results, discuss their relationship to other probes of large-scale structure and sketch our plans to extend these analyses in the future.

2. The Angular Dipoles

In this section we will consider dipole moments of the angular counts of galaxies. In Sec. 2.1 we state the model for the matter density field, and for the distribution of galaxies. In Sec. 2.2 we calculate the probability distribution function (pdf) for a single dipole moment, and in Sec. 2.3 we calculate the pdf for a collection of dipoles. We obtain expressions for the likelihood as a function of the normalisation parameters of the model. In Sec. 2.4 we apply these results to the IRAS dipoles, and also consider some alternative models.

2.1 - The Matter Density Field and the Galaxies

We assume that the Fourier components of the density field are drawn from a Gaussian distribution with $\langle\Delta(\mathbf{k})\Delta^*(\mathbf{k})\rangle = P(k)$. The field

$$\Delta(\mathbf{r}) \equiv \frac{\varrho(\mathbf{r}) - \langle\varrho\rangle}{\langle\varrho\rangle} = \frac{V_u}{(2\pi)^3}\int d^3k\, \Delta(\mathbf{k})\, e^{-i\mathbf{k}\cdot\mathbf{r}} \qquad (2.1)$$

is therefore a statistically homogeneous and isotropic random field which is periodic in the large but arbitrary volume V_u.

Writing $P(k) = P_0 F(k)$, the specific form we adopt is a fit to the CDM spectrum with initially adiabatic fluctuations and with $\Omega_0 = 1$ and $h = 0.5$ as given by Bond and Efstathiou (1984):

$$F(k) = k \{1 + [11.55 k + (5.70 k)^{3/2} + (3.24 k)^2]^{1.25} \}^{-1.6}. \quad (2.2)$$

The unit of length here and throughout is 1 Mpc/h, or the distance corresponding to a recession velocity of 100 km/s.

We express the normalization in terms of the variance of density in a 'top-hat' sphere of radius a = 8 Mpc/h:

$$\sigma_\varrho^2 \equiv \langle \left(\frac{\Delta M}{M}\right)^2 \rangle = \frac{1}{(2\pi)^3} \int d^3k \, P(k) \, |U(ka)|^2 \quad (2.3)$$

where $U(ka) = 3 j_1(ka)/(ka)$, and $j_1(y) \equiv \sin y/y^2 - \cos y/y$.

We assume that there is a universal luminosity function $\Phi(L)$ and that the galaxies 'fairly sample' the linearly biased density field $1 + \Delta_{galaxies} = 1 + b\Delta$. Thus, if we describe this point process by occupation numbers $n(\mathbf{r}_i, L_j)$ for microscopic cells in $\mathbf{r}$ and L space which are 0 or 1 if the volume element $d^3r \, dL$ with labels i, j is empty or contains 1 particle respectively, then the probability of occupation is

$$P(\text{galaxy in cell } \mathbf{r}_i, L_j) = \langle n(\mathbf{r}_i, L_j) \rangle \propto (1 + b\Delta(\mathbf{r}_i)) \cdot \Phi(L_j). \quad (2.4)$$

The inclusion of the parameter b is motivated by current thinking about biased models for galaxy formation in which production of galaxies is modulated by longer wavelength modes. Various schemes have been proposed to effect biasing, but the property shared by all of these is that for small amplitude waves one obtains a linear bias. More generally one would let b be a function of luminosity or some other attribute of the galaxies, but we will not explore that possibility.

The 'universal luminosity function' assumption according to which luminosities are assigned purely at random is highly idealised, and is almost certainly unrealistic. This assumption is important in determining the statistical noise in the observations. If, as seems at least plausible, environmental effects systematically perturb the luminosities of galaxies, then one would expect the simple 'Poissonian' estimate used below to underestimate the true noise. We have made tests to see if this is important by analysing the differences between dipoles obtained from IRAS and optical samples with similar depth. The constraint is unfortunately rather weak, and while we found no strong indication of excess fluctuations, we would warn against too literally interpreting dipoles for which the shot noise term is appreciable.

2.2 - Modeling a Single Dipole

Given a flux limited catalogue which lists positions and fluxes we can construct a dimensionless dipole moment which we define to be

$$\mathbf{D} = \frac{3 \sum w(S_q)\, \hat{\mathbf{r}}_q}{\int_{S_{\min}}^{S_{\max}} dS\, N(\mathrm{N})\, w(S)} \tag{2.5}$$

where the sum is over all galaxies with flux S in the range $S_{\max} > S \geq S_{\min}$. $N(S)$ is defined such that $\int_S^\infty N(S)\, dS = N_0(S/S_{\lim})^{-3/2}$, where N_0 is the total *expected* number of galaxies brighter than $S_{\lim}$ (the flux limit of the catalogue) in the entire sky. The choice of weighting function $w(S)$ is at the disposal of the observer.

In terms of the occupation numbers defined above,

$$\mathbf{D} = \frac{3 \sum_{\mathrm{i,j}} w(\mathrm{L_j}/4\pi r_i^2)\, \hat{\mathbf{r}}_i\, n(\mathbf{r}_i, L_j)}{\int dS\, N(S)\, w(\mathrm{S})} \tag{2.6}$$

where the sum is now over microcells. We now write **D** as the sum of two terms; the first we call the 'theoretical' dipole $\mathbf{D}^{\mathrm{th}}$, which is the value that would be obtained in a world where galaxies are so numerous that they approach the continuum limit, while the second $\mathbf{D}^{\mathrm{sn}}$ represents the sampling 'shot noise' error arising from the discreteness of the galaxies. For the former we have

$$\mathbf{D}^{\mathrm{th}} = \langle \mathbf{D} \rangle = \frac{3 \sum_{i,j} w(L_j/4\pi r_i^2)\, \hat{\mathbf{r}}\, \langle n(\mathbf{r}_i, L_j) \rangle}{\int dS\, N(S)\, \mathrm{w}(S)}\,. \tag{2.7}$$

Using 2.4, converting the sums to integrals, and shifting the spatial origin so the observer lives at $\mathbf{r}_0$, gives

$$\mathbf{D}^{\mathrm{th}}(\mathbf{r}_0) = 3\, b \int d^3r\, W(r)\, \Delta(\mathbf{r}_0 + \mathbf{r})\, \hat{\mathbf{r}}, \tag{2.8}$$

where the radial window is

$$W(r) \propto \int_{4\pi r^2 S_{\min}}^{4\pi r^2 S_{\max}} dL\, w(L/4\pi r^2)\, \Phi(L), \tag{2.9}$$

where $\Phi(L)dL$ is the luminosity function, and we normalise $W(r)$ so that $\int d^3r\, W(r) = 1$.

From the convolution theorem we find

$$\mathbf{D}^{\text{th}}(\mathbf{k}) = W^*(k)\, \hat{\mathbf{k}}\, \Delta(\mathbf{k}) \tag{2.10}$$

where

$$W^*(k) \equiv -12\pi i \int dr\, r^2\, W(\mathrm{r})\, j_1(kr). \tag{2.11}$$

From Parseval's theorem we find:

$$\langle \mathbf{D}^{\text{th}} \cdot \mathbf{D}^{\text{th}} \rangle \equiv \frac{1}{V_u} \int d^3r_0\, \mathbf{D}^{\text{th}} \cdot \mathbf{D}^{\text{th}} = \frac{b^2}{(2\pi)^3} \int d^3k\, P(k)\, \mathrm{W}(k) \mathrm{W}^*(k). \tag{2.12}$$

To obtain the total variance we must add the shot noise variance, which turns out to be

$$\langle \mathbf{D}^{\text{sn}} \cdot \mathbf{D}^{\text{sn}} \rangle = \frac{9}{N_0} \frac{\int_{S_{\min}}^{S_{max}} dS\, S^{-5/2} \int_{S_{\min}}^{S_{\max}} dS\, w^2(S)\, S^{-5/2}}{\left(\int_{S_{\min}}^{S_{\max}} dS\, w(S)\, \mathrm{S}^{-5/2} \right)^2}. \tag{2.13}$$

The final result then is that the observation defined by equation 2.5 is a single sample of a zero-mean, statistically isotropic, Gaussian vector field **D** with variance

$$\langle \mathbf{D} \cdot \mathbf{D} \rangle = \langle \mathbf{D}^{\text{th}} \cdot \mathbf{D}^{\text{th}} \rangle + \langle \mathbf{D}^{\text{sn}} \cdot \mathbf{D}^{\text{sn}} \rangle. \tag{2.14}$$

2.3 - Modeling N-dipoles

One can learn more from the data by splitting it into flux bins; and in the limiting case of very fine bins one would be using full information contained in the run of the dipole with flux.

The multivariate probability distribution of a set of M measurements of Gaussian variables a_i is

$$P(a_1, a_2, \ldots a_M)\, d^M a = \frac{d^M a}{\sqrt{(2\pi)^M |A|}} \exp - \frac{1}{2} \sum_i \sum_j a_i\, A_{ij}^{-1}\, a_j. \tag{2.15}$$

where $A_{ij} \equiv \langle a_i a_j \rangle$. Here we have N dipoles, defined by N different weighting functions, each with 3 components labelled α, β etc. By virtue of the assumed spherical symmetry of the sample we have $A_{(i\alpha),(j\beta)} = \delta_{\alpha\beta} C_{ij}$ with

$$C_{ij} \equiv \langle D^{\rm th}_{i\alpha}(\mathbf{r}) D^{\rm th}_{j\alpha}(\mathbf{r})\rangle + \langle D^{\rm sn}_{i\alpha}(\mathbf{r}) D^{\rm sn}_{j\alpha}(\mathbf{r})\rangle, \quad (2.16)$$

where no summation over α is implied, and where

$$\langle D^{\rm sn}_{i\alpha}(\mathbf{r}) D^{\rm th}_{j\beta}(\mathbf{r})\rangle = \delta_{\alpha\beta} \frac{1}{3} \frac{b^2}{(2\pi)^3} \int d^3k \, P(k) \, W_i(k) \, W^*_j(k) \quad (2.17)$$

and

$$\langle D^{\rm th}_{i\alpha}(\mathbf{r}) D^{\rm sn}_{j\beta}(\mathbf{r})\rangle = \frac{3\delta_{\alpha\beta}}{N_0} \frac{\int dS\, N(S) \int dS\, N(S)\, w_i(S)\, w_j(S)}{\int dS\, N(S)\, w_i(S) \int dS\, N(S)\, w_j(S)} . \quad (2.18)$$

The final probability distribution for the $D_{i\alpha}$ can then be written as

$$P(D_{i\alpha}) d^{3N} D = \frac{d^{3N}D}{(2\pi)^{3N/2} |C|^{3/2}} \exp - \frac{1}{2} \sum_i \sum_j C^{-1}_{ij} \sum_\alpha D_{i\alpha} D_{j\alpha} . \quad (2.19)$$

This formula contains, via 2.17, the normalisation parameter $b\sigma_\varrho$, and gives, on inserting actual measurements, the likelihood function $L(b\sigma_\varrho) = P(\text{data}\,|\,b\sigma_\varrho)$. We will also look at the χ^2 statistic that appears in the argument of the exponential.

2.4 - IRAS Dipoles

As our data source we use a colour-selected IRAS sample (Meurs and Harmon 1988; Harmon, Lahav, and Meurs 1987) at 60μ, which is based on version II of the IRAS *Point Source Catalog*. The advantages of this sample are its good sky coverage (more than 75%) and its "objective" selection of galaxies, free from association with optical galaxies. For this IRAS sample we use a double power law luminosity function at 60μ (Lawrence *et al.* 1986):

$$\Phi(L) dL = C L^{-\alpha} [1 + \frac{L}{L_* \beta}]^{-\beta} dL \quad (2.20)$$

with $\alpha = 1.8$, $\beta = 1.7$ and $L_* = 1.58 \times 10^{10} h^{-2} L_\odot$.

Fig. 2.1a shows the IRAS radial window functions (eq. (2.9)-(2.10)) for flux limits 0.7 — 10.0 Jy. The solid line corresponds to a number weighted window ($w(S) = 1$) and the dashed line corresponds to a flux weighted window ($w(S) = S$). The ordinate chosen is $r^3 W(r)$, which is proportional to the net weight per logarithmic distance interval, and this peaks at about 10000 km/sec.

Fig. 2.1b shows the corresponding k-space window functions (eq. (2.11)). The ordinate shows $k^2 W^2(k)$ for the number and flux weighted windows and

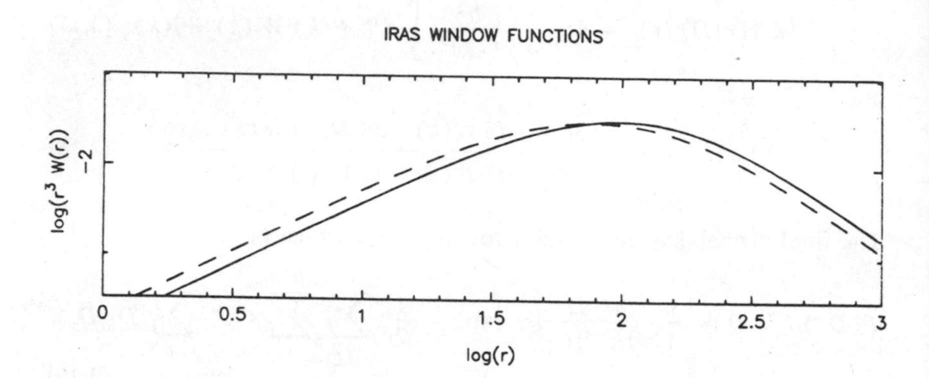

FIG. 2.1a. IRAS radial window functions (weighted by the cube of the distance), as calculated from the IRAS luminosity function (eq. 2.9) for a flux range 0.7-10.0 Jy. The curves correspond to number weighted (solid line) and flux weighted (dashed line) windows. The plot indicates that in a uniform universe a galaxy with a flux in this range has the highest probability to be found at a distance of about 100 Mpc/h. Note the similarity of the two curves.

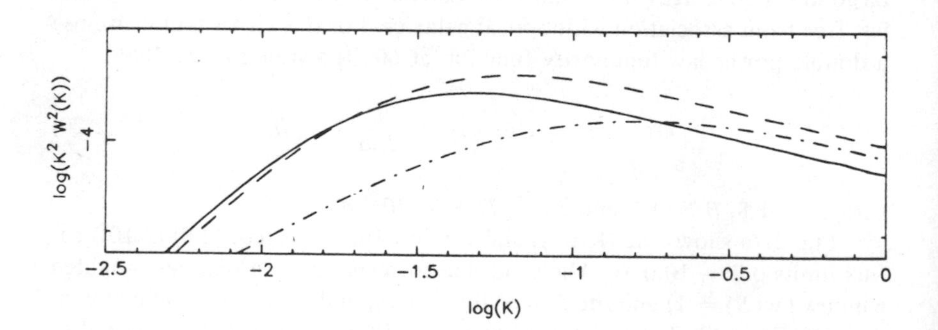

FIG. 2.1b. The Fourier transform of the curves in Fig. 2.1a according to eq. 2.11 (weighted by the square of the wave-number). The dashed-dotted line shows $k\,F(k)$, where $F(k)$ is the CDM power-spectrum (eq. 2.2). The curves show the contribution to the integral 2.12, for the theoretical dipole.

the dashed-dotted line shows $k\ F(k)$, where $F(k)$ is the CDM power spectrum given in (2.2). By plotting these expressions we can see the contribution of various wavelengths to the integral (2.12). It is clear from these figures that for this sample, the two weighting schemes give very similar windows, and therefore probe very similar depths. Consequently, measuring both dipoles will give little more information than either one alone.

Windows can however be constructed which do give better resolution in depth, and which therefore have greater power to discriminate between theories. Fig. 2.2 shows the windows for 5 IRAS differential flux slices, where the slices boundaries are 0.7, 1.2, 2.0, 3.5, 5.9, 10.0 Jy (see also Table 2.2). The solid curve corresponds to the fainter slice (0.7-1.2 Jy). The other curves are shifted to smaller distances in increasing order of their fluxes.

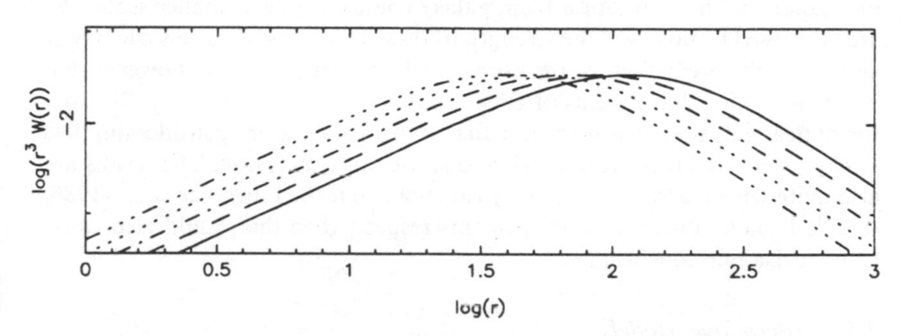

FIG. 2.2. IRAS radial window function for 5 differential flux slices (see Table 2.2). The solid line corresponds to the faintest slice, which samples the most distant volume. The other curves are shifted to smaller distances in increasing order of their fluxes.

Details of the construction of the dipoles are given in Kaiser and Lahav (1988). The results are given in columns 1-6 of Table 2.1. In column 7 we show the predicted dipoles assuming $b\sigma_{\varrho} = 1$. The predictions and the observations are quite similar. Both increase with increasing flux, but the predictions do so more rapidly than the data. This is interesting, since it is an indication that the data would fit better to a model with somewhat more power at larger scales than the CDM model - though in order to fit the normalisation condition for 8 Mpc/h spheres the spectrum would have to turn up again at small scales. To see quantitatively whether these dipoles are consistent with CDM we calculated the statistic $\chi^2 \equiv \Sigma_i \Sigma_j C_{ij}^{-1} \Sigma_\alpha D_{i\alpha} D_{j\alpha}$. This should be roughly Gaussian distributed with mean $3N$ (= 15 in this case) and variance $6N$. The result was 15.0 exactly, clearly a bit of a fluke, but equally clearly showing that the dipoles are quite compatible with CDM.

TABLE 2.1
FLUX WEIGHTED IRAS DIPOLES

S_{min}	S_{max}	D_x	D_y	D_z	D_{obs}	D_{pred}
0.7	1.2	− 0.008	− 0.131	0.049	0.140	0.110
1.2	2.0	− 0.003	− 0.103	0.089	0.136	0.154
2.0	3.5	0.053	− 0.179	0.147	0.238	0.212
3.5	5.9	− 0.102	− 0.182	0.081	0.224	0.295
5.9	10.0	− 0.177	− 0.090	0.168	0.260	0.401

We have also calculated the likelihood versus $b\sigma_\varrho$ as shown by the strong solid line in Fig. 2.3. The ML solution is $b\sigma_\varrho = 1.1$, which agrees well with the 'canonical' normalisation from galaxy counts at much smaller scale. We are impressed by how well the strength of these measures of large-scale power agree with the predictions of the normalised CDM model. Note however that we are assuming that the rms of IRAS counts in 8 Mpc/h spheres is like that for optical galaxies. If it turns out that IRAS galaxies are considerably less strongly clustered then we may wish to revise our conclusion. The results are also quite sensitive to the luminosity function, so if the Lawrence *et al.* (1986) sample is badly unrepresentative in this respect, then this would also cause us to revise our conclusions.

2.5 - *Alternative Models*

An alternative to the random Gaussian fields considered so far is to model the perturbation as a single localised overdense region. Figure 2.2 is very useful here as it shows graphically how changing the distance of the region producing the dipole affects the run of **D** with flux. If the perturbing region is situated at $r < 20$ Mpc/h then the high flux dipole should be 4.5 times that for the lowest flux, whereas we observe a factor 2 difference. Setting the perturbation ar $r = 60$ Mpc/h gives equal dipole for these flux bins, and for $r > 200$ Mpc/h the low flux dipole is 8 times larger than the highest flux bin - quite different from that observed. The best fitting single distance perturber lies at $r \simeq 40$ Mpc/h. Our result differs from the conclusion of Villumsen and Strauss (1987) who used the same 'delta-function' model and obtained a best fit for a perturber at 17.5 Mpc/h. They used the Meiksin and Davis (1986) sample which is based on different selection criteria and has poorer sky coverage than that used here, so perhaps the difference in results stems from this. Our result is also somewhat in disagreement with the result of Lahav, Rowan-Robinson, and Lynden-Bell (1988), who concluded that both the IRAS and optical dipoles are generated on scales smaller than 4000 km/sec. Again, their samples

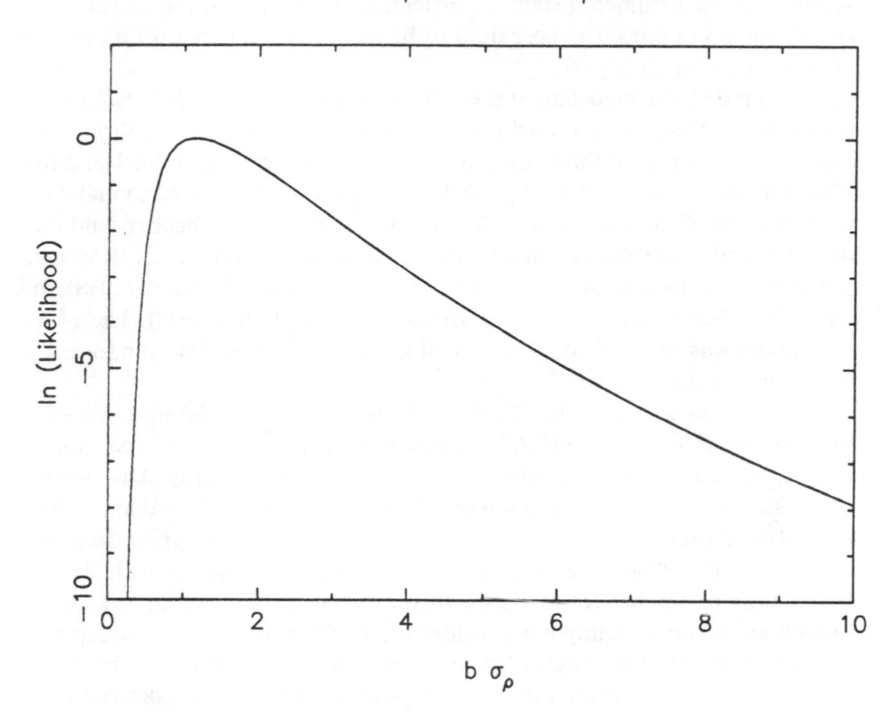

FIG. 2.3. The likelihood function versus $b\sigma_\varrho$ (where b is the bias parameter and σ_ϱ is the normalization), using 5 IRAS flux dipoles (see Table 2.2). The curve peaks at $b\sigma_\varrho = 1.1$.

and model (a "shell model") are different, and that may be a cause for discrepancy.

Our preferred distance $r \simeq 40$ Mpc/h is reminiscent of the Great Attractor picture. We have made a more detailed calculation of these 5 dipoles for a spherically symmetric r^{-2} density enhancement profile centred at 40 Mpc/h and which induces a retardation of the Hubble flow at our position of 600 km/s in the direction of the Local Group - microwave background motion. If we assume a b of unity the dipoles come out too small, but with $b = 1.5$ the predicted dipole amplitudes are in very good agreement with those observed. The directions of the dipoles are in poorer agreement - clearly, a single attractor model predicts that all dipoles should be parallel - but it is encouraging that we see at least rough agreement in the scale of the dominant perturbation in our vicinity as determined from these very different techniques. The value of $\chi^2 \equiv \Sigma(D_{obs} - D_{pred})^2/D_{sn}^2$ was 13, with 15 degrees of freedom. We experimented with other distances and found that χ^2 was disappointingly insensitive to the assumed distance, particularly if b was allowed to vary too, so we would not consider these data to be in strong conflict with a distance even as small as 20 Mpc/h.

Inspired by the modeling of the velocity field by Lynden-Bell *et al.* (1988) we made a 3-component model comprising the Great Attractor, Virgo, and a third component. All three components had the same r^{-2} profile. The direction and velocity induced by the third component were adjusted to make the total velocity of the Local Group agree with the microwave background motion. The optimum value for the distance to the third component was 18 Mpc/h, in accord with the idea of a Local Anomaly of Faber and Burstein (1988) and a possible effect of the Local Void (Lynden-Bell and Lahav 1988). The χ^2 for this model was reduced to 10, but again, the distance to this component is not very well determined.

To summarise then, the CDM model with our favoured normalisation predicts rms values for the IRAS dipoles roughly like those observed - the χ^2 is spot on, and the ML normalisation matches almost perfectly. This is quite impressive since the dipoles probe a much larger length scale than that at which the normalisation is imposed, so the outcome was by no means a foregone conclusion. If we look carefully at the run of dipole amplitude with flux we find a weak preference for models with somewhat more power at large scale, though we would certainly not consider the CDM model to be unacceptable on these grounds. The observed dipoles are compatible with those predicted by simple models motivated by the picture of the density field derived from quite independent studies of the peculiar velocity field, though clearly the real universe is more complicated than any of these simple models.

3. Peculiar Velocities

We now consider dipole moments of the line of sight peculiar velocity field - often called 'bulk flow solutions'. The analysis here is formally very similar to that of the previous section. One important difference, however, is that while the angular counts constrain the product $b\sigma_\varrho$, the velocity data constrain the *mass* fluctuations σ_ϱ alone.

The velocity field $\mathbf{v}(\mathbf{r})$ and $\Delta(\mathbf{r})$ are connected by the equation of continuity $\nabla \cdot \mathbf{v} = -\partial\Delta/\partial t$, or for $\Omega_0 = 1$, in terms of Fourier components

$$\mathbf{v}(\mathbf{k}) = \frac{-iH_0\mathbf{k}}{k^2}\,\Delta(\mathbf{k}). \tag{3.1}$$

This formula gives the velocity of a particle with respect to the uniformly expanding frame defined by the matter at large distances. Operationally this can be defined to be the frame in which the microwave background dipole anisotropy vanishes.

3.1 - Window Function for a Velocity Dipole

Let us assume that we are given a catalogue of direction vectors $\hat{\mathbf{r}}_q$, distance estimates and associated distance error estimates r_q, σ_q and redshifts corrected to the microwave frame z_q. An estimate of the line of sight peculiar velocity of the qth galaxy is $u_q \equiv cz_q - r_q$. We can construct a distance weighted dipole of the u_q which is analogous to equation 2.5:

$$\mathbf{U} \equiv \frac{3 \Sigma\, w(r_q)\, \hat{\mathbf{r}}_q\, u_q}{\int d^3r\, n(r)\, w(r)}. \tag{3.2}$$

As with the angular dipoles, the weighting scheme $w(r)$ is as yet unspecified. Note that for the case of uniform Hubble expansion, **U** vanishes, and for a pure 'bulk flow', **U** simply measures the flow velocity.

Neglecting for the moment the uncertainty in the distance estimates, and assuming that the galaxies observed are approximately spherically distributed around the sky with a radially varying number density $n(r)$, we can write this 'theoretical' **U** (see 2.7) as a convolution of the peculiar velocity field **v**:

$$\mathbf{U}^{\rm th}(\mathbf{r}_0) = 3 \int d^3r\, W(r)\, (\hat{\mathbf{r}}.\mathbf{v}(\mathbf{r}_0 + \mathbf{r}))\, \hat{\mathbf{r}}, \tag{3.3}$$

where the window $W(r)$ is simply a normalised version of the radial weight distribution $n(r)w(r)$:

$$W(r) = \frac{n(r)\, w(r)}{\int d^3r\, n(r)\, w(r)}. \tag{3.4}$$

Applying the convolution theorem we obtain

$$\langle \mathbf{U}^{\rm th} \cdot \mathbf{U}^{\rm th} \rangle = \frac{H_0^2}{(2\pi)^3} \int d^3k\, W^2(k)\, \mathrm{P}(k), \tag{3.5}$$

where

$$W(k) = \frac{12\pi}{k} \int dr\, r^2\, W(r)\, j_2(kr), \tag{3.6}$$

and we have defined $j_2(x) \equiv 1/2 \int_{-1}^{1} d\mu\, \mu^2\, e^{ix\mu}$ (note however that this differs from the usual definition of the spherical Bessel function).

With 3-dimensional velocity data, one would compute

$$\mathbf{U}(\mathbf{r}_0) = \int d^3r\, W(r)\, \mathbf{v}(\mathbf{r}_0 + \mathbf{r}),$$

and the corresponding k-space window would be the standard Fourier integral rather than the j_2 integral above. The main difference is that the use of line of sight velocities increases the acceptance to high-k modes, and thereby increases the variance (Kaiser 1988).

As before, we must include the 'shot noise' velocity that would arise even in a universe with zero peculiar velocity. If the rms error in the line of sight peculiar velocity u for a galaxy at estimated distance r is $\sigma(r)$, then the shot-noise variance in the dipole is

$$\langle \mathbf{U}^{sn} \cdot \mathbf{U}^{sn} \rangle = 9 \int d^3r \, W^2(r) \, \sigma^2(r). \tag{3.7}$$

If the error is dominated by scatter in the 'Tully-Fisher' relation, or whatever analogue thereof is used, then the fractional error is constant with $\sigma(r) = fr$, where $f \simeq 23\%$ for the best methods, and if an inverse variance weighting scheme is used, this becomes

$$\langle \mathbf{U}^{sn} \cdot \mathbf{U}^{sn} \rangle = \frac{9f^2}{\Sigma r^{-2}} \, . \tag{3.8}$$

Just as for the dipole moments in Sec. 2, if we work in 1st order perturbation theory then the shot noise variance 3.7 and the theoretical variance 3.5 simply add to give the final total variance.

We can construct the probability density function for a set of N bulk flow solutions (defined by N different weighting functions $w(r)$) just as in Sec. 2. First one must construct the window $W(k)$ for each shell. One then proceeds to calculate the theoretical covariance matrix elements:

$$\langle U_{i\alpha} U_{j\beta} \rangle^{th} = \frac{1}{3} \frac{H_0^2 \delta_{\alpha\beta}}{(2\pi)^3} \int d^3k \, P(k) \, W_i(k) \, W_j^*(k). \tag{3.9}$$

Add to these the shot noise matrix elements

$$\langle U_{i\alpha} U_{j\beta} \rangle^{sn} = 3f^2 \delta_{ij} \delta_{\alpha\beta} / \sum r^{-2}, \tag{3.10}$$

which we have again taken to be diagonal since we will not consider overlapping samples. Finally, after inverting the resulting matrix $C_{ij} \equiv \langle U_{i\alpha} U_{j\alpha} \rangle^{th} + \langle U_{i\alpha} U_{j\alpha} \rangle^{sn}$, we have

$$P(U_{i\alpha}) d^{3N}U = \frac{d^{3N}U}{(2\pi)^{3N/2} |C|^{3/2}} \exp - \frac{1}{2} \sum_i \sum_j C_{ij}^{-1} \sum_\alpha U_{i\alpha} U_{j\alpha} \tag{3.11}$$

(see equation 2.19).

3.2 - *The Data vs Predictions*

Dressler *et al.* (1987) have given a streaming solution of the form 3.3, for the galaxies with estimated distances $r < 60$ Mpc/h, with a weighting function which, for large r, varies inversely as the variance; $w(r) \propto 1/r^2$. This weighting scheme minimises the shot noise fluctuations, which seems a desirable property, but does not necessarily give the maximum signal to noise or maximal cosmologically significant information. The radial distribution of number of galaxies and weight are shown in Figures 3.1 and 3.2. These figures are very informative; note how the weight is concentrated at small distances; this results in a much broader window function in k-space than if one imagined the weight to be uniformly distributed in a sphere of radius 60 Mpc/h. Indeed, it is apparent that with so little weight at large distances, the choice of outer boundary is essentially irrelevant. We stress that in order to model a real observation it is necessary to know how the weight in the solution is distributed in radius. If a theoretical window (Gaussian sphere or whatever) is adopted with a scale length tied to the outer boundary of the sample then the results are likely to be misleading.

In a more recent paper, Lynden-Bell *et al.* (1988) have given bulk flow solutions for a variety of shells around us. In Table 3.1 we give the rms predictions for streaming solutions with $1/r^2$ weighting and with upper and lower limits to estimated distances shown. The predictions given correspond to an 'unbiased' normalisation $\sigma_\varrho = 1$. We have also included in this table the prediction for the Local Group motion, which we model as the motion of a point.

TABLE 3.1
FIVE *BULK FLOW* SOLUTIONS AND THEIR EXPECTATION VALUES

Sample	U_x	U_y	U_z	$\lvert U\rvert$	U_{th}	U_{sn}	U_{tot}	
1	− 20	− 532	277	600	990	— —	990	local group
2	362	− 486	112	616	431	90	440	E's 0-3200
3	194	− 132	− 3	235	268	251	376	E's 3200-8000
4	61	− 349	− 13	355	627	31	628	AHMST 0-3000
5	139	− 398	15	420	590	29	591	E's + AHMST < 3200

If we consider a biased model with $\sigma_\varrho = 0.7$ for instance, then the Local Group motion lies right on the rms prediction; the distant sample is still a little lower than the prediction; but the solution for the galaxies in the range 0-32 Mpc/h is now about twice the rms prediction, which is starting to look embarrassing. Before condemning this model on the basis of this single observation one should pause and consider the following points. First, one should

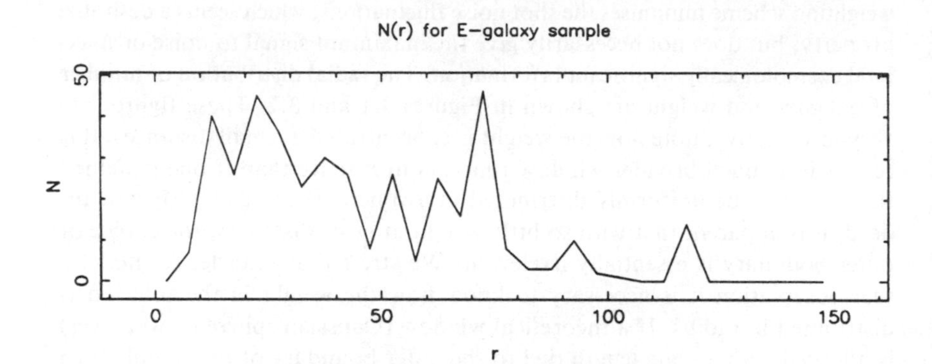

FIG. 3.1. The radial distribution of the sample of elliptical galaxies of Lynden-Bell *et al.* (1988). Each galaxy is equally weighted.

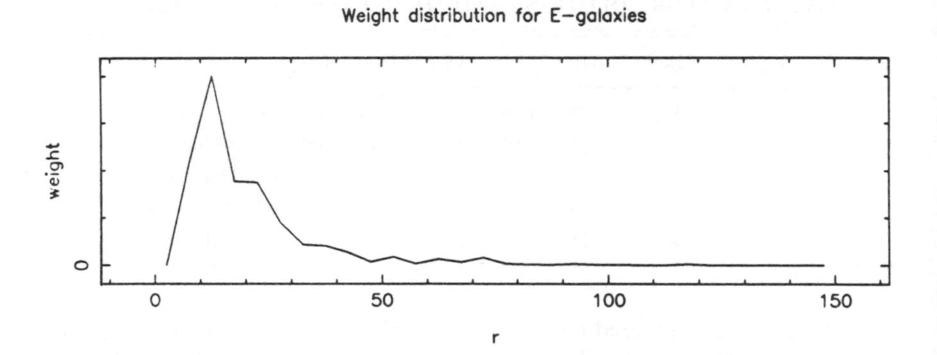

FIG. 3.2. The radial distribution of ellipticals, where each galaxy is weighted by the inverse of its variance. Since the variance increases as the square of the distance, the distribution is peaked at a much smaller distance than the unweighted distribution in Fig. 3.1.

note that the elliptical galaxies give the largest motion for this distance range (616 km/s), while the solution of Aaronson *et al.* (1982, hereafter AHMST) for their nearby spiral sample gives the much smaller value of 355 km/s (line 4 of Table 3.1), as do the results of de Vaucouleurs and Peters. Second, one should ask how the division between the shells was obtained; in our statistical model we implicity assume that such parameters are set *a priori,* but clearly that is a fiction and there is the possibility for bias here. Third, but closely related to the second point, one should ask oneself how many quasi-independent tests of the model could have been constructed, and whether to have found one 2-sigma vector is really unacceptable. We have also included as the 5th line in Table 3.1 a weighted combination of the ellipticals and the spirals which has $|\mathbf{U}| = 420$ km/s.

It is interesting to note that the 'far field' elliptical sample has a motion which is not detected above the noise level. The initial response to these observations was that they ruled out CDM because they demonstrated coherent flows on very large scales (e.g. Vittorio *et al.* 1986; Bond 1987). With the data separated into near and far field, we can see clearly that even the unbiased CDM model predicts an rms velocity which is now, if anything, *larger* than that observed in the far field. It is only fair to add that there have been other claims of much larger motions for distant spirals, and quite possibly these may be problematic for theories like CDM, in which the rms velocity falls inversely as the sample depth at large distances. If these observations hold up then this would be very exciting.

As with the angular dipoles, an interesting quantitative way to see whether these data, taken together, are consistent with the models is to calculate χ^2 (i.e. twice the argument of the exponential in 3.11). We took 3 dipoles as the data. The first is the Local Group motion; the second is the weighted combination of the near field ellipticals and the AHMST spirals; the third is the far field ellipticals. We consider these to be the best data available for the task. With our preferred normalisation $\sigma_\varrho = 0.6$, χ^2 came out to be 8.7, whereas the expectation is 9. Again, the fact that this is so close to the expectation can only be a fluke, but this does not alter the conclusion that these data are compatible with CDM.

In Figure 3.3 we show the likelihood function $L(\sigma_\varrho)$ using again the 3 dipoles. The most likely value $\sigma_\varrho \simeq 0.6$ agrees very well with our canonical biased CDM model, and even a smaller amplitude of $\sigma_\varrho \simeq 0.5$ would not be ruled out. The form of these functions is very similar to those obtained from the angular dipoles. It is important to note that the likelihood falls quite slowly for large values of σ_ϱ - reflecting the fact that it is easier to reconcile a low observation with a high rms prediction than *vice versa* - so we would not exclude models with somewhat more power at large scales than CDM. As with the dipole results however, we are struck by how well the predictions of the CDM model agree with the observations.

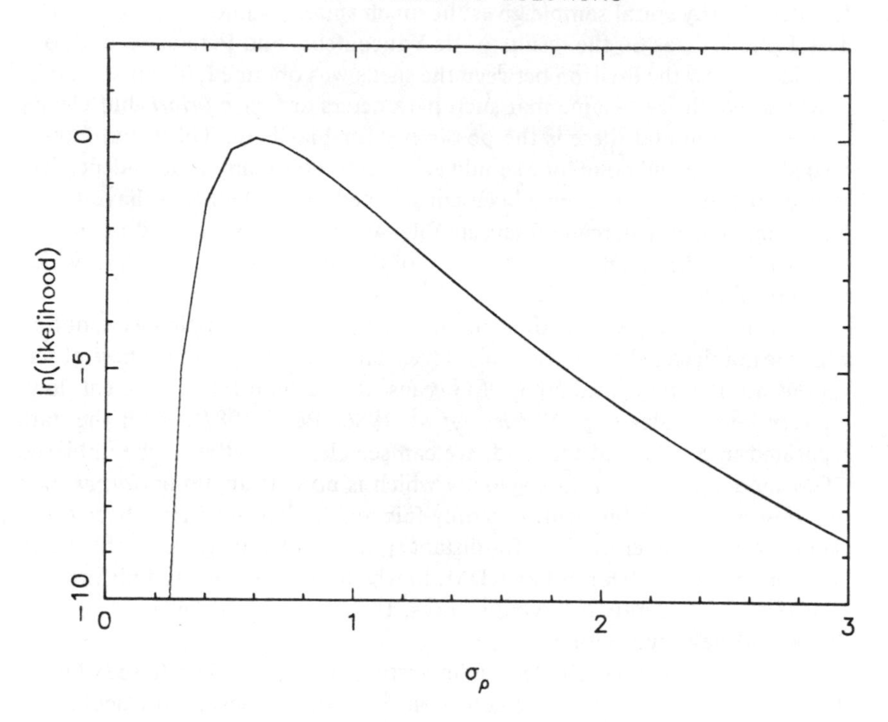

FIG. 3.3. The likelihood function versus the normalization parameter σ_ϱ, using 3 velocity vectors: the motion of the Local Group, a motion of a nearby (< 32 Mpc/h) shell of ellipticals and spirals, and a distant (32-80 Mpc/h) shell of ellipticals. The most likely value is $\sigma_\varrho = 0.6$. Note the asymmetry of the likelihood function.

4. Angular Dipoles, Bulk Flows, and the Bias Parameter

In the previous two sections we looked at the angular dipoles and bulk flow solutions and asked whether they were compatible with the CDM predictions. Here we wish to use these data together to test for the correlations expected between these vectors under the hypothesis that structure is growing by gravitational instability, and to determine the bias parameter b which appears as a constant of proportionality.

The basic relation between the peculiar velocity and the peculiar acceleration is (Peebles 1980):

$$\mathbf{v} = H_0 f(\Omega_0) \mathbf{g} \tag{4.1}$$

where $f(\Omega_0) \simeq \Omega_0^{0.6}$ and

$$\mathbf{g} = \frac{1}{4\pi} \int d^3r \, \Delta(r) \, \frac{\mathbf{r}}{r^3} \, . \tag{4.2}$$

The test, which we understand was originally due to Gott, has most commonly been applied (e.g. Yahil *et al.* 1980, Davis and Huchra 1982, Yahil *et al.* 1986, Lahav 1987) using the flux weighted angular dipole and the velocity of the Local Group. The idea here is that since both flux and acceleration vary as $1/r^2$, the flux weighted dipole should look very much like equaton 4.2, or ly with Δ_{galaxies} in place of the mass fluctuation Δ, and give an estimate of ɔ times) the acceleration g.

The limitations of this test arise from the flux limits that must realistically be imposed and from the shot noise resulting from the finite number of galaxies. With no flux limits, the window function for a flux weighted dipole would vary as $1/k$ for all wavelength. We see from Figure 2.1b that this ideal is not realised in practice for the IRAS sample, and that $kW(k)$ is strongly dependent on wavelength. The lower flux limit is unavoidable. The upper limit could in principle be increased, but the price paid is that the shot noise becomes unacceptably large. The result then is that a realistic dipole will not sample the acceleration from all shells equally, and consequently any constraint on b will depend on where one assumes the acceleration originates. Here we will obtain a best estimate of b, and of the statistical uncertainty, under the hypothesis that the fluctuations are Gaussian with the CDM spectrum.

We first consider the traditional application of the test using $\mathbf{D}_{\text{F}}$ and $\mathbf{U}_{\text{LG}}$. We find the uncertainty to be quite large. The reasons for this are twofold. In the CDM-world there are big fluctuations in the distance from which the acceleration arises. Because the shape of the window function differs from the idealised flux weighted $1/k$ form this converts to a big uncertainty in b. Second, because one is attempting to measure the acceleration acting on a point, the graininess of the nearby galaxies inevitably gives a big shot noise term. Additionally, there is the underlying worry as to how well the Local Group motion is described by linear theory, though that is much harder to quantify.

In Sec. 4.2 we look at a modified test that gives reduced uncertainty. The basic idea here is that by giving less weight to the high flux galaxies the shot noise will be reduced. Now, this will also attenuate the acceptance of the window to high-k modes still further, and this will degrade the correlation of this dipole with the Local Group motion, so little, if any, gain in performance will be achieved. However, the correlation with other bulk-flow solutions (which also have window functions with high-k modes strongly attenuated) will increase, so if we use these in place of the Local Group motion we should get a more precise test. Crudely, the aim is to find a low noise dipole which measures the acceleration acting on the Local Supercluster, or some other large region for which the peculiar velocity has been measured. The specific problem we solve is: given a particular bulk-flow solution, what is the flux-weighting scheme that results in the optimum dipole for estimating b. This question of what is the 'best' dipole to measure the acceleration was discussed by Lynden-Bell and Lahav (1988). Clearly any such optimisation is model dependent; the solution given here is contingent on the assumption of Gaussian fluctuations. By using many bulk-flow solutions, a whole family of tests can be constructed which are independent to the extent that the windows sample different regions of k-space. Finally, in Sec. 4.3 we apply all tests simultaneously by means of multivariate likelihood analysis to obtain our best estimate of b.

4.1 - Flux Weighted Dipole and the Local Group Velocity

The joint probability distribution $P(\mathbf{D}_F, \mathbf{U}_{LG})$ is a 6-variate zero-mean Gaussian with covariance matrix elements

$$C_{11} = \frac{1}{3} \langle \mathbf{U}_{LG} \cdot \mathbf{U}_{LG} \rangle = \frac{H_0^2}{3(2\pi)^3} \int d^3k \, P(k) \, W_U(k) \, W_U^*(k)$$

$$C_{12} = \frac{1}{3} \langle \mathbf{U}_{LG} \cdot \mathbf{D}_F \rangle = \frac{bH_0}{3(2\pi)^3} \int d^3k \, P(k) \, W_U(k) \, W_D^*(k) \quad (4.3)$$

$$C_{22} = \frac{1}{3} \langle \mathbf{D}_F \cdot \mathbf{D}_F \rangle = \frac{b^2}{3(2\pi)^3} \int d^3k \, P(k) \, W_D(k) \, W_D^*(k) + \text{shot noise}$$

where $W_D(k)$ is given by equation (2.12), with the shot noise variance given by equation (2.15), and, treating the local Group as a point, we have $W_U = 1/k$. We ignore shot noise for the Local Group velocity.

The likelihood function $L(b) \equiv P(\mathbf{D}_F, \mathbf{U}_{LG} | b, ...)$ is shown in Figure 4.1 for the IRAS flux dipole given in Table 2.1 and for a Local Group velocity of 600 km/sec towards galactic coordinates ($l = 268°$, $b = 27°$). We have set $\sigma_\varrho = 0.7$. The curve peaks at $b_{max} = 1.5$. For $\sigma_\varrho = 1.0$ we find $b_{max} = 1.3$. These estimates are very similar to those obtained in Sec. 2.5 by matching the 5-dipole IRAS amplitudes with the predictions in the Great Attractor model.

An illuminating way of quantifying the uncertainty in b in the method is to use the conditional probability $P(\mathbf{U}_{LG} | \mathbf{D}_F, ...)$ which is Gaussian with mean

$$\bar{\mathbf{U}}_{LG} = a\mathbf{D}_F, \quad (4.4)$$

where

$$a = \langle \mathbf{U}_{LG} \cdot \mathbf{D}_F \rangle \, / \, \langle \mathbf{D}_F^2 \rangle \quad (4.5)$$

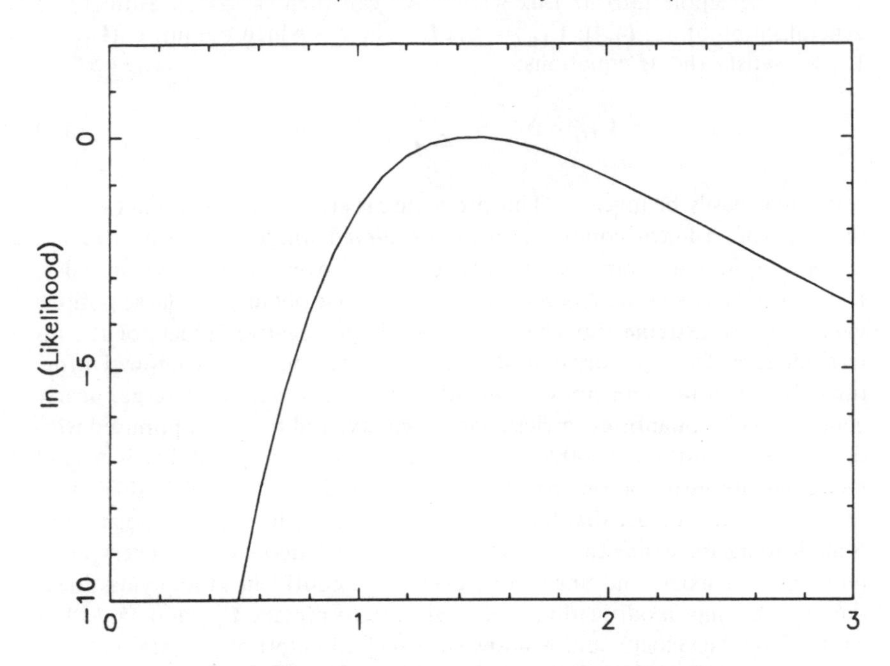

FIG. 4.1. The likelihood function versus the bias parameter b, using the velocity vector of the Local Group and an IRAS flux dipole. The normalization is set to be $\sigma_Q = 0.7$. The curve peaks at $b = 1.5$, but it is rather broad.

and with variance

$$\langle (\mathbf{U}_{\mathrm{LG}} - \bar{\mathbf{U}}_{\mathrm{LG}})^2 \rangle = \langle 3\mathbf{U}_{\mathrm{LG}}^2 \rangle \left[1 - \frac{\langle \mathbf{U}_{\mathrm{LG}} \cdot \mathbf{D}_{\mathrm{F}} \rangle^2}{\langle \mathbf{U}_{\mathrm{LG}}^2 \rangle \langle \mathbf{D}_{\mathrm{F}}^2 \rangle} \right]. \tag{4.6}$$

We can say that a measurement of a flux-dipole $\mathbf{D}_{\mathrm{F}}$ *predicts* a velocity $a\mathbf{D}_{\mathrm{F}}$ with uncertainty given by (4.6). A good estimator would be one that has $E_U^2 \equiv \langle (\mathbf{U}_{\mathrm{LG}} - \bar{\mathbf{U}}_{\mathrm{LG}})^2 \rangle \,/\, \langle \mathbf{U}_{\mathrm{LG}}^2 \rangle \ll 1$. Unfortunately, with this choice of dipoles the coupling is quite weak because of the poor shape of W_D and the shot noise. For the IRAS dipole we find $E_U = 35\%$.

We will now consider some modifications which result in a more tightly coupled combination of dipoles.

4.2 - *Alternative Schemes*

One possibility is to vary the flux weighting scheme. As in Section 2 we can bin the dipole into M flux shells. We can then try as an estimator a generalisation of eq. (4.4): $\bar{\mathbf{U}}_{LG} = \Sigma_i a_i \mathbf{D}_i$. The a_i's which minimise $\langle(\mathbf{U}_{LG} - \bar{\mathbf{U}}_{LG})^2\rangle$ satisfy the M equations:

$$\langle \mathbf{U}_{LG} \cdot \mathbf{D}_j \rangle = \sum_i a_i \langle \mathbf{D}_i \cdot \mathbf{D}_j \rangle \tag{4.7}$$

which may easily be inverted. This procedure is effectively an attempt to synthesise, from a linear combination of the curved windows shown in Figure 2.2, a window which varies approximately as $1/k$ over a wide range of scales. If one artificially sets the shot noise to zero, then one obtains very large positive values for the extreme flux shells and very large negative values for the intermediate shells. This does indeed synthesise a well behaved window. Unfortunately, when the shot noise is included, this substraction of large, nearly equal but noisy quantities carries a large penalty, and the new optimised window performs little better than that using $\mathbf{D}_F$. Another possibility is to add more velocity information. As we have discussed, the potential advantages of this modification are that the shot noise is much reduced, since no attempt is made to measure the near field acceleration which derives from a very small number of galaxies, and one can be even more confident in applying linear theory. The only modifications necessary are to replace $\mathbf{U}_{LG}$ and $W_U(k)$ in eq. (4.3) by the velocity and window of a shell of elliptical or spiral galaxies given in Table 3.1. Rather than present the results for a number of quasi-independent bulk-flows, we prefer to proceed directly to the more general likelihood function which uses all the data together rather than special linear combinations.

4.3 - *Multi-dipole Likelihood Function*

We use 5 IRAS flux dipoles (Table 2.2) and 3-velocity shells (from Table 3.1) to construct the likelihood function.

$$L(b) = P(\mathbf{D}_1, \ldots, \mathbf{D}_5, \mathbf{U}_1, \ldots, \mathbf{U}_3 | b \ldots). \tag{4.8}$$

As $\mathbf{U}_1$ we use the Local Group velocity (though because this has a strong correlation matrix element to the noisy high flux dipole it enters the likelood function with little weight), as $\mathbf{U}_2$ we use the combined velocity of the nearby shell (ellipticals and spirals) and as $\mathbf{U}_3$ we use the velocity of the distant ellipticals shell (3200 — 8000 km/sec). Figure 4.2 shows the likelihood function (4.8) for $\sigma_\varrho = 0.7$. The curve peaks at $b_{max} = 1.6$, and the location of the peaks is insensitive to σ_ϱ. See how narrow the curve is, in particular when compared

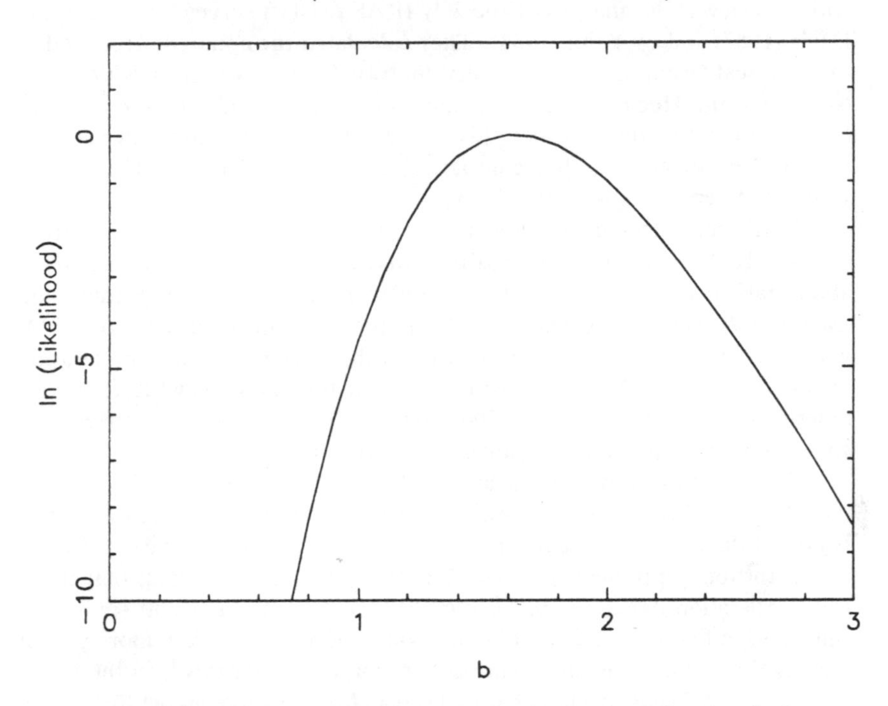

FIG. 4.2. The likelihood function versus the bias parameter b, using the 3 velocity vectors (of Fig. 3.3) and 5 IRAS flux dipoles with $\sigma_\varrho = 0.7$. The curve peaks at $b = 1.6$. Note that the curve is much narrower than the curve of Fig. 4.1.

to Figs. 4.2 and 5.3. This demonstrates how likelihood analysis, which uses all the data simultaneously, has considerably reduced the statistical uncertainty.

5. 'Gravity' Vector from Redshift Surveys

In the previous sections we have used angular dipoles of the galaxy counts to estimate the acceleration acting on the Local Group and on larger regions, and we have used the run of dipole with flux to get a handle on the distance from which this acceleration derives. A more or less consistent picture has emerged; the run of **D** with flux agrees approximately with that predicted in the Great Attractor model, and comparing the acceleration with the Local Group velocity (or the more sophisticated test of Sec. 4.3) gives consistent estimates of $b \simeq 1.5$. In addition, the amplitudes of both angular dipoles and peculiar velocities are quite consistent with CDM.

An important piece of evidence which, at first sight, seems to conflict with this view is the analysis of the 2 Jy IRAS redshift survey by Strauss and Davis (1987) and by Yahil (1988). They calculated the apparent acceleration of the Local Group using the density contrast field as seen in *redshift* space (see Davis and Huchra 1982 for a description of the method). They found: (i) that integrating out in shells, this vector converged to a constant value by $r = 40$ mpc/h; and (ii) that comparing this acceleration with $\mathbf{U}_{LG}$ gave a value of b very close to unity.

With regard to the 'convergence radius', the lack of acceleration from $r > 40$ Mpc/h is clearly incompatible with the Great Attractor picture where about half of the total acceleration arises from these distances. Strauss and Davis (1987) claimed that what they saw conformed to the predictions of CDM-world. A simple calculation reveals that in this model, the rms acceleration arising from $r > 40$ Mpc/h is about 1/3 of the total rms, or, when expressed in terms of the induced velocity, 300 σ_ϱ km/s. Thus the apparent convergence seen would actually be quite embarassing for CDM.

We will show below that the apparent conflict between these observations (and the theory) is quite simply resolved once allowance is made for the perturbation of the clustering pattern in redshift space. The mathematical details of this distortion are presented in Sec. 5.1. We then show (Sec. 5.2) that if the true acceleration did converge for $r < 40$ Mpc/h then one would see in redshift space a fictitious and substantial negative acceleration from more distant regions. In so far as the apparent acceleration converges this hypothesis can be discounted. We then show that in a Great Attractor-like model the net apparent acceleration for 80 Mpc/h $> r >$ 40 Mpc/h is small - much like that observed - even though about 1/2 of the true acceleration arises from this region.

Turning to Gaussian models like CDM, we shall show that if we Fourier analyse the density field we find that each Fourier mode gives a contribution to the apparent acceleration vector which is proportional to the true acceleration, but with an amplitude which is wavelength dependent. This can be analysed in terms of a window function for this observation, which is quite analogous to those used in preceding sections. In Sec. 5.2.3 we construct this window function, and use this to calculate the uncertainty in the estimate of b under the assumption of CDM. We find that the uncertainty is larger than that obtained in Sec. 4 using only angular coordinates and fluxes. This is in part because of the distortion of the clustering pattern and in part because of the large shot noise. This latter source of uncertainty is made particularly large by the choice of a 'semi volume limited' weighting scheme. We discuss ways in which the test can be improved.

5.1 - Distortion of Clustering in Redshift Space

The mapping from real space coordinates $\mathbf{r}$ to redshift space s is just

$$\mathbf{s} = \mathbf{r}\left(1 + \frac{\hat{\mathbf{r}} \cdot (\mathbf{v}(\mathbf{r}) - \mathbf{v}(\mathbf{0}))}{r}\right) \tag{5.1}$$

where $\mathbf{v}(\mathbf{0})$ is the observer's velocity relative to the cosmic frame and it is assumed that $H_0 = 1$. From this one obtains (Kaiser 1987) for the linearised apparent density perturbation in redshift space:

$$\Delta_s(\mathbf{r}) = \Delta_r(\mathbf{r}) - \left(2 + \frac{d\ln\phi}{d\ln r}\right)\frac{\hat{\mathbf{r}} \cdot (\mathbf{v}(\mathbf{r}) - \mathbf{v}(\mathbf{0}))}{r} - \hat{\mathbf{r}} \cdot \frac{d\mathbf{v}(\mathbf{r})}{dr}. \tag{5.2}$$

The origin of these terms can be understood as follows. If $\mathbf{v}$ were identically zero then redshifts would exactly measure distances and we would have just the first term $\Delta_s = \Delta_r$. The other terms arise because peculiar velocities distort the pattern of clustering in redshift space and introduce spurious apparent density enhancements. Consider the second term in equation 5.2. From 5.1 we see that $2\hat{\mathbf{r}} \cdot (\mathbf{v}(\mathbf{r}) - \mathbf{v}(\mathbf{0}))/r$ is just twice the fractional radial displacement of a galaxy in redshift space, so this factor gives the amount by which a volume element is transversely compressed or rarefied assuming $\mathbf{v}(\mathbf{r})$ to be constant over the volume element. If, for instance, redshift underestimates the distance to the galaxies in some region, this causes an apparent density enhancement, relative to the density we would measure for the same galaxies with real distance estimates. The logarithmic derivative of the selection function comes in because, in computing Δ_s we compare the number of galaxies with that predicted from the selection function, but evaluated at a perturbed position. (It is interesting to note that if nature were so generous as to provide a population of galaxies with $\Phi(L) \propto L^{-2}$, so $\phi(r) \propto r^{-2}$, this correction terms vanishes. Were this the case one would also be spared the need to collect redshifts, since the sum in equation 5.3 below is then independent of the redshifts. Unfortunately, for real galaxies this does not seem to be the case.)

Finally, the third term is necessary to allow for gradients of the peculiar velocity field, which act to compress or rarefy the volume element in the radial direction.

It is not difficult to modify the analysis to treat a semi-volume limited survey. Such a survey is obtained by taking a redshift-dependent flux limit; $S(s) = S_{\rm lim}$ if $s > s_{\rm i}$ and $S(s) = (s/s_{\rm i})^{-2}S_{\rm lim}$ if $s < s_{\rm i}$. The result is simply to replace in equation 5.1 the factor $2 + d\ln\phi/d\ln r$ by its value at the boundary of the magnitude limited region.

An interesting consequence of this is that by choosing the boundary appropriately, the second term in equation 5.2 can be made to vanish throughout the volume-limited region. If one had a much deeper redshift survey this might appear to be a very nice way to get around the problems of redshift space distortions, since provided the volume limited region encompasses the source of our acceleration the 3rd term in 5.2 can easily be corrected for (it simply causes the 'acceleration' measured in redshift space to exceed the true acceleration by a constant factor). Unfortunately, for a 2 Jy survey this would limit the volume limited regime to $r < 20$ Mpc/h which is uninterestingly small.

5.2 - Apparent Acceleration Vector in Redshift Space

Define the 'acceleration' vector:

$$\mathbf{v}_s = \frac{H_0}{4\pi n_1} \sum_i \frac{1}{\phi(s_i)} \frac{\hat{\mathbf{s}}_i}{s_i^2} \tag{5.3}$$

just as in Davis and Huchra (1982), and in Strauss and Davis (1987), and where it is implied that we take $\phi = \phi(s_1)$ in the volume-limited region. Taking the limits of the sum to be spheres in redshift space, and setting $H_0 = 1$ for convenience gives

$$\mathbf{v}_s = \frac{1}{4\pi} \int_{s_{\min}}^{s_{\max}} d^3s \, \frac{\hat{\mathbf{s}}}{s^2} \, \Delta_s(\mathbf{s}) + \text{shot noise}$$

$$= \frac{1}{4\pi} \int_{s_{\min}}^{s_{\max}} d^3r \, \frac{\hat{\mathbf{r}}}{r^2} \left(\Delta_r(\mathbf{r}) - \left(2 + \frac{d\ln\phi}{d\ln r}\right) \frac{\hat{\mathbf{r}} \cdot (\mathbf{v}(\mathbf{r}) - \mathbf{v}(\mathbf{0}))}{r} - \right.$$

$$\left. - \hat{\mathbf{r}} \cdot \frac{d\mathbf{v}(\mathbf{r})}{dr} \right) + \text{shot noise} \tag{5.4}$$

We will presently do the usual decomposition into plane waves, construct the window function for $\mathbf{v}_s$ and proceed as before. There are, however, important conclusions that can be drawn from equation 5.4 directly.

5.2.1 - Apparent acceleration seen by a rocket borne observer

Consider an observer who has a locally generated peculiar motion $\mathbf{v}(\mathbf{0})$ (perhaps generated by a rocket, or by the gravitational influence of nearby matter), but who resides in a universe which is uniform and at rest with respect to the microwave background on larger scales. Outside of the local perturbation, Δ_r and $\mathbf{v}(\mathbf{r})$ vanish, so this observer calculates

$$\mathbf{v}_s = \frac{1}{3} \mathbf{v}(\mathbf{0}) \int \frac{dr}{r} \left(2 + \frac{d\ln\phi}{d\ln r}\right)$$

$$= \frac{1}{3} \mathbf{v}(\mathbf{0}) \left[\left(2 + \frac{d\ln\phi}{d\ln r}\right)_{s_1} \ln \frac{s_1}{s_{\min}} + \ln \frac{\phi(s_{\max}) s_{\max}{}^2}{\phi(s_1) s_1^2} \right] . \tag{5.5}$$

Thus, the motion of the observer induces a spurious $\mathbf{v}_s$ vector which is of order the true velocity and contains logarithmic divergences. Note that this result is independent of the true value of Ω_0 - it simply results from calculating the 1st order change in equation 5.3 when one changes the inertial frame to which the redshifts are referred. It is hoped that this simple example makes clear that the simple estimator (equation 5.3) cannot in general give a reliable estimate of the acceleration vector. It should also make clear the futility of trying to correct equation 5.3 for Virgo infall for instance, when there are much larger sources of error coming from our uncorrected motion with respect to the galaxies at greater distances.

It is interesting to compare the predicted run of $\mathbf{v}_s$ for this simple 'rocket' model with that observed. Strauss and Davis (1987) found that, integrating out in redshift shells, the acceleration converges by about 4000 km/s, there being negligible contribution from the magnitude limited region from 4000 to 8000 km/s. While no details of the luminosity function adopted by Strauss and Davis (1987) are given, they do tell us that the selection function falls by a factor 10 going from 4000 to 8000 km/s, so using this in equation 5.5 we would predict that in this shell they should have found a *negative* contribution to the dipole of about 1/3 of the observer's velocity, or about 150 km/s since with their correction for Virgo infall the redshifts for distant galaxies used are those that would be seen by an observer moving at 450 km/s with respect to the microwave frame. We do not wish to go into the question of exactly how reliably the data exclude or constrain any such fall in $\mathbf{v}_s$, but it is interesting to note that if one were convinced that $\mathbf{v}_s$ had really converged then the one hypothesis that can be excluded immediately is that we live in a world where the true acceleration converges! One of the many hypotheses that *would* be allowed is that we live in a world with velocities generated by perturbations on a scale much larger than the survey region.

5.2.2 - Apparent acceleration in the Great Attractor-like model

With a specific model for the velocity field one can predict the radial run of $\mathbf{v}_s$ using equation 5.4. The heavy line in Figure 5.1 shows the apparent acceleration build up with increasing redshift for a single attractor model centred at $r = 40$ Mpc/h with an r^{-2} density contrast run producing an infall of 600 km/s at our position (see Lynden-Bell *et al.* 1987). The calculation assumes $b = 1$. One finds that the overall $\mathbf{v}_s$ agrees quite well with the true $\mathbf{v}$ - though this is not generally the case. The light continuous line, the dotted-dashed line and the dotted line correspond to the 3 terms in eq. 5.4 in the order they appear. The growth of $\mathbf{v}_s$ with r is quite different; the appearance is that the result has converged by about 40 Mpc/h, and if anything, there appears to be a net repulsion from the back of the perturbation. The behaviour of $\mathbf{v}_s$ in this model is qualitatively similar to that observed, particularly in the

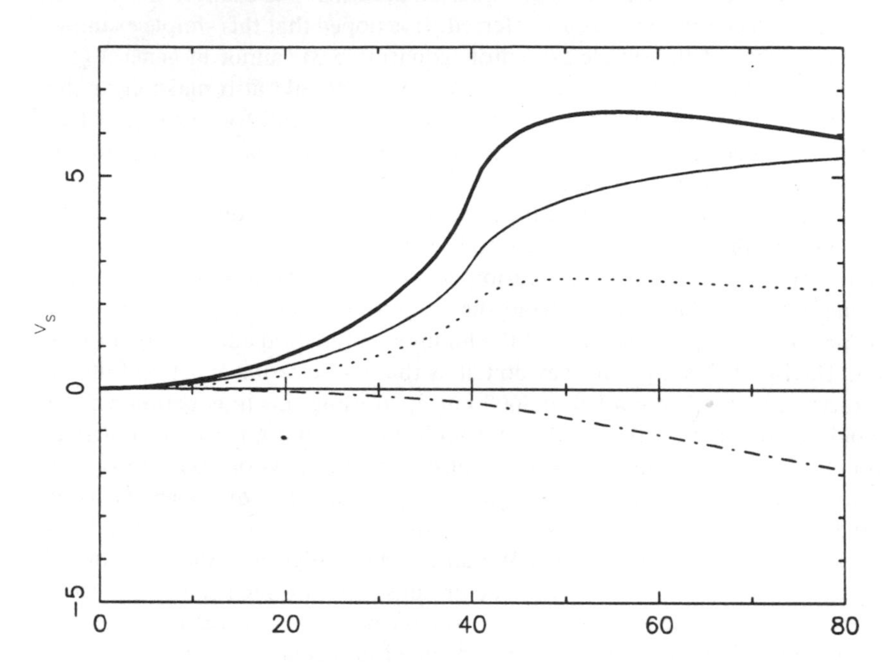

FIG. 5.1. The heavy solid line shows how the apparent acceleration of the Local Group (in 100 km/sec units) builds up with distance (in Mpc/h) for a single attractor model placed at 40 Mpc/h (see text). The light continuous line, the dotted-dashed line and the dotted line correspond to the three terms of eq. 5.4, in the order they appear. In particular, the light continuous line shows the true acceleration.

far field, so there is clearly no conflict between the redshift space picture and the Great Attractor-like model once allowance is made for peculiar velocities.

5.2.3 - Apparent acceleration vector in Gaussian models

While one cannot simply equate $\mathbf{v}_s$ and the true acceleration, it is still a potentially useful statistic. Just as for the observations discussed in the previous sections, that defined by equation 5.3 corresponds to the convolution of the density field with a particular window function, and so we can proceed much as before to calculate the corresponding window function, and calculate the correlation between $\mathbf{v}_s$ and other vectors such as the Local Group motion. For any particular model, we can then try to compensate for any systematic bias introduced by the distortion, obtain an improved estimate of b, and quantify the uncertainty.

Doing the usual plane-wave decomposition: $\Delta(r) = \Delta_k e^{i\mathbf{k}.\mathbf{r}}$, $\mathbf{v}(\mathbf{r}) = \hat{\mathbf{k}} v_k e^{i\mathbf{k}.\mathbf{r}}$, with $\Delta_k = (bk/i)v_k$, and performing the integral in 5.3 results in

$$v_s(k) = [b\, W_r(k) + W_s(k)]\, v(k) + \text{shot noise} \tag{5.6}$$

where

$$W_r(k) = j_0(ks_{\text{min}}) - j_0(ks_{\text{max}}) \tag{5.7}$$

and

$$W_s(k) = j_2(ks_{\text{min}}) - j_2(ks_{\text{max}}) + \int_{s_{\text{min}}}^{s_{\text{max}}} \frac{dr}{r} \left(2 + \frac{d\ln\phi}{d\ln r}\right)\left(\frac{1}{3} - j_2(kr)\right) \tag{5.8}$$

where, as before

$$j_2 \equiv \frac{\sin(kr)}{kr} + 2\left[\frac{\cos(kr)}{(kr)^2} - \frac{\sin(kr)}{(kr)^3}\right].$$

We have evaluated these windows for the selection function derived from the Lawrence *et al.* (1986) luminosity function. Figure 5.2 shows the window functions W_r (light continuous), W_s (dashed line), and their sum (heavy line). The actual window differs considerably from the desired window, which in these plots would be the horizontal line $W(k) = 1$. Were the actual window function simply offset vertically this would present no problem, since this would bias the result by a multiplicative factor which can easily be corrected for. The problem revealed by Figure 5.2 is that the actual window for the $\mathbf{v}_s$ estimate is wavelength dependent, so for a realistic power spectrum, this reduces the correlation between $\mathbf{v}_s$ *and* $\mathbf{v}$.

For a quantitative comparison of theory and observation we must of course include the noise in the measurement. We find for the shot noise term

$$\langle \mathbf{v}_s \cdot \mathbf{v}_s \rangle^{\text{sn}} = \frac{1}{N} \int_{s_{\text{min}}}^{s_{\text{max}}} dr\, r^2\, \phi(r) \times$$

$$\left[\int_{s_{\text{min}}}^{s_1} \frac{dr}{r^2 \phi(r_1)} + \int_{s_1}^{s_{\text{max}}} \frac{dr}{r^2 \phi(r)}\right], \tag{5.9}$$

where s_1 is the boundary of the volume limited region and N is the total number of galaxies in the sample (*not* the number in the semi-volume limited

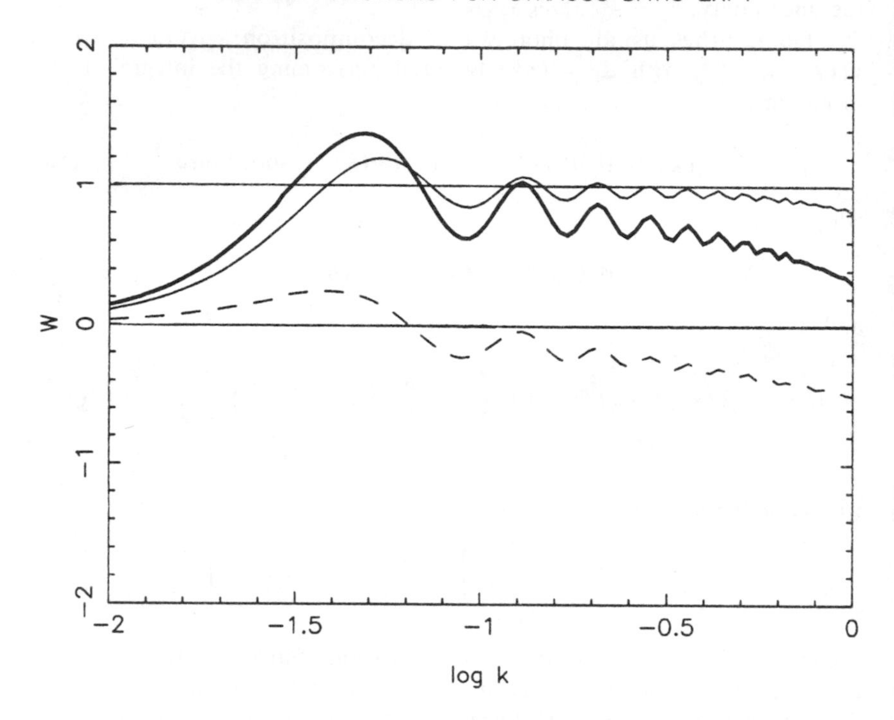

FIG. 5.2. The window functions in real space (light continuous line) and in redshift space (dashed line) and their sum (heavy line). The desired window, which represents the true gravity, is the horizontal line.

sample). This diverges with the lower redshift cuf-off. Taking $s_{\min} = 500$ km/s, which seems reasonable, we get $\langle \mathbf{v} \cdot \mathbf{v} \rangle \simeq (340 \text{ km/s})^2$, so the observed signal is only about a factor 2 larger than the rms noise.

We are now in the position to use the $\mathbf{v}_s$ of Strauss and Davis (1987) and the Local Group motion $\mathbf{U}_{LG}$ to esimate b, and, perhaps more importantly, give a realistic estimate of the uncertainty. A good indication of the performance of this test can be obtained by looking at the probability distribution for $\mathbf{U}_{LG}$ conditional on an observation $\mathbf{v}_s$ as used in Sec. 4. With $b = 1$ one finds $\bar{\mathbf{U}}_{LG} \simeq 0.8\mathbf{v}_s$, but the scatter about this is quite large:

$$\langle (\mathbf{U}_{LG} - \bar{\mathbf{U}}_{LG})^2 \rangle \simeq 0.37 \, \langle \mathbf{U}_{LG}^2 \rangle, \tag{5.10}$$

which, with our favoured normalisation parameter $\sigma_\varrho = 0.7$, gives an rms fluctuation about $\bar{\mathbf{U}}_{LG}$ of about 450 km/s. Since the observed value is about

600 km/s, the total statistical noise in this test is considerable, and arises roughly equally from shot noise and distortion effects. Figure 5.3 shows the likelihood function for b using these data. Not surprisingly it is very broad, though it is interesting that it peaks at a very similar value to that obtained from the angular dipoles in the previous section.

While the performance of this estimator of b (or of Ω_0) is rather poor when compared with the estimate of Sec. 4, there are however several ways to improve the test. The shot noise is large both because of the use of a semi-volume limiting flux cut-off and because the measurement samples the acceleration right up to the observer. Provided one has a good luminosity function estimate, there is no reason for making a semi-volume limited catalogue - unless of course one particlarly wishes to exclude intrinsically faint galaxies, and by not doing this one decreases the shot noise considerably (though tests we have made using alternative forms for the luminosity function suggest that the results

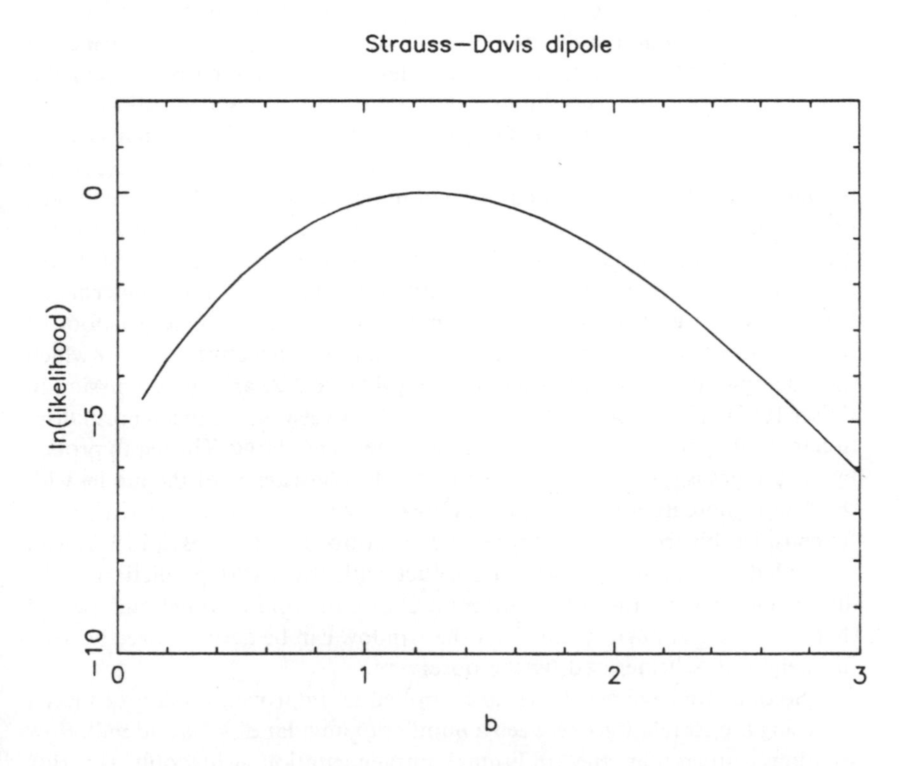

FIG. 5.3. The likelihood function versus the bias parameter b, using the IRAS redshift dipole. The curve peaks at $b = 1.3$. Compare the breadth of this curve to the curves in Figs. 4.1 and 4.2.

are then more sensitive to the choice of luminosity function adopted). With regard to sampling the local acceleration, just as discussed in Sec. 4, there are various advantages in calculating an acceleration which largely ignores the local regions, and which correlates well with the motion of the Local Supercluster, or some other extended region, rather than with the Local Group motion.

Finally, just as in Sec. 4, the formalism we have constructed allows one in principle to use all of the information in the dipoles as a function of distance and luminosity. Ultimately, we hope that by extension of the techniques illustrated here much useful cosmological information can be recovered from the distorted, but 3-dimensional, view of the world provided by redshift surveys.

6. Summary

We have concentrated fairly arbitrarily on *dipole* moments of the line of sight peculiar velocities, and of the counts of galaxies both in flux-angle space and in redshift space. Our choice was motivated in part by common use of these statistics in previous works; in part by the desire to probe density fluctuations on the largest scale, and also because of the special significance of the correlations between the various dipoles. We can summarise our conclusions drawn from this very limited description of the data as follows.

One major goal was to see if the angular dipoles and the peculiar velocity data are consistent with the CDM-world predictions. We were very impressed (though somewhat disappointed) to find in Sec. 2 and Sec. 3 that the model predictions fit very nicely with the observations; the χ^2 statistics were perfectly acceptable, and the likelihood analysis using these probes of large-scale structure gave essentially identical normalisation parameters as those determined from considerations of clustering on much smaller scales. One offshoot of the analysis of the peculiar velocity data is that the effective depth to which these samples probe is quite small (see e.g. Figure 3.2, and the discussion in Kaiser 1988). This means that the 'lever arm' between the 8 Mpc/h normalisation scale and the scale probed by the velocities is not large. The depth probed by the dipoles is greater, and consequently the consistency of the model with the data is more impressive. Our conclusions regarding the velocity data differ considerably from earlier studies (e.g. Vittorio *et al.* 1986, Bond 1987) who concluded that these data are in conflict with the CDM predictions. The discrepancy can be traced to a different choice of window function. We feel that there is no ambiguity, and that the window can be derived directly from the weighting scheme used by the observers.

Second, we have developed and applied an improved version of Gott's test using the correlations between a number of angular dipoles and bulk flow solutions rather than the traditional implementation using only the flux weighted dipole and Local Group. We obtain a value for the bias parameter for IRAS galaxies of $b \simeq 1.5$, corresponding to an Ω_0 of around 0.4. Our analysis differs from previous applications of this type of test in that we make a maximum likelihood estimate of b, contingent on the assumption of the CDM spectrum. The ML estimate of b should be fairly insensitive to the specific spectrum assumed, but the model does play an important role in that it enables us to determine the statistical uncertainty in b. We have applied a very similar analysis to the determination of b (or Ω) from the 2 Jy redshift survey. We find that the statistical uncertainty here is much larger, but improvements are possible.

Third, while our main results assume for the model of the density field the random Gaussian CDM Δ field, we have interspersed these calculations with calculations based on other simpler models using one or more spherical perturbations. These models give a more graphic picture of the density and velocity fields and provide an important link to other studies which have used this type of model extensively. We find a broad degree of consistency between the various observations regarding the distribution in distance of the major components. One way we illustrated this (Sec. 2.5) was to use a single component attractor model centred at 40 Mpc/h, and with amplitude adjusted to give the Local Group microwave background motion. This model reproduces quite nicely the run of angular dipoles with flux, and gave an acceptable chi-squared. Following Lynden-Bell's modeling of the velocity data, we then went on to show that a 3-component model comprising the Great Attractor, Virgo and a third component to make up the microwave velocity provided an even better fit to the dipoles. The distance to the third component was kept as a free parameter, and adjusting gave a best fitting distance of 18 Mpc/h, so this seems to accord with the idea of a 'Local Anomaly' in the velocity field (see Faber and Burstein 1988 and Lynden-Bell and Lahav 1988). Unfortunately, we found that the angular dipole data did not very tightly constrain the depth from which the dominant acceleration derives. We also showed (Sec. 5.2.2) that the simple 1-component model (in which roughly 300 km/s of our peculiar motion derives from inhomogeneity at distances between about 40 and 80 Mpc/h) also reproduces the apparent *lack* of acceleration seen at these distances in redshift space, once the distortion of the clustering in redshift space by peculiar velocities is taken into account. Indeed, the redshift results would seem to preclude the possibility that nearly all of our acceleration derives from smaller distance (Sec. 5.2.1).

These simple models provide at best a very broad brush description of reality; all we claim here is that these very different data sets are all at least *consistent* with models in which a substantial part of our peculiar motion (roughly 300 km/s or so) derives from $r > 40$ Mpc/h. While each line of evidence appears questionable when subjected to close scrutiny, it is encouraging to see a farily coherent picture emerging. On the theoretical side, we presented the simple result that in CDM-world the rms velocity driven by perturbations at $r > 40$ Mpc/h comes to $300\sigma_\varrho$ km/s. This provides a simple way to see that the scale and amplitude of large-scale structure suggested by the data are in reasonable accord with this theoretical model.

7. Discussion

In Sec. 7.1 we discuss the relation of the results obtained above to other probes of large-scale structure; in Sec. 7.2 we consider the question of normalisation; and in Sec. 7.3 we discuss the assumption that $\Omega = 1$, and some implications for galaxy formation. Finally, in Sec. 7.4 we consider the outlook for future extensions of the type of observations described above.

7.1 - Other Probes of Large-Scale Structure

There are various other probes of large-scale structure that provide important independent tests of models like CDM. On very large scales there are constraints on the level of the diffuse X-ray backround (Shafer 1983) and from the isotropy of the microwave background (Kaiser and Silk 1986 and references therein). Mostly, these have given null results, and have been compatible with the rather low values predicted in CDM-world: on large scales the rms fluctuation amplitude varies as λ^{-2}, falling to around 10^{-4} at the Hubble length scale. A notable exception to these null results is the microwave anisotropy detection claimed by Lasenby (1988) and his collaborators. This anisotropy is about an order of magnitude larger than that predicted in CDM-world with our choice of normalisation. If this is confirmed at a higher level of significance, then this would be a very strong indication of much more power at large scales than provided by CDM.

Another very important quantitative probe of large-scale structure is provided by the clustering analysis of the Abell cluster catalogue. There are two lines of evidence here; on one hand we have the angular correlation function estimates from the complete catalogue (now augmented by the southern clusters), and on the other we have the 3-dimensional clustering studies of the much shallower subsample for which redshifts are available. The advantages and disadvantages of these approaches are to a large extent complementary; the formal error bars on the redshift survey estimates of ξ are large and the signal falls below noise at a lag of at most $\simeq$ 30 Mpc/h. Note that for a clustered population the realistic error bars are larger than those for a Poissonian distribution, roughly by a factor $1 + 4\pi n J_3$, (see e.g. Kaiser 1986), so the realistic error bars are several times larger than the Poissonian formula at this scale. More graphically, it is apparent from the pair plots of Bahcall and Soneira (1983) that the number of independent clumps is quite small, so it is unlikely that we are seeing a large enough sample to ensure representative results. The formal errors for the deeper angular surveys (Hauser and Peebles 1973) are smaller, and we can be fairly confident that we are seeing a representative sample, but there are other sources of error or bias that might creep in here. Taken at face value the estimates obtained appear to be consistent and give a correlation length of around 25 Mpc/h. This correlation strength is much greater that that predicted in CDM. In this theory the clustering amplitude *is* predicted to be greatly enhanced over the amplitude of the mass correlation function (e.g. Kaiser 1984), but the predicted correlation length is still only around 14 Mpc/h, almost a factor of two below that observed.

Before concluding that CDM is ruled out it is worth critically examining these observations. One possible problem is the spurious clustering that should arise because of Abell's selection criteria. Since he defined clusters to be regions on the sky where the projected density of galaxies was particularly high, one would expect to find an enhancement of pairs of clusters at small angular separation where the physical neighbours of one cluster are counted as members of the other and vice versa. The result is a spurious contribution to the correlation function as a function of the separation of the clusters perpendicular to the line of sight, so this effect would produce an elongation of the clustering pattern along the redshift axis as is seen in the redshift survey. This effect has been explored in some detail by Dekel (1988). To model this effect we have calculated the spurious clustering for a set of spherically symmetric clusters which are laid down in a Poissonian manner, and which have profiles modeled on the cluster-galaxy cross-correlation function. After projecting this onto the sky and selecting an apparent richness limited cluster sample, we get a correlation function which is proportional to the angular cluster-galaxy cross-correlation function, and which falls to unity for projected separation of about 7 Mpc/h (though this is quite sensitive to the assumed cross-correlation function, the cluster richness distribution etc., and is therefore quite uncertain). This strength of correlation would not strongly perturb the estimate of the correlation length from the shallow redshift survey, but would be expected to strongly contaminate the deeper angular surveys. This is very worrying, since it is on these samples that we rely for statistically significant results.

Another source of spurious clustering is expected to arise from the manner in which Abell determined the background density for each plate. Lucey (1983) points out that the background correction becomes large for the deeper samples, so one would expect to find an excess of clusters on plates where the background correction is abnormally low and *vice versa.* Since Abell gives few details of how this background was derived it is difficult to model this effect. It is worth noting, however, that both this and the projection effect mentioned above would be predicted to give a scaling with depth R which is intermediate between that expected for true spatial clustering; $w(\theta) \propto R^{-2}$, and that for a distance independent angular pattern $w \propto R^0$, and so would be quite difficult to detect from scaling tests on the deep samples.

Estimating effects like these accurately is very difficult, and it is easy to think of numerous other sources of spurious clustering, or indeed to speculate on how Abell might have judiciously corrected for such effects. An alternative approach is to use the redshift survey data directly to see if the clustering pattern is really isotropic, and if not, to correct for non-spatial clustering. Bahcall and Soneira (1983) presented plots of the excess pair counts as a function of redshift separation for fixed bins of angular separation and showed that these fall at large Δz. This was held to be evidence that we are seeing true spatial clustering. However, since the pair counts for a uniform distribution also fall off at large Δz, this conclusion is questionable. If one looks at the correlation function which is given by the ratio of the excess pair counts to that for a random distribution (Sutherland 1988), then this shows no sign of falling, and this would suggest a very strong component of non-spatial clustering. Puzzlingly, the effect is seen even for large angular separation of 20 degrees or so, and this would not be predicted by either of the effects mentioned above. Sutherland has tried to correct for this contamination in a manner similar to that used in quasar clustering studies where the 'raw' value of $1 + \xi$ is divided by the mean $1 + \xi$ for a bin with the same perpendicular separation but much larger Δz. The corrected ξ thus obtained looks quite spherically symmetric, has a much lower amplitude, and appears to be quite compatible with the CDM predictions. It is still questionable whether Sutherland's corrected estimate is representative; it is, after all, still based on the redshift survey sample, but his results certainly cast a shadow of doubt over the robustness of the conventional estimates. Sutherland and Efstathiou (1988) have subsequently tested for the reality of the clustering by comparing estimates of $w(\theta)$ derived from inter and intra-plate pairs respectively. They find a significant discrepancy, casting further doubt over the reality of the apparent cluster clustering, so we feel it is premature to discard CDM to the scrap heap as yet. Cluster-clustering does seem a promising area of vulnerability for CDM, and the prospect of a large X-ray selected cluster sample from ROSAT is very exciting.

7.2 - *Normalisation*

Our normalisation of the fluctuations in galaxy counts is fairly direct and conventional. Another commonly used way to quote the normalisation is in terms of the J_3 integral. Our choice in very similar, but uses a different integral over the 2-point function (see e.g. Peebles 1980, Sec. 36). One might question however, our assumption that the 2-point function for IRAS galaxies is the same as for optical galaxies. If it were shown convincingly that the correlation length is much smaller then we would consider the dipoles to be evidence for more power at large scale. Another minor problem is the assumption that linear theory is applicable for the 8 Mpc/h normalisation. If this were shown to be badly in error we would again have to revise our conclusions.

Our choice of normalisation $\sigma_\varrho \simeq 0.6$—$0.7$ for the mass fluctuations is more controversial. It has come to our notice (Efstathiou, private communication) that the normalisation used for N-body experiments is roughly $\sigma_\varrho = 0.35$. This is about a factor 2 smaller than our preferred value, so with their choice of normalisation the peculiar velocities would be quite a problem for CDM. The results of the N-body experiments look in many ways quite realistic, and a very interesting question is whether with our normalisation the results would be unacceptable. It seems unlikely that considerations of galaxy clustering would usefully discriminate since to a large degree one can play off a larger amplitude mass fluctuation against a smaller level of biasing. The most effective way to discriminate between these different normalisation options is through comparison of peculiar velocity observations.

If we assume a normalisation which is a factor 2 larger than used in the numerical experiments then we have to reinterpret the results using the red-shift scaling $z \rightarrow 2z + 1$, so a snapshot of the particle positions at the 'present' would be interpreted as a snapshot of $z = 1$ for instance, and the scaling for physical velocities $V \rightarrow \sqrt{2}V$, so all rotation velocities would increase by about 40%. It is arguable that in many respects these changes would be beneficial; the epoch of formation of halos that one would wish to identify with galaxies and perhaps quasars would be pushed back to much higher red-shift, and there is some indication that the velocity dispersion of clusters in the simulations are on the low side (White *et al.* 1987), so this would also improve. The Press-Schechter (1974) formula for the distribution of haloes provides a direct link between properties of non-linear clusters at the present epoch and smaller objects of a similar density contrast at early times. One infers from the CDM spectrum that *if* 1% of the present universe is in rich clusters with mean separation 50 Mpc/h and rotation velocities 1000 km/s, then since the rms fluctuations rise by about a factor 5 going to a mass scale 1000 times smaller, we should see at $1 + z \simeq 5$ about 1% of the total mass in objects with mean separation 5 Mpc/h and rotation velocities $\sqrt{1+z}(M/M_{\text{cluster}})^{1/3} \times 1000$ km/s, or $\simeq 220$ km/s. These look very much like the properties of real galaxies. Perhaps improved studies of rich clusters will give lower rotation velocities, but if this value is appropriate then this would imply quite an early epoch for the formation of objects with abundance and internal velocities like real galaxies. Other observational tests that would be very useful to resolve the normalisation question are studies of high velocity dispersion clusters at higher redshift from X-ray studies and possibly from gravitational lensing, since the rapid evolution expected for these object is quite sensitive to the normalisation adopted.

7.3 - *Biasing Galaxy Formation*

We have assumed throughout that the galaxies trace a biased field which has fluctuations which are amplified relative to the underlying mass fluctuations. The main motivation for this is that if Ω really is unity then some sort of biasing of galaxy formation towards dense regions seems to be necessary in order to reconcile the high L/M of clusters compared to the L/M for the universe as a whole. We now discuss the plausibility or otherwise of biased galaxy formation. The discussion here is largely quantitative and any discussion of the physics of galaxy formation is necessarily speculative. We will make the fairly conventional assumption that galaxies form in the dark matter potential wells, and this at least is amenable to quantitative calculations. Here we will use the Press-Schechter (1974) formalism for the evolution of dark matter condensations in CDM-world (Cole and Kaiser 1988).

The basic idea that we shall appeal to is that the long wavelength perturbations present in the CDM spectrum modulate the formation of halos of smaller mass scale. (We shall use the word halo here to denote dark matter condensations in general). If one accepts that to a useful approximation one can associate the halos which have just formed at any epoch with overdense regions in the initial state (e.g. Press and Schechter 1974, Bardeen *et al.* 1986) then the effect of long wavelength modes is simply to modulate the collapse times of the halos - perturbations will turn around and virialise earlier in a region subject to a positive long wavelength perturbation.

This type of analysis has been applied to rich clusters (Kaiser 1984), which it seems appropriate to identify with very massive halos forming at the present epoch. Since, in CDM and other hierarchical pictures the mass scale of clustering increases as time proceeds, the modulation of collapse times by long waves converts to a spatial modulation of the number density of clusters selected according to some minimum mass threshold. This calculation gives a very strong amplification of cluster-clustering, though as we have discussed, this may still be too small. This analysis, where we effectively take a 'snapshot' and analyse the spatial distribution of clusters at that instant of time, can readily be extended to lower mass halos using the Press-Schechter (1974) or analogous formulae, and one finds that the bias decreases in strength, with a negative bias for objects at the low mass end of the spectrum. The borderline unbiased objects are those where $M^2 n(M)$ peaks, so these objects contain the greatest fraction of the mass of the universe per log interval of cluster mass.

Whether this 'snapshot' calculation is relevant to galaxies depends on how one imagines galaxies form. One not unreasonable picture for the formation of disk galaxies is that halos with roughly flat rotation curves form containing gas which is heated to the virial temperature on collapse. As time goes on the gas can cool from progressively larger radii and this gas contracts within the dark halo to form the disk. Aspects of this type of formation picture have been explored by many authors (e.g. White and Rees 1987, Fall and Efstathiou 1980, Gunn 1982) and have become something of a paradigm. An appealing feature of this model in the context of CDM is that with the gas density implied by baryosynthesis the gas in a $V_{\rm rot}$ = 220 km/s halo can just cool by the present at a radius of about 100 kpc, so would plausibly make a reasonable disk after a factor 10 decrease in radius. In this picture, galaxy formation would be an ongoing process, the 'snapshot' result should be applicable, and so one would expect at best a rather weak positive bias for the most massive spirals and a negative bias for small spirals. Worse still, since the effect of a positive long-wavelength 'swell' is to increase the characteristic virial temperature of the halos, this acts to impede the cooling, and results in a net antibias for the total amount of gas which has cooled, and therefore most naturally in a net antibias for the total number of stars. A closely related problem is that if one approximates the cooling time - temperature dependence as a power law, $t_{cool} \propto \varrho^{-1} T^{\beta}$, and if one assumes that the gas is initially distributed like the dark matter, then one finds that the mass of gas which has cooled by the present varies with the halo temperature as $M_{cool} \propto T^{(3-\beta)/2}$. Taking $\beta \simeq 1$ as appropriate for galactic temperatures, one would naively predict a luminosity - rotation velocity relation $L \propto V^2$ unlike the observed relation $L \propto V^4$. The relation between this result and the antibiasing becomes apparent when one notes that the halo mass is proportional to the cube of the velocity, so if one has $L \propto V^{\gamma}$, with $\gamma < 3$, then L/M is a decreasing function of mass. Thus, in a region subject to a positive long wave perturbation so the characteristic halo mass is increased, one would expect the net L/M to be decreased. Finally, one would also expect an additional antibias to arise in this model if the gas is stripped from the halo in a dense region like a rich cluster. All in all, it is very hard to see how a strong positive bias would result in this kind of picture.

Another type of scenario in which the 'snapshot' calculation might be relevant is if galaxy formation is abruptly terminated, perhaps by some kind of negative feedback. If this occurs while the number density of 'galaxies' is still rapidly rising, or equivalently, when only a very small fraction of the mass is in these objects, then the 'galaxies' would be quite analogous to rich clusters at the present epoch and a strong positive bias seems much more promising. The weakest feature of this scenario is that no obvious candidate exists for the hypothetical feedback process which is invoked solely to effect the bias and reconcile the theoretical prejudice for a closed universe with observations.

Without a feedback to abruptly switch off galaxy formation it seems inappropriate to apply the 'snapshot' calculation. To be sure, at any particular instant one can find a population of objects which are strongly biased towards regions where the long-wave swell is positive, but this is only because in regions which are underdense on large scales the analogous objects are destined to collapse just a little later. If there is to be a positive bias then it must be that the delayed collapse somehow reduces the luminosity of the galaxy, but how might such an effect arise?

Before embarking on what might otherwise seem to be a course of idle speculation it is illuminating to consider for a moment what one would expect to see in an unbiased universe - i.e. a universe in which the stellar luminosity generated in a dark matter condensation is independent of the time of formation and is simply proportional to the mass. When CDM first became popular it was argued that such a 'what-you-see-is-what-you-get' model would nicely reproduce the observed luminosity - velocity relations for galaxies (e.g. Blumenthal *et al.* 1984). The argument is that on galaxy scales the spectral index $n \equiv d\log P(k)/d\log k$ is close to -2, so for the '1-sigma' perturbations, for instance, we have $\Delta\varrho/\varrho \propto M^{-1/6}$. The final density of the system at collapse varies as the cube of the initial perturbation amplitude, or as $M^{-1/2}$. Invoking the virial theorem then gives $M \propto V^4$. While the resemblance of this to the observed $L - V$ relation is encouraging, when one looks at the expected scatter about the mean $M - V$ relation the result is less pleasant. The internal velocity varies as $\sqrt{\nu}$, where ν is the initial overdensity in units of the rms. Consider, for instance, the 1 and 2-sigma perturbations on a given mass scale. These have by assumption the same L, but have V^4 differing by a factor 4. A little more quantitatively, one can translate the Gaussian distribution for ν into the scatter for V at constant L. One finds that 2.5 log V^4 has variance 2.2, which is roughly 5 times larger than the scatter of 0.45 magnitudes for the Tully-Fisher relation. Since this presumably contains substantial observational uncertainty we see that the intrinsic $M - V$ correlation for Gaussian fluctuations is much broader than the intrinsic $L - V$ relation for real galaxies. Perhaps, as suggested by Faber (1982) the different morphological types are stratified along lines of constant ν, possibly because of systematic dependence of the angular momentum on ν. If angular momentum, or more generally some factor depending on the shape of the perturbation were the discriminator for morphological type, then one would still have a potential problem. Long wavelength modes have a negligible effect on the shape of much smaller perturbations, so the galaxies would still be unbiased by such modes. If we take the objects at constant ν and subject them to a positive swell they will collapse earlier and have a higher V^4 at the same L, and consequently, using these obiects to determine the distance scale, one would infer a smaller Hubble constant using galaxies in dense environments as compared to field galaxies, and as far as we know no such effect is seen.

This 'unbiased' model, therefore, seems to predict that we should see a broad scatter about the $L - V$ relation, and systematic environmental dependence of the $L - V$ intercept, essentially because the random and systematic variations in collapse time expected in this model will change V but not L. There are two possible ways to resolve this dilemma. First, we could stick with $L \propto M$ and argue that the V we observe for stars is not the V for the dark halo which is what we have used above. Second, one can keep $V_{\rm stars} \simeq V_{\rm halo}$, and assume that in perturbations with the same mass, the efficiency of star formation is greater the earlier is the collapse time. Following this option we are led back (though now on rather firmer empirical grounds) to ask: why should an earlier collapse time give a brighter galaxy?

A positive long wave swell will increase the density, the virial temperature, the pressure, and the collapse redshift. Higher density tends to increase the cooling rate, but the increase in temperature nearly balances this, so the ratio of cooling time to dynamical time is hardly altered by a long wave swell. More promising is the idea that one can form more stars per unit mass of gas in a deeper potential well. Such a dependence has been suggested by Larson (1974), and the idea has more recently been revived by Dekel and Silk (1986) in the context of CDM. Such a dependence might plausibly arise if supernovae are efficient at expelling gas from shallow potential wells. A related possibility is that the increased pressure might modify the initial stellar mass function and perhaps reduce the fraction of massive stars, resulting in more moderate mass stars being produced before the supernovae blow off the gas. Yet another possibility is that speeding up the collapse allows more stars to form before the supernovae go off.

Whatever physics is operating, if we take seriously the idea that it is the velocity dispersion of the potential well which determines the luminosity, and not the amount of gas available, then a positive bias seems inevitable. Quite how strong this bias is depends on when galaxies are assumed to form. Using the Press-Schechter formalism (with the normalisation proposed in Sec. 1) one can calculate the number density of objects with $V_{rot} > 220$ km/s, for instance. This reaches a maximum around $(1 + z) = 3$, and is biased up and down by long wavelength modes with a b value of 1.4-1.5. If one uses the simple spherical collapse model to estimate the initial density contrast for recently virialised systems such as rich clusters, then one obtains about a factor 2 enhancement for the L/M of such systems relative to the global L/M. Thus, the bias we find is substantial, though these simple estimates are somewhat lower than seem to be required to reproduce the virial analysis.

7.4 - *Future Prospects*

As the data we have looked at have failed to falsify CDM, it is interesting to consider the outlook for more powerful tests. One possibility would be to repeat the analysis with much deeper samples. The fluctuation amplitude predicted in CDM does fall quite fast as one probes deeper. Unfortunately, the fundamental statistical errors fall more slowly, so quite rapidly one will reach the point where the CDM predictions fall below the noise level, and then only models which produce much larger fluctuations than CDM can be effectively tested. For the angular dipoles, the predicted dipole falls asymptotically with depth as R^{-2}, whereas the shot noise falls as $R^{-3/2}$. For the velocity, the rms prediction falls as R^{-1}, and the noise falls as $R^{-1/2}$. Since the observations we have considered are at best a few sigma detections and agree with the CDM predictions, it is clear that the possibility for extension is unlikely to be fruitful.

A more hopeful outlook is to use a more complete description of the data, rather than using just the dipole moments. The formalism described here can readly be extended to a more or less complete description of the data (subject only to the constraint that one use only representations of the data that are well described by linear theory). The more general analysis for the peculiar velocity field can be found in Kaiser (1988), and we are currently developing the analysis for counts in cells of galaxies. While the mathematics can be straightforwardly generalised, the problem is deciding which representation of the data provides the best compromise between loss of information and the requirement that linear theory be applicable.

Generalising to e.g. counts in cells rather than dipoles enables us to extract more information from the currently available data, but at the price of probing smaller scales. This also solves the problem of how to deal with the zone of avoidance or catalogue boundaries. For deeper samples this type of analysis will enable us to sample many independent volumes, similar to the volumes sampled by the dipoles we have used here, and that will greatly improve the power of the test. For deeper peculiar velocity samples the analogous benefit will be much less because of the increase in uncertainty.

There is one very important benefit to be obtained from deeper peculiar velocity studies. These will enable the importance of systematic errors due to environmental effects to be quantified. The peculiar velocities of galaxies other than our own are determined rather indirectly; one assumes that the Tully-Fisher relation or whatever is universal. Systematic dependence of internal properties of galaxies on environment must be important at some level, and will tend to inflate the estimates of the peculiar velocities inferred. Some tests for these effects can be made within the currently available data, but the most convincing test would be to determine the 'velocities' of a sample of much more distant clusters since one would then expect any real peculiar velocities to be negligible (though this betrays some theoretical prejudice on the part of the authors), and one would then measure the real total statistical uncertainty. Such a program would be arduous, but would not require a full sky survey.

Major increases in the data available will take some time. On a shorter timescale we might anticipate revisions to the luminosity function or perhaps the appropriate normalisation, which would change our conclusions. For Gaussian models the theoretical formalism provides a promising methodology for analysing the large-scale structure data base. The theoretical questions that seem to us most interesting here are: Does the structure appear to be growing by gravitational instability? If so, how strongly biased are the various types of galaxies? How do the data constrain the space of theoretical models for the initial power spectrum?

REFERENCES

Aaronson, M., Huchra, J., Mould, J., Schechter, P.L., and Tully, R.B. 1982. *Ap J.* **258**, 64. (AHMST)
Bahcall, N.A. and Soneira, R.M. 1983. *Ap J.* **270,** 20.
Bardeen, J.M., Bond, J.R., Kaiser, N., and Szalay, A., 1986. *Ap J.* **304,** 15.
Blumenthal, G.R., Faber, S.M., Primack, J.R., and Rees, M.J. 1984. *Nature.* **311,** 517.
Bond, J.R., 1987. in *Nearly Normal Galaxies, The Eighth Santa-Cruz Summer Workshop.* ed. S.M. Faber, p. 388. New York: Springer-Verlag.
Bond, J.R. and Efstathiou, G., 1984. *Ap J.* **285,** L45.
Cole, S. and Kaiser, N. 1988. in preparation.
Davis, M. and Huchra, J. 1982. *Ap J.* **254,** 437.
Dekel, A. 1988. this volume.
Dekel, A. and Silk, J. 1986. *Ap J.* **303,** 39.
Dressler, A., Faber, S.M., Burstein, D., Davies, R.L., Lynden-Bell, D., Terlevich, R.J., and Wegner, G. 1987. *Ap. J.* **313,** L37.
Faber, S.M. 1982. in *Astrophysical Cosmology.* ed. H.A. Bruck, G.V. Coyne, and M.S. Longair. Vatican City State: Pontifical Academy, p. 191.
Faber, S.M. and Burstein, D. 1988. this volume.
Fall, M.S. and Efstathiou, G. 1980. *MN.* **193,** 189.
Gunn, J. 1982. in *Astrophysical Cosmology.* ed. H.A. Bruck, G.V. Coyne, and M.S. Longair. Vatican City State: Pontifical Academy, p. 233.
Harmon, R.T., Lahav, O., and Meurs, E.J.A. 1987. *MN.* **228,** 5p.
Hauser, M.G. and Peebles, P.J.E. 1973. *Ap J.* **185,** 757.
Kaiser, N. 1984. *Ap J.* **284,** L9.
_____. 1986. *MN.* **219,** 795.
_____. 1987. *MN.* **227,** 1.
_____. 1988. *MN.* **231,** 149.
Kaiser, N. and Lahav, O. 1988. preprint.
Kaiser, N. and Silk, J. 1986. *Nature.* **324,** 529.
Lahav, O. 1987. *MN.* **225,** 213.
Lahav, O., Rowan-Robinson, M., and Lynden-Bell. 1988. *MN.* **234**, 677.
Larson, R.B. 1974. *MN.* **169,** 229.
Lasenby, A. 1988. this volume.
Lawrence, A., Walker, D., Rowan-Robinson, M., Leech, K.J., and Penston, M.V. 1986. *MN.* **219,** 687.
Lucey, J.R. 1983. *MN.* **204,** 33.
Lynden-Bell, D. and Lahav, O. 1988. this volume.

Lynden-Bell, D., Faber, S.M., Burstein, D., Davies, R.L., Dressler, A., Terlevich, R.J., and Wegner, G. 1988. *Ap J.* **326**, 19.

Meiksin, A. and Davis, M. 1986. *AJ.* **91,** 191.

Meurs, E.J.A. and Harmon, R.T. 1988. preprint.

Peebles, P.J.E. 1980. *The Large Scale Structure of The Universe.* Princeton: Princeton University Press.

Press, W.H. and Schechter, P. 1974. *Ap J.* **187,** 425.

Rowan-Robinson, M., Walker, D., Chester, T., Soifer, T., and Fairclough, J. 1986. *MN.* **219,** 273.

Shafer, R.A. 1983. Ph.D. thesis, University of Maryland.

Strauss, M.A. and Davis, M. 1988. in *Large-Scale Structures of the Universe,* eds. J. Audouze, M.-C. Pelletan, and A. Szalay. Dordrecht: Kluwer Academic Publishers, p. 191.

Sutherland, W. 1988. *MN.* **234,** 159.

Sutherland, W. and Efstathiou, G. 1988. in preparation.

Villumsen, J.V. and Strauss, M.A. 1987. *Ap J.* **322,** 37.

Vittorio, N., Juszkiewicz, R., and Davis, M. 1986. *Nature.* **323,** 132.

Webster, A., 1976. *MN.* **175,** 71.

White, S.D.M. and Rees, M.J. 1987. *MN.* **183,** 341.

White, S.D.M., Frenk, C.S., Davis, M., and Efstathiou, G. 1987. *Ap J.* **313,** 505.

Yahil, A. 1988. this volume.

Yahil, A., Sandage, A., and Tammann, G.A., 1980. *Ap J.* **242,** 448.

Yahil, A., Walker, D., and Rowan-Robinson, M. 1986. *Ap J.* **301,** L1.

CAN SCALE INVARIANCE BE BROKEN IN INFLATION?

J. R. BOND
CIAR Cosmology Program, Canadian Institute for Theoretical Physics, Toronto

D. SALOPEK
Physics Department, University of Toronto

and

J. M. BARDEEN
Physics Department, University of Washington

Abstract

Scale invariant fluctuation spectra are the most natural outcomes of inflation. In this paper we explore how difficult it is to break scale invariance naturally. It can be done when there is more than one scalar field. We sketch the effects that varying the expansion rate, structure of the potential surface and the curvature coupling constant have on the quantum fluctuation spectra. We conclude that: (1) power laws different than scale invariant are very difficult to realize in inflation; (2) designing mountains of extra power added on top of an underlying scale invariant spectrum is also difficult; and (3) double inflation leading to a mountain leveling off at a scale invariant high amplitude plateau at long wavelengths is generic, but to tune the cliff rising up to the plateau to lie in an interesting wavelength range, a special choice of initial conditions and/or scalar field potentials is required.

1. Introduction

In this paper, we discuss the difficulties of designing a density fluctuation spectrum to order using physical processes occurring during an initial inflation epoch. We denote the two-point functions in momentum (comoving wavenumber) space for homogeneous and isotropic random fields by, for example for the density, $\mathbf{P}_\varrho(k) \equiv (k^3/(2\pi^2)) \langle |(\delta\varrho/\varrho)(k)|^2\rangle$, the power per logarithmic interval of comoving wavenumber k. If the fluctuations are Gaussian then this is all that is required to completely specify the random field. In the inflation picture, quantum fluctuations in scalar fields ϕ generated during the inflation epoch become density fluctuations. If the spatial eigenmodes have zero occupation number and mode-mode couplings can be ignored (linearity is expected since fluctuations are apparently quite small at recombination), then these quantum fluctuations will be like zero point oscillations of sound waves in a crystal — in a Gaussian ground state. Curiously, the long wavelength sound modes of a zero temperature crystal have a Zeldovich scale invariant spectrum, $\mathbf{P}_\varrho \propto k^{3+n}$ with $n = 1$, the generic outcome of inflation. However, the way a Zeldovich spectrum arises during inflation depends critically upon the way scalar perturbations in the expanding universe suffer 'Hubble drag' which damps the oscillations for wavelengths smaller than a Hubble distance and freezes out their amplitude for longer wavelengths.

Inflation occurs if the comoving Hubble distance $(Ha)^{-1}$ decreases with time, *i.e.*, $\ddot{a} > 0$; the equation of state of the homogeneous background universe must then obey $p/\varrho < -1/3$, where p is the total pressure and ϱ is the total energy density. This is realizable for a homogeneous scalar field ϕ self-interacting through a potential $V(\phi)$ for which $p = \dot{\phi}^2/2 - V(\phi)$ and $\varrho = \dot{\phi}^2/2 + V(\phi)$, provided the energy density is potential dominated for some period and over some patch of space. As inflation proceeds, $(Ha)^{-1}$ sweeps in, encompassing ever smaller comoving length scales, arresting causal communication across waves with $k^{-1} > (Ha)^{-1}$.

In the inflation picture, a region of size smaller than the Planck length is inflated by a factor $\sim e^{60+N_I}$ to a region (much) larger than the current Hubble length $H_0^{-1} = 3000\ \mathrm{h}^{-1}$ Mpc provided the number of e-foldings during inflation is $N_I \gtrsim 70$. Without inflation, the region would have expanded by only a factor $\sim e^{60}$ to less than a micron across by the current time — an aspect of the 'horizon' problem which inflation so successfully solves.

If only one scalar field drives inflation approximate scale invariance, $\mathbf{P}_\phi(k) \approx$ constant, for the spectrum of scalar field fluctuations is the natural outcome. This translates into scale invariant fluctuations for the gravitational potential fluctuations $\mathbf{P}_\Phi(k) \approx$ constant for adiabatic scalar perturbations, leading to a Zeldovich spectrum for the density. The Poisson-Newton equation $a^{-2}\nabla^2\Phi = 4\pi G\delta\varrho$ for the gravitational potential in an infinite expanding background remains an exact relation in general relativistic perturbation theory if $\delta\varrho$ is the density perturbation in the comoving frame, and $-\Phi$ is Bardeen's (1980) gauge invariant Φ_H variable.

For isocurvature perturbations, the ϕ fluctuations become fluctuations in one of the species of particles present, for example, in axions, $\mathbf{P}_\varrho(k) \propto \mathbf{P}_\mathrm{f}(k)$, where ϱ_A is the axion mass density (*isocurvature CDM perturbations*) or, possibly, in baryons, $\mathbf{P}_{n_B}(k) \propto \mathbf{P}_\mathrm{f}(k)$, where n_B is the baryon number density (*isocurvature baryon perturbations*). In these cases, the total density and gravitational potential fluctuations vanish, $\mathbf{P}_{\varrho} = \mathbf{P}_{\Phi} = 0$, at the time when either the axion mass or the baryon number is generated — provided it happens after inflation has ended. The power law $n = -3$ corresponds to scale invariance in these isocurvature cases for the initial density perturbations in the nonrelativistic matter. We refer to a scalar field giving rise to an isocurvature spectrum as an *isocon* in keeping with the terminology *inflaton* for the scalar driving inflation which gives rise to adiabatic fluctuations.

Although very gentle deviations from scale invariance are expected in inflation it is rather difficult to arrange drastic modifications. Here we are interested in exploring the requirements for obtaining power laws different from $n = 1$ for adiabatic perturbations or $n = -3$ for isocurvature perturbations over extended regimes of k-space, or mountains of extra power built on top of an underlying scale invariant spectrum. More generic are two scale invariant spectra of different amplitudes, the larger at longer wavelengths, corresponding to regimes when two different scalar fields drive inflation, with a matching region (a ramp) that depends upon the details of how one scalar takes over from the other. This is double inflation. Tuning the location of the ramp or the mountain or the power law regime to lie in a specific k range requires rather precise conditions to be imposed on the inflationary model. For example, to put modified power at the scale of clusters implies that some special physics was operating at a redshift $z \sim e^{122}$ involving either finely tuning initial conditions for the background scalars or the potential parameters. We discuss these possibilities in more detail in Sec. 2, reviewing work reported in Bardeen, Bond, and Salopek (1987, 1988 [BBS1, BBS2]).

2. Response of Scalar Fields to Potential Changes

There are only a few parameters at our disposal if we wish to modify the spectrum of fluctuations that comes out of inflation. The linearized equations obeyed by the perturbations in the fluctuations $\delta\phi_j(k, t)$ of fields $\phi_j(\vec{x}, t)$, $j = 1, \ldots, N$, about a homogeneous average $\bar{\phi}_j(t)$ are:

$$\ddot{\delta\phi}_j + 3H\dot{\delta\phi}_j + (k^2 / a^2)\, \delta\phi_j + \sum_i (m_{ij}^2 + 12\xi H^2 \delta_{ij})\, \delta\phi_i \quad (2.1a)$$

$$+\Gamma_j \dot{\delta\phi}_j + (\dot{\bar{\phi}}_j \delta\Gamma_j) + \dot{\bar{\phi}}_j \dot{h}/2 + \bar{\phi}_j \xi \delta R \quad (2.1b)$$

$$= 0.$$

Equation (2.1) is coupled to equations describing the evolution of the gravitational metric and the perturbed total (field plus radiation) energy and momentum density. There are also equations describing the evolution of the background fields, the expansion factor (Friedmann equation) and the background radiation energy density. Details are given in BBS1, BBS2.

For this discussion we shall ignore the terms in the second line, which involve such factors as the rate at which field energy is dissipated into radiation (Γ_j) and the growth of the field perturbations due to gravitational instability (perturbed metric $h = h^b_b$ in the synchronous gauge and perturbed curvature δR). These operate primarily when the perturbations are well outside the Hubble distance and the shape of the fluctuation spectrum is set.

When the wave is still causally connected, $k \gg H\, a$, with momenta high compared with masses $k \gg |m|a$, a WKB solution of equation (2.1) exhibits the rapid oscillation appropriate for massless scalars with a general decline due to the Hubble drag: $\delta\phi_j \approx (2k)^{-1/2} a^{-1} \exp[-i \int dtk/a]$, where the positive frequency solution has been selected. $\phi_j(k, t)$ has real $\mathrm{Re}(\delta\phi_j)$ and imaginary $\mathrm{Im}(\delta\phi_j)$ parts which must be calculated separately by solving two sets of evolution equations (2.1). The power spectrum for ϕ_j is given by

$$\mathbf{P}\phi_j\,(k, t) = \frac{k^3}{2\pi^2}\,([\mathrm{Re}(\delta\phi_j(k, t))]^2 + [\mathrm{Im}(\delta\phi_j(k, t))]^2)\,(1 + 2\bar{n}_{jk}). \quad (2.2)$$

Here $\bar{n}_{jk}$ is the average occupation number of mode k for the scalar field ϕ_j. At horizon crossing ($k = H\, a$) the interior WKB solution referred to above would give $\mathbf{P}\phi_j = [H/(2\pi)]^2\,(1 + 2\bar{n}_{jk})$, relating the amplitude to the Hawking temperature $H/(2\pi)$.

To modify the outcome of inflation — assuming linearization is valid which in some cases is debatable — we are only free to modify (1) the occupation number $\bar{n}_{jk}$, (2) the Hubble parameter $H(t) = \dot{a}/a$, which enters in the Hubble drag term $-3H\dot{\delta\phi_j}$, (3) the effective mass matrix $m^2_{ij} \equiv \partial^2 V/\partial\phi_i\partial\phi_j$, or (4) the curvature coupling constant. ξ enters as an effective term in the mass matrix; it may be positive or negative. It is 0 for minimally coupled fields, 1/6 for conformally coupled fields, and is subject to renormalization so the value could evolve.

It is usually argued that the modes which are currently within our horizon would have had such large values of k/a relative to any characteristic energy scale describing mode occupation that $\bar{n}_{jk}$ can be taken to be zero, so that only zero point fluctuations contribute to the spectrum. Since we are dealing with sub-Planck scale physics it is by no means clear that this assumption is valid. If $\bar{n}_{jk}$ is nonzero then the fluctuation spectrum would be unlikely to be scale invariant, reflecting instead whatever physics would determine the primordial occupation of modes, and the fluctuations would be non-Gaussian.

An approximate solution to equation (2.1) can be used to illustrate the effect of varying $H(t)$ and $m^2(t)$ histories (ignoring the off-diagonal m^2_{ij}, $i \neq j$ terms): outside the horizon $k^{-1} < (Ha)^{-1}$, but before growth due to the perturbed metric occurs, the power spectrum is

$$\mathbf{P}\phi(k) \approx [H(t_k)/(2\pi)]^2 \exp[-\int_{t_k}^{t} dt\, \mathrm{H}(t)\,(3 + n(t))], \quad (2.3a)$$

$$n(t) \equiv \mathrm{Re}\,[-3(1 - 4m^2(t)/(9H^2(t)))^{1/2}], \quad (2.3b)$$

$$\varrho_\phi \propto \exp[-\int_{t_k}^{t} dt\, \mathrm{H}(t)\,(3 + n(t))], \quad (2.3c)$$

$$a_k \equiv a(t_k) \equiv k/H. \quad (2.3d)$$

The time t_k and expansion factor a_k when a wavenumber k crosses the horizon are given by eq. (2.3d). Equation (2.3a) is valid provided n >0 at t_k, and remains valid even if n subsequently vanishes ($2m/3H > 1$). If n is zero at t_k, the prefactor $H^2 t_k$ should be replaced by $H^3(t_k)/m(t_k)$. Note that if n is constant, then the exponential term in (2.3a) is simply $[k/(H(t_k)a)]^{3+n}$, a power law in k if H is constant. A number of interesting cases follow from this result.

2.1 - Scale Invariance

If $m^2 \approx 0$, then the spectrum is scale invariant if H is independent of the expansion factor a. The value of a when k first 'crosses' the horizon is a_k. Since $H^2 = (8\pi G/3)\,[\Sigma_j \dot{\phi}_j^2 + V]$, judicious choice of V can lead to structure in $\mathbf{P}_\phi$. However, the most likely case over the ~ 6 orders of magnitude observable k range and the corresponding 14 e-foldings in a is that V will fall gently, leading to only slight deviations from scale invariance, with just a little more power on large scales than on small.

We can distinguish two types of behaviour, depending upon whether the field is driving inflation (in which case it is the inflaton) or its Hubble drag is driven by another field (in which case it is an isocon). The *inflaton* leads to gravitational metric fluctuations Φ proportional to fluctuations in the gauge invariant variable ζ introduced by Bardeen, Steinhardt, and Turner (1983, see also BBS1) which has the advantage of being independent of the spacelike hypersurface upon which the perturbation is measured: $\mathbf{P}_\Phi \propto \mathbf{P}_\zeta \sim (3H(a_k)/\dot{\phi})^2\, \mathbf{P}_\phi(k, a_k)/H^2$. For an *isocon,* for which we take the axion as the generic case, the fluctuations in the axion mass density after the axion mass is generated is $\mathbf{P}_{\varrho A} \propto \mathbf{P}_\phi$. In both cases, approximate scale invariance is the outcome.

An *isocon* which preferentially dissipates into baryons rather than antibaryons can lead to *isocurvature baryon perturbations,* although the δn_B spectrum will be scale invariant if the ϕ spectrum is. Such spectra can be ruled out by their large angle CMB anisotropies which are predicted to be enormous (Efstathiou and Bond 1986, see also Efstathiou 1988).

2.2 - Double Inflation

If we allow for more than one degree of freedom for our scalar fields, then marvelous mountain ranges, valleys and moguls can be envisaged for potential space. One's intuition regarding motion in this space can be utilized, except that the rolling fields are subject to Hubble drag ($-3H\dot{\phi}_j$) which can result in a terminal velocity (slow-rollover), a phenomenon fundamental to the realization of inflation. Fluctuations $\delta\phi$ are small spreads in the field about the rolling background value $\bar{\phi}$. Fluctuations within the horizon at the specific epoch are still oscillating, while those outside have their shapes frozen in although the amplitudes may change with time. For some purposes it is useful to include those waves outside the horizon with the background field, to allow for a gentle variation from place to place (Starobinsky 1983).

If there are a number of flat directions in potential space, inflation could be a complicated process, modifying the H profile with time and also the form of the mass matrix m^2_{ij}. This will certainly map onto structure in the fluctuation spectrum. However, to transform specific features in potential space to features in a particular range in k-space, we must arrange for the fields to pass through this V structure at a specific range of a values.

Consider first the case of 2 scalars having a potential forming a broad valley with the valley minimum line very gently sloping down towards the origin with somewhat steeper walls rising away from the minimum. This configuration leads to double inflation (Starobinsky 1985, Kofman and Linde 1987, Silk and Turner 1987, BBS1, BBS2). If the 2-dimensional field begins high enough on the wall away from the origin, it will be potential dominated by the wall part, and experience inflation with a large value of H. The field will roll down towards the valley minimum, oscillating in one direction ($m^2 > 0$ so the power in the field in this direction damps away as $a^{-(3+n)}$ as above), while continuing to roll down towards the origin in the other direction with a lower value of H. The field in the second direction is all that is left after the end of inflation. Since it experiences first the high H for long waves as they leave the horizon, then the low H value for short waves, the plateau structure referred to in Sec. 1 is generic. The ramp between the two levels will depend upon the specific form of the potential. To arrange for the location of the ramp to be tuned to an astrophysically interesting scale, the initial location of the field matters. If the field remains near the valley minimum for too long a period then the spectrum within the current Hubble length will only reflect the low value of H and be effectively scale invariant.

Another procedure to adopt is to litter the valley minimum with moguls that will impose structure in the fluctuations of the field direction along the valley minimum. Bardeen (1988) has analyzed this case for a specific choice of mogul potential which leads to non-Gaussian fluctuations. The resulting spectrum becomes independent of initial conditions of the background field, but completely dependent upon mogul emplacement.

2.3 - *Mountains*

Choosing m^2 positive or negative can give spectra rising or falling with increasing k, but at the expense of exponential decreases or increases in both the background field energy density and in the fluctuation power, assuming inflation is continuing.

If m^2 is fixed, the drop of H with time eventually leads to n reaching zero. The field oscillates coherently with an energy density averaged over an oscillation period decreasing as $\varrho_\phi \sim a^{-3}$ like nonrelativistic matter, with fractional fluctuations $\mathbf{P}_\phi/\varrho_\phi$ being constant. This is the mechanism by which the axion, once it attains its mass when the temperature of the universe is about 200 MeV, behaves like cold nonrelativistic matter. However this occurs *after* inflation, so the a^{-3} law is not devastating.

To have a field whose fluctuations are still of interest we can only have m^2 positive or negative over a limited regime *during* inflation. To shape a mountain of power on top of a scale invariant spectrum, we would want m^2 to begin at 0 to ensure scale invariance at the longest wavelengths, to become positive for the rise, then negative for the drop, finally returning to zero to maintain scale invariance on short wavelengths. However, the scale invariant long wave structure will first decline in amplitude due to the $m^2 > 0$, then rise in amplitude due to $m^2 < 0$. It will require a restrictive class of choices to ensure the long wave part is not too high. Having m^2 become negative first will tend to give the plateau plus ramp structure of double inflation.

2.4 - *Power Laws*

One way to get extended power laws over some k range is to arrange for n to be constant over the associated range in a_k. This would be possible if m^2 scales with H^2. The natural way for this to occur is to make use of non-zero curvature coupling constants ξ, for then the local spectral index for isocons is $n = -3(1 - 16\xi/3)^{1/2}$. The severe price to pay is the $\sim a^{-(3+n)}$ fall of the energy density in the field throughout inflation. Only for $n = -2$ might we envisage getting the perturbation strength back by relative growth compared with the radiation, and this requires the very special choice $\xi = 5/48$. Even if such a value were to arise for some obscure reason, it would be subject to renormalization. It is easier to contemplate ξ negative with $n < -3$, falling to short wavelengths. We are currently exploring the $\xi<0$ region and find some interesting features.

Another way to get power laws has been to invoke power law inflation (Lucchin and Matarrese 1985), with the expansion going as $a \sim t^q$, with $q > 1$ to ensure $\ddot{a} > 0$. This necessitates an equation of state yielding fixed $p/\varrho = -1 + 2/(3q)$. For the scalar field which drives power law inflation, $\mathbf{P}_\phi \sim (H/2\pi)^2 \sim k^{(3+n)}$, where $n = -(3q-1)/(q-1) = 3(1-p/\varrho)/(1+3p/\varrho)$.

This is also the k-dependence of $\mathbf{P}_\zeta$, hence the density fluctuations subsequently generated would have the power law index $n_\varrho = n + 4$. This always gives $n \leq -3$, $n_\varrho \leq 1$, if inflation is realized, hence there is more power at large scales than the Zeldovich spectrum. A disadvantage of such spectra is that the redshifts of cluster and galaxy formation will not be well separated for power laws with $n_\varrho \lesssim 0$. Further if n is too negative large angle CMB anisotropies become too large. With an interaction potential of the specific form $V(\phi) = V_0 \exp(-4(\pi/q)^{1/2} \phi/m_\mathrm{P})$ for the inflaton, power law inflation could be realized, with n related to q as given above. Power law inflation could also drive an isocon to develop power law isocurvature perturbations. For example, a Goldstone boson such as the axion would have $\mathbf{P}_\phi \sim k^{3+n}$, with n also as given above. The axion density perturbations developed once the axion mass is generated would have the same power n; such isocurvature CDM spectra can be strongly ruled out by large angle CMB anisotropies (Efstathiou and Bond 1986, Efstathiou 1988).

From this discussion there seems to be little hope that the $-1 \lesssim n \lesssim 0$ power law isocurvature baryon spectra advocated by Peebles (1988) will arise within the inflationary paradigm. In any case, these models would have $\Omega = 1$, and smaller values, Ω 0.2 to 0.4, are preferred.

3. Specific Realizations of Broken Scale Invariance

Consider two scalar fields interacting through a chaotic inflation potential containing quadratic and quartic terms:

$$V(\phi_1, \phi_2) = \frac{(m_2^2 + \xi_2 R)\phi_2^2}{2} + \frac{\lambda_2 \phi_2^4}{4} + \frac{(m_1^2 + \xi_1 R)\phi_1^2}{2} + \frac{\lambda_1 \phi_1^4}{4} - \frac{\nu \phi_1^2 \phi_2^2}{2}\,. \tag{3.1}$$

If $\nu = 0$, and $\xi_1 = \xi_2 = 0$, then inflation is likely to be double inflation: If, $\lambda \gg \lambda_2$, first ϕ_1 dominates H, then ϕ_2. The effective m^2 of the first field ϕ_1 is always positive. Hence during the second phase its final amplitude would inflate away to an exponentially small value. $\mathbf{P}_{\phi 2}$ develops the ramp plus plateau structure.

The position of the ramp in k-space is controlled by the initial value $\phi_2(t_i)$. In the spirit of chaotic inflation as originally proposed by Linde (1983), $\phi_2(t_i)$ should fluctuate in space on comoving scales similar to the comoving scales on which ϕ_1 fluctuates, with amplitudes which should range up to $\frac{1}{4}\lambda_2\phi_2^4 \sim m_\mathrm{P}^4$. Given that the initial values of ϕ_1 allow the first stage of inflation to take place in some region, the inhomogeneities in ϕ_2 will inflate away (if ϕ_2 is not rough on arbitrarily small scales compared with the initial Planck scale). The value of ϕ_2 at a given location is frozen until the end of the first stage of inflation. Only if the frozen value of ϕ_2 is very precisely a certain value near $4.3 m_\mathrm{P}$ will the ramp on the final perturbation spectrum be at an astrophysically interesting scale. If initial values of ϕ_2 do indeed range over $|\phi_2(t_i)| \lesssim \lambda_2^{-1/4} m_\mathrm{P}$ (Linde 1983) and if, as is necessary for the amplitude of the density perturbations, $\lambda_2 \lesssim 10^{-14}$, then the desired range of $\phi_2(t_i)$ would be a tiny fraction of its possible range. Although the probability distribution of $\phi_2(t_i)$ over this range is unknown, it will undoubtedly be necessary to 'fine tune' the initial conditions — as well as the potential parameters — to place the ramp in the desired location, making this version of double inflation rather unattractive in our view.

In any case, making the ramp interestingly large to generate structure will result in a high plateau giving a high amplitude scale invariant spectrum for microwave background fluctuations which would violate the stringent bounds set by the Soviet RELICT experiment discussed *e.g.*, by Bond (1988).

Adding the coupling $\nu \neq 0$ in (3.1) does not aid matters appreciably. Even though one can arrange ν so that the effective m^2 vanishes in the potential valley minimum, Hubble drag forces the background field to lie outside the trough with $m^2 > 0$ always. This is true whether the field starts in the minimum (so there is no double inflation) or outside of it (so there is). Thus ϕ_1 declines exponentially quickly, becoming dynamically unimportant. The $\mathbf{P}_{\phi 2}$ that is left is still of the ramp plus plateau form.

To get enough e-foldings of inflation with a potential for ϕ_2 of the form eq. (3.1), it is necessary that the field start at a value several times the Planck mass. However, the Lagrangian term involving the Ricci scalar R becomes $\mathbf{L}_G = (m_{\mathrm{P}}^2/(16\pi) - \xi\phi^2/_2)R$, which can change sign if ϕ is too large. To avoid this catastrophe it is necessary that $\xi_2 \lesssim 0.002$; *i.e.*, that the field be effectively minimally coupled (or have $\xi_2 < 0$). A similar restriction would hold for ξ_1 if we were contemplating large initial values of ϕ_1 as well. Even if ξ_1 did not have to be zero, unless it is quite negative the potential gives a positive effective mass $m_{11}^2 = m_1^2 + 12\xi_1 H^2 + 3\lambda_1\phi_1^2 - \nu\phi_2^2$: the ϕ_1 fluctuations again inflate away.

The addition of cubic interaction terms to the potential (3.1) can give $m^2 < 0$ over a short range. Subsequently however m^2 becomes positive and the ϕ_1 fluctuations inflate away. Bardeen (1988) shows that before they inflate away they could induce non-Gaussian fluctuations in the field driving inflation, ϕ_2. In another approach to the generation of non-Gaussian perturbations, Grinstein and Wise (1987) considered a universe with massive axions which were a very small component of the dark matter, $\Omega_A \ll 1$, but which had very large amplitude fluctuations.

If a (non-chaotic) 'new inflation' potential is chosen, *e.g.*, of form $\lambda_2(\phi_2^2 - \sigma_2^2)^2/4$, with the initial value of the field near the origin, we would still have to tune $\phi_2(t_i)$ to get the ramp in the right place. In this case it is also not clear why ϕ_2 would start close enough to the origin for inflation to be feasible (Masenko, Unruh, and Wald 1984, but see Albrecht and Brandenberger 1985 who partially address this objection).

Starobinsky (1983) showed that the conformal anomaly of massless scalar fields (nonzero trace of the stress-energy tensor due to quantum effects) is o order R^2, where R is the scalar curvature, and this might drive inflation.

However the conformal anomaly terms that were most likely to appear in typical theories were shown to be unlikely to lead to inflationary behaviour. Starobinsky (1985) now considers the R^2 terms independently of the conformal anomaly, parameterizing the gravitational Lagrangian $\mathbf{L}_G = (m_P^2 16\pi)(R + R^2/(6M^2))$, by a small mass scale $M \ll m_P$. Another massive scalar field, the "scalaron," was introduced to get this mass scale $M \sim 10^{-5} m_P$. The Friedman equation is now significantly modified over the form with no R^2 term in the Lagrangian. Kofman, Linde and Starobinsky (1985) show that double inflation may follow provided that $M > (\lambda/6\pi)^{1/2}\, m_P$. Both the scalar field potential and the R^2 term drive the initial inflationary epoch, followed by an era when the scalar field dominates. Again a ramp plus plateau spectrum remains the generic outcome.

Kofman, Linde and Einasto (1987) suggested that the transition to the second phase might proceed by quantum tunnelling through a potential barrier, as in Guth's original 'old' inflation model, leading to bubbles being generated, superimposed upon a universe smoothed during the first inflationary epoch. Such fluctuations would certainly be non-Gaussian, but it would be extremely difficult to arrange for their amplitude to be just right to be useful to explain the large-scale texture we observe now.

We can conclude that no compelling case now exists either for a natural way to break scale invariance either with a mountain-like spectrum built above a scale invariant base, or for power laws differing appreciably from the scale invariant form.

Acknowlegements

We would like to thank L. Kofman, A. Starobinsky, R, Fakir, M. Mijic, and B. Unruh for informative discussions. The hospitality of J. Monaghan at Monash University while some of this paper was written is gratefully acknowledged by J.R.B. J.M.B. was supported by DOE grant DE-AC0G-81ER40048 at U.W., J.R.B. by a Canadian Institute for Advanced Research Fellowship and a Sloan Foundation Fellowship. J.R.B. and D.S.S. were also supported by the NSERC of Canada.

REFERENCES

Albrecht, A. and Brandenberger, R. 1985. *Phys Rev. D.* **31**, 1225.

Bardeen, J.M. 1980. *Phys Rev D.* **22**, 1882.

_____. 1988. in preparation.

Bardeen, J.M., Steinhardt, P.J., and Turner, M.S. 1983. *Phys Rev D.* **28**, 679.

Bardeen, J.M., Bond, J.R. and Salopek, D. 1987. *Proc. Second Canadian Conference on General Relativity and Relativistic Astrophysics.* ed. A. Coley and C. Dyer. Singapore: World Scientific. [BBS1]

_____. 1988. in preparation. [BBS2]

Bond, J.R. 1988. in *Large-Scale Structures of the Universe.* eds. J. Audouze, M.-C. Pelletan, and A. Szalay. Dordrecht: Kluwer Academic Publishers, p. 93.

Efstathiou, G. 1988, this volume.

Efstathiou, G. and Bond, J.R. 1986. *MN.* **218,** 103.

Grinstein, B. and Wise, M. 1987. preprint.

Kofman, L.A. and Linde, A.D. 1987. *Nuc Phys.* B**282,** 555.

Kofman, L.A., Linde, A.D., and Starobinsky, A.A. 1985. *Phys Lett.* **157B,** 361.

Kofman, L.A., Linde, A.D., and Einasto, J. 1987. *Nature.* **326,** 48.

Linde, A.D. 1983. *Phys Lett.* **129B,** 177.

Lucchin, F. and Matarrese, S. 1985. *Phys Rev D.* **32,** 1316.

Masenko, G., Unruh, W., and Wald, R. 1985. *Phys Rev D.* **31,** 273.

Peebles, P.J.E. 1988, this volume.

Starobinsky, A.A. 1983. In *Quantum Gravity.* p. 103. ed. M.A. Markov and P. West. New York: Plenum.

_____. 1985. *JETP Lett.* **42,** 152.

LARGE-SCALE ANISOTROPY OF THE COSMIC BACKGROUND RADIATION

NICOLA VITTORIO

Dipartimento di Fisica, Università dell'Aquila

ABSTRACT

In this paper we discuss a number of related issues concerning the large-scale anisotropy of the cosmic background radiation (CBR). We present cold dark matter model predictions for the peculiar acceleration field and for the CBR large-scale anisotropy. We discuss the number and the angular sizes of the hot spots expected in the CBR temperature distribution, under the assumption of Gaussian cosmic background fluctuations. We show that comparing the available CBR observations on large angular scales may constrain the density fluctuation spectral index. We also present an analysis of the Davies *et al.* (1987) data on CBR anisotropy and a Monte Carlo simulation of the experiment.

1. INTRODUCTION

The cosmic background radiation (CBR) temperature anisotropies expected in initially slightly inhomogeneous cosmological models have proved to be one of the most stringent constraints on different galaxy formation scenarios. In fact, CBR anisotropy studies bypass many of the uncertainties associated with the non linear models of small-scale galaxy clustering. Perhaps the most fundamental aspect of the CBR is that it probes epochs (~ 700,000 years after the Big Bang) and scales ($\gtrsim$ 200 Mpc) of the universe that are otherwise inaccessible.

In particular, the study of the large scale CBR temperature distribution is very important for at least two reasons. On one hand, the large sky coverage, provided by balloon and satellite experiments, ensures we observe a significant sample of the sky. One the other hand, the observational upper limits to the CBR temperature anisotropy can be interpreted independently of the presence or absence of late reheating of the intergalactic medium. In fact, the gravitational potential fluctuations responsible for the CBR temperature anisotropies (Sachs and Wolfe 1967) are independent of the location of the last scattering surface.

A detection of large scale CBR temperature fluctuations would be, of course, of fundamental interest, not only because it would confirm the current ideas on the origin and the evolution of the large-scale structure of the universe, but also because it would provide a direct measure of the amplitude of the initial density fluctuations in the framework of the linear theory. This, in turn, would constrain both the epoch of galaxy formation and the properties of the dark matter in the universe (possible kinds of weakly interacting particle, mass, lifetime, etc.).

The difficulty in detecting CBR temperature fluctuations has raised the question of devising the best observational strategy. For this goal, it is necessary to calculate for different models, not only rms values for the CBR temperature fluctuations on a given angular scale, but also the pattern expected for the CBR temperature distribution. It is thus possible to evaluate in a given cosmological scenario, for a given observational configuration, the probability of detecting CBR temperature fluctuations at a certain level, and then, to define the best experimental configuration and observational strategy.

The standard inflationary scenario (see, e.g., Turner 1987 for a recent review) predicts a flat universe and adiabatic scale invariant density fluctuations. Attention has been drawn in the last years to a very specific model for the formation and evolution of the large-scale structure of the universe, where the mass density is at the present dominated by cold dark matter (CDM; for a review see Blumenthal *et al.* 1984). This model combines the inflationary predictions with the observational constraints derived by small-scale galaxy clustering, if galaxies are biased to form in the highest peaks of the density field. But in this case there could be a problem with the predicted large-scale peculiar velocity field (Vittorio, Juzkiewicz, and Davis 1986; Juzkiewicz and Bertschinger 1988; but see also Kaiser 1988).

Here we discuss some related issues concerning the large-scale anisotropy of the CBR. The CDM model predictions for the large-scale peculiar acceleration field are presented in Section 2. The expected large-scale pattern of the CBR is discussed in Section 3. Observational upper limits and CDM model predictions for CBR temperature fluctuations are reviewed in Section 4 and Section 5 respectively. An analysis of the Davies *et al.* data (1987) and a Monte Carlo simulation of the experiment is presented in Section 6. Finally, a brief summary is given in Section 7.

2. CBR Dipole Anisotropy and Peculiar Acceleration Field

The dipole anisotropy is the only unambigously detected anisotropy of the CBR. Its amplitude ($\sim 10^{-3}$) is indicative of our motion relative to the comoving frame and implies a Local Group peculiar velocity of ~ 600 km s^{-1}, in a direction which is 45° away from the Virgo Cluster (for a recent review see Lubin and Villela 1987). Since the Local Group is falling into Virgo with a velocity of ~ 250 km s^{-1} (see, e.g., Yahil, 1985), the Virgo cluster as a whole moves relative to the CBR, with a velocity of ~ 500 km s^{-1}, in the general direction of the Hydra-Centaurus Supercluster. Galaxy formation scenarios can be tested by checking if the expected gravitational field of typical mass concentrations is sufficient to generate the observed Virgo Cluster peculiar velocity.

To compare the observations with the model predictions it is important to determine the minimum depth of a sample of galaxies necessary to define the comoving frame. If these galaxies trace the Hubble flow, the peculiar velocities of the Local Group relative to the sample and to the comoving frame should coincide. Otherwise, a coherent motion of the sample as a whole relative to the CBR is implied. From a given sample of galaxies, one can, for example, evaluate the acceleration exerted on the Local Group by all the galaxies in the sample. In order to verify that all the material outside the sample does not exert a significant acceleration on the Local Group, it is usual to calculate the cumulative acceleration exerted by galaxies in nested subsamples of increasing radii. If the cumulative acceleration converges to a finite value, then the inhomogeneities responsible for the dipole anisotropy should be contained in the sample and the sample itself should be at rest relative to the CBR.

The IRAS Point Source Catalogue provides for the first time a galaxy sample uniformly selected and with a nearly complete sky coverage. The depth of the IRAS sample is estimated to be $\lesssim$ 200 Mpc. The analysis of this sample has proven to be a powerful probe of the Hubble flow field (Yahil *et al.* 1986, Meiksin and Davis 1985, Villumsen and Strauss 1987, Strauss and Davis 1987). It has been shown that the distribution of galaxies in the catalogue exhibits a dipole anisotropy in reasonable agreement with the direction of the CBR dipole anisotropy to within ~ 20° to 30°.

If the IRAS galaxies trace the mass distribution, the gravitational acceleration exerted on the Local Group is proportional to (Yahil *et al.*, 1986)

$$\vec{d} = \frac{3}{4\pi} \int_0^\infty \delta(\vec{x}) \frac{\vec{x}}{x^3} \Phi(x, R) d^3x, \tag{1}$$

where $\delta(\vec{x})$ is the density fluctuation field and the window function

$$\Phi(x, R) = [1 + x^2/(2.4R^2)]^{-2.4}$$

describes the geometry of the sample. The parameter R gives the effective depth of the sample.

The rms values of $|\vec{d}|$ and of the bulk motion of the sample relative to be CBR are [Juszkiewicz, Vittorio, and Wyse (1988)]:

$$D^2(r, R) = \frac{9}{2\pi^2} \int_0^\infty dk\, P_s(k) W_g^2(k, R) \tag{2}$$

and

$$v^2(r, R) = \frac{1}{2\pi^2} \int_0^\infty dk P_s(k)\, W_v^2(k, R), \tag{3}$$

where $P_s(k)$ is the power spectrum smoothed on scale r: $P_s(k) = P(k) exp(-k^2r^2)$. For the explicit expression for W_g and W_v we refer to Juszkiewicz, Vittorio, and Wyse (1988). In the simple case of a volume limited sample, $W_g^2(k, R) = 1 - j_0(kR)$ and $W_v^2(k, R) = 3j_1(kR)/(kR)$, where j_0 and j_1 are the spherical Bessel functions. This shows that W_g acts as a high pass filter and supresses the contribution of long wave ($kR \rightarrow 0$) density perturbations. On the contrary, W_v acts as a low pass filter, suppresses the small ($kR \rightarrow \infty$) wavelength perturbations. So D and v provide us with information about the density inhomogeneities.

In a CDM dominated universe the density fluctuation spectrum is conveniently described by the following fitting formula: $P(k) = A\, k^n\, [1 + 6.8\, k / \Omega_0 + 72\, (k / \Omega_0)^{1.5} + 16\, (k / \Omega_0)^2]^{-2}$ (e.g., Davis *et al.* 1985). Here n is the primordial spectral index, Ω_0 is the density parameter, the wavenumber k is measured in Mpc^{-1} and the Hubble constant is assumed to be $H_0 = 50$ km s^{-1}/Mpc. The constant A defines the initial amplitude of the density fluctuations. The spectrum tilts down from the primordial slope ($\propto k^n$) at low k, to k^{n-4} for high k, as a consequence of the reduced growth of fluctuations entering the horizon during the radiation dominated era (see, e.g., Blumenthal *et al.* 1984). The variance of the density fluctuations is defined as (Peebles 1980)

$$\sigma_0^2(R) = \frac{1}{2\pi^2} \int_0^\infty dk\, k^2\, P(k) W_m^2(kR). \tag{4}$$

The window function $W_m(kR) = 3j_1(kR)/(kR)$ is introduced because of the averaging over a sharp edged sphere. The quantity R is the typical size of the proto-object, whose rms density fluctuation is $\sigma_0(R)$. The *a priori* unknown amplitude of the primordial density fluctuation spectrum, i.e. the constant A, is usually fixed by requiring $\sigma_0(8h^{-1}Mpc) = b^{-1}$. The biasing factor b is the ratio of the galaxy counts-to-mass fluctuations on the scale of 8 h^{-1} Mpc. The results below have been calculated with b = 1. If light does not trace mass they must be rescaled by b^{-1}.

Figure 1 shows D(r,R) and v(r,R) as a function of R, for a flat CDM dominated universe with scale invariant density fluctuations. Due to the lack of large-scale power in the model, D(r,R) converges at large scales to a finite value, which implies that density inhomogeneities on scales $\lesssim$ 100 Mpc are

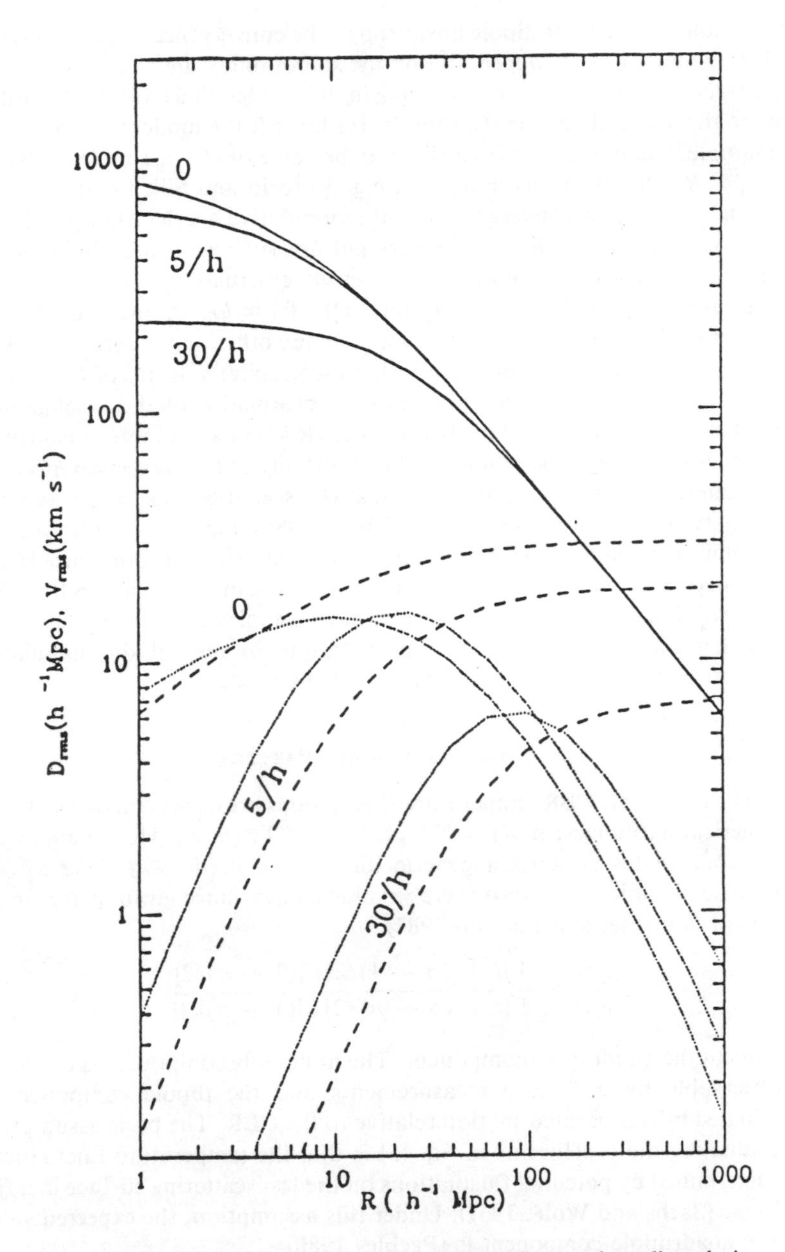

FIG. 1. Rms values of the density moment (dashed lines) and peculiar velocity (continuous line) of a sample defined by the window function $\Phi(x, R)$ (see text). Labels refer to different small-scale cutoffs. The dotted lines show $(d|\vec{d}|/dlog_{10}R)$ vs. R to determine the distance to the dominant fraction of the accelerating material. (From Juszkiewicz, Vittorio, and Wyse 1988).

responsible for the CBR dipole ansiotropy. The convergence values, however, differ by a factor ~ 4, depending on the smoothing scale r. This shows the importance of the small-scale clustering in this model. Since v(r,R) is sensitive only to the material outside the sample, for large R the model predictions for the sample bulk motion are insensitive to the value of r. The simple analytical law $v \propto R^{-1}$ holds in this limit (see, e.g. Vittorio and Silk 1985).

The criterion of convergence for the cumulative acceleration provides a condition which is definitely necessary but not sufficient to exclude the existence of large-scale, large amplitude density fluctuations. In the quite extreme case in which $P(k) \propto k^{-1}$, in fact, $D(r,R) \propto \log R$, and it may easily deceive the observer as being convergent. On the other hand, density fluctuations on large-scale are so important that the peculiar velocity of the sample is formally infinite. However, the additional information on the misalignment angle between the apex of the CBR and the IRAS galaxy dipole anisotropies can help to disentangle the problem. The probability of having a given misalignment angle in different galaxy formation scenarios has been recently investigated (Juszkiewicz, Vittorio, and Wyse 1988). Figure 2 shows, for a flat CDM universe, limits to the misalignment angle, at different confidence levels, as a function of the depth of the sample. If the sample has a depth ~ 200 Mpc, there is a probability of $\sim 95\%$ of having a misalignment angle $< 40°$, given that the amplitude of the dipole anisotropy and of the cumulative acceleration are equal to the rms values of the model.

3. CBR Large-Scale Pattern

The large scale CBR temperature fluctuations are conveninetly expanded in spherical harmonics: $\Delta(\hat{n}) = \Sigma_{l=2}^{\infty} \Sigma_{m=-l}^{m=+l} a_l^m \, Y_l^m(\theta, \phi)$. Here θ and ϕ are polar coordinates defining a generic direction $\hat{n}$ in the sky. The a_l^m are stochastic variables, with zero average value and variance given by (see, e.g., Fabbri, Matarrese, and Lucchin 1987):

$$\frac{|a_l|^2}{|a_2|^2} = \frac{\Gamma[l + (n - 1)/2]\Gamma[(9 - n)/2]}{\Gamma[l + (5 - n)/2]\Gamma[(3 - n)/2]} \tag{5}$$

in units of the quadrupole component. The monopole component is of course unobservable by difference measurements and the dipole component is dominated by our peculiar motion relative to the CBR. The basic assumption in evaluating the coefficients in Eq. (5) is that the temperature fluctuations are determined by potential fluctuations on the last scattering surface in a flat universe (Sachs and Wolfe 1967). Under this assumption, the expected value of the quadrupole component is (Peebles 1980)

$$|a_2|^2 = \frac{A}{16} \left(\frac{H_0}{c}\right)^{3+n} \frac{\Gamma(3 - n)\Gamma(\frac{3+n}{2})}{[\Gamma(\frac{4-n}{2})]^2\Gamma(\frac{9-n}{2})} . \tag{6}$$

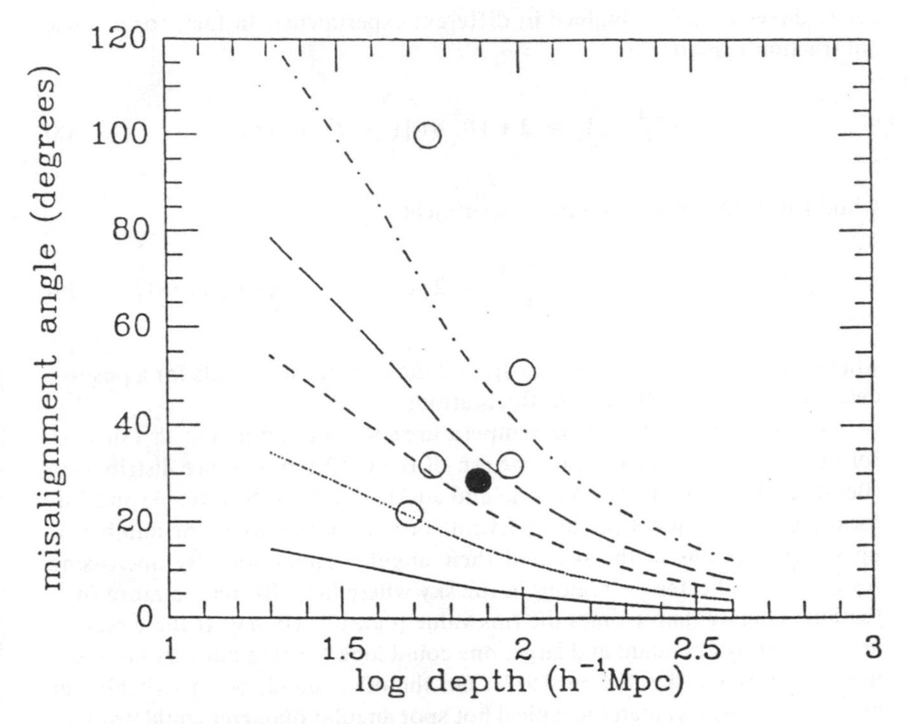

FIG. 2. Confidence levels on the maximum misalignment angle expected in a flat cold dark matter dominated universe: 5% (continuous line), 25%, 50%, 75%, 95% (dotted dashed lines). (From Juszkiewicz, Vittorio, and Wyse 1988).

Here A is the overall amplitude of the density fluctuations and n is the density fluctuation spectral index. The statistics of the CBR temperature field are completely described by the two point angular correlation function $C(\alpha) \equiv \langle \Delta(\hat{n}_1) \cdot \Delta(\hat{n}_2) \rangle$, where $\alpha = \cos^{-1}(\hat{n}_1 \cdot \hat{n}_2)$. It is usual to take into account the finite angular resolution of the antenna by modeling the antenna beam with a Gaussian profile of FWHM $= 2.35\ \sigma$. Then, the smoothed correlation function is (Scaramella and Vittorio 1988)

$$C(\alpha, \sigma) = \frac{1}{4\pi} \sum_{l=2}^{\infty} |a_l|^2 (2l + 1) P_l (cos\ \alpha)\ exp \left\{ -\left[\left(l + \frac{1}{2}\right)\sigma\right]^2 \right\} \tag{7}$$

The effect of the beam, as expected, acts as a low pass filter, which severely attenuates harmonics of order $l \gg 1/\sigma$. Knowing the shape of the correlation function allows one to derive from the observations the amplitude of the rms temperature fluctuations, i.e. $C^{1/2}(0, \sigma)$, as a function of n, and, then, to relate directly results obtained in different experiments. In fact, for a single subtraction experiment,

$$\left(\frac{\Delta T}{T}\Big|_{obs}\right)^2 \geq 2\, C(0, \sigma)\, [1 - R(\alpha, \sigma)], \tag{8}$$

while for a double substraction experiment,

$$\left(\frac{\Delta T}{T}\Big|_{obs}\right)^2 \geq C(0, \sigma) \left[\frac{3}{2} - 2\, R(\alpha, \sigma) + \frac{1}{2}\, R(2\alpha, \sigma)\right] \tag{9}$$

where $R(\alpha, \sigma) \equiv [C(\alpha, \sigma)/C(0, \sigma)]$, and the equality sign holds for a positive detection of CBR temperature fluctuations.

The knowledge of the CBR temperature correlation function also suffices for investigating the large-scale pattern of the CBR temperature distribution (Bond and Efstathiou 1987, Vittorio and Juskiewicz 1987, Scaramella and Vittorio 1988). Two quantities are relevant for the observations: the number of upcrossing regions in the sky and their angular dimension. By upcrossing regions one indicates the regions in the sky where the CBR temperature fluctuation is higher than ν times the rms value [i.e., $C^{1/2}(0, \sigma)$]. If these regions are sufficiently abundant and large, one could look for rare but very hot spots in the microwave sky (Sazhin 1985). On the other hand, beam-switching at an angular scale less than the typical hot spot angular diameter could produce a strong reduction in any detectable anisotropy. So, the knowledge of the typical hot spot angular diameter can at least be a guide in designing the observational configuration.

For $\nu \gg 1$, the number of hot spots is well approximated, by (see, e.g., Scaramella and Vittorio 1988):

$$N_{>\nu} = N_*(\sigma, n)\, \nu\, exp(-\frac{\nu^2}{2}). \tag{10}$$

The function $N_*(\sigma, n)$ is plotted in Figure 3a as function of σ, for different values of the spectral index n. The number of upcrossing regions scales as σ^{-2} for very steep density fluctuation power spectra. This dependence flattens out for negative spectral indices. In fact, lowering n reduces the small scale (relative to the large scale) power. Eventually the finite antenna beam size has pratically no effect. If the primordial fluctuations have a Zeldovich spectrum, the number of upcrossing regions expected in all the sky has an analytical expression (Vittorio and Juszkiewicz 1987):

$$N_{>\nu} = \frac{650}{\sigma^2} \frac{\nu\, e^{-\nu^2/2}}{[-ln\, 2\sigma + 3.78]} \quad . \tag{11}$$

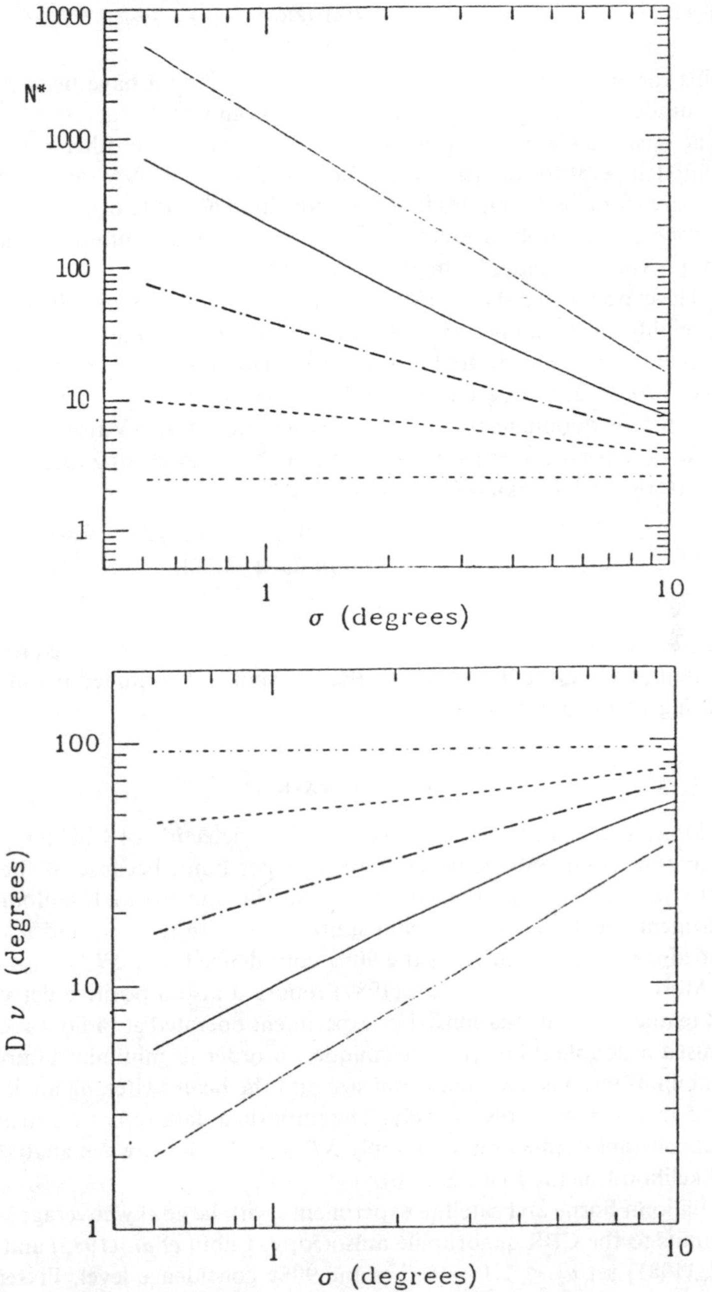

FIG. 3. The quantity N_* (3a), and the hot spot average diameter times the threshold ν (3b), vs. the antenna beam size σ, for different values of the spectral index: $n = 3$ (dotted), $n = 1$ (solid), $n = -1$ (dashed), and $n = -2.9$ (dotted—dashed). (From Scaramella and Vittorio 1988).

In this formula σ is measured in degrees. As it should have been expected, this number depends only upon the antenna beam size. In fact, this is the only characteristic scale which is introduced in observing the otherwise scale invariant temperature distribution. The dependence of $N_{>\nu}$ on the antenna beam size (for $n \gtrsim 0$) implies that hot spots in the CBR temperature distribution appear as unresolved sources: in fact, their number continuously increases on improving the antenna angular resolution.

The expected angular diameter of the upcrossing regions is inversely proportional to their number and depends on the threshold as ν^{-1} (for $\nu \gg 1$). The quantity $D \cdot \nu$ is plotted in Figure 3b. For fixed ν, D increases linearly with σ for $n \simeq 3$, but flattens out to 90° (formally for $\nu = 1$) when only the quadrupole is dominant (i.e., $n < -2$) (Scaramella and Vittorio 1988). For a Zeldovich spectrum, the angular diameter of the upcrossing regions is given by (Vittorio and Juszkiewicz 1987):

$$D = \frac{5.6}{\nu} \sigma[-ln\ 2\sigma + 3.78]. \tag{12}$$

Here σ is measured in degrees. Again, as should have been expected for a scale invariant process, the dimension of the hot spot is determined mainly by the smearing of the antenna beam.

4. Observations

Melchiorri *et al.* (1981) reported a positive detection of CBR temperature fluctuation, commonly considered as an upper limit, because of the uncertainties in possible galactic contamination. This far infrared, balloon borne experiment involved a single beam subtraction with $\alpha = 6°$ and $\sigma = 2°.2$. The deduced upper limit is, at the 90% confidence level, $\Delta T/T < 4 \cdot 10^{-5}$.

More recently Davies *et al.* (1987) reported also a positive detection of CBR temperature fluctuations. The experiment operated at radio wavelengths and used a double subtraction technique, in order to minimize atmospheric contaminations. The antenna beam size and the beamswitching angle were $\sigma = 3^o.5$ and $\alpha = 8^o.2$, respectively. The published data refer to a strip of the sky at constant declination, and imply $\Delta T/T = 3 \cdot 10^{-5}$, when analysed with the likelihood method (see Sec. 6).

Balloon borne and satellite experiments with large sky coverage give upper limits to the CBR quadrupole anisotropy. Lubin *et al.* (1985) and Fixsen *et al.* (1981) set $a_2 < 1.1 \cdot 10^{-4}$ at the 90% confidence level. Fixsen *et al.* placed also an upper limit to the CBR correlation function: $C(10^o < \alpha < 180^o, \sigma) < 1.37 \cdot 10^{-9}$. More recently, the RELIC satellite borne experiment (Klypin *et al.* 1986) set $C(20^o, \sigma) < 5.5 \cdot 10^{-10}$ and $a_2 < 4.75 \cdot 10^{-5}$ at the 95% confidence level. Assuming a scale invariant density fluctuation power spectrum, this limit is even more severe: $a_2 < 2.54 \cdot 10^{-5}$. The RELIC experiment sets an upper limit to the amplitude of the octupole component: $a_3 < 9.4 \cdot 10^{-5}$.

In order to have a first comparison among these different experiments we proceed as follows (Scaramella and Vittorio 1988). For a given value of the primordial spectral index, Eq. (8) or (9) allows us to derive from an observational upper limit to (a positive detection of) $\Delta T/T$, an upper limit to (a measure of) $C(0, \sigma)$. Since different experiments operate with different antennas, given a value for $C(0, \sigma)$, Eq. (7) can be used to evaluate $C(0, \sigma')$ for any value of the beam size σ'. So, for each n, one can have four values of $C(0, 3^o.5)$: three upper limits (determined by the Melchiorri *et al.* 1981, Fixsen *et al.* 1982, and Klypin *et al.* 1986 experiments) and one detection (from the Davies *et al.* 1987 experiment). Comparing these different values of CBR temperature variances (see Figure 4) shows that the Davies *et al.* (1987) detection is consistent with the Fixsen *et al.* (1981) and the RELIC upper limits only if $n > 1$. This comparison is fairly independent of the actual value of the density parameter Ω_0, as for $0.2 \lesssim \Omega_0 < 1$, the angles involved are smaller than that subtended by the curvature radius, $\sim 30^o \Omega_0^{1/2}$ (Peebles 1981, 1982).

With a similar strategy, one can predict the quadrupole anisotropy implied by the values of $C(0, \sigma)$, for a given spectral index, deduced from the Melchiorri *et al.* (1981), Fixsen *et al.* (1981), and Davies *et al.* (1987) experiments (see Figure 5). The latter experiment predicts a quadrupole which is higher than that derived by the Melchiorri *et al.* (1981) experiment for any value of the spectral index n, and it is consistent with the RELIC upper limit only if $n > 1.5$.

5. CDM Model Predictions

In this Section we discuss the large scale CBR temperature anisotropies expected in a CDM dominated universe, where the primordial power spectrum [see Sec. 2] is not necessarily scale invariant. Observational upper limits to the CBR anisotropy can provide bounds on n for a given Ω_0. On intermediate angular scale ($1^o < \alpha < 10^o$), the expected CBR temperature anisotropy, as it would be measured in a single subtraction experiment, is (Vittorio, Matarrese, and Lucchin 1988)

$$\frac{\delta T}{T}\Big|^2_{rms}(\alpha, \sigma) = A \, \frac{F^2(\Omega_0^{-1} - 1)}{\Omega_0^2} \, \frac{2}{\pi^{3/2}} \, \frac{\Gamma(\frac{3-n}{2})}{\Gamma(2 - \frac{n}{2})} \left(\frac{H_0 \Omega_0}{2c}\right)^{n+3} \sigma^{1-n}$$

$$\times \sum_{m=1}^{m=\infty} \frac{(-1)^{(m-1)}}{(m!)^2} \, \Gamma\left(\frac{2m + n - 1}{2}\right) \left(\frac{\alpha}{2\sigma}\right)^{2m} . \qquad (13)$$

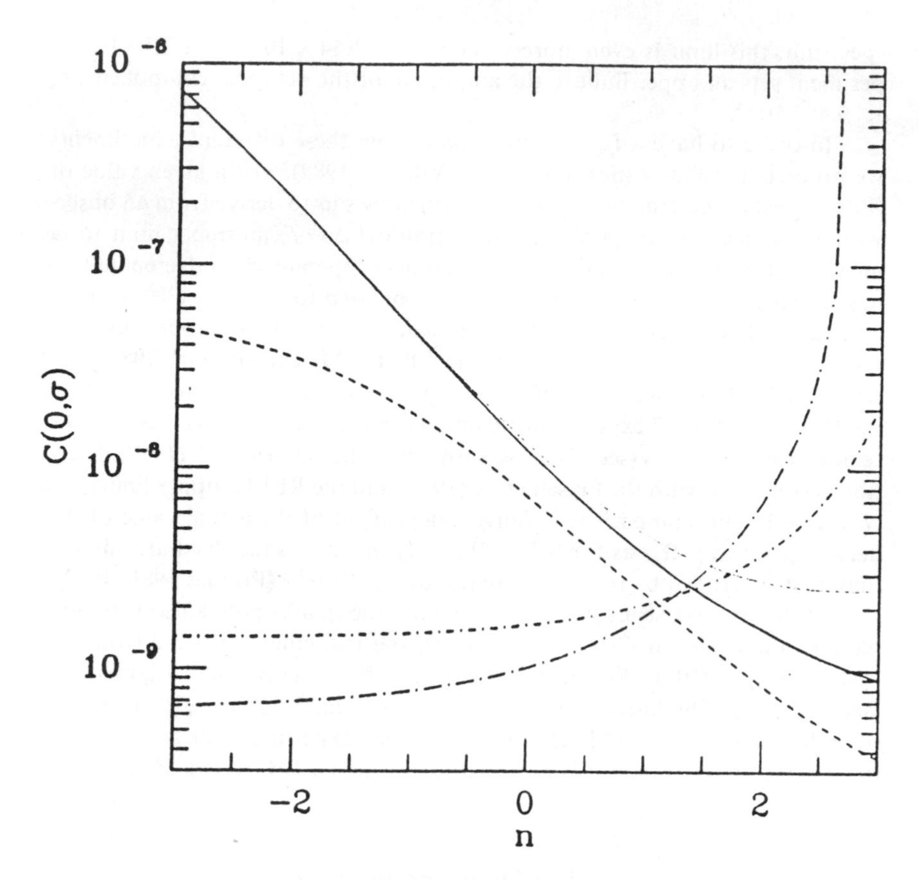

FIG. 4. CBR temperature variance derived from the Melchiorri *et al.* 1981 (dashed), Davies *et al.* 1987 (solid), Fixsen *et al.* 1987 (short dash—dotted), and the RELIC (long dash—dotted) experiments vs. the spectral index. (From Scaramella and Vittorio 1988).

The function $F(y) = 2y/[5 + 15y^{-1} + 15\sqrt{1+y}\,y^{-3/2} ln\,(\sqrt{1+y} - \sqrt{y})]$ takes into account the reduced growth of fluctuations in an open universe (Peebles 1981) and the constant A is the overall amplitude of the density fluctuations. If $n=1$, the scale invariance of the spectrum reflects the fact that the expected CBR anisotropy depends only on the ratio α/σ. Figure 6 shows the rms value of the CBR temperature anisotropy, expected in this case in a flat universe (Vittorio and Silk 1985). The two fields of view have an angular radius $\sim \sigma$ and are separated by an angle α. The expected anisotropy is bigger for large values of the ratio α/σ. At small values of α/σ the two antennas are essentially looking to the same region of the sky and the signal is strongly suppressed because of the difference procedure. In Figure 6 the continuous lines

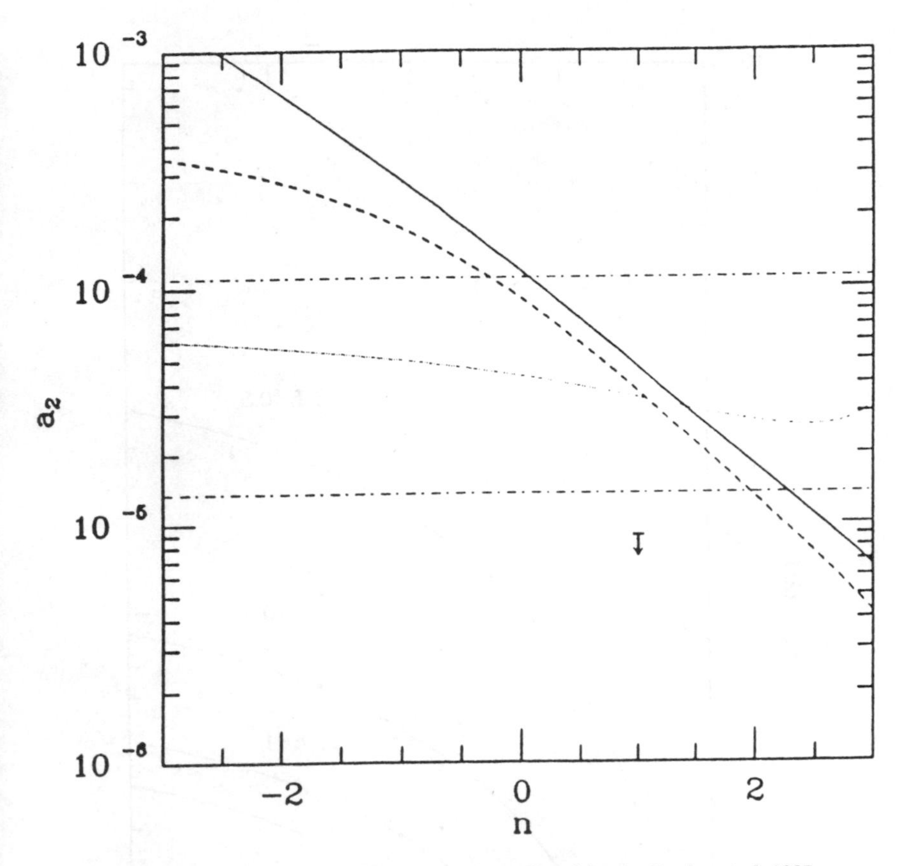

FIG. 5. The rms quadrupole component predicted by the Davies *et al.* 1987 experiment (solid) and the upper limits inferred from the Melchiorri *et al.* 1981 (dashed) and Fixsen *et al.* 1981 (dotted) experiments vs. the spectral index. The upper limits from Lubin *et al.* 1985, and Fixsen *et al.* 1981 (upper dot—dashed), and the RELIC (lower dot—dashed) experiments are also shown. (From Scaramella and Vittorio 1988).

(labeled c) are the predictions for a CDM dominated universe with scale invariant density fluctuations normalized as described in Sect. 2. If $n \neq 1$, at fixed α/σ, the anisotropy varies as $\sigma^{(1-n)/2}$. A good fit to the numerical result, for $H_0 = 50$ km s^{-1} Mpc^{-1}, is given by (Vittorio, Matarrese, and Lucchin 1988)

$$\frac{\delta T}{T}\Big|_{rms}(6^o, 2^o.2) \simeq 6 \cdot 10^{-5+(0.23\Omega_0-1.15)n}\,\Omega_0^{-1.15}. \tag{14}$$

Consistency with the Melchiorri *et al.* (1981) upper limit requires $n > 0.2$ and $n > 0.6$ for $\Omega_0 = 1$ and $\Omega_0 = 0.4$, respectively. For a scale-invariant spec-

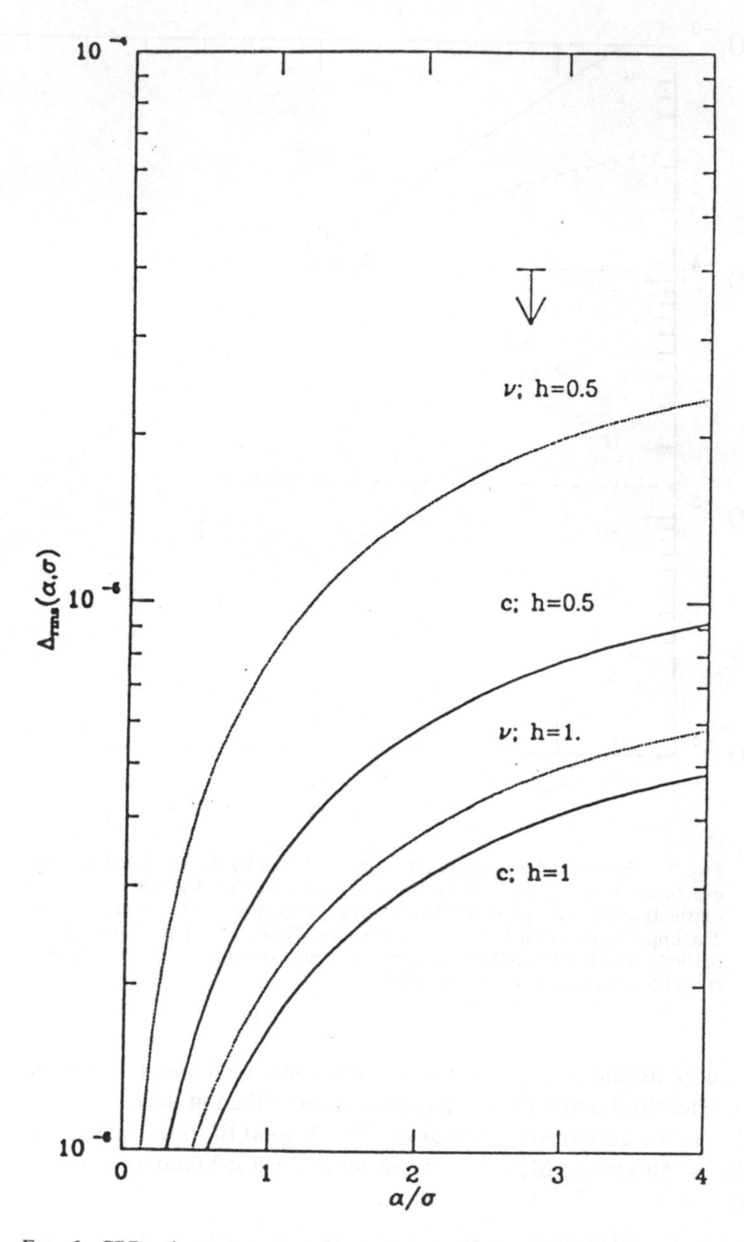

FIG. 6. CBR anisotropy expected in a CDM dominated universe for h = 0.5 (heavy continuous line) and for h = 1.0 (light continuous line); h is the Hubble constant in units of 100 km s^{-1}/Mpc. The arrow refers to the Melchiorri *et al.* 1981 upper limit. Dotted lines refer to a massive neutrino dominated universe. (From Vittorio and Silk 1985).

trum and $\Omega_0 > 0.4$ there is no conflict with the observed upper limit (see also Vittorio and Silk 1984).

The calculation of the intermediate scale anisotropy in Eq. (13) neglects the global space curvature. For the quadrupole anisotropy this approximation is not valid any longer, and, because of this, we restrict ourselves to the flat case. In a flat CDM dominated universe, where light traces the mass, the rms value of the quadrupole component is (Vittorio, Matarrese, and Lucchin 1988)

$$|a_2| = 1.3 \cdot 10^{-(4+1.285n)}. \quad (15)$$

This implies $n > 0$, in order to be consistent with the observational upper limit set by the Lubin *et al.* (1985) and Fixsen *et al.* (1981). Consistency with the RELIC upper limit requires $n > 0.35$. The amplitude of the higher multipoles are (Vittorio, Matarrese, and Lucchin 1988)

$$|a_l| = 1.3 \cdot 10^{-(4+1.285n)} \cdot \sqrt{\frac{\Gamma[l + (n-1)/2]}{\Gamma[(3+n)/2]}} \sqrt{\frac{\Gamma[(9-n)/2]}{\Gamma[l + (5-n)/2]}} \cdot \quad (16)$$

The RELIC upper limit to the octupole anisotropy constrains the primordial spectral index to be greater than 0.15.

6. Simulating a Microwave Anisotropy Experiment

The analysis of CBR anisotropy data with the likelihood method (see, e.g., Kaiser and Lasenby 1987) requires an explicit guess for the functional form of the CBR temperature correlation function. Davies *et al.* (1987) assumed a Gaussian functional form: $C(\alpha, \sigma) = C(0, \sigma)\, exp\, \{-\alpha^2/[2(2\sigma^2 + \theta_c^2)]\}$, with a sky intrinsic coherence angle of $\theta_c \simeq 4^o$, and found $C(0, \sigma) = 3.7 \cdot 10^{-5}$ (corresponding to $\Delta T/T(8^o.2, 3^o.5) = 3 \cdot 10^{-5}$).

The Davies *et al.* (1987) published data have also been analysed using the correlation function given in Eq. (7), for primordial spectral indices $-3 < n < 3$ (Vittorio *et al.* 1988). Figure 7 shows the value of the rms CBR temperature fluctuation, $C_{0M}^{1/2}$, obtained with the maximum likelihood method, and the likelihood ratio vs. the primordial spectral index n. For spectral indices between 0 and 1, the likelihood ratio shows a peak value $\sim$ 10. A similar value was found by Davies *et al.* (1987) analysing the data with the Gaussian correlation function. So, in the framework of the likelihood analysis, white noise or scale-invariant density fluctuation spectra are also consistent with the Davies *et al.* (1987) data, producing best fit values of $C_{0M}^{1/2} = 10^{-4}$ and $C_{0M}^{1/2} = 5.9 \cdot 10^{-5}$, respectively.

Unfortunately, in the present case, the likelihood analysis provides only the best estimates for the parameters of the model, but does not provide a

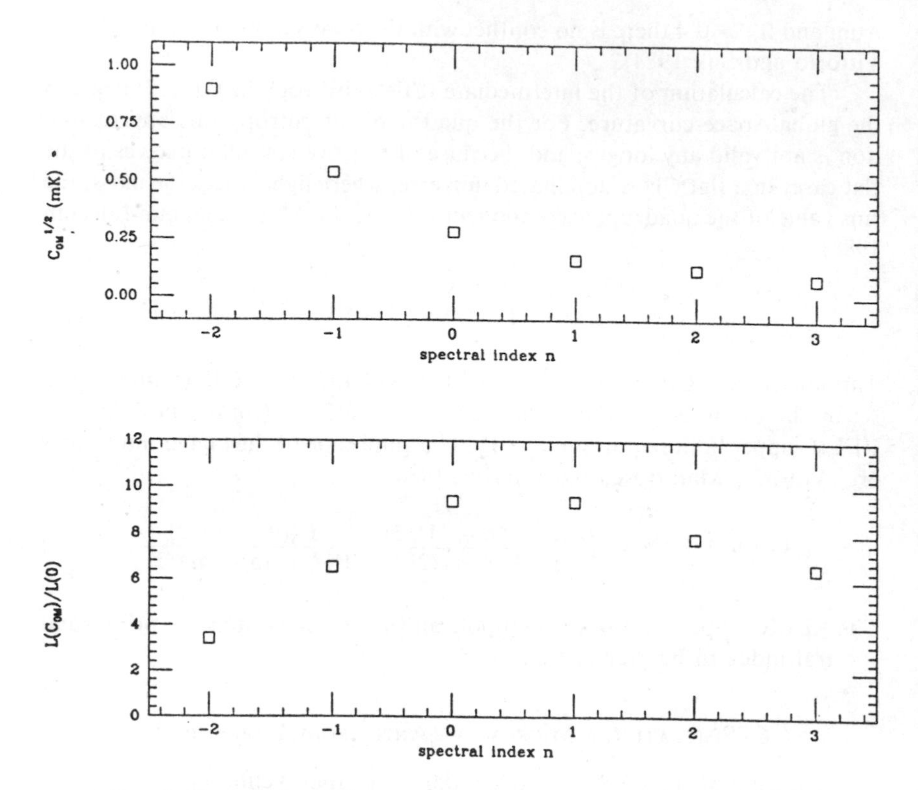

FIG. 7. The quantity $C_{0M}^{1/2}$, and the likelihood ratio vs. the perturbation spectral index assumed in analysing the Davies *et al.* 1987 data. (From Vittorio, de Bernardis, Masi, and Scaramella 1988).

confidence interval for them. For evaluating the probability distribution for the parameters one can simulate the observations of a theoretical sky, including effects as, e.g., receiver noise, modulation geometry, beam pattern, etc. This seems also a promising method for fairly comparing theory and observations, since it takes into account all the important experimental effects otherwise neglected in a purely theoretical analysis.

The Davies *et al.* (1987) experiment has been simulated in the following way (Vittorio *et al.* 1988). Two hundred different realizations of a strip of the theoretical sky have been generated. Each of these strips is $\sim 160^o$ wide, 20° thick, and has a resolution of $\sim 0^o.8$. They are realizations of the theoretical ensemble described by Eq. (7), where temperature fluctuations are determined by the Sachs and Wolfe effect: the density fluctuations are assumed to be Gaussian distributed, adiabatic, and scale-invariant. The rms temperature fluctuation of the theoretical ensemble is fixed to be $5.9 \cdot 10^{-5}$, as consistently found analysing the Davies *et al.* (1987) data for $n = 1$. Each strip is sampled as was done in the actual experiment, so a data set of 70 points per strip is generated. The receiver noise is simulated by adding to the theoretical temperature fluctuations a white noise of amplitude (rms) ~ 0.22 mK, the amplitude of error bars in the Davies *et al.* (1987) data.

Each data set has been analysed as the real data, obtaining values for $C_{0M}^{1/2}$ and likelihood ratios. Figure 8 shows the frequency of the results over the 200 realizations. If the temperature fluctuations are determined by scale invariant density fluctuations, the simulation of the experiment by Davies *et al.* (1987) shows that a likelihood ratio > 10 should be expected in 25% of the realizations. However the probability of having a likelihood ratio between 10 and 30 and $C_{0M}^{1/2} = 5.9\ (\pm 5.9)\ 10^{-5}$ is only 5%.

7. Summary

We presented CDM model predictions for the peculiar acceleration exerted on the Local Group by the material inside a galaxy sample of depth R. The convergence of the peculiar acceleration, exerted by the material contained

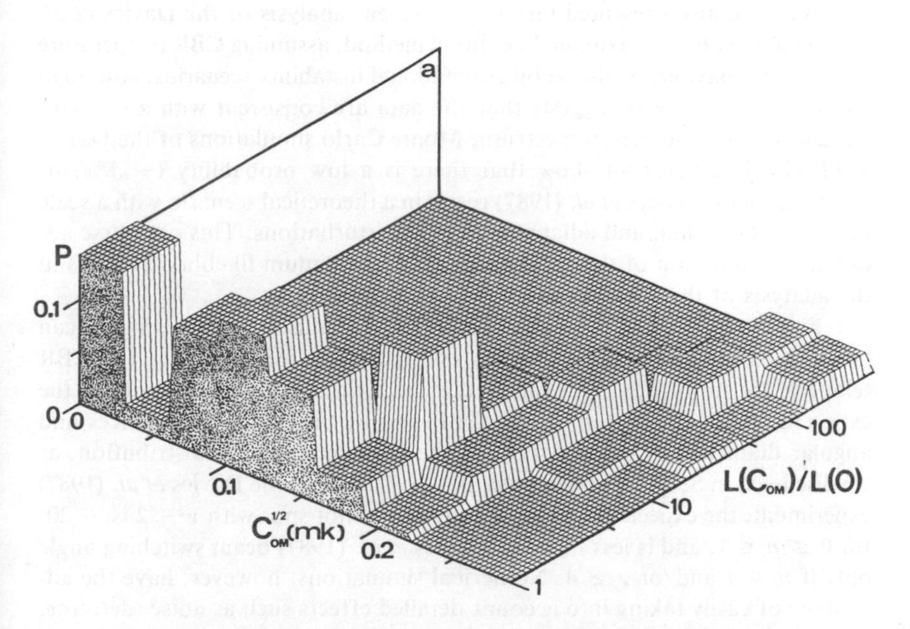

Fig. 8. Results of Monte-Carlo simulations of the Davies *et al.* 1987 experiment: the frequency of the results is plotted vs. the estimated $C_{0M}^{1/2}$ and the likelihood ratio, $L(C_{OM})/L(0)$. (From Vittorio, de Bernardis, Masi, and Scaramella 1988).

in nested subsamples, to a finite value is only a necessary condition for the absence of large-scale, large amplitude density inhomogeneities. To confidently assess this absence, we need to independently estimate the peculiar velocity of the considered sample relative to the comoving frame. We consider a flat CDM dominated universe, where light traces the mass, and we discuss the convergence of the peculiar acceleration. We also show that in this model the expected misalignment angle is less than 40° (95% confidence level) for a galaxy sample comparable with the IRAS sample.

Comparing among themselves different experiments can provide constraints on the primordial spectral index. In particular, if the Davies *et al.* (1987) experiment really provides a measure of the primordial density fluctuations and the RELIC result sets a realistic upper limit to the quadrupole anisotropy, then there is evidence for n to be greater than unity, if the density fluctuation spectrum on large scales is a single power law.

We have discussed the large scale CBR anisotropies expected in a CDM dominated universe and we have shown that the observational upper limits to the CBR anisotropy again constrain the value of the primordial spectral index. The tightest constraint comes from the RELIC upper limit to the quadrupole anisotropy which implies, for a flat CDM universe where light traces the mass, that $n > 0.35$.

We have also presented results of a recent analysis of the Davies *et al.* (1987) data, with the maximum likelihood method, assuming CBR temperature correlation functions predicted by gravitational instability scenarios. Although the likelihood analysis suggests that the data are consistent with a scale invariant density fluctuation spectrum, Monte Carlo simulations of the Davies *et al.* (1987) experiment show that there is a low probability ($\sim$ 5%) of reproducing the Davies *et al.* (1987) result in a theoretical scenario with a scale invariant, Gaussian, and adiabatic density perturbations. This of course addresses the question of the confidence of the maximum likelihood results in the analysis of the CBR data.

Finally, Monte Carlo simulations of the CBR anisotropy experiments can be used to define the experimental configuration for searching for CBR temperature anisotropy. This can be done as a first approach by studying the expected large-scale pattern of the CBR and by predicting abundances and angular diameters of the hot spots in the CBR temperature distribution, as we discussed in Sec. 3. For example, if $\sigma = 3^o.5$, as in the Davies *et al.* (1987) experiment, the expected angular diameter of a hot spot with $\nu \sim 2$ is $\sim 20^o$ for $0 \lesssim n \lesssim 1$, and is less than the Davies *et al.* (1987) beam switching angle only if $n > 1$ and/or $\nu \gtrsim 4$. Numerical simulations, however, have the advantage of easily taking into account detailed effects such as noise (detector, atmosphere, instrumentation, etc), sky coverage, modulation geometry, beam pattern, etc. Therefore, they seem to be extremely promising, from one side, for fairly comparing theory and observations and properly constraining theoretical models, and, from the other side, for assessing the best observational strategy for a positive detection of CBR temperature anisotropies.

ACKNOWLEDGEMENTS

The results presented at this meeting have been obtained in a collaborative effort with many friends. I am particularly indebted to Paolo de Bernardis, Roman Juszkiewicz, Francesco Lucchin, Silvia Masi, Sabino Matarrese, Roberto Scaramella, and Rosemary Wyse. I am also indebted to Alfonso Cavaliere for discussions and comments on this manuscript.

REFERENCES

Bardeen, J.M., Bond, J.R., Kaiser, N., Szalay, A.S. 1986. *Ap J.* **300**, 15.
Blumenthal, G., Faber, S., Primack, J., and Rees, M. 1984. *Nature.* **301**, 584.
Bond, J. and Efstathiou, G. 1987. *MN.* **226**, 655.
Davis, M., Efstathiou, G., Frenk, C., and White, S.D.M. 1985. *Ap J.* **292**, 371.
Davis, M. and Peebles, P.J.E. 1983. *Ap J.* **267**, 465.
Davies, R.D., Lasenby, A.N., Watson, R.A., Daintree, E.J., Hopkins, J., Beckman, J., Sanchez-Almeida, J., and Rebolo, R. 1987. *Nature.* **326,** 462.
Fabbri, R., Lucchin, F., and Matarrese, S., 1987. *Ap J.* **315,** 1.
Fixsen, D.J., Cheng, E.S., and Wilkinson, D.T. 1981. *Phys Rev Lett.* **44,** 1563.
Juszkiewicz, R. and Bertschinger, E. 1988, *Ap J Letters.* **344,** L59.
Juszkiewicz, R., Vittorio, N., and Wyse, R.M. 1988. in preparation.
Kaiser, N. 1988. preprint.
Kaiser, N., and Lasenby, A.N. 1987. preprint.
Klypin, A., Sazhin, M., Strukov, A., and Skulachev, D. 1986. *Pisma Astron. Zh.* in press.
Lubin, P. and Villela, T. 1987. in *Proceedings of the E. Fermi Summer School, Confrontation Bewteen Theories and Observations in Cosmology.* ed. F. Melchiorri. Varenna.
Lubin, P., Villela, T., Epstein, G., and Smoot, G. 1985. *Ap J Letters.* **298,** 1.
Meiksin, A., and Davis, M. 1985. *AJ.* **91,** 191.
Melchiorri, F., Melchiorri, B., Ceccarelli, C., and Pietranera, L. 1981. *Ap J.* **250,** L1.
Peebles, P.J.E. 1980. *Large-Scale Structure of the Universe.* Princeton: Princeton University Press.
_____. 1981. *Ap J. Letters.* **263,** L119.
_____. 1982. *Ap J.* **259,** 442.
Sachs, R.W. and Wolfe, A.M. 1967. *Ap J.* **147,** 73.
Sazhin, M.V., 1985. *MN.* **216,** 25p.
Scaramella, R. and Vittorio, N. 1988. *Ap J Letters.* in press.
Strauss, M. and Davis M. 1988. in *Large-Scale Structures of the Universe.* eds. J. Audouze, M.-C. Pelletan, and A. Szalay. Dordrecht: Kluwer Academic Publishers, p. 191.
Turner, M.S. 1987. *Lecture at the E. Fermi Summer School, Confrontation between Theories and Observations in Cosmology.* ed. F. Melchiorri. Varenna, Italy.
Villumsen, J. and Strauss, M. 1987. *Ap J.* **322,** 37.
Vittorio, N. and Silk, J. 1984. *Ap J Letters.* **285,** L39.
_____. J. 1985. *Ap J Letters.* **293,** L1.

Vittorio, N. and Juszkiewicz, R. 1987. *Ap J Letters.* **314,** L29.
Vittorio, N., Matarrese, S., and Lucchin, F. 1988. *Ap J.* **328**, 69.
Vittorio, N., Juszkiewicz, R., and Davis, M. 1986. *Nature.* **323,** 132.
Vittorio, N., de Bernardis, P., Masi, S., and Scaramella, R. 1988. *Ap J.* in press.
Yahil, A. 1985. in *The Virgo Cluster of Galaxies.* eds. O.G. Richter and B. Binggeli. Munich: ESO, p. 359.
Yahil, A., Walker, D., and Rowan-Robinson, M. 1986. *Ap J Letters.* **301,** L1.

PROBING COSMIC DENSITY FLUCTUATION SPECTRA

J. R. BOND

CIAR Cosmology Program, Canadian Institute for Theoretical Astrophysics, Toronto

ABSTRACT

The most popular approach to structure formation in the universe is to assume perturbations at early times form a homogeneous and isotropic Gaussian random field characterized by a density perturbation spectrum $\mathbf{P}_\varrho(k) \equiv (k^3/(2\pi^2)) \langle|(\delta\varrho/\varrho)(k)|^2\rangle$. In this paper I summarize the status of a wide variety of tests based on the theory of Gaussian random fields and approximate analytic treatments of dynamics which probe different wavenumber ranges of the $\mathbf{P}_\varrho(k)$ function : essentially every region of k-space is covered by at least one test. Granted complete freedom in the choice of $\mathbf{P}_\varrho(k)$, there is currently no clear need for non-Gaussian statistics. If we regard the interpretation of ξ_{cc} and cluster patches determined from the Abell catalogue as biased, the Great Attractor hypothesis as being an oversimplification of a complex large-scale velocity field, and the Tenerife $\Delta T/T$ observation at 8° as non-primordial, then the required spectrum looks very promising for the *minimal* $\Omega = 1$ cold dark matter (CDM) model, with structure growing from scale invariant adiabatic initial conditions. If these interpretations of the observations stand, it is not clear whether the $\mathbf{P}_\varrho(k)$ suggested by *all* of the data can arise in any natural setting.

1. INTRODUCTION

There are currently two main techniques that can be applied to the study of structure formation on large scales in models based on Gaussian initial conditions. One can evolve realizations of the linear initial conditions by N-body (and/or hydrodynamic) techniques, measuring various observables in the realizations to confront with observations. The alternative is the approach adopted here: dynamical evolution is approximated by the Zeldovich solution, using it to map the Gaussian statistics of the filtered fluctuations in unperturbed Lagrangian space to non-Gaussian statistics in Eulerian space. Abundances and correlation functions of selected classes of points chosen to correspond to classes of cosmic objects are then directly computed by ensemble averaging various statistical 'operators'. The formalism to treat the statistics of density peaks in the unevolved Gaussian random field is given in Bardeen *et al.* (1986, hereafter BBKS) and is extended by Bardeen *et al.* (1988, hereafter BBRS). See also Bond (1986, 1987a,b,c, 1988) and Bardeen, Bond, and Efstathiou (1987, hereafter BBE). The inclusion of Zeldovich dynamics in this formalism was given by Bond and Couchman (1987, 1988).

For such calculations to be feasible it is necessary to identify the complex objects that we can observe with points conditioned by relatively simple criteria, *e.g.*, with peaks of the Gaussian random density field smoothed over a scale R_f roughly corresponding to the mass of the objects, with the selection function $P(\mathrm{obj}|\nu, \nu_b)$ depending only upon simple parameters like the relative height ν of the peak (in units of $\sigma_\varrho(R_f)$, where $\sigma_\varrho(R_f)$ is the (linear) amplitude of the rms density fluctuations), and perhaps the height ν_b of a lower resolution 'background' (BBKS) field smoothed on $R_b > R_f$. For example, P might be unity if $\nu\sigma_\varrho(R_f)$ is above some critical value appropriate to the collapse of the R_f-scale peak *and* $\nu_b\sigma_\varrho(R_b)$ is below some threshold associated with merging, being zero otherwise. The $R_b \to R_f$ limit of this is similar to the Press and Schechter (1974) formalism for estimating the mass function of cosmic objects from a fluctuation spectrum. A variant of this approach appropriate to peaks is used in Section 2.1 and applied to the study of rare events in Section 2.2.

Estimations of correlation function evolution can be obtained by moving the peaks using the Zeldovich approximation: the position at time t of a peak initially at position t is $x(r, t) = r - a(t)[2H_0^{-2}/3]\,\nabla\Phi$ for $\Omega = 1$ universes, where Φ is the (linear and time independent) gravitational potential field, and a is the scale factor. The assumption in this model is that the peaks are primarily generated by short waves and follow the bulk flow of the mass described by $x(r, t)$, smoothed on some $R_b > R_f$ scale. The collapse of highly asymmetric peaks could also be followed using the Zeldovich approximation. However, if the peak is relatively spherically symmetric, as is the case if $\nu \gtrsim 2.5$, the Zeldovich approximation fails badly. In this case, a spherical approximation like the top hat model would do better: I adopt this in Section 2 to describe localized collapses.

In Section 2 I summarize the status of some of the quantities which can be calculated in the statistical theory. Table 1 lists a variety of tests of the fluctuation spectrum, some of which probe the shape of the spectrum over an extended range in k-space, and all of which probe the amplitude of the spectrum in some k-band. In this paper I emphasize calculations appropriate to Great Attractors: the velocity field in their presence and how probable they

TABLE 1
DIRECT PROBES OF THE FLUCTUATION SPECTRUM

PROBES	Wavenumber Range k^{-1} (h^{-1} Mpc)	Local Power Law Index	CDM Status
Shape and Amp Tests			
galaxy halos	0.2-1	$-2 \lesssim n \lesssim -1$	+
$\xi_{gg}(r)$, $w_{gg}(\theta)$ power law	0.5-10	$-2 \lesssim n \lesssim -1$	+
$w_{gg}(\theta)$ break	10-20	$n \gtrsim -1/2$	+
ξ_{cg}	5-20	$n \gtrsim -1/2$	+
ξ_{cc}	5-100	$n \sim -1$ (?)	—
ξ_{cg}, ξ_{cc} anisotropies	5-20	$n \lesssim -1/2$ (?)	+
Void/Wall Texture	5-15	$n \lesssim -1/2$	+ (?)
Local Flow	3-5		+
Large-Scale Flow	10-40 (wishful?)		?
Amplitude Tests			
Lyα clouds	0.04-0.1	$4 \lesssim \sigma_\varrho(.05) \lesssim 6$	+
gal form. epoch z_{gf}	0.2-7	$3 \lesssim \sigma_\varrho(.4) \lesssim 6$?	+
cl abund. $n_{cluster}$	4-7	$0.4 \lesssim \sigma_\varrho(5) \lesssim 1$	+
rare gal events *e.g.*, n_{GA}	8-20	$\sigma_\varrho(15) \sim .2$?	—
rare cl events $n_{cl\ patch}$	~ 20+	$\sigma_\varrho(25) \sim .1$?	—
CMB Test			
small $\theta \sim 2' - 20'$	3-30		+
interm $\theta \sim 2^o - 10^o$	200-1000		—
large θ (*e.g.*, RELICT)	700-6000		+
BN distortion	0.001-0.01		— (?)

Notes: The range of wavenumbers listed above is a crude indication of where the tests based on Gaussian statistical techniques are applicable. N-body studies can go in closer to collapsed structures with accuracy. The Gaussian tests listed are direct probes only for a hierarchical theory for which dynamical evolution on the scales in question is relatively mild, and are not applicable, for example, to highly nonlinear portions of $n = 0$ spectra. Other interpretations of some 'tests' are certainly possible. In particular the models of Lyα clouds (Rees 1986, Bond, Szalay, and Silk 1988) and the high redshift dust source (Bond, Carr, and Hogan 1988) of the CMB spectral distortion (Matsumoto *et al.* 1987) are certainly controversial. While texture comparisons remain visual tests, Gaussian field methods are at best qualitative indicators. Although galaxy halo formation really requires N-body studies (Frenk *et al.* 1985, Quinn, Salmon, and Zurek 1986), a good indication of the final shape is found by considering the cross-correlation $\xi_{pk,\varrho}$, describing the statistically averaged shape of matter that can infall onto a galaxy. $n = 0$ spectra lead to halos with profiles that are too steep. The $n > -1/2$ limits from w_{cg} and the w_{gg} break are not very serious, but are meant to indicate that $n = -1$ will *not* do. I expect that the w_{gg} constraint on n is stronger. Similarly the $n \lesssim -1/2$ limits indicate $n \gtrsim 0$ will definitely not do.

are. I mainly discuss these tests within the context of the $\Omega = 1$ CDM model, with an amplitude parameterized by a biasing factor b_g (BBKS, Bond 1986). I take $b_g = 1.44$ following Bond (1987a,b); this compares with the value $b_g \approx 2.5$ used by Davis *et al.* (1985) and $b_g = 1.7$ used by BBE. Kaiser and Lahav (1988) give evidence for $b_g \approx 1.5$ from the streaming data in these proceedings; Kaiser and Cole (1988) have also used a similar value.

The data probing shorter scales than for clusters, $k^{-1} \lesssim 5\ h^{-1}$ Mpc, seem to agree reasonably well with the predictions of this theory. Some of the large-scale tests suggest extra power might be required to explain them. A reasonable phenomenological approach is to add the extra power to a CDM base. This could arise physically if we were free to alter the initial conditions from the scale invariant form predicted in the standard inflation model. How difficult this is to arrange naturally is discussed by Bond, Salopek and Bardeen in these proceedings. There are three fiducial cases we consider: (1) CDM + plateau. with the initial spectrum having a ramp beginning at some wavenumber k_R which levels off to a scale invariant high amplitude plateau at some smaller wavenumber k_P. Such a spectrum could arise in 'double inflation'; (2) CDM + mountain, with the power falling down to either the standard CDM value for $k < k_P$, or even below it. A simple idealization of this which makes the calculations easy, CDM + spike, is to add a δ-function of extra power at a specific wavenumber: $\mathbf{P}_\varrho = \mathbf{P}_\varrho\,[CDM] + \sigma^2_{\varrho,P} k_P \delta(k - k_P)$: we show some of its effects in Section 2; (3) Initial spectra of fluctuations with arbitrary power laws. An example used in Section 2 is isocurvature baryon perturbations in an open universe with $n = -1$ fluctuations in the entropy-per-baryon ($n = -3$ is scale invariant), as advocated by Peebles (1987, 1988). The transfer function for such universes naturally imprints features at the horizon scale when the relativistic and non-relativistic matter have equal density. A mountain of power is generated followed by a precipitous $n = 3$ dropoff at low k. A fourth approach is to assume scale invariance but change the constituents of the universe, thereby modifying the transfer function which maps the initial fluctuation spectrum from the very early universe to the post-recombination one (BBE). None of these spectra can have power in the gravitational potential fluctuations above the power on very large scales: they cannot look like CDM + mountain; extra large-scale power looks more like the CDM + plateau model.

2. Direct Probes of the Fluctuation Spectrum

2.1 - Abundances Using a Spherical Peak Model

With the statistical theory one can estimate the abundances of objects ranging from dwarf galaxies and Lyα clouds through galaxies, groups, and clusters, given models for the selection functions $P(\text{object}|\nu, R_f)$. Although these abundances qualitatively agree with observations for the CDM theory, for a more precise test we would ideally like an expression for the differential mass function $n(>\delta_*, M)d\ln M$ for objects with overdensity δ (relative to the background) above δ_* (i.e., belonging to a 'catalogue' of contrast δ_*) and with mass between M and $M + dM$. As we emphasized in BBKS, no satisfactory theoretical derivation of a semi-analytic form for $n(>\delta_*, M)d\ln M$ exists, and I believe this conclusion still remains valid, although many people, including myself, have explored the consequences of simple ansatzes for the mass function. A better approach is to calibrate $n(>\delta_*, M)$ for many different catalogue thresholds δ_* using N-body experiments. Although a beginning has been made to such a calibration (Efstathiou and Rees 1987), the high M (rare events) and low M (destruction through merging) ends have not been well determined.

The two main analytic techniques utilize either the Press-Schechter (1974, hereafter PS) ansatz or the theory of peaks of a Gaussian field. I prefer the latter (Bond 1987a) which I think is better motivated physically, although most authors seem to prefer the former (*e.g.*, Schaeffer and Silk 1985, Coles and Kaiser 1988, Efstathiou and Rees 1988, Narayan and White 1988). Here I shall emphasize the peak formulation, and adopt Gaussian smoothing of the field. (Although top hat smoothing seems better motivated physically, problems arise since the short wavelength limit of the top hat filter of the density fluctuation spectrum only drops as a power $\sim (kR_{TH})^{-4}$; the top-hat-filtered CDM random density field is then only once differentiable and has an infinite density of peaks of tiny radius for a flicker noise unfiltered spectrum at high k). The ideal approach from the point of view of the theory of Gaussian fields is to consider peaks on a hierarchy of resolution scales $\{R_{f,s}\}$. In the limit in which the resolution scales become densely packed, a reasonable (although not rigorously derivable) expression for $n(>\delta_*, M)$ is

$$n(>\delta_*, M)d\ln M = \mathrm{N}_{pk}(\nu\delta_*, R_f)\ \frac{d\nu_{\delta_*}}{d\ln M}\ d\ln M\,, \qquad (2.1)$$

the number density of peaks on the scale R_{fs} with overdensities in excess of δ_* which do not have overdensities when smoothed on the 'background' scale $R_{f,s+1} = R_{fs} + dR_f$ as large as δ_*. Here $\nu_{\delta*}$ is the height of R_f-scale peaks which have $\delta = \delta_*$ at the redshift at which $n(>\delta_*, M)$ is determined. $N_{pk}(\nu)$ is the differential peak density (BBKS).

To evaluate this expression the parameters of the smoothed Gaussian field ν and R_f must be related to the (possibly nonlinear) overdensity δ and the mass M by adopting a simplified model of the dynamics of collapse of the matter surrounding the peak. Here I shall use the spherical top hat model. For universes with $\Omega = \Omega_{nr} = 1$, where Ω_{nr} is the density in non-relativistic particles, the relation between $\nu\sigma_\varrho$ and the actual nonlinear overdensity is accurately given by

$$1 + \delta \approx (1 - \nu\sigma_\varrho/f_c)^{-f_c},\ f_c = 1.68647,\ \nu\sigma_\varrho(z) < f_\upsilon \approx 1.606, \tag{2.2a}$$

$$X(r, t) = ar/(1 + \delta)^{1/3}, \tag{2.2b}$$

$$V(r, t) = Har(1 - \nu\sigma_\varrho/f_c)^{f_c/3-1}(1 - \nu\sigma_\varrho/f_{ta}),\ f_{ta} = f_c/(1 + f_c/3) = 1.08 \tag{2.2c}$$

The only failure in these expressions is the velocity V near turnaround, which occurs when $\nu\sigma_\varrho = 1.06$ in the exact model rather than at 1.08. Suppose collapse ceases once the overdensity rises to a value δ_υ with a corresponding value of $f_\upsilon \equiv \nu\sigma_\varrho$ obtained from eq. (2.2a). After this redshift, the subsequent evolution of the overdensity is given by the $\sim (1 + z)^{-3}$ law:

$$1 + \delta = (1 + \delta_\nu)\,(\nu\sigma_\varrho(z)/f_\nu)^3,\ \nu\sigma_\varrho(z) > f_\upsilon. \tag{2.3}$$

The 'classical' value for the virialized density of the top hat model with $\Omega_{nr} = 1$ is $\delta_\upsilon = 170$, obtained with $f_\upsilon = 1.606$, the value quoted in eq. (2.2a).

The relation between the initial comoving top hat radius of the cloud required to determine the mass M, the height ν of the peak and the filtering radius R_f is complex and indeed time dependent as infall occurs. In Bond (1987a), I adopted the mass associated with a Gaussian profile of comoving radius $r_s \equiv f_s(\nu, R_f)R_f$:

$$M = (2\pi)^{3\text{-}2}\,\bar{\varrho}_0 r_s^3(R_f, \nu), \quad \text{with} \quad r_s \sim 1.5R_f, \tag{2.4}$$

where $\bar{\varrho}_0$ is the current background density, and argued that $1 \lesssim r_s/R_f \lesssim 1.7$, with higher values appropriate for higher ν peaks and smaller filtering radii. As a simple compromise I chose a ν-independent value $f_s = 1.5$. The faint end of the PS fit to the numerical results for CDM given by Efstathiou and Rees (1987) indicate $f_s \sim 1.3$ might be a better choice. With these approximations for M and δ, eq.(2.1) becomes

$$n(> \delta_*, M)d \ln M = N_{pk}(\nu_{\delta*}, R_f)\nu_{\delta*}\frac{\gamma^2 R_f^2}{R_*^2}\,d \ln M,\ \langle k^2\rangle \equiv 3\gamma_f^2/R_*^2. \tag{2.5}$$

The quantities γ and R_* introduced by BBKS are related to the average $\langle k^2\rangle$ of the density spectrum as given in eq.(2.5). N_{Pk} is replaced by $2\exp[-\nu_{\delta*}^2/2]/(2\pi)^{1/2}\,[(2\pi)^{3/2}r_s^3]^{-1}$ in the PS theory, where f_s is typically chosen to be 1 rather than the 1.5 adopted here.

Fig. 1 illustrates how eq.(2.5) for 'virialized' structures with $\delta > 170$ compares with the data for the Gott-Turner groups and Abell clusters. The main

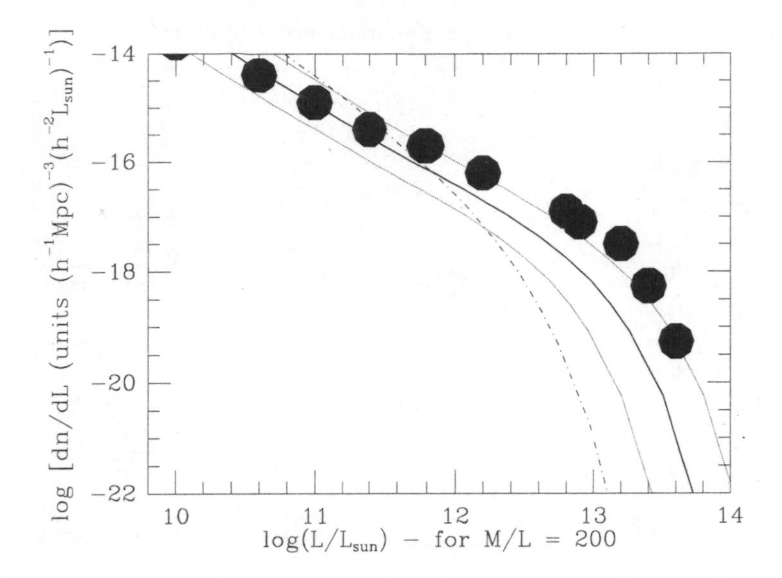

FIG. 1. The luminosity function computed for the CDM theory with biasing factor $b_g = 1.44$ and $f_s = 1.5$, assuming a fixed M/L ratio for the groups and clusters. The heavy solid line corresponds to $M/L = 200$, and the dotted lines have $M/L = 400$ (left) and 100 (right). The dot-dash line is the PS result (without the factor of 2 fudge usually included) with $f_s = 1$; $f_s = 1.5$ would shift it over toward the peak curve, although the shape would remain different. The black dots are data for the Gott-Turner groups and Abell clusters. The data for the groups is not particularly reliable.

conclusion to be drawn is that the CDM model does not obviously fail to reproduce the hierarchy of observed virialized structures in the universe. The more precise group analysis of Nolthenius and White (1987) using N-body realizations of the CDM model gives even better fits to the group abundances they determine from the CfA redshift survey.

2.2 - Abundances of Rare Events: Great Attractors and Cluster Patches

In this section, I estimate the abundances of very large-scale objects with relatively small overdensities $1 \lesssim \delta_* \lesssim 5$ which have not yet turned around. Equations (2.2) and (2.5) would still be applicable. Thus $\delta = 0.8$ if $\nu\sigma_\varrho = 0.5$, and $\delta = 3.5$ if $\nu\sigma_\varrho = 1$. Figure 2 illustrates how $n(> \delta_*, M)$ falls to high mass for varying δ_*. Note that since N_{pk} drops at low ν, for small M the mass function with $\delta > 1$ can fall below that with $\delta > 5$. The PS formalism suffers from this as well although not as severely. However, the high M values should provide a reasonable estimate of the rare event density; this will be true whether the PS or peak technique is used since the behaviour on

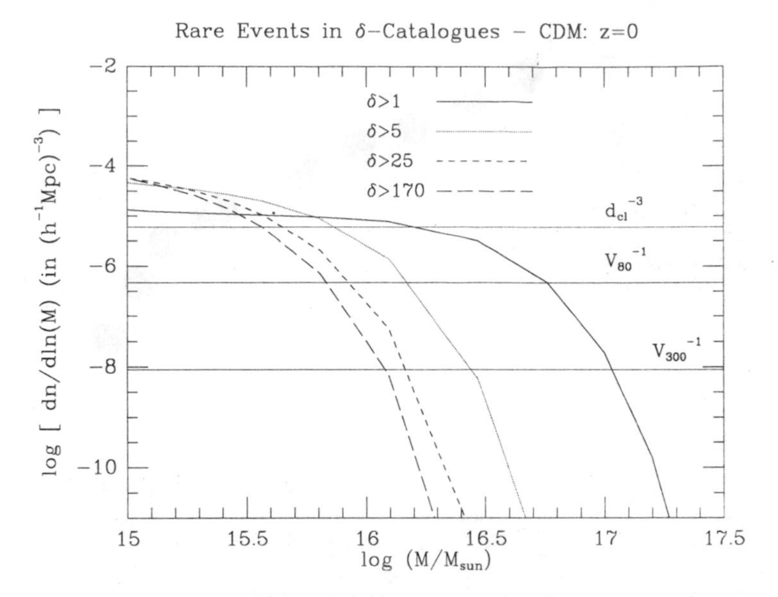

FIG. 2. The mass function $n(> \delta_*, M)$ derived from eq.(2.5) for the CDM model with $b_g = 1.44$ and $f_s = 1.5$ for the values of δ_* shown. For comparison the density of ($R \geq 1$) Abell clusters, $d_{cl}^{-3} = (55\ \mathrm{h}^{-1}\ \mathrm{Mpc})^{-3}$, and the densities corresponding to one object in volume limited samples of radius $R_{sample} = 80$ and 300 h^{-1} Mpc are indicated, corresponding roughly to the elliptical survey region and to Tully's cluster sample.

the rare event side is dominated by the $\exp[-\nu_{\delta_*}^2/2]$ tail characteristic of Gaussians: when it begins to dominate the fall is dramatic. A difficulty of applying this technique is that abundances of objects of a given mass will then be very sensitive to the choice of f_s.

If a Great Attractor (GA) is identified with a peak of the Gaussian field with overdensity $\sim \delta_*$ and 'mass' in a logarithmic range $d \ln M$ about M, then the probability of finding a GA in a volume-limited sample of radius R_{sample} is

$$P(GA) = \frac{4\pi}{3} R_{\mathrm{sample}}^3\, n(> \delta_*, M).$$

For the elliptical survey, I take $R_{\mathrm{sample}} = 80\ \mathrm{h}^{-1}$ Mpc. The mass excess is $\sim 5 \times 10^{16}\ \mathrm{M}_\odot$ for the GA parameters presented in this conference by Faber and Burstein (1988). According to Figure 2, for the CDM model (with biasing factor $b_g = 1.44$) there would be about one $\nu\sigma_\varrho = 0.6$ object in the survey, but certainly no $\nu\sigma_\varrho = 1$ objects. In Section 2.6 I show that if the $\sim 600\ \mathrm{km\ s}^{-1}$ flow experienced by Tully's Local (Coma-Sculptor) Cloud —

of which the Local Group is a small part — were entirely due to a GA, then $\nu\sigma_\varrho \approx 1$ would be required. However if about half of this velocity is due to other pullers and pushers (*e.g.*, the Local Void) which just happen to align the flow pattern in the Centaurus flow direction, then $\nu\sigma_\varrho \approx 0.5$ would suffice: the GA would not necessarily be too rare for the CDM model. This is sensitive to f_s: $f_s = 1$ would give a probability about a percent; hence to be more precise requires calibration of the M — R_f relation. The $f_s = 1.5$ choice probably errs on the pro-CDM side.

These rare events arise from high ν peaks which are approximately spherical, justifying our use of the top hat model. With $f_s = 1.5$, 5×10^{16} $M_\odot$ corresponds to $R_f \sim 12$ h^{-1} Mpc for which $\sigma_\varrho = 0.17$, hence ν is 6 for $\nu\sigma_\varrho = 1$, and 3 for $\nu\sigma_\varrho = 0.5$. The three eigenvalues of the shear tensor for high ν are approximately (Bond 1987b) $\lambda_1 \approx (\nu\sigma_\varrho/3)(1 + 3e_\upsilon)$, $\lambda_2 \approx \nu\sigma_\varrho/3$ and $\lambda_3 \approx (\nu\sigma_\varrho/3)(1 - 3e_\upsilon)$. For given ν, the average value of e_υ is $\langle e_\upsilon | \nu \rangle \approx (3\pi/40)^{1/2}\nu^{-1}$; hence $3e_\upsilon \sim 0.25$ for $\nu = 6$ and 0.5 for $\nu = 3$, with non-negligible deviations from a spherical flow predicted in the latter case. However, the spherical δ — ν relation still provides a reasonable estimate. In a pancake approximation (which will not be accurate for such small values of e_υ), $\delta \approx \nu\sigma_\varrho[1 - \nu\sigma_\varrho(1/3 + e_\upsilon)]^{-1}$: the overdensities predicted are not very disimilar from the spherical model.

The cluster patches advocated by Tully could also arise from Gaussian random fields. The large biasing factor of clusters implies that a great increase in the number of clusters will occur for relatively modest overdensities:

$$1 + \delta_{cl} \sim (1 + \delta_\varrho) \exp[(b_{cl} - 1)\nu\sigma_\varrho] \exp[-(b_{cl} - 1)^2 \sigma_\varrho^2/2]. \quad (2.6)$$

For clusters in the CDM model $b_{cl} \approx 6$, hence $\nu\sigma_\varrho = 0.5$ leads to $\delta_{cl} \approx 20$, and $\nu\sigma_\varrho = 1$ gives $\delta_{cl} \approx 670$. In a sample of radius $R_{sample} = 300$ h^{-1} Mpc there would be about one $\nu\sigma_\varrho = 0.5$ peak on scale $R_f \approx 20$ h^{-1} Mpc; over $R_f \approx 40$ h^{-1} Mpc, there would not even be one of these patches in the observable universe. By contrast consider the isocurvature baryon model with $\Omega = \Omega_B = 0.4$ normalized by mass traces light ($b_g = 1$ for J_3 normalization at 10 h^{-1} Mpc). In this case $b_{cl} \sim 4$. Over 300 h^{-1} Mpc there would be 3 regions with the cluster density enhanced by 7 in patches of $R_f = 40$ h^{-1} Mpc. This model would have quite a few $\nu\sigma_\varrho = 1$ regions within the elliptical sample volume as well.

2.3 - Two-point Correlation Functions

The correlation function for galaxies and clusters ξ_{gg} and ξ_{cc} and the cross correlation between clusters and galaxies ξ_{cg} can be computed assuming the above models for the objects as peaks of the smoothed Gaussian random field. (See BBKS, BBE and BBRS for details). The theoretical predictions for the

CDM theory are in agreement with the observed ξ_{gg} and the new determination of ξ_{cg} found by Lilje and Efstathiou (1988), but the CDM ξ_{cc} is definitely too low relative to the Bahcall and Soneira (1983) result. The shape of the correlation functions on large scales is very much conditioned by the shape of the fluctuation spectrum for CDM as it steepens towards the primordial Zeldovich shape as one passes from cluster scales where the local power law index of the fluctuation spectrum is $n \approx -1$ to $n = 1$ at large scales. The reported $\xi_{cc} \sim r^{-2}$ shape suggests a $n = -1$ ramp beyond cluster scale, delaying the drop to $n = 1$ to beyond $\sim$ 100 h^{-1} Mpc. However, the Groth-Peebles (1977) break at $\theta \sim 3^o$ in the angular correlation function $w_{gg}(\theta)$ evaluated from the Shane-Wirtanen catalogue of the Lick survey, which has been confirmed by Maddox and Efstathiou (1988) in their Southern Sky Catalogue, implies such a power law extension is incompatible with w_{gg} (Bond and Couchman 1987). Further, the amplitude of ξ_{cg} would have to be more similar to the original Seldner and Peebles (1977) value than the smaller one determined by Lilje and Efstathiou (1988). Thus if ξ_{cc} persists in maintaining its current power law form to $\sim$ 50 h^{-1} Mpc then it appears it cannot be compatible with a biasing picture based on Gaussian initial fluctuations.

I now consider the effect of extra power in a simplistic model which mimics the CDM + mountain spectrum and is somewhat similar to the isocurvature baryon spectrum. If a pulse of power is added at wavenumber k_P, then added to the usual scale invariant ξ_{ij} will be a term $b_i b_j \sigma_\varrho^2(k_P) \sin(k_P r)/(k_P r)$, where i and j can be c or g. The amplitude of this extra piece would then be approximately constant, of magnitude $\sim 0.3(\sigma_\varrho(k_P)/0.1)^2$ for clusters, out to $\sqrt{3} k_P^{-1}$. The addition of such a term could therefore mimic the large ξ_{cc} obtained, assuming we are not overly sold on the reliability of the r^{-2} power law. However such a term might already be ruled out. Lilje and Efstathiou (1988) find no evidence for such behaviour in ξ_{cg} whether or not they remove large-scale gradients from the Lick catalogue. Although if k_P^{-1} is large enough this contribution to w_{gg} would not have shown up in the Groth-Peebles determination since large-scale gradients are subtracted, the Maddox-Efstathiou determination does not require gradient subtraction: they find no evidence for such a term. These are both high precision tests and even the low $\xi_{cg} \sim 0.07$ and $\xi_{gg} \sim 0.02$ levels predicted could probably have been picked up.

2.4 - Two-point Function Anisotropies

The Binggeli (1982) effect (see also Struble and Peebles 1985, Rhee and Katgert 1987) that clusters are preferentially aligned along their long axis, and the Argyres *et al.* (1986) effect that galaxies are also preferentially found in the plane of the long axis of clusters out to $\sim$ 15 — 20 h^{-1} Mpc is predicted by the theory of Gaussian statistics (Bond 1986, 1987c). The constraint imposed by requiring a cluster to be oriented along a specified axis forces the density waves that had to constructively interfere at the cluster peak to also preferentially constructively interfere along the long axis compared with other directions. For fluctuation spectra shallower than $n = 0$, long waves as well as short ones build the cluster, hence the effect can persist weakly out to large distances from the collapsed cluster. This coherent wave bunching will make the overdensities of clusters, of galaxies and of the mass field largest along the long axis and smallest along the short axis for a given radius. In addition the same effect will tend to align the long axes of the clusters so that they point toward each other. Further, the orientation of a central smaller scale peak will be correlated with the imposed cluster scale orientation, especially if the two scales are not inordinately disparate. A cD galaxy might correspond to Gaussian smoothing of $R_f \sim (1 - 2)\ h^{-1}$ Mpc, which is not very far from the $\sim 5\ h^{-1}$ Mpc appropriate for clusters, suggesting a plausible origin of the Carter and Metcalfe (1981) effect that cD galaxies tend to be aligned with their host clusters.

The critical question is whether these effects are quantitatively large enough to explain the observed effects and whether they can be used to differentiate among possible fluctuation spectra. Little alignment will occur if the spectral index is $n > 0$: $n = 0$ spectra are uncorrelated and $n = 1$ spectra are anticorrelated; such indices can be ruled out over the range that alignments are observed.

If ξ for oriented clusters is expanded in spherical harmonics then, in the linear regime, the nonzero components will be monopoles ($\ell = 0$) for ξ_{cg} and ξ_{cc}, the usual correlation function when no account is taken of the orientation; a quadrupole ($\ell = 2$) for ξ_{cg}, and a quadrupole and octupole ($\ell = 4$) for ξ_{cc}. Nonlinear corrections to the statistics or the dynamics add all higher even multipoles, but these will be unimportant in the far field. For clusters the far field is $r \gtrsim 15\ h^{-1}$ Mpc. If the mass correlation function is a power law $\xi \sim r^{-\gamma}$ in the far field then $\xi_{l=2} \sim r^{-(\gamma+2)}$ falls off faster, being negligible at distances where the monopole component can still be large. Nonetheless the anisotropy amplitude can be quite large to many times the filtering radius, similar to the alignment levels reported. (See BBRS for detailed calculations.)

2.5 - *Large-Scale Streaming Velocities with Great Attractors and Repulsors*

The most direct and probably the best method to relate the data on large-scale streaming velocities to theory is to average the velocity over some selection function, forming a bulk velocity v_{bulk}, which has a distribution $P(v_{bulk}|R_f)$ depending upon the specific choice of selection function, *e.g.*, a Gaussian of scale R_f. The work discussed by Kaiser and Lahav (1988) and Szalay (1988) considerably extends this approach and places it on a firm theoretical foundation. The technique is unbiased in that our location is taken to be random. Alternatively one can use the data in an *a posteriori* fashion to recognize that we may not be in a typical region and condition the velocity field determination by the constraint that there may be large entities such as a GA surrounding us. One can compute conditional probabilities $P(v(r)|GA)$, $P(v(r)|GA$, Local Void) *etc.* The probability of the observed velocity field would then be $P(v(r)|GA) \times P(GA)$, shifting the question of large-scale streaming velocities to one of rare event probability $P(GA)$ for a given fluctuation spectrum, as described in Sec. 2.2.

Within linear theory, $P(v(r)|GA)$ is a Gaussian distribution. Assume that the GA can be modelled by a region smoothed with a Gaussian filter R_{ga} with linear perturbation amplitude $\nu_{ga}\sigma_{ga}$. Requiring the point to be a peak of the density field does not change the result shown here very much. The regions whose velocities we are interested in computing are assumed to be smoothed over a Gaussian scale R_c, which would typically be much smaller than R_{ga}. The mean velocity of such R_c-smoothed clouds at position r_2 subject to the constraint that a GA is located at position r_1 is given by (see BBKS, Appendix D, for the derivational technique)

$$\langle \mathbf{v}(\mathbf{r}_2)|\nu_{ga}(\mathbf{r}_1)\rangle = \hat{\mathbf{r}}\, \nu_{ga}\, \xi_v'(r; R_h)[H_D\sigma_{\varrho,ga}]^{-1} \quad (2.7a)$$

$$= -\frac{H_D\mathbf{r}}{3}\langle \frac{\Delta M}{M}(< r; R_h)|\nu_{ga}\rangle, \; \mathbf{r} = \mathbf{r}_2 - \mathbf{r}_1, \quad (2.7b)$$

$$\xi_v(r) \equiv \langle \mathbf{v}(\mathbf{r}_2)\cdot \mathbf{v}(\mathbf{r}_1)\rangle, \; \xi_v' \equiv d\xi_v/dr = -H_D^2 J_{3\varrho}(r; R_h)/r^2, \quad (2.7c)$$

$$R_h \equiv [(R_{ga}^2 + R_c^2)/2]^{1/2}, \; \sigma_{\varrho,ga} \equiv \sigma_\varrho(R_{ga}), \; H_D \equiv \dot{D}/D. \quad (2.7d)$$

The cross correlation of the velocity field with an asemble of Great Attractors is $\xi_{\mathrm{vg\dot{a}}}$. Here, the velocity correlation function ξ_v and $J_{3\varrho}(r) \equiv \int_0^r \xi_\varrho r^2\, dr$ are to be smoothed over the intermediate filter R_h. $D(t)$ describes the linear growth of perturbations (Peebles 1980, Sec. II) and $= a(t)$ in a $\Omega = \Omega_{nr} = 1$ universe. Equation (2.7b) relates the mean velocity to the mean mass excess $\Delta M/M$ within the radius r from a the GA. The dispersion in the direction perpendicular and parallel to the line joining the point r_2 to GA are

$$\langle(\Delta \mathbf{v}_\perp(\mathbf{r}))^2\rangle^{1/2} = (2/3)^{1/2}\sigma_v, \; \Delta\mathbf{v} \equiv \mathbf{v} - \langle \mathbf{v}|\nu_{ga}\rangle, \; \sigma_v \equiv \sigma_v(R_c), \quad (2.8a)$$

$$\langle(\Delta \mathrm{v}_\parallel(r))^2\rangle^{1/2} = [\sigma_v^2/3 - (\xi_v'(r)/(H_D\sigma_{\varrho,ga}))^2]^{1/2}. \quad (2.8b)$$

The 3-dimensional velocity dispersion of the clouds in the absence of a GA is σ_v evaluated using R_c. It is not very sensitive to the specific value of R_c provided it is small enough, since the velocity spectrum peaks at $\sim 3\ h^{-1}$ Mpc for CDM; this turns out to be similar to the coherence scale for velocities of field points smoothed on galactic scales, which defines the typical coherent flow unit. The behaviour of the mean velocities and the dispersions is shown in Fig. 3 for the CDM and isocurvature baryon models. Here I took R_c = 3.2 h^{-1} Mpc, motivated by the above discussion. The choice $R_{ga} \sim 15\ h^{-1}$ Mpc is arbitrary; $R_{ga} \sim 12\ h^{-1}$ Mpc might be a better choice to get the mass excess to be $\sim 5 \times 10^{16}\ M_\odot$ if $f_s = 1.5$ as assumed in Sec. 2.2.

Note that eq. (2.7b) bears some similarity to the phenomenological formula for the GA velocity field adopted by Faber and Burstein (1988). For a power law correlation $\xi_\varrho \sim r^{-\gamma}$, $\gamma = 3 + n$ if $n \lesssim 0$, where n is the spectral index, $3J_{3_\varrho}/r^3$ will be $(1 - \gamma/3)^{-1}\ \xi_\varrho(r)$, for $r \gg R_h$, with a 'core radius' of order R_h. Apparently $n = n_A - 2$, where their far field velocity falloff power is n_A, fit to be 1.7. Useful calibration of eq.(2.7) using their result is difficult as it is quite sensitive to the core radius which could be considerably shorter than the linear value if the GA has already turned around. A crude but simple estimate of how the perturbed (Eulerian) position $\mathbf{x}_2$ has evolved from the unperturbed (Lagrangian) position $\mathbf{r}_2$ is to adopt the spherical top hat model given by eq.(2.2): the distance contracts according to $\mathbf{x}_2 = (\mathbf{r}_2 - \mathbf{r}_1)/(1 + \delta)^{1/3} + \mathbf{x}_1$, where $\delta = \Delta M/M$ is the mean overdensity in the interior; *e.g.*, δ = 3.5 implies the significant contraction $x_2 - x_1 \approx 0.6(r_2 - r_1)$ over the core radius. At our distance there would not have been as much collapse of comoving space.

The interpretation of Fig. 3a is the following: If the Local Group velocity toward the GA direction must reach 580 km s^{-1}, then it could be obtained coherently using $\langle v|GA\rangle$ alone, provided the GA has $\delta \sim 3.5$; or it could be only partially coherent if the GA has $\delta \sim 0.8$, with the GA contributing $\sim$ 300 km s^{-1}, the remaining $\sim$ 300 km s^{-1} being provided by the unconstrained part of the density field: such a value is within the 1σ upward fluctuation indicated by Fig. 3a. However, the objects around us which are doing the pulling and pushing should be identifiable. An example would be the Local Void, which lies above the plane of the Local Supercluster and in the direction opposite to the Centaurus cluster. The velocity field given by eq.(2.7) then points away from the void since $\nu < 0$ — voids are repulsive. As a crude estimate of the sort of amplitude we might expect, take our distance from the Local Void centre as 20 h^{-1} Mpc and let it have an 'overdensity' $\nu\sigma_\varrho = -0.5$, smoothed over 10 h^{-1} Mpc. With these parameters we would get a contribution of 320 km s^{-1} pointing away from the void, with a component directed below the supergalactic plane, and presumably a residual component in the GA direction.

2.6 - *Are Mountains of Extra Power Called For?*

In this paper we have seen that judicious interpretations of the observations (and the rather low b_g and high f_s choices) might make all of the data compatible with the CDM model. However this cannot be the case if the

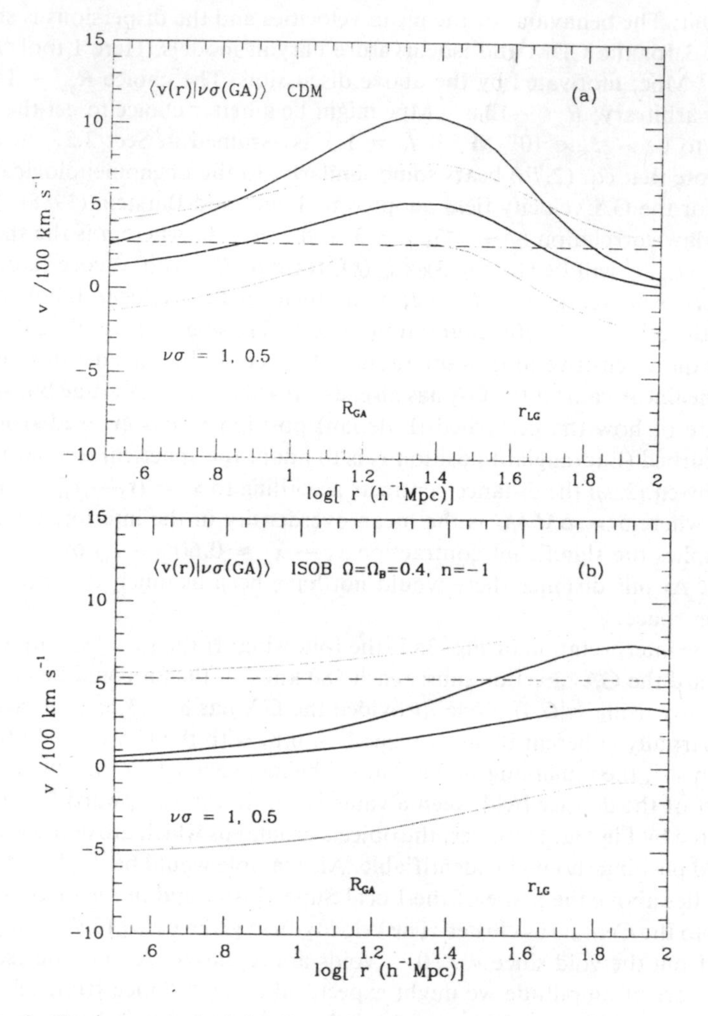

FIG. 3. The mean velocity $\langle v|GA\rangle$ (solid curves) is shown for the CDM model with $b_g = 1.44$ (3a) and the $\Omega = \Omega_B = 0.4$ isocurvature baryon model with $b_g = 1$ and $n = -1$ (3b) for two choices of height $\nu_{ga}\sigma_{ga} = 0.5$ and 1 (upper solid curve), corresponding to overdensities δ_{ga} in the spherical model of Sec. 2.2 of 0.8 and 3.5. The Gaussian smoothing radii are $R_{ga} = 15\ h^{-1}$ Mpc for the GA and $R_c = 3.2\ h^{-1}$ Mpc — the coherence length for velocities smoothed over galaxy scales, 0.35 h^{-1} Mpc, in the CDM model — for the clouds whose flow velocity we wish to determine; thus $R_h = 10.8\ h^{-1}$ Mpc. With such a large R_{ga}, the top hat 'GA' would encompass the Centaurus cluster at $\sim 30\ h^{-1}$ Mpc from us as well as the GA at 42 h^{-1} Mpc. The R_c scale corresponds roughly to the scale of our 'Coma-Sculptor' cloud, whose flow is relatively coherent as Faber, Burstein and Tully pointed out. The dotted curves are the $\pm 1\sigma$ variations about the mean velocity in the parallel direction (toward the GA) for the $\nu_{ga}\sigma_{ga} = 0.5$ case. The variations are similar for the $\nu_{ga}\sigma_{ga} = 1$ case. The dispersion in the perpendicular direction is indicated by the dot-dash curve.

Tenerife experiment (Davies *et al.* 1987) is really seeing primordial CMB anisotropies at the 4×10^{-5} level in a beam of size $\theta_{fwhm} = 8^o$. If we accept the need for extra power at large scales indicated by this experiment, yet still require compatibility with the large angle RELICT experiment of Strukov *et al.* (1988), a mountain of extra power is the required solution, the ramp plus plateau possibility being ruled out by RELICT. The idealized sharp mountain at wavenumber k_P — a spike — must be obtained by initial condition modification, not by transfer function modification of scale invariant initial conditions. Such a spike gives an enhanced ξ_{cc}, impressive cluster patches and increases the number of entities of the Great Attractor type, but might already be incompatible with w_{gg} and ξ_{cg}. It would add (in quadrature) a term to the 3D *rms* velocity dispersion given by $\sigma_{vP} \approx 1000 k_{P2}^{-1}\, \sigma_{\varrho P}/0.1$ km s^{-1}, where $k_{P2}^{-1} \equiv k_P^{-1}/$ 100 h^{-1} Mpc. If k_P^{-1} is large enough, the CMB anisotropy would just be due to the Sachs Wolfe effect. The spike would add (again in quadrature) an *rms* single beam temperature fluctuation, smoothed over a beam profile with Gaussian dispersion $\theta_s = 0.425\theta_{fwhm}$, given by (for $\Omega_{nr} = 1$ universes)

$$\frac{\Delta T}{T}(\theta_{fwhm}) \approx \frac{1}{2}(Hk_P^{-1})^2 \sigma_{\varrho P} e^{-(\ell\theta_s)^2/3} \tag{2.9}$$

$$\sim 6 \times 10^{-5}[k_{P2}^{-1}]^2 \; \frac{\sigma_{\varrho P}}{0.1} \; \exp\left[-k_{P2}^2 \left(\frac{\theta_{fwhm}}{3.9^o}\right)^2\right],$$

$$\textit{where } \ell_P = k_P\chi_0 \approx 60[k_{P2}^{-1}]^{-1} \approx (1^o)^{-1}k_{P2}.$$

Here χ_0 is the comoving distance to the last scattering surface. (See Bond 1987a for the treatment of the beam smearing.) Choosing $k_P^{-1} \sim 300$ h^{-1} Mpc with $\sigma_{\varrho P} \sim 0.1$ could therefore give the reported Tenerife level. I would also guess that such a pulse might be compatible with the RELICT data, since there would be little pixel-to-pixel correlation, and the associated angular power spectrum for the spike (per logarithmic interval of angular wavenumber ℓ), $\ell^2 C_\ell = (\ell/\ell_P)^2 (1 - \ell^2/\ell_P^2)^{-1/2}$ for $\ell < \ell_P$, vanishing for larger ℓ, would have little power over the ℓ-range that the RELICT experiment is most sensitive to. (However, the effective beam smearing for RELICT is similar to that for the Tenerife experiment.) In conclusion, if we accept the need for large-scale power, a mountain of extra power from ~ 25 h^{-1} Mpc to ~ 300 h^{-1} Mpc might be compatible with all of the data as currently stated. However, there is apparently no natural physical mechanism to add such power.

Acknowledgement

This work was supported by a Canadian Institute for Advanced Research Fellowship, a Sloan Foundation Fellowship and the NSERC of Canada.

REFERENCES

Argyres, P.C., Groth, E.J., and Peebles, P.J.E. 1986. *AJ.* **91,** 471.

Bahcall, N. and Soneira, R. 1983. *Ap J.* **270,** 70.

Bardeen, J.M., Bond, J.R., Kaiser, N., and Szalay, A.S. 1986. *Ap J.* **304,** 15 [BBKS].

Bardeen, J.M. *et al.* 1988. in preparation. [BBRS].

Bardeen, J.M., Bond, J.R., and Efstathiou, G. 1987. *Ap J.* **321,** 28. [BBE].

Binggeli, B. 1982. *AA.* **107,** 338.

Bond, J.R. 1986. in *Galaxy Distances and Deviations from the Hubble Flow.* eds. B. F. Madore and R. B. Tully. Dordrecht: Reidel, p. 255.

_____. 1987a. in *The Early Universe.* ed. W. G. Unruh. Dordrecht: Reidel.

_____. 1987b. in *Cosmology Particle Physics.* ed. I. Hinchcliffe. Singapore: World Scientific.

_____. 1987c. In *Nearly Normal Galaxies.* ed. S. Faber. New York: Springer-Verlag.

_____. 1988. in *Large-Scale Structures of the Universe.* eds. J. Audouze, M.-C. Pelletan, and A. Szalay. Dordrecht: Kluwer Academic Publishers, p. 93.

Bond, J.R. and Couchman, H. 1987. *Proceedings of the Second Canadian Conference on General Relativity and Relativistic Astrophysics.* eds. A. Coley and C. Dyer. Singapore: World Scientific.

_____. 1988. in preparation.

Bond, J.R., Carr, B.J., and Hogan, C. 1988. Preprint.

Bond, J.R., Szalay, A.S., and Silk, J. 1988. *Ap J.* **324,** 627.

Carter, D. and Metcalfe, N. 1981. *MN.* **191,** 325.

Coles, S. and Kaiser, N. 1988. in preparation.

Davies, R.D., Lasenby, A.L., Watson, R.A., Daintree, E.J., Hopkins, J., Beckman, J., Sanchez-Almeida, J., and Rebolo, R. 1987. *Nature.* **326,** 462.

Davis, M., Efstahiou, G., Frenk, C.S., and White, S.D.M. 1985. *Ap J.* **292,** 371.

Efstathiou, G. and Rees, M. 1987. *MN.* **230,** 5P.

Efstathiou, G., Frenk, C.S., White, S.D.M. and Davis, M. 1988. preprint.

Faber, S., and Burstein, D. 1988, this volume.

Frenk, C.S., White, S.D.M., Davis, M., and Efstathiou, G. 1985. *Nature.* **317,** 595.

Groth, E.J., and Peebles, P.J.E. 1977. *Ap J.* **217,** 385.

Kaiser, N., and Lahav, O. 1988, this volume.

Lilje, P., and Efstathiou, G. 1988. *MN.* **231,** 635.

Maddox, S., Efstathiou, G. and Loveday, J. 1988. in *Large-Scale Structures of the Universe.* eds. J. Audouze, M.-C. Pelletan, and A. Szalay. Dordrecht: Kluwer Academic Publishers, p. 151.

Matsumoto, T., Hayakawa, S., Matsuo, H., Murakami, H., Sato, S., Lange, A.E. and Richards, P.L. 1987. preprint.
Narayan, R. and White, S.D.M. 1988. preprint.
Nolthenius, R. and White, S.D.M. 1987. *MN.* **225,** 505.
Peebles, P.J.E. 1980. *The Large Scale Structure of the Universe.* Princeton: Princeton University Press.
_____. 1987. *Ap J.* **277,** L1.
_____. 1988. this volume.
Press, W.H. and Schechter, P. 1974. *Ap J.* **187,** 425.
Quinn, P.J., Salmon, J.K. and Zurek, W.H. 1986. *Nature.* **322,** 392.
Rees, M. 1986. *MN.* **218,** 25P.
Rhee, G. and Katgert, P. 1987. *AA.* **183,** 217.
Schaeffer, R., and Silk, J. 1985. *Ap J.* **292,** 319.
Seldner, M., and Peebles, P.J.E. 1977. *Ap J.* **215,** 703.
Struble, M.F., and Peebles, P.J.E. 1985. *AJ.* **90,** 582.
Strukov, I.A., Skulachev, D.P. and Klypin, A.A. 1988, in *Large-Scale Structures of the Universe.* eds. J. Audouze, M.-C. Pelletan, and A. Szalay. Dordrecht: Kluwer Academic Publishers, p. 27.
Szalay, A. 1988, this volume.

Matsumoto, T., Hayakawa, S., Matsuo, H., Murakami, H., Sato, S., Lange, A.E., and Richards, P.L. 1987, preprint.
Narayan, R. and White, S.D.M. 1988, preprint.
Nolthenius, R. and White, S.D.M. 1987, MN 226, 505.
Peebles, P.J.E. 1980, The Large-Scale Structure of the Universe. Princeton: Princeton University Press.
——. 1984, ApJ 277, L1.
——. 1988, this volume.
Press, W.H. and Schechter, P. 1974, ApJ 187, 425.
O[illegible] J.P., [illegible] 1986, Nature 322, 702.
Rees, M. [illegible] MN 218, 25P.
[illegible] 1987, [illegible] 183, 21.
[illegible] 1985, ApJ 292, 319.
[illegible] M., and Peebles, P.J.E. 1985, ApJ 298, 701.
[illegible] 1985, [illegible] 41, 90, 38.
[illegible] Skulachev, D.P., and Klypin, A.A. 1988, in Large Scale Structures of the Universe, eds. J. Audouze, M.-C. Pelletan, and A. Szalay, (Dordrecht: Kluwer Academic Publishers), p. 2[illegible]
Szalay, A. 1988, this volume.

VI

THEORY: PHENOMENOLOGICAL

N-BODY SIMULATIONS OF A UNIVERSE DOMINATED BY COLD DARK MATTER

MARC DAVIS
Departments of Astronomy and Physics, University of California, Berkeley

and

GEORGE EFSTATHIOU
Institute of Astronomy, Cambridge

Abstract

There exist a variety of observational constraints that can be used to test and reject theories of large-scale structure in the universe. Many of these tests require N-body simulations to follow the nonlinear clustering evolution of the models. In this review we describe how the standard cold dark matter (CDM) model compares to the available constraints. The theory is parameterized by two free parameters: the initial amplitude of the perturbation spectrum, and the horizon scale at the epoch of equality of the radiation and matter density. With judicious choice of these parameters (the latter of which is fixed by the Hubble constant), the CDM theory matches an impressive list of observations, but is inconsistent with the reports of clustering on scales $\gtrsim$ 5000 km/s. We describe recent simulations which show that the distribution of luminous galaxies in a CDM universe will be more strongly clustered than the mass distribution. The amplitude of the "bias" is sufficiently large to reconcile observations with the high cosmological density ($\Omega = 1$) assumed in the CDM model.

1. Introduction

Enormous progress has been made in the last several years toward understanding the development of large-scale structure in the universe. The theoretical underpinning of this progress is the presumption that the universe underwent an early inflationary episode, which provides the only known explanation for the near flatness of the universe, and also prescribes an initial spectrum of density perturbations. The current standard model is the minimal theory expected from any inflationary universe. The density perturbations arising from inflation are assumed to be random phase Gaussian noise with a scale invariant adiabatic spectrum. The universe is assumed to have the critical cosmological density parameter ($\Omega = 1$) and to be dominated by some form of cold dark matter (CDM), presumably a weakly interacting particle. This allows us to have a flat universe while still remaining compatible with the standard theory of primordial nucleosynthesis. The overall amplitude of the fluctuation spectrum is sensitive to unknown details of the field theory, so at present we adjust this free parameter to match observations. Subsequent evolution of these adiabatic density perturbations during the radiation dominated phase (Peebles 1982, Blumenthal and Primack 1983) leads to a unique spectrum at the epoch of recombination. These linear perturbations will evolve via gravitational instability into the nonlinear structure observed at present. The goal of N-body studies is to directly simulate the clustering into this nonlinear phase. Because the theory is so completely specified there is always the chance that it can be falsified if it fails to match observations.

We will describe the current effort of our "gang of four" collaboration (Davis, Efstathiou, Frenk and White), but will not attempt a review of all our past work (see White *et al.* 1987 and references therein). Most of our projects have been focused on direct comparison of the models to a variety of observational facts, which we summarize in Table 1. Any acceptable theory of large-scale structure should pass the tests listed in Table 1, although it must be remembered that some of the observations lack precision and others (*e.g.* filamentary structure) are rather qualitative and subjective. CDM has become the focus of attention because it passes most of the tests so well. Perhaps the model will eventually be falsified, but we have been unable to do so convincingly after four years of trying.

Setting the Hubble constant is equivalent to setting the horizon scale at the epoch of equality of radiation and matter density, which determines the peak in the perturbation spectrum. We presume that $H_0 = 50$ km s^{-1} Mpc^{-1}, because this value provides a good match to large-scale structure. The CDM model works well for smaller values of H_0, but fails to give sufficient large-scale structure for $H_0 = 100$. If in fact H_0 eventually settles in the range of 100 km s^{-1} Mpc^{-1}, the minimal CDM model is dead, quite apart from problems of the age constraints (see Bardeen *et al.* 1987, for a discussion of various non-standard CDM models). The amplitude of the initial perturbations is set by matching the observed strength of galaxy correlations $\xi_{gg}(r)$ (point 6 in Table 1). With no further adjustable parameters, the model must match the tests described in Table 1. The cosmological constant could be used as a further parameter, but we shall assume $\Lambda = 0$.

TABLE 1
WHAT A *GOOD* THEORY SHOULD EXPLAIN

On the scale of galaxies:

1. Galaxy abundance as a function of luminosity $\phi(L)$
2. Flatness of galaxy rotation curves at v_{rot} = 150 — 250 km/s
3. Luminosity-rotation correlation ($L \propto \sigma^4$)
4. Proper specific angular momentum λ and internal distribution
5. Morphology-environment correlation

On intermediate scales:

6. Amplitude and slope of $\xi_{gg}(r)$
7. Amplitude and slope of relative pair velocities σ_{12}(r)
8. Plausible bias mechanism

On larger scales:

9. Proper "filamentary frothiness" in galaxy clustering
10. Large voids, apparently empty
11. Correct abundance of rich clusters
12. Amplitude and shape of $\xi_{gc}(r)$
13. Sufficiently enhanced amplitude of $\xi_{cc}(r)$
14. Sufficient coherence in large-scale flows
15. Consistency with IRAS dipole direction $\Delta\theta(r)$

As a function of redshift:

16. Evolution in abundance of QSO's, AGN's, radio sources
17. Age of galactic disks
18. Change in populations of galaxies in clusters
19. Redshift of "galaxy" formation

2. Small-Scale Structure

Let us discuss the tests one by one. The tests on galaxy size scales are detailed by Frenk *et al.* (1985, 1988). We evolved a series of simulations of comoving size 14 Mpc, so that galaxy halos are comprised of many particles. The halos in these models form with flat rotation curves (we measure $v_c(r) = (GM(r)/r)^{1/2}$) in exactly the range to compare to real galaxies, so item 2 falls out immediately. These halos extend for several hundred kpc, and the CDM model is quite inconsistent with halos truncated at ~ 50 kpc (*e.g.*

Little and Tremaine 1987). The model can thus be tested by measuring the extent of halos via radial velocity measurements of faint satellites or by the X-ray emission of hot gas.

From the models we can measure $N(v_c)$, the number density of halos with circular velocity $\gtrsim v_c$. From the observed galaxy luminosity distribution function (Felten 1985) we infer $N(L)$, the number density of galaxies with luminosity $\gtrsim L$. If we assume each halo to be associated with one galaxy and luminosity to be a monotonic function of circular velocity, then by matching $N(v_c) = N(L)$ we derive the functional form of $L(v_c)$, with results shown in Figure 1a. For comparison, we show the Tully-Fisher relation for spirals (dashed line), the Faber-Jackson relation for ellipticals (dotted line) and a data point for dwarf irregulars in Virgo from Bothun *et al.* (1984; see Frenk *et al.* 1988 for further details concerning these comparisons). The halos in our simulations must harbour galaxies of all types, so it is encouraging that our predicted $L - v_c$ relation lies between the observed relations for spirals and ellipticals and passes close to the data point for dwarfs. This agreement is especially noteworthy because we had no adjustable parameters in deriving our prediction.

Assuming that the 'light' associated with each halo may be computed as described above, we can determine a "mass-to-light" ratio for each halo. The results are shown in Figure 1b. Note that this is M/L measured at a surface

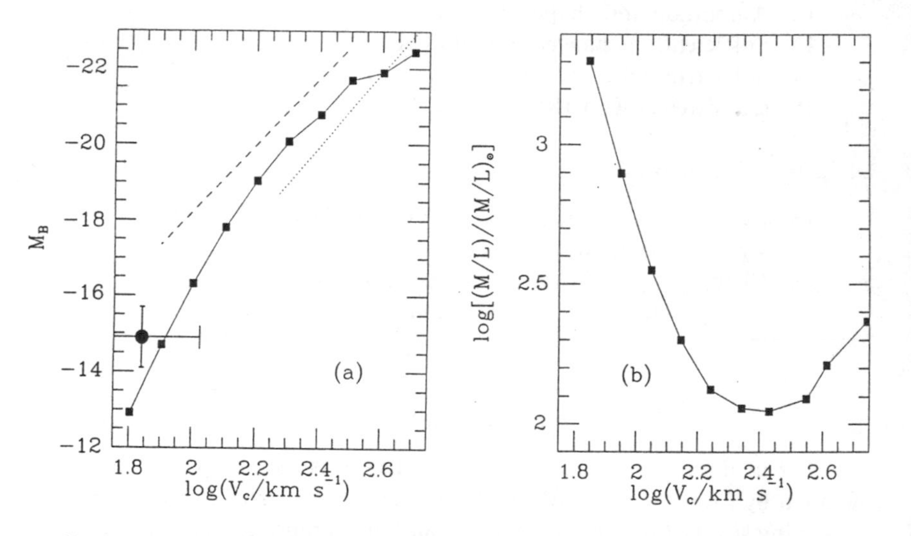

FIG. 1. (a) The Tully-Fisher relation for halos in the simulations. The "magnitude" of the "galaxy" associated with each halo is plotted against characteristic velocity. The dashed and dotted lines give the observed relations for spirals and ellipticals respectively. The point with error bars refers to dwarf irregulars in the Virgo cluster. (b) Predicted mass-to-light ratios in solar units for the "galaxies" associated with the halos in our simulations.

of density contrast of approximately 200. At this large radius, most halos with $V_c \simeq 200$ km/s have $M/L \simeq 100$ and for halos with shallower potential wells the M/L value is expected to be very large. It is not that the fainter galaxies have extraordinarily massive halos, but rather that their star formation efficiency is likely to be less efficient, as discussed by Dekel and Silk (1986). The upturn in M/L for larger V_c is a consequence of our assumption that each halo is associated with only one luminous galaxy; some of our halos have $v_c > 400$km/s and are uncharacteristic of even the brightest spirals. While some of these systems may correspond to the halos of bright ellipticals, others appear too massive to be associated with any single galaxy. Rather, they must be identified with the merged halos of groups and clusters of galaxies. This problem could be overcome by simulating the dissipative formation of galaxies, though we describe below an approximate prescription for assigning several galaxies to a common halo.

Angular momentum of galaxies has been discussed in detail by Barnes and Efstathiou (1987). Sufficient angular momentum is generated via tidal torques in the CDM model; the formation of centrifugally supported spiral galaxies would follow if gas collapsed dissipatively by a factor of ~ 10 in radius within the dark halos (Fall and Efstathiou 1980). The smaller specific angular momentum observed in the luminous portions of elliptical galaxies may result from angular momentum redistribution during the highly aspherical collapse and merging which occurs in the denser regions in the CDM models. In fact, in the simulations we find that halos grow both by slow accretion of diffuse material and by violent merging of subunits. Presumably a galactic disk will form under quiet conditions of gradual accretion, but disks are likely to be disrupted by strong mergers. The maximum ages of white dwarf stars in the disk of our own galaxy suggest it has not undergone a strong merger since $Z \sim 1$. The models demonstrate that strong merging for $Z < 1$ occurs mostly in the denser regions of the models, i.e. the group centers. It seems therefore that CDM provides a framework for the well established clustering differences of early versus late type galaxies (item 5 of Table 1). This statement can be better quantified by further analysis of our existing simulations.

3. Intermediate-Scale Clustering

We use the observed amplitude of galaxy clustering to set the initial amplitude of the perturbation spectrum. However, this does not guarantee that $\xi(r)$ will have a power law form with the observed slope. Furthermore we must determine the relationship of the galaxy distribution to the underlying mass distribution. It is well known that observations imply $\Omega \approx 0.2$ if galaxies trace the mass, so to meet the inflationary imperative that $\Omega = 1$, luminous galaxies must be a biased tracer of mass. Understanding the mechanism by which galaxies become a biased mass tracer is fundamental in any model of large-scale structure. The usual assumption is that bright galaxies are associated with rare peaks in the Gaussian noise field (Bardeen *et al.* 1986). In our initial studies we put the bias in "by hand", assuming galaxies were associated with the peaks of 2.5σ fluctuations in the initial density field. This procedure worked remarkably well but had no strong physical justification, although possible astrophysical biasing mechanisms have been discussed (e.g. Rees 1985, Dekel and Silk 1986). The absence of a compelling physical explanation of the needed bias has always been an unattractive feature of the CDM model.

However in our recent studies we have shown that halos, particularly those with $v_c \geq 200$ km/s are "naturally biased" by a purely gravitational effect. In the CDM model, deep potential wells are assembled late, when there are substantial fluctuations on larger scale. A protovoid will act as a section of an open uiverse and a protocluster will act as a section of a closed universe; the growth rate of linear density fluctuations will be enhanced in the dense regions and it will be depressed in the underdense regions. The net result is that denser regions are more efficient at condensing material into deep potential well objects, so a bias is guaranteed. The bias is expected to be less pronounced for smaller v_c because these objects are assembled earlier, when the amplitude of the large-scale structure is smaller and the growth rate of linear perturbations more nearly equal in the over and underdense regions.

This work is described by White *et al.* (1988). In this study we required more dynamic range than in our previous work because we needed to resolve individual galaxies as condensations of multiple points in a volume sufficiently large to measure galaxy clustering. We have begun therefore to use Cray supercomputers and can now evolve models of 262144 particles in potential grids of size 128^3. With the PPPM code our force softening length is 1/600th size of the periodic cube, and each model requires approximately 20 Cray hours to expand a factor of 8. In Figure 2 are presented slices of a snapshot of a model evolved in a cube of comoving size 3200 km/s. The mass of each particle is $7.0 \times 10^{10}\ M_\odot$ which is sufficient to resolve virialized halos with $v_c = 100$ km/s into approximately 10 particles.

The simulations are purely gravitational and so do not allow us to study the behavior of the dissipative gas from which galaxies must form. We have, therefore, included galaxy formation and merging in a way which, although plausible, remains somewhat *ad hoc*. At various stages during the evolution of a model we locate the most strongly bound particle in each dark matter halo. These are labelled "galaxies" and are assigned a circular speed at the radius of a sphere centered on each "galaxy" of mean overdensity 1000 times the present critical density. We then adopt simple algorithms to model galaxy mergers and to avoid multiple galaxy formation within each halo. The results are not especially sensitive to these procedures which are described in White *et al.* (1988). In Figures 2b and 2c we show the distribution of "galaxies" with $v_c > 100$ km/s and $v_c > 200$ km/s respectively. Note that the galaxies trace the ridges of high density in the matter distribution and are very deficient in the low density regions.

The spatial autocorrelation functions $\xi(r)$ for the mass distribution and for galaxies with $v_c > 100$ km/s and $v_c > 200$ km/s of this model are shown in Figure 3. The galaxy autocorrelation functions are steeper than that of the mass and have a larger amplitude. Over the range of separations shown in Figure 3 the mean correlation enhancement is ~ 1.3 for $v_c > 100$ km/s and ~ 2.5 for $v_c > 200$ km/s. When correlations are computed in redshift space with velocity broadening consistent with observed peculiar velocities, the slope of the correlations flattens to a power law of approximately $\gamma = 1.8$.

The amplitude of the correlations for "galaxies" with $v_c > 200$ km/s is a factor of 2 below that observed in magnitude limited galaxy catalogs; a sufficient level of natural bias occurs for "galaxies" with $v_c > 250$ km/s. The model predicts that the strength of clustering depends on the asymptotic circular velocity v_c of a galaxy, and, therefore, on the luminosity by the Tully-Fisher relation. There is little direct evidence for this in present samples, though good statistics are available for only a narrow range of luminosities. To some degree, however, galaxy morphology is correlated with v_c, and it is quite apparent that the strength of galaxy clustering is strongly dependent on morphology (e.g. Giovanelli *et al.* 1986, Davis and Geller 1976). Further study will hopefully provide a more quantitative measure of this point.

Previously the amplitude of the initial spectrum was adjusted to give the correct correlations of 2.5σ peaks, but now we must readjust the amplitude so that the naturally biased galaxy distribution has the correct correlation behavior. The correlation of the bias strength with the depth of the halo potential well makes this readjustment somewhat uncertain at present.

4. Large-Scale Clustering

N-body simulations in a box larger than 10000 km/s necessarily sacrifice small-scale resolution; the particle masses exceed the mass of a galaxy and it becomes necessary to use the statistical prescription to generate the bias. Large-scale clustering properties of a CDM universe are described in White *et al.* (1987). Items 9-11 of Table 1 were shown to be consistent with CDM, but items 12-14 presented problems.

We were able to reproduce the void in Boötes without difficulty if we sampled the galaxy distribution dilutely, as Kirshner *et al.* (1987) had sampled Boötes. However, we commented that in the CDM model Boötes sized voids would be very rare events, if completely devoid of galaxies. A new full survey of *IRAS* selected galaxies in the Boötes region (Strauss and Huchra 1988) found 3 galaxies within the void region where 11 were expected. These galaxies

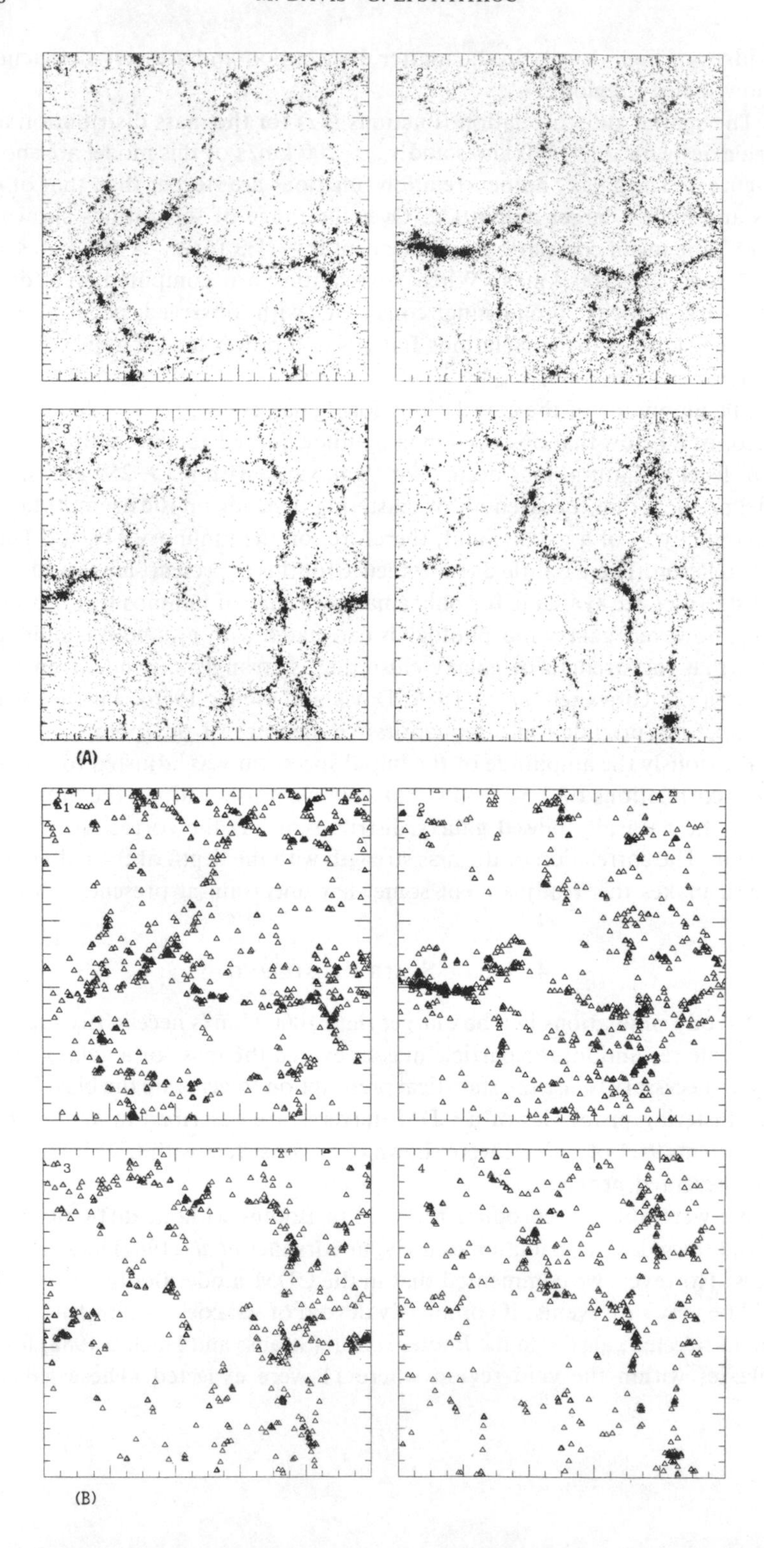
1
2
3
4
(A)
1
2
3
4
(B)

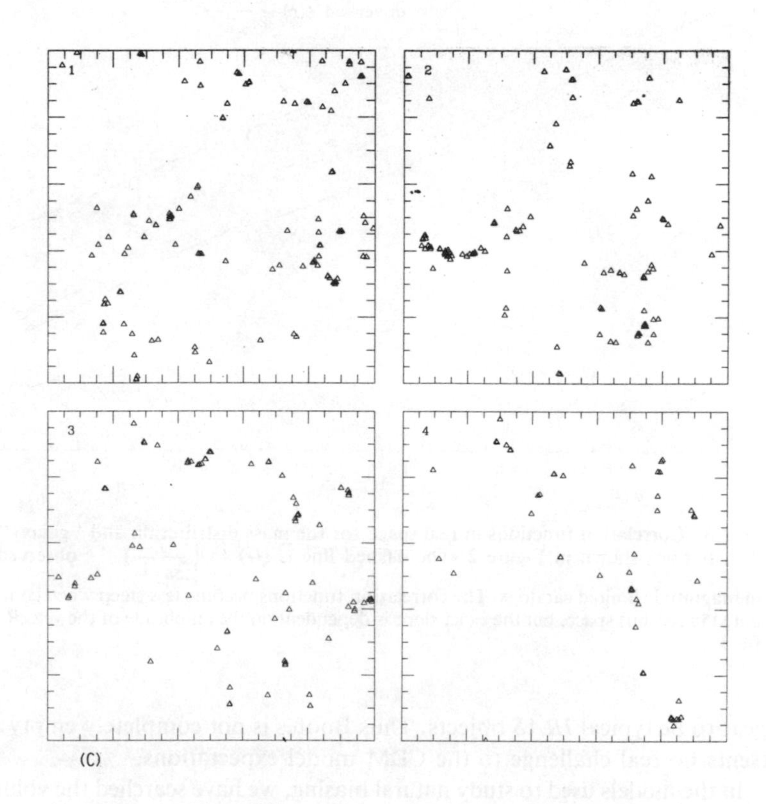

FIG. 2. (a) Four consecutive slices, each $^1/_8$ the thickness of the cube for a simulation of 262144 points in a volume 3200 km/s in length. The epoch is $z = 0$. All points are plotted. (b) The same four slices, but now only points at the center of potential wells of depth $\geq$ 100 km/s are plotted. (c) The same four slices, but now only points centered on potential wells of depth $\geq$ 200 km/s are plotted.

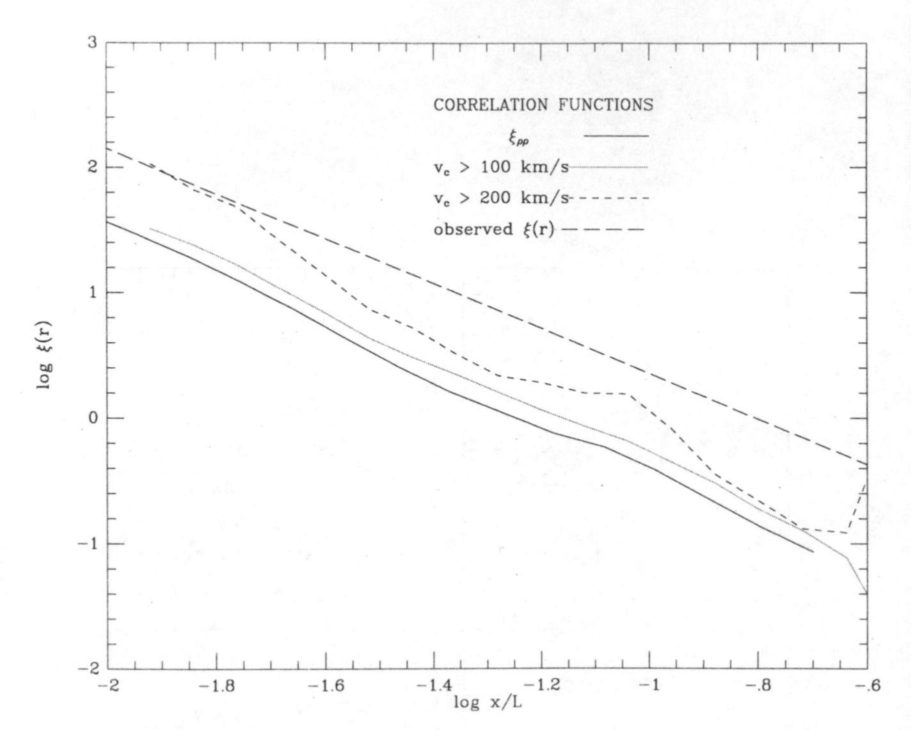

FIG. 3. Correlation functions in real space for the mass distribution and "galaxy" distributions shown in Figure 2. The dashed line is $\xi(r) = (\frac{r}{5h^{-1}})^{-1.8}$ observed in magnitude limited catalogs. The correlation functions become less steep when computed in redshift space, but the exact slope is dependent on the amplitude of the velocity field.

appear to be typical *IRAS* objects. Thus Boötes is not completely empty and presents no real challenge to the CDM model expectations.

In the models used to study natural biasing, we have searched the volumes of length size 2500 km/s to find the largest sphere devoid of all "galaxies" with $v_c > 100$ km/s. Plotted in Figure 4 is the average mass density as a function of distance from these void centers. The void radius was approximately 20% the size of the simulation, and the figure shows that the mass density within the voids is typically 20% of the mean. The voids are not empty, but they are certainly of low density, and they are not expected to be "filled" with faint galaxies. These voids are small compared to the size of those reported in the galaxy distribution, but they are as large as could be expected within simulations of this size.

White *et al.* (1987) demonstrated the observed abundance of the Abell clusters was compatible with CDM, but that the model cluster correlation func-

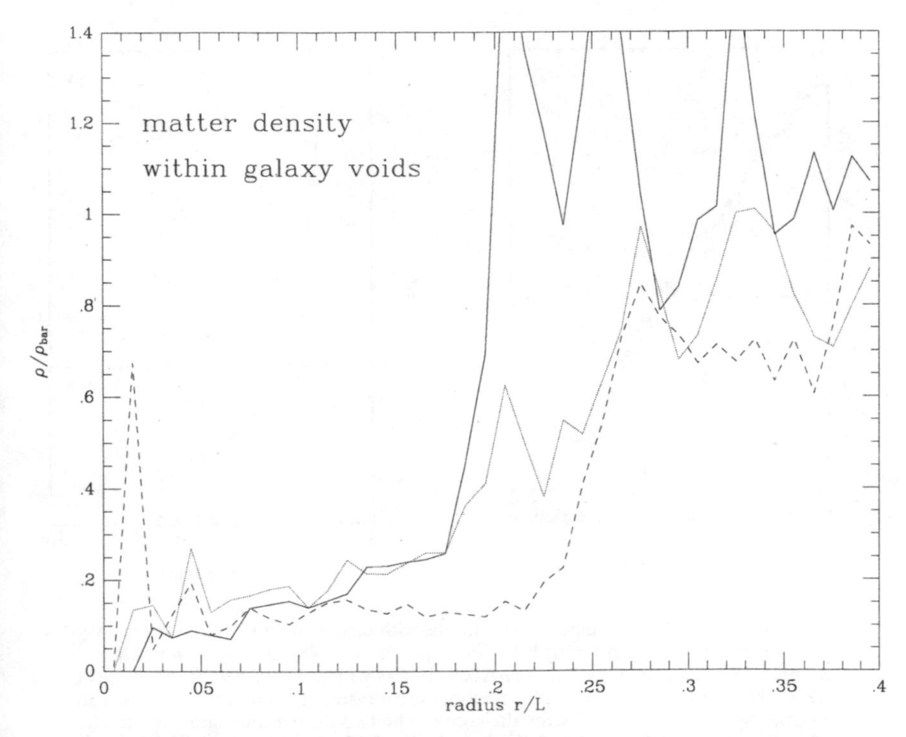

FIG. 4. The matter density distribution within the largest voids in each of 3 simulations of size $L = 2500$ km/s. We plot the density relative to the mean as a function of radius from the center of the void.

tions, $\xi_{cc}(r)$ and $\xi_{cg}(r)$, were of insufficient amplitude compared to observations (Bahcall and Soneira 1983, Seldner and Peebles 1977). This has always been the chief deficiency of the CDM model, and it is, therefore, of considerable interest to reconfirm the observational measures of cluster correlations.

Recently Lilje and Efstathiou (1988) have reanalyzed $\xi_{cg}(r)$ using redshifts for 204 Abell clusters. Using the redshift information and modern estimates of the galaxy luminosity function, they found a correlation amplitude a factor of two smaller than previously reported; furthermore, they showed that $\xi_{cg}(r)$ could not be reliably determined on scales $\gtrsim 20h^{-1}$ Mpc. Figure 5 shows a plot comparing $\xi_{cg}(r)$ from the models of White *et al.* (1987) and the recent measurements of Lilje and Efstathiou (1988). No parameters were used to scale one result to the other. The amplitude and shape are reasonably well reproduced by the model; there is no indication of any serious discrepancy.

The question of $\xi_{cc}(r)$ has been much discussed at this conference and the problem of projection effects, raised by Sutherland (1988) and presented here by Kaiser and Lahav (1988), must be studied further. The CDM predic-

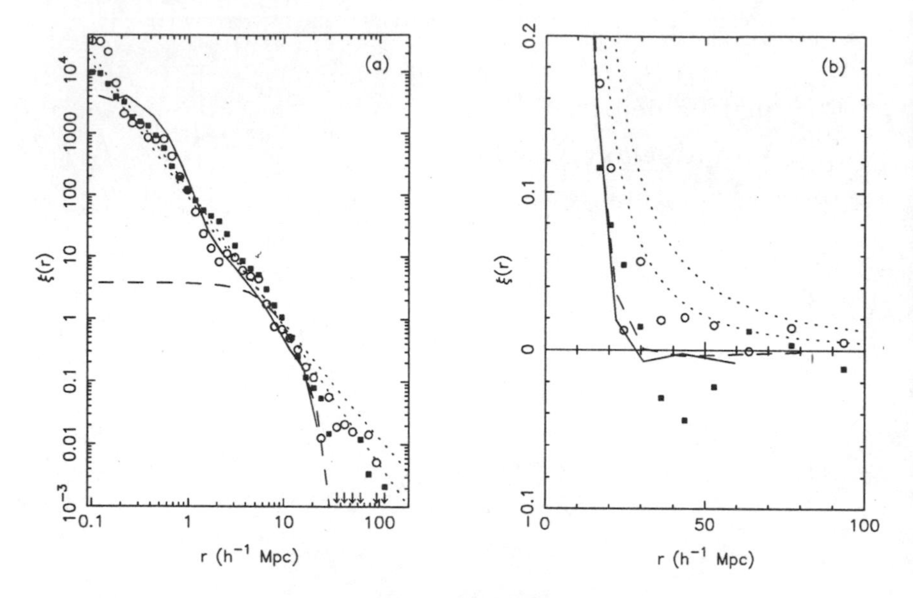

FIG. 5. Comparison of the biased CDM model with observations of the cluster-galaxy cross correlation function (from Lilje and Efstathiou 1988). Symbols show the ξ_{cg} determined from a cross-correlation of Abell clusters with richness class $R \geq 1$ against galaxies in the Lick catalogue. Solid symbols show results for the north galactic cap and open symbols show results for the south. The two dotted lines show the model $\xi_{cg}(r) = (r/8.8\mathrm{h}^{-1}\ \mathrm{Mpc})^{-2.2}$ and $\xi_{cg}(r) = (r/6\mathrm{h}^{-1}\ \mathrm{Mpc})^{-2.3} + (r/7\mathrm{h}^{-1}\ \mathrm{Mpc})^{-1}$. The solid lines show ξ_{cg} for the CDM model as determined from N-body simulations while the dashed line shows an analytic prediction of ξ_{cg} using the statistical techniques described in Bardeen *et al.* (1987). Figure 5(a) shows these results on a log-log plot, while Figure 5(b) shows the results at large scales on a linear-linear plot.

tion is quite definite; there should not be substantial clustering of Abell clusters on scales larger than ~ 5000 km/s. The key issue here concerns reliability of the cluster catalogues. In our view, it would be premature to discard the CDM model on this account until more observational work has been done.

More recent considerations are related to the question of large-scale velocity flows in the universe. In principle, this is an easy calculation for models of structure with random phase initial conditions because linear theory should apply on the scales in question. CDM is not compatible with large amplitude flows with coherence lengths of order 5000 km/s. Superficially, it would seem that the coherence length of the velocity flows observed by Dressler *et al.* (1987) are considerably larger than predicted by the CDM model. However, it is important to assess the relative weights assigned to galaxies as a function of their distance from the observer; the more distant galaxies will carry much less weight than nearby galaxies so the effective "window function" for a bulk flow solution may be considerably smaller than the depth of the sample (Kaiser 1988).

Careful statistical comparisons of the elliptical galaxy data, together with results on the IRAS and optical dipoles have been described at this meeting by Kaiser and Lahav (1988). They find that a biased CDM model is compatible with these data, though there are indications of more power at large scales; their maximum likelihood analysis favours a larger amplitude for the CDM power spectrum than we have used in our simulations by a factor of about 2, but the data do not seem incompatible with our lower value.

Vittorio (1988) has shown that the misalignment angle between the microwave dipole direction and the IRAS gravity direction presents a strong test of the large-scale power spectrum, provided the IRAS galaxies at least approximately trace the mass distribution (item 15 in Table 1). The observed misalignment is small at modest distances, which is consistent with CDM but perhaps is a problem for the isocurvature baryon model.

5. Clustering Evolution

Based on the simulations we can certainly predict the evolutionary history of the correlation functions, but a suitable data set is not yet available for comparison. The evolutionary behavior of individual objects is much more difficult because we must include additional physics into simulations to begin to describe the dissipation of the baryons. The CDM halo models have been criticized as implying galaxy formation at too low a redshift, but as Baron and White (1987) have shown, galaxy formation could have continued to low redshifts ($z \sim 1 - 2$) without conflicting with observational searches for primeval galaxies. Constraints on the redshift of galaxy formation could offer an important constraint on the amplitude of the CDM fluctuation spectrum and could perhaps supply a decisive test of the theory, but at present the observational limits seem weak.

We also remark that there is no conflict between the standard CDM model and observations of bright quasars at $z \gtrsim 4$ (Efstathiou and Rees 1988). The comoving density of luminous quasars ($L > 10^{47}$ erg/s between $z = 2 - 4$) is small ($N_Q \simeq 10^{-6} h^{-3}$ Mpc^{-3}), and can be associated with proto-galactic nuclei in the CDM model if quasars are short lived and radiate at about the Eddington limit.

In some respects the kind of ongoing galaxy formation expected in the CDM model seems to us to be a positive virtue. For example, point 18 in Table 1 is another indicator that galaxies, as recently as $z = 0.4$, often have different stellar populations, indicative of more active star formation than at the present (*e.g.* Dressler and Gunn 1983). The CDM model predicts that galaxies at $z \sim 1$ should show an enhanced merger rate. More work clearly needs to be done on the expected evolution of galaxies, but certainly the view with the Space Telescope should be illuminating.

6. Conclusions

The CDM model is attractive because it fits in naturally with our current ideas of the early universe. The model has been tested against a large number of observations, yet there are no clear-cut discrepancies. The most serious problem comes from observations of the cluster-cluster correlation function, which may indicate a requirement for more power at large scales.

We have been motivated to investigate the CDM model in so much detail precisely because it contains so few parameters. Although we haven't been able to rule it out yet, we can at least point to the sort of observations and further investigations which need to be done. These include:

(i) *Tests of biased galaxy formation.* Our "natural biasing" mechanism makes a clear prediction that galaxies associated with halos of high v_c should be more strongly clustered than those with low v_c. This can be tested in a variety of ways, *e.g.* by comparing the correlation functions for faint and bright galaxies.

(ii) *Formation of galaxies.* With our choice of amplitude for the CDM spectrum, most halos with $v_c \sim 250$km/s form between redshifts $z \approx 3$—1. Such a low redshift of galaxy formation can be tested by primeval galaxy searches and by studies of spectral evolution (*e.g.* the amplitude of the 4000 Å break, Hamilton 1985). If massive protogalaxies were abundant at $z \gtrsim 3$, this would require either a revision of our choice of amplitude for the CDM spectrum, or perhaps indicate that the CDM theory is wrong.

(iii) *Clustering on large scales.* The CDM model makes a precise prediction of the amplitude of the cluster-cluster correlation function which is clearly at odds with the results of Bahcall and Soneira (1983). We have played down the discrepancy because we don't really know how reliable the Abell catalog is. However, in the near future, we should have a new X-ray selected cluster catalog from the ROSAT satellite which should be objective and complete. It will be fascinating to see whether the CDM model passes this test. The new deep galaxy survey constructed at Cambridge (Maddox, Efstathiou and Loveday 1988) also promises to shed some light on this problem as discussed at this meeting.

(iv) *Large-scale streaming motions.* As mentioned by several speakers at this meeting, the CDM model predicts a noisy velocity field with a relatively small coherence length; but this is not in obvious conflict with the available data. Further studies of large-scale streaming motions are necessary.

(v) *Extent of galaxy halos.* In the CDM model, we would expect galaxy halos to extend to large radii with approximately isothermal radial profiles. It should be possible to test this idea, either though X-ray observations or from kinematic studies.

(vi) *Anisotropies and spectral distortions of the cosmic background radiation.* Anisotropies in the cosmic microwave background would provide a clean test of the CDM model. Detailed predictions have been made by Bond and Efstathiou (1987). The results are within the accessible range for well designed experiments but are below the sensitivity levels of current experiments. The reported detection on scales of 8° discussed by Lasenby (1988) is too large to be compatible with the CDM model; we await with great interest to see whether this result is confirmed. We also mention the distortion of the backround radiation spectrum reported by Matsumoto *et al.* (1988). This perhaps indicates the presence of a substantial quantity of dust at redshifts $z \sim 40$, which would be difficult to reconcile with the CDM model.

Doubtless the reader will be able to add other items to this list. Even if the CDM model is eventually proved wrong, it is at least a good approximation to the truth!

Acknowledgements

We thank our collaborators, Carlos Frenk and Simon White, for allowing us to discuss our joint work. We have received financial support from NATO, the SERC, and the NSF.

REFERENCES

Bahcall, N. and Soneira, R. 1983. *Ap J.* **270,** 20.
Bardeen, J.M., Bond, J.R., Kaiser, N., and Szalay, A.S. 1986. *Ap J.* **304,** 15.
Bardeen, J.M., Bond, J.R., and Efstathiou, G. 1987. *Ap J.* **321,** 28.
Barnes, J. and Efstathiou., G. 1987. *Ap J.* **319,** 573.
Baron, E. and White, S.D.M. 1987. *Ap J.* **322,** 585.
Blumenthal, G.R. and Primack, J.R. 1983. in *Fourth Workshop on Grand Unification.* p. 256. ed. H.A. Weldon, P. Langacker, and P.J. Steinhardt. Boston: Birkhauser.
Bond, J.R. and Efstathiou, G. 1987. *MN.* **226,** 655.
Bothun, G.D., Aaronson, M., Schommer, R., Huchra, J., and Mould, J. 1984. *Ap J.* **278,** 475.
Davis M. and Geller, M.J. 1976. *Ap J.* **208,** 13.
Dekel, A. and Silk, J. 1986. *Ap J.* **303,** 39.
Dressler, A. and Gunn, J.E. 1983. *Ap J.* **270,** 7.
Dressler, A., Faber, S.M., Burstein, D., Davies, R.L., Lynden-Bell, D., Terlevich, R.J., and Wegner, G. 1987. *Ap J.* **313,** 42.
Efstathiou, G. and Rees, M.J. 1988. *MN.* **230,** 5p.
Fall, S.M. and Efstathiou, G. 1980. *MN.* **193,** 189.
Felten, J.E. 1985. *Comm Astr Sp Sci.* **11,** 53.
Frenk, C.S., White, S.D.M., Efstathiou, G., and Davis, M. 1985. *Nature.* **317,** 595.
Frenk, C.S., White, S.D.M., Davis, M., and Efstathiou, G. 1988. *Ap J.,* **327,** 507.
Giovanelli, R., Haynes, M.P., and Chincarini, G.L. 1986. *Ap J.* **300,** 77.
Hamilton, D. 1985. *Ap J.* **297,** 371.
Kaiser, N. 1988. *MN.* **231,** 149.
Kaiser, N. and Lahav, O. 1988, this volume.
Kirshner, R., Oemler, A. Schechter, P., and Shectman, S. 1987. *Ap J.* **314,** 493.
Lasenby, A. 1988, this volume.
Lilje, P. and Efstathiou, G. 1988. *MN.* **231,** 635.
Little, B. and Tremaine, S. 1987. *Ap J.* **320,** 493.
Maddox, S.J., Efstathiou, G., and Loveday, J. 1988. in *Large-Scale Structures of the Universe.* eds. J. Audouze, M.-C. Pelletan, and A. Szalay. Dordrecht: Kluwer Academic Publishers, p. 151.
Matsumoto, T., Hayakawa, S., Matsuo, H., Murakami, H., Sato, S., Lange, A.E., and Richards, P.L. 1988. *Ap J.,* **329,** 567.
Peebles, P.J.E. 1982. *Ap J Letters.* **263,** L1.
Rees, M.J. 1985. *MN.* **213,** 75p.

Seldner, M. and Peebles, P.J.E. 1977. *Ap J.* **215**, 703.
Strauss, M. and Huchra, J. 1988. *AJ.,* **95,** 1602.
Sutherland, W. 1988. *MN.* **234**, 159.
Vittorio, N. 1988. this volume.
White, S.D.M., Frenk, C.S., Davis, M., and Efstathiou, G. 1987. *Ap J.* **313,** 505.
White, S.D.M., Davis, M., Efstathiou, G., and Frenk, C.S. 1988. *Nature.* **330**, 451.

THE STATISTICS OF THE DISTRIBUTIONS OF GALAXIES, MASS AND PECULIAR VELOCITIES

P. J. E. PEEBLES
Joseph Henry Laboratories, Princeton University

1. Introduction

The galaxy peculiar velocity field seems to have two noteworthy features: the rms value is large, perhaps $\sim$ 600 to 1000 km sec^{-1}, and the coherence length is large, $r_c \gtrsim 10\ h^{-1}$ Mpc ($H = 100\ h$ km sec^{-1} Mpc^{-1}). The first is based on the observation that the velocity of the Local Group relative to the rest frame defined by the microwave background radiation is 600 km sec^{-1} (Wilkinson 1986). Since we are in a relatively calm part of space, it seems likely that the rms velocity is no smaller than our motion. This velocity is considerably greater than the deviations from Hubble's law observed in our neighborhood (Sandage 1986): at distances $hr \sim 6$ Mpc the scatter is no more than about 150 km sec^{-1} (Peebles 1988). The standard interpretation is that a large part of our peculiar motion is a component that varies only slowly with position. Possible examples of this large-scale component are seen in the Rubin-Ford effect (Rubin *et al.* 1976) and in the Great Attractor discussed in these proceedings by Faber and Burstein (1988).

A convenient measure of the length scales over which the peculiar velocity field, $\mathbf{v}(\mathbf{r})$, varies is provided by the statistic

$$<(v_1 - v_2)^2> = \delta v(r)^2. \quad (1)$$

The average is supposed to be taken over a fair sample of luminous ($L > L_*$) pairs of galaxies with separations in a small range around r. We expect that the velocities, $\mathbf{v}_1$ and $\mathbf{v}_2$, of the two galaxies are uncorrelated in the limit of large separation so

$$\delta v(r \rightarrow \infty)^2 = 2<v^2>.$$

The observations lead us to suspect that $\delta v(r)$ decreases with decreasing r, reflecting the fact that the two galaxies tend to have a common component of velocity that cancels out the difference $\mathbf{v}_1$-$\mathbf{v}_2$. A coherence length for the velocity field is the separation r_c at which δv^2 is down from its value at $r \rightarrow \infty$ by a factor of 2:

$$\delta v(r_c) = \delta v(r \rightarrow \infty)/2^{1/2}. \tag{2}$$

At this separation the peculiar velocities of the two galaxies can be imagined to have on average equal contributions from a part common to the two galaxies and a part uncorrelated between the two galaxies.

At small r, $\delta v(r)$ is observed to be (Davis and Peebles 1983, Peebles 1984, Fig. 6)

$$\delta v = 3^{1/2}(340 \pm 40)\,(hr_{\mathrm{Mpc}})^{0.13 \pm 0.04}\ \mathrm{km\ sec}^{-1},$$
$$0.1 \lesssim hr \lesssim 1\ \mathrm{Mpc}. \tag{3}$$

The factor $3^{1/2}$ corrects from line-of-sight to three-dimensional velocities under the assumption of isotropy. The fact that δv in this equation is greater than the local noise in the Hubble flow is, I think, a reflection of the fact that we are in an unusually calm region. The fact that δv at $hr = 1$ Mpc is less than 600 km sec^{-1} indicates that the coherence length r_c is greater than 1 h^{-1} Mpc, as we would expect from the Rubin-Ford and Great Attractor effects.

If the peculiar velocity field is produced by gravity then $\delta v(r)$ tells us something about the character of the mass distribution. I propose to consider here whether our constraints on $\delta v(r)$ are consistent with what would be expected if mass were distributed in the same way as galaxies. This question was first considered by Clutton-Brock and Peebles (1981); the discussion in the next section is an update. I will argue that the assumption that galaxies trace mass looks phenomenologically promising although, as is well known, it does force us to the inelegant assumption that the mean mass density is about 20% of the critical Einstein-de Sitter value. In Section 3 I present some generally negative comments on the biasing assumption required to reconcile the galaxy position and velocity observations with the Einstein-de Sitter density.

2. Dynamics under the Assumption that Galaxies Trace Mass

The statistic $\delta v(r)$ is determined by, among other things, the dimensionless mass autocorrelation function

$$\xi(r) = \langle (\varrho(\mathbf{x}) - \langle\varrho\rangle)\,(\varrho(\mathbf{x} + \mathbf{r}) - \langle\varrho\rangle)\rangle / \langle\varrho\rangle^2, \tag{4}$$

where $\varrho(\mathbf{x})$ is the mass density, $<\varrho>$ the mean value. The autocorrelation function is the Fourier transform of the power spectrum, P(k):

$$\xi(r) = \int d^3k \, P(k) \sin kr/kr. \tag{5}$$

The rms velocity difference $\delta v(r)$ at small separations is determined by the condition that the relative acceleration $\sim v^2/r$ be balanced on average by gravity, for otherwise the galaxy clustering pattern would dissolve in less than a Hubble time. If the mass two—and three—point correlation functions agree with the galaxy functions then the balance condition is (Davis and Peebles 1983, eq. [44])

$$<(\mathbf{v}_1 - \mathbf{v}_2)^2> = KQ(Hr)^2 \, \xi(r) \, \Omega, \tag{6}$$

where (Davis and Peebles 1983; Groth and Peebles 1986)

$$\xi = (r_0/r)^\gamma, \; hr_0 = 5.4 \text{ Mpc}, \; \gamma = 1.77; \tag{7}$$

the amplitude of the three-point correlation function is $Q \sim 0.7$; the mean mass density is $\varrho = \Omega$ times the critical Einstein-de Sitter value; the galaxy pair separation is r; and the dimensionless factor is $K(1.77) \cong 12$. Equation (6) agrees with the observations (eq. [3]) if the density parameter is

$$\Omega = 0.2. \tag{8}$$

This value is adopted in the following.

The energy equation (Irvine 1965) fixes the galaxy rms peculiar velocity in terms of the mass autocorrelation function, as was first discussed by Fall (1975). A convenient form of the energy equation is (Peebles 1980, Sec. 74)

$$<v^2> = \frac{6}{7} \, \Omega H^2 \, J_2, \; J_2 = \int_0^\infty r dr \, \xi(r). \tag{9}$$

The best estimate of J_2 seems to be that of Clutton-Brock and Peebles (1981) from the Lick catalog:

$$J_2 = 164 \, e^{\pm 0.15} \, h^{-2} \text{ Mpc}^2. \tag{10}$$

With $\Omega = 0.2$ this gives

$$<v^2>^{1/2} = 530 \, e^{\pm 0.08} \text{ km sec}^{-1}. \tag{11}$$

Since the motion of the Local Group relative to the microwave background is about 600 km sec^{-1} (Wilkinson 1986) equation (11) seems reasonable.

Equation (1) is

$$\delta v(r)^2 = 2<v^2> - 2 <\mathbf{v}_1 \cdot \mathbf{v}_2>. \tag{12}$$

At large r we can use perturbation theory to find $<\mathbf{v}_1 \cdot \mathbf{v}_2>$ (Peebles 1980, Sec. 72):

$$<\mathbf{v}_1 \cdot \mathbf{v}_2> = 4\pi(Hf(\Omega))^2 \int_0^\infty dk\ P(k) \sin kr/kr,\ f \cong \Omega^{0.6}. \tag{13}$$

This with equation (5) for P(k) and equation (11) for $<v^2>$ shows how δv is expected to behave at large r.

To get a rough approximation to the integral in equation (13) let us note that on scales $20 \lesssim hr \lesssim 40$ Mpc $\xi(r)$ is small so the power spectrum is nearly flat, $P(k) \cong P_0$. In the approximation $P(k) = P_0$ equation (5) says

$$J_3 = \int_0^r r^2 dr\ \xi(r) \cong 2\pi^2 P_0, \tag{14}$$

and equation (13) is

$$<\mathbf{v}_1 \cdot \mathbf{v}_2> \cong (Hf)^2 J_3/r. \tag{15}$$

The estimate of J_3 from the Lick catalog is (Clutton-Brock and Peebles 1981)

$$J_3 = 596\ e^{\pm 0.2}\ h^{-3}\ \text{Mpc}^3. \tag{16}$$

Equations (11), (12), (15) and (16) yield

$$\delta v^2 = (750)^2 - (1300)^2/hr_{\text{Mpc}}\ \text{km}^2\ \text{sec}^{-2}. \tag{17}$$

In this approximation the coherence length (eq. [2]) is

$$r_c \sim 6\ h^{-1}\ \text{Mpc}. \tag{18}$$

The coherence length may be underestimated because J_3 in equation (16) assumes $\xi = 0$ at $r \geq 25\ h^{-1}$ Mpc. Because of the volume factor $r^2 dr$ the integral J_3 is very sensitive to the value of ξ at large r (and the value of J_3 in eq. [9] is less sensitive), so that a small positive tail of ξ at larger r could appreciably increase r_c. Examples of this effect, which would be expected in the baryon isocurvature model, are shown in Peebles (1987).

We have no direct estimates of r_c from the observations, though it may be possible to get useful constraints from the sample of Burstein *et al.* (1987). Until such results are available we can get some feeling for the velocity field expected under the present assumptions by considering the mean velocity averaged through a spherical window of radius r:

$$\mathbf{v}_a = \int_r \mathbf{v}\, dV/V. \tag{19}$$

The mean square value of v_a is (eq. [15], or eqs. [13] and [16] of Clutton-Brock and Peebles 1981)

$$<v_a^2> = \frac{6}{5}\,(Hf(\Omega))^2 J_3/r. \tag{20}$$

With $r = 40\ h^{-1}$ Mpc, $\Omega = 0.2$, and J_3 from equation (16) we get

$$<v_a^2>^{1/2} = 160 \text{ km sec}^{-1},\ r = 40\ h^{-1} \text{ Mpc}. \tag{21}$$

In an observation like that of Faber and Burstein (1988) one would move the sphere around to find a local maximum of $|\mathbf{v}_a|$; one would not have to look hard to find a 2 σ fluctuation so one would expect to see

$$v_a \sim 300 \text{ km sec}^{-1}, \tag{22}$$

at the depth $r = 40\ h^{-1}$ Mpc of the Great Attractor. This is a factor of about two below what Faber *et al.* (1987) observe. Given the rather large uncertainties in the observations of the Great Attractor and of J_3, I consider this to be tolerably reasonable agreement. R. Juszkiewicz is working on a more detailed check of this point.

3. Comments on the Biasing Picture

As is discussed in these proceedings by Kaiser and Lahav (1988), we could reconcile the velocity field observations with $\Omega = 1$ by assuming that mass clusters less strongly than do galaxies. This is a powerful concept and, as Efstathiou (1988) describes in these proceedings, it follows in a natural way in the scale-invariant cold dark matter model. There are of course problems. The most immediate, to my mind, is that the bias factor would have to be remarkably insensitive to the length scale. If equation (6) for δv^2 were biased because r_0^γ is overestimated by a factor $b^2 \sim 5$, then b^2 would have to be nearly constant at $0.01 \lesssim hr \lesssim 1$ Mpc, because the observed δv (eq. [3]) varies with r about as expected from equation (6). Since $\xi(r)$ varies by more than three orders of magnitude at $0.01 \lesssim hr \lesssim 1$ Mpc, it seems doubtful that one would have expected b^2 to be so nearly constant (unless $b^2 = 1$).

The statistic J_3 is mainly sensitive to $\xi(r)$ ar $r \sim 10\ h^{-1}$ Mpc. If $\xi(r)$ were correctly estimated from the galaxy distribution on this scale, and $\Omega = 1$, then we would expect, instead of equation (22), $v_a \sim 850$ km sec^{-1}. This would say that the Great Attractor is unusually weak, which would seem a little surprising. But the alternative assumption, that the bias factor has held nearly constant from 10 kpc to 10 Mpc separation, would also seem surprising.

4. Conclusion

The point of this discussion is that, under the assumption that mass clusters like galaxies, and with suitable adjustment of one parameter, the mean mass density, the predicted character of the galaxy peculiar velocity field is reasonably close to the observations. This assumption is not very popular because it would require that the mean mass density be less than the critical Einstein-de Sitter value. On the other hand, it seems to me that the fact that the assumption can be fitted to the observations in a simple and direct way is a considerable virtue, and that it therefore merits further close attention.

This research was supported in part by the National Science Foundation of the United States.

REFERENCES

Burstein, D., Davies, R., Dressler, A., Faber, S., Stone, R., Lynden-Bell, D., Terlevich, R., and Wegner, G. 1987. *Ap J.* **64,** 601.

Clutton-Brock, M. and Peebles, P.J.E. 1981. *AJ.* **86,** 1115.

Davis, M. and Peebles, P.J.E. 1983. *Ap J.* **267,** 465.

Efstathiou, G. 1988. this volume.

Faber, S. and Burstein, D. 1988. this volume.

Faber, S., Dressler, A., Davies, R., Burstein, D., Lynden-Bell, D., Terlevich, R., and Wegner, G. 1987. in *Nearly Normal Galaxies.* p. 175. ed. S.M. Faber. New York: Springer Verlag.

Fall, S.M. 1975. MN. **172,** 23p.

Groth, E.J. and Peebles, P.J.E. 1986. *Ap J.* **310,** 507.

Irvine, W.M. 1965. *Ann. Phys.* (N.Y.), **32,** 322.

Kaiser, N. and Lahav, O. 1988. this volume.

Peebles, P.J.E. 1980. *The Large-Scale Structure of the Universe.* Princeton: Princeton University Press.

_____. 1984. *Science.* **224,** 1385.

_____. 1987. *Nature.* **327,** 210.

_____. 1988. *Ap J.* in the press.

Rubin, V.C., Ford, W.K., Thonnard, N., Roberts M.S., and Graham, J.A. 1976. *AJ.* **81,** 687.

Sandage. A.R. 1986. *Ap J.* **307,** 1.

Wilkinson, D.T. 1986. *Science.* **232,** 1517.

THEORETICAL IMPLICATIONS OF SUPERCLUSTERING

AVISHAI DEKEL
Racah Institute of Physics, The Hebrew University of Jerusalem

ABSTRACT

I report here on three studies of the superclustering of clusters.

M. J. West, A. Oemler Jr. and myself have found, using N-body simulations, that the alignment of clusters of galaxies with the surrounding galaxy distribution is a useful discriminant between certain Gaussian cosmologies within the gravitational instability picture. The observed alignment of the Shane-Wirtanen galaxy counts with Abell clusters requires either a coherence length in the initial fluctuation spectrum on a scale of a few tens of megaparsecs, as in pancake scenarios or a flat ($n \leq -2$) power spectrum on such large scales. It is in conflict with hierarchical scenarios such as the Cold Dark Matter (CDM) model, which indicates a general weakness in 'filamentary structure' in such models.

G. R. Blumenthal, J. R. Primack and myself have estimated the effect of projection contamination in the Abell catalog on the cluster-cluster correlation function. A simple empirical test where pairs of small angular separations are excluded indicates only a $\sim$ 50% effect in amplitude near the correlation length. A statistical estimate relating the cluster auto-correlation function to the cross-correlation function of clusters and galaxies suggests that at least one of these functions is inconsistent with the conventional models of Gaussian fluctuations in an $\Omega = 1$ universe, such as standard CDM or pancake scenarios.

D. Weinberg, J. P. Ostriker and myself have found that superclustering in a generic explosion scenario substantially exceeds that in flat, Gaussian models. The two points where three shells intersect are found to be the natural sites for the formation of rich clusters. A simple toy model is used to study the spatial distribution of rich clusters in an explosion picture with a random distribution of seeds. The shell topology gives rise to a correlation function which is close to the observed power-law, with richer clusters having stronger correlations. The correlation amplitude is larger than observed by a factor of $\sim$ 2, but a natural anti-correlation of seeds reduces the superclustering to the observed level. Typical shell radii of a few tens of megaparsecs produce the observed number density of Abell clusters. Percolation superclustering analysis, cluster void probabilities, and topology tests confirm strong high-order correlations. Thus, the superclustering in certain explosion models is fairly consistent with the available observational data.

1. Introduction

This paper describes parts of three separate long-term projects. They address different implications of superclustering on the theory of the formation of large-scale structure in the universe.

The first project is an observational and theoretical effort by M. J. West, A. Oemler, Jr. and myself to study the structure and dynamics of rich clusters in N-body simulations of certain competing cosmological scenarios, in comparison with observed Abell clusters. Although clusters are non-linear systems, they might preserve some tracers of the initial conditions which led to their formation. What makes them appealing for such a study is that while they are relatively easy to identify and study, they are young dynamically and involve mostly gravity in their formation and evolution. We have found, contrary to what we had hoped for, that the light distibution in the observed central regions of clusters is similar in most aspects to the mass distribution of the simulated clusters in most of the scenarios. This general result tells us useful things, such as how efficient violent relaxation is in erasing the initial conditions, and how well the light profile traces the mass profile in clusters, but it means that the inner parts of clusters are not very helpful in trying to distinguish between cosmologies.

The outer regions of clusters are more promising, however. We find, for example, that subclustering at radii $\sim 5\ h^{-1}$ *Mpc* around clusters is sensitive to initial conditions and we suggest observations in that direction. Here, I will describe another property which we find useful for cosmology — the alignment of clusters with the surrounding galaxy distribution. We find that the observed large-scale alignment of the Shane-Wirtanen galaxy counts with Abell clusters (see Peebles 1988) indicates either a coherence length in the spectrum on a scale of a few tens of megaparsecs, as in pancake scenarios or a flat ($n \leq -2$) power spectrum on such large scales. The observed alignment is in conflict with hierarchical scenarios such as the Cold Dark Matter (CDM) model, which seems to reflect a general deficiency of large-scale 'filamentary' structure in CDM.

My second contribution in collaboration with G. R. Blumenthal and J. R. Primack, attempts to evaluate the validity of the excessive two-point correlation function of Abell clusters, which provides constraints on the formation of very large-scale structure. We estimate in two different ways the effect of contamination of Abell clusters by foreground or background galaxies on the cluster-cluster correlations (an effect suggested by W. Sutherland and N. Kaiser). We find the effect to be weak. Relating the cluster-galaxy correlation function and the effect of contamination on the cluster-cluster correlation function, we conclude that the inconsistency with conventional Gaussian models, such as CDM in an $\Omega = 1$ universe, does not go way so easily.

We studied a possible scenario which is capable of reproducing the observed superclustering (and the streaming motions on very large scales) within the framework of Gaussian fluctuations with a Harrison-Zeldovich spectrum. This is an open ($\Omega < 1$) hybrid model of CDM and baryons, which carries more large-scale power in the spectrum of fluctuations but is not associated with high-order correlations in the linear regime.

Based on a non-Gaussian cosmological model a collaboration, which started with S. Saarinen and B. J. Carr, is going on with D. Weinberg and J. P. Ostriker. We study the formation of large-scale structure in the explosion picture where positive-energy perturbations sweep matter onto high-density shells. The pairs of knots where three shells intersect are found to be deep potential wells which are the natural sites for the formation of rich galaxy clusters. We use a simple toy model to study the spatial distribution of rich clusters in a generic type of explosion scenario, concentrating on statistical measures of superclustering and especially on the cluster-cluster correlation function. The toy model, parameterized by the distribution of shell radii and the filling factor, places spherical shells at random and identifies each intersection point as a cluster.

The two-point correlation function, which results from the shell topology despite the random distribution of seeds, is a power-law consistent with observations, with the richer clusters forming at the intersections of bigger shells and so having stronger correlations. However, the correlation amplitude in our simple models is larger than observed by a factor $\sim 2 - 3$. The superclustering must be reduced by an anti-correlation of the explosion seeds — an effect which has a natural origin. Typical shell radii $\sim 25 - 50h^{-1}$ *Mpc* and filling factors $\sim 0.3 - 0.6$ are required to produce the observed number density of clusters. A toy model with a power-law radius distribution $\propto R^{-4}$ for $R < R_{max}$, can reproduce the richness distribution of clusters in the Abell catalog. Percolation supercluster statistics, cluster void probabilities, and topology tests confirm the presence of strong high-order correlations, reflecting the non-Gaussian nature of the superclustering. Thus, superclustering in the explosion scenario substantially exceeds that in the $\Omega = 1$ Gaussian models and is quite distinguishable from them. We conclude that certain explosion models are consistent with the available observational data.

2. Alignment of Clusters with their Surroundings: a Problem for CDM?

2.1 - Introduction

As an attempt to constrain the range of viable cosmogonic scenarios, rich clusters of galaxies have been studied in a series of papers by Michael J. West, Augustus Oemler Jr. and myself. We have compared N-body simulations of clusters formed from different cosmological initial conditions with observations. In paper I (West, Dekel, and Oemler 1987), we studied density profiles and velocity dispersion profiles. We found good agreement between the simulated mass profiles and the observed light profiles, indicating that the radial distribution of light and mass inside clusters follow each other, but the profiles were found to be poor discriminants between cosmological scenarios. In Paper II (West, Oemler, and Dekel 1988), we focused on substructures in and around clusters. We found the subclustering in the inner $\sim 3\ h^{-1}\ Mpc$ to be poorly correlated with the initial conditions, probably because of violent relaxation, but subclustering in the vicinity of rich clusters is a discriminant which might be useful. In Paper III (West, Dekel and Oemler 1988) we study alignments and ellipticities. I will describe here only one part of this work — an effect which has been found to provide a useful discriminant between competing cosmological scenarios. This is the alignment of clusters with the surrounding galaxy distribution.

Most clusters of galaxies are elongated to a degree which allows a relatively unambiguous definition of a major axis for each cluster. The position angles of first ranked cluster galaxies show a strong tendency to be aligned with the major axis of their parent cluster (Sastry 1986, Dressler 1980, Carter and Metcalfe 1980, Binggeli 1982, Struble 1987, Tucker and Peterson 1987). On larger scales, up to $\sim 30\ h^{-1}\ Mpc$, the clusters themselves exhibit a statistical tendency to be aligned with the lines connecting them to neighboring clusters (Binggeli 1982). Although a subsequent study (Struble and Peebles 1985) found the evidence for such an alignment to be weaker, more recent work (Rhee and Katgert 1987) gives more support to the effect.

In a recent study of a possibly related effect on intermediate scales, Argyres *et al.* (1986, hereafter AGPS) and Lambas, Groth, and Peebles (1988, hereafter LGP) have found that Shane-Wirtanen galaxy counts in the regions surrounding Abell clusters tend to be systematically higher along the direction defined by the cluster major axis (or by the position angle of its first ranked galaxy). This suggests a correlation between the orientation of rich clusters and the large-scale galaxy distribution which extends to at least $15h^{-1}$ *Mpc*. Taken together, these various observations indicate the alignment of structure on scales from several tens of kiloparsecs on up to several tens of megaparsecs. These alignments are one aspect of what is sometimes referred to as 'filamentary structure' in the distribution of galaxies. Accounting for these observations presents a formidable challenge to any theory of the formation of structure in the universe.

The cosmogonic scenarios which we address in this work all assume pure gravitational clustering from small-amplitude, Gaussian density fluctuations. Depending on the exact form of the initial spectrum of fluctuations, the nature of the dark matter and the cosmological model, structure may have formed either by clustering bottom-up from small to large scales, or via fragmentation top-down from large to small scales. In CDM, for instance, the sequence proceeds bottom-up, with nearly simultaneous collapse of systems on galactic scales (Peebles 1982, Blumental and Primack 1983). In the case of hot dark matter, like neutrinos, small-scale fluctuations would have been erased and thus large-scale perturbations would have been the first to collapse, resulting in the formation of flattened superclusters ('pancakes'), followed by subsequent fragmentation to galaxies and clusters (Zeldovich 1970; Doroshkevich, Shandarin, and Saar 1978). Hybrids of these two extreme senarios are also possible, as might result from the presence of different types of perturbations or different types of dark matter (e.g. Dekel 1983, 1984a, 1984b; Dekel and Aarseth 1984), or if the universe underwent more than one inflationary phase (Silk and Turner 1987, Turner *et al.* 1987).

Qualitative arguments suggest that elongated and aligned clusters could arise in various cosmogonies. The flattening cannot be accounted for by rotation (e.g. Rood *et al.* 1972, Dressler 1980, Noonan 1980); anisotropic velocity dispersion is required. In general, there is high probability for any initial perturbation to be aspherical (Doroshkevich 1970, Bardeen *et al.* 1986). Any initial asphericity is amplified during gravitational collapse (Lin, Mestel and Shu 1965) and some of it is preserved later on (e.g. Aarseth and Binney 1978). It would seem natural to expect flattened clusters which tend to be correlated with one another in the pancake scenario, where large-scale collapse to sheets and filaments is expected to occur first. Hierarchical scenarios are expected in general to show weaker large-scale alignments, with decreasing strength for steeper spectra (larger power index n). The CDM spectrum, for example, is expected to yield certain filaments on galactic scales, where the spectrum is flat ($n \simeq -2$) and objects of different scales collapse almost simultaneously, but not on scales of $\sim 10\ h^{-1}$ *Mpc* and up, where the spectrum is steeper ($n \simeq 0$) (Nusser and Dekel 1988). Nevertheless, the situation might be more complex. For instance, elongation and alignment might arise in hierarchical clustering as a result of tidal interactions between neighboring protoclusters (Binney and Silk 1979, Palmer 1981, 1983). Even though this has been found by N-body simulations (Dekel, West, and Aarseth 1984, Dekel 1984b) not to be enough to explain the Binggeli alignment, it might still produce some alignments on smaller scales. Also, simulations of cluster formation in hierarchical clustering (e.g. White 1976, Cavaliere *et al.* 1986) have shown that subclustering during the cluster collapse is also capable of producing an overall flattening of the galaxy distribution which persists long after the collapse. These effects on the resultant alignments have to be quantified.

The simulations used in this study are described brifly in Sec. 2.2. In Sec. 2.3 the alignment is measured in the simulated clusters of the different scenarios and compared with the observations of AGPS. Our conclusions are in Sec. 2.4.

2.2 - *The Simulations*

In order to generate high-resolution cluster simulations beginning from realistic cosmological initial conditions, a novel approach for stretching the dynamical range was used, in which the clusters were simulated in two separate stages. In step I, low-resolution, large-scale cosmological simulations were performed in order to find the locations where protoclusters formed for a given random realization of the initial conditions. These results were then used to generate the initial conditions for step II, in which high-resolution simulations of individual clusters and their surroundings were performed. Such an approach makes it possible to study the detailed properties of rich clusters with sufficient resolution to detect systematic differences which may arise due to the different initial conditions.

The desired initial fluctuation spectra for the different cosmologies were generated using a method based on the approximation of Zeldovich (1970) for describing the linear evolution of fluctuations. The spectra considered here have the general power-law form

$$\langle |\delta_{\vec{k}}|^2 \rangle \propto k^n \tag{2.1}$$

over certain intervals of wave number k. Simulations were performed for the following five initial spectra: a) a pancake scenario with $n = 0$ on large scales and no fluctuations below a critical coherence wavelength, b) a hybrid scenario where a large-scale component ($n = 0$) with a coherence length as in the pancake scenario is combined with a small-scale component ($n = 0$) whose amplitude was one-half that of the large-scale component at the coherence length, and c), d), and e) three hierarchical clustering scenarios with power spectrum indices $n = 0, -1$, and -2 respectively. The CDM spectrum can be approximated near the relevant scales for rich clusters by either the $n = 0$ or $n = -1$ models, while $n = -2$ is more appropriate for CDM on the scale of galaxies. All of these simulations assumed an Einstein-de Sitter universe

($\Omega = 1$). (The $n = 0$ simulations were also repeated for the case of an open universe, $\Omega_0 = 0.15$, but the clusters there ended up relatively isolated so that there were not enough particles in the regions surrounding the clusters to allow meaningful alignment analysis.) The large-scale simulations of step I were performed with $\sim$ 4000 equal-mass particles using a comoving version of a direct N-body code (Aarseth 1985). Four random realizations were performed for each of the different theoretical scenarios. The stages of the simulations that correspond to the present epoch were determined by matching the slope of the two-point correlation function, $\xi(r)$, with the slope $\gamma = 1.8$ of the observed galaxy correlation function, ignoring biasing. Equating the correlation length of the simulations with the claimed value for galaxies, $r_0 \simeq 5h^{-1}$ *Mpc* (Davis and Peebles 1983; although see Oemler *et al.* 1988), then sets the scaling from simulation to physical units. With this scaling, the diameter of the simulated volumes corresponds to $\sim 100h^{-1}$ *Mpc*, and the coherence length in the pancake and hybrid scenarios results in superclusters of roughly $30h^{-1}$ *Mpc* in diameter. Rich clusters were then identified in these large-scale simulations using a simple group-finding algorithm described in Paper I. The five richest clusters found by this procedure in each simulation, with a mean number density of $\simeq 6 \times 10^{-6}\,(h^{-1}\,Mpc)^{-3}$, were assumed to correspond to Abell clusters of richness $R \geq 1$.

Having identified the rich clusters in each simulation of step I, new simulations were then performed using the same initial fluctuations, but now modeling smaller volumes centered on the locations of each of the clusters (a total of 20 clusters per cosmogonic scenario). The radius of these volumes was 45% that of the large-scale simulations. These high-resolution simulations of individual clusters were run using a non-comoving version of the Aarseth code, with $\sim$ 1000 equal-mass particles. With the adopted scaling each particle in these simulations should correspond roughly to an L_* galaxy.

In order to study the simulated clusters in a manner similar to that used by observers, three orthogonal projected views of each cluster have been used. A cubic volume with origin at the center of the simulated volume was cut to ensure equal depth along the line of sight. The cluster center in each case was then taken as the location of the density maximum of the projected distribution of particles within this cube, as determined by an iterative count procedure using square grid cells. Several representative clusters formed in the different cosmological scenarios are shown in Figure 2.1.

2.3 - Statistical Analysis of Alignment

In order to compare the simulations with available observational results, the same test used by AGPS to analyse the observational catalogs has been applied here to the simulated clusters. Each projected view of each simulated cluster was assigned a redshift drawn at random from the cluster redshifts in

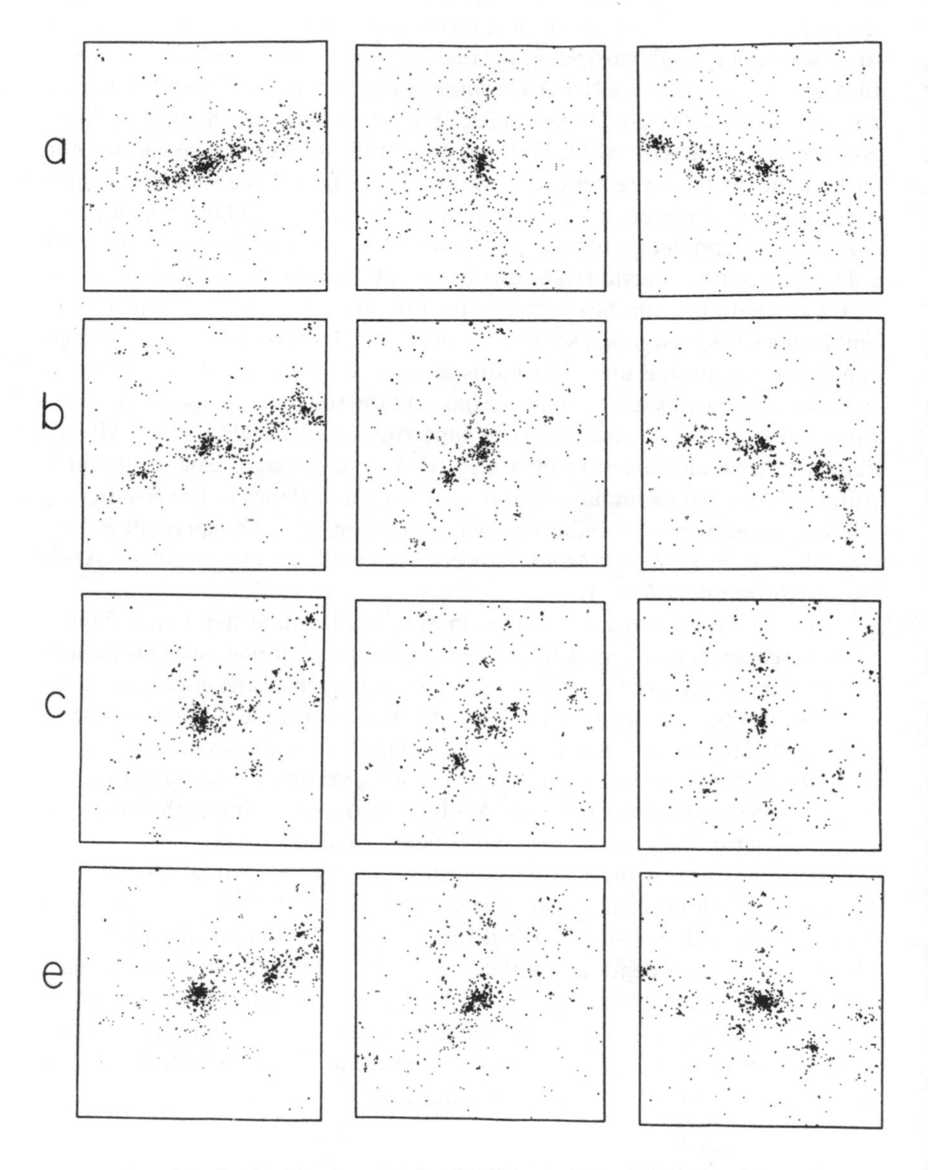

FIG. 2.1. Projected galaxy distribution in typical simulated clusters formed in the different cosmological scenarios. Labels denote: (a) pancake scenario (b) hybrid scenario; (c), and (e) hierarchical clustering scenarios with $n = 0$ and -2 respectively. Each box is 30 h^{-1} *Mpc* on a side. Each particle roughly represents an L_* galaxy. The large-scale fluctuations were chosen from the same set of random numbers so that the clusters in the different scenarios are directly comparable.

the sample of 137 clusters with $z \leq 0.1$ and $b \geq 40$ which comprised the data set used in the study of AGPS (83 of the clusters in their sample were taken from the earlier study of Struble and Peebles 1985). Whereas AGPS determined the major axis of each cluster by eye, from either the galaxy distribution on the palomar Sky Survey prints or from the position angle of the brightest cluster galaxy, the major axis of each of the simulated clusters was found here from the moments of inertia of all particles within a distance corresponding to a projected angular separation $\Theta = 0.25°$ from the cluster center. Then, the mean surface density of particles in different angular and radial bins around each cluster was measured. Following AGPS, four circular rings of radius Θ centered on each cluster were used: (0.25°, 0.5°), (0.5°, 1°), (1°, 2°) and (2°, 4°). [AGPS also used a bin (4°, 8°) which was not possible to use here because of the limited size of our simulations. They detected no clear signal of alignments in that bin anyway.] For comparison, an angle of $\Theta = 4°$ subtends a distance of $\sim 17\ h^{-1}\ Mpc$ at a redshift of $z = 0.1$ (for $\Omega = 1$). The relative position angle ϕ of each particle was then defined as the angle between the radius vector to the particle and the cluster major axis. Next, for each radial bin, Θ, the excess surface number density of particles, $\eta(\phi)$, was determined by first counting particles in bins of 10° in ϕ, and then subtracting the mean surface density of particles within the given radial bin. Once this was done for all simulated clusters, the means and standard deviations of $\eta(\phi)$ were computed for all clusters of a given cosmological scenario. The results for some of our scenarios are presented in Fig. 2.2. For comparison the results of AGPS are reproduced in Fig. 2.3.

For the pancake scenario (Fig. 2.2a), the particle counts show a clear tendency to be systematically higher along the direction defined by the cluster major axis; this effect extending to $\Theta = 4°$ and perhaps more. Moderate alignment is detected in the hybrid scenario and in the case n = −2. No alignment is detected for n = 0 or n = 1.

To provide a more quantitative measure AGPS fit the $\eta(\phi)$ data for each Θ bin to a function of the form $a\ cos(2\phi)$, by defining for each cluster, i,

$$a_i(\Theta) = \frac{2}{9} \sum_{k=1}^{9} \eta_i(\Theta, \phi_k) cos(2\phi_k), \tag{2.2}$$

where $\phi_k = \pi(2k - 1)/36$ radians. Then, the mean value for all clusters, $a(\Theta)$, was computed, along with its standard deviation $\sigma(\Theta)$. As a measure of the significance of the alignment, a χ^2 is defined by

$$\chi^2 = \sum_{j=1}^{m} \left(\frac{a(\Theta_j)}{\sigma(\Theta_j)} \right)^2, \tag{2.3}$$

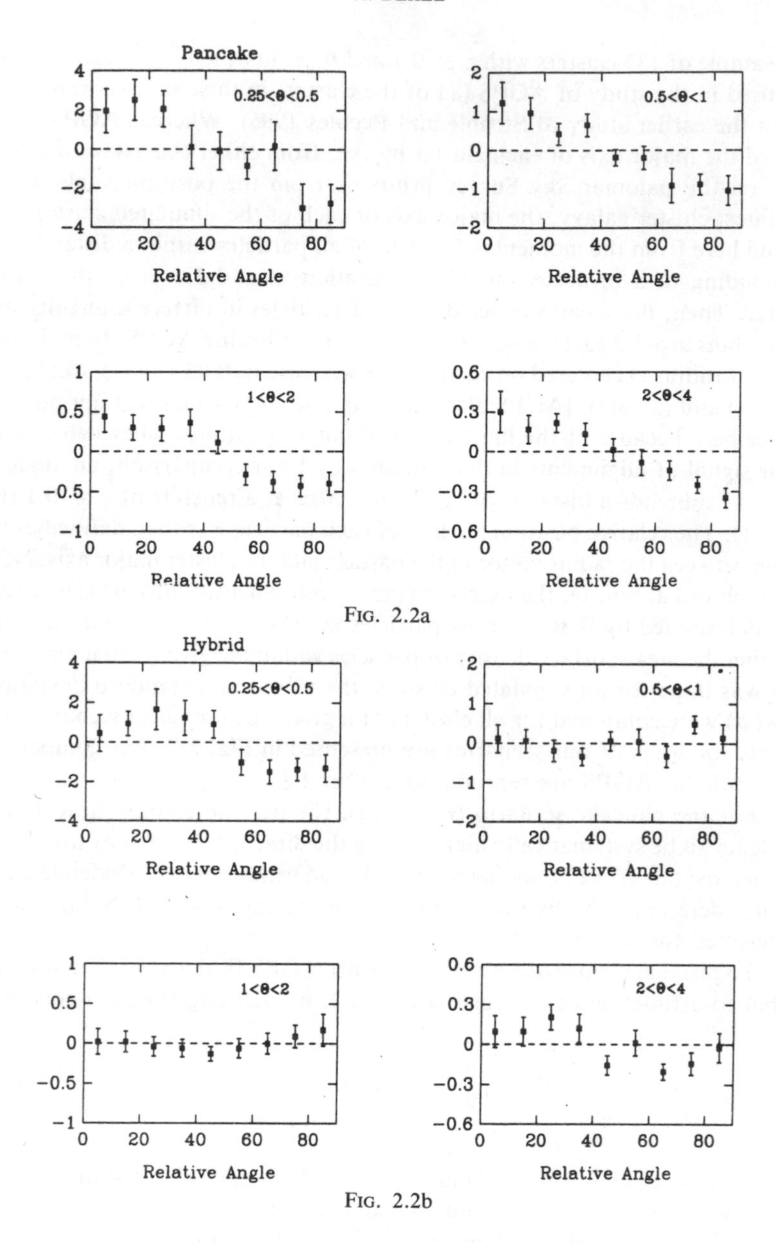

FIG. 2.2a

FIG. 2.2b

FIG. 2.2. Alignment in the simulated clusters of the different scenarios. Shown is the surface density of galaxies, η, as a function of the azimuthal angle ϕ as measured from the cluster major axis ($\phi = 0$), for different radial bins (Θ): (a) pancake scenario; (c), (d) and (e) hierarchical scenarios with $n = 0$, -1, and -2 respectively.

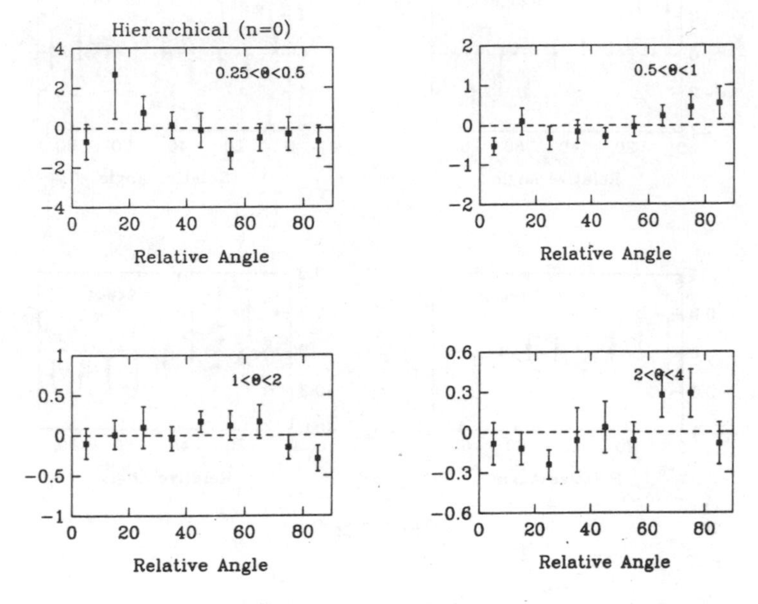

FIG. 2.2c

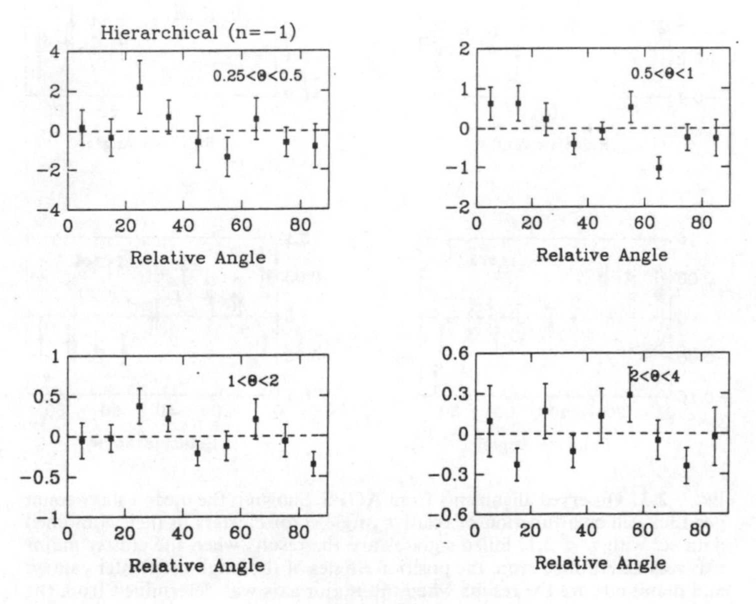

FIG. 2.2d

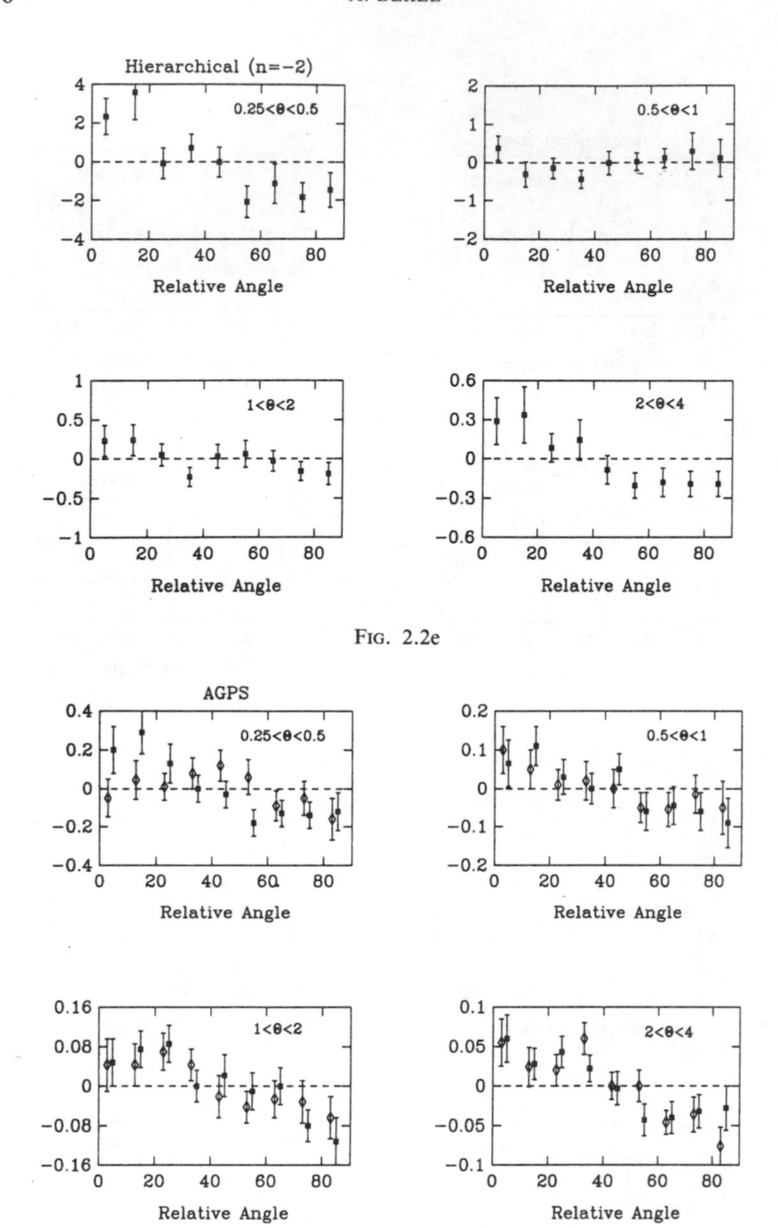

FIG. 2.2e

FIG. 2.3. Observed alignments from AGPS. Shown is the mean galaxy count per Lick cell as a function of relative angle ϕ for clusters in their combined data set with $z < 0.1$. Filled squares are the results when the cluster major axis was determined from the position angles of the brightest cluster galaxy, and diamonds are the results when the major axis was determined from the galaxy distribution within the inner regions of the cluster.

where m is the number of Θ bins. From their sample of 137 clusters (the combined data sets 1 + 2 in their paper), AGPS find $\chi^2 = 21.3$ when the cluster major axes were determined from their brightest galaxies, and $\chi^2 = 13.9$ when the galaxy distribution in the inner regions of the cluster was used. LGP find $\chi^2 = 11.5$ and 17.1, respectively. This statistic should approximate the usual χ^2 distribution with m degrees of freedom (in this case $m = 5$), i.e. $\chi^2 \simeq 5$ for no significant alignment. Because the probability of having measured the above values of χ^2 in the absence of any true alignments is exceedingly small, this is a strong signal of large-scale alignment of clusters with the surrounding galaxy distribution.

This same analysis has been applied here to the results of the simulated clusters shown in Fig. 2.2. Values of χ^2 found for each of the different scenarios are listed in Table 1, along with the corresponding probabilities of having measured such a value of χ^2 in the absence of any true alignments (here $m = 4$ instead of 5). The *only* scenarios which yield results comparable or more significant than those observed are the panacake, hybrid, and n = −2 scenarios.

TABLE 1

χ^2 VALUES AND CORRESPONDING PROBABILITIES

Scenario	χ^2	$P(\chi^2)$
pancake	43.77	< 0.001
hybrid	11.47	0.002
hierarchical n = 0	6.44	0.184
hierarchical n = −1	5.03	0.294
hierarchical n = −2	18.44	<0.001
AGPS (observed north)		
brightest galaxy	21.3	0.0008
clusters core	13.9	0.02
LGP (observed south)		
z ≤ 0.055	11.5	0.042
z ≤ 0.075	17.1	0.0042

2.4 - Conclusion

The orientations of clusters with respect to the galaxy distribution on larger scales were found to provide a very interesting test. A very pronounced tendency for alignments is found in the pancake scenario, on scales up to at least $10h^{-1}$ *Mpc*. Alignments were seen in the $n = -2$ case and in the hybrid case. No detectable alignments were found for $n = 0$ or $n = -1$ hierarchical clustering scenarios, which approximate CDM on cluster scales. (The ellipticities of the main bodies of the simulated clusters, on the other hand, were found to be independent of the initial conditions and therefore not very useful for cosmology.)

That the pancake scenario leads to alignments between clusters and the large-scale galaxy distribution is not too surprising, since the coherence length in the initial fluctuation spectrum results in the formation of large-scale filaments and sheets with which the clusters are associated. Likewise, it would seem that the lack of such alignments in the $n = 0$ and $n = -1$ hierarchical clustering scenarios reflects the dominance of small-scale fluctuations which results in a different sequence of formation in which clusters collapse and virialize before supercluster formation, and are not much correlated with them. These results are in agreement with the earlier results of Dekel, West, and Aarseth (1984) concerning cluster-cluster alignment, and indicate that tidal forces do not have as significant an effect on the final shapes and orientations of clusters as naive estimates might suggest. That a positive, though weaker, signal of alignment is detected in the $n = -2$ case and in the hybrid model reflects the important contribution of large-scale fluctuations there, which both induce tidal effects and make the collapses of clusters and superclusters follow each other in quicker succession so that there is a cross-talk between them.

The observational evidence for large-scale alignments of the galaxy distribution around Abell clusters found by AGPS and LGP is found to be consistent with that predicted by the pancake scenarios or the $n \leq -2$ hierarchical models. The hierarchical models with $n = 0$ and $n = -1$, which approximate CDM on the relevant scales, reproduce no detectable alignments, in clear conflict with the observations. [The fact that analytic estimates based on Gaussian fluctuations (J.R. Bond, private communication) do seem to predict some sort of alignments in CDM might be explained by the different definition used for 'alignment' which is not directly comparable to the observed property.]

I would like to argue that this alignment is a measure of one aspect of the general concept sometimes referred to as 'filamentary' structure. Our results indicate that standard CDM does have difficulties in reproducing the filamentary distribution of galaxies on scales beyond $10h^{-1}$ *Mpc* (see also Dekel 1984b; Dekel, West and Aarseth 1984). It warns us that the large-scale filaments seen in some of the N-body simulations of CDM (White *et al.* 1987) might be artifacts of the numerical procedure, as has been argued recently by P.J.E. Peebles (private communication).

More work is needed to determine whether or not similar alignments could also be produced by models which are not based solely on gravitational instability; for example, we are in a process of studying alignments in the explosion scenario.

3. Contamination of the Cluster Correlation Function: is the Superclustering Problem Real?

3.1 - Introduction

Perhaps the strongest clue for the presence of very large-scale structure in the universe is provided by the spatial distribution of rich clusters of galaxies. The two-point correlation function of Abell clusters (Abell 1958), $\xi_{cc}(r)$, is about twenty times stronger than the galaxy-galaxy correlation function, and it remains positive out to about 50 h^{-1} *Mpc* (see Bahcall 1988a for a review). These strong correlations are in clear conflict with the traditional cosmogonic scenario which assume Gaussian initial density fluctuations in an Einstein-deSitter ($\Omega = 1$) universe (e.g. Barnes *et al.* 1985). In particular, the 'standard' cold dark matter scenario (CDM), when normalized to fit either the distribution of galaxies or the isotropy of the microwave background, fails to reproduce the observed ξ_{cc} (Bardeen, Bond, and Efstathiou 1987; Batuski, Melott, and Burns 1987; White *et al.* 1987). CDM predicts that ξ_{cc} should become negative beyond $\sim 20\ h^{-2}$ *Mpc*, because if clusters form at high density peaks then $\xi_{cc}(r) \propto \xi(r)$ at large separations (Kaiser 1984), and the matter two-point correlation function, $\xi(r)$, is predicted to go negative at $\sim 20\ (\Omega h^2)^{-1}$ *Mpc* in a CDM universe.

The important theoretical implications of the apparently strong cluster correlations, which I will partly address in the next section, motivate first a careful evaluation of the reality of the observed result. The limited number of clusters in the redshift surveys leads to quite large uncertainties in ξ_{cc} (e.g. Ling, Frenk, and Barrow 1986). But the most serious questions concerning the validity of the measured ξ_{cc} as an indicator for real large-scale clustering arise because of possible systematic selection effects. Sutherland (1988), whose work has been discussed in this meeting by Nick Kaiser, has raised the possibility that the high amplitude of ξ_{cc} is mostly due to the effect of contamination intrinsic to the Abell catalog (or similar cluster catalogs). The effect is overcounting pairs of small angular separations as a result of mutual contamination of the galaxy counts in the one cluster by galaxies of the other. The general contamination in the Abell catalog by foreground and background galaxies is a well-known problem; it has previously been estimated, for example, by Lucey (1983) using Monte Carlo realizations. This effect, Sutherland argues, is responsible for the anisotropy detected by Bahcall, Soneira and Burgett (1986) in the distribution of cluster pair separations in redshift space. When they plot separation in redshift against angular separation (i.e. projected on the sky), they find a strong excess of pairs with small angular separations, mostly below 10 h^{-1} *Mpc*. While these authors blame the effect on cluster peculiar velocities on the order of 2000 km s^{-1}, namely the so called 'fingers of God', Sutherland argues that the anisotropy extends to too high redshift separations, and is a natural result of the projection effect as estimated by him.

We are investigating the contamination effect on ξ_{cc} of the northern Abell sample. I describe below preliminary results which ignore the effect of cluster correlation functions of order ≥ 2.

First, a brief description of the Abell procedure in classifying the clusters in his catalog. For any cluster which he detected on the Palomar Sky Survey plates, he estimated the distance based on the magnitude, m_{10}, of the 10th brightest galaxy, which is assumed to be a standard candle. Using this distance he counted galaxies within a circle of a radius which corresponds to $r_A = 1.5\ h^{-1}\ Mpc$ about the cluster center. Galaxies are included only if their magnitude falls in the band $(m_3,\ m_3 + 2)$, where m_3 is the magnitude of the third brightest galaxy. The background count in a reference field on each plate, chosen to be away from any cluster, has been subtracted out. The clusters were classified into richness classes $R = 0,1,\ \ldots\ 5$ if their net count fell in the range 30-49, 50-79, ... 300 - respectively. The clusters were also assigned a distance class $D = 1,2,\ \ldots\ 6$ according to m_{10}. The outer boundary of $D = 4$ is at about 300 $h^{-1}\ Mpc$ ($z \simeq 0.1$) and the $D = 6$ clusters extend to $\simeq 600\ h^{-1}\ Mpc$. The richness classification is assumed to be independent of distance out to $D = 6$, and the catalog is assumed to be complete for $R \geq 1$. There are 102 clusters in the $R \geq 1$, $D \leq 4$ sample, which, together with the cluster redshifts, is the main source for the current direct estimates of ξ_{cc}.

3.2 - *Simple Empirical Estimate*

To obtain the most simple estimate we compare the correlation function of the 102 clusters in the redshift sample as is, with that obtained eliminating the pairs with small projected separations — those which are blamed for the correlation excess.

We first calculate $\xi_{cc}(r)$ as in Bahcall and Soneira (1983, hereafter BS). This is shown in Figure 3.1. It can be approximated by

$$\xi_{cc}(r) = (r/r_o)^{-\gamma},\ r_o = 25\ h^{-1}\ Mpc,\ \gamma = 1.8 \qquad (3.1)$$

(also Klypin and Kopylov 1983). Then, we eliminate from the pair counts, both in the Abell sample and in the appropriate Poissonian sample which we use as a reference, all pairs that are of projected separation less than 10 $h^{-1}\ Mpc$. The resultant correlation function is also shown in Fig. 3.1. The difference between the two correlation functions is small; less than a factor of 25% in amplitude on any scale.

This empirical test might suffer from over-simplification, but, nevertheless, it does indicate that the projection effect is quite small, unlike what is expected based on the estimate of Sutherland.

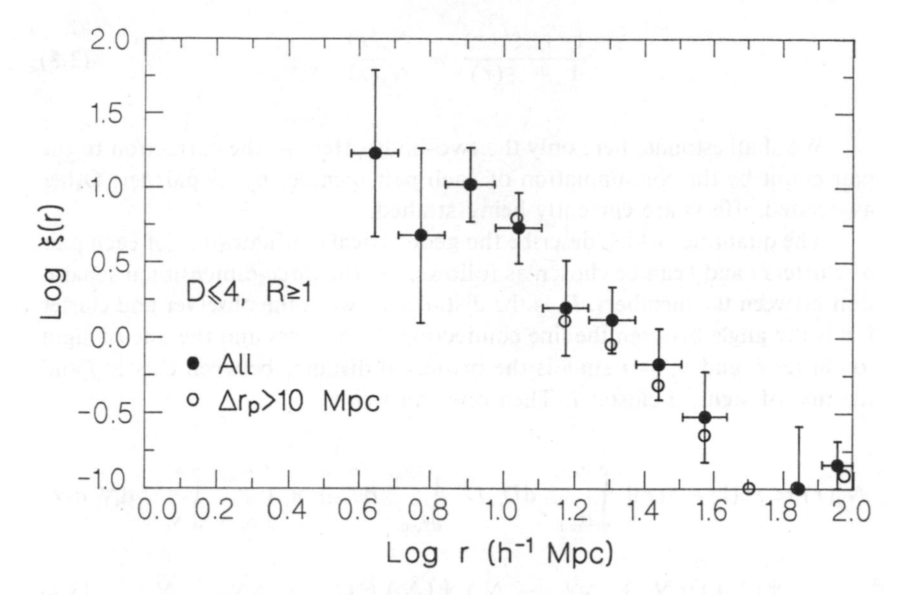

FIG. 3.1. Empirical test: The cluster-cluster correlation function in redshift space of the 102 $R \geq 1$, $D \leq 4$ Abell clusters (error bars) is compared with the correlation function of the same sample from which all pairs of angular separations less than 10 h^{-1} *Mpc* were excluded.

3.3 - *Statistical Estimate of the Two-Body Effect*

We now wish to estimate the expected effect of contamination on ξ_{cc}, for assumed number density profiles of galaxies in extended halos around the clusters.

Write the desired cluster two-point correlation function of the "pure" clusters using the integral definition

$$1 + \xi(r) = N_p(r)/N_{pp}(r), \tag{3.2}$$

where $N_p(r)$ is the number of cluster pairs with separation in the interval $(r, r + dr)$, and $N_{pp}(r)$ is the expected number of pairs in an equivalent Poisson distribution. Denote the analogous quantities as derived from the contaminated catalog by $\xi'(r)$, $N_p'(r)$ and $N_{pp}'(r)$ respectively. We wish to calculate $[1 + \xi'(r)]/[1 + \xi(r)]$.

The Poisson pair counts scale with the total number of pairs within the volume. Most pairs are of large angular separation, on the order of the volume size R, which are not expected to be affected by contamination. Thus, the ratio $N_{pp}'(r)/N_{pp}(r)$ is much closer to unity than the ratio $N_p'(r)/N_p(r)$ for $r \ll R$, and the desired effect is therefore given, to first order, by

$$\frac{1 + \xi'(r)}{1 + \xi(r)} = \frac{N_p'(r)}{N_p(r)}. \quad (3.3)$$

We shall estimate here only the two-body effect — the correction to the pair count by the contamination of each pair member by its partner. Other associated effects are currently being studied.

The quantities which describe the geometrical configuration of each pair of clusters i and j can be chosen as follows: r is the three-dimensional separation between the members, D_i is the distance between the observer and cluster i, θ is the angle between the line connecting the clusters and the line of sight to cluster i, and $x_{ij} = r \sin\theta$ is the projected distance between cluster j and the line of sight to cluster i. Then one can write

$$N_p'(r) \propto r^2[1 + \xi(r)] \int_{D_{\min}}^{D_{\max}} \mathrm{d}D_i\, D_i^2 \int_{\theta_{\min}}^{\theta_{\max}} \mathrm{d}\theta \sin\theta \int_{N_i=0}^{\infty} \int_{N_j=0}^{\infty} \mathrm{d}N_i\, \mathrm{d}N_j$$

$$\Psi(N_i)\, \Theta(N_i + \Delta N_{ij} - N_c)\, \Psi(N_j)\, \Theta(N_j + \Delta N_{ji} - N_c). \quad (3.4)$$

The r-dependent term in front of the integrals is the same for $N_p(r)$ and it therefore drops out from the ratio $N_p'(r)/N_p(r)$. The function $\Psi(N)$ is the multiplicity function of Abell clusters such that $\Psi(N)\mathrm{d}N$ is the number density of clusters with Abell count of galaxies in the range $(N, N + \mathrm{d}N)$. $N_c = 50$ is the critical count which borders between richness classes $R = 0$ and $R = 1$. The contribution to the counts in cluster i by galaxies of cluster j is given by the correction term ΔN_{ij}. $\Theta(N)$ is the usual step function; only pairs in which the "contaminated" richness of each cluster, $N + \Delta N$, is larger than N_c, qualify to be included in the pair count.

The correction term can be written as a product of three terms:

$$\Delta N_{ij}(D_{ij}, x_{ij}, N_j) = D(D_{ij})\, \sigma(x_{ij})\, N_j. \quad (3.5)$$

The distance dependence is given as a function of D_{ij}, the ratio of the distance to cluster j projected on the line of sight to i and D_i. We assume a universal shape for the projected density profile around each cluster, $\sigma(x)$, which is weighted by the Abell number count N.

The distance dependence of the correction term is calculated from

$$D(D_{ij}) = D_{ij}^2\, \pi r_A^2 \int_{L_2 D_{ij}^2}^{L_3 D_{ij}^2} \mathrm{d}L\, \Phi(L)/\bar{n}, \quad (3.6)$$

where $\Phi(L)$ is the galaxy luminosity function and L_3 and L_2 are the intrinsic luminosities corresponding to the magnitudes m_3 and $m_3 + 2$ which define the Abell band. The varying lower limit of the integral takes into account the fact that some foreground galaxies, which are intrinsically fainter than L_2, are included in the Abell counts as if they were brighter than L_2, because their distance is overestimated when they are assumed to belong to the cluster under counting.

The projected number density profile of each cluster, assuming spherical symmetry, is related to the three-dimensional number density profile, $n(r)$, via

$$\sigma(x)\mathrm{N} = \int_x^{R_{max}} dr\, n(r)\, \mathrm{r}\, (r^2 - x^2)^{-1/2}. \tag{3.7}$$

For the profile $n(r)$ we use a most crucial observational input: the cluster-galaxy correlation function, $\xi_{cg}(r)$, of Abell clusters of richness $R \geq 1$ and galaxies from the Lick counts. A possible realization of ξ_{cg} is an ensemble of randomly distributed clusters of richness $R \geq 1$, with density profiles weighted by their richness,

$$n(N, r) = \bar{n}\, \xi_{cg}(r)\, \mathrm{N}/\langle N\rangle, \tag{3.8}$$

out to a large halo radius, R_{max}. Here $\langle N\rangle$ is the mean richness

$$\langle N\rangle = \int_{N_c}^{\infty} \mathrm{d}N\, \Psi(N)\, N \Big/ \int_{N_c}^{\infty} \mathrm{d}N\, \Psi(N), \tag{3.9}$$

and $\bar{n}$ is the mean number density of galaxies with luminosities in the Abell band,

$$\bar{n} = \int_{L_2}^{L_3} \mathrm{d}L\, \Phi(L). \tag{3.10}$$

Thus, we need three observational inputs: $\Phi(L)$, $\Psi(N)$, and $\xi_{cg}(r)$ with R_{max}. For the galaxies we adopt a Schechter Luminosity function,

$$\Phi(L) = \frac{\Phi_*}{L_*}\left(\frac{L}{L_*}\right)^{-\alpha} e^{-L/L_*}. \tag{3.11}$$

Following the Abell definition of the luminosity band one obtains for $\alpha = 1.3$

$$L_2 = 0.28\, L_*, \quad L_3 = 1.75\, L_*. \tag{3.12}$$

Then, in Eq. (3.10), $\bar{n} = 1.07\Phi_*$. We adopt the Schechter parameters

$$\alpha = 1.3, \quad \Phi_* = 0.01\ (h^{-1}\, Mpc)^{-3}. \tag{3.13}$$

A more uncertain input is the Abell cluster multiplicity function $\Psi(N)$ in the vicinity of $N = N_c$. The steepness of this function just below N_c determines how many clusters have real richnesses just below N_c, that can easily be upgraded by contamination into richness class $R = 1$. Figure 3.2 shows the distribution of cluster richnesses in the Abell catalog, binned in intervals of $\Delta N = 5$. There is an obvious bend near N_c which may reflect the onset of the catalog incompleteness below N_c. We use different fits to $\Psi(N)$ of the $D \leq 4$ sample, using power-laws of the form

$$\Psi(N) \propto (N/N_c)^{-\psi}. \tag{3.14}$$

A power-law with $\psi = 3.5$ is a reasonable fit in the range $40 \leq N \leq 60$ about N_c. It serves as an upper limit on smaller scales, where the distribution of groups in the catalog of Gott and Turner (1977) can be fitted by $\psi = 2$ (see the next section). A power of $\psi = 5$ provides an upper limit on the possible power below N_c. We consider only clusters of $N \geq 30$ (i.e. $R \geq 0$).

For the cluster-galaxy correlation function we test two alternative observational results. The old result of Seldner and Peebles (1977, hereafter SP)

$$\xi_{cg}(r) = (r/7h^{-1}\,Mpc)^{-2.5} + (r/12h^{-1}\,Mpc)^{-1.7}, \tag{3.15}$$

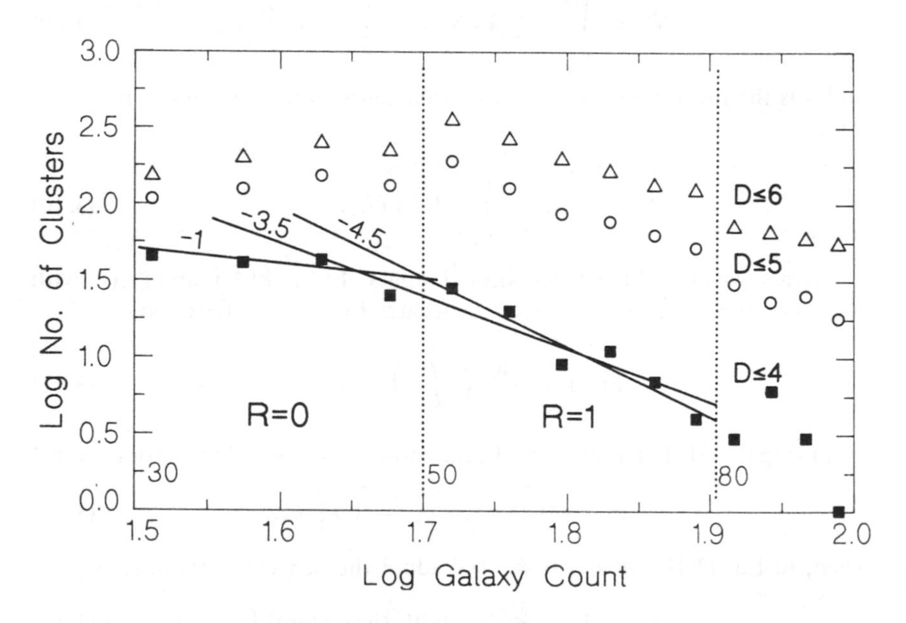

FIG. 3.2. Multiplicity function of Abell clusters in bins of 5 galaxies.

and the recent result of Lilje and Efstathiou (1988, hereafter LE)

$$\xi_{cg}(r) = (r/8.8h^{-1}\,Mpc)^{-2.2}. \tag{3.16}$$

The ξ_{cg} of SP has strong power on large scales, indicating extended halos of galaxies around clusters. This ξ_{cg} is in conflict with 'standard' CDM. The LE result, which is different mostly because of a difference in the adopted luminosity function for the Lick galaxy counts, indicates more limited cluster halos, in better agreement with CDM. In each case we tried two alternative maximum radii for the halos around the clusters, $R_{max} = 25\,h^{-1}\,Mpc$ and $50\,h^{-1}\,Mpc$. Recall that the mean separation between $R \geq 1$ neighboring Abell clusters is $55\,h^{-1}\,Mpc$, so the assumed values for R_{max} should lead to an overestimate of the contamination effect.

The correction to the observed $\xi_{cc}(r)$ of the contaminated catalog (BS) is calculated by evaluating the multiple integral (3.4) under the assumed $\xi_{cg}(r)$, $\Psi(N)$ and $\Phi(L)$. The resultant 'pure' $\xi_{cc}(r)$ are shown in Figure 3.3 for the various choices of parameters.

If the ξ_{cg} of LE is adopted, the contamination hardly affects ξ_{cc}. Even with the maximal cluster multiplicity function ($\psi = 5$) and halo extent ($R_{max} = 50\,h^{-1}\,Mpc$), the effect on ξ_{cc} at $r = 25\,h^{-1}\,Mpc$ is weaker than the 1σ upper limit of 25% - effect indicated by our empirical test of the previous section. If we adopt the ξ_{cg} of SP instead, the effect is somewhat stronger. For the reasonable choice of parameters ($\psi \leq 3.5$ below N_c and $R_{max} = 25\,h^{-1}\,Mpc$) the correction is still on the order of 25%. The effect becomes large only when the extreme parameters are used; for $\psi = 5$ and $R_{max} = 50\,h^{-1}\,Mpc$, the correction factor in ξ_{cc} is of order 4 at $25\,h^{-1}\,Mpc$. This latter result, which we view as an overestimate of the two-body effect, is comparable to the estimate claimed by Sutherland.

3.4 - Conclusion

The contamination of the Abell catalog by foreground and background galaxies systematically enhances the cluster-cluster correlation by preferentially introducing cluster pairs of small angular separations. It is very important to estimate this effect because of the strong implications of the high ξ_{cc} on the conventional theories of structure formation, and in particular on standard CDM.

Our empirical test, where we simply eliminated pairs of projected separation less than $10\,h^{-1}\,Mpc$, showed a weak effect; $\xi_{cc}/\xi'_{cc} \simeq 0.75$ at $25\,h^{-1}\,Mpc$.

The analytic estimate of the effect depends mostly on the assumed cluster-galaxy correlation function. The recent ξ_{cg} estimate of Lilje and Efstathiou (1988) yields $\xi_{cc}/\xi'_{cc} \simeq 0.75$ in agreement with the empirical test. The old ξ_{cg}

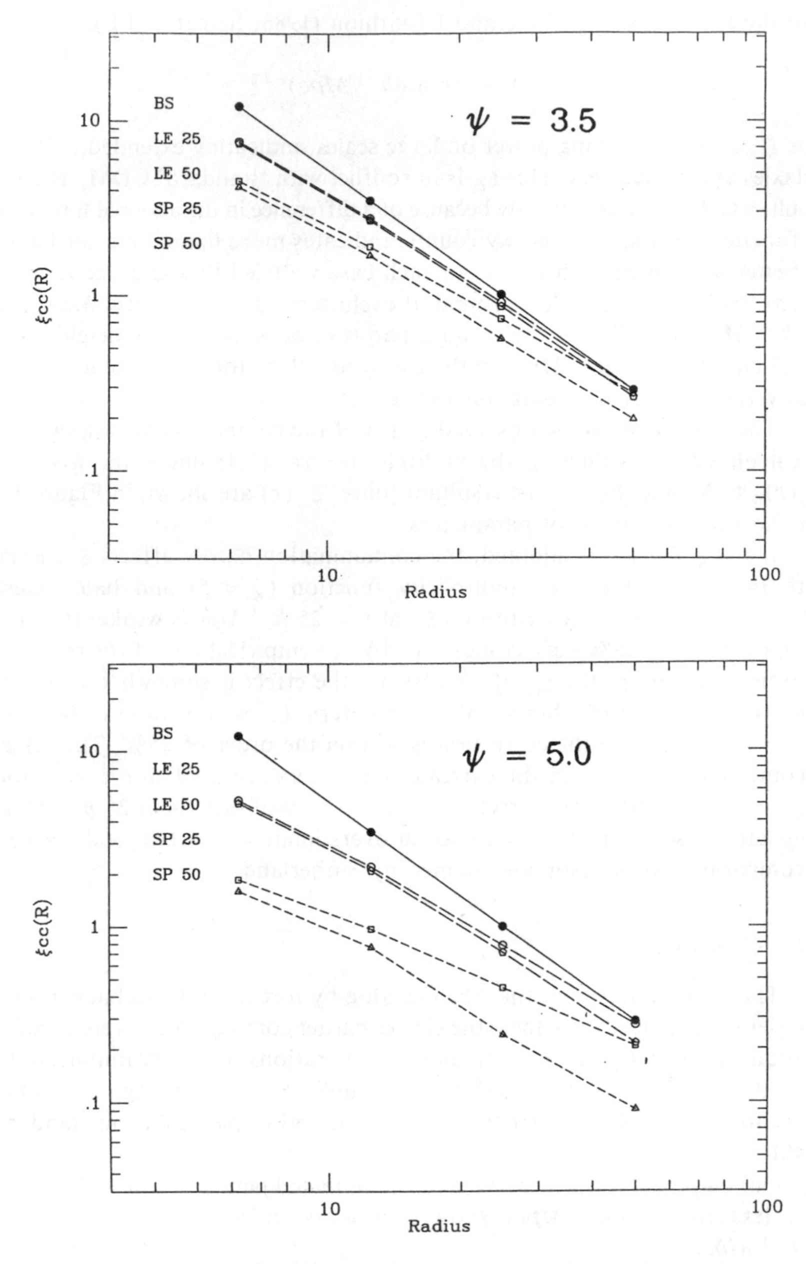

FIG. 3.3. Corrected cluster-cluster correlation function for the cluster-galaxy correlation function of Lilje and Efstathiou (1988, LE) or of Seldner and Peebles (1977, SP), with $R_{max} = 25$ or $50\ h^{-1}\ Mpc$. The assumed cluster multiplicity function is a power-law with $\psi = 3.5$ (top) and $\psi = 5$ (bottom).

estimate of Seldner and Peebles (1977) can give rise to corrections of $\xi_{cc}/\xi'_{cc} \leq 0.5$, but only if the cluster halos are assumed to extend to 50 h^{-1} *Mpc* (!) and the cluster multiplicity function is assumed to be steep ($\psi = 5$) for $R = 0$ clusters. The former is an over-estimate because it assumes that the clusters extend beyond the half mean separation between them. The latter means that clusters are missing from the catalog in a rapidly growing rate as N decreases, starting just below $N_c = 50$.

If the result of the empirical test for ξ_{cc} is to be taken as is, the extreme result based on the ξ_{cg} of SP, with very extended halos and a very steep cluster multiplicity function, can be excluded based on the analytic estimate.

The theoretical implication is that ξ_{cc} and ξ_{cg} cannot simultaneously be compatible with the predictions of CDM (and similar traditional scenarios). If one of these functions is low enough in amplitude, the other must be too high. The high ξ_{cg} of SP is by itself incompatible with CDM (independently of whether the associated ξ_{cc} may or may not be compatible with CDM). The lower ξ_{cg} of LE is in better agreement with CDM, but the corrected ξ_{cc} is still in conflict with the theory. However, projection effects from three-point correlations might be stronger.

4. Superclustering in the Explosion Scenario: One Possible Solution?

4.1 - Introduction

Since all the cluster samples studied so far yield a similar correlation function, and there is no convincing evidence that it has been severely overestimated, it would be worth while to adopt the observed excess of ξ_{cc} as a working hypothesis and seek a theoretical explanation.

Within the framework of Gaussian fluctuations, the attemps to get more power on large scales have baryons playing an important role in an open universe (e.g. Dekel 1984a) or in a universe with a large cosmological constant. The hybrid scenario of baryons and CDM (Bardeen, Bond, and Efstathiou 1987; Blumenthal, Dekel, and Primack 1988) and the baryon isocurvature scenario (Peebles 1987), both have substantial power on scales of 50 h^{-1} *Mpc*. However, in addition to giving up the simplicity of the Einstein-deSitter cosmology, these scenarios may run into problems with the observed isotropy of the microwave background, so they require either a finely tuned cosmological constant or *ad hoc* reionization at $z \sim 100$.

The difficulties of the Gaussian models on large scales motivate a serious consideration of non-Gaussian scenarios. One possibility is the cosmic string model, where clusters form by accretion onto string loops, which have non-Gaussian correlations because they are chopped from the same parent loop (Turok 1985; Primack, Blumenthal, and Dekel 1986; Scherrer 1987). This picture has some promising features, but detailed predictions await more accurate studies of string-loop fragmentation and loop velocities, since the current results from numerical simulations are in conflict with one another (compare, for example, Albrecht and Turok 1985 with Bennet and Bouchet 1988).

Here I will focus on the explosion scenario. I will describe below the surprising result that a *random* distribution of shells could produce the observed clustering of clusters which form at the vertices where three shells intersect. This is a part of a study of the formation of large-scale structure in generic explosion models, which is being carried out by David Weinberg, Jeremiah P. Ostriker and myself [Weinberg, Dekel and Ostriker 1988 (Paper I); Weinberg, Ostriker and Dekel 1988 (Paper II)]. I wish to refer to Bahcall (1988b) and to Geller (M. Geller, private comm.), who have referred in general to a similar model of cluster formation primarily on the basis of observational considerations.

In this picture, positive energy perturbations drive material away from "seeds", sweeping gas into dense, expanding shells that cool and fragment into galaxies (Ostriker and Cowie 1981; Ikeuchi 1981). Explosions of supermassive stars could act as seeds (also Carr and Ikeuchi 1985). Alternatively, the long-wavelength radiation produced by superconducting cosmic strings could sweep plasma into shells which could be a few tens of Mpc's in size (Ostriker, Thompson, and Witten 1987; 1988). In fact, the scenario does not require explosive energy at all — negative density fluctuations grow by gravity alone into expanding voids that have structure similar to other cosmological blast waves (Bertschinger 1983, 1985; Ostriker and McKee 1988).

The explosion scenario naturally accounts for the "bubbles" in the galaxy distribution (e.g. de Lapparent, Geller, and Huchra 1986). Galaxy redshift surveys suggest that shells of radius 10 to 30 h^{-1} *Mpc* with a filling factor of order unity may dominate the structure, such that the dynamical interactions of shells must play an important role in the development of clusters, superclusters, and voids. Saarinen, Dekel, and Carr (1987) have simulated clustering in a universe of interacting shells, and found that the explosion scenario may be able to account for the observed galaxy distribution. In Paper I, we extend this study to a detailed investigation of shell interactions, focusing on two- and three-shell collisions. The work of Paper II, some of which I will describe here, grew out of this dynamical investigation and it explores one of the explosion scenario's most striking predictions — a distribution of clusters with strong signatures of high-order superclustering on scales of ~ 50 h^{-1} *Mpc*.

The explosion seeds produce shells that expand, enclosing an ever-increasing fraction of space, and eventually overlap. Gravitational instabilities grow slowly on isolated shells (White and Ostriker 1988), so shell interactions are essential to the formation of massive galaxy clusters. When two shells cross they interesect in an overdense ring, which accretes matter from the shells;

the two voids "push" matter into the circular wall where the shells overlap and from this wall into the ring (Figure 4.1). Rich clusters then form when a *third* shell intersects this ring in two points — "knots". (When three shells overlap they intersect in two, and only two, points, independent of their relative sizes!) The knots are the sites of deep potential minima that accumulate the surrounding matter (Figure 4.2). The process is somewhat analogous to the formation of structure in the pancake scenario (Zeldovich 1970), where the matter collapses to flat "sheets", then flows towards the lines of intersection, and along those to their points of intersection.

These results suggest a simple, geometrical toy model for the spatial distribution of clusters in the explosion scenario. For a given distribution of

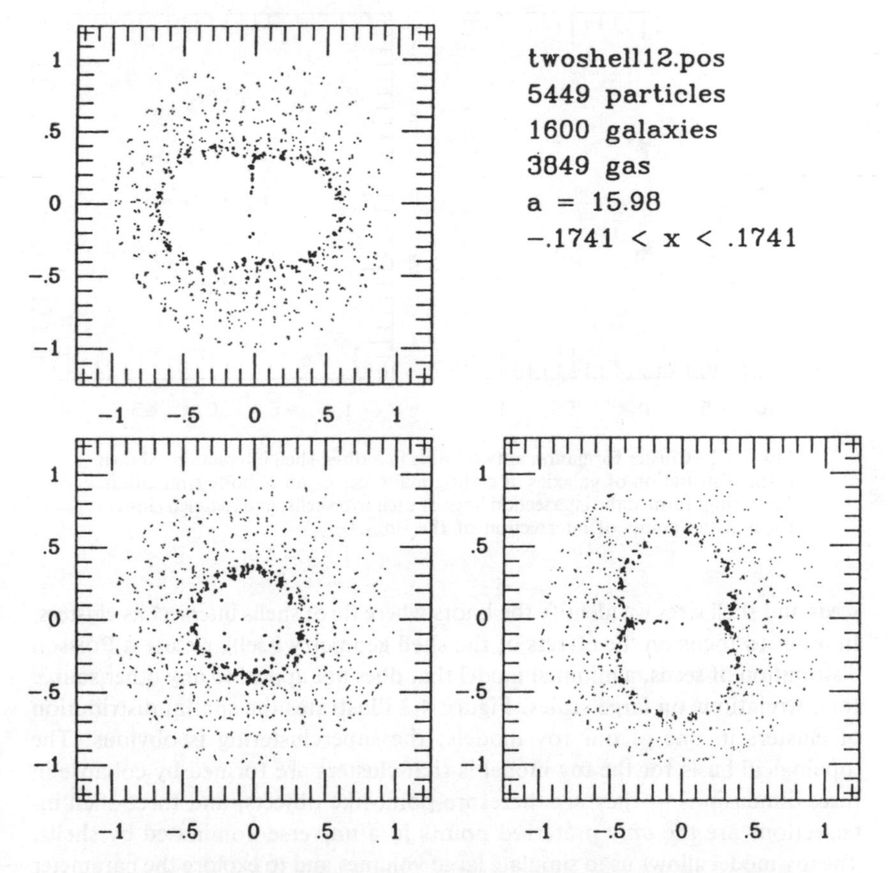

FIG. 4.1. Ring formation in a two-shell interaction. Shown is the distribution of galaxies (and background gas) in orthogonal slices of an N-body simulations. The overlapping wall has been evacutaed into a dense ring.

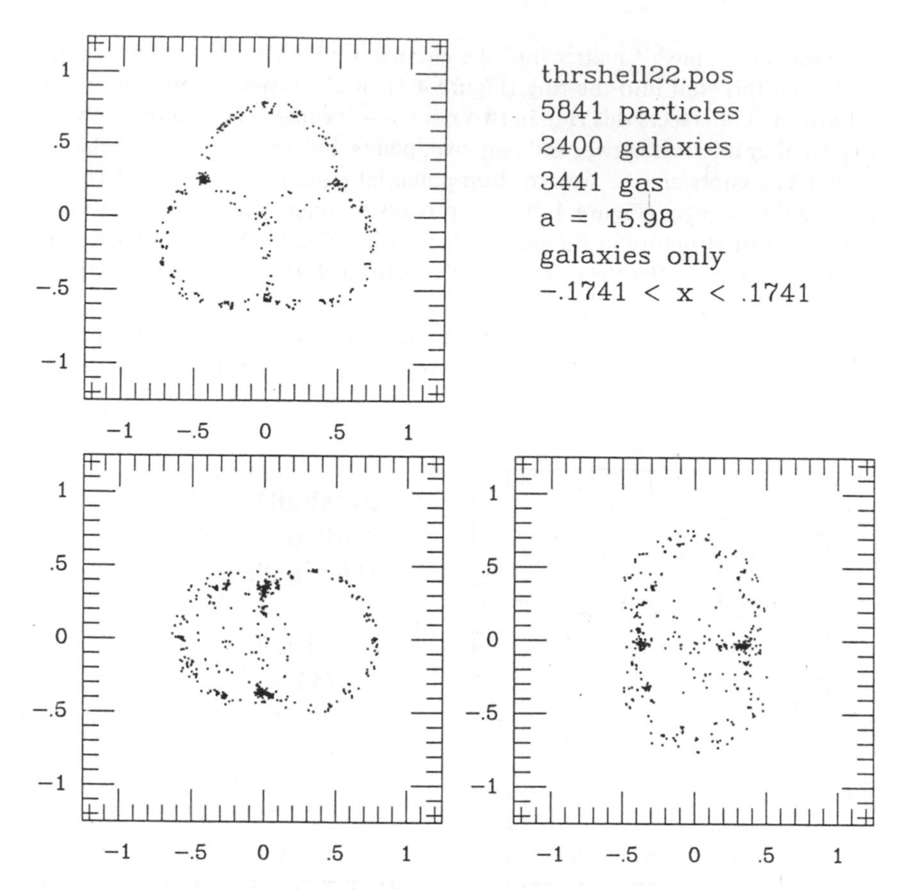

FIG. 4.2. Cluster formation at two knots in a three-shell interaction. Shown is the distribution of galaxies in orthogonal slices of an N-body simulation. Three rings form at the intersection lines of each two shells, and two rich clusters form at the points of intersection of the rings.

seeds and shell sizes we identify the knots where three shells intersect as clusters. In order to focus on the effects of the shell geometry itself, we use a Poisson distribution of seeds, a minimal model that does not appeal to any other source for correlations on large scales. Figure 4.3 illustrates the spatial distribution of clusters in one of our toy models; the superclustering is obvious. The topological basis for the toy model is that clusters are formed by collapse in three dimensions — they are therefore point-like objects, and three-shell intersections are the *only* preferred points in a universe dominated by shells. The toy model allows us to simulate large volumes and to explore the parameter space of the explosion scenario in order to identify the range of models that seems most promising.

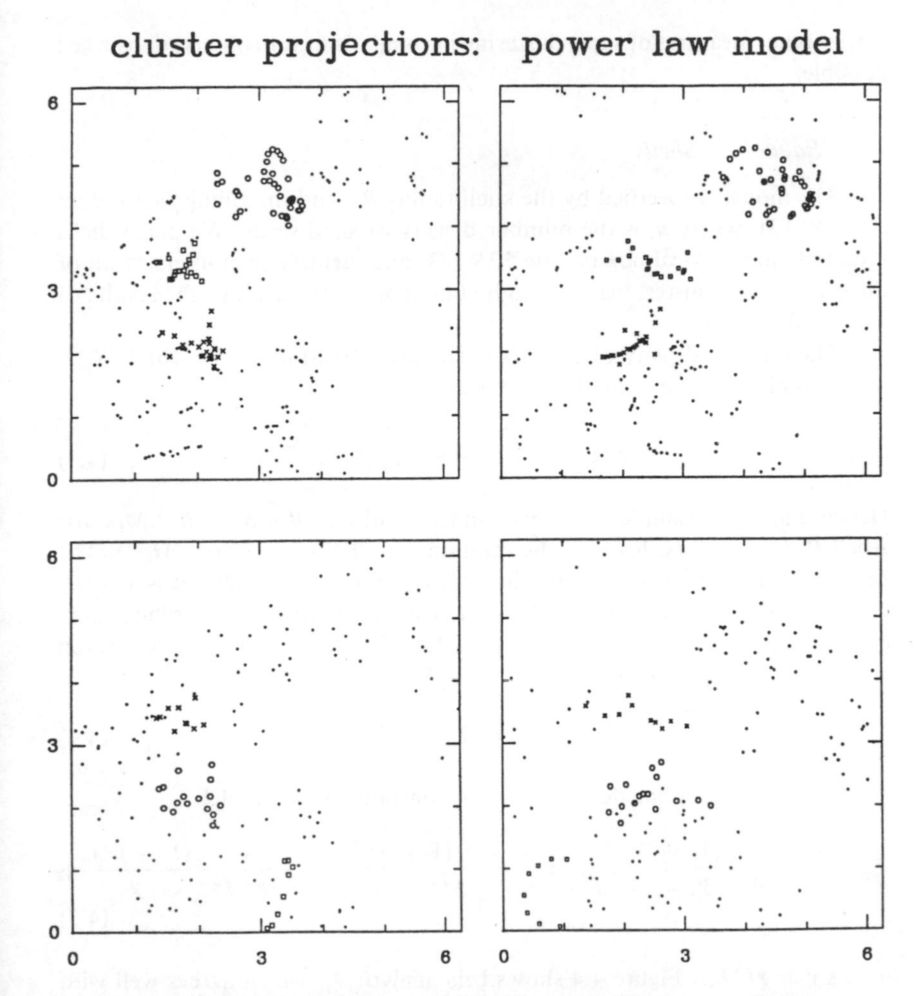

FIG. 4.3. Projections of cluster positions in a toy model. Top: a realization of the power law model. Bottom: the merged version of the same realization. The models have $\beta = -4.5$, $f = 0.3$, and the merger radius is $R_{mer} = R_{max}/5$. Left and right panels are projected along different axes. Hexagons, crosses, and squares indicate members of the three largest superclusters, identified by percolation at an overdensity of 20. Tick marks are in units of R_{max}.

I will discuss here only some highlights of Paper II, which includes a very detailed investigation of several toy models. In Sec. 4.2 I describe the simplest version of equal-size shells, focusing on the number density of clusters and ξ_{cc}. In Sec. 4.3 I address one of our more realistic models that assumes a spectrum of shell sizes, where we also compute the cluster mass function and look for trends of correlation strength with richness. Finally, in Sec. 4.4 I summarize other measures of superclustering in comparison with observations when possible.

4.2 - *Equal Size Shells*

The model is specified by the shell radius R_{sh} and the filling factor $f \equiv n_s(4\pi R_{sh}^3/3)$, where n_s is the number density of shell seeds. We place shells randomly in a cubical box of side $50R_{sh}/3$, and identify each intersection of three shells as a cluster, using periodic boundaries. We average the results of eight realizations.

The number density of clusters is (Kulsrud 1988) $n_c = (9\pi/16)\, f^3\, R_{sh}^{-3}$, which implies a mean neighbor separation

$$\bar{d} \equiv n_c^{-1/3} \simeq 0.83\; R_{sh}\, f^{-1}. \tag{4.1}$$

Demanding, for example, that this equals the observed $\bar{d}$ of 55 h^{-1} *Mpc* for Abell $R \geq 1$ clusters, leads to the relation $f = 0.45(R_{sh}/30\ h^{-1}\ Mpc)$. The points in Figure 4.4 show the resulting $\xi_{cc}(r)$ in our toy model. It is sharply truncated at $r = 2\ R_{sh}$, as expected. At smaller separations it is remarkably close to a power law, with a slope $\gamma \sim 1.5$. The correlation length is given to within a few percent by

$$r_0 \simeq 1.6\; R_{sh}\, f^{-0.85}. \tag{4.2}$$

Kulsrud (1988) has derived for the constant radius model

$$\xi_{cc}(y) = \frac{3}{4f}\,\frac{1}{y} + \frac{2}{\pi^2 f^2}\,\frac{(1-y^2)\,(1+3y^2)}{y^2} + \frac{1}{3\pi^2 f^3}\,\frac{(1-y^2)^2}{y}, \tag{4.3}$$

where $y \equiv r/2R_{sh}$. Figure 4.4 shows this analytic ξ_{cc} which agrees well with the numerical results. The probability of finding a cluster in a randomly chosen volume is $\propto f^3$, because it requires three shells. Given a cluster, a formation of a cluster elsewhere on one of the parent shells requires two additional shells, so it contributes a term $\propto f^2/f^3$. This term is $\propto r^{-2}$ at $r < R_{sh}$ because of the two-dimensional nature of the parent shell where the other cluster forms. Similarly, formation of a cluster on one of the three rings requires only one additional shell and it therefore contributes a term $\propto f/f^3$, which is $\propto r^{-2}$ at small r. The guaranteed presence of the twin cluster contributes a term $\propto 1/f^3$, which dominates only at very low f. Thus, the three terms in eq. 4.3 come from five-, four- and three-shell configurations respectively.

The two-point correlations have a maximum range of $2R_{sh}$. Higher-order correlations, however, can extend beyond $2R_{sh}$. For example, a connected

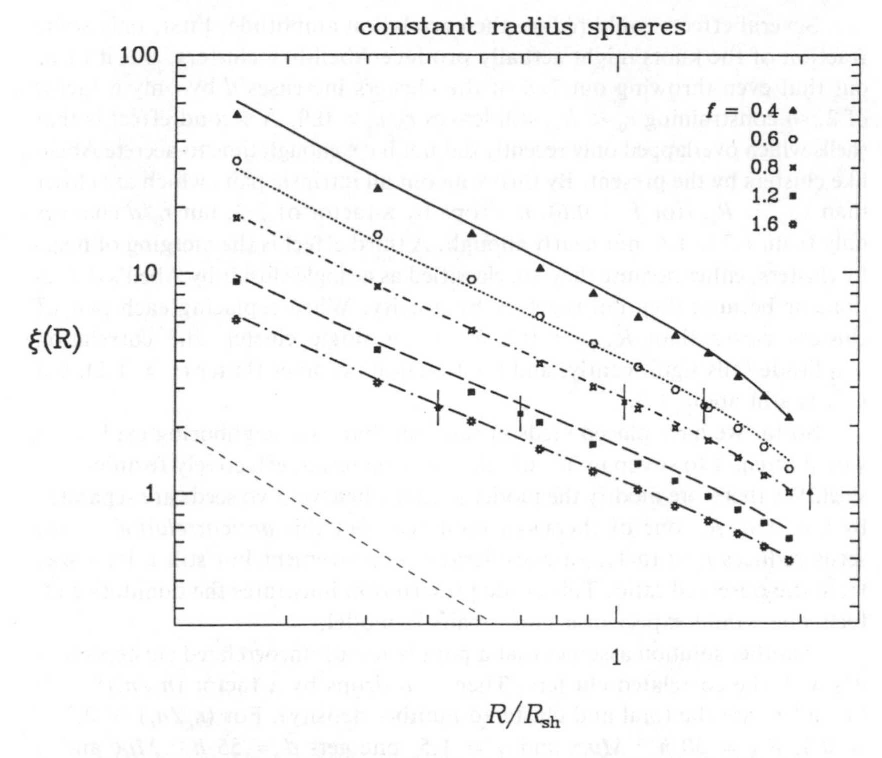

FIG. 4.4. Correlation function of the constant radius model. Points are average results of eight realizations for each of the indicated filling factors f. Curves show the analytic formula of Kulsrud (1988) for the same filling factors. Dashed line at lower left is an $r^{-1.8}$ power law. Small points with vertical bars mark the mean neighbor separations $\bar{d}$.

chain of shells can produce clusters with positive N-point correlations out to $2(N-1)\,R_{sh}$.

How does the model compare with observations? The power law form is encouraging, and while the slope $\gamma \simeq 1.5$ is somewhat shallow, it is acceptable given the uncertainties. The correlation amplitude, however, is too high. In all cases the ratio $r_0/\bar{d}$ exceeds the observed value of 0.5 by at least a factor of three, as can be seen in Figure 4.4, where vertical bars mark $\bar{d}$. Take the fact that the observed ξ_{cc} is still a power law (and certainly positive) at r_0, which requires $r_0 < 2R_{sh}$ because correlations vanish beyond the shell diameter. By equation (4.2), this condition implies $f > 0.77$. The lower limit on f, by equations (4.1) and (4.2), forces

$$r_0/\bar{d} = 1.93 f^{0.15} > 1.85, \qquad (4.4)$$

in *clear conflict* with the observed value of 0.5.

Several effects could reduce the correlation amplitude. First, only some fraction of the knots might actually produce Abell-like clusters. But it turns out that even throwing out 7/8 of the clusters increases $\bar{d}$ by only a factor of 2, so constraining $r_0 < R_{sh}$ still leaves $r_0/\bar{d} > 0.9$. A second effect is that shells which overlapped only recently did not have enough time to accrete Abell-like clusters by the present. By throwing out all intrinsic pairs which are closer than $s_{min} = R_{sh}$ (for $f = 0.6$), n_c drops by a factor of 2.3, but $r_0/\bar{d}$ changes only from 1.7 to 1.6, not nearly enough. A third effect is the merging of near-by clusters, either because they are classified as a single cluster by Abell's definition, or because they fall together by gravity. When replacing each pair of clusters closer than $R_{mer} = 0.2\ R_{sh}$ by a single cluster, the correlation amplitude falls significantly, and the function becomes flatter ($\gamma \simeq 1.2$), but $r_0/\bar{d}$ is still about 1.3.

So far we have placed seeds at random, but two neighboring explosions would attempt to sweep up mostly the same material, effectively forming one shell. We therefore modify the model so that whenever two seeds are separated by less than R_{sh} one of them is eliminated. But this *anticorrelation* of the seeds reduces $r_0/\bar{d}$ to 1.2, a considerable improvement but still a long way from the observed ratio. This crude prescription illustrates the qualitative effects one would expect in a more realistic model.

Another solution assumes that a population of uncorrelated clusters coexists with the correlated clusters. Then $r_0/\bar{d}$ drops by a factor $(n_c/n_t)^{(6-\gamma)/3\gamma}$ (n_t and n_c are the total and clustered number density). For $(n_c/n_t) = 0.3$, $f = 0.3$, $R_{sh} = 30\ h^{-1}\ Mpc$, and $\gamma = 1.5$, one gets $\bar{d} = 55\ h^{-1}\ Mpc$ and $r_0 = 27\ h^{-1}\ Mpc$, consistent with the observed ξ_{cc}. But note that the random clusters must outnumber the correlated clusters, which would require the *ad hoc* addition of large density fluctuations to the model.

In summary, while the simplest model is attractive in magically producing correlated clusters from uncorrelated shells with approximately the correct functional form, it is clearly incorrect, and no simple modification that we have considered can rescue it. The model fails at low filling factors because ξ_{cc} tends to run into the cutoff at $2R_{sh}$ while its amplitude is still too high, and it fails at filling factors above the limit imposed by the positivity of ξ_{cc} at r_0 because the dimensionless correlation length is too strong.

4.3 - A Range of Shell Sizes

The presence of very large shells allows ξ_{cc} to continue beyond the typical shell diameter, giving hope for improvement. The models require new parameters, but they produce a spectrum of cluster masses that can be tested against additional observational constraints. The surface density of mass swept up by the shell is proportional to its radius, so the cluster mass is roughly proportional to the multiple of the three radii of the interacting shells. Assuming that M/L is constant, we can compare the form of the model mass function to the observed luminosity (or multiplicity) function of clusters, and study the dependence of correlation strength on cluster richness.

I will focus here on a model with a power-law distribution for shell radii, with a sharp cutoff at some maximum. The superconducting cosmic string model predicts such a distribution, with an upper cutoff due to the last generation of shells that can cool and form galaxies before Compton cooling becomes ineffective (Ostriker and Thompson 1987). Something similar might also arise in a model where density fluctuations with a scale-free power spectrum collapse, cool, and convert some fraction of their energy into explosive blast waves.

The probability for a shell to have a radius R is taken to be $P(R) \propto R^{\beta}$ for $R \leq R_{max}$. When $\beta = -4$, shells in equal logarithmic bins of radius occupy an equal volume. The value $\beta = -4.5$ is predicted by the superconducting cosmic model for shells of radius $R \geq 5$ Mpc (Ostriker, Thompson and Witten 1988). In computational realizations we must have a lower cutoff R_{min} as well. The total filling factor, for $\beta \leq -4$, diverges as R_{min} goes to zero, so here f corresponds to shells with $1/2\ R_{max} < R < R_{max}$ only. We eliminate shells that happen to be placed entirely within larger shells. The model is thus specified by β, f, and R_{max}. The dynamic range is $R_{max}/R_{min} = 8$. Because clusters of mass $M/M_{max} = R_1 R_2 R_3/R^3_{max} > 1/8$ cannot have parent shells smaller than R_{min}, such a simulation correctly represents the cluster distribution for masses greater than $M_{max}/8$.

4.3.1 - *Mass-Luminosity Function*

Bahcall (1979) finds that the distribution of cluster richnesses in the Abell catalog (see Fig. 3.2) and the group catalog of Gott and Turner (1977), is well described by the function

$$\Psi_c(L) = \Psi_0(L/L_0)^{-2} \exp(-L/L_0) \tag{4.5}$$

$$L_0 = 2.5 \cdot 10^{12}\ h^{-2}\ L_{\odot}, \qquad \Psi_0 = 5.3 \cdot 10^{-6}\ (h^{-1}\ Mpc)^{-3}\ L_0^{-1}.$$

Here L represents the total luminosity within the Abell radius, and $\Psi_c(L)dL$ is the number density of clusters in the range $(L, L + dL)$. The minimum luminosity of an $R = 1$ cluster is $1.9 \cdot 10^{12}\ h^{-2}\ L_{\odot}$, about L_0. The complete redshift sample of $D \leq 4$ clusters in the northern Abell catalog (Hoessel, Gunn, and Thuan 1980, HGT) contains 102 clusters of richness $R \geq 1$, implying a mean number density $n_c = 6 \cdot 10^{-6}\ (h^{-1}\ Mpc)^{-3}$. The corresponding mean neighbor separation is $d = 55\ h^{-1}\ Mpc$.

Figure 4.5 shows the mass function of model clusters. The cutoff in radius at R_{max} induces the exponential decline at high masses. As long as the filling

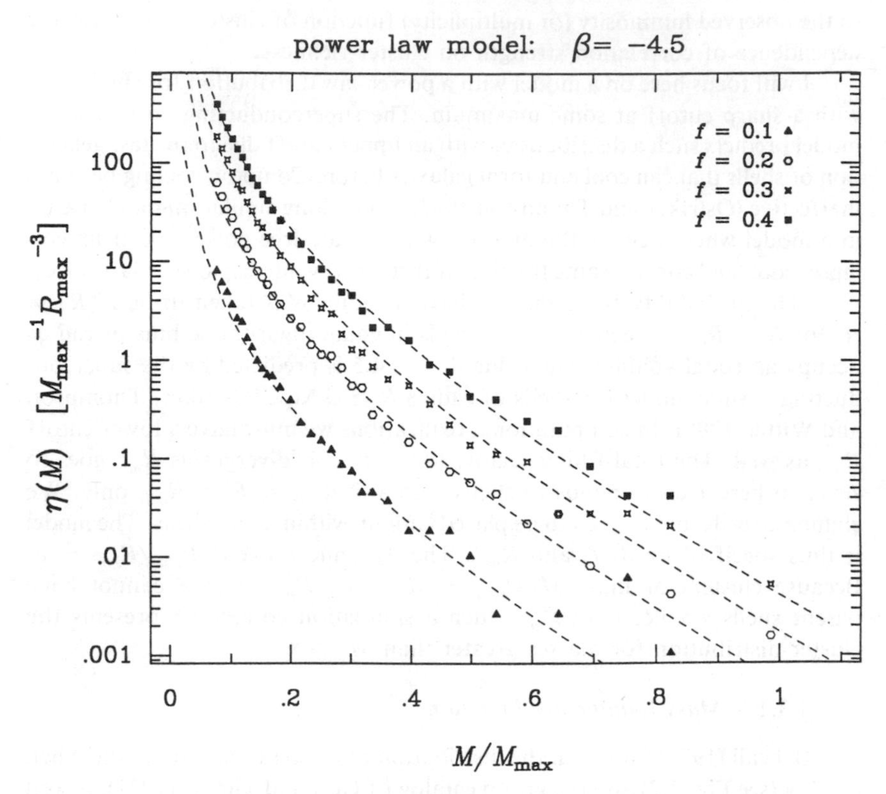

FIG. 4.5. Mass function of the power law model with $\beta = -4.5$, filling factors of 0.1, 0.2, 0.3, and 0.4. Points are average results of eight realizations. Dashed lines are Schechter-like functions fit to the numerical results, with characteristic $M_* = 0.14\ M_{max}$.

factor is low, n scales with f^3 just as in the constant radius case. The dashed lines show the fit of a function $\Psi(M) = f^3\ A(M/M_*)^{-2}\ exp(-M/M_*)\ M_{max}^{-1}\ R_{max}^{-3}$. We allowed A and M_* to assume their best fit values, but the curves differ only by the f^3 factor. The best fit parameters for $\beta = -4.5$ are $A = 2066$ and $M_* = 0.14\ M_{max}$. The value of M_* is insensitive to the assumed behavior at small masses because most of the data points lie in the exponential cutoff region. The range $-4.5 < \beta < -3.5$ gives an acceptable mass function within the uncertainties. The mass function near M_* is insensitive to our prescription for eliminating shells because only a few shells of this size are eliminated. Shells with small radii begin to fill space, and the mass function depends on our approximation to the action of sweeping, so it is not appropriate to compare the model with poor groups. However, the power law model naturally accounts for the observation that the cluster $\Psi(L)$ smoothly extends the luminosity function of rich groups, since the transition from three-shell processes to ring and shell fragmentation that give rise to smaller groups is a gradual one.

We can identify objects of mass $M > M_*$ as Abell clusters of richness $R \geq 1$. For the above mass function, $\bar{d}_* = 0.29\ R_{max} f^{-1}$ for $\beta = -4.5$. This is a direct generalization of equation (4.1). A similar integration of the cluster luminosity function yields $\bar{d}_0 = 58\ h^{-1}\ Mpc$ for the mean separation of clusters with $L > L_0$. Using the best fit values of A and M_*, the filling factor required to make $\bar{d}_* = \bar{d}_0$ is $f = 0.25\ (R_{max}/50\ h^{-1}\ Mpc)$.

4.3.2 - *Correlation Function*

Figure 4.6 shows ξ_{cc} for $f = 0.3$, and its richness dependence. The correlation function at small distances is again nearly a power law. Instead of cutting off sharply at $2R_{sh}$, it rolls over gradually towards $2R_{max}$. We determine γ by a least squares fit of $(r/r_0)^{-\gamma}$ in the range $10 > \xi_{cc} > 0.5$. The correlation length r_0 determined in this way typically exceeds by 5 - 10% the radius at which ξ_{cc} itself is unity. The amplitude of ξ_{cc} is higher for richer clusters, in qualitative agreement with the observed trend, but in our simulation $r_0/\bar{d}$ becomes somewhat smaller for more massive clusters. The dependence of ξ_{cc} on f is similar to that in the constant radius model; only for $f \geq 0.3$ does the power law extend relatively unchanged to $\xi_{cc} = 1$. The constraint that γ should be $\sim 1.5 - 2$ imposes $f \geq 0.3$. But the constraint that r_0 should be $\approx 0.5\bar{d}$ is not satisfied for any $f \geq 0.1$. For $f \geq 0.3$ the model r_0 is a factor of two or more too large. For small f, on the other hand, r_0 becomes greater than R_{max}, which is unacceptable because ξ_{cc} is observed to be still positive at $2r_0 \simeq 50\ h^{-1}\ Mpc$.

On balance, the parameter values $\beta = -4.5$ and $f = 0.3$ come closest to meeting the observational constraints on ξ_{cc}. The best fit slope for $M \geq M_*$ clusters is -1.75. ξ_{cc} does fall below the power law for $\xi_{cc} \leq 1$, but the observations are uncertain enough in this regime for the drop to be acceptable. The correlation length is $1.5\ R_{max}$, in some disagreement with the positivity condition $r_0 < R_{max}$. The most serious discrepancy is that $r_0/\bar{d} = 1.5$, about a factor of three higher than observed.

4.3.3 - *Modified Power Law Models*

Merging clusters considerably improves the agreement with observations (Figure 4.7). With $R_{mer} = R_{max}/5$, $r_0/\bar{d}$ drops to 0.95 and $\gamma \simeq 1.5$. $r_0/\bar{d}$ thus remains about a factor of two too large, and r_0 still slightly exceeds R_{max}. We have tested various extreme prescriptions for assigning mass to the merged clusters and found negligible effects.

Finally, we anticorrelate the seeds as before. One can turn strongly overlapping shells into a single shell that conserves their combined energy ($\propto R^5$),

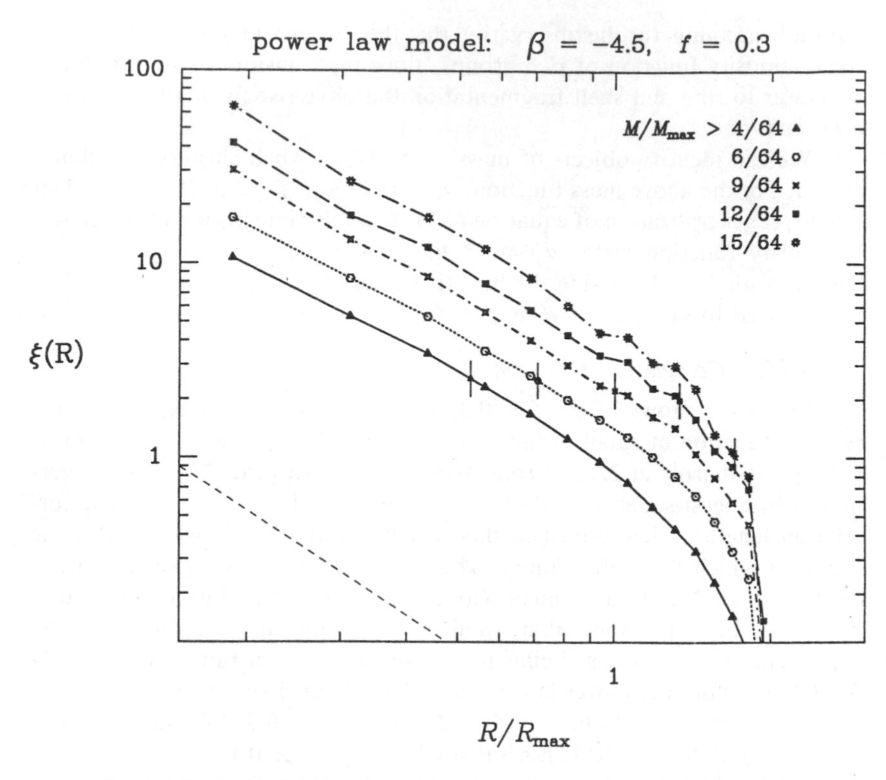

FIG. 4.6. Correlation function of the power law model, with $\beta = -4.5$ and f = 0.3. Different symbols indicate different mass cutoffs — the correlation function is evaluated for all clusters exceeding the specified mass. Abell clusters of $R \geq 1$ correspond to $M/M_{max} \geq 0.14$. Vertical bars mark the mean neighbor separations.

or volume, or surface area. But our simpler prescription of eliminating seeds that lie inside larger shells differs from more elaborate merger procedures only at the 10 — 20% level. Now $\gamma \simeq 2.2$, $r_0 \simeq 1.4\ R_{sh}$, n_c drops by nearly a factor of five, so $r_0/\bar{d} = 0.81$ — a great improvement over the standard model. The mass function retains the same form and characteristic mass, just reduced by a constant factor. Mergers turn out to have only a minimal impact on the anticorrelated model; $r_0/\bar{d} = 0.74$. In all three mdified versions of the power law model, $r_0/\bar{d}$ is nearly independent of mass, in agreement with the observed trend.

In summary, although none of the power law models accurately satisfies all of the observational constraints, they are an improvement over models with a single shell size. The mass function of models with $\beta \sim -4$ agrees well with the form of the observed luminosity function, and the values of f and R_{max}

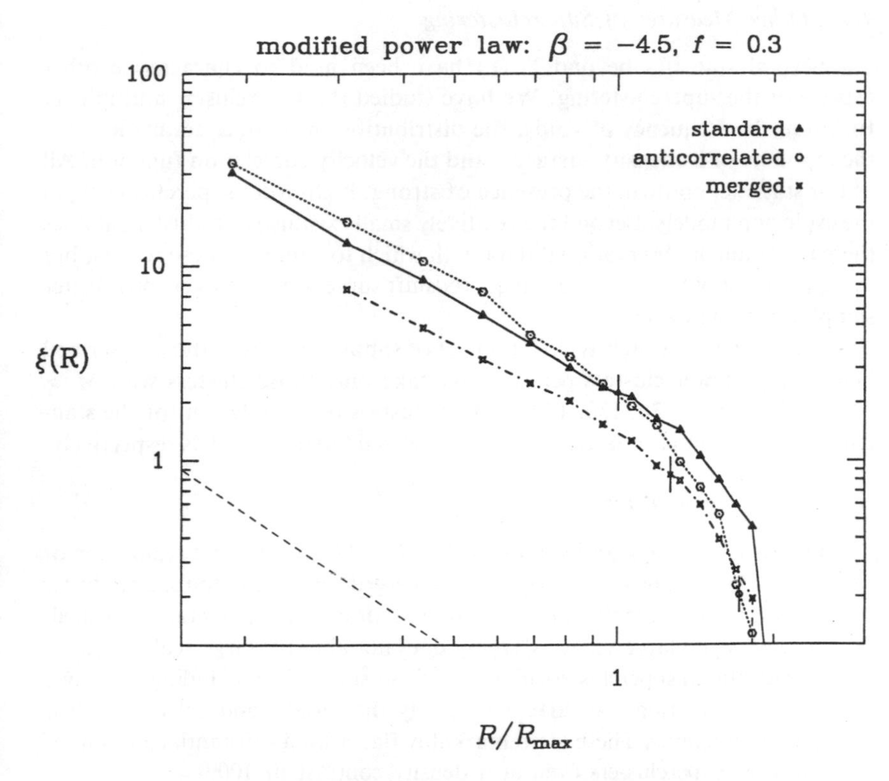

FIG. 4.7. Correlation functions of modified power law models, $\beta = -4.5$, f = 0.3 and clusters with $M/M_{max} > 9/64$. Triangles (solid line) show the standard model. Hexagons show the anticorrelated model, where seeds lying inside larger shells are eliminated. Crosses show the merged model, in which cluster groups smaller than $R_{mer} = R_{max}/5$ are merged into single clusters. Vertical bars mark the mean neighbor separations (near the lower right corner for the anticorrelated model).

needed to reproduce the observed number density of clusters are quite plausible. In the constant radius case even the minimal constraint $r_0 < 2R_{sh}$ forces $r_0/\bar{d} \geq 1.85$. For power law models the separation at which correlations vanish, $2R_{max}$, significantly exceeds the diameter of a typical cluster's parent shells, so the effects of the cutoff are less harsh. The most successful model, $\beta = -4.5, f = 0.3$, with anticorrelated seeds and mergers of close groups, has $r_0/\bar{d} \approx 0.75$, only about 50% larger than observed.

We note in passing that superconducting cosmic strings might produce elongated ellipsoidal bubbles rather than shells because the strings have substantial peculiar velocities. Since elongated structures create correlations at large distances, changing shells to ellipsoids might reduce the remaining problems.

4.4 - *Other Measures of Superclustering*

Several statistics beyond $\xi_{cc}(r)$ have been used to characterize other aspects of the superclustering. We have studied the supercluster multiplicity function, the frequency of voids, the distribution of number counts in cells, the topology of isodensity surfaces, and the velocity correlation function. All of our statistics confirm the presence of strong, high-order superclustering in the explosion models. Beyond the relatively small existing supercluster catalogs there is not much observational data with which to compare these results, but the situation should improve as larger redshift surveys of homogeneous cluster samples become available.

We have used a slightly different set of simulations from those described before, with fewer clusters per run. We take only those clusters with $M > M_*$, leaving about 230, 130, 120, and 95 clusters per simulation for the standard, merged, anticorrelated, and anticorrelated/merged models respectively.

4.4.1 - *Superclusters*

Figure 4.3 displays projections of the $M > M_*$ clusters in a realization of the standard power law model (top) and its merged version (bottom). The cluster distribution appears highly non-random, with dramatic filamentary and shell-like structures and large voids. Merging does not alter the large-scale features. We have identified superclusters in our models using a cluster-finding technique based on "percolation" at different density thresholds, and calculated their multiplicity functions. These are remarkably flat, with a substantial amount of mass in large superclusters even at a density contrast of 100.

In the analysis of Batuski, Melott and Burns (1987, BMB) of a sample of 225 $R \geq 0$ Abell clusters of $\varrho/\bar{\varrho} = 2.8$, two superclusters, with 38 and 36 members respectively, contain one third of all the clusters in the sample. The explosion models are clearly capable of producing such large superclusters. All of the models appear roughly consistent with the BMB data, although the constant radius and the standard power law models show perhaps too mucn superclustering. Bahcall and Soneira (1984, BS4) provide four separate catalogs of superclusters from the sample of 104 Abell clusters, at overdensities of 20, 40, 100, and 400. At an overdensity of 20 a single supercluster, whose central concentration is like the Corona Borealis supercluster, contains nearly 15% of all the clusters in the survey. Again the explosion models can easily produce such superclusters. In fact the three modified power law models all predict that $\sim$ 15% of the clusters should be in groups of multiplicity 9 or larger, in agreement with the BSA result. Given the small size of the observational sample, the merged power law model seems quite consistent with the BS4 multiplicity function.

BMB compare observational results to numerical simulations of several initially Gaussian models. The ξ_{cc} of all these Gaussian models are weaker than observed, the opposite problem from our explosion models. BMB then find less large-scale superclustering in their models than in the observations, with only 20 - 30% of clusters in groups larger than 10 at $\varrho/\bar{\varrho} = 2.8$. The multiplicity functions of the Gaussian models are declining at this point — (there are fewer clusters in the > 10 bin than in the 5 — 10 bin) whereas the multiplicity functions of the explosion models are still rising. It seems that the differences between the explosion models and the Gaussian models would become still more evident at larger multiplicities and higher density contrasts. The only Gaussian models that can reproduce the observed ξ_{cc} appeal to an open cosmology dominated by baryons or hot dark matter, but these would also fail in reproducing very big superclusters.

Superclustering, as measured by the multiplicity function, depends on high-order correlations in addition to the two-point correlations. We expect strong high-order correlations in the explosion models, as a result of multiple-shell interactions; four shells, for example, give rise to n-point correlation functions up to $n = 6$. Elongated superclusters can form along chains of overlapping shells. While ξ_{cc} vanishes beyond $2R_{sh}$, higher-order correlations can extend beyond the shell diameter! This is in contrast to the Gaussian models whose statistical properties are completely specified by their power spectrum or by its Fourier transform, the two-point correlation function; all higher order correlations vanish except where they are created by non-linear evolution.

4.4.2 - *Voids*

The void probability function $P_0(V)$, the probability that a randomly placed volume V is empty, depends on correlation functions of all orders. A Poisson distribution with number density n, for example, has no other correlations, yielding the familiar $P_0 = \exp(-nV)$. For a Gaussian distribution only the one and two-point correlations are non-zero, so $P_0 = \exp(-nV + n^2V^2\langle\xi\rangle/2)$.

In Figure 4.8 we plot P_0 as a function of nV, the expected number of clusters in a sphere of volume V. There are big differences between the various explosion models — we show here only the results for the merged power law model, which has the lowest P_0 of all. The differences between the explosion models and a Poisson distribution are enormous, an order of magnitude by $nV = 4$, and 3—4 orders of magnitude by $nV = 8$. We have analyzed the simulated cluster sample of BMB for CDM and neutrinos in the same way. The most weakly clustered of our models has void probabilities that are substantially higher than the most strongly clustered Gaussian models, an open universe dominated by CDM.

The void probability function is therefore a potentially powerful tool for testing theoretical models. One needs large samples to measure low values of P_0, however. A sample of N clusters contains N/nV independent volumes of size V, so one can measure probabilities down to $\sim nV/N$. Bahcall and Soneira

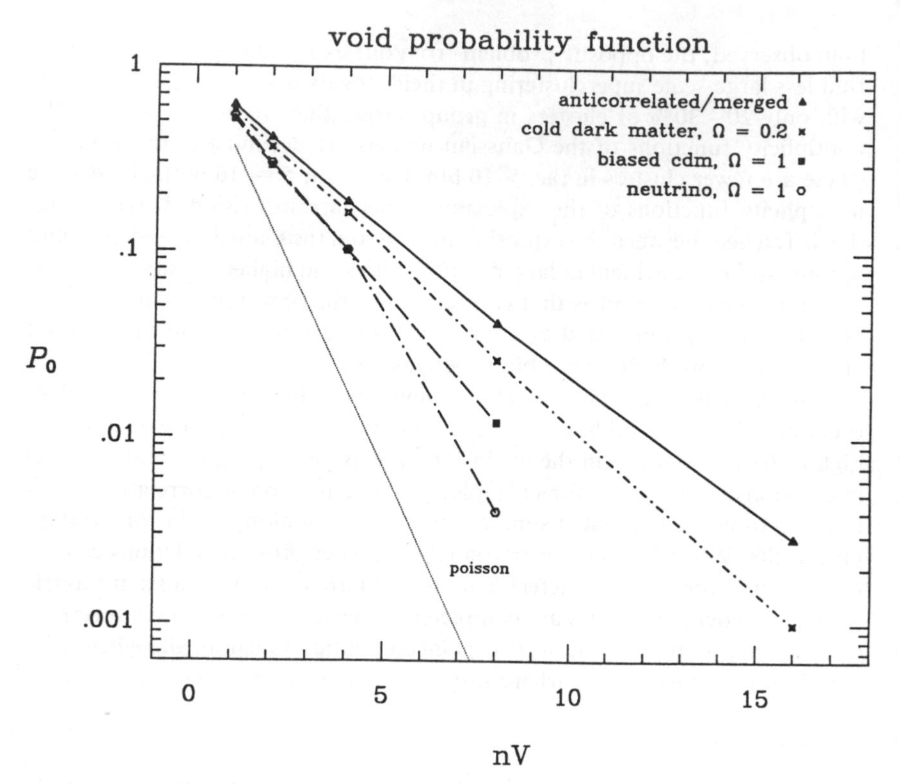

FIG. 4.8. Void probability function for the anticorrelated/merged explosion model in comparison with Gaussian models (as simulated by Melott).

(1982) have reported the existence of a very large void in the Abell cluster distribution, with a volume $\sim 10^6\, h^{-3}$ Mpc^3 and $nV \sim 6$. The probability of finding such a void is about 0.001, 0.01 and 0.1 in the Poisson, CDM and explosion models respectively. However, because Bahcall and Soneira search for a large empty volume rather than place randomly centered spheres, the statistical significance of this void is unclear.

We also generalize P_0 to the number count distribution, the probability $P_N(nV)$ that a randomly placed volume V contains N particles, and we make the appropriate predictions (paper II).

4.4.3 - *Topology*

The method was introduced by Gott, Melott, and Dicknson (1986) and discussed at length by Weinberg, Gott, and Melott (1987, hereafter WGM). Gott, Weinberg, and Melott (1987), and Melott, Weinberg, and Gott (1988) have applied this technique to the galaxy distributions of Gaussian models. After smoothing, we draw contour surfaces at a range of threshold densities

(ν = the number of standard deviations from the mean) and measure the genus of the surface. The genus G_s is the difference between the number of holes and the number of isolated regions. The number of "holes", in the sense of donut holes, is the maximum number of closed curves that can be drawn on the surface without dividing it further. At high or low ν contours typically break into isolated bags surrounding local peaks or valleys, so they have a negative genus. At intermediate thresholds contours tend to be multiply connected, "sponge-like", and therefore have positive genus.

For random phase distributions the genus is symmetric in ν because high and low density regions are statistically indistinguishable (e.g. Hamilton, Gott, and Weinberg 1986). The transition between isolated clusters or voids and sponge-like contours occurs at $\nu = \pm 1$. Non-random phase models generally produce an asymmetric $G_s(\nu)$ (WGM); this asymmetry can serve as an important distinguishing feature between Gaussian and non-Gaussian models. The amplitude of the $G_s(\nu)$ curve reflects the amount of structure present in the cluster distribution on scales of the smoothing length, and we can use this amplitude to define an "effective index" n_{eff} that can be compared to other models with a power law spectrum.

All the explosion models display definite asymmetry in $G_s(\nu)$. Isolated superclusters appear over an expanded range of ν with G_s crossing zero at $\nu \sim 0.7$; isolated voids begin to appear only at $\nu \leq -1.2$, and for the constant-radius models the curve peaks below the median density. The minimum at positive ν tends to be deeper than that at negative ν, indicating more isolated superclusters than isolated voids. The asymmetry is opposite to that of "bubble" models in which mass points lie uniformly on walls surrounding empty voids (Figure 14 of WGM). Rather, these cluster distributions show the "meatball" topology seen in isolated cluster distributions (WGM, Figure 19). The shape of the $G_s(\nu)$ curves is very similar for all the models, but the amplitudes are different; they indicate for the modified power law models $n_{eff} = 0.1, 0.25$, and 0.4 for the merged, anticorrelated, and anticorrelated/merged cases respectively. The other models show a negative index; the $f = 1.0$ model, in particular, has very few holes, with $n_{eff} = -1.5$.

Preliminary results from an analysis of 155 Abell clusters by Gott and collaborators are consistent with a random phase distribution and an index $n_{eff} \sim 0.5$ (Gott, private communication), but the sample contains only 4 "holes", so the results are not strong enough to rule out any of these models.

4.4.4 - *Velocities*

Because clusters form on expanding shells, they should acquire peculiar velocities. We estimate the peculiar velocities by making the simplifying assumptions that the peculiar velocity of each shell is a fixed fraction of its Hubble velocity, each shell contributes an equal amount of mass to the cluster, and the cluster conserves the net momentum of this matter. For shells that overlap substantially, the velocity calculated in this way approaches the velocity of the intersection point itself. We calculate the peculiar velocity correlation function $\langle \vec{v}(\vec{x}) \cdot \vec{v}(\vec{x} + \vec{s}) \rangle$. It is positive and decreasing at separations s smaller than the typical shell radius R, and it turns negative at $s \simeq R$. Nearby clusters tend to share the peculiar velocity of their common shells, so their velocity vectors point in the same direction. This turns to an anticorrelation at $s > R$ because many pairs in this separation range lie on opposite sides of the same shell. More cluster pairs share common shells when the filling factor is low, hence a deeper anticorrelation shows up. Any correlations beyond the shell diameter disappear in the noise. This pattern of positive velocity correlations at small separations turning to *anticorrelations* at larger separations should be a distinctive signature of the explosion models. If clusters formed instead by gravitational collapse of density fluctuations, we would more likely find nearby clusters falling towards each other, producing an anticorrelation at small s.

The rms peculiar velocity is about 2/3 the shell peculiar velocity for the constant radius models and 1/3 the R_{max} peculiar velocity for the power law model. The peculiar expansion velocities of shells depend on details of the cosmological scenario. Shells that sweep up all matter in a flat universe approach a self-similar state (Bertschinger 1983, Ikeuchi, Tomisaka, and Ostriker 1983) where the peculiar expansion velocity is 20% of the Hubble velocity. However, peculiar velocities will eventually decay in an open universe, and in a universe dominated by collisionless dark matter whose gravity will slow a shell down until the dark matter inside catches up and crosses it. It appears that the typical cluster velocities will be too low to produce the ~ 1000 km s^{-1} broadening suggested by the anisotropy of pair separations in the redshift sample (Bahcall, Soneira, and Burgett 1986). The large velocities on large scales is currently the most severe problem for the explosion picture.

In summary, all our statistical measures demonstrate the impressive degree of large-scale structure in the explosion models. Each of these superclustering statistics depends on the various orders of correlation functions in a different way, so they can provide at least partially independent tests of the models. All of the explosion models show substantially more superclustering than the with traditional Gaussian models. The merged power law model agrees quite well with the observed supercluster multiplicity function. The other models are reasonably consistent with the multiplicity data, except that they tend to have fewer isolated clusters at the higher density thresholds. No observational analyses are available for the void probability function or number counts, though these are not difficult to measure in principle. The supercluster-void topology of all the explosion models displays a "meatball" asymmetry. This asymmetry does not appear in the current observational data, but larger cluster samples will be needed before this becomes a statistically significant discrepancy. Cluster peculiar velocities are expected to be fairly small, but they should show a systematic pattern of correlation at close separations and an anticorrelation at separations between R_{sh} and $2R_{sh}$.

5. Conclusion

The three different projects described above are meant to demonstrate how unique the system of rich clusters of galaxies is in constraining the theories of the formation of large-scale structure. These objects could be detective out to cosmological distances of order $z \sim 1$, and their clustering process must be dominated by gravity, so they should make ideal tracers of large-scale initial conditions and dynamics. Although the study of internal properties of clusters yielded little evidence for any specific initial conditions, the immediate vicinity of clusters, on scales of 1 — 10 h^{-1} *Mpc*, and its relationship to the central parts, does contain very interesting cosmological signatures. But of crucial importance is the superclustering of clusters, which indicates non-trivial structure on scales of 10 — 100 h^{-1} *Mpc*. While the galaxy two-point correlation function vanishes in the noise at about 10 h^{-1} *Mpc*, the cluster correlations provide strong statistical evidence for non-trivial structure on larger scales. Combined with the presence of filamentary structure and big voids in the distribution of galaxies on scales of a few tens of megaparsecs, and the intriguing coherent motions on similar scales, the cluster distribution indicates detectable dynamical evolution of structure on very large scales, on the order of 5% of the present horizon, and it has the potential of probing the structure on even larger scales.

Our analysis of the systematic effect of contamination in the Abell catalog indicates that although the effect is interesting the superclustering is not a fluke — it might be real and should be taken seriously.

The very large-scale structure is not a natural outcome of the standard theories which are based on Gaussian, scale-invariant initial fluctuations in an Einstein-de Sitter universe. The repair of the Gaussian models requires either an open cosmological model, or a non-zero cosmological constant. Alternatively, the explosion scenario offers a natural source of non-Gaussian fluctuations, which can reproduce and easily over-produce the observed superclustering in a generic way.

What makes the cluster distribution in the explosion scenario so powerful is that it emerges straightforwardly from generic topological considerations. This allows the use of simple toy models which provide strong constraints on the possible nature of the model. For example, the required sizes of $\sim$ 30 h^{-1} *Mpc* for the biggest shells indicate explosions which are non-trivial energetically. Such explosions could not be seeded by supernovae in a single galaxy; they must originate in something like detonations, or superconducting cosmic strings. The model provides quantitative predictions which will be testable with larger redshift samples of clusters.

The moral is that we are in great need for larger, homogeneous, redshift samples of rich clusters of galaxies. The Abell catalog to distance class $D = 6$, and its southern counterpart are there, waiting for cluster redshifts to be measured. Cluster catalogs based on their X-ray emission will also be very useful; they would suffer even less from possible contamination because the cluster cores are smaller in X-rays, and they will provide independent evidence for the degree of superclustering. On smaller scales, more detailed studies of the galaxy distribution a few megaparsecs away from the centers of nearby clusters also have the promise of providing very useful cosmological information.

Acknowledgements

I am deeply indebted to my collaborators on the various projects described here: G.R. Blumenthal, A. Oemler Jr., J.P. Ostriker, J.R. Primack, D. Weinberg and M.J. West.

REFERENCES

Aarseth, S.J. 1985. In *Multiple Time Scales*. ed. J.U. Brackbill and B.I. Cohen. p. 377. New York: Academic.
Abell, G. 1958. *Ap J Suppl*. **3,** 211.
Albrecht, A. and Turok, N. 1985. *Phys Rev Letters*. **54,** 1868.
Argyres, P.C., Groth, E.J., Peebles, P.J.E., and Struble, M.F. 1986. *AJ*. **91,** 471. (AGPS).
Bahcall, N.A. 1979. *Ap J*. **232,** 689.
_____. 1988. in *Large-Scale Structures of the Universe,* eds. J. Audouze, M.C. Pelletan, and A. Szalay. Dordrecht: Kluwer Academic Publishers, p. 229.
Bahcall, N.A. and Soneira, R.M. 1982 *Ap J*. **262,** 419.
_____. 1983. *Ap J*. **270,** 20 (BS).
_____. 1984. *Ap J*. **277,** 27 (BS4).
Bahcall, N.A., Soneira, R.M., and Burgett, W.S. 1986 *Ap J*. **311,** 15.
Bardeen, J.M., Bond, J.R., and Efstathiou, G. 1987. *Ap J*. **321,** 28.
Bardeen, J.M., Bond, J.R., Kaiser, N., and Szalay, A.S. 1986. *Ap J*. **304,** 15.
Barnes, J., Dekel, A., Efstathiou, G. and Frenk, C.S. 1985. *Ap J*. **295,** 368.
Batuski, D.J. and Burns, J.O. 1985. *AJ*. **90,** 1413.
Batuski, D.J., Melott, A.L., and Burns, J.O. 1987 *Ap J*. **322,** 48. (BMB).
Bennet, D. and Bouchet, F. 1988. *Phys Rev Letters*. **60,** 257.
Bertschinger, E.W. 1983. *Ap J*. **268,** 17.
_____. 1985. *Ap J*. **295,** 1.
Binggeli, B. *AA*. **107,** 338.
Binney, J. and Silk, J. 1979. *MN*. **188,** 273.
Blumenthal, G.R. and Primack, J.R. 1983. in *Fourth Workshop on Grand Unification*. ed. H.A. Weldon, P. Langacker, and P.J. Steinhardt. p. 256. Boston: Birkhauser.
Blumenthal, G.R., Dekel, A. and Primack, J.R. 1988. *Ap J*. **326,** 539.
Carr, B. and Ikeuchi, S. 1985. *MN*. **213,** 497.
Carter, D. and Metcalfe, N. 1981. *MN*. **191,** 325.
Cavaliere, A., Santangelo, P., Tarquini, G., and Vittorio, N. 1986. *Ap J*. **305,** 651.
Davis, M. and Peebles, P.J.E. 1983. *Ap J*. **267,** 465.
Dekel, A. 1983. *Ap J*. **264,** 373.
_____. 1984a. *Ap J*. **284,** 445.
_____. 1984b. in *Eighth Johns Hopkins Workshop on Current Problems in Particle Theory*. eds. G. Domokos and S. Koveski-Domokos. p. 191. Singapore: World Scientific.
Dekel, A. and Aarseth, S.J. 1984. *Ap J*. **238,** 1.

Dekel, A., West, M.J., and Aarseth, S.J. 1984. *Ap J.* **279,** 1.
de Lapparent, V., Geller, M., and Huchra, J. 1986. *Ap J Letters.* **302,** L1.
Doroshkevich, A.G. 1970. *Astrophysica.* **6,** 320.
Doroshkevich, A.G., Shandarin, S.F., and Saar, E. 1978. *MN.* **184,** 643.
Dressler, A. 1980. *Ap J.* **236,** 351.
Gott, J.R. and Turner, E.L. 1977. *Ap J.* **216,** 357.
Gott, J.R., Melott, A.L., and Dickinson, M. 1986. *Ap J.* **306,** 341.
Gott, J.R., Weinberg, D.W., and Melott, A.L. 1987. *Ap J.* **319,** 1.
Hamilton, A.J.S., Gott, J.R., and Weinberg, D.H. 1986. *Ap J.* **309,** 1.
Hoessel, J., Gunn, J.E., and Thuan, T.X. 1980. *Ap J.* **241,** 486. (HGT).
Ikeuchi, S. 1981. *PASJ.* **33,** 221.
Ikeuchi, S., Tomisaka, K., and Ostriker, J.P. 1983. *Ap J.* **265,** 583.
Klypin, A.A. and Kopylov, A.I. 1983. *Sov Astron Letters.* **9,** 41.
Kulsrud, R. 1988, in preparation.
Lambas, D.G., Groth, E.J. and Peebles, P.J.E. 1988, *AJ.* **95,** 975 (LGP).
Lilje, P.B. and Efstathiou, G. 1988. *MN.* **231,** 635 (LE).
Lin, C.C., Mestel, L., and Shu, F.H. 1965. *Ap J.* **142,** 1431.
Ling, E.N., Frenk, C.S., and Barrow, J.D. 1986. *MN.* **223,** 21p.
Lukey, J.R. 1983. *MN.* **204,** 33.
Melott, A.L., Weinberg, D.W., and Gott, J.R. 1988. *Ap J.,* in press.
Noonan, T. 1980. *Ap J.* **238,** 793.
Nusser, A. and Dekel, A. 1988. in preparation.
Oemler, A., Schechter, P.L., Shectman, S.A., and Kirshner, R.P. 1988, in preparation.
Ostriker, J.P. and Cowie, L.L. 1981. *Ap J Letters.* **243,** L127.
Ostriker, J.P., and McKee, C.M. 1988. *Rev Mod Phys.* **60,** 1.
Ostriker. J.P., Thompson, C., and Witten, E. 1986. *Phys Lett B.* **280,** 231.
_____. 1988. *Rev Mod Phys.,* in preparation.
Palmer, P.L. 1981. *MN.* **197,** 721.
_____. 1983. *MN.* **202,** 561.
Peebles P.J.E. 1982. *Ap J Letters.* **263,** L1.
_____. 1987. *Ap J Letters.* **315,** L73.
_____. 1988. this volume.
Primack, J.R., Blumenthal, G.R. and Dekel, A. 1986. in *Galaxy Distances and Deviations from Universal Expansion.* eds. B.F. Madore and R.B. Tully. Dordrecht: Reidel, p. 265.
Rhee, G.F.R.N. and Katgert, P. 1987. *AA.* **183,** 217.
Rood, H.J., Page, T.L., Kintner, E.C., and King, I.R. 1972. *Ap J.* **175,** 627.
Saarinen, S., Dekel, A. and Carr, B. 1987. *Nature.* **325,** 598.
Sastry, G.N. 1986. *PASP.* **80,** 252.
Scherrer, R. 1987, preprint.
Seldner, M. and Peebles, P.J.E. 1977. *Ap J.* **215,** 703 (SE).
Silk, J. and Turner, M.S. 1987, preprint.

Struble, M.F. 1987. *Ap J.* **323,** 468.
Struble, M.F. and Peebles, P.J.E. 1985. *AJ.* **90,** 582.
Sutherland, W. 1988. *MN.* **234**, 159.
Tucker, G.S. and Peterson, J.B. 1987, preprint.
Turner, M.S., Villumsen, J.V., Vittorio, N., and Silk, J. 1987, preprint.
Turok, N. 1985. *Phys Rev Letters.* **55,** 1801.
Weinberg, D.H., Dekel, A., and Ostriker, J.P. 1988, in preparation. (Paper I).
Weinberg, D.H., Ostriker, J.P., and Dekel, A. 1988, in preparation. (Paper II).
Weinberg, D.H., Gott, J.R., and Melott, A.L. 1987. *Ap J.* **321,** 2. (WGM).
West, M.J., Dekel, A., and Oemler, A.Jr. 1987. *Ap J.* **316,** 1. (Paper I).
_____. 1988. *Ap J.* in preparation. (Paper III).
West, M.J., Oemler, A., and Dekel, A. 1988. *Ap J.* in press. (Paper II).
White, S.D.M. 1976. *MN.* **177,** 717.
White, S.D.M., Frenk, C., Davis, M., and Efstathiou, G. 1987. *Ap J.* **313,** 505.
White, S.D.M. and Ostriker, J.P. 1988, in preparation.
Zeldovich, Ya.B. 1970. *AA.* **5,** 84.

VII

PROPERTIES OF GALAXIES AT HIGH Z AND LOW Z

RECENT OBSERVATIONS OF DISTANT MATTER: DIRECT CLUES TO BIRTH AND EVOLUTION

DAVID C. KOO
Space Telescope Science Institute, Baltimore

and

Lick Observatory, Board of Studies in Astronomy and Astrophysics

Abstract

Highlights of recent deep observations of field galaxies, clusters of galaxies, radio galaxies, quasar absorption lines, and quasars are used to illustrate our progress since the 1981 Vatican Conference on Astrophysical Cosmology and to review the current status of evidence for evolution in their intrinsic properties and large-scale clustering. The birth and ages of galaxies can be explored directly by exploiting these classes of objects to search for primeval galaxies.

1. Introducion

Except for the microwave background measurements, most of the other observations presented in this volume are equivalent to snapshots, many quite detailed, of the universe as it appears now, and reach distances that only encompass less than 1% of the accessible volume. In contrast, although the data become much coarser as we peer farther into the past to higher redshifts and fainter limits, the resultant glimpses of the early history of the universe provide, in principle, direct evidence for changes in the contents, properties, and distribution of matter. Even the birth process may be visible.

These faint observations are also important as powerful constraints to various cosmological scenarios, including the popular Cold Dark Matter (CDM) theory, which has been so successful in explaining many of the known correlations among a variety of galaxy properties and their clustering and motions, all with a critical density of $\Omega = 1$ for the universe. Of interest for faint observations, this theory predicts that the bulk of star formation should have occurred at small redshifts, typically from $z = 0.5$ to perhaps 2 or so, a range that is quite accessible within an $\Omega = 1$ universe. For example, assuming little correction for the shape of the spectra (i.e., a spectral index of -1) and no evolution in luminosity for most galaxies (i.e., those similar to spiral galaxies with active star formation), the predicted brightness of a typical galaxy of $M_v = -22$ (Hubble constant of 50 km $\sec^{-1}$ Mpc^{-1} assumed, so the universe is about 13 Gyr old) should be around $V = 20.6$ at $z = 0.5$, 22.2 at $z = 1$, and 23.9 at $z = 2$. A 4-m telescope equipped with modern CCD detectors would image to these limits easily and should even yield redshifts in several hours, if emission lines are strong, as expected with active star formation.

Since the last Vatican Conference on Astrophysical Cosmology in 1981 (Brück *et al.* 1982, henceforth VAC) when CCDs were still a novelty, tremendous progress has been made. A handful of the more recent and exciting (at least tantalizing) discoveries will be highlighted here for five classes of objects receiving the most attention in observations of distant matter; field galaxies, clusters of galaxies, radio galaxies, quasar absorption lines, and quasars themselves. For each I will touch on the current status of our view of their evolution, on the presence and perhaps evolution of large-scale clustering at lookback times that are a significant fraction of the age of the universe ($z > 0.1$), and on new candidates for primeval galaxies.

2. Field Galaxies

2.1 - Evolution

At the VAC Gunn in his review on this subject wrote as follows: (1) on counts and colors: "There is no evidence for evolution at J^+ mag brighter than about 23" and "the agreement among various workers is not very good", where J^+ refers to a photographic broadband blue; (2) on redshifts: he discussed his pioneering survey of 58 optically selected field galaxies to $B \sim 20$ that had a median redshift consistent with no evolution, but also an unexplained high redshift tail; (3) on the 4000 Å break: "It is quite clear that one does see the expected stellar evolution in ellipticals as one looks to larger... redshifts," demonstrated with an example at $z = 0.75$.

Today the data quantity is vastly superior, but our picture is still far from final: (1) on counts: the depth has reached almost 10 times fainter (Tyson 1988), with divergence of the no-evolution model and data by about a factor of 10 (Fig. 1); even if the claim for no significant field galaxy evolution in luminosity and color to $z \sim 0.8$ from the multicolor work of Loh and Spillar (1986) is not accepted, the nature of the strong color evolution seen in the multicolor data of Tyson (1988) and Koo (1986*b*) remains ambiguous but suggestive of

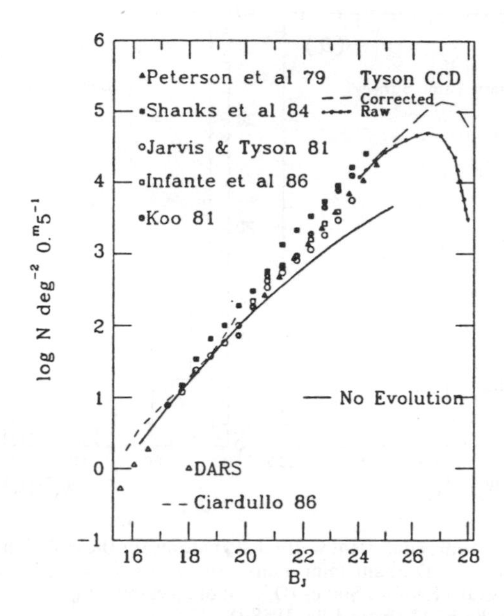

FIG. 1. Recent published galaxy number counts reduced to a photographic blue band. The solid curve represents the best estimate no-evolution model (see Ellis 1988*b* for details and references).

more extensive star formation at moderate redshifts z = 0.4 to 2; (2) on redshifts: nearly 600 redshifts to limits ~ 5x fainter than that mentioned above are now available from two independent surveys. Both are consistent in suggesting that luminosity evolution has indeed been slight, if any, since z ~ 0.4, but that many galaxies do show signs of extensive star formation, either through very ultraviolet colors (Koo and Kron 1988*b*) or from the overall increase in the strengths of [O II] emission lines (Ellis 1988*a*), as seen in Figure 2. The high redshift tail mentioned by Gunn above is not confirmed (Koo 1985); (3) on the 4000 Å break: among four groups all working to limits of z ~0.8, Hamilton (1985) and Oke(1983) both claim no evidence for evolution while Spinrad (1986) and Dressler and Gunn (1988) see a decrease in the amplitude of the break consistent with "expected evolution" (see Figure 3).

2.2 - *Clustering Evolution*

This area of research remains largely unexplored and among existing data, inconclusive if not inconsistent. Based on 4-m photographic surveys to B ~ 24, Koo and Szalay (1984) and Pritchet and Infante (1986) show consistent amplitudes for the two-point autocorrelation function that suggest little cluster-

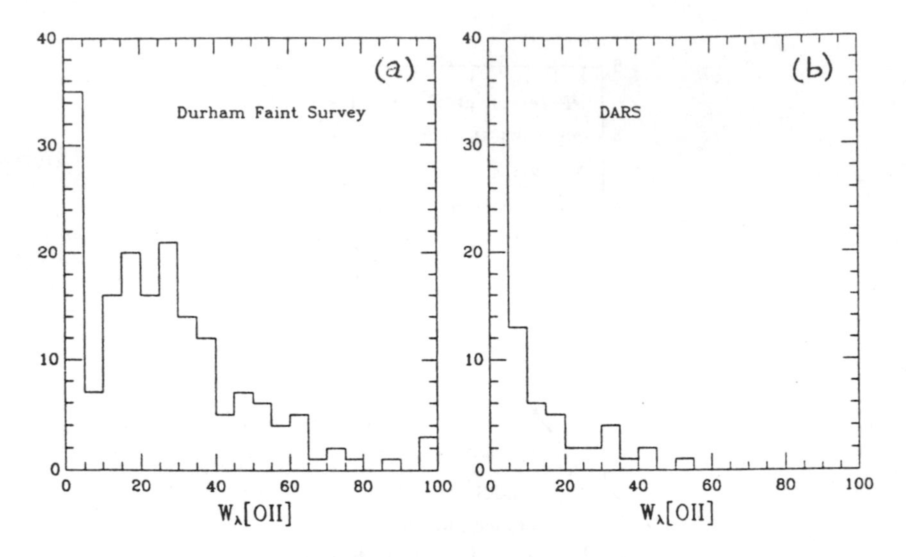

FIG. 2. Rest-frame equivalent width distributions for the emission line [O II] 3727 Å in (a) the Durham faint galaxy survey to $B_j = 21.5$, and (b) the Durham-Australia Redshift Survey (DARS) of a nearby sample of field galaxies complete to $B_J = 17$ (from Ellis 1988*a*).

ing evolution, whereas Stevenson *et al.* (1985) measure amplitudes nearly a factor of two lower, implying that clustering has increased significantly since larger redshifts. Based on field-to-field fluctuations being twice Poisson in the CCD data of Tyson and 5 times Poisson as reported by Ellis (1987) for photographic data, Koo (1988) notes that these values are entirely consistent with a simple extrapolation of the -0.8 power-law slope found for bright galaxies and of the no-evolution extension of the scaling by number density of counts. Using spectral redshifts (Ellis 1987) or multicolor estimates of redshifts (Loh 1988) to measure the spatial correlation amplitude at $z \sim 0.5$, both report values consistent with lower amplitudes and hence clustering evolution. On larger scales ($\sim$ 100 Mpc), simple histograms of the redshifts in some fields suggest the presence of large-scale voids and perhaps sheet-like superclusters (Ellis 1987; Koo and Kron 1988*b*; see Fig. 4), but before we accapt their reality the significance of these fluctuations needs to be carefully compared to realistic simulations, since even the CDM model may naturally produce such features (White *et al.* 1987) in a fraction of independent samples.

2.3 - Primeval Galaxies

The prospect of detecting primeval galaxies, i.e. those undergoing their initial star formation, is an area of research pioneered nearly two decades ago

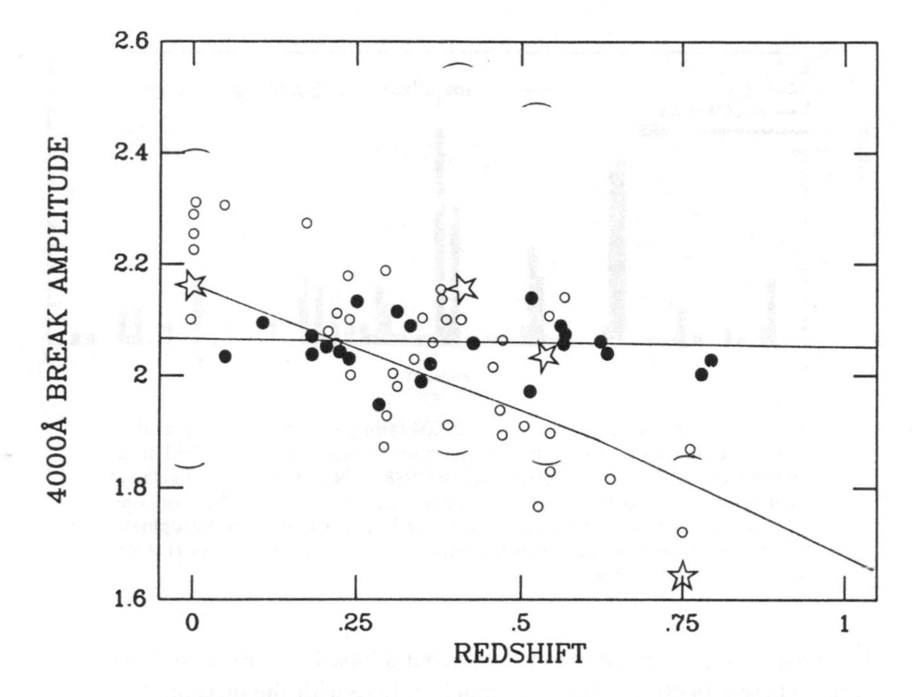

FIG. 3. Observations of the 4000 Å break amplitude versus redshift from Spinrad (1986) in open circles, Hamilton (1985) in solid circles, and Dressler (1987) in stars that represent the median and brackets that enclose the central two-thirds of each sample. The sloping solid line is the prediction for evolving ellipticals; the horizontal line is for reference (see Spinrad 1986 for more details).

by Partridge and Peebles (1967), but it has received much attention in the last few years, though virtually ignored at the VAC. In addition to the largely negative results reviewed by Davis (1980) and Koo (1986*a*), three recent surveys for field primeval galaxies deserve attention: (1) an extremely deep survey for Lyman-α to R $\sim$ 26 limits has been completed by Pritchet and Hartwick (1987), with negative results that begin to severely constrain traditional models (see Figure 5); more recent explorations of the extended formation times and perhaps gradual merging of subcomponents may, however, allow the CDM model to survive (Baron and White 1987); (2) preliminary results from comparing a very deep ultraviolet CCD image (see Majewski 1988) covering a 10 square arcmin field to a blue photographic image reaching B $\sim$ 24 show no candidates with z > 3.5. Any such candidate would be recognized by its invisibility in the optical UV, due to the expected drop of flux at the Lyman continuum break at 912 Å; (3) about 20% of the B > 26 objects break up

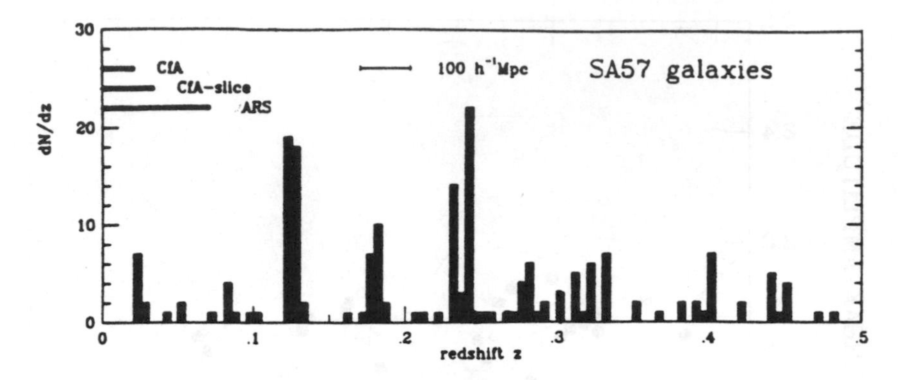

FIG. 4. Redshift distribution of about 200 faint galaxies, some fainter than 22nd mag, in a 40 arcmin diameter region in the Selected Area 57 field at the North Galactic Pole from Koo and Kron (1988*b*). Note the dearth of galaxies centered near z = 0.15 as well as the peaks at z = 0.125 and 0.24. For comparison, the ranges of redshifts covered by the 15th mag Center for Astrophysics (CfA) surveys and the 17th mag Durham Australia Redshift Survey (DARS, same as in Figure 2) are shown.

into blobs when observed in excellent ground-based conditions of 0.89 arcsec seeing (Tyson 1988), a result very much in tune with the merging formation scenarios of at least the CDM model.

3. CLUSTERS OF GALAXIES

3.1 - Evolution

Gunn in his VAC review mentions the initiation with Dressler of spectroscopic investigations of the Butcher-Oemler effect, in which distant (z ~ 0.4) compact rich clusters exhibit a larger fraction of blue galaxies in their cores than their present-day counterparts (Butcher and Oemler 1978). By now, several groups have contributed to the accumulation of several hundred spectra for perhaps a dozen or so clusters (see Dressler and Gunn 1988 for a recent review) with important results. First, these data largely support the membership of enough blue galaxies to confirm the photometric effect, as redefined (Butcher and Oemler 1984; see Figure 6). Of more interest, unlike most cluster spirals today which have spectra consistent with reasonably continuous star formation, spectra of distant cluster galaxies often show strong enough Balmer absorption lines, with weak, if any, accompanying emission lines, to suggest bursts of star formation (about 1% to 10% by mass) a billion years or so prior to the epoch being observed (see Figure 7). Moreover, some

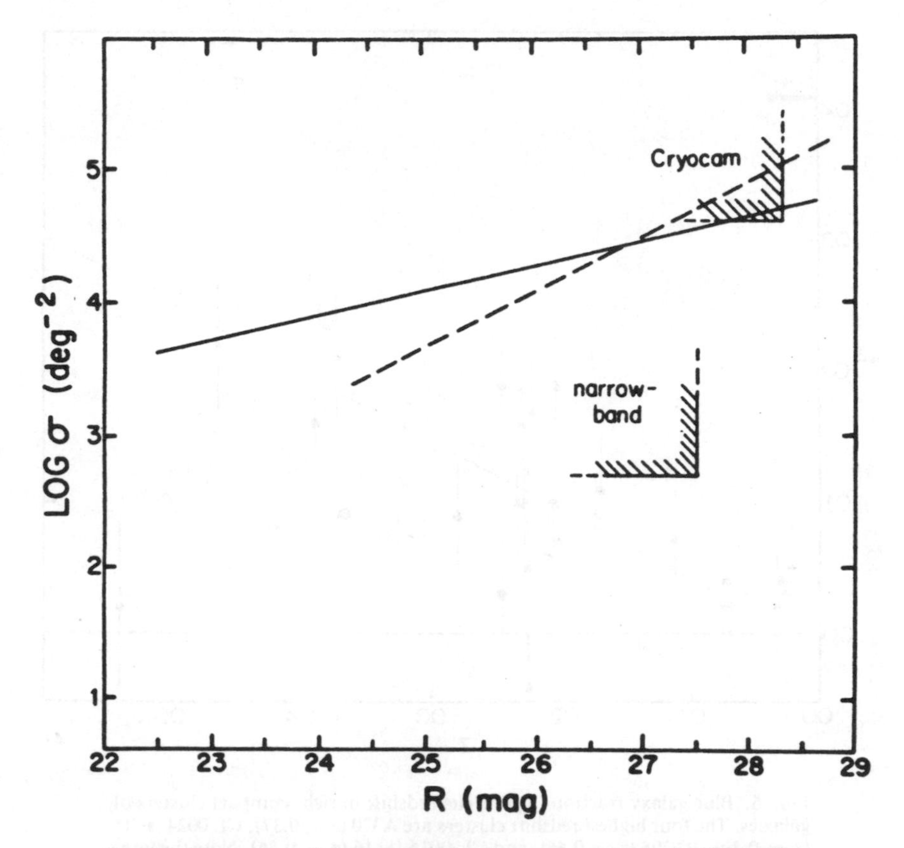

FIG. 5. Comparison of observational limits on primeval galaxies (PG) set by Pritchet and Hartwick (1987) with theoretical predictions. The solid line represents the predictions of a variety of models by Meier (1976). The dashed line represents models of PGs at z = 5 by Davis (1980), modified as described by Koo (1986*a*). Their observed limits correspond to less than one PG per CCD field at 6800 Å for the narrow-band filter observations, and less than one object per 272 arcsec^2 for the Cryocam observations. The sense of the limits is to exclude PGs whose properties lie to the upper left of the plotted points. The limits are plotted for pure Lyman-α sources with angular extent less than approximately 2.″5 (Cryocam) and less than approximately 1.″5 (narrowband). No correction for the sampling in redshift space has been made in this diagram (taken from Pritchet and Hartwick 1987; see their text for details).

evidence for activity and evolution in otherwise very red cluster galaxies at these redshifts is also found, either by changes in the 4000 Å break (Gunn and Dressler 1988) or by ultraviolet excesses (MacLaren *et al.* 1988, Couch and Sharples 1987). The underlying physical mechanisms remain uncertain, but range from cluster-gas ram-pressure induced star-formation to galaxy

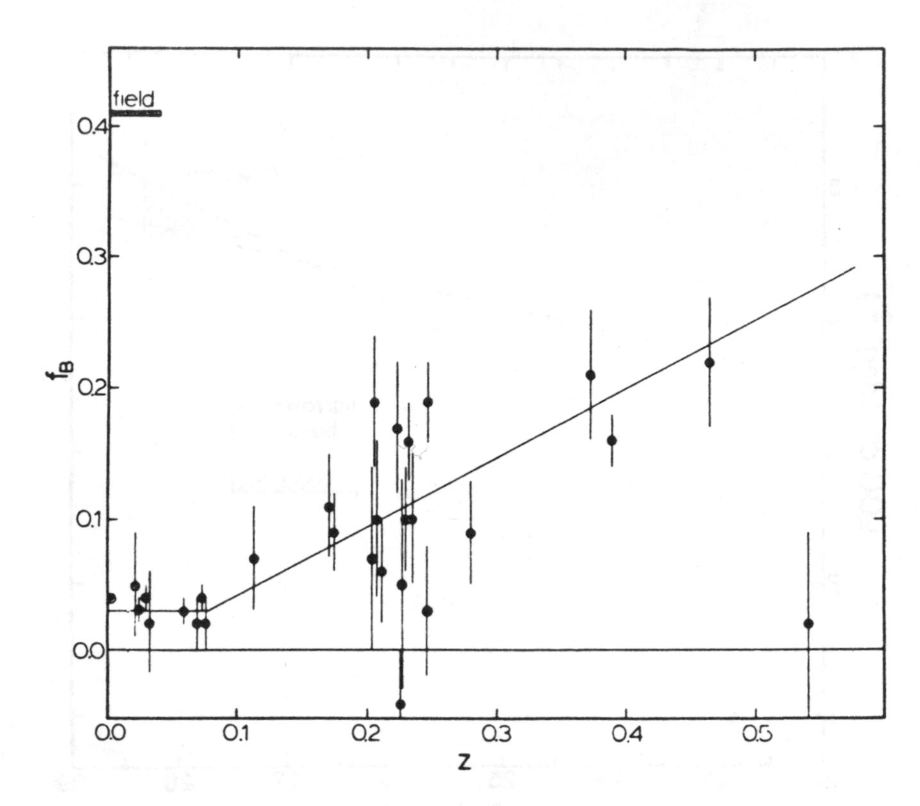

FIG. 6. Blue galaxy fraction (f_B) versus redshift in rich, compact clusters of galaxies. The four highest redshift clusters are A370 (z = 0.37), CL 0024 + 16 (z = 0.39), 3C 295 (z = 0.46), and CL 0016 + 16 (z = 0.54). Note the large dispersion of f_B at high redshifts. The "field" refers to local field galaxies; the solid line is an eye-drawn trend of increasing f_B with redshift claimed by Butcher and Oemler (1984). This figure is a simplified version of their Figure 3.

interactions; whatever the cause, different clusters at the same redshift may show different average properties, hinting that the evolutionary clocks are set cluster by cluster rather than by universal effects. As previously mentioned, field galaxies also show excess star formation activity at the same redshifts, but this vital relationship of evolution and environment has yet to be studied in detail.

3.2 - Cluster-Cluster Evolution

This area of research is still in its infancy but is likely to blossom over the next five years, as more systematic searches for distant clusters are undertaken (Gunn *et al.* 1986, Ellis 1988*b*). These will be among the first surveys to yield

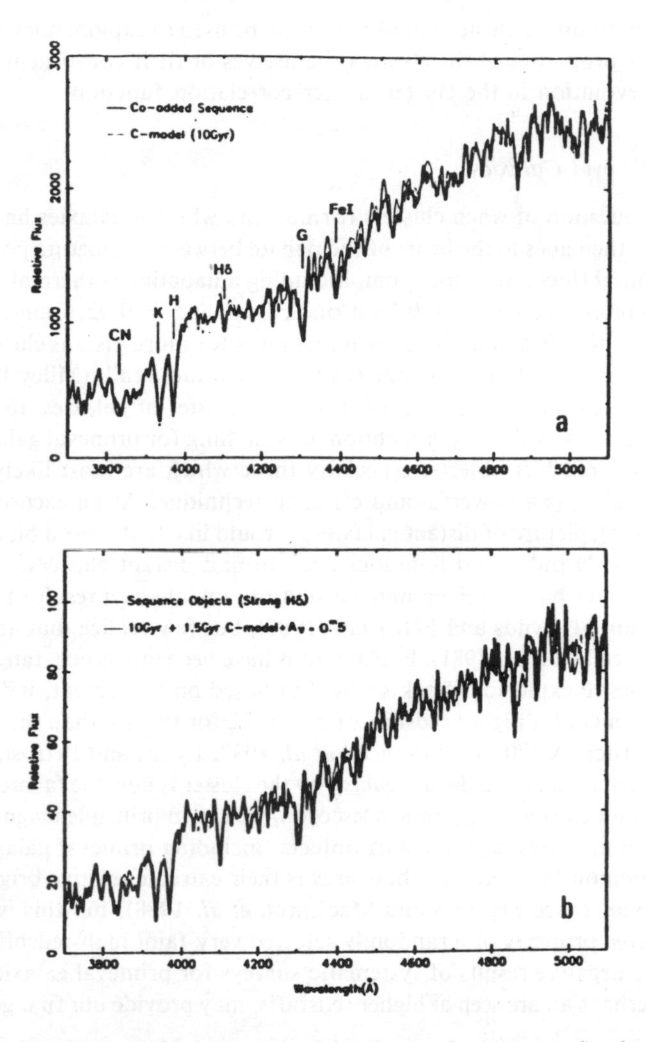

FIG. 7. (a) The coadded spectrum of the 62 normal sequence (red) galaxies identified in AC 103, AC 114, and AC 118, plotted on a rest-wavelength scale. Prominent absorption features are identified. For comparison, the spectrum of a standard old population model (10 Gyr c-model with a prompt 1Gyr initial burst) is superimposed upon the observed spectrum. The agreement between the two is seen to be excellent; (b) the coadded spectrum of the 11 sequence galaxies with noticeably stronger Hδ absorption. The comparison spectrum (light line) is the combination of a burst model spectrum and the old population model spectrum shown in (a) added together so as to contribute equal amounts of light at λ_{rest} = 4000 Å and reddened assuming A_V = 0.5 mag. Note that the poor match between the observed and model spectrum in the vicinity of Hβ (λ4861) is due to a strong night-sky emission line falling on this feature in the observed spectra of many of the Hδ-strong objects (from Couch and Sharples 1987).

reasonably complete cluster samples that can be used to explore not only evolution in the properties of the clusters themselves or their constituent galaxies, but also evolution in the cluster-cluster correlation function.

3.3 - Primeval Galaxies

The question of when clusters formed and whether galaxies had already existed by then goes to the heart of the debate between competing possibilities for the initial fluctuation spectrum, including adiabatic, isothermal, and now CDM theories. We know of at least one cluster at $z = 0.92$, found optically by Gunn *et al.* (1986) and there are many hints for more distant clusters from objects at higher redshifts. As one example, bent radio tails (Miley 1987) suggest the presence of gas, perhaps from a rich cluster of galaxies, to redshifts of nearly two. As will be later mentioned, searching for primeval galaxies near known high redshift objects, especially those which are most likely to exist within clusters, is a powerful and efficient technique. As an excuse to show an interesting picture of distant galaxies, I would like to digress a bit and mention the highly publicized luminous arcs around distant clusters.

These arcs have received much attention, mainly as a result of the AAS presentation by Lynds and Petrosian (1986), but few realize that an arc was first reported by Hoag (1981). Explanations have been numerous, ranging from light echoes to explosion shock shells, but based on two recent, independent measurements of a higher redshift of $z = 0.72$ for the arc than the $z = 0.37$ for the cluster, A 370, itself (Soucail *et al.* 1987, Lynds and Petrosian 1988), gravitational lensing of a distant galaxy by the cluster is now the favored theory. Using distant clusters as gigantic telescopes, we can in principle magnify otherwise very faint, very high redshift objects, including primeval galaxies. One of the common properties of these arcs is their extreme relative brightness in the ultraviolet (see Figure 8 and MacLaren *et al.* 1988), but this is perhaps the expected property of a randomly selected very-faint high-redshift galaxy. Given the negative results of systematic surveys for primeval galaxies, serendipity, perhaps an arc seen at higher redshifts, may provide our first good case.

4. Radio Galaxies

4.1 - Evolution

At the VAC the claim was made that "the fact, ... that strong extragalactic radio sources ... if they are not QSOs, are uniquely and without exception ... ellipticals and S0's." Based on this assumption, van der Laan and Windhorst not only suggested an increase of a factor of 100 in density by $z = 0.8$ in radio galaxies, but also suggested the detection of color evolution, since many

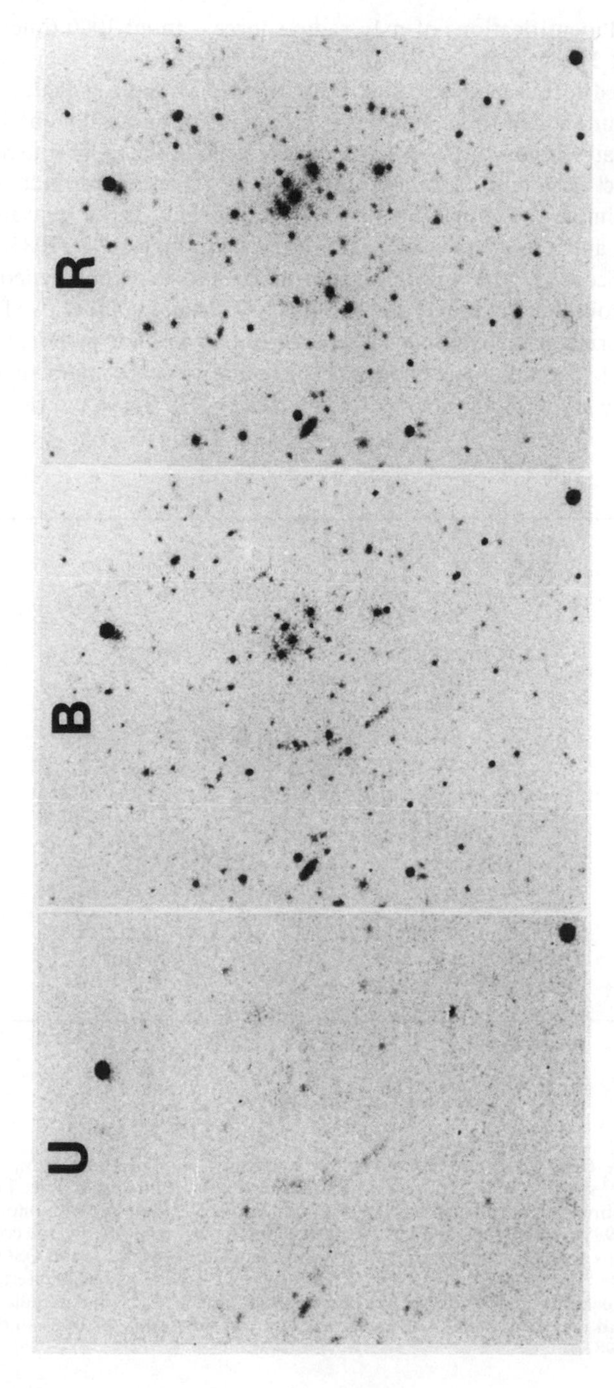

FIG. 8. Pictures of cluster 0024 + 16 at z = 0.39 showing a possible three piece "luminous arc". North is to the top, east to the left. The size of the images is ~ 130 arc sec E-W and ~ 160 arc sec N-S; at 6.3 Mpc per arc sec at the cluster redshift and $H_0 = 50$ km sec^{-1} Mpc^{-1} and $q_0 = 0.5$, the arc is ~ 240 kpc from the cluster center. The images are produced from plates taken with the 4 m telescope at Kitt Peak National Observatory. The red (R), blue (B), and ultraviolet (U) plates were exposed for 60m, 60m and 150m, respectively; taken in 1975, 1980, and 1983, respectively; and correspond to rest-frame 4400 Å, 3350 Å, and 2600 Å, respectively, if the arc is at the cluster redshift. Recent observations of an arc in the cluster A370 favor the theory that an arc is a background galaxy gravitationally lensed by the rich cluster.

of the optical identifications of mJy radio sources, about 1000 times fainter than the 3CR Jy level catalog, were quite blue. Furthermore, they suggested a cutoff at redshifts z between 3 to 5 for these mJy radio galaxies.

Today our view of radio galaxies has changed dramatically and in many ways complicates any picture for evolution. First of all, the radio source counts show a distinct flattening of the slope at low fluxes (Figure 9), which is inconsistent with simple extrapolations of the traditional classes of quasars, elliptical galaxies, and a few very weak radio spirals (Windhorst *et al.* 1985, Danese *et al.* 1987). Secondly, the view that radio galaxies form a homogeneous class has been demolished. With over 100 redshifts of the mJy sources, we find that many of the brighter blue radio galaxies are not star-forming ellipticals at high redshifts $z > 1$, but rather morphologically peculiar, perhaps interacting, galaxies (see Figure 10) at moderate redshifts typically $z \sim 0.2$ to 0.6 (Kron *et al.*

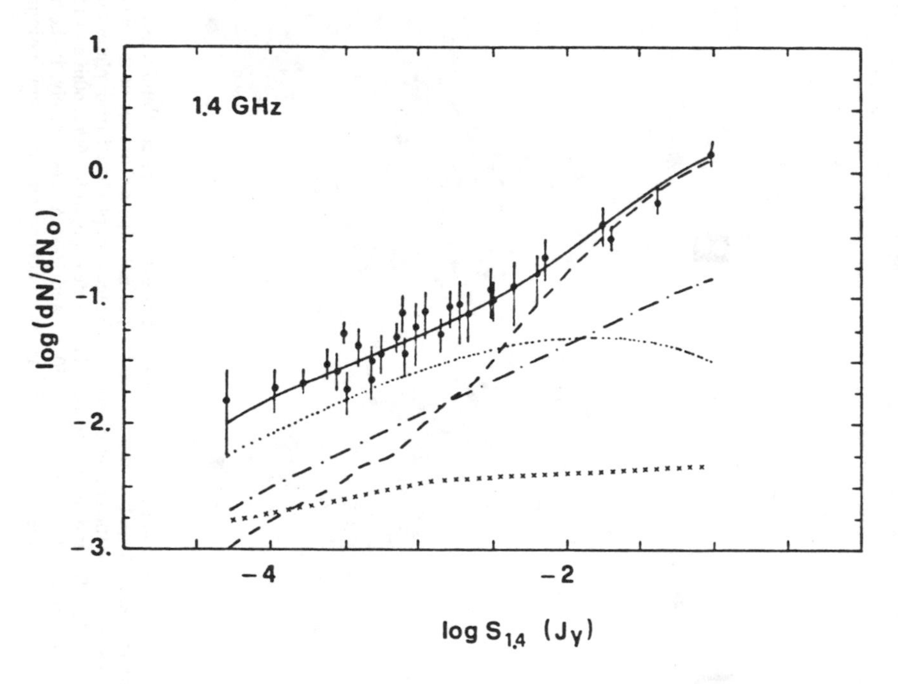

FIG. 9. Deep counts at 1.4 GHz in relative differential form ($dN_0 = 150 S_{1.4}^{-2.5}$ sr^{-1} Jy^{-1} such that an Euclidian rise would be a horizontal line. The data points are those listed by Condon and Mitchell (1984) and Windhorst *et al.* (1985). The dashed and the dot-dashed lines show, respectively, the contributions of steep-spectrum and "flat"-spectrum ellipticals, S0's, and QSOs; the dotted line displays the expected counts from evolving starbust/interacting galaxies. The crosses display the counts of unevolving spirals and irregulars. The solid line is the sum of all the above contributions (taken from Danese *et al.* 1987).

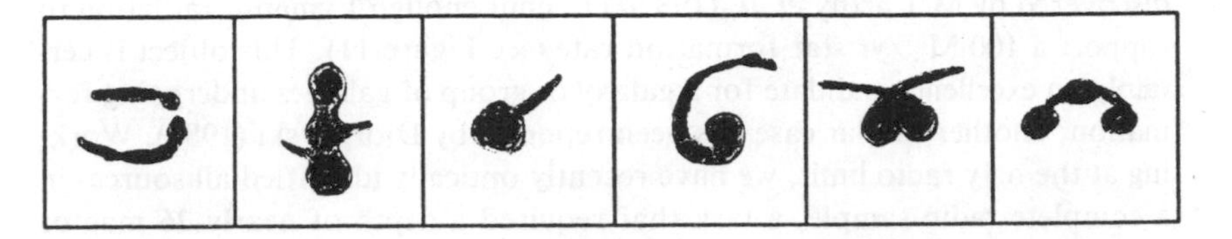

FIG. 10. Hand-drawn pictures of several optical identifications of mJy radio galaxies displaying diverse morphologies suggestive of close interactions (see Kron *et al.* 1985 for more details).

1985, Windhorst *et al.* 1987). This view of radio galaxies belonging to different classes occurs even at the 3CR level, where vastly improved CCD imaging by, e.g., Hutchings (1987) or Heckman *et al.* (1986) shows that strong radio galaxies also frequently possess peculiar morphologies, including the mysterious tendency for alignment between optical images and radio jets among distant sources (McCarthy *et al.* 1987*b*; Chambers *et al.* 1987). Until this diversity is well studied and understood, any claims for evolution, either of the underlying optical or infrared hosts in luminosity and color (see e.g. Spinrad 1986 or Lilly and Longair 1984), or of the radio population (Peacock 1985; Windhorst 1984), should be treated with caution.

4.2 - *Clustering Evolution*

Since radio sources can be detected to very high redshifts without the usual optical bias, they may also serve as powerful tracers of large-scale clustering with time, assuming of course that their typical environments are not changing at different epochs. Until one reaches at least the mJy level of around 50 sources or so per square degree (where typical separations are around 10 arcmin or a few Mpc at $z = 0.5$ and higher), any clustering amplitudes extrapolated from that of galaxy-galaxy correlations found today would be difficult to detect. Of course, most high redshift radio sources might reside in clusters and thus possess higher correlation amplitudes, as found among clusters today. At present several similar redshifts, hinting of large-scale clustering, are already quite common among these deep samples (Kron *et al.* 1985), but a more detailed analysis awaits improved redshift coverage.

4.3 - *Primeval Galaxies*

Despite the disappointing results of not finding the 20th mag blue mJy radio galaxies to be ellipticals undergoing extensive star formation at high redshift, as suggested by Katgert *et al.* (1979), radio galaxies have recently provided some of the best primeval galaxy candidates. Most exciting is the giant (10″ or 100 kpc) gas cloud at $z = 1.82$, identified with a 3CR source, and discovered by McCarthy *et al.* (1987*a*) to emit enough Lyman-α radiation to support a 100 $M_\odot$/yr star formation rate (see Figure 11). This object is certainly an excellent candidate for a galaxy or group of galaxies undergoing formation; another similar case has been reported by Djorgovski (1988). Working at the mJy radio limit, we have recently optically identified all sources in a complete radio sample, a task that required a depth of nearly 26 mag or so (Windhorst *et al.* 1987). What is tantalizing are the dozen or more faintest candidates, for they appear to be higher-redshift counterparts to the confirmed 3CR gas clouds. So far, spectroscopy has not been successful in yielding any redshifts on any of these "fuzzballs", but this and other samples should soon provide more examples of these giant Lyman-α gas clouds. In contrast to these radio sources Lilly (1988) has discovered a radio galaxy at z ~ 3.4 and argues that it contains evolved stars and is ~ 2 Gyr old. If so, at least some galaxies not associated with quasars formed at z ~ 4.5 or earlier.

5. QSO Absorption Lines

5.1 - *Evolution*

These features in the spectra of QSOs, which occur in various classes with the narrow Lyman-α lines most studied, were barely mentioned at the VAC but are now standard topics at many meetings; recently an entire conference was devoted to this very active area of research (Blades *et al.* 1987). These

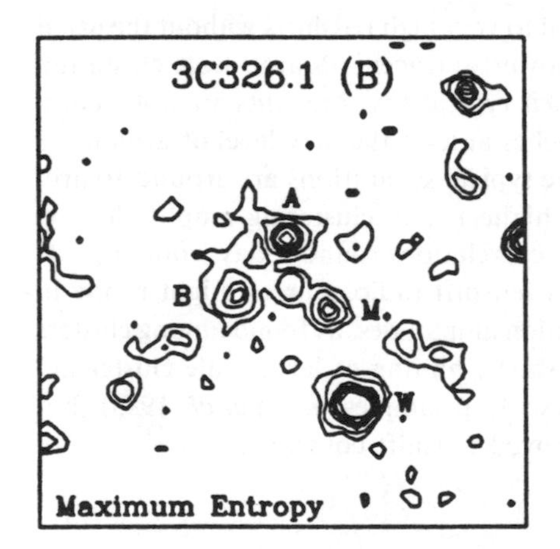

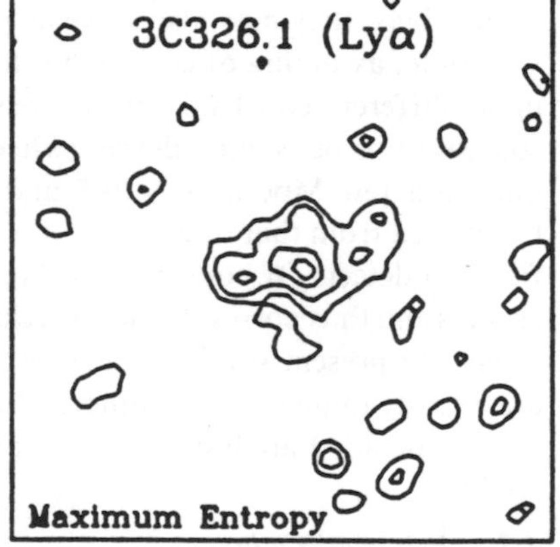

Fig. 11. Contour maps of a 100 kpc gas cloud primeval galaxy candidate, 3C 326.1 (z = 1.825): left panel represents the broad-band blue (B), which is free of any line emission; right panel a 60 Å narrow-band filter, centered on the redshifted Lyman-α. Both fields are 37.4 arcsec on each side, with North to the top, East to the left. The data were obtained with the KPNO 4-m telescope and a CCD, and processed with a Maximum Entropy algorithm (taken from Djorgovski 1988).

lines, observable to the highest redshifts, provide unique probes of the intergalactic medium as well as otherwise invisible gaseous structures that may not even be associated with luminous matter. Along with the counts of radio sources and QSOs, these absorption features give some of the least ambiguous pieces of evidence for evolution in the universe (at least with standard cosmologies). The increase with redshift per comoving volume of the number of narrow Lyman-α lines has been put on more solid footing (greater than 3σ above no-evolution predictions) with, e.g., the high precision work of Murdoch *et al.* (1986), as seen in Figure 12; using observations with IUE, this

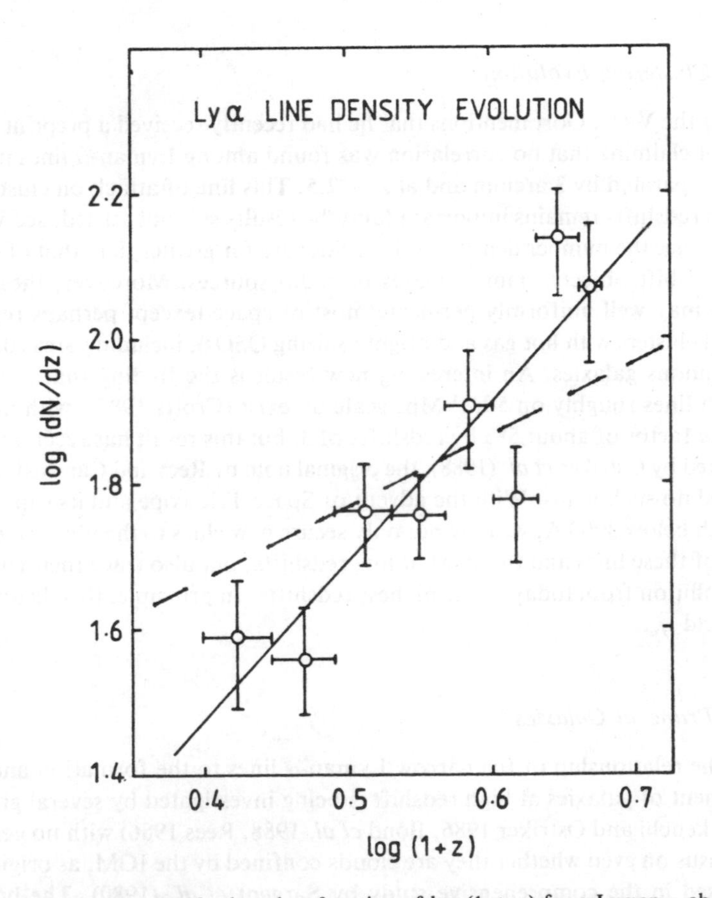

FIG. 12. Log dN/dz plotted as function of log $(1 + z)$ for a Lyman-α absorption line sample with equivalent widths $\gtrsim 0.32$ Å. The vertical bars are 1σ errors, and the horizontal bars represent chosen ranges of Δz. The solid line drawn is the Maximum Likelihood fit to the 277 lines of the Lyman-α sample (taken from Murdoch *et al.* 1986); for $q_0 \lesssim 0.5$, the no-evolution predictions are shallower than the dashed line.

evolution has been found to continue to redshifts z ~ 1 (Jenkins *et al.* 1987). An interesting result from the Murdoch *et al.* (1986) work is the confirmation of previous claims that the density of lines decrease near the QSO; this "inverse effect" or "proximity effect" is presumably the result of the increased ionization in vast volumes (radii of several Mpc) surrounding bright QSOs (but see Tytler 1987 for a contrary view). This effect places constraints on the properties of the IGM at high redshifts (Bajtlik *et al.* 1988). In addition, a recent 15 times improvement of the Gunn and Peterson (1965) measurement of the lack of neutral hydrogen in the IGM at high redshifts by Steidel and Sargent (1987) shows that the IGM must still be highly ionized beyond z = 3.

5.2 - Clustering Evolution

At the VAC, Oort mentions that he had recently received a preprint from Sargent claiming that no correlation was found among Lyman-α lines in two QSOs separated by 3 arcmin and at z ~ 2.5. This line of attack on clustering at high redshifts remains important (and the results still not settled; see Webb 1987), since the number density of these lines are far greater than that of other high redshift objects, namely QSOs or radio sources. Moreover, these gas clouds may well uniformly permeate most of space (except perhaps regions around clusters with hot gas and bright ionizing QSOs), including areas devoid of luminous galaxies. An interesting new result is the finding that "voids" of such lines roughly on 5 h^{-1} Mpc scale do exist (Crotts 1987), with maybe a filling factor of about 5% by redshifts of 3, but this result has recently been contested by Ostriker *et al.* (1988); the original note by Rees and Carswell (1987) claimed no such voids. With the advent of Space Telescope and its capability to reach below 3000 Å, we may not only secure new clues to the physical properties of these lines and the IGM at low redshifts, but also trace their clustering evolution from today to the highest redshifts. In principle, this diagnostic can yield q_o.

5.3 - Primeval Galaxies

The relationship of the narrow Lyman-α lines to the formation and environment of galaxies at high redshift is being investigated by several groups (e.g., Ikeuchi and Ostriker 1986, Bond *et al.* 1988, Rees 1986) with no general consensus on even whether they are clouds confined by the IGM, as originally proposed in the comprehensive study by Sargent *et al.* (1980). The broad, damped Lyman-α absorption lines and those from heavy metals, on the other hand, are proposed to be directly related to galaxies, probably as gaseous extensions of galaxy disks. Since such features are found to cover 20% of the sky in one sample, whereas simple predictions suggest 4%, these disks were either larger in the past or more numerous (Smith *et al.* 1986, Wolfe *et al.* 1986). Using narrow band filters tuned to Lyman-α at the redshift of the damped lines, Smith *et al.* (1987) have searched for the underlying or nearby galaxies. Their results constrain the star formation rate of these high redshift, possibly primeval spiral galaxies, to less than 1 $M_{\odot}$/yr, far less than the 100 $M_{\odot}$/yr values of the giant clouds mentioned in the previous section and perhaps even less than that from our Milky Way. This approach to finding primeval galaxies is attractive in that neutral hydrogen gas is known to exist in the search area at a specific and very high redshift.

6. QSOs

6.1 - Evolution

In their review of QSO evolution, Schmidt and Green (VAC) in the first sentence say: "The space density of quasars increases steeply with distance out to a redshift of three at least." Later, based mainly on the failure to find high redshift QSOs by Osmer (1982), they state "the number of observable quasars beyond a redshift of 3.5 is an order of magnitude lower than that predicted on the basis of a smooth extrapolation from lower redshifts", i.e., a high redshift cutoff exists that may signal an important epoch in the history of galaxy and QSO formation. At that time, only one useable QSO fainter than B = 20 had a firm redshift, and although the picture drawn by Schmidt and Green (1983), in which more luminous QSOs were evolving faster in density (luminosity-dependent density evolution), did fit the available data well, radically different physical scenarios, such as one in which the total volume density of QSOs has been approximately constant but their luminosities were brighter in the past (pure luminosity evolution) could not be distinguished (Mathez 1976, Cheny and Rowan-Robinson 1981), as demonstrated in Figure 13.

Today about 250 redshifts or more of such faint QSOs are available, mainly from the fiber-optic spectroscopic survey of the Durham group (Boyle *et al.* 1987). Combined with data from several other groups, a simple scenario is emerging that describes the overall changes in the luminosity function to the highest redshifts (see Figure 14). Using the break in the shape of the luminosity function as a fiducial marker of evolution, we find that QSOs evolved, as an ensemble, mainly in luminosity with a *gradual decline* in density. In other words, there was not a much larger density of QSO objects (or events) in the past and there is no evidence for an abrupt cutoff at redshifts near 3 to 4. Moreover, there are insufficient numbers of observed QSOs to ionize the IGM as required by the Gunn-Peterson test (Shapiro 1986). Although one

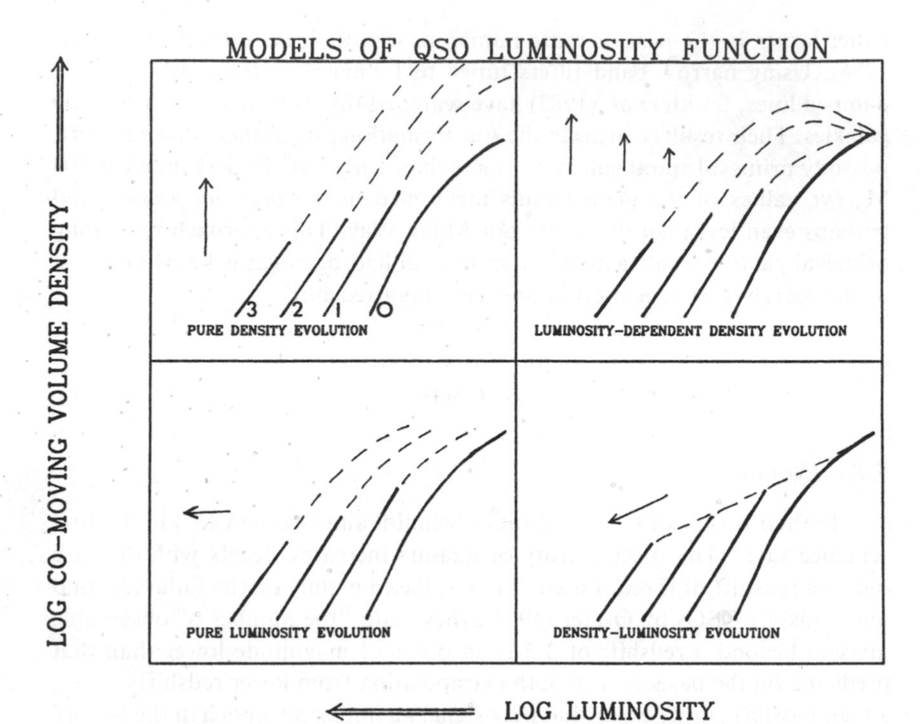

FIG. 13. Rough representations of several proposed models of QSO luminosity-function evolution. The solid lines for redshifts $z = 0, 1, 2,$ and 3 show the observed luminosity function of QSOs (and Seyfert 1s) to a limit of $B \sim 20$; note the lack of discrimination among the models. The dashed lines show the predictions for samples complete to magnitudes fainter than 20. The arrows show the direction of evolution of the luminosity function towards higher redshift; the luminosity-dependent density evolution model of Schmidt and Green (1983) predicts larger density increases for brighter objects.

interpretation is that brighter QSOs die off faster after their birth at high redshifts (Koo and Kron 1988a), alternatives which modify birth and death rates and luminosity history are equally acceptable to the extent that the continuity equation is satisfied (Cavaliere *et al.* 1982). Both the finding that many low-redshift QSOs appear to be activated by interactions (Stockton 1982, Hutchings 1987) and the claim that QSOs change from lower to higher density environments with redshift (Yee and Green 1987) argue against any picture in which QSOs were all formed in the distant past.

6.2 - *Clustering Evolution*

Although the clustering history of QSOs may be complicated by evolution in their sites of formation, QSOs remain one of the few available probes

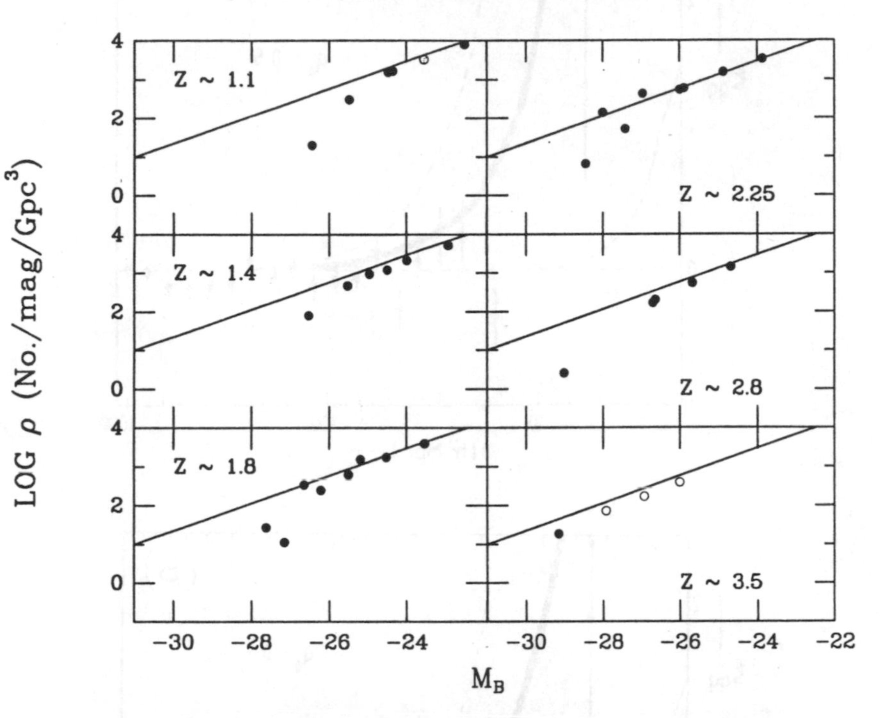

FIG. 14. Each panel shows the observed comoving volume density of QSOs at different redshifts near 1 and larger. A Hubble constant of 50 km sec^{-1} Mpc^{-1}, a critical $q_0 = 0.5$ and a spectral index of $\alpha = -1$ have been assumed in the calculation of absolute magnitudes. Open circles show values corresponding to the detection of one QSO, though none were found. The solid lines serve as fiducials. Note that the main trend in the evolution of the apparent break of the luminosity function is for higher redshift QSOs to be brighter and if anything, systematically fewer with no abrupt cutoff at even high redshifts. Details of the data from several groups and a density-luminosity model that fits can be found in Koo and Kron (1988*a*).

at high redshift. Except for a handful of pairs and triplets of QSOs that suggest possible association in superclusters (Oort *et al.* 1981), Woltjer and Setti at the VAC had little else and thus stated: "the data are certainly inadequate to establish fully clustering of quasars." Today the situation has improved but we are still at the 3σ level, though we can expect substantial gains in this area of research as more surveys achieve spectroscopic completeness. Two recent examples of positive detection include a 170 QSO sample that shows strong correlations at less than 10 Mpc (Shanks *et al.* 1987) as seen in Figure 15 and

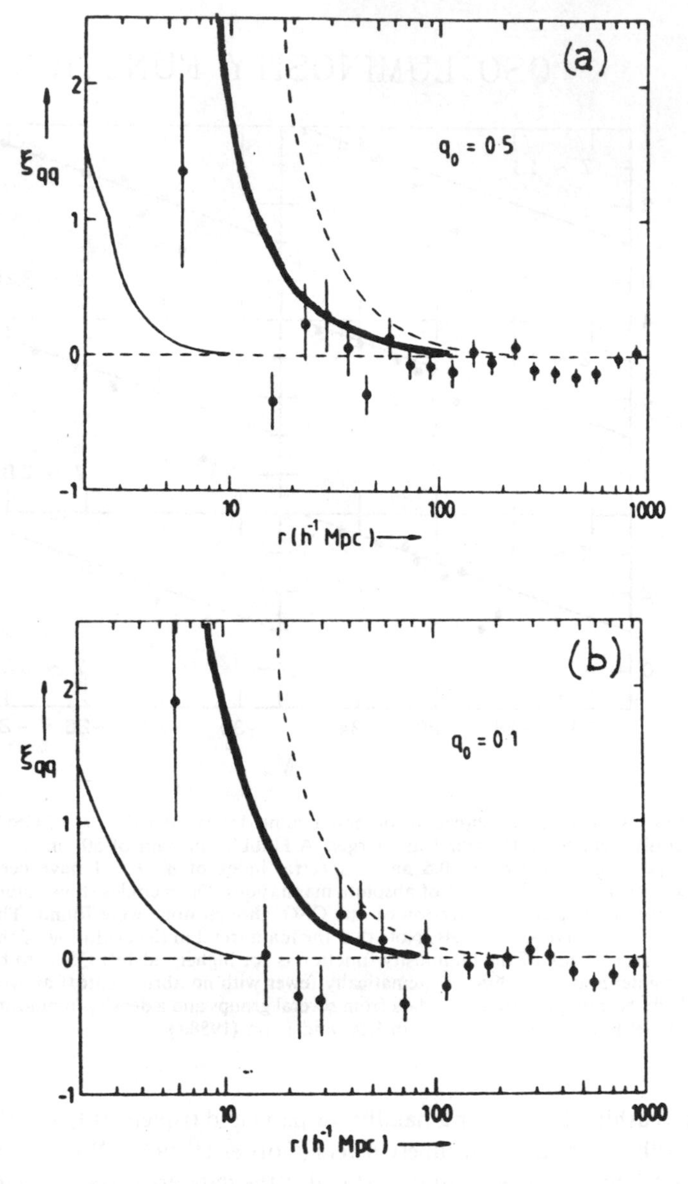

FIG. 15. (a) The QSO two-point correlation function ξ_{qq} for a complete sample of 172 QSOs assuming a value of $q_0 = 0.5$. QSO separation, r, is measured in comoving coordinates. The thin solid line is the expected galaxy correlation function at z = 1.5 based on the 'stable' clustering model. The thick solid line and dashed line are the predicted rich cluster correlation functions at z = 1.5 based on the 'stable' and 'comoving' models respectively. (b) As (a) for the $q_0 = 0.1$ model (from Shanks *et al.* 1987).

the claim by Shaver (1987) for a significant excess of QSOs on scales larger than superclusters at redshifts $z < 0.5$ in the direction of the Local Group motion.

6.3 - *Primeval Galaxy Candidates*

Although the redshift 4 barrier was first toppled by Warren *et al.* (1987*a*) nearly a year ago with a $z = 4.01$ QSO, a flurry of five new cases has recently appeared, including the record holder of $z = 4.43$ by Warren *et al.* (1987*b*) and one at $z = 4.40$ found accidentally by McCarthy *et al.* (1988), while observing a $z = 0.71$ radio galaxy. These push the epoch of formation, for at least some objects, beyond $z = 4$ and support the QSO evolution picture drawn above in which no abrupt cutoff of QSOs has yet been found. Moreover, these QSOs provide new high-redshift beacons to possible primeval candidates as have already been found in two lower redshift cases. On their first attempt using narrow-band imaging Djorgovski *et al.* (1985) found a $z = 3.22$ Lyman-α fuzzy object, presumably a galaxy (though possibly one with some non-thermal activity or interaction with the nearby QSO) (see Figure 16). Another one at $z = 3.27$ was found near the interesting QSO gravitational lens pair, MG 2016 + 112, by Schneider *et al.* (1986, 1987). In its perversity, the universe has not yielded another one despite deeper surveys of over four dozen other high redshift QSOs (see Djorgovski 1988)!

7. Summary

7.1 - *Evolution*

Beyond redshifts of $z \sim 0.4$, evolution in number density, luminosity, or spectral properties has been observed with reasonable confidence in all classes of high redshift objects. The coherent picture that emerges is the importance of gas in this evolution, as the fuel for active nuclei that produces radio galaxies and QSOs, as the source itself in the case of the QSO absorption lines, and as revealed by increased star formation activity in field galaxies and cluster galaxies. The physical mechanisms for the observed evolution remain poorly understood, largely because the major theoretical advancements have been results from N-body simulations using different fluctuation spectra with perhaps some biasing. Admittedly, dynamics that rely solely on gravity may play an important role in controlling evolution through interactions and mergers, but complex inelastic gas processes are likely to dominate more and more as we look further back in time when less gas had already undergone a conversion into stars. The explosive galaxy formation theory of Ostriker and Cowie (1981) and Ikeuchi (1981) shows how gas may even affect large-scale

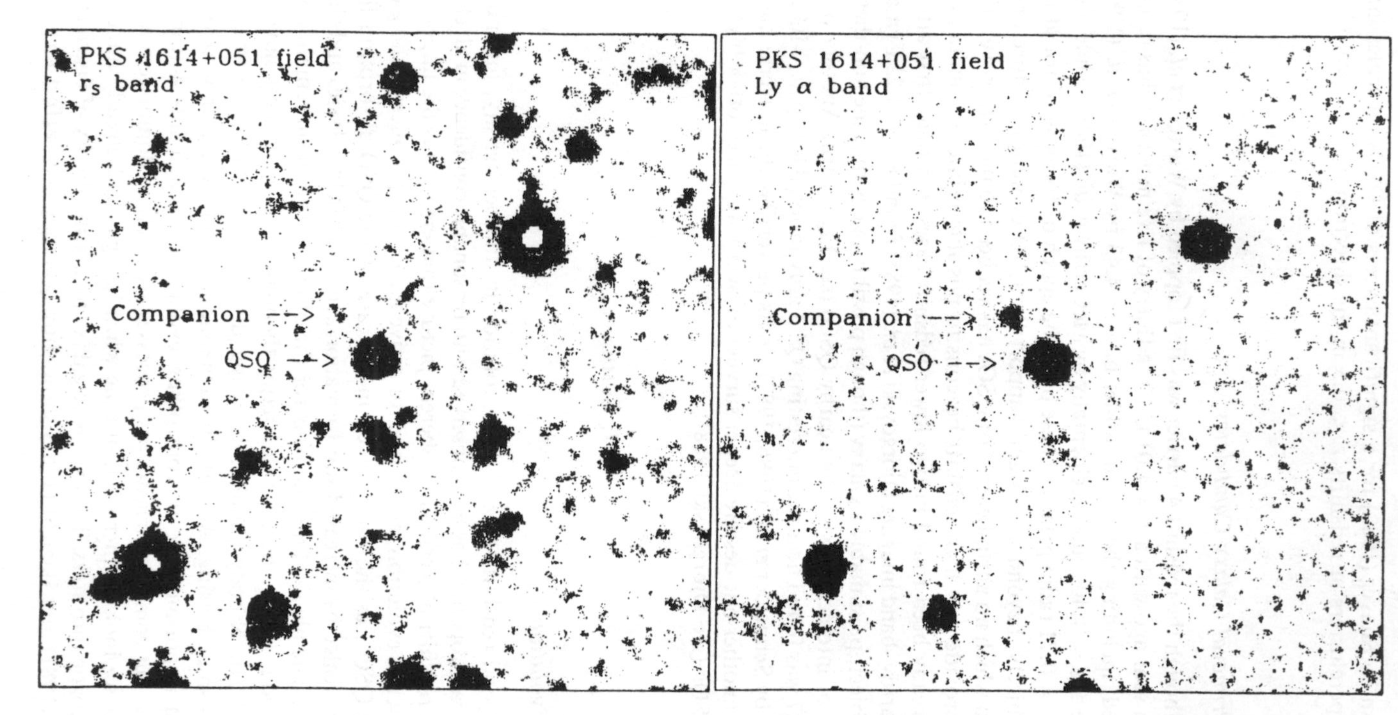

FIG. 16. CCD pictures (North at top; East to left; 73 arcsec each side) of the region around the radio QSO, PKS 1614 + 051 at $z = 3.218$, taken with a broadband red filter (r) and with a narrow band filter that brackets Lyman-α redshifted by $z \sim 3.22$. A possible bridge of Lyman-α emission appears to connect the QSO and its companion, presumably a galaxy interacting with it (from Djorgovski *et al.* 1987).

structure; similarly, bright QSOs can ionize vast volumes of the surrounding IGM. The incorporation of gas into some of the N-body codes has already begun (Quinn, private communication) and will undoubtedly provide new insights and understanding of the rapid influx of new observational data in the years to come.

7.2 - *Clustering*

Several 3σ level detections of clustering at large redshifts have already been reported for QSOs, Lyman-α lines, and field galaxies. Despite its importance as a probe of various cosmological scenarios and perhaps even q_o itself (assuming the comoving clustering scale is approximately invariant over most of the later life of the universe), the research area of tracking the evolution of clustering on all scales is very much in its infancy. With the advent of several fiber-optic spectrograph systems, especially those designed for use in dedicated redshift programs of faint objects, we can expect tremendous progress in this field over the next decade.

7.3 - *Primeval Galaxies*

Starting with the Golden Age of primeval galaxy searches in the 60's, when Partridge and Peebles (1967) proposed that galaxies at birth would be red (high redshift) and fuzzy (with star-formation occurring at maximum extent of the protogalactic gas cloud), continuing to a Renaissance era in the 70's when the dissipative galaxy formation models of Larson (1974) inspired Meier (1976) to suggest that primeval galaxies might instead be blue (low redshift) and compact (star formation occurring after collapse), we are now in the 80's with CDM models predicting that they might be blue (low redshift) but fuzzy (with blobs continually merging) or with recent IRAS observations hinting that they might even be red (due to dust obscuration) and compact (as galaxies undergoing bursts of star formation). We appear to be on the verge of a new era, when primeval galaxies will finally be discovered. Several good candidates have recently been found among high redshift radio galaxies and QSOs, with many promising avenues yet to be explored. Overall, the clues point to the oldest galaxies being formed at redshifts beyond 3 to 4.

These searches are motivated by the hope of pinning down the age of galaxies and the epoch of galaxy formation, as well as constraining the subsequent processes that control the evolution of star formation rate, initial mass function, chemical enrichment, gas dynamics, structure formation, etc. A reasonable concern is that our underlying assumption of coeval formation of all galaxies is wrong. If so, we can no longer rely on special classes of objects, such as the reddest galaxies, strong emission line galaxies, cluster galaxies, QSO absorption lines, or QSOs to reveal the history of "normal" galaxy formation. In the worst case, perhaps even a direct attack through a redshift study of extremely faint field galaxies will not yield definitive answers. On the other hand, much larger ground based telescopes will soon be a reality; larger CCDs should be forthcoming; deep infrared imaging and spectroscopy are progressing rapidly; Hubble Space telescope may yet fly; etc. Even if we do not reach the Holy Grail of watching galaxies being born, the search itself will be exciting and full of surprises. "Look afar and see the end from the beginning" (Fortune Cookie 1983).

REFERENCES

Bajtlik, S., Duncan, R.C., and Ostriker, J.P. 1988. *Ap J.* **327,** 570.

Baron, E. and White, S.D.M. 1987. *Ap J.* **322,** 585.

Blades, J.C., Norman, C., and Turnshek, D.A. 1987. *Proceedings of STScI Workshop on QSO Absorption Lines: Probing the Universe.*

Bond, J.R., Szalay, A., and Silk, J. 1988. *Ap J.***324,** 627.

Boyle, B.J., Fong, R., Shanks, T., and Peterson, B.A. 1987. *MN.* **227,** 717.

Brück, H.A., Coyne, G.V., and Longair, M.S. (eds.) 1982. *Astrophysical Cosmology.* Vatican City State: Pontifical Academy of Sciences. (VAC).

Butcher, H. and Oemler, A. 1978. *Ap J.* **219,** 18.

_____. 1984. *Ap J.* **285,** 426.

Cavaliere, A., Giallongo, E., Messina, A., and Vagnetti, F. 1982. *AA.* **114,** L1.

Chambers, K., Miley, G., and van Breugel, W. 1987. *Nature,* **329**, 604.

Cheny, J.E. and Rowan-Robinson, M. 1981. *MN.* **195,** 497.

Condon, J.J. and Mitchell, K.J. 1984. AJ. **89,** 610.

Couch, W.J. and Sharples, R.M. 1987. *MN.* **229,** 423.

Crotts, A.P.S. 1987. *MN.* **228,** 41p.

Danese, L., DeZotti, G., Franceschini, A., and Toffolatti, L. 1987. *Ap J Letters.* **318,** L15.

Davis, M. 1980. in *IAU 92, Objects of High Redshift.* eds. G.O. Abell and P.J.E. Peebles. p. 57. Dordrecht: Reidel.

Djorgovski, S. 1988. in *Towards Understanding Galaxies at Large Redshifts.* eds. R.G. Kron and A. Renzini. p. 259. Dordrecht: Reidel.

Djorgovski, S., Spinrad, H., McCarthy, P., and Strauss, M. 1985. *Ap J Letters.* **299,** L1.

Djorgovski, S., Strauss, M., Perly, R., Spinrad, H., and McCarthy, P. 1987. *AJ.* **93,** 1318.

Dressler, A. 1987. in *Nearly Normal Galaxies.* ed. S.M. Faber. p. 265. New York: Springer-Verlag.

Dressler, A. and Gunn, J.E. 1988. in *Large-Scale Structures of the Universe,* eds. J. Audouze, M.-C. Pelletan, and A. Szalay. Dordrecht: Kluwer Academic Pubblisher, p. 311.

Ellis, R.S. 1987. in *IAU Symp. 124, Observational Cosmology.* eds. A. Hewitt, G. Burbidge, and L.Z. Fang., p. 367. Dordrecht: Reidel.

_____. 1988*a*. in *Towards Understanding Galaxies at large Redshifts.* eds. R.G. Kron and A. Renzini, p. 147. Dordrecht: Reidel.

_____. 1988*b*. in *High Redshift and Primeval Galaxies.* eds. J. Bergeron, D. Kunth, and B. Rocca-Volmerange.

Fortune Cookie. 1983, eaten with inspiration by the author.

Gunn, J.E. and Dressler, A. 1988. in *Towards Understanding Galaxies at Large Redshifts.* eds. R.G. Kron and A. Renzini. p. 227. Dordrecht: Reidel.
Gunn, J.E. and Peterson, B.A. 1965. *Ap J.* **142,** 1633.
Gunn, J.E., Hoessel, J., and Oke, J.B. 1986. *Ap J.* **306,** 30.
Hamilton, D. 1985. *Ap J.* **297,** 371.
Heckman, T.M., Smith, E.P., Baum, S.A., van Breugel, W.J.M., Miley, G.K., Illingworth, G.D., Bothun, G.D., and Balick, B. 1986. *Ap J.* **311,** 526.
Hoag, A.A. 1981. *BAAS.* **13,** 799.
Hutchings, J.B. 1987. *Ap J.* **320,** 122.
Ikeuchi, S. 1981. *PASJ.* **33,** 211.
Ikeuchi, S. and Ostriker, J. 1986. *Ap J.* **301,** 522.
Jenkins, E.B., Caulet, A., Wamstecker, W., Blades, J.C., Morton, D.C., and York, D.G. 1987. in *QSO Absorption Lines: Probing the Universe.* eds. J.C. Blades, C. Norman, and D.A. Turnshek. p. 34. (STScI).
Katgert, P., de Ruiter, H.R., and van der Laan, H. 1979. *Nature.* **280,** 20.
Koo, D.C. 1985. *AJ.* **90,** 418.
_____. 1986*a*. in *The Spectral Evolution of Galaxies.* eds. C. Chiosi and A. Renzini. p. 419. Dordrecht: Reidel.
_____. 1986*b*. *Ap J.* **311,** 651.
_____. 1988. in *Large-Scale Structures of the Universe,* eds. J. Audouze, M.-C. Pelletan, and A. Szalay. Dordrecht: Kluwer Academic Publishers, p. 221.
Koo, D.C. and Kron, R.G. 1988*a*. *Ap J.* **325,** 92.
_____. 1988*b*. in *Towards Understanding Galaxies at Large Redshifts.* eds. R.G. Kron and A. Renzini. p. 209. Dordrecht: Reidel.
Koo, D.C. and Szalay, A.S. 1984. *Ap J.* **282,** 390.
Kron, R.G., Koo, D.C., and Windhorst, R.A. 1985. *AA.* **146,** 38.
Larson, R.B. 1974. *MN.* **166,** 585.
Lilly, S.J. 1988. *Ap J.* **333,** 161.
Lilly, S.J. and Longair, M.S. 1984. *MN.* **211,** 833.
Loh, E.D. 1988. in *Large-Scale Structures of the Universe,* eds. J. Audouze, M.-C. Pelletan, and A. Szalay. Dordrecht: Kluwer Academic Publishers, p. 529.
Loh, E.D. and Spillar, E.J. 1986. *Ap J Letters.* **307,** L1.
Lynds, R. and Petrosian, V. 1986. *BAAS.* **18,** 1014.
_____. 1988, preprint.
MacLaren, I., Ellis, R.S., and Couch, W.J. 1988. *MN.* **230,** 249.
Majewski, S.R. 1988. in *Towards Understanding Galaxies at Large Redshifts.* eds. R.G. Kron and A. Renzini. p. 203. Dordrecht: Reidel.
Mathez, G. 1976. *AA.* **53,** 15.
McCarthy, P.J., Spinrad, H., Djorgovski, S., Strauss, M.A., van Breugel, W., and Liebert, J. 1987a. *Ap J Letters.* **319,** L39.
McCarthy, P.J., van Breugel, W., Spinrad, H., and Djorgovski, S. 1987b. *Ap J Letters.* **321,** L29.

McCarthy, P.J., Dickinson, M., Filippenko, A.V., Spinrad, H., and van Breugel, W.J.M. 1988. *Ap J Letters.* **328**, L29.
Meier, D.L. 1976. *Ap J.* **207**, 343.
Miley, G. 1987. in *IAU Symp. 124 Observational Cosmology.* eds. A. Hewitt, G. Burbidge, and L.Z. Fang. p. 267. Dordrecht: Reidel.
Murdoch, H.S., Hunstead, R.W., Pettini, M., and Blades, J.C. 1986. *Ap J.* **309,** 19.
Oke, J.B. 1983. in *Clusters and Groups of Galaxies.* eds. F. Mardiorossian, G. Giuricin, and M. Mesetti. p. 99. Dordrecht: Reidel.
Oort, J.H., Arp, H., and de Ruiter, H. 1981. *AA.* **95,** 7.
Osmer, P.S. 1982. *Ap J.* **253,** 28.
Ostriker, J.P., Bajtlik, S., and Duncan, R.C. 1988. *Ap J Letters.* **327**, L35.
Ostriker, J.P. and Cowie, L.L. 1981. *Ap J Letters.* **243,** L127.
Partridge , R.B. and Peebles, P.J.E. 1967. *Ap. J.* **147,** 868.
Peacock, J.A. 1985. *MN.* **217,** 601.
Pritchet, C.J. and Hartwick, F.D.A. 1987. *Ap. J.* **320,** , 464.
Pritchet, C. and Infante, L. 1986. *AJ.* **91,** 1.
Rees, M. 1986. *MN.* **218,** 25p.
Rees, M. and Carswell, R. 1987. *MN.* **224,** 13p.
Sargent, W.L.W., Young, P.J., Boksenberg, A., and Tytler, D. 1980. *Ap J.* **42,** 41.
Schmidt, M. and Green, R.F. 1983. *Ap J.* **269,** 352.
Schneider, D., Gunn, J.E., Turner, E., Lawrence, C., Hewitt, J., Schmidt, M., and Burke, B. 1986. *AJ.* **91,** 991.
Schneider, D., Gunn, J.E. Turner, E., Lawrence, C., Schmidt, M., and Burke, B. 1987. *AJ.* **94,** 12.
Shanks, T., Fong, R., Boyle, B.J., and Peterson, B.A. 1987. *MN.* **227,** 739.
Shapiro, P.R. 1986. *PASP.* **98,** 1014.
Shaver, P.A. 1987. *Nature.* **326,** 773.
Smith, H.E., Cohen, R.D., and Bradley, S. 1986. *Ap J.* **310,** 583.
Smith, H.E., Cohen, R.D., and Burns, J.E. 1987. in *QSO Absorption Lines: Probing the Universe.* eds. J.C. Blades, C. Norman, and D.A. Turnshek. p. 148. (STScI).
Soucail, G., Mellier, Y., Fort, B., Mathez, G., and d'Odorico, S. 1987. *IAU Circular No. 4482.*
Spinrad, H. 1986. *PASP.* **98,** 269.
Steidel, C.C. and Sargent, W.L.W. 1987. *Ap J Letters.* **318,** L11.
Stevenson, P.R.F., Shanks, T., Fong, R., and MacGillivray, H.T. 1985. *MN.* **213,** 953.
Stockton, A. 1982. *Ap J.* **257**, 33.
Tyson, J.A. 1988, preprint.
Tytler, D. 1987. *Ap J.* **321,** 69.

Warren, S.J., Hewett, P.C., Irwin, M.J., McMahon, R.G., Bridgeland, M.T., Bunclark, P.S., and Kibblewhite, E.J. 1987a. *Nature.* **325,** 131.
Warren, S.J., Hewett, P.C., Osmer, P.S., and Irwin, M.J. 1987b. *Nature.* **330,** 453.
Webb, J.K. 1987. in *IAU Symp. 124 Observational Cosmology.* eds. A. Hewitt, G. Burbidge, and L.Z. Fang. p. 803. Dordrecht: Reidel.
White, S.D.M., Frenk, C.S., Davis, M., and Efstathiou, G. 1987. *Ap J.* **313,** 505.
Windhorst, R.A. 1984. *Ph. D. thesis.* Leiden University.
Windhorst, R.A., Dressler, A., and Koo, D.C. 1987. *BAAS.* **18,** 1006.
Windhorst, R.A., Miley, G.A., Owen, F.N., Kron, R.G., and Koo, D.C. 1985. *Ap J.* **289,** 494.
Wolfe, A., Turnshek, D., Smith, H.E., and Cohen, R.D. 1986. *Ap J Suppl.* **61,** 249.
Yee, H.K.C. and Green, R.F. 1987. *Ap J.* **319,** 28.

FIELD AND CLUSTER GALAXIES: DO THEY DIFFER DYNAMICALLY?

VERA C. RUBIN
Department of Terrestrial Magnetism,
Carnegie Institution of Washington

1. Introduction

Astronomers are optimists by nature. To decipher the large-scale structure of the universe from observations of bits and pieces of galaxies made at vast distances demands optimism. And the results discussed in this volume, the initial successes in mapping large-scale motions in the universe, show that our optimism is warranted.

However, we should not be so impressed by our success that we fail to assess our procedures critically. Between the observations and the conclusions lie many assumptions, some of which we continually examine: Malmquist biases, extinction corrections, sampling effects. I examine a different question here: Do galaxies of similar morphology have similar dynamical properties, regardless of their local environments and their evolutionary histories? Over fifty years ago Hubble and Humason (1931) noted that galaxies are segregated by types, depending upon their location. More recently Dressler (1980a) quantified this separation by Hubble types: 80% of field galaxies are spirals, but as few as 15% of cluster galaxies are spirals, and these are most often found in the cluster periphery. Depending predominantly on the local density, clustering properties differ (Davis and Geller 1976), angular correlations differ (Haynes and Giovannelli 1988), luminosity functions differ (Sandage *et al.* 1985), and dwarf fraction differs (Sharp, Jones, and Jones 1988), among other properties.

Is it then reasonable to expect that the Tully-Fisher and the Faber-Jackson relations have identical zero points and identical slopes at all locations across the sky and in depth? Are the rotation curves the same whether we study galaxies in the field or in clusters or with optical telescopes or with radio telescopes?

For parts of these questions answers exist. It is now well-established that some spiral galaxies in clusters have HI disks which are deficient in neutral hydrogen (Chamaraux *et al.* 1980, Haynes *et al.* 1985) compared to their cohorts in the field, and that in Virgo this deficiency decreases as a function of distance from M87 (van Gorkom and Kotanyi 1985), decreases as a function of later Hubble type, is a function of orbital parameters (Dressler 1986), and is accompanied by a truncation and an asymmetry related to the direction to M87 (Warmels 1985). Surprisingly, the CO diameters, normalized by HI diameters, are larger near the core of the cluster (Kenney and Young 1986), making it possible that the HI profile width can be artificially reduced if the HI disk does not extend as far as the peak of the rotation curve (Stauffer, Kenney, and Young 1986). Distributions and fluxes of HII regions (Kennecutt *et al.* 1984) of cluster and field spirals are generally indistinguishable. Optical spectra of the inner parts (for the few galaxies with measured velocities) appear normal (Chincarini and de Souza 1985), but no sample prior to the one discussed here has rotation curves for cluster galaxies which extend to the outer regions of spiral disks. We do not yet know if the maximum rotation velocity of the optical disk is never, sometimes, often, or always located beyond the limits of the neutral hydrogen disk in HI deficient galaxies. This is a large ignorance.

It is fair to assume, I think, that dynamical differences between field and cluster spirals are not large, or variations would already be known. But even subtle differences are important if we misinterpret small variations among galaxies as large-scale velocities. There are numerous occasions in the history of astronomy when small differences which were overlooked have led to large misunderstandings. All studies of large-scale motions are based on the premise that we understand galaxy properties so well that we can define parameters or correlations that remain unchanged as a function of environment, and that we can use these correlations to estimate the distance and hence the expected velocity of recession for any galaxy. In this contribution, I examine the rotation properties of galaxies in spiral-rich clusters, and in the compact Hickson groups, and ask if they differ as a function of position in the cluster, and from field spirals.

2. Rotation Curves for Galaxies in Spiral-Rich Clusters

2.1 - Forms of Rotation Curves

Although over 100 spiral galaxies now have accurate emission line rotation curves extending virtually to the limits of the optical disk, few cluster galaxies have comparable rotation curves. Data necessary to carry out a comparison of optical rotation curves of field and cluster galaxies are woefully incomplete. Over the past few years Rubin *et al.* (1988a) have observed optical rotation curves for about 20 Sa, Sb, Sc, and Irr galaxies in four spiral-rich clusters:

Cancer, Peg I, Hercules, and DC 1842-63 (Dressler 1980b). While individually many of the rotation curves appear normal, a study of their mass forms [i.e., log(V^2R) interior to R versus log R] by Burstein *et al.* (1986) indicates that statistically they differ subtly from the mass forms defined by field galaxies. Particularly, rotation curves with steeply rising inner velocities and rising outer velocities (mass types I) are absent from the cluster sample.

Of the cluster galaxies, 13 have morphology that can be called normal. Ten have rotation curves of normal form; only seven have rotation curves of normal amplitude. Rotation curves for some cluster galaxies are shown in Figure 1. Two-thirds of the galaxies of normal morphology have rotation curves of normal form; most of the galaxies of peculiar morphology have rotation curves which are peculiar. Thus, while we can generalize that normal galaxies have normal rotation curves and peculiar galaxies have peculiar rotation curves, the counter examples are equally important. The most common abnormality is a falling rotation curve, which is observed for about 1/3 of the cluster galaxies, of which about half have normal morphology.

A good correlation is found (Whitmore *et al.* 1988) between the outer gradients of the rotation curves and the distances from the centers of the clusters, in the sense that galaxies near the cluster core tend to have falling rotation curves, while galaxies farther from the cluster center [and the field galaxies studied earlier (Rubin *et al.* 1985)] have flat or rising rotation curves (Fig. 2). The outer gradient is defined (Whitmore 1984) as the percentage increase in the rotation velocity from $R=0.4R_{25}$ to $R=0.8R_{25}$, normalized to Vmax. While the sample is small, especially when divided into Hubble types, and the scatter large, gradients which are zero or negative (i.e., falling rotation curves) occur only in galaxies observed at small central distances, while large gradients are seen only in galaxies at large central distances. It is interesting that, although the *range* of values for the outer gradient in the field sample is as large as the range of values observed in the cluster sample, 6 of the 16 cluster galaxies (with well determined rotation curves) have zero or negative outer gradients, compared with only 7 of the 50 or so field spirals. Moreover, while for the field sample the outer gradients clump about zero or slightly positive values, for the cluster galaxies the values spread uniformly over the entire range. Many cluster spirals have rotation properties which resemble those of field spirals, but there is the additional small population of spirals near the cluster cores whose dynamics are unlike field spirals. The cluster galaxies must have had a more diverse evolutionary history than the field galaxies to establish these differences.

The ratio of mass-to-red luminosity ($M/L=V^2R/L_R$) interior to R is shown as a function of radial distance in the galaxy in Figure 3 for several cluster galaxies. The galaxies are arranged in order of the distance from the cluster centers, $R_{cluster}$. All cluster galaxies, except the few closest to the cluster centers, exhibit integral M/L ratios which increase with galaxy radius. A

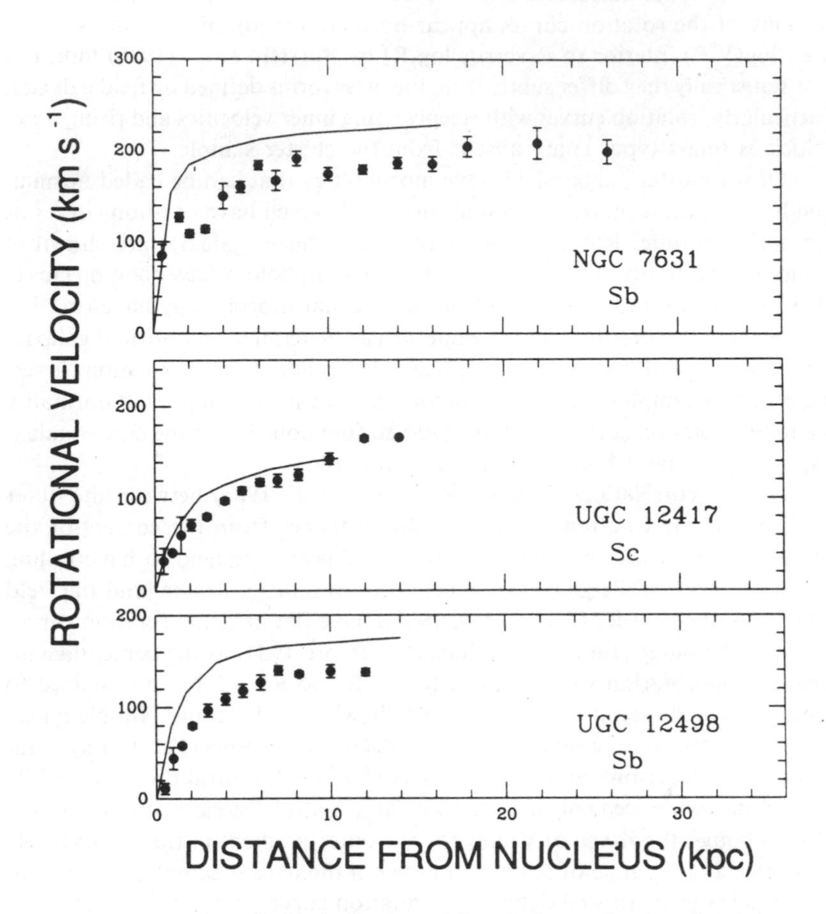

FIG. 1a. Rotation velocities in the plane of the galaxies as a function of nuclear distance for 3 galaxies in the Peg I cluster, compared with the rotation curves predicted by field spirals of corresponding morphology and luminosity. The mean velocity and the 1σ error bars are indicated for all measurements in each radial bin. The cluster rotation curves we have observed are generally low compared to their field cohorts.

luminous disk of constant M/L ratio, the assumption made in models which deconvolve the mass into disk and halo, is a horizontal line on Figure 3. This is not an acceptable fit to the observed M/L variation, and hence is not a suitable mass model for any program galaxy beyond about 0.3 Mpc from the cluster center. Cluster galaxies beyond this distance require a dark halo.

I show in Figure 4 the gradient in M/L [defined as:

$$M/L(0.8R/R_{25})/M/L(0.1R/R_{25})$$

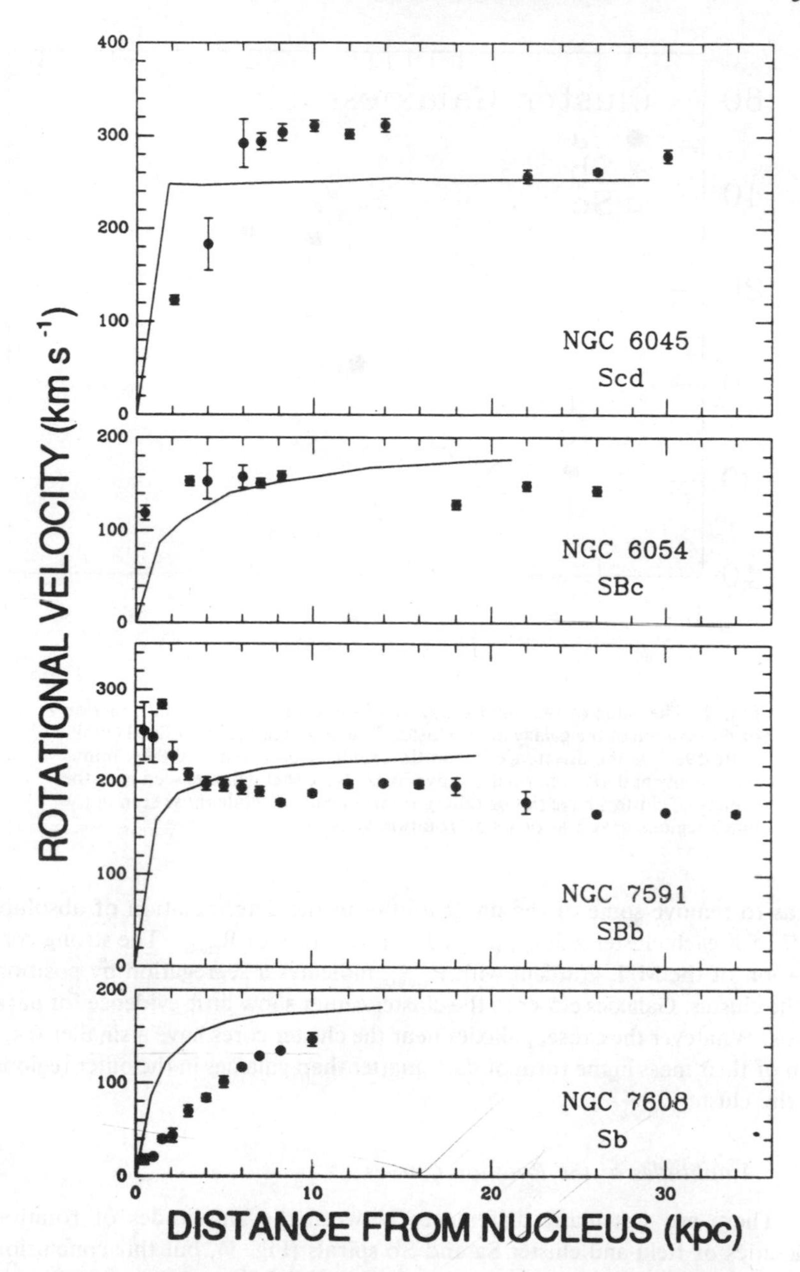

FIG. 1b. Rotation velocities in the plane of the galaxies as a function of nuclear distance for 2 galaxies in Hercules (NGC 6045 and NGC 6054) and 2 galaxies in the Peg I cluster, compared with the rotation curves predicted by field spirals of corresponding morphology and luminosity. Note the peculiar forms for some of the rotation curves.

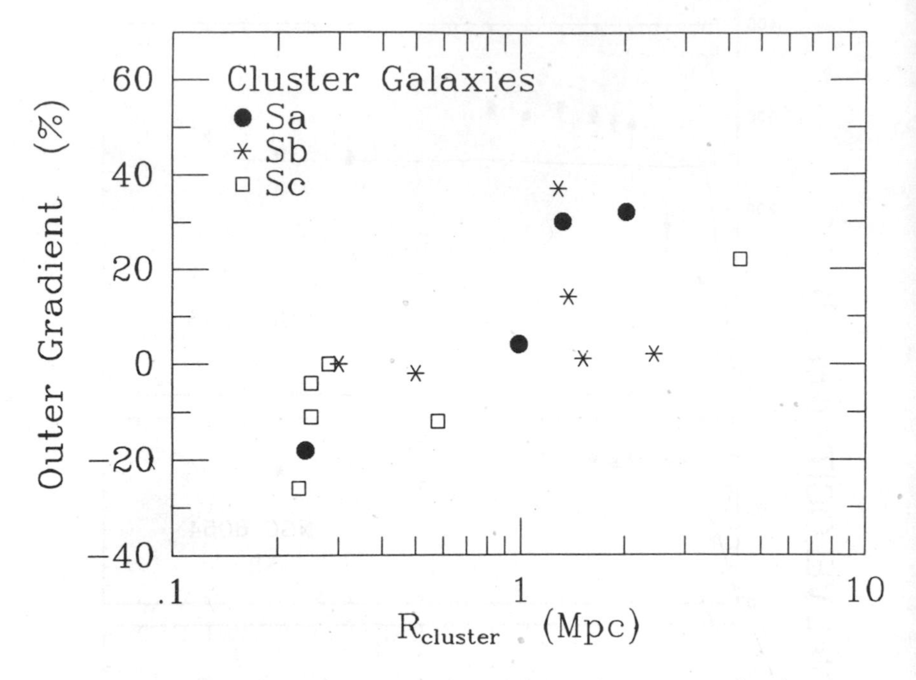

FIG. 2. The value of the outer gradient of the rotation curve as a function of the position of the galaxy in the cluster. The outer gradient is defined (Whitmore 1984) as the difference in rotation velocity observed at $0.8R_{25}$ minus the velocity at $0.4R_{25}$, normalized by Vmax. Note that galaxies seen near the centers of clusters have flat or falling rotation curves, while those seen in the outer regions have flat or rising rotation velocities.

so as to remove some of the uncertainties in the determination of absolute M/L] for each cluster galaxy, plotted as a function of $R_{cluster}$. The strong correlation of the M/L gradient with $R_{cluster}$ indicates a segregation by position in the cluster. Galaxies closer to the cluster center show little evidence for dark halos. Whatever the cause, galaxies near the cluster cores have a smaller fraction of their mass in the form of dark matter than galaxies in the outer regions of the clusters.

2.2 - Amplitudes of the Rotation Curves

There are systematic differences between the amplitudes of rotation velocities of field and cluster Sa and Sb spirals (Fig. 1), but this conclusion is of low statistical weight due to the small sample size and the possibility of systematic differences in magnitude scale compared with the field galaxies. We place each cluster galaxy at its mean cluster distance, and use its observed Vmax to predict an absolute magnitude, based on field galaxies of the same

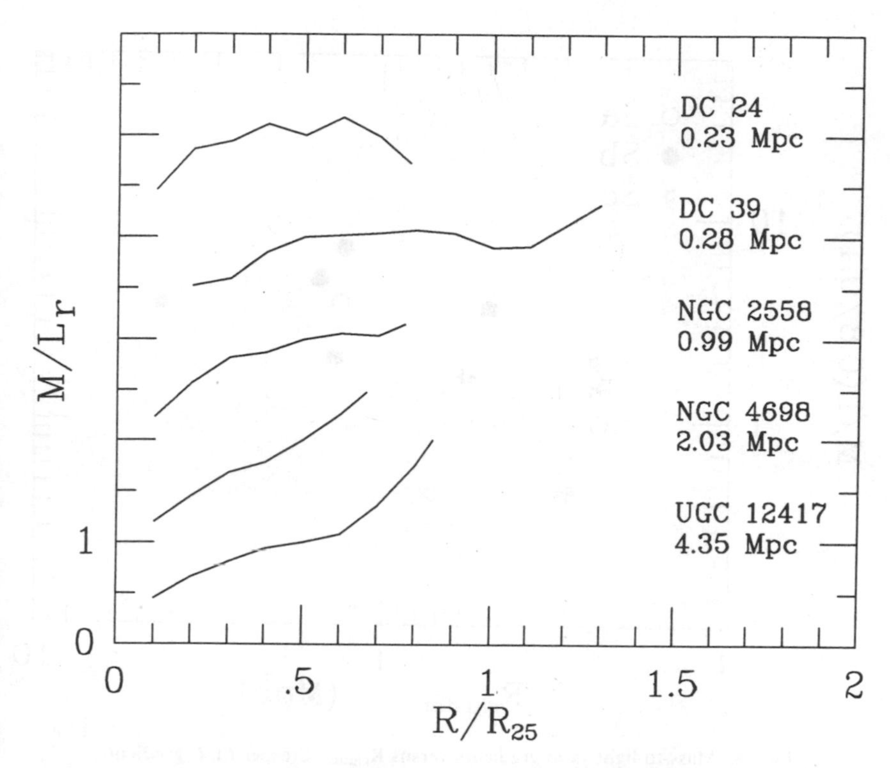

FIG. 3. Mass-to-red light ratios (normalized to 1 at $0.5R_{25}$) versus the distance from the center of the galaxy; galaxies are arranged in order of increasing distances from the centers of the clusters.

Hubble type; alternatively, we use its photometric absolute magnitude to predict a rotation curve and the value of Vmax. All Sa and Sb cluster galaxies have rotation curves which fall below the rotation curve predicted (Rubin *et al.* 1985) for a field galaxy of equivalent Hubble type and luminosity. An equivalent statement is that the Tully-Fisher relation has a different zero-point for the cluster and field samples.

This result, if supported by more extensive rotation curve data, has implications for the evaluation of the Hubble constant and for large-scale motions. However, we are aware of the numerous systematic effects which can enter the cluster analysis (Rubin *et al.* 1988a). Different selection procedures for the field and cluster samples may make the field synthetic rotation curves inappropriate for the cluster sample. Cluster galaxy apparent magnitudes, most from our CCD photometry, may be systematically too bright compared with the Zwicky values for field spirals. Our data indicate that internal extinction in cluster spirals may be smaller than internal extinction in the field counter-

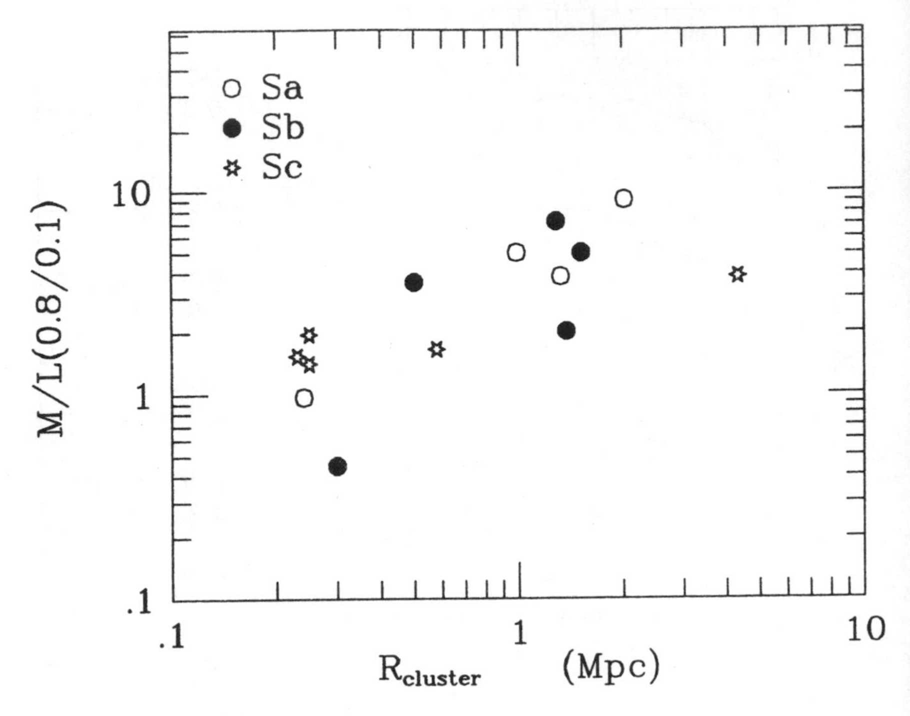

FIG. 4. Mass-to-light ratio gradients versus $R_{cluster}$. Steeper M/L gradients, evidence of massive halos, are found in galaxies in the outer regions of the clusters.

parts, a suggestion previously advanced by van den Bergh (1984) for cluster spirals and by Sandage and Tammann (1981) for Sa galaxies. Or each cluster may have a motion of 600 to 800 km/s away from the observer with respect to the smooth Hubble flow. This is expected for DC 1852-63, which is located near the apex of the microwave background dipole and not far on the sky or in velocity from the Great Attractor. However, velocities of approach, not recession, are predicted on this model for the Cancer and Peg I clusters. This is discussed further in Rubin (1988).

2.3 - *Comparison of Optical Studies and 21-cm Observations of Clusters Spirals*

In many observations and discussions of the Tully-Fisher relation, optical observations of field spirals are combined with 21-cm observations of cluster spirals. Many calibrators are field spirals. It is an unsolved puzzle why the optical data, generally for field galaxies, generally with blue magnitudes, sometimes show a steeper slope and a separation by Hubble type in a plot of luminosity versus log Vmax, compared with the 21-cm data, which are generally for cluster galaxies, often with H magnitudes.

The determination of the slope for the Tully-Fisher relation has been the subject of many studies (Tully and Fisher 1977, Roberts 1978, de Vaucouleurs *et al.* 1982, Richter and Huchtmeier 1984, Rubin *et al.* 1985, Kraan-Korteweg *et al.* 1987). While the Rubin field sample gives a slope of 10, other samples, predominantly cluster spirals, show slopes generally in the range from 5 to 8. Our small cluster sample also indicates shallower slopes, 6.1 for Sb's, 8.8 for Sc's. Perhaps the difference is simply the result of mixing two populations; field spirals with a steep slope and cluster spirals with a shallower slope.

Kraan-Korteweg *et al.* (1987) suggest that the large number of very bright galaxies in the Rubin field sample may increase the slope in the Tully-Fisher relation. But removing those with luminosities greater than $10^{11}L_{\odot}$ (i.e., the top 30%) reduces the slope for the field sample insignificantly.

To compare Vmax optical with the 21-cm profile width, we need a galaxy-by-galaxy comparison for HI deficient galaxies. Only then can we answer the question, To the limits of the optical disks what are the rotation properties of galaxies which have tidally truncated HI disks? In a pioneering effort Chincarini and de Souza (1985) obtained optical rotation curves for 8 spirals in Virgo known to be deficient in HI, and showed them to be normal. However, the limited extent of the optical spectra are not adequate to answer the above question.

Data do not yet exist to investigate directly such an effect, but I attempt a first look. I plot in Figure 5 the Fisher-Tully relation for Virgo spirals from the recent study of Kraan-Korteweg (1987), along with the best fit she determines. Vmax values come from integrated 21-cm profiles. Only galaxies with measured deficiency values are shown; this leaves 6 galaxies classified Sa or Sab, and 25 classified Sbc, Sc, or Scd. Although the sample suffers from the segregation usual in samples with only a small range in luminosity for each Hubble type, with Sd and Im at lowest magnitudes and Sa and Sab at highest, all galaxies define the same broad Tully-Fisher line. However, of the 8 galaxies with deficiencies over 0.80 (i.e., HI deficient by $\log^{-1} 0.8$ compared with field galaxies), 5 are classified Sab, 3 are Sc. All Sa's lie far from the mean line, as they will if Vmax for these galaxies is too small. The case is far from proven, but is suggestive and points a direction for future observations. It will be important to learn the optical rotation properties of the 5 Sab galaxies.

Only eight cluster galaxies in our sample have measured HI deficiencies (Giovanelli and Haynes 1985). The sample size is disappointingly small, yet there is a trend of HI deficiency to increase with increasing depression of the cluster rotation curves from the template rotation curves (Figure 6). That is, cluster spirals whose rotation velocities lie lowest when compared with field galaxies of the same morphology and luminosity have the largest measured HI deficiencies. Following the discussion at the Study Week, and in an effort to examine

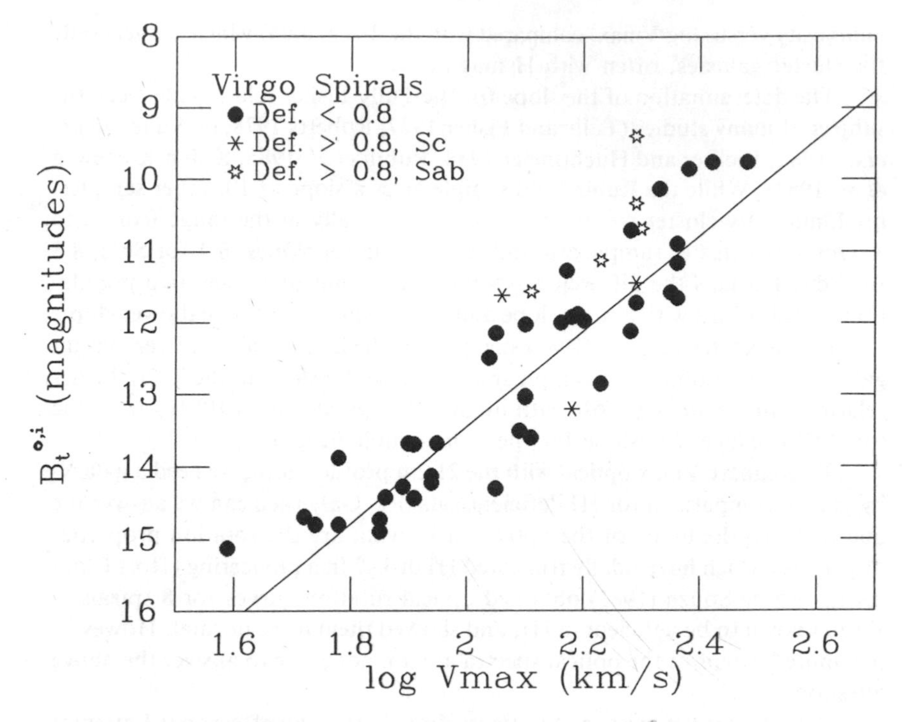

FIG. 5. The Tully-Fisher diagram for spirals in the Virgo cluster (from Kraan-Korteweg *et al.* 1987) which have a measured deficiency (Giovanelli and Haynes 1985). Although the scatter is large, there is no separation by Hubble type. However, Sab galaxies with the largest HI deficiencies lie highest with respect to the mean line copied from Kraan-Korteweg *et al.* See text for details.

this effect for a larger sample, Burstein and Rubin are presently calculating deficiency values for the field and cluster spirals in the Aaronson *et al.* (1986) sample, to see if these deficiency values correlate with the inferred deviations from a smooth Hubble flow. If these two parameters correlate, (and there is as yet no evidence that they do), this would offer evidence that deviant dynamical properties are an alternative explanation for large-scale motions.

3. Dynamics of Spirals in Denser Environments

Spiral galaxies located in large clusters, especially late type spirals, could be anomalies: they may inhabit outlying regions, even if seen near the centers by projection effects, or they may represent a population which has maintained a normal spiral morphology by lack of interaction with other cluster members. In order to resolve questions concerning differences between spiral galaxies isolated in the field and those located in truly crowded regions, we have ex-

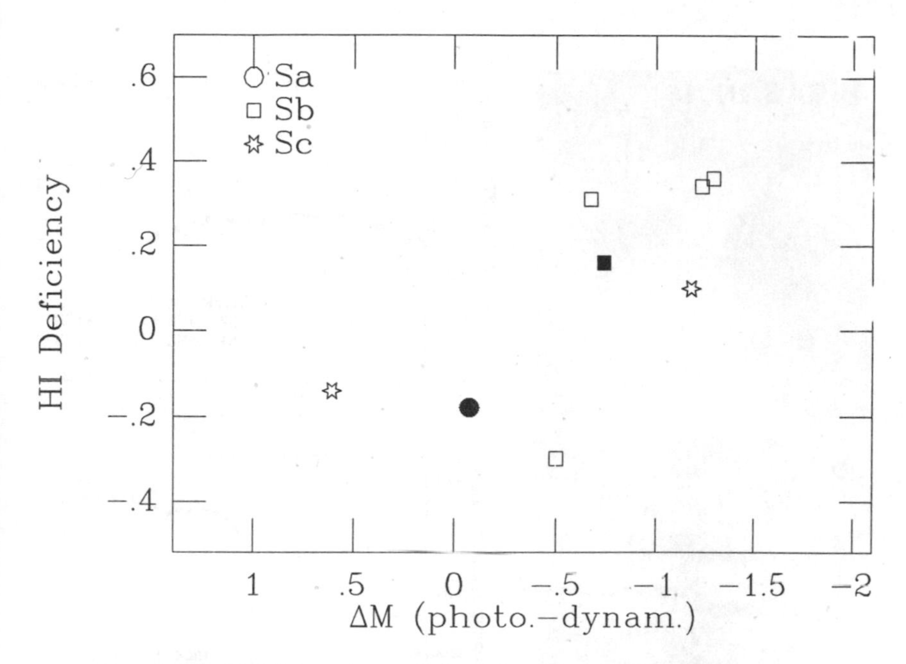

FIG. 6. The HI deficiency for 8 galaxies in the Cancer and Peg I clusters, plotted against $\Delta M_{photo\text{-}dyn}$ (i.e., the difference between the photometrically determined absolute magnitude and that determined from the rotation curve). A positive value of ΔM means that the cluster rotation velocity is low compared with that of a field galaxy of corresponding Hubble type and luminosity. Open Symbols, Peg I, filled symbols, Cancer.

tended our observations to study the dynamical properties of spirals located in the compact groups of galaxies catalogued by Hickson (1982). If these are bound groups, then the spirals inhabit regions of density as high as those near the cores of large clusters.

Compact groups pose an intriguing question for galaxy environmentalists. Crossing times are short; mergers and tidal effects would destroy the galaxies in times much less than a Hubble time. Yet morphological and spectral evidence indicate that many of these groups are dynamical entities. Rubin *et al.* (1988b) are currently completing a study of 40 spirals and 10 elliptical and S0 galaxies in 12 Hickson groups. Although many of the galaxies have a relatively normal morphology, most of them have rotation patterns which are at least moderately abnormal. Characteristic abnormalities are shown in Figures 7 and 8. Hickson 16a, an SBab spiral, and Hickson 100a, an Sb galaxy, have rotational velocities which are dissimilar on the two sides of the major axis, although each side separately would probably not be recognized as abnormal. A continuous warping might produce such an effect but, especially in H100a, the

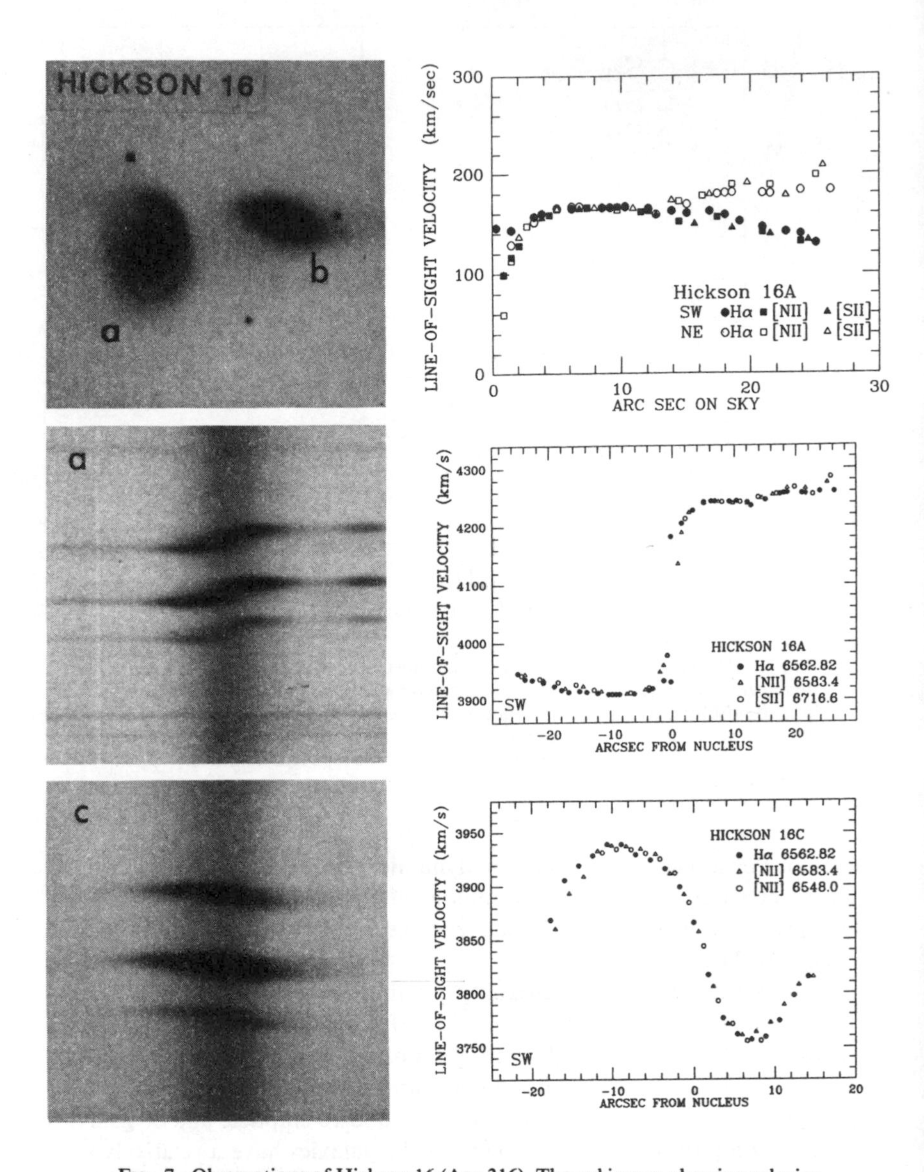

FIG. 7. Observations of Hickson 16 (Arp 316). The red image, showing galaxies a and b, comes from a Kitt Peak 36-inch CCD frame taken by J. Young. Major axis spectra of H16a (SBab) and H16c (Im) were taken with the Palomar 200-inch double spectrograph; the scale and dispersion are 0.8A/pix and 0.59''/pix. Integration times were 60 and 70 minutes, respectively. Note the lack of agreement between the SW and the NE rotation velocities in H16a beyond 14'', corresponding to the interarm region, and the sinusoidal rotation curve for H16c.

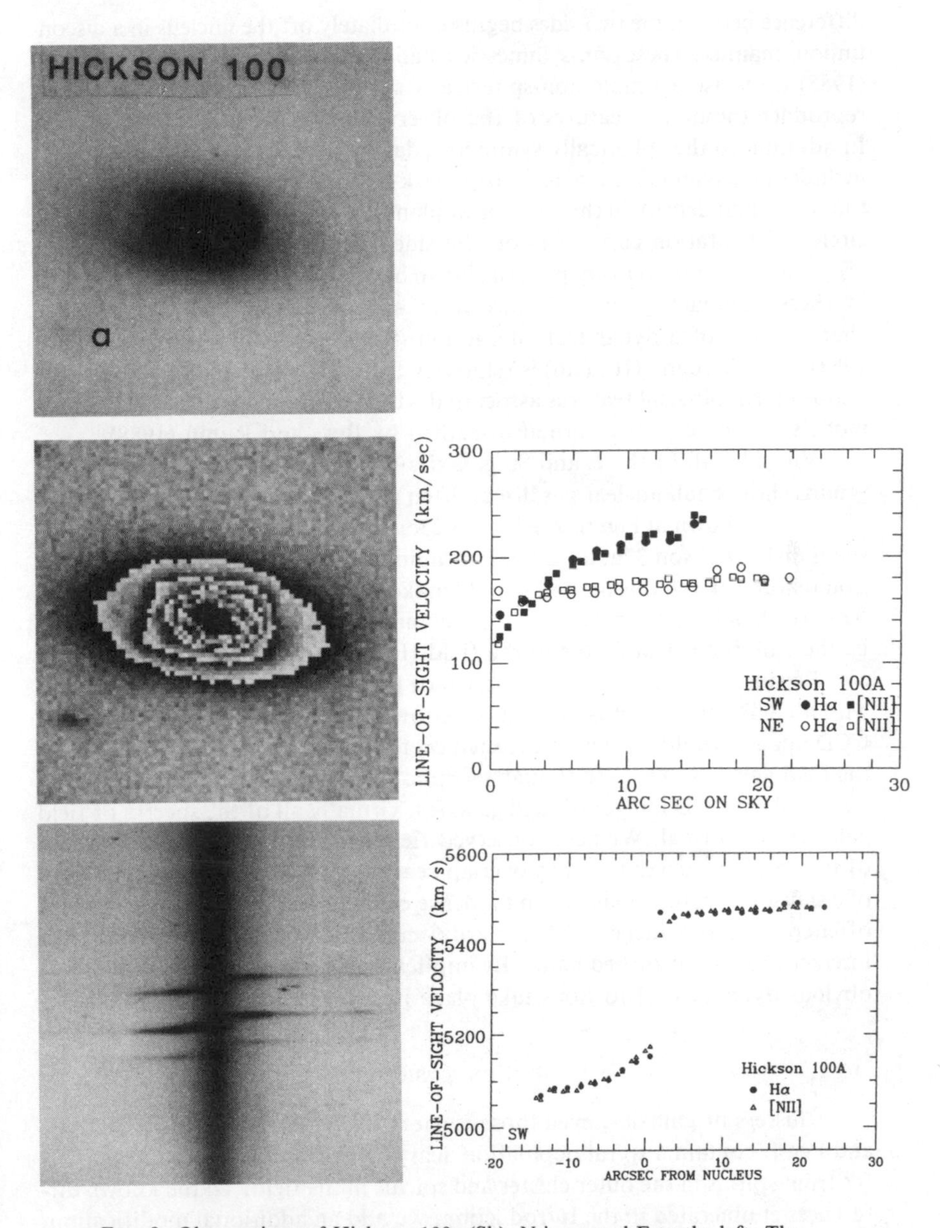

FIG. 8. Observations of Hickson 100a (Sb), N is up and E to the left. The red image comes from a KPNO 36-inch CCD frame taken by J. Young. Note the curious structure to the E, and the boxy intermediate isophotes. Velocities obtained from the Palomar 200-inch spectrum (scale and dispersion as in Fig. 7; integration time 60 min.) show a strong SW/NE asymmetry starting just beyond the nucleus.

difference between the two sides begins immediately off the nucleus in a discontinuous manner. These non-symmetric rotation curves have led Boss and Rubin (1988) to devise a simple non-spherically symmetric mass model which can reproduce the major features of the observed asymmetrical rotation curve. In addition to the spherically symmetric dark halo with $\varrho \propto R^{-2}$, this model includes an asymmetric $\varrho \propto R^{-1} \cos \theta$ dark halo. In the adopted model, contours of equal density in the equatorial plane are non-concentric approximate circles; the rotation curve rises on the side where the density is higher.

The sinusoidal velocity pattern shown by H16c is also often seen in spirals in these compact groups. It has been suggested that this pattern is a characteristic of a dynamical interaction (Schweizer 1982, Rubin and Ford 1983), even though H16c (Im) is relatively featureless and exhibits few of the major morphological features associated with a recent interaction. Halo mass models for these systems are also studied by Boss and Rubin (1988).

Virtually all of the E and S0 galaxies we have observed in the Hickson groups show small nuclear gas disks, often with high rotation velocities. Two of these are shown in Figure 9. Hickson 23c, an S0, contains a counter-rotating small disk; Hickson 37a, an E7, has a rapidly rotating disk with strong emission features. Even this limited sample makes it clear that gas disks in E's and S0's are ubiquitous for galaxies in these compact groups. Gas disks may equally be the rule for E's and S0's in the field (Phillips *et al.* 1986).

Because the images and the spectra of the Hickson galaxies are obtained with a CCD detector, it is difficult to know if the weak features seen on the CCD images would have been detected on the earlier photographic images of the field galaxies. Yet even though morphological peculiarities might now be detectable in CCD images of field galaxies, virtually all of the spectra of field galaxies are normal. We never observed field galaxies in which the two sides of the rotation curves failed to overlap reasonably well, nor rotation curves of a sinusoidal nature. Galaxies in the dense environment of the compact groups offer convincing evidence of dynamical peculiarities, probably related to tidal interactions and disturbed halos. By implication, it is likely that smaller, less obvious dynamical distortions take place in galaxies located in clusters.

4. Conclusions

Clusters of galaxies, even loose clusters like Cancer, Hercules (A2151), and Peg I, contain a spiral population near their cores, whose properties differ from spirals in the outer cluster and spirals in the field. To the known differences enumerated in the Introduction, we add an additional modification. Spiral galaxies located near the cluster cores have falling rotation curves. This suggests that only in the dense cluster cores are properties sufficiently extreme that they can modify the mass distributions of the halos. Halos of spirals in cluster cores may have been stripped by gravitational interactions with other

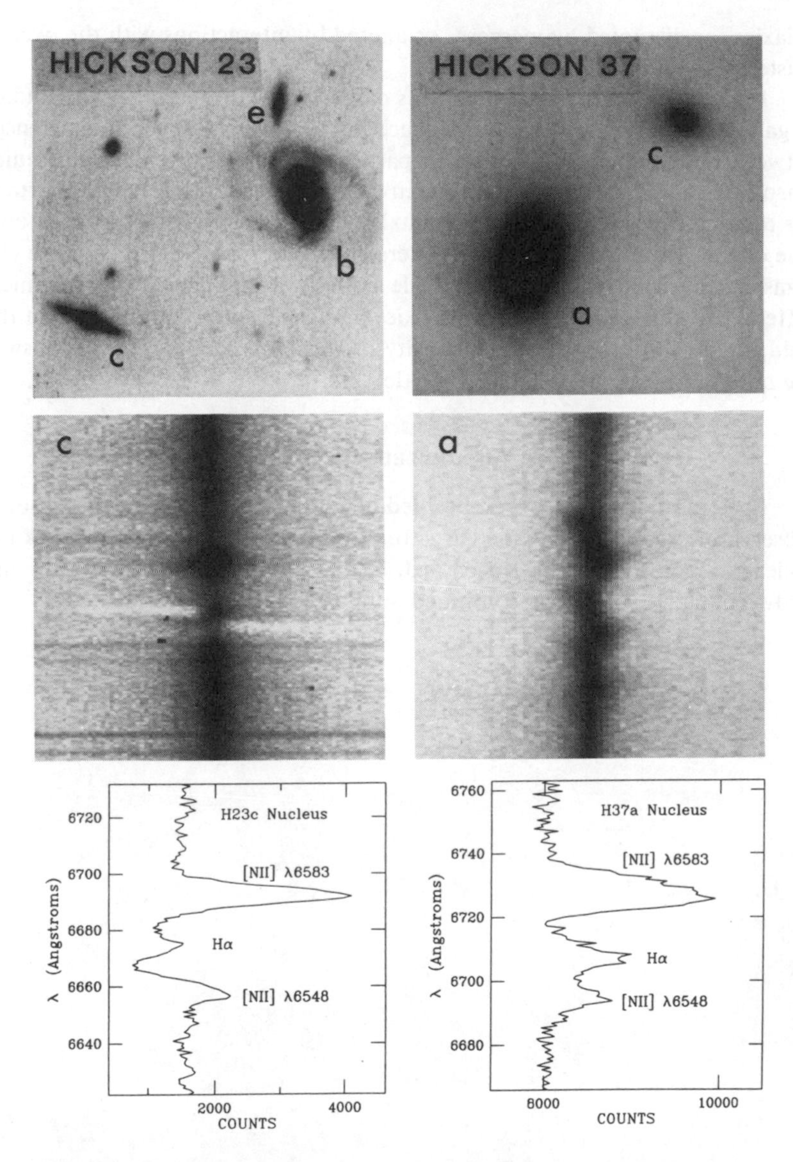

FIG. 9. Red images and spectra of Hickson 23c (S0) and Hickson 37a (E7). The images were taken with the KPNO 36-inch telescope (H23 by J. Young), and the spectra with the Palomar 200-inch double spectrograph (dispersion and scale as in Fig. 7; integration times 60 min. and 33 min.). The spectra are displayed after removal of 80% of the galaxy continuum radiation. Note strong Hα absorption, and the counter-rotating [NII] along the major axis of H23c. In H37a, the rapid rotation in Hα and [NII] is observed in position angle 312°, the position angle joining H37a and H37c, but 36° from the measured major axis of H37a.

galaxies, or disrupted by mergers, or altered by interactions with the overall cluster gravitational field.

It is not difficult to reconcile this conclusion with recent 21-cm studies of galaxies in clusters which have produced little evidence for major differences between field and cluster spirals. In particular, from their extensive studies Aaronson *et al.* (1986) point out that environmental effects are not important for their conclusions because the galaxies studied are generally located outside of the cluster cores, and are generally of late type. Although these circumstances will aid studies of large-scale motions by minimizing any dynamical differences which exist between galaxies in cluster cores and galaxies in the field, they will not satisfy the curiosity of those of us who want to answer the question posed by this paper's title.

Acknowledgements

Spectra and images were obtained at Palomar and Kitt Peak National Observatories, and I thank the Directors for observing time. I also thank my colleagues: Dave Burstein, Kent Ford, Deidre Hunter, Brad Whitmore, and Judy Young, for for their continued support and collaboration.

REFERENCES

Aaronson, M., Bothun, G., Mould, J., Huchra, J., Schommer, R. A., and Cornell, M. E. 1986. *Ap J.* **303,** 536.

Boss, A. P. and Rubin, V. C. 1988, in preparation.

Burstein, D., Rubin, V. C., Ford, W. K., and Whitmore, B. C. 1986. *Ap. J Letters.* **305,** L11.

Chamaraux, P., Balkowski, C., and Gerard, E. 1980. *Ap J.* **83,** 38.

Chincarini, G. and de Souza, R. 1985. *AA.* **153,** 218.

Davis, M. and Geller, M. J. 1976. *Ap J.* **208,** 13.

de Vaucouleurs, G., Buta, R., Bottinelli, L., Gouguenheim, L., and Paturel, G. 1982. *Ap J.* **254,** 8.

Dressler, A. 1980a. *Ap J.* **236,** 351.

_____. 1980b. *Ap J.* **42,** 569.

_____. 1986. *Ap J.* **301,** 35.

Giovanelli, R. and Haynes, M. P. 1985. *Ap J.* **292,** 404.

Haynes, M. P. and Giovanelli, R. 1988, this volume.

Haynes, M. P., Giovanelli, R., and Chincarini, G. L. 1985. *Ann Rev Astr Ap.* **22,** 445.

Hickson, P. 1982. *Ap J.* **255,** 382.

Hubble, E. and Humason, M. L. 1931. *Ap J.* **74,** 43.

Kennecutt, R. C., Bothun, G. D., and Schommer, R. A. 1984. *AJ.* **89,** 1279.

Kenney, J. D. and Young, J. 1986. *Ap J Letters.* **301,** L13.

Kraan-Korteweg, R. C., Cameron, L. C., and Tammann, G. A. 1987, preprint.

Phillips, M. M., Jenkins, C. R., Dopita, M. A., Sadler, E. M., and Binette, L. 1986. *A J.* **91,** 1062

Richter, O. G. and Huchtmeier, W. K. 1984. *AA.* **132,** 253.

Roberts, M. S. 1978. *AJ.* **83,** 1026.

Rubin, V. C. 1988, this volume.

Rubin, V. C., Burstein, D., Ford, W. K., and Thonnard, N. 1985. *Ap J.* **289,** 81.

Rubin, V. C. and Ford, W. K. 1983. *Ap J.* **271,** 556.

Rubin, V. C., Hunter, D., and Ford, W. K. 1988a, in preparation.

Rubin, V. C. Whitmore, B. C., and Ford, W. K. 1988b. *Ap J.* **333,** 522.

Sandage, A., Binggeli, B., and Tammann, G. A. 1985. *AJ.* **90,** 1759.

Sandage, A. and Tammann, G. A. 1981. *A Revised Shapley-Ames Catalog of Bright Galaxies,* Washington, D.C.: Carnegie Institution of Washington.

Schweizer, F. 1982. *AJ.* **252,** 455.

Sharp, N. A., Jones, B. J. T., and Jones, J. E. 1988. *MN.* **185,** 457.

Stauffer, J. B., Kenney, J. D., and Young, J. S. 1986. *AJ.* **91,** 1286.

Tully, R. B. and Fisher, J. R. 1977. *AA.* **54,** 661.
van den Bergh, S. 1984. *AJ.* **89,** 608.
van Gorkom, J. and Kotanyi, C. 1985. in *The Virgo Cluster.* p. 61. ed. O.G. Richter and Binggeli, B. Munich: ESO.
Warmels, R. H. 1985. in *The Virgo Cluster.* p. 51. ed. O.G. Richter and Binggeli, B. Munch: ESO.
Whitmore, B. C. 1984. *Ap J.* **278,** 61.
Whitmore, B. C., Forbes, D. A., and Rubin, V. C. 1988. *Ap J.* **333**, 542.

VIII

PROSPECTS FOR THE FUTURE

PROSPECTS FOR THE FUTURE OR THE OPTIMISTIC COSMOLOGIST

AVISHAI DEKEL
Racah Institute of Physics, The Hebrew University of Jerusalem

Believe me, I did not invent this title. The organizing committee is to be blamed. I find it almost impossible to say something intelligent summarizing a workshop of such a team of leading experts. Moreover, how can one talk of prospects for the future if we can't even predict trivia such as whether we are doomed to fall into the Great Attractor or not. We might avoid it, of course, if $\Omega < 1$. (We asked a taxi driver in Rome how should we spell Great Attractor in Italian and he replied: "You mean Grande Attrattore? Like Sofia Loren?"). And then, what does 'optimism' mean in this context? Do we really hope to reach the ultimate 'theory of everything'? Understanding everything might make it all very boring for us at the end. Perhaps more exciting is the intellectual struggle associated with the challenges set by conflicting evidence for our theories?

Trying to meet the challenge anyway, I consulted some of my friends. I must quote George Blumenthal, who advised me as follows: "I can think of only three reasonable approaches: (1) Speak doubletalk, or Yiddish. No one will understand and nobody will cite you for stupidity; (2) Discuss upcoming observations and where they may lead theory. This is the safe choice; (3) Make an outrageous prediction. People will laugh and think you are joking, but this is your chance to be the Zwicky of the 1980s if you are right". Unfortunately, I can speak neither doubletalk nor Yiddish, but I will try to follow his other two suggestions. Since every participant is going to contribute to the summary of this meeting, I will be brief and will not make any attempt to be complete or even objective. This is, by all considerations, my own biased view.

The safe way to make a prediction is by extrapolation from the past into the future. When I look into the evolution of our field in the recent past, the one thing which strikes me most is the poor causal connection between theory and observation. This is especially surprising in a field which we all regard as very phenomenological. Let me demonstrate my point with a few examples. Consider first some of the *observational* results which were the major issues of discussion in this workshop, and ask in each case whether the result was predicted by theory, or even whether the observational effort was motivated *a priori* by theory.

Most of results concerning the *spatial distribution* of galaxies were neither predicted nor motivated by theory. The extensive study of galaxy correlations by Peebles and co-workers was certainly guided by general theoretical considerations, but I do not think that the discovery of a power-law correlation function has been predicted *a priori* by any specific theory. Superclusters and voids were found independently of theory. Some exceptions were the discoveries of 'filaments' and 'bubbles', which were predicted by theories of pancakes and explosions, respectively. The correlation of galaxy type with environment had not been predicted *a priori* either (and it is not well understood even today).

The discovery of large-scale *streaming motions* — the theme of this meeting which has been thoroughly disscussed by Burstein, Faber, Lynden-Bell, Mould, and Rubin — was (and still is) a big surprise for all theories.

The discovery of *superclustering* of clusters of galaxies, as measured for example by the cluster-cluster correlation function discussed here by Neta Bahcall and myself, was neither predicted nor motivated by theory.

The study of angular *temperature fluctuations* in the microwave background radiation as described by Anthony Lasenby was motivated by theory, but the current demanding upper limits were certainly not expected *a priori*.

Consider, in turn, some of the major *theoretical* ideas concerning the formation of large-scale structure, and ask whether they were motivated by observations. Causality is somewhat more apparent here.

The theory of *inflation* was motivated by very general observational evidence such as the causality indicated by the large-scale isotropy of the microwave background.

The idea of *non-baryonic* dark matter was partly motivated by observations. The fact that the observed isotropy of the microwave background is hard to explain in a baryonic universe provided some motivation. But the main argument for it is still wishful thinking. The theory of primordial nucleosynthesis and the measured abundances of He and D put an upper limit on the mean density of baryons, $\Omega_b \leq 0.2$. Then, the theoretical desire to have a flat universe with $\Omega = 1$ and $\Lambda = 0$, motivated by Inflation, suggests non-baryonic dark matter. As for the existence of candidates, the neutrino-dominated scenario had been somewhat motivated by the false alarm concerning the detection of $\sim 30\ eV$ mass for the electron neutrino in the early eighties, but our current 'standard' Cold Dark Matter Model (CDM) was (and still is) a purely theoretical speculation (ask Jim Peebles, George Blumenthal, or Joel Primack).

Neither the idea of *Cosmic Explosions* nor the galaxy formation scenario based on *Cosmic Strings* were motivated by observations.

The newly revived scenarios of an open universe where *baryons* play an important role, both the isocurvature version which has recently been suggested by Jim Peebles and the hybrid versions which I have been working on during the years, were motivated by the evidence for very large-scale structure; first the cluster-cluster correlations and then the streaming velocities.

The *Decaying Particle Cosmology* was motivated by the wish to reconcile the dynamical evidence for $\Omega \sim 0.2$ with the theoretical desire for $\Omega = 1$. Certainly not by experimental particle physics.

The popular idea of *biased galaxy formation* was motivated by the same desire. But recall that this idea first emerged in the context of CDM, trying to reconcile the predicted correlation length with the correlation length observed for galaxies.

The study of *Gaussian Processes*, which was the basis for some of what Dick Bond and Alex Szalay told us here, has been initiated by the classical work of Doroshkevich, and has been pursued in order to provide a semi-analytic tool for comparing theories with observations. The same is true of other efforts made in developing various statistics to quantify this comparison, such as the maximum likelihood method discussed here by Nick Kaiser.

The theoretical study of the *microwave* fluctuations, some of which has been described here by Nicola Vittorio, is guided, to a certain extent, by the null discovery of such fluctuations on any angular scale.

Going through the above examples, it is perhaps not a big surprise that our field is led by observations, but it is quite surprising to realize what a weak effect the theoretical ideas actually have on the observational discoveries. There is some influence in the opposite direction, but even here it is surprising to notice that some of the major theoretical efforts were not motivated by observations.

The prospects for the future are easy to predict in this case. More of the same!

Let me summarize first some of the *observational developments* which are expected in the near future, and the major questions which they hopefully answer.

Galaxy redshift surveys. I am told by Margaret Geller and John Huchra that we should expect complete wide-angle surveys like the slices which they have shown us here, i.e. out to a magnitude limit of 15.5 (an effective depth of $\sim 100\ h^{-1}\ Mpc$ for L_* galaxies), in seven years. Deeper surveys to $m \leq 17.5$ ($\sim 250\ h^{-1}\ Mpc$ for L_*) in cones of $\sim 100\ deg^2$ are expected in five years, and 20th magnitude surveys will come next. The number of IRAS galaxies for which redshifts are measured is expected to be doubled soon. One can pose at least three important questions which should motivate such surveys: (a) Does the galaxy-galaxy correlation function go negative at $\sim 20\ h^{-1}\ Mpc$

or beyond? This is crucial for CDM and more generally for the standard assumption of scale-invariant Zeldovich spectrum of initial fluctuations; (b) Can one quantify the spatial distribution of galaxies in terms of statistically meaningful measures of filamentary structure, alignments, topology, etc.? (c) Are we approaching a 'fair sample'?

Velocities and redshift-independent distances. Martha Haynes promises that Tully-Fisher distances to spirals will be extended from the current ~ 150 h^{-1} *Mpc* in the Perseus-Pisces region to $\sim 300\ h^{-1}$ *Mpc* within one to three years. What one really needs though is a radio relescope like Arecibo in the southern hemisphere, and lots of optical time. Distances to ellipticals using the revised Faber-Jackson method can be, and should be, extended beyond the current 60 h^{-1} *Mpc* limit. A common desire we all share, of course, is to refine our current distance indicators, discover new ones, and hopefully understand the internal physical correlations of galaxy properties on which they are based, but I am quite skeptical about the chances of making significant progress on this front in the near future. On the other hand, the ability to extract the large-scale patterns of the velocity field by applying linear-theory iterations to a uniform redshift sample, as demonstrated so impressively at this meeting by Amos Yahil and Marc Davis using the IRAS sample, is a very promising new approach. After exploring the motion towards the Great Attractor, one can expect detailed mapping of the velocity field in the Perseus-Pisces region, and in larger volumes.

I share the worry expressed here by some of us that we cannot regard the current samples as "fair" samples as long as the significant features we detect, like superclusters, voids and streaming motions, are on scales comparable to the volume sampled itself. I can see how the recent evolution from the old notion of 'Virgo Infall' to the current idea of 'Grande Attrattore' will lead next to a discovery of 'Grande Repulsore' in the opposite side of the sky, and then to something which we will probably call in the next Vatican Workshop 'Attrattore di Tutti Gli Attrattori" (see Figure 1), and so on.

After pursuing so impressively the 'geographical approach' of mapping the velocity field and modeling it with a simple model, we should proceed to *a priori* statistical analysis which would enable quantitative comparison with theory. The use of multi-parameter models, even if they are very descriptive and they provide a great fit to the data, is limited. The difference is analogous to the difference between discovering America and understanding plate tectonics. A simple example of a useful measurable quantity is the mean ('bulk') velocity in a given volume(s). Most interesting is the coherence length of the velocity field, and efforts will be made to analyse the velocity-velocity correlation function. This analysis is complicated by the fact that only the radial velocities are available, but I expect that several techniques will be developed to make its application useful.

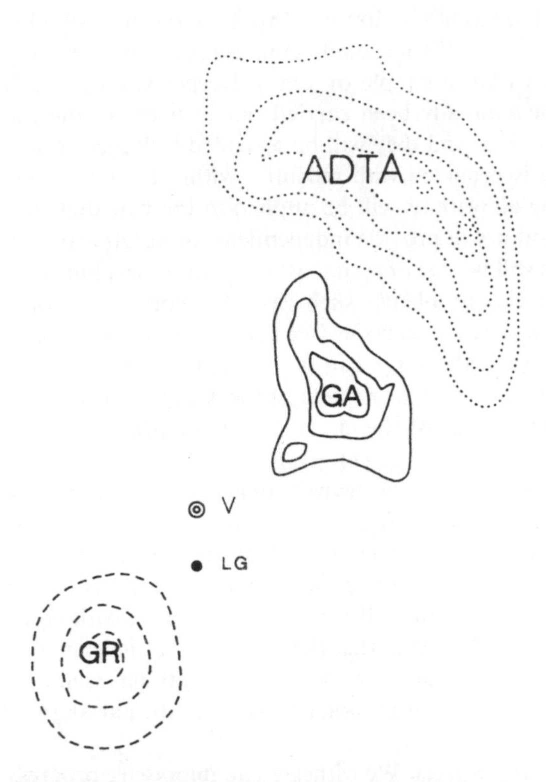

FIG. 1. Do we have a fair sample? The current geographical notion of *Grande Attrattore,* which has replaced the old model of *Virgo Infall,* is probably not the final word. Next, one might expect the discovery of an analogous *Grande Repulsore* — a negative perturbation somewhere in the opposite side of the sky. Then, when our samples expand further, we shall naturally discover something bigger and more distant, which we might naturally name *Attrattore di Tutti Gli Attrattori,* and so on.

Cluster-clustering. The 64 h^{-1} thousand dollar question is whether the cluster-cluster correlation function is, indeed, as high as it seems to be based on the studied sub-samples of the Abell catalog. Is the correlation length as large as half the mean separation between neighbors ($r_0/\bar{d} = 0.5$), as argued by Neta Bahcall, or is it significantly smaller? The current large correlation length is in clear conflict with all current theories that are based on Gaussian, scale-invariant fluctuations in an Einstein-deSitter flat universe, but a value of $r_0/\bar{d} \simeq 0.1 - 0.2$ would be compatible with 'standard' CDM, for example. Significant progress is expected on this issue. The southern Abell catalog has been completed recently and it allows a new, immediate analysis of the angular cluster correlation function, $w(\theta)$. One can expect to have redshifts for all $R \geq 1$, $D \leq 4$ ($\leq 300\ h^{-1}$ Mpc) southern Abell clusters, a sample comparable to the northern redshift sample which has been analysed by Bahcall and Soneira, within a couple of years. Deeper surveys, to $D \leq 6$ ($\leq 600\ h^{-1}$ Mpc), have already been carried out in narrow cones as described by Geller and Huchra, and they will be extended to larger areas until the whole Abell catalog is provided with redshifts within the next decade. Automatic cluster-finding algorithms will be applied to the new digitized sky survey of Cambridge and it will provide independent, more objective cluster catalogs. Such catalogs will be free of systematic human biases, but they will still suffer from systematic problems such as the contamination by projected foreground/background galaxies (see contributions by Kaiser and by myself). Most promising in the short run is an upcoming survey of X-ray clusters. Because of the smaller angular extent of the X-ray emitting regions, this catalog will be practically free of the above contamination effect, and will provide more reliable measure of $\xi_{cc}(r)$.

Fluctuations in the microwave background. Anthony Lasenby promises that we shall reach the level of $\delta T/T \sim 6 \times 10^{-6}$ on angular scales of $30''$ — $4'$ within one year, and $\delta T/T \sim 3 \times 10^{-6}$ on scales of $5'$ — $2°$ within five years. We seem to be very close now to upper limits on the various scales at the level of $\delta T/T < 10^{-5}$. If we actually get there with no believable detection, and convince ourselves that the limits are statistically significant, it will become a very severe challenge to most conventional scenarios, and it might require a revolution in our general approach to the paradigm of gravitational instability.

High redshift objects. We witnessed an impressive progress in the last few years, as described by David Koo, and we can expect more in the future. The homogeneous surveys of quasars and absorption clouds have the potential of providing a new dimension to our view of the formation of large scale structure — the time evolution. The interpretation of the data, however, is not trivial. A study of the time evolution of clustering, for example, will require a decomposition of the effects of luminosity/density evolution of the sources themselves and the gravitational clustering process. I do believe that if the data improve, and if we are clever enough, we will be able to extract useful constraints separately on the physics of the sources and on the cosmology which they are embedded in.

Predicting observations is relatively easy. Let's proceed now to the harder part of the prospects for the future, namely, trying to predict how the various *theories* will develop. This is going to be almost pure speculation. In these days of crisis in the stock exchange markets, let me use an analog — the 'Dow Jones' index of popularity for cosmological scenarios. Figure 2 shows first my view of the evolution of this index in the last decade.

We ended the seventies with only *baryons* in mind, but realized around the turn of the decade, facing the isotropy of the microwave background and

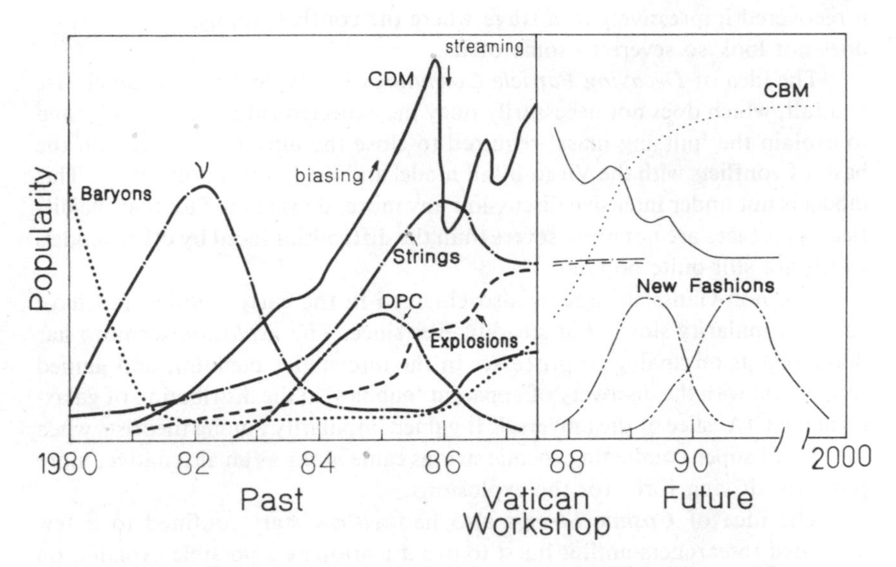

FIG. 2. Evolution of the 'Dow Jones' popularity index for theories of the formation of structure in the universe; before the Vatican Workshop and after it. See the text for details.

nucleosynthesis constraints, that non-baryonic dark matter may be required. Only a small group of non-baryonic skeptics were left to keep the baryonic models on low fire.

Neutrinos became the highlight of the early eighties, associated with the revival of the Zeldovich pancake picture, but then this scenario lost popularity drastically because of the difficulties it has in forming galaxies in time without over-clustering on larger scales. Two little positive jumps in the popularity of neutrinos occurred recently: the idea of anti-biasing helped easing some of the difficulties, and the discovery of large-scale streaming velocities reminded us that we need more power on large scales in the spectrum of fluctuations. The neutrino-dominated cosmology might provide the required large-scale power, especially if $\Omega < 1$.

Cold Dark Matter took over in about 1983 and has been the 'hot'theory since. Its main appeal is in its predictive power, which has been pursued very impressively by the N-body simulations, some of which were described by Marc Davis. Among the major milestones on its way to become the current 'standard' model one can mention the continuous great success in explaining many properties of galaxies, the problem of too weak clustering which was answered by the natural idea of 'biased galaxy formation', and recently the discovery of streaming motions on large scales. This discovery was at first interpreted as a fatal problem for CDM, and caused a drastic drop in its popularity, but it recovered impressively to a stage where the conflict, though it still exists, does not look so severe to some of us.

The idea of *Decaying Particle Cosmology* is a typical case of quick rise and fall, which does not necessarily obey the expected rules. It was suggested to explain the 'missing mass' required to close the universe, but fell on the basis of conflicts with the Virgo-infall model and the flat rotation curves. This model is not under intensive discussion any more, despite the fact that the difficulties it faces are not more severe than the difficulties faced by other models which are still quite popular.

The non-Gaussian models also emerged in the early eighties and have gained popularity slowly but steadily ever since. The *explosion* scenario has developed as an analog to processes in the interstellar medium, and gained recognition with the discovery of apparent 'bubbles' in the distribution of galaxies in the CfA 'slice of the universe'. It gained popularity among theorists when the idea of super-conducting cosmic strings came along as an alternative, more powerful driving force for the explosions.

The idea of *Cosmic Strings* also had a slow start confined to a few dedicated researchers until it burst to our attention as a possible explanation for non-Gaussian clustering of galaxies and clusters. It gained popularity in some circles at an outrageous rate until Jim Peebles brought the community back to its senses with two privately circulated 'screeds' of embarrassing questions concerning galaxy formation in the string picture as compared to observations.

This is my evaluation of where the various thoeries stand at present. How will they do in the future? This brings me to the *outrageous predictions*.

I take seriously the difficulties of CDM to cope with the following problems: (a) the very large-scale structure reflected by the cluster clustering and the large coherence length of the streaming towards the Great Attractor; (b) the filamentary structure on scales beyond $10\ h^{-1}\ Mpc$; (c) the late galaxy formation and the fragility of disks; (d) the finite extent of the halo of the Galaxy; (e) the lack of actual, existing candidate particle. I therefore predict a steady decrease in popularity for CDM. This is an outrageous prediction because the current popularity gradient, as reflected by the general attitude of the participants of this workshop, is certainly still positive.

Neutrinos will probably maintain their moderately-low level of popularity. The shrinking experimental upper limits on the mass of the electron neutrino leave the tau neutrino as the only viable neutrino candidate which can still close the universe (need $m_\nu \sim 30\ eV$). The predicted clusters in the neutrino scenario are somewhat too big and too hot even if galaxy formation in them was subject to anti-biasing. Massive halos around dwarf galaxies, if confirmed, pose another argument against neutrinos as the dark matter candidate there.

Cosmic Strings, I am afraid, might go down in popularity too. In addition to the difficulties raised by Jim Peebles concerning galaxy formation seeded by strings, I am confident that it will be realized soon that our understanding of the basic processes involved in the string scenario, and in particular the fragmentation of strings, is only poorly understood. The fragmentation pattern in the string simulations is extremely sensitive to the way the intercommuting process, and the resultant nasty kinks that form and propagate along the strings, are treated numerically. The different simulations by the different groups show very different results. This will add to the complexity of the string scenario, and will cause a continuous damage to its popularity. Enthusiasm is expected, at least for a while, with regard to a new type of strings — *Fundamental Strings* — which is an attempt to relate a cosmological string scenario to the concept of super-strings. The parameters of this model seem to be predicted by theory (e.g. $G_\mu \sim 10^{-3}$) but its ability to make a reasonable galaxy formation scenario is still to be investigated.

The *Explosion* model, on the other hand, is in a sense more immune against loss of popularity because it is backed by powerful forces that can match almost any observation with an ingenious theoretical idea. It has been found, as I briefly described in my earlier contribution, that clusters supercluster naturally in a generic explosion scenario, significantly more than in any of the other scenarios under discussion, and in better agreement with observations. It remains to be seen, however, how one can explain the large streaming motions in the explosion picture.

The existing reservoir of theoretical ingenuity, which has given rise to fancy ideas like decaying particles, cosmic strings, cosmic explosions and cosmic explosions generated by super-conducting cosmic strings etc. etc., promises that we will see more new theories which will emerge in order to face new conflicting evidence, attract attention for a while, and then decay the way the decaying particles did, perhaps to burst again when needed.

My outrageous prediction, which is not so outrageous for me if you consider my work in the past, is that the dark matter will eventually prove to be *baryonic*. After all, aren't baryons the only dark matter candidate which is known to exist? Take for example planets as the dark matter building blocks. Baryons, in an open universe, guarantee sufficent power on large scales to explain both the streaming motions and the superclustering (or 'super-pancaking'). The baryonic universe has started regaining popularity these days since Jim Peebles proposed the isocurvature model. I predict that baryonic models will go strong in the near future. Still, one should not give up so easily the nice features of CDM in explaining galaxies, which is basically an argument for an effective logarithmic slope of $n \simeq -2$ for the spectrum of fluctuations on galactic scales. The ideal scenario might, therefore, be a baryonic universe with a CDM-like spectrum of fluctuations on galactic scales. The isocurvature model comes close. Also, baryonic matter would behave as 'cold' matter if the gas had collapsed into objects (e.g. black holes) before galactic scales entered the horizon, giving rise to an exciting scenario of 'Cold Baryonic Matter'.

Despite my attempts to indicate the prospects for the future, I do not recommend we rush to buy stocks based on my speculations. Judging from the rapid evolution of ideas in the recent past and the current state of relative confusion in our field, one must admit that the future is mostly unpredictable. Nevertheless, this might be a good reason for 'optimism'.

FIG. 3. I must admit that I did something useful during the long discussions we had. I tried to sketch amateur cartoons of some of the participants at their best moments. It matches the high spirit of the workshop, but I have mixed feelings about how it would fit in these serious proceedings. Anyway, since the editors encouraged me to put in at least one example and they decided to take the responsibility, here it is. It is left for the reader to decide who this character is.

COMMENT

In order to recapture, in a somewhat systematic way, the lively and creative exchanges that took place during the Study Week and to preserve the personal enthusiasm that characterized much of the week's work, the Scientific Organizing Committee decided that at the end of the week, following the summary by Avishai Dekel, each participant would be asked to present for a period of time, not to exceed five minutes, whatever they wished in terms of final thoughts concerning the Study Week. Since the organizers had designed the meeting as a research week with the hope of opening new horizons of discovery or, at least, defining more accurately the questions to be addressed, a great deal of importance was attached to this final exercise of the week. From the recording a written transcript was provided to each participant for personal editing and the results are presented in the following pages.

The order of speakers was determined by drawing names from a hat, — John Huchra's best western sombrero, to be precise. This session was presided over by George V. Coyne, whose remarks to each as they began their small discourse have been, except for fragments retraceable from their remarks, discretely suppressed.

George V. Coyne, S.J.

COMMENT

[illegible]

[illegible]

SUMMARY COMMENTS

DAVID BURSTEIN

I would like to make two points. The first is that the prospects for getting peculiar velocities in the future seem somewhat limited in the following sense. We are currently stuck with distance predictors that give us, at best, accuracies of about 15%. This means that, for an individual object at a distance of 10,000 km s^{-1}, you can easily measure peculiar velocities of 3000 km s^{-1} or more. That doesn't get us very far. It seems to me that the only prospect for getting peculiar velocities on much larger scales is to look at groups of galaxies, groups of groups, and to try to beat down the statistics in that region. Even that game will not get us much past about 10,000 or 15,000 km s^{-1}, and when I looked to see if we could measure peculiar velocities at, say 50,000 km s^{-1}, the future actually looks pretty dim with present techniques. We simply do not have the accuracy and we do not have the number of galaxies that will be necessary to measure peculiar velocities on the very largest scales. That is my first point.

My second point is that we should come away from this meeting with a clear desideratum to map out the plane of our Galaxy. It is apparent that, for whatever reason, our universe is created in such a way that, by our current view, much of that matter that is nearby (within 5,000-6,000 km s^{-1}) is obscured by galactic extinction. I do not think it is a coincidence that Perseus-Pisces dives into the Galactic Plane at a longitude of 120°, and the Great Attractor dives into the plane at a longitude of 310°, situated on opposite parts of the sky. The chances that galactic obscuration could do this are not negligible. We could have just as well been oriented so that our Galactic Plane would have totally obscured the large-scale structure that Brent Tully has identified.

So, the question in my mind is to what extent are the large-scale structures we see molded by this large dark band, and Zone of Avoidance that exists in our Galaxy?

NETA BAHCALL

The discussions held this week reveal the great progress made over the last several years in the field of the large-scale structure and motion in the universe. The existence of large-scale structure as a common phenomenon is now clearly established. Observations of the spatial distribution of galaxies, clusters, and even quasars, reveal a consistent picture of the existence of large-scale structure to scales of $\sim$ 50-100h^{-1}Mpc. Three dimensional maps of the galaxy and cluster distribution show the existence of large overdense regions surrounding voids, or underdense regions, that may extend to 50h^{-1}Mpc or more. Specific quantitative tools such as the correlation function and the "sponginess" measure of the distribution have been applied to describe the nature of the structure. Some of these results were discussed here this week. The correlation function analysis yields an average strength and scale of the correlated structure, showing positive galaxy correlations to $\sim$ 10h^{-1}Mpc, and considerably stronger correlations to larger scales for galaxy clusters. While the correlation analysis provides important information regarding the structure, additional quantitative measures are needed in order to describe the shape, extent, and nature of the distribution on large scales.

Observations over the last decade preceded theories in revealing structure and motion in the universe. New theories quickly followed and, as we heard summarized, are still in the process of development. It is difficult to predict at this point which of the theories — baryonic models, cold dark matter, explosion, or something else — will provide a consistent picture of the universe. Additional observations of structure and motion, combined with a quantitative description of the results, will be available over the next several years. These will surely provide new limits and ideas on theoretical models.

The use of clusters of galaxies as a tracer of the large-scale structure has proved to be effective, revealing the largest-scale structures yet observed. The cluster correlation function, with its large amplitude and scale-length, has stimulated new ideas such as biased galaxy formation and large structures surrounding clusters, as well as encouraging suggestions of cosmic strings as a possible source of cosmic fluctuations. However, the strong cluster correlation function has also created a problem for some models that cannot produce as strong a clustering and as much power on large scales as suggested by the cluster observations. While the correlation amplitude may be uncertain by about 30% due to various effects (see article above), it is still an order of magnitude stronger than the galaxy correlations, and stronger than estimated for example by cold dark matter simulations. New and improved cluster samples such as will be available in the future from digitized sky surveys and from X-ray cluster surveys should be able to reduce the uncertainty on the amplitude and scale of the cluster correlation, as well as determine whether weak correlations exist on scales $\gtrsim$50h^{-1}Mpc.

The use of quasars as tracers of the large-scale structure was discussed only briefly this week. I expect that this new tool will yield important results, including the dependence of the correlations, or structure, on redshift. Current results indicate strong quasar correlations, close to the cluster correlation strength, that also extend to nearly $\sim 100h^{-1}$ Mpc (for $z \lesssim 2$). (No significant correlations are detected at $z > 2$ in the current samples.) Larger complete samples are expected to be analyzed in the next few years; these should yield a better understanding of the evolution of the large-scale structure and the role of quasars in these early structures.

Peculiar velocities on large scales have been discussed in detail during the week. It is important to continue the investigation of the possible motion of clusters in superclusters, summarized earlier, by using accurate distance indicators to these clusters. If large motions are indeed detected, the quantitative nature of the motion will yield important information regarding the masses of superclusters and/or possible non-gravitational effects such as explosions.

GEORGE EFSTATHIOU

I think that perhaps one of the reasons that we're having this session has to do with the Church's love for confession. If I had arranged it, each of us in turn would have confessed our biases and prejudices to Father Coyne at the start of the meeting. During the meeting, he would assess our performance and at the end he could hand out our penance in the form of camera ready sheets. The more "sins" you'd committed, the more camera ready sheets you would get!

We have been asked for our personal view of the meeting. It isn't sensible to confess in public, but I will admit to a couple of things that have bothered me. I have been frustrated by Jim Peebles' remarks of filaments in the N-body simulations. It goes against his intuition that a cold dark matter simulation should show fairly large-scale filaments. Is there a mistake in the initial conditions, or in the N-body dynamics? I understand Jim's point, but it isn't clear to me that anything is wrong. I am frustrated because I can't answer a query like this until I have interrogated my computer. Jim's remarks are qualitative so it isn't even clear what would constitute a satisfactory answer. In any case, there is an interesting problem to investigate. The second point concerns comparisons with the "standard cold dark matter" model. Several of us have compared the CDM model to observations and concluded that it is O.K. But I have noticed that at least three different values have been assumed for the amplitude of the power-spectrum; so we aren't all discussing the same CDM model after all, although it appears that way to others. Let's be precise about the amplitude and make sure that everyone knows the value that we are using.

One of the nicest things about coming to a meeting like this is discussing work before it has been published. That's very valuable. You show people graphs, you say "look at this" and "look at that". You get a good idea of what is going on. It has been really useful to discuss the results of the big galaxy survey that my students are working on in Cambridge. Hopefully, we should be able to tackle some of the important issues raised at this meeting.

I thank the organisers of this very successful workshop and I am especially grateful to Father Coyne for granting me absolution.

RICHARD BOND

As someone who inhabits the space of theories, and only sometimes the space of reality my interest in what I've been hearing this week is to learn how I should limit the regions of theory space I regularly haunt. In Avishai Dekel's stock market language I would avoid spreading my portfolio over too many models, and invest heavily in a winner. There is one model that we've all invested heavily in and that's the biased cold dark matter theory. We should emphasize strongly that it is not only just a pretty theory but it is a minimal theory, in the sense that we start with an inflationary model and one of the simplest assumptions for the nature of the dark matter. There has been *no other* viable mechanisms proposed but inflation to smooth out the universe. I'm sure that you all came away with the strong belief that inflation implies scale invariance after my talk in which I twisted and turned to try to break scale invariance. This means that, within the context of the theoretical models based upon a quantum origin for fluctuations, the isocurvature baryon models with the non-scale-invariant power laws advocated by Jim Peebles are extremely difficult to arrange, and are at least as unlikely as the extra power deviants that I was talking about.

We should, therefore, adopt the view that the biased cold dark matter model requires very firm evidence against it for us to completely reject it. What we've seen this week is a fantastic development of cosmography, the subject of making maps of the universe, and the maps are extending out from us to far distant lands. But it is only the most distant and the most hidden that gives the standard biased CDM model the most difficulty: the most hidden being the Great Attractor and the most distant being the cluster patches that Brent Tully tells us we should take very seriously. If so, we must have power which is of an enormous amplitude on large scales ($\delta \sim 0.1$ at ~ 50—$100h^{-1}$ Mpc), and the cluster-cluster correlation function is just one aspect of that. I think the Henry-Huchra-Postman cluster survey provides even more compelling evidence for very large-scale patchiness of the clusters in redshift space. It is hard to imagine how Abell could have biased his cluster selection in the redshift direction. Making sure that these cluster islands really exist is one of the most important things for us to show in the next number of years. On the other hand, I would say that, since we don't know how much of what Brent Tully is seeing is due to a biasing problem in the cluster catalogue, then we should not be surprised if these objects do not survive in new cluster catalogues independent of Abell's eyes.

Given that there are all sorts of exciting possibilities out there indicating very large-scale structure, it is still important for us to keep exploring theoretical space in order to find those new rising theories that Avishai pointed to in his forecast of trends in theory. Perhaps hints of very large-scale structures at high redshift from the distribution of Lyα clouds may force us into the very radical. However, we should bear in mind that, as with the space of theories, the space of current observational interpretations is not always synonymous with the space of reality.

As a final point: The most important thing that I'll be looking for over the next few years is confirmation of Anthony Lasenby's large angle microwave background anisotropy at a few degrees, which would destroy the scale invariant cold dark matter theory if the anisotropies are shown to be primordial.

SANDRA FABER

Since I was assigned to be one of the conference summarizers, I have tried to think out a more encompassing set of remarks. Actually, I think Avishai's summary was very much in the flavour of my own thoughts. I thought it was a brilliant summary, Avishai.

The great advances at this meeting, for me, were three. First, there are George Efstathiou's new galaxy-galaxy and galaxy-cluster correlation functions, which I had not seen before, and the implications that they have for cold dark matter. Secondly, I commend the magnificent work that has been done by Amos Yahil and Mark Davis and their collaborators in interpreting the IRAS survey. When I first heard about the IRAS work — I think it was at the Princeton cold dark matter meeting over a year ago — I was nery negative about it for all the reasons that various people have raised. I've now completely turned around on this, and I think it has been well worth all the effort that has been invested. It might not be the ultimate answer, but it's certainly opened our eyes to a new way of studying the universe. I see three things about it that are really significant.

Looking at those velocity flow maps, one realizes the importance of being freed from over-simplified infall flow models. This has been obvious to our group as we have tried to map our own velocity information and been forced to put in so many extra parameters. We're very conscious of the inelegance of this. It really is wonderful to have maps that don't have parameters like that and can be compared one-for-one with every observed galaxy.

Secondly, I think that the IRAS results as they now stand are on the edge of confirming once and for all that the velocities we see *are* due to density inhomogeneities. This is something we have all believed, but it's very nice to see it actually demonstrated. I think the demonstration will go down as one of the fundamental experiments in astrophysics, because it's the crucial information that we needed for the gravitational instability picture.

Thirdly, I think that the IRAS approach still needs work. Right now, the predicted motion of the Local Group is too close to Virgo, and in fact, I think that's exactly the direction one would have picked from just looking at the *visible* galaxies in Brent's catalogue of nearby galaxies. We know though that the Local Group doesn't move quite in that direction — it moves more toward the Great Attractor. So it's clear to me that something as yet is still missing from the data base; there needs to be more mass in the Great Attractor. My hope is that filling in the galactic plane better by interpolating rationally from one side of the plane to the other will fill in the missing link. In the longer term we need to survey the plane better, perhaps with 21 cm or X-rays.

The third advance for me at this meeting was the more detailed N-body simulations for biased galaxy formation and cold dark matter. I would say a gratifying point was seeing for the first time all the correlations between different kinds of galaxies, their velocities of rotation, masses, and so on, as a function of density and environment. Qualitatively, these effects were all predicted before. We knew, just from simple linear statistics, what kinds of galaxies would form and where in the CDM picture. Still, it was nice to see that demonstrated more precisely and more quantitatively.

A major question, though, is the question of exactly how good the biasing is. I would have liked to have heard more discussion about this. I asked Mark Davis and George Efstathiou the following question. I said, "if you observe your simulations the way observers observe the universe and measure the virial masses of little groups, attach a certain mass per galaxy that way and then infer the mass of the entire volume — this is the way we measure the mass density of the universe traditionally — how far short of the total does that inferred mass actually fall?" And the answer was, it falls short by a factor of 5, they said, which is exactly the factor we need in order to make Ω equal to unity. So that's very reassuring.

In this model now you can look and see where that missing mass is, and the answer they gave was that some of it is in the voids; and then we have these very high mass-to-light ratios of dwarf galaxies. That's where the rest of it is. The voids I can understand, obviously, and we've got to do something drastically different from what we're now doing in order to find the matter that's in the voids that never made galaxies. But I am puzzled about the matter around the dwarf galaxies. Anything that's around a galaxy to me is potentially fair game for discovery. Why haven't we discovered it so far? Where are the baryons that are associated with that matter? Have they simply not fallen into those galaxies yet? Perhaps one of these gentlemen will reply to that, but let me summarize by saying simply that looking at these simulations in more detail and trying to figure out clever ways of observing the missing mass might be very informative.

I will briefly enumerate three observational questions for the future. First, I believe that the future of large-scale motion work lies in two regions: in Perseus-Pisces, where we see very interesting structures and there is a hope of correlating motions with the morphology; and in the Great Attractor, where we badly need to figure out what the backside of this entity looks like and whether all the drama is merely on the front side. If so, we've already seen all there is, in which case it's not such a large or interesting structure. Second, I believe we badly need an unbiased and believable cluster catalogue to help us answer whether or not any of the cluster correlations remain positive beyond 3,000 kilometers per second. Third, I believe we are on the verge of seeing cosmic microwave background fluctuations. Even the foreground patch of the Great Attractor would have produced an observable fluctuation in the microwave background. I'd like to see emphasis on actually making maps and finding real features instead of being content with just RMS measurements. However, I'm sure the people in this field already feel this way.

My overall assessment of theory at this meeting is that cold dark matter is still alive and well, and I would not have put the gradient at zero as Avishai Dekel did; I would have put it positive.

JEREMY MOULD

Astronomers are divided into many tribes and there are three main tribes represented here. The first tribe is the tribe of observers. It's a very primitive tribe. We are given to gathering on mountain tops and worshipping the stars and the moon, and recently we were told that we should also worship the Great Attractor. There was some concern about this because it was difficult for some of us to see it. The second tribe is the tribe of model makers, and in this context the function of the model makers is to determine the peculiar velocities, from magnitudes and velocities that the observers measure. The third tribe is the tribe of theorists and the function of the theorists is to deduce the initial fluctuation spectrum from the peculiar velocities that we find from the models.

So what is there left that hasn't been talked about for each of the tribes to take home to their own camps? Well, the observers' tribe learned that there were good and poor data, and clearly we want to concentrate on the good material. Of course, this is not a fixed data set. We cannot simply confine ourselves to the galaxies in the Second Reference Catalogue; we have to measure those discrepant erroneous diameters and get them right.

A lesson for the parameterized model maker to take home, I think, is that now that the number of parameters is getting large, we need to know what the error bars are on the ten parameters. When you minimize chi-squared with ten parameters, you get out a ten by ten error matrix and you want to know what the covariance between, for example, the Virgo infall and the Great Attractor infall is, so that we know how much freedom there is to trade off one against the other.

I cannot offer any suggestions as to what the theorist tribe might take home, apart from noting that Avishai Dekel thought that they should wait and see what we came up with next.

An impression I have from the unparameterized models which, I agree with Sandy Faber, were the big new advance at this meeting is that, apart from Virgo, Hydra-Centaurus was probably the largest entity responsible for our local acceleration. The reason I got that impression was that apparently the acceleration vector has converged early as you move out in that direction towards Virgo and Hydra-Cen; that was also the appearance, I thought, of Amos Yahil's models.

At this meeting I think the question, is there a Great Attractor?, has been replaced by the question, do the IRAS density maps predict the observed velocities? In other words, does light trace mass? And if not, do we need to add extra attractors? In addition, we should, as Nick Kaiser said, worry about the uniqueness of these models. Perhaps we do need to add more biasing to the model, put more mass or less in the peaks.

Based on Marc Davis' results I think a tentative answer to the question: do the IRAS density maps predict the observed velocities? is yes. Remember, most of the outflowing test particles that we've seen in our own Tully-Fisher work are in the foreground of the Hydra-Cen mass distribution, at 1.8 Virgo distances, to introduce a new unit of distance. One is actually in it, at 2.8 Virgo distances. That's the Hydra cluster and we see zero peculiar velocity for it, plus or minus a large number. Clearly, as has been said before and was said at IAU Symposium 130 in Hungary, we have to try harder to get behind the Hydra-Cen mass distribution where inflow is predicted. Remember too, that it is not sufficient to observe galaxies in the 4500 kilometer per second peak discovered in Dressler's survey, because we found the Cen 45 subcluster peak was mostly peculiar velocity. We have to step out further in distance and that's particularly a problem for the Parkes Tully-Fisher observations, because we are right at the limit of the telescope. But if we had a few more distant clusters, the picture would be a lot clearer. I join the plea for a southern hemisphere Arecibo.

So, finally, to reiterate a bit, I think the future in this particular subfield lies in perturbing the IRAS map that Amos Yahil and Marc Davis have shown us, and in minimizing chi-squared between the observed and predicted velocity distributions. If we have to stick with the parameterized models, I think we should try pulling the Great Attractor in, placing it between the 3000 km/s and 4500 km/s peaks, and spreading it out, moving it over the zone of avoidance in the Pavo-Indus region. In the meantime, we will try to provide a lot more data to fully constrain the problem.

BRENT TULLY

I would like to dwell on the issue of the morphology of large-scale structure. My approach has been rather phenomenological, more so than that of almost anyone else in this room. George Efstathiou gives me a hard time because he is obstreperous, but I contend that the approach is a valid one in an area where we are so incredibly ignorant.

Jim Peebles delights in warning us about connecting the dots. We could be grossly misled. But a few pictures may be pushing us in surprising directions: a lizard in Perseus, a stick man centered on the Coma Cluster, a vast plane in Pisces-Cetus. These phenomena could not have been found by statistical investigations, because we would never have asked the right questions.

Consider that possible plane in Pisces-Cetus extending over 0.1c. There are reasonable reservations about whether it exists. I am relatively confident that it is real because the one-dimensional two-point correlation test provides a strong confirmation. But the situation will be clarified soon enough, when the new redshift surveys that extend to this region are analyzed.

There *can* be more to connect-the-dot patterns than a pretty picture. If truth is beauty, might beauty portend truth?

JOHN HUCHRA

What has impressed me the most at this meeting, a situation which is especially apparent at conferences discussing problems at the frontiers of science, is the basic conflict between theory and observations which seems to continue no matter how advanced the theory is nor how complete the observations are. At this meeting the theorists seems to be running a little rampant again and it's about time for those of us who are observers to rein them in a little and bring them at least a little closer to reality.

One annoying problem I want to point out, also mentioned earlier by Sandy Faber and George Efstathiou, is that we keep forgetting to lay out properly our ground rules and definitions. I don't mean definitions on viewgraphs but rather the definitions of those things that we are trying to actually go out and look at or measure. A good example for the observers is the definition of a cluster of galaxies. What exactly is a cluster? We had better have a good definition of what one is before we really start talking about the distribution of these somethat messy things (or complaining about the inability of theories to describe the statistics of their spatial distribution)! We need to get our house in order before either the theorists or observers can make any progress on such problems.

Now I want to play a role that would have been called a few weeks ago "snake-oil salesman," but to continue the analogy used by Avishai Dekel and Nick Kaiser this morning, this role might now be more likened to "stockbroker." I think the time has come for the observers to start concentrating on the big project needed to make any headway on large-scale-structure problems: a Digital Sky Survey. Let me not say too many bad things about photographic plates, but those of you in this room who have taken some know that they are not exactly the greatest detectors in the world. We *now* have much better detectors available and even better ones are coming in the next year or so. It is not beyond our reach to start projects to map the sky with galaxies to 19th or 20th magnitude over the next 4 or 5 years.

I would like to make a strong plea that we consider a national or even international effort to do a digital sky survey with CCD's and 1 arc sec pixels. For those of you who are not observers, 1 arc sec pixels are about what is needed to identify galaxies to 20th magnitude. The integration time should be set to produce ~ a few percent photometry at 20th. Such a project could cover a hemisphere, would contain roughly a terrabyte of data, and would occupy the photometric dark time on a 4-meter class telescope for about 5 years. If you wish to go a little less faint or do somewhat poorer photometry, you could go faster, or even better yet, cover the sky in more than one color.

People have objected to the amount of storage such a survey might take, but there are answers to that problem. With current technology a terrabyte will fit on roughly 50 laser disks, but that number will come down. I have seen a preliminary announcement from a German company for optical disk technology with sub-micron spot sizes; this could bring the storage requirement down to ten disks or even a single disk. Imagine having a digital version of the Palomar Sky Survey in your briefcase!

Such a CCD survey has amazing advantages. National centers and computer networks could allow almost instant access to the digital data to astronomers all over the world. CCD's can work into the red or near infrared and thus minimize the galactic extinction problem. You can easily do all those things we all like to do to galaxies — get their profiles, diameter, inclinations, position angles, you name it, from a homogeneous data base. You could identify x-ray or infrared sources. You could also (I hope!) figure out how to pick out galaxy clusters in a reproducible fashion. That's what I'd like to sell to you. Let's do it!

ANTHONY LASENBY

As a member of the primitive tribe of observers, I must say I've learnt a lot over this past week. It has been very beneficial for me. I am particularly glad to have learnt about the three-dimensional structure in some detail. I now think I have some feeling as to where the various clusters and superclusters are and am able to some extent to picture them in my mind. But of course as one continues with this basic programme of finding out the real details of what we have arounds us, it continues to look more and more complex, as one should have expected it to. One worry I have on large-scale structure is that perhaps we are not yet using the right statistic for evaluating it. The two-point correlation function is good since we understand (or are beginning to understand) how to put errors on it, and it is independent of the density of the fields, but I am not sure that it is adequate to quantify what we are seeing in terms of complex structure. Perhaps a new statistic — multifractals perhaps, though doubt has already been expressed about these — will be necessary to encode the higher order correlation functions which contain the information we need.

On the question of velocities, it is clear that there do exist large-scale motions, and I think the agreement between the data sets which Dave Burstien has been showing is very impressive. However, perhaps it is rather too early as yet to say that they are definitely caused by a single object — the Great Attractor. It would be nice to see some exploration of the range of models which could give rise to a similar quality of fit, and of course a definitive test will be possible when we are able to look *beyond* the Great Attractor, and see if infall is occuring there.

I really liked seeing an attempt to determine all the components of the velocity field. As several people have said, the diagrams shown by Amos Yahil and Mark Davis were very impressive, particularly Amos' diagrams of the self-consistent velocity and density fields. Even if one cannot believe all the details, because of problems with shot noise for example, this is still an extremely nice blend of observations and dynamical theory.

One detail I'd like to see cleared up is what Brent Tully was saying about H_0 and Malmquist bias, which is a subject we don't seem to have returned to very much this week. If H_0 is really as high as 95 km s^{-1} Mpc $^{-1}$, it seems it would represent a crushing blow for cold dark matter theory, so that for this and many other reasons the accurate determination of H_0 is still of course a top priority. On this topic, let me put in another quick 'plug' for microwave background observations. The determination of H_0 via observations of the Sunyaev-Zeldovich effect in clusters of galaxies is a possibility that has been discussed for some years now. With the advent of the enhanced Cambridge 5-km telescope, due to start operation in late 1988, for the first time it will be possible to make accurate maps of the Sunyaev-Zeldovich decrement in several galaxy clusters, with reasonable integration times. Thus useful bounds on H_0, set via this technique, will be possible within a few years.

Now H_0 is critical for CDM. On this question of critical numbers I'd like to support what Avishai Dekel was saying with respect to the tendency to take a single number and perhaps too quickly treat it as having demolished a theory (or provided great support for it). I think our theories now, like nature itself, are becoming complex enough that in some cases we don't immediately know what they are telling us. Time has to be spent working out consequences. For instance, in the meeting this week we have seen that the bulk velocities observed are not inconsistent with cold dark matter after all. The two can be reconciled. As a theory I think that cold dark matter is looking more complete and impressive all the time, and one is particularly struck by its predictive power from only a few assumptions.

On the microwave background, as Nick Kaiser was saying during the week, there has been a sort of 'golden age' up to now, where one could represent the confrontation between theory and experiment in terms of a single number — the $\Delta T/T$ on a given scale. However, if the Tenerife 'detection' is really of intrinsic anisotropy then detailed comparison on a 'morphological' level will start becoming possible (and necessary). Again a 'critical number' approach may be too simplistic. However, if one is still talking about critical numbers, and our detection turns out *not* to be of intrinsic anisotropy, so that one has to dig to still smaller fluctuation levels, then I'm glad to report that 10^{-5} will be enough. I was speaking with Marc Davis yesterday, and he said that if we could get the $\Delta T/T$ limits below 10^{-5}, then we'll have eliminated everything but cold dark matter, and he promised not to invent any alternative theories. So I'd like that written into the record!

To be realistic, in summary, on our microwave background fluctuations, we, Rod Davies, myself and colleagues, are the first observers to find fluctuations in a situation where we can actually go back and look in detail at what we've found. There is already some evidence for the reality of the fluctuations in the agreement of two independent experiments on different angular scales (our 5° and 8° experiments). However, the 'clincher' has obviously got to be to prove that the frequency dependence of what we have found has the form appropriate to a black body perturbation. Thus experiments at higher (and lower) frequencies will be vital, and we fully intend to carry these out and discover definitively the nature of the fluctuations we see.

MARTHA HAYNES

I'd like to return to the analogy that Avishai made about plate tectonics and cartography. I believe that we are at a very exciting point, and I think we will be giving Avishai Dekel a lot more information in the next few years that will help him develop his theory of plate tectonics.

One of the important points that we have learned recently, since the IAU Symposium 130 in Hungary in particular, is that there seems to be continuing convergence of ideas among us even when we look at very different data sets and pursue them in different ways. Although perhaps the details are not all ironed out yet, I find it encouraging that a lot of the pictures of large-scale structure that are emerging are similar. For example, when we take the southern part of Pisces-Perseus and the CfA slice, the structures are remarkably similar when they are displayed in a similar manner. When we consider how our ideas have modified in the last five or ten years, I think this convergence of thought means that we are certainly on the right track. That's true not only for observations but for theory. It seems to me that while we haven't ruled out all theories and we haven't ruled in some of them, cold dark matter makes some nice preditions that we can now go out and test. Maybe we'll rule cold dark matter theories out, but at least we have to be on some of the right tracks.

At the same time, I think we've also shown the importance of pursuing the subject from a variety of approaches. I can say that sometimes similarity in structure exists, but in some places, Pisces-Perseus just doesn't look like anything else. We have also to worry, therefore, that our little part of the universe is not like everywhere else, and locally at least we certainly have evidence now that we inhabit an anomaly. We should really wonder then whether our view of the universe would be different if we were sitting on the outskirts of Perseus, for example.

It is unfortunate at this point that while we have pictures of superclusters like Pisces-Perseus, Coma-A1367, and Hercules, we really can't measure distances to better than about 15%, and that just is not good enough. But, within the next few years, we ought to be able to do better.

I'd like to bring us back to the cartography to take a look at an analogy that we might keep in mind, and that's the story of the first "discovery" of the Pacific Ocean. Verrazano was looking to sail to India, not to discover a continent. While exploring the East Coast of the United States, he sailed past the outer banks of North Carolina Cape Hatteras, and he discovered the Pacific Ocean; at least, he thought he did. He looked over the narrow cape, saw another body of water, and said, "Ah, we can follow that to India. I don't have to do anything else. I can sail back". Now, he should have seen the land on the other side of the spit and the bay, but perhaps it was foggy that day, or perhaps he had to finish in time for a coffee break (as I am supposed to).

This little story does teach us something, that we should worry when the scales that we are studying are about the same size as our sample, and so I'll just leave with that warning. But, on the other hand, I know that, as Brent Tully has said, we will be continuing to extend our scales and we'll probably find more interesting structures and motions on larger scales too.

AMOS YAHIL

I want to concentrate on the rather narrow area that I myself have worked on, and to try and discuss some prospects for the future. I tend to go along with what Jeremy Mould has said, that the initial comparison between the peculiar velocity field predicted from the IRAS density distribution and the observed velocity field, as shown by various people here, is encouraging, but that, in detail, there is a lot more work left to be done. I think there is work to be done on both the side of density and velocity, and I want briefly to outline what I think are the major tasks in the next, two, three, or five years.

Let me start with things that we can already do, but have not yet done, on the density side. One objective is to put in the contribution of the early-type galaxies in a more realistic way. You have seen the sort of simple-minded, quick-fix algorithm of double counting which Marc Davis showed. We really need to supplement, indeed to replace it, by a careful analysis of the ratio of optical to infrared galaxies. Perhaps then we can get a better idea of the total mass distribution, and improve the calculation of gravity. That can be done with existing data.

The second thing we can do with existing data is to put in nonlinear corrections. This has so far not been done, but we have some ideas, and hopefully can report on them within a short time. I think this is also a doable problem.

One can always improve by obtaining more observations. We are ourselves pushing the IRAS measurements down to five degress from the galactic plane. I also agree with the suggestion of Sandy Faber, that we not replace what is left—whether it is five or ten degrees—by a simple homogeneous distribution, but by a more sophisticated interpolation. That is a nice idea, and it can improve our predicted velocities considerably.

I am hoping that another observational handle on density will come if ROSAT gives us a list of elliptical galaxies which we might use in a similar way to the IRAS spiral galaxies, but that is into the future.

On the side of the velocity distribution, we need to expand on what has already been done. One of the nice things about that is that many people can get those peculiar velocities in many different ways. My prediction is that many observers, both present here and ones not at this meeting, will now be scrambling to get as many peculiar velocities as they can. The body of data on peculiar velocities will then begin to grow very, very rapidly. I agree with what Martha Haynes said earlier, that the direction of Perseus-Pisces is an extremely interesting first shot, because we have relatively little information in that direction, and the density perturbation there is comparable to the Great Attractor in Hydra-Centaurus. So that is a place to go first.

Now, given that we are going to make all those improvements in the density determination, and obtain more peculiar velocities, we can begin to do a careful one-on-one comparison between the prediction of the theory and the observed peculiar velocity field. This can be done in a very, very detailed way, and will give us primarily two things.

First, we will have a quantitative test of biasing. I am stretching my neck out here saying that we will overcome all the observational difficulties, and actually be able to confront theory. This goes back to a point which Jim Gunn made in his summary talk at IAU Symposium 117 on *Dark Matter in the Universe* at Princeton two years ago, after the initial IRAS results were presented. He said that, if you have biasing, you violate the principle of superposition, and therefore vectors add up in different ways, and velocity fields get distorted. So, if you find good agreement between the predicted and observed velocity fields, you are fairly confident that biasing is minimal. On the other hand, if you do not find agreement, you may be able to find a prescription by which you re-scale the density perturbations, so as to obtain a better predicted velocity field, and then you have discovered biasing. I think that is a project that we should continue to pursue, and hopefully, when the data improve, will be able to accomplish.

Secondly, if we do find that everything fits together, we can realistically seek the cosmological density parameter, Ω_0. Again, I will be optimistic, and say that perhaps we can do that in a reliable way. So, my hope is that by the time of the next Vatican meeting, in another five or six years, they can shut us all up in the Sistine Chapel, and not let us out until we have decided on the value of Ω_0.

VERA RUBIN

To capture the spirit of the meeting, and in the fashion of the times, I present a List. *The Prettiest Pictures:* Yahil's galaxy plots showing gravitationally induced velocity vectors in the supergalactic plane, Tully's tinker-toy clusters, and Koo's faint UV arc. *The Biggest Failure:* using the microphones. *The Biggest IF:* IF light traces mass. *The Most Overworked Words:* toy, tophat, generic. *Questions I Wish I Had Asked, to Burstein, Faber, Mould, and Lynden-Bell:* Assuming a motion due to the Great Attractor, why is the decrease in dispersion in the Aaronson, Huchra, and Mould spiral sample so much more dramatic than for the Faber-Burstein elliptical sample? Does it arise from different sky coverage, or do spirals better map the large-scale flow than do ellipticals?

First Disappointing Statement: Geller's opening remarks that photometric surveys are not accurate enough to do meaningful statistics. *Most Beautiful Doodles:* Alex Szalay's. *Most Trivial Statistic:* every left-handed participant sat next to a left-handed participant. *Best Food:* every meal. *Second Disappointment:* unease concerning possible biases in the Abell catalogue. *What I Was Pleased To Learn:* Martha Hayne's impressive evidence for luminosity function differences in regions of different densities and how well Marc Davis's N-body simulations show such properties.

What I Was Not Really Surprised To Learn: questions concerning the reality of a Virgo infall. *What I Would Most Like to Learn:* the cause of the motion with respect to the microwave background and its relation to the inferred motion toward the Great Attractor; is it only gravity? *Least Esoteric Question:* how should we properly define groups, clusters, and superclusters, for they will surely in the future play the role that galaxies now play in defining large-scale structure and motions.

Most Optimistic View of The Great Attractor: probably at present the best detailed description of our motion, but unfortunate in: 1) being located close to the zone of avoidance, the zone made unobservable by the Milky Way; and 2) precariously near the velocity limit of the observed sample. *Big Question:* how will more observations change the model of the Great Attractor?

Most Thanks: to all the participants for their thoughtful contributions and spontaneous contributions, their agreements and disagreements all carried out cheerfully and in the spirit of advancing our understanding, even though we did not answer the 31 questions asked at the outset of the meeting. *Overall Conclusion:* it should be no surprise to the participants to hear that I support the concept that large-scale motions exist. A variety of observations of several samples now show that large-scale motions do exist. The interpretation of velocity structure may be affected by large-scale differences among galaxies. These must be sorted out by future observations.

I opened the meeting with the 16th century sky map of Centaurus. Perhaps a map of the world at that time would have been equally appropriate. When we compare 16th to 20th century world maps, and 16th and 20th century sky maps, we see an enormous change. I presume that the cosmography of the universe will change equally much in the next few hundred years.

JIM PEEBLES

There are several points in our discussion where it seems to me that we have reached a crisis point, such that relatively small further advances in theory and observation might be expected to yield a considerable improvement in our understanding. Here is my list of the hottest crises.

The main topic of our meeting is the large-scale peculiar velocity field. I remind you that the big news from the Burstein *et al.* study was not the magnitude of the velocity but rather the suggestion that the coherence length over which it varies may be as large as 4000 km s^{-1}. If this were so then as Dick Bond, Marc Davis and others have noted, it would be a considerable embarrassment for the scale-invariant cold dark matter model. Faber and Burstein have presented a thorough and closely reasoned case for their interpretation, which we certainly must take seriously. But the case would be stronger still, if a model with a small coherence length consistent with inflation had the benefit of a similar degree of attention and adjustment and been found to be wanting. An important start in this direction has been taken by Kaiser, as is described in these proceedings, but I have the impression that his method of weighting reduces sensitivity to the Burstein *et al.* picture, that the peculiar velocity has an appreciable component that varies over scales $\sim$ 4000 km sec^{-1}. I think we understand the issues here, and that we may even be converging on a resolution, but am not so sure which way the resolution will go, large coherence length or small.

The second crisis point is the interpretation of the tendency of the galaxy space distribution to show linear features. There was a time when I was skeptical of this effect, but there is no believer like a reformed sinner. I am convinced by the redshift maps, and by the statistical tendency of elliptical galaxies to line up with the large-scale galaxy distribution, that the effect is real and crying for an explanation.

The numerical simulations of the scale-invariant cold dark matter model shown by George Efstathiou tend to have remarkable linear structures that are strikingly similar to what is observed. If this impression is confirmed by closer checks it will be a dramatic triumph for this model (and for the general class of gravitational instability pictures with roughly similar power spectra). I am a little uneasy about the effect, however, because we seem to have in these simulations an example of pattern formation without a source for the pattern. In Benard convection rolls one can see the effect of the container. In pancaking in hot dark matter one can see the effect of the hard high frequency cutoff of the power spectrum. Before we buy the linear structures seen in the N-body model simulations with slowly varying power spectra (flatter than k^{-3}) we are owed an explanation of where these patterns come from.

The explosion picture predicted the existence of linear features in the galaxy space distribution, as the remnants of the ridges piled up by the explosions. That is an impressive success. However, I am also impressed by an apparent problem for a simple explosion picture. We know that the local sheet of galaxies, at distances less than about 900 km sec^{-1}, is moving very nearly uniformly at 600 km sec^{-1} relative to the rest frame defined by the microwave background. The local galaxies off the sheet are moving at very nearly the same peculiar velocity. How does one explain this in an explosion picture?

My third crisis has to do with the shape of the galaxy two-point correlation function, ξ. At small separations ξ is well approximated as a power law, $\xi \propto r^{-\gamma}$, $\gamma = 1.77$. At separation, $r \sim 10h^{-1}$ Mpc (H = 100 h km sec^{-1} Mpc^{-1}), γ has a feature, a local excess over the power law, followed by an increase of slope. (Beyond that, at hr $\gtrsim$ 15 Mpc, ξ has not yet been reliably measured.) This feature was discovered by Ed Groth and myself some ten years ago, but the evidence for it has been critically debated only recently as a result of stimulating discussions by Margaret Geller and her colleagues. The outcome is that I am not convinced of the reality of the feature in the galaxy distribution at the Lick depth. Even so, it was with more than a little relief that I learned of the preliminary results of George Efstathiou and his colleagues, that seem to reveal the same feature in the deeper Cambridge APM survey.

What is the significance of this feature in ξ at hr $\sim$ 10 Mpc? A clue might be that the feature appears where $\xi(r)$ passes through unity. This suggests to me that the feature may have something to do with the transition from the highly non-linear character of the density fluctuations on small scales to the linear character of the density fluctuations observed on large scales; but I know of no convincing application of this idea (including my own).

N-body models with power law initial power spectra tend to make the slope of the mass autocorrelation function, $\xi_{\varrho\varrho}$, at $\xi_{\varrho\varrho} \sim 1$ much steeper than the slope of the galaxy two-point correlation function at $\xi \sim 1$. The unwanted steep slope is the result of the tendency of newly forming levels of the clustering hierarchy to collapse in a nearly radial way. I do not think that this radial collapse describes what is happening in the Local Supercluster. One can think of several ways to avoid radial collapse: assume mass does not cluster like galaxies, or assume the initial power spectrum is not a pure power law or, perhaps, assume the initial density fluctuations are not Gaussian. The scale-invariant cold dark matter model is an example of the first possibility. I gather that the N-body model simulations of this model described by George Efstathiou could be used to predict the expected shape of the galaxy two-point correlation function at $\xi \sim 1$ in this picture. We will await the results with considerable interest.

Another possibility is that the initial power spectrum of the mass distribution is not well approximated as a pure power law. An example is the baryonic isocurvature model, where the power spectrum of the baryon distribution develops a spike at a wavelength $\lambda \sim 100$ Mpc. The spike adds to the mass autocorrelation function a broad shoulder, with width $\sim \lambda$, which is at least in the wanted direction. Perhaps when we next meet I will be able to give an assessment of the prospects for this idea.

The final crisis is the value of the density parameter, $\Omega = \varrho / \varrho_{\text{crit}}$. The elegant answer is $\Omega = 1$. The dynamical estimates pretty consistently indicate $\Omega \sim 0.1$ to 0.3. As the quality and variety of the dynamical tests have improved I have started to take seriously the possibility that the universe knows something we do not know, that Ω may be less than unity. The tension between our belief in the reasonableness of the argument for $\Omega = 1$ and our wish to accept the observations will be an interesting case study for the sociology of science.

If $\Omega = 1$ then galaxies have to be more strongly clustered than is mass. Also, if the dark mass has negligible pressure, then there is an upper bound on the redshift at which the sites of galaxy formation can have been determined, because mass concentrations exist and these concentrations grow by gravitational instability, drawing in galaxies and mass and so tending to erase the original segregation of mass from galaxies. The redshift bound seems to be particularly tight in the scale-invariant cold dark matter model. My impression of the numerical simulations of this model is that galaxies are being assembled at redshifts less than unity. If so, I think this is a crisis for the model because, as David Koo described for us, there already is pretty good observational evidence for the existence of well-developed galaxies at redshifts greater than unity. It's reasonable to hope that we will see a resolution of the conflict between early and late galaxy formation in the not too distant future, because the observations of galaxies at redshifts $z \gtrsim 1$ are improving rapidly. This is a result of the dramatic improvement of optical and, very recently, panoramic infrared detectors, and of the very impressive work of the observers. The Ω puzzle is more complicated, in many ways; it is likely to be with us for some time.

MARC DAVIS

During the course of several long sleepless nights this week I perused the volume from the last Vatican Study Week, *Astrophysical Cosmology,* and I was intrigued to read Martin Rees' opening introduction. He addressed seven points for consideration at that workshop. They start with such questions as whether we know the Hubble constant to a precision better than 25%, and they end with the question of whether comprehensible physical processes at very early times can account for the overall homogeneity but small-scale roughness of the universe.

That week Steven Weinberg told us about the fantastic notion of inflation, baryogenesis and other exotica that were currently happening in the world of grand unification physics. It was all very new, very young, and the notion of fluctuation generation was unresolved. Also at that meeting Sandy Faber argued strongly and very impressively that the flat rotation curves and the observed scaling behavior of galaxies implied that we could understand quite a few properties of galaxies if the effective spectral index of perturbations was $n = -2$. I remember giving her a hard time about that, because I was concerned particularly with the problems such a spectral index would cause on large scales.

Inflation, of course, had just been invented. Cold dark matter wasn't invented until a year later. After cold dark matter was invented it was obvious that spectral index $n_{eff} = -2$ was going to be the case on galaxy scales, but that we wouldn't have a problem with large scales. So Sandy Faber's contribution was really prescient. Massive neutrinos were in vogue at that meeting; Dennis Sciama spoke about them, but even then I note in the proceedings that we had serious questions about whether that model would possibly work. Massive neutrino models basically died the following year.

Now cold dark matter is the standard model and it's a matter of taste whether you like the model. I happen to like it because it is so specific and so minimal, as Dick Bond has emphasized, and therefore we can make all types of predictions and it's a great model to try to shoot down. Now, it is really all a house of cards, because not only does the model fail if we have 600 km/s coherent bulk flows on a large-scale or if the cluster-custer correlation function is sufficiently large, but it fails just as well if we can find convincing evidence of rotation curves that fall at large radii. If we start to see edges of galactic halos, the theory dies. The rotation curves in our models extend hundreds of kiloparsecs and are essentially flat over this range. There's no way around that in the simulations. The theory will also die if the Hubble constant settles in at 100 km s^{-1} Mpc^{-1}. I think the proponents of CDM got away awfully easily this week as there was little discussion of the actual value of H_o. To teach myself a little bit more about the Hubble constant, I've been teaching a seminar to undergraduates this semester and I've learned considerably more about the situation. There are many remaining problems but there is a lot of consistency in the different measurements of the calibrators of H_o, and I don't think the calibrators are off by a factor of 2. I think Brent Tully is right. 95 is a lot better estimate than 50. So I don't know what to do at the moment.

I would like to mention one thing about Avishai Dekel's Dow-Jones prediction. I can see that he is a careful market watcher, but I think he missed one blip in the CDM history. In particular, the biased CDM model was issued simultaneously with the original model, because the theory didn't work at all without the bias; the first announcement of success for the theory came with the bias. However, as Gary Steigman and others emphasized, this invoked the tooth fairy too many times because the bias had no physical motivation. Without a physical understanding of the bias, the CDM stock was oversold, and the theory was not all that convincing. I personally feel that it was the natural bias, invented only last year, that really raised the stock of CDM up to its present level, and I agree that it is still climbing (I recommend 'buy').

Six years from now, at perhaps the next Vatican Study Week, my prediction, for what it is worth, is that CDM will have passed the observational tests on large scale. The clusters-cluster correlations will not be insurmountable and the drift velocities will be acceptable, but H_o is still going to be a problem. At that time, we are going to wonder how nature could be so perverse as to make a theory seem so consistent with the observations and yet clearly be wrong.

DONALD LYNDEN-BELL

As someone who believes in the importance of conventions I also think that society needs people who are prepared to break them; and Brent Tully's cosmography and bravery in exploring the forbidden regions, beset by the ogres of observational selection, I both admire and deprecate.

Now, Avishai Dekel told us about the future of the stock that is most closely conventional. As a conventionalist, I would like to back it. I think the stock is very down at present. It's very depressed, like my voice. The ordinary matter is the only matter we know of and of course it's very bad to predict from the only thing you know, but I like to do it nevertheless. I think we'll need a guillotine for all those who predicted the cosmological abundance. I think something is very wrong there, because omega is just a little greater than 1, and all matter is baryonic.

Now, I thought one of the objects of these brief remarks was to explain what we were going to do when we went away. As usual, I stand criticized by Sandy Faber, so I'll go away and allow for the mass associated with the seen mass and not just the seen mass in trying to calculate the optical dipole. I'll also extrapolate over the galactic plane as she suggests, and I'll calculate the nearby part of the dipole from actual luminosity rather than from angular-diameter squared, because if it's true that a lot of the dipole comes from nearby; then we know about all those galaxies.

I have been greatly impressed, as I was when we wrote that paper on the Great Attractor, that the velocity field is well-delineated by the Great Attractor by and large. I have also been very impressed, as have others, by the IRAS density maps. Their gravity field, I think, is at the present time not well-represented by the Great Attractor model, though the velocity field *is* and that raises a doubt. If you correct the picture from Huchra's catalogue for the incompleteness, you find that between Virgo and the Great Attractor there is an offset of the center of mass. This includes material only within 15 degrees of the supergalactic plane. If you now go off the supergalactic plane and try and plot density following those regions which seem to be the strongest on the plane of the sky, you find even further complications. I think, for instance, that the region in Capricorn may in practice be as important or even more important than the Great Attractor.

NICOLA VITTORIO

Many ideas have been discussed during this week and I feel I am going back home with a lot of homework to do. A crucial issue for comparing theory and observations concerns having a fair sample of the universe. As Peebles commented, the fairness of the sample depends on the quantity to be measured. In the case of the large-scale peculiar velocity field it is important to be careful, since, as we heard, the accelerating material seems to be just outside the Seven Samurai galaxy sample.

In my opinion the consistency of the CDM model with the large-scale flow is not a settled issue. The problem is the scale at which the flow occurs. Only if the Seven Samurai sample really shows that the drift occurs on scales as small as $15h^{-1}$ Mpc, will a CDM model, where light traces the mass, fulfill the observational constraints.

Undoubtedly, cold dark matter provides a very specific model for the formation and evolution of the large-scale structure. The natural biasing that Marc Davis discussed here seems to be an automatic way of introducing a bias in the galaxy distribution. Also, a natural biasing seems to be quite model independent, as it rests mainly on the gravitational growth of structure in a hierarchical scenario. Unfortunately, I must add, I am not convinced that a biased cold dark matter model satisfies the large-scale drifts constraints.

For the cosmic microwave background it is, of course, crucial to confirm the Davies *et al.* result and to provide tighter bounds on the temperature anisotropy. The RELIC II satellite is very promising in this respect.

BILL STOEGER

Well, I have been listening from the fringe, so to speak. I am not sure where the fringe is. It's somewhere between the outer edge of the Great Attractor and the last scattering surface. I have been amazed at the progress that has been made in mapping what I would call our cosmological neighborhood, the region of intermediate scale to which we have access with precise astronomical technique. However, I have a few worries, which have been reflected upon already by other people here which are superimposed on my amazement at what has been accomplished. One of them is that the structures that we are finding are just about the size, or a little bit bigger perhaps, than the sample volume we have been using. Another concern is selection effects. We are, as a matter of fact, coming to understand the important selection effects much better than before, and developing much more uniform ways of dealing with galaxy samples. But there still seems to be a lot more to be done in terms of dealing with this crucial issue.

Then, there is the whole problem which Nicola Vittorio and Donald Lynden-Bell mentioned: the fair sampling hypothesis. At what distances are we going to be able to say that we have a fair sample of the galaxy distribution in the universe? This is a question which worries me and a number of the people I work with. As we extend our more precise astronomical observations to larger and larger distances, we find structures, and possibly even coherent flows, on those larger scales. Where will this end? Are we approaching distance scales, as maybe Tony Tyson's deep galaxy count work and radio source counts would indicate, at which we shall be able to say that we do have spatial homogeneity on scales above between 200 and 500 Mpc? I hope so!

That leads me to the second half of my remarks, which is related more closely to the last scattering surface than it is to the Great Attractor. The issue is: How does what we have been talking about here in this Study Week fit in with our use of cosmological models, or *a* cosmological model, on larger scales? When we employ a Friedmann-Robertson-Walker (FRW) model to characterize or describe the universe — or an almost-FRW model — we must always specify the particular length scale below which it does not work. We already have a good idea about what the lower limit for that length scale is. Certainly, from what we have seen, FRW or almost-FRW models, do not adequately describe our universe on scales below 200 Mpc. On such scales, the universe is still too inhomogeneous. Above what length scale can we be sure that an almost-FRW model fits the universe? We do not yet know.

Some of the work that I have been doing with George Ellis and with others is aimed at giving precision to what is an almost-FRW model and what is not. We still do not have a workable criterion. But I think that the one thing that gives us hope is in the confrontation between the data provided by the more precise observational measurements of traditional astronomy (with which we can map our cosmological neighborhood) and the data we obtain from microwave background observations, which tell us about the epochs at last scattering and before. If we really believe that the microwave background, in its orthodox interpretation, is smooth enough to indicate that on some length scale in the past there was almost-homogeneity, then surely we can construct an argument to demonstrate that there is almost-homogeneity now on some very large length scale, even though it is much larger than we can deal with astronomically at present.

Relevant to this particular point, I believe that it behooves theorists and cosmological background observers alike to show that the smoothness of the background we have seen up to now is smooth enough to support an almost-FRW model of the universe on scales above some large value. That still has not been demonstrated. If you go back into the literature to see why people think using an FRW model is valid in the first place, you find that the strongest arguments rest on what is known as the Ehlers-Geren-Sachs theorem, which is outlined in its most accessible form in Hawking and Ellis: *The Large Scale Structure of Space-Time.* We really have to go back and re-examine (and certainly strengthen) that theorem (by proving it in its "almost form") in order to confirm the view that at least on some very large scale the almost-smoothness of the microwave background justified our use of FRW, or almost-FRW. These large scales on which we may have spatial homogeneity are linked to the intermediate scales we have been talking about, through models of galaxy formation, and the consistency they impose between these two types of observations — the astronomical and the microwave-background. What are the connections between the progenitors of the intermediate-scale structures we are seeing, like the Great Attractor and the surrounding galaxies affected by it, and the fluctuations (or lack of fluctuations) in the microwave background?

NICK KAISER

I would like to say a few words about the present status of the cold dark matter model in the light of what we have heard at this meeting.

The biggest problem is the Hubble constant. Most theorists want *H* of 50 (or less) in order to get the most pronounced large-scale structure, yet we have heard that most data point to a larger value. The problem of what are the real uncertainties here is very involved; however, one cannot help but feel somewhat ostrich-like in simply hoping that this problem will just go away.

If we brush this problem aside then the predictions of cold dark matter (CDM) on intermediate scales (2-10 Mpc/h say) look good, but given that one has the initial amplitude and bias as free parameters these successes are less impressive. In order to really test the models we must go to much larger scales. Fortunately, models like CDM are very cooperative here. If we stick to linear theory, we can simply write down the probability distribution for counts of galaxies in flux, redshift and angle space, and for peculiar velocities. Perhaps the most pressing questions here are; Is the divergence of the velocity field proportional to the underdensity of galaxies? This is expected if the structure is growing by gravitational instability - even if the galaxies are biased. What is the bias factor *b* which appears as the constant of proportionality? Are the fluctuations consistent with the cold dark matter power spectrum?

My feeling is that we are now seeing a fairly consistent picture emerging if all the different observations are taken with a pinch of salt. The comparison of peculiar velocities and the various dipole moments at least suggest a positive answer to the first question, and seem to give consistent and reasonable values for the bias parameter. When we come to the question of whether the spectrum of fluctuations agrees, things are more complicated. Provided we only look at these very 'broad brush' statistics, the distribution with depth looks quite compatible with the CDM prediction, and at least with the normalisation I prefer, the amplitude looks compatible. It seems that rumours of the death of CDM at the hands of these data were greatly exaggerated. Whether this will remain true when we look at the data in more detail remains a very challenging open question.

MARGARET GELLER

It will come as no surprise that I've been hearing many others mention the points on my list. I can thus be very brief.

I, like many others, have been very much impressed by the IRAS results. They demonstrate the power of a nearly uniform all-sky survey, where the selection biases are perhaps better understood than they are in the optical. In fact, I was struck by the difficulty of comparing the IRAS results with current optical catalogues. This problem brings me to the need John Huchra emphasized: we're going to have to obtain well-calibrated photometric catalogues in the optical. There is already the promise of progress toward this goal with the APM survey for the South and with the scanning of the new Palomar Sky Survey in the North. In the long run, we will probably need a digital survey to address all the problems.

One issue which particularly concerns me is the comparison between the scale of the largest structures and the extent of our redshift surveys. It's sobering that every survey we do seems to contain a structure as big as the survey. Thus it's rather difficult to derive statistics we can trust. The errors and uncertainties in the statistics we generally use are probably larger than we have thought.

We have to know the frequency of structures like the void in Boötes. Voids of this size appear to be common; most surveys which can contain them do. From redshift surveys, we don't have very good (or any) limits on structures which are intermediate in scale between Boötes (6,000 km s^{-1}/H_o) and gigaparsec scales. We would like to know the distribution of these very large structures and to have some statistics which are sensitive to the high order properties of the galaxy distribution. Perhaps we should be measuring the distribution of sizes of voids or the properties of the sheets. Of course, the first is a tall order when the surveys aren't big enough to contain more than one of the largest known voids!

Another statistical issue, also affected by problems in catalogues, is the relationship between the distribution of rich clusters and the distribution of individual galaxies. The cold dark matter models cannot match the current results. Certainly the new digitized sky surveys with objectively derived cluster catalogues will help us to make some progress in understanding clusters of galaxies as tracers of the large-scale matter distribution. However, I think that we will not really understand the relationship until redshift surveys for clusters and for individual galaxies overlap sufficiently that we can examine the relationship directly. There's hope that within ten years we will actually see such measures. We will be able to compare what we expect physically with what we see.

One certainty is that none of us are going to have any problems finding something to do. We'll all probably be surprised at some of the things we find.

DAVID KOO

Well, my worst fear has been borne out. I was the last one. Fortunately, not all my points have been expressed. Let me start by mentioning the conclusions of the 1981 Vatican Study Week, in which Longair emphasized the need for systematic surveys and here I quote: "...particularly profitable, I would suggest: further detailed studies of nearby galaxies, deep surveys of large areas of sky, the velocities and spectra of large samples of galaxies, the relation of active galaxies to the properties of the stellar and gaseous component of their parent galaxies and to their environments, measurement of fluctuations in the microwave background radiation with high precision" and "studies of how the properties of galaxies and the large-scale distribution of galaxies and diffuse matter change with cosmic epoch". Except for perhaps the relationship of AGNs to host galaxies, I was extremely impressed how the relatively small number of observers here have covered so well almost all these areas with great advances; and probably just as remarkable has been the work of the theorists in terms of their energy, creativity, and sophistication in handling the large amounts of new data. Although I had come fully expecting to be convinced of the Great Attractor, I leave wondering whether some of the observed peculiar motions result more from repulsion by large pockets of under-densities, locally and perhaps beyond 5,000 km per second.

On the theoretical side I came expecting to see convincing evidence that the standard biased cold dark matter theory is inadequate to explain all this new data but I leave somewhat amazed by how well it actually does work. On the other hand, I am looking forward to future improvements by my next door neighbour here (A. Lasenby) in exploring the microwave background and am hoping that his detection really is positive on the large scales.

I also came hoping to see much discussion on different and quantitative diagnostics of the large-scale structure, especially of the voids, but I think that despite the mathematical rigorousness of PC's (visually blind alien from the Pisces-Cetus supercluster introduced by G. Efstathiou in the discussions) view of the world, the universe is really more complex and beautiful, or ugly if you're a theorist, than we can imagine with numbers. We should not forget the power of Eccle's human eye-brain pattern recognizer, with which we notice arcs around the clusters, bubbles or filaments in the galaxy distribution, different Hubble types of galaxies, maybe spectral features not much discussed here, and unusual classes of objects, such as quasars, blasars, AGNs, etc... Basically, much of what we learn results from recognizing deviations from the norm. Along these lines, I, like very many others, was extremely impressed by the flow diagrams of Amos Yahil. I guess that's because I am an observer, but, unlike others who were impressed by the attraction seen in certain areas, I was really amazed by where I thought I could see outflows from the voids. A simple question is: are we actually seeing the voids pushing? This point is important, for if they are pushing away, they must be true underdensities, and not just regions with lots of hidden matter.

Along the theme of cartography and maps that have been expressed by many here, I felt like an outsider exploring the frontiers, instead of mapping all the little cities on the East Coast. Our domain encompasses the more distant but 99% of the Universe that was largely ignored, perhaps appropriately so for this workshop. Besides penetrating new continents or even the poles, our approach, though crude, may even include drilling holes into the earth, probing the ocean bottom, or sending up balloons and rockets.

One point that I'd like to emphasize is the role of gas. Although gravity is very easy to model, gas is where the action is, especially in much of the detected evolution beyond redshift of 0.1. Gas is intimately linked to much of what is observable; to the morphology, physical parameters, and formation of galaxies, active galactic nuclei, the intergalactic medium, and quasar absorption lines; and to even the large-scale motions discussed in this workshop.

After all, what would be the effects of bulk flow among huge volumes of gas, either today or especially at early epochs before stars and galaxies were formed (and before simple dynamics are applicable)?

The future is very promising for us explorers. We'll soon have space telescope, large 10 meter or maybe 16 meter telescopes on the ground, vast imaging and spectroscopic surveys with large-area CCD's, and new adaptive and active optics that provide high resolution. Infrared imaging with arrays has already begun, giving us new eyes to explore the distant past.

In closing, since I am the last one, I wish to offer great thanks to the Vatican for a delightful week, more busy than free, but full of wonderful food, interesting science, great company, and very warm hospitality.

INDEX

A Brief Guide to **Getting the Most** from this Book

1 Read the Book

Feature	Description	Benefit
Applications Using Real-World Data	From the chapter and section openers through the examples and exercises, interesting applications from nearly every discipline, supported by up-to-date real-world data, are included in every section.	Ever wondered how you'll use algebra? This feature will show you how algebra can solve real problems.
Detailed Worked-Out Examples	Examples are clearly written and provide step-by-step solutions. No steps are omitted, and key steps are thoroughly explained to the right of the mathematics.	The blue annotations will help you to understand the solutions by providing the reason why the algebraic steps are true.
Explanatory Voice Balloons	Voice balloons help to demystify algebra. They translate algebraic language into plain English, clarify problem-solving procedures, and present alternative ways of understanding.	Does math ever look foreign to you? This feature translates math into everyday English.
Great Question!	Answers to students' questions offer suggestions for problem solving, point out common errors to avoid, and provide informal hints and suggestions.	This feature should help you not to feel anxious or threatened when asking questions in class.
Achieving Success	The book's Achieving Success boxes offer strategies for success in learning algebra.	Follow these suggestions to help achieve your full academic potential in mathematics.

2 Work the Problems

Feature	Description	Benefit
Check Point Examples	Each example is followed by a similar problem, called a Check Point, that offers you the opportunity to work a similar exercise. Answers to all Check Points are provided in the answer section.	You learn best by doing. You'll solidify your understanding of worked examples if you try a similar problem right away to be sure you understand what you've just read.
Concept and Vocabulary Checks	These short-answer questions, mainly fill-in-the blank and true/false items, assess your understanding of the definitions and concepts presented in each section.	It is difficult to learn algebra without knowing its special language. These exercises test your understanding of the vocabulary and concepts.
Extensive and Varied Exercise Sets	An abundant collection of exercises is included in an Exercise Set at the end of each section. Exercises are organized within several categories. Practice Exercises follow the same order as the section's worked examples. Practice PLUS Exercises contain more challenging problems that often require you to combine several skills or concepts.	The parallel order of the Practice Exercises lets you refer to the worked examples and use them as models for solving these problems. Practice PLUS provides you with ample opportunity to dig in and develop your problem-solving skills.

3 Review for Quizzes and Tests

Feature	Description	Benefit
Mid-Chapter Check Points	Near the midway point in the chapter, an integrated set of review exercises allows you to review the skills and concepts you learned separately over several sections.	Combining exercises from the first half of the chapter gives you a comprehensive review before you continue on.
Chapter Review Charts	Each chapter contains a review chart that summarizes the definitions and concepts in every section of the chapter, complete with examples.	Review this chart and you'll know the most important material in the chapter.
Chapter Tests	Each chapter contains a practice test with problems that cover the important concepts in the chapter. Take the test, check your answers, and then watch the Chapter Test Prep Videos.	You can use the Chapter Test to determine whether you have mastered the material covered in the chapter.
Chapter Test Prep Videos	These videos contain worked-out solutions to every exercise in each chapter test.	These videos let you review any exercises you miss on the chapter test.
Lecture Series on DVD	These interactive lecture videos highlight key examples from every section of the textbook. A new interface allows easy navigation to sections, objectives, and examples.	These videos let you review each objective from the textbook that you need extra help on.

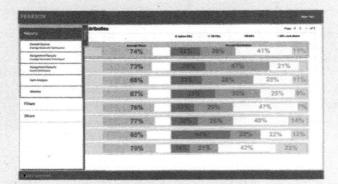

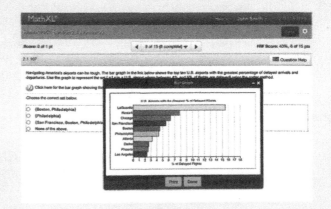

PEARSON ALWAYS LEARNING

Robert F. Blitzer

Intermediate Algebra for College Students

Fourth Custom Edition for Queensborough Community College of CUNY

Taken from:
Intermediate Algebra for College Students, Seventh Edition
by Robert F. Blitzer

Cover Art: Courtesy of Pearson Learning Solutions

Taken from:

Intermediate Algebra for College Students, Seventh Edition
by Robert F. Blitzer

New York, NY 10013

This special edition published in cooperation with Pearson Education, Inc.

Pearson Education, Inc., 330 Hudson Street, New York, New York 10013
A Pearson Education Company
www.pearsoned.com

Printed in the United States of America

1 17

000200010272091820

EJ

ISBN 10: 1-323-76341-4
ISBN 13: 978-1-323-76341-4

Table of Contents

3 Systems of Linear Equations 177

4 Inequalities and Problem Solving 253

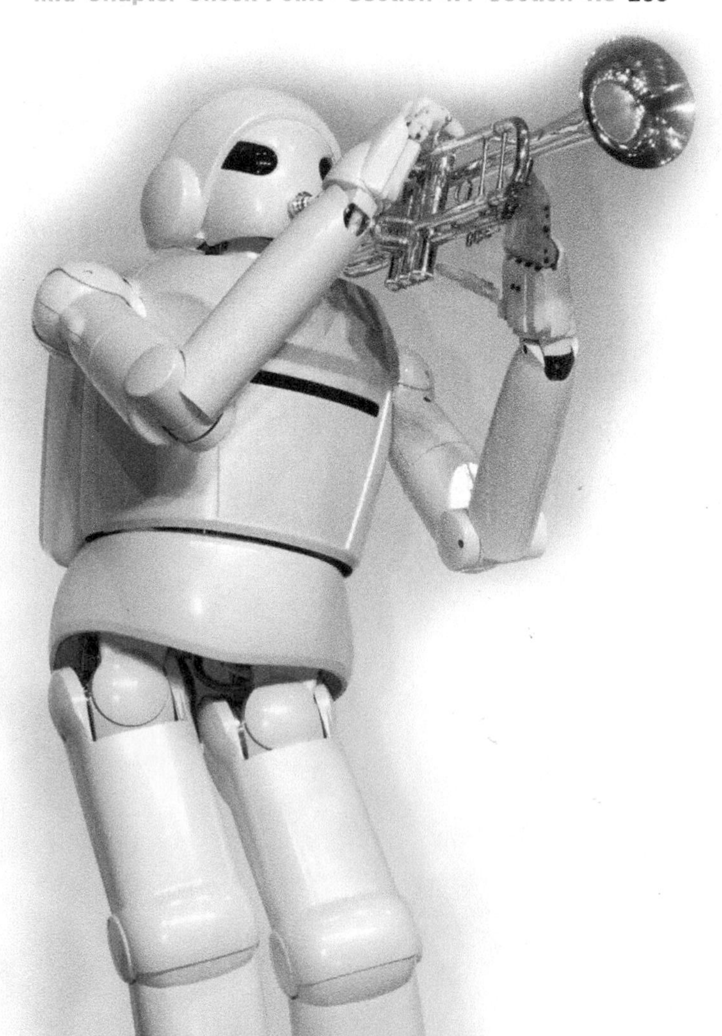

5 Polynomials, Polynomial Functions, and Factoring 313

6 Rational Expressions, Functions, and Equations 403

Preface

Intermediate Algebra for College Students, Seventh Edition, provides comprehensive, in-depth coverage of the topics required in a one-term course in intermediate algebra. The book is written for college students who have had a course in introductory algebra or who have demonstrated proficiency in the objectives of such a course. I wrote the book to help diverse students with different backgrounds and career plans succeed in intermediate algebra. *Intermediate Algebra for College Students,* Seventh Edition, has two primary goals:

1. To help students acquire a solid foundation in the skills and concepts of intermediate algebra needed for success in future STEM and non-STEM directed math courses.
2. To show students how algebra can model and solve authentic real-world problems.

One major obstacle in the way of achieving these goals is the fact that very few students actually read their textbook. This has been a regular source of frustration for me and for my colleagues in the classroom. Anecdotal evidence gathered over years highlights two basic reasons students give when asked why they do not take advantage of their textbook:

- "I'll never use this information."
- "I can't follow the explanations."

I've written every page of the Seventh Edition with the intent of eliminating these two objections. The ideas and tools I've used to do so are described in the features that follow. These features and their benefits are highlighted for the student in "A Brief Guide to Getting the Most from This Book," which appears inside the front cover.

What's New in the Seventh Edition?

- **New Applications and Real-World Data.** The Seventh Edition contains 131 new or revised worked-out examples and exercises based on updated and new data sets. Many of these applications involve topics relevant to college students and newsworthy items. Among topics of interest to college students, you'll find new and updated data sets describing student loan debt (Section 1.1, Example 4), tuition costs (Section 1.4, Example 7; Exercise Set 1.4, Exercises 67–68), dormitory charges (Exercise Set 11.2, Exercises 65–66), the number of hours college students study, by major (Exercise Set 3.3, Exercises 39–40), sleep hours of college students (Exercise Set 3.2, Exercise 10), college students and video games (Chapter 6 opener, Section 6.7 opener; Section 6.7, Example 3), and self-assesment of physical health by college freshmen (Chapter 7 Review, Exercise 85). Among newsworthy items, new applications range from the frivolous (Hamachiphobia: correlating those who won't try sushi and those who don't approve of marriage equality: Exercise Set 9.6, Exercise 41) to the weighty (mentally ill adults in the United States: Chapter 9 Review Exercise 90). Other new and updated applications address climate change (Section 2.4, Example 10 and Check Point 10), the war on drugs and nonviolent inmates in federal prisons (Exercise Set 3.1, Exercise 94; Exercise Set 10.5, Exercise 63), and the changing U.S. population by race/ethnicity (Section 11.2, Example 3).

- **New Blitzer Bonus Videos with Assessment.** The Blitzer Bonus features throughout the textbook have been turned into animated videos that are built into the MyMathLab course. These videos help students make visual connections to algebra and the world around them. Assignable exercises have been created within the MyMathLab course to assess conceptual understanding and mastery. These videos and exercises can be turned into a media assignment within the Blitzer MyMathLab course.
- **Updated Learning Guide.** Organized by the textbook's learning objectives, this updated Learning Guide helps students learn how to make the most of their textbook for test preparation. Projects are now included to give students an opportunity to discover and reinforce the concepts in an active learning environment and are ideal for group work in class.
- **Updated Graphing Calculator Screens.** All screens have been updated using the TI-84 Plus C.

What's New in the Blitzer Developmental Mathematics Series?

Two new textbooks and MyMathLab courses have been added to the series:

- ***Developmental Mathematics,*** First Edition, is intended for a course sequence covering prealgebra, introductory algebra, and intermediate algebra. The text provides a solid foundation in arithmetic and algebra.
- ***Pathways to College Mathematics,*** First Edition, provides a general survey of topics to prepare STEM and non-STEM students for success in a variety of college math courses, including college algebra, statistics, liberal arts mathematics, quantitative reasoning, finite mathematics, and mathematics for education majors. The prerequisite is basic math or prealgebra.
- ***MyMathLab with Integrated Review*** courses are also available for select Blitzer titles. These MyMathLab courses provide the full suite of resources for the core textbook, but also add in study aids and skills check assignments keyed to the prerequisite topics that students need to know, helping them quickly get up to speed.

What Familiar Features Have Been Retained in the Seventh Edition of *Intermediate Algebra for College Students*?

- **Learning Objectives.** Learning objectives, framed in the context of a student question (What am I supposed to learn?), are clearly stated at the beginning of each section. These objectives help students recognize and focus on the section's most important ideas. The objectives are restated in the margin at their point of use.
- **Chapter-Opening and Section-Opening Scenarios.** Every chapter and every section open with a scenario presenting a unique application of mathematics in students' lives outside the classroom. These scenarios are revisited in the course of the chapter or section in an example, discussion, or exercise.
- **Innovative Applications.** A wide variety of interesting applications, supported by up-to-date, real-world data, are included in every section.
- **Detailed Worked-Out Examples.** Each example is titled, making the purpose of the example clear. Examples are clearly written and provide students with detailed step-by-step solutions. No steps are omitted and each step is thoroughly explained to the right of the mathematics.

- **Explanatory Voice Balloons.** Voice balloons are used in a variety of ways to demystify mathematics. They translate algebraic ideas into everyday English, help clarify problem-solving procedures, present alternative ways of understanding concepts, and connect problem solving to concepts students have already learned.
- **Check Point Examples.** Each example is followed by a similar matched problem, called a Check Point, offering students the opportunity to test their understanding of the example by working a similar exercise. The answers to the Check Points are provided in the answer section.
- **Concept and Vocabulary Checks.** This feature offers short-answer exercises, mainly fill-in-the-blank and true/false items, that assess students' understanding of the definitions and concepts presented in each section. The Concept and Vocabulary Checks appear as separate features preceding the Exercise Sets.
- **Extensive and Varied Exercise Sets.** An abundant collection of exercises is included in an Exercise Set at the end of each section. Exercises are organized within eight category types: Practice Exercises, Practice Plus Exercises, Application Exercises, Explaining the Concepts, Critical Thinking Exercises, Technology Exercises, Review Exercises, and Preview Exercises. This format makes it easy to create well-rounded homework assignments. The order of the Practice Exercises is exactly the same as the order of the section's worked examples. This parallel order enables students to refer to the titled examples and their detailed explanations to achieve success working the Practice Exercises.
- **Practice Plus Problems.** This category of exercises contains more challenging practice problems that often require students to combine several skills or concepts. With an average of ten Practice Plus problems per Exercise Set, instructors are provided with the option of creating assignments that take Practice Exercises to a more challenging level.
- **Mid-Chapter Check Points.** At approximately the midway point in each chapter, an integrated set of Review Exercises allows students to review and assimilate the skills and concepts they learned separately over several sections.
- **Graphing and Functions.** Graphing is introduced in Chapter 1 and functions are introduced in Chapter 2, with an integrated graphing functional approach emphasized throughout the book. Graphs and functions that model data appear in nearly every section and Exercise Set. Examples and exercises use graphs of functions to explore relationships between data and to provide ways of visualizing a problem's solution. Because functions are the core of this course, students are repeatedly shown how functions relate to equations and graphs.
- **Integration of Technology Using Graphic and Numerical Approaches to Problems.** Side-by-side features in the technology boxes connect algebraic solutions to graphic and numerical approaches to problems. Although the use of graphing utilities is optional, students can use the explanatory voice balloons to understand different approaches to problems even if they are not using a graphing utility in the course.
- **Great Question!** This feature presents a variety of study tips in the context of students' questions. Answers to questions offer suggestions for problem solving, point out common errors to avoid, and provide informal hints and suggestions. As a secondary benefit, this feature should help students not to feel anxious or threatened when asking questions in class.
- **Achieving Success.** The Achieving Success boxes at the end of many sections offer strategies for persistence and success in college mathematics courses.
- **Chapter Review Grids.** Each chapter contains a review chart that summarizes the definitions and concepts in every section of the chapter. Examples that illustrate these key concepts are also included in the chart.

- **End-of-Chapter Materials.** A comprehensive collection of Review Exercises for each of the chapter's sections follows the review grid. This is followed by a Chapter Test that enables students to test their understanding of the material covered in the chapter. Beginning with Chapter 2, each chapter concludes with a comprehensive collection of mixed Cumulative Review Exercises.
- **Blitzer Bonuses.** These enrichment essays provide historical, interdisciplinary, and otherwise interesting connections to the algebra under study, showing students that math is an interesting and dynamic discipline.
- **Discovery.** Discover for Yourself boxes, found throughout the text, encourage students to further explore algebraic concepts. These explorations are optional and their omission does not interfere with the continuity of the topic under consideration.

I hope that my passion for teaching, as well as my respect for the diversity of students I have taught and learned from over the years, is apparent throughout this new edition. By connecting algebra to the whole spectrum of learning, it is my intent to show students that their world is profoundly mathematical, and indeed, π is in the sky.

Robert Blitzer

BREAKTHROUGH
To improving results

Resources for Success

MyMathLab for the Blitzer Developmental Algebra Series

MyMathLab is available to accompany Pearson's market-leading text offerings. This text's flavor and approach are tightly integrated throughout the accompanying MyMathLab course, giving students a consistent tone, voice, and teaching method that make learning the material as seamless as possible.

Section Lecture and Chapter Test Prep Videos

An **updated** video program provides a multitude of resources for students. Section Lecture videos walk students through the concepts from every section of the text in a fresh, modern presentation format. Chapter Test Prep videos walk students through the solution of every problem in the text's Chapter Tests, giving students video resources when they might need it most.

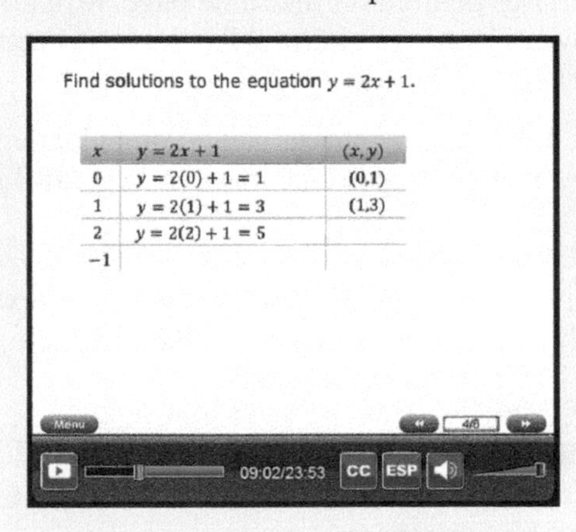

Find solutions to the equation $y = 2x + 1$.
x
$y = 2x + 1$
(x, y)
0
$y = 2(0) + 1 = 1$
(0,1)
1
$y = 2(1) + 1 = 3$
(1,3)
2
$y = 2(2) + 1 = 5$
−1
Menu
09:02/23:53
CC
ESP

Blitzer Bonus Videos

NEW! Animated videos have been created to mirror the Blitzer Bonus features throughout the textbook. Blitzer Bonus features in the text provide interesting real-world connections to the mathematical topics at hand, conveying Bob Blitzer's signature style to engage students. These new Blitzer Bonus videos will help students to connect the topics to the world around them in a visual way. Corresponding assignable exercises in MyMathLab are also available, allowing these new videos to be turned into a media assignment to truly ensure that students have understood what they've watched.

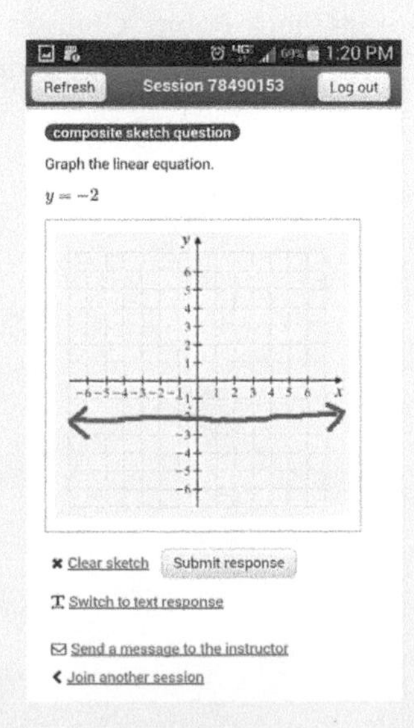

Learning Catalytics

Integrated into MyMathLab, the Learning Catalytics feature uses students' devices in the classroom for an engagement, assessment, and classroom intelligence system that gives instructors real-time feedback on student learning. Learning Catalytics contains Pearson-created content for developmental math topics that allows you to take advantage of this exciting technology immediately.

Student Success Modules

These modules are integrated within the MyMathLab course to help students succeed in college courses and prepare for future professions.

www.mymathlab.com

BREAKTHROUGH
To improving results

Resources for Success

Instructor Resources

Annotated Instructor's Edition

This version of the text contains answers to exercises printed on the same page, with graphing answers in a special Graphing Answer Section at the back of the text.

The following resources can be downloaded from www.pearsonhighered.com or in MyMathLab.

PowerPoint® Lecture Slides

Fully editable slides correlated with the textbook include definitions, key concepts, and examples for use in a lecture setting.

Instructor's Solutions Manual

This manual includes fully worked-out solutions to all text exercises.

Instructor's Resource Manual

This manual includes a Mini-Lecture, Skill Builder, and Additional Exercises for every section of the text. It also includes Chapter Test forms, as well as Cumulative and Final Exams, with answers.

TestGen®

TestGen® (www.pearsoned.com/testgen) enables instructors to build, edit, print, and administer tests using a computerized bank of questions developed to cover all the objectives of the text.

Student Resources

The following additional resources are available to support student success:

Learning Guide

UPDATED! Organized by learning objectives, the Learning Guide helps students make the most of their textbook and prepare for tests. Now updated to include projects, students will have the opportunity to discover and reinforce the concepts in an active learning environment. These projects are ideal for group work in class. The Learning Guide is available in MyMathLab, and available as a printed supplement.

Video Lecture Series

Available in MyMathLab, the video program covers every section in the text, providing students with a video tutor at home, in lab, or on the go. The program includes Section Lecture Videos and Chapter Test Prep videos.

Student Solutions Manual

This manual provides detailed, worked-out solutions to odd-numbered section exercises, plus all Check Points, Review/Preview Exercises, Mid-Chapter Check Points, Chapter Reviews, Chapter Tests, and Cumulative Reviews.

www.mymathlab.com

Acknowledgments

An enormous benefit of authoring a successful series is the broad-based feedback I receive from the students, dedicated users, and reviewers. Every change to this edition is the result of their thoughtful comments and suggestions. I would like to express my appreciation to all the reviewers, whose collective insights form the backbone of this revision. In particular, I would like to thank the following people for reviewing *Intermediate Algebra for College Students.*

Cindy Adams, *San Jacinto College*
Gwen P. Aldridge, *Northwest Mississippi Community College*
Ronnie Allen, *Central New Mexico Community College*
Dr. Simon Aman, *Harry S. Truman College*
Howard Anderson, *Skagit Valley College*
John Anderson, *Illinois Valley Community College*
Michael H. Andreoli, *Miami Dade College – North Campus*
Michele Bach, *Kansas City Kansas Community College*
Jana Barnard, *Angelo State University*
Rosanne Benn, *Prince George's Community College*
Christine Brady, *Suffolk County Community College*
Gale Brewer, *Amarillo College*
Carmen Buhler, *Minneapolis Community & Technical College*
Warren J. Burch, *Brevard College*
Alice Burstein, *Middlesex Community College*
Edie Carter, *Amarillo College*
Jerry Chen, *Suffolk County Community College*
Sandra Pryor Clarkson, *Hunter College*
Sally Copeland, *Johnson County Community College*
Valerie Cox, *Calhoun Community College*
Carol Curtis, *Fresno City College*
Robert A. Davies, *Cuyahoga Community College*
Deborah Detrick, *Kansas City Kansas Community College*
Jill DeWitt, *Baker College of Muskegon*
Ben Divers, Jr., *Ferrum College*
Irene Doo, *Austin Community College*
Charles C. Edgar, *Onondaga Community College*
Karen Edwards, *Diablo Valley College*
Scott Fallstrom, *MiraCosta College*
Elise Fischer, *Johnson County Community College*
Susan Forman, *Bronx Community College*
Wendy Fresh, *Portland Community College*
Jennifer Garnes, *Cuyahoga Community College*
Gary Glaze, *Eastern Washington University*
Jay Graening, *University of Arkansas*
Robert B. Hafer, *Brevard College*
Andrea Hendricks, *Georgia Perimeter College*
Donald Herrick, *Northern Illinois University*
Beth Hooper, *Golden West College*
Sandee House, *Georgia Perimeter College*
Tracy Hoy, *College of Lake County*
Laura Hoye, *Trident Community College*
Margaret Huddleston, *Schreiner University*
Marcella Jones, *Minneapolis Community & Technical College*
Shelbra B. Jones, *Wake Technical Community College*
Sharon Keenee, *Georgia Perimeter College*
Regina Keller, *Suffolk County Community College*
Gary Kersting, *North Central Michigan College*
Dennis Kimzey, *Rogue Community College*
Kandace Kling, *Portland Community College*

Gray Knippenberg, *Lansing Community College*
Mary Kochler, *Cuyahoga Community College*
Scot Leavitt, *Portland Community College*
Robert Leibman, *Austin Community College*
Jennifer Lempke, *North Central Michigan College*
Ann M. Loving, *J. Sargent Reynolds Community College*
Kent MacDougall, *Temple College*
Jean-Marie Magnier, *Springfield Technical Community College*
Hank Martel, *Broward College*
Kim Martin, *Southeastern Illinois College*
John Robert Martin, *Tarrant County College*
Lisa McMillen, *Baker College of Auburn Hills*
Irwin Metviner, *State University of New York at Old Westbury*
Jean P. Millen, *Georgia Perimeter College*
Lawrence Morales, *Seattle Central Community College*
Morteza Shafii-Mousavi, *Indiana University South Bend*
Lois Jean Nieme, *Minneapolis Community & Technical College*
Allen R. Newhart, *Parkersburg Community College*
Karen Pain, *Palm Beach State College*
Peg Pankowski, *Community College of Allegheny County – South Campus*
Robert Patenaude, *College of the Canyons*
Matthew Peace, *Florida Gateway College*
Dr. Bernard J. Piña, *New Mexico State University – Doña Ana Community College*
Jill Rafael, *Sierra College*
James Razavi, *Sierra College*
Christopher Reisch, *The State University of New York at Buffalo*
Nancy Ressler, *Oakton Community College*
Katalin Rozsa, *Mesa Community College*
Haazim Sabree, *Georgia Perimeter College*
Chris Schultz, *Iowa State University*
Shannon Schumann, *University of Phoenix*
Barbara Sehr, *Indiana University Kokomo*
Brian Smith, *Northwest Shoals Community College*
Gayle Smith, *Lane Community College*
Dick Spangler, *Tacoma Community College*
Janette Summers, *University of Arkansas*
Robert Thornton, *Loyola University*
Lucy C. Thrower, *Francis Marion College*
Mary Thurow, *Minneapolis Community & Tech College*
Richard Townsend, *North Carolina Central University*
Cindie Wade, *St. Clair County Community College*
Andrew Walker, *North Seattle Community College*
Kathryn Wetzel, *Amarillo College*

Additional acknowledgments are extended to Dan Miller and Kelly Barber for preparing the solutions manuals and the new Learning Guide; Brad Davis, for preparing the answer section and serving as accuracy checker; the codeMantra formatting team for the book's brilliant paging; Brian Morris and Kevin Morris at Scientific Illustrators, for superbly illustrating the book; and Francesca Monaco, project manager, and Kathleen Manley, production editor, whose collective talents kept every aspect of this complex project moving through its many stages.

I would like to thank my editors at Pearson, Dawn Giovanniello and Megan Tripp, who guided and coordinated the book from manuscript through production. Thanks to Beth Paquin and Studio Montage for the quirky cover and interior design. Finally, thanks to marketing manager Alicia Frankel for your innovative marketing efforts, and to the entire Pearson sales force, for your confidence and enthusiasm about the book.

To the Student

The bar graph shows some of the qualities that students say make a great teacher.

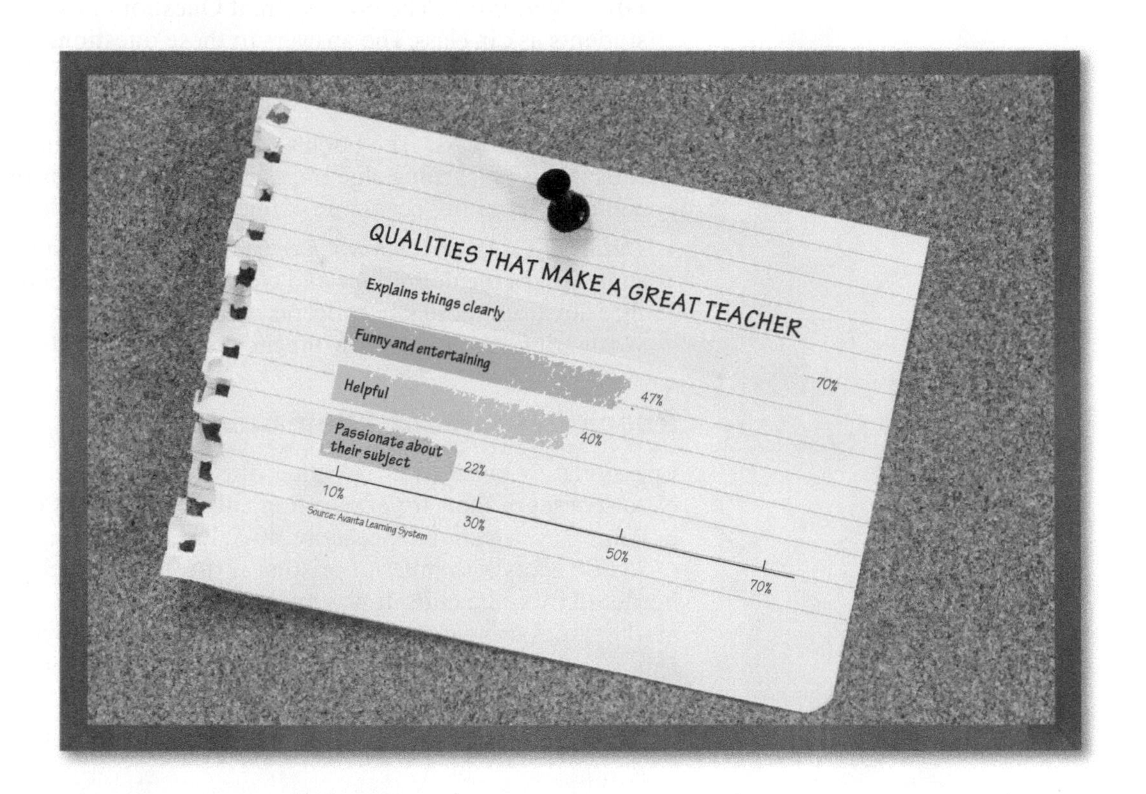

It was my goal to incorporate each of the qualities that make a great teacher throughout the pages of this book.

Explains Things Clearly

I understand that your primary purpose in reading *Intermediate Algebra for College Students* is to acquire a solid understanding of the required topics in your algebra course. In order to achieve this goal, I've carefully explained each topic. Important definitions and procedures are set off in boxes, and worked-out examples that present solutions in a step-by-step manner appear in every section. Each example is followed by a similar matched problem, called a Check Point, for you to try so that you can actively participate in the learning process as you read the book. (Answers to all Check Points appear in the back of the book.)

Funny/Entertaining

Who says that an algebra textbook can't be entertaining? From our quirky cover to the photos in the chapter and section openers, prepare to expect the unexpected. I hope some of the book's enrichment essays, called Blitzer Bonuses, will put a smile on your face from time to time.

Helpful

I designed the book's features to help you acquire knowledge of intermediate algebra, as well as to show you how algebra can solve authentic problems that apply to your life. These helpful features include:

- ***Explanatory Voice Balloons:*** Voice balloons are used in a variety of ways to make math less intimidating. They translate algebraic language into everyday English, help clarify problem-solving procedures, present alternative ways of understanding concepts, and connect new concepts to concepts you have already learned.
- ***Great Question!:*** The book's Great Question! boxes are based on questions students ask in class. The answers to these questions give suggestions for problem solving, point out common errors to avoid, and provide informal hints and suggestions.
- ***Achieving Success:*** The book's Achieving Success boxes give you helpful strategies for success in learning algebra, as well as suggestions that can be applied for achieving your full academic potential in future college coursework.
- ***Detailed Chapter Review Charts:*** Each chapter contains a review chart that summarizes the definitions and concepts in every section of the chapter. Examples that illustrate these key concepts are also included in the chart. Review these summaries and you'll know the most important material in the chapter!

Passionate about Their Subject

I passionately believe that no other discipline comes close to math in offering a more extensive set of tools for application and development of your mind. I wrote the book in Point Reyes National Seashore, 40 miles north of San Francisco. The park consists of 75,000 acres with miles of pristine surf-washed beaches, forested ridges, and bays bordered by white cliffs. It was my hope to convey the beauty and excitement of mathematics using nature's unspoiled beauty as a source of inspiration and creativity. Enjoy the pages that follow as you empower yourself with the algebra needed to succeed in college, your career, and your life.

Regards,
Bob
Robert Blitzer

About the Author

Bob Blitzer is a native of Manhattan and received a Bachelor of Arts degree with dual majors in mathematics and psychology (minor: English literature) from the City College of New York. His unusual combination of academic interests led him toward a Master of Arts in mathematics from the University of Miami and a doctorate in behavioral sciences from Nova University. Bob's love for teaching mathematics was nourished for nearly 30 years at Miami Dade College, where he received numerous teaching awards, including Innovator of the Year from the League for Innovations in the Community College and an endowed chair based on excellence in the classroom. In addition to *Intermediate Algebra for College Students*, Bob has written textbooks covering developmental mathematics, introductory algebra, college algebra, algebra and trigonometry, precalculus, and liberal arts mathematics, all published by Pearson. When not secluded in his Northern California writer's cabin, Bob can be found hiking the beaches and trails of Point Reyes National Seashore, and tending to the chores required by his beloved entourage of horses, chickens, and irritable roosters.

CHAPTER 1

Algebra, Mathematical Models, and Problem Solving

How would your lifestyle change if a gallon of gas cost $9.15? Of if the price of a staple such as milk were $15? That's how much those products would cost if their prices had increased at the same rate as college tuition has increased since 1980. If this trend continues, what can we expect in the 2020s and beyond?

We can answer this question by representing data for tuition and fees at U.S. colleges mathematically. With such representations, called *mathematical models*, we can gain insights and predict what might occur in the future on a variety of issues, ranging from college costs to a possible Social Security doomsday, and even the changes that occur as we age.

Here's where you'll find these applications:

- Mathematical models involving college costs appear as Example 7 in Section 1.4 and Exercises 67–68 in Exercise Set 1.4.
- The insecurities of Social Security are explored in Exercise 78 in the Review Exercises.
- Some surprising changes that occur with aging appear as Example 2 in Section 1.1, Exercises 89–92 in Exercise Set 1.1, and Exercises 53–56 in Exercise Set 1.3.

SECTION

1.1 Algebraic Expressions, Real Numbers, and Interval Notation

What am I supposed to learn?

After studying this section, you should be able to:

1. Translate English phrases into algebraic expressions.
2. Evaluate algebraic expressions.
3. Use mathematical models.
4. Recognize the sets that make up the real numbers.
5. Use set-builder notation.
6. Use the symbols $\in$ and $\notin$.
7. Use inequality symbols.
8. Use interval notation.

As we get older, do we mellow out or become more neurotic? In this section, you will learn how the special language of algebra describes your world, including our improving emotional health with age.

Algebraic Expressions

Algebra uses letters, such as x and y, to represent numbers. If a letter is used to represent various numbers, it is called a **variable**. For example, imagine that you are basking in the sun on the beach. We can let x represent the number of minutes that you can stay in the sun without burning with no sunscreen. With a number 6 sunscreen, exposure time without burning is six times as long, or 6 times x. This can be written $6 \cdot x$, but it is usually expressed as $6x$. Placing a number and a letter next to one another indicates multiplication.

Notice that $6x$ combines the number 6 and the variable x using the operation of multiplication. A combination of variables and numbers using the operations of addition, subtraction, multiplication, or division, as well as powers or roots, is called an **algebraic expression**. Here are some examples of algebraic expressions:

$$x + 6, \quad x - 6, \quad 6x, \quad \frac{x}{6}, \quad 3x + 5, \quad x^2 - 3, \quad \sqrt{x} + 7.$$

Is every letter in algebra a variable? No. Some letters stand for a particular number. Such a letter is called a **constant**. For example, let $d =$ the number of days in a week. The letter d represents just one number, namely 7, and is a constant.

1 Translate English phrases into algebraic expressions.

Translating English Phrases into Algebraic Expressions

Problem solving in algebra involves translating English phrases into algebraic expressions. Here is a list of words and phrases for the four basic operations:

Addition	Subtraction	Multiplication	Division
sum	difference	product	quotient
plus	minus	times	divide
increased by	decreased by	of (used with fractions)	per
more than	less than	twice	ratio

EXAMPLE 1 Translating English Phrases into Algebraic Expressions

Write each English phrase as an algebraic expression. Let x represent the number.

a. Nine less than six times a number

b. The quotient of five and a number, increased by twice the number

Great Question!

Why is it so important to work each of the book's Check Points?

You learn best by doing. Do not simply look at the worked examples and conclude that you know how to solve them. To be sure that you understand the worked examples, try each Check Point. Check your answer in the answer section before continuing your reading. Expect to read this book with pencil and paper handy to work the Check Points.

Solution

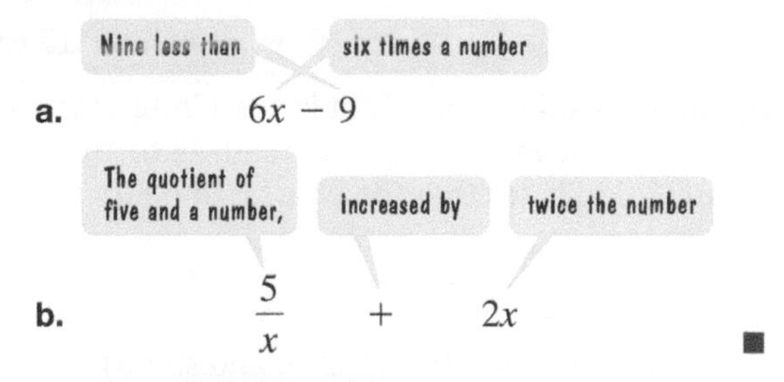

■

CHECK POINT 1 Write each English phrase as an algebraic expression. Let x represent the number.

a. Five more than 8 times a number

b. The quotient of a number and 7, decreased by twice the number

2 Evaluate algebraic expressions.

Evaluating Algebraic Expressions

Evaluating an algebraic expression means to find the value of the expression for a given value of the variable.

EXAMPLE 2 Evaluating an Algebraic Expression

We opened the section with a comment about our improving emotional health with age. A test measuring neurotic traits, such as anxiety and hostility, indicates that people may become less neurotic as they get older. **Figure 1.1** shows the average level of neuroticism, on a scale of 0 to 50, for persons at various ages.

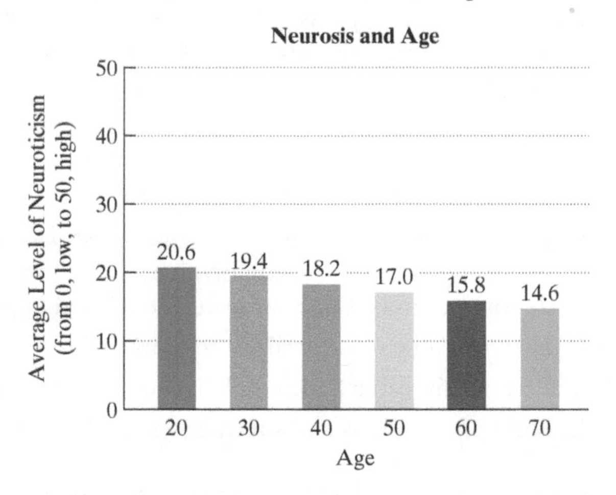

Figure 1.1

Source: L. M. Williams, "The Mellow Years? Neural Basis of Improving Emotional Stability over Age," *The Journal of Neuroscience*, June 14, 2006.

The algebraic expression $23 - 0.12x$ describes the average neurotic level for people who are x years old. Evaluate the expression for $x = 80$. Describe what the answer means in practical terms.

Solution We begin by substituting 80 for x. Because $x = 80$, we will be finding the average neurotic level at age 80.

$$23 - 0.12x$$

Replace x with 80.

$$= 23 - 0.12(80) = 23 - 9.6 = 13.4$$

Thus, at age 80, the average level of neuroticism on a scale of 0 to 50 is 13.4. ■

CHECK POINT 2 Evaluate the expression from Example 2, $23 - 0.12x$, for $x = 10$. Describe what the answer means in practical terms.

Many algebraic expressions involve *exponents*. For example, the algebraic expression

$$46x^2 + 541x + 17{,}650$$

approximates student-loan debt in the United States, in dollars, x years after 2000. The expression x^2 means $x \cdot x$, and is read "x to the second power" or "x squared." The exponent, 2, indicates that the base, x, appears as a factor two times.

Exponential Notation

If n is a counting number (1, 2, 3, and so on),

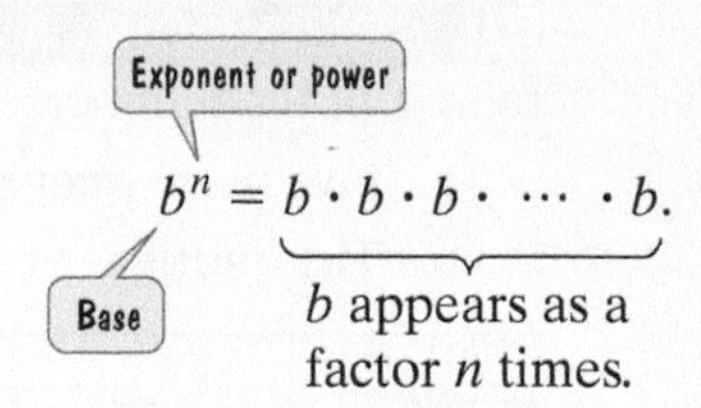

$$b^n = \underbrace{b \cdot b \cdot b \cdot \cdots \cdot b}_{b \text{ appears as a factor } n \text{ times.}}.$$

b^n is read "the nth power of b" or "b to the nth power." Thus, the nth power of b is defined as the product of n factors of b. The expression b^n is called an **exponential expression**. Furthermore, $b^1 = b$.

Using Technology

You can use a calculator to evaluate exponential expressions. For example, to evaluate 2^4, press the following keys:

Many Scientific Calculators

2 [y^x] 4 [=]

Many Graphing Calculators

2 [∧] 4 [ENTER]

Although calculators have special keys to evaluate powers of ten and to square bases, you can always use one of the sequences shown here.

For example,

$$8^2 = 8 \cdot 8 = 64, \quad 5^3 = 5 \cdot 5 \cdot 5 = 125, \quad \text{and} \quad 2^4 = 2 \cdot 2 \cdot 2 \cdot 2 = 16.$$

Many algebraic expressions involve more than one operation. Evaluating an algebraic expression without a calculator involves carefully applying the following order of operations agreement:

The Order of Operations Agreement

1. Perform operations within the innermost parentheses and work outward. If the algebraic expression involves a fraction, treat the numerator and the denominator as if they were each enclosed in parentheses.
2. Evaluate all exponential expressions.
3. Perform multiplications and divisions as they occur, working from left to right.
4. Perform additions and subtractions as they occur, working from left to right.

EXAMPLE 3 Evaluating an Algebraic Expression

Evaluate $7 + 5(x - 4)^3$ for $x = 6$.

Solution

$7 + 5(x - 4)^3 = 7 + 5(6 - 4)^3$ — Replace x with 6.

$= 7 + 5(2)^3$ — First work inside parentheses: $6 - 4 = 2$.

$= 7 + 5(8)$ — Evaluate the exponential expression: $2^3 = 2 \cdot 2 \cdot 2 = 8$.

$= 7 + 40$ — Multiply: $5(8) = 40$.

$= 47$ — Add. ■

Great Question!

What am I supposed to do with the worked examples?

Study the step-by-step solutions in the examples. Reading the solutions slowly and with great care will prepare you for success with the Check Points and the exercises in the Exercise Sets.

✓ CHECK POINT 3 Evaluate $8 + 6(x - 3)^2$ for $x = 13$.

3 Use mathematical models.

Formulas and Mathematical Models

An **equation** is formed when an equal sign is placed between two algebraic expressions. One aim of algebra is to provide a compact, symbolic description of the world. These descriptions involve the use of *formulas*. A **formula** is an equation that uses variables to express a relationship between two or more quantities. Here is an example of a formula:

$$C = \frac{5}{9}(F - 32).$$

Celsius temperature | is | $\frac{5}{9}$ of | the difference between Fahrenheit temperature and 32°.

The process of finding formulas to describe real-world phenomena is called **mathematical modeling**. Such formulas, together with the meaning assigned to the variables, are called **mathematical models**. We often say that these formulas model, or describe, the relationships among the variables.

EXAMPLE 4 Modeling Student-Loan Debt

College students are graduating with the highest debt burden in history. **Figure 1.2** shows the mean, or average, student-loan debt in the United States for five selected graduating years from 2001 through 2013.

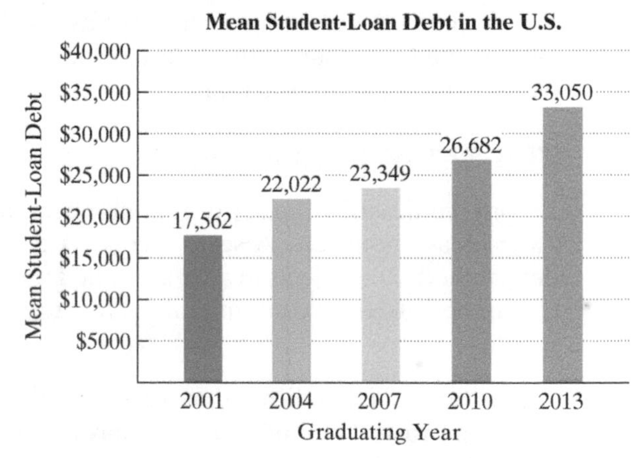

Figure 1.2
Source: Pew Research Center

The formula

$$D = 46x^2 + 541x + 17{,}650$$

models mean student-loan debt, D, in dollars, x years after 2000.

a. Use the formula to find mean student-loan debt for college students who graduated in 2013.

b. Does the mathematical model underestimate or overestimate mean student-loan debt for 2013 shown in **Figure 1.2**? By how much?

Solution

a. Because 2013 is 13 years after 2000, we substitute 13 for x in the given formula. Then we use the order of operations to find D, the mean student-loan debt for the graduating class of 2013.

$D = 46x^2 + 541x + 17{,}650$ This is the given mathematical model.

$D = 46(13)^2 + 541(13) + 17{,}650$ Replace each occurrence of x with 13.

$D = 46(169) + 541(13) + 17{,}650$ Evaluate the exponential expression: $13^2 = 13 \cdot 13 = 169$.

$D = 7774 + 7033 + 17{,}650$ Multiply from left to right: $46(169) = 7774$ and $541(13) = 7033$.

$D = 32{,}457$ Add.

The formula indicates that the mean student-loan debt for college students who graduated in 2013 was $32,457.

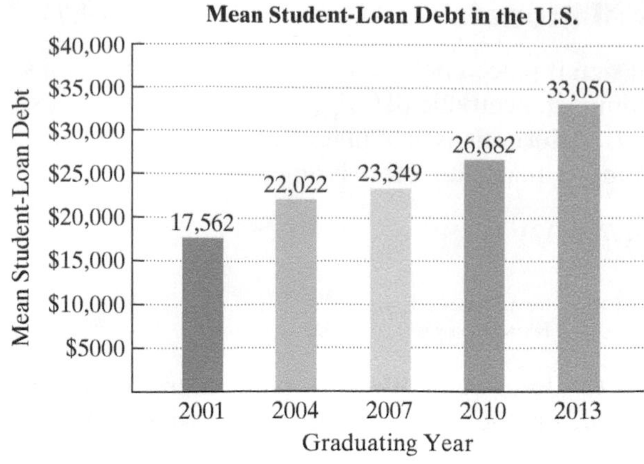

Figure 1.2 (repeated)

b. The mean student-loan debt for 2013 given in **Figure 1.2** is \$33,050. The model value, \$32,457, is less than the actual data value, \$33,050. Thus, the mathematical model underestimates mean student-loan debt for 2013 by

$$\$33{,}050 - \$32{,}457,$$

or by \$593. ■

✓ CHECK POINT 4

a. Use the formula in Example 4 to find mean student-loan debt for college students who graduated in 2010.

b. Does the mathematical model underestimate or overestimate mean student-loan debt for 2010 shown in **Figure 1.2**? By how much?

Sometimes a mathematical model gives an estimate that is not a good approximation or is extended to include values of the variable that do not make sense. In these cases, we say that **model breakdown** has occurred. Models that accurately describe data for the past ten years might not serve as reliable predictions for what can reasonably be expected to occur in the future. Model breakdown can occur when formulas are extended too far into the future.

4 Recognize the sets that make up the real numbers.

The Set of Real Numbers

Before we describe the set of real numbers, let's be sure you are familiar with some basic ideas about sets. A **set** is a collection of objects whose contents can be clearly determined. The objects in a set are called the **elements** of the set. For example, the set of numbers used for counting can be represented by

$$\{1, 2, 3, 4, 5, \ldots\}.$$

The braces, { }, indicate that we are representing a set. This form of representation, called the **roster method**, uses commas to separate the elements of the set. The three dots after the 5, called an *ellipsis*, indicate that there is no final element and that the listing goes on forever.

Three common sets of numbers are the *natural numbers*, the *whole numbers*, and the *integers*.

Great Question!

Can I use symbols other than braces to indicate sets in the roster method?

No. Grouping symbols such as parentheses, ( ), and square brackets, [ ], are not used to represent sets in the roster method. Furthermore, only commas are used to separate the elements of a set. Separators such as colons or semicolons are not used.

Natural Numbers, Whole Numbers, and Integers

The Set of Natural Numbers

$$\{1, 2, 3, 4, 5, \ldots\}$$

These are the numbers that we use for counting.

The Set of Whole Numbers

$$\{0, 1, 2, 3, 4, 5, \ldots\}$$

The set of whole numbers includes 0 and the natural numbers.

The Set of Integers

$$\{\ldots, -5, -4, -3, -2, -1, 0, 1, 2, 3, 4, 5, \ldots\}$$

The set of integers includes the negatives of the natural numbers and the whole numbers.

5 Use set-builder notation.

A set can also be written in **set-builder notation**. In this notation, the elements of the set are described, but not listed. Here is an example:

$$\{x \mid x \text{ is a natural number less than 6}\}.$$

The set of all x | such that | x is a natural number less than 6.

The same set is written using the roster method as

$$\{1, 2, 3, 4, 5\}.$$

6 Use the symbols $\in$ and $\notin$.

The symbol $\in$ is used to indicate that a number or object is in a particular set. The symbol $\in$ is read "is an element of." Here is an example:

$$7 \in \{1, 2, 3, 4, 5, \ldots\}.$$

7 | is an element of | the set of natural numbers.

The symbol $\notin$ is used to indicate that a number or object is not in a particular set. The symbol $\notin$ is read "is not an element of." Here is an example:

$$\frac{1}{2} \notin \{1, 2, 3, 4, 5, \ldots\}.$$

$\frac{1}{2}$ | is not an element of | the set of natural numbers.

EXAMPLE 5 Using the Symbols $\in$ and $\notin$

Determine whether each statement is true or false:

a. $100 \in \{x \mid x \text{ is an integer}\}$ **b.** $20 \notin \{5, 10, 15\}$.

Solution

a. Because 100 is an integer, the statement

$$100 \in \{x \mid x \text{ is an integer}\}$$

is true. The number 100 is an element of the set of integers.

b. Because 20 is not an element of $\{5, 10, 15\}$, the statement $20 \notin \{5, 10, 15\}$ is true. ■

✓ CHECK POINT 5 Determine whether each statement is true or false:

a. $13 \in \{x \mid x \text{ is an integer}\}$ **b.** $6 \notin \{7, 8, 9, 10\}$.

Another common set is the set of *rational numbers*. Each of these numbers can be expressed as an integer divided by a nonzero integer.

Rational Numbers

The set of **rational numbers** is the set of all numbers that can be expressed as a quotient of two integers, with the denominator not 0.

$$\left\{\frac{a}{b} \,\middle|\, a \text{ and } b \text{ are integers and } b \neq 0\right\}$$

This means that b is not equal to zero.

Three examples of rational numbers are

$$\frac{1}{4}, \quad \frac{-2}{3}, \quad \text{and } 5 = \frac{5}{1}.$$

$a = 1$, $b = 4$; $a = -2$, $b = 3$; $a = 5$, $b = 1$

Can you see that integers are also rational numbers because they can be written in terms of division by 1?

Rational numbers can be expressed in fraction or decimal notation. To express the fraction $\frac{a}{b}$ as a decimal, divide the denominator, b, into the numerator, a. In decimal notation, rational numbers either terminate (stop) or have a digit, or block of digits, that repeats. For example,

$$\frac{3}{8} = 3 \div 8 = 0.375 \quad \text{and} \quad \frac{7}{11} = 7 \div 11 = 0.6363\ldots = 0.\overline{63}.$$

The decimal stops: it is a terminating decimal.

This is a repeating decimal. The bar is written over the repeating part.

Some numbers cannot be expressed as terminating or repeating decimals. An example of such a number is $\sqrt{2}$, the square root of 2. The number $\sqrt{2}$ is a number that can be squared to give 2. No terminating or repeating decimal can be squared to get 2. However, some approximations have squares that come close to 2. We use the symbol $\approx$, which means "is approximately equal to."

- $\sqrt{2} \approx 1.4$ because $(1.4)^2 = (1.4)(1.4) = 1.96$.
- $\sqrt{2} \approx 1.41$ because $(1.41)^2 = (1.41)(1.41) = 1.9881$.
- $\sqrt{2} \approx 1.4142$ because $(1.4142)^2 = (1.4142)(1.4142) = 1.99996164$.

$\sqrt{2}$ is an example of an *irrational number.*

Irrational Numbers

The set of **irrational numbers** is the set of numbers whose decimal representations neither terminate nor repeat. Irrational numbers cannot be expressed as quotients of integers.

Examples of irrational numbers include

$$\sqrt{3} \approx 1.73205 \quad \text{and} \quad \pi(\text{pi}) \approx 3.141593.$$

Not all square roots are irrational. For example, $\sqrt{25} = 5$ because $5^2 = 5 \cdot 5 = 25$. Thus, $\sqrt{25}$ is a natural number, a whole number, an integer, and a rational number $\left(\sqrt{25} = \frac{5}{1}\right)$.

The set of *real numbers* is formed by combining the sets of rational numbers and irrational numbers. Thus, every real number is either rational or irrational, as shown in **Figure 1.3**.

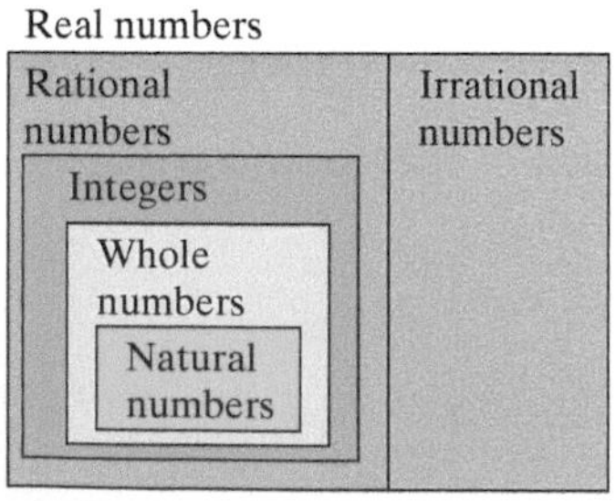

Real numbers
Rational numbers
Integers
Whole numbers
Natural numbers
Irrational numbers

Figure 1.3 Every real number is either rational or irrational.

Real Numbers

The set of **real numbers** is the set of numbers that are either rational or irrational:

$$\{x \mid x \text{ is rational or } x \text{ is irrational}\}.$$

The Real Number Line

The **real number line** is a graph used to represent the set of real numbers. An arbitrary point, called the **origin**, is labeled 0. Select a point to the right of 0 and label it 1. The distance from 0 to 1 is called the **unit distance**. Numbers to the right of the origin are **positive** and numbers to the left of the origin are **negative**. The real number line is shown in **Figure 1.4**.

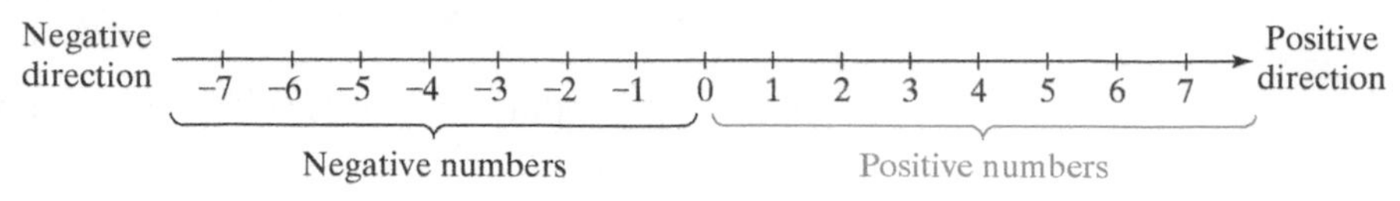

Figure 1.4 The real number line

Great Question!

How did you locate $\sqrt{2}$ as a precise point on the number line in Figure 1.5?

We used a right triangle with two legs of length 1. The remaining side has a length measuring $\sqrt{2}$.

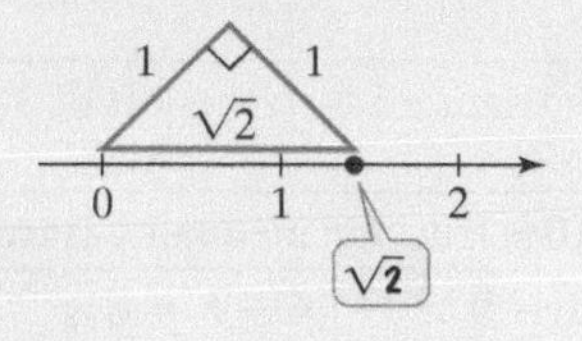

We'll have lots more to say about right triangles later in the book.

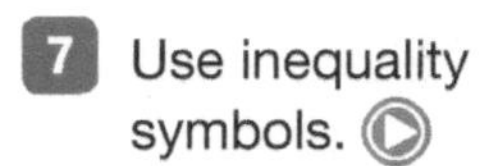

7 Use inequality symbols.

Real numbers are **graphed** on a number line by placing a dot at the correct location for each number. The integers are easiest to locate. In **Figure 1.5**, we've graphed six rational numbers and three irrational numbers on a real number line.

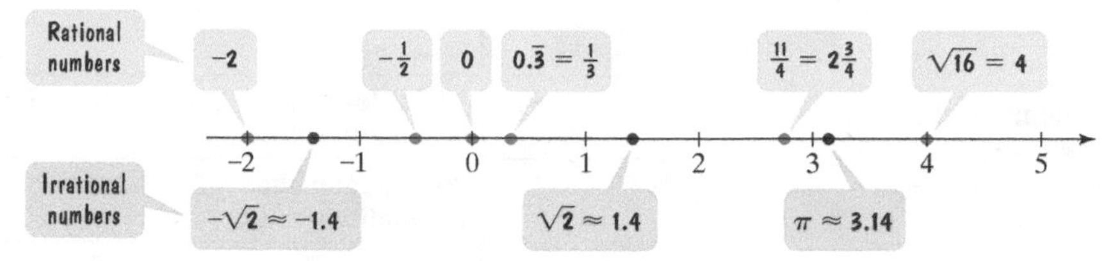

Figure 1.5 Graphing numbers on a real number line

Every real number corresponds to a point on the number line and every point on the number line corresponds to a real number. We say that there is a **one-to-one correspondence** between all the real numbers and all points on a real number line.

Ordering the Real Numbers

On the real number line, the real numbers increase from left to right. The lesser of two real numbers is the one farther to the left on a number line. The greater of two real numbers is the one farther to the right on a number line.

Look at the number line in **Figure 1.6**. The integers -4 and -1 are graphed.

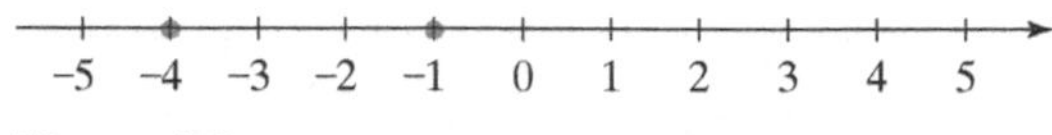

Figure 1.6

Observe that -4 is to the left of -1 on the number line. This means that -4 is less than -1.

$$-4 < -1$$

-4 is less than -1 because -4 is to the left of -1 on the number line.

In **Figure 1.6**, we can also observe that -1 is to the right of -4 on the number line. This means that -1 is greater than -4.

$$-1 > -4$$

-1 is greater than -4 because -1 is to the right of -4 on the number line.

The symbols $<$ and $>$ are called **inequality symbols**. These symbols always point to the lesser of the two real numbers when the inequality statement is true.

-4 is less than -1. $\quad -4 < -1 \quad$ The symbol points to -4, the lesser number.

-1 is greater than -4. $\quad -1 > -4 \quad$ The symbol still points to -4, the lesser number.

The symbols $<$ and $>$ may be combined with an equal sign, as shown in the following table:

Symbol	Meaning	Examples	Explanation
$a \le b$	a is less than or equal to b.	$2 \le 9$ $9 \le 9$	Because $2 < 9$ Because $9 = 9$
$b \ge a$	b is greater than or equal to a.	$9 \ge 2$ $2 \ge 2$	Because $9 > 2$ Because $2 = 2$

This inequality is true if either the $<$ part or the $=$ part is true.

This inequality is true if either the $>$ part or the $=$ part is true.

EXAMPLE 6 Using Inequality Symbols

Write out the meaning of each inequality. Then determine whether the inequality is true or false.

a. $-5 < -1$ **b.** $6 > -2$ **c.** $-6 \le 3$ **d.** $10 \ge 10$ **e.** $-9 \ge 6$

Solution The solution is illustrated by the number line in **Figure 1.7**.

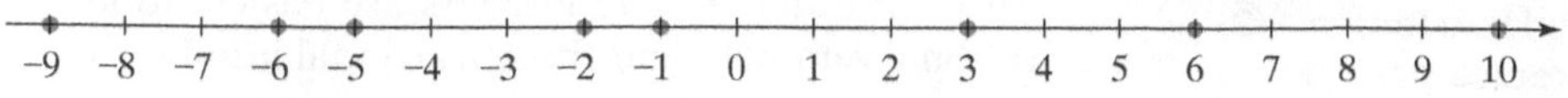

Figure 1.7

−9 −8 −7 −6 −5 −4 −3 −2 −1 0 1 2 3 4 5 6 7 8 9 10

Inequality	Meaning
a. $-5 < -1$	"-5 is less than -1." Because -5 is to the left of -1 on the number line, the inequality is true.
b. $6 > -2$	"6 is greater than -2." Because 6 is to the right of -2 on the number line, the inequality is true.
c. $-6 \le 3$	"-6 is less than or equal to 3." Because $-6 < 3$ is true (-6 is to the left of 3 on the number line), the inequality is true.
d. $10 \ge 10$	"10 is greater than or equal to 10." Because $10 = 10$ is true, the inequality is true.
e. $-9 \ge 6$	"-9 is greater than or equal to 6." Because neither $-9 > 6$ nor $-9 = 6$ is true, the inequality is false.

Great Question!

Can similar English phrases have different algebraic representations?

Yes. Here are three similar English phrases that have very different translations:

- 7 minus a number: $7 - x$
- 7 less than a number: $x - 7$
- 7 is less than a number: $7 < x$

Think carefully about what is expressed in English before you translate into the language of algebra.

CHECK POINT 6 Write out the meaning of each inequality. Then determine whether the inequality is true or false.

a. $-8 < -2$

b. $7 > -3$

c. $-1 \le -4$

d. $5 \ge 5$

e. $2 \ge -14$

8 Use interval notation.

Interval Notation

Some sets of real numbers can be represented using **interval notation**. Suppose that a and b are two real numbers such that $a < b$.

Interval Notation	Graph
The **open interval** (a, b) represents the set of real numbers between, but not including, a and b. $$(a, b) = \{x \mid a < x < b\}$$ x is greater than a ($a < x$) and x is less than b ($x < b$).	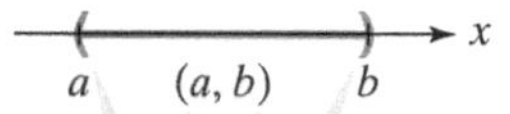 a (a, b) b x The parentheses in the graph and in interval notation indicate that a and b, the endpoints, are excluded from the interval.
The **closed interval** $[a, b]$ represents the set of real numbers between, and including, a and b. $$[a, b] = \{x \mid a \le x \le b\}$$ x is greater than or equal to a ($a \le x$) and x is less than or equal to b ($x \le b$).	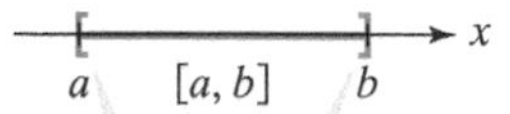 a $[a, b]$ b x The square brackets in the graph and in interval notation indicate that a and b, the endpoints, are included in the interval.
The **infinite interval** (a, ∞) represents the set of real numbers that are greater than a. $$(a, \infty) = \{x \mid x > a\}$$ The infinity symbol does not represent a real number. It indicates that the interval extends indefinitely to the right.	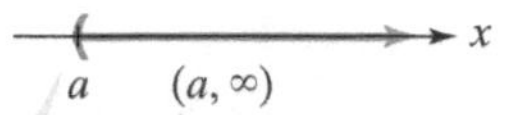 a (a, ∞) x The parenthesis indicates that a is excluded from the interval.
The **infinite interval** $(-\infty, b]$ represents the set of real numbers that are less than or equal to b. $$(-\infty, b] = \{x \mid x \le b\}$$ The negative infinity symbol indicates that the interval extends indefinitely to the left.	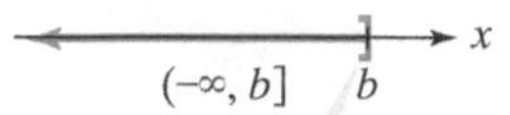 $(-\infty, b]$ b x The square bracket indicates that b is included in the interval.

Parentheses and Brackets in Interval Notation

Parentheses indicate endpoints that are not included in an interval. Square brackets indicate endpoints that are included in an interval. Parentheses are always used with ∞ or $-\infty$.

Table 1.1 lists nine possible types of intervals used to describe sets of real numbers.

Table 1.1 Intervals on the Real Number Line

Let a and b be real numbers such that $a < b$.

Interval Notation	Set-Builder Notation	Graph
(a, b)	$\{x \mid a < x < b\}$	
$[a, b]$	$\{x \mid a \le x \le b\}$	
$[a, b)$	$\{x \mid a \le x < b\}$	
$(a, b]$	$\{x \mid a < x \le b\}$	
(a, ∞)	$\{x \mid x > a\}$	
$[a, \infty)$	$\{x \mid x \ge a\}$	
$(-\infty, b)$	$\{x \mid x < b\}$	
$(-\infty, b]$	$\{x \mid x \le b\}$	
$(-\infty, \infty)$	$\{x \mid x$ is a real number$\}$ or $\mathbb{R}$ (set of real numbers)	

Achieving Success

Practice! Practice! Practice!

The way to learn algebra is by seeing solutions to examples and **doing exercises.** This means working the Check Points and the assigned exercises in the Exercise Sets. There are no alternatives. It's easy to read a solution, or watch your professor solve an example, and believe you know what to do. However, learning algebra requires that you actually **perform solutions by yourself**. Get in the habit of working exercises every day. The more time you spend solving exercises, the easier the process becomes. It's okay to take a short break after class, but start reviewing and working the assigned homework as soon as possible.

EXAMPLE 7 Interpreting Interval Notation

Express each interval in set-builder notation and graph:

a. $(-1, 4]$ **b.** $[2.5, 4]$ **c.** $(-4, \infty)$.

Solution

a. $(-1, 4] = \{x \mid -1 < x \le 4\}$

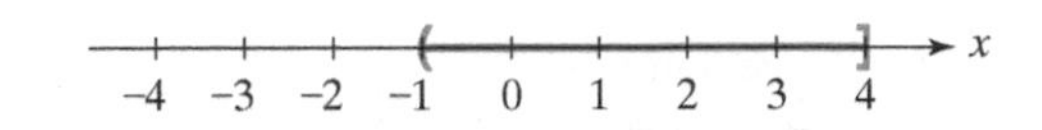

b. $[2.5, 4] = \{x \mid 2.5 \le x \le 4\}$

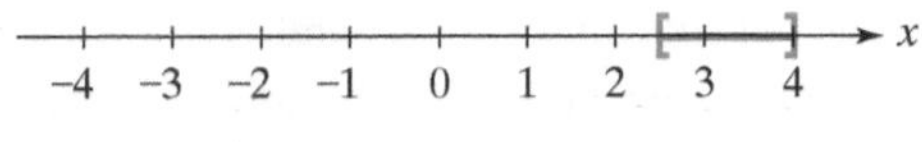

c. $(-4, \infty) = \{x \mid x > -4\}$

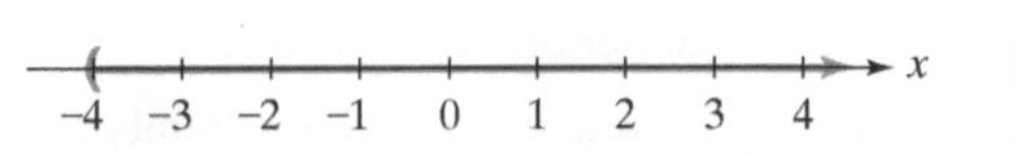

■

CHECK POINT 7 Express each interval in set-builder notation and graph:

a. $[-2, 5)$ **b.** $[1, 3.5]$ **c.** $(-\infty, -1)$.

CONCEPT AND VOCABULARY CHECK

Fill in each blank so that the resulting statement is true.

1. A letter that represents a variety of different numbers is called a/an ____________.
2. A combination of numbers, letters that represent numbers, and operation symbols is called an algebraic ____________.
3. If n is a counting number, b^n, read ____________, indicates that there are n factors of b. The number b is called the ____________ and the number n is called the ____________.
4. A statement that expresses a relationship between two or more variables, such as $C = \frac{5}{9}(F - 32)$, is called a/an ____________. The process of finding such statements to describe real-world phenomena is called mathematical ____________. Such statements, together with the meaning assigned to the variables, are called mathematical ____________.
5. The set $\{1, 2, 3, 4, 5, \ldots\}$ is called the set of ____________ numbers.
6. The set $\{0, 1, 2, 3, 4, 5, \ldots\}$ is called the set of ____________ numbers.
7. The set $\{\ldots, -4, -3, -2, -1, 0, 1, 2, 3, 4, \ldots\}$ is called the set of ____________.
8. The set of numbers in the form $\frac{a}{b}$, where a and b belong to the set in statement 7 above and $b \neq 0$, is called the set of ____________ numbers.
9. The set of numbers whose decimal representations are neither terminating nor repeating is called the set of ____________ numbers.
10. Every real number is either a/an ____________ number or a/an ____________ number.
11. The notation $2 < 5$ means that 2 is to the ____________ of 5 on a number line.
12. In interval notation, $[2, 5)$ represents the set of real numbers between ____________ and ____________, including ____________ but not including ____________.
13. In interval notation, $(-2, \infty)$ represents the set of real numbers ____________ -2.
14. In interval notation, $(-\infty, -1]$ represents the set of real numbers ____________ -1.

1.1 EXERCISE SET

MyMathLab®

Practice Exercises

In Exercises 1–14, write each English phrase as an algebraic expression. Let x represent the number.

1. Five more than a number
2. A number increased by six
3. Four less than a number
4. Nine less than a number
5. Four times a number
6. Twice a number
7. Ten more than twice a number
8. Four more than five times a number
9. The difference of six and half of a number
10. The difference of three and half of a number
11. Two less than the quotient of four and a number
12. Three less than the quotient of five and a number
13. The quotient of three and the difference of five and a number
14. The quotient of six and the difference of ten and a number

In Exercises 15–26, evaluate each algebraic expression for the given value or values of the variable(s).

15. $7 + 5x$, for $x = 10$
16. $8 + 6x$, for $x = 5$
17. $6x - y$, for $x = 3$ and $y = 8$
18. $8x - y$, for $x = 3$ and $y = 4$
19. $x^2 + 3x$, for $x = \frac{1}{3}$
20. $x^2 + 2x$ for $x = \frac{1}{2}$
21. $x^2 - 6x + 3$, for $x = 7$
22. $x^2 - 7x + 4$, for $x = 8$
23. $4 + 5(x - 7)^3$, for $x = 9$
24. $6 + 5(x - 6)^3$, for $x = 8$
25. $x^2 - 3(x - y)$, for $x = 8$ and $y = 2$
26. $x^2 - 4(x - y)$, for $x = 8$ and $y = 3$

In Exercises 27–34, use the roster method to list the elements in each set.

27. $\{x \mid x$ is a natural number less than $5\}$
28. $\{x \mid x$ is a natural number less than $4\}$
29. $\{x \mid x$ is an integer between -8 and $-3\}$
30. $\{x \mid x$ is an integer between -7 and $-2\}$
31. $\{x \mid x$ is a natural number greater than $7\}$
32. $\{x \mid x$ is a natural number greater than $9\}$
33. $\{x \mid x$ is an odd whole number less than $11\}$
34. $\{x \mid x$ is an odd whole number less than $9\}$

In Exercises 35–48, use the meaning of the symbols $\in$ and $\notin$ to determine whether each statement is true or false.

35. $7 \in \{x \mid x \text{ is an integer}\}$

36. $9 \in \{x \mid x \text{ is an integer}\}$

37. $7 \in \{x \mid x \text{ is a rational number}\}$

38. $9 \in \{x \mid x \text{ is a rational number}\}$

39. $7 \in \{x \mid x \text{ is an irrational number}\}$

40. $9 \in \{x \mid x \text{ is an irrational number}\}$

41. $3 \notin \{x \mid x \text{ is an irrational number}\}$

42. $5 \notin \{x \mid x \text{ is an irrational number}\}$

43. $\frac{1}{2} \notin \{x \mid x \text{ is a rational number}\}$

44. $\frac{1}{4} \notin \{x \mid x \text{ is a rational number}\}$

45. $\sqrt{2} \notin \{x \mid x \text{ is a rational number}\}$

46. $\pi \notin \{x \mid x \text{ is a rational number}\}$

47. $\sqrt{2} \notin \{x \mid x \text{ is a real number}\}$

48. $\pi \notin \{x \mid x \text{ is a real number}\}$

In Exercises 49–64, write out the meaning of each inequality. Then determine whether the inequality is true or false.

49. $-6 < -2$

50. $-7 < -3$

51. $5 > -7$

52. $3 > -8$

53. $0 < -4$

54. $0 < -5$

55. $-4 \le 1$

56. $-5 \le 1$

57. $-2 \le -6$

58. $-3 \le -7$

59. $-2 \le -2$

60. $-3 \le -3$

61. $-2 \ge -2$

62. $-3 \ge -3$

63. $2 \le -\frac{1}{2}$

64. $4 \le -\frac{1}{2}$

In Exercises 65–78, express each interval in set-builder notation and graph the interval on a number line.

65. $(1, 6]$

66. $(-2, 4]$

67. $[-5, 2)$

68. $[-4, 3)$

69. $[-3, 1]$

70. $[-2, 5]$

71. $(2, \infty)$

72. $(3, \infty)$

73. $[-3, \infty)$

74. $[-5, \infty)$

75. $(-\infty, 3)$

76. $(-\infty, 2)$

77. $(-\infty, 5.5)$

78. $(-\infty, 3.5]$

Practice PLUS

By definition, an "and" statement is true only when the statements before and after the "and" connective are both true. Use this definition to determine whether each statement in Exercises 79–88 is true or false.

79. $0.\overline{3} > 0.3$ and $-10 < 4 + 6$.

80. $0.6 < 0.\overline{6}$ and $2 \cdot 5 \le 4 + 6$.

81. $12 \in \{1, 2, 3, \dots\}$ and $\{3\} \in \{1, 2, 3, 4\}$

82. $17 \in \{1, 2, 3, \dots\}$ and $\{4\} \in \{1, 2, 3, 4, 5\}$

83. $\left(\frac{2}{5} + \frac{3}{5}\right) \in \{x \mid x \text{ is a natural number}\}$ and the value of $9x^2(x + 11) - 9(x + 11)x^2$, for $x = 100$, is 0.

84. $\left(\frac{14}{19} + \frac{5}{19}\right) \in \{x \mid x \text{ is a natural number}\}$ and the value of $12x^2(x + 10) - 12(x + 10)x^2$, for $x = 50$, is 0.

85. $\{x \mid x \text{ is an integer between } -3 \text{ and } 0\} = \{-3, -2, -1, 0\}$ and $-\pi > -3.5$.

86. $\{x \mid x \text{ is an integer between } -4 \text{ and } 0\} = \{-4, -3, -2, -1, 0\}$ and $-\frac{\pi}{2} > -2.3$.

87. Twice the sum of a number and three is represented by $2x + 3$ and $-1{,}100{,}000 \in \{x \mid x \text{ is an integer}\}$

88. Three times the sum of a number and five is represented by $3x + 5$ and $-4{,}500{,}000 \in \{x \mid x \text{ is an integer}\}$

Application Exercises

We opened the section with a comment about our improving emotional health with age. We also saw an example indicating people become less neurotic as they get older. How can this be explained? One theory is that key centers of the brain tend to create less resistance to feelings of happiness as we age. The graph shows the average resistance to happiness, on a scale of 0 (no resistance) to 8 (completely resistant), for persons at various ages.

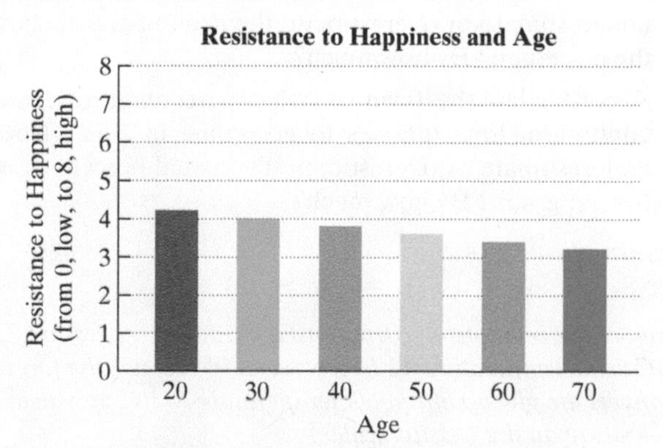

Source: L. M. Williams, "The Mellow Years? Neural Basis for Improving Stability over Age." THE JOURNAL OF NEUROSCIENCE, June 14, 2006

The data in the graph above can be modeled by the formula

$$R = 4.6 - 0.02x,$$

where R represents the average resistance to happiness, on a scale of 0 to 8, for a person who is x years old. Use this formula to solve Exercises 89–92.

89. According to the formula, what is the average resistance to happiness at age 20?

(In Exercises 90–92, refer to the formula and the graph at the bottom of the previous page.)

90. According to the formula, what is the average resistance to happiness at age 30?

91. What is the difference between the average resistance to happiness at age 30 and at age 50?

92. What is the difference between the average resistance to happiness at age 20 and at age 70?

The majority of American adults use their smartphones to go online. The bar graph shows the percentage of Americans using phones to go online from 2009 through 2013.

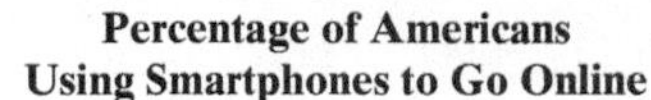

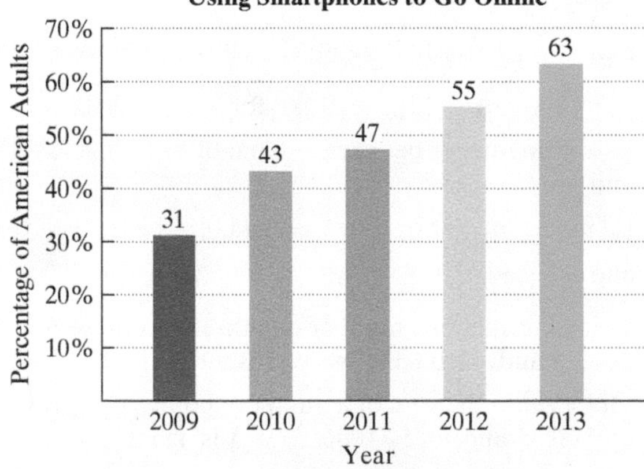

Source: Pew Internet and American Life Project

The data in the graph can be modeled by the formula

$$S = 32 + 8.7x - 0.3x^2,$$

where S represents the percentage of American adults using smartphones to go online x years after 2009. Use this formula to solve Exercises 93–94.

93. According to the formula, what percentage of American adults used smartphones to go online in 2013? Does this underestimate or overestimate the actual percent shown by the bar graph? By how much?

94. According to the formula, what percentage of American adults used smartphones to go online in 2012? Does this underestimate or overestimate the actual percent shown by the bar graph? By how much?

The formula

$$C = \frac{5}{9}(F - 32)$$

expresses the relationship between Fahrenheit temperature, F, and Celsius temperature, C. In Exercises 95–96, use the formula to convert the given Fahrenheit temperature to its equivalent temperature on the Celsius scale.

95. 50°F

96. 86°F

A football was kicked vertically upward from a height of 4 feet with an initial speed of 60 feet per second. The formula

$$h = 4 + 60t - 16t^2$$

describes the ball's height above the ground, h, in feet, t seconds after it was kicked. Use this formula to solve Exercises 97–98.

97. What was the ball's height 2 seconds after it was kicked?

98. What was the ball's height 3 seconds after it was kicked?

Explaining the Concepts

Achieving Success

An effective way to understand something is to explain it to someone else. You can do this by using the Explaining the Concepts exercises that ask you to respond with verbal or written explanations. Speaking about a new concept uses a different part of your brain than thinking about the concept. Explaining new ideas verbally will quickly reveal any gaps in your understanding. It will also help you to remember new concepts for longer periods of time.

99. What is a variable?

100. What is an algebraic expression? Give an example with your explanation.

101. If n is a natural number, what does b^n mean? Give an example with your explanation.

102. What does it mean when we say that a formula models real-world phenomena?

103. What is model breakdown?

104. What is a set?

105. Describe the roster method for representing a set.

106. What are the natural numbers?

107. What are the whole numbers?

108. What are the integers?

109. Describe the rational numbers.

110. Describe the difference between a rational number and an irrational number.

111. What are the real numbers?

112. What is set-builder notation?

113. Describe the meanings of the symbols $\in$ and $\notin$. Provide an example showing the correct use of each symbol.

114. What is the real number line?

115. If you are given two real numbers, explain how to determine which one is the lesser.

116. In interval notation, what does a parenthesis signify? What does a bracket signify?

Critical Thinking Exercises

Make Sense? *In Exercises 117–120, determine whether each statement makes sense or does not make sense, and explain your reasoning.*

117. My mathematical model describes the data for the past ten years extremely well, so it will serve as an accurate prediction for what will occur in 2050.

118. My calculator will not display the value of 13^{1500}, so the algebraic expression $4x^{1500} - 3x + 7$ cannot be evaluated for $x = 13$ even without a calculator.

119. Regardless of what real numbers I substitute for x and y, I will always obtain zero when evaluating $2x^2y - 2yx^2$.

120. A model that describes the number of smartphone users x years after 2007 cannot be used to estimate the number in 2007.

In Exercises 121–124, determine whether each statement is true or false. If the statement is false, make the necessary change(s) to produce a true statement.

121. Every rational number is an integer.

122. Some whole numbers are not integers.

123. Some rational numbers are not positive.

124. Some irrational numbers are negative.

125. A bird lover visited a pet shop where there were twice 4 and 20 parrots. The bird lover purchased $\frac{1}{7}$ of the birds. English, of course, can be ambiguous and "twice 4 and 20" can mean $2(4 + 20)$ or $2 \cdot 4 + 20$. Explain how the conditions of the situation determine if "twice 4 and 20" means $2(4 + 20)$ or $2 \cdot 4 + 20$.

In Exercises 126–127, insert parentheses to make each statement true.

126. $2 \cdot 3 + 3 \cdot 5 = 45$

127. $8 + 2 \cdot 4 - 3 = 10$

128. Between which two consecutive integers is $-\sqrt{26}$? Do not use a calculator.

Preview Exercises

Exercises 129–131 will help you prepare for the material covered in the next section.

129. There are two real numbers whose distance is five units from zero on a real number line. What are these numbers?

130. Simplify: $\dfrac{16 + 3(2)^4}{12 - (10 - 6)}$.

131. Evaluate $2(3x + 5)$ and $6x + 10$ for $x = 4$.

SECTION

1.2 Operations with Real Numbers and Simplifying Algebraic Expressions

What am I supposed to learn?

After studying this section, you should be able to:

1. Find a number's absolute value.
2. Add real numbers.
3. Find opposites.
4. Subtract real numbers.
5. Multiply real numbers.
6. Evaluate exponential expressions.
7. Divide real numbers.
8. Use the order of operations.
9. Use commutative, associative, and distributive properties.
10. Simplify algebraic expressions.

College students have money to spend, often courtesy of Mom and Dad. The most sophisticated college marketers, from American Eagle to Apple to Red Bull, are increasingly turning to social media focused on students' wants and needs. In 2011, there were nearly 10,000 student reps on U.S. campuses facebooking, tweeting, and partying their way to selling things vital and specific to college students. Call it New World College Marketing 101.

How much do college students have to spend? In this section's Exercise Set (see Exercises 147–148), you'll be working with a model that addresses this question. In order to use the model, we need to review operations with real numbers, our focus of this section.

Absolute Value

Absolute value is used to describe how to operate with positive and negative numbers.

Geometric Meaning of Absolute Value

The **absolute value** of a real number a, denoted by $|a|$, is the distance from 0 to a on the number line. This distance is always taken to be nonnegative.

1 Find a number's absolute value.

EXAMPLE 1 Finding Absolute Value

Find the absolute value:

a. $|-4|$ **b.** $|3.5|$ **c.** $|0|$.

Solution The solution is illustrated in **Figure 1.8**.

a. $|-4| = 4$ The absolute value of −4 is 4 because −4 is 4 units from 0.

b. $|3.5| = 3.5$ The absolute value of 3.5 is 3.5 because 3.5 is 3.5 units from 0.

c. $|0| = 0$ The absolute value of 0 is 0 because 0 is 0 units from itself.

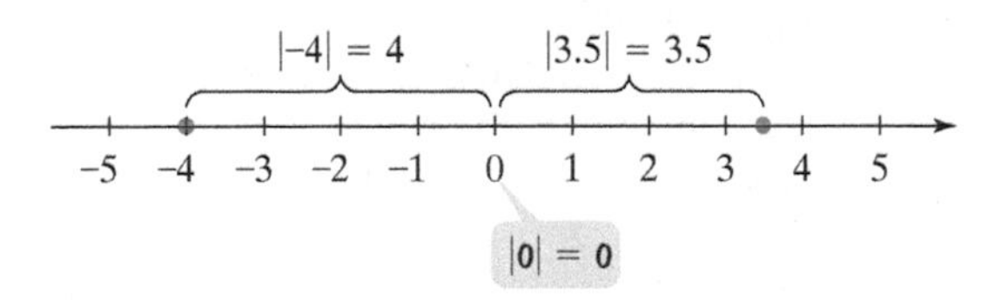

Figure 1.8 ■

Can you see that the absolute value of a real number is either positive or zero? Zero is the only real number whose absolute value is 0:

$$|0| = 0.$$

The absolute value of a real number is never negative.

✓ CHECK POINT 1 Find the absolute value:

a. $|-6|$ **b.** $|4.5|$ **c.** $|0|$.

2 Add real numbers.

Adding Real Numbers

Table 1.2 reviews how to add real numbers.

Table 1.2 Adding Real Numbers

Rule	Example
To add two real numbers with the same sign, add their absolute values. Use the common sign as the sign of the sum.	$(-7) + (-4) = -(\|-7\| + \|-4\|)$ $= -(7 + 4)$ $= -11$
To add two real numbers with different signs, subtract the smaller absolute value from the greater absolute value. Use the sign of the number with the greater absolute value as the sign of the sum.	$7 + (-15) = -(\|-15\| - \|7\|)$ $= -(15 - 7)$ $= -8$

EXAMPLE 2 Adding Real Numbers

Add:

a. $-12 + (-14)$ **b.** $-0.3 + 0.7$ **c.** $-\frac{3}{4} + \frac{1}{2}$.

Solution

a. $-12 + (-14) = -26$

Add absolute values: $12 + 14 = 26$.

Use the common sign.

b. $-0.3 + 0.7 = 0.4$

Subtract absolute values: $0.7 - 0.3 = 0.4$.

Use the sign of the number with the greater absolute value. The sign of the sum is assumed to be positive.

c. $-\frac{3}{4} + \frac{1}{2} = -\frac{1}{4}$

Subtract absolute values: $\frac{3}{4} - \frac{1}{2} = \frac{3}{4} - \frac{2}{4} = \frac{1}{4}$.

Use the sign of the number with the greater absolute value. ■

CHECK POINT 2 Add:

a. $-10 + (-18)$ **b.** $-0.2 + 0.9$ **c.** $-\frac{3}{5} + \frac{1}{2}$.

If one of two numbers being added is zero, the sum is the other number. For example,

$$-3 + 0 = -3 \quad \text{and} \quad 0 + 2 = 2.$$

In general,

$$a + 0 = a \quad \text{and} \quad 0 + a = a.$$

We call 0 the **identity element of addition** or the **additive identity**. Thus, the additive identity can be deleted from a sum.

3 Find opposites.

Numbers with different signs but the same absolute value are called **opposites** or **additive inverses**. For example, 3 and -3 are additive inverses. When additive inverses are added, their sum is 0. For example,

$$3 + (-3) = 0 \quad \text{and} \quad -3 + 3 = 0.$$

Inverse Property of Addition

The sum of a real number and its additive inverse is 0, the additive identity.

$$a + (-a) = 0 \quad \text{and} \quad (-a) + a = 0$$

The symbol "$-$" is used to name the opposite, or additive inverse, of a. When a is a negative number, $-a$, its opposite, is positive. For example, if a is -4, its opposite is 4. Thus,

$$-(-4) = 4.$$

The opposite of -4 is 4.

In general, if a is any real number,

$$-(-a) = a.$$

EXAMPLE 3 Finding Opposites

Find $-x$ if

a. $x = -6$ **b.** $x = \frac{1}{2}$.

Solution

a. If $x = -6$, then $-x = -(-6) = 6$. The opposite of -6 is 6.

b. If $x = \frac{1}{2}$, then $-x = -\frac{1}{2}$. The opposite of $\frac{1}{2}$ is $-\frac{1}{2}$. ■

✓ CHECK POINT 3 Find $-x$ if

a. $x = -8$ **b.** $x = \frac{1}{3}$.

We can define the absolute value of the real number a using opposites, without referring to a number line. The algebraic definition of the absolute value of a is given as follows:

Definition of Absolute Value

$$|a| = \begin{cases} a & \text{if } a \geq 0 \\ -a & \text{if } a < 0 \end{cases}$$

If a is nonnegative (that is, $a \geq 0$), the absolute value of a is the number itself: $|a| = a$. For example,

$$|5| = 5 \qquad |\pi| = \pi \qquad \left|\frac{1}{3}\right| = \frac{1}{3} \qquad |0| = 0$$

Zero is the only number whose absolute value is 0.

If a is a negative number (that is, $a < 0$), the absolute value of a is the opposite of a: $|a| = -a$. This makes the absolute value positive. For example,

$$|-3| = -(-3) = 3 \qquad |-\pi| = -(-\pi) = \pi \qquad \left|-\frac{1}{3}\right| = -\left(-\frac{1}{3}\right) = \frac{1}{3}.$$

This middle step is usually omitted.

4 Subtract real numbers.

Subtracting Real Numbers

Subtraction of real numbers is defined in terms of addition.

Definition of Subtraction

If a and b are real numbers,

$$a - b = a + (-b).$$

To subtract a real number, add its opposite or additive inverse.

Thus, to subtract real numbers,

1. Change the subtraction to addition.
2. Change the sign of the number being subtracted.
3. Add, using one of the rules for adding numbers with the same sign or different signs.

EXAMPLE 4 Subtracting Real Numbers

Subtract:

a. $6 - 13$ **b.** $5.1 - (-4.2)$ **c.** $-\frac{11}{3} - \left(-\frac{4}{3}\right)$.

Solution

a. $6 - 13 = 6 + (-13) = -7$

Change the subtraction to addition. Replace 13 with its opposite.

b. $5.1 - (-4.2) = 5.1 + 4.2 = 9.3$

Change the subtraction to addition. Replace −4.2 with its opposite.

c. $-\frac{11}{3} - \left(-\frac{4}{3}\right) = -\frac{11}{3} + \frac{4}{3} = -\frac{7}{3}$

Change the subtraction to addition. Replace $-\frac{4}{3}$ with its opposite. ■

✓ **CHECK POINT 4** Subtract:

a. $7 - 10$ **b.** $4.3 - (-6.2)$ **c.** $-\frac{4}{5} - \left(-\frac{1}{5}\right)$.

5 Multiply real numbers.

Multiplying Real Numbers

You can think of multiplication as repeated addition or subtraction that starts at 0. For example,

$$3(-4) = 0 + (-4) + (-4) + (-4) = -12$$

The numbers have different signs and the product is negative.

and

$$(-3)(-4) = 0 - (-4) - (-4) - (-4) = 0 + 4 + 4 + 4 = 12$$

The numbers have the same sign and the product is positive.

Table 1.3 reviews how to multiply real numbers.

Table 1.3 Multiplying Real Numbers

Rule	Example
The product of two real numbers with different signs is found by multiplying their absolute values. The product is negative.	$7(-5) = -35$
The product of two real numbers with the same sign is found by multiplying their absolute values. The product is positive.	$(-6)(-11) = 66$
The product of 0 and any real number is 0: $a \cdot 0 = 0$ and $0 \cdot a = 0$.	$-17(0) = 0$
If no number is 0, a product with an odd number of negative factors is found by multiplying their absolute values. The product is negative.	$-2(-3)(-5) = -30$ Three (odd) negative factors
If no number is 0, a product with an even number of negative factors is found by multiplying absolute values. The product is positive.	$-2(3)(-5) = 30$ Two (even) negative factors

6 Evaluate exponential expressions.

Because exponents indicate repeated multiplication, rules for multiplying real numbers can be used to evaluate exponential expressions.

EXAMPLE 5 Evaluating Exponential Expressions

Evaluate:

a. $(-6)^2$ **b.** -6^2 **c.** $(-5)^3$ **d.** $\left(-\frac{2}{3}\right)^4$.

Solution

a. $(-6)^2 = (-6)(-6) = 36$

Base is −6. Same signs give positive product.

b. $-6^2 = -(6 \cdot 6) = -36$

Base is 6. The negative is not inside parentheses and is not taken to the second power.

c. $(-5)^3 = (-5)(-5)(-5) = -125$

An odd number of negative factors gives a negative product.

d. $\left(-\frac{2}{3}\right)^4 = \left(-\frac{2}{3}\right)\left(-\frac{2}{3}\right)\left(-\frac{2}{3}\right)\left(-\frac{2}{3}\right) = \frac{16}{81}$

Base is $-\frac{2}{3}$. An even number of negative factors gives a positive product. ■

CHECK POINT 5 Evaluate:

a. $(-5)^2$ **b.** -5^2 **c.** $(-4)^3$ **d.** $\left(-\frac{3}{5}\right)^4$.

---

7 Divide real numbers.

Dividing Real Numbers

If a and b are real numbers and b is not 0, then the quotient of a and b is defined as follows:

$$a \div b = a \cdot \frac{1}{b} \quad \text{or} \quad \frac{a}{b} = a \cdot \frac{1}{b}.$$

Thus, to find the quotient of a and b, we can divide by b or multiply by $\frac{1}{b}$. The nonzero real numbers b and $\frac{1}{b}$ are called **reciprocals**, or **multiplicative inverses**, of one another. When reciprocals are multiplied, their product is 1:

$$b \cdot \frac{1}{b} = 1.$$

Because division is defined in terms of multiplication, the sign rules for dividing numbers are the same as the sign rules for multiplying them.

Dividing Real Numbers

The quotient of two numbers with different signs is negative. The quotient of two numbers with the same sign is positive. The quotient is found by dividing absolute values.

EXAMPLE 6 Dividing Real Numbers

Divide:

a. $\frac{20}{-5}$ **b.** $-\frac{3}{4} \div \left(-\frac{5}{9}\right)$.

Solution

a. $\frac{20}{-5} = -4$

Divide absolute values: $\frac{20}{5} = 4$.

Different signs: negative quotient

b. $-\frac{3}{4} \div \left(-\frac{5}{9}\right) = \frac{27}{20}$

Divide absolute values: $\frac{3}{4} \div \frac{5}{9} = \frac{3}{4} \cdot \frac{9}{5} = \frac{27}{20}$.

Same sign: positive quotient ■

✓ CHECK POINT 6 Divide:

a. $\frac{32}{-4}$ **b.** $-\frac{2}{3} \div \left(-\frac{5}{4}\right)$.

We must be careful with division when 0 is involved. Zero divided by any nonzero real number is 0. For example,

$$\frac{0}{-5} = 0.$$

Can you see why $\frac{0}{-5}$ must be 0? The definition of division tells us that

$$\frac{0}{-5} = 0 \cdot \left(-\frac{1}{5}\right)$$

and the product of 0 and any real number is 0. By contrast, what happens if we divide −5 by 0. The answer must be a number that, when multiplied by 0, gives −5. However, any number multiplied by 0 is 0. Thus, we cannot divide −5, or any other real number, by 0.

Division by Zero

Division by zero is not allowed; it is undefined. A real number can never have a denominator of 0.

8 Use the order of operations.

Order of Operations

The rules for order of operations can be applied to positive and negative real numbers. Recall that if no grouping symbols are present, we

- Evaluate exponential expressions.
- Multiply and divide, from left to right.
- Add and subtract, from left to right.

EXAMPLE 7 Using the Order of Operations

Simplify: $4 - 7^2 + 8 \div 2(-3)^2$.

Solution

$$
\begin{aligned}
&4 - 7^2 + 8 \div 2(-3)^2 \\
&= 4 - 49 + 8 \div 2(9) && \text{Evaluate exponential expressions: } 7^2 = 7 \cdot 7 = 49 \text{ and } (-3)^2 = (-3)(-3) = 9. \\
&= 4 - 49 + 4(9) && \text{Divide: } 8 \div 2 = 4. \\
&= 4 - 49 + 36 && \text{Multiply: } 4(9) = 36. \\
&= -45 + 36 && \text{Subtract: } 4 - 49 = 4 + (-49) = -45. \\
&= -9 && \text{Add.}
\end{aligned}
$$

✓ CHECK POINT 7 Simplify: $3 - 5^2 + 12 \div 2(-4)^2$.

If an expression contains grouping symbols, we perform operations within these symbols first. Common grouping symbols are parentheses, brackets, and braces. Other grouping symbols include fraction bars, absolute value symbols, and radical symbols, such as square root signs ($\sqrt{\ }$).

EXAMPLE 8 Using the Order of Operations

Simplify: $\dfrac{13 - 3(-2)^4}{3 - (6 - 10)}$.

Solution Simplify the numerator and the denominator separately. Then divide.

$$
\begin{aligned}
&\frac{13 - 3(-2)^4}{3 - (6 - 10)} \\
&= \frac{13 - 3(16)}{3 - (-4)} && \text{Evaluate the exponential expression in the numerator: } (-2)^4 = (-2)(-2)(-2)(-2) = 16. \\
& && \text{Subtract inside parentheses in the denominator: } 6 - 10 = 6 + (-10) = -4. \\
&= \frac{13 - 48}{7} && \text{Multiply in the numerator: } 3(16) = 48. \\
& && \text{Subtract in the denominator: } 3 - (-4) = 3 + 4 = 7. \\
&= \frac{-35}{7} && \text{Subtract in the numerator: } 13 - 48 = 13 + (-48) = -35. \\
&= -5 && \text{Divide.}
\end{aligned}
$$

✓ CHECK POINT 8 Simplify: $\dfrac{4 + 3(-2)^3}{2 - (6 - 9)}$.

9 Use commutative, associative, and distributive properties.

The Commutative, Associative, and Distributive Properties

Basic algebraic properties enable us to write *equivalent algebraic expressions.* Two algebraic expressions that have the same value for all replacements are called **equivalent algebraic expressions.** In Section 1.4, you will use such expressions to solve equations.

In arithmetic, when two numbers are added or multiplied, the order in which the numbers are written does not affect the answer. These facts are called **commutative properties.**

The Commutative Properties

Let a and b represent real numbers, variables, or algebraic expressions.

Addition: $a + b = b + a$

Multiplication: $ab = ba$

Changing order when adding or multiplying does not affect a sum or product.

EXAMPLE 9 Using the Commutative Properties

Write an algebraic expression equivalent to $3x + 7$ using each of the commutative properties.

Solution

Commutative of Addition

$$3x + 7 = 7 + 3x$$

Change the order of the addition.

Commutative of Multiplication

$$3x + 7 = x \cdot 3 + 7$$

Change the order of the multiplication. ■

✓ CHECK POINT 9 Write an algebraic expression equivalent to $4x + 9$ using each of the commutative properties.

The **associative properties** enable us to form equivalent expressions by regrouping.

The Associative Properties

Let a, b, and c represent real numbers, variables, or algebraic expressions.

Addition: $(a + b) + c = a + (b + c)$

Multiplication: $(ab)c = a(bc)$

Changing grouping when adding or multiplying does not affect a sum or product.

Blitzer Bonus

The Associative Property and the English Language

In the English language, phrases can take on different meanings depending on the way the words are associated with commas. Here are three examples.

- Woman, without her man, is nothing.
  Woman, without her, man is nothing.
- What's the latest dope?
  What's the latest, dope?
- Population of Amsterdam broken down by age and sex
  Population of Amsterdam, broken down by age and sex

EXAMPLE 10 Using the Associative Properties

Use an associative property to write an equivalent expression and simplify:

a. $7 + (3 + x)$ **b.** $-6(5x)$.

Solution

a. $7 + (3 + x) = (7 + 3) + x = 10 + x$

b. $-6(5x) = (-6 \cdot 5)x = -30x$ ■

✓ CHECK POINT 10 Use an associative property to write an equivalent expression and simplify:

a. $6 + (12 + x)$ **b.** $-7(4x)$.

The **distributive property** allows us to rewrite the product of a number and a sum as the sum of two products.

The Distributive Property

Let a, b, and c represent real numbers, variables, or algebraic expressions.

$$a(b + c) = ab + ac$$

Multiplication distributes over addition.

Great Question!

What's the most important thing I should keep in mind when using the distributive property?

When using a distributive property to remove parentheses, be sure to multiply *each term* inside the parentheses by the factor outside.

Incorrect

$5(3y + 7) = 5 \cdot 3y + 7$
$= 15y + 7$

EXAMPLE 11 Using the Distributive Property

Use the distributive property to write an equivalent expression:

$$-2(3x + 5).$$

Solution

$$-2(3x + 5) = -2 \cdot 3x + (-2) \cdot 5 = -6x + (-10) = -6x - 10 \quad \blacksquare$$

☑ **CHECK POINT 11** Use the distributive property to write an equivalent expression: $-4(7x + 2)$.

Table 1.4 shows a number of other forms of the distributive property.

Table 1.4 Other Forms of the Distributive Property

Property	Meaning	Example
$a(b - c) = ab - ac$	Multiplication distributes over subtraction.	$5(4x - 3) = 5 \cdot 4x - 5 \cdot 3$ $= 20x - 15$
$a(b + c + d) = ab + ac + ad$	Multiplication distributes over three or more terms in parentheses.	$4(x + 10 + 3y)$ $= 4x + 4 \cdot 10 + 4 \cdot 3y$ $= 4x + 40 + 12y$
$(b + c)a = ba + ca$	Multiplication on the right distributes over addition (or subtraction).	$(x + 7)9 = x \cdot 9 + 7 \cdot 9$ $= 9x + 63$

10 Simplify algebraic expressions.

Combining Like Terms and Simplifying Algebraic Expressions

The **terms** of an algebraic expression are those parts that are separated by addition. For example, consider the algebraic expression

$$7x - 9y + z - 3,$$

which can be expressed as

$$7x + (-9y) + z + (-3).$$

This expression contains four terms, namely $7x$, $-9y$, z, and -3.

The numerical part of a term is called its **coefficient**. In the term $7x$, the 7 is the coefficient. If a term containing one or more variables is written without a coefficient, the coefficient is understood to be 1. Thus, z means $1z$. If a term is a constant, its coefficient is that constant. Thus, the coefficient of the constant term -3 is -3.

$$7x + (-9y) + z + (-3)$$

Coefficient is 7. Coefficient is −9. Coefficient is 1; z means $1z$. Coefficient is −3.

The parts of each term that are multiplied are called the **factors** of the term. The factors of the term $7x$ are 7 and x.

Like terms are terms that have exactly the same variable factors. For example, $3x$ and $7x$ are like terms. The distributive property in the form

$$ba + ca = (b + c)a$$

enables us to add or subtract like terms. For example,

$$3x + 7x = (3 + 7)x = 10x$$
$$7y^2 - y^2 = 7y^2 - 1y^2 = (7 - 1)y^2 = 6y^2.$$

This process is called **combining like terms.**

An algebraic expression is **simplified** when grouping symbols have been removed and like terms have been combined.

Great Question!

What is the bottom line for combining like terms?

To combine like terms mentally, add or subtract the coefficients of the terms. Use this result as the coefficient of the terms' variable factor(s).

EXAMPLE 12 Simplifying an Algebraic Expression

Simplify: $7x + 12x^2 + 3x + x^2$.

Solution

$7x + 12x^2 + 3x + x^2$

$= (7x + 3x) + (12x^2 + x^2)$ Rearrange terms and group like terms using commutative and associative properties. This step is often done mentally.

$= (7 + 3)x + (12 + 1)x^2$ Apply the distributive property.

$= 10x + 13x^2$ Simplify. Because $10x$ and $13x^2$ are not like terms, this is the final answer.

Using the commutative property of addition, we can write this simplified expression as $13x^2 + 10x$. ■

CHECK POINT 12 Simplify: $3x + 14x^2 + 11x + x^2$.

EXAMPLE 13 Simplifying an Algebraic Expression

Simplify: $4(7x - 3) - 10x$.

Solution

$4(7x - 3) - 10x$

$= 4 \cdot 7x - 4 \cdot 3 - 10x$ Use the distributive property to remove the parentheses.

$= 28x - 12 - 10x$ Multiply.

$= (28x - 10x) - 12$ Group like terms.

$= (28 - 10)x - 12$ Apply the distributive property.

$= 18x - 12$ Simplify. ■

CHECK POINT 13 Simplify: $8(2x - 5) - 4x$.

It is not uncommon to see algebraic expressions with parentheses preceded by a negative sign or subtraction. An expression of the form $-(b + c)$ can be simplified as follows:

$$-(b + c) = -1(b + c) = (-1)b + (-1)c = -b + (-c) = -b - c.$$

Do you see a fast way to obtain the simplified expression on the right? **If a negative sign or a subtraction symbol appears outside parentheses, drop the parentheses and change the sign of every term within the parentheses.** For example,

$$-(3x^2 - 7x - 4) = -3x^2 + 7x + 4.$$

EXAMPLE 14 Simplifying an Algebraic Expression

Simplify: $8x + 2[5 - (x - 3)]$.

Solution

$8x + 2[5 - (x - 3)]$

$= 8x + 2[5 - x + 3]$ Drop parentheses and change the sign of each term in parentheses: $-(x - 3) = -x + 3$.

$= 8x + 2[8 - x]$ Simplify inside brackets: $5 + 3 = 8$.

$= 8x + 16 - 2x$ Apply the distributive property: $2[8 - x] = 2 \cdot 8 - 2x = 16 - 2x$.

$= (8x - 2x) + 16$ Group like terms.

$= (8 - 2)x + 16$ Apply the distributive property.

$= 6x + 16$ Simplify. ■

✓ CHECK POINT 14 Simplify: $6 + 4[7 - (x - 2)]$.

CONCEPT AND VOCABULARY CHECK

In items 1–7, state whether each result is a positive number, a negative number, or 0. Do not actually perform the computation.

1. $-16 + (-30)$ ____________
2. $-16 + 16$ ____________
3. $-(-16)$ ____________
4. $|-16|$ ____________
5. $-16 - (-30)$ ____________
6. $(-16)^3$ ____________
7. $(-16)(-30)$ ____________

In items 8–9, use the choices below to fill in each blank:

add subtract multiply divide.

8. To simplify $4 - 49 + 8 \div 4 \cdot 2$, first ____________.
9. To simplify $15 - [3 - (-1)] + 12 \div 2 \cdot 3$, first ____________.

In the remaining items, fill in each blank so that the resulting statement is true.

10. $|a|$, called the ____________ of a, represents the distance from ____________ to ____________ on the number line.
11. If $a \geq 0$, then $|a| =$ ____________. If $a < 0$, then $|a| =$ ____________.
12. $a + (-a) =$ ____________: The sum of a real number and its additive ____________ is ____________, the additive ____________.
13. If a and b are real numbers, the commutative property of addition states that $a + b =$ ____________.
14. If a, b, and c are real numbers, the associative property of multiplication states that ____________ $= a(bc)$.
15. If a, b, and c are real numbers, the distributive property states that $a(b + c) =$ ____________.
16. An algebraic expression is ____________ when parentheses have been removed and like terms have been combined.

1.2 EXERCISE SET MyMathLab®

Practice Exercises

In Exercises 1–12, find each absolute value.

1. $|-7|$
2. $|-10|$
3. $|4|$
4. $|13|$
5. $|-7.6|$
6. $|-8.3|$
7. $\left|\frac{\pi}{2}\right|$
8. $\left|\frac{\pi}{3}\right|$
9. $|-\sqrt{2}|$
10. $|-\sqrt{3}|$
11. $-\left|-\frac{2}{5}\right|$
12. $-\left|-\frac{7}{10}\right|$

In Exercises 13–28, add as indicated.

13. $-3 + (-8)$
14. $-5 + (-10)$
15. $-14 + 10$
16. $-15 + 6$
17. $-6.8 + 2.3$
18. $-7.9 + 2.4$
19. $\frac{11}{15} + \left(-\frac{3}{5}\right)$
20. $\frac{7}{10} + \left(-\frac{4}{5}\right)$
21. $-\frac{2}{9} - \frac{3}{4}$
22. $-\frac{3}{5} - \frac{4}{7}$
23. $-3.7 + (-4.5)$
24. $-6.2 + (-5.9)$
25. $0 + (-12.4)$
26. $0 + (-15.3)$
27. $12.4 + (-12.4)$
28. $15.3 + (-15.3)$

In Exercises 29–34, find $-x$ for the given value of x.

29. $x = 11$
30. $x = 13$
31. $x = -5$
32. $x = -9$
33. $x = 0$
34. $x = -\sqrt{2}$

In Exercises 35–46, subtract as indicated.

35. $3 - 15$
36. $4 - 20$
37. $8 - (-10)$
38. $7 - (-13)$
39. $-20 - (-5)$
40. $-30 - (-10)$
41. $\frac{1}{4} - \frac{1}{2}$
42. $\frac{1}{10} - \frac{2}{5}$
43. $-2.3 - (-7.8)$
44. $-4.3 - (-8.7)$
45. $0 - (-\sqrt{2})$
46. $0 - (-\sqrt{3})$

In Exercises 47–58, multiply as indicated.

47. $9(-10)$
48. $8(-10)$
49. $(-3)(-11)$
50. $(-7)(-11)$
51. $\frac{15}{13}(-1)$
52. $\frac{11}{13}(-1)$
53. $-\sqrt{2}\cdot 0$
54. $-\sqrt{3}\cdot 0$
55. $(-4)(-2)(-1)$
56. $(-5)(-3)(-2)$
57. $2(-3)(-1)(-2)(-4)$
58. $3(-2)(-1)(-5)(-3)$

In Exercises 59–70, evaluate each exponential expression.

59. $(-10)^2$
60. $(-8)^2$
61. -10^2
62. -8^2
63. $(-2)^3$
64. $(-3)^3$
65. $(-1)^4$
66. $(-4)^4$
67. $(-1)^{33}$
68. $(-1)^{35}$
69. $-\left(-\frac{1}{2}\right)^3$
70. $-\left(-\frac{1}{4}\right)^3$

In Exercises 71–82, divide as indicated or state that the division is undefined.

71. $\frac{12}{-4}$
72. $\frac{30}{-5}$
73. $\frac{-90}{-2}$
74. $\frac{-55}{-5}$
75. $\frac{0}{-4.6}$
76. $\frac{0}{-5.3}$
77. $-\frac{4.6}{0}$
78. $-\frac{5.3}{0}$
79. $-\frac{1}{2} \div \left(-\frac{7}{9}\right)$
80. $-\frac{1}{2} \div \left(-\frac{3}{5}\right)$
81. $6 \div \left(-\frac{2}{5}\right)$
82. $8 \div \left(-\frac{2}{9}\right)$

In Exercises 83–100, use the order of operations to simplify each expression.

83. $4(-5) - 6(-3)$
84. $8(-3) - 5(-6)$
85. $3(-2)^2 - 4(-3)^2$
86. $5(-3)^2 - 2(-2)^2$
87. $8^2 - 16 \div 2^2 \cdot 4 - 3$
88. $10^2 - 100 \div 5^2 \cdot 2 - 3$
89. $\frac{5\cdot 2 - 3^2}{[3^2 - (-2)]^2}$
90. $\frac{10 \div 2 + 3\cdot 4}{(12 - 3\cdot 2)^2}$
91. $8 - 3[-2(2 - 5) - 4(8 - 6)]$
92. $8 - 3[-2(5 - 7) - 5(4 - 2)]$
93. $\frac{2(-2) - 4(-3)}{5 - 8}$
94. $\frac{6(-4) - 5(-3)}{9 - 10}$
95. $\frac{(5 - 6)^2 - 2|3 - 7|}{89 - 3\cdot 5^2}$
96. $\frac{12 \div 3\cdot 5|2^2 + 3^2|}{7 + 3 - 6^2}$
97. $15 - \sqrt{3 - (-1)} + 12 \div 2\cdot 3$
98. $17 - |5 - (-2)| + 12 \div 2\cdot 3$
99. $20 + 1 - \sqrt{10^2 - (5 + 1)^2(-2)}$
100. $24 \div \sqrt{3\cdot(5 - 2)} \div [-1 - (-3)]^2$

In Exercises 101–104, write an algebraic expression equivalent to the given expression using each of the commutative properties.

101. $4x + 10$
102. $5x + 30$
103. $7x - 5$
104. $3x - 7$

In Exercises 105–110, use an associative property to write an algebraic expression equivalent to each expression and simplify.

105. $4 + (6 + x)$

106. $12 + (3 + x)$

107. $-7(3x)$

108. $-10(5x)$

109. $-\frac{1}{3}(-3y)$

110. $-\frac{1}{4}(-4y)$

In Exercises 111–116, use the distributive property to write an equivalent expression.

111. $3(2x + 5)$ **112.** $5(4x + 7)$

113. $-7(2x + 3)$ **114.** $-9(3x + 2)$

115. $-(3x - 6)$ **116.** $-(6x - 3)$

In Exercises 117–130, simplify each algebraic expression.

117. $7x + 5x$ **118.** $8x + 10x$

119. $6x^2 - x^2$ **120.** $9x^2 - x^2$

121. $6x + 10x^2 + 4x + 2x^2$

122. $9x + 5x^2 + 3x + 4x^2$

123. $8(3x - 5) - 6x$

124. $7(4x - 5) - 8x$

125. $5(3y - 2) - (7y + 2)$

126. $4(5y - 3) - (6y + 3)$

127. $7 - 4[3 - (4y - 5)]$

128. $6 - 5[8 - (2y - 4)]$

129. $18x^2 + 4 - [6(x^2 - 2) + 5]$

130. $14x^2 + 5 - [7(x^2 - 2) + 4]$

Practice PLUS

In Exercises 131–138, write each English phrase as an algebraic expression. Then simplify the expression. Let x represent the number.

131. A number decreased by the sum of the number and four

132. A number decreased by the difference between eight and the number

133. Six times the product of negative five and a number

134. Ten times the product of negative four and a number

135. The difference between the product of five and a number and twice the number

136. The difference between the product of six and a number and negative two times the number

137. The difference between eight times a number and six more than three times the number

138. Eight decreased by three times the sum of a number and six

Application Exercises

The number line shows the approval ratings by Americans of five selected countries. Use this information to solve Exercises 139–146.

Approval Ratings by Americans of Selected Countries

50: Very Favorable
0: Neutral
−50: Very Unfavorable

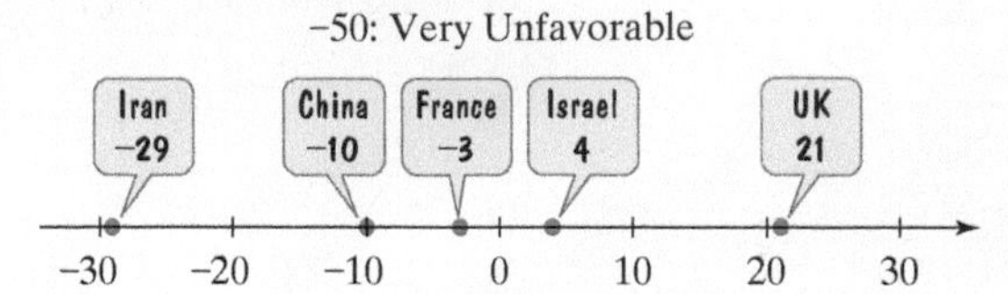

Source: www.thechicagocouncil.org

139. What is the combined approval rating of the UK and Iran?

140. What is the combined approval rating of Israel and China?

141. What is the difference between the approval rating of the UK and the approval rating of Iran?

142. What is the difference between the approval rating of Israel and the approval rating of China?

143. By how much does the approval rating of France exceed the approval rating of China?

144. By how much does the approval rating of France exceed the approval rating of Iran?

145. What is the average of the approval ratings for China, France, and Israel?

146. What is the average of the approval ratings for Iran, China, and the UK?

We opened the section by noting that college students have money to spend, often courtesy of Mom and Dad. The bar graph shows discretionary spending on non-essentials, excluding food and college costs, by full- and part-time college students in the United States for four selected years from 2007 through 2013.

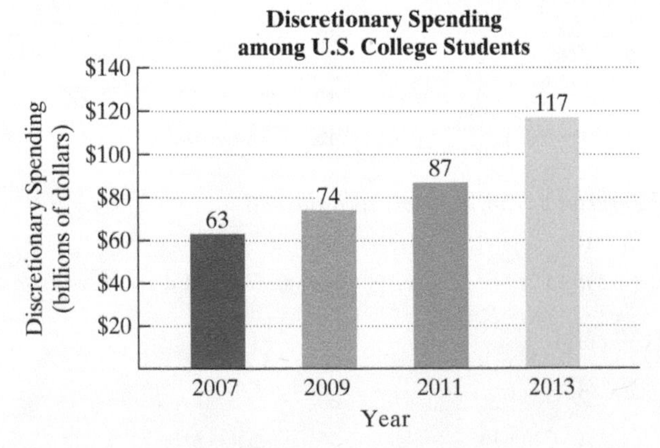

Source: Alloy Media and Marketing

The data in the graph can be modeled by the formula

$$D = 1.2x^2 + 1.6(x + 40),$$

where D represents discretionary spending among U.S. college students, in billions of dollars, x years after 2007. Use this formula to solve Exercises 147–148.

147. According to the formula on the previous page, how much money did college students spend in 2013? Does the model underestimate or overestimate the amount of discretionary spending displayed by the bar graph? By how much?

148. According to the formula on the previous page, how much money did college students spend in 2011? Does the model underestimate or overestimate the amount of discretionary spending displayed by the bar graph? By how much?

149. You had $10,000 to invest. You put x dollars in a safe, government-insured certificate of deposit paying 5% per year. You invested the remainder of the money in noninsured corporate bonds paying 12% per year. Your total interest earned at the end of the year is given by the algebraic expression

$$0.05x + 0.12(10{,}000 - x).$$

a. Simplify the algebraic expression.

b. Use each form of the algebraic expression to determine your total interest earned at the end of the year if you invested $6000 in the safe, government-insured certificate of deposit.

150. It takes you 50 minutes to get to campus. You spend t minutes walking to the bus stop and the rest of the time riding the bus. Your walking rate is 0.06 mile per minute and the bus travels at a rate of 0.5 mile per minute. The total distance walking and traveling by bus is given by the algebraic expression

$$0.06t + 0.5(50 - t).$$

a. Simplify the algebraic expression.

b. Use each form of the algebraic expression to determine the total distance that you travel if you spend 20 minutes walking to the bus stop.

Explaining the Concepts

151. What is the meaning of $|a|$ in terms of a number line?

152. Explain how to add two numbers with the same sign. Give an example with your explanation.

153. Explain how to add two numbers with different signs. Give an example with your explanation.

154. What are opposites, or additive inverses? What happens when finding the sum of a number and its opposite?

155. Explain how to subtract real numbers.

156. Explain how to multiply two numbers with different signs. Give an example with your explanation.

157. Explain how to multiply two numbers with the same sign. Give an example with your explanation.

158. Explain how to determine the sign of a product that involves more than two numbers.

159. Explain how to divide real numbers.

160. Why is $\frac{0}{4} = 0$, although $\frac{4}{0}$ is undefined?

161. What are equivalent algebraic expressions?

162. State a commutative property and give an example of how it is used to write equivalent algebraic expressions.

163. State an associative property and give an example of how it is used to write equivalent algebraic expressions.

164. State a distributive property and give an example of how it is used to write equivalent algebraic expressions.

165. What are the terms of an algebraic expression? How can you tell if terms are like terms?

166. What does it mean to simplify an algebraic expression?

167. If a negative sign appears outside parentheses, explain how to simplify the expression. Give an example.

Critical Thinking Exercises

Make Sense? *In Exercises 168–171, determine whether each statement makes sense or does not make sense, and explain your reasoning.*

168. My mathematical model, although it contains an algebraic expression that is not simplified, describes the data perfectly well, so it will describe the data equally well when simplified.

169. Subtraction actually means the addition of an additive inverse.

170. The terms $13x^2$ and $10x$ both contain the variable x, so I can combine them to obtain $23x^3$.

171. There is no number in front of the term x, so this means that the term has no coefficient.

In Exercises 172–176, determine whether each statement is true or false. If the statement is false, make the necessary change(s) to produce a true statement.

172. $16 \div 4 \cdot 2 = 16 \div 8 = 2$

173. $6 - 2(4 + 3) = 4(4 + 3) = 4(7) = 28$

174. $5 + 3(x - 4) = 8(x - 4) = 8x - 32$

175. $-x - x = -x + (-x) = 0$

176. $x - 0.02(x + 200) = 0.98x - 4$

In Exercises 177–178, insert parentheses to make each statement true.

177. $8 - 2 \cdot 3 - 4 = 14$

178. $2 \cdot 5 - \frac{1}{2} \cdot 10 \cdot 9 = 45$

179. Simplify: $\dfrac{9[4 - (1 + 6)] - (3 - 9)^2}{5 + \dfrac{12}{5 - \dfrac{6}{2 + 1}}}$.

Review Exercises

From here on, each Exercise Set will contain three review exercises. It is important to review previously covered topics to improve your understanding of the topics and to help maintain your mastery of the material. If you are not certain how to solve a review exercise, turn to the section and the worked-out example given in parentheses at the end of each exercise.

180. Write the following English phrase as an algebraic expression: "The quotient of ten and a number, decreased by four times the number." Let x represent the number. (Section 1.1, Example 1)

181. Evaluate $10 + 2(x - 5)^4$ for $x = 7$. (Section 1.1, Example 3)

182. Graph $(-5, \infty)$ on a number line. (Section 1.1, Example 7)

Preview Exercises

Exercises 183–185 will help you prepare for the material covered in the next section.

183. If $y = 4 - x^2$, find the value of y that corresponds to values of x for each integer starting with -3 and ending with 3.

184. If $y = 1 - x^2$, find the value of y that corresponds to values of x for each integer starting with -3 and ending with 3.

185. If $y = |x + 1|$, find the value of y that corresponds to values of x for each integer starting with -4 and ending with 2.

SECTION

1.3 Graphing Equations

What am I supposed to learn?

After studying this section, you should be able to:

1. Plot points in the rectangular coordinate system.
2. Graph equations in the rectangular coordinate system.
3. Use the rectangular coordinate system to visualize relationships between variables.
4. Interpret information about a graphing utility's viewing rectangle or table.

The beginning of the seventeenth century was a time of innovative ideas and enormous intellectual progress in Europe. English theatergoers enjoyed a succession of exciting new plays by Shakespeare. William Harvey proposed the radical notion that the heart was a pump for blood rather than the center of emotion. Galileo, with his new-fangled invention called the telescope, supported the theory of Polish astronomer Copernicus that the sun, not the Earth, was the center of the solar system. Monteverdi was writing the world's first grand operas. French mathematicians Pascal and Fermat invented a new field of mathematics called probability theory.

Into this arena of intellectual electricity stepped French aristocrat René Descartes (1596–1650). Descartes (pronounced "day cart"), propelled by the creativity surrounding him, developed a new branch of mathematics that brought together algebra and geometry in a unified way—a way that visualized numbers as points on a graph, equations as geometric figures, and geometric figures as equations. This new branch of mathematics, called *analytic geometry*, established Descartes as one of the founders of modern thought and among the most original mathematicians and philosophers of any age. We begin this section by looking at Descartes's deceptively simple idea, called the **rectangular coordinate system** or (in his honor) the **Cartesian coordinate system.**

1 Plot points in the rectangular coordinate system.

Points and Ordered Pairs

Descartes used two number lines that intersect at right angles at their zero points, as shown in **Figure 1.9.** The horizontal number line is the ***x*-axis.** The vertical number line is the ***y*-axis.** The point of intersection of these axes is their zero points, called the **origin.**

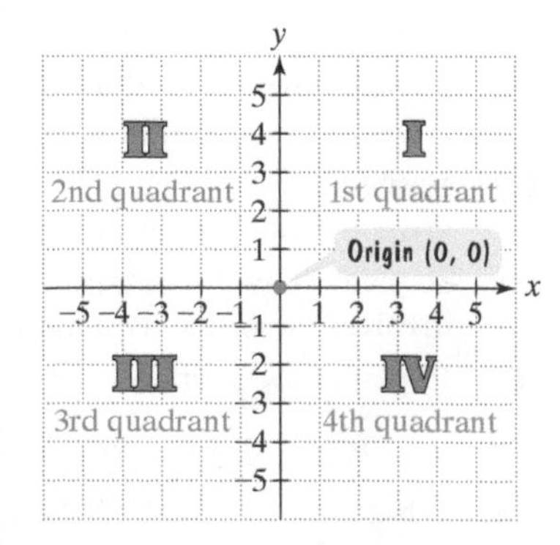

Figure 1.9 The rectangular coordinate system

Great Question!

What's the significance of the word "ordered" when describing a pair of real numbers?

The phrase *ordered pair* is used because order is important. The order in which coordinates appear makes a difference in a point's location. This is illustrated in **Figure 1.10**.

Positive numbers are shown to the right and above the origin. Negative numbers are shown to the left and below the origin. The axes divide the plane into four quarters, called **quadrants**. The points located on the axes are not in any quadrant.

Each point in the rectangular coordinate system corresponds to an **ordered pair** of real numbers, (x, y). Examples of such pairs are $(-5, 3)$ and $(3, -5)$. The first number in each pair, called the **x-coordinate**, denotes the distance and direction from the origin along the x-axis. The second number, called the **y-coordinate**, denotes vertical distance and direction along a line parallel to the y-axis or along the y-axis itself.

Figure 1.10 shows how we **plot**, or locate, the points corresponding to the ordered pairs $(-5, 3)$ and $(3, -5)$. We plot $(-5, 3)$ by going 5 units from 0 to the left along the x-axis. Then we go 3 units up parallel to the y-axis. We plot $(3, -5)$ by going 3 units from 0 to the right along the x-axis and 5 units down parallel to the y-axis. The phrase "the points corresponding to the ordered pairs $(-5, 3)$ and $(3, -5)$" is often abbreviated as "the points $(-5, 3)$ and $(3, -5)$."

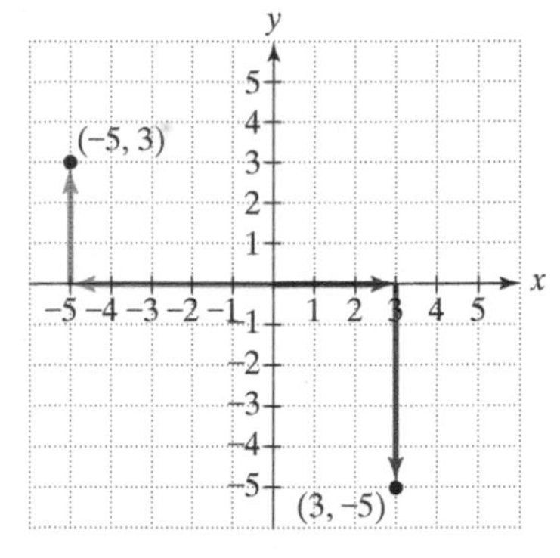

Figure 1.10 Plotting $(-5, 3)$ and $(3, -5)$

Great Question!

Isn't it easy to confuse the notation for an ordered pair and the notation for an open interval? For example, how do I know if $(-4, 5)$ represents a point in the rectangular coordinate system or the set of numbers between, but not including, -4 and 5?

The only way to tell the difference between ordered-pair notation and open-interval notation involves the situation in which the notation is used. If the situation involves plotting points, $(-4, 5)$ is an ordered pair. If the situation involves a set of numbers along a number line, $(-4, 5)$ is an open interval of numbers.

EXAMPLE 1 Plotting Points in the Rectangular Coordinate System

Plot the points: $A(-4, 5)$, $B(3, -4)$, $C(-5, 0)$, $D(-4, -2)$, $E(0, 3.5)$, and $F(0, 0)$.

Solution See **Figure 1.11**. We move from the origin and **plot** the points in the following way:

$A(-4, 5)$: 4 units left, 5 units up

$B(3, -4)$: 3 units right, 4 units down

$C(-5, 0)$: 5 units left, 0 units up or down

$D(-4, -2)$: 4 units left, 2 units down

$E(0, 3.5)$: 0 units right or left, 3.5 units up

$F(0, 0)$: 0 units right or left, 0 units up or down

Notice that the origin is represented by (0, 0).

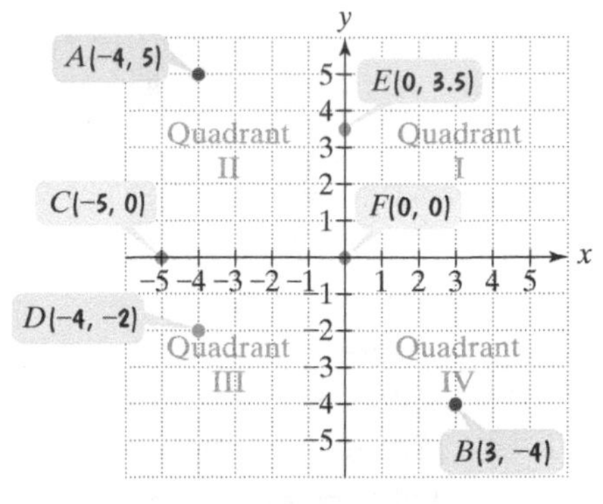

Figure 1.11 Plotting points ■

CHECK POINT 1 Plot the points: $A(2, 5)$, $B(-1, 3)$, $C(-1.5, -4.5)$, and $D(0, -2)$.

2 Graph equations in the rectangular coordinate system.

Graphs of Equations

A relationship between two quantities can be expressed as an **equation in two variables**, such as

$$y = 4 - x^2.$$

A **solution of an equation in two variables**, x and y, is an ordered pair of real numbers with the following property: When the x-coordinate is substituted for x and the y-coordinate is substituted for y in the equation, we obtain a true statement. For example, consider the equation $y = 4 - x^2$ and the ordered pair $(3, -5)$. When 3 is substituted for x and -5 is substituted for y, we obtain the statement $-5 = 4 - 3^2$, or $-5 = 4 - 9$, or $-5 = -5$. Because this statement is true, the ordered pair $(3, -5)$ is a solution of the equation $y = 4 - x^2$. We also say that $(3, -5)$ **satisfies** the equation.

We can generate as many ordered-pair solutions as desired to $y = 4 - x^2$ by substituting numbers for x and then finding the corresponding values for y. For example, suppose we let $x = 3$:

Start with x. Compute y. Form the ordered pair (x, y).

x	$y = 4 - x^2$	Ordered Pair (x, y)
3	$y = 4 - 3^2 = 4 - 9 = -5$	$(3, -5)$

Let $x = 3$. $(3, -5)$ is a solution of $y = 4 - x^2$.

The **graph of an equation in two variables** is the set of all points whose coordinates satisfy the equation. One method for graphing such equations is the **point-plotting method**. First, we find several ordered pairs that are solutions of the equation. Next, we plot these ordered pairs as points in the rectangular coordinate system. Finally, we connect the points with a smooth curve or line. This often gives us a picture of all ordered pairs that satisfy the equation.

EXAMPLE 2 Graphing an Equation Using the Point-Plotting Method

Graph $y = 4 - x^2$. Select integers for x, starting with -3 and ending with 3.

Solution For each value of x, we find the corresponding value for y.

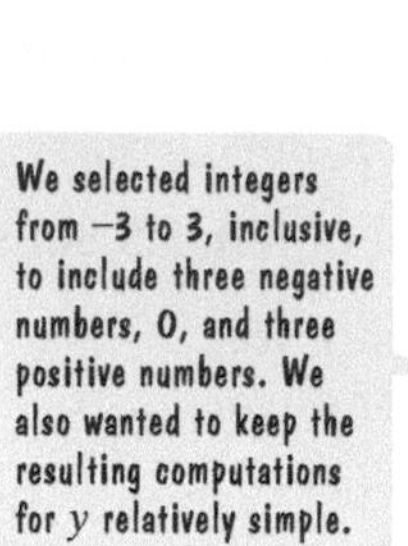

We selected integers from −3 to 3, inclusive, to include three negative numbers, 0, and three positive numbers. We also wanted to keep the resulting computations for y relatively simple.

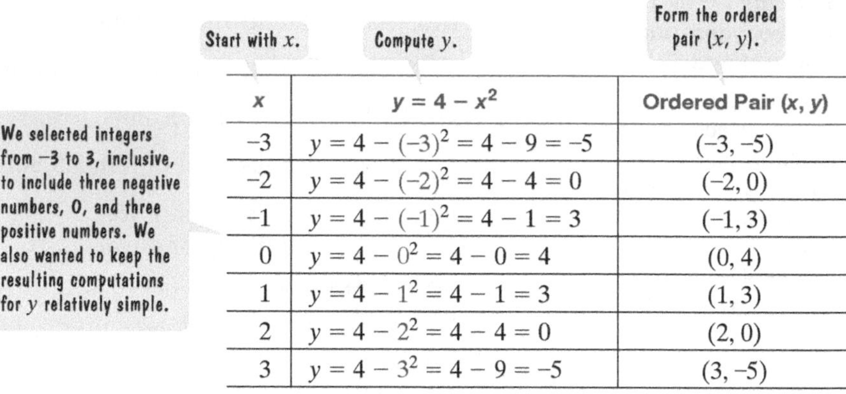

Start with x. Compute y. Form the ordered pair (x, y).

x	$y = 4 - x^2$	Ordered Pair (x, y)
-3	$y = 4 - (-3)^2 = 4 - 9 = -5$	$(-3, -5)$
-2	$y = 4 - (-2)^2 = 4 - 4 = 0$	$(-2, 0)$
-1	$y = 4 - (-1)^2 = 4 - 1 = 3$	$(-1, 3)$
0	$y = 4 - 0^2 = 4 - 0 = 4$	$(0, 4)$
1	$y = 4 - 1^2 = 4 - 1 = 3$	$(1, 3)$
2	$y = 4 - 2^2 = 4 - 4 = 0$	$(2, 0)$
3	$y = 4 - 3^2 = 4 - 9 = -5$	$(3, -5)$

Figure 1.12 The graph of $y = 4 - x^2$

Now we plot the seven points and join them with a smooth curve, as shown in **Figure 1.12**. The graph of $y = 4 - x^2$ is a curve where the part of the graph to the right of the y-axis is a reflection of the part to the left of it and vice versa. The arrows on the left and the right of the curve indicate that it extends indefinitely in both directions. ■

CHECK POINT 2 Graph $y = 1 - x^2$. Select integers for x, starting with -3 and ending with 3.

EXAMPLE 3 Graphing an Equation Using the Point-Plotting Method

Graph $y = |x|$. Select integers for x, starting with -3 and ending with 3.

Solution For each value of x, we find the corresponding value for y.

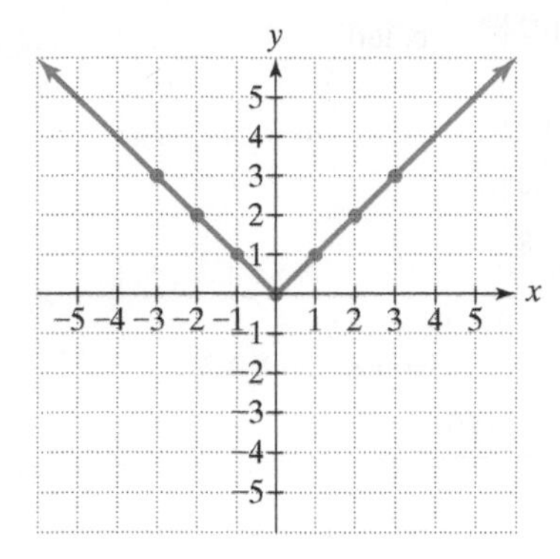

Figure 1.13 The graph of $y = |x|$

x	$y = \|x\|$	Ordered Pair (x, y)
-3	$y = \|-3\| = 3$	$(-3, 3)$
-2	$y = \|-2\| = 2$	$(-2, 2)$
-1	$y = \|-1\| = 1$	$(-1, 1)$
0	$y = \|0\| = 0$	$(0, 0)$
1	$y = \|1\| = 1$	$(1, 1)$
2	$y = \|2\| = 2$	$(2, 2)$
3	$y = \|3\| = 3$	$(3, 3)$

We plot the points and connect them, resulting in the graph shown in **Figure 1.13**. The graph is V-shaped and centered at the origin. For every point (x, y) on the graph, the point $(-x, y)$ is also on the graph. This shows that the absolute value of a positive number is the same as the absolute value of its opposite. ■

✓ **CHECK POINT 3** Graph $y = |x + 1|$. Select integers for x, starting with -4 and ending with 2.

3 Use the rectangular coordinate system to visualize relationships between variables.

Applications

The rectangular coordinate system allows us to visualize relationships between two variables by associating any equation in two variables with a graph. Graphs in the rectangular coordinate system can also be used to tell a story.

EXAMPLE 4 Telling a Story with a Graph

Too late for that flu shot now! It's only 8 A.M. and you're feeling lousy. Fascinated by the way that algebra models the world (your author is projecting a bit here), you construct a graph showing your body temperature from 8 A.M. through 3 P.M. You decide to let x represent the number of hours after 8 A.M. and y your body temperature at time x. The graph is shown in **Figure 1.14**. The symbol ⸸ on the y-axis shows that there is a break in values between 0 and 98. Thus, the first tick mark on the y-axis represents a temperature of 98°F.

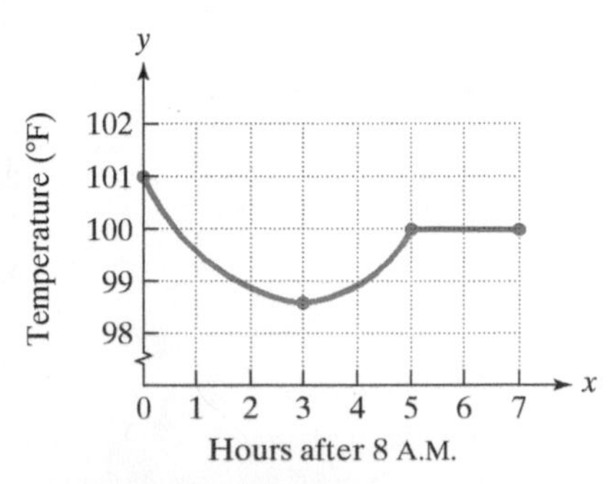

Figure 1.14 Body temperature from 8 A.M. through 3 P.M.

a. What is your temperature at 8 A.M.?

b. During which period of time is your temperature decreasing?

c. Estimate your minimum temperature during the time period shown. How many hours after 8 A.M. does this occur? At what time does this occur?

d. During which period of time is your temperature increasing?

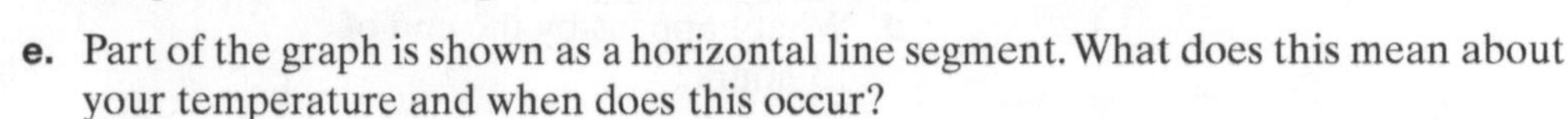

e. Part of the graph is shown as a horizontal line segment. What does this mean about your temperature and when does this occur?

Solution

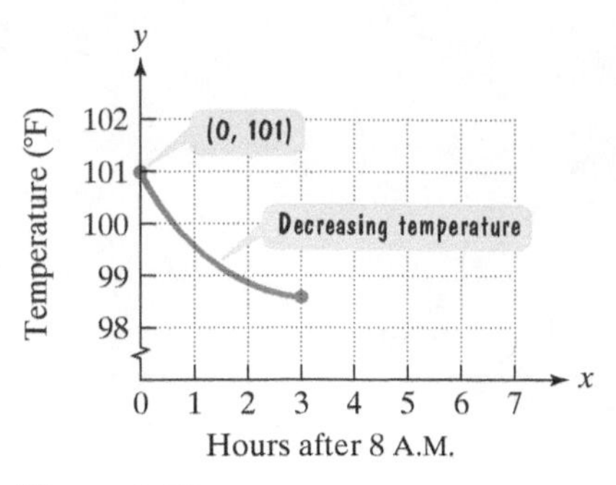

Figure 1.15

a. Because x is the number of hours after 8 A.M., your temperature at 8 A.M. corresponds to $x = 0$. Locate 0 on the horizontal axis and look at the point on the graph above 0. **Figure 1.15** shows that your temperature at 8 A.M. is 101°F.

b. Your temperature is decreasing when the graph falls from left to right. This occurs between $x = 0$ and $x = 3$, also shown in **Figure 1.15**. Because x represents the number of hours after 8 A.M., your temperature is decreasing between 8 A.M. and 11 A.M.

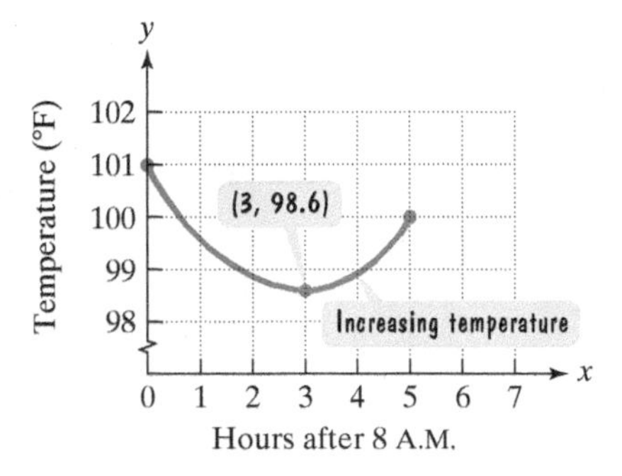

Figure 1.16

c. Your minimum temperature can be found by locating the lowest point on the graph. This point lies above 3 on the horizontal axis, shown in **Figure 1.16**. The y-coordinate of this point falls more than midway between 98 and 99, at approximately 98.6. The lowest point on the graph, (3, 98.6), shows that your minimum temperature, 98.6°F, occurs 3 hours after 8 A.M., at 11 A.M.

d. Your temperature is increasing when the graph rises from left to right. This occurs between $x = 3$ and $x = 5$, shown in **Figure 1.16**. Because x represents the number of hours after 8 A.M., your temperature is increasing between 11 A.M. and 1 P.M.

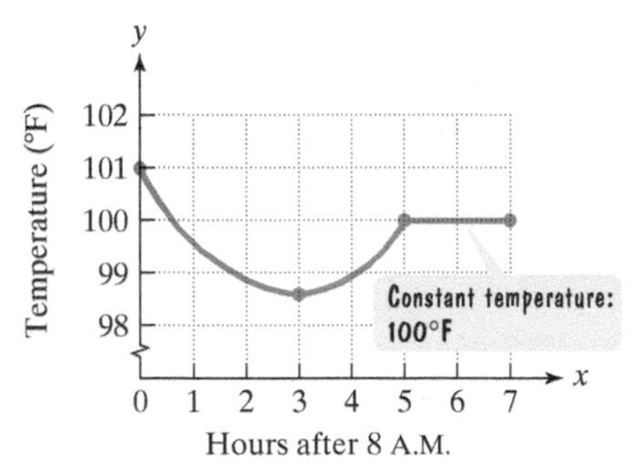

Figure 1.17

e. The horizontal line segment shown in **Figure 1.17** indicates that your temperature is neither increasing nor decreasing. Your temperature remains the same, 100°F, between $x = 5$ and $x = 7$. Thus, your temperature is at a constant 100°F between 1 P.M. and 3 P.M. ■

✓ **CHECK POINT 4** When a physician injects a drug into a patient's muscle, the concentration of the drug in the body, measured in milligrams per 100 milliliters, depends on the time elapsed after the injection, measured in hours. **Figure 1.18** shows the graph of drug concentration over time, where x represents hours after the injection and y represents the drug concentration at time x.

a. During which period of time is the drug concentration increasing?

b. During which period of time is the drug concentration decreasing?

c. What is the drug's maximum concentration and when does this occur?

d. What happens by the end of 13 hours?

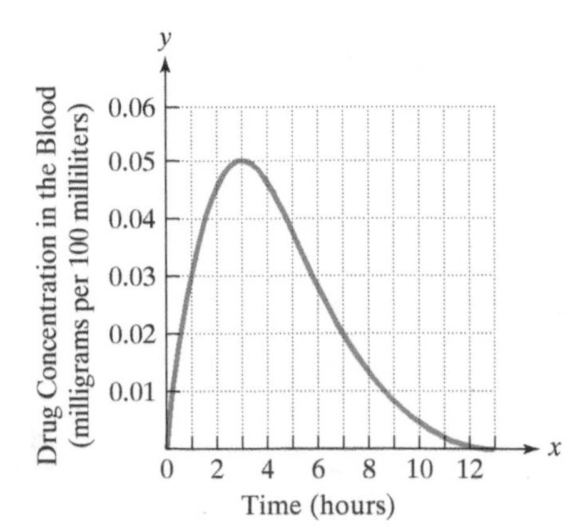

Figure 1.18

Graphing Equations and Creating Tables Using a Graphing Utility

4 Interpret information about a graphing utility's viewing rectangle or table.

Graphing calculators and graphing software packages for computers are referred to as **graphing utilities** or graphers. A graphing utility is a powerful tool that quickly generates the graph of an equation in two variables. **Figures 1.19(a)** and **1.19(b)** show two such graphs for the equations in Examples 2 and 3.

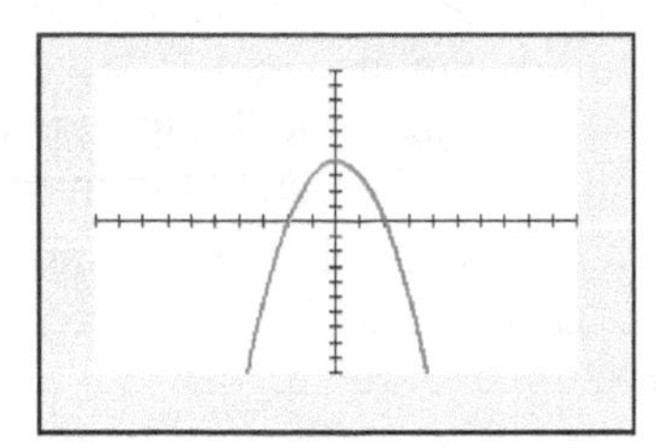

Figure 1.19(a) The graph of $y = 4 - x^2$

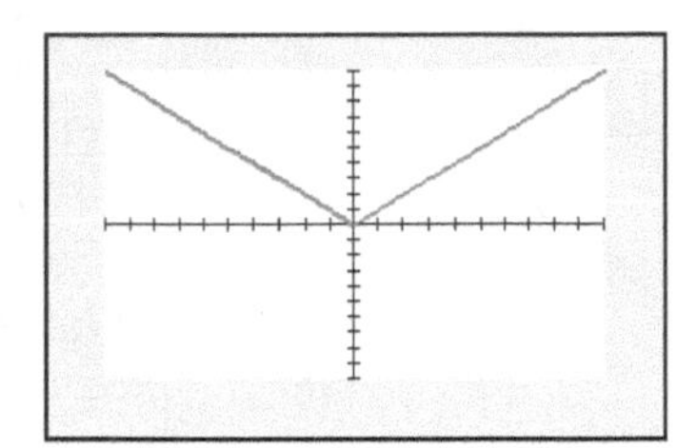

Figure 1.19(b) The graph of $y = |x|$

Great Question!

I'm not using a graphing calculator, so should I just skip to the end of this section?

Even if you are not using a graphing utility in the course, read this part of the section. Knowing about viewing rectangles will enable you to understand the graphs that we display in the technology boxes throughout the book.

What differences do you notice between these graphs and the graphs that we drew by hand? They do seem a bit "jittery." Arrows do not appear on the left and right ends of the graphs. Furthermore, numbers are not given along the axes. For both graphs in **Figure 1.19**, the x-axis extends from -10 to 10 and the y-axis also extends from -10 to 10. The distance represented by each consecutive tick mark is one unit. We say that the **viewing rectangle**, or the **viewing window**, is $[-10, 10, 1]$ by $[-10, 10, 1]$.

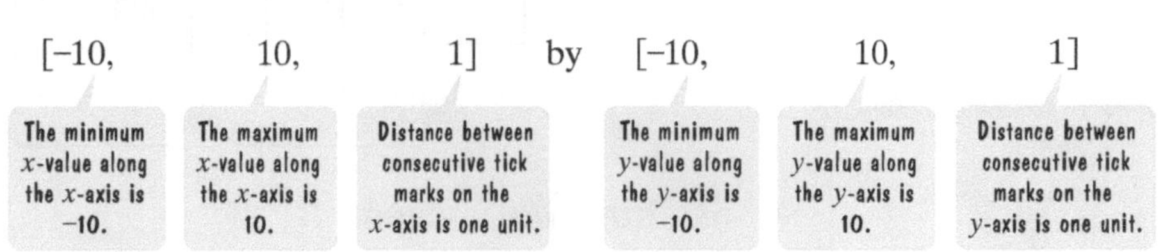

To graph an equation in x and y using a graphing utility, enter the equation and specify the size of the viewing rectangle. The size of the viewing rectangle sets minimum and maximum values for both the x- and y-axes. Enter these values, as well as the values between consecutive tick marks, on the respective axes. The $[-10, 10, 1]$ by $[-10, 10, 1]$ viewing rectangle used in **Figure 1.19** is called the **standard viewing rectangle**.

EXAMPLE 5 Understanding the Viewing Rectangle

What is the meaning of a $[-2, 3, 0.5]$ by $[-10, 20, 5]$ viewing rectangle?

Solution We begin with $[-2, 3, 0.5]$, which describes the x-axis. The minimum x-value is -2 and the maximum x-value is 3. The distance between consecutive tick marks is 0.5.

Next, consider $[-10, 20, 5]$, which describes the y-axis. The minimum y-value is -10 and the maximum y-value is 20. The distance between consecutive tick marks is 5.

Figure 1.20 illustrates a $[-2, 3, 0.5]$ by $[-10, 20, 5]$ viewing rectangle. To make things clearer, we've placed numbers by each tick mark. These numbers do not appear on the axes when you use a graphing utility to graph an equation. ■

Figure 1.20 A $[-2, 3, 0.5]$ by $[-10, 20, 5]$ viewing rectangle

✓ **CHECK POINT 5** What is the meaning of a $[-100, 100, 50]$ by $[-100, 100, 10]$ viewing rectangle? Create a figure like the one in **Figure 1.20** that illustrates this viewing rectangle.

On many graphing utilities, the display screen is five-eights as high as it is wide. By using a square setting, you can equally space the x and y tick marks. (This does not occur in the standard viewing rectangle.) Graphing utilities can also *zoom in* and *zoom out*. When you zoom in, you see a smaller portion of the graph, but you do so in greater detail. When you zoom out, you see a larger portion of the graph. Thus, zooming out may help you to develop a better understanding of the overall character of the graph. With practice, you will become more comfortable with graphing equations in two variables using your graphing utility. You will also develop a better sense of the size of the viewing rectangle that will reveal needed information about a particular graph.

Graphing utilities can also be used to create tables showing solutions of equations in two variables. Use the Table Setup function to choose the starting value of x and to input the increment, or change, between the consecutive x-values. The corresponding y-values are calculated based on the equation(s) in two variables in the Y= screen. In **Figure 1.21**, we used a TI-84 Plus C to create a table for $y = 4 - x^2$ and $y_2 = |x|$, the equations in Examples 2 and 3.

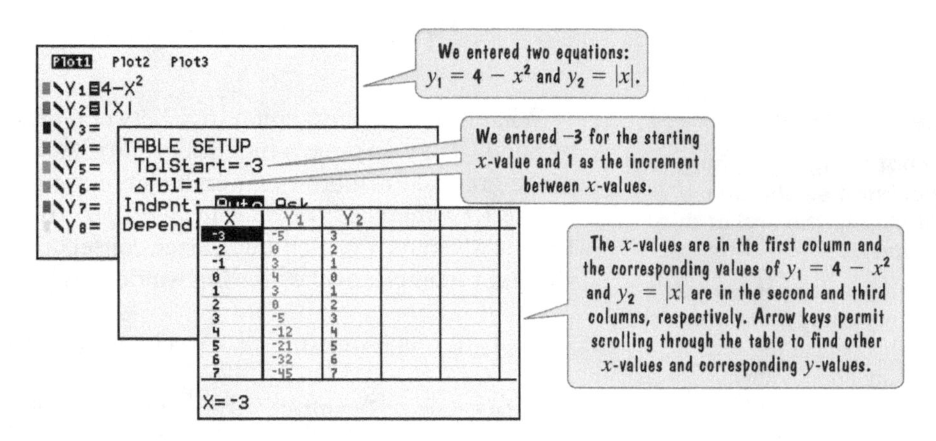

Figure 1.21 Creating a table for $y_1 = 4 - x^2$ and $y_2 = |x|$

Achieving Success

Check! Check! Check!

After completing each Check Point or odd-numbered exercise, compare your answer with the one given in the answer section at the back of the book. To make this process more convenient, place a Post-it® or some other marker at the appropriate page of the answer section. If your answer is different from the one given in the answer section, try to figure out your mistake. Then correct the error. If you cannot determine what went wrong, show your work to your professor. **By recording each step neatly and using as much paper as you need,** your professor will find it easier to determine where you had trouble.

CONCEPT AND VOCABULARY CHECK

Fill in each blank so that the resulting statement is true.

1. In the rectangular coordinate system, the horizontal number line is called the ______________.
2. In the rectangular coordinate system, the vertical number line is called the ______________.
3. In the rectangular coordinate system, the point of intersection of the horizontal axis and the vertical axis is called the ______________.
4. The axes of the rectangular coordinate system divide the plane into regions, called ______________. There are ______________ of these regions.
5. The first number in an ordered pair such as (8, 3) is called the ______________. The second number in such an ordered pair is called the ______________.
6. The ordered pair (4, 19) is a/an ______________ of the equation $y = x^2 + 3$ because when 4 is substituted for x and 19 is substituted for y, we obtain a true statement. We also say that (4, 19) ______________ the equation.

1.3 EXERCISE SET

MyMathLab®

Practice Exercises

In Exercises 1–10, plot the given point in a rectangular coordinate system.

1. $(1, 4)$ **2.** $(2, 5)$

3. $(-2, 3)$ **4.** $(-1, 4)$

5. $(-3, -5)$ **6.** $(-4, -2)$

7. $(4, -1)$ **8.** $(3, -2)$

9. $(-4, 0)$ **10.** $(0, -3)$

Graph each equation in Exercises 11–26. Let $x = -3, -2, -1, 0, 1, 2,$ and 3.

11. $y = x^2 - 4$ **12.** $y = x^2 - 9$

13. $y = x - 2$ **14.** $y = x + 2$

15. $y = 2x + 1$ **16.** $y = 2x - 4$

17. $y = -\frac{1}{2}x$ **18.** $y = -\frac{1}{2}x + 2$

19. $y = |x| + 1$ **20.** $y = |x| - 1$

21. $y = 2|x|$ **22.** $y = -2|x|$

23. $y = -x^2$ **24.** $y = -\frac{1}{2}x^2$

25. $y = x^3$ **26.** $y = x^3 - 1$

In Exercises 27–30, match the viewing rectangle with the correct figure. Then label the tick marks in the figure to illustrate this viewing rectangle.

27. $[-5, 5, 1]$ by $[-5, 5, 1]$

28. $[-10, 10, 2]$ by $[-4, 4, 2]$

29. $[-20, 80, 10]$ by $[-30, 70, 10]$

30. $[-40, 40, 20]$ by $[-1000, 1000, 100]$

a.

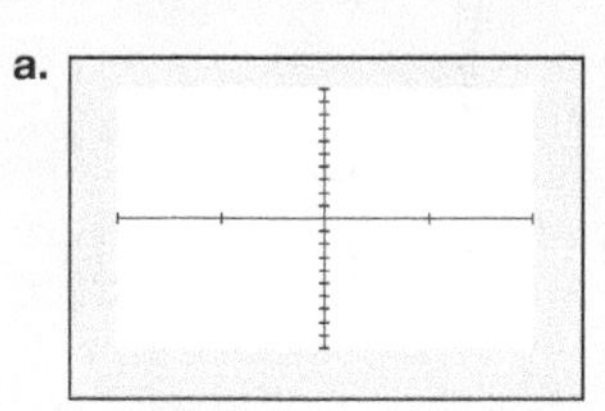

b.

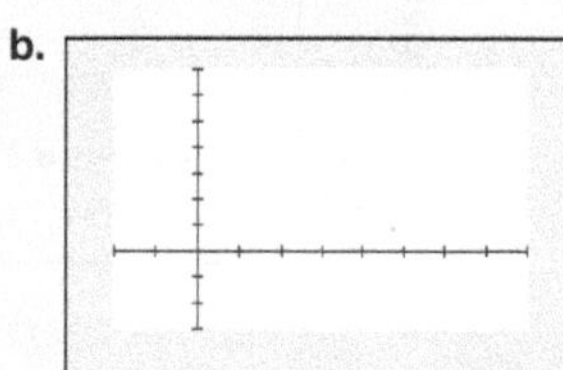

c.

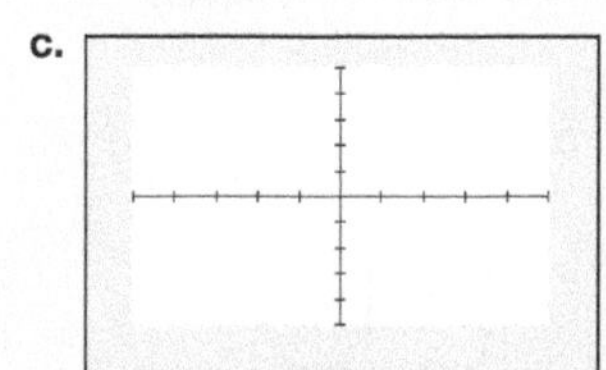

d.

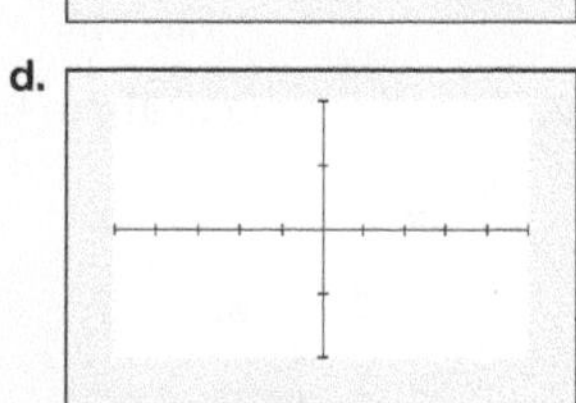

The table of values was generated by a graphing utility with a TABLE feature. Use the table to solve Exercises 31–38.

X	Y1	Y2
-3	9	5
-2	4	4
-1	1	3
0	0	2
1	1	1
2	4	0
3	9	-1
4	16	-2
5	25	-3
6	36	-4
7	49	-5

X=-3

31. Which equation corresponds to Y_2 in the table?

a. $y_2 = x + 8$ **b.** $y_2 = x - 2$
c. $y_2 = 2 - x$ **d.** $y_2 = 1 - 2x$

32. Which equation corresponds to Y_1 in the table?

a. $y_1 = -3x$ **b.** $y_1 = x^2$
c. $y_1 = -x^2$ **d.** $y_1 = 2 - x$

33. Does the graph of Y_2 pass through the origin?

34. Does the graph of Y_1 pass through the origin?

35. At which point does the graph of Y_2 cross the x-axis?

36. At which point does the graph of Y_2 cross the y-axis?

37. At which points do the graphs of Y_1 and Y_2 intersect?

38. For which values of x is $Y_1 = Y_2$?

Practice PLUS

In Exercises 39–42, write each English sentence as an equation in two variables. Then graph the equation.

39. The y-value is four more than twice the x-value.

40. The y-value is the difference between four and twice the x-value.

41. The y-value is three decreased by the square of the x-value.

42. The y-value is two more than the square of the x-value.

In Exercises 43–46, graph each equation.

43. $y = 5$ (Let $x = -3, -2, -1, 0, 1, 2,$ and 3.)

44. $y = -1$ (Let $x = -3, -2, -1, 0, 1, 2,$ and 3.)

45. $y = \frac{1}{x}$ (Let $x = -2, -1, -\frac{1}{2}, -\frac{1}{3}, \frac{1}{3}, \frac{1}{2}, 1,$ and 2.)

46. $y = -\frac{1}{x}$ (Let $x = -2, -1, -\frac{1}{2}, -\frac{1}{3}, \frac{1}{3}, \frac{1}{2}, 1,$ and 2.)

Application Exercises

The line graph shows the percentage of the U.S. population using the Internet from 2007 through 2013. Use the graph to solve Exercises 47–52.

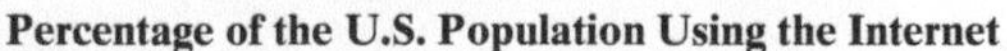

Percentage of the U.S. Population Using the Internet

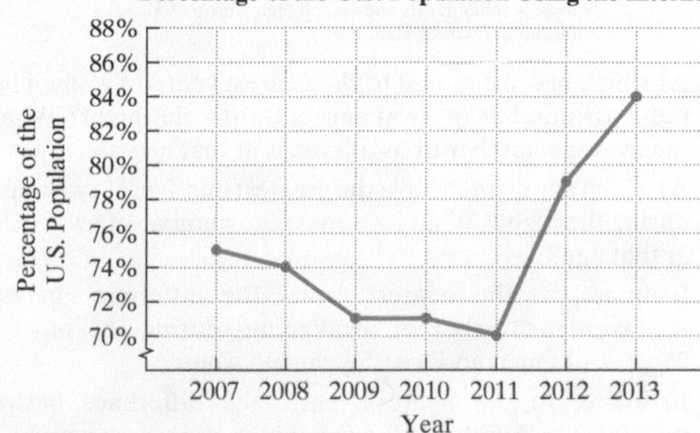

Source: International Telecommunication Union

(In Exercises 47–52, be sure to refer to the graph at the bottom of the previous page.)

47. For the period shown, in which year did the greatest percentage of the U.S. population use the Internet? What percentage of the population used the Internet in that year?

48. For the period shown, in which year did the least percentage of the U.S. population use the Internet? What percentage of the population used the Internet in that year?

49. For the period shown, between which two years did the percentage of the U.S. population using the Internet remain constant? What percentage of the population used the Internet in those years?

50. For the period shown, between which two years did the percentage of the U.S. population using the Internet increase most rapidly? What was the increase in the percentage using the Internet between those years?

51. For the period shown, between which two years did the percentage of the U.S. population using the Internet decrease most rapidly? What was the decrease in the percentage using the Internet between those years?

52. Between 2007 and 2013, what was the increase in the percentage of the U.S. population using the Internet?

Contrary to popular belief, older people do not need less sleep than younger adults. However, the line graphs show that they awaken more often during the night. The numerous awakenings are one reason why some elderly individuals report that sleep is less restful than it had been in the past. Use the line graphs to solve Exercises 53–56.

Average Number of Awakenings During the Night, by Age and Gender

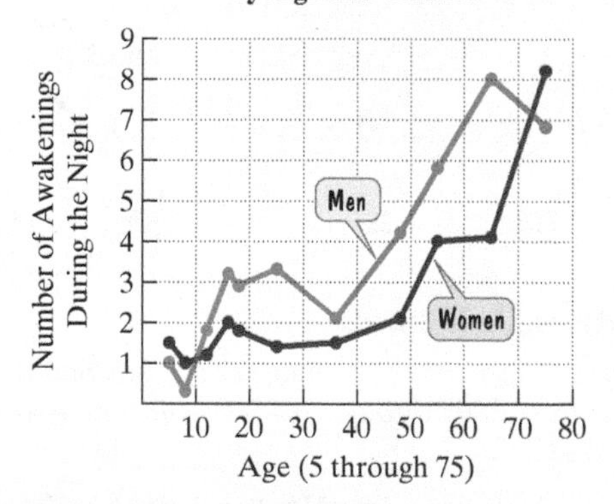

Source: Stephen Davis and Joseph Palladino, *Psychology*, 5th Edition, Prentice Hall, 2007

53. At which age, estimated to the nearest year, do women have the least number of awakenings during the night? What is the average number of awakenings at that age?

54. At which age do men have the greatest number of awakenings during the night? What is the average number of awakenings at that age?

55. Estimate, to the nearest tenth, the difference between the average number of awakenings during the night for 25-year-old men and for 25-year-old women.

56. Estimate, to the nearest tenth, the difference between the average number of awakenings during the night for 18-year-old men and for 18-year-old women.

In Exercises 57–60, match the story with the correct figure. The figures are labeled (a), (b), (c), and (d).

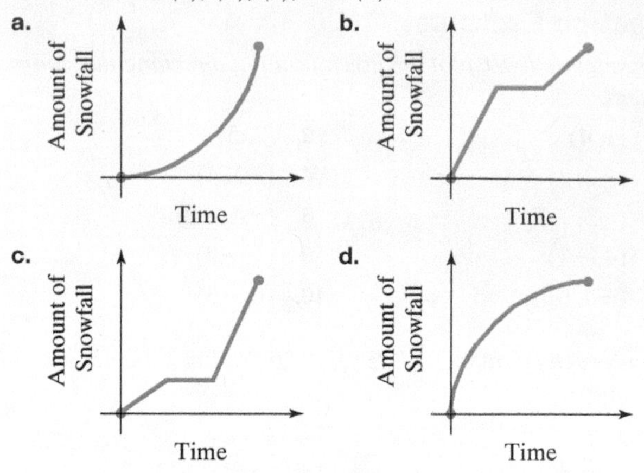

57. As the blizzard got worse, the snow fell harder and harder.

58. The snow fell more and more softly.

59. It snowed hard, but then it stopped. After a short time, the snow started falling softly.

60. It snowed softly, and then it stopped. After a short time, the snow started falling hard.

In Exercises 61–64, select the graph that best illustrates each story.

61. An airplane flew from Miami to San Francisco.

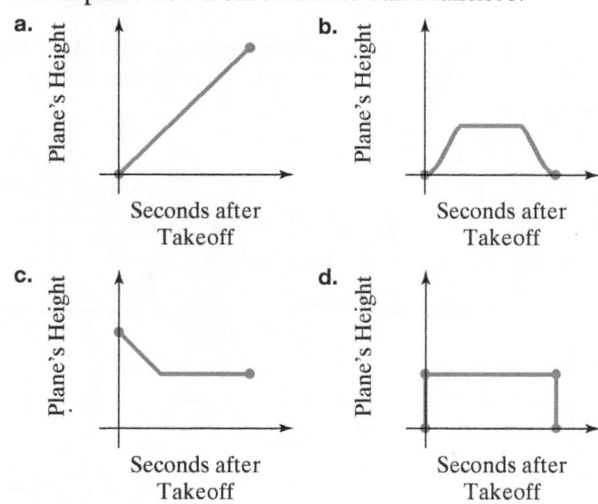

62. At noon, you begin to breathe in.

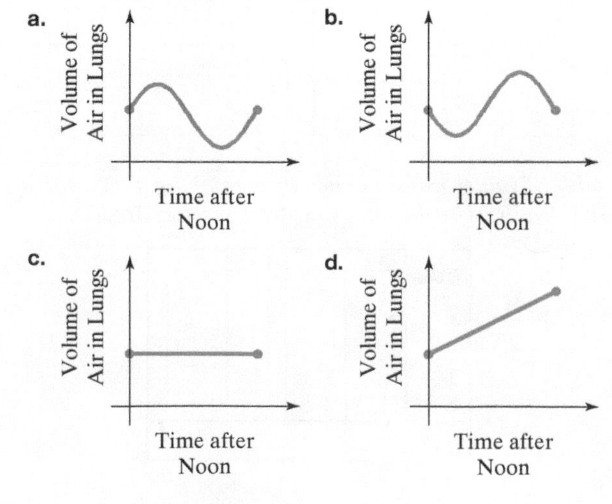

63. Measurements are taken of a person's height from birth to age 100.

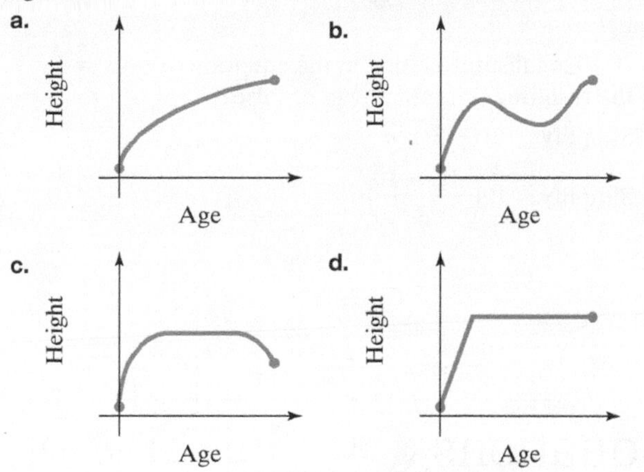

64. You begin your bike ride by riding down a hill. Then you ride up another hill. Finally, you ride along a level surface before coming to a stop.

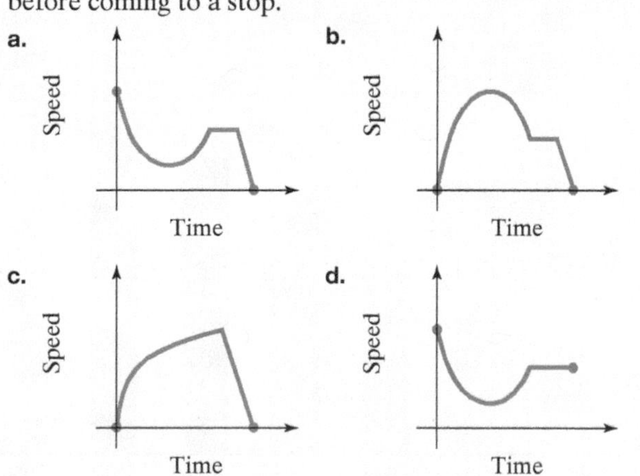

Explaining the Concepts

65. What is the rectangular coordinate system?

66. Explain how to plot a point in the rectangular coordinate system. Give an example with your explanation.

67. Explain why $(5, -2)$ and $(-2, 5)$ do not represent the same point.

68. Explain how to graph an equation in the rectangular coordinate system.

69. What does a $[-20, 2, 1]$ by $[-4, 5, 0.5]$ viewing rectangle mean?

In Exercises 70–71, write a story, or description, to match each title and graph.

70. **Checking Account Balance** 71. **Hair Length**

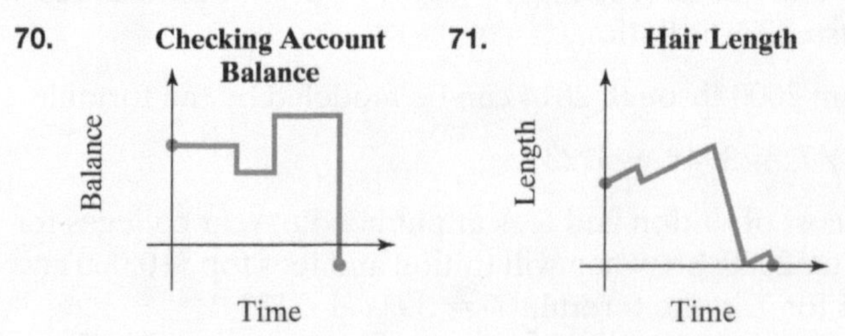

Technology Exercise

72. Use a graphing utility to verify each of your hand-drawn graphs in Exercises 11–26. Experiment with the viewing rectangle to make the graph displayed by the graphing utility resemble your hand-drawn graph as much as possible.

Critical Thinking Exercises

Make Sense? *In Exercises 73–76, determine whether each statement makes sense or does not make sense, and explain your reasoning.*

73. The rectangular coordinate system provides a geometric picture of what an equation in two variables looks like.

74. There is something wrong with my graphing utility because it is not displaying numbers along the x- and y-axes.

75. A horizontal line is not a graph that tells the story of the number of calories that I burn throughout the day.

76. I told my story with a graph, so I can be confident that there is a mathematical model that perfectly describes the graph's data.

In Exercises 77–80, determine whether each statement is true or false. If the statement is false, make the necessary change(s) to produce a true statement.

77. If the product of a point's coordinates is positive, the point must be in quadrant I.

78. If a point is on the x-axis, it is neither up nor down, so $x = 0$.

79. If a point is on the y-axis, its x-coordinate must be 0.

80. The ordered pair $(2, 5)$ satisfies $3y - 2x = -4$.

The graph shows the costs at a parking garage that allows cars to be parked for up to ten hours per day. Closed dots indicate that points belong to the graph and open dots indicate that points are not part of the graph. Use the graph to solve Exercises 81–82.

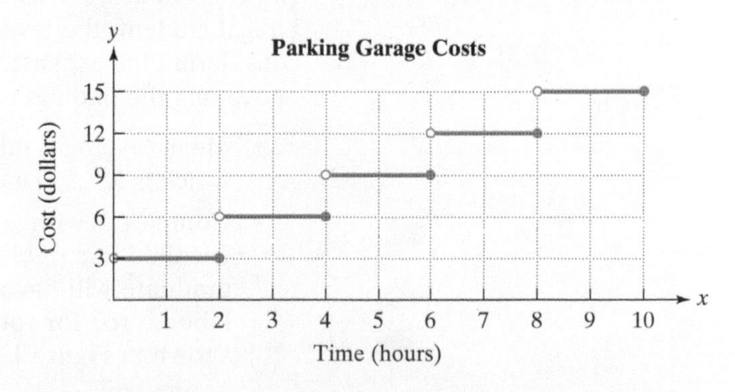

81. You park your car at the garage for four hours on Tuesday and five hours on Wednesday. What are the total parking garage costs for the two days?

82. On Thursday, you paid $12 for parking at the garage. Describe how long your car was parked.

Review Exercises

83. Find the absolute value: $|-14.3|$. (Section 1.2, Example 1)

84. Simplify: $[12 - (13 - 17)] - [9 - (6 - 10)]$. (Section 1.2, Examples 7 and 8)

85. Simplify: $6x - 5(4x + 3) - 10$. (Section 1.2, Example 13)

Preview Exercises

Exercises 86–88 will help you prepare for the material covered in the next section.

86. If -9 is substituted for x in the equation $4x - 3 = 5x + 6$, is the resulting statement true or false?

87. Simplify: $13 - 3(x + 2)$.

88. Simplify: $10\left(\frac{3x + 1}{2}\right)$.

SECTION

1.4 Solving Linear Equations

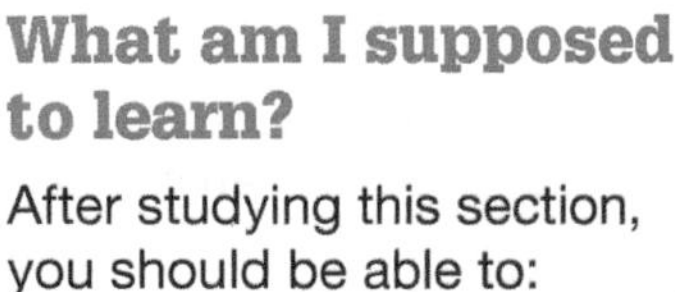

What am I supposed to learn?

After studying this section, you should be able to:

1. Solve linear equations.
2. Recognize identities, conditional equations, and inconsistent equations.
3. Solve applied problems using mathematical models.

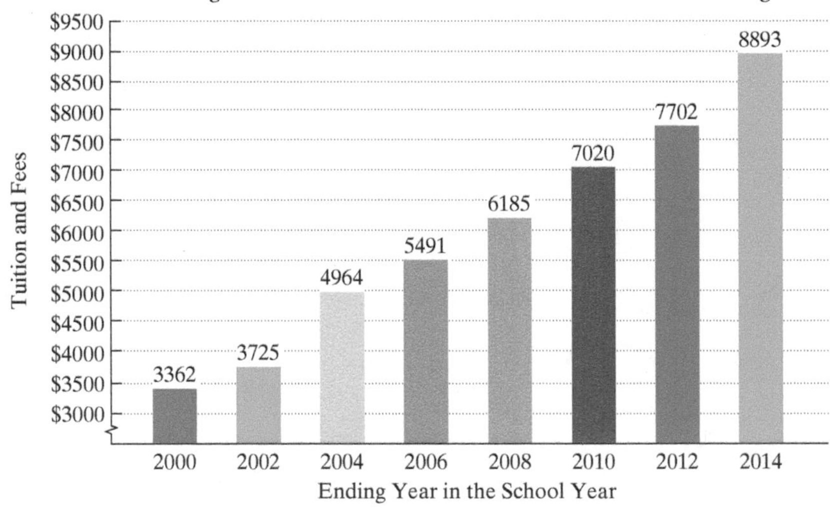

Figure 1.22

Source: The College Board

"Questions have intensified about whether going to college is worthwhile," says *Education Pays*, released by the College Board Advocacy & Policy Center. "For the typical student, the investment pays off very well over the course of a lifetime, even considering the expense."

Among the findings in *Education Pays*:

- Mean (average) full-time earnings with a bachelor's degree in 2012 were \$60,159, which is \$27,529 more than high school graduates.
- Compared with a high school graduate, a four-year college graduate who enrolled in a public university at age 18 will break even by age 33. The college graduate will have earned enough by then to compensate for being out of the labor force for four years and for borrowing enough to pay tuition and fees, shown in **Figure 1.22**, adjusted for inflation.

The rising college costs from 2000 through 2014 can be modeled by the formula

$$T = 394x + 3123,$$

where T represents the average cost of tuition and fees at public four-year colleges for the school year ending x years after 2000. So, when will tuition and fees top \$10,000 and reach \$10,215? Substitute 10,215 for T in the formula $T = 394x + 3123$:

$$10{,}215 = 394x + 3123.$$

Our goal is to determine the value of x, the number of years after 2000, when tuition and fees will reach \$10,215. Notice that the exponent on the variable in $10{,}215 = 394x + 3123$ is 1. In this section, we will study how to determine the value of x in such equations. With this skill, you will be able to use certain mathematical models, such as the model for tuition and fees at public colleges, to project what might occur in the future.

1 Solve linear equations.

Solving Linear Equations in One Variable

We begin with a general definition of a linear equation in one variable.

Definition of a Linear Equation

A **linear equation in one variable x** is an equation that can be written in the form

$$ax + b = 0,$$

where a and b are real numbers, and $a \neq 0$ (a is not equal to 0).

An example of a linear equation in one variable is

$$4x + 12 = 0.$$

Solving an equation in x involves determining all values of x that result in a true statement when substituted into the equation. Such values are **solutions**, or **roots**, of the equation. For example, substitute -3 for x in $4x + 12 = 0$. We obtain

$$4(-3) + 12 = 0, \quad \text{or} \quad -12 + 12 = 0.$$

This simplifies to the true statement $0 = 0$. Thus, -3 is a solution of the equation $4x + 12 = 0$. We also say that -3 **satisfies** the equation $4x + 12 = 0$, because when we substitute -3 for x, a true statement results. The set of all such solutions is called the equation's **solution set**. For example, the solution set of the equation $4x + 12 = 0$ is $\{-3\}$.

Two or more equations that have the same solution set are called **equivalent equations**. For example, the equations

$$4x + 12 = 0 \quad \text{and} \quad 4x = -12 \quad \text{and} \quad x = -3$$

are equivalent equations because the solution set for each is $\{-3\}$. To solve a linear equation in x, we transform the equation into an equivalent equation one or more times. Our final equivalent equation should be of the form

$$x = \text{a number}.$$

The solution set of this equation is the set consisting of the number.

To generate equivalent equations, we will use the following properties:

The Addition and Multiplication Properties of Equality

The Addition Property of Equality

The same real number or algebraic expression may be added to both sides of an equation without changing the equation's solution set.

$$a = b \text{ and } a + c = b + c \text{ are equivalent equations.}$$

The Multiplication Property of Equality

The same nonzero real number may multiply both sides of an equation without changing the equation's solution set.

$$a = b \text{ and } ac = bc \text{ are equivalent equations as long as } c \neq 0.$$

Because subtraction is defined in terms of addition, the addition property also lets us subtract the same number from both sides of an equation without changing the equation's solution set. Similarly, because division is defined in terms of multiplication, the multiplication property of equality can be used to divide both sides of an equation by the same nonzero number to obtain an equivalent equation.

Table 1.5 illustrates how these properties are used to isolate x to obtain an equation of the form x = a number.

Great Question!

Have I solved an equation if I obtain something like $-x = 3$?

No. Your final equivalent equation should not be of the form

$-x$ = a number.

We're not finished. A negative sign should not precede the variable.

Isolate x by multiplying or dividing both sides of this equation by -1.

Table 1.5 Using Properties of Equality to Solve Linear Equations

	Equation	How to Isolate x	Solving the Equation	The Equation's Solution Set
These equations are solved using the Addition Property of Equality.	$x - 3 = 8$	Add 3 to both sides.	$x - 3 + 3 = 8 + 3$ $x = 11$	$\{11\}$
	$x + 7 = -15$	Subtract 7 from both sides.	$x + 7 - 7 = -15 - 7$ $x = -22$	$\{-22\}$
These equations are solved using the Multiplication Property of Equality.	$6x = 30$	Divide both sides by 6 (or multiply both sides by $\frac{1}{6}$).	$\frac{6x}{6} = \frac{30}{6}$ $x = 5$	$\{5\}$
	$\frac{x}{5} = 9$	Multiply both sides by 5.	$5 \cdot \frac{x}{5} = 5 \cdot 9$ $x = 45$	$\{45\}$

EXAMPLE 1 Solving a Linear Equation

Solve and check: $2x + 3 = 17$.

Solution Our goal is to obtain an equivalent equation with x isolated on one side and a number on the other side.

$$2x + 3 = 17 \quad \text{This is the given equation.}$$

$$2x + 3 - 3 = 17 - 3 \quad \text{Subtract 3 from both sides.}$$

$$2x = 14 \quad \text{Simplify.}$$

$$\frac{2x}{2} = \frac{14}{2} \quad \text{Divide both sides by 2.}$$

$$x = 7 \quad \text{Simplify.}$$

Now we check the proposed solution, 7, by replacing x with 7 in the original equation.

$$2x + 3 = 17 \quad \text{This is the original equation.}$$

$$2 \cdot 7 + 3 \stackrel{?}{=} 17 \quad \text{Substitute 7 for } x\text{. The question mark indicates that we do not yet know if the two sides are equal.}$$

$$14 + 3 \stackrel{?}{=} 17 \quad \text{Multiply: } 2 \cdot 7 = 14.$$

$$17 = 17 \quad \text{Add: } 14 + 3 = 17.$$

This statement is true.

Because the check results in a true statement, we conclude that the solution of the given equation is 7, or the solution set is $\{7\}$. ■

CHECK POINT 1 Solve and check: $4x + 5 = 29$.

Great Question!

What are the differences between what I'm supposed to do with algebraic expressions and algebraic equations?

We simplify algebraic expressions. We solve algebraic equations. Notice the differences between the procedures:

Simplifying an Algebraic Expression

Simplify: $3(x - 7) - (5x - 11)$.

This is not an equation. There is no equal sign.

Solution
$$\begin{aligned} &3(x - 7) - (5x - 11) \\ &= 3x - 21 - 5x + 11 \\ &= (3x - 5x) + (-21 + 11) \\ &= -2x + (-10) \\ &= -2x - 10 \end{aligned}$$

Stop! Further simplification is not possible. Avoid the common error of setting $-2x - 10$ equal to 0.

Solving an Algebraic Equation

Solve: $3(x - 7) - (5x - 11) = 14$.

This is an equation. There is an equal sign.

Solution
$$\begin{aligned} 3(x - 7) - (5x - 11) &= 14 \\ 3x - 21 - 5x + 11 &= 14 \\ -2x - 10 &= 14 \\ -2x - 10 + 10 &= 14 + 10 \\ -2x &= 24 \\ \frac{-2x}{-2} &= \frac{24}{-2} \\ x &= -12 \end{aligned}$$

Add 10 to both sides.

Divide both sides by -2.

The solution set is $\{-12\}$.

Here is a step-by-step procedure for solving a linear equation in one variable. Not all of these steps are necessary to solve every equation.

Solving a Linear Equation

1. Simplify the algebraic expression on each side by removing grouping symbols and combining like terms.
2. Collect all the variable terms on one side and all the numbers, or constant terms, on the other side.
3. Isolate the variable and solve.
4. Check the proposed solution in the original equation.

EXAMPLE 2 Solving a Linear Equation

Solve and check: $2x - 7 + x = 3x + 1 + 2x$.

Solution

Step 1. Simplify the algebraic expression on each side.

$$2x - 7 + x = 3x + 1 + 2x$$ This is the given equation.

$$3x - 7 = 5x + 1$$ Combine like terms: $2x + x = 3x$ and $3x + 2x = 5x$.

Step 2. Collect variable terms on one side and constant terms on the other side. We will collect variable terms on the left by subtracting $5x$ from both sides. We will collect the numbers on the right by adding 7 to both sides.

$$3x - 5x - 7 = 5x - 5x + 1$$ Subtract $5x$ from both sides.

$$-2x - 7 = 1$$ Simplify.

$$-2x - 7 + 7 = 1 + 7$$ Add 7 to both sides.

$$-2x = 8$$ Simplify.

Discover for Yourself

Solve the equation in Example 2 by collecting terms with the variable on the right and constant terms on the left. What do you observe?

Step 3. Isolate the variable and solve. We isolate x in $-2x = 8$ by dividing both sides by -2.

$$\frac{-2x}{-2} = \frac{8}{-2} \qquad \text{Divide both sides by } -2.$$

$$x = -4 \qquad \text{Simplify.}$$

Step 4. Check the proposed solution in the original equation. Substitute -4 for x in the original equation.

$$\begin{aligned} 2x - 7 + x &= 3x + 1 + 2x && \text{This is the original equation.} \\ 2(-4) - 7 + (-4) &\stackrel{?}{=} 3(-4) + 1 + 2(-4) && \text{Substitute } -4 \text{ for } x. \\ -8 - 7 + (-4) &\stackrel{?}{=} -12 + 1 + (-8) && \text{Multiply: } 2(-4) = -8,\ 3(-4) = -12, \text{ and } 2(-4) = -8. \\ -15 + (-4) &\stackrel{?}{=} -11 + (-8) && \text{Add or subtract from left to right: } -8 - 7 = -15 \text{ and } -12 + 1 = -11. \\ -19 &= -19 && \text{Add.} \end{aligned}$$

The true statement $-19 = -19$ verifies that -4 is the solution, or the solution set is $\{-4\}$. ■

CHECK POINT 2 Solve and check: $2x - 12 + x = 6x - 4 + 5x$.

EXAMPLE 3 Solving a Linear Equation

Solve and check: $4(2x + 1) - 29 = 3(2x - 5)$.

Solution

Step 1. Simplify the algebraic expression on each side.

$$\begin{aligned} 4(2x + 1) - 29 &= 3(2x - 5) && \text{This is the given equation.} \\ 8x + 4 - 29 &= 6x - 15 && \text{Use the distributive property.} \\ 8x - 25 &= 6x - 15 && \text{Simplify.} \end{aligned}$$

Step 2. Collect variable terms on one side and constant terms on the other side. We will collect the variable terms on the left by subtracting $6x$ from both sides. We will collect the numbers on the right by adding 25 to both sides.

$$\begin{aligned} 8x - 6x - 25 &= 6x - 6x - 15 && \text{Subtract } 6x \text{ from both sides.} \\ 2x - 25 &= -15 && \text{Simplify.} \\ 2x - 25 + 25 &= -15 + 25 && \text{Add 25 to both sides.} \\ 2x &= 10 && \text{Simplify.} \end{aligned}$$

Step 3. Isolate the variable and solve. We isolate x by dividing both sides by 2.

$$\frac{2x}{2} = \frac{10}{2} \qquad \text{Divide both sides by 2.}$$

$$x = 5 \qquad \text{Simplify.}$$

Step 4. Check the proposed solution in the original equation. Substitute 5 for x in the original equation.

$$\begin{aligned} 4(2x + 1) - 29 &= 3(2x - 5) && \text{This is the original equation.} \\ 4(2 \cdot 5 + 1) - 29 &\stackrel{?}{=} 3(2 \cdot 5 - 5) && \text{Substitute 5 for } x. \\ 4(11) - 29 &\stackrel{?}{=} 3(5) && \text{Simplify inside parentheses: } 2 \cdot 5 + 1 = 10 + 1 = 11 \text{ and } 2 \cdot 5 - 5 = 10 - 5 = 5. \\ 44 - 29 &\stackrel{?}{=} 15 && \text{Multiply: } 4(11) = 44 \text{ and } 3(5) = 15. \\ 15 &= 15 && \text{Subtract.} \end{aligned}$$

The true statement $15 = 15$ verifies that 5 is the solution, or the solution set is $\{5\}$. ■

CHECK POINT 3 Solve and check: $2(x-3)-17=13-3(x+2)$.

Using Technology

Numeric and Graphic Connections

In many algebraic situations, technology provides numeric and visual insights into problem solving. For example, you can use a graphing utility to check the solution of a linear equation, giving numeric and geometric meaning to the solution. Enter each side of the equation separately under y_1 and y_2. Then use the table or the graphs to locate the x-value for which the y-values are the same. This x-value is the solution.

Let's verify our work in Example 3 and show that 5 is the solution of

$$4(2x+1)-29=3(2x-5).$$

Enter $y_1 = 4(2x+1) - 29$ in the $\boxed{y=}$ screen.

Enter $y_2 = 3(2x-5)$ in the $\boxed{y=}$ screen.

Numeric Check

Display a table for y_1 and y_2.

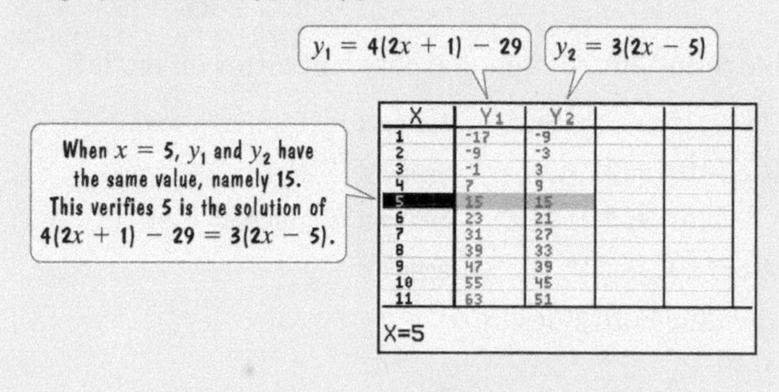

X	Y1	Y2
1	-17	-9
2	-9	-3
3	-1	3
4	7	9
5	15	15
6	23	21
7	31	27
8	39	33
9	47	39
10	55	45
11	63	51

Graphic Check

Display graphs for y_1 and y_2 and use the intersection feature. The solution is the x-coordinate of the intersection point.

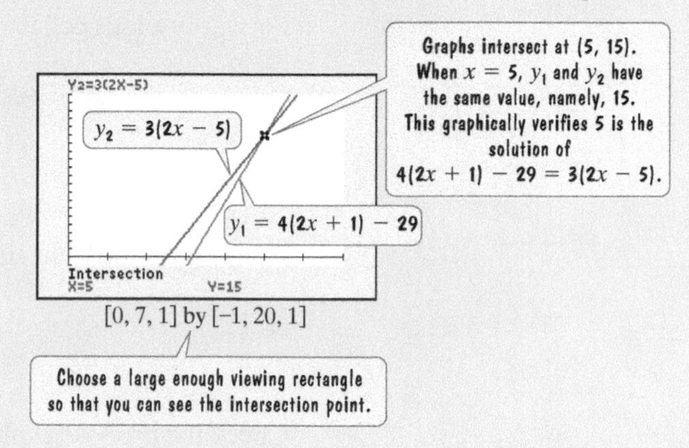

Y2=3(2X-5)
y₂ = 3(2x − 5)
y₁ = 4(2x + 1) − 29
Intersection X=5 Y=15
Graphs intersect at (5, 15). When x = 5, y₁ and y₂ have the same value, namely, 15. This graphically verifies 5 is the solution of 4(2x + 1) − 29 = 3(2x − 5).
Choose a large enough viewing rectangle so that you can see the intersection point.

[0, 7, 1] by [−1, 20, 1]

Linear Equations with Fractions

Equations are easier to solve when they do not contain fractions. How do we remove fractions from an equation? We begin by multiplying both sides of the equation by the least common denominator (LCD) of any fractions in the equation. The least common denominator is the smallest number that all denominators will divide into. Multiplying every term on both sides of the equation by the least common denominator will eliminate the fractions in the equation. Example 4 shows how we "clear an equation of fractions."

EXAMPLE 4 Solving a Linear Equation Involving Fractions

Solve: $\dfrac{2x+5}{5}+\dfrac{x-7}{2}=\dfrac{3x+1}{2}.$

Solution The denominators are 5, 2, and 2. The smallest number that is divisible by 5, 2, and 2 is 10. We begin by multiplying both sides of the equation by 10, the least common denominator.

$$\frac{2x+5}{5}+\frac{x-7}{2}=\frac{3x+1}{2}$$ This is the given equation.

$$10\left(\frac{2x+5}{5}+\frac{x-7}{2}\right)=10\left(\frac{3x+1}{2}\right)$$ Multiply both sides by 10.

$$\frac{10}{1}\cdot\left(\frac{2x+5}{5}\right)+\frac{10}{1}\cdot\left(\frac{x-7}{2}\right)=\frac{10}{1}\cdot\left(\frac{3x+1}{2}\right)$$ Use the distributive property and multiply each term by 10.

$$\frac{\overset{2}{\cancel{10}}}{1}\cdot\left(\frac{2x+5}{\underset{1}{\cancel{5}}}\right)+\frac{\overset{5}{\cancel{10}}}{1}\cdot\left(\frac{x-7}{\underset{1}{\cancel{2}}}\right)=\frac{\overset{5}{\cancel{10}}}{1}\cdot\left(\frac{3x+1}{\underset{1}{\cancel{2}}}\right)$$ Divide out common factors in each multiplication.

$$2(2x+5)+5(x-7)=5(3x+1)$$ The fractions are now cleared.

At this point, we have an equation similar to those we have previously solved. Use the distributive property to begin simplifying each side.

$$4x+10+5x-35=15x+5$$ Use the distributive property.

$$9x-25=15x+5$$ Combine like terms on the left side: $4x+5x=9x$ and $10-35=-25$.

For variety, let's collect variable terms on the right and constant terms on the left.

$$9x-9x-25=15x-9x+5$$ Subtract $9x$ from both sides.

$$-25=6x+5$$ Simplify.

$$-25-5=6x+5-5$$ Subtract 5 from both sides.

$$-30=6x$$ Simplify.

Isolate x on the right side by dividing both sides by 6.

$$\frac{-30}{6}=\frac{6x}{6}$$ Divide both sides by 6.

$$-5=x$$ Simplify.

Check the proposed solution in the original equation. Substitute -5 for x in the original equation. You should obtain $-7=-7$. This true statement verifies that -5 is the solution, or the solution set is $\{-5\}$. ■

CHECK POINT 4 Solve: $\frac{x+5}{7}+\frac{x-3}{4}=\frac{5}{14}$.

---

2 Recognize identities, conditional equations, and inconsistent equations.

Types of Equations

Equations can be placed into categories that depend on their solution sets.

An equation that is true for all real numbers for which both sides are defined is called an **identity**. An example of an identity is

$$x+3=x+2+1.$$

Every number plus 3 is equal to that number plus 2 plus 1. Therefore, the solution set to this equation is the set of all real numbers. This set is written either as

$$\{x\,|\,x \text{ is a real number}\} \quad\text{or}\quad (-\infty,\infty) \quad\text{or}\quad \mathbb{R}.$$

An equation that is not an identity, but that is true for at least one real number, is called a **conditional equation**. The equation $2x+3=17$ is an example of a conditional equation. The equation is not an identity and is true only if x is 7.

Great Question!

What's the bottom line on all the vocabulary associated with types of equations?

If you are concerned by the vocabulary of equation types, keep in mind that there are three possible situations. We can state these situations informally as follows:

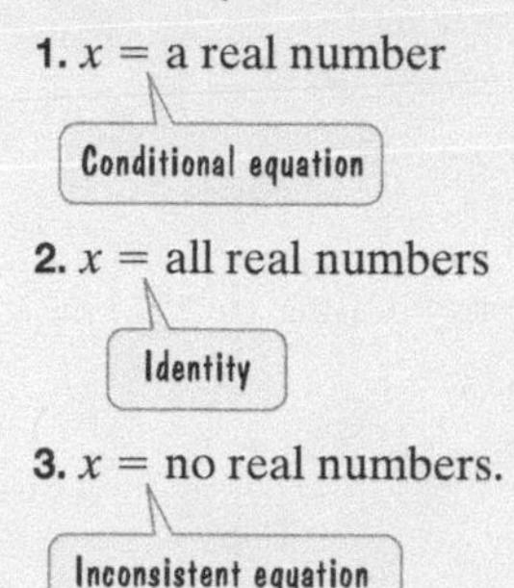

1. x = a real number
Conditional equation
2. x = all real numbers
Identity
3. x = no real numbers.
Inconsistent equation

An **inconsistent equation** is an equation that is not true for even one real number. An example of an inconsistent equation is

$$x = x + 7.$$

There is no number that is equal to itself plus 7. The equation $x = x + 7$ has no solution. Its solution set is written either as

$$\{\ \} \quad \text{or} \quad \varnothing.$$

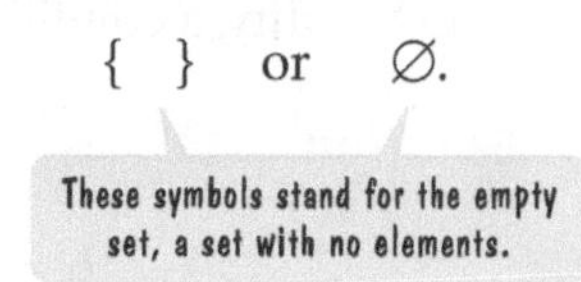

These symbols stand for the empty set, a set with no elements.

If you attempt to solve an identity or an inconsistent equation, you will eliminate the variable. A true statement such as $6 = 6$ or a false statement such as $2 = 3$ will be the result. **If a true statement results, the equation is an identity that is true for all real numbers. If a false statement results, the equation is an inconsistent equation with no solution.**

EXAMPLE 5 Categorizing an Equation

Solve and determine whether the equation

$$2(x + 1) = 2x + 3$$

is an identity, a conditional equation, or an inconsistent equation.

Solution Begin by applying the distributive property on the left side. We obtain

$$2x + 2 = 2x + 3.$$

Does something look strange about $2x + 2 = 2x + 3$? Can doubling a number and increasing the product by 2 give the same result as doubling the same number and increasing the product by 3? No. Let's continue solving the equation by subtracting $2x$ from both sides of $2x + 2 = 2x + 3$.

$$2x - 2x + 2 = 2x - 2x + 3$$

$$2 = 3$$

Keep reading. $2 = 3$ is not the solution.

The original equation is equivalent to the statement $2 = 3$, which is false for every value of x. The equation is inconsistent and has no solution. You can express this by writing "no solution" or using one of the symbols for the empty set, { } or ∅. ■

Using Technology

Graphic Connections

How can technology visually reinforce the fact that the equation

$$2(x + 1) = 2x + 3$$

has no solution? Enter $y_1 = 2(x + 1)$ and $y_2 = 2x + 3$. The graphs of y_1 and y_2 appear to be parallel lines with no intersection point. This supports our conclusion that $2(x + 1) = 2x + 3$ is an inconsistent equation with no solution.

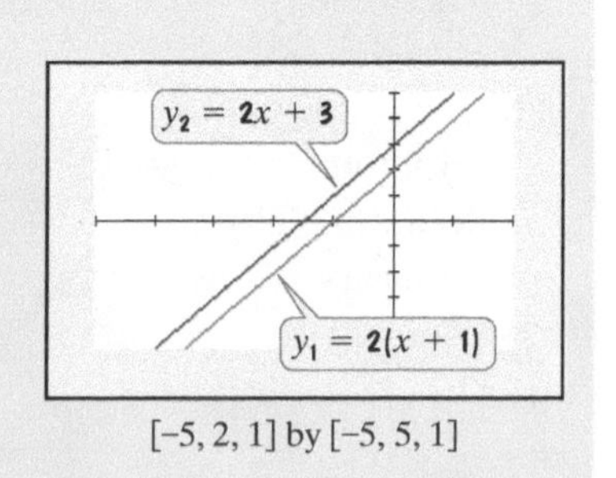

[−5, 2, 1] by [−5, 5, 1]

CHECK POINT 5 Solve and determine whether the equation

$$4x - 7 = 4(x - 1) + 3$$

is an identity, a conditional equation, or an inconsistent equation.

EXAMPLE 6 Categorizing an Equation

Solve and determine whether the equation

$$4x + 6 = 6(x + 1) - 2x$$

is an identity, a conditional equation, or an inconsistent equation.

Solution

$4x + 6 = 6(x + 1) - 2x$	This is the given equation.
$4x + 6 = 6x + 6 - 2x$	Apply the distributive property on the right side.
$4x + 6 = 4x + 6$	Combine like terms on the right side: $6x - 2x = 4x$.

Can you see that the equation $4x + 6 = 4x + 6$ is true for every value of x? Let's continue solving the equation by subtracting $4x$ from both sides.

$$4x - 4x + 6 = 4x - 4x + 6$$

$$6 = 6$$

Keep reading. $6 = 6$ is not the solution.

The original equation is equivalent to the statement $6 = 6$, which is true for every value of x. The equation is an identity, and all real numbers are solutions. You can express this by writing "all real numbers" or using one of the following notations:

$$\{x \mid x \text{ is a real number}\} \quad \text{or} \quad (-\infty, \infty) \quad \text{or} \quad \mathbb{R}. \blacksquare$$

Using Technology

Numeric Connections

A graphing utility's TABLE feature can be used to numerically verify that the solution set of

$$4x + 6 = 6(x + 1) - 2x$$

is the set of all real numbers.

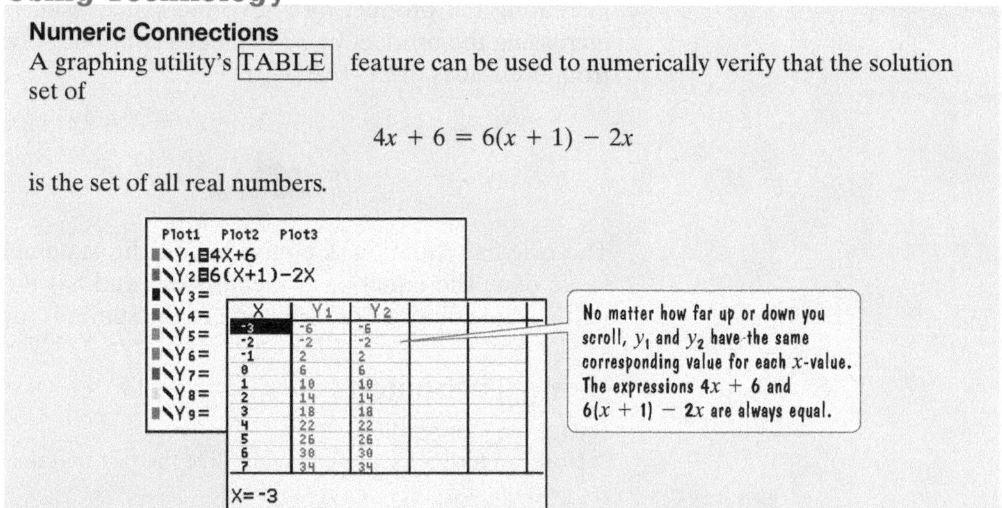

✓ CHECK POINT 6 Solve and determine whether the equation

$$7x + 9 = 9(x + 1) - 2x$$

is an identity, a conditional equation, or an inconsistent equation.

---

3 Solve applied problems using mathematical models.

Applications

We opened both the chapter and this section with a discussion of skyrocketing college costs. In our next example, we use a formula that models these costs. The example shows how the procedure for solving linear equations can be used to find the value of a variable in a mathematical model.

EXAMPLE 7 The Cost of Attending a Public College

The formula

$$T = 394x + 3123$$

models the average cost of tuition and fees, T, at public four-year colleges for the school year ending x years after 2000. When will tuition and fees reach \$10,215?

Solution We are interested in when tuition and fees will reach \$10,215. We substitute 10,215 for T in the formula and solve for x, the number of years after 2000.

$T = 394x + 3123$	This is the given formula.
$10{,}215 = 394x + 3123$	Replace T with 10,215.
$10{,}215 - 3123 = 394x + 3123 - 3123$	Subtract 3123 from both sides.
$7092 = 394x$	Simplify.
$\frac{7092}{394} = \frac{394x}{394}$	Divide both sides by 394.
$18 = x$	Simplify.

The model indicates that for the school year ending 18 years after 2000, or the school year ending 2018, tuition and fees will reach \$10,215. ■

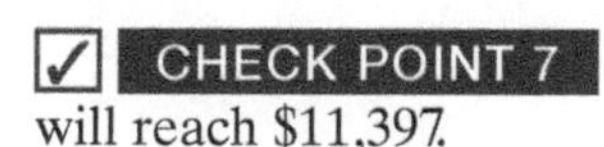

Use the formula in Example 7 to find when tuition and fees will reach \$11,397.

Achieving Success

Because concepts in mathematics build on each other, **it is extremely important that you complete all homework assignments.** This requires more than attempting a few of the assigned exercises. When it comes to assigned homework, you need to do four things and to do these things consistently throughout any math course:

1. Attempt to work every assigned problem.
2. Check your answers.
3. Correct your errors.
4. Ask for help with the problems you have attempted, but do not understand.

Having said this, **don't panic at the length of the Exercise Sets.** You are not expected to work all, or even most, of the problems. Your professor will provide guidance on which exercises to work by assigning those problems that are consistent with the goals and objectives of your course.

CONCEPT AND VOCABULARY CHECK

Fill in each blank so that the resulting statement is true.

1. An equation in the form $ax + b = 0$, $a \neq 0$, such as $3x + 17 = 0$, is called a/an ____________ equation in one variable.
2. Two or more equations that have the same solution set are called ____________ equations.
3. The addition property of equality states that if $a = b$, then $a + c =$ ____________.
4. The multiplication property of equality states that if $a = b$ and $c \neq 0$, then $ac =$ ____________.
5. The first step in solving $7 + 3(x - 2) = 2x + 10$ is to ____________.

6. The equation

$$\frac{x}{4} = 2 + \frac{x-3}{3}$$

can be solved by multiplying both sides by the ____________ of the fractions, which is ____________.

7. In solving an equation, if you eliminate the variable and obtain a false statement such as $2 = 3$, the equation is a/an ____________ equation. The solution set can be expressed using the symbol ____________.

8. In solving an equation, if you eliminate the variable and obtain a true statement such as $6 = 6$, the equation is a/an ____________. The solution set can be expressed using interval notation as ____________.

1.4 EXERCISE SET MyMathLab®

Practice Exercises

In Exercises 1–24, solve and check each linear equation.

1. $5x + 3 = 18$
2. $3x + 8 = 50$
3. $6x - 3 = 63$
4. $5x - 8 = 72$
5. $14 - 5x = -41$
6. $25 - 6x = -83$
7. $11x - (6x - 5) = 40$
8. $5x - (2x - 8) = 35$
9. $2x - 7 = 6 + x$
10. $3x + 5 = 2x + 13$
11. $7x + 4 = x + 16$
12. $8x + 1 = x + 43$
13. $8y - 3 = 11y + 9$
14. $5y - 2 = 9y + 2$
15. $3(x - 2) + 7 = 2(x + 5)$
16. $2(x - 1) + 3 = x - 3(x + 1)$
17. $3(x - 4) - 4(x - 3) = x + 3 - (x - 2)$
18. $2 - (7x + 5) = 13 - 3x$
19. $16 = 3(x - 1) - (x - 7)$
20. $5x - (2x + 2) = x + (3x - 5)$
21. $7(x + 1) = 4[x - (3 - x)]$
22. $2[3x - (4x - 6)] = 5(x - 6)$
23. $\frac{1}{2}(4z + 8) - 16 = -\frac{2}{3}(9z - 12)$
24. $\frac{3}{4}(24 - 8z) - 16 = -\frac{2}{3}(6z - 9)$

In Exercises 25–38, solve each equation.

25. $\frac{x}{3} = \frac{x}{2} - 2$
26. $\frac{x}{5} = \frac{x}{6} + 1$
27. $20 - \frac{x}{3} = \frac{x}{2}$
28. $\frac{x}{5} - \frac{1}{2} = \frac{x}{6}$
29. $\frac{3x}{5} = \frac{2x}{3} + 1$
30. $\frac{x}{2} = \frac{3x}{4} + 5$
31. $\frac{3x}{5} - x = \frac{x}{10} - \frac{5}{2}$
32. $2x - \frac{2x}{7} = \frac{x}{2} + \frac{17}{2}$
33. $\frac{x+3}{6} = \frac{2}{3} + \frac{x-5}{4}$
34. $\frac{x+1}{4} = \frac{1}{6} + \frac{2-x}{3}$
35. $\frac{x}{4} = 2 + \frac{x-3}{3}$
36. $5 + \frac{x-2}{3} = \frac{x+3}{8}$
37. $\frac{x+1}{3} = 5 - \frac{x+2}{7}$
38. $\frac{3x}{5} - \frac{x-3}{2} = \frac{x+2}{3}$

In Exercises 39–50, solve each equation. Then state whether the equation is an identity, a conditional equation, or an inconsistent equation.

39. $5x + 9 = 9(x + 1) - 4x$
40. $4x + 7 = 7(x + 1) - 3x$
41. $3(y + 2) = 7 + 3y$
42. $4(y + 5) = 21 + 4y$
43. $10x + 3 = 8x + 3$
44. $5x + 7 = 2x + 7$
45. $\frac{1}{2}(6z + 20) - 8 = 2(z - 4)$

46. $\frac{1}{3}(6z + 12) = \frac{1}{5}(20z + 30) - 8$

47. $-4x - 3(2 - 2x) = 7 + 2x$

48. $3x - 3(2 - x) = 6(x - 1)$

49. $y + 3(4y + 2) = 6(y + 1) + 5y$

50. $9y - 3(6 - 5y) = y - 2(3y + 9)$

In Exercises 51–54, use the Y= *screen to write the equation being solved. Then use the table to solve the equation.*

51.

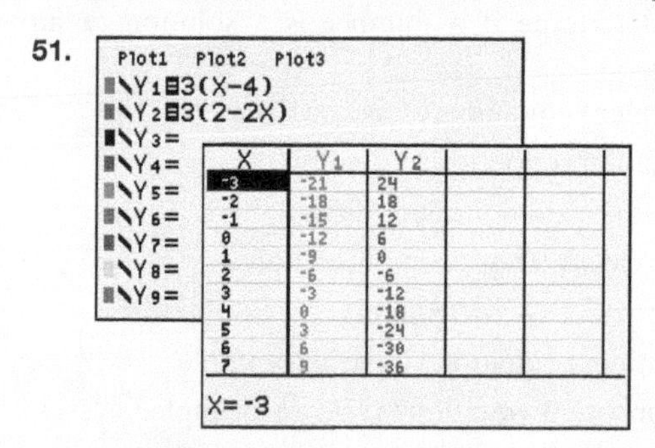

52.

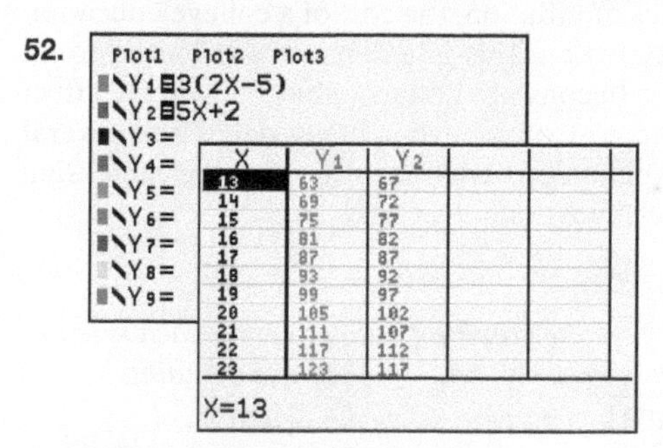

53.

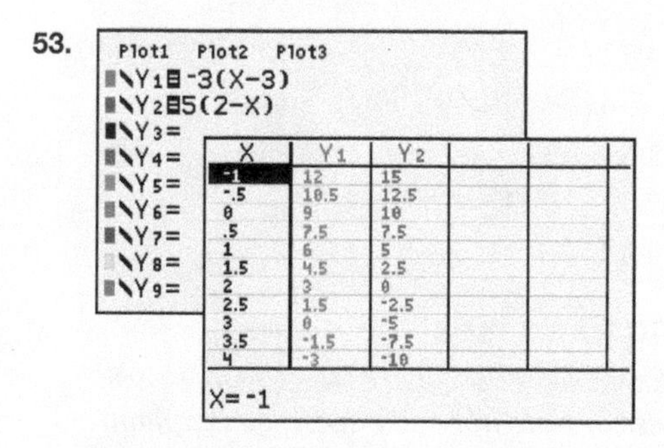

54.

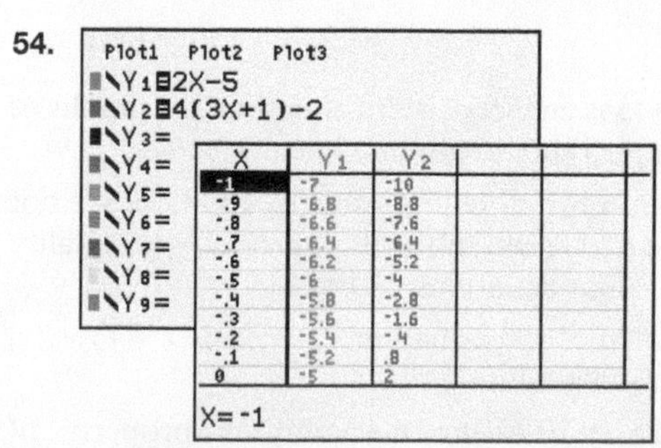

Practice PLUS

55. Evaluate $x^2 - x$ for the value of x satisfying $4(x - 2) + 2 = 4x - 2(2 - x)$.

56. Evaluate $x^2 - x$ for the value of x satisfying $2(x - 6) = 3x + 2(2x - 1)$.

57. Evaluate $x^2 - (xy - y)$ for x satisfying $\frac{3(x + 3)}{5} = 2x + 6$ and y satisfying $-2y - 10 = 5y + 18$.

58. Evaluate $x^2 - (xy - y)$ for x satisfying $\frac{13x - 6}{4} = 5x + 2$ and y satisfying $5 - y = 7(y + 4) + 1$.

In Exercises 59–66, solve each equation.

59. $[(3 + 6)^2 \div 3] \cdot 4 = -54x$

60. $2^3 - [4(5 - 3)^3] = -8x$

61. $5 - 12x = 8 - 7x - [6 \div 3(2 + 5^3) + 5x]$

62. $2(5x + 58) = 10x + 4(21 \div 3.5 - 11)$

63. $0.7x + 0.4(20) = 0.5(x + 20)$

64. $0.5(x + 2) = 0.1 + 3(0.1x + 0.3)$

65. $4x + 13 - \{2x - [4(x - 3) - 5]\} = 2(x - 6)$

66. $-2\{7 - [4 - 2(1 - x) + 3]\} = 10 - [4x - 2(x - 3)]$

Application Exercises

The bar graph shows the average cost of tuition and fees at private four-year colleges in the United States.

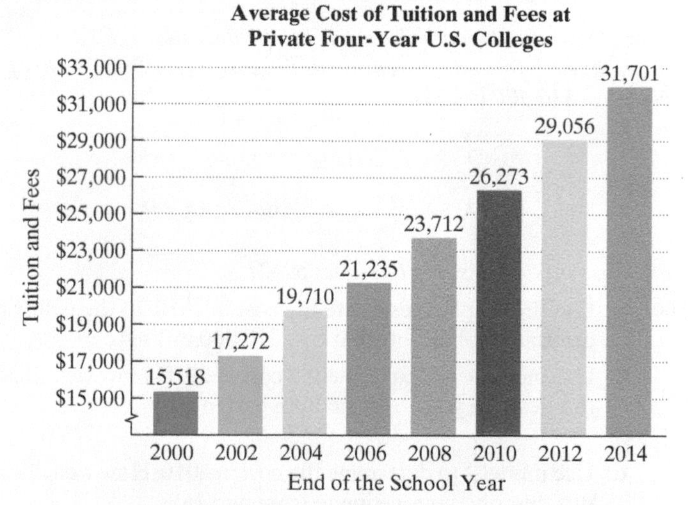

Source: The College Board

Here are two mathematical models for the data shown in the graph. In each formula, T represents the average cost of tuition and fees at private U.S. colleges for the school year ending x years after 2000.

Model 1 $T = 1157x + 14{,}961$

Model 2 $T = 21x^2 + 862x + 15{,}552$

Use this information to solve Exercises 67–68.

67. a. Use each model to find the average cost of tuition and fees at private U.S. colleges for the school year ending in 2014. By how much does each model underestimate or overestimate the actual cost shown for the school year ending in 2014?

b. Use model 1 to determine when tuition and fees at private four-year colleges will average $36,944.

68. **a.** Use each model at the bottom of the previous page to find the average cost of tuition and fees at private U.S. colleges for the school year ending in 2012. By how much does each model underestimate or overestimate the actual cost shown for the school year ending in 2012?

b. Use model 1 to determine when tuition and fees at private four-year colleges will average $39,258.

The line graph shows the cost of inflation. What cost $10,000 in 1984 would cost the amount shown by the graph in subsequent years.

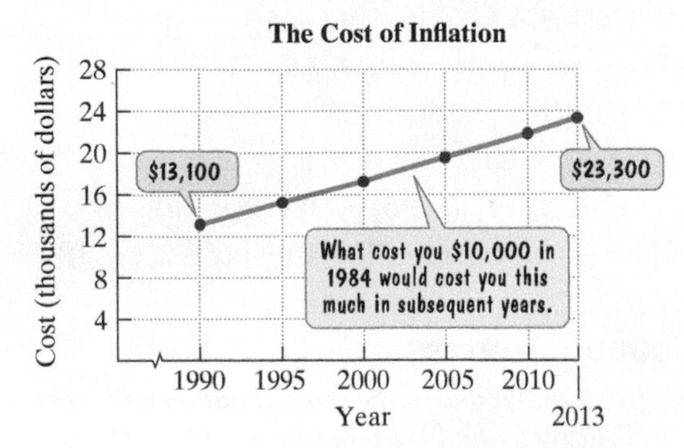

Source: U.S. Bureau of Labor Statistics

Here are two mathematical models for the data shown by the graph. In each formula, C represents the cost x years after 1990 of what cost $10,000 in 1984.

Model 1 $C = 442x + 12{,}969$

Model 2 $C = 2x^2 + 390x + 13{,}126$

Use these models to solve Exercises 69–74.

69. **a.** Use the graph to estimate the cost in 2010, to the nearest thousand dollars, of what cost $10,000 in 1984.

b. Use model 1 to determine the cost in 2010. How well does this describe your estimate from part (a)?

c. Use model 2 to determine the cost in 2010. How well does this describe your estimate from part (a)?

70. **a.** Use the graph to estimate the cost in 2000, to the nearest thousand dollars, of what cost $10,000 in 1984.

b. Use model 1 to determine the cost in 2000. How well does this describe your estimate from part (a)?

c. Use model 2 to determine the cost in 2000. How well does this describe your estimate from part (a)?

71. Which model is a better description for the cost in 1990 of what cost $10,000 in 1984? Does this model underestimate or overestimate the cost shown by the graph? By how much?

72. Which model is a better description for the cost in 2013 of what cost $10,000 in 1984? Does this model underestimate or overestimate the cost shown by the graph? By how much?

73. Use model 1 to determine in which year the cost will be $26,229 for what cost $10,000 in 1984.

74. Use model 1 to determine in which year the cost will be $25,345 for what cost $10,000 in 1984.

Explaining the Concepts

75. What is a linear equation in one variable? Give an example of this type of equation.

76. What does it mean to solve an equation?

77. How do you determine if a number is a solution of an equation?

78. What are equivalent equations? Give an example.

79. What is the addition property of equality?

80. What is the multiplication property of equality?

81. Explain how to clear an equation of fractions.

82. What is an identity? Give an example.

83. What is a conditional equation? Give an example.

84. What is an inconsistent equation? Give an example.

85. Despite low rates of inflation, the cost of a college education continues to skyrocket. This is a departure from the trend during the 1970s: In constant dollars (which negate the effect of inflation), the cost of college actually decreased several times. What explanations can you offer for the increasing cost of a college education?

Technology Exercises

In Exercises 86–89, use your graphing utility to enter each side of the equation separately under y_1 and y_2. Then use the utility's TABLE *or* GRAPH *feature to solve the equation.*

86. $5x + 2(x - 1) = 3x + 10$

87. $2x + 3(x - 4) = 4x - 7$

88. $3(2x + 11) = 3(5 + x)$

89. $\frac{2x - 1}{3} - \frac{x - 5}{6} = \frac{x - 3}{4}$

Critical Thinking Exercises

Make Sense? *In Exercises 90–93, determine whether each statement makes sense or does not make sense, and explain your reasoning.*

90. Because $x = x + 5$ is an inconsistent equation, the graphs of $y = x$ and $y = x + 5$ should not intersect.

91. Because subtraction is defined in terms of addition, it's not necessary to state a separate subtraction property of equality to generate equivalent equations.

92. The number 3 satisfies the equation $7x + 9 = 9(x + 1) - 2x$, so $\{3\}$ is the equation's solution set.

93. I can solve $-2x = 10$ using the addition property of equality.

In Exercises 94–97, determine whether each statement is true or false. If the statement is false, make the necessary change(s) to produce a true statement.

94. The equation $-7x = x$ has no solution.

95. The equations $\frac{x}{x-4} = \frac{4}{x-4}$ and $x = 4$ are equivalent.

96. The equations $3y - 1 = 11$ and $3y - 7 = 5$ are equivalent.

97. If a and b are any real numbers, then $ax + b = 0$ always has only one number in its solution set.

98. Solve for x: $ax + b = c$.

99. Write three equations that are equivalent to $x = 5$.

100. If x represents a number, write an English sentence about the number that results in an inconsistent equation.

101. Find b such that $\frac{7x+4}{b} + 13 = x$ will have a solution set given by $\{-6\}$.

Review Exercises

In Exercises 102–103, perform the indicated operations.

102. $-\frac{1}{5} - \left(-\frac{1}{2}\right)$ (Section 1.2, Example 4)

103. $4(-3)(-1)(-5)$ (Section 1.2, Examples in **Table 1.3**)

104. Graph $y = x^2 - 4$. Let $x = -3, -2, -1, 0, 1, 2,$ and 3. (Section 1.3, Example 2)

Preview Exercises

Exercises 105–107 will help you prepare for the material covered in the next section.

105. Let x represent a number.

a. Write an equation in x that describes the following conditions:

Four less than three times the number is 32.

b. Solve the equation and determine the number.

106. Let x represent the number of countries in the world that are not free. The number of free countries exceeds the number of not-free countries by 44. Write an algebraic expression that represents the number of free countries.

107. You purchase a new car for \$20,000. Each year the value of the car decreases by \$2500. Write an algebraic expression that represents the car's value, in dollars, after x years.

MID-CHAPTER CHECK POINT Section 1.1 – Section 1.4

What You Know: We reviewed a number of topics from introductory algebra, including the real numbers and their representations on number lines. We used interval notation to represent sets of real numbers, where parentheses indicate endpoints that are not included and square brackets indicate endpoints that are included. We performed operations with real numbers and applied the order-of-operations agreement to expressions containing more than one operation. We used commutative, associative, and distributive properties to simplify algebraic expressions. We used the rectangular coordinate system to represent ordered pairs of real numbers and graph equations in two variables. Finally, we solved linear equations, including equations with fractions. We saw that some equations have no solution, whereas others have all real numbers as solutions.

In Exercises 1–14, simplify the expression or solve the equation, whichever is appropriate.

1. $-5 + 3(x + 5)$

2. $-5 + 3(x + 5) = 2(3x - 4)$

3. $3[7 - 4(5 - 2)]$

4. $\frac{x-3}{5} - 1 = \frac{x-5}{4}$

5. $\frac{-2^4 + (-2)^2}{-4 - (2 - 2)}$

6. $7x - [8 - 3(2x - 5)]$

7. $3(2x - 5) - 2(4x + 1) = -5(x + 3) - 2$

8. $3(2x - 5) - 2(4x + 1) - 5(x + 3) - 2$

9. $-4^2 \div 2 + (-3)(-5)$

10. $3x + 1 - (x - 5) = 2x - 4$

11. $\frac{3x}{4} - \frac{x}{3} + 1 = \frac{4x}{5} - \frac{3}{20}$

12. $(6 - 9)(8 - 12) \div \frac{5^2 + 4 \div 2}{8^2 - 9^2 + 8}$

13. $4x - 2(1 - x) = 3(2x + 1) - 5$

14. $\frac{3[4 - 3(-2)^2]}{2^2 - 2^4}$

In Exercises 15–16, express each interval in set-builder notation and graph.

15. $[-2, 0)$ **16.** $(-\infty, 0]$

In Exercises 17–19, graph each equation in a rectangular coordinate system.

17. $y = 2x - 1$ **18.** $y = 1 - |x|$ **19.** $y = x^2 + 2$

In Exercises 20–23, determine whether each statement is true or false.

20. $-\left|-\frac{\sqrt{3}}{5}\right| = -\frac{\sqrt{3}}{5}$

21. $\{x \mid x \text{ is a negative integer greater than } -4\} = \{-4, -3, -2, -1\}$

22. $-17 \notin \{x \mid x \text{ is a rational number}\}$

23. $-128 \div (2 \cdot 4) > (-128 \div 2) \cdot 4$

SECTION

1.5 Problem Solving and Using Formulas

1969

2013

What am I supposed to learn?

After studying this section, you should be able to:

1. Solve algebraic word problems using linear equations.
2. Solve a formula for a variable.

Many changes occurred from 1969 to 2013, including the shifting goals of college students. Compared to 1969, college freshmen in 2013 had making money on their minds. In this section, you will learn a problem-solving strategy that uses linear equations to model the changing attitudes of college freshmen.

Problem Solving with Linear Equations

1 Solve algebraic word problems using linear equations.

We have seen that a model is a mathematical representation of a real-world situation. In this section, we will be solving problems that are presented in English. This means that we must obtain models by translating from the ordinary language of English into the language of algebraic equations. To translate, however, we must understand the English prose and be familiar with the forms of algebraic language. Here are some general steps we will follow in solving word problems:

Strategy for Solving Word Problems

Step 1. Read the problem carefully several times until you can state in your own words what is given and what the problem is looking for. Let x (or any variable) represent one of the unknown quantities in the problem.

Step 2. If necessary, write expressions for any other unknown quantities in the problem in terms of x.

Step 3. Write an equation in x that models the verbal conditions of the problem.

Step 4. Solve the equation and answer the problem's question.

Step 5. Check the solution *in the original wording* of the problem, not in the equation obtained from the words.

Great Question!

Why are word problems important?

There is great value in reasoning through the steps for solving a word problem. This value comes from the problem-solving skills that you will attain and is often more important than the specific problem or its solution.

EXAMPLE 1 Education Pays Off

The graph in **Figure 1.23** at the top of the next page shows average yearly earnings in the United States by highest educational attainment.

The average yearly salary of a man with a bachelor's degree exceeds that of a man with an associate's degree by $25 thousand. The average yearly salary of a man with a master's degree exceeds that of a man with an associate's degree by $45 thousand. Combined, three men with each of these degrees earn $214 thousand. Find the average yearly salary of men with each of these levels of education.

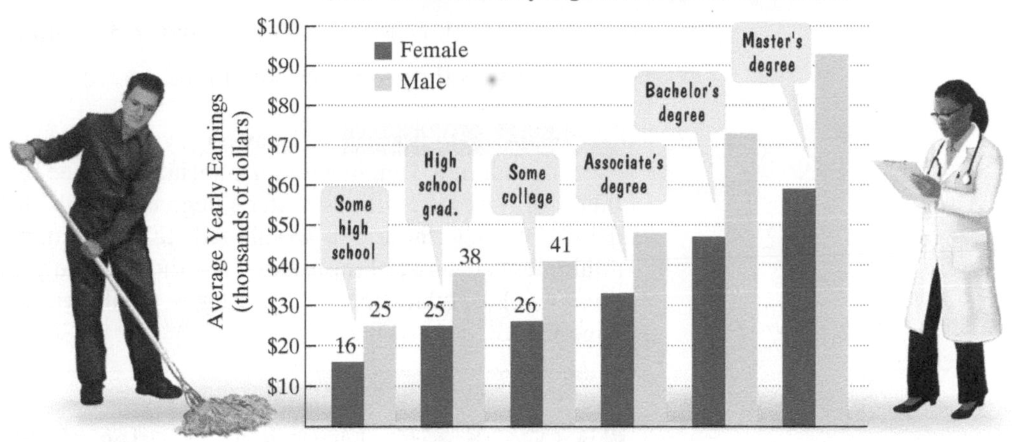

Figure 1.23
Source: U.S. Census Bureau (2012 data)

Solution

Step 1. Let x represent one of the unknown quantities. We know something about salaries of men with bachelor's degrees and master's degrees: They exceed the salary of a man with an associate's degree by \$25 thousand and \$45 thousand, respectively. We will let

x = the average yearly salary of a man with an associate's degree (in thousands of dollars).

Step 2. Represent other unknown quantities in terms of x. Because a man with a bachelor's degree earns \$25 thousand more than a man with an associate's degree, let

$x + 25$ = the average yearly salary of a man with a bachelor's degree.

Because a man with a master's degree earns \$45 thousand more than a man with an associate's degree, let

$x + 45$ = the average yearly salary of a man with a master's degree.

Step 3. Write an equation in x that models the conditions. Combined, three men with each of these degrees earn \$214 thousand.

Salary: associate's degree	plus	salary: bachelor's degree	plus	salary: master's degree	equals	\$214 thousand.
x	$+$	$(x + 25)$	$+$	$(x + 45)$	$=$	214

Step 4. Solve the equation and answer the question.

$$x + (x + 25) + (x + 45) = 214$$ This is the equation that models the problem's conditions.

$$3x + 70 = 214$$ Remove parentheses, regroup, and combine like terms.

$$3x = 144$$ Subtract 70 from both sides.

$$x = 48$$ Divide both sides by 3.

Because we isolated the variable in the model and obtained $x = 48$,

average salary with an associate's degree $= x = 48$

average salary with a bachelor's degree $= x + 25 = 48 + 25 = 73$

average salary with a master's degree $= x + 45 = 48 + 45 = 93$.

Men with associate's degrees average \$48 thousand per year, men with bachelor's degrees average \$73 thousand per year, and men with master's degrees average \$93 thousand per year.

Great Question!

Example 1 involves using the word "exceeds" to represent two of the unknown quantities. Can you help me write algebraic expressions for quantities described using "exceeds"?

Modeling with the word *exceeds* can be a bit tricky. It's helpful to identify the smaller quantity. Then add to this quantity to represent the larger quantity. For example, suppose that Tim's height exceeds Tom's height by a inches. Tom is the shorter person. If Tom's height is represented by x, then Tim's height is represented by $x + a$.

Step 5. Check the proposed solution in the original wording of the problem. The problem states that combined, three men with each of these educational attainments earn \$214 thousand. Using the salaries we determined in step 4, the sum is

\$48 thousand + \$73 thousand + \$93 thousand, or \$214 thousand,

which satisfies the problem's conditions. ■

CHECK POINT 1 The average yearly salary of a woman with a bachelor's degree exceeds that of a woman with an associate's degree by \$14 thousand. The average yearly salary of a woman with a master's degree exceeds that of a woman with an associate's degree by \$26 thousand. Combined, three women with each of these educational attainments earn \$139 thousand. Find the average yearly salary of women with each of these levels of education. (These salaries are illustrated by the bar graph on the previous page.)

EXAMPLE 2 Modeling Attitudes of College Freshmen

Your author teaching math in 1969

Researchers have surveyed college freshmen every year since 1969. **Figure 1.24** shows that attitudes about some life goals have changed dramatically over the years. In particular, the freshmen class of 2013 was more interested in making money than the freshmen of 1969 had been. In 1969, 42% of first-year college students considered "being well-off financially" essential or very important. For the period from 1969 through 2013, this percentage increased by approximately 0.9 each year. If this trend continues, by which year will all college freshmen consider "being well-off financially" essential or very important?

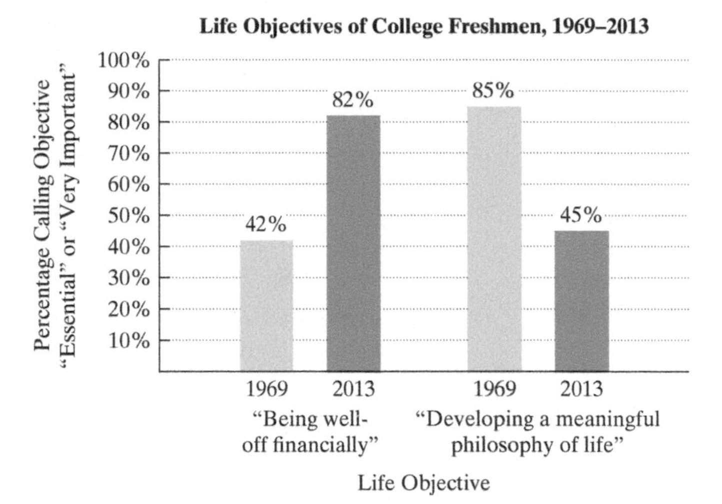

Figure 1.24

Source: Higher Education Research Institute

Solution

Step 1. Let x represent one of the unknown quantities. We are interested in the year when all college freshmen, or 100% of the freshmen, will consider this life objective essential or very important. Let

x = the number of years after 1969 when all freshmen will consider "being well-off financially"essential or very important.

Step 2. Represent other unknown quantities in terms of x. There are no other unknown quantities to find, so we can skip this step.

Step 3. Write an equation in x that models the conditions.

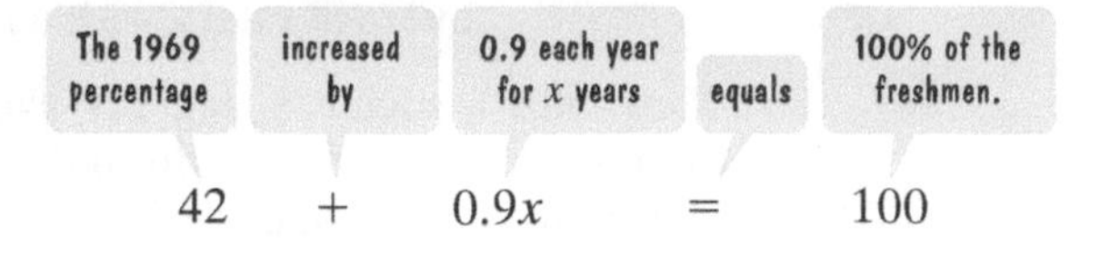

The 1969 percentage
increased by
0.9 each year for x years
equals
100% of the freshmen.
42 + 0.9x = 100

Step 4. Solve the equation and answer the question.

$42 + 0.9x = 100$	This is the equation that models the problem's conditions.
$42 - 42 + 0.9x = 100 - 42$	Subtract 42 from both sides.
$0.9x = 58$	Simplify.
$\frac{0.9x}{0.9} = \frac{58}{0.9}$	Divide both sides by 0.9.
$x = 64.\overline{4} \approx 64$	Simplify and round to the nearest whole number.

Using current trends, by approximately 64 years after 1969, or in 2033, all freshmen will consider "being well-off financially" essential or very important.

Step 5. Check the proposed solution in the original wording of the problem. The problem states that all freshmen (100%, represented by 100 using the model) will consider the objective essential or very important. Does this approximately occur if we increase the 1969 percentage, 42%, by 0.9 each year for 64 years, our proposed solution?

$$42 + 0.9(64) = 42 + 57.6 = 99.6 \approx 100$$

This verifies that using trends shown in **Figure 1.24**, all first-year college students will consider the objective essential or very important approximately 64 years after 1969. ■

Great Question!

I notice that the equation in Example 2, 42 + 0.9x = 100, contains a decimal. Can I clear an equation of decimals much like I cleared equations of fractions?

- You can clear an equation of decimals by multiplying each side by a power of 10. The exponent on 10 will be equal to the greatest number of digits to the right of any decimal point in the equation.
- Multiplying a decimal number by 10^n has the effect of moving the decimal point n places to the right.

Example

$$42 + 0.9x = 100$$

The greatest number of digits to the right of any decimal point in the equation is 1. Multiply each side by 10^1, or 10.

$$\begin{aligned} 10(42 + 0.9x) &= 10(100) \\ 10(42) + 10(0.9x) &= 10(100) \\ 420 + 9x &= 1000 \\ 420 - 420 + 9x &= 1000 - 420 \\ 9x &= 580 \\ \frac{9x}{9} &= \frac{580}{9} \\ x &= 64.\overline{4} \approx 64 \end{aligned}$$

It is not a requirement to clear decimals before solving an equation. Compare this solution to the one in step 4 of Example 2. Which method do you prefer?

✓ CHECK POINT 2 **Figure 1.24** on the previous page shows that the freshmen class of 2013 was less interested in developing a philosophy of life than the freshmen of 1969 had been. In 1969, 85% of the freshmen considered this objective essential or very important. Since then, this percentage has decreased by approximately 0.9 each year. If this trend continues, by which year will only 25% of college freshmen consider "developing a meaningful philosophy of life" essential or very important?

EXAMPLE 3 Modeling Options for a Toll

The toll to a bridge costs \$7. Commuters who use the bridge frequently have the option of purchasing a monthly discount pass for \$30. With the discount pass, the toll is reduced to \$4. For how many bridge crossings per month will the total monthly cost without the discount pass be the same as the total monthly cost with the discount pass?

Solution

Step 1. Let x represent one of the unknown quantities. Let

$$x = \text{the number of bridge crossings per month.}$$

Step 2. Represent other unknown quantities in terms of x. There are no other unknown quantities, so we can skip this step.

Step 3. Write an equation in x that models the conditions. The monthly cost without the discount pass is the toll, \$7, times the number of bridge crossings per month, x. The monthly cost with the discount pass is the cost of the pass, \$30, plus the toll, \$4, times the number of bridge crossings per month, x.

The monthly cost without the discount pass | must equal | the monthly cost with the discount pass.

$$7x = 30 + 4x$$

Step 4. Solve the equation and answer the question.

$$7x = 30 + 4x$$ This is the equation that models the problem's conditions.

$$3x = 30$$ Subtract $4x$ from both sides.

$$x = 10$$ Divide both sides by 3.

Because x represents the number of bridge crossings per month, the total monthly cost without the discount pass will be the same as the total monthly cost with the discount pass for 10 bridge crossings per month.

Step 5. Check the proposed solution in the original wording of the problem. The problem states that the monthly cost without the discount pass should be the same as the monthly cost with the discount pass. Let's see if they are the same with 10 bridge crossings per month.

$$\text{Cost without the discount pass} = \$7(10) = \$70$$

Cost of the pass | Toll

$$\text{Cost with the discount pass} = \$30 + \$4(10) = \$30 + \$40 = \$70$$

With 10 bridge crossings per month, both options cost \$70 for the month. Thus the proposed solution, 10 bridge crossings, satisfies the problem's conditions. ■

Using Technology

Numeric and Graphic Connections

We can use a graphing utility to numerically or graphically verify our work in Example 3.

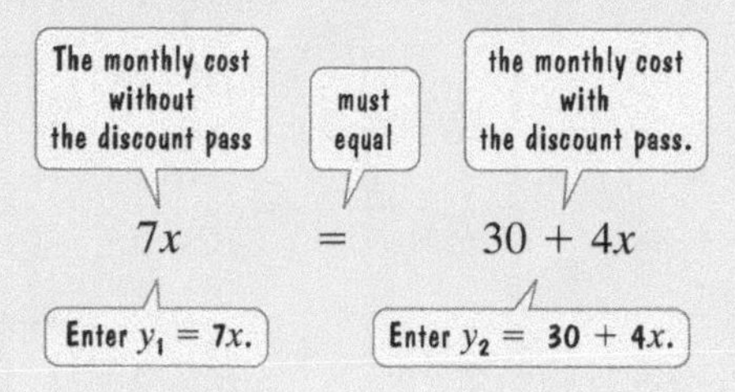

Numeric Check

Display a table for y_1 and y_2.

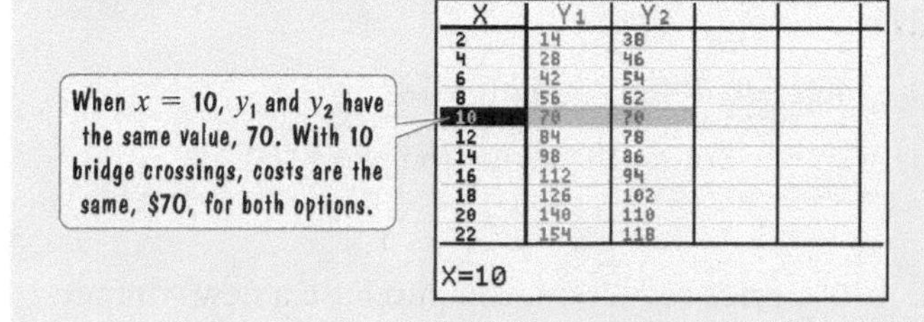

Graphic Check

Display graphs for y_1 and y_2. Use the intersection feature.

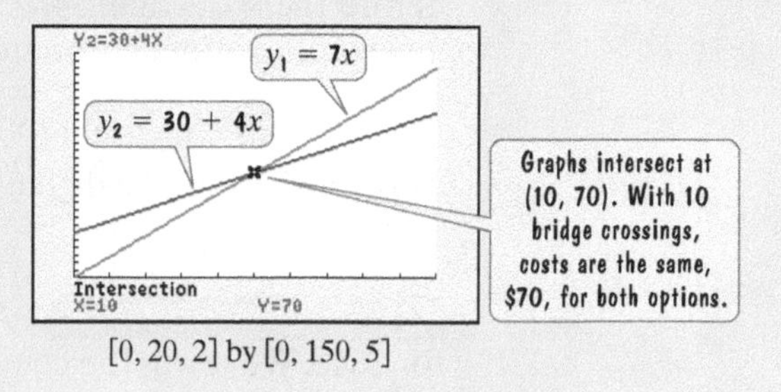

[0, 20, 2] by [0, 150, 5]

☑ **CHECK POINT 3** The toll to a bridge costs $5. Commuters who use the bridge frequently have the option of purchasing a monthly discount pass for $40. With the discount pass, the toll is reduced to $3. For how many bridge crossings per month will the total monthly cost without the discount pass be the same as the total monthly cost with the discount pass?

EXAMPLE 4 A Price Reduction on a Digital Camera

Your local computer store is having a terrific sale on digital cameras. After a 40% price reduction, you purchase a digital camera for $276. What was the camera's price before the reduction?

Solution

Step 1. Let x represent one of the unknown quantities. We will let

$$x = \text{the original price of the digital camera prior to the reduction.}$$

Step 2. Represent other unknown quantities in terms of x. There are no other unknown quantities to find, so we can skip this step.

Step 3. Write an equation in x that models the conditions. The camera's original price minus the 40% reduction is the reduced price, $276.

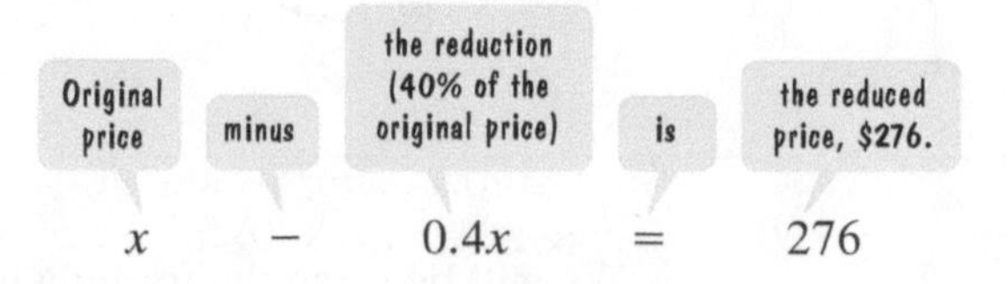

Great Question!

Why is the 40% reduction written as $0.4x$ in Example 4?

- 40% is written 0.40 or 0.4.
- "Of" represents multiplication, so 40% of the original price is $0.4x$.

Notice that the original price, x, reduced by 40% is $x - 0.4x$ and *not* $x - 0.4$.

Step 4. Solve the equation and answer the question.

$$x - 0.4x = 276$$ This is the equation that models the problem's conditions.

$$0.6x = 276$$ Combine like terms: $x - 0.4x = 1x - 0.4x = (1 - 0.4)x = 0.6x$.

$$\frac{0.6x}{0.6} = \frac{276}{0.6}$$ Divide both sides by 0.6.

$$x = 460$$ Simplify: $0.6\overline{)276.0}$ = 460.

The digital camera's price before the reduction was $460.

Step 5. Check the proposed solution in the original wording of the problem. The price before the reduction, $460, minus the 40% reduction should equal the reduced price given in the original wording, $276:

$$460 - 40\% \text{ of } 460 = 460 - 0.4(460) = 460 - 184 = 276.$$

This verifies that the digital camera's price before the reduction was $460. ■

✓ **CHECK POINT 4** After a 30% price reduction, you purchase a new computer for $840. What was the computer's price before the reduction?

Solving geometry problems usually requires a knowledge of basic geometric ideas and formulas. Formulas for area, perimeter, and volume are given in **Table 1.6.**

Table 1.6 Common Formulas for Area, Perimeter, and Volume

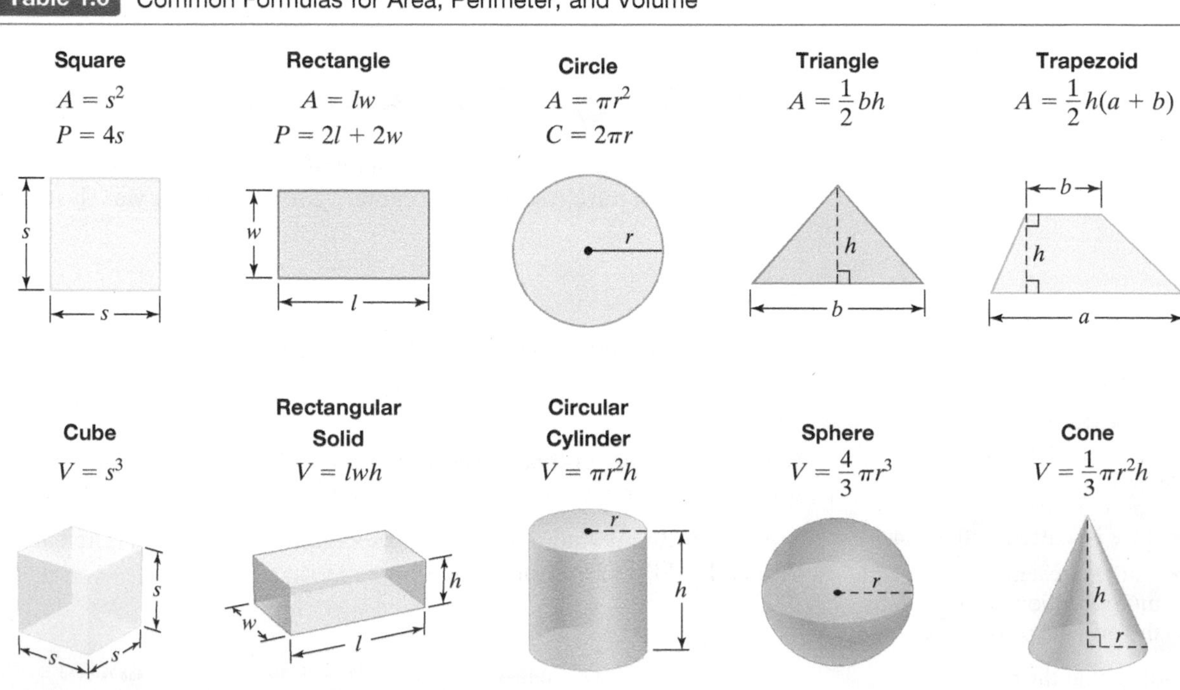

Square
$A = s^2$
$P = 4s$
Rectangle
$A = lw$
$P = 2l + 2w$
Circle
$A = \pi r^2$
$C = 2\pi r$
Triangle
$A = \frac{1}{2}bh$
Trapezoid
$A = \frac{1}{2}h(a + b)$
Cube
$V = s^3$
Rectangular Solid
$V = lwh$
Circular Cylinder
$V = \pi r^2 h$
Sphere
$V = \frac{4}{3}\pi r^3$
Cone
$V = \frac{1}{3}\pi r^2 h$

We will be using the formula for the perimeter of a rectangle, $P = 2l + 2w$, in our next example. The formula states that a rectangle's perimeter is the sum of twice its length and twice its width.

EXAMPLE 5 Finding the Dimensions of an American Football Field

The length of an American football field is 200 feet more than the width. If the perimeter of the field is 1040 feet, what are its dimensions?

Solution

Step 1. Let x represent one of the unknown quantities. We know something about the length; the length is 200 feet more than the width. We will let

$$x = \text{the width.}$$

Step 2. Represent other unknown quantities in terms of x. Because the length is 200 feet more than the width, let

$$x + 200 = \text{the length.}$$

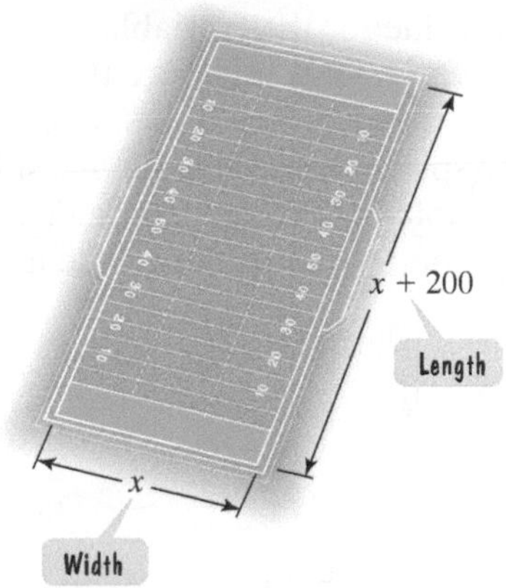

Figure 1.25 illustrates an American football field and its dimensions.

Figure 1.25 An American football field

Step 3. Write an equation in x that models the conditions. Because the perimeter of the field is 1040 feet,

Twice the length	plus	twice the width	is	the perimeter.
$2(x + 200)$	$+$	$2x$	$=$	$1040.$

Step 4. Solve the equation and answer the question.

$$2(x + 200) + 2x = 1040$$ This is the equation that models the problem's conditions.

$$2x + 400 + 2x = 1040$$ Apply the distributive property.

$$4x + 400 = 1040$$ Combine like terms: $2x + 2x = 4x$.

$$4x = 640$$ Subtract 400 from both sides.

$$x = 160$$ Divide both sides by 4.

Thus,

$$\text{width} = x = 160.$$

$$\text{length} = x + 200 = 160 + 200 = 360.$$

The dimensions of an American football field are 160 feet by 360 feet. (The 360-foot length is usually described as 120 yards.)

Step 5. Check the proposed solution in the original wording of the problem. The perimeter of the football field using the dimensions that we found is

$$2(360 \text{ feet}) + 2(160 \text{ feet}) = 720 \text{ feet} + 320 \text{ feet} = 1040 \text{ feet}.$$

Because the problem's wording tells us that the perimeter is 1040 feet, our dimensions are correct. ■

Great Question!

Should I draw pictures like Figure 1.25 when solving geometry problems?

When solving word problems, particularly problems involving geometric figures, drawing a picture of the situation is often helpful. Label x on your drawing and, where appropriate, label other parts of the drawing in terms of x.

✓ **CHECK POINT 5** The length of a rectangular basketball court is 44 feet more than the width. If the perimeter of the basketball court is 288 feet, what are its dimensions?

---

2 Solve a formula for a variable.

Solving a Formula for One of Its Variables

We know that solving an equation is the process of finding the number (or numbers) that make the equation a true statement. All of the equations we have solved contained only one letter, x.

By contrast, formulas contain two or more letters, representing two or more variables. An example is the formula for the perimeter of a rectangle:

$$2l + 2w = P.$$

We say that this formula is solved for the variable P because P is alone on one side of the equation and the other side does not contain a P.

Solving a formula for a variable means using the addition and multiplication properties of equality to rewrite the formula so that the variable is isolated on one side of the equation. It does not mean obtaining a numerical value for that variable.

To solve a formula for one of its variables, treat that variable as if it were the only variable in the equation. Think of the other variables as if they were numbers. Use the addition property of equality to isolate all terms with the specified variable on one side of the equation and all terms without the specified variable on the other side. Then use the multiplication property of equality to get the specified variable alone. The next example shows how to do this.

EXAMPLE 6 Solving a Formula for a Variable

Solve the formula $2l + 2w = P$ for l.

Solution First, isolate $2l$ on the left by subtracting $2w$ from both sides. Then solve for l by dividing both sides by 2.

We need to isolate l.

$$2l + 2w = P$$ This is the given formula.

$$2l + 2w - 2w = P - 2w$$ Isolate $2l$ by subtracting $2w$ from both sides.

$$2l = P - 2w$$ Simplify.

$$\frac{2l}{2} = \frac{P - 2w}{2}$$ Solve for l by dividing both sides by 2.

$$l = \frac{P - 2w}{2}$$ Simplify. ■

✓ CHECK POINT 6 Solve the formula $2l + 2w = P$ for w.

EXAMPLE 7 Solving a Formula for a Variable

Circular Cylinder

$V = \pi r^2 h$

Table 1.6 on page 60 shows that the volume of a circular cylinder is given by the formula

$$V = \pi r^2 h,$$

where r is the radius of the circle at either end and h is the height. Solve this formula for h.

Solution Our goal is to get h by itself on one side of the formula. There is only one term with h, $\pi r^2 h$, and it is already isolated on the right side. We isolate h on the right by dividing both sides by πr^2.

We need to isolate h.

$$V = \pi r^2 h$$ This is the given formula.

$$\frac{V}{\pi r^2} = \frac{\pi r^2 h}{\pi r^2}$$ Isolate h by dividing both sides by πr^2.

$$\frac{V}{\pi r^2} = h$$ Simplify: $\frac{\pi r^2 h}{\pi r^2} = \frac{\pi r^2}{\pi r^2} \cdot h = 1h = h$.

Equivalently,

$$h = \frac{V}{\pi r^2}. \blacksquare$$

CHECK POINT 7 The volume of a rectangular solid is the product of its length, width, and height:

$$V = lwh.$$

Solve this formula for h.

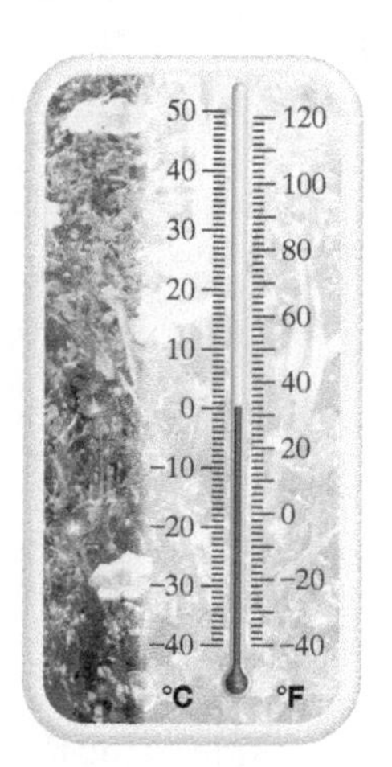

The Celsius scale is on the left and the Fahrenheit scale is on the right.

You'll be leaving the cold of winter for a vacation to Hawaii. CNN International reports a temperature in Hawaii of 30°C. Should you pack a winter coat? You can convert from Celsius temperature, C, to Fahrenheit temperature, F, using the formula

$$F = \frac{9}{5}C + 32.$$

A temperature of 30°C corresponds to a Fahrenheit temperature of

$$F = \frac{9}{5} \cdot 30 + 32 = \frac{9}{\cancel{5}_1} \cdot \frac{\cancel{30}^6}{1} + 32 = 54 + 32 = 86,$$

or a balmy 86°F. (Don't pack the coat.)

Visitors to the United States are more likely to be familiar with the Celsius temperature scale. For them, a useful formula is one that can be used to convert from Fahrenheit to Celsius. In Example 8, you will see how to obtain such a formula.

EXAMPLE 8 Solving a Formula for a Variable

Solve the formula

$$F = \frac{9}{5}C + 32$$

for C.

Solution We begin by multiplying both sides of the formula by 5 to clear the fraction. Then we isolate the variable C.

$$F = \frac{9}{5}C + 32 \qquad \text{This is the given formula.}$$

$$5F = 5\left(\frac{9}{5}C + 32\right) \qquad \text{Multiply both sides by 5.}$$

$$5F = 5 \cdot \frac{9}{5}C + 5 \cdot 32 \qquad \text{Apply the distributive property.}$$

We need to isolate C.

$$5F = 9C + 160 \qquad \text{Simplify.}$$

$$5F - 160 = 9C + 160 - 160 \qquad \text{Subtract 160 from both sides.}$$

$$5F - 160 = 9C \qquad \text{Simplify.}$$

$$\frac{5F - 160}{9} = \frac{9C}{9} \qquad \text{Divide both sides by 9.}$$

$$\frac{5F - 160}{9} = C \qquad \text{Simplify.}$$

Using the distributive property, we can express $5F - 160$ as $5(F - 32)$. Thus,

$$C = \frac{5F - 160}{9} = \frac{5(F - 32)}{9}.$$

This formula, used to convert from Fahrenheit to Celsius, is usually given as

$$C = \frac{5}{9}(F - 32). \blacksquare$$

CHECK POINT 8 The formula

$$\frac{W}{2} - 3H = 53$$

models the recommended weight, W, in pounds, for a male, where H represents his height, in inches, over 5 feet. Solve this formula for W.

EXAMPLE 9 Solving a Formula for a Variable That Occurs Twice

The formula

$$A = P + Prt$$

describes the amount, A, that a principal of P dollars is worth after t years when invested at a simple annual interest rate, r. Solve this formula for P.

Solution Notice that all the terms with P already occur on the right side of the formula.

We need to isolate P.

$$A = P + Prt$$

We can use the distributive property in the form $ab + ac = a(b + c)$ to convert the two occurrences of P into one.

$$A = P + Prt$$ This is the given formula.

$$A = P(1 + rt)$$ Use the distributive property to obtain a single occurrence of P.

$$\frac{A}{1 + rt} = \frac{P(1 + rt)}{1 + rt}$$ Divide both sides by $1 + rt$.

$$\frac{A}{1 + rt} = P$$ Simplify: $\frac{P(1 + rt)}{1(1 + rt)} = \frac{P}{1} = P.$

Equivalently,

$$P = \frac{A}{1 + rt}. \blacksquare$$

Great Question!

Can I solve $A = P + Prt$ for P by subtracting Prt from both sides and writing

$$A - Prt = P?$$

No. When a formula is solved for a specified variable, that variable must be isolated on one side. The variable P occurs on both sides of

$$A - Prt = P.$$

CHECK POINT 9 Solve the formula $P = C + MC$ for C.

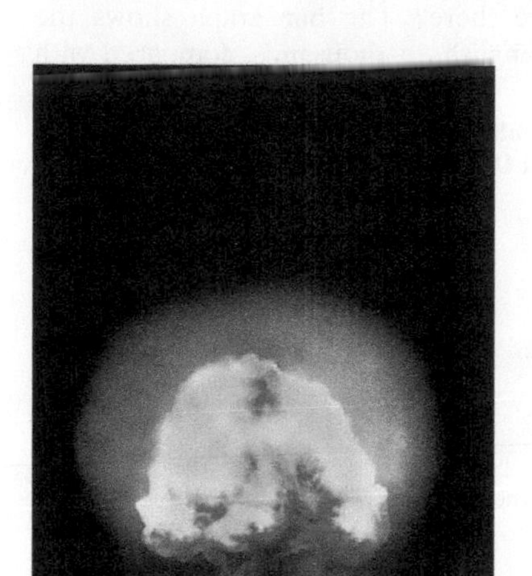

Blitzer Bonus

Einstein's Famous Formula: $E = mc^2$

One of the most famous formulas in the world is $E = mc^2$, formulated by Albert Einstein. Einstein showed that any form of energy has mass and that mass itself has an associated energy that can be released if the matter is destroyed. In this formula, E represents energy, in ergs, m represents mass, in grams, and c represents the speed of light. Because light travels at 30 billion centimeters per second, the formula indicates that 1 gram of mass will produce 900 billion billion ergs of energy.

Einstein's formula implies that the mass of a golf ball could provide the daily energy needs of the metropolitan Boston area. Mass and energy are equivalent, and the transformation of even a tiny amount of mass releases an enormous amount of energy. If this energy is released suddenly, a destructive force is unleashed, as in an atom bomb. When the release is gradual and controlled, the energy can be used to generate power.

The theoretical results implied by Einstein's formula $E = mc^2$ have not been realized because scientists have not yet developed a way of converting a mass completely to energy. Japan's devastating earthquake–tsunami crisis rekindled debate over safety problems associated with nuclear energy.

Achieving Success

Do not expect to solve every word problem immediately. As you read each problem, underline the important parts. It's a good idea to read the problem at least twice. Be persistent, but use the **"Ten Minutes of Frustration" Rule**. If you have exhausted every possible means for solving a problem and you are still bogged down, stop after ten minutes. Put a question mark by the exercise and move on. When you return to class, ask your professor for assistance.

CONCEPT AND VOCABULARY CHECK

Fill in each blank so that the resulting statement is true.

1. According to the U.S. Office of Management and Budget, the 2011 budget for defense exceeded the budget for education by \$658.6 billion. If x represents the budget for education, in billions of dollars, the budget for defense can be represented by ____________.
2. In 2000, 31% of U.S. adults viewed a college education as essential for success. For the period from 2000 through 2010, this percentage increased by approximately 2.4 each year. The percentage of U.S. adults who viewed a college education as essential for success x years after 2000 can be represented by ____________.
3. A text message plan costs \$4 per month plus \$0.15 per text. The monthly cost for x text messages can be represented by ____________.
4. I purchased a computer after a 15% price reduction. If x represents the computer's original price, the reduced price can be represented by ____________.
5. Solving a formula for a variable means rewriting the formula so that the variable is ____________.
6. The first step in solving $IR + Ir = E$ for I is to obtain a single occurrence of I using the ____________ property.

1.5 EXERCISE SET

MyMathLab®

Practice Exercises

Use the five-step strategy for solving word problems to find the number or numbers described in Exercises 1–10.

1. When five times a number is decreased by 4, the result is 26. What is the number?
2. When two times a number is decreased by 3, the result is 11. What is the number?
3. When a number is decreased by 20% of itself, the result is 20. What is the number?
4. When a number is decreased by 30% of itself, the result is 28. What is the number?
5. When 60% of a number is added to the number, the result is 192. What is the number?
6. When 80% of a number is added to the number, the result is 252. What is the number?

7. 70% of what number is 224?

8. 70% of what number is 252?

9. One number exceeds another by 26. The sum of the numbers is 64. What are the numbers?

10. One number exceeds another by 24. The sum of the numbers is 58. What are the numbers?

Practice PLUS

In Exercises 11–16, find all values of x satisfying the given conditions.

11. $y_1 = 13x - 4$, $y_2 = 5x + 10$, and y_1 exceeds y_2 by 2.

12. $y_1 = 10x + 6$, $y_2 = 12x - 7$, and y_1 exceeds y_2 by 3.

13. $y_1 = 10(2x - 1)$, $y_2 = 2x + 1$, and y_1 is 14 more than 8 times y_2.

14. $y_1 = 9(3x - 5)$, $y_2 = 3x - 1$, and y_1 is 51 less than 12 times y_2.

15. $y_1 = 2x + 6$, $y_2 = x + 8$, $y_3 = x$, and the difference between 3 times y_1 and 5 times y_2 is 22 less than y_3.

16. $y_1 = 2.5$, $y_2 = 2x + 1$, $y_3 = x$, and the difference between 2 times y_1 and 3 times y_2 is 8 less than 4 times y_3.

Application Exercises

In Exercises 17–18, use the five-step strategy for solving word problems.

17. Some animals stick around much longer than their species' life expectancy would predict. The bar graph shows three critters that lived to a ripe old age.

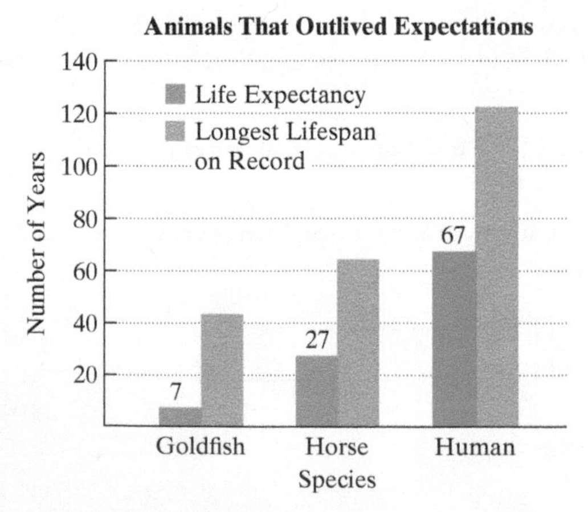

Source: Listomania, HarperCollins

The longest lifespan of a horse exceeded that of a goldfish by 21 years. The longest lifespan of a human exceeded that of a goldfish by 79 years. Combined, the longest lifespan on record for a goldfish, a horse, and a human was 229 years. Determine the longest lifespan on record for a goldfish, a horse, and a human.

18. How many words are there? The bar graph shows the number of words in English, in thousands, compared with four other languages.

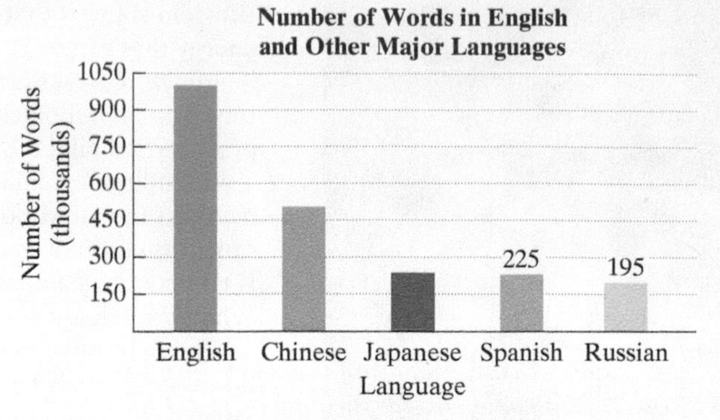

Source: Global Language Monitor

The number of words in English exceeds the number of words in Japanese by 767 thousand. The number of words in Chinese (various dialects) exceeds the number of words in Japanese by 268 thousand. Combined, these three languages contain 1731 thousand words. Determine the number of words, in thousands, for each of these three languages.

Solve Exercises 19–22 using the fact that the sum of the measures of the three angles of a triangle is 180°.

19. In a triangle, the measure of the first angle is twice the measure of the second angle. The measure of the third angle is 8° less than the measure of the second angle. What is the measure of each angle?

20. In a triangle, the measure of the first angle is three times the measure of the second angle. The measure of the third angle is 35° less than the measure of the second angle. What is the measure of each angle?

21. In a triangle, the measures of the three angles are consecutive integers. What is the measure of each angle?

22. In a triangle, the measures of the three angles are consecutive even integers. What is the measure of each angle?

Even as Americans increasingly view a college education as essential for success, many believe that a college education is becoming less available to qualified students. Exercises 23–24 are based on the data displayed by the graph.

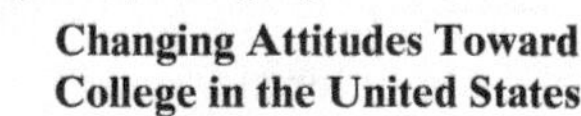

Changing Attitudes Toward College in the United States

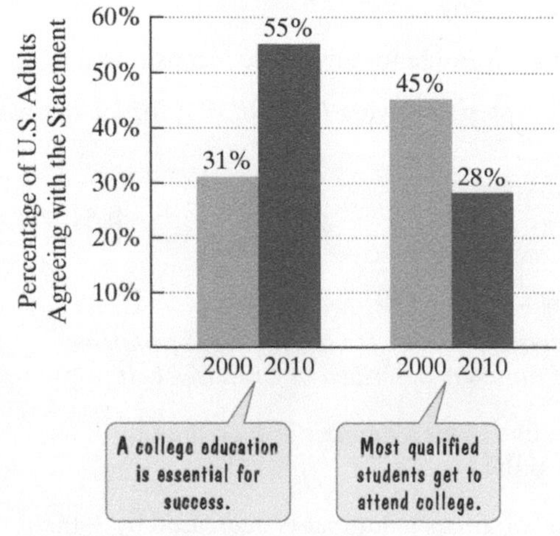

Source: Public Agenda

23. In 2000, 31% of U.S. adults viewed a college education as essential for success. For the period 2000 through 2010, the percentage viewing a college education as essential for success increased on average by approximately 2.4 each year. If this trend continues, by which year will 67% of all American adults view a college education as essential for success?

24. In 2000, 45% of U.S. adults believed that most qualified students get to attend college. For the period from 2000 through 2010, the percentage who believed that a college education is available to most qualified students decreased by approximately 1.7 each year. If this trend continues, by which year will only 11% of all American adults believe that most qualified students get to attend college?

The line graph indicates that in 1960, 23% of U.S. taxes came from corporate income tax. For the period from 1960 through 2010, this percentage decreased by approximately 0.28 each year. Use this information to solve Exercises 25–26.

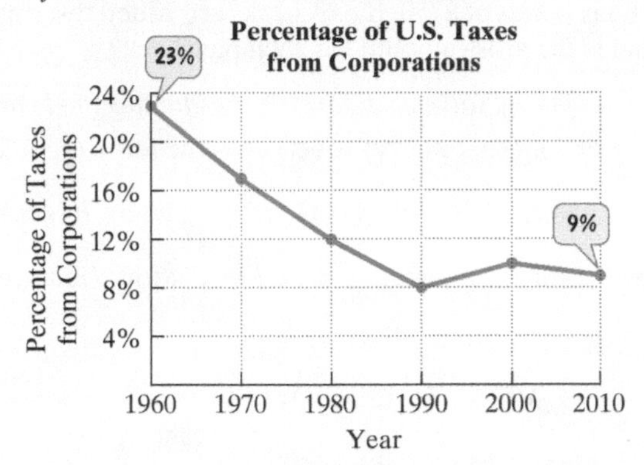

Source: Office of Management and Budget

25. If this trend continues, by which year will corporations pay zero taxes? Round to the nearest year.

26. If this trend continues, by which year will the percentage of federal tax receipts coming from the corporate income tax drop to 5%? Round to the nearest year.

On average, every minute of every day, 158 babies are born. The bar graph represents the results of a single day of births, deaths, and population increase worldwide. Exercises 27–28 are based on the information displayed by the graph.

Daily Growth of World Population

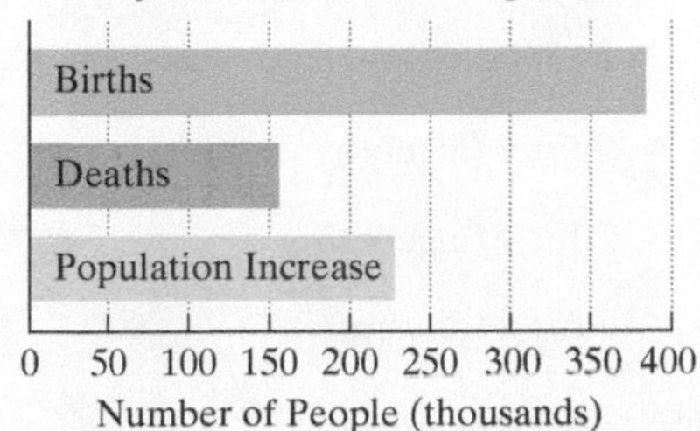

Source: James Henslin, *Sociology*, Eleventh Edition, Pearson, 2012.

27. Each day, the number of births in the world is 84 thousand less than three times the number of deaths.
    a. If the population increase in a single day is 228 thousand, determine the number of births and deaths per day.
    b. If the population increase in a single day is 228 thousand, by how many millions of people does the worldwide population increase each year? Round to the nearest million.
    c. Based on your answer to part (b), approximately how many years does it take for the population of the world to increase by an amount greater than the entire U.S. population (320 million)?

28. Each day, the number of births in the world exceeds twice the number of deaths by 72 thousand.
    a. If the population increase in a single day is 228 thousand, determine the number of births and deaths per day.
    b. If the population increase in a single day is 228 thousand, by how many millions of people does the worldwide population increase each year? Round to the nearest million.
    c. Based on your answer to part (b), approximately how many years does it take for the population of the world to increase by an amount greater than the entire U.S. population (320 million)?

29. The bus fare in a city is $1.25. People who use the bus have the option of purchasing a monthly discount pass for $15.00. With the discount pass, the fare is reduced to $0.75. Determine the number of times in a month the bus must be used so that the total monthly cost without the discount pass is the same as the total monthly cost with the discount pass.

30. You are choosing between two health clubs. Club A offers membership for a fee of $40 plus a monthly fee of $25. Club B offers membership for a fee of $15 plus a monthly fee of $30. After how many months will the total cost at each health club be the same? What will be the total cost for each club?

31. A discount pass for a bridge costs $30 per month. The toll for the bridge is normally $5.00, but it is reduced to $3.50 for people who have purchased the discount pass. Determine the number of times in a month the bridge must be crossed so that the total monthly cost without the discount pass is the same as the total monthly cost with the discount pass.

32. You are choosing between two cellphone plans. Data Plan A has a monthly fee of $40 with a charge of $15 per gigabyte (GB). Data Plan B has a monthly fee of $30 with a charge of $20 per GB. For how many GB of data will the costs for the two data plans be the same?

33. In 2008, there were 13,300 students at college A, with a projected enrollment increase of 1000 students per year. In the same year, there were 26,800 students at college B, with a projected enrollment decline of 500 students per year.
    a. According to these projections, when will the colleges have the same enrollment? What will be the enrollment in each college at that time?
    b. Use the following table to numerically check your work in part (a). What equations were entered for y_1 and y_2 to obtain this table?

X	Y_1	Y_2		
7	20300	23300		
8	21300	22800		
9	22300	22300		
10	23300	21800		
11	24300	21300		
12	25300	20800		
13	26300	20300		
14	27300	19800		
15	28300	19300		
16	29300	18800		
17	30300	18300		

X=7

34. In 2000, the population of Greece was 10,600,000, with projections of a population decrease of 28,000 people per year. In the same year, the population of Belgium was 10,200,000, with projections of a population decrease of 12,000 people per year. (*Source:* United Nations) According to these projections, when will the two countries have the same population? What will be the population at that time?

35. After a 20% reduction, you purchase a television for \$336. What was the television's price before the reduction?

36. After a 30% reduction, you purchase a dictionary for \$30.80. What was the dictionary's price before the reduction?

37. Including 8% sales tax, an inn charges \$162 per night. Find the inn's nightly cost before the tax is added.

38. Including 5% sales tax, an inn charges \$252 per night. Find the inn's nightly cost before the tax is added.

Exercises 39–40 involve markup, the amount added to the dealer's cost of an item to arrive at the selling price of that item.

39. The selling price of a refrigerator is \$584. If the markup is 25% of the dealer's cost, what is the dealer's cost of the refrigerator?

40. The selling price of a scientific calculator is \$15. If the markup is 25% of the dealer's cost, what is the dealer's cost of the calculator?

41. A rectangular soccer field is twice as long as it is wide. If the perimeter of the soccer field is 300 yards, what are its dimensions?

42. A rectangular swimming pool is three times as long as it is wide. If the perimeter of the pool is 320 feet, what are its dimensions?

43. The length of the rectangular tennis court at Wimbledon is 6 feet longer than twice the width. If the court's perimeter is 228 feet, what are the court's dimensions?

44. The length of a rectangular pool is 6 meters less than twice the width. If the pool's perimeter is 126 meters, what are its dimensions?

45. The rectangular painting in the figure shown measures 12 inches by 16 inches and contains a frame of uniform width around the four edges. The perimeter of the rectangle formed by the painting and its frame is 72 inches. Determine the width of the frame.

46. The rectangular swimming pool in the figure shown measures 40 feet by 60 feet and contains a path of uniform width around the four edges. The perimeter of the rectangle formed by the pool and the surrounding path is 248 feet. Determine the width of the path.

40 feet
60 feet
x
x

47. For an international telephone call, a telephone company charges \$0.43 for the first minute, \$0.32 for each additional minute, and a \$2.10 service charge. If the cost of a call is \$5.73, how long did the person talk?

48. A job pays an annual salary of \$33,150, which includes a holiday bonus of \$750. If paychecks are issued twice a month, what is the gross amount for each paycheck?

In Exercises 49–74, solve each formula for the specified variable.

49. $A = lw$ for l (geometry)

50. $A = lw$ for w (geometry)

51. $A = \frac{1}{2}bh$ for b (geometry)

52. $A = \frac{1}{2}bh$ for h (geometry)

53. $I = Prt$ for P (finance)

54. $I = Prt$ for t (finance)

55. $T = D + pm$ for p (finance)

56. $P = C + MC$ for M (finance)

57. $A = \frac{1}{2}h(a + b)$ for a (geometry)

58. $A = \frac{1}{2}h(a + b)$ for b (geometry)

59. $V = \frac{1}{3}\pi r^2 h$ for h (geometry)

60. $V = \frac{1}{3}\pi r^2 h$ for r^2 (geometry)

61. $y - y_1 = m(x - x_1)$ for m (algebra)

62. $y_2 - y_1 = m(x_2 - x_1)$ for m (algebra)

63. $V = \frac{d_1 - d_2}{t}$ for d_1 (physics)

64. $z = \frac{x - u}{s}$ for x (statistics)

65. $Ax + By = C$ for x (algebra)

66. $Ax + By = C$ for y (algebra)

67. $s = \frac{1}{2}at^2 + vt$ for v (physics)

68. $s = \frac{1}{2}at^2 + vt$ for a (physics)

69. $L = a + (n - 1)d$ for n (algebra)

70. $L = a + (n - 1)d$ for d (algebra)

71. $A = 2lw + 2lh + 2wh$ for l (geometry)

72. $A = 2lw + 2lh + 2wh$ for h (geometry)

73. $IR + Ir = E$ for I (physics)

74. $A = \frac{x_1 + x_2 + x_n}{n}$ for n (statistics)

Explaining the Concepts

75. In your own words, describe a step-by-step approach for solving algebraic word problems.

76. Write an original word problem that can be solved using a linear equation. Then solve the problem.

77. Explain what it means to solve a formula for a variable.

78. Did you have difficulties solving some of the problems that were assigned in this Exercise Set? Discuss what you did if this happened to you. Did your course of action enhance your ability to solve algebraic word problems?

79. The bar graph in **Figure 1.23** on page 55 indicates a gender gap in pay at all levels of education. What explanations can you offer for this phenomenon?

Technology Exercises

80. Use a graphing utility to numerically or graphically verify your work in any one exercise from Exercises 29–32. For assistance on how to do this, refer to the Using Technology box on page 59.

81. The formula $y = 31 + 2.4x$ models the percentage of U.S. adults, y, viewing a college education as essential for success x years after 2000. Graph the formula in a [0, 20, 2] by [0, 100, 10] viewing rectangle. Then use the TRACE or ZOOM feature to verify your answer in Exercise 23.

82. In Exercises 25–26, we saw that in 1960, 23% of U.S. taxes came from corporate income tax, decreasing by approximately 0.28% per year.

a. Write a formula that models the percentage of U.S. taxes from corporations, y, x years after 1960.

b. Enter the formula from part (a) as y_1 in your graphing utility. Then use either a table for y_1 or a graph of y_1 to numerically or graphically verify your answer to Exercise 25 or Exercise 26.

Critical Thinking Exercises

Make Sense? *In Exercises 83–86, determine whether each statement makes sense or does not make sense, and explain your reasoning.*

83. I solved the formula for one of its variables, so now I have a numerical value for that variable.

84. Reasoning through a word problem often increases problem-solving skills in general.

85. The hardest part in solving a word problem is writing the equation that models the verbal conditions.

86. When traveling in Europe, the most useful form of the two Celsius-Fahrenheit conversion formulas is the formula used to convert from Fahrenheit to Celsius.

In Exercises 87–90, determine whether each statement is true or false. If the statement is false, make the necessary change(s) to produce a true statement.

87. If $I = prt$, then $t = I - pr$.

88. If $y = \frac{kx}{z}$, then $z = \frac{kx}{y}$.

89. It is not necessary to use the distributive property to solve $P = C + MC$ for C.

90. An item's price, x, reduced by $\frac{1}{3}$ is modeled by $x - \frac{1}{3}$.

91. The price of a dress is reduced by 40%. When the dress still does not sell, it is reduced by 40% of the reduced price. If the price of the dress after both reductions is \$72, what was the original price?

92. Suppose that we agree to pay you 8¢ for every problem in this chapter that you solve correctly and fine you 5¢ for every problem done incorrectly. If at the end of 26 problems we do not owe each other any money, how many problems did you solve correctly?

93. It was wartime when the Ricardos found out Mrs. Ricardo was pregnant. Ricky Ricardo was drafted and made out a will, deciding that \$14,000 in a savings account was to be divided between his wife and his child-to-be. Rather strangely, and certainly with gender bias, Ricky stipulated that if the child were a boy, he would get twice the amount of the mother's portion. If it were a girl, the mother would get twice the amount the girl was to receive. We'll never know what Ricky was thinking, for (as fate would have it) he did not return from war. Mrs. Ricardo gave birth to twins—a boy and a girl. How was the money divided?

94. A thief steals a number of rare plants from a nursery. On the way out, the thief meets three security guards, one after another. To each security guard, the thief is forced to give one-half the plants that he still has, plus 2 more. Finally, the thief leaves the nursery with 1 lone palm. How many plants were originally stolen?

95. Solve for C: $V = C - \frac{C - S}{L}N$.

Review Exercises

96. Express in set-builder notation and graph: $(-4, 0]$. (Section 1.1, Example 7)

97. Simplify: $\frac{(2 + 4)^2 + (-1)^5}{12 \div 2 \cdot 3 - 3}$. (Section 1.2, Example 8)

98. Solve: $\frac{2x}{3} - \frac{8}{3} = x$. (Section 1.4, Example 4)

Preview Exercises

Exercises 99–101 will help you prepare for the material covered in the next section.

99. In parts (a) and (b), complete each statement.

a. $b^4 \cdot b^3 = (b \cdot b \cdot b \cdot b)(b \cdot b \cdot b) = b^?$

b. $b^5 \cdot b^5 = (b \cdot b \cdot b \cdot b \cdot b)(b \cdot b \cdot b \cdot b \cdot b) = b^?$

c. Generalizing from parts (a) and (b), what should be done with the exponents when multiplying exponential expressions with the same base?

100. In parts (a) and (b), complete each statement.

a. $\frac{b^7}{b^3} = \frac{\cancel{b} \cdot \cancel{b} \cdot \cancel{b} \cdot b \cdot b \cdot b \cdot b}{\cancel{b} \cdot \cancel{b} \cdot \cancel{b}} = b^?$

b. $\frac{b^8}{b^2} = \frac{\cancel{b} \cdot \cancel{b} \cdot b \cdot b \cdot b \cdot b \cdot b \cdot b}{\cancel{b} \cdot \cancel{b}} = b^?$

c. Generalizing from parts (a) and (b), what should be done with the exponents when dividing exponential expressions with the same base?

101. Simplify: $\frac{1}{\left(-\frac{1}{2}\right)^3}$.

SECTION

1.6

Properties of Integral Exponents

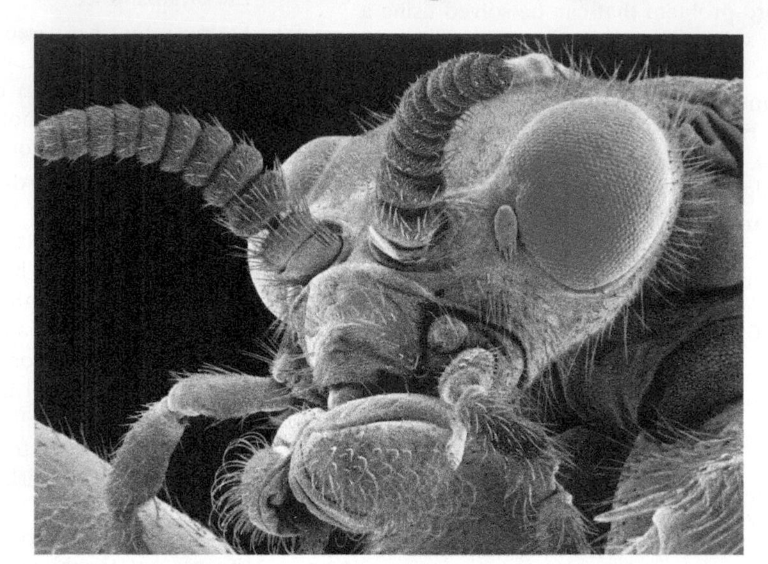

What am I supposed to learn?

After studying this section, you should be able to:

1. Use the product rule.
2. Use the quotient rule.
3. Use the zero-exponent rule.
4. Use the negative-exponent rule.
5. Use the power rule.
6. Find the power of a product.
7. Find the power of a quotient.
8. Simplify exponential expressions.

Our opening photo shows the head of a fly as seen under an electronic microscope. Some electronic microscopes can view objects that are less than 10^{-4} meter, or 0.0001 meter, in size. In this section, we'll make sense of the negative exponent in 10^{-4}, as we turn to integral exponents and their properties.

The Product and Quotient Rules

1 Use the product rule.

We have seen that exponents are used to indicate repeated multiplication. Now consider the multiplication of two exponential expressions, such as $b^4 \cdot b^3$. We are multiplying 4 factors of b and 3 factors of b. We have a total of 7 factors of b:

4 factors of b | 3 factors of b

$$b^4 \cdot b^3 = (b \cdot b \cdot b \cdot b)(b \cdot b \cdot b) = b^7.$$

Total: 7 factors of b

The product is exactly the same if we add the exponents:

$$b^4 \cdot b^3 = b^{4+3} = b^7.$$

This suggests the following rule:

The Product Rule

$$b^m \cdot b^n = b^{m+n}$$

When multiplying exponential expressions with the same base, add the exponents. Use this sum as the exponent of the common base.

EXAMPLE 1 Using the Product Rule

Multiply each expression using the product rule:

a. $b^8 \cdot b^{10}$ **b.** $(6x^4y^3)(5x^2y^7)$.

Solution

a. $b^8 \cdot b^{10} = b^{8+10} = b^{18}$

b. $(6x^4y^3)(5x^2y^7)$

$$= 6 \cdot 5 \cdot x^4 \cdot x^2 \cdot y^3 \cdot y^7$$

Use the associative and commutative properties. This step can be done mentally.

$$= 30x^{4+2}y^{3+7}$$
$$= 30x^6y^{10} \quad \blacksquare$$

✓ CHECK POINT 1 Multiply each expression using the product rule:

a. $b^6 \cdot b^5$ **b.** $(4x^3y^4)(10x^2y^6)$.

---

2 Use the quotient rule.

Now, consider the division of two exponential expressions, such as the quotient of b^7 and b^3. We are dividing 7 factors of b by 3 factors of b.

$$\frac{b^7}{b^3} = \frac{b \cdot b \cdot b \cdot b \cdot b \cdot b \cdot b}{b \cdot b \cdot b} = \boxed{\frac{b \cdot b \cdot b}{b \cdot b \cdot b}} \cdot b \cdot b \cdot b \cdot b = 1 \cdot b \cdot b \cdot b \cdot b = b^4$$

This factor is equal to 1.

The quotient is exactly the same if we subtract the exponents:

$$\frac{b^7}{b^3} = b^{7-3} = b^4.$$

This suggests the following rule:

The Quotient Rule

$$\frac{b^m}{b^n} = b^{m-n}, \quad b \neq 0$$

When dividing exponential expressions with the same nonzero base, subtract the exponent in the denominator from the exponent in the numerator. Use this difference as the exponent of the common base.

EXAMPLE 2 Using the Quotient Rule

Divide each expression using the quotient rule:

a. $\dfrac{(-2)^7}{(-2)^4}$ **b.** $\dfrac{30x^{12}y^9}{5x^3y^7}$.

Solution

a. $\dfrac{(-2)^7}{(-2)^4} = (-2)^{7-4} = (-2)^3$ or -8

$(-2)^3 = (-2)(-2)(-2) = -8$

b. $\dfrac{30x^{12}y^9}{5x^3y^7} = \dfrac{30}{5} \cdot \dfrac{x^{12}}{x^3} \cdot \dfrac{y^9}{y^7} = 6x^{12-3}y^{9-7} = 6x^9y^2 \quad \blacksquare$

✓ CHECK POINT 2 Divide each expression using the quotient rule:

a. $\dfrac{(-3)^6}{(-3)^3}$ **b.** $\dfrac{27x^{14}y^8}{3x^3y^5}$.

---

3 Use the zero-exponent rule.

Zero as an Exponent

A nonzero base can be raised to the 0 power. The quotient rule can be used to help determine what zero as an exponent should mean. Consider the quotient of b^4 and b^4, where b is not zero. We can determine this quotient in two ways.

$$\frac{b^4}{b^4} = 1 \qquad \frac{b^4}{b^4} = b^{4-4} = b^0$$

Any nonzero expression divided by itself is 1.

Use the quotient rule and subtract exponents.

This means that b^0 must equal 1.

The Zero-Exponent Rule

If b is any real number other than 0,

$$b^0 = 1.$$

EXAMPLE 3 Using the Zero-Exponent Rule

Use the zero-exponent rule to simplify each expression:

a. 8^0 **b.** $(-6)^0$ **c.** -6^0 **d.** $5x^0$ **e.** $(5x)^0$.

Solution

a. $8^0 = 1$ — Any nonzero number raised to the 0 power is 1.

b. $(-6)^0 = 1$

c. $-6^0 = -(6^0) = -1$ — Only 6 is raised to the 0 power.

d. $5x^0 = 5 \cdot 1 = 5$ — Only x is raised to the 0 power.

e. $(5x)^0 = 1$ — The entire expression, $5x$, is raised to the 0 power. ■

✓ CHECK POINT 3 Use the zero-exponent rule to simplify each expression:

a. 7^0 **b.** $(-5)^0$ **c.** -5^0 **d.** $10x^0$ **e.** $(10x)^0$.

4 Use the negative-exponent rule.

Negative Integers as Exponents

A nonzero base can be raised to a negative power. The quotient rule can be used to help determine what a negative integer as an exponent should mean. Consider the quotient of b^3 and b^5, where b is not zero. We can determine this quotient in two ways.

$$\frac{b^3}{b^5} = \frac{\cancel{b} \cdot \cancel{b} \cdot \cancel{b}}{\cancel{b} \cdot \cancel{b} \cdot \cancel{b} \cdot b \cdot b} = \frac{1}{b^2} \qquad \frac{b^3}{b^5} = b^{3-5} = b^{-2}$$

After dividing common factors, we have two factors of b in the denominator.

Use the quotient rule and subtract exponents.

Notice that $\frac{b^3}{b^5}$ equals both b^{-2} and $\frac{1}{b^2}$. This means that b^{-2} must equal $\frac{1}{b^2}$. This example is a special case of the **negative-exponent rule**.

The Negative-Exponent Rule

If b is any real number other than 0 and n is a natural number, then

$$b^{-n} = \frac{1}{b^n}.$$

EXAMPLE 4 Using the Negative-Exponent Rule

Use the negative-exponent rule to write each expression with a positive exponent. Simplify, if possible:

a. 9^{-2} **b.** $(-2)^{-5}$ **c.** $\frac{1}{6^{-2}}$ **d.** $7x^{-5}y^2$.

Solution

a. $9^{-2} = \frac{1}{9^2} = \frac{1}{81}$

b. $(-2)^{-5} = \frac{1}{(-2)^5} = \frac{1}{(-2)(-2)(-2)(-2)(-2)} = \frac{1}{-32} = -\frac{1}{32}$

Only the sign of the exponent, −5, changes. The base, −2, does not change sign.

c. $\frac{1}{6^{-2}} = \frac{1}{\frac{1}{6^2}} = 1 \cdot \frac{6^2}{1} = 6^2 = 36$

d. $7x^{-5}y^2 = 7 \cdot \frac{1}{x^5} \cdot y^2 = \frac{7y^2}{x^5}$ ■

✓ **CHECK POINT 4** Use the negative-exponent rule to write each expression with a positive exponent. Simplify, if possible:

a. 5^{-2} **b.** $(-3)^{-3}$ **c.** $\frac{1}{4^{-2}}$ **d.** $3x^{-6}y^4$.

In Example 4 and Check Point 4, did you notice that

$$\frac{1}{6^{-2}} = 6^2 \quad \text{and} \quad \frac{1}{4^{-2}} = 4^2?$$

In general, if a negative exponent appears in a denominator, an expression can be written with a positive exponent using

$$\frac{1}{b^{-n}} = b^n.$$

Negative Exponents in Numerators and Denominators

If b is any real number other than 0 and n is a natural number, then

$$b^{-n} = \frac{1}{b^n} \quad \text{and} \quad \frac{1}{b^{-n}} = b^n.$$

When a negative number appears as an exponent, switch the position of the base (from numerator to denominator or from denominator to numerator) and make the exponent positive. The sign of the base does not change.

EXAMPLE 5 Using Negative Exponents

Write each expression with positive exponents only. Then simplify, if possible:

a. $\frac{5^{-3}}{4^{-2}}$ **b.** $\frac{1}{6x^{-4}}$.

Solution

a. $\frac{5^{-3}}{4^{-2}} = \frac{4^2}{5^3} = \frac{4 \cdot 4}{5 \cdot 5 \cdot 5} = \frac{16}{125}$ Switch the position of each base to the other side of the fraction bar and change the sign of the exponent.

b. $\frac{1}{6x^{-4}} = \frac{x^4}{6}$ Switch the position of x to the other side of the fraction bar and change -4 to 4.

Don't switch the position of 6. It is not affected by a negative exponent.

■

CHECK POINT 5 Write each expression with positive exponents only. Then simplify, if possible:

a. $\frac{7^{-2}}{4^{-3}}$ **b.** $\frac{1}{5x^{-2}}$.

---

5 Use the power rule.

The Power Rule for Exponents (Powers to Powers)

The next property of exponents applies when an exponential expression is raised to a power. Here is an example:

$$(b^2)^4.$$

The exponential expression b^2 is raised to the fourth power.

There are 4 factors of b^2. Thus,

$$(b^2)^4 = b^2 \cdot b^2 \cdot b^2 \cdot b^2 = b^{2+2+2+2} = b^8.$$

Add exponents when multiplying with the same base.

We can obtain the answer, b^8, by multiplying the exponents:

$$(b^2)^4 = b^{2 \cdot 4} = b^8.$$

This suggests the following rule:

The Power Rule (Powers to Powers)

$$(b^m)^n = b^{mn}$$

When an exponential expression is raised to a power, multiply the exponents. Place the product of the exponents on the base and remove the parentheses.

EXAMPLE 6 Using the Power Rule (Powers to Powers)

Simplify each expression using the power rule:

a. $(x^6)^4$ **b.** $(y^5)^{-3}$ **c.** $(b^{-4})^{-2}$.

Solution

a. $(x^6)^4 = x^{6 \cdot 4} = x^{24}$

b. $(y^5)^{-3} = y^{5(-3)} = y^{-15} = \frac{1}{y^{15}}$

c. $(b^{-4})^{-2} = b^{(-4)(-2)} = b^8$ ■

CHECK POINT 6 Simplify each expression using the power rule:

a. $(x^5)^3$ **b.** $(y^7)^{-2}$ **c.** $(b^{-3})^{-4}$.

6 Find the power of a product.

The Products-to-Powers Rule for Exponents

The next property of exponents applies when we are raising a product to a power. Here is an example:

$$(2x)^4.$$

The product $2x$ is raised to the fourth power.

There are four factors of $2x$. Thus,

$$(2x)^4 = 2x \cdot 2x \cdot 2x \cdot 2x = 2 \cdot 2 \cdot 2 \cdot 2 \cdot x \cdot x \cdot x \cdot x = 2^4x^4.$$

We can obtain the answer, 2^4x^4, by raising each factor within the parentheses to the fourth power:

$$(2x)^4 = 2^4x^4.$$

This suggests the following rule:

Products to Powers

$$(ab)^n = a^nb^n$$

When a product is raised to a power, raise each factor to that power.

EXAMPLE 7 Using the Products-to-Powers Rule

Simplify each expression using the products-to-powers rule:

a. $(6x)^3$ **b.** $(-2y^2)^4$ **c.** $(-3x^{-1}y^3)^{-2}$.

Solution

a. $(6x)^3 = 6^3x^3$ Raise each factor to the third power.

$= 216x^3$ Simplify: $6^3 = 6 \cdot 6 \cdot 6 = 216$.

b. $(-2y^2)^4 = (-2)^4(y^2)^4$ Raise each factor to the fourth power.

$= (-2)^4y^{2 \cdot 4}$ To raise an exponential expression to a power, multiply exponents: $(b^m)^n = b^{mn}$.

$= 16y^8$ Simplify: $(-2)^4 = (-2)(-2)(-2)(-2) = 16$.

c. $(-3x^{-1}y^3)^{-2} = (-3)^{-2}(x^{-1})^{-2}(y^3)^{-2}$ Raise each factor to the -2 power.

$= (-3)^{-2}x^{(-1)(-2)}y^{3(-2)}$ Use $(b^m)^n = b^{mn}$ on the second and third factors.

$= (-3)^{-2}x^2y^{-6}$ Simplify.

$= \dfrac{1}{(-3)^2} \cdot x^2 \cdot \dfrac{1}{y^6}$ Apply $b^{-n} = \dfrac{1}{b^n}$ to the first and last factors.

$= \dfrac{x^2}{9y^6}$ Simplify: $(-3)^2 = (-3)(-3) = 9$. ■

CHECK POINT 7 Simplify each expression using the products-to-powers rule:

a. $(2x)^4$ **b.** $(-3y^2)^3$ **c.** $(-4x^5y^{-1})^{-2}$.

7 Find the power of a quotient.

The Quotients-to-Powers Rule for Exponents

The following rule is used to raise a quotient to a power:

Quotients to Powers

If b is a nonzero real number, then

$$\left(\frac{a}{b}\right)^n = \frac{a^n}{b^n}.$$

When a quotient is raised to a power, raise the numerator to that power and divide by the denominator to that power.

EXAMPLE 8 Using the Quotients-to-Powers Rule

Simplify each expression using the quotients-to-powers rule:

a. $\left(\frac{x^2}{4}\right)^3$ **b.** $\left(\frac{2x^3}{y^{-4}}\right)^5$ **c.** $\left(\frac{x^3}{y^2}\right)^{-4}$.

Great Question!

In Example 8, why didn't you start by simplifying inside parentheses?

When simplifying exponential expressions, the first step should be to simplify inside parentheses. In Example 8, the expressions inside parentheses have different bases and cannot be simplified. This is why we begin with the quotients-to-powers rule.

Solution

a. $\left(\frac{x^2}{4}\right)^3 = \frac{(x^2)^3}{4^3} = \frac{x^{2\cdot 3}}{4\cdot 4\cdot 4} = \frac{x^6}{64}$ Cube the numerator and the denominator.

b. $\left(\frac{2x^3}{y^{-4}}\right)^5 = \frac{(2x^3)^5}{(y^{-4})^5}$ Raise the numerator and the denominator to the fifth power.

$= \frac{2^5(x^3)^5}{(y^{-4})^5}$ Raise each factor in the numerator to the fifth power.

$= \frac{2^5 \cdot x^{3\cdot 5}}{y^{(-4)(5)}}$ Multiply exponents in both powers-to-powers expressions: $(b^m)^n = b^{mn}$.

$= \frac{32x^{15}}{y^{-20}}$ Simplify.

$= 32x^{15}y^{20}$ Move y to the other side of the fraction bar and change -20 to 20: $\frac{1}{b^{-n}} = b^n$.

c. $\left(\frac{x^3}{y^2}\right)^{-4} = \frac{(x^3)^{-4}}{(y^2)^{-4}}$ Raise the numerator and the denominator to the -4 power.

$= \frac{x^{3(-4)}}{y^{2(-4)}}$ Multiply exponents in both powers-to-powers expressions: $(b^m)^n = b^{mn}$.

$= \frac{x^{-12}}{y^{-8}}$ Simplify.

$= \frac{y^8}{x^{12}}$ Move each base to the other side of the fraction bar and make each exponent positive. ■

✓ CHECK POINT 8 Simplify each expression using the quotients-to-powers rule:

a. $\left(\frac{x^5}{4}\right)^3$ **b.** $\left(\frac{2x^{-3}}{y^2}\right)^4$ **c.** $\left(\frac{x^{-3}}{y^4}\right)^{-5}$.

8 Simplify exponential expressions.

Simplifying Exponential Expressions

Properties of exponents are used to simplify exponential expressions. An exponential expression is **simplified** when

- No parentheses appear.
- No powers are raised to powers.
- Each base occurs only once.
- No negative or zero exponents appear.

Simplifying Exponential Expressions

	Example
1. If necessary, remove parentheses by using $(ab)^n = a^n b^n$ or $\left(\frac{a}{b}\right)^n = \frac{a^n}{b^n}$.	$(xy)^3 = x^3y^3$
2. If necessary, simplify powers to powers by using $(b^m)^n = b^{mn}$.	$(x^4)^3 = x^{4\cdot 3} = x^{12}$
3. If necessary, be sure that each base appears only once by using $b^m \cdot b^n = b^{m+n}$ or $\frac{b^m}{b^n} = b^{m-n}$.	$x^4 \cdot x^3 = x^{4+3} = x^7$
4. If necessary, rewrite exponential expressions with zero powers as 1 ($b^0 = 1$). Furthermore, write the answer with positive exponents by using $b^{-n} = \frac{1}{b^n}$ or $\frac{1}{b^{-n}} = b^n$.	$\frac{x^5}{x^8} = x^{5-8} = x^{-3} = \frac{1}{x^3}$

The following example shows how to simplify exponential expressions. Throughout the example, assume that no variable in a denominator is equal to zero.

EXAMPLE 9 Simplifying Exponential Expressions

Simplify:

a. $(-2xy^{-14})(-3x^4y^5)^3$ **b.** $\left(\frac{25x^2y^4}{-5x^6y^{-8}}\right)^2$ **c.** $\left(\frac{x^{-4}y^7}{2}\right)^{-5}$.

Solution

a. $(-2xy^{-14})(-3x^4y^5)^3$

$= (-2xy^{-14})(-3)^3(x^4)^3(y^5)^3$ Cube each factor in the second parentheses.

$= (-2xy^{-14})(-27)x^{12}y^{15}$ Multiply the exponents when raising a power to a power: $(x^4)^3 = x^{4\cdot 3} = x^{12}$ and $(y^5)^3 = y^{5\cdot 3} = y^{15}$.

$= (-2)(-27)x^{1+12}y^{-14+15}$ Mentally rearrange factors and multiply like bases by adding the exponents.

$= 54x^{13}y$ Simplify.

b. $\left(\dfrac{25x^2y^4}{-5x^6y^{-8}}\right)^2$ The expression inside parentheses contains more than one occurrence of each base and can be simplified. Begin with this simplification.

$= (-5x^{2-6}y^{4-(-8)})^2$ Simplify inside the parentheses. Subtract the exponents when dividing.

$= (-5x^{-4}y^{12})^2$ Simplify.

$= (-5)^2(x^{-4})^2(y^{12})^2$ Square each factor in parentheses.

$= 25x^{-8}y^{24}$ Multiply the exponents when raising a power to a power: $(x^{-4})^2 = x^{-4(2)} = x^{-8}$ and $(y^{12})^2 = y^{12\cdot 2} = y^{24}$.

$= \dfrac{25y^{24}}{x^8}$ Simplify x^{-8} using $b^{-n} = \dfrac{1}{b^n}$.

c. $\left(\dfrac{x^{-4}y^7}{2}\right)^{-5}$

$= \dfrac{(x^{-4}y^7)^{-5}}{2^{-5}}$ Raise the numerator and the denominator to the -5 power.

$= \dfrac{(x^{-4})^{-5}(y^7)^{-5}}{2^{-5}}$ Raise each factor in the numerator to the -5 power.

$= \dfrac{x^{20}y^{-35}}{2^{-5}}$ Multiply the exponents when raising a power to a power: $(x^{-4})^{-5} = x^{-4(-5)} = x^{20}$ and $(y^7)^{-5} = y^{7(-5)} = y^{-35}$.

$= \dfrac{2^5x^{20}}{y^{35}}$ Move each base with a negative exponent to the other side of the fraction bar and make each negative exponent positive.

$= \dfrac{32x^{20}}{y^{35}}$ Simplify: $2^5 = 2\cdot 2\cdot 2\cdot 2\cdot 2 = 32$. ■

✓ CHECK POINT 9 Simplify:

a. $(-3x^{-6}y)(-2x^3y^4)^2$ **b.** $\left(\dfrac{10x^3y^5}{5x^6y^{-2}}\right)^2$ **c.** $\left(\dfrac{x^3y^5}{4}\right)^{-3}$.

Great Question!

Simplifying exponential expressions seems to involve lots of steps. Are there common errors I can avoid along the way?

Yes. Here's a list. The first column has the correct simplification. The second column contains common errors you should try to avoid.

Correct	Incorrect	Description of Error
$b^3\cdot b^4 = b^7$	$b^3\cdot b^4 = b^{12}$	The exponents should be added, not multiplied.
$3^2\cdot 3^4 = 3^6$	$3^2\cdot 3^4 = 9^6$	The common base should be retained, not multiplied.
$\dfrac{5^{16}}{5^4} = 5^{12}$	$\dfrac{5^{16}}{5^4} = 5^4$	The exponents should be subtracted, not divided.
$(4a)^3 = 64a^3$	$(4a)^3 = 4a^3$	Both factors should be cubed.
$b^{-n} = \dfrac{1}{b^n}$	$b^{-n} = -\dfrac{1}{b^n}$	Only the exponent should change sign.
$(a+b)^{-1} = \dfrac{1}{a+b}$	$(a+b)^{-1} = \dfrac{1}{a} + \dfrac{1}{b}$	The exponent applies to the entire expression $a + b$.

CONCEPT AND VOCABULARY CHECK

Fill in each blank so that the resulting statement is true.

1. The product rule for exponents states that $b^m \cdot b^n =$ ________. When multiplying exponential expressions with the same base, ________ the exponents.
2. The quotient rule for exponents states that $\frac{b^m}{b^n} =$ ________, $b \neq 0$. When dividing exponential expressions with the same nonzero base, ________ the exponents.
3. If $b \neq 0$, then $b^0 =$ ________.
4. The negative-exponent rule states that $b^{-n} =$ ________, $b \neq 0$.
5. True or false: $5^{-2} = -5^2$ ________
6. Negative exponents in denominators can be evaluated using $\frac{1}{b^{-n}} =$ ________, $b \neq 0$.
7. True or false: $\frac{1}{8^{-2}} = 8^2$ ________

1.6 EXERCISE SET

MyMathLab®

Practice Exercises

In Exercises 1–14, multiply using the product rule.

1. $b^4 \cdot b^7$
2. $b^5 \cdot b^9$
3. $x \cdot x^3$
4. $x \cdot x^4$
5. $2^3 \cdot 2^2$
6. $2^4 \cdot 2^2$
7. $3x^4 \cdot 2x^2$
8. $5x^3 \cdot 3x^2$
9. $(-2y^{10})(-10y^2)$
10. $(-4y^8)(-8y^4)$
11. $(5x^3y^4)(20x^7y^8)$
12. $(4x^5y^6)(20x^7y^4)$
13. $(-3x^4y^0z)(-7xyz^3)$
14. $(-9x^3yz^4)(-5xy^0z^2)$

In Exercises 15–24, divide using the quotient rule.

15. $\frac{b^{12}}{b^3}$
16. $\frac{b^{25}}{b^5}$
17. $\frac{15x^9}{3x^4}$
18. $\frac{18x^{11}}{3x^4}$
19. $\frac{x^9y^7}{x^4y^2}$
20. $\frac{x^9y^{12}}{x^2y^6}$
21. $\frac{50x^2y^7}{5xy^4}$
22. $\frac{36x^{12}y^4}{4xy^2}$
23. $\frac{-56a^{12}b^{10}c^8}{7ab^2c^4}$
24. $\frac{-66a^9b^7c^6}{6a^3bc^2}$

In Exercises 25–34, use the zero-exponent rule to simplify each expression.

25. 6^0
26. 9^0
27. $(-4)^0$
28. $(-2)^0$
29. -4^0
30. -2^0
31. $13y^0$
32. $17y^0$
33. $(13y)^0$
34. $(17y)^0$

In Exercises 35–52, write each expression with positive exponents only. Then simplify, if possible.

35. 3^{-2}
36. 4^{-2}
37. $(-5)^{-2}$
38. $(-7)^{-2}$
39. -5^{-2}
40. -7^{-2}
41. x^2y^{-3}
42. x^3y^{-4}
43. $8x^{-7}y^3$
44. $9x^{-8}y^4$
45. $\frac{1}{5^{-3}}$
46. $\frac{1}{2^{-5}}$
47. $\frac{1}{(-3)^{-4}}$
48. $\frac{1}{(-2)^{-4}}$
49. $\frac{x^{-2}}{y^{-5}}$
50. $\frac{x^{-3}}{y^{-7}}$
51. $\frac{a^{-4}b^7}{c^{-3}}$
52. $\frac{a^{-3}b^8}{c^{-2}}$

In Exercises 53–58, simplify each expression using the power rule.

53. $(x^6)^{10}$
54. $(x^3)^2$
55. $(b^4)^{-3}$
56. $(b^8)^{-3}$
57. $(7^{-4})^{-5}$
58. $(9^{-4})^{-5}$

In Exercises 59–72, simplify each expression using the products-to-powers rule.

59. $(4x)^3$
60. $(2x)^5$
61. $(-3x^7)^2$
62. $(-4x^9)^2$
63. $(2xy^2)^3$
64. $(3x^2y)^4$

65. $(-3x^2y^5)^2$

66. $(-3x^4y^6)^2$

67. $(-3x^{-2})^{-3}$

68. $(-2x^{-4})^{-3}$

69. $(5x^3y^{-4})^{-2}$

70. $(7x^2y^{-5})^{-2}$

71. $(-2x^{-5}y^4z^2)^{-4}$

72. $(-2x^{-4}y^5z^3)^{-4}$

In Exercises 73–84, simplify each expression using the quotients-to-powers rule.

73. $\left(\frac{2}{x}\right)^4$

74. $\left(\frac{y}{2}\right)^5$

75. $\left(\frac{x^3}{5}\right)^2$

76. $\left(\frac{x^4}{6}\right)^2$

77. $\left(-\frac{3x}{y}\right)^4$

78. $\left(-\frac{2x}{y}\right)^5$

79. $\left(\frac{x^4}{y^2}\right)^6$

80. $\left(\frac{x^5}{y^3}\right)^6$

81. $\left(\frac{x^3}{y^{-4}}\right)^3$

82. $\left(\frac{x^4}{y^{-2}}\right)^3$

83. $\left(\frac{a^{-2}}{b^3}\right)^{-4}$

84. $\left(\frac{a^{-3}}{b^5}\right)^{-4}$

In Exercises 85–116, simplify each exponential expression.

85. $\frac{x^3}{x^9}$

86. $\frac{x^6}{x^{10}}$

87. $\frac{20x^3}{-5x^4}$

88. $\frac{10x^5}{-2x^6}$

89. $\frac{16x^3}{8x^{10}}$

90. $\frac{15x^2}{3x^{11}}$

91. $\frac{20a^3b^8}{2ab^{13}}$

92. $\frac{72a^5b^{11}}{9ab^{17}}$

93. $x^3 \cdot x^{-12}$

94. $x^4 \cdot x^{-12}$

95. $(2a^5)(-3a^{-7})$

96. $(4a^2)(-2a^{-5})$

97. $\left(-\frac{1}{4}x^{-4}y^5z^{-1}\right)(-12x^{-3}y^{-1}z^4)$

98. $\left(-\frac{1}{3}x^{-5}y^4z^6\right)(-18x^{-2}y^{-1}z^{-7})$

99. $\frac{6x^2}{2x^{-8}}$

100. $\frac{12x^5}{3x^{-10}}$

101. $\frac{x^{-7}}{x^3}$

102. $\frac{x^{-10}}{x^4}$

103. $\frac{30x^2y^5}{-6x^8y^{-3}}$

104. $\frac{24x^2y^{13}}{-2x^5y^{-2}}$

105. $\frac{-24a^3b^{-5}c^5}{-3a^{-6}b^{-4}c^{-7}}$

106. $\frac{-24a^2b^{-2}c^8}{-8a^{-5}b^{-1}c^{-3}}$

107. $\left(\frac{x^3}{x^{-5}}\right)^2$

108. $\left(\frac{x^4}{x^{-11}}\right)^3$

109. $\left(\frac{-15a^4b^2}{5a^{10}b^{-3}}\right)^3$

110. $\left(\frac{-30a^{14}b^8}{10a^{17}b^{-2}}\right)^3$

111. $\left(\frac{3a^{-5}b^2}{12a^3b^{-4}}\right)^0$

112. $\left(\frac{4a^{-5}b^3}{12a^3b^{-5}}\right)^0$

113. $\left(\frac{x^{-5}y^8}{3}\right)^{-4}$

114. $\left(\frac{x^6y^{-7}}{2}\right)^{-3}$

115. $\left(\frac{20a^{-3}b^4c^5}{-2a^{-5}b^{-2}c}\right)^{-2}$

116. $\left(\frac{-2a^{-4}b^3c^{-1}}{3a^{-2}b^{-5}c^{-2}}\right)^{-4}$

Practice PLUS

In Exercises 117–124, simplify each exponential expression.

117. $\frac{9y^4}{x^{-2}} + \left(\frac{x^{-1}}{y^2}\right)^{-2}$

118. $\frac{7x^3}{y^{-9}} + \left(\frac{x^{-1}}{y^3}\right)^{-3}$

119. $\left(\frac{3x^4}{y^{-4}}\right)^{-1}\left(\frac{2x}{y^2}\right)^3$

120. $\left(\frac{2^{-1}x^{-2}y}{x^4y^{-1}}\right)^{-2}\left(\frac{xy^{-3}}{x^{-3}y}\right)^3$

121. $(-4x^3y^{-5})^{-2}(2x^{-8}y^{-5})$

122. $(-4x^{-4}y^5)^{-2}(-2x^5y^{-6})$

123. $\frac{(2x^2y^4)^{-1}(4xy^3)^{-3}}{(x^2y)^{-5}(x^3y^2)^4}$

124. $\frac{(3x^3y^2)^{-1}(2x^2y)^{-2}}{(xy^2)^{-5}(x^2y^3)^3}$

Application Exercises

The formula

$$A = 1000 \cdot 2^t$$

models the population, A, of aphids in a field of potato plants after t weeks. Use this formula to solve Exercises 125–126.

125. **a.** What is the present aphid population?
b. What will the aphid population be in 4 weeks?
c. What was the aphid population 3 weeks ago?

126. **a.** What is the present aphid population?
b. What will the aphid population be in 3 weeks?
c. What was the aphid population 2 weeks ago?

A rumor about algebra CDs that you can listen to as you sleep, allowing you to awaken refreshed and algebraically empowered, is spreading among the students in your math class. The formula

$$N = \frac{25}{1 + 24 \cdot 2^{-t}}$$

models the number of people in the class, N, who have heard the rumor after t minutes. Use this formula to solve Exercises 127–128.

127. **a.** How many people in the class started the rumor?
b. How many people in the class have heard the rumor after 4 minutes?

128. **a.** How many people in the class started the rumor?
b. How many people in the class, rounded to the nearest whole number, have heard the rumor after 6 minutes?

Use the graph of the rumor model to solve Exercises 129–132.

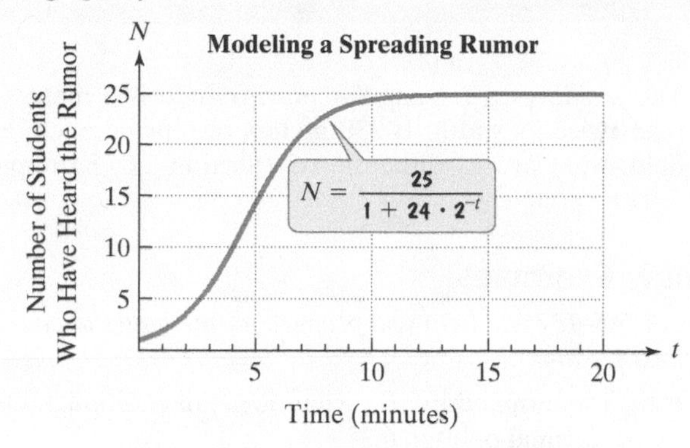

129. Identify your answers to Exercise 127, parts (a) and (b), as points on the graph.

130. Identify your answers to Exercise 128, parts (a) and (b), as points on the graph.

131. Which one of the following best describes the rate of growth of the rumor as shown by the graph?

a. The number of people in the class who heard the rumor grew steadily over time.

b. The number of people in the class who heard the rumor remained constant over time.

c. The number of people in the class who heard the rumor increased slowly at the beginning, but this rate of increase continued to escalate over time.

d. The number of people in the class who heard the rumor increased quite rapidly at the beginning, but this rate of increase eventually slowed down, ultimately limited by the number of students in the class.

132. Use the graph to determine how many people in the class eventually heard the rumor.

The astronomical unit (AU) is often used to measure distances within the solar system. One AU is equal to the average distance between Earth and the sun, or 92,955,630 miles. The distance, d, of the nth planet from the sun is modeled by the formula

$$d = \frac{3(2^{n-2}) + 4}{10},$$

where d is measured in astronomical units. Use this formula to solve Exercises 133–136.

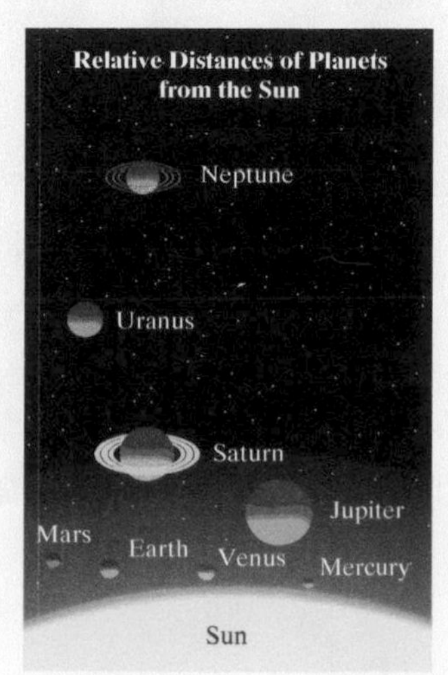

133. Substitute 1 for n and find the distance between Mercury and the sun.

134. Substitute 2 for n and find the distance between Venus and the sun.

135. How much farther from the sun is Jupiter than Earth?

136. How much farther from the sun is Uranus than Earth?

Explaining the Concepts

137. Explain the product rule for exponents. Use $b^2 \cdot b^3$ in your explanation.

138. Explain the quotient rule for exponents. Use $\frac{b^8}{b^2}$ in your explanation.

139. Explain how to find any nonzero number to the 0 power.

140. Explain the negative-exponent rule and give an example.

141. Explain the power rule for exponents. Use $(b^2)^3$ in your explanation.

142. Explain how to simplify an expression that involves a product raised to a power. Give an example.

143. Explain how to simplify an expression that involves a quotient raised to a power. Give an example.

144. How do you know if an exponential expression is simplified?

Technology Exercise

145. Enter the rumor formula

$$N = \frac{25}{1 + 24 \cdot 2^{-t}}$$

in your graphing utility as

$y_1 = 25$ [÷] [(] 1 [+] 24 [×] 2 [∧] [(−)] [x] [)].

Then use a table for y_1 to numerically verify your answers to Exercise 127 or 128.

Critical Thinking Exercises

Make Sense? *In Exercises 146–149, determine whether each statement makes sense or does not make sense, and explain your reasoning.*

146. The properties $(ab)^n = a^n b^n$ and $\left(\frac{a}{b}\right)^n = \frac{a^n}{b^n}$ are like distributive properties of powers over multiplication and division.

147. If 7^{-2} is raised to the third power, the result is a number between 0 and 1.

148. There are many exponential expressions that are equal to $25x^{12}$, such as $(5x^6)^2$, $(5x^3)(5x^9)$, $25(x^3)^9$, and $5^2(x^2)^6$.

149. The expression $\frac{a^n}{b^0}$ is undefined because division by 0 is undefined.

In Exercises 150–157, determine whether each statement is true or false. If the statement is false, make the necessary change(s) to produce a true statement.

150. $2^2 \cdot 2^4 = 2^8$

151. $5^6 \cdot 5^2 = 25^8$

152. $2^3 \cdot 3^2 = 6^5$

153. $\frac{1}{(-2)^3} = 2^{-3}$

154. $\frac{2^8}{2^{-3}} = 2^5$

155. $2^4 + 2^5 = 2^9$

156. $2000.002 = (2 \times 10^3) + (2 \times 10^{-3})$

157. $40{,}000.04 = (4 \times 10^4) + (4 \times 10^{-2})$

In Exercises 158–161, simplify the expression. Assume that all variables used as exponents represent integers and that all other variables represent nonzero real numbers.

158. $x^{n-1} \cdot x^{3n+4}$

159. $(x^{-4n} \cdot x^n)^{-3}$

160. $\left(\frac{x^{3-n}}{x^{6-n}}\right)^{-2}$

161. $\left(\frac{x^n y^{3n+1}}{y^n}\right)^3$

Review Exercises

162. Graph $y = 2x - 1$ in a rectangular coordinate system. Let $x = -3, -2, -1, 0, 1, 2,$ and 3. (Section 1.3, Example 2)

163. Solve $Ax + By = C$ for y. (Section 1.5, Example 6)

164. The length of a rectangular playing field is 5 meters less than twice its width. If 230 meters of fencing enclose the field, what are its dimensions? (Section 1.5, Example 5)

Preview Exercises

Exercises 165–167 will help you prepare for the material covered in the next section.

165. If 6.2 is multiplied by 10^3, what does this multiplication do to the decimal point in 6.2?

166. If 8.5 is multiplied by 10^{-2}, what does this multiplication do to the decimal point in 8.5?

167. Write each computation as a single power of 10. Then evaluate this exponential expression.

a. $10^9 \times 10^{-4}$ **b.** $\frac{10^4}{10^{-2}}$

SECTION

1.7

Scientific Notation

What am I supposed to learn?

After studying this section, you should be able to:

1. Convert from scientific to decimal notation.
2. Convert from decimal to scientific notation.
3. Perform computations with scientific notation.
4. Use scientific notation to solve problems.

People who complain about paying their income tax can be divided into two types: men and women. Perhaps we can quantify the complaining by examining the data in **Figure 1.26**. The bar graphs show the U.S. population, in millions, and the total amount we paid in federal taxes, in trillions of dollars, from 2004 through 2012.

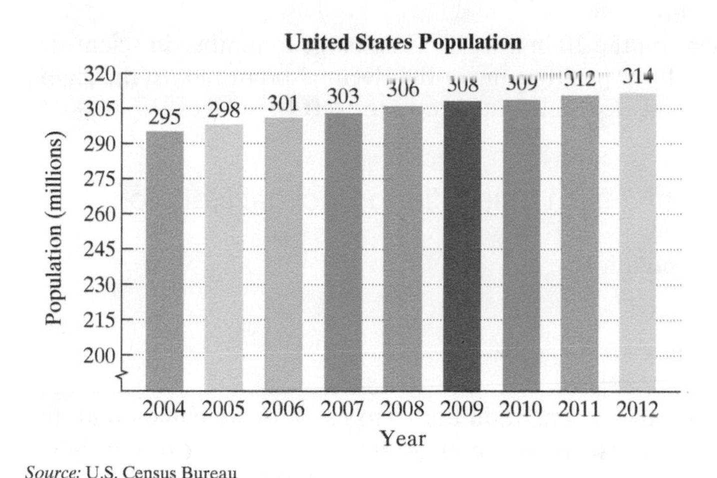

Source: U.S. Census Bureau

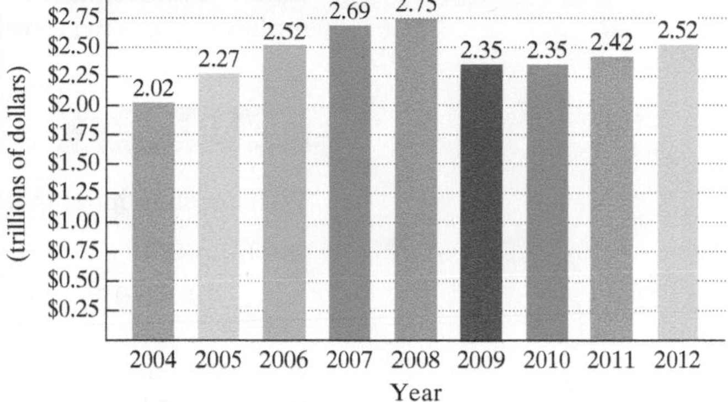

Source: Internal Revenue Service

Figure 1.26 Population and total tax collections in the United States

The bar graph on the right shows that in 2012 total tax collections were \$2.52 trillion. How can we place this amount in the proper perspective? If the total tax collections were evenly divided among all Americans, how much would each citizen pay in taxes?

In this section, you will learn to use exponents to provide a way of putting large and small numbers in perspective. Using this skill, we will explore the per capita tax for some of the years shown in **Figure 1.26**.

1 Convert from scientific to decimal notation.

Scientific Notation

We have seen that in 2012 total tax collections were \$2.52 trillion. Because a trillion is 10^{12} (see **Table 1.7**), this amount can be expressed as

$$2.52 \times 10^{12}.$$

The number 2.52×10^{12} is written in a form called *scientific notation*.

Table 1.7 Names of Large Numbers

10^2	hundred
10^3	thousand
10^6	million
10^9	billion
10^{12}	trillion
10^{15}	quadrillion
10^{18}	quintillion
10^{21}	sextillion
10^{24}	septillion
10^{27}	octillion
10^{30}	nonillion
10^{100}	googol

Scientific Notation

A number is written in **scientific notation** when it is expressed in the form

$$a \times 10^n,$$

where the absolute value of a is greater than or equal to 1 and less than 10 $(1 \le |a| < 10)$, and n is an integer.

It is customary to use the multiplication symbol, ×, rather than a dot, when writing a number in scientific notation.

Converting from Scientific to Decimal Notation

Here are two examples of numbers in scientific notation:

$$6.4 \times 10^5 \quad \text{means} \quad 640{,}000.$$
$$2.17 \times 10^{-3} \quad \text{means} \quad 0.00217.$$

Do you see that the number with the positive exponent is relatively large and the number with the negative exponent is relatively small?

We can use n, the exponent on the 10 in $a \times 10^n$, to change a number in scientific notation to decimal notation. If n is **positive**, move the decimal point in a to the **right** n places. If n is **negative**, move the decimal point in a to the **left** $|n|$ places.

EXAMPLE 1 Converting from Scientific to Decimal Notation

Write each number in decimal notation:

a. 6.2×10^7 **b.** -6.2×10^7

c. 2.019×10^{-3} **d.** -2.019×10^{-3}.

Solution In each case, we use the exponent on the 10 to move the decimal point. In parts (a) and (b), the exponent is positive, so we move the decimal point to the right. In parts (c) and (d), the exponent is negative, so we move the decimal point to the left.

a. $6.2 \times 10^7 = 62{,}000{,}000$

$n = 7$ Move the decimal point 7 places to the right.

b. $-6.2 \times 10^7 = -62{,}000{,}000$

$n = 7$ Move the decimal point 7 places to the right.

c. $2.019 \times 10^{-3} = 0.002019$

$n = -3$ Move the decimal point $|-3|$ places, or 3 places, to the left.

d. $-2.019 \times 10^{-3} = -0.002019$

$n = -3$ Move the decimal point $|-3|$ places, or 3 places, to the left. ■

✓ CHECK POINT 1 Write each number in decimal notation:

a. -2.6×10^9 **b.** 3.017×10^{-6}.

2 Convert from decimal to scientific notation.

Converting from Decimal to Scientific Notation

To convert from decimal notation to scientific notation, we reverse the procedure of Example 1.

Converting from Decimal to Scientific Notation

Write the number in the form $a \times 10^n$.

- Determine a, the numerical factor. Move the decimal point in the given number to obtain a number whose absolute value is between 1 and 10, including 1.
- Determine n, the exponent on 10^n. The absolute value of n is the number of places the decimal point was moved. The exponent n is positive if the decimal point was moved to the left, negative if the decimal point was moved to the right, and 0 if the decimal point was not moved.

EXAMPLE 2 Converting from Decimal Notation to Scientific Notation

Write each number in scientific notation:

a. 34,970,000,000,000 **b.** −34,970,000,000,000

c. 0.0000000000802 **d.** −0.0000000000802.

Using Technology

You can use your calculator's [EE] (enter exponent) or [EXP] key to convert from decimal to scientific notation. Here is how it's done for 0.0000000000802.

Many Scientific Calculators

Keystrokes

.0000000000802 [EE] [=]

Display

8.02 − 11

Many Graphing Calculators

Use the mode setting for scientific notation.

Keystrokes

.0000000000802 [ENTER]

Display

8.02E − 11

Solution

a. $34{,}970{,}000{,}000{,}000 = 3.497 \times 10^{13}$

Move the decimal point to get a number whose absolute value is between 1 and 10.

The decimal point was moved 13 places to the left, so $n = 13$.

b. $-34{,}970{,}000{,}000{,}000 = -3.497 \times 10^{13}$

c. $0.0000000000802 = 8.02 \times 10^{-11}$

Move the decimal point to get a number whose absolute value is between 1 and 10.

The decimal point was moved 11 places to the right, so $n = -11$.

d. $-0.0000000000802 = -8.02 \times 10^{-11}$ ■

Great Question!

In scientific notation, which numbers have positive exponents and which have negative exponents?

If the absolute value of a number is greater than 10, it will have a positive exponent in scientific notation. If the absolute value of a number is less than 1, it will have a negative exponent in scientific notation.

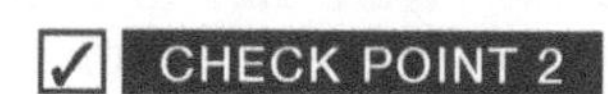

Write each number in scientific notation:

a. 5,210,000,000 **b.** −0.00000006893.

EXAMPLE 3 What a Difference a Day Makes in Scientific Notation

What's happening to the planet's 6.9 billion people today? A very partial answer is provided by the bar graph in **Figure 1.27**.

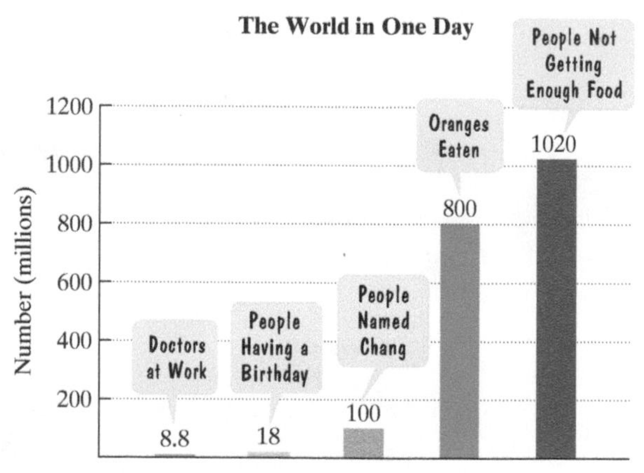

Figure 1.27

Source: Fullman et al., *Look Now*, DK Publishing, 2010

Express the number of oranges eaten in the world in one day in scientific notation.

Solution Because a million is 10^6, the number of oranges eaten in a day can be expressed as

$$800 \times 10^6.$$

This factor is not between 1 and 10, so the number is not in scientific notation.

The voice balloon indicates that we need to convert 800 to scientific notation.

$$800 \times 10^6 = (8 \times 10^2) \times 10^6 = 8 \times (10^2 \times 10^6) = 8 \times 10^{2+6} = 8 \times 10^8$$

There are 8×10^8 oranges eaten in the world in one day. ■

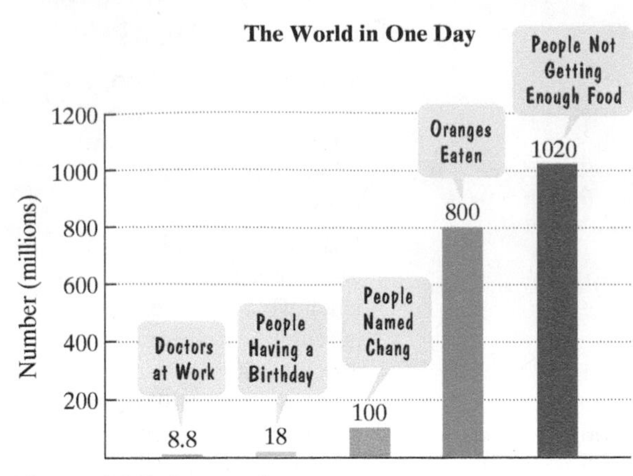

Figure 1.27 (repeated)

Great Question!

Is there more than one way to describe a large number, such as the number of oranges eaten in a day?

Yes. Many of the large numbers you encounter in newspapers, magazines, and online are expressed in millions (10^6), billions (10^9), and trillions (10^{12}). We can use exponential properties to describe these large numbers, such as the number of oranges eaten in a day.

$$800 \times 10^6 = 8 \times 10^8 = 0.8 \times 10^9$$

There are 800 million oranges eaten.

This expresses the number of oranges in scientific notation.

There are $\frac{4}{5}$ (0.8) of a of a billion oranges eaten.

CHECK POINT 3 Use **Figure 1.27** to express the number of people in the world who have a birthday today in scientific notation.

3 Perform computations with scientific notation.

Computations with Scientific Notation

Properties of exponents are used to perform computations with numbers that are expressed in scientific notation.

Computations with Numbers in Scientific Notation

Multiplication

$$(a \times 10^n)(b \times 10^m) = (a \times b) \times 10^{n+m}$$

Add the exponents on 10 and multiply the other parts of the numbers separately.

Division

$$\frac{a \times 10^n}{b \times 10^m} = \left(\frac{a}{b}\right) \times 10^{n-m}$$

Subtract the exponents on 10 and divide the other parts of the numbers separately.

After the computation is completed, the answer may require an adjustment before it is back in scientific notation.

EXAMPLE 4 Computations with Scientific Notation

Perform the indicated computations, writing the answers in scientific notation:

a. $(6.1 \times 10^5)(4 \times 10^{-9})$ **b.** $\dfrac{1.8 \times 10^4}{3 \times 10^{-2}}$.

Solution

a. $(6.1 \times 10^5)(4 \times 10^{-9})$

$= (6.1 \times 4) \times (10^5 \times 10^{-9})$ Regroup factors.

$= 24.4 \times 10^{5+(-9)}$ Add the exponents on 10 and multiply the other parts.

$= 24.4 \times 10^{-4}$ Simplify.

$= (2.44 \times 10^1) \times 10^{-4}$ Convert 24.4 to scientific notation: $24.4 = 2.44 \times 10^1$.

$= 2.44 \times 10^{-3}$ $10^1 \times 10^{-4} = 10^{1+(-4)} = 10^{-3}$

b. $\frac{1.8 \times 10^4}{3 \times 10^{-2}} = \left(\frac{1.8}{3}\right) \times \left(\frac{10^4}{10^{-2}}\right)$ Regroup factors.

$= 0.6 \times 10^{4-(-2)}$ Subtract the exponents on 10 and divide the other parts.

$= 0.6 \times 10^6$ Simplify: $4 - (-2) = 4 + 2 = 6$.

$= (6 \times 10^{-1}) \times 10^6$ Convert 0.6 to scientific notation: $0.6 = 6 \times 10^{-1}$.

$= 6 \times 10^5$ $0.6 = 6 \times 10^{-1}$. ■

Using Technology

$(6.1 \times 10^5)(4 \times 10^{-9})$
on a Calculator:

Many Scientific Calculators

6.1 [EE] 5 [×] 4 [EE] 9 [+/−] [=]

Display

2.44 − 03

Many Graphing Calculators

6.1 [EE] 5 [×] 4 [EE] [(−)] 9 [ENTER]

Display (in scientific notation mode)

2.44E − 3

✓ **CHECK POINT 4** Perform the indicated computations, writing the answers in scientific notation:

a. $(7.1 \times 10^5)(5 \times 10^{-7})$

b. $\frac{1.2 \times 10^6}{3 \times 10^{-3}}$.

4 Use scientific notation to solve problems.

Applications: Putting Numbers in Perspective

In the section opener, we saw that in 2012 the U.S. government collected \$2.52 trillion in taxes. Example 5 shows how we can use scientific notation to comprehend the meaning of a number such as 2.52 trillion.

EXAMPLE 5 Tax per Capita

In 2012, the U.S. government collected 2.52×10^{12} dollars in taxes. At that time, the U.S. population was approximately 314 million, or 3.14×10^8. If the total tax collections were evenly divided among all Americans, how much would each citizen pay? Express the answer in decimal notation, rounded to the nearest dollar.

Solution The amount that we would each pay, or the tax per capita, is the total amount collected, 2.52×10^{12}, divided by the number of Americans, 3.14×10^8.

$$\frac{2.52 \times 10^{12}}{3.14 \times 10^8} = \left(\frac{2.52}{3.14}\right) \times \left(\frac{10^{12}}{10^8}\right) \approx 0.8025 \times 10^{12-8} = 0.8025 \times 10^4 = 8025$$

To obtain an answer in decimal notation, it is not necessary to express this number in scientific notation.

Move the decimal point 4 places to the right.

If total tax collections were evenly divided, we would each pay approximately $8025 in taxes. ■

☑ CHECK POINT 5 In 2011, the U.S. government collected 2.42×10^{12} dollars in taxes. At that time, the U.S. population was approximately 312 million or 3.12×10^8. Find the per capita tax, rounded to the nearest dollar, in 2011.

Many problems in algebra involve motion. Suppose that you ride your bike at an average speed of 12 miles per hour. What distance do you cover in 2 hours? Your distance is the product of your speed and the time that you travel:

$$\frac{12 \text{ miles}}{\cancel{\text{hour}}} \times 2 \cancel{\text{hours}} = 24 \text{ miles}.$$

Your distance is 24 miles. Notice how the hour units cancel. The distance is expressed in miles.

In general, the distance covered by any moving body is the product of its average speed, or rate, and its time in motion.

A Formula for Motion

$$d = rt$$

Distance equals rate times time.

EXAMPLE 6 Using the Motion Formula

Light travels at a rate of approximately 1.86×10^5 miles per second. It takes light 5×10^2 seconds to travel from the sun to Earth. What is the distance between Earth and the sun?

Solution

$d = rt$	Use the motion formula.
$d = (1.86 \times 10^5) \times (5 \times 10^2)$	Substitute the given values.
$d = (1.86 \times 5) \times (10^5 \times 10^2)$	Rearrange factors.
$d = 9.3 \times 10^7$	Add the exponents on 10 and multiply the other parts.

The distance between Earth and the sun is approximately 9.3×10^7 miles, or 93 million miles. ■

☑ CHECK POINT 6 A futuristic spacecraft traveling at 1.55×10^3 miles per hour takes 20,000 hours (about 833 days) to travel from Venus to Mercury. What is the distance from Venus to Mercury?

Achieving Success

Do not wait until the last minute to study for an exam. Cramming is a high-stress activity that forces your brain to make a lot of weak connections. No wonder crammers tend to forget everything they learned minutes after taking a test.

Preparing for Tests Using the Book

- Study the appropriate sections from the review chart in the Chapter Summary. The chart contains definitions, concepts, procedures, and examples. Review this chart and you'll know the most important material in each section!
- Work the assigned exercises from the Review Exercises. The Review Exercises contain the most significant problems for each of the chapter's sections.
- Find a quiet place to take the Chapter Test. Do not use notes, index cards, or any other resources. Check your answers and ask your professor to review any exercises you missed.

CONCEPT AND VOCABULARY CHECK

Fill in each blank so that the resulting statement is true.

1. A positive number is written in scientific notation when it is expressed in the form $a \times 10^n$, where a is ______________ and n is a/an ______________.
2. True or false: 7×10^4 is written in scientific notation. ______________.
3. True or false: 70×10^3 is written in scientific notation. ______________

1.7 EXERCISE SET MyMathLab®

Practice Exercises

In Exercises 1–14, write each number in decimal notation without the use of exponents.

1. 3.8×10^2
2. 9.2×10^2
3. 6×10^{-4}
4. 7×10^{-5}
5. -7.16×10^6
6. -8.17×10^6
7. 1.4×10^0
8. 2.4×10^0
9. 7.9×10^{-1}
10. 6.8×10^{-1}
11. -4.15×10^{-3}
12. -3.14×10^{-3}
13. -6.00001×10^{10}
14. -7.00001×10^{10}

In Exercises 15–30, write each number in scientific notation.

15. 32,000
16. 64,000
17. 638,000,000,000,000,000
18. 579,000,000,000,000,000
19. −317
20. −326
21. −5716
22. −3829
23. 0.0027
24. 0.0083
25. −0.00000000504
26. −0.00000000405
27. 0.007
28. 0.005
29. 3.14159
30. 2.71828

In Exercises 31–50, perform the indicated computations. Write the answers in scientific notation. If necessary, round the decimal factor in your scientific notation answer to two decimal places.

31. $(3 \times 10^4)(2.1 \times 10^3)$
32. $(2 \times 10^4)(4.1 \times 10^3)$
33. $(1.6 \times 10^{15})(4 \times 10^{-11})$
34. $(1.4 \times 10^{15})(3 \times 10^{-11})$
35. $(6.1 \times 10^{-8})(2 \times 10^{-4})$
36. $(5.1 \times 10^{-8})(3 \times 10^{-4})$
37. $(4.3 \times 10^8)(6.2 \times 10^4)$
38. $(8.2 \times 10^8)(4.6 \times 10^4)$
39. $\dfrac{8.4 \times 10^8}{4 \times 10^5}$
40. $\dfrac{6.9 \times 10^8}{3 \times 10^5}$
41. $\dfrac{3.6 \times 10^4}{9 \times 10^{-2}}$
42. $\dfrac{1.2 \times 10^4}{2 \times 10^{-2}}$
43. $\dfrac{4.8 \times 10^{-2}}{2.4 \times 10^6}$
44. $\dfrac{7.5 \times 10^{-2}}{2.5 \times 10^6}$
45. $\dfrac{2.4 \times 10^{-2}}{4.8 \times 10^{-6}}$
46. $\dfrac{1.5 \times 10^{-2}}{3 \times 10^{-6}}$
47. $\dfrac{480{,}000{,}000{,}000}{0.00012}$
48. $\dfrac{282{,}000{,}000{,}000}{0.00141}$
49. $\dfrac{0.00072 \times 0.003}{0.00024}$
50. $\dfrac{66{,}000 \times 0.001}{0.003 \times 0.002}$

Practice PLUS

In Exercises 51–58, solve each equation. Express the solution in scientific notation.

51. $(2 \times 10^{-5})x = 1.2 \times 10^9$
52. $(3 \times 10^{-2})x = 1.2 \times 10^4$
53. $\dfrac{x}{2 \times 10^8} = -3.1 \times 10^{-5}$

54. $\frac{x}{5 \times 10^{11}} = -2.9 \times 10^{-3}$

55. $x - (7.2 \times 10^{18}) = 9.1 \times 10^{18}$

56. $x - (5.3 \times 10^{-16}) = 8.4 \times 10^{-16}$

57. $(-1.2 \times 10^{-3})x = (1.8 \times 10^{-4})(2.4 \times 10^{6})$

58. $(-7.8 \times 10^{-4})x = (3.9 \times 10^{-7})(6.8 \times 10^{5})$

Application Exercises

The graph shows the net worth, in billions of dollars, of the five richest Americans in 2014. Each bar also indicates the source of each person's wealth. Use 10^9 for one billion and the figures shown to solve Exercises 59–62. Express all answers in scientific notation.

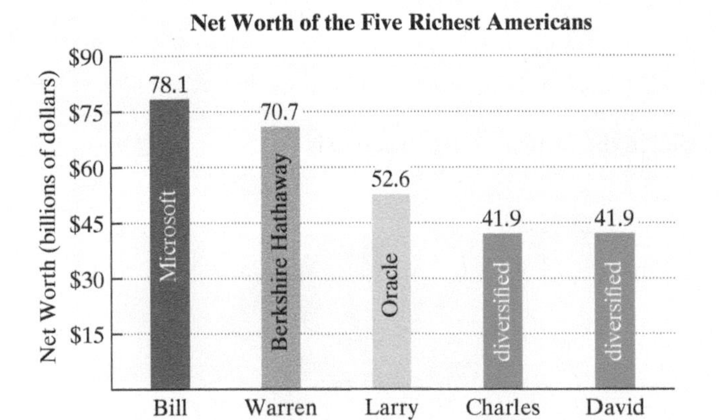

Source: Forbes

59. How much is Bill Gates worth?

60. How much is Warren Buffet worth?

61. By how much does Larry Ellison's worth exceed that of Charles Koch?

62. What is the combined worth of the Koch brothers (Charles and David)?

Our ancient ancestors hunted for their meat and expended a great deal of energy chasing it down. Today, our animal protein is raised in cages and on feedlots, delivered in great abundance nearly to our door. Use the numbers shown below to solve Exercises 63–66. Use 10^6 for one million and 10^9 for one billion.

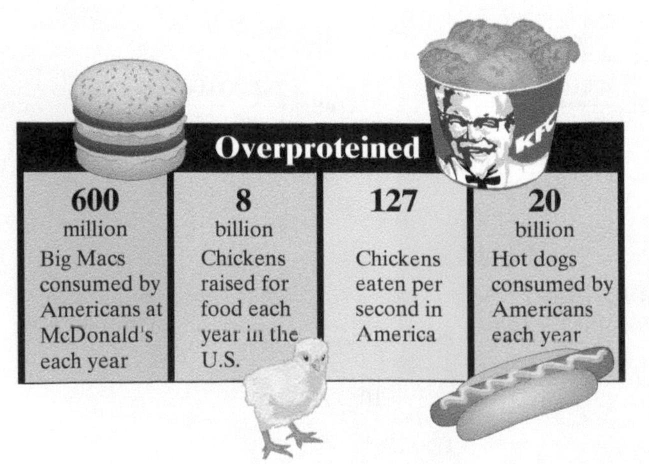

Overproteined
600
million
Big Macs consumed by Americans at McDonald's each year
8
billion
Chickens raised for food each year in the U.S.
127
Chickens eaten per second in America
20
billion
Hot dogs consumed by Americans each year

Source: TIME Magazine

In Exercises 63–64, use 300 million, or 3×10^8, for an approximation of the U.S. population. Express answers in decimal notation, rounded, if necessary, to the nearest whole number.

63. Find the number of hot dogs consumed by each American in a year.

64. If the consumption of Big Macs was divided evenly among all Americans, how many Big Macs would we each consume in a year?

In Exercises 65–66, use the fact that there are approximately 3.2×10^7 seconds in a year.

65. How many chickens are raised for food each second in the United States? Express the answer in scientific and decimal notations.

66. How many chickens are eaten per year in the United States? Express the answer in scientific notation.

The graph shows the cost, in billions of dollars, and the enrollment, in millions of people, for various federal social programs for a recent year. Use the numbers shown to solve Exercises 67–69.

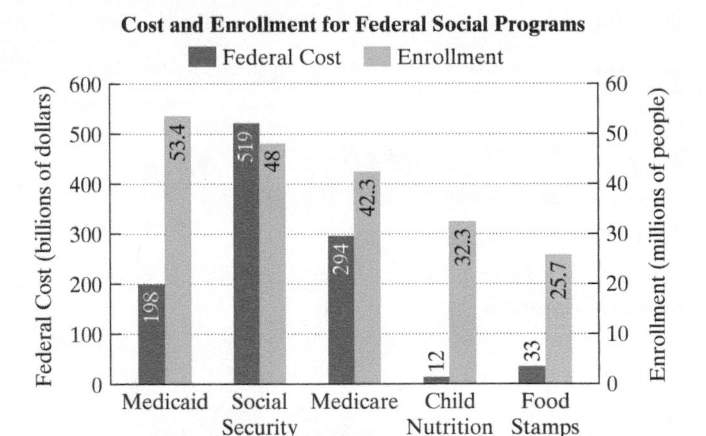

Source: Office of Management and Budget

67. a. What was the average per person benefit for Social Security? Express the answer in scientific notation and in decimal notation, rounded to the nearest dollar.

b. What was the average monthly per person benefit, rounded to the nearest dollar, for Social Security?

68. a. What was the average per person benefit for the food stamps program? Express the answer in scientific notation and in decimal notation, rounded to the nearest dollar.

b. What was the average monthly per person benefit, rounded to the nearest dollar, for the food stamps program?

69. Medicaid provides health insurance for the poor. Medicare provides health insurance for people 65 and older, as well as younger people who are disabled. Which program provides the greater per person benefit? By how much, rounded to the nearest dollar?

70. The area of Alaska is approximately 3.66×10^8 acres. The state was purchased in 1867 from Russia for \$7.2 million. What price per acre, to the nearest cent, did the United States pay Russia?

71. The mass of one oxygen molecule is 5.3×10^{-23} gram. Find the mass of 20,000 molecules of oxygen. Express the answer in scientific notation.

72. The mass of one hydrogen atom is 1.67×10^{-24} gram. Find the mass of 80,000 hydrogen atoms. Express the answer in scientific notation.

73. In Exercises 65–66, we used 3.2×10^7 as an approximation for the number of seconds in a year. Convert 365 days (one year) to hours, to minutes, and, finally, to seconds, to determine precisely how many seconds there are in a year. Express the answer in scientific notation.

Explaining the Concepts

74. How do you know if a number is written in scientific notation?

75. Explain how to convert from scientific to decimal notation and give an example.

76. Explain how to convert from decimal to scientific notation and give an example.

77. Describe one advantage of expressing a number in scientific notation over decimal notation.

Technology Exercises

78. Use a calculator to check any three of your answers in Exercises 1–14.

79. Use a calculator to check any three of your answers in Exercises 15–30.

80. Use a calculator with an EE or EXP key to check any four of your computations in Exercises 31–50. Display the result of the computation in scientific notation.

Critical Thinking Exercises

Make Sense? *In Exercises 81–84, determine whether each statement makes sense or does not make sense, and explain your reasoning.*

81. For a recent year, total tax collections in the United States were $\$2.02 \times 10^7$.

82. I just finished reading a book that contained approximately 1.04×10^5 words.

83. If numbers in the form $a \times 10^n$ are listed from least to greatest, values of a need not appear from least to greatest.

84. When expressed in scientific notation, 58 million and 58 millionths have exponents on 10 with the same absolute value.

In Exercises 85–89, determine whether each statement is true or false. If the statement is false, make the necessary change(s) to produce a true statement.

85. $534.7 = 5.347 \times 10^3$

86. $\dfrac{8 \times 10^{30}}{4 \times 10^{-5}} = 2 \times 10^{25}$

87. $(7 \times 10^5) + (2 \times 10^{-3}) = 9 \times 10^2$

88. $(4 \times 10^3) + (3 \times 10^2) = 43 \times 10^2$

89. The numbers 8.7×10^{25}, 1.0×10^{26}, 5.7×10^{26}, and 3.7×10^{27} are listed from least to greatest.

In Exercises 90–91, perform the indicated additions. Write the answers in scientific notation.

90. $5.6 \times 10^{13} + 3.1 \times 10^{13}$

91. $8.2 \times 10^{-16} + 4.3 \times 10^{-16}$

92. Our hearts beat approximately 70 times per minute. Express in scientific notation how many times the heart beats over a lifetime of 80 years. Round the decimal factor in your scientific notation answer to two decimal places.

93. Give an example of a number where there is no advantage in using scientific notation over decimal notation.

Review Exercises

94. Simplify: $9(10x - 4) - (5x - 10)$. (Section 1.2, Example 14)

95. Solve: $\dfrac{4x - 1}{10} = \dfrac{5x + 2}{4} - 4$. (Section 1.4, Example 4)

96. Simplify: $(8x^4y^{-3})^{-2}$. (Section 1.6, Example 7)

Preview Exercises

Exercises 97–99 will help you prepare for the material covered in the first section of the next chapter.

97. Here are two sets of ordered pairs:

set 1: $\{(1, 5), (2, 5)\}$
set 2: $\{(5, 1), (5, 2)\}$

In which set is each x-coordinate paired with only one y-coordinate?

98. Evaluate $r^3 - 2r^2 + 5$ for $r = -5$.

99. Evaluate $5x + 7$ for $x = a + h$.

Chapter 1 Summary

Definitions and Concepts	Examples

Section 1.1 Algebraic Expressions, Real Numbers, and Interval Notation

Letters that represent numbers are called variables.

An algebraic expression is a combination of variables, numbers, and operation symbols. English phrases can be translated into algebraic expressions:

- Addition: sum, plus, increased by, more than
- Subtraction: difference, minus, decreased by, less than
- Multiplication: product, times, of, twice
- Division: quotient, divide, per, ratio

Translate: Six less than the product of a number and five.

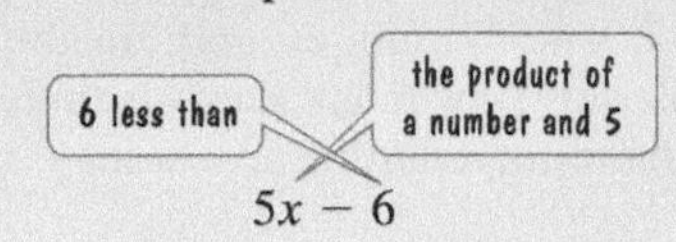

$$5x - 6$$

Many algebraic expressions contain exponents. If b is a natural number, b^n, the nth power of b, is the product of n factors of b.

Furthermore, $b^1 = b$.

Evaluating an algebraic expression means to find the value of the expression for a given value of the variable.

Evaluate $6 + 5(x - 10)^3$ for $x = 12$.

$$\begin{aligned} &6 + 5(12 - 10)^3 \\ &= 6 + 5 \cdot 2^3 \\ &= 6 + 5 \cdot 8 \\ &= 6 + 40 = 46 \end{aligned}$$

An equation is a statement that two expressions are equal. Formulas are equations that express relationships among two or more variables. Mathematical modeling is the process of finding formulas to describe real-world phenomena. Such formulas, together with the meaning assigned to the variables, are called mathematical models. The formulas are said to model, or describe, the relationships among the variables.

The formula

$$h = -16t^2 + 200t + 4$$

models the height, h, in feet, of fireworks t seconds after launch. What is the height after 2 seconds?

$$\begin{aligned} h &= -16(2)^2 + 200(2) + 4 \\ &= -16(4) + 200(2) + 4 \\ &= -64 + 400 + 4 = 340 \end{aligned}$$

The height is 340 feet.

A set is a collection of objects, called elements, enclosed in braces. The roster method uses commas to separate the elements of the set. Set-builder notation describes the elements of a set, but does not list them. The symbol $\in$ means that a number or object is in a set; $\notin$ means that a number or object is not in a set. The set of real numbers is the set of all numbers that can be represented by points on the number line. Sets that make up the real numbers include

- Natural numbers: $\{1, 2, 3, 4, \ldots\}$
- Whole numbers: $\{0, 1, 2, 3, 4, \ldots\}$
- Integers: $\{\ldots, -4, -3, -2, -1, 0, 1, 2, 3, 4, \ldots\}$
- Rational numbers: $\left\{\frac{a}{b} \middle| a \text{ and } b \text{ are integers and } b \neq 0\right\}$
- Irrational numbers: $\{x \mid x \text{ is a real number and } x \text{ is not a rational number}\}$.

In decimal form, rational numbers terminate or repeat.

In decimal form, irrational numbers do neither.

- Use the roster method to list the elements of $\{x \mid x \text{ is a natural number less than } 6\}$.

Solution

$$\{1, 2, 3, 4, 5\}$$

- True or false: $\sqrt{2} \notin \{x \mid x \text{ is a rational number}\}$.

Solution

The statement is true: $\sqrt{2}$ is not a rational number.

The decimal form of $\sqrt{2}$ neither terminates nor repeats. Thus, $\sqrt{2}$ is an irrational number.

Definitions and Concepts	Examples

Section 1.1 Algebraic Expressions, Real Numbers, and Interval Notation (continued)

For any two real numbers, a and b, a is less than b if a is to the left of b on the number line.

Inequality Symbols

$<$: is less than
$>$: is greater than
$\leq$: is less than or equal to
$\geq$: is greater than or equal to

- $-1 < 5$, or -1 is less than 5, is true because -1 is to the left of 5 on a number line.
- $-3 \geq 7$, -3 is greater than or equal to 7, is false. Neither $-3 > 7$ nor $-3 = 7$ is true.

In interval notation, parentheses indicate endpoints that are not included in an interval. Square brackets indicate endpoints that are included in an interval. Parentheses are always used with ∞ or $-\infty$.

- $(-2, 1] = \{x \mid -2 < x \leq 1\}$

  −4 −3 −2 −1 0 1 2 3 4 x

- $[-2, \infty) = \{x \mid x \geq -2\}$

  −4 −3 −2 −1 0 1 2 3 4 x

Section 1.2 Operations with Real Numbers and Simplifying Algebraic Expressions

Absolute Value

$$|a| = \begin{cases} a & \text{if } a \geq 0 \\ -a & \text{if } a < 0 \end{cases}$$

The opposite, or additive inverse, of a is $-a$. When a is a negative number, $-a$ is positive.

- $|6.03| = 6.03$
- $|0| = 0$
- $|-4.9| = -(-4.9) = 4.9$

Adding Real Numbers

To add two numbers with the same sign, add their absolute values and use their common sign. To add two numbers with different signs, subtract the smaller absolute value from the greater absolute value and use the sign of the number with the greater absolute value.

- $-4.1 + (-6.2) = -10.3$
- $-30 + 25 = -5$
- $12 + (-8) = 4$

Subtracting Real Numbers

$$a - b = a + (-b)$$

$$-\frac{3}{4} - \left(-\frac{1}{2}\right) = -\frac{3}{4} + \frac{1}{2} = -\frac{3}{4} + \frac{2}{4} = -\frac{1}{4}$$

Multiplying and Dividing Real Numbers

The product or quotient of two numbers with the same sign is positive and with different signs is negative. If no number is 0, a product with an even number of negative factors is positive and a product with an odd number of negative factors is negative. Division by 0 is undefined.

- $2(-6)(-1)(-5) = -60$

  Three (odd) negative factors give a negative product.

- $(-2)^3 = (-2)(-2)(-2) = -8$
- $-\frac{1}{3}\left(-\frac{2}{5}\right) = \frac{2}{15}$
- $\frac{-14}{2} = -7$

Definitions and Concepts	Examples

Section 1.2 Operations with Real Numbers and Simplifying Algebraic Expressions (continued)

Order of Operations

1. Perform operations within grouping symbols, starting with the innermost grouping symbols. Grouping symbols include parentheses, brackets, fraction bars, absolute value symbols, and square root signs.
2. Evaluate exponential expressions.
3. Multiply and divide from left to right.
4. Add and subtract from left to right.

Simplify: $\frac{6(8-10)^3+(-2)}{(-5)^2(-2)}$.

$$= \frac{6(-2)^3+(-2)}{(-5)^2(-2)} = \frac{6(-8)+(-2)}{25(-2)}$$

$$= \frac{-48+(-2)}{-50} = \frac{-50}{-50} = 1$$

Basic Algebraic Properties

Commutative: $a + b = b + a$

$ab = ba$

Associative: $(a + b) + c = a + (b + c)$

$(ab)c = a(bc)$

Distributive: $a(b + c) = ab + ac$

$a(b - c) = ab - ac$

$(b + c)a = ba + ca$

- Commutative

$$3x + 5 = 5 + 3x = 5 + x \cdot 3$$

- Associative

$$-4(6x) = (-4 \cdot 6)x = -24x$$

- Distributive

$$-4(9x + 3) = -4(9x) + (-4) \cdot 3$$
$$= -36x + (-12)$$
$$= -36x - 12$$

Simplifying Algebraic Expressions

Terms are separated by addition. Like terms have the same variable factors and are combined using the distributive property. An algebraic expression is simplified when grouping symbols have been removed and like terms have been combined.

Simplify: $7(3x - 4) - (10x - 5)$.

$$= 21x - 28 - 10x + 5$$
$$= 21x - 10x - 28 + 5$$
$$= 11x - 23$$

Section 1.3 Graphing Equations

The rectangular coordinate system consists of a horizontal number line, the x-axis, and a vertical number line, the y-axis, intersecting at their zero points, the origin. Each point in the system corresponds to an ordered pair of real numbers (x, y). The first number in the pair is the x-coordinate; the second number is the y-coordinate.

Plot: (4, 2), (−3, 4), (−5, −4), and (4, −3).

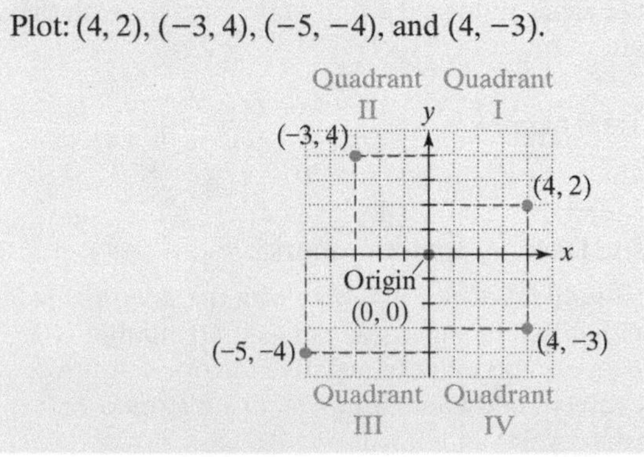

An ordered pair is a solution of an equation in two variables if replacing the variables by the corresponding coordinates results in a true statement. The ordered pair is said to satisfy the equation. The graph of the equation is the set of all points whose coordinates satisfy the equation. One method for graphing an equation is to plot ordered-pair solutions and connect them with a smooth curve or line.

Graph: $y = x^2 - 1$.

x	$y = x^2 - 1$
−2	$(-2)^2 - 1 = 3$
−1	$(-1)^2 - 1 = 0$
0	$0^2 - 1 = -1$
1	$1^2 - 1 = 0$
2	$2^2 - 1 = 3$

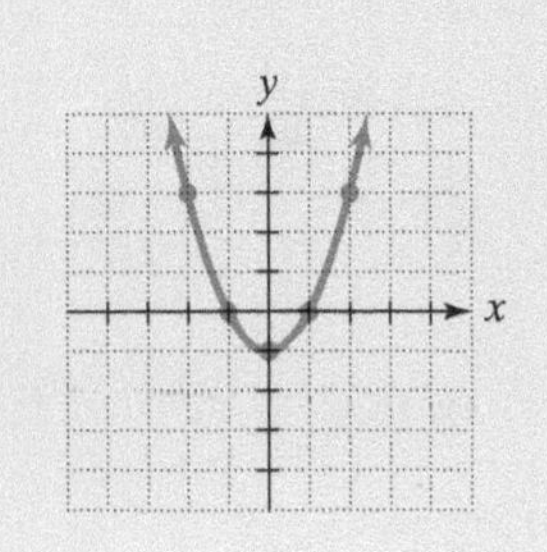

Definitions and Concepts	Examples

Section 1.4 Solving Linear Equations

A linear equation in one variable can be written in the form $ax + b = 0, a \neq 0$. A solution is a value of the variable that makes the equation a true statement. The set of all such solutions is the equation's solution set. Equivalent equations have the same solution set. To solve a linear equation,

1. Simplify each side.
2. Collect variable terms on one side and constant terms on the other side.
3. Isolate the variable and solve.
4. Check the proposed solution in the original equation.

Solve: $4(x-5) = 2x - 14$.

$$\begin{aligned} 4x - 20 &= 2x - 14 \\ 4x - 2x - 20 &= 2x - 2x - 14 \\ 2x - 20 &= -14 \\ 2x - 20 + 20 &= -14 + 20 \\ 2x &= 6 \\ \frac{2x}{2} &= \frac{6}{2} \\ x &= 3 \end{aligned}$$

Checking gives $-8 = -8$, so 3 is the solution, or $\{3\}$ is the solution set.

Equations Containing Fractions

Multiply both sides (all terms) by the least common denominator. This clears the equation of fractions.

Solve: $\frac{x-2}{5} + \frac{x+2}{2} = \frac{x+4}{3}$.

$$\begin{aligned} 30\left(\frac{x-2}{5} + \frac{x+2}{2}\right) &= 30\left(\frac{x+4}{3}\right) \\ 6(x-2) + 15(x+2) &= 10(x+4) \\ 6x - 12 + 15x + 30 &= 10x + 40 \\ 21x + 18 &= 10x + 40 \\ 11x &= 22 \\ x &= 2 \end{aligned}$$

Checking gives $2 = 2$, so 2 is the solution, or $\{2\}$ is the solution set.

Types of Equations

An equation that is true for all real numbers, $(-\infty, \infty)$, is called an identity. When solving an identity, the variable is eliminated and a true statement, such as $3 = 3$, results. An equation that is not true for even one real number is called an inconsistent equation. A false statement, such as $3 = 7$, results when solving such an equation, whose solution set is $\varnothing$, the empty set. A conditional equation is not an identity, but is true for at least one real number.

Solve: $4x + 5 = 4(x + 2)$.

$$\begin{aligned} 4x + 5 &= 4x + 8 \\ 5 &= 8, \quad \text{false} \end{aligned}$$

The inconsistent equation has no solution: $\varnothing$.

Solve: $5x - 4 = 5(x + 1) - 9$.

$$\begin{aligned} 5x - 4 &= 5x + 5 - 9 \\ 5x - 4 &= 5x - 4 \\ -4 &= -4, \quad \text{true} \end{aligned}$$

All real numbers satisfy the identity: $(-\infty, \infty)$.

Definitions and Concepts	Examples

Section 1.5 Problem Solving and Using Formulas

Strategy for Solving Algebraic Word Problems

1. Let x represent one of the unknown quantities.
2. Represent other unknown quantities in terms of x.
3. Write an equation that models the conditions.
4. Solve the equation and answer the question.
5. Check the proposed solution in the original wording of the problem.

After a 60% reduction, a suit sold for \$32. What was the original price?

Let x = the original price.

Original price minus 60% reduction = reduced price

$$x - 0.6x = 32$$
$$0.4x = 32$$
$$\frac{0.4x}{0.4} = \frac{32}{0.4}$$
$$x = 80$$

The original price was \$80. Check this amount using the first sentence in the problem's conditions.

To solve a formula for a variable, use the steps for solving a linear equation and isolate that variable on one side of the equation.

Solve for r: $E = I(R + r)$.

$$E = IR + Ir$$ (We need to isolate r.)
$$E - IR = Ir$$
$$\frac{E - IR}{I} = r$$

Section 1.6 Properties of Integral Exponents

The Product Rule

$$b^m \cdot b^n = b^{m+n}$$

$$(-3x^{10})(5x^{20}) = -3 \cdot 5x^{10+20} = -15x^{30}$$

The Quotient Rule

$$\frac{b^m}{b^n} = b^{m-n}, b \neq 0$$

$$\frac{5x^{20}}{10x^{10}} = \frac{5}{10} \cdot x^{20-10} = \frac{x^{10}}{2}$$

Zero and Negative Exponents

$$b^0 = 1, b \neq 0$$

$$b^{-n} = \frac{1}{b^n} \text{ and } \frac{1}{b^{-n}} = b^n, b \neq 0$$

- $(3x)^0 = 1$
- $3x^0 = 3 \cdot 1 = 3$
- $\frac{2^{-3}}{4^{-2}} = \frac{4^2}{2^3} = \frac{16}{8} = 2$

Power Rule

$$(b^m)^n = b^{mn}$$

$$(x^5)^{-4} = x^{5(-4)} = x^{-20} = \frac{1}{x^{20}}$$

Products to Powers

$$(ab)^n = a^n b^n$$

$$(5x^3y^{-4})^{-2} = 5^{-2} \cdot (x^3)^{-2} \cdot (y^{-4})^{-2} = 5^{-2}x^{-6}y^8 = \frac{y^8}{5^2x^6} = \frac{y^8}{25x^6}$$

Definitions and Concepts	Examples
Section 1.6 Properties of Integral Exponents (continued)	
Quotients to Powers $$\left(\frac{a}{b}\right)^n = \frac{a^n}{b^n}$$	$$\left(\frac{2}{x^3}\right)^{-4} = \frac{2^{-4}}{(x^3)^{-4}} = \frac{2^{-4}}{x^{-12}} = \frac{x^{12}}{2^4} = \frac{x^{12}}{16}$$
An exponential expression is simplified when • No parentheses appear. • No powers are raised to powers. • Each base occurs only once. • No negative or zero exponents appear.	Simplify: $\frac{-5x^{-3}y^2}{-20x^2y^{-6}}$. $$= \frac{-5}{-20} \cdot x^{-3-2} \cdot y^{2-(-6)} = \frac{1}{4}x^{-5}y^8 = \frac{y^8}{4x^5}$$
Section 1.7 Scientific Notation	
A number in scientific notation is expressed in the form $$a \times 10^n,$$ where $\lvert a \rvert$ is greater than or equal to 1 and less than 10, and n is an integer.	Write in decimal notation: 3.8×10^{-3}. $$3.8 \times 10^{-3} = .0038 = 0.0038$$ Write in scientific notation: 26,000. $$26{,}000 = 2.6 \times 10^4$$
Computations with Numbers in Scientific Notation $$(a \times 10^n)(b \times 10^m) = (a \times b) \times 10^{n+m}$$ $$\frac{a \times 10^n}{b \times 10^m} = \left(\frac{a}{b}\right) \times 10^{n-m}$$	$$(8 \times 10^3)(5 \times 10^{-8}) = 8 \cdot 5 \times 10^{3+(-8)} = 40 \times 10^{-5} = (4 \times 10^1) \times 10^{-5} = 4 \times 10^{-4}$$

CHAPTER 1 REVIEW EXERCISES

1.1 *In Exercises 1–3, write each English phrase as an algebraic expression. Let x represent the number.*

1. Ten less than twice a number

2. Four more than the product of six and a number

3. The quotient of nine and a number, increased by half of the number

In Exercises 4–6, evaluate each algebraic expression for the given value or values of the variable.

4. $x^2 - 7x + 4$, for $x = 10$

5. $6 + 2(x - 8)^3$, for $x = 11$

6. $x^4 - (x - y)$, for $x = 2$ and $y = 1$

In Exercises 7–8, use the roster method to list the elements in each set.

7. $\{x \mid x$ is a natural number less than $3\}$

8. $\{x \mid x$ is an integer greater than -4 and less than $2\}$

In Exercises 9–11, determine whether each statement is true or false.

9. $0 \in \{x \mid x$ is a natural number$\}$

10. $-2 \in \{x \mid x$ is a rational number$\}$

11. $\frac{1}{3} \notin \{x \mid x$ is an irrational number$\}$

In Exercises 12–14, write out the meaning of each inequality. Then determine whether the inequality is true or false.

12. $-5 < 2$

13. $-7 \geq -3$

14. $-7 \leq -7$

15. On average, smartphone users in the United States look at their phones more than 150 times per day, spending approximately 2.5 hours on their phones. The bar graph shows the rise of iPhone sales, in millions, from 2007 through 2013.

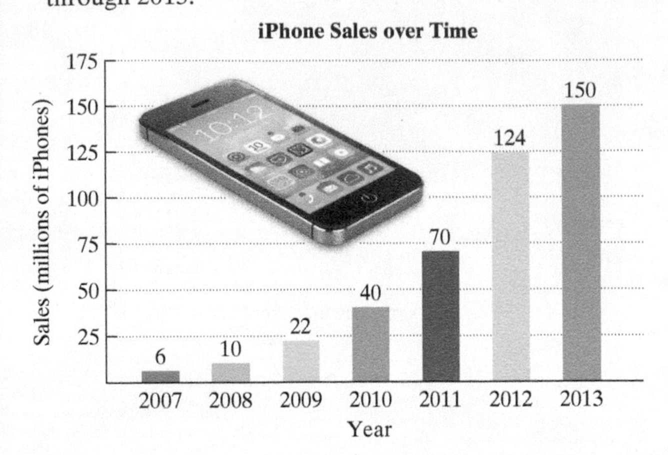

Source: Statista

The data can be modeled by the formula

$$S = 4x^2 + 0.7x + 5,$$

where S represents iPhone sales, in millions of units, x years after 2007. Does the model underestimate or overestimate sales shown for 2013? By how much?

In Exercises 16–18, express each interval in set-builder notation and graph the interval on a number line.

16. $(-2, 3]$

17. $[-1.5, 2]$

18. $(-1, \infty)$

1.2 *In Exercises 19–21, find each absolute value.*

19. $|-9.7|$ **20.** $|5.003|$ **21.** $|0|$

In Exercises 22–33, perform the indicated operation.

22. $-2.4 + (-5.2)$ **23.** $-6.8 + 2.4$

24. $-7 - (-20)$ **25.** $(-3)(-20)$

26. $-\frac{3}{5} - \left(-\frac{1}{2}\right)$ **27.** $\left(\frac{2}{7}\right)\left(-\frac{3}{10}\right)$

28. $4(-3)(-2)(-10)$ **29.** $(-2)^4$

30. -2^5 **31.** $-\frac{2}{3} \div \frac{8}{5}$

32. $\frac{-35}{-5}$ **33.** $\frac{54.6}{-6}$

34. Find $-x$ if $x = -7$.

In Exercises 35–41, simplify each expression.

35. $-11 - [-17 + (-3)]$ **36.** $\left(-\frac{1}{2}\right)^3 \cdot 2^4$

37. $-3[4 - (6 - 8)]$ **38.** $8^2 - 36 \div 3^2 \cdot 4 - (-7)$

39. $\frac{(-2)^4 + (-3)^2}{2^2 - (-21)}$ **40.** $\frac{(7-9)^3 - (-4)^2}{2 + 2(8) \div 4}$

41. $4 - (3 - 8)^2 + 3 \div 6 \cdot 4^2$

In Exercises 42–46, simplify each algebraic expression.

42. $5(2x - 3) + 7x$

43. $5x + 7x^2 - 4x + 2x^2$

44. $3(4y - 5) - (7y + 2)$

45. $8 - 2[3 - (5x - 1)]$

46. $6(2x - 3) - 5(3x - 2)$

1.3 *In Exercises 47–49, plot the given point in a rectangular coordinate system.*

47. $(-1, 3)$ **48.** $(2, -5)$ **49.** $(0, -6)$

In Exercises 50–53, graph each equation. Let $x = -3, -2, -1, 0, 1, 2,$ and 3.

50. $y = 2x - 2$ **51.** $y = x^2 - 3$

52. $y = x$ **53.** $y = |x| - 2$

54. What does a $[-20, 40, 10]$ by $[-5, 5, 1]$ viewing rectangle mean? Draw axes with tick marks and label the tick marks to illustrate this viewing rectangle.

The caseload of Alzheimer's disease in the United States is expected to explode as baby boomers head into their later years. The graph shows the percentage of Americans with the disease, by age. Use the graph to solve Exercises 55–57.

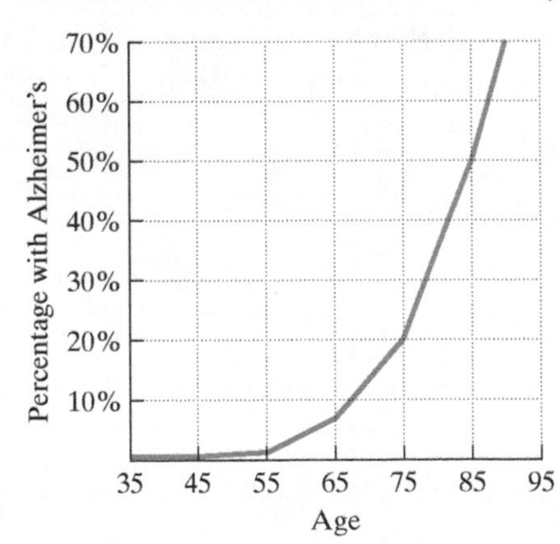

Source: U.S. Centers for Disease Control

55. What percentage of Americans who are 75 have Alzheimer's disease?

56. What age represents 50% prevalence of Alzheimer's disease?

57. Describe the trend shown by the graph.

58. Select the graph that best illustrates the following description: A train pulls into a station and lets off its passengers.

a.

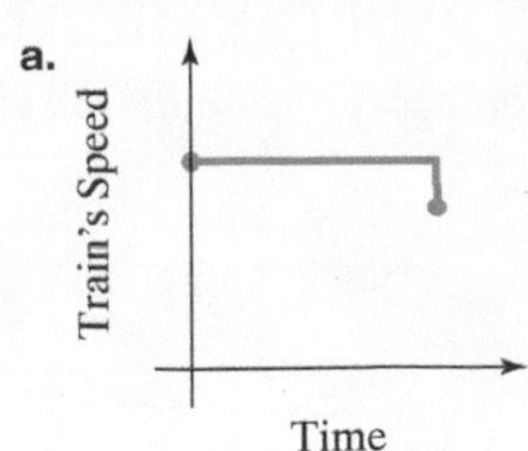

b.

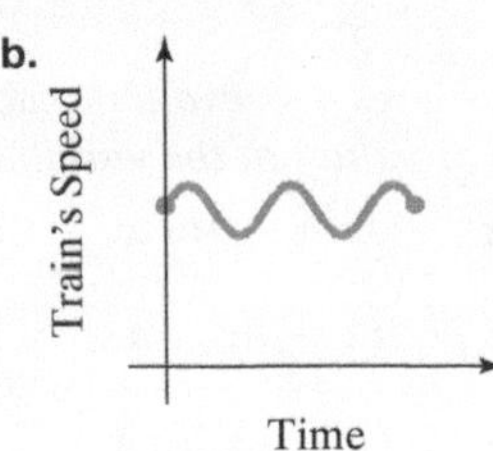

c.

Train's Speed

Time

d.

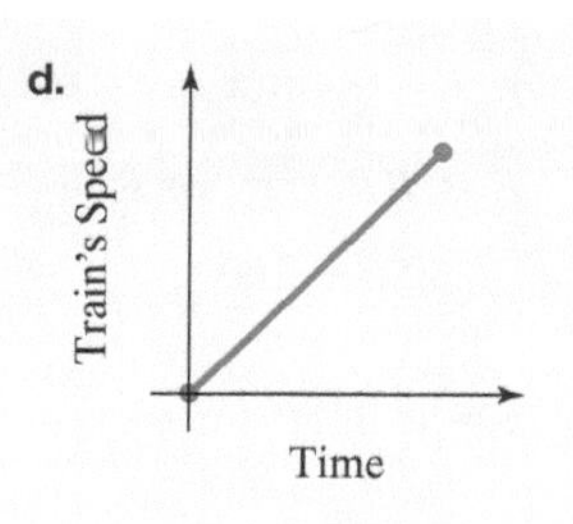

1.4 *In Exercises 59–64, solve and check each linear equation.*

59. $2x - 5 = 7$

60. $5x + 20 = 3x$

61. $7(x - 4) = x + 2$

62. $1 - 2(6 - x) = 3x + 2$

63. $2(x - 4) + 3(x + 5) = 2x - 2$

64. $2x - 4(5x + 1) = 3x + 17$

In Exercises 65–69, solve each equation.

65. $\frac{2x}{3} = \frac{x}{6} + 1$

66. $\frac{x}{2} - \frac{1}{10} = \frac{x}{5} + \frac{1}{2}$

67. $\frac{2x}{3} = 6 - \frac{x}{4}$

68. $\frac{x}{4} = 2 + \frac{x - 3}{3}$

69. $\frac{3x + 1}{3} - \frac{13}{2} = \frac{1 - x}{4}$

In Exercises 70–74, solve each equation. Then state whether the equation is an identity, a conditional equation, or an inconsistent equation.

70. $7x + 5 = 5(x + 3) + 2x$

71. $7x + 13 = 4x - 10 + 3x + 23$

72. $7x + 13 = 3x - 10 + 2x + 23$

73. $4(x - 3) + 5 = x + 5(x - 2)$

74. $(2x - 3)2 - 3(x + 1) = (x - 2)4 - 3(x + 5)$

75. The bar graph shows the changing face of America's federal tax returns.

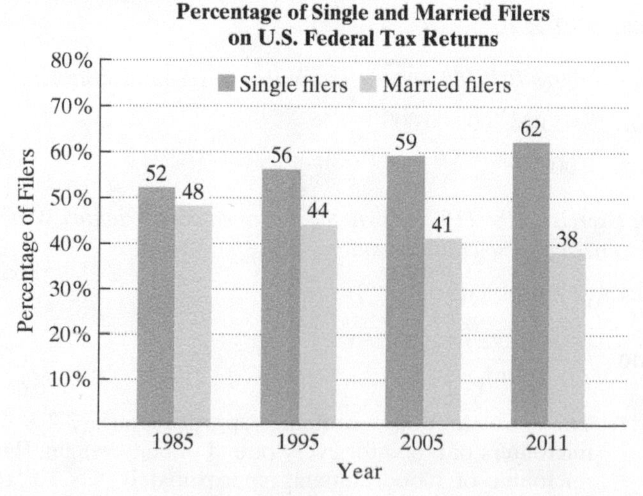

Source: IRS

The data can be modeled by the formula

$$M = -0.4x + 48,$$

where M represents the percentage of married filers x years after 1985.

a. Does the model underestimate or overestimate the percentage of married filers in 2005? By how much?

b. According to the model, when will 34% of federal tax returns be submitted by married filers?

1.5 *In Exercises 76–82, use the five-step strategy for solving word problems.*

76. Although you want to choose a career that fits your interests and abilities, it is good to have an idea of what jobs pay when looking at career options. The bar graph shows the average yearly earnings of full-time employed college graduates with only a bachelor's degree based on their college major.

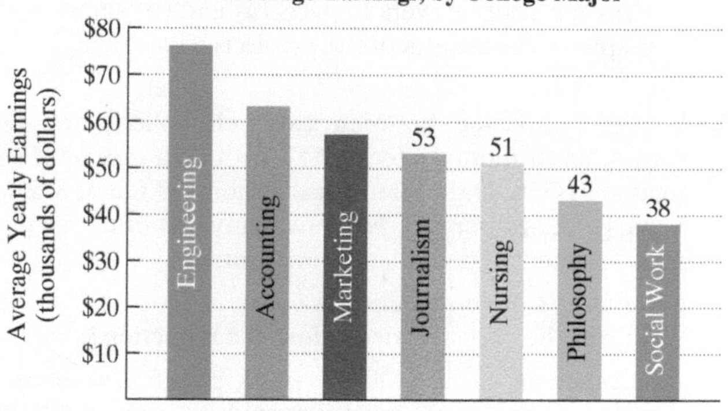

Source: Arthur J. Keown, *Personal Finance*, Pearson.

The average yearly earnings of engineering majors exceeds the earnings of marketing majors by $19 thousand. The average yearly earnings of accounting majors exceeds the earnings of marketing majors by $6 thousand. Combined, the average yearly earnings for these three college majors is $196 thousand. Determine the average yearly earnings, in thousands of dollars, for each of these three college majors.

77. One angle of a triangle measures 10° more than the second angle. The measure of the third angle is twice the sum of the measures of the first two angles. Determine the measure of each angle.

78. Without changes, the graphs show projections for the amount being paid in Social Security benefits and the amount going into the system. All data are expressed in billions of dollars.

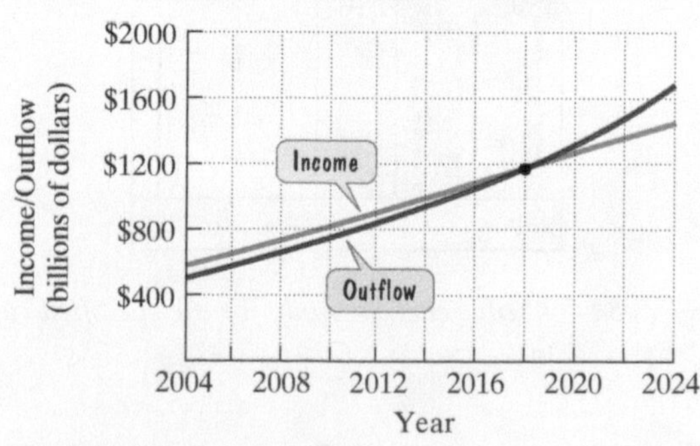

Source: 2004 Social Security Trustees Report

(Be sure to refer to the information at the bottom of the previous page.)

a. In 2004, the Social Security system's income was \$575 billion, projected to increase at an average rate of \$43 billion per year. In which year will the system's income be \$1177 billion?

b. The data for the Social Security system's outflow can be modeled by the formula

$$B = 0.07x^2 + 47.4x + 500,$$

where B represents the amount paid in benefits, in billions of dollars, x years after 2004. According to this model, what will be the amount paid in benefits for the year you determined in part (a)? Round to the nearest billion dollars.

c. How are your answers to parts (a) and (b) shown by the graphs at the bottom of the previous page?

79. You are choosing between two cellphone plans. Data Plan A has a monthly fee of \$52 with a charge of \$18 per gigabyte (GB). Data Plan B has a monthly fee of \$32 with a charge of \$22 per GB. For how many GB of data will the costs for the two data plans be the same?

80. After a 20% price reduction, a cordless phone sold for \$48. What was the phone's price before the reduction?

81. A salesperson earns \$300 per week plus 5% commission of sales. How much must be sold to earn \$800 in a week?

82. The length of a rectangular field is 6 yards less than triple the width. If the perimeter of the field is 340 yards, what are its dimensions?

83. In 2015, there were 14,100 students at college A, with a projected enrollment increase of 1500 students per year. In the same year, there were 41,700 students at college B, with a projected enrollment decline of 800 students per year.

a. Let x represent the number of years after 2015. Write, but do not solve, an equation that can be used to find how many years after 2015 the colleges will have the same enrollment.

b. The following table is based on your equation in part (a). Y_1 represents one side of the equation and Y_2 represents the other side of the equation. Use the table to answer these questions: In which year will the colleges have the same enrollment? What will be the enrollment in each college at that time?

X	Y1	Y2		
7	24600	36100		
8	26100	35300		
9	27600	34500		
10	29100	33700		
11	30600	32900		
12	32100	32100		
13	33600	31300		
14	35100	30500		
15	36600	29700		
16	38100	28900		
17	39600	28100		

X=7

In Exercises 84–89, solve each formula for the specified variable.

84. $V = \frac{1}{3}Bh$ for h

85. $y - y_1 = m(x - x_1)$ for x

86. $E = I(R + r)$ for R

87. $C = \frac{5F - 160}{9}$ for F

88. $s = vt + gt^2$ for g

89. $T = gr + gvt$ for g

1.6 *In Exercises 90–104, simplify each exponential expression. Assume that no denominators are 0.*

90. $(-3x^7)(-5x^6)$

91. x^2y^{-5}

92. $\frac{3^{-2}x^4}{y^{-7}}$

93. $(x^3)^{-6}$

94. $(7x^3y)^2$

95. $\frac{16y^3}{-2y^{10}}$

96. $(-3x^4)(4x^{-11})$

97. $\frac{12x^7}{4x^{-3}}$

98. $\frac{-10a^5b^6}{20a^{-3}b^{11}}$

99. $(-3xy^4)(2x^2)^3$

100. $2^{-2} + \frac{1}{2}x^0$

101. $(5x^2y^{-4})^{-3}$

102. $(3x^4y^{-2})(-2x^5y^{-3})$

103. $\left(\frac{3xy^3}{5x^{-3}y^{-4}}\right)^2$

104. $\left(\frac{-20x^{-2}y^3}{10x^5y^{-6}}\right)^{-3}$

1.7 *In Exercises 105–106, write each number in decimal notation.*

105. 7.16×10^6

106. 1.07×10^{-4}

In Exercises 107–108, write each number in scientific notation.

107. $-41{,}000{,}000{,}000{,}000$

108. 0.00809

In Exercises 109–110, perform the indicated computations. Write the answers in scientific notation.

109. $(4.2 \times 10^{13})(3 \times 10^{-6})$

110. $\frac{5 \times 10^{-6}}{20 \times 10^{-8}}$

111. The human body contains approximately 3.2×10^4 microliters of blood for every pound of body weight. Each microliter of blood contains approximately 5×10^6 red blood cells. Express in scientific notation the approximate number of red blood cells in the body of a 180-pound person.

CHAPTER 1 TEST

Step-by-step test solutions are found on the Chapter Test Prep Videos available in MyMathLab® or on YouTube (search "BlitzerInterAlg7e" and click on "Channels").

1. Write the following English phrase as an algebraic expression:

   Five less than the product of a number and four.

   Let x represent the number.
2. Evaluate $8 + 2(x - 7)^4$ for $x = 10$.
3. Use the roster method to list the elements in the set:

   $\{x \mid x \text{ is a negative integer greater than } -5\}$.
4. Determine whether the following statement is true or false:

   $\frac{1}{4} \notin \{x \mid x \text{ is a natural number}\}$.
5. Write out the meaning of the inequality $-3 > -1$. Then determine whether the inequality is true or false.

In Exercises 6–7, express each interval in set-builder notation and graph the interval on a number line.

6. $[-3, 2)$
7. $(-\infty, -1]$
8. One possible reason for the explosion of college tuition involves the decrease in government aid per student. In 2001, higher-education revenues per student averaged \$8500. The bar graph shows government aid per U.S. college student from 2005 through 2012. (All figures are adjusted for inflation and expressed in 2012 dollars.)

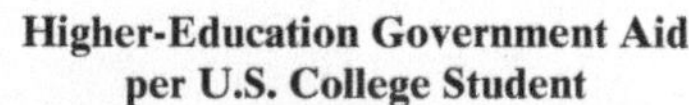

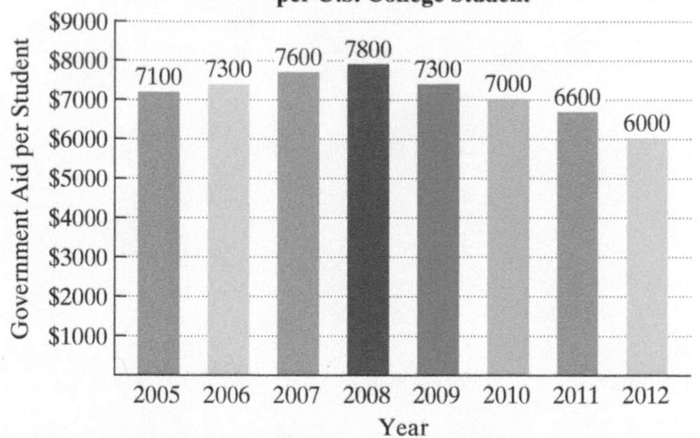

Source: State Higher Education Executive Officers Association

The mathematical model

$$G = -82x^2 + 410x + 7079$$

describes government aid per college student, G, x years after 2005. Does the model underestimate or overestimate aid per student in 2011? By how much?

9. Find the absolute value: $|-17.9|$.

In Exercises 10–14, perform the indicated operation.

10. $-10.8 + 3.2$
11. $-\frac{1}{4} - \left(-\frac{1}{2}\right)$
12. $2(-3)(-1)(-10)$
13. $-\frac{1}{4}\left(-\frac{1}{2}\right)$
14. $\frac{-27.9}{-9}$

In Exercises 15–20, simplify each expression.

15. $24 - 36 \div 4 \cdot 3$
16. $(5^2 - 2^4) + [9 \div (-3)]$
17. $\frac{(8 - 10)^3 - (-4)^2}{2 + 8(2) \div 4}$
18. $7x - 4(3x + 2) - 10$
19. $5(2y - 6) - (4y - 3)$
20. $9x - [10 - 4(2x - 3)]$
21. Plot $(-2, -4)$ in a rectangular coordinate system.
22. Graph $y = x^2 - 4$ in a rectangular coordinate system.

In Exercises 23–25, solve each equation. If the solution set is $\varnothing$ or $(-\infty, \infty)$, classify the equation as an inconsistent equation or an identity.

23. $3(2x - 4) = 9 - 3(x + 1)$
24. $\frac{2x - 3}{4} = \frac{x - 4}{2} - \frac{x + 1}{4}$
25. $3(x - 4) + x = 2(6 + 2x)$

In Exercises 26–30, use the five-step strategy for solving word problems.

26. Find two numbers such that the second number is 3 more than twice the first number and the sum of the two numbers is 72.
27. You bought a new car for \$13,805. Its value is decreasing by \$1820 per year. After how many years will its value be \$4705?
28. The toll to a bridge costs \$8. Commuters who use the bridge frequently have the option of purchasing a monthly discount pass for \$45. With the discount pass, the toll is reduced to \$5. For how many crossings per month will the monthly cost without the discount pass be the same as the monthly cost with the discount pass?
29. After a 60% reduction, a jacket sold for \$20. What was the jacket's price before the reduction?
30. The length of a rectangular field exceeds the width by 260 yards. If the perimeter of the field is 1000 yards, what are its dimensions?

In Exercises 31–32, solve each formula for the specified variable.

31. $V = \frac{1}{3}lwh$ for h
32. $Ax + By = C$ for y

In Exercises 33–37, simplify each exponential expression.

33. $(-2x^5)(7x^{-10})$

34. $(-8x^{-5}y^{-3})(-5x^2y^{-5})$

35. $\dfrac{-10x^4y^3}{-40x^{-2}y^6}$

36. $(4x^{-5}y^2)^{-3}$

37. $\left(\dfrac{-6x^{-5}y}{2x^3y^{-4}}\right)^{-2}$

38. Write in decimal notation: 3.8×10^{-6}.

39. Write in scientific notation: 407,000,000,000.

40. Divide and write the answer in scientific notation:

$$\frac{4 \times 10^{-3}}{8 \times 10^{-7}}.$$

41. In 2010, world population was approximately 6.9×10^9. By some projections, world population will double by 2080. Express the population at that time in scientific notation.

CHAPTER 2

Functions and Linear Functions

A vast expanse of open water at the top of our world was once covered with ice. The melting of the Arctic ice caps has forced polar bears to swim as far as 40 miles, causing them to drown in significant numbers. Such deaths were rare in the past.

There is strong scientific consensus that human activities are changing the Earth's climate. Scientists now believe that there is a striking correlation between atmospheric carbon dioxide concentration and global temperature. As both of these variables increase at significant rates, there are warnings of a planetary emergency that threatens to condemn coming generations to a catastrophically diminished future.*

In this chapter, you'll learn to approach our climate crisis mathematically by creating formulas, called functions, that model data for average global temperature and carbon dioxide concentration over time. Understanding the concept of a function will give you a new perspective on many situations, ranging from climate change to using mathematics in a way that is similar to making a movie.

*Sources: Al Gore, *An Inconvenient Truth*, Rodale, 2006; *Time*, April 3, 2006; *Rolling Stone*, September 26, 2013

Here's where you'll find these applications:

- Mathematical models involving climate change are developed in Example 10 and Check Point 10 in Section 2.4.
- Using mathematics in a way that is similar to making a movie is discussed in the Blitzer Bonus on page 148.

SECTION

2.1

Introduction to Functions 

What am I supposed to learn?

After studying this section, you should be able to:

1. Find the domain and range of a relation.
2. Determine whether a relation is a function.
3. Evaluate a function.

Relations

Forbes magazine published a list of the highest-paid TV celebrities between June 2013 and June 2014. The results are shown in **Figure 2.1**.

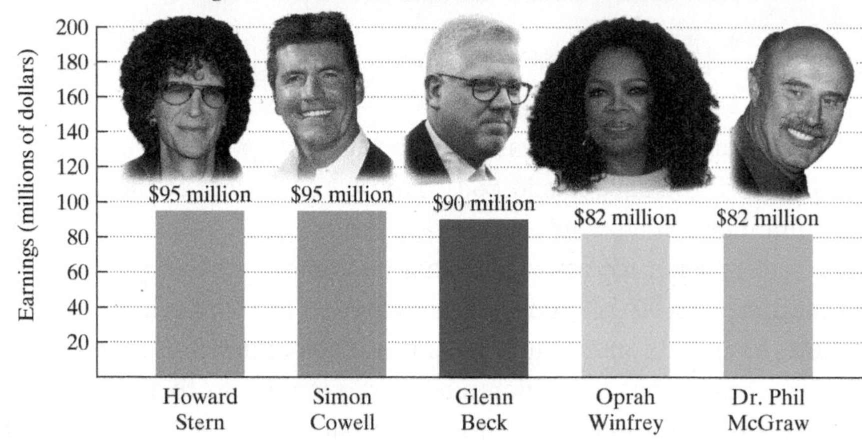

Figure 2.1

Source: Forbes

The graph indicates a correspondence between a TV celebrity and that person's earnings, in millions of dollars. We can write this correspondence using a set of ordered pairs:

{(Stern, 95), (Cowell, 95), (Beck, 90), (Winfrey, 82), (McGraw, 82)}.

These braces indicate we are representing a set.

The mathematical term for a set of ordered pairs is a *relation*.

Definition of a Relation

A **relation** is any set of ordered pairs. The set of all first components of the ordered pairs is called the **domain** of the relation and the set of all second components is called the **range** of the relation.

1 Find the domain and range of a relation.

EXAMPLE 1 Finding the Domain and Range of a Relation

Find the domain and range of the relation:

{(Stern, 95), (Cowell, 95), (Beck, 90), (Winfrey, 82), (McGraw, 82)}.

Solution The domain is the set of all first components. Thus, the domain is

{Stern, Cowell, Beck, Winfrey, McGraw}.

The range is the set of all second components. Thus, the range is

{95, 90, 82}.

Although Stern and Cowell both earned $95 million, it is not necessary to list 95 twice.

Although Winfrey and McGraw both earned $82 million, it is not necessary to list 82 twice.

■

✓ CHECK POINT 1 Find the domain and range of the relation:

$$\{(0, 9.1), (10, 6.7), (20, 10.7), (30, 13.2), (40, 21.2)\}.$$

As you worked Check Point 1, did you wonder if there was a rule that assigned the "inputs" in the domain to the "outputs" in the range? For example, for the ordered pair (30, 13.2), how does the output 13.2 depend on the input 30? The ordered pair is based on the data in **Figure 2.2(a)**, which shows the percentage of first-year U.S. college students claiming no religious affiliation.

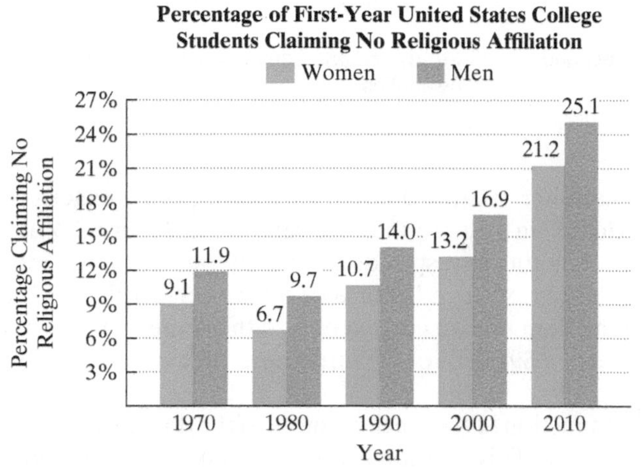

Figure 2.2(a) Data for women and men

Source: John Macionis, *Sociology*, Fourteenth Edition, Pearson, 2012.

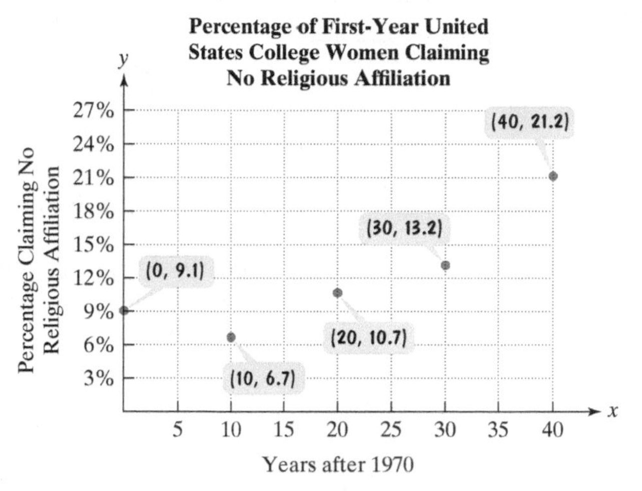

Figure 2.2(b) Visually representing the relation for the women's data

In **Figure 2.2(b)**, we used the data for college women to create the following ordered pairs:

(years after 1970, percentage of first-year college women claiming no religious affiliation).

Consider, for example, the ordered pair (30, 13.2).

(30, 13.2)

30 years after 1970, or in 2000, — 13.2% of first-year college women claimed no religious affiliation.

The five points in **Figure 2.2(b)** visually represent the relation formed from the women's data. Another way to visually represent the relation is as follows:

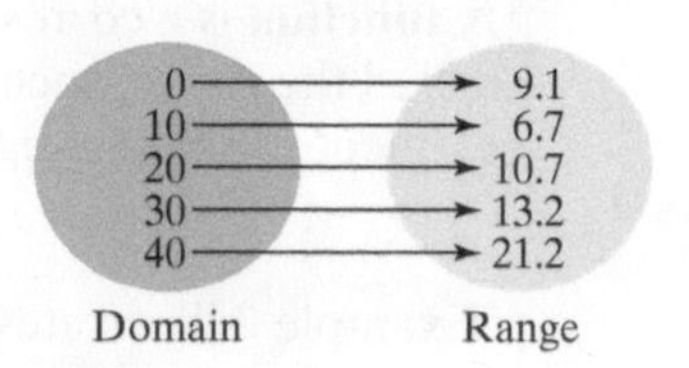

2 Determine whether a relation is a function.

Table 2.1 Highest-Paid TV Celebrities

Celebrity	Earnings (millions of dollars)
Stern	95
Cowell	95
Beck	90
Winfrey	82
McGraw	82

Functions

Table 2.1, based on our earlier discussion, shows the highest-paid TV celebrities and their earnings between June 2013 and June 2014, in millions of dollars. We've used this information to define two relations.

Figure 2.3(a) shows a correspondence between celebrities and their earnings. **Figure 2.3(b)** shows a correspondence between earnings and celebrities.

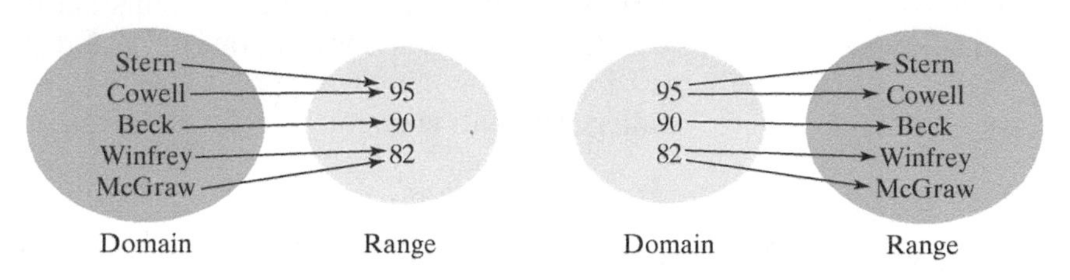

Figure 2.3(a) Celebrities correspond to earnings.

Figure 2.3(b) Earnings correspond to celebrities.

A relation in which each member of the domain corresponds to exactly one member of the range is a **function**. Can you see that the relation in **Figure 2.3(a)** is a function? Each celebrity in the domain corresponds to exactly one earning amount in the range. If we know the celebrity, we can be sure of his or her earnings. Notice that more than one element in the domain can correspond to the same element in the range: Stern and Cowell both earned \$95 million. Winfrey and McGraw both earned \$82 million.

Is the relation in **Figure 2.3(b)** a function? Does each member of the domain correspond to precisely one member of the range? This relation is not a function because there are members of the domain that correspond to two different members of the range:

(95, Stern) (95, Cowell) (82, Winfrey) (82, McGraw).

The member of the domain 95 corresponds to both Stern and Cowell. The member of the domain 82 corresponds to both Winfrey and McGraw. If we know that the earnings are \$95 million or \$82 million, we cannot be sure of the celebrity. Because **a function is a relation in which no two ordered pairs have the same first component and different second components**, the ordered pairs (95, Stern) and (95, Cowell) are not ordered pairs of a function. Similarly, (82, Winfrey) and (82, McGraw) are not ordered pairs of a function.

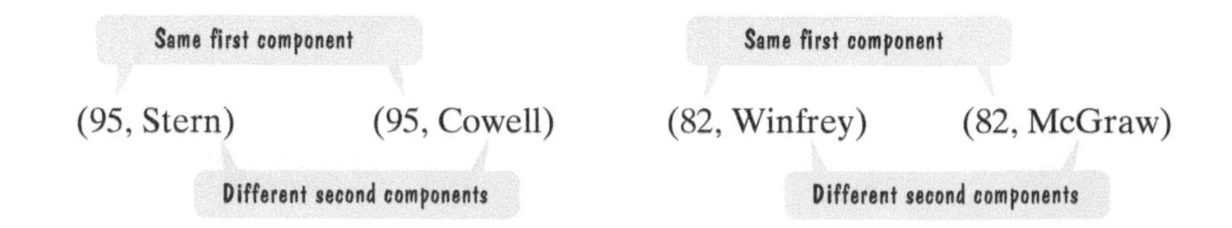

Definition of a Function

A **function** is a correspondence from a first set, called the **domain**, to a second set, called the **range**, such that each element in the domain corresponds to *exactly one* element in the range.

Example 2 illustrates that not every correspondence between sets is a function.

EXAMPLE 2 Determining Whether a Relation Is a Function

Determine whether each relation is a function:

a. $\{(1, 5), (2, 5), (3, 7), (4, 8)\}$ **b.** $\{(5, 1), (5, 2), (7, 3), (8, 4)\}$.

Solution We begin by making a figure for each relation that shows the domain and the range (**Figure 2.4**).

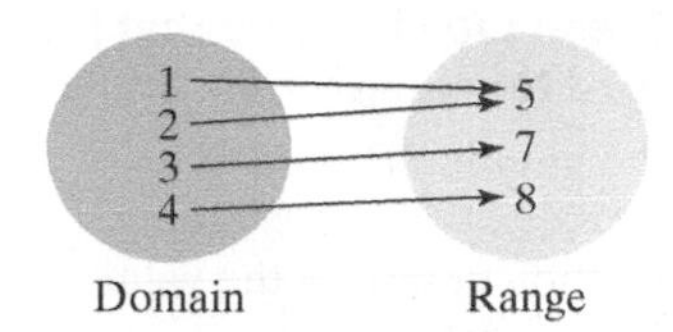

Figure 2.4(a)

a. **Figure 2.4(a)** shows that every element in the domain corresponds to exactly one element in the range. The element 1 in the domain corresponds to the element 5 in the range. Furthermore, 2 corresponds to 5, 3 corresponds to 7, and 4 corresponds to 8. No two ordered pairs in the given relation have the same first component and different second components. Thus, the relation is a function.

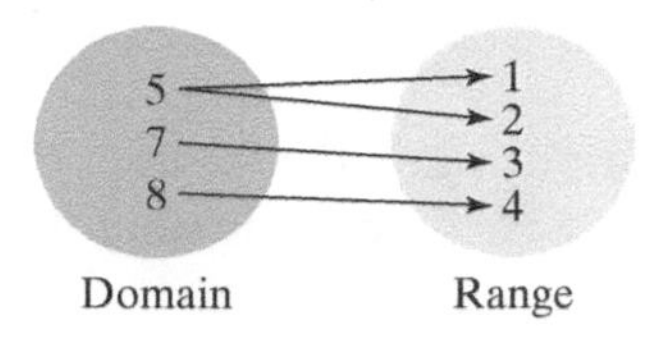

Figure 2.4(b)

b. **Figure 2.4(b)** shows that 5 corresponds to both 1 and 2. If any element in the domain corresponds to more than one element in the range, the relation is not a function. This relation is not a function because two ordered pairs have the same first component and different second components.

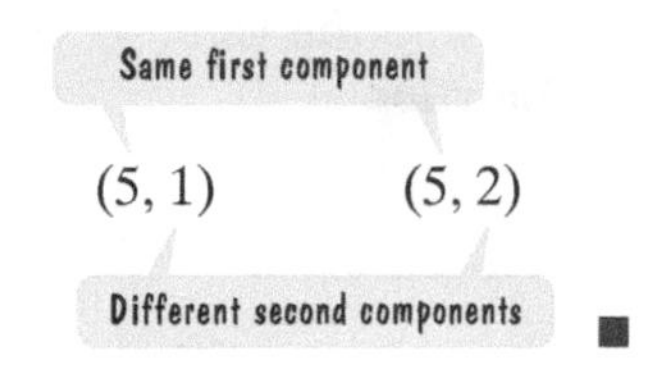

■

Great Question!

If I reverse a function's components, will this new relation be a function?

If a relation is a function, reversing the components in each of its ordered pairs may result in a relation that is not a function.

Look at **Figure 2.4(a)** again. The fact that 1 and 2 in the domain correspond to the same number, 5, in the range does not violate the definition of a function. **A function can have two different first components with the same second component.** By contrast, a relation is not a function when two different ordered pairs have the same first component and different second components. Thus, the relation in Example 2(b) is not a function.

✓ **CHECK POINT 2** Determine whether each relation is a function:

a. $\{(1, 2), (3, 4), (5, 6), (5, 7)\}$

b. $\{(1, 2), (3, 4), (6, 5), (7, 5)\}$.

3 Evaluate a function. 

Functions as Equations and Function Notation

Functions are usually given in terms of equations rather than as sets of ordered pairs. For example, here is an equation that models the percentage of first-year college women claiming no religious affiliation as a function of time:

$$y = 0.014x^2 - 0.24x + 8.8.$$

The variable x represents the number of years after 1970. The variable y represents the percentage of first-year college women claiming no religious affiliation. The variable y is a function of the variable x. For each value of x, there is one and only one value of y. The variable x is called the **independent variable** because it can be assigned any value from the domain. Thus, x can be assigned any nonnegative integer representing the number of years after 1970. The variable y is called the **dependent variable** because its value depends on x. The percentage claiming no religious affiliation depends on the number of years after 1970. The value of the dependent variable, y, is calculated after selecting a value for the independent variable, x.

If an equation in x and y gives one and only one value of y for each value of x, then the variable y is a function of the variable x. When an equation represents a function, the function is often named by a letter such as f, g, h, F, G, or H. Any letter can be used to name a function. Suppose that f names a function. Think of the domain as the set of the function's inputs and the range as the set of the function's outputs. As shown in **Figure 2.5**, the input is represented by x and the output by $f(x)$. The special notation $f(x)$, read "f of x" or "f at x," represents the **value of the function at the number** x.

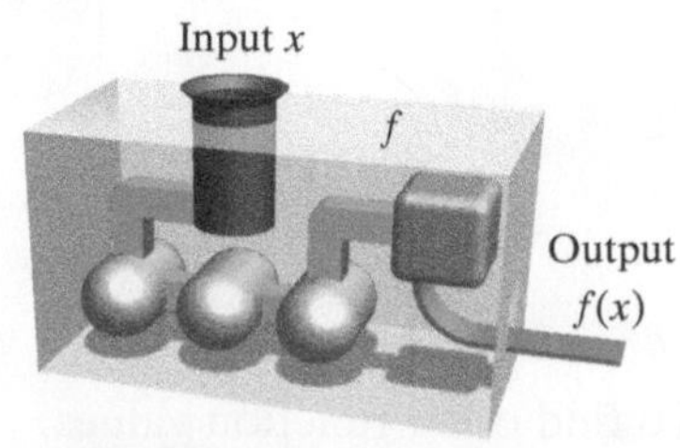

Figure 2.5 A "function machine" with inputs and outputs

Let's make this clearer by considering a specific example. We know that the equation

$$y = 0.014x^2 - 0.24x + 8.8$$

defines y as a function of x. We'll name the function f. Now, we can apply our new function notation.

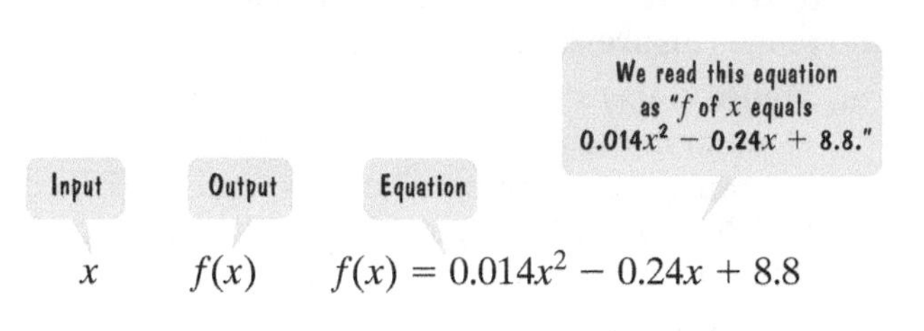

Great Question!

Doesn't $f(x)$ indicate that I need to multiply f and x?

The notation $f(x)$ does *not* mean "f times x." The notation describes the value of the function at x.

Suppose we are interested in finding $f(30)$, the function's output when the input is 30. To find the value of the function at 30, we substitute 30 for x. We are **evaluating the function** at 30.

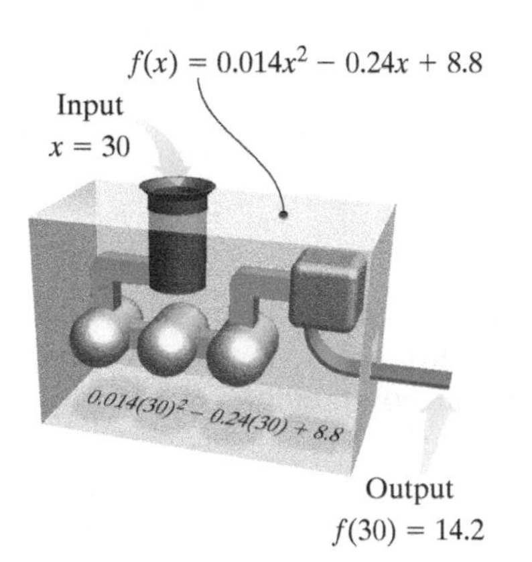

Figure 2.6 A function machine at work

$$f(x) = 0.014x^2 - 0.24x + 8.8$$ This is the given function.

$$f(30) = 0.014(30)^2 - 0.24(30) + 8.8$$ Replace each occurrence of x with 30.

$$= 0.014(900) - 0.24(30) + 8.8$$ Evaluate the exponential expression: $30^2 = 30 \cdot 30 = 900$.

$$= 12.6 - 7.2 + 8.8$$ Perform the multiplications.

$$f(30) = 14.2$$ Subtract and add from left to right.

The statement $f(30) = 14.2$, read "f of 30 equals 14.2," tells us that the value of the function at 30 is 14.2. When the function's input is 30, its output is 14.2. **Figure 2.6** illustrates the input and output in terms of a function machine.

$$f(30) = 14.2$$

30 years after 1970, or in 2000, — 14.2% of first-year college women claimed no religious affiliation.

We have seen that in 2000, 13.2% actually claimed nonaffiliation, so our function that models the data slightly overestimates the percent for 2000.

Using Technology

Graphing utilities can be used to evaluate functions. The screens on the right show the evaluation of

$$f(x) = 0.014x^2 - 0.24x + 8.8$$

at 30 on a TI-84 Plus C graphing calculator. The function f is named Y_1.

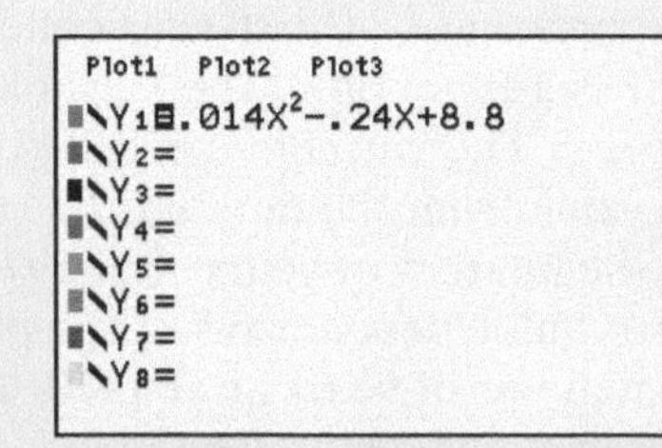

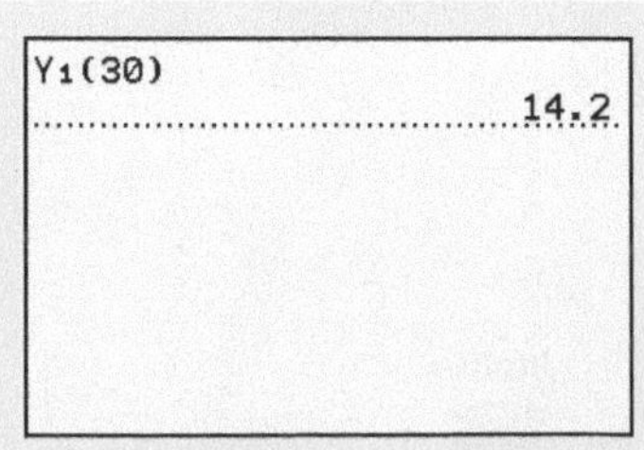

We used $f(x) = 0.014x^2 - 0.24x + 8.8$ to find $f(30)$. To find other function values, such as $f(40)$ or $f(55)$, substitute the specified input value, 40 or 55, for x in the function's equation.

If a function is named f and x represents the independent variable, the notation $f(x)$ corresponds to the y-value for a given x. Thus,

$$f(x) = 0.014x^2 - 0.24x + 8.8 \quad \text{and} \quad y = 0.014x^2 - 0.24x + 8.8$$

define the same function. This function may be written as

$$y = f(x) = 0.014x^2 - 0.24x + 8.8.$$

EXAMPLE 3 Using Function Notation

Find the indicated function value:

a. $f(4)$ for $f(x) = 2x + 3$

b. $g(-2)$ for $g(x) = 2x^2 - 1$

c. $h(-5)$ for $h(r) = r^3 - 2r^2 + 5$

d. $F(a + h)$ for $F(x) = 5x + 7$.

Solution

a.

$f(x) = 2x + 3$ — This is the given function.

$f(4) = 2 \cdot 4 + 3$ — To find f of 4, replace x with 4.

$= 8 + 3$ — Multiply: $2 \cdot 4 = 8$.

$f(4) = 11$ — Add.

f of 4 is 11.

b.

$g(x) = 2x^2 - 1$ — This is the given function.

$g(-2) = 2(-2)^2 - 1$ — To find g of -2, replace x with -2.

$= 2(4) - 1$ — Evaluate the exponential expression: $(-2)^2 = 4$.

$= 8 - 1$ — Multiply: $2(4) = 8$.

$g(-2) = 7$ — Subtract.

g of −2 is 7.

c.

$h(r) = r^3 - 2r^2 + 5$ — The function's name is h and r represents the independent variable.

$h(-5) = (-5)^3 - 2(-5)^2 + 5$ — To find h of -5, replace each occurrence of r with -5.

$= -125 - 2(25) + 5$ — Evaluate exponential expressions.

$= -125 - 50 + 5$ — Multiply.

$h(-5) = -170$ — $-125 - 50 = -175$ and $-175 + 5 = -170$.

h of −5 is −170.

d.

$F(x) = 5x + 7$ — This is the given function.

$F(a + h) = 5(a + h) + 7$ — Replace x with $a + h$.

$F(a + h) = 5a + 5h + 7$ — Apply the distributive property. ■

F of $a + h$ is $5a + 5h + 7$.

✓ **CHECK POINT 3** Find the indicated function value:

a. $f(6)$ for $f(x) = 4x + 5$

b. $g(-5)$ for $g(x) = 3x^2 - 10$

c. $h(-4)$ for $h(r) = r^2 - 7r + 2$

d. $F(a + h)$ for $F(x) = 6x + 9$.

Great Question!

In Example 3 and Check Point 3, finding some of the function values involved evaluating exponential expressions. Can't this be a bit tricky when such functions are evaluated at negative numbers?

Yes. Be particularly careful if there is a term with a coefficient of −1. Notice the following differences:

$$f(x) = -x^2 \qquad\qquad g(x) = (-x)^2$$

Replace x with −4.   Replace x with −4.

$$f(-4) = -(-4)^2 = -16 \qquad\qquad g(-4) = (-(-4))^2 = 4^2 = 16$$

Functions Represented by Tables and Function Notation

Function notation can be applied to functions that are represented by tables.

EXAMPLE 4 Using Function Notation

Function f is defined by the following table:

x	$f(x)$
−2	5
−1	0
0	3
1	1
2	4

a. Explain why the table defines a function.

b. Find the domain and the range of the function.

Find the indicated function value:

c. $f(-1)$

d. $f(0)$

e. Find x such that $f(x) = 4$.

Solution

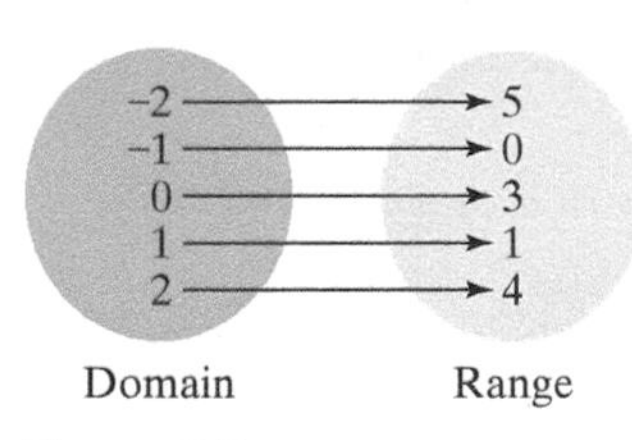

Figure 2.7

a. Values in the first column of the table make up the domain, or input values. Values in the second column of the table make up the range, or output values. We see that every element in the domain corresponds to exactly one element in the range, shown in **Figure 2.7**. Therefore, the relation given by the table is a function.

The voice balloons pointing to appropriate parts of the table illustrate the solution to parts (b)–(e).

x	$f(x)$
−2	5
−1	0
0	3
1	1
2	4

c. $f(-1) = 0$: When the input is −1, the output is 0.

d. $f(0) = 3$: When the input is 0, the output is 3.

e. $f(x) = 4$ when $x = 2$: The output, $f(x)$, is 4 when the input, x, is 2.

b. The domain is the set of inputs: $\{-2, -1, 0, 1, 2\}$.

b. The range is the set of outputs: $\{5, 0, 3, 1, 4\}$.

■

CHECK POINT 4 Function g is defined by the following table:

x	$g(x)$
0	3
1	0
2	1
3	2
4	3

a. Explain why the table defines a function.

b. Find the domain and the range of the function.

Find the indicated function value:

c. $g(1)$

d. $g(3)$

e. Find x such that $g(x) = 3$.

Achieving Success

Check out the *Learning Guide* that accompanies this textbook.

Benefits of using the *Learning Guide* include:

- It will help you become better organized. This includes organizing your class notes, assigned homework, quizzes, and tests.
- It will enable you to use your textbook more efficiently.
- It will bring together the learning tools for this course, including the textbook, the Video Lecture Series, and the PowerPoint Presentation.
- It will help increase your study skills.
- It will help you prepare for the chapter tests.

Ask your professor about the availability of this textbook supplement.

CONCEPT AND VOCABULARY CHECK

Fill in each blank so that the resulting statement is true.

1. Any set of ordered pairs is called a/an __________. The set of all first components of the ordered pairs is called the __________. The set of all second components of the ordered pairs is called the __________.
2. A set of ordered pairs in which each member of the set of first components corresponds to exactly one member of the set of second components is called a/an __________.
3. The notation $f(x)$ describes the value of __________ at __________.
4. If $h(r) = -r^2 + 4r - 7$, we can find $h(-2)$ by replacing each occurrence of __________ by __________.

2.1 EXERCISE SET

MyMathLab®

Practice Exercises

In Exercises 1–8, determine whether each relation is a function. Give the domain and range for each relation.

1. $\{(1, 2), (3, 4), (5, 5)\}$
2. $\{(4, 5), (6, 7), (8, 8)\}$
3. $\{(3, 4), (3, 5), (4, 4), (4, 5)\}$
4. $\{(5, 6), (5, 7), (6, 6), (6, 7)\}$
5. $\{(-3, -3), (-2, -2), (-1, -1), (0, 0)\}$
6. $\{(-7, -7), (-5, -5), (-3, -3), (0, 0)\}$
7. $\{(1, 4), (1, 5), (1, 6)\}$
8. $\{(4, 1), (5, 1), (6, 1)\}$

In Exercises 9–24, find the indicated function values.

9. $f(x) = x + 1$
   a. $f(0)$ **b.** $f(5)$ **c.** $f(-8)$
   d. $f(2a)$ **e.** $f(a + 2)$
10. $f(x) = x + 3$
   a. $f(0)$ **b.** $f(5)$ **c.** $f(-8)$
   d. $f(2a)$ **e.** $f(a + 2)$

11. $g(x) = 3x - 2$
a. $g(0)$
b. $g(-5)$
c. $g\left(\frac{2}{3}\right)$
d. $g(4b)$
e. $g(b + 4)$

12. $g(x) = 4x - 3$
a. $g(0)$
b. $g(-5)$
c. $g\left(\frac{3}{4}\right)$
d. $g(5b)$
e. $g(b + 5)$

13. $h(x) = 3x^2 + 5$
a. $h(0)$
b. $h(-1)$
c. $h(4)$
d. $h(-3)$
e. $h(4b)$

14. $h(x) = 2x^2 - 4$
a. $h(0)$
b. $h(-1)$
c. $h(5)$
d. $h(-3)$
e. $h(5b)$

15. $f(x) = 2x^2 + 3x - 1$
a. $f(0)$
b. $f(3)$
c. $f(-4)$
d. $f(b)$
e. $f(5a)$

16. $f(x) = 3x^2 + 4x - 2$
a. $f(0)$
b. $f(3)$
c. $f(-5)$
d. $f(b)$
e. $f(5a)$

17. $f(x) = (-x)^3 - x^2 - x + 7$
a. $f(0)$
b. $f(2)$
c. $f(-2)$
d. $f(1) + f(-1)$

18. $f(x) = (-x)^3 - x^2 - x + 10$
a. $f(0)$
b. $f(2)$
c. $f(-2)$
d. $f(1) + f(-1)$

19. $f(x) = \dfrac{2x - 3}{x - 4}$
a. $f(0)$
b. $f(3)$
c. $f(-4)$
d. $f(-5)$
e. $f(a + h)$
f. Why must 4 be excluded from the domain of f?

20. $f(x) = \dfrac{3x - 1}{x - 5}$
a. $f(0)$
b. $f(3)$
c. $f(-3)$
d. $f(10)$
e. $f(a + h)$
f. Why must 5 be excluded from the domain of f?

21.

x	$f(x)$
−4	3
−2	6
0	9
2	12
4	15

a. $f(-2)$
b. $f(2)$
c. For what value of x is $f(x) = 9$?

22.

x	$f(x)$
−5	4
−3	8
0	12
3	16
5	20

a. $f(-3)$
b. $f(3)$
c. For what value of x is $f(x) = 12$?

23.

x	$h(x)$
−2	2
−1	1
0	0
1	1
2	2

a. $h(-2)$
b. $h(1)$
c. For what values of x is $h(x) = 1$?

24.

x	$h(x)$
−2	−2
−1	−1
0	0
1	−1
2	−2

a. $h(-2)$
b. $h(1)$
c. For what values of x is $h(x) = -1$?

Practice PLUS

In Exercises 25–26, let $f(x) = x^2 - x + 4$ and $g(x) = 3x - 5$.

25. Find $g(1)$ and $f(g(1))$.
26. Find $g(-1)$ and $f(g(-1))$.

In Exercises 27–28, let f and g be defined by the following table:

x	$f(x)$	$g(x)$
−2	6	0
−1	3	4
0	−1	1
1	−4	−3
2	0	−6

27. Find $\sqrt{f(-1) - f(0)} - [g(2)]^2 + f(-2) \div g(2) \cdot g(-1)$.

28. Find $|f(1) - f(0)| - [g(1)]^2 + g(1) \div f(-1) \cdot g(2)$.

In Exercises 29–30, find $f(-x) - f(x)$ for the given function f. Then simplify the expression.

29. $f(x) = x^3 + x - 5$
30. $f(x) = x^2 - 3x + 7$

In Exercises 31–32, each function is defined by two equations. The equation in the first row gives the output for negative numbers in the domain. The equation in the second row gives the output for nonnegative numbers in the domain. Find the indicated function values.

31. $f(x) = \begin{cases} 3x + 5 & \text{if } x < 0 \\ 4x + 7 & \text{if } x \geq 0 \end{cases}$
a. $f(-2)$
b. $f(0)$
c. $f(3)$
d. $f(-100) + f(100)$

32. $f(x) = \begin{cases} 6x - 1 & \text{if } x < 0 \\ 7x + 3 & \text{if } x \geq 0 \end{cases}$

  a. $f(-3)$

  b. $f(0)$

  c. $f(4)$

  d. $f(-100) + f(100)$

Application Exercises

The Corruption Perceptions Index uses perceptions of the general public, business people, and risk analysts to rate countries by how likely they are to accept bribes. The ratings are on a scale from 0 to 10, where higher scores represent less corruption. The graph shows the corruption ratings for the world's least corrupt and most corrupt countries. (The rating for the United States is 7.6.) Use the graph to solve Exercises 33–34.

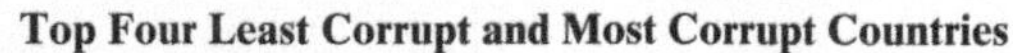

Top Four Least Corrupt and Most Corrupt Countries

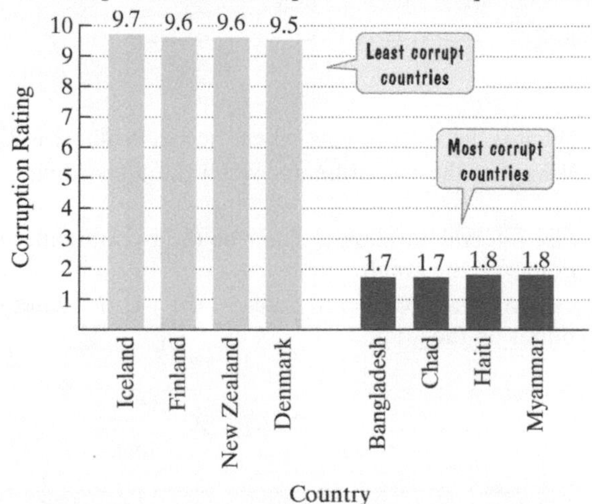

Source: Transparency International, *Corruption Perceptions Index*

33. a. Write a set of four ordered pairs in which each of the least corrupt countries corresponds to a corruption rating. Each ordered pair should be in the form

(country, corruption rating).

  b. Is the relation in part (a) a function? Explain your answer.

  c. Write a set of four ordered pairs in which corruption ratings for the least corrupt countries correspond to countries. Each ordered pair should be in the form

(corruption rating, country).

  d. Is the relation in part (c) a function? Explain your answer.

34. a. Write a set of four ordered pairs in which each of the most corrupt countries corresponds to a corruption rating. Each ordered pair should be in the form

(country, corruption rating).

  b. Is the relation in part (a) a function? Explain your answer.

  c. Write a set of four ordered pairs in which corruption ratings for the most corrupt countries correspond to countries. Each ordered pair should be in the form

(corruption rating, country).

  d. Is the relation in part (c) a function? Explain your answer.

Explaining the Concepts

35. What is a relation? Describe what is meant by its domain and its range.

36. Explain how to determine whether a relation is a function. What is a function?

37. Does $f(x)$ mean f times x when referring to function f? If not, what does $f(x)$ mean? Provide an example with your explanation.

38. For people filing a single return, federal income tax is a function of adjusted gross income because for each value of adjusted gross income there is a specific tax to be paid. By contrast, the price of a house is not a function of the lot size on which the house sits because houses on same-sized lots can sell for many different prices.

  a. Describe an everyday situation between variables that is a function.

  b. Describe an everyday situation between variables that is not a function.

Critical Thinking Exercises

Make Sense? *In Exercises 39–42, determine whether each statement makes sense or does not make sense, and explain your reasoning.*

39. Today's temperature is a function of the time of day.

40. My height is a function of my age.

41. Although I presented my function as a set of ordered pairs, I could have shown the correspondences using a table or using points plotted in a rectangular coordinate system.

42. My function models how the chance of divorce depends on the number of years of marriage, so the range is $\{x \mid x \text{ is the number of years of marriage}\}$.

In Exercises 43–48, determine whether each statement is true or false. If the statement is false, make the necessary change(s) to produce a true statement.

43. All relations are functions.

44. No two ordered pairs of a function can have the same second components and different first components.

Using the tables that define f and g, determine whether each statement in Exercises 45–48 is true or false.

x	$f(x)$
-4	-1
-3	-2
-2	-3
-1	-4

x	$g(x)$
-1	-4
-2	-3
-3	-2
-4	-1

45. The domain of $f =$ the range of f.

46. The range of $f =$ the domain of g.

47. $f(-4) - f(-2) = 2$

48. $g(-4) + f(-4) = 0$

49. If $f(x) = 3x + 7$, find $\dfrac{f(a + h) - f(a)}{h}$.

50. Give an example of a relation with the following characteristics: The relation is a function containing two ordered pairs. Reversing the components in each ordered pair results in a relation that is not a function.

51. If $f(x + y) = f(x) + f(y)$ and $f(1) = 3$, find $f(2)$, $f(3)$, and $f(4)$. Is $f(x + y) = f(x) + f(y)$ for all functions?

Review Exercises

52. Simplify: $24 \div 4[2 - (5 - 2)]^2 - 6$ (Section 1.2, Example 7)

53. Simplify: $\left(\dfrac{3x^2y^{-2}}{y^3}\right)^{-2}$. (Section 1.6, Example 9)

54. Solve: $\dfrac{x}{3} = \dfrac{3x}{5} + 4$. (Section 1.4, Example 4)

Preview Exercises

Exercises 55–57 will help you prepare for the material covered in the next section.

55. Graph $y = 2x$. Select integers for x, starting with -2 and ending with 2.

56. Graph $y = 2x + 4$. Select integers for x, starting with -2 and ending with 2.

57. Use the following graph to solve this exercise.

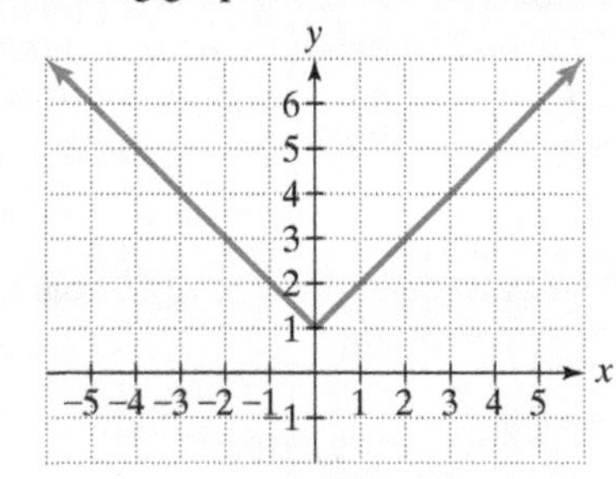

a. What is the y-coordinate when the x-coordinate is 2?

b. What are the x-coordinates when the y-coordinate is 4?

c. Use interval notation to describe the x-coordinates of all points on the graph.

d. Use interval notation to describe the y-coordinates of all points on the graph.

SECTION

2.2 Graphs of Functions

Magnified 6000 times, this color-scanned image shows a T-lymphocyte blood cell (green) infected with the HIV virus (red). Depletion of the number of T cells causes destruction of the immune system.

What am I supposed to learn?

After studying this section, you should be able to:

1. Graph functions by plotting points.
2. Use the vertical line test to identify functions.
3. Obtain information about a function from its graph.
4. Identify the domain and range of a function from its graph.

The number of T cells in a person with HIV is a function of time after infection. In this section, we'll analyze the graph of this function, using the rectangular coordinate system to visualize what functions look like.

Graphs of Functions

The **graph of a function** is the graph of its ordered pairs. For example, the graph of $f(x) = 2x$ is the set of points (x, y) in the rectangular coordinate system satisfying $y = 2x$. Similarly, the graph of $g(x) = 2x + 4$ is the set of points (x, y) in the rectangular coordinate system satisfying the equation $y = 2x + 4$. In the next example, we graph both of these functions in the same rectangular coordinate system.

1 Graph functions by plotting points.

EXAMPLE 1 Graphing Functions

Graph the functions $f(x) = 2x$ and $g(x) = 2x + 4$ in the same rectangular coordinate system. Select integers for x, starting with -2 and ending with 2.

Solution We begin by setting up a partial table of coordinates for each function. Then, we plot the five points in each table and connect them, as shown in **Figure 2.8**. The graph of each function is a straight line. Do you see a relationship between the two graphs? The graph of g is the graph of f shifted vertically up by 4 units.

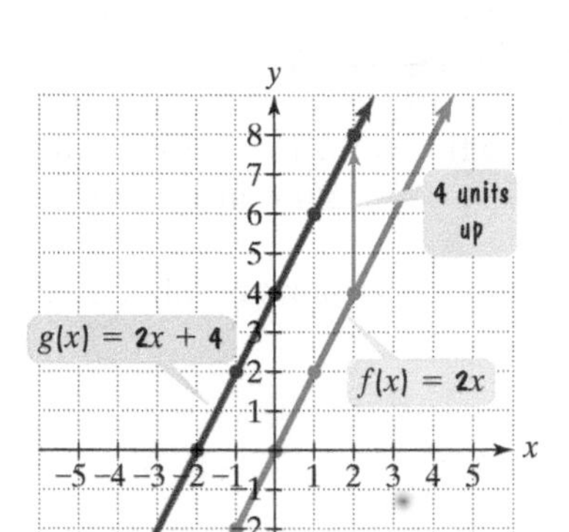

4 units up
$g(x) = 2x + 4$
$f(x) = 2x$

Figure 2.8

x	$f(x) = 2x$	(x, y) or $(x, f(x))$
-2	$f(-2) = 2(-2) = -4$	$(-2, -4)$
-1	$f(-1) = 2(-1) = -2$	$(-1, -2)$
0	$f(0) = 2 \cdot 0 = 0$	$(0, 0)$
1	$f(1) = 2 \cdot 1 = 2$	$(1, 2)$
2	$f(2) = 2 \cdot 2 = 4$	$(2, 4)$

x	$g(x) = 2x + 4$	(x, y) or $(x, g(x))$
-2	$g(-2) = 2(-2) + 4 = 0$	$(-2, 0)$
-1	$g(-1) = 2(-1) + 4 = 2$	$(-1, 2)$
0	$g(0) = 2 \cdot 0 + 4 = 4$	$(0, 4)$
1	$g(1) = 2 \cdot 1 + 4 = 6$	$(1, 6)$
2	$g(2) = 2 \cdot 2 + 4 = 8$	$(2, 8)$

Choose x. Compute $f(x)$ by evaluating f at x. Form the ordered pair. Choose x. Compute $g(x)$ by evaluating g at x. Form the ordered pair.

■

The graphs in Example 1 are straight lines. All functions with equations of the form $f(x) = mx + b$ graph as straight lines. Such functions, called **linear functions**, will be discussed in detail in Sections 2.4–2.5.

Using Technology

We can use a graphing utility to check the tables and the graphs in Example 1 for the functions

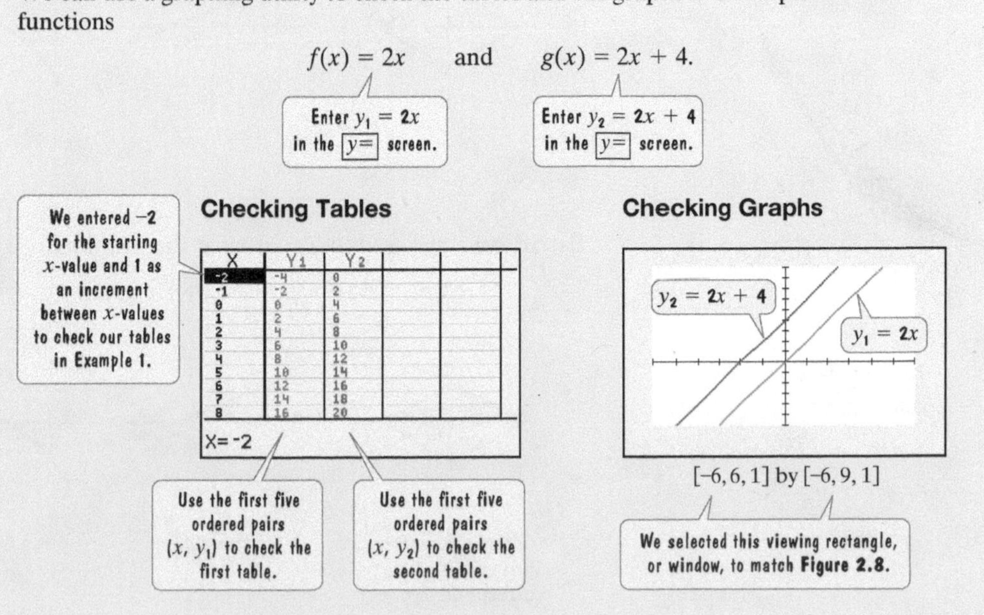

$f(x) = 2x$ and $g(x) = 2x + 4$.
Enter $y_1 = 2x$ in the y= screen.
Enter $y_2 = 2x + 4$ in the y= screen.
Checking Tables
Checking Graphs
We entered -2 for the starting x-value and 1 as an increment between x-values to check our tables in Example 1.
X=-2
$y_2 = 2x + 4$
$y_1 = 2x$
[−6, 6, 1] by [−6, 9, 1]
Use the first five ordered pairs (x, y_1) to check the first table.
Use the first five ordered pairs (x, y_2) to check the second table.
We selected this viewing rectangle, or window, to match **Figure 2.8**.

✓ CHECK POINT 1 Graph the functions $f(x) = 2x$ and $g(x) = 2x - 3$ in the same rectangular coordinate system. Select integers for x, starting with -2 and ending with 2. How is the graph of g related to the graph of f?

2 Use the vertical line test to identify functions.

The Vertical Line Test

Not every graph in the rectangular coordinate system is the graph of a function. The definition of a function specifies that no value of x can be paired with two or more different values of y. Consequently, if a graph contains two or more different points with the same first coordinate, the graph cannot represent a function. This is illustrated in **Figure 2.9**. Observe that points sharing a common first coordinate are vertically above or below each other.

This observation is the basis of a useful test for determining whether a graph defines y as a function of x. The test is called the **vertical line test.**

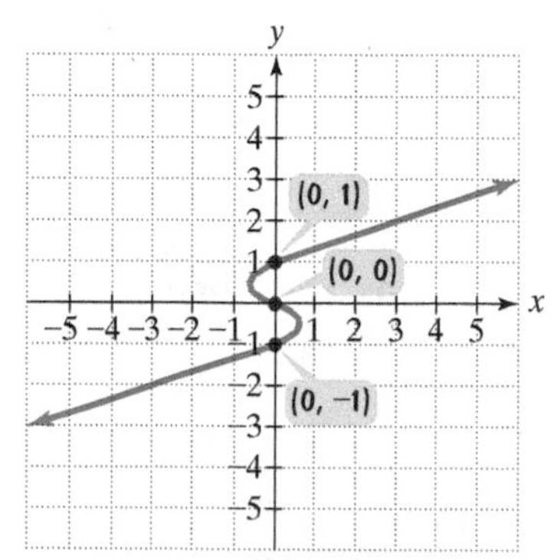

Figure 2.9 y is not a function of x because 0 is paired with three values of y, namely, 1, 0, and -1.

The Vertical Line Test for Functions

If any vertical line intersects a graph in more than one point, the graph does not define y as a function of x.

EXAMPLE 2 Using the Vertical Line Test

Use the vertical line test to identify graphs in which y is a function of x.

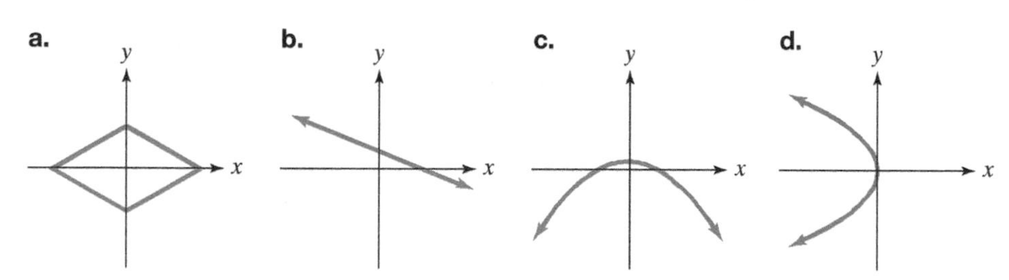

Solution y is a function of x for the graphs in (b) and (c).

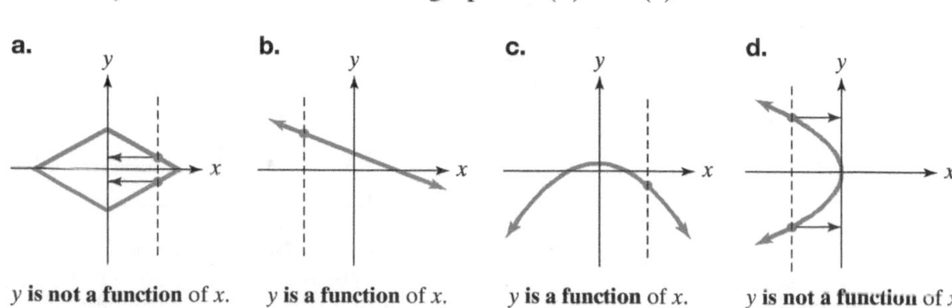

a. **y is not a function** of x. Two values of y correspond to one x-value.

b. **y is a function** of x.

c. **y is a function** of x.

d. **y is not a function** of x. Two values of y correspond to one x-value. ■

CHECK POINT 2 Use the vertical line test to identify graphs in which y is a function of x.

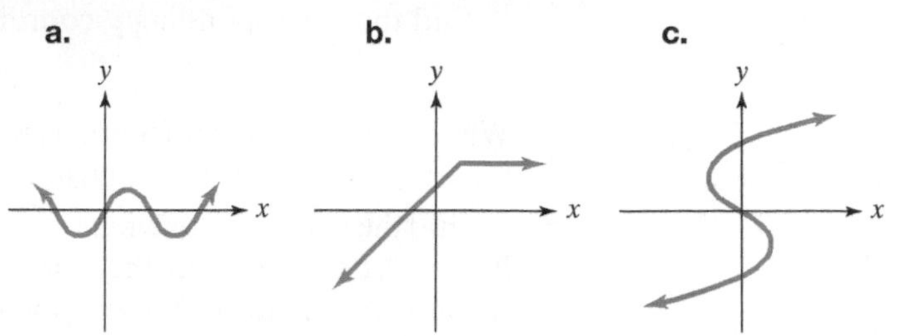

3 Obtain information about a function from its graph.

Obtaining Information from Graphs

You can obtain information about a function from its graph. At the right or left of a graph, you will often find closed dots, open dots, or arrows.

- A closed dot indicates that the graph does not extend beyond this point and the point belongs to the graph.
- An open dot indicates that the graph does not extend beyond this point and the point does not belong to the graph.
- An arrow indicates that the graph extends indefinitely in the direction in which the arrow points.

EXAMPLE 3 Analyzing the Graph of a Function

The human immunodeficiency virus, or HIV, infects and kills helper T cells. Because T cells stimulate the immune system to produce antibodies, their destruction disables the body's defenses against other pathogens. By counting the number of T cells that remain active in the body, the progression of HIV can be monitored. The fewer helper T cells, the more advanced the disease. Without the drugs that are now used to inhibit the progression of the virus, **Figure 2.10** shows a graph that is used to monitor the average progression of the disease. The number of T cells, $f(x)$, is a function of time after infection, x.

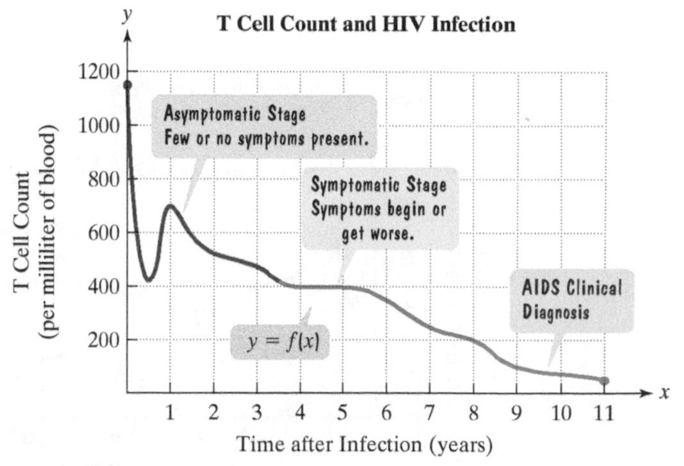

Figure 2.10

Source: Human Sexuality by B. Pruitt, et al. © 1997 Prentice Hall

a. Explain why f represents the graph of a function.

b. Use the graph to find $f(8)$.

c. For what value of x is $f(x) = 350$?

d. Describe the general trend shown by the graph.

Solution

a. No vertical line can be drawn that intersects the graph of f in **Figure 2.10** more than once. By the vertical line test, f represents the graph of a function.

b. To find $f(8)$, or f of 8, we locate 8 on the x-axis. **Figure 2.11** shows the point on the graph of f for which 8 is the first coordinate. From this point, we look to the y-axis to find the corresponding y-coordinate. We see that the y-coordinate is 200. Thus,

$$f(8) = 200.$$

When the time after infection is 8 years, the T cell count is 200 cells per milliliter of blood. (AIDS clinical diagnosis is given at a T cell count of 200 or below.)

c. To find the value of x for which $f(x) = 350$, we approximately locate 350 on the y-axis. **Figure 2.12** shows that there is one point on the graph of f for which 350 is the second coordinate. From this point, we look to the x-axis to find the corresponding x-coordinate. We see that the x-coordinate is 6. Thus,

$$f(x) = 350 \text{ for } x = 6.$$

A T cell count of 350 occurs 6 years after infection.

d. **Figure 2.13** uses voice balloons to describe the general trend shown by the graph.

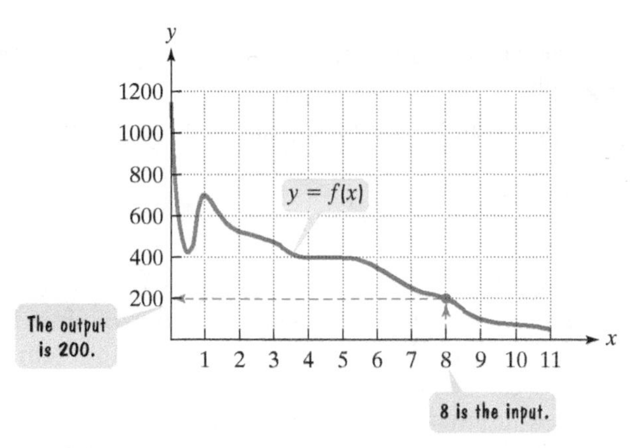

Figure 2.11 Finding $f(8)$

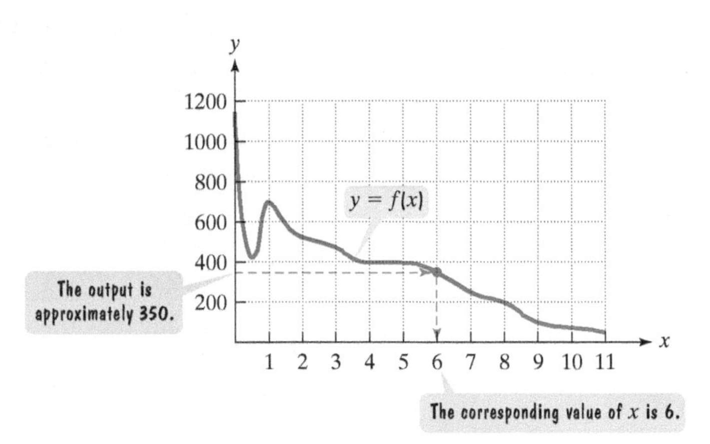

Figure 2.12 Finding x for which $f(x) = 350$

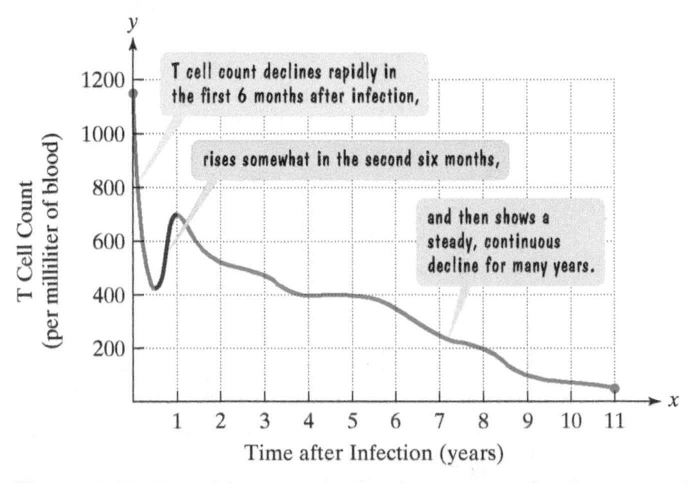

Figure 2.13 Describing changing T cell count over time in a person infected with HIV ■

✓ CHECK POINT 3

a. Use the graph of f in **Figure 2.10** on page 117 to find $f(5)$.

b. For what value of x is $f(x) = 100$?

c. Estimate the minimum T cell count during the asymptomatic stage.

4 Identify the domain and range of a function from its graph.

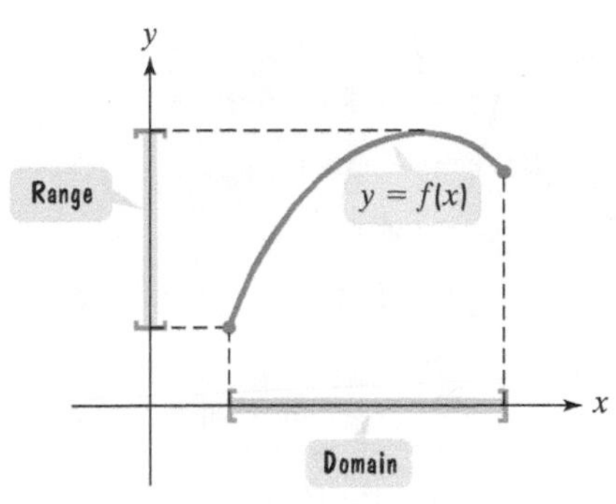
Range
$y = f(x)$
Domain

Figure 2.14 Domain and range of f

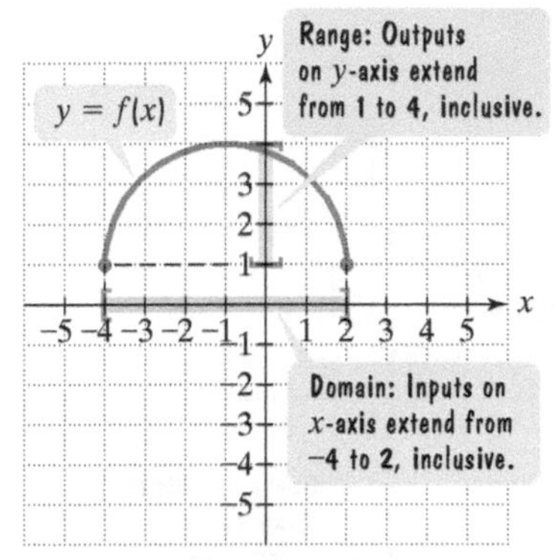
$y = f(x)$
Range: Outputs on y-axis extend from 1 to 4, inclusive.
Domain: Inputs on x-axis extend from −4 to 2, inclusive.

Figure 2.15 Domain and range of f

Identifying Domain and Range from a Function's Graph

Figure 2.14 illustrates how the graph of a function is used to determine the function's domain and its range.

Domain: set of inputs

Found on the x-axis

Range: set of outputs

Found on the y-axis

Let's apply these ideas to the graph of the function shown in **Figure 2.15**. To find the domain, look for all the inputs on the x-axis that correspond to points on the graph. Can you see that they extend from −4 to 2, inclusive? Using interval notation, the function's domain can be represented as follows:

$$[-4, 2].$$

The square brackets indicate −4 and 2 are included. Note the square brackets on the x-axis in **Figure 2.15**.

To find the range, look for all the outputs on the y-axis that correspond to points on the graph. They extend from 1 to 4, inclusive. Using interval notation, the function's range can be represented as follows:

$$[1, 4].$$

The square brackets indicate 1 and 4 are included. Note the square brackets on the y-axis in **Figure 2.15**.

EXAMPLE 4 Identifying the Domain and Range of a Function from Its Graph

Use the graph of each function to identify its domain and its range.

a.

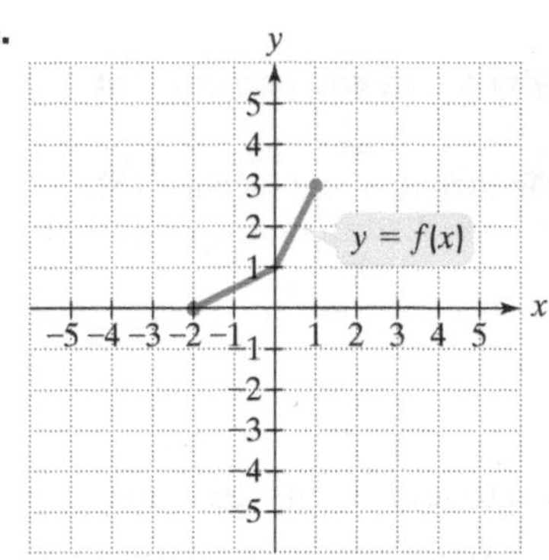
$y = f(x)$

b.

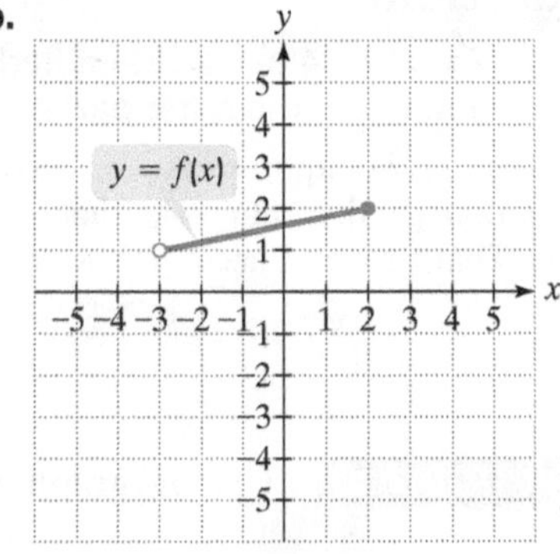
$y = f(x)$

c.

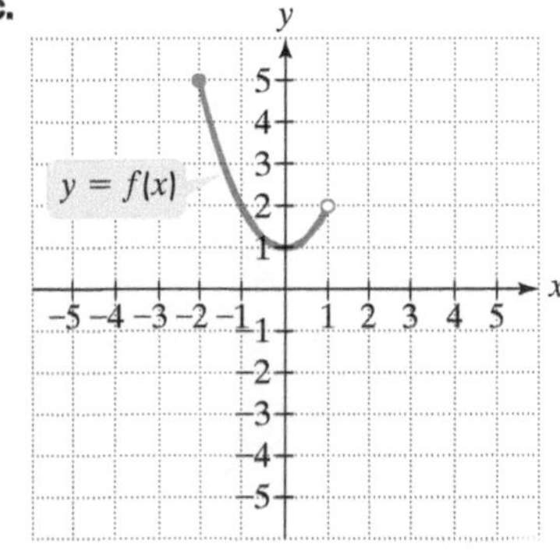
$y = f(x)$

d.

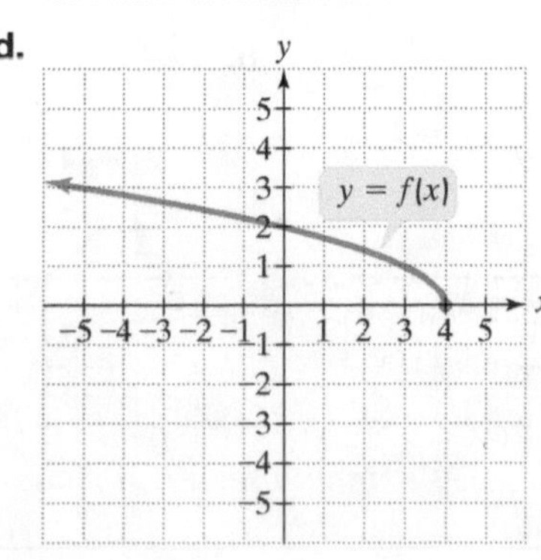
$y = f(x)$

e.

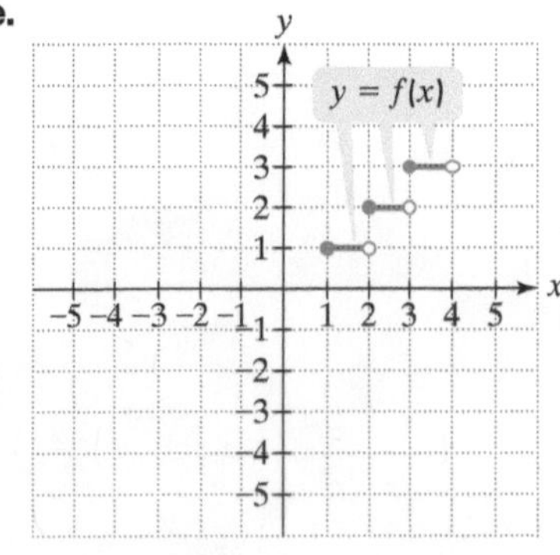
$y = f(x)$

Solution For the graph of each function, the domain is highlighted in purple on the x-axis and the range is highlighted in green on the y-axis.

a.

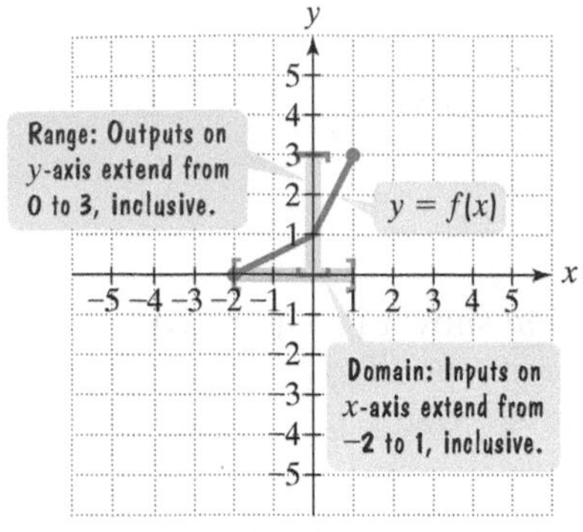

Domain = [−2, 1]; Range = [0, 3]

b.

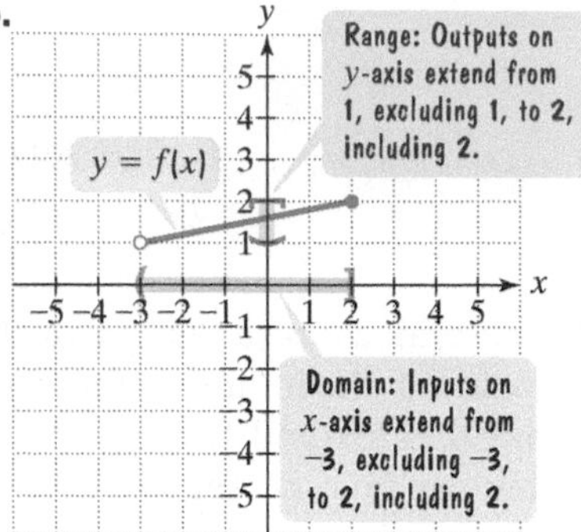

Domain = (−3, 2]; Range = (1, 2]

c.

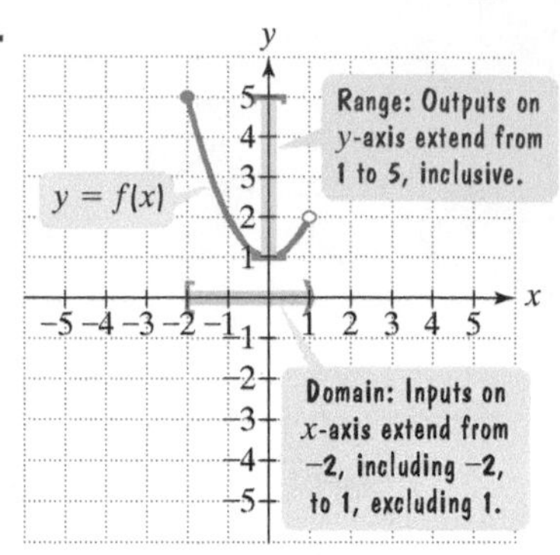

Range: Outputs on
y-axis extend from
1 to 5, inclusive.
y = f(x)
Domain: Inputs on
x-axis extend from
−2, including −2,
to 1, excluding 1.

Domain = [−2, 1); Range = [1, 5]

d.

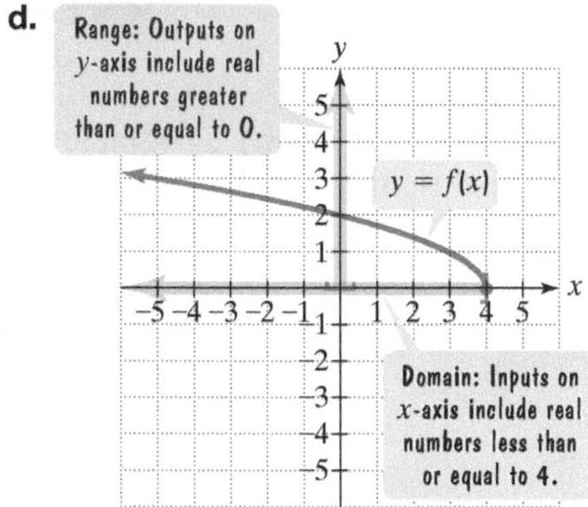

Domain = (−∞, 4]; Range = [0, ∞)

e.

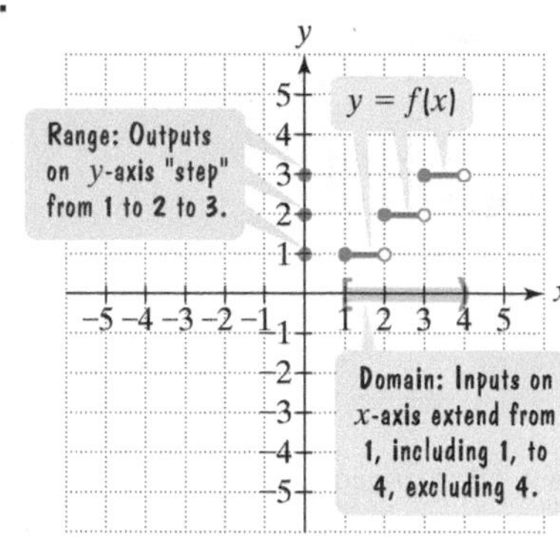

Domain = [1, 4); Range = {1, 2, 3} ■

Great Question!

The range in Example 4(e) was identified as {1, 2, 3}. Why didn't you use interval notation like you did in the other parts of Example 4?

Interval notation is not appropriate for describing a set of distinct numbers such as {1, 2, 3}. Interval notation, [1, 3], would mean that numbers such as 1.5 and 2.99 are in the range, but they are not. That's why we used the roster method.

✓ CHECK POINT 4 Use the graph of each function to identify its domain and its range.

a.

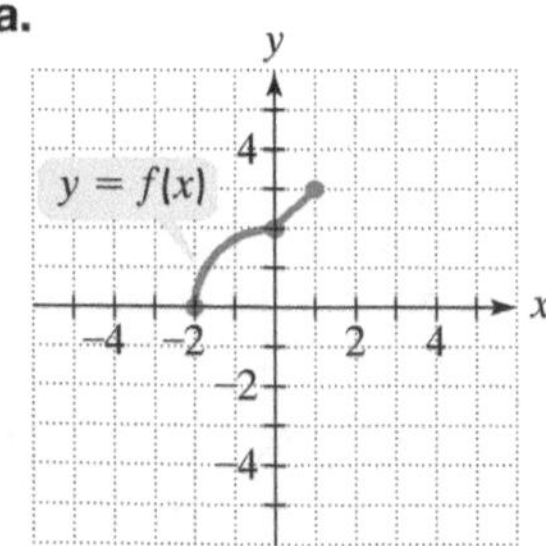

b.

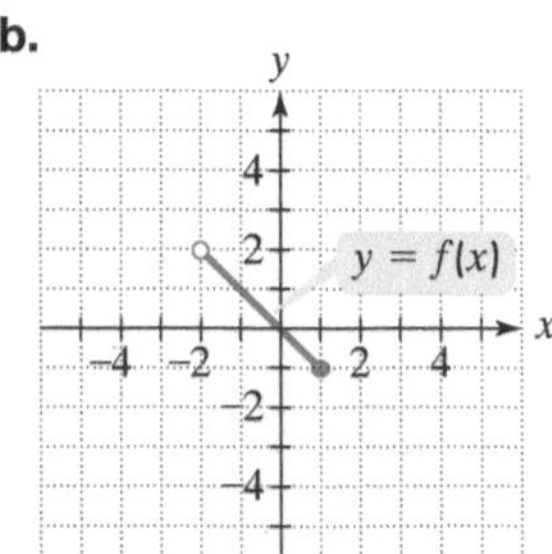

c.

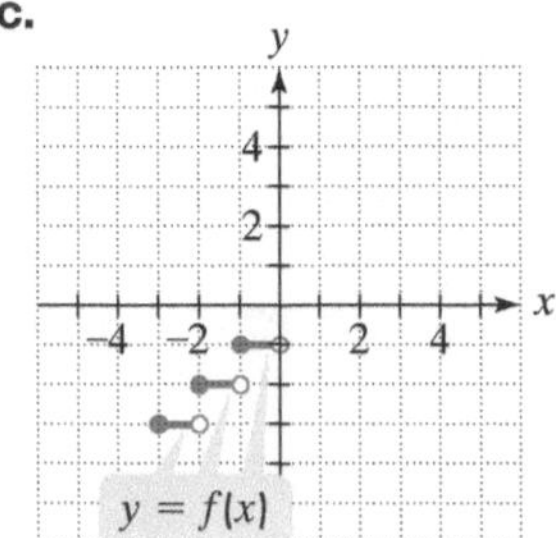

CONCEPT AND VOCABULARY CHECK

Fill in each blank so that the resulting statement is true.

1. The graph of a function is the graph of its ____________.
2. If any vertical line intersects a graph ____________, the graph does not define y as a/an ____________ of x.
3. The shaded set of numbers shown on the x-axis can be expressed in interval notation as ____________. This set represents the function's ____________.

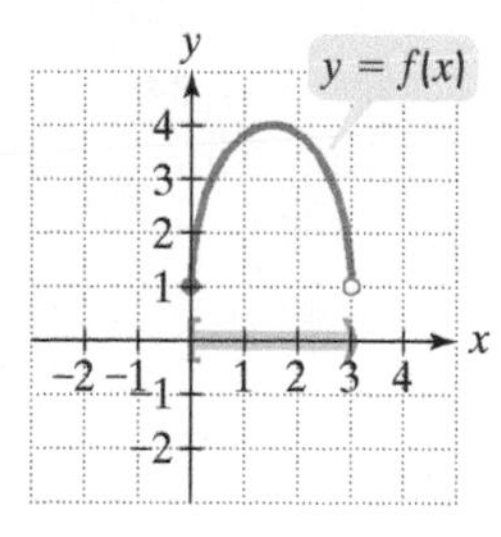

4. The shaded set of numbers shown on the y-axis can be expressed in interval notation as ____________. This set represents the function's ____________.

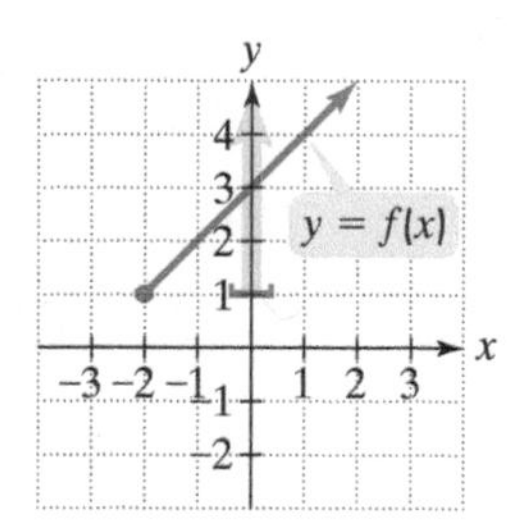

2.2 EXERCISE SET

MyMathLab®

Practice Exercises

In Exercises 1–10, graph the given functions, f and g, in the same rectangular coordinate system. Select integers for x, starting with −2 and ending with 2. Once you have obtained your graphs, describe how the graph of g is related to the graph of f.

1. $f(x) = x, g(x) = x + 3$
2. $f(x) = x, g(x) = x - 4$
3. $f(x) = -2x, g(x) = -2x - 1$
4. $f(x) = -2x, g(x) = -2x + 3$
5. $f(x) = x^2, g(x) = x^2 + 1$
6. $f(x) = x^2, g(x) = x^2 - 2$
7. $f(x) = |x|, g(x) = |x| - 2$
8. $f(x) = |x|, g(x) = |x| + 1$
9. $f(x) = x^3, g(x) = x^3 + 2$
10. $f(x) = x^3, g(x) = x^3 - 1$

In Exercises 11–18, use the vertical line test to identify graphs in which y is a function of x.

11.

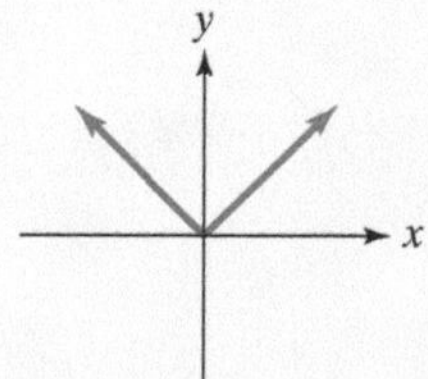

12.

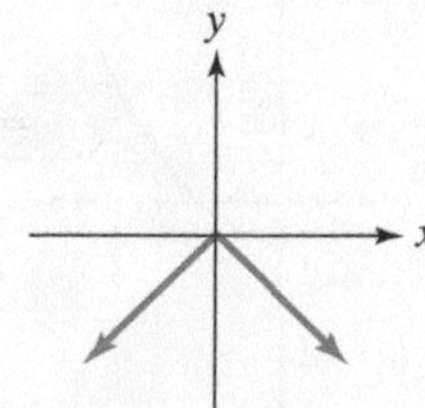

13.

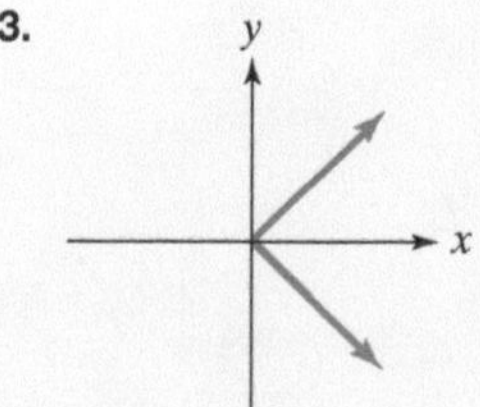

14.

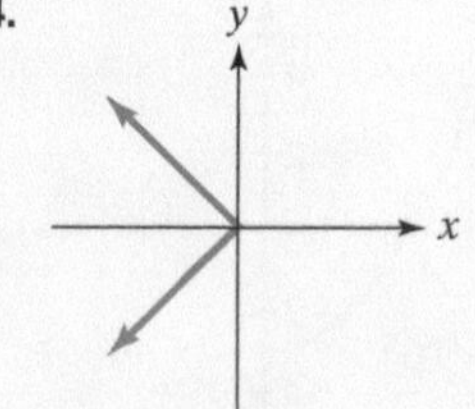

15.

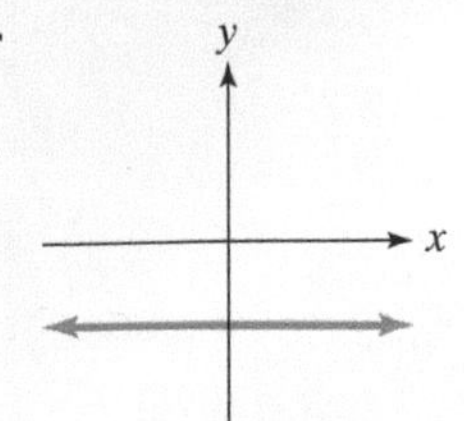

16.

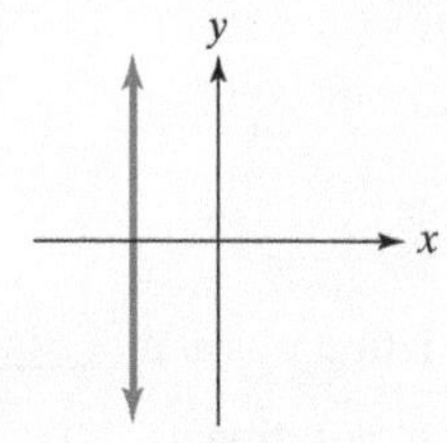

17.

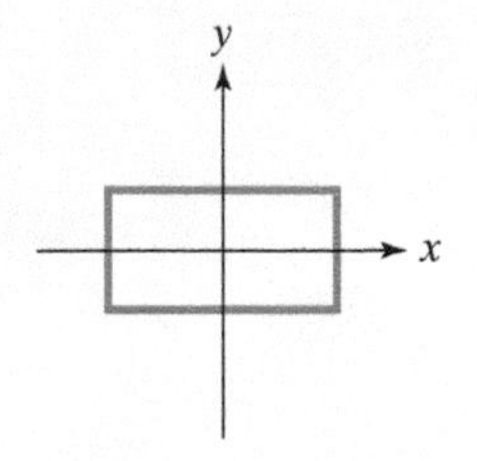

18.

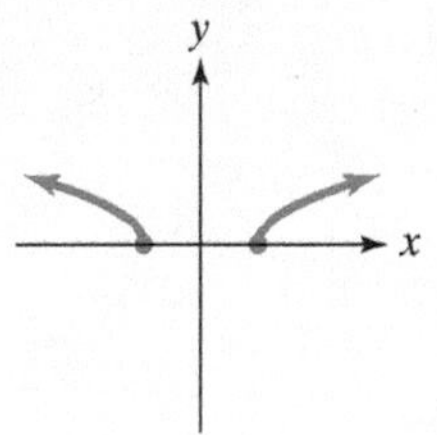

In Exercises 19–24, use the graph of f to find each indicated function value.

19. $f(-2)$

20. $f(2)$

21. $f(4)$

22. $f(-4)$

23. $f(-3)$

24. $f(-1)$

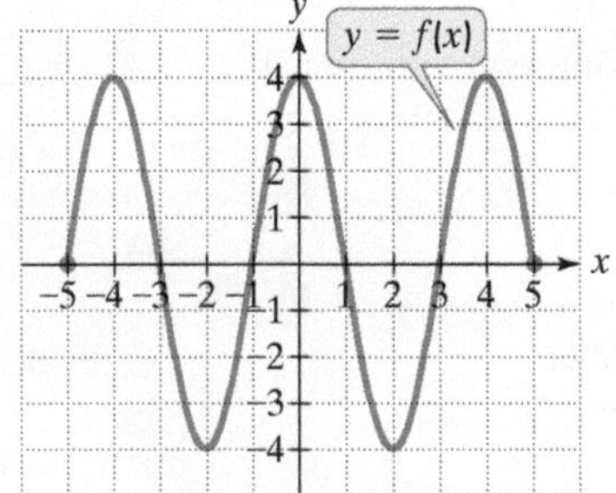

Use the graph of g to solve Exercises 25–30.

25. Find $g(-4)$.

26. Find $g(2)$.

27. Find $g(-10)$.

28. Find $g(10)$.

29. For what value of x is $g(x) = 1$?

30. For what value of x is $g(x) = -1$?

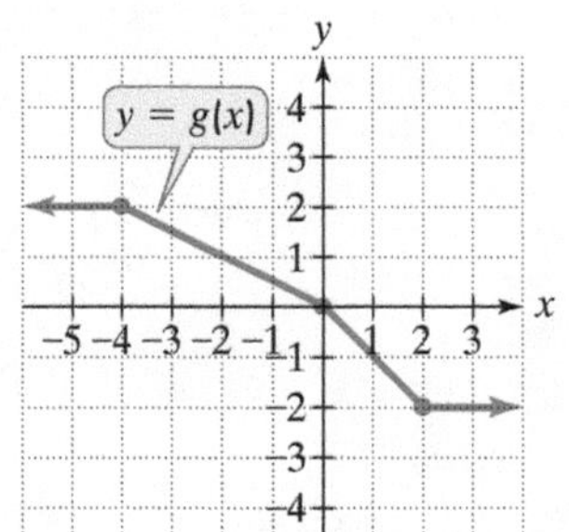

In Exercises 31–40, use the graph of each function to identify its domain and its range.

31.

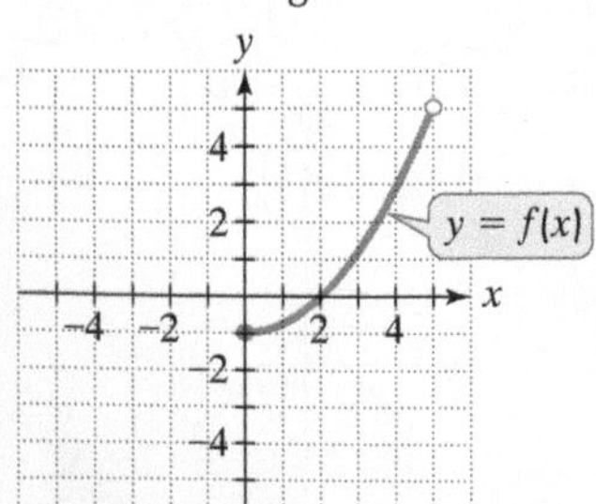

32.

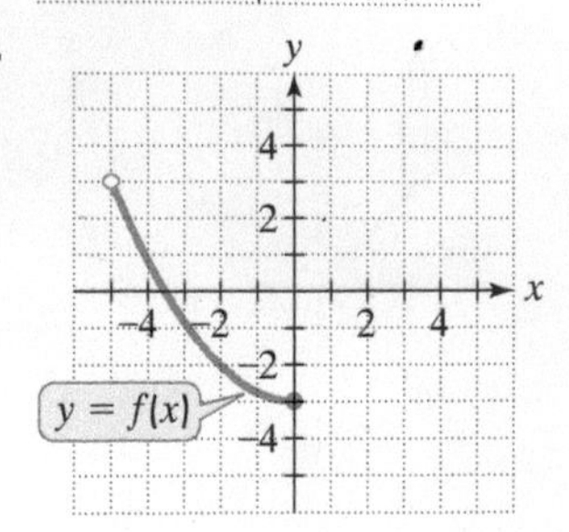

33.

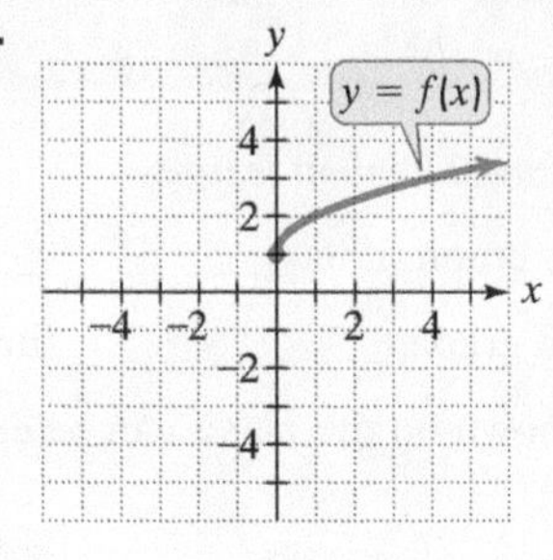

34.

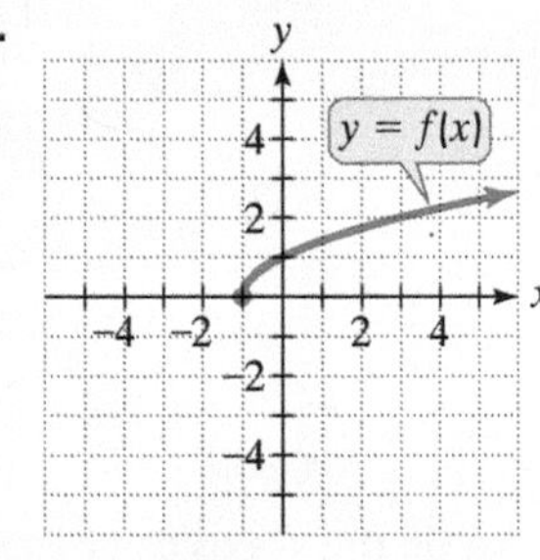

35.

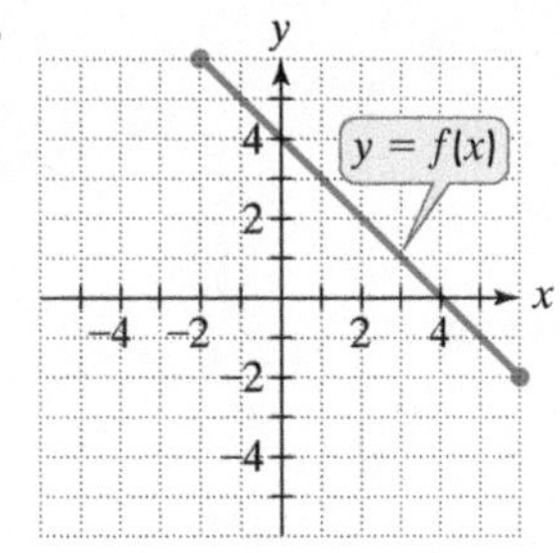

36.

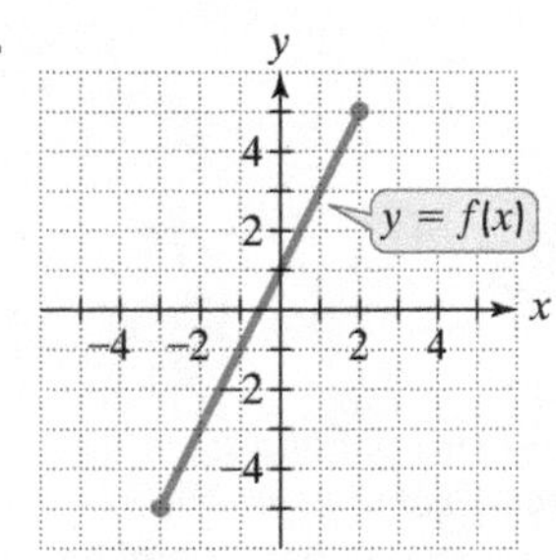

37.

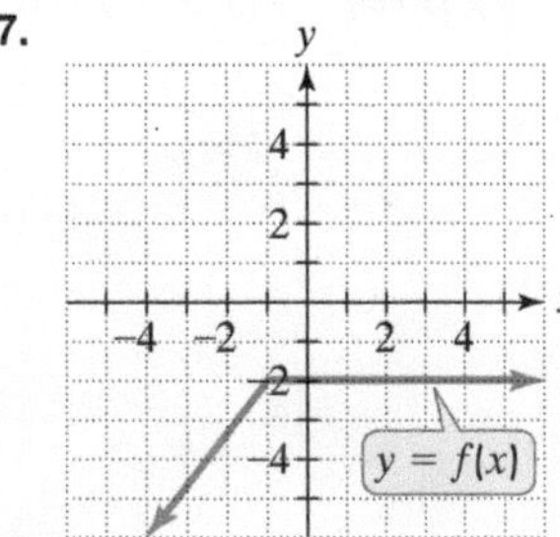

38.

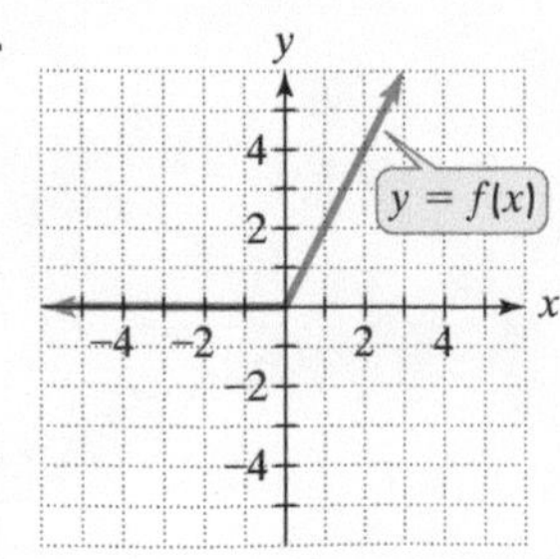

39.

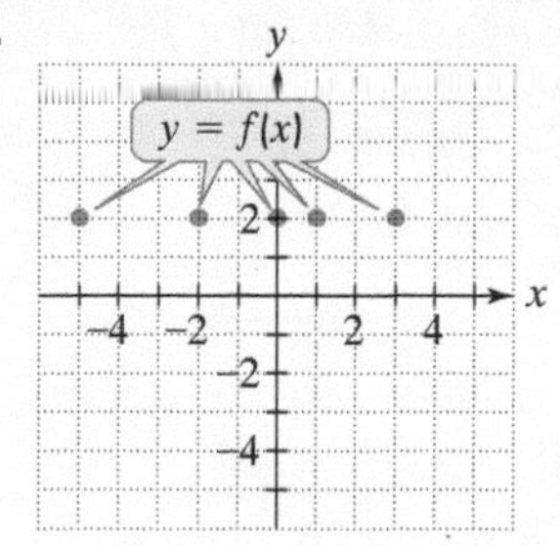

40.

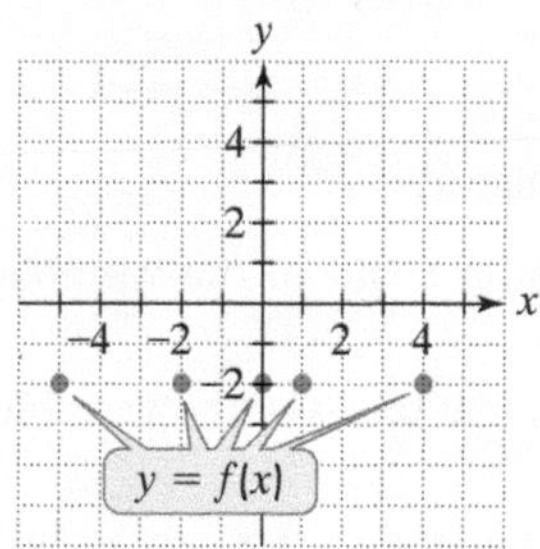

Practice PLUS

41. Use the graph of f to determine (a) through (h) at the top of the next column. Where applicable, use interval notation.

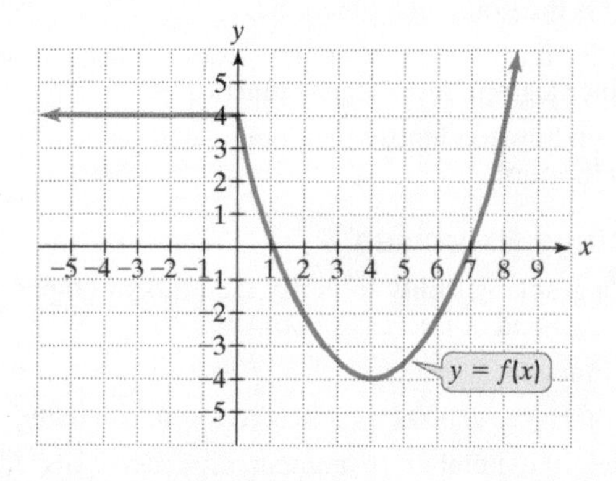

a. the domain of f

b. the range of f

c. $f(-3)$

d. the values of x for which $f(x) = -2$

e. the points where the graph of f crosses the x-axis

f. the point where the graph of f crosses the y-axis

g. values of x for which $f(x) < 0$

h. Is $f(-8)$ positive or negative?

42. Use the graph of f to determine each of the following. Where applicable, use interval notation.

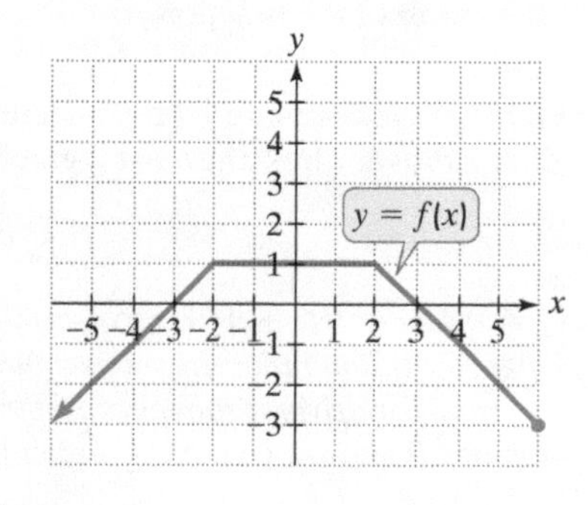

a. the domain of f

b. the range of f

c. $f(-4)$

d. the values of x for which $f(x) = -3$

e. the points where the graph of f crosses the x-axis

f. the point where the graph of f crosses the y-axis

g. values of x for which $f(x) > 0$

h. Is $f(-2)$ positive or negative?

Application Exercises

The wage gap is used to compare the status of women's earnings relative to men's. The wage gap is expressed as a percent and is calculated by dividing the median, or middlemost, annual earnings for women by the median annual earnings for men. The bar graph shows the wage gap for selected years from 1980 through 2010.

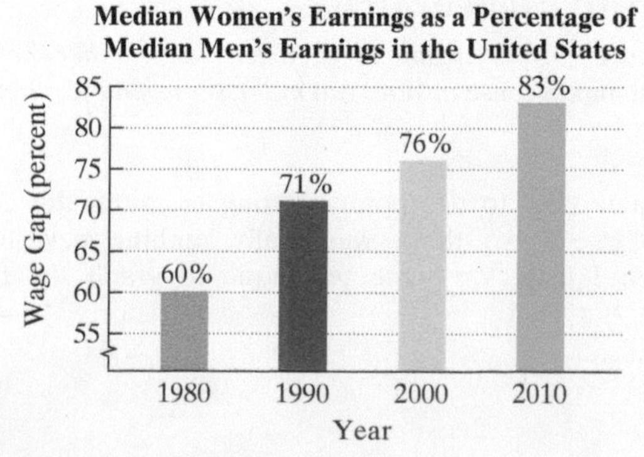

Source: U.S. Bureau of Labor Statistics

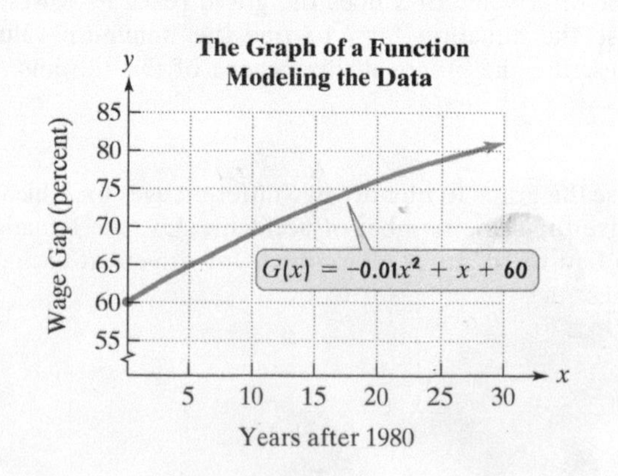

(Refer to the graphs at the bottom of the previous page.) The function $G(x) = -0.01x^2 + x + 60$ *models the wage gap, as a percent, x years after 1980. The graph of function G is shown to the right of the actual data. Use this information to solve Exercises 43–44.*

43. **a.** Find and interpret $G(30)$. Identify this information as a point on the graph of the function.

b. Does $G(30)$ overestimate or underestimate the actual data shown by the bar graph? By how much?

44. **a.** Find and interpret $G(10)$. Identify this information as a point on the graph of the function.

b. Does $G(10)$ overestimate or underestimate the actual data shown by the bar graph? By how much?

The function $f(x) = 0.4x^2 - 36x + 1000$ *models the number of accidents, f(x), per 50 million miles driven as a function of a driver's age, x, in years, for drivers from ages 16 through 74, inclusive. The graph of f is shown. Use the equation for f to solve Exercises 45–48.*

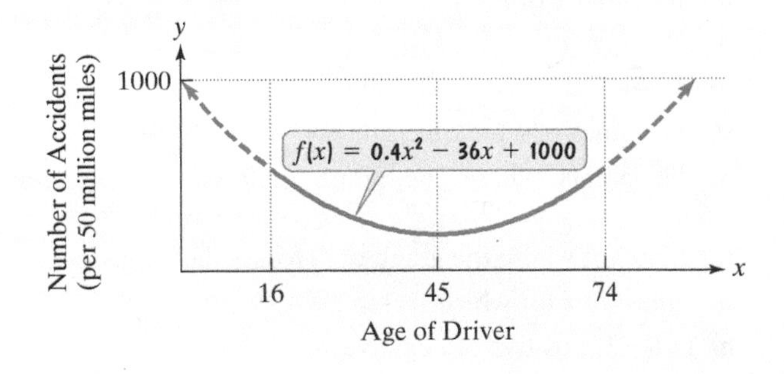

45. Find and interpret $f(20)$. Identify this information as a point on the graph of f.

46. Find and interpret $f(50)$. Identify this information as a point on the graph of f.

47. For what value of x does the graph reach its lowest point? Use the equation for f to find the minimum value of y. Describe the practical significance of this minimum value.

48. Use the graph to identify two different ages for which drivers have the same number of accidents. Use the equation for f to find the number of accidents for drivers at each of these ages.

The figure shows the cost of mailing a first-class letter, f(x), as a function of its weight, x, in ounces, for weights not exceeding 3.5 ounces. Use the graph to solve Exercises 49–52.

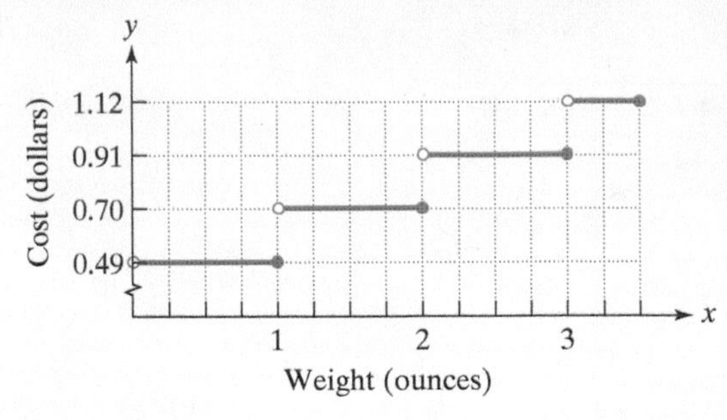

49. Find $f(3)$. What does this mean in terms of the variables in this situation?

50. Find $f(3.5)$. What does this mean in terms of the variables in this situation?

51. What is the cost of mailing a letter that weighs 1.5 ounces?

52. What is the cost of mailing a letter that weighs 1.8 ounces?

Explaining the Concepts

53. What is the graph of a function?

54. Explain how the vertical line test is used to determine whether a graph represents a function.

55. Explain how to identify the domain and range of a function from its graph.

Technology Exercises

56. Use a graphing utility to verify the pairs of graphs that you drew by hand in Exercises 1–10.

57. The function

$$f(x) = -0.00002x^3 + 0.008x^2 - 0.3x + 6.95$$

models the number of annual physician visits, $f(x)$, by a person of age x. Graph the function in a $[0, 100, 5]$ by $[0, 40, 2]$ viewing rectangle. What does the shape of the graph indicate about the relationship between one's age and the number of annual physician visits? Use the TRACE or minimum function capability to find the coordinates of the minimum point on the graph of the function. What does this mean?

Critical Thinking Exercises

Make Sense? *In Exercises 58–61, determine whether each statement makes sense or does not make sense, and explain your reasoning.*

58. I knew how to use point plotting to graph the equation $y = x^2 - 1$, so there was really nothing new to learn when I used the same technique to graph the function $f(x) = x^2 - 1$.

59. The graph of my function revealed aspects of its behavior that were not obvious by just looking at its equation.

60. I graphed a function showing how paid vacation days depend on the number of years a person works for a company. The domain was the number of paid vacation days.

61. I graphed a function showing how the number of annual physician visits depends on a person's age. The domain was the number of annual physician visits.

In Exercises 62–67, determine whether each statement is true or false. If the statement is false, make the necessary change(s) to produce a true statement.

62. The graph of every line is a function.

63. A horizontal line can intersect the graph of a function in more than one point.

Use the graph of f to determine whether each statement in Exercises 64–67 is true or false.

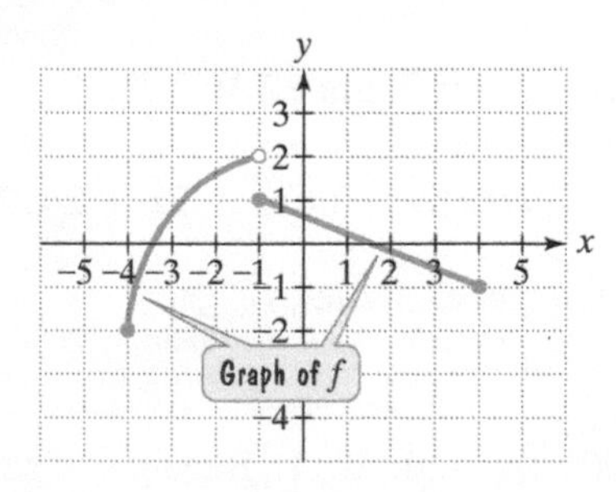

64. The domain of f is $[-4, 4]$.

65. The range of f is $[-2, 2]$.

66. $f(-1) - f(4) = 2$

67. $f(0) = 2.1$

In Exercises 68–69, let f be defined by the following graph.

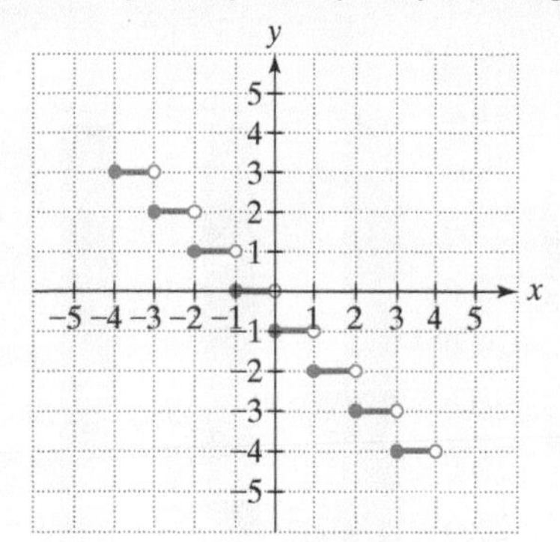

68. Find
$\sqrt{f(-1.5) + f(-0.9)} - [f(\pi)]^2 + f(-3) \div f(1) \cdot f(-\pi)$.

69. Find
$\sqrt{f(-2.5) - f(1.9)} - [f(-\pi)]^2 + f(-3) \div f(1) \cdot f(\pi)$.

Review Exercises

70. Is $\{(1, 1), (2, 2), (3, 3), (4, 4)\}$ a function? (Section 2.1, Example 2)

71. Solve: $12 - 2(3x + 1) = 4x - 5$. (Section 1.4, Example 3)

72. The length of a rectangle exceeds 3 times the width by 8 yards. If the perimeter of the rectangle is 624 yards, what are its dimensions? (Section 1.5, Example 5)

Preview Exercises

Exercises 73–75 will help you prepare for the material covered in the next section.

73. If $f(x) = \frac{4}{x-3}$, why must 3 be excluded from the domain of f?

74. If $f(x) = x^2 + x$ and $g(x) = x - 5$, find $f(4) + g(4)$.

75. Simplify: $-2.6x^2 + 49x + 3994 - (-0.6x^2 + 7x + 2412)$.

SECTION

2.3 The Algebra of Functions

What am I supposed to learn?

After studying this section, you should be able to:

1. Find the domain of a function.
2. Use the algebra of functions to combine functions and determine domains.

We're born. We die. **Figure 2.16** at the top of the next page quantifies these statements by showing the number of births and deaths in the United States from 2000 through 2011.

In this section, we look at these data from the perspective of functions. By considering the yearly change in the U.S. population, you will see that functions can be subtracted using procedures that will remind you of combining algebraic expressions.

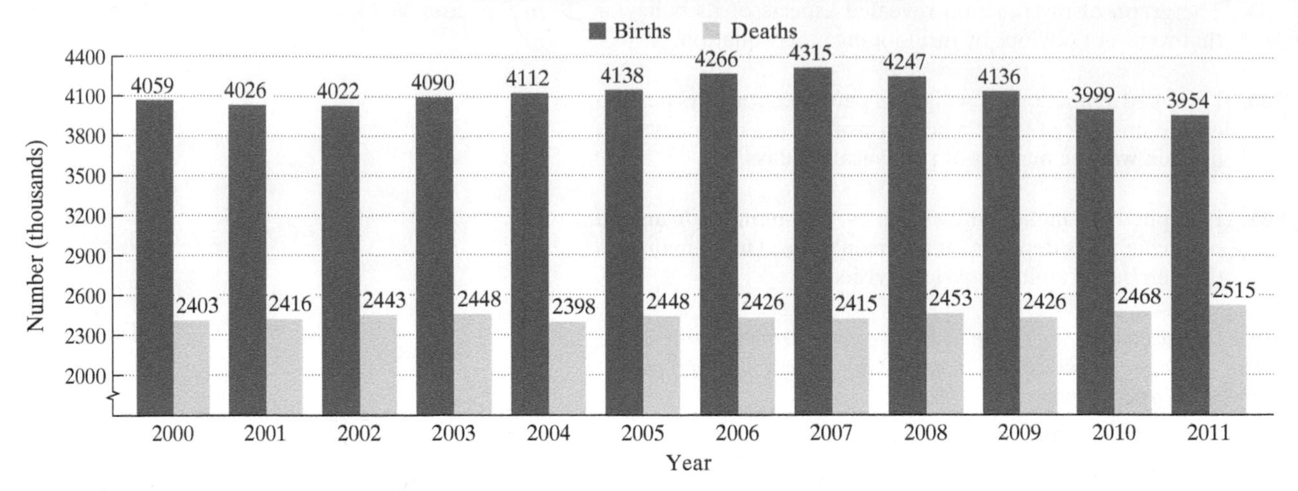

Figure 2.16

Source: U.S. Department of Health and Human Services

1 Find the domain of a function.

The Domain of a Function

We begin with two functions that model the data in **Figure 2.16**.

$$B(x) = -2.6x^2 + 49x + 3994 \qquad D(x) = -0.6x^2 + 7x + 2412.$$

Number of births, $B(x)$, in thousands, x years after 2000

Number of deaths, $D(x)$, in thousands, x years after 2000

The years in **Figure 2.16** extend from 2000 through 2011. Because x represents the number of years after 2000,

$$\text{Domain of } B = \{0, 1, 2, 3, \ldots, 11\}$$

and

$$\text{Domain of } D = \{0, 1, 2, 3, \ldots, 11\}.$$

Functions that model data often have their domains explicitly given with the function's equation. However, for most functions, only an equation is given and the domain is not specified. In cases like this, the domain of a function f is the largest set of real numbers for which the value of $f(x)$ is a real number. For example, consider the function

$$f(x) = \frac{1}{x-3}.$$

Because division by 0 is undefined (and not a real number), the denominator, $x - 3$, cannot be 0. Thus, x cannot equal 3. The domain of the function consists of all real numbers other than 3, represented by

$$\text{Domain of } f = (-\infty, 3) \text{ or } (3, \infty).$$

Great Question!

Can you explain how $(-\infty, 3)$ or $(3, \infty)$ represents all real numbers other than 3?

The following voice balloons should help:

$(-\infty, 3)$ or $(3, \infty)$.

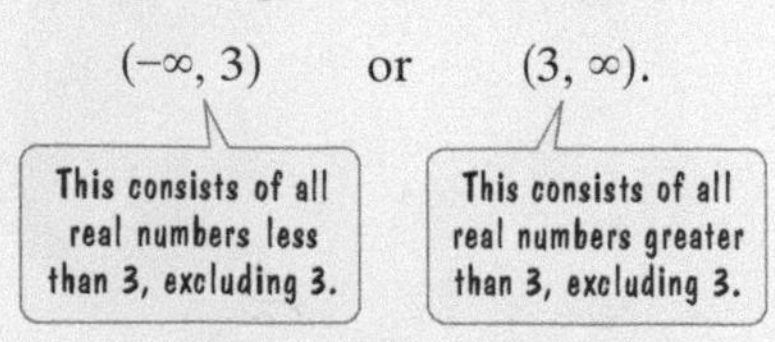

In Chapter 7, we will be studying square root functions such as

$$g(x) = \sqrt{x}.$$

The equation tells us to take the square root of x. Because only nonnegative numbers have square roots that are real numbers, the expression under the square root sign, x, must be nonnegative. Thus,

$$\text{Domain of } g = [0, \infty).$$

Using Technology

Graphic Connections

You can graph a function and visually determine its domain. Look for inputs on the x-axis that correspond to points on the graph. For example, consider

$g(x) = \sqrt{x}$, or $y = \sqrt{x}$.

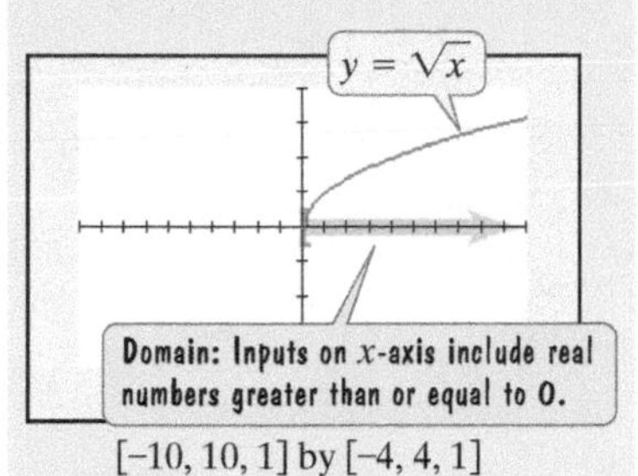

$y = \sqrt{x}$
Domain: Inputs on x-axis include real numbers greater than or equal to 0.

[−10, 10, 1] by [−4, 4, 1]

This verifies that $[0, \infty)$ is the domain.

Finding a Function's Domain

If a function f does not model data or verbal conditions, its domain is the largest set of real numbers for which the value of $f(x)$ is a real number. Exclude from a function's domain real numbers that cause division by zero and real numbers that result in a square root of a negative number.

EXAMPLE 1 Finding the Domain of a Function

Find the domain of each function:

a. $f(x) = 3x + 2$ **b.** $g(x) = \dfrac{3x + 2}{x + 1}$.

Solution

a. The function $f(x) = 3x + 2$ contains neither division nor a square root. For every real number, x, the algebraic expression $3x + 2$ is a real number. Thus, the domain of f is the set of all real numbers.

$$\text{Domain of } f = (-\infty, \infty).$$

b. The function $g(x) = \dfrac{3x + 2}{x + 1}$ contains division. Because division by 0 is undefined, we must exclude from the domain the value of x that causes $x + 1$ to be 0. Thus, x cannot equal -1.

$$\text{Domain of } g = (-\infty, -1) \text{ or } (-1, \infty). \blacksquare$$

✓ **CHECK POINT 1** Find the domain of each function:

a. $f(x) = \dfrac{1}{2}x + 3$

b. $g(x) = \dfrac{7x + 4}{x + 5}$.

---

2 Use the algebra of functions to combine functions and determine domains.

The Algebra of Functions

We can combine functions using addition, subtraction, multiplication, and division by performing operations with the algebraic expressions that appear on the right side of the equations. For example, the functions $f(x) = 2x$ and $g(x) = x - 1$ can be combined to form the sum, difference, product, and quotient of f and g. Here's how it's done:

For each function, $f(x) = 2x$ and $g(x) = x - 1$.

Sum: $f + g$
$$(f + g)(x) = f(x) + g(x) = 2x + (x - 1) = 3x - 1$$

Difference: $f - g$
$$(f - g)(x) = f(x) - g(x) = 2x - (x - 1) = 2x - x + 1 = x + 1$$

Product: fg
$$(fg)(x) = f(x) \cdot g(x) = 2x(x - 1) = 2x^2 - 2x$$

Quotient: $\dfrac{f}{g}$
$$\left(\frac{f}{g}\right)(x) = \frac{f(x)}{g(x)} = \frac{2x}{x - 1}, x \neq 1.$$

The domain for each of these functions consists of all real numbers that are common to the domains of f and g. In the case of the quotient function $\dfrac{f(x)}{g(x)}$, we must remember not to divide by 0, so we add the further restriction that $g(x) \neq 0$.

The Algebra of Functions: Sum, Difference, Product, and Quotient of Functions

Let f and g be two functions. The **sum** $f + g$, the **difference** $f - g$, the **product** fg, and the **quotient** $\frac{f}{g}$ are functions whose domains are the set of all real numbers common to the domains of f and g, defined as follows:

1. Sum: $(f + g)(x) = f(x) + g(x)$
2. Difference: $(f - g)(x) = f(x) - g(x)$
3. Product: $(fg)(x) = f(x) \cdot g(x)$
4. Quotient: $\left(\frac{f}{g}\right)(x) = \frac{f(x)}{g(x)}$, provided $g(x) \neq 0$.

EXAMPLE 2 Using the Algebra of Functions

Let $f(x) = x^2 - 3$ and $g(x) = 4x + 5$. Find each of the following:

a. $(f + g)(x)$

b. $(f + g)(3)$.

Solution

a. $(f + g)(x) = f(x) + g(x) = (x^2 - 3) + (4x + 5) = x^2 + 4x + 2$

Thus,

$$(f + g)(x) = x^2 + 4x + 2.$$

b. We find $(f + g)(3)$ by substituting 3 for x in the equation for $f + g$.

$$(f + g)(x) = x^2 + 4x + 2$$ This is the equation for $f + g$.

Substitute 3 for x.

$$(f + g)(3) = 3^2 + 4 \cdot 3 + 2 = 9 + 12 + 2 = 23$$ ■

✓ **CHECK POINT 2** Let $f(x) = 3x^2 + 4x - 1$ and $g(x) = 2x + 7$. Find each of the following:

a. $(f + g)(x)$ **b.** $(f + g)(4)$.

EXAMPLE 3 Using the Algebra of Functions

Let $f(x) = \frac{4}{x}$ and $g(x) = \frac{3}{x + 2}$. Find each of the following:

a. $(f - g)(x)$

b. the domain of $f - g$.

Solution

a. $(f - g)(x) = f(x) - g(x) = \frac{4}{x} - \frac{3}{x+2}$

(In Chapter 6, we will discuss how to perform the subtraction with these algebraic fractions. Perhaps you remember how to do so from your work in introductory algebra. For now, we will leave these fractions in the form shown.)

b. The domain of $f - g$ is the set of all real numbers that are common to the domain of f and the domain of g. Thus, we must find the domains of f and g. We will do so for f first.

Note that $f(x) = \frac{4}{x}$ is a function involving division. Because division by 0 is undefined, x cannot equal 0.

The function $g(x) = \frac{3}{x+2}$ is also a function involving division. Because division by 0 is undefined, x cannot equal -2.

To be in the domain of $f - g$, x must be in both the domain of f and the domain of g. This means that $x \neq 0$ and $x \neq -2$. **Figure 2.17** shows these excluded values on a number line.

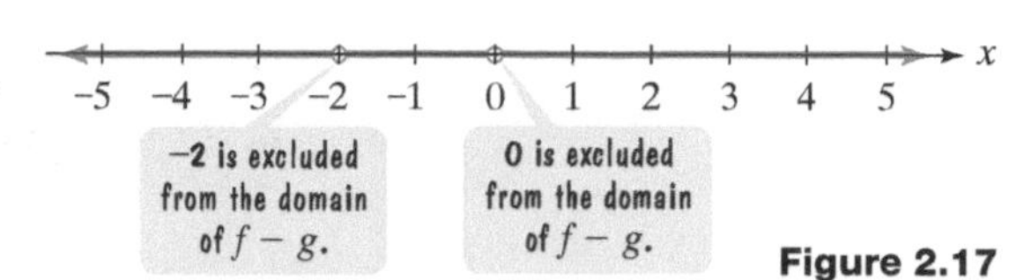
x
-5 -4 -3 -2 -1 0 1 2 3 4 5
−2 is excluded from the domain of f − g.
0 is excluded from the domain of f − g.

Figure 2.17

The number line in **Figure 2.17** is helpful in writing each of the three intervals that describe the domain of $f - g$.

Domain of $f - g = (-\infty, -2)$ or $(-2, 0)$ or $(0, \infty)$

All real numbers less than −2, excluding −2

All real numbers between −2 and 0, excluding −2 and excluding 0

All real numbers greater than 0, excluding 0

■

Great Question!

I find the interval notation in Example 3(b), where you excluded −2 and 0 from the domain of $f - g$, somewhat confusing. Is there another way to describe such a domain?

You could use set-builder notation:

$$\text{Domain of } f - g = \{x \mid x \text{ is a real number and } x \neq -2 \text{ and } x \neq 0\}.$$

Ask your professor if set-builder notation is an acceptable alternative for describing domains that exclude two or more numbers. The downside: We'll be using interval notation throughout the book to indicate both domains and ranges.

✓ **CHECK POINT 3** Let $f(x) = \frac{5}{x}$ and $g(x) = \frac{7}{x-8}$. Find each of the following:

a. $(f - g)(x)$

b. the domain of $f - g$.

EXAMPLE 4 Using the Algebra of Functions

Let $f(x) = x^2 + x$ and $g(x) = x - 5$. Find each of the following:

a. $(f + g)(4)$

b. $(f - g)(x)$ and $(f - g)(-3)$

c. $\left(\frac{f}{g}\right)(x)$ and $\left(\frac{f}{g}\right)(7)$

d. $(fg)(-2)$.

$f(x) = x^2 + x$
$g(x) = x - 5$
The given functions in Example 4 (repeated)

Solution

a. We can find $(f + g)(4)$ using $f(4)$ and $g(4)$.

$$f(x) = x^2 + x \qquad g(x) = x - 5$$
$$f(4) = 4^2 + 4 = 20 \qquad g(4) = 4 - 5 = -1$$

Thus,

$$(f + g)(4) = f(4) + g(4) = 20 + (-1) = 19.$$

We can also find $(f + g)(4)$ by first finding $(f + g)(x)$ and then substituting 4 for x:

$(f + g)(x) = f(x) + g(x)$ This is the definition of the sum $f + g$.
$= (x^2 + x) + (x - 5)$ Substitute the given functions.
$= x^2 + 2x - 5.$ Simplify.

Using $(f + g)(x) = x^2 + 2x - 5$, we have

$$(f + g)(4) = 4^2 + 2 \cdot 4 - 5 = 16 + 8 - 5 = 19.$$

b. $(f - g)(x) = f(x) - g(x)$ This is the definition of the difference $f - g$.
$= (x^2 + x) - (x - 5)$ Substitute the given functions.
$= x^2 + x - x + 5$ Remove parentheses and change the sign of each term in the second set of parentheses.
$= x^2 + 5$ Simplify.

Using $(f - g)(x) = x^2 + 5$, we have

$$(f - g)(-3) = (-3)^2 + 5 = 9 + 5 = 14.$$

c. $\left(\frac{f}{g}\right)(x) = \frac{f(x)}{g(x)}$ This is the definition of the quotient $\frac{f}{g}$.
$= \frac{x^2 + x}{x - 5}$ Substitute the given functions.

Using $\left(\frac{f}{g}\right)(x) = \frac{x^2 + x}{x - 5}$, we have

$$\left(\frac{f}{g}\right)(7) = \frac{7^2 + 7}{7 - 5} = \frac{56}{2} = 28.$$

d. We can find $(fg)(-2)$ using the fact that

$$(fg)(-2) = f(-2) \cdot g(-2).$$

$$f(x) = x^2 + x \qquad g(x) = x - 5$$
$$f(-2) = (-2)^2 + (-2) = 4 - 2 = 2 \qquad g(-2) = -2 - 5 = -7$$

Thus,

$$(fg)(-2) = f(-2) \cdot g(-2) = 2(-7) = -14.$$

We could also have found $(fg)(-2)$ by multiplying $f(x) \cdot g(x)$ and then substituting -2 into the product. We will discuss how to multiply expressions such as $x^2 + x$ and $x - 5$ in Chapter 5. ■

CHECK POINT 4 Let $f(x) = x^2 - 2x$ and $g(x) = x + 3$. Find each of the following:

a. $(f + g)(5)$

b. $(f - g)(x)$ and $(f - g)(-1)$

c. $\left(\frac{f}{g}\right)(x)$ and $\left(\frac{f}{g}\right)(7)$

d. $(fg)(-4)$.

EXAMPLE 5 Applying the Algebra of Functions

We opened the section with functions that model the number of births and deaths in the United States from 2000 through 2011:

$$B(x) = -2.6x^2 + 49x + 3994 \qquad D(x) = -0.6x^2 + 7x + 2412$$

Number of births, $B(x)$, in thousands, x years after 2000

Number of deaths, $D(x)$, in thousands, x years after 2000

a. Write a function that models the change in U.S. population for the years from 2000 through 2011.
b. Use the function from part (a) to find the change in U.S. population in 2008.
c. Does the result in part (b) overestimate or underestimate the actual population change in 2008 obtained from the data in **Figure 2.16** on page 126? By how much?

Solution

a. The change in population is the number of births minus the number of deaths. Thus, we will find the difference function, $B - D$.

$$\begin{aligned}
&(B - D)(x) \\
&= B(x) - D(x) \\
&= (-2.6x^2 + 49x + 3994) - (-0.6x^2 + 7x + 2412) && \text{Substitute the given functions.} \\
&= -2.6x^2 + 49x + 3994 + 0.6x^2 - 7x - 2412 && \text{Remove parentheses and change the sign of each term in the second set of parentheses.} \\
&= (-2.6x^2 + 0.6x^2) + (49x - 7x) + (3994 - 2412) && \text{Group like terms.} \\
&= -2x^2 + 42x + 1582 && \text{Combine like terms.}
\end{aligned}$$

The function

$$(B - D)(x) = -2x^2 + 42x + 1582$$

models the change in U.S. population, in thousands, x years after 2000.

b. Because 2008 is 8 years after 2000, we substitute 8 for x in the difference function $(B - D)(x)$.

$$\begin{aligned}
(B - D)(x) &= -2x^2 + 42x + 1582 && \text{Use the difference function } B - D. \\
(B - D)(8) &= -2(8)^2 + 42(8) + 1582 && \text{Substitute 8 for } x. \\
&= -2(64) + 42(8) + 1582 && \text{Evaluate the exponential expression: } 8^2 = 64. \\
&= -128 + 336 + 1582 && \text{Perform the multiplications.} \\
&= 1790 && \text{Add from left to right.}
\end{aligned}$$

We see that $(B - D)(8) = 1790$. The model indicates that there was a population increase of 1790 thousand, or approximately 1,790,000 people, in 2008.

c. The data for 2008 in **Figure 2.16** on page 126 show 4247 thousand births and 2453 thousand deaths.

$$\begin{aligned}
\text{population change} &= \text{births} - \text{deaths} \\
&= 4247 - 2453 = 1794
\end{aligned}$$

The actual population increase was 1794 thousand, or 1,794,000. Our model gave us an increase of 1790 thousand. Thus, the model underestimates the actual increase by 1794 − 1790, or 4 thousand people. ■

Achieving Success

Ask! Ask! Ask!

Do not be afraid to ask questions in class. Your professor may not realize that a concept is unclear until you raise your hand and ask a question. Other students who have problems asking questions in class will be appreciative that you have spoken up. Be polite and professional, but ask as many questions as required.

✓ **CHECK POINT 5** Use the birth and death models from Example 5.

a. Write a function that models the total number of births and deaths in the United States for the years from 2000 through 2011.

b. Use the function from part (a) to find the total number of births and deaths in the United States in 2003.

c. Does the result in part (b) overestimate or underestimate the actual number of total births and deaths in 2003 obtained from the data in **Figure 2.16** on page 126? By how much?

CONCEPT AND VOCABULARY CHECK

Fill in each blank so that the resulting statement is true.

1. We exclude from a function's domain real numbers that cause division by ________.
2. We exclude from a function's domain real numbers that result in a square root of a/an ________ number.
3. $(f + g)(x) =$ ________
4. $(f - g)(x) =$ ________
5. $(fg)(x) =$ ________
6. $\frac{f}{g}(x) =$ ________, provided ________ $\neq 0$
7. The domain of $f(x) = 5x + 7$ consists of all real numbers, represented in interval notation as ________.
8. The domain of $g(x) = \frac{3}{x - 2}$ consists of all real numbers except 2, represented in interval notation as $(-\infty, 2)$ or ________.
9. The domain of $h(x) = \frac{1}{x} + \frac{7}{x - 3}$ consists of all real numbers except 0 and 3, represented in interval notation as $(-\infty, 0)$ or ________ or ________.

2.3 EXERCISE SET

MyMathLab®

Practice Exercises

In Exercises 1–10, find the domain of each function.

1. $f(x) = 3x + 5$
2. $f(x) = 4x + 7$
3. $g(x) = \frac{1}{x + 4}$
4. $g(x) = \frac{1}{x + 5}$
5. $f(x) = \frac{2x}{x - 3}$
6. $f(x) = \frac{4x}{x - 2}$
7. $g(x) = x + \frac{3}{5 - x}$
8. $g(x) = x + \frac{7}{6 - x}$
9. $f(x) = \frac{1}{x + 7} + \frac{3}{x - 9}$
10. $f(x) = \frac{1}{x + 8} + \frac{3}{x - 10}$

In Exercises 11–16, find $(f + g)(x)$ and $(f + g)(5)$.

11. $f(x) = 3x + 1, g(x) = 2x - 6$
12. $f(x) = 4x + 2, g(x) = 2x - 9$
13. $f(x) = x - 5, g(x) = 3x^2$
14. $f(x) = x - 6, g(x) = 2x^2$
15. $f(x) = 2x^2 - x - 3, g(x) = x + 1$
16. $f(x) = 4x^2 - x - 3, g(x) = x + 1$
17. Let $f(x) = 5x$ and $g(x) = -2x - 3$. Find $(f + g)(x)$, $(f - g)(x)$, $(fg)(x)$, and $\left(\frac{f}{g}\right)(x)$.

18. Let $f(x) = -4x$ and $g(x) = -3x + 5$. Find $(f + g)(x)$, $(f - g)(x)$, $(fg)(x)$, and $\left(\frac{f}{g}\right)(x)$.

In Exercises 19–30, for each pair of functions, f and g, determine the domain of f + g.

19. $f(x) = 3x + 7, g(x) = 9x + 10$

20. $f(x) = 7x + 4, g(x) = 5x - 2$

21. $f(x) = 3x + 7, g(x) = \frac{2}{x - 5}$

22. $f(x) = 7x + 4, g(x) = \frac{2}{x - 6}$

23. $f(x) = \frac{1}{x}, g(x) = \frac{2}{x - 5}$

24. $f(x) = \frac{1}{x}, g(x) = \frac{2}{x - 6}$

25. $f(x) = \frac{8x}{x - 2}, g(x) = \frac{6}{x + 3}$

26. $f(x) = \frac{9x}{x - 4}, g(x) = \frac{7}{x + 8}$

27. $f(x) = \frac{8x}{x - 2}, g(x) = \frac{6}{2 - x}$

28. $f(x) = \frac{9x}{x - 4}, g(x) = \frac{7}{4 - x}$

29. $f(x) = x^2, g(x) = x^3$

30. $f(x) = x^2 + 1, g(x) = x^3 - 1$

In Exercises 31–50, let

$$f(x) = x^2 + 4x \quad \text{and} \quad g(x) = 2 - x.$$

Find each of the following.

31. $(f + g)(x)$ and $(f + g)(3)$

32. $(f + g)(x)$ and $(f + g)(4)$

33. $f(-2) + g(-2)$

34. $f(-3) + g(-3)$

35. $(f - g)(x)$ and $(f - g)(5)$

36. $(f - g)(x)$ and $(f - g)(6)$

37. $f(-2) - g(-2)$

38. $f(-3) - g(-3)$

39. $(fg)(-2)$

40. $(fg)(-3)$

41. $(fg)(5)$

42. $(fg)(6)$

43. $\left(\frac{f}{g}\right)(x)$ and $\left(\frac{f}{g}\right)(1)$

44. $\left(\frac{f}{g}\right)(x)$ and $\left(\frac{f}{g}\right)(3)$

45. $\left(\frac{f}{g}\right)(-1)$

46. $\left(\frac{f}{g}\right)(0)$

47. The domain of $f + g$

48. The domain of $f - g$

49. The domain of $\frac{f}{g}$

50. The domain of fg

Practice PLUS

Use the graphs of f and g below to solve Exercises 51–58.

51. Find $(f + g)(-3)$.

52. Find $(g - f)(-2)$.

53. Find $(fg)(2)$.

54. Find $\left(\frac{g}{f}\right)(3)$.

55. Find the domain of $f + g$.

56. Find the domain of $\frac{f}{g}$.

57. Graph $f + g$.

58. Graph $f - g$.

Use the table defining f and g to solve Exercises 59–62.

x	f(x)	g(x)
−2	5	0
−1	3	−2
0	−2	4
1	−6	−3
2	0	1

59. Find $(f + g)(1) - (g - f)(-1)$.

60. Find $(f + g)(-1) - (g - f)(0)$.

61. Find $(fg)(-2) - \left[\left(\frac{f}{g}\right)(1)\right]^2$.

62. Find $(fg)(2) - \left[\left(\frac{g}{f}\right)(0)\right]^2$.

Application Exercises

The bar graph shows the population of the United States, in millions, for seven selected years.

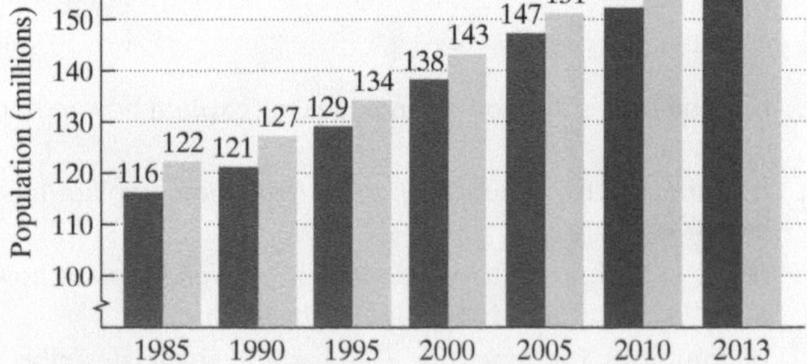

Source: U.S. Census Bureau

Here are two functions that model the data in the bar graph at the bottom of the previous page:

$M(x) = 1.5x + 115$ — Male U.S. population, $M(x)$, in millions, x years after 1985

$F(x) = 1.4x + 121.$ — Female U.S. population, $F(x)$, in millions, x years after 1985

Use the functions above to solve Exercises 63–65.

63. **a.** Write a function that models the total U.S. population for the years shown in the bar graph.
b. Use the function from part (a) to find the total U.S. population in 2010.
c. Does the result in part (b) overestimate or underestimate the actual total U.S. population in 2010 shown by the bar graph? By how much?

64. **a.** Write a function that models the difference between the female U.S. population and the male U.S. population for the years shown in the bar graph.
b. Use the function from part (a) to find how many more women than men there were in the U.S. population in 2005.
c. How well does the result in part (b) describe the actual difference between the female and male population in 2005 shown by the bar graph?

65. **a.** Write a function that models the ratio of men to women in the U.S. population for the years shown in the bar graph.
b. Use the function from part (a) to find the ratio of men to women, correct to three decimal places, in 2000.
c. Does the result in part (b) overestimate or underestimate the actual ratio of men to women in 2000 shown by the bar graph? By how much?

66. A company that sells radios has yearly fixed costs of \$600,000. It costs the company \$45 to produce each radio. Each radio will sell for \$65. The company's costs and revenue are modeled by the following functions:

$C(x) = 600{,}000 + 45x$ — This function models the company's costs.

$R(x) = 65x$ — This function models the company's revenue.

Find and interpret $(R - C)(20{,}000)$, $(R - C)(30{,}000)$, and $(R - C)(40{,}000)$.

Explaining the Concepts

67. If a function is defined by an equation, explain how to find its domain.

68. If equations for functions f and g are given, explain how to find $f + g$.

69. If the equations of two functions are given, explain how to obtain the quotient function and its domain.

70. If equations for functions f and g are given, describe two ways to find $(f - g)(3)$.

Technology Exercises

In Exercises 71–74, graph each of the three functions in the same $[-10, 10, 1]$ by $[-10, 10, 1]$ viewing rectangle.

71. $y_1 = 2x + 3$
$y_2 = 2 - 2x$
$y_3 = y_1 + y_2$

72. $y_1 = x - 4$
$y_2 = 2x$
$y_3 = y_1 - y_2$

73. $y_1 = x$
$y_2 = x - 4$
$y_3 = y_1 \cdot y_2$

74. $y_1 = x^2 - 2x$
$y_2 = x$
$y_3 = \dfrac{y_1}{y_2}$

75. In Exercise 74, use the TRACE feature to trace along y_3. What happens at $x = 0$? Explain why this occurs.

Critical Thinking Exercises

Make Sense? *In Exercises 76–79, determine whether each statement makes sense or does not make sense, and explain your reasoning.*

76. There is an endless list of real numbers that cannot be included in the domain of $f(x) = \sqrt{x}$.

77. I used a function to model data from 1990 through 2015. The independent variable in my model represented the number of years after 1990, so the function's domain was $\{0, 1, 2, 3, \ldots, 25\}$.

78. If I have equations for functions f and g, and 3 is in both domains, then there are always two ways to determine $(f + g)(3)$.

79. I have two functions. Function f models total world population x years after 2000 and function g models population of the world's more-developed regions x years after 2000. I can use $f - g$ to determine the population of the world's less-developed regions for the years in both functions' domains.

In Exercises 80–83, determine whether each statement is true or false. If the statement is false, make the necessary change(s) to produce a true statement.

80. If $(f + g)(a) = 0$, then $f(a)$ and $g(a)$ must be opposites, or additive inverses.

81. If $(f - g)(a) = 0$, then $f(a)$ and $g(a)$ must be equal.

82. If $\left(\dfrac{f}{g}\right)(a) = 0$, then $f(a)$ must be 0.

83. If $(fg)(a) = 0$, then $f(a)$ must be 0.

Review Exercises

84. Solve for b: $R = 3(a + b)$. (Section 1.5, Example 6)

85. Solve: $3(6 - x) = 3 - 2(x - 4)$. (Section 1.4, Example 3)

86. If $f(x) = 6x - 4$, find $f(b + 2)$. (Section 2.1, Example 3)

Preview Exercises

Exercises 87–89 will help you prepare for the material covered in the next section.

87. Consider $4x - 3y = 6$.

a. What is the value of x when $y = 0$?

b. What is the value of y when $x = 0$?

88. a. Graph $y = 2x + 4$. Select integers for x from -3 to 1, inclusive.

b. At what point does the graph cross the x-axis?

c. At what point does the graph cross the y-axis?

89. Solve for y: $5x + 3y = -12$.

MID-CHAPTER CHECK POINT Section 2.1–Section 2.3

What You Know: We learned that a function is a relation in which no two ordered pairs have the same first component and different second components. We represented functions as equations and used function notation. We graphed functions and applied the vertical line test to identify graphs of functions. We determined the domain and range of a function from its graph, using inputs on the x-axis for the domain and outputs on the y-axis for the range. Finally, we developed an algebra of functions to combine functions and determine their domains.

In Exercises 1–6, determine whether each relation is a function. Give the domain and range for each relation.

1. $\{(2, 6), (1, 4), (2, -6)\}$

2. $\{(0, 1), (2, 1), (3, 4)\}$

3.

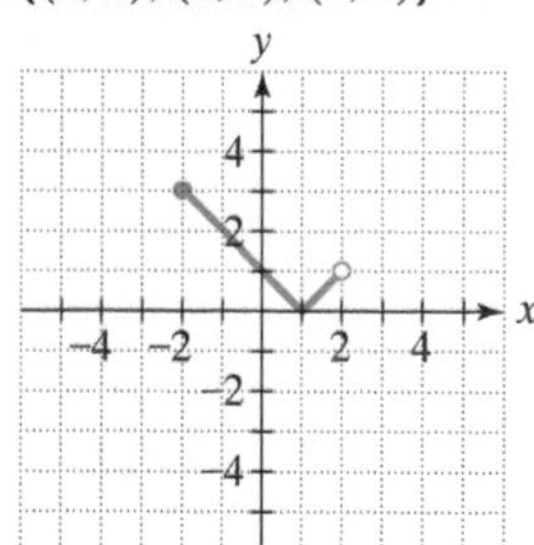

4.

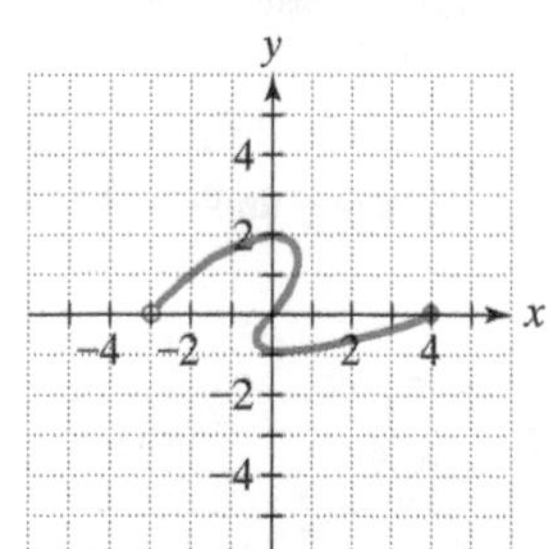

5.

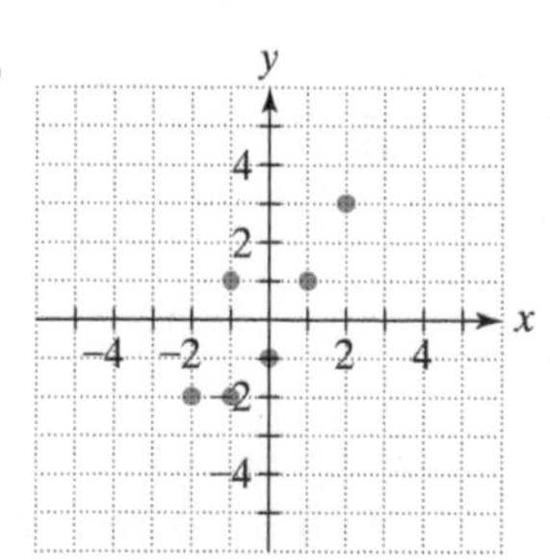

6.

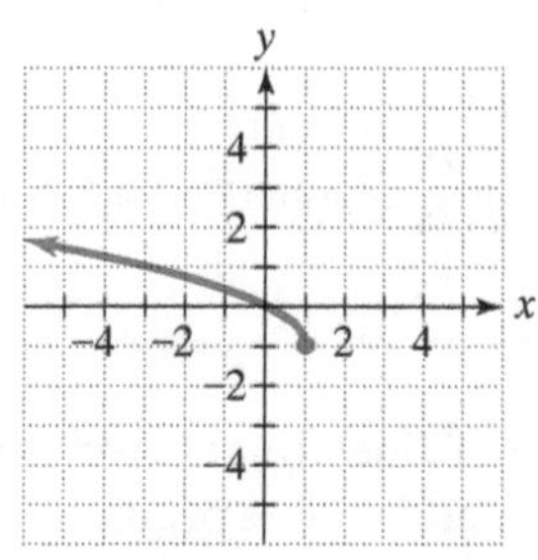

Use the graph of f to solve Exercises 7–12.

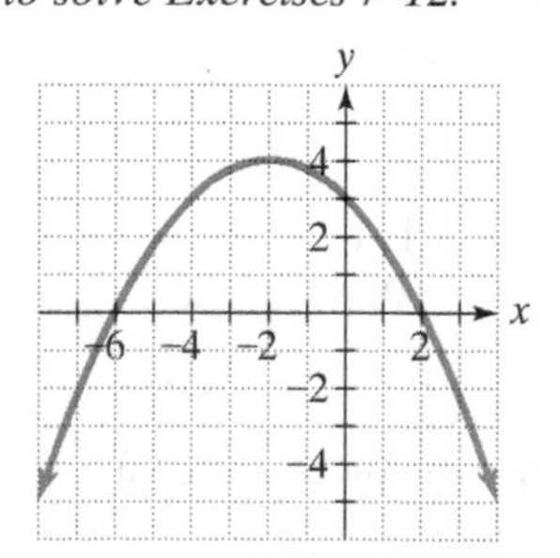

7. Explain why f represents the graph of a function.

8. Use the graph to find $f(-4)$.

9. For what value or values of x is $f(x) = 4$?

10. For what value or values of x is $f(x) = 0$?

11. Find the domain of f.

12. Find the range of f.

In Exercises 13–14, find the domain of each function.

13. $f(x) = (x + 2)(x - 2)$

14. $g(x) = \dfrac{1}{(x + 2)(x - 2)}$

In Exercises 15–22, let

$$f(x) = x^2 - 3x + 8 \text{ and } g(x) = -2x - 5.$$

Find each of the following.

15. $f(0) + g(-10)$

16. $f(-1) - g(3)$

17. $f(a) + g(a + 3)$

18. $(f + g)(x)$ and $(f + g)(-2)$

19. $(f - g)(x)$ and $(f - g)(5)$

20. $(fg)(-1)$

21. $\left(\dfrac{f}{g}\right)(x)$ and $\left(\dfrac{f}{g}\right)(-4)$

22. The domain of $\dfrac{f}{g}$

SECTION

2.4 Linear Functions and Slope

What am I supposed to learn?

After studying this section, you should be able to:

1. Use intercepts to graph a linear function in standard form.
2. Compute a line's slope.
3. Find a line's slope and y-intercept from its equation.
4. Graph linear functions in slope-intercept form.
5. Graph horizontal or vertical lines.
6. Interpret slope as rate of change.
7. Find a function's average rate of change.
8. Use slope and y-intercept to model data.

Is there a relationship between literacy and child mortality? As the percentage of adult females who are literate increases, does the mortality of children under age five decrease? **Figure 2.18** is based on data from the United Nations and indicates that this is, indeed, the case. Each point in the figure represents one country.

Data presented in a visual form as a set of points is called a **scatter plot**. Also shown in **Figure 2.18** is a line that passes through or near the points. A line that best fits the data points in a scatter plot is called a **regression line**. By writing the equation of this line, we can obtain a model of the data and make predictions about child mortality based on the percentage of literate adult females in a country.

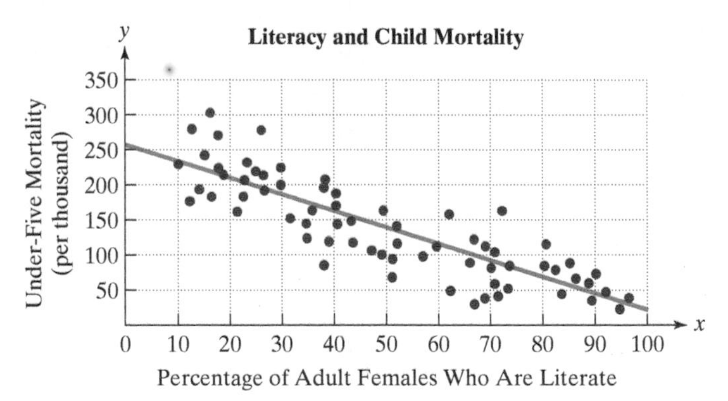

Figure 2.18

Source: United Nations Development Programme

Data often fall on or near a line. In the remainder of this chapter, we will use equations to model such data and make predictions. We begin with a discussion of graphing linear functions using intercepts.

1 Use intercepts to graph a linear function in standard form.

Graphing Using Intercepts

The equation of the regression line in **Figure 2.18** is

$$y = -2.39x + 254.47.$$

The variable x represents the percentage of adult females in a country who are literate. The variable y represents child mortality, per thousand, for children under five in that country. Using function notation, we can rewrite the equation as

$$f(x) = -2.39x + 254.47.$$

A function such as this, whose graph is a straight line, is called a **linear function**. There is another way that we can write the function's equation

$$y = -2.39x + 254.47.$$

We will collect the x- and y-terms on the left side. This is done by adding $2.39x$ to both sides:

$$2.39x + y = 254.47.$$

The form of this equation is $Ax + By = C$.

$$2.39x \quad + \quad y = 254.47$$

A, the coefficient of x, is 2.39.

B, the coefficient of y, is 1.

C, the constant on the right, is 254.47.

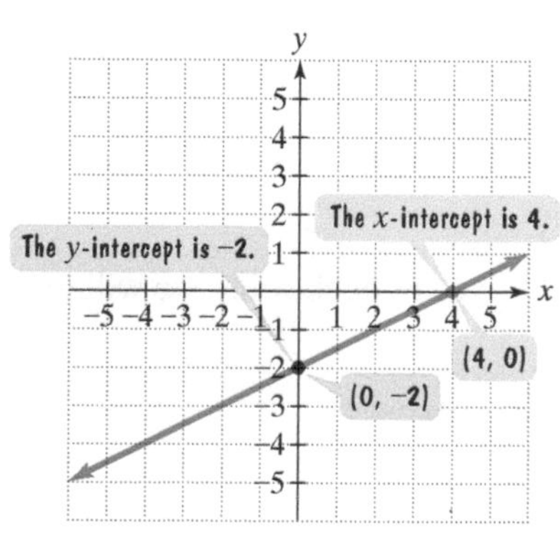

Figure 2.19 The graph of $2x - 4y = 8$

All equations of the form $Ax + By = C$ are straight lines when graphed, as long as A and B are not both zero. Such an equation is called the **standard form of the equation of a line**. To graph equations of this form, we will use two important points: the **intercepts**.

An ***x*-intercept** of a graph is the x-coordinate of a point where the graph intersects the x-axis. For example, look at the graph of $2x - 4y = 8$ in **Figure 2.19**. The graph crosses the x-axis at $(4, 0)$. Thus, the x-intercept is 4. **The *y*-coordinate corresponding to an *x*-intercept is always zero.**

A ***y*-intercept** of a graph is the y-coordinate of a point where the graph intersects the y-axis. The graph of $2x - 4y = 8$ in **Figure 2.19** shows that the graph crosses the y-axis at $(0, -2)$. Thus, the y-intercept is -2. **The *x*-coordinate corresponding to a *y*-intercept is always zero.**

Great Question!

Are single numbers the only way to represent intercepts? Can ordered pairs also be used?

Mathematicians tend to use two ways to describe intercepts. Did you notice that we are using single numbers? If a graph's x-intercept is a, it passes through the point $(a, 0)$. If a graph's y-intercept is b, it passes through the point $(0, b)$.

Some books state that the x-intercept is the *point* $(a, 0)$ and the x-intercept is *at a* on the x-axis. Similarly, the y-intercept is the *point* $(0, b)$ and the y-intercept is *at b* on the y-axis. In these descriptions, the intercepts are the actual points where a graph crosses the axes.

Although we'll describe intercepts as single numbers, we'll immediately state the point on the x- or y-axis that the graph passes through. Here's the important thing to keep in mind:

x-intercept: The corresponding y-coordinate is 0.
y-intercept: The corresponding x-coordinate is 0.

When graphing using intercepts, it is a good idea to use a third point, a checkpoint, before drawing the line. A checkpoint can be obtained by selecting a value for either variable, other than 0, and finding the corresponding value for the other variable. The checkpoint should lie on the same line as the x- and y-intercepts. If it does not, recheck your work and find the error.

Using Intercepts to Graph $Ax + By = C$

1. Find the x-intercept. Let $y = 0$ and solve for x.
2. Find the y-intercept. Let $x = 0$ and solve for y.
3. Find a checkpoint, a third ordered-pair solution.
4. Graph the equation by drawing a line through the three points.

EXAMPLE 1 Using Intercepts to Graph a Linear Equation

Graph: $4x - 3y = 6$.

Solution

Step 1. Find the x-intercept. Let $y = 0$ and solve for x.

$$4x - 3 \cdot 0 = 6 \quad \text{Replace } y \text{ with 0 in } 4x - 3y = 6.$$

$$4x = 6 \quad \text{Simplify.}$$

$$x = \frac{6}{4} = \frac{3}{2} \quad \text{Divide both sides by 4.}$$

The x-intercept is $\frac{3}{2}$, so the line passes through $\left(\frac{3}{2}, 0\right)$ or $(1.5, 0)$.

Step 2. Find the y-intercept. Let $x = 0$ and solve for y.

$$4 \cdot 0 - 3y = 6 \quad \text{Replace } x \text{ with 0 in } 4x - 3y = 6.$$

$$-3y = 6 \quad \text{Simplify.}$$

$$y = -2 \quad \text{Divide both sides by } -3.$$

The y-intercept is -2, so the line passes through $(0, -2)$.

Step 3. Find a checkpoint, a third ordered-pair solution. For our checkpoint, we will let $x = 1$ and find the corresponding value for y.

$$4x - 3y = 6 \quad \text{This is the given equation.}$$

$$4 \cdot 1 - 3y = 6 \quad \text{Substitute 1 for } x.$$

$$4 - 3y = 6 \quad \text{Simplify.}$$

$$-3y = 2 \quad \text{Subtract 4 from both sides.}$$

$$y = -\frac{2}{3} \quad \text{Divide both sides by } -3.$$

The checkpoint is the ordered pair $\left(1, -\frac{2}{3}\right)$.

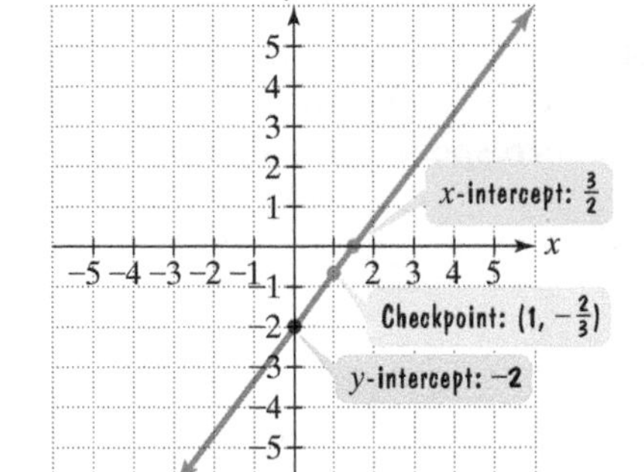

Figure 2.20 The graph of $4x - 3y = 6$

Step 4. Graph the equation by drawing a line through the three points. The three points in **Figure 2.20** lie along the same line. Drawing a line through the three points results in the graph of $4x - 3y = 6$. ■

Using Technology

You can use a graphing utility to graph equations of the form $Ax + By = C$. Begin by solving the equation for y. For example, to graph $4x - 3y = 6$, solve the equation for y.

$$4x - 3y = 6 \quad \text{This is the equation to be graphed.}$$

$$4x - 4x - 3y = -4x + 6 \quad \text{Add } -4x \text{ to both sides.}$$

$$-3y = -4x + 6 \quad \text{Simplify.}$$

$$\frac{-3y}{-3} = \frac{-4x + 6}{-3} \quad \text{Divide both sides by } -3.$$

$$y = \frac{4}{3}x - 2 \quad \text{Simplify.}$$

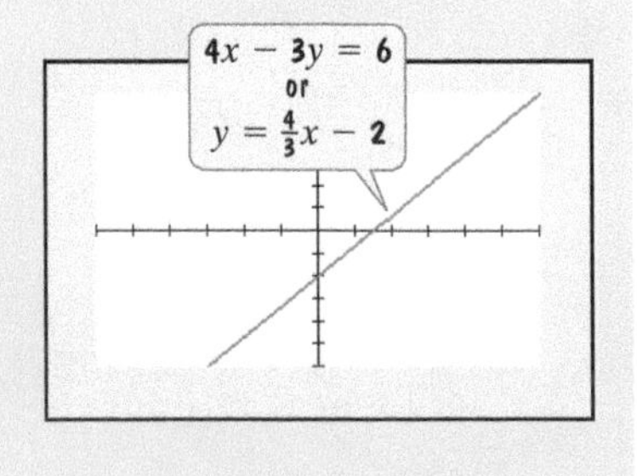

This is the equation to enter in your graphing utility. The graph of $y = \frac{4}{3}x - 2$, or, equivalently, $4x - 3y = 6$ is shown above in a $[-6, 6, 1]$ by $[-6, 6, 1]$ viewing rectangle.

CHECK POINT 1 Graph: $3x - 2y = 6$.

2 Compute a line's slope.

The Slope of a Line

Mathematicians have developed a useful measure of the steepness of a line, called the **slope** of the line. Slope compares the vertical change (the **rise**) to the horizontal change (the **run**) when moving from one fixed point to another along the line. To calculate the slope of a line, we use a ratio that compares the change in y (the rise) to the change in x (the run).

Definition of Slope

The **slope** of the line through the distinct points (x_1, y_1) and (x_2, y_2) is

$$\frac{\text{Change in } y}{\text{Change in } x} = \frac{\text{Rise}}{\text{Run}} = \frac{y_2 - y_1}{x_2 - x_1},$$

Rise — Vertical change
Run — Horizontal change

where $x_2 - x_1 \neq 0$.

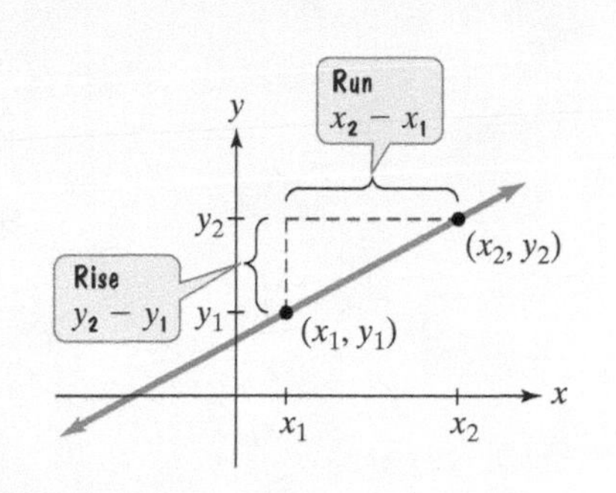

It is common notation to let the letter m represent the slope of a line. The letter m is used because it is the first letter of the French verb *monter*, meaning "to rise," or "to ascend."

EXAMPLE 2 Using the Definition of Slope

Find the slope of the line passing through each pair of points:

a. $(-3, -4)$ and $(-1, 6)$ **b.** $(-1, 3)$ and $(-4, 5)$.

Solution

a. Let $(x_1, y_1) = (-3, -4)$ and $(x_2, y_2) = (-1, 6)$. The slope is obtained as follows:

$$m = \frac{\text{Change in } y}{\text{Change in } x} = \frac{y_2 - y_1}{x_2 - x_1} = \frac{6 - (-4)}{-1 - (-3)} = \frac{6 + 4}{-1 + 3} = \frac{10}{2} = 5.$$

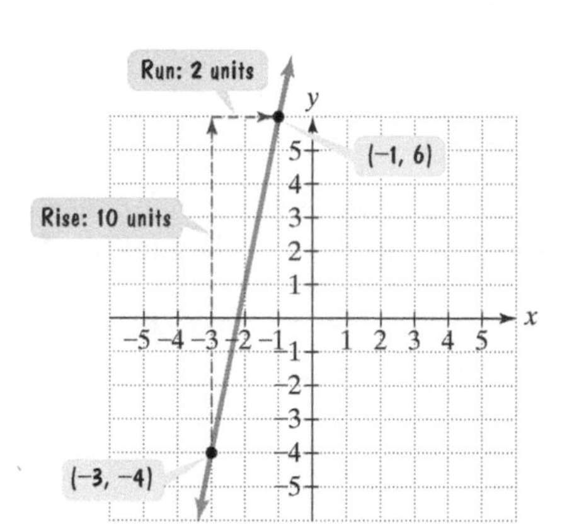

Figure 2.21 Visualizing a slope of 5

The situation is illustrated in **Figure 2.21**. The slope of the line is 5, or $\frac{10}{2}$. For every vertical change, or rise, of 10 units, there is a corresponding horizontal change, or run, of 2 units. The slope is positive and the line rises from left to right.

Great Question!

When using the definition of slope, how do I know which point to call (x_1, y_1) and which point to call (x_2, y_2)?

When computing slope, it makes no difference which point you call (x_1, y_1) and which point you call (x_2, y_2). If we let $(x_1, y_1) = (-1, 6)$ and $(x_2, y_2) = (-3, -4)$, the slope is still 5:

$$m = \frac{y_2 - y_1}{x_2 - x_1} = \frac{-4 - 6}{-3 - (-1)} = \frac{-10}{-2} = 5.$$

However, you should not subtract in one order in the numerator $(y_2 - y_1)$ and then in a different order in the denominator $(x_1 - x_2)$.

$$\frac{-4 - 6}{-1 - (-3)} = \frac{-10}{2} = -5. \quad \text{Incorrect! The slope is not } -5.$$

b. To find the slope of the line passing through $(-1, 3)$ and $(-4, 5)$, we can let $(x_1, y_1) = (-1, 3)$ and $(x_2, y_2) = (-4, 5)$. The slope is computed as follows:

$$m = \frac{\text{Change in } y}{\text{Change in } x} = \frac{y_2 - y_1}{x_2 - x_1} = \frac{5 - 3}{-4 - (-1)} = \frac{2}{-3} = -\frac{2}{3}.$$

The situation is illustrated in **Figure 2.22**. The slope of the line is $-\frac{2}{3}$. For every vertical change of -2 units (2 units down), there is a corresponding horizontal change of 3 units. The slope is negative and the line falls from left to right.

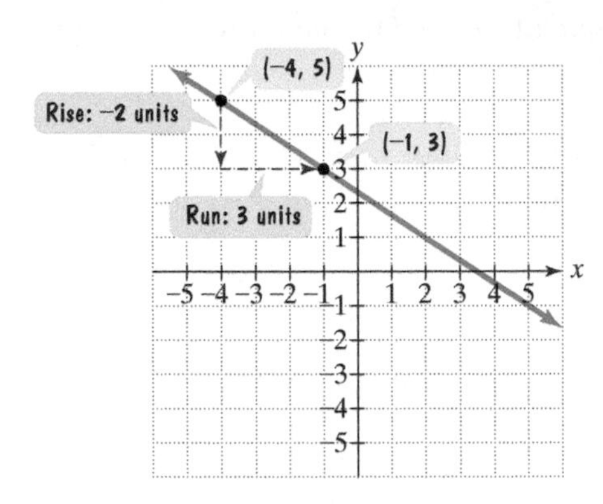

Figure 2.22 Visualizing a slope of $-\frac{2}{3}$ ■

☑ CHECK POINT 2 Find the slope of the line passing through each pair of points:

a. $(-3, 4)$ and $(-4, -2)$ **b.** $(4, -2)$ and $(-1, 5)$.

Example 2 illustrates that a line with a positive slope is rising from left to right and a line with a negative slope is falling from left to right. By contrast, a horizontal line neither rises nor falls and has a slope of zero. A vertical line has no horizontal change, so $x_2 - x_1 = 0$ in the formula for slope. Because we cannot divide by zero, the slope of a vertical line is undefined. This discussion is summarized in **Table 2.2.**

Table 2.2 Possibilities for a Line's Slope

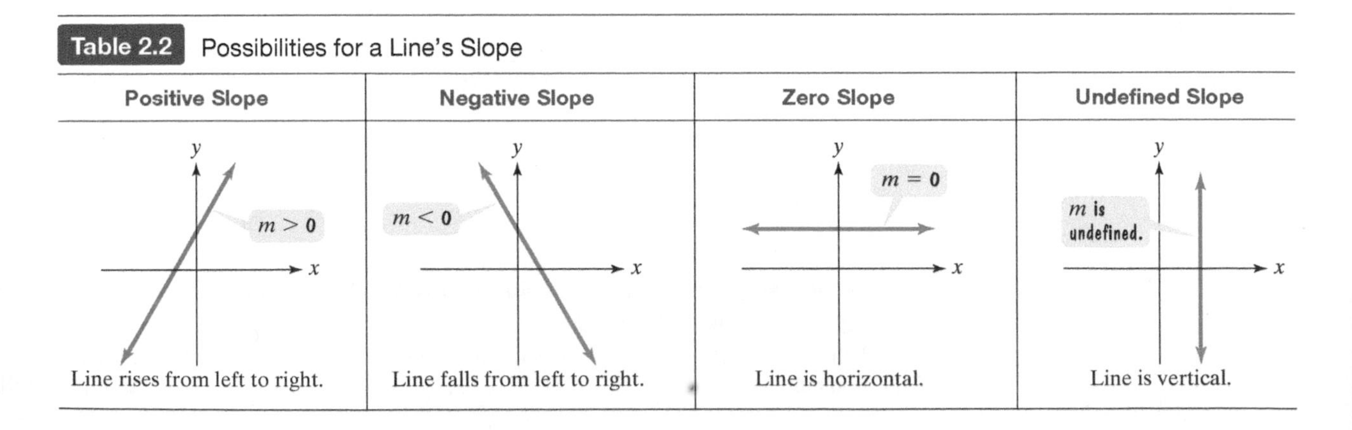

Positive Slope	Negative Slope	Zero Slope	Undefined Slope
Line rises from left to right.	Line falls from left to right.	Line is horizontal.	Line is vertical.

Great Question!

Is it OK to say that a vertical line has no slope?

Always be clear in the way you use language, especially in mathematics. For example, it's not a good idea to say that a line has "no slope." This could mean that the slope is zero or that the slope is undefined.

The Slope-Intercept Form of the Equation of a Line

We opened this section with a linear function that models child mortality as a function of literacy. The function's equation can be expressed as

$$y = -2.39x + 254.47 \quad \text{or} \quad f(x) = -2.39x + 254.47.$$

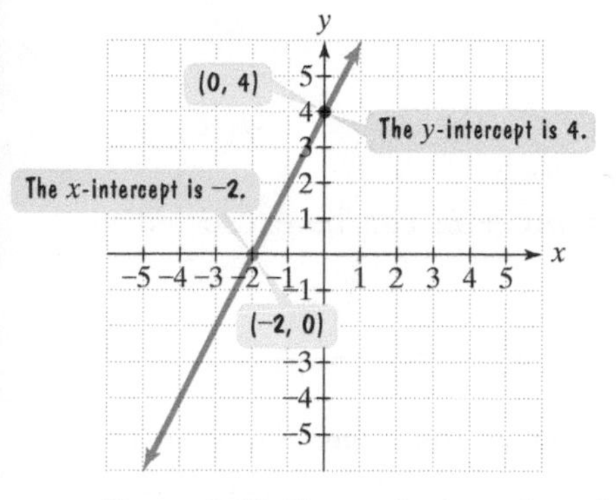

Figure 2.23 The graph of $y = 2x + 4$

What is the significance of −2.39, the x-coefficient, or of 254.47, the constant term? To answer this question, let's look at an equation in the same form with simpler numbers. In particular, consider the equation $y = 2x + 4$.

Figure 2.23 shows the graph of $y = 2x + 4$. Verify that the x-intercept is −2 by setting y equal to 0 and solving for x. Similarly, verify that the y-intercept is 4 by setting x equal to 0 and solving for y.

Now that we have two points on the line, we can calculate the slope of the graph of $y = 2x + 4$.

$$\text{Slope} = \frac{\text{Change in } y}{\text{Change in } x} = \frac{4 - 0}{0 - (-2)} = \frac{4}{2} = 2$$

We see that the slope of the line is 2, the same as the coefficient of x in the equation $y = 2x + 4$. The y-intercept is 4, the same as the constant in the equation $y = 2x + 4$.

$$y = 2x + 4$$

The slope is 2. The y-intercept is 4.

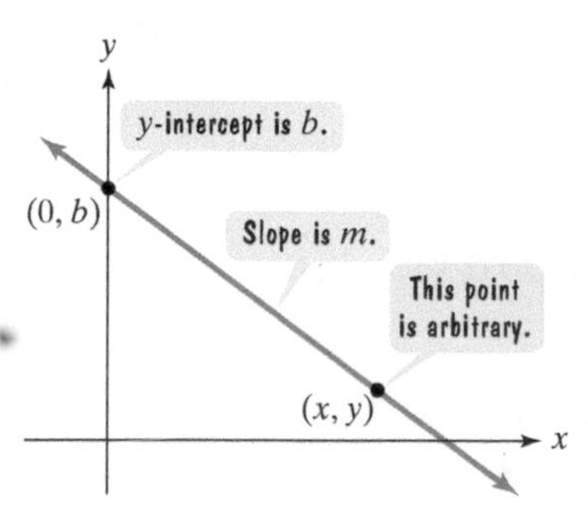

Figure 2.24 A line with slope m and y-intercept b

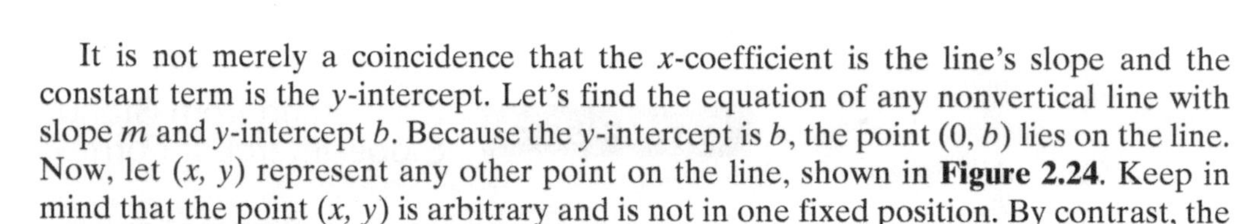

It is not merely a coincidence that the x-coefficient is the line's slope and the constant term is the y-intercept. Let's find the equation of any nonvertical line with slope m and y-intercept b. Because the y-intercept is b, the point $(0, b)$ lies on the line. Now, let (x, y) represent any other point on the line, shown in **Figure 2.24**. Keep in mind that the point (x, y) is arbitrary and is not in one fixed position. By contrast, the point $(0, b)$ is fixed.

Regardless of where the point (x, y) is located, the steepness of the line in **Figure 2.24** remains the same. Thus, the ratio for slope stays a constant m. This means that for all points along the line,

$$m = \frac{\text{Change in } y}{\text{Change in } x} = \frac{y - b}{x - 0} = \frac{y - b}{x}.$$

We can clear the fraction by multiplying both sides by x, the least common denominator.

$m = \dfrac{y - b}{x}$	This is the slope of the line in Figure 2.24.
$mx = \dfrac{y - b}{x} \cdot x$	Multiply both sides by x.
$mx = y - b$	Simplify: $\dfrac{y - b}{\not{x}} \cdot \not{x} = y - b$.
$mx + b = y - b + b$	Add b to both sides and solve for y.
$mx + b = y$	Simplify.

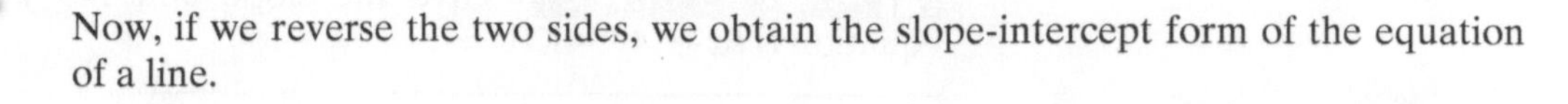

Now, if we reverse the two sides, we obtain the slope-intercept form of the equation of a line.

3 Find a line's slope and y-intercept from its equation.

Great Question!

Which are the variables and which are the constants in the slope-intercept equation $y = mx + b$?

The variables in $y = mx + b$ vary in different ways. The values for slope, m, and y-intercept, b, vary from one line's equation to another. However, they remain constant in the equation of a single line. By contrast, the variables x and y represent the infinitely many points, (x, y), on a single line. Thus, these variables vary in both the equation of a single line, as well as from one equation to another.

Slope-Intercept Form of the Equation of a Line

The **slope-intercept form of the equation** of a nonvertical line with slope m and y-intercept b is

$$y = mx + b.$$

The slope-intercept form of a line's equation, $y = mx + b$, can be expressed in function notation by replacing y with $f(x)$:

$$f(x) = mx + b.$$

We have seen that functions in this form are called **linear functions**. Thus, in the equation of a linear function, the x-coefficient is the line's slope and the constant term is the y-intercept. Here are two examples:

$$y = 2x - 4$$

The slope is 2. The y-intercept is -4.

$$f(x) = \frac{1}{2}x + 2.$$

The slope is $\frac{1}{2}$. The y-intercept is 2.

If a linear function's equation is in standard form, $Ax + By = C$, do you see how we can identify the line's slope and y-intercept? Solve the equation for y and convert to slope-intercept form.

EXAMPLE 3 Converting from Standard Form to Slope-Intercept Form

Give the slope and the y-intercept for the line whose equation is

$$5x + 3y = -12.$$

Solution We convert $5x + 3y = -12$ to slope-intercept form by solving the equation for y. In this form, the coefficient of x is the line's slope and the constant term is the y-intercept.

$$5x + 3y = -12$$ This is the given equation in standard form, $Ax + By = C$.

Our goal is to isolate y.

$$5x - 5x + 3y = -5x - 12$$ Add $-5x$ to both sides.

$$3y = -5x - 12$$ Simplify.

$$\frac{3y}{3} = \frac{-5x - 12}{3}$$ Divide both sides by 3.

$$y = -\frac{5}{3}x - 4$$ Divide each term in the numerator by 3.

The slope is $-\frac{5}{3}$. The y-intercept is -4. ■

✓ **CHECK POINT 3** Give the slope and the y-intercept for the line whose equation is $8x - 4y = 20$.

4 Graph linear functions in slope-intercept form.

If a linear function's equation is in slope-intercept form, we can use the y-intercept and the slope to obtain its graph.

Graphing $y = mx + b$ Using the Slope and y-Intercept

1. Plot the point containing the y-intercept on the y-axis. This is the point $(0, b)$.
2. Obtain a second point using the slope, m. Write m as a fraction, and use rise over run, starting at the point on the y-axis, to plot this point.
3. Use a straightedge to draw a line through the two points. Draw arrowheads at the ends of the line to show that the line continues indefinitely in both directions.

EXAMPLE 4 Graphing by Using the Slope and y-Intercept

Graph the line whose equation is $y = 3x - 4$.

Solution The equation $y = 3x - 4$ is in the form $y = mx + b$. The slope, m, is the coefficient of x. The y-intercept, b, is the constant term.

$$y = 3x + (-4)$$

The slope is 3. The y-intercept is -4.

Now that we have identified the slope and the y-intercept, we use the three-step procedure to graph the equation.

Step 1. Plot the point containing the y-intercept on the y-axis. The y-intercept is -4. We plot the point $(0, -4)$, shown in **Figure 2.25(a)**.

Step 2. Obtain a second point using the slope, m. Write m as a fraction, and use rise over run, starting at the point on the y-axis, to plot this point. We express the slope, 3, as a fraction.

$$m = \frac{3}{1} = \frac{\text{Rise}}{\text{Run}}$$

We plot the second point on the line by starting at $(0, -4)$, the first point. Based on the slope, we move 3 units *up* (the rise) and 1 unit to the *right* (the run). This puts us at a second point on the line, $(1, -1)$, shown in **Figure 2.25(b)**.

Step 3. Use a straightedge to draw a line through the two points. The graph of $y = 3x - 4$ is shown in **Figure 2.25(c)**.

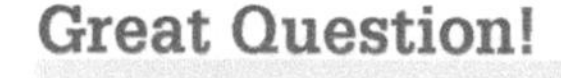

If the slope is an integer, such as 3, why should I express it as $\frac{3}{1}$ for graphing purposes?

Writing the slope, m, as a fraction allows you to identify the rise (the fraction's numerator) and the run (the fraction's denominator).

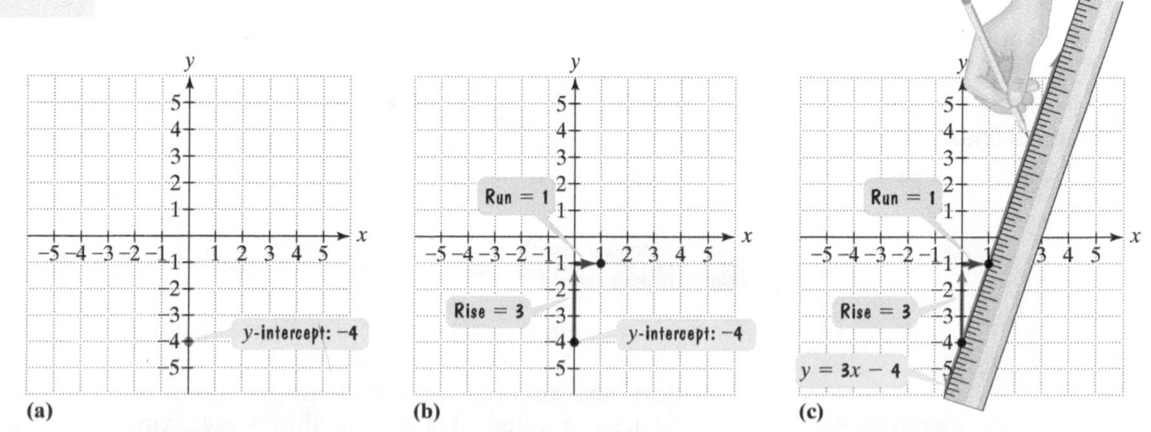

Figure 2.25 Graphing $y = 3x - 4$ using the y-intercept and slope ■

CHECK POINT 4 Graph the line whose equation is $y = 4x - 3$.

EXAMPLE 5 Graphing by Using the Slope and y-Intercept

Graph the linear function: $f(x) = -\frac{3}{2}x + 2.$

Solution The equation of the line is in the form $f(x) = mx + b$. We can find the slope, m, by identifying the coefficient of x. We can find the y-intercept, b, by identifying the constant term.

$$f(x) = -\frac{3}{2}x + 2$$

The slope is $-\frac{3}{2}$. The y-intercept is 2.

Now that we have identified the slope, $-\frac{3}{2}$, and the y-intercept, 2, we use the three-step procedure to graph the equation.

Step 1. Plot the point containing the y-intercept on the y-axis. The y-intercept is 2. We plot $(0, 2)$, shown in **Figure 2.26**.

Step 2. Obtain a second point using the slope, m. Write m as a fraction, and use rise over run, starting at the point on the y-axis, to plot this point. The slope, $-\frac{3}{2}$, is already written as a fraction.

$$m = -\frac{3}{2} = \frac{-3}{2} = \frac{\text{Rise}}{\text{Run}}$$

We plot the second point on the line by starting at $(0, 2)$, the first point. Based on the slope, we move 3 units *down* (the rise) and 2 units to the *right* (the run). This puts us at a second point on the line, $(2, -1)$, shown in **Figure 2.26**.

Step 3. Use a straightedge to draw a line through the two points. The graph of the linear function $f(x) = -\frac{3}{2}x + 2$ is shown as a blue line in **Figure 2.26**. ■

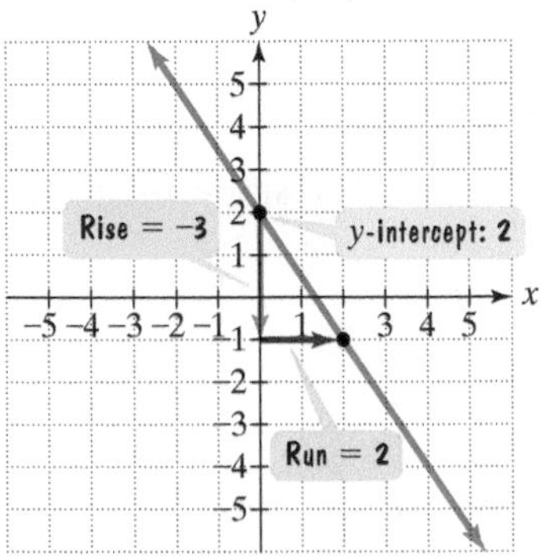

Rise = −3
y-intercept: 2
Run = 2

Figure 2.26 The graph of $f(x) = -\frac{3}{2}x + 2$

Discover for Yourself

Obtain a second point in Example 5 by writing the slope as follows:

$$m = \frac{3}{-2} = \frac{\text{Rise}}{\text{Run}}.$$

$-\frac{3}{2}$ can be expressed as $\frac{-3}{2}$ or $\frac{3}{-2}$.

Obtain a second point in **Figure 2.26** by moving *up* 3 units and to the *left* 2 units, starting at $(0, 2)$. What do you observe once you draw the line?

✓ **CHECK POINT 5** Graph the linear function: $f(x) = -\frac{2}{3}x.$

---

5 Graph horizontal or vertical lines.

Equations of Horizontal and Vertical Lines

If a line is horizontal, its slope is zero: $m = 0$. Thus, the equation $y = mx + b$ becomes $y = b$, where b is the y-intercept. All horizontal lines have equations of the form $y = b$.

EXAMPLE 6 Graphing a Horizontal Line

Graph $y = -4$ in the rectangular coordinate system.

Solution All ordered pairs that are solutions of $y = -4$ have a value of y that is always -4. Any value can be used for x. In the table on the right, we have selected three of the possible values for x: -2, 0, and 3. The table shows that three ordered pairs that are solutions of $y = -4$ are $(-2, -4)$, $(0, -4)$, and $(3, -4)$. Drawing a line that passes through the three points gives the horizontal line shown in **Figure 2.27**. ■

x	$y = -4$	(x, y)
−2	−4	(−2, −4)
0	−4	(0, −4)
3	−4	(3, −4)

For all choices of x, y is a constant −4.

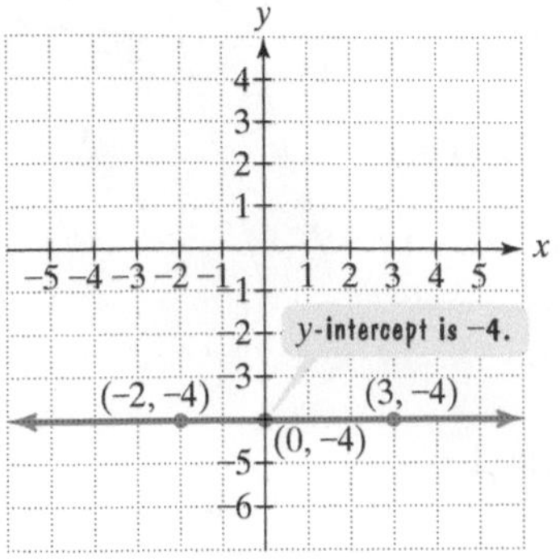

y-intercept is −4.
(−2, −4)
(3, −4)
(0, −4)

Figure 2.27 The graph of $y = -4$ or $f(x) = -4$

CHECK POINT 6 Graph $y = 3$ in the rectangular coordinate system.

Equation of a Horizontal Line

A horizontal line is given by an equation of the form

$$y = b,$$

where b is the y-intercept.

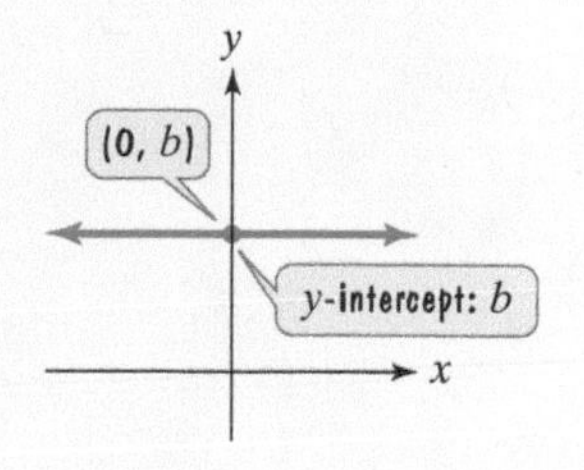

Because any vertical line can intersect the graph of a horizontal line $y = b$ only once, a horizontal line is the graph of a function. Thus, we can express the equation $y = b$ as $f(x) = b$. This linear function is often called a **constant function**.

Next, let's see what we can discover about the graph of an equation of the form $x = a$ by looking at an example.

EXAMPLE 7 Graphing a Vertical Line

Graph the linear equation: $x = 2$.

Solution All ordered pairs that are solutions of $x = 2$ have a value of x that is always 2. Any value can be used for y. In the table on the right, we have selected three of the possible values for y: -2, 0, and 3. The table shows that three ordered pairs that are solutions of $x = 2$ are $(2, -2)$, $(2, 0)$, and $(2, 3)$. Drawing a line that passes through the three points gives the vertical line shown in **Figure 2.28**. ■

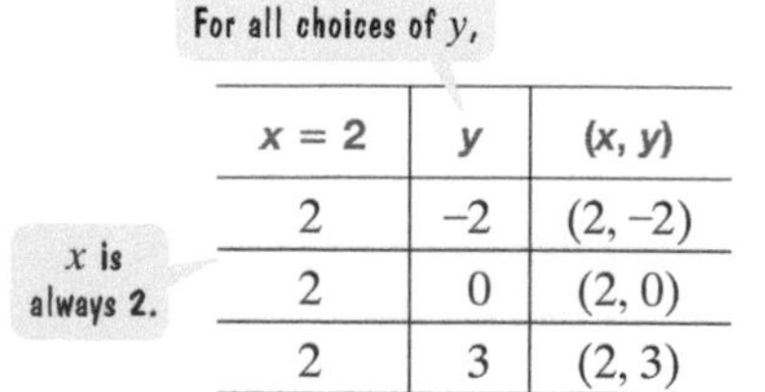

$x = 2$	y	(x, y)
2	-2	$(2, -2)$
2	0	$(2, 0)$
2	3	$(2, 3)$

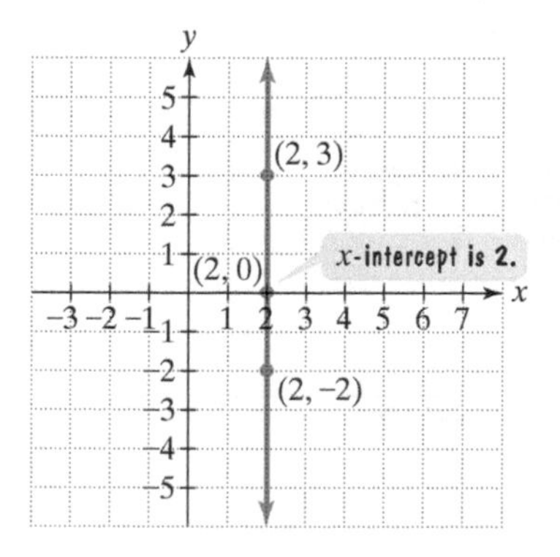

Figure 2.28 The graph of $x = 2$

Equation of a Vertical Line

A vertical line is given by an equation of the form

$$x = a,$$

where a is the x-intercept.

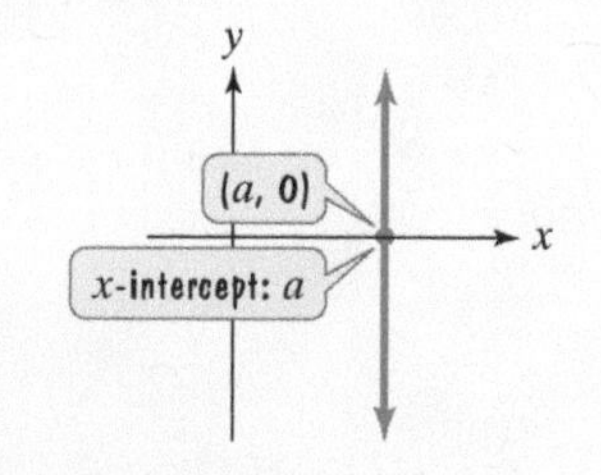

Does a vertical line represent the graph of a linear function? No. Look at the graph of $x = 2$ in **Figure 2.28**. A vertical line drawn through $(2, 0)$ intersects the graph infinitely many times. This shows that infinitely many outputs are associated with the input 2. **No vertical line is a linear function.**

CHECK POINT 7 Graph the linear equation: $x = -3$.

Great Question!

How can equations of horizontal and vertical lines fit into the standard form of a line's equation, $Ax + By = C$, which contains two variables?

The linear equations in Examples 6 and 7, $y = -4$ and $x = 2$, each show only one variable.

- $y = -4$ means $0x + 1y = -4$.
- $x = 2$ means $1x + 0y = 2$.

6 Interpret slope as rate of change.

Slope as Rate of Change

Slope is defined as the ratio of a change in y to a corresponding change in x. It describes how fast y is changing with respect to x. For a linear function, slope may be interpreted as the rate of change of the dependent variable per unit change in the independent variable.

Our next example shows how slope can be interpreted as a rate of change in an applied situation. When calculating slope in applied problems, keep track of the units in the numerator and the denominator.

EXAMPLE 8 Slope as a Rate of Change: Garbage Overload

The line graph in **Figure 2.29** shows the average waste, in pounds, produced per person per day in the United States for selected years from 1960 through 2012. Find the slope of the steepest line segment. Describe what this slope represents.

Average Waste Produced per Person per Day in the U.S.

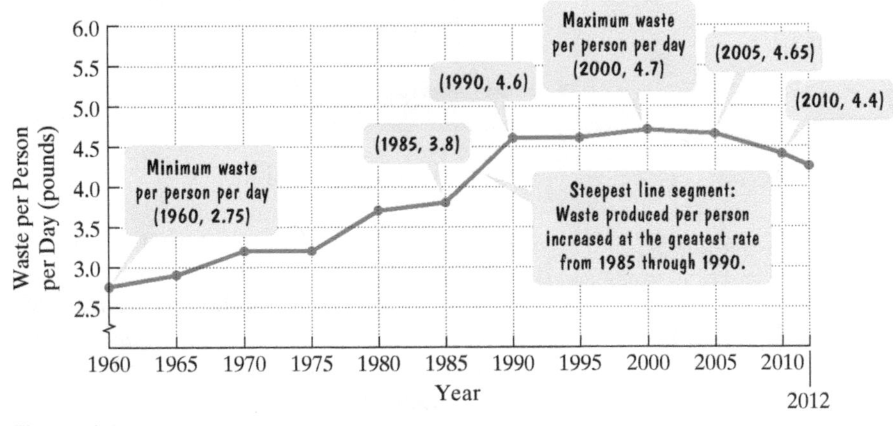

Figure 2.29

Source: Environmental Protection Agency

Solution We will let x represent a year and y the average waste produced per person per day for that year. The two points shown on the steepest line segment have the following coordinates:

$$(1985, 3.8) \quad \text{and} \quad (1990, 4.6).$$

In 1985, Americans produced 3.8 pounds of waste per day.

In 1990, Americans produced 4.6 pounds of waste per day.

Now we compute the slope.

$$m = \frac{\text{Change in } y}{\text{Change in } x} = \frac{4.6 - 3.8}{1990 - 1985} = \frac{0.8}{5} = 0.16$$

The unit in the numerator is *pounds per day*.

The unit in the denominator is *year*.

The slope indicates that from 1985 through 1990, waste production for each American increased by 0.16 pound per day each year. The rate of change per person is an increase of 0.16 pound per day per year. ■

✓ CHECK POINT 8 Use the graph in **Figure 2.29** to find the slope of the line segment that is decreasing the most over a five-year period. Describe what this slope represents.

7 Find a function's average rate of change.

The Average Rate of Change of a Function

If the graph of a function is not a straight line, the **average rate of change** between any two points is the slope of the line containing the two points. For example, **Figure 2.30** shows the graph of a particular man's height, in inches, as a function of his age, in years. Two points on the graph are labeled (13, 57) and (18, 76). At age 13, this man was 57 inches tall, and at age 18, he was 76 inches tall.

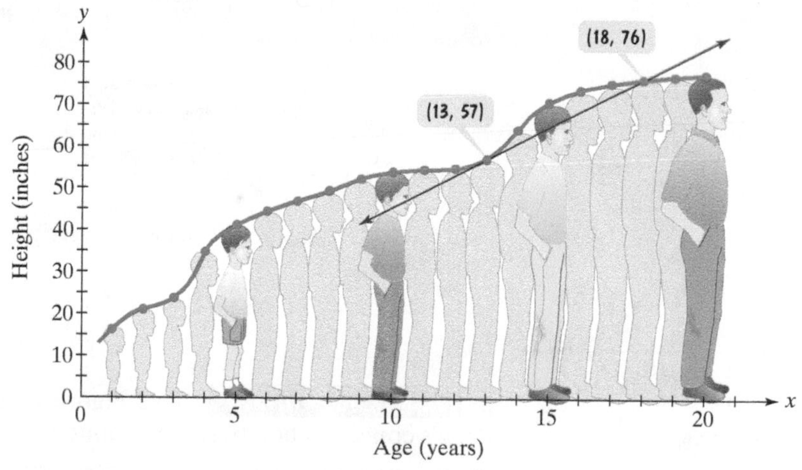

Figure 2.30 Height as a function of age

The man's average growth rate between ages 13 and 18 is the slope of the line containing (13, 57) and (18, 76):

$$m = \frac{\text{Change in } y}{\text{Change in } x} = \frac{76 - 57}{18 - 13} = \frac{19}{5} = 3\frac{4}{5}.$$

This man's average rate of change, or average growth rate, from age 13 to age 18 was $3\frac{4}{5}$, or 3.8, inches per year.

For any function, $y = f(x)$, the slope of the line between any two points on its graph is the **average change in y per unit change in x.**

EXAMPLE 9 Finding the Average Rate of Change

When a person receives a drug injected into a muscle, the concentration of the drug in the body, measured in milligrams per 100 milliliters, is a function of the time elapsed after the injection, measured in hours. **Figure 2.31** shows the graph of such a function, where x represents hours after the injection and $f(x)$ is the drug's concentration at time x. Find the average rate of change in the drug's concentration between 3 and 7 hours.

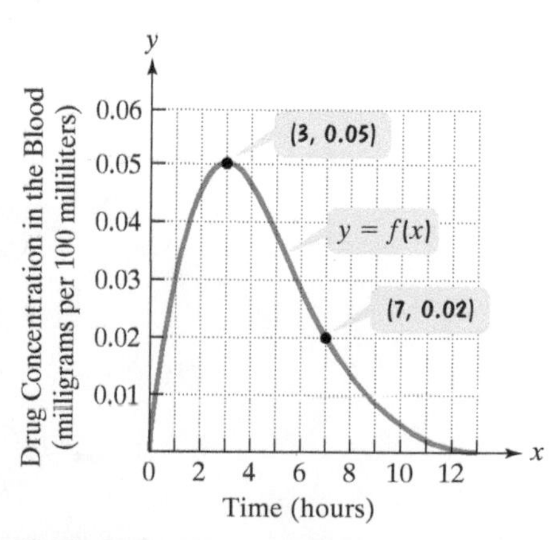

Figure 2.31 Concentration of a drug as a function of time

Solution At 3 hours, the drug's concentration is 0.05 and at 7 hours, the concentration is 0.02. The average rate of change in its concentration between 3 and 7 hours is the slope of the line connecting the points (3, 0.05) and (7, 0.02).

$$m = \frac{\text{Change in } y}{\text{Change in } x} = \frac{0.02 - 0.05}{7 - 3} = \frac{-0.03}{4} = -0.0075$$

The average rate of change is -0.0075. This means that the drug's concentration is decreasing at an average rate of 0.0075 milligram per 100 milliliters per hour. ■

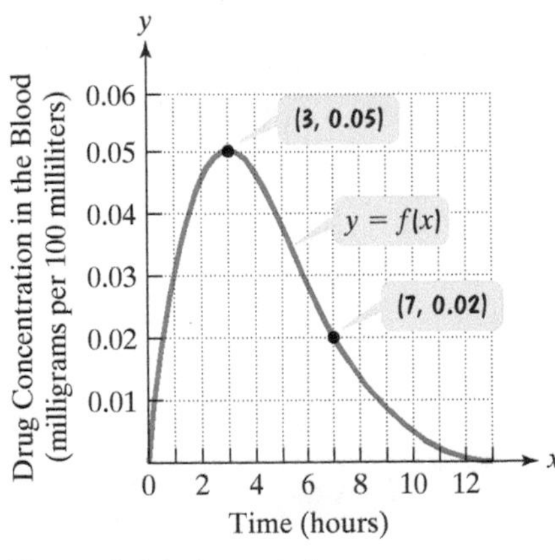

Figure 2.31 (repeated)

Great Question!

Can you clarify how you determine the units that you use when describing slope as a rate of change?

Units used to describe x and y tend to "pile up" when expressing the rate of change of y with respect to x. The unit used to express the rate of change of y with respect to x is

the unit used to describe y **per** the unit used to describe x.

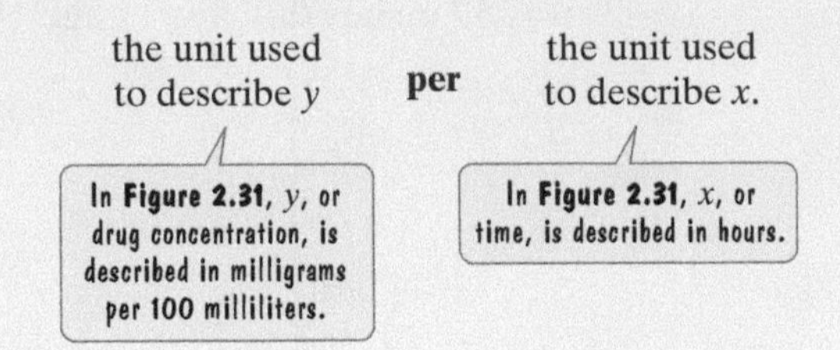

In Figure 2.31, y, or drug concentration, is described in milligrams per 100 milliliters.
In Figure 2.31, x, or time, is described in hours.

In Example 9, the rate of change is described in terms of milligrams per 100 milliliters per hour.

☑ CHECK POINT 9 Use **Figure 2.31** to find the average rate of change in the drug's concentration between 1 hour and 3 hours.

Blitzer Bonus

How Calculus Studies Change

Take a rapid sequence of still photographs of a moving scene and project them onto a screen at thirty shots a second or faster. Our eyes see the results as continuous motion. The small difference between one frame and the next cannot be detected by the human visual system. The idea of calculus likewise regards continuous motion as made up of a sequence of still configurations. Calculus masters the mystery of movement by "freezing the frame" of a continuous changing process, instant by instant. For example, **Figure 2.32** shows a male's changing height over intervals of time. Over the period of time from P to D, his average rate of growth is his change in height—that is, his height at time D minus his height at time P—divided by the change in time from P to D. This is the slope of line PD.

The lines PD, PC, PB, and PA shown in **Figure 2.32** have slopes that show average growth rates for successively shorter periods of time. Calculus makes these time frames so small that they approach a single point—that is, a single instant in time. This point is shown as point P in **Figure 2.32**. The slope of the line that touches the graph at P gives the male's growth rate at one instant in time, P.

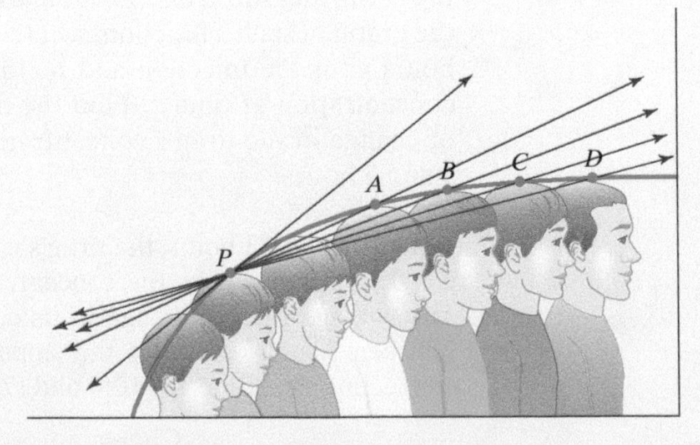

Figure 2.32 Analyzing continuous growth over intervals of time and at an instant in time

8 Use slope and y-intercept to model data.

Modeling Data with the Slope-Intercept Form of the Equation of a Line

Our planet has been heating up for more than a century. Most experts have concluded that the increase in the amount of carbon dioxide in the atmosphere has been at least partly responsible for the increase in global surface temperature. The bar graph in **Figure 2.33(a)** shows the average global temperature for selected years from 1900 through 2014. The data are displayed as a set of four points in the scatter plot in **Figure 2.33(b)**.

Also shown on the scatter plot in **Figure 2.33(b)** is a line that passes through or near the four points. Linear functions are useful for modeling data that fall on or near a line. Example 10 illustrates how we can use the equation $y = mx + b$ to obtain a model for the data and make predictions about climate change in the future.

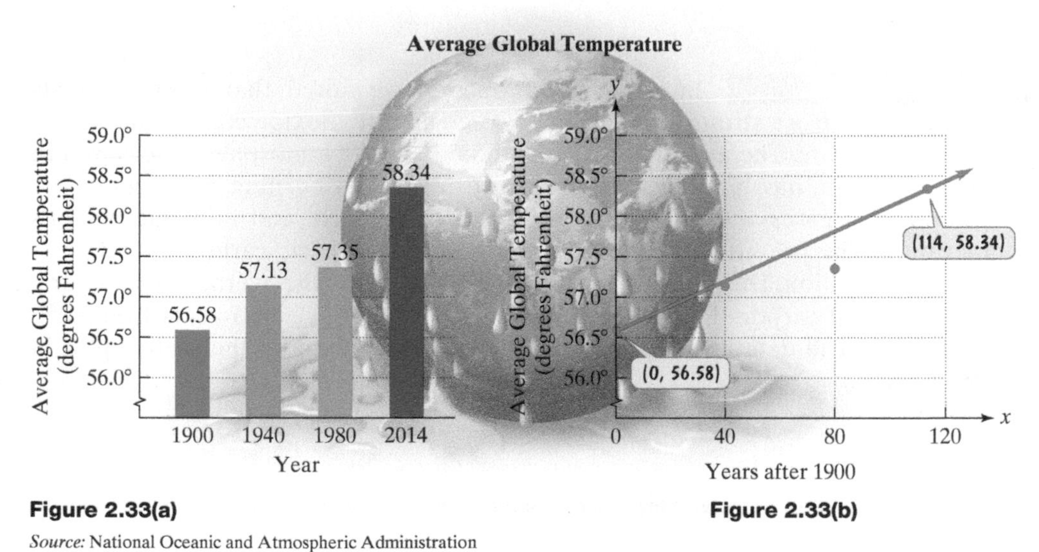

Figure 2.33(a)

Source: National Oceanic and Atmospheric Administration

Figure 2.33(b)

EXAMPLE 10 Modeling with the Slope-Intercept Form of the Equation

a. Use the scatter plot in **Figure 2.33(b)** to find a function in the form $T(x) = mx + b$ that models average global temperature, $T(x)$, in degrees Fahrenheit, x years after 1900.

b. If the trend shown by the data continues, use the model to project the average global temperature in 2050.

Solution

a. We will use the line passing through the points (0, 56.58) and (114, 58.34) to obtain a model. We need values for m, the slope, and b, the y-intercept.

$$T(x) = mx + b$$

$$m = \frac{\text{Change in } y}{\text{Change in } x}$$

$$m = \frac{58.34 - 56.58}{114 - 0} \approx 0.02$$

The point (0, 56.58) lies on the line, so the y-intercept is 56.58: $b = 56.58$.

Average global temperature, $T(x)$, in degrees Fahrenheit (°F), x years after 1900 can be modeled by the linear function

$$T(x) = 0.02x + 56.58.$$

The slope, approximately 0.02, indicates an increase in average global temperature of 0.02° Fahrenheit per year from 1900 through 2014.

b. Now let's use the function $T(x) = 0.02x + 56.58$ to project average global temperature in 2050. Because 2050 is 150 years after 1900, substitute 150 for x and evaluate the function at 150.

$$T(150) = 0.02(150) + 56.58 = 3 + 56.58 = 59.58$$

Our model projects an average global temperature of 59.58°F in 2050. ■

CHECK POINT 10 In our chapter opener, we noted that most scientists believe that there is a strong correlation between carbon dioxide concentration and global temperature. The amount of carbon dioxide in the atmosphere, measured in parts per million, has been increasing as a result of the burning of oil and coal. The buildup of gases and particles is believed to trap heat and raise the planet's temperature. The pre-industrial concentration of atmospheric carbon dioxide was 280 parts per million. The bar graph in **Figure 2.34(a)** shows the average atmospheric concentration of carbon dioxide, in parts per million, for selected years from 1950 through 2010. The data are displayed as a set of four points in the scatter plot in **Figure 2.34(b)**.

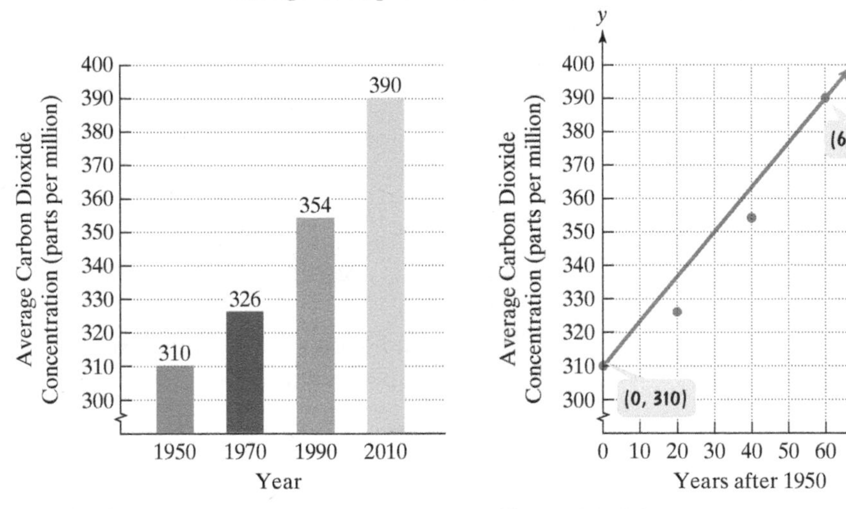

Figure 2.34(a)

Source: National Oceanic and Atmospheric Administration

Figure 2.34(b)

a. Use the line in the scatter plot in **Figure 2.34(b)** to find a function in the form $C(x) = mx + b$ that models average atmospheric concentration of carbon dioxide, $C(x)$, in parts per million, x years after 1950. Round m to two decimal places.

b. If the trend shown by the data continues, use your model from part (a) to project the average atmospheric concentration of carbon dioxide in 2050.

CONCEPT AND VOCABULARY CHECK

Fill in each blank so that the resulting statement is true.

1. Data presented in a visual form as a set of points is called a/an ____________. A line that best fits this set of points is called a/an ____________ line.
2. The equation $Ax + By = C$, where A and B are not both zero, is called the ____________ form of the equation of a line.
3. The x-coordinate of a point where a graph crosses the x-axis is called a/an ____________. The y-coordinate of such a point is always ____________.
4. The y-coordinate of a point where a graph crosses the y-axis is called a/an ____________. The x-coordinate of such a point is always ____________.
5. The slope, m, of a line through the distinct points (x_1, y_1) and (x_2, y_2) is given by the formula $m =$ ____________.
6. If a line rises from left to right, the line has ____________ slope.
7. If a line falls from left to right, the line has ____________ slope.
8. The slope of a horizontal line is ____________.
9. The slope of a vertical line is ____________.
10. The slope-intercept form of the equation of a line is ____________, where m represents the ____________ and b represents the ____________.
11. In order to graph the line whose equation is $y = \frac{2}{5}x + 3$, begin by plotting the point ____________. From this point, we move ____________ units up (the rise) and ____________ units to the right (the run).
12. The graph of the equation $y = 3$ is a/an ____________ line.
13. The graph of the equation $x = -2$ is a/an ____________ line.
14. The slope of the line through the distinct points (x_1, y_1) and (x_2, y_2) can be interpreted as the rate of change in ____________ with respect to ____________.

2.4 EXERCISE SET

MyMathLab®

Practice Exercises

In Exercises 1–14, use intercepts and a checkpoint to graph each linear function.

1. $x + y = 4$
2. $x + y = 2$
3. $x + 3y = 6$
4. $2x + y = 4$
5. $6x - 2y = 12$
6. $6x - 9y = 18$
7. $3x - y = 6$
8. $x - 4y = 8$
9. $x - 3y = 9$
10. $2x - y = 5$
11. $2x = 3y + 6$
12. $3x = 5y - 15$
13. $6x - 3y = 15$
14. $8x - 2y = 12$

In Exercises 15–26, find the slope of the line passing through each pair of points or state that the slope is undefined. Then indicate whether the line through the points rises, falls, is horizontal, or is vertical.

15. $(2, 4)$ and $(3, 8)$
16. $(3, 1)$ and $(5, 4)$
17. $(-1, 4)$ and $(2, 5)$
18. $(-3, -2)$ and $(2, 5)$
19. $(2, 5)$ and $(-1, 5)$
20. $(-6, -3)$ and $(4, -3)$
21. $(-7, 1)$ and $(-4, -3)$
22. $(2, -1)$ and $(-6, 3)$
23. $(-7, -4)$ and $(-3, 6)$
24. $(-3, -4)$ and $(-1, 6)$
25. $\left(\frac{7}{2}, -2\right)$ and $\left(\frac{7}{2}, \frac{1}{4}\right)$
26. $\left(\frac{3}{2}, -6\right)$ and $\left(\frac{3}{2}, \frac{1}{6}\right)$

In Exercises 27–28, find the slope of each line.

27.

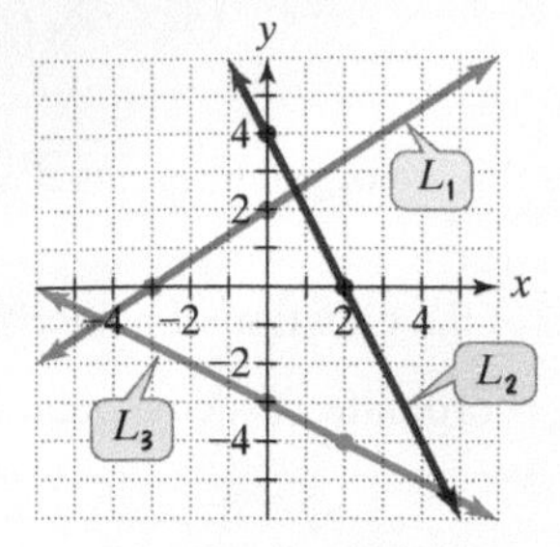

28.

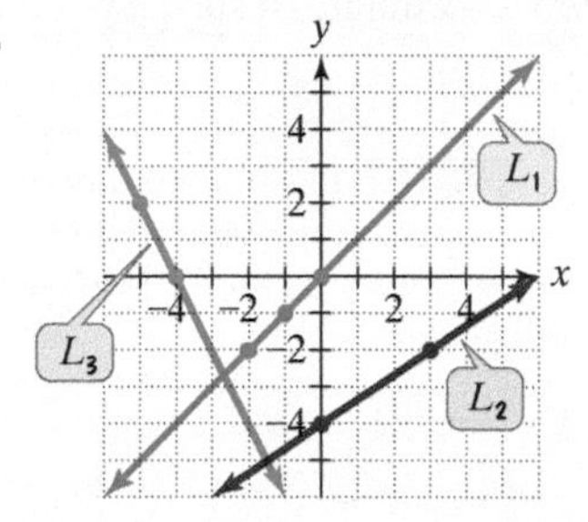

In Exercises 29–40, give the slope and y-intercept of each line whose equation is given. Then graph the linear function.

29. $y = 2x + 1$

30. $y = 3x + 2$

31. $y = -2x + 1$

32. $y = -3x + 2$

33. $f(x) = \frac{3}{4}x - 2$

34. $f(x) = \frac{3}{4}x - 3$

35. $f(x) = -\frac{3}{5}x + 7$

36. $f(x) = -\frac{2}{5}x + 6$

37. $y = -\frac{1}{2}x$ **38.** $y = -\frac{1}{3}x$

39. $y = -\frac{1}{2}$ **40.** $y = -\frac{1}{3}$

In Exercises 41–48,

a. *Rewrite the given equation in slope-intercept form by solving for y.*

b. *Give the slope and y-intercept.*

c. *Use the slope and y-intercept to graph the linear function.*

41. $2x + y = 0$

42. $3x + y = 0$

43. $5y = 4x$

44. $4y = 3x$

45. $3x + y = 2$

46. $2x + y = 4$

47. $5x + 3y = 15$

48. $7x + 2y = 14$

In Exercises 49–62, graph each equation in a rectangular coordinate system.

49. $y = 3$	**50.** $y = 5$	**51.** $f(x) = -2$
52. $f(x) = -4$	**53.** $3y = 18$	**54.** $5y = -30$
55. $f(x) = 2$	**56.** $f(x) = 1$	**57.** $x = 5$
58. $x = 4$	**59.** $3x = -12$	**60.** $4x = -12$
61. $x = 0$	**62.** $y = 0$	

Practice PLUS

In Exercises 63–66, find the slope of the line passing through each pair of points or state that the slope is undefined. Assume that all variables represent positive real numbers. Then indicate whether the line through the points rises, falls, is horizontal, or is vertical.

63. $(0, a)$ and $(b, 0)$

64. $(-a, 0)$ and $(0, -b)$

65. (a, b) and $(a, b + c)$

66. $(a - b, c)$ and $(a, a + c)$

In Exercises 67–68, give the slope and y-intercept of each line whose equation is given. Assume that $B \neq 0$.

67. $Ax + By = C$

68. $Ax = By - C$

In Exercises 69–70, find the value of y if the line through the two given points is to have the indicated slope.

69. $(3, y)$ and $(1, 4)$, $m = -3$

70. $(-2, y)$ and $(4, -4)$, $m = \frac{1}{3}$

In Exercises 71–72, graph each linear function.

71. $3x - 4f(x) = 6$

72. $6x - 5f(x) = 20$

73. If one point on a line is $(3, -1)$ and the line's slope is -2, find the y-intercept.

74. If one point on a line is $(2, -6)$ and the line's slope is $-\frac{3}{2}$, find the y-intercept.

Use the figure to make the lists in Exercises 75–76.

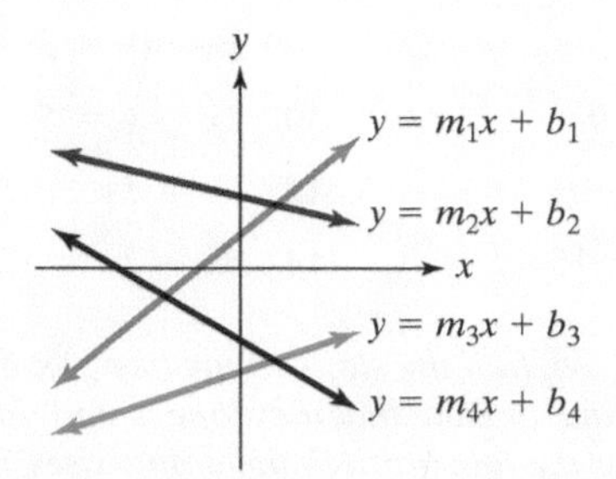

75. List the slopes m_1, m_2, m_3, and m_4 in order of decreasing size.

76. List the y-intercepts b_1, b_2, b_3, and b_4 in order of decreasing size.

Application Exercises

In Exercises 77–80, a linear function that models data is described. Find the slope of each model. Then describe what this means in terms of the rate of change of the dependent variable per unit change in the independent variable.

77. The linear function $f(x) = 55.7x + 60.1$ models the number of smartphones sold in the United States, $f(x)$, in millions, x years after 2006. (*Source:* Newsweek, April 11, 2011)

78. The linear function $f(x) = 2x + 10$ models the amount, $f(x)$, in billions of dollars, that the drug industry spent on marketing information about drugs to doctors x years after 2000. (*Source:* IMS Health)

79. The linear function $f(x) = -0.52x + 24.7$ models the percentage of U.S. adults who smoked cigarettes, $f(x)$, x years after 1997. (*Source:* National Center for Health Statistics)

80. The linear function $f(x) = -0.28x + 1.7$ models the percentage of U.S. taxpayers who were audited by the IRS, $f(x)$, x years after 1996. (*Source:* IRS)

Divorce rates are typically higher for couples who marry in their teens. The graph shows the percentage of marriages ending in divorce by wife's age at marriage. Use the information shown to solve Exercises 81–82.

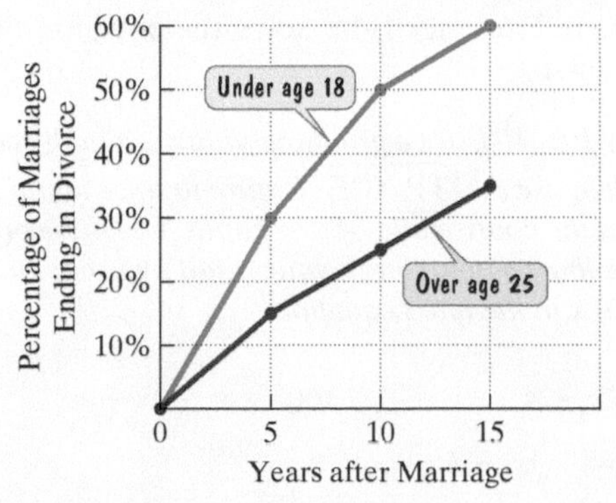

Source: National Center for Health Statistics

81. **a.** What percentage of marriages in which the wife is under 18 when she marries end in divorce within the first five years?

b. What percentage of marriages in which the wife is under 18 when she marries end in divorce within the first ten years?

c. Find the average rate of change in the percentage of marriages ending in divorce between five and ten years of marriage in which the wife is under 18 when she marries.

82. **a.** What percentage of marriages in which the wife is over age 25 when she marries end in divorce within the first five years?

b. What percentage of marriages in which the wife is over age 25 when she marries end in divorce within the first ten years?

c. Find the average rate of change in the percentage of marriages ending in divorce between five and ten years of marriage in which the wife is over age 25 when she marries.

83. Shown, again, is the scatter plot that indicates a relationship between the percentage of adult females in a country who are literate and the mortality of children under five. Also shown is a line that passes through or near the points.

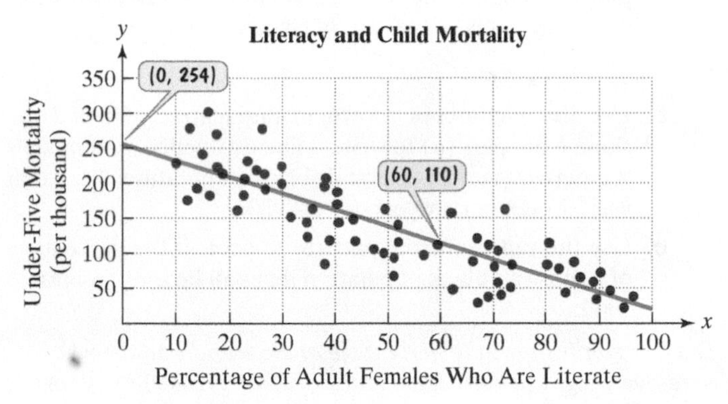

Source: United Nations Development Programme

a. According to the graph, what is the y-intercept of the line? Describe what this represents in this situation.

b. Use the coordinates of the two points shown to compute the slope of the line. Describe what this means in terms of the rate of change.

c. Use the y-intercept from part (a) and the slope from part (b) to write a linear function that models child mortality, $f(x)$, per thousand, for children under five in a country where x% of adult women are literate.

d. Use the function from part (c) to predict the mortality rate of children under five in a country where 50% of adult females are literate.

84. A wave of immigration from the Caribbean, Africa, and Latin America is reshaping America's black population. In 2013, approximately one in 11 blacks in America was foreign-born. The figure is projected to rise to one in six by 2060. The bar graph on the left shows the percentage of the U.S. black population that was foreign-born in four selected years from 1980 through 2013. The data are displayed as a set of four points in the scatter plot on the right.

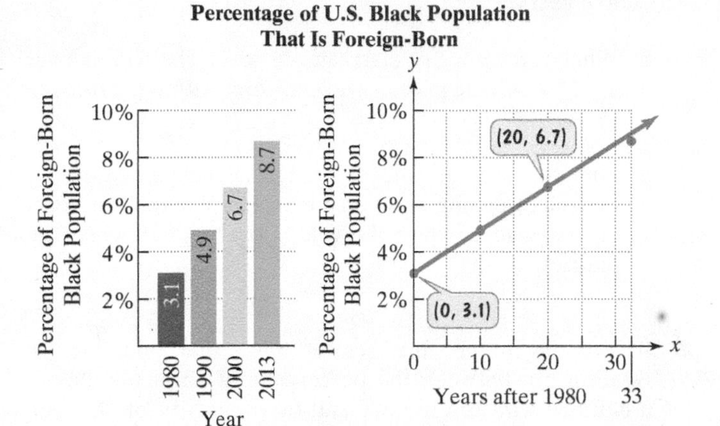

Source: Pew Research Center

a. Use the coordinates of the two points shown in the scatter plot to compute the slope of the line. Describe what this means in terms of the rate of change.

b. Use the y-intercept shown in the scatter plot and the slope from part (a) to write a linear function that models the percentage of America's black population, $P(x)$, that was foreign-born x years after 1980.

c. Use the function from part (b) to project the percentage of America's black population that will be foreign-born in 2060.

The bar graph shows the percentage of the world's adults who were overweight (a body mass index of 25 or greater) by gender, for two selected years.

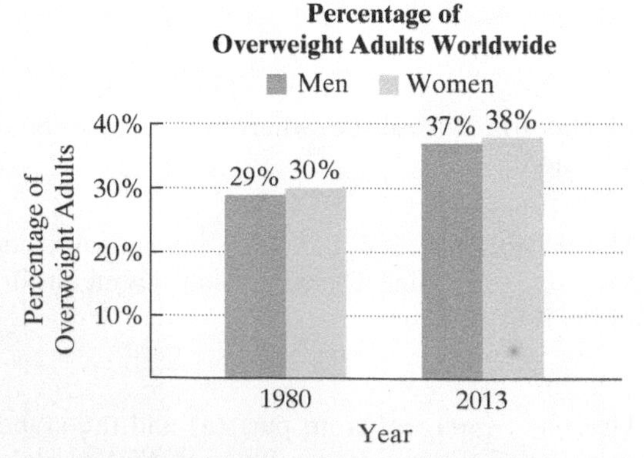

Source: TheLancet.com

In Exercises 85–86, find a linear function in slope-intercept form that models the given description. Each function should model the percentage of overweight men or women worldwide, $P(x)$, x years after 1980.

85. In 1980, 29% of adult men were overweight. This has increased at an average rate of approximately 0.24% per year since then.

86. In 1980, 30% of adult women were overweight. This has increased at an average rate of approximately 0.24% per year since then.

Explaining the Concepts

87. What is a scatter plot?

88. What is a regression line?

89. What is the standard form of the equation of a line?

90. What is an x-intercept of a graph?

91. What is a y-intercept of a graph?

92. If you are given the standard form of the equation of a line, explain how to find the x-intercept.

93. If you are given the standard form of the equation of a line, explain how to find the y-intercept.

94. What is the slope of a line?

95. Describe how to calculate the slope of a line passing through two points.

96. What does it mean if the slope of a line is zero?

97. What does it mean if the slope of a line is undefined?

98. Describe how to find the slope of a line whose equation is given.

99. Describe how to graph a line using the slope and y-intercept. Provide an original example with your description.

100. Describe the graph of $y = b$.

101. Describe the graph of $x = a$.

102. If the graph of a function is not a straight line, explain how to find the average rate of change between two points.

103. Take another look at the scatter plot in Exercise 83. Although there is a relationship between literacy and child mortality, we cannot conclude that increased literacy causes child mortality to decrease. Offer two or more possible explanations for the data in the scatter plot.

Technology Exercises

104. Use a graphing utility to verify any three of your hand-drawn graphs in Exercises 1–14. Solve the equation for y before entering it.

In Exercises 105–108, use a graphing utility to graph each linear function. Then use the TRACE *feature to trace along the line and find the coordinates of two points. Use these points to compute the line's slope. Check your result by using the coefficient of x in the line's equation.*

105. $y = 2x + 4$

106. $y = -3x + 6$

107. $f(x) = -\frac{1}{2}x - 5$

108. $f(x) = \frac{3}{4}x - 2$

Critical Thinking Exercises

Make Sense? *In Exercises 109–112, determine whether each statement makes sense or does not make sense, and explain your reasoning.*

109. The graph of my linear function at first rose from left to right, reached a maximum point, and then fell from left to right.

110. A linear function that models tuition and fees at public four-year colleges from 2000 through 2010 has negative slope.

111. The function $S(x) = 49{,}100x + 1700$ models the average salary for a college professor, $S(x)$, x years after 2000.

112. The federal minimum wage was \$5.15 per hour from 1997 through 2006, so $f(x) = 5.15$ models the minimum wage, $f(x)$, in dollars, for the domain $\{1997, 1998, 1999, \ldots, 2006\}$.

In Exercises 113–116, determine whether each statement is true or false. If the statement is false, make the necessary change(s) to produce a true statement.

113. A linear function with nonnegative slope has a graph that rises from left to right.

114. Every line in the rectangular coordinate system has an equation that can be expressed in slope-intercept form.

115. The graph of the linear function $5x + 6y = 30$ is a line passing through the point $(6, 0)$ with slope $-\frac{5}{6}$.

116. The graph of $x = 7$ in the rectangular coordinate system is the single point $(7, 0)$.

In Exercises 117–118, find the coefficients that must be placed in each shaded area so that the function's graph will be a line satisfying the specified conditions.

117. $\blacksquare x + \blacksquare y = 12$; x-intercept $= -2$; y-intercept $= 4$

118. $\blacksquare x + \blacksquare y = 12$; y-intercept $= -6$; slope $= \frac{1}{2}$

119. For the linear function

$$f(x) = mx + b,$$

a. Find $f(x_1 + x_2)$.

b. Find $f(x_1) + f(x_2)$.

c. Is $f(x_1 + x_2) = f(x_1) + f(x_2)$?

Review Exercises

120. Simplify: $\left(\frac{4x^2}{y^{-3}}\right)^2$. (Section 1.6, Example 9)

121. Multiply and write the answer in scientific notation:

$$(8 \times 10^{-7})(4 \times 10^3).$$

(Section 1.7, Example 4)

122. Simplify: $5 - [3(x - 4) - 6x]$.
(Section 1.2, Example 14)

Preview Exercises

Exercises 123–125 will help you prepare for the material covered in the next section.

123. Write the equation $y - 5 = 7(x + 4)$ in slope-intercept form.

124. Write the equation $y + 3 = -\frac{7}{3}(x - 1)$ in slope-intercept form.

125. The equation of a line is $x + 4y - 8 = 0$.

a. Write the equation in slope-intercept form and determine the slope.

b. The product of the line's slope in part (a) and the slope of a second line is -1. What is the slope of the second line?

SECTION

2.5 The Point-Slope Form of the Equation of a Line

What am I supposed to learn?

After studying this section, you should be able to:

1. Use the point-slope form to write equations of a line.
2. Model data with linear functions and make predictions.
3. Find slopes and equations of parallel and perpendicular lines.

If present trends continue, is it possible that our descendants could live to be 200 years of age? To answer this question, we need to develop a function that models life expectancy by birth year. In this section, you will learn to use another form of a line's equation to obtain functions that model data.

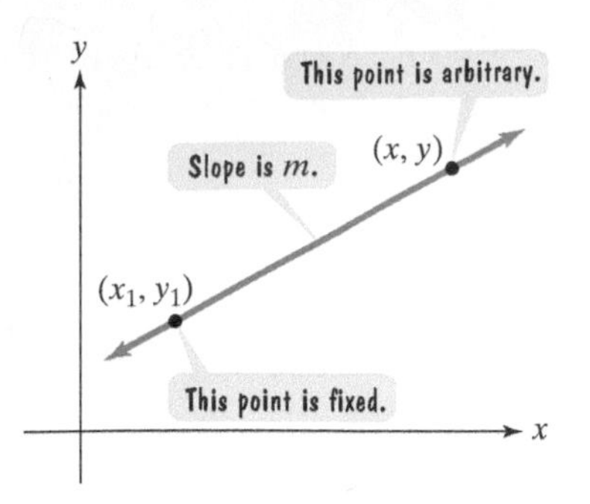

Figure 2.35 A line passing through (x_1, y_1) with slope m

Point-Slope Form

We can use the slope of a line to obtain another useful form of the line's equation. Consider a nonvertical line that has slope m and contains the point (x_1, y_1). Now, let (x, y) represent any other point on the line, shown in **Figure 2.35**. Keep in mind that the point (x, y) is arbitrary and is not in one fixed position. By contrast, the point (x_1, y_1) is fixed.

Regardless of where the point (x, y) is located, the steepness of the line in **Figure 2.35** remains the same. Thus, the ratio for slope stays a constant m. This means that for all points (x, y) along the line

$$m = \frac{\text{Change in } y}{\text{Change in } x} = \frac{y - y_1}{x - x_1}.$$

We can clear the fraction by multiplying both sides by $x - x_1$, the least common denominator, where $x - x_1 \neq 0$.

$$m = \frac{y - y_1}{x - x_1}$$ This is the slope of the line in Figure 2.35.

$$m(x - x_1) = \frac{y - y_1}{x - x_1} \cdot (x - x_1)$$ Multiply both sides by $x - x_1$.

$$m(x - x_1) = y - y_1$$ Simplify: $\frac{y - y_1}{\cancel{x - x_1}} \cdot \cancel{(x - x_1)} = y - y_1$.

Great Question!

When using $y - y_1 = m(x - x_1)$, for which variables do I substitute numbers?

When writing the point-slope form of a line's equation, you will never substitute numbers for x and y. You will substitute values for x_1, y_1, and m.

Now, if we reverse the two sides, we obtain the **point-slope form** of the equation of a line.

Point-Slope Form of the Equation of a Line

The **point-slope form of the equation** of a nonvertical line with slope m that passes through the point (x_1, y_1) is

$$y - y_1 = m(x - x_1).$$

For example, the point-slope form of the equation of the line passing through (1, 5) with slope 2 ($m = 2$) is

$$y - 5 = 2(x - 1).$$

1 Use the point-slope form to write equations of a line.

Using the Point-Slope Form to Write a Line's Equation

If we know the slope of a line and a point not containing the y-intercept through which the line passes, the point-slope form is the equation that we should use. Once we have obtained this equation, it is customary to solve for y and write the equation in slope-intercept form. Examples 1 and 2 illustrate these ideas.

EXAMPLE 1 Writing the Point-Slope Form and the Slope-Intercept Form

Write the point-slope form and the slope-intercept form of the equation of the line with slope 7 that passes through the point $(-4, 5)$.

Solution We begin with the point-slope form of the equation of a line with $m = 7$, $x_1 = -4$, and $y_1 = 5$.

$$y - y_1 = m(x - x_1)$$ This is the point-slope form of the equation.

$$y - 5 = 7[x - (-4)]$$ Substitute the given values.

$$y - 5 = 7(x + 4)$$ We now have the point-slope form of the equation of the given line.

Now we solve $y - 5 = 7(x + 4)$, the point-slope form, for y and write an equivalent equation in slope-intercept form ($y = mx + b$).

We need to isolate y.

$$y - 5 = 7(x + 4)$$ This is the point-slope form of the equation.

$$y - 5 = 7x + 28$$ Use the distributive property.

$$y = 7x + 33$$ Add 5 to both sides.

The slope-intercept form of the line's equation is $y = 7x + 33$. Using function notation, the equation is $f(x) = 7x + 33$. ■

✓ **CHECK POINT 1** Write the point-slope form and the slope-intercept form of the equation of the line with slope -2 that passes through the point $(4, -3)$.

EXAMPLE 2 Writing the Point-Slope Form and the Slope-Intercept Form

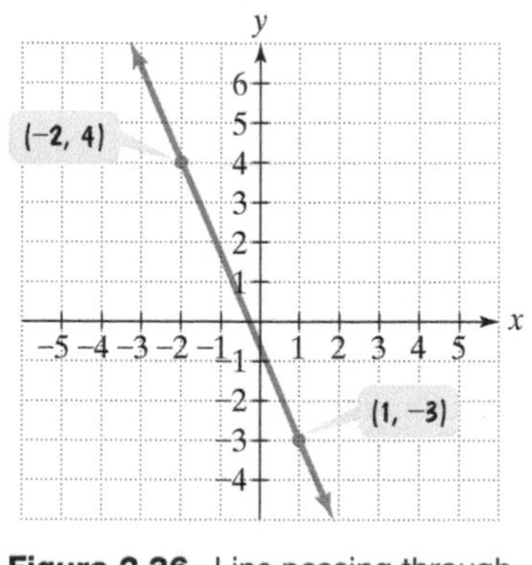

Figure 2.36 Line passing through $(1, -3)$ and $(-2, 4)$

A line passes through the points $(1, -3)$ and $(-2, 4)$. (See **Figure 2.36**.) Find an equation of the line

a. in point-slope form. **b.** in slope-intercept form.

Solution

a. To use the point-slope form, we need to find the slope. The slope is the change in the y-coordinates divided by the corresponding change in the x-coordinates.

$$m = \frac{4 - (-3)}{-2 - 1} = \frac{7}{-3} = -\frac{7}{3}$$ This is the definition of slope using $(1, -3)$ and $(-2, 4)$.

We can take either point on the line to be (x_1, y_1). Let's use $(x_1, y_1) = (1, -3)$. Now, we are ready to write the point-slope form of the equation.

$$y - y_1 = m(x - x_1)$$ This is the point-slope form of the equation.

$$y - (-3) = -\frac{7}{3}(x - 1)$$ Substitute: $(x_1, y_1) = (1, -3)$ and $m = -\frac{7}{3}$.

$$y + 3 = -\frac{7}{3}(x - 1)$$ Simplify.

This equation is the point-slope form of the equation of the line shown in **Figure 2.36**.

b. Now, we solve this equation for y and write an equivalent equation in slope-intercept form ($y = mx + b$).

We need to isolate y.

$$y + 3 = -\frac{7}{3}(x - 1)$$ This is the point-slope form of the equation.

$$y + 3 = -\frac{7}{3}x + \frac{7}{3}$$ Use the distributive property.

$$y = -\frac{7}{3}x - \frac{2}{3}$$ Subtract 3 from both sides: $\frac{7}{3} - 3 = \frac{7}{3} - \frac{9}{3} = -\frac{2}{3}$.

This equation is the slope-intercept form of the equation of the line shown in **Figure 2.36**. Using function notation, the equation is $f(x) = -\frac{7}{3}x - \frac{2}{3}$. ■

Discover for Yourself

If you are given two points on a line, you can use either point for (x_1, y_1) when you write the point-slope form of its equation. Rework Example 2 using $(-2, 4)$ for (x_1, y_1). Once you solve for y, you should obtain the same slope-intercept form of the equation as the one shown in the last line of the solution to Example 2.

✓ CHECK POINT 2 A line passes through the points $(6, -3)$ and $(2, 5)$. Find an equation of the line

a. in point-slope form.

b. in slope-intercept form.

Here is a summary of the various forms for equations of lines:

Equations of Lines

1. Standard form: $Ax + By = C$
2. Slope-intercept form: $y = mx + b$ or $f(x) = mx + b$
3. Horizontal line: $y = b$
4. Vertical line: $x = a$
5. Point-slope form: $y - y_1 = m(x - x_1)$

In Examples 1 and 2, we eventually wrote a line's equation in slope-intercept form, or in function notation. But where do we start our work?

Starting with $y = mx + b$	Starting with $y - y_1 = m(x - x_1)$
Begin with the slope-intercept form if you know • The slope of the line and the y-intercept or • Two points on the line, one of which contains the y-intercept.	Begin with the point-slope form if you know • The slope of the line and a point on the line not containing the y-intercept or • Two points on the line, neither of which contains the y-intercept.

2 Model data with linear functions and make predictions.

Applications

We have seen that linear functions are useful for modeling data that fall on or near a line. For example, the bar graph in **Figure 2.37(a)** gives the life expectancy for American men and women born in the indicated year. The data for the men are displayed as a set of six points in the scatter plot in **Figure 2.37(b)**.

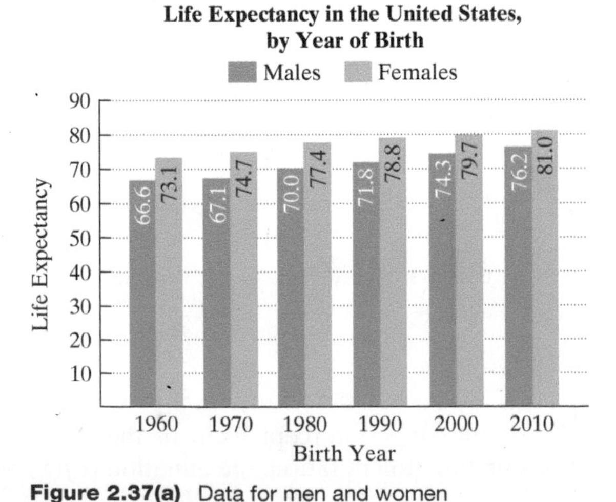

Figure 2.37(a) Data for men and women

Source: National Center for Health Statistics

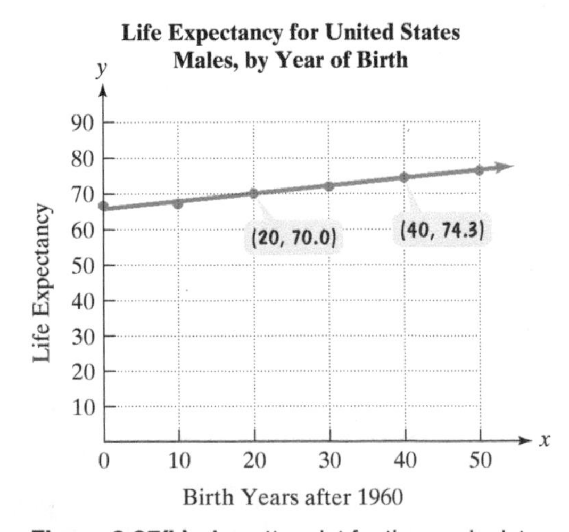

Figure 2.37(b) A scatter plot for the men's data

Also shown on the scatter plot in **Figure 2.37(b)** is a line that passes through or near the six points. By writing the equation of this line, we can obtain a model for life expectancy and make predictions about how long American men will live in the future.

EXAMPLE 3 Modeling Life Expectancy

Write the slope-intercept form of the equation of the line shown in **Figure 2.37(b)**. Use the equation to predict the life expectancy of an American man born in 2020.

Solution The line in **Figure 2.37(b)** passes through (20, 70.0) and (40, 74.3). We start by finding its slope.

$$m = \frac{\text{Change in } y}{\text{Change in } x} = \frac{74.3 - 70.0}{40 - 20} = \frac{4.3}{20} = 0.215$$

The slope indicates that for each subsequent birth year, a man's life expectancy is increasing by 0.215 year.

Now we write the line's equation in slope-intercept form.

$$y - y_1 = m(x - x_1)$$ Begin with the point-slope form.

$$y - 70.0 = 0.215(x - 20)$$ Either ordered pair can be (x_1, y_1). Let $(x_1, y_1) = (20, 70.0)$. From above, $m = 0.215$.

$$y - 70.0 = 0.215x - 4.3$$ Apply the distributive property: $0.215(20) = 4.3$.

$$y = 0.215x + 65.7$$ Add 70 to both sides and solve for y.

A linear function that models life expectancy, $f(x)$, for American men born x years after 1960 is

$$f(x) = 0.215x + 65.7.$$

Now let's use this function to predict the life expectancy of an American man born in 2020. Because 2020 is 60 years after 1960, substitute 60 for x and evaluate the function at 60.

$$f(60) = 0.215(60) + 65.7 = 78.6$$

Our model predicts that American men born in 2020 will have a life expectancy of 78.6 years. ■

Using Technology

You can use a graphing utility to obtain a model for a scatter plot in which the data points fall on or near a straight line. After entering the data for men in **Figure 2.37(a)**, a graphing utility displays a scatter plot of the data and the regression line, that is, the line that best fits the data.

[0, 50, 10] by [0, 90, 10]

Also displayed is the regression line's equation.

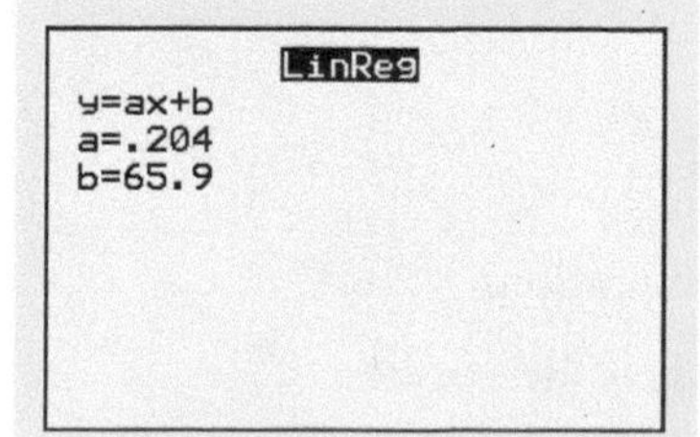

CHECK POINT 3 The data for the life expectancy for American women are displayed as a set of six points in the scatter plot in **Figure 2.38**. Also shown is a line that passes through or near the six points. Use the data points labeled by the voice balloons to write the slope-intercept form of the equation of this line. Round the slope to two decimal places. Then use the linear function to predict the life expectancy of an American woman born in 2020.

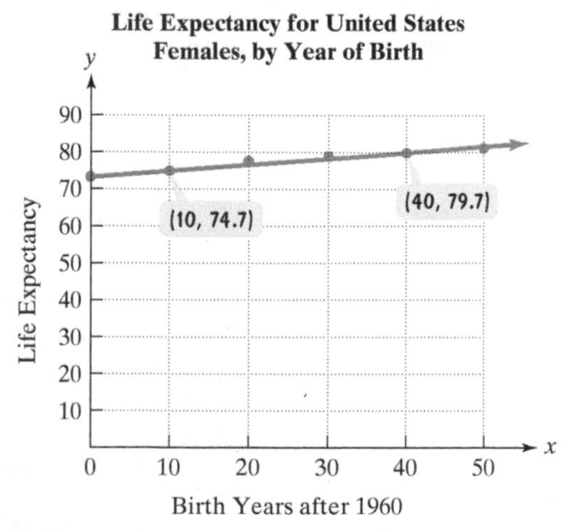

Figure 2.38

Blitzer Bonus

Predicting Your Own Life Expectancy

The models in Example 3 and Check Point 3 do not take into account your current health, lifestyle, and family history, all of which could increase or decrease your life expectancy. Thomas Perls at Boston University Medical School, who studies centenarians, developed a much more detailed model for life expectancy at livingto100.com. The model takes into account everything from your stress level to your sleep habits and gives you the exact age it predicts you will live to.

3 Find slopes and equations of parallel and perpendicular lines.

Parallel and Perpendicular Lines

Two nonintersecting lines that lie in the same plane are **parallel.** If two lines do not intersect, the ratio of the vertical change to the horizontal change is the same for each line. Because two parallel lines have the same "steepness," they must have the same slope.

Slope and Parallel Lines

1. If two nonvertical lines are parallel, then they have the same slope.
2. If two distinct nonvertical lines have the same slope, then they are parallel.
3. Two distinct vertical lines, both with undefined slopes, are parallel.

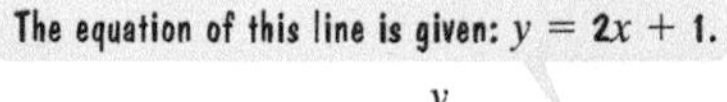

The equation of this line is given: y = 2x + 1.

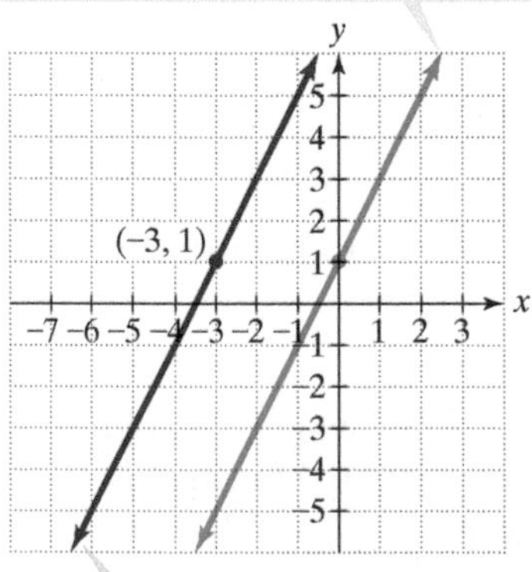

Figure 2.39

EXAMPLE 4 Writing Equations of a Line Parallel to a Given Line

Write an equation of the line passing through $(-3, 1)$ and parallel to the line whose equation is $y = 2x + 1$. Express the equation in point-slope form and slope-intercept form.

Solution The situation is illustrated in **Figure 2.39**. We are looking for the equation of the red line shown on the left. How do we obtain this equation? Notice that the line passes through the point $(-3, 1)$. Using the point-slope form of the line's equation, we have $x_1 = -3$ and $y_1 = 1$.

$$y - y_1 = m(x - x_1)$$

$y_1 = 1$ $\quad x_1 = -3$

Now the only thing missing from the equation of the red line is m, the slope. Do we know anything about the slope of either line in **Figure 2.39**? The answer is yes; we know the slope of the blue line on the right, whose equation is given.

$$y = 2x + 1$$

The slope of the blue line on the right in **Figure 2.39** is 2.

Parallel lines have the same slope. Because the slope of the blue line is 2, the slope of the red line, the line whose equation we must write, is also 2: $m = 2$. We now have values for x_1, y_1, and m for the red line.

$$y - y_1 = m(x - x_1)$$

$y_1 = 1$ $\quad m = 2$ $\quad x_1 = -3$

The point-slope form of the red line's equation is

$$y - 1 = 2[x - (-3)] \text{ or}$$
$$y - 1 = 2(x + 3).$$

Solving for y, we obtain the slope-intercept form of the equation.

$$y - 1 = 2x + 6 \quad \text{Apply the distributive property.}$$
$$y = 2x + 7 \quad \text{Add 1 to both sides. This is the slope-intercept form, } y = mx + b, \text{ of the equation. Using function notation, the equation is } f(x) = 2x + 7. \ \blacksquare$$

✓ CHECK POINT 4 Write an equation of the line passing through $(-2, 5)$ and parallel to the line whose equation is $y = 3x + 1$. Express the equation in point-slope form and slope-intercept form.

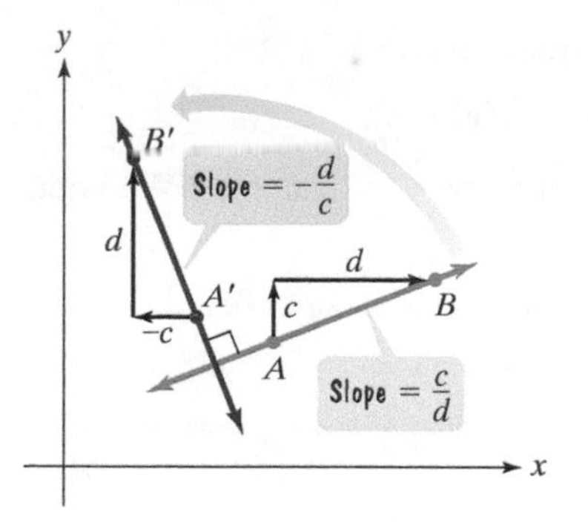

Figure 2.40 Slopes of perpendicular lines

Two lines that intersect at a right angle (90°) are said to be **perpendicular**, shown in **Figure 2.40**. The relationship between the slopes of perpendicular lines is not as obvious as the relationship between parallel lines. **Figure 2.40** shows line AB, with slope $\frac{c}{d}$. Rotate line AB through 90° counterclockwise to obtain line $A'B'$, perpendicular to line AB. The figure indicates that the rise and the run of the new line are reversed from the original line, but the run is now negative. This means that the slope of the new line is $-\frac{d}{c}$. Notice that the product of the slopes of the two perpendicular lines is -1:

$$\left(\frac{c}{d}\right)\left(-\frac{d}{c}\right) = -1.$$

This relationship holds for all nonvertical perpendicular lines and is summarized in the following box:

Slope and Perpendicular Lines

1. If two nonvertical lines are perpendicular, then the product of their slopes is -1.
2. If the product of the slopes of two lines is -1, then the lines are perpendicular.
3. A horizontal line having zero slope is perpendicular to a vertical line having undefined slope.

An equivalent way of stating this relationship between two nonvertical lines is to say that **one line is perpendicular to another line if its slope is the *negative reciprocal* of the slope of the other line.** For example, if a line has slope 5, any line having slope $-\frac{1}{5}$ is perpendicular to it. Similarly, if a line has slope $-\frac{3}{4}$, any line having slope $\frac{4}{3}$ is perpendicular to it.

EXAMPLE 5 Writing Equations of a Line Perpendicular to a Given Line

a. Find the slope of any line that is perpendicular to the line whose equation is $x + 4y = 8$.

b. Write the equation of the line passing through $(3, -5)$ and perpendicular to the line whose equation is $x + 4y = 8$. Express the equation in point-slope form and slope-intercept form.

Solution

a. We begin by writing the equation of the given line, $x + 4y = 8$, in slope-intercept form. Solve for y.

$$x + 4y = 8$$ This is the given equation.

$$4y = -x + 8$$ To isolate the y-term, subtract x from both sides.

$$y = -\frac{1}{4}x + 2$$ Divide both sides by 4.

Slope is $-\frac{1}{4}$.

The given line has slope $-\frac{1}{4}$. Any line perpendicular to this line has a slope that is the negative reciprocal of $-\frac{1}{4}$. Thus, the slope of any perpendicular line is 4.

b. Let's begin by writing the point-slope form of the perpendicular line's equation. Because the line passes through the point $(3, -5)$, we have $x_1 = 3$ and $y_1 = -5$. In part (a), we determined that the slope of any line perpendicular to $x + 4y = 8$ is 4, so the slope of this particular perpendicular line must be 4: $m = 4$.

$$y - y_1 = m(x - x_1)$$

$y_1 = -5$ $m = 4$ $x_1 = 3$

The point-slope form of the perpendicular line's equation is

$$y - (-5) = 4(x - 3) \quad \text{or} \quad y + 5 = 4(x - 3).$$

How can we express this equation in slope-intercept form, $y = mx + b$? We need to solve for y.

$y + 5 = 4(x - 3)$ — This is the point-slope form of the line's equation.

$y + 5 = 4x - 12$ — Apply the distributive property.

$y = 4x - 17$ — Subtract 5 from both sides of the equation and solve for y.

The slope-intercept form of the perpendicular line's equation is

$$y = 4x - 17 \quad \text{or} \quad f(x) = 4x - 17. \ \blacksquare$$

✓ CHECK POINT 5

a. Find the slope of any line that is perpendicular to the line whose equation is $x + 3y = 12$.

b. Write the equation of the line passing through $(-2, -6)$ and perpendicular to the line whose equation is $x + 3y = 12$. Express the equation in point-slope form and slope-intercept form.

Achieving Success

Organizing and creating your own compact chapter summaries can reinforce what you know and help with the retention of this information. Imagine that your professor will permit two index cards of notes (3 by 5; front and back) on all exams. Organize and create such a two-card summary for the test on this chapter. Begin by determining what information you would find most helpful to include on the cards. Take as long as you need to create the summary. Based on how effective you find this strategy, you may decide to use the technique to help prepare for future exams.

CONCEPT AND VOCABULARY CHECK

Fill in each blank so that the resulting statement is true.

1. The point-slope form of the equation of a nonvertical line with slope m that passes through the point (x_1, y_1) is ______________.
2. Two parallel lines have ______________ slopes.
3. The product of the slopes of two nonvertical perpendicular lines is ______________.
4. The negative reciprocal of 5 is ______________.
5. The negative reciprocal of $-\frac{3}{5}$ is ______________.
6. The slope of the line whose equation is $y = -4x + 3$ is ______________. The slope of any line parallel to $y = -4x + 3$ is ______________.
7. The slope of the line whose equation is $y = \frac{1}{2}x - 5$ is ______________. The slope of any line perpendicular to $y = \frac{1}{2}x - 5$ is ______________.

2.5 EXERCISE SET

MyMathLab®

Practice Exercises

Write the point-slope form of the equation of the line satisfying each of the conditions in Exercises 1–28. Then use the point-slope form to write the slope-intercept form of the equation in function notation.

1. Slope $= 3$, passing through $(2, 5)$
2. Slope $= 4$, passing through $(3, 1)$
3. Slope $= 5$, passing through $(-2, 6)$
4. Slope $= 8$, passing through $(-4, 1)$
5. Slope $= -4$, passing through $(-3, -2)$
6. Slope $= -6$, passing through $(-2, -4)$
7. Slope $= -5$, passing through $(-2, 0)$
8. Slope $= -4$, passing through $(0, -3)$
9. Slope $= -1$, passing through $\left(-2, -\frac{1}{2}\right)$
10. Slope $= -1$, passing through $\left(-\frac{1}{4}, -4\right)$
11. Slope $= \frac{1}{4}$, passing through the origin
12. Slope $= \frac{1}{5}$, passing through the origin
13. Slope $= -\frac{2}{3}$, passing through $(6, -4)$
14. Slope $= -\frac{2}{5}$, passing through $(15, -4)$
15. Passing through $(6, 3)$ and $(5, 2)$
16. Passing through $(1, 3)$ and $(2, 4)$
17. Passing through $(-2, 0)$ and $(0, 4)$
18. Passing through $(2, 0)$ and $(0, -1)$
19. Passing through $(-6, 13)$ and $(-2, 5)$
20. Passing through $(-3, 2)$ and $(2, -8)$
21. Passing through $(1, 9)$ and $(4, -2)$
22. Passing through $(4, -8)$ and $(8, -3)$
23. Passing through $(-2, -5)$ and $(3, -5)$
24. Passing through $(-1, -4)$ and $(3, -4)$
25. Passing through $(7, 8)$ with x-intercept $= 3$
26. Passing through $(-4, 5)$ with y-intercept $= -3$
27. x-intercept $= 2$ and y-intercept $= -1$
28. x-intercept $= -2$ and y-intercept $= 4$

In Exercises 29–44, the equation of a line is given. Find the slope of a line that is **a.** *parallel to the line with the given equation; and* **b.** *perpendicular to the line with the given equation.*

29. $y = 5x$
30. $y = 3x$
31. $y = -7x$
32. $y = -9x$
33. $y = \dfrac{1}{2}x + 3$
34. $y = \dfrac{1}{4}x - 5$
35. $y = -\dfrac{2}{5}x - 1$
36. $y = -\dfrac{3}{7}x - 2$
37. $4x + y = 7$
38. $8x + y = 11$
39. $2x + 4y = 8$
40. $3x + 2y = 6$
41. $2x - 3y = 5$
42. $3x - 4y = -7$
43. $x = 6$
44. $y = 9$

In Exercises 45–48, write an equation for line L in point-slope form and slope-intercept form.

45.

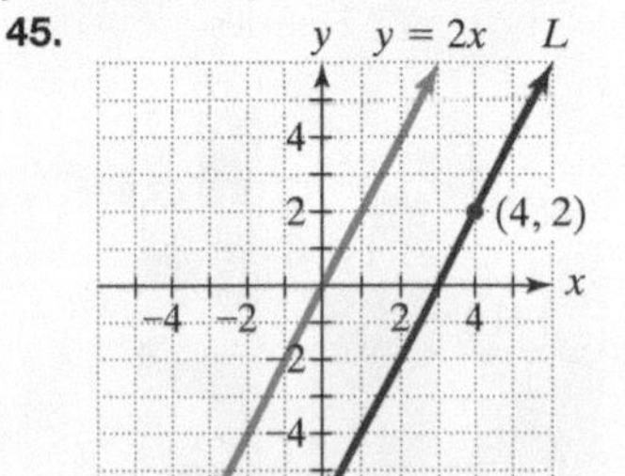

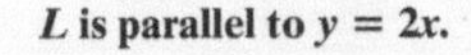
L is parallel to $y = 2x$.

46.

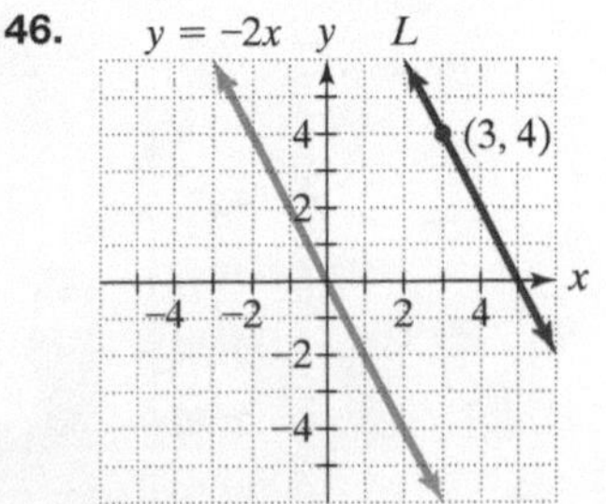

L is parallel to $y = -2x$.

47.

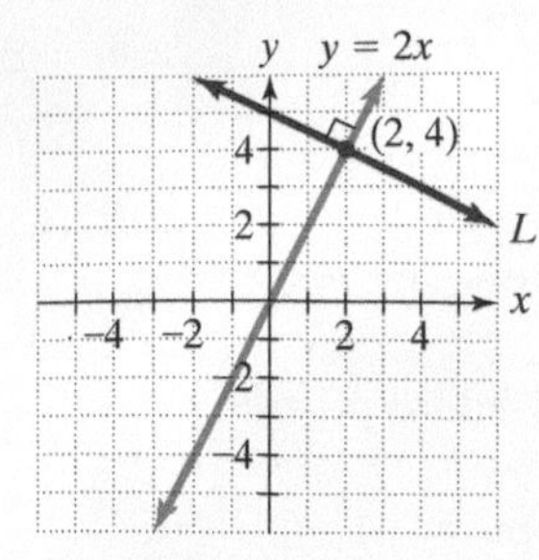

L is perpendicular to $y = 2x$.

48.

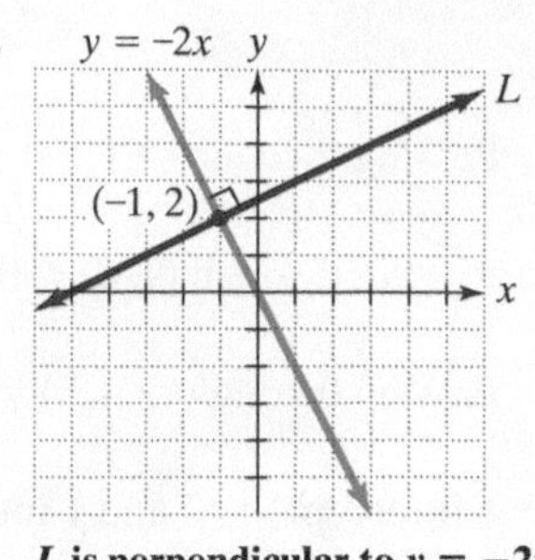

L is perpendicular to $y = -2x$.

In Exercises 49–56, use the given conditions to write an equation for each line in point-slope form and slope-intercept form.

49. Passing through $(-8, -10)$ and parallel to the line whose equation is $y = -4x + 3$

50. Passing through $(-2, -7)$ and parallel to the line whose equation is $y = -5x + 4$

51. Passing through $(2, -3)$ and perpendicular to the line whose equation is $y = \frac{1}{5}x + 6$

52. Passing through $(-4, 2)$ and perpendicular to the line whose equation is $y = \frac{1}{3}x + 7$

53. Passing through $(-2, 2)$ and parallel to the line whose equation is $2x - 3y = 7$

54. Passing through $(-1, 3)$ and parallel to the line whose equation is $3x - 2y = 5$

55. Passing through $(4, -7)$ and perpendicular to the line whose equation is $x - 2y = 3$

56. Passing through $(5, -9)$ and perpendicular to the line whose equation is $x + 7y = 12$

Practice PLUS

In Exercises 57–64, write the slope-intercept form of the equation of a function f whose graph satisfies the given conditions.

57. The graph of f passes through $(-1, 5)$ and is perpendicular to the line whose equation is $x = 6$.

58. The graph of f passes through $(-2, 6)$ and is perpendicular to the line whose equation is $x = -4$.

59. The graph of f passes through $(-6, 4)$ and is perpendicular to the line that has an x-intercept of 2 and a y-intercept of -4.

60. The graph of f passes through $(-5, 6)$ and is perpendicular to the line that has an x-intercept of 3 and a y-intercept of -9.

61. The graph of f is perpendicular to the line whose equation is $3x - 2y = 4$ and has the same y-intercept as this line.

62. The graph of f is perpendicular to the line whose equation is $4x - y = 6$ and has the same y-intercept as this line.

63. The graph of f is the graph of $g(x) = 4x - 3$ shifted down 2 units.

64. The graph of f is the graph of $g(x) = 2x - 5$ shifted up 3 units.

65. What is the slope of a line that is parallel to the line whose equation is $Ax + By = C, B \neq 0$?

66. What is the slope of a line that is perpendicular to the line whose equation is $Ax + By = C, A \neq 0$ and $B \neq 0$?

Application Exercises

Americans are getting married later in life, or not getting married at all. In 2010, more than half of Americans ages 25 through 29 were unmarried. The bar graph shows the percentage of never-married men and women in this age group for four selected years. The data are displayed as two sets of four points each, one scatter plot for the percentage of never-married American men and one for the percentage of never-married American women. Also shown for each scatter plot is a line that passes through or near the four points. Use these lines to solve Exercises 67–68 on the next page.

Percentage of United States Population Never Married, Ages 25–29

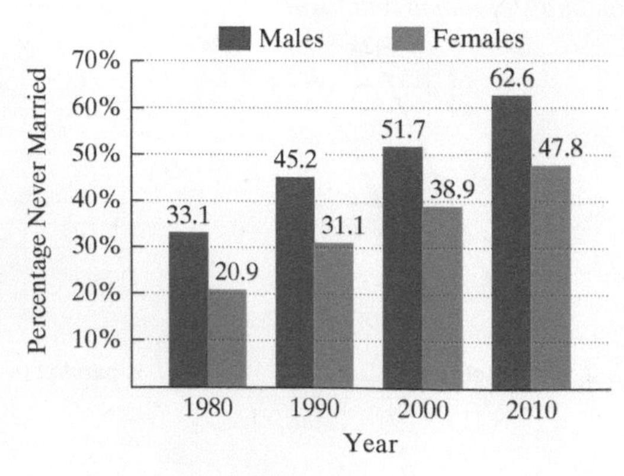

Source: U.S. Census Bureau

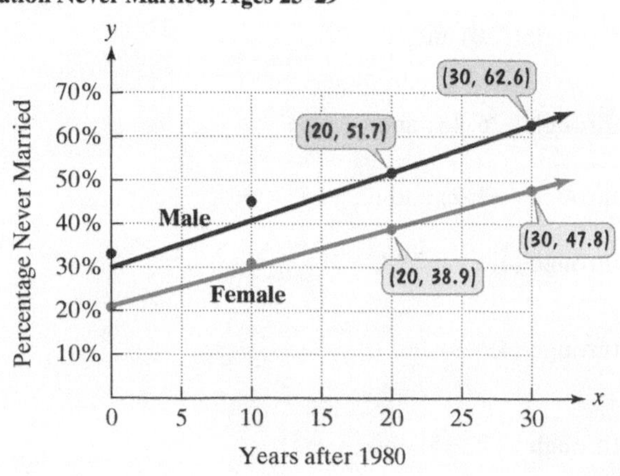

67. In this exercise, you will use the blue line for the women shown on the scatter plot on the previous page to develop a model for the percentage of never-married American females ages 25–29.

a. Use the two points whose coordinates are shown by the voice balloons to find the point-slope form of the equation of the line that models the percentage of never-married American females ages 25–29, y, x years after 1980.

b. Write the equation from part (a) in slope-intercept form. Use function notation.

c. Use the linear function to predict the percentage of never-married American females, ages 25–29, in 2020.

68. In this exercise, you will use the red line for the men shown on the scatter plot on the previous page to develop a model for the percentage of never-married American males ages 25–29.

a. Use the two points whose coordinates are shown by the voice balloons to find the point-slope form of the equation of the line that models the percentage of never-married American males ages 25–29, y, x years after 1980.

b. Write the equation from part (a) in slope-intercept form. Use function notation.

c. Use the linear function to predict the percentage of never-married American males, ages 25–29, in 2015.

Phone Fight! *Some of the smartphone giants, such as Motorola, Panasonic, Microsoft, and Apple, are suing each other, charging competitors with stealing their patents. Do more phones mean more lawyers? Think about this as you create linear models for the number of smartphones sold (Exercise 69) and the number of infringement filings by smartphone companies (Exercise 70).*

69. The bar graph shows the number of smartphones sold in the United States from 2004 through 2010.

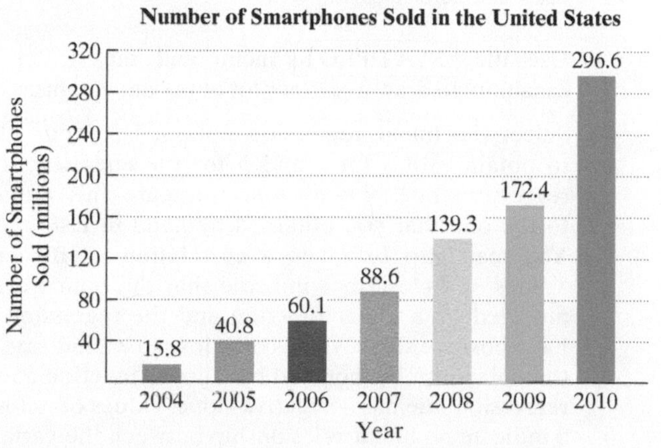

Source: *Newsweek*, April 11, 2011

a. Let x represent the number of years after 2004 and let y represent the number of smartphones sold, in millions, in the United States. Create a scatter plot that displays the data as a set of seven points in a rectangular coordinate system.

b. Draw a line through the two points that show the number of smartphones sold in 2005 and 2010. Use the coordinates of these points to write the line's equation in point-slope form and slope-intercept form.

c. Use the slope-intercept form of the equation from part (b) to project the number of smartphones sold in the United States in 2015.

70. The bar graph shows the number of U.S. lawsuits by smartphone companies for patent infringement from 2004 through 2010.

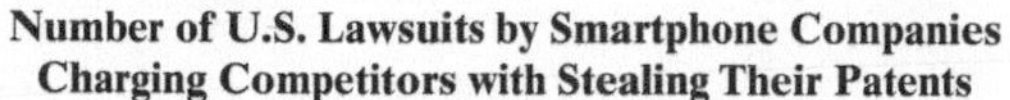

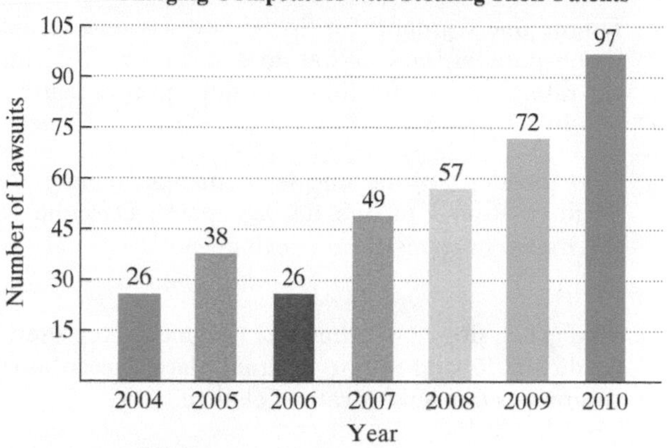

Source: *Newsweek*, April 11, 2011

a. Let x represent the number of years after 2004 and let y represent the number of lawsuits. Create a scatter plot that displays the data as a set of seven points in a rectangular coordinate system.

b. Draw a line through the two points that show the number of lawsuits in 2007 and 2010. Use the coordinates of these points to write the line's equation in point-slope form and slope-intercept form.

c. Use the slope-intercept form of the equation from part (b) to project the number of lawsuits by smartphone companies for patent infringement in 2016.

In 2007, the U.S. government faced the prospect of paying out more and more in Social Security, Medicare, and Medicaid benefits. The line graphs show the costs of these entitlement programs, in billions of dollars, from 2007 through 2016 (projected). Use this information to solve Exercises 71–72 on the next page.

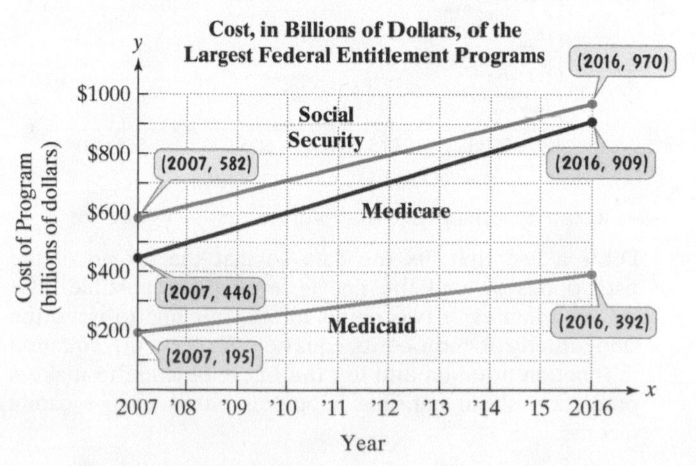

Source: U.S. Congressional Budget Office

(In Exercises 71–72, refer to the line graphs at the bottom of the previous page.)

71. **a.** Find the slope of the line segment representing Social Security. Round to one decimal place. Describe what this means in terms of rate of change.

b. Find the slope of the line segment representing Medicare. Round to one decimal place. Describe what this means in terms of rate of change.

c. Do the line segments for Social Security and Medicare lie on parallel lines? What does this mean in terms of the rate of change for these entitlement programs?

72. **a.** Find the slope of the line segment representing Social Security. Round to one decimal place. Describe what this means in terms of rate of change.

b. Find the slope of the line segment representing Medicaid. Round to one decimal place. Describe what this means in terms of rate of change.

c. Do the line segments for Social Security and Medicaid lie on parallel lines? What does this mean in terms of the rate of change for these entitlement programs?

73. Just as money doesn't buy happiness for individuals, the two don't necessarily go together for countries either. However, the scatter plot does show a relationship between a country's annual per capita income and the percentage of people in that country who call themselves "happy."

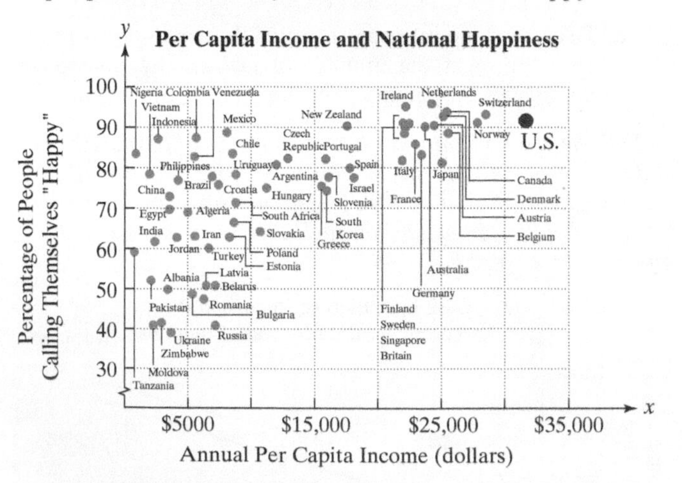

Source: Richard Layard, *Happiness: Lessons from a New Science*, Penguin, 2005

Draw a line that fits the data so that the spread of the data points around the line is as small as possible. Use the coordinates of two points along your line to write the slope-intercept form of its equation. Express the equation in function notation and use the linear function to make a prediction about national happiness based on per capita income.

Explaining the Concepts

74. Describe how to write the equation of a line if its slope and a point along the line are known.

75. Describe how to write the equation of a line if two points along the line are known.

76. If two lines are parallel, describe the relationship between their slopes.

77. If two lines are perpendicular, describe the relationship between their slopes.

78. If you know a point on a line and you know the equation of a line parallel to this line, explain how to write the line's equation.

79. In Example 3 on page 159, we developed a model that predicted American men born in 2020 will have a life expectancy of 78.6 years. Describe something that might occur that would make this prediction inaccurate.

Technology Exercises

80. The lines whose equations are $y = \frac{1}{3}x + 1$ and $y = -3x - 2$ are perpendicular because the product of their slopes, $\frac{1}{3}$ and -3, respectively, is -1.

a. Use a graphing utility to graph the equations in a $[-10, 10, 1]$ by $[-10, 10, 1]$ viewing rectangle. Do the lines appear to be perpendicular?

b. Now use the zoom square feature of your utility. Describe what happens to the graphs. Explain why this is so.

81. **a.** Use the statistical menu of your graphing utility to enter the seven data points shown in the scatter plot that you drew in Exercise 69(a).

b. Use the STAT PLOT menu and the scatter plot capability to draw a scatter plot of the data points.

c. Select the linear regression option. Use your utility to obtain values for a and b for the equation of the regression line, $y = ax + b$. Compare this equation to the one that you obtained by hand in Exercise 69. You may also be given a **correlation coefficient**, r. Values of r close to 1 indicate that the points can be modeled by a linear function and the regression line has a positive slope. Values of r close to -1 indicate that the points can be modeled by a linear function and the regression line has a negative slope. Values of r close to 0 indicate no linear relationship between the variables. In this case, a linear model does not accurately describe the data.

d. Use the appropriate sequence (consult your manual) to graph the regression equation on top of the points in the scatter plot.

82. Repeat Exercise 81 using the seven data points shown in the scatter plot that you drew in Exercise 70(a).

Critical Thinking Exercises

Make Sense? In Exercises 83–86, determine whether each statement makes sense or does not make sense, and explain your reasoning.

83. I can use any two points in a scatter plot to write the point-slope form of the equation of the line through those points. However, the other data points in the scatter plot might not fall on, or even near, this line.

84. I have linear functions that model changes for men and women over the same time period. The functions have the same slope, so their graphs are parallel lines, indicating that the rate of change for men is the same as the rate of change for women.

85. Some of the steel girders in this photo of the Eiffel Tower appear to be perpendicular. I can verify my observation by determining that their slopes are negative reciprocals.

86. When writing equations of lines, it's always easiest to begin by writing the point-slope form of the equation.

In Exercises 87–90, determine whether each statement is true or false. If the statement is false, make the necessary change(s) to produce a true statement.

87. The standard form of the equation of a line passing through $(-3, -1)$ and perpendicular to the line whose equation is $y = -\frac{2}{5}x - 4$ is $5x - 2y = -13$.

88. If I change the subtraction signs to addition signs in $y - 12 = 8(x - 2)$, the y-intercept of the corresponding graph will change from -4 to 4.

89. $y - 5 = 2(x - 1)$ is an equation of a line passing through $(4, 11)$.

90. The function $\{(-1, 4), (3, 6), (5, 7), (11, 10)\}$ can be described using $y - 7 = \frac{1}{2}(x - 5)$ with a domain of $\{-1, 3, 5, 11\}$.

91. Determine the value of B so that the line whose equation is $By = 8x - 1$ has slope -2.

92. Determine the value of A so that the line whose equation is $Ax + y = 2$ is perpendicular to the line containing the points $(1, -3)$ and $(-2, 4)$.

93. Consider a line whose x-intercept is -3 and whose y-intercept is -6. Provide the missing coordinate for the following two points that lie on this line: $(-40, \quad)$ and $(\quad, -200)$.

94. Prove that the equation of a line passing through $(a, 0)$ and $(0, b)$ $(a \neq 0, b \neq 0)$ can be written in the form $\frac{x}{a} + \frac{y}{b} = 1$. Why is this called the *intercept form* of a line?

Review Exercises

95. If $f(x) = 3x^2 - 8x + 5$, find $f(-2)$. (Section 2.1, Example 3)

96. If $f(x) = x^2 - 3x + 4$ and $g(x) = 2x - 5$, find $(fg)(-1)$. (Section 2.3, Example 4)

97. The sum of the angles of a triangle is 180°. Find the three angles of a triangle if one angle is 20° greater than the smallest angle and the third angle is twice the smallest angle. (Section 1.5, Example 1)

Preview Exercises

Exercises 98–100 will help you prepare for the material covered in the first section of the next chapter.

98. **a.** Does $(-5, -6)$ satisfy $2x - y = -4$?

b. Does $(-5, -6)$ satisfy $3x - 5y = 15$?

99. Graph $y = -x - 1$ and $4x - 3y = 24$ in the same rectangular coordinate system. At what point do the graphs intersect?

100. Solve: $7x - 2(-2x + 4) = 3$.

Chapter 2 Summary

Definitions and Concepts	Examples

Section 2.1 Introduction to Functions

A relation is any set of ordered pairs. The set of first components of the ordered pairs is the domain and the set of second components is the range. A function is a relation in which each member of the domain corresponds to exactly one member of the range. No two ordered pairs of a function can have the same first component and different second components.

The domain of the relation $\{(1, 2), (3, 4), (3, 7)\}$ is $\{1, 3\}$. The range is $\{2, 4, 7\}$. The relation is not a function: 3, in the domain, corresponds to both 4 and 7 in the range.

If a function is defined by an equation, the notation $f(x)$, read "f of x" or "f at x," describes the value of the function at the number, or input, x.

If $f(x) = 7x - 5$, then

$$\begin{aligned} f(a+2) &= 7(a+2) - 5 \\ &= 7a + 14 - 5 \\ &= 7a + 9. \end{aligned}$$

Section 2.2 Graphs of Functions

The graph of a function is the graph of its ordered pairs.

The Vertical Line Test for Functions
If any vertical line intersects a graph in more than one point, the graph does not define y as a function of x.

At the left or right of a function's graph, you will often find closed dots, open dots, or arrows. A closed dot shows that the graph ends and the point belongs to the graph. An open dot shows that the graph ends and the point does not belong to the graph. An arrow indicates that the graph extends indefinitely.

The graph of a function can be used to determine the function's domain and its range. To find the domain, look for all the inputs on the x-axis that correspond to points on the graph. To find the range, look for all the outputs on the y-axis that correspond to points on the graph.

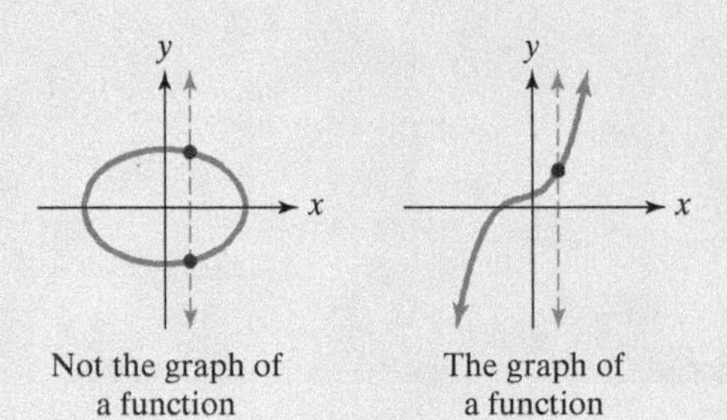

Not the graph of a function

The graph of a function

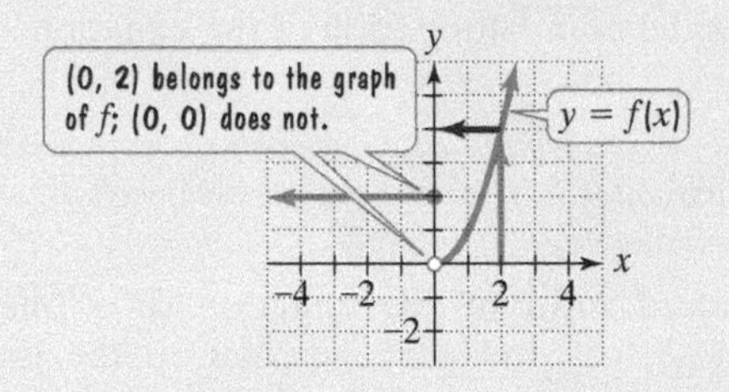

To find $f(2)$, locate 2 on the x-axis. The graph shows $f(2) = 4$.
Domain of $f = (-\infty, \infty)$
Range of $f = (0, \infty)$

Section 2.3 The Algebra of Functions

A Function's Domain
If a function f does not model data or verbal conditions, its domain is the largest set of real numbers for which the value of $f(x)$ is a real number. Exclude from a function's domain real numbers that cause division by zero and real numbers that result in a square root of a negative number.

$$f(x) = 7x + 13$$

Domain of $f = (-\infty, \infty)$

$$g(x) = \frac{7x}{12 - x}$$

Domain of $g = (-\infty, 12)$ or $(12, \infty)$

Definitions and Concepts | **Examples**

Section 2.3 The Algebra of Functions (continued)

The Algebra of Functions

Let f and g be two functions. The sum $f + g$, the difference $f - g$, the product fg, and the quotient $\frac{f}{g}$ are functions whose domains are the set of all real numbers common to the domains of f and g, defined as follows:

1. Sum: $(f + g)(x) = f(x) + g(x)$
2. Difference: $(f - g)(x) = f(x) - g(x)$
3. Product: $(fg)(x) = f(x) \cdot g(x)$
4. Quotient: $\left(\frac{f}{g}\right)(x) = \frac{f(x)}{g(x)}, g(x) \neq 0.$

Let $f(x) = x^2 + 2x$ and $g(x) = 4 - x$.

- $(f + g)(x) = (x^2 + 2x) + (4 - x) = x^2 + x + 4$

  $(f + g)(-2) = (-2)^2 + (-2) + 4 = 4 - 2 + 4 = 6$
- $(f - g)(x) = (x^2 + 2x) - (4 - x) = x^2 + 2x - 4 + x$

  $= x^2 + 3x - 4$

  $(f - g)(5) = 5^2 + 3 \cdot 5 - 4 = 25 + 15 - 4 = 36$
- $(fg)(1) = f(1) \cdot g(1) = (1^2 + 2 \cdot 1)(4 - 1)$

  $= 3(3) = 9$
- $\left(\frac{f}{g}\right)(x) = \frac{x^2 + 2x}{4 - x}, x \neq 4$

  $\left(\frac{f}{g}\right)(3) = \frac{3^2 + 2 \cdot 3}{4 - 3} = \frac{9 + 6}{1} = 15$

Section 2.4 Linear Functions and Slope

Data presented in a visual form as a set of points is called a scatter plot. A line that best fits the data points is called a regression line.

A function whose graph is a straight line is called a linear function. All linear functions can be written in the form $f(x) = mx + b$.

$f(x) = 3x + 10$ is a linear function.

$g(x) = 3x^2 + 10$ is not a linear function.

If a graph intersects the x-axis at $(a, 0)$, then a is an x-intercept. If a graph intersects the y-axis at $(0, b)$, then b is a y-intercept. The standard form of the equation of a line,

$$Ax + By = C,$$

can be graphed using intercepts and a checkpoint.

Graph using intercepts: $4x + 3y = 12$.

x-intercept: (Set $y = 0$.)

$4x = 12$

$x = 3$ — Line passes through $(3, 0)$.

y-intercept: (Set $x = 0$.)

$3y = 12$

$y = 4$ — Line passes through $(0, 4)$.

Checkpoint: Let $x = 2$.

$4 \cdot 2 + 3y = 12$

$8 + 3y = 12$

$3y = 4$

$y = \frac{4}{3}$

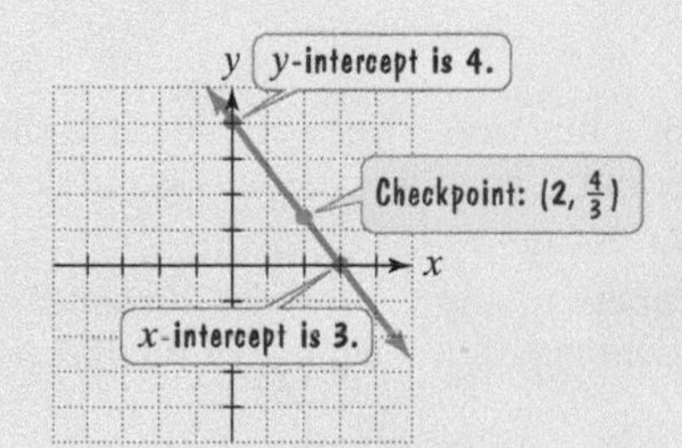

Definitions and Concepts | **Examples**

Section 2.4 Linear Functions and Slope (continued)

The slope, m, of the line through the points (x_1, y_1) and (x_2, y_2) is

$$m = \frac{y_2 - y_1}{x_2 - x_1}, \quad x_2 - x_1 \neq 0.$$

If the slope is positive, the line rises from left to right. If the slope is negative, the line falls from left to right. The slope of a horizontal line is 0. The slope of a vertical line is undefined.

For points $(-7, 2)$ and $(3, -4)$, the slope of the line through the points is

$$m = \frac{\text{Change in } y}{\text{Change in } x} = \frac{-4 - 2}{3 - (-7)} = \frac{-6}{10} = -\frac{3}{5}.$$

The slope is negative, so the line falls.

For points $(2, -5)$ and $(2, 16)$, the slope of the line through the points is

$$m = \frac{\text{Change in } y}{\text{Change in } x} = \frac{16 - (-5)}{2 - 2} = \frac{21}{0}.$$

undefined

The slope is undefined, so the line is vertical.

The slope-intercept form of the equation of a nonvertical line with slope m and y-intercept b is

$$y = mx + b.$$

Using function notation, the equation is

$$f(x) = mx + b.$$

Graph: $f(x) = -\frac{3}{4}x + 1.$

Slope is $-\frac{3}{4}$. | y-intercept is 1.

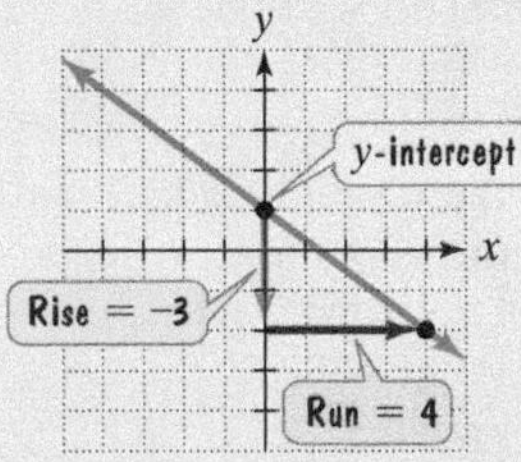

Horizontal and Vertical Lines

The graph of $y = b$, or $f(x) = b$, is a horizontal line. The y-intercept is b. The linear function $f(x) = b$ is called a constant function.

The graph of $x = a$ is a vertical line. The x-intercept is a. A vertical line is not a linear function.

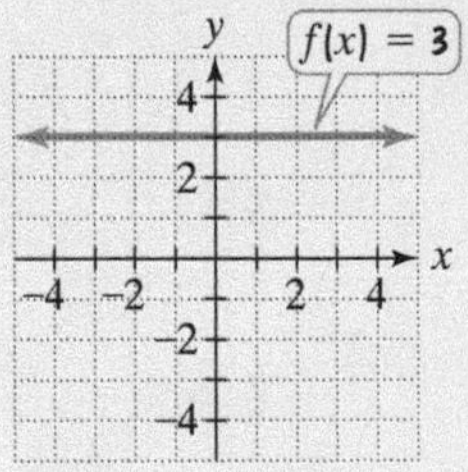

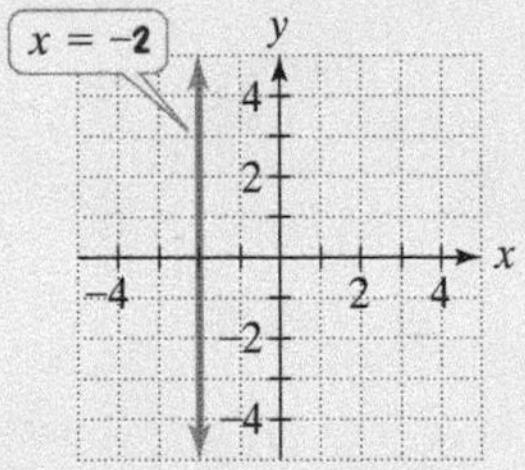

If the graph of a function is not a straight line, the average rate of change between any two points is the slope of the line containing the two points.

For a linear function, slope is the rate of change of the dependent variable per unit change of the independent variable.

The function

$$p(t) = -0.59t + 80.75$$

slope

models the percentage, $p(t)$, of Americans smoking cigarettes t years after 1900. The slope, -0.59, shows that the percentage of smokers is decreasing by 0.59% per year.

Definitions and Concepts	Examples

Section 2.5 The Point-Slope Form of the Equation of a Line

The point-slope form of the equation of a nonvertical line with slope m that passes through the point (x_1, y_1) is

$$y - y_1 = m(x - x_1).$$

Slope $= -4$, passing through $(-1, 5)$

$m = -4$ $x_1 = -1$ $y_1 = 5$

The point-slope form of the line's equation is

$$y - 5 = -4[x - (-1)].$$

Simplify:

$$y - 5 = -4(x + 1).$$

To write the point-slope form of the line passing through two points, begin by using the points to compute the slope, m. Use either given point as (x_1, y_1) and write the point-slope equation:

$$y - y_1 = m(x - x_1).$$

Solving this equation for y gives the slope-intercept form of the line's equation.

Write equations in point-slope form and in slope-intercept form of the line passing through $(4, 1)$ and $(3, -2)$.

$$m = \frac{-2 - 1}{3 - 4} = \frac{-3}{-1} = 3$$

Using $(4, 1)$ as (x_1, y_1), the point-slope form of the equation is

$$y - 1 = 3(x - 4).$$

Solve for y to obtain the slope-intercept form.

$$y - 1 = 3x - 12$$
$$y = 3x - 11$$

In function notation,

$$f(x) = 3x - 11.$$

Nonvertical parallel lines have the same slope. If the product of the slopes of two lines is -1, then the lines are perpendicular. One line is perpendicular to another line if its slope is the negative reciprocal of the slope of the other. A horizontal line having zero slope is perpendicular to a vertical line having undefined slope.

Write equations in point-slope form and in slope-intercept form of the line passing through $(2, -1)$

x_1 y_1

and perpendicular to $y = -\frac{1}{5}x + 6$.

slope

The slope, m, of the perpendicular line is 5, the negative reciprocal of $-\frac{1}{5}$.

$$y - (-1) = 5(x - 2)$$ Point-slope form of the equation

$$y + 1 = 5(x - 2)$$
$$y + 1 = 5x - 10$$
$$y = 5x - 11 \quad \text{or} \quad f(x) = 5x - 11$$

Slope-intercept form of the equation

CHAPTER 2 REVIEW EXERCISES

2.1 *In Exercises 1–3, determine whether each relation is a function. Give the domain and range for each relation.*

1. $\{(3, 10), (4, 10), (5, 10)\}$
2. $\{(1, 12), (2, 100), (3, \pi), (4, -6)\}$
3. $\{(13, 14), (15, 16), (13, 17)\}$

In Exercises 4–5, find the indicated function values.

4. $f(x) = 7x - 5$
   a. $f(0)$ b. $f(3)$ c. $f(-10)$
   d. $f(2a)$ e. $f(a + 2)$
5. $g(x) = 3x^2 - 5x + 2$
   a. $g(0)$ b. $g(5)$ c. $g(-4)$
   d. $g(b)$ e. $g(4a)$

2.2 *In Exercises 6–7, graph the given functions, f and g, in the same rectangular coordinate system. Select integers for x, starting with −2 and ending with 2. Once you have obtained your graphs, describe how the graph of g is related to the graph of f.*

6. $f(x) = x^2, \quad g(x) = x^2 - 1$
7. $f(x) = |x|, \quad g(x) = |x| + 2$

In Exercises 8–13, use the vertical line test to identify graphs in which y is a function of x.

8.
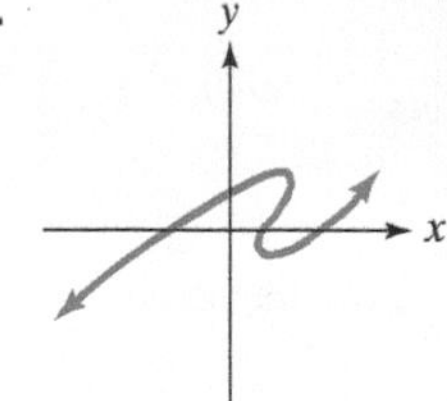

9.
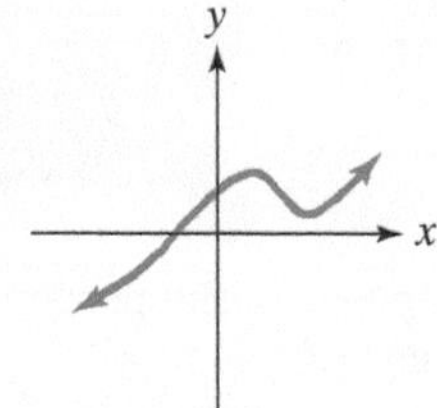

10.
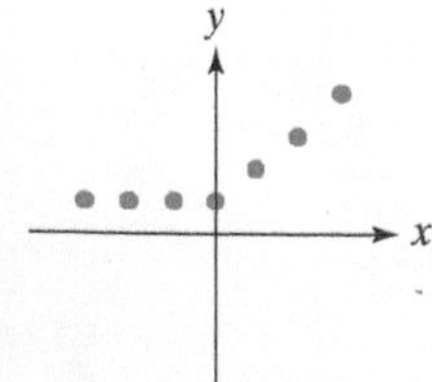

11.
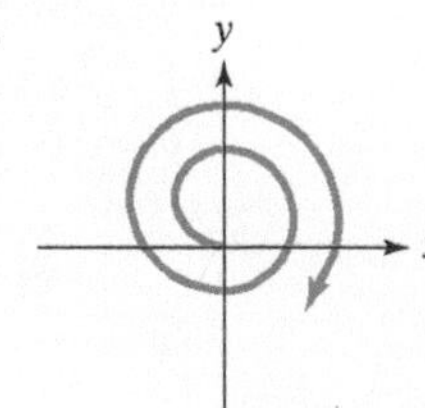

12.
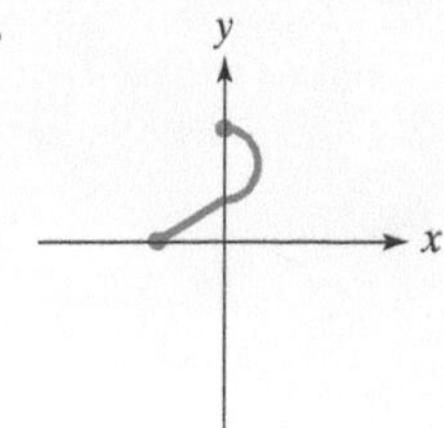

13.
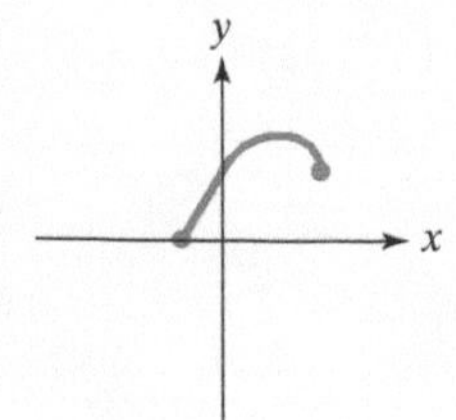

Use the graph of f to solve Exercises 14–18.

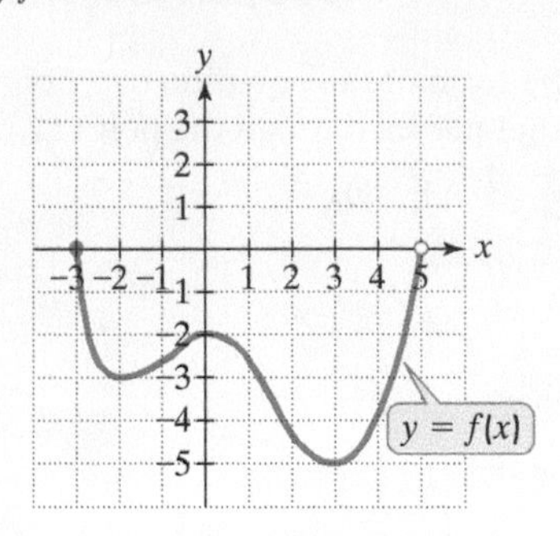

14. Find $f(-2)$.
15. Find $f(0)$.
16. For what value of x is $f(x) = -5$?
17. Find the domain of f.
18. Find the range of f.
19. The graph shows the height, in meters, of an eagle in terms of its time, in seconds, in flight.

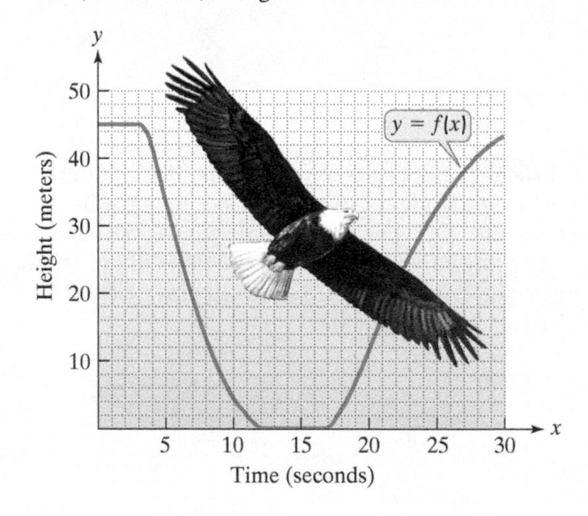

   a. Use the graph to explain why the eagle's height is a function of its time in flight.
   b. Find $f(15)$. Describe what this means in practical terms.
   c. What is a reasonable estimate of the eagle's maximum height?
   d. For what values of x is $f(x) = 20$? Describe what this means in practical terms.
   e. Use the graph of the function to write a description of the eagle's flight.

2.3 *In Exercises 20–22, find the domain of each function.*

20. $f(x) = 7x - 3$
21. $g(x) = \dfrac{1}{x + 8}$
22. $f(x) = x + \dfrac{3x}{x - 5}$

In Exercises 23–24, find **a.** $(f + g)(x)$ *and* **b.** $(f + g)(3)$.

23. $f(x) = 4x - 5, g(x) = 2x + 1$

24. $f(x) = 5x^2 - x + 4, g(x) = x - 3$

In Exercises 25–26, for each pair of functions, f and g, determine the domain of f + g.

25. $f(x) = 3x + 4, g(x) = \frac{5}{4 - x}$

26. $f(x) = \frac{7x}{x + 6}, g(x) = \frac{4}{x + 1}$

In Exercises 27–34, let

$$f(x) = x^2 - 2x \quad \text{and} \quad g(x) = x - 5.$$

Find each of the following.

27. $(f + g)(x)$ and $(f + g)(-2)$

28. $f(3) + g(3)$

29. $(f - g)(x)$ and $(f - g)(1)$

30. $f(4) - g(4)$

31. $(fg)(-3)$

32. $\left(\frac{f}{g}\right)(x)$ and $\left(\frac{f}{g}\right)(4)$

33. The domain of $f - g$

34. The domain of $\frac{f}{g}$

2.4 *In Exercises 35–37, use intercepts and a checkpoint to graph each linear function.*

35. $x + 2y = 4$

36. $2x - 3y = 12$

37. $4x = 8 - 2y$

In Exercises 38–41, find the slope of the line passing through each pair of points or state that the slope is undefined. Then indicate whether the line through the points rises, falls, is horizontal, or is vertical.

38. $(5, 2)$ and $(2, -4)$

39. $(-2, 3)$ and $(7, -3)$

40. $(3, 2)$ and $(3, -1)$

41. $(-3, 4)$ and $(-1, 4)$

In Exercises 42–44, give the slope and y-intercept of each line whose equation is given. Then graph the linear function.

42. $y = 2x - 1$

43. $f(x) = -\frac{1}{2}x + 4$

44. $y = \frac{2}{3}x$

In Exercises 45–47, rewrite the equation in slope-intercept form. Give the slope and y-intercept.

45. $2x + y = 4$

46. $-3y = 5x$

47. $5x + 3y = 6$

In Exercises 48–52, graph each equation in a rectangular coordinate system.

48. $y = 2$

49. $7y = -21$

50. $f(x) = -4$

51. $x = 3$

52. $2x = -10$

53. The function $f(t) = -0.27t + 70.45$ models record time, $f(t)$, in seconds, for the women's 400-meter run t years after 1900. What is the slope of this model? Describe what this means in terms of rate of change.

54. The stated intent of the 1994 "don't ask, don't tell" policy was to reduce the number of discharges of gay men and lesbians from the military. Nearly 14,000 active-duty gay servicemembers were dismissed under the policy, which officially ended in 2011, after 18 years. The line graph shows the number of discharges under "don't ask, don't tell" from 1994 through 2010.

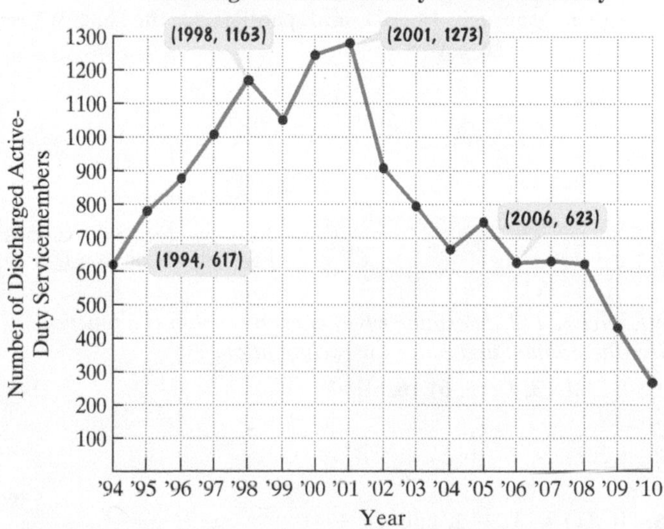

Source: General Accountability Office

a. Find the average rate of change, rounded to the nearest whole number, from 1994 through 1998. Describe what this means.

b. Find the average rate of change, rounded to the nearest whole number, from 2001 through 2006. Describe what this means.

55. The graph shows that a linear function describes the relationship between Fahrenheit temperature, F, and Celsius temperature, C.

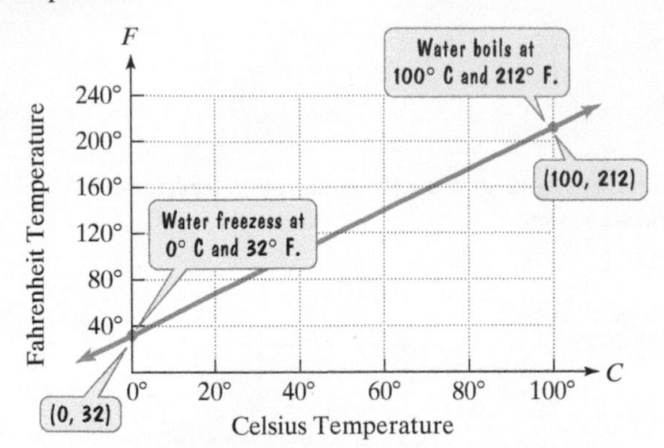

a. Use the points labeled by the voice balloons to find a function in the form $F = mC + b$ that expresses Fahrenheit temperature, F, in terms of Celsius temperature, C.

b. Use the function from part (a) to find the Fahrenheit temperature when the Celsius temperature is 30°.

2.5 *In Exercises 56–59, use the given conditions to write an equation for each line in point-slope form and in slope-intercept form.*

56. Passing through $(-3, 2)$ with slope -6

57. Passing through $(1, 6)$ and $(-1, 2)$

58. Passing through $(4, -7)$ and parallel to the line whose equation is $3x + y = 9$

59. Passing through $(-2, 6)$ and perpendicular to the line whose equation is $y = \frac{1}{3}x + 4$

60. The bar graph shows the average age at which men in the United States married for the first time from 2009 through 2013. The data are displayed as five points in a scatter plot. Also shown is a line that passes through or near the points.

Men's Average Age of First Marriage

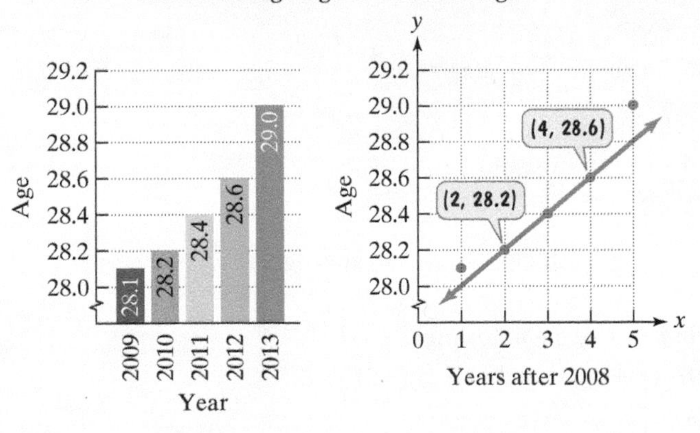

Source: U.S. Census Bureau

a. Use the two points whose coordinates are shown by the voice balloons to find the point-slope form of the equation of the line that models men's average age of first marriage, y, x years after 2008.

b. Write the equation from part (a) in slope-intercept form. Use function notation.

c. If trends shown from 2009 through 2013 continue, use the linear function to project men's average age of first marriage in 2020.

CHAPTER 2 TEST

Step-by-step test solutions are found on the Chapter Test Prep Videos available in MyMathLab® or on YouTube (search "BlitzerInterAlg7e" and click on "Channels").

In Exercises 1–2, determine whether each relation is a function. Give the domain and range for each relation.

1. $\{(1, 2), (3, 4), (5, 6), (6, 6)\}$

2. $\{(2, 1), (4, 3), (6, 5), (6, 6)\}$

3. If $f(x) = 3x - 2$, find $f(a + 4)$.

4. If $f(x) = 4x^2 - 3x + 6$, find $f(-2)$.

5. Graph $f(x) = x^2 - 1$ and $g(x) = x^2 + 1$ in the same rectangular coordinate system. Select integers for x, starting with -2 and ending with 2. Once you have obtained your graphs, describe how the graph of g is related to the graph of f.

In Exercises 6–7, identify the graph or graphs in which y is a function of x.

6. **7.**

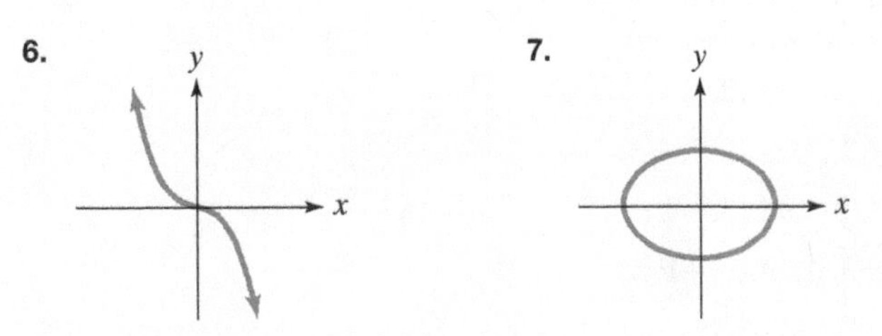

Use the graph of f to solve Exercises 8–11.

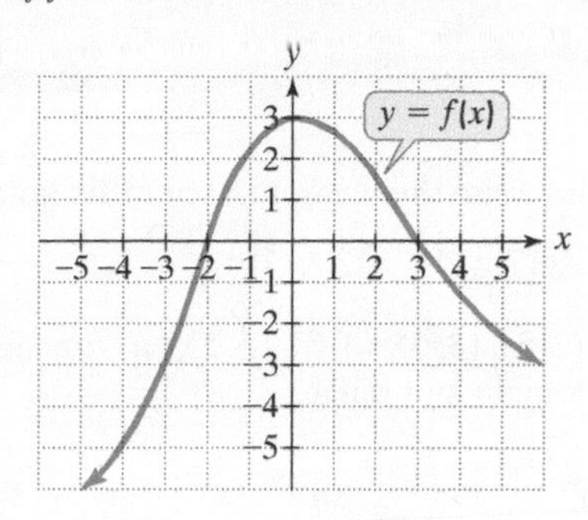

8. Find $f(6)$.
9. List two values of x for which $f(x) = 0$.
10. Find the domain of f.
11. Find the range of f.
12. Find the domain of $f(x) = \dfrac{6}{10 - x}$.

In Exercises 13–17, let

$$f(x) = x^2 + 4x \quad \textit{and} \quad g(x) = x + 2.$$

Find each of the following.

13. $(f + g)(x)$ and $(f + g)(3)$
14. $(f - g)(x)$ and $(f - g)(-1)$
15. $(fg)(-5)$
16. $\left(\dfrac{f}{g}\right)(x)$ and $\left(\dfrac{f}{g}\right)(2)$
17. The domain of $\dfrac{f}{g}$

In Exercises 18–20, graph each linear function.

18. $4x - 3y = 12$
19. $f(x) = -\dfrac{1}{3}x + 2$
20. $f(x) = 4$

In Exercises 21–22, find the slope of the line passing through each pair of points or state that the slope is undefined. Then indicate whether the line through the points rises, falls, is horizontal, or is vertical.

21. $(5, 2)$ and $(1, 4)$
22. $(4, 5)$ and $(4, -5)$

The function $V(t) = 3.6t + 140$ models the number of Super Bowl viewers, $V(t)$, in millions, t years after 1995. Use the model to solve Exercises 23–24.

23. Find $V(10)$. Describe what this means in terms of the variables in the model.
24. What is the slope of this model? Describe what this means in terms of rate of change.

In Exercises 25–27, use the given conditions to write an equation for each line in point-slope form and slope-intercept form.

25. Passing through $(-1, -3)$ and $(4, 2)$
26. Passing through $(-2, 3)$ and perpendicular to the line whose equation is $y = -\frac{1}{2}x - 4$
27. Passing through $(6, -4)$ and parallel to the line whose equation is $x + 2y = 5$
28. In the United States, it is illegal to drive with a blood alcohol concentration of 0.08 or higher. (A blood alcohol concentration of 0.08 indicates 0.08% alcohol in the blood.) The scatter plot shows the relationship between the number of one-ounce beers consumed per hour by a 200-pound person and that person's corresponding blood alcohol concentration. Also shown is a line that passes through or near the data points.

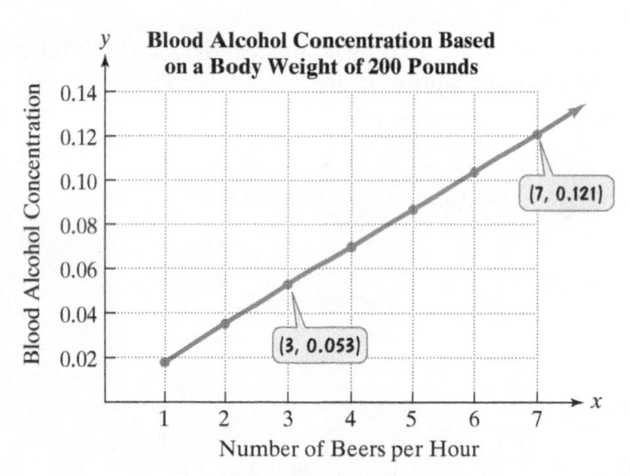

Source: National Highway Traffic Safety Administration

a. Use the two points whose coordinates are shown by the voice balloons to find the point-slope form of the equation of the line that models the blood alcohol concentration of a 200-pound person, y, who consumes x one-ounce beers per hour.
b. Write the equation from part (a) in slope-intercept form. Use function notation.
c. Use the linear function to predict the blood alcohol concentration of a 200-pound person who consumes 8 one-ounce beers in an hour. (Effects at this level include major impairment of all physical and mental functions.)

CUMULATIVE REVIEW EXERCISES (CHAPTERS 1–2)

1. Use the roster method to list the elements in the set:

$$\{x \mid x \text{ is a whole number less than } 4\}.$$

2. Determine whether the following statement is true or false:

$$\pi \notin \{x \mid x \text{ is an irrational number}\}.$$

In Exercises 3–4, use the order of operations to simplify each expression.

3. $\dfrac{8 - 3^2 \div 9}{|-5| - [5 - (18 \div 6)]^2}$

4. $4 - (2 - 9)^0 + 3^2 \div 1 + 3$

5. Simplify: $3 - [2(x - 2) - 5x]$.

In Exercises 6–8, solve each equation. If the solution set is $\varnothing$ or $\mathbb{R}$, classify the equation as an inconsistent equation or an identity.

6. $2 + 3x - 4 = 2(x - 3)$

7. $4x + 12 - 8x = -6(x - 2) + 2x$

8. $\dfrac{x - 2}{4} = \dfrac{2x + 6}{3}$

9. After a 20% reduction, a computer sold for \$1800. What was the computer's price before the reduction?

10. Solve for t: $A = p + prt$.

In Exercises 11–12, simplify each exponential expression.

11. $(3x^4y^{-5})^{-2}$

12. $\left(\dfrac{3x^2y^{-4}}{x^{-3}y^2}\right)^2$

13. Multiply and write the answer in scientific notation:

$$(7 \times 10^{-8})(3 \times 10^2).$$

14. Is $\{(1, 5), (2, 5), (3, 5), (4, 5), (6, 5)\}$ a function? Give the relation's domain and range.

15. Graph $f(x) = |x| - 1$ and $g(x) = |x| + 2$ in the same rectangular coordinate system. Select integers for x, starting with -2 and ending with 2. Once you have obtained your graphs, describe how the graph of g is related to the graph of f.

16. Find the domain of $f(x) = \dfrac{1}{15 - x}$.

17. If $f(x) = 3x^2 - 4x + 2$ and $g(x) = x^2 - 5x - 3$, find $(f - g)(x)$ and $(f - g)(-1)$.

In Exercises 18–19, graph each linear function.

18. $f(x) = -2x + 4$

19. $x - 2y = 6$

20. Write equations in point-slope form and slope-intercept form for the line passing through $(3, -5)$ and parallel to the line whose equation is $y = 4x + 7$.

CHAPTER 3

Systems of Linear Equations

Held in a different country every five years, with contributions from more than 120 countries, the World Expo is a showcase of cool gizmos, a multinational take on what the future might be. Who needs a human friend when you can connect with an adorable robot that recognizes you, engages in (meaningful?) conversation in any language, takes orders, plays a mean trumpet, and even does algebra? And what do the entrepreneurs who create these robots and other global visions want to do? Generate profit, of course. In this chapter, you'll learn how algebra models every business venture, from a kid selling lemonade to companies producing innovative "personal partner" robots.

Here's where you'll find this application:

- Functions of business, including modeling business ventures with profit functions, are discussed in Section 3.2, pages 201–203.

SECTION

3.1 Systems of Linear Equations in Two Variables

What am I supposed to learn?

After studying this section, you should be able to:

1. Determine whether an ordered pair is a solution of a system of linear equations.
2. Solve systems of linear equations by graphing.
3. Solve systems of linear equations by substitution.
4. Solve systems of linear equations by addition.
5. Select the most efficient method for solving a system of linear equations.
6. Identify systems that do not have exactly one ordered-pair solution.

"My wife and I were happy for twenty years. Then we met."

—Henny Youngman

"Getting married for sex is like buying a 747 for the free peanuts."

—Jeff Foxworthy

Despite an endless array of jokes about marriage, **Figure 3.1** indicates that fewer U.S. adults are getting married. Our changing marital status can be analyzed using a pair of linear models in two variables.

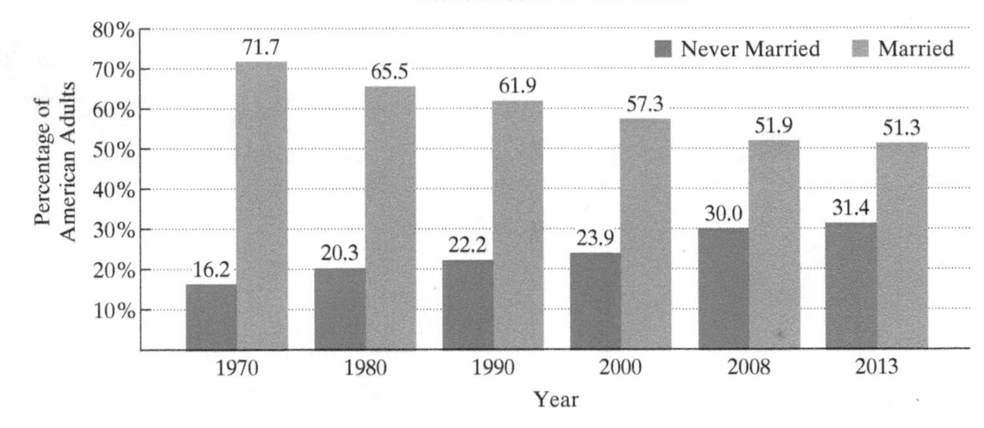

Figure 3.1

Source: U.S. Census Bureau

In the first three sections of this chapter, you will learn to model your world with two equations in two variables and three equations in three variables. The methods you learn for solving these systems provide the foundation for solving complex problems involving thousands of equations containing thousands of variables. In this section's Exercise Set (Exercise 91), you will apply these methods to analyze America's changing marital status.

1 Determine whether an ordered pair is a solution of a system of linear equations.

Systems of Linear Equations and Their Solutions

We have seen that all equations in the form $Ax + By = C$ are straight lines when graphed. Two such equations are called a **system of linear equations**, or a **linear system**. A **solution of a system of linear equations** is an ordered pair that satisfies both equations in the system. For example, (3, 4) satisfies the system

$$\begin{cases} x + y = 7 & (3 + 4 \text{ is, indeed, } 7.) \\ x - y = -1. & (3 - 4 \text{ is, indeed, } -1.) \end{cases}$$

Thus, (3, 4) satisfies both equations and is a solution of the system. The solution can be described by saying that $x = 3$ and $y = 4$. The solution can also be described using set notation. The solution set of the system is $\{(3, 4)\}$—that is, the set consisting of the ordered pair $(3, 4)$.

A system of linear equations can have exactly one solution, no solution, or infinitely many solutions. We begin with systems with exactly one solution.

EXAMPLE 1 Determining Whether Ordered Pairs Are Solutions of a System

Consider the system:

$$\begin{cases} x + 2y = -7 \\ 2x - 3y = 0. \end{cases}$$

Determine if each ordered pair is a solution of the system:

a. $(-3, -2)$ **b.** $(1, -4)$.

Solution

a. We begin by determining whether $(-3, -2)$ is a solution. Because -3 is the x-coordinate and -2 is the y-coordinate of $(-3, -2)$, we replace x with -3 and y with -2.

$$\begin{aligned} x + 2y &= -7 \\ -3 + 2(-2) &\stackrel{?}{=} -7 \\ -3 + (-4) &\stackrel{?}{=} -7 \\ -7 &= -7, \quad \text{true} \end{aligned} \qquad \begin{aligned} 2x - 3y &= 0 \\ 2(-3) - 3(-2) &\stackrel{?}{=} 0 \\ -6 - (-6) &\stackrel{?}{=} 0 \\ -6 + 6 &\stackrel{?}{=} 0 \\ 0 &= 0, \quad \text{true} \end{aligned}$$

The pair $(-3, -2)$ satisfies both equations: It makes each equation true. Thus, the ordered pair is a solution of the system.

b. To determine whether $(1, -4)$ is a solution, we replace x with 1 and y with -4.

$$\begin{aligned} x + 2y &= -7 \\ 1 + 2(-4) &\stackrel{?}{=} -7 \\ 1 + (-8) &\stackrel{?}{=} -7 \\ -7 &= -7, \quad \text{true} \end{aligned} \qquad \begin{aligned} 2x - 3y &= 0 \\ 2 \cdot 1 - 3(-4) &\stackrel{?}{=} 0 \\ 2 - (-12) &\stackrel{?}{=} 0 \\ 2 + 12 &\stackrel{?}{=} 0 \\ 14 &= 0, \quad \text{false} \end{aligned}$$

The pair $(1, -4)$ fails to satisfy *both* equations: It does not make both equations true. Thus, the ordered pair is not a solution of the system. ■

✓ CHECK POINT 1 Consider the system:

$$\begin{cases} 2x + 5y = -24 \\ 3x - 5y = 14. \end{cases}$$

Determine if each ordered pair is a solution of the system:

a. $(-7, -2)$ **b.** $(-2, -4)$.

---

2 Solve systems of linear equations by graphing.

Solving Linear Systems by Graphing

The solution of a system of two linear equations in two variables can be found by graphing both of the equations in the same rectangular coordinate system. For a system with one solution, **the coordinates of the point of intersection give the system's solution.**

Great Question!

Can I use a rough sketch on scratch paper to solve a linear system by graphing?

No. When solving linear systems by graphing, neatly drawn graphs are essential for determining points of intersection.

- Use rectangular coordinate graph paper.
- Use a ruler or straightedge.
- Use a pencil with a sharp point.

Solving Systems of Two Linear Equations in Two Variables, x and y, by Graphing

1. Graph the first equation.
2. Graph the second equation on the same set of axes.
3. If the lines intersect at a point, determine the coordinates of this point of intersection. The ordered pair is the solution to the system.
4. Check the solution in both equations.

EXAMPLE 2 Solving a Linear System by Graphing

Solve by graphing:

$$\begin{cases} y = -x - 1 \\ 4x - 3y = 24. \end{cases}$$

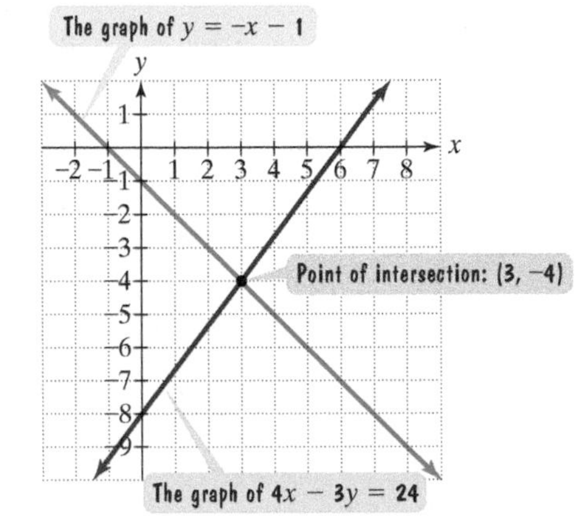

Figure 3.2

Solution

Step 1. Graph the first equation. We use the y-intercept and the slope to graph $y = -x - 1$.

$$y = -x - 1$$

The slope is −1. The y-intercept is −1.

The graph of the linear function is shown as a blue line in **Figure 3.2**.

Step 2. Graph the second equation on the same axes. We use intercepts to graph $4x - 3y = 24$.

x-intercept (Set $y = 0$.)	y-intercept (Set $x = 0$.)
$4x - 3 \cdot 0 = 24$	$4 \cdot 0 - 3y = 24$
$4x = 24$	$-3y = 24$
$x = 6$	$y = -8$

The x-intercept is 6, so the line passes through (6, 0). The y-intercept is −8, so the line passes through (0, −8). The graph of $4x - 3y = 24$ is shown as a red line in **Figure 3.2**.

Step 3. Determine the coordinates of the intersection point. This ordered pair is the system's solution. Using **Figure 3.2**, it appears that the lines intersect at (3, −4). The "apparent" solution of the system is (3, −4).

Step 4. Check the solution in both equations.

Check (3, −4) in $y = -x - 1$:	Check (3, −4) in $4x - 3y = 24$:
$-4 \stackrel{?}{=} -3 - 1$	$4(3) - 3(-4) \stackrel{?}{=} 24$
$-4 = -4$, true	$12 + 12 \stackrel{?}{=} 24$
	$24 = 24$, true

Because both equations are satisfied, (3, −4) is the solution and {(3, −4)} is the solution set. ■

Using Technology

A graphing utility can be used to solve the system in Example 2. Solve each equation for y, if necessary, graph the equations, and use the intersection feature. The utility displays the solution (3, −4) as $x = 3$, $y = -4$.

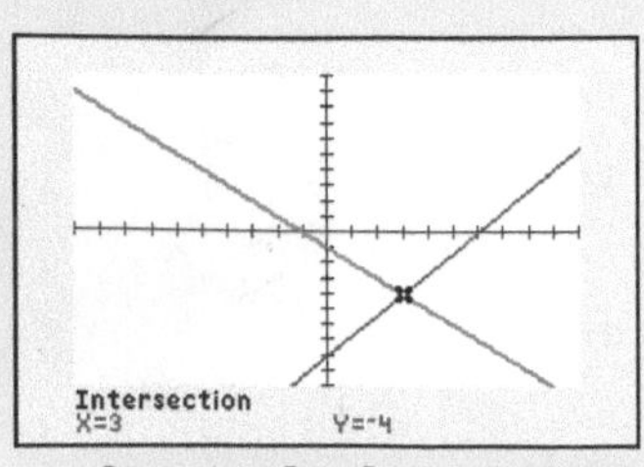

[−10, 10, 1] by [−10, 10, 1]

CHECK POINT 2 Solve by graphing:

$$\begin{cases} y = -2x + 6 \\ 2x - y = -2. \end{cases}$$

3 Solve systems of linear equations by substitution.

Eliminating a Variable Using the Substitution Method

Finding the solution to a linear system by graphing equations may not be easy to do. For example, a solution of $\left(-\frac{2}{3}, \frac{157}{29}\right)$ would be difficult to "see" as an intersection point on a graph.

Let's consider a method that does not depend on finding a system's solution visually: the substitution method. This method involves converting the system to one equation in one variable by an appropriate substitution.

Great Question!

In the first step of the substitution method, how do I know which variable to isolate and in which equation?

You can choose both the variable and the equation. If possible, solve for a variable whose coefficient is 1 or −1 to avoid working with fractions.

Solving Linear Systems by Substitution

1. Solve either of the equations for one variable in terms of the other. (If one of the equations is already in this form, you can skip this step.)
2. Substitute the expression found in step 1 into the other equation. This will result in an equation in one variable.
3. Solve the equation containing one variable.
4. Back-substitute the value found in step 3 into one of the original equations. Simplify and find the value of the remaining variable.
5. Check the proposed solution in both of the system's given equations.

EXAMPLE 3 Solving a System by Substitution

Solve by the substitution method:

$$\begin{cases} y = -2x + 4 \\ 7x - 2y = 3. \end{cases}$$

Solution

Step 1. Solve either of the equations for one variable in terms of the other. This step has already been done for us. The first equation, $y = -2x + 4$, is solved for y in terms of x.

Step 2. Substitute the expression from step 1 into the other equation. We substitute the expression $-2x + 4$ for y in the second equation:

$$y = \boxed{-2x + 4} \qquad 7x - 2\boxed{y} = 3. \qquad \text{Substitute } -2x + 4 \text{ for } y.$$

This gives us an equation in one variable, namely

$$7x - 2(-2x + 4) = 3.$$

The variable y has been eliminated.

Step 3. Solve the resulting equation containing one variable.

$$\begin{aligned} 7x - 2(-2x + 4) &= 3 && \text{This is the equation containing one variable.} \\ 7x + 4x - 8 &= 3 && \text{Apply the distributive property.} \\ 11x - 8 &= 3 && \text{Combine like terms.} \\ 11x &= 11 && \text{Add 8 to both sides.} \\ x &= 1 && \text{Divide both sides by 11.} \end{aligned}$$

Step 4. Back-substitute the obtained value into one of the original equations. We now know that the x-coordinate of the solution is 1. To find the y-coordinate, we back-substitute the x-value into either original equation. We will use

$$y = -2x + 4.$$

Substitute 1 for x.

$$y = -2 \cdot 1 + 4 = -2 + 4 = 2$$

With $x = 1$ and $y = 2$, the proposed solution is $(1, 2)$.

Step 5. Check the proposed solution in both of the system's given equations. Replace x with 1 and y with 2.

$$\begin{aligned} y &= -2x + 4 \\ 2 &\stackrel{?}{=} -2 \cdot 1 + 4 \\ 2 &\stackrel{?}{=} -2 + 4 \\ 2 &= 2, \quad \text{true} \end{aligned} \qquad \begin{aligned} 7x - 2y &= 3 \\ 7(1) - 2(2) &\stackrel{?}{=} 3 \\ 7 - 4 &\stackrel{?}{=} 3 \\ 3 &= 3, \quad \text{true} \end{aligned}$$

The pair (1, 2) satisfies both equations. The solution is (1, 2) and the system's solution set is {(1, 2)}. ■

✓ CHECK POINT 3 Solve by the substitution method:

$$\begin{cases} y = 3x - 7 \\ 5x - 2y = 8. \end{cases}$$

EXAMPLE 4 Solving a System by Substitution

Solve by the substitution method:

$$\begin{cases} 5x + 2y = 1 \\ x - 3y = 7. \end{cases}$$

Solution

Step 1. Solve either of the equations for one variable in terms of the other. We begin by isolating one of the variables in either of the equations. By solving for x in the second equation, which has a coefficient of 1, we can avoid fractions.

$$x - 3y = 7 \quad \text{This is the second equation in the given system.}$$

$$x = 3y + 7 \quad \text{Solve for } x \text{ by adding } 3y \text{ to both sides.}$$

Step 2. Substitute the expression from step 1 into the other equation. We substitute $3y + 7$ for x in the first equation.

$$x = \boxed{3y + 7} \qquad 5\boxed{x} + 2y = 1$$

This gives us an equation in one variable, namely

$$5(3y + 7) + 2y = 1.$$

The variable x has been eliminated.

Step 3. Solve the resulting equation containing one variable.

$$\begin{aligned} 5(3y + 7) + 2y &= 1 && \text{This is the equation containing one variable.} \\ 15y + 35 + 2y &= 1 && \text{Apply the distributive property.} \\ 17y + 35 &= 1 && \text{Combine like terms.} \\ 17y &= -34 && \text{Subtract 35 from both sides.} \\ y &= -2 && \text{Divide both sides by 17.} \end{aligned}$$

Step 4. Back-substitute the obtained value into one of the original equations. We back-substitute -2 for y into one of the original equations to find x. Let's use both equations to show that we obtain the same value for x in either case.

Using the first equation:

$$\begin{aligned} 5x + 2y &= 1 \\ 5x + 2(-2) &= 1 \\ 5x - 4 &= 1 \\ 5x &= 5 \\ x &= 1 \end{aligned}$$

Using the second equation:

$$\begin{aligned} x - 3y &= 7 \\ x - 3(-2) &= 7 \\ x + 6 &= 7 \\ x &= 1 \end{aligned}$$

With $x = 1$ and $y = -2$, the proposed solution is $(1, -2)$.

Great Question!

When I back-substitute the value for one of the variables, which equation should I use?

The equation from step 1, in which one variable is expressed in terms of the other, is equivalent to one of the original equations. It is often easiest to back-substitute the obtained value into this equation to find the value of the other variable. After obtaining both values, get into the habit of checking the ordered-pair solution in *both* equations of the system.

Step 5. Check. Take a moment to show that $(1, -2)$ satisfies both given equations. The solution is $(1, -2)$ and the solution set is $\{(1, -2)\}$. ■

CHECK POINT 4 Solve by the substitution method:

$$\begin{cases} 3x + 2y = 4 \\ 2x + y = 1. \end{cases}$$

---

4 Solve systems of linear equations by addition.

Eliminating a Variable Using the Addition Method

The substitution method is most useful if one of the given equations has an isolated variable. A third method for solving a linear system is the addition method. Like the substitution method, the addition method involves eliminating a variable and ultimately solving an equation containing only one variable. However, this time we eliminate a variable by adding the equations.

For example, consider the following system of linear equations:

$$\begin{cases} 3x - 4y = 11 \\ -3x + 2y = -7. \end{cases}$$

When we add these two equations, the x-terms are eliminated. This occurs because the coefficients of the x-terms, 3 and -3, are opposites (additive inverses) of each other:

$$\begin{cases} 3x - 4y = 11 \\ \underline{-3x + 2y = -7} \end{cases}$$

$$\text{Add: } 0x - 2y = 4$$

$$-2y = 4$$

$$y = -2.$$

The sum is an equation in one variable.

Divide both sides by -2 and solve for y.

Now we can back-substitute -2 for y into one of the original equations to find x. It does not matter which equation you use; you will obtain the same value for x in either case. If we use either equation, we can show that $x = 1$ and the solution $(1, -2)$ satisfies both equations in the system.

When we use the addition method, we want to obtain two equations whose sum is an equation containing only one variable. The key step is to **obtain, for one of the variables, coefficients that differ only in sign**. To do this, we may need to multiply one or both equations by some nonzero number so that the coefficients of one of the variables, x or y, become opposites. Then when the two equations are added, this variable will be eliminated.

EXAMPLE 5 Solving a System by the Addition Method

Solve by the addition method:

$$\begin{cases} 3x + 4y = -10 \\ 5x - 2y = 18. \end{cases}$$

Solution We must rewrite one or both equations in equivalent forms so that the coefficients of the same variable (either x or y) are opposites of each other. Consider the terms in y in each equation, that is, $4y$ and $-2y$. To eliminate y, we can multiply each term of the second equation by 2 and then add equations.

$$\begin{cases} 3x + 4y = -10 \\ 5x - 2y = 18 \end{cases} \xrightarrow[\text{Multiply by 2.}]{\text{No change}} \begin{cases} 3x + 4y = -10 \\ \underline{10x - 4y = 36} \end{cases}$$

$$\text{Add: } 13x + 0y = 26$$

$$13x = 26$$

$$x = 2$$

Divide both sides by 13 and solve for x.

Great Question!

Isn't the addition method also called the elimination method?

Although the addition method is also known as the elimination method, variables are eliminated when using both the substitution and addition methods. The name *addition method* specifically tells us that the elimination of a variable is accomplished by adding two equations.

Thus, $x = 2$. We back-substitute this value into either one of the given equations. We'll use the first one.

$$
\begin{aligned}
3x + 4y &= -10 && \text{This is the first equation in the given system.}\\
3(2) + 4y &= -10 && \text{Substitute 2 for } x.\\
6 + 4y &= -10 && \text{Multiply.}\\
4y &= -16 && \text{Subtract 6 from both sides.}\\
y &= -4 && \text{Divide both sides by 4.}
\end{aligned}
$$

We see that $x = 2$ and $y = -4$. The ordered pair $(2, -4)$ can be shown to satisfy both equations in the system. Consequently, the solution is $(2, -4)$ and the solution set is $\{(2, -4)\}$. ■

Solving Linear Systems by Addition

1. If necessary, rewrite both equations in the form $Ax + By = C$.
2. If necessary, multiply either equation or both equations by appropriate nonzero numbers so that the sum of the x-coefficients or the sum of the y-coefficients is 0.
3. Add the equations in step 2. The sum will be an equation in one variable.
4. Solve the equation in one variable.
5. Back-substitute the value obtained in step 4 into either of the given equations and solve for the other variable.
6. Check the solution in both of the original equations.

✓ CHECK POINT 5 Solve by the addition method:

$$
\begin{cases} 4x - 7y = -16 \\ 2x + 5y = 9. \end{cases}
$$

EXAMPLE 6 Solving a System by the Addition Method

Solve by the addition method:

$$
\begin{cases} 7x = 5 - 2y \\ 3y = 16 - 2x. \end{cases}
$$

Solution

Step 1. Rewrite both equations in the form $Ax + By = C$. We first arrange the system so that variable terms appear on the left and constants appear on the right. We obtain

$$
\begin{cases} 7x + 2y = 5 & \text{Add } 2y \text{ to both sides of the first equation.} \\ 2x + 3y = 16. & \text{Add } 2x \text{ to both sides of the second equation.} \end{cases}
$$

Step 2. If necessary, multiply either equation or both equations by appropriate numbers so that the sum of the x-coefficients or the sum of the y-coefficients is 0. We can eliminate x or y. Let's eliminate y by multiplying the first equation by 3 and the second equation by -2.

$$
\begin{cases} 7x + 2y = 5 \\ 2x + 3y = 16 \end{cases}
\begin{array}{c} \xrightarrow{\text{Multiply by 3.}} \\ \xrightarrow{\text{Multiply by } -2.} \end{array}
\begin{cases} 21x + 6y = 15 \\ \underline{-4x - 6y = -32} \end{cases}
$$

Step 3. Add the equations.

$$
\begin{aligned}
\text{Add: } 17x + 0y &= -17\\
17x &= -17
\end{aligned}
$$

Step 4. Solve the equation in one variable. We solve $17x = -17$ by dividing both sides by 17.

$$\frac{17x}{17} = \frac{-17}{17} \quad \text{Divide both sides by 17.}$$

$$x = -1 \quad \text{Simplify.}$$

Step 5. Back-substitute and find the value of the other variable. We can back-substitute -1 for x into either one of the given equations. We'll use the second one.

$$3y = 16 - 2x \quad \text{This is the second equation in the given system.}$$
$$3y = 16 - 2(-1) \quad \text{Substitute } -1 \text{ for } x.$$
$$3y = 16 + 2 \quad \text{Multiply.}$$
$$3y = 18 \quad \text{Add.}$$
$$y = 6 \quad \text{Divide both sides by 3.}$$

We found that $x = -1$ and $y = 6$. The proposed solution is $(-1, 6)$.

Step 6. Check. Take a moment to show that $(-1, 6)$ satisfies both given equations. The solution is $(-1, 6)$ and the solution set is $\{(-1, 6)\}$. ■

✓ CHECK POINT 6 Solve by the addition method:

$$\begin{cases} 3x = 2 - 4y \\ 5y = -1 - 2x. \end{cases}$$

Some linear systems have solutions that are not integers. If the value of one variable turns out to be a "messy" fraction, back-substitution might lead to cumbersome arithmetic. If this happens, you can return to the original system and use the addition method a second time to find the value of the other variable.

EXAMPLE 7 Solving a System by the Addition Method

Solve by the addition method:

$$\begin{cases} \frac{x}{2} - 5y = 32 \\ \frac{3x}{2} - 7y = 45. \end{cases}$$

Solution

Step 1. Rewrite both equations in the form $Ax + By = C$. Although each equation is already in this form, the coefficients of x are not integers. There is less chance for error if the coefficients for x and y in $Ax + By = C$ are integers. Consequently, we begin by clearing fractions. Multiply both sides of each equation by 2.

$$\begin{cases} \frac{x}{2} - 5y = 32 \xrightarrow{\text{Multiply by 2.}} \\ \frac{3x}{2} - 7y = 45 \xrightarrow{\text{Multiply by 2.}} \end{cases} \begin{cases} x - 10y = 64 \\ 3x - 14y = 90 \end{cases}$$

Step 2. If necessary, multiply either equation or both equations by appropriate numbers so that the sum of the x-coefficients or the sum of the y-coefficients is 0. We will eliminate x. Multiply the first equation with integral coefficients by -3 and leave the second equation unchanged.

$$\begin{cases} x - 10y = 64 \xrightarrow{\text{Multiply by } -3.} \\ 3x - 14y = 90 \xrightarrow{\text{No change}} \end{cases} \begin{cases} -3x + 30y = -192 \\ 3x - 14y = 90 \end{cases}$$

Step 3. Add the equations.

$$\text{Add: } 0x + 16y = -102$$
$$16y = -102$$

Step 4. Solve the equation in one variable. We solve $16y = -102$ by dividing both sides by 16.

$$\frac{16y}{16} = \frac{-102}{16} \quad \text{Divide both sides by 16.}$$

$$y = -\frac{102}{16} = -\frac{51}{8} \quad \text{Simplify.}$$

Step 5. Back-substitute and find the value of the other variable. Back-substitution of $-\frac{51}{8}$ for y into either of the given equations results in cumbersome arithmetic. Instead, let's use the addition method on the system with integral coefficients from step 1 to find the value of x. Thus, we eliminate y by multiplying the first equation by -7 and the second equation by 5.

$$\begin{cases} x - 10y = 64 \\ 3x - 14y = 90 \end{cases} \xrightarrow[\text{Multiply by 5.}]{\text{Multiply by } -7.} \begin{cases} -7x + 70y = -448 \\ 15x - 70y = 450 \end{cases}$$

$$\text{Add:} \quad 8x = 2$$

$$x = \frac{2}{8} = \frac{1}{4}$$

We found that $x = \frac{1}{4}$ and $y = -\frac{51}{8}$. The proposed solution is $\left(\frac{1}{4}, -\frac{51}{8}\right)$.

Step 6. Check. For this system, a calculator is helpful in showing that $\left(\frac{1}{4}, -\frac{51}{8}\right)$ satisfies both of the original equations of the system. The solution is $\left(\frac{1}{4}, -\frac{51}{8}\right)$ and the solution set is $\left\{\left(\frac{1}{4}, -\frac{51}{8}\right)\right\}$. ■

✓ CHECK POINT 7 Solve by the addition method:

$$\begin{cases} \frac{3x}{2} - 2y = \frac{5}{2} \\ x - \frac{5y}{2} = -\frac{3}{2}. \end{cases}$$

5 Select the most efficient method for solving a system of linear equations.

Comparing the Three Solution Methods

The following chart compares the graphing, substitution, and addition methods for solving systems of linear equations in two variables. With increased practice, you will find it easier to select the best method for solving a particular linear system.

Comparing Solution Methods

Method	Advantages	Disadvantages
Graphing	You can see the solutions.	If the solutions do not involve integers or are too large to be seen on the graph, it's impossible to tell exactly what the solutions are.
Substitution	Gives exact solutions. Easy to use if a variable is on one side by itself.	Solutions cannot be seen. Introduces extensive work with fractions when no variable has a coefficient of 1 or -1.
Addition	Gives exact solutions. Easy to use if no variable has a coefficient of 1 or -1.	Solutions cannot be seen.

6 Identify systems that do not have exactly one ordered-pair solution.

Linear Systems Having No Solution or Infinitely Many Solutions

We have seen that a system of linear equations in two variables represents a pair of lines. The lines either intersect at one point, are parallel, or are identical. Thus, there are three possibilities for the number of solutions to a system of two linear equations.

The Number of Solutions to a System of Two Linear Equations

The number of solutions to a system of two linear equations in two variables is given by one of the following. (See **Figure 3.3.**)

Number of Solutions	What This Means Graphically
Exactly one ordered-pair solution	The two lines intersect at one point.
No solution	The two lines are parallel.
Infinitely many solutions	The two lines are identical.

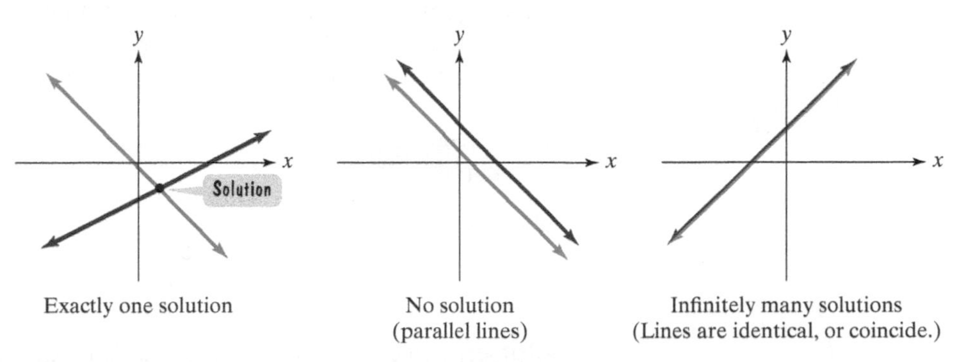

Figure 3.3 Possible graphs for a system of two linear equations in two variables

A linear system with no solution is called an **inconsistent system**. If you attempt to solve such a system by substitution or addition, you will eliminate both variables. A false statement, such as $0 = 6$, will be the result.

EXAMPLE 8 A System with No Solution

Solve the system:

$$\begin{cases} 3x - 2y = 6 \\ 6x - 4y = 18. \end{cases}$$

Solution Because no variable is isolated, we will use the addition method. To obtain coefficients of x that differ only in sign, we multiply the first equation by -2.

$$\begin{cases} 3x - 2y = 6 \xrightarrow{\text{Multiply by } -2.} \\ 6x - 4y = 18 \xrightarrow{\text{No change}} \end{cases} \begin{cases} -6x + 4y = -12 \\ \underline{6x - 4y = 18} \end{cases}$$

$$\text{Add:} \quad 0 = 6$$

There are no values of x and y for which $0 = 6$. No values of x and y satisfy $0x + 0y = 6$.

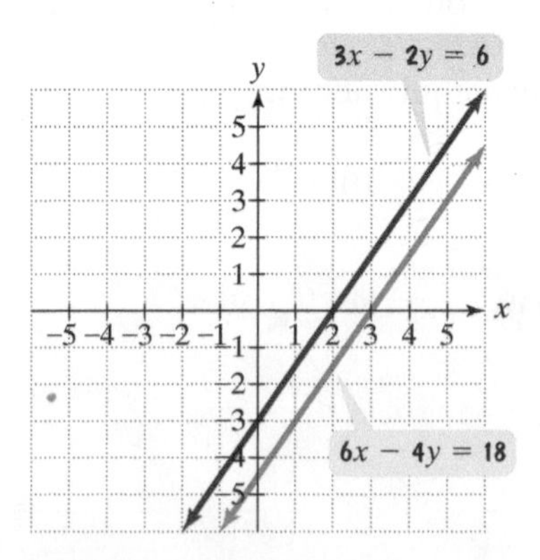

Figure 3.4 The graph of an inconsistent system

The false statement $0 = 6$ indicates that the system is inconsistent and has no solution. The solution set is the empty set, $\varnothing$. ■

The lines corresponding to the two equations in Example 8 are shown in **Figure 3.4**. The lines are parallel and have no point of intersection.

Discover for Yourself

Show that the graphs of $3x - 2y = 6$ and $6x - 4y = 18$ must be parallel lines by solving each equation for y. What is the slope and the y-intercept for each line? What does this mean? If a linear system is inconsistent, what must be true about the slopes and the y-intercepts for the system's graphs?

✓ CHECK POINT 8 Solve the system:

$$\begin{cases} 5x - 2y = 4 \\ -10x + 4y = 7. \end{cases}$$

A linear system that has at least one solution is called a **consistent system.** Lines that intersect and lines that coincide both represent consistent systems. If the lines coincide, then the consistent system has infinitely many solutions, represented by every point on the coinciding lines.

The equations in a linear system with infinitely many solutions are called **dependent.** If you attempt to solve such a system by substitution or addition, you will eliminate both variables. However, a true statement, such as $10 = 10$, will be the result.

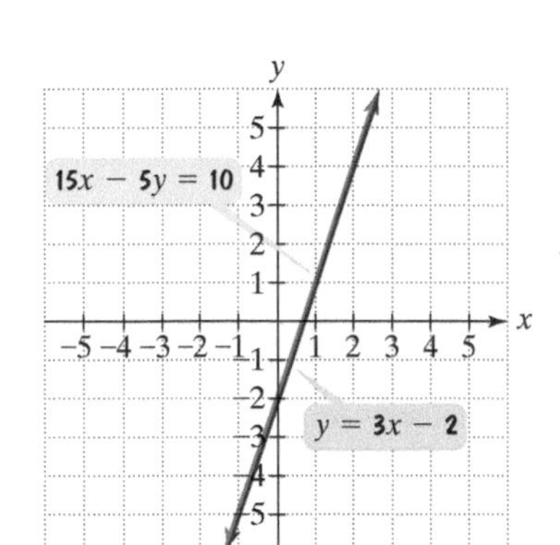

Figure 3.5 The graph of a system with infinitely many solutions

EXAMPLE 9 A System with Infinitely Many Solutions

Solve the system:

$$\begin{cases} y = 3x - 2 \\ 15x - 5y = 10. \end{cases}$$

Solution Because the variable y is isolated in $y = 3x - 2$, the first equation, we can use the substitution method. We substitute the expression for y into the second equation.

$$y = \boxed{3x - 2} \qquad 15x - 5\boxed{y} = 10 \quad \text{Substitute } 3x - 2 \text{ for } y.$$

$$15x - 5(3x - 2) = 10 \quad \text{The substitution results in an equation in one variable.}$$

$$15x - 15x + 10 = 10 \quad \text{Apply the distributive property.}$$

$$10 = 10 \quad \text{Simplify.}$$

This statement is true for all values of x and y.

In our final step, both variables have been eliminated and the resulting statement, $10 = 10$, is true. This true statement indicates that the system has infinitely many solutions. The solution set consists of all points (x, y) lying on either of the coinciding lines, $y = 3x - 2$ or $15x - 5y = 10$, as shown in **Figure 3.5**.

We express the solution set for the system in one of two equivalent ways:

$$\{(x, y) \mid y = 3x - 2\} \quad \text{or} \quad \{(x, y) \mid 15x - 5y = 10\}.$$

The set of all ordered pairs (x, y) such that $y = 3x - 2$ — The set of all ordered pairs (x, y) such that $15x - 5y = 10$ ■

Great Question!

The system in Example 9 has infinitely many solutions. Does that mean that any ordered pair of numbers is a solution?

No. Although the system in Example 9 has infinitely many solutions, this does not mean that any ordered pair of numbers you can form will be a solution. The ordered pair (x, y) must satisfy one of the system's equations, $y = 3x - 2$ or $15x - 5y = 10$, and there are infinitely many such ordered pairs. Because the graphs are coinciding lines, the ordered pairs that are solutions of one of the equations are also solutions of the other equation.

✓ CHECK POINT 9 Solve the system:

$$\begin{cases} x = 4y - 8 \\ 5x - 20y = -40. \end{cases}$$

Achieving Success

Take some time to consider how you are doing in this course. Check your performance by answering the following questions:

- Are you attending all lectures?
- For each hour of class time, are you spending at least two hours outside of class completing all homework assignments, checking answers, correcting errors, and using all resources to get the help that you need?
- Are you reviewing for quizzes and tests?
- Are you reading the textbook? In all college courses, you are responsible for the information in the text, whether or not it is covered in class.
- Are you keeping an organized notebook? Does each page have the appropriate section number from the text on top? Do the pages contain examples your instructor works during lecture and other relevant class notes? Have you included your worked-out homework exercises? Do you keep a special section for graded exams?
- Are you analyzing your mistakes and learning from your errors?
- Are there ways you can improve how you are doing in the course?

CONCEPT AND VOCABULARY CHECK

Fill in each blank so that the resulting statement is true.

1. A solution to a system of linear equations in two variables is an ordered pair that ________________.
2. When solving a system of linear equations by graphing, the system's solution is determined by locating ________________.
3. When solving

$$\begin{cases} 3x - 2y = 5 \\ y = 3x - 3 \end{cases}$$

by the substitution method, we obtain $x = \frac{1}{3}$, so the solution set is ________________.

4. When solving

$$\begin{cases} 2x + 10y = 9 \\ 8x + 5y = 7 \end{cases}$$

by the addition method, we can eliminate y by multiplying the second equation by ________________ and then adding the equations.

5. When solving

$$\begin{cases} 4x - 3y = 15 \\ 3x - 2y = 10 \end{cases}$$

by the addition method, we can eliminate y by multiplying the first equation by 2 and the second equation by ________________ and then adding the equations.

6. When solving

$$\begin{cases} 12x - 21y = 24 \\ 4x - 7y = 7 \end{cases}$$

by the addition method, we obtain $0 = 3$, so the solution set is ________________. The linear system is a/an ________________ system. If you attempt to solve such a system by graphing, you will obtain two lines that are ________________.

7. When solving

$$\begin{cases} x = 3y + 2 \\ 5x - 15y = 10 \end{cases}$$

by the substitution method, we obtain $10 = 10$, so the solution set is ________________. The equations in this system are called ________________. If you attempt to solve such a system by graphing, you will obtain two lines that ________________.

3.1 EXERCISE SET

MyMathLab®

Practice Exercises

In Exercises 1–6, determine whether the given ordered pair is a solution of the system.

1. $(7, -5)$ $\begin{cases} x - y = 12 \\ x + y = 2 \end{cases}$

2. $(-3, 1)$ $\begin{cases} x - y = -4 \\ 2x + 10y = 4 \end{cases}$

3. $(2, -1)$ $\begin{cases} 3x + 4y = 2 \\ 2x + 5y = 1 \end{cases}$

4. $(4, 2)$ $\begin{cases} 2x - 5y = -2 \\ 3x + 4y = 18 \end{cases}$

5. $(5, -3)$ $\begin{cases} y = 2x - 13 \\ 4x + 9y = -7 \end{cases}$

6. $(-3, -4)$ $\begin{cases} y = 3x + 5 \\ 5x - 2y = -7 \end{cases}$

In Exercises 7–24, solve each system by graphing. Identify systems with no solution and systems with infinitely many solutions, using set notation to express their solution sets.

7. $\begin{cases} x + y = 4 \\ x - y = 2 \end{cases}$

8. $\begin{cases} x + y = 6 \\ x - y = -4 \end{cases}$

9. $\begin{cases} 2x + y = 4 \\ y = 4x + 1 \end{cases}$

10. $\begin{cases} x + 2y = 4 \\ y = -2x - 1 \end{cases}$

11. $\begin{cases} 3x - 2y = 6 \\ x - 4y = -8 \end{cases}$

12. $\begin{cases} 4x + y = 4 \\ 3x - y = 3 \end{cases}$

13. $\begin{cases} 2x + 3y = 6 \\ 4x = -6y + 12 \end{cases}$

14. $\begin{cases} 3x - 3y = 6 \\ 2x = 2y + 4 \end{cases}$

15. $\begin{cases} y = 2x - 2 \\ y = -5x + 5 \end{cases}$

16. $\begin{cases} y = -x + 1 \\ y = 3x + 5 \end{cases}$

17. $\begin{cases} 3x - y = 4 \\ 6x - 2y = 4 \end{cases}$

18. $\begin{cases} 2x - y = -4 \\ 4x - 2y = 6 \end{cases}$

19. $\begin{cases} 2x + y = 4 \\ 4x + 3y = 10 \end{cases}$

20. $\begin{cases} 4x - y = 9 \\ x - 3y = 16 \end{cases}$

21. $\begin{cases} x - y = 2 \\ y = 1 \end{cases}$

22. $\begin{cases} x + 2y = 1 \\ x = 3 \end{cases}$

23. $\begin{cases} 3x + y = 3 \\ 6x + 2y = 12 \end{cases}$

24. $\begin{cases} 2x - 3y = 6 \\ 4x - 6y = 24 \end{cases}$

In Exercises 25–42, solve each system by the substitution method. Identify inconsistent systems and systems with dependent equations, using set notation to express their solution sets.

25. $\begin{cases} x + y = 6 \\ y = 2x \end{cases}$

26. $\begin{cases} x + y = 10 \\ y = 4x \end{cases}$

27. $\begin{cases} 2x + 3y = 9 \\ x = y + 2 \end{cases}$

28. $\begin{cases} 3x - 4y = 18 \\ y = 1 - 2x \end{cases}$

29. $\begin{cases} y = -3x + 7 \\ 5x - 2y = 8 \end{cases}$

30. $\begin{cases} x = 3y + 8 \\ 2x - y = 6 \end{cases}$

31. $\begin{cases} 4x + y = 5 \\ 2x - 3y = 13 \end{cases}$

32. $\begin{cases} x - 3y = 3 \\ 3x + 5y = -19 \end{cases}$

33. $\begin{cases} x - 2y = 4 \\ 2x - 4y = 5 \end{cases}$

34. $\begin{cases} x - 3y = 6 \\ 2x - 6y = 5 \end{cases}$

35. $\begin{cases} 2x + 5y = -4 \\ 3x - y = 11 \end{cases}$

36. $\begin{cases} 2x + 5y = 1 \\ -x + 6y = 8 \end{cases}$

37. $\begin{cases} 2(x - 1) - y = -3 \\ y = 2x + 3 \end{cases}$

38. $\begin{cases} x + y - 1 = 2(y - x) \\ y = 3x - 1 \end{cases}$

39. $\begin{cases} \frac{x}{4} - \frac{y}{4} = -1 \\ x + 4y = -9 \end{cases}$

40. $\begin{cases} \frac{x}{6} - \frac{y}{2} = \frac{1}{3} \\ x + 2y = -3 \end{cases}$

41. $\begin{cases} y = \frac{2}{5}x - 2 \\ 2x - 5y = 10 \end{cases}$

42. $\begin{cases} y = \frac{1}{3}x + 4 \\ 3y = x + 12 \end{cases}$

In Exercises 43–58, solve each system by the addition method. Identify inconsistent systems and systems with dependent equations, using set notation to express their solution sets.

43. $\begin{cases} x + y = 7 \\ x - y = 3 \end{cases}$

44. $\begin{cases} 2x + y = 3 \\ x - y = 3 \end{cases}$

45. $\begin{cases} 12x + 3y = 15 \\ 2x - 3y = 13 \end{cases}$

46. $\begin{cases} 4x + 2y = 12 \\ 3x - 2y = 16 \end{cases}$

47. $\begin{cases} x + 3y = 2 \\ 4x + 5y = 1 \end{cases}$

48. $\begin{cases} x + 2y = -1 \\ 2x - y = 3 \end{cases}$

49. $\begin{cases} 6x - y = -5 \\ 4x - 2y = 6 \end{cases}$

50. $\begin{cases} x - 2y = 5 \\ 5x - y = -2 \end{cases}$

51. $\begin{cases} 3x - 5y = 11 \\ 2x - 6y = 2 \end{cases}$

52. $\begin{cases} 4x - 3y = 12 \\ 3x - 4y = 2 \end{cases}$

53. $\begin{cases} 2x - 5y = 13 \\ 5x + 3y = 17 \end{cases}$

54. $\begin{cases} 4x + 5y = -9 \\ 6x - 3y = -3 \end{cases}$

55. $\begin{cases} 2x + 6y = 8 \\ 3x + 9y = 12 \end{cases}$

56. $\begin{cases} x - 3y = -6 \\ 3x - 9y = 9 \end{cases}$

57. $\begin{cases} 2x - 3y = 4 \\ 4x + 5y = 3 \end{cases}$

58. $\begin{cases} 4x - 3y = 8 \\ 2x - 5y = -14 \end{cases}$

In Exercises 59–82, solve each system by the method of your choice. Identify inconsistent systems and systems with dependent equations, using set notation to express solution sets.

59. $\begin{cases} 3x - 7y = 1 \\ 2x - 3y = -1 \end{cases}$

60. $\begin{cases} 2x - 3y = 2 \\ 5x + 4y = 51 \end{cases}$

61. $\begin{cases} x = y + 4 \\ 3x + 7y = -18 \end{cases}$

62. $\begin{cases} y = 3x + 5 \\ 5x - 2y = -7 \end{cases}$

63. $\begin{cases} 9x + \frac{4y}{3} = 5 \\ 4x - \frac{y}{3} = 5 \end{cases}$

64. $\begin{cases} \frac{x}{6} - \frac{y}{5} = -4 \\ \frac{x}{4} - \frac{y}{6} = -2 \end{cases}$

65. $\begin{cases} \frac{1}{4}x - \frac{1}{9}y = \frac{2}{3} \\ \frac{1}{2}x - \frac{1}{3}y = 1 \end{cases}$

66. $\begin{cases} \frac{1}{16}x - \frac{3}{4}y = -1 \\ \frac{3}{4}x + \frac{5}{2}y = 11 \end{cases}$

67. $\begin{cases} x = 3y - 1 \\ 2x - 6y = -2 \end{cases}$

68. $\begin{cases} x = 4y - 1 \\ 2x - 8y = -2 \end{cases}$

69. $\begin{cases} y = 2x + 1 \\ y = 2x - 3 \end{cases}$

70. $\begin{cases} y = 2x + 4 \\ y = 2x - 1 \end{cases}$

71. $\begin{cases} 0.4x + 0.3y = 2.3 \\ 0.2x - 0.5y = 0.5 \end{cases}$

72. $\begin{cases} 0.2x - y = -1.4 \\ 0.7x - 0.2y = -1.6 \end{cases}$

73. $\begin{cases} 5x - 40 = 6y \\ 2y = 8 - 3x \end{cases}$

74. $\begin{cases} 4x - 24 = 3y \\ 9y = 3x - 1 \end{cases}$

75. $\begin{cases} 3(x + y) = 6 \\ 3(x - y) = -36 \end{cases}$

76. $\begin{cases} 4(x - y) = -12 \\ 4(x + y) = -20 \end{cases}$

77. $\begin{cases} 3(x - 3) - 2y = 0 \\ 2(x - y) = -x - 3 \end{cases}$

78. $\begin{cases} 5x + 2y = -5 \\ 4(x + y) = 6(2 - x) \end{cases}$

79. $\begin{cases} x + 2y - 3 = 0 \\ 12 = 8y + 4x \end{cases}$

80. $\begin{cases} 2x - y - 5 = 0 \\ 10 = 4x - 2y \end{cases}$

81. $\begin{cases} 3x + 4y = 0 \\ 7x = 3y \end{cases}$

82. $\begin{cases} 5x + 8y = 20 \\ 4y = -5x \end{cases}$

Practice PLUS

In Exercises 83–84, solve each system by the method of your choice.

83. $\begin{cases} \dfrac{x+2}{2} - \dfrac{y+4}{3} = 3 \\ \dfrac{x+y}{5} = \dfrac{x-y}{2} - \dfrac{5}{2} \end{cases}$

84. $\begin{cases} \dfrac{x-y}{3} = \dfrac{x+y}{2} - \dfrac{1}{2} \\ \dfrac{x+2}{2} - 4 = \dfrac{y+4}{3} \end{cases}$

In Exercises 85–86, solve each system for x and y, expressing either value in terms of a or b, if necessary. Assume that $a \neq 0$ and $b \neq 0$.

85. $\begin{cases} 5ax + 4y = 17 \\ ax + 7y = 22 \end{cases}$

86. $\begin{cases} 4ax + by = 3 \\ 6ax + 5by = 8 \end{cases}$

87. For the linear function $f(x) = mx + b$, $f(-2) = 11$ and $f(3) = -9$. Find m and b.

88. For the linear function $f(x) = mx + b$, $f(-3) = 23$ and $f(2) = -7$. Find m and b.

Use the graphs of the linear functions to solve Exercises 89–90.

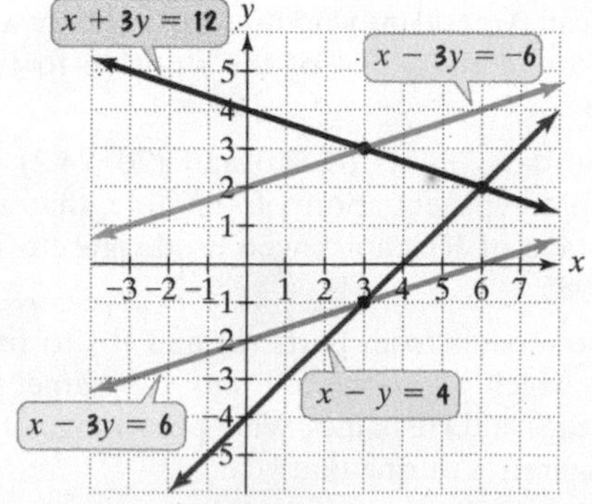

89. Write the linear system whose solution set is $\{(6, 2)\}$. Express each equation in the system in slope-intercept form.

90. Write the linear system whose solution set is $\varnothing$. Express each equation in the system in slope-intercept form.

Application Exercises

91. We opened the section with a bar graph that showed fewer U.S. adults are getting married. (See **Figure 3.1** on page 178.) The data can be modeled by the following system of linear equations:

$$\begin{cases} -3x + 10y = 160 & \text{Percentage of never-married American adults, } y, x \text{ years after 1970} \\ x + 2y = 142. & \text{Percentage of married American adults, } y, x \text{ years after 1970} \end{cases}$$

a. Use these models to determine the year, rounded to the nearest year, when the percentage of never-married adults will be the same as the percentage of married adults. For that year, approximately what percentage of Americans, rounded to the nearest percent, will belong to each group?

b. How is your approximate solution from part (a) shown by the following graphs?

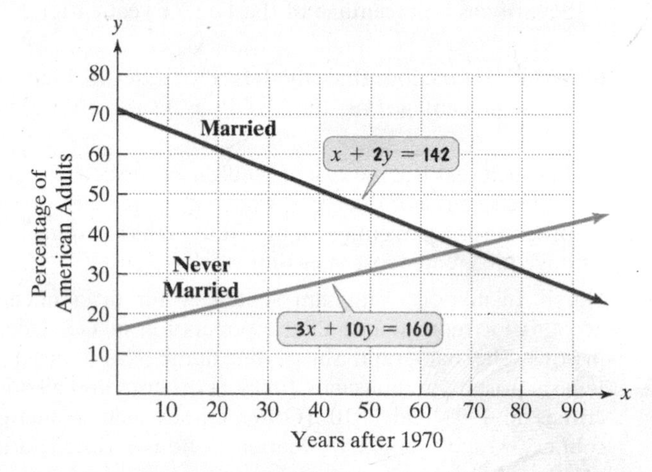

92. The graph shows that from 2000 through 2006, Americans unplugged land lines and switched to cellphones.

Number of Cellphone and Land-Line Customers in the United States

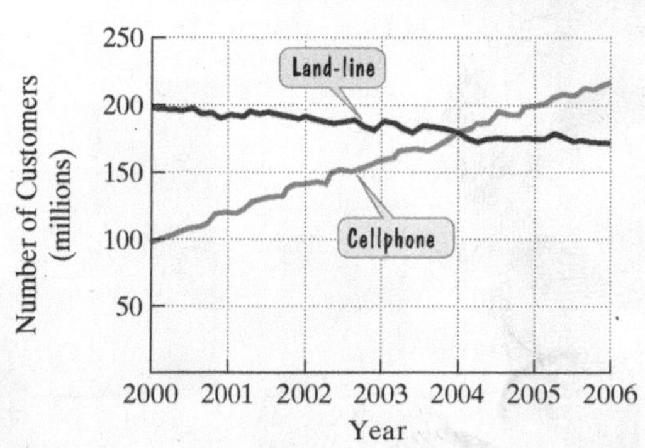

Source: Federal Communications Commission

a. Use the graphs to estimate the point of intersection. In what year was the number of cellphone and land-line customers the same? How many millions of customers were there for each?

b. The function $4.3x + y = 198$ models the number of land-line customers, in millions, x years after 2000.

The function $y = 19.8x + 98$ models the number of cellphone customers, in millions, x years after 2000. Use these models to determine the year, rounded to the nearest year, when the number of cellphone and land-line customers was the same. According to the models, how many millions of customers, rounded to the nearest ten million, were there for each?

c. How well do the models in part (b) describe the point of intersection of the graphs that you estimated in part (a)?

93. Although Social Security is a problem, some projections indicate that there's a much bigger time bomb ticking in the federal budget, and that's Medicare. In 2000, the cost of Social Security was 5.48% of the gross domestic product, increasing by 0.04% of the GDP per year. In 2000, the cost of Medicare was 1.84% of the gross domestic product, increasing by 0.17% of the GDP per year. (*Source*: Congressional Budget Office)

a. Write a function that models the cost of Social Security as a percentage of the GDP x years after 2000.

b. Write a function that models the cost of Medicare as a percentage of the GDP x years after 2000.

c. In which year will the cost of Medicare and Social Security be the same? For that year, what will be the cost of each program as a percentage of the GDP? Which program will have the greater cost after that year?

94. Harsh, mandatory minimum sentences for drug offenses account for more than half the population in U.S. federal prisons. The bar graph shows the number of inmates in federal prisons, in thousands, for drug offenses and all other crimes in 1998 and 2010. (Other crimes include murder, robbery, fraud, burglary, weapons offenses, immigration offenses, racketeering, and perjury.)

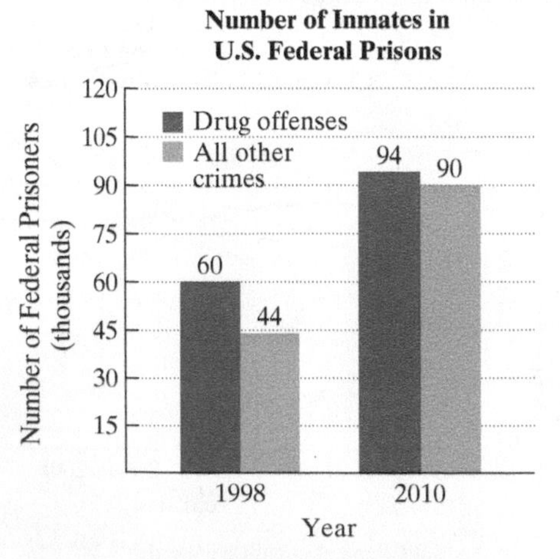

Source: Bureau of Justice Statistics

a. In 1998, there were 60 thousand inmates in federal prisons for drug offenses. For the period shown by the graph, this number increased by approximately 2.8 thousand inmates per year. Write a function that models the number of inmates, y, in thousands, for drug offenses x years after 1998.

b. In 1998, there were 44 thousand inmates in federal prisons for all crimes other than drug offenses. For the period shown by the graph, this number increased by approximately 3.8 thousand inmates per year. Write a function that models the number of inmates, y, in thousands, for all crimes other than drug offenses x years after 1998.

c. Use the models from parts (a) and (b) to determine in which year the number of federal inmates for drug offenses was the same as the number of federal inmates for all other crimes. How many inmates were there for drug offenses and for all other crimes in that year?

The bar graph shows the percentage of Americans who used cigarettes, by ethnicity, in 1985 and 2005. For each of the groups shown, cigarette use has been linearly decreasing. Use this information to solve Exercises 95–96.

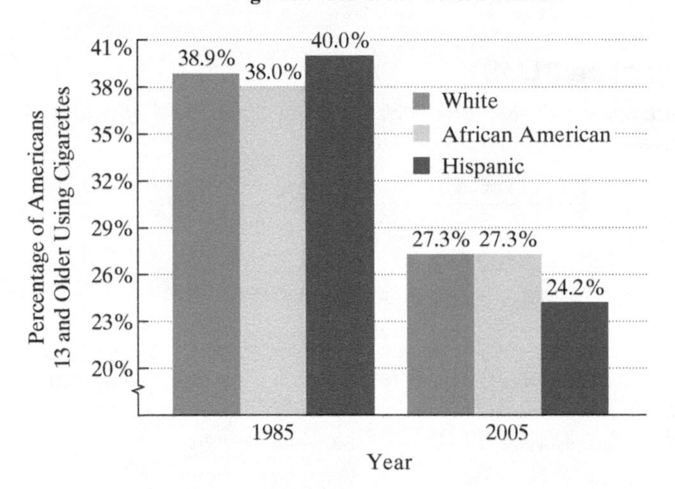

Source: Department of Health and Human Services

95. In this exercise, let x represent the number of years after 1985 and let y represent the percentage of Americans in one of the groups shown who used cigarettes.

a. Use the data points (0, 38) and (20, 27.3) to find the slope-intercept equation of the line that models the percentage of African Americans who used cigarettes, y, x years after 1985. Round the value of the slope m to two decimal places.

b. Use the data points (0, 40) and (20, 24.2) to find the slope-intercept equation of the line that models the percentage of Hispanics who used cigarettes, y, x years after 1985.

c. Use the models from parts (a) and (b) to find the year during which cigarette use was the same for African Americans and Hispanics. What percentage of each group used cigarettes during that year?

96. In this exercise, let x represent the number of years after 1985 and let y represent the percentage of Americans in one of the groups shown who used cigarettes.

a. Use the data points (0, 38.9) and (20, 27.3) to find the slope-intercept equation of the line that models the percentage of whites who used cigarettes, y, x years after 1985.

b. Use the data points (0, 40) and (20, 24.2) to find the slope-intercept equation of the line that models the percentage of Hispanics who used cigarettes, y, x years after 1985.

c. Use the models from parts (a) and (b) to find the year, to the nearest whole year, during which cigarette use was the same for whites and Hispanics. What percentage of each group, to the nearest percent, used cigarettes during that year?

An important application of systems of equations arises in connection with supply and demand. As the price of a product increases, the demand for that product decreases. However, at higher prices, suppliers are willing to produce greater quantities of the product. Exercises 97–98 involve supply and demand.

97. A chain of electronics stores sells hand-held color televisions. The weekly demand and supply models are given as follows:

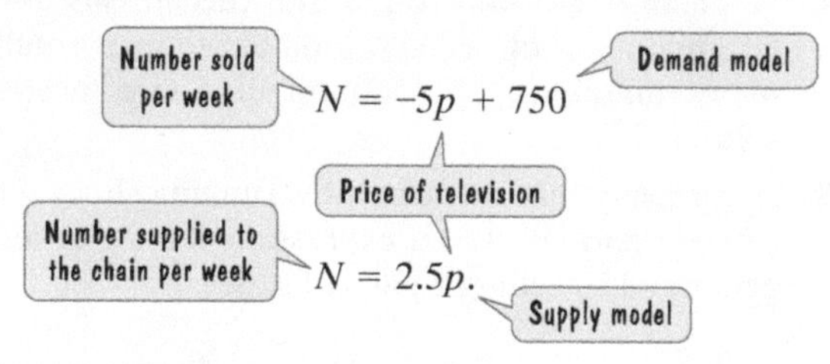

a. How many hand-held color televisions can be sold and supplied at \$120 per television?

b. Find the price at which supply and demand are equal. At this price, how many televisions can be supplied and sold each week?

98. At a price of p dollars per ticket, the number of tickets to a rock concert that can be sold is given by the demand model $N = -25p + 7800$. At a price of p dollars per ticket, the number of tickets that the concert's promoters are willing to make available is given by the supply model $N = 5p + 6000$.

a. How many tickets can be sold and supplied for \$50 per ticket?

b. Find the ticket price at which supply and demand are equal. At this price, how many tickets will be supplied and sold?

Explaining the Concepts

99. What is a system of linear equations? Provide an example with your description.

100. What is a solution of a system of linear equations?

101. Explain how to determine if an ordered pair is a solution of a system of linear equations.

102. Explain how to solve a system of linear equations by graphing.

103. Explain how to solve a system of equations using the substitution method. Use $y = 3 - 3x$ and $3x + 4y = 6$ to illustrate your explanation.

104. Explain how to solve a system of equations using the addition method. Use $5x + 8y = -1$ and $3x + y = 7$ to illustrate your explanation.

105. When is it easier to use the addition method rather than the substitution method to solve a system of equations?

106. When using the addition or substitution method, how can you tell if a system of linear equations has no solution? What is the relationship between the graphs of the two equations?

107. When using the addition or substitution method, how can you tell if a system of linear equations has infinitely many solutions? What is the relationship between the graphs of the two equations?

Technology Exercise

108. Verify your solutions to any five exercises from Exercises 7–24 by using a graphing utility to graph the two equations in the system in the same viewing rectangle. Then use the intersection feature to display the solution.

Critical Thinking Exercises

Make Sense? *In Exercises 109–112, determine whether each statement makes sense or does not make sense, and explain your reasoning.*

109. Even if a linear system has a solution set involving fractions, such as $\{(\frac{8}{11}, \frac{43}{11})\}$, I can use graphs to determine if the solution set is reasonable.

110. If I add the equations on the right and solve the resulting equation for x, I will obtain the x-coordinate of the intersection point of the lines represented by the equations on the left.

$$\begin{cases} 4x - 6y = 1 \\ 3x + 5y = -8 \end{cases} \longrightarrow \begin{cases} 20x - 30y = 5 \\ 18x + 30y = -8 \end{cases}$$

111. In the previous chapter, we developed models for life expectancy, y, for U.S. men and women born x years after 1960:

$$\begin{cases} y = 0.22x + 65.7 & \text{Men} \\ y = 0.17x + 72.9. & \text{Women} \end{cases}$$

The system indicates that life expectancy for men is increasing at a faster rate than for women, so if these trends continue, life expectancies for men and women will be the same for some future birth year.

112. Here are two models that describe winning times for the Olympic 400-meter run, y, in seconds, x years after 1968:

$$\begin{cases} y = -0.02433x + 44.43 & \text{Men} \\ y = -0.08883x + 50.86. & \text{Women} \end{cases}$$

The system indicates that winning times have been decreasing more rapidly for women than for men, so if these trends continue, there will be a year when winning times for men and women are the same.

In Exercises 113–116, determine whether each statement is true or false. If the statement is false, make the necessary change(s) to produce a true statement.

113. The addition method cannot be used to eliminate either variable in a system of two equations in two variables.

114. The solution set of the system

$$\begin{cases} 5x - y = 1 \\ 10x - 2y = 2 \end{cases}$$

is $\{(2, 9)\}$.

115. A system of linear equations can have a solution set consisting of precisely two ordered pairs.

116. The solution set of the system

$$\begin{cases} y = 4x - 3 \\ y = 4x + 5 \end{cases}$$

is the empty set.

117. Determine a and b so that $(2, 1)$ is a solution of this system:

$$\begin{cases} ax - by = 4 \\ bx + ay = 7. \end{cases}$$

118. Write a system of equations having $\{(-2, 7)\}$ as a solution set. (More than one system is possible.)

119. Solve the system for x and y in terms of a_1, b_1, c_1, a_2, b_2, and c_2:

$$\begin{cases} a_1x + b_1y = c_1 \\ a_2x + b_2y = c_2. \end{cases}$$

Review Exercises

120. Solve: $6x = 10 + 5(3x - 4)$. (Section 1.4, Example 3)

121. Simplify: $(4x^2y^4)^2(-2x^5y^0)^3$. (Section 1.6, Example 9)

122. If $f(x) = x^2 - 3x + 7$, find $f(-1)$. (Section 2.1, Example 3)

Preview Exercises

Exercises 123–125 will help you prepare for the material covered in the next section.

123. The formula $I = Pr$ is used to find the simple interest, I, earned for one year when the principal, P, is invested at an annual interest rate, r. Write an expression for the total interest earned on a principal of x dollars at a rate of 15% ($r = 0.15$) and a principal of y dollars at a rate of 7% ($r = 0.07$).

124. A chemist working on a flu vaccine needs to obtain 50 milliliters of a 30% sodium-iodine solution. How many milliliters of sodium-iodine are needed in the solution?

125. A company that manufactures running shoes sells them at \$80 per pair. Write an expression for the revenue that is generated by selling x pairs of shoes.

SECTION

3.2 Problem Solving and Business Applications Using Systems of Equations

What am I supposed to learn?

After studying this section, you should be able to:

1. Solve problems using systems of equations.
2. Use functions to model revenue, cost, and profit, and perform a break-even analysis.

Driving through your neighborhood, you see kids selling lemonade. Would it surprise you to know that this activity can be analyzed using functions and systems of equations? By doing so, you will view profit and loss in the business world in a new way. In this section, we use systems of equations to solve problems and model business ventures.

A Strategy for Solving Word Problems Using Systems of Equations

When we solved problems in Chapter 1, we let x represent a quantity that was unknown. Problems in this section involve two unknown quantities. We will let x and y represent these quantities. We then translate from the verbal conditions of the problem into a *system* of linear equations.

1 Solve problems using systems of equations.

EXAMPLE 1 Solving a Problem Involving Energy Efficiency of Building Materials

A heat-loss survey by an electric company indicated that a wall of a house containing 40 square feet of glass and 60 square feet of plaster lost 1920 Btu (British thermal units) of heat. A second wall containing 10 square feet of glass and 100 square feet of plaster lost 1160 Btu of heat. Determine the heat lost per square foot for the glass and for the plaster.

Solution

Step 1. Use variables to represent unknown quantities.

Let x = the heat lost per square foot for the glass.

Let y = the heat lost per square foot for the plaster.

Step 2. Write a system of equations that models the problem's conditions. The heat loss for each wall is the heat lost by the glass plus the heat lost by the plaster. One wall containing 40 square feet of glass and 60 square feet of plaster lost 1920 Btu of heat.

Heat lost by the glass + heat lost by the plaster = total heat lost.

$$\begin{pmatrix}\text{Number}\\ \text{of ft}^2\end{pmatrix} \cdot \begin{pmatrix}\text{heat lost}\\ \text{per ft}^2\end{pmatrix} + \begin{pmatrix}\text{number}\\ \text{of ft}^2\end{pmatrix} \cdot \begin{pmatrix}\text{heat lost}\\ \text{per ft}^2\end{pmatrix} = \text{total heat lost.}$$

$$40 \cdot x + 60 \cdot y = 1920$$

A second wall containing 10 square feet of glass and 100 square feet of plaster lost 1160 Btu of heat.

Heat lost by the glass + heat lost by the plaster = total heat lost.

$$\begin{pmatrix}\text{Number}\\ \text{of ft}^2\end{pmatrix} \cdot \begin{pmatrix}\text{heat lost}\\ \text{per ft}^2\end{pmatrix} + \begin{pmatrix}\text{number}\\ \text{of ft}^2\end{pmatrix} \cdot \begin{pmatrix}\text{heat lost}\\ \text{per ft}^2\end{pmatrix} = \text{total heat lost.}$$

$$10 \cdot x + 100 \cdot y = 1160$$

Step 3. Solve the system and answer the problem's question. The system

$$\begin{cases} 40x + 60y = 1920 \\ 10x + 100y = 1160 \end{cases}$$

can be solved by addition. We'll multiply the second equation by -4 and then add equations to eliminate x.

$$\begin{cases} 40x + 60y = 1920 \xrightarrow{\text{No change}} \\ 10x + 100y = 1160 \xrightarrow{\text{Multiply by } -4.} \end{cases} \begin{cases} 40x + 60y = 1920 \\ \underline{-40x - 400y = -4640} \end{cases}$$

$$\text{Add:} \quad -340y = -2720$$

$$y = \frac{-2720}{-340} = 8$$

Now we can find the value of x by back-substituting 8 for y in either of the system's equations.

$10x + 100y = 1160$	We'll use the second equation.
$10x + 100(8) = 1160$	Back-substitute 8 for y.
$10x + 800 = 1160$	Multiply.
$10x = 360$	Subtract 800 from both sides.
$x = 36$	Divide both sides by 10.

We see that $x = 36$ and $y = 8$. Because x represents heat lost per square foot for the glass and y for the plaster, the glass lost 36 Btu of heat per square foot and the plaster lost 8 Btu per square foot.

Step 4. Check the proposed solution in the original wording of the problem. The problem states that the wall with 40 square feet of glass and 60 square feet of plaster lost 1920 Btu.

$$40(36) + 60(8) = 1440 + 480 = 1920 \text{ Btu of heat}$$

Proposed solution is 36 Btu per ft² for glass and 8 Btu per ft² for plaster.

Our proposed solution checks with the first statement. The problem also states that the wall with 10 square feet of glass and 100 square feet of plaster lost 1160 Btu.

$$10(36) + 100(8) = 360 + 800 = 1160 \text{ Btu of heat}$$

Our proposed solution also checks with the second statement. ■

✓ **CHECK POINT 1** University of Arkansas researchers discovered that we underestimate the number of calories in restaurant meals. The next time you eat out, take the number of calories you think you ate and double it. The researchers concluded that this number should be a more accurate estimate. The actual number of calories in one portion of hamburger and fries and two portions of fettuccine Alfredo is 4240. The actual number of calories in two portions of hamburger and fries and one portion of fettuccine Alfredo is 3980. Find the actual number of calories in each of these dishes. (*Source: Consumer Reports*, January/February, 2007)

Test Your Calorie I.Q.

Hamburger and Fries
Average guess:
777 calories

Fettuccine Alfredo
Average guess:
704 calories

Next, we will solve problems involving investments, mixtures, and motion with systems of equations. We will continue using our four-step problem-solving strategy. We will also use tables to help organize the information in the problems.

Dual Investments with Simple Interest

Simple interest involves interest calculated only on the amount of money that we invest, called the **principal**. The formula $I = Pr$ is used to find the simple interest, I, earned for one year when the principal, P, is invested at an annual interest rate, r. Dual investment problems involve different amounts of money in two or more investments, each paying a different rate.

EXAMPLE 2 Solving a Dual Investment Problem

Your grandmother needs your help. She has \$50,000 to invest. Part of this money is to be invested in noninsured bonds paying 15% annual interest. The rest of this money is to be invested in a government-insured certificate of deposit paying 7% annual interest. She told you that she requires \$6000 per year in extra income from both of these investments. How much money should be placed in each investment?

Solution

Step 1. Use variables to represent unknown quantities.

Let x = the amount invested in the 15% noninsured bonds.

Let y = the amount invested in the 7% certificate of deposit.

Step 2. Write a system of equations that models the problem's conditions. Because Grandma has \$50,000 to invest,

The amount invested at 15% | plus | the amount invested at 7% | equals | \$50,000.

$$x \quad + \quad y \quad = \quad 50{,}000$$

Furthermore, Grandma requires \$6000 in total interest. We can use a table to organize the information in the problem and obtain a second equation.

	Principal (amount invested) ×	Interest rate =	Interest earned
15% Investment	x	0.15	$0.15x$
7% Investment	y	0.07	$0.07y$

The interest for the two investments combined must be $6000.

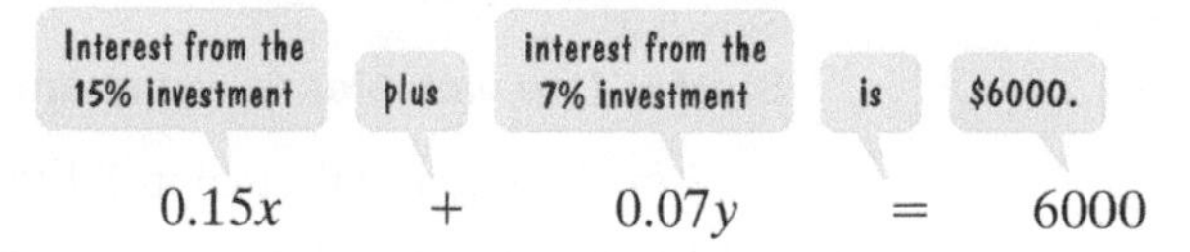

$$0.15x + 0.07y = 6000$$

Step 3. Solve the system and answer the problem's question. The system

$$\begin{cases} x + y = 50{,}000 \\ 0.15x + 0.07y = 6000 \end{cases}$$

can be solved by substitution or addition. Substitution works well because both variables in the first equation have coefficients of 1. Addition also works well; if we multiply the first equation by -0.15 or -0.07, adding equations will eliminate a variable. We will use addition.

$$\begin{cases} x + y = 50{,}000 \\ 0.15x + 0.07y = 6000 \end{cases} \xrightarrow[\text{No change}]{\text{Multiply by } -0.07.} \begin{cases} -0.07x - 0.07y = -3500 \\ 0.15x + 0.07y = 6000 \end{cases}$$

$$\text{Add: } 0.08x = 2500$$

$$x = \frac{2500}{0.08}$$

$$x = 31{,}250$$

Because x represents the amount that should be invested at 15%, Grandma should place $31,250 in 15% noninsured bonds. Now we can find y, the amount that she should place in the 7% certificate of deposit. We do so by back-substituting 31,250 for x in either of the system's equations.

$x + y = 50{,}000$ We'll use the first equation.

$31{,}250 + y = 50{,}000$ Back-substitute 31,250 for x.

$y = 18{,}750$ Subtract 31,250 from both sides.

Because $x = 31{,}250$ and $y = 18{,}750$, Grandma should invest $31,250 at 15% and $18,750 at 7%.

Step 4. Check the proposed answers in the original wording of the problem. Has Grandma invested $50,000?

$$\$31{,}250 + \$18{,}750 = \$50{,}000$$

Yes, all her money was placed in the dual investments. Can she count on $6000 interest? The interest earned on $31,250 at 15% is ($31,250)(0.15), or $4687.50. The interest earned on $18,750 at 7% is ($18,750)(0.07), or $1312.50. The total interest is $4687.50 + $1312.50, or $6000, exactly as it should be. You've made your grandmother happy. (Now if you would just visit her more often . . .) ■

✓ **CHECK POINT 2** You inherited $5000 with the stipulation that for the first year the money had to be invested in two funds paying 9% and 11% annual interest. How much did you invest at each rate if the total interest earned for the year was $487?

Problems Involving Mixtures

Chemists and pharmacists often have to change the concentration of solutions and other mixtures. In these situations, the amount of a particular ingredient in the solution or mixture is expressed as a percentage of the total.

EXAMPLE 3 Solving a Mixture Problem

A chemist working on a flu vaccine needs to mix a 10% sodium-iodine solution with a 60% sodium-iodine solution to obtain 50 milliliters of a 30% sodium-iodine solution. How many milliliters of the 10% solution and of the 60% solution should be mixed?

Solution

Step 1. Use variables to represent unknown quantities.

Let x = the number of milliliters of the 10% solution to be used in the mixture.

Let y = the number of milliliters of the 60% solution to be used in the mixture.

Step 2. Write a system of equations that models the problem's conditions. The situation is illustrated in **Figure 3.6**. The chemist needs 50 milliliters of a 30% sodium-iodine solution. We form a table that shows the amount of sodium-iodine in each of the three solutions.

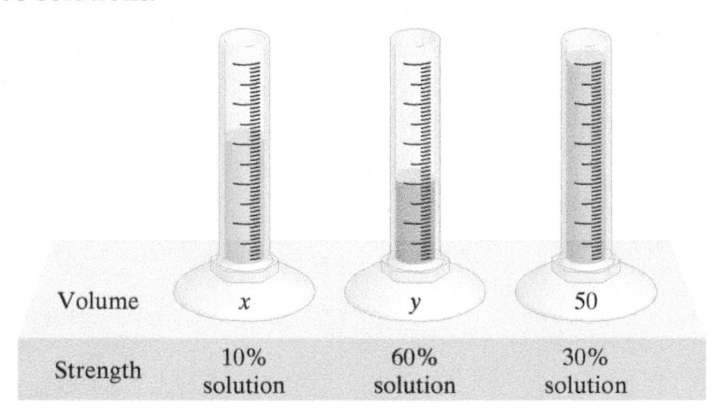
Volume
x
y
50
Strength
10% solution
60% solution
30% solution

Figure 3.6

Solution	Number of Milliliters	×	Percent of Sodium-Iodine	=	Amount of Sodium-Iodine
10% Solution	x		$10\% = 0.1$		$0.1x$
60% Solution	y		$60\% = 0.6$		$0.6y$
30% Mixture	50		$30\% = 0.3$		$0.3(50) = 15$

The chemist needs to obtain a 50-milliliter mixture.

The number of milliliters used of the 10% solution | plus | the number of milliliters used of the 60% solution | must equal | 50 milliliters.

$$x + y = 50$$

The 50-milliliter mixture must be 30% sodium-iodine. The amount of sodium-iodine must be 30% of 50, or $(0.3)(50) = 15$ milliliters.

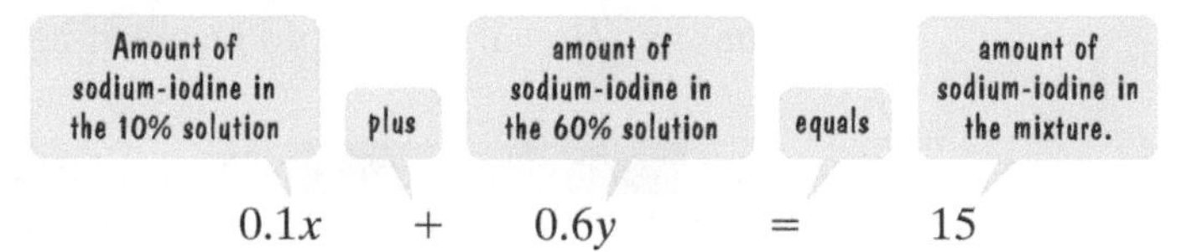
Amount of sodium-iodine in the 10% solution
plus
amount of sodium-iodine in the 60% solution
equals
amount of sodium-iodine in the mixture.

$$0.1x + 0.6y = 15$$

Step 3. Solve the system and answer the problem's question. The system

$$\begin{cases} x + y = 50 \\ 0.1x + 0.6y = 15 \end{cases}$$

can be solved by substitution or addition. Let's use substitution. The first equation can easily be solved for x or y. Solving for y, we obtain $y = 50 - x$.

$$y = \boxed{50 - x} \qquad 0.1x + 0.6\boxed{y} = 15$$

We substitute $50 - x$ for y in the second equation. This gives us an equation in one variable.

$$\begin{aligned} 0.1x + 0.6(50 - x) &= 15 && \text{This equation contains one variable, } x. \\ 0.1x + 30 - 0.6x &= 15 && \text{Apply the distributive property.} \\ -0.5x + 30 &= 15 && \text{Combine like terms.} \\ -0.5x &= -15 && \text{Subtract 30 from both sides.} \\ x = \frac{-15}{-0.5} &= 30 && \text{Divide both sides by } -0.5. \end{aligned}$$

Back-substituting 30 for x in either of the system's equations ($x + y = 50$ is easier to use) gives $y = 20$. Because x represents the number of milliliters of the 10% solution and y the number of milliliters of the 60% solution, the chemist should mix 30 milliliters of the 10% solution with 20 milliliters of the 60% solution.

Step 4. Check the proposed solution in the original wording of the problem. The problem states that the chemist needs 50 milliliters of a 30% sodium-iodine solution. The amount of sodium-iodine in this mixture is 0.3(50), or 15 milliliters. The amount of sodium-iodine in 30 milliliters of the 10% solution is 0.1(30), or 3 milliliters. The amount of sodium-iodine in 20 milliliters of the 60% solution is $0.6(20) = 12$ milliliters. The amount of sodium-iodine in the two solutions used in the mixture is 3 milliliters + 12 milliliters, or 15 milliliters, exactly as it should be. ■

✓ CHECK POINT 3 A chemist needs to mix a 12% acid solution with a 20% acid solution to obtain 160 ounces of a 15% acid solution. How many ounces of each of the acid solutions must be used?

Great Question!

Are there similarities between dual investment problems and mixture problems?

Problems involving dual investments and problems involving mixtures are both based on the same idea: The total amount times the rate gives the amount.

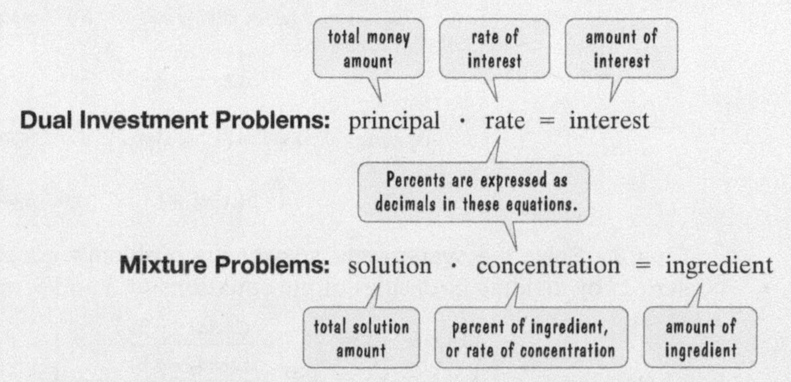
total money amount
rate of interest
amount of interest
Dual Investment Problems: principal · rate = interest
Percents are expressed as decimals in these equations.
Mixture Problems: solution · concentration = ingredient
total solution amount
percent of ingredient, or rate of concentration
amount of ingredient

Our dual investment problem involved mixing two investments. Our mixture problem involved mixing two liquids. The equations in these problems are obtained from similar conditions:

Dual Investment Problems **Mixture Problems**

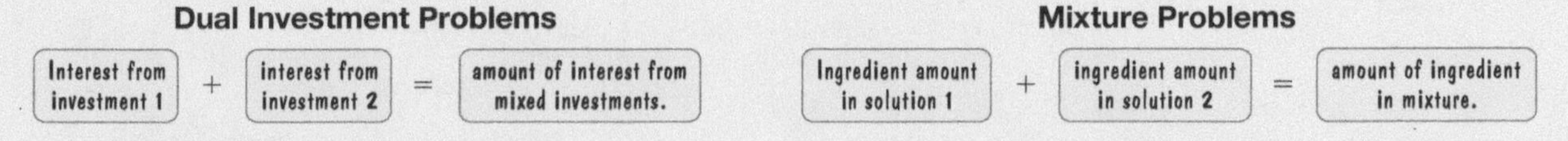
Interest from investment 1
+
interest from investment 2
=
amount of interest from mixed investments.
Ingredient amount in solution 1
+
ingredient amount in solution 2
=
amount of ingredient in mixture.

Being aware of the similarities between dual investment and mixture problems should make you a better problem solver in a variety of situations that involve mixtures.

Problems Involving Motion

We have seen that the distance, d, covered by any moving body is the product of its average rate, r, and its time in motion, t:

$$d = rt. \quad \text{Distance equals rate times time.}$$

Wind and water current have the effect of increasing or decreasing a traveler's rate.

Great Question!

Is it always necessary to use x and y to represent a problem's variables?

No. Select letters that help you remember what the variables represent. For example, in Example 4, you may prefer using p and w rather than x and y:

p = plane's average rate in still air

w = wind's average rate.

EXAMPLE 4 Solving a Motion Problem

When a small airplane flies with the wind, it can travel 450 miles in 3 hours. When the same airplane flies in the opposite direction against the wind, it takes 5 hours to fly the same distance. Find the average rate of the plane in still air and the average rate of the wind.

Solution

Step 1. Use variables to represent unknown quantities.

Let x = the average rate of the plane in still air.

Let y = the average rate of the wind.

Step 2. Write a system of equations that models the problem's conditions. As it travels with the wind, the plane's rate is increased. The net rate is its rate in still air, x, plus the rate of the wind, y, given by the expression $x + y$. As it travels against the wind, the plane's rate is decreased. The net rate is its rate in still air, x, minus the rate of the wind, y, given by the expression $x - y$. Here is a chart that summarizes the problem's information and includes the increased and decreased rates.

	Rate	×	Time	=	Distance
Trip with the Wind	$x + y$		3		$3(x + y)$
Trip against the Wind	$x - y$		5		$5(x - y)$

The problem states that the distance in each direction is 450 miles. We use this information to write our system of equations.

The distance of the trip with the wind is 450 miles.

$$3(x + y) = 450$$

The distance of the trip against the wind is 450 miles.

$$5(x - y) = 450$$

Step 3. Solve the system and answer the problem's question. We can simplify the system by dividing both sides of the equations by 3 and 5, respectively.

$$\begin{cases} 3(x + y) = 450 \\ 5(x - y) = 450 \end{cases} \xrightarrow[\text{Divide by 5.}]{\text{Divide by 3.}} \begin{cases} x + y = 150 \\ x - y = 90 \end{cases}$$

Solve the system on the right by the addition method.

$$\begin{cases} x + y = 150 \\ \underline{x - y = \ \ 90} \end{cases}$$

$$\text{Add: } 2x \quad = 240$$

$$x = 120 \quad \text{Divide both sides by 2.}$$

Back-substituting 120 for x in either of the system's equations gives $y = 30$. Because $x = 120$ and $y = 30$, the average rate of the plane in still air is 120 miles per hour and the average rate of the wind is 30 miles per hour.

Step 4. Check the proposed solution in the original wording of the problem. The problem states that the distance in each direction is 450 miles. The average rate of the plane with the wind is $120 + 30 = 150$ miles per hour. In 3 hours, it travels $150 \cdot 3$, or 450 miles, which checks with the stated condition. Furthermore, the average rate of the plane against the wind is $120 - 30 = 90$ miles per hour. In 5 hours, it travels $90 \cdot 5 = 450$ miles, which is the stated distance. ■

✓ CHECK POINT 4 With the current, a motorboat can travel 84 miles in 2 hours. Against the current, the same trip takes 3 hours. Find the average rate of the boat in still water and the average rate of the current.

---

2 Use functions to model revenue, cost, and profit, and perform a break-even analysis.

Functions of Business: Break-Even Analysis

Suppose that a company produces and sells x units of a product. Its *revenue* is the money generated by selling x units of the product. Its *cost* is the cost of producing x units of the product.

Revenue and Cost Functions

A company produces and sells x units of a product.

Revenue Function

$$R(x) = (\text{price per unit sold})x$$

Cost Function

$$C(x) = \text{fixed cost} + (\text{cost per unit produced})x$$

The point of intersection of the graphs of the revenue and cost functions is called the **break-even point**. The x-coordinate of the point reveals the number of units that a company must produce and sell so that money coming in, the revenue, is equal to money going out, the cost. The y-coordinate of the break-even point gives the amount of money coming in and going out. Example 5 illustrates the use of the substitution method in determining a company's break-even point.

EXAMPLE 5 Finding a Break-Even Point

Technology is now promising to bring light, fast, and beautiful wheelchairs to millions of disabled people. A company is planning to manufacture these radically different wheelchairs. Fixed cost will be \$500,000 and it will cost \$400 to produce each wheelchair. Each wheelchair will be sold for \$600.

a. Write the cost function, C, of producing x wheelchairs.

b. Write the revenue function, R, from the sale of x wheelchairs.

c. Determine the break-even point. Describe what this means.

Solution

a. The cost function is the sum of the fixed cost and variable cost.

Fixed cost of \$500,000 | plus | Variable cost: \$400 for each chair produced

$$C(x) = 500{,}000 + 400x$$

b. The revenue function is the money generated from the sale of x wheelchairs.

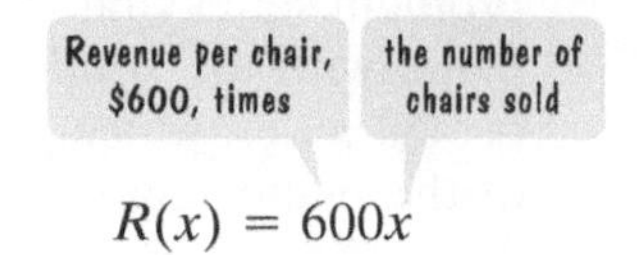

$$R(x) = 600x$$

c. The break-even point occurs where the graphs of C and R intersect. Thus, we find this point by solving the system

$$\begin{cases} C(x) = 500{,}000 + 400x \\ R(x) = 600x \end{cases} \quad \text{or} \quad \begin{cases} y = 500{,}000 + 400x \\ y = 600x. \end{cases}$$

Using the substitution method, we can substitute $600x$ for y in the first equation.

$600x = 500{,}000 + 400x$ Substitute $600x$ for y in $y = 500{,}000 + 400x$.

$200x = 500{,}000$ Subtract $400x$ from both sides.

$x = 2500$ Divide both sides by 200.

Back-substituting 2500 for x in either of the system's equations (or functions), we obtain

$$R(2500) = 600(2500) = 1{,}500{,}000.$$

We used $R(x) = 600x$.

The break-even point is (2500, 1,500,000). This means that the company will break even if it produces and sells 2500 wheelchairs. At this level, the money coming in is equal to the money going out: $1,500,000. ■

Figure 3.7 shows the graphs of the revenue and cost functions for the wheelchair business. Similar graphs and models apply no matter how small or large a business venture may be.

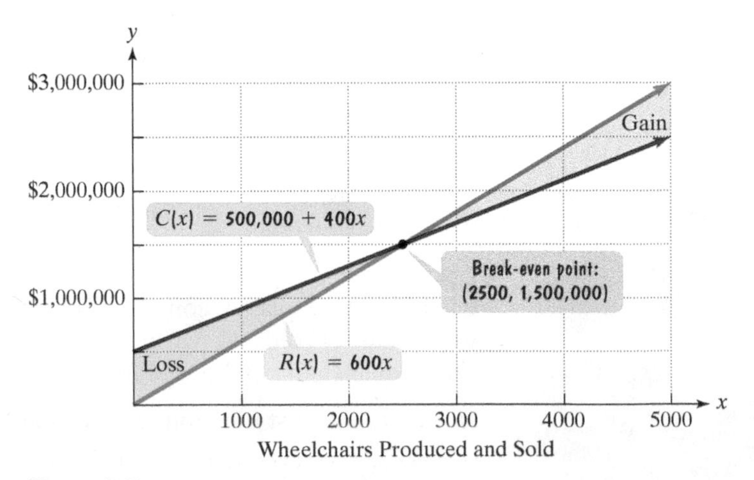

Figure 3.7

The intersection point confirms that the company breaks even by producing and selling 2500 wheelchairs. Can you see what happens for $x < 2500$? The red cost graph lies above the blue revenue graph. The cost is greater than the revenue and the business is losing money. Thus, if they sell fewer than 2500 wheelchairs, the result is a *loss*. By contrast, look at what happens for $x > 2500$. The blue revenue graph lies above the red cost graph. The revenue is greater than the cost and the business is making money. Thus, if they sell more than 2500 wheelchairs, the result is a *gain*.

✓ CHECK POINT 5 A company that manufactures running shoes has a fixed cost of $300,000. Additionally, it costs $30 to produce each pair of shoes. The shoes are sold at $80 per pair.

a. Write the cost function, C, of producing x pairs of running shoes.

b. Write the revenue function, R, from the sale of x pairs of running shoes.

c. Determine the break-even point. Describe what this means.

What does every entrepreneur, from a kid selling lemonade to Mark Zuckerberg, want to do? Generate profit, of course. The *profit* made is the money taken in, or the revenue, minus the money spent, or the cost. This relationship between revenue and cost allows us to define the *profit function,* $P(x)$.

The Profit Function

The profit, $P(x)$, generated after producing and selling x units of a product is given by the **profit function**

$$P(x) = R(x) - C(x),$$

where R and C are the revenue and cost functions, respectively.

EXAMPLE 6 Writing a Profit Function

Use the revenue and cost functions for the wheelchair business in Example 5,

$$R(x) = 600x \quad \text{and} \quad C(x) = 500{,}000 + 400x,$$

to write the profit function for producing and selling x wheelchairs.

Solution The profit function is the difference between the revenue function and the cost function.

$$\begin{aligned} P(x) &= R(x) - C(x) && \text{This is the definition of the profit function.} \\ &= 600x - (500{,}000 + 400x) && \text{Substitute the given functions.} \\ &= 600x - 500{,}000 - 400x && \text{Distribute } -1 \text{ to each term in parentheses.} \\ &= 200x - 500{,}000 && \text{Simplify: } 600x - 400x = 200x. \end{aligned}$$

The profit function is $P(x) = 200x - 500{,}000$. ■

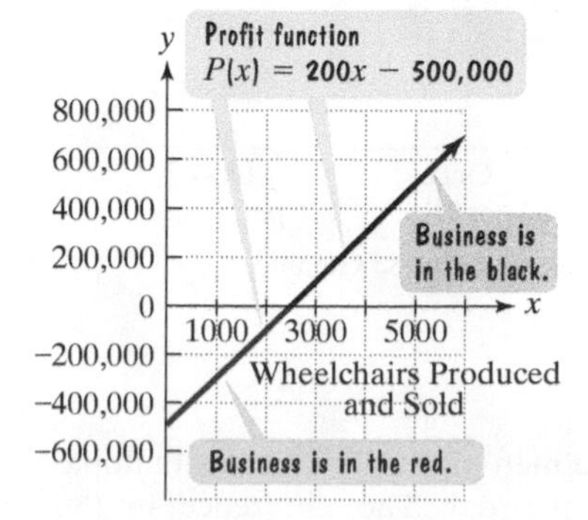

Figure 3.8

The graph of the profit function for the wheelchair business, $P(x) = 200x - 500{,}000$, is shown in **Figure 3.8**. The red portion lies below the x-axis and shows a loss when fewer than 2500 wheelchairs are sold. The business is "in the red." The black portion lies above the x-axis and shows a gain when more than 2500 wheelchairs are sold. The wheelchair business is "in the black."

✓ CHECK POINT 6 Use the revenue and cost functions that you obtained in Check Point 5 to write the profit function for producing and selling x pairs of running shoes.

Achieving Success

Avoid asking for help on a problem that you have not thought about. This is basically asking your instructor to do the work for you. First try solving the problem on your own!

CONCEPT AND VOCABULARY CHECK

Fill in each blank so that the resulting statement is true.

1. If computers sell for \$1180 per unit and hard drives sell for \$125 per unit, the revenue from x computers and y hard drives can be represented by ____________.
2. The combined yearly interest for x dollars invested at 12% and y dollars invested at 9% is represented by ____________.
3. The total amount of acid in x milliliters of a 9% acid solution and y milliliters of a 60% acid solution is represented by ____________.
4. If x represents the average rate of a plane in still air and y represents the average rate of the wind, the plane's rate with the wind is represented by ____________ and the plane's rate against the wind is represented by ____________.
5. If $x + y$ represents a motorboat's rate with the current, in miles per hour, its distance after 4 hours is represented by ____________.
6. A company's ____________ function is the money generated by selling x units of its product. The difference between this function and the company's cost function is called its ____________ function.
7. A company has a graph that shows the money it generates by selling x units of its product. It also has a graph that shows its cost of producing x units of its product. The point of intersection of these graphs is called the company's ____________.

3.2 EXERCISE SET MyMathLab®

Practice Exercises

In Exercises 1–4, let x represent one number and let y represent the other number. Use the given conditions to write a system of equations. Solve the system and find the numbers.

1. The sum of two numbers is 7. If one number is subtracted from the other, the result is −1. Find the numbers.
2. The sum of two numbers is 2. If one number is subtracted from the other, the result is 8. Find the numbers.
3. Three times a first number decreased by a second number is 1. The first number increased by twice the second number is 12. Find the numbers.
4. The sum of three times a first number and twice a second number is 8. If the second number is subtracted from twice the first number, the result is 3. Find the numbers.

In Exercises 5–8, cost and revenue functions for producing and selling x units of a product are given. Cost and revenue are expressed in dollars.

a. *Find the number of units that must be produced and sold to break even. At this level, what is the dollar amount coming in and going out?*

b. *Write the profit function from producing and selling x units of the product.*

5. $C(x) = 25{,}500 + 15x$
   $R(x) = 32x$
6. $C(x) = 15{,}000 + 12x$
   $R(x) = 32x$
7. $C(x) = 105x + 70{,}000$
   $R(x) = 245x$
8. $C(x) = 1.2x + 1500$
   $R(x) = 1.7x$

Application Exercises

In Exercises 9–40, use the four-step strategy to solve each problem.

9. The current generation of college students grew up playing interactive online games, and many continue to play in college. The bar graph shows the percentage of U.S. college students playing online games, by gender.

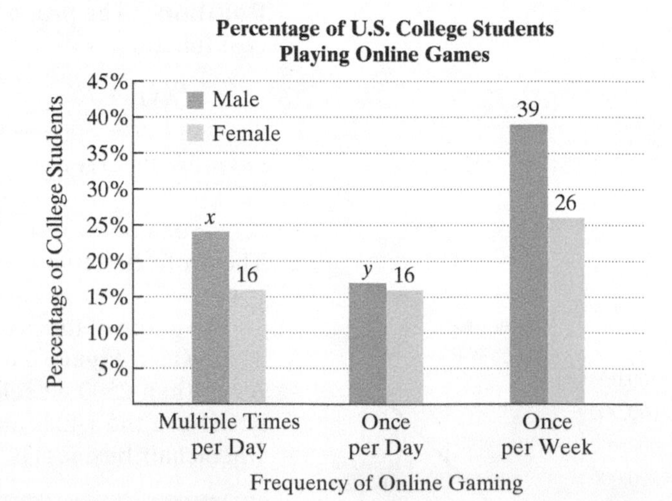

Source: statista.com

A total of 41% of college men play online games multiple times per day or once per day. The difference in the percentage who play multiple times per day and once per day is 7%. Find the percentage of college men who play online games multiple times per day and the percentage of college men who play online games once per day.

10. A number of studies have emphasized the importance of sleep for students' success in their academic performance. The bar graph shows the actual sleep hours and the sleep hours to function best for U.S. college students.

Sleep Hours of U.S. College Students

Actual sleep hours
Sleep hours to function best

Sleep Hours	Actual sleep hours	Sleep hours to function best
[3, 5)	5	2
[5, 7)	x	12
[7, 9)	50	66
[9, 11)	y	20

Percentage of College Students: 10%, 20%, 30%, 40%, 50%, 60%, 70%, 80%

Sleep Hours

Source: Chiang et al., "The Effects of Sleep on Academic Performance and Job Performance," *College Student Journal*, Spring 2014

A total of 45% of college students sleep between 5 and 7 hours or between 9 and 11 hours. The difference in the percentage who sleep between 5 and 7 hours and between 9 and 11 hours is 41%. Find the percentage of college students who sleep between 5 and 7 hours and the percentage of college students who sleep between 9 and 11 hours.

11. One week a computer store sold a total of 36 computers and external hard drives. The revenue from these sales was $27,710. If computers sold for $1180 per unit and hard drives for $125 per unit, how many of each did the store sell?

12. There were 180 people at a civic club fundraiser. Members paid $4.50 per ticket and nonmembers paid $8.25 per ticket. If total receipts amounted to $1222.50, how many members and how many nonmembers attended the fundraiser?

13. You invested $7000 in two accounts paying 6% and 8% annual interest. If the total interest earned for the year was $520, how much was invested at each rate?

14. You invested $11,000 in stocks and bonds, paying 5% and 8% annual interest. If the total interest earned for the year was $730, how much was invested in stocks and how much was invested in bonds?

15. You invested money in two funds. Last year, the first fund paid a dividend of 9% and the second a dividend of 3%, and you received a total of $900. This year, the first fund paid a 10% dividend and the second only 1%, and you received a total of $860. How much money did you invest in each fund?

16. You invested money in two funds. Last year, the first fund paid a dividend of 8% and the second a dividend of 5%, and you received a total of $1330. This year, the first fund paid a 12% dividend and the second only 2%, and you received a total of $1500. How much money did you invest in each fund?

17. Things did not go quite as planned. You invested $20,000, part of it in a stock with a 12% annual return. However, the rest of the money suffered a 5% loss. If the total annual income from both investments was $1890, how much was invested at each rate?

18. Things did not go quite as planned. You invested $30,000, part of it in a stock with a 14% annual return. However, the rest of the money suffered a 6% loss. If the total annual income from both investments was $200, how much was invested at each rate?

19. A wine company needs to blend a California wine with a 5% alcohol content and a French wine with a 9% alcohol content to obtain 200 gallons of wine with a 7% alcohol content. How many gallons of each kind of wine must be used?

20. A jeweler needs to mix an alloy with a 16% gold content and an alloy with a 28% gold content to obtain 32 ounces of a new alloy with a 25% gold content. How many ounces of each of the original alloys must be used?

21. For thousands of years, gold has been considered one of Earth's most precious metals. One hundred percent pure gold is 24-karat gold, which is too soft to be made into jewelry. In the United States, most gold jewelry is 14-karat gold, approximately 58% gold. If 18-karat gold is 75% gold and 12-karat gold is 50% gold, how much of each should be used to make a 14-karat gold bracelet weighing 300 grams?

22. In the "Peanuts" cartoon shown, solve the problem that is sending Peppermint Patty into an agitated state. How much cream and how much milk, to the nearest hundredth of a gallon, must be mixed together to obtain 50 gallons of cream that contains 12.5% butterfat?

23. The manager of a candystand at a large multiplex cinema has a popular candy that sells for \$1.60 per pound. The manager notices a different candy worth \$2.10 per pound that is not selling well. The manager decides to form a mixture of both types of candy to help clear the inventory of the more expensive type. How many pounds of each kind of candy should be used to create a 75-pound mixture selling for \$1.90 per pound?

24. A grocer needs to mix raisins at \$2.00 per pound with granola at \$3.25 per pound to obtain 10 pounds of a mixture that costs \$2.50 per pound. How many pounds of raisins and how many pounds of granola must be used?

25. A coin purse contains a mixture of 15 coins in nickels and dimes. The coins have a total value of \$1.10. Determine the number of nickels and the number of dimes in the purse.

26. A coin purse contains a mixture of 15 coins in dimes and quarters. The coins have a total value of \$3.30. Determine the number of dimes and the number of quarters in the purse.

27. When a small plane flies with the wind, it can travel 800 miles in 5 hours. When the plane flies in the opposite direction, against the wind, it takes 8 hours to fly the same distance. Find the rate of the plane in still air and the rate of the wind.

28. When a plane flies with the wind, it can travel 4200 miles in 6 hours. When the plane flies in the opposite direction, against the wind, it takes 7 hours to fly the same distance. Find the rate of the plane in still air and the rate of the wind.

29. A boat's crew rowed 16 kilometers downstream, with the current, in 2 hours. The return trip upstream, against the current, covered the same distance, but took 4 hours. Find the crew's rowing rate in still water and the rate of the current.

30. A motorboat traveled 36 miles downstream, with the current, in 1.5 hours. The return trip upstream, against the current, covered the same distance, but took 2 hours. Find the boat's rate in still water and the rate of the current.

31. With the current, you can canoe 24 miles in 4 hours. Against the same current, you can canoe only $\frac{3}{4}$ of this distance in 6 hours. Find your rate in still water and the rate of the current.

32. With the current, you can row 24 miles in 3 hours. Against the same current, you can row only $\frac{2}{3}$ of this distance in 4 hours. Find your rowing rate in still water and the rate of the current.

33. A student has two test scores. The difference between the scores is 12 and the mean, or average, of the scores is 80. What are the two test scores?

34. A student has two test scores. The difference between the scores is 8 and the mean, or average, of the scores is 88. What are the two test scores?

In Exercises 35–36, an isosceles triangle containing two angles with equal measure is shown. The degree measure of each triangle's three interior angles and an exterior angle is represented with variables. Find the measure of the three interior angles.

35.

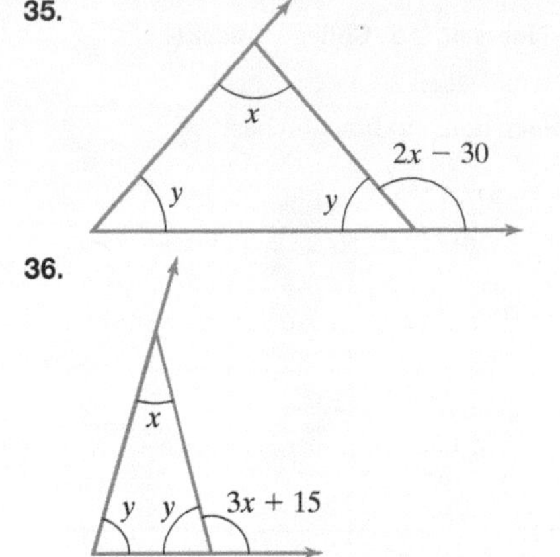

36.

37. A rectangular lot whose perimeter is 220 feet is fenced along three sides. An expensive fencing along the lot's length costs \$20 per foot, and an inexpensive fencing along the two side widths costs only \$8 per foot. The total cost of the fencing along the three sides comes to \$2040. What are the lot's dimensions?

38. A rectangular lot whose perimeter is 260 feet is fenced along three sides. An expensive fencing along the lot's length costs \$16 per foot, and an inexpensive fencing along the two side widths costs only \$5 per foot. The total cost of the fencing along the three sides comes to \$1780. What are the lot's dimensions?

39. A new restaurant is to contain two-seat tables and four-seat tables. Fire codes limit the restaurant's maximum occupancy to 56 customers. If the owners have hired enough servers to handle 17 tables of customers, how many of each kind of table should they purchase?

40. A hotel has 200 rooms. Those with kitchen facilities rent for \$100 per night and those without kitchen facilities rent for \$80 per night. On a night when the hotel was completely occupied, revenues were \$17,000. How many of each type of room does the hotel have?

The figure shows the graphs of the cost and revenue functions for a company that manufactures and sells small radios. Use the information in the figure to solve Exercises 41–46.

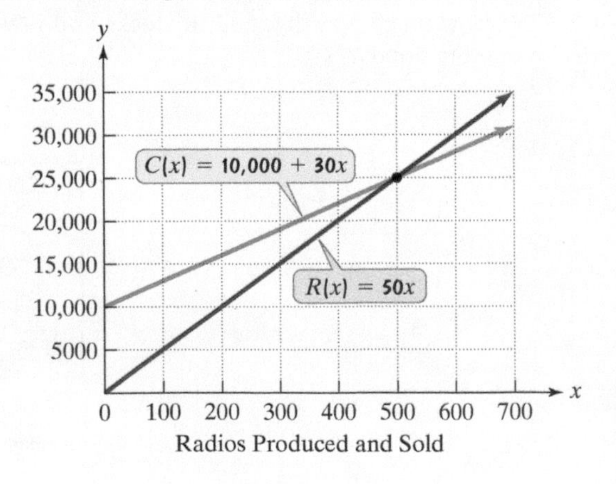

41. How many radios must be produced and sold for the company to break even?

42. More than how many radios must be produced and sold for the company to have a profit?

43. Use the formulas shown in the voice balloons to find $R(200) - C(200)$. Describe what this means for the company.

44. Use the formulas shown in the voice balloons to find $R(300) - C(300)$. Describe what this means for the company.

45. a. Use the formulas shown in the voice balloons to write the company's profit function, P, from producing and selling x radios.
    b. Find the company's profit if 10,000 radios are produced and sold.

46. a. Use the formulas shown in the voice balloons to write the company's profit function, P, from producing and selling x radios.
    b. Find the company's profit if 20,000 radios are produced and sold.

Exercises 47–50 describe a number of business ventures. For each exercise,

a. *Write the cost function, C.*

b. *Write the revenue function, R.*

c. *Determine the break-even point. Describe what this means.*

47. A company that manufactures small canoes has a fixed cost of \$18,000. It costs \$20 to produce each canoe. The selling price is \$80 per canoe. (In solving this exercise, let x represent the number of canoes produced and sold.)

48. A company that manufactures bicycles has a fixed cost of \$100,000. It costs \$100 to produce each bicycle. The selling price is \$300 per bike. (In solving this exercise, let x represent the number of bicycles produced and sold.)

49. You invest in a new play. The cost includes an overhead of \$30,000, plus production costs of \$2500 per performance. A sold-out performance brings in \$3125. (In solving this exercise, let x represent the number of sold-out performances.)

50. You invested \$30,000 and started a business writing greeting cards. Supplies cost 2¢ per card and you are selling each card for 50¢. (In solving this exercise, let x represent the number of cards produced and sold.)

Explaining the Concepts

51. Describe the conditions in a problem that enable it to be solved using a system of linear equations.

52. Write a word problem that can be solved by translating to a system of linear equations. Then solve the problem.

53. Describe a revenue function for a business venture.

54. Describe a cost function for a business venture. What are the two kinds of costs that are modeled by this function?

55. What is the profit function for a business venture and how is it determined?

56. Describe the break-even point for a business.

57. The law of supply and demand states that, in a free market economy, a commodity tends to be sold at its equilibrium price. At this price, the amount that the seller will supply is the same amount that the consumer will buy. Explain how graphs can be used to determine the equilibrium price.

58. Many students hate mixture problems and decide to ignore them, stating, "I'll just skip that one on the test." If you share this opinion, describe what you find particularly unappealing about this kind of problem.

Technology Exercises

In Exercises 59–60, graph the revenue and cost functions in the same viewing rectangle. Then use the intersection feature to determine the break-even point.

59. $R(x) = 50x, \quad C(x) = 20x + 180$

60. $R(x) = 92.5x, \quad C(x) = 52x + 1782$

61. Use the procedure in Exercises 59–60 to verify your work for any one of the break-even points that you found in Exercises 47–50.

Critical Thinking Exercises

Make Sense? *In Exercises 62–65, determine whether each statement makes sense or does not make sense, and explain your reasoning.*

62. A system of linear equations can be used to model and compare the fees charged by two different taxicab companies.

63. I should mix 6 liters of a 50% acid solution with 4 liters of a 25% acid solution to obtain 10 liters of a 75% acid solution.

64. If I know the perimeter of this rectangle and triangle, each in the same unit of measure, I can use a system of linear equations to determine values for x and y.

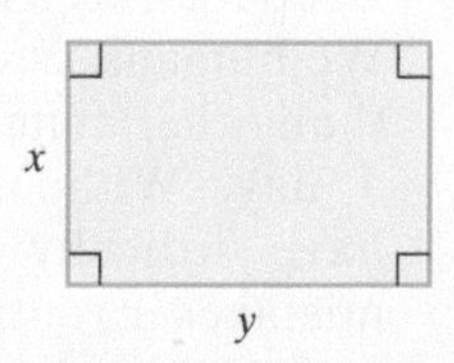

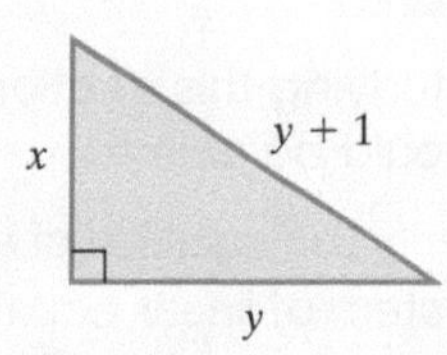

65. You told me that you flew against the wind from Miami to Seattle in 7 hours and, at the same time, your friend flew with the wind from Seattle to Miami in only 5.6 hours. You have not given me enough information to determine the average rate of the wind.

66. The radiator in your car contains 4 gallons of antifreeze and water. The mixture is 45% antifreeze. How much of this mixture should be drained and replaced with pure antifreeze in order to have a 60% antifreeze solution? Round to the nearest tenth of a gallon.

67. A marching band has 52 members, and there are 24 in the pom-pom squad. They wish to form several hexagons and squares like those diagrammed below. Can it be done with no people left over?

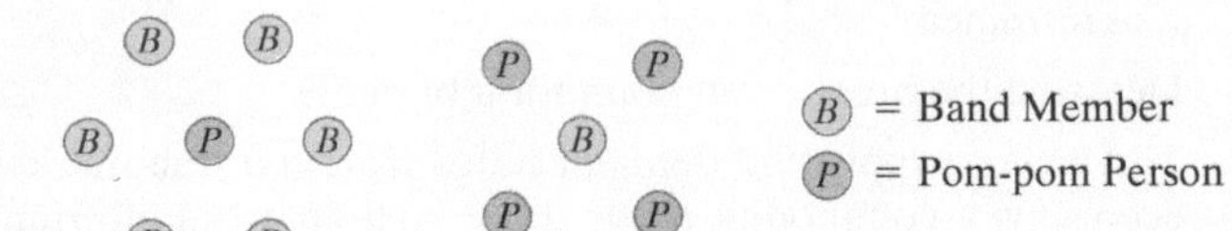

68. A boy has as many brothers as he has sisters. Each of his sisters has twice as many brothers as she has sisters. How many boys and girls are in this family?

69. When entering your test score into a computer, your professor accidently reversed the two digits. This error reduced your score by 36 points. Your professor told you that the sum of the digits of your actual score was 14, corrected the error, and agreed to give you extra credit if you could determine the actual score without looking back at the test. What was your actual test score? (*Hint*: Let $t =$ the tens-place digit of your actual score and let $u =$ the units-place digit of your actual score. Thus, $10t + u$ represents your actual test score.)

70. A dealer paid a total of $67 for mangos and avocados. The mangos were sold at a profit of 20% on the dealer's cost, but the avocados started to spoil, resulting in a selling price of a 2% loss on the dealer's cost. The dealer made a profit of $8.56 on the total transaction. How much did the dealer pay for the mangos and for the avocados?

Review Exercises

In Exercises 71–72, use the given conditions to write an equation for each line in point-slope form and slope-intercept form.

71. Passing through $(-2, 5)$ and $(-6, 13)$ (Section 2.5, Example 2)

72. Passing through $(-3, 0)$ and parallel to the line whose equation is $-x + y = 7$ (Section 2.5, Example 4)

73. Find the domain of $g(x) = \frac{x-2}{3-x}$. (Section 2.3, Example 1)

Preview Exercises

Exercises 74–76 will help you prepare for the material covered in the next section.

74. If $x = 3$, $y = 2$, and $z = -3$, does the ordered triple (x, y, z) satisfy the equation $2x - y + 4z = -8$?

75. Consider the following equations:

$$\begin{cases} 5x - 2y - 4z = 3 & \text{Equation 1} \\ 3x + 3y + 2z = -3. & \text{Equation 2} \end{cases}$$

Eliminate z by copying Equation 1, multiplying Equation 2 by 2, and then adding the equations.

76. Write an equation involving a, b, and c based on the following description:

When the value of x in $y = ax^2 + bx + c$ is 4, the value of y is 1682.

SECTION

3.3 Systems of Linear Equations in Three Variables

What am I supposed to learn?

After studying this section, you should be able to:

1. Verify the solution of a system of linear equations in three variables.
2. Solve systems of linear equations in three variables.
3. Identify inconsistent and dependent systems.
4. Solve problems using systems in three variables.

All animals sleep, but the length of time they sleep varies widely: Cattle sleep for only a few minutes at a time. We humans seem to need more sleep than other animals, up to eight hours a day. Without enough sleep, we have difficulty concentrating, make mistakes in routine tasks, lose energy, and feel bad-tempered. There is a relationship between hours of sleep and death rate per year per 100,000 people. How many hours of sleep will put you in the group with the minimum death rate? In this section, we will answer this question by solving a system of linear equations with more than two variables.

1 Verify the solution of a system of linear equations in three variables.

Systems of Linear Equations in Three Variables and Their Solutions

An equation such as $x + 2y - 3z = 9$ is called a *linear equation in three variables*. In general, any equation of the form

$$Ax + By + Cz = D,$$

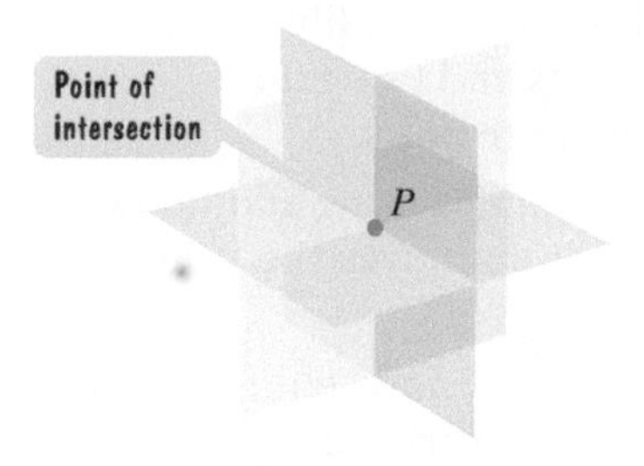

Figure 3.9

where A, B, C, and D are real numbers such that A, B, and C are not all 0, is a **linear equation in three variables: x, y, and z.** The graph of this linear equation in three variables is a plane in three-dimensional space.

The process of solving a system of three linear equations in three variables is geometrically equivalent to finding the point of intersection (assuming that there is one) of three planes in space (see **Figure 3.9**). A **solution** of a system of linear equations in three variables is an ordered triple of real numbers that satisfies all equations in the system. The **solution set** of the system is the set of all its solutions.

EXAMPLE 1 Determining Whether an Ordered Triple Satisfies a System

Show that the ordered triple $(-1, 2, -2)$ is a solution of the system:

$$\begin{cases} x + 2y - 3z = 9. \\ 2x - y + 2z = -8. \\ -x + 3y - 4z = 15. \end{cases}$$

Solution Because -1 is the x-coordinate, 2 is the y-coordinate, and -2 is the z-coordinate of $(-1, 2, -2)$, we replace x with -1, y with 2, and z with -2 in each of the three equations.

$$\begin{aligned} x + 2y - 3z &= 9 \\ -1 + 2(2) - 3(-2) &\stackrel{?}{=} 9 \\ -1 + 4 + 6 &\stackrel{?}{=} 9 \\ 9 &= 9, \quad \text{true} \end{aligned}$$

$$\begin{aligned} 2x - y + 2z &= -8 \\ 2(-1) - 2 + 2(-2) &\stackrel{?}{=} -8 \\ -2 - 2 - 4 &\stackrel{?}{=} -8 \\ -8 &= -8, \quad \text{true} \end{aligned}$$

$$\begin{aligned} -x + 3y - 4z &= 15 \\ -(-1) + 3(2) - 4(-2) &\stackrel{?}{=} 15 \\ 1 + 6 + 8 &\stackrel{?}{=} 15 \\ 15 &= 15, \quad \text{true} \end{aligned}$$

The ordered triple $(-1, 2, -2)$ satisfies the three equations: It makes each equation true. Thus, the ordered triple is a solution of the system. ■

CHECK POINT 1 Show that the ordered triple $(-1, -4, 5)$ is a solution of the system:

$$\begin{cases} x - 2y + 3z = 22 \\ 2x - 3y - z = 5 \\ 3x + y - 5z = -32. \end{cases}$$

2 Solve systems of linear equations in three variables.

Solving Systems of Linear Equations in Three Variables by Eliminating Variables

The method for solving a system of linear equations in three variables is similar to that used on systems of linear equations in two variables. We use addition to eliminate any variable, reducing the system to two equations in two variables. Once we obtain a system of two equations in two variables, we use addition or substitution to eliminate a variable. The result is a single equation in one variable. We solve this equation to get the value of the remaining variable. Other variable values are found by back-substitution.

Great Question!

When solving a linear system in three variables, which variable should I eliminate first?

It does not matter which variable you eliminate first, as long as you eliminate the same variable in two different pairs of equations.

Solving Linear Systems in Three Variables by Eliminating Variables

1. Reduce the system to two equations in two variables. This is usually accomplished by taking two different pairs of equations and using the addition method to eliminate the same variable from both pairs.
2. Solve the resulting system of two equations in two variables using addition or substitution. The result is an equation in one variable that gives the value of that variable.
3. Back-substitute the value of the variable found in step 2 into either of the equations in two variables to find the value of the second variable.
4. Use the values of the two variables from steps 2 and 3 to find the value of the third variable by back-substituting into one of the original equations.
5. Check the proposed solution in each of the original equations.

EXAMPLE 2 Solving a System in Three Variables

Solve the system:

$$\begin{cases} 5x - 2y - 4z = 3 & \text{Equation 1} \\ 3x + 3y + 2z = -3 & \text{Equation 2} \\ -2x + 5y + 3z = 3. & \text{Equation 3} \end{cases}$$

Solution There are many ways to proceed. Because our initial goal is to reduce the system to two equations in two variables, **the central idea is to take two different pairs of equations and eliminate the same variable from both pairs.**

Step 1. Reduce the system to two equations in two variables. We choose any two equations and use the addition method to eliminate a variable. Let's eliminate z using Equations 1 and 2. We do so by multiplying Equation 2 by 2. Then we add equations.

$$\begin{matrix} \text{(Equation 1)} \\ \text{(Equation 2)} \end{matrix} \begin{cases} 5x - 2y - 4z = 3 \xrightarrow{\text{No change}} \\ 3x + 3y + 2z = -3 \xrightarrow{\text{Multiply by 2.}} \end{cases} \begin{cases} 5x - 2y - 4z = 3 \\ \underline{6x + 6y + 4z} = \underline{-6} \end{cases}$$

$$\text{Add: } 11x + 4y = -3 \quad \text{Equation 4}$$

Now we must eliminate the *same* variable from another pair of equations. We can eliminate z using Equations 2 and 3. First, we multiply Equation 2 by -3. Next, we multiply Equation 3 by 2. Finally, we add equations.

$$\begin{matrix} \text{(Equation 2)} \\ \text{(Equation 3)} \end{matrix} \begin{cases} 3x + 3y + 2z = -3 \xrightarrow{\text{Multiply by } -3.} \\ -2x + 5y + 3z = 3 \xrightarrow{\text{Multiply by 2.}} \end{cases} \begin{cases} -9x - 9y - 6z = 9 \\ \underline{-4x + 10y + 6z} = \underline{6} \end{cases}$$

$$\text{Add: } -13x + y = 15 \quad \text{Equation 5}$$

Equations 4 and 5 give us a system of two equations in two variables.

Step 2. Solve the resulting system of two equations in two variables. We will use the addition method to solve Equations 4 and 5 for x and y. To do so, we multiply Equation 5 by -4 and add this to Equation 4.

$$\begin{matrix} \text{(Equation 4)} \\ \text{(Equation 5)} \end{matrix} \begin{cases} 11x + 4y = -3 \xrightarrow{\text{No change}} \\ -13x + y = 15 \xrightarrow{\text{Multiply by } -4.} \end{cases} \begin{cases} 11x + 4y = -3 \\ \underline{52x - 4y} = \underline{-60} \end{cases}$$

$$\text{Add: } 63x = -63$$

$$x = -1 \quad \text{Divide both sides by 63.}$$

Step 3. Use back-substitution in one of the equations in two variables to find the value of the second variable. We back-substitute -1 for x in either Equation 4 or 5 to find the value of y. We will use Equation 5.

$$\begin{aligned} -13x + y &= 15 && \text{Equation 5} \\ -13(-1) + y &= 15 && \text{Substitute } -1 \text{ for } x. \\ 13 + y &= 15 && \text{Multiply.} \\ y &= 2 && \text{Subtract 13 from both sides.} \end{aligned}$$

Step 4. Back-substitute the values found for two variables into one of the original equations to find the value of the third variable. We can now use any one of the original equations and back-substitute the values of x and y to find the value for z. We will use Equation 2.

$$\begin{aligned} 3x + 3y + 2z &= -3 && \text{Equation 2} \\ 3(-1) + 3(2) + 2z &= -3 && \text{Substitute } -1 \text{ for } x \text{ and 2 for } y. \\ 3 + 2z &= -3 && \text{Multiply and then add: } 3(-1) + 3(2) = -3 + 6 = 3. \\ 2z &= -6 && \text{Subtract 3 from both sides.} \\ z &= -3 && \text{Divide both sides by 2.} \end{aligned}$$

With $x = -1$, $y = 2$, and $z = -3$, the proposed solution is the ordered triple $(-1, 2, -3)$.

Step 5. Check. Check the proposed solution, $(-1, 2, -3)$, by substituting the values for x, y, and z into each of the three original equations. These substitutions yield three true statements. Thus, the solution is $(-1, 2, -3)$ and the solution set is $\{(-1, 2, -3)\}$. ■

✓ **CHECK POINT 2** Solve the system:

$$\begin{cases} x + 4y - z = 20 \\ 3x + 2y + z = 8 \\ 2x - 3y + 2z = -16. \end{cases}$$

In some examples, one of the variables is missing from a given equation. In this case, the missing variable should be eliminated from the other two equations, thereby making it possible to omit one of the elimination steps. We illustrate this idea in Example 3.

EXAMPLE 3 Solving a System of Equations with a Missing Term

Solve the system:

$$\begin{cases} x + z = 8 & \text{Equation 1} \\ x + y + 2z = 17 & \text{Equation 2} \\ x + 2y + z = 16. & \text{Equation 3} \end{cases}$$

Solution

Step 1. Reduce the system to two equations in two variables. Because Equation 1 contains only x and z, we can omit one of the elimination steps by eliminating y using Equations 2 and 3. This will give us two equations in x and z. To eliminate y using Equations 2 and 3, we multiply Equation 2 by -2 and add Equation 3.

$$\begin{matrix} \text{(Equation 2)} & x + y + 2z = 17 & \xrightarrow{\text{Multiply by } -2.} & -2x - 2y - 4z = -34 \\ \text{(Equation 3)} & x + 2y + z = 16 & \xrightarrow{\text{No change}} & \underline{x + 2y + z = 16} \\ & & \text{Add:} & -x \quad\quad - 3z = -18 \quad \text{Equation 4} \end{matrix}$$

Equation 4 and the given Equation 1 provide us with a system of two equations in two variables:

$$\begin{cases} x + z = 8 & \text{Equation 1} \\ -x - 3z = -18. & \text{Equation 4} \end{cases}$$

Step 2. Solve the resulting system of two equations in two variables. We will solve Equations 1 and 4 for x and z.

$$\begin{cases} x + z = 8 & \text{Equation 1} \\ -x - 3z = -18 & \text{Equation 4} \end{cases}$$

$$\begin{aligned} \text{Add: } -2z &= -10 \\ z &= 5 \quad \text{Divide both sides by } -2. \end{aligned}$$

Step 3. Use back-substitution in one of the equations in two variables to find the value of the second variable. To find x, we back-substitute 5 for z in either Equation 1 or 4. We will use Equation 1.

$$\begin{aligned} x + z &= 8 && \text{Equation 1} \\ x + 5 &= 8 && \text{Substitute 5 for } z. \\ x &= 3 && \text{Subtract 5 from both sides.} \end{aligned}$$

Step 4. Back-substitute the values found for two variables into one of the original equations to find the value of the third variable. To find y, we back-substitute 3 for x and 5 for z into Equation 2, $x + y + 2z = 17$, or Equation 3, $x + 2y + z = 16$. We can't use Equation 1, $x + z = 8$, because y is missing in this equation. We will use Equation 2.

$$\begin{aligned} x + y + 2z &= 17 && \text{Equation 2} \\ 3 + y + 2(5) &= 17 && \text{Substitute 3 for } x \text{ and 5 for } z. \\ y + 13 &= 17 && \text{Multiply and add.} \\ y &= 4 && \text{Subtract 13 from both sides.} \end{aligned}$$

We found that $z = 5$, $x = 3$, and $y = 4$. Thus, the proposed solution is the ordered triple $(3, 4, 5)$.

Step 5. Check. Substituting 3 for x, 4 for y, and 5 for z into each of the three original equations yields three true statements. Consequently, the solution is $(3, 4, 5)$ and the solution set is $\{(3, 4, 5)\}$. ■

✓ CHECK POINT 3 Solve the system:

$$\begin{cases} 2y - z = 7 \\ x + 2y + z = 17 \\ 2x - 3y + 2z = -1. \end{cases}$$

---

3 Identify inconsistent and dependent systems.

Inconsistent and Dependent Systems

A system of three linear equations in three variables represents three planes. The three planes need not intersect at one point. The planes may have no common point of intersection and represent an **inconsistent system** with no solution. **Figure 3.10** illustrates some of the geometric possibilities for inconsistent systems.

Three planes are parallel with no common intersection point. | Two planes are parallel with no common intersection point. | Planes intersect two at a time. There is no intersection point common to all three planes.

Figure 3.10 Three planes may have no common point of intersection.

If you attempt to solve an inconsistent system algebraically, at some point in the solution process you will eliminate all three variables. A false statement, such as $0 = -10$, will be the result. For example, consider the system

$$\begin{cases} 2x + 5y + z = 12 & \text{Equation 1} \\ x - 2y + 4z = -10 & \text{Equation 2} \\ -3x + 6y - 12z = 20. & \text{Equation 3} \end{cases}$$

Suppose we reduce the system to two equations in two variables by eliminating x. To eliminate x using Equations 2 and 3, we multiply Equation 2 by 3 and add Equation 3:

$$\begin{cases} x - 2y + 4z = -10 \\ -3x + 6y - 12z = 20 \end{cases} \begin{matrix} \xrightarrow{\text{Multiply by 3.}} \\ \xrightarrow{\text{No change}} \end{matrix} \begin{cases} 3x - 6y + 12z = -30 \\ -3x + 6y - 12z = 20 \end{cases}$$

$$\text{Add:} \quad 0 = -10$$

There are no values of x, y, and z for which $0 = -10$. The false statement $0 = -10$ indicates that the system is inconsistent and has no solution. The solution set is the empty set, $\varnothing$.

We have seen that a linear system that has at least one solution is called a **consistent system**. Planes that intersect at one point and planes that intersect at infinitely many points both represent consistent systems. **Figure 3.11** illustrates two different cases of three planes that intersect at infinitely many points. The equations in these linear systems with infinitely many solutions are called **dependent**.

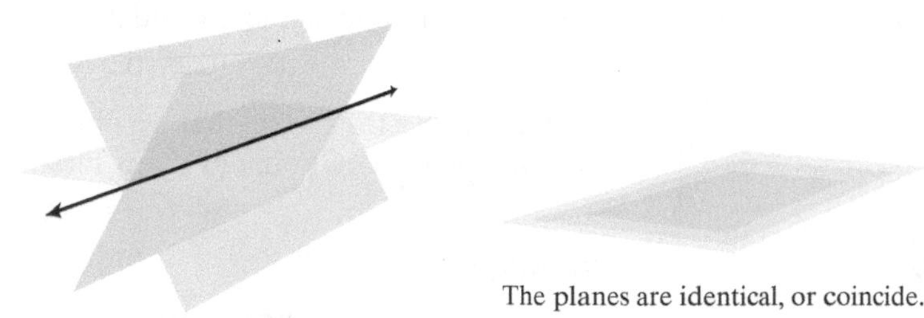

The planes intersect along a common line.

The planes are identical, or coincide.

Figure 3.11 Three planes may intersect at infinitely many points.

If you attempt to solve a system with dependent equations algebraically, at some point in the solution process you will eliminate all three variables. A true statement, such as $0 = 0$, will be the result. If this occurs as you are solving a linear system, simply state that the equations are dependent.

4 Solve problems using systems in three variables.

Applications

Systems of equations may allow us to find models for data without using a graphing utility. Three data points that do not lie on or near a line determine the graph of a function of the form

$$y = ax^2 + bx + c, a \neq 0.$$

Such a function is called a **quadratic function.** If $a > 0$, its graph is shaped like a bowl, making it ideal for modeling situations in which values of y are decreasing and then increasing. In Chapter 8, we'll have lots of interesting things to tell you about quadratic functions and their graphs.

The process of determining a function whose graph contains given points is called **curve fitting**. In our next example, we fit the curve whose equation is $y = ax^2 + bx + c$ to three data points. Using a system of equations, we find values for a, b, and c.

EXAMPLE 4 Modeling Data Relating Sleep and Death Rate

In a study relating sleep and death rate, the following data were obtained. Use the function $y = ax^2 + bx + c$ to model the data.

x (Average Number of Hours of Sleep)	y (Death Rate per Year per 100,000 Males)
4	1682
7	626
9	967

Using Technology

The graph of

$$y = 104.5x^2 - 1501.5x + 6016$$

is displayed in a [3, 12, 1] by [500, 2000, 100] viewing rectangle. The minimum function feature shows that the lowest point on the graph is approximately (7.2, 622.5). Men who average 7.2 hours of sleep are in the group with the lowest death rate, approximately 622.5 deaths per 100,000 males.

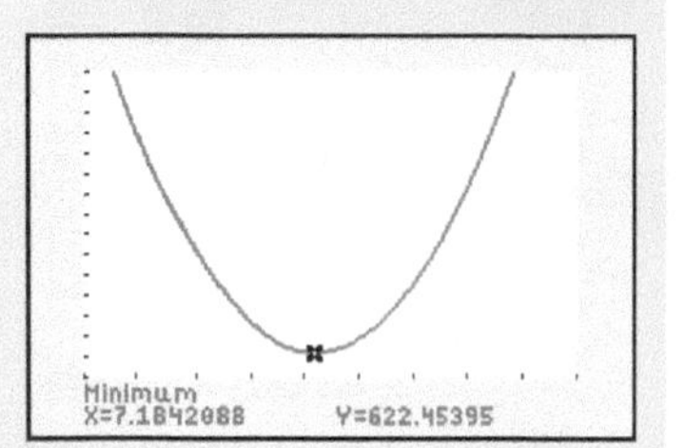

Solution We need to find values for a, b, and c in $y = ax^2 + bx + c$. We can do so by solving a system of three linear equations in a, b, and c. We obtain the three equations by using the values of x and y from the data as follows:

$y = ax^2 + bx + c$ Use the quadratic function to model the data.

When $x = 4$, $y = 1682$: When $x = 7$, $y = 626$: When $x = 9$, $y = 967$:

$$\begin{cases} 1682 = a \cdot 4^2 + b \cdot 4 + c \\ 626 = a \cdot 7^2 + b \cdot 7 + c \\ 967 = a \cdot 9^2 + b \cdot 9 + c \end{cases} \quad \text{or} \quad \begin{cases} 16a + 4b + c = 1682 \\ 49a + 7b + c = 626 \\ 81a + 9b + c = 967. \end{cases}$$

The easiest way to solve this system is to eliminate c from two pairs of equations, obtaining two equations in a and b. Solving this system gives $a = 104.5$, $b = -1501.5$, and $c = 6016$. We now substitute the values for a, b, and c into $y = ax^2 + bx + c$. The function that models the given data is

$$y = 104.5x^2 - 1501.5x + 6016. \blacksquare$$

We can use the model that we obtained in Example 4 to find the death rate of males who average, say, 6 hours of sleep. First, write the model in function notation:

$$f(x) = 104.5x^2 - 1501.5x + 6016.$$

Substitute 6 for x:

$$f(6) = 104.5(6)^2 - 1501.5(6) + 6016 = 769.$$

According to the model, the death rate for males who average 6 hours of sleep is 769 deaths per 100,000 males.

☑ **CHECK POINT 4** Find the quadratic function $y = ax^2 + bx + c$ whose graph passes through the points (1, 4), (2, 1), and (3, 4).

Problems involving three unknowns can be solved using the same strategy for solving problems with two unknown quantities. You can let x, y, and z represent the unknown quantities. We then translate from the verbal conditions of the problem to a system of three equations in three variables. Problems of this type are included in the Exercise Set that follows.

Achieving Success

Think about finding a tutor.

If you're attending all lectures, taking good class notes, reading the textbook, and doing all the assigned homework, but still having difficulty in a math course, you might want to find a tutor. Many on-campus learning centers and math labs have trained people available to help you. Sometimes a TA who has previously taught the course is available. **Make sure the tutor is both good at math and familiar with the particular course you're taking.** Bring your textbook, class notes, the problems you've done, and information about course policy and tests to each meeting with your tutor. That way he or she can be sure the tutoring sessions address your exact needs.

CONCEPT AND VOCABULARY CHECK

Fill in each blank so that the resulting statement is true.

1. A solution of a system of linear equations in three variables is an ordered ________________ of real numbers that satisfies all/some of the equations in the system.

Circle the correct choice.

2. Consider the following system:

$$\begin{cases} x + y - z = -1 & \text{Equation 1} \\ 2x - 2y - 5z = 7 & \text{Equation 2} \\ 4x + y - 2z = 7. & \text{Equation 3} \end{cases}$$

We can eliminate x from Equations 1 and 2 by multiplying Equation 1 by ________________ and adding equations. We can eliminate x from Equations 1 and 3 by multiplying Equation 1 by ________________ and adding equations.

3. Consider the following system:

$$\begin{cases} x + y + z = 2 & \text{Equation 1} \\ 2x - 3y = 3 & \text{Equation 2} \\ 10y - z = 12. & \text{Equation 3} \end{cases}$$

Equation 2 does not contain the variable ________________. To obtain a second equation that does not contain this variable, we can ________________.

4. A function of the form $y = ax^2 + bx + c, a \neq 0$, is called a/an ________________ function.

5. The process of determining a function whose graph contains given points is called ________________.

3.3 EXERCISE SET

MyMathLab®

Practice Exercises

In Exercises 1–4, determine if the given ordered triple is a solution of the system.

1. $(2, -1, 3)$
$$\begin{cases} x + y + z = 4 \\ x - 2y - z = 1 \\ 2x - y - z = -1 \end{cases}$$

2. $(5, -3, -2)$
$$\begin{cases} x + y + z = 0 \\ x + 2y - 3z = 5 \\ 3x + 4y + 2z = -1 \end{cases}$$

3. $(4, 1, 2)$
$$\begin{cases} x - 2y = 2 \\ 2x + 3y = 11 \\ y - 4z = -7 \end{cases}$$

4. $(-1, 3, 2)$
$$\begin{cases} x - 2z = -5 \\ y - 3z = -3 \\ 2x - z = -4 \end{cases}$$

Solve each system in Exercises 5–22. If there is no solution or if there are infinitely many solutions and a system's equations are dependent, so state.

5. $$\begin{cases} x + y + 2z = 11 \\ x + y + 3z = 14 \\ x + 2y - z = 5 \end{cases}$$

6. $$\begin{cases} 2x + y - 2z = -1 \\ 3x - 3y - z = 5 \\ x - 2y + 3z = 6 \end{cases}$$

7. $$\begin{cases} 4x - y + 2z = 11 \\ x + 2y - z = -1 \\ 2x + 2y - 3z = -1 \end{cases}$$

8. $$\begin{cases} x - y + 3z = 8 \\ 3x + y - 2z = -2 \\ 2x + 4y + z = 0 \end{cases}$$

9. $$\begin{cases} 3x + 2y - 3z = -2 \\ 2x - 5y + 2z = -2 \\ 4x - 3y + 4z = 10 \end{cases}$$

10. $$\begin{cases} 2x + 3y + 7z = 13 \\ 3x + 2y - 5z = -22 \\ 5x + 7y - 3z = -28 \end{cases}$$

11. $$\begin{cases} 2x - 4y + 3z = 17 \\ x + 2y - z = 0 \\ 4x - y - z = 6 \end{cases}$$

12. $$\begin{cases} x + z = 3 \\ x + 2y - z = 1 \\ 2x - y + z = 3 \end{cases}$$

13. $$\begin{cases} 2x + y = 2 \\ x + y - z = 4 \\ 3x + 2y + z = 0 \end{cases}$$

14. $$\begin{cases} x + 3y + 5z = 20 \\ y - 4z = -16 \\ 3x - 2y + 9z = 36 \end{cases}$$

15. $$\begin{cases} x + y = -4 \\ y - z = 1 \\ 2x + y + 3z = -21 \end{cases}$$

16. $\begin{cases} x + y = 4 \\ x + z = 4 \\ y + z = 4 \end{cases}$

17. $\begin{cases} 2x + y + 2z = 1 \\ 3x - y + z = 2 \\ x - 2y - z = 0 \end{cases}$

18. $\begin{cases} 3x + 4y + 5z = 8 \\ x - 2y + 3z = -6 \\ 2x - 4y + 6z = 8 \end{cases}$

19. $\begin{cases} 5x - 2y - 5z = 1 \\ 10x - 4y - 10z = 2 \\ 15x - 6y - 15z = 3 \end{cases}$

20. $\begin{cases} x + 2y + z = 4 \\ 3x - 4y + z = 4 \\ 6x - 8y + 2z = 8 \end{cases}$

21. $\begin{cases} 3(2x + y) + 5z = -1 \\ 2(x - 3y + 4z) = -9 \\ 4(1 + x) = -3(z - 3y) \end{cases}$

22. $\begin{cases} 7z - 3 = 2(x - 3y) \\ 5y + 3z - 7 = 4x \\ 4 + 5z = 3(2x - y) \end{cases}$

In Exercises 23–26, find the quadratic function $y = ax^2 + bx + c$ whose graph passes through the given points.

23. $(-1, 6), (1, 4), (2, 9)$

24. $(-2, 7), (1, -2), (2, 3)$

25. $(-1, -4), (1, -2), (2, 5)$

26. $(1, 3), (3, -1), (4, 0)$

In Exercises 27–28, let x represent the first number, y the second number, and z the third number. Use the given conditions to write a system of equations. Solve the system and find the numbers.

27. The sum of three numbers is 16. The sum of twice the first number, 3 times the second number, and 4 times the third number is 46. The difference between 5 times the first number and the second number is 31. Find the three numbers.

28. The following is known about three numbers: Three times the first number plus the second number plus twice the third number is 5. If 3 times the second number is subtracted from the sum of the first number and 3 times the third number, the result is 2. If the third number is subtracted from the sum of 2 times the first number and 3 times the second number, the result is 1. Find the numbers.

Practice PLUS

Solve each system in Exercises 29–30.

29. $\begin{cases} \dfrac{x+2}{6} - \dfrac{y+4}{3} + \dfrac{z}{2} = 0 \\ \dfrac{x+1}{2} + \dfrac{y-1}{2} - \dfrac{z}{4} = \dfrac{9}{2} \\ \dfrac{x-5}{4} + \dfrac{y+1}{3} + \dfrac{z-2}{2} = \dfrac{19}{4} \end{cases}$

30. $\begin{cases} \dfrac{x+3}{2} - \dfrac{y-1}{2} + \dfrac{z+2}{4} = \dfrac{3}{2} \\ \dfrac{x-5}{2} + \dfrac{y+1}{3} - \dfrac{z}{4} = -\dfrac{25}{6} \\ \dfrac{x-3}{4} - \dfrac{y+1}{2} + \dfrac{z-3}{2} = -\dfrac{5}{2} \end{cases}$

In Exercises 31–32, find the equation of the quadratic function $y = ax^2 + bx + c$ whose graph is shown. Select three points whose coordinates appear to be integers.

31.

32.

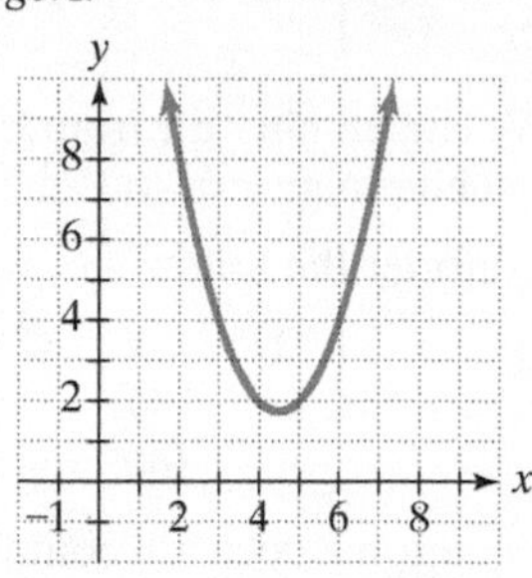

In Exercises 33–34, solve each system for (x, y, z) in terms of the nonzero constants a, b, and c.

33. $\begin{cases} ax - by - 2cz = 21 \\ ax + by + cz = 0 \\ 2ax - by + cz = 14 \end{cases}$

34. $\begin{cases} ax - by + 2cz = -4 \\ ax + 3by - cz = 1 \\ 2ax + by + 3cz = 2 \end{cases}$

Application Exercises

35. The bar graph shows the percentage of U.S. parents willing to pay for all, some, or none of their child's college education.

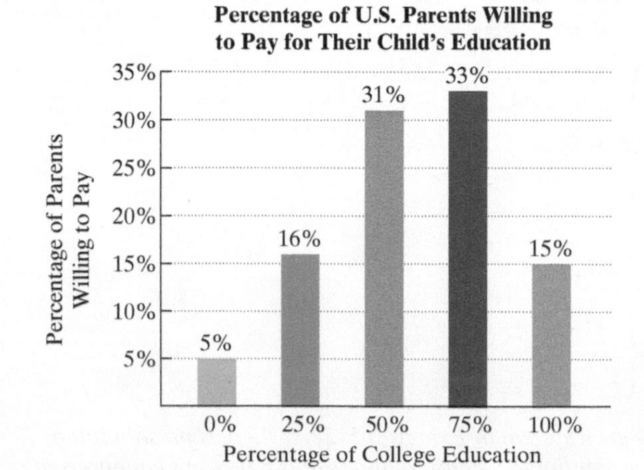

Source: Student Monitor LLC

a. Use the bars that represent none, half, and all of a college education. Represent the data for each bar as an ordered pair (x, y), where x is the percentage of college education and y is the percentage of parents willing to pay for that percent of their child's education.

b. The three data points in part (a) can be modeled by the quadratic function $y = ax^2 + bx + c$, where $a < 0$. Substitute each ordered pair into this function, one ordered pair at a time, and write a system of linear equations in three variables that can be used to find values for a, b, and c. It is not necessary to solve the system.

36. How much time do you spend on hygiene/grooming in the morning (including showering, washing face and hands, brushing teeth, shaving, and applying makeup)? The bar graph shows the time American adults spend on morning grooming.

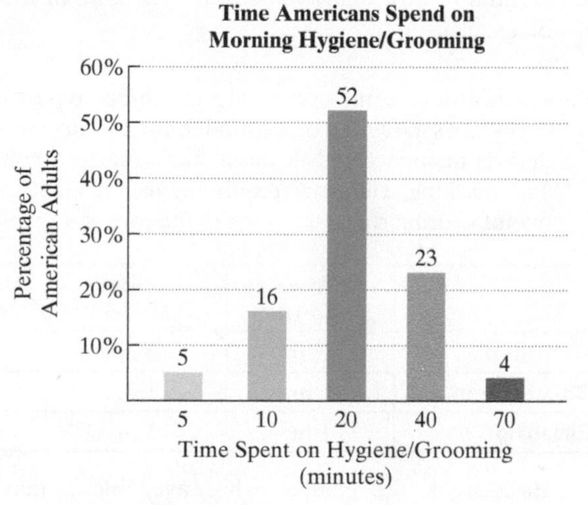

Source: Hygiene Matters Report, 2011

a. Write the data for 10 minutes, 20 minutes, and 40 minutes as ordered pairs (x, y), where x is the time spent on morning grooming and y is the percentage of American adults spending that much time on grooming.

b. The three data points in part (a) can be modeled by the quadratic function $y = ax^2 + bx + c$, where $a < 0$. Substitute each ordered pair into this function, one ordered pair at a time, and write a system of linear equations in three variables that can be used to find values for a, b, and c. It is not necessary to solve the system.

37. You throw a ball straight up from a rooftop. The ball misses the rooftop on its way down and eventually strikes the ground. A mathematical model can be used to describe the ball's height above the ground, y, after x seconds. Consider the following data.

x, seconds after the ball is thrown	y, ball's height, in feet, above the ground
1	224
3	176
4	104

a. Find the quadratic function $y = ax^2 + bx + c$ whose graph passes through the given points.

b. Use the function in part (a) to find the value for y when $x = 5$. Describe what this means.

38. A mathematical model can be used to describe the relationship between the number of feet a car travels once the brakes are applied, y, and the number of seconds the car is in motion after the brakes are applied, x. A research firm collects the data shown below.

x, seconds in motion after brakes are applied	y, feet car travels once the brakes are applied
1	46
2	84
3	114

a. Find the quadratic function $y = ax^2 + bx + c$ whose graph passes through the given points.

b. Use the function in part (a) to find the value for y when $x = 6$. Describe what this means.

In Exercises 39–46, use the four-step strategy to solve each problem. Use x, y, and z to represent unknown quantities. Then translate from the verbal conditions of the problem to a system of three equations in three variables.

39. The bar graph shows the average number of hours U.S. college students study per week for seven selected majors. Study times are rounded to the nearest hour.

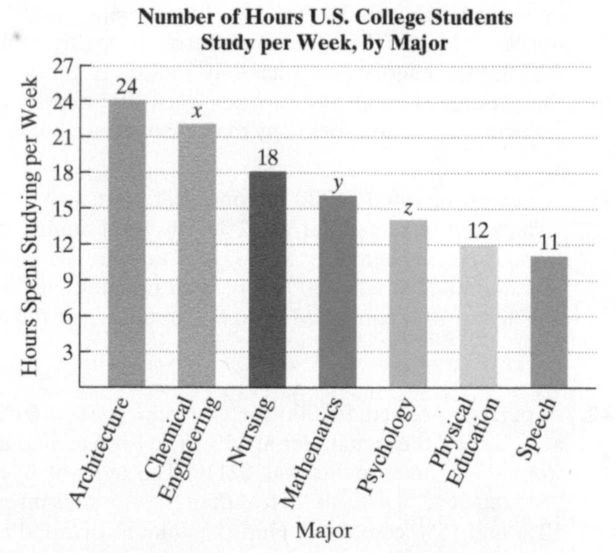

Source: statista.com

The combined weekly study time for students majoring in chemical engineering, mathematics, and psychology is 52 hours. The difference between weekly study time for chemical engineering majors and math majors is 6 hours. The difference between weekly study time for chemical engineering majors and psychology majors is 8 hours. Find the average number of hours per week that chemical engineering majors, mathematics majors, and psychology majors spend studying.

40. The bar graph shows the average number of hours U.S. college students study per week for seven selected majors. Study times are rounded to the nearest hour.

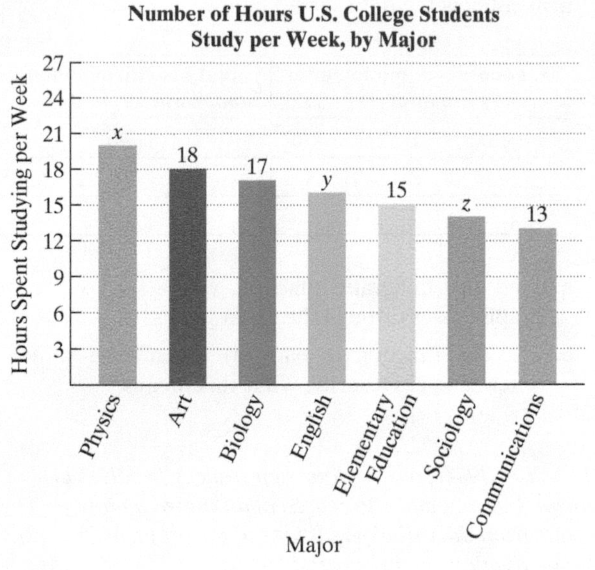

Source: statista.com

The combined weekly study time for students majoring in physics, English, and sociology is 50 hours. The difference between weekly study time for physics majors and English majors is 4 hours. The difference between weekly study time for physics majors and sociology majors is 6 hours. Find the average number of hours per week that physics majors, English majors, and sociology majors spend studying.

41. A person invested \$6700 for one year, part at 8%, part at 10%, and the remainder at 12%. The total annual income from these investments was \$716. The amount of money invested at 12% was \$300 more than the amounts invested at 8% and 10% combined. Find the amount invested at each rate.

42. A person invested \$17,000 for one year, part at 10%, part at 12%, and the remainder at 15%. The total annual income from these investments was \$2110. The amount of money invested at 12% was \$1000 less than the amounts invested at 10% and 15% combined. Find the amount invested at each rate.

43. At a college production of *Streetcar Named Desire*, 400 tickets were sold. The ticket prices were \$8, \$10, and \$12, and the total income from ticket sales was \$3700. How many tickets of each type were sold if the combined number of \$8 and \$10 tickets sold was 7 times the number of \$12 tickets sold?

44. A certain brand of razor blades comes in packages of 6, 12, and 24 blades, costing \$2, \$3, and \$4 per package, respectively. A store sold 12 packages containing a total of 162 razor blades and took in \$35. How many packages of each type were sold?

45. Three foods have the following nutritional content per ounce.

	Calories	Protein (in grams)	Vitamin C (in milligrams)
Food A	40	5	30
Food B	200	2	10
Food C	400	4	300

If a meal consisting of the three foods allows exactly 660 calories, 25 grams of protein, and 425 milligrams of vitamin C, how many ounces of each kind of food should be used?

46. A furniture company produces three types of desks: a children's model, an office model, and a deluxe model. Each desk is manufactured in three stages: cutting, construction, and finishing. The time requirements for each model and manufacturing stage are given in the following table.

	Children's Model	Office Model	Deluxe Model
Cutting	2 hr	3 hr	2 hr
Construction	2 hr	1 hr	3 hr
Finishing	1 hr	1 hr	2 hr

Each week the company has available a maximum of 100 hours for cutting, 100 hours for construction, and 65 hours for finishing. If all available time must be used, how many of each type of desk should be produced each week?

Explaining the Concepts

47. What is a system of linear equations in three variables?

48. How do you determine whether a given ordered triple is a solution of a system of linear equations in three variables?

49. Describe in general terms how to solve a system in three variables.

50. Describe what happens when using algebraic methods to solve an inconsistent system.

51. Describe what happens when using algebraic methods to solve a system with dependent equations.

Technology Exercises

52. Does your graphing utility have a feature that allows you to solve linear systems by entering coefficients and constant terms? If so, use this feature to verify the solutions to any five exercises that you worked by hand from Exercises 5–16.

53. Verify your results in Exercises 23–26 by using a graphing utility to graph the quadratic function. Trace along the curve and convince yourself that the three points given in the exercise lie on the function's graph.

Critical Thinking Exercises

Make Sense? *In Exercises 54–57, determine whether each statement makes sense or does not make sense, and explain your reasoning.*

54. Solving a system in three variables, I found that $x = 3$ and $y = -1$. Because z represents a third variable, z cannot equal 3 or -1.

55. A system of linear equations in three variables, x, y, and z, cannot contain an equation in the form $y = mx + b$.

56. I'm solving a three-variable system in which one of the given equations has a missing term, so it will not be necessary to use any of the original equations twice when I reduce the system to two equations in two variables.

57. Because the percentage of the U.S. population that was foreign-born decreased from 1910 through 1970 and then increased after that, a quadratic function of the form $f(x) = ax^2 + bx + c$, rather than a linear function of the form $f(x) = mx + b$, should be used to model the data.

In Exercises 58–61, determine whether each statement is true or false. If the statement is false, make the necessary change(s) to produce a true statement.

58. The ordered triple (2, 15, 14) is the only solution of the equation $x + y - z = 3$.

59. The equation $x - y - z = -6$ is satisfied by $(2, -3, 5)$.

60. If two equations in a system are $x + y - z = 5$ and $x + y - z = 6$, then the system must be inconsistent.

61. An equation with four variables, such as $x + 2y - 3z + 5w = 2$, cannot be satisfied by real numbers.

62. In the following triangle, the degree measures of the three interior angles and two of the exterior angles are represented with variables. Find the measure of each interior angle.

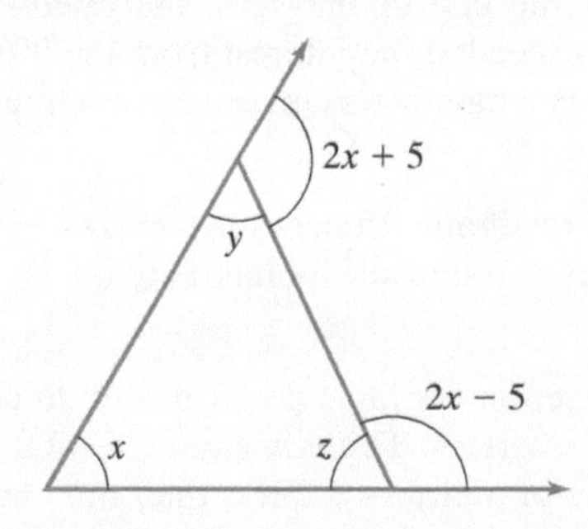

63. A modernistic painting consists of triangles, rectangles, and pentagons, all drawn so as to not overlap or share sides. Within each rectangle are drawn 2 red roses, and each pentagon contains 5 carnations. How many triangles, rectangles, and pentagons appear in the painting if the painting contains a total of 40 geometric figures, 153 sides of geometric figures, and 72 flowers?

64. Two blocks of wood having the same length and width are placed on the top and bottom of a table, as shown in (a). Length A measures 32 centimeters. The blocks are rearranged as shown in (b). Length B measures 28 centimeters. Determine the height of the table.

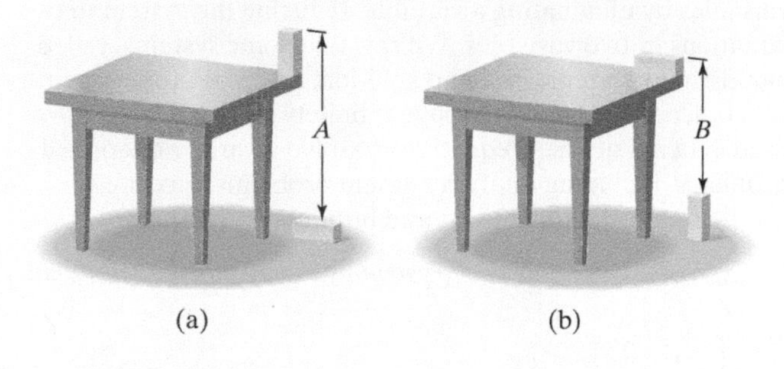

Review Exercises

In Exercises 65–67, graph each linear function.

65. $f(x) = -\frac{3}{4}x + 3$ (Section 2.4, Example 5)

66. $-2x + y = 6$ (Section 2.4, Example 1)

67. $f(x) = -5$ (Section 2.4, Example 6)

Preview Exercises

Exercises 68–70 will help you prepare for the material covered in the next section.

68. Solve the system:

$$\begin{cases} x + 2y = -1 \\ y = 1. \end{cases}$$

What makes it fairly easy to find the solution?

69. Solve the system:

$$\begin{cases} x + y + 2z = 19 \\ y + 2z = 13 \\ z = 5. \end{cases}$$

What makes it fairly easy to find the solution?

70. Consider the following array of numbers:

$$\begin{bmatrix} 1 & 2 & -1 \\ 4 & -3 & -15 \end{bmatrix}.$$

Rewrite the array as follows: Multiply each number in the top row by -4 and add this product to the corresponding number in the bottom row. Do not change the numbers in the top row.

MID-CHAPTER CHECK POINT Section 3.1–Section 3.3

What You Know: We learned to solve systems of linear equations. We solved systems in two variables by graphing, by the substitution method, and by the addition method. We solved systems in three variables by eliminating a variable, reducing the system to two equations in two variables. We saw that some systems, called inconsistent systems, have no solution, whereas other systems, called dependent systems, have infinitely many solutions. We used systems of linear equations to solve a variety of applied problems, including dual investment problems, mixture problems, motion problems, and business problems.

In Exercises 1–8, solve each system by the method of your choice.

1. $\begin{cases} x = 3y - 7 \\ 4x + 3y = 2 \end{cases}$

2. $\begin{cases} 3x + 4y = -5 \\ 2x - 3y = 8 \end{cases}$

3. $\begin{cases} \frac{2x}{3} + \frac{y}{5} = 6 \\ \frac{x}{6} - \frac{y}{2} = -4 \end{cases}$

4. $\begin{cases} y = 4x - 5 \\ 8x - 2y = 10 \end{cases}$

5. $\begin{cases} 2x + 5y = 3 \\ 3x - 2y = 1 \end{cases}$

6. $\begin{cases} \frac{x}{12} - y = \frac{1}{4} \\ 4x - 48y = 16 \end{cases}$

7. $\begin{cases} 2x - y + 2z = -8 \\ x + 2y - 3z = 9 \\ 3x - y - 4z = 3 \end{cases}$

8. $\begin{cases} x - 3z = -5 \\ 2x - y + 2z = 16 \\ 7x - 3y - 5z = 19 \end{cases}$

In Exercises 9–10, solve each system by graphing.

9. $\begin{cases} 2x - y = 4 \\ x + y = 5 \end{cases}$

10. $\begin{cases} y = x + 3 \\ y = -\frac{1}{2}x \end{cases}$

11. A company is planning to manufacture PDAs (personal digital assistants). The fixed cost will be \$400,000 and it will cost \$20 to produce each PDA. Each PDA will be sold for \$100.

a. Write the cost function, C, of producing x PDAs.

b. Write the revenue function, R, from the sale of x PDAs.

c. Write the profit function, P, from producing and selling x PDAs.

d. Determine the break-even point. Describe what this means.

In Exercises 12–18, solve each problem.

12. Roses sell for \$3 each and carnations for \$1.50 each. If a mixed bouquet of 20 flowers consisting of roses and carnations costs \$39, how many of each type of flower is in the bouquet?

13. You invested \$15,000 in two funds paying 5% and 6% annual interest. At the end of the year, the total interest from these investments was \$837. How much was invested at each rate?

14. The manager of a gardening center needs to mix a plant food that is 13% nitrogen with one that is 18% nitrogen to obtain 50 gallons of a plant food that is 16% nitrogen. How many gallons of each of the plant foods must be used?

15. With the current, you can row 9 miles in 2 hours. Against the current, your return trip takes 6 hours. Find your average rowing rate in still water and the average rate of the current.

16. You invested \$8000 in two funds paying 2% and 5% annual interest. At the end of the year, the interest from the 5% investment exceeded the interest from the 2% investment by \$85. How much money was invested at each rate?

17. Find the quadratic function $y = ax^2 + bx + c$ whose graph passes through the points $(-1, 0)$, $(1, 4)$, and $(2, 3)$.

18. A coin collection contains a mixture of 26 coins in nickels, dimes, and quarters. The coins have a total value of \$4.00. The number of quarters is 2 less than the number of nickels and dimes combined. Determine the number of nickels, the number of dimes, and the number of quarters in the collection.

SECTION

3.4 Matrix Solutions to Linear Systems

What am I supposed to learn?

After studying this section, you should be able to:

1. Write the augmented matrix for a linear system.
2. Perform matrix row operations.
3. Use matrices to solve linear systems in two variables.
4. Use matrices to solve linear systems in three variables.
5. Use matrices to identify inconsistent and dependent systems.

In 2014, women made up 47% of the U.S. workforce. Although the number of women entering the workforce has increased, the data below show they are still outnumbered by men in many careers, including science and technology professions. They also continue to earn less than men for the same job and hold fewer positions of power. In 2014, there were only 23 female CEOs of *Fortune 500* companies.

Gender Breakdown of Various Careers in the United States (2014 data)

	Computer Occupations	Lawyers	CEOs of *Fortune 500* Companies	Teachers	Physicians and Surgeons	Registered Nurses
Men	74%	67%	95.4%	25%	65%	11%
Women	26%	33%	4.6%	75%	35%	89%

Source: U.S. Bureau of Labor Statistics

The 12 numbers inside the brackets are arranged in two rows and six columns. This rectangular array of 12 numbers, arranged in rows and columns and placed in brackets, is an example of a **matrix** (plural: **matrices**). The numbers inside the brackets are called **elements** of the matrix. Matrices are used to display information and to solve systems of linear equations.

1 Write the augmented matrix for a linear system.

Augmented Matrices

A matrix gives us a shortened way of writing a system of equations. The first step in solving a system of linear equations using matrices is to write the *augmented matrix*. An **augmented matrix** has a vertical bar separating the columns of the matrix into two groups. The coefficients of each variable are placed to the left of the vertical line and the constants are placed to the right. If any variable is missing, its coefficient is 0. Here are two examples.

System of Linear Equations / **Augmented Matrix**

$$\begin{aligned} x + 3y &= 5 \\ 2x - y &= -4 \end{aligned} \qquad \left[\begin{array}{cc|c} 1 & 3 & 5 \\ 2 & -1 & -4 \end{array}\right]$$

$$\begin{aligned} 3x + 4y &= 19 \\ 2y + 3z &= 8 \\ 4x - 5z &= 7 \end{aligned} \qquad \left[\begin{array}{ccc|c} 3 & 4 & 0 & 19 \\ 0 & 2 & 3 & 8 \\ 4 & 0 & -5 & 7 \end{array}\right]$$

Our goal in solving a linear system using matrices is to produce a matrix with 1s down the diagonal from upper left to lower right on the left side of the vertical bar, called the **main diagonal**, and 0s below the 1s. In general, the matrix will be one of the following forms.

This is the desired form for systems with two equations.

$$\left[\begin{array}{cc|c} 1 & a & b \\ 0 & 1 & c \end{array}\right] \qquad \left[\begin{array}{ccc|c} 1 & a & b & c \\ 0 & 1 & d & e \\ 0 & 0 & 1 & f \end{array}\right]$$

This is the desired form for systems with three equations.

The last row of these matrices gives us the value of one variable. The values of the other variables can then be found by back-substitution.

2 Perform matrix row operations.

Matrix Row Operations

A matrix with 1s down the main diagonal and 0s below the 1s is said to be in **row-echelon form**. How do we produce a matrix in this form? We use **row operations** on the augmented matrix. These row operations are just like what you did when solving a linear system by the addition method. The difference is that we no longer write the variables, usually represented by x, y, and z.

Matrix Row Operations

The following row operations produce matrices that represent systems with the same solution set:

1. Two rows of a matrix may be interchanged. This is the same as interchanging two equations in a linear system.
2. The elements in any row may be multiplied by a nonzero number. This is the same as multiplying both sides of an equation by a nonzero number.
3. The elements in any row may be multiplied by a nonzero number, and these products may be added to the corresponding elements in any other row. This is the same as multiplying an equation by a nonzero number and then adding equations to eliminate a variable.

Two matrices are **row equivalent** if one can be obtained from the other by a sequence of row operations.

Great Question!

Can you clarify what I'm supposed to do to find $kR_i + R_j$? Which row do I work with and which row do I replace?

When performing the row operation

$$kR_i + R_j$$

you use row i to find the products. However, **elements in row i do not change. It is the elements in row j that change:** Add k times the elements in row i to the corresponding elements in row j. Replace elements in row j by these sums.

Each matrix row operation in the preceding box can be expressed symbolically as follows:

1. Interchange the elements in the ith and jth rows: $R_i \leftrightarrow R_j$.
2. Multiply each element in the ith row by k: kR_i.
3. Add k times the elements in row i to the corresponding elements in row j: $kR_i + R_j$.

EXAMPLE 1 Performing Matrix Row Operations

Use the matrix

$$\left[\begin{array}{ccc|c} 3 & 18 & -12 & 21 \\ 1 & 2 & -3 & 5 \\ -2 & -3 & 4 & -6 \end{array}\right]$$

and perform each indicated row operation:

a. $R_1 \leftrightarrow R_2$ **b.** $\frac{1}{3}R_1$ **c.** $2R_2 + R_3$.

Solution

a. The notation $R_1 \leftrightarrow R_2$ means to interchange the elements in row 1 and row 2. This results in the row-equivalent matrix

$$\left[\begin{array}{rrr|r} 1 & 2 & -3 & 5 \\ 3 & 18 & -12 & 21 \\ -2 & -3 & 4 & -6 \end{array}\right].$$

This was row 2; now it's row 1.

This was row 1; now it's row 2.

b. The notation $\frac{1}{3}R_1$ means to multiply each element in row 1 by $\frac{1}{3}$. This results in the row-equivalent matrix

$$\left[\begin{array}{cccc|c} \frac{1}{3}(3) & \frac{1}{3}(18) & \frac{1}{3}(-12) & \frac{1}{3}(21) \\ 1 & 2 & -3 & 5 \\ -2 & -3 & 4 & -6 \end{array}\right] = \left[\begin{array}{rrr|r} 1 & 6 & -4 & 7 \\ 1 & 2 & -3 & 5 \\ -2 & -3 & 4 & -6 \end{array}\right].$$

c. The notation $2R_2 + R_3$ means to add 2 times the elements in row 2 to the corresponding elements in row 3. Replace the elements in row 3 by these sums. First, we find 2 times the elements in row 2, namely, 1, 2, −3, and 5:

$$2(1) \text{ or } 2, \quad 2(2) \text{ or } 4, \quad 2(-3) \text{ or } -6, \quad 2(5) \text{ or } 10.$$

Now we add these products to the corresponding elements in row 3. Although we use row 2 to find the products, row 2 does not change. It is the elements in row 3 that change, resulting in the row-equivalent matrix

Replace row 3 by the sum of itself and 2 times row 2.

$$\left[\begin{array}{ccc|c} 3 & 18 & -12 & 21 \\ 1 & 2 & -3 & 5 \\ -2+2=0 & -3+4=1 & 4+(-6)=-2 & -6+10=4 \end{array}\right] = \left[\begin{array}{rrr|r} 3 & 18 & -12 & 21 \\ 1 & 2 & -3 & 5 \\ 0 & 1 & -2 & 4 \end{array}\right].$$ ■

☑ **CHECK POINT 1** Use the matrix

$$\left[\begin{array}{rrr|r} 4 & 12 & -20 & 8 \\ 1 & 6 & -3 & 7 \\ -3 & -2 & 1 & -9 \end{array}\right]$$

and perform each indicated row operation:

a. $R_1 \leftrightarrow R_2$

b. $\frac{1}{4}R_1$

c. $3R_2 + R_3$.

---

3 Use matrices to solve linear systems in two variables.

Solving Linear Systems in Two Variables Using Matrices

The process that we use to solve linear systems using matrix row operations is often called **Gaussian elimination**, after the German mathematician Carl Friedrich Gauss (1777–1855). At the top of the next page are the steps used in solving linear systems in two variables with matrices.

Solving Linear Systems in Two Variables Using Matrices

1. Write the augmented matrix for the system.
2. Use matrix row operations to simplify the matrix to a row-equivalent matrix in row-echelon form, with 1s down the main diagonal from upper left to lower right, and a 0 below the 1 in the first column.

$$\left[\begin{array}{cc|c} 1 & * & * \\ * & * & * \end{array}\right] \rightarrow \left[\begin{array}{cc|c} 1 & * & * \\ 0 & * & * \end{array}\right] \rightarrow \left[\begin{array}{cc|c} 1 & * & * \\ 0 & 1 & * \end{array}\right]$$

Get 1 in the upper left-hand corner.

Use the 1 in the first column to get 0 below it.

Get 1 in the second row, second column position.

3. Write the system of linear equations corresponding to the matrix from step 2 and use back-substitution to find the system's solution.

EXAMPLE 2 Using Matrices to Solve a Linear System

Use matrices to solve the system:

$$\begin{cases} 4x - 3y = -15 \\ x + 2y = -1. \end{cases}$$

Solution

Step 1. Write the augmented matrix for the system.

Linear System	Augmented Matrix
$\begin{cases} 4x - 3y = -15 \\ x + 2y = -1 \end{cases}$	$\left[\begin{array}{cc\|c} 4 & -3 & -15 \\ 1 & 2 & -1 \end{array}\right]$

Step 2. Use matrix row operations to simplify the matrix to row-echelon form, with 1s down the main diagonal from upper left to lower right, and a 0 below the 1 in the first column. Our first step in achieving this goal is to get 1 in the top position of the first column.

We want 1 in this position.

$$\left[\begin{array}{cc|c} 4 & -3 & -15 \\ 1 & 2 & -1 \end{array}\right]$$

To get 1 in this position, we interchange row 1 and row 2: $R_1 \leftrightarrow R_2$.

$$\left[\begin{array}{cc|c} 1 & 2 & -1 \\ 4 & -3 & -15 \end{array}\right]$$

This was row 2; now it's row 1.

This was row 1; now it's row 2.

Now we want a 0 below the 1 in the first column.

We want 0 in this position.

$$\left[\begin{array}{cc|c} 1 & 2 & -1 \\ 4 & -3 & -15 \end{array}\right]$$

Let's get a 0 where there is now a 4. If we multiply the top row of numbers by -4 and add these products to the second row of numbers, we will get 0 in this position: $-4R_1 + R_2$. *We change only row 2.*

Replace row 2 by $-4R_1 + R_2$.

$$\left[\begin{array}{cc|c} 1 & 2 & -1 \\ -4(1) + 4 & -4(2) + (-3) & -4(-1) + (-15) \end{array}\right] = \left[\begin{array}{cc|c} 1 & 2 & -1 \\ 0 & -11 & -11 \end{array}\right]$$

We move on to the second column. We want 1 in the second row, second column.

We want 1 in this position.

$$\left[\begin{array}{cc|c} 1 & 2 & -1 \\ 0 & -11 & -11 \end{array}\right]$$

To get 1 in the desired position, we multiply -11 by its multiplicative inverse, or reciprocal, $-\frac{1}{11}$. Therefore, we multiply all the numbers in the second row by $-\frac{1}{11}$: $-\frac{1}{11}R_2$.

$$-\tfrac{1}{11}R_2 \quad \left[\begin{array}{cc|c} 1 & 2 & -1 \\ -\frac{1}{11}(0) & -\frac{1}{11}(-11) & -\frac{1}{11}(-11) \end{array}\right] = \left[\begin{array}{cc|c} 1 & 2 & -1 \\ 0 & 1 & 1 \end{array}\right]$$

We now have the desired matrix in row-echelon form, with 1s down the main diagonal and a 0 below the 1 in the first column.

Step 3. Write the system of linear equations corresponding to the matrix from step 2 and use back-substitution to find the system's solution. The system represented by the matrix from step 2 is

$$\left[\begin{array}{cc|c} 1 & 2 & -1 \\ 0 & 1 & 1 \end{array}\right] \rightarrow \begin{cases} 1x + 2y = -1 \\ 0x + 1y = 1 \end{cases} \quad \text{or} \quad \begin{cases} x + 2y = -1 & (1) \\ y = 1. & (2) \end{cases}$$

We immediately see from Equation 2 that the value for y is 1. To find x, we back-substitute 1 for y in Equation 1.

$$\begin{aligned} x + 2y &= -1 && \text{Equation 1} \\ x + 2 \cdot 1 &= -1 && \text{Substitute 1 for } y. \\ x + 2 &= -1 && \text{Multiply.} \\ x &= -3 && \text{Subtract 2 from both sides.} \end{aligned}$$

With $x = -3$ and $y = 1$, the proposed solution is $(-3, 1)$. Take a moment to show that $(-3, 1)$ satisfies both equations. The solution is $(-3, 1)$ and the solution set is $\{(-3, 1)\}$. ■

✓ CHECK POINT 2

Use matrices to solve the system:

$$\begin{cases} 2x - y = -4 \\ x + 3y = 5. \end{cases}$$

---

4 Use matrices to solve linear systems in three variables.

Solving Linear Systems in Three Variables Using Matrices

Gaussian elimination is also used to solve a system of linear equations in three variables. Most of the work involves using matrix row operations to obtain a matrix with 1s down the main diagonal and 0s below the 1s in the first and second columns.

Solving Linear Systems in Three Variables Using Matrices

1. Write the augmented matrix for the system.
2. Use matrix row operations to simplify the matrix to a row-equivalent matrix in row-echelon form, with 1s down the main diagonal from upper left to lower right, and 0s below the 1s in the first and second columns.

$$\left[\begin{array}{ccc|c} 1 & * & * & * \\ * & * & * & * \\ * & * & * & * \end{array}\right] \rightarrow \left[\begin{array}{ccc|c} 1 & * & * & * \\ 0 & * & * & * \\ 0 & * & * & * \end{array}\right] \rightarrow \left[\begin{array}{ccc|c} 1 & * & * & * \\ 0 & 1 & * & * \\ 0 & * & * & * \end{array}\right] \rightarrow \left[\begin{array}{ccc|c} 1 & * & * & * \\ 0 & 1 & * & * \\ 0 & 0 & * & * \end{array}\right] \rightarrow \left[\begin{array}{ccc|c} 1 & * & * & * \\ 0 & 1 & * & * \\ 0 & 0 & 1 & * \end{array}\right]$$

Get 1 in the upper left-hand corner.
Use the 1 in the first column to get 0s below it.
Get 1 in the second row, second column position.
Use the 1 in the second column to get 0 below it.
Get 1 in the third row, third column position.

3. Write the system of linear equations corresponding to the matrix from step 2 and use back-substitution to find the system's solution.

EXAMPLE 3 Using Matrices to Solve a Linear System

Use matrices to solve the system:

$$\begin{cases} 3x + y + 2z = 31 \\ x + y + 2z = 19 \\ x + 3y + 2z = 25. \end{cases}$$

Solution

Step 1. Write the augmented matrix for the system.

Linear System **Augmented Matrix**

$$\begin{cases} 3x + y + 2z = 31 \\ x + y + 2z = 19 \\ x + 3y + 2z = 25 \end{cases} \qquad \left[\begin{array}{ccc|c} 3 & 1 & 2 & 31 \\ 1 & 1 & 2 & 19 \\ 1 & 3 & 2 & 25 \end{array}\right]$$

Step 2. Use matrix row operations to simplify the matrix to row-echelon form, with 1s down the main diagonal from upper left to lower right, and 0s below the 1s in the first and second columns. Our first step in achieving this goal is to get 1 in the top position of the first column.

We want 1 in this position.

$$\left[\begin{array}{ccc|c} 3 & 1 & 2 & 31 \\ 1 & 1 & 2 & 19 \\ 1 & 3 & 2 & 25 \end{array}\right]$$

To get 1 in this position, we interchange row 1 and row 2: $R_1 \leftrightarrow R_2$.
We could also interchange row 1 and row 3 to get 1 in the upper left-hand corner.

$$\left[\begin{array}{ccc|c} 1 & 1 & 2 & 19 \\ 3 & 1 & 2 & 31 \\ 1 & 3 & 2 & 25 \end{array}\right]$$

This was row 2; now it's row 1.

This was row 1; now it's row 2.

Now we want to get 0s below the 1 in the first column.

We want 0 in these positions.

$$\left[\begin{array}{ccc|c} 1 & 1 & 2 & 19 \\ 3 & 1 & 2 & 31 \\ 1 & 3 & 2 & 25 \end{array}\right]$$

To get a 0 where there is now a 3, multiply the top row of numbers by -3 and add these products to the second row of numbers: $-3R_1 + R_2$. To get a 0 in the bottom of the first column where there is now a 1, multiply the top row of numbers by -1 and add these products to the third row of numbers: $-1R_1 + R_3$. Although we are using row 1 to find the products, the numbers in row 1 do not change.

Replace row 2 by $-3R_1 + R_2$.

Replace row 3 by $-1R_1 + R_3$.

$$\left[\begin{array}{ccc|c} 1 & 1 & 2 & 19 \\ -3(1)+3 & -3(1)+1 & -3(2)+2 & -3(19)+31 \\ -1(1)+1 & -1(1)+3 & -1(2)+2 & -1(19)+25 \end{array}\right] = \left[\begin{array}{ccc|c} 1 & 1 & 2 & 19 \\ 0 & -2 & -4 & -26 \\ 0 & 2 & 0 & 6 \end{array}\right]$$

We want 1 in this position.

We move on to the second column. To get 1 in the desired position, we multiply -2 by its reciprocal, $-\frac{1}{2}$. Therefore, we multiply all the numbers in the second row by $-\frac{1}{2}$: $-\frac{1}{2}R_2$.

$-\frac{1}{2}R_2$

$$\left[\begin{array}{ccc|c} 1 & 1 & 2 & 19 \\ -\frac{1}{2}(0) & -\frac{1}{2}(-2) & -\frac{1}{2}(-4) & -\frac{1}{2}(-26) \\ 0 & 2 & 0 & 6 \end{array}\right] = \left[\begin{array}{ccc|c} 1 & 1 & 2 & 19 \\ 0 & 1 & 2 & 13 \\ 0 & 2 & 0 & 6 \end{array}\right]$$

We want 0 in this position.

We are not yet done with the second column. The voice balloon shows that we want to get a 0 where there is now a 2. If we multiply the second row of numbers by -2 and add these products to the third row of numbers, we will get 0 in this position: $-2R_2 + R_3$. Although we are using the numbers in row 2 to find the products, the numbers in row 2 do not change.

Replace row 3 by $-2R_2 + R_3$.

$$\left[\begin{array}{ccc|c} 1 & 1 & 2 & 19 \\ 0 & 1 & 2 & 13 \\ -2(0)+0 & -2(1)+2 & -2(2)+0 & -2(13)+6 \end{array}\right] = \left[\begin{array}{ccc|c} 1 & 1 & 2 & 19 \\ 0 & 1 & 2 & 13 \\ 0 & 0 & -4 & -20 \end{array}\right]$$

We want 1 in this position.

We move on to the third column. To get 1 in the desired position, we multiply -4 by its reciprocal, $-\frac{1}{4}$. Therefore, we multiply all the numbers in the third row by $-\frac{1}{4}$: $-\frac{1}{4}R_3$.

$$-\tfrac{1}{4}R_3 \quad \left[\begin{array}{ccc|c} 1 & 1 & 2 & 19 \\ 0 & 1 & 2 & 13 \\ -\frac{1}{4}(0) & -\frac{1}{4}(0) & -\frac{1}{4}(-4) & -\frac{1}{4}(-20) \end{array}\right] = \left[\begin{array}{ccc|c} 1 & 1 & 2 & 19 \\ 0 & 1 & 2 & 13 \\ 0 & 0 & 1 & 5 \end{array}\right]$$

We now have the desired matrix in row-echelon form, with 1s down the main diagonal and 0s below the 1s in the first and second columns.

Using Technology

Most graphing utilities can convert an augmented matrix to row-echelon form, with 1s down the main diagonal and 0s below the 1s. However, row-echelon form is not unique. Your graphing utility might give a row-echelon form different from the one you obtained by hand. However, all row-echelon forms for a given system's augmented matrix produce the same solution to the system. Enter the augmented matrix and name it A. Then use the REF (row-echelon form) command on matrix A.

Step 3. Write the system of linear equations corresponding to the matrix from step 2 and use back-substitution to find the system's solution. The system represented by the matrix from step 2 is

$$\left[\begin{array}{ccc|c} 1 & 1 & 2 & 19 \\ 0 & 1 & 2 & 13 \\ 0 & 0 & 1 & 5 \end{array}\right] \rightarrow \begin{cases} 1x + 1y + 2z = 19 \\ 0x + 1y + 2z = 13 \\ 0x + 0y + 1z = 5 \end{cases} \text{ or } \begin{cases} x + y + 2z = 19 & 1 \\ y + 2z = 13. & 2 \\ z = 5 & 3 \end{cases}$$

We immediately see from Equation 3 that the value for z is 5. To find y, we back-substitute 5 for z in the second equation.

$$\begin{aligned} y + 2z &= 13 && \text{Equation 2} \\ y + 2(5) &= 13 && \text{Substitute 5 for } z. \\ y &= 3 && \text{Solve for } y. \end{aligned}$$

Finally, back-substitute 3 for y and 5 for z in the first equation.

$$\begin{aligned} x + y + 2z &= 19 && \text{Equation 1} \\ x + 3 + 2(5) &= 19 && \text{Substitute 3 for } y \text{ and 5 for } z. \\ x + 13 &= 19 && \text{Multiply and add.} \\ x &= 6 && \text{Subtract 13 from both sides.} \end{aligned}$$

The solution of the original system is (6, 3, 5) and the solution set is {(6, 3, 5)}. Check to see that the solution satisfies all three equations in the given system. ■

✓ CHECK POINT 3

Use matrices to solve the system:

$$\begin{cases} 2x + y + 2z = 18 \\ x - y + 2z = 9 \\ x + 2y - z = 6. \end{cases}$$

Modern supercomputers are capable of solving systems with more than 600,000 variables. The augmented matrices for such systems are huge, but the solution using matrices is exactly like what we did in Example 3. Work with the augmented matrix, one column at a time. Get 1s down the main diagonal from upper left to lower right, and 0s below the 1s.

5 Use matrices to identify inconsistent and dependent systems.

Inconsistent Systems and Systems with Dependent Equations

When solving a system using matrices, you might obtain a matrix with a row in which the numbers to the left of the vertical bar are all zeros, but a nonzero number appears on the right. In such a case, the system is inconsistent and has no solution. For example, a system of equations that yields the following matrix is an inconsistent system:

$$\left[\begin{array}{cc|c} 1 & -2 & 3 \\ 0 & 0 & -4 \end{array}\right].$$

The second row of the matrix represents the equation $0x + 0y = -4$, which is false for all values of x and y.

If you obtain a matrix in which a 0 appears across an entire row, the system contains dependent equations and has infinitely many solutions. This row of zeros represents $0x + 0y = 0$ or $0x + 0y + 0z = 0$. These equations are satisfied by infinitely many ordered pairs or triples.

CONCEPT AND VOCABULARY CHECK

Fill in each blank so that the resulting statement is true.

1. A rectangular array of numbers, arranged in rows and columns and placed in brackets, is called a/an ____________. The numbers inside the brackets are called ____________.

2. The augmented matrix for the system

$$\begin{cases} 3x - 2y = -6 \\ 4x + 5y = -8 \end{cases}$$

is $\left[\begin{array}{cc|c} \_ & \_ & \_ \\ \_ & \_ & \_ \end{array}\right]$.

3. The augmented matrix for the system

$$\begin{cases} 2x + y + 4z = -4 \\ 3x \quad\;\; + z = 1 \\ 4x + 3y + z = 8 \end{cases}$$

is $\left[\begin{array}{ccc|c} \_ & \_ & \_ & \_ \\ \_ & \_ & \_ & \_ \\ \_ & \_ & \_ & \_ \end{array}\right]$.

4. Consider the matrix

$$\left[\begin{array}{cc|c} 2 & -1 & 5 \\ 4 & -2 & 7 \end{array}\right].$$

We can obtain 1 in the shaded position if we multiply all the numbers in the ____________ row by ____________.

5. Consider the matrix

$$\left[\begin{array}{ccc|c} 1 & 1 & -1 & -2 \\ -3 & -4 & 2 & 4 \\ 2 & 1 & 1 & 6 \end{array}\right].$$

We can obtain 0 in the position shaded by a rectangle if we multiply the top row of numbers by ____________ and add these products to the ____________ row of numbers. We can obtain 0 in the position shaded by an oval if we multiply the top row of numbers by ____________ and add these products to the ____________ row of numbers.

6. True or false: Two columns of a matrix may be interchanged to form an equivalent matrix. ____________

7. True or false: The matrix

$$\left[\begin{array}{cc|c} 1 & 2 & 5 \\ 0 & 0 & 8 \end{array}\right]$$

represents an inconsistent system. ____________

3.4 EXERCISE SET

MyMathLab®

Practice Exercises

In Exercises 1–14, perform each matrix row operation and write the new matrix.

1. $\left[\begin{array}{cc|c} 2 & 2 & 5 \\ 1 & -\frac{3}{2} & 5 \end{array}\right] R_1 \leftrightarrow R_2$

2. $\left[\begin{array}{cc|c} -6 & 9 & 4 \\ 1 & -\frac{3}{2} & 4 \end{array}\right] R_1 \leftrightarrow R_2$

3. $\left[\begin{array}{cc|c} -6 & 8 & -12 \\ 3 & 5 & -2 \end{array}\right] -\frac{1}{6}R_1$

4. $\left[\begin{array}{cc|c} -2 & 3 & -10 \\ 4 & 2 & 5 \end{array}\right] -\frac{1}{2}R_1$

5. $\left[\begin{array}{cc|c} 1 & -3 & 5 \\ 2 & 6 & 4 \end{array}\right] -2R_1 + R_2$

6. $\left[\begin{array}{cc|c} 1 & -3 & 1 \\ 2 & 1 & -5 \end{array}\right] -2R_1 + R_2$

7. $\left[\begin{array}{cc|c} 1 & -\frac{3}{2} & \frac{7}{2} \\ 3 & 4 & 2 \end{array}\right] -3R_1 + R_2$

8. $\left[\begin{array}{cc|c} 1 & -\frac{2}{5} & \frac{3}{4} \\ 4 & 2 & -1 \end{array}\right] -4R_1 + R_2$

9. $\left[\begin{array}{ccc|c} 2 & -6 & 4 & 10 \\ 1 & 5 & -5 & 0 \\ 3 & 0 & 4 & 7 \end{array}\right] \frac{1}{2}R_1$

10. $\left[\begin{array}{ccc|c} 3 & -12 & 6 & 9 \\ 1 & -4 & 4 & 0 \\ 2 & 0 & 7 & 4 \end{array}\right] \frac{1}{3}R_1$

11. $\left[\begin{array}{ccc|c} 1 & -3 & 2 & 0 \\ 3 & 1 & -1 & 7 \\ 2 & -2 & 1 & 3 \end{array}\right] -3R_1 + R_2$

12. $\left[\begin{array}{ccc|c} 1 & -1 & 5 & -6 \\ 3 & 3 & -1 & 10 \\ 1 & 3 & 2 & 5 \end{array}\right] -3R_1 + R_2$

13. $\left[\begin{array}{ccc|c} 1 & 1 & -1 & 6 \\ 2 & -1 & 1 & -3 \\ 3 & -1 & -1 & 4 \end{array}\right] \begin{array}{l} \\ -2R_1 + R_2 \\ -3R_1 + R_3 \end{array}$

14. $\left[\begin{array}{ccc|c} 1 & 2 & 1 & 2 \\ -2 & -1 & 2 & 5 \\ 1 & 3 & -2 & -8 \end{array}\right] \begin{array}{l} \\ 2R_1 + R_2 \\ -1R_1 + R_3 \end{array}$

In Exercises 15–38, solve each system using matrices. If there is no solution or if there are infinitely many solutions and a system's equations are dependent, so state.

15. $\begin{cases} x + y = 6 \\ x - y = 2 \end{cases}$

16. $\begin{cases} x + 2y = 11 \\ x - y = -1 \end{cases}$

17. $\begin{cases} 2x + y = 3 \\ x - 3y = 12 \end{cases}$

18. $\begin{cases} 3x - 5y = 7 \\ x - y = 1 \end{cases}$

19. $\begin{cases} 5x + 7y = -25 \\ 11x + 6y = -8 \end{cases}$

20. $\begin{cases} 3x - 5y = 22 \\ 4x - 2y = 20 \end{cases}$

21. $\begin{cases} 4x - 2y = 5 \\ -2x + y = 6 \end{cases}$

22. $\begin{cases} -3x + 4y = 12 \\ 6x - 8y = 16 \end{cases}$

23. $\begin{cases} x - 2y = 1 \\ -2x + 4y = -2 \end{cases}$

24. $\begin{cases} 3x - 6y = 1 \\ 2x - 4y = \frac{2}{3} \end{cases}$

25. $\begin{cases} x + y - z = -2 \\ 2x - y + z = 5 \\ -x + 2y + 2z = 1 \end{cases}$

26. $\begin{cases} x - 2y - z = 2 \\ 2x - y + z = 4 \\ -x + y - 2z = -4 \end{cases}$

27. $\begin{cases} x + 3y = 0 \\ x + y + z = 1 \\ 3x - y - z = 11 \end{cases}$

28. $\begin{cases} 3y - z = -1 \\ x + 5y - z = -4 \\ -3x + 6y + 2z = 11 \end{cases}$

29. $\begin{cases} 2x + 2y + 7z = -1 \\ 2x + y + 2z = 2 \\ 4x + 6y + z = 15 \end{cases}$

30. $\begin{cases} 3x + 2y + 3z = 3 \\ 4x - 5y + 7z = 1 \\ 2x + 3y - 2z = 6 \end{cases}$

31. $\begin{cases} x + y + z = 6 \\ x - z = -2 \\ y + 3z = 11 \end{cases}$

32. $\begin{cases} x + y + z = 3 \\ -y + 2z = 1 \\ -x + z = 0 \end{cases}$

33. $\begin{cases} x - y + 3z = 4 \\ 2x - 2y + 6z = 7 \\ 3x - y + 5z = 14 \end{cases}$

34. $\begin{cases} 3x - y + 2z = 4 \\ -6x + 2y - 4z = 1 \\ 5x - 3y + 8z = 0 \end{cases}$

35. $\begin{cases} x - 2y + z = 4 \\ 5x - 10y + 5y = 20 \\ -2x + 4y - 2z = -8 \end{cases}$

36. $\begin{cases} x - 3y + z = 2 \\ 4x - 12y + 4z = 8 \\ -2x + 6y - 2z = -4 \end{cases}$

37. $\begin{cases} x + y = 1 \\ y + 2z = -2 \\ 2x - z = 0 \end{cases}$

38. $\begin{cases} x + 3y = 3 \\ y + 2z = -8 \\ x - z = 7 \end{cases}$

Practice PLUS

In Exercises 39–40, write the system of linear equations represented by the augmented matrix. Use w, x, y, and z for the variables. Once the system is written, use back-substitution to find its solution set, {(w, x, y, z)}.

39. $\left[\begin{array}{cccc|c} 1 & -1 & 1 & 1 & 3 \\ 0 & 1 & -2 & -1 & 0 \\ 0 & 0 & 1 & 6 & 17 \\ 0 & 0 & 0 & 1 & 3 \end{array}\right]$

40. $\left[\begin{array}{cccc|c} 1 & 2 & -1 & 0 & 2 \\ 0 & 1 & 1 & -2 & -3 \\ 0 & 0 & 1 & -1 & -2 \\ 0 & 0 & 0 & 1 & 3 \end{array}\right]$

In Exercises 41–42, perform each matrix row operation and write the new matrix.

41. $$\left[\begin{array}{rrrr|r} 1 & -1 & 1 & 1 & 3 \\ 0 & 1 & -2 & -1 & 0 \\ 2 & 0 & 3 & 4 & 11 \\ 5 & 1 & 2 & 4 & 6 \end{array}\right] \begin{array}{l} \\ \\ -2R_1 + R_3 \\ -5R_1 + R_4 \end{array}$$

42. $$\left[\begin{array}{rrrr|r} 1 & -5 & 2 & -2 & 4 \\ 0 & 1 & -3 & -1 & 0 \\ 3 & 0 & 2 & -1 & 6 \\ -4 & 1 & 4 & 2 & -3 \end{array}\right] \begin{array}{l} \\ \\ -3R_1 + R_3 \\ 4R_1 + R_4 \end{array}$$

In Exercises 43–44, solve each system using matrices. You will need to use matrix row operations to obtain matrices like those in Exercises 39 and 40, with 1s down the main diagonal and 0s below the 1s. Express the solution set as $\{(w, x, y, z)\}$.

43. $$\begin{cases} w + x + y + z = 4 \\ 2w + x - 2y - z = 0 \\ w - 2x - y - 2z = -2 \\ 3w + 2x + y + 3z = 4 \end{cases}$$

44. $$\begin{cases} w + x + y + z = 5 \\ w + 2x - y - 2z = -1 \\ w - 3x - 3y - z = -1 \\ 2w - x + 2y - z = -2 \end{cases}$$

Application Exercises

45. A ball is thrown straight upward. The graph shows the ball's height, $s(t)$, in feet, after t seconds.

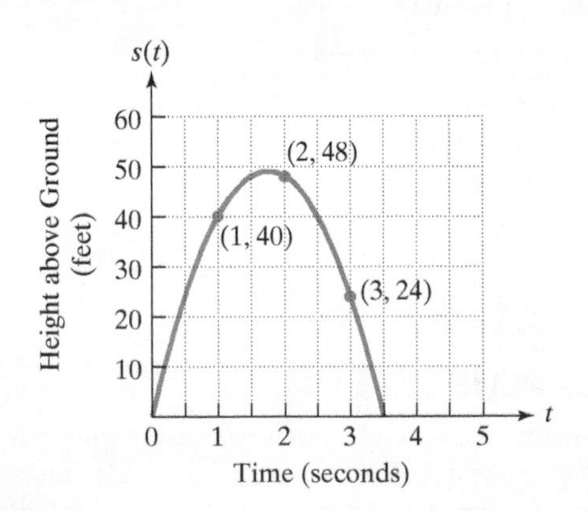

a. Find the quadratic function

$$s(t) = at^2 + bt + c$$

whose graph passes through the three points labeled on the graph. Solve the system of linear equations involving a, b, and c using matrices.

b. Find and interpret $s(3.5)$. Identify your solution as a point on the graph shown.

46. A football is kicked straight upward. The graph shows the football's height, $s(t)$, in feet, after t seconds.

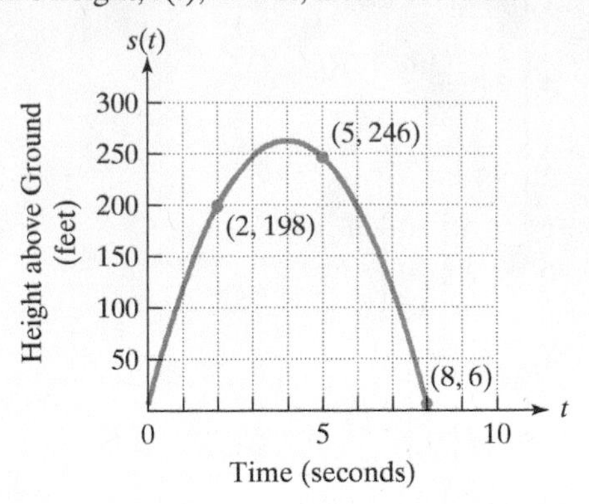

a. Find the quadratic function

$$s(t) = at^2 + bt + c$$

whose graph passes through the three points labeled on the graph. Solve the system of linear equations involving a, b, and c using matrices.

b. Find and interpret $s(7)$. Identify your solution as a point on the graph shown.

Write a system of linear equations in three variables to solve Exercises 47–48. Then use matrices to solve the system.

Exercises 47–48 are based on a Time/CNN telephone poll that included never-married single women between the ages of 18 and 49 and never-married single men between the ages of 18 and 49. The circle graphs show the results for one of the questions in the poll.

If You Couldn't Find the Perfect Mate, Would You Marry Someone Else?

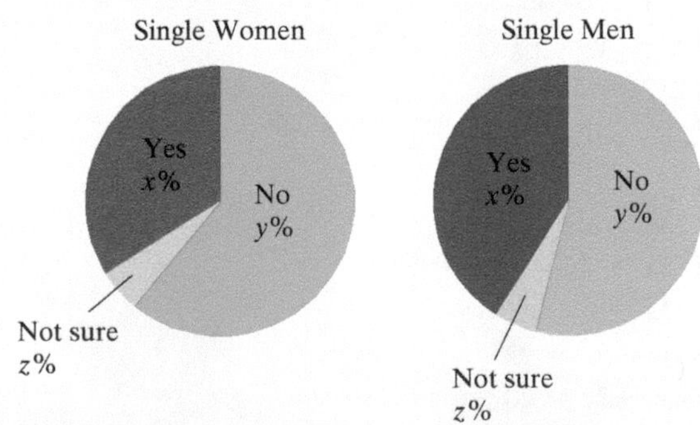

Source: Time/CNN

47. For single women in the poll, the percentage who said no exceeded the combined percentages for those who said yes and those who said not sure by 22%. If the percentage who said yes is doubled, it is 7% more than the percentage who said no. Find the percentage of single women who responded yes, no, and not sure.

48. For single men in the poll, the percentage who said no exceeded the combined percentages for those who said yes and those who said not sure by 8%. If the percentage who said yes is doubled, it is 28% more than the percentage who said no. Find the percentage of single men who responded yes, no, and not sure.

Explaining the Concepts

49. What is a matrix?
50. Describe what is meant by the augmented matrix of a system of linear equations.
51. In your own words, describe each of the three matrix row operations. Give an example of each of the operations.
52. Describe how to use matrices and row operations to solve a system of linear equations.
53. When solving a system using matrices, how do you know if the system has no solution?
54. When solving a system using matrices, how do you know if the system has infinitely many solutions?

Technology Exercises

55. Most graphing utilities can perform row operations on matrices. Consult the owner's manual for your graphing utility to learn proper keystroke sequences for performing these operations. Then duplicate the row operations of any three exercises that you solved from Exercises 3–12.
56. If your graphing utility has a REF (row-echelon form) command, use this feature to verify your work with any five systems from Exercises 15–38.
57. A matrix with 1s down the main diagonal and 0s in every position *above and below* each 1 is said to be in **reduced row-echelon form.**

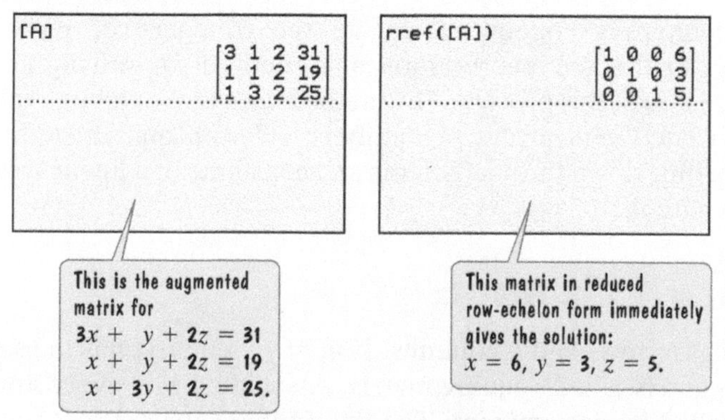
[A]
[3 1 2 31]
[1 1 2 19]
[1 3 2 25]
rref([A])
[1 0 0 6]
[0 1 0 3]
[0 0 1 5]
This is the augmented matrix for
3x + y + 2z = 31
x + y + 2z = 19
x + 3y + 2z = 25.
This matrix in reduced row-echelon form immediately gives the solution:
x = 6, y = 3, z = 5.

If your graphing utility has a RREF (reduced row-echelon form) command, use this feature to verify your work with any five systems from Exercises 15–38.

Critical Thinking Exercises

Make Sense? *In Exercises 58–61, determine whether each statement makes sense or does not make sense, and explain your reasoning.*

58. Matrix row operations remind me of what I did when solving a linear system by the addition method, although I no longer write the variables.
59. When I use matrices to solve linear systems, the only arithmetic involves multiplication or a combination of multiplication and addition.
60. When I use matrices to solve linear systems, I spend most of my time using row operations to express the system's augmented matrix in row-echelon form.
61. Using row operations on an augmented matrix, I obtain a row in which 0s appear to the left of the vertical bar, but 6 appears on the right, so the system I'm working with has infinitely many solutions.

In Exercises 62–65, determine whether each statement is true or false. If the statement is false, make the necessary change(s) to produce a true statement.

62. A matrix row operation such as $-\frac{4}{5}R_1 + R_2$ is not permitted because of the negative fraction.
63. The augmented matrix for the system

$$\begin{cases} x - 3y = 5 \\ y - 2z = 7 \\ 2x + z = 4 \end{cases} \text{ is } \left[\begin{array}{cc|c} 1 & -3 & 5 \\ 1 & -2 & 7 \\ 2 & 1 & 4 \end{array}\right].$$

64. In solving a linear system of three equations in three variables, we begin with the augmented matrix and use row operations to obtain a row-equivalent matrix with 0s down the diagonal from left to right and 1s below each 0.
65. The row operation $kR_i + R_j$ indicates that it is the elements in row i that change.
66. The vitamin content per ounce for three foods is given in the following table.

	Milligrams per Ounce		
	Thiamin	Riboflavin	Niacin
Food A	3	7	1
Food B	1	5	3
Food C	3	8	2

a. Use matrices to show that no combination of these foods can provide exactly 14 milligrams of thiamin, 32 milligrams of riboflavin, and 9 milligrams of niacin.

b. Use matrices to describe in practical terms what happens if the riboflavin requirement is increased by 5 milligrams and the other requirements stay the same.

Review Exercises

67. If $f(x) = -3x + 10$, find $f(2a - 1)$. (Section 2.1, Example 3)
68. If $f(x) = 3x$ and $g(x) = 2x - 3$, find $(fg)(-1)$. (Section 2.3, Example 4)
69. Simplify: $\frac{-4x^8y^{-12}}{12x^{-3}y^{24}}$. (Section 1.6, Example 9)

Preview Exercises

Exercises 70–72 will help you prepare for the material covered in the next section. Simplify the expression in each exercise.

70. $2(-5) - (-3)(4)$
71. $\frac{2(-5) - 1(-4)}{5(-5) - 6(-4)}$
72. $2(-30 - (-3)) - 3(6 - 9) + (-1)(1 - 15)$

SECTION

3.5 Determinants and Cramer's Rule

What am I supposed to learn?

After studying this section, you should be able to:

1. Evaluate a second-order determinant.
2. Solve a system of linear equations in two variables using Cramer's rule.
3. Evaluate a third-order determinant.
4. Solve a system of linear equations in three variables using Cramer's rule.

As cyberspace absorbs more and more of our work, play, shopping, and socializing, where will it all end? Which activities will still be offline in 2025?

A portion of Charles Babbage's unrealized Difference Engine

Our technologically transformed lives can be traced back to the English inventor Charles Babbage (1791–1871). Babbage knew of a method for solving linear systems called *Cramer's rule*, in honor of the Swiss geometer Gabriel Cramer (1704–1752). Cramer's rule was simple, but involved numerous multiplications for large systems. Babbage designed a machine, called the "difference engine," that consisted of toothed wheels on shafts for performing these multiplications. Despite the fact that only one-seventh of the functions ever worked, Babbage's invention demonstrated how complex calculations could be handled mechanically. In 1944, scientists at IBM used the lessons of the difference engine to create the world's first computer.

Those who invented computers hoped to relegate the drudgery of repeated computation to a machine. In this section, we look at a method for solving linear systems that played a critical role in this process. The method uses real numbers, called *determinants*, that are associated with arrays of numbers. As with matrix methods, solutions are obtained by writing down the coefficients and constants of a linear system and performing operations with them.

1 Evaluate a second-order determinant.

The Determinant of a 2 × 2 Matrix

A matrix of **order $m \times n$** has m rows and n columns. If $m = n$, a matrix has the same number of rows as columns and is called a **square matrix**. Associated with every square matrix is a real number, called its **determinant**. The determinant for a 2×2 square matrix is defined as follows:

Great Question!

What does the definition of a determinant mean? What am I supposed to do?

To evaluate a second-order determinant, find the difference of the product of the two diagonals.

$$\begin{vmatrix} a_1 & b_1 \\ a_2 & b_2 \end{vmatrix} = a_1b_2 - a_2b_1$$

Definition of the Determinant of a 2 × 2 Matrix

The determinant of the matrix $\begin{bmatrix} a_1 & b_1 \\ a_2 & b_2 \end{bmatrix}$ is denoted by $\begin{vmatrix} a_1 & b_1 \\ a_2 & b_2 \end{vmatrix}$ and is defined by

$$\begin{vmatrix} a_1 & b_1 \\ a_2 & b_2 \end{vmatrix} = a_1b_2 - a_2b_1.$$

We also say that the **value** of the **second-order determinant** $\begin{vmatrix} a_1 & b_1 \\ a_2 & b_2 \end{vmatrix}$ is $a_1b_2 - a_2b_1$.

Example 1 illustrates that the determinant of a matrix may be positive or negative. The determinant can also have 0 as its value.

EXAMPLE 1 Evaluating the Determinant of a 2 × 2 Matrix

Evaluate the determinant of each of the following matrices:

a. $\begin{bmatrix} 5 & 6 \\ 7 & 3 \end{bmatrix}$

b. $\begin{bmatrix} 2 & 4 \\ -3 & -5 \end{bmatrix}$.

Discover for Yourself

Write and then evaluate three determinants, one whose value is positive, one whose value is negative, and one whose value is 0.

Solution We multiply and subtract as indicated.

a. $\begin{vmatrix} 5 & 6 \\ 7 & 3 \end{vmatrix} = 5 \cdot 3 - 7 \cdot 6 = 15 - 42 = -27$ The value of the second-order determinant is -27.

b. $\begin{vmatrix} 2 & 4 \\ -3 & -5 \end{vmatrix} = 2(-5) - (-3)(4) = -10 + 12 = 2$ The value of the second-order determinant is 2. ■

CHECK POINT 1 Evaluate the determinant of each of the following matrices:

a. $\begin{bmatrix} 10 & 9 \\ 6 & 5 \end{bmatrix}$ b. $\begin{bmatrix} 4 & 3 \\ -5 & -8 \end{bmatrix}$.

2 Solve a system of linear equations in two variables using Cramer's rule.

Solving Systems of Linear Equations in Two Variables Using Determinants

Determinants can be used to solve a linear system in two variables. In general, such a system appears as

$$\begin{cases} a_1x + b_1y = c_1 \\ a_2x + b_2y = c_2. \end{cases}$$

Let's first solve this system for x using the addition method. We can solve for x by eliminating y from the equations. Multiply the first equation by b_2 and the second equation by $-b_1$. Then add the two equations:

$$\begin{cases} a_1x + b_1y = c_1 \xrightarrow{\text{Multiply by } b_2.} \\ a_2x + b_2y = c_2 \xrightarrow{\text{Multiply by } -b_1.} \end{cases} \begin{cases} a_1b_2x + b_1b_2y = c_1b_2 \\ -a_2b_1x - b_1b_2y = -c_2b_1 \end{cases}$$

$$\text{Add: } (a_1b_2 - a_2b_1)x = c_1b_2 - c_2b_1$$

$$x = \frac{c_1b_2 - c_2b_1}{a_1b_2 - a_2b_1}$$

Because

$$\begin{vmatrix} c_1 & b_1 \\ c_2 & b_2 \end{vmatrix} = c_1b_2 - c_2b_1 \quad \text{and} \quad \begin{vmatrix} a_1 & b_1 \\ a_2 & b_2 \end{vmatrix} = a_1b_2 - a_2b_1,$$

we can express our answer for x as the quotient of two determinants:

$$x = \frac{\begin{vmatrix} c_1 & b_1 \\ c_2 & b_2 \end{vmatrix}}{\begin{vmatrix} a_1 & b_1 \\ a_2 & b_2 \end{vmatrix}}.$$

Similarly, we could use the addition method to solve our system for y, again expressing y as the quotient of two determinants. This method of using determinants to solve the linear system, called **Cramer's rule**, is summarized in the box on the next page.

Solving a Linear System in Two Variables Using Determinants

Cramer's Rule

If

$$\begin{cases} a_1x + b_1y = c_1 \\ a_2x + b_2y = c_2 \end{cases}$$

then

$$x = \frac{\begin{vmatrix} c_1 & b_1 \\ c_2 & b_2 \end{vmatrix}}{\begin{vmatrix} a_1 & b_1 \\ a_2 & b_2 \end{vmatrix}} \quad \text{and} \quad y = \frac{\begin{vmatrix} a_1 & c_1 \\ a_2 & c_2 \end{vmatrix}}{\begin{vmatrix} a_1 & b_1 \\ a_2 & b_2 \end{vmatrix}},$$

where

$$\begin{vmatrix} a_1 & b_1 \\ a_2 & b_2 \end{vmatrix} \neq 0.$$

Here are some helpful tips when solving

$$\begin{cases} a_1x + b_1y = c_1 \\ a_2x + b_2y = c_2 \end{cases}$$

using determinants:

1. Three different determinants are used to find x and y. The determinants in the denominators for x and y are identical. The determinants in the numerators for x and y differ. In abbreviated notation, we write

$$x = \frac{D_x}{D} \quad \text{and} \quad y = \frac{D_y}{D}, \text{ where } D \neq 0.$$

2. The elements of D, the determinant in the denominator, are the coefficients of the variables in the system.

$$D = \begin{vmatrix} a_1 & b_1 \\ a_2 & b_2 \end{vmatrix}$$

3. D_x, the determinant in the numerator of x, is obtained by replacing the x-coefficients, in D, a_1 and a_2, with the constants on the right sides of the equations, c_1 and c_2.

$$D = \begin{vmatrix} a_1 & b_1 \\ a_2 & b_2 \end{vmatrix} \quad \text{and} \quad D_x = \begin{vmatrix} c_1 & b_1 \\ c_2 & b_2 \end{vmatrix}$$

Replace the column with a_1 and a_2 with the constants c_1 and c_2 to get D_x.

4. D_y, the determinant in the numerator for y, is obtained by replacing the y-coefficients, in D, b_1 and b_2, with the constants on the right sides of the equations, c_1 and c_2.

$$D = \begin{vmatrix} a_1 & b_1 \\ a_2 & b_2 \end{vmatrix} \quad \text{and} \quad D_y = \begin{vmatrix} a_1 & c_1 \\ a_2 & c_2 \end{vmatrix}$$

Replace the column with b_1 and b_2 with the constants c_1 and c_2 to get D_y

EXAMPLE 2 Using Cramer's Rule to Solve a Linear System

Use Cramer's rule to solve the system:

$$\begin{cases} 5x - 4y = 2 \\ 6x - 5y = 1. \end{cases}$$

Solution Because

$$x = \frac{D_x}{D} \quad \text{and} \quad y = \frac{D_y}{D},$$

we will set up and evaluate the three determinants D, D_x, and D_y.

$\begin{cases} 5x - 4y = 2 \\ 6x - 5y = 1 \end{cases}$

The system we are interested in solving (repeated so that you dont't have to look back)

1. D, the determinant in both denominators, consists of the x- and y-coefficients.

$$D = \begin{vmatrix} 5 & -4 \\ 6 & -5 \end{vmatrix} = (5)(-5) - (6)(-4) = -25 + 24 = -1$$

Because this determinant is not zero, we continue to use Cramer's rule to solve the system.

2. D_x, the determinant in the numerator for x, is obtained by replacing the x-coefficients in D, 5 and 6, by the constants on the right sides of the equations, 2 and 1.

$$D_x = \begin{vmatrix} 2 & -4 \\ 1 & -5 \end{vmatrix} = (2)(-5) - (1)(-4) = -10 + 4 = -6$$

3. D_y, the determinant in the numerator for y, is obtained by replacing the y-coefficients in D, -4 and -5, by the constants on the right sides of the equations, 2 and 1.

$$D_y = \begin{vmatrix} 5 & 2 \\ 6 & 1 \end{vmatrix} = (5)(1) - (6)(2) = 5 - 12 = -7$$

4. Thus,

$$x = \frac{D_x}{D} = \frac{-6}{-1} = 6 \quad \text{and} \quad y = \frac{D_y}{D} = \frac{-7}{-1} = 7.$$

As always, the ordered pair (6, 7) should be checked by substituting these values into the original equations. The solution is (6, 7) and the solution set is {(6, 7)}. ■

CHECK POINT 2 Use Cramer's rule to solve the system:

$$\begin{cases} 5x + 4y = 12 \\ 3x - 6y = 24. \end{cases}$$

---

3 Evaluate a third-order determinant.

The Determinant of a 3 × 3 Matrix

The determinant for a 3 × 3 matrix is defined in terms of second-order determinants:

Definition of the Determinant of a 3 × 3 Matrix

A third-order determinant is defined by

$$\begin{vmatrix} a_1 & b_1 & c_1 \\ a_2 & b_2 & c_2 \\ a_3 & b_3 & c_3 \end{vmatrix} = a_1\begin{vmatrix} b_2 & c_2 \\ b_3 & c_3 \end{vmatrix} - a_2\begin{vmatrix} b_1 & c_1 \\ b_3 & c_3 \end{vmatrix} + a_3\begin{vmatrix} b_1 & c_1 \\ b_2 & c_2 \end{vmatrix}.$$

Subtract.

Add.

Each a on the right comes from the first column.

$$\begin{vmatrix} a_1 & b_1 & c_1 \\ a_2 & b_2 & c_2 \\ a_3 & b_3 & c_3 \end{vmatrix} =$$

$$a_1\begin{vmatrix} b_2 & c_2 \\ b_3 & c_3 \end{vmatrix} - a_2\begin{vmatrix} b_1 & c_1 \\ b_3 & c_3 \end{vmatrix} + a_3\begin{vmatrix} b_1 & c_1 \\ b_2 & c_2 \end{vmatrix}$$

The definition of a third-order determinant (repeated)

Here are some tips that should be helpful when evaluating the determinant of a 3×3 matrix:

Evaluating the Determinant of a 3 × 3 Matrix

1. Each of the three terms in the definition of a third-order determinant contains two factors—a numerical factor and a second-order determinant.
2. The numerical factor in each term is an element from the first column of the third-order determinant.
3. The minus sign precedes the second term.
4. The second-order determinant that appears in each term is obtained by crossing out the row and the column containing the numerical factor.

$$a_1\begin{vmatrix} b_2 & c_2 \\ b_3 & c_3 \end{vmatrix} - a_2\begin{vmatrix} b_1 & c_1 \\ b_3 & c_3 \end{vmatrix} + a_3\begin{vmatrix} b_1 & c_1 \\ b_2 & c_2 \end{vmatrix}$$

$$\downarrow \qquad \downarrow \qquad \downarrow$$

$$\begin{vmatrix} \not{a_1} & \not{b_1} & \not{c_1} \\ \not{a_2} & b_2 & c_2 \\ \not{a_3} & b_3 & c_3 \end{vmatrix} \begin{vmatrix} \not{a_1} & b_1 & c_1 \\ \not{a_2} & \not{b_2} & \not{c_2} \\ \not{a_3} & b_3 & c_3 \end{vmatrix} \begin{vmatrix} \not{a_1} & b_1 & c_1 \\ \not{a_2} & b_2 & c_2 \\ \not{a_3} & \not{b_3} & \not{c_3} \end{vmatrix}$$

The **minor** of an element is the determinant that remains after deleting the row and column of that element. For this reason, we call this method **expansion by minors.**

EXAMPLE 3 Evaluating the Determinant of a 3 × 3 Matrix

Evaluate the determinant of the following matrix:

$$\begin{bmatrix} 4 & 1 & 0 \\ -9 & 3 & 4 \\ -3 & 8 & 1 \end{bmatrix}.$$

Solution We know that each of the three terms in the determinant contains a numerical factor and a second-order determinant. The numerical factors are from the first column of the given matrix. They are shown in red in the following matrix:

$$\begin{bmatrix} 4 & 1 & 0 \\ -9 & 3 & 4 \\ -3 & 8 & 1 \end{bmatrix}.$$

We find the minor for each numerical factor by deleting the row and column of that element.

$$\begin{bmatrix} \not{4} & \not{1} & \not{0} \\ -\not{9} & 3 & 4 \\ -\not{3} & 8 & 1 \end{bmatrix} \quad \begin{bmatrix} \not{4} & 1 & 0 \\ -\not{9} & \not{3} & \not{4} \\ -\not{3} & 8 & 1 \end{bmatrix} \quad \begin{bmatrix} \not{4} & 1 & 0 \\ -\not{9} & 3 & 4 \\ -\not{3} & \not{8} & \not{1} \end{bmatrix}$$

The minor for 4 is $\begin{vmatrix} 3 & 4 \\ 8 & 1 \end{vmatrix}$. The minor for −9 is $\begin{vmatrix} 1 & 0 \\ 8 & 1 \end{vmatrix}$. The minor for −3 is $\begin{vmatrix} 1 & 0 \\ 3 & 4 \end{vmatrix}$.

Now we have three numerical factors, 4, −9, and −3, and three second-order determinants. We multiply each numerical factor by its second-order determinant to find the three terms of the third-order determinant:

$$4\begin{vmatrix} 3 & 4 \\ 8 & 1 \end{vmatrix}, \quad -9\begin{vmatrix} 1 & 0 \\ 8 & 1 \end{vmatrix}, \quad -3\begin{vmatrix} 1 & 0 \\ 3 & 4 \end{vmatrix}.$$

Using Technology

A graphing utility can be used to evaluate the determinant of a matrix. Enter the matrix and call it *A*. Then use the determinant command. The screen below verifies our result in Example 3.

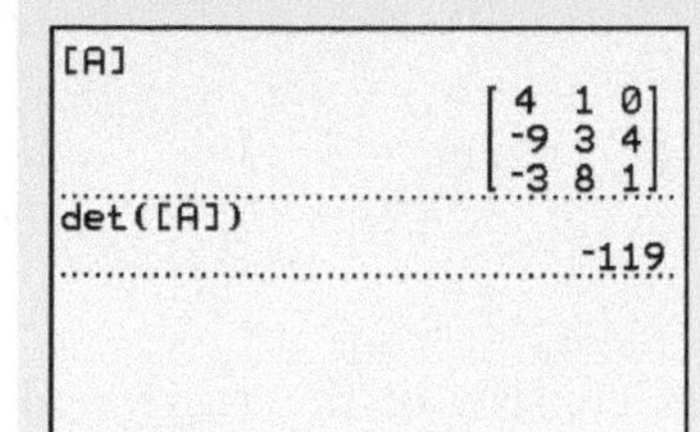

Based on the preceding definition, we subtract the second term from the first term and add the third term.

Don't forget to supply the minus sign.

$$\begin{vmatrix} 4 & 1 & 0 \\ -9 & 3 & 4 \\ -3 & 8 & 1 \end{vmatrix} = 4\begin{vmatrix} 3 & 4 \\ 8 & 1 \end{vmatrix} - (-9)\begin{vmatrix} 1 & 0 \\ 8 & 1 \end{vmatrix} - 3\begin{vmatrix} 1 & 0 \\ 3 & 4 \end{vmatrix}$$ Begin by evaluating the three second-order determinants.

$$= 4(3 \cdot 1 - 8 \cdot 4) + 9(1 \cdot 1 - 8 \cdot 0) - 3(1 \cdot 4 - 3 \cdot 0)$$

$$= 4(3 - 32) + 9(1 - 0) - 3(4 - 0)$$ Multiply within parentheses.

$$= 4(-29) + 9(1) - 3(4)$$ Subtract within parentheses.

$$= -116 + 9 - 12$$ Multiply.

$$= -119$$ Add and subtract as indicated. ■

✓ CHECK POINT 3 Evaluate the determinant of the following matrix:

$$\begin{bmatrix} 2 & 1 & 7 \\ -5 & 6 & 0 \\ -4 & 3 & 1 \end{bmatrix}.$$

4 Solve a system of linear equations in three variables using Cramer's rule.

Solving Systems of Linear Equations in Three Variables Using Determinants

Cramer's rule can be applied to solving systems of linear equations in three variables. The determinants in the numerator and denominator of the quotients determining each variable are third-order determinants.

Solving Three Equations in Three Variables Using Determinants

Cramer's Rule

If

$$\begin{cases} a_1x + b_1y + c_1z = d_1 \\ a_2x + b_2y + c_2z = d_2 \\ a_3x + b_3y + c_3z = d_3, \end{cases}$$

then

$$x = \frac{D_x}{D}, \quad y = \frac{D_y}{D}, \quad \text{and} \quad z = \frac{D_z}{D}.$$

These four third-order determinants are given by

$$D = \begin{vmatrix} a_1 & b_1 & c_1 \\ a_2 & b_2 & c_2 \\ a_3 & b_3 & c_3 \end{vmatrix}$$ These are the coefficients of the variables *x*, *y*, and *z*. $D \neq 0$

$$D_x = \begin{vmatrix} d_1 & b_1 & c_1 \\ d_2 & b_2 & c_2 \\ d_3 & b_3 & c_3 \end{vmatrix}$$ Replace *x*-coefficients in *D* with the constants on the right of the three equations.

$$D_y = \begin{vmatrix} a_1 & d_1 & c_1 \\ a_2 & d_2 & c_2 \\ a_3 & d_3 & c_3 \end{vmatrix}$$ Replace *y*-coefficients in *D* with the constants on the right of the three equations.

$$D_z = \begin{vmatrix} a_1 & b_1 & d_1 \\ a_2 & b_2 & d_2 \\ a_3 & b_3 & d_3 \end{vmatrix}.$$ Replace *z*-coefficients in *D* with the constants on the right of the three equations.

EXAMPLE 4 Using Cramer's Rule to Solve a Linear System in Three Variables

Use Cramer's rule to solve:

$$\begin{cases} x + 2y - z = -4 \\ x + 4y - 2z = -6 \\ 2x + 3y + z = 3. \end{cases}$$

Solution Because

$$x = \frac{D_x}{D}, \quad y = \frac{D_y}{D}, \quad \text{and} \quad z = \frac{D_z}{D},$$

we need to set up and evaluate four determinants.

Step 1. Set up the determinants.

1. D, the determinant in all three denominators, consists of the x-, y-, and z-coefficients.

$$D = \begin{vmatrix} 1 & 2 & -1 \\ 1 & 4 & -2 \\ 2 & 3 & 1 \end{vmatrix}$$

2. D_x, the determinant in the numerator for x, is obtained by replacing the x-coefficients in D, 1, 1, and 2, with the constants on the right sides of the equations, -4, -6, and 3.

$$D_x = \begin{vmatrix} -4 & 2 & -1 \\ -6 & 4 & -2 \\ 3 & 3 & 1 \end{vmatrix}$$

3. D_y, the determinant in the numerator for y, is obtained by replacing the y-coefficients in D, 2, 4, and 3, with the constants on the right sides of the equations, -4, -6, and 3.

$$D_y = \begin{vmatrix} 1 & -4 & -1 \\ 1 & -6 & -2 \\ 2 & 3 & 1 \end{vmatrix}$$

4. D_z, the determinant in the numerator for z, is obtained by replacing the z-coefficients in D, -1, -2, and 1, with the constants on the right sides of the equations, -4, -6, and 3.

$$D_z = \begin{vmatrix} 1 & 2 & -4 \\ 1 & 4 & -6 \\ 2 & 3 & 3 \end{vmatrix}$$

Step 2. Evaluate the four determinants.

$$D = \begin{vmatrix} 1 & 2 & -1 \\ 1 & 4 & -2 \\ 2 & 3 & 1 \end{vmatrix} = 1\begin{vmatrix} 4 & -2 \\ 3 & 1 \end{vmatrix} - 1\begin{vmatrix} 2 & -1 \\ 3 & 1 \end{vmatrix} + 2\begin{vmatrix} 2 & -1 \\ 4 & -2 \end{vmatrix}$$

$$= 1(4 + 6) - 1(2 + 3) + 2(-4 + 4)$$

$$= 1(10) - 1(5) + 2(0) = 5$$

Using the same technique to evaluate each determinant, we obtain

$$D_x = -10, \quad D_y = 5, \quad \text{and} \quad D_z = 20.$$

Great Question!

Can I use D, the determinant in each denominator, to find the determinants in the numerators?

To find D_x, D_y, and D_z, you'll need to apply the evaluation process for a 3×3 determinant three times. The values of D_x, D_y, and D_z cannot be obtained from the numbers that occur in the computation of D.

Step 3. Substitute these four values and solve the system.

$$x = \frac{D_x}{D} = \frac{-10}{5} = -2$$

$$y = \frac{D_y}{D} = \frac{5}{5} = 1$$

$$z = \frac{D_z}{D} = \frac{20}{5} = 4$$

The ordered triple $(-2, 1, 4)$ can be checked by substitution into the original three equations. The solution is $(-2, 1, 4)$ and the solution set is $\{(-2, 1, 4)\}$. ■

CHECK POINT 4 Use Cramer's rule to solve the system:

$$\begin{cases} 3x - 2y + z = 16 \\ 2x + 3y - z = -9 \\ x + 4y + 3z = 2. \end{cases}$$

Great Question!

What should I do if D, the determinant in the denominator of Cramer's rule, is zero?

If $D = 0$, the system is inconsistent or contains dependent equations. Use a method other than Cramer's rule to determine the solution set.

Although we have focused on applying determinants to solve linear systems, they have other applications, some of which we consider in the Exercise Set that follows.

CONCEPT AND VOCABULARY CHECK

Fill in each blank so that the resulting statement is true.

1. $\begin{vmatrix} 5 & 4 \\ 2 & 3 \end{vmatrix} = \_\cdot\_ - \_\cdot\_ = \_ - \_ = \_$

The value of this second-order ______________ is ______________.

2. Using Cramer's rule to solve

$$\begin{cases} x + y = 8 \\ x - y = -2, \end{cases}$$

we obtain

$$x = \frac{\begin{vmatrix} \_ & \_ \\ \_ & \_ \end{vmatrix}}{\begin{vmatrix} \_ & \_ \\ \_ & \_ \end{vmatrix}} \quad \text{and} \quad y = \frac{\begin{vmatrix} \_ & \_ \\ \_ & \_ \end{vmatrix}}{\begin{vmatrix} \_ & \_ \\ \_ & \_ \end{vmatrix}}.$$

3. $\begin{vmatrix} 3 & 2 & 1 \\ 4 & 3 & 1 \\ 5 & 1 & 1 \end{vmatrix} = 3\begin{vmatrix} \_ & \_ \\ \_ & \_ \end{vmatrix} - 4\begin{vmatrix} \_ & \_ \\ \_ & \_ \end{vmatrix} + 5\begin{vmatrix} \_ & \_ \\ \_ & \_ \end{vmatrix}$

4. Using Cramer's rule to solve

$$\begin{cases} 3x + y + 4z = -8 \\ 2x + 3y - 2z = 11 \\ x - 3y - 2z = 4 \end{cases}$$

for y, we obtain

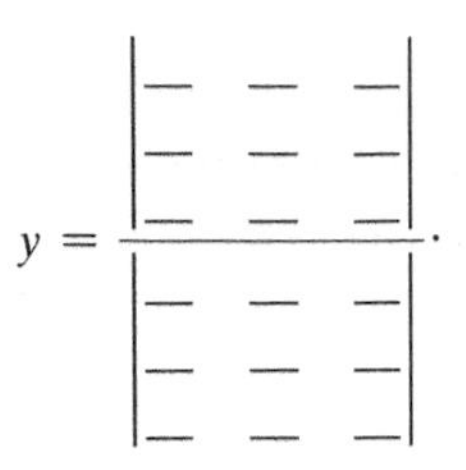

3.5 EXERCISE SET MyMathLab®

Practice Exercises

Evaluate each determinant in Exercises 1–10.

1. $\begin{vmatrix} 5 & 7 \\ 2 & 3 \end{vmatrix}$

2. $\begin{vmatrix} 4 & 8 \\ 5 & 6 \end{vmatrix}$

3. $\begin{vmatrix} -4 & 1 \\ 5 & 6 \end{vmatrix}$

4. $\begin{vmatrix} 7 & 9 \\ -2 & -5 \end{vmatrix}$

5. $\begin{vmatrix} -7 & 14 \\ 2 & -4 \end{vmatrix}$

6. $\begin{vmatrix} 1 & 3 \\ -8 & 2 \end{vmatrix}$

7. $\begin{vmatrix} -5 & -1 \\ -2 & -7 \end{vmatrix}$

8. $\begin{vmatrix} \frac{1}{5} & \frac{1}{6} \\ -6 & 5 \end{vmatrix}$

9. $\begin{vmatrix} \frac{1}{2} & \frac{1}{2} \\ \frac{1}{8} & -\frac{3}{4} \end{vmatrix}$

10. $\begin{vmatrix} \frac{2}{3} & \frac{1}{3} \\ -\frac{1}{2} & \frac{3}{4} \end{vmatrix}$

For Exercises 11–26, use Cramer's rule to solve each system or use another method to determine that the system is inconsistent or contains dependent equations.

11. $\begin{cases} x + y = 7 \\ x - y = 3 \end{cases}$

12. $\begin{cases} 2x + y = 3 \\ x - y = 3 \end{cases}$

13. $\begin{cases} 12x + 3y = 15 \\ 2x - 3y = 13 \end{cases}$

14. $\begin{cases} x - 2y = 5 \\ 5x - y = -2 \end{cases}$

15. $\begin{cases} 4x - 5y = 17 \\ 2x + 3y = 3 \end{cases}$

16. $\begin{cases} 3x + 2y = 2 \\ 2x + 2y = 3 \end{cases}$

17. $\begin{cases} x - 3y = 4 \\ 3x - 4y = 12 \end{cases}$

18. $\begin{cases} 2x - 9y = 5 \\ 3x - 3y = 11 \end{cases}$

19. $\begin{cases} 3x - 4y = 4 \\ 2x + 2y = 12 \end{cases}$

20. $\begin{cases} 3x = 7y + 1 \\ 2x = 3y - 1 \end{cases}$

21. $\begin{cases} 2x = 3y + 2 \\ 5x = 51 - 4y \end{cases}$

22. $\begin{cases} y = -4x + 2 \\ 2x = 3y + 8 \end{cases}$

23. $\begin{cases} 3x = 2 - 3y \\ 2y = 3 - 2x \end{cases}$

24. $\begin{cases} x + 2y - 3 = 0 \\ 12 = 8y + 4x \end{cases}$

25. $\begin{cases} 4y = 16 - 3x \\ 6x = 32 - 8y \end{cases}$

26. $\begin{cases} 2x = 7 + 3y \\ 4x - 6y = 3 \end{cases}$

Evaluate each determinant in Exercises 27–32.

27. $\begin{vmatrix} 3 & 0 & 0 \\ 2 & 1 & -5 \\ 2 & 5 & -1 \end{vmatrix}$

28. $\begin{vmatrix} 4 & 0 & 0 \\ 3 & -1 & 4 \\ 2 & -3 & 5 \end{vmatrix}$

29. $\begin{vmatrix} 3 & 1 & 0 \\ -3 & 4 & 0 \\ -1 & 3 & -5 \end{vmatrix}$

30. $\begin{vmatrix} 2 & -4 & 2 \\ -1 & 0 & 5 \\ 3 & 0 & 4 \end{vmatrix}$

31. $\begin{vmatrix} 1 & 1 & 1 \\ 2 & 2 & 2 \\ -3 & 4 & -5 \end{vmatrix}$

32. $\begin{vmatrix} 1 & 2 & 3 \\ 2 & 2 & -3 \\ 3 & 2 & 1 \end{vmatrix}$

In Exercises 33–40, use Cramer's rule to solve each system.

33. $\begin{cases} x + y + z = 0 \\ 2x - y + z = -1 \\ -x + 3y - z = -8 \end{cases}$

34. $\begin{cases} x - y + 2z = 3 \\ 2x + 3y + z = 9 \\ -x - y + 3z = 11 \end{cases}$

35. $\begin{cases} 4x - 5y - 6z = -1 \\ x - 2y - 5z = -12 \\ 2x - y = 7 \end{cases}$

36. $\begin{cases} x - 3y + z = -2 \\ x + 2y = 8 \\ 2x - y = 1 \end{cases}$

37. $\begin{cases} x + y + z = 4 \\ x - 2y + z = 7 \\ x + 3y + 2z = 4 \end{cases}$

38. $\begin{cases} 2x + 2y + 3z = 10 \\ 4x - y + z = -5 \\ 5x - 2y + 6z = 1 \end{cases}$

39. $\begin{cases} x + 2z = 4 \\ 2y - z = 5 \\ 2x + 3y = 13 \end{cases}$

40. $\begin{cases} 3x + 2z = 4 \\ 5x - y = -4 \\ 4y + 3z = 22 \end{cases}$

Practice PLUS

In Exercises 41–42, evaluate each determinant.

41. $\begin{vmatrix} \begin{vmatrix} 3 & 1 \\ -2 & 3 \end{vmatrix} & \begin{vmatrix} 7 & 0 \\ 1 & 5 \end{vmatrix} \\ \begin{vmatrix} 3 & 0 \\ 0 & 7 \end{vmatrix} & \begin{vmatrix} 9 & -6 \\ 3 & 5 \end{vmatrix} \end{vmatrix}$

42. $\begin{vmatrix} \begin{vmatrix} 5 & 0 \\ 4 & -3 \end{vmatrix} & \begin{vmatrix} -1 & 0 \\ 0 & -1 \end{vmatrix} \\ \begin{vmatrix} 7 & -5 \\ 4 & 6 \end{vmatrix} & \begin{vmatrix} 4 & 1 \\ -3 & 5 \end{vmatrix} \end{vmatrix}$

In Exercises 43–44, write the system of linear equations for which Cramer's rule yields the given determinants.

43. $D = \begin{vmatrix} 2 & -4 \\ 3 & 5 \end{vmatrix}$, $D_x = \begin{vmatrix} 8 & -4 \\ -10 & 5 \end{vmatrix}$

44. $D = \begin{vmatrix} 2 & -3 \\ 5 & 6 \end{vmatrix}$, $D_x = \begin{vmatrix} 8 & -3 \\ 11 & 6 \end{vmatrix}$

In Exercises 45–48, solve each equation for x.

45. $\begin{vmatrix} -2 & x \\ 4 & 6 \end{vmatrix} = 32$

46. $\begin{vmatrix} x+3 & -6 \\ x-2 & -4 \end{vmatrix} = 28$

47. $\begin{vmatrix} 1 & x & -2 \\ 3 & 1 & 1 \\ 0 & -2 & 2 \end{vmatrix} = -8$

48. $\begin{vmatrix} 2 & x & 1 \\ -3 & 1 & 0 \\ 2 & 1 & 4 \end{vmatrix} = 39$

Application Exercises

Determinants are used to find the area of a triangle whose vertices are given by three points in a rectangular coordinate system. The area of a triangle with vertices (x_1, y_1), (x_2, y_2), and (x_3, y_3) is

$$\text{Area} = \pm\frac{1}{2}\begin{vmatrix} x_1 & y_1 & 1 \\ x_2 & y_2 & 1 \\ x_3 & y_3 & 1 \end{vmatrix},$$

where the $\pm$ symbol indicates that the appropriate sign should be chosen to yield a positive area. Use this information to work Exercises 49–50.

49. Use determinants to find the area of the triangle whose vertices are $(3, -5)$, $(2, 6)$, and $(-3, 5)$.

50. Use determinants to find the area of the triangle whose vertices are $(1, 1)$, $(-2, -3)$, and $(11, -3)$.

Determinants are used to show that three points lie on the same line (are collinear). If $\begin{vmatrix} x_1 & y_1 & 1 \\ x_2 & y_2 & 1 \\ x_3 & y_3 & 1 \end{vmatrix} = 0$, *then the points (x_1, y_1), (x_2, y_2), and (x_3, y_3) are collinear. If the determinant does not equal 0, then the points are not collinear. Use this information to work Exercises 51–52.*

51. Are the points $(3, -1)$, $(0, -3)$, and $(12, 5)$ collinear?

52. Are the points $(-4, -6)$, $(1, 0)$, and $(11, 12)$ collinear?

Determinants are used to write an equation of a line passing through two points. An equation of the line passing through the distinct points (x_1, y_1) and (x_2, y_2) is given by

$$\begin{vmatrix} x & y & 1 \\ x_1 & y_1 & 1 \\ x_2 & y_2 & 1 \end{vmatrix} = 0.$$

Use this information to work Exercises 53–54.

53. Use the determinant to write an equation for the line passing through $(3, -5)$ and $(-2, 6)$. Then expand the determinant, expressing the line's equation in slope-intercept form.

54. Use the determinant to write an equation for the line passing through $(-1, 3)$ and $(2, 4)$. Then expand the determinant, expressing the line's equation in slope-intercept form.

Explaining the Concepts

55. Explain how to evaluate a second-order determinant.

56. Describe the determinants D, D_x, and D_y in terms of the coefficients and constants in a system of two equations in two variables.

57. Explain how to evaluate a third-order determinant.

58. When expanding a determinant by minors, when is it necessary to supply a minus sign?

59. Without going into too much detail, describe how to solve a linear system in three variables using Cramer's rule.

60. In applying Cramer's rule, what does it mean if $D = 0$?

61. The process of solving a linear system in three variables using Cramer's rule can involve tedious computation. Is there a way of speeding up this process, perhaps using Cramer's rule to find the value for only one of the variables? Describe how this process might work, presenting a specific example with your description. Remember that your goal is still to find the value for each variable in the system.

62. If you could use only one method to solve linear systems in three variables, which method would you select? Explain why this is so.

Technology Exercises

63. Use the feature of your graphing utility that evaluates the determinant of a square matrix to verify any five of the determinants that you evaluated by hand in Exercises 1–10 or Exercises 27–32.

64. What is the fastest method for solving a linear system with your graphing utility?

Critical Thinking Exercises

Make Sense? *In Exercises 65–68, determine whether each statement makes sense or does not make sense, and explain your reasoning.*

65. I'm solving a linear system using a determinant that contains two rows and three columns.

66. I can speed up the tedious computations required by Cramer's rule by using the value of D to determine the value of D_x.

67. When using Cramer's rule to solve a linear system, the number of determinants that I set up and evaluate is the same as the number of variables in the system.

68. Using Cramer's rule to solve a linear system, I found the value of D_x to be zero, so the system is inconsistent.

In Exercises 69–72, determine whether each statement is true or false. If the statement is false, make the necessary change(s) to produce a true statement.

69. Only one 2×2 determinant is needed to evaluate

$$\begin{vmatrix} 2 & 3 & -2 \\ 0 & 1 & 3 \\ 0 & 4 & -1 \end{vmatrix}.$$

70. If $D = 0$, then every variable has a value of 0.

71. Because there are different determinants in the numerators of x and y, if a system is solved using Cramer's rule, x and y cannot have the same value.

72. Using Cramer's rule, we use $\frac{D}{D_y}$ to get the value of y.

73. What happens to the value of a second-order determinant if the two columns are interchanged?

74. Consider the system

$$\begin{cases} a_1x + b_1y = c_1 \\ a_2x + b_2y = c_2. \end{cases}$$

Use Cramer's rule to prove that if the first equation of the system is replaced by the sum of the two equations, the resulting system has the same solution as the original system.

75. Show that the equation of a line through (x_1, y_1) and (x_2, y_2) is given by the following determinant.

$$\begin{vmatrix} x & y & 1 \\ x_1 & y_1 & 1 \\ x_2 & y_2 & 1 \end{vmatrix} = 0$$

Review Exercises

76. Solve: $6x - 4 = 2 + 6(x - 1)$. (Section 1.4, Example 6)

77. Solve for y: $-2x + 3y = 7$. (Section 1.5, Example 6)

78. Solve: $\frac{4x + 1}{3} = \frac{x - 3}{6} + \frac{x + 5}{6}$. (Section 1.4, Example 4)

Preview Exercises

Exercises 79–81 will help you prepare for the material covered in the first section of the next chapter.

79. Solve: $\frac{x + 3}{4} = \frac{x - 2}{3} + \frac{1}{4}$.

80. Solve: $-2x - 4 = x + 5$.

81. Use interval notation to describe values of x for which $2(x + 4)$ is greater than $2x + 3$.

Chapter 3 Summary

Definitions and Concepts	Examples

Section 3.1 Systems of Linear Equations in Two Variables

A system of linear equations in two variables, x and y, consists of two equations that can be written in the form $Ax + By = C$. The solution set is the set of all ordered pairs that satisfy both equations. Using the graphing method, a solution of a linear system is a point common to the graphs of both equations in the system.

Solve by graphing: $\begin{cases} 2x - y = 6 \\ x + y = 6. \end{cases}$

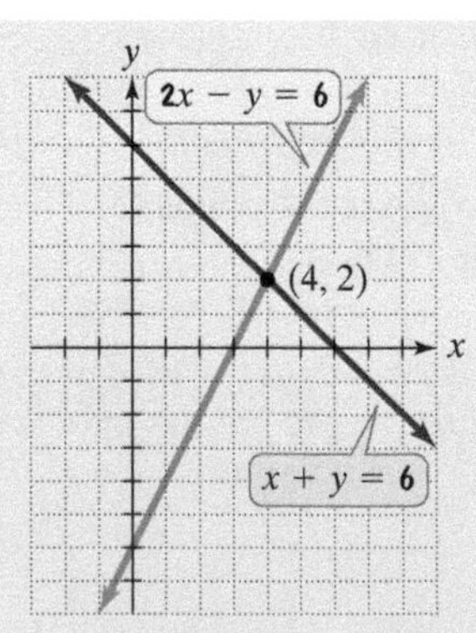

The intersection point gives the solution: (4, 2). The solution set is {(4, 2)}.

Definitions and Concepts | **Examples**

Section 3.1 Systems of Linear Equations in Two Variables (continued)

To solve a linear system by the substitution method,

1. Solve either equation for one variable in terms of the other.
2. Substitute the expression for that variable into the other equation.
3. Solve the equation in one variable.
4. Back-substitute the value of the variable into one of the original equations and find the value of the other variable.
5. Check the proposed solution in both equations.

Solve by the substitution method:

$$\begin{cases} y = 2x - 3 \\ 4x - 3y = 5. \end{cases}$$

Substitute $2x - 3$ for y in the second equation.

$$\begin{aligned} 4x - 3(2x - 3) &= 5 \\ 4x - 6x + 9 &= 5 \\ -2x + 9 &= 5 \\ -2x &= -4 \\ x &= 2 \end{aligned}$$

Find y. Substitute 2 for x in $y = 2x - 3$.

$$y = 2(2) - 3 = 4 - 3 = 1$$

The ordered pair (2, 1) checks. The solution is (2, 1) and {(2, 1)} is the solution set.

To solve a linear system by the addition method,

1. Write the equations in $Ax + By = C$ form.
2. Multiply one or both equations by nonzero numbers so that coefficients of one variable are opposites.
3. Add equations.
4. Solve the resulting equation for the variable.
5. Back-substitute the value of the variable into either original equation and find the value of the remaining variable.
6. Check the proposed solution in both equations.

Solve by the addition method:

$$\begin{cases} 2x + y = 10 \\ 3x + 4y = 25. \end{cases}$$

Eliminate y. Multiply the first equation by -4.

$$\begin{aligned} &\begin{cases} -8x - 4y = -40 \\ 3x + 4y = 25 \end{cases} \\ \text{Add: } &-5x = -15 \\ &x = 3 \end{aligned}$$

Find y. Back-substitute 3 for x. Use the first equation, $2x + y = 10$.

$$\begin{aligned} 2(3) + y &= 10 \\ 6 + y &= 10 \\ y &= 4 \end{aligned}$$

The ordered pair (3, 4) checks. The solution is (3, 4) and {(3, 4)} is the solution set.

A linear system with at least one solution is a consistent system. A system that has no solution, with $\varnothing$ as its solution set, is an inconsistent system. A linear system with infinitely many solutions has dependent equations. Solving inconsistent systems by substitution or addition leads to a false statement, such as $0 = 3$. Solving systems with dependent equations leads to a true statement, such as $7 = 7$.

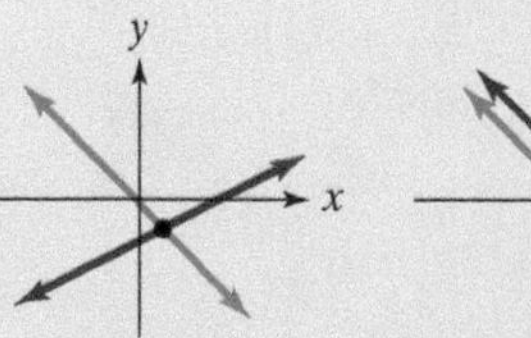

One solution: consistent

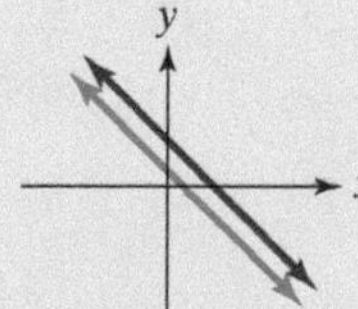

No solution: inconsistent

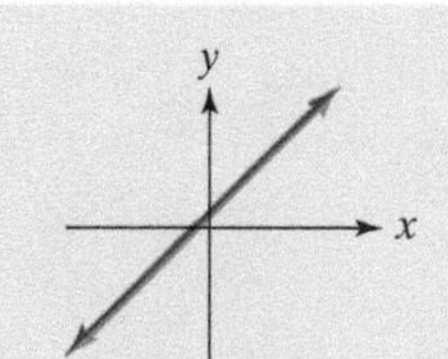

Infinitely many solutions: dependent and consistent

Definitions and Concepts	Examples

Section 3.2 Problem Solving and Business Applications Using Systems of Equations

A Problem-Solving Strategy

1. Use variables, usually x and y, to represent unknown quantities.
2. Write a system of equations describing the problem's conditions.
3. Solve the system and answer the problem's question.
4. Check proposed answers in the problem's wording.

You invested \$14,000 in two funds paying 7% and 9% interest. Total year-end interest was \$1180. How much was invested at each rate?

Let x = amount invested at 7% and

y = amount invested at 9%.

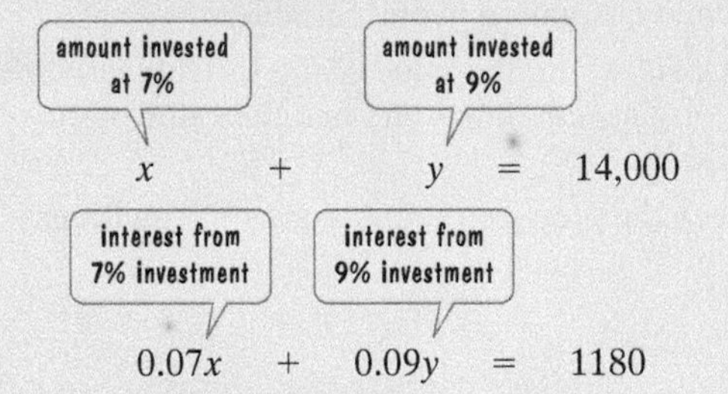

$$x + y = 14{,}000$$

$$0.07x + 0.09y = 1180$$

Solving by substitution or addition, $x = 4000$ and $y = 10{,}000$. Thus, \$4000 was invested at 7% and \$10,000 at 9%.

Functions of Business

A company produces and sells x units of a product.

Revenue Function

$$R(x) = (\text{price per unit sold})x$$

Cost Function

$$C(x) = \text{fixed cost} + (\text{cost per unit produced})x$$

Profit Function

$$P(x) = R(x) - C(x)$$

The point of intersection of the graphs of R and C is the break-even point. The x-coordinate of the point reveals the number of units that a company must produce and sell so that the money coming in, the revenue, is equal to the money going out, the cost. The y-coordinate gives the amount of money coming in and going out.

A company that manufactures lamps has a fixed cost of \$80,000 and it costs \$20 to produce each lamp. Lamps are sold for \$70.

a. Write the cost function.

$$C(x) = 80{,}000 + 20x$$

Fixed cost

Variable cost: \$20 per lamp

b. Write the revenue function.

$$R(x) = 70x$$

Revenue per lamp, \$70, times number of lamps sold

c. Find the break-even point.

Solve

$$\begin{cases} y = 80{,}000 + 20x \\ y = 70x \end{cases}$$

by substitution. Solving

$$70x = 80{,}000 + 20x$$

yields $x = 1600$. Back-substituting, $y = 112{,}000$. The break-even point is (1600, 112,000): The company breaks even if it sells 1600 lamps. At this level, money coming in equals money going out: \$112,000.

Definitions and Concepts	Examples

Section 3.3 Systems of Linear Equations in Three Variables

A system of linear equations in three variables, x, y, and z, consists of three equations of the form $Ax + By + Cz = D$. The solution set is the set consisting of the ordered triple that satisfies all three equations. The solution represents the point of intersection of three planes in space.

Is $(2, -1, 3)$ a solution of

$$\begin{cases} 3x + 5y - 2z = -5 \\ 2x + 3y - z = -2 \\ 2x + 4y + 6z = 18? \end{cases}$$

Replace x with 2, y with -1, and z with 3. Using the first equation, we obtain:

$$3 \cdot 2 + 5(-1) - 2(3) \stackrel{?}{=} -5$$
$$6 - 5 - 6 \stackrel{?}{=} -5$$
$$-5 = -5, \quad \text{true}$$

The ordered triple $(2, -1, 3)$ satisfies the first equation. In a similar manner, you can verify that it satisfies the other two equations and is a solution.

To solve a linear system in three variables by eliminating variables,

1. Reduce the system to two equations in two variables.
2. Solve the resulting system of two equations in two variables.
3. Use back-substitution in one of the equations in two variables to find the value of the second variable.
4. Back-substitute the values for two variables into one of the original equations to find the value of the third variable.
5. Check.

If all variables are eliminated and a false statement results, the system is inconsistent and has no solution. If a true statement results, the system contains dependent equations and has infinitely many solutions.

Solve

$$\begin{cases} 2x + 3y - 2z = 0 & \text{Equation 1} \\ x + 2y - z = 1 & \text{Equation 2} \\ 3x - y + z = -15. & \text{Equation 3} \end{cases}$$

Add Equations 2 and 3 to eliminate z.

$$4x + y = -14 \quad \text{Equation 4}$$

Eliminate z again. Multiply Equation 3 by 2 and add to Equation 1.

$$8x + y = -30 \quad \text{Equation 5}$$

Multiply Equation 4 by -1 and add to Equation 5.

$$\begin{cases} -4x - y = 14 \\ 8x + y = -30 \end{cases}$$
$$\text{Add: } 4x = -16$$
$$x = -4$$

Substitute -4 for x in Equation 4.

$$4(-4) + y = -14$$
$$y = 2$$

Substitute -4 for x and 2 for y in Equation 3.

$$3(-4) - 2 + z = -15$$
$$-14 + z = -15$$
$$z = -1$$

Checking verifies that $(-4, 2, -1)$ is the solution and $\{(-4, 2, -1)\}$ is the solution set.

Definitions and Concepts	Examples

Section 3.3 Systems of Linear Equations in Three Variables (continued)

Curve Fitting

Curve fitting is determining a function whose graph contains given points. Three points that do not lie on a line determine the graph of a quadratic function

$$y = ax^2 + bx + c.$$

Use the three given points to create a system of three equations. Solve the system to find a, b, and c.

Find the quadratic function $y = ax^2 + bx + c$ whose graph passes through the points $(-1, 2)$, $(1, 8)$, and $(2, 14)$.

Use $y = ax^2 + bx + c$.

When $x = -1, y = 2$: $\quad 2 = a(-1)^2 + b(-1) + c$

When $x = 1, y = 8$: $\quad 8 = a \cdot 1^2 + b \cdot 1 + c$

When $x = 2, y = 14$: $\quad 14 = a \cdot 2^2 + b \cdot 2 + c$

Solving

$$\begin{cases} a - b + c = 2 \\ a + b + c = 8 \\ 4a + 2b + c = 14, \end{cases}$$

$a = 1, b = 3$, and $c = 4$. The quadratic function, $y = ax^2 + bx + c$, is $y = x^2 + 3x + 4$.

Section 3.4 Matrix Solutions to Linear Systems

A matrix is a rectangular array of numbers. The augmented matrix of a linear system is obtained by writing the coefficients of each variable, a vertical bar, and the constants of the system.

$$\begin{cases} x + 4y = 9 \\ 3x + y = 5 \end{cases}$$

The augmented matrix is $\left[\begin{array}{cc|c} 1 & 4 & 9 \\ 3 & 1 & 5 \end{array}\right]$.

The following row operations produce matrices that represent systems with the same solution. Two matrices are row equivalent if one can be obtained from the other by a sequence of these row operations.

1. Interchange the elements in the ith and jth rows: $R_i \leftrightarrow R_j$.
2. Multiply each element in the ith row by k: kR_i.
3. Add k times the elements in row i to the corresponding elements in row j: $kR_i + R_j$.

Find the result of the row operation $-4R_1 + R_2$:

$$\left[\begin{array}{ccc|c} 1 & 0 & -2 & 5 \\ 4 & -1 & 2 & 6 \\ 3 & -7 & 9 & 10 \end{array}\right].$$

Add -4 times the elements in row 1 to the corresponding elements in row 2.

$$\left[\begin{array}{ccc|c} 1 & 0 & -2 & 5 \\ -4(1) + 4 & -4(0) + (-1) & -4(-2) + 2 & -4(5) + 6 \\ 3 & -7 & 9 & 10 \end{array}\right]$$

$$= \left[\begin{array}{ccc|c} 1 & 0 & -2 & 5 \\ 0 & -1 & 10 & -14 \\ 3 & -7 & 9 & 10 \end{array}\right]$$

Solving Linear Systems Using Matrices

1. Write the augmented matrix for the system.
2. Use matrix row operations to simplify the matrix to row-echelon form, with 1s down the main diagonal from upper left to lower right, and 0s below the 1s.
3. Write the system of linear equations corresponding to the matrix from step 2 and use back-substitution to find the system's solution.

If you obtain a matrix with a row containing 0s to the left of the vertical bar and a nonzero number on the right, the system is inconsistent. If 0s appear across an entire row, the system contains dependent equations.

Solve using matrices:

$$\begin{cases} 3x + y = 5 \\ x + 4y = 9. \end{cases}$$

$$\left[\begin{array}{cc|c} 3 & 1 & 5 \\ 1 & 4 & 9 \end{array}\right] \xrightarrow{R_1 \leftrightarrow R_2} \left[\begin{array}{cc|c} 1 & 4 & 9 \\ 3 & 1 & 5 \end{array}\right]$$

$$\xrightarrow{-3R_1 + R_2} \left[\begin{array}{cc|c} 1 & 4 & 9 \\ 0 & -11 & -22 \end{array}\right] \xrightarrow{-\frac{1}{11}R_2} \left[\begin{array}{cc|c} 1 & 4 & 9 \\ 0 & 1 & 2 \end{array}\right]$$

$$\longrightarrow \begin{cases} x + 4y = 9 \\ y = 2. \end{cases}$$

When $y = 2$, $x + 4 \cdot 2 = 9$, so $x = 1$. The solution is $(1, 2)$ and the solution set is $\{(1, 2)\}$.

Definitions and Concepts	Examples

Section 3.5 Determinants and Cramer's Rule

A square matrix has the same number of rows as columns. A determinant is a real number associated with a square matrix. The determinant is denoted by placing vertical bars about the array of numbers. The value of a second-order determinant is

$$\begin{vmatrix} a_1 & b_1 \\ a_2 & b_2 \end{vmatrix} = a_1b_2 - a_2b_1.$$

Evaluate:

$$\begin{vmatrix} 2 & -1 \\ 3 & 4 \end{vmatrix} = 2(4) - 3(-1) = 8 + 3 = 11.$$

Cramer's Rule for Two Linear Equations in Two Variables
If

$$\begin{cases} a_1x + b_1y = c_1 \\ a_2x + b_2y = c_2, \end{cases}$$

then

$$x = \frac{\begin{vmatrix} c_1 & b_1 \\ c_2 & b_2 \end{vmatrix}}{\begin{vmatrix} a_1 & b_1 \\ a_2 & b_2 \end{vmatrix}} = \frac{D_x}{D} \text{ and } y = \frac{\begin{vmatrix} a_1 & c_1 \\ a_2 & c_2 \end{vmatrix}}{\begin{vmatrix} a_1 & b_1 \\ a_2 & b_2 \end{vmatrix}} = \frac{D_y}{D}, D \neq 0.$$

If $D = 0$, the system is inconsistent or contains dependent equations. Use a method other than Cramer's rule to determine the solution set.

Solve by Cramer's rule:

$$\begin{cases} 5x + 3y = 7 \\ -x + 2y = 9. \end{cases}$$

$$D = \begin{vmatrix} 5 & 3 \\ -1 & 2 \end{vmatrix} = 5(2) - (-1)(3) = 10 + 3 = 13$$

$$D_x = \begin{vmatrix} 7 & 3 \\ 9 & 2 \end{vmatrix} = 7 \cdot 2 - 9 \cdot 3 = 14 - 27 = -13$$

$$D_y = \begin{vmatrix} 5 & 7 \\ -1 & 9 \end{vmatrix} = 5(9) - (-1)(7) = 45 + 7 = 52$$

$$x = \frac{D_x}{D} = \frac{-13}{13} = -1, \quad y = \frac{D_y}{D} = \frac{52}{13} = 4$$

The solution is $(-1, 4)$ and the solution set is $\{(-1, 4)\}$.

The value of a third-order determinant is

$$\begin{vmatrix} a_1 & b_1 & c_1 \\ a_2 & b_2 & c_2 \\ a_3 & b_3 & c_3 \end{vmatrix}$$

$$= a_1\begin{vmatrix} b_2 & c_2 \\ b_3 & c_3 \end{vmatrix} - a_2\begin{vmatrix} b_1 & c_1 \\ b_3 & c_3 \end{vmatrix} + a_3\begin{vmatrix} b_1 & c_1 \\ b_2 & c_2 \end{vmatrix}.$$

Each second-order determinant is called a minor.

Evaluate:

$$\begin{vmatrix} 1 & -2 & 1 \\ 3 & 1 & -2 \\ 5 & 5 & 3 \end{vmatrix}$$

$$= 1\begin{vmatrix} 1 & -2 \\ 5 & 3 \end{vmatrix} - 3\begin{vmatrix} -2 & 1 \\ 5 & 3 \end{vmatrix} + 5\begin{vmatrix} -2 & 1 \\ 1 & -2 \end{vmatrix}$$

$$= 1(3 - (-10)) - 3(-6 - 5) + 5(4 - 1)$$

$$= 1(13) - 3(-11) + 5(3)$$

$$= 13 + 33 + 15 = 61.$$

Definitions and Concepts	Examples

Section 3.5 Determinants and Cramer's Rule (continued)

Cramer's Rule for Three Linear Equations in Three Variables

If

$$\begin{cases} a_1x + b_1y + c_1z = d_1 \\ a_2x + b_2y + c_2z = d_2 \\ a_3x + b_3y + c_3z = d_3, \end{cases}$$

then

$$x = \frac{D_x}{D}, \quad y = \frac{D_y}{D}, \quad z = \frac{D_z}{D}.$$

$$D = \begin{vmatrix} a_1 & b_1 & c_1 \\ a_2 & b_2 & c_2 \\ a_3 & b_3 & c_3 \end{vmatrix} \neq 0, \quad D_x = \begin{vmatrix} d_1 & b_1 & c_1 \\ d_2 & b_2 & c_2 \\ d_3 & b_3 & c_3 \end{vmatrix}$$

$$D_y = \begin{vmatrix} a_1 & d_1 & c_1 \\ a_2 & d_2 & c_2 \\ a_3 & d_3 & c_3 \end{vmatrix}, \quad D_z = \begin{vmatrix} a_1 & b_1 & d_1 \\ a_2 & b_2 & d_2 \\ a_3 & b_3 & d_3 \end{vmatrix}$$

If $D = 0$, the system is inconsistent or contains dependent equations. Use a methed other than Cramer's rule to determine the solution set.

Solve by Cramer's rule:

$$\begin{cases} x - 2y + z = 4 \\ 3x + y - 2z = 3 \\ 5x + 5y + 3z = -8. \end{cases}$$

$$D = \begin{vmatrix} 1 & -2 & 1 \\ 3 & 1 & -2 \\ 5 & 5 & 3 \end{vmatrix} = 61$$

This evaluation is shown on the bottom of page 247.

$$D_x = \begin{vmatrix} 4 & -2 & 1 \\ 3 & 1 & -2 \\ -8 & 5 & 3 \end{vmatrix} = 61$$

$$D_y = \begin{vmatrix} 1 & 4 & 1 \\ 3 & 3 & -2 \\ 5 & -8 & 3 \end{vmatrix} = -122$$

$$D_z = \begin{vmatrix} 1 & -2 & 4 \\ 3 & 1 & 3 \\ 5 & 5 & -8 \end{vmatrix} = -61$$

$$x = \frac{D_x}{D} = \frac{61}{61} = 1, \quad y = \frac{D_y}{D} = \frac{-122}{61} = -2,$$

$$z = \frac{D_z}{D} = \frac{-61}{61} = -1.$$

The solution is $(1, -2, -1)$ and the solution set is $\{(1, -2, -1)\}$.

CHAPTER 3 REVIEW EXERCISES

3.1 *In Exercises 1–2, determine whether the given ordered pair is a solution of the system.*

1. $(4, 2)$
$$\begin{cases} 2x - 5y = -2 \\ 3x + 4y = 4 \end{cases}$$

2. $(-5, 3)$
$$\begin{cases} -x + 2y = 11 \\ y = -\frac{x}{3} + \frac{4}{3} \end{cases}$$

In Exercises 3–6, solve each system by graphing. Identify systems with no solution and systems with infinitely many solutions, using set notation to express their solution sets.

3. $\begin{cases} x + y = 5 \\ 3x - y = 3 \end{cases}$

4. $\begin{cases} 3x - 2y = 6 \\ 6x - 4y = 12 \end{cases}$

5. $\begin{cases} y = \frac{3}{5}x - 3 \\ 2x - y = -4 \end{cases}$

6. $\begin{cases} y = -x + 4 \\ 3x + 3y = -6 \end{cases}$

In Exercises 7–13, solve each system by the substitution method or the addition method. Identify systems with no solution and systems with infinitely many solutions, using set notation to express their solution sets.

7. $\begin{cases} 2x - y = 2 \\ x + 2y = 11 \end{cases}$

8. $\begin{cases} y = -2x + 3 \\ 3x + 2y = -17 \end{cases}$

9. $\begin{cases} 3x + 2y = -8 \\ 2x + 5y = 2 \end{cases}$

10. $\begin{cases} 5x - 2y = 14 \\ 3x + 4y = 11 \end{cases}$

11. $\begin{cases} y = 4 - x \\ 3x + 3y = 12 \end{cases}$

12. $\begin{cases} \frac{x}{8} + \frac{3y}{4} = \frac{19}{8} \\ -\frac{x}{2} + \frac{3y}{4} = \frac{1}{2} \end{cases}$

13. $\begin{cases} x - 2y + 3 = 0 \\ 2x - 4y + 7 = 0 \end{cases}$

3.2 *In Exercises 14–18, use the four-step strategy to solve each problem.*

14. An appliance store is having a sale on small TVs and stereos. One day a salesperson sells 3 of the TVs and 4 stereos for \$2530. The next day the salesperson sells 4 of the same TVs and 3 of the same stereos for \$2510. What are the prices of a TV and a stereo?

15. You invested \$9000 in two funds paying 4% and 7% annual interest. At the end of the year, the total interest from these investments was \$555. How much was invested at each rate?

16. A chemist needs to mix a solution that is 34% silver nitrate with one that is 4% silver nitrate to obtain 100 milliliters of a mixture that is 7% silver nitrate. How many milliliters of each of the solutions must be used?

17. When a plane flies with the wind, it can travel 2160 miles in 3 hours. When the plane flies in the opposite direction, against the wind, it takes 4 hours to fly the same distance. Find the rate of the plane in still air and the rate of the wind.

18. The perimeter of a rectangular tabletop is 34 feet. The difference between 4 times the length and 3 times the width is 33 feet. Find the dimensions.

The cost and revenue functions for producing and selling x units of a new graphing calculator are

$$C(x) = 22{,}500 + 40x \quad \text{and} \quad R(x) = 85x.$$

Use these functions to solve Exercises 19–21.

19. Find the loss or the gain from selling 400 graphing calculators.

20. Determine the break-even point. Describe what this means.

21. Write the profit function, P, from producing and selling x graphing calculators.

22. A company is planning to manufacture computer desks. The fixed cost will be $60,000 and it will cost $200 to produce each desk. Each desk will be sold for $450.

  a. Write the cost function, C, of producing x desks.

  b. Write the revenue function, R, from the sale of x desks.

  c. Determine the break-even point. Describe what this means.

3.3

23. Is $(-3, -2, 5)$ a solution of the system

$$\begin{cases} x + y + z = 0 \\ 2x - 3y + z = 5 \\ 4x + 2y + 4z = 3? \end{cases}$$

Solve each system in Exercises 24–26 by eliminating variables using the addition method. If there is no solution or if there are infinitely many solutions and a system's equations are dependent, so state.

24. $$\begin{cases} 2x - y + z = 1 \\ 3x - 3y + 4z = 5 \\ 4x - 2y + 3z = 4 \end{cases}$$

25. $$\begin{cases} x + 2y - z = 5 \\ 2x - y + 3z = 0 \\ 2y + z = 1 \end{cases}$$

26. $$\begin{cases} 3x - 4y + 4z = 7 \\ x - y - 2z = 2 \\ 2x - 3y + 6z = 5 \end{cases}$$

27. Find the quadratic function $y = ax^2 + bx + c$ whose graph passes through the points (1, 4), (3, 20), and (−2, 25).

28. **20th Century Death** The greatest cause of death in the 20th century was disease, killing 1390 million people. The bar graph shows the five leading causes of death in that century, excluding disease.

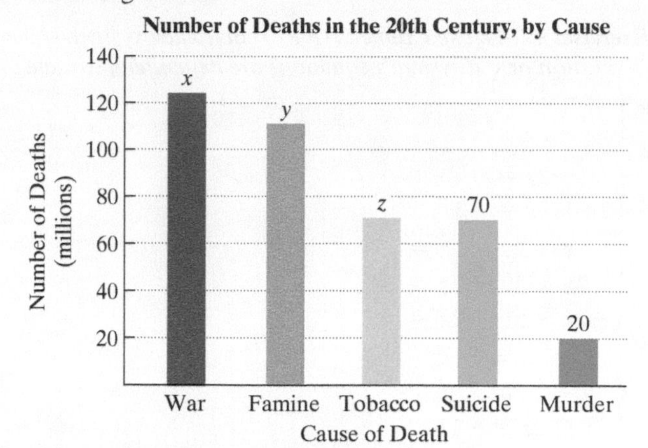

Source: Centers for Disease Control and Prevention

War, famine, and tobacco combined resulted in 306 million deaths. The difference between the number of deaths from war and famine was 13 million. The difference between the number of deaths from war and tobacco was 53 million. Find the number of 20th century deaths from war, famine, and tobacco.

3.4

In Exercises 29–32, perform each matrix row operation and write the new matrix.

29. $$\left[\begin{array}{cc|c} 1 & -8 & 3 \\ 0 & 7 & -14 \end{array}\right] \; \frac{1}{7}R_2$$

30. $$\left[\begin{array}{cc|c} 1 & -3 & 1 \\ 2 & 1 & -5 \end{array}\right] \; -2R_1 + R_2$$

31. $$\left[\begin{array}{ccc|c} 2 & -2 & 1 & -1 \\ 1 & 2 & -1 & 2 \\ 6 & 4 & 3 & 5 \end{array}\right] \; \frac{1}{2}R_1$$

32. $$\left[\begin{array}{ccc|c} 1 & 2 & 2 & 2 \\ 0 & 1 & -1 & 2 \\ 0 & 5 & 4 & 1 \end{array}\right] \; -5R_2 + R_3$$

In Exercises 33–36, solve each system using matrices. If there is no solution or if a system's equations are dependent, so state.

33. $$\begin{cases} x + 4y = 7 \\ 3x + 5y = 0 \end{cases}$$

34. $$\begin{cases} 2x - 3y = 8 \\ -6x + 9y = 4 \end{cases}$$

35. $$\begin{cases} x + 2y + 3z = -5 \\ 2x + y + z = 1 \\ x + y - z = 8 \end{cases}$$

36. $$\begin{cases} x - 2y + z = 0 \\ y - 3z = -1 \\ 2y + 5z = -2 \end{cases}$$

3.5

In Exercises 37–40, evaluate each determinant.

37. $$\begin{vmatrix} 3 & 2 \\ -1 & 5 \end{vmatrix}$$

38. $$\begin{vmatrix} -2 & -3 \\ -4 & -8 \end{vmatrix}$$

39. $\begin{vmatrix} 2 & 4 & -3 \\ 1 & -1 & 5 \\ -2 & 4 & 0 \end{vmatrix}$

40. $\begin{vmatrix} 4 & 7 & 0 \\ -5 & 6 & 0 \\ 3 & 2 & -4 \end{vmatrix}$

In Exercises 41–44, use Cramer's rule to solve each system. If there is no solution or if a system's equations are dependent, so state.

41. $\begin{cases} x - 2y = 8 \\ 3x + 2y = -1 \end{cases}$

42. $\begin{cases} 7x + 2y = 0 \\ 2x + y = -3 \end{cases}$

43. $\begin{cases} x + 2y + 2z = 5 \\ 2x + 4y + 7z = 19 \\ -2x - 5y - 2z = 8 \end{cases}$

44. $\begin{cases} 2x + y = -4 \\ y - 2z = 0 \\ 3x - 2z = -11 \end{cases}$

45. Use the quadratic function $y = ax^2 + bx + c$ to model the following data:

x (Age of a Driver)	y (Average Number of Automobile Accidents per Day in the United States)
20	400
40	150
60	400

Use Cramer's rule to determine values for a, b, and c. Then use the model to write a statement about the average number of automobile accidents in which 30-year-old drivers and 50-year-old drivers are involved daily.

CHAPTER 3 TEST

Step-by-step test solutions are found on the Chapter Test Prep Videos available in MyMathLab® or on YouTube (search "BlitzerInterAlg7e" and click on "Channels").

1. Solve by graphing

$$\begin{cases} x + y = 6 \\ 4x - y = 4. \end{cases}$$

In Exercises 2–4, solve each system by the substitution method or the addition method. Identify systems with no solution and systems with infinitely many solutions, using set notation to express their solution sets.

2. $\begin{cases} 5x + 4y = 10 \\ 3x + 5y = -7 \end{cases}$

3. $\begin{cases} x = y + 4 \\ 3x + 7y = -18 \end{cases}$

4. $\begin{cases} 4x = 2y + 6 \\ y = 2x - 3 \end{cases}$

In Exercises 5–8, solve each problem.

5. In a new development, 50 one- and two-bedroom condominiums were sold. Each one-bedroom condominium sold for \$120 thousand and each two-bedroom condominium sold for \$150 thousand. If sales totaled \$7050 thousand, how many of each type of unit was sold?

6. You invested \$9000 in two funds paying 6% and 7% annual interest. At the end of the year, the total interest from these investments was \$610. How much was invested at each rate?

7. You need to mix a 6% peroxide solution with a 9% peroxide solution to obtain 36 ounces of an 8% peroxide solution. How many ounces of each of the solutions must be used?

8. A paddleboat on the Mississippi River travels 48 miles downstream, with the current, in 3 hours. The return trip, against the current, takes the paddleboat 4 hours. Find the boat's rate in still water and the rate of the current.

Use this information to solve Exercises 9–11: A company is planning to produce and sell a new line of computers. The fixed cost will be \$360,000 and it will cost \$850 to produce each computer. Each computer will be sold for \$1150.

9. Write the cost function, C, of producing x computers.

10. Write the revenue function, R, from the sale of x computers.

11. Determine the break-even point. Describe what this means.

12. The cost and revenue functions for producing and selling x units of a toaster oven are

$$C(x) = 40x + 350{,}000 \quad \text{and} \quad R(x) = 125x.$$

Write the profit function, P, from producing and selling x toaster ovens.

13. Solve by eliminating variables using the addition method:

$$\begin{cases} x + y + z = 6 \\ 3x + 4y - 7z = 1 \\ 2x - y + 3z = 5. \end{cases}$$

14. Perform the indicated matrix row operation and write the new matrix.

$$\left[\begin{array}{ccc|c} 1 & 0 & -4 & 5 \\ 6 & -1 & 2 & 10 \\ 2 & -1 & 4 & -3 \end{array}\right] \quad -6R_1 + R_2$$

In Exercises 15–16, solve each system using matrices.

15. $\begin{cases} 2x + y = 6 \\ 3x - 2y = 16 \end{cases}$

16. $\begin{cases} x - 4y + 4z = -1 \\ 2x - y + 5z = 6 \\ -x + 3y - z = 5 \end{cases}$

In Exercises 17–18, evaluate each determinant.

17. $\begin{vmatrix} -1 & -3 \\ 7 & 4 \end{vmatrix}$

18. $\begin{vmatrix} 3 & 4 & 0 \\ -1 & 0 & -3 \\ 4 & 2 & 5 \end{vmatrix}$

In Exercises 19–20, use Cramer's rule to solve each system.

19. $\begin{cases} 4x - 3y = 14 \\ 3x - y = 3 \end{cases}$

20. $\begin{cases} 2x + 3y + z = 2 \\ 3x + 3y - z = 0 \\ x - 2y - 3z = 1 \end{cases}$

CUMULATIVE REVIEW EXERCISES (CHAPTERS 1–3)

1. Simplify: $\dfrac{6(8 - 10)^3 + (-2)}{(-5)^2(-2)}$.

2. Simplify: $7x - [5 - 2(4x - 1)]$.

In Exercises 3–5, solve each equation.

3. $5 - 2(3 - x) = 2(2x + 5) + 1$

4. $\dfrac{3x}{5} + 4 = \dfrac{x}{3}$

5. $3x - 4 = 2(3x + 2) - 3x$

6. For a summer sales job, you are choosing between two pay arrangements: a weekly salary of \$200 plus 5% commission on sales, or a straight 15% commission. For how many dollars of sales will the earnings be the same regardless of the pay arrangement?

7. Simplify: $\dfrac{-5x^6y^{-10}}{20x^{-2}y^{20}}$.

8. If $f(x) = -4x + 5$, find $f(a + 2)$.

9. Find the domain of $f(x) = \dfrac{4}{x + 3}$.

10. If $f(x) = 2x^2 - 5x + 2$ and $g(x) = x^2 - 2x + 3$, find $(f - g)(x)$ and $(f - g)(3)$.

In Exercises 11–12, graph each linear function.

11. $f(x) = -\dfrac{2}{3}x + 2$

12. $2x - y = 6$

In Exercises 13–14, use the given conditions to write an equation for each line in point-slope form and slope-intercept form.

13. Passing through $(2, 4)$ and $(4, -2)$

14. Passing through $(-1, 0)$ and parallel to the line whose equation is $3x + y = 6$

In Exercises 15–16, solve each system by eliminating variables using the addition method.

15. $\begin{cases} 3x + 12y = 25 \\ 2x - 6y = 12 \end{cases}$

16. $\begin{cases} x + 3y - z = 5 \\ -x + 2y + 3z = 13 \\ 2x - 5y - z = -8 \end{cases}$

17. If two pads of paper and 19 pens are sold for \$5.40 and 7 of the same pads and 4 of the same pens sell for \$6.40, find the cost of one pad and one pen.

18. Evaluate:

$$\begin{vmatrix} 0 & 1 & -2 \\ -7 & 0 & -4 \\ 3 & 0 & 5 \end{vmatrix}.$$

19. Solve using matrices:

$$\begin{cases} 2x + 3y - z = -1 \\ x + 2y + 3z = 2 \\ 3x + 5y - 2x = -3. \end{cases}$$

20. Solve using Cramer's rule (determinants):

$$\begin{cases} 3x + 4y = -1 \\ -2x + y = 8. \end{cases}$$

CHAPTER 4

Inequalities and Problem Solving

You are in Yosemite National Park in California, surrounded by evergreen forests, alpine meadows, and sheer walls of granite. The beauty of soaring cliffs, plunging waterfalls, gigantic trees, rugged canyons, mountains, and valleys is overwhelming. This is so different from where you live and attend college, a region in which grasslands predominate. What variables affect whether regions are forests, grasslands, or deserts, and what kinds of mathematical models are used to describe the incredibly diverse land that forms the surface of our planet?

Here's where you'll find this application:

- The role that temperature and precipitation play in determining whether regions are forests, grasslands, or deserts can be modeled using inequalities in two variables. You will use these models and their graphs in Example 4 of Section 4.4.

SECTION

4.1 Solving Linear Inequalities

What am I supposed to learn?

After studying this section, you should be able to:

1. Solve linear inequalities.
2. Recognize inequalities with no solution or all real numbers as solutions.
3. Solve applied problems using linear inequalities.

You can go online and obtain a list of cellphone companies and their monthly price plans. You've chosen a plan that has a monthly fee of \$80 that includes 1500 voice minutes, 4 GB of data, with a charge of \$0.08 per text (regular, photo, or video). Suppose you are limited by how much money you can spend for the month: You can spend at most \$100. If we let x represent the number of text messages in a month, we can write an inequality that describes the given conditions:

The monthly fee of \$80	plus	the charge of \$0.08 per text for x text messages	must be less than or equal to	\$100.
80	$+$	$0.08x$	$\leq$	100.

Using the commutative property of addition, we can express this inequality as

$$0.08x + 80 \leq 100.$$

Placing an inequality symbol between a linear expression $(mx + b)$ and a constant results in a *linear inequality in one variable*. In this section, we will study how to solve linear inequalities such as the one shown above. **Solving an inequality** is the process of finding the set of numbers that make the inequality a true statement. These numbers are called the **solutions** of the inequality and we say that they **satisfy** the inequality. The set of all solutions is called the **solution set** of the inequality. We will use interval notation to represent these solution sets.

1. Solve linear inequalities.

Solving Linear Inequalities in One Variable

We know that a linear equation in x can be expressed as $ax + b = 0$. A **linear inequality in x** can be written in one of the following forms: $ax + b < 0$, $ax + b \leq 0$, $ax + b > 0$, $ax + b \geq 0$. In each form, $a \neq 0$.

Back to our question from above: How many text messages can you send in a month if you can spend at most \$100? We answer the question by solving the linear inequality

$$0.08x + 80 \leq 100$$

for x. The solution procedure is nearly identical to that for solving the equation

$$0.08x + 80 = 100.$$

Our goal is to get x by itself on the left side. We do this by first subtracting 80 from both sides to isolate $0.08x$:

$$0.08x + 80 \leq 100 \quad \text{This is the given inequality.}$$

$$0.08x + 80 - 80 \leq 100 - 80 \quad \text{Subtract 80 from both sides.}$$

$$0.08x \leq 20. \quad \text{Simplify.}$$

Finally, we isolate x from $0.08x$ by dividing both sides of the inequality by 0.08:

$$\frac{0.08x}{0.08} \le \frac{20}{0.08} \quad \text{Divide both sides by 0.08.}$$

$$x \le 250. \quad \text{Simplify.}$$

With at most $100 per month to spend, you can send no more than 250 text messages each month.

We started with the inequality $0.08x + 80 \le 100$ and obtained the inequality $x \le 250$ in the final step. Both of these inequalities have the same solution set. Inequalities such as these, with the same solution set, are said to be **equivalent**.

We isolated x from $0.08x$ by dividing both sides of $0.08x \le 20$ by 0.08, a positive number. Let's see what happens if we divide both sides of an inequality by a negative number. Consider the inequality $10 < 14$. Divide 10 and 14 by -2:

$$\frac{10}{-2} = -5 \quad \text{and} \quad \frac{14}{-2} = -7.$$

Because -5 lies to the right of -7 on the number line, -5 is greater than -7:

$$-5 > -7.$$

Notice that the direction of the inequality symbol is reversed:

$$10 < 14$$

$$\downarrow$$

$$-5 > -7.$$

Dividing by -2 changes the direction of the inequality symbol.

In general, **when we multiply or divide both sides of an inequality by a negative number, the direction of the inequality symbol is reversed.** When we reverse the direction of the inequality symbol, we say that we change the *sense* of the inequality.

We can isolate a variable in a linear inequality the same way we can isolate a variable in a linear equation. The following properties are used to create equivalent inequalities:

Properties of Inequalities

Property	The Property in Words	Example
The Addition Property of Inequality If $a < b$, then $a + c < b + c$. If $a < b$, then $a - c < b - c$.	If the same quantity is added to or subtracted from both sides of an inequality, the resulting inequality is equivalent to the original one.	$2x + 3 < 7$ Subtract 3: $2x + 3 - 3 < 7 - 3.$ Simplify: $2x < 4.$
The Positive Multiplication Property of Inequality If $a < b$ and c is positive, then $ac < bc$. If $a < b$ and c is positive, then $\frac{a}{c} < \frac{b}{c}$.	If we multiply or divide both sides of an inequality by the same positive quantity, the resulting inequality is equivalent to the original one.	$2x < 4$ Divide by 2: $\frac{2x}{2} < \frac{4}{2}.$ Simplify: $x < 2.$
The Negative Multiplication Property of Inequality If $a < b$ and c is negative, then $ac > bc$. If $a < b$ and c is negative, then $\frac{a}{c} > \frac{b}{c}$.	If we multiply or divide both sides of an inequality by the same negative quantity and reverse the direction of the inequality symbol, the resulting inequality is equivalent to the original one.	$-4x < 20$ Divide by -4 and change the sense of the inequality: $\frac{-4x}{-4} > \frac{20}{-4}.$ Simplify: $x > -5.$

If an inequality does not contain fractions, it can be solved using the following procedure. (In Example 3, we will see how to clear fractions.) Notice, again, how similar this procedure is to the procedure for solving a linear equation.

Solving a Linear Inequality

1. Simplify the algebraic expression on each side.
2. Use the addition property of inequality to collect all the variable terms on one side and all the constant terms on the other side.
3. Use the multiplication property of inequality to isolate the variable and solve. Change the sense of the inequality when multiplying or dividing both sides by a negative number.
4. Express the solution set in interval notation and graph the solution set on a number line.

EXAMPLE 1 Solving a Linear Inequality

Solve and graph the solution set on a number line:

$$3x - 5 > -17.$$

Solution

Step 1. Simplify each side. Because each side is already simplified, we can skip this step.

Step 2. Collect variable terms on one side and constant terms on the other side. The variable term, $3x$, is already on the left side of $3x - 5 > -17$. We will collect constant terms on the right side by adding 5 to both sides.

$$3x - 5 > -17 \quad \text{This is the given inequality.}$$

$$3x - 5 + 5 > -17 + 5 \quad \text{Add 5 to both sides.}$$

$$3x > -12 \quad \text{Simplify.}$$

Step 3. Isolate the variable and solve. We isolate the variable, x, by dividing both sides by 3. Because we are dividing by a positive number, we do not reverse the direction of the inequality symbol.

$$\frac{3x}{3} > \frac{-12}{3} \quad \text{Divide both sides by 3.}$$

$$x > -4 \quad \text{Simplify.}$$

Step 4. Express the solution set in interval notation and graph the set on a number line. The solution set consists of all real numbers that are greater than -4. The interval notation for this solution set is $(-4, \infty)$. The graph of the solution set is shown as follows:

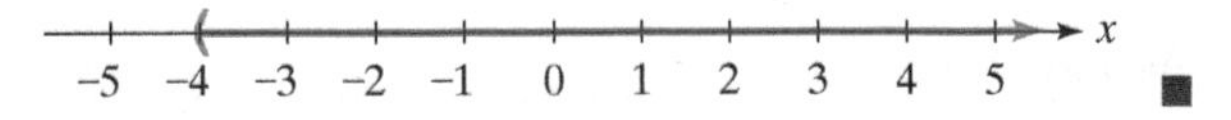

■

Discover for Yourself

As a partial check, select one number from the solution set of $3x - 5 > -17$, the inequality in Example 1. Substitute that number into the original inequality. Perform the resulting computations. You should obtain a true statement.

Is it possible to perform a partial check using a number that is not in the solution set? What should happen in this case? Try doing this.

✓ CHECK POINT 1 Solve and graph the solution set on a number line: $4x - 3 > -23$.

EXAMPLE 2 Solving a Linear Inequality

Solve and graph the solution set on a number line:

$$-2x - 4 > x + 5.$$

Solution

Step 1. Simplify each side. Because each side is already simplified, we can skip this step.

Great Question!

Can I solve the inequality in Example 2 by isolating the variable on the right?

Yes, you can solve

$$-2x - 4 > x + 5$$

by isolating x on the right side. Add $2x$ to both sides.

$$-2x - 4 + 2x > x + 5 + 2x$$
$$-4 > 3x + 5$$

Now subtract 5 from both sides.

$$-4 - 5 > 3x + 5 - 5$$
$$-9 > 3x$$

Finally, divide both sides by 3.

$$\frac{-9}{3} > \frac{3x}{3}$$
$$-3 > x$$

This last inequality means the same thing as $x < -3$.

Step 2. Collect variable terms on one side and constant terms on the other side. We will collect variable terms on the left and constant terms on the right.

$-2x - 4 > x + 5$	This is the given inequality.
$-2x - 4 - x > x + 5 - x$	Subtract x from both sides.
$-3x - 4 > 5$	Simplify.
$-3x - 4 + 4 > 5 + 4$	Add 4 to both sides.
$-3x > 9$	Simplify.

Step 3. Isolate the variable and solve. We isolate the variable, x, by dividing both sides by -3. Because we are dividing by a negative number, we must reverse the direction of the inequality symbol.

$\frac{-3x}{-3} < \frac{9}{-3}$	Divide both sides by -3 and change the sense of the inequality.
$x < -3$	Simplify.

Step 4. Express the solution set in interval notation and graph the set on a number line. The solution set consists of all real numbers that are less than -3. The interval notation for this solution set is $(-\infty, -3)$. The graph of the solution set is shown as follows:

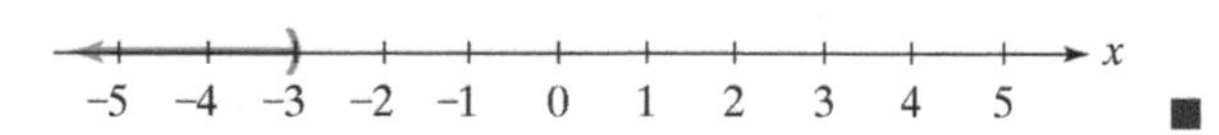

■

Using Technology

Numeric and Graphic Connections

You can use a graphing utility to check the solution set of a linear inequality. Enter each side of the inequality separately under y_1 and y_2. Then use the table or the graphs. To use the table, first locate the x-value for which the y-values are the same. Then scroll up or down to locate x values for which y_1 is greater than y_2 or for which y_1 is less than y_2. To use the graphs, locate the intersection point and then find the x-values for which the graph of y_1 lies above the graph of y_2 ($y_1 > y_2$) or for which the graph of y_1 lies below the graph of y_2 ($y_1 < y_2$).

Let's verify our work in Example 2 and show that $(-\infty, -3)$ is the solution set of

$$-2x - 4 > x + 5.$$

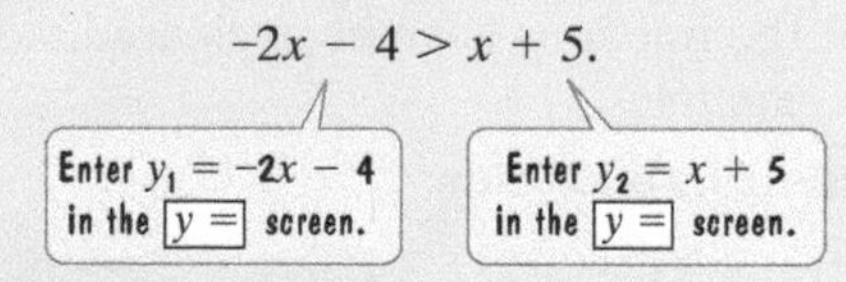

We are looking for values of x for which y_1 is greater than y_2.

Numeric Check

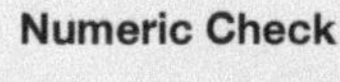

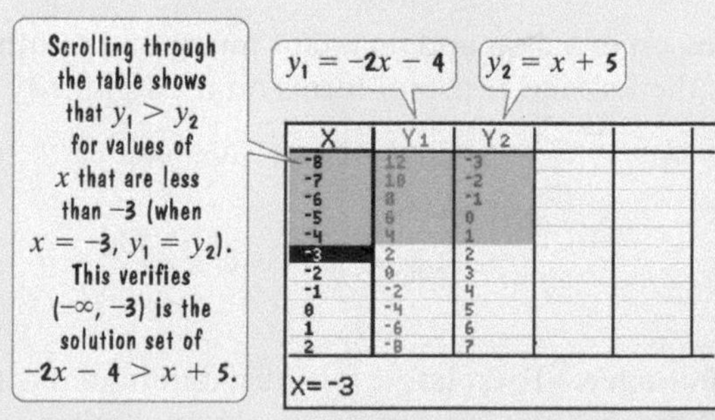

Graphic Check

Display the graphs for y_1 and y_2. Use the intersection feature. The solution set is the set of x-values for which the graph of y_1 lies above the graph of y_2.

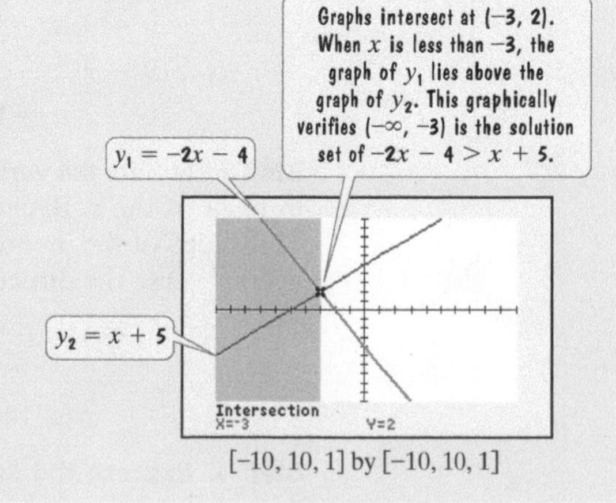

Graphs intersect at $(-3, 2)$. When x is less than -3, the graph of y_1 lies above the graph of y_2. This graphically verifies $(-\infty, -3)$ is the solution set of $-2x - 4 > x + 5$.
$y_1 = -2x - 4$
$y_2 = x + 5$
Intersection
X=-3 Y=2

[−10, 10, 1] by [−10, 10, 1]

☑ CHECK POINT 2 Solve and graph the solution set: $3x + 1 > 7x - 15$.

If an inequality contains fractions, begin by multiplying both sides by the least common denominator. This will clear the inequality of fractions.

EXAMPLE 3 Solving a Linear Inequality Containing Fractions

Solve and graph the solution set on a number line:

$$\frac{x+3}{4} \geq \frac{x-2}{3} + \frac{1}{4}.$$

Solution The denominators are 4, 3, and 4. The least common denominator is 12. We begin by multiplying both sides of the inequality by 12.

$$\frac{x+3}{4} \geq \frac{x-2}{3} + \frac{1}{4}$$ This is the given inequality.

$$12\left(\frac{x+3}{4}\right) \geq 12\left(\frac{x-2}{3} + \frac{1}{4}\right)$$ Multiply both sides by 12. Multiplying by a positive number preserves the sense of the inequality.

$$\frac{12}{1} \cdot \frac{x+3}{4} \geq \frac{12}{1} \cdot \frac{x-2}{3} + \frac{12}{1} \cdot \frac{1}{4}$$ Multiply each term by 12. Use the distributive property on the right side.

$$\frac{\overset{3}{\cancel{12}}}{1} \cdot \frac{x+3}{\underset{1}{\cancel{4}}} \geq \frac{\overset{4}{\cancel{12}}}{1} \cdot \frac{x-2}{\underset{1}{\cancel{3}}} + \frac{\overset{3}{\cancel{12}}}{1} \cdot \frac{1}{\underset{1}{\cancel{4}}}$$ Divide out common factors in each multiplication.

$$3(x+3) \geq 4(x-2) + 3$$ The fractions are now cleared.

Now that the fractions have been cleared, we follow the four steps that we used in the previous examples.

Step 1. Simplify each side.

$$3(x+3) \geq 4(x-2) + 3$$ This is the inequality with the fractions cleared.

$$3x + 9 \geq 4x - 8 + 3$$ Use the distributive property.

$$3x + 9 \geq 4x - 5$$ Simplify.

Step 2. Collect variable terms on one side and constant terms on the other side. We will collect variable terms on the left and constant terms on the right.

$$3x + 9 - 4x \geq 4x - 5 - 4x$$ Subtract $4x$ from both sides.

$$-x + 9 \geq -5$$ Simplify.

$$-x + 9 - 9 \geq -5 - 9$$ Subtract 9 from both sides.

$$-x \geq -14$$ Simplify.

Step 3. Isolate the variable and solve. To isolate x, we must eliminate the negative sign in front of the x. Because $-x$ means $-1x$, we can do this by multiplying (or dividing) both sides of the inequality by -1. We are multiplying by a negative number. Thus, we must reverse the direction of the inequality symbol.

$$(-1)(-x) \leq (-1)(-14)$$ Multiply both sides by -1 and change the sense of the inequality.

$$x \leq 14$$ Simplify.

Step 4. Express the solution set in interval notation and graph the set on a number line. The solution set consists of all real numbers that are less than or equal to 14. The interval notation for this solution set is $(-\infty, 14]$. The graph of the solution set is shown as follows:

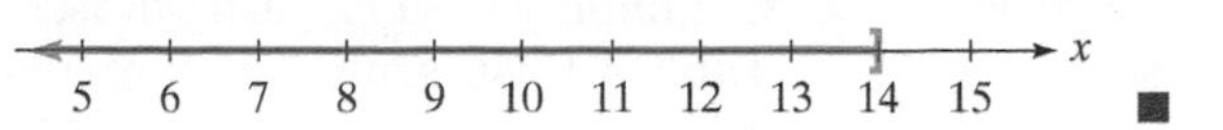

■

CHECK POINT 3 Solve and graph the solution set on a number line: $\frac{x-4}{2} \geq \frac{x-2}{3} + \frac{5}{6}$.

2 Recognize inequalities with no solution or all real numbers as solutions.

Inequalities with Unusual Solution Sets

We have seen that some equations have no solution. This is also true for some inequalities. An example of such an inequality is

$$x > x + 1.$$

There is no number that is greater than itself plus 1. This inequality has no solution and its solution set is $\varnothing$, the empty set.

By contrast, some inequalities are true for all real numbers. An example of such an inequality is

$$x < x + 1.$$

Every real number is less than itself plus 1. The solution set is $(-\infty, \infty)$.

If you attempt to solve an inequality that has no solution, you will eliminate the variable and obtain a false statement, such as $0 > 1$. If you attempt to solve an inequality that is true for all real numbers, you will eliminate the variable and obtain a true statement, such as $0 < 1$.

EXAMPLE 4 Solving Linear Inequalities

Solve each inequality:

a. $2(x + 4) > 2x + 3$

b. $x + 7 \leq x - 2$.

Solution

a.

$2(x + 4) > 2x + 3$	This is the given inequality.
$2x + 8 > 2x + 3$	Apply the distributive property.
$2x + 8 - 2x > 2x + 3 - 2x$	Subtract $2x$ from both sides.
$8 > 3$	Simplify. The statement $8 > 3$ is true.

The inequality $8 > 3$ is true for all values of x. Because this inequality is equivalent to the original inequality, the original inequality is true for all real numbers. The solution set is $(-\infty, \infty)$.

b.

$x + 7 \leq x - 2$	This is the given inequality.
$x + 7 - x \leq x - 2 - x$	Subtract x from both sides.
$7 \leq -2$	Simplify. The statement $7 \leq -2$ is false.

The inequality $7 \leq -2$ is false for all values of x. Because this inequality is equivalent to the original inequality, the original inequality has no solution. The solution set is $\varnothing$. ■

Using Technology

Graphic Connections

The graphs of

$$y_1 = 2(x + 4) \text{ and } y_2 = 2x + 3$$

are parallel lines. The graph of y_1 is always above the graph of y_2. Every value of x satisfies the inequality $y_1 > y_2$. Thus, the solution set of the inequality

$$2(x + 4) > 2x + 3$$

is $(-\infty, \infty)$.

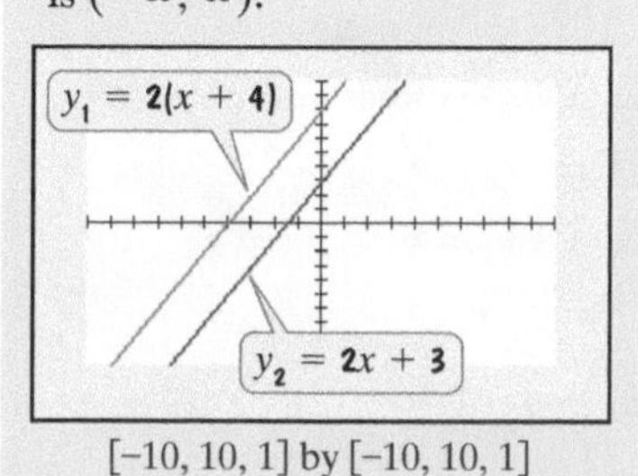

$[-10, 10, 1]$ by $[-10, 10, 1]$

CHECK POINT 4 Solve each inequality:

a. $3(x + 1) > 3x + 2$

b. $x + 1 \leq x - 1$.

3 Solve applied problems using linear inequalities.

Applications

Commonly used English phrases such as "at least" and "at most" indicate inequalities. **Table 4.1** lists sentences containing these phrases and their algebraic translations into inequalities.

Table 4.1 English Sentences and Inequalities

English Sentence	Inequality
x is at least 5.	$x \geq 5$
x is at most 5.	$x \leq 5$
x is between 5 and 7.	$5 < x < 7$
x is no more than 5.	$x \leq 5$
x is no less than 5.	$x \geq 5$

Our next example shows how to use an inequality to select the better deal when considering two pricing options. We use our strategy for solving word problems, translating from the verbal conditions of the problem to a linear inequality.

EXAMPLE 5 Selecting the Better Deal

Acme Car rental agency charges $4 a day plus $0.15 per mile, whereas Interstate rental agency charges $20 a day and $0.05 per mile. How many miles must be driven to make the daily cost of an Acme rental a better deal than an Interstate rental?

Solution

Step 1. Let x represent one of the unknown quantities. We are looking for the number of miles that must be driven in a day to make Acme the better deal. Thus,

let x = the number of miles driven in a day.

Step 2. Represent other unknown quantities in terms of x. We are not asked to find another quantity, so we can skip this step.

Step 3. Write an inequality in x that models the conditions. Acme is a better deal than Interstate if the daily cost of Acme is less than the daily cost of Interstate.

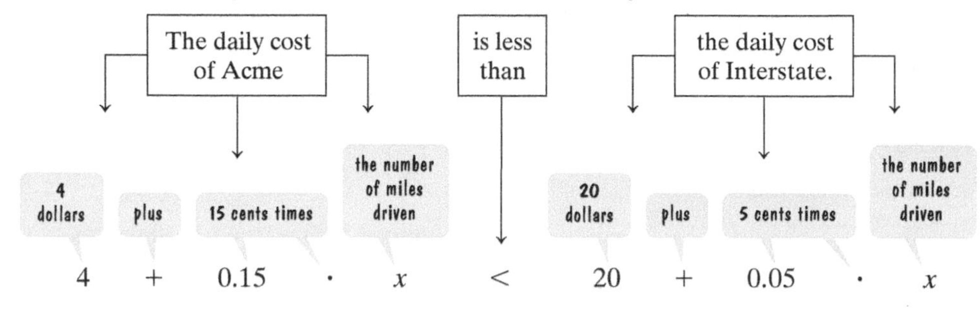

Step 4. Solve the inequality and answer the question.

$$4 + 0.15x < 20 + 0.05x$$ This is the inequality that models the verbal conditions.

$$4 + 0.15x - 0.05x < 20 + 0.05x - 0.05x$$ Subtract 0.05x from both sides.

$$4 + 0.1x < 20$$ Simplify.

$$4 + 0.1x - 4 < 20 - 4$$ Subtract 4 from both sides.

$$0.1x < 16$$ Simplify.

$$\frac{0.1x}{0.1} < \frac{16}{0.1}$$ Divide both sides by 0.1.

$$x < 160$$ Simplify.

Thus, driving fewer than 160 miles per day makes Acme the better deal.

Using Technology

Graphic Connections

The graphs of the daily cost models for the car rental agencies

$$y_1 = 4 + 0.15x$$
$$\text{and } y_2 = 20 + 0.05x$$

are shown in a [0, 300, 10] by [0, 40, 4] viewing rectangle. The graphs intersect at (160, 28). To the left of $x = 160$, the graph of Acme's daily cost lies below that of Interstate's daily cost. This shows that for fewer than 160 miles per day, Acme offers the better deal.

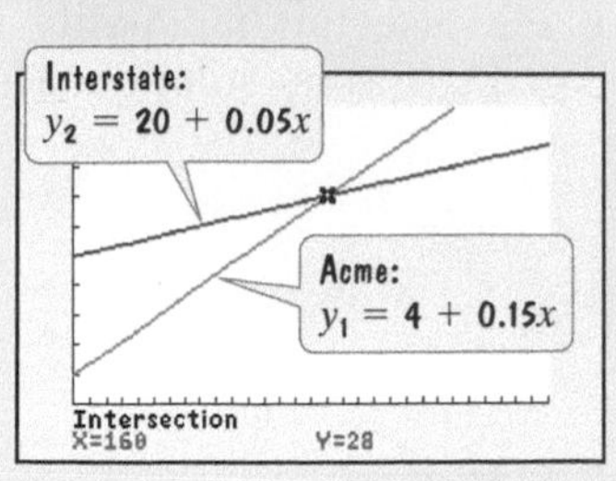

Step 5. Check the proposed solution in the original wording of the problem. One way to do this is to take a mileage less than 160 miles per day to see if Acme is the better deal. Suppose that 150 miles are driven in a day.

$$\text{Cost for Acme} = 4 + 0.15(150) = 26.50$$
$$\text{Cost for Interstate} = 20 + 0.05(150) = 27.50$$

Acme has a lower daily cost, making Acme the better deal. ■

✓ **CHECK POINT 5** A car can be rented from Basic Rental for $260 per week with no extra charge for mileage. Continental charges $80 per week plus 25 cents for each mile driven to rent the same car. How many miles must be driven in a week to make the rental cost for Basic Rental a better deal than Continental's?

Achieving Success

Learn from your mistakes. Being human means making mistakes. By finding and understanding your errors, you will become a better math student.

Source of Error	Remedy
Not Understanding a Concept	Review the concept by finding a similar example in your textbook or class notes. Ask your professor questions to help clarify the concept.
Skipping Steps	Show clear step-by-step solutions. Detailed solution procedures help organize your thoughts and enhance understanding. Doing too many steps mentally often results in preventable mistakes.
Carelessness	Write neatly. Not being able to read your own math writing leads to errors. Avoid writing in pen so you won't have to put huge marks through incorrect work.

"You can achieve your goal if you persistently pursue it."

—Cha Sa-Soon, a 68-year-old South Korean woman who passed her country's written driver's-license exam on her 950th try (*Source: Newsweek*)

CONCEPT AND VOCABULARY CHECK

Fill in each blank so that the resulting statement is true.

1. The addition property of inequality states that if $a < b$, then $a + c$ ________.
2. The positive multiplication property of inequality states that if $a < b$ and c is positive, then ac ________.
3. The negative multiplication property of inequality states that if $a < b$ and c is negative, then ac ________.
4. The linear inequality $-3x - 4 > 5$ can be solved by first ________ to both sides and then ________ both sides by ________, which changes the ________ of the inequality symbol from ________ to ________.
5. In solving an inequality, if you eliminate the variable and obtain a false statement such as $7 < -2$, the solution set is ________.
6. In solving an inequality, if you eliminate the variable and obtain a true statement such as $8 > 3$, the solution set is ________.
7. The algebraic translation of "x is at least 7" is ________.
8. The algebraic translation of "x is at most 7" is ________.
9. The algebraic translation of "x is no more than 7" is ________.
10. The algebraic translation of "x is no less than 7" is ________.

4.1 EXERCISE SET

MyMathLab®

Practice Exercises

In Exercises 1–32, solve each linear inequality. Other than ∅, graph the solution set on a number line.

1. $5x + 11 < 26$
2. $2x + 5 < 17$
3. $3x - 8 \geq 13$
4. $8x - 2 \geq 14$
5. $-9x \geq 36$
6. $-5x \leq 30$
7. $8x - 11 \leq 3x - 13$
8. $18x + 45 \leq 12x - 8$
9. $4(x + 1) + 2 \geq 3x + 6$
10. $8x + 3 > 3(2x + 1) + x + 5$
11. $2x - 11 < -3(x + 2)$
12. $-4(x + 2) > 3x + 20$
13. $1 - (x + 3) \geq 4 - 2x$
14. $5(3 - x) \leq 3x - 1$
15. $\frac{x}{4} - \frac{1}{2} \leq \frac{x}{2} + 1$
16. $\frac{3x}{10} + 1 \geq \frac{1}{5} - \frac{x}{10}$
17. $1 - \frac{x}{2} > 4$
18. $7 - \frac{4}{5}x < \frac{3}{5}$
19. $\frac{x-4}{6} \geq \frac{x-2}{9} + \frac{5}{18}$
20. $\frac{4x-3}{6} + 2 \geq \frac{2x-1}{12}$
21. $4(3x - 2) - 3x < 3(1 + 3x) - 7$
22. $3(x - 8) - 2(10 - x) < 5(x - 1)$
23. $8(x + 1) \leq 7(x + 5) + x$
24. $4(x - 1) \geq 3(x - 2) + x$
25. $3x < 3(x - 2)$
26. $5x < 5(x - 3)$
27. $7(x + 4) - 13 < 12 + 13(3 + x)$
28. $-3[7x - (2x - 3)] > -2(x + 1)$
29. $6 - \frac{2}{3}(3x - 12) \leq \frac{2}{5}(10x + 50)$
30. $\frac{2}{7}(7 - 21x) - 4 > 10 - \frac{3}{11}(11x - 11)$
31. $3[3(x + 5) + 8x + 7] + 5[3(x - 6) - 2(3x - 5)] < 2(4x + 3)$
32. $5[3(2 - 3x) - 2(5 - x)] - 6[5(x - 2) - 2(4x - 3)] < 3x + 19$
33. Let $f(x) = 3x + 2$ and $g(x) = 5x - 8$. Find all values of x for which $f(x) > g(x)$.
34. Let $f(x) = 2x - 9$ and $g(x) = 5x + 4$. Find all values of x for which $f(x) > g(x)$.
35. Let $f(x) = \frac{2}{5}(10x + 15)$ and $g(x) = \frac{1}{4}(8 - 12x)$. Find all values of x for which $g(x) \leq f(x)$.
36. Let $f(x) = \frac{3}{5}(10x - 15) + 9$ and $g(x) = \frac{3}{8}(16 - 8x) - 7$. Find all values of x for which $g(x) \leq f(x)$.
37. Let $f(x) = 1 - (x + 3) + 2x$. Find all values of x for which $f(x)$ is at least 4.
38. Let $f(x) = 2x - 11 + 3(x + 2)$. Find all values of x for which $f(x)$ is at most 0.

Practice PLUS

In Exercises 39–40, solve each linear inequality and graph the solution set on a number line.

39. $2(x + 3) > 6 - \{4[x - (3x - 4) - x] + 4\}$
40. $3(4x - 6) < 4 - \{5x - [6x - (4x - (3x + 2))]\}$

In Exercises 41–42, write an inequality with x isolated on the left side that is equivalent to the given inequality.

41. $ax + b > c$; Assume $a < 0$.
42. $\frac{ax + b}{c} > b$; Assume $a > 0$ and $c < 0$.

In Exercises 43–44, use the graphs of y_1 and y_2 to solve each inequality.

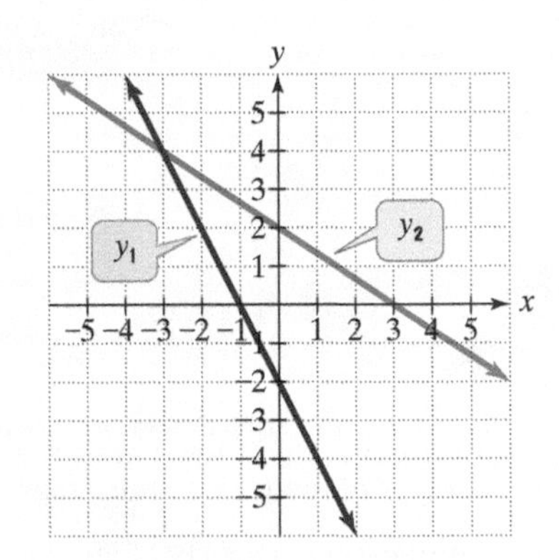

43. $y_1 \geq y_2$
44. $y_1 \leq y_2$

In Exercises 45–46, use the table of values for the linear functions y_1 and y_2 to solve each inequality.

X	Y1	Y2		
-1.8	9.4	7.4		
-1.7	9.1	7.6		
-1.6	8.8	7.8		
-1.5	8.5	8		
-1.4	8.2	8.2		
-1.3	7.9	8.4		
-1.2	7.6	8.6		
-1.1	7.3	8.8		
-1	7	9		
-.9	6.7	9.2		
-.8	6.4	9.4		

X=-1.8

45. $y_1 < y_2$
46. $y_1 > y_2$

Application Exercises

The graphs show that the three components of love, namely passion, intimacy, and commitment, progress differently over time. Passion peaks early in a relationship and then declines. By contrast, intimacy and commitment build gradually. Use the graphs to solve Exercises 47–54.

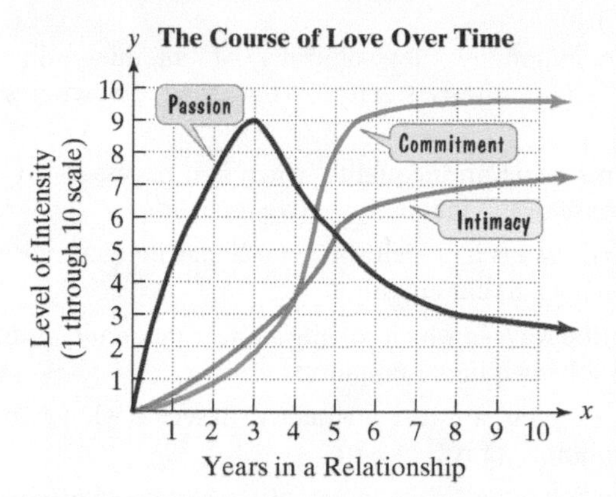

Source: R.J. Sternberg, A Triangular Theory of Love, *Psychological Review, 93,* 119–135.

47. Use interval notation to write an inequality that expresses for which years in a relationship intimacy is greater than commitment.

48. Use interval notation to write an inequality that expresses for which years in a relationship passion is greater than or equal to intimacy.

49. What is the relationship between passion and intimacy on the interval $[5, 7)$?

50. What is the relationship between intimacy and commitment on the interval $[4, 7)$?

51. What is the relationship between passion and commitment on the interval $(6, 8)$?

52. What is the relationship between passion and commitment on the interval $(7, 9)$?

53. What is the maximum level of intensity for passion? After how many years in a relationship does this occur?

54. After approximately how many years do levels of intensity for commitment exceed the maximum level of intensity for passion?

Diversity Index. *What is the chance that the next person I meet will be different from me? The diversity index, from 0 (no diversity) to 100, measures the chance that two randomly selected people are a different race or ethnicity. The diversity index in the United States varies widely from region to region, from as high as 81 in Hawaii to as low as 11 in Vermont. The bar graph at the top of the next column shows the national diversity index for the United States for four years in the period from 1980 through 2010. Exercises 55–56 are based on the data in the graph.*

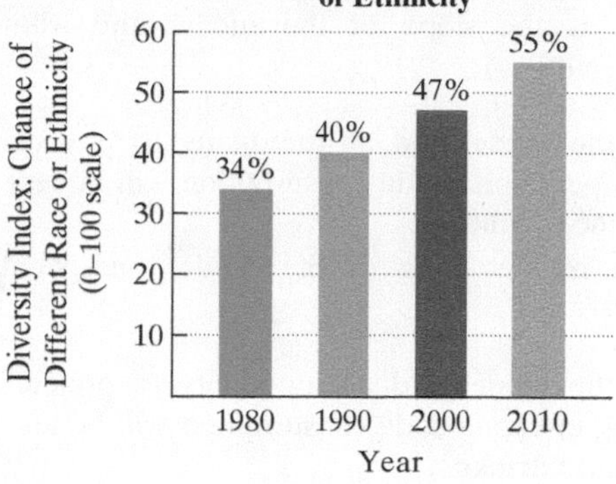

Source: USA Today

55. The data in the graph can be modeled by the formula

$$D = 0.7x + 34,$$

where D is the national diversity index for the United States x years after 1980.

a. How well does the formula model the national diversity index for 2010?

b. Use an inequality to determine in which years the national diversity index will exceed 62.

56. The data in the graph can be modeled by the formula

$$D = 0.7x + 34,$$

where D is the national diversity index for the United States x years after 1980.

a. Does the formula underestimate or overestimate the national diversity index for 2000? By how many points?

b. Use an inequality to determine in which years the national diversity index will exceed 69.

Drinking Less Soda. *The bar graph shows annual per-capita consumption of 12-ounce servings of carbonated soda in the United States for three selected years.*

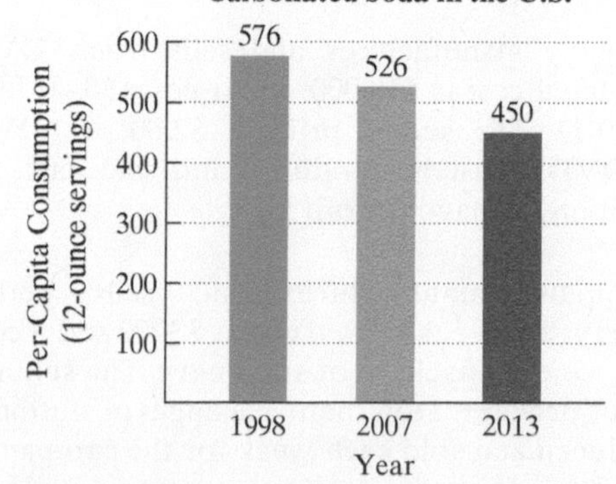

Source: Beverage Digest

The data displayed by the graph can be modeled by the formula

$$S = -8x + 583,$$

where S is per-capita consumption of carbonated soda (12-ounce servings) x years after 1998 (the year of U.S. peak soda consumption). Use this model to solve Exercises 57–58.

(In Exercises 57–58, refer to the model and graph at the bottom of the previous page.)

57. **a.** What is the slope of the model and what does it represent?

b. Use the model and an inequality to project in which years per-capita soda consumption will be less than 415 12-ounce drinks.

58. **a.** What is the slope of the model and what does it represent?

b. Use the model and an inequality to project in which years per-capita soda consumption will be less than 383 12-ounce drinks.

In Exercises 59–66, use the strategy for solving word problems, translating from the verbal conditions of the problem to a linear inequality.

59. A truck can be rented from Basic Rental for \$50 a day plus \$0.20 per mile. Continental charges \$20 per day plus \$0.50 per mile to rent the same truck. How many miles must be driven in a day to make the rental cost for Basic Rental a better deal than Continental's?

60. You are choosing between two landline telephone plans. Plan A has a monthly fee of \$15 with a charge of \$0.08 per minute for all calls. Plan B has a monthly fee of \$3 with a charge of \$0.12 per minute for all calls. How many calling minutes in a month make plan A the better deal?

61. A city commission has proposed two tax bills. The first bill requires that a homeowner pay \$1800 plus 3% of the assessed home value in taxes. The second bill requires taxes of \$200 plus 8% of the assessed home value. What price range of home assessment would make the first bill a better deal for the homeowner?

62. A local bank charges \$8 per month plus 5¢ per check. The credit union charges \$2 per month plus 8¢ per check. How many checks should be written each month to make the credit union a better deal?

63. A company manufactures and sells blank DVDs. The weekly fixed cost is \$10,000 and it costs \$0.40 to produce each DVD. The selling price is \$2.00 per DVD. How many DVDs must be produced and sold each week for the company to have a profit?

64. A company manufactures and sells personalized stationery. The weekly fixed cost is \$3000 and it costs \$3.00 to produce each package of stationery. The selling price is \$5.50 per package. How many packages of stationery must be produced and sold each week for the company to have a profit?

65. An elevator at a construction site has a maximum capacity of 3000 pounds. If the elevator operator weighs 200 pounds and each cement bag weighs 70 pounds, how many bags of cement can be safely lifted on the elevator in one trip?

66. An elevator at a construction site has a maximum capacity of 2500 pounds. If the elevator operator weighs 160 pounds and each cement bag weighs 60 pounds, how many bags of cement can be safely lifted on the elevator in one trip?

Explaining the Concepts

67. When graphing the solutions of an inequality, what does a parenthesis signify? What does a bracket signify?

68. When solving an inequality, when is it necessary to change the sense of the inequality? Give an example.

69. Describe ways in which solving a linear inequality is similar to solving a linear equation.

70. Describe ways in which solving a linear inequality is different from solving a linear equation.

71. When solving a linear inequality, describe what happens if the solution set is $(-\infty, \infty)$.

72. When solving a linear inequality, describe what happens if the solution set is $\varnothing$.

73. What is the slope of the model in Exercise 55 or Exercise 56? What does this represent in terms of the 0-to-100 national diversity index?

Technology Exercises

In Exercises 74–75, solve each inequality using a graphing utility. Graph each side separately. Then determine the values of x for which the graph on the left side lies above the graph on the right side.

74. $-3(x - 6) > 2x - 2$

75. $-2(x + 4) > 6x + 16$

76. Use a graphing utility's TABLE feature to verify your work in Exercises 74–75.

Use the same technique employed in Exercises 74–75 to solve each inequality in Exercises 77–78. In each case, what conclusion can you draw? What happens if you try solving the inequalities algebraically?

77. $12x - 10 > 2(x - 4) + 10x$

78. $2x + 3 > 3(2x - 4) - 4x$

79. A bank offers two checking account plans. Plan A has a base service charge of \$4.00 per month plus 10¢ per check. Plan B charges a base service charge of \$2.00 per month plus 15¢ per check.

a. Write models for the total monthly costs for each plan if x checks are written.

b. Use a graphing utility to graph the models in the same [0, 50, 10] by [0, 10, 1] viewing rectangle.

c. Use the graphs (and the intersection feature) to determine for what number of checks per month plan A will be better than plan B.

d. Verify the result of part (c) algebraically by solving an inequality.

Critical Thinking Exercises

Make Sense? *In Exercises 80–83, determine whether each statement makes sense or does not make sense, and explain your reasoning.*

80. I began the solution of $5 - 3(x + 2) > 10x$ by simplifying the left side, obtaining $2x + 4 > 10x$.

81. I have trouble remembering when to reverse the direction of an inequality symbol, so I avoid this difficulty by collecting variable terms on an appropriate side.

82. If you tell me that three times a number is less than two times that number, it's obvious that no number satisfies this condition, and there is no need for me to write and solve an inequality.

83. Whenever I solve a linear inequality in which the coefficients of the variable on each side are the same, the solution set is $\varnothing$ or $(-\infty, \infty)$.

In Exercises 84–87, determine whether each statement is true or false. If the statement is false, make the necessary change(s) to produce a true statement.

84. The inequality $3x > 6$ is equivalent to $2 > x$.

85. The smallest real number in the solution set of $2x > 6$ is 4.

86. If x is at least 7, then $x > 7$.

87. The inequality $-3x > 6$ is equivalent to $-2 > x$.

88. Find a so that the solution set of $ax + 4 \leq -12$ is $[8, \infty)$.

89. What's wrong with this argument? Suppose x and y represent two real numbers, where $x > y$.

$2 > 1$	This is a true statement.
$2(y - x) > 1(y - x)$	Multiply both sides by $y - x$.
$2y - 2x > y - x$	Use the distributive property.
$y - 2x > -x$	Subtract y from both sides.
$y > x$	Add $2x$ to both sides.

The final inequality, $y > x$, is impossible because we were initially given $x > y$.

Review Exercises

90. If $f(x) = x^2 - 2x + 5$, find $f(-4)$. (Section 2.1, Example 3)

91. Solve the system:

$$\begin{cases} 2x - y - z = -3 \\ 3x - 2y - 2z = -5 \\ -x + y + 2z = 4. \end{cases}$$

(Section 3.3, Example 2)

92. Simplify: $\left(\frac{2x^4y^{-2}}{4xy^3}\right)^3$. (Section 1.6, Example 9)

Preview Exercises

Exercises 93–95 will help you prepare for the material covered in the next section.

93. Consider the sets $A = \{1, 2, 3, 4\}$ and $B = \{3, 4, 5, 6, 7\}$.

a. Write the set consisting of elements common to both set A and set B.

b. Write the set consisting of elements that are members of set A or of set B or of both sets.

94. a. Solve: $x - 3 < 5$.

b. Solve: $2x + 4 < 14$.

c. Give an example of a number that satisfies the inequality in part (a) and the inequality in part (b).

d. Give an example of a number that satisfies the inequality in part (a), but not the inequality in part (b).

95. a. Solve: $2x - 6 \geq -4$.

b. Solve: $5x + 2 \geq 17$.

c. Give an example of a number that satisfies the inequality in part (a) and the inequality in part (b).

d. Give an example of a number that satisfies the inequality in part (a), but not the inequality in part (b).

SECTION

4.2 Compound Inequalities

What am I supposed to learn?

After studying this section, you should be able to:

1. Find the intersection of two sets.
2. Solve compound inequalities involving *and*.
3. Find the union of two sets.
4. Solve compound inequalities involving *or*.

Among the U.S. presidents, more have had dogs than cats in the White House. Sets of presidents with dogs, cats, dogs and cats, dogs or cats, and neither, with the number of presidents belonging to each of these sets, are shown in **Figure 4.1**.

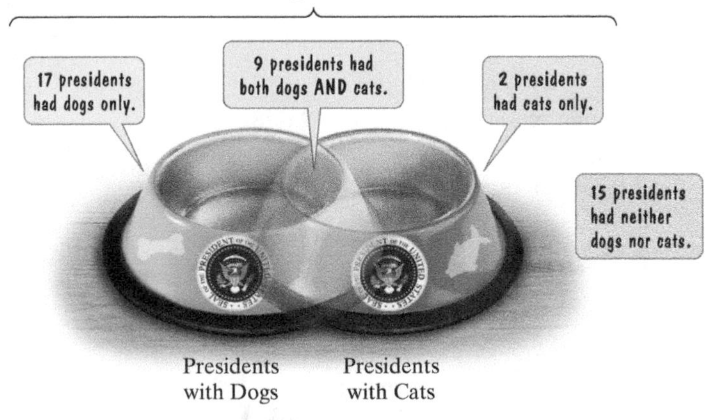
17 + 9 + 2 = 28 presidents had dogs OR cats.
17 presidents had dogs only.
9 presidents had both dogs AND cats.
2 presidents had cats only.
15 presidents had neither dogs nor cats.
Presidents with Dogs
Presidents with Cats

Figure 4.1 Presidential pets in the White House
Source: factmonster.com

Figure 4.1 indicates that 9 presidents had dogs *and* cats in the White House. Furthermore, 28 presidents had dogs *or* cats in the White House. In this section, we'll focus on sets joined with the word *and* and the word *or*. The difference is that we'll be using solution sets of inequalities rather than sets of presidents with pets.

A **compound inequality** is formed by joining two inequalities with the word *and* or the word *or*.

Examples of Compound Inequalities

- $x - 3 < 5$ and $2x + 4 < 14$
- $3x - 5 \leq 13$ or $5x + 2 > -3$

Compound inequalities illustrate the importance of the words *and* and *or* in mathematics, as well as in everyday English.

1 Find the intersection of two sets.

Compound Inequalities Involving *And*

If A and B are sets, we can form a new set consisting of all elements that are in both A and B. This set is called the *intersection* of the two sets.

Definition of the Intersection of Sets

The **intersection** of sets A and B, written $A \cap B$, is the set of elements common to both set A **and** set B. This definition can be expressed in set-builder notation as follows:

$$A \cap B = \{x | x \in A \text{ and } x \in B\}.$$

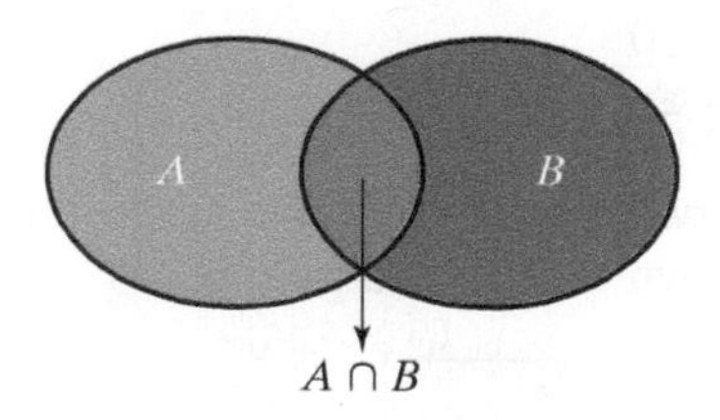

Figure 4.2 Picturing the intersection of two sets

Figure 4.2 shows a useful way of picturing the intersection of sets A and B. The figure indicates that $A \cap B$ contains those elements that belong to both A and B at the same time.

EXAMPLE 1 Finding the Intersection of Two Sets

Find the intersection: $\{7, 8, 9, 10, 11\} \cap \{6, 8, 10, 12\}$.

Solution The elements common to $\{7, 8, 9, 10, 11\}$ and $\{6, 8, 10, 12\}$ are 8 and 10. Thus,

$$\{7, 8, 9, 10, 11\} \cap \{6, 8, 10, 12\} = \{8, 10\}. \blacksquare$$

✓ **CHECK POINT 1** Find the intersection: $\{3, 4, 5, 6, 7\} \cap \{3, 7, 8, 9\}$.

2 Solve compound inequalities involving *and*.

A number is a **solution of a compound inequality formed by the word *and*** if it is a solution of both inequalities. For example, the solution set of the compound inequality

$$x \leq 6 \quad \text{and} \quad x \geq 2$$

is the set of values of x that satisfy both $x \leq 6$ and $x \geq 2$. Thus, the solution set is the intersection of the solution sets of the two inequalities.

What are the numbers that satisfy both $x \leq 6$ and $x \geq 2$? These numbers are easier to see if we graph the solution set to each inequality on a number line. These graphs are shown in **Figure 4.3**. The intersection is shown in the third graph.

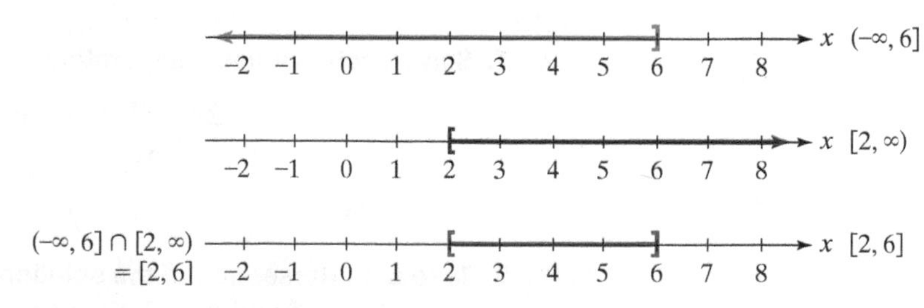

Figure 4.3 Numbers satisfying both $x \leq 6$ and $x \geq 2$

The numbers common to both sets are those that are less than or equal to 6 and greater than or equal to 2. This set in interval notation is $[2, 6]$.

Here is a procedure for finding the solution set of a compound inequality containing the word *and*.

Solving Compound Inequalities Involving *AND*

1. Solve each inequality separately.
2. Graph the solution set to each inequality on a number line and take the intersection of these solution sets. This intersection appears as the portion of the number line that the two graphs have in common.

EXAMPLE 2 Solving a Compound Inequality with *And*

Solve: $x - 3 < 5$ and $2x + 4 < 14$.

Solution

Step 1. Solve each inequality separately.

$$\begin{array}{rcr} x - 3 < 5 & \text{and} & 2x + 4 < 14 \\ x < 8 & & 2x < 10 \\ & & x < 5 \end{array}$$

Step 2. Take the intersection of the solution sets of the two inequalities. We graph the solution sets of $x < 8$ and $x < 5$. The intersection is shown in the third graph.

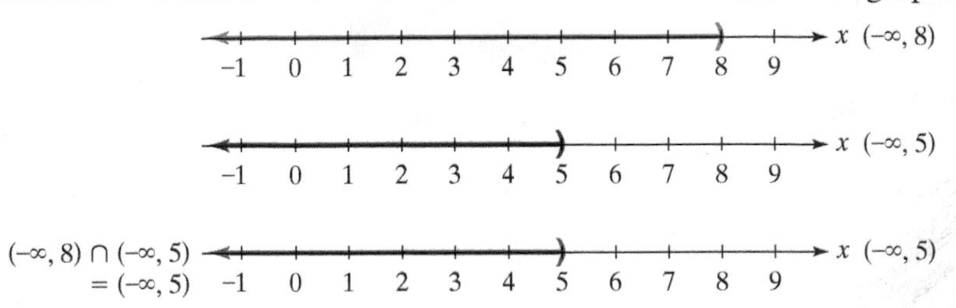

The numbers common to both sets are those that are less than 5. The solution set in interval notation is $(-\infty, 5)$. Take a moment to check that any number in $(-\infty, 5)$ satisfies both of the original inequalities. ■

✓ **CHECK POINT 2** Solve: $x + 2 < 5$ and $2x - 4 < -2$.

EXAMPLE 3 Solving a Compound Inequality with *And*

Solve: $2x - 7 > 3$ and $5x - 4 < 6$.

Solution

Step 1. Solve each inequality separately.

$$\begin{array}{rcr} 2x - 7 > 3 & \text{and} & 5x - 4 < 6 \\ 2x > 10 & & 5x < 10 \\ x > 5 & & x < 2 \end{array}$$

Step 2. Take the intersection of the solution sets of the two inequalities. We graph the solution sets of $x > 5$ and $x < 2$. We use these graphs to find their intersection.

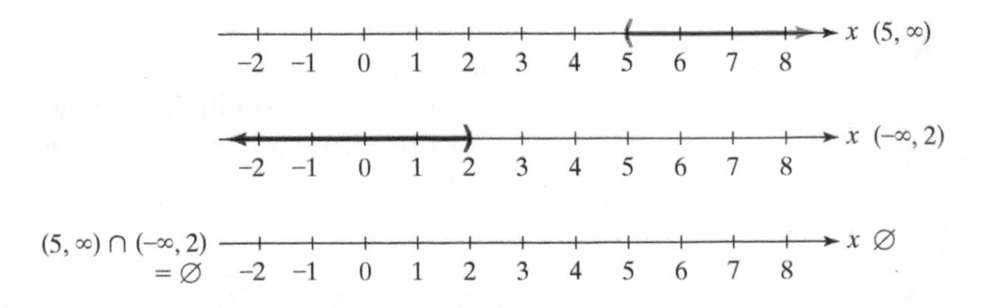

There is no number that is both greater than 5 and at the same time less than 2. Thus, the solution set is the empty set, $\varnothing$. ■

✓ **CHECK POINT 3** Solve: $4x - 5 > 7$ and $5x - 2 < 3$.

If $a < b$, the compound inequality

$$a < x \text{ and } x < b$$

can be written in the shorter form

$$a < x < b.$$

For example, the compound inequality

$$-3 < 2x + 1 \text{ and } 2x + 1 < 3$$

can be abbreviated

$$-3 < 2x + 1 < 3.$$

The word *and* does not appear when the inequality is written in the shorter form, although it is implied. The shorter form enables us to solve both inequalities at once. By performing the same operations on all three parts of the inequality, our goal is to **isolate x in the middle**.

EXAMPLE 4 Solving a Compound Inequality

Solve and graph the solution set:

$$-3 < 2x + 1 \leq 3.$$

Solution We would like to isolate x in the middle. We can do this by first subtracting 1 from all three parts of the compound inequality. Then we isolate x from $2x$ by dividing all three parts of the inequality by 2.

$$-3 < 2x + 1 \leq 3 \quad \text{This is the given inequality.}$$

$$-3 - 1 < 2x + 1 - 1 \leq 3 - 1 \quad \text{Subtract 1 from all three parts.}$$

$$-4 < 2x \leq 2 \quad \text{Simplify.}$$

$$\frac{-4}{2} < \frac{2x}{2} \leq \frac{2}{2} \quad \text{Divide each part by 2.}$$

$$-2 < x \leq 1 \quad \text{Simplify.}$$

The solution set consists of all real numbers greater than -2 and less than or equal to 1, represented by $(-2, 1]$ in interval notation. The graph is shown as follows:

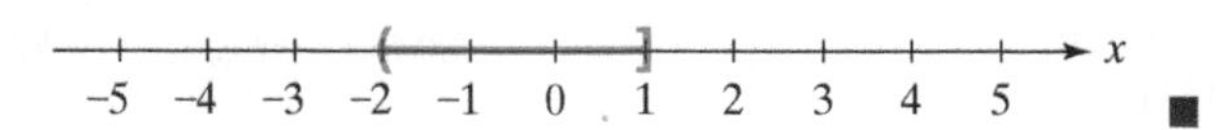

■

Using Technology

Numeric and Graphic Connections

Let's verify our work in Example 4 and show that $(-2, 1]$ is the solution set of $-3 < 2x + 1 \leq 3$.

Numeric Check

To check numerically, enter $y_1 = 2x + 1$.

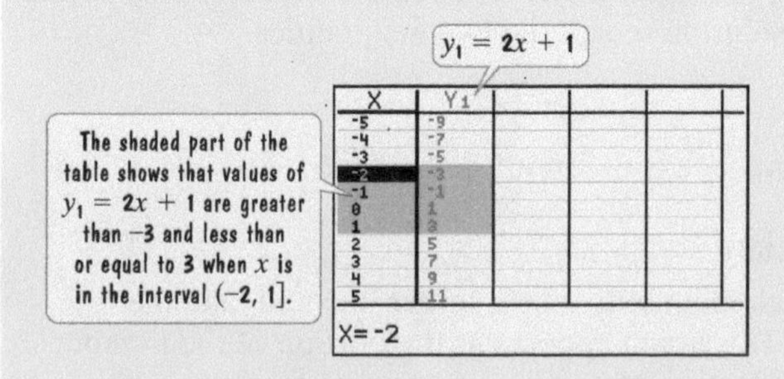

y1 = 2x + 1
The shaded part of the table shows that values of y1 = 2x + 1 are greater than −3 and less than or equal to 3 when x is in the interval (−2, 1].
X Y1
-5 -9
-4 -7
-3 -5
-2 -3
-1 -1
0 1
1 3
2 5
3 7
4 9
5 11
X=-2

Graphic Check

To check graphically, graph each part of

$$-3 < 2x + 1 \leq 3.$$

Enter $y_1 = -3$. Enter $y_2 = 2x + 1$. Enter $y_3 = 3$.

The figure shows that the graph of $y_2 = 2x + 1$ lies above the graph of $y_1 = -3$ and on or below the graph of $y_3 = 3$ when x is in the interval $(-2, 1]$.

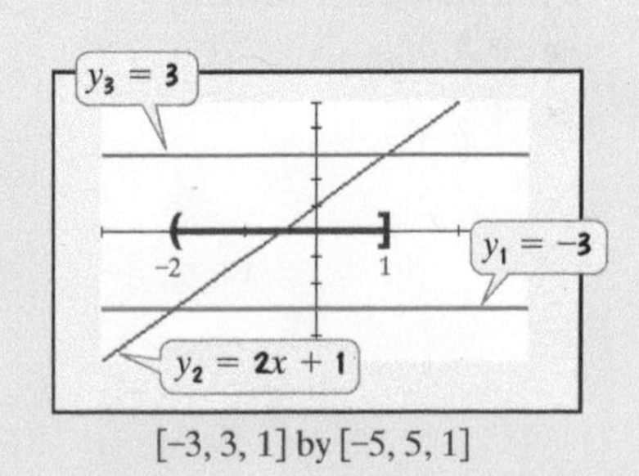

$[-3, 3, 1]$ by $[-5, 5, 1]$

CHECK POINT 4 Solve and graph the solution set: $1 \le 2x + 3 < 11$.

3 Find the union of two sets.

Compound Inequalities Involving *Or*

Another set that we can form from sets A and B consists of elements that are in A or B or in both sets. This set is called the *union* of the two sets.

Definition of the Union of Sets

The **union** of sets A and B, written $A \cup B$, is the set of elements that are members of set A **or** of set B or of both sets. This definition can be expressed in set-builder notation as follows:

$$A \cup B = \{x \mid x \in A \text{ or } x \in B\}.$$

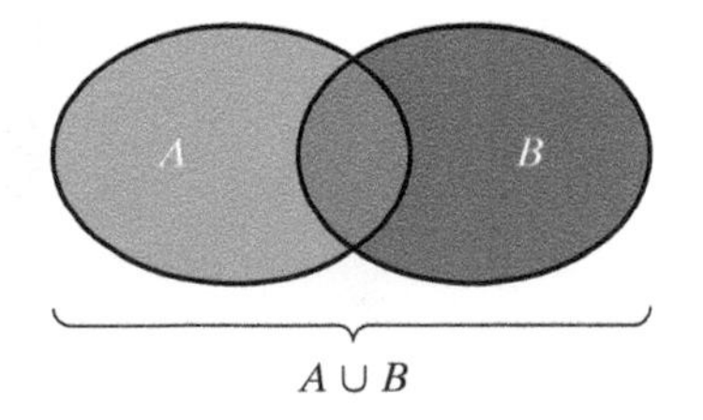

Figure 4.4 Picturing the union of two sets

Figure 4.4 shows a useful way of picturing the union of sets A and B. The figure indicates that $A \cup B$ is formed by joining the sets together.

We can find the union of set A and set B by listing the elements of set A. Then, we include any elements of set B that have not already been listed. Enclose all elements that are listed with braces. This shows that the union of two sets is also a set.

Great Question!

How can I use the words *union* and *intersection* to help me distinguish between these two operations?

Union, as in a marriage union, suggests joining things, or uniting them. Intersection, as in the intersection of two crossing streets, brings to mind the area common to both, suggesting things that overlap.

EXAMPLE 5 Finding the Union of Two Sets

Find the union: $\{7, 8, 9, 10, 11\} \cup \{6, 8, 10, 12\}$.

Solution To find $\{7, 8, 9, 10, 11\} \cup \{6, 8, 10, 12\}$, start by listing all the elements from the first set, namely 7, 8, 9, 10, and 11. Now list all the elements from the second set that are not in the first set, namely 6 and 12. The union is the set consisting of all these elements. Thus,

$$\{7, 8, 9, 10, 11\} \cup \{6, 8, 10, 12\} = \{6, 7, 8, 9, 10, 11, 12\}.$$

Although 8 and 10 appear in both sets, do not list 8 and 10 twice. ■

CHECK POINT 5 Find the union: $\{3, 4, 5, 6, 7\} \cup \{3, 7, 8, 9\}$.

4 Solve compound inequalities involving *or*.

A number is a **solution of a compound inequality formed by the word *or*** if it is a solution of either inequality. Thus, the solution set of a compound inequality formed by the word *or* is the union of the solution sets of the two inequalities.

Solving Compound Inequalities Involving *OR*

1. Solve each inequality separately.
2. Graph the solution set to each inequality on a number line and take the union of these solution sets. This union appears as the portion of the number line representing the total collection of numbers in the two graphs.

EXAMPLE 6 Solving a Compound Inequality with *Or*

Solve: $2x - 3 < 7$ or $35 - 4x \leq 3$.

Solution

Step 1. Solve each inequality separately.

$$\begin{aligned} 2x - 3 &< 7 \quad \text{or} \quad & 35 - 4x &\leq 3 \\ 2x &< 10 & -4x &\leq -32 \\ x &< 5 & x &\geq 8 \end{aligned}$$

Step 2. Take the union of the solution sets of the two inequalities. We graph the solution sets of $x < 5$ and $x \geq 8$. We use these graphs to find their union.

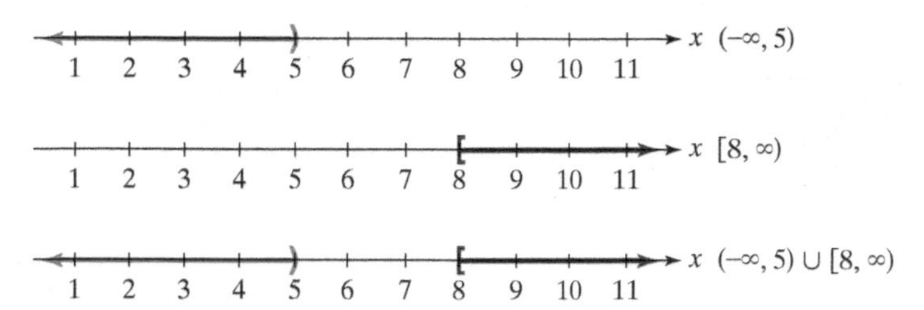

The solution set consists of all numbers that are less than 5 or greater than or equal to 8. The solution set in interval notation is $(-\infty, 5) \cup [8, \infty)$. There is no shortcut way to express this union when interval notation is used. ■

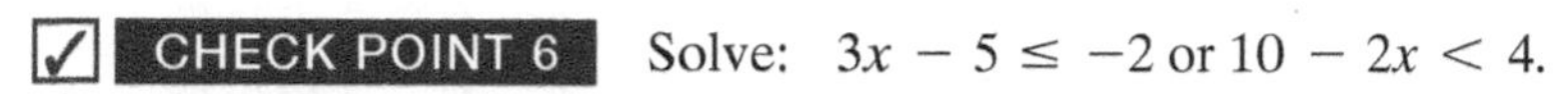

CHECK POINT 6 Solve: $3x - 5 \leq -2$ or $10 - 2x < 4$.

EXAMPLE 7 Solving a Compound Inequality with *Or*

Solve: $3x - 5 \leq 13$ or $5x + 2 > -3$.

Solution

Step 1. Solve each inequality separately.

$$\begin{aligned} 3x - 5 &\leq 13 \quad \text{or} \quad & 5x + 2 &> -3 \\ 3x &\leq 18 & 5x &> -5 \\ x &\leq 6 & x &> -1 \end{aligned}$$

Step 2. Take the union of the solution sets of the two inequalities. We graph the solution sets of $x \leq 6$ and $x > -1$. We use these graphs to find their union.

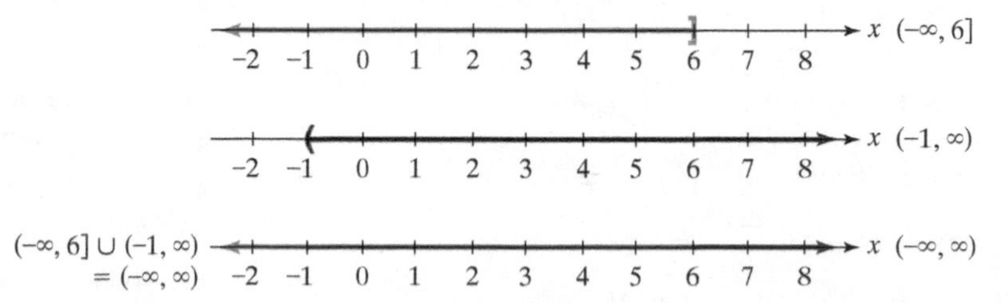

Because all real numbers are either less than or equal to 6 or greater than −1 or both, the union of the two sets fills the entire number line. Thus, the solution set in interval notation is $(-\infty, \infty)$. Any real number that you select will satisfy at least one of the original inequalities. ■

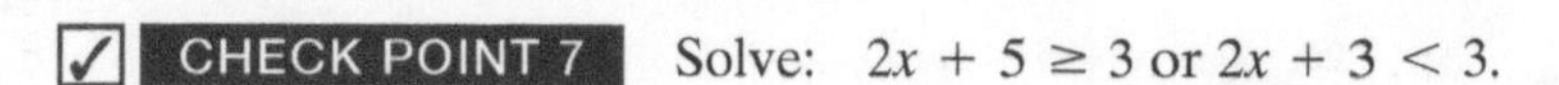

CHECK POINT 7 Solve: $2x + 5 \geq 3$ or $2x + 3 < 3$.

CONCEPT AND VOCABULARY CHECK

Fill in each blank so that the resulting statement is true.

1. The set of elements common to both set A and set B is called the ________________ of sets A and B, and is symbolized by ________________.
2. The set of elements that are members of set A or set B or of both sets is called the ________________ of sets A and B, and is symbolized by ________________.
3. The set of elements common to both $(-\infty, 9)$ and $(-\infty, 12)$ is ________________.
4. The set of elements in $(-\infty, 9)$ or $(-\infty, 12)$ or in both sets is ________________.
5. The way to solve $-7 < 3x - 4 \leq 5$ is to isolate x in the ________________.

4.2 EXERCISE SET

MyMathLab®

Practice Exercises

In Exercises 1–6, find the intersection of the sets.

1. $\{1, 2, 3, 4\} \cap \{2, 4, 5\}$
2. $\{1, 3, 7\} \cap \{2, 3, 8\}$
3. $\{1, 3, 5, 7\} \cap \{2, 4, 6, 8, 10\}$
4. $\{0, 1, 3, 5\} \cap \{-5, -3, -1\}$
5. $\{a, b, c, d\} \cap \varnothing$
6. $\{w, y, z\} \cap \varnothing$

In Exercises 7–24, solve each compound inequality. Use graphs to show the solution set to each of the two given inequalities, as well as a third graph that shows the solution set of the compound inequality. Except for the empty set, express the solution set in interval notation.

7. $x > 3$ and $x > 6$
8. $x > 2$ and $x > 4$
9. $x \leq 5$ and $x \leq 1$
10. $x \leq 6$ and $x \leq 2$
11. $x < 2$ and $x \geq -1$
12. $x < 3$ and $x \geq -1$
13. $x > 2$ and $x < -1$
14. $x > 3$ and $x < -1$
15. $5x < -20$ and $3x > -18$
16. $3x \leq 15$ and $2x > -6$
17. $x - 4 \leq 2$ and $3x + 1 > -8$
18. $3x + 2 > -4$ and $2x - 1 < 5$
19. $2x > 5x - 15$ and $7x > 2x + 10$
20. $6 - 5x > 1 - 3x$ and $4x - 3 > x - 9$
21. $4(1 - x) < -6$ and $\frac{x - 7}{5} \leq -2$
22. $5(x - 2) > 15$ and $\frac{x - 6}{4} \leq -2$
23. $x - 1 \leq 7x - 1$ and $4x - 7 < 3 - x$
24. $2x + 1 > 4x - 3$ and $x - 1 \geq 3x + 5$

In Exercises 25–32, solve each inequality and graph the solution set on a number line. Express the solution set in interval notation.

25. $6 < x + 3 < 8$
26. $7 < x + 5 < 11$
27. $-3 \leq x - 2 < 1$
28. $-6 < x - 4 \leq 1$
29. $-11 < 2x - 1 \leq -5$
30. $3 \leq 4x - 3 < 19$
31. $-3 \leq \frac{2x}{3} - 5 < -1$
32. $-6 \leq \frac{x}{2} - 4 < -3$

In Exercises 33–38, find the union of the sets.

33. $\{1, 2, 3, 4\} \cup \{2, 4, 5\}$
34. $\{1, 3, 7, 8\} \cup \{2, 3, 8\}$
35. $\{1, 3, 5, 7\} \cup \{2, 4, 6, 8, 10\}$
36. $\{0, 1, 3, 5\} \cup \{2, 4, 6\}$
37. $\{a, e, i, o, u\} \cup \varnothing$
38. $\{e, m, p, t, y\} \cup \varnothing$

In Exercises 39–54, solve each compound inequality. Use graphs to show the solution set to each of the two given inequalities, as well as a third graph that shows the solution set of the compound inequality. Express the solution set in interval notation.

39. $x > 3$ or $x > 6$
40. $x > 2$ or $x > 4$
41. $x \leq 5$ or $x \leq 1$
42. $x \leq 6$ or $x \leq 2$
43. $x < 2$ or $x \geq -1$
44. $x < 3$ or $x \geq -1$
45. $x \geq 2$ or $x < -1$
46. $x \geq 3$ or $x < -1$
47. $3x > 12$ or $2x < -6$
48. $3x < 3$ or $2x > 10$
49. $3x + 2 \leq 5$ or $5x - 7 \geq 8$
50. $2x - 5 \leq -11$ or $5x + 1 \geq 6$
51. $4x + 3 < -1$ or $2x - 3 \geq -11$
52. $2x + 1 < 15$ or $3x - 4 \geq -1$
53. $-2x + 5 > 7$ or $-3x + 10 > 2x$
54. $16 - 3x \geq -8$ or $13 - x > 4x + 3$
55. Let $f(x) = 2x + 3$ and $g(x) = 3x - 1$. Find all values of x for which $f(x) \geq 5$ and $g(x) > 11$.
56. Let $f(x) = 4x + 5$ and $g(x) = 3x - 4$. Find all values of x for which $f(x) \geq 5$ and $g(x) \leq 2$.

57. Let $f(x) = 3x - 1$ and $g(x) = 4 - x$. Find all values of x for which $f(x) \le -1$ or $g(x) \le$ [illegible].

58. Let $f(x) = 2x - 5$ and $g(x) = 3 - x$. Find all values of x for which $f(x) \geq 3$ or $g(x) < 0$.

Practice PLUS

In Exercises 59–60, write an inequality with x isolated in the middle that is equivalent to the given inequality. Assume $a > 0, b > 0$, and $c > 0$.

59. $-c < ax - b < c$

60. $-2 < \dfrac{ax - b}{c} < 2$

In Exercises 61–62, use the graphs of y_1, y_2, and y_3 to solve each compound inequality.

61. $-3 \le 2x - 1 \le 5$

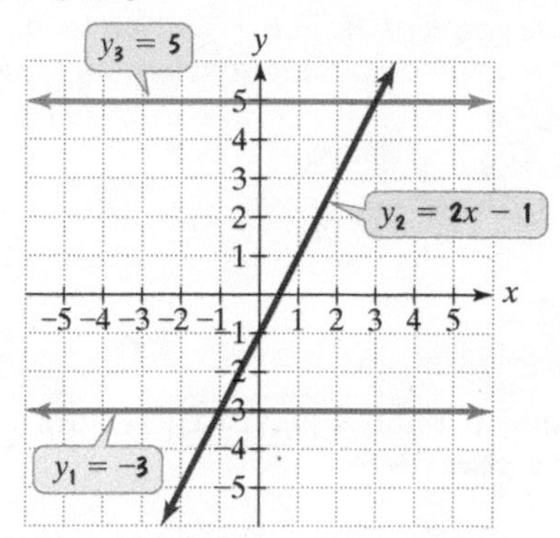

62. $x - 2 < 2x - 1 < x + 2$

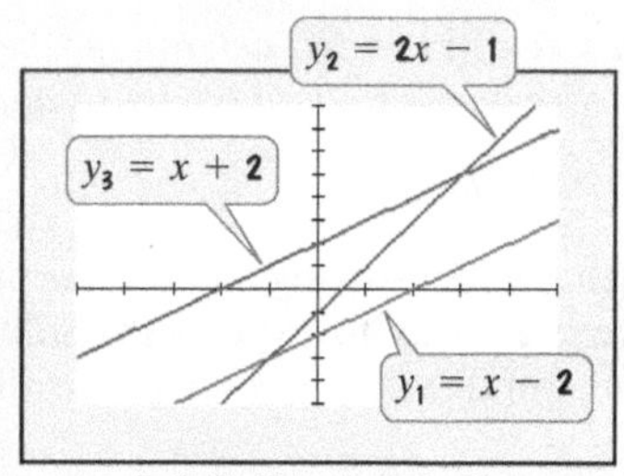

[−5, 5, 1] by [−5, 8, 1]

63. Solve $x - 2 < 2x - 1 < x + 2$, the inequality in Exercise 62, using algebraic methods. (*Hint:* Rewrite the inequality as $2x - 1 > x - 2$ and $2x - 1 < x + 2$.)

64. Use the hint given in Exercise 63 to solve $x \le 3x - 10 \le 2x$.

In Exercises 65–66, use the table to solve each inequality.

65. $-2 \le 5x + 3 < 13$

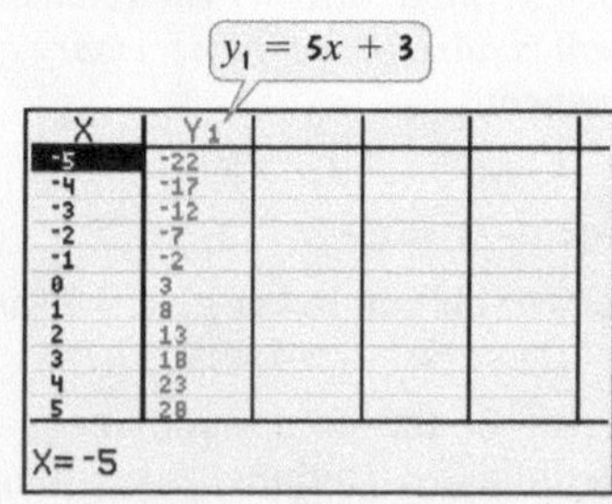

66. $-3 < 2x - 5 \le 3$

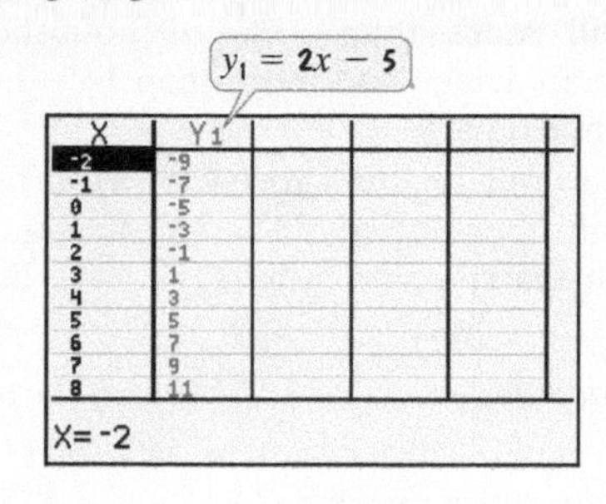

In Exercises 67–68, use the roster method to find the set of negative integers that are solutions of each inequality.

67. $5 - 4x \ge 1$ and $3 - 7x < 31$

68. $-5 < 3x + 4 \le 16$

Application Exercises

In more U.S. marriages, spouses have different faiths. The bar graph shows the percentage of households with an interfaith marriage in 1988 and 2012. Also shown is the percentage of households in which a person of faith is married to someone with no religion.

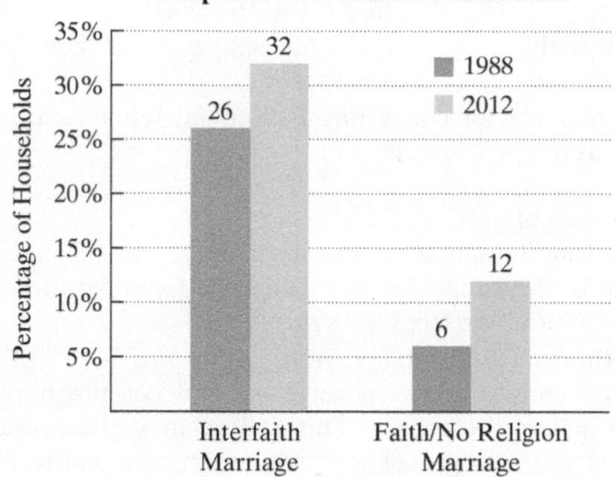

Source: General Social Survey, University of Chicago

The formula

$$I = \frac{1}{4}x + 26$$

models the percentage of U.S. households with an interfaith marriage, I, x years after 1988. The formula

$$N = \frac{1}{4}x + 6$$

models the percentage of U.S. households in which a person of faith is married to someone with no religion, N, x years after 1988. Use these models to solve Exercises 69–70.

69. a. In which years will more than 33% of U.S. households have an interfaith marriage?

b. In which years will more than 14% of U.S. households have a person of faith married to someone with no religion?

c. Based on your answers to parts (a) and (b), in which years will more than 33% of households have an interfaith marriage and more than 14% have a faith/no religion marriage?

d. Based on your answers to parts (a) and (b), in which years will more than 33% of households have an interfaith marriage or more than 14% have a faith/no religion marriage?

(Refer to the models and graph at the bottom of the previous page.)

70. a. In which years will more than 34% of U.S. households have an interfaith marriage?

b. In which years will more than 15% of U.S. households have a person of faith married to someone with no religion?

c. Based on your answers to parts (a) and (b), in which years will more than 34% of households have an interfaith marriage and more than 15% have a faith/no religion marriage?

d. Based on your answers to parts (a) and (b), in which years will more than 34% of households have an interfaith marriage or more than 15% have a faith/no religion marriage?

71. The formula for converting Celsius temperature, C, to Fahrenheit temperature, F, is

$$F = \frac{9}{5}C + 32.$$

If Fahrenheit temperature ranges from 41° to 50°, inclusive, what is the range for Celsius temperature? Use interval notation to express this range.

72. The formula for converting Fahrenheit temperature, F, to Celsius temperature, C, is

$$C = \frac{5}{9}(F - 32).$$

If Celsius temperature ranges from 15° to 35°, inclusive, what is the range for the Fahrenheit temperature? Use interval notation to express this range.

73. On the first four exams, your grades are 70, 75, 87, and 92. There is still one more exam, and you are hoping to earn a B in the course. This will occur if the average of your five exam grades is greater than or equal to 80 and less than 90. What range of grades on the fifth exam will result in earning a B? Use interval notation to express this range.

74. On the first four exams, your grades are 82, 75, 80, and 90. There is still a final exam, and it counts as two grades. You are hoping to earn a B in the course: This will occur if the average of your six exam grades is greater than or equal to 80 and less than 90. What range of grades on the final exam will result in earning a B? Use interval notation to express this range.

75. The toll to a bridge is \$3.00. A three-month pass costs \$7.50 and reduces the toll to \$0.50. A six-month pass costs \$30 and permits crossing the bridge for no additional fee. How many crossings per three-month period does it take for the three-month pass to be the best deal?

76. Parts for an automobile repair cost \$175. The mechanic charges \$34 per hour. If you receive an estimate for at least \$226 and at most \$294 for fixing the car, what is the time interval that the mechanic will be working on the job?

Explaining the Concepts

77. Describe what is meant by the intersection of two sets. Give an example.

78. Explain how to solve a compound inequality involving *and*.

79. Why is $1 < 2x + 3 < 9$ a compound inequality? What are the two inequalities and what is the word that joins them?

80. Explain how to solve $1 < 2x + 3 < 9$.

81. Describe what is meant by the union of two sets. Give an example.

82. Explain how to solve a compound inequality involving *or*.

Technology Exercises

In Exercises 83–86, solve each inequality using a graphing utility. Graph each of the three parts of the inequality separately in the same viewing rectangle. The solution set consists of all values of x for which the graph of the linear function in the middle lies between the graphs of the constant functions on the left and the right.

83. $1 < x + 3 < 9$

84. $-1 < \frac{x + 4}{2} < 3$

85. $1 \le 4x - 7 \le 3$

86. $2 \le 4 - x \le 7$

87. Use a graphing utility's TABLE feature to verify your work in Exercises 83–86.

Critical Thinking Exercises

Make Sense? *In Exercises 88–91, determine whether each statement makes sense or does not make sense, and explain your reasoning.*

88. I've noticed that when solving some compound inequalities with *or*, there is no way to express the solution set using a single interval, but this does not happen with *and* compound inequalities.

89. Compound inequalities with *and* have solutions that satisfy both inequalities, whereas compound inequalities with *or* have solutions that satisfy at least one of the inequalities.

90. I'm considering the compound inequality $x < 8$ and $x \ge 8$, so the solution set is $\varnothing$.

91. I'm considering the compound inequality $x < 8$ or $x \ge 8$, so the solution set is $(-\infty, \infty)$.

In Exercises 92–95, determine whether each statement is true or false. If the statement is false, make the necessary change(s) to produce a true statement.

92. $(-\infty, -1] \cap [-4, \infty) = [-4, -1]$

93. $(-\infty, 3) \cup (-\infty, -2) = (-\infty, -2)$

94. The union of two sets can never give the same result as the intersection of those same two sets.

95. The solution set of the compound inequality $x < a$ and $x > a$ is the set of all real numbers excluding a.

96. Solve and express the solution set in interval notation: $-7 \le 8 - 3x \le 20$ and $-7 < 6x - 1 < 41$.

The graphs of $f(x) = \sqrt{4 - x}$ and $g(x) = \sqrt{x + 1}$ are shown in a $[-3, 10, 1]$ by $[-2, 5, 1]$ viewing rectangle.

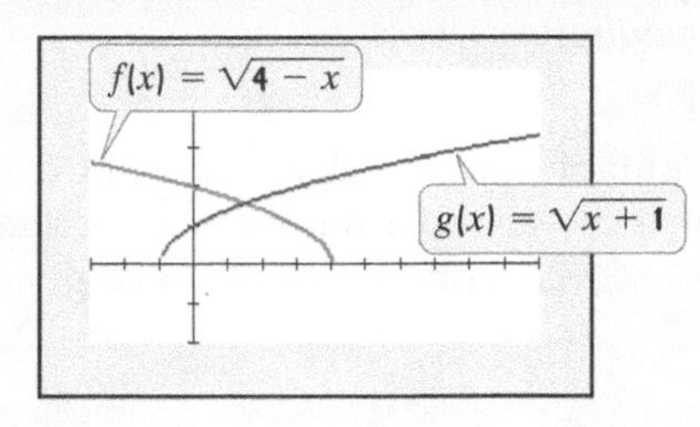

In Exercises 97–100, use the graphs and interval notation to express the domain of the given function.

97. The domain of f

98. The domain of g

99. The domain of $f + g$

100. The domain of $\frac{f}{g}$

101. At the end of the day, the change machine at a laundrette contained at least \$3.20 and at most \$5.45 in nickels, dimes, and quarters. There were 3 fewer dimes than twice the number of nickels and 2 more quarters than twice the number of nickels. What was the least possible number and the greatest possible number of nickels?

Review Exercises

102. If $f(x) = x^2 - 3x + 4$ and $g(x) = 2x - 5$, find $(g - f)(x)$ and $(g - f)(-1)$. (Section 2.3, Example 4)

103. Use function notation to write the equation of the line passing through $(4, 2)$ and perpendicular to the line whose equation is $4x - 2y = 8$. (Section 2.5, Example 5)

104. Simplify: $4 - [2(x - 4) - 5]$. (Section 1.2, Example 14)

Preview Exercises

Exercises 105–107 will help you prepare for the material covered in the next section.

105. Find all values of x satisfying $1 - 4x = 3$ or $1 - 4x = -3$.

106. Find all values of x satisfying $3x - 1 = x + 5$ or $3x - 1 = -(x + 5)$.

107. **a.** Substitute -5 for x and determine whether -5 satisfies $|2x + 3| \geq 5$.

b. Does 0 satisfy $|2x + 3| \geq 5$?

SECTION

4.3 Equations and Inequalities Involving Absolute Value

What am I supposed to learn?

After studying this section, you should be able to:

1. Solve absolute value equations.
2. Solve absolute value inequalities of the form $|u| < c$.
3. Solve absolute value inequalities of the form $|u| > c$.
4. Recognize absolute value inequalities with no solution or all real numbers as solutions.
5. Solve problems using absolute value inequalities.

What activities do you dread? Reading math textbooks with bottle-cap designs painted on brick walls on the cover? (Be kind!) No, that's not America's most-dreaded activity. In a random sample of 1000 U.S. adults, 46% of those questioned responded, "Public speaking."

Numerical information, such as the percentage of adults who dread public speaking, is often given with a margin of error. Inequalities involving absolute value are used to describe errors in polling, as well as errors of measurement in manufacturing, engineering, science, and other fields. In this section, you will learn to solve equations and inequalities containing absolute value. With these skills, you will be able to analyze data on the activities U.S. adults say they dread.

1 Solve absolute value equations.

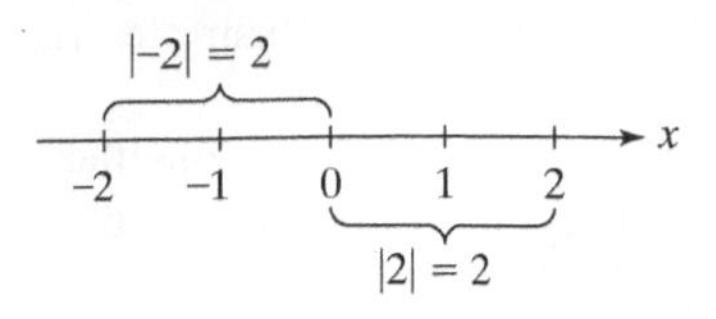

Figure 4.5 If $|x| = 2$, then $x = 2$ or $x = -2$.

Equations Involving Absolute Value

We have seen that the absolute value of a, denoted $|a|$, is the distance from 0 to a on a number line. Now consider **absolute value equations**, such as

$$|x| = 2.$$

This means that we must determine real numbers whose distance from the origin on a number line is 2. **Figure 4.5** shows that there are two numbers such that $|x| = 2$, namely, 2 and -2. We write $x = 2$ or $x = -2$. This observation can be generalized as follows:

Rewriting an Absolute Value Equation Without Absolute Value Bars

If c is a positive real number and u represents any algebraic expression, then $|u| = c$ is equivalent to $u = c$ or $u = -c$.

EXAMPLE 1 Solving an Equation Involving Absolute Value

Solve: $|2x - 3| = 11$.

Solution

$|2x - 3| = 11$ This is the given equation.

$2x - 3 = 11$ or $2x - 3 = -11$ Rewrite the equation without absolute value bars: $|u| = c$ is equivalent to $u = c$ or $u = -c$. In this case $u = 2x - 3$.

$2x = 14$ $2x = -8$ Add 3 to both sides of each equation.

$x = 7$ $x = -4$ Divide both sides of each equation by 2.

Check 7:	Check −4:	
$\|2x - 3\| = 11$	$\|2x - 3\| = 11$	This is the original equation.
$\|2(7) - 3\| \stackrel{?}{=} 11$	$\|2(-4) - 3\| \stackrel{?}{=} 11$	Substitute the proposed solutions.
$\|14 - 3\| \stackrel{?}{=} 11$	$\|-8 - 3\| \stackrel{?}{=} 11$	Perform operations inside the absolute value bars.
$\|11\| \stackrel{?}{=} 11$	$\|-11\| \stackrel{?}{=} 11$	
$11 = 11$, true	$11 = 11$, true	These true statements indicate that 7 and −4 are solutions.

The solutions are -4 and 7. We can also say that the solution set is $\{-4, 7\}$. ■

Using Technology

Graphic Connections

You can use a graphing utility to verify the solution set of an absolute value equation. Consider, for example,

$$|2x - 3| = 11.$$

Graph $y_1 = |2x - 3|$ and $y_2 = 11$. The graphs are shown in a $[-10, 10, 1]$ by $[-1, 15, 1]$ viewing rectangle. The x-coordinates of the intersection points are -4 and 7, verifying that $\{-4, 7\}$ is the solution set.

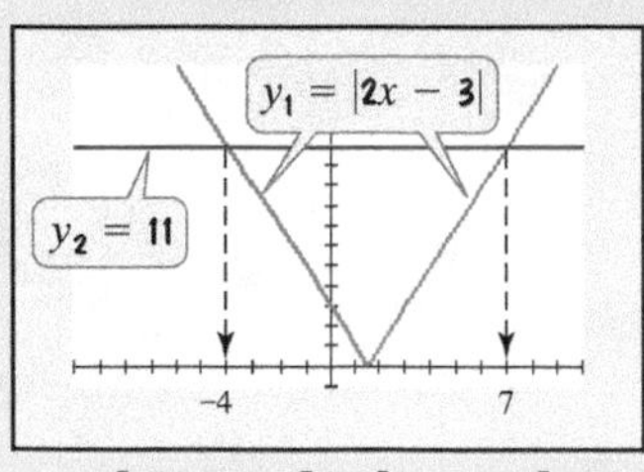

$[-10, 10, 1]$ by $[-1, 15, 1]$

✓ CHECK POINT 1 Solve: $|2x - 1| = 5$.

EXAMPLE 2 Solving an Equation Involving Absolute Value

Solve: $5|1 - 4x| - 15 = 0$.

Solution

$5|1 - 4x| - 15 = 0$ This is the given equation.

We need to isolate $|1 - 4x|$, the absolute value expression.

$5|1 - 4x| = 15$ Add 15 to both sides.

$|1 - 4x| = 3$ Divide both sides by 5.

$1 - 4x = 3$ or $1 - 4x = -3$ Rewrite $|u| = c$ as $u = c$ or $u = -c$.

$-4x = 2$ $-4x = -4$ Subtract 1 from both sides of each equation.

$x = -\frac{1}{2}$ $x = 1$ Divide both sides of each equation by -4.

Take a moment to check $-\frac{1}{2}$ and 1, the proposed solutions, in the original equation, $5|1 - 4x| - 15 = 0$. In each case, you should obtain the true statement $0 = 0$. The solutions are $-\frac{1}{2}$ and 1, and the solution set is $\left\{-\frac{1}{2}, 1\right\}$. ■

CHECK POINT 2 Solve: $2|1 - 3x| - 28 = 0$.

The absolute value of a number is never negative. Thus, if u is an algebraic expression and c is a negative number, then $|u| = c$ has no solution. For example, the equation $|3x - 6| = -2$ has no solution because $|3x - 6|$ cannot be negative. The solution set is $\varnothing$, the empty set.

The absolute value of 0 is 0. Thus, if u is an algebraic expression and $|u| = 0$, the solution is found by solving $u = 0$. For example, the solution of $|x - 2| = 0$ is obtained by solving $x - 2 = 0$. The solution is 2 and the solution set is {2}.

Some equations have two absolute value expressions, such as

$$|3x - 1| = |x + 5|.$$

These absolute value expressions are equal when the expressions inside the absolute value bars are equal to or opposites of each other.

Rewriting an Absolute Value Equation with Two Absolute Values without Absolute Value Bars

If $|u| = |v|$, then $u = v$ or $u = -v$.

EXAMPLE 3 Solving an Absolute Value Equation with Two Absolute Values

Solve: $|3x - 1| = |x + 5|$.

Solution We rewrite the equation without absolute value bars.

$$|u| = |v| \quad \text{means} \quad u = v \quad \text{or} \quad u = -v$$

$$|3x - 1| = |x + 5| \quad \text{means} \quad 3x - 1 = x + 5 \quad \text{or} \quad 3x - 1 = -(x + 5).$$

We now solve the two equations that do not contain absolute value bars.

$$\begin{aligned} 3x - 1 &= x + 5 & \text{or} \quad 3x - 1 &= -(x + 5) \\ 2x - 1 &= 5 & 3x - 1 &= -x - 5 \\ 2x &= 6 & 4x - 1 &= -5 \\ x &= 3 & 4x &= -4 \\ & & x &= -1 \end{aligned}$$

Take a moment to complete the solution process by checking the two proposed solutions in the original equation. The solutions are -1 and 3, and the solution set is $\{-1, 3\}$. ■

CHECK POINT 3 Solve: $|2x - 7| = |x + 3|$.

---

2 Solve absolute value inequalities of the form $|u| < c$.

Inequalities Involving Absolute Value

Absolute value can also arise in inequalities. Consider, for example,

$$|x| < 2.$$

This means that the distance from 0 to x on a number line is less than 2, as shown in **Figure 4.6**.

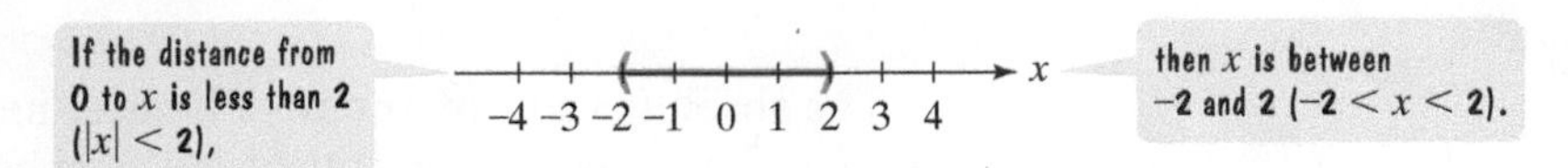

Figure 4.6

Generalizing from the observations in the voice balloons gives us a method for solving inequalities of the form $|u| < c$, where c is a positive number. This method involves rewriting the given inequality without absolute value bars.

Solving Absolute Value Inequalities of the Form $|u| < c$

If c is a positive real number and u represents any algebraic expression, then

$$|u| < c \text{ is equivalent to } -c < u < c.$$

This rule is valid if $<$ is replaced by $\leq$.

EXAMPLE 4 Solving an Absolute Value Inequality of the Form $|u| < c$

Solve and graph the solution set on a number line:

$$|x - 4| < 3.$$

Solution We rewrite the inequality without absolute value bars.

$$|u| < c \quad \text{means} \quad -c < u < c.$$

$$|x - 4| < 3 \quad \text{means} \quad -3 < x - 4 < 3.$$

We solve the compound inequality by adding 4 to all three parts.

$$-3 < x - 4 < 3$$
$$-3 + 4 < x - 4 + 4 < 3 + 4$$
$$1 < x < 7$$

The solution set is all real numbers greater than 1 and less than 7, denoted in interval notation by $(1, 7)$. The graph of the solution set is shown as follows:

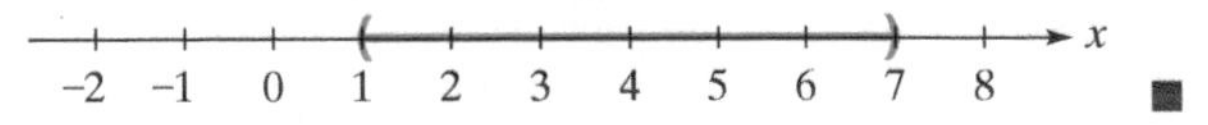

■

We can use the rectangular coordinate system to visualize the solution set of

$$|x - 4| < 3.$$

Figure 4.7 shows the graphs of $f(x) = |x - 4|$ and $g(x) = 3$. The solution set of $|x - 4| < 3$ consists of all values of x for which the blue graph of f lies below the red graph of g. These x-values make up the interval $(1, 7)$, which is the solution set.

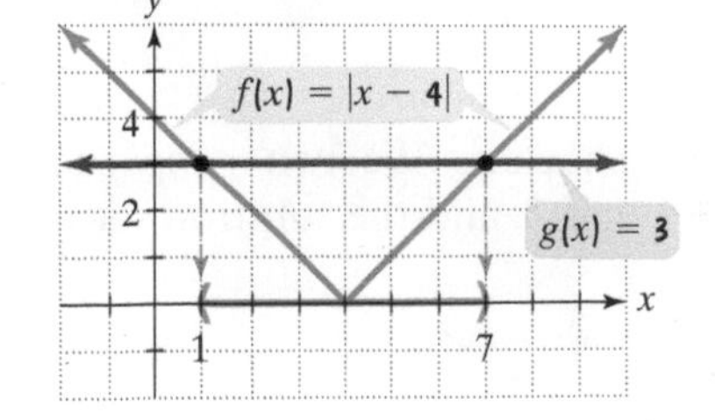

Figure 4.7 The solution set of $|x - 4| < 3$ is $(1, 7)$.

✓ CHECK POINT 4 Solve and graph the solution set on a number line: $|x - 2| < 5$.

Before rewriting an absolute value inequality without absolute value bars, isolate the absolute value expression on one side of the inequality.

EXAMPLE 5 Solving an Absolute Value Inequality

Solve and graph the solution set on a number line: $-2|3x + 5| + 7 \geq -13$.

Solution

$-2|3x + 5| + 7 \geq -13$ This is the given inequality.

We need to isolate $|3x + 5|$, the absolute value expression.

$-2|3x + 5| + 7 - 7 \geq -13 - 7$ Subtract 7 from both sides.

$-2|3x + 5| \geq -20$ Simplify.

$\frac{-2|3x + 5|}{-2} \leq \frac{-20}{-2}$ Divide both sides by -2 and change the sense of the inequality.

$|3x + 5| \leq 10$ Simplify.

$-10 \leq 3x + 5 \leq 10$ Rewrite without absolute value bars: $|u| \leq c$ means $-c \leq u \leq c$.

Now we need to isolate x in the middle.

$-10 - 5 \leq 3x + 5 - 5 \leq 10 - 5$ Subtract 5 from all three parts.

$-15 \leq 3x \leq 5$ Simplify.

$\frac{-15}{3} \leq \frac{3x}{3} \leq \frac{5}{3}$ Divide each part by 3.

$-5 \leq x \leq \frac{5}{3}$ Simplify.

The solution set is $\left[-5, \frac{5}{3}\right]$ in interval notation. The graph is shown as follows:

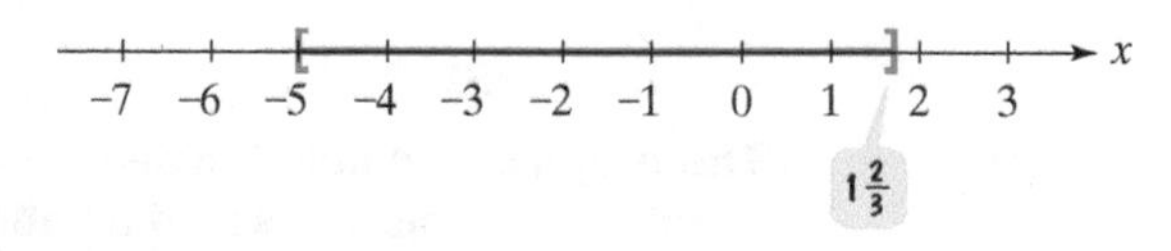

■

CHECK POINT 5 Solve and graph the solution set on a number line: $-3|5x - 2| + 20 \geq -19$.

---

3 Solve absolute value inequalities of the form $|u| > c$.

Now let's consider absolute value inequalities with greater than symbols, such as

$$|x| > 2.$$

This means that the distance from 0 to x on a number line is greater than 2, as shown in **Figure 4.8**.

If the distance from 0 to x is greater than 2 ($|x| > 2$),
-4 -3 -2 -1 0 1 2 3 4
x
then x is less than -2 or greater than 2 ($x < -2$ or $x > 2$).

Figure 4.8

Generalizing from the observations in the voice balloons gives us a method for solving inequalities of the form $|u| > c$, where c is a positive number. This method once again involves rewriting the given inequality without absolute value bars.

Solving Absolute Value Inequalities of the Form $|u| > c$

If c is a positive real number and u represents any algebraic expression, then

$$|u| > c \text{ is equivalent to } u < -c \text{ or } u > c.$$

This rule is valid if $>$ is replaced by $\geq$.

EXAMPLE 6 Solving an Absolute Value Inequality of the Form $|u| \geq c$

Solve and graph the solution set on a number line:

$$|2x + 3| \geq 5.$$

Solution We rewrite the inequality without absolute value bars.

$$|u| \geq c \quad \text{means} \quad u \leq -c \quad \text{or} \quad u \geq c.$$

$$|2x + 3| \geq 5 \quad \text{means} \quad 2x + 3 \leq -5 \quad \text{or} \quad 2x + 3 \geq 5.$$

We solve this compound inequality by solving each of these inequalities separately. Then we take the union of their solution sets.

$2x + 3 \leq -5$	or	$2x + 3 \geq 5$	These are the inequalities without absolute value bars.
$2x \leq -8$		$2x \geq 2$	Subtract 3 from both sides.
$x \leq -4$		$x \geq 1$	Divide both sides by 2.

The solution set consists of all numbers that are less than or equal to -4 or greater than or equal to 1. The solution set in interval notation is $(-\infty, -4] \cup [1, \infty)$. The graph of the solution set is shown as follows:

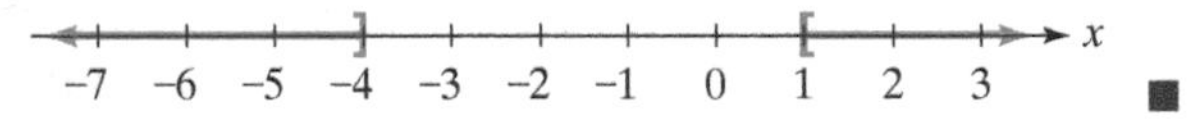

■

Great Question!

The graph of the solution set in Example 6 consists of two intervals. When is the graph of the solution set of an absolute value inequality a single interval and when is it divided into two intervals?

If u is a linear expression and $c > 0$, the graph of the solution set for $|u| > c$ will be divided into two intervals whose union cannot be represented as a single interval. The graph of the solution set for $|u| < c$ will be a single interval. Avoid the common error of rewriting $|u| > c$ as $-c < u > c$.

✓ CHECK POINT 6 Solve and graph the solution set on a number line: $|2x - 5| \geq 3$.

Great Question!

Please cut to the chase. What do I need to know when rewriting absolute value equations and inequalities without absolute value bars?

Here's a brief summary. If $c > 0$,

- $|u| = c$ is equivalent to $u = c$ or $u = -c$.
- $|u| < c$ is equivalent to $-c < u < c$.
- $|u| > c$ is equivalent to $u < -c$ or $u > c$.

4 Recognize absolute value inequalities with no solution or all real numbers as solutions.

Absolute Value Inequalities with Unusual Solution Sets

We have been working with $|u| < c$ and $|u| > c$, where c is a positive number. Now let's see what happens to these inequalities if c is a negative number. Consider, for example, $|x| < -2$. Because $|x|$ always has a value that is greater than or equal to 0, there is no number whose absolute value is less than -2. The inequality $|x| < -2$ has no solution. The solution set is $\varnothing$.

Now consider the inequality $|x| > -2$. Because $|x|$ is never negative, all numbers have an absolute value greater than -2. All real numbers satisfy the inequality $|x| > -2$. The solution set is $(-\infty, \infty)$.

Absolute Value Inequalities with Unusual Solution Sets

If u is an algebraic expression and c is a negative number,

1. The inequality $|u| < c$ has no solution.
2. The inequality $|u| > c$ is true for all real numbers for which u is defined.

5 Solve problems using absolute value inequalities.

Applications

We opened this section with this question:

What activities do you dread?

In a random sample of 1000 U.S. adults, 46% of those questioned responded, "Public speaking." The problem is that this is a single random sample. Do 46% of adults in the entire U.S. population dread public speaking?

If you look at the results of a poll like the one in **Figure 4.9**, you will observe that a margin of error is reported.

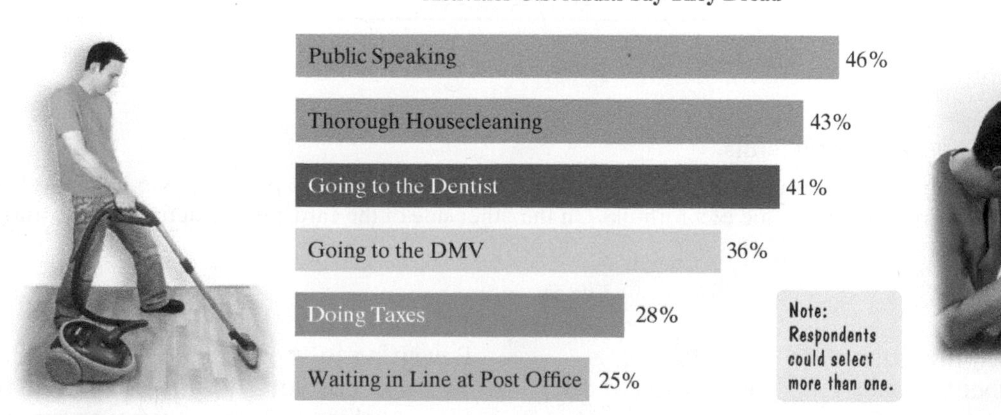

Figure 4.9

Source: TNS survey of 1000 adults, March 2010, Margin of error: ±3.2%

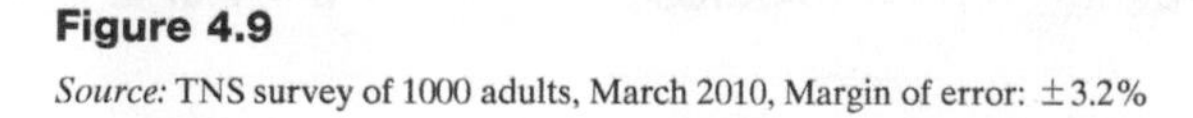

The margin of error in the dreaded-activities poll is ±3.2%. This means that the actual percentage of U.S. adults who dread public speaking is at most 3.2% greater than or less than 46%. If x represents the percentage of U.S. adults in the population who dread public speaking, then the poll's margin of error can be expressed as an absolute value inequality:

$$|x - 46| \leq 3.2.$$

EXAMPLE 7 Analyzing a Poll's Margin of Error

The inequality

$$|x - 28| \leq 3.2$$

describes the percentage of U.S. adults in the population who dread doing taxes. Solve the inequality and interpret the solution.

Solution We rewrite the inequality without absolute value bars.

|u| ≤ c
means
−c ≤ u ≤ c.

$$|x - 28| \leq 3.2 \quad \text{means} \quad -3.2 \leq x - 28 \leq 3.2.$$

We solve the compound inequality by adding 28 to all three parts.

$$-3.2 \leq x - 28 \leq 3.2$$ This is $|x - 28| \leq 3.2$ without absolute value bars.

$$-3.2 + 28 \leq x - 28 + 28 \leq 3.2 + 28$$ Add 28 to all three parts.

$$24.8 \leq x \leq 31.2$$ Simplify. The solution is [24.8, 31.2].

The percentage of U.S. adults in the population who dread doing taxes is somewhere between a low of 24.8% and a high of 31.2%. Notice that these percents are 3.2% above and below the given 28% in **Figure 4.9** on the previous page, and that 3.2% is the poll's margin of error. ■

✓ CHECK POINT 7 Solve the inequality:

$$|x - 41| \leq 3.2.$$

Interpret the solution in terms of the information in **Figure 4.9** on the previous page.

Achieving Success

Use index cards to help learn new terms.

Many of the terms, notations, and formulas used in this book will be new to you. Buy a pack of 3 × 5 index cards. On each card, list a new vocabulary word, symbol, or title of a formula. On the other side of the card, put the definition or formula. Here are four examples:

Effective Index Cards

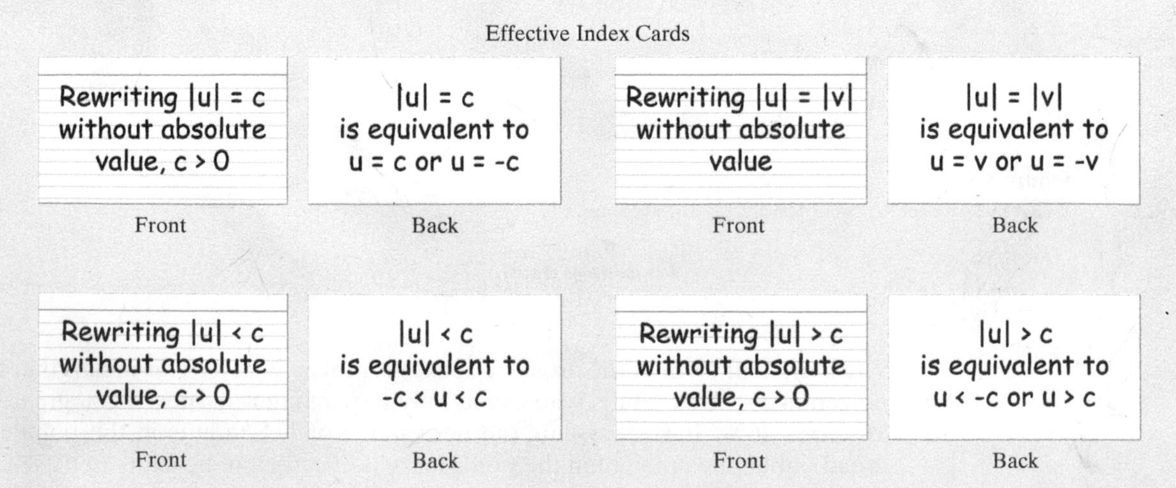

Rewriting |u| = c without absolute value, c > 0
|u| = c is equivalent to u = c or u = -c
Rewriting |u| = |v| without absolute value
|u| = |v| is equivalent to u = v or u = -v
Front
Back
Front
Back
Rewriting |u| < c without absolute value, c > 0
|u| < c is equivalent to -c < u < c
Rewriting |u| > c without absolute value, c > 0
|u| > c is equivalent to u < -c or u > c
Front
Back
Front
Back

Review these cards frequently. Use the cards to quiz yourself and prepare for exams.

CONCEPT AND VOCABULARY CHECK

Fill in each blank so that the resulting statement is true.

1. If $c > 0$, $|u| = c$ is equivalent to $u =$ __________ or $u =$ __________.
2. $|u| = |v|$ is equivalent to $u =$ __________ or $u =$ __________.
3. If $c > 0$, $|u| < c$ is equivalent to __________ $< u <$ __________.
4. If $c > 0$, $|u| > c$ is equivalent to $u <$ __________ or $u >$ __________.
5. $|u| < c$ has no solution if c __________ 0.
6. $|u| > c$ is true for all real numbers if c __________ 0.

Match each absolute value equation or inequality in the left column with an equivalent statement from the right column.

7. $|3x - 1| = 5$
8. $|3x - 1| \geq 5$
9. $|3x - 1| \leq 5$
10. $|3x - 1| = |x|$
11. $|3x - 1| < -5$
12. $|3x - 1| > -5$

A. $-5 \leq 3x - 1 \leq 5$
B. $3x - 1 = x$ or $3x - 1 = -x$
C. $3x - 1 = 5$ or $3x - 1 = -5$
D. $\varnothing$
E. $3x - 1 \leq -5$ or $3x - 1 \geq 5$
F. $(-\infty, \infty)$

4.3 EXERCISE SET

MyMathLab®

Practice Exercises

In Exercises 1–38, find the solution set for each equation.

1. $|x| = 8$
2. $|x| = 6$
3. $|x - 2| = 7$
4. $|x + 1| = 5$
5. $|2x - 1| = 7$
6. $|2x - 3| = 11$
7. $\left|\frac{4x - 2}{3}\right| = 2$
8. $\left|\frac{3x - 1}{5}\right| = 1$
9. $|x| = -8$
10. $|x| = -6$
11. $|x + 3| = 0$
12. $|x + 2| = 0$
13. $2|y + 6| = 10$
14. $3|y + 5| = 12$
15. $3|2x - 1| = 21$
16. $2|3x - 2| = 14$
17. $|6y - 2| + 4 = 32$
18. $|3y - 1| + 10 = 25$
19. $7|5x| + 2 = 16$
20. $7|3x| + 2 = 16$
21. $|x + 1| + 5 = 3$
22. $|x + 1| + 6 = 2$
23. $|4y + 1| + 10 = 4$
24. $|3y - 2| + 8 = 1$
25. $|2x - 1| + 3 = 3$
26. $|3x - 2| + 4 = 4$
27. $|5x - 8| = |3x + 2|$
28. $|4x - 9| = |2x + 1|$
29. $|2x - 4| = |x - 1|$
30. $|6x| = |3x - 9|$
31. $|2x - 5| = |2x + 5|$
32. $|3x - 5| = |3x + 5|$
33. $|x - 3| = |5 - x|$
34. $|x - 3| = |6 - x|$
35. $|2y - 6| = |10 - 2y|$
36. $|4y + 3| = |4y + 5|$
37. $\left|\frac{2x}{3} - 2\right| = \left|\frac{x}{3} + 3\right|$
38. $\left|\frac{x}{2} - 2\right| = \left|x - \frac{1}{2}\right|$

In Exercises 39–74, solve and graph the solution set on a number line.

39. $|x| < 3$
40. $|x| < 5$
41. $|x - 2| < 1$
42. $|x - 1| < 5$
43. $|x + 2| \leq 1$
44. $|x + 1| \leq 5$
45. $|2x - 6| < 8$
46. $|3x + 5| < 17$
47. $|x| > 3$
48. $|x| > 5$
49. $|x + 3| > 1$
50. $|x - 2| > 5$
51. $|x - 4| \geq 2$
52. $|x - 3| \geq 4$
53. $|3x - 8| > 7$
54. $|5x - 2| > 13$
55. $|2(x - 1) + 4| \leq 8$
56. $|3(x - 1) + 2| \leq 20$

57. $\left|\frac{2x+6}{3}\right| < 2$

58. $\left|\frac{3x-3}{4}\right| < 6$

59. $\left|\frac{2x+2}{4}\right| \geq 2$

60. $\left|\frac{3x-3}{9}\right| \geq 1$

61. $\left|3 - \frac{2x}{3}\right| > 5$

62. $\left|3 - \frac{3x}{4}\right| > 9$

63. $|x - 2| < -1$

64. $|x - 3| < -2$

65. $|x + 6| > -10$

66. $|x + 4| > -12$

67. $|x + 2| + 9 \leq 16$

68. $|x - 2| + 4 \leq 5$

69. $2|2x - 3| + 10 > 12$

70. $3|2x - 1| + 2 > 8$

71. $-4|1 - x| < -16$

72. $-2|5 - x| < -6$

73. $3 \leq |2x - 1|$

74. $9 \leq |4x + 7|$

75. Let $f(x) = |5 - 4x|$. Find all values of x for which $f(x) = 11$.

76. Let $f(x) = |2 - 3x|$. Find all values of x for which $f(x) = 13$.

77. Let $f(x) = |3 - x|$ and $g(x) = |3x + 11|$. Find all values of x for which $f(x) = g(x)$.

78. Let $f(x) = |3x + 1|$ and $g(x) = |6x - 2|$. Find all values of x for which $f(x) = g(x)$.

79. Let $g(x) = |-1 + 3(x + 1)|$. Find all values of x for which $g(x) \leq 5$.

80. Let $g(x) = |-3 + 4(x + 1)|$. Find all values of x for which $g(x) \leq 3$.

81. Let $h(x) = |2x - 3| + 1$. Find all values of x for which $h(x) > 6$.

82. Let $h(x) = |2x - 4| - 6$. Find all values of x for which $h(x) > 18$.

Practice PLUS

83. When 3 times a number is subtracted from 4, the absolute value of the difference is at least 5. Use interval notation to express the set of all real numbers that satisfy this condition.

84. When 4 times a number is subtracted from 5, the absolute value of the difference is at most 13. Use interval notation to express the set of all real numbers that satisfy this condition.

In Exercises 85–86, solve each inequality. Assume that $a > 0$ and $c > 0$.

85. $|ax + b| < c$

86. $|ax + b| \geq c$

In Exercises 87–88, use the graph of $f(x) = |4 - x|$ to solve each equation or inequality.

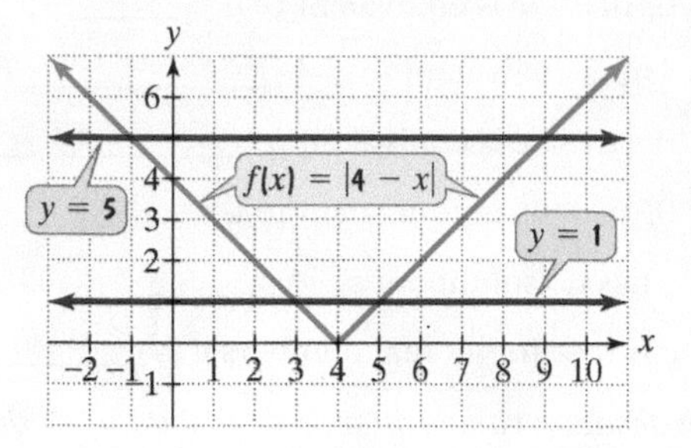

87. $|4 - x| = 1$

88. $|4 - x| < 5$

In Exercises 89–90, use the table to solve each inequality.

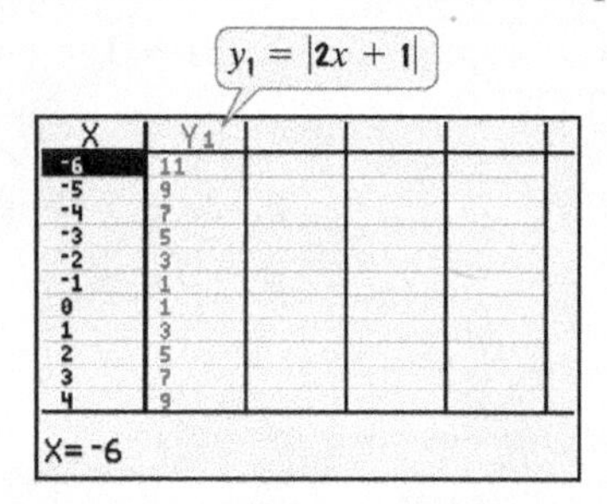

89. $|2x + 1| \leq 3$

90. $|2x + 1| \geq 3$

Application Exercises

How to Blow Your Job Interview. *The data in the bar graph are from a random survey of 1910 job interviewers. The graph shows the top interviewer turnoffs and the percentage of surveyed interviewers who were offended by each of these behaviors. In Exercises 91–92, let x represent the actual percentage of interviewers in the entire population of job interviewers.*

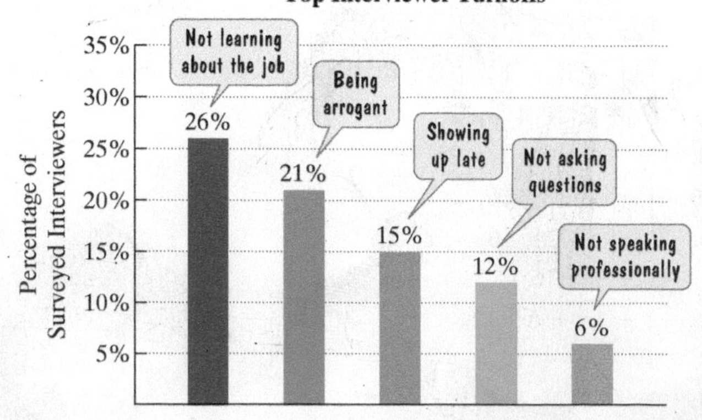

Source: Scott Erker, PhD., and Kelli Buczynski, "Are You Failing the Interview? 2009 Survey of Global Interviewing Practices and Perceptions." Development Dimensions International.

91. Solve the inequality: $|x - 21| \leq 3$. Interpret the solution in terms of the information in the graph. What is the margin of error?

92. Solve the inequality: $|x - 15| \le 3$. Interpret the solution in terms of the information in the graph at the bottom of the previous page. What is the margin of error?

93. The inequality $|T - 57| \le 7$ describes the range of monthly average temperature, T, in degrees Fahrenheit, for San Francisco, California. Solve the inequality and interpret the solution.

94. The inequality $|T - 50| \le 22$ describes the range of monthly average temperature, T, in degrees Fahrenheit, for Albany, New York. Solve the inequality and interpret the solution.

The specifications for machine parts are given with tolerance limits that describe a range of measurements for which the part is acceptable. In Exercises 95–96, x represents the length of a machine part, in centimeters. The tolerance limit is 0.01 centimeter.

95. Solve: $|x - 8.6| \le 0.01$. If the length of the machine part is supposed to be 8.6 centimeters, interpret the solution.

96. Solve: $|x - 9.4| \le 0.01$. If the length of the machine part is supposed to be 9.4 centimeters, interpret the solution.

97. If a coin is tossed 100 times, we would expect approximately 50 of the outcomes to be heads. It can be demonstrated that a coin is unfair if h, the number of outcomes that result in heads, satisfies $\left|\frac{h - 50}{5}\right| \ge 1.645$. Describe the number of outcomes that result in heads that determine an unfair coin that is tossed 100 times.

Explaining the Concepts

98. Explain how to solve an equation containing one absolute value expression.

99. Explain why the procedure that you described in Exercise 98 does not apply to the equation $|x - 5| = -3$. What is the solution set of this equation?

100. Describe how to solve an absolute value equation with two absolute values.

101. Describe how to solve an absolute value inequality of the form $|u| < c$ for $c > 0$.

102. Explain why the procedure that you described in Exercise 101 does not apply to the inequality $|x - 5| < -3$. What is the solution set of this inequality?

103. Describe how to solve an absolute value inequality of the form $|u| > c$ for $c > 0$.

104. Explain why the procedure that you described in Exercise 103 does not apply to the inequality $|x - 5| > -3$. What is the solution set of this inequality?

Technology Exercises

In Exercises 105–107, solve each equation using a graphing utility. Graph each side separately in the same viewing rectangle. The solutions are the x-coordinates of the intersection points.

105. $|x + 1| = 5$

106. $|3(x + 4)| = 12$

107. $|2x - 3| = |9 - 4x|$

In Exercises 108–110, solve each inequality using a graphing utility. Graph each side separately in the same viewing rectangle. The solution set consists of all values of x for which the graph of the left side lies below the graph of the right side.

108. $|2x + 3| < 5$

109. $\left|\frac{2x - 1}{3}\right| < \frac{5}{3}$

110. $|x + 4| < -1$

In Exercises 111–113, solve each inequality using a graphing utility. Graph each side separately in the same viewing rectangle. The solution set consists of all values of x for which the graph of the left side lies above the graph of the right side.

111. $|2x - 1| > 7$

112. $|0.1x - 0.4| + 0.4 > 0.6$

113. $|x + 4| > -1$

114. Use a graphing utility to verify the solution sets for any five equations or inequalities that you solved by hand in Exercises 1–74.

Critical Thinking Exercises

Make Sense? *In Exercises 115–118, determine whether each statement makes sense or does not make sense, and explain your reasoning.*

115. I solved $|x - 2| = 5$ by rewriting the equation as $x - 2 = 5$ or $x + 2 = 5$.

116. I solved $|x - 2| > 5$ by rewriting the inequality as $-5 < x - 2 > 5$.

117. Because the absolute value of any expression is never less than a negative number, I can immediately conclude that the inequality $|2x - 5| - 9 < -4$ has no solution.

118. I'll win the contest if I can complete the crossword puzzle in 20 minutes plus or minus 5 minutes, so my winning time, x, is modeled by $|x - 20| \le 5$.

In Exercises 119–122, determine whether each statement is true or false. If the statement is false, make the necessary change(s) to produce a true statement.

119. All absolute value equations have two solutions.

120. The equation $|x| = -6$ is equivalent to $x = 6$ or $x = -6$.

121. Values of -5 and 5 satisfy $|x| = 5$, $|x| \le 5$, and $|x| \ge -5$.

122. The absolute value of any linear expression is greater than 0 for all real numbers except the number for which the expression is equal to 0.

123. Write an absolute value inequality for which the interval shown is the solution.

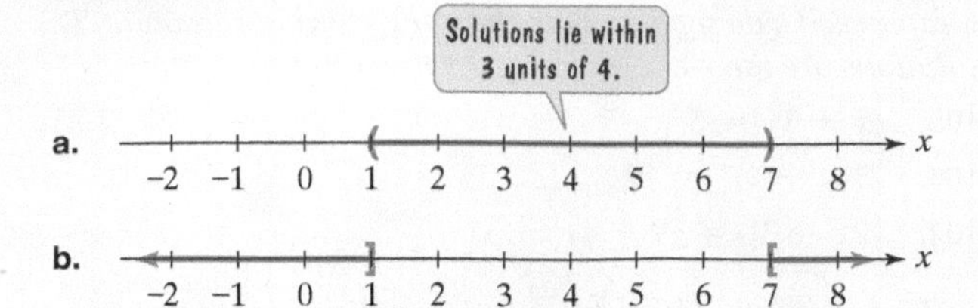

124. The percentage, p, of defective products manufactured by a company is given by $|p - 0.3\%| \le 0.2\%$. If 100,000 products are manufactured and the company offers a \$5 refund for each defective product, describe the company's cost for refunds.

125. Solve: $|2x + 5| = 3x + 4$.

Review and Preview Exercises

Exercises 126–128 will enable you to review graphing linear functions. In addition, they will help you prepare for the material covered in the next section. In each exercise, graph the linear function.

126. $3x - 5y = 15$ (Section 2.4, Example 1)

127. $f(x) = -\frac{2}{3}x$ (Section 2.4, Example 5)

128. $f(x) = -2$ (Section 2.4, Example 6)

MID-CHAPTER CHECK POINT Section 4.1–Section 4.3

What You Know: We learned to solve linear inequalities, expressing solution sets in interval notation. We know that it is necessary to change the sense of an inequality when multiplying or dividing both sides by a negative number. We solved compound inequalities with *and* by finding the intersection of solution sets and with *or* by finding the union of solution sets. Finally, we solved equations and inequalities involving absolute value by carefully rewriting the given equation or inequality without absolute value bars. For positive values of c, we wrote $|u| = c$ as $u = c$ or $u = -c$. We wrote $|u| < c$ as $-c < u < c$, and we wrote $|u| > c$ as $u < -c$ or $u > c$.

In Exercises 1–18, solve each inequality or equation.

1. $4 - 3x \ge 12 - x$
2. $5 \le 2x - 1 < 9$
3. $|4x - 7| = 5$
4. $-10 - 3(2x + 1) > 8x + 1$
5. $2x + 7 < -11$ or $-3x - 2 < 13$
6. $|3x - 2| \le 4$
7. $|x + 5| = |5x - 8|$
8. $5 - 2x \ge 9$ and $5x + 3 > -17$
9. $3x - 2 > -8$ or $2x + 1 < 9$
10. $\frac{x}{2} + 3 \le \frac{x}{3} + \frac{5}{2}$
11. $\frac{2}{3}(6x - 9) + 4 > 5x + 1$
12. $|5x + 3| > 2$
13. $7 - \left|\frac{x}{2} + 2\right| \le 4$
14. $5(x - 2) - 3(x + 4) \ge 2x - 20$
15. $\frac{x + 3}{4} < \frac{1}{3}$
16. $5x + 1 \ge 4x - 2$ and $2x - 3 > 5$
17. $3 - |2x - 5| = -6$
18. $3 + |2x - 5| = -6$

In Exercises 19–22, solve each problem.

19. A car rental agency rents a certain car for \$40 per day with unlimited mileage or \$24 per day plus \$0.20 per mile. How far can a customer drive this car per day for the \$24 option to cost no more than the unlimited mileage option?

20. To receive a B in a course, you must have an average of at least 80% but less than 90% on five exams. Your grades on the first four exams were 95%, 79%, 91%, and 86%. What range of grades on the fifth exam will result in a B for the course?

21. A retiree requires an annual income of at least \$9000 from an investment paying 7.5% annual interest. How much should the retiree invest to achieve the desired return?

22. A company that manufactures compact discs has fixed monthly overhead costs of \$60,000. Each disc costs \$0.18 to produce and sells for \$0.30. How many discs should be produced and sold each month for the company to have a profit of at least \$30,000?

SECTION

4.4 Linear Inequalities in Two Variables 

What am I supposed to learn?

After studying this section, you should be able to:

1. Graph a linear inequality in two variables.
2. Use mathematical models involving linear inequalities.
3. Graph a system of linear inequalities.

This book was written in Point Reyes National Seashore, 40 miles north of San Francisco. The park consists of 75,000 acres with miles of pristine surf-washed beaches, forested ridges, and bays bordered by white cliffs.

Like your author, many people are kept inspired and energized surrounded by nature's unspoiled beauty. In this section, you will see how systems of inequalities model whether a region's natural beauty manifests itself in forests, grasslands, or deserts.

Linear Inequalities in Two Variables and Their Solutions

We have seen that equations in the form $Ax + By = C$ are straight lines when graphed. If we change the symbol $=$ to $>$, $<$, $\geq$, or $\leq$, we obtain a **linear inequality in two variables**. Some examples of linear inequalities in two variables are $x + y > 2$, $3x - 5y \leq 15$, and $2x - y < 4$.

A **solution of an inequality in two variables**, x and y, is an ordered pair of real numbers with the following property: When the x-coordinate is substituted for x and the y-coordinate is substituted for y in the inequality, we obtain a true statement. For example, (3, 2) is a solution of the inequality $x + y > 1$. When 3 is substituted for x and 2 is substituted for y, we obtain the true statement $3 + 2 > 1$, or $5 > 1$. Because there are infinitely many pairs of numbers that have a sum greater than 1, the inequality $x + y > 1$ has infinitely many solutions. Each ordered-pair solution is said to **satisfy** the inequality. Thus, (3, 2) satisfies the inequality $x + y > 1$.

1 Graph a linear inequality in two variables.

The Graph of a Linear Inequality in Two Variables

We know that the graph of an equation in two variables is the set of all points whose coordinates satisfy the equation. Similarly, the **graph of an inequality in two variables** is the set of all points whose coordinates satisfy the inequality.

Let's use **Figure 4.10** to get an idea of what the graph of a linear inequality in two variables looks like. Part of the figure shows the graph of the linear equation $x + y = 2$. The line divides the points in the rectangular coordinate system into three sets. First, there is the set of points along the line, satisfying $x + y = 2$. Next, there is the set of points in the green region above the line. Points in the green region satisfy the linear inequality $x + y > 2$. Finally, there is the set of points in the purple region below the line. Points in the purple region satisfy the linear inequality $x + y < 2$.

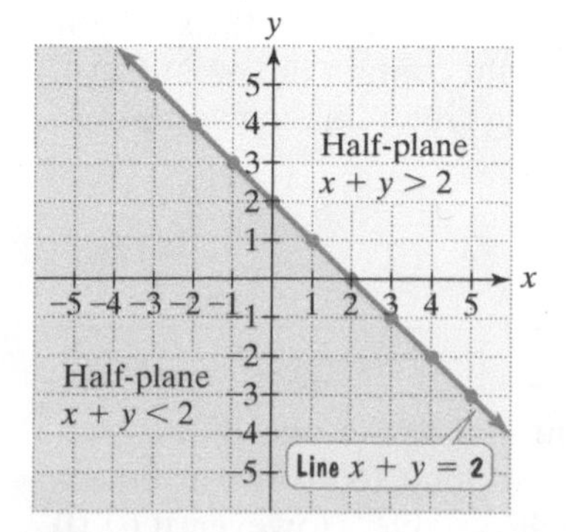

Figure 4.10

A **half-plane** is the set of all the points on one side of a line. In **Figure 4.10**, the green region is a half-plane. The purple region is also a half-plane. A half-plane is the graph of a linear inequality that involves $>$ or $<$. The graph of a linear inequality that involves $\geq$ or $\leq$ is a half-plane and a line. A solid line is used to show that a line is part of a graph. A dashed line is used to show that a line is not part of a graph.

Graphing a Linear Inequality in Two Variables

1. Replace the inequality symbol with an equal sign and graph the corresponding linear equation. Draw a solid line if the original inequality contains a $\le$ or $\ge$ symbol. Draw a dashed line if the original inequality contains a $<$ or $>$ symbol.
2. Choose a test point from one of the half-planes. (Do not choose a point on the line.) Substitute the coordinates of the test point into the inequality.
3. If a true statement results, shade the half-plane containing this test point. If a false statement results, shade the half-plane not containing this test point.

EXAMPLE 1 Graphing a Linear Inequality in Two Variables

Graph: $2x - 3y \ge 6$.

Solution

Step 1. Replace the inequality symbol by = and graph the linear equation. We need to graph $2x - 3y = 6$. We can use intercepts to graph this line.

We set $y = 0$ to find the x-intercept:	**We set $x = 0$ to find the y-intercept:**
$2x - 3y = 6$	$2x - 3y = 6$
$2x - 3 \cdot 0 = 6$	$2 \cdot 0 - 3y = 6$
$2x = 6$	$-3y = 6$
$x = 3.$	$y = -2.$

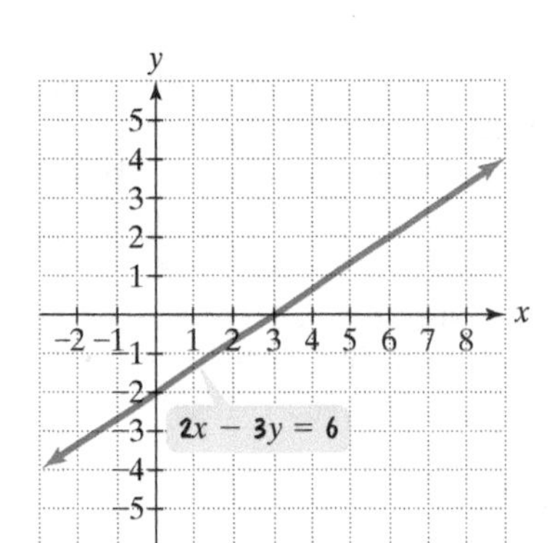

Figure 4.11 Preparing to graph $2x - 3y \ge 6$

The x-intercept is 3, so the line passes through $(3, 0)$. The y-intercept is -2, so the line passes through $(0, -2)$. Using the intercepts, the line is shown in **Figure 4.11** as a solid line. This is because the inequality $2x - 3y \ge 6$ contains a $\ge$ symbol, in which equality is included.

Step 2. Choose a test point from one of the half-planes and not from the line. Substitute its coordinates into the inequality. **Figure 4.11** shows that the line $2x - 3y = 6$ divides the plane into three parts—the line itself and two half-planes. The points in one half-plane satisfy $2x - 3y > 6$. The points in the other half-plane satisfy $2x - 3y < 6$. We need to find which half-plane belongs to the solution of $2x - 3y \ge 6$. To do so, we test a point from either half-plane. The origin, $(0, 0)$, is the easiest point to test.

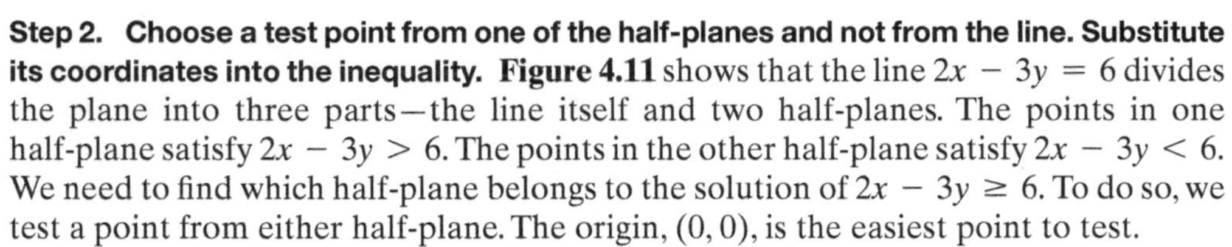

$2x - 3y \ge 6$	This is the given inequality.
$2 \cdot 0 - 3 \cdot 0 \overset{?}{\ge} 6$	Test (0, 0) by substituting 0 for x and 0 for y.
$0 - 0 \overset{?}{\ge} 6$	Multiply.
$0 \ge 6$	This statement is false.

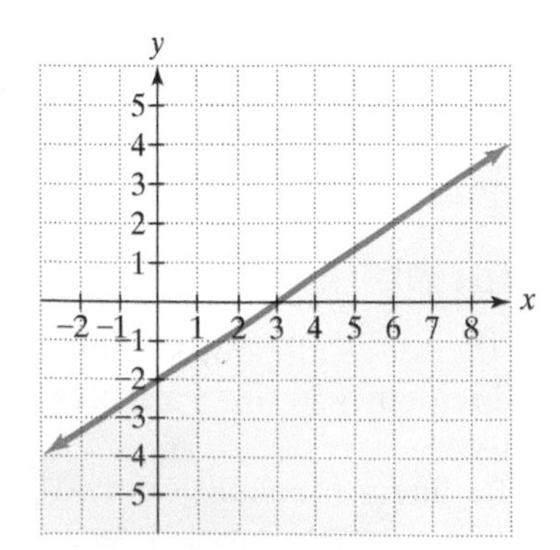

Figure 4.12 The graph of $2x - 3y \ge 6$

Step 3. If a false statement results, shade the half-plane not containing the test point. Because 0 is not greater than or equal to 6, the test point, $(0, 0)$, is not part of the solution set. Thus, the half-plane below the solid line $2x - 3y = 6$ is part of the solution set. The solution set is the line and the half-plane that does not contain the point $(0, 0)$, indicated by shading this half-plane. The graph is shown using green shading and a blue line in **Figure 4.12**. ■

✓ **CHECK POINT 1** Graph: $4x - 2y \ge 8$.

When graphing a linear inequality, choose a test point that lies in one of the half-planes and *not on the line dividing the half-planes*. The test point $(0, 0)$ is convenient because it is easy to calculate when 0 is substituted for each variable. However, if $(0, 0)$ lies on the dividing line and not in a half-plane, a different test point must be selected.

EXAMPLE 2 Graphing a Linear Inequality in Two Variables

Graph: $y > -\frac{2}{3}x$.

Solution

Step 1. Replace the inequality symbol by = and graph the linear equation. Because we are interested in graphing $y > -\frac{2}{3}x$, we begin by graphing $y = -\frac{2}{3}x$. We can use the slope and the y-intercept to graph this linear function.

$$y = -\frac{2}{3}x + 0$$

Slope $= \frac{-2}{3} = \frac{\text{rise}}{\text{run}}$ y-intercept $= 0$

The y-intercept is 0, so the line passes through (0, 0). Using the y-intercept and the slope, the line is shown in **Figure 4.13** as a dashed line. This is because the inequality $y > -\frac{2}{3}x$ contains a > symbol, in which equality is not included.

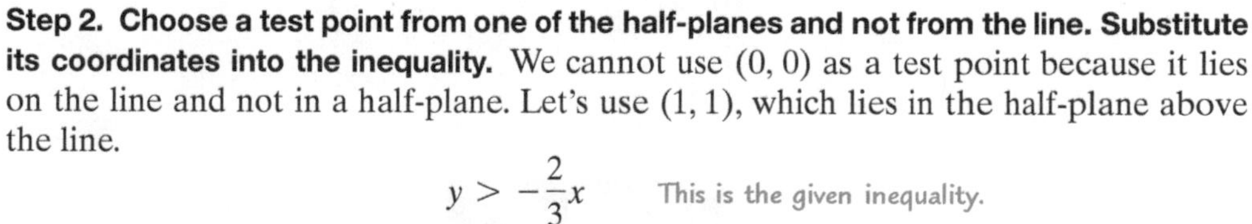

Figure 4.13 The graph of $y > -\frac{2}{3}x$

Step 2. Choose a test point from one of the half-planes and not from the line. Substitute its coordinates into the inequality. We cannot use (0, 0) as a test point because it lies on the line and not in a half-plane. Let's use (1, 1), which lies in the half-plane above the line.

$$y > -\frac{2}{3}x \quad \text{This is the given inequality.}$$

$$1 \overset{?}{>} -\frac{2}{3}\cdot 1 \quad \text{Test (1, 1) by substituting 1 for } x \text{ and 1 for } y.$$

$$1 > -\frac{2}{3} \quad \text{This statement is true.}$$

Step 3. If a true statement results, shade the half-plane containing the test point. Because 1 is greater than $-\frac{2}{3}$, the test point (1, 1) is part of the solution set. All the points on the same side of the line $y = -\frac{2}{3}x$ as the point (1, 1) are members of the solution set. The solution set is the half-plane that contains the point (1, 1), indicated by shading this half-plane. The graph is shown using green shading and a dashed blue line in **Figure 4.13**. ■

Using Technology

Most graphing utilities can graph inequalities in two variables with the SHADE feature. The procedure varies by model, so consult your manual. For most graphing utilities, you must first solve for y if it is not already isolated. The figure shows the graph of $y > -\frac{2}{3}x$. Most displays do not distinguish between dashed and solid boundary lines.

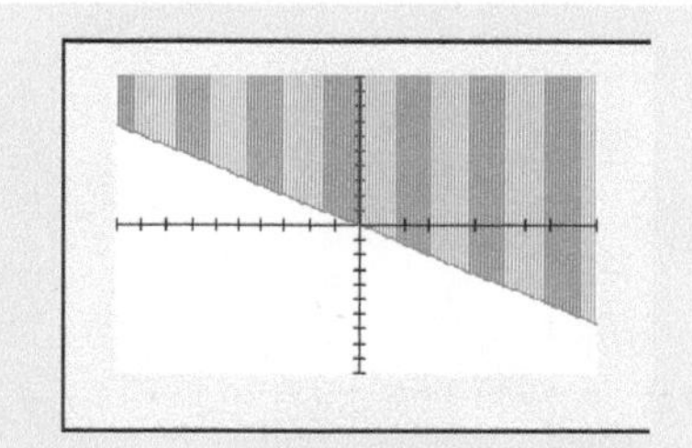

✓ **CHECK POINT 2** Graph: $y > -\frac{3}{4}x$.

Graphing Linear Inequalities without Using Test Points

You can graph inequalities in the form $y > mx + b$ or $y < mx + b$ without using test points. The inequality symbol indicates which half-plane to shade.

- If $y > mx + b$, shade the half-plane above the line $y = mx + b$.
- If $y < mx + b$, shade the half-plane below the line $y = mx + b$.

Observe how this is illustrated in **Figure 4.13**. The graph of $y > -\frac{2}{3}x$ is the half-plane above the line $y = -\frac{2}{3}x$.

Great Question!

When is it important to use test points to graph linear inequalities?

Continue using test points to graph inequalities in the form $Ax + By > C$ or $Ax + By < C$. The graph of $Ax + By > C$ can lie above or below the line given by $Ax + By = C$, depending on the value of B. The same comment applies to the graph of $Ax + By < C$.

It is also not necessary to use test points when graphing inequalities involving half-planes on one side of a vertical or a horizontal line.

For the Vertical Line $x = a$:	For the Horizontal Line $y = b$:
• If $x > a$, shade the half-plane to the right of $x = a$.	• If $y > b$, shade the half-plane above $y = b$.
• If $x < a$, shade the half-plane to the left of $x = a$.	• If $y < b$, shade the half-plane below $y = b$.

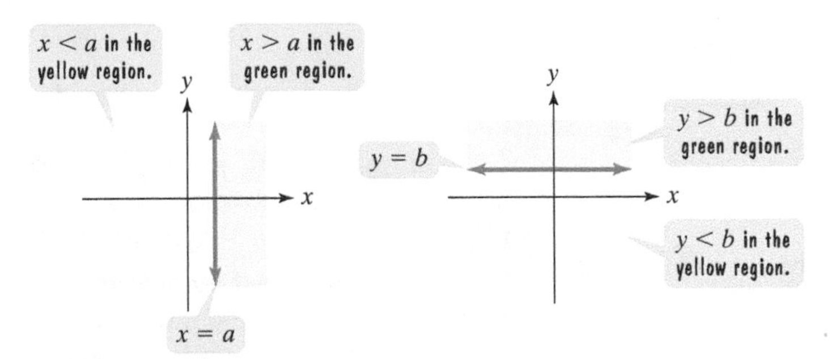

x < a in the yellow region.
x > a in the green region.
y
x
x = a
y = b
y
x
y > b in the green region.
y < b in the yellow region.

EXAMPLE 3 Graphing Inequalities without Using Test Points

Graph each inequality in a rectangular coordinate system:

a. $y \le -3$ **b.** $x > 2$.

Solution

a. $y \le -3$

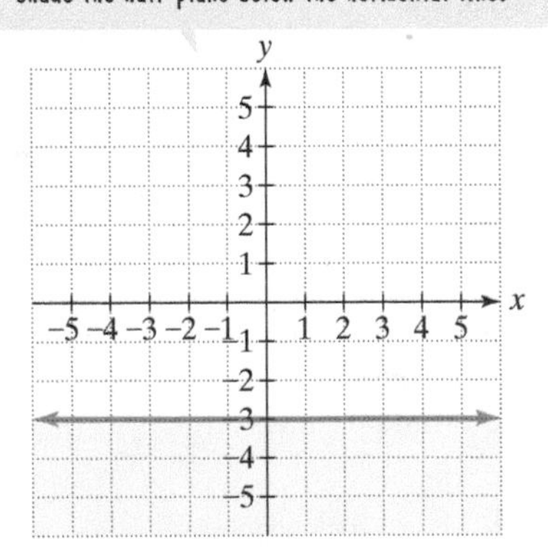

Graph y = −3, a horizontal line with y-intercept −3. The line is solid because equality is included in y ≤ −3. Because of the less than part of ≤, shade the half-plane below the horizontal line.
y
x

b. $x > 2$

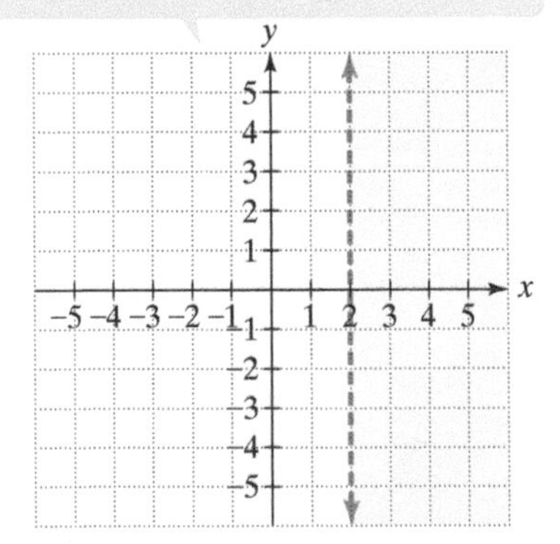

Graph x = 2, a vertical line with x-intercept 2. The line is dashed because equality is not included in x > 2. Because of >, the greater than symbol, shade the half-plane to the right of the vertical line.
y
x

■

CHECK POINT 3 Graph each inequality in a rectangular coordinate system:

a. $y > 1$ **b.** $x \le -2$.

2 Use mathematical models involving linear inequalities.

Modeling with Systems of Linear Inequalities

Just as two or more linear equations make up a system of linear equations, two or more linear inequalities make up a **system of linear inequalities**. A **solution of a system of linear inequalities** in two variables is an ordered pair that satisfies each inequality in the system.

EXAMPLE 4 Forests, Grasslands, Deserts, and Systems of Inequalities

Temperature and precipitation affect whether or not trees and forests can grow. At certain levels of precipitation and temperature, only grasslands and deserts will exist. **Figure 4.14** shows three kinds of regions—deserts, grasslands, and forests—that result from various ranges of temperature, T, and precipitation, P.

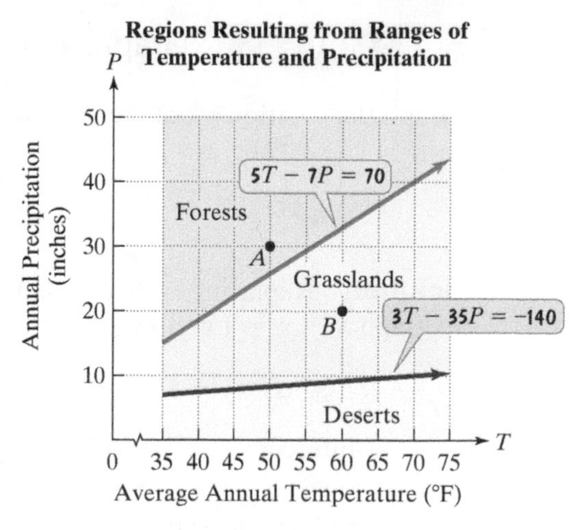

Figure 4.14

Source: Albert Miller et al., *Elements of Meteorology*. © 1983 Merrill Publishing Company

Systems of inequalities can be used to model where forests, grasslands, and deserts occur. Because these regions occur when the average annual temperature, T, is 35°F or greater, each system contains the inequality $T \geq 35$.

Forests occur if	Grasslands occur if	Deserts occur if
$\begin{cases} T \geq 35 \\ 5T - 7P < 70. \end{cases}$	$\begin{cases} T \geq 35 \\ 5T - 7P \geq 70 \\ 3T - 35P \leq -140. \end{cases}$	$\begin{cases} T \geq 35 \\ 3T - 35P > -140. \end{cases}$

Show that point A in **Figure 4.14** is a solution of the system of inequalities that models where forests occur.

Solution Point A has coordinates (50, 30). This means that if a region has an average annual temperature of 50°F and an average annual precipitation of 30 inches, a forest occurs. We can show that (50, 30) satisfies the system of inequalities for forests by substituting 50 for T and 30 for P in each inequality in the system.

$$T \geq 35$$
$$50 \geq 35, \quad \text{true}$$

$$5T - 7P < 70$$
$$5 \cdot 50 - 7 \cdot 30 \stackrel{?}{<} 70$$
$$250 - 210 \stackrel{?}{<} 70$$
$$40 < 70, \quad \text{true}$$

The coordinates (50, 30) make each inequality true. Thus, (50, 30) satisfies the system for forests. ■

✓ **CHECK POINT 4** Show that point B in **Figure 4.14** is a solution of the system of inequalities that models where grasslands occur.

3 Graph a system of linear inequalities.

Graphing Systems of Linear Inequalities

The **solution set of a system of linear inequalities in two variables** is the set of all ordered pairs that satisfy each inequality in the system. Thus, to graph a system of inequalities in two variables, begin by graphing each individual inequality in the same rectangular coordinate system. Then find the region, if there is one, that is common to every graph in the system. This region of intersection gives a picture of the system's solution set.

EXAMPLE 5 Graphing a System of Linear Inequalities

Graph the solution set of the system:

$$\begin{cases} x - y < 1 \\ 2x + 3y \geq 12. \end{cases}$$

Solution Replacing each inequality symbol with an equal sign indicates that we need to graph $x - y = 1$ and $2x + 3y = 12$. We can use intercepts to graph these lines.

$x - y = 1$		$2x + 3y = 12$
x-intercept: $x - 0 = 1$ $x = 1$ The line passes through (1, 0).	Set $y = 0$ in each equation.	x-intercept: $2x + 3 \cdot 0 = 12$ $2x = 12$ $x = 6$ The line passes through (6, 0).
y-intercept: $0 - y = 1$ $-y = 1$ $y = -1$ The line passes through (0, −1).	Set $x = 0$ in each equation.	y-intercept: $2 \cdot 0 + 3y = 12$ $3y = 12$ $y = 4$ The line passes through (0, 4).

Now we are ready to graph the solution set of the system of linear inequalities.

Graph $x - y < 1$. The blue line, $x - y = 1$, is dashed: Equality is not included in $x - y < 1$. Because (0, 0) makes the inequality true ($0 - 0 < 1$, or $0 < 1$, is true), shade the half-plane containing (0, 0) in yellow.

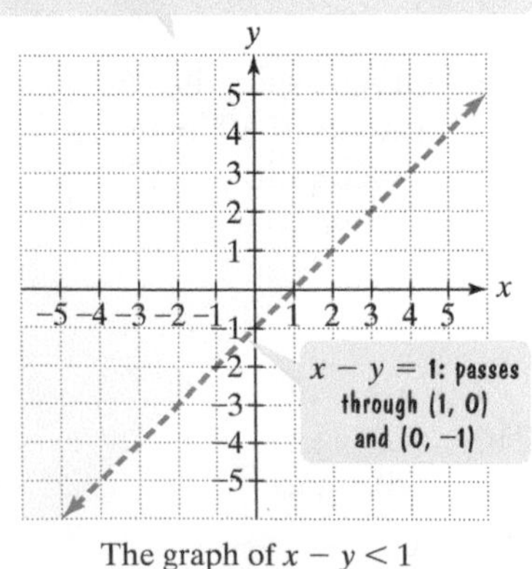

The graph of $x - y < 1$

Add the graph of $2x + 3y \geq 12$. The red line, $2x + 3y = 12$, is solid: Equality is included in $2x + 3y \geq 12$. Because (0, 0) makes the inequality false ($2 \cdot 0 + 3 \cdot 0 \geq 12$, or $0 \geq 12$, is false), shade the half-plane not containing (0, 0) using green vertical shading.

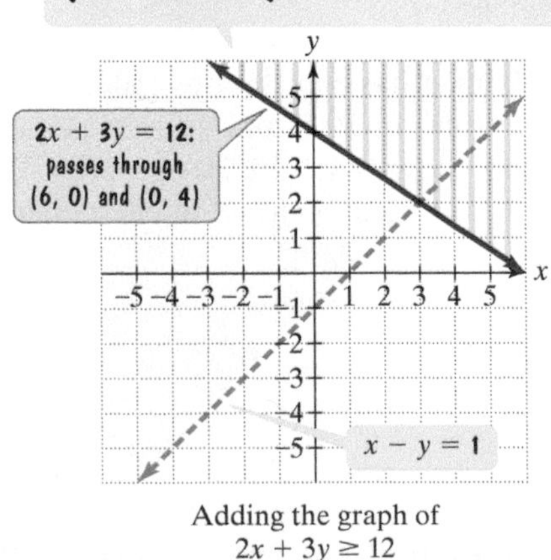

Adding the graph of $2x + 3y \geq 12$

The solution set of the system is graphed as the intersection (the overlap) of the two half-planes. This is the region in which the yellow shading and the green vertical shading overlap.

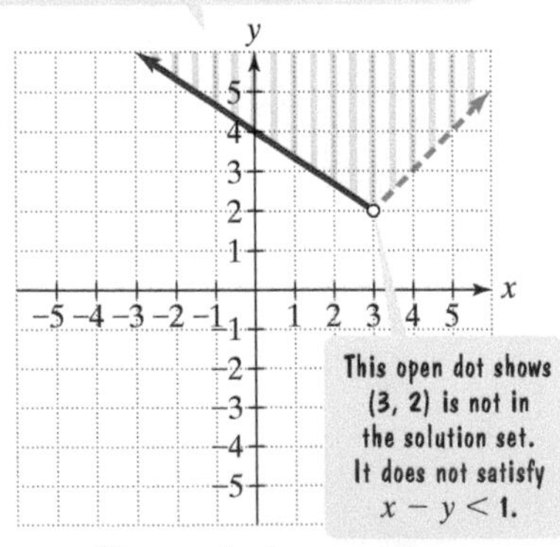

The graph of $x - y < 1$ and $2x + 3y \geq 12$

■

✓ **CHECK POINT 5** Graph the solution set of the system:

$$\begin{cases} x - 3y < 6 \\ 2x + 3y \geq -6. \end{cases}$$

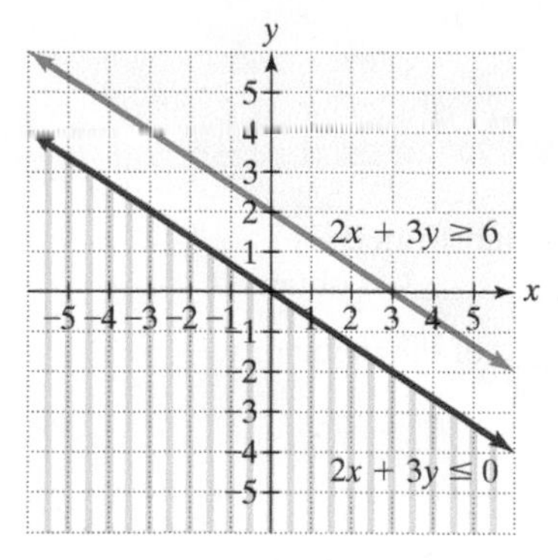

Figure 4.15 A system of inequalities with no solution

A system of inequalities has no solution if there are no points in the rectangular coordinate system that simultaneously satisfy each inequality in the system. For example, the system

$$\begin{cases} 2x + 3y \geq 6 \\ 2x + 3y \leq 0, \end{cases}$$

whose separate graphs are shown in **Figure 4.15**, has no overlapping region. Thus, the system has no solution. The solution set is $\varnothing$, the empty set.

EXAMPLE 6 Graphing a System of Inequalities

Graph the solution set of the system:

$$\begin{cases} x - y < 2 \\ -2 \leq x < 4 \\ y < 3. \end{cases}$$

Solution We begin by graphing $x - y < 2$, the first given inequality. The line $x - y = 2$ has an x-intercept of 2 and a y-intercept of -2. The test point $(0, 0)$ makes the inequality $x - y < 2$ true. The graph of $x - y < 2$ is shown in **Figure 4.16**.

Now, let's consider the second given inequality, $-2 \leq x < 4$. Replacing the inequality symbols by $=$, we obtain $x = -2$ and $x = 4$, graphed as red vertical lines in **Figure 4.17**. The line $x = 4$ is not included. Because x is between -2 and 4, we shade the region between the vertical lines. We must intersect this region with the yellow region in **Figure 4.16**. The resulting region is shown in yellow and green vertical shading in **Figure 4.17**.

Finally, let's consider the third given inequality, $y < 3$. Replacing the inequality symbol by $=$, we obtain $y = 3$, which graphs as a horizontal line. Because of the less than symbol in $y < 3$, the graph consists of the half-plane below the line $y = 3$. We must intersect this half-plane with the region in **Figure 4.17**. The resulting region is shown in yellow and green vertical shading in **Figure 4.18**. This region represents the graph of the solution set of the given system.

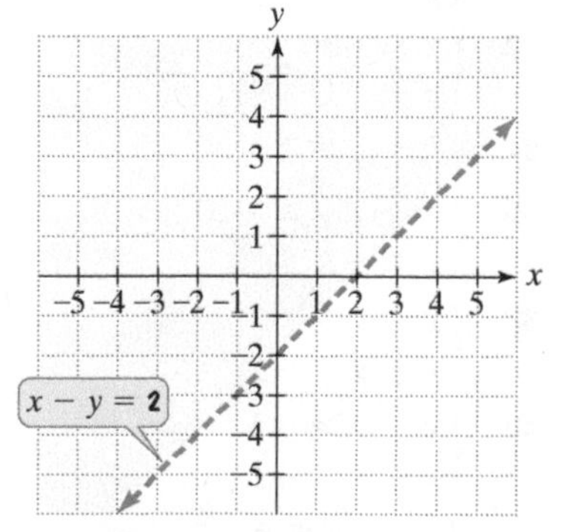

Figure 4.16 The graph of $x - y < 2$

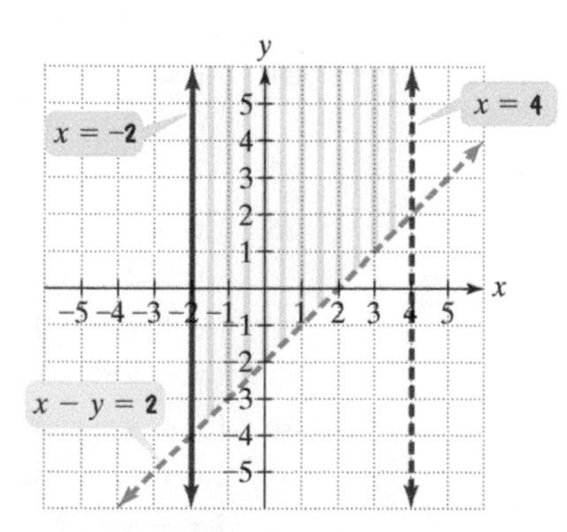

Figure 4.17 The graph of $x - y < 2$ and $-2 \leq x < 4$

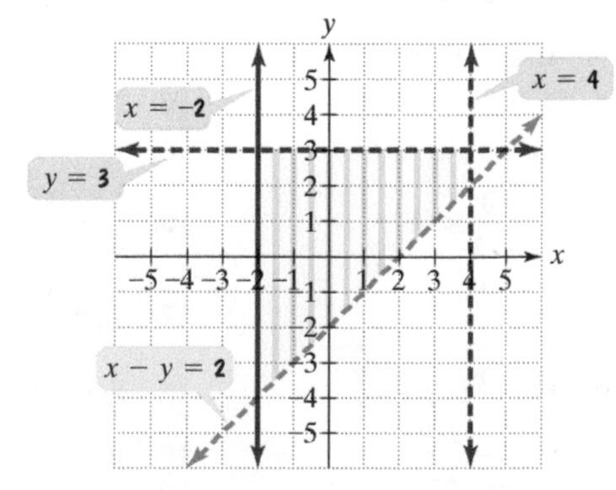

Figure 4.18 The graph of $x - y < 2$ and $-2 \leq x < 4$ and $y < 3$ ■

CHECK POINT 6 Graph the solution set of the system:

$$\begin{cases} x + y < 2 \\ -2 \leq x < 1 \\ y > -3. \end{cases}$$

CONCEPT AND VOCABULARY CHECK

Fill in each blank so that the resulting statement is true.

1. The ordered pair (3, 2) is a/an ______________ of the inequality $x + y > 1$ because when 3 is substituted for ______________ and 2 is substituted for ______________, the true statement ______________ is obtained.
2. The set of all points that satisfy a linear inequality in two variables is called the ______________ of the inequality.
3. The set of all points on one side of a line is called a/an ______________.
4. True or false: The graph of $2x - 3y > 6$ includes the line $2x - 3y = 6$. ______________
5. True or false: The graph of the linear equation $2x - 3y = 6$ is used to graph the linear inequality $2x - 3y > 6$.

   ____
6. True or false: When graphing $4x - 2y \geq 8$, to determine which side of the line to shade, choose a test point on $4x - 2y = 8$.

   ____
7. The solution set of the system

$$\begin{cases} x - y < 1 \\ 2x + 3y \geq 12 \end{cases}$$

   is the set of ordered pairs that satisfy ______________ and ______________.
8. True or false: The graph of the solution set of the system

$$\begin{cases} x - 3y < 6 \\ 2x + 3y \geq -6 \end{cases}$$

   includes the intersection point of $x - 3y = 6$ and $2x + 3y = -6$. ______________

4.4 EXERCISE SET

MyMathLab®

Practice Exercises

In Exercises 1–22, graph each inequality.

1. $x + y \geq 3$
2. $x + y \geq 2$
3. $x - y < 5$
4. $x - y < 6$
5. $x + 2y > 4$
6. $2x + y > 6$
7. $3x - y \leq 6$
8. $x - 3y \leq 6$
9. $\frac{x}{2} + \frac{y}{3} < 1$
10. $\frac{x}{4} + \frac{y}{2} < 1$
11. $y > \frac{1}{3}x$
12. $y > \frac{1}{4}x$
13. $y \leq 3x + 2$
14. $y \leq 2x - 1$
15. $y < -\frac{1}{4}x$
16. $y < -\frac{1}{3}x$
17. $x \leq 2$
18. $x \leq -4$
19. $y > -4$
20. $y > -2$
21. $y \geq 0$
22. $x \leq 0$

In Exercises 23–46, graph the solution set of each system of inequalities or indicate that the system has no solution.

23. $\begin{cases} 3x + 6y \leq 6 \\ 2x + y \leq 8 \end{cases}$
24. $\begin{cases} x - y \geq 4 \\ x + y \leq 6 \end{cases}$
25. $\begin{cases} 2x - 5y \leq 10 \\ 3x - 2y > 6 \end{cases}$
26. $\begin{cases} 2x - y \leq 4 \\ 3x + 2y > -6 \end{cases}$
27. $\begin{cases} y > 2x - 3 \\ y < -x + 6 \end{cases}$
28. $\begin{cases} y < -2x + 4 \\ y < x - 4 \end{cases}$
29. $\begin{cases} x + 2y \leq 4 \\ y \geq x - 3 \end{cases}$
30. $\begin{cases} x + y \leq 4 \\ y \geq 2x - 4 \end{cases}$
31. $\begin{cases} x \leq 2 \\ y \geq -1 \end{cases}$
32. $\begin{cases} x \leq 3 \\ y \leq -1 \end{cases}$
33. $-2 \leq x < 5$
34. $-2 < y \leq 5$
35. $\begin{cases} x - y \leq 1 \\ x \geq 2 \end{cases}$
36. $\begin{cases} 4x - 5y \geq -20 \\ x \geq -3 \end{cases}$
37. $\begin{cases} x + y > 4 \\ x + y < -1 \end{cases}$
38. $\begin{cases} x + y > 3 \\ x + y < -2 \end{cases}$
39. $\begin{cases} x + y > 4 \\ x + y > -1 \end{cases}$
40. $\begin{cases} x + y > 3 \\ x + y > -2 \end{cases}$
41. $\begin{cases} x - y \leq 2 \\ x \geq -2 \\ y \leq 3 \end{cases}$
42. $\begin{cases} 3x + y \leq 6 \\ x \geq -2 \\ y \leq 4 \end{cases}$
43. $\begin{cases} x \geq 0 \\ y \geq 0 \\ 2x + 5y \leq 10 \\ 3x + 4y \leq 12 \end{cases}$
44. $\begin{cases} x \geq 0 \\ y \geq 0 \\ 2x + y \leq 4 \\ 2x - 3y \leq 6 \end{cases}$
45. $\begin{cases} 3x + y \leq 6 \\ 2x - y \leq -1 \\ x \geq -2 \\ y \leq 4 \end{cases}$
46. $\begin{cases} 2x + y \leq 6 \\ x + y \geq 2 \\ 1 \leq x \leq 2 \\ y \leq 3 \end{cases}$

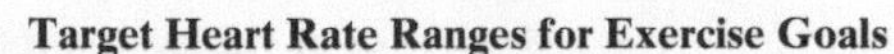

Practice PLUS

In Exercises 47–48, write each sentence as a linear inequality in two variables. Then graph the inequality.

47. The y-variable is at least 4 more than the product of -2 and the x-variable.

48. The y-variable is at least 2 more than the product of -3 and the x-variable.

In Exercises 49–50, write the given sentences as a system of linear inequalities in two variables. Then graph the system.

49. The sum of the x-variable and the y-variable is at most 4. The y-variable added to the product of 3 and the x-variable does not exceed 6.

50. The sum of the x-variable and the y-variable is at most 3. The y-variable added to the product of 4 and the x-variable does not exceed 6.

In Exercises 51–52, rewrite each inequality in the system without absolute value bars. Then graph the rewritten system in rectangular coordinates.

51. $\begin{cases} |x| \leq 2 \\ |y| \leq 3 \end{cases}$

52. $\begin{cases} |x| \leq 1 \\ |y| \leq 2 \end{cases}$

*The graphs of solution sets of systems of inequalities involve finding the intersection of the solution sets of two or more inequalities. By contrast, in Exercises 53–54 you will be graphing the **union** of the solution sets of two inequalities.*

53. Graph the union of $y > \frac{3}{2}x - 2$ and $y < 4$.

54. Graph the union of $x - y \geq -1$ and $5x - 2y \leq 10$.

Without graphing, in Exercises 55–58, determine if each system has no solution or infinitely many solutions.

55. $\begin{cases} 3x + y < 9 \\ 3x + y > 9 \end{cases}$

56. $\begin{cases} 6x - y \leq 24 \\ 6x - y > 24 \end{cases}$

57. $\begin{cases} 3x + y \leq 9 \\ 3x + y \geq 9 \end{cases}$

58. $\begin{cases} 6x - y \leq 24 \\ 6x - y \geq 24 \end{cases}$

Application Exercises

Maximum heart rate, H, in beats per minute is a function of age, a, modeled by the formula

$$H = 220 - a,$$

where $10 \leq a \leq 70$. The bar graph at the top of the next column shows the target heart rate ranges for four types of exercise goals in terms of maximum heart rate.

Target Heart Rate Ranges for Exercise Goals

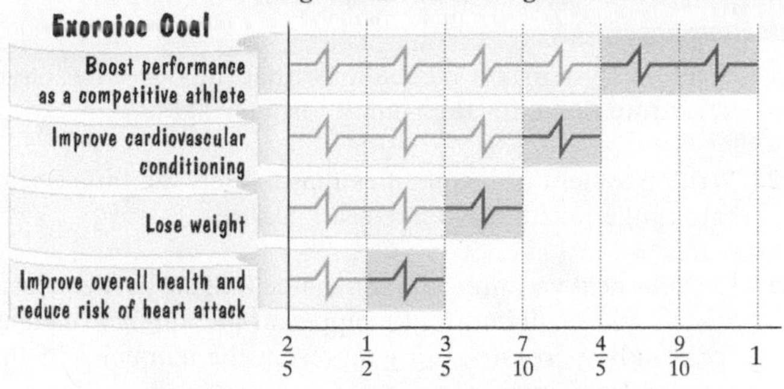

Source: *Vitality* Magazine

In Exercises 59–62, systems of inequalities will be used to model three of the target heart rate ranges shown in the bar graph. We begin with the target heart rate range for cardiovascular conditioning, modeled by the following system of inequalities:

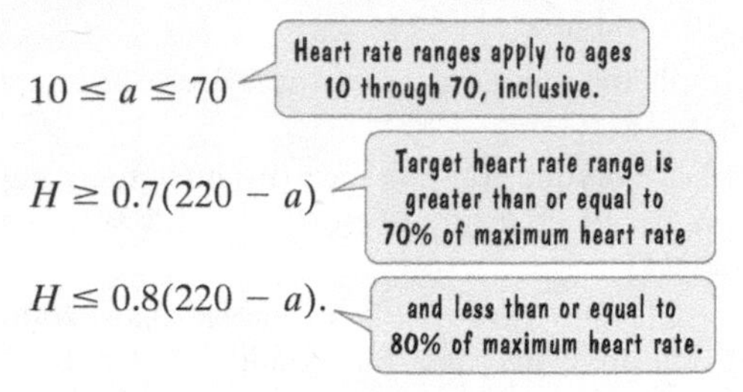

10 ≤ a ≤ 70
Heart rate ranges apply to ages 10 through 70, inclusive.
H ≥ 0.7(220 − a)
Target heart rate range is greater than or equal to 70% of maximum heart rate
H ≤ 0.8(220 − a).
and less than or equal to 80% of maximum heart rate.

The graph of this system is shown in the figure. Use the graph to solve Exercises 59–60.

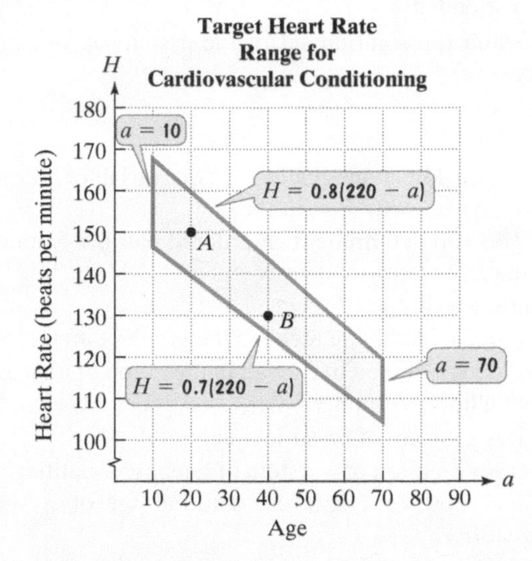

59. a. What are the coordinates of point A and what does this mean in terms of age and heart rate?

b. Show that point A is a solution of the system of inequalities.

60. a. What are the coordinates of point B and what does this mean in terms of age and heart rate?

b. Show that point B is a solution of the system of inequalities.

(In Exercises 61–62, refer to the models and graphs on the previous page.)

61. Write a system of inequalities that models the target heart rate range for the goal of losing weight.

62. Write a system of inequalities that models the target heart rate range for improving overall health.

63. On your next vacation, you will divide lodging between large resorts and small inns. Let x represent the number of nights spent in large resorts. Let y represent the number of nights spent in small inns.

a. Write a system of inequalities that models the following conditions:

You want to stay at least 5 nights. At least one night should be spent at a large resort. Large resorts average \$200 per night and small inns average \$100 per night. Your budget permits no more than \$700 for lodging.

b. Graph the solution set of the system of inequalities in part (a).

c. Based on your graph in part (b), how many nights could you spend at a large resort and still stay within your budget?

64. a. An elevator can hold no more than 2000 pounds. If children average 80 pounds and adults average 160 pounds, write a system of inequalities that models when the elevator holding x children and y adults is overloaded.

b. Graph the solution set of the system of inequalities in part (a).

Explaining the Concepts

65. What is a linear inequality in two variables? Provide an example with your description.

66. How do you determine if an ordered pair is a solution of an inequality in two variables, x and y?

67. What is a half-plane?

68. What does a solid line mean in the graph of an inequality?

69. What does a dashed line mean in the graph of an inequality?

70. Explain how to graph $x - 2y < 4$.

71. What is a system of linear inequalities?

72. What is a solution of a system of linear inequalities?

73. Explain how to graph the solution set of a system of inequalities.

74. What does it mean if a system of linear inequalities has no solution?

Technology Exercises

Graphing utilities can be used to shade regions in the rectangular coordinate system, thereby graphing an inequality in two variables. Read the section of the user's manual for your graphing utility that describes how to shade a region. Then use your graphing utility to graph the inequalities in Exercises 75–78.

75. $y \le 4x + 4$

76. $y \ge \frac{2}{3}x - 2$

77. $2x + y \le 6$

78. $3x - 2y \ge 6$

79. Does your graphing utility have any limitations in terms of graphing inequalities? If so, what are they?

80. Use a graphing utility with a SHADE feature to verify any five of the graphs that you drew by hand in Exercises 1–22.

81. Use a graphing utility with a SHADE feature to verify any five of the graphs that you drew by hand for the systems in Exercises 23–46.

Critical Thinking Exercises

Make Sense? *In Exercises 82–85, determine whether each statement makes sense or does not make sense, and explain your reasoning.*

82. When graphing a linear inequality, I should always use (0, 0) as a test point because it's easy to perform the calculations when 0 is substituted for each variable.

83. If you want me to graph $x < 3$, you need to tell me whether to use a number line or a rectangular coordinate system.

84. When graphing $3x - 4y < 12$, it's not necessary for me to graph the linear equation $3x - 4y = 12$ because the inequality contains a $<$ symbol, in which equality is not included.

85. Linear inequalities can model situations in which I'm interested in purchasing two items at different costs, I can spend no more than a specified amount on both items, and I want to know how many of each item I can purchase.

In Exercises 86–89, determine whether each statement is true or false. If the statement is false, make the necessary change(s) to produce a true statement.

86. The graph of $3x - 5y < 10$ consists of a dashed line and a shaded half-plane below the line.

87. The graph of $y \ge -x + 1$ consists of a solid line that rises from left to right and a shaded half-plane above the line.

88. The ordered pair $(-2, 40)$ satisfies the following system:

$$\begin{cases} y \ge 9x + 11 \\ 13x + y > 14. \end{cases}$$

89. For the graph of $y < x - 3$, the points $(0, -3)$ and $(8, 5)$ lie on the graph of the corresponding linear equation, but neither point is a solution of the inequality.

90. Write a linear inequality in two variables whose graph is shown.

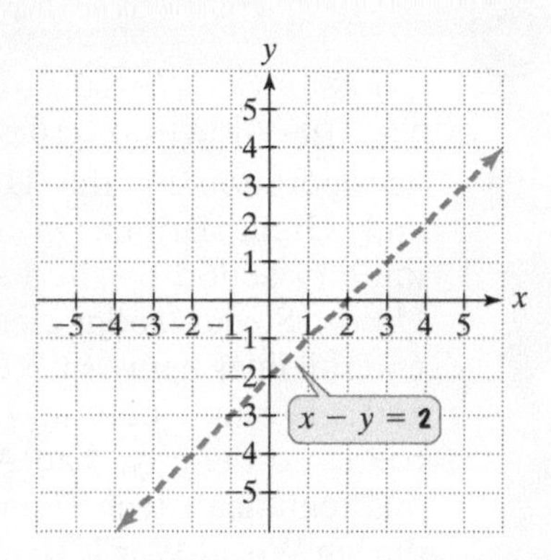

In Exercises 91–92, write a system of inequalities for each graph.

91.

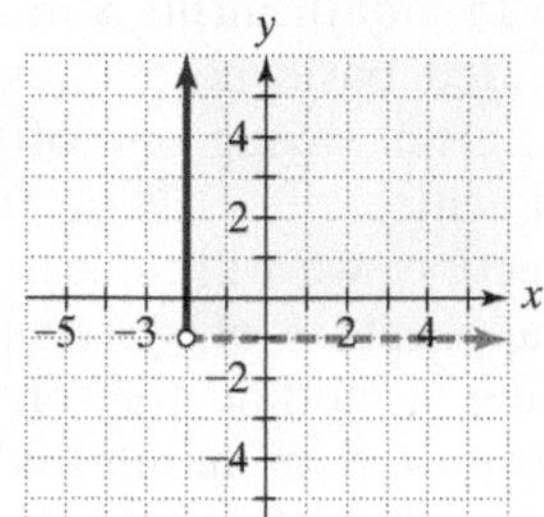

92.

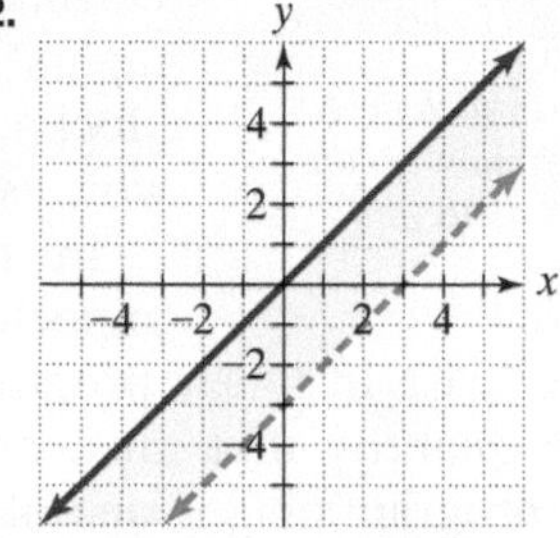

93. Write a linear inequality in two variables satisfying the following conditions: The points $(-3, -8)$ and $(4, 6)$ lie on the graph of the corresponding linear equation and each point is a solution of the inequality. The point $(1, 1)$ is also a solution.

94. Write a system of inequalities whose solution set includes every point in the rectangular coordinate system.

95. Sketch the graph of the solution set for the following system of inequalities:

$$\begin{cases} y \geq nx + b \ (n < 0, b > 0) \\ y \leq mx + b \ (m > 0, b > 0). \end{cases}$$

Review Exercises

96. Solve using matrices:

$$\begin{cases} 3x - y = 8 \\ x - 5y = -2. \end{cases}$$

(Section 3.4, Example 2)

97. Solve by graphing:

$$\begin{cases} y = 3x - 2 \\ y = -2x + 8. \end{cases}$$

(Section 3.1, Example 2)

98. Evaluate:

$$\begin{vmatrix} 8 & 2 & -1 \\ 3 & 0 & 5 \\ 6 & -3 & 4 \end{vmatrix}.$$

(Section 3.5, Example 3)

Preview Exercises

Exercises 99–101 will help you prepare for the material covered in the next section.

99. a. Graph the solution set of the system:

$$\begin{cases} x + y \geq 6 \\ x \leq 8 \\ y \leq 5. \end{cases}$$

b. List the points that form the corners of the graphed region in part (a).

c. Evaluate $3x + 2y$ at each of the points obtained in part (b).

100. a. Graph the solution set of the system:

$$\begin{cases} x \geq 0 \\ y \geq 0 \\ 3x - 2y \leq 6 \\ y \leq -x + 7. \end{cases}$$

b. List the points that form the corners of the graphed region in part (a).

c. Evaluate $2x + 5y$ at each of the points obtained in part (b).

101. Bottled water and medical supplies are to be shipped to survivors of an earthquake by plane. The bottled water weighs 20 pounds per container and medical kits weigh 10 pounds per kit. Each plane can carry no more than 80,000 pounds. If x represents the number of bottles of water to be shipped per plane and y represents the number of medical kits per plane, write an inequality that models each plane's 80,000-pound weight restriction.

SECTION

4.5 Linear Programming

What am I supposed to learn?

After studying this section, you should be able to:

1. Write an objective function modeling a quantity that must be maximized or minimized.
2. Use inequalities to model limitations in a situation.
3. Use linear programming to solve problems.

West Berlin children at Tempelhof airport watch fleets of U.S. airplanes bringing in supplies to circumvent the Soviet blockade. The airlift began June 28, 1948, and continued for 15 months.

The Berlin Airlift (1948–1949) was an operation by the United States and Great Britain in response to military action by the former Soviet Union: Soviet troops closed all roads and rail lines between West Germany and Berlin, cutting off supply routes to the city. The Allies used a mathematical technique developed during World War II to maximize the quantities of supplies transported. During the 15-month airlift, 278,228 flights provided basic necessities to blockaded Berlin, saving one of the world's great cities.

In this section, we will look at an important application of systems of linear inequalities. Such systems arise in **linear programming**, a method for solving problems in which a particular quantity that must be maximized or minimized is limited by other factors. Linear programming is one of the most widely used tools in management science. It helps businesses allocate resources to manufacture products in a way that will maxmize profit. Linear programming accounts for more than 50% and perhaps as much as 90% of all computing time used for management decisions in business. The Allies used linear programming to save Berlin.

1 Write an objective function modeling a quantity that must be maximized or minimized.

Objective Functions in Linear Programming

Many problems involve quantities that must be maximized or minimized. Businesses are interested in maximizing profit. A relief operation in which bottled water and medical kits are shipped to earthquake survivors needs to maximize the number of survivors helped by this shipment. An **objective function** is an algebraic expression in two or more variables describing a quantity that must be maximized or minimized.

EXAMPLE 1 Writing an Objective Function

Bottled water and medical supplies are to be shipped to survivors of an earthquake by plane. Each container of bottled water will serve 10 people and each medical kit will aid 6 people. If x represents the number of bottles of water to be shipped and y represents the number of medical kits, write the objective function that models the number of people that can be helped.

Solution Because each bottle of water serves 10 people and each medical kit aids 6 people, we have

The number of people helped | is | 10 times the number of bottles of water | plus | 6 times the number of medical kits.

$$= \quad 10x \quad + \quad 6y.$$

Using z to represent the number of people helped, the objective function is

$$z = 10x + 6y.$$

Unlike the functions that we have seen so far, the objective function is an equation in three variables. For a value of x and a value of y, there is one and only one value of z. Thus, z is a function of x and y. ■

✓ CHECK POINT 1 A company manufactures bookshelves and desks for computers. Let x represent the number of bookshelves manufactured daily and y the number of desks manufactured daily. The company's profits are \$25 per bookshelf and \$55 per desk. Write the objective function that models the company's total daily profit, z, from x bookshelves and y desks. (Check Points 2 through 4 are related to this situation, so keep track of your answers.)

---

2 Use inequalities to model limitations in a situation.

Constraints in Linear Programming

Ideally, the number of earthquake survivors helped in Example 1 should increase without restriction so that every survivor receives water and medical supplies. However, the planes that ship these supplies are subject to weight and volume restrictions. In linear programming problems, such restrictions are called **constraints**. Each constraint is expressed as a linear inequality. The list of constraints forms a system of linear inequalities.

EXAMPLE 2 Writing a Constraint

Each plane can carry no more than 80,000 pounds. The bottled water weighs 20 pounds per container and each medical kit weighs 10 pounds. Let x represent the number of bottles of water to be shipped and y the number of medical kits. Write an inequality that models this constraint.

Solution Because each plane can carry no more than 80,000 pounds, we have

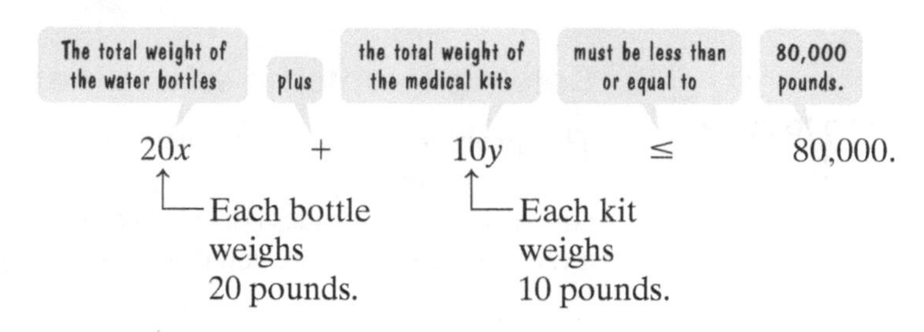
The total weight of the water bottles
plus
the total weight of the medical kits
must be less than or equal to
80,000 pounds.
20x
+
10y
≤
80,000.
Each bottle weighs 20 pounds.
Each kit weighs 10 pounds.

The plane weight constraint is modeled by the inequality

$$20x + 10y \leq 80{,}000. \blacksquare$$

✓ CHECK POINT 2 To maintain high quality, the company in Check Point 1 should not manufacture more than a total of 80 bookshelves and desks per day. Write an inequality that models this constraint.

In addition to a weight constraint on its cargo, each plane has a limited amount of space in which to carry supplies. Example 3 demonstrates how to express this constraint.

EXAMPLE 3 Writing a Constraint

Each plane can carry a total volume of supplies that does not exceed 6000 cubic feet. Each water bottle is 1 cubic foot and each medical kit also has a volume of 1 cubic foot. With x still representing the number of water bottles and y the number of medical kits, write an inequality that models this second constraint.

Solution Because each plane can carry a volume of supplies that does not exceed 6000 cubic feet, we have

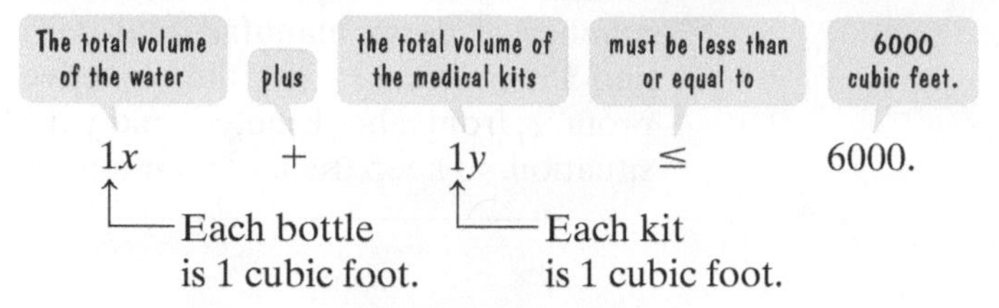

The plane volume constraint is modeled by the inequality $x + y \leq 6000$. ■

In summary, here's what we have described so far in this aid-to-earthquake-survivors situation:

$$z = 10x + 6y$$

This is the objective function modeling the number of people helped with x bottles of water and y medical kits.

$$\begin{cases} 20x + 10y \leq 80{,}000 \\ x + y \leq 6000. \end{cases}$$

These are the constraints based on each plane's weight and volume limitations.

✓ CHECK POINT 3 To meet customer demand, the company in Check Point 1 must manufacture between 30 and 80 bookshelves per day, inclusive. Furthermore, the company must manufacture at least 10 and no more than 30 desks per day. Write an inequality that models each of these sentences. Then summarize what you have described about this company by writing the objective function for its profits and the three constraints.

---

3 Use linear programming to solve problems.

Solving Problems with Linear Programming

The problem in the earthquake situation described previously is to maximize the number of survivors who can be helped, subject to each plane's weight and volume constraints. The process of solving this problem is called *linear programming*, based on a theorem that was proven during World War II.

Solving a Linear Programming Problem

Let $z = ax + by$ be an objective function that depends on x and y. Furthermore, z is subject to a number of linear constraints on x and y. If a maximum or minimum value of z exists, it can be determined as follows:

1. Graph the system of inequalities representing the constraints.
2. Find the value of the objective function at each corner, or **vertex**, of the graphed region. The maximum and minimum of the objective function occur at one or more of the corner points.

EXAMPLE 4 Solving a Linear Programming Problem

Determine how many bottles of water and how many medical kits should be sent on each plane to maximize the number of earthquake survivors who can be helped.

Solution We must maximize $z = 10x + 6y$ subject to the following constraints:

$$\begin{cases} 20x + 10y \leq 80{,}000 \\ x + y \leq 6000. \end{cases}$$

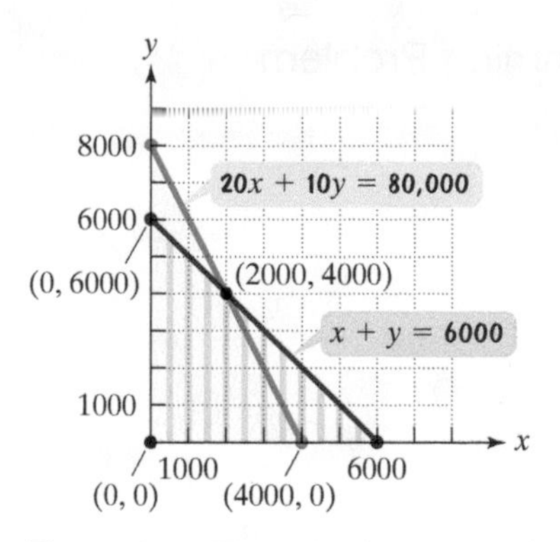

Figure 4.19 The region in quadrant I representing the constraints $20x + 10y \le 80{,}000$ and $x + y \le 6000$

Step 1. Graph the system of inequalities representing the constraints. Because x (the number of bottles of water per plane) and y (the number of medical kits per plane) must be nonnegative, we need to graph the system of inequalities in quadrant I and its boundary only.

To graph the inequality $20x + 10y \le 80{,}000$, we graph the equation $20x + 10y = 80{,}000$ as a solid blue line (**Figure 4.19**). Setting $y = 0$, the x-intercept is 4000 and setting $x = 0$, the y-intercept is 8000. Using $(0, 0)$ as a test point, the inequality is satisfied, so we shade below the blue line, as shown in yellow in **Figure 4.19.**

Now we graph $x + y \le 6000$ by first graphing $x + y = 6000$ as a solid red line. Setting $y = 0$, the x-intercept is 6000. Setting $x = 0$, the y-intercept is 6000. Using $(0, 0)$ as a test point, the inequality is satisfied, so we shade below the red line, as shown using green vertical shading in **Figure 4.19.**

We use the addition method to find where the lines $20x + 10y = 80{,}000$ and $x + y = 6000$ intersect.

$$\begin{cases} 20x + 10y = 80{,}000 \xrightarrow{\text{No change}} \\ x + y = 6000 \xrightarrow{\text{Multiply by } -10.} \end{cases} \quad \begin{cases} 20x + 10y = 80{,}000 \\ -10x - 10y = -60{,}000 \end{cases}$$

$$\text{Add:} \quad 10x = 20{,}000$$

$$x = 2000$$

Back-substituting 2000 for x in $x + y = 6000$, we find $y = 4000$, so the intersection point is $(2000, 4000)$.

The system of inequalities representing the constraints is shown by the region in which the yellow shading and the green vertical shading overlap in **Figure 4.19**. The graph of the system of inequalities is shown again in **Figure 4.20**. The red and blue line segments are included in the graph.

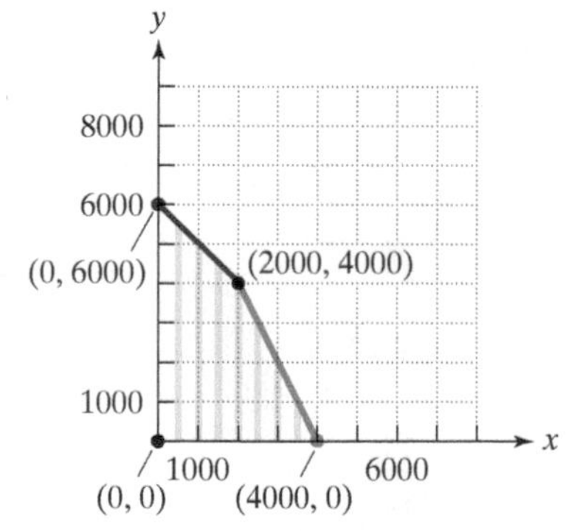

Figure 4.20

Step 2. Find the value of the objective function at each corner of the graphed region. The maximum and minimum of the objective function occur at one or more of the corner points. We must evaluate the objective function, $z = 10x + 6y$, at the four corners, or vertices, of the region in **Figure 4.20**.

Corner (x, y)	Objective Function $z = 10x + 6y$
$(0, 0)$	$z = 10(0) + 6(0) = 0$
$(4000, 0)$	$z = 10(4000) + 6(0) = 40{,}000$
$(2000, 4000)$	$z = 10(2000) + 6(4000) = 44{,}000$ ← maximum
$(0, 6000)$	$z = 10(0) + 6(6000) = 36{,}000$

Thus, the maximum value of z is 44,000 and this occurs when $x = 2000$ and $y = 4000$. In practical terms, this means that the maximum number of earthquake survivors who can be helped with each plane shipment is 44,000. This can be accomplished by sending 2000 water bottles and 4000 medical kits per plane. ■

✓ **CHECK POINT 4** For the company in Check Points 1–3, how many bookshelves and how many desks should be manufactured per day to obtain maximum profit? What is the maximum daily profit?

EXAMPLE 5 Solving a Linear Programming Problem

Find the maximum value of the objective function

$$z = 2x + y$$

subject to the following constraints:

$$\begin{cases} x \geq 0, y \geq 0 \\ x + 2y \leq 5 \\ x - y \leq 2. \end{cases}$$

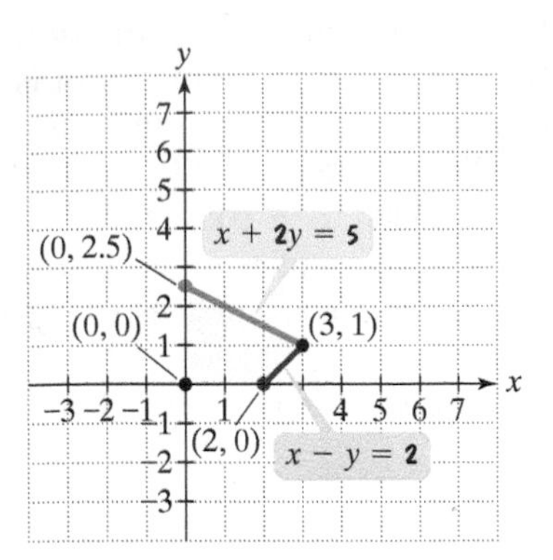

Figure 4.21 The graph of $x + 2y \leq 5$ and $x - y \leq 2$ in quadrant I

Solution We begin by graphing the region in quadrant I ($x \geq 0, y \geq 0$) formed by the constraints. The graph is shown in **Figure 4.21**.

Now we evaluate the objective function at the four vertices of this region.

Objective function: $z = 2x + y$

At (0, 0): $z = 2 \cdot 0 + 0 = 0$

At (2, 0): $z = 2 \cdot 2 + 0 = 4$

At (3, 1): $z = 2 \cdot 3 + 1 = 7$ Maximum value of z

At (0, 2.5): $z = 2 \cdot 0 + 2.5 = 2.5$

Thus, the maximum value of z is 7, and this occurs when $x = 3$ and $y = 1$. ■

We can see why the objective function in Example 5 has a maximum value that occurs at a vertex by solving the equation for y.

$$z = 2x + y$$ This is the objective function of Example 5.

$$y = -2x + z$$ Solve for y. Recall that the slope-intercept form of the equation of a line is $y = mx + b$.

Slope = −2 y-intercept = z

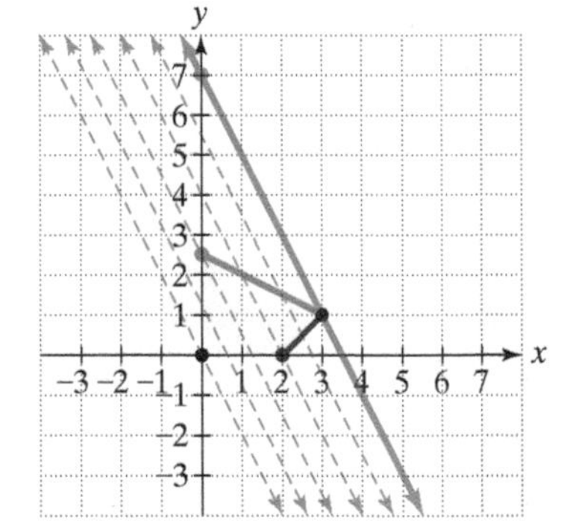

Figure 4.22 The line with slope −2 with the greatest y-intercept that intersects the shaded region passes through one of the vertices of the region.

In this form, z represents the y-intercept of the objective function. The equation describes infinitely many parallel lines (one for each value of z), each with slope −2. The process in linear programming involves finding the maximum z-value for all lines that intersect the region determined by the constraints. Of all the lines whose slope is −2, we're looking for the one with the greatest y-intercept that intersects the given region. As we see in **Figure 4.22**, such a line will pass through one (or possibly more) of the vertices of the region.

✓ CHECK POINT 5 Find the maximum value of the objective function $z = 3x + 5y$ subject to the constraints $x \geq 0, y \geq 0, x + y \geq 1, x + y \leq 6$.

CONCEPT AND VOCABULARY CHECK

Fill in each blank so that the resulting statement is true.

1. A method for finding the maximum or minimum value of a quantity that is subject to various limitations is called ________________.

2. An algebraic expression in two or more variables describing a quantity that must be maximized or minimized is called a/an ________________ function.

3. A system of linear inequalities is used to represent restrictions, or ________________, on a function that must be maximized or minimized. Using the graph of such a system of inequalities, the maximum and minimum values of the function occur at one or more of the ________________ points.

4.5 EXERCISE SET

MyMathLab®

Practice Exercises

In Exercises 1–4, find the value of the objective function at each corner of the graphed region. What is the maximum value of the objective function? What is the minimum value of the objective function?

1. Objective Function
   $z = 5x + 6y$

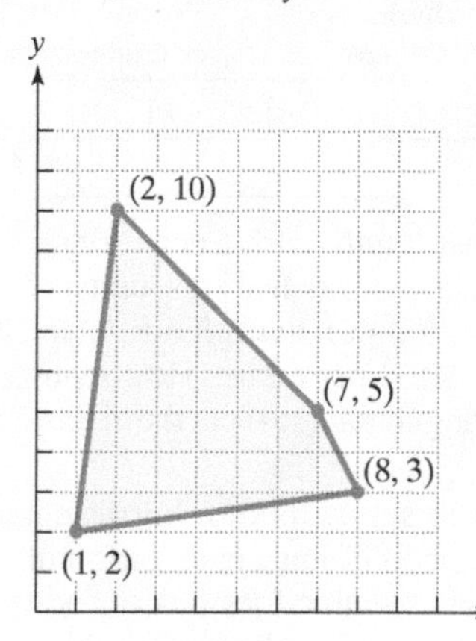

2. Objective Function
   $z = 3x + 2y$

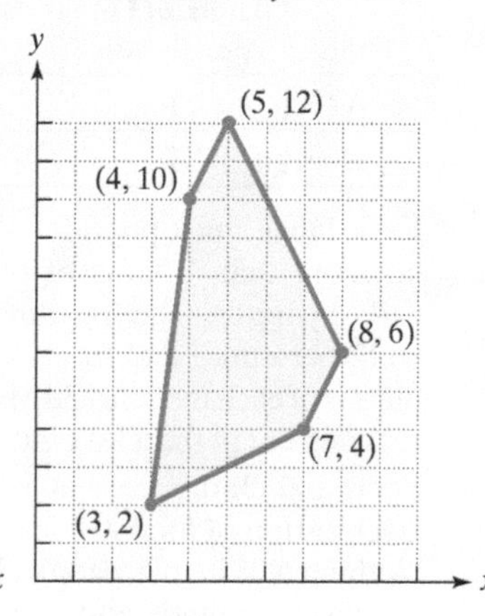

3. Objective Function
   $z = 40x + 50y$

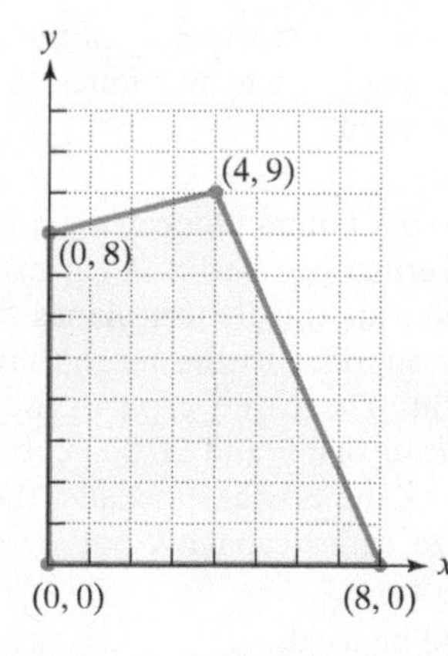

4. Objective Function
   $z = 30x + 45y$

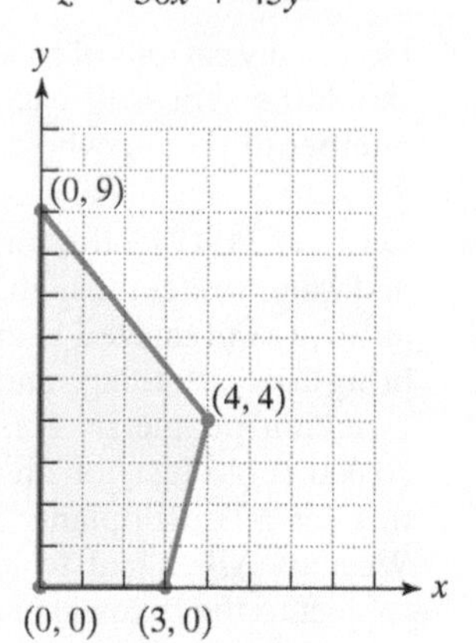

In Exercises 5–14, an objective function and a system of linear inequalities representing constraints are given.

a. *Graph the system of inequalities representing the constraints.*

b. *Find the value of the objective function at each corner of the graphed region.*

c. *Use the values in part (b) to determine the maximum value of the objective function and the values of x and y for which the maximum occurs.*

5. Objective Function $z = 3x + 2y$
   Constraints $\begin{cases} x \geq 0, y \geq 0 \\ 2x + y \leq 8 \\ x + y \geq 4 \end{cases}$

6. Objective Function $z = 2x + 3y$
   Constraints $\begin{cases} x \geq 0, y \geq 0 \\ 2x + y \leq 8 \\ 2x + 3y \leq 12 \end{cases}$

7. Objective Function $z = 4x + y$
   Constraints $\begin{cases} x \geq 0, y \geq 0 \\ 2x + 3y \leq 12 \\ x + y \geq 3 \end{cases}$

8. Objective Function $z = x + 6y$
   Constraints $\begin{cases} x \geq 0, y \geq 0 \\ 2x + y \leq 10 \\ x - 2y \geq -10 \end{cases}$

9. Objective Function $z = 3x - 2y$
   Constraints $\begin{cases} 1 \leq x \leq 5 \\ y \geq 2 \\ x - y \geq -3 \end{cases}$

10. Objective Function $z = 5x - 2y$
    Constraints $\begin{cases} 0 \leq x \leq 5 \\ 0 \leq y \leq 3 \\ x + y \geq 2 \end{cases}$

11. Objective Function $z = 4x + 2y$
    Constraints $\begin{cases} x \geq 0, y \geq 0 \\ 2x + 3y \leq 12 \\ 3x + 2y \leq 12 \\ x + y \geq 2 \end{cases}$

12. Objective Function $z = 2x + 4y$
    Constraints $\begin{cases} x \geq 0, y \geq 0 \\ x + 3y \geq 6 \\ x + y \geq 3 \\ x + y \leq 9 \end{cases}$

13. Objective Function $z = 10x + 12y$
    Constraints $\begin{cases} x \geq 0, y \geq 0 \\ 2x + y \leq 10 \\ 2x + 3y \leq 18 \end{cases}$

14. Objective Function $z = 5x + 6y$
    Constraints $\begin{cases} x \geq 0, y \geq 0 \\ 2x + y \geq 10 \\ x + 2y \geq 10 \\ x + y \leq 10 \end{cases}$

Application Exercises

15. A television manufacturer makes rear-projection and plasma televisions. The profit per unit is \$125 for the rear-projection televisions and \$200 for the plasma televisions.

    a. Let $x =$ the number of rear-projection televisions manufactured in a month and $y =$ the number of plasma televisions manufactured in a month. Write the objective function that models the total monthly profit.

b. The manufacturer is bound by the following constraints:
- Equipment in the factory allows for making at most 450 rear-projection televisions in one month.
- Equipment in the factory allows for making at most 200 plasma televisions in one month.
- The cost to the manufacturer per unit is $600 for the rear-projection televisions and $900 for the plasma televisions. Total monthly costs cannot exceed $360,000.

Write a system of three inequalities that models these constraints.

c. Graph the system of inequalities in part (b). Use only the first quadrant and its boundary, because *x* and *y* must both be nonnegative.

d. Evaluate the objective function for total monthly profit at each of the five vertices of the graphed region. [The vertices should occur at (0, 0), (0, 200), (300, 200), (450, 100), and (450, 0).]

e. Complete the missing portions of this statement: The television manufacturer will make the greatest profit by manufacturing ___ rear-projection televisions each month and ___ plasma televisions each month. The maximum monthly profit is $_____.

16. a. A student earns $15 per hour for tutoring and $10 per hour as a teacher's aide. Let $x =$ the number of hours each week spent tutoring and $y =$ the number of hours each week spent as a teacher's aide. Write the objective function that models total weekly earnings.

b. The student is bound by the following constraints:
- To have enough time for studies, the student can work no more than 20 hours a week.
- The tutoring center requires that each tutor spend at least three hours a week tutoring.
- The tutoring center requires that each tutor spend no more than eight hours a week tutoring.

Write a system of three inequalities that models these constraints.

c. Graph the system of inequalities in part (b). Use only the first quadrant and its boundary, because *x* and *y* are nonnegative.

d. Evaluate the objective function for total weekly earnings at each of the four vertices of the graphed region. [The vertices should occur at (3, 0), (8, 0), (3, 17), and (8, 12).]

e. Complete the missing portions of this statement: The student can earn the maximum amount per week by tutoring for ___ hours per week and working as a teacher's aide for ___ hours per week. The maximum amount that the student can earn each week is $___.

Use the two steps for solving a linear programming problem, given in the box on page 300, to solve the problems in Exercises 17–23.

17. A manufacturer produces two models of mountain bicycles. The times (in hours) required for assembling and painting each model are given in the following table:

	Model A	Model B
Assembling	5	4
Painting	2	3

The maximum total weekly hours available in the assembly department and the paint department are 200 hours and 108 hours, respectively. The profits per unit are $25 for model A and $15 for model B. How many of each type should be produced to maximize profit?

18. A large institution is preparing lunch menus containing foods A and B. The specifications for the two foods are given in the following table:

Food	Units of Fat per Ounce	Units of Carbohydrates per Ounce	Units of Protein per Ounce
A	1	2	1
B	1	1	1

Each lunch must provide at least 6 units of fat per serving, no more than 7 units of protein, and at least 10 units of carbohydrates. The institution can purchase food A for $0.12 per ounce and food B for $0.08 per ounce. How many ounces of each food should a serving contain to meet the dietary requirement at the least cost?

19. Food and clothing are shipped to survivors of a hurricane. Each carton of food will feed 12 people, while each carton of clothing will help 5 people. Each 20-cubic-foot box of food weighs 50 pounds and each 10-cubic-foot box of clothing weighs 20 pounds. The commercial carriers transporting food and clothing are bound by the following constraints:
- The total weight per carrier cannot exceed 19,000 pounds.
- The total volume must be less than 8000 cubic feet.

How many cartons of food and how many cartons of clothing should be sent with each plane shipment to maximize the number of people who can be helped?

20. On June 24, 1948, the former Soviet Union blocked all land and water routes through East Germany to Berlin. A gigantic airlift was organized using American and British planes to bring food, clothing, and other supplies to the more than 2 million people in West Berlin. The cargo capacity was 30,000 cubic feet for an American plane and 20,000 cubic feet for a British plane. To break the Soviet blockade, the Western Allies had to maximize cargo capacity, but were subject to the following restrictions:
- No more than 44 planes could be used.
- The larger American planes required 16 personnel per flight, double that of the requirement for the British planes. The total number of personnel available could not exceed 512.
- The cost of an American flight was $9000 and the cost of a British flight was $5000. Total weekly costs could not exceed $300,000.

Find the number of American planes and the number of British planes that were used to maximize cargo capacity.

21. A theater is presenting a program on drinking and driving for students and their parents. The proceeds will be donated to a local alcohol information center. Admission is $2.00 for parents and $1.00 for students. However, the situation has two constraints: The theater can hold no more than 150 people and every two parents must bring at least one student. How many parents and students should attend to raise the maximum amount of money?

22. You are about to take a test that contains computation problems worth 6 points each and word problems worth 10 points each. You can do a computation problem in 2 minutes and a word problem in 4 minutes. You have 40 minutes to take the test and may answer no more than 12 problems. Assuming you answer all the problems attempted correctly, how many of each type of problem must you do to maximize your score? What is the maximum score?

23. In 1978, a ruling by the Civil Aeronautics Board allowed Federal Express to purchase larger aircraft. Federal Express's options included 20 Boeing 727s that United Airlines was retiring and/or the French-built Dassault Fanjet Falcon 20. To aid in their decision, executives at Federal Express analyzed the following data:

	Boeing 727	Falcon 20
Direct Operating Cost	\$1400 per hour	\$500 per hour
Payload	42,000 pounds	6000 pounds

Federal Express was faced with the following constraints:

- Hourly operating cost was limited to \$35,000.
- Total payload had to be at least 672,000 pounds.
- Only twenty 727s were available.

Given the constraints, how many of each kind of aircraft should Federal Express have purchased to maximize the number of aircraft?

Explaining the Concepts

24. What kinds of problems are solved using the linear programming method?

25. What is an objective function in a linear programming problem?

26. What is a constraint in a linear programming problem? How is a constraint represented?

27. In your own words, describe how to solve a linear programming problem.

28. Describe a situation in your life in which you would like to maximize something, but are limited by at least two constraints. Can linear programming be used in this situation? Explain your answer.

Critical Thinking Exercises

Make Sense? *In Exercises 29–32, determine whether each statement makes sense or does not make sense, and explain your reasoning.*

29. In order to solve a linear programming problem, I use the graph representing the constraints and the graph of the objective function.

30. I use the coordinates of each vertex from my graph representing the constraints to find the values that maximize or minimize an objective function.

31. I need to be able to graph systems of linear inequalities in order to solve linear programming problems.

32. An important application of linear programming for businesses involves maximizing profit.

33. Suppose that you inherit \$10,000. The will states how you must invest the money. Some (or all) of the money must be invested in stocks and bonds. The requirements are that at least \$3000 be invested in bonds, with expected returns of \$0.08 per dollar, and at least \$2000 be invested in stocks, with expected returns of \$0.12 per dollar. Because the stocks are medium risk, the final stipulation requires that the investment in bonds should never be less than the investment in stocks. How should the money be invested so as to maximize your expected returns?

34. Consider the objective function $z = Ax + By$ ($A > 0$ and $B > 0$) subject to the following constraints: $2x + 3y \le 9$, $x - y \le 2$, $x \ge 0$, and $y \ge 0$. Prove that the objective function will have the same maximum value at the vertices $(3, 1)$ and $(0, 3)$ if $A = \frac{2}{3}B$.

Review Exercises

35. Simplify: $(2x^4y^3)(3xy^4)^3$. (Section 1.6, Example 9)

36. Solve for L: $3P = \dfrac{2L - W}{4}$. (Section 1.5, Example 8)

37. If $f(x) = x^3 + 2x^2 - 5x + 4$, find $f(-1)$. (Section 2.1, Example 3)

Preview Exercises

Exercises 38–40 will help you prepare for the material covered in the first section of the next chapter.

In Exercises 38–39, simplify each algebraic expression.

38. $(-9x^3 + 7x^2 - 5x + 3) + (13x^3 + 2x^2 - 8x - 6)$

39. $(7x^3 - 8x^2 + 9x - 6) - (2x^3 - 6x^2 - 3x + 9)$

40. The figures show the graphs of two functions.

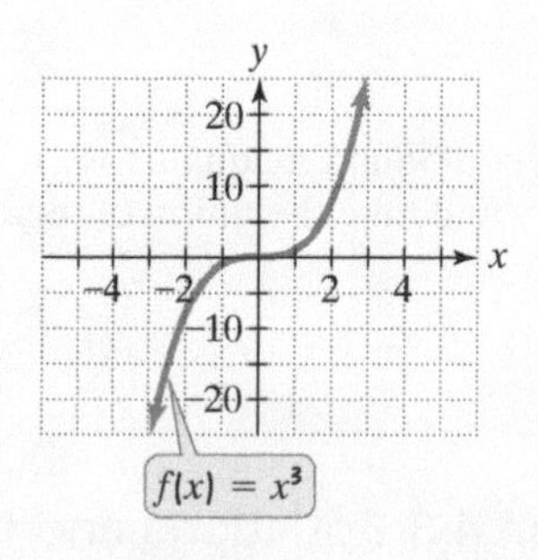

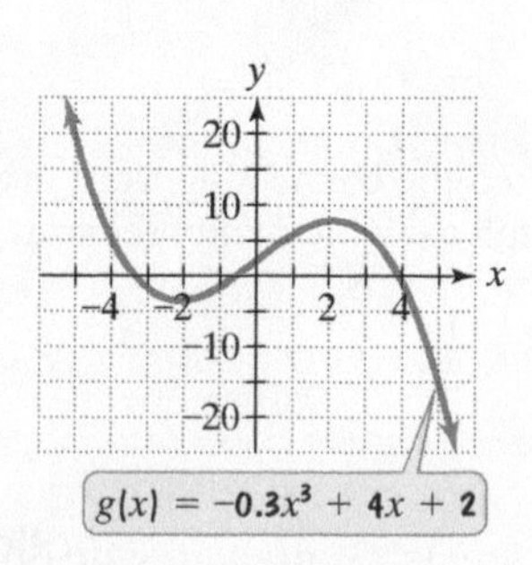

a. Which function, f or g, has a graph that rises to the left and falls to the right?

b. Which function, f or g, has a graph that falls to the left and rises to the right?

Chapter 4 Summary

Definitions and Concepts	Examples

Section 4.1 Solving Linear Inequalities

A linear inequality in one variable can be written in the form $ax + b < 0$, $ax + b \leq 0$, $ax + b > 0$, or $ax + b \geq 0$. The set of all numbers that make the inequality a true statement is its solution set, represented using interval notation.

Solving a Linear Inequality

1. Simplify each side.
2. Collect variable terms on one side and constant terms on the other side.
3. Isolate the variable and solve.

If an inequality is multiplied or divided by a negative number, the direction of the inequality symbol must be reversed.

Solve: $2(x + 3) - 5x \leq 15$.

$$2x + 6 - 5x \leq 15$$
$$-3x + 6 \leq 15$$
$$-3x \leq 9$$
$$\frac{-3x}{-3} \geq \frac{9}{-3}$$
$$x \geq -3$$

Solution set: $[-3, \infty)$

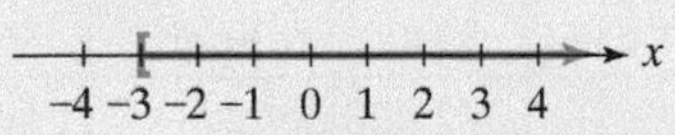

Section 4.2 Compound Inequalities

Intersection ($\cap$) and Union ($\cup$)

$A \cap B$ is the set of elements common to both set A and set B.
$A \cup B$ is the set of elements that are members of set A or set B or of both sets.

$\{1, 3, 5, 7\} \cap \{5, 7, 9, 11\} = \{5, 7\}$
$\{1, 3, 5, 7\} \cup \{5, 7, 9, 11\} = \{1, 3, 5, 7, 9, 11\}$

A compound inequality is formed by joining two inequalities with the word *and* or *or*.

When the connecting word is *and*, graph each inequality separately and take the intersection of their solution sets.

Solve: $x + 1 > 3$ and $x + 4 \leq 8$.

$x > 2$ and $x \leq 4$

-4 -3 -2 -1 0 1 2 3 4

Solution set: $(2, 4]$

The compound inequality $a < x < b$ means $a < x$ and $x < b$. Solve by isolating the variable in the middle.

Solve: $-1 < \dfrac{2x + 1}{3} \leq 2$.

$-3 < 2x + 1 \leq 6$ Multiply by 3.
$-4 < 2x \leq 5$ Subtract 1.
$-2 < x \leq \frac{5}{2}$ Divide by 2.

Solution set: $\left(-2, \frac{5}{2}\right]$

-4 -3 -2 -1 0 1 2 3 4

When the connecting word in a compound inequality is *or*, graph each inequality separately and take the union of their solution sets.

Solve: $x - 2 > -3$ or $2x \leq -6$.

$x > -1$ or $x \leq -3$

Solution set: $(-\infty, -3] \cup (-1, \infty)$

-6 -5 -4 -3 -2 -1 0 1 2

Section 4.3 Equations and Inequalities Involving Absolute Value

Absolute Value Equations

1. If $c > 0$, then $|u| = c$ means $u = c$ or $u = -c$.
2. If $c < 0$, then $|u| = c$ has no solution.
3. If $c = 0$, then $|u| = 0$ means $u = 0$.

Solve: $|2x - 7| = 3$.

$2x - 7 = 3$ or $2x - 7 = -3$
$2x = 10$ $\quad$ $2x = 4$
$x = 5$ $\quad$ $x = 2$

The solution set is $\{2, 5\}$.

Definitions and Concepts	Examples

Section 4.3 Equations and Inequalities Involving Absolute Value (continued)

Absolute Value Equations with Two Absolute Value Bars

If $|u| = |v|$, then $u = v$ or $u = -v$.

Solve: $|x - 6| = |2x + 1|$.

$$\begin{aligned} x - 6 &= 2x + 1 \quad \text{or} \quad & x - 6 &= -(2x + 1) \\ -x - 6 &= 1 & x - 6 &= -2x - 1 \\ -x &= 7 & 3x - 6 &= -1 \\ x &= -7 & 3x &= 5 \\ & & x &= \frac{5}{3} \end{aligned}$$

The solutions are -7 and $\frac{5}{3}$, and the solution set is $\left\{-7, \frac{5}{3}\right\}$.

Solving Absolute Value Inequalities

If c is a positive number,

1. $|u| < c$ is equivalent to $-c < u < c$.
2. $|u| > c$ is equivalent to $u < -c$ or $u > c$.

In each case, the absolute value inequality is rewritten as an equivalent compound inequality without absolute value bars.

Solve: $|x - 4| < 3$.

$$-3 < x - 4 < 3$$
$$1 < x < 7 \quad \text{Add 4.}$$

The solution set is $(1, 7)$.

0 1 2 3 4 5 6 7 8 x

Solve: $\left|\frac{x}{3} - 1\right| \geq 2$.

$$\frac{x}{3} - 1 \leq -2 \quad \text{or} \quad \frac{x}{3} - 1 \geq 2.$$
$$x - 3 \leq -6 \quad \text{or} \quad x - 3 \geq 6 \quad \text{Multiply by 3.}$$
$$x \leq -3 \quad \text{or} \quad x \geq 9 \quad \text{Add 3.}$$

The solution set is $(-\infty, -3] \cup [9, \infty)$.

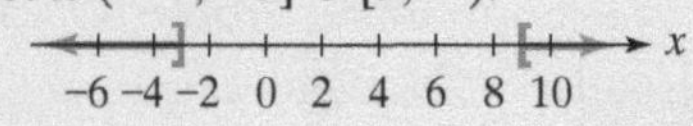

Absolute Value Inequalities with Unusual Solution Sets

If c is a negative number,

1. $|u| < c$ has no solution.
2. $|u| > c$ is true for all real numbers for which u is defined.

- $|x - 4| < -3$ has no solution. The solution set is $\varnothing$.
- $|3x + 6| > -12$ is true for all real numbers. The solution set is $(-\infty, \infty)$.

Section 4.4 Linear Inequalities in Two Variables

If the equal sign in $Ax + By = C$ is replaced with an inequality symbol, the result is a linear inequality in two variables. Its graph is the set of all points whose coordinates satisfy the inequality. To obtain the graph,

1. Replace the inequality symbol with an equal sign and graph the boundary line. Use a solid line for $\leq$ or $\geq$ and a dashed line for $<$ or $>$.
2. Choose a test point not on the line and substitute its coordinates into the inequality.
3. If a true statement results, shade the half-plane containing the test point. If a false statement results, shade the half-plane not containing the test point.

Graph: $x - 2y \leq 4$.

1. Graph $x - 2y = 4$. Use a solid line because the inequality symbol is $\leq$.
2. Test $(0, 0)$.
$$x - 2y \leq 4$$
$$0 - 2 \cdot 0 \overset{?}{\leq} 4$$
$$0 \leq 4, \quad \text{true}$$
3. The inequality is true. Shade the half-plane containing $(0, 0)$.

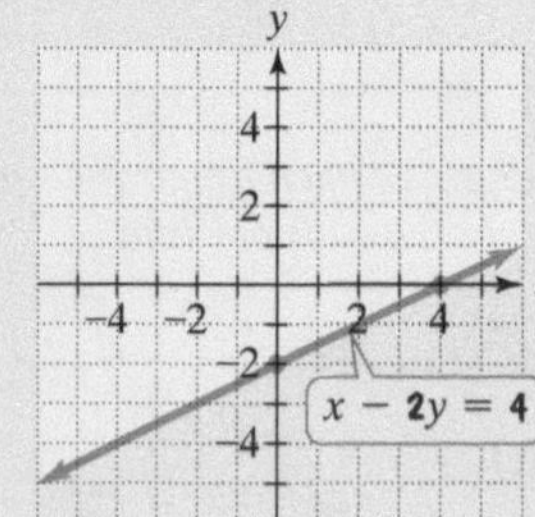

Definitions and Concepts	Examples

Section 4.4 Linear Inequalities in Two Variables (continued)

Two or more linear inequalities make up a system of linear inequalities. A solution is an ordered pair satisfying all inequalities in the system. To graph a system of inequalities, graph each inequality in the system. The overlapping region, if there is one, represents the solutions of the system. If there is no overlapping region, the system has no solution.

Graph the solutions of the system:

$$\begin{cases} y \le -2x \\ x - y \ge 3. \end{cases}$$

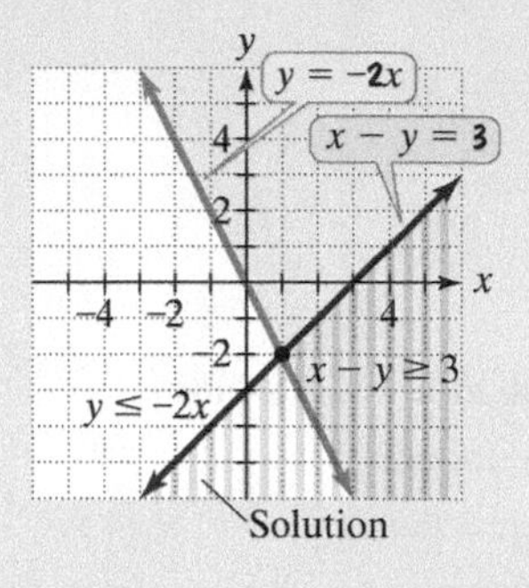

Section 4.5 Linear Programming

Linear programming is a method for solving problems in which a particular quantity that must be maximized or minimized is limited. An objective function is an algebraic expression in three variables modeling a quantity that must be maximized or minimized. Constraints are restrictions, expressed as linear inequalities.

Solving a Linear Programming Problem

1. Graph the system of inequalities representing the constraints.
2. Find the value of the objective function at each corner, or vertex, of the graphed region. The maximum and minimum of the objective function occur at one or more vertices.

Find the maximum value of the objective function $z = 3x + 2y$ subject to the following constraints: $x \ge 0, y \ge 0, 2x + 3y \le 18, 2x + y \le 10$.

1. Graph the system of inequalities representing the constraints.

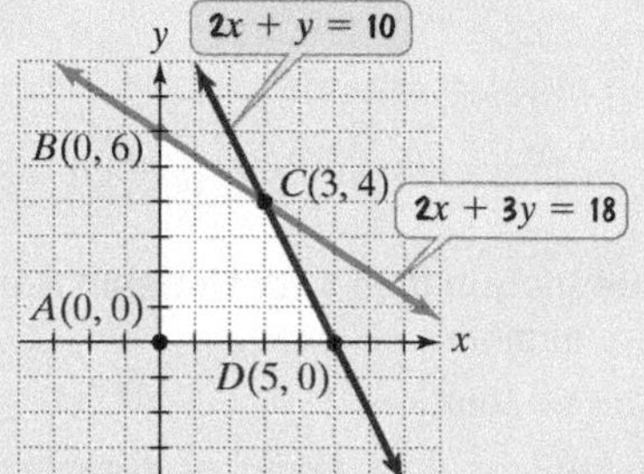

2. Evaluate the objective function at each vertex.

Vertex	$z = 3x + 2y$
$A(0, 0)$	$z = 3(0) + 2(0) = 0$
$B(0, 6)$	$z = 3(0) + 2(6) = 12$
$C(3, 4)$	$z = 3(3) + 2(4) = 17$
$D(5, 0)$	$z = 3(5) + 2(0) = 15$

The maximum value of the objective function is 17.

CHAPTER 4 REVIEW EXERCISES

4.1 *In Exercises 1–6, solve each linear inequality. Other than ∅, graph the solution set on a number line.*

1. $-6x + 3 \le 15$
2. $6x - 9 \ge -4x - 3$
3. $\frac{x}{3} - \frac{3}{4} - 1 > \frac{x}{2}$
4. $6x + 5 > -2(x - 3) - 25$
5. $3(2x - 1) - 2(x - 4) \ge 7 + 2(3 + 4x)$
6. $2x + 7 \le 5x - 6 - 3x$

7. A person can choose between two charges on a checking account. The first method involves a fixed cost of \$11 per month plus 6¢ for each check written. The second method involves a fixed cost of \$4 per month plus 20¢ for each check written. How many checks should be written to make the first method a better deal?

8. A salesperson earns \$500 per month plus a commission of 20% of sales. Describe the sales needed to receive a total income that exceeds \$3200 per month.

4.2 *In Exercises 9–12, let $A = \{a, b, c\}$, $B = \{a, c, d, e\}$, and $C = \{a, d, f, g\}$. Find the indicated set.*

9. $A \cap B$

10. $A \cap C$

11. $A \cup B$

12. $A \cup C$

In Exercises 13–23, solve each compound inequality. Other than $\varnothing$, graph the solution set on a number line.

13. $x \leq 3$ and $x < 6$

14. $x \leq 3$ or $x < 6$

15. $-2x < -12$ and $x - 3 < 5$

16. $5x + 3 \leq 18$ and $2x - 7 \leq -5$

17. $2x - 5 > -1$ and $3x < 3$

18. $2x - 5 > -1$ or $3x < 3$

19. $x + 1 \leq -3$ or $-4x + 3 < -5$

20. $5x - 2 \leq -22$ or $-3x - 2 > 4$

21. $5x + 4 \geq -11$ or $1 - 4x \geq 9$

22. $-3 < x + 2 \leq 4$

23. $-1 \leq 4x + 2 \leq 6$

24. To receive a B in a course, you must have an average of at least 80% but less than 90% on five exams. Your grades on the first four exams were 95%, 79%, 91%, and 86%. What range of grades on the fifth exam will result in a B for the course? Use interval notation to express this range.

4.3 *In Exercises 25–28, find the solution set for each equation.*

25. $|2x + 1| = 7$

26. $|3x + 2| = -5$

27. $2|x - 3| - 7 = 10$

28. $|4x - 3| = |7x + 9|$

In Exercises 29–33, solve each absolute value inequality. Other than $\varnothing$, graph the solution set on a number line.

29. $|2x + 3| \leq 15$

30. $\left|\dfrac{2x + 6}{3}\right| > 2$

31. $|2x + 5| - 7 < -6$

32. $-4|x + 2| + 5 \leq -7$

33. $|2x - 3| + 4 \leq -10$

34. Approximately 90% of the population sleeps h hours daily, where h is modeled by the inequality $|h - 6.5| \leq 1$. Write a sentence describing the range for the number of hours that most people sleep. Do *not* use the phrase "absolute value" in your description.

4.4 *In Exercises 35–40, graph each inequality in a rectangular coordinate system.*

35. $3x - 4y > 12$

36. $x - 3y \leq 6$

37. $y \leq -\frac{1}{2}x + 2$

38. $y > \frac{3}{5}x$

39. $x \leq 2$

40. $y > -3$

In Exercises 41–49, graph the solution set of each system of inequalities or indicate that the system has no solution.

41. $\begin{cases} 2x - y \leq 4 \\ x + y \geq 5 \end{cases}$

42. $\begin{cases} y < -x + 4 \\ y > x - 4 \end{cases}$

43. $-3 \leq x < 5$

44. $-2 < y \leq 6$

45. $\begin{cases} x \geq 3 \\ y \leq 0 \end{cases}$

46. $\begin{cases} 2x - y > -4 \\ x \geq 0 \end{cases}$

47. $\begin{cases} x + y \leq 6 \\ y \geq 2x - 3 \end{cases}$

48. $\begin{cases} 3x + 2y \geq 4 \\ x - y \leq 3 \\ x \geq 0, y \geq 0 \end{cases}$

49. $\begin{cases} 2x - y > 2 \\ 2x - y < -2 \end{cases}$

4.5

50. Find the value of the objective function $z = 2x + 3y$ at each corner of the graphed region shown. What is the maximum value of the objective function? What is the minimum value of the objective function?

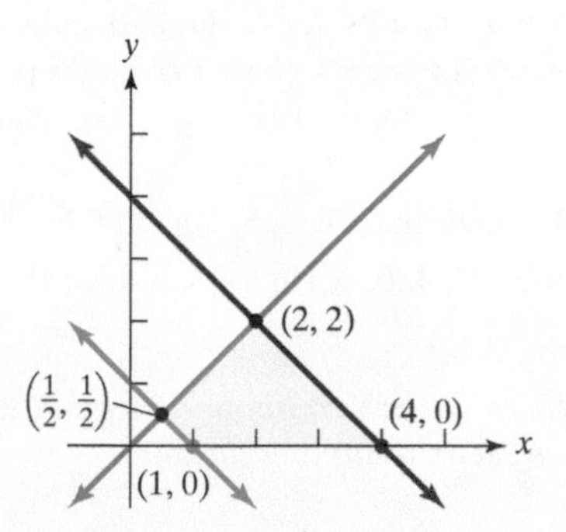

In Exercises 51–53, graph the region determined by the constraints. Then find the maximum value of the given objective function, subject to the constraints.

51. Objective Function $z = 2x + 3y$
Constraints $\begin{cases} x \geq 0, y \geq 0 \\ x + y \leq 8 \\ 3x + 2y \geq 6 \end{cases}$

52. Objective Function $z = x + 4y$
Constraints $\begin{cases} 0 \leq x \leq 5, 0 \leq y \leq 7 \\ x + y \geq 3 \end{cases}$

53. Objective Function $z = 5x + 6y$

Constraints $\begin{cases} x \geq 0, y \geq 0 \\ y \leq x \\ 2x + y \leq 12 \\ 2x + 3y \geq 6 \end{cases}$

54. A paper manufacturing company converts wood pulp to writing paper and newsprint. The profit on a unit of writing paper is $500 and the profit on a unit of newsprint is $350.

a. Let x represent the number of units of writing paper produced daily. Let y represent the number of units of newsprint produced daily. Write the objective function that models total daily profit.

b. The manufacturer is bound by the following constraints:
- Equipment in the factory allows for making at most 200 units of paper (writing paper and newsprint) in a day.
- Regular customers require at least 10 units of writing paper and at least 80 units of newsprint daily.

Write a system of inequalities that models these constraints.

c. Graph the inequalities in part (b). Use only the first quadrant, because x and y must both be positive. (*Suggestion*: Let each unit along the x- and y-axes represent 20.)

d. Evaluate the objective function at each of the three vertices of the graphed region.

e. Complete the missing portions of this statement: The company will make the greatest profit by producing ___ units of writing paper and ___ units of newsprint each day. The maximum daily profit is $________.

55. A manufacturer of lightweight tents makes two models whose specifications are given in the following table.

	Cutting Time per Tent	Assembly Time per Tent
Model A	0.9 hour	0.8 hour
Model B	1.8 hours	1.2 hours

Each month, the manufacturer has no more than 864 hours of labor available in the cutting department and at most 672 hours in the assembly division. The profits come to $25 per tent for model A and $40 per tent for model B. How many of each should be manufactured monthly to maximize the profit?

CHAPTER 4 TEST

Step-by-step test solutions are found on the Chapter Test Prep Videos available in MyMathLab® or on YouTube (search "BlitzerInterAlg7e" and click on "Channels").

In Exercises 1–2, solve and graph the solution set on a number line.

1. $3(x + 4) \geq 5x - 12$

2. $\frac{x}{6} + \frac{1}{8} \leq \frac{x}{2} - \frac{3}{4}$

3. You are choosing between two landline phone plans. Plan A charges $25 per month for unlimited calls. Plan B has a monthly fee of $13 with a charge of $0.06 per minute. How many minutes of calls in a month make plan A the better deal?

4. Find the intersection: $\{2, 4, 6, 8, 10\} \cap \{4, 6, 12, 14\}$.

5. Find the union: $\{2, 4, 6, 8, 10\} \cup \{4, 6, 12, 14\}$.

In Exercises 6–10, solve each compound inequality. Other than ∅, graph the solution set on a number line.

6. $2x + 4 < 2$ and $x - 3 > -5$

7. $x + 6 \geq 4$ and $2x + 3 \geq -2$

8. $2x - 3 < 5$ or $3x - 6 \leq 4$

9. $x + 3 \leq -1$ or $-4x + 3 < -5$

10. $-3 \leq \frac{2x + 5}{3} < 6$

In Exercises 11–12, find the solution set for each equation.

11. $|5x + 3| = 7$

12. $|6x + 1| = |4x + 15|$

In Exercises 13–14, solve and graph the solution set on a number line.

13. $|2x - 1| < 7$

14. $|2x - 3| \geq 5$

15. The inequality $|b - 98.6| > 8$ describes a person's body temperature, b, in degrees Fahrenheit, when hyperthermia (extremely high body temperature) or hypothermia (extremely low body temperature) occurs. Solve the inequality and interpret the solution.

In Exercises 16–18, graph each inequality in a rectangular coordinate system.

16. $3x - 2y < 6$

17. $y \geq \frac{1}{2}x - 1$

18. $y \leq -1$

In Exercises 19–21, graph the solution set of each system of inequalities.

19. $\begin{cases} x + y \geq 2 \\ x - y \geq 4 \end{cases}$

20. $\begin{cases} 3x + y \leq 9 \\ 2x + 3y \geq 6 \\ x \geq 0, y \geq 0 \end{cases}$

21. $-2 < x \leq 4$

22. Find the maximum value of the objective function $z = 3x + 5y$ subject to the following constraints: $x \geq 0, y \geq 0, x + y \leq 6, x \geq 2$.

23. A manufacturer makes two types of jet skis, regular and deluxe. The profit on a regular jet ski is \$200 and the profit on the deluxe model is \$250. To meet customer demand, the company must manufacture at least 50 regular jet skis per week and at least 75 deluxe models. To maintain high quality, the total number of both models of jet skis manufactured by the company should not exceed 150 per week. How many jet skis of each type should be manufactured per week to obtain maximum profit? What is the maximum weekly profit?

CUMULATIVE REVIEW EXERCISES (CHAPTERS 1–4)

In Exercises 1–2, solve each equation.

1. $5(x + 1) + 2 = x - 3(2x + 1)$

2. $\dfrac{2(x + 6)}{3} = 1 + \dfrac{4x - 7}{3}$

3. Simplify: $\dfrac{-10x^2y^4}{15x^7y^{-3}}$.

4. If $f(x) = x^2 - 3x + 4$, find $f(-3)$ and $f(2a)$.

5. If $f(x) = 3x^2 - 4x + 1$ and $g(x) = x^2 - 5x - 1$, find $(f - g)(x)$ and $(f - g)(2)$.

6. Use function notation to write the equation of the line passing through $(2, 3)$ and perpendicular to the line whose equation is $y = 2x - 3$.

In Exercises 7–10, graph each equation or inequality in a rectangular coordinate system.

7. $f(x) = 2x + 1$

8. $y > 2x$

9. $2x - y \geq 6$

10. $f(x) = -1$

11. Solve the system:

$$\begin{cases} 3x - y + z = -15 \\ x + 2y - z = 1. \\ 2x + 3y - 2z = 0 \end{cases}$$

12. Solve using matrices:

$$\begin{cases} 2x - y = -4 \\ x + 3y = 5. \end{cases}$$

13. Evaluate: $\begin{vmatrix} 4 & 3 \\ -1 & -5 \end{vmatrix}$.

14. A motel with 60 rooms charges \$90 per night for rooms with kitchen facilities and \$80 per night for rooms without kitchen facilities. When all rooms are occupied, the nightly revenue is \$5260. How many rooms of each kind are there?

15. Which of the following are functions?

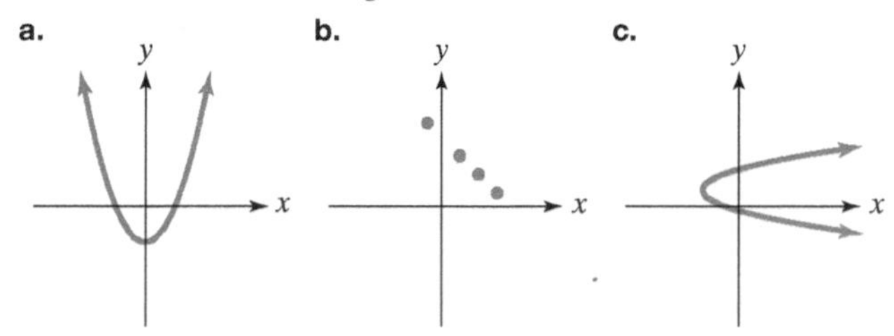

In Exercises 16–20, solve and graph the solution set on a number line.

16. $\dfrac{x}{4} - \dfrac{3}{4} - 1 \leq \dfrac{x}{2}$

17. $2x + 5 \leq 11$ and $-3x > 18$

18. $x - 4 \geq 1$ or $-3x + 1 \geq -5 - x$

19. $|2x + 3| \leq 17$

20. $|3x - 8| > 7$

CHAPTER 5

Polynomials, Polynomial Functions, and Factoring

Just one hundred years ago, there were at least 100,000 wild tigers. By 2010, the estimated world tiger population was 3200. Without conservation efforts, tigers could disappear from the wild by 2022.

Here's where you'll find this application:

A function that models this scenario is

$$f(x) = 0.76x^3 - 30x^2 - 882x + 37{,}807,$$

where $f(x)$ is the world tiger population x years after 1970. The algebraic expression on the right side contains variables to powers that are whole numbers and is an example of a polynomial. Much of what we do in algebra involves polynomials and polynomial functions. The vanishing tiger model, with an accompanying graph, is developed in Exercises 65–68 in Exercise Set 5.1.

SECTION

5.1 Introduction to Polynomials and Polynomial Functions

What am I supposed to learn?

After studying this section, you should be able to:

1. Use the vocabulary of polynomials.
2. Evaluate polynomial functions.
3. Determine end behavior.
4. Add polynomials.
5. Subtract polynomials.

In 1980, U.S. doctors diagnosed 41 cases of a rare form of cancer, Kaposi's sarcoma, that involved skin lesions, pneumonia, and severe immunological deficiencies. All cases involved gay men ranging in age from 26 to 51. By the end of 2011, approximately 1.2 million Americans, straight and gay, male and female, old and young, were infected with the HIV virus.

Modeling AIDS-related data and making predictions about the epidemic's havoc is serious business. **Figure 5.1** shows the number of AIDS cases diagnosed in the United States from 1983 through 2011.

Basketball player Magic Johnson (1959–) tested positive for HIV in 1991.

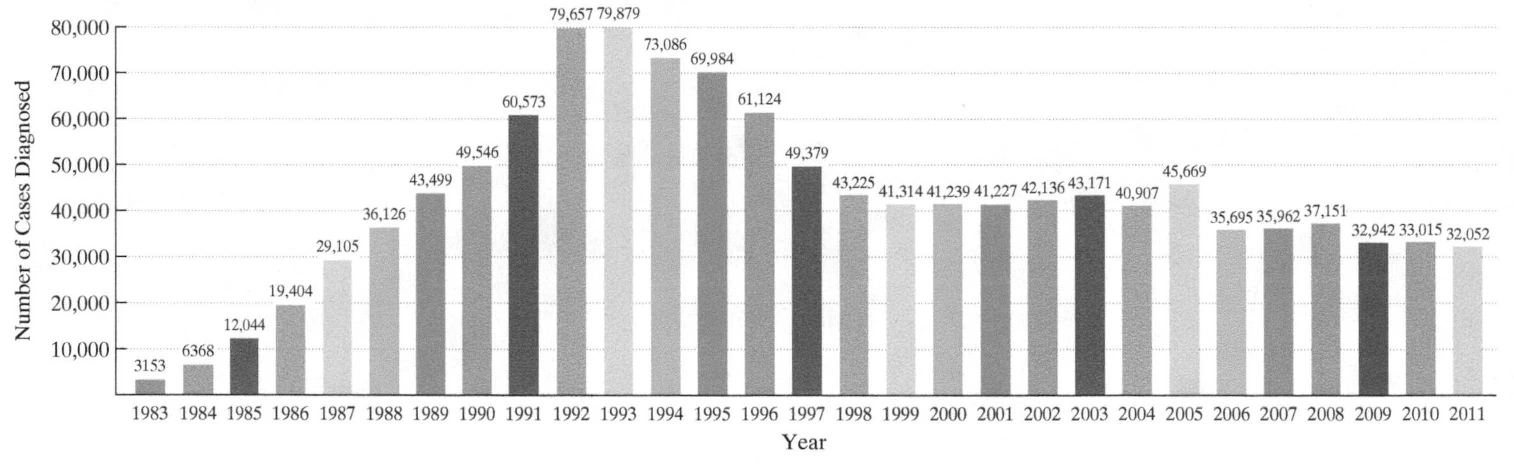

Figure 5.1

Source: U.S. Department of Health and Human Services

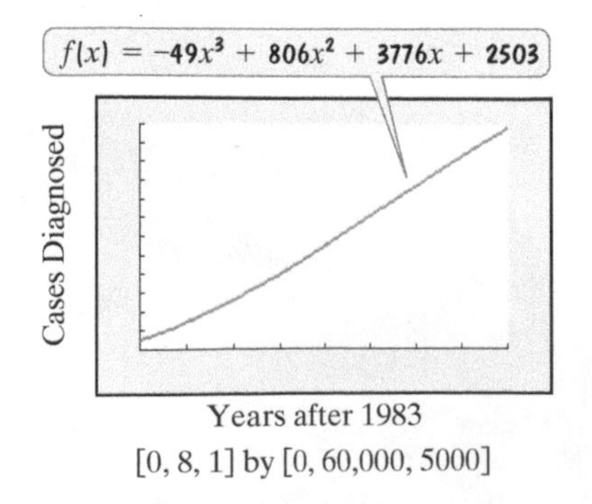

Figure 5.2 The graph of a function modeling the number of AIDS cases from 1983 through 1991

Changing circumstances and unforeseen events can result in models for AIDS-related data that are not particularly useful over long periods of time. For example, the function

$$f(x) = -49x^3 + 806x^2 + 3776x + 2503$$

models the number of AIDS cases diagnosed in the United States x years after 1983. The model was obtained using a portion of the data shown in **Figure 5.1**, namely cases diagnosed from 1983 through 1991, inclusive. **Figure 5.2** shows the graph of f from 1983 through 1991.

The voice balloon in **Figure 5.2** displays the algebraic expression used to define f. This algebraic expression is an example of a *polynomial*. A **polynomial** is a single term or the sum of two or more terms containing variables with whole-number exponents. Functions containing polynomials are used in such diverse areas as science, business, medicine, psychology, and sociology. In this section, we present basic ideas about polynomials and polynomial functions. We will use our knowledge of combining like terms to find sums and differences of polynomials.

1 Use the vocabulary of polynomials.

How We Describe Polynomials

Consider the polynomial

$$-49x^3 + 806x^2 + 3776x + 2503.$$

This polynomial contains four terms. It is customary to write the terms in the order of descending powers of the variable. This is the **standard form** of a polynomial.

Some polynomials contain only one variable. Each term of such a polynomial in x is of the form ax^n. If $a \neq 0$, the **degree** of ax^n is n. For example, the degree of the term $-49x^3$ is 3.

Great Question!

Why doesn't the constant 0 have a degree?

We can express 0 in many ways, including $0x$, $0x^2$, and $0x^3$. It is impossible to assign a single exponent on the variable. This is why 0 has no defined degree.

The Degree of ax^n

If $a \neq 0$ and n is a whole number, the degree of ax^n is n. The degree of a nonzero constant is 0. The constant 0 has no defined degree.

Here is the polynomial modeling AIDS cases and the degree of each of its four terms:

$$-49x^3 + 806x^2 + 3776x + 2503$$

degree 3 | degree 2 | degree 1 | degree of nonzero constant: 0

Notice that the exponent on x for the term $3776x$ is understood to be 1: $3776x^1$. For this reason, the degree of $3776x$ is 1. You can think of 2503 as $2503x^0$; thus, its degree is 0.

A polynomial is simplified when it contains no grouping symbols and no like terms. A simplified polynomial that has exactly one term is called a **monomial**. A **binomial** is a simplified polynomial that has two terms. A **trinomial** is a simplified polynomial with three terms. Simplified polynomials with four or more terms have no special names.

Some polynomials contain two or more variables. Here is an example of a polynomial in two variables, x and y:

$$7x^2y^3 - 17x^4y^2 + xy - 6y^2 + 9.$$

A polynomial in two variables, x and y, contains the sum of one or more monomials of the form ax^ny^m. The constant a is the **coefficient**. The exponents, n and m, represent whole numbers. The **degree of the term** ax^ny^m is the sum of the exponents of the variables, $n + m$.

The **degree of a polynomial** is the greatest degree of any term of the polynomial. If there is precisely one term of the greatest degree, it is called the **leading term**. Its coefficient is called the **leading coefficient**.

EXAMPLE 1 Using the Vocabulary of Polynomials

Determine the coefficient of each term, the degree of each term, the degree of the polynomial, the leading term, and the leading coefficient of the polynomial

$$7x^2y^3 - 17x^4y^2 + xy - 6y^2 + 9.$$

Solution Consider each term of the polynomial $7x^2y^3 - 17x^4y^2 + xy - 6y^2 + 9$.

Term	Coefficient	Degree (Sum of Exponents on the Variables)
$7x^2y^3$	7	$2 + 3 = 5$
$-17x^4y^2$	-17	$4 + 2 = 6$
xy	1	$1 + 1 = 2$
$-6y^2$	-6	$0 + 2 = 2$
9	9	$0 + 0 = 0$

Think of xy as $1x^1y^1$.

Think of $-6y^2$ as $-6x^0y^2$.

Think of 9 as $9x^0y^0$.

The degree of the polynomial is the greatest degree of any term of the polynomial, which is 6. The leading term is the term of the greatest degree, which is $-17x^4y^2$. Its coefficient, -17, is the leading coefficient. ■

✓ CHECK POINT 1 Determine the coefficient of each term, the degree of each term, the degree of the polynomial, the leading term, and the leading coefficient of the polynomial

$$8x^4y^5 - 7x^3y^2 - x^2y - 5x + 11.$$

If a polynomial contains three or more variables, the degree of a term is the sum of the exponents of all the variables. Here is an example of a polynomial in three variables, x, y, and z:

The coefficients are $\frac{1}{4}$, -2, 6, and 5.

$$\frac{1}{4}xy^2z^4 - 2xyz + 6x^2 + 5.$$

Degree: $1 + 2 + 4 = 7$ | Degree: $1 + 1 + 1 = 3$ | Degree: 2 | Degree: 0

The degree of this polynomial is the greatest degree of any term of the polynomial, which is 7.

2 Evaluate polynomial functions.

Polynomial Functions

The expression $4x^3 - 5x^2 + 3$ is a polynomial. If we write

$$f(x) = 4x^3 - 5x^2 + 3,$$

then we have a **polynomial function**. In a polynomial function, the expression that defines the function is a polynomial. How do we evaluate a polynomial function? Use substitution, just as we did to evaluate other functions in Chapter 2.

EXAMPLE 2 Evaluating a Polynomial Function

The polynomial function

$$f(x) = -49x^3 + 806x^2 + 3776x + 2503$$

models the number of AIDS cases diagnosed in the United States, $f(x)$, x years after 1983, where $0 \le x \le 8$. Find $f(6)$ and describe what this means in practical terms.

Solution To find $f(6)$, or f of 6, we replace each occurrence of x in the function's formula with 6.

$f(x) = -49x^3 + 806x^2 + 3776x + 2503$	This is the given function.
$f(6) = -49(6)^3 + 806(6)^2 + 3776(6) + 2503$	Replace each occurrence of x with 6.
$= -49(216) + 806(36) + 3776(6) + 2503$	Evaluate exponential expressions.
$= -10{,}584 + 29{,}016 + 22{,}656 + 2503$	Multiply.
$= 43{,}591$	Add.

Thus, $f(6) = 43{,}591$. According to the model, this means that 6 years after 1983, in 1989, there were 43,591 AIDS cases diagnosed in the United States. (The actual number, shown in **Figure 5.1**, is 43,499.) ■

Using Technology

Once each occurrence of x in $f(x) = -49x^3 + 806x^2 + 3776x + 2503$ is replaced with 6, the resulting computation can be performed using a scientific calculator or a graphing calculator.

$$-49(6)^3 + 806(6)^2 + 3776(6) + 2503$$

Many Scientific Calculators

49 [+/−] [×] 6 [y^x] 3 [+] 806 [×] 6 [y^x] 2 [+] 3776 [×] 6 [+] 2503 [=]

Many Graphing Calculators

[(−)] 49 [×] 6 [^] 3 [+] 806 [×] 6 [^] 2 [+] 3776 [×] 6 [+] 2503 [ENTER]

The display should be 43591. This number can also be obtained by using a graphing utility's feature that evaluates a function or by using its table feature.

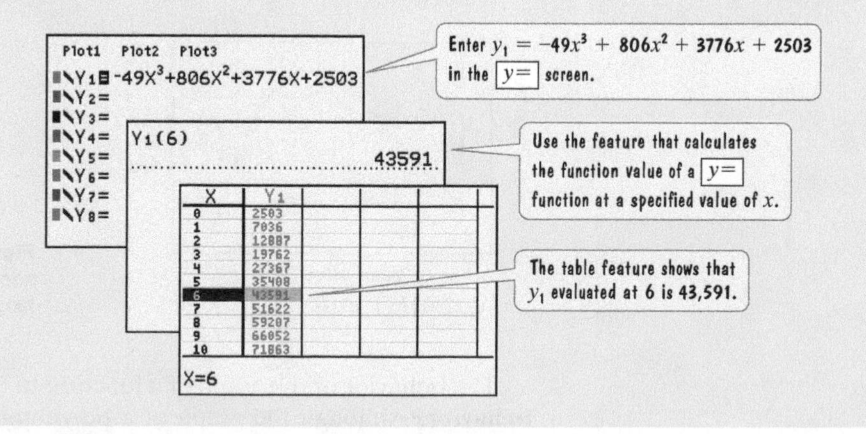

☑ **CHECK POINT 2** For the polynomial function
$$f(x) = 4x^3 - 3x^2 - 5x + 6,$$
find $f(2)$.

Smooth, Continuous Graphs

Polynomial functions of degree 2 or higher have graphs that are *smooth* and *continuous*. By **smooth**, we mean that the graph contains only rounded curves with no sharp corners. By **continuous**, we mean that the graph has no breaks and can be drawn without lifting your pencil from the rectangular coordinate system. These ideas are illustrated in **Figure 5.3**.

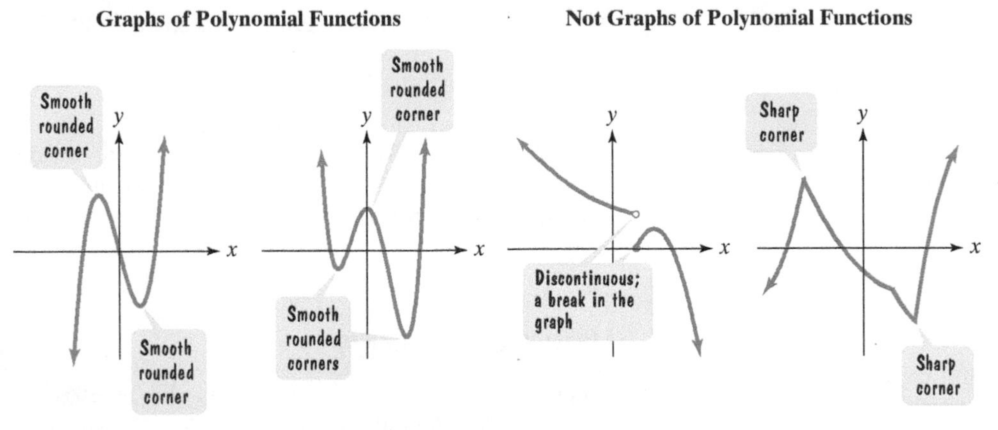

Figure 5.3 Recognizing graphs of polynomial functions

3 Determine end behavior.

End Behavior of Polynomial Functions

Figure 5.4 shows the graph of the function

$$f(x) = -49x^3 + 806x^2 + 3776x + 2503,$$

which models U.S. AIDS cases from 1983 through 1991. Look at what happens to the graph when we extend the year up through 2005. By year 21 (2004), the values of y are negative and the function no longer models AIDS cases. We've added an arrow to the graph at the far right to emphasize that it continues to decrease without bound. This far-right *end behavior* of the graph is one reason that this function is inappropriate for modeling AIDS cases into the future.

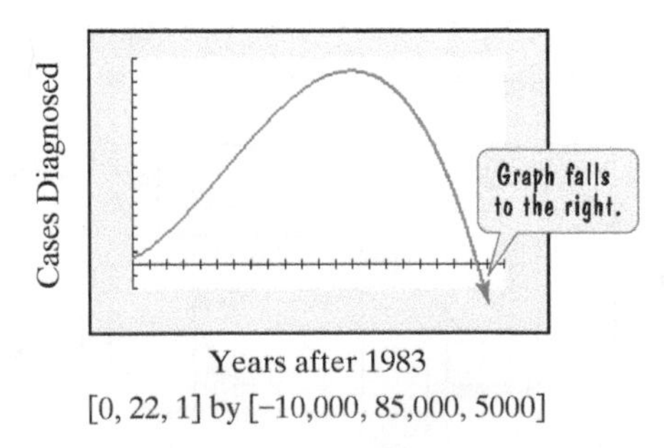

Figure 5.4 By extending the viewing rectangle, we see that y is eventually negative and the function no longer models the number of AIDS cases.

The behavior of the graph of a function to the far left or the far right is called its **end behavior**. Although the graph of a polynomial function may have intervals where it increases and intervals where it decreases, the graph will eventually rise or fall without bound as it moves far to the left or far to the right.

How can you determine whether the graph of a polynomial function goes up or down at each end? **The end behavior depends upon the leading term.** In particular, the sign of the leading coefficient and the degree of the polynomial reveal the graph's end behavior. With regard to end behavior, only the leading term—that is, the term of the greatest degree—counts, as summarized by the **Leading Coefficient Test**.

The Leading Coefficient Test

As x increases or decreases without bound, the graph of a polynomial function eventually rises or falls. In particular,

1. For odd-degree polynomials:

   If the leading coefficient is positive, the graph falls to the left and rises to the right. (↙, ↗)

   If the leading coefficient is negative, the graph rises to the left and falls to the right. (↖, ↘)

2. For even-degree polynomials:

   If the leading coefficient is positive, the graph rises to the left and rises to the right. (↖, ↗)

   If the leading coefficient is negative, the graph falls to the left and falls to the right. (↙, ↘)

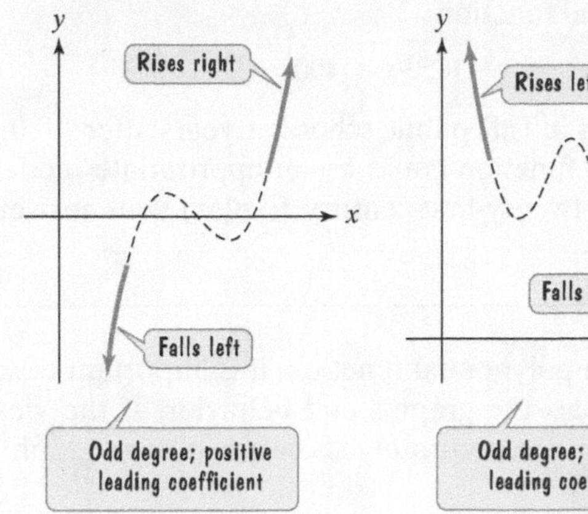

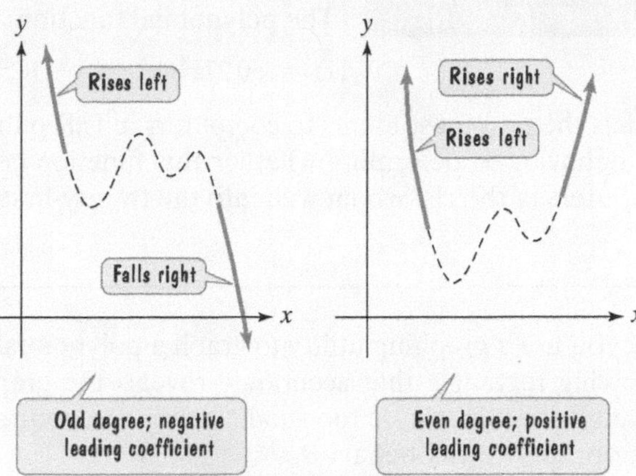

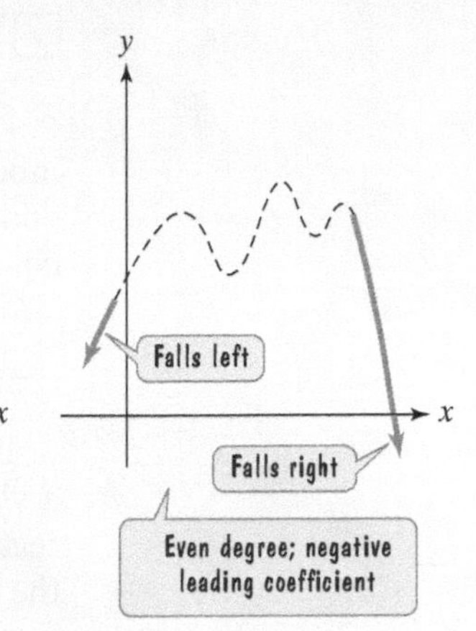

Great Question!

What's the bottom line on the Leading Coefficient Test?

Odd-degree polynomial functions have graphs with opposite behavior at each end. Even-degree polynomial functions have graphs with the same behavior at each end.

EXAMPLE 3 Using the Leading Coefficient Test

Use the Leading Coefficient Test to determine the end behavior of the graph of

$$f(x) = x^3 + 3x^2 - x - 3.$$

Solution We begin by identifying the sign of the leading coefficient and the degree of the polynomial.

$$f(x) = x^3 + 3x^2 - x - 3$$

The leading coefficient, 1, is positive. The degree of the polynomial, 3, is odd.

The degree of the function f is 3, which is odd. Odd-degree polynomial functions have graphs with opposite behavior at each end. The leading coefficient, 1, is positive. Thus, the graph falls to the left and rises to the right (↙, ↗). The graph of f is shown in **Figure 5.5**. ■

Rises right

Falls left

Figure 5.5 The graph of $f(x) = x^3 + 3x^2 - x - 3$

☑ CHECK POINT 3 Use the Leading Coefficient Test to determine the end behavior of the graph of $f(x) = x^4 - 4x^2$.

EXAMPLE 4 Using the Leading Coefficient Test

Use end behavior to explain why

$$f(x) = -49x^3 + 806x^2 + 3776x + 2503$$

is only an appropriate model for AIDS cases for a limited time period.

Solution We begin by identifying the sign of the leading coefficient and the degree of the polynomial.

$$f(x) = -49x^3 + 806x^2 + 3776x + 2503$$

The leading coefficient, −49, is negative. The degree of the polynomial, 3, is odd.

The degree of f in $f(x) = -49x^3 + 806x^2 + 3776x + 2053$ is 3, which is odd. Odd-degree polynomial functions have graphs with opposite behavior at each end. The leading coefficient, -49, is negative. Thus, the graph rises to the left and falls to the right (↖, ↘). The fact that the graph falls to the right indicates that at some point the number of AIDS cases will be negative, an impossibility. If a function has a graph that decreases without bound over time, it will not be capable of modeling nonnegative phenomena over long time periods. Model breakdown will eventually occur. ■

☑ CHECK POINT 4 The polynomial function

$$f(x) = -0.27x^3 + 9.2x^2 - 102.9x + 400$$

models the ratio of students to computers in U.S. public schools x years after 1980. Use end behavior to determine whether this function could be an appropriate model for computers in the classroom well into the twenty-first century. Explain your answer.

If you use a graphing utility to graph a polynomial function, it is important to select a viewing rectangle that accurately reveals the graph's end behavior. If the viewing rectangle, or window, is too small, it may not accurately show a complete graph with the appropriate end behavior.

EXAMPLE 5 Using the Leading Coefficient Test

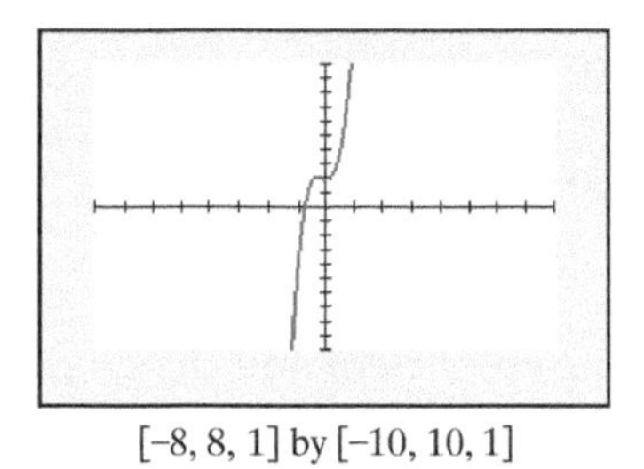

$[-8, 8, 1]$ by $[-10, 10, 1]$

Figure 5.6

The graph of $f(x) = -x^4 + 8x^3 + 4x^2 + 2$ was obtained with a graphing utility using a $[-8, 8, 1]$ by $[-10, 10, 1]$ viewing rectangle. The graph is shown in **Figure 5.6**. Is this a complete graph that shows the end behavior of the function?

Solution We begin by identifying the sign of the leading coefficient and the degree of the polynomial.

$$f(x) = -x^4 + 8x^3 + 4x^2 + 2$$

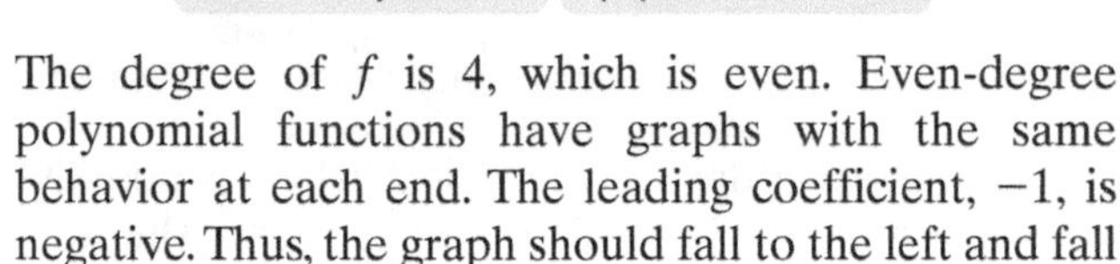
The leading coefficient, −1, is negative.
The degree of the polynomial, 4, is even.

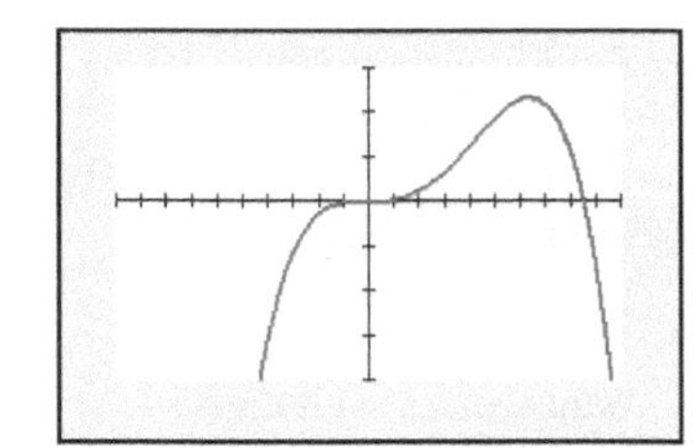

$[-10, 10, 1]$ by $[-1000, 750, 250]$

Figure 5.7

The degree of f is 4, which is even. Even-degree polynomial functions have graphs with the same behavior at each end. The leading coefficient, -1, is negative. Thus, the graph should fall to the left and fall to the right (↙, ↘). The graph in **Figure 5.6** is falling to the left, but it is not falling to the right. Therefore, the graph is not complete enough to show end behavior. A more complete graph of the function is shown in a larger viewing rectangle in **Figure 5.7**. ■

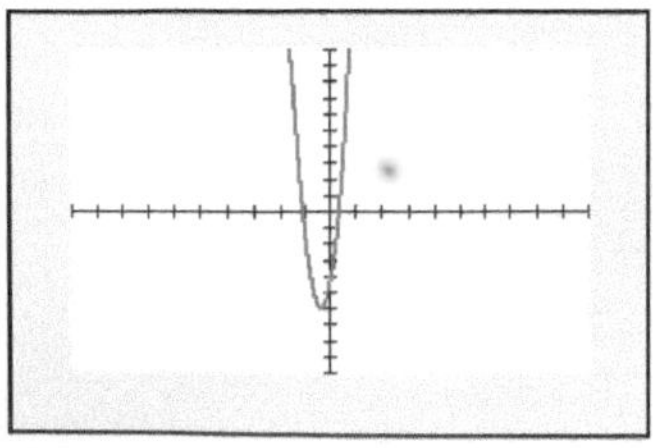

Figure 5.8

☑ CHECK POINT 5 The graph of $f(x) = x^3 + 13x^2 + 10x - 4$ is shown in a standard viewing rectangle in **Figure 5.8**. Use the Leading Coefficient Test to determine whether this is a complete graph that shows the end behavior of the function. Explain your answer.

---

4 Add polynomials.

Adding Polynomials

Polynomials are added by combining like terms. Here are two examples that illustrate the use of the distributive property in adding monomials and combining like terms:

$$-9x^3 + 13x^3 = (-9 + 13)x^3 = 4x^3$$

Add coefficients and keep the same variable factor(s).

$$-7x^3y^2 + 4x^3y^2 = (-7 + 4)x^3y^2 = -3x^3y^2.$$

EXAMPLE 6 Adding Polynomials

Add: $(-6x^3 + 5x^2 + 4) + (2x^3 + 7x^2 - 10)$.

Solution

$(-6x^3 + 5x^2 + 4) + (2x^3 + 7x^2 - 10)$

$= -6x^3 + 5x^2 + 4 + 2x^3 + 7x^2 - 10$ Remove the parentheses. Like terms are shown in the same color.

$= \underbrace{-6x^3 + 2x^3} + \underbrace{5x^2 + 7x^2} + \underbrace{4 - 10}$ Rearrange the terms so that like terms are adjacent.

$= -4x^3 + 12x^2 - 6$ Combine like terms.

$= -4x^3 + 12x^2 - 6$ This is the same sum as above, written more concisely. ■

✓ CHECK POINT 6 Add: $(-7x^3 + 4x^2 + 3) + (4x^3 + 6x^2 - 13)$.

EXAMPLE 7 Adding Polynomials

Add: $(5x^3y - 4x^2y - 7y) + (2x^3y + 6x^2y - 4y - 5)$.

Solution

$(5x^3y - 4x^2y - 7y) + (2x^3y + 6x^2y - 4y - 5)$ The given problem involves adding polynomials in two variables.

$= 5x^3y - 4x^2y - 7y + 2x^3y + 6x^2y - 4y - 5$ Remove the parentheses. Like terms are shown in the same color.

$= \underbrace{5x^3y + 2x^3y} - \underbrace{4x^2y + 6x^2y} - \underbrace{7y - 4y} - \underbrace{5}$ Rearrange the terms so that like terms are adjacent.

$= 7x^3y + 2x^2y - 11y - 5$ Combine like terms. ■

Polynomials can be added by arranging like terms in columns. Then combine like terms, column by column. Here's the solution to Example 7 using columns and a vertical format:

$$\begin{array}{r} 5x^3y - 4x^2y - 7y \phantom{{}-5} \\ 2x^3y + 6x^2y - 4y - 5 \\ \hline 7x^3y + 2x^2y - 11y - 5 \end{array}.$$

✓ CHECK POINT 7 Add: $(7xy^3 - 5xy^2 - 3y) + (2xy^3 + 8xy^2 - 12y - 9)$.

5 Subtract polynomials.

Subtracting Polynomials

We subtract real numbers by adding the opposite, or additive inverse, of the number being subtracted. For example,

$$8 - 3 = 8 + (-3) = 5.$$

We can apply this idea to polynomial subtraction. For example,

Change the subtraction to addition.

$$(6x^2 + 5x - 3) - (4x^2 - 2x - 7) = (6x^2 + 5x - 3) + (-4x^2 + 2x + 7)$$

Replace $4x^2 - 2x - 7$ with its opposite, or additive inverse.

$= 6x^2 + 5x - 3 - 4x^2 + 2x + 7$ Remove the parentheses.

$= 6x^2 - 4x^2 + 5x + 2x - 3 + 7$ Rearrange the terms.

$= 2x^2 + 7x + 4$ Combine like terms.

Great Question!

How do I find the opposite, or additive inverse, of the polynomial being subtracted?

To write the additive inverse, change the sign of each term of the polynomial.

Subtracting Polynomials

To subtract one polynomial from another, add the opposite, or additive inverse, of the polynomial being subtracted.

Great Question!

Can I use a vertical format to subtract polynomials, like we did with addition?

Yes. Here's the solution to Example 8 with a vertical format. Notice that you still distribute the negative sign, thereby adding the opposite.

$$\begin{array}{r} 7x^3 - 8x^2 + 9x - 6 \\ -(2x^3 - 6x^2 - 3x + 9) \\ \hline 7x^3 - 8x^2 + 9x - 6 \\ + \; -2x^3 + 6x^2 + 3x - 9 \\ \hline 5x^3 - 2x^2 + 12x - 15 \end{array}$$

EXAMPLE 8 Subtracting Polynomials

Subtract: $(7x^3 - 8x^2 + 9x - 6) - (2x^3 - 6x^2 - 3x + 9)$.

Solution

$(7x^3 - 8x^2 + 9x - 6) - (2x^3 - 6x^2 - 3x + 9)$

$= (7x^3 - 8x^2 + 9x - 6) + (-2x^3 + 6x^2 + 3x - 9)$ Add the opposite of the polynomial being subtracted.

$= 7x^3 - 8x^2 + 9x - 6 - 2x^3 + 6x^2 + 3x - 9$ Remove the parentheses. Like terms are shown in the same color.

$= \underbrace{7x^3 - 2x^3}_{5x^3} \underbrace{- 8x^2 + 6x^2}_{-2x^2} + \underbrace{9x + 3x}_{+12x} \underbrace{- 6 - 9}_{-15}$ Rearrange terms.

$= 5x^3 - 2x^2 + 12x - 15$ Combine like terms. ■

✓ CHECK POINT 8 Subtract: $(14x^3 - 5x^2 + x - 9) - (4x^3 - 3x^2 - 7x + 1)$.

EXAMPLE 9 Subtracting Polynomials

Subtract $-2x^5y^2 - 3x^3y + 7$ from $3x^5y^2 - 4x^3y - 3$.

Solution

$(3x^5y^2 - 4x^3y - 3) - (-2x^5y^2 - 3x^3y + 7)$

$= (3x^5y^2 - 4x^3y - 3) + (2x^5y^2 + 3x^3y - 7)$ Add the opposite of the polynomial being subtracted.

$= 3x^5y^2 - 4x^3y - 3 + 2x^5y^2 + 3x^3y - 7$ Remove the parentheses. Like terms are shown in the same color.

$= \underbrace{3x^5y^2 + 2x^5y^2}_{5x^5y^2} \underbrace{- 4x^3y + 3x^3y}_{-x^3y} \underbrace{- 3 - 7}_{-10}$ Rearrange terms.

$= 5x^5y^2 - x^3y - 10$ Combine like terms. ■

Great Question!

When I subtract one polynomial from another, as in Example 9, how can I tell which polynomial is being subtracted?

Be careful of the order in Example 9. For example, subtracting 2 from 5 is equivalent to $5 - 2$. In general, subtracting B from A means $A - B$. The order of the resulting algebraic expression is not the same as the order in English.

✓ CHECK POINT 9 Subtract $-7x^2y^5 - 4xy^3 + 2$ from $6x^2y^5 - 2xy^3 - 8$.

Achieving Success

Form a study group with other students in your class. Working in small groups often serves as an excellent way to learn and reinforce new material. Set up helpful procedures and guidelines for the group. "Talk" math by discussing and explaining the concepts and exercises to one another.

CONCEPT AND VOCABULARY CHECK

Fill in each blank so that the resulting statement is true.

1. A polynomial is a single term or the sum of two or more terms containing variables with exponents that are ________ numbers.
2. It is customary to write the terms of a polynomial in the order of descending powers of the variable. This is called the ________ form of a polynomial.
3. A simplified polynomial that has exactly one term is called a/an ________.
4. A simplified polynomial that has two terms is called a/an ________.
5. A simplified polynomial that has three terms is called a/an ________.
6. If $a \neq 0$, the degree of ax^n is ________.
7. If $a \neq 0$, the degree of $ax^n y^m$ is ________.
8. The degree of a polynomial is the ________ degree of all the terms of the polynomial. If there is precisely one term of the greatest degree, it is called the ________ term. Its coefficient is called the ________ coefficient.
9. True or false: Polynomial functions of degree 2 or higher have graphs that contain only rounded curves with no sharp corners. ________
10. True or false: Polynomial functions of degree 2 or higher have breaks in their graphs. ________
11. The behavior of the graph of a polynomial function to the far left or the far right is called its ________ behavior, which depends upon the ________ term.
12. The graph of $f(x) = x^3$ ________ to the left and ________ to the right.
13. The graph of $f(x) = -x^3$ ________ to the left and ________ to the right.
14. The graph of $f(x) = x^2$ ________ to the left and ________ to the right.
15. The graph of $f(x) = -x^2$ ________ to the left and ________ to the right.
16. True or false: Odd-degree polynomial functions have graphs with opposite behavior at each end. ________
17. True or false: Even-degree polynomial functions have graphs with the same behavior at each end. ________
18. Terms of a polynomial that contain the same variables raised to the same powers are called ________ terms.

Add or subtract, if possible.

19. $-7x^3 + 4x^3$
20. $-4x^3y + x^3y$
21. $x^5 + x^5$
22. $7x^5y^2 - (-3x^5y^2)$
23. $12xy^2 - 12y^2$

5.1 EXERCISE SET

MyMathLab®

Practice Exercises

In Exercises 1–10, determine the coefficient of each term, the degree of each term, the degree of the polynomial, the leading term, and the leading coefficient of the polynomial.

1. $-x^4 + x^2$

2. $x^3 - 4x^2$

3. $5x^3 + 7x^2 - x + 9$

4. $11x^3 - 6x^2 + x + 3$

5. $3x^2 - 7x^4 - x + 6$

6. $2x^2 - 9x^4 - x + 5$

7. $x^3y^2 - 5x^2y^7 + 6y^2 - 3$

8. $12x^4y - 5x^3y^7 - x^2 + 4$

9. $x^5 + 3x^2y^4 + 7xy + 9x - 2$

10. $3x^6 + 4x^4y^4 - x^3y + 4x^2 - 5$

In Exercises 11–20, let

$$f(x) = x^2 - 5x + 6 \quad \text{and} \quad g(x) = 2x^3 - x^2 + 4x - 1.$$

Find the indicated function values.

11. $f(3)$ **12.** $f(4)$ **13.** $f(-1)$

14. $f(-2)$ **15.** $g(3)$ **16.** $g(2)$

17. $g(-2)$ **18.** $g(-3)$

19. $g(0)$ **20.** $f(0)$

In Exercises 21–24, identify which graphs are not those of polynomial functions.

21.

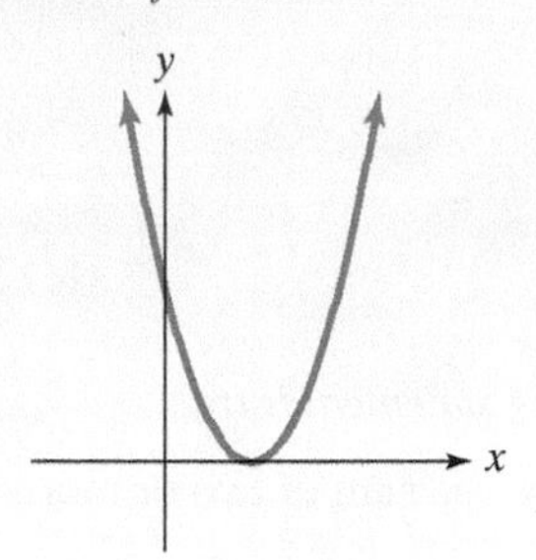

22.

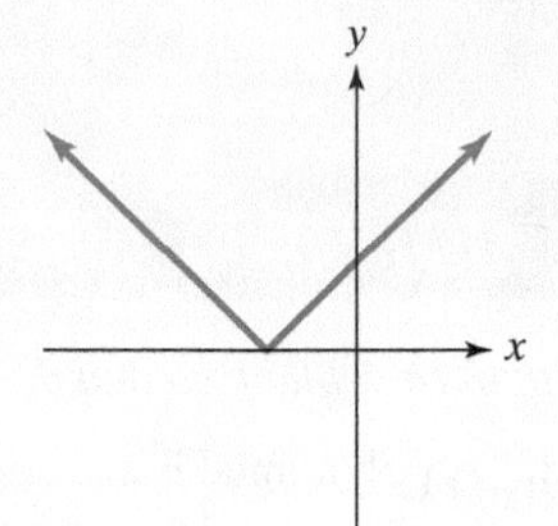

23.

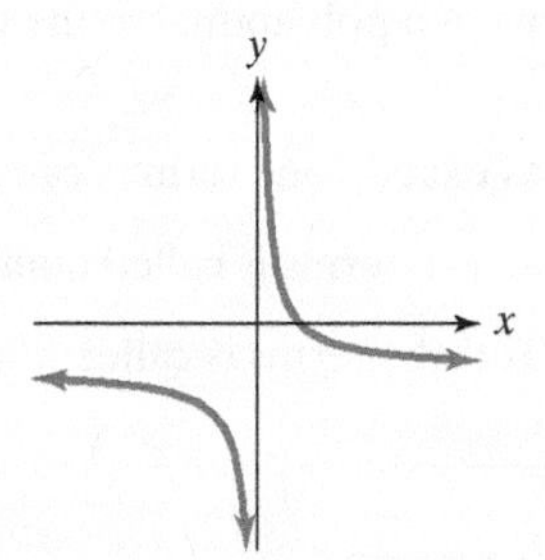

24.

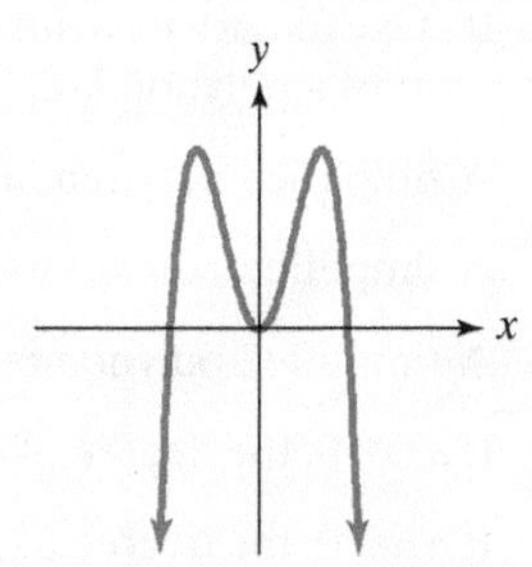

In Exercises 25–28, use the Leading Coefficient Test to determine the end behavior of the graph of the given polynomial function. Then use this end behavior to match the polynomial function with its graph. [The graphs are labeled (a) through (d).]

25. $f(x) = -x^4 + x^2$

26. $f(x) = x^3 - 4x^2$

27. $f(x) = x^2 - 6x + 9$

28. $f(x) = -x^3 - x^2 + 5x - 3$

a.

b.

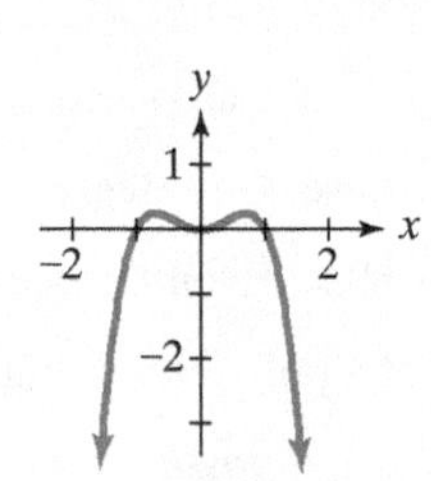

c.

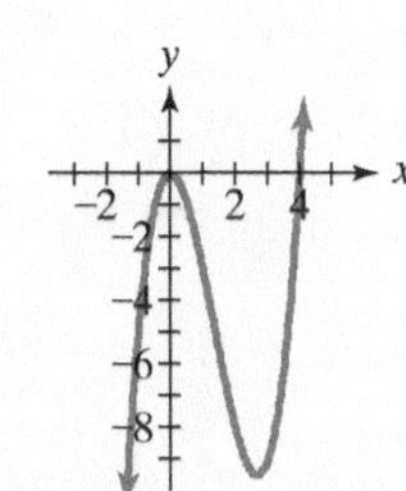

d.

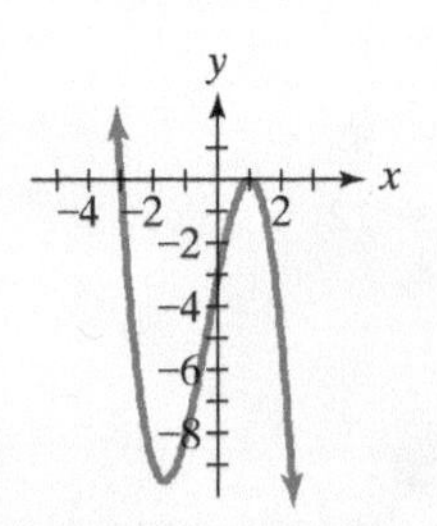

In Exercises 29–40, add the polynomials. Assume that all variable exponents represent whole numbers.

29. $(-6x^3 + 5x^2 - 8x + 9) + (17x^3 + 2x^2 - 4x - 13)$

30. $(-7x^3 + 6x^2 - 11x + 13) + (19x^3 - 11x^2 + 7x - 17)$

31. $\left(\frac{2}{5}x^4 + \frac{2}{3}x^3 + \frac{5}{8}x^2 + 7\right) + \left(-\frac{4}{5}x^4 + \frac{1}{3}x^3 - \frac{1}{4}x^2 - 7\right)$

32. $\left(\frac{1}{5}x^4 + \frac{1}{3}x^3 + \frac{3}{8}x^2 + 6\right) + \left(-\frac{3}{5}x^4 + \frac{2}{3}x^3 - \frac{1}{2}x^2 - 6\right)$

33. $(7x^2y - 5xy) + (2x^2y - xy)$

34. $(-4x^2y + xy) + (7x^2y + 8xy)$

35. $(5x^2y + 9xy + 12) + (-3x^2y + 6xy + 3)$

36. $(8x^2y + 12xy + 14) + (-2x^2y + 7xy + 4)$

37. $(9x^4y^2 - 6x^2y^2 + 3xy) + (-18x^4y^2 - 5x^2y - xy)$

38. $(10x^4y^2 - 3x^2y^2 + 2xy) + (-16x^4y^2 - 4x^2y - xy)$

39. $(x^{2n} + 5x^n - 8) + (4x^{2n} - 7x^n + 2)$

40. $(6y^{2n} + y^n + 5) + (3y^{2n} - 4y^n - 15)$

In Exercises 41–50, subtract the polynomials. Assume that all variable exponents represent whole numbers.

41. $(17x^3 - 5x^2 + 4x - 3) - (5x^3 - 9x^2 - 8x + 11)$

42. $(18x^3 - 2x^2 - 7x + 8) - (9x^3 - 6x^2 - 5x + 7)$

43. $(13y^5 + 9y^4 - 5y^2 + 3y + 6) - (-9y^5 - 7y^3 + 8y^2 + 11)$

44. $(12y^5 + 7y^4 - 3y^2 + 6y + 7) - (-10y^5 - 8y^3 + 3y^2 + 14)$

45. $(x^3 + 7xy - 5y^2) - (6x^3 - xy + 4y^2)$

46. $(x^4 - 7xy - 5y^3) - (6x^4 - 3xy + 4y^3)$

47. $(3x^4y^2 + 5x^3y - 3y) - (2x^4y^2 - 3x^3y - 4y + 6x)$

48. $(5x^4y^2 + 6x^3y - 7y) - (3x^4y^2 - 5x^3y - 6y + 8x)$

49. $(7y^{2n} + y^n - 4) - (6y^{2n} - y^n - 1)$

50. $(8x^{2n} + x^n - 4) - (9x^{2n} - x^n - 2)$

51. Subtract $-5a^2b^4 - 8ab^2 - ab$ from $3a^2b^4 - 5ab^2 + 7ab$.

52. Subtract $-7a^2b^4 - 8ab^2 - ab$ from $13a^2b^4 - 17ab^2 + ab$.

53. Subtract $-4x^3 - x^2y + xy^2 + 3y^3$ from $x^3 + 2x^2y - y^3$.

54. Subtract $-6x^3 + x^2y - xy^2 + 2y^3$ from $x^3 + 2xy^2 - y^3$.

Practice PLUS

55. Add $6x^4 - 5x^3 + 2x$ to the difference between $4x^3 + 3x^2 - 1$ and $x^4 - 2x^2 + 7x - 3$.

56. Add $5x^4 - 2x^3 + 7x$ to the difference between $2x^3 + 5x^2 - 3$ and $-x^4 - x^2 - x - 1$.

57. Subtract $9x^2y^2 - 3x^2 - 5$ from the sum of $-6x^2y^2 - x^2 - 1$ and $5x^2y^2 + 2x^2 - 1$.

58. Subtract $6x^2y^3 - 2x^2 - 7$ from the sum of $-5x^2y^3 + 3x^2 - 4$ and $4x^2y^3 - 2x^2 - 6$.

In Exercises 59–64, let

$$f(x) = -3x^3 - 2x^2 - x + 4$$
$$g(x) = x^3 - x^2 - 5x - 4$$
$$h(x) = -2x^3 + 5x^2 - 4x + 1.$$

Find the indicated function, function value, or polynomial.

59. $(f - g)(x)$ and $(f - g)(-1)$

60. $(g - h)(x)$ and $(g - h)(-1)$

61. $(f + g - h)(x)$ and $(f + g - h)(-2)$

62. $(g + h - f)(x)$ and $(g + h - f)(-2)$

63. $2f(x) - 3g(x)$

64. $-2g(x) - 3h(x)$

Application Exercises

As we noted in the chapter opener, experts fear that without conservation efforts, tigers could disappear from the wild by 2022. Just one hundred years ago, there were at least 100,000 wild tigers. By 2010, the estimated world tiger population was 3200. The bar graph shows the estimated world tiger population for selected years from 1970 through 2010. Also shown is a polynomial function, with its graph, that models the data. Use this information to solve Exercises 65–68.

Vanishing Tigers

Estimated World Tiger Population

Year	World Tiger Population
1970	37,500
1980	28,000
1990	12,500
2000	6300
2010	3200

Source: World Wildlife Fund

Graph of a Polynomial Model for the Data

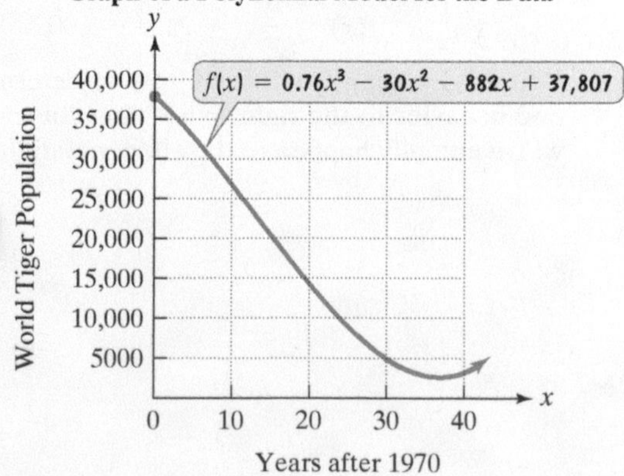

(In Exercises 65–68, be sure to refer to the information on vanishing tigers at the bottom of the previous page.)

65. **a.** Find and interpret $f(40)$. Identify this information as a point on the graph of f.

b. Does $f(40)$ overestimate or underestimate the actual data shown by the bar graph? By how much?

66. **a.** Find and interpret $f(10)$. Identify this information as a point on the graph of f.

b. Does $f(10)$ overestimate or underestimate the actual data shown by the bar graph? By how much?

67. Use the Leading Coefficient Test to determine the end behavior to the right for the graph of f. Will this function be useful in modeling the world tiger population if conservation efforts to save wild tigers fail? Explain your answer.

68. Use the Leading Coefficient Test to determine the end behavior to the right for the graph of f. Might this function be useful in modeling the world tiger population if conservation efforts to save wild tigers are successful? Explain your answer.

69. The common cold is caused by a rhinovirus. After x days of invasion by the viral particles, the number of particles in our bodies, $f(x)$, in billions, can be modeled by the polynomial function

$$f(x) = -0.75x^4 + 3x^3 + 5.$$

Use the Leading Coefficient Test to determine the graph's end behavior to the right. What does this mean about the number of viral particles in our bodies over time?

70. The polynomial function

$$f(x) = -0.87x^3 + 0.35x^2 + 81.62x + 7684.94$$

models the number of thefts, $f(x)$, in thousands, in the United States x years after 1987. Will this function be useful in modeling the number of thefts over an extended period of time? Explain your answer.

71. A herd of 100 elk is introduced to a small island. The number of elk, $f(x)$, after x years is modeled by the polynomial function

$$f(x) = -x^4 + 21x^2 + 100.$$

Use the Leading Coefficient Test to determine the graph's end behavior to the right. What does this mean about what will eventually happen to the elk population?

Explaining the Concepts

72. What is a polynomial?

73. Explain how to determine the degree of each term of a polynomial.

74. Explain how to determine the degree of a polynomial.

75. Explain how to determine the leading coefficient of a polynomial.

76. What is a polynomial function?

77. What do we mean when we describe the graph of a polynomial function as smooth and continuous?

78. What is meant by the end behavior of a polynomial function?

79. Explain how to use the Leading Coefficient Test to determine the end behavior of a polynomial function.

80. Why is a polynomial function of degree 3 with a negative leading coefficient not appropriate for modeling nonnegative real-world phenomena over a long period of time?

81. Explain how to add polynomials.

82. Explain how to subtract polynomials.

83. In a favorable habitat and without natural predators, a population of reindeer is introduced to an island preserve. The reindeer population t years after their introduction is modeled by the polynomial function

$$f(t) = -0.125t^5 + 3.125t^4 + 4000.$$

Discuss the growth and decline of the reindeer population. Describe the factors that might contribute to this population model.

Technology Exercises

Write a polynomial function that imitates the end behavior of each graph in Exercises 84–87. The dashed portions of the graphs indicate that you should focus only on imitating the left and right end behavior of the graph. You can be flexible about what occurs between the left and right ends. Then use your graphing utility to graph the polynomial function and verify that you imitated the end behavior shown in the given graph.

84.

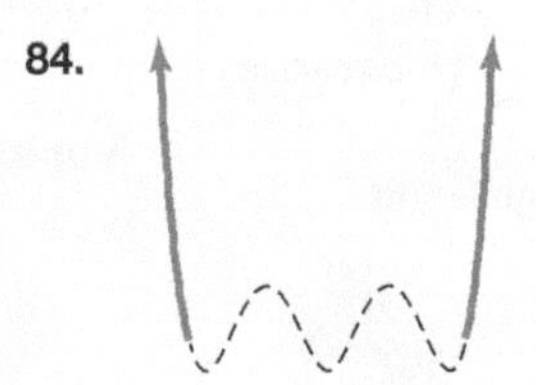

85.

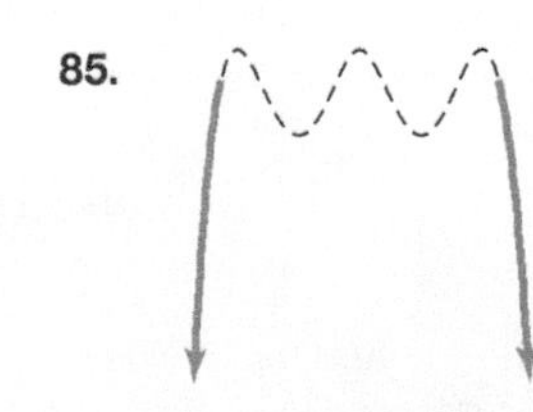

86. **87.**

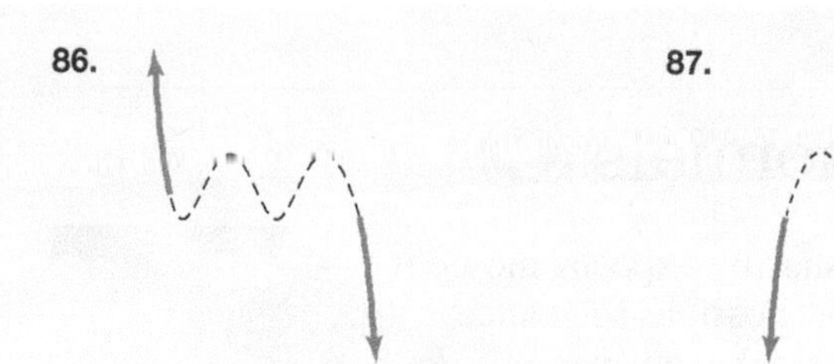

In Exercises 88–91, use a graphing utility with a viewing rectangle large enough to show end behavior to graph each polynomial function.

88. $f(x) = x^3 + 13x^2 + 10x - 4$

89. $f(x) = -2x^3 + 6x^2 + 3x - 1$

90. $f(x) = -x^4 + 8x^3 + 4x^2 + 2$

91. $f(x) = -x^5 + 5x^4 - 6x^3 + 2x + 20$

In Exercises 92–93, use a graphing utility to graph f and g in the same viewing rectangle. Then use the ZOOM OUT *feature to show that f and g have identical end behavior.*

92. $f(x) = x^3 - 6x + 1, \quad g(x) = x^3$

93. $f(x) = -x^4 + 2x^3 - 6x, \quad g(x) = -x^4$

Critical Thinking Exercises

Make Sense? *In Exercises 94–97, determine whether each statement makes sense or does not make sense, and explain your reasoning.*

94. Many English words have prefixes with meanings similar to those used to describe polynomials, such as *monologue*, *binocular*, and *tricuspid*.

95. I can determine a polynomial's leading coefficient by inspecting the coefficient of the first term.

96. When I'm trying to determine end behavior, it's the coefficient of the first term of a polynomial function written in standard form that I should inspect.

97. When I rearrange the terms of a polynomial, it's important that I move the sign in front of a term with that term.

In Exercises 98–101, determine whether each statement is true or false. If the statement is false, make the necessary change(s) to produce a true statement.

98. $4x^3 + 7x^2 - 5x + \frac{2}{x}$ is a polynomial containing four terms.

99. If two polynomials of degree 2 are added, the sum must be a polynomial of degree 2.

100. $(x^2 - 7x) - (x^2 - 4x) = -11x$ for all values of x.

101. All terms of a polynomial are monomials.

In Exercises 102–103, perform the indicated operations. Assume that exponents represent whole numbers.

102. $(x^{2n} - 3x^n + 5) + (4x^{2n} - 3x^n - 4) - (2x^{2n} - 5x^n - 3)$

103. $(y^{3n} - 7y^{2n} + 3) - (-3y^{3n} - 2y^{2n} - 1) + (6y^{3n} - y^{2n} + 1)$

104. From what polynomial must $4x^2 + 2x - 3$ be subtracted to obtain $5x^2 - 5x + 8$?

Review Exercises

105. Solve: $9(x - 1) = 1 + 3(x - 2)$. (Section 1.4, Example 3)

106. Graph: $2x - 3y < -6$. (Section 4.4, Example 1)

107. Write the point-slope form and slope-intercept form of equations of a line passing through the point $(-2, 5)$ and parallel to the line whose equation is $3x - y = 9$. (Section 2.5, Example 4)

Preview Exercises

Exercises 108–110 will help you prepare for the material covered in the next section.

108. Multiply: $(2x^3y^2)(5x^4y^7)$.

109. Use the distributive property to multiply: $2x^4(8x^4 + 3x)$.

110. Simplify and express the polynomial in standard form:

$$3x(x^2 + 4x + 5) + 7(x^2 + 4x + 5).$$

SECTION

5.2 Multiplication of Polynomials

What am I supposed to learn?

After studying this section, you should be able to:

1. Multiply monomials.
2. Multiply a monomial and a polynomial.
3. Multiply polynomials when neither is a monomial.
4. Use FOIL in polynomial multiplication.
5. Square binomials.
6. Multiply the sum and difference of two terms.
7. Find the product of functions.
8. Use polynomial multiplication to evaluate functions.

Can that be Axl, your author's yellow lab, sharing a special moment with a baby chick? And if it is (it is), what possible relevance can this have to multiplying polynomials? An answer is promised before you reach the Exercise Set. For now, let's begin by reviewing how to multiply monomials, a skill that you will apply in every polynomial multiplication problem.

Old Dog ... New Chicks

1 Multiply monomials.

Multiplying Monomials

To multiply monomials, begin by multiplying the coefficients. Then multiply the variables. Use the product rule for exponents to multiply the variables: Retain the variable and add the exponents.

EXAMPLE 1 Multiplying Monomials

Multiply:

a. $(5x^3y^4)(-6x^7y^8)$ **b.** $(4x^3y^2z^5)(2x^5y^2z^4)$.

Solution

a.

$$(5x^3y^4)(-6x^7y^8) = 5(-6)x^3 \cdot x^7 \cdot y^4 \cdot y^8$$ Rearrange factors. This step is usually done mentally.

$$= -30x^{3+7}y^{4+8}$$ Multiply coefficients and add exponents.

$$= -30x^{10}y^{12}$$ Simplify.

b.

$$(4x^3y^2z^5)(2x^5y^2z^4) = 4 \cdot 2 \cdot x^3 \cdot x^5 \cdot y^2 \cdot y^2 \cdot z^5 \cdot z^4$$ Rearrange factors.

$$= 8x^{3+5}y^{2+2}z^{5+4}$$ Multiply coefficients and add exponents.

$$= 8x^8y^4z^9$$ Simplify. ■

✓ CHECK POINT 1 Multiply:

a. $(6x^5y^7)(-3x^2y^4)$ **b.** $(10x^4y^3z^6)(3x^6y^3z^2)$.

2 Multiply a monomial and a polynomial.

Multiplying a Monomial and a Polynomial That Is Not a Monomial

We use the distributive property to multiply a monomial and a polynomial that is not a monomial. For example,

$$3x^2(2x^3 + 5x) = 3x^2 \cdot 2x^3 + 3x^2 \cdot 5x = 3 \cdot 2x^{2+3} + 3 \cdot 5x^{2+1} = 6x^5 + 15x^3.$$

Monomial Binomial Multiply coefficients and add exponents.

To multiply a monomial and a polynomial, multiply each term of the polynomial by the monomial. Once the monomial factor is distributed, we multiply the resulting monomials using the procedure shown in Example 1.

EXAMPLE 2 Multiplying a Monomial and a Trinomial

Multiply:

a. $4x^3(6x^5 - 2x^2 + 3)$ **b.** $5x^3y^4(2x^7y - 6x^4y^3 - 3)$.

Solution

a. $4x^3(6x^5 - 2x^2 + 3) = 4x^3 \cdot 6x^5 - 4x^3 \cdot 2x^2 + 4x^3 \cdot 3$ Use the distributive property.

$= 24x^8 - 8x^5 + 12x^3$ Multiply coefficients and add exponents.

b. $5x^3y^4(2x^7y - 6x^4y^3 - 3)$

$= 5x^3y^4 \cdot 2x^7y - 5x^3y^4 \cdot 6x^4y^3 - 5x^3y^4 \cdot 3$ Use the distributive property.

$= 10x^{10}y^5 - 30x^7y^7 - 15x^3y^4$ Multiply coefficients and add exponents. ■

✓ **CHECK POINT 2** Multiply:

a. $6x^4(2x^5 - 3x^2 + 4)$

b. $2x^4y^3(5xy^6 - 4x^3y^4 - 5)$.

---

3 Multiply polynomials when neither is a monomial.

Multiplying Polynomials When Neither Is a Monomial

How do we multiply two polynomials if neither is a monomial? For example, consider

$$(3x + 7)(x^2 + 4x + 5).$$

Binomial Trinomial

One way to perform this multiplication is to distribute $3x$ throughout the trinomial

$$3x(x^2 + 4x + 5)$$

and 7 throughout the trinomial

$$7(x^2 + 4x + 5).$$

Then combine the like terms that result.

Multiplying Polynomials When Neither Is a Monomial

Multiply each term of one polynomial by each term of the other polynomial. Then combine like terms.

EXAMPLE 3 Multiplying a Binomial and a Trinomial

Multiply: $(3x + 7)(x^2 + 4x + 5)$.

Solution

$(3x + 7)(x^2 + 4x + 5)$

$= 3x(x^2 + 4x + 5) + 7(x^2 + 4x + 5)$ Multiply the trinomial by each term of the binomial.

$= 3x \cdot x^2 + 3x \cdot 4x + 3x \cdot 5 + 7x^2 + 7 \cdot 4x + 7 \cdot 5$ Use the distributive property.

$= 3x^3 + 12x^2 + 15x + 7x^2 + 28x + 35$ Multiply monomials: Multiply coefficients and add exponents.

$= 3x^3 + 19x^2 + 43x + 35$ Combine like terms: $12x^2 + 7x^2 = 19x^2$ and $15x + 28x = 43x$. ■

CHECK POINT 3 Multiply: $(3x + 2)(2x^2 - 2x + 1)$.

Another method for solving Example 3 is to use a vertical format similar to that used for multiplying whole numbers.

$$\begin{array}{r} x^2 + 4x + 5 \\ 3x + 7 \\ \hline 7x^2 + 28x + 35 \\ 3x^3 + 12x^2 + 15x \quad\quad \\ \hline 3x^3 + 19x^2 + 43x + 35 \end{array}$$

Write like terms in the same column.

$7(x^2 + 4x + 5)$

$3x(x^2 + 4x + 5)$

Combine like terms.

EXAMPLE 4 Multiplying a Binomial and a Trinomial

Multiply: $(2x^2y + 3y)(5x^4y - 4x^2y + y)$.

Solution

$(2x^2y + 3y)(5x^4y - 4x^2y + y)$

$= 2x^2y(5x^4y - 4x^2y + y) + 3y(5x^4y - 4x^2y + y)$ Multiply the trinomial by each term of the binomial.

$= 2x^2y \cdot 5x^4y - 2x^2y \cdot 4x^2y + 2x^2y \cdot y + 3y \cdot 5x^4y - 3y \cdot 4x^2y + 3y \cdot y$ Use the distributive property.

$= 10x^6y^2 - 8x^4y^2 + 2x^2y^2 + 15x^4y^2 - 12x^2y^2 + 3y^2$ Multiply coefficients and add exponents.

$= 10x^6y^2 + 7x^4y^2 - 10x^2y^2 + 3y^2$ Combine like terms:

$-8x^4y^2 + 15x^4y^2 = 7x^4y^2$

$2x^2y^2 - 12x^2y^2 = -10x^2y^2$. ■

CHECK POINT 4 Multiply: $(4xy^2 + 2y)(3xy^4 - 2xy^2 + y)$.

4 Use FOIL in polynomial multiplication.

The Product of Two Binomials: FOIL

Frequently we need to find the product of two binomials. One way to perform this multiplication is to distribute each term in the first binomial throughout the second binomial. For example, we can find the product of the binomials $7x + 2$ and $4x + 5$ as follows:

$$(7x + 2)(4x + 5) = 7x(4x + 5) + 2(4x + 5)$$
$$= 7x(4x) + 7x(5) + 2(4x) + 2(5)$$
$$= 28x^2 + 35x + 8x + 10.$$

Distribute $7x$ over $4x + 5$.

Distribute 2 over $4x + 5$.

We'll combine these like terms later. For now, our interest is in how to obtain *each* of these four terms.

We can also find the product of $7x + 2$ and $4x + 5$ using a method called FOIL, which is based on our work shown above. Any two binomials can be quickly multiplied using the FOIL method, in which **F** represents the product of the **first** terms in each binomial, **O** represents the product of the **outside** terms, **I** represents the product of the two **inside** terms, and **L** represents the product of the **last**, or second, terms in each binomial. For example, we can use the FOIL method to find the product of the binomials $7x + 2$ and $4x + 5$ as follows:

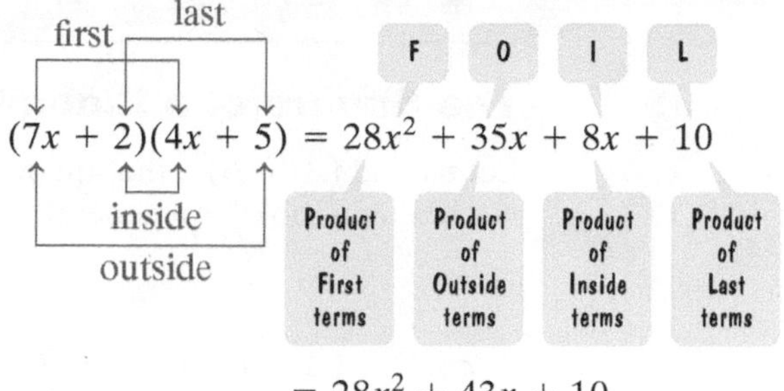

$$(7x + 2)(4x + 5) = 28x^2 + 35x + 8x + 10$$

$$= 28x^2 + 43x + 10$$ Combine like terms.

In general, here's how to use the FOIL method to find the product of $ax + b$ and $cx + d$:

Using the FOIL Method to Multiply Binomials

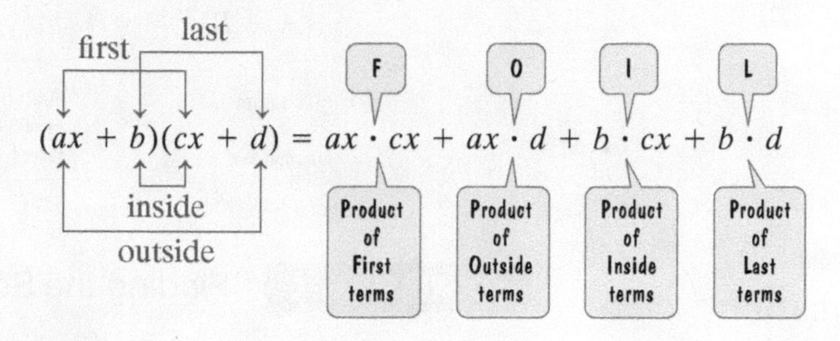

$$(ax + b)(cx + d) = ax \cdot cx + ax \cdot d + b \cdot cx + b \cdot d$$

EXAMPLE 5 Using the FOIL Method

Multiply:

a. $(x + 3)(x + 2)$ **b.** $(3x + 5y)(x - 2y)$ **c.** $(5x^3 - 6)(4x^3 - x)$.

Solution

last, first, F, O, I, L, inside, outside

a. $(x + 3)(x + 2) = x \cdot x + x \cdot 2 + 3 \cdot x + 3 \cdot 2$

$= x^2 + 2x + 3x + 6$

$= x^2 + 5x + 6$ Combine like terms.

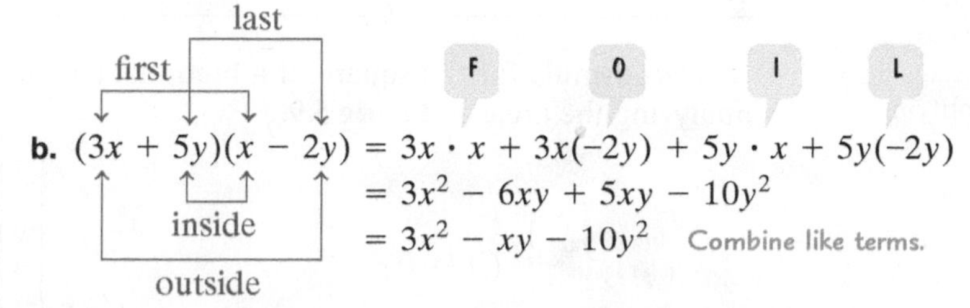

b. $(3x + 5y)(x - 2y) = 3x \cdot x + 3x(-2y) + 5y \cdot x + 5y(-2y)$

$= 3x^2 - 6xy + 5xy - 10y^2$

$= 3x^2 - xy - 10y^2$ Combine like terms.

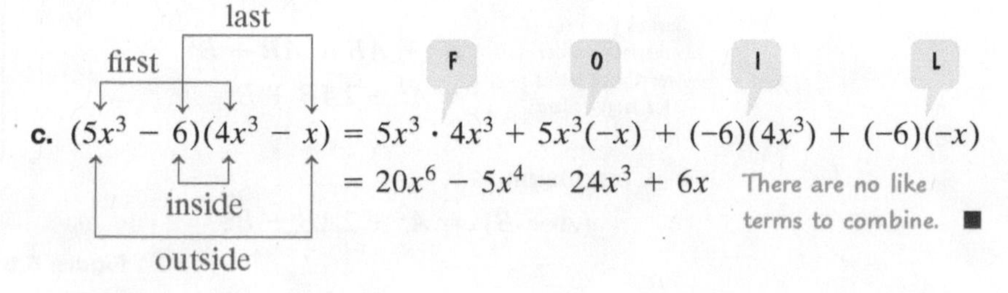

c. $(5x^3 - 6)(4x^3 - x) = 5x^3 \cdot 4x^3 + 5x^3(-x) + (-6)(4x^3) + (-6)(-x)$

$= 20x^6 - 5x^4 - 24x^3 + 6x$ There are no like terms to combine. ■

✓ CHECK POINT 5 Multiply:

a. $(x + 5)(x + 3)$ b. $(7x + 4y)(2x - y)$ c. $(4x^3 - 5)(x^3 - 3x)$.

5 Square binomials.

The Square of a Binomial

Let us find $(A + B)^2$, the square of a binomial sum. To do so, we begin with the FOIL method and look for a general rule.

$$(A + B)^2 = (A + B)(A + B) = \overset{F}{A \cdot A} + \overset{O}{A \cdot B} + \overset{I}{A \cdot B} + \overset{L}{B \cdot B}$$
$$= A^2 + 2AB + B^2$$

This result implies the following rule, which is often called a **special-product formula**:

The Square of a Binomial Sum

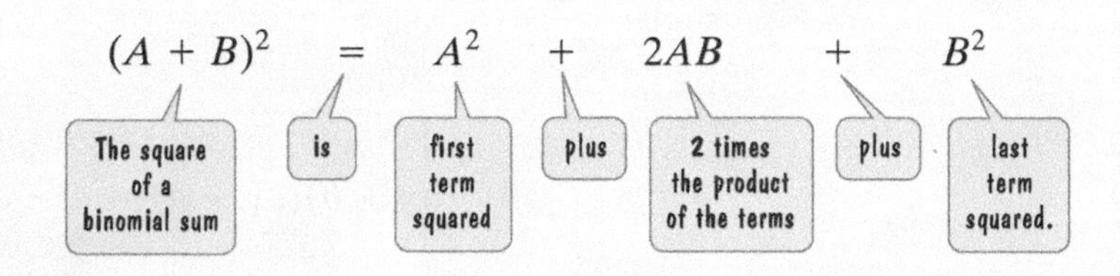

Great Question!

When finding $(x + 5)^2$, why can't I just write $x^2 + 5^2$, or $x^2 + 25$?

Caution! The square of a sum is *not* the sum of the squares.

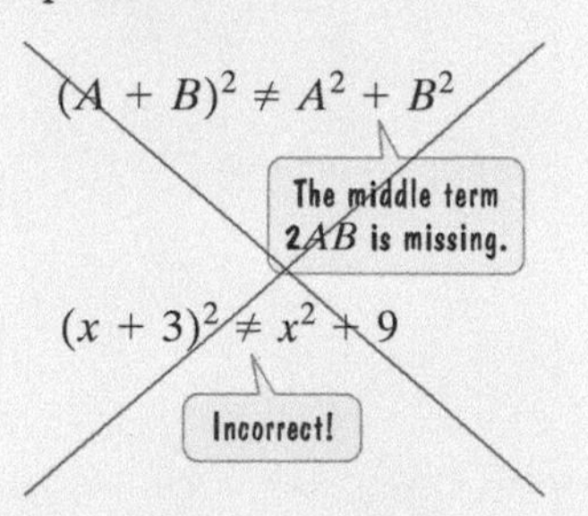

$(A + B)^2 \neq A^2 + B^2$
The middle term 2AB is missing.
$(x + 3)^2 \neq x^2 + 9$
Incorrect!

Show that $(x + 5)^2$ and $x^2 + 25$ are not equal by substituting 3 for x in each expression and simplifying.

EXAMPLE 6 Finding the Square of a Binomial Sum

Multiply:

a. $(x + 5)^2$ b. $(3x + 2y)^2$.

Solution Use the special-product formula shown.

$(A + B)^2 = \quad A^2 \quad + \quad 2AB \quad + \quad B^2$

	(First Term)^2	+	2 · Product of the Terms	+	(Last Term)^2	= Product
a. $(x + 5)^2$ =	x^2	+	$2 \cdot x \cdot 5$	+	5^2	$= x^2 + 10x + 25$
b. $(3x + 2y)^2$ =	$(3x)^2$	+	$2 \cdot 3x \cdot 2y$	+	$(2y)^2$	$= 9x^2 + 12xy + 4y^2$

■

✓ CHECK POINT 6 Multiply:

a. $(x + 8)^2$ b. $(4x + 5y)^2$.

The formula for the square of a binomial sum can be interpreted geometrically by analyzing the areas in **Figure 5.9**.

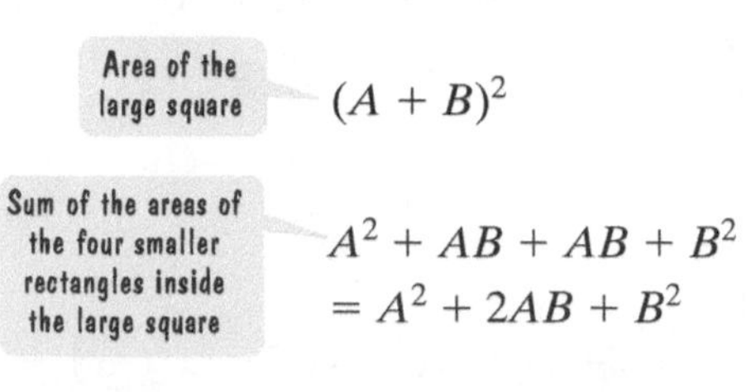

Area of the large square
$(A + B)^2$
Sum of the areas of the four smaller rectangles inside the large square
$A^2 + AB + AB + B^2$
$= A^2 + 2AB + B^2$

Conclusion:

$(A + B)^2 = A^2 + 2AB + B^2$

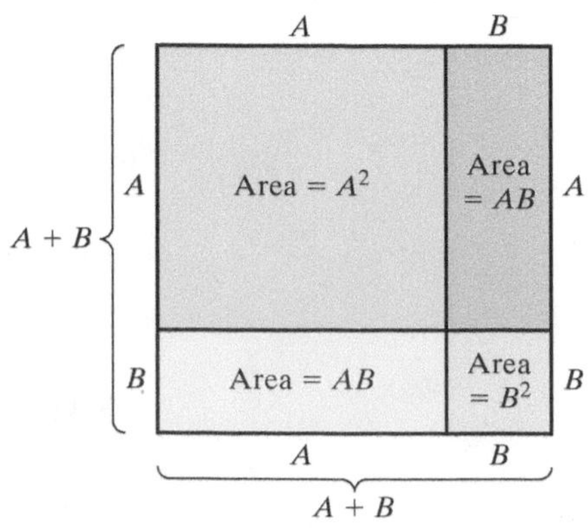

Figure 5.9

A similar pattern occurs for $(A - B)^2$, the square of a binomial difference. Using the FOIL method on $(A - B)^2$, we obtain the following rule:

The Square of a Binomial Difference

$$(A - B)^2 = A^2 - 2AB + B^2$$

The square of a binomial difference is first term squared minus 2 times the product of the terms plus last term squared.

EXAMPLE 7 Finding the Square of a Binomial Difference

Multiply:

a. $(x - 8)^2$ **b.** $\left(\frac{1}{2}x - 4y^3\right)^2$.

Solution Use the special-product formula shown.

$(A - B)^2 = \quad A^2 \quad - \quad 2AB \quad + \quad B^2$

	(First Term)^2	$-$	2 · Product of the Terms	$+$	(Last Term)^2	= Product
a. $(x - 8)^2 =$	x^2	$-$	$2 \cdot x \cdot 8$	$+$	8^2	$= x^2 - 16x + 64$
b. $\left(\frac{1}{2}x - 4y^3\right)^2 =$	$\left(\frac{1}{2}x\right)^2$	$-$	$2 \cdot \frac{1}{2}x \cdot 4y^3$	$+$	$(4y^3)^2$	$= \frac{1}{4}x^2 - 4xy^3 + 16y^6$

■

CHECK POINT 7 Multiply:

a. $(x - 5)^2$ **b.** $(2x - 6y^4)^2$.

6 Multiply the sum and difference of two terms.

Multiplying the Sum and Difference of Two Terms

We can use the FOIL method to multiply $A + B$ and $A - B$ as follows:

F O I L

$$(A + B)(A - B) = A^2 - AB + AB - B^2 = A^2 - B^2.$$

Notice that the outside and inside products have a sum of 0 and the terms cancel. The FOIL multiplication provides us with a quick rule for multiplying the sum and difference of two terms, which is another example of a special-product formula.

The Product of the Sum and Difference of Two Terms

$$(A + B)(A - B) = A^2 - B^2$$

The product of the sum and the difference of the same two terms is the square of the first term minus the square of the second term.

EXAMPLE 8 Finding the Product of the Sum and Difference of Two Terms

Multiply:

a. $(x + 8)(x - 8)$ **b.** $(9x + 5y)(9x - 5y)$ **c.** $(6a^2b - 3b)(6a^2b + 3b)$.

Solution Use the special-product formula shown.

$$(A + B)(A - B) = A^2 - B^2$$

First term squared − Second term squared = Product

a. $(x + 8)(x - 8) = x^2 - 8^2 = x^2 - 64$

b. $(9x + 5y)(9x - 5y) = (9x)^2 - (5y)^2 = 81x^2 - 25y^2$

c. $(6a^2b - 3b)(6a^2b + 3b) = (6a^2b)^2 - (3b)^2 = 36a^4b^2 - 9b^2$ ■

✓ CHECK POINT 8 Multiply:

a. $(x + 3)(x - 3)$

b. $(5x + 7y)(5x - 7y)$

c. $(5ab^2 - 4a)(5ab^2 + 4a)$.

Special products can sometimes be used to find the products of certain trinomials, as illustrated in Example 9.

EXAMPLE 9 Using the Special Products

Multiply:

a. $(7x + 5 + 4y)(7x + 5 - 4y)$ b. $(3x + y + 1)^2$.

Solution

a. By grouping the first two terms within each set of parentheses, we can find the product using the form for the sum and difference of two terms.

$$(A + B) \cdot (A - B) = A^2 - B^2$$

$$[(7x + 5) + 4y] \cdot [(7x + 5) - 4y] = (7x + 5)^2 - (4y)^2$$
$$= (7x)^2 + 2 \cdot 7x \cdot 5 + 5^2 - (4y)^2$$
$$= 49x^2 + 70x + 25 - 16y^2$$

b. We can group the terms so that the formula for the square of a binomial can be applied.

$$(A + B)^2 = A^2 + 2 \cdot A \cdot B + B^2$$

$$[(3x + y) + 1]^2 = (3x + y)^2 + 2 \cdot (3x + y) \cdot 1 + 1^2$$
$$= 9x^2 + 6xy + y^2 + 6x + 2y + 1$$ ■

Discover for Yourself

Group $(3x + y + 1)^2$ as $[3x + (y + 1)]^2$. Verify that you get the same product as we obtained in Example 9(b).

✓ CHECK POINT 9 Multiply:

a. $(3x + 2 + 5y)(3x + 2 - 5y)$

b. $(2x + y + 3)^2$.

---

7 Find the product of functions.

Multiplication of Polynomial Functions

In Chapter 2, we developed an algebra of functions, defining the product of functions f and g as follows:

$$(fg)(x) = f(x) \cdot g(x).$$

Now that we know how to multiply polynomials, we can find the product of functions.

EXAMPLE 10 Using the Algebra of Functions

Let $f(x) = x - 5$ and $g(x) = x - 2$. Find:

a. $(fg)(x)$ **b.** $(fg)(1)$.

Solution

a. $(fg)(x) = f(x) \cdot g(x)$ This is the definition of the product function, fg.

$= (x - 5)(x - 2)$ Substitute the given functions.

F O I L

$= x^2 - 2x - 5x + 10$ Multiply by the FOIL method.

$= x^2 - 7x + 10$ Combine like terms.

Thus,

$$(fg)(x) = x^2 - 7x + 10.$$

b. We use the product function to find $(fg)(1)$—that is, the value of the function fg at 1. Replace x with 1.

$$(fg)(1) = 1^2 - 7 \cdot 1 + 10 = 4 \quad \blacksquare$$

Example 10 involved linear and quadratic functions.

$f(x) = x - 5$ $g(x) = x - 2$ $(fg)(x) = x^2 - 7x + 10$

These are linear functions of the form $f(x) = mx + b$.

This is a quadratic function of the form $f(x) = ax^2 + bx + c$.

All three of these functions are polynomial functions. A linear function is a first-degree polynomial function. A quadratic function is a second-degree polynomial function.

CHECK POINT 10 Let $f(x) = x - 3$ and $g(x) = x - 7$. Find:

a. $(fg)(x)$

b. $(fg)(2)$.

8 Use polynomial multiplication to evaluate functions.

If you are given a function, f, calculus can reveal how it is changing at any instant in time. The algebraic expression

$$\frac{f(a + h) - f(a)}{h}$$

plays an important role in this process. Our work with polynomial multiplication can be used to evaluate the numerator of this expression.

EXAMPLE 11 Using Polynomial Multiplication to Evaluate Functions

Given $f(x) = x^2 - 7x + 3$, find and simplify each of the following:

a. $f(a + 4)$ **b.** $f(a + h) - f(a)$.

Solution

a. We find $f(a + 4)$, read "f at a plus 4," by replacing x with $a + 4$ each time that x appears in the polynomial.

$$f(x) = x^2 - 7x + 3$$

Replace x with $a + 4$. Replace x with $a + 4$. Replace x with $a + 4$. Copy the 3. There is no x in this term.

$$f(a+4) = (a+4)^2 - 7(a+4) + 3$$

$= a^2 + 8a + 16 - 7a - 28 + 3$ Multiply as indicated.

$= a^2 + a - 9$ Combine like terms: $8a - 7a = a$ and $16 - 28 + 3 = -9$.

b. To find $f(a+h) - f(a)$, we first replace each occurrence of x in $f(x) = x^2 - 7x + 3$ with $a + h$ and then replace each occurrence of x with a. Then we perform the resulting operations and simplify.

This is $f(a+h)$. Use $f(x) = x^2 - 7x + 3$ and replace x with $a + h$.

This is $f(a)$. Use $f(x) = x^2 - 7x + 3$ and replace x with a.

$$f(a+h) - f(a) = \boxed{(a+h)^2 - 7(a+h) + 3} - (a^2 - 7a + 3)$$

$= (a^2 + 2ah + h^2 - 7a - 7h + 3) - (a^2 - 7a + 3)$ Perform the multiplications required by $f(a+h)$.

$= (a^2 + 2ah + h^2 - 7a - 7h + 3) + (-a^2 + 7a - 3)$ Add the opposite of the polynomial being subtracted. Like terms are shown in the same color.

$= a^2 - a^2 - 7a + 7a + 3 - 3 + 2ah + h^2 - 7h$ Group like terms.

$= 2ah + h^2 - 7h$ Simplify. Observe that $a^2 - a^2 = 0$, $-7a + 7a = 0$, and $3 - 3 = 0$. ■

CHECK POINT 11 Given $f(x) = x^2 - 5x + 4$, find and simplify each of the following:

a. $f(a+3)$ **b.** $f(a+h) - f(a)$.

Blitzer Bonus

Labrador Retrievers and Polynomial Multiplication

The color of a Labrador retriever is determined by its pair of genes. A single gene is inherited at random from each parent. The black-fur gene, B, is dominant. The yellow-fur gene, Y, is recessive. This means that labs with at least one black-fur gene (BB or BY) have black coats. Only labs with two yellow-fur genes (YY) have yellow coats.

Axl, your author's yellow lab, inherited his genetic makeup from two black BY parents.

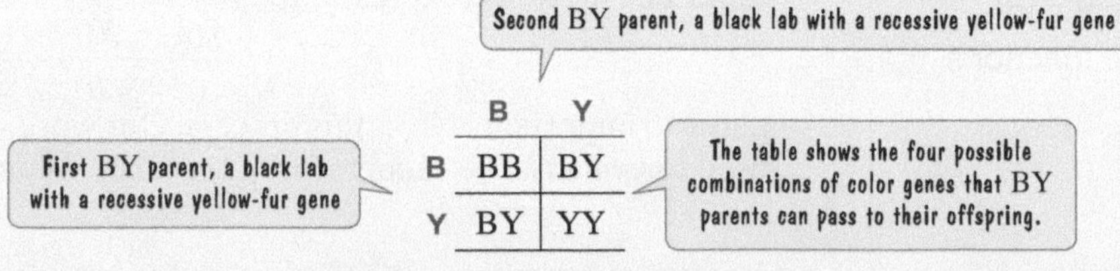
Second BY parent, a black lab with a recessive yellow-fur gene
First BY parent, a black lab with a recessive yellow-fur gene
The table shows the four possible combinations of color genes that BY parents can pass to their offspring.

	B	Y
B	BB	BY
Y	BY	YY

Because YY is one of four possible outcomes, the probability that a yellow lab like Axl will be the offspring of these black parents is $\frac{1}{4}$.

The probabilities suggested by the table can be modeled by the expression $\left(\frac{1}{2}B + \frac{1}{2}Y\right)^2$.

$$\left(\frac{1}{2}B + \frac{1}{2}Y\right)^2 = \left(\frac{1}{2}B\right)^2 + 2\left(\frac{1}{2}B\right)\left(\frac{1}{2}Y\right) + \left(\frac{1}{2}Y\right)^2$$

$$= \frac{1}{4}BB + \frac{1}{2}BY + \frac{1}{4}YY$$

The probability of a black lab with two dominant black genes is $\frac{1}{4}$.

The probability of a black lab with a recessive yellow gene is $\frac{1}{2}$.

The probability of a yellow lab with two recessive yellow genes is $\frac{1}{4}$.

CONCEPT AND VOCABULARY CHECK

Fill in each blank so that the resulting statement is true.

1. To multiply the variables in monomials, retain each variable and ____________ the exponents.
2. To multiply $7x^3(4x^5 - 8x^2 + 6)$, use the ____________ property to multiply each term of the trinomial ____________ by the monomial ____________.
3. To multiply $(5x + 3)(x^2 + 8x + 7)$, begin by multiplying each term of $x^2 + 8x + 7$ by ____________. Then multiply each term of $x^2 + 8x + 7$ by ____________. Then combine ____________ terms.
4. When using the FOIL method to find $(x + 7)(3x + 5)$, the product of the first terms is ____________, the product of the outside terms is ____________, the product of the inside terms is ____________, and the product of the last terms is ____________.
5. $(A + B)^2 =$ ____________. The square of a binomial sum is the first term ____________ plus 2 times the ____________ plus the last term ____________.
6. $(A - B)^2 =$ ____________. The square of a binomial difference is the first term squared ____________ 2 times the ____________ ____________ the last term squared.

plus or minus?

7. $(A + B)(A - B) =$ ____________. The product of the sum and difference of the same two terms is the square of the first term ____________ the square of the second term.
8. If $f(x) = x^2 - 4x + 7$, we find $f(a + h)$ by replacing each occurrence of ____________ with ____________.

5.2 EXERCISE SET

MyMathLab®

Practice Exercises

Throughout the practice exercises, assume that any variable exponents represent whole numbers.

In Exercises 1–8, multiply the monomials.

1. $(3x^2)(5x^4)$
2. $(4x^2)(6x^4)$
3. $(3x^2y^4)(5xy^7)$
4. $(6x^4y^2)(3x^7y)$
5. $(-3xy^2z^5)(2xy^7z^4)$
6. $(11x^2yz^4)(-3xy^5z^6)$
7. $(-8x^{2n}y^{n-5})\left(-\frac{1}{4}x^ny^3\right)$
8. $(-9x^{3n}y^{n-3})\left(-\frac{1}{3}x^ny^2\right)$

In Exercises 9–22, multiply the monomial and the polynomial.

9. $4x^2(3x + 2)$
10. $5x^2(6x + 7)$
11. $2y(y^2 - 5y)$
12. $3y(y^2 - 4y)$
13. $5x^3(2x^5 - 4x^2 + 9)$
14. $6x^3(3x^5 - 5x^2 + 7)$
15. $4xy(7x + 3y)$
16. $5xy(8x + 3y)$
17. $3ab^2(6a^2b^3 + 5ab)$
18. $5ab^2(10a^2b^3 + 7ab)$
19. $-4x^2y(3x^4y^2 - 7xy^3 + 6)$
20. $-3x^2y(10x^2y^4 - 2xy^3 + 7)$
21. $-4x^n\left(3x^{2n} - 5x^n + \frac{1}{2}x\right)$
22. $-10x^n\left(4x^{2n} - 3x^n + \frac{1}{5}x\right)$

In Exercises 23–34, find each product using either a horizontal or a vertical format.

23. $(x - 3)(x^2 + 2x + 5)$
24. $(x + 4)(x^2 - 5x + 8)$
25. $(x - 1)(x^2 + x + 1)$
26. $(x - 2)(x^2 + 2x + 4)$
27. $(a - b)(a^2 + ab + b^2)$
28. $(a + b)(a^2 - ab + b^2)$
29. $(x^2 + 2x - 1)(x^2 + 3x - 4)$
30. $(x^2 - 2x + 3)(x^2 + x + 1)$
31. $(x - y)(x^2 - 3xy + y^2)$
32. $(x - y)(x^2 - 4xy + y^2)$
33. $(xy + 2)(x^2y^2 - 2xy + 4)$
34. $(xy + 3)(x^2y^2 - 2xy + 5)$

In Exercises 35–54, use the FOIL method to multiply the binomials.

35. $(x + 4)(x + 7)$
36. $(x + 5)(x + 8)$
37. $(y + 5)(y - 6)$
38. $(y + 5)(y - 8)$

39. $(5x + 3)(2x + 1)$
40. $(4x + 3)(5x + 1)$
41. $(3y - 4)(2y - 1)$
42. $(5y - 2)(3y - 1)$
43. $(3x - 2)(5x - 4)$
44. $(2x - 3)(4x - 5)$
45. $(x - 3y)(2x + 7y)$
46. $(3x - y)(2x + 5y)$
47. $(7xy + 1)(2xy - 3)$
48. $(3xy - 1)(5xy + 2)$
49. $(x - 4)(x^2 - 5)$
50. $(x - 5)(x^2 - 3)$
51. $(8x^3 + 3)(x^2 - 5)$
52. $(7x^3 + 5)(x^2 - 2)$
53. $(3x^n - y^n)(x^n + 2y^n)$
54. $(5x^n - y^n)(x^n + 4y^n)$

In Exercises 55–68, multiply using one of the rules for the square of a binomial.

55. $(x + 3)^2$
56. $(x + 4)^2$
57. $(y - 5)^2$
58. $(y - 6)^2$
59. $(2x + y)^2$
60. $(4x + y)^2$
61. $(5x - 3y)^2$
62. $(3x - 4y)^2$
63. $(2x^2 + 3y)^2$
64. $(4x^2 + 5y)^2$
65. $(4xy^2 - xy)^2$
66. $(5xy^2 - xy)^2$
67. $(a^n + 4b^n)^2$
68. $(3a^n - b^n)^2$

In Exercises 69–82, multiply using the rule for the product of the sum and difference of two terms.

69. $(x + 4)(x - 4)$
70. $(x + 5)(x - 5)$
71. $(5x + 3)(5x - 3)$
72. $(3x + 2)(3x - 2)$
73. $(4x + 7y)(4x - 7y)$
74. $(8x + 7y)(8x - 7y)$
75. $(y^3 + 2)(y^3 - 2)$
76. $(y^3 + 3)(y^3 - 3)$
77. $(1 - y^5)(1 + y^5)$
78. $(2 - y^5)(2 + y^5)$
79. $(7xy^2 - 10y)(7xy^2 + 10y)$
80. $(3xy^2 - 4y)(3xy^2 + 4y)$
81. $(5a^n - 7)(5a^n + 7)$
82. $(10b^n - 3)(10b^n + 3)$

In Exercises 83–94, find each product.

83. $[(2x + 3) + 4y][(2x + 3) - 4y]$
84. $[(3x + 2) + 5y][(3x + 2) - 5y]$
85. $(x + y + 3)(x + y - 3)$
86. $(x + y + 4)(x + y - 4)$
87. $(5x + 7y - 2)(5x + 7y + 2)$
88. $(7x + 5y - 2)(7x + 5y + 2)$
89. $[5y + (2x + 3)][5y - (2x + 3)]$
90. $[8y + (3x + 2)][8y - (3x + 2)]$
91. $(x + y + 1)^2$
92. $(x + y + 2)^2$
93. $(x + 1)(x - 1)(x^2 + 1)$
94. $(x + 2)(x - 2)(x^2 + 4)$
95. Let $f(x) = x - 2$ and $g(x) = x + 6$. Find each of the following.
    a. $(fg)(x)$
    b. $(fg)(-1)$
    c. $(fg)(0)$
96. Let $f(x) = x - 4$ and $g(x) = x + 10$. Find each of the following.
    a. $(fg)(x)$
    b. $(fg)(-1)$
    c. $(fg)(0)$
97. Let $f(x) = x - 3$ and $g(x) = x^2 + 3x + 9$. Find each of the following.
    a. $(fg)(x)$
    b. $(fg)(-2)$
    c. $(fg)(0)$
98. Let $f(x) = x + 3$ and $g(x) = x^2 - 3x + 9$. Find each of the following.
    a. $(fg)(x)$
    b. $(fg)(-2)$
    c. $(fg)(0)$

In Exercises 99–102, find each of the following and simplify:

a. $f(a + 2)$ b. $f(a + h) - f(a)$.

99. $f(x) = x^2 - 3x + 7$
100. $f(x) = x^2 - 4x + 9$
101. $f(x) = 3x^2 + 2x - 1$
102. $f(x) = 4x^2 + 5x - 1$

Practice PLUS

In Exercises 103–112, perform the indicated operation or operations.

103. $(3x + 4y)^2 - (3x - 4y)^2$
104. $(5x + 2y)^2 - (5x - 2y)^2$
105. $(5x - 7)(3x - 2) - (4x - 5)(6x - 1)$
106. $(3x + 5)(2x - 9) - (7x - 2)(x - 1)$
107. $(2x + 5)(2x - 5)(4x^2 + 25)$
108. $(3x + 4)(3x - 4)(9x^2 + 16)$

109. $(x - 1)^3$

110. $(x - 2)^3$

111. $\dfrac{(2x - 7)^5}{(2x - 7)^3}$

112. $\dfrac{(5x - 3)^6}{(5x - 3)^4}$

Application Exercises

In Exercises 113–114, find the area of the large rectangle in two ways:

a. *Find the sum of the areas of the four smaller rectangles.*

b. *Multiply the length and the width of the large rectangle using the FOIL method. Compare this product with your answer to part (a).*

113.

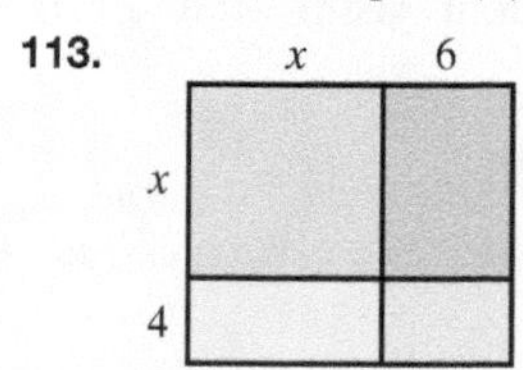

114.

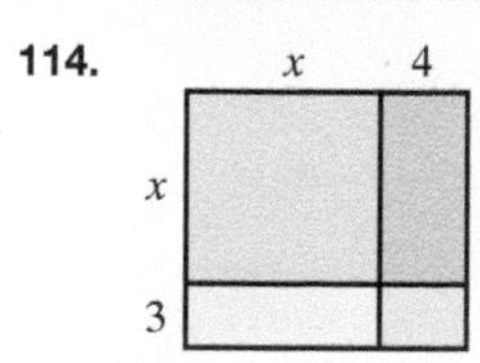

In Exercises 115–116, express each polynomial in standard form—that is, in descending powers of x.

a. *Write a polynomial that represents the area of the large rectangle.*

b. *Write a polynomial that represents the area of the small, unshaded rectangle.*

c. *Write a polynomial that represents the area of the shaded blue region.*

115.

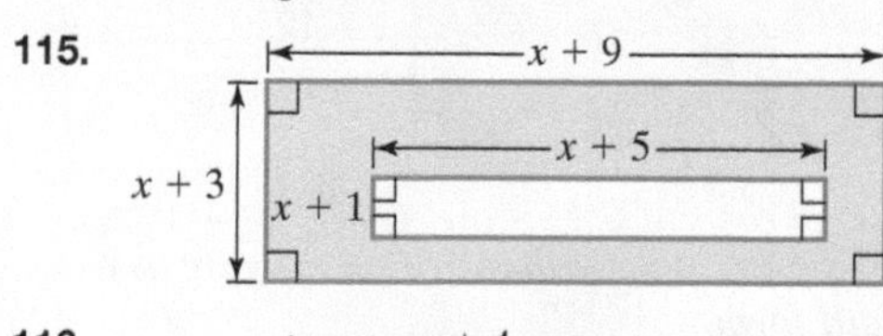

116.

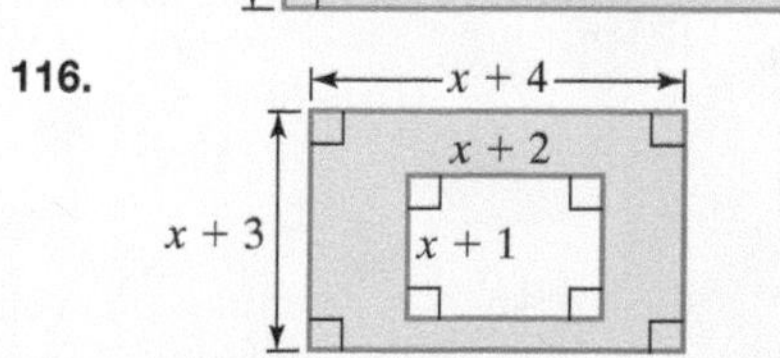

In Exercises 117–118, express each polynomial in standard form.

a. *Write a polynomial that represents the area of the rectangular base of the open box.*

b. *Write a polynomial that represents the volume of the open box.*

117.

118.

x

$8 - 2x$

$5 - 2x$

119. A popular model of carry-on luggage has a length that is 10 inches greater than its depth. Airline regulations require that the sum of the length, width, and depth cannot exceed 40 inches. These conditions, with the assumption that this sum *is* 40 inches, can be modeled by a function that gives the volume of the luggage, V, in cubic inches, in terms of its depth, x, in inches.

Volume = depth · length · width: 40 − (depth + length)

$$V(x) = x \cdot (x + 10) \cdot [40 - (x + x + 10)]$$

$$V(x) = x(x + 10)(30 - 2x)$$

a. Perform the multiplications in the formula for $V(x)$ and express the formula in standard form.

b. Use the function's formula from part (a) and the Leading Coefficient Test to determine the end behavior of its graph.

c. Does the end behavior to the right make this function useful in modeling the volume of carry-on luggage as its depth continues to increase?

d. Use the formula from part (a) to find $V(10)$. Describe what this means in practical terms.

e. The graph of the function modeling the volume of carry-on luggage is shown below. Identify your answer from part (d) as a point on the graph.

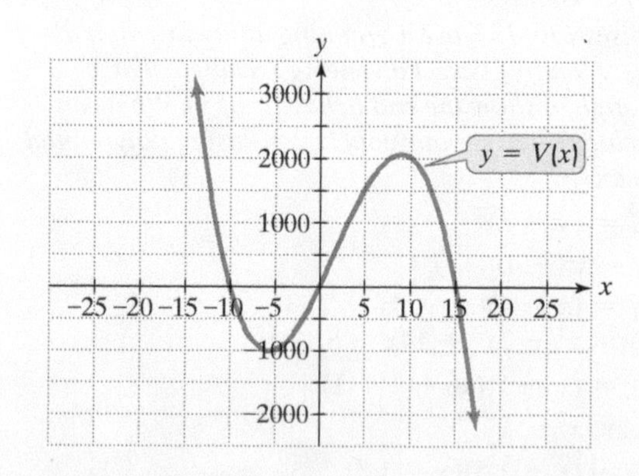

f. Use the graph to describe a realistic domain, x, for the volume function, where x represents the depth of the carry-on luggage. Use interval notation to express this realistic domain.

120. Before working this exercise, be sure that you have read the Blitzer Bonus on page 336. The table shows the four combinations of color genes that a YY yellow lab and a BY black lab can pass to their offspring.

	B	Y
Y	BY	YY
Y	BY	YY

a. How many combinations result in a yellow lab with two recessive yellow genes? What is the probability of a yellow lab?

b. How many combinations result in a black lab with a recessive yellow gene? What is the probability of a black lab?

c. Find the product of Y and $\frac{1}{2}$B + $\frac{1}{2}$Y. How does this product model the probabilities that you determined in parts (a) and (b)?

Explaining the Concepts

121. Explain how to multiply monomials. Give an example.

122. Explain how to multiply a monomial and a polynomial that is not a monomial. Give an example.

123. Explain how to multiply a binomial and a trinomial.

124. What is the FOIL method and when is it used? Give an example of the method.

125. Explain how to square a binomial sum. Give an example.

126. Explain how to square a binomial difference. Give an example.

127. Explain how to find the product of the sum and difference of two terms. Give an example with your explanation.

128. How can the graph of function fg be obtained from the graphs of functions f and g?

129. Explain how to find $f(a + h) - f(a)$ for a given function f.

Technology Exercises

In Exercises 130–133, use a graphing utility to graph the functions y_1 and y_2. Select a viewing rectangle that is large enough to show the end behavior of y_2. What can you conclude? Verify your conclusions using polynomial multiplication.

130. $y_1 = (x - 2)^2$
$y_2 = x^2 - 4x + 4$

131. $y_1 = (x - 4)(x^2 - 3x + 2)$
$y_2 = x^3 - 7x^2 + 14x - 8$

132. $y_1 = (x - 1)(x^2 + x + 1)$
$y_2 = x^3 - 1$

133. $y_1 = (x + 1.5)(x - 1.5)$
$y_2 = x^2 - 2.25$

134. Graph $f(x) = x + 4$, $g(x) = x - 2$, and the product function, fg, in a $[-6, 6, 1]$ by $[-10, 10, 1]$ viewing rectangle. Trace along the curves and show that $(fg)(1) = f(1) \cdot g(1)$.

Critical Thinking Exercises

Make Sense? *In Exercises 135–138, determine whether each statement makes sense or does not make sense, and explain your reasoning.*

135. Knowing the difference between factors and terms is important: In $(3x^2y)^2$, I can distribute the exponent 2 on each factor, but in $(3x^2 + y)^2$, I cannot do the same thing on each term.

136. I used the FOIL method to find the product of $x + 5$ and $x^2 + 2x + 1$.

137. Instead of using the formula for the square of a binomial sum, I prefer to write the binomial sum twice and then apply the FOIL method.

138. Special-product formulas have patterns that make their multiplications quicker than using the FOIL method.

In Exercises 139–142, determine whether each statement is true or false. If the statement is false, make the necessary change(s) to produce a true statement.

139. If f is a polynomial function, then

$$f(a + h) - f(a) = f(a) + f(h) - f(a) = f(h).$$

140. $(x - 5)^2 = x^2 - 5x + 25$

141. $(x + 1)^2 = x^2 + 1$

142. Suppose a square garden has an area represented by $9x^2$ square feet. If one side is made 7 feet longer and the other side is made 2 feet shorter, then the trinomial that represents the area of the larger garden is $9x^2 + 15x - 14$ square feet.

143. Express the area of the plane figure shown as a polynomial in standard form.

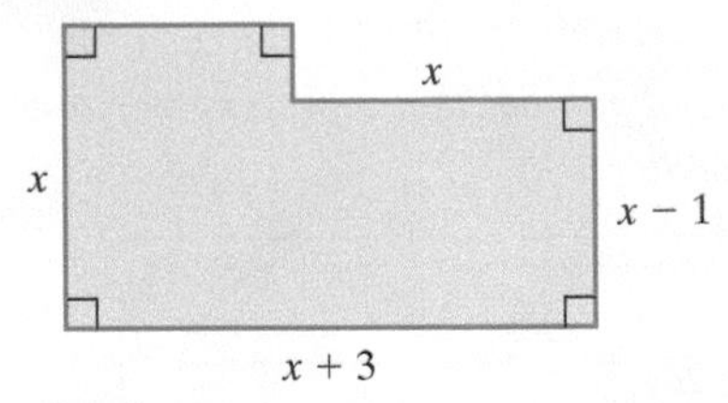

In Exercises 144–145, represent the volume of each figure as a polynomial in standard form.

144.

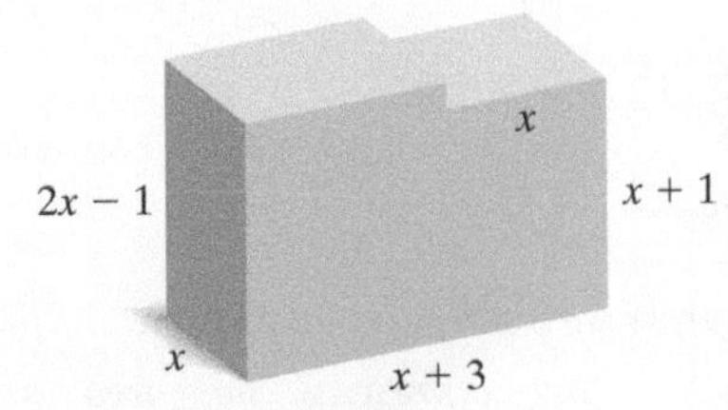

145.

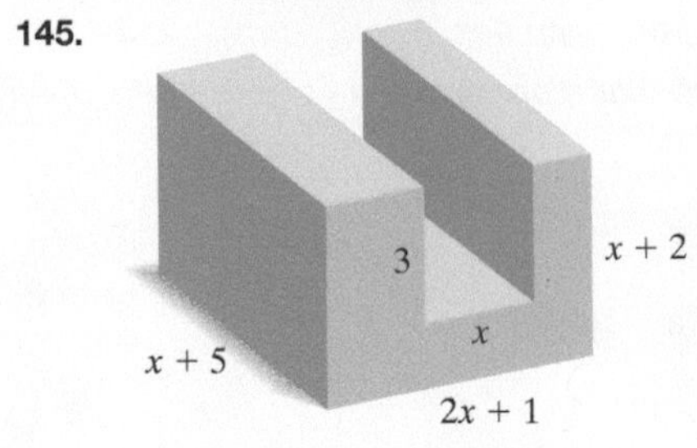

146. Simplify: $(y^n + 2)(y^n - 2) - (y^n - 3)^2$.

147. The product of two consecutive odd integers is 22 less than the square of the greater integer. Find the integers.

Review Exercises

148. Solve: $|3x + 4| \geq 10$. (Section 4.3, Example 6)

149. Solve: $2 - 6x \leq 20$. (Section 4.1, Example 2)

150. Write in scientific notation: 8,034,000,000. (Section 1.7, Example 2)

Preview Exercises

Exercises 151–153 will help you prepare for the material covered in the next section.

151. Replace each boxed question mark with a polynomial that results in the given product.

a. $3x^3 \cdot \boxed{?} = 9x^5$

b. $2x^3y^2 \cdot \boxed{?} = 12x^5y^4$

In Exercises 152–153, a polynomial is given in factored form. Use multiplication to find the product of the factors.

152. $(x - 5)(x^2 + 3)$

153. $(x + 4)(3x - 2y)$

SECTION

5.3 Greatest Common Factors and Factoring by Grouping

What am I supposed to learn?

After studying this section, you should be able to:

1. Factor out the greatest common factor of a polynomial.
2. Factor out a common factor with a negative coefficient.
3. Factor by grouping.

A two-year-old girl is asked, "Do you have a sister?" She answers, "Yes." "What is your sister's name?" "Hannah." Asked if Hannah has a sister, the two-year-old replies, "No." The child can go in the direction from self to sister but she cannot reverse this direction and move from sister back to self.

As our intellects develop, we learn to reverse the direction of our thinking. Reversibility of thought is found throughout algebra. For example, we can multiply polynomials and show that

$$7x(3x + 4) = 21x^2 + 28x.$$

We can also reverse $7x(3x + 4) = 21x^2 + 28x$ and express the resulting polynomial as

$$21x^2 + 28x = 7x(3x + 4).$$

Factoring a polynomial consisting of the sum of monomials means finding an equivalent expression that is a product.

Factoring $21x^2 + 28x$

Sum of monomials

Equivalent expression that is a product

$$21x^2 + 28x = 7x(3x + 4)$$

The factors of $21x^2 + 28x$ are $7x$ and $3x + 4$.

In this chapter, we will be factoring over the set of integers, meaning that the coefficients in the factors are integers. Polynomials that cannot be factored using integer coefficients are called **prime polynomials over the set of integers.**

1 Factor out the greatest common factor of a polynomial.

Factoring Out the Greatest Common Factor

In any factoring problem, the first step is to look for the *greatest common factor*. The **greatest common factor**, abbreviated GCF, is an expression with the greatest coefficient and of the highest degree that divides each term of the polynomial. Can you see that $7x$ is the greatest common factor of $21x^2 + 28x$? 7 is the greatest integer that divides both 21 and 28. Furthermore, x is the greatest power of x that divides x^2 and x.

The variable part of the greatest common factor always contains the *smallest* power of a variable that appears in all terms of the polynomial. For example, consider the polynomial

$$21x^2 + 28x.$$

x^1, or x, is the variable raised to the smallest exponent.

We see that x is the variable part of the greatest common factor, $7x$.

When factoring a monomial from a polynomial, determine the greatest common factor of all terms in the polynomial. Sometimes there may not be a GCF other than 1. When a GCF other than 1 exists, we use the following procedure:

Factoring a Monomial from a Polynomial

1. Determine the greatest common factor of all terms in the polynomial.
2. Express each term as the product of the GCF and its other factor.
3. Use the distributive property to factor out the GCF.

EXAMPLE 1 Factoring Out the Greatest Common Factor

Factor: $21x^2 + 28x$.

Solution The GCF of the two terms of the polynomial is $7x$.

$$21x^2 + 28x$$

$$= 7x(3x) + 7x(4)$$ Express each term as the product of the GCF and its other factor.

$$= 7x(3x + 4)$$ Factor out the GCF.

We can check this factorization by multiplying $7x$ and $3x + 4$, obtaining the original polynomial as the answer. ■

CHECK POINT 1 Factor: $20x^2 + 30x$.

EXAMPLE 2 Factoring Out the Greatest Common Factor

Factor:

a. $9x^5 + 15x^3$ **b.** $16x^2y^3 - 24x^3y^4$ **c.** $12x^5y^4 - 4x^4y^3 + 2x^3y^2$.

Solution

a. First, determine the greatest common factor.

3 is the greatest integer that divides 9 and 15.

$$9x^5 + 15x^3$$

x^3 is the variable raised to the smallest exponent.

The GCF of the two terms of the polynomial is $3x^3$.

$$9x^5 + 15x^3$$

$$= 3x^3 \cdot 3x^2 + 3x^3 \cdot 5$$ Express each term as the product of the GCF and its other factor.

$$= 3x^3(3x^2 + 5)$$ Factor out the GCF.

b. Begin by determining the greatest common factor.

8 is the greatest integer that divides 16 and 24.

$$16x^2y^3 - 24x^3y^4$$

The variables raised to the smallest exponents are x^2 and y^3.

The GCF of the two terms of the polynomial is $8x^2y^3$.

$$16x^2y^3 - 24x^3y^4$$

$$= 8x^2y^3 \cdot 2 - 8x^2y^3 \cdot 3xy$$ Express each term as the product of the GCF and its other factor.

$$= 8x^2y^3(2 - 3xy)$$ Factor out the GCF.

c. First, determine the greatest common factor of the three terms.

2 is the greatest integer that divides 12, 4, and 2.

$$12x^5y^4 - 4x^4y^3 + 2x^3y^2$$

The variables raised to the smallest exponents are x^3 and y^2.

The GCF is $2x^3y^2$.

$$12x^5y^4 - 4x^4y^3 + 2x^3y^2$$

$$= 2x^3y^2 \cdot 6x^2y^2 - 2x^3y^2 \cdot 2xy + 2x^3y^2 \cdot 1$$ Express each term as the product of the GCF and its other factor.

You can obtain the factors shown in black by dividing each term of the given polynomial by $2x^3y^2$, the GCF.

$$\frac{12x^5y^4}{2x^3y^2} = 6x^2y^2 \qquad \frac{4x^4y^3}{2x^3y^2} = 2xy \qquad \frac{2x^3y^2}{2x^3y^2} = 1$$

$$= 2x^3y^2(6x^2y^2 - 2xy + 1)$$ Factor out the GCF. ■

Because factoring reverses the process of multiplication, all factorizations can be checked by multiplying. Take a few minutes to check each of the three factorizations in Example 2. Use the distributive property to multiply the factors. This should give the original polynomial.

✓ CHECK POINT 2 Factor:

a. $9x^4 + 21x^2$

b. $15x^3y^2 - 25x^4y^3$

c. $16x^4y^5 - 8x^3y^4 + 4x^2y^3$.

---

2 Factor out a common factor with a negative coefficient.

When the leading coefficient of a polynomial is negative, it is often desirable to factor out a common factor with a negative coefficient. The common factor is the GCF preceded by a negative sign.

EXAMPLE 3 Using a Common Factor with a Negative Coefficient

Factor: $-3x^3 + 12x^2 - 15x$.

Solution The GCF is $3x$. Because the leading coefficient, -3, is negative, we factor out a common factor with a negative coefficient. We will factor out the opposite of the GCF, or $-3x$.

$$-3x^3 + 12x^2 - 15x$$

$$= -3x(x^2) - 3x(-4x) - 3x(5)$$ Express each term as the product of the common factor and its other factor.

$$= -3x(x^2 - 4x + 5)$$ Factor out the opposite of the GCF. ■

✓ CHECK POINT 3 Factor out a common factor with a negative coefficient: $-2x^3 + 10x^2 - 6x$.

Factoring by Grouping

Up to now, we have factored a monomial from a polynomial. By contrast, in our next example, the greatest common factor of the polynomial is a binomial.

EXAMPLE 4 Factoring Out the Greatest Common Binomial Factor

Factor:

a. $2(x - 7) + 9a(x - 7)$ **b.** $5y(a - b) - (a - b)$.

Solution Let's identify the common binomial factor in each part of the problem.

$$2(x - 7) + 9a(x - 7) \qquad 5y(a - b) - (a - b)$$

The GCF, a binomial, is $x - 7$. The GCF, a binomial, is $a - b$.

We factor out each common binomial factor as follows.

a. $2(x - 7) + 9a(x - 7)$

$= (x - 7)2 + (x - 7)9a$ — This step, usually omitted, shows each term as the product of the GCF and its other factor, in that order.

$= (x - 7)(2 + 9a)$ — Factor out the GCF.

b. $5y(a - b) - (a - b)$

$= 5y(a - b) - 1(a - b)$ — Write $-(a - b)$ as $-1\,(a - b)$ to aid in the factoring.

$= (a - b)(5y - 1)$ — Factor out the GCF. ■

✓ **CHECK POINT 4** Factor:

a. $3(x - 4) + 7a(x - 4)$

b. $7x(a + b) - (a + b)$.

---

3 Factor by grouping.

Some polynomials have only a greatest common factor of 1. However, by a suitable grouping of the terms, it still may be possible to factor. This process, called **factoring by grouping**, is illustrated in Example 5.

EXAMPLE 5 Factoring by Grouping

Factor: $x^3 - 5x^2 + 3x - 15$.

Solution There is no factor other than 1 common to all four terms. However, we can group terms that have a common factor:

$$\boxed{x^3 - 5x^2} + \boxed{3x - 15}$$

Common factor is x^2. Common factor is 3.

Discover for Yourself

In Example 5, group the terms as follows:

$$x^2 + 3x - 5x^2 - 15.$$

Factor out the common factor from each group and complete the factoring process. Describe what happens. What can you conclude?

We now factor the given polynomial as follows.

$x^3 - 5x^2 + 3x - 15$

$= (x^3 - 5x^2) + (3x - 15)$ — Group terms with common factors.

$= x^2(x - 5) + 3(x - 5)$ — Factor out the greatest common factor from the grouped terms. The remaining two terms have $x - 5$ as a common binomial factor.

$= (x - 5)(x^2 + 3).$ — Factor out the GCF.

Thus, $x^3 - 5x^2 + 3x - 15 = (x - 5)(x^2 + 3)$. Check the factorization by multiplying the right side of the equation using the FOIL method. Because the factorization is correct, you should obtain the original polynomial. ■

✓ **CHECK POINT 5** Factor: $x^3 - 4x^2 + 5x - 20$.

Factoring by Grouping

1. Group terms that have a common monomial factor. There will usually be two groups. Sometimes the terms must be rearranged.
2. Factor out the common monomial factor from each group.
3. Factor out the remaining common binomial factor (if one exists).

EXAMPLE 6 Factoring by Grouping

Factor: $3x^2 + 12x - 2xy - 8y$.

Solution There is no factor other than 1 common to all four terms. However, we can group terms that have a common factor:

$$\boxed{3x^2 + 12x} + \boxed{-2xy - 8y}.$$

Common factor is $3x$: $3x^2 + 12x = 3x(x + 4)$.

Use $-2y$, rather than $2y$, as the common factor: $-2xy - 8y = -2y(x + 4)$. In this way, the common binomial factor, $x + 4$, appears.

The voice balloons illustrate that it is sometimes necessary to use a factor with a negative coefficient to obtain a common binomial factor for the two groupings. We now factor the given polynomial as follows:

$$\begin{aligned} &3x^2 + 12x - 2xy - 8y \\ &= (3x^2 + 12x) + (-2xy - 8y) && \text{Group terms with common factors.} \\ &= 3x(x + 4) - 2y(x + 4) && \text{Factor out the common factors from the grouped terms.} \\ &= (x + 4)(3x - 2y). && \text{Factor out the GCF.} \end{aligned}$$

Thus, $3x^2 + 12x - 2xy - 8y = (x + 4)(3x - 2y)$. Using the commutative property of multiplication, the factorization can also be expressed as $(3x - 2y)(x + 4)$. Verify the factorization by showing that, regardless of the order, FOIL multiplication gives the original polynomial. ■

✓ **CHECK POINT 6** Factor: $4x^2 + 20x - 3xy - 15y$.

Factoring by grouping sometimes requires that the terms be rearranged before the groupings are made. For example, consider the polynomial

$$3x^2 - 8y + 12x - 2xy.$$

The first two terms have no common factor other than 1. We must rearrange the terms and try a different grouping. Example 6 showed one such rearrangement of two groupings.

Achieving Success

When using your professor's office hours, show up prepared. If you are having difficulty with a concept or problem, bring your work so that your instructor can determine where you are having trouble. If you miss a lecture, read the appropriate section in the textbook, borrow class notes, and attempt the assigned homework before your office visit. Because this text has video lectures for every section, you might find it helpful to view them in MyMathLab to review the material you missed. It is not realistic to expect your professor to rehash all or part of a class lecture during office hours.

CONCEPT AND VOCABULARY CHECK

Fill in each blank so that the resulting statement is true.

1. The process of writing a polynomial containing the sum of monomials as a product is called ______________.
2. An expression with the greatest coefficient and of the highest degree that divides each term of a polynomial is called the ______________. The variable part of this expression contains the ______________ power of a variable that appears in all terms of the polynomial.

3. True or false: The factorization of $12x^4 + 21x^2$ is $3x^2 \cdot 4x^2 + 3x^2 \cdot 7$. ____________
4. We factor $-2x^3 + 10x^2 - 6x$ by factoring out ____________.
5. True or false: The factorization of $x^3 - 4x^2 + 5x - 20$ is $x^2(x - 4) + 5(x - 4)$. ____________

5.3 EXERCISE SET

MyMathLab®

Practice Exercises

Throughout the practice exercises, assume that any variable exponents represent whole numbers.

In Exercises 1–22, factor the greatest common factor from each polynomial.

1. $10x^2 + 4x$
2. $12x^2 + 9x$
3. $y^2 - 4y$
4. $y^2 - 7y$
5. $x^3 + 5x^2$
6. $x^3 + 7x^2$
7. $12x^4 - 8x^2$
8. $20x^4 - 8x^2$
9. $32x^4 + 2x^3 + 8x^2$
10. $9x^4 + 18x^3 + 6x^2$
11. $4x^2y^3 + 6xy$
12. $6x^3y^2 + 9xy$
13. $30x^2y^3 - 10xy^2$
14. $27x^2y^3 - 18xy^2$
15. $12xy - 6xz + 4xw$
16. $14xy - 10xz + 8xw$
17. $15x^3y^6 - 9x^4y^4 + 12x^2y^5$
18. $15x^4y^6 - 3x^3y^5 + 12x^4y^4$
19. $25x^3y^6z^2 - 15x^4y^4z^4 + 25x^2y^5z^3$
20. $49x^4y^3z^5 - 70x^3y^5z^4 + 35x^4y^4z^3$
21. $15x^{2n} - 25x^n$
22. $12x^{3n} - 9x^{2n}$

In Exercises 23–34, factor out the negative of the greatest common factor.

23. $-4x + 12$
24. $-5x + 20$
25. $-8x - 48$
26. $-7x - 63$
27. $-2x^2 + 6x - 14$
28. $-2x^2 + 8x - 12$
29. $-5y^2 + 40x$
30. $-9y^2 + 45x$
31. $-4x^3 + 32x^2 - 20x$
32. $-5x^3 + 50x^2 - 10x$
33. $-x^2 - 7x + 5$
34. $-x^2 - 8x + 8$

In Exercises 35–44, factor the greatest common binomial factor from each polynomial.

35. $4(x + 3) + a(x + 3)$
36. $5(x + 4) + a(x + 4)$
37. $x(y - 6) - 7(y - 6)$
38. $x(y - 9) - 5(y - 9)$
39. $3x(x + y) - (x + y)$
40. $7x(x + y) - (x + y)$
41. $4x^2(3x - 1) + 3x - 1$
42. $6x^2(5x - 1) + 5x - 1$
43. $(x + 2)(x + 3) + (x - 1)(x + 3)$
44. $(x + 4)(x + 5) + (x - 1)(x + 5)$

In Exercises 45–68, factor by grouping.

45. $x^2 + 3x + 5x + 15$
46. $x^2 + 2x + 4x + 8$
47. $x^2 + 7x - 4x - 28$
48. $x^2 + 3x - 5x - 15$
49. $x^3 - 3x^2 + 4x - 12$
50. $x^3 - 2x^2 + 5x - 10$
51. $xy - 6x + 2y - 12$
52. $xy - 5x + 9y - 45$
53. $xy + x - 7y - 7$
54. $xy + x - 5y - 5$
55. $10x^2 - 12xy + 35xy - 42y^2$
56. $3x^2 - 6xy + 5xy - 10y^2$
57. $4x^3 - x^2 - 12x + 3$
58. $3x^3 - 2x^2 - 6x + 4$
59. $x^2 - ax - bx + ab$
60. $x^2 + ax - bx - ab$
61. $x^3 - 12 - 3x^2 + 4x$
62. $2x^3 - 10 + 4x^2 - 5x$
63. $ay - by + bx - ax$
64. $cx - dx + dy - cy$
65. $ay^2 + 2by^2 - 3ax - 6bx$
66. $3a^2x + 6a^2y - 2bx - 4by$
67. $x^ny^n + 3x^n + y^n + 3$
68. $x^ny^n - x^n + 2y^n - 2$

Practice PLUS

In Exercises 69–78, factor each polynomial.

69. $ab - c - ac + b$

70. $ab - 3c - ac + 3b$

71. $x^3 - 5 + 4x^3y - 20y$

72. $x^3 - 2 + 3x^3y - 6y$

73. $2y^7(3x - 1)^5 - 7y^6(3x - 1)^4$

74. $3y^9(3x - 2)^7 - 5y^8(3x - 2)^6$

75. $ax^2 + 5ax - 2a + bx^2 + 5bx - 2b$

76. $ax^2 + 3ax - 11a + bx^2 + 3bx - 11b$

77. $ax + ay + az - bx - by - bz + cx + cy + cz$

78. $ax^2 + ay^2 - az^2 + bx^2 + by^2 - bz^2 + cx^2 + cy^2 - cz^2$

Application Exercises

79. A ball is thrown straight upward. The function

$$f(t) = -16t^2 + 40t$$

describes the ball's height above the ground, $f(t)$, in feet, t seconds after it is thrown.

a. Find and interpret $f(2)$.

b. Find and interpret $f(2.5)$.

c. Factor the polynomial $-16t^2 + 40t$ and write the function in factored form.

d. Use the factored form of the function to find $f(2)$ and $f(2.5)$. Do you get the same answers as you did in parts (a) and (b)? If so, does this prove that your factorization is correct? Explain.

80. An explosion causes debris to rise vertically. The function

$$f(t) = -16t^2 + 72t$$

describes the height of the debris above the ground, $f(t)$, in feet, t seconds after the explosion.

a. Find and interpret $f(2)$.

b. Find and interpret $f(4.5)$.

c. Factor the polynomial $-16t^2 + 72t$ and write the function in factored form.

d. Use the factored form of the function to find $f(2)$ and $f(4.5)$. Do you get the same answers as you did in parts (a) and (b)? If so, does this prove that your factorization is correct? Explain.

81. Your computer store is having an incredible sale. The price on one model is reduced by 40%. Then the sale price is reduced by another 40%. If x is the computer's original price, the sale price can be represented by

$$(x - 0.4x) - 0.4(x - 0.4x).$$

a. Factor out $(x - 0.4x)$ from each term. Then simplify the resulting expression.

b. Use the simplified expression from part (a) to answer these questions. With a 40% reduction followed by a 40% reduction, is the computer selling at 20% of its original price? If not, at what percentage of the original price is it selling?

82. Your local electronics store is having an end-of-the-year sale. The price on a plasma television had been reduced by 30%. Now the sale price is reduced by another 30%. If x is the television's original price, the sale price can be represented by

$$(x - 0.3x) - 0.3(x - 0.3x).$$

a. Factor out $(x - 0.3x)$ from each term. Then simplify the resulting expression.

b. Use the simplified expression from part (a) to answer these questions. With a 30% reduction followed by a 30% reduction, is the television selling at 40% of its original price? If not, at what percentage of the original price is it selling?

Exercises 83–84 involve compound interest. **Compound interest** *is interest computed on your original savings as well as on any accumulated interest.*

83. After 2 years, the balance, A, in an account with principal P and interest rate r compounded annually is given by the formula

$$A = P + Pr + (P + Pr)r.$$

Use factoring by grouping to express the formula as $A = P(1 + r)^2$.

84. After 3 years, the balance, A, in an account with principal P and interest rate r compounded annually is given by the formula

$$A = P(1 + r)^2 + P(1 + r)^2r.$$

Use factoring by grouping to express the formula as $A = P(1 + r)^3$.

85. The area of the skating rink with semicircular ends shown is $A = \pi r^2 + 2rl$. Express the area, A, in factored form.

86. The amount of sheet metal needed to manufacture a cylindrical tin can, that is, its surface area, S, is $S = 2\pi r^2 + 2\pi rh$. Express the surface area, S, in factored form.

Explaining the Concepts

87. What is factoring?

88. If a polynomial has a greatest common factor other than 1, explain how to find its GCF.

89. Using an example, explain how to factor out the greatest common factor of a polynomial.

90. Suppose that a polynomial contains four terms and can be factored by grouping. Explain how to obtain the factorization.

91. Use two different groupings to factor

$$ac - ad + bd - bc$$

in two ways. Then explain why the two factorizations are the same.

92. Write a sentence that uses the word *factor* as a noun. Then write a sentence that uses the word *factor* as a verb.

Technology Exercises

In Exercises 93–96, use a graphing utility to graph the function on each side of the equation in the same viewing rectangle. Use end behavior to show a complete picture of the polynomial function on the left side. Do the graphs coincide? If so, this means that the polynomial on the left side has been factored correctly. If not, factor the polynomial correctly and then use your graphing utility to verify the factorization.

93. $x^2 - 4x = x(x - 4)$

94. $x^2 - 2x + 5x - 10 = (x - 2)(x - 5)$

95. $x^2 + 2x + x + 2 = x(x + 2) + 1$

96. $x^3 - 3x^2 + 4x - 12 = (x^2 + 4)(x - 3)$

Critical Thinking Exercises

Make Sense? *In Exercises 97–100, determine whether each statement makes sense or does not make sense, and explain your reasoning.*

97. After I've factored a polynomial, my answer cannot always be checked by multiplication.

98. The word *greatest* in greatest common factor is helpful because it tells me to look for the greatest power of a variable appearing in all terms.

99. Although $20x^3$ appears as a term in both $20x^3 + 8x^2$ and $20x^3 + 10x$, I'll need to factor $20x^3$ in different ways to obtain each polynomial's factorization.

100. You grouped the polynomial's terms using different groupings than I did, yet we both obtained the same factorization.

In Exercises 101–104, determine whether each statement is true or false. If the statement is false, make the necessary change(s) to produce a true statement.

101. Because the GCF of $9x^3 + 6x^2 + 3x$ is $3x$, it is not necessary to write the 1 when $3x$ is factored from the last term.

102. Some polynomials with four terms, such as $x^3 + x^2 + 4x - 4$, cannot be factored by grouping.

103. The polynomial $28x^3 - 7x^2 + 36x - 9$ can be factored by grouping terms as follows:

$$(28x^3 + 36x) + (-7x^2 - 9).$$

104. $x^2 - 2$ is a factor of $2 - 50x - x^2 + 25x^3$.

In Exercises 105–107, factor each polynomial. Assume that all variable exponents represent whole numbers.

105. $x^{4n} + x^{2n} + x^{3n}$

106. $3x^{3m}y^m - 6x^{2m}y^{2m}$

107. $8y^{2n+4} + 16y^{2n+3} - 12y^{2n}$

In Exercises 108–109, write a polynomial that fits the given description. Do not use a polynomial that appeared in this section or in the Exercise Set.

108. The polynomial has three terms and can be factored using a greatest common factor that has both a negative coefficient and a variable.

109. The polynomial has four terms and can be factored by grouping.

Review Exercises

110. Solve by Cramer's rule:

$$\begin{cases} 3x - 2y = 8 \\ 2x - 5y = 10. \end{cases}$$

(Section 3.5, Example 2)

111. Determine whether each relation is a function.

a. $\{(0, 5), (3, -5), (5, 5), (7, -5)\}$

b. $\{(1, 2), (3, 4), (5, 5), (5, 6)\}$ (Section 2.1, Example 2)

112. The length of a rectangle is 2 feet greater than twice its width. If the rectangle's perimeter is 22 feet, find the length and width. (Section 1.5, Example 5)

Preview Exercises

Exercises 113–115 will help you prepare for the material covered in the next section. In each exercise, replace the boxed question mark with an integer that results in the given product. Some trial and error may be necessary.

113. $(x + 3)(x + \boxed{?}) = x^2 + 7x + 12$

114. $(x - \boxed{?})(x - 12) = x^2 - 14x + 24$

115. $(x + 3y)(x - \boxed{?}y) = x^2 - 4xy - 21y^2$

SECTION

5.4 Factoring Trinomials

What am I supposed to learn?

After studying this section, you should be able to:

1. Factor a trinomial whose leading coefficient is 1.
2. Factor using a substitution.
3. Factor a trinomial whose leading coefficient is not 1.
4. Factor trinomials by grouping.

A great deal of trial and error is involved in finding your way out of this maze. Trial and error play an important role in problem solving and can be helpful in leading to correct solutions. In this section, you will use trial and error to factor trinomials following a problem-solving process that is not very different from learning to traverse the maze.

1 Factor a trinomial whose leading coefficient is 1.

Factoring a Trinomial Whose Leading Coefficient Is 1

In Section 5.2, we used the FOIL method to multiply two binomials. The product was often a trinomial. The following are some examples:

Factored Form	F	O	I	L	Trinomial Form
$(x+3)(x+4)$	$= x^2$	$+ 4x$	$+ 3x$	$+ 12$	$= x^2 + 7x + 12$
$(x-3)(x-4)$	$= x^2$	$- 4x$	$- 3x$	$+ 12$	$= x^2 - 7x + 12$
$(x+3)(x-5)$	$= x^2$	$- 5x$	$+ 3x$	$- 15$	$= x^2 - 2x - 15$

Observe that each trinomial is of the form $x^2 + bx + c$, where the coefficient of the squared term is 1. Our goal in the first part of this section is to start with the trinomial form and, assuming that it is factorable, return to the factored form.

The first FOIL multiplication shown in our list indicates that

$$(x+3)(x+4) = x^2 + 7x + 12.$$

Let's reverse the sides of this equation:

$$x^2 + 7x + 12 = (x+3)(x+4).$$

We can use $x^2 + 7x + 12 = (x+3)(x+4)$ to make several important observations about the factors on the right side.

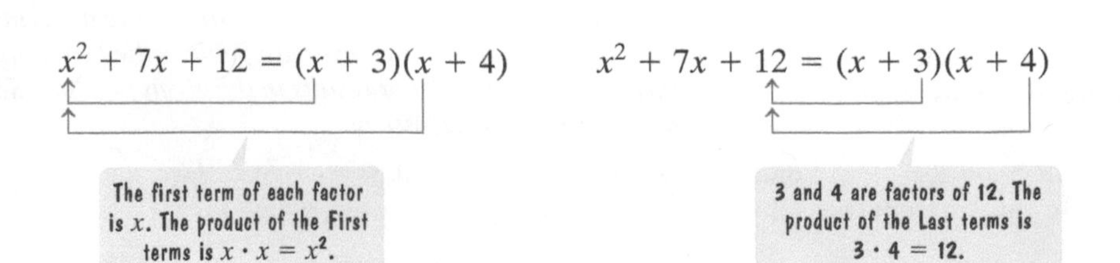

x² + 7x + 12 = (x + 3)(x + 4)
The first term of each factor is x. The product of the First terms is x · x = x².
x² + 7x + 12 = (x + 3)(x + 4)
3 and 4 are factors of 12. The product of the Last terms is 3 · 4 = 12.

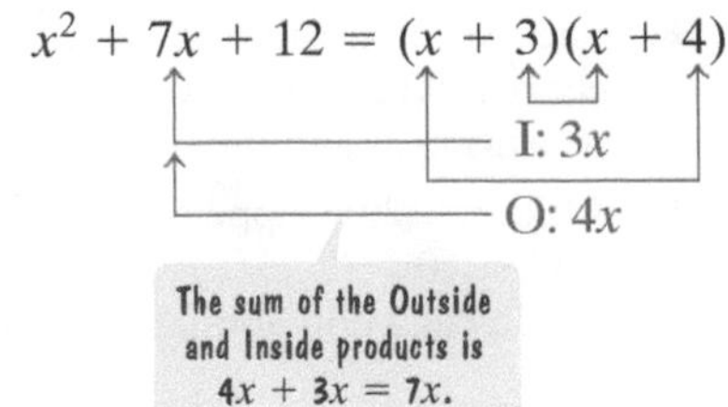

x² + 7x + 12 = (x + 3)(x + 4)
I: 3x
O: 4x
The sum of the Outside and Inside products is 4x + 3x = 7x.

These observations provide us with a procedure for factoring $x^2 + bx + c$.

A Strategy for Factoring $x^2 + bx + c$

1. Enter x as the first term of each factor.

$$(x \quad)(x \quad) = x^2 + bx + c$$

2. List pairs of factors of the constant c.
3. Try various combinations of these factors. Select the combination in which the sum of the Outside and Inside products is equal to bx.

$$(x + \square)(x + \square) = x^2 + bx + c$$

I

O

Sum of O + I

4. Check your work by multiplying the factors using the FOIL method. You should obtain the original trinomial.

If none of the possible combinations yield an Outside product and an Inside product whose sum is equal to bx, the trinomial cannot be factored using integers and is called **prime** over the set of integers.

EXAMPLE 1 Factoring a Trinomial Whose Leading Coefficient Is 1

Factor: $x^2 + 5x + 6$.

Solution

Step 1. Enter x as the first term of each factor.

$$x^2 + 5x + 6 = (x \quad)(x \quad)$$

Step 2. List pairs of factors of the constant, 6.

Factors of 6	6, 1	3, 2	−6, −1	−3, −2

Step 3. Try various combinations of these factors. The correct factorization of $x^2 + 5x + 6$ is the one in which the sum of the Outside and Inside products is equal to $5x$. Here is a list of the possible factorizations.

Possible Factorizations of $x^2 + 5x + 6$	Sum of Outside and Inside Products (Should Equal $5x$)
$(x + 6)(x + 1)$	$x + 6x = 7x$
$(x + 3)(x + 2)$	$2x + 3x = 5x$
$(x - 6)(x - 1)$	$-x - 6x = -7x$
$(x - 3)(x - 2)$	$-2x - 3x = -5x$

This is the required middle term.

Thus,

$$x^2 + 5x + 6 = (x + 3)(x + 2).$$

Step 4. Check this result by multiplying the right side using the FOIL method. You should obtain the original trinomial. Because of the commutative property, the factorization can also be expressed as

$$x^2 + 5x + 6 = (x + 2)(x + 3). \quad ■$$

In factoring a trinomial of the form $x^2 + bx + c$, you can speed things up by listing the factors of c and then finding their sums. We are interested in a sum of b. For example, in factoring $x^2 + 5x + 6$, we are interested in the factors of 6 whose sum is 5.

Factors of 6	6, 1	3, 2	−6, −1	−3, −2
Sum of Factors	7	5	−7	−5

This is the desired sum.

Thus, $x^2 + 5x + 6 = (x + 3)(x + 2)$.

Using Technology

Numeric and Graphic Connections

If a polynomial contains one variable, a graphing utility can be used to check its factorization. For example, the factorization in Example 1 can be checked graphically or numerically.

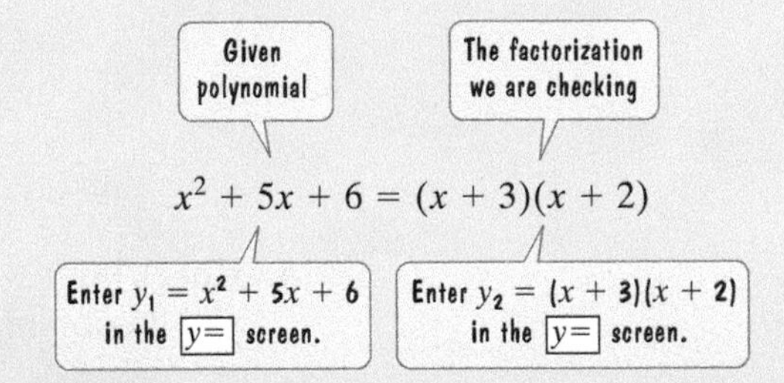

Given polynomial
The factorization we are checking
$x^2 + 5x + 6 = (x + 3)(x + 2)$
Enter $y_1 = x^2 + 5x + 6$ in the y= screen.
Enter $y_2 = (x + 3)(x + 2)$ in the y= screen.

Numeric Check

Use the TABLE feature.

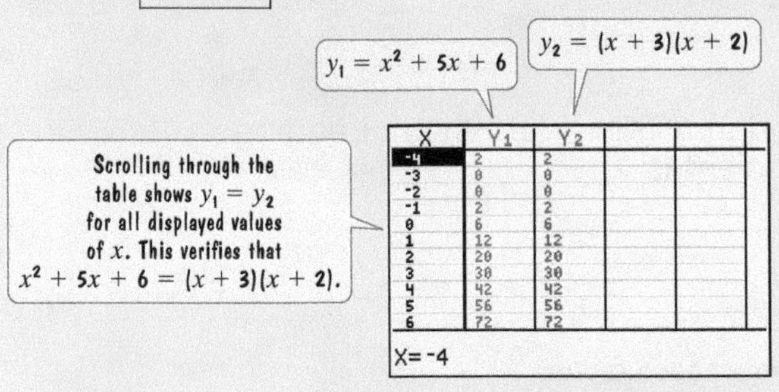

$y_1 = x^2 + 5x + 6$
$y_2 = (x + 3)(x + 2)$
Scrolling through the table shows $y_1 = y_2$ for all displayed values of x. This verifies that $x^2 + 5x + 6 = (x + 3)(x + 2)$.
X Y1 Y2
-4 2 2
-3 0 0
-2 0 0
-1 2 2
0 6 6
1 12 12
2 20 20
3 30 30
4 42 42
5 56 56
6 72 72
X=-4

Graphic Check

Use the GRAPH feature to display graphs for y_1 and y_2.

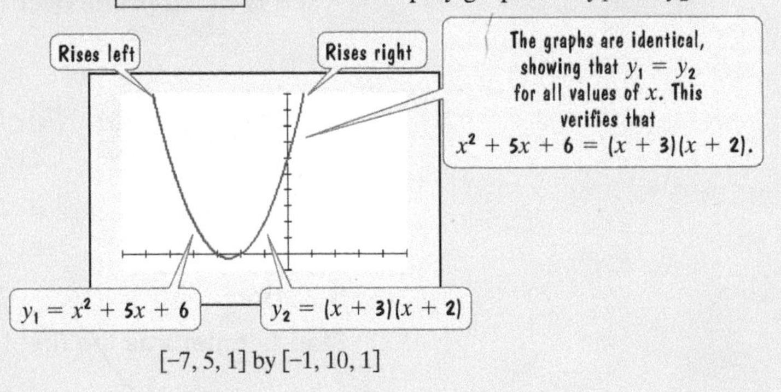

Rises left
Rises right
The graphs are identical, showing that $y_1 = y_2$ for all values of x. This verifies that $x^2 + 5x + 6 = (x + 3)(x + 2)$.
$y_1 = x^2 + 5x + 6$
$y_2 = (x + 3)(x + 2)$

[−7, 5, 1] by [−1, 10, 1]

Notice that the graph of the quadratic function is shaped like a bowl. The graph of the even-degree quadratic function exhibits the same behavior at each end, rising to the left and rising to the right (↖, ↗).

✓ **CHECK POINT 1** Factor: $x^2 + 6x + 8$.

EXAMPLE 2 Factoring a Trinomial Whose Leading Coefficient Is 1

Factor: $x^2 - 14x + 24$.

Solution

Step 1. Enter x as the first term of each factor.

$$x^2 - 14x + 24 = (x \quad\ )(x \quad\ )$$

To find the second term of each factor, we must find two integers whose product is 24 and whose sum is −14.

Step 2. List pairs of factors of the constant, 24. Because the desired sum, −14, is negative, we will list only the negative pairs of factors of 24.

Negative Factors of 24	−24, −1	−12, −2	−8, −3	−6, −4

Step 3. Try various combinations of these factors. We are interested in the factors whose sum is -14.

Negative Factors of 24	−24, −1	−12, −2	−8, −3	−6, −4
Sum of Factors	−25	−14	−11	−10

This is the desired sum.

Thus, $x^2 - 14x + 24 = (x - 12)(x - 2)$. ■

Great Question!

Is there a way to eliminate some of the combinations of factors for $x^2 + bx + c$ when c is positive?

Yes. To factor $x^2 + bx + c$ when c is positive, find two numbers with the same sign as the middle term.

$$x^2 + 5x + 6 = (x + 3)(x + 2)$$

Same signs

$$x^2 - 14x + 24 = (x - 12)(x - 2)$$

Same signs

✓ CHECK POINT 2 Factor: $x^2 - 9x + 20$.

EXAMPLE 3 Factoring a Trinomial Whose Leading Coefficient Is 1

Factor: $y^2 + 7y - 60$.

Solution

Step 1. Enter y as the first term of each factor.

$$y^2 + 7y - 60 = (y \quad)(y \quad)$$

To find the second term of each factor, we must find two integers whose product is -60 and whose sum is 7.

Steps 2 and 3. List pairs of factors of the constant, -60, and try various combinations of these factors. Because the desired sum, 7, is positive, the positive factor of -60 must be farther from 0 than the negative factor is. Thus, we will only list pairs of factors of -60 in which the positive factor has the larger absolute value.

Some Factors of −60	60, −1	30, −2	20, −3	15, −4	12, −5	10, −6
Sum of Factors	59	28	17	11	7	4

This is the desired sum.

Thus, $y^2 + 7y - 60 = (y + 12)(y - 5)$. ■

Great Question!

Is there a way to eliminate some of the combinations of factors for $x^2 + bx + c$ when c is negative?

Yes. To factor $x^2 + bx + c$ when c is negative, find two numbers with opposite signs whose sum is the coefficient of the middle term.

$$y^2 + 7y - 60 = (y + 12)(y - 5)$$

Negative

Opposite signs

✓ CHECK POINT 3 Factor: $y^2 + 19y - 66$.

EXAMPLE 4 Factoring a Trinomial in Two Variables

Factor: $x^2 - 4xy - 21y^2$.

Solution

Step 1. Enter x as the first term of each factor. Because the last term of the trinomial contains y^2, the second term of each factor must contain y.

$$x^2 - 4xy - 21y^2 = (x \quad ?y)(x \quad ?y)$$

The question marks indicate that we are looking for the coefficients of y in each factor. To find these coefficients, we must find two integers whose product is -21 and whose sum is -4.

Steps 2 and 3. List pairs of factors of the coefficient of the last term, -21, and try various combinations of these factors. We are interested in the factors whose sum is -4.

Factors of −21	1, −21	3, −7	−1, 21	−3, 7
Sum of Factors	−20	−4	20	4

This is the desired sum.

Thus, $x^2 - 4xy - 21y^2 = (x + 3y)(x - 7y)$ or $(x - 7y)(x + 3y)$.

Step 4. Verify the factorization using the FOIL method.

$$(x + 3y)(x - 7y) = x^2 - 7xy + 3xy - 21y^2 = x^2 - 4xy - 21y^2$$

Because the product of the factors is the original polynomial, the factorization is correct. ■

✓ CHECK POINT 4 Factor: $x^2 - 5xy + 6y^2$.

Can every trinomial be factored? The answer is no. For example, consider

$$x^2 + x - 5 = (x \qquad)(x \qquad).$$

To find the second term of each factor, we must find two integers whose product is -5 and whose sum is 1. Because no such integers exist, $x^2 + x - 5$ cannot be factored. This trinomial is prime.

To factor some polynomials, more than one technique must be used. **Always begin by trying to factor out the greatest common factor.** A polynomial is **factored completely** when it is written as the product of prime polynomials.

Great Question!

Can my graphing utility verify that a polynomial is factored completely?

The technology box on page 352 shows how to use a graphing utility to check a polynomial's factorization graphically or numerically. The graphing utility verifies that the factorization is equivalent to the original polynomial. However, a graphing utility cannot verify that the factorization is complete.

EXAMPLE 5 Factoring Completely

Factor: $8x^3 - 40x^2 - 48x$.

Solution The GCF of the three terms of the polynomial is $8x$. We begin by factoring out $8x$. Then we factor the remaining trinomial.

$$8x^3 - 40x^2 - 48x$$
$$= 8x(x^2 - 5x - 6) \quad \text{Factor out the GCF.}$$
$$= 8x(x \quad)(x \quad) \quad \text{Begin factoring } x^2 - 5x - 6. \text{ Find two integers whose product is } -6 \text{ and whose sum is } -5.$$
$$= 8x(x - 6)(x + 1) \quad \text{The integers are } -6 \text{ and } 1.$$

Thus,

$$8x^3 - 40x^2 - 48x = 8x(x - 6)(x + 1).$$

Be sure to include the GCF in the factorization.

You can check this factorization by multiplying the binomials using the FOIL method. Then use the distributive property and multiply each term in this product by $8x$. Try doing this now. Because the factorization is correct, you should obtain the original polynomial. ■

☑ CHECK POINT 5 Factor: $3x^3 - 15x^2 - 42x$.

Some trinomials, such as $-x^2 + 5x + 6$, have a leading coefficient of -1. Because it is easier to factor a trinomial with a positive leading coefficient, begin by factoring out -1. For example,

$$-x^2 + 5x + 6 = -1(x^2 - 5x - 6) = -(x - 6)(x + 1).$$

2 Factor using a substitution.

In some trinomials, the highest power is greater than 2, and the exponent in one of the terms is half that of the other term. By letting u equal the variable to the smaller power, the trinomial can be written in a form that makes its possible factorization more obvious. Here are some examples:

Given Trinomial	Substitution	New Trinomial
$x^6 - 8x^3 + 15$ or $(x^3)^2 - 8x^3 + 15$	$u = x^3$	$u^2 - 8u + 15$
$x^4 - 8x^2 - 9$ or $(x^2)^2 - 8x^2 - 9$	$u = x^2$	$u^2 - 8u - 9$

In each case, we factor the given trinomial by working with the new trinomial on the right. If a factorization is found, we replace all occurrences of u in the factorization with the substitution shown in the middle column.

EXAMPLE 6 Factoring by Substitution

Factor: $x^6 - 8x^3 + 15$.

Solution Notice that the exponent on x^3 is half that of the exponent on x^6. We will let u equal the variable to the power that is half of 6. Thus, let $u = x^3$.

$$(x^3)^2 - 8x^3 + 15 \quad \text{This is the given polynomial, with } x^6 \text{ written as } (x^3)^2.$$
$$= u^2 - 8u + 15 \quad \text{Let } u = x^3. \text{ Rewrite the trinomial in terms of } u.$$
$$= (u-5)(u-3) \quad \text{Factor.}$$
$$= (x^3-5)(x^3-3) \quad \text{Now substitute } x^3 \text{ for } u.$$

Thus, the given trinomial can be factored as

$$x^6 - 8x^3 + 15 = (x^3 - 5)(x^3 - 3).$$

Check this result using FOIL multiplication on the right. ■

✓ CHECK POINT 6 Factor: $x^6 - 7x^3 + 10$.

---

3 Factor a trinomial whose leading coefficient is not 1.

Factoring a Trinomial Whose Leading Coefficient Is Not 1

How do we factor a trinomial such as $5x^2 - 14x + 8$? Notice that the leading coefficient is 5. We must find two binomials whose product is $5x^2 - 14x + 8$. The product of the First terms must be $5x^2$:

$$(5x \quad)(x \quad).$$

From this point on, the factoring strategy is exactly the same as the one we use to factor a trinomial whose leading coefficient is 1.

A Strategy for Factoring $ax^2 + bx + c$

Assume, for the moment, that there is no greatest common factor.

1. Find two First terms whose product is ax^2:

$$(\square x + \quad)(\square x + \quad) = ax^2 + bx + c.$$

2. Find two Last terms whose product is c:

$$(\square x + \square)(\square x + \square) = ax^2 + bx + c.$$

3. By trial and error, perform steps 1 and 2 until the sum of the Outside product and Inside product is bx:

$$(\square x + \square)(\square x + \square) = ax^2 + bx + c.$$

I

O

Sum of O + I

If no such combinations exist, the polynomial is prime.

Great Question!

Should I feel discouraged if it takes me a while to get the correct factorization?

The *error* part of the factoring strategy plays an important role in the process. If you do not get the correct factorization the first time, this is not a bad thing. This error is often helpful in leading you to the correct factorization.

EXAMPLE 7 Factoring a Trinomial Whose Leading Coefficient Is Not 1

Factor: $5x^2 - 14x + 8$.

Solution

Step 1. Find two First terms whose product is $5x^2$.

$$5x^2 - 14x + 8 = (5x \quad)(x \quad)$$

Step 2. Find two Last terms whose product is 8. The number 8 has pairs of factors that are either both positive or both negative. Because the middle term, $-14x$, is negative, both factors must be negative. The negative factorizations of 8 are $-1(-8)$ and $-2(-4)$.

Step 3. Try various combinations of these factors. The correct factorization of $5x^2 - 14x + 8$ is the one in which the sum of the Outside and Inside products is equal to $-14x$. Here is a list of the possible factorizations:

Possible Factorizations of $5x^2 - 14x + 8$	Sum of Outside and Inside Products (Should Equal $-14x$)
$(5x - 1)(x - 8)$	$-40x - x = -41x$
$(5x - 8)(x - 1)$	$-5x - 8x = -13x$
$(5x - 2)(x - 4)$	$-20x - 2x = -22x$
$(5x - 4)(x - 2)$	$-10x - 4x = -14x$

This is the required middle term.

Thus,

$$5x^2 - 14x + 8 = (5x - 4)(x - 2).$$

Show that this factorization is correct by multiplying the factors using the FOIL method. You should obtain the original trinomial. ■

Great Question!

When factoring trinomials, must I list every possible factorization before getting the correct one?

With practice, you will find that you don't have to list every possible factorization of the trinomial. As you practice factoring, you will be able to narrow down the list of possible factorizations to just a few. When it comes to factoring, practice makes perfect.

✓ **CHECK POINT 7** Factor: $3x^2 - 20x + 28$.

EXAMPLE 8 Factoring a Trinomial Whose Leading Coefficient Is Not 1

Factor: $8x^6 - 10x^5 - 3x^4$.

Solution The GCF of the three terms of the polynomial is x^4. We begin by factoring out x^4.

$$8x^6 - 10x^5 - 3x^4 = x^4(8x^2 - 10x - 3)$$

Now we factor the remaining trinomial, $8x^2 - 10x - 3$.

Step 1. Find two First terms whose product is $8x^2$.

$$8x^2 - 10x - 3 \stackrel{?}{=} (8x \quad)(x \quad)$$
$$8x^2 - 10x - 3 \stackrel{?}{=} (4x \quad)(2x \quad)$$

Step 2. Find two Last terms whose product is -3. The possible factorizations are $1(-3)$ and $-1(3)$.

Step 3. Try various combinations of these factors. The correct factorization of $8x^2 - 10x - 3$ is the one in which the sum of the Outside and Inside products is equal to $-10x$. At the top of the next page is a list of the possible factorizations.

Great Question!

I space out reading your long lists of possible factorizations. Are there any rules for shortening these lists?

Here are some suggestions for reducing the list of possible factorizations for $ax^2 + bx + c$.

1. If b is relatively small, avoid the larger factors of a.
2. If c is positive, the signs in both binomial factors must match the sign of b.
3. If the trinomial has no common factor, no binomial factor can have a common factor.
4. Reversing the signs in the binomial factors reverses the sign of bx, the middle term.

Possible Factorizations of $8x^2 - 10x - 3$	Sum of Outside and Inside Products (Should Equal $-10x$)
$(8x + 1)(x - 3)$	$-24x + x = -23x$
$(8x - 3)(x + 1)$	$8x - 3x = 5x$
$(8x - 1)(x + 3)$	$24x - x = 23x$
$(8x + 3)(x - 1)$	$-8x + 3x = -5x$
$(4x + 1)(2x - 3)$	$-12x + 2x = -10x$
$(4x - 3)(2x + 1)$	$4x - 6x = -2x$
$(4x - 1)(2x + 3)$	$12x - 2x = 10x$
$(4x + 3)(2x - 1)$	$-4x + 6x = 2x$

This is the required middle term.

The factorization of $8x^2 - 10x - 3$ is $(4x + 1)(2x - 3)$. Now we include the GCF in the complete factorization of the given polynomial. Thus,

$$8x^6 - 10x^5 - 3x^4 = x^4(8x^2 - 10x - 3) = x^4(4x + 1)(2x - 3).$$

This is the complete factorization with the GCF, x^4, included. ■

✓ CHECK POINT 8 Factor: $6x^6 + 19x^5 - 7x^4$.

We have seen that not every trinomial can be factored. For example, consider

$$6x^2 + 14x + 7 = (6x + \square)(x + \square)$$
$$6x^2 + 14x + 7 = (3x + \square)(2x + \square).$$

The possible factors for the last term are 1 and 7. However, regardless of how these factors are placed in the boxes shown, the sum of the Outside and Inside products is not equal to $14x$. Thus, the trinomial $6x^2 + 14x + 7$ cannot be factored and is prime.

EXAMPLE 9 Factoring a Trinomial in Two Variables

Factor: $3x^2 - 13xy + 4y^2$.

Solution

Step 1. Find two First terms whose product is $3x^2$.

$$3x^2 - 13xy + 4y^2 = (3x \quad ?y)(x \quad ?y)$$

The question marks indicate that we are looking for the coefficients of y in each factor. **Steps 2 and 3. List pairs of factors of the coefficient of the last term, 4, and try various combinations of these factors.** The correct factorization is the one in which the sum of the Outside and Inside products is equal to $-13xy$. Because of the negative coefficient, -13, we will consider only the negative pairs of factors of 4. The possible factorizations are $-1(-4)$ and $-2(-2)$.

Possible Factorizations of $3x^2 - 13xy + 4y^2$	Sum of Outside and Inside Products (Should Equal $-13xy$)
$(3x - y)(x - 4y)$	$-12xy - xy = -13xy$
$(3x - 4y)(x - y)$	$-3xy - 4xy = -7xy$
$(3x - 2y)(x - 2y)$	$-6xy - 2xy = -8xy$

This is the required middle term.

Thus,

$$3x^2 - 13xy + 4y^2 = (3x - y)(x - 4y). \quad ■$$

CHECK POINT 9 Factor: $2x^2 - 7xy + 3y^2$.

EXAMPLE 10 Factoring by Substitution

Factor: $6y^4 + 13y^2 + 6$.

Solution Notice that the exponent on y^2 is half that of the exponent on y^4. We will let u equal the variable to the smaller power. Thus, let $u = y^2$.

$$6(y^2)^2 + 13y^2 + 6$$ This is the given polynomial, with y^4 written as $(y^2)^2$.

$$= 6u^2 + 13u + 6$$ Let $u = y^2$. Rewrite the trinomial in terms of u.

$$= (3u + 2)(2u + 3)$$ Factor the trinomial.

$$= (3y^2 + 2)(2y^2 + 3)$$ Now substitute y^2 for u.

Therefore, $6y^4 + 13y^2 + 6 = (3y^2 + 2)(2y^2 + 3)$. Check using FOIL multiplication. ■

CHECK POINT 10 Factor: $3y^4 + 10y^2 - 8$.

4 Factor trinomials by grouping.

Factoring Trinomials by Grouping

A second method for factoring $ax^2 + bx + c$, $a \neq 1$, is called the **grouping method.** This method involves both trial and error, as well as grouping. The trial and error in factoring $ax^2 + bx + c$ depends upon finding two numbers, p and q, for which $p + q = b$. Then we factor $ax^2 + px + qx + c$ using grouping.

Let's see how this works by looking at a particular factorization:

$$15x^2 - 7x - 2 = (3x - 2)(5x + 1).$$

If we multiply using FOIL on the right, we obtain

$$(3x - 2)(5x + 1) = 15x^2 + 3x - 10x - 2.$$

In this case, the desired numbers, p and q, are $p = 3$ and $q = -10$. Compare these numbers to ac and b in the given polynomial.

$a = 15$ $b = -7$ $c = -2$

$$15x^2 - 7x - 2$$

$ac = 15(-2) = -30$

Can you see that p and q, 3 and -10, are factors of ac, or -30? Furthermore, p and q have a sum of b, namely -7. By expressing the middle term, $-7x$, in terms of p and q, we can factor by grouping as follows:

$$15x^2 - 7x - 2$$

$$= 15x^2 + (3x - 10x) - 2$$ Rewrite $-7x$ as $3x - 10x$.

$$= (15x^2 + 3x) + (-10x - 2)$$ Group terms.

$$= 3x(5x + 1) - 2(5x + 1)$$ Factor from each group.

$$= (5x + 1)(3x - 2).$$ Factor out $5x + 1$, the common binomial factor.

Factoring $ax^2 + bx + c$ Using Grouping ($a \neq 1$)

1. Multiply the leading coefficient, a, and the constant, c.
2. Find the factors of ac whose sum is b.
3. Rewrite the middle term, bx, as a sum or difference using the factors from step 2.
4. Factor by grouping.

EXAMPLE 11 Factoring a Trinomial by Grouping

Factor by grouping: $12x^2 - 5x - 2$.

Solution The trinomial is of the form $ax^2 + bx + c$.

$$12x^2 - 5x - 2$$

$a = 12$ $b = -5$ $c = -2$

Step 1. Multiply the leading coefficient, a, and the constant, c. Using $a = 12$ and $c = -2$,

$$ac = 12(-2) = -24.$$

Step 2. Find the factors of ac whose sum is b. We want the factors of -24 whose sum is b, or -5. The factors of -24 whose sum is -5 are -8 and 3.

Step 3. Rewrite the middle term, $-5x$, as a sum or difference using the factors from step 2, -8 and 3.

$$12x^2 - 5x - 2 = 12x^2 - 8x + 3x - 2$$

Step 4. Factor by grouping.

$$= (12x^2 - 8x) + (3x - 2) \quad \text{Group terms.}$$
$$= 4x(3x - 2) + 1(3x - 2) \quad \text{Factor from each group.}$$
$$= (3x - 2)(4x + 1) \quad \text{Factor out } 3x - 2\text{, the common binomial factor.}$$

Thus,

$$12x^2 - 5x - 2 = (3x - 2)(4x + 1). \blacksquare$$

Discover for Yourself

In step 2, we found that the desired numbers were -8 and 3. We wrote $-5x$ as $-8x + 3x$. What happens if we write $-5x$ as $3x - 8x$? Use factoring by grouping on

$$12x^2 - 5x - 2$$
$$= 12x^2 + 3x - 8x - 2.$$

Describe what happens.

✓ CHECK POINT 11 Factor by grouping: $8x^2 - 22x + 5$.

CONCEPT AND VOCABULARY CHECK

Fill in each blank so that the resulting statement is true.

1. A polynomial is factored ______________ when it is written as a product of prime polynomials.
2. We begin the process of factoring a polynomial by first factoring out the ______________, assuming that there is one other than 1.
3. $x^2 + 8x + 15 = (x + 3)(x$ ______________$)$
4. $x^2 - 9x + 20 = (x - 5)(x$ ______________$)$
5. $x^2 + 13x - 48 = (x - 3)(x$ ______________$)$
6. $x^2 + 17xy - 60y^2 = (x + 20y)(x$ ______________$)$
7. $2x^3 - 30x^2 - 108x =$ ______________ $(x^2 - 15x - 54) =$ ______________ $(x + 3)(x$ ______________$)$
8. $3x^2 - 14x + 11 = (x - 1)(3x$ ______________$)$
9. $6x^2 + 13x - 63 = (3x - 7)($______________$)$
10. $63x^2 - 89xy + 30y^2 = (9x - 5y)(7x$ ______________$)$

5.4 EXERCISE SET

Practice Exercises

In Exercises 1–30, factor each trinomial, or state that the trinomial is prime. Check each factorization using FOIL multiplication.

1. $x^2 + 5x + 6$
2. $x^2 + 10x + 9$
3. $x^2 + 8x + 12$
4. $x^2 + 8x + 15$
5. $x^2 + 9x + 20$
6. $x^2 + 11x + 24$
7. $y^2 + 10y + 16$
8. $y^2 + 9y + 18$
9. $x^2 - 8x + 15$
10. $x^2 - 5x + 6$
11. $y^2 - 12y + 20$
12. $y^2 - 25y + 24$
13. $a^2 + 5a - 14$
14. $a^2 + a - 12$
15. $x^2 + x - 30$
16. $x^2 + 14x - 32$
17. $x^2 - 3x - 28$
18. $x^2 - 4x - 21$
19. $y^2 - 5y - 36$
20. $y^2 - 3y - 40$
21. $x^2 - x + 7$
22. $x^2 + 3x + 8$
23. $x^2 - 9xy + 14y^2$
24. $x^2 - 8xy + 15y^2$
25. $x^2 - xy - 30y^2$
26. $x^2 - 3xy - 18y^2$
27. $x^2 + xy + y^2$
28. $x^2 - xy + y^2$
29. $a^2 - 18ab + 80b^2$
30. $a^2 - 18ab + 45b^2$

In Exercises 31–38, factor completely.

31. $3x^2 + 3x - 18$
32. $4x^2 - 4x - 8$
33. $2x^3 - 14x^2 + 24x$
34. $2x^3 + 6x^2 + 4x$
35. $3y^3 - 15y^2 + 18y$
36. $4y^3 + 12y^2 - 72y$
37. $2x^4 - 26x^3 - 96x^2$
38. $3x^4 + 54x^3 + 135x^2$

In Exercises 39–44, factor by introducing an appropriate substitution.

39. $x^6 - x^3 - 6$
40. $x^6 + x^3 - 6$
41. $x^4 - 5x^2 - 6$
42. $x^4 - 4x^2 - 5$
43. $(x + 1)^2 + 6(x + 1) + 5$ (Let $u = x + 1$.)
44. $(x + 1)^2 + 8(x + 1) + 7$ (Let $u = x + 1$.)

In Exercises 45–68, use the method of your choice to factor each trinomial, or state that the trinomial is prime. Check each factorization using FOIL multiplication.

45. $3x^2 + 8x + 5$
46. $2x^2 + 9x + 7$
47. $5x^2 + 56x + 11$
48. $5x^2 - 16x + 3$
49. $3y^2 + 22y - 16$
50. $5y^2 + 33y - 14$
51. $4y^2 + 9y + 2$
52. $8y^2 + 10y + 3$
53. $10x^2 + 19x + 6$
54. $6x^2 + 19x + 15$
55. $8x^2 - 18x + 9$
56. $4x^2 - 27x + 18$
57. $6y^2 - 23y + 15$
58. $16y^2 - 6y - 27$
59. $6y^2 + 14y + 3$
60. $4y^2 + 22y - 5$
61. $3x^2 + 4xy + y^2$
62. $2x^2 + 3xy + y^2$
63. $6x^2 - 7xy - 5y^2$
64. $6x^2 - 5xy - 6y^2$
65. $15x^2 - 31xy + 10y^2$
66. $15x^2 + 11xy - 14y^2$
67. $3a^2 - ab - 14b^2$
68. $15a^2 - ab - 6b^2$

In Exercises 69–82, factor completely.

69. $15x^3 - 25x^2 + 10x$
70. $10x^3 + 24x^2 + 14x$
71. $24x^4 + 10x^3 - 4x^2$
72. $15x^4 - 39x^3 + 18x^2$
73. $15y^5 - 2y^4 - y^3$
74. $10y^5 - 17y^4 + 3y^3$
75. $24x^2 + 3xy - 27y^2$
76. $12x^2 + 10xy - 8y^2$
77. $6a^2b - 2ab - 60b$
78. $8a^2b + 34ab - 84b$
79. $12x^2y - 34xy^2 + 14y^3$
80. $12x^2y - 46xy^2 + 14y^3$
81. $13x^3y^3 + 39x^3y^2 - 52x^3y$
82. $4x^3y^5 + 24x^2y^5 - 64xy^5$

In Exercises 83–92, factor by introducing an appropriate substitution.

83. $2x^4 - x^2 - 3$

84. $5x^4 + 2x^2 - 3$

85. $2x^6 + 11x^3 + 15$

86. $2x^6 + 13x^3 + 15$

87. $2y^{10} + 7y^5 + 3$

88. $5y^{10} + 29y^5 - 42$

89. $5(x + 1)^2 + 12(x + 1) + 7$ (Let $u = x + 1$.)

90. $3(x + 1)^2 - 5(x + 1) + 2$ (Let $u = x + 1$.)

91. $2(x - 3)^2 - 5(x - 3) - 7$

92. $3(x - 2)^2 - 5(x - 2) - 2$

Practice PLUS

In Exercises 93–100, factor completely.

93. $x^2 - 0.5x + 0.06$

94. $x^2 + 0.3x - 0.04$

95. $x^2 - \frac{3}{49} + \frac{2}{7}x$

96. $x^2 - \frac{6}{25} + \frac{1}{5}x$

97. $acx^2 - bcx + adx - bd$

98. $acx^2 - bcx - adx + bd$

99. $-4x^5y^2 + 7x^4y^3 - 3x^3y^4$

100. $-5x^4y^3 + 7x^3y^4 - 2x^2y^5$

101. If $(fg)(x) = 3x^2 - 22x + 39$, find f and g.

102. If $(fg)(x) = 4x^2 - x - 5$, find f and g.

In Exercises 103–104, a large rectangle formed by a number of smaller rectangles is shown. Factor the sum of the areas of the smaller rectangles to determine the dimensions of the large rectangle.

103.

x^2	x^2	x
x	x	1
x	x	1
x	x	1

104.

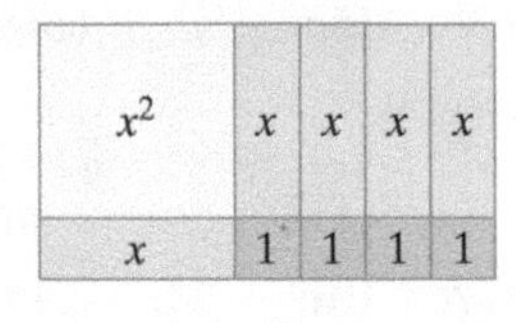

Application Exercises

105. A diver jumps directly upward from a board that is 32 feet high. The function

$$f(t) = -16t^2 + 16t + 32$$

describes the diver's height above the water, $f(t)$, in feet, after t seconds.

a. Find and interpret $f(1)$.

b. Find and interpret $f(2)$.

c. Factor the expression for $f(t)$ and write the function in completely factored form.

d. Use the factored form of the function to find $f(1)$ and $f(2)$.

106. The function $V(x) = 3x^3 - 2x^2 - 8x$ describes the volume, $V(x)$, in cubic inches, of the box shown whose height is x inches.

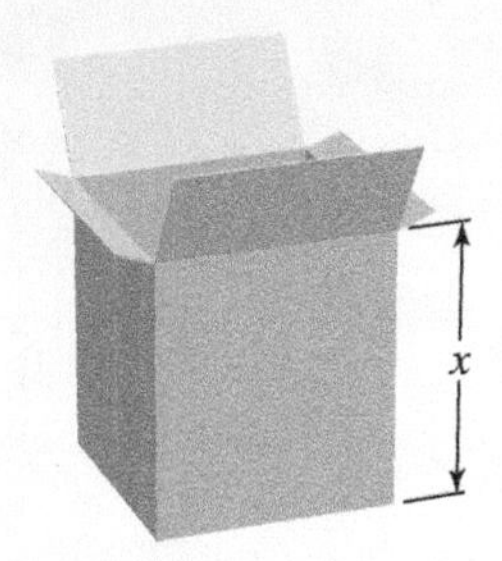

a. Find and interpret $V(4)$.

b. Factor the expression for $V(x)$ and write the function in completely factored form.

c. Use the factored form of the function to find $V(4)$ and $V(5)$.

107. Find the area of the large rectangle shown below in two ways.

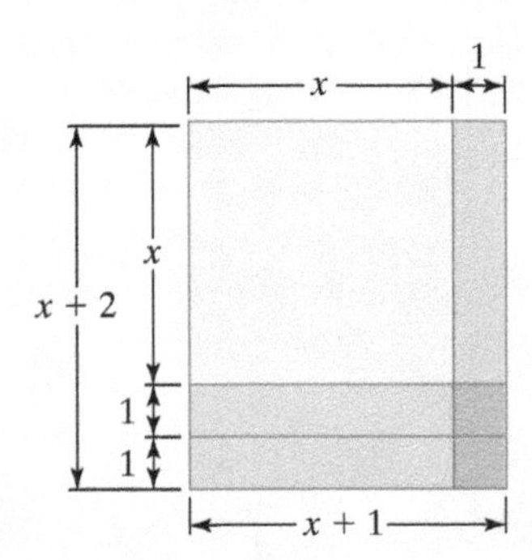

a. Find the sum of the areas of the six smaller rectangles and squares.

b. Express the area of the large rectangle as the product of its length and width.

c. Explain how the figure serves as a geometric model for the factorization of the sum that you wrote in part (a).

108. If x represents a positive integer, factor $x^3 + 3x^2 + 2x$ to show that the trinomial represents the product of three consecutive integers.

Explaining the Concepts

109. Explain how to factor $x^2 + 8x + 15$.

110. Give two helpful suggestions for factoring $x^2 - 5x + 6$.

111. In factoring $x^2 + bx + c$, describe how the last term in each binomial factor is related to b and to c.

112. Describe the first thing that you should try doing when factoring a polynomial.

113. What does it mean to factor completely?

114. Explain how to factor $x^6 - 7x^3 + 10$ by substitution.

115. Is it possible to factor $x^6 - 7x^3 + 10$ without using substitution? How might this be done?

116. Explain how to factor $2x^2 - x - 1$.

Technology Exercises

In Exercises 117–120, use a graphing utility to graph the function on each side of the equation in the same viewing rectangle. Use end behavior to show a complete picture of the polynomial function on the left side. Do the graphs coincide? If so, this means that the polynomial on the left side has been factored correctly. If not, factor the polynomial correctly and then use your graphing utility to verify the factorization.

117. $x^2 + 7x + 12 = (x + 4)(x + 3)$

118. $x^2 - 7x + 6 = (x - 2)(x - 3)$

119. $6x^3 + 5x^2 - 4x = x(3x + 4)(2x - 1)$

120. $x^4 - x^2 - 20 = (x^2 + 5)(x^2 - 4)$

121. Use the TABLE feature of a graphing utility to verify any two of your factorizations in Exercises 39–44.

Critical Thinking Exercises

Make Sense? *In Exercises 122–125, determine whether each statement makes sense or does not make sense, and explain your reasoning.*

122. Although $(x + 2)(x - 5)$ is the same as $(x - 5)(x + 2)$, the factorization $(2 - x)(2 + x)$ is not the same as $-(x - 2)(x + 2)$.

123. I'm often able to use an incorrect factorization to lead me to the correct factorization.

124. My graphing calculator showed the same graph for $y_1 = 20x^3 - 70x^2 + 60x$ and $y_2 = 10x(2x^2 - 7x + 6)$, so I can conclude that the complete factorization of $20x^3 - 70x^2 + 60x$ is $10x(2x^2 - 7x + 6)$.

125. First factoring out the greatest common factor makes it easier for me to determine how to factor the remaining factor, assuming that it is not prime.

In Exercises 126–129, determine whether each statement is true or false. If the statement is false, make the necessary change(s) to produce a true statement.

126. Once a GCF is factored from $6y^6 - 19y^5 + 10y^4$, the remaining trinomial factor is prime.

127. One factor of $8y^2 - 51y + 18$ is $8y - 3$.

128. We can immediately tell that $6x^2 - 11xy - 10y^2$ is prime because 11 is a prime number and the polynomial contains two variables.

129. A factor of $12x^2 - 19xy + 5y^2$ is $4x - y$.

In Exercises 130–131, find all integers b so that the trinomial can be factored.

130. $4x^2 + bx - 1$

131. $3x^2 + bx + 5$

In Exercises 132–137, factor each polynomial. Assume that all variable exponents represent whole numbers.

132. $9x^{2n} + x^n - 8$

133. $4x^{2n} - 9x^n + 5$

134. $a^{2n+2} - a^{n+2} - 6a^2$

135. $b^{2n+2} + 3b^{n+2} - 10b^2$

136. $3c^{n+2} - 10c^{n+1} + 3c^n$

137. $2d^{n+2} - 5d^{n+1} + 3d^n$

Review Exercises

138. Solve: $-2x \le 6$ and $-2x + 3 < -7$. (Section 4.2, Example 2)

139. Solve the system:

$$\begin{cases} 2x - y - 2z = -1 \\ x - 2y - z = 1 \\ x + y + z = 4. \end{cases}$$

(Section 3.3, Example 2)

140. Factor: $4x^3 + 8x^2 - 5x - 10$. (Section 5.3, Example 5)

Preview Exercises

Exercises 141–143 will help you prepare for the material covered in the next section. In each exercise, factor the polynomial. (You'll soon be learning techniques that will shorten the factoring process.)

141. $x^2 + 14x + 49$

142. $x^2 - 8x + 16$

143. $x^2 - 25$ (or $x^2 + 0x - 25$)

MID-CHAPTER CHECK POINT Section 5.1–Section 5.4

What You Know: We learned the vocabulary of polynomials and observed the smooth, continuous graphs of polynomial functions. We used the Leading Coefficient Test to describe the end behavior of these graphs. We learned to add, subtract, and multiply polynomials. We used a number of fast methods for finding products of polynomials, including the FOIL method for multiplying binomials, special-product formulas for squaring binomials $[(A + B)^2 = A^2 + 2AB + B^2; (A - B)^2 = A^2 - 2AB + B^2]$, and a special-product formula for the product of the sum and difference of two terms $[(A + B)(A - B) = A^2 - B^2]$. We learned to factor out a polynomial's greatest common factor and to use grouping to factor polynomials with more than three terms. We factored polynomials with three terms, beginning with trinomials with leading coefficient 1 and moving on to $ax^2 + bx + c$, with $a \neq 1$. We saw that the factoring process should begin by looking for a GCF and, if there is one, factoring it out first.

In Exercises 1–18, perform the indicated operations.

1. $(-8x^3 + 6x^2 - x + 5) - (-7x^3 + 2x^2 - 7x - 12)$

2. $(6x^2yz^4)\left(-\frac{1}{3}x^5y^2z\right)$

3. $5x^2y\left(6x^3y^2 - 7xy - \frac{2}{5}\right)$

4. $(3x - 5)(x^2 + 3x - 8)$

5. $(x^2 - 2x + 1)(2x^2 + 3x - 4)$

6. $(x^2 - 2x + 1) - (2x^2 + 3x - 4)$

7. $(6x^3y - 11x^2y - 4y) + (-11x^3y + 5x^2y - y - 6)$
$-(-x^3y + 2y - 1)$

8. $(2x + 5)(4x - 1)$

9. $(2xy - 3)(5xy + 2)$

10. $(3x - 2y)(3x + 2y)$

11. $(3xy + 1)(2x^2 - 3y)$

12. $(7x^3y + 5x)(7x^3y - 5x)$

13. $3(x + h)^2 - 2(x + h) + 5 - (3x^2 - 2x + 5)$

14. $(x^2 - 3)^2$

15. $(x^2 - 3)(x^3 + 5x + 2)$

16. $(2x + 5y)^2$

17. $(x + 6 + 3y)(x + 6 - 3y)$

18. $(x + y + 5)^2$

In Exercises 19–30, factor completely, or state that the polynomial is prime.

19. $x^2 - 5x - 24$

20. $15xy + 5x + 6y + 2$

21. $5x^2 + 8x - 4$

22. $35x^2 + 10x - 50$

23. $9x^2 - 9x - 18$

24. $10x^3y^2 - 20x^2y^2 + 35x^2y$

25. $18x^2 + 21x + 5$

26. $12x^2 - 9xy - 16x + 12y$

27. $9x^2 - 15x + 4$

28. $3x^6 + 11x^3 + 10$

29. $25x^3 + 25x^2 - 14x$

30. $2x^4 - 6x - x^3y + 3y$

SECTION

5.5 Factoring Special Forms

What am I supposed to learn?

After studying this section, you should be able to:

1. Factor the difference of two squares.
2. Factor perfect square trinomials.
3. Use grouping to obtain the difference of two squares.
4. Factor the sum or difference of two cubes.

Bees use honeycombs to store honey and house larvae. They construct honey storage cells from wax. Each cell has the shape of a six-sided figure whose sides are all the same length and whose angles all have the same measure, called a regular hexagon. The cells fit together perfectly, preventing dirt or predators from entering. Squares or equilateral triangles would fit equally well, but regular hexagons provide the largest storage space for the amount of wax used.

In this section, we develop factoring techniques by reversing the formulas for special products discussed in Section 5.2. Like the construction of honeycombs, these factorizations can be visualized by perfectly fitting together "cells" of squares and rectangles to form larger rectangles.

Factoring the Difference of Two Squares

A method for factoring the difference of two squares is obtained by reversing the special product for the sum and difference of two terms.

1 Factor the difference of two squares.

The Difference of Two Squares

If A and B are real numbers, variables, or algebraic expressions, then

$$A^2 - B^2 = (A + B)(A - B).$$

In words: The difference of the squares of two terms factors as the product of a sum and a difference of those terms.

EXAMPLE 1 Factoring the Difference of Two Squares

Factor:

a. $9x^2 - 100$ **b.** $36y^6 - 49x^4$.

Solution We must express each term as the square of some monomial. Then we use the formula for factoring $A^2 - B^2$.

a. $9x^2 - 100 = (3x)^2 - 10^2 = (3x + 10)(3x - 10)$

$A^2 - B^2 = (A + B)(A - B)$

b. $36y^6 - 49x^4 = (6y^3)^2 - (7x^2)^2 = (6y^3 + 7x^2)(6y^3 - 7x^2)$ ■

In order to apply the factoring formula for $A^2 - B^2$, each term must be the square of an integer or a polynomial.

- A number that is the square of an integer is called a **perfect square**. For example, 100 is a perfect square because $100 = 10^2$.
- Any exponential expression involving a perfect-square coefficient and variables to even powers is a perfect square. For example, $100y^6$ is a perfect square because $100y^6 = (10y^3)^2$.

Great Question!

You mentioned that because 100 = 10^2, 100 is a perfect square. What are some other perfect squares that I should recognize?

It's helpful to identify perfect squares. Here are 16 perfect squares, each printed in boldface.

1 $= 1^2$	**25** $= 5^2$	**81** $= 9^2$	**169** $= 13^2$
4 $= 2^2$	**36** $= 6^2$	**100** $= 10^2$	**196** $= 14^2$
9 $= 3^2$	**49** $= 7^2$	**121** $= 11^2$	**225** $= 15^2$
16 $= 4^2$	**64** $= 8^2$	**144** $= 12^2$	**256** $= 16^2$

CHECK POINT 1 Factor:

a. $16x^2 - 25$ **b.** $100y^6 - 9x^4$.

Be careful when determining whether or not to apply the factoring formula for the difference of two squares.

Prime Over the Integers

- $x^2 - 5$ — 5 is not a perfect square.
- $x^7 - 25$ — 7 is an odd power. x^7 is not the square of any integer power of x.

Factorable

- $1 - x^6y^4$ — Even powers. Perfect square: $1 = 1^2$. Perfect square: $x^6y^4 = (x^3y^2)^2$

When factoring, always check first for common factors. If there are common factors, factor out the GCF and then factor the resulting polynomial.

EXAMPLE 2 Factoring Out the GCF and Then Factoring the Difference of Two Squares

Factor: $3y - 3x^6y^5$.

Solution The GCF of the two terms of the polynomial is $3y$. We begin by factoring out $3y$.

$$3y - 3x^6y^5 = 3y(1 - x^6y^4) = 3y[1^2 - (x^3y^2)^2] = 3y(1 + x^3y^2)(1 - x^3y^2)$$

Factor out the GCF. $A^2 - B^2 = (A + B)(A - B)$ ■

CHECK POINT 2 Factor: $6y - 6x^2y^7$.

We have seen that a polynomial is factored completely when it is written as the product of prime polynomials. To be sure that you have factored completely, check to see whether any factors with more than one term in the factored polynomial can be factored further. If so, continue factoring.

EXAMPLE 3 A Repeated Factorization

Factor completely: $81x^4 - 16$.

Solution

$$81x^4 - 16 = (9x^2)^2 - 4^2$$ Express as the difference of two squares.

$$= (9x^2 + 4)(9x^2 - 4)$$ The factors are the sum and difference of the expressions being squared.

$$= (9x^2 + 4)[(3x)^2 - 2^2]$$ The factor $9x^2 - 4$ is the difference of two squares and can be factored.

$$= (9x^2 + 4)(3x + 2)(3x - 2)$$ The factors of $9x^2 - 4$ are the sum and difference of the expressions being squared. ■

Great Question!

Why isn't factoring $81x^4 - 16$ as $(9x^2 + 4)(9x^2 - 4)$ a complete factorization?

The second factor, $9x^2 - 4$, is itself a difference of two squares and can be factored.

Are you tempted to further factor $9x^2 + 4$, the sum of two squares, in Example 3? Resist the temptation! **The sum of two squares, $A^2 + B^2$, with no common factor other than 1 is a prime polynomial.**

CHECK POINT 3 Factor completely: $16x^4 - 81$.

In our next example, we begin with factoring by grouping. We can then factor further using the difference of two squares.

EXAMPLE 4 Factoring Completely

Factor completely: $x^3 + 5x^2 - 9x - 45$.

Solution

$$x^3 + 5x^2 - 9x - 45$$

$$= (x^3 + 5x^2) + (-9x - 45)$$ Group terms with common factors.

$$= x^2(x + 5) - 9(x + 5)$$ Factor out the common factor from each group.

$$= (x + 5)(x^2 - 9)$$ Factor out $x + 5$, the common binomial factor, from both terms.

$$= (x + 5)(x + 3)(x - 3)$$ Factor $x^2 - 3^2$, the difference of two squares. ■

✓ CHECK POINT 4 Factor completely: $x^3 + 7x^2 - 4x - 28$.

In Examples 1–4, we used the formula for factoring the difference of two squares. Although we obtained the formula by reversing the special product for the sum and difference of two terms, it can also be obtained geometrically.

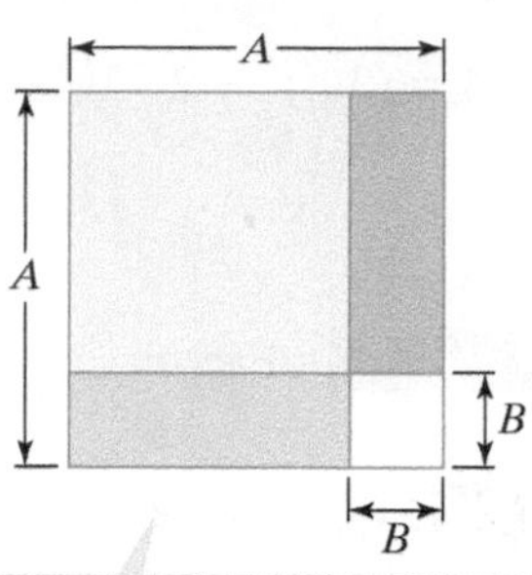

The combined area of the three purple rectangles is $A^2 - B^2$.

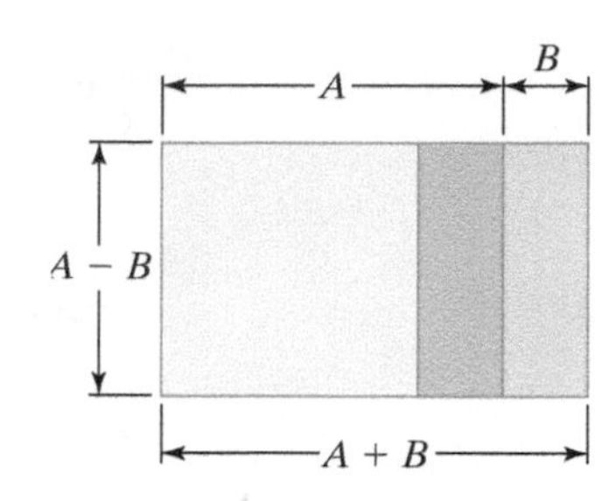

We've rearranged the three purple rectangles. Their combined area is $(A + B)(A - B)$.

Because the three purple rectangles make up the same combined area in both figures,

$$A^2 - B^2 = (A + B)(A - B).$$

2 Factor perfect square trinomials.

Factoring Perfect Square Trinomials

Our next factoring technique is obtained by reversing the special products for squaring binomials. The trinomials that are factored using this technique are called **perfect square trinomials.**

Factoring Perfect Square Trinomials

Let A and B be real numbers, variables, or algebraic expressions.

1. $A^2 + 2AB + B^2 = (A + B)^2$ (Same sign)

2. $A^2 - 2AB + B^2 = (A - B)^2$ (Same sign)

The two items in the box show that perfect square trinomials, $A^2 + 2AB + B^2$ and $A^2 - 2AB + B^2$, come in two forms: one in which the coefficient of the middle term is positive and one in which the coefficient of the middle term is negative. Here's how to recognize a perfect square trinomial:

1. The first and last terms are squares of monomials or integers.
2. The middle term is twice the product of the expressions being squared in the first and last terms.

EXAMPLE 5 Factoring Perfect Square Trinomials

Factor:

a. $x^2 + 14x + 49$ **b.** $4x^2 + 12xy + 9y^2$ **c.** $9y^4 - 12y^2 + 4$.

Solution

a. $x^2 + 14x + 49 = x^2 + 2 \cdot x \cdot 7 + 7^2 = (x + 7)^2$ The middle term has a positive sign.

$A^2 + 2AB + B^2 = (A + B)^2$

b. We suspect that $4x^2 + 12xy + 9y^2$ is a perfect square trinomial because $4x^2 = (2x)^2$ and $9y^2 = (3y)^2$. The middle term can be expressed as twice the product of $2x$ and $3y$.

$$4x^2 + 12xy + 9y^2 = (2x)^2 + 2 \cdot 2x \cdot 3y + (3y)^2 = (2x + 3y)^2$$

$$A^2 + 2AB + B^2 = (A + B)^2$$

c. $9y^4 - 12y^2 + 4 = (3y^2)^2 - 2 \cdot 3y^2 \cdot 2 + 2^2 = (3y^2 - 2)^2$

$$A^2 - 2AB + B^2 = (A - B)^2$$

The middle term has a negative sign. ■

CHECK POINT 5 Factor:

a. $x^2 + 6x + 9$

b. $16x^2 + 40xy + 25y^2$

c. $4y^4 - 20y^2 + 25$.

---

3 Use grouping to obtain the difference of two squares.

Using Special Forms When Factoring by Grouping

If a polynomial contains four terms, try factoring by grouping. In the next example, we group the terms to obtain the difference of two squares. One of the squares is a perfect square trinomial.

EXAMPLE 6 Using Grouping to Obtain the Difference of Two Squares

Factor: $x^2 - 8x + 16 - y^2$.

Solution

$x^2 - 8x + 16 - y^2$

$= (x^2 - 8x + 16) - y^2$ Group as a perfect square trinomial minus y^2 to obtain a difference of two squares.

$= (x - 4)^2 - y^2$ Factor the perfect square trinomial.

$= (x - 4 + y)(x - 4 - y)$ Factor the difference of two squares. The factors are the sum and difference of the expressions being squared. ■

CHECK POINT 6 Factor: $x^2 + 10x + 25 - y^2$.

EXAMPLE 7 Using Grouping to Obtain the Difference of Two Squares

Factor: $a^2 - b^2 + 10b - 25$.

Solution Grouping into two groups of two terms does not result in a common binomial factor. Let's look for a perfect square trinomial. Can you see that the perfect square trinomial is the expression being subtracted from a^2?

$a^2 - b^2 + 10b - 25$

$= a^2 - (b^2 - 10b + 25)$ Factor out -1 and group as $a^2 -$ (perfect square trinomial) to obtain a difference of two squares.

$= a^2 - (b - 5)^2$ Factor the perfect square trinomial.

$= [a + (b - 5)][a - (b - 5)]$ Factor the difference of squares. The factors are the sum and difference of the expressions being squared.

$= (a + b - 5)(a - b + 5)$ Simplify. ■

✓ CHECK POINT 7 Factor: $a^2 - b^2 + 4b - 4$.

4 Factor the sum or difference of two cubes.

Factoring the Sum or Difference of Two Cubes

Here are two multiplications that lead to factoring formulas for the sum of two cubes and the difference of two cubes:

$$\begin{aligned}(A + B)(A^2 - AB + B^2) &= A(A^2 - AB + B^2) + B(A^2 - AB + B^2)\\ &= A^3 - A^2B + AB^2 + A^2B - AB^2 + B^3\\ &= A^3 + B^3\end{aligned}$$

The product results in the sum of two cubes.

Combine like terms: $-A^2B + A^2B = 0$ and $AB^2 - AB^2 = 0$.

and

$$\begin{aligned}(A - B)(A^2 + AB + B^2) &= A(A^2 + AB + B^2) - B(A^2 + AB + B^2)\\ &= A^3 + A^2B + AB^2 - A^2B - AB^2 - B^3\\ &= A^3 - B^3.\end{aligned}$$

The product results in the difference of two cubes.

Combine like terms: $A^2B - A^2B = 0$ and $AB^2 - AB^2 = 0$.

By reversing the two sides of these equations, we obtain formulas that allow us to factor a sum or difference of two cubes. These formulas should be memorized.

Factoring the Sum or Difference of Two Cubes

1. Factoring the Sum of Two Cubes

$$A^3 + B^3 = (A + B)(A^2 - AB + B^2)$$

Same signs Opposite signs

2. Factoring the Difference of Two Cubes

$$A^3 - B^3 = (A - B)(A^2 + AB + B^2)$$

Same signs Opposite signs

EXAMPLE 8 Factoring the Sum of Two Cubes

Factor:

a. $x^3 + 125$ **b.** $x^6 + 64y^3$.

Solution We must express each term as the cube of some monomial. Then we use the formula for factoring $A^3 + B^3$.

a. $x^3 + 125 = x^3 + 5^3 = (x + 5)(x^2 - x \cdot 5 + 5^2) = (x + 5)(x^2 - 5x + 25)$

$A^3 + B^3 = (A + B)(A^2 - AB + B^2)$

b. $x^6 + 64y^3 = (x^2)^3 + (4y)^3 = (x^2 + 4y)[(x^2)^2 - x^2 \cdot 4y + (4y)^2]$

$A^3 + B^3 = (A + B)(A^2 - AB + B^2)$

$= (x^2 + 4y)(x^4 - 4x^2y + 16y^2)$ ■

Great Question!

What are some cubes that I should be able to identify?

When factoring the sum or difference of cubes, it is helpful to recognize the following cubes:

$$\begin{aligned}1 &= 1^3\\ 8 &= 2^3\\ 27 &= 3^3\\ 64 &= 4^3\\ 125 &= 5^3\\ 216 &= 6^3\\ 1000 &= 10^3.\end{aligned}$$

✓ CHECK POINT 8 Factor:

a. $x^3 + 27$

b. $x^6 + 1000y^3$.

EXAMPLE 9 Factoring the Difference of Two Cubes

Factor:

a. $x^3 - 216$ **b.** $8 - 125x^3y^3$.

Solution We must express each term as the cube of some monomial. Then we use the formula for factoring $A^3 - B^3$.

a. $x^3 - 216 = x^3 - 6^3 = (x - 6)(x^2 + x \cdot 6 + 6^2) = (x - 6)(x^2 + 6x + 36)$

$A^3 - B^3 = (A - B)(A^2 + AB + B^2)$

b. $8 - 125x^3y^3 = 2^3 - (5xy)^3 = (2 - 5xy)[2^2 + 2 \cdot 5xy + (5xy)^2]$

$A^3 - B^3 = (A - B)(A^2 + AB + B^2)$

$= (2 - 5xy)(4 + 10xy + 25x^2y^2)$ ■

✓ CHECK POINT 9 Factor:

a. $x^3 - 8$

b. $1 - 27x^3y^3$.

Great Question!

The formulas for factoring $A^3 + B^3$ and $A^3 - B^3$ are difficult to remember and easy to confuse. Can you help me out?

A Cube of SOAP

When factoring sums or differences of cubes, observe the sign patterns.

Same signs

$A^3 + B^3 = (A + B)(A^2 - AB + B^2)$

Opposite signs | Always positive

Same signs

$A^3 - B^3 = (A - B)(A^2 + AB + B^2)$

Opposite signs | Always positive

The word *SOAP* is a way to remember these patterns:

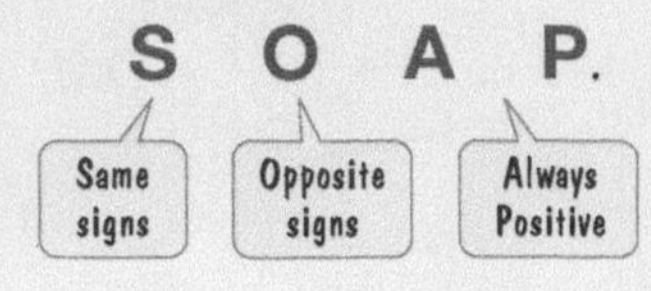

CONCEPT AND VOCABULARY CHECK

Fill in each blank so that the resulting statement is true.

1. The formula for factoring the difference of two squares is $A^2 - B^2 =$ ________.
2. A formula for factoring a perfect square trinomial is $A^2 + 2AB + B^2 =$ ________.
3. A formula for factoring a perfect square trinomial is $A^2 - 2AB + B^2 =$ ________.
4. The formula for factoring the sum of two cubes is $A^3 + B^3 =$ ________.
5. The formula for factoring the difference of two cubes is $A^3 - B^3 =$ ________.
6. $16x^2 - 49 = ($ ________ $+ 7)($ ________ $- 7)$
7. $a^2 - (b + 3)^2 = [a +$ ________ $][a -$ ________ $]$
8. $x^2 - 14x + 49 = (x$ ________ $)^2$
9. $16x^2 + 40xy + 25y^2 = ($ ________ $+ 5y)^2$
10. $x^3 + 27 = (x$ ________ $)(x^2$ ________ $+ 9)$
11. $x^3 - 1000 = (x$ ________ $)(x^2 + 10x$ ________ $)$
12. True or false: $x^2 - 10$ is the difference of two perfect squares. ________
13. True or false: $x^2 + 8x + 16$ is a perfect square trinomial. ________
14. True or false: $x^2 - 5x + 25$ is a perfect square trinomial. ________
15. True or false: $x^6 + 1000y^3$ is the sum of two cubes. ________
16. True or false: $x^3 - 100$ is the difference of two cubes. ________

5.5 EXERCISE SET

MyMathLab®

Practice Exercises

In Exercises 1–22, factor each difference of two squares. Assume that any variable exponents represent whole numbers.

1. $x^2 - 4$
2. $x^2 - 16$
3. $9x^2 - 25$
4. $4x^2 - 9$
5. $9 - 25y^2$
6. $16 - 49y^2$
7. $36x^2 - 49y^2$
8. $64x^2 - 25y^2$
9. $x^2y^2 - 1$
10. $x^2y^2 - 100$
11. $9x^4 - 25y^6$
12. $25x^4 - 9y^6$
13. $x^{14} - y^4$
14. $x^4 - y^{10}$
15. $(x - 3)^2 - y^2$
16. $(x - 6)^2 - y^2$
17. $a^2 - (b - 2)^2$
18. $a^2 - (b - 3)^2$
19. $x^{2n} - 25$
20. $x^{2n} - 36$
21. $1 - a^{2n}$
22. $4 - b^{2n}$

In Exercises 23–48, factor completely, or state that the polynomial is prime.

23. $2x^3 - 8x$
24. $2x^3 - 72x$
25. $50 - 2y^2$
26. $72 - 2y^2$
27. $8x^2 - 8y^2$
28. $6x^2 - 6y^2$
29. $2x^3y - 18xy$
30. $2x^3y - 32xy$
31. $a^3b^2 - 49ac^2$
32. $4a^3c^2 - 16ax^2y^2$
33. $5y - 5x^2y^7$
34. $2y - 2x^6y^3$
35. $8x^2 + 8y^2$
36. $6x^2 + 6y^2$
37. $x^2 + 25y^2$

38. $x^2 + 36y^2$
39. $x^4 - 16$
40. $x^4 - 1$
41. $81x^4 - 1$
42. $1 - 81x^4$
43. $2x^5 - 2xy^4$
44. $3x^5 - 3xy^4$
45. $x^3 + 3x^2 - 4x - 12$
46. $x^3 + 3x^2 - 9x - 27$
47. $x^3 - 7x^2 - x + 7$
48. $x^3 - 6x^2 - x + 6$

In Exercises 49–64, factor any perfect square trinomials, or state that the polynomial is prime.

49. $x^2 + 4x + 4$
50. $x^2 + 2x + 1$
51. $x^2 - 10x + 25$
52. $x^2 - 14x + 49$
53. $x^4 - 4x^2 + 4$
54. $x^4 - 6x^2 + 9$
55. $9y^2 + 6y + 1$
56. $4y^2 + 4y + 1$
57. $64y^2 - 16y + 1$
58. $25y^2 - 10y + 1$
59. $x^2 - 12xy + 36y^2$
60. $x^2 + 16xy + 64y^2$
61. $x^2 - 8xy + 64y^2$
62. $x^2 - 9xy + 81y^2$
63. $9x^2 + 48xy + 64y^2$
64. $16x^2 - 40xy + 25y^2$

In Exercises 65–74, factor by grouping to obtain the difference of two squares.

65. $x^2 - 6x + 9 - y^2$
66. $x^2 - 12x + 36 - y^2$
67. $x^2 + 20x + 100 - x^4$
68. $x^2 + 16x + 64 - x^4$
69. $9x^2 - 30x + 25 - 36y^2$
70. $25x^2 - 20x + 4 - 81y^2$
71. $x^4 - x^2 - 2x - 1$
72. $x^4 - x^2 - 6x - 9$
73. $z^2 - x^2 + 4xy - 4y^2$
74. $z^2 - x^2 + 10xy - 25y^2$

In Exercises 75–94, factor using the formula for the sum or difference of two cubes.

75. $x^3 + 64$
76. $x^3 + 1$
77. $x^3 - 27$
78. $x^3 - 1000$
79. $8y^3 + 1$
80. $27y^3 + 1$
81. $125x^3 - 8$
82. $27x^3 - 8$
83. $x^3y^3 + 27$
84. $x^3y^3 + 64$
85. $64x - x^4$
86. $216x - x^4$
87. $x^6 + 27y^3$
88. $x^6 + 8y^3$
89. $125x^6 - 64y^6$
90. $125x^6 - y^6$
91. $x^9 + 1$
92. $x^9 - 1$
93. $(x - y)^3 - y^3$
94. $x^3 + (x + y)^3$

Practice PLUS

In Exercises 95–104, factor completely.

95. $0.04x^2 + 0.12x + 0.09$
96. $0.09x^2 - 0.12x + 0.04$
97. $8x^4 - \frac{x}{8}$
98. $27x^4 + \frac{x}{27}$
99. $x^6 - 9x^3 + 8$
100. $x^6 + 9x^3 + 8$
101. $x^8 - 15x^4 - 16$
102. $x^8 + 15x^4 - 16$
103. $x^5 - x^3 - 8x^2 + 8$
104. $x^5 - x^3 + 27x^2 - 27$
105. The figure shows four purple rectangles that fit together to form a large square.

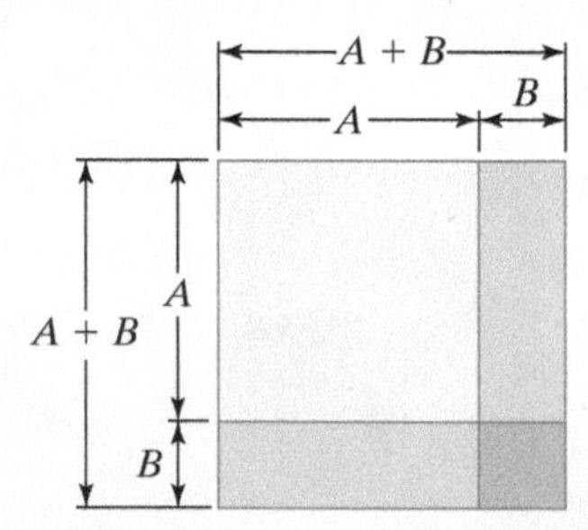

a. Express the area of the large square in terms of one of its sides, $A + B$.

b. Write an expression for the area of each of the four rectangles that form the large square.

c. Use the sum of the areas from part (b) to write a second expression for the area of the large square.

d. Set the expression from part (c) equal to the expression from part (a). What factoring technique have you established?

Application Exercises

In Exercises 106–109, find the formula for the area of the shaded blue region and express it in factored form.

106.

107.

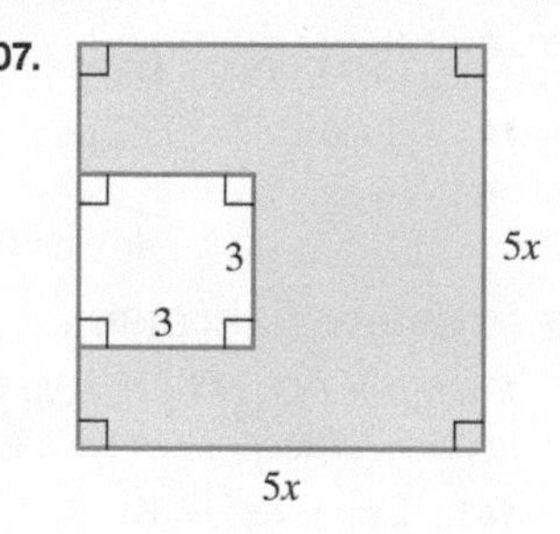

108.

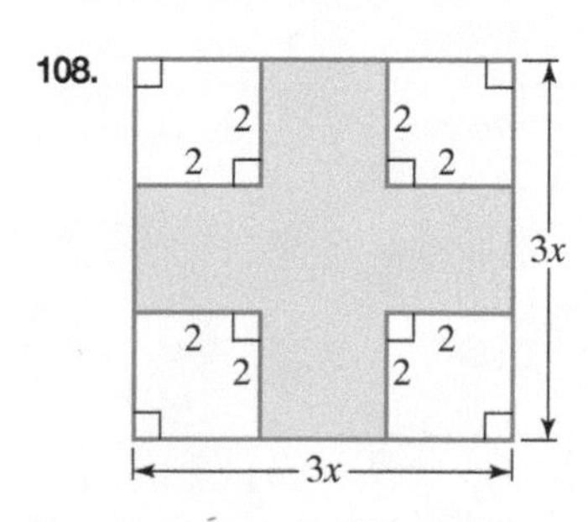

109.

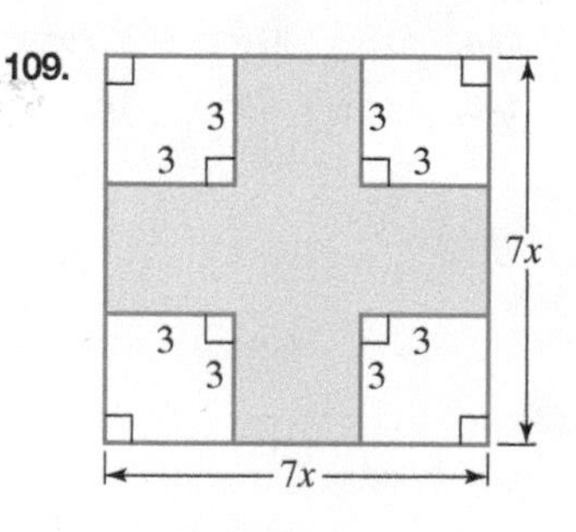

In Exercises 110–111, find the formula for the volume of the region outside the smaller rectangular solid and inside the larger rectangular solid. Then express the volume in factored form.

110.

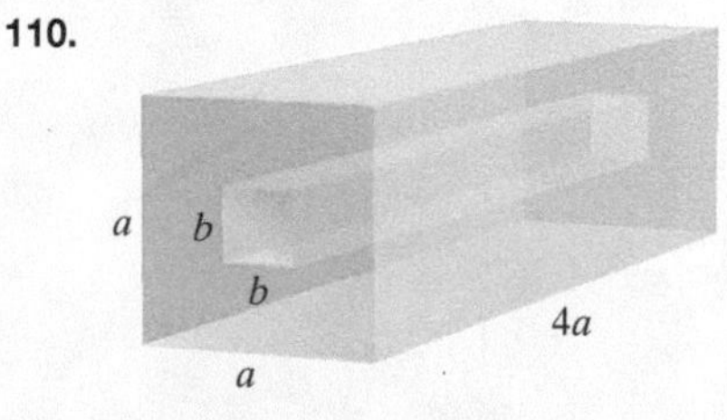

111.

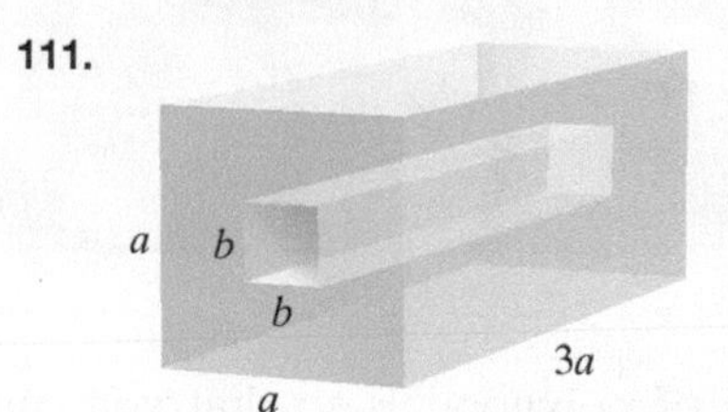

Explaining the Concepts

112. Explain how to factor the difference of two squares. Provide an example with your explanation.

113. What is a perfect square trinomial and how is it factored?

114. Explain how to factor $x^2 - y^2 + 8x - 16$. Should the expression be grouped into two groups of two terms? If not, why not, and what sort of grouping should be used?

115. Explain how to factor $x^3 + 1$.

Technology Exercises

In Exercises 116–123, use a graphing utility to graph the function on each side of the equation in the same viewing rectangle. Use end behavior to show a complete picture of the polynomial function on the left side. Do the graphs coincide? If so, this means that the polynomial on the left side has been factored correctly. If not, factor the polynomial correctly and then use your graphing utility to verify the factorization.

116. $9x^2 - 4 = (3x + 2)(3x - 2)$

117. $x^2 + 4x + 4 = (x + 4)^2$

118. $9x^2 + 12x + 4 = (3x + 2)^2$

119. $25 - (x^2 + 4x + 4) = (x + 7)(x - 3)$

120. $(2x + 3)^2 - 9 = 4x(x + 3)$

121. $(x - 3)^2 + 8(x - 3) + 16 = (x - 1)^2$

122. $x^3 - 1 = (x - 1)(x^2 - x + 1)$

123. $(x + 1)^3 + 1 = (x + 1)(x^2 + x + 1)$

124. Use the TABLE feature of a graphing utility to verify any two of your factorizations in Exercises 67–68 or 77–78.

Critical Thinking Exercises

Make Sense? *In Exercises 125–128, determine whether each statement makes sense or does not make sense, and explain your reasoning.*

125. Although I can factor the difference of squares and perfect square trinomials using trial and error associated with FOIL, recognizing these special forms shortens the process.

126. Although $x^3 + 2x^2 - 5x - 6$ can be factored as $(x + 1)(x + 3)(x - 2)$, I have not yet learned techniques to obtain this factorization.

127. I factored $4x^2 - 100$ completely and obtained $(2x + 10)(2x - 10)$.

128. You told me that the area of a square is represented by $9x^2 + 12x + 4$ square inches, so I factored and concluded that the length of one side must be $3x + 2$ inches.

In Exercises 129–132, determine whether each statement is true or false. If the statement is false, make the necessary change(s) to produce a true statement.

129. $9x^2 + 15x + 25 = (3x + 5)^2$

130. $x^3 - 27 = (x - 3)(x^2 + 6x + 9)$

131. $x^3 - 64 = (x - 4)^3$

132. $4x^2 - 121 = (2x - 11)^2$

In Exercises 133–136, factor each polynomial completely. Assume that any variable exponents represent whole numbers.

133. $y^3 + x + x^3 + y$

134. $36x^{2n} - y^{2n}$

135. $x^{3n} + y^{12n}$

136. $4x^{2n} + 20x^n y^m + 25y^{2m}$

137. Factor $x^6 - y^6$ first as the difference of squares and then as the difference of cubes. From these two factorizations, determine a factorization for $x^4 + x^2y^2 + y^4$.

In Exercises 138–139, find all integers k so that the trinomial is a perfect square trinomial.

138. $kx^2 + 8xy + y^2$

139. $64x^2 - 16x + k$

Review Exercises

140. Solve: $2x + 2 \geq 12$ and $\frac{2x - 1}{3} \leq 7$. (Section 4.2, Example 2)

141. Solve using matrices:

$$\begin{cases} 3x - 2y = -8 \\ x + 6y = 4. \end{cases}$$

(Section 3.4, Example 2)

142. Factor: $3x^2 + 21x - xy - 7y$. (Section 5.3, Example 6)

Preview Exercises

Exercises 143–145 will help you prepare for the material covered in the next section. In each exercise, factor completely.

143. $2x^3 + 8x^2 + 8x$

144. $5x^3 - 40x^2y + 35xy^2$

145. $9b^2x + 9b^2y - 16x - 16y$

SECTION

5.6 A General Factoring Strategy

What am I supposed to learn?

After studying this section, you should be able to:

1 Use a general strategy for factoring polynomials.

Successful problem solving involves understanding the problem, devising a plan for solving it, and then carrying out the plan. In this section, you will learn a step-by-step strategy that provides a plan and direction for solving factoring problems.

1 Use a general strategy for factoring polynomials.

A Strategy for Factoring Polynomials

It is important to practice factoring a wide variety of polynomials so that you can quickly select the appropriate technique. The polynomial is factored completely when all its polynomial factors, except possibly for monomial factors, are prime. Because of the commutative property, the order of the factors does not matter.

Here is a general strategy for factoring polynomials:

A Strategy for Factoring a Polynomial

1. If there is a common factor, factor out the GCF or factor out a common factor with a negative coefficient.
2. Determine the number of terms in the polynomial and try factoring as follows:
   a. If there are two terms, can the binomial be factored by using one of the following special forms?
      Difference of two squares: $A^2 - B^2 = (A + B)(A - B)$
      Sum of two cubes: $A^3 + B^3 = (A + B)(A^2 - AB + B^2)$
      Difference of two cubes: $A^3 - B^3 = (A - B)(A^2 + AB + B^2)$

b. If there are three terms, is the trinomial a perfect square trinomial? If so, factor by using one of the following special forms:

$$A^2 + 2AB + B^2 = (A + B)^2$$
$$A^2 - 2AB + B^2 = (A - B)^2.$$

If the trinomial is not a perfect square trinomial, try factoring by trial and error or grouping.

c. If there are four or more terms, try factoring by grouping.

3. Check to see if any factors with more than one term in the factored polynomial can be factored further. If so, factor completely.

Remember to check the factored form by multiplying or by using the TABLE or GRAPH feature of a graphing utility.

The following examples and those in the Exercise Set are similar to the previous factoring problems. However, these factorizations are not all of the same type. They are intentionally mixed to promote the development of a general factoring strategy.

Using Technology

Graphic Connections

The polynomial functions $y_1 = 2x^3 + 8x^2 + 8x$ and $y_2 = 2x(x + 2)^2$ have identical graphs. This verifies that

$$2x^3 + 8x^2 + 8x = 2x(x + 2)^2.$$

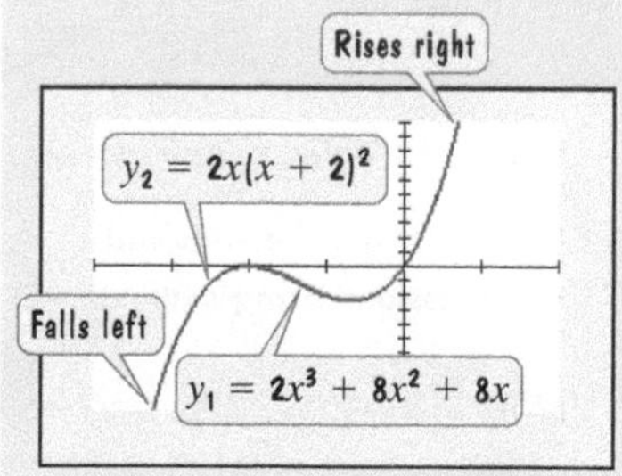

[−4, 2, 1] by [−10, 10, 1]

The degree of y_1 is 3, which is odd. Odd-degree polynomial functions have graphs with opposite behavior at each end. The leading coefficient, 2, is positive. The graph should fall to the left and rise to the right (↙, ↗). The viewing rectangle used is complete enough to show this end behavior.

EXAMPLE 1 Factoring a Polynomial

Factor: $2x^3 + 8x^2 + 8x$.

Solution

Step 1. If there is a common factor, factor out the GCF. Because $2x$ is common to all terms, we factor it out.

$$2x^3 + 8x^2 + 8x = 2x(x^2 + 4x + 4) \quad \text{Factor out the GCF.}$$

Step 2. Determine the number of terms and factor accordingly. The factor $x^2 + 4x + 4$ has three terms and is a perfect square trinomial. We factor using $A^2 + 2AB + B^2 = (A + B)^2$.

$$2x^3 + 8x^2 + 8x = 2x(x^2 + 4x + 4)$$
$$= 2x(x^2 + 2 \cdot x \cdot 2 + 2^2)$$

$A^2 + 2AB + B^2$

$$= 2x(x + 2)^2 \quad A^2 + 2AB + B^2 = (A + B)^2$$

Step 3. Check to see if factors can be factored further. In this problem, they cannot. Thus,

$$2x^3 + 8x^2 + 8x = 2x(x + 2)^2. \blacksquare$$

✓ CHECK POINT 1 Factor: $3x^3 - 30x^2 + 75x$.

EXAMPLE 2 Factoring a Polynomial

Factor: $4x^2y - 16xy - 20y$.

Solution

Step 1. If there is a common factor, factor out the GCF. Because $4y$ is common to all terms, we factor it out.

$$4x^2y - 16xy - 20y = 4y(x^2 - 4x - 5) \quad \text{Factor out the GCF.}$$

Step 2. Determine the number of terms and factor accordingly. The factor $x^2 - 4x - 5$ has three terms, but it is not a perfect square trinomial. We factor it using trial and error.

$$4x^2y - 16xy - 20y = 4y(x^2 - 4x - 5) = 4y(x + 1)(x - 5)$$

Step 3. Check to see if factors can be factored further. In this case, they cannot, so we have factored completely. ■

☑ **CHECK POINT 2** Factor: $3x^2y - 12xy - 36y$.

EXAMPLE 3 Factoring a Polynomial

Factor: $9b^2x - 16y - 16x + 9b^2y$.

Solution

Step 1. If there is a common factor, factor out the GCF. Other than 1 or -1, there is no common factor.

Step 2. Determine the number of terms and factor accordingly. There are four terms. We try factoring by grouping. Notice that the first and last terms have a common factor of $9b^2$ and the two middle terms have a common factor of -16. Thus, we begin by rearranging the terms.

$$9b^2x - 16y - 16x + 9b^2y$$

$$= (9b^2x + 9b^2y) + (-16x - 16y)$$ Rearrange terms and group terms with common factors.

$$= 9b^2(x + y) - 16(x + y)$$ Factor from each group.

$$= (x + y)(9b^2 - 16)$$ Factor out the common binomial factor, $x + y$.

Step 3. Check to see if factors can be factored further. We note that $9b^2 - 16$ is the difference of two squares, $(3b)^2 - 4^2$, so we continue factoring.

$$9b^2x - 16y - 16x + 9b^2y$$

$$= (x + y)[(3b)^2 - 4^2]$$ Express $9b^2 - 16$ as the difference of squares.

$$= (x + y)(3b + 4)(3b - 4)$$ The factors of $9b^2 - 16$ are the sum and difference of the expressions being squared. ■

☑ **CHECK POINT 3** Factor: $16a^2x - 25y - 25x + 16a^2y$.

EXAMPLE 4 Factoring a Polynomial

Factor: $x^2 - 25a^2 + 8x + 16$.

Solution

Step 1. If there is a common factor, factor out the GCF. Other than 1 or -1, there is no common factor.

Step 2. Determine the number of terms and factor accordingly. There are four terms. We try factoring by grouping. Grouping into two groups of two terms does not result in a common binomial factor. Let's try grouping as a difference of squares.

$$x^2 - 25a^2 + 8x + 16$$

$$= (x^2 + 8x + 16) - 25a^2$$ Rearrange terms and group as a perfect square trinomial minus $25a^2$ to obtain a difference of squares.

$$= (x + 4)^2 - (5a)^2$$ Factor the perfect square trinomial.

$$= (x + 4 + 5a)(x + 4 - 5a)$$ Factor the difference of squares. The factors are the sum and difference of the expressions being squared.

Step 3. Check to see if factors can be factored further. In this case, they cannot, so we have factored completely. ■

☑ CHECK POINT 4 Factor: $x^2 - 36a^2 + 20x + 100$.

EXAMPLE 5 Factoring a Polynomial

Factor: $3x^{10} + 3x$.

Solution

Step 1. If there is a common factor, factor out the GCF. Because $3x$ is common to both terms, we factor it out.

$$3x^{10} + 3x = 3x(x^9 + 1) \quad \text{Factor out the GCF.}$$

Step 2. Determine the number of terms and factor accordingly. The factor $x^9 + 1$ has two terms. This binomial can be expressed as $(x^3)^3 + 1^3$, so it can be factored as the sum of two cubes.

$$\begin{aligned} 3x^{10} + 3x &= 3x(x^9 + 1) \\ &= 3x[(x^3)^3 + 1^3] = 3x(x^3 + 1)[(x^3)^2 - x^3 \cdot 1 + 1^2] \end{aligned}$$

$A^3 + B^3 = (A + B)(A^2 - AB + B^2)$

$$= 3x(x^3 + 1)(x^6 - x^3 + 1) \quad \text{Simplify.}$$

Step 3. Check to see if factors can be factored further. We note that $x^3 + 1$ is the sum of two cubes, $x^3 + 1^3$, so we continue factoring.

$$\begin{aligned} &3x^{10} + 3x \\ &= 3x(x^3 + 1)(x^6 - x^3 + 1) \end{aligned}$$

This is our factorization in the previous step.

$A^3 + B^3 = (A + B)(A^2 - AB + B^2)$

Factor completely by factoring $x^3 + 1^3$, the sum of cubes.

$$= 3x(x + 1)(x^2 - x + 1)(x^6 - x^3 + 1) \quad ■$$

☑ CHECK POINT 5 Factor: $x^{10} + 512x$. *Hint*: $512 = 8^3$.

Achieving Success

In the next chapter, you will see how mathematical models involving quotients of polynomials describe environmental issues. Factoring is an essential skill for working with such models. **Success in mathematics cannot be achieved without a complete understanding of factoring.** Be sure to work all the assigned exercises in Exercise Set 5.6 so that you can apply each of the factoring techniques discussed in this chapter. The more deeply you force your brain to think about factoring by working many exercises, the better will be your chances of achieving success in the next chapter and in future mathematics courses.

CONCEPT AND VOCABULARY CHECK

Here is a list of the factoring techniques that we have discussed.

a. Factoring out the GCF

b. Factoring out the negative of the GCF

c. Factoring by grouping

d. Factoring trinomials by trial and error or grouping

e. Factoring the difference of two squares

$A^2 - B^2 = (A + B)(A - B)$

f. Factoring perfect square trinomials

$A^2 + 2AB + B^2 = (A + B)^2$

$A^2 - 2AB + B^2 = (A - B)^2$

g. Factoring the sum of two cubes

$A^3 + B^3 = (A + B)(A^2 - AB + B^2)$

h. Factoring the difference of two cubes

$A^3 - B^3 = (A - B)(A^2 + AB + B^2)$

Fill in each blank by writing the letter of the technique (a through h) for factoring the polynomial.

1. $-4x^2 + 20x$ _______

2. $9x^2 - 16$ _______

3. $125x^3 - 1$ _______

4. $x^2 + 7x + xy + 7y$ _______

5. $x^2 - 4x - 5$ _______

6. $x^2 + 4x + 4$ _______

7. $3x^6 + 6x$ _______

8. $x^9 + 1$ _______

5.6 EXERCISE SET

MyMathLab®

Practice Exercises

In Exercises 1–68, factor completely, or state that the polynomial is prime.

1. $x^3 - 16x$

2. $x^3 - x$

3. $3x^2 + 18x + 27$

4. $8x^2 + 40x + 50$

5. $81x^3 - 3$

6. $24x^3 - 3$

7. $x^2y - 16y + 32 - 2x^2$

8. $12x^2y - 27y - 4x^2 + 9$

9. $4a^2b - 2ab - 30b$

10. $32y^2 - 48y + 18$

11. $ay^2 - 4a - 4y^2 + 16$

12. $ax^2 - 16a - 2x^2 + 32$

13. $11x^5 - 11xy^2$

14. $4x^9 - 400x$

15. $4x^5 - 64x$

16. $7x^5 - 7x$

17. $x^3 - 4x^2 - 9x + 36$

18. $x^3 - 5x^2 - 4x + 20$

19. $2x^5 + 54x^2$

20. $3x^5 + 24x^2$

21. $3x^4y - 48y^5$

22. $32x^4y - 2y^5$

23. $12x^3 + 36x^2y + 27xy^2$

24. $18x^3 + 48x^2y + 32xy^2$

25. $x^2 - 12x + 36 - 49y^2$

26. $x^2 - 10x + 25 - 36y^2$

27. $4x^2 + 25y^2$

28. $16x^2 + 49y^2$

29. $12x^3y - 12xy^3$

30. $9x^2y^2 - 36y^2$

31. $6bx^2 + 6by^2$

32. $6x^2 - 66$

33. $x^4 - xy^3 + x^3y - y^4$

34. $x^3 - xy^2 + x^2y - y^3$

35. $x^2 - 4a^2 + 12x + 36$

36. $x^2 - 49a^2 + 14x + 49$

37. $5x^3 + x^6 - 14$

38. $6x^3 + x^6 - 16$

39. $4x - 14 + 2x^3 - 7x^2$

40. $3x^3 + 8x + 9x^2 + 24$
41. $54x^3 - 16y^3$
42. $54x^3 - 250y^3$
43. $x^2 + 10x - y^2 + 25$
44. $x^2 + 6x - y^2 + 9$
45. $x^8 - y^8$
46. $x^8 - 1$
47. $x^3y - 16xy^3$
48. $x^3y - 100xy^3$
49. $x + 8x^4$
50. $x + 27x^4$
51. $16y^2 - 4y - 2$
52. $32y^2 + 4y - 6$
53. $14y^3 + 7y^2 - 10y$
54. $5y^3 - 45y^2 + 70y$
55. $27x^2 + 36xy + 12y^2$
56. $125x^2 + 50xy + 5y^2$
57. $12x^3 + 3xy^2$
58. $3x^4 + 27x^2$
59. $x^6y^6 - x^3y^3$
60. $x^3 - 2x^2 - x + 2$
61. $(x + 5)(x - 3) + (x + 5)(x - 7)$
62. $(x + 4)(x - 9) + (x + 4)(2x - 3)$
63. $a^2(x - y) + 4(y - x)$
64. $b^2(x - 3) + c^2(3 - x)$
65. $(c + d)^4 - 1$
66. $(c + d)^4 - 16$
67. $p^3 - pq^2 + p^2q - q^3$
68. $p^3 - pq^2 - p^2q + q^3$

Practice PLUS

In Exercises 69–80, factor completely.

69. $x^4 - 5x^2y^2 + 4y^4$
70. $x^4 - 10x^2y^2 + 9y^4$
71. $(x + y)^2 + 6(x + y) + 9$
72. $(x - y)^2 - 8(x - y) + 16$
73. $(x - y)^4 - 4(x - y)^2$
74. $(x + y)^4 - 100(x + y)^2$
75. $2x^2 - 7xy^2 + 3y^4$
76. $3x^2 + 5xy^2 + 2y^4$
77. $x^3 - y^3 - x + y$
78. $x^3 + y^3 + x^2 - y^2$
79. $x^6y^3 + x^3 - 8x^3y^3 - 8$
80. $x^6y^3 - x^3 + x^3y^3 - 1$

Application Exercises

In Exercises 81–86,

a. *Write an expression for the area of the shaded blue region.*
b. *Write the expression in factored form.*

81.
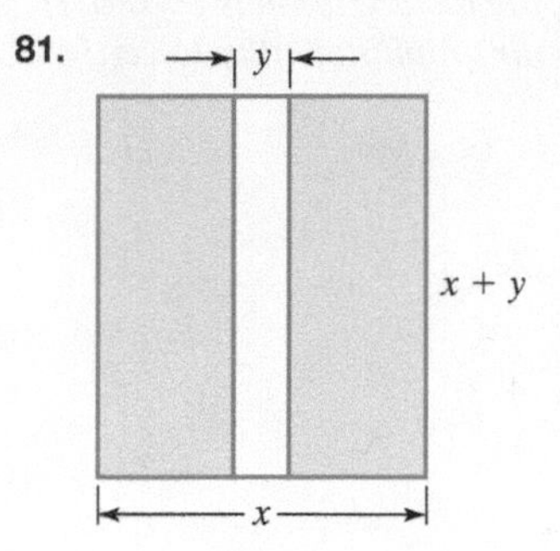

82.
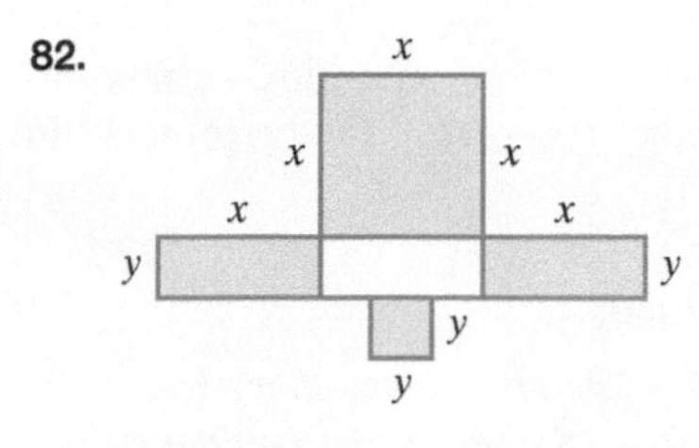

83.
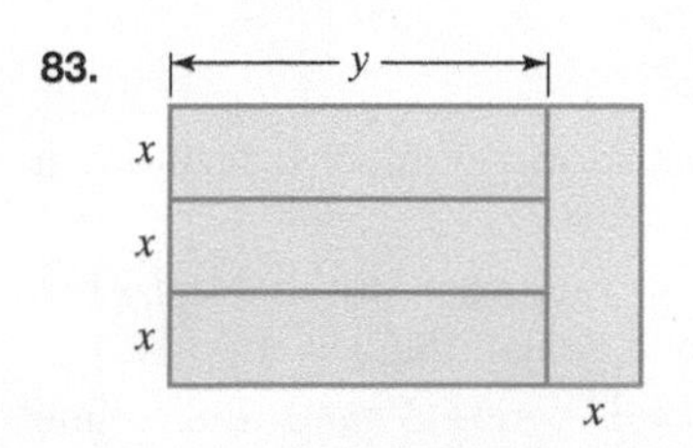

84.
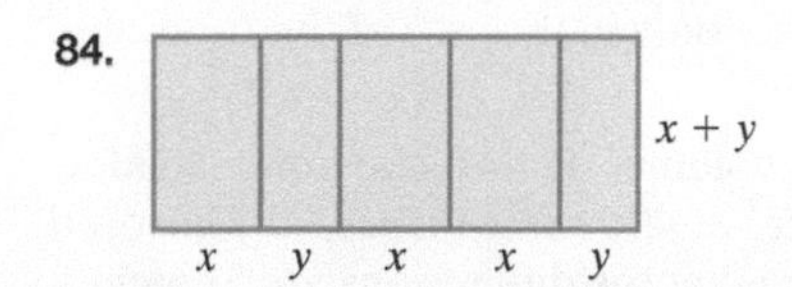

85.

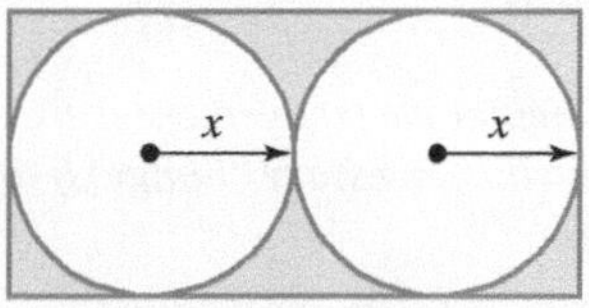

86.
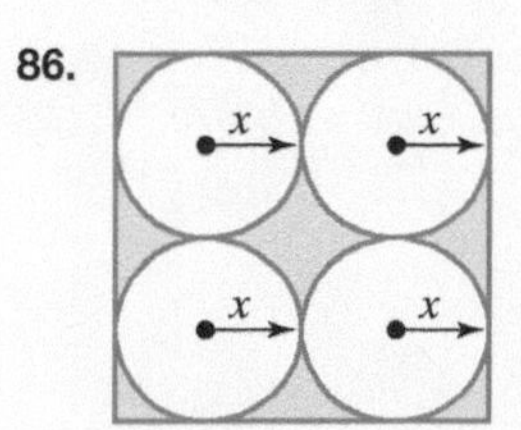

Explaining the Concepts

87. Describe a strategy that can be used to factor polynomials.
88. Describe some of the difficulties in factoring polynomials. What suggestions can you offer to overcome these difficulties?

Technology Exercises

In Exercises 89–92, use a graphing utility to graph the function on each side of the equation in the same viewing rectangle. Use end behavior to show a complete picture of the polynomial on the left side. Do the graphs coincide? If so, the factorization is correct. If not, factor correctly and then use your graphing utility to verify the factorization.

89. $4x^2 - 12x + 9 = (4x - 3)^2$

90. $2x^3 + 10x^2 - 2x - 10 = 2(x + 5)(x^2 + 1)$

91. $x^4 - 16 = (x^2 + 4)(x + 2)(x - 2)$

92. $x^3 + 1 = (x + 1)^3$

93. Use the [TABLE] feature of a graphing utility to verify any two of your complete factorizations in Exercises 15–20.

Critical Thinking Exercises

Make Sense? *In Exercises 94–97, determine whether each statement makes sense or does not make sense, and explain your reasoning.*

94. It takes a great deal of practice to get good at factoring a wide variety of polynomials.

95. Multiplying polynomials is relatively mechanical, but factoring often requires a great deal of thought.

96. The factorable trinomial $4x^2 + 8x + 3$ and the prime trinomial $4x^2 + 8x + 1$ are in the form $ax^2 + bx + c$, but $b^2 - 4ac$ is a perfect square only in the case of the factorable trinomial.

97. You told me that the volume of a rectangular solid is represented by $5x^3 + 30x^2 + 40x$ cubic inches, so I factored completely and concluded that the dimensions are $5x$ inches, $x + 2$ inches, and $x + 5$ inches.

In Exercises 98–101, determine whether each statement is true or false. If the statement is false, make the necessary change(s) to produce a true statement.

98. $x^4 - 16$ is factored completely as $(x^2 + 4)(x^2 - 4)$.

99. The trinomial $x^2 - 4x - 4$ is a prime polynomial.

100. $x^2 + 36 = (x + 6)^2$

101. $x^3 - 64 = (x + 4)(x^2 + 4x - 16)$

In Exercises 102–104, factor completely. Assume that variable exponents represent whole numbers.

102. $x^{2n+3} - 10x^{n+3} + 25x^3$

103. $3x^{n+2} - 13x^{n+1} + 4x^n$

104. $x^{4n+1} - xy^{4n}$

105. In certain circumstances, the sum of two perfect squares can be factored by adding and subtracting the same perfect square. For example,

$$x^4 + 4 = x^4 + 4x^2 + 4 - 4x^2. \quad \text{Add and subtract } 4x^2.$$

Use this first step to factor $x^4 + 4$.

106. Express $x^3 + x + 2x^4 + 4x^2 + 2$ as the product of two polynomials of degree 2.

Review Exercises

107. Solve: $\frac{3x - 1}{5} + \frac{x + 2}{2} = -\frac{3}{10}$.

(Section 1.4, Example 4)

108. Simplify: $(4x^3y^{-1})^2(2x^{-3}y)^{-1}$.

(Section 1.6, Example 9)

109. Evaluate: $\begin{vmatrix} 0 & -3 & 2 \\ 1 & 5 & 3 \\ -2 & 1 & 4 \end{vmatrix}$.

(Section 3.5, Example 3)

Preview Exercises

Exercises 110–112 will help you prepare for the material covered in the next section.

110. Evaluate $(2x + 3)(x - 4)$ in your head for $x = 4$.

111. Evaluate $-16(t - 6)(t + 4)$ in your head for $t = 6$.

112. Express as an equivalent equation with a factored trinomial on the left side and zero on the right side:

$$x^2 + (x + 7)^2 = (x + 8)^2.$$

SECTION

5.7 Polynomial Equations and Their Applications

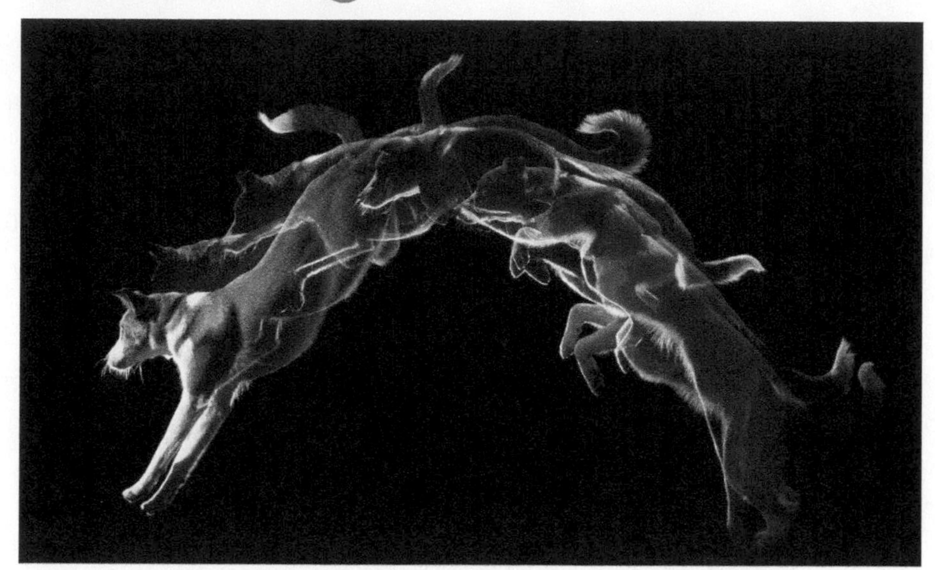

What am I supposed to learn?

After studying this section, you should be able to:

1. Solve quadratic equations by factoring.
2. Solve higher-degree polynomial equations by factoring.
3. Solve problems using polynomial equations.

Motion and change are the very essence of life. Moving air brushes against our faces; rain falls on our heads; birds fly past us; plants spring from the earth, grow, and then die; and rocks thrown upward reach a maximum height before falling to the ground. In this section, you will use quadratic functions and factoring strategies to model and visualize motion. Analyzing the where and when of moving objects involves equations in which the highest exponent on the variable is 2, called *quadratic equations*.

The Standard Form of a Quadratic Equation

We begin by defining a quadratic equation.

Definition of a Quadratic Equation

A **quadratic equation** in x is an equation that can be written in the **standard form**

$$ax^2 + bx + c = 0,$$

where a, b, and c are real numbers, with $a \neq 0$. A quadratic equation in x is also called a **second-degree polynomial equation** in x.

Here is an example of a quadratic equation in standard form:

$$x^2 - 12x + 27 = 0.$$

$a = 1$ $b = -12$ $c = 27$

1 Solve quadratic equations by factoring.

Solving Quadratic Equations by Factoring

We can factor the left side of the quadratic equation $x^2 - 12x + 27 = 0$. We obtain $(x - 3)(x - 9) = 0$. If a quadratic equation has zero on one side and a factored expression on the other side, it can be solved using the **zero-product principle**.

The Zero-Product Principle

If the product of two algebraic expressions is zero, then at least one of the factors is equal to zero.

$$\text{If } AB = 0, \text{ then } A = 0 \text{ or } B = 0.$$

For example, consider the equation $(x - 3)(x - 9) = 0$. According to the zero-product principle, this product can be zero only if at least one of the factors is zero. We set each individual factor equal to zero and solve the resulting equations for x.

$$(x - 3)(x - 9) = 0$$

$$x - 3 = 0 \quad \text{or} \quad x - 9 = 0$$

$$x = 3 \qquad\qquad x = 9$$

The solutions of the original quadratic equation, $x^2 - 12x + 27 = 0$, are 3 and 9. The solution set is $\{3, 9\}$.

Solving a Quadratic Equation by Factoring

1. If necessary, rewrite the equation in the standard form $ax^2 + bx + c = 0$, moving all terms to one side, thereby obtaining zero on the other side.
2. Factor completely.
3. Apply the zero-product principle, setting each factor containing a variable equal to zero.
4. Solve the equations in step 3.
5. Check the solutions in the original equation.

EXAMPLE 1 Solving a Quadratic Equation by Factoring

Solve: $2x^2 - 5x = 12$.

Solution

Step 1. Move all terms to one side and obtain zero on the other side. Subtract 12 from both sides and write the equation in standard form.

$$2x^2 - 5x - 12 = 12 - 12$$

$$2x^2 - 5x - 12 = 0$$

Step 2. Factor.

$$(2x + 3)(x - 4) = 0$$

Steps 3 and 4. Set each factor equal to zero and solve the resulting equations.

$$2x + 3 = 0 \quad \text{or} \quad x - 4 = 0$$

$$2x = -3 \qquad\qquad x = 4$$

$$x = -\frac{3}{2}$$

Step 5. Check the solutions in the original equation.

Check $-\frac{3}{2}$:

$$2x^2 - 5x = 12$$
$$2\left(-\frac{3}{2}\right)^2 - 5\left(-\frac{3}{2}\right) \stackrel{?}{=} 12$$
$$2\left(\frac{9}{4}\right) - 5\left(-\frac{3}{2}\right) \stackrel{?}{=} 12$$
$$\frac{9}{2} + \frac{15}{2} \stackrel{?}{=} 12$$
$$\frac{24}{2} \stackrel{?}{=} 12$$
$$12 = 12, \text{ true}$$

Check 4:

$$2x^2 - 5x = 12$$
$$2(4)^2 - 5(4) \stackrel{?}{=} 12$$
$$2(16) - 5(4) \stackrel{?}{=} 12$$
$$32 - 20 \stackrel{?}{=} 12$$
$$12 = 12, \text{ true}$$

The solutions are $-\frac{3}{2}$ and 4, and the solution set is $\{-\frac{3}{2}, 4\}$. ■

CHECK POINT 1 Solve: $2x^2 - 9x = 5$.

Great Question!

After factoring a polynomial, should I set each factor equal to zero?

No. Do not confuse factoring a polynomial with solving a quadratic equation by factoring.

Factoring a Polynomial

Factor: $2x^2 - 5x - 12$.

Solving a Quadratic Equation

Solve: $2x^2 - 5x - 12 = 0$.

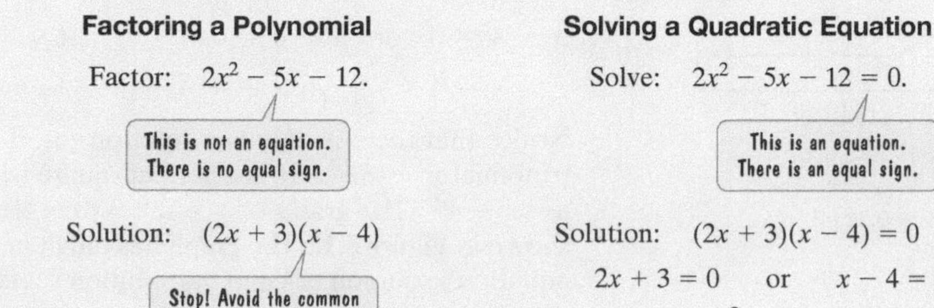
This is not an equation. There is no equal sign.
This is an equation. There is an equal sign.

Solution: $(2x + 3)(x - 4)$

Stop! Avoid the common error of setting each factor equal to zero.

Solution: $(2x + 3)(x - 4) = 0$

$2x + 3 = 0$ or $x - 4 = 0$

$x = -\frac{3}{2}$ $\quad$ $x = 4$

The solution set is $\left\{-\frac{3}{2}, 4\right\}$.

There is an important relationship between a quadratic equation in standard form, such as

$$2x^2 - 5x - 12 = 0$$

and a quadratic function, such as

$$y = 2x^2 - 5x - 12.$$

The solutions of $ax^2 + bx + c = 0$ correspond to the x-intercepts of the graph of the quadratic function $y = ax^2 + bx + c$. For example, you can visualize the solutions of $2x^2 - 5x - 12 = 0$ by looking at the x-intercepts of the graph of the quadratic function $y = 2x^2 - 5x - 12$. The graph, shaped like a bowl, is shown in **Figure 5.10**. The solutions of the equation $2x^2 - 5x - 12 = 0$, $-\frac{3}{2}$ and 4, appear as the graph's x-intercepts.

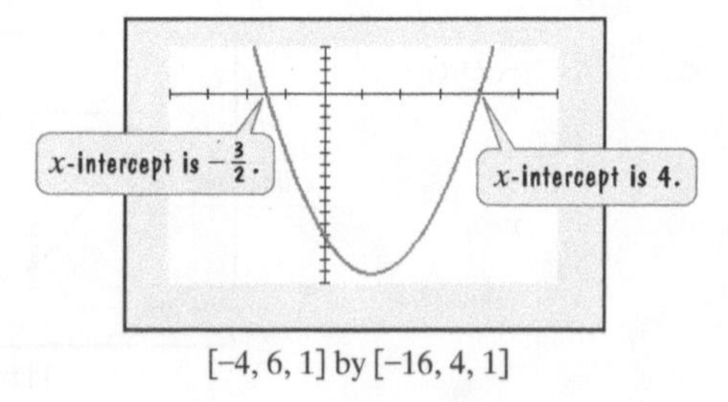
x-intercept is $-\frac{3}{2}$.
x-intercept is 4.
[−4, 6, 1] by [−16, 4, 1]

Figure 5.10

EXAMPLE 2 Solving Quadratic Equations by Factoring

Solve:

a. $5x^2 = 20x$ **b.** $x^2 + 4 = 8x - 12$ **c.** $(x - 7)(x + 5) = -20$.

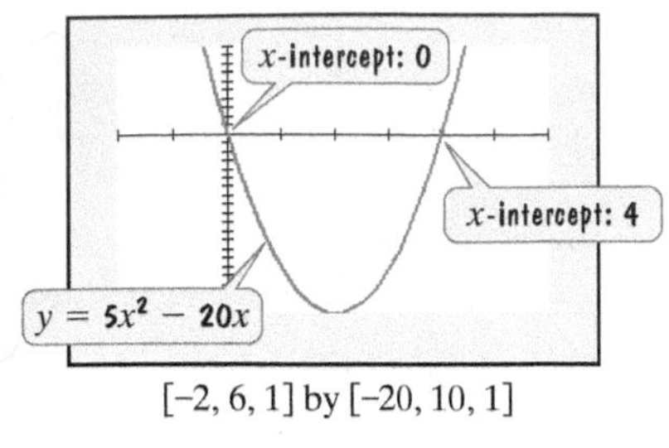

[−2, 6, 1] by [−20, 10, 1]

Figure 5.11 The solution set of $5x^2 = 20x$, or $5x^2 - 20x = 0$, is $\{0, 4\}$.

Solution

a.

$5x^2 = 20x$	This is the given equation.
$5x^2 - 20x = 0$	Subtract 20x from both sides and write the equation in standard form.
$5x(x - 4) = 0$	Factor.
$5x = 0$ or $x - 4 = 0$	Set each factor equal to 0.
$x = 0$ $\quad x = 4$	Solve the resulting equations.

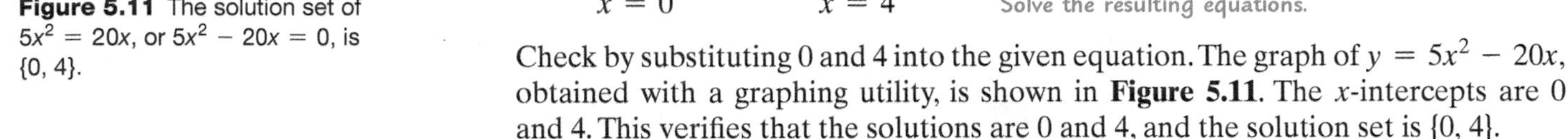

Check by substituting 0 and 4 into the given equation. The graph of $y = 5x^2 - 20x$, obtained with a graphing utility, is shown in **Figure 5.11**. The x-intercepts are 0 and 4. This verifies that the solutions are 0 and 4, and the solution set is $\{0, 4\}$.

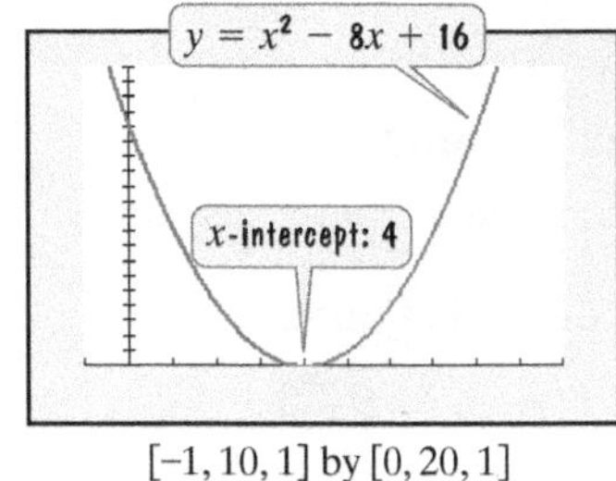

[−1, 10, 1] by [0, 20, 1]

Figure 5.12 The solution set of $x^2 + 4 = 8x - 12$, or $x^2 - 8x + 16 = 0$, is $\{4\}$.

b.

$x^2 + 4 = 8x - 12$	This is the given equation.
$x^2 - 8x + 16 = 0$	Write the equation in standard form by subtracting 8x and adding 12 on both sides.
$(x - 4)(x - 4) = 0$	Factor.
$x - 4 = 0$ or $x - 4 = 0$	Set each factor equal to 0.
$x = 4$ $\quad x = 4$	Solve the resulting equations.

Notice that there is only one solution (or, if you prefer, a repeated solution). The trinomial $x^2 - 8x + 16$ is a perfect square trinomial that could have been factored as $(x - 4)^2$. The graph of $y = x^2 - 8x + 16$, obtained with a graphing utility, is shown in **Figure 5.12**. The graph has only one x-intercept at 4. This verifies that the equation's solution is 4 and the solution set is $\{4\}$.

c. Be careful! Although the left side of $(x - 7)(x + 5) = -20$ is factored, we cannot use the zero-product principle. Why not? The right side of the equation is not 0. So we begin by multiplying the factors on the left side of the equation. Then we add 20 to both sides to obtain 0 on the right side.

$(x - 7)(x + 5) = -20$	This is the given equation.
$x^2 - 2x - 35 = -20$	Use the FOIL method to multiply on the left side.
$x^2 - 2x - 15 = 0$	Add 20 to both sides.
$(x + 3)(x - 5) = 0$	Factor.
$x + 3 = 0$ or $x - 5 = 0$	Set each factor equal to 0.
$x = -3$ $\quad x = 5$	Solve the resulting equations.

Check by substituting -3 and 5 into the given equation. The graph of $y = x^2 - 2x - 15$, obtained with a graphing utility, is shown in **Figure 5.13**. The x-intercepts are -3 and 5. This verifies that the solutions are -3 and 5, and the solution set is $\{-3, 5\}$.

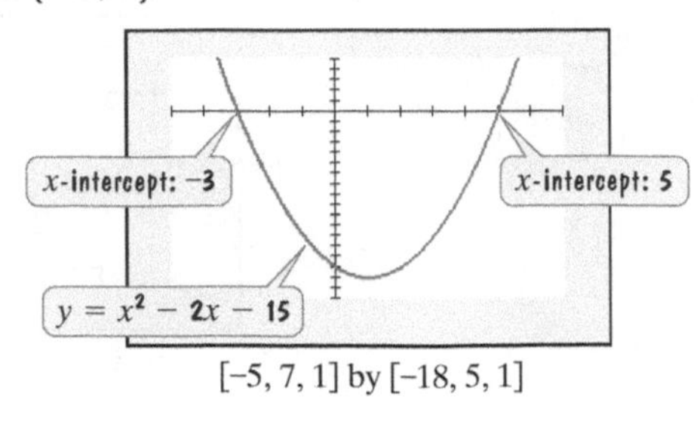

[−5, 7, 1] by [−18, 5, 1]

Figure 5.13 The solution set of $(x - 7)(x + 5) = -20$, or $x^2 - 2x - 15 = 0$, is $\{-3, 5\}$. ■

Great Questions!

In Example 2(a), can I simplify $5x^2 = 20x$ by dividing both sides by x? In Example 2(c), can I solve $(x - 7)(x + 5) = -20$ by using the zero-product principle in the first step?

No and no. Avoid the following errors:

$$5x^2 = 20x$$
$$\frac{5x^2}{x} = \frac{20x}{x}$$
$$5x = 20$$
$$x = 4$$

Never divide both sides of an equation by x. Division by zero is undefined and x may be zero. Indeed, the solutions for this equation (Example 2a) are 0 and 4. Dividing both sides by x does not permit us to find both solutions.

$$(x - 7)(x + 5) = -20$$
$$x - 7 = -20 \text{ or } x + 5 = -20$$
$$x = -13 \text{ or } x = -25$$

The zero-product principle cannot be used because the right side of the equation is not equal to 0.

CHECK POINT 2 Solve:

a. $3x^2 = 2x$

b. $x^2 + 7 = 10x - 18$

c. $(x - 2)(x + 3) = 6$.

---

2 Solve higher-degree polynomial equations by factoring.

Polynomial Equations

A **polynomial equation** is the result of setting two polynomials equal to each other. The equation is in **standard form** if one side is 0 and the polynomial on the other side is in standard form, that is, in descending powers of the variable. The **degree of a polynomial equation** is the same as the highest degree of any term in the equation. Here are examples of three polynomial equations:

$$3x + 5 = 14 \qquad 2x^2 + 7x = 4 \qquad x^3 + x^2 = 4x + 4.$$

This equation is of degree 1 because 1 is the highest degree.

This equation is of degree 2 because 2 is the highest degree.

This equation is of degree 3 because 3 is the highest degree.

Notice that a polynomial equation of degree 1 is a linear equation. A polynomial equation of degree 2 is a quadratic equation.

Some polynomial equations of degree 3 or higher can be solved by moving all terms to one side, thereby obtaining 0 on the other side. Once the equation is in standard form, factor and then set each factor equal to 0.

EXAMPLE 3 Solving a Polynomial Equation by Factoring

Solve by factoring: $x^3 + x^2 = 4x + 4$.

Solution

Step 1. Move all terms to one side and obtain zero on the other side. Subtract $4x$ and subtract 4 from both sides.

$$x^3 + x^2 - 4x - 4 = 4x + 4 - 4x - 4$$
$$x^3 + x^2 - 4x - 4 = 0$$

Step 2. Factor. Use factoring by grouping. Group terms that have a common factor.

$$\boxed{x^3 + x^2} + \boxed{-4x - 4} = 0$$

Common factor is x^2. Common factor is -4.

Using Technology

Numeric Connections

A graphing utility's TABLE feature can be used to numerically verify that $\{-2, -1, 2\}$ is the solution set of

$$x^3 + x^2 = 4x + 4.$$

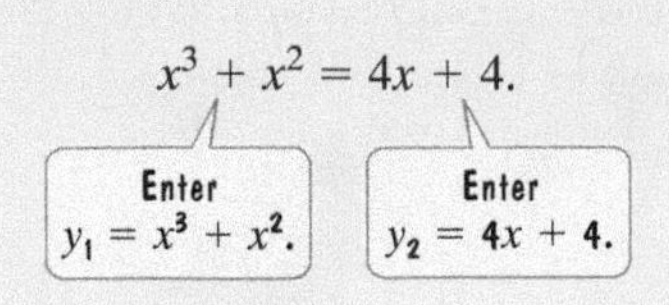
Enter $y_1 = x^3 + x^2$.
Enter $y_2 = 4x + 4$.

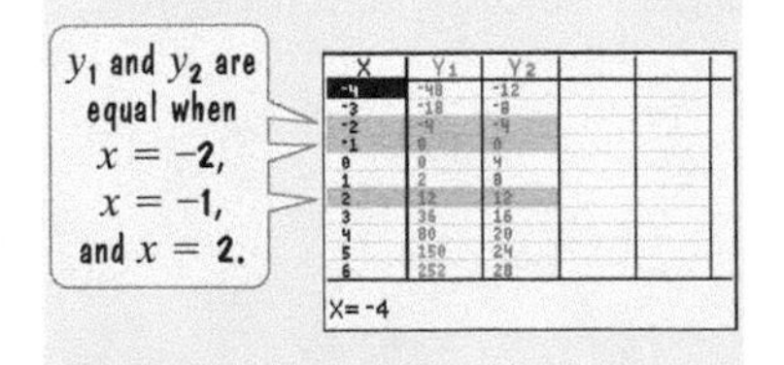
y_1 and y_2 are equal when $x = -2$, $x = -1$, and $x = 2$.
X=-4

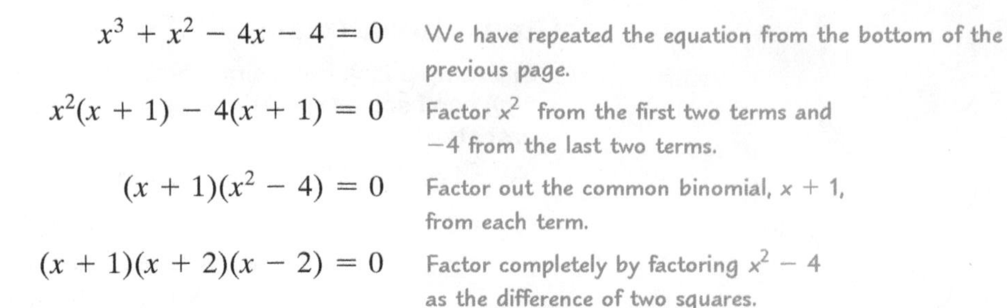

$$x^3 + x^2 - 4x - 4 = 0$$ We have repeated the equation from the bottom of the previous page.

$$x^2(x + 1) - 4(x + 1) = 0$$ Factor x^2 from the first two terms and -4 from the last two terms.

$$(x + 1)(x^2 - 4) = 0$$ Factor out the common binomial, $x + 1$, from each term.

$$(x + 1)(x + 2)(x - 2) = 0$$ Factor completely by factoring $x^2 - 4$ as the difference of two squares.

Steps 3 and 4. Set each factor equal to zero and solve the resulting equations.

$$x + 1 = 0 \quad \text{or} \quad x + 2 = 0 \quad \text{or} \quad x - 2 = 0$$
$$x = -1 \qquad\qquad x = -2 \qquad\qquad x = 2$$

Step 5. Check the solutions in the original equation. Check the three solutions, -1, -2, and 2, by substituting them into the original equation. Can you verify that the solutions are -1, -2, and 2, and the solution set is $\{-2, -1, 2\}$? ■

Using Technology

Graphic Connections

You can use a graphing utility to check the solutions to $x^3 + x^2 - 4x - 4 = 0$. Graph $y = x^3 + x^2 - 4x - 4$, as shown on the right. Is the graph complete? Because the degree, 3, is odd, and the leading coefficient, 1, is positive, it should fall to the left and rise to the right (↙, ↗). The graph shows this end behavior and is therefore complete. The x-intercepts are -2, -1, and 2, corresponding to the equation's solutions.

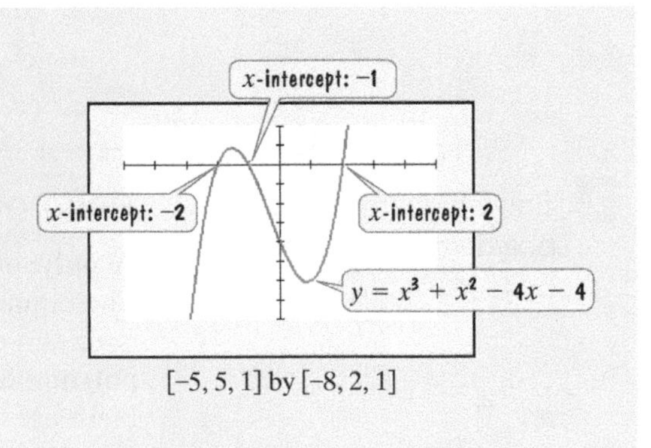
x-intercept: −1
x-intercept: −2
x-intercept: 2
y = x³ + x² − 4x − 4

[−5, 5, 1] by [−8, 2, 1]

Discover for Yourself

Suggest a method involving intersecting graphs that can be used with a graphing utility to verify that $\{-2, -1, 2\}$ is the solution set of

$$x^3 + x^2 = 4x + 4.$$

Apply this method to verify the solution set.

✓ CHECK POINT 3 Solve by factoring: $2x^3 + 3x^2 = 8x + 12$.

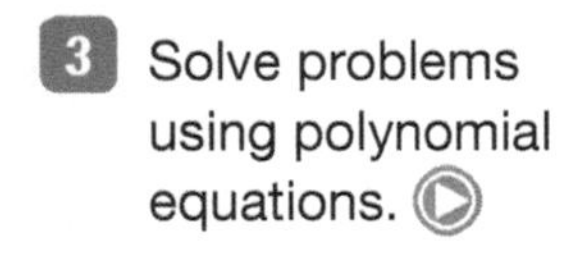

3 Solve problems using polynomial equations.

Applications of Polynomial Equations

Solving polynomial equations by factoring can be used to answer questions about variables contained in mathematical models.

EXAMPLE 4 Modeling Motion

You throw a ball straight up from a rooftop 384 feet high with an initial speed of 32 feet per second. The function

$$s(t) = -16t^2 + 32t + 384$$

describes the ball's height above the ground, $s(t)$, in feet, t seconds after you throw it. The ball misses the rooftop on its way down and eventually strikes the ground. How long will it take for the ball to hit the ground?

Solution The ball hits the ground when $s(t)$, its height above the ground, is 0 feet. Thus, we substitute 0 for $s(t)$ in the given function and solve for t.

$$s(t) = -16t^2 + 32t + 384$$ This is the function that models the ball's height.

$$0 = -16t^2 + 32t + 384$$ Substitute 0 for $s(t)$.

$$0 = -16(t^2 - 2t - 24)$$ Factor out -16, the negative of the GCF.

$$0 = -16(t - 6)(t + 4)$$ Factor $t^2 - 2t - 24$, the trinomial.

Do not set the constant, -16, equal to zero: $-16 \neq 0$.

$$t - 6 = 0 \quad \text{or} \quad t + 4 = 0$$ Set each variable factor equal to 0.

$$t = 6 \qquad\qquad t = -4$$ Solve for t.

Because we begin describing the ball's height at $t = 0$, we discard the solution $t = -4$. The ball hits the ground after 6 seconds. ■

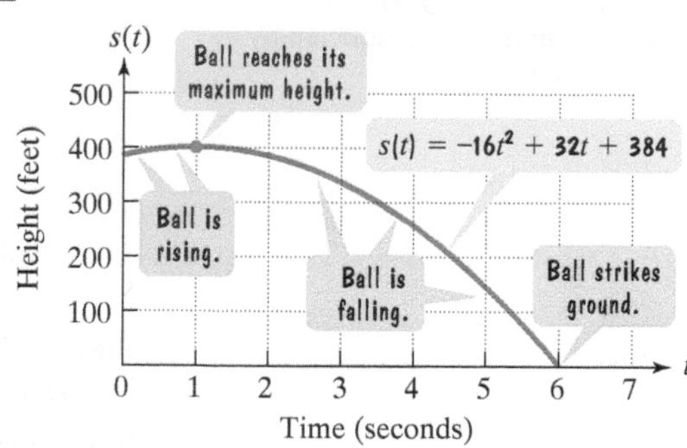

Figure 5.14

Figure 5.14 shows the graph of the function $s(t) = -16t^2 + 32t + 384$. The horizontal axis is labeled t, for the ball's time in motion. The vertical axis is labeled $s(t)$, for the ball's height above the ground at time t. Because time and height are both positive, the function is graphed in quadrant I only.

The graph visually shows what we discovered algebraically: The ball hits the ground after 6 seconds. The graph also reveals that the ball reaches its maximum height, 400 feet, after 1 second. Then the ball begins to fall.

✓ **CHECK POINT 4** Use the function $s(t) = -16t^2 + 32t + 384$ to determine when the ball's height is 336 feet. Identify your meaningful solution as a point on the graph in **Figure 5.14**.

In our next example, we use our five-step strategy for solving word problems.

EXAMPLE 5 Solving a Problem Involving Landscape Design

A rectangular garden measures 80 feet by 60 feet. A large path of uniform width is to be added along both shorter sides and one longer side of the garden. The landscape designer doing the work wants to double the garden's area with the addition of this path. How wide should the path be?

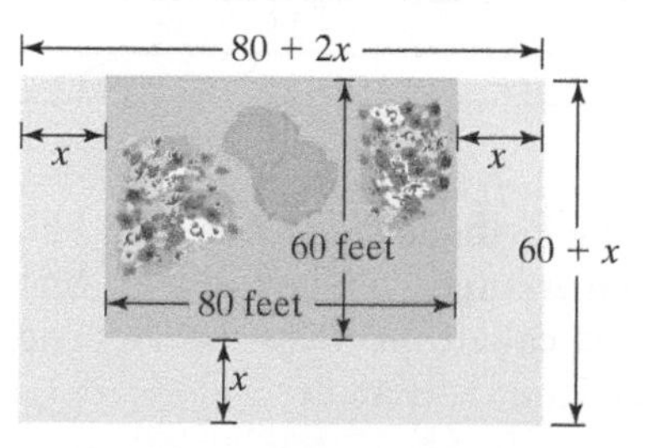

Figure 5.15 The garden's area is to be doubled by adding the path.

Solution

Step 1. Let x represent one of the unknown quantities. We will let

$$x = \text{the width of the path.}$$

The situation is illustrated in **Figure 5.15**. The figure shows the original 80-by-60 foot rectangular garden and the path of width x added along both shorter sides and one longer side.

Step 2. Represent other unknown quantities in terms of x. Because the path is added along both shorter sides and one longer side, **Figure 5.15** shows that

$$80 + 2x = \text{the length of the new, expanded rectangle}$$

$$60 + x = \text{the width of the new, expanded rectangle.}$$

Step 3. Write an equation that models the conditions. The area of the rectangle must be doubled by the addition of the path.

The area, or length times width, of the new, expanded rectangle | must be | twice that of | the area of the garden.

$$(80 + 2x)(60 + x) = 2 \cdot 80 \cdot 60$$

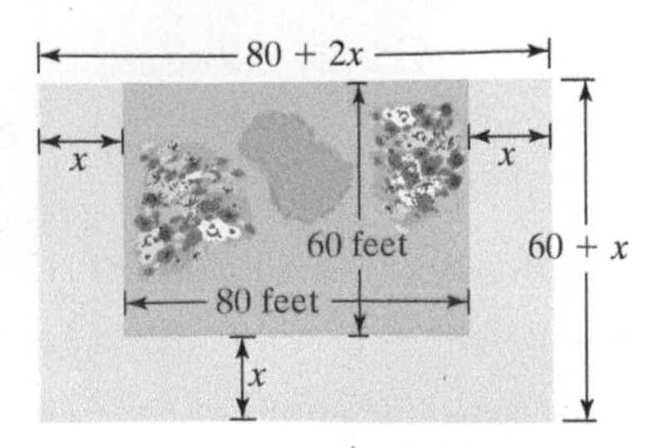

Figure 5.15 (repeated) The garden's area is to be doubled by adding the path.

Step 4. Solve the equation and answer the question.

$(80 + 2x)(60 + x) = 2 \cdot 80 \cdot 60$	This is the equation that models the problem's conditions.
$4800 + 200x + 2x^2 = 9600$	Multiply. Use FOIL on the left side.
$2x^2 + 200x - 4800 = 0$	Subtract 9600 from both sides and write the equation in standard form.
$2(x^2 + 100x - 2400) = 0$	Factor out 2, the GCF.
$2(x - 20)(x + 120) = 0$	Factor the trinomial.
$x - 20 = 0$ or $x + 120 = 0$	Set each variable factor equal to 0.
$x = 20$ or $x = -120$	Solve for x.

The path cannot have a negative width. Because -120 is geometrically impossible, we use $x = 20$. The width of the path should be 20 feet.

Step 5. Check the proposed solution in the original wording of the problem. Has the landscape architect doubled the garden's area with the 20-foot-wide path? The area of the garden is 80 feet times 60 feet, or 4800 square feet. Because $80 + 2x$ and $60 + x$ represent the length and width of the expanded rectangle,

$$80 + 2x = 80 + 2 \cdot 20 = 120 \text{ feet is the expanded rectangle's length.}$$

$$60 + x = 60 + 20 = 80 \text{ feet is the expanded rectangle's width.}$$

The area of the expanded rectangle is 120 feet times 80 feet, or 9600 square feet. This is double the area of the garden, 4800 square feet, as specified by the problem's conditions. ■

✓ **CHECK POINT 5** A rectangular garden measures 16 feet by 12 feet. A path of uniform width is to be added so as to surround the entire garden. The landscape artist doing the work wants the garden and path to cover an area of 320 square feet. How wide should the path be?

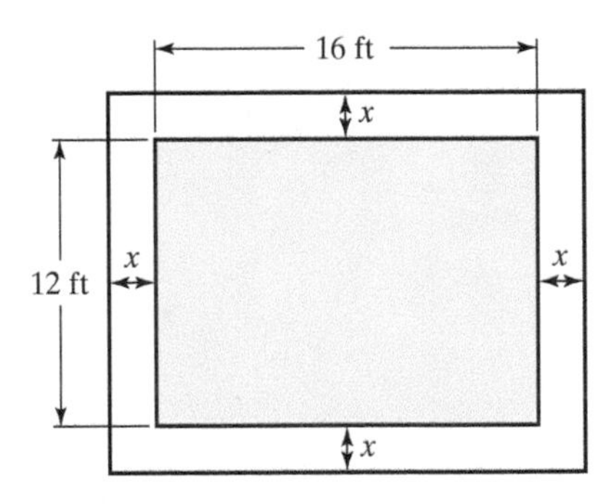

The solution to our next problem relies on knowing the **Pythagorean Theorem**. The theorem relates the lengths of the three sides of a **right triangle**, a triangle with one angle measuring 90°. The side opposite the 90° angle is called the **hypotenuse**. The other sides are called **legs**. The legs form the two sides of the right angle.

The Pythagorean Theorem

The sum of the squares of the lengths of the legs of a right triangle equals the square of the length of the hypotenuse.

If the legs have lengths a and b, and the hypotenuse has length c, then

$$a^2 + b^2 = c^2.$$

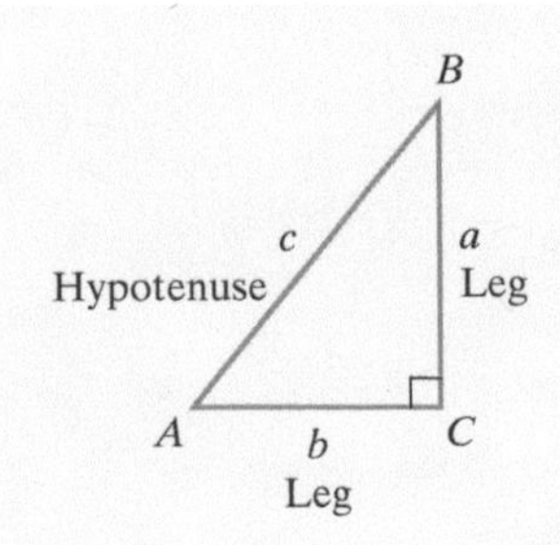

EXAMPLE 6 Using the Pythagorean Theorem to Obtain a Polynomial Equation

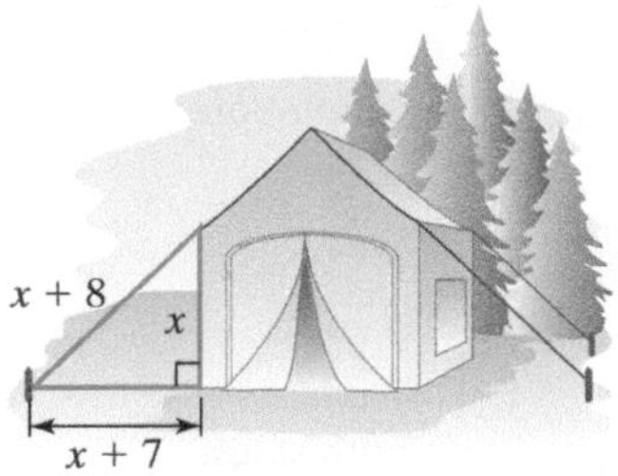

Figure 5.16 shows a tent with wires attached to help stabilize it. The length of each wire is 8 feet greater than the distance from the ground to where it is attached to the tent. The distance from the base of the tent to where the wire is anchored exceeds this height by 7 feet. Find the length of each wire used to stabilize the tent.

$x + 8$ x $x + 7$

Figure 5.16

Solution **Figure 5.16** shows a right triangle. The lengths of the legs are x and $x + 7$. The length of the hypotenuse, $x + 8$, represents the length of the wire. We use the Pythagorean Theorem to find this length.

$$\text{leg}^2 + \text{leg}^2 = \text{hypotenuse}^2$$

$$x^2 + (x+7)^2 = (x+8)^2$$ This is the equation arising from the Pythagorean Theorem.

$$x^2 + x^2 + 14x + 49 = x^2 + 16x + 64$$ Square $x + 7$ and $x + 8$.

$$2x^2 + 14x + 49 = x^2 + 16x + 64$$ Combine like terms: $x^2 + x^2 = 2x^2$.

$$x^2 - 2x - 15 = 0$$ Subtract $x^2 + 16x + 64$ from both sides and write the quadratic equation in standard form.

$$(x-5)(x+3) = 0$$ Factor the trinomial.

$$x - 5 = 0 \quad \text{or} \quad x + 3 = 0$$ Set each factor equal to 0.

$$x = 5 \qquad\qquad x = -3$$ Solve for x.

Because x represents the distance from the ground to where the wire is attached, x cannot be negative. Thus, we only use $x = 5$. **Figure 5.16** shows that the length of the wire is $x + 8$ feet. The length of the wire is $5 + 8$ feet, or 13 feet.

We can check to see that the lengths of the three sides of the right triangle, x, $x + 7$, and $x + 8$, satisfy the Pythagorean Theorem when $x = 5$. The lengths are 5 feet, 12 feet, and 13 feet.

$$\text{leg}^2 + \text{leg}^2 = \text{hypotenuse}^2$$

$$5^2 + 12^2 \stackrel{?}{=} 13^2$$

$$25 + 144 \stackrel{?}{=} 169$$

$$169 \stackrel{?}{=} 169, \text{ true}$$ ■

$x + 1$ $x + 2$ x

Figure 5.17

✓ CHECK POINT 6 A guy wire is attached to a tree to help it grow straight. The situation is illustrated in **Figure 5.17**. The length of the wire is 2 feet greater than the distance from the base of the tree to the stake. Find the length of the wire.

Achieving Success

Be sure to use the chapter test prep video on YouTube to prepare for each chapter test. These videos contain worked-out solutions to every exercise in the chapter test and let you review any exercises you miss.

Are you using any of the other textbook supplements for help and additional study? These include:

- The Student Solutions Manual. This contains fully worked solutions to the odd-numbered section exercises plus all Check Points, Concept and Vocabulary Checks, Review/Preview Exercises, Mid-Chapter Check Points, Chapter Reviews, Chapter Tests, and Cumulative Reviews.
- Lecture Videos on DVD. These are keyed to each section of the text and contain short video clips of an instructor working key text examples and exercises.
- MyMathLab is a text-specific online course.

CONCEPT AND VOCABULARY CHECK

Fill in each blank so that the resulting statement is true.

1. An equation that can be written in the standard form $ax^2 + bx + c = 0$, $a \neq 0$, is called a/an ____________ equation.
2. The zero-product principle states that if $AB = 0$, then ____________.
3. The solutions of $ax^2 + bx + c = 0$ correspond to the ____________ for the graph of $y = ax^2 + bx + c$.
4. The equation $5x^2 = 20x$ can be written in standard form by ____________ on both sides.
5. The equation $x^2 + 4 = 8x - 12$ can be written in standard form by ____________ and ____________ on both sides.
6. The result of setting two polynomials equal to each other is called a/an ____________ equation. The equation is in standard form if one side is ____________ and the polynomial on the other side is in standard form, or in ____________ powers of the variable. The degree of the equation is the ____________ degree of any term in the equation.
7. A triangle with one angle measuring 90° is called a/an ____________ triangle. The side opposite the 90° angle is called the ____________. The other sides are called ____________.
8. The Pythagorean Theorem states that in any ____________ triangle, the sum of the squares of the lengths of the ____________ equals ____________.

5.7 EXERCISE SET MyMathLab®

Practice Exercises

In Exercises 1–36, use factoring to solve each quadratic equation. Check by substitution or by using a graphing utility and identifying x-intercepts.

1. $x^2 + x - 12 = 0$
2. $x^2 - 2x - 15 = 0$
3. $x^2 + 6x = 7$
4. $x^2 - 4x = 45$
5. $3x^2 + 10x - 8 = 0$
6. $2x^2 - 5x - 3 = 0$
7. $5x^2 = 8x - 3$
8. $7x^2 = 30x - 8$
9. $3x^2 = 2 - 5x$
10. $5x^2 = 2 + 3x$
11. $x^2 = 8x$
12. $x^2 = 4x$
13. $3x^2 = 5x$
14. $2x^2 = 5x$
15. $x^2 + 4x + 4 = 0$
16. $x^2 + 6x + 9 = 0$
17. $x^2 = 14x - 49$
18. $x^2 = 12x - 36$
19. $9x^2 = 30x - 25$
20. $4x^2 = 12x - 9$
21. $x^2 - 25 = 0$
22. $x^2 - 49 = 0$
23. $9x^2 = 100$
24. $4x^2 = 25$
25. $x(x - 3) = 18$
26. $x(x - 4) = 21$
27. $(x - 3)(x + 8) = -30$
28. $(x - 1)(x + 4) = 14$
29. $x(x + 8) = 16(x - 1)$
30. $x(x + 9) = 4(2x + 5)$
31. $(x + 1)^2 - 5(x + 2) = 3x + 7$
32. $(x + 1)^2 = 2(x + 5)$
33. $x(8x + 1) = 3x^2 - 2x + 2$
34. $2x(x + 3) = -5x - 15$
35. $\frac{x^2}{18} + \frac{x}{2} + 1 = 0$
36. $\frac{x^2}{4} - \frac{5x}{2} + 6 = 0$

In Exercises 37–46, use factoring to solve each polynomial equation. Check by substitution or by using a graphing utility and identifying x-intercepts.

37. $x^3 + 4x^2 - 25x - 100 = 0$
38. $x^3 - 2x^2 - x + 2 = 0$
39. $x^3 - x^2 = 25x - 25$
40. $x^3 + 2x^2 = 16x + 32$
41. $3x^4 - 48x^2 = 0$
42. $5x^4 - 20x^2 = 0$
43. $x^4 - 4x^3 + 4x^2 = 0$
44. $x^4 - 6x^3 + 9x^2 = 0$
45. $2x^3 + 16x^2 + 30x = 0$
46. $3x^3 - 9x^2 - 30x = 0$

In Exercises 47–50, determine the x-intercepts of the graph of each quadratic function. Then match the function with its graph, labeled (a)–(d). Each graph is shown in a $[-10, 10, 1]$ *by* $[-10, 10, 1]$ *viewing rectangle.*

47. $y = x^2 - 6x + 8$

48. $y = x^2 - 2x - 8$

49. $y = x^2 + 6x + 8$

50. $y = x^2 + 2x - 8$

a.

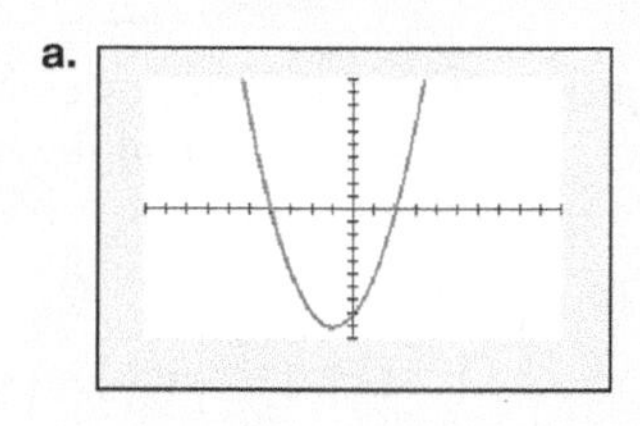

b.

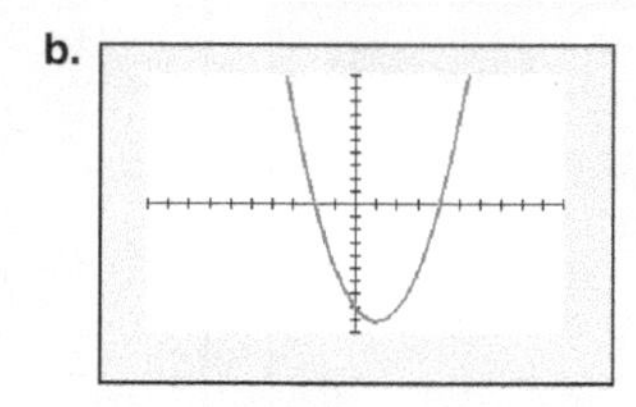

c.

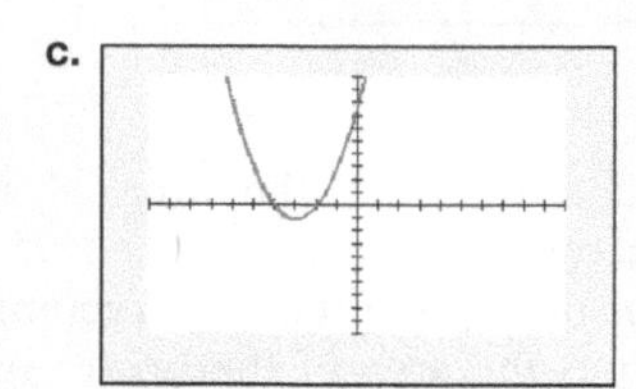

d.

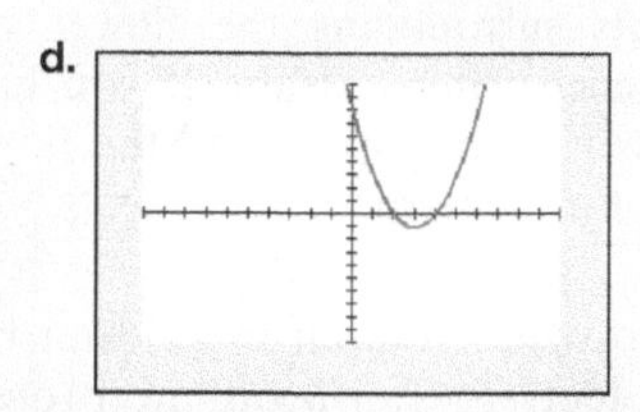

Practice PLUS

In Exercises 51–54, solve each polynomial equation.

51. $x(x + 1)^3 - 42(x + 1)^2 = 0$

52. $x(x - 2)^3 - 35(x - 2)^2 = 0$

53. $-4x[x(3x - 2) - 8](25x^2 - 40x + 16) = 0$

54. $-7x[x(2x - 5) - 12](9x^2 + 30x + 25) = 0$

In Exercises 55–58, find all values of c satisfying the given conditions.

55. $f(x) = x^2 - 4x - 27$ and $f(c) = 5$.

56. $f(x) = 5x^2 - 11x + 6$ and $f(c) = 4$.

57. $f(x) = 2x^3 + x^2 - 8x + 2$ and $f(c) = 6$.

58. $f(x) = x^3 + 4x^2 - x + 6$ and $f(c) = 10$.

In Exercises 59–62, find all numbers satisfying the given conditions.

59. The product of the number decreased by 1 and increased by 4 is 24.

60. The product of the number decreased by 6 and increased by 2 is 20.

61. If 5 is subtracted from 3 times the number, the result is the square of 1 less than the number.

62. If the square of the number is subtracted from 61, the result is the square of 1 more than the number.

In Exercises 63–64, list all numbers that must be excluded from the domain of the given function.

63. $f(x) = \dfrac{3}{x^2 + 4x - 45}$

64. $f(x) = \dfrac{7}{x^2 - 3x - 28}$

Application Exercises

A gymnast dismounts the uneven parallel bars at a height of 8 feet with an initial upward velocity of 8 feet per second. The function

$$s(t) = -16t^2 + 8t + 8$$

describes the height of the gymnast's feet above the ground, s(t), in feet, t seconds after dismounting. The graph of the function is shown, with unlabeled tick marks along the horizontal axis. Use the function to solve Exercises 65–66.

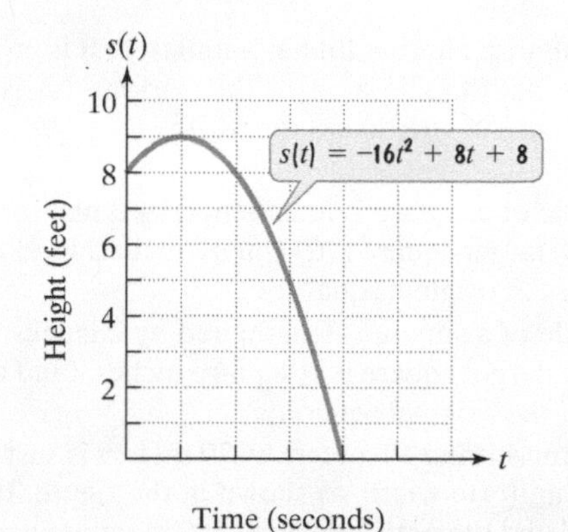

65. How long will it take the gymnast to reach the ground? Use this information to provide a number on each tick mark along the horizontal axis in the figure shown.

66. When will the gymnast be 8 feet above the ground? Identify the solution(s) as one or more points on the graph.

In a round-robin chess tournament, each player is paired with every other player once. The function

$$f(x) = \frac{x^2 - x}{2}$$

models the number of chess games, $f(x)$, that must be played in a round-robin tournament with x chess players. Use this function to solve Exercises 67–68.

67. In a round-robin chess tournament, 21 games were played. How many players were entered in the tournament?

68. In a round-robin chess tournament, 36 games were played. How many players were entered in the tournament?

The graph of the quadratic function in Exercises 67–68 is shown. Use the graph to solve Exercises 69–70.

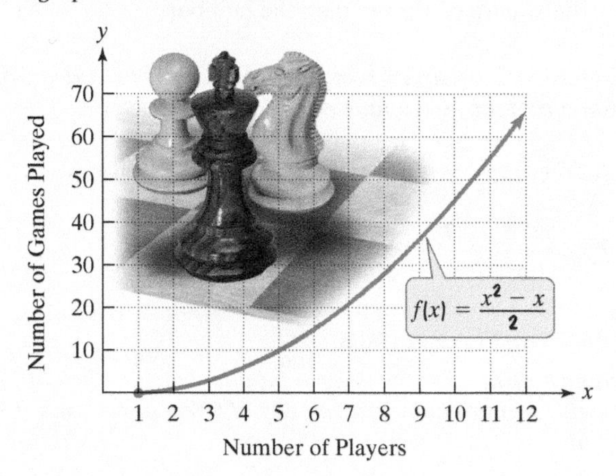

69. Identify your solution to Exercise 67 as a point on the graph.

70. Identify your solution to Exercise 68 as a point on the graph.

71. The length of a rectangular sign is 3 feet longer than the width. If the sign's area is 54 square feet, find its length and width.

72. A rectangular parking lot has a length that is 3 yards greater than the width. The area of the parking lot is 180 square yards. Find the length and the width.

73. Each side of a square is lengthened by 3 inches. The area of this new, larger square is 64 square inches. Find the length of a side of the original square.

74. Each side of a square is lengthened by 2 inches. The area of this new, larger square is 36 square inches. Find the length of a side of the original square.

75. A pool measuring 10 meters by 20 meters is surrounded by a path of uniform width, as shown in the figure. If the area of the pool and the path combined is 600 square meters, what is the width of the path?

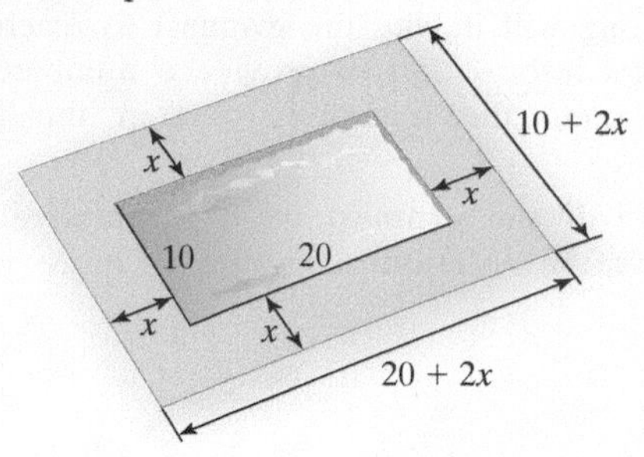

76. A vacant rectangular lot is being turned into a community vegetable garden measuring 15 meters by 12 meters. A path of uniform width is to surround the garden. If the area of the lot is 378 square meters, find the width of the path surrounding the garden.

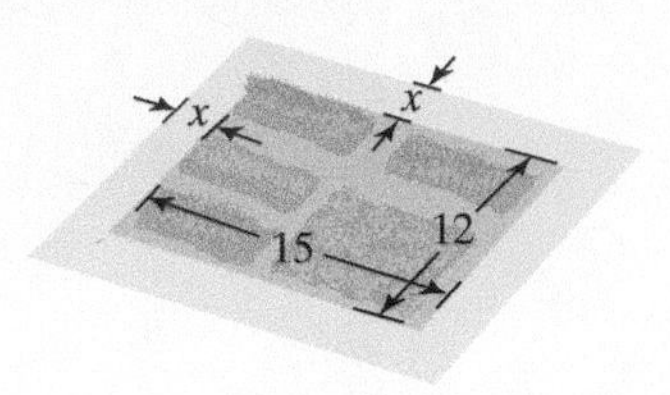

77. As part of a landscaping project, you put in a flower bed measuring 10 feet by 12 feet. You plan to surround the bed with a uniform border of low-growing plants.

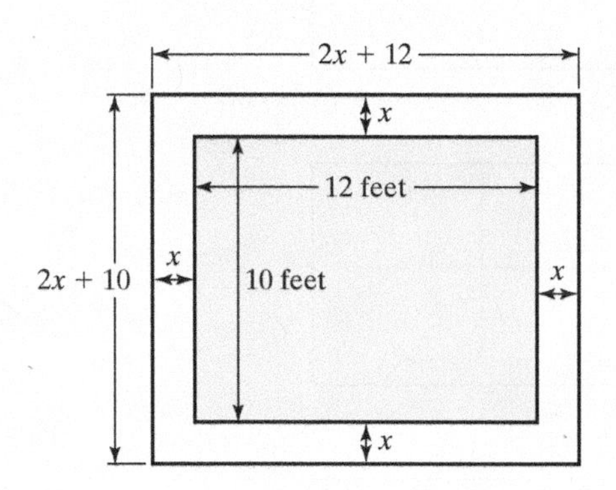

a. Write a polynomial that describes the area of the uniform border that surrounds your flower bed. *Hint*: The area of the border is the area of the large rectangle shown in the figure minus the area of the flower bed.

b. The low growing plants surrounding the flower bed require 1 square foot each when mature. If you have 168 of these plants, how wide a strip around the flower bed should you prepare for the border?

78. As part of a landscaping project, you put in a flower bed measuring 20 feet by 30 feet. To finish off the project, you are putting in a uniform border of pine bark around the outside of the rectangular garden. You have enough pine bark to cover 336 square feet. How wide should the border be?

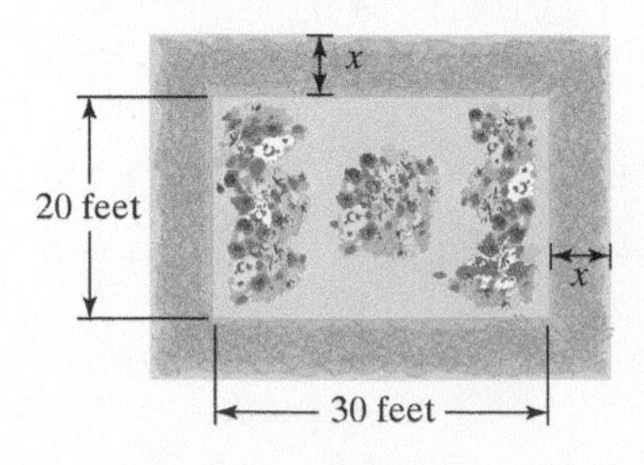

79. A machine produces open boxes using square sheets of metal. The figure illustrates that the machine cuts equal-sized squares measuring 2 inches on a side from the corners and then shapes the metal into an open box by turning up the sides. If each box must have a volume of 200 cubic inches, find the length and width of the open box.

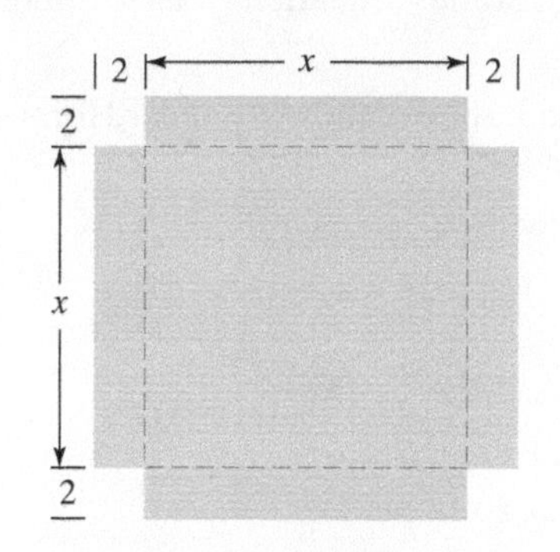

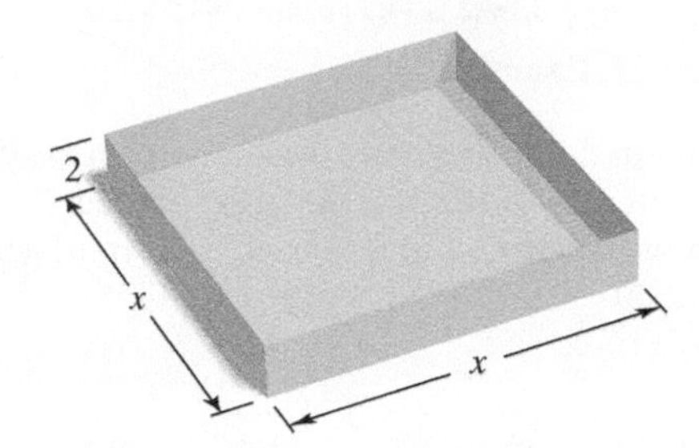

80. A machine produces open boxes using square sheets of metal. The machine cuts equal-sized squares measuring 3 inches on a side from the corners and then shapes the metal into an open box by turning up the sides. If each box must have a volume of 75 cubic inches, find the length and width of the open box.

81. The rectangular floor of a closet is divided into two right triangles by drawing a diagonal, as shown in the figure. One leg of the right triangle is 2 feet more than twice the other leg. The hypotenuse is 13 feet. Determine the closet's length and width.

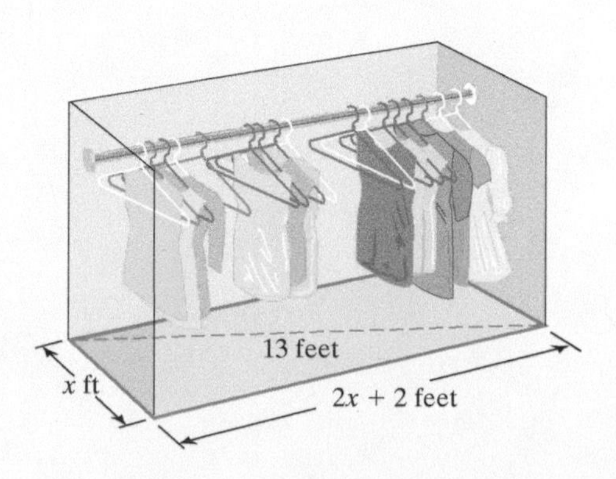

82. A piece of wire measuring 20 feet is attached to a telephone pole as a guy wire. The distance along the ground from the bottom of the pole to the end of the wire is 4 feet greater than the height where the wire is attached to the pole. How far up the pole does the guy wire reach?

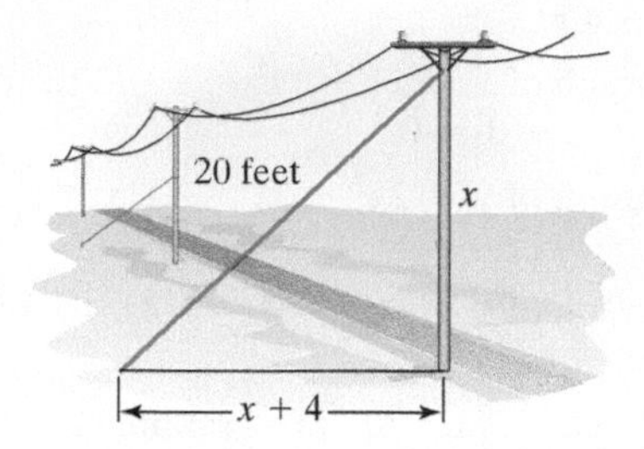

83. A tree is supported by a wire anchored in the ground 15 feet from its base. The wire is 4 feet longer than the height that it reaches on the tree. Find the length of the wire.

84. A tree is supported by a wire anchored in the ground 5 feet from its base. The wire is 1 foot longer than the height that it reaches on the tree. Find the length of the wire.

Explaining the Concepts

85. What is a quadratic equation?

86. What is the zero-product principle?

87. Explain how to solve $x^2 - x = 6$.

88. Describe the relationship between the solutions of a quadratic equation and the graph of the corresponding quadratic function.

89. What is a polynomial equation? When is it in standard form?

90. What is the degree of a polynomial equation? What are polynomial equations of degree 1 and degree 2, respectively, called?

91. Explain how to solve $x^3 + x^2 = x + 1$.

92. If something is thrown straight up, or possibly dropped, describe a situation in which it is important to know how long it will take the object to hit the ground or possibly the water.

93. A toy rocket is launched vertically upward. Using a quadratic equation, we find that the rocket will reach a height of 220 feet at 2.5 seconds and again at 5.5 seconds. How can this be?

94. Describe a situation in which a landscape designer might use polynomials and polynomial equations.

95. In your own words, state the Pythagorean Theorem.

Technology Exercises

In Exercises 96–99, use a graphing utility with a viewing rectangle large enough to show end behavior to graph each polynomial function. Then use the x-intercepts for the graph to solve the polynomial equation. Check by substitution.

96. Use the graph of $y = x^2 + 3x - 4$ to solve $x^2 + 3x - 4 = 0$.

97. Use the graph of $y = x^3 + 3x^2 - x - 3$ to solve $x^3 + 3x^2 - x - 3 = 0$.

98. Use the graph of $y = 2x^3 - 3x^2 - 11x + 6$ to solve $2x^3 - 3x^2 - 11x + 6 = 0$.

99. Use the graph of $y = -x^4 + 4x^3 - 4x^2$ to solve $-x^4 + 4x^3 - 4x^2 = 0$.

100. Use the TABLE feature of a graphing utility to verify the solution sets for any two equations in Exercises 31–32 or 39–40.

Critical Thinking Exercises

Make Sense? *In Exercises 101–104, determine whether each statement makes sense or does not make sense, and explain your reasoning.*

101. I'm working with a quadratic function that describes the length of time a ball has been thrown into the air and its height above the ground, and I find the function's graph more meaningful than its equation.

102. I set the quadratic equation $2x^2 - 5x = 12$ equal to zero and obtained $2x^2 - 5x = 0$.

103. Because some trinomials are prime, some quadratic equations cannot be solved by factoring.

104. I'm looking at a graph with one x-intercept, so it must be the graph of a linear function or a vertical line.

In Exercises 105–108, determine whether each statement is true or false. If the statement is false, make the necessary change(s) to produce a true statement.

105. Quadratic equations solved by factoring always have two different solutions.

106. If $4x(x^2 + 49) = 0$, then

$$4x = 0 \quad \text{or} \quad x^2 + 49 = 0$$
$$x = 0 \quad \text{or} \quad x = 7 \quad \text{or} \quad x = -7.$$

107. If -4 is a solution of $7y^2 + (2k - 5)y - 20 = 0$, then k must equal 14.

108. Some quadratic equations have more than two solutions.

109. Write a quadratic equation in standard form whose solutions are -3 and 7.

110. Solve: $|x^2 + 2x - 36| = 12$.

Review Exercises

111. Solve: $|3x - 2| = 8$. (Section 4.3, Example 1)

112. Simplify: $3(5 - 7)^2 + \sqrt{16} + 12 \div (-3)$. (Section 1.2, Example 7)

113. You invested \$3000 in two accounts paying 5% and 8% annual interest. If the total interest earned for the year is \$189, how much was invested at each rate? (Section 3.2, Example 2)

Preview Exercises

Exercises 114–116 will help you prepare for the material covered in the first section of the next chapter.

114. If $f(x) = \dfrac{120x}{100 - x}$, find $f(20)$.

115. Find the domain of $f(x) = \dfrac{4}{x - 2}$.

116. Factor the numerator and the denominator. Then simplify by dividing out the common factor in the numerator and the denominator.

$$\frac{x^2 - 7x - 18}{2x^2 + 3x - 2}$$

Chapter 5 Summary

Definitions and Concepts	Examples

Section 5.1 Introduction to Polynomials and Polynomial Functions

A polynomial is a single term or the sum of two or more terms containing variables with whole-number exponents. A monomial is a polynomial with exactly one term; a binomial has exactly two terms; a trinomial has exactly three terms. If $a \neq 0$, the degree of ax^n is n and the degree of ax^ny^m is $n + m$. The degree of a nonzero constant is 0. The constant 0 has no defined degree. The degree of a polynomial is the greatest degree of any term. The leading term is the term of greatest degree. Its coefficient is called the leading coefficient.

$$7x^3y - 4x^5y^4 - 2x^4y$$

Degree is $3 + 1 = 4$. Degree is $5 + 4 = 9$. Degree is $4 + 1 = 5$.

The degree of the polynomial is 9. The leading term is $-4x^5y^4$. The leading coefficient is -4. This polynomial is a trinomial.

In a polynomial function, the expression that defines the function is a polynomial. Polynomial functions have graphs that are smooth and continuous. The behavior of the graph of a polynomial function to the far left or the far right is called its end behavior.

The Leading Coefficient Test

1. Odd-degree polynomial functions have graphs with opposite behavior at each end. If the leading coefficient is positive, the graph falls to the left and rises to the right. If the leading coefficient is negative, the graph rises to the left and falls to the right.
2. Even-degree polynomial functions have graphs with the same behavior at each end. If the leading coefficient is positive, the graph rises to the left and rises to the right. If the leading coefficient is negative, the graph falls to the left and falls to the right.

Describe the end behavior of the graph of each polynomial function:

- $f(x) = -2x^3 + 3x^2 + 11x - 6$

  The degree, 3, is odd. The leading coefficient, -2, is negative. The graph rises to the left and falls to the right.

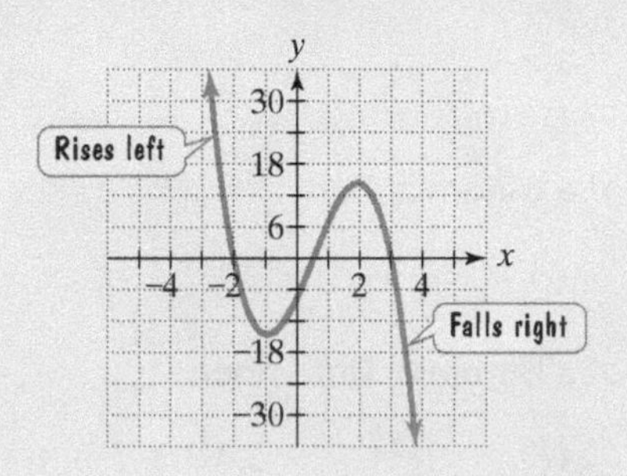

- $f(x) = x^2 - 4$

  The degree, 2, is even. The leading coefficient, 1, is positive. The graph rises to the left and rises to the right.

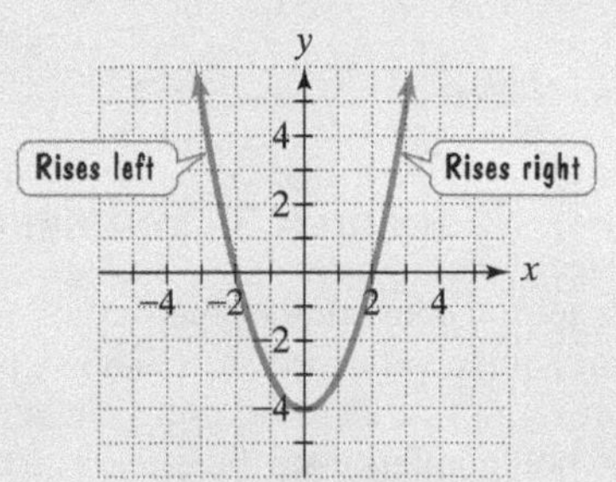

To add polynomials, add like terms.

$$\begin{aligned}&(6x^3y + 5x^2y - 7y) + (-9x^3y + x^2y + 6y)\\ &= (6x^3y - 9x^3y) + (5x^2y + x^2y) + (-7y + 6y)\\ &= -3x^3y + 6x^2y - y\end{aligned}$$

To subtract two polynomials, add the opposite, or additive inverse, of the polynomial being subtracted.

$$\begin{aligned}&(5y^3 - 9y^2 - 4) - (3y^3 - 12y^2 - 5)\\ &= (5y^3 - 9y^2 - 4) + (-3y^3 + 12y^2 + 5)\\ &= (5y^3 - 3y^3) + (-9y^2 + 12y^2) + (-4 + 5)\\ &= 2y^3 + 3y^2 + 1\end{aligned}$$

Definitions and Concepts	Examples

Section 5.2 Multiplication of Polynomials

To multiply monomials, multiply coefficients and add exponents.	$(-2x^2y^4)(-3x^3y)$ $= (-2)(-3)x^{2+3}y^{4+1} = 6x^5y^5$
To multiply a monomial and a polynomial, multiply each term of the polynomial by the monomial.	$7x^2y(4x^3y^5 - 2xy - 1)$ $= 7x^2y \cdot 4x^3y^5 - 7x^2y \cdot 2xy - 7x^2y \cdot 1$ $= 28x^5y^6 - 14x^3y^2 - 7x^2y$
To multiply polynomials if neither is a monomial, multiply each term of one by each term of the other.	$(x^3 + 2x)(5x^2 - 3x + 4)$ $= x^3(5x^2 - 3x + 4) + 2x(5x^2 - 3x + 4)$ $= 5x^5 - 3x^4 + 4x^3 + 10x^3 - 6x^2 + 8x$ $= 5x^5 - 3x^4 + 14x^3 - 6x^2 + 8x$
The FOIL method may be used when multiplying two binomials: First terms multiplied. Outside terms multiplied. Inside terms multiplied. Last terms multiplied.	F O I L $(5x - 3y)(2x + y) = 5x \cdot 2x + 5x \cdot y + (-3y) \cdot 2x + (-3y) \cdot y$ $= 10x^2 + 5xy - 6xy - 3y^2$ $= 10x^2 - xy - 3y^2$
The Square of a Binomial Sum $(A + B)^2 = A^2 + 2AB + B^2$	$(x^2 + 6)^2 = (x^2)^2 + 2 \cdot x^2 \cdot 6 + 6^2$ $= x^4 + 12x^2 + 36$
The Square of a Binomial Difference $(A - B)^2 = A^2 - 2AB + B^2$	$(4x - 5)^2 = (4x)^2 - 2 \cdot 4x \cdot 5 + 5^2$ $= 16x^2 - 40x + 25$
The Product of the Sum and Difference of Two Terms $(A + B)(A - B) = A^2 - B^2$	• $(3x + 7y)(3x - 7y) = (3x)^2 - (7y)^2$ $= 9x^2 - 49y^2$ • $[(x + 2) - 4y][(x + 2) + 4y]$ $= (x + 2)^2 - (4y)^2 = x^2 + 4x + 4 - 16y^2$

Section 5.3 Greatest Common Factors and Factoring by Grouping

Factoring a polynomial consisting of the sum of monomials means finding an equivalent expression that is a product. Polynomials that cannot be factored using integer coefficients are called prime polynomials over the integers. The greatest common factor, GCF, is an expression that divides every term of the polynomial. The GCF is the product of the largest common numerical factor and the variable of lowest degree common to every term of the polynomial. To factor a monomial from a polynomial, express each term as the product of the GCF and its other factor. Then use the distributive property to factor out the GCF. When the leading coefficient of a polynomial is negative, it is often desirable to factor out a common factor with a negative coefficient.	Factor: $4x^4y^2 - 12x^2y^3 + 20xy^2$. (GCF is $4xy^2$.) $= 4xy^2 \cdot x^3 - 4xy^2 \cdot 3xy + 4xy^2 \cdot 5$ $= 4xy^2(x^3 - 3xy + 5)$ Factor: $-25x^3 + 10x^2 - 15x$. (Use $-5x$ as a common factor.) $= -5x(5x^2) - 5x(-2x) - 5x(3)$ $= -5x(5x^2 - 2x + 3)$

Definitions and Concepts	Examples

Section 5.3 Greatest Common Factors and Factoring by Grouping (continued)

Definitions and Concepts	Examples
To factor by grouping, factor out the GCF from each group. Then factor out the remaining factor.	$xy + 7x - 2y - 14$ $= x(y + 7) - 2(y + 7)$ $= (y + 7)(x - 2)$

Section 5.4 Factoring Trinomials

Definitions and Concepts	Examples
To factor a trinomial of the form $x^2 + bx + c$, find two numbers whose product is c and whose sum is b. The factorization is $(x + \text{one number})(x + \text{other number})$.	Factor: $x^2 + 9x + 20$. Find two numbers whose product is 20 and whose sum is 9. The numbers are 4 and 5. $x^2 + 9x + 20 = (x + 4)(x + 5)$
In some trinomials, the highest power is greater than 2, and the exponent in one of the terms is half that of the other term. Factor by introducing a substitution. Let u equal the variable to the smaller power.	Factor: $x^6 - 7x^3 + 12$. $= (x^3)^2 - 7x^3 + 12$ $= u^2 - 7u + 12$ Let $u = x^3$. $= (u - 4)(u - 3)$ $= (x^3 - 4)(x^3 - 3)$
To factor $ax^2 + bx + c$ by trial and error, try various combinations of factors of ax^2 and c until a middle term of bx is obtained for the sum of the outside and inside products.	Factor: $2x^2 + 7x - 15$. Factors of $2x^2$: $2x, x$ Factors of -15: 1 and -15, -1 and 15, 3 and -5, -3 and 5 $(2x - 3)(x + 5)$ Sum of outside and inside products should equal $7x$. $10x - 3x = 7x$ Thus, $2x^2 + 7x - 15 = (2x - 3)(x + 5)$.
To factor $ax^2 + bx + c$ by grouping, find the factors of ac whose sum is b. Write bx as a sum or difference using these factors. Then factor by grouping.	Factor: $2x^2 + 7x - 15$. Find the factors of $2(-15)$, or -30, whose sum is 7. They are 10 and -3. $2x^2 + 7x - 15$ $= 2x^2 + 10x - 3x - 15$ $= 2x(x + 5) - 3(x + 5) = (x + 5)(2x - 3)$

Section 5.5 Factoring Special Forms

Definitions and Concepts	Examples
The Difference of Two Squares $A^2 - B^2 = (A + B)(A - B)$	$16x^2 - 9y^2$ $= (4x)^2 - (3y)^2 = (4x + 3y)(4x - 3y)$
Perfect Square Trinomials $A^2 + 2AB + B^2 = (A + B)^2$ $A^2 - 2AB + B^2 = (A - B)^2$	• $x^2 + 20x + 100 = x^2 + 2 \cdot x \cdot 10 + 10^2 = (x + 10)^2$ • $9x^2 - 30x + 25 = (3x)^2 - 2 \cdot 3x \cdot 5 + 5^2 = (3x - 5)^2$
Sum or Difference of Cubes $A^3 + B^3 = (A + B)(A^2 - AB + B^2)$ $A^3 - B^3 = (A - B)(A^2 + AB + B^2)$	$125x^3 - 8 = (5x)^3 - 2^3$ $= (5x - 2)[(5x)^2 + 5x \cdot 2 + 2^2]$ $= (5x - 2)(25x^2 + 10x + 4)$

Definitions and Concepts	Examples

Section 5.5 Factoring Special Forms (continued)

When using factoring by grouping, terms can sometimes be grouped to obtain the difference of two squares. One of the squares is a perfect square trinomial.

$$\underbrace{x^2 + 18x + 81} - 25y^2$$
$$= (x + 9)^2 - (5y)^2$$
$$= (x + 9 + 5y)(x + 9 - 5y)$$

Section 5.6 A General Factoring Strategy

A Factoring Strategy

1. Factor out the GCF or a common factor with a negative coefficient.
2. **a.** If two terms, try
$$A^2 - B^2 = (A + B)(A - B)$$
$$A^3 + B^3 = (A + B)(A^2 - AB + B^2)$$
$$A^3 - B^3 = (A - B)(A^2 + AB + B^2).$$
   b. If three terms, try
$$A^2 + 2AB + B^2 = (A + B)^2$$
$$A^2 - 2AB + B^2 = (A - B)^2.$$
   If not a perfect square trinomial, try trial and error or grouping.
   c. If four terms, try factoring by grouping.
3. See if any factors can be factored further.

Factor: $3x^4 + 12x^3 - 3x^2 - 12x$.
The GCF is $3x$.

$$3x^4 + 12x^3 - 3x^2 - 12x$$
$$= 3x(x^3 + 4x^2 - x - 4)$$

Four terms: Try grouping.

$$= 3x[x^2(x + 4) - 1(x + 4)]$$
$$= 3x(x + 4)(x^2 - 1)$$

This can be factored further.

$$= 3x(x + 4)(x + 1)(x - 1)$$

Section 5.7 Polynomial Equations and Their Applications

A quadratic equation in x can be written in the standard form

$$ax^2 + bx + c = 0, a \neq 0.$$

A polynomial equation is the result of setting two polynomials equal to each other. The equation is in standard form if one side is 0 and the polynomial on the other side is in standard form, that is, in descending powers of the variable. In standard form, its degree is the highest degree of any term in the equation. A polynomial equation of degree 1 is a linear equation and of degree 2 a quadratic equation. Some polynomial equations can be solved by writing the equation in standard form, factoring, and then using the zero-product principle: If a product is 0, then at least one of the factors is equal to 0.

Solve: $5x^2 + 7x = 6$.

$$5x^2 + 7x - 6 = 0$$
$$(5x - 3)(x + 2) = 0$$
$$5x - 3 = 0 \quad \text{or} \quad x + 2 = 0$$
$$5x = 3 \qquad\qquad x = -2$$
$$x = \frac{3}{5}$$

The solutions are -2 and $\frac{3}{5}$, and the solution set is $\left\{-2, \frac{3}{5}\right\}$. (The solutions are the x-intercepts of the graph of $y = 5x^2 + 7x - 6$.)

CHAPTER 5 REVIEW EXERCISES

5.1 *In Exercises 1–2, determine the coefficient of each term, the degree of each term, the degree of the polynomial, the leading term, and the leading coefficient of the polynomial.*

1. $-5x^3 + 7x^2 - x + 2$
2. $8x^4y^2 - 7xy^6 - x^3y$
3. If $f(x) = x^3 - 4x^2 + 3x - 1$, find $f(-2)$.

4. The bar graph shows the number of record daily high temperatures set across the United States, by decade.

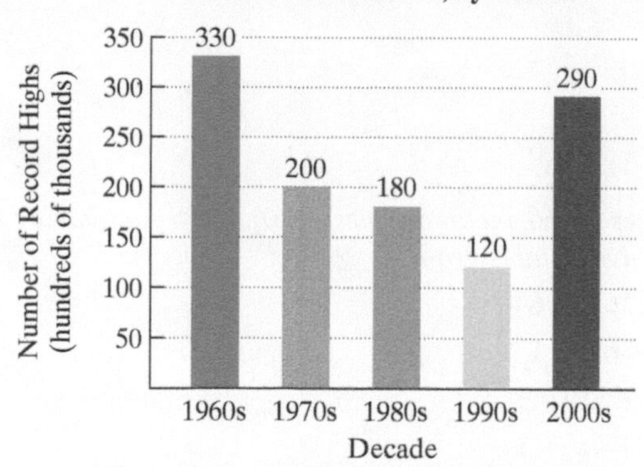

Source: University Corporation for Atmospheric Research, UCAR MAGAZINE

The data can be modeled by the function

$$f(x) = 10x^3 - 200x^2 + 1230x - 2016,$$

where $f(x)$ is the number of record daily high temperatures in the United States, in hundreds of thousands, in decade x, with $x = 6 = 1960\text{s}$, $7 = 1970\text{s}$, $8 = 1980\text{s}$, $9 = 1990\text{s}$, and $10 = 2000\text{s}$.

a. Find and interpret $f(10)$.

b. Does your answer in part (a) underestimate or overestimate the number of record highs shown by the graph? By how much?

In Exercises 5–8, use the Leading Coefficient Test to determine the end behavior of the graph of the given polynomial function. Then use this end behavior to match the polynomial function with its graph. [The graphs are labeled (a) through (d).]

5. $f(x) = -x^3 + x^2 + 2x$

6. $f(x) = x^6 - 6x^4 + 9x^2$

7. $f(x) = x^5 - 5x^3 + 4x$

8. $f(x) = -x^4 + 1$

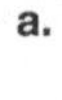
a.

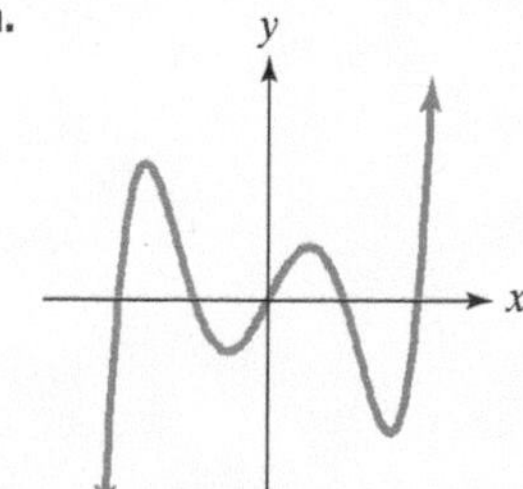

b.

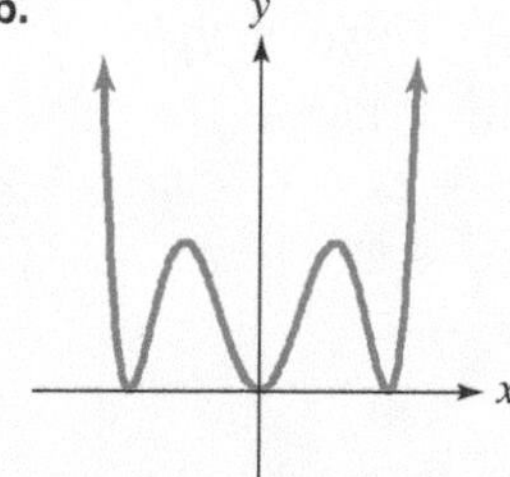

c.

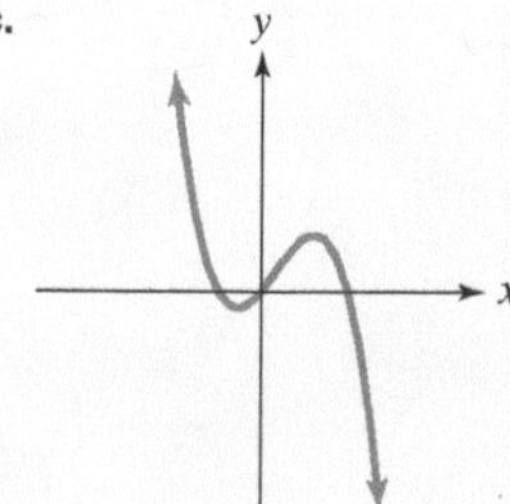

d.

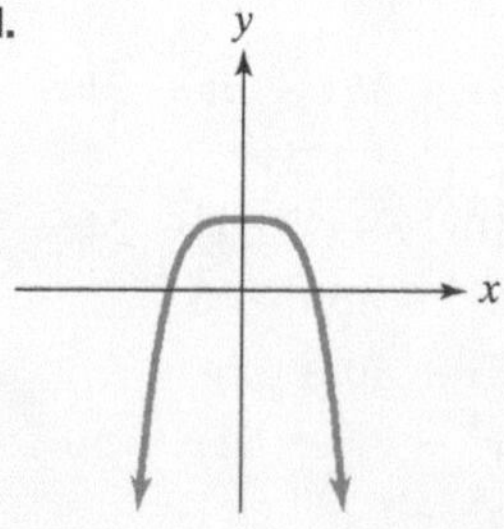

9. The Brazilian Amazon rain forest is the world's largest tropical rain forest, with some of the greatest biodiversity of any region. In 2009, the number of trees cut down in the Amazon dropped to its lowest level in 20 years. The line graph shows the number of square kilometers cleared from 2001 through 2009.

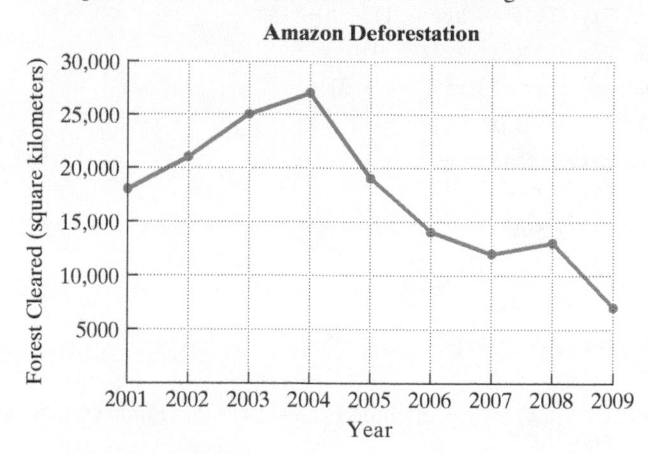

Source: Brazil's National Institute for Space Research

The data in the line graph can be modeled by the following third- and fourth-degree polynomial functions:

$$f(x) = 158x^3 - 2845x^2 + 12{,}926x + 7175$$
$$g(x) = -17x^4 + 508x^3 - 5180x^2 + 18{,}795x + 2889.$$

Amazon deforestation, in square kilometers, x years after 2000

a. Use the Leading Coefficient Test to determine the end behavior to the right for the graph of f.

b. Assume that the rate at which the Amazon rain forest is being cut down continues to decline. Based on your answer to part (a), will f be useful in modeling Amazon deforestation over an extended period of time? Explain your answer.

c. Use the Leading Coefficient Test to determine the end behavior to the right for the graph of g.

d. Assume that the rate at which the Amazon rain forest is being cut down continues to decline. Based on your answer to part (c), will g be useful in modeling Amazon deforestation over an extended period of time? Explain your answer.

In Exercises 10–11, add the polynomials.

10. $(-8x^3 + 5x^2 - 7x + 4) + (9x^3 - 11x^2 + 6x - 13)$

11. $(7x^3y - 13x^2y - 6y) + (5x^3y + 11x^2y - 8y - 17)$

In Exercises 12–13, subtract the polynomials.

12. $(7x^3 - 6x^2 + 5x - 11) - (-8x^3 + 4x^2 - 6x - 3)$

13. $(4x^3y^2 - 7x^3y - 4) - (6x^3y^2 - 3x^3y + 4)$

14. Subtract $-2x^3 - x^2y + xy^2 + 7y^3$ from $x^3 + 4x^2y - y^3$.

5.2 *In Exercises 15–27, multiply the polynomials.*

15. $(4x^2yz^5)(-3x^4yz^2)$

16. $6x^3\left(\frac{1}{3}x^5 - 4x^2 - 2\right)$

17. $7xy^2(3x^4y^2 - 5xy - 1)$

18. $(2x + 5)(3x^2 + 7x - 4)$

19. $(x^2 + x - 1)(x^2 + 3x + 2)$

20. $(4x - 1)(3x - 5)$

21. $(3xy - 2)(5xy + 4)$

22. $(3x + 7y)^2$

23. $(x^2 - 5y)^2$

24. $(2x + 7y)(2x - 7y)$

25. $(3xy^2 - 4x)(3xy^2 + 4x)$

26. $[(x + 3) + 5y][(x + 3) - 5y]$

27. $(x + y + 4)^2$

28. Let $f(x) = x - 3$ and $g(x) = 2x + 5$. Find $(fg)(x)$ and $(fg)(-4)$.

29. Let $f(x) = x^2 - 7x + 2$. Find each of the following and simplify:

a. $f(a - 1)$

b. $f(a + h) - f(a)$.

5.3 *In Exercises 30–33, factor the greatest common factor from each polynomial.*

30. $16x^3 + 24x^2$

31. $2x - 36x^2$

32. $21x^2y^2 - 14xy^2 + 7xy$

33. $18x^3y^2 - 27x^2y$

In Exercises 34–35, factor out a common factor with a negative coefficient.

34. $-12x^2 + 8x - 48$

35. $-x^2 - 11x + 14$

In Exercises 36–38, factor by grouping.

36. $x^3 - x^2 - 2x + 2$

37. $xy - 3x - 5y + 15$

38. $5ax - 15ay + 2bx - 6by$

5.4 *In Exercises 39–47, factor each trinomial completely, or state that the trinomial is prime.*

39. $x^2 + 8x + 15$

40. $x^2 + 16x - 80$

41. $x^2 + 16xy - 17y^2$

42. $3x^3 - 36x^2 + 33x$

43. $3x^2 + 22x + 7$

44. $6x^2 - 13x + 6$

45. $5x^2 - 6xy - 8y^2$

46. $6x^3 + 5x^2 - 4x$

47. $2x^2 + 11x + 15$

In Exercises 48–51, factor by introducing an appropriate substitution.

48. $x^6 + x^3 - 30$

49. $x^4 - 10x^2 - 39$

50. $(x + 5)^2 + 10(x + 5) + 24$

51. $5x^6 + 17x^3 + 6$

5.5 *In Exercises 52–55, factor each difference of two squares.*

52. $4x^2 - 25$

53. $1 - 81x^2y^2$

54. $x^8 - y^6$

55. $(x - 1)^2 - y^2$

In Exercises 56–60, factor any perfect square trinomials, or state that the polynomial is prime.

56. $x^2 + 16x + 64$

57. $9x^2 - 6x + 1$

58. $25x^2 + 20xy + 4y^2$

59. $49x^2 + 7x + 1$

60. $25x^2 - 40xy + 16y^2$

In Exercises 61–62, factor by grouping to obtain the difference of two squares.

61. $x^2 + 18x + 81 - y^2$

62. $z^2 - 25x^2 + 10x - 1$

In Exercises 63–65, factor using the formula for the sum or difference of two cubes.

63. $64x^3 + 27$

64. $125x^3 - 8$

65. $x^3y^3 + 1$

5.6 *In Exercises 66–90, factor completely, or state that the polynomial is prime.*

66. $15x^2 + 3x$

67. $12x^4 - 3x^2$

68. $20x^4 - 24x^3 + 28x^2 - 12x$

69. $x^3 - 15x^2 + 26x$

70. $-2y^4 + 24y^3 - 54y^2$

71. $9x^2 - 30x + 25$

72. $5x^2 - 45$

73. $2x^3 - x^2 - 18x + 9$

74. $6x^2 - 23xy + 7y^2$

75. $2y^3 + 12y^2 + 18y$

76. $x^2 + 6x + 9 - 4a^2$

77. $8x^3 - 27$

78. $x^5 - x$

79. $x^4 - 6x^2 + 9$

80. $x^2 + xy + y^2$

81. $4a^3 + 32$

82. $x^4 - 81$

83. $ax + 3bx - ay - 3by$

84. $27x^3 - 125y^3$

85. $10x^3y + 22x^2y - 24xy$

86. $6x^6 + 13x^3 - 5$

87. $2x + 10 + x^2y + 5xy$

88. $y^3 + 2y^2 - 25y - 50$

89. $a^8 - 1$

90. $9(x - 4) + y^2(4 - x)$

In Exercises 91–92,

a. *Write an expression for the area of the shaded blue region.*

b. *Write the expression in factored form.*

91.

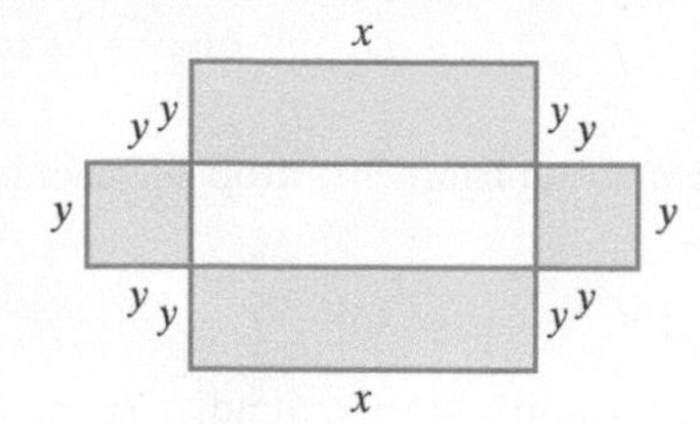

92.

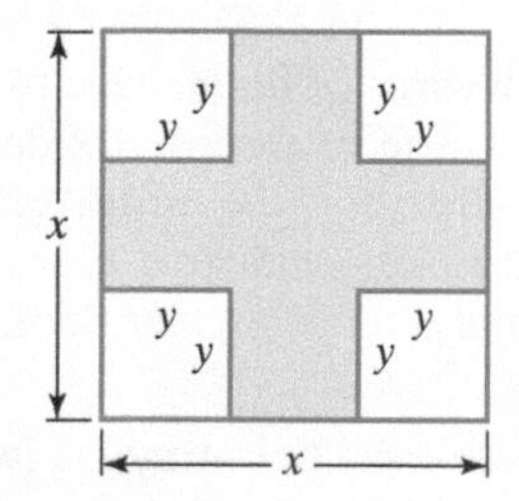

5.7 *In Exercises 93–97, use factoring to solve each polynomial equation.*

93. $x^2 + 6x + 5 = 0$

94. $3x^2 = 22x - 7$

95. $(x + 3)(x - 2) = 50$

96. $3x^2 = 12x$

97. $x^3 + 5x^2 = 9x + 45$

98. A model rocket is launched from the top of a cliff 144 feet above sea level. The function

$$s(t) = -16t^2 + 128t + 144$$

describes the rocket's height above the water, $s(t)$, in feet, t seconds after it is launched. The rocket misses the edge of the cliff on its way down and eventually lands in the ocean. How long will it take for the rocket to hit the water?

99. How much distance do you need to bring your car to a complete stop? A function used by those who study automobile safety is

$$d(x) = \frac{x^2}{20} + x,$$

where $d(x)$ is the stopping distance, in feet, for a car traveling at x miles per hour.

a. If it takes you 40 feet to come to a complete stop, how fast was your car traveling?

b. The graph of the quadratic function that models stopping distance is shown. Identify your solution from part (a) as a point on the graph.

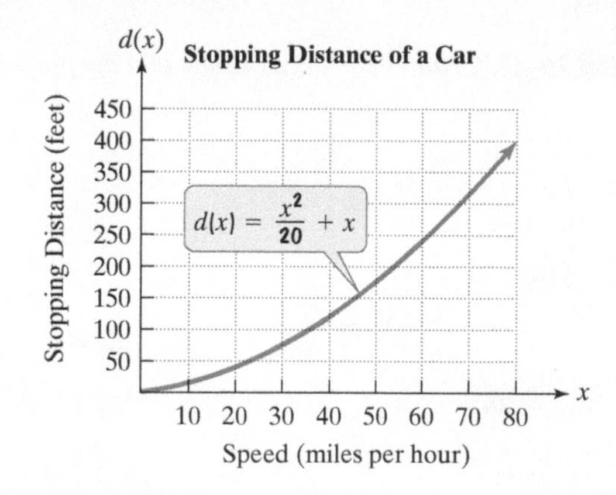

c. Describe the trend shown by the graph.

100. The length of a rectangular sign is 3 feet longer than the width. If the sign has space for 54 square feet of advertising, find its length and its width.

101. A painting measuring 10 inches by 16 inches is surrounded by a frame of uniform width. If the combined area of the painting and frame is 280 square inches, determine the width of the frame.

102. A lot is in the shape of a right triangle. The longer leg of the triangle is 20 yards longer than twice the length of the shorter leg. The hypotenuse is 30 yards longer than twice the length of the shorter leg. What are the lengths of the three sides?

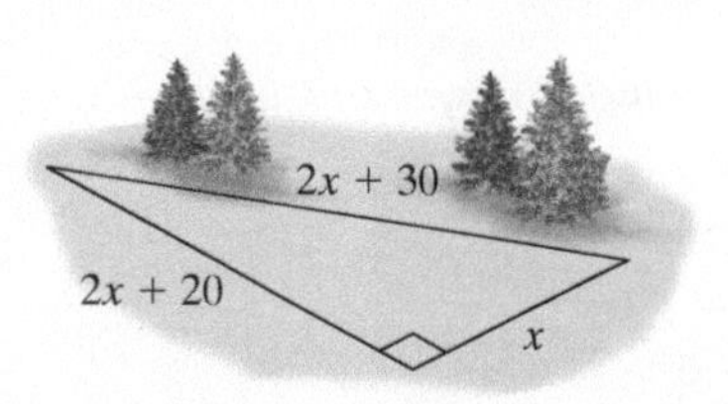

CHAPTER 5 TEST

Step-by-step test solutions are found on the Chapter Test Prep Videos available in MyMathLab® or on YouTube (search "BlitzerInterAlg7e" and click on "Channels").

In Exercises 1–2, give the degree and the leading coefficient of the polynomial.

1. $7x - 5 + x^2 - 6x^3$

2. $4xy^3 + 7x^4y^5 - 3xy^4$

3. If $f(x) = 3x^3 + 5x^2 - x + 6$, find $f(0)$ and $f(-2)$.

In Exercises 4–5, use the Leading Coefficient Test to describe the end behavior of the graph of the polynomial function.

4. $f(x) = -16x^2 + 160x$

5. $f(x) = 4x^3 + 12x^2 - x - 3$

In Exercises 6–13, perform the indicated operations.

6. $(4x^3y - 19x^2y - 7y) + (3x^3y + x^2y + 6y - 9)$

7. $(6x^2 - 7x - 9) - (-5x^2 + 6x - 3)$

8. $(-7x^3y)(-5x^4y^2)$

9. $(x - y)(x^2 - 3xy - y^2)$

10. $(7x - 9y)(3x + y)$

11. $(2x - 5y)(2x + 5y)$
12. $(4y - 7)^2$
13. $[(x + 2) + 3y][(x + 2) - 3y]$
14. Let $f(x) = x + 2$ and $g(x) = 3x - 5$. Find $(fg)(x)$ and $(fg)(-5)$.
15. Let $f(x) = x^2 - 5x + 3$. Find $f(a + h) - f(a)$ and simplify.

In Exercises 16–33, factor completely, or state that the polynomial is prime.

16. $14x^3 - 15x^2$
17. $81y^2 - 25$
18. $x^3 + 3x^2 - 25x - 75$
19. $25x^2 - 30x + 9$
20. $x^2 + 10x + 25 - 9y^2$
21. $x^4 + 1$
22. $y^2 - 16y - 36$
23. $14x^2 + 41x + 15$
24. $5x^3 - 5$
25. $12x^2 - 3y^2$
26. $12x^2 - 34x + 10$
27. $3x^4 - 3$
28. $x^8 - y^8$
29. $12x^2y^4 + 8x^3y^2 - 36x^2y$
30. $x^6 - 12x^3 - 28$
31. $x^4 - 2x^2 - 24$
32. $12x^2y - 27xy + 6y$
33. $y^4 - 3y^3 + 2y^2 - 6y$

In Exercises 34–37, solve each polynomial equation.

34. $3x^2 = 5x + 2$
35. $(5x + 4)(x - 1) = 2$
36. $15x^2 - 5x = 0$
37. $x^3 - 4x^2 - x + 4 = 0$
38. A baseball is thrown straight up from a rooftop 448 feet high. The function

$$s(t) = -16t^2 + 48t + 448$$

describes the ball's height above the ground, $s(t)$, in feet, t seconds after it is thrown. How long will it take for the ball to hit the ground?

39. An architect is allowed 15 square yards of floor space to add a small bedroom to a house. Because of the room's design in relationship to the existing structure, the width of the rectangular floor must be 7 yards less than two times the length. Find the length and width of the rectangular floor that the architect is permitted.
40. Find the lengths of the three sides of the right triangle in the figure shown.

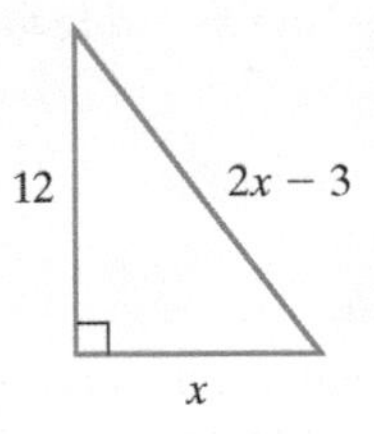

CUMULATIVE REVIEW EXERCISES (CHAPTERS 1–5)

In Exercises 1–7, solve each equation, inequality, or system of equations.

1. $8(x + 2) - 3(2 - x) = 4(2x + 6) - 2$
2. $\begin{cases} 2x + 4y = -6 \\ x = 2y - 5 \end{cases}$
3. $\begin{cases} 2x - y + 3z = 0 \\ 2y + z = 1 \\ x + 2y - z = 5 \end{cases}$
4. $2x + 4 < 10$ and $3x - 1 > 5$
5. $|2x - 5| \geq 9$
6. $2x^2 = 7x - 5$
7. $2x^3 + 6x^2 = 20x$
8. Solve for x: $x = \dfrac{ax + b}{c}$.
9. Use function notation to write the equation of the line passing through $(-2, -3)$ and $(2, 5)$.
10. In a campuswide election for student government president, 2800 votes were cast for the two candidates. If the winner had 160 more votes than the loser, how many votes were cast for each candidate?

In Exercises 11–13, graph each equation or inequality in a rectangular coordinate system.

11. $f(x) = -\dfrac{1}{3}x + 1$
12. $4x - 5y < 20$
13. $y \leq -1$
14. Simplify: $-\dfrac{8x^3y^6}{16x^9y^{-4}}$.
15. Write in scientific notation: 0.0000706.

In Exercises 16–17, perform the indicated operations.

16. $(3x^2 - y)^2$
17. $(3x^2 - y)(3x^2 + y)$

In Exercises 18–20, factor completely.

18. $x^3 - 3x^2 - 9x + 27$
19. $x^6 - x^2$
20. $14x^3y^2 - 28x^4y^2$

CHAPTER 6

Rational Expressions, Functions, and Equations

Is the newest route to college through video games? A few collegiate gaming factoids:

- Hundreds of colleges have formed video-gaming teams and some have made gaming a varsity sport. Top Nexus destroyers can receive thousands of dollars in e-sports scholarships.
- College e-sports are predicted to galvanize the video-game industry. The numbers are impressive: In 2014, an estimated 10,000 students competed in the video game *Collegiate StarLeague*. More than 130,000 viewers watched the *League of Legends* college finals.
- Video-game companies, recognizing a great market opportunity when they see one, are helping to underwrite the college gaming explosion. In 2015, Riot Games, creator of *League of Legends*, offered $360,000 in scholarship money to gifted players. Blizzard Entertainment ran a 64-team collegiate tournament when they introduced *Heroes of the Storm* (dubbed Heroes of the Dorm). ESPN televised the finals and each member of the winning team received up to three years of tuition. Teams from over 650 colleges tried to qualify.

But wait, there's much more! We are on the cusp of a revolution in the way people play video games. Virtual reality headsets will enable players to experience video games as immersive three-dimensional environments, basically putting the gamer inside the game.

In this chapter, you will see how functions involving fractional expressions provide insights into phenomena as diverse as virtual reality video games, the aftereffects of candy on your mouth, the cost of environmental cleanup, the relationship between heart rate and life span, and even our ongoing processes of learning and forgetting.

Here's where you'll find these applications:

- Virtual reality video games: Section 6.7 opener and Section 6.7, Example 3
- Sugar and your mouth: Exercise Set 6.1, Exercises 109–112
- Environmental cleanup: Section 6.1, Example 1, and Section 6.6, Example 6
- Heart rate and life span: Exercise Set 6.8, Exercises 29–32
- Learning and forgetting: Exercise Set 6.6, Exercises 53–60.

SECTION

6.1

Rational Expressions and Functions: Multiplying and Dividing

What am I supposed to learn?

After studying this section, you should be able to:

1. Evaluate rational functions.
2. Find the domain of a rational function.
3. Interpret information given by the graph of a rational function.
4. Simplify rational expressions.
5. Multiply rational expressions.
6. Divide rational expressions.

Environmental scientists and municipal planners often make decisions using **cost-benefit models**. These mathematical models estimate the cost of removing a pollutant from the atmosphere as a function of the percentage of pollutant removed. What kinds of functions describe the cost of reducing environmental pollution? In this section, we introduce this new category of functions, called *rational functions*.

Rational Expressions

A **rational expression** consists of a polynomial divided by a nonzero polynomial. Here are some examples of rational expressions:

$$\frac{120x}{100-x}, \quad \frac{2x+1}{2x^2-x-1}, \quad \frac{3x^2+12xy-15y^2}{6x^3-6xy^2}.$$

1 Evaluate rational functions.

Rational Functions

A **rational function** is a function defined by a formula that is a rational expression.

EXAMPLE 1 Using a Cost-Benefit Model

The rational function

$$f(x) = \frac{120x}{100-x}$$

models the cost, $f(x)$, in thousands of dollars, to remove $x\%$ of the pollutants that a city has discharged into a lake. Find and interpret each of the following:

a. $f(20)$ **b.** $f(80)$.

Solution We use substitution to evaluate a rational function, just as we did to evaluate other functions in Chapter 2.

a. $f(x) = \frac{120x}{100-x}$ This is the given rational function.

$f(20) = \frac{120(20)}{100-20}$ To find f of 20, replace x with 20.

$= \frac{2400}{80} = 30$ Perform the indicated operations.

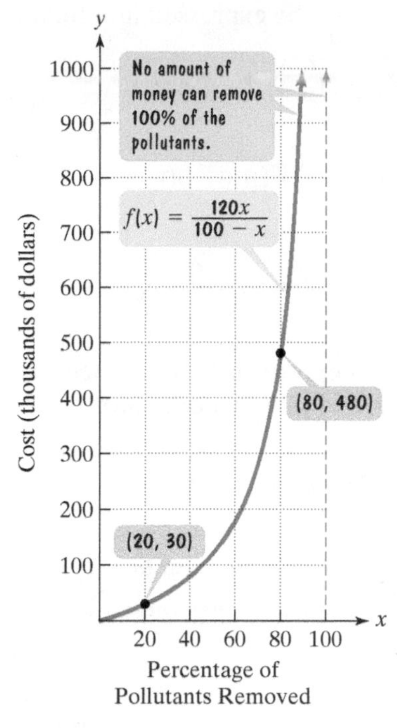

Figure 6.1

Thus, $f(20) = 30$. This means that the cost to remove 20% of the lake's pollutants is \$30 thousand. **Figure 6.1** illustrates the solution by the point (20, 30) on the graph of the rational function.

b. $f(x) = \dfrac{120x}{100 - x}$ This is the given rational function.

$f(80) = \dfrac{120(80)}{100 - 80} = \dfrac{9600}{20} = 480$ To find f of 80, replace x with 80.

Thus, $f(80) = 480$. This means that the cost to remove 80% of the lake's pollutants is \$480 thousand. **Figure 6.1** illustrates the solution by the point (80, 480) on the graph of the cost-benefit model. The graph illustrates that costs rise steeply as the percentage of pollutants removed increases. ■

✓ CHECK POINT 1 Use the rational function in Example 1 to find and interpret: **a.** $f(40)$ **b.** $f(60)$. Identify each solution as a point on the graph of the rational function in **Figure 6.1**.

---

2 Find the domain of a rational function.

The Domain of a Rational Function

Does the cost-benefit model

$$f(x) = \frac{120x}{100 - x}$$

indicate that the city can clean up its lake completely? To do this, the city must remove 100% of the pollutants. The problem is that the rational function is undefined for $x = 100$.

$$f(x) = \frac{120x}{100 - x}$$

If $x = 100$, the value of the denominator is 0.

Notice how the graph of the rational function in **Figure 6.1** approaches, but never touches, the dashed green vertical line whose equation is $x = 100$. The graph continues to rise more and more steeply, visually showing the escalating costs. By never touching the dashed vertical line, the graph illustrates that no amount of money will be enough to remove all pollutants from the lake.

In Chapter 2, we learned to exclude from a function's domain real numbers that cause division by zero. Thus, the **domain of a rational function** is the set of all real numbers except those for which the denominator is zero. We can find the domain by determining when the denominator is zero. For the cost-benefit model, the denominator is zero when $x = 100$. Furthermore, for this model, negative values of x and values of x greater than 100 are not meaningful. The domain of the function is [0, 100) and excludes 100.

Inspection can sometimes be used to find a rational function's domain. Here are two examples.

$$f(x) = \frac{4}{x - 2} \qquad g(x) = \frac{x - 3}{(x + 1)(x - 1)}$$

This numerator can be zero, so there is no need to exclude 3 from the domain.

This denominator would equal zero if $x = 2$.

This factor would equal zero if $x = -1$.

This factor would equal zero if $x = 1$.

The domain of f, which excludes 2, can be expressed in interval notation:

$$\text{Domain of } f = (-\infty, 2) \cup (2, \infty).$$

$$g(x) = \frac{x - 3}{(x + 1)(x - 1)}$$

This factor would equal zero if $x = -1$.

This factor would equal zero if $x = 1$.

Function g (repeated)

Likewise, the domain of g, which excludes both -1 and 1, can be expressed in interval notation:

$$\text{Domain of } g = (-\infty, -1) \cup (-1, 1) \cup (1, \infty).$$

EXAMPLE 2 Finding the Domain of a Rational Function

Find the domain of f if

$$f(x) = \frac{2x + 1}{2x^2 - x - 1}.$$

Solution The domain of f is the set of all real numbers except those for which the denominator is zero. We can identify such numbers by setting the denominator equal to zero and solving for x.

$$2x^2 - x - 1 = 0 \quad \text{Set the denominator equal to 0.}$$

$$(2x + 1)(x - 1) = 0 \quad \text{Factor.}$$

$$2x + 1 = 0 \quad \text{or} \quad x - 1 = 0 \quad \text{Set each factor equal to 0.}$$

$$2x = -1 \qquad x = 1 \quad \text{Solve the resulting equations.}$$

$$x = -\frac{1}{2}$$

Because $-\frac{1}{2}$ and 1 make the denominator zero, these are the values to exclude. Thus,

$$\text{Domain of } f = \left(-\infty, -\frac{1}{2}\right) \cup \left(-\frac{1}{2}, 1\right) \cup (1, \infty). \quad ■$$

Using Technology

We can use the TABLE feature of a graphing utility (with ΔTbl = .5) to verify our work with

$$f(x) = \frac{2x + 1}{2x^2 - x - 1}.$$

Enter

y_1 = (2 x + 1) ÷ (2 x ^ 2 − x − 1)

and press TABLE.

On some calculators, it may be necessary to press the right arrow key to exit the exponent after entering the exponent, 2.

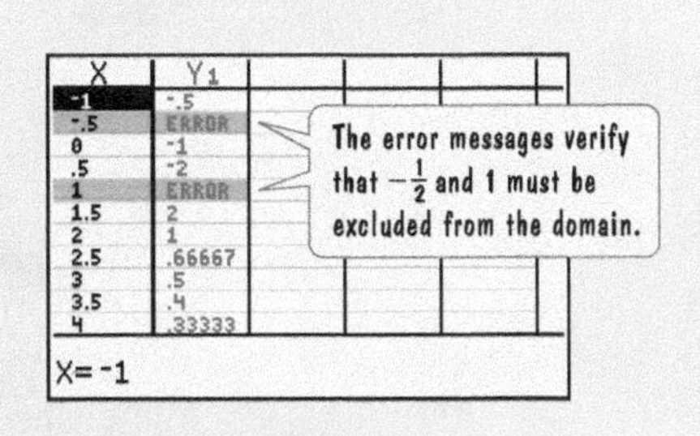

The error messages verify that $-\frac{1}{2}$ and 1 must be excluded from the domain.

CHECK POINT 2 Find the domain of f if

$$f(x) = \frac{x - 5}{2x^2 + 5x - 3}.$$

Does every rational function have values to exclude? The answer is no. For example, consider

$$f(x) = \frac{2}{x^2 + 1}.$$

No real-number values of x cause this denominator to equal zero.

The domain of f consists of the set of all real numbers. Recall that this set can be expressed in three different ways:

$$\text{Domain of } f = \{x \mid x \text{ is a real number}\} \text{ or } \mathbb{R} \text{ or } (-\infty, \infty).$$

3 Interpret information given by the graph of a rational function.

What Graphs of Rational Functions Look Like

In everyday speech, a continuous process is one that goes on without interruption and without abrupt changes. In mathematics, a continuous function has much the same meaning. The graph of a continuous function does not have interrupting breaks, such as holes, gaps, or jumps. Thus, the graph of a continuous function can be drawn without lifting a pencil off the paper.

Most rational functions are not continuous functions. For example, the graph of the rational function

$$f(x) = \frac{2x + 1}{2x^2 - x - 1}$$

is shown in **Figure 6.2**. In Example 2, we excluded $-\frac{1}{2}$ and 1 from this function's domain. Unlike the graph of a polynomial function, this graph has two breaks in it—one at each of the excluded values. At $-\frac{1}{2}$, a hole in the graph appears. The graph is composed of two distinct branches. Each branch approaches, but never touches, the dashed vertical line drawn at 1, the other excluded value.

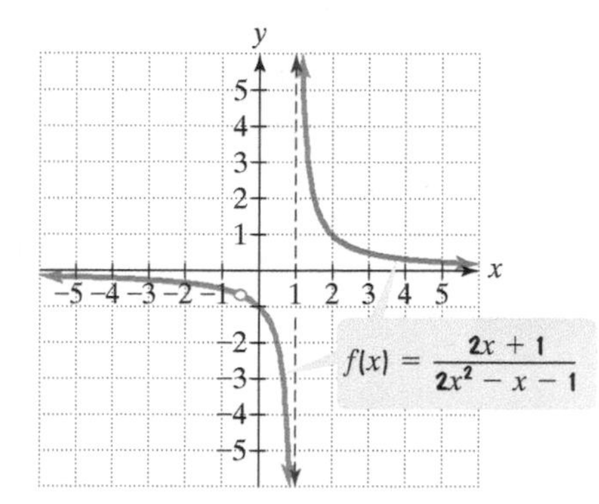

Figure 6.2

A vertical line that the graph of a function approaches, but does not touch, is said to be a **vertical asymptote** of the graph. In **Figure 6.2**, the line $x = 1$ is a vertical asymptote of the graph of f. A rational function may have no vertical asymptotes, one vertical asymptote, or several vertical asymptotes. The graph of a rational function never intersects a vertical asymptote.

Unlike the graph of a polynomial function, the graph of the rational function in **Figure 6.2** does not go up or down at each end. At the far left and the far right, the graph is getting closer to, but not actually reaching, the x-axis. This shows that as x increases or decreases without bound, the function values are approaching 0. The line $y = 0$ (that is, the x-axis) is a **horizontal asymptote** of the graph. Many, but not all, rational functions have horizontal asymptotes.

For the remainder of this section, we will focus on the rational expressions that define rational functions. Operations with these expressions should remind you of those performed in arithmetic with fractions.

Using Technology

When using a graphing utility to graph a rational function, remember to enclose both the numerator and denominator in parentheses. The graph of the rational function

$$y = \frac{2x + 1}{2x^2 - x - 1}$$

is shown in a $[-6.6, 6.6, 1]$ by $[-6.6, 6.6, 1]$ viewing rectangle.

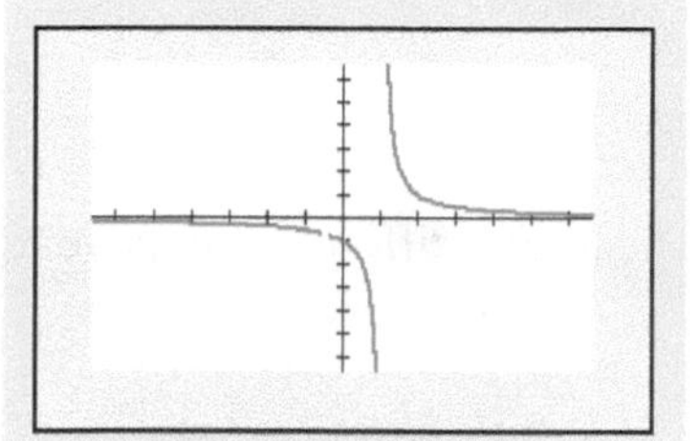

Some graphing utilities incorrectly connect the two pieces of the graph shown above, resulting in a nearly vertical line appearing on the screen. If your graphing utility does this, we recommend that you use DOT mode. Consult your owner's manual.

4 Simplify rational expressions.

Simplifying Rational Expressions

A rational expression is **simplified** if its numerator and denominator have no common factors other than 1 or −1. The following procedure can be used to simplify rational expressions:

Simplifying Rational Expressions

1. Factor the numerator and the denominator completely.
2. Divide both the numerator and the denominator by any common factors.

EXAMPLE 3 Simplifying a Rational Expression

Simplify: $\dfrac{x^2 + 4x + 3}{x + 1}$.

Solution

$$\frac{x^2 + 4x + 3}{x + 1} = \frac{(x + 1)(x + 3)}{1(x + 1)}$$ Factor the numerator and denominator.

$$= \frac{\cancel{(x + 1)}(x + 3)}{1\cancel{(x + 1)}}$$ Divide out the common factor, $x + 1$.

$$= x + 3 \quad \blacksquare$$

Simplifying a rational expression can change the numbers that make it undefined. For example, we just showed that

$$\frac{x^2 + 4x + 3}{x + 1} = x + 3.$$

This is undefined for $x = -1$.

The simplified form is defined for all real numbers.

Thus, to equate the two expressions, we must restrict the values of x in the simplified expression to exclude -1. We can write

$$\frac{x^2 + 4x + 3}{x + 1} = x + 3, x \neq -1.$$

Without this restriction, the expressions are not equal. The slight difference between them is illustrated using the graphs of

$$f(x) = \frac{x^2 + 4x + 3}{x + 1} \quad \text{and} \quad g(x) = x + 3$$

in **Figure 6.3**.

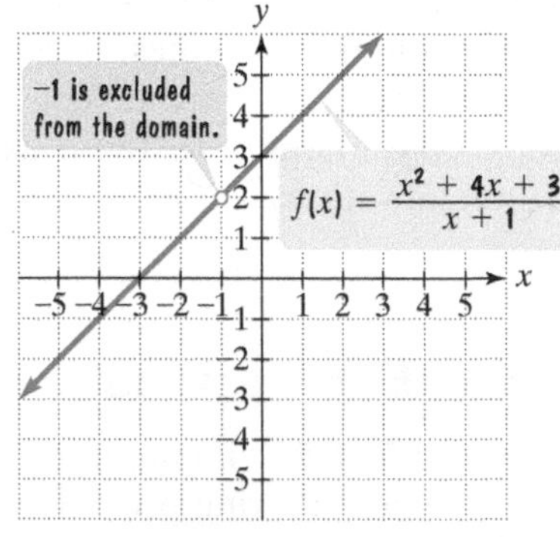

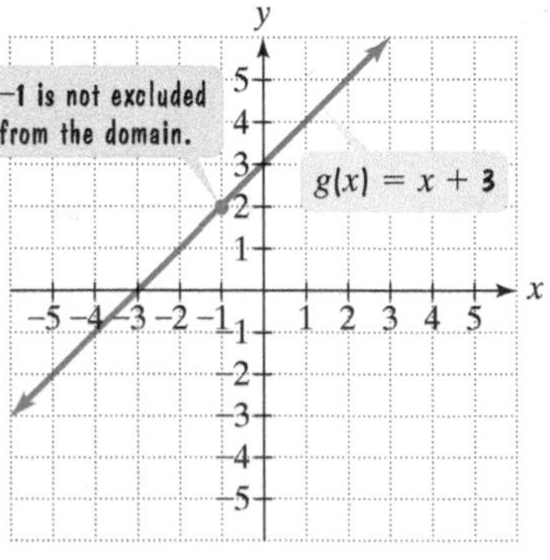

Figure 6.3 Visualizing the difference between $\dfrac{x^2 + 4x + 3}{x + 1}$ and $x + 3$

Hereafter, we will assume that a simplified rational expression is equal to the original rational expression for all real numbers except those for which either denominator is 0.

☑ CHECK POINT 3 Simplify: $\dfrac{x^2 + 7x + 10}{x + 2}$.

EXAMPLE 4 Simplifying Rational Expressions

Simplify:

a. $\dfrac{x^2 - 7x - 18}{2x^2 + 3x - 2}$ b. $\dfrac{3x^2 + 12xy - 15y^2}{6x^3 - 6xy^2}$.

Solution

a.
$$\frac{x^2 - 7x - 18}{2x^2 + 3x - 2} = \frac{(x-9)(x+2)}{(2x-1)(x+2)}$$
Factor the numerator and denominator.
$$= \frac{(x-9)\cancel{(x+2)}}{(2x-1)\cancel{(x+2)}}$$
Divide out the common factor, $x + 2$.
$$= \frac{x-9}{2x-1}$$

b.
$$\frac{3x^2 + 12xy - 15y^2}{6x^3 - 6xy^2} = \frac{3(x^2 + 4xy - 5y^2)}{6x(x^2 - y^2)}$$
Factor out the GCF in the numerator and denominator.
$$= \frac{3(x+5y)(x-y)}{6x(x+y)(x-y)}$$
Factor the numerator and denominator completely.
$$= \frac{\overset{1}{\cancel{3}}(x+5y)\cancel{(x-y)}}{\underset{2}{\cancel{6}}x(x+y)\cancel{(x-y)}}$$
Divide out the common factor, $3(x - y)$.
$$= \frac{x+5y}{2x(x+y)}$$
It is not necessary to carry out the multiplication in the denominator.

■

CHECK POINT 4 Simplify:

a. $\dfrac{x^2 - 2x - 15}{3x^2 + 8x - 3}$ b. $\dfrac{3x^2 + 9xy - 12y^2}{9x^3 - 9xy^2}$.

Great Question!

Can I simplify $\dfrac{x+4}{x}$ by dividing the numerator and the denominator by x?

No. When simplifying rational expressions, you can only divide out, or cancel, *factors* common to the numerator and denominator. **It is incorrect to divide out common terms from the numerator and denominator.**

Incorrect!

$$\frac{\cancel{x}+4}{\cancel{x}} = 4 \qquad \frac{x^2 - \cancel{4}}{\cancel{4}} = x^2 - 1 \qquad \frac{\overset{x}{\cancel{x^2}} - \overset{3}{\cancel{9}}}{\underset{1}{\cancel{x}} - \underset{1}{\cancel{3}}} = x - 3$$

The first two expressions have no common factors in their numerators and denominators. Only when expressions are multiplied can they be factors. **If you can't factor, then don't try to cancel.** The third rational expression can be simplified as follows:

Correct

$$\frac{x^2 - 9}{x - 3} = \frac{(x+3)\cancel{(x-3)}}{1\cancel{(x-3)}} = x + 3.$$

Divide out the common factor, $x - 3$.

5 Multiply rational expressions.

Multiplying Rational Expressions

The product of two rational expressions is the product of their numerators divided by the product of their denominators. For example,

$$\frac{x^2}{y+3} \cdot \frac{x+5}{y-7} = \frac{x^2(x+5)}{(y+3)(y-7)}.$$

Multiply numerators.

Multiply denominators.

Here is a step-by-step procedure for multiplying rational expressions. Before multiplying, divide out any factors common to both a numerator and a denominator.

Multiplying Rational Expressions

1. Factor all numerators and denominators completely.
2. Divide numerators and denominators by common factors.
3. Multiply the remaining factors in the numerators and multiply the remaining factors in the denominators.

EXAMPLE 5 Multiplying Rational Expressions

Multiply: $\frac{x+3}{x-4} \cdot \frac{x^2-2x-8}{x^2-9}$.

Solution

$$\frac{x+3}{x-4} \cdot \frac{x^2-2x-8}{x^2-9}$$

$$= \frac{1(x+3)}{1(x-4)} \cdot \frac{(x-4)(x+2)}{(x+3)(x-3)}$$ Factor all numerators and denominators completely.

$$= \frac{1\cancel{(x+3)}}{1\cancel{(x-4)}} \cdot \frac{\cancel{(x-4)}(x+2)}{\cancel{(x+3)}(x-3)}$$ Divide numerators and denominators by common factors.

$$= \frac{x+2}{x-3}$$ Multiply the remaining factors in the numerators and in the denominators. ■

✓ CHECK POINT 5 Multiply: $\frac{x+4}{x-7} \cdot \frac{x^2-4x-21}{x^2-16}$.

Some rational expressions contain factors in the numerator and denominator that are opposites, or additive inverses. Here is an example of such an expression:

$$\frac{(2x+5)(2x-5)}{3(5-2x)}.$$

The factors $2x - 5$ and $5 - 2x$ are opposites. They differ only in sign.

Although you can factor out -1 from the numerator or the denominator and then divide out the common factor, there is an even faster way to simplify this rational expression.

Simplifying Rational Expressions with Opposite Factors in the Numerator and Denominator

The quotient of two polynomials that have opposite signs and are additive inverses is -1.

For example,

$$\frac{(2x+5)(2x-5)}{3(5-2x)} = \frac{(2x+5)\overset{(-1)}{\cancel{(2x-5)}}}{3\cancel{(5-2x)}} = \frac{-(2x+5)}{3} \text{ or } -\frac{2x+5}{3} \text{ or } \frac{-2x-5}{3}.$$

Factoring out −1 is done mentally: $\frac{(2x+5)(-1)\cancel{(5-2x)}}{3\cancel{(5-2x)}}$.

EXAMPLE 6 Multiplying Rational Expressions

Multiply: $\frac{5x+5}{7x-7x^2} \cdot \frac{2x^2+x-3}{4x^2-9}$.

Solution

$$\frac{5x+5}{7x-7x^2} \cdot \frac{2x^2+x-3}{4x^2-9}$$

$$= \frac{5(x+1)}{7x(1-x)} \cdot \frac{(2x+3)(x-1)}{(2x+3)(2x-3)}$$

Factor all numerators and denominators completely.

$$= \frac{5(x+1)}{7x\cancel{(1-x)}} \cdot \frac{\cancel{(2x+3)}\overset{(-1)}{\cancel{(x-1)}}}{\cancel{(2x+3)}(2x-3)}$$

Divide numerators and denominators by common factors. Because $1-x$ and $x-1$ are opposites, their quotient is -1.

$$= \frac{-5(x+1)}{7x(2x-3)} \text{ or } -\frac{5(x+1)}{7x(2x-3)}$$

Multiply the remaining factors in the numerators and in the denominators. ■

CHECK POINT 6 Multiply: $\frac{4x+8}{6x-3x^2} \cdot \frac{3x^2-4x-4}{9x^2-4}$.

6 Divide rational expressions.

Dividing Rational Expressions

The quotient of two rational expressions is the product of the first expression and the multiplicative inverse, or reciprocal, of the second expression. The reciprocal is found by interchanging the numerator and the denominator of the expression.

Dividing Rational Expressions

If P, Q, R, and S are polynomials, where $Q \neq 0$, $R \neq 0$, and $S \neq 0$, then

$$\frac{P}{Q} \div \frac{R}{S} = \frac{P}{Q} \cdot \frac{S}{R} = \frac{PS}{QR}.$$

Change division to multiplication.

Replace $\frac{R}{S}$ with its reciprocal by interchanging numerator and denominator.

Thus, **we find the quotient of two rational expressions by inverting the divisor and multiplying**. For example,

$$\frac{x}{7} \div \frac{6}{y} = \frac{x}{7} \cdot \frac{y}{6} = \frac{xy}{42}.$$

Change the division to multiplication.

Replace $\frac{6}{y}$ with its reciprocal by interchanging numerator and denominator.

Great Question!

If I'm multiplying or dividing rational expressions and an expression appears without a denominator, what should I do?

When performing operations with rational expressions, if a rational expression is written without a denominator, it is helpful to write the expression with a denominator of 1. In Example 7(a), we wrote

$$4x^2 - 25 \text{ as } \frac{4x^2 - 25}{1}.$$

EXAMPLE 7 Dividing Rational Expressions

Divide:

a. $(4x^2 - 25) \div \frac{2x + 5}{14}$ **b.** $\frac{x^2 + 3x - 10}{2x} \div \frac{x^2 - 5x + 6}{x^2 - 3x}$.

Solution

a. $(4x^2 - 25) \div \frac{2x + 5}{14}$

$= \frac{4x^2 - 25}{1} \div \frac{2x + 5}{14}$ Write $4x^2 - 25$ with a denominator of 1.

$= \frac{4x^2 - 25}{1} \cdot \frac{14}{2x + 5}$ Invert the divisor and multiply.

$= \frac{(2x + 5)(2x - 5)}{1} \cdot \frac{14}{1(2x + 5)}$ Factor.

$= \frac{\cancel{(2x + 5)}(2x - 5)}{1} \cdot \frac{14}{1\cancel{(2x + 5)}}$ Divide the numerator and denominator by the common factor, $2x + 5$.

$= 14(2x - 5)$ Multiply the remaining factors in the numerators and in the denominators.

b. $\frac{x^2 + 3x - 10}{2x} \div \frac{x^2 - 5x + 6}{x^2 - 3x}$

$= \frac{x^2 + 3x - 10}{2x} \cdot \frac{x^2 - 3x}{x^2 - 5x + 6}$ Invert the divisor and multiply.

$= \frac{(x + 5)(x - 2)}{2x} \cdot \frac{x(x - 3)}{(x - 3)(x - 2)}$ Factor.

$= \frac{(x + 5)\cancel{(x - 2)}}{2\cancel{x}} \cdot \frac{\cancel{x}\cancel{(x - 3)}}{\cancel{(x - 3)}\cancel{(x - 2)}}$ Divide numerators and denominators by common factors.

$= \frac{x + 5}{2}$ Multiply the remaining factors in the numerators and in the denominators. ■

✓ CHECK POINT 7 Divide:

a. $(9x^2 - 49) \div \frac{3x - 7}{9}$

b. $\frac{x^2 - x - 12}{5x} \div \frac{x^2 - 10x + 24}{x^2 - 6x}$.

Achieving Success

Analyze the errors you make on quizzes and tests. For each error, write out the correct solution along with a description of the concept needed to solve the problem correctly. Do your mistakes indicate gaps in understanding concepts or do you at times believe that you are just not a good test taker? Are you repeatedly making the same kinds of mistakes on tests? Keeping track of errors should increase your understanding of the material, resulting in improved test scores.

CONCEPT AND VOCABULARY CHECK

Fill in each blank so that the resulting statement is true.

1. A rational expression consists of a/an _______________ divided by a nonzero _______________.
2. The domain of a rational function is the set of all real numbers except those for which the denominator is _______________.
3. A vertical line that the graph of a function approaches, but does not touch, is said to be a vertical _______________ of the graph.
4. A horizontal line that the graph of a rational function approaches at the far left or the far right is called a horizontal _______________ of the graph.
5. We simplify a rational expression by _______________ the numerator and the denominator completely. Then we divide the numerator and the denominator by any _______________.
6. The rational expression $\frac{(x+3)(x+5)}{x+3}$ simplifies to _______________ if $x \neq -3$.
7. True or false: $\frac{x^2-9}{9}$ simplifies to $x^2 - 1$. _______________
8. The rational expression $\frac{x-5}{5-x}$ simplifies to _______________.
9. The product of two rational expressions is the product of their _______________ divided by the product of their _______________.
10. The quotient of two rational expressions is the product of the first expression and the _______________ of the second:

$$\frac{P}{Q} \div \frac{R}{S} = \frac{P}{Q} \cdot \frac{\_}{\_} = \frac{\_}{\_}, Q \neq 0, R \neq 0, S \neq 0.$$

11. $\frac{x}{7} \cdot \frac{x}{10} = \frac{}{\_}$
12. $\frac{x}{7} \div \frac{x}{10} = \frac{}{\_}$

6.1 EXERCISE SET

MyMathLab®

Practice Exercises

In Exercises 1–6, use the given rational function to find the indicated function values. If a function value does not exist, so state.

1. $f(x) = \frac{x^2-9}{x+3}$; $f(-2), f(0), f(5)$
2. $f(x) = \frac{x^2-16}{x+4}$; $f(-2), f(0), f(5)$
3. $f(x) = \frac{x^2-2x-3}{4-x}$; $f(-1), f(4), f(6)$
4. $f(x) = \frac{x^2-3x-4}{3-x}$; $f(-1), f(3), f(5)$
5. $g(t) = \frac{2t^3-5}{t^2+1}$; $g(-1), g(0), g(2)$
6. $g(t) = \frac{2t^3-1}{t^2+4}$; $g(-1), g(0), g(2)$

In Exercises 7–16, find the domain of the given rational function.

7. $f(x) = \frac{x-2}{x-5}$
8. $f(x) = \frac{x-3}{x-6}$
9. $f(x) = \frac{x-4}{(x-1)(x+3)}$
10. $f(x) = \frac{x-5}{(x-2)(x+4)}$
11. $f(x) = \frac{2x}{(x+5)^2}$
12. $f(x) = \frac{2x}{(x+7)^2}$
13. $f(x) = \frac{3x}{x^2-8x+15}$
14. $f(x) = \frac{3x}{x^2-13x+36}$
15. $f(x) = \frac{(x-1)^2}{3x^2-2x-8}$
16. $f(x) = \frac{(x-1)^2}{4x^2-13x+3}$

The graph of a rational function, f, is shown in the figure. Use the graph to solve Exercises 17–26.

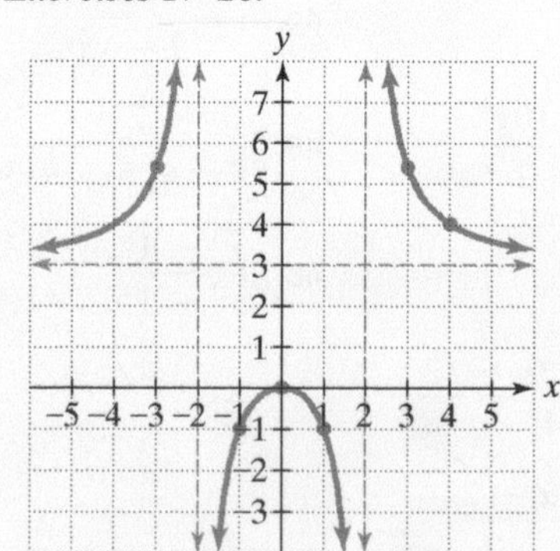

(In Exercises 17–26, use the graph on the previous page.)

17. Find $f(4)$.

18. Find $f(1)$.

19. What is the domain of f? What is the range of f?

20. What are the equations of the vertical asymptotes of the graph of f?

21. Describe the end behavior of the graph at the far left. What is the equation of the horizontal asymptote?

22. Describe the end behavior of the graph at the far right. What is the equation of the horizontal asymptote?

23. Explain how the graph shows that $f(-2)$ does not exist.

24. Explain how the graph shows that $f(2)$ does not exist.

25. How can you tell that this is not the graph of a polynomial function?

26. List two real numbers that are not function values of f.

In Exercises 27–50, simplify each rational expression. If the rational expression cannot be simplified, so state.

27. $\frac{x^2 - 4}{x - 2}$

28. $\frac{x^2 - 25}{x - 5}$

29. $\frac{x + 2}{x^2 - x - 6}$

30. $\frac{x + 1}{x^2 - 2x - 3}$

31. $\frac{4x + 20}{x^2 + 5x}$

32. $\frac{5x + 30}{x^2 + 6x}$

33. $\frac{4y - 20}{y^2 - 25}$

34. $\frac{6y - 42}{y^2 - 49}$

35. $\frac{3x - 5}{25 - 9x^2}$

36. $\frac{5x - 2}{4 - 25x^2}$

37. $\frac{y^2 - 49}{y^2 - 14y + 49}$

38. $\frac{y^2 - 9}{y^2 - 6y + 9}$

39. $\frac{x^2 + 7x - 18}{x^2 - 3x + 2}$

40. $\frac{x^2 - 4x - 5}{x^2 + 5x + 4}$

41. $\frac{3x + 7}{3x + 10}$

42. $\frac{2x + 3}{2x + 5}$

43. $\frac{x^2 - x - 12}{16 - x^2}$

44. $\frac{x^2 - 7x + 12}{9 - x^2}$

45. $\frac{x^2 + 3xy - 10y^2}{3x^2 - 7xy + 2y^2}$

46. $\frac{x^2 + 2xy - 3y^2}{2x^2 + 5xy - 3y^2}$

47. $\frac{x^3 - 8}{x^2 - 4}$

48. $\frac{x^3 - 1}{x^2 - 1}$

49. $\frac{x^3 + 4x^2 - 3x - 12}{x + 4}$

50. $\frac{x^3 - 2x^2 + x - 2}{x - 2}$

In Exercises 51–72, multiply as indicated.

51. $\frac{x - 3}{x + 7} \cdot \frac{3x + 21}{2x - 6}$

52. $\frac{x - 2}{x + 3} \cdot \frac{2x + 6}{5x - 10}$

53. $\frac{x^2 - 49}{x^2 - 4x - 21} \cdot \frac{x + 3}{x}$

54. $\frac{x^2 - 25}{x^2 - 3x - 10} \cdot \frac{x + 2}{x}$

55. $\frac{x^2 - 9}{x^2 - x - 6} \cdot \frac{x^2 + 5x + 6}{x^2 + x - 6}$

56. $\frac{x^2 - 1}{x^2 - 4} \cdot \frac{x^2 - 5x + 6}{x^2 - 2x - 3}$

57. $\frac{x^2 + 4x + 4}{x^2 + 8x + 16} \cdot \frac{(x + 4)^3}{(x + 2)^3}$

58. $\frac{x^2 - 2x + 1}{x^2 - 4x + 4} \cdot \frac{(x - 2)^3}{(x - 1)^3}$

59. $\frac{8y + 2}{y^2 - 9} \cdot \frac{3 - y}{4y^2 + y}$

60. $\frac{6y + 2}{y^2 - 1} \cdot \frac{1 - y}{3y^2 + y}$

61. $\frac{y^3 - 8}{y^2 - 4} \cdot \frac{y + 2}{2y}$

62. $\frac{y^2 + 6y + 9}{y^3 + 27} \cdot \frac{1}{y + 3}$

63. $(x - 3) \cdot \frac{x^2 + x + 1}{x^2 - 5x + 6}$

64. $(x + 1) \cdot \frac{x + 2}{x^2 + 7x + 6}$

65. $\frac{x^2 + xy}{x^2 - y^2} \cdot \frac{4x - 4y}{x}$

66. $\frac{x^2 - y^2}{x} \cdot \frac{x^2 + xy}{x + y}$

67. $\frac{x^2 + 2xy + y^2}{x^2 - 2xy + y^2} \cdot \frac{4x - 4y}{3x + 3y}$

68. $\frac{2x^2 - 3xy - 2y^2}{3x^2 - 4xy + y^2} \cdot \frac{3x^2 - 2xy - y^2}{x^2 + xy - 6y^2}$

69. $\frac{4a^2 + 2ab + b^2}{2a + b} \cdot \frac{4a^2 - b^2}{8a^3 - b^3}$

70. $\frac{27a^3 - 8b^3}{b^2 - b - 6} \cdot \frac{bc - b - 3c + 3}{3ac - 2bc - 3a + 2b}$

71. $\frac{10z^2 + 13z - 3}{3z^2 - 8z + 5} \cdot \frac{2z^2 - 3z - 2z + 3}{25z^2 - 10z + 1} \cdot \frac{15z^2 - 28z + 5}{4z^2 - 9}$

72. $\frac{2z^2 - 2z - 12}{z^2 - 49} \cdot \frac{4z^2 - 1}{2z^2 + 5z + 2} \cdot \frac{2z^2 - 13z - 7}{2z^2 - 7z + 3}$

In Exercises 73–90, divide as indicated.

73. $\frac{x + 5}{7} \div \frac{4x + 20}{9}$

74. $\frac{x + 1}{3} \div \frac{3x + 3}{7}$

75. $\frac{4}{y - 6} \div \frac{40}{7y - 42}$

76. $\frac{7}{y - 5} \div \frac{28}{3y - 15}$

77. $\frac{x^2 - 2x}{15} \div \frac{x - 2}{5}$

78. $\frac{x^2 - x}{15} \div \frac{x - 1}{5}$

79. $\dfrac{y^2-25}{2y-2} \div \dfrac{y^2+10y+25}{y^2+4y-5}$

80. $\dfrac{y^2+y}{y^2-4} \div \dfrac{y^2-1}{y^2+5y+6}$

81. $(x^2-16) \div \dfrac{x^2+3x-4}{x^2+4}$

82. $(x^2+4x-5) \div \dfrac{x^2-25}{x+7}$

83. $\dfrac{y^2-4y-21}{y^2-10y+25} \div \dfrac{y^2+2y-3}{y^2-6y+5}$

84. $\dfrac{y^2+4y-21}{y^2+3y-28} \div \dfrac{y^2+14y+48}{y^2+4y-32}$

85. $\dfrac{8x^3-1}{4x^2+2x+1} \div \dfrac{x-1}{(x-1)^2}$

86. $\dfrac{x^2-9}{x^3-27} \div \dfrac{x^2+6x+9}{x^2+3x+9}$

87. $\dfrac{x^2-4y^2}{x^2+3xy+2y^2} \div \dfrac{x^2-4xy+4y^2}{x+y}$

88. $\dfrac{xy-y^2}{x^2+2x+1} \div \dfrac{2x^2+xy-3y^2}{2x^2+5xy+3y^2}$

89. $\dfrac{x^4-y^8}{x^2+y^4} \div \dfrac{x^2-y^4}{3x^2}$

90. $\dfrac{(x-y)^3}{x^3-y^3} \div \dfrac{x^2-2xy+y^2}{x^2-y^2}$

Practice PLUS

In Exercises 91–98, perform the indicated operation or operations.

91. $\dfrac{x^3-4x^2+x-4}{2x^3-8x^2+x-4} \cdot \dfrac{2x^3+2x^2+x+1}{x^4-x^3+x^2-x}$

92. $\dfrac{y^3+y^2+yz^2+z^2}{y^3+y+y^2+1} \cdot \dfrac{y^3+y+y^2z+z}{2y^2+2yz-yz^2-z^3}$

93. $\dfrac{ax-ay+3x-3y}{x^3+y^3} \div \dfrac{ab+3b+ac+3c}{xy-x^2-y^2}$

94. $\dfrac{a^3+b^3}{ac-ad-bc+bd} \div \dfrac{ab-a^2-b^2}{ac-ad+bc-bd}$

95. $\dfrac{a^2b+b}{3a^2-4a-20} \cdot \dfrac{a^2+5a}{2a^2+11a+5} \div \dfrac{ab^2}{6a^2-17a-10}$

96. $\dfrac{a^2-8a+15}{2a^3-10a^2} \cdot \dfrac{2a^2+3a}{3a^3-27a} \div \dfrac{14a+21}{a^2-6a-27}$

97. $\dfrac{a-b}{4c} \div \left(\dfrac{b-a}{c} \div \dfrac{a-b}{c^2}\right)$

98. $\left(\dfrac{a-b}{4c} \div \dfrac{b-a}{c}\right) \div \dfrac{a-b}{c^2}$

In Exercises 99–102, find $\dfrac{f(a+h)-f(a)}{h}$ *and simplify.*

99. $f(x) = 7x - 4$

100. $f(x) = -3x + 5$

101. $f(x) = x^2 - 5x + 3$

102. $f(x) = 3x^2 - 4x + 7$

In Exercises 103–104, let

$$f(x) = \frac{(x+2)^2}{1-2x} \quad \text{and} \quad g(x) = \frac{x+2}{2x-1}.$$

103. Find $\left(\dfrac{f}{g}\right)(x)$ and the domain of $\dfrac{f}{g}$.

104. Find $\left(\dfrac{g}{f}\right)(x)$ and the domain of $\dfrac{g}{f}$.

Application Exercises

The rational function

$$f(x) = \frac{130x}{100-x}$$

models the cost, $f(x)$, in millions of dollars, to inoculate x% of the population against a particular strain of flu. The graph of the rational function is shown. Use the function's equation to solve Exercises 105–108.

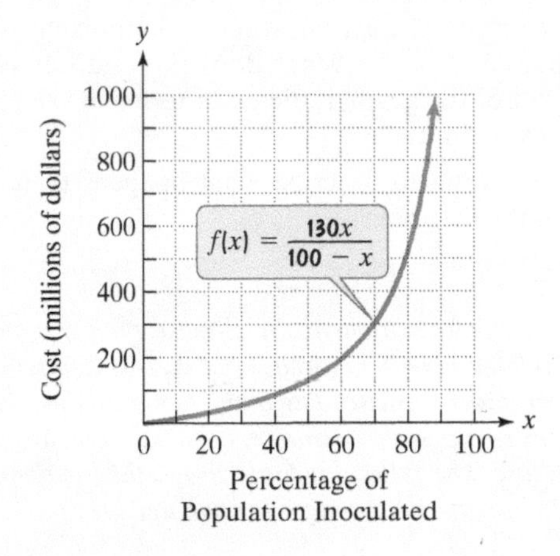

105. Find and interpret $f(60)$. Identify your solution as a point on the graph.

106. Find and interpret $f(80)$. Identify your solution as a point on the graph.

107. What value of x must be excluded from the rational function's domain? Does the cost model indicate that we can inoculate all of the population against the flu? Explain.

108. What happens to the cost as x approaches 100%? How is this shown by the graph? Explain what this means.

The function

$$f(x) = \frac{6.5x^2 - 20.4x + 234}{x^2 + 36}$$

models the pH level, $f(x)$, of the human mouth x minutes after a person eats food containing sugar. The graph of this function is shown in the figure. Use the function's equation or graph, as specified, to solve Exercises 109–112.

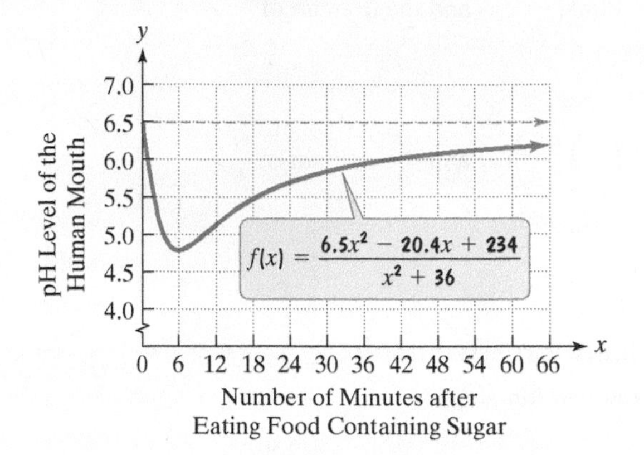

109. After eating sugar, when is the pH level the lowest? Use the function's equation to determine the pH level, to the nearest tenth, at this time.

110. Use the graph to obtain a reasonable estimate, to the nearest tenth, of the pH level of the human mouth 42 minutes after a person eats food containing sugar.

111. According to the graph, what is the normal pH level of the human mouth? What does the end behavior at the far right of the graph indicate in terms of the mouth's pH level over time?

112. Use the graph to describe what happens to the pH level during the first hour.

Among all deaths from a particular disease, the percentage that is smoking related (21–39 cigarettes per day) is a function of the disease's ***incidence ratio****. The incidence ratio describes the number of times more likely smokers are than nonsmokers to die from the disease. The following table shows the incidence ratios for heart disease and lung cancer for two age groups.*

Incidence Ratios

	Heart Disease	Lung Cancer
Ages 55–64	1.9	10
Ages 65–74	1.7	9

Source: Alexander M. Walker, *Observation and Inference*, Epidemiology Resources Inc., 1991.

For example, the incidence ratio of 9 in the table means that smokers between the ages of 65 and 74 are 9 times more likely than nonsmokers in the same age group to die from lung cancer.

The rational function

$$P(x) = \frac{100(x - 1)}{x}$$

models the percentage of smoking-related deaths among all deaths from a disease, $P(x)$, in terms of the disease's incidence ratio, x. The graph of the rational function is shown. Use this function to solve Exercises 113–116.

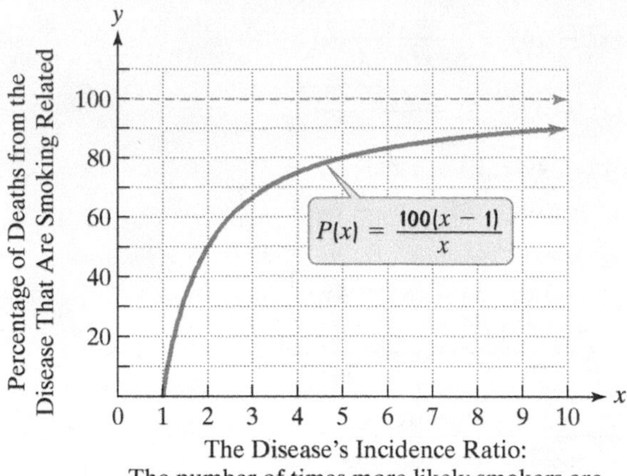

113. Find $P(10)$. Describe what this means in terms of the incidence ratio, 10, given in the table. Identify your solution as a point on the graph.

114. Find $P(9)$. Round to the nearest percent. Describe what this means in terms of the incidence ratio, 9, given in the table. Identify your solution as a point on the graph.

115. What is the horizontal asymptote of the graph? Describe what this means about the percentage of deaths caused by smoking with increasing incidence ratios.

116. According to the model and its graph, is there a disease for which all deaths are caused by smoking? Explain your answer.

Explaining the Concepts

117. What is a rational expression? Give an example with your explanation.

118. What is a rational function? Provide an example.

119. What is the domain of a rational function?

120. If you are given the equation of a rational function, explain how to determine its domain.

121. Describe two ways the graph of a rational function differs from the graph of a polynomial function.

122. What is a vertical asymptote?

123. What is a horizontal asymptote?

124. Explain how to simplify a rational expression.

125. Explain how to simplify a rational expression with opposite factors in the numerator and the denominator.

126. Explain how to multiply rational expressions.

127. Explain how to divide rational expressions.

128. Although your friend has a family history of heart disease, he smokes, on average, 25 cigarettes per day. He sees the table showing incidence ratios for heart disease (see Exercises 113–116) and feels comfortable that they are less than 2, compared to 9 and 10 for lung cancer. He claims that all family deaths have been from heart disease and decides not to give up smoking. Use the given function and its graph to describe some additional information not given in the table that might influence his decision.

Technology Exercises

In Exercises 129–132, determine if the multiplication or division has been performed correctly by graphing the function on each side of the equation in the same viewing rectangle. If the graphs do not coincide, correct the expression on the right side and then verify your correction using the graphing utility.

129. $\dfrac{x^2+x}{3x}\cdot\dfrac{6x}{x+1}=2x$

130. $\dfrac{x^3-25x}{x^2-3x-10}\cdot\dfrac{x+2}{x}=x+5$

131. $\dfrac{x^2-9}{x+4}\div\dfrac{x-3}{x+4}=x-3$

132. $(x-5)\div\dfrac{2x^2-11x+5}{4x^2-1}=2x-1$

133. Use the TABLE feature of a graphing utility to verify the domains that you determined for any two functions in Exercises 7–16.

134. **a.** Graph $f(x)=\dfrac{x^2-x-2}{x-2}$ and $g(x)=x+1$ in the same viewing rectangle. What do you observe?

b. Simplify the formula in the definition of function f. Do f and g represent exactly the same function? Explain.

c. Display the graphs of f and g separately in the utility's viewing rectangle. TRACE along each of the curves until you get to $x=2$. What difference do you observe? What does this mean?

Critical Thinking Exercises

Make Sense? *In Exercises 135–138, determine whether each statement makes sense or does not make sense, and explain your reasoning.*

135. I cannot simplify, multiply, or divide rational expressions without knowing how to factor polynomials.

136. I simplified $\dfrac{2(x+2)-5(x+1)}{(x+2)(x+1)}$ by dividing the numerator and the denominator by $(x+2)(x+1)$.

137. The values to exclude from the domain of $f(x)=\dfrac{x-3}{x-7}$ are 3 and 7.

138. When performing the division

$$\frac{3x}{x+2}\div\frac{(x+2)^2}{x-4},$$

I began by dividing the numerator and the denominator by the common factor, $x+2$.

In Exercises 139–142, determine whether each statement is true or false. If the statement is false, make the necessary change(s) to produce a true statement.

139. $\dfrac{x^2-25}{x-5}=x-5$

140. $\dfrac{x^2+7}{7}=x^2+1$

141. The domain of $f(x)=\dfrac{7}{x(x-3)+5(x-3)}$ is $(-\infty,3)\cup(3,\infty)$.

142. The restrictions on the values of x when performing the division

$$\frac{f(x)}{g(x)}\div\frac{h(x)}{k(x)}$$

are $g(x)\neq 0$, $k(x)\neq 0$, and $h(x)\neq 0$.

143. Graph: $f(x)=\dfrac{x^2-x-2}{x-2}$.

144. Simplify: $\dfrac{6x^{3n}+6x^{2n}y^n}{x^{2n}-y^{2n}}$.

145. Divide: $\dfrac{y^{2n}-1}{y^{2n}+3y^n+2}\div\dfrac{y^{2n}+y^n-12}{y^{2n}-y^n-6}$.

146. Solve for x in terms of a and write the resulting rational expression in simplified form. Include any necessary restrictions on a.

$$a^2(x-1)=4(x-1)+5a+10$$

Review Exercises

147. Graph: $4x-5y\geq 20$. (Section 4.4, Example 1)

148. Multiply: $(2x-5)(x^2-3x-6)$. (Section 5.2, Example 3)

149. Simplify: $\left(\dfrac{ab^{-3}c^{-4}}{4a^5b^{10}c^{-3}}\right)^{-2}$. (Section 1.6, Example 9)

Preview Exercises

Exercises 150–152 will help you prepare for the material covered in the next section. In each exercise, perform the indicated operation. Where possible, reduce the answer to its lowest terms.

150. $\dfrac{7}{10}-\dfrac{3}{10}$

151. $\dfrac{1}{2}+\dfrac{2}{3}$

152. $\dfrac{7}{15}-\dfrac{3}{10}$

SECTION

6.2 Adding and Subtracting Rational Expressions

What am I supposed to learn?

After studying this section, you should be able to:

1. Add rational expressions with the same denominator.
2. Subtract rational expressions with the same denominator.
3. Find the least common denominator.
4. Add and subtract rational expressions with different denominators.
5. Add and subtract rational expressions with opposite denominators.

Did you know that people in California are at far greater risk of injury or death from drunk drivers than from earthquakes? According to the U.S. Bureau of Justice Statistics, half the arrests for driving under the influence of alcohol involve drivers ages 25 through 34. The rational function

$$f(x) = \frac{27{,}725(x - 14)}{x^2 + 9} - 5x$$

models the number of arrests for driving under the influence, $f(x)$, per 100,000 drivers, as a function of a driver's age, x.

The formula for function f involves the subtraction of two expressions. It is possible to perform this subtraction and express the function's formula as a single rational expression. In this section, we will draw on your experience from arithmetic to add and subtract rational expressions. We return to the driving-under-the-influence model and its graph in Exercise 83 in the section's Exercise Set.

1 Add rational expressions with the same denominator.

Addition and Subtraction When Denominators Are the Same

To add rational numbers having the same denominators, such as $\frac{2}{9}$ and $\frac{5}{9}$, we add the numerators and place the sum over the common denominator:

$$\frac{2}{9} + \frac{5}{9} = \frac{2+5}{9} = \frac{7}{9}.$$

We add rational expressions with the same denominator in an identical manner.

Adding Rational Expressions with Common Denominators

If $\frac{P}{R}$ and $\frac{Q}{R}$ are rational expressions, then

$$\frac{P}{R} + \frac{Q}{R} = \frac{P+Q}{R}.$$

To add rational expressions with the same denominator, add numerators and place the sum over the common denominator. If possible, factor and simplify the result.

EXAMPLE 1 Adding Rational Expressions When Denominators Are the Same

Add: $\frac{x^2 + 2x - 2}{x^2 + 3x - 10} + \frac{5x + 12}{x^2 + 3x - 10}$.

Solution

$$\frac{x^2 + 2x - 2}{x^2 + 3x - 10} + \frac{5x + 12}{x^2 + 3x - 10}$$

$$= \frac{x^2 + 2x - 2 + 5x + 12}{x^2 + 3x - 10}$$

Add numerators. Place this sum over the common denominator.

$$= \frac{x^2 + 7x + 10}{x^2 + 3x - 10}$$

Combine like terms: $2x + 5x = 7x$ and $-2 + 12 = 10$.

$$= \frac{(x + 2)\cancel{(x + 5)}}{(x - 2)\cancel{(x + 5)}}$$

Factor and simplify by dividing out the common factor, $x + 5$.

$$= \frac{x + 2}{x - 2} \quad \blacksquare$$

☑ **CHECK POINT 1** Add: $\frac{x^2 - 5x - 15}{x^2 + 5x + 6} + \frac{2x + 5}{x^2 + 5x + 6}$.

The following box shows how to subtract rational expressions with the same denominator:

2 Subtract rational expressions with the same denominator.

Subtracting Rational Expressions with Common Denominators

If $\frac{P}{R}$ and $\frac{Q}{R}$ are rational expressions, then

$$\frac{P}{R} - \frac{Q}{R} = \frac{P - Q}{R}.$$

To subtract rational expressions with the same denominator, subtract numerators and place the difference over the common denominator. If possible, factor and simplify the result.

EXAMPLE 2 Subtracting Rational Expressions When Denominators Are the Same

Subtract: $\frac{3y^3 - 5x^3}{x^2 - y^2} - \frac{4y^3 - 6x^3}{x^2 - y^2}$.

Solution

$$\frac{3y^3 - 5x^3}{x^2 - y^2} - \frac{4y^3 - 6x^3}{x^2 - y^2} = \frac{3y^3 - 5x^3 - (4y^3 - 6x^3)}{x^2 - y^2}$$

Subtract numerators and include parentheses to indicate that both terms are subtracted. Place this difference over the common denominator.

$$= \frac{3y^3 - 5x^3 - 4y^3 + 6x^3}{x^2 - y^2}$$

Remove parentheses and then distribute the minus to change the sign of each term.

$$= \frac{x^3 - y^3}{x^2 - y^2}$$

Combine like terms: $-5x^3 + 6x^3 = x^3$ and $3y^3 - 4y^3 = -y^3$.

$$= \frac{\cancel{(x - y)}(x^2 + xy + y^2)}{(x + y)\cancel{(x - y)}}$$

Factor and simplify by dividing out the common factor, $x - y$.

$$= \frac{x^2 + xy + y^2}{x + y} \quad \blacksquare$$

Great Question!

When subtracting a numerator containing more than one term, do I just subtract the first term?

No. When a numerator is being subtracted, be sure to **subtract every term in that expression.**

The − sign applies to the entire numerator, $4y^3 - 6x^3$.

Insert parentheses to indicate this.

The sign of every term of $4y^3 - 6x^3$ changes.

$$\frac{3y^3 - 5x^3}{x^2 - y^2} - \frac{4y^3 - 6x^3}{x^2 - y^2} = \frac{3y^3 - 5x^3 - (4y^3 - 6x^3)}{x^2 - y^2} = \frac{3y^3 - 5x^3 - 4y^3 + 6x^3}{x^2 - y^2}$$

The entire numerator of the second rational expression must be subtracted. Avoid the common error of subtracting only the first term.

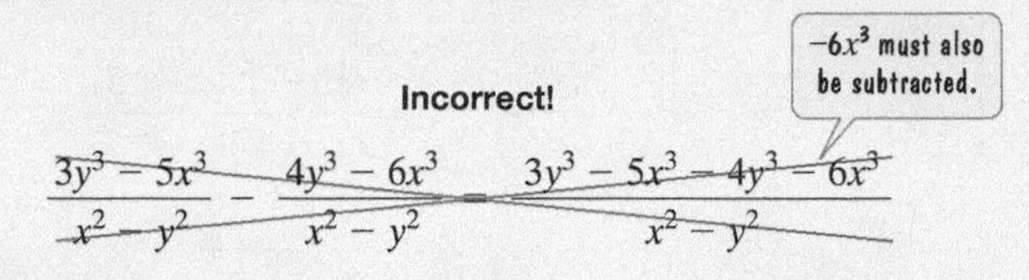

✓ CHECK POINT 2 Subtract: $\frac{5x - y}{x^2 - y^2} - \frac{4x - 2y}{x^2 - y^2}$.

3 Find the least common denominator.

Finding the Least Common Denominator

We can gain insight into adding rational expressions with different denominators by looking closely at what we do when adding fractions with different denominators. For example, suppose that we want to add $\frac{1}{2}$ and $\frac{2}{3}$. We must first write the fractions with the same denominator. We look for the smallest number that contains both 2 and 3 as factors. This number, 6, is then used as the *least common denominator*, or LCD.

The **least common denominator** of several rational expressions is a polynomial consisting of the product of all prime factors in the denominators, with each factor raised to the greatest power of its occurrence in any denominator.

Finding the Least Common Denominator

1. Factor each denominator completely.
2. List the factors of the first denominator.
3. Add to the list in step 2 any factors of the second denominator that do not appear in the list.
4. Form the product of each different factor from the list in step 3. This product is the least common denominator.

EXAMPLE 3 Finding the Least Common Denominator

Find the LCD of

$$\frac{3}{10x^2} \text{ and } \frac{7}{15x}.$$

Solution

Step 1. Factor each denominator completely.

$$10x^2 = 5 \cdot 2x^2 \quad (\text{or } 5 \cdot 2 \cdot x \cdot x)$$
$$15x = 5 \cdot 3x$$

Great Question!

What's the relationship between the powers in the factored denominators and the LCD?

After factoring denominators completely, the LCD can be determined by taking each factor to the highest power it appears in any factorization. The *Mathematics Teacher* magazine accused the LCD of keeping up with the Joneses. The LCD wants everything (all of the factors) the other denominators have.

Step 2. List the factors of the first denominator.

$$5, 2, x^2 \quad (\text{or } 5, 2, x, x)$$

Step 3. Add any unlisted factors from the second denominator. Two factors from $5 \cdot 3x$ are already in our list. These factors include 5 and x. We add the unlisted factor, 3, to our list. We have

$$3, 5, 2, x^2.$$

Step 4. The least common denominator is the product of all factors in the final list. Thus,

$$3 \cdot 5 \cdot 2 \cdot x^2,$$

or $30x^2$, is the least common denominator. ■

✓ CHECK POINT 3 Find the LCD of

$$\frac{7}{6x^2} \text{ and } \frac{2}{9x}.$$

EXAMPLE 4 Finding the Least Common Denominator

Find the LCD of

$$\frac{9}{7x^2 + 28x} \text{ and } \frac{11}{x^2 + 8x + 16}.$$

Solution

Step 1. Factor each denominator completely.

$$7x^2 + 28x = 7x(x + 4)$$
$$x^2 + 8x + 16 = (x + 4)^2$$

Step 2. List the factors of the first denominator.

$$7, x, (x + 4)$$

Step 3. Add any unlisted factors from the second denominator. The second denominator is $(x + 4)^2$, or $(x + 4)(x + 4)$. One factor of $x + 4$ is already in our list, but the other factor is not. We add a second factor of $x + 4$ to the list. We have

$$7, x, (x + 4), (x + 4).$$

Step 4. The least common denominator is the product of all factors in the final list. Thus,

$$7x(x + 4)(x + 4), \text{ or } 7x(x + 4)^2,$$

is the least common denominator. ■

✓ CHECK POINT 4 Find the LCD of

$$\frac{7}{5x^2 + 15x} \text{ and } \frac{9}{x^2 + 6x + 9}.$$

---

4 Add and subtract rational expressions with different denominators.

Addition and Subtraction When Denominators Are Different

Finding the least common denominator for two (or more) rational expressions is the first step needed to add or subtract the expressions. For example, to add $\frac{1}{2}$ and $\frac{2}{3}$, we first determine that the LCD is 6. Then we write each fraction in terms of the LCD.

$$\frac{1}{2}+\frac{2}{3}=\frac{1}{2}\cdot\frac{3}{3}+\frac{2}{3}\cdot\frac{2}{2}$$

Multiply the numerator and denominator of each fraction by whatever extra factors are required to form 6, the LCD.

$\frac{3}{3}=1$ and $\frac{2}{2}=1$. Multiplying by 1 does not change a fraction's value.

$$=\frac{3}{6}+\frac{4}{6}$$

Perform the required multiplications.

$$=\frac{3+4}{6}=\frac{7}{6}$$

Add numerators. Place this sum over the LCD. Simplify.

We follow the same steps in adding or subtracting rational expressions with different denominators.

Adding and Subtracting Rational Expressions That Have Different Denominators

1. Find the LCD of the rational expressions.
2. Rewrite each rational expression as an equivalent expression whose denominator is the LCD. To do so, multiply the numerator and the denominator of each rational expression by any factor(s) needed to convert the denominator into the LCD.
3. Add or subtract numerators, placing the resulting expression over the LCD.
4. If possible, simplify the resulting rational expression.

EXAMPLE 5 Adding Rational Expressions with Different Denominators

Add: $\frac{3}{10x^2}+\frac{7}{15x}$.

Solution

Step 1. Find the least common denominator. In Example 3, we found that the LCD for these rational expressions is $30x^2$.

Step 2. Write equivalent expressions with the LCD as denominators. We must rewrite each rational expression with a denominator of $30x^2$.

$$\frac{3}{10x^2}\cdot\frac{3}{3}=\frac{9}{30x^2} \qquad \frac{7}{15x}\cdot\frac{2x}{2x}=\frac{14x}{30x^2}$$

Multiply the numerator and denominator by 3 to get $30x^2$, the LCD.

Multiply the numerator and denominator by $2x$ to get $30x^2$, the LCD.

Because $\frac{3}{3}=1$ and $\frac{2x}{2x}=1$, we are not changing the value of either rational expression, only its appearance. In summary, we have

$$\frac{3}{10x^2}+\frac{7}{15x}$$

The LCD is $30x^2$.

$$=\frac{3}{10x^2}\cdot\frac{3}{3}+\frac{7}{15x}\cdot\frac{2x}{2x}$$

Write equivalent expressions with the LCD.

$$=\frac{9}{30x^2}+\frac{14x}{30x^2}.$$

Perform the required multiplications.

Steps 3 and 4. Add numerators, putting this sum over the LCD. Simplify, if possible.

$$=\frac{9+14x}{30x^2} \text{ or } \frac{14x+9}{30x^2}$$

The numerator is prime and further simplification is not possible. ■

Great Question!

Can't I just simplify things and add rational expressions by adding numerators and adding denominators?

No. It is incorrect to add rational expressions by adding numerators and adding denominators. Avoid this common error.

Incorrect!

$$\frac{3}{10x^2}+\frac{7}{15x}=\frac{3+7}{10x^2+15x}=\frac{10}{10x^2+15x}$$

✓ **CHECK POINT 5** Add: $\frac{7}{6x^2} + \frac{2}{9x}$.

EXAMPLE 6 Adding Rational Expressions with Different Denominators

Add: $\frac{x}{x-3} + \frac{x-1}{x+3}$.

Solution

Step 1. Find the least common denominator. Begin by factoring the denominators.

$$x - 3 = 1(x - 3)$$
$$x + 3 = 1(x + 3)$$

The factors of the first denominator are 1 and $x - 3$. The only factor from the second denominator that is unlisted is $x + 3$. Thus, the least common denominator is $1(x - 3)(x + 3)$, or $(x - 3)(x + 3)$.

Step 2. Write equivalent expressions with the LCD as denominators.

$$\frac{x}{x-3} + \frac{x-1}{x+3}$$

$$= \frac{x(x+3)}{(x-3)(x+3)} + \frac{(x-1)(x-3)}{(x-3)(x+3)}$$

Multiply each numerator and denominator by the extra factor required to form $(x - 3)(x + 3)$, the LCD.

Steps 3 and 4. Add numerators, putting this sum over the LCD. Simplify, if possible.

$$= \frac{x(x+3) + (x-1)(x-3)}{(x-3)(x+3)}$$

$$= \frac{x^2 + 3x + x^2 - 4x + 3}{(x-3)(x+3)}$$

Perform the multiplications using the distributive property and FOIL.

$$= \frac{2x^2 - x + 3}{(x-3)(x+3)}$$

Combine like terms: $x^2 + x^2 = 2x^2$ and $3x - 4x = -x$.

The numerator is prime and further simplification is not possible. ■

✓ **CHECK POINT 6** Add: $\frac{x}{x-4} + \frac{x-2}{x+4}$.

EXAMPLE 7 Subtracting Rational Expressions with Different Denominators

Subtract: $\frac{x-1}{x^2+x-6} - \frac{x-2}{x^2+4x+3}$.

Solution

Step 1. Find the least common denominator. Begin by factoring the denominators.

$$x^2 + x - 6 = (x + 3)(x - 2)$$
$$x^2 + 4x + 3 = (x + 3)(x + 1)$$

The factors of the first denominator are $x + 3$ and $x - 2$. The only factor from the second denominator that is unlisted is $x + 1$. Thus, the least common denominator is $(x + 3)(x - 2)(x + 1)$.

Step 2. Write equivalent expressions with the LCD as denominators.

$$\frac{x-1}{x^2+x-6} - \frac{x-2}{x^2+4x+3}$$

$$= \frac{x-1}{(x+3)(x-2)} - \frac{x-2}{(x+3)(x+1)}$$

Factor denominators. The LCD is $(x+3)(x-2)(x+1)$.

$$= \frac{(x-1)(x+1)}{(x+3)(x-2)(x+1)} - \frac{(x-2)(x-2)}{(x+3)(x-2)(x+1)}$$

Multiply each numerator and denominator by the extra factor required to form $(x+3)(x-2)(x+1)$, the LCD.

Steps 3 and 4. Subtract numerators, putting this difference over the LCD. Simplify, if possible.

$$= \frac{(x-1)(x+1)-(x-2)(x-2)}{(x+3)(x-2)(x+1)}$$

$$= \frac{x^2-1-(x^2-4x+4)}{(x+3)(x-2)(x+1)}$$

Perform the multiplications in the numerator. Don't forget the parentheses.

$$= \frac{x^2-1-x^2+4x-4}{(x+3)(x-2)(x+1)}$$

Remove parentheses and change the sign of each term in parentheses.

$$= \frac{4x-5}{(x+3)(x-2)(x+1)}$$

Combine like terms.

The numerator is prime and further simplification is not possible. ■

✓ CHECK POINT 7 Subtract: $\frac{2x-3}{x^2-5x+6} - \frac{x+4}{x^2-2x-3}$.

EXAMPLE 8 Adding and Subtracting Rational Expressions with Different Denominators

Perform the indicated operations:

$$\frac{3y+2}{y-5} + \frac{4}{3y+4} - \frac{7y^2+24y+28}{3y^2-11y-20}.$$

Solution

Step 1. Find the least common denominator. Begin by factoring the denominators.

$$y-5 = 1(y-5)$$
$$3y+4 = 1(3y+4)$$
$$3y^2-11y-20 = (3y+4)(y-5)$$

The factors of the first denominator are 1 and $y-5$. The only factor from the second denominator that is unlisted is $3y+4$. Adding this factor to our list, we have 1, $y-5$, and $3y+4$. We have listed all factors from the third denominator. Thus, the least common denominator is $1(y-5)(3y+4)$, or $(y-5)(3y+4)$.

Step 2. Write equivalent expressions with the LCD as denominators.

$$\frac{3y+2}{y-5} + \frac{4}{3y+4} - \frac{7y^2+24y+28}{3y^2-11y-20}$$

$$= \frac{3y+2}{y-5} + \frac{4}{3y+4} - \frac{7y^2+24y+28}{(3y+4)(y-5)}$$

Factor denominators. The LCD is $(y-5)(3y+4)$.

$$= \frac{(3y+2)(3y+4)}{(y-5)(3y+4)} + \frac{4(y-5)}{(y-5)(3y+4)} - \frac{7y^2+24y+28}{(3y+4)(y-5)}$$

Multiply the first two numerators and denominators by the extra factor required to form the LCD.

Steps 3 and 4. Add and subtract numerators, putting this result over the LCD. Simplify, if possible.

$$= \frac{(3y+2)(3y+4) + 4(y-5) - (7y^2 + 24y + 28)}{(y-5)(3y+4)}$$

$$= \frac{9y^2 + 18y + 8 + 4y - 20 - 7y^2 - 24y - 28}{(y-5)(3y+4)}$$

Perform multiplications. Remove parentheses and change the sign of each term.

$$= \frac{2y^2 - 2y - 40}{(y-5)(3y+4)}$$

Combine like terms: $9y^2 - 7y^2 = 2y^2$, $18y + 4y - 24y = -2y$, and $8 - 20 - 28 = -40$.

$$= \frac{2(y^2 - y - 20)}{(y-5)(3y+4)}$$

Factor out the GCF in the numerator.

$$= \frac{2(y+4)\cancel{(y-5)}}{\cancel{(y-5)}(3y+4)}$$

Factor completely and simplify.

$$= \frac{2(y+4)}{3y+4} \quad ■$$

CHECK POINT 8 Perform the indicated operations:

$$\frac{y-1}{y-2} + \frac{y-6}{y^2-4} - \frac{y+1}{y+2}.$$

---

5 Add and subtract rational expressions with opposite denominators.

In some situations, we need to add or subtract rational expressions with denominators that are opposites, or additive inverses. Multiply the numerator and the denominator of either of the rational expressions by -1. Then they will have the same denominators.

EXAMPLE 9 Adding Rational Expressions When Denominators Are Opposites

Add: $\dfrac{4x - 16y}{x - 5y} + \dfrac{x - 6y}{5y - x}$.

Solution

$$\frac{4x - 16y}{x - 5y} + \frac{x - 6y}{5y - x}$$

The denominators, $x - 5y$ and $5y - x$, are opposites, or additive inverses.

$$= \frac{4x - 16y}{x - 5y} + \frac{(-1)}{(-1)} \cdot \frac{x - 6y}{5y - x}$$

Multiply the numerator and denominator of the second rational expression by -1.

$$= \frac{4x - 16y}{x - 5y} + \frac{-x + 6y}{-5y + x}$$

Perform the multiplications by -1.

$$= \frac{4x - 16y}{x - 5y} + \frac{-x + 6y}{x - 5y}$$

Rewrite $-5y + x$ as $x - 5y$. Both rational expressions have the same denominator.

$$= \frac{4x - 16y - x + 6y}{x - 5y}$$

Add numerators. Place this sum over the common denominator.

$$= \frac{3x - 10y}{x - 5y}$$

Combine like terms. ■

CHECK POINT 9 Add: $\dfrac{4x - 7y}{x - 3y} + \dfrac{x - 2y}{3y - x}$.

CONCEPT AND VOCABULARY CHECK

Fill in each blank so that the resulting statement is true.

1. $\frac{P}{R} + \frac{Q}{R} =$ ________: To add rational expressions with the same denominator, add ________ and place the sum over the ________.

2. $\frac{P}{R} - \frac{Q}{R} =$ ________: To subtract rational expressions with the same denominator, subtract ________ and place the difference over the ________.

3. $\frac{x}{3} - \frac{5 - y}{3} =$ ________

4. When adding or subtracting rational expressions with denominators that are opposites, or additive inverses, multiply either expression by ________ to obtain a common denominator.

5. The first step in finding the least common denominator of $\frac{7}{5x^2 + 15x}$ and $\frac{9}{x^2 + 6x + 9}$ is to ________.

6. Consider the following subtraction problem:
$$\frac{x - 1}{x^2 + x - 6} - \frac{x - 2}{x^2 + 4x + 3}.$$
The factors of the first denominator are ________. The factors of the second denominator are ________. The LCD is ________.

7. An equivalent expression for $\frac{7}{15x}$ with a denominator of $30x^2$ can be obtained by multiplying the numerator and denominator by ________.

8. An equivalent expression for $\frac{3y + 2}{y - 5}$ with a denominator of $(3y + 4)(y - 5)$ can be obtained by multiplying the numerator and denominator by ________.

9. An equivalent expression for $\frac{x - 2y}{3y - x}$ with a denominator of $x - 3y$ can be obtained by multiplying the numerator and denominator by ________.

6.2 EXERCISE SET

MyMathLab®

Practice Exercises

In Exercises 1–16, perform the indicated operations. These exercises involve addition and subtraction when denominators are the same. Simplify the result, if possible.

1. $\frac{2}{9x} + \frac{4}{9x}$
2. $\frac{11}{6x} + \frac{4}{6x}$
3. $\frac{x}{x - 5} + \frac{9x + 3}{x - 5}$
4. $\frac{x}{x - 3} + \frac{11x + 5}{x - 3}$
5. $\frac{x^2 - 2x}{x^2 + 3x} + \frac{x^2 + x}{x^2 + 3x}$
6. $\frac{x^2 + 7x}{x^2 - 5x} + \frac{x^2 - 4x}{x^2 - 5x}$
7. $\frac{y^2}{y^2 - 9} + \frac{9 - 6y}{y^2 - 9}$
8. $\frac{y^2}{y^2 - 25} + \frac{25 - 10y}{y^2 - 25}$
9. $\frac{3x}{4x - 3} - \frac{2x - 1}{4x - 3}$
10. $\frac{3x}{7x - 4} - \frac{2x - 1}{7x - 4}$
11. $\frac{x^2 - 2}{x^2 + 6x - 7} - \frac{19 - 4x}{x^2 + 6x - 7}$
12. $\frac{x^2 + 6x + 2}{x^2 + x - 6} - \frac{2x - 1}{x^2 + x - 6}$
13. $\frac{20y^2 + 5y + 1}{6y^2 + y - 2} - \frac{8y^2 - 12y - 5}{6y^2 + y - 2}$
14. $\frac{y^2 + 3y - 6}{y^2 - 5y + 4} - \frac{4y - 4 - 2y^2}{y^2 - 5y + 4}$
15. $\frac{2x^3 - 3y^3}{x^2 - y^2} - \frac{x^3 - 2y^3}{x^2 - y^2}$
16. $\frac{4y^3 - 3x^3}{y^2 - x^2} - \frac{3y^3 - 2x^3}{y^2 - x^2}$

In Exercises 17–28, find the least common denominator of the rational expressions.

17. $\frac{11}{25x^2}$ and $\frac{14}{35x}$

18. $\frac{7}{15x^2}$ and $\frac{9}{24x}$

19. $\frac{2}{x-5}$ and $\frac{3}{x^2-25}$

20. $\frac{2}{x+3}$ and $\frac{5}{x^2-9}$

21. $\frac{7}{y^2-100}$ and $\frac{13}{y(y-10)}$

22. $\frac{7}{y^2-4}$ and $\frac{15}{y(y+2)}$

23. $\frac{8}{x^2-16}$ and $\frac{x}{x^2-8x+16}$

24. $\frac{3}{x^2-25}$ and $\frac{x}{x^2-10x+25}$

25. $\frac{7}{y^2-5y-6}$ and $\frac{y}{y^2-4y-5}$

26. $\frac{3}{y^2-y-20}$ and $\frac{y}{2y^2+7y-4}$

27. $\frac{7y}{2y^2+7y+6}, \frac{3}{y^2-4}$, and $\frac{-7y}{2y^2-3y-2}$

28. $\frac{5y}{y^2-9}, \frac{8}{y^2+6y+9}$, and $\frac{-5y}{2y^2+5y-3}$

In Exercises 29–66, perform the indicated operations. These exercises involve addition and subtraction when denominators are different. Simplify the result, if possible.

29. $\frac{3}{5x^2}+\frac{10}{x}$

30. $\frac{7}{2x^2}+\frac{4}{x}$

31. $\frac{4}{x-2}+\frac{3}{x+1}$

32. $\frac{2}{x-3}+\frac{7}{x+2}$

33. $\frac{3x}{x^2+x-2}+\frac{2}{x^2-4x+3}$

34. $\frac{7x}{x^2+2x-8}+\frac{3}{x^2-3x+2}$

35. $\frac{x-6}{x+5}+\frac{x+5}{x-6}$

36. $\frac{x-2}{x+7}+\frac{x+7}{x-2}$

37. $\frac{3x}{x^2-25}-\frac{4}{x+5}$

38. $\frac{8x}{x^2-16}-\frac{5}{x+4}$

39. $\frac{3y+7}{y^2-5y+6}-\frac{3}{y-3}$

40. $\frac{2y+9}{y^2-7y+12}-\frac{2}{y-3}$

41. $\frac{x^2-6}{x^2+9x+18}-\frac{x-4}{x+6}$

42. $\frac{x^2-39}{x^2+3x-10}-\frac{x-7}{x-2}$

43. $\frac{4x+1}{x^2+7x+12}+\frac{2x+3}{x^2+5x+4}$

44. $\frac{3x-2}{x^2-x-6}+\frac{4x-3}{x^2-9}$

45. $\frac{x+4}{x^2-x-2}-\frac{2x+3}{x^2+2x-8}$

46. $\frac{2x+1}{x^2-7x+6}-\frac{x+3}{x^2-5x-6}$

47. $4+\frac{1}{x-3}$

48. $7+\frac{1}{x-5}$

49. $\frac{y-7}{y^2-16}+\frac{7-y}{16-y^2}$

50. $\frac{y-3}{y^2-25}+\frac{y-3}{25-y^2}$

51. $\frac{x+7}{3x+6}+\frac{x}{4-x^2}$

52. $\frac{x+5}{4x+12}+\frac{x}{9-x^2}$

53. $\frac{2x}{x-4}+\frac{64}{x^2-16}-\frac{2x}{x+4}$

54. $\frac{x}{x-3}+\frac{x+2}{x^2-2x-3}-\frac{4}{x+1}$

55. $\frac{5x}{x^2-y^2}-\frac{7}{y-x}$

56. $\frac{9x}{x^2-y^2}-\frac{10}{y-x}$

57. $\frac{3}{5x+6}-\frac{4}{x-2}+\frac{x^2-x}{5x^2-4x-12}$

58. $\frac{x-1}{x^2+2x+1}-\frac{3}{2x-2}+\frac{x}{x^2-1}$

59. $\frac{3x-y}{x^2-9xy+20y^2}+\frac{2y}{x^2-25y^2}$

60. $\frac{x+2y}{x^2+4xy+4y^2}-\frac{2x}{x^2-4y^2}$

61. $\frac{3x}{x^2-4}+\frac{5x}{x^2+x-2}-\frac{3}{x^2-4x+4}$

62. $\frac{1}{x} + \frac{4}{x^2 - 4} - \frac{2}{x^2 - 2x}$

63. $\frac{6a + 5b}{6a^2 + 5ab - 4b^2} - \frac{a + 2b}{9a^2 - 16b^2}$

64. $\frac{5a - b}{a^2 + ab - 2b^2} - \frac{3a + 2b}{a^2 + 5ab - 6b^2}$

65. $\frac{1}{m^2 + m - 2} - \frac{3}{2m^2 + 3m - 2} + \frac{2}{2m^2 - 3m + 1}$

66. $\frac{5}{2m^2 - 5m - 3} + \frac{3}{2m^2 + 5m + 2} - \frac{1}{m^2 - m - 6}$

Practice PLUS

In Exercises 67–74, perform the indicated operations. Simplify the result, if possible.

67. $\left(\frac{2x + 3}{x + 1} \cdot \frac{x^2 + 4x - 5}{2x^2 + x - 3}\right) - \frac{2}{x + 2}$

68. $\frac{1}{x^2 - 2x - 8} \div \left(\frac{1}{x - 4} - \frac{1}{x + 2}\right)$

69. $\left(2 - \frac{6}{x + 1}\right)\left(1 + \frac{3}{x - 2}\right)$

70. $\left(4 - \frac{3}{x + 2}\right)\left(1 + \frac{5}{x - 1}\right)$

71. $\left(\frac{1}{x + h} - \frac{1}{x}\right) \div h$

72. $\left(\frac{5}{x - 5} - \frac{2}{x + 3}\right) \div (3x + 25)$

73. $\left(\frac{1}{a^3 - b^3} \cdot \frac{ac + ad - bc - bd}{1}\right) - \frac{c - d}{a^2 + ab + b^2}$

74. $\frac{ab}{a^2 + ab + b^2} + \left(\frac{ac - ad - bc + bd}{ac - ad + bc - bd} \div \frac{a^3 - b^3}{a^3 + b^3}\right)$

75. If $f(x) = \frac{2x - 3}{x + 5}$ and $g(x) = \frac{x^2 - 4x - 19}{x^2 + 8x + 15}$, find $(f - g)(x)$ and the domain of $f - g$.

76. If $f(x) = \frac{2x - 1}{x^2 + x - 6}$ and $g(x) = \frac{x + 2}{x^2 + 5x + 6}$, find $(f - g)(x)$ and the domain of $f - g$.

Application Exercises

You plan to drive from Miami, Florida, to Atlanta, Georgia. Your trip involves approximately 470 miles of travel in Florida and 250 miles in Georgia. The speed limit is 70 miles per hour in Florida and 65 miles per hour in Georgia. If you average x miles per hour over these speed limits, the total driving time, T(x), in hours, is given by the function

$$T(x) = \frac{470}{x + 70} + \frac{250}{x + 65}.$$

The graph of T is shown in the figure. Use the function's equation to solve Exercises 77–82.

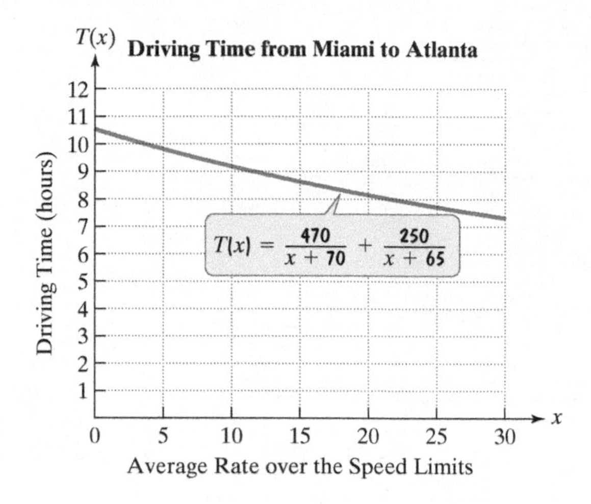

77. Find and interpret $T(0)$. Round to the nearest hour. Identify your solution as a point on the graph.

78. Find and interpret $T(5)$. Round to the nearest hour. Identify your solution as a point on the graph.

79. Find a simplified form of $T(x)$ by adding the rational expressions in the function's formula. Then use this form of the function to find $T(0)$.

80. Find a simplified form of $T(x)$ by adding the rational expressions in the function's formula. Then use this form of the function to find $T(5)$.

81. Use the graph to answer this question. If you want the driving time to be 9 hours, how much over the speed limits do you need to drive? Round to the nearest mile per hour. Does this seem like a realistic driving time for the trip? Explain your answer.

82. Use the graph to answer this question. If you want the driving time to be 8 hours, how much over the speed limits do you need to drive? Round to the nearest mile per hour. Does this seem like a realistic driving time for the trip? Explain your answer.

83. In the section opener, we saw that the rational function

$$f(x) = \frac{27{,}725(x - 14)}{x^2 + 9} - 5x$$

models the number of arrests, $f(x)$, per 100,000 drivers, for driving under the influence of alcohol as a function of a driver's age, x. The graph of f for an appropriate domain is shown in the figure.

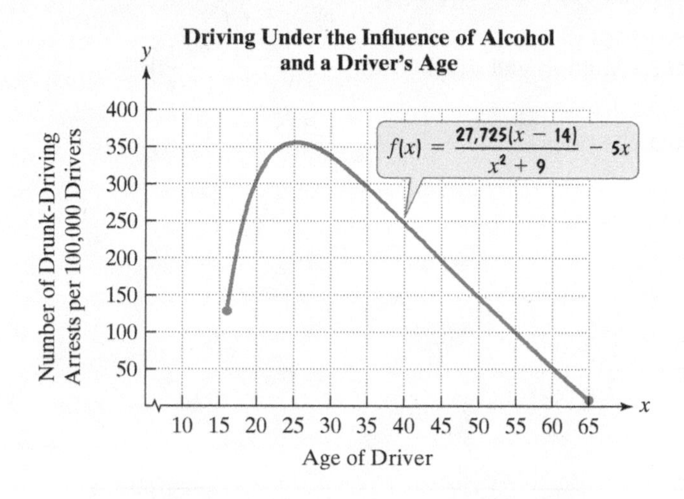

a. Use the function's equation to find and interpret $f(20)$. Round to the nearest whole number. Identify your solution as a point on the graph.

b. Find a simplified form of $f(x)$ by subtracting the rational expressions in the function's formula and writing the equation as a single rational expression.

c. Use the graph to determine the age, to the nearest five years, that corresponds to the greatest number of arrests. Then use the form of the function that you obtained in part (b) to determine the number of arrests, per 100,000 drivers, for this age group. Round to the nearest whole number.

In Exercises 84–85, express the perimeter of each rectangle as a single rational expression.

84.

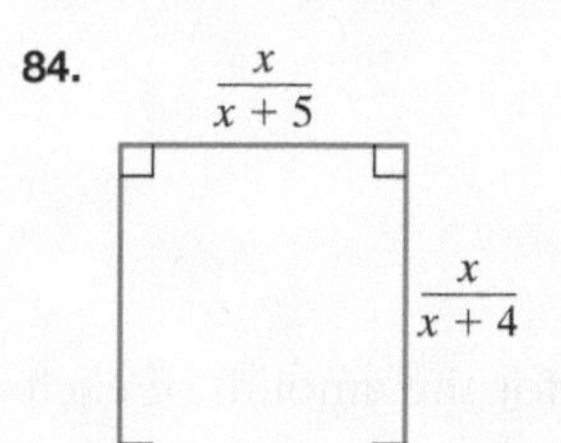

85.

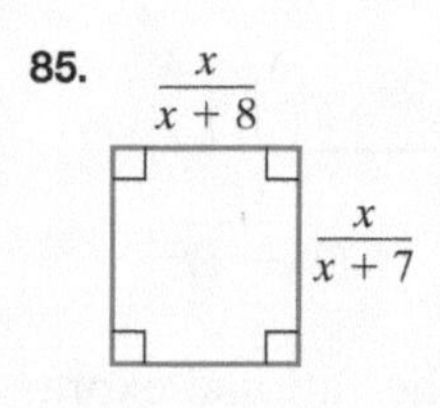

Explaining the Concepts

86. Explain how to add rational expressions when denominators are the same. Give an example with your explanation.

87. Explain how to subtract rational expressions when denominators are the same. Give an example with your explanation.

88. Explain how to find the least common denominator for denominators of $x^2 - 100$ and $x^2 - 20x + 100$.

89. Explain how to add rational expressions that have different denominators. Use $\frac{5}{x+1} + \frac{3}{x+4}$ in your explanation.

90. Explain how to add rational expressions when denominators are opposites. Use an example to support your explanation.

Explain the error in Exercises 91–92. Then rewrite the right side of the equation to correct the error that now exists.

91. $\frac{1}{a} + \frac{1}{b} = \frac{1}{a+b}$

92. $\frac{1}{x} + \frac{3}{7} = \frac{4}{x+7}$

Critical Thinking Exercises

Make Sense? *In Exercises 93–96, determine whether each statement makes sense or does not make sense, and explain your reasoning.*

93. When a numerator is being subtracted, I find that inserting parentheses helps me to distribute the negative sign to every term.

94. The reason I can rewrite rational expressions with a common denominator is that 1 is the multiplicative identity.

95. The fastest way for me to add $\frac{5}{x-7} + \frac{3}{7-x}$ is by using $(x - 7)(7 - x)$ as the LCD.

96. Although $\frac{2x^3 + 11x^2}{x+3} + \frac{5x^3 + 4x^2}{x+3}$ looks more complicated than $\frac{2}{x+3} + \frac{5}{x-3}$, it takes me more steps to perform the less complicated-looking addition.

In Exercises 97–100, determine whether each statement is true or false. If the statement is false, make the necessary change(s) to produce a true statement.

97. $\frac{2}{x+3} + \frac{3}{x+4} = \frac{5}{2x+7}$

98. $\frac{a}{b} + \frac{a}{c} = \frac{a}{b+c}$

99. $6 + \frac{1}{x} = \frac{7}{x}$

100. $\frac{1}{x+3} + \frac{x+3}{2} = \frac{1}{\cancel{(x+3)}} + \frac{\cancel{(x+3)}}{2} = 1 + \frac{1}{2} = \frac{3}{2}$

In Exercises 101–103, perform the indicated operations.

101. $\dfrac{1}{x^n - 1} - \dfrac{1}{x^n + 1} - \dfrac{1}{x^{2n} - 1}$

102. $\left(1 - \dfrac{1}{x}\right)\left(1 - \dfrac{1}{x+1}\right)\left(1 - \dfrac{1}{x+2}\right)\left(1 - \dfrac{1}{x+3}\right)$

103. $(x - y)^{-1} + (x - y)^{-2}$

Review Exercises

104. Simplify: $\left(\dfrac{3x^2y^{-2}}{y^3}\right)^{-2}$. (Section 1.6, Example 9)

105. Solve: $|3x - 1| \leq 14$. (Section 4.3, Example 4)

106. Factor completely: $50x^3 - 18x$. (Section 5.5, Example 2)

Preview Exercises

Exercises 107–109 will help you prepare for the material covered in the next section.

107. Multiply and simplify: $x^2y^2\left(\dfrac{1}{x} + \dfrac{y}{x^2}\right)$.

108. Multiply and simplify: $x(x + h)\left(\dfrac{1}{x + h} - \dfrac{1}{x}\right)$.

109. Divide: $\dfrac{x^2 - 1}{x^2} \div \dfrac{x^2 - 4x + 3}{x^2}$.

SECTION

6.3 Complex Rational Expressions

What am I supposed to learn?

After studying this section, you should be able to:

1. Simplify complex rational expressions by multiplying by 1.
2. Simplify complex rational expressions by dividing.

One area in finance of great interest to us ordinary folks when we buy a new car is the amount of each monthly payment. If P is the principal, or the amount borrowed, i is the monthly interest rate, and n is the number of monthly payments, then the amount, A, of each monthly payment is

$$A = \frac{Pi}{1 - \dfrac{1}{(1 + i)^n}}.$$

Do you notice anything unusual about the rational expression for the amount of each payment? It has a separate rational expression in its denominator.

$$\frac{Pi}{1 - \dfrac{1}{(1 + i)^n}}$$

Numerator

Main fraction bar

Denominator

A separate rational expression occurs in the denominator.

Complex rational expressions, also called **complex fractions**, have numerators or denominators containing one or more rational expressions. Here is another example of such an expression:

Numerator

$$\frac{\frac{1}{x}+\frac{y}{x^2}}{\frac{1}{y}+\frac{x}{y^2}}.$$

Main fraction bar

Denominator

In this section, we study two methods for simplifying complex rational expressions.

1 Simplify complex rational expressions by multiplying by 1.

Simplifying Complex Rational Expressions by Multiplying by 1

One method for simplifying a complex rational expression is to find the least common denominator of all the rational expressions in its numerator and denominator. Then multiply each term in its numerator and denominator by this least common denominator. Because we are multiplying by a form of 1, we will obtain an equivalent expression that does not contain fractions in the numerator or denominator.

Simplifying a Complex Rational Expression by Multiplying by 1 in the Form $\frac{\text{LCD}}{\text{LCD}}$

1. Find the LCD of all rational expressions within the complex rational expression.
2. Multiply both the numerator and the denominator of the complex rational expression by this LCD.
3. Use the distributive property and multiply each term in the numerator and denominator by this LCD. Simplify each term. No fractional expressions should remain within the numerator or denominator of the main fraction.
4. If possible, factor and simplify.

EXAMPLE 1 Simplifying a Complex Rational Expression

Simplify:

$$\frac{\frac{1}{x}+\frac{y}{x^2}}{\frac{1}{y}+\frac{x}{y^2}}.$$

Solution The denominators in the complex rational expression are x, x^2, y, and y^2. The LCD is x^2y^2. Multiply both the numerator and the denominator of the complex rational expression by x^2y^2.

$$\frac{\frac{1}{x}+\frac{y}{x^2}}{\frac{1}{y}+\frac{x}{y^2}} = \frac{x^2y^2}{x^2y^2}\cdot\frac{\left(\frac{1}{x}+\frac{y}{x^2}\right)}{\left(\frac{1}{y}+\frac{x}{y^2}\right)}$$

Multiply the numerator and the denominator by x^2y^2.

$$= \frac{x^2y^2 \cdot \frac{1}{x} + x^2y^2 \cdot \frac{y}{x^2}}{x^2y^2 \cdot \frac{1}{y} + x^2y^2 \cdot \frac{x}{y^2}}$$

Use the distributive property on $\frac{x^2y^2}{x^2y^2} \cdot \frac{\left(\frac{1}{x}+\frac{y}{x^2}\right)}{\left(\frac{1}{y}+\frac{x}{y^2}\right)}$.

In all four rational expressions, we have divided numerators and denominators by common boxed factors.

$$= \frac{xy^2 + y^3}{x^2y + x^3}$$

Simplify.

$$= \frac{y^2(x+y)}{x^2(y+x)}$$

Factor and simplify.

$$= \frac{y^2}{x^2} \quad \blacksquare$$

CHECK POINT 1 Simplify:

$$\frac{\frac{x}{y} - 1}{\frac{x^2}{y^2} - 1}.$$

Great Question!

It's been a while since I've seen negative exponents. Can you refresh my memory and tell me what they mean?

In Section 1.6, we introduced the negative-exponent rule:

$$b^{-n} = \frac{1}{b^n}, b \neq 0.$$

See pages 72–74 if you need to review negative integers as exponents.

Complex rational expressions are often written with negative exponents. For example,

$$\frac{x^{-1} + x^{-2}y}{y^{-1} + xy^{-2}} \quad \text{means} \quad \frac{\frac{1}{x} + \frac{y}{x^2}}{\frac{1}{y} + \frac{x}{y^2}}.$$

This is the expression that we simplified in Example 1. If an expression contains negative exponents, first rewrite it as an equivalent expression with positive exponents. Then simplify by multiplying the numerator and the denominator by the LCD.

EXAMPLE 2 Simplifying a Complex Rational Expression

Simplify:

$$\frac{\frac{1}{x+h} - \frac{1}{x}}{h}.$$

Solution The denominators in the complex rational expression are $x + h$ and x. The LCD is $x(x + h)$. Multiply both the numerator and the denominator of the complex rational expression by $x(x + h)$.

$$\frac{\frac{1}{x+h} - \frac{1}{x}}{h} = \frac{x(x+h)}{x(x+h)} \cdot \frac{\left(\frac{1}{x+h} - \frac{1}{x}\right)}{h}$$

Multiply the numerator and the denominator by $x(x + h)$.

$$= \frac{x(x+h) \cdot \frac{1}{x+h} - x(x+h) \cdot \frac{1}{x}}{x(x+h)h}$$

Use the distributive property in the numerator to multiply every term by the LCD.

$$= \frac{x - (x+h)}{x(x+h)h}$$

Simplify: $x(x+h) \cdot \frac{1}{(x+h)} = x$ and $x(x+h) \cdot \frac{1}{x} = x + h$.

$$= \frac{x - x - h}{x(x+h)h}$$

Remove parentheses and change the sign of each term.

$$= \frac{-h}{x(x+h)h} \quad \text{Simplify: } x - x - h = -h.$$

$$= \frac{-\cancel{h}}{x(x+h)\cancel{h}} \quad \text{Divide the numerator and the denominator by the common factor, } h.$$

$$= -\frac{1}{x(x+h)}$$ ■

✓ CHECK POINT 2 Simplify:

$$\frac{\frac{1}{x+7} - \frac{1}{x}}{7}.$$

---

2 Simplify complex rational expressions by dividing.

Simplifying Complex Rational Expressions by Dividing

A second method for simplifying a complex rational expression is to combine its numerator into a single rational expression and combine its denominator into a single rational expression. Then perform the division by inverting the denominator and multiplying.

Simplifying a Complex Rational Expression by Dividing

1. If necessary, add or subtract to get a single rational expression in the numerator.
2. If necessary, add or subtract to get a single rational expression in the denominator.
3. Perform the division indicated by the main fraction bar: Invert the denominator of the complex rational expression and multiply.
4. If possible, simplify.

EXAMPLE 3 Simplifying a Complex Rational Expression

Simplify:

$$\frac{\frac{x+1}{x} + \frac{x+1}{x-1}}{\frac{x+2}{x} - \frac{2}{x-1}}.$$

Solution

Step 1. Add to get a single rational expression in the numerator.

$$\frac{x+1}{x} + \frac{x+1}{x-1}$$

The LCD is $x(x-1)$.

$$= \frac{(x+1)(x-1)}{x(x-1)} + \frac{x(x+1)}{x(x-1)} = \frac{(x+1)(x-1) + x(x+1)}{x(x-1)} = \frac{x^2 - 1 + x^2 + x}{x(x-1)} = \frac{2x^2 + x - 1}{x(x-1)}$$

Step 2. Subtract to get a single rational expression in the denominator.

$$\frac{x+2}{x} - \frac{2}{x-1}$$

The LCD is $x(x-1)$.

$$= \frac{(x+2)(x-1)}{x(x-1)} - \frac{2x}{x(x-1)} = \frac{(x+2)(x-1) - 2x}{x(x-1)} = \frac{x^2 + x - 2 - 2x}{x(x-1)} = \frac{x^2 - x - 2}{x(x-1)}$$

Steps 3 and 4. Perform the division indicated by the main fraction bar: Invert and multiply. If possible, simplify.

$$\frac{\frac{x+1}{x}+\frac{x+1}{x-1}}{\frac{x+2}{x}-\frac{2}{x-1}} = \frac{\frac{2x^2+x-1}{x(x-1)}}{\frac{x^2-x-2}{x(x-1)}}$$

These are the single rational expressions from steps 1 and 2.

$$= \frac{2x^2+x-1}{x(x-1)} \cdot \frac{x(x-1)}{x^2-x-2} = \frac{(2x-1)\cancel{(x+1)}}{\cancel{x(x-1)}} \cdot \frac{\cancel{x(x-1)}}{(x-2)\cancel{(x+1)}} = \frac{2x-1}{x-2}$$

Invert and multiply.

■

☑ CHECK POINT 3 Simplify:

$$\frac{\frac{x+1}{x-1}-\frac{x-1}{x+1}}{\frac{x-1}{x+1}+\frac{x+1}{x-1}}.$$

Which of the two methods do you prefer? Let's try them both in Example 4.

EXAMPLE 4 Simplifying a Complex Rational Expression: Comparing Methods

Simplify:

$$\frac{1-x^{-2}}{1-4x^{-1}+3x^{-2}}.$$

Solution First rewrite the expression without negative exponents.

$$\frac{1-x^{-2}}{1-4x^{-1}+3x^{-2}} = \frac{1-\frac{1}{x^2}}{1-\frac{4}{x}+\frac{3}{x^2}}$$

The negative exponents affect only the variables and not the constants.

Method 1 Multiplying by 1

$$\frac{1-\frac{1}{x^2}}{1-\frac{4}{x}+\frac{3}{x^2}} = \frac{x^2}{x^2}\cdot\frac{\left(1-\frac{1}{x^2}\right)}{\left(1-\frac{4}{x}+\frac{3}{x^2}\right)}$$

Multiply the numerator and denominator by x^2, the LCD of all fractions.

$$= \frac{x^2\cdot 1 - x^2\cdot\frac{1}{x^2}}{x^2\cdot 1 - x^2\cdot\frac{4}{x} + x^2\cdot\frac{3}{x^2}}$$

Apply the distributive property.

$$= \frac{x^2-1}{x^2-4x+3}$$

Simplify.

$$= \frac{(x+1)\cancel{(x-1)}}{(x-3)\cancel{(x-1)}}$$

Factor and simplify.

$$= \frac{x+1}{x-3}$$

Method 2 Dividing

$$\frac{1-\frac{1}{x^2}}{1-\frac{4}{x}+\frac{3}{x^2}} = \frac{\frac{x^2}{x^2}-\frac{1}{x^2}}{\frac{x^2}{x^2}-\frac{4}{x}\cdot\frac{x}{x}+\frac{3}{x^2}}$$

Get a single rational expression in the numerator and in the denominator.

$$= \frac{\frac{x^2-1}{x^2}}{\frac{x^2-4x+3}{x^2}}$$

$$= \frac{x^2-1}{x^2}\cdot\frac{x^2}{x^2-4x+3}$$

Invert and multiply.

$$= \frac{(x+1)\cancel{(x-1)}}{x^2}\cdot\frac{x^2}{\cancel{(x-1)}(x-3)}$$

Factor and simplify.

$$= \frac{x+1}{x-3}$$

■

CHECK POINT 4 Simplify by the method of your choice:

$$\frac{1 - 4x^{-2}}{1 - 7x^{-1} + 10x^{-2}}.$$

Achieving Success

According to the Ebbinghaus retention model, you forget 50% of processed information within one hour of leaving the classroom. You lose 60% to 70% within 24 hours. After 30 days, 70% is gone. Reviewing and rewriting class notes is an effective way to counteract this phenomenon. At the very least, read your lecture notes at the end of each day. The more you engage with the material, the more you retain.

CONCEPT AND VOCABULARY CHECK

Fill in each blank so that the resulting statement is true.

1. A rational expression whose numerator or denominator or both contains rational expressions is called a/an ____________ rational expression or a/an ____________ fraction.

2. $$\frac{\frac{7}{x} + \frac{5}{x^2}}{\frac{5}{x} + 1} = \frac{x^2}{x^2} \cdot \frac{\left(\frac{7}{x} + \frac{5}{x^2}\right)}{\left(\frac{5}{x} + 1\right)} = \frac{x^2 \cdot \frac{7}{x} + x^2 \cdot \frac{5}{x^2}}{x^2 \cdot \frac{5}{x} + x^2 \cdot 1} = \frac{\_\_ + \_\_}{\_\_ + \_\_}$$

3. $$\frac{\frac{1}{x+3} - \frac{1}{x}}{3} = \frac{x(x+3)}{x(x+3)} \cdot \frac{\left(\frac{1}{x+3} - \frac{1}{x}\right)}{3} = \frac{x(x+3) \cdot \_\_\_\_ - x(x+3) \cdot \_\_}{3x(x+3)} = \frac{\_\_ - (\_\_\_)}{3x(x+3)} = \frac{\_\_}{3x(x+3)} = \_\_\_\_\_\_$$

6.3 EXERCISE SET

MyMathLab®

Practice Exercises

In Exercises 1–40, simplify each complex rational expression by the method of your choice.

1. $\dfrac{4 + \frac{2}{x}}{1 - \frac{3}{x}}$

2. $\dfrac{5 - \frac{2}{x}}{3 + \frac{1}{x}}$

3. $\dfrac{\frac{3}{x} + \frac{x}{3}}{\frac{x}{3} - \frac{3}{x}}$

4. $\dfrac{\frac{x}{5} - \frac{5}{x}}{\frac{1}{5} + \frac{1}{x}}$

5. $\dfrac{\frac{1}{x} + \frac{1}{y}}{\frac{1}{x} - \frac{1}{y}}$

6. $\dfrac{\frac{x}{y} + \frac{1}{x}}{\frac{y}{x} + \frac{1}{x}}$

7. $\dfrac{8x^{-2} - 2x^{-1}}{10x^{-1} - 6x^{-2}}$

8. $\dfrac{12x^{-2} - 3x^{-1}}{15x^{-1} - 9x^{-2}}$

9. $\dfrac{\frac{1}{x-2}}{1 - \frac{1}{x-2}}$

10. $\dfrac{\frac{1}{x+2}}{1 + \frac{1}{x+2}}$

11. $\dfrac{\frac{1}{x+5} - \frac{1}{x}}{5}$

12. $\dfrac{\frac{1}{x+6} - \frac{1}{x}}{6}$

13. $\dfrac{\frac{4}{x+4}}{\frac{1}{x+4} - \frac{1}{x}}$

14. $\dfrac{\frac{7}{x+7}}{\frac{1}{x+7} - \frac{1}{x}}$

15. $\dfrac{\frac{1}{x-1} + 1}{\frac{1}{x+1} - 1}$

16. $\dfrac{\frac{1}{x+1} - 1}{\frac{1}{x-1} + 1}$

17. $\dfrac{x^{-1} + y^{-1}}{(x+y)^{-1}}$

18. $(x^{-1} + y^{-1})^{-1}$

19. $\dfrac{\frac{x+2}{x-2} - \frac{x-2}{x+2}}{\frac{x-2}{x+2} + \frac{x+2}{x-2}}$

20. $\dfrac{\frac{x+1}{x-1} - \frac{x-1}{x+1}}{\frac{x-1}{x+1} + \frac{x+2}{x-1}}$

21. $\dfrac{\frac{2}{x^3y} + \frac{5}{xy^4}}{\frac{5}{x^3y} - \frac{3}{xy}}$

22. $\dfrac{\frac{3}{xy^2} + \frac{2}{x^2y}}{\frac{1}{x^2y} + \frac{2}{xy^3}}$

23. $\dfrac{\dfrac{3}{x+2}-\dfrac{3}{x-2}}{\dfrac{5}{x^2-4}}$

24. $\dfrac{\dfrac{3}{x+1}-\dfrac{3}{x-1}}{\dfrac{5}{x^2-1}}$

25. $\dfrac{3a^{-1}+3b^{-1}}{4a^{-2}-9b^{-2}}$

26. $\dfrac{5a^{-1}-2b^{-1}}{25a^{-2}-4b^{-2}}$

27. $\dfrac{\dfrac{4x}{x^2-4}-\dfrac{5}{x-2}}{\dfrac{2}{x-2}+\dfrac{3}{x+2}}$

28. $\dfrac{\dfrac{2}{x+3}+\dfrac{5x}{x^2-9}}{\dfrac{4}{x+3}+\dfrac{2}{x-3}}$

29. $\dfrac{\dfrac{2y}{y^2+4y+3}}{\dfrac{1}{y+3}+\dfrac{2}{y+1}}$

30. $\dfrac{\dfrac{5y}{y^2-5y+6}}{\dfrac{3}{y-3}+\dfrac{2}{y-2}}$

31. $\dfrac{\dfrac{2}{a^2}-\dfrac{1}{ab}-\dfrac{1}{b^2}}{\dfrac{1}{a^2}-\dfrac{3}{ab}+\dfrac{2}{b^2}}$

32. $\dfrac{\dfrac{2}{b^2}-\dfrac{5}{ab}-\dfrac{3}{a^2}}{\dfrac{2}{b^2}+\dfrac{7}{ab}+\dfrac{3}{a^2}}$

33. $\dfrac{\dfrac{2x}{x^2-25}+\dfrac{1}{3x-15}}{\dfrac{5}{x-5}+\dfrac{3}{4x-20}}$

34. $\dfrac{\dfrac{7x}{2x-2}+\dfrac{x}{x^2-1}}{\dfrac{4}{x+1}-\dfrac{1}{3x+3}}$

35. $\dfrac{\dfrac{3}{x+2y}-\dfrac{2y}{x^2+2xy}}{\dfrac{3y}{x^2+2xy}+\dfrac{5}{x}}$

36. $\dfrac{\dfrac{1}{x^3-y^3}}{\dfrac{1}{x-y}-\dfrac{1}{x^2+xy+y^2}}$

37. $\dfrac{\dfrac{2}{m^2-3m+2}+\dfrac{2}{m^2-m-2}}{\dfrac{2}{m^2-1}+\dfrac{2}{m^2+4m+3}}$

38. $\dfrac{\dfrac{m}{m^2-9}-\dfrac{2}{m^2-4m+4}}{\dfrac{3}{m^2-5m+6}+\dfrac{m}{m^2+m-6}}$

39. $\dfrac{\dfrac{2}{a^2+2a-8}+\dfrac{1}{a^2+5a+4}}{\dfrac{1}{a^2-5a+6}+\dfrac{2}{a^2-a-2}}$

40. $\dfrac{\dfrac{3}{a^2+10a+25}-\dfrac{1}{a^2-a-2}}{\dfrac{4}{a^2+6a+5}-\dfrac{2}{a^2+3a-10}}$

Practice PLUS

In Exercises 41–46, perform the indicated operations. Simplify the result, if possible.

41. $\dfrac{\dfrac{x-1}{x^2-4}}{1+\dfrac{1}{x-2}}-\dfrac{1}{x-2}$

42. $\dfrac{\dfrac{x-3}{x^2-16}}{1+\dfrac{1}{x-4}}-\dfrac{1}{x-4}$

43. $\dfrac{3}{1-\dfrac{3}{3+x}}-\dfrac{3}{\dfrac{3}{3-x}-1}$

44. $\dfrac{5}{1-\dfrac{5}{5+x}}-\dfrac{5}{\dfrac{5}{5-x}-1}$

45. $\dfrac{x}{1-\dfrac{1}{1+\dfrac{1}{x}}}$

46. $\dfrac{\dfrac{1}{x+1}}{x-\dfrac{1}{x+\dfrac{1}{x}}}$

In Exercises 47–48, let $f(x)=\dfrac{1+x}{1-x}$.

47. Find $f\left(\dfrac{1}{x+3}\right)$ and simplify.

48. Find $f\left(\dfrac{1}{x-6}\right)$ and simplify.

In Exercises 49–50, use the given rational function to find and simplify

$$\frac{f(a+h)-f(a)}{h}.$$

49. $f(x)=\dfrac{3}{x}$

50. $f(x)=\dfrac{1}{x^2}$

Application Exercises

51. How much are your monthly payments on a loan? If P is the principal, or amount borrowed, i is the monthly interest rate (as a decimal), and n is the number of monthly payments, then the amount, A, of each monthly payment is

$$A=\frac{Pi}{1-\dfrac{1}{(1+i)^n}}.$$

a. Simplify the complex rational expression for the amount of each payment.

b. You purchase a \$20,000 automobile at 1% monthly interest to be paid over 48 months. How much do you pay each month? Use the simplified rational expression from part (a) and a calculator. Round to the nearest dollar.

52. The average rate on a round-trip commute having a one-way distance d is given by the complex rational expression

$$\frac{2d}{\frac{d}{r_1}+\frac{d}{r_2}},$$

in which r_1 and r_2 are the rates on the outgoing and return trips, respectively.

a. Simplify the complex rational expression.

b. Find your average rate if you drive to campus averaging 30 miles per hour and return home on the same route averaging 40 miles per hour. Use the simplified rational expression from part (a).

53. If three resistors with resistances $R_1, R_2,$ and R_3 are connected in parallel, their combined resistance, R, is given by the formula

$$R = \frac{1}{\frac{1}{R_1}+\frac{1}{R_2}+\frac{1}{R_3}}.$$

R_1 R_2 R_3

Simplify the complex rational expression on the right side of the formula. Then find R, to the nearest hundredth of an ohm, when R_1 is 4 ohms, R_2 is 8 ohms, and R_3 is 12 ohms.

54. A camera lens has a measurement called its focal length, f. When an object is in focus, its distance from the lens, p, and its image distance from the lens, q, satisfy the formula

$$f = \frac{1}{\frac{1}{p}+\frac{1}{q}}.$$

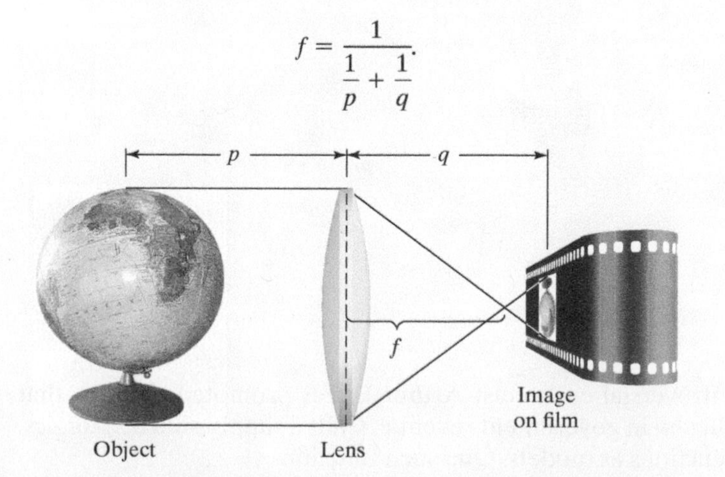

Simplify the complex rational expression on the right side of the formula.

Explaining the Concepts

55. What is a complex rational expression? Give an example with your explanation.

56. Describe two ways to simplify $\frac{\frac{2}{x}+\frac{2}{y}}{\frac{2}{x}-\frac{2}{y}}$.

57. Which method do you prefer for simplifying complex rational expressions? Why?

58. Of the four complex rational expressions in Exercises 51–54, which one do you find most useful? Explain how you might use this expression in a practical situation.

Technology Exercises

In Exercises 59–62, use a graphing utility to determine if the simplification is correct by graphing the function on each side of the equation in the same viewing rectangle. If the graphs do not coincide, correct the expression on the right side and then verify your correction using the graphing utility.

59. $\frac{x-\frac{1}{2x+1}}{1-\frac{x}{2x+1}} = 2x-1$

60. $\frac{\frac{1}{x}+1}{\frac{1}{x}} = 2$

61. $\frac{\frac{1}{x}+\frac{1}{3}}{\frac{1}{3x}} = x+\frac{1}{3}$

62. $\frac{\frac{x}{3}}{\frac{2}{x+1}} = \frac{3(x+1)}{2}$

Critical Thinking Exercises

Make Sense? *In Exercises 63–66, determine whether each statement makes sense or does not make sense, and explain your reasoning.*

63. I simplified

$$\frac{\frac{1+3x}{xy}}{5+4y}$$

by multiplying the numerator by xy.

64. By noticing that

$$\frac{\frac{1}{x+7}-\frac{1}{x}}{7}$$

repeats x and 7 twice, it's fairly easy to simplify the complex fraction in my head without showing all the steps.

65. I simplified

$$\frac{3 - \frac{6}{x+5}}{1 + \frac{7}{x-4}}$$

by multiplying by 1 and obtained $\frac{3 - 6(x-4)}{1 + 7(x+5)}$.

66. Before simplifying $\frac{1 - x^{-2}}{1 - 5x^{-3}}$, I wrote the complex fraction without negative exponents as

$$\frac{1 - \frac{1}{x^2}}{1 - \frac{1}{5x^3}}.$$

67. Simplify:

$$\frac{\frac{x+h}{x+h+1} - \frac{x}{x+1}}{h}.$$

68. Simplify:

$$x + \cfrac{1}{x + \cfrac{1}{x + \cfrac{1}{x}}}.$$

69. If $f(x) = \frac{1}{x+1}$, find $f(f(a))$ and simplify.

70. Let x represent the first of two consecutive integers. Find a simplified expression that represents the reciprocal of the sum of the reciprocals of the two integers.

Review Exercises

71. Solve: $x^2 + 27 = 12x$. (Section 5.7, Example 2)

72. Multiply: $(4x^2 - y)^2$. (Section 5.2, Example 7)

73. Solve: $-4 < 3x - 7 < 8$. (Section 4.2, Example 4)

Preview Exercises

Exercises 74–76 will help you prepare for the material covered in the next section.

74. Simplify: $\frac{8x^4y^5}{4x^3y^2}$.

75. Divide 737 by 21 without using a calculator. Write the answer as

$$\text{quotient} + \frac{\text{remainder}}{\text{divisor}}.$$

76. Simplify: $6x^2 + 3x - (6x^2 - 4x)$.

SECTION

6.4 Division of Polynomials

What am I supposed to learn?

After studying this section, you should be able to:

1. Divide a polynomial by a monomial.
2. Use long division to divide by a polynomial containing more than one term.

During the 1980s, the controversial economist Arthur Laffer promoted the idea that tax *increases* lead to a *reduction* in government revenue. Called supply-side economics, the theory uses rational functions as models. One such function

$$f(x) = \frac{80x - 8000}{x - 110}, \quad 30 \le x \le 100$$

models the government tax revenue, $f(x)$, in tens of billions of dollars, as a function of the tax rate, x. The graph of the rational function is shown in **Figure 6.4**. The graph shows tax revenue decreasing quite dramatically as the tax rate increases. At a tax rate of (gasp) 100%, the government takes all our money and no one has an incentive to work. With no income earned, zero dollars in tax revenue is generated.

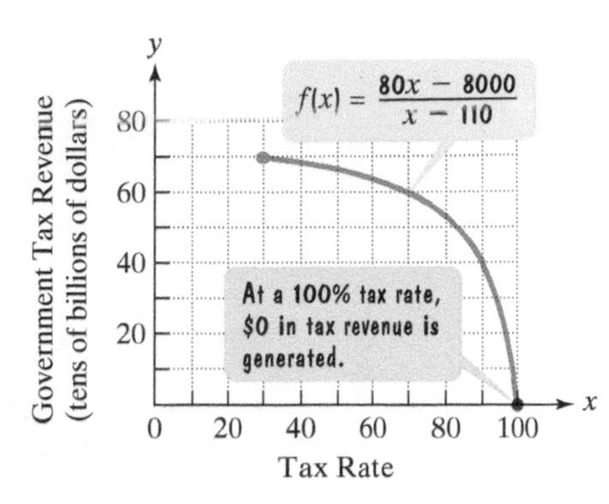

Figure 6.4

Like all rational functions, the Laffer model consists of the quotient of polynomials. Although the rational expression in the model cannot be simplified, it is possible to perform the division, thereby expressing the function in another form. In this section, you will learn to divide polynomials.

1 Divide a polynomial by a monomial.

Dividing a Polynomial by a Monomial

We have seen that to divide monomials, we divide the coefficients and subtract the exponents when bases are the same. For example,

$$\frac{25x^{12}}{5x^4} = \frac{25}{5}x^{12-4} = 5x^8 \quad \text{and} \quad \frac{60x^4y^3}{-30x^2y} = \frac{60}{-30}x^{4-2}y^{3-1} = -2x^2y^2.$$

How do we divide a polynomial that is not a monomial by a monomial? We divide each term of the polynomial by the monomial. For example,

$$\frac{10x^8 + 15x^6}{5x^3} = \frac{10x^8}{5x^3} + \frac{15x^6}{5x^3} = \frac{10}{5}x^{8-3} + \frac{15}{5}x^{6-3} = 2x^5 + 3x^3.$$

Polynomial dividend

Monomial divisor

Divide the first term by $5x^3$.

Divide the second term by $5x^3$.

Dividing a Polynomial That Is Not a Monomial by a Monomial

To divide a polynomial by a monomial, divide each term of the polynomial by the monomial.

EXAMPLE 1 Dividing a Polynomial by a Monomial

Divide: $(15x^3 - 5x^2 + x + 5) \div (5x)$.

Solution

$$\frac{15x^3 - 5x^2 + x + 5}{5x}$$ Rewrite the division in a vertical format.

$$= \frac{15x^3}{5x} - \frac{5x^2}{5x} + \frac{x}{5x} + \frac{5}{5x}$$ Divide each term of the polynomial by the monomial.

$$= 3x^2 - x + \frac{1}{5} + \frac{1}{x}$$ Simplify each quotient. ■

CHECK POINT 1 Divide: $(16x^3 - 32x^2 + 2x + 4) \div 4x$.

EXAMPLE 2 Dividing a Polynomial by a Monomial

Divide $8x^4y^5 - 10x^4y^3 + 12x^2y^3$ by $4x^3y^2$.

Solution

$$\frac{8x^4y^5 - 10x^4y^3 + 12x^2y^3}{4x^3y^2}$$ Express the division of $8x^4y^5 - 10x^4y^3 + 12x^2y^3$ by $4x^3y^2$ in a vertical format.

$$= \frac{8x^4y^5}{4x^3y^2} - \frac{10x^4y^3}{4x^3y^2} + \frac{12x^2y^3}{4x^3y^2}$$ Divide each term of the polynomial by the monomial.

$$= 2xy^3 - \frac{5}{2}xy + \frac{3y}{x}$$ Simplify each quotient. ■

☑ CHECK POINT 2 Divide $15x^4y^5 - 5x^3y^4 + 10x^2y^2$ by $5x^2y^3$.

2 Use long division to divide by a polynomial containing more than one term.

Dividing by a Polynomial Containing More Than One Term

We now look at division by a polynomial containing more than one term, such as

$$x - 2\overline{)x^2 - 14x + 24}.$$

Divisor has two terms and is a binomial.

The polynomial dividend has three terms and is a trinomial.

When a divisor has more than one term, the four steps used to divide whole numbers—**divide, multiply, subtract, bring down the next term**—form the repetitive procedure for polynomial long division.

EXAMPLE 3 Dividing a Polynomial by a Binomial

Divide $x^2 - 14x + 24$ by $x - 2$.

Solution The following steps illustrate how polynomial division is very similar to numerical division.

$$x - 2\overline{)x^2 - 14x + 24}$$

Arrange the terms of the dividend ($x^2 - 14x + 24$) and the divisor ($x - 2$) in descending powers of x.

$$\begin{array}{r} x \\ x - 2\overline{)x^2 - 14x + 24} \end{array}$$

DIVIDE x^2 (the first term in the dividend) by x (the first term in the divisor): $\frac{x^2}{x} = x$. Align like terms.

$x(x - 2) = x^2 - 2x$

$$\begin{array}{r} x \\ x - 2\overline{)x^2 - 14x + 24} \\ x^2 - 2x \end{array}$$

MULTIPLY each term in the divisor ($x - 2$) by x, aligning terms of the product under like terms in the dividend.

$$\begin{array}{r} x \\ x - 2\overline{)x^2 - 14x + 24} \\ \ominus x^2 \oplus 2x \\ \hline -12x \end{array}$$

Change signs of the polynomial being subtracted.

SUBTRACT $x^2 - 2x$ from $x^2 - 14x$ by changing the sign of each term in the lower expression and adding.

$$\begin{array}{r} x \\ x - 2\overline{)x^2 - 14x + 24} \\ x^2 - 2x \downarrow \\ \hline -12x + 24 \end{array}$$

BRING DOWN 24 from the original dividend and add algebraically to form a new dividend.

$$\begin{array}{r} x - 12 \\ x - 2\overline{)x^2 - 14x + 24} \\ x^2 - 2x \\ \hline -12x + 24 \end{array}$$

Find the second term of the quotient. DIVIDE the first term of $-12x + 24$ by x, the first term of the divisor: $\frac{-12x}{x} = -12$.

$$\begin{array}{r} x - 12 \\ x - 2\overline{)x^2 - 14x + 24} \\ \underline{x^2 - 2x} \\ -12x + 24 \\ \underline{\overset{\oplus}{-}12x \overset{\ominus}{+} 24} \\ 0 \end{array}$$

$-12(x - 2) = -12x + 24$

MULTIPLY the divisor $(x - 2)$ by -12, aligning under like terms in the new dividend. Then subtract to obtain the remainder of 0.

Remainder

The quotient is $x - 12$. Because the remainder is 0, we can conclude that $x - 2$ is a factor of $x^2 - 14x + 24$ and

$$\frac{x^2 - 14x + 24}{x - 2} = x - 12. \blacksquare$$

After performing polynomial long division, the answer can be checked. Find the product of the divisor and the quotient, and add the remainder. If the result is the dividend, the answer to the division problem is correct. For example, let's check our work in Example 3.

$$\frac{x^2 - 14x + 24}{x - 2} = x - 12$$

Dividend; Divisor; Quotient to be checked

Multiply the divisor and the quotient, and add the remainder, 0:

$$(x - 2)(x - 12) + 0 = x^2 - 12x - 2x + 24 + 0 = x^2 - 14x + 24.$$

Divisor; Quotient; Remainder; This is the dividend.

Because we obtained the dividend, the quotient is correct.

CHECK POINT 3 Divide $3x^2 - 14x + 16$ by $x - 2$.

Before considering additional examples, let's summarize the general procedure for dividing by a polynomial that contains more than one term.

Long Division of Polynomials

1. Arrange the terms of both the dividend and the divisor in descending powers of any variable.
2. **Divide** the first term in the dividend by the first term in the divisor. The result is the first term of the quotient.
3. **Multiply** every term in the divisor by the first term in the quotient. Write the resulting product beneath the dividend with like terms lined up.
4. **Subtract** the product from the dividend.
5. **Bring down** the next term in the original dividend and write it next to the remainder to form a new dividend.
6. Use this new expression as the dividend and repeat this process until the remainder can no longer be divided. This will occur when the degree of the remainder (the highest exponent on a variable in the remainder) is less than the degree of the divisor.

In our next long division, we will obtain a nonzero remainder.

EXAMPLE 4 Long Division of Polynomials

Divide $4 - 5x - x^2 + 6x^3$ by $3x - 2$.

Solution In order to divide $4 - 5x - x^2 + 6x^3$ by $3x - 2$, we begin by writing the dividend in descending powers of x.

$$4 - 5x - x^2 + 6x^3 = 6x^3 - x^2 - 5x + 4$$

$$\begin{array}{r} 2x^2 \phantom{{}- 5x + 4} \\ 3x - 2\overline{)6x^3 - x^2 - 5x + 4} \\ \underline{6x^3 - 4x^2} \phantom{{}- 5x + 4} \\ 3x^2 - 5x \phantom{{}+ 4} \end{array}$$

$2x^2(3x - 2) = 6x^3 - 4x^2$

Change signs of the polynomial being subtracted.

Divide: $\frac{6x^3}{3x} = 2x^2$.

Multiply: $2x^2(3x - 2) = 6x^3 - 4x^2$.

Subtract $6x^3 - 4x^2$ from $6x^3 - x^2$, and bring down $-5x$.

Now we divide $3x^2$ by $3x$ to obtain x, multiply x and the divisor, and subtract.

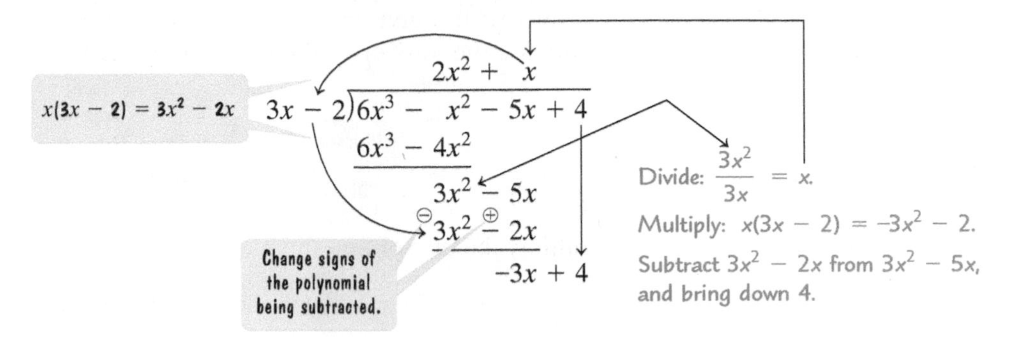

Now we divide $-3x$ by $3x$ to obtain -1, multiply -1 and the divisor, and subtract.

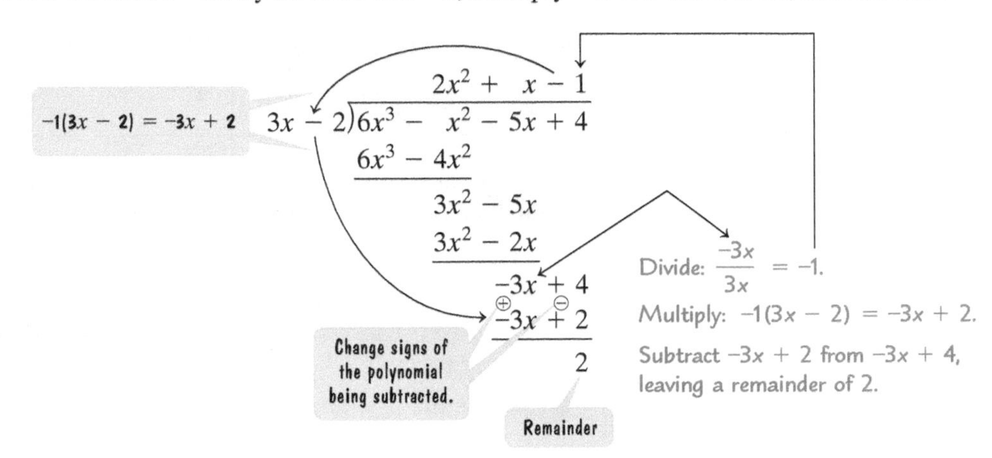

The quotient is $2x^2 + x - 1$ and the remainder is 2. When there is a nonzero remainder, as in this example, list the quotient, plus the remainder above the divisor. Thus,

$$\frac{6x^3 - x^2 - 5x + 4}{3x - 2} = \underbrace{2x^2 + x - 1}_{\text{Quotient}} + \frac{2}{3x - 2}.$$

Remainder above divisor

Check this result by showing that the product of the divisor and the quotient,

$$(3x - 2)(2x^2 + x - 1),$$

plus the remainder, 2, is the dividend, $6x^3 - x^2 - 5x + 4$. ■

CHECK POINT 4 Divide $-9 + 7x - 4x^2 + 4x^3$ by $2x - 1$.

If a power of x is missing in either a dividend or a divisor, add that power of x with a coefficient of 0 and then divide. In this way, like terms will be aligned as you carry out the long division.

EXAMPLE 5 Long Division of Polynomials

Divide $6x^4 + 5x^3 + 3x - 5$ by $3x^2 - 2x$.

Solution We write the dividend, $6x^4 + 5x^3 + 3x - 5$, as $6x^4 + 5x^3 + 0x^2 + 3x - 5$ to keep all like terms aligned.

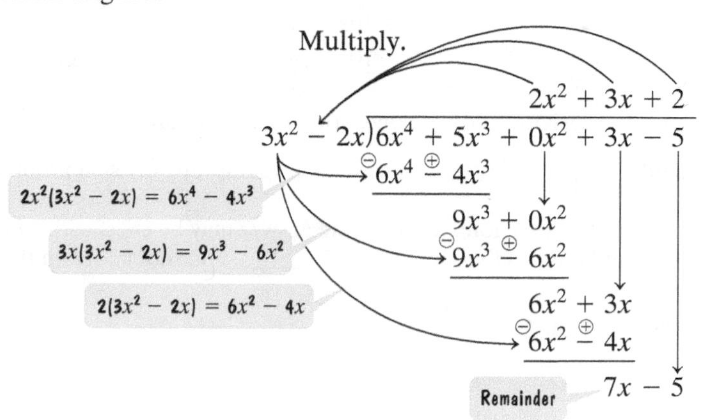

The division process is finished because the degree of $7x - 5$, which is 1, is less than the degree of the divisor $3x^2 - 2x$, which is 2.

We see that the answer is

$$\frac{6x^4 + 5x^3 + 3x - 5}{3x^2 - 2x} = 2x^2 + 3x + 2 + \frac{7x - 5}{3x^2 - 2x}.$$ ■

✓ CHECK POINT 5 Divide $2x^4 + 3x^3 - 7x - 10$ by $x^2 - 2x$.

CONCEPT AND VOCABULARY CHECK

Fill in each blank so that the resulting statement is true.

1. To divide $16x^3 - 32x^2 + 2x + 4$ by $4x$, divide each term of ________ by ________.
2. Consider the following long division problem:

   $x + 4\overline{)6x - 4 + 2x^3}.$

   We begin the division process by rewriting the dividend as ________.
3. Consider the following long division problem:

   $3x - 1\overline{)6x^3 + 7x^2 + 12x - 5}.$

   We begin the division process by dividing ________ by ________. We obtain ________. We write this result above ________ in the dividend.
4. In the following long division problem, the first step has been completed:

   $$\begin{array}{r} 2x^2 \\ 5x - 2\overline{)10x^3 + 6x^2 - 9x + 10} \end{array}$$

   The next step is to multiply ________ and ________. We obtain ________. We write this result below ________.

5. In the following long division problem, the first two steps have been completed:

$$\begin{array}{r} 2x \\ 3x - 5\overline{)6x^2 + 8x - 4.} \\ \underline{6x^2 - 10x} \end{array}$$

The next step is to subtract ______________ from ______________. We obtain ______________. Then we bring down ______________ and form the new dividend ______________.

6. In the following long division problem, most of the steps have been completed:

$$\begin{array}{r} 3x - 5 \\ 2x + 1\overline{)6x^2 - 7x + 4.} \\ \underline{6x^2 + 3x} \\ -10x + 4 \\ \underline{-10x - 5} \\ ? \end{array}$$

Completing the step designated by the question mark, we obtain ______________. Thus, the quotient is ______________

and the remainder is ______________. The answer to this long division problem is ______________.

7. After performing polynomial long division, the answer may be checked by multiplying the ______________ by the ______________, and then adding the ______________. You should obtain the ______________.

6.4 EXERCISE SET

MyMathLab®

Practice Exercises

In Exercises 1–12, divide the polynomial by the monomial.

1. $\dfrac{25x^7 - 15x^5 + 10x^3}{5x^3}$

2. $\dfrac{49x^7 - 28x^5 + 14x^3}{7x^3}$

3. $\dfrac{18x^3 + 6x^2 - 9x - 6}{3x}$

4. $\dfrac{25x^3 + 50x^2 - 40x - 10}{5x}$

5. $(28x^3 - 7x^2 - 16x) \div (4x^2)$

6. $(70x^3 - 10x^2 - 14x) \div (7x^2)$

7. $(25x^8 - 50x^7 + 3x^6 - 40x^5) \div (-5x^5)$

8. $(18x^7 - 9x^6 + 20x^5 - 10x^4) \div (-2x^4)$

9. $(18a^3b^2 - 9a^2b - 27ab^2) \div (9ab)$

10. $(12a^2b^2 + 6a^2b - 15ab^2) \div (3ab)$

11. $(36x^4y^3 - 18x^3y^2 - 12x^2y) \div (6x^3y^3)$

12. $(40x^4y^3 - 20x^3y^2 - 50x^2y) \div (10x^3y^3)$

In Exercises 13–36, divide as indicated. Check at least five of your answers by showing that the product of the divisor and the quotient, plus the remainder, is the dividend.

13. $(x^2 + 8x + 15) \div (x + 5)$

14. $(x^2 + 3x - 10) \div (x - 2)$

15. $(x^3 - 2x^2 - 5x + 6) \div (x - 3)$

16. $(x^3 + 5x^2 + 7x + 2) \div (x + 2)$

17. $(x^2 - 7x + 12) \div (x - 5)$

18. $(2x^2 + x - 9) \div (x - 2)$

19. $(2x^2 + 13x + 5) \div (2x + 3)$

20. $(8x^2 + 6x - 25) \div (4x + 9)$

21. $(x^3 + 3x^2 + 5x + 4) \div (x + 1)$

22. $(x^3 + 6x^2 - 2x + 3) \div (x - 1)$

23. $(4y^3 + 12y^2 + 7y - 3) \div (2y + 3)$

24. $(6y^3 + 7y^2 + 12y - 5) \div (3y - 1)$

25. $(9x^3 - 3x^2 - 3x + 4) \div (3x + 2)$

26. $(2x^3 + 13x^2 + 9x - 6) \div (2x + 3)$

27. $(4x^3 - 6x - 11) \div (2x - 4)$

28. $(2x^3 + 6x - 4) \div (x + 4)$

29. $(4y^3 - 5y) \div (2y - 1)$

30. $(6y^3 - 5y) \div (2y - 1)$

31. $(4y^4 - 17y^2 + 14y - 3) \div (2y - 3)$

32. $(2y^4 - y^3 + 16y^2 - 4) \div (2y - 1)$

33. $(4x^4 + 3x^3 + 4x^2 + 9x - 6) \div (x^2 + 3)$

34. $(3x^5 - x^3 + 4x^2 - 12x - 8) \div (x^2 - 2)$

35. $(15x^4 + 3x^3 + 4x^2 + 4) \div (3x^2 - 1)$

36. $(18x^4 + 9x^3 + 3x^2) \div (3x^2 + 1)$

In Exercises 37–40, find a simplified expression for $\left(\frac{f}{g}\right)(x)$.

37. $f(x) = 8x^3 - 38x^2 + 49x - 10,$
$g(x) = 4x - 1$

38. $f(x) = 2x^3 - 9x^2 - 17x + 39,$
$g(x) = 2x - 3$

39. $f(x) = 2x^4 - 7x^3 + 7x^2 - 9x + 10,$
$g(x) = 2x - 5$

40. $f(x) = 4x^4 + 6x^3 + 3x - 1,$
$g(x) = 2x^2 + 1$

Practice PLUS

In Exercises 41–50, divide as indicated.

41. $\dfrac{x^4 + y^4}{x + y}$

42. $\dfrac{x^5 + y^5}{x + y}$

43. $\dfrac{3x^4 + 5x^3 + 7x^2 + 3x - 2}{x^2 + x + 2}$

44. $\dfrac{x^4 - x^3 - 7x^2 - 7x - 2}{x^2 - 3x - 2}$

45. $\dfrac{4x^3 - 3x^2 + x + 1}{x^2 + x + 1}$

46. $\dfrac{x^4 - x^2 + 1}{x^2 + x + 1}$

47. $\dfrac{x^5 - 1}{x^2 - x + 2}$

48. $\dfrac{5x^5 - 7x^4 + 3x^3 - 20x^2 + 28x - 12}{x^3 - 4}$

49. $\dfrac{4x^3 - 7x^2y - 16xy^2 + 3y^3}{x - 3y}$

50. $\dfrac{12x^3 - 19x^2y + 13xy^2 - 10y^3}{4x - 5y}$

In Exercises 51–52, find $\left(\frac{f-g}{h}\right)(x)$ *and the domain of* $\frac{f-g}{h}$.

51. $f(x) = 3x^3 + 4x^2 - x - 4, g(x) = -5x^3 + 22x^2 - 28x - 12,$
$h(x) = 4x + 1$

52. $f(x) = x^3 + 9x^2 - 6x + 25, g(x) = -3x^3 + 2x^2 - 14x + 5,$
$h(x) = 2x + 4$

In Exercises 53–54, solve each equation for x in terms of a and simplify.

53. $ax + 2x + 4 = 3a^3 + 10a^2 + 6a$

54. $ax - 3x + 6 = a^3 - 6a^2 + 11a$

Application Exercises

In the section opener, we saw that

$$f(x) = \frac{80x - 8000}{x - 110}, \quad 30 \le x \le 100$$

models the government tax revenue, f(x), in tens of billions of dollars, as a function of the tax rate percentage, x. Use this function to solve Exercises 55–58. Round to the nearest ten billion dollars.

55. Find and interpret $f(30)$. Identify the solution as a point on the graph of the function in **Figure 6.4** on page 439.

56. Find and interpret $f(70)$. Identify the solution as a point on the graph of the function in **Figure 6.4** on page 439.

57. Rewrite the function by using long division to perform

$$(80x - 8000) \div (x - 110).$$

Then use this new form of the function to find $f(30)$. Do you obtain the same answer as you did in Exercise 55? Which form of the function do you find easier to use?

58. Rewrite the function by using long division to perform

$$(80x - 8000) \div (x - 110).$$

Then use this new form of the function to find $f(70)$. Do you obtain the same answer as you did in Exercise 56? Which form of the function do you find easier to use?

Explaining the Concepts

59. Explain how to divide a polynomial that is not a monomial by a monomial. Give an example.
60. In your own words, explain how to divide by a polynomial containing more than one term. Use $\frac{x^2+4}{x+2}$ in your explanation.
61. When performing polynomial long division, explain when to stop dividing.
62. After performing polynomial long division, explain how to check the answer.
63. When performing polynomial long division with missing terms, explain the advantage of writing the missing terms with zero coefficients.
64. The idea of supply-side economics is that an increase in the tax rate may actually reduce government revenue. What explanation can you offer for this theory?

Technology Exercises

In Exercises 65–67, use a graphing utility to determine if the division has been performed correctly. Graph the function on each side of the equation in the same viewing rectangle. If the graphs do not coincide, correct the expression on the right side by using polynomial long division. Then verify your correction using the graphing utility.

65. $(6x^2 + 16x + 8) \div (3x + 2) = 2x + 4$
66. $(4x^3 + 7x^2 + 8x + 20) \div (2x + 4) = 2x^2 - \frac{1}{2}x + 3$
67. $(3x^4 + 4x^3 - 32x^2 - 5x - 20) \div (x + 4) = 3x^3 - 8x^2 + 5$
68. Use the TABLE feature of a graphing utility to verify any two division results that you obtained in Exercises 13–36.

Critical Thinking Exercises

Make Sense? *In Exercises 69–72, determine whether each statement makes sense or does not make sense, and explain your reasoning.*

69. When performing the division

$$(2x^3 + 13x^2 + 9x - 6) \div (2x + 3),$$

I mentally cover up the +3, the second term of the binomial divisor, before dividing into the dividend.
70. When performing the division $(x^5 + 1) \div (x + 1)$, there's no need for me to follow all the steps involved in polynomial long division because I can work the problem in my head and see that the quotient must be $x^4 + 1$.
71. Because of exponential properties, the degree of the quotient must be the difference between the degree of the dividend and the degree of the divisor.
72. When performing the division $(x^3 + 1) \div (x + 2)$, the purpose of rewriting $x^3 + 1$ as $x^3 + 0x^2 + 0x + 1$ is to keep all like terms aligned.

In Exercises 73–76, determine whether each statement is true or false. If the statement is false, make the necessary change(s) to produce a true statement.

73. All long-division problems can be done by the alternative method of factoring the dividend and canceling identical factors in the dividend and the divisor.
74. Polynomial long division always shows that the answer is a polynomial.
75. The long division process should be continued until the degree of the remainder is the same as the degree of the divisor.
76. If a polynomial long-division problem results in a remainder that is zero, then the divisor is a factor of the dividend.

In Exercises 77–78, divide as indicated.

77. $(x^{3n} - 4x^{2n} - 2x^n - 12) \div (x^n - 5)$
78. $(x^{3n} + 1) \div (x^n + 1)$
79. When $2x^2 - 7x + 9$ is divided by a polynomial, the quotient is $2x - 3$ and the remainder is 3. Find the polynomial.
80. Find k so that the remainder is 0: $(20x^3 + 23x^2 - 10x + k) \div (4x + 3)$.

Review Exercises

81. Solve: $|2x - 3| > 4$. (Section 4.3, Example 6)
82. Write 40,610,000 in scientific notation. (Section 1.7, Example 2)
83. Simplify: $2x - 4[x - 3(2x + 1)]$. (Section 1.2, Example 14)

Preview Exercises

Exercises 84–86 will help you prepare for the material covered in the next section.

84. **a.** Divide: $\frac{5x^3 + 6x + 8}{x + 2}$.

b. Find the sum of the numbers in each column, designated by $\square$, in the following array of numbers. Then describe the relationship between the four numbers in the bottom row of the array and your answer to the division problem in part (a).

−2	5	0	6	8
		−10	20	−52
	5	$\square$	$\square$	$\square$

85. a. Divide: $\dfrac{3x^3 - 4x^2 + 2x - 1}{x + 1}$.

b. Find the sum of the numbers in each column, designated by □, in the following array of numbers. Then describe the relationship between the four numbers in the bottom row of the array and your answer to the division problem in part (a).

$$\begin{array}{r|rrrr} -1 & 3 & -4 & 2 & -1 \\ & & -3 & 7 & -9 \\ \hline & 3 & \square & \square & \square \end{array}$$

86. Divide $2x^3 - 3x^2 - 11x + 6$ by $x - 3$. Use your answer to factor $2x^3 - 3x^2 - 11x + 6$ completely.

MID-CHAPTER CHECK POINT Section 6.1–Section 6.4

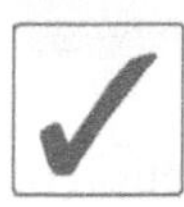

What You Know: We learned that the domain of a rational function is the set of all real numbers except those for which the denominator is zero. We saw that graphs of rational functions have vertical asymptotes or breaks at these excluded values. We learned to simplify rational expressions by dividing the numerator and the denominator by common factors. We performed a variety of operations with rational expressions, including multiplication, division, addition, and subtraction. We used two methods (multiplying by 1 and dividing) to simplify complex rational expressions. Finally, we used long division when dividing by a polynomial with more than one term.

1. Simplify: $\dfrac{x^2 - x - 6}{x^2 + 3x - 18}$.

In Exercises 2–19, perform the indicated operation(s) and, if possible, simplify.

2. $\dfrac{2x^2 - 8x - 11}{x^2 + 3x - 4} + \dfrac{x^2 + 14x - 13}{x^2 + 3x - 4}$

3. $\dfrac{x^3 - 27}{4x^2 - 4x} \cdot \dfrac{4x}{x - 3}$

4. $5 + \dfrac{7}{x - 2}$

5. $\dfrac{x - \dfrac{4}{x + 6}}{\dfrac{1}{x + 6} + x}$

6. $(2x^4 - 13x^3 + 17x^2 + 18x - 24) \div (x - 4)$

7. $\dfrac{x^3y - y^3x}{x^2y - xy^2}$

8. $(28x^8y^3 - 14x^6y^2 + 3x^2y^2) \div (7x^2y)$

9. $\dfrac{2x - 1}{x + 6} - \dfrac{x + 3}{x - 2}$

10. $\dfrac{3}{x - 2} - \dfrac{2}{x + 2} - \dfrac{x}{x^2 - 4}$

11. $\dfrac{3x^2 - 7x - 6}{3x^2 - 13x - 10} \div \dfrac{2x^2 - x - 1}{4x^2 - 18x - 10}$

12. $\dfrac{3}{7 - x} + \dfrac{x - 2}{x - 7}$

13. $(6x^4 - 3x^3 - 11x^2 + 2x + 4) \div (3x^2 - 1)$

14. $\dfrac{5 + \dfrac{2}{x}}{3 - \dfrac{1}{x}}$

15. $\dfrac{x}{x^2 - 7x + 6} - \dfrac{x}{x^2 - 2x - 24}$

16. $\dfrac{\dfrac{3}{x + 1} + \dfrac{4}{x}}{\dfrac{4}{x}}$

17. $\dfrac{x^2 - x - 6}{x + 1} \div \left(\dfrac{x^2 - 9}{x^2 - 1} \cdot \dfrac{x - 1}{x + 3}\right)$

18. $(64x^3 + 4) \div (4x + 2)$

19. $\dfrac{x + 1}{x^2 + x - 2} - \dfrac{1}{x^2 - 3x + 2} + \dfrac{2x}{x^2 - 4}$

20. Find the domain of $f(x) = \dfrac{5x - 10}{x^2 + 5x - 14}$. Then simplify the function's equation.

SECTION

6.5 Synthetic Division and the Remainder Theorem

What am I supposed to learn?

After studying this section, you should be able to:

1. Divide polynomials using synthetic division.
2. Evaluate a polynomial function using the Remainder Theorem.
3. Show that a number is a solution of a polynomial equation using the Remainder Theorem.

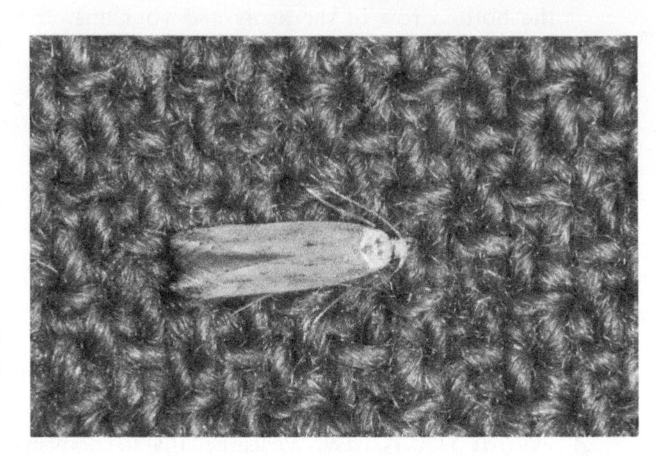

A moth has moved into your closet. She appeared in your bedroom at night, but somehow her relatively stout body escaped your clutches. Within a few weeks, swarms of moths in your tattered wardrobe suggest that Mama Moth was in the family way. There must be at least 200 critters nesting in every crevice of your clothing.

Two hundred plus moth-tykes from one female moth—is this possible? Indeed it is. The number of eggs, $f(x)$, in a female moth is a function of her abdominal width, x, in millimeters, modeled by

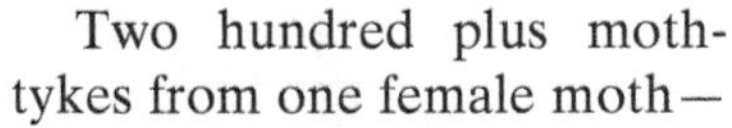

$$f(x) = 14x^3 - 17x^2 - 16x + 34, \quad 1.5 \leq x \leq 3.5.$$

Because there are 200 moths feasting on your favorite sweaters, Mama's abdominal width can be estimated by finding the solutions of the polynomial equation

$$14x^3 - 17x^2 - 16x + 34 = 200.$$

With mathematics present even in your quickly disappearing attire, we move from rags to a shortcut for long division, called *synthetic division*. In the Exercise Set (Exercise 41), you will use this shortcut to find Mama Moth's abdominal width.

1 Divide polynomials using synthetic division.

Dividing Polynomials Using Synthetic Division

We can use **synthetic division** to divide polynomials if the divisor is of the form $x - c$. This method provides a quotient more quickly than long division. Let's compare the two methods showing $x^3 + 4x^2 - 5x + 5$ divided by $x - 3$.

Long Division

$$\begin{array}{r} x^2 + 7x + 16 \\ x - 3\overline{)x^3 + 4x^2 - 5x + 5} \\ \ominus x^3 \oplus 3x^2 \qquad\qquad \\ \hline 7x^2 - 5x \qquad \\ \ominus 7x^2 \oplus 21x \qquad \\ \hline 16x + 5 \\ \ominus 16x \oplus 48 \\ \hline 53 \end{array}$$

Quotient: $x^2 + 7x + 16$

Dividend: $x^3 + 4x^2 - 5x + 5$

Divisor: $x - c$; $c = 3$

Remainder: 53

Synthetic Division

$$\begin{array}{r|rrrr} 3 & 1 & 4 & -5 & 5 \\ & & 3 & 21 & 48 \\ \hline & 1 & 7 & 16 & 53 \end{array}$$

Notice the relationship between the polynomials in the long division process and the numbers that appear in synthetic division.

These are the coefficients of the dividend $x^3 + 4x^2 - 5x + 5$.

The divisor is $x - 3$. This is 3, or c, in $x - c$.

$$\begin{array}{r|rrrr} 3 & 1 & 4 & -5 & 5 \\ & & 3 & 21 & 48 \\ \hline & 1 & 7 & 16 & 53 \end{array}$$

These are the coefficients of the quotient $x^2 + 7x + 16$.

This is the remainder.

Now let's look at the steps involved in synthetic division.

Synthetic Division To divide a polynomial by $x - c$:

Example

1. Arrange polynomials in descending powers, with a 0 coefficient for any missing term.

$$x - 3\overline{)x^3 + 4x^2 - 5x + 5}$$

2. Write c for the divisor, $x - c$. To the right, write the coefficients of the dividend.

$$\begin{array}{r|rrrr} 3 & 1 & 4 & -5 & 5 \end{array}$$

3. Write the leading coefficient of the dividend on the bottom row.

$$\begin{array}{r|rrrr} 3 & 1 & 4 & -5 & 5 \\ & \downarrow & & & \\ \hline & 1 & & & \end{array}$$

Bring down 1.

4. Multiply c (in this case, 3) times the value just written on the bottom row. Write the product in the next column in the second row.

$$\begin{array}{r|rrrr} 3 & 1 & 4 & -5 & 5 \\ & & 3 & & \\ \hline & 1 & & & \end{array}$$

Multiply by 3: $3 \cdot 1 = 3$.

5. Add the values in this new column, writing the sum in the bottom row.

$$\begin{array}{r|rrrr} 3 & 1 & 4 & -5 & 5 \\ & & 3 & & \\ \hline & 1 & 7 & & \end{array}$$

Add.

6. Repeat this series of multiplications and additions until all columns are filled in.

$$\begin{array}{r|rrrr} 3 & 1 & 4 & -5 & 5 \\ & & 3 & 21 & \\ \hline & 1 & 7 & 16 & \end{array}$$

Add.

Multiply by 3: $3 \cdot 7 = 21$.

$$\begin{array}{r|rrrr} 3 & 1 & 4 & -5 & 5 \\ & & 3 & 21 & 48 \\ \hline & 1 & 7 & 16 & 53 \end{array}$$

Add.

Multiply by 3: $3 \cdot 16 = 48$.

7. Use the numbers in the last row to write the quotient, plus the remainder above the divisor. **The degree of the first term of the quotient is one less than the degree of the first term of the dividend.** The final value in this row is the remainder.

Written from 1 7 16 53 the last row of the synthetic division

$$x - 3\overline{)x^3 + 4x^2 - 5x + 5} \quad \text{quotient: } x^2 + 7x + 16 + \frac{53}{x-3}$$

EXAMPLE 1 Using Synthetic Division

Use synthetic division to divide $5x^3 + 6x + 8$ by $x + 2$.

Solution The divisor must be in the form $x - c$. Thus, we write $x + 2$ as $x - (-2)$. This means that $c = -2$. Writing a 0 coefficient for the missing x^2-term in the dividend, we can express the division as follows:

$$x - (-2)\overline{)5x^3 + 0x^2 + 6x + 8}.$$

Now we are ready to set up the problem so that we can use synthetic division.

Use the coefficients of the dividend $5x^3 + 0x^2 + 6x + 8$ in descending powers of x.

This is c in $x - (-2)$.

$$\begin{array}{r|rrrr} -2 & 5 & 0 & 6 & 8 \end{array}$$

We begin the synthetic division process by bringing down 5. This is followed by a series of multiplications and additions.

1. Bring down 5.

$$\begin{array}{r|rrrr} -2 & 5 & 0 & 6 & 8 \\ & & & & \\ \hline & 5 & & & \end{array}$$

2. Multiply: $-2(5) = -10$.

$$\begin{array}{r|rrrr} -2 & 5 & 0 & 6 & 8 \\ & & -10 & & \\ \hline & 5 & & & \end{array}$$

Multiply 5 by −2.

3. Add: $0 + (-10) = -10$.

$$\begin{array}{r|rrrr} -2 & 5 & 0 & 6 & 8 \\ & & -10 & & \\ \hline & 5 & -10 & & \end{array}$$

Add.

4. Multiply: $-2(-10) = 20$.

$$\begin{array}{r|rrrr} -2 & 5 & 0 & 6 & 8 \\ & & -10 & 20 & \\ \hline & 5 & -10 & & \end{array}$$

Multiply −10 by −2.

5. Add: $6 + 20 = 26$.

$$\begin{array}{r|rrrr} -2 & 5 & 0 & 6 & 8 \\ & & -10 & 20 & \\ \hline & 5 & -10 & 26 & \end{array}$$

Add.

6. Multiply: $-2(26) = -52$.

$$\begin{array}{r|rrrr} -2 & 5 & 0 & 6 & 8 \\ & & -10 & 20 & -52 \\ \hline & 5 & -10 & 26 & \end{array}$$

Multiply 26 by −2.

7. Add: $8 + (-52) = -44$.

$$\begin{array}{r|rrrr} -2 & 5 & 0 & 6 & 8 \\ & & -10 & 20 & -52 \\ \hline & 5 & -10 & 26 & -44 \end{array}$$

Add.

The numbers in the last row represent the coefficients of the quotient and the remainder. The degree of the first term of the quotient is one less than that of the dividend. Because the degree of the dividend, $5x^3 + 6x + 8$, is 3, the degree of the quotient is 2. This means that the 5 in the last row represents $5x^2$.

$$\begin{array}{r|rrrr} -2 & 5 & 0 & 6 & 8 \\ & & -10 & 20 & -52 \\ \hline & 5 & -10 & 26 & -44 \end{array}$$

The quotient is $5x^2 - 10x + 26$.

The remainder is −44.

Thus,

$$x + 2\overline{)5x^3 + 6x + 8} \quad \text{quotient: } 5x^2 - 10x + 26 - \frac{44}{x + 2}$$

■

CHECK POINT 1 Use synthetic division to divide $x^3 - 7x - 6$ by $x + 2$.

---

2 Evaluate a polynomial function using the Remainder Theorem.

The Remainder Theorem

We have seen that the answer to a long division problem can be checked: Find the product of the divisor and the quotient and add the remainder. The result should be the dividend. If the divisor is $x - c$, we can express this idea symbolically:

$$f(x) = (x - c)q(x) + r.$$

Dividend | Divisor | Quotient | The remainder, r, is a constant when dividing by $x - c$.

Now let's evaluate f at c.

$$f(c) = (c - c)q(c) + r$$ Find $f(c)$ by letting $x = c$ in $f(x) = (x - c)q(x) + r$. This will give an expression for r.

$$f(c) = 0 \cdot q(c) + r$$ $c - c = 0$

$$f(c) = r$$ $0 \cdot q(c) = 0$ and $0 + r = r$.

What does this last equation mean? If a polynomial is divided by $x - c$, the remainder is the value of the polynomial at c. This result is called the **Remainder Theorem**.

The Remainder Theorem

If the polynomial $f(x)$ is divided by $x - c$, then the remainder is $f(c)$.

Example 2 shows how we can use the Remainder Theorem to evaluate a polynomial function at 2. Rather than substituting 2 for x, we divide the function by $x - 2$. The remainder is $f(2)$.

EXAMPLE 2 Using the Remainder Theorem to Evaluate a Polynomial Function

Given $f(x) = x^3 - 4x^2 + 5x + 3$, use the Remainder Theorem to find $f(2)$.

Solution By the Remainder Theorem, if $f(x)$ is divided by $x - 2$, then the remainder is $f(2)$. We'll use synthetic division to divide.

$$\begin{array}{r|rrrr} 2 & 1 & -4 & 5 & 3 \\ & & 2 & -4 & 2 \\ \hline & 1 & -2 & 1 & 5 \end{array}$$

Remainder (5)

The remainder, 5, is the value of $f(2)$. Thus, $f(2) = 5$. We can verify that this is correct by evaluating $f(2)$ directly. Using $f(x) = x^3 - 4x^2 + 5x + 3$, we obtain

$$f(2) = 2^3 - 4 \cdot 2^2 + 5 \cdot 2 + 3 = 8 - 16 + 10 + 3 = 5. \blacksquare$$

CHECK POINT 2 Given $f(x) = 3x^3 + 4x^2 - 5x + 3$, use the Remainder Theorem to find $f(-4)$.

---

3 Show that a number is a solution of a polynomial equation using the Remainder Theorem.

If the polynomial $f(x)$ is divided by $x - c$ and the remainder is zero, then $f(c) = 0$. This means that c is a solution of the polynomial equation $f(x) = 0$.

EXAMPLE 3 Using the Remainder Theorem

Show that 3 is a solution of the equation

$$2x^3 - 3x^2 - 11x + 6 = 0.$$

Then solve the polynomial equation.

Solution One way to show that 3 is a solution is to substitute 3 for x in the equation and obtain 0. An easier way is to use synthetic division and the Remainder Theorem.

Proposed solution

$$\begin{array}{r|rrrr} 3 & 2 & -3 & -11 & 6 \\ & & 6 & 9 & -6 \\ \hline & 2 & 3 & -2 & 0 \end{array}$$

Remainder

$$x - 3\overline{)2x^3 - 3x^2 - 11x + 6} \quad \text{quotient: } 2x^2 + 3x - 2$$

Equivalently,
$$2x^3 - 3x^2 - 11x + 6 = (x - 3)(2x^2 + 3x - 2).$$

Because the remainder is 0, the polynomial has a value of 0 when $x = 3$. Thus, 3 is a solution of the given equation.

The synthetic division also shows that $x - 3$ divides the polynomial with a zero remainder. Thus, $x - 3$ is a factor of the polynomial, as shown to the right of the synthetic division. The other factor is the quotient found in the last row of the synthetic division. Now we can solve the polynomial equation.

$2x^3 - 3x^2 - 11x + 6 = 0$ This is the given equation.

$(x - 3)(2x^2 + 3x - 2) = 0$ Factor using the result from the synthetic division.

$(x - 3)(2x - 1)(x + 2) = 0$ Factor the trinomial.

$x - 3 = 0$ or $2x - 1 = 0$ or $x + 2 = 0$ Set each factor equal to 0.

$x = 3$ $\quad x = \frac{1}{2}$ $\quad x = -2$ Solve for x.

The solutions are -2, $\frac{1}{2}$, and 3, and the solution set is $\{-2, \frac{1}{2}, 3\}$. ■

Using Technology

Graphic Connections

Because the solution set of

$$2x^3 - 3x^2 - 11x + 6 = 0$$

is $\{-2, \frac{1}{2}, 3\}$, this implies that the polynomial function

$$f(x) = 2x^3 - 3x^2 - 11x + 6$$

has x-intercepts at -2, $\frac{1}{2}$, and 3. This is verified by the graph of f.

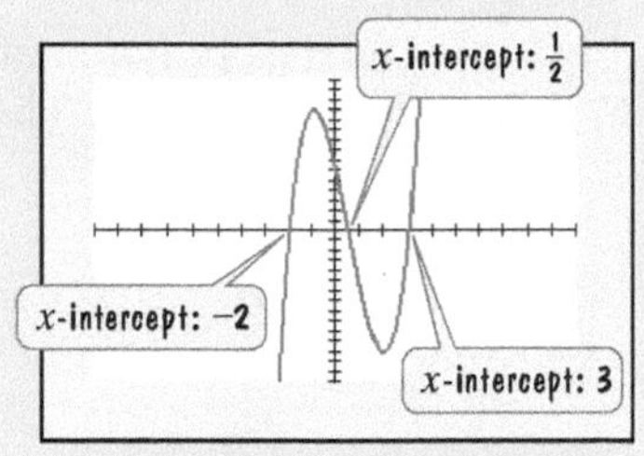

[−10, 10, 1] by [−15, 15, 1]

✓ **CHECK POINT 3** Use synthetic division to show that -1 is a solution of the equation

$$15x^3 + 14x^2 - 3x - 2 = 0.$$

Then solve the polynomial equation.

CONCEPT AND VOCABULARY CHECK

Fill in each blank so that the resulting statement is true.

1. To divide $x^3 + 5x^2 - 7x + 1$ by $x - 4$ using synthetic division, the first step is to write

   ____⌋ ____ ____ ____ ____.

2. To divide $4x^3 - 8x - 2$ by $x + 5$ using synthetic division, the first step is to write

   ____⌋ ____ ____ ____ ____.

3. True or false:

$$\begin{array}{r|rrrr} -1 & 3 & -4 & 2 & -1 \\ & & -3 & 7 & -9 \\ \hline & 3 & -7 & 9 & -10 \end{array} \quad \text{means} \quad \frac{3x^3 - 4x^2 + 2x - 1}{x + 1} = 3x^2 - 7x + 9 - \frac{10}{x + 1}.$$ ________

4. The Remainder Theorem states that if the polynomial $f(x)$ is divided by $x - c$, then the remainder is ________.

6.5 EXERCISE SET MyMathLab®

Practice Exercises

In Exercises 1–18, divide using synthetic division. In the first two exercises, begin the process as shown.

1. $(2x^2 + x - 10) \div (x - 2)$ $\quad 2\rfloor \;\; 2 \quad 1 \quad -10$
2. $(x^2 + x - 2) \div (x - 1)$ $\quad 1\rfloor \;\; 1 \quad 1 \quad -2$
3. $(3x^2 + 7x - 20) \div (x + 5)$
4. $(5x^2 - 12x - 8) \div (x + 3)$
5. $(4x^3 - 3x^2 + 3x - 1) \div (x - 1)$
6. $(5x^3 - 6x^2 + 3x + 11) \div (x - 2)$
7. $(6x^5 - 2x^3 + 4x^2 - 3x + 1) \div (x - 2)$
8. $(x^5 + 4x^4 - 3x^2 + 2x + 3) \div (x - 3)$
9. $(x^2 - 5x - 5x^3 + x^4) \div (5 + x)$
10. $(x^2 - 6x - 6x^3 + x^4) \div (6 + x)$
11. $(3x^3 + 2x^2 - 4x + 1) \div \left(x - \frac{1}{3}\right)$
12. $(2x^4 - x^3 + 2x^2 - 3x + 1) \div \left(x - \frac{1}{2}\right)$
13. $\dfrac{x^5 + x^3 - 2}{x - 1}$
14. $\dfrac{x^7 + x^5 - 10x^3 + 12}{x + 2}$
15. $\dfrac{x^4 - 256}{x - 4}$
16. $\dfrac{x^7 - 128}{x - 2}$
17. $\dfrac{2x^5 - 3x^4 + x^3 - x^2 + 2x - 1}{x + 2}$
18. $\dfrac{x^5 - 2x^4 - x^3 + 3x^2 - x + 1}{x - 2}$

In Exercises 19–26, use synthetic division and the Remainder Theorem to find the indicated function value.

19. $f(x) = 2x^3 - 11x^2 + 7x - 5;\ f(4)$
20. $f(x) = x^3 - 7x^2 + 5x - 6;\ f(3)$
21. $f(x) = 3x^3 - 7x^2 - 2x + 5;\ f(-3)$
22. $f(x) = 4x^3 + 5x^2 - 6x - 4;\ f(-2)$
23. $f(x) = x^4 + 5x^3 + 5x^2 - 5x - 6;\ f(3)$
24. $f(x) = x^4 - 5x^3 + 5x^2 + 5x - 6;\ f(2)$
25. $f(x) = 2x^4 - 5x^3 - x^2 + 3x + 2;\ f\left(-\frac{1}{2}\right)$
26. $f(x) = 6x^4 + 10x^3 + 5x^2 + x + 1;\ f\left(-\frac{2}{3}\right)$

In Exercises 27–32, use synthetic division to show that the number given to the right of each equation is a solution of the equation. Then solve the polynomial equation.

27. $x^3 - 4x^2 + x + 6 = 0;\ \ -1$
28. $x^3 - 2x^2 - x + 2 = 0;\ \ -1$
29. $2x^3 - 5x^2 + x + 2 = 0;\ \ 2$
30. $2x^3 - 3x^2 - 11x + 6 = 0;\ \ -2$
31. $6x^3 + 25x^2 - 24x + 5 = 0;\ \ -5$
32. $3x^3 + 7x^2 - 22x - 8 = 0;\ \ -4$

Practice PLUS

In Exercises 33–36, use the graph or the table to determine a solution of each equation. Use synthetic division to verify that this number is a solution of the equation. Then solve the polynomial equation.

33. $x^3 + 2x^2 - 5x - 6 = 0$

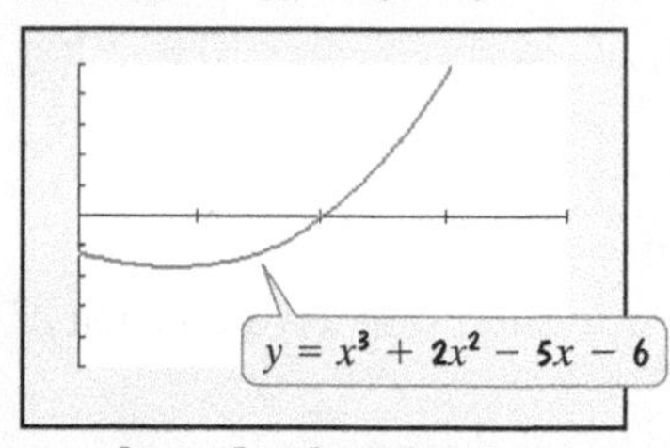

[0, 4, 1] by [−25, 25, 5]

34. $2x^3 + x^2 - 13x + 6 = 0$

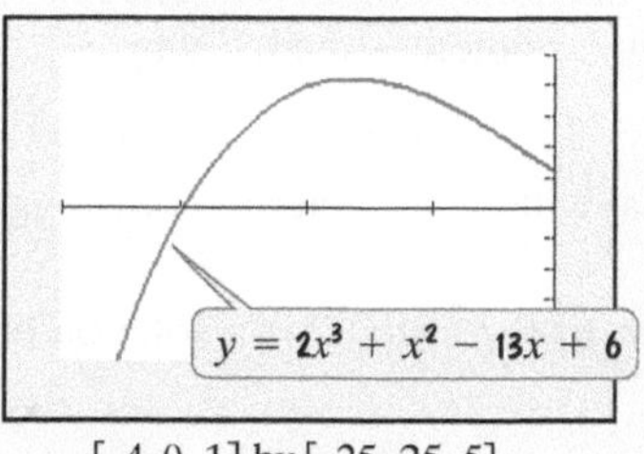

[−4, 0, 1] by [−25, 25, 5]

35. $6x^3 - 11x^2 + 6x - 1 = 0$

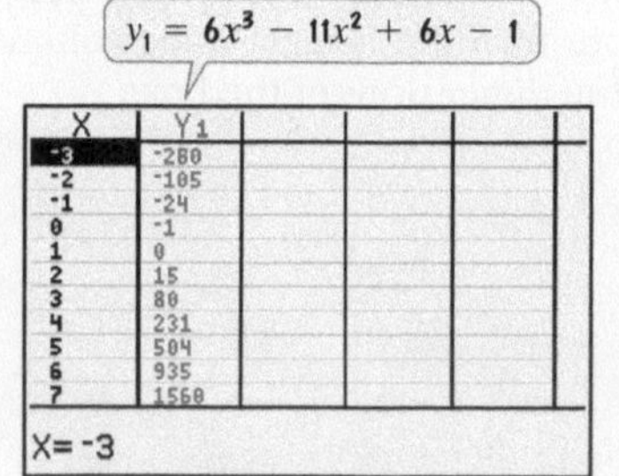

X	Y1
-3	-280
-2	-105
-1	-24
0	-1
1	0
2	15
3	80
4	231
5	504
6	935
7	1560

36. $2x^3 + 11x^2 - 7x - 6 = 0$

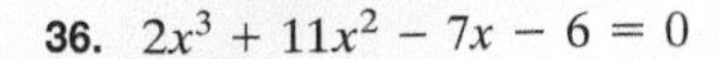

$y_1 = 2x^3 + 11x^2 - 7x - 6$

X	Y_1
-3	60
-2	36
-1	10
0	-6
1	0
2	40
3	126
4	270
5	484
6	780
7	1170

X=-3

In Exercises 37–38, perform the given operations.

37. $(22x - 24 + 7x^3 - 29x^2 + 4x^4)(x + 4)^{-1}$

38. $(9 - x^2 + 6x + 2x^3)(x + 1)^{-1}$

In Exercises 39–40, write a polynomial that represents the length of each rectangle.

39.

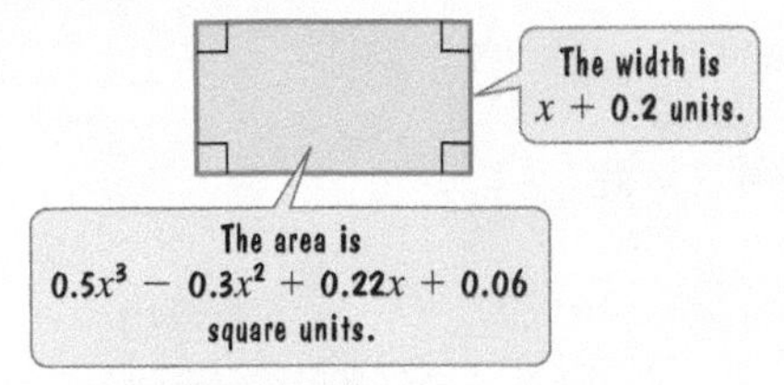
The width is $x + 0.2$ units.
The area is $0.5x^3 - 0.3x^2 + 0.22x + 0.06$ square units.

40.

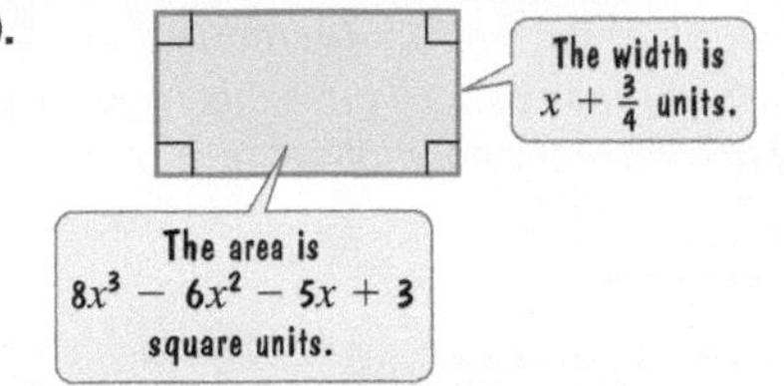
The width is $x + \frac{3}{4}$ units.
The area is $8x^3 - 6x^2 - 5x + 3$ square units.

Application Exercises

41. a. Use synthetic division to show that 3 is a solution of the polynomial equation

$$14x^3 - 17x^2 - 16x - 177 = 0.$$

b. Use the solution from part (a) to solve this problem. The number of eggs, $f(x)$, in a female moth is a function of her abdominal width, x, in millimeters, modeled by

$$f(x) = 14x^3 - 17x^2 - 16x + 34.$$

What is the abdominal width when there are 211 eggs?

42. a. Use synthetic division to show that 2 is a solution of the polynomial equation

$$2h^3 + 14h^2 - 72 = 0.$$

b. Use the solution from part (a) to solve this problem. The width of a rectangular box is twice the height and the length is 7 inches more than the height. If the volume is 72 cubic inches, find the dimensions of the box.

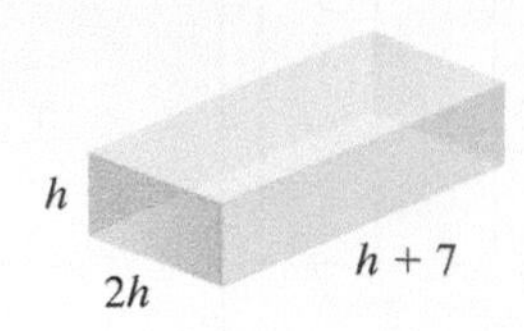
h
$2h$
$h + 7$

Explaining the Concepts

43. Explain how to perform synthetic division. Use the division problem

$$(2x^3 - 3x^2 - 11x + 7) \div (x - 3)$$

to support your explanation.

44. State the Remainder Theorem.

45. Explain how the Remainder Theorem can be used to find $f(-6)$ if $f(x) = x^4 + 7x^3 + 8x^2 + 11x + 5$. What advantage is there to using the Remainder Theorem in this situation rather than evaluating $f(-6)$ directly?

46. Explain how the Remainder Theorem and synthetic division can be used to determine whether -4 is a solution of the following equation:

$$5x^3 + 22x^2 + x - 28 = 0.$$

Technology Exercise

47. For each equation that you solved in Exercises 27–32, use a graphing utility to graph the polynomial function on the left side of the equation. Use end behavior to obtain a complete graph. Then use the graph's x-intercepts to verify your solutions.

Critical Thinking Exercises

Make Sense? *In Exercises 48–51, determine whether each statement makes sense or does not make sense, and explain your reasoning.*

48. There are certain kinds of polynomial divisions that I can perform using long division, but not using synthetic division.

49. Every time I divide polynomials using synthetic division, I am using a highly condensed form of the long division procedure where omitting the variables and exponents does not involve the loss of any essential data.

50. The only nongraphic method that I have for evaluating a function at a given value is to substitute that value into the function's equation.

51. The Remainder Theorem gives me a method for factoring certain polynomials that I could not factor using the factoring strategies from the previous chapter.

52. Synthetic division is a process for dividing a polynomial by $x - c$. The coefficient of x is 1. How might synthetic division be used if you are dividing by $2x - 4$?

53. Use synthetic division to show that 5 is a solution of

$$x^4 - 4x^3 - 9x^2 + 16x + 20 = 0.$$

Then solve the polynomial equation. *Hint:* Use factoring by grouping when working with the quotient factor.

Review Exercises

54. Solve: $4x + 3 - 13x - 7 < 2(3 - 4x)$. (Section 4.1, Example 3)

55. Solve: $2x(x + 3) + 6(x - 3) = -28$. (Section 5.7, Example 2)

56. Solve by Cramer's rule:

$$\begin{cases} 7x - 6y = 17 \\ 3x + y = 18. \end{cases}$$

(Section 3.5, Example 2)

Preview Exercises

Exercises 57–59 will help you prepare for the material covered in the next section. In each exercise, find the LCD of the rational expressions.

57. $\frac{x+4}{2x}, \frac{x+20}{3x}$

58. $\frac{x}{3}, \frac{9}{x}, 4$

59. $\frac{2x}{x-3}, \frac{6}{x+3}, -\frac{28}{x^2-9}$

SECTION

6.6 Rational Equations

What am I supposed to learn?

After studying this section, you should be able to:

1. Solve rational equations.
2. Solve problems involving rational functions that model applied situations.

The lake is in one of the city's favorite parks and the time has come to clean it up. Voters in the city have committed \$80 thousand for the cleanup. We know that

$$f(x) = \frac{120x}{100 - x}$$

models the cost, $f(x)$, in thousands of dollars, to remove x% of the lake's pollutants. What percentage of the pollutants can be removed for \$80 thousand?

To determine the percentage, we use the given cost-benefit model. Voters have committed \$80 thousand, so substitute 80 for $f(x)$:

$$80 = \frac{120x}{100 - x}.$$

This equation contains a rational expression.

Now we have to solve the equation and find the value for x. This variable represents the percentage of the pollutants that can be removed for \$80 thousand.

A **rational equation**, also called a **fractional equation**, is an equation containing one or more rational expressions. The equation shown above is an example of a rational equation. Do you see that there is a variable in a denominator? This is a characteristic of many rational equations. In this section, you will learn a procedure for solving such equations.

1 Solve rational equations.

Solving Rational Equations

We have seen that the LCD is used to add and subtract rational expressions. By contrast, when solving rational equations, **the LCD is used as a multiplier that clears an equation of fractions.**

EXAMPLE 1 Solving a Rational Equation

Solve: $\frac{x+4}{2x} + \frac{x+20}{3x} = 3.$

Solution Notice that the variable x appears in both denominators. We must avoid any values of the variable that make a denominator zero.

$$\frac{x+4}{2x} + \frac{x+20}{3x} = 3$$

This denominator would equal zero if $x = 0$. This denominator would equal zero if $x = 0$.

We see that x cannot equal zero.

The denominators are $2x$ and $3x$. The least common denominator is $6x$. We begin by multiplying both sides of the equation by $6x$. We will also write the restriction that x cannot equal zero to the right of the equation.

$$\frac{x+4}{2x} + \frac{x+20}{3x} = 3, x \neq 0 \quad \text{This is the given equation.}$$

$$6x\left(\frac{x+4}{2x} + \frac{x+20}{3x}\right) = 6x \cdot 3 \quad \text{Multiply both sides by } 6x\text{, the LCD.}$$

$$\frac{6x}{1}\cdot\frac{x+4}{2x} + \frac{6x}{1}\cdot\frac{x+20}{3x} = 18x \quad \text{Use the distributive property.}$$

$$3(x+4) + 2(x+20) = 18x \quad \text{Divide out common factors in the multiplications.}$$

Observe that the equation is now cleared of fractions.

$$3x + 12 + 2x + 40 = 18x \quad \text{Use the distributive property.}$$

$$5x + 52 = 18x \quad \text{Combine like terms.}$$

$$52 = 13x \quad \text{Subtract } 5x \text{ from both sides.}$$

$$4 = x \quad \text{Divide both sides by 13.}$$

The proposed solution, 4, is not part of the restriction $x \neq 0$. It should check in the original equation.

Check 4:

$$\frac{x+4}{2x} + \frac{x+20}{3x} = 3$$

$$\frac{4+4}{2\cdot 4} + \frac{4+20}{3\cdot 4} \stackrel{?}{=} 3$$

$$\frac{8}{8} + \frac{24}{12} \stackrel{?}{=} 3$$

$$1 + 2 \stackrel{?}{=} 3$$

$$3 = 3, \quad \text{true}$$

This true statement verifies that the solution is 4 and the solution set is $\{4\}$. ■

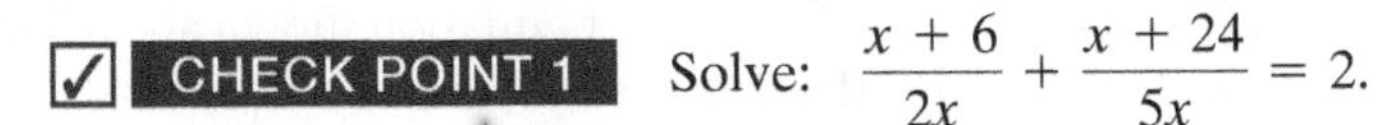

CHECK POINT 1 Solve: $\frac{x+6}{2x} + \frac{x+24}{5x} = 2$.

The following steps may be used to solve a rational equation:

Solving Rational Equations

1. List restrictions on the variable. Avoid any values of the variable that make a denominator zero.
2. Clear the equation of fractions by multiplying both sides by the LCD of all rational expressions in the equation.
3. Solve the resulting equation.
4. Reject any proposed solution that is in the list of restrictions on the variable. Check other proposed solutions in the original equation.

EXAMPLE 2 Solving a Rational Equation

Solve: $\frac{x+1}{x+10} = \frac{x-2}{x+4}$.

Solution

Step 1. List restrictions on the variable.

$$\frac{x+1}{x+10} = \frac{x-2}{x+4}$$

This denominator would equal zero if $x = -10$. This denominator would equal zero if $x = -4$.

The restrictions are $x \neq -10$ and $x \neq -4$.

Step 2. Multiply both sides by the LCD. The denominators are $x + 10$ and $x + 4$. Thus, the LCD is $(x + 10)(x + 4)$.

$$\frac{x+1}{x+10} = \frac{x-2}{x+4}, \quad x \neq -10, x \neq -4 \qquad \text{This is the given equation.}$$

$$(x+10)(x+4)\cdot\left(\frac{x+1}{x+10}\right) = (x+10)(x+4)\cdot\left(\frac{x-2}{x+4}\right) \qquad \text{Multiply both sides by the LCD.}$$

$$(x+4)(x+1) = (x+10)(x-2) \qquad \text{Simplify.}$$

Step 3. Solve the resulting equation.

$$(x+4)(x+1) = (x+10)(x-2) \qquad \text{This is the equation cleared of fractions.}$$
$$x^2 + 5x + 4 = x^2 + 8x - 20 \qquad \text{Use FOIL multiplication on each side.}$$
$$5x + 4 = 8x - 20 \qquad \text{Subtract } x^2 \text{ from both sides.}$$
$$-3x + 4 = -20 \qquad \text{Subtract } 8x \text{ from both sides.}$$
$$-3x = -24 \qquad \text{Subtract 4 from both sides.}$$
$$x = 8 \qquad \text{Divide both sides by } -3.$$

Step 4. Check the proposed solution in the original equation. The proposed solution, 8, is not part of the restriction that $x \neq -10$ and $x \neq -4$. Substitute 8 for x in the given equation. You should obtain the true statement $\frac{1}{2} = \frac{1}{2}$. The solution is 8 and the solution set is $\{8\}$. ■

CHECK POINT 2 Solve: $\frac{x-3}{x+1} = \frac{x-2}{x+6}$.

Great Question!

Earlier in the chapter, I learned how to add and subtract rational expressions. In this section, I'm solving equations that contain addition and subtraction of rational expressions. Can you do a side-by-side comparison of the two procedures?

We simplify rational expressions. We solve rational equations. Notice the differences between the procedures.

Simplifying a Rational Expression	Solving a Rational Equation
Simplify: $\frac{9}{4x} - \frac{5}{2x} - \frac{3}{4}$. This is not an equation. There is no equal sign.	Solve: $\frac{9}{4x} - \frac{5}{2x} = \frac{3}{4}$. This is an equation. There is an equal sign.
Solution The LCD is $4x$. Rewrite each expression with this LCD and retain the LCD. $\frac{9}{4x} - \frac{5}{2x} - \frac{3}{4}$ $= \frac{9}{4x} - \frac{5\cdot 2}{2x\cdot 2} - \frac{3\cdot x}{4\cdot x}$ $= \frac{9}{4x} - \frac{10}{4x} - \frac{3x}{4x}$ $= \frac{9-10-3x}{4x} = \frac{-1-3x}{4x}$	**Solution** The LCD is $4x$. Multiply both sides by this LCD and clear the fractions. $4x\left(\frac{9}{4x} - \frac{5}{2x}\right) = 4x\cdot\frac{3}{4}$ $\frac{4x}{1}\cdot\frac{9}{4x} - \frac{4x}{1}\cdot\frac{5}{2x} = 4x\cdot\frac{3}{4}$ $9 - 10 = 3x$ $-1 = 3x$ $-\frac{1}{3} = x$ The solution set is $\left\{-\frac{1}{3}\right\}$.

You only eliminate the denominators when solving a rational equation with an equal sign. You should never begin by eliminating the denominators when simplifying a rational expression involving addition or subtraction with no equal sign.

EXAMPLE 3 Solving a Rational Equation

Solve: $\frac{x}{x-3} = \frac{3}{x-3} + 9.$

Solution

Step 1. List restrictions on the variable.

$$\frac{x}{x-3} = \frac{3}{x-3} + 9$$

These denominators are zero if $x = 3$.

The restriction is $x \neq 3$.

Step 2. Multiply both sides by the LCD. The LCD is $x - 3$.

$$\frac{x}{x-3} = \frac{3}{x-3} + 9, \quad x \neq 3$$ This is the given equation.

$$(x-3) \cdot \frac{x}{x-3} = (x-3)\left(\frac{3}{x-3} + 9\right)$$ Multiply both sides by the LCD.

$$\cancel{(x-3)} \cdot \frac{x}{\cancel{x-3}} = \cancel{(x-3)} \cdot \frac{3}{\cancel{x-3}} + 9(x-3)$$ Use the distributive property on the right side.

$$x = 3 + 9(x-3)$$ Simplify.

Step 3. Solve the resulting equation.

$$x = 3 + 9(x-3)$$ This is the equation cleared of fractions.

$$x = 3 + 9x - 27$$ Use the distributive property on the right side.

$$x = 9x - 24$$ Combine numerical terms.

$$-8x = -24$$ Subtract $9x$ from both sides.

$$x = 3$$ Divide both sides by -8.

Step 4. Check proposed solutions. The proposed solution, 3, is *not* a solution because of the restriction that $x \neq 3$. Notice that 3 makes both of the denominators zero in the original equation. There is no solution for this equation. The solution set is $\varnothing$, the empty set. ■

Great Question!

Cut to the chase: When do I get rid of proposed solutions in rational equations?

Reject any proposed solution that causes any denominator in a rational equation to equal 0.

✓ **CHECK POINT 3** Solve: $\frac{8x}{x+1} = 4 - \frac{8}{x+1}.$

Examples 4 and 5 involve rational equations that become quadratic equations after clearing fractions.

EXAMPLE 4 Solving a Rational Equation

Solve: $\frac{x}{3} + \frac{9}{x} = 4.$

Solution

Step 1. List restrictions on the variable.

$$\frac{x}{3} + \frac{9}{x} = 4$$

This denominator would equal zero if $x = 0$.

The restriction is $x \neq 0$.

Step 2. Multiply both sides by the LCD. The denominators are 3 and x. Thus, the LCD is $3x$.

$$\frac{x}{3} + \frac{9}{x} = 4, \quad x \neq 0 \qquad \text{This is the given equation.}$$

$$3x\left(\frac{x}{3} + \frac{9}{x}\right) = 3x \cdot 4 \qquad \text{Multiply both sides by the LCD.}$$

$$3x \cdot \frac{x}{3} + 3x \cdot \frac{9}{x} = 12x \qquad \text{Use the distributive property on the left side.}$$

$$x^2 + 27 = 12x \qquad \text{Simplify.}$$

Using Technology

Graphic Connections

The graphs of

$$y_1 = \frac{x}{3} + \frac{9}{x}$$

and

$$y_2 = 4$$

have two intersection points. The x-coordinates of the points are 3 and 9. This verifies that

$$\frac{x}{3} + \frac{9}{x} = 4$$

has both 3 and 9 as solutions.

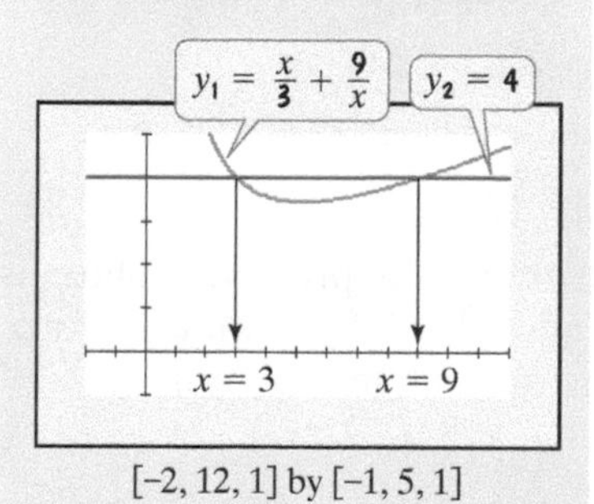

[−2, 12, 1] by [−1, 5, 1]

Step 3. Solve the resulting equation. Can you see that we have a quadratic equation? Write the equation in standard form and solve for x.

$$x^2 + 27 = 12x \qquad \text{This is the equation cleared of fractions.}$$

$$x^2 - 12x + 27 = 0 \qquad \text{Subtract } 12x \text{ from both sides.}$$

$$(x - 9)(x - 3) = 0 \qquad \text{Factor.}$$

$$x - 9 = 0 \quad \text{or} \quad x - 3 = 0 \qquad \text{Set each factor equal to 0.}$$

$$x = 9 \qquad x = 3 \qquad \text{Solve the resulting equations.}$$

Step 4. Check proposed solutions in the original equation. The proposed solutions, 9 and 3, are not part of the restriction that $x \neq 0$. Substitute 9 for x, and then 3 for x, in the given equation. In each case, you should obtain the true statement $4 = 4$. The solutions are 3 and 9, and the solution set is {3, 9}. ■

CHECK POINT 4 Solve: $\frac{x}{2} + \frac{12}{x} = 5$.

EXAMPLE 5 Solving a Rational Equation

Solve: $\frac{2x}{x - 3} + \frac{6}{x + 3} = -\frac{28}{x^2 - 9}$.

Solution

Step 1. List restrictions on the variable. By factoring denominators, it makes it easier to see values that make the denominators zero.

$$\frac{2x}{x - 3} + \frac{6}{x + 3} = -\frac{28}{(x + 3)(x - 3)}$$

This denominator is zero if $x = 3$. This denominator is zero if $x = -3$. This denominator is zero if $x = -3$ or $x = 3$.

The restrictions are $x \neq -3$ and $x \neq 3$.

Step 2. Multiply both sides by the LCD. The LCD is $(x+3)(x-3)$.

$$\frac{2x}{x-3}+\frac{6}{x+3}=-\frac{28}{(x+3)(x-3)},\quad x\neq -3, x\neq 3$$ This is the given equation with a denominator factored.

$$(x+3)(x-3)\left(\frac{2x}{x-3}+\frac{6}{x+3}\right)=(x+3)(x-3)\left(-\frac{28}{(x+3)(x-3)}\right)$$ Multiply both sides by the LCD.

$$(x+3)\cancel{(x-3)}\cdot\frac{2x}{\cancel{x-3}}+\cancel{(x+3)}(x-3)\cdot\frac{6}{\cancel{x+3}}=\cancel{(x+3)}\cancel{(x-3)}\left(-\frac{28}{\cancel{(x+3)}\cancel{(x-3)}}\right)$$ Use the distributive property on the left side.

$$2x(x+3)+6(x-3)=-28$$ Simplify.

Step 3. Solve the resulting equation.

$2x(x+3)+6(x-3)=-28$	This is the equation cleared of fractions.
$2x^2+6x+6x-18=-28$	Use the distributive property twice on the left side.
$2x^2+12x-18=-28$	Combine like terms.
$2x^2+12x+10=0$	Add 28 to both sides and write the quadratic equation in standard form.
$2(x^2+6x+5)=0$	Factor out the GCF.
$2(x+5)(x+1)=0$	Factor the trinomial.
$x+5=0$ or $x+1=0$	Set each variable factor equal to 0.
$x=-5$ $\quad x=-1$	Solve for x.

Step 4. Check the proposed solutions in the original equation. The proposed solutions, -5 and -1, are not part of the restriction that $x\neq-3$ and $x\neq 3$. Substitute -5 for x, and then -1 for x, in the given equation. The resulting true statements verify that -5 and -1 are the solutions, and $\{-5,-1\}$ is the solution set. ■

✓ **CHECK POINT 5** Solve: $\dfrac{3}{x-3}+\dfrac{5}{x-4}=\dfrac{x^2-20}{x^2-7x+12}$.

---

2 Solve problems involving rational functions that model applied situations.

Applications of Rational Equations

Solving rational equations can be used to answer questions about variables contained in rational functions.

EXAMPLE 6 Using a Cost-Benefit Model

The function

$$f(x)=\frac{120x}{100-x}$$

models the cost, $f(x)$, in thousands of dollars, to remove x% of a lake's pollutants. If voters commit \$80 thousand for this project, what percentage of the pollutants can be removed?

Solution Substitute 80, the cost in thousands of dollars, for $f(x)$ and solve the resulting rational equation for x.

$$80 = \frac{120x}{100 - x} \quad \text{The LCD is } 100 - x.$$

$$(100 - x)80 = \cancel{(100 - x)} \cdot \frac{120x}{\cancel{100 - x}} \quad \text{Multiply both sides by the LCD.}$$

$$80(100 - x) = 120x \quad \text{Simplify.}$$

$$8000 - 80x = 120x \quad \text{Use the distributive property on the left side.}$$

$$8000 = 200x \quad \text{Add } 80x \text{ to both sides.}$$

$$40 = x \quad \text{Divide both sides by 200.}$$

If voters commit $80 thousand, 40% of the lake's pollutants can be removed. ■

☑ **CHECK POINT 6** Use the cost-benefit model in Example 6 to answer this question: If voters in the city commit $120 thousand for the project, what percentage of the lake's pollutants can be removed?

Achieving Success

Assuming that you have done very well preparing for an exam, **there are certain things you can do that will make you a better test taker**.

- Get a good sleep the night before the exam.
- Have a good breakfast that balances protein, carbohydrates, and fruit.
- Just before the exam, briefly review the relevant material in the chapter summary.
- Bring everything you need to the exam, including two pencils, an eraser, scratch paper (if permitted), a calculator (if you're allowed to use one), water, and a watch.
- Survey the entire exam quickly to get an idea of its length.
- Read the directions to each problem carefully. Make sure that you have answered the specific question asked.
- Work the easy problems first. Then return to the hard problems you are not sure of. Doing the easy problems first will build your confidence. If you get bogged down on any one problem, you may not be able to complete the exam and receive credit for the questions you can easily answer.
- Attempt every problem. There may be partial credit even if you do not obtain the correct answer.
- Work carefully. Show your step-by-step solutions neatly. Check your work and answers.
- Watch the time. Pace yourself and be aware of when half the time is up. Determine how much of the exam you have completed. This will indicate if you're moving at a good pace or need to speed up. Prepare to spend more time on problems worth more points.
- Never turn in a test early. Use every available minute you are given for the test. If you have extra time, double check your arithmetic and look over your solutions.

CONCEPT AND VOCABULARY CHECK

Fill in each blank so that the resulting statement is true.

1. We clear a rational equation of fractions by multiplying both sides by the __________ of all rational expressions in the equation.
2. We reject any proposed solution of a rational equation that causes a denominator to equal __________.

3. The first step in solving

$$\frac{4}{x} + \frac{1}{2} = \frac{5}{x}$$

is to multiply both sides by ________________.

4. The first step in solving

$$\frac{x-6}{x+5} = \frac{x-3}{x+1}$$

is to multiply both sides by ________________.

5. The restrictions on the variable in the rational equation

$$\frac{1}{x-2} - \frac{2}{x+4} = \frac{2x-1}{x^2+2x-8}$$

are ________________ and ________________.

6.
$$\frac{5}{x+4} + \frac{3}{x+3} = \frac{12x+9}{(x+4)(x+3)}$$

$$(x+4)(x+3)\left(\frac{5}{x+4} + \frac{3}{x+3}\right) = (x+4)(x+3)\left(\frac{12x+9}{(x+4)(x+3)}\right)$$

The resulting equation cleared of fractions is ________________.

7. True or false: A rational equation can have no solution. ________________

6.6 EXERCISE SET

MyMathLab®

Practice Exercises

In Exercises 1–34, solve each rational equation. If an equation has no solution, so state.

1. $\frac{1}{x} + 2 = \frac{3}{x}$

2. $\frac{1}{x} - 3 = \frac{4}{x}$

3. $\frac{5}{x} + \frac{1}{3} = \frac{6}{x}$

4. $\frac{4}{x} + \frac{1}{2} = \frac{5}{x}$

5. $\frac{x-2}{2x} + 1 = \frac{x+1}{x}$

6. $\frac{7x-4}{5x} = \frac{9}{5} - \frac{4}{x}$

7. $\frac{3}{x+1} = \frac{5}{x-1}$

8. $\frac{6}{x+3} = \frac{4}{x-3}$

9. $\frac{x-6}{x+5} = \frac{x-3}{x+1}$

10. $\frac{x+2}{x+10} = \frac{x-3}{x+4}$

11. $\frac{x+6}{x+3} = \frac{3}{x+3} + 2$

12. $\frac{3x+1}{x-4} = \frac{6x+5}{2x-7}$

13. $1 - \frac{4}{x+7} = \frac{5}{x+7}$

14. $5 - \frac{2}{x-5} = \frac{3}{x-5}$

15. $\frac{4x}{x+2} + \frac{2}{x-1} = 4$

16. $\frac{3x}{x+1} + \frac{4}{x-2} = 3$

17. $\frac{8}{x^2-9} + \frac{4}{x+3} = \frac{2}{x-3}$

18. $\frac{32}{x^2-25} = \frac{4}{x+5} + \frac{2}{x-5}$

19. $x + \frac{7}{x} = -8$

20. $x + \frac{6}{x} = -7$

21. $\frac{6}{x} - \frac{x}{3} = 1$

22. $\frac{x}{2} - \frac{12}{x} = 1$

23. $\frac{x+6}{3x-12} = \frac{5}{x-4} + \frac{2}{3}$

24. $\frac{1}{5x+5} = \frac{3}{x+1} - \frac{7}{5}$

25. $\frac{1}{x-1} + \frac{1}{x+1} = \frac{2}{x^2-1}$

26. $\frac{1}{x-2} + \frac{1}{x+2} = \frac{4}{x^2-4}$

27. $\frac{5}{x+4} + \frac{3}{x+3} = \frac{12x+19}{x^2+7x+12}$

28. $\frac{2x-1}{x^2+2x-8} + \frac{2}{x+4} = \frac{1}{x-2}$

29. $\frac{4x}{x+3} - \frac{12}{x-3} = \frac{4x^2+36}{x^2-9}$

30. $\frac{2}{x+3} - \frac{5}{x+1} = \frac{3x+5}{x^2+4x+3}$

31. $\frac{4}{x^2+3x-10} + \frac{1}{x^2+9x+20} = \frac{2}{x^2+2x-8}$

32. $\frac{4}{x^2+3x-10} - \frac{1}{x^2+x-6} = \frac{3}{x^2-x-12}$

33. $\frac{3y}{y^2+5y+6} + \frac{2}{y^2+y-2} = \frac{5y}{y^2+2y-3}$

34. $\frac{y-1}{y^2-4} + \frac{y}{y^2-y-2} = \frac{2y-1}{y^2+3y+2}$

In Exercises 35–38, a rational function g is given. Find all values of a for which g(a) is the indicated value.

35. $g(x) = \frac{x}{2} + \frac{20}{x}; g(a) = 7$

36. $g(x) = \frac{x}{4} + \frac{5}{x}; g(a) = 3$

37. $g(x) = \frac{5}{x+2} + \frac{25}{x^2+4x+4}; g(a) = 20$

38. $g(x) = \frac{3x-2}{x+1} + \frac{x+2}{x-1}; g(a) = 4$

Practice PLUS

In Exercises 39–46, solve or simplify, whichever is appropriate.

39. $\frac{x+2}{x^2-x} - \frac{6}{x^2-1}$

40. $\frac{x+3}{x^2-x} - \frac{8}{x^2-1}$

41. $\frac{x+2}{x^2-x} - \frac{6}{x^2-1} = 0$

42. $\frac{x+3}{x^2-x} - \frac{8}{x^2-1} = 0$

43. $\frac{1}{x^3-8} + \frac{3}{(x-2)(x^2+2x+4)} = \frac{2}{x^2+2x+4}$

44. $\frac{2}{x^3-1} + \frac{4}{(x-1)(x^2+x+1)} = -\frac{1}{x^2+x+1}$

45. $\frac{1}{x^3-8} + \frac{3}{(x-2)(x^2+2x+4)} - \frac{2}{x^2+2x+4}$

46. $\frac{2}{x^3-1} + \frac{4}{(x-1)(x^2+x+1)} + \frac{1}{x^2+x+1}$

In Exercises 47–48, find all values of a for which $f(a) = g(a) + 1$.

47. $f(x) = \frac{x+2}{x+3}, g(x) = \frac{x+1}{x^2+2x-3}$

48. $f(x) = \frac{4}{x-3}, g(x) = \frac{10}{x^2+x-12}$

In Exercises 49–50, find all values of a for which $(f+g)(a) = h(a)$.

49. $f(x) = \frac{5}{x-4}, g(x) = \frac{3}{x-3}, h(x) = \frac{x^2-20}{x^2-7x+12}$

50. $f(x) = \frac{6}{x+3}, g(x) = \frac{2x}{x-3}, h(x) = -\frac{28}{x^2-9}$

Application Exercises

The function

$$f(x) = \frac{250x}{100-x}$$

models the cost, f(x), in millions of dollars, to remove x% of a river's pollutants. Use this function to solve Exercises 51–52.

51. If the government commits \$375 million for this project, what percentage of the pollutants can be removed?

52. If the government commits \$750 million for this project, what percentage of the pollutants can be removed?

In an experiment about memory, students in a language class are asked to memorize 40 vocabulary words in Latin, a language with which they are not familiar. After studying the words for one day, students are tested each day thereafter to see how many words they remember. The class average is then found. The function

$$f(x) = \frac{5x+30}{x}$$

models the average number of Latin words remembered by the students, f(x), after x days. The graph of the rational function is shown. Use the function to solve Exercises 53–56.

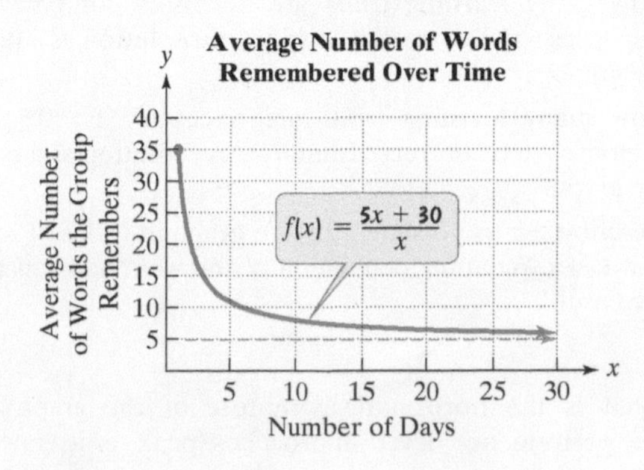

(In Exercises 53–56, be sure to refer to the rational function $f(x) = \frac{5x + 30}{x}$ *and its graph on the previous page.)*

53. After how many days do the students remember 8 words? Identify your solution as a point on the graph.

54. After how many days do the students remember 7 words? Identify your solution as a point on the graph.

55. What is the horizontal asymptote of the graph? Describe what this means about the average number of Latin words remembered by the students over an extended period of time.

56. According to the graph, between which two days do students forget the most? Describe the trend shown by the memory function's graph.

Rational functions can be used to model learning. Many of these functions model the proportion of correct responses as a function of the number of trials of a particular task. One such model, called a learning curve, is

$$f(x) = \frac{0.9x - 0.4}{0.9x + 0.1},$$

where f(x) is the proportion of correct responses after x trials. If f(x) = 0, there are no correct responses. If f(x) = 1, all responses are correct. The graph of the rational function is shown. Use the function to solve Exercises 57–60.

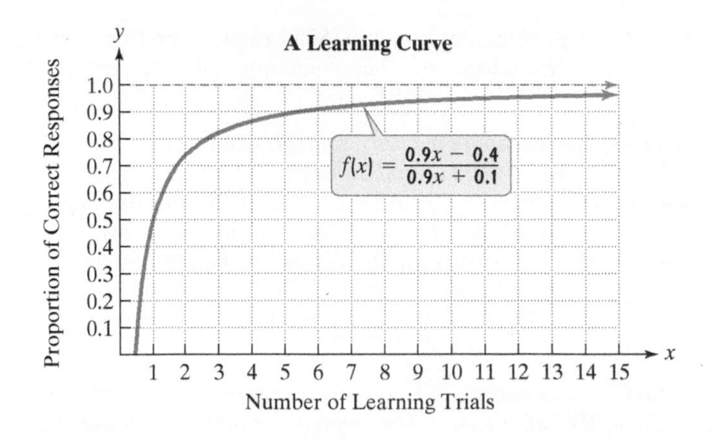

57. How many learning trials are necessary for 0.95 of the responses to be correct? Identify your solution as a point on the graph.

58. How many learning trials are necessary for 0.5 of the responses to be correct? Identify your solution as a point on the graph.

59. Describe the trend shown by the graph in terms of learning new tasks. What happens initially and what happens as time increases?

60. What is the horizontal asymptote of the graph? Once the performance level approaches peak efficiency, what effect does additional practice have on performance? Describe how this is shown by the graph.

61. A company wants to increase the 10% peroxide content of its product by adding pure peroxide (100% peroxide). If x liters of pure peroxide are added to 500 liters of its 10% solution, the concentration, $C(x)$, of the new mixture is given by

$$C(x) = \frac{x + 0.1(500)}{x + 500}.$$

How many liters of pure peroxide should be added to produce a new product that is 28% peroxide?

62. Suppose that x liters of pure acid are added to 200 liters of a 35% acid solution.

a. Write a function that gives the concentration, $C(x)$, of the new mixture. (*Hint*: See Exercise 61.)

b. How many liters of pure acid should be added to produce a new mixture that is 74% acid?

Explaining the Concepts

63. What is a rational equation?

64. Explain how to solve a rational equation.

65. Explain how to find restrictions on the variable in a rational equation.

66. Why should restrictions on the variable in a rational equation be listed before you begin solving the equation?

67. Describe similarities and differences between the procedures needed to solve the following problems:

Add: $\frac{2}{x} + \frac{3}{4}.$ Solve: $\frac{2}{x} + \frac{3}{4} = 1.$

68. Rational functions model learning and forgetting. Use the graphs in Exercises 53–56 and in Exercises 57–60 to describe one similarity and one difference between learning and forgetting over time.

69. Does the graph of the learning curve shown in Exercises 57–60 indicate that practice makes perfect? Explain. Does this have anything to do with what psychologists call "the curse of perfection"?

Technology Exercises

In Exercises 70–74, use a graphing utility to solve each rational equation. Graph each side of the equation in the given viewing rectangle. The solution is the first coordinate of the point(s) of intersection. Check by direct substitution.

70. $\frac{x}{2} + \frac{x}{4} = 6$; $[-5, 10, 1]$ by $[-5, 10, 1]$

71. $\frac{50}{x} = 2x$; $[-10, 10, 1]$ by $[-20, 20, 2]$

72. $x + \frac{6}{x} = -5$; $[-10, 10, 1]$ by $[-7, 10, 1]$

73. $\frac{2}{x} = x + 1$; $[-5, 5, 1]$ by $[-5, 5, 1]$

74. $\frac{3}{x} - \frac{x + 21}{3x} = \frac{5}{3}$; $[-5, 5, 1]$ by $[-5, 5, 1]$

Critical Thinking Exercises

Make Sense? *In Exercises 75–78, determine whether each statement makes sense or does not make sense, and explain your reasoning.*

75. I must have made an error if a rational equation produces no solution.

76. I added two rational expressions and found the solution set.

77. I can solve the equation $\frac{40}{x} = \frac{15}{x - 20}$ by multiplying both sides by the LCD or by setting the cross product of 40 and $x - 20$ equal to the cross product of 15 and x.

78. I'm solving a rational equation that became a quadratic equation, so my rational equation will have two solutions.

In Exercises 79–82, determine whether each statement is true or false. If the statement is false, make the necessary change(s) to produce a true statement.

79. Once a rational equation is cleared of fractions, all solutions of the resulting equation are also solutions of the rational equation.

80. We find

$$\frac{4}{x} - \frac{2}{x + 1}$$

by multiplying each term by the LCD, $x(x + 1)$, thereby clearing fractions.

81. All real numbers satisfy the equation $\frac{7}{x} - \frac{2}{x} = \frac{5}{x}$.

82. In order to find a number to add to the numerator and denominator of $\frac{3}{16}$ to result in $\frac{1}{2}$, we could solve the following rational equation:

$$\frac{3 + x}{16 + x} = \frac{1}{2}.$$

83. Solve: $\left(\frac{1}{x + 1} + \frac{x}{1 - x}\right) \div \left(\frac{x}{x + 1} - \frac{1}{x - 1}\right) = -1.$

84. Solve: $\left|\frac{x + 1}{x + 8}\right| = \frac{2}{3}.$

85. Write an original rational equation that has no solution.

86. Solve

$$\left(\frac{4}{x - 1}\right)^2 + 2\left(\frac{4}{x - 1}\right) + 1 = 0$$

by introducing the substitution $u = \frac{4}{x - 1}$.

Review Exercises

87. Graph the solution set:

$$\begin{cases} x + 2y \geq 2 \\ x - y \geq -4. \end{cases}$$

(Section 4.4, Example 5)

88. Solve:

$$\frac{x - 4}{2} - \frac{1}{5} = \frac{7x + 1}{20}.$$

(Section 1.4, Example 4)

89. Solve for F:

$$C = \frac{5F - 160}{9}.$$

(Section 1.5, Example 8)

Preview Exercises

Exercises 90–92 will help you prepare for the material covered in the next section.

90. Solve for p: $qf + pf = pq$.

91. Solve: $\frac{40}{x} + \frac{40}{x + 30} = 2.$

92. A plane flies at an average rate of 450 miles per hour. It can travel 980 miles with the wind in the same amount of time as it travels 820 miles against the wind. Solve the equation

$$\frac{980}{450 + x} = \frac{820}{450 - x}$$

to find the average rate of the wind, x, in miles per hour.

SECTION

6.7 Formulas and Applications of Rational Equations

What am I supposed to learn?

After studying this section, you should be able to:

1. Solve a formula with a rational expression for a variable.
2. Solve business problems involving average cost.
3. Solve problems involving time in motion.
4. Solve problems involving work.

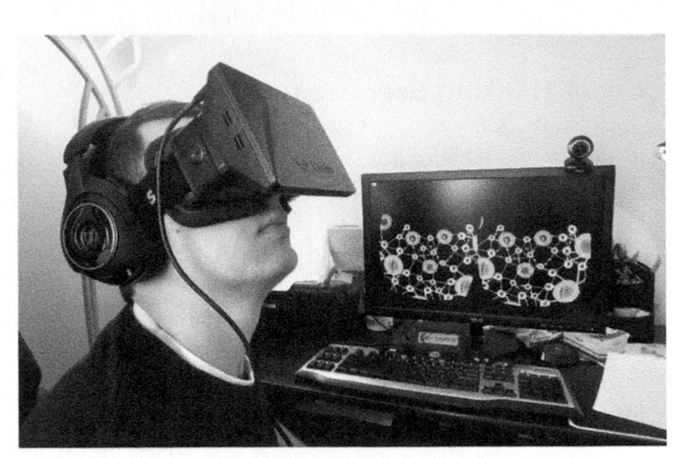

The current generation of college students grew up playing interactive online games and many continue to play in school. Hundreds of colleges have formed organized gaming teams, many as campus clubs. Enter the Oculus Rift, revolutionizing the way people experience gaming. The Oculus Rift is a virtual reality headset that enables users to experience video games as immersive three-dimensional environments. Basically, it puts the gamer inside the game.

The cost of manufacturing Oculus Rift headsets can be modeled by rational functions. In this section, you will see that high production levels of the Oculus Rift can keep the price of this amazing invention low, perhaps making this the device that brings home virtual reality to reality.

1 Solve a formula with a rational expression for a variable.

Solving a Formula for a Variable

Formulas and mathematical models frequently contain rational expressions. We solve for a specified variable using the procedure for solving rational equations. The goal is to get the specified variable alone on one side of the equation. To do so, collect all terms with this variable on one side and all other terms on the other side. It is sometimes necessary to factor out the variable you are solving for.

EXAMPLE 1 Solving for a Variable in a Formula

The formula

$$S = \frac{C}{1 - r}$$

describes a product's selling price, S, in terms of its cost to the seller, C, and its markup rate, r, in decimal form.

a. Solve the formula for r.

b. What is the markup rate on a textbook costing the campus bookstore \$140 and selling for \$200?

Solution

a. Our goal is to isolate the variable r.

$$S = \frac{C}{1 - r}$$ We need r by itself on one side of the formula.

We begin by multiplying both sides by the least common denominator, $1 - r$, to clear the equation of fractions.

$$S = \frac{C}{1-r} \qquad \text{This is the given formula.}$$

$$(1-r)S = (1-r)\left(\frac{C}{1-r}\right) \qquad \text{Multiply both sides by } 1 - r\text{, the LCD.}$$

$$S - Sr = C$$

Use the distributive property on the left side. On the right side, $(1-r)\left(\frac{C}{1-r}\right) = C$.

Observe that the formula is now cleared of fractions. The term with r, the specified variable, is already on the left side of the equation. To isolate this term, subtract S from both sides.

$$S - Sr = C \qquad \text{This is the equation cleared of fractions.}$$

$$-Sr = C - S \qquad \text{Subtract } S \text{ from both sides.}$$

$$\frac{-Sr}{-S} = \frac{C-S}{-S} \qquad \text{Divide both sides by } -S \text{ and solve for } r.$$

$$r = \frac{C-S}{-S} \qquad \text{Simplify.}$$

Great Question!

Now that we've solved the formula for r, it looks weird with a negative sign in the denominator. Is there another way to write this formula?

Yes. Multiply the numerator and the denominator by -1.

$$r = \frac{C-S}{-S} = \frac{(-1)}{(-1)} \cdot \frac{C-S}{-S} = \frac{-C+S}{S} = \frac{S-C}{S}$$

b. Now we find the markup rate on a textbook costing the campus bookstore \$140 and selling for \$200.

$$r = \frac{C-S}{-S} \qquad C = 140,\ S = 200 \qquad \text{This is the formula solved for } r \text{ from part (a).}$$

$$r = \frac{140-200}{-200} = \frac{-60}{-200} = \frac{3}{10} = 0.3 = 30\%$$

The bookstore's markup rate is 30%. ■

CHECK POINT 1 Solve for x: $a = \frac{b}{x+2}$.

EXAMPLE 2 Solving for a Variable in a Formula

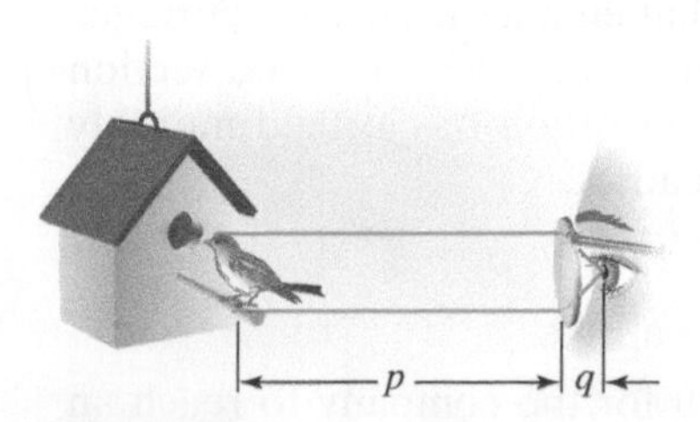

Figure 6.5

If you wear glasses, did you know that each lens has a measurement called its focal length, f? When an object is in focus, its distance from the lens, p, and the distance from the lens to your retina, q, satisfy the formula

$$\frac{1}{p} + \frac{1}{q} = \frac{1}{f}.$$

(See **Figure 6.5**.) Solve this formula for p.

Solution Our goal is to isolate the variable p. We begin by multiplying both sides by the least common denominator, pqf, to clear the equation of fractions.

$$\frac{1}{p} + \frac{1}{q} = \frac{1}{f}$$ This is the given formula.

$$pqf\left(\frac{1}{p} + \frac{1}{q}\right) = pqf\left(\frac{1}{f}\right)$$ Multiply both sides by pqf, the LCD.

$$pqf\left(\frac{1}{p}\right) + pqf\left(\frac{1}{q}\right) = pqf\left(\frac{1}{f}\right)$$ Use the distributive property on the left side.

$$qf + pf = pq$$ Simplify.

Observe that the formula is now cleared of fractions. Collect terms with p, the specified variable, on one side of the equation. To do so, subtract pf from both sides.

$$qf + pf = pq$$ This is the equation cleared of fractions.

$$qf = pq - pf$$ Subtract pf from both sides.

$$qf = p(q - f)$$ Factor out p, the specified variable.

$$\frac{qf}{q - f} = \frac{p(q - f)}{q - f}$$ Divide both sides by $q - f$ and solve for p.

$$\frac{qf}{q - f} = p$$ Simplify. ■

CHECK POINT 2 Solve $\frac{1}{x} + \frac{1}{y} = \frac{1}{z}$ for x.

---

2 Solve business problems involving average cost.

Business Problems Involving Average Cost

We have seen that the cost function for a business is the sum of its fixed and variable costs:

$$C(x) = (\text{fixed cost}) + cx$$

Cost per unit times the number of units produced, x

The **average cost** per unit for a company to produce x units is the sum of its fixed and variable costs divided by the number of units produced. The **average cost function** is a rational function that is denoted by $\overline{C}$. Thus,

Cost of producing x units: fixed plus variable costs

$$\overline{C}(x) = \frac{(\text{fixed cost}) + cx}{x}.$$

Number of units produced

EXAMPLE 3 Putting the Video-Game Player Inside the Game

We return to the Oculus Rift, a virtual reality headset that enables users to experience video games as immersive three-dimensional environments described in the section opener. Suppose the company that manufactures this invention has a fixed monthly cost of $600,000 and that it costs $50 to produce each headset.

a. Write the cost function, C, of producing x headsets.

b. Write the average cost function, $\overline{C}$, of producing x headsets.

c. How many headsets must be produced each month for the company to reach an average cost of $350 per headset?

Solution

a. The cost function, C, is the sum of the fixed cost and the variable costs.

$$C(x) = 600{,}000 + 50x$$

Fixed cost is \$600,000.

Variable cost: \$50 for each headset

b. The average cost function, $\overline{C}$, is the sum of fixed and variable costs divided by the number of headsets produced.

$$\overline{C}(x) = \frac{600{,}000 + 50x}{x}$$

c. We are interested in the company's production level that results in an average cost of \$350 per headset. Substitute 350, the average cost, for $\overline{C}(x)$ and solve the resulting rational equation for x.

$$350 = \frac{600{,}000 + 50x}{x} \quad \text{Substitute 350 for } \overline{C}(x).$$

$$350x = 600{,}000 + 50x \quad \text{Multiply both sides by the LCD, } x.$$

$$300x = 600{,}000 \quad \text{Subtract } 50x \text{ from both sides.}$$

$$x = 2000 \quad \text{Divide both sides by 300.}$$

The company must produce 2000 headsets each month for an average cost of \$350 per headset. ■

Figure 6.6 shows the graph of the average cost function in Example 3. As the production level increases, the average cost of producing each headset decreases. The horizontal asymptote, $y = 50$, is also shown in the figure. This means that the more headsets produced each month, the closer the average cost per system for the company comes to \$50. The least possible cost per headset is approaching \$50. Competitively low prices take place with high production levels, posing a major problem for small businesses.

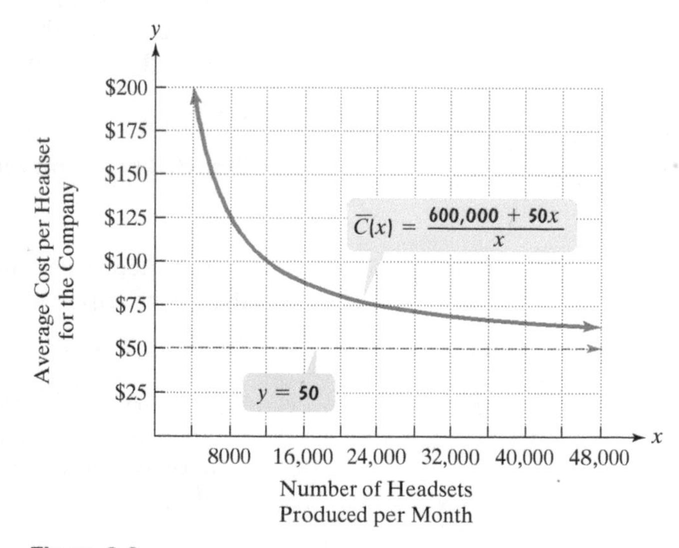

Figure 6.6

✓ CHECK POINT 3 A company is planning to manufacture wheelchairs that are light, fast, and beautiful. Fixed monthly cost will be \$500,000 and it will cost \$400 to produce each radically innovative chair.

a. Write the cost function, C, of producing x wheelchairs.

b. Write the average cost function, $\overline{C}$, of producing x wheelchairs.

c. How many wheelchairs must be produced each month for the company to reach an average cost of \$450 per chair?

3 Solve problems involving time in motion.

Problems Involving Motion

We have seen that the distance, d, covered by any moving body is the product of its average rate, r, and its time in motion, t: $d = rt$. Rational expressions appear in motion problems when the conditions of the problem involve the time traveled. We can obtain an expression for t, the time traveled, by dividing both sides of $d = rt$ by r.

$$d = rt \quad \text{Distance equals rate times time.}$$

$$\frac{d}{r} = \frac{rt}{r} \quad \text{Divide both sides by } r.$$

$$\frac{d}{r} = t \quad \text{Simplify.}$$

Time in Motion

$$t = \frac{d}{r}$$

$$\text{Time traveled} = \frac{\text{Distance traveled}}{\text{Rate of travel}}$$

EXAMPLE 4 A Motion Problem Involving Time

You commute to work a distance of 40 miles and return on the same route at the end of the day. Your average rate on the return trip is 30 miles per hour faster than your average rate on the outgoing trip. If the round trip takes 2 hours, what is your average rate on the outgoing trip to work?

Solution

Step 1. Let x represent one of the unknown quantities. Let

x = the rate on the outgoing trip.

Step 2. Represent other unknown quantities in terms of x. Because the average rate on the return trip is 30 miles per hour faster than the average rate on the outgoing trip, let

$x + 30$ = the rate on the return trip.

Step 3. Write an equation that models the conditions. By reading the problem again, we discover that the crucial idea is that the time for the round trip is 2 hours. Thus, the time on the outgoing trip plus the time on the return trip is 2 hours.

	Distance	Rate	Time = $\frac{\text{Distance}}{\text{Rate}}$
Outgoing Trip	40	x	$\frac{40}{x}$
Return Trip	40	$x + 30$	$\frac{40}{x+30}$

The sum of these times is 2 hours.

We are now ready to write an equation that models the problem's conditions.

Time on the outgoing trip plus time on the return trip equals 2 hours.

$$\frac{40}{x} + \frac{40}{x+30} = 2$$

Step 4. Solve the equation and answer the question.

$$\frac{40}{x} + \frac{40}{x+30} = 2$$ This is the equation that models the problem's conditions.

$$x(x+30)\left(\frac{40}{x} + \frac{40}{x+30}\right) = 2x(x+30)$$ Multiply both sides by the LCD, $x(x+30)$.

$$\cancel{x}(x+30)\cdot\frac{40}{\cancel{x}} + x\cancel{(x+30)}\cdot\frac{40}{\cancel{x+30}} = 2x^2 + 60x$$ Use the distributive property on each side.

$$40(x+30) + 40x = 2x^2 + 60x$$ Simplify.

$$40x + 1200 + 40x = 2x^2 + 60x$$ Use the distributive property.

$$80x + 1200 = 2x^2 + 60x$$ Combine like terms: $40x + 40x = 80x$.

$$0 = 2x^2 - 20x - 1200$$ Subtract $80x + 1200$ from both sides.

$$0 = 2(x^2 - 10x - 600)$$ Factor out the GCF.

$$0 = 2(x-30)(x+20)$$ Factor completely.

$$x - 30 = 0 \quad \text{or} \quad x + 20 = 0$$ Set each variable factor equal to 0.

$$x = 30 \qquad\qquad x = -20$$ Solve for x.

Using Technology

Graphic Connections

The graph of the rational function for the time for the round trip,

$$f(x) = \frac{40}{x} + \frac{40}{x+30},$$

is shown in a [0, 60, 30] by [0, 10, 1] viewing rectangle.

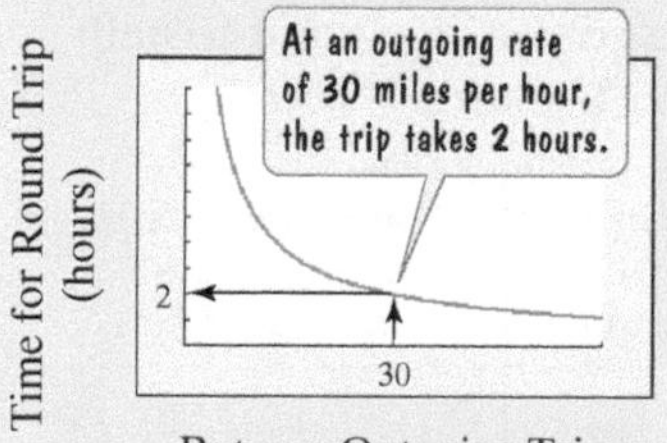

[0, 60, 30] by [0, 10, 1]

The graph is falling from left to right. This shows that the time for the round trip decreases as the rate increases.

Because x represents the rate on the outgoing trip, we reject the negative value, -20. The rate on the outgoing trip is 30 miles per hour. At an outgoing rate of 30 miles per hour, the round trip should take 2 hours.

Step 5. Check the proposed solution in the original wording of the problem. Does the round trip take 2 hours? Because the rate on the return trip is 30 miles per hour faster than the rate on the outgoing trip, the rate on the return trip is $30 + 30$, or 60 miles per hour.

$$\text{Time on the outgoing trip} = \frac{\text{Distance}}{\text{Rate}} = \frac{40}{30} = \frac{4}{3}\text{ hours}$$

$$\text{Time on the return trip} = \frac{\text{Distance}}{\text{Rate}} = \frac{40}{60} = \frac{2}{3}\text{ hour}$$

The total time for the round trip is $\frac{4}{3} + \frac{2}{3} = \frac{6}{3}$, or 2 hours. This checks with the original conditions of the problem. ■

✓ **CHECK POINT 4** After riding at a steady speed for 40 miles, a bicyclist had a flat tire and walked 5 miles to a repair shop. The cycling rate was 4 times faster than the walking rate. If the time spent cycling and walking was 5 hours, at what rate was the cyclist riding?

---

4 Solve problems involving work.

Problems Involving Work

You are thinking of designing your own Web site. You estimate that it will take 15 hours to do the job. In 1 hour, $\frac{1}{15}$ of the job is completed. In 2 hours, $\frac{2}{15}$ of the job is completed. In 3 hours, the fractional part of the job done is $\frac{3}{15}$, or $\frac{1}{5}$. In x hours, the fractional part of the job that you can complete is $\frac{x}{15}$.

Your friend, who has experience developing Web sites, took 10 hours working on her own to design an impressive site. You wonder about the possibility of working together. How long would it take both of you to design your Web site?

Problems involving work usually have two (or more) people or machines working together to complete a job. The amount of time it takes each person to do the job working alone is frequently known. The question deals with how long it will take both people working together to complete the job.

In work problems, **the number 1 represents one whole job completed.** For example, the completion of your Web site is represented by 1. Equations in work problems are often based on the following condition:

Fractional part of the job done by the first person + fractional part of the job done by the second person = 1 (one whole job completed).

EXAMPLE 5 Solving a Problem Involving Work

You can design a Web site in 15 hours. Your friend can design the same site in 10 hours. How long will it take to design the Web site if you both work together?

Solution

Step 1. Let x represent one of the unknown quantities. Let

$x =$ the time, in hours, for you and your friend working together to design the Web site.

Step 2. Represent other unknown quantities in terms of x. Because there are no other unknown quantities, we can skip this step.

Step 3. Write an equation that models the conditions. We construct a table to help find the fractional parts of the task completed by you and your friend in x hours.

	Fractional part of job completed in 1 hour	Time working together	Fractional part of job completed in x hours
You (You can design the site in 15 hours.)	$\frac{1}{15}$	x	$\frac{x}{15}$
Your friend (Your friend can design the site in 10 hours.)	$\frac{1}{10}$	x	$\frac{x}{10}$

Fractional part of the job done by you + fractional part of the job done by your friend = one whole job.

$$\frac{x}{15} + \frac{x}{10} = 1$$

Step 4. Solve the equation and answer the question.

$$\frac{x}{15} + \frac{x}{10} = 1$$ This is the equation that models the problem's conditions.

$$30\left(\frac{x}{15} + \frac{x}{10}\right) = 30 \cdot 1$$ Multiply both sides by 30, the LCD.

$$30 \cdot \frac{x}{15} + 30 \cdot \frac{x}{10} = 30$$ Use the distributive property on the left side.

$$2x + 3x = 30$$ Simplify: $\frac{\overset{2}{\cancel{30}}}{1} \cdot \frac{x}{\underset{1}{\cancel{15}}} = 2x$ and $\frac{\overset{3}{\cancel{30}}}{1} \cdot \frac{x}{\underset{1}{\cancel{10}}} = 3x$.

$$5x = 30$$ Combine like terms.

$$x = 6$$ Divide both sides by 5.

If you both work together, you can design your Web site in 6 hours.

Step 5. Check the proposed solution in the original wording of the problem. Will you both complete the job in 6 hours? Because you can design the site in 15 hours, in 6 hours, you can complete $\frac{6}{15}$, or $\frac{2}{5}$, of the job. Because your friend can design the site in 10 hours, in 6 hours, she can complete $\frac{6}{10}$, or $\frac{3}{5}$, of the job. Notice that $\frac{2}{5} + \frac{3}{5} = 1$, which represents the completion of the entire job, or one whole job. ■

Great Question!

Is there an equation that models all the work problems in this section?
Yes. Let

a = the time it takes person A to do a job working alone, and

b = the time it takes person B to do the same job working alone.

If x represents the time it takes for A and B to complete the entire job working together, then the situation can be modeled by the rational equation

$$\frac{x}{a} + \frac{x}{b} = 1.$$

✓ **CHECK POINT 5** A new underwater tunnel is being built using tunnel-boring machines that begin at opposite ends of the tunnel. One tunnel-boring machine can complete the tunnel in 18 months. A faster machine can tunnel to the other side in 9 months. If both machines start at opposite ends and work at the same time, in how many months will the tunnel be finished?

EXAMPLE 6 Solving a Problem Involving Work

After designing your Web site, you and your friend decide to go into business setting up sites for others. With lots of practice, you can now work together and design a modest site in 4 hours. Your friend is still a faster worker. Working alone, you require 6 more hours than she does to design a site for a client. How many hours does it take your friend to design a Web site if she works alone?

Solution

Step 1. Let x represent one of the unknown quantities. Let

x = the time, in hours, for your friend to design a Web site working alone.

Step 2. Represent other unknown quantities in terms of x. Because you require 6 more hours than your friend to do the job, let

$x + 6$ = the time, in hours, for you to design a Web site working alone.

Step 3. Write an equation that models the conditions. Working together, you and your friend can complete the job in 4 hours. We construct a table to find the fractional part of the task completed by you and your friend in 4 hours.

	Fractional part of job completed in 1 hour	Time working together	Fractional part of job completed in 4 hours
Your friend	$\frac{1}{x}$	4	$\frac{4}{x}$
You	$\frac{1}{x+6}$	4	$\frac{4}{x+6}$

Your friend can design the site in x hours.

You can design the site in $x + 6$ hours.

Because you can both complete the job in 4 hours,

Fractional part of the job done by your friend + fractional part of the job done by you = one whole job.

$$\frac{4}{x} + \frac{4}{x+6} = 1.$$

Step 4. Solve the equation and answer the question.

$$\frac{4}{x} + \frac{4}{x+6} = 1$$ This is the equation that models the problem's conditions.

$$x(x+6)\left(\frac{4}{x} + \frac{4}{x+6}\right) = x(x+6) \cdot 1$$ Multiply both sides by $x(x + 6)$, the LCD.

$$\cancel{x}(x+6) \cdot \frac{4}{\cancel{x}} + x\cancel{(x+6)} \cdot \frac{4}{\cancel{(x+6)}} = x^2 + 6x$$ Use the distributive property on each side.

$$4(x+6) + 4x = x^2 + 6x$$ Simplify.

$$4x + 24 + 4x = x^2 + 6x$$ Use the distributive property.

$$8x + 24 = x^2 + 6x$$ Combine like terms: $4x + 4x = 8x$.

$$0 = x^2 - 2x - 24$$ Subtract $8x + 24$ from both sides and write the quadratic equation in standard form.

$$0 = (x-6)(x+4)$$ Factor.

$$x - 6 = 0 \quad \text{or} \quad x + 4 = 0$$ Set each factor equal to 0.

$$x = 6 \qquad\qquad x = -4$$ Solve for x.

Because x represents the time for your friend to design a Web site working alone, we reject the negative value, -4. Your friend can design a Web site in 6 hours.

Step 5. Check the proposed solution in the original wording of the problem. Will you both complete the job in 4 hours? Working alone, your friend takes 6 hours. Because you require 6 more hours than your friend, you require 12 hours to complete the job on your own.

In 4 hours, your friend completes $\frac{4}{6} = \frac{2}{3}$ of the job.
In 4 hours, you complete $\frac{4}{12} = \frac{1}{3}$ of the job.

Notice that $\frac{2}{3} + \frac{1}{3} = 1$, which represents the completion of the entire job, or the design of one Web site. ■

☑ **CHECK POINT 6** An experienced carpenter can panel a room 3 times faster than an apprentice can. Working together, they can panel the room in 6 hours. How long would it take each person working alone to do the job?

CONCEPT AND VOCABULARY CHECK

Fill in each blank so that the resulting statement is true.

1. The first step in solving $\frac{1}{x} + \frac{1}{y} = \frac{1}{z}$ for z is to multiply both sides of the equation by the least common denominator, which is ________.
2. The cost function for a business, $C(x)$, is the sum of the ________ cost and the ________ costs.
3. The average cost function for a company, $\overline{C}(x)$, is its cost function divided by the ________.
4. The formula $t = \frac{d}{r}$ states that time traveled is ________ divided by ________.
5. In work problems, the number ________ represents the whole job completed.
6. If you can complete a job in 19 hours, the fractional part of the job that you can complete in x hours is represented by ________.

6.7 EXERCISE SET

MyMathLab®

Practice Exercises

In Exercises 1–14, solve each formula for the specified variable.

1. $\frac{V_1}{V_2} = \frac{P_2}{P_1}$ for P_1 (chemistry)
2. $\frac{V_1}{V_2} = \frac{P_2}{P_1}$ for V_2 (chemistry)
3. $\frac{1}{p} + \frac{1}{q} = \frac{1}{f}$ for f (optics)
4. $\frac{1}{p} + \frac{1}{q} = \frac{1}{f}$ for q (optics)
5. $P = \frac{A}{1 + r}$ for r (investment)
6. $S = \frac{a}{1 - r}$ for r (mathematics)
7. $F = \frac{Gm_1m_2}{d^2}$ for m_1 (physics)
8. $F = \frac{Gm_1m_2}{d^2}$ for m_2 (physics)
9. $z = \frac{x - \bar{x}}{s}$ for x (statistics)
10. $z = \frac{x - \bar{x}}{s}$ for s (statistics)
11. $I = \frac{E}{R + r}$ for R (electronics)
12. $I = \frac{E}{R + r}$ for r (electronics)
13. $f = \frac{f_1f_2}{f_1 + f_2}$ for f_1 (optics)
14. $f = \frac{f_1f_2}{f_1 + f_2}$ for f_2 (optics)

Application Exercises

The figure shows the graph of the average cost function for the company described in Check Point 3 that manufactures wheelchairs. Use the graph to solve Exercises 15–18.

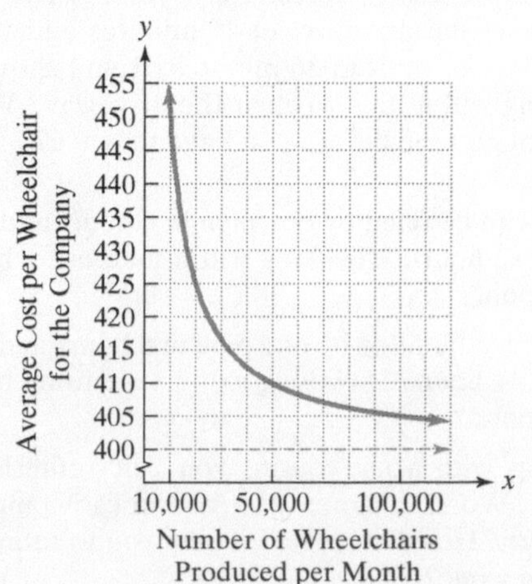

15. How many wheelchairs must be produced each month for the company to have an average cost of \$410 per chair?
16. How many wheelchairs must be produced each month for the company to have an average cost of \$425 per chair?
17. What is the equation of the horizontal asymptote shown by the dashed green line? What is the meaning of the horizontal asymptote as production level increases?
18. Describe the end behavior of the graph at the far right. Is there a production level that results in an average cost of \$400 per chair? Explain your answer.
19. A company is planning to manufacture mountain bikes. Fixed monthly cost will be \$100,000 and it will cost \$100 to produce each bicycle.
    a. Write the cost function, C, of producing x mountain bikes.
    b. Write the average cost function, $\overline{C}$, of producing x mountain bikes.
    c. How many mountain bikes must be produced each month for the company to have an average cost of \$300 per bike?
20. A company is planning to manufacture small canoes. Fixed monthly cost will be \$20,000 and it will cost \$20 to produce each canoe.
    a. Write the cost function, C, of producing x canoes.
    b. Write the average cost function, $\overline{C}$, of producing x canoes.
    c. How many canoes must be produced each month for the company to have an average cost of \$40 per canoe?

It's vacation time. You drive 90 miles along a scenic highway and then take a 5-mile run along a hiking trail. Your driving rate is nine times that of your running rate. The graph shows the total time you spend driving and running, $f(x)$, as a function of your running rate, x. Use the graph of this rational function to solve Exercises 21–24.

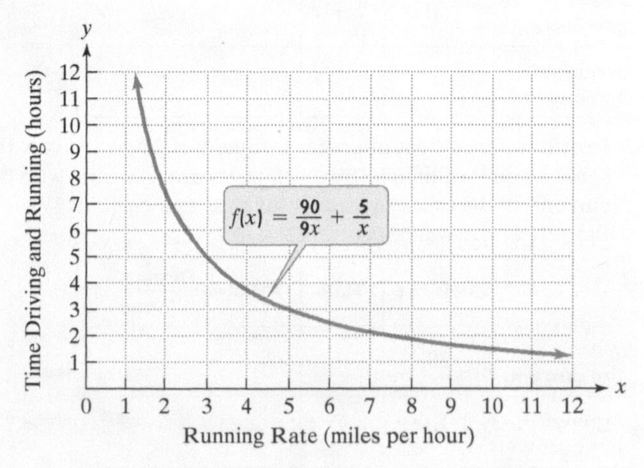

(Be sure to use the graph of the rational function at the bottom of the previous page to solve Exercises 21–24.)

21. If the total time for driving and running is 3 hours, what is your running rate?

22. If the total time for driving and running is 5 hours, what is your running rate?

23. Describe the behavior of the graph as x approaches 0. What does this show about the time driving and running as your running rate is close to zero miles per hour (a stupefied crawl)?

24. The graph is falling from left to right. What does this show?

Use a rational equation to solve Exercises 25–36. Each exercise is a problem involving motion.

25. A car can travel 300 miles in the same amount of time it takes a bus to travel 180 miles. If the rate of the bus is 20 miles per hour slower than the rate of the car, find the average rate for each.

26. A passenger train can travel 240 miles in the same amount of time it takes a freight train to travel 160 miles. If the rate of the freight train is 20 miles per hour slower than the rate of the passenger train, find the average rate of each.

27. You ride your bike to campus a distance of 5 miles and return home on the same route. Going to campus, you ride mostly downhill and average 9 miles per hour faster than on your return trip home. If the round trip takes one hour and ten minutes—that is, $\frac{7}{6}$ hours—what is your average rate on the return trip?

28. An engine pulls a train 140 miles. Then a second engine, whose average rate is 5 miles per hour faster than the first engine, takes over and pulls the train 200 miles. The total time required for both engines is 9 hours. Find the average rate of each engine.

29. In still water, a boat averages 7 miles per hour. It takes the same amount of time to travel 20 miles downstream, with the current, as 8 miles upstream, against the current. What is the rate of the water's current?

	Distance	Rate	Time = $\frac{\text{Distance}}{\text{Rate}}$
With the current	20	$7 + x$	
Against the current	8	$7 - x$	

These times are equal.

30. In still water, a boat averages 8 miles per hour. It takes the same amount of time to travel 30 miles downstream, with the current, as 18 miles upstream, against the current. What is the rate of the water's current?

	Distance	Rate	Time = $\frac{\text{Distance}}{\text{Rate}}$
With the current	30	$8 + x$	
Against the current	18	$8 - x$	

These times are equal.

31. The rate of the jet stream is 100 miles per hour. Traveling with the jet stream, an airplane can fly 2400 miles in the same amount of time as it takes to fly 1600 miles against the jet stream. What is the airplane's average rate in calm air?

32. The wind is blowing at an average rate of 10 miles per hour. Riding with the wind, a bicyclist can cycle 75 miles in the same amount of time it takes to cycle 15 miles against the wind. What is the bicyclist's average rate in calm air?

33. A moving sidewalk at an airport glides at a rate of 1.8 feet per second. Walking on the moving sidewalk, you travel 100 feet forward in the same time it takes to travel 40 feet in the opposite direction. Find your walking speed on a nonmoving sidewalk.

34. A moving sidewalk at an airport glides at a rate of 1.8 feet per second. Walking on the moving sidewalk, you travel 105 feet forward in the same time it takes to travel 50 feet in the opposite direction. Find your walking speed on a nonmoving sidewalk. Round to the nearest tenth.

35. Two runners, one averaging 8 miles per hour and the other 6 miles per hour, start at the same place and run along the same trail. The slower runner arrives at the end of the trail a half hour after the faster runner. How far did each person run?

36. Two sailboats, one averaging 20 miles per hour and the other 18 miles per hour, start at the same place and follow the same course. The slower boat arrives at the end of the course $\frac{1}{6}$ of an hour after the faster sailboat. How far did each boat travel?

Use a rational equation to solve Exercises 37–48. Each exercise is a problem involving work.

37. You promised your parents that you would wash the family car. You have not started the job and they are due home in 20 minutes. You can wash the car in 45 minutes and your sister claims she can do it in 30 minutes. If you work together, how long will it take to do the job? Will this give you enough time before your parents return?

38. You must leave for campus in half an hour, or you will be late for class. Unfortunately, you are snowed in. You can shovel the driveway in 45 minutes and your brother claims he can do it in 36 minutes. If you shovel together, how long will it take to clear the driveway? Will this give you enough time before you have to leave?

39. A pool can be filled by one pipe in 6 hours and by a second pipe in 12 hours. How long will it take using both pipes to fill the pool?

40. A pond can be filled by one pipe in 8 hours and by a second pipe in 24 hours. How long will it take using both pipes to fill the pond?

41. Working with your cousin, you can refinish a table in 3 hours. Working alone, your cousin can complete the job in 4 hours. How long would it take you to refinish the table working alone?

42. Working with your cousin, you can split a cord of firewood in 5 hours. Working alone, your cousin can complete the job in 7 hours. How long would it take you to split the firewood working alone?

43. An earthquake strikes and an isolated area is without food or water. Three crews arrive. One can dispense needed supplies in 20 hours, a second in 30 hours, and a third in 60 hours. How long will it take all three crews working together to dispense food and water?

44. A hurricane strikes and a rural area is without food or water. Three crews arrive. One can dispense needed supplies in 10 hours, a second in 15 hours, and a third in 20 hours. How long will it take all three crews working together to dispense food and water?

45. An office has an old copying machine and a new one. Working together, it takes both machines 6 hours to make all the copies of the annual financial report. Working alone, it takes the old copying machine 5 hours longer than the new one to make all the copies of the report. How long would it take the new copying machine to make all the copies working alone?

46. A demolition company wants to build a brick wall to hide from public view the area where they store wrecked cars. Working together, an experienced bricklayer and an apprentice can build the wall in 12 hours. Working alone, it takes the apprentice 10 hours longer than the experienced bricklayer to do the job. How long would it take the experienced bricklayer to build the wall working alone?

47. A faucet can fill a sink in 5 minutes. It takes twice as long for the drain to empty the sink. How long will it take to fill the sink if the drain is open and the faucet is on?

48. A pool can be filled by a pipe in 3 hours. It takes 3 times as long for another pipe to empty the pool. How long will it take to fill the pool if both pipes are open?

Exercises 49–56 contain a variety of problems. Use a rational equation to solve each exercise.

49. What number multiplied by the numerator and added to the denominator of $\frac{4}{5}$ makes the resulting fraction equivalent to $\frac{3}{2}$?

50. What number multiplied by the numerator and subtracted from the denominator of $\frac{9}{11}$ makes the resulting fraction equivalent to $-\frac{12}{5}$?

51. The sum of 2 times a number and twice its reciprocal is $\frac{20}{3}$. Find the number(s).

52. If 2 times the reciprocal of a number is subtracted from 3 times the number, the difference is 1. Find the number(s).

53. You have 35 hits in 140 times at bat. Your batting average is $\frac{35}{140}$, or 0.25. How many consecutive hits must you get to increase your batting average to 0.30?

54. You have 30 hits in 120 times at bat. Your batting average is $\frac{30}{120}$, or 0.25. How many consecutive hits must you get to increase your batting average to 0.28?

55. If one pipe can fill a pool in a hours and a second pipe can fill the pool in b hours, write a formula for the time, x, in terms of a and b, for the number of hours it takes both pipes, working together, to fill the pool.

56. If one pipe can fill a pool in a hours and a second pipe can empty the pool in b hours, write a formula for the time, x, in terms of a and b, for the number of hours it takes to fill the pool with both of these pipes open.

Explaining the Concepts

57. Without showing the details, explain how to solve the formula

$$\frac{1}{R} = \frac{1}{R_1} + \frac{1}{R_2}$$

for R_1. (The formula is used in electronics.)

58. Explain how to find the average cost function for a business.

59. How does the average cost function illustrate a problem for small businesses?

60. What is the relationship among time traveled, distance traveled, and rate of travel?

61. If you know how many hours it takes for you to do a job, explain how to find the fractional part of the job you can complete in x hours.

62. If you can do a job in 6 hours and your friend can do the same job in 3 hours, explain how to find how long it takes to complete the job working together. It is not necessary to solve the problem.

63. When two people work together to complete a job, describe one factor that can result in more or less time than the time given by the rational equations we have been using.

Technology Exercises

64. For Exercises 19–20, use a graphing utility to graph the average cost function described by the problem's conditions. Then [TRACE] along the curve and find the point that visually shows the solution in part (c).

65. For Exercises 45–46, use a graphing utility to graph the function representing the sum of the fractional parts of the job done by the two machines or the two people. Then [TRACE] along the curve and find the point that visually shows the problem's solution.

66. A boat can travel 10 miles per hour in still water. The boat travels 24 miles upstream, against the current, and then 24 miles downstream, with the current.

  a. Let x = the rate of the current. Write a function in terms of x that models the total time for the boat to travel upstream and downstream.

  b. Use a graphing utility to graph the rational function in part (a).

  c. [TRACE] along the curve and determine the current's rate if the trip's time is 5 hours. Then verify this result algebraically.

Critical Thinking Exercises

Make Sense? *Read this excerpt from an advertisement for bikes with aerodynamic coverings.*

> Our high-performance bicycles have the aerodynamic design of custom racing bikes, but are practical for everyday riding. The aerodynamic covering will increase your average speed by 10 miles per hour. Cyclists using our bicycles, versus bikes without aerodynamic coverings, reduced time on a 75-mile test run by 2 hours.

Now you are interested in finding the average rate of the bikes with the aerodynamic coverings on the 75-mile test run. With this goal in mind, determine whether each statement in Exercises 67–70 makes sense or does not make sense, and explain your reasoning.

67. I decided to organize the critical information from the advertisement in a table with the following entries:

	Distance	Rate	Time
With covering	75	$x + 10$	$\frac{75}{x+10}$
Without covering	75	x	$\frac{75}{x}$

68. The ad stated that bikes with coverings reduced time on the 75-mile test run by 2 hours, so I used my table from Exercise 67 and modeled this condition with the rational equation

$$\frac{75}{x} = \frac{75}{x+10} - 2.$$

69. The equation in x that modeled the conditions had a positive and a negative value for x, so I rejected the negative solution.

70. My professor verified that 15 is the correct value for x in the equation modeling the conditions, so I used my table from Exercise 67 and concluded that the average rate of the covered bikes on the 75-mile test run was 15 miles per hour.

In Exercises 71–74, determine whether each statement is true or false. If the statement is false, make the necessary change(s) to produce a true statement.

71. As production level increases, the average cost for a company to produce each unit of its product also increases.

72. To solve $qf + pf = pq$ for p, subtract qf from both sides and then divide by f.

73. If you plan a theater trip that costs \$300 to rent a limousine and \$25 per ticket, the cost per person, $f(x)$, for a group of x people is modeled by the rational function

$$f(x) = \frac{300 + 25x}{x}.$$

74. If you can clean the house in 3 hours and your sloppy friend can completely mess it up in 6 hours, then $\frac{x}{3} - \frac{x}{6} = 1$ can be used to find how long it takes to clean the house if you both "work" together.

75. Solve $\frac{1}{s} = f + \frac{1-f}{p}$ for f.

76. A new schedule for a train requires it to travel 351 miles in $\frac{1}{4}$ hour less time than before. To accomplish this, the rate of the train must be increased by 2 miles per hour. What should the average rate of the train be so that it can keep on the new schedule?

77. It takes Mr. Todd 4 hours longer to prepare an order of pies than it takes Mrs. Lovett. They bake together for 2 hours when Mrs. Lovett leaves. Mr. Todd takes 7 additional hours to complete the work. Working alone, how long does it take Mrs. Lovett to prepare the pies?

Review Exercises

78. Factor: $x^2 + 4x + 4 - 9y^2$. (Section 5.6, Example 4)

79. Solve using matrices:

$$\begin{cases} 2x + 5y = -5 \\ x + 2y = -1. \end{cases}$$

(Section 3.4, Example 2)

80. Solve the system:

$$\begin{cases} x + y + z = 4 \\ 2x + 5y = 1 \\ x - y - 2x = 0. \end{cases}$$

(Section 3.3, Example 3)

Preview Exercises

Exercises 81–83 will help you prepare for the material covered in the next section.

81. **a.** If $y = kx^2$, find the value of k using $x = 2$ and $y = 64$.

b. Substitute the value for k into $y = kx^2$ and write the resulting equation.

c. Use the equation from part (b) to find y when $x = 5$.

82. **a.** If $y = \frac{k}{x}$, find the value of k using $x = 8$ and $y = 12$.

b. Substitute the value for k into $y = \frac{k}{x}$ and write the resulting equation.

c. Use the equation from part (b) to find y when $x = 3$.

83. If $S = \frac{kA}{P}$, find the value of k using $A = 60{,}000$, $P = 40$, and $S = 12{,}000$.

SECTION

6.8 Modeling Using Variation

What am I supposed to learn?

After studying this section, you should be able to:

1. Solve direct variation problems.
2. Solve inverse variation problems.
3. Solve combined variation problems.
4. Solve problems involving joint variation.

Have you ever wondered how telecommunication companies estimate the number of phone calls expected per day between two cities? The formula

$$C = \frac{0.02P_1P_2}{d^2}$$

shows that the daily number of phone calls, C, increases as the populations of the cities, P_1 and P_2, in thousands, increase, and decreases as the distance, d, between the cities increases.

Certain formulas occur so frequently in applied situations that they are given special names. Variation formulas show how one quantity changes in relation to other quantities. Quantities can vary *directly, inversely*, or *jointly*. In this section, we look at situations that can be modeled by each of these kinds of variation.

1 Solve direct variation problems.

Direct Variation

When you swim underwater, the pressure in your ears depends on the depth at which you are swimming. The formula

$$p = 0.43d$$

describes the water pressure, p, in pounds per square inch, at a depth of d feet. We can use this linear function to determine the pressure in your ears at various depths.

In each case, use $p = 0.43d$:

If $d = 20$, $p = 0.43(20) = 8.6$. At a depth of 20 feet, water pressure is 8.6 pounds per square inch.

Doubling the depth doubles the pressure.

If $d = 40$, $p = 0.43(40) = 17.2$. At a depth of 40 feet, water pressure is 17.2 pounds per square inch.

Doubling the depth doubles the pressure.

If $d = 80$, $p = 0.43(80) = 34.4$. At a depth of 80 feet, water pressure is 34.4 pounds per square inch.

The formula $p = 0.43d$ illustrates that water pressure is a constant multiple of your underwater depth. If your depth is doubled, the pressure is doubled; if your depth is tripled, the pressure is tripled; and so on. Because of this, the pressure in your ears is said to **vary directly** as your underwater depth. The **equation of variation** is

$$p = 0.43d.$$

Generalizing, we obtain the following statement:

Direct Variation

If a situation is described by an equation in the form

$$y = kx,$$

where k is a nonzero constant, we say that **y varies directly as x** or **y is directly proportional to x**. The number k is called the **constant of variation** or the **constant of proportionality**.

Can you see that **the direct variation equation, $y = kx$, is a special case of the linear function $y = mx + b$**? When $m = k$ and $b = 0$, $y = mx + b$ becomes $y = kx$. Thus, the slope of a direct variation equation is k, the constant of variation. Because b, the y-intercept, is 0, the graph of a direct variation equation is a line passing through the origin. This is illustrated in **Figure 6.7**, which shows the graph of $p = 0.43d$: Water pressure varies directly as depth.

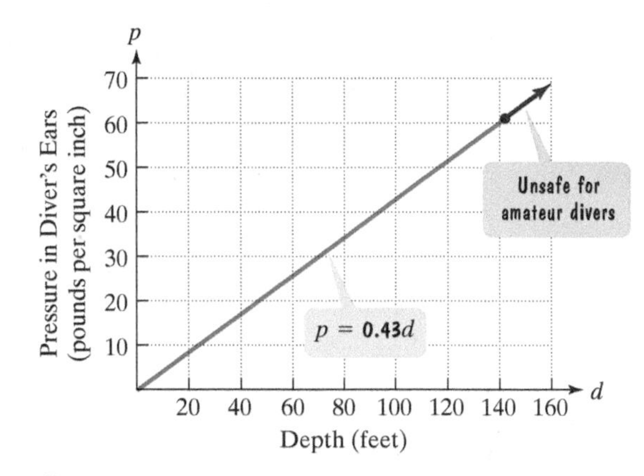

Figure 6.7 Water pressure at various depths

Problems involving direct variation can be solved using the following procedure. This procedure applies to direct variation problems, as well as to the other kinds of variation problems that we will discuss.

Solving Variation Problems

1. Write an equation that models the given English statement.
2. Substitute the given pair of values into the equation in step 1 and find the value of k, the constant of variation.
3. Substitute the value of k into the equation in step 1.
4. Use the equation from step 3 to answer the problem's question.

EXAMPLE 1 Solving a Direct Variation Problem

Many areas of Northern California depend on the snowpack of the Sierra Nevada mountain range for their water supply. The volume of water produced from melting snow varies directly as the volume of snow. Meteorologists have determined that 250 cubic centimeters of snow will melt to 28 cubic centimeters of water. How much water does 1200 cubic centimeters of melting snow produce?

Solution

Step 1. Write an equation. We know that *y varies directly as x* is expressed as

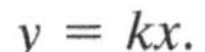

$$y = kx.$$

By changing letters, we can write an equation that models the following English statement: Volume of water, W, varies directly as volume of snow, S.

$$W = kS$$

Step 2. Use the given values to find *k*. We are told that 250 cubic centimeters of snow will melt to 28 cubic centimeters of water. Substitute 28 for W and 250 for S in the direct variation equation. Then solve for k.

$W = kS$	Volume of water varies directly as volume of melting snow.
$28 = k(250)$	250 cubic centimeters of snow melt to 28 cubic centimeters of water.
$\frac{28}{250} = \frac{k(250)}{250}$	Divide both sides by 250.
$0.112 = k$	Simplify.

Step 3. Substitute the value of *k* into the equation.

$W = kS$	This is the equation from step 1.
$W = 0.112S$	Replace k, the constant of variation, with 0.112.

Step 4. Answer the problem's question. How much water does 1200 cubic centimeters of melting snow produce? Substitute 1200 for S in $W = 0.112S$ and solve for W.

$W = 0.112S$	Use the equation from step 3.
$W = 0.112(1200)$	Substitute 1200 for S.
$W = 134.4$	Multiply.

A snowpack measuring 1200 cubic centimeters will produce 134.4 cubic centimeters of water. ■

✓ **CHECK POINT 1** The number of gallons of water, W, used when taking a shower varies directly as the time, t, in minutes, in the shower. A shower lasting 5 minutes uses 30 gallons of water. How much water is used in a shower lasting 11 minutes?

The direct variation equation $y = kx$ is a linear function. If $k > 0$, then the slope of the line is positive. Consequently, as x increases, y also increases.

A direct variation situation can involve variables to higher powers. For example, y can vary directly as x^2 ($y = kx^2$) or as x^3 ($y = kx^3$).

Direct Variation with Powers

y* varies directly as the *n*th power of *x if there exists some nonzero constant k such that

$$y = kx^n.$$

We also say that ***y* is directly proportional to the *n*th power of *x*.**

Direct variation with exponents that are whole numbers is modeled by polynomial functions. In our next example, the graph of the variation equation is the graph of a quadratic function.

EXAMPLE 2 Solving a Direct Variation Problem

The distance, s, that a body falls from rest varies directly as the square of the time, t, of the fall. If skydivers fall 64 feet in 2 seconds, how far will they fall in 4.5 seconds?

Solution

Step 1. Write an equation. We know that y *varies directly as the square of* x is expressed as

$$y = kx^2.$$

By changing letters, we can write an equation that models the following English statement: Distance, s, varies directly as the square of time, t, of the fall.

$$s = kt^2$$

Step 2. Use the given values to find k. Skydivers fall 64 feet in 2 seconds. Substitute 64 for s and 2 for t in the direct variation equation. Then solve for k.

$s = kt^2$ Distance varies directly as the square of time.

$64 = k \cdot 2^2$ Skydivers fall 64 feet in 2 seconds.

$64 = 4k$ Simplify: $2^2 = 4$.

$\frac{64}{4} = \frac{4k}{4}$ Divide both sides by 4.

$16 = k$ Simplify.

Step 3. Substitute the value of k into the equation.

$s = kt^2$ Use the equation from step 1.

$s = 16t^2$ Replace k, the constant of variation, with 16.

Step 4. Answer the problem's question. How far will the skydivers fall in 4.5 seconds? Substitute 4.5 for t in $s = 16t^2$ and solve for s.

$$s = 16(4.5)^2 = 16(20.25) = 324$$

Thus, in 4.5 seconds, the skydivers will fall 324 feet. ■

We can express the variation equation from Example 2 in function notation, writing

$$s(t) = 16t^2.$$

The distance that a body falls from rest is a function of the time, t, of the fall. The graph of this quadratic function is shown in **Figure 6.8**. The graph increases rapidly from left to right, showing the effects of the acceleration of gravity.

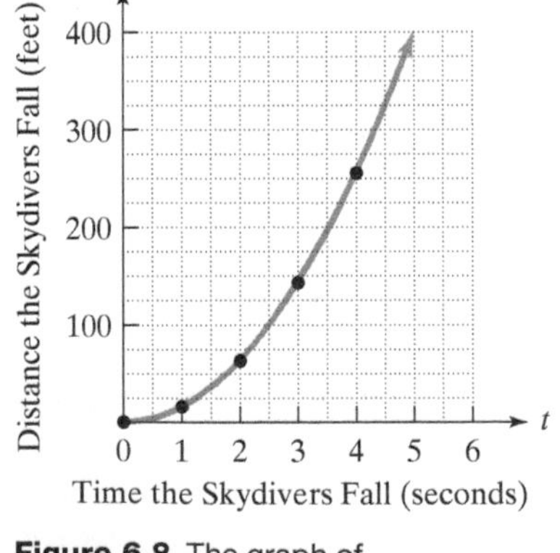

Figure 6.8 The graph of $s(t) = 16t^2$

CHECK POINT 2 The distance required to stop a car varies directly as the square of its speed. If it requires 200 feet to stop a car traveling 60 miles per hour, how many feet are required to stop a car traveling 100 miles per hour?

---

2 Solve inverse variation problems.

Inverse Variation

The distance from San Francisco to Los Angeles is 420 miles. The time that it takes to drive from San Francisco to Los Angeles depends on the average rate at which one drives and is given by

$$\text{Time} = \frac{420}{\text{Rate}}.$$

For example, if you average 30 miles per hour, the time for the drive is

$$\text{Time} = \frac{420}{30} = 14,$$

or 14 hours. If you average 50 miles per hour, the time for the drive is

$$\text{Time} = \frac{420}{50} = 8.4,$$

or 8.4 hours. As your rate (or speed) increases, the time for the trip decreases and vice versa. This is illustrated by the graph in **Figure 6.9**.

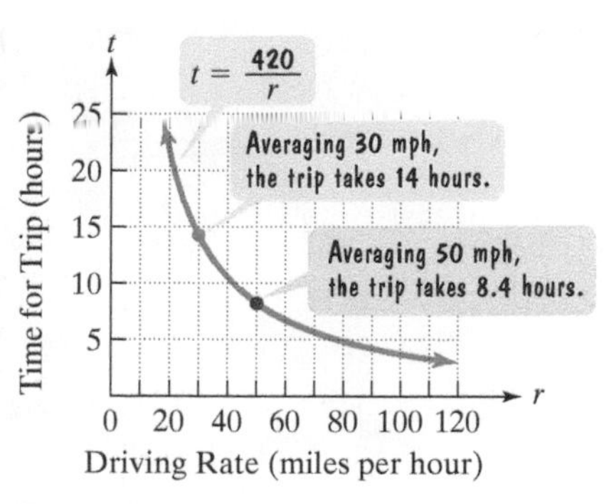

Figure 6.9

We can express the time for the San Francisco–Los Angeles trip using t for time and r for rate:

$$t = \frac{420}{r}.$$

This equation is an example of an **inverse variation** equation. Time, t, **varies inversely** as rate, r. When two quantities vary inversely, one quantity increases as the other decreases, and vice versa.

Generalizing, we obtain the following statement:

Inverse Variation

If a situation is described by an equation in the form

$$y = \frac{k}{x},$$

where k is a nonzero constant, we say that **y varies inversely as x** or **y is inversely proportional to x.** The number k is called the **constant of variation.**

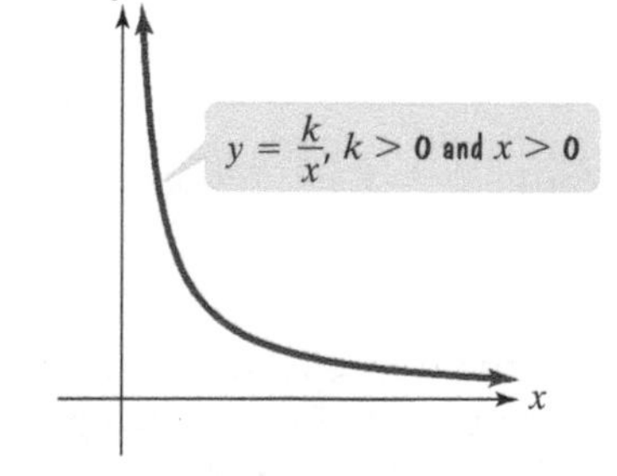

Figure 6.10 The graph of the inverse variation equation

Notice that **the inverse variation equation**

$$y = \frac{k}{x}, \quad \text{or} \quad f(x) = \frac{k}{x},$$

is a rational function. For $k > 0$ and $x > 0$, the graph of the function takes on the shape shown in **Figure 6.10**.

We use the same procedure to solve inverse variation problems as we did to solve direct variation problems. Example 3 illustrates this procedure.

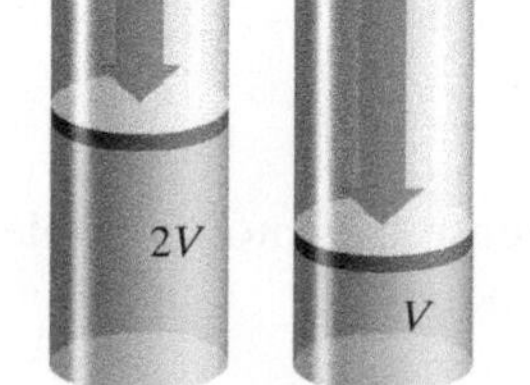

Doubling the pressure halves the volume.

EXAMPLE 3 Solving an Inverse Variation Problem

When you use a spray can and press the valve at the top, you decrease the pressure of the gas in the can. This decrease of pressure causes the volume of the gas in the can to increase. Because the gas needs more room than is provided in the can, it expands in spray form through the small hole near the valve. In general, if the temperature is constant, the pressure, P, of a gas in a container varies inversely as the volume, V, of the container. The pressure of a gas sample in a container whose volume is 8 cubic inches is 12 pounds per square inch. If the sample expands to a volume of 22 cubic inches, what is the new pressure of the gas?

Solution

Step 1. Write an equation. We know that *y varies inversely as x* is expressed as

$$y = \frac{k}{x}.$$

By changing letters, we can write an equation that models the following English statement: The pressure, P, of a gas in a container varies inversely as the volume, V.

$$P = \frac{k}{V}$$

Step 2. Use the given values to find k. The pressure of a gas sample in a container whose volume is 8 cubic inches is 12 pounds per square inch. Substitute 12 for P and 8 for V in the inverse variation equation. Then solve for k.

$$P = \frac{k}{V}$$ Pressure varies inversely as volume.

$$12 = \frac{k}{8}$$ The pressure in an 8-cubic-inch container is 12 pounds per square inch.

$$12 \cdot 8 = \frac{k}{8} \cdot 8$$ Multiply both sides by 8.

$$96 = k$$ Simplify.

Step 3. Substitute the value of k into the equation.

$$P = \frac{k}{V}$$ Use the equation from step 1.

$$P = \frac{96}{V}$$ Replace k, the constant of variation, with 96.

Step 4. Answer the problem's question. We need to find the pressure when the volume expands to 22 cubic inches. Substitute 22 for V and solve for P.

$$P = \frac{96}{V} = \frac{96}{22} = 4\frac{4}{11}$$

When the volume is 22 cubic inches, the pressure of the gas is $4\frac{4}{11}$ pounds per square inch. ■

☑ **CHECK POINT 3** The length of a violin string varies inversely as the frequency of its vibrations. A violin string 8 inches long vibrates at a frequency of 640 cycles per second. What is the frequency of a 10-inch string?

3 Solve combined variation problems.

Combined Variation

In **combined variation**, direct variation and inverse variation occur at the same time. For example, as the advertising budget, A, of a company increases, its monthly sales, S, also increase. Monthly sales vary directly as the advertising budget:

$$S = kA.$$

By contrast, as the price of the company's product, P, increases, its monthly sales, S, decrease. Monthly sales vary inversely as the price of the product:

$$S = \frac{k}{P}.$$

We can combine these two variation equations into one equation:

$$S = \frac{kA}{P}.$$ Monthly sales, S, vary directly as the advertising budget, A, and inversely as the price of the product, P.

The following example illustrates an application of combined variation.

EXAMPLE 4 Solving a Combined Variation Problem

The owners of Rollerblades Now determined that the monthly sales, S, of its skates vary directly as its advertising budget, A, and inversely as the price of the skates, P. When \$60,000 is spent on advertising and the price of the skates is \$40, the monthly sales are 12,000 pairs of rollerblades.

a. Write an equation of variation that models this situation.

b. Determine monthly sales if the amount of the advertising budget is increased to \$70,000.

Solution

a. Write an equation:

$$S = \frac{kA}{P}.$$ Translate "sales vary directly as the advertising budget and inversely as the skates' price."

Use the given values to find k.

$$12{,}000 = \frac{k(60{,}000)}{40}$$ When \$60,000 is spent on advertising ($A = 60{,}000$) and the price is \$40 ($P = 40$), monthly sales are 12,000 units ($S = 12{,}000$).

$$12{,}000 = k \cdot 1500$$ Divide 60,000 by 40.

$$\frac{12{,}000}{1500} = \frac{k \cdot 1500}{1500}$$ Divide both sides of the equation by 1500.

$$8 = k$$ Simplify.

Therefore, the equation of variation that models monthly sales is

$$S = \frac{8A}{P}.$$ Substitute 8 for k in $S = \frac{kA}{P}$.

b. The advertising budget is increased to \$70,000, so $A = 70{,}000$. The skates' price is still \$40, so $P = 40$.

$$S = \frac{8A}{P}$$ This is the equation from part (a).

$$S = \frac{8(70{,}000)}{40}$$ Substitute 70,000 for A and 40 for P.

$$S = 14{,}000$$ Simplify.

With a \$70,000 advertising budget and \$40 price, the company can expect to sell 14,000 pairs of rollerblades in a month (up from 12,000). ■

✓ **CHECK POINT 4** The number of minutes needed to solve an Exercise Set of variation problems varies directly as the number of problems and inversely as the number of people working to solve the problems. It takes 4 people 32 minutes to solve 16 problems. How many minutes will it take 8 people to solve 24 problems?

4 Solve problems involving joint variation.

Joint Variation

Joint variation is a variation in which a variable varies directly as the product of two or more other variables. Thus, the equation $y = kxz$ is read "y varies jointly as x and z."

Joint variation plays a critical role in Isaac Newton's formula for gravitation:

$$F = G\frac{m_1 m_2}{d^2}.$$

The formula states that the force of gravitation, F, between two bodies varies jointly as the product of their masses, m_1 and m_2, and inversely as the square of the distance between them, d. (G is the gravitational constant.) The formula indicates that gravitational force exists between any two objects in the universe, increasing as the distance between the bodies decreases. One practical result is that the pull of the moon on the oceans is greater on the side of Earth closer to the moon. This gravitational imbalance is what produces tides.

EXAMPLE 5 Modeling Centrifugal Force

The centrifugal force, C, of a body moving in a circle varies jointly with the radius of the circular path, r, and the body's mass, m, and inversely with the square of the time, t, it takes to move about one full circle. A 6-gram body moving in a circle with radius 100 centimeters at a rate of 1 revolution in 2 seconds has a centrifugal force of 6000 dynes. Find the centrifugal force of an 18-gram body moving in a circle with radius 100 centimeters at a rate of 1 revolution in 3 seconds.

Solution

$C = \frac{krm}{t^2}$	Translate "Centrifugal force, C, varies jointly with radius, r, and mass, m, and inversely with the square of time, t."
$6000 = \frac{k(100)(6)}{2^2}$	A 6-gram body ($m = 6$) moving in a circle with radius 100 centimeters ($r = 100$) at 1 revolution in 2 seconds ($t = 2$) has a centrifugal force of 6000 dynes ($C = 6000$).
$6000 = 150k$	Simplify: $\frac{100(6)}{2^2} = \frac{600}{4} = 150$.
$40 = k$	Divide both sides by 150 and solve for k.
$C = \frac{40rm}{t^2}$	Substitute 40 for k in the model for centrifugal force.
$C = \frac{40(100)(18)}{3^2}$	Find centrifugal force, C, of an 18-gram body ($m = 18$) moving in a circle with radius 100 centimeters ($r = 100$) at 1 revolution in 3 seconds ($t = 3$).
$= 8000$	Simplify.

The centrifugal force is 8000 dynes. ■

CHECK POINT 5 The volume of a cone, V, varies jointly as its height, h, and the square of its radius, r. A cone with a radius measuring 6 feet and a height measuring 10 feet has a volume of 120π cubic feet. Find the volume of a cone having a radius of 12 feet and a height of 2 feet.

CONCEPT AND VOCABULARY CHECK

Fill in each blank so that the resulting statement is true.

1. y varies directly as x can be modeled by the equation _______________, where k is called the _______________.
2. y varies directly as the nth power of x can be modeled by the equation _______________.
3. y varies inversely as x can be modeled by the equation _______________.
4. y varies directly as x and inversely as z can be modeled by the equation _______________.
5. y varies jointly as x and z can be modeled by the equation _______________.
6. In the equation $S = \frac{8A}{P}$, S varies _______________ as A and _______________ as P.
7. In the equation $C = \frac{0.02P_1P_2}{d^2}$, C varies _______________ as P_1 and P_2 and _______________ as the square of d.

6.8 EXERCISE SET

MyMathLab®

Practice Exercises

Use the four-step procedure for solving variation problems given on page 480 to solve Exercises 1–10.

1. y varies directly as x. $y = 65$ when $x = 5$. Find y when $x = 12$.
2. y varies directly as x. $y = 45$ when $x = 5$. Find y when $x = 13$.
3. y varies inversely as x. $y = 12$ when $x = 5$. Find y when $x = 2$.
4. y varies inversely as x. $y = 6$ when $x = 3$. Find y when $x = 9$.
5. y varies directly as x and inversely as the square of z. $y = 20$ when $x = 50$ and $z = 5$. Find y when $x = 3$ and $z = 6$.
6. a varies directly as b and inversely as the square of c. $a = 7$ when $b = 9$ and $c = 6$. Find a when $b = 4$ and $c = 8$.
7. y varies jointly as x and z. $y = 25$ when $x = 2$ and $z = 5$. Find y when $x = 8$ and $z = 12$.
8. C varies jointly as A and T. $C = 175$ when $A = 2100$ and $T = 4$. Find C when $A = 2400$ and $T = 6$.
9. y varies jointly as a and b, and inversely as the square root of c. $y = 12$ when $a = 3$, $b = 2$, and $c = 25$. Find y when $a = 5$, $b = 3$, and $c = 9$.
10. y varies jointly as m and the square of n, and inversely as p. $y = 15$ when $m = 2$, $n = 1$, and $p = 6$. Find y when $m = 3$, $n = 4$, and $p = 10$.

Practice PLUS

In Exercises 11–20, write an equation that expresses each relationship. Then solve the equation for y.

11. x varies jointly as y and z.
12. x varies jointly as y and the square of z.
13. x varies directly as the cube of z and inversely as y.
14. x varies directly as the cube root of z and inversely as y.
15. x varies jointly as y and z and inversely as the square root of w.
16. x varies jointly as y and z and inversely as the square of w.
17. x varies jointly as z and the sum of y and w.
18. x varies jointly as z and the difference between y and w.
19. x varies directly as z and inversely as the difference between y and w.
20. x varies directly as z and inversely as the sum of y and w.

Application Exercises

Use the four-step procedure for solving variation problems given on page 480 to solve Exercises 21–28.

21. An alligator's tail length, T, varies directly as its body length, B. An alligator with a body length of 4 feet has a tail length of 3.6 feet. What is the tail length of an alligator whose body length is 6 feet?

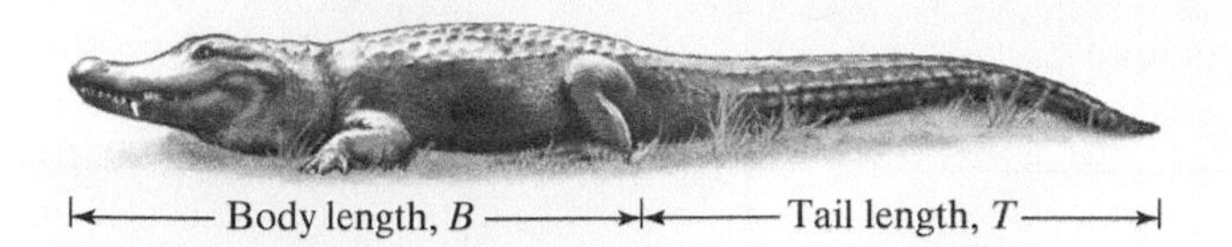

22. An object's weight on the moon, M, varies directly as its weight on Earth, E. Neil Armstrong, the first person to step on the moon on July 20, 1969, weighed 360 pounds on Earth (with all of his equipment on) and 60 pounds on the moon. What is the moon weight of a person who weighs 186 pounds on Earth?

23. The height that a ball bounces varies directly as the height from which it was dropped. A tennis ball dropped from 12 inches bounces 8.4 inches. From what height was the tennis ball dropped if it bounces 56 inches?

24. The distance that a spring will stretch varies directly as the force applied to the spring. A force of 12 pounds is needed to stretch a spring 9 inches. What force is required to stretch the spring 15 inches?

25. If all men had identical body types, their weight would vary directly as the cube of their height. Shown below is Robert Wadlow, who reached a record height of 8 feet 11 inches (107 inches) before his death at age 22. If a man who is 5 feet 10 inches tall (70 inches) with the same body type as Mr. Wadlow weighs 170 pounds, what was Robert Wadlow's weight shortly before his death?

26. On a dry asphalt road, a car's stopping distance varies directly as the square of its speed. A car traveling at 45 miles per hour can stop in 67.5 feet. What is the stopping distance for a car traveling at 60 miles per hour?

27. The figure shows that a bicyclist tips the cycle when making a turn. The angle B, formed by the vertical direction and the bicycle, is called the banking angle. The banking angle varies inversely as the cycle's turning radius. When the turning radius is 4 feet, the banking angle is 28°. What is the banking angle when the turning radius is 3.5 feet?

28. The water temperature of the Pacific Ocean varies inversely as the water's depth. At a depth of 1000 meters, the water temperature is 4.4° Celsius. What is the water temperature at a depth of 5000 meters?

Heart rates and life spans of most mammals can be modeled using inverse variation. The bar graph shows the average heart rate and the average life span of five mammals. You will use the data to solve Exercises 29–30.

Heart Rate and Life Span

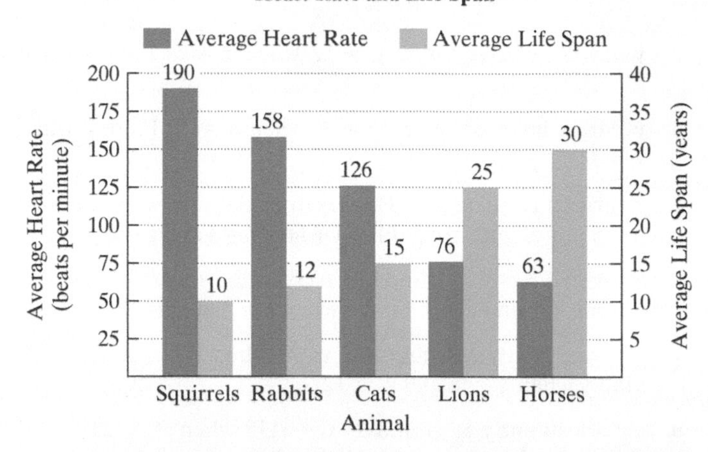

Source: The Carnegie Library of Pittsburgh, THE HANDY SCIENCE ANSWER BOOK, Centennial Edition. © 2003

29. a. A mammal's average life span, L, in years, varies inversely as its average heart rate, R, in beats per minute. Use the data shown for horses to write the equation that models this relationship.

b. Is the inverse variation equation in part (a) an exact model or an approximate model for the data shown for lions?

c. Elephants have an average heart rate of 27 beats per minute. Determine their average life span.

30. a. A mammal's average life span, L, in years, varies inversely as its average heart rate, R, in beats per minute. Use the data shown for cats at the bottom of the previous page to write the equation that models this relationship.

b. Is the inverse variation equation in part (a) an exact model or an approximate model for the data shown for squirrels?

c. Mice have an average heart rate of 634 beats per minute. Determine their average life span, rounded to the nearest year.

The figure shows the graph of the inverse variation model that you wrote in Exercise 29(a) or Exercise 30(a). Use the graph of this rational function to solve Exercises 31–32.

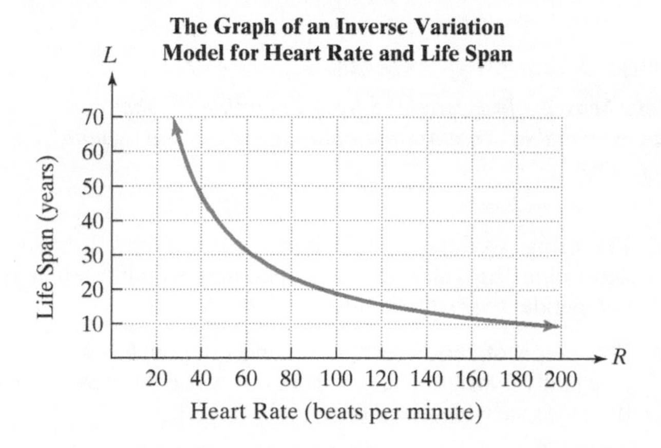

31. a. If a mammal has a life span of 20 years, use the graph to estimate its heart rate, rounded to the nearest 10 beats per minute.

b. Use the inverse variation equation that you wrote in Exercise 29(a) or Exercise 30(a) to determine the heart rate, rounded to the nearest beat per minute, for a mammal with a life span of 20 years.

c. The bar graph at the bottom of the previous page uses two bars to display the data for horses. How is this information shown on the graph of the inverse variation model?

32. a. If a mammal has a life span of 50 years, use the graph to estimate its heart rate, rounded to the nearest 10 beats per minute.

b. Use the inverse variation equation that you wrote in Exercise 29(a) or Exercise 30(a) to determine the heart rate, rounded to the nearest beat per minute, for a mammal with a life span of 50 years.

c. The bar graph at the bottom of the previous page uses two bars to display the data for lions. How is this information shown on the graph of the inverse variation model?

Continue to use the four-step procedure for solving variation problems given on page 480 to solve Exercises 33–40.

33. Radiation machines, used to treat tumors, produce an intensity of radiation that varies inversely as the square of the distance from the machine. At 3 meters, the radiation intensity is 62.5 milliroentgens per hour. What is the intensity at a distance of 2.5 meters?

34. The illumination provided by a car's headlight varies inversely as the square of the distance from the headlight. A car's headlight produces an illumination of 3.75 footcandles at a distance of 40 feet. What is the illumination when the distance is 50 feet?

35. Body-mass index, or BMI, takes both weight and height into account when assessing whether an individual is underweight or overweight. BMI varies directly as one's weight, in pounds, and inversely as the square of one's height, in inches. In adults, normal values for the BMI are between 20 and 25, inclusive. Values below 20 indicate that an individual is underweight and values above 30 indicate that an individual is obese. A person who weighs 180 pounds and is 5 feet, or 60 inches, tall has a BMI of 35.15. What is the BMI, to the nearest tenth, for a 170 pound person who is 5 feet 10 inches tall. Is this person overweight?

36. One's intelligence quotient, or IQ, varies directly as a person's mental age and inversely as that person's chronological age. A person with a mental age of 25 and a chronological age of 20 has an IQ of 125. What is the chronological age of a person with a mental age of 40 and an IQ of 80?

37. The heat loss of a glass window varies jointly as the window's area and the difference between the outside and inside temperatures. A window 3 feet wide by 6 feet long loses 1200 Btu per hour when the temperature outside is 20° colder than the temperature inside. Find the heat loss through a glass window that is 6 feet wide by 9 feet long when the temperature outside is 10° colder than the temperature inside.

38. Kinetic energy varies jointly as the mass and the square of the velocity. A mass of 8 grams and velocity of 3 centimeters per second has a kinetic energy of 36 ergs. Find the kinetic energy for a mass of 4 grams and velocity of 6 centimeters per second.

39. Sound intensity varies inversely as the square of the distance from the sound source. If you are in a movie theater and you change your seat to one that is twice as far from the speakers, how does the new sound intensity compare to that of your original seat?

40. Many people claim that as they get older, time seems to pass more quickly. Suppose that the perceived length of a period of time is inversely proportional to your age. How long will a year seem to be when you are three times as old as you are now?

41. The average number of daily phone calls, C, between two cities varies jointly as the product of their populations, P_1 and P_2, and inversely as the square of the distance, d, between them.

a. Write an equation that expresses this relationship.

b. The distance between San Francisco (population: 777,000) and Los Angeles (population: 3,695,000) is 420 miles. If the average number of daily phone calls between the cities is 326,000, find the value of k to two decimal places and write the equation of variation.

c. Memphis (population: 650,000) is 400 miles from New Orleans (population: 490,000). Find the average number of daily phone calls, to the nearest whole number, between these cities.

42. The force of wind blowing on a window positioned at a right angle to the direction of the wind varies jointly as the area of the window and the square of the wind's speed. It is known that a wind of 30 miles per hour blowing on a window measuring 4 feet by 5 feet exerts a force of 150 pounds. During a storm with winds of 60 miles per hour, should hurricane shutters be placed on a window that measures 3 feet by 4 feet and is capable of withstanding 300 pounds of force?

43. The table shows the values for the current, I, in an electric circuit and the resistance, R, of the circuit.

I (amperes)	0.5	1.0	1.5	2.0	2.5	3.0	4.0	5.0
R (ohms)	12	6.0	4.0	3.0	2.4	2.0	1.5	1.2

a. Graph the ordered pairs in the table of values, with values of I along the x-axis and values of R along the y-axis. Connect the eight points with a smooth curve.

b. Does current vary directly or inversely as resistance? Use your graph and explain how you arrived at your answer.

c. Write an equation of variation for I and R, using one of the ordered pairs in the table to find the constant of variation. Then use your variation equation to verify the other seven ordered pairs in the table.

Explaining the Concepts

44. What does it mean if two quantities vary directly?

45. In your own words, explain how to solve a variation problem.

46. What does it mean if two quantities vary inversely?

47. Explain what is meant by combined variation. Give an example with your explanation.

48. Explain what is meant by joint variation. Give an example with your explanation.

In Exercises 49–50, describe in words the variation shown by the given equation.

49. $z = \dfrac{k\sqrt{x}}{y^2}$

50. $z = kx^2\sqrt{y}$

51. We have seen that the daily number of phone calls between two cities varies jointly as their populations and inversely as the square of the distance between them. This model, used by telecommunication companies to estimate the line capacities needed among various cities, is called the *gravity model.* Compare the model to Newton's formula for gravitation on page 486 and describe why the name *gravity model* is appropriate.

Technology Exercises

52. Use a graphing utility to graph any three of the variation equations in Exercises 21–28. Then TRACE along each curve and identify the point that corresponds to the problem's solution.

Critical Thinking Exercises

Make Sense? *In Exercises 53–56, determine whether each statement makes sense or does not make sense, and explain your reasoning.*

53. I'm using an inverse variation equation and I need to determine the value of the dependent variable when the independent variable is zero.

54. The graph of this direct variation equation has a positive constant of variation and shows one variable increasing as the other variable decreases.

55. When all is said and done, it seems to me that direct variation equations are special kinds of linear functions and inverse variation equations are special kinds of rational functions.

56. Using the language of variation, I can now state the formula for the area of a trapezoid, $A = \frac{1}{2}h(b_1 + b_2)$, as, "A trapezoid's area varies jointly with its height and the sum of its bases."

57. In a hurricane, the wind pressure varies directly as the square of the wind velocity. If wind pressure is a measure of a hurricane's destructive capacity, what happens to this destructive power when the wind speed doubles?

58. The heat generated by a stove element varies directly as the square of the voltage and inversely as the resistance. If the voltage remains constant, what needs to be done to triple the amount of heat generated?

59. Galileo's telescope brought about revolutionary changes in astronomy. A comparable leap in our ability to observe the universe took place as a result of the Hubble Space Telescope. The space telescope can see stars and galaxies whose brightness is $\frac{1}{50}$ of the faintest objects now observable using ground-based telescopes. Use the fact that the brightness of a point source, such as a star, varies inversely as the square of its distance from an observer to show that the space telescope can see about seven times farther than a ground-based telescope.

Review Exercises

60. Evaluate:

$$\begin{vmatrix} -1 & 2 \\ 3 & -4 \end{vmatrix}.$$

(Section 3.5, Example 1)

61. Factor completely:

$$x^2y - 9y - 3x^2 + 27.$$

(Section 5.6, Example 3)

62. Find the degree of

$$7xy + x^2y^2 - 5x^3 - 7.$$

(Section 5.1, Example 1)

Preview Exercises

Exercises 63–65 will help you prepare for the material covered in the first section of the next chapter.

63. If $f(x) = \sqrt{3x + 12}$, find $f(-1)$.

64. If $f(x) = \sqrt{3x + 12}$, find $f(8)$.

65. Use the graph of $f(x) = \sqrt{3x + 12}$ to identify the function's domain and its range.

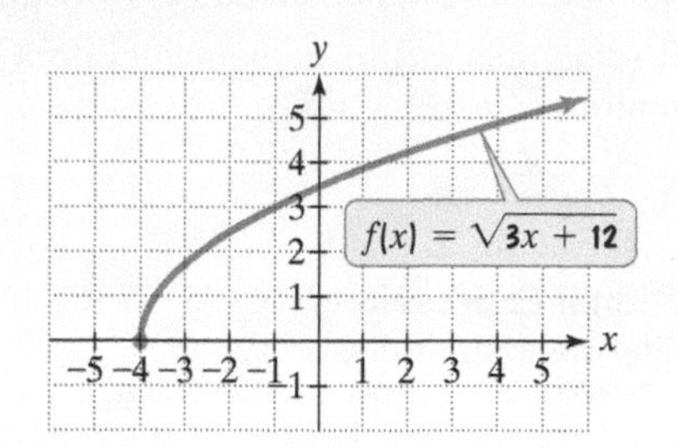

Chapter 6 Summary

Definitions and Concepts	Examples

Section 6.1 Rational Expressions and Functions: Multiplying and Dividing

A rational expression consists of a polynomial divided by a nonzero polynomial. A rational function is defined by a formula that is a rational expression. The domain of a rational function is the set of all real numbers except those for which the denominator is zero. Graphs of rational functions often contain disconnected branches. The graphs often approach, but do not touch, vertical or horizontal lines, called vertical asymptotes and horizontal asymptotes, respectively.

Let $f(x) = \dfrac{x + 2}{x^2 + x - 6}$. Find the domain of f.

$$x^2 + x - 6 = 0$$
$$(x + 3)(x - 2) = 0$$
$$x + 3 = 0 \quad \text{or} \quad x - 2 = 0$$
$$x = -3 \qquad\qquad x = 2$$

These values of x make the denominator zero. Exclude them from the domain.

Domain of $f = (-\infty, -3) \cup (-3, 2) \cup (2, \infty)$

The branches of the graph of f break at -3 and 2.

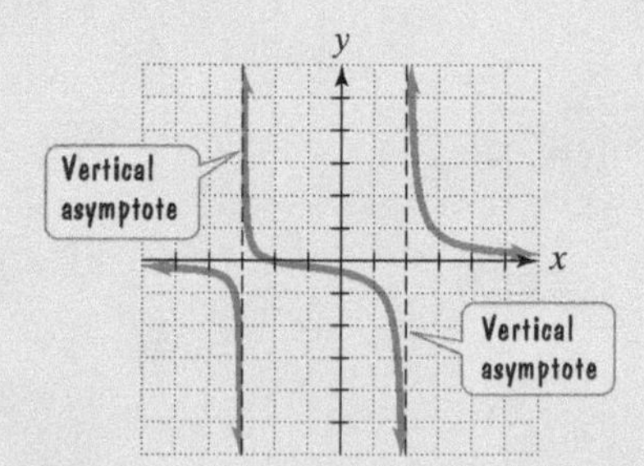

Simplifying Rational Expressions

1. Factor the numerator and the denominator completely.
2. Divide both the numerator and the denominator by any common factors.

Simplify: $\dfrac{4x + 28}{x^2 - 49}$.

$$\frac{4x + 28}{x^2 - 49} = \frac{4\cancel{(x + 7)}}{\cancel{(x + 7)}(x - 7)} = \frac{4}{x - 7}$$

Definitions and Concepts	Examples

Section 6.1 Rational Expressions and Functions: Multiplying and Dividing (continued)

Multiplying Rational Expressions

1. Factor completely.
2. Divide numerators and denominators by common factors.
3. Multiply remaining factors in the numerators and in the denominators.

$$\frac{x^2+3x-18}{x^2-3x}\cdot\frac{x^2}{x^2-36}$$
$$=\frac{(x+6)(x-3)}{x(x-3)}\cdot\frac{x\cdot x}{(x+6)(x-6)}$$
$$=\frac{x}{x-6}$$

Dividing Rational Expressions

Invert the divisor and multiply.

$$\frac{5y+15}{(y+5)^2}\div\frac{y^2-9}{y+5}$$
$$=\frac{5y+15}{(y+5)^2}\cdot\frac{y+5}{y^2-9}=\frac{5(y+3)}{(y+5)(y+5)}\cdot\frac{(y+5)}{(y+3)(y-3)}$$
$$=\frac{5}{(y+5)(y-3)}$$

Section 6.2 Adding and Subtracting Rational Expressions

Adding or Subtracting Rational Expressions

If the denominators are the same, add or subtract the numerators. Place the result over the common denominator. If the denominators are different, write all rational expressions with the least common denominator. The LCD is a polynomial consisting of the product of all prime factors in the denominators, with each factor raised to the greatest power of its occurrence in any denominator. Once all rational expressions are written in terms of the LCD, add or subtract as described above. Simplify the result, if possible.

$$\frac{x+1}{2x-2}-\frac{2x}{x^2+2x-3}$$
$$=\frac{x+1}{2(x-1)}-\frac{2x}{(x-1)(x+3)}$$ LCD is $2(x-1)(x+3)$.
$$=\frac{(x+1)(x+3)}{2(x-1)(x+3)}-\frac{2x\cdot 2}{2(x-1)(x+3)}$$
$$=\frac{x^2+4x+3-4x}{2(x-1)(x+3)}=\frac{x^2+3}{2(x-1)(x+3)}$$

Section 6.3 Complex Rational Expressions

Complex rational expressions have numerators or denominators containing one or more rational expressions. Complex rational expressions can be simplified by multiplying the numerator and the denominator by the LCD. They can also be simplified by obtaining single expressions in the numerator and denominator and then dividing.

Simplify: $\dfrac{\frac{x+3}{x}}{x-\frac{9}{x}}$.

Multiplying by the LCD, x:

$$\frac{x}{x}\cdot\frac{\left(\frac{x+3}{x}\right)}{\left(x-\frac{9}{x}\right)}=\frac{x+3}{x^2-9}=\frac{(x+3)}{(x+3)(x-3)}=\frac{1}{x-3}$$

Simplifying by dividing:

$$\frac{\frac{x+3}{x}}{x-\frac{9}{x}}=\frac{\frac{x+3}{x}}{\frac{x}{1}\cdot\frac{x}{x}-\frac{9}{x}}=\frac{\frac{x+3}{x}}{\frac{x^2-9}{x}}=\frac{x+3}{x}\cdot\frac{x}{x^2-9}$$

LCD is x.

$$=\frac{x+3}{x}\cdot\frac{x}{(x+3)(x-3)}=\frac{1}{x-3}$$

Definitions and Concepts	Examples

Section 6.4 Division of Polynomials

To divide a polynomial by a monomial, divide each term of the polynomial by the monomial.

$$\frac{9x^3y^4 - 12x^3y^2 - 7x^2y^2}{3xy^2}$$

$$= \frac{9x^3y^4}{3xy^2} - \frac{12x^3y^2}{3xy^2} - \frac{7x^2y^2}{3xy^2} = 3x^2y^2 - 4x^2 - \frac{7x}{3}$$

To divide by a polynomial containing more than one term, use long division. If necessary, arrange the dividend in descending powers of the variable. If a power of a variable is missing, add that power with a coefficient of 0. Repeat the four steps of the long-division process—divide, multiply, subtract, bring down the next term—until the degree of the remainder is less than the degree of the divisor.

Divide: $(2x^3 - x^2 - 7) \div (x - 2)$.

$$\begin{array}{r} 2x^2 + 3x + 6 \\ x - 2\overline{)2x^3 - x^2 + 0x - 7} \\ \underline{2x^3 - 4x^2} \qquad\qquad \\ 3x^2 + 0x \qquad \\ \underline{3x^2 - 6x} \qquad \\ 6x - 7 \\ \underline{6x - 12} \\ 5 \end{array}$$

The answer is $2x^2 + 3x + 6 + \frac{5}{x - 2}$.

Section 6.5 Synthetic Division and the Remainder Theorem

A shortcut to long division, called synthetic division, can be used to divide a polynomial by a binomial of the form $x - c$.

Here is the division problem shown above using synthetic division: $(2x^3 - x^2 - 7) \div (x - 2)$.

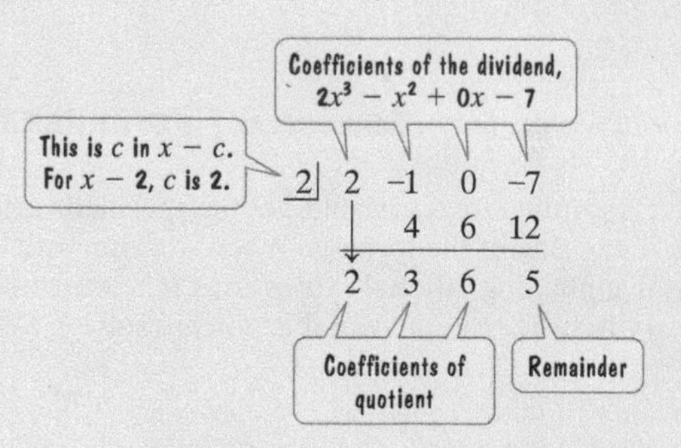

The answer is $2x^2 + 3x + 6 + \frac{5}{x - 2}$.

The Remainder Theorem
If the polynomial $f(x)$ is divided by $x - c$, then the remainder is $f(c)$. This can be used to evaluate a polynomial function at c. Rather than substituting c for x, divide the function by $x - c$. The remainder is $f(c)$.

If $f(x) = x^4 - 3x^2 - 2x + 5$, use the Remainder Theorem to find $f(-2)$.

$$\begin{array}{r|rrrrr} -2 & 1 & 0 & -3 & -2 & 5 \\ & & -2 & 4 & -2 & 8 \\ \hline & 1 & -2 & 1 & -4 & 13 \end{array}$$

Remainder

Thus, $f(-2) = 13$.

Definitions and Concepts	Examples

Section 6.6 Rational Equations

A rational equation is an equation containing one or more rational expressions.

Solving Rational Equations

1. List restrictions on the variable.
2. Clear fractions by multiplying both sides by the LCD.
3. Solve the resulting equation.
4. Reject any proposed solution in the list of restrictions. Check other proposed solutions in the original equation.

Solve: $\frac{7x}{x^2-4}+\frac{5}{x-2}=\frac{2x}{x^2-4}$.

$$\frac{7x}{(x+2)(x-2)}+\frac{5}{x-2}=\frac{2x}{(x+2)(x-2)}$$

Denominators would equal 0 if $x=-2$ or $x=2$. Restrictions: $x\neq-2$ and $x\neq 2$.

LCD is $(x+2)(x-2)$.

$$(x+2)(x-2)\left[\frac{7x}{(x+2)(x-2)}+\frac{5}{x-2}\right]=(x+2)(x-2)\cdot\frac{2x}{(x+2)(x-2)}$$

$$7x+5(x+2)=2x$$
$$7x+5x+10=2x$$
$$12x+10=2x$$
$$10=-10x$$
$$-1=x$$

The proposed solution, -1, is not part of the restriction $x\neq-2$ and $x\neq 2$. It checks. The solution is -1 and the solution set is $\{-1\}$.

Section 6.7 Formulas and Applications of Rational Equations

To solve a formula for a variable, get the specified variable alone on one side of the formula. When working with formulas containing rational expressions, it is sometimes necessary to factor out the variable you are solving for.

Solve: $\frac{e}{E}=\frac{r}{r+R}$ for r.

$$E(r+R)\cdot\frac{e}{E}=E(r+R)\cdot\frac{r}{r+R}$$ LCD is $E(r+R)$.

$$e(r+R)=Er$$
$$er+eR=Er$$
$$eR=Er-er$$
$$eR=(E-e)r$$
$$\frac{eR}{E-e}=r$$

The average cost function, $\overline{C}$, for a business is a rational function representing the average cost for the company to produce each unit of its product. The function consists of the sum of fixed and variable costs divided by the number of units produced. As production level increases, the average cost to produce each unit of a product decreases.

A company has a fixed cost of \$80,000 monthly and it costs \$20 to produce each unit of its product. Its cost function, C, of producing x units is

$$C(x)=80{,}000+20x.$$

Its average cost function, $\overline{C}$, is

$$\overline{C}(x)=\frac{80{,}000+20x}{x}.$$

Definitions and Concepts	Examples

Section 6.7 Formulas and Applications of Rational Equations (continued)

Problems involving time in motion and problems involving work translate into rational equations. Motion problems involving time are solved using

$$\text{Time traveled} = \frac{\text{Distance traveled}}{\text{Rate of travel}}.$$

Work problems are solved using the following condition:

$$\boxed{\text{Fraction of job done by first}} + \boxed{\text{fraction of job done by second}} = \boxed{1}.$$

It takes a cyclist who averages 16 miles per hour in still air the same time to travel 48 miles with the wind as 16 miles against the wind. What is the wind's rate?

$$x = \text{wind's rate}$$

$$16 + x = \text{cyclist's rate with wind}$$

$$16 - x = \text{cyclist's rate against wind}$$

	Distance	Rate	Time = $\frac{\text{Distance}}{\text{Rate}}$
With wind	48	$16 + x$	$\frac{48}{16 + x}$
Against wind	16	$16 - x$	$\frac{16}{16 - x}$

These times are equal.

$$(16 + x)(16 - x) \cdot \frac{48}{16 + x} = \frac{16}{16 - x} \cdot (16 + x)(16 - x)$$

$$48(16 - x) = 16(16 + x)$$

Solving this equation, $x = 8$. The wind's rate is 8 miles per hour.

Section 6.8 Modeling Using Variation

English Statement	Equation
y varies directly as x. y is directly proportional to x.	$y = kx$
y varies directly as x^n. y is directly proportional to x^n.	$y = kx^n$
y varies inversely as x. y is inversely proportional to x.	$y = \frac{k}{x}$
y varies inversely as x^n. y is inversely proportional to x^n.	$y = \frac{k}{x^n}$
y varies directly as x and inversely as z.	$y = \frac{kx}{z}$
y varies jointly as x and z.	$y = kxz$

Solving Variation Problems

1. Write an equation that models the given English statement.
2. Substitute the pair of values into the equation in step 1 and find k.
3. Substitute k into the equation in step 1.
4. Use the equation in step 3 to answer the problem's question.

The time that it takes you to drive a certain distance varies inversely as your driving rate. Averaging 40 miles per hour, it takes you 10 hours to drive the distance. How long would the trip take averaging 50 miles per hour?

1. $t = \frac{k}{r}$ Time, t, varies inversely as rate, r.

2. It takes 10 hours at 40 miles per hour.

$$10 = \frac{k}{40}$$

$$k = 10(40) = 400$$

3. $t = \frac{400}{r}$

4. How long at 50 miles per hour? Substitute 50 for r.

$$t = \frac{400}{50} = 8$$

It takes 8 hours at 50 miles per hour.

CHAPTER 6 REVIEW EXERCISES

6.1

1. If $f(x) = \dfrac{x^2 + 2x - 3}{x^2 - 4}$, find the following function values. If a function value does not exist, so state.

   a. $f(4)$ b. $f(0)$

   c. $f(-2)$ d. $f(-3)$

In Exercises 2–3, find the domain of the given rational function.

2. $f(x) = \dfrac{x - 6}{(x - 3)(x + 4)}$

3. $f(x) = \dfrac{x + 2}{x^2 + x - 2}$

In Exercises 4–8, simplify each rational expression. If the rational expression cannot be simplified, so state.

4. $\dfrac{5x^3 - 35x}{15x^2}$

5. $\dfrac{x^2 + 6x - 7}{x^2 - 49}$

6. $\dfrac{6x^2 + 7x + 2}{2x^2 - 9x - 5}$

7. $\dfrac{x^2 + 4}{x^2 - 4}$

8. $\dfrac{x^3 - 8}{x^2 - 4}$

In Exercises 9–15, multiply or divide as indicated.

9. $\dfrac{5x^2 - 5}{3x + 12} \cdot \dfrac{x + 4}{x - 1}$

10. $\dfrac{2x + 5}{4x^2 + 8x - 5} \cdot \dfrac{4x^2 - 4x + 1}{x + 1}$

11. $\dfrac{x^2 - 9x + 14}{x^3 + 2x^2} \cdot \dfrac{x^2 - 4}{x^2 - 4x + 4}$

12. $\dfrac{1}{x^2 + 8x + 15} \div \dfrac{3}{x + 5}$

13. $\dfrac{x^2 + 16x + 64}{2x^2 - 128} \div \dfrac{x^2 + 10x + 16}{x^2 - 6x - 16}$

14. $\dfrac{y^2 - 16}{y^3 - 64} \div \dfrac{y^2 - 3y - 18}{y^2 + 5y + 6}$

15. $\dfrac{x^2 - 4x + 4 - y^2}{2x^2 - 11x + 15} \cdot \dfrac{x^4y}{x - 2 + y} \div \dfrac{x^3y - 2x^2y - x^2y^2}{3x - 9}$

16. Deer are placed into a newly acquired habitat. The deer population over time is modeled by a rational function whose graph is shown in the figure. Use the graph to answer each of the following questions.

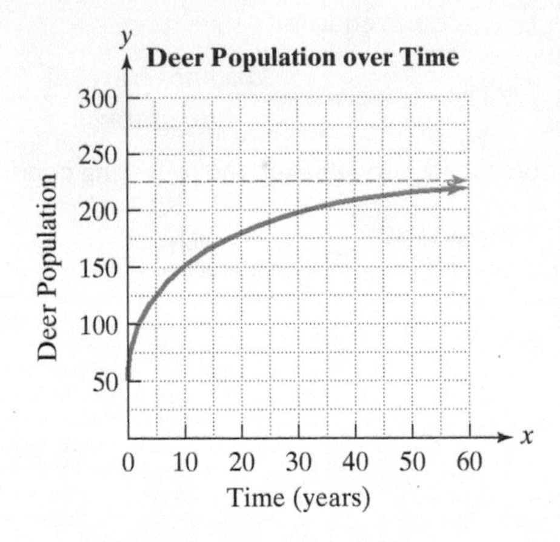

   a. How many deer were introduced into the habitat?

   b. What is the pop.ulation after 10 years?

   c. What is the equation of the horizontal asymptote shown in the figure? What does this mean in terms of the deer population?

6.2

In Exercises 17–19, add or subtract as indicated. Simplify the result, if possible.

17. $\dfrac{4x + 1}{3x - 1} + \dfrac{8x - 5}{3x - 1}$

18. $\dfrac{2x - 7}{x^2 - 9} - \dfrac{x - 4}{x^2 - 9}$

19. $\dfrac{4x^2 - 11x + 4}{x - 3} - \dfrac{x^2 - 4x + 10}{x - 3}$

In Exercises 20–21, find the least common denominator of the rational expressions.

20. $\dfrac{7}{9x^3}$ and $\dfrac{5}{12x}$

21. $\dfrac{x + 7}{x^2 + 2x - 35}$ and $\dfrac{x}{x^2 + 9x + 14}$

In Exercises 22–28, perform the indicated operations. Simplify the result, if possible.

22. $\dfrac{1}{x} + \dfrac{2}{x - 5}$

23. $\frac{2}{x^2 - 5x + 6} + \frac{3}{x^2 - x - 6}$

24. $\frac{x - 3}{x^2 - 8x + 15} + \frac{x + 2}{x^2 - x - 6}$

25. $\frac{3x^2}{9x^2 - 16} - \frac{x}{3x + 4}$

26. $\frac{y}{y^2 + 5y + 6} - \frac{2}{y^2 + 3y + 2}$

27. $\frac{x}{x + 3} + \frac{x}{x - 3} - \frac{9}{x^2 - 9}$

28. $\frac{3x^2}{x - y} + \frac{3y^2}{y - x}$

6.3 *In Exercises 29–34, simplify each complex rational expression.*

29. $\frac{\frac{3}{x} - 3}{\frac{8}{x} - 8}$

30. $\frac{\frac{5}{x} + 1}{1 - \frac{25}{x^2}}$

31. $\frac{3 - \frac{1}{x + 3}}{3 + \frac{1}{x + 3}}$

32. $\frac{\frac{4}{x + 3}}{\frac{2}{x - 2} - \frac{1}{x^2 + x - 6}}$

33. $\frac{\frac{2}{x^2 - x - 6} + \frac{1}{x^2 - 4x + 3}}{\frac{3}{x^2 + x - 2} - \frac{2}{x^2 + 5x + 6}}$

34. $\frac{x^{-2} + x^{-1}}{x^{-2} - x^{-1}}$

6.4 *In Exercises 35–36, divide the polynomial by the monomial.*

35. $(15x^3 - 30x^2 + 10x - 2) \div (5x^2)$

36. $(36x^4y^3 + 12x^2y^3 - 60x^2y^2) \div (6xy^2)$

In Exercises 37–40, divide as indicated.

37. $(6x^2 - 5x + 5) \div (2x + 3)$

38. $(10x^3 - 26x^2 + 17x - 13) \div (5x - 3)$

39. $(x^6 + 3x^5 - 2x^4 + x^2 - 3x + 2) \div (x - 2)$

40. $(4x^4 + 6x^3 + 3x - 1) \div (2x^2 + 1)$

6.5 *In Exercises 41–43, divide using synthetic division.*

41. $(4x^3 - 3x^2 - 2x + 1) \div (x + 1)$

42. $(3x^4 - 2x^2 - 10x - 20) \div (x - 2)$

43. $(x^4 + 16) \div (x + 4)$

In Exercises 44–45, use synthetic division and the Remainder Theorem to find the indicated function value.

44. $f(x) = 2x^3 - 5x^2 + 4x - 1;\ f(2)$

45. $f(x) = 3x^4 + 7x^3 + 8x^2 + 2x + 4;\ f\left(-\frac{1}{3}\right)$

In Exercises 46–47, use synthetic division to determine whether or not the number given to the right of each equation is a solution of the equation.

46. $2x^3 - x^2 - 8x + 4 = 0;\ -2$

47. $x^4 - x^3 - 7x^2 + x + 6 = 0;\ 4$

48. Use synthetic division to show that $\frac{1}{2}$ is a solution of

$$6x^3 + x^2 - 4x + 1 = 0.$$

Then solve the polynomial equation.

6.6 *In Exercises 49–55, solve each rational equation. If an equation has no solution, so state.*

49. $\frac{3}{x} + \frac{1}{3} = \frac{5}{x}$

50. $\frac{5}{3x + 4} = \frac{3}{2x - 8}$

51. $\frac{1}{x - 5} - \frac{3}{x + 5} = \frac{6}{x^2 - 25}$

52. $\frac{x + 5}{x + 1} - \frac{x}{x + 2} = \frac{4x + 1}{x^2 + 3x + 2}$

53. $\frac{2}{3} - \frac{5}{3x} = \frac{1}{x^2}$

54. $\frac{2}{x - 1} = \frac{1}{4} + \frac{7}{x + 2}$

55. $\frac{2x + 7}{x + 5} - \frac{x - 8}{x - 4} = \frac{x + 18}{x^2 + x - 20}$

56. The function

$$f(x) = \frac{4x}{100 - x}$$

models the cost, $f(x)$, in millions of dollars, to remove x% of pollutants from a river due to pesticide runoff from area farms. What percentage of the pollutants can be removed for $16 million?

6.7 *In Exercises 57–61, solve each formula for the specified variable.*

57. $P = \frac{R - C}{n}$ for C

58. $\frac{P_1V_1}{T_1} = \frac{P_2V_2}{T_2}$ for T_1

59. $T = \frac{A - P}{Pr}$ for P

60. $\frac{1}{R} = \frac{1}{R_1} + \frac{1}{R_2}$ for R

61. $I = \frac{nE}{R + nr}$ for n

62. A company is planning to manufacture affordable graphing calculators. Fixed monthly cost will be \$50,000, and it will cost \$25 to produce each calculator.

a. Write the cost function, C, of producing x graphing calculators.

b. Write the average cost function, $\overline{C}$, of producing x graphing calculators.

c. How many graphing calculators must be produced each month for the company to have an average cost of \$35 per graphing calculator?

63. After riding at a steady rate for 60 miles, a bicyclist had a flat tire and walked 8 miles to a repair shop. The cycling rate was 3 times faster than the walking rate. If the time spent cycling and walking was 7 hours, at what rate was the cyclist riding?

64. The current of a river moves at 3 miles per hour. It takes a boat a total of 3 hours to travel 12 miles upstream, against the current, and return the same distance traveling downstream, with the current. What is the boat's rate in still water?

65. Working alone, two people can clean their house in 3 hours and 6 hours, respectively. They have agreed to clean together so that they can finish in time to watch a TV program that begins in $1\frac{1}{2}$ hours. How long will it take them to clean the house working together? Can they finish before the program starts?

66. Working together, two crews can clear snow from the city's streets in 20 hours. Working alone, the faster crew requires 9 hours less time than the slower crew. How many hours would it take each crew to clear the streets working alone?

67. An inlet faucet can fill a small pond in 60 minutes. The pond can be emptied by an outlet pipe in 80 minutes. You begin filling the empty pond. By accident, the outlet pipe that empties the pond is left open. Under these conditions, how long will it take for the pond to fill?

6.8 *Solve the variation problems in Exercises 68–73.*

68. A company's profit varies directly as the number of products it sells. The company makes a profit of \$1175 on the sale of 25 products. What is the company's profit when it sells 105 products?

69. The distance that a body falls from rest varies directly as the square of the time of the fall. If skydivers fall 144 feet in 3 seconds, how far will they fall in 10 seconds?

70. The pitch of a musical tone varies inversely as its wavelength. A tone has a pitch of 660 vibrations per second and a wavelength of 1.6 feet. What is the pitch of a tone that has a wavelength of 2.4 feet?

71. The loudness of a stereo speaker, measured in decibels, varies inversely as the square of your distance from the speaker. When you are 8 feet from the speaker, the loudness is 28 decibels. What is the loudness when you are 4 feet from the speaker?

72. The time required to assemble computers varies directly as the number of computers assembled and inversely as the number of workers. If 30 computers can be assembled by 6 workers in 10 hours, how long would it take 5 workers to assemble 40 computers?

73. The volume of a pyramid varies jointly as its height and the area of its base. A pyramid with a height of 15 feet and a base with an area of 35 square feet has a volume of 175 cubic feet. Find the volume of a pyramid with a height of 20 feet and a base with an area of 120 square feet.

CHAPTER 6 TEST

Step-by-step test solutions are found on the Chapter Test Prep Videos available in MyMathLab® or on YouTube (search "BlitzerInterAlg7e" and click on "Channels").

1. Find the domain of $f(x) = \frac{x^2 - 2x}{x^2 - 7x + 10}$. Then simplify the right side of the function's equation.

In Exercises 2–11, perform the indicated operations. Simplify where possible.

2. $\frac{x^2}{x^2 - 16} \cdot \frac{x^2 + 7x + 12}{x^2 + 3x}$

3. $\frac{x^3 + 27}{x^2 - 1} \div \frac{x^2 - 3x + 9}{x^2 - 2x + 1}$

4. $\frac{x^2 + 3x - 10}{x^2 + 4x + 3} \cdot \frac{x^2 + x - 6}{x^2 + 10x + 25} \cdot \frac{x + 1}{x - 2}$

5. $\frac{x^2 - 6x - 16}{x^3 + 3x^2 + 2x} \cdot (x^2 - 3x - 4) \div \frac{x^2 - 7x + 12}{3x}$

6. $\frac{x^2 - 5x - 2}{6x^2 - 11x - 35} - \frac{x^2 - 7x + 5}{6x^2 - 11x - 35}$

7. $\frac{x}{x + 3} + \frac{5}{x - 3}$

8. $\frac{2}{x^2 - 4x + 3} + \frac{3x}{x^2 + x - 2}$

9. $\frac{5x}{x^2 - 4} - \frac{2}{x^2 + x - 2}$

10. $\frac{x - 4}{x - 5} - \frac{3}{x + 5} - \frac{10}{x^2 - 25}$

11. $\frac{1}{10 - x} + \frac{x - 1}{x - 10}$

In Exercises 12–13, simplify each rational expression.

12. $\dfrac{\frac{x}{4} - \frac{1}{x}}{1 + \frac{x + 4}{x}}$

13. $\dfrac{\frac{1}{x} - \frac{3}{x + 2}}{\frac{2}{x^2 + 2x}}$

In Exercises 14–16, divide as indicated.

14. $(12x^4y^3 + 16x^2y^3 - 10x^2y^2) \div (4x^2y)$

15. $(9x^3 - 3x^2 - 3x + 4) \div (3x + 2)$

16. $(3x^4 + 2x^3 - 8x + 6) \div (x^2 - 1)$

17. Divide using synthetic division:

$$(3x^4 + 11x^3 - 20x^2 + 7x + 35) \div (x + 5).$$

18. Given that

$$f(x) = x^4 - 2x^3 - 11x^2 + 5x + 34,$$

use synthetic division and the Remainder Theorem to find $f(-2)$.

19. Use synthetic division to decide whether -2 is a solution of $2x^3 - 3x^2 - 11x + 6 = 0$.

In Exercises 20–21, solve each rational equation.

20. $\frac{x}{x + 4} = \frac{11}{x^2 - 16} + 2$

21. $\frac{x + 1}{x^2 + 2x - 3} - \frac{1}{x + 3} = \frac{1}{x - 1}$

22. Park rangers introduce 50 elk into a wildlife preserve. The function

$$f(t) = \frac{250(3t + 5)}{t + 25}$$

models the elk population, $f(t)$, after t years. How many years will it take for the population to increase to 125 elk?

23. Solve for a: $R = \frac{as}{a + s}$.

24. A company is planning to manufacture portable satellite radio players. Fixed monthly cost will be $300,000 and it will cost $10 to produce each player.

a. Write the cost function, C, of producing x players.

b. Write the average cost function, $\overline{C}$, of producing x players.

c. How many portable satellite radio players must be produced each month for the company to have an average cost of $25 per player?

25. It takes one pipe 3 hours to fill a pool and a second pipe 4 hours to drain the pool. The pool is empty and the first pipe begins to fill it. The second pipe is accidently left open, so the water is also draining out of the pool. Under these conditions, how long will it take to fill the pool?

26. A motorboat averages 20 miles per hour in still water. It takes the boat the same amount of time to travel 3 miles with the current as it does to travel 2 miles against the current. What is the current's rate?

27. The intensity of light received at a source varies inversely as the square of the distance from the source. A particular light has an intensity of 20 footcandles at 15 feet. What is the light's intensity at 10 feet?

CUMULATIVE REVIEW EXERCISES (CHAPTERS 1–6)

In Exercises 1–5, solve each equation, inequality, or system.

1. $2x + 5 \le 11$ and $-3x > 18$
2. $2x^2 = 7x + 4$
3. $\begin{cases} 4x + 3y + 3z = 4 \\ 3x + 2z = 2 \\ 2x - 5y = -4 \end{cases}$
4. $|3x - 4| \le 10$
5. $\dfrac{x}{x-8} + \dfrac{6}{x-2} = \dfrac{x^2}{x^2 - 10x + 16}$
6. Solve for s: $I = \dfrac{2R}{w + 2s}$.
7. Solve by graphing: $\begin{cases} 2x - y = 4 \\ x + y = 5. \end{cases}$
8. Use function notation to write the equation of the line with slope -3 and passing through the point $(1, -5)$.

In Exercises 9–11, graph each equation, inequality, or system in a rectangular coordinate system.

9. $y = |x| + 2$
10. $\begin{cases} y \ge 2x - 1 \\ x \ge 1 \end{cases}$
11. $2x - y < 4$

In Exercises 12–15, perform the indicated operations.

12. $[(x + 2) + 3y][(x + 2) - 3y]$
13. $\dfrac{2x^2 + x - 1}{2x^2 - 9x + 4} \div \dfrac{6x + 15}{3x^2 - 12x}$
14. $\dfrac{3x}{x^2 - 9x + 20} - \dfrac{5}{2x - 8}$
15. $(3x^2 + 10x + 10) \div (x + 2)$

In Exercises 16–17, factor completely.

16. $xy - 6x + 2y - 12$
17. $24x^3y + 16x^2y - 30xy$
18. A baseball is thrown straight up from a height of 64 feet. The function
$$s(t) = -16t^2 + 48t + 64$$
describes the ball's height above the ground, $s(t)$, in feet, t seconds after it is thrown. How long will it take for the ball to hit the ground?
19. The local cable television company offers two deals. Basic cable service with one movie channel costs \$35 per month. Basic service with two movie channels costs \$45 per month. Find the charge for the basic cable service and the charge for each movie channel.
20. A rectangular garden 10 feet wide and 12 feet long is surrounded by a rock border of uniform width. The area of the garden and rock border combined is 168 square feet. What is the width of the rock border?

CHAPTER 7

Radicals, Radical Functions, and Rational Exponents

Can mathematical models be created for events that appear to involve random behavior, such as stock market fluctuations or air turbulence? Chaos theory, a new frontier of mathematics, offers models and computer-generated images that reveal order and underlying patterns where only the erratic and the unpredictable had been observed. Because most behavior is chaotic, the computer has become a canvas that looks more like the real world than anything previously seen. Magnified portions of these computer images yield repetitions of the original structure, as well as new and unexpected patterns. The computer generates these visualizations of chaos by plotting large numbers of points for functions whose domains are nonreal numbers involving the square root of negative one.

Here's where you'll find this application:

- We define $\sqrt{-1}$ in Section 7.7 and hint at chaotic possibilities in the Blitzer Bonus on page 569. If you are intrigued by how the operations of nonreal numbers in Section 7.7 reveal that the world is not random [rather, the underlying patterns are far more intricate than we had previously assumed], we suggest reading *Chaos* by James Gleick, published by Penguin Books.

SECTION

7.1 Radical Expressions and Functions 

What am I supposed to learn?

After studying this section, you should be able to:

1. Evaluate square roots.
2. Evaluate square root functions.
3. Find the domain of square root functions.
4. Use models that are square root functions.
5. Simplify expressions of the form $\sqrt{a^2}$.
6. Evaluate cube root functions.
7. Simplify expressions of the form $\sqrt[3]{a^3}$.
8. Find even and odd roots.
9. Simplify expressions of the form $\sqrt[n]{a^n}$.

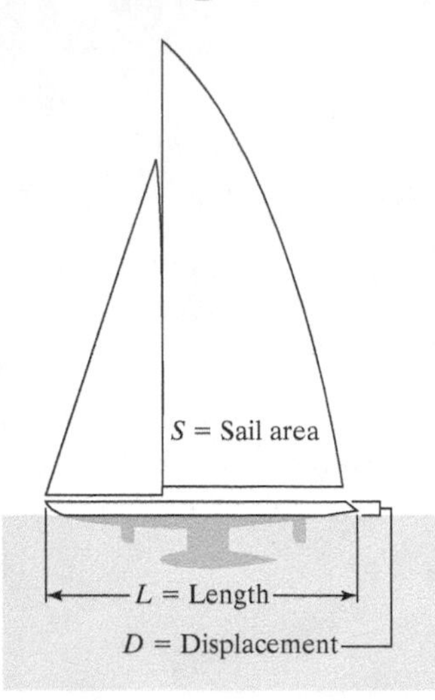

The America's Cup is the supreme event in ocean sailing. Competition is fierce and the costs are huge. Competitors look to mathematics to provide the critical innovation that can make the difference between winning and losing. The basic dimensions of competitors' yachts must satisfy an inequality containing square roots and cube roots:

$$L + 1.25\sqrt{S} - 9.8\sqrt[3]{D} \leq 16.296.$$

In the inequality, L is the yacht's length, in meters, S is its sail area, in square meters, and D is its displacement, in cubic meters.

In this section, we introduce a new category of expressions and functions that contain roots. You will see why square root functions are used to describe phenomena that are continuing to grow but whose growth is leveling off.

Square Roots

1 Evaluate square roots.

From our earlier work with exponents, we are aware that the square of 5 and the square of −5 are both 25:

$$5^2 = 25 \quad \text{and} \quad (-5)^2 = 25.$$

The reverse operation of squaring a number is finding the *square root* of the number. For example,

- One square root of 25 is 5 because $5^2 = 25$.
- Another square root of 25 is −5 because $(-5)^2 = 25$.

In general, **if $b^2 = a$, then b is a square root of a.**

The symbol $\sqrt{}$ is used to denote the *positive* or *principal square root* of a number. For example,

- $\sqrt{25} = 5$ because $5^2 = 25$ and 5 is positive.
- $\sqrt{100} = 10$ because $10^2 = 100$ and 10 is positive.

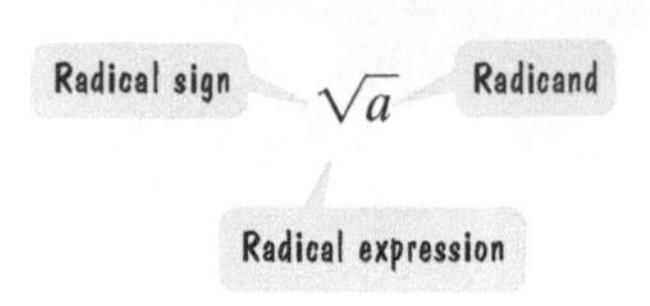

The symbol $\sqrt{}$ that we use to denote the principal square root is called a **radical sign.** The number under the radical sign is called the **radicand.** Together we refer to the radical sign and its radicand as a **radical expression.**

Definition of the Principal Square Root

If a is a nonnegative real number, the nonnegative number b such that $b^2 = a$, denoted by $b = \sqrt{a}$, is the **principal square root** of a.

The symbol $-\sqrt{\;}$ is used to denote the negative square root of a number. For example,

- $-\sqrt{25} = -5$ because $(-5)^2 = 25$ and -5 is negative.
- $-\sqrt{100} = -10$ because $(-10)^2 = 100$ and -10 is negative.

EXAMPLE 1 Evaluating Square Roots

Evaluate:

a. $\sqrt{81}$ **b.** $-\sqrt{9}$ **c.** $\sqrt{\frac{4}{49}}$

d. $\sqrt{0.0064}$ **e.** $\sqrt{36 + 64}$ **f.** $\sqrt{36} + \sqrt{64}$.

Solution

a. $\sqrt{81} = 9$ The principal square root of 81 is 9 because $9^2 = 81$.

b. $-\sqrt{9} = -3$ The negative square root of 9 is -3 because $(-3)^2 = 9$.

c. $\sqrt{\frac{4}{49}} = \frac{2}{7}$ The principal square root of $\frac{4}{49}$ is $\frac{2}{7}$ because $\left(\frac{2}{7}\right)^2 = \frac{4}{49}$.

d. $\sqrt{0.0064} = 0.08$ The principal square root of 0.0064 is 0.08 because $(0.08)^2 = (0.08)(0.08) = 0.0064$.

e. $\sqrt{36 + 64} = \sqrt{100}$ Simplify the radicand.

$= 10$ Take the principal square root of 100, which is 10.

f. $\sqrt{36} + \sqrt{64} = 6 + 8$ $\sqrt{36} = 6$ because $6^2 = 36$. $\sqrt{64} = 8$ because $8^2 = 64$.

$= 14$ ■

Great Question!

Is $\sqrt{a + b}$ equal to $\sqrt{a} + \sqrt{b}$?

No. In Example 1, parts (e) and (f), observe that $\sqrt{36 + 64}$ is not equal to $\sqrt{36} + \sqrt{64}$. In general,

$$\sqrt{a + b} \neq \sqrt{a} + \sqrt{b}$$

and

$$\sqrt{a - b} \neq \sqrt{a} - \sqrt{b}.$$

✓ **CHECK POINT 1** Evaluate:

a. $\sqrt{64}$ **b.** $-\sqrt{49}$ **c.** $\sqrt{\frac{16}{25}}$

d. $\sqrt{0.0081}$ **e.** $\sqrt{9 + 16}$ **f.** $\sqrt{9} + \sqrt{16}$.

Let's see what happens to the radical expression $\sqrt{x}$ if x is a negative number. Is the square root of a negative number a real number? For example, consider $\sqrt{-25}$. Is there a real number whose square is -25? No. Thus, $\sqrt{-25}$ is not a real number. In general, **a square root of a negative number is not a real number.**

2 Evaluate square root functions.

Square Root Functions

Because each nonnegative real number, x, has precisely one principal square root, $\sqrt{x}$, there is a **square root function** defined by

$$f(x) = \sqrt{x}.$$

The domain of this function is $[0, \infty)$. We can graph $f(x) = \sqrt{x}$ by selecting nonnegative real numbers for x. It is easiest to choose perfect squares, numbers that have rational square roots. **Table 7.1** shows five such choices for x and the calculations for the corresponding outputs. We plot these ordered pairs as points in the rectangular coordinate system and connect the points with a smooth curve. The graph of $f(x) = \sqrt{x}$ is shown in **Figure 7.1**.

Table 7.1

x	$f(x) = \sqrt{x}$	(x, y) or $(x, f(x))$
0	$f(0) = \sqrt{0} = 0$	(0, 0)
1	$f(1) = \sqrt{1} = 1$	(1, 1)
4	$f(4) = \sqrt{4} = 2$	(4, 2)
9	$f(9) = \sqrt{9} = 3$	(9, 3)
16	$f(16) = \sqrt{16} = 4$	(16, 4)

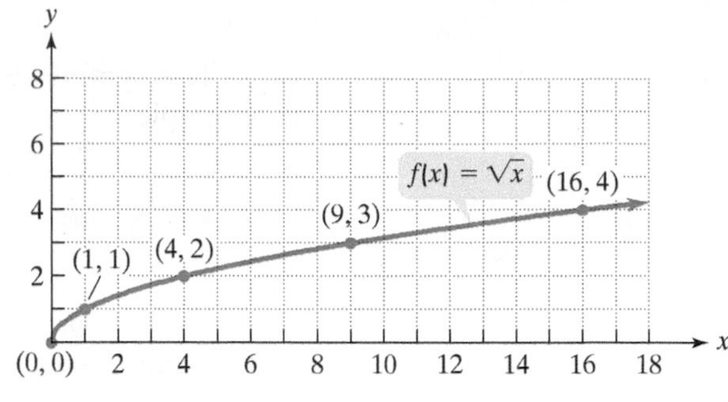

Figure 7.1 The graph of the square root function $f(x) = \sqrt{x}$

Is it possible to choose values of x for **Table 7.1** that are not squares of integers, or perfect squares? Yes. For example, we can let $x = 3$. Thus, $f(3) = \sqrt{3}$. Because 3 is not a perfect square, $\sqrt{3}$ is an irrational number, one that cannot be expressed as a quotient of integers. We can use a calculator to find a decimal approximation of $\sqrt{3}$.

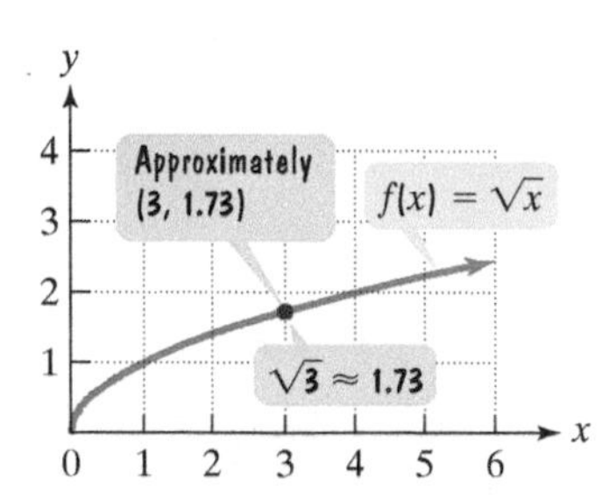

Figure 7.2 Visualizing $\sqrt{3}$ as a point on the graph of $f(x) = \sqrt{x}$

Many Scientific Calculators	Many Graphing Calculators
3 [√]	[√] 3 [ENTER]

Rounding the displayed number to two decimal places, $\sqrt{3} \approx 1.73$. This information is shown visually as a point, approximately (3, 1.73), on the graph of $f(x) = \sqrt{x}$ in **Figure 7.2**.

To evaluate a square root function, we use substitution, just as we have done to evaluate other functions.

EXAMPLE 2 Evaluating Square Root Functions

For each function, find the indicated function value:

a. $f(x) = \sqrt{5x - 6}$; $f(2)$ **b.** $g(x) = -\sqrt{64 - 8x}$; $g(-3)$.

Solution

a. $f(2) = \sqrt{5 \cdot 2 - 6}$ Substitute 2 for x in $f(x) = \sqrt{5x - 6}$.

$= \sqrt{4} = 2$ Simplify the radicand and take the square root.

b. $g(-3) = -\sqrt{64 - 8(-3)}$ Substitute -3 for x in $g(x) = -\sqrt{64 - 8x}$.

$= -\sqrt{88} \approx -9.38$ Simplify the radicand: $64 - 8(-3) = 64 - (-24) = 64 + 24 = 88$. Then use a calculator to approximate $\sqrt{88}$. ■

✓ CHECK POINT 2 For each function, find the indicated function value:

a. $f(x) = \sqrt{12x - 20}$; $f(3)$

b. $g(x) = -\sqrt{9 - 3x}$; $g(-5)$.

3 Find the domain of square root functions.

We have seen that the domain of a function f is the largest set of real numbers for which the value of $f(x)$ is a real number. Because only nonnegative numbers have real square roots, the domain of a square root function is the set of real numbers for which the radicand is nonnegative.

EXAMPLE 3 Finding the Domain of a Square Root Function

Find the domain of

$$f(x) = \sqrt{3x + 12}.$$

Solution The domain is the set of real numbers, x, for which the radicand, $3x + 12$, is nonnegative. We set the radicand greater than or equal to 0 and solve the resulting inequality.

$$3x + 12 \geq 0$$
$$3x \geq -12$$
$$x \geq -4$$

The domain of f is $[-4, \infty)$. ■

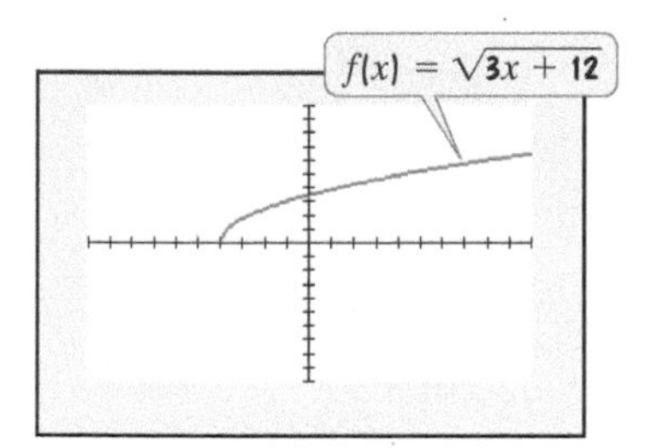

Figure 7.3

Figure 7.3 shows the graph of $f(x) = \sqrt{3x + 12}$ in a $[-10, 10, 1]$ by $[-10, 10, 1]$ viewing rectangle. The graph appears only for $x \geq -4$, verifying $[-4, \infty)$ as the domain. Can you see how the graph also illustrates this square root function's range? The graph only appears for nonnegative values of y. Thus, the range is $[0, \infty)$.

CHECK POINT 3 Find the domain of

$$f(x) = \sqrt{9x - 27}.$$

---

4 Use models that are square root functions.

The graph of the square root function $f(x) = \sqrt{x}$ is increasing from left to right. However, the rate of increase is slowing down as the graph moves to the right. This is why square root functions are often used to model growing phenomena with growth that is leveling off.

EXAMPLE 4 Modeling with a Square Root Function

By 2011, the amount of "clutter," including commercials and plugs for other shows, had increased to the point where an "hour-long" drama on cable TV was 44.8 minutes. The graph in **Figure 7.4** shows the average number of nonprogram minutes in an hour of prime-time cable television. Although the minutes of clutter grew from 1996 through 2011, the growth was leveling off. The data can be modeled by the function

$$M(x) = 0.7\sqrt{x} + 12.5,$$

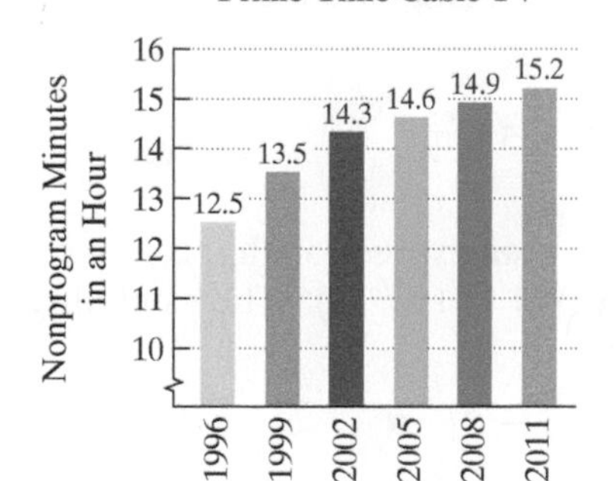

Figure 7.4

Source: The Nielsen Media Company, Monitor-Plus

where $M(x)$ is the average number of nonprogram minutes in an hour of prime-time cable x years after 1996. According to the model, in 2002, how many cluttered minutes disrupted cable TV action in an hour? Round to the nearest tenth of a minute. What is the difference between the actual data and the number of minutes that you obtained?

Solution Because 2002 is 6 years after 1996, we substitute 6 for x and evaluate the function at 6.

$$M(x) = 0.7\sqrt{x} + 12.5 \quad \text{Use the given function.}$$
$$M(6) = 0.7\sqrt{6} + 12.5 \quad \text{Substitute 6 for } x.$$
$$\approx 14.2 \quad \text{Use a calculator.}$$

The model indicates that there were approximately 14.2 nonprogram minutes in an hour of prime-time cable in 2002. **Figure 7.4** shows 14.3 minutes, so the difference is $14.3 - 14.2$, or 0.1 minute. ■

CHECK POINT 4 If the trend from 1996 through 2011 continues, use the square root function in Example 4 to project how many cluttered minutes, rounded to the nearest tenth, there will be in an hour in 2017.

5 Simplify expressions of the form $\sqrt{a^2}$.

Simplifying Expressions of the Form $\sqrt{a^2}$

You may think that $\sqrt{a^2} = a$. However, this is not necessarily true. Consider the following examples:

$$\sqrt{4^2} = \sqrt{16} = 4$$
$$\sqrt{(-4)^2} = \sqrt{16} = 4.$$

The result is not −4, but rather the absolute value of −4, or 4.

Using Technology

Graphic Connections

The graphs of $f(x) = \sqrt{x^2}$ and $g(x) = |x|$ are shown in a $[-10, 10, 1]$ by $[-2, 10, 1]$ viewing rectangle. The graphs are the same. Thus,

$$\sqrt{x^2} = |x|.$$

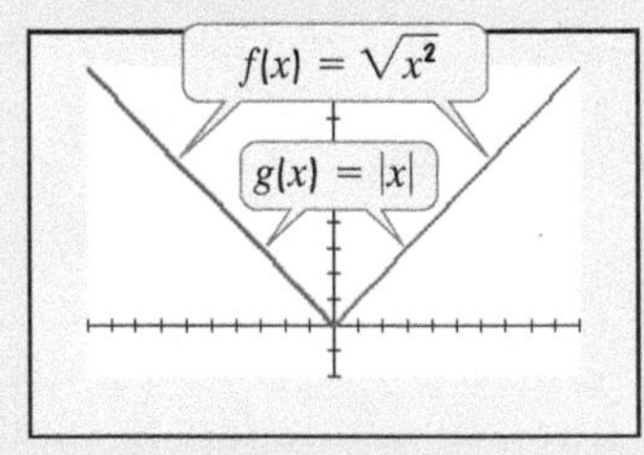

Here is a rule for simplifying expressions of the form $\sqrt{a^2}$:

Simplifying $\sqrt{a^2}$

For any real number a,

$$\sqrt{a^2} = |a|.$$

In words, the principal square root of a^2 is the absolute value of a.

EXAMPLE 5 Simplifying Radical Expressions

Simplify each expression:

a. $\sqrt{(-6)^2}$ **b.** $\sqrt{(x+5)^2}$ **c.** $\sqrt{25x^6}$ **d.** $\sqrt{x^2 - 4x + 4}$.

Solution The principal square root of an expression squared is the absolute value of that expression. In parts (a) and (b), we are given squared radicands. In parts (c) and (d), it will first be necessary to express the radicand as an expression that is squared.

a. $\sqrt{(-6)^2} = |-6| = 6$

b. $\sqrt{(x+5)^2} = |x+5|$

c. To simplify $\sqrt{25x^6}$, first write $25x^6$ as an expression that is squared: $25x^6 = (5x^3)^2$. Then simplify.

$$\sqrt{25x^6} = \sqrt{(5x^3)^2} = |5x^3| \text{ or } 5|x^3|$$

d. To simplify $\sqrt{x^2 - 4x + 4}$, first write $x^2 - 4x + 4$ as an expression that is squared by factoring the perfect square trinomial: $x^2 - 4x + 4 = (x-2)^2$. Then simplify.

$$\sqrt{x^2 - 4x + 4} = \sqrt{(x-2)^2} = |x-2| \quad ■$$

✓ **CHECK POINT 5** Simplify each expression:

a. $\sqrt{(-7)^2}$

b. $\sqrt{(x+8)^2}$

c. $\sqrt{49x^{10}}$

d. $\sqrt{x^2 - 6x + 9}$.

In some situations, we are told that no radicands involve negative quantities raised to even powers. When the expression being squared is nonnegative, it is not necessary to use absolute value when simplifying $\sqrt{a^2}$. For example, assuming that no radicands contain negative quantities that are squared,

$$\sqrt{x^6} = \sqrt{(x^3)^2} = x^3$$
$$\sqrt{25x^2 + 10x + 1} = \sqrt{(5x+1)^2} = 5x + 1.$$

6 Evaluate cube root functions.

Cube Roots and Cube Root Functions

Finding the square root of a number reverses the process of squaring a number. Similarly, finding the cube root of a number reverses the process of cubing a number. For example, $2^3 = 8$, and so the cube root of 8 is 2. The notation that we use is $\sqrt[3]{8} = 2$.

Definition of the Cube Root of a Number

The **cube root** of a real number a is written $\sqrt[3]{a}$.

$$\sqrt[3]{a} = b \quad \text{means that} \quad b^3 = a.$$

Great Question!

Should I know the cube roots of certain numbers by heart?

Some cube roots occur so frequently that you might want to memorize them.

$$\sqrt[3]{1} = 1$$
$$\sqrt[3]{8} = 2$$
$$\sqrt[3]{27} = 3$$
$$\sqrt[3]{64} = 4$$
$$\sqrt[3]{125} = 5$$
$$\sqrt[3]{216} = 6$$
$$\sqrt[3]{1000} = 10$$

For example,

$$\sqrt[3]{64} = 4 \quad \text{because} \quad 4^3 = 64.$$
$$\sqrt[3]{-27} = -3 \quad \text{because} \quad (-3)^3 = -27.$$

In contrast to square roots, the cube root of a negative number is a real number. All real numbers have cube roots. The cube root of a positive number is positive. The cube root of a negative number is negative.

Because every real number, x, has precisely one cube root, $\sqrt[3]{x}$, there is a **cube root function** defined by

$$f(x) = \sqrt[3]{x}.$$

The domain of this function is the set of all real numbers. We can graph $f(x) = \sqrt[3]{x}$ by selecting perfect cubes, numbers that have rational cube roots, for x. **Table 7.2** shows five such choices for x and the calculations for the corresponding outputs. We plot these ordered pairs as points in the rectangular coordinate system and connect the points with a smooth curve. The graph of $f(x) = \sqrt[3]{x}$ is shown in **Figure 7.5**.

Table 7.2

x	$f(x) = \sqrt[3]{x}$	(x, y) or $(x, f(x))$
-8	$f(-8) = \sqrt[3]{-8} = -2$	$(-8, -2)$
-1	$f(-1) = \sqrt[3]{-1} = -1$	$(-1, -1)$
0	$f(0) = \sqrt[3]{0} = 0$	$(0, 0)$
1	$f(1) = \sqrt[3]{1} = 1$	$(1, 1)$
8	$f(8) = \sqrt[3]{8} = 2$	$(8, 2)$

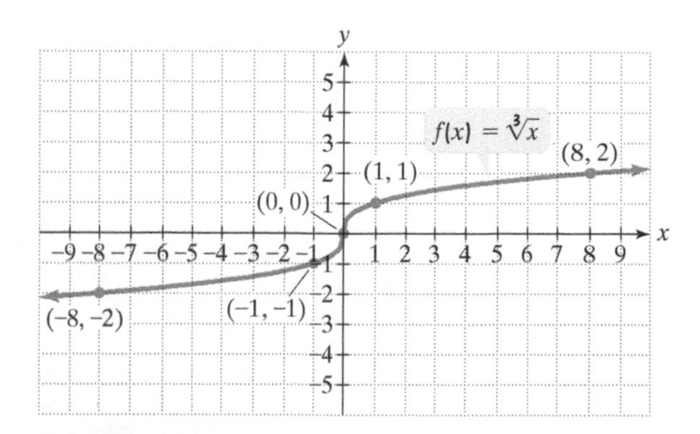

Figure 7.5 The graph of the cube root function $f(x) = \sqrt[3]{x}$

Notice that both the domain and the range of $f(x) = \sqrt[3]{x}$ are the set of all real numbers, $(-\infty, \infty)$.

EXAMPLE 6 Evaluating Cube Root Functions

For each function, find the indicated function value:

a. $f(x) = \sqrt[3]{x - 2}$; $f(127)$

b. $g(x) = \sqrt[3]{8x - 8}$; $g(-7)$.

Solution

a. $f(x) = \sqrt[3]{x-2}$ — This is the given function.

$f(127) = \sqrt[3]{127-2}$ — Substitute 127 for x.

$= \sqrt[3]{125}$ — Simplify the radicand.

$= 5$ — $\sqrt[3]{125} = 5$ because $5^3 = 125$.

b. $g(x) = \sqrt[3]{8x-8}$ — This is the given function.

$g(-7) = \sqrt[3]{8(-7)-8}$ — Substitute -7 for x.

$= \sqrt[3]{-64}$ — Simplify the radicand: $8(-7) - 8 = -56 - 8 = -64$.

$= -4$ — $\sqrt[3]{-64} = -4$ because $(-4)^3 = -64$. ■

✓ CHECK POINT 6 For each function, find the indicated function value:

a. $f(x) = \sqrt[3]{x-6}$; $f(33)$

b. $g(x) = \sqrt[3]{2x+2}$; $g(-5)$.

---

7 Simplify expressions of the form $\sqrt[3]{a^3}$.

Because the cube root of a positive number is positive and the cube root of a negative number is negative, absolute value is not needed to simplify expressions of the form $\sqrt[3]{a^3}$.

Simplifying $\sqrt[3]{a^3}$

For any real number a,

$$\sqrt[3]{a^3} = a.$$

In words, the cube root of any expression cubed is that expression.

EXAMPLE 7 Simplifying a Cube Root

Simplify: $\sqrt[3]{-64x^3}$.

Solution Begin by expressing the radicand as an expression that is cubed: $-64x^3 = (-4x)^3$. Then simplify.

$$\sqrt[3]{-64x^3} = \sqrt[3]{(-4x)^3} = -4x$$

We can check our answer by cubing $-4x$:

$$(-4x)^3 = (-4)^3x^3 = -64x^3.$$

By obtaining the original radicand, we know that our simplification is correct. ■

✓ CHECK POINT 7 Simplify: $\sqrt[3]{-27x^3}$.

---

8 Find even and odd roots.

Even and Odd *n*th Roots

Up to this point, we have focused on square roots and cube roots. Other radical expressions have different roots. For example, the fifth root of a, written $\sqrt[5]{a}$, is the number b for which $b^5 = a$. Thus,

$$\sqrt[5]{32} = 2 \quad \text{because} \quad 2^5 = 2\cdot2\cdot2\cdot2\cdot2 = 32.$$

The radical expression $\sqrt[n]{a}$ represents the ***n*th root** of a. The number n is called the **index**. An index of 2 represents a square root and is not written. An index of 3 represents a cube root.

If the index n in $\sqrt[n]{a}$ is an odd number, a root is said to be an **odd root**. A cube root is an odd root. Other odd roots have the same characteristics as cube roots.

- Every real number has exactly one real root when n is odd. An odd root of a positive number is positive and an odd root of a negative number is negative.

$$3^5 = 3 \cdot 3 \cdot 3 \cdot 3 \cdot 3 = 243, \text{ so the fifth root of 243 is 3.}$$

$$(-3)^5 = (-3)(-3)(-3)(-3)(-3) = -243, \text{ so the fifth root of } -243 \text{ is } -3.$$

- The (odd) nth root of a, $\sqrt[n]{a}$, is the number b for which $b^n = a$.

$$\sqrt[5]{243} = 3 \qquad \sqrt[5]{-243} = -3$$

$3^5 = 243$ $\qquad$ $(-3)^5 = -243$

Great Question!

Should I know the higher roots of certain numbers by heart?

Some higher even and odd roots occur so frequently that you might want to memorize them.

Fourth Roots	Fifth Roots
$\sqrt[4]{1} = 1$	$\sqrt[5]{1} = 1$
$\sqrt[4]{16} = 2$	$\sqrt[5]{32} = 2$
$\sqrt[4]{81} = 3$	$\sqrt[5]{243} = 3$
$\sqrt[4]{256} = 4$	
$\sqrt[4]{625} = 5$	

If the index n in $\sqrt[n]{a}$ is an even number, a root is said to be an **even root**. A square root is an even root. Other even roots have the same characteristics as square roots.

- Every positive real number has two real roots when n is even. One root is positive and one is negative.

$$2^4 = 2 \cdot 2 \cdot 2 \cdot 2 = 16 \quad \text{and} \quad (-2)^4 = (-2)(-2)(-2)(-2) = 16,$$

so both 2 and -2 are fourth roots of 16.

- The positive root, called the **principal *n*th root** and represented by $\sqrt[n]{a}$, is the nonnegative number b for which $b^n = a$. The symbol $-\sqrt[n]{a}$ is used to denote the negative nth root.

$$\sqrt[4]{16} = 2 \qquad -\sqrt[4]{16} = -2$$

$2^4 = 16$ $\qquad$ $(-2)^4 = 16$

- **An even root of a negative number is not a real number.**

$$\sqrt[4]{-16} \text{ is not a real number.}$$

EXAMPLE 8 Finding Even and Odd Roots

Find the indicated root, or state that the expression is not a real number:

a. $\sqrt[4]{81}$ **b.** $-\sqrt[4]{81}$ **c.** $\sqrt[4]{-81}$ **d.** $\sqrt[5]{-32}$.

Solution

a. $\sqrt[4]{81} = 3$ — The principal fourth root of 81 is 3 because $3^4 = 3 \cdot 3 \cdot 3 \cdot 3 = 81$.

b. $-\sqrt[4]{81} = -3$ — The negative fourth root of 81 is -3 because $(-3)^4 = (-3)(-3)\ (-3)(-3) = 81$.

c. $\sqrt[4]{-81}$ is not a real number because the index, 4, is even and the radicand, -81, is negative. No real number can be raised to the fourth power to give a negative result such as -81. Real numbers to even powers can only result in nonnegative numbers.

d. $\sqrt[5]{-32} = -2$ because $(-2)^5 = (-2)(-2)(-2)(-2)(-2) = -32$. An odd root of a negative real number is always negative. ■

CHECK POINT 8 Find the indicated root, or state that the expression is not a real number:

a. $\sqrt[4]{16}$ **b.** $-\sqrt[4]{16}$ **c.** $\sqrt[4]{-16}$ **d.** $\sqrt[5]{-1}$.

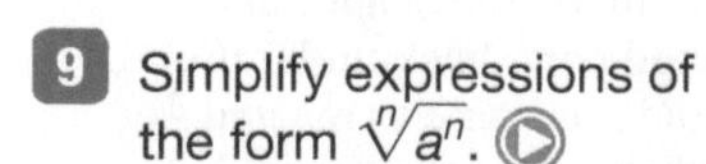
9 Simplify expressions of the form $\sqrt[n]{a^n}$.

Simplifying Expressions of the Form $\sqrt[n]{a^n}$

We have seen that

$$\sqrt{a^2} = |a| \quad \text{and} \quad \sqrt[3]{a^3} = a.$$

Expressions of the form $\sqrt[n]{a^n}$ can be simplified in the same manner. Unless a is known to be nonnegative, absolute value notation is needed when n is even. When the index is odd, absolute value bars are not used.

Simplifying $\sqrt[n]{a^n}$

For any real number a,

1. If n is even, $\sqrt[n]{a^n} = |a|$.
2. If n is odd, $\sqrt[n]{a^n} = a$.

EXAMPLE 9 Simplifying Radical Expressions

Simplify:

a. $\sqrt[4]{(x-3)^4}$ b. $\sqrt[5]{(2x+7)^5}$ c. $\sqrt[6]{(-5)^6}$.

Solution Each expression involves the nth root of a radicand raised to the nth power. Thus, each radical expression can be simplified. Absolute value bars are necessary in parts (a) and (c) because the index, n, is even.

a. $\sqrt[4]{(x-3)^4} = |x-3|$ $\quad$ $\sqrt[n]{a^n} = |a|$ if n is even.

b. $\sqrt[5]{(2x+7)^5} = 2x+7$ $\quad$ $\sqrt[n]{a^n} = a$ if n is odd.

c. $\sqrt[6]{(-5)^6} = |-5| = 5$ $\quad$ $\sqrt[n]{a^n} = |a|$ if n is even. ■

CHECK POINT 9 Simplify:

a. $\sqrt[4]{(x+6)^4}$ b. $\sqrt[5]{(3x-2)^5}$ c. $\sqrt[6]{(-8)^6}$.

Achieving Success

The Secret of Math Success

What's the secret of math success? The bar graph in **Figure 7.6** shows that Japanese teachers and students are more likely than their American counterparts to believe that the key to doing well in math is working hard. Americans tend to think that either you have mathematical intelligence or you don't. Alan Bass, author of *Math Study Skills* (Pearson Education, 2008), strongly disagrees with this American perspective:

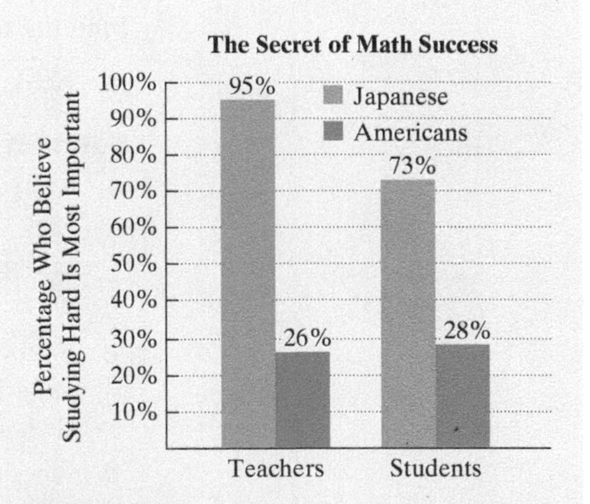

Figure 7.6

Source: Wade and Tavris, *Psychology*, Ninth Edition, Pearson, 2008.

"Human beings are easily intelligent enough to understand the basic principles of math. I cannot repeat this enough, but I'll try… **Poor performance in math is not due to a lack of intelligence!** The fact is that the key to success in math is in taking an intelligent approach. Students come up to me and say, 'I'm just not good at math.' Then I ask to see their class notebooks and they show me a chaotic mess of papers jammed into a folder. In math, that's a lot like taking apart your car's engine, dumping the heap of disconnected parts back under the hood, and then going to a mechanic to ask why it won't run. Students come to me and say, 'I'm just not good at math.' Then I ask them about their study habits and they say, 'I have to do all my studying on the weekend.' In math, that's a lot like trying to do all your eating or sleeping on the weekends and wondering why you're so tired and hungry all the time. **How you approach math is much more important than how smart you are.**"

—Alan Bass

CONCEPT AND VOCABULARY CHECK

Fill in each blank so that the resulting statement is true.

1. The symbol $\sqrt{}$ is used to denote the nonnegative, or ____________, square root of a number.
2. $\sqrt{64} = 8$ because ____________ $= 64$.
3. The domain of $f(x) = \sqrt{x}$ is ____________.
4. The domain of $f(x) = \sqrt{5x - 20}$ can be found by solving the inequality ____________.
5. For any real number a, $\sqrt{a^2} =$ ____________.
6. $\sqrt[3]{1000} = 10$ because ____________ $= 1000$.
7. $\sqrt[3]{-125} = -5$ because ____________ $= -125$.
8. For any real number a, $\sqrt[3]{a^3} =$ ____________.
9. The domain of $f(x) = \sqrt[3]{x}$ is ____________.
10. The radical expression $\sqrt[n]{a}$ represents the ____________ root of a. The number n is called the ____________.
11. If n is even, $\sqrt[n]{a^n} =$ ____________. If n is odd, $\sqrt[n]{a^n} =$ ____________.
12. True or false: $-\sqrt{25}$ is a real number. ____________
13. True or false: $\sqrt{-25}$ is a real number. ____________
14. True or false: $\sqrt[3]{-1}$ is a real number. ____________
15. True or false: $\sqrt[4]{-1}$ is a real number. ____________

7.1 EXERCISE SET

Practice Exercises

In Exercises 1–20, evaluate each expression, or state that the expression is not a real number.

1. $\sqrt{36}$
2. $\sqrt{16}$
3. $-\sqrt{36}$
4. $-\sqrt{16}$
5. $\sqrt{-36}$
6. $\sqrt{-16}$
7. $\sqrt{\frac{1}{25}}$
8. $\sqrt{\frac{1}{49}}$
9. $-\sqrt{\frac{9}{16}}$
10. $-\sqrt{\frac{4}{25}}$
11. $\sqrt{0.81}$
12. $\sqrt{0.49}$
13. $-\sqrt{0.04}$
14. $-\sqrt{0.64}$
15. $\sqrt{25 - 16}$
16. $\sqrt{144 + 25}$
17. $\sqrt{25} - \sqrt{16}$
18. $\sqrt{144} + \sqrt{25}$
19. $\sqrt{16 - 25}$
20. $\sqrt{25 - 144}$

In Exercises 21–26, find the indicated function values for each function. If necessary, round to two decimal places. If the function value is not a real number and does not exist, so state.

21. $f(x) = \sqrt{x - 2};\quad f(18), f(3), f(2), f(-2)$
22. $f(x) = \sqrt{x - 3};\quad f(28), f(4), f(3), f(-1)$
23. $g(x) = -\sqrt{2x + 3};\quad g(11), g(1), g(-1), g(-2)$
24. $g(x) = -\sqrt{2x + 1};\quad g(4), g(1), g\left(-\frac{1}{2}\right), g(-1)$
25. $h(x) = \sqrt{(x - 1)^2};\quad h(5), h(3), h(0), h(-5)$
26. $h(x) = \sqrt{(x - 2)^2};\quad h(5), h(3), h(0), h(-5)$

In Exercises 27–32, find the domain of each square root function. Then use the domain to match the radical function with its graph. [The graphs are labeled (a) through (f) and are shown in [−10, 10, 1] by [−10, 10, 1] viewing rectangles below and on the next page.]

27. $f(x) = \sqrt{x - 3}$
28. $f(x) = \sqrt{x + 2}$
29. $f(x) = \sqrt{3x + 15}$
30. $f(x) = \sqrt{3x - 15}$
31. $f(x) = \sqrt{6 - 2x}$
32. $f(x) = \sqrt{8 - 2x}$

a.

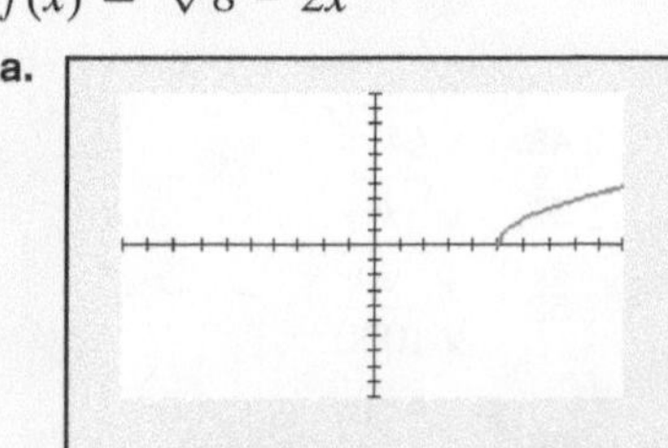

b.

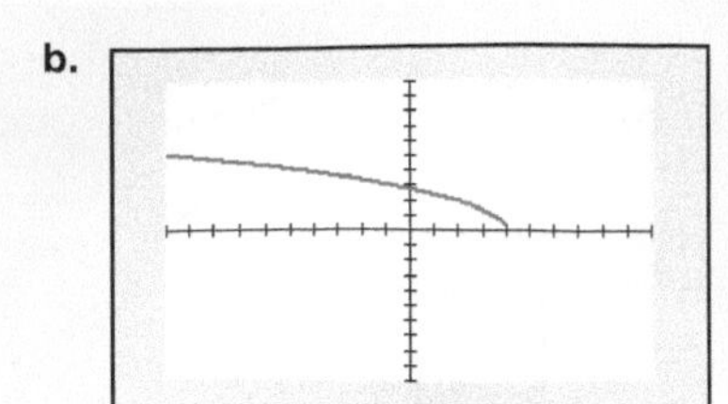

c.

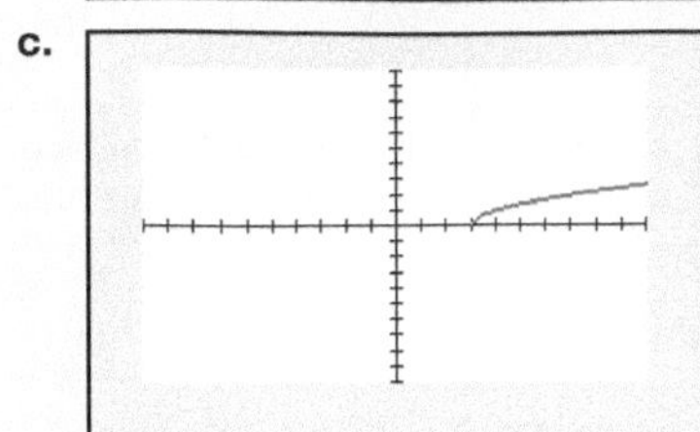

d.

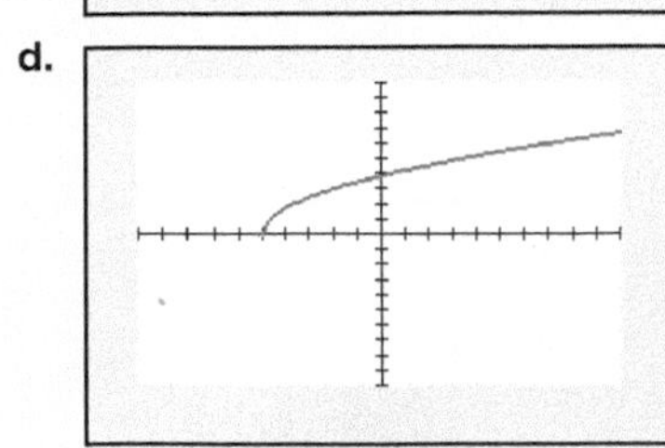

e.

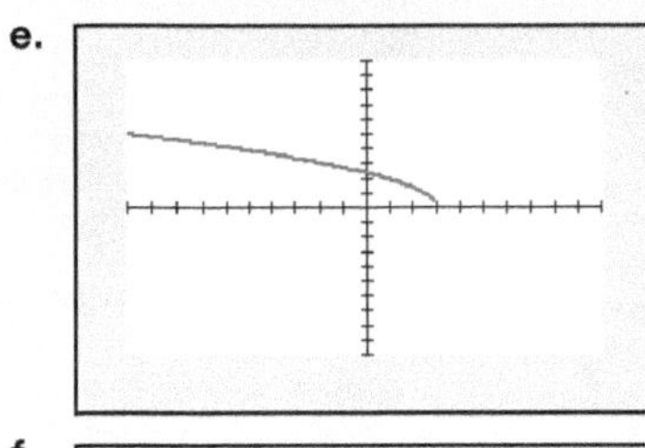

f.

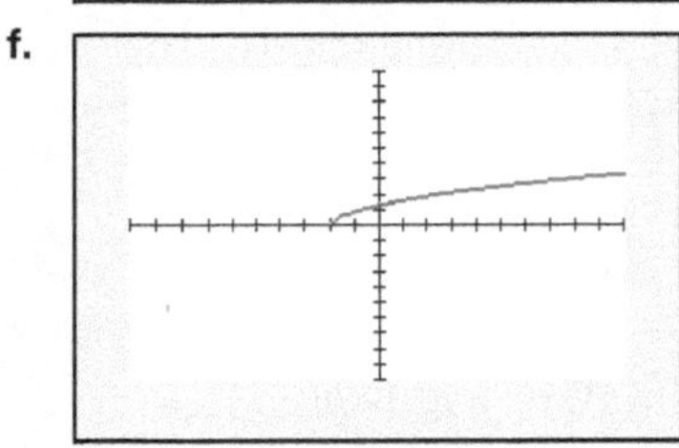

In Exercises 33–46, simplify each expression.

33. $\sqrt{5^2}$ **34.** $\sqrt{7^2}$

35. $\sqrt{(-4)^2}$ **36.** $\sqrt{(-10)^2}$

37. $\sqrt{(x-1)^2}$ **38.** $\sqrt{(x-2)^2}$

39. $\sqrt{36x^4}$ **40.** $\sqrt{81x^4}$

41. $-\sqrt{100x^6}$

42. $-\sqrt{49x^6}$

43. $\sqrt{x^2+12x+36}$

44. $\sqrt{x^2+14x+49}$

45. $-\sqrt{x^2-8x+16}$

46. $-\sqrt{x^2-10x+25}$

In Exercises 47–54, find each cube root.

47. $\sqrt[3]{27}$ **48.** $\sqrt[3]{64}$

49. $\sqrt[3]{-27}$ **50.** $\sqrt[3]{-64}$

51. $\sqrt[3]{\frac{1}{125}}$ **52.** $\sqrt[3]{\frac{1}{1000}}$

53. $\sqrt[3]{\frac{-27}{1000}}$ **54.** $\sqrt[3]{\frac{-8}{125}}$

In Exercises 55–58, find the indicated function values for each function.

55. $f(x) = \sqrt[3]{x-1}$; $f(28), f(9), f(0), f(-63)$

56. $f(x) = \sqrt[3]{x-3}$; $f(30), f(11), f(2), f(-122)$

57. $g(x) = -\sqrt[3]{8x-8}$; $g(2), g(1), g(0)$

58. $g(x) = -\sqrt[3]{2x+1}$; $g(13), g(0), g(-63)$

In Exercises 59–76, find the indicated root, or state that the expression is not a real number.

59. $\sqrt[4]{1}$ **60.** $\sqrt[5]{1}$

61. $\sqrt[4]{16}$ **62.** $\sqrt[4]{81}$

63. $-\sqrt[4]{16}$ **64.** $-\sqrt[4]{81}$

65. $\sqrt[4]{-16}$ **66.** $\sqrt[4]{-81}$

67. $\sqrt[5]{-1}$ **68.** $\sqrt[7]{-1}$

69. $\sqrt[6]{-1}$ **70.** $\sqrt[8]{-1}$

71. $-\sqrt[4]{256}$ **72.** $-\sqrt[4]{10{,}000}$

73. $\sqrt[6]{64}$ **74.** $\sqrt[5]{32}$

75. $-\sqrt[5]{32}$ **76.** $-\sqrt[6]{64}$

In Exercises 77–90, simplify each expression. Include absolute value bars where necessary.

77. $\sqrt[3]{x^3}$ **78.** $\sqrt[5]{x^5}$

79. $\sqrt[4]{y^4}$ **80.** $\sqrt[6]{y^6}$

81. $\sqrt[3]{-8x^3}$ **82.** $\sqrt[3]{-125x^3}$

83. $\sqrt[3]{(-5)^3}$ **84.** $\sqrt[3]{(-6)^3}$

85. $\sqrt[4]{(-5)^4}$ **86.** $\sqrt[6]{(-6)^6}$

87. $\sqrt[4]{(x+3)^4}$ **88.** $\sqrt[4]{(x+5)^4}$

89. $\sqrt[5]{-32(x-1)^5}$

90. $\sqrt[5]{-32(x-2)^5}$

Practice PLUS

In Exercises 91–94, complete each table and graph the given function. Identify the function's domain and range.

91. $f(x) = \sqrt{x} + 3$

x	$f(x) = \sqrt{x} + 3$
0	
1	
4	
9	

92. $f(x) = \sqrt{x} - 2$

x	$f(x) = \sqrt{x} - 2$
0	
1	
4	
9	

93. $f(x) = \sqrt{x - 3}$

x	$f(x) = \sqrt{x - 3}$
3	
4	
7	
12	

94. $f(x) = \sqrt{4 - x}$

x	$f(x) = \sqrt{4 - x}$
−5	
0	
3	
4	

In Exercises 95–98, find the domain of each function.

95. $f(x) = \dfrac{\sqrt[3]{x}}{\sqrt{30 - 2x}}$

96. $f(x) = \dfrac{\sqrt[3]{x}}{\sqrt{80 - 5x}}$

97. $f(x) = \dfrac{\sqrt{x - 1}}{\sqrt{3 - x}}$

98. $f(x) = \dfrac{\sqrt{x - 2}}{\sqrt{7 - x}}$

In Exercises 99–100, evaluate each expression.

99. $\sqrt[3]{\sqrt[4]{16} + \sqrt{625}}$

100. $\sqrt[3]{\sqrt{\sqrt{169} + \sqrt{9}} + \sqrt{\sqrt[3]{1000} + \sqrt[3]{216}}}$

Application Exercises

101. The function $f(x) = 2.9\sqrt{x} + 20.1$ models the median height, $f(x)$, in inches, of boys who are x months of age. The graph of f is shown.

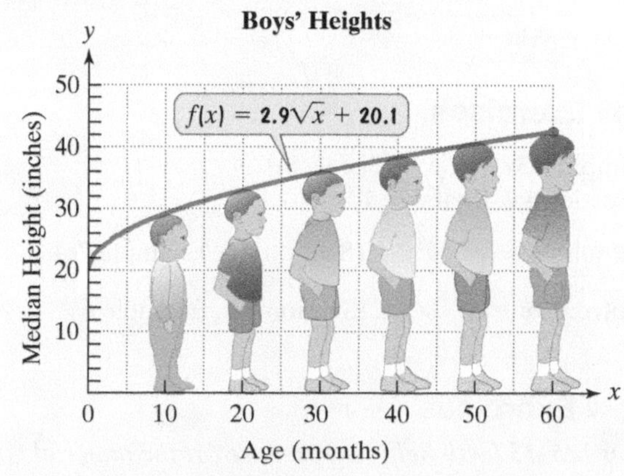

Source: Laura Walther Nathanson, THE PORTABLE PEDIATRICIAN FOR PARENTS, © 1994 Harper Perennial

a. According to the model, what is the median height of boys who are 48 months, or four years, old? Use a calculator and round to the nearest tenth of an inch. The actual median height for boys at 48 months is 40.8 inches. Does the model overestimate or underestimate the actual height? By how much?

b. Use the model to find the average rate of change, in inches per month, between birth and 10 months. Round to the nearest tenth.

c. Use the model to find the average rate of change, in inches per month, between 50 and 60 months. Round to the nearest tenth. How does this compare with your answer in part (b)? How is this difference shown by the graph?

102. The function $f(x) = 3.1\sqrt{x} + 19$ models the median height, $f(x)$, in inches, of girls who are x months of age. The graph of f is shown.

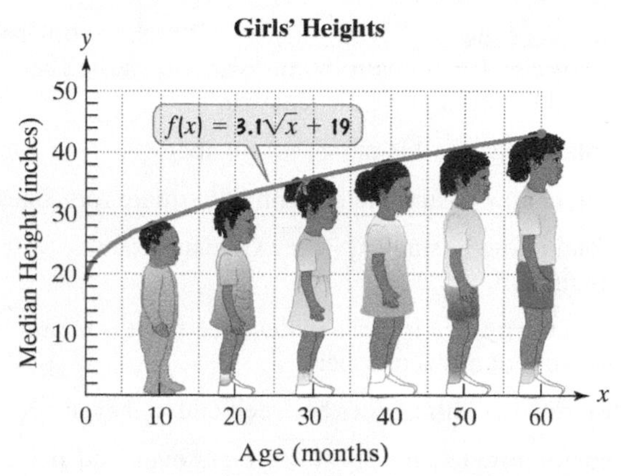

Source: Laura Walther Nathanson, THE PORTABLE PEDIATRICIAN FOR PARENTS© 1994 Harper Perennial

a. According to the model, what is the median height of girls who are 48 months, or four years, old? Use a calculator and round to the nearest tenth of an inch. The actual median height for girls at 48 months is 40.2 inches. Does the model overestimate or underestimate the actual height? By how much?

b. Use the model to find the average rate of change, in inches per month, between birth and 10 months. Round to the nearest tenth.

c. Use the model to find the average rate of change, in inches per month, between 50 and 60 months. Round to the nearest tenth. How does this compare with your answer in part (b)? How is this difference shown by the graph?

Police use the function $f(x) = \sqrt{20x}$ to estimate the speed of a car, $f(x)$, in miles per hour, based on the length, x, in feet, of its skid marks upon sudden braking on a dry asphalt road. Use the function to solve Exercises 103–104.

103. A motorist is involved in an accident. A police officer measures the car's skid marks to be 245 feet long. Estimate the speed at which the motorist was traveling before braking. If the posted speed limit is 50 miles per hour and the motorist tells the officer he was not speeding, should the officer believe him? Explain.

104. A motorist is involved in an accident. A police officer measures the car's skid marks to be 45 feet long. Use the function described at the bottom of the previous page to estimate the speed at which the motorist was traveling before braking. If the posted speed limit is 35 miles per hour and the motorist tells the officer she was not speeding, should the officer believe her? Explain.

Explaining the Concepts

105. What are the square roots of 36? Explain why each of these numbers is a square root.

106. What does the symbol $\sqrt{\ }$ denote? Which of your answers in Exercise 105 is given by this symbol? Write the symbol needed to obtain the other answer.

107. Explain why $\sqrt{-1}$ is not a real number.

108. Explain how to find the domain of a square root function.

109. Explain how to simplify $\sqrt{a^2}$. Give an example with your explanation.

110. Explain why $\sqrt[3]{8}$ is 2. Then describe what is meant by the cube root of a real number.

111. Describe two differences between odd and even roots.

112. Explain how to simplify $\sqrt[n]{a^n}$ if n is even and if n is odd. Give examples with your explanations.

113. Explain the meaning of the words *radical*, *radicand*, and *index*. Give an example with your explanation.

114. Describe the trend in a boy's growth from birth through five years, shown in the graph for Exercise 101. Why is a square root function a useful model for the data?

Technology Exercises

115. Use a graphing utility to graph $y_1 = \sqrt{x}$, $y_2 = \sqrt{x+4}$, and $y_3 = \sqrt{x-3}$ in the same $[-5, 10, 1]$ by $[0, 6, 1]$ viewing rectangle. Describe one similarity and one difference that you observe among the graphs. Use the word *shift* in your response.

116. Use a graphing utility to graph $y = \sqrt{x}$, $y = \sqrt{x} + 4$, and $y = \sqrt{x} - 3$ in the same $[-1, 10, 1]$ by $[-10, 10, 1]$ viewing rectangle. Describe one similarity and one difference that you observe among the graphs.

117. Use a graphing utility to graph $f(x) = \sqrt{x}$, $g(x) = -\sqrt{x}$, $h(x) = \sqrt{-x}$, and $k(x) = -\sqrt{-x}$ in the same $[-10, 10, 1]$ by $[-4, 4, 1]$ viewing rectangle. Use the graphs to describe the domains and the ranges of functions f, g, h, and k.

118. Use a graphing utility to graph $y_1 = \sqrt{x^2}$ and $y_2 = -x$ in the same viewing rectangle.

a. For what values of x is $\sqrt{x^2} = -x$?

b. For what values of x is $\sqrt{x^2} \neq -x$?

Critical Thinking Exercises

Make Sense? *In Exercises 119–122, determine whether each statement makes sense or does not make sense, and explain your reasoning.*

119. $\sqrt[4]{(-8)^4}$ cannot be positive 8 because the power and the index cancel each other.

120. If I am given any real number, that number has exactly one odd root and two even roots.

121. I need to restrict the domains of radical functions with even indices, but these restrictions are not necessary when indices are odd.

122. Using my calculator, I determined that $5^5 = 3125$, so 5 must be the fifth root of 3125.

In Exercises 123–126, determine whether each statement is true or false. If the statement is false, make the necessary change(s) to produce a true statement.

123. The domain of $f(x) = \sqrt[3]{x-4}$ is $[4, \infty)$.

124. If n is odd and b is negative, then $\sqrt[n]{b}$ is not a real number.

125. If $x = -2$, then $\sqrt{x^6} = x^3$.

126. The expression $\sqrt[n]{4}$ represents increasingly larger numbers for $n = 2, 3, 4, 5, 6$, and so on.

127. Write a function whose domain is $(-\infty, 5]$.

128. Let $f(x) = \sqrt{x-3}$ and $g(x) = \sqrt{x+1}$. Find the domain of $f + g$ and $\frac{f}{g}$.

129. Simplify: $\sqrt{(2x+3)^{10}}$.

In Exercises 130–131, graph each function by hand. Then describe the relationship between the function that you graphed and the graph of $f(x) = \sqrt{x}$.

130. $g(x) = \sqrt{x} + 2$

131. $h(x) = \sqrt{x+3}$

Review Exercises

132. Simplify: $3x - 2[x - 3(x+5)]$. (Section 1.2, Example 14)

133. Simplify: $(-3x^{-4}y^3)^{-2}$. (Section 1.6, Example 7c)

134. Solve: $|3x - 4| > 11$. (Section 4.3, Example 6)

Preview Exercises

Exercises 135–137 will help you prepare for the material covered in the next section. In each exercise, use properties of exponents to simplify the expression. Be sure that no negative exponents appear in your simplified expression. (If you have forgotten how to simplify an exponential expression, see the box on page 77.)

135. $(2^3x^5)(2^4x^{-6})$

136. $\dfrac{32x^2}{16x^5}$

137. $(x^{-2}y^3)^4$

SECTION

7.2

What am I supposed to learn?

After studying this section, you should be able to:

1. Use the definition of $a^{\frac{1}{n}}$.
2. Use the definition of $a^{\frac{m}{n}}$.
3. Use the definition of $a^{-\frac{m}{n}}$.
4. Simplify expressions with rational exponents.
5. Simplify radical expressions using rational exponents.

Rational Exponents

The Galápagos Islands are a chain of volcanic islands lying 600 miles west of Ecuador. They are famed for over 5000 species of plants and animals, including a rare flightless cormorant, marine iguanas, and giant tortoises weighing more than 600 pounds. Early in 2001, the plants and wildlife that live in the Galápagos were at risk from a massive oil spill that flooded 150,000 gallons of toxic fuel into one of the world's most fragile ecosystems. The long-term danger of the accident is that fuel sinking to the ocean floor will destroy algae that is vital to the food chain. Any imbalance in the food chain could threaten the rare Galápagos plant and animal species that have evolved for thousands of years in isolation with little human intervention.

Marine iguana of the Galápagos Islands

At risk on these ecologically vulnerable islands are unique flora and fauna that helped to inspire Charles Darwin's theory of evolution. Darwin made an enormous collection of the island's plant species. The function

$$f(x) = 29x^{\frac{1}{3}}$$

models the number of plant species, $f(x)$, on the various islands of the Galápagos in terms of the area, x, in square miles, of a particular island. But x to the *what* power? How can we interpret the information given by this function? In this section, we turn our attention to rational exponents such as $\frac{1}{3}$ and their relationship to roots of real numbers.

Defining Rational Exponents

We define rational exponents so that their properties are the same as the properties for integer exponents. For example, suppose that $x = 7^{\frac{1}{3}}$. We know that exponents are multiplied when an exponential expression is raised to a power. For this to be true,

$$x^3 = \left(7^{\frac{1}{3}}\right)^3 = 7^{\frac{1}{3}\cdot 3} = 7^1 = 7.$$

We see that $x^3 = 7$. This means that x is the number whose cube is 7. Thus, $x = \sqrt[3]{7}$. Remember that we began with $x = 7^{\frac{1}{3}}$. This means that

$$7^{\frac{1}{3}} = \sqrt[3]{7}.$$

We can generalize $7^{\frac{1}{3}} = \sqrt[3]{7}$ with the following definition:

1 Use the definition of $a^{\frac{1}{n}}$.

The Definition of $a^{\frac{1}{n}}$

If $\sqrt[n]{a}$ represents a real number and $n \geq 2$ is an integer, then

$$a^{\frac{1}{n}} = \sqrt[n]{a}.$$

The denominator of the rational exponent is the radical's index.

If n is even, a must be nonnegative. If n is odd, a can be any real number.

Using Technology

This graphing utility screen shows that $\sqrt{64} = 8$ and $64^{\frac{1}{2}} = 8$.

```
√64
                 8
64^1/2
                 8
```

EXAMPLE 1 Using the Definition of $a^{\frac{1}{n}}$

Use radical notation to rewrite each expression. Simplify, if possible:

a. $64^{\frac{1}{2}}$ **b.** $(-125)^{\frac{1}{3}}$ **c.** $(6x^2y)^{\frac{1}{5}}$.

Solution

a. $64^{\frac{1}{2}} = \sqrt{64} = 8$

The denominator is the index.

b. $(-125)^{\frac{1}{3}} = \sqrt[3]{-125} = -5$

c. $(6x^2y)^{\frac{1}{5}} = \sqrt[5]{6x^2y}$ ■

CHECK POINT 1 Use radical notation to rewrite each expression. Simplify, if possible:

a. $25^{\frac{1}{2}}$ **b.** $(-8)^{\frac{1}{3}}$ **c.** $(5xy^2)^{\frac{1}{4}}$.

In our next example, we begin with radical notation and rewrite the expression with rational exponents.

The radical's index becomes the exponent's denominator.

$$\sqrt[n]{a} = a^{\frac{1}{n}}$$

The radicand becomes the base.

EXAMPLE 2 Using the Definition of $a^{\frac{1}{n}}$

Rewrite with rational exponents:

a. $\sqrt[5]{13ab}$ **b.** $\sqrt[7]{\frac{xy^2}{17}}$.

Solution Parentheses are needed to show that the entire radicand becomes the base.

a. $\sqrt[5]{13ab} = (13ab)^{\frac{1}{5}}$

The index is the exponent's denominator.

b. $\sqrt[7]{\frac{xy^2}{17}} = \left(\frac{xy^2}{17}\right)^{\frac{1}{7}}$ ■

CHECK POINT 2 Rewrite with rational exponents:

a. $\sqrt[4]{5xy}$ b. $\sqrt[5]{\frac{a^3b}{2}}$.

---

2 Use the definition of $a^{\frac{m}{n}}$.

Can rational exponents have numerators other than 1? The answer is yes. If the numerator is some other integer, we still want to multiply exponents when raising a power to a power. For this reason,

$$a^{\frac{2}{3}} = \left(a^{\frac{1}{3}}\right)^2 \quad \text{and} \quad a^{\frac{2}{3}} = (a^2)^{\frac{1}{3}}.$$

This means $(\sqrt[3]{a})^2$. This means $\sqrt[3]{a^2}$.

Thus,

$$a^{\frac{2}{3}} = (\sqrt[3]{a})^2 = \sqrt[3]{a^2}.$$

Do you see that the denominator, 3, of the rational exponent is the same as the index of the radical? The numerator, 2, of the rational exponent serves as an exponent in each of the two radical forms. We generalize these ideas with the following definition:

The Definition of $a^{\frac{m}{n}}$

If $\sqrt[n]{a}$ represents a real number, $\frac{m}{n}$ is a positive rational number reduced to lowest terms, and $n \geq 2$ is an integer, then

$$a^{\frac{m}{n}} = (\sqrt[n]{a})^m.$$

First take the nth root of a.

and

$$a^{\frac{m}{n}} = \sqrt[n]{a^m}.$$

First raise a to the m power.

The first form of the definition shown in the box involves taking the root first. This form is often preferable because smaller numbers are involved. Notice that the rational exponent consists of two parts, indicated by the following voice balloons:

The numerator is the exponent.

$$a^{\frac{m}{n}} = (\sqrt[n]{a})^m.$$

The denominator is the radical's index.

EXAMPLE 3 Using the Definition of $a^{\frac{m}{n}}$

Use radical notation to rewrite each expression and simplify:

a. $1000^{\frac{2}{3}}$ b. $16^{\frac{3}{2}}$ c. $-32^{\frac{3}{5}}$.

Solution

a. $(1000)^{\frac{2}{3}} = (\sqrt[3]{1000})^2 = 10^2 = 100$

The denominator of $\frac{2}{3}$ is the root and the numerator is the exponent.

b. $16^{\frac{3}{2}} = (\sqrt{16})^3 = 4^3 = 64$

c. $-32^{\frac{3}{5}} = -(\sqrt[5]{32})^3 = -2^3 = -8$

The base is 32 and the negative sign is not affected by the exponent.

■

Using Technology

Here are the calculator keystroke sequences for $1000^{\frac{2}{3}}$:

Many Scientific Calculators

1000 [y^x] [(] 2 [÷] 3 [)] [=]

Many Graphing Calculators

1000 [^] [(] 2 [÷] 3 [)] [ENTER]

✓ **CHECK POINT 3** Use radical notation to rewrite each expression and simplify:

a. $8^{\frac{4}{3}}$ **b.** $25^{\frac{3}{2}}$ **c.** $-81^{\frac{3}{4}}$.

In our next example, we begin with radical notation and rewrite the expression with rational exponents. When changing from radical form to exponential form, the index becomes the denominator of the rational exponent.

EXAMPLE 4 Using the Definition of $a^{\frac{m}{n}}$

Rewrite with rational exponents:

a. $\sqrt[3]{7^5}$ **b.** $(\sqrt[4]{13xy})^9$.

Solution

a. $\sqrt[3]{7^5} = 7^{\frac{5}{3}}$

The index is the exponent's denominator.

b. $(\sqrt[4]{13xy})^9 = (13xy)^{\frac{9}{4}}$ ■

✓ **CHECK POINT 4** Rewrite with rational exponents:

a. $\sqrt[3]{6^4}$ **b.** $(\sqrt[5]{2xy})^7$.

---

3 Use the definition of $a^{-\frac{m}{n}}$.

Can a rational exponent be negative? Yes. The way that negative rational exponents are defined is similar to the way that negative integer exponents are defined.

The Definition of $a^{-\frac{m}{n}}$

If $a^{\frac{m}{n}}$ is a nonzero real number, then

$$a^{-\frac{m}{n}} = \frac{1}{a^{\frac{m}{n}}}.$$

EXAMPLE 5 Using the Definition of $a^{-\frac{m}{n}}$

Rewrite each expression with a positive exponent. Simplify, if possible:

a. $36^{-\frac{1}{2}}$ **b.** $125^{-\frac{1}{3}}$ **c.** $16^{-\frac{3}{4}}$ **d.** $(7xy)^{-\frac{4}{7}}$.

Solution

a. $36^{-\frac{1}{2}} = \frac{1}{36^{\frac{1}{2}}} = \frac{1}{\sqrt{36}} = \frac{1}{6}$

b. $125^{-\frac{1}{3}} = \frac{1}{125^{\frac{1}{3}}} = \frac{1}{\sqrt[3]{125}} = \frac{1}{5}$

c. $16^{-\frac{3}{4}} = \frac{1}{16^{\frac{3}{4}}} = \frac{1}{(\sqrt[4]{16})^3} = \frac{1}{2^3} = \frac{1}{8}$

d. $(7xy)^{-\frac{4}{7}} = \frac{1}{(7xy)^{\frac{4}{7}}}$ ■

Using Technology

Here are the calculator keystroke sequences for $16^{-\frac{3}{4}}$:

Many Scientific Calculators

16 [y^x] [(] 3 [+/−] [÷] 4 [)] [=]

Many Graphing Calculators

16 [^] [(] [(−)] 3 [÷] 4 [)] [ENTER]

CHECK POINT 5 Rewrite each expression with a positive exponent. Simplify, if possible:

a. $100^{-\frac{1}{2}}$

b. $8^{-\frac{1}{3}}$

c. $32^{-\frac{3}{5}}$

d. $(3xy)^{-\frac{5}{9}}$.

Properties of Rational Exponents

The same properties apply to rational exponents as to integer exponents. The following is a summary of these properties:

Properties of Rational Exponents

If m and n are rational exponents, and a and b are real numbers for which the following expressions are defined, then

1. $b^m \cdot b^n = b^{m+n}$ When multiplying exponential expressions with the same base, add the exponents. Use this sum as the exponent of the common base.
2. $\frac{b^m}{b^n} = b^{m-n}$ When dividing exponential expressions with the same base, subtract the exponents. Use this difference as the exponent of the common base.
3. $(b^m)^n = b^{mn}$ When an exponential expression is raised to a power, multiply the exponents. Place the product of the exponents on the base and remove the parentheses.
4. $(ab)^n = a^n b^n$ When a product is raised to a power, raise each factor to that power and multiply.
5. $\left(\frac{a}{b}\right)^n = \frac{a^n}{b^n}$ When a quotient is raised to a power, raise the numerator to that power and divide by the denominator to that power.

4 Simplify expressions with rational exponents.

We can use these properties to simplify exponential expressions with rational exponents. As with integer exponents, an expression with rational exponents is **simplified** when:

- No parentheses appear.
- No powers are raised to powers.
- Each base occurs only once.
- No negative or zero exponents appear.

Great Question!

Because $6 \cdot 6 = 36$, should I multiply 6 and 6 when simplifying $6^{\frac{1}{7}} \cdot 6^{\frac{4}{7}}$?

No. To simplify $6^{\frac{1}{7}} \cdot 6^{\frac{4}{7}}$, **do not multiply the numerical bases.**

Incorrect!

~~$6^{\frac{1}{7}} \cdot 6^{\frac{4}{7}} = 36^{\frac{1}{7}+\frac{4}{7}}$~~

EXAMPLE 6 Simplifying Expressions with Rational Exponents

Simplify:

a. $6^{\frac{1}{7}} \cdot 6^{\frac{4}{7}}$

b. $\frac{32x^{\frac{1}{2}}}{16x^{\frac{3}{4}}}$

c. $\left(8.3^{\frac{3}{4}}\right)^{\frac{2}{3}}$

d. $\left(x^{-\frac{2}{5}}y^{\frac{1}{3}}\right)^{\frac{1}{2}}$.

Solution

a. $6^{\frac{1}{7}} \cdot 6^{\frac{4}{7}} = 6^{\frac{1}{7}+\frac{4}{7}}$ To multiply with the same base, add exponents.

$= 6^{\frac{5}{7}}$ Simplify: $\frac{1}{7}+\frac{4}{7}=\frac{5}{7}$.

b. $\dfrac{32x^{\frac{1}{2}}}{16x^{\frac{3}{4}}} = \dfrac{32}{16}x^{\frac{1}{2}-\frac{3}{4}}$ — Divide coefficients. To divide with the same base, subtract exponents.

$= 2x^{\frac{2}{4}-\frac{3}{4}}$ — Write exponents in terms of the LCD, 4.

$= 2x^{-\frac{1}{4}}$ — Subtract: $\frac{2}{4} - \frac{3}{4} = -\frac{1}{4}$.

$= \dfrac{2}{x^{\frac{1}{4}}}$ — Rewrite with a positive exponent: $a^{-\frac{m}{n}} = \dfrac{1}{a^{\frac{m}{n}}}$.

c. $\left(8.3^{\frac{3}{4}}\right)^{\frac{2}{3}} = 8.3^{\left(\frac{3}{4}\right)\left(\frac{2}{3}\right)}$ — To raise a power to a power, multiply exponents.

$= 8.3^{\frac{1}{2}}$ — Multiply: $\frac{3}{4}\cdot\frac{2}{3} = \frac{6}{12} = \frac{1}{2}$.

d. $\left(x^{-\frac{2}{5}}y^{\frac{1}{3}}\right)^{\frac{1}{2}} = \left(x^{-\frac{2}{5}}\right)^{\frac{1}{2}}\left(y^{\frac{1}{3}}\right)^{\frac{1}{2}}$ — To raise a product to a power, raise each factor to the power.

$= x^{-\frac{1}{5}}y^{\frac{1}{6}}$ — Multiply: $-\frac{2}{5}\cdot\frac{1}{2} = -\frac{1}{5}$ and $\frac{1}{3}\cdot\frac{1}{2} = \frac{1}{6}$.

$= \dfrac{y^{\frac{1}{6}}}{x^{\frac{1}{5}}}$ — Rewrite with positive exponents. ■

CHECK POINT 6 Simplify:

a. $7^{\frac{1}{2}}\cdot 7^{\frac{1}{3}}$ **b.** $\dfrac{50x^{\frac{1}{3}}}{10x^{\frac{4}{3}}}$ **c.** $\left(9.1^{\frac{2}{5}}\right)^{\frac{3}{4}}$ **d.** $\left(x^{-\frac{3}{5}}y^{\frac{1}{4}}\right)^{\frac{1}{3}}$.

5 Simplify radical expressions using rational exponents.

Using Rational Exponents to Simplify Radical Expressions

Some radical expressions can be simplified using rational exponents. We will use the following procedure:

Simplifying Radical Expressions Using Rational Exponents

1. Rewrite each radical expression as an exponential expression with a rational exponent.
2. Simplify using properties of rational exponents.
3. Rewrite in radical notation if rational exponents still appear.

EXAMPLE 7 Simplifying Radical Expressions Using Rational Exponents

Use rational exponents to simplify. Assume that all variables represent nonnegative numbers.

a. $\sqrt[10]{x^5}$ **b.** $\sqrt[3]{27a^{15}}$ **c.** $\sqrt[4]{x^6y^2}$ **d.** $\sqrt{x}\cdot\sqrt[3]{x}$ **e.** $\sqrt[3]{\sqrt{x}}$

Solution

a. $\sqrt[10]{x^5} = x^{\frac{5}{10}}$ — Rewrite as an exponential expression.

$= x^{\frac{1}{2}}$ — Simplify the exponent.

$= \sqrt{x}$ — Rewrite in radical notation.

b. $\sqrt[3]{27a^{15}} = (27a^{15})^{\frac{1}{3}}$ — Rewrite as an exponential expression.

$= 27^{\frac{1}{3}}(a^{15})^{\frac{1}{3}}$ — Raise each factor in parentheses to the $\frac{1}{3}$ power.

$= \sqrt[3]{27}\cdot a^{15\left(\frac{1}{3}\right)}$ — To raise a power to a power, multiply exponents.

$= 3a^5$ — $\sqrt[3]{27} = 3$. Multiply exponents: $15\cdot\frac{1}{3} = 5$.

c. $\sqrt[4]{x^6y^2} = (x^6y^2)^{\frac{1}{4}}$ Rewrite as an exponential expression.

$= (x^6)^{\frac{1}{4}}(y^2)^{\frac{1}{4}}$ Raise each factor in parentheses to the $\frac{1}{4}$ power.

$= x^{\frac{6}{4}}y^{\frac{2}{4}}$ To raise powers to powers, multiply.

$= x^{\frac{3}{2}}y^{\frac{1}{2}}$ Simplify.

$= (x^3y)^{\frac{1}{2}}$ $a^nb^n = (ab)^n$

$= \sqrt{x^3y}$ Rewrite in radical notation.

d. $\sqrt{x}\cdot\sqrt[3]{x} = x^{\frac{1}{2}}\cdot x^{\frac{1}{3}}$ Rewrite as exponential expressions.

$= x^{\frac{1}{2}+\frac{1}{3}}$ To multiply with the same base, add exponents.

$= x^{\frac{3}{6}+\frac{2}{6}}$ Write exponents in terms of the LCD, 6.

$= x^{\frac{5}{6}}$ Add: $\frac{3}{6}+\frac{2}{6}=\frac{5}{6}$.

$= \sqrt[6]{x^5}$ Rewrite in radical notation.

e. $\sqrt[3]{\sqrt{x}} = \sqrt[3]{x^{\frac{1}{2}}}$ Write the radicand as an exponential expression.

$= \left(x^{\frac{1}{2}}\right)^{\frac{1}{3}}$ Write the entire expression in exponential form.

$= x^{\frac{1}{6}}$ To raise powers to powers, multiply: $\frac{1}{2}\cdot\frac{1}{3}=\frac{1}{6}$.

$= \sqrt[6]{x}$ Rewrite in radical notation. ■

✓ **CHECK POINT 7** Use rational exponents to simplify:

a. $\sqrt[6]{x^3}$ **b.** $\sqrt[3]{8a^{12}}$ **c.** $\sqrt[8]{x^4y^2}$

d. $\frac{\sqrt{x}}{\sqrt[3]{x}}$ **e.** $\sqrt{\sqrt[3]{x}}$.

CONCEPT AND VOCABULARY CHECK

Fill in each blank so that the resulting statement is true.

1. $36^{\frac{1}{2}} = \sqrt{\_} = \_$
2. $8^{\frac{1}{3}} = \sqrt[3]{\_} = \_$
3. $a^{\frac{1}{n}} = \_\_\_$
4. $16^{\frac{3}{4}} = (\sqrt[4]{\_})^3 = (\_)^3 = \_$
5. $a^{\frac{m}{n}} = \_\_\_\_\_\_$
6. $\sqrt[3]{2^5} = 2^{-}$
7. $16^{-\frac{3}{2}} = \frac{1}{\_\_\_} = \frac{1}{(\sqrt{\_})^{-}} = \frac{1}{(\_)^{-}} = \frac{1}{\_}$

7.2 EXERCISE SET MyMathLab®

Practice Exercises

In Exercises 1–20, use radical notation to rewrite each expression. Simplify, if possible.

1. $49^{\frac{1}{2}}$
2. $100^{\frac{1}{2}}$
3. $(-27)^{\frac{1}{3}}$
4. $(-64)^{\frac{1}{3}}$
5. $-16^{\frac{1}{4}}$
6. $-81^{\frac{1}{4}}$
7. $(xy)^{\frac{1}{3}}$
8. $(xy)^{\frac{1}{4}}$
9. $(2xy^3)^{\frac{1}{5}}$
10. $(3xy^4)^{\frac{1}{5}}$
11. $81^{\frac{3}{2}}$
12. $25^{\frac{3}{2}}$
13. $125^{\frac{2}{3}}$
14. $1000^{\frac{2}{3}}$
15. $(-32)^{\frac{3}{5}}$
16. $(-27)^{\frac{2}{3}}$
17. $27^{\frac{2}{3}} + 16^{\frac{3}{4}}$
18. $4^{\frac{5}{2}} - 8^{\frac{2}{3}}$
19. $(xy)^{\frac{4}{7}}$
20. $(xy)^{\frac{4}{9}}$

In Exercises 21–38, rewrite each expression with rational exponents.

21. $\sqrt{7}$
22. $\sqrt{13}$
23. $\sqrt[3]{5}$
24. $\sqrt[3]{6}$
25. $\sqrt[5]{11x}$
26. $\sqrt[5]{13x}$
27. $\sqrt{x^3}$
28. $\sqrt{x^5}$
29. $\sqrt[5]{x^3}$
30. $\sqrt[7]{x^4}$
31. $\sqrt[5]{x^2y}$
32. $\sqrt[7]{xy^3}$
33. $(\sqrt{19xy})^3$
34. $(\sqrt{11xy})^3$
35. $(\sqrt[6]{7xy^2})^5$
36. $(\sqrt[8]{9x^2y})^5$
37. $2x\sqrt[3]{y^2}$
38. $4x\sqrt[5]{y^2}$

In Exercises 39–54, rewrite each expression with a positive rational exponent. Simplify, if possible.

39. $49^{-\frac{1}{2}}$
40. $9^{-\frac{1}{2}}$

41. $27^{-\frac{1}{3}}$
42. $125^{-\frac{1}{3}}$
43. $16^{-\frac{3}{4}}$
44. $81^{-\frac{5}{4}}$
45. $8^{-\frac{2}{3}}$
46. $32^{-\frac{4}{5}}$
47. $\left(\frac{8}{27}\right)^{-\frac{1}{3}}$
48. $\left(\frac{8}{125}\right)^{-\frac{1}{3}}$
49. $(-64)^{-\frac{2}{3}}$
50. $(-8)^{-\frac{2}{3}}$
51. $(2xy)^{-\frac{7}{10}}$
52. $(4xy)^{-\frac{4}{7}}$
53. $5xz^{-\frac{1}{3}}$
54. $7xz^{-\frac{1}{4}}$

In Exercises 55–78, use properties of rational exponents to simplify each expression. Assume that all variables represent positive numbers.

55. $3^{\frac{3}{4}} \cdot 3^{\frac{1}{4}}$
56. $5^{\frac{2}{3}} \cdot 5^{\frac{1}{3}}$
57. $\dfrac{16^{\frac{3}{4}}}{16^{\frac{1}{4}}}$
58. $\dfrac{100^{\frac{3}{4}}}{100^{\frac{31}{4}}}$
59. $x^{\frac{1}{2}} \cdot x^{\frac{1}{3}}$
60. $x^{\frac{1}{2}} \cdot x^{\frac{2}{3}}$
61. $\dfrac{x^{\frac{4}{5}}}{x^{\frac{1}{5}}}$
62. $\dfrac{x^{\frac{3}{7}}}{x^{\frac{1}{7}}}$
63. $\dfrac{x^{\frac{1}{3}}}{x^{\frac{3}{4}}}$
64. $\dfrac{x^{\frac{1}{4}}}{x^{\frac{3}{5}}}$
65. $\left(5^{\frac{2}{3}}\right)^3$
66. $\left(3^{\frac{4}{5}}\right)^5$
67. $\left(y^{-\frac{2}{3}}\right)^{\frac{1}{4}}$
68. $\left(y^{-\frac{3}{4}}\right)^{\frac{1}{6}}$
69. $\left(2x^{\frac{1}{5}}\right)^5$
70. $\left(2x^{\frac{1}{4}}\right)^4$
71. $(25x^4y^6)^{\frac{1}{2}}$
72. $(125x^9y^6)^{\frac{1}{3}}$
73. $\left(x^{\frac{1}{2}}y^{-\frac{3}{5}}\right)^{\frac{1}{2}}$
74. $\left(x^{\frac{1}{4}}y^{-\frac{2}{5}}\right)^{\frac{1}{3}}$
75. $\dfrac{3^{\frac{1}{2}} \cdot 3^{\frac{3}{4}}}{3^{\frac{1}{4}}}$
76. $\dfrac{5^{\frac{3}{4}} \cdot 5^{\frac{1}{2}}}{5^{\frac{1}{4}}}$
77. $\dfrac{\left(3y^{\frac{1}{4}}\right)^3}{y^{\frac{1}{12}}}$
78. $\dfrac{\left(2y^{\frac{1}{5}}\right)^4}{y^{\frac{3}{10}}}$

In Exercises 79–112, use rational exponents to simplify each expression. If rational exponents appear after simplifying, write the answer in radical notation. Assume that all variables represent positive numbers.

79. $\sqrt[8]{x^2}$
80. $\sqrt[10]{x^2}$
81. $\sqrt[3]{8a^6}$
82. $\sqrt[3]{27a^{12}}$
83. $\sqrt[5]{x^{10}y^{15}}$
84. $\sqrt[5]{x^{15}y^{20}}$
85. $\left(\sqrt[3]{xy}\right)^{18}$
86. $\left(\sqrt[3]{xy}\right)^{21}$
87. $\sqrt[10]{(3y)^2}$
88. $\sqrt[12]{(3y)^2}$
89. $\left(\sqrt[6]{2a}\right)^4$
90. $\left(\sqrt[8]{2a}\right)^6$
91. $\sqrt[9]{x^6y^3}$
92. $\sqrt[4]{x^2y^6}$
93. $\sqrt{2} \cdot \sqrt[3]{2}$
94. $\sqrt{3} \cdot \sqrt[3]{3}$
95. $\sqrt[5]{x^2} \cdot \sqrt{x}$
96. $\sqrt[7]{x^2} \cdot \sqrt{x}$
97. $\sqrt[4]{a^2b} \cdot \sqrt[3]{ab}$
98. $\sqrt[6]{ab^2} \cdot \sqrt[3]{a^2b}$
99. $\dfrac{\sqrt[4]{x}}{\sqrt[5]{x}}$
100. $\dfrac{\sqrt[3]{x}}{\sqrt[4]{x}}$
101. $\dfrac{\sqrt[3]{y^2}}{\sqrt[6]{y}}$
102. $\dfrac{\sqrt[5]{y^2}}{\sqrt[10]{y^3}}$
103. $\sqrt[4]{\sqrt{x}}$
104. $\sqrt[5]{\sqrt{x}}$
105. $\sqrt{\sqrt{x^2y}}$
106. $\sqrt{\sqrt{xy^2}}$
107. $\sqrt[4]{\sqrt[3]{2x}}$
108. $\sqrt[5]{\sqrt[3]{2x}}$
109. $\left(\sqrt[4]{x^3y^5}\right)^{12}$
110. $\left(\sqrt[5]{x^4y^2}\right)^{20}$
111. $\dfrac{\sqrt[4]{a^5b^5}}{\sqrt{ab}}$
112. $\dfrac{\sqrt[4]{a^3b^3}}{\sqrt{ab}}$

Practice PLUS

In Exercises 113–116, use the distributive property or the FOIL method to perform each multiplication.

113. $x^{\frac{1}{3}}\left(x^{\frac{1}{3}} - x^{\frac{2}{3}}\right)$
114. $x^{-\frac{1}{4}}\left(x^{\frac{9}{4}} - x^{\frac{1}{4}}\right)$
115. $\left(x^{\frac{1}{2}} - 3\right)\left(x^{\frac{1}{2}} + 5\right)$
116. $\left(x^{\frac{1}{3}} - 2\right)\left(x^{\frac{1}{3}} + 6\right)$

In Exercises 117–120, factor out the greatest common factor from each expression.

117. $6x^{\frac{1}{2}} + 2x^{\frac{3}{2}}$
118. $8x^{\frac{1}{4}} + 4x^{\frac{5}{4}}$
119. $15x^{\frac{1}{3}} - 60x$
120. $7x^{\frac{1}{3}} - 70x$

In Exercises 121–124, simplify each expression. Assume that all variables represent positive numbers.

121. $(49x^{-2}y^4)^{-\frac{1}{2}}\left(xy^{\frac{1}{2}}\right)$
122. $(8x^{-6}y^3)^{\frac{1}{3}}\left(x^{\frac{5}{6}}y^{-\frac{1}{3}}\right)^6$
123. $\left(\dfrac{x^{-\frac{5}{4}}y^{\frac{1}{3}}}{x^{-\frac{3}{4}}}\right)^{-6}$
124. $\left(\dfrac{x^{\frac{1}{2}}y^{-\frac{7}{4}}}{y^{-\frac{5}{4}}}\right)^{-4}$

Application Exercises

The Galápagos Islands, lying 600 miles west of Ecuador, are famed for their extraordinary wildlife. The function

$$f(x) = 29x^{\frac{1}{3}}$$

models the number of plant species, $f(x)$, on the various islands of the Galápagos chain in terms of the area, x, in square miles, of a particular island. Use the function to solve Exercises 125–126.

125. How many species of plants are on a Galápagos island that has an area of 8 square miles?

126. How many species of plants are on a Galápagos island that has an area of 27 square miles?

The function

$$f(x) = 70x^{\frac{3}{4}}$$

models the number of calories per day, $f(x)$, a person needs to maintain life in terms of that person's weight, x, in kilograms. (1 kilogram is approximately 2.2 pounds.) Use this model and a calculator to solve Exercises 127–128. Round answers to the nearest calorie.

127. How many calories per day does a person who weighs 80 kilograms (approximately 176 pounds) need to maintain life?

128. How many calories per day does a person who weighs 70 kilograms (approximately 154 pounds) need to maintain life?

The way that we perceive the temperature on a cold day depends on both air temperature and wind speed. The windchill is what the air temperature would have to be with no wind to achieve the same chilling effect on the skin. In 2002, the National Weather Service issued new windchill temperatures, shown in the table below. (One reason for this new windchill index is that the wind speed is now calculated at 5 feet, the average height of the human body's face, rather than 33 feet, the height of the standard anemometer, an instrument that calculates wind speed.)

New Windchill Temperature Index

Wind Speed (miles per hour)	Air Temperature (°F)											
	30	25	20	15	10	5	0	−5	−10	−15	−20	−25
5	25	19	13	7	1	−5	−11	−16	−22	−28	−34	−40
10	21	15	9	3	−4	−10	−16	−22	−28	−35	−41	−47
15	19	13	6	0	−7	−13	−19	−26	−32	−39	−45	−51
20	17	11	4	−2	−9	−15	−22	−29	−35	−42	−48	−55
25	16	9	3	−4	−11	−17	−24	−31	−37	−44	−51	−58
30	15	8	1	−5	−12	−19	−26	−33	−39	−46	−53	−60
35	14	7	0	−7	−14	−21	−27	−34	−41	−48	−55	−62
40	13	6	−1	−8	−15	−22	−29	−36	−43	−50	−57	−64
45	12	5	−2	−9	−16	−23	−30	−37	−44	−51	−58	−65
50	12	4	−3	−10	−17	−24	−31	−38	−45	−52	−60	−67
55	11	4	−3	−11	−18	−25	−32	−39	−46	−54	−61	−68
60	10	3	−4	−11	−19	−26	−33	−40	−48	−55	−62	−69

Frostbite occurs in 15 minutes or less.

Source: National Weather Service

The windchill temperatures shown in the table can be calculated using

$$C = 35.74 + 0.6215t - 35.74\sqrt[25]{v^4} + 0.4275t\sqrt[25]{v^4},$$

in which C is the windchill, in degrees Fahrenheit, t is the air temperature, in degrees Fahrenheit, and v is the wind speed, in miles per hour. Use the formula to solve Exercises 129–132.

129. a. Rewrite the equation for calculating windchill temperatures using rational exponents.

b. Use the form of the equation in part (a) and a calculator to find the windchill temperature, to the nearest degree, when the air temperature is 25°F and the wind speed is 30 miles per hour.

130. a. Rewrite the equation for calculating windchill temperatures using rational exponents.

b. Use the form of the equation in part (a) and a calculator to find the windchill temperature, to the nearest degree, when the air temperature is 35°F and the wind speed is 15 miles per hour.

131. a. Substitute 0 for t in the equation with rational exponents from Exercise 129(a) and write a function $C(v)$ that gives the windchill temperature as a function of wind speed for an air temperature of 0°F.

b. Find and interpret $C(25)$. Use a calculator and round to the nearest degree.

c. Identify your solution to part (b) on the graph shown.

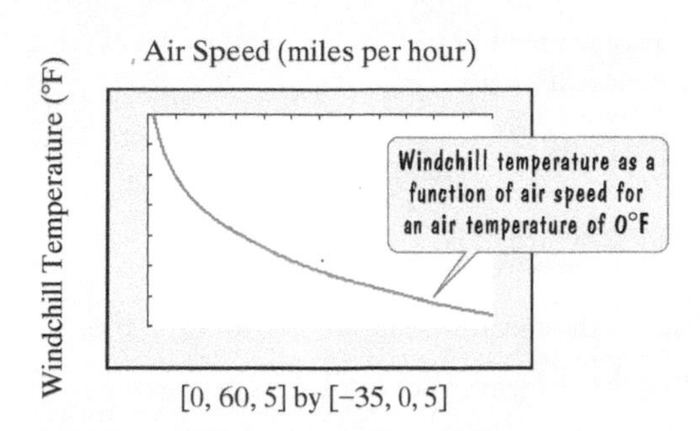

[0, 60, 5] by [−35, 0, 5]

132. a. Substitute 30 for t in the equation with rational exponents from Exercise 130(a) and write a function $C(v)$ that gives the windchill temperature as a function of wind speed for an air temperature of 30°F. Simplify the function's formula so that it contains exactly two terms.

b. Find and interpret $C(40)$. Use a calculator and round to the nearest degree.

c. Identify your solution to part (b) on the graph shown.

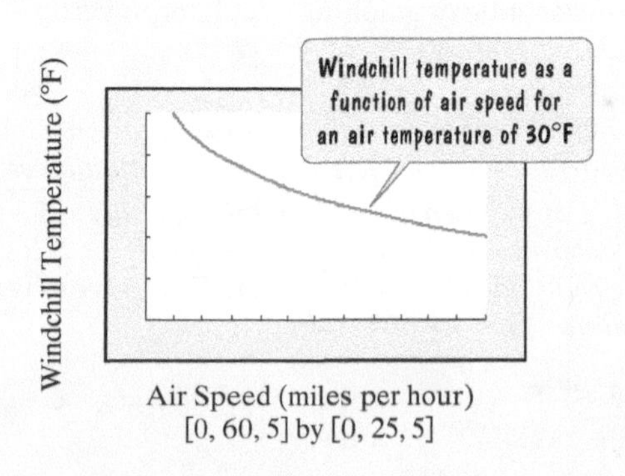

[0, 60, 5] by [0, 25, 5]

Your job is to determine whether or not yachts are eligible for the America's Cup, the supreme event in ocean sailing. The basic dimensions of competitors' yachts must satisfy

$$L + 1.25\sqrt{S} - 9.8\sqrt[3]{D} \le 16.296,$$

where L is the yacht's length, in meters, S is its sail area, in square meters, and D is its displacement, in cubic meters. Use this information to solve Exercises 133–134.

133. **a.** Rewrite the inequality using rational exponents.

b. Use your calculator to determine if a yacht with length 20.85 meters, sail area 276.4 square meters, and displacement 18.55 cubic meters is eligible for the America's Cup.

134. **a.** Rewrite the inequality using rational exponents.

b. Use your calculator to determine if a yacht with length 22.85 meters, sail area 312.5 square meters, and displacement 22.34 cubic meters is eligible for the America's Cup.

Explaining the Concepts

135. What is the meaning of $a^{\frac{1}{n}}$? Give an example to support your explanation.

136. What is the meaning of $a^{\frac{m}{n}}$? Give an example.

137. What is the meaning of $a^{-\frac{m}{n}}$? Give an example.

138. Explain why $a^{\frac{1}{n}}$ is negative when n is odd and a is negative. What happens if n is even and a is negative? Why?

139. In simplifying $36^{\frac{3}{2}}$, is it better to use $a^{\frac{m}{n}} = \sqrt[n]{a^m}$ or $a^{\frac{m}{n}} = (\sqrt[n]{a})^m$? Explain.

140. How can you tell if an expression with rational exponents is simplified?

141. Explain how to simplify $\sqrt[3]{x} \cdot \sqrt{x}$.

142. Explain how to simplify $\sqrt[3]{\sqrt{x}}$.

Technology Exercises

143. Use a scientific or graphing calculator to verify your results in Exercises 15–18.

144. Use a scientific or graphing calculator to verify your results in Exercises 45–50.

Exercises 145–147 show a number of simplifications, not all of which are correct. Enter the left side of each equation as y_1 and the right side as y_2. Then use your graphing utility's TABLE *feature to determine if the simplification is correct. If it is not, correct the right side and use the* TABLE *feature to verify your simplification.*

145. $x^{\frac{3}{5}} \cdot x^{-\frac{1}{10}} = x^{\frac{1}{2}}$

146. $\left(x^{-\frac{1}{2}} \cdot x^{\frac{3}{4}}\right)^{-2} = x^{\frac{1}{2}}$

147. $\dfrac{x^{\frac{1}{4}}}{x^{\frac{1}{2}} \cdot x^{-\frac{3}{4}}} = \dfrac{1}{x^{\frac{1}{2}}}$

Critical Thinking Exercises

Make Sense? *In Exercises 148–151, determine whether each statement makes sense or does not make sense, and explain your reasoning.*

148. By adding the exponents, I simplified $7^{\frac{1}{2}} \cdot 7^{\frac{1}{2}}$ and obtained 49.

149. When I use the definition for $a^{\frac{m}{n}}$, I usually prefer to first raise a to the m power because smaller numbers are involved.

150. There's no question that $(-64)^{\frac{1}{3}} = -64^{\frac{1}{3}}$, so I can conclude that $(-64)^{\frac{1}{2}} = -64^{\frac{1}{2}}$.

151. I checked the following simplification and every step is correct:

$$\begin{aligned} 5\left(4 - 5^{\frac{1}{2}}\right) &= 5 \cdot 4 - 5 \cdot 5^{\frac{1}{2}} \\ &= 20 - 25^{\frac{1}{2}} \\ &= 20 - \sqrt{25} \\ &= 20 - 5 = 15. \end{aligned}$$

In Exercises 152–155, determine whether each statement is true or false. If the statement is false, make the necessary change(s) to produce a true statement.

152. If n is odd, then $(-b)^{\frac{1}{n}} = -b^{\frac{1}{n}}$.

153. $(a + b)^{\frac{1}{n}} = a^{\frac{1}{n}} + b^{\frac{1}{n}}$

154. $8^{-\frac{2}{3}} = -4$

155. $4^{-3.5} = \dfrac{1}{128}$

156. A mathematics professor recently purchased a birthday cake for her son with the inscription

$$\text{Happy}\left(2^{\frac{5}{2}} \cdot 2^{\frac{3}{4}} \div 2^{\frac{1}{4}}\right)\text{th Birthday.}$$

How old is the son?

157. The birthday boy in Exercise 156, excited by the inscription on the cake, tried to wolf down the whole thing. Professor Mom, concerned about the possible metamorphosis of her son into a blimp, exclaimed, "Hold on! It is your birthday, so why not take $\dfrac{8^{-\frac{4}{3}} + 2^{-2}}{16^{-\frac{3}{4}} + 2^{-1}}$ of the cake? I'll eat half of what's left over." How much of the cake did the professor eat?

158. Simplify: $\left[3 + \left(27^{\frac{2}{3}} + 32^{\frac{2}{5}}\right)\right]^{\frac{3}{2}} - 9^{\frac{1}{2}}$.

159. Find the domain of $f(x) = (x - 3)^{\frac{1}{2}}(x + 4)^{-\frac{1}{2}}$.

Review Exercises

160. Write the equation of the linear function whose graph passes through (5, 1) and (4, 3). (Section 2.5, Example 2)

161. Graph $y \le -\frac{3}{2}x + 3$. (Section 4.4, Example 2)

162. Solve by Cramer's rule:

$$\begin{cases} 5x - 3y = 3 \\ 7x + y = 25. \end{cases}$$

(Section 3.5, Example 2)

Preview Exercises

Exercises 163–165 will help you prepare for the material covered in the next section.

163. **a.** Find $\sqrt{16} \cdot \sqrt{4}$.

b. Find $\sqrt{16 \cdot 4}$.

c. Based on your answers to parts (a) and (b), what can you conclude?

164. **a.** Use a calculator to approximate $\sqrt{300}$ to two decimal places.

b. Use a calculator to approximate $10\sqrt{3}$ to two decimal places.

c. Based on your answers to parts (a) and (b), what can you conclude?

165. Simplify: **a.** $\sqrt[3]{x^{21}}$ **b.** $\sqrt[6]{y^{24}}$.

SECTION

7.3 Multiplying and Simplifying Radical Expressions

What am I supposed to learn?

After studying this section, you should be able to:

1. Use the product rule to multiply radicals.
2. Use factoring and the product rule to simplify radicals.
3. Multiply radicals and then simplify.

Mirror II (1963), George Tooker. Addison Gallery, Phillips Academy, MA. © George Tooker.

We opened this book with a model that described our improving emotional health as we age. Unfortunately, not everything gets better. The aging process is also accompanied by a number of physical transformations, including changes in vision that require glasses for reading, the onset of wrinkles and sagging skin, and a decrease in heart response. A change in heart response occurs fairly early; after 20, our hearts become less adept at accelerating in response to exercise. In this section's Exercise Set (Exercises 95–96), you will see how a radical function models changes in heart function throughout the aging process, as we turn to multiplying and simplifying radical expressions.

1 Use the product rule to multiply radicals.

The Product Rule for Radicals

A rule for multiplying radicals can be generalized by comparing $\sqrt{25}\cdot\sqrt{4}$ and $\sqrt{25\cdot 4}$. Notice that

$$\sqrt{25}\cdot\sqrt{4} = 5\cdot 2 = 10 \quad\text{and}\quad \sqrt{25\cdot 4} = \sqrt{100} = 10.$$

Because we obtain 10 in both situations, the original radical expressions must be equal. That is,

$$\sqrt{25}\cdot\sqrt{4} = \sqrt{25\cdot 4}.$$

This result is a special case of the **product rule for radicals** that can be generalized as follows:

The Product Rule for Radicals

If $\sqrt[n]{a}$ and $\sqrt[n]{b}$ are real numbers, then

$$\sqrt[n]{a}\cdot\sqrt[n]{b} = \sqrt[n]{ab}.$$

The product of two nth roots is the nth root of the product of the radicands.

Great Question!

Can I use the product rule to simplify radicals with different indices, such as $\sqrt{x}\cdot\sqrt[3]{x}$?

No. The product rule can be used only when the radicals have the same index. If indices differ, rational exponents can be used, as in $\sqrt{x}\cdot\sqrt[3]{x}$, which was Example 7(d) in the previous section.

EXAMPLE 1 Using the Product Rule for Radicals

Multiply:

a. $\sqrt{3}\cdot\sqrt{7}$ **b.** $\sqrt{x+7}\cdot\sqrt{x-7}$ **c.** $\sqrt[3]{7}\cdot\sqrt[3]{9}$ **d.** $\sqrt[8]{10x}\cdot\sqrt[8]{8x^4}$.

Solution In each problem, the indices are the same. Thus, we multiply by multiplying the radicands.

a. $\sqrt{3}\cdot\sqrt{7} = \sqrt{3\cdot 7} = \sqrt{21}$

b. $\sqrt{x+7}\cdot\sqrt{x-7} = \sqrt{(x+7)(x-7)} = \sqrt{x^2-49}$

This is not equal to $\sqrt{x^2} - \sqrt{49}$.

c. $\sqrt[3]{7}\cdot\sqrt[3]{9} = \sqrt[3]{7\cdot 9} = \sqrt[3]{63}$

d. $\sqrt[8]{10x}\cdot\sqrt[8]{8x^4} = \sqrt[8]{10x\cdot 8x^4} = \sqrt[8]{80x^5}$ ■

CHECK POINT 1 Multiply:

a. $\sqrt{5}\cdot\sqrt{11}$ **b.** $\sqrt{x+4}\cdot\sqrt{x-4}$

c. $\sqrt[3]{6}\cdot\sqrt[3]{10}$ **d.** $\sqrt[7]{2x}\cdot\sqrt[7]{6x^3}$.

---

2 Use factoring and the product rule to simplify radicals.

Using Factoring and the Product Rule to Simplify Radicals

In Chapter 5, we saw that a number that is the square of an integer is a **perfect square**. For example, 100 is a perfect square because $100 = 10^2$. A number is a **perfect cube** if it is the cube of an integer. Thus, 125 is a perfect cube because $125 = 5^3$. In general, a number is a **perfect *n*th power** if it is the nth power of an integer. Thus, p is a perfect nth power if there is an integer q such that $p = q^n$.

A radical of index n is **simplified** when its radicand has no factors other than 1 that are perfect nth powers. For example, $\sqrt{300}$ is not simplified because it can be expressed as $\sqrt{100\cdot 3}$ and 100 is a perfect square. We can use the product rule in the form

$$\sqrt[n]{ab} = \sqrt[n]{a}\cdot\sqrt[n]{b}$$

to simplify $\sqrt[n]{ab}$ when a or b is a perfect nth power. Consider $\sqrt{300}$. To simplify, we factor 300 so that one of its factors is the greatest perfect square possible.

Using Technology

You can use a calculator to provide numerical support that $\sqrt{300} = 10\sqrt{3}$. First find an approximation for $\sqrt{300}$:

300 [√] ≈ 17.32

or

[√] 300 [ENTER] ≈ 17.32.

Now find an aproximation for $10\sqrt{3}$:

10 [×] 3 [√] ≈ 17.32

or

10 [√] 3 [ENTER] ≈ 17.32.

Correct to two decimal places,

$\sqrt{300} \approx 17.32$ and
$10\sqrt{3} \approx 17.32$.

This verifies that

$\sqrt{300} = 10\sqrt{3}$.

Use this technique to support the numerical results for the answers in this section. *Caution:* A simplified radical does not mean a decimal approximation.

$$\sqrt{300} = \sqrt{100 \cdot 3}$$ Factor 300: 100 is the greatest perfect square factor.

$$= \sqrt{100} \cdot \sqrt{3}$$ Use the product rule: $\sqrt[n]{ab} = \sqrt[n]{a} \cdot \sqrt[n]{b}$.

$$= 10\sqrt{3}$$ Write $\sqrt{100}$ as 10. We read $10\sqrt{3}$ as "ten times the square root of three."

Simplifying Radical Expressions by Factoring

A radical expression whose index is n is **simplified** when its radicand has no factors that are perfect nth powers. To simplify, use the following procedure:

1. Write the radicand as the product of two factors, one of which is the greatest perfect nth power.
2. Use the product rule to take the nth root of each factor.
3. Find the nth root of the perfect nth power.

EXAMPLE 2 Simplifying Radicals by Factoring

Simplify by factoring:

a. $\sqrt{75}$ **b.** $\sqrt[3]{54}$ **c.** $\sqrt[5]{64}$ **d.** $\sqrt{500xy^2}$.

Solution

a. $\sqrt{75} = \sqrt{25 \cdot 3}$ 25 is the greatest perfect square that is a factor of 75.

$= \sqrt{25} \cdot \sqrt{3}$ Take the square root of each factor: $\sqrt[n]{ab} = \sqrt[n]{a} \cdot \sqrt[n]{b}$.

$= 5\sqrt{3}$ Write $\sqrt{25}$ as 5.

b. $\sqrt[3]{54} = \sqrt[3]{27 \cdot 2}$ 27 is the greatest perfect cube that is a factor of 54: $27 = 3^3$.

$= \sqrt[3]{27} \cdot \sqrt[3]{2}$ Take the cube root of each factor: $\sqrt[n]{ab} = \sqrt[n]{a} \cdot \sqrt[n]{b}$.

$= 3\sqrt[3]{2}$ Write $\sqrt[3]{27}$ as 3.

c. $\sqrt[5]{64} = \sqrt[5]{32 \cdot 2}$ 32 is the greatest perfect fifth power that is a factor of 64: $32 = 2^5$.

$= \sqrt[5]{32} \cdot \sqrt[5]{2}$ Take the fifth root of each factor: $\sqrt[n]{ab} = \sqrt[n]{a} \cdot \sqrt[n]{b}$.

$= 2\sqrt[5]{2}$ Write $\sqrt[5]{32}$ as 2.

d. $\sqrt{500xy^2} = \sqrt{100y^2 \cdot 5x}$ $100y^2$ is the greatest perfect square that is a factor of $500xy^2$: $100y^2 = (10y)^2$.

$= \sqrt{100y^2} \cdot \sqrt{5x}$ Factor into two radicals.

$= 10|y|\sqrt{5x}$ Take the square root of $100y^2$. ■

✓ CHECK POINT 2 Simplify by factoring:

a. $\sqrt{80}$ **b.** $\sqrt[3]{40}$
c. $\sqrt[4]{32}$ **d.** $\sqrt{200x^2y}$.

EXAMPLE 3 Simplifying a Radical Function

If

$$f(x) = \sqrt{2x^2 + 4x + 2},$$

express the function, f, in simplified form.

Solution Begin by factoring the radicand. The GCF is 2. Simplification is possible if we obtain a factor that is a perfect square.

$f(x) = \sqrt{2x^2 + 4x + 2}$	This is the given function.
$= \sqrt{2(x^2 + 2x + 1)}$	Factor out the GCF.
$= \sqrt{2(x + 1)^2}$	Factor the perfect square trinomial: $A^2 + 2AB + B^2 = (A + B)^2$.
$= \sqrt{2} \cdot \sqrt{(x + 1)^2}$	Take the square root of each factor. The factor $(x + 1)^2$. is a perfect square.
$= \sqrt{2}\lvert x + 1\rvert$	Take the square root of $(x + 1)^2$.

In simplified form,

$$f(x) = \sqrt{2}|x + 1|. \blacksquare$$

Using Technology

Graphic Connections

The graphs of

$$f(x) = \sqrt{2x^2 + 4x + 2}, \quad g(x) = \sqrt{2}|x + 1|, \quad \text{and} \quad h(x) = \sqrt{2}(x + 1)$$

are shown in **Figure 7.7** in three separate $[-5, 5, 1]$ by $[-5, 5, 1]$ viewing rectangles. The graphs in **Figure 7.7 (a)** and **(b)** are identical. This verifies that our simplification in Example 3 is correct: $\sqrt{2x^2 + 4x + 2} = \sqrt{2}|x + 1|$. Now compare the graphs in **Figure 7.7 (a)** and **(c)**. Can you see that they are not the same? This illustrates the importance of not leaving out absolute value bars:

$$\sqrt{2x^2 + 4x + 2} \neq \sqrt{2}(x + 1).$$

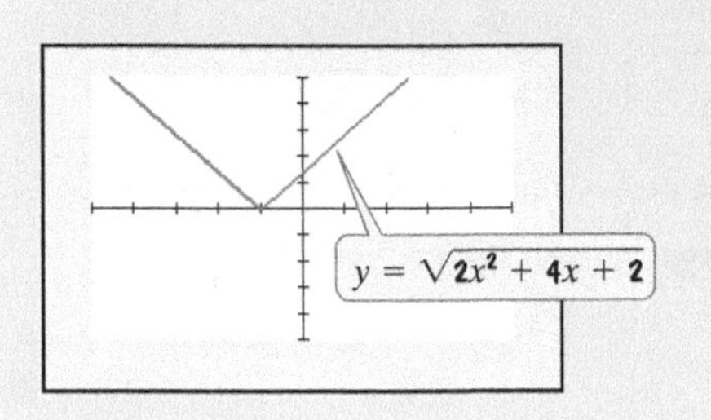

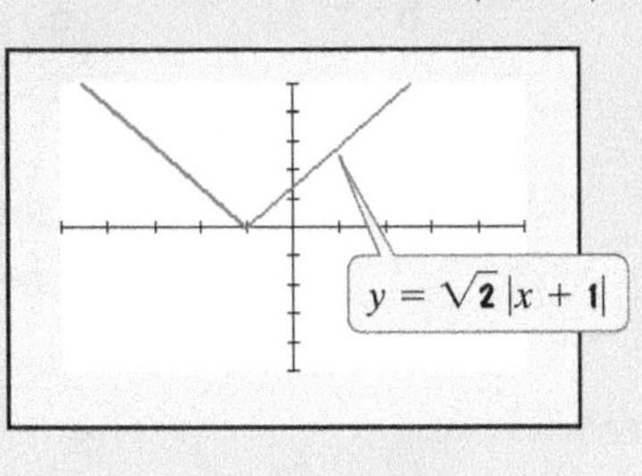

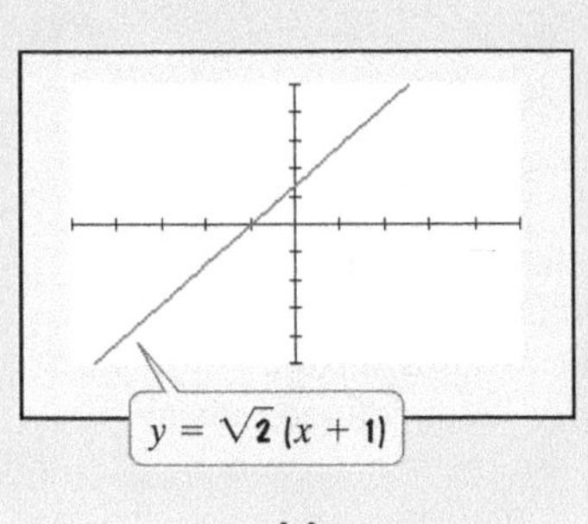

Figure 7.7 (a) **(b)** **(c)**

CHECK POINT 3 If $f(x) = \sqrt{3x^2 - 12x + 12}$, express the function, f, in simplified form.

For the remainder of this chapter, in situations that do not involve functions, we will **assume that no radicands involve negative quantities raised to even powers. Based upon this assumption, absolute value bars are not necessary when taking even roots.**

Simplifying When Variables to Even Powers in a Radicand Are Nonnegative Quantities

For any nonnegative real number a,

$$\sqrt[n]{a^n} = a.$$

In simplifying an nth root, how do we find variable factors in the radicand that are perfect nth powers? **The perfect nth powers have exponents that are divisible by n.** Simplification is possible by observation or by using rational exponents. Here are some examples:

- $\sqrt{x^6} = \sqrt{(x^3)^2} = x^3$ or $\sqrt{x^6} = (x^6)^{\frac{1}{2}} = x^3$

  6 is divisible by the index, 2. Thus, x^6 is a perfect square.

- $\sqrt[3]{y^{21}} = \sqrt[3]{(y^7)^3} = y^7$ or $\sqrt[3]{y^{21}} = (y^{21})^{\frac{1}{3}} = y^7$

  21 is divisible by the index, 3. Thus, y^{21} is a perfect cube.

- $\sqrt[6]{z^{24}} = \sqrt[6]{(z^4)^6} = z^4$ or $\sqrt[6]{z^{24}} = (z^{24})^{\frac{1}{6}} = z^4$.

  24 is divisible by the index, 6. Thus, z^{24} is a perfect 6th power.

EXAMPLE 4 Simplifying a Radical by Factoring

Simplify: $\sqrt{x^5y^{13}z^7}$.

Solution We write the radicand as the product of the greatest perfect square factor and another factor. Because the index is 2, variables that have exponents that are divisible by 2 are part of the perfect square factor. We use the greatest exponents that are divisible by 2.

Discover for Yourself

Square the answer in Example 4 and show that it is correct. If it is a square root, you should obtain the given radicand, $x^5y^{13}z^7$.

$$\begin{aligned}\sqrt{x^5y^{13}z^7} &= \sqrt{x^4 \cdot x \cdot y^{12} \cdot y \cdot z^6 \cdot z} && \text{Use the greatest even power of each variable.}\\ &= \sqrt{(x^4y^{12}z^6)(xyz)} && \text{Group the perfect square factors.}\\ &= \sqrt{x^4y^{12}z^6} \cdot \sqrt{xyz} && \text{Factor into two radicals.}\\ &= x^2y^6z^3\sqrt{xyz} && \sqrt{x^4y^{12}z^6} = (x^4y^{12}z^6)^{\frac{1}{2}} = x^2y^6z^3 \quad \blacksquare\end{aligned}$$

☑ CHECK POINT 4 Simplify: $\sqrt{x^9y^{11}z^3}$.

EXAMPLE 5 Simplifying a Radical by Factoring

Simplify: $\sqrt[3]{32x^8y^{16}}$.

Solution We write the radicand as the product of the greatest perfect cube factor and another factor. Because the index is 3, variables that have exponents that are divisible by 3 are part of the perfect cube factor. We use the greatest exponents that are divisible by 3.

$$\begin{aligned}\sqrt[3]{32x^8y^{16}} &= \sqrt[3]{8 \cdot 4 \cdot x^6 \cdot x^2 \cdot y^{15} \cdot y} && \text{Identify perfect cube factors.}\\ &= \sqrt[3]{(8x^6y^{15})(4x^2y)} && \text{Group the perfect cube factors.}\\ &= \sqrt[3]{8x^6y^{15}} \cdot \sqrt[3]{4x^2y} && \text{Factor into two radicals.}\\ &= 2x^2y^5\sqrt[3]{4x^2y} && \sqrt[3]{8} = 2 \text{ and } \sqrt[3]{x^6y^{15}} = (x^6y^{15})^{\frac{1}{3}} = x^2y^5. \quad \blacksquare\end{aligned}$$

☑ CHECK POINT 5 Simplify: $\sqrt[3]{40x^{10}y^{14}}$.

EXAMPLE 6 Simplifying a Radical by Factoring

Simplify: $\sqrt[5]{64x^3y^7z^{29}}$.

Solution We write the radicand as the product of the greatest perfect 5th power and another factor. Because the index is 5, variables that have exponents that are divisible by 5 are part of the perfect fifth factor. We use the greatest exponents that are divisible by 5.

$$\begin{aligned}\sqrt[5]{64x^3y^7z^{29}} &= \sqrt[5]{32 \cdot 2 \cdot x^3 \cdot y^5 \cdot y^2 \cdot z^{25} \cdot z^4} && \text{Identify perfect fifth factors.}\\ &= \sqrt[5]{(32y^5z^{25})(2x^3y^2z^4)} && \text{Group the perfect fifth factors.}\\ &= \sqrt[5]{32y^5z^{25}} \cdot \sqrt[5]{2x^3y^2z^4} && \text{Factor into two radicals.}\\ &= 2yz^5\sqrt[5]{2x^3y^2z^4} && \sqrt[5]{32} = 2 \text{ and } \sqrt[5]{y^5z^{25}} = (y^5z^{25})^{\frac{1}{5}} = yz^5. \quad \blacksquare\end{aligned}$$

☑ CHECK POINT 6 Simplify: $\sqrt[5]{32x^{12}y^2z^8}$.

---

3 Multiply radicals and then simplify.

Multiplying and Simplifying Radicals

We have seen how to use the product rule when multiplying radicals with the same index. Sometimes after multiplying, we can simplify the resulting radical.

EXAMPLE 7 Multiplying Radicals and Then Simplifying

Multiply and simplify:

a. $\sqrt{15} \cdot \sqrt{3}$ **b.** $7\sqrt[3]{4} \cdot 5\sqrt[3]{6}$ **c.** $\sqrt[4]{8x^3y^2} \cdot \sqrt[4]{8x^5y^3}$.

Great Question!

When should I write an expression under one radical and when should I separate the radicals?

- Use $\sqrt[n]{a} \cdot \sqrt[n]{b} = \sqrt[n]{ab}$, writing under one radical, when *multiplying*.
- Use $\sqrt[n]{ab} = \sqrt[n]{a} \cdot \sqrt[n]{b}$, factoring into two radicals, when *simplifying*.

Solution

a. $\sqrt{15} \cdot \sqrt{3} = \sqrt{15 \cdot 3}$ Use the product rule.

$= \sqrt{45} = \sqrt{9 \cdot 5}$ 9 is the greatest perfect square factor of 45.

$= \sqrt{9} \cdot \sqrt{5} = 3\sqrt{5}$

b. $7\sqrt[3]{4} \cdot 5\sqrt[3]{6} = 35\sqrt[3]{4 \cdot 6}$ Use the product rule.

$= 35\sqrt[3]{24} = 35\sqrt[3]{8 \cdot 3}$ 8 is the greatest perfect cube factor of 24.

$= 35\sqrt[3]{8} \cdot \sqrt[3]{3} = 35 \cdot 2 \cdot \sqrt[3]{3}$

$= 70\sqrt[3]{3}$

c. $\sqrt[4]{8x^3y^2} \cdot \sqrt[4]{8x^5y^3} = \sqrt[4]{8x^3y^2 \cdot 8x^5y^3}$ Use the product rule.

$= \sqrt[4]{64x^8y^5}$ Multiply.

$= \sqrt[4]{16 \cdot 4 \cdot x^8 \cdot y^4 \cdot y}$ Identify perfect fourth factors.

$= \sqrt[4]{(16x^8y^4)(4y)}$ Group the perfect fourth factors.

$= \sqrt[4]{16x^8y^4} \cdot \sqrt[4]{4y}$ Factor into two radicals.

$= 2x^2y\sqrt[4]{4y}$ $\sqrt[4]{16} = 2$ and $\sqrt[4]{x^8y^4} = (x^8y^4)^{\frac{1}{4}} = x^2y$. ■

CHECK POINT 7 Multiply and simplify:

a. $\sqrt{6} \cdot \sqrt{2}$

b. $10\sqrt[3]{16} \cdot 5\sqrt[3]{2}$

c. $\sqrt[4]{4x^2y} \cdot \sqrt[4]{8x^6y^3}$.

CONCEPT AND VOCABULARY CHECK

Fill in each blank so that the resulting statement is true.

1. If $\sqrt[n]{a}$ and $\sqrt[n]{b}$ are real numbers, then $\sqrt[n]{a} \cdot \sqrt[n]{b} =$ _______.
2. $\sqrt[3]{7} \cdot \sqrt[3]{11} = \sqrt[3]{\_}$
3. $\sqrt[3]{8 \cdot 5} = \sqrt[3]{\_} \cdot \sqrt[3]{\_} =$ _______
4. $\sqrt{5(x+1)^2} =$ _______
5. Variable factors in a radicand that are perfect *n*th powers can be simplified by observation. For example, $\sqrt{x^{10}} =$ ______, $\sqrt[3]{x^{12}} =$ ______, and $\sqrt[6]{x^{30}} =$ ______.

7.3 EXERCISE SET MyMathLab®

Practice Exercises

In Exercises 1–20, use the product rule to multiply.

1. $\sqrt{3} \cdot \sqrt{5}$
2. $\sqrt{7} \cdot \sqrt{5}$
3. $\sqrt[3]{2} \cdot \sqrt[3]{9}$
4. $\sqrt[3]{5} \cdot \sqrt[3]{4}$
5. $\sqrt[4]{11} \cdot \sqrt[4]{3}$
6. $\sqrt[5]{9} \cdot \sqrt[5]{3}$
7. $\sqrt{3x} \cdot \sqrt{11y}$
8. $\sqrt{5x} \cdot \sqrt{11y}$
9. $\sqrt[5]{6x^3} \cdot \sqrt[5]{4x}$
10. $\sqrt[4]{6x^2} \cdot \sqrt[4]{3x}$
11. $\sqrt{x+3} \cdot \sqrt{x-3}$
12. $\sqrt{x+6} \cdot \sqrt{x-6}$
13. $\sqrt[6]{x-4} \cdot \sqrt[6]{(x-4)^4}$
14. $\sqrt[6]{x-5} \cdot \sqrt[6]{(x-5)^4}$
15. $\sqrt{\frac{2x}{3}} \cdot \sqrt{\frac{3}{2}}$
16. $\sqrt{\frac{2x}{5}} \cdot \sqrt{\frac{5}{2}}$

17. $\sqrt[4]{\frac{x}{7}} \cdot \sqrt[4]{\frac{3}{y}}$

18. $\sqrt[4]{\frac{x}{3}} \cdot \sqrt[4]{\frac{7}{y}}$

19. $\sqrt[7]{7x^2y} \cdot \sqrt[7]{11x^3y^2}$

20. $\sqrt[9]{12x^2y^3} \cdot \sqrt[9]{3x^3y^4}$

In Exercises 21–32, simplify by factoring.

21. $\sqrt{50}$

22. $\sqrt{27}$

23. $\sqrt{45}$

24. $\sqrt{28}$

25. $\sqrt{75x}$

26. $\sqrt{40x}$

27. $\sqrt[3]{16}$

28. $\sqrt[3]{54}$

29. $\sqrt[3]{27x^3}$

30. $\sqrt[3]{250x^3}$

31. $\sqrt[3]{-16x^2y^3}$

32. $\sqrt[3]{-32x^2y^3}$

In Exercises 33–38, express the function, f, in simplified form. Assume that x can be any real number.

33. $f(x) = \sqrt{36(x+2)^2}$

34. $f(x) = \sqrt{81(x-2)^2}$

35. $f(x) = \sqrt[3]{32(x+2)^3}$

36. $f(x) = \sqrt[3]{48(x-2)^3}$

37. $f(x) = \sqrt{3x^2 - 6x + 3}$

38. $f(x) = \sqrt{5x^2 - 10x + 5}$

In Exercises 39–60, simplify by factoring. Assume that all variables in a radicand represent positive real numbers and no radicands involve negative quantities raised to even powers.

39. $\sqrt{x^7}$

40. $\sqrt{x^5}$

41. $\sqrt{x^8y^9}$

42. $\sqrt{x^6y^7}$

43. $\sqrt{48x^3}$

44. $\sqrt{40x^3}$

45. $\sqrt[3]{y^8}$

46. $\sqrt[3]{y^{11}}$

47. $\sqrt[3]{x^{14}y^3z}$

48. $\sqrt[3]{x^3y^{17}z^2}$

49. $\sqrt[3]{81x^8y^6}$

50. $\sqrt[3]{32x^9y^{17}}$

51. $\sqrt[3]{(x+y)^5}$

52. $\sqrt[3]{(x+y)^4}$

53. $\sqrt[5]{y^{17}}$

54. $\sqrt[5]{y^{18}}$

55. $\sqrt[5]{64x^6y^{17}}$

56. $\sqrt[5]{64x^7y^{16}}$

57. $\sqrt[4]{80x^{10}}$

58. $\sqrt[4]{96x^{11}}$

59. $\sqrt[4]{(x-3)^{10}}$

60. $\sqrt[4]{(x-2)^{14}}$

In Exercises 61–82, multiply and simplify. Assume that all variables in a radicand represent positive real numbers and no radicands involve negative quantities raised to even powers.

61. $\sqrt{12} \cdot \sqrt{2}$

62. $\sqrt{3} \cdot \sqrt{6}$

63. $\sqrt{5x} \cdot \sqrt{10y}$

64. $\sqrt{8x} \cdot \sqrt{10y}$

65. $\sqrt{12x} \cdot \sqrt{3x}$

66. $\sqrt{20x} \cdot \sqrt{5x}$

67. $\sqrt{50xy} \cdot \sqrt{4xy^2}$

68. $\sqrt{5xy} \cdot \sqrt{10xy^2}$

69. $2\sqrt{5} \cdot 3\sqrt{40}$

70. $3\sqrt{15} \cdot 5\sqrt{6}$

71. $\sqrt[3]{12} \cdot \sqrt[3]{4}$

72. $\sqrt[4]{4} \cdot \sqrt[4]{8}$

73. $\sqrt{5x^3} \cdot \sqrt{8x^2}$

74. $\sqrt{2x^7} \cdot \sqrt{12x^4}$

75. $\sqrt[3]{25x^4y^2} \cdot \sqrt[3]{5xy^{12}}$

76. $\sqrt[3]{6x^7y} \cdot \sqrt[3]{9x^4y^{12}}$

77. $\sqrt[4]{8x^2y^3z^6} \cdot \sqrt[4]{2x^4yz}$

78. $\sqrt[4]{4x^2y^3z^3} \cdot \sqrt[4]{8x^4yz^6}$

79. $\sqrt[5]{8x^4y^6z^2} \cdot \sqrt[5]{8xy^7z^4}$

80. $\sqrt[5]{8x^4y^3z^3} \cdot \sqrt[5]{8xy^9z^8}$

81. $\sqrt[3]{x-y} \cdot \sqrt[3]{(x-y)^7}$

82. $\sqrt[3]{x-6} \cdot \sqrt[3]{(x-6)^7}$

Practice PLUS

In Exercises 83–90, simplify each expression. Assume that all variables in a radicand represent positive real numbers and no radicands involve negative quantities raised to even powers.

83. $-2x^2y\left(\sqrt[3]{54x^3y^7z^2}\right)$

84. $\frac{-x^2y^7}{2}\left(\sqrt[3]{-32x^4y^9z^7}\right)$

85. $-3y\left(\sqrt[5]{64x^3y^6}\right)$

86. $-4x^2y^7\left(\sqrt[5]{-32x^{11}y^{17}}\right)$

87. $\left(-2xy^2\sqrt{3x}\right)\left(xy\sqrt{6x}\right)$

88. $\left(-5x^2y^3z\sqrt{2xyz}\right)\left(-x^4z\sqrt{10xz}\right)$

89. $\left(2x^2y\sqrt[4]{8xy}\right)\left(-3xy^2\sqrt[4]{2x^2y^3}\right)$

90. $\left(5a^2b\sqrt[4]{8a^2b}\right)\left(4ab\sqrt[4]{4a^3b^2}\right)$

Application Exercises

The function

$$d(x) = \sqrt{\frac{3x}{2}}$$

models the distance, d(x), in miles, that a person x feet high can see to the horizon. Use this function to solve Exercises 91–92.

91. The pool deck on a cruise ship is 72 feet above the water. How far can passengers on the pool deck see? Write the answer in simplified radical form. Then use the simplified radical form and a calculator to express the answer to the nearest tenth of a mile.

92. The captain of a cruise ship is on the star deck, which is 120 feet above the water. How far can the captain see? Write the answer in simplified radical form. Then use the simplified radical form and a calculator to express the answer to the nearest tenth of a mile.

Paleontologists use the function

$$W(x) = 4\sqrt{2x}$$

to estimate the walking speed of a dinosaur, $W(x)$, in feet per second, where x is the length, in feet, of the dinosaur's leg. The graph of W is shown in the figure. Use this information to solve Exercises 93–94.

Dinosaur Walking Speeds

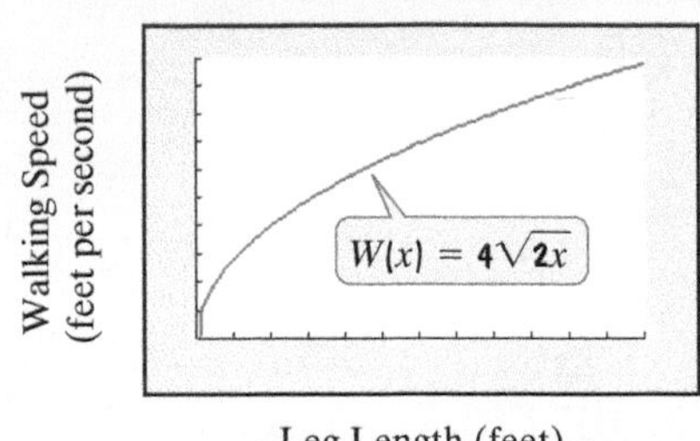

Leg Length (feet)
[0, 12, 1] by [0, 20, 2]

93. What is the walking speed of a dinosaur whose leg length is 6 feet? Use the function's equation and express the answer in simplified radical form. Then use the function's graph to estimate the answer to the nearest foot per second.

94. What is the walking speed of a dinosaur whose leg length is 10 feet? Use the function's equation and express the answer in simplified radical form. Then use the function's graph to estimate the answer to the nearest foot per second.

Your **cardiac index** *is your heart's output, in liters of blood per minute, divided by your body's surface area, in square meters. The cardiac index, $C(x)$, can be modeled by*

$$C(x) = \frac{7.644}{\sqrt[4]{x}}, \quad 10 \le x \le 80,$$

where x is an individual's age, in years. The graph of the function is shown. Use the function to solve Exercises 95–96.

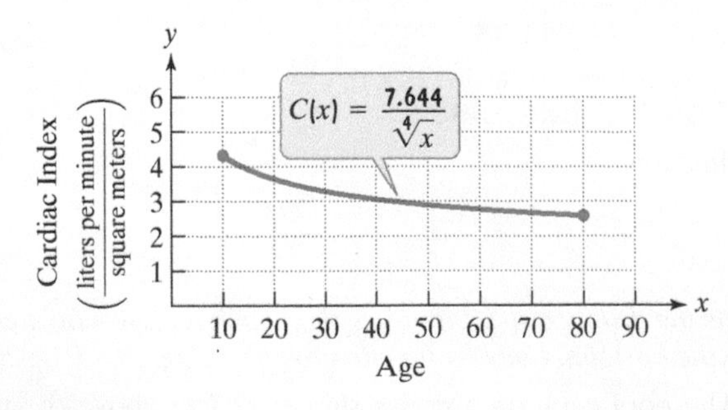

95. a. Find the cardiac index of a 32-year-old. Express the denominator in simplified radical form and reduce the fraction.

b. Use the form of the answer in part (a) and a calculator to express the cardiac index to the nearest hundredth. Identify your solution as a point on the graph.

96. a. Find the cardiac index of an 80-year-old. Express the denominator in simplified radical form and reduce the fraction.

b. Use the form of the answer in part (a) and a calculator to express the cardiac index to the nearest hundredth. Identify your solution as a point on the graph.

Explaining the Concepts

97. What is the product rule for radicals? Give an example to show how it is used.

98. Explain why $\sqrt{50}$ is not simplified. What do we mean when we say a radical expression is simplified?

99. In simplifying an nth root, explain how to find variable factors in the radicand that are perfect nth powers.

100. Without showing all the details, explain how to simplify $\sqrt[3]{16x^{14}}$.

101. As you get older, what would you expect to happen to your heart's output? Explain how this is shown in the graph for Exercises 95–96. Is this trend taking place progressively more rapidly or more slowly over the entire interval? What does this mean about this aspect of aging?

Technology Exercises

102. Use a calculator to provide numerical support for your simplifications in Exercises 21–24 and 27–28. In each case, find a decimal approximation for the given expression. Then find a decimal approximation for your simplified expression. The approximations should be the same.

In Exercises 103–106, determine if each simplification is correct by graphing the function on each side of the equation with your graphing utility. Use the given viewing rectangle. The graphs should be the same. If they are not, correct the right side of the equation and then use your graphing utility to verify the simplification.

103. $\sqrt{x^4} = x^2$; [0, 5, 1] by [0, 20, 1]

104. $\sqrt{8x^2} = 4x\sqrt{2}$; [−5, 5, 1] by [−5, 20, 1]

105. $\sqrt{3x^2 - 6x + 3} = (x - 1)\sqrt{3}$; [−5, 5, 1] by [−5, 5, 1]

106. $\sqrt[3]{2x} \cdot \sqrt[3]{4x^2} = 4x$; [−10, 10, 1] by [−10, 10, 1]

Critical Thinking Exercises

Make Sense? *In Exercises 107–110, determine whether each statement makes sense or does not make sense, and explain your reasoning.*

107. Because the product rule for radicals applies when $\sqrt[n]{a}$ and $\sqrt[n]{b}$ are real numbers, I can use it to find $\sqrt[3]{16} \cdot \sqrt[3]{-4}$, but not to find $\sqrt{8} \cdot \sqrt{-2}$.

108. I multiply nth roots by taking the nth root of the product of the radicands.

109. I need to know how to factor a trinomial to simplify $\sqrt{x^2 - 10x + 25}$.

110. I know that I've simplified a radical expression when it contains a single radical.

In Exercises 111–114, determine whether each statement is true or false. If the statement is false, make the necessary change(s) to produce a true statement.

111. $2\sqrt{5}\cdot 6\sqrt{5} = 12\sqrt{5}$

112. $\sqrt[3]{4}\cdot\sqrt[3]{4} = 4$

113. $\sqrt{12} = 2\sqrt{6}$

114. $\sqrt[3]{3^{15}} = 243$

115. If a number is tripled, what happens to its square root?

116. What must be done to a number so that its cube root is tripled?

117. If $f(x) = \sqrt[3]{2x}$ and $(fg)(x) = 2x$, find $g(x)$.

118. Graph $f(x) = \sqrt{(x-1)^2}$ by hand.

Review Exercises

119. Solve: $2x - 1 \le 21$ and $2x + 2 \ge 12$. (Section 4.2, Example 2)

120. Solve:

$$\begin{cases} 5x + 2y = 2 \\ 4x + 3y = -4. \end{cases}$$

(Section 3.1, Example 6)

121. Factor: $64x^3 - 27$. (Section 5.5, Example 9)

Preview Exercises

Exercises 122–124 will help you prepare for the material covered in the next section.

122. a. Simplify: $21x + 10x$.

b. Simplify: $21\sqrt{2} + 10\sqrt{2}$.

123. a. Simplify: $4x - 12x$.

b. Simplify: $4\sqrt[3]{2} - 12\sqrt[3]{2}$.

124. Simplify: $\dfrac{\sqrt[4]{7y^5}}{\sqrt[4]{x^{12}}}$.

SECTION

7.4 Adding, Subtracting, and Dividing Radical Expressions

What am I supposed to learn?

After studying this section, you should be able to:

1. Add and subtract radical expressions.
2. Use the quotient rule to simplify radical expressions.
3. Use the quotient rule to divide radical expressions.

The future is now: You have the opportunity to explore the cosmos in a starship traveling near the speed of light. The experience will enable you to understand the mysteries of the universe firsthand, transporting you to unimagined levels of knowing and being. The downside: According to Einstein's theory of relativity, close to the speed of light, your aging rate relative to friends on Earth is nearly zero. You will return from your two-year journey to an unknown futuristic world. In this section's Exercise Set (Exercise 86), we provide an expression involving radical division that models your return to this unrecognizable world. To make sense of the model, we turn to various operations with radicals, including addition, subtraction, and division.

1 Add and subtract radical expressions.

Adding and Subtracting Radical Expressions

We know that like terms have exactly the same variable factors and can be combined. For example,

$$7x + 6x = (7 + 6)x = 13x.$$

Two or more radical expressions that have the same indices *and* the same radicands are called **like radicals**. Like radicals are combined in exactly the same way that we combine like terms. For example,

$$7\sqrt{11} + 6\sqrt{11} = (7 + 6)\sqrt{11} = 13\sqrt{11}.$$

7 square roots of 11 plus 6 square roots of 11 result in 13 square roots of 11.

EXAMPLE 1 Adding and Subtracting Like Radicals

Simplify (add or subtract) by combining like radical terms:

a. $7\sqrt{2} + 8\sqrt{2}$

b. $\sqrt[3]{5} - 4x\sqrt[3]{5} + 8\sqrt[3]{5}$

c. $8\sqrt[6]{5x} - 5\sqrt[6]{5x} + 4\sqrt[3]{5x}$.

Solution

a. $7\sqrt{2} + 8\sqrt{2}$

$= (7 + 8)\sqrt{2}$ Apply the distributive property.

$= 15\sqrt{2}$ Simplify.

b. $\sqrt[3]{5} - 4x\sqrt[3]{5} + 8\sqrt[3]{5}$

$= (1 - 4x + 8)\sqrt[3]{5}$ Apply the distributive property.

$= (9 - 4x)\sqrt[3]{5}$ Simplify.

c. $8\sqrt[6]{5x} - 5\sqrt[6]{5x} + 4\sqrt[3]{5x}$

$= (8 - 5)\sqrt[6]{5x} + 4\sqrt[3]{5x}$ Apply the distributive property to the two terms with like radicals.

$= 3\sqrt[6]{5x} + 4\sqrt[3]{5x}$ The indices, 6 and 3, differ. These are not like radicals and cannot be combined. ■

✓ CHECK POINT 1 Simplify by combining like radical terms:

a. $8\sqrt{13} + 2\sqrt{13}$

b. $9\sqrt[3]{7} - 6x\sqrt[3]{7} + 12\sqrt[3]{7}$

c. $7\sqrt[4]{3x} - 2\sqrt[4]{3x} + 2\sqrt[3]{3x}$.

In some cases, radical expressions can be combined once they have been simplified. For example, to add $\sqrt{2}$ and $\sqrt{8}$, we can write $\sqrt{8}$ as $\sqrt{4 \cdot 2}$ because 4 is a perfect square factor of 8.

$$\sqrt{2} + \sqrt{8} = \sqrt{2} + \sqrt{4 \cdot 2} = 1\sqrt{2} + 2\sqrt{2} = (1 + 2)\sqrt{2} = 3\sqrt{2}$$

Always begin by simplifying radical terms. This makes it possible to identify and combine any like radicals.

EXAMPLE 2 Combining Radicals That First Require Simplification

Simplify by combining like radical terms, if possible:

a. $7\sqrt{18} + 5\sqrt{8}$ **b.** $4\sqrt{27x} - 8\sqrt{12x}$ **c.** $7\sqrt{3} - 2\sqrt{5}$.

Solution

a. $7\sqrt{18} + 5\sqrt{8}$

$= 7\sqrt{9\cdot 2} + 5\sqrt{4\cdot 2}$ Factor the radicands using the greatest perfect square factors.

$= 7\sqrt{9}\cdot\sqrt{2} + 5\sqrt{4}\cdot\sqrt{2}$ Take the square root of each factor.

$= 7\cdot 3\cdot\sqrt{2} + 5\cdot 2\cdot\sqrt{2}$ $\sqrt{9} = 3$ and $\sqrt{4} = 2$.

$= 21\sqrt{2} + 10\sqrt{2}$ Multiply.

$= (21 + 10)\sqrt{2}$ Apply the distributive property.

$= 31\sqrt{2}$ Simplify.

b. $4\sqrt{27x} - 8\sqrt{12x}$

$= 4\sqrt{9\cdot 3x} - 8\sqrt{4\cdot 3x}$ Factor the radicands using the greatest perfect square factors.

$= 4\sqrt{9}\cdot\sqrt{3x} - 8\sqrt{4}\cdot\sqrt{3x}$ Take the square root of each factor.

$= 4\cdot 3\cdot\sqrt{3x} - 8\cdot 2\cdot\sqrt{3x}$ $\sqrt{9} = 3$ and $\sqrt{4} = 2$.

$= 12\sqrt{3x} - 16\sqrt{3x}$ Multiply.

$= (12 - 16)\sqrt{3x}$ Apply the distributive property.

$= -4\sqrt{3x}$ Simplify.

c. $7\sqrt{3} - 2\sqrt{5}$ cannot be simplified. The radical expressions have different radicands, namely 3 and 5, and are not like terms. ■

✓ **CHECK POINT 2** Simplify by combining like radical terms, if possible:

a. $3\sqrt{20} + 5\sqrt{45}$

b. $3\sqrt{12x} - 6\sqrt{27x}$

c. $8\sqrt{5} - 6\sqrt{2}$.

EXAMPLE 3 Adding and Subtracting with Higher Indices

Simplify by combining like radical terms, if possible:

a. $2\sqrt[3]{16} - 4\sqrt[3]{54}$ **b.** $5\sqrt[3]{xy^2} + \sqrt[3]{8x^4y^5}$.

Solution

a. $2\sqrt[3]{16} - 4\sqrt[3]{54}$

$= 2\sqrt[3]{8\cdot 2} - 4\sqrt[3]{27\cdot 2}$ Factor the radicands using the greatest perfect cube factors.

$= 2\sqrt[3]{8}\cdot\sqrt[3]{2} - 4\sqrt[3]{27}\cdot\sqrt[3]{2}$ Take the cube root of each factor.

$= 2\cdot 2\cdot\sqrt[3]{2} - 4\cdot 3\cdot\sqrt[3]{2}$ $\sqrt[3]{8} = 2$ and $\sqrt[3]{27} = 3$.

$= 4\sqrt[3]{2} - 12\sqrt[3]{2}$ Multiply.

$= (4 - 12)\sqrt[3]{2}$ Apply the distributive property.

$= -8\sqrt[3]{2}$ Simplify.

b. $5\sqrt[3]{xy^2} + \sqrt[3]{8x^4y^5}$

$= 5\sqrt[3]{xy^2} + \sqrt[3]{(8x^3y^3)xy^2}$ Factor the second radicand using the greatest perfect cube factor.

$= 5\sqrt[3]{xy^2} + \sqrt[3]{8x^3y^3} \cdot \sqrt[3]{xy^2}$ Take the cube root of each factor.

$= 5\sqrt[3]{xy^2} + 2xy\sqrt[3]{xy^2}$ $\sqrt[3]{8} = 2$ and $\sqrt[3]{x^3y^3} = (x^3y^3)^{\frac{1}{3}} = xy$.

$= (5 + 2xy)\sqrt[3]{xy^2}$ Apply the distributive property. ■

CHECK POINT 3 Simplify by combining like radical terms, if possible:

a. $3\sqrt[3]{24} - 5\sqrt[3]{81}$

b. $5\sqrt[3]{x^2y} + \sqrt[3]{27x^5y^4}$.

Dividing Radical Expressions

We have seen that the root of a product is the product of the roots. The root of a quotient can also be expressed as the quotient of roots. Here is an example:

$$\sqrt{\frac{64}{4}} = \sqrt{16} = 4 \quad \text{and} \quad \frac{\sqrt{64}}{\sqrt{4}} = \frac{8}{2} = 4.$$

This expression is the square root of a quotient.

This expression is the quotient of two square roots.

The two procedures produce the same result, 4. This is a special case of the **quotient rule for radicals.**

2 Use the quotient rule to simplify radical expressions.

The Quotient Rule for Radicals

If $\sqrt[n]{a}$ and $\sqrt[n]{b}$ are real numbers and $b \neq 0$, then

$$\sqrt[n]{\frac{a}{b}} = \frac{\sqrt[n]{a}}{\sqrt[n]{b}}.$$

The *nth* root of a quotient is the quotient of the *nth* roots of the numerator and denominator.

We know that a radical is simplified when its radicand has no factors other than 1 that are perfect *nth* powers. The quotient rule can be used to simplify some radicals. Keep in mind that all variables in radicands represent positive real numbers.

EXAMPLE 4 Using the Quotient Rule to Simplify Radicals

Simplify using the quotient rule:

a. $\sqrt[3]{\frac{16}{27}}$ **b.** $\sqrt{\frac{x^2}{25y^6}}$ **c.** $\sqrt[4]{\frac{7y^5}{x^{12}}}$.

Solution We simplify each expression by taking the roots of the numerator and the denominator. Then we use factoring to simplify the resulting radicals, if possible.

a. $\sqrt[3]{\frac{16}{27}} = \frac{\sqrt[3]{16}}{\sqrt[3]{27}} = \frac{\sqrt[3]{8 \cdot 2}}{3} = \frac{\sqrt[3]{8} \cdot \sqrt[3]{2}}{3} = \frac{2\sqrt[3]{2}}{3}$

b. $\sqrt{\frac{x^2}{25y^6}} = \frac{\sqrt{x^2}}{\sqrt{25y^6}} = \frac{x}{5(y^6)^{\frac{1}{2}}} = \frac{x}{5y^3}$

Try to do this step mentally.

c. $\sqrt[4]{\frac{7y^5}{x^{12}}} = \frac{\sqrt[4]{7y^5}}{\sqrt[4]{x^{12}}} = \frac{\sqrt[4]{y^4 \cdot 7y}}{\sqrt[4]{x^{12}}} = \frac{y\sqrt[4]{7y}}{x^3}$ ■

CHECK POINT 4 Simplify using the quotient rule:

a. $\sqrt[3]{\frac{24}{125}}$ b. $\sqrt{\frac{9x^3}{y^{10}}}$ c. $\sqrt[3]{\frac{8y^7}{x^{12}}}$.

By reversing the two sides of the quotient rule, we obtain a procedure for dividing radical expressions.

3 Use the quotient rule to divide radical expressions.

Dividing Radical Expressions

If $\sqrt[n]{a}$ and $\sqrt[n]{b}$ are real numbers and $b \neq 0$, then

$$\frac{\sqrt[n]{a}}{\sqrt[n]{b}} = \sqrt[n]{\frac{a}{b}}.$$

To divide two radical expressions with the same index, divide the radicands and retain the common index.

Great Question!

When should I write a quotient under a single radical and when should I use separate radicals for a quotient's numerator and denominator?

- Use
$$\frac{\sqrt[n]{a}}{\sqrt[n]{b}} = \sqrt[n]{\frac{a}{b}},$$
writing under one radical (as in Example 5), when *dividing*.
- Use
$$\sqrt[n]{\frac{a}{b}} = \frac{\sqrt[n]{a}}{\sqrt[n]{b}},$$
with separate radicals for the numerator and the denominator (as in Example 4), when *simplifying* a quotient.

EXAMPLE 5 Dividing Radical Expressions

Divide and, if possible, simplify:

a. $\frac{\sqrt{48x^3}}{\sqrt{6x}}$ b. $\frac{\sqrt{45xy}}{2\sqrt{5}}$ c. $\frac{\sqrt[3]{16x^5y^2}}{\sqrt[3]{2xy^{-1}}}$.

Solution In each part of this problem, the indices in the numerator and the denominator are the same. Perform each division by dividing the radicands and retaining the common index.

a. $\frac{\sqrt{48x^3}}{\sqrt{6x}} = \sqrt{\frac{48x^3}{6x}} = \sqrt{8x^2} = \sqrt{4x^2 \cdot 2} = \sqrt{4x^2} \cdot \sqrt{2} = 2x\sqrt{2}$

b. $\frac{\sqrt{45xy}}{2\sqrt{5}} = \frac{1}{2} \cdot \sqrt{\frac{45xy}{5}} = \frac{1}{2} \cdot \sqrt{9xy} = \frac{1}{2} \cdot 3\sqrt{xy}$ or $\frac{3\sqrt{xy}}{2}$

c. $\frac{\sqrt[3]{16x^5y^2}}{\sqrt[3]{2xy^{-1}}} = \sqrt[3]{\frac{16x^5y^2}{2xy^{-1}}}$ Divide the radicands and retain the common index.

$= \sqrt[3]{8x^{5-1}y^{2-(-1)}}$ Divide factors in the radicand. Subtract exponents on common bases.

$= \sqrt[3]{8x^4y^3}$ Simplify.

$= \sqrt[3]{(8x^3y^3)x}$ Factor using the greatest perfect cube factor.

$= \sqrt[3]{8x^3y^3} \cdot \sqrt[3]{x}$ Factor into two radicals.

$= 2xy\sqrt[3]{x}$ Simplify. ■

CHECK POINT 5 Divide and, if possible, simplify:

a. $\frac{\sqrt{40x^5}}{\sqrt{2x}}$ b. $\frac{\sqrt{50xy}}{2\sqrt{2}}$ c. $\frac{\sqrt[3]{48x^7y}}{\sqrt[3]{6xy^{-2}}}$.

CONCEPT AND VOCABULARY CHECK

Fill in each blank so that the resulting statement is true.

1. $5\sqrt{3} + 8\sqrt{3} = (\_ + \_)\sqrt{3} = \_\_\_\_$
2. $\sqrt{27} - \sqrt{12} = \sqrt{\_\cdot 3} - \sqrt{\_\cdot 3} = \_\sqrt{3} - \_\sqrt{3} = \_$
3. $\sqrt[3]{54} + \sqrt[3]{16} = \sqrt[3]{\_\cdot 2} + \sqrt[3]{\_\cdot 2} = \_\sqrt[3]{2} + \_\sqrt[3]{2} = \_\_\_\_$
4. If $\sqrt[n]{a}$ and $\sqrt[n]{b}$ are real numbers and $b \neq 0$, the quotient rule for radicals states that
$$\sqrt[n]{\frac{a}{b}} = \_\_\_.$$
5. $\sqrt[3]{\frac{8}{27}} = \frac{\sqrt[3]{\_}}{\sqrt[3]{\_}} = \frac{\_}{\_}$
6. If $x > 0$,
$$\frac{\sqrt{72x^3}}{\sqrt{2x}} = \sqrt{\frac{\_\_}{\_\_}} = \sqrt{\_\_\_} = \_.$$

7.4 EXERCISE SET

MyMathLab®

Practice Exercises

In this Exercise Set, assume that all variables represent positive real numbers.

In Exercises 1–10, add or subtract as indicated.

1. $8\sqrt{5} + 3\sqrt{5}$
2. $7\sqrt{3} + 2\sqrt{3}$
3. $9\sqrt[3]{6} - 2\sqrt[3]{6}$
4. $9\sqrt[3]{7} - 4\sqrt[3]{7}$
5. $4\sqrt[5]{2} + 3\sqrt[5]{2} - 5\sqrt[5]{2}$
6. $6\sqrt[5]{3} + 2\sqrt[5]{3} - 3\sqrt[5]{3}$
7. $3\sqrt{13} - 2\sqrt{5} - 2\sqrt{13} + 4\sqrt{5}$
8. $8\sqrt{17} - 5\sqrt{19} - 6\sqrt{17} + 4\sqrt{19}$
9. $3\sqrt{5} - \sqrt[3]{x} + 4\sqrt{5} + 3\sqrt[3]{x}$
10. $6\sqrt{7} - \sqrt[3]{x} + 2\sqrt{7} + 5\sqrt[3]{x}$

In Exercises 11–28, add or subtract as indicated. You will need to simplify terms to identify the like radicals.

11. $\sqrt{3} + \sqrt{27}$
12. $\sqrt{5} + \sqrt{20}$
13. $7\sqrt{12} + \sqrt{75}$
14. $5\sqrt{12} + \sqrt{75}$
15. $3\sqrt{32x} - 2\sqrt{18x}$
16. $5\sqrt{45x} - 2\sqrt{20x}$
17. $5\sqrt[3]{16} + \sqrt[3]{54}$
18. $3\sqrt[3]{24} + \sqrt[3]{81}$
19. $3\sqrt{45x^3} + \sqrt{5x}$
20. $8\sqrt{45x^3} + \sqrt{5x}$
21. $\sqrt[3]{54xy^3} + y\sqrt[3]{128x}$
22. $\sqrt[3]{24xy^3} + y\sqrt[3]{81x}$
23. $\sqrt[3]{54x^4} - \sqrt[3]{16x}$
24. $\sqrt[3]{81x^4} - \sqrt[3]{24x}$
25. $\sqrt{9x - 18} + \sqrt{x - 2}$
26. $\sqrt{4x - 12} + \sqrt{x - 3}$
27. $2\sqrt[3]{x^4y^2} + 3x\sqrt[3]{xy^2}$
28. $4\sqrt[3]{x^4y^2} + 5x\sqrt[3]{xy^2}$

In Exercises 29–44, simplify using the quotient rule.

29. $\sqrt{\frac{11}{4}}$
30. $\sqrt{\frac{19}{25}}$
31. $\sqrt[3]{\frac{19}{27}}$
32. $\sqrt[3]{\frac{11}{64}}$
33. $\sqrt{\frac{x^2}{36y^8}}$
34. $\sqrt{\frac{x^2}{144y^{12}}}$
35. $\sqrt{\frac{8x^3}{25y^6}}$
36. $\sqrt{\frac{50x^3}{81y^8}}$
37. $\sqrt[3]{\frac{x^4}{8y^3}}$
38. $\sqrt[3]{\frac{x^5}{125y^3}}$
39. $\sqrt[3]{\frac{50x^8}{27y^{12}}}$
40. $\sqrt[3]{\frac{81x^8}{8y^{15}}}$
41. $\sqrt[4]{\frac{9y^6}{x^8}}$
42. $\sqrt[4]{\frac{13y^7}{x^{12}}}$
43. $\sqrt[5]{\frac{64x^{13}}{y^{20}}}$
44. $\sqrt[5]{\frac{64x^{14}}{y^{15}}}$

In Exercises 45–66, divide and, if possible, simplify.

45. $\dfrac{\sqrt{40}}{\sqrt{5}}$ **46.** $\dfrac{\sqrt{200}}{\sqrt{10}}$

47. $\dfrac{\sqrt[3]{48}}{\sqrt[3]{6}}$ **48.** $\dfrac{\sqrt[3]{54}}{\sqrt[3]{2}}$

49. $\dfrac{\sqrt{54x^3}}{\sqrt{6x}}$ **50.** $\dfrac{\sqrt{72x^3}}{\sqrt{2x}}$

51. $\dfrac{\sqrt{x^5y^3}}{\sqrt{xy}}$ **52.** $\dfrac{\sqrt{x^7y^6}}{\sqrt{x^3y^2}}$

53. $\dfrac{\sqrt{200x^3}}{\sqrt{10x^{-1}}}$ **54.** $\dfrac{\sqrt{500x^3}}{\sqrt{10x^{-1}}}$

55. $\dfrac{\sqrt{48a^8b^7}}{\sqrt{3a^{-2}b^{-3}}}$ **56.** $\dfrac{\sqrt{54a^7b^{11}}}{\sqrt{3a^{-4}b^{-2}}}$

57. $\dfrac{\sqrt{72xy}}{2\sqrt{2}}$ **58.** $\dfrac{\sqrt{50xy}}{2\sqrt{2}}$

59. $\dfrac{\sqrt[3]{24x^3y^5}}{\sqrt[3]{3y^2}}$ **60.** $\dfrac{\sqrt[3]{250x^5y^3}}{\sqrt[3]{2x^3}}$

61. $\dfrac{\sqrt[4]{32x^{10}y^8}}{\sqrt[4]{2x^2y^{-2}}}$

62. $\dfrac{\sqrt[5]{96x^{12}y^{11}}}{\sqrt[5]{3x^2y^{-2}}}$ **63.** $\dfrac{\sqrt[3]{x^2+5x+6}}{\sqrt[3]{x+2}}$

64. $\dfrac{\sqrt[3]{x^2+7x+12}}{\sqrt[3]{x+3}}$

65. $\dfrac{\sqrt[3]{a^3+b^3}}{\sqrt[3]{a+b}}$

66. $\dfrac{\sqrt[3]{a^3-b^3}}{\sqrt[3]{a-b}}$

Practice PLUS

In Exercises 67–76, perform the indicated operations.

67. $\dfrac{\sqrt{32}}{5}+\dfrac{\sqrt{18}}{7}$ **68.** $\dfrac{\sqrt{27}}{2}+\dfrac{\sqrt{75}}{7}$

69. $3x\sqrt{8xy^2}-5y\sqrt{32x^3}+\sqrt{18x^3y^2}$

70. $6x\sqrt{3xy^2}-4x^2\sqrt{27xy}-5\sqrt{75x^5y}$

71. $5\sqrt{2x^3}+\dfrac{30x^3\sqrt{24x^2}}{3x^2\sqrt{3x}}$

72. $7\sqrt{2x^3}+\dfrac{40x^3\sqrt{150x^2}}{5x^2\sqrt{3x}}$

73. $2x\sqrt{75xy}-\dfrac{\sqrt{81xy^2}}{\sqrt{3x^{-2}y}}$

74. $5\sqrt{8x^2y^3}-\dfrac{9x^2\sqrt{64y}}{3x\sqrt{2y^{-2}}}$

75. $\dfrac{15x^4\sqrt[3]{80x^3y^2}}{5x^3\sqrt[3]{2x^2y}}-\dfrac{75\sqrt[3]{5x^3y}}{25\sqrt[3]{x^{-1}}}$

76. $\dfrac{16x^4\sqrt[3]{48x^3y^2}}{8x^3\sqrt[3]{3x^2y}}-\dfrac{20\sqrt[3]{2x^3y}}{4\sqrt[3]{x^{-1}}}$

In Exercises 77–80, find $\left(\frac{f}{g}\right)(x)$ and the domain of $\frac{f}{g}$. Express each quotient function in simplified form.

77. $f(x)=\sqrt{48x^5},\ g(x)=\sqrt{3x^2}$

78. $f(x)=\sqrt{x^2-25},\ g(x)=\sqrt{x+5}$

79. $f(x)=\sqrt[3]{32x^6},\ g(x)=\sqrt[3]{2x^2}$

80. $f(x)=\sqrt[3]{2x^6},\ g(x)=\sqrt[3]{16x}$

Application Exercises

Exercises 81–84 involve the perimeter and area of various geometric figures. Refer to ***Table 1.6*** *on page 60 if you've forgotten any of the formulas.*

In Exercises 81–82, find the perimeter and area of each rectangle. Express answers in simplified radical form.

81.

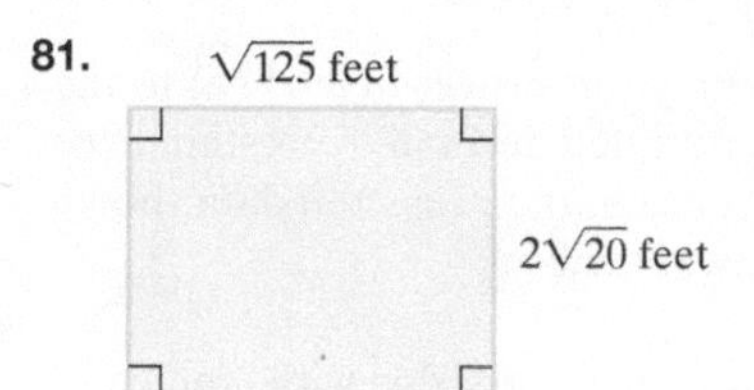

82.

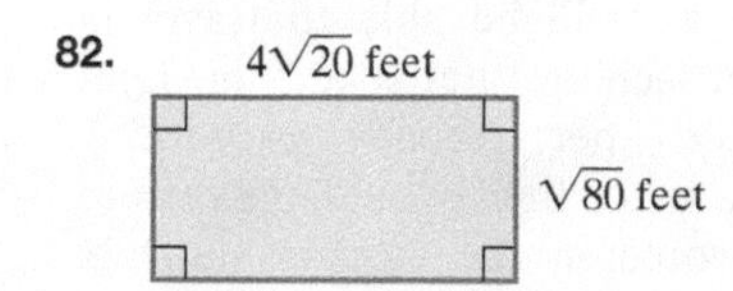

83. Find the perimeter of the triangle in simplified radical form.

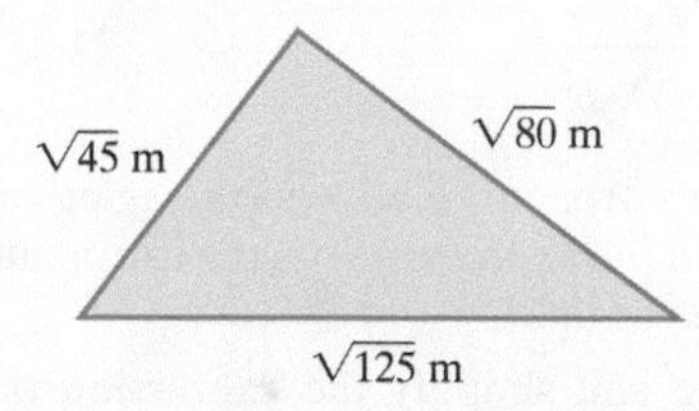

84. Find the area of the trapezoid in simplified radical form.

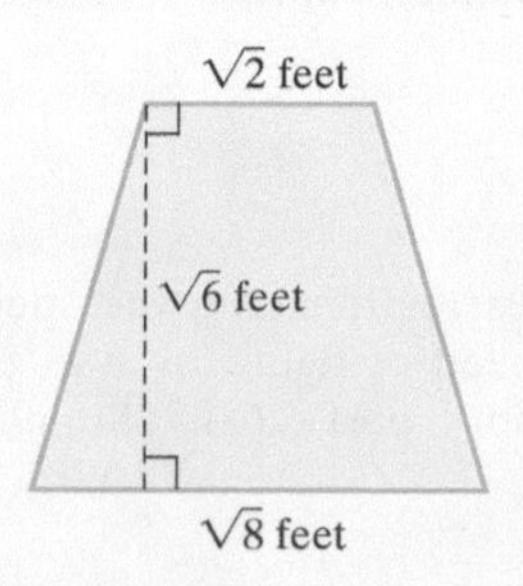

85. America is getting older. The graph shows the elderly U.S. population for ages 65–84 and for ages 85 and older in 2010, with projections for 2020 and beyond.

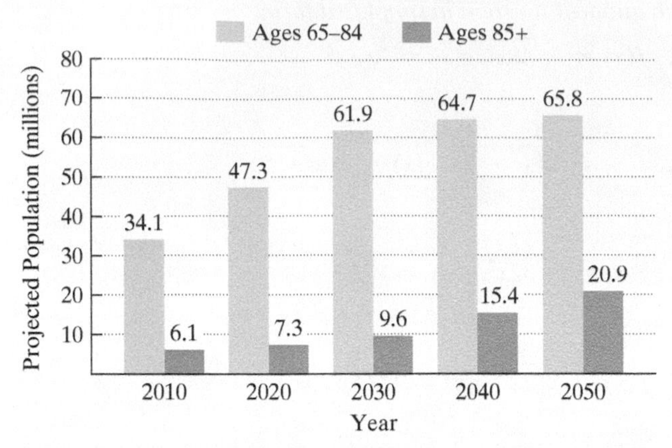

Source: U.S. Census Bureau

The function $f(x) = 5\sqrt{x} + 34.1$ models the projected number of Americans ages 65–84, $f(x)$, in millions, x years after 2010.

a. Use the function to find $f(40) - f(10)$. Express this difference in simplified radical form. What does this simplified radical represent?

b. Use a calculator and write your answer in part (a) to the nearest tenth. Does this rounded decimal overestimate or underestimate the difference in the projected data shown by the bar graph? By how much?

86. What does travel in space have to do with radicals? Imagine that in the future we will be able to travel in starships at velocities approaching the speed of light (approximately 186,000 miles per second). According to Einstein's theory of relativity, time would pass more quickly on Earth than it would in the moving starship. The radical expression

$$R_f \frac{\sqrt{c^2 - v^2}}{\sqrt{c^2}}$$

gives the aging rate of an astronaut relative to the aging rate of a friend, R_f, on Earth. In the expression, v is the astronaut's velocity and c is the speed of light.

a. Use the quotient rule and simplify the expression that shows your aging rate relative to a friend on Earth. Working in a step-by-step manner, express your aging rate as

$$R_f\sqrt{1 - \left(\frac{v}{c}\right)^2}.$$

b. You are moving at velocities approaching the speed of light. Substitute c, the speed of light, for v in the simplified expression from part (a). Simplify completely. Close to the speed of light, what is your aging rate relative to a friend on Earth? What does this mean?

Explaining the Concepts

87. What are like radicals? Give an example with your explanation.

88. Explain how to add like radicals. Give an example with your explanation.

89. If only like radicals can be combined, why is it possible to add $\sqrt{2}$ and $\sqrt{8}$?

90. Explain how to simplify a radical expression using the quotient rule. Provide an example.

91. Explain how to divide radical expressions with the same index.

92. In Exercise 85, use the data displayed by the bar graph to explain why we used a square root function to model the projected population for the 65–84 age group, but not for the 85+ group.

Technology Exercises

93. Use a calculator to provide numerical support for any four exercises that you worked from Exercises 1–66 that do not contain variables. Begin by finding a decimal approximation for the given expression. Then find a decimal approximation for your answer. The two decimal approximations should be the same.

In Exercises 94–96, determine if each operation is performed correctly by graphing the function on each side of the equation with your graphing utility. Use the given viewing rectangle. The graphs should be the same. If they are not, correct the right side of the equation and then use your graphing utility to verify the correction.

94. $\sqrt{4x} + \sqrt{9x} = 5\sqrt{x}$

$[0, 5, 1]$ by $[0, 10, 1]$

95. $\sqrt{16x} - \sqrt{9x} = \sqrt{7x}$

$[0, 5, 1]$ by $[0, 5, 1]$

96. $x\sqrt{8} + x\sqrt{2} = x\sqrt{10}$

$[-5, 5, 1]$ by $[-15, 15, 1]$

Critical Thinking Exercises

Make Sense? *In Exercises 97–100, determine whether each statement makes sense or does not make sense, and explain your reasoning.*

97. I divide nth roots by taking the nth root of the quotient of the radicands.

98. The unlike radicals $3\sqrt{2}$ and $5\sqrt{3}$ remind me of the unlike terms $3x$ and $5y$ that cannot be combined by addition or subtraction.

99. I simplified the terms of $3\sqrt[3]{81} + 2\sqrt[3]{54}$, and then I was able to identify and add the like radicals.

100. Without using a calculator, it's easier for me to estimate the decimal value of $\sqrt{72} + \sqrt{32} + \sqrt{18}$ by first simplifying.

In Exercises 101–104, determine whether each statement is true or false. If the statement is false, make the necessary change(s) to produce a true statement.

101. $\sqrt{5} + \sqrt{5} = \sqrt{10}$

102. $4\sqrt{3} + 5\sqrt{3} = 9\sqrt{6}$

103. If any two radical expressions are completely simplified, they can then be combined through addition or subtraction.

104. $\dfrac{\sqrt{-8}}{\sqrt{2}} = \sqrt{\dfrac{-8}{2}} = \sqrt{-4} = -2$

105. If an irrational number is decreased by $2\sqrt{18} - \sqrt{50}$, the result is $\sqrt{2}$. What is the number?

106. Simplify: $\dfrac{\sqrt{20}}{3} + \dfrac{\sqrt{45}}{4} - \sqrt{80}$.

107. Simplify: $\dfrac{6\sqrt{49xy}\sqrt{ab^2}}{7\sqrt{36x^{-3}y^{-5}}\sqrt{a^{-9}b^{-1}}}$.

Review Exercises

108. Solve: $2(3x - 1) - 4 = 2x - (6 - x)$.
(Section 1.4, Example 3)

109. Factor: $x^2 - 8xy + 12y^2$.
(Section 5.4, Example 4)

110. Add: $\dfrac{2}{x^2 + 5x + 6} + \dfrac{3x}{x^2 + 6x + 9}$.
(Section 6.2, Example 6)

Preview Exercises

Exercises 111–113 will help you prepare for the material covered in the next section.

111. a. Multiply: $7(x + 5)$.
b. Multiply: $\sqrt{7}(x + \sqrt{5})$.

112. a. Multiply: $(x + 5)(6x + 3)$.
b. Multiply: $(\sqrt{2} + 5)(6\sqrt{2} + 3)$.

113. Multiply and simplify:
$$\frac{10y}{\sqrt[5]{4x^3y}} \cdot \frac{\sqrt[5]{8x^2y^4}}{\sqrt[5]{8x^2y^4}}.$$

MID-CHAPTER CHECK POINT Section 7.1–Section 7.4

What You Know: We learned to find roots of numbers. We saw that the domain of a square root function is the set of real numbers for which the radicand is nonnegative. We learned to simplify radical expressions, using $\sqrt[n]{a^n} = |a|$ if n is even and $\sqrt[n]{a^n} = a$ if n is odd. The definition $a^{\frac{m}{n}} = (\sqrt[n]{a})^m = \sqrt[n]{a^m}$ connected rational exponents and radicals. Finally, we performed various operations with radicals, including multiplication, addition, subtraction, and division.

In Exercises 1–23, simplify the given expression or perform the indicated operation(s) and, if possible, simplify. Assume that all variables represent positive real numbers.

1. $\sqrt{100} - \sqrt[3]{-27}$
2. $\sqrt{8x^5y^7}$
3. $3\sqrt[3]{4x^2} + 2\sqrt[3]{4x^2}$
4. $(3\sqrt[3]{4x^2})(2\sqrt[3]{4x^2})$
5. $27^{\frac{2}{3}} + (-32)^{\frac{3}{5}}$
6. $(64x^3y^{\frac{1}{4}})^{\frac{1}{3}}$
7. $5\sqrt{27} - 4\sqrt{48}$
8. $\sqrt{\dfrac{500x^3}{4y^4}}$
9. $\dfrac{x}{\sqrt[4]{x}}$
10. $\sqrt[3]{54x^5}$
11. $\dfrac{\sqrt[3]{160}}{\sqrt[3]{2}}$
12. $\sqrt[5]{\dfrac{x^{10}}{y^{20}}}$
13. $\dfrac{(x^{\frac{2}{3}})^2}{(x^{\frac{1}{4}})^3}$
14. $\sqrt[6]{x^6y^4}$
15. $\sqrt[7]{(x-2)^3} \cdot \sqrt[7]{(x-2)^6}$
16. $\sqrt[4]{32x^{11}y^{17}}$
17. $4\sqrt[3]{16} + 2\sqrt[3]{54}$
18. $\dfrac{\sqrt[7]{x^4y^9}}{\sqrt[7]{x^{-5}y^7}}$
19. $(-125)^{-\frac{2}{3}}$
20. $\sqrt{2} \cdot \sqrt[3]{2}$
21. $\sqrt[3]{\dfrac{32x}{y^4}} \cdot \sqrt[3]{\dfrac{2x^2}{y^2}}$
22. $\sqrt{32xy^2} \cdot \sqrt{2x^3y^5}$
23. $4x\sqrt{6x^4y^3} - 7y\sqrt{24x^6y}$

In Exercises 24–25, find the domain of each function.

24. $f(x) = \sqrt{30 - 5x}$
25. $g(x) = \sqrt[3]{3x - 15}$

SECTION

7.5 Multiplying with More Than One Term and Rationalizing Denominators

What am I supposed to learn?

After studying this section, you should be able to:

1. Multiply radical expressions with more than one term.
2. Use polynomial special products to multiply radicals.
3. Rationalize denominators containing one term.
4. Rationalize denominators containing two terms.
5. Rationalize numerators.

The late Charles Schulz, creator of the "Peanuts" comic strip, transfixed 350 million readers worldwide with the joys and angst of his hapless Charlie Brown and Snoopy, a romantic self-deluded beagle. In 18,250 comic strips that spanned nearly 50 years, mathematics was often featured. Is the discussion of radicals shown on the left the real thing, or is it just an algebraic scam? By the time you complete this section on multiplying and dividing radicals, you will be able to decide.

1 Multiply radical expressions with more than one term.

Multiplying Radical Expressions with More Than One Term

Radical expressions with more than one term are multiplied in much the same way that polynomials with more than one term are multiplied. Example 1 uses the distributive property and the FOIL method to perform multiplications.

EXAMPLE 1 Multiplying Radicals

Multiply:

a. $\sqrt{7}(x+\sqrt{2})$ **b.** $\sqrt[3]{x}(\sqrt[3]{6}-\sqrt[3]{x^2})$ **c.** $(5\sqrt{2}+2\sqrt{3})(4\sqrt{2}-3\sqrt{3})$.

Solution

a. $\sqrt{7}(x+\sqrt{2})$

$= \sqrt{7}\cdot x + \sqrt{7}\cdot\sqrt{2}$ Use the distributive property.

$= x\sqrt{7} + \sqrt{14}$ Multiply the radicals.

b. $\sqrt[3]{x}(\sqrt[3]{6}-\sqrt[3]{x^2})$

$= \sqrt[3]{x}\cdot\sqrt[3]{6} - \sqrt[3]{x}\cdot\sqrt[3]{x^2}$ Use the distributive property.

$= \sqrt[3]{6x} - \sqrt[3]{x^3}$ Multiply the radicals: $\sqrt[n]{a}\cdot\sqrt[n]{b} = \sqrt[n]{ab}$.

$= \sqrt[3]{6x} - x$ Simplify: $\sqrt[3]{x^3} = x$.

first, last, inside, outside

c. $(5\sqrt{2}+2\sqrt{3})(4\sqrt{2}-3\sqrt{3})$ Use the FOIL method.

F O I L

$= (5\sqrt{2})(4\sqrt{2}) + (5\sqrt{2})(-3\sqrt{3}) + (2\sqrt{3})(4\sqrt{2}) + (2\sqrt{3})(-3\sqrt{3})$

$= 20\cdot 2 - 15\sqrt{6} + 8\sqrt{6} - 6\cdot 3$ Multiply. Note that $\sqrt{2}\cdot\sqrt{2}=\sqrt{4}=2$ and $\sqrt{3}\cdot\sqrt{3}=\sqrt{9}=3$.

$= 40 - 15\sqrt{6} + 8\sqrt{6} - 18$ Complete the multiplications.

$= (40-18) + \left(-15\sqrt{6}+8\sqrt{6}\right)$ Group like terms. Try to do this step mentally.

$= 22 - 7\sqrt{6}$ Combine numerical terms and like radicals. ■

✓ **CHECK POINT 1** Multiply:

a. $\sqrt{6}\left(x+\sqrt{10}\right)$

b. $\sqrt[3]{y}\left(\sqrt[3]{y^2}-\sqrt[3]{7}\right)$

c. $\left(6\sqrt{5}+3\sqrt{2}\right)\left(2\sqrt{5}-4\sqrt{2}\right)$.

---

2 Use polynomial special products to multiply radicals.

Some radicals can be multiplied using the special products for multiplying polynomials.

EXAMPLE 2 Using Special Products to Multiply Radicals

Multiply:

a. $\left(\sqrt{3}+\sqrt{7}\right)^2$ **b.** $\left(\sqrt{7}+\sqrt{3}\right)\left(\sqrt{7}-\sqrt{3}\right)$ **c.** $\left(\sqrt{a}-\sqrt{b}\right)\left(\sqrt{a}+\sqrt{b}\right)$.

Solution Use the special-product formulas shown.

$(A+B)^2 = A^2 + 2\cdot A\cdot B + B^2$

a. $(\sqrt{3}+\sqrt{7})^2 = (\sqrt{3})^2 + 2\cdot\sqrt{3}\cdot\sqrt{7} + (\sqrt{7})^2$ Use the special product for $(A+B)^2$.

$= 3 + 2\sqrt{21} + 7$ Multiply the radicals.

$= 10 + 2\sqrt{21}$ Simplify.

$(A+B)\cdot(A-B) = A^2 - B^2$

b. $(\sqrt{7}+\sqrt{3})(\sqrt{7}-\sqrt{3}) = (\sqrt{7})^2 - (\sqrt{3})^2$ Use the special product for $(A+B)(A-B)$.

$= 7 - 3$ Simplify: $\left(\sqrt{a}\right)^2 = a$.

$= 4$

$(A-B)\cdot(A+B) = A^2 - B^2$

c. $(\sqrt{a}-\sqrt{b})(\sqrt{a}+\sqrt{b}) = (\sqrt{a})^2 - (\sqrt{b})^2 = a - b$ ■

Radical expressions that involve the sum and difference of the same two terms are called **conjugates**. For example,

$$\sqrt{7}+\sqrt{3} \quad \text{and} \quad \sqrt{7}-\sqrt{3}$$

are conjugates of each other. Parts (b) and (c) of Example 2 illustrate that the product of two radical expressions need not be a radical expression:

$$(\sqrt{7}+\sqrt{3})(\sqrt{7}-\sqrt{3})=4$$
$$(\sqrt{a}-\sqrt{b})(\sqrt{a}+\sqrt{b})=a-b.$$

The product of conjugates does not contain a radical.

Later in this section, we will use conjugates to simplify quotients.

CHECK POINT 2 Multiply:

a. $(\sqrt{5}+\sqrt{6})^2$

b. $(\sqrt{6}+\sqrt{5})(\sqrt{6}-\sqrt{5})$

c. $(\sqrt{a}-\sqrt{7})(\sqrt{a}+\sqrt{7})$.

3 Rationalize denominators containing one term.

Rationalizing Denominators Containing One Term

You can use a calculator to compare the approximate values for $\frac{1}{\sqrt{3}}$ and $\frac{\sqrt{3}}{3}$. The two approximations are the same. This is not a coincidence:

$$\frac{1}{\sqrt{3}}=\frac{1}{\sqrt{3}}\cdot\boxed{\frac{\sqrt{3}}{\sqrt{3}}}=\frac{\sqrt{3}}{\sqrt{9}}=\frac{\sqrt{3}}{3}.$$

Any nonzero number divided by itself is 1. Multiplication by 1 does not change the value of $\frac{1}{\sqrt{3}}$.

This process involves rewriting a radical expression as an equivalent expression in which the denominator no longer contains any radicals. The process is called **rationalizing the denominator**. When the denominator contains a single radical with an nth root, **multiply the numerator and the denominator by a radical of index n that produces a perfect nth power in the denominator's radicand.**

EXAMPLE 3 Rationalizing Denominators

Rationalize each denominator:

a. $\frac{\sqrt{5}}{\sqrt{6}}$

b. $\sqrt[3]{\frac{7}{25}}$.

Solution

a. If we multiply the numerator and the denominator of $\frac{\sqrt{5}}{\sqrt{6}}$ by $\sqrt{6}$, the denominator becomes $\sqrt{6}\cdot\sqrt{6}=\sqrt{36}=6$. The denominator's radicand, 36, is a perfect square. The denominator no longer contains a radical. Therefore, we multiply by 1, choosing $\frac{\sqrt{6}}{\sqrt{6}}$ for 1.

$$\frac{\sqrt{5}}{\sqrt{6}}=\frac{\sqrt{5}}{\sqrt{6}}\cdot\frac{\sqrt{6}}{\sqrt{6}}$$

Multiply the numerator and denominator by $\sqrt{6}$ to remove the radical in the denominator.

$$=\frac{\sqrt{30}}{\sqrt{36}}$$

Multiply numerators and multiply denominators. The denominator's radicand, 36, is a perfect square.

$$=\frac{\sqrt{30}}{6}$$

Simplify: $\sqrt{36}=6$.

b. Using the quotient rule, we can express $\sqrt[3]{\frac{7}{25}}$ as $\frac{\sqrt[3]{7}}{\sqrt[3]{25}}$. We have cube roots, so we want the denominator's radicand to be a perfect cube. Right now, the denominator's radicand is 25 or 5^2. We know that $\sqrt[3]{5^3} = 5$. If we multiply the numerator and the denominator of $\frac{\sqrt[3]{7}}{\sqrt[3]{25}}$ by $\sqrt[3]{5}$, the denominator becomes

$$\sqrt[3]{25} \cdot \sqrt[3]{5} = \sqrt[3]{5^2} \cdot \sqrt[3]{5} = \sqrt[3]{5^3} = 5.$$

The denominator's radicand, 5^3, is a perfect cube. The denominator no longer contains a radical. Therefore, we multiply by 1, choosing $\frac{\sqrt[3]{5}}{\sqrt[3]{5}}$ for 1.

$$\sqrt[3]{\frac{7}{25}} = \frac{\sqrt[3]{7}}{\sqrt[3]{25}}$$ Use the quotient rule and rewrite as the quotient of radicals.

$$= \frac{\sqrt[3]{7}}{\sqrt[3]{5^2}}$$ Write the denominator's radicand as an exponential expression.

$$= \frac{\sqrt[3]{7}}{\sqrt[3]{5^2}} \cdot \frac{\sqrt[3]{5}}{\sqrt[3]{5}}$$ Multiply numerator and denominator by $\sqrt[3]{5}$ to remove the radical in the denominator.

$$= \frac{\sqrt[3]{35}}{\sqrt[3]{5^3}}$$ Multiply numerators and denominators. The denominator's radicand, 5^3, is a perfect cube.

$$= \frac{\sqrt[3]{35}}{5}$$ Simplify: $\sqrt[3]{5^3} = 5$. ■

Great Question!

What exactly does rationalizing a denominator do to an irrational number in the denominator?

Rationalizing a numerical denominator makes that denominator a rational number.

✓ CHECK POINT 3 Rationalize each denominator:

a. $\frac{\sqrt{3}}{\sqrt{7}}$ **b.** $\sqrt[3]{\frac{2}{9}}$.

Example 3 showed that it is helpful to express the denominator's radicand using exponents. In this way, we can easily find the extra factor or factors needed to produce a perfect nth power. For example, suppose that $\sqrt[5]{8}$ appears in the denominator. We want a perfect fifth power. By expressing $\sqrt[5]{8}$ as $\sqrt[5]{2^3}$, we would multiply the numerator and denominator by $\sqrt[5]{2^2}$ because

$$\sqrt[5]{2^3} \cdot \sqrt[5]{2^2} = \sqrt[5]{2^5} = 2.$$

EXAMPLE 4 Rationalizing Denominators

Rationalize each denominator:

a. $\sqrt{\frac{3x}{5y}}$ **b.** $\frac{\sqrt[3]{x}}{\sqrt[3]{36y}}$ **c.** $\frac{10y}{\sqrt[5]{4x^3y}}$.

Solution By examining each denominator, you can determine how to multiply by 1. For the square root, we must produce exponents of 2 in the radicand. For the cube root, we need exponents of 3, and for the fifth root, we want exponents of 5.

Examine the denominators to determine how to remove each radical.

- $\sqrt{5y}$ — Multiply by $\sqrt{5y}$: $\sqrt{5y}\cdot\sqrt{5y}=\sqrt{25y^2}=5y$.
- $\sqrt[3]{36y}$ or $\sqrt[3]{6^2y}$ — Multiply by $\sqrt[3]{6y^2}$: $\sqrt[3]{6^2y}\cdot\sqrt[3]{6y^2}=\sqrt[3]{6^3y^3}=6y$.
- $\sqrt[5]{4x^3y}$ or $\sqrt[5]{2^2x^3y}$ — Multiply by $\sqrt[5]{2^3x^2y^4}$: $\sqrt[5]{2^2x^3y}\cdot\sqrt[5]{2^3x^2y^4}=\sqrt[5]{2^5x^5y^5}=2xy$.

a. $$\sqrt{\frac{3x}{5y}}=\frac{\sqrt{3x}}{\sqrt{5y}}=\frac{\sqrt{3x}}{\sqrt{5y}}\cdot\frac{\sqrt{5y}}{\sqrt{5y}}=\frac{\sqrt{15xy}}{\sqrt{25y^2}}=\frac{\sqrt{15xy}}{5y}$$

Multiply by 1. $25y^2$ is a perfect square.

b. $$\frac{\sqrt[3]{x}}{\sqrt[3]{36y}}=\frac{\sqrt[3]{x}}{\sqrt[3]{6^2y}}=\frac{\sqrt[3]{x}}{\sqrt[3]{6^2y}}\cdot\frac{\sqrt[3]{6y^2}}{\sqrt[3]{6y^2}}=\frac{\sqrt[3]{6xy^2}}{\sqrt[3]{6^3y^3}}=\frac{\sqrt[3]{6xy^2}}{6y}$$

Multiply by 1. 6^3y^3 is a perfect cube.

c. $$\frac{10y}{\sqrt[5]{4x^3y}}=\frac{10y}{\sqrt[5]{2^2x^3y}}=\frac{10y}{\sqrt[5]{2^2x^3y}}\cdot\frac{\sqrt[5]{2^3x^2y^4}}{\sqrt[5]{2^3x^2y^4}}$$

Multiply by 1.

$$=\frac{10y\sqrt[5]{2^3x^2y^4}}{\sqrt[5]{2^5x^5y^5}}=\frac{10y\sqrt[5]{8x^2y^4}}{2xy}=\frac{5\sqrt[5]{8x^2y^4}}{x}$$

$2^5x^5y^5$ is a perfect 5th power.

Simplify: Divide numerator and denominator by $2y$. ■

CHECK POINT 4 Rationalize each denominator:

a. $\sqrt{\frac{2x}{7y}}$ **b.** $\frac{\sqrt[3]{x}}{\sqrt[3]{9y}}$ **c.** $\frac{6x}{\sqrt[5]{8x^2y^4}}$.

4 Rationalize denominators containing two terms.

Rationalizing Denominators Containing Two Terms

How can we rationalize a denominator if the denominator contains two terms with one or more square roots? **Multiply the numerator and the denominator by the conjugate of the denominator.** Here are three examples of such expressions:

- $\frac{8}{3\sqrt{2}+4}$ — The conjugate of the denominator is $3\sqrt{2}-4$.
- $\frac{2+\sqrt{5}}{\sqrt{6}-\sqrt{3}}$ — The conjugate of the denominator is $\sqrt{6}+\sqrt{3}$.
- $\frac{h}{\sqrt{x+h}-\sqrt{x}}$ — The conjugate of the denominator is $\sqrt{x+h}+\sqrt{x}$.

The product of the denominator and its conjugate is found using the formula

$$(A+B)(A-B)=A^2-B^2.$$

The simplified product will not contain a radical.

EXAMPLE 5 Rationalizing a Denominator Containing Two Terms

Rationalize the denominator: $\frac{8}{3\sqrt{2}+4}$.

Solution The conjugate of the denominator is $3\sqrt{2} - 4$. If we multiply the numerator and the denominator by $3\sqrt{2} - 4$, the simplified denominator will not contain a radical. Therefore, we multiply by 1, choosing $\frac{3\sqrt{2} - 4}{3\sqrt{2} - 4}$ for 1.

$$\frac{8}{3\sqrt{2}+4} = \frac{8}{3\sqrt{2}+4} \cdot \frac{3\sqrt{2}-4}{3\sqrt{2}-4}$$ Multiply by 1.

$$= \frac{8(3\sqrt{2}-4)}{(3\sqrt{2})^2 - 4^2}$$ $(A + B)(A - B) = A^2 - B^2$

$$= \frac{8(3\sqrt{2}-4)}{18-16}$$ Leave the numerator in factored form. This helps simplify, if possible. $(3\sqrt{2})^2 = 9 \cdot 2 = 18$

$$= \frac{8(3\sqrt{2}-4)}{2}$$ This expression can still be simplified.

$$= \frac{\overset{4}{\cancel{8}}(3\sqrt{2}-4)}{\underset{1}{\cancel{2}}}$$ Divide the numerator and denominator by 2.

$$= 4(3\sqrt{2}-4) \quad \text{or} \quad 12\sqrt{2} - 16$$ ■

CHECK POINT 5 Rationalize the denominator: $\frac{18}{2\sqrt{3}+3}$.

EXAMPLE 6 Rationalizing a Denominator Containing Two Terms

Rationalize the denominator: $\frac{2+\sqrt{5}}{\sqrt{6}-\sqrt{3}}$.

Solution The conjugate of the denominator is $\sqrt{6} + \sqrt{3}$. Multiplication of both the numerator and denominator by $\sqrt{6} + \sqrt{3}$ will rationalize the denominator. This will produce a rational number in the denominator.

$$\frac{2+\sqrt{5}}{\sqrt{6}-\sqrt{3}} = \frac{2+\sqrt{5}}{\sqrt{6}-\sqrt{3}} \cdot \frac{\sqrt{6}+\sqrt{3}}{\sqrt{6}+\sqrt{3}}$$ Multiply by 1.

F O I L

$$= \frac{2\sqrt{6} + 2\sqrt{3} + \sqrt{5}\cdot\sqrt{6} + \sqrt{5}\cdot\sqrt{3}}{(\sqrt{6})^2 - (\sqrt{3})^2}$$ Use FOIL in the numerator and $(A - B)(A + B) = A^2 - B^2$ in the denominator.

$$= \frac{2\sqrt{6} + 2\sqrt{3} + \sqrt{30} + \sqrt{15}}{6-3}$$

$$= \frac{2\sqrt{6} + 2\sqrt{3} + \sqrt{30} + \sqrt{15}}{3}$$ Further simplification is not possible. ■

CHECK POINT 6 Rationalize the denominator: $\frac{3+\sqrt{7}}{\sqrt{5}-\sqrt{2}}$.

5 Rationalize numerators.

Rationalizing Numerators

We have seen that square root functions are often used to model growing phenomena with growth that is leveling off. **Figure 7.8** shows a male's height as a function of his age. The pattern of his growth suggests modeling with a square root function.

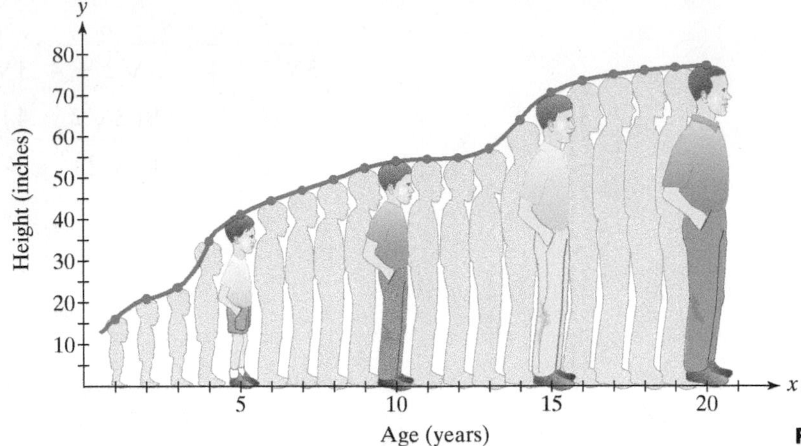

Figure 7.8

If we use $f(x) = \sqrt{x}$ to model height, $f(x)$, at age x, the expression

$$\frac{f(a+h)-f(a)}{h} = \frac{\sqrt{a+h}-\sqrt{a}}{h}$$

describes the man's average growth rate from age a to age $a + h$. Can you see that this expression is not defined if $h = 0$? However, to explore the man's average growth rates for successively shorter periods of time, we need to know what happens to the expression as h takes on values that get closer and closer to 0.

What happens to growth near the instant in time that the man is age a? The question is answered in calculus by **rationalizing the numerator**. The procedure is similar to rationalizing the denominator. **To rationalize a numerator, multiply by 1 to eliminate the radical in the *numerator*.**

EXAMPLE 7 Rationalizing a Numerator

Rationalize the numerator:

$$\frac{\sqrt{a+h}-\sqrt{a}}{h}.$$

Solution The conjugate of the numerator is $\sqrt{a+h}+\sqrt{a}$. If we multiply the numerator and the denominator by $\sqrt{a+h}+\sqrt{a}$, the simplified numerator will not contain radicals. Therefore, we multiply by 1, choosing $\frac{\sqrt{a+h}+\sqrt{a}}{\sqrt{a+h}+\sqrt{a}}$ for 1.

$$\frac{\sqrt{a+h}-\sqrt{a}}{h} = \frac{\sqrt{a+h}-\sqrt{a}}{h}\cdot\frac{\sqrt{a+h}+\sqrt{a}}{\sqrt{a+h}+\sqrt{a}}$$ Multiply by 1.

$$= \frac{(\sqrt{a+h})^2-(\sqrt{a})^2}{h(\sqrt{a+h}+\sqrt{a})}$$ $(A-B)(A+B)=A^2-B^2$ Leave the denominator in factored form.

$$= \frac{a+h-a}{h(\sqrt{a+h}+\sqrt{a})}$$ $(\sqrt{a+h})^2 = a+h$ and $(\sqrt{a})^2 = a$.

$$= \frac{h}{h(\sqrt{a+h}+\sqrt{a})}$$ Simplify the numerator.

$$= \frac{1}{\sqrt{a+h}+\sqrt{a}}$$ Simplify by dividing the numerator and the denominator by h. ■

CHECK POINT 7 Rationalize the numerator: $\frac{\sqrt{x+3}-\sqrt{x}}{3}$.

Achieving Success

Do you get to choose your seat during an exam? If so, select a desk that minimizes distractions and puts you in the right frame of mind. Many students can focus better if they do not sit near friends. If possible, sit near a window, next to a wall, or in the front row. In these locations, you can glance up from time to time without looking at another student's work.

CONCEPT AND VOCABULARY CHECK

Fill in each blank so that the resulting statement is true.

1. Consider the following multiplication problem:

$$(7\sqrt{5}+3\sqrt{2})(10\sqrt{5}-6\sqrt{2}).$$

Using the FOIL method, the product of the first terms is __________, the product of the outside terms is __________, the product of the inside terms is __________, and the product of the last terms is __________.

2. $(\sqrt{10}+\sqrt{5})(\sqrt{10}-\sqrt{5}) = (\_\_)^2 - (\_\_)^2 = \_\_ - \_\_ = \_\_$

3. The process of rewriting a radical expression as an equivalent expression in which the denominator no longer contains any radicals is called __________.

4. The number $\frac{\sqrt{7}}{\sqrt{5}}$ can be rewritten without a radical in the denominator by multiplying the numerator and denominator by __________.

5. The number $\sqrt[3]{\frac{2}{3}}$ can be rewritten without a radical in the denominator by multiplying the numerator and the denominator by __________.

6. The conjugate of $7\sqrt{2}+5$ is __________.

7. The number $\frac{2\sqrt{6}+\sqrt{5}}{3\sqrt{6}-\sqrt{5}}$ can be rewritten without a radical in the denominator by multiplying the numerator and denominator by __________.

7.5 EXERCISE SET

MyMathLab®

Practice Exercises

In this Exercise Set, assume that all variables represent positive real numbers.

In Exercises 1–38, multiply as indicated. If possible, simplify any radical expressions that appear in the product.

1. $\sqrt{2}(x+\sqrt{7})$
2. $\sqrt{5}(x+\sqrt{3})$
3. $\sqrt{6}(7-\sqrt{6})$
4. $\sqrt{3}(5-\sqrt{3})$
5. $\sqrt{3}(4\sqrt{6}-2\sqrt{3})$
6. $\sqrt{6}(4\sqrt{6}-3\sqrt{2})$
7. $\sqrt[3]{2}(\sqrt[3]{6}+4\sqrt[3]{5})$
8. $\sqrt[3]{3}(\sqrt[3]{6}+7\sqrt[3]{4})$
9. $\sqrt[3]{x}(\sqrt[3]{16x^2}-\sqrt[3]{x})$
10. $\sqrt[3]{x}(\sqrt[3]{24x^2}-\sqrt[3]{x})$
11. $(5+\sqrt{2})(6+\sqrt{2})$
12. $(7+\sqrt{2})(8+\sqrt{2})$
13. $(6+\sqrt{5})(9-4\sqrt{5})$
14. $(4+\sqrt{5})(10-3\sqrt{5})$
15. $(6-3\sqrt{7})(2-5\sqrt{7})$
16. $(7-2\sqrt{7})(5-3\sqrt{7})$
17. $(\sqrt{2}+\sqrt{7})(\sqrt{3}+\sqrt{5})$

18. $(\sqrt{3}+\sqrt{2})(\sqrt{10}+\sqrt{11})$
19. $(\sqrt{2}-\sqrt{7})(\sqrt{3}-\sqrt{5})$
20. $(\sqrt{3}-\sqrt{2})(\sqrt{10}-\sqrt{11})$
21. $(3\sqrt{2}-4\sqrt{3})(2\sqrt{2}+5\sqrt{3})$
22. $(3\sqrt{5}-2\sqrt{3})(4\sqrt{5}+5\sqrt{3})$
23. $(\sqrt{3}+\sqrt{5})^2$
24. $(\sqrt{2}+\sqrt{7})^2$
25. $(\sqrt{3x}-\sqrt{y})^2$
26. $(\sqrt{2x}-\sqrt{y})^2$
27. $(\sqrt{5}+7)(\sqrt{5}-7)$
28. $(\sqrt{6}+2)(\sqrt{6}-2)$
29. $(2-5\sqrt{3})(2+5\sqrt{3})$
30. $(3-5\sqrt{2})(3+5\sqrt{2})$
31. $(3\sqrt{2}+2\sqrt{3})(3\sqrt{2}-2\sqrt{3})$
32. $(4\sqrt{3}+3\sqrt{2})(4\sqrt{3}-3\sqrt{2})$
33. $(3-\sqrt{x})(2-\sqrt{x})$
34. $(4-\sqrt{x})(3-\sqrt{x})$
35. $(\sqrt[3]{x}-4)(\sqrt[3]{x}+5)$
36. $(\sqrt[3]{x}-3)(\sqrt[3]{x}+7)$
37. $(x+\sqrt[3]{y^2})(2x-\sqrt[3]{y^2})$
38. $(x-\sqrt[5]{y^3})(2x+\sqrt[5]{y^3})$

In Exercises 39–64, rationalize each denominator.

39. $\frac{\sqrt{2}}{\sqrt{5}}$
40. $\frac{\sqrt{7}}{\sqrt{3}}$
41. $\sqrt{\frac{11}{x}}$
42. $\sqrt{\frac{6}{x}}$
43. $\frac{9}{\sqrt{3y}}$
44. $\frac{12}{\sqrt{3y}}$
45. $\frac{1}{\sqrt[3]{2}}$
46. $\frac{1}{\sqrt[3]{3}}$
47. $\frac{6}{\sqrt[3]{4}}$
48. $\frac{10}{\sqrt[3]{5}}$
49. $\sqrt[3]{\frac{2}{3}}$
50. $\sqrt[3]{\frac{3}{4}}$
51. $\frac{4}{\sqrt[3]{x}}$
52. $\frac{7}{\sqrt[3]{x}}$
53. $\sqrt[3]{\frac{2}{y^2}}$
54. $\sqrt[3]{\frac{5}{y^2}}$
55. $\frac{7}{\sqrt[3]{2x^2}}$
56. $\frac{10}{\sqrt[3]{4x^2}}$
57. $\sqrt[3]{\frac{2}{xy^2}}$
58. $\sqrt[3]{\frac{3}{xy^2}}$
59. $\frac{3}{\sqrt[4]{x}}$
60. $\frac{5}{\sqrt[4]{x}}$
61. $\frac{6}{\sqrt[5]{8x^3}}$
62. $\frac{10}{\sqrt[5]{16x^2}}$
63. $\frac{2x^2y}{\sqrt[5]{4x^2y^4}}$
64. $\frac{3xy^2}{\sqrt[5]{8xy^3}}$

In Exercises 65–74, simplify each radical expression and then rationalize the denominator.

65. $\frac{9}{\sqrt{3x^2y}}$
66. $\frac{25}{\sqrt{5x^2y}}$
67. $-\sqrt{\frac{75a^5}{b^3}}$
68. $-\sqrt{\frac{150a^3}{b^5}}$
69. $\sqrt{\frac{7m^2n^3}{14m^3n^2}}$
70. $\sqrt{\frac{5m^4n^6}{15m^3n^4}}$
71. $\frac{3}{\sqrt[4]{x^5y^3}}$
72. $\frac{5}{\sqrt[4]{x^2y^7}}$
73. $\frac{12}{\sqrt[3]{-8x^5y^8}}$
74. $\frac{15}{\sqrt[3]{-27x^4y^{11}}}$

In Exercises 75–92, rationalize each denominator. Simplify, if possible.

75. $\frac{8}{\sqrt{5}+2}$
76. $\frac{15}{\sqrt{6}+1}$
77. $\frac{13}{\sqrt{11}-3}$
78. $\frac{17}{\sqrt{10}-2}$
79. $\frac{6}{\sqrt{5}+\sqrt{3}}$
80. $\frac{12}{\sqrt{7}+\sqrt{3}}$
81. $\frac{\sqrt{a}}{\sqrt{a}-\sqrt{b}}$
82. $\frac{\sqrt{b}}{\sqrt{a}-\sqrt{b}}$
83. $\frac{25}{5\sqrt{2}-3\sqrt{5}}$
84. $\frac{35}{5\sqrt{2}-3\sqrt{5}}$
85. $\frac{\sqrt{5}+\sqrt{3}}{\sqrt{5}-\sqrt{3}}$
86. $\frac{\sqrt{11}-\sqrt{5}}{\sqrt{11}+\sqrt{5}}$
87. $\frac{\sqrt{x}+1}{\sqrt{x}+3}$
88. $\frac{\sqrt{x}-2}{\sqrt{x}-5}$
89. $\frac{5\sqrt{3}-3\sqrt{2}}{3\sqrt{2}-2\sqrt{3}}$
90. $\frac{2\sqrt{6}+\sqrt{5}}{3\sqrt{6}-\sqrt{5}}$
91. $\frac{2\sqrt{x}+\sqrt{y}}{\sqrt{y}-2\sqrt{x}}$
92. $\frac{3\sqrt{x}+\sqrt{y}}{\sqrt{y}-3\sqrt{x}}$

In Exercises 93–104, rationalize each numerator. Simplify, if possible.

93. $\sqrt{\frac{3}{2}}$
94. $\sqrt{\frac{5}{3}}$
95. $\frac{\sqrt[3]{4x}}{\sqrt[3]{y}}$
96. $\frac{\sqrt[3]{2x}}{\sqrt[3]{y}}$
97. $\frac{\sqrt{x}+3}{\sqrt{x}}$
98. $\frac{\sqrt{x}+4}{\sqrt{x}}$
99. $\frac{\sqrt{a}+\sqrt{b}}{\sqrt{a}-\sqrt{b}}$
100. $\frac{\sqrt{a}-\sqrt{b}}{\sqrt{a}+\sqrt{b}}$
101. $\frac{\sqrt{x+5}-\sqrt{x}}{5}$

102. $\dfrac{\sqrt{x+7}-\sqrt{x}}{7}$

103. $\dfrac{\sqrt{x}+\sqrt{y}}{x^2-y^2}$

104. $\dfrac{\sqrt{x}-\sqrt{y}}{x^2-y^2}$

Practice PLUS

In Exercises 105–112, add or subtract as indicated. Begin by rationalizing denominators for all terms in which denominators contain radicals.

105. $\sqrt{2}+\dfrac{1}{\sqrt{2}}$ **106.** $\sqrt{5}+\dfrac{1}{\sqrt{5}}$

107. $\sqrt[3]{25}-\dfrac{15}{\sqrt[3]{5}}$ **108.** $\sqrt[4]{8}-\dfrac{20}{\sqrt[4]{2}}$

109. $\sqrt{6}-\sqrt{\dfrac{1}{6}}+\sqrt{\dfrac{2}{3}}$

110. $\sqrt{15}-\sqrt{\dfrac{5}{3}}+\sqrt{\dfrac{3}{5}}$

111. $\dfrac{2}{\sqrt{2}+\sqrt{3}}+\sqrt{75}-\sqrt{50}$

112. $\dfrac{5}{\sqrt{2}+\sqrt{7}}-2\sqrt{32}+\sqrt{28}$

113. Let $f(x)=x^2-6x-4$. Find $f(3-\sqrt{13})$.

114. Let $f(x)=x^2+4x-2$. Find $f(-2+\sqrt{6})$.

115. Let $f(x)=\sqrt{9+x}$. Find $f(3\sqrt{5})\cdot f(-3\sqrt{5})$.

116. Let $f(x)=x^2$. Find $f(\sqrt{a+1}-\sqrt{a-1})$.

Application Exercises

117. The early Greeks believed that the most pleasing of all rectangles were **golden rectangles**, whose ratio of width to height is

$$\frac{w}{h}=\frac{2}{\sqrt{5}-1}.$$

The Parthenon at Athens fits into a golden rectangle once the triangular pediment is reconstructed.

Rationalize the denominator of the golden ratio. Then use a calculator and find the ratio of width to height, correct to the nearest hundredth, in golden rectangles.

118. In the "Peanuts" cartoon shown in the section opener on page 542, Woodstock appears to be working steps mentally. Fill in the missing steps that show how to go from $\dfrac{7\sqrt{2\cdot2\cdot3}}{6}$ to $\dfrac{7}{3}\sqrt{3}$.

In Exercises 119–120, write expressions for the perimeter and area of each figure. Then simplify these expressions. Assume that all measures are given in inches.

119.

$\sqrt{8}+1$

$\sqrt{8}-1$

120.

$2\sqrt{3}+\sqrt{2}$

$2\sqrt{3}+\sqrt{2}$

The Pythagorean Theorem for right triangles tells us that the length of the hypotenuse is the square root of the sum of the squares of the lengths of the legs. In Exercises 121–122, use the Pythagorean Theorem to find the length of each hypotenuse in simplified radical form. Assume that all measures are given in inches.

121.

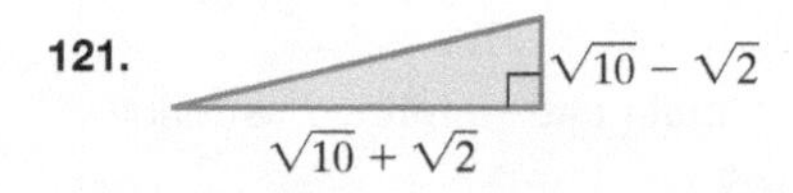

122.

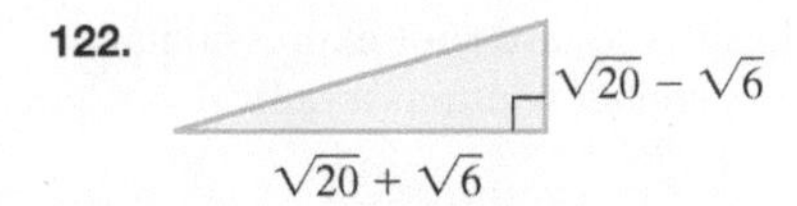

Explaining the Concepts

123. Explain how to perform this multiplication: $\sqrt{2}(\sqrt{7}+\sqrt{10})$.

124. Explain how to perform this multiplication: $(2+\sqrt{3})(4+\sqrt{3})$.

125. Explain how to perform this multiplication: $(2+\sqrt{3})^2$.

126. What are conjugates? Give an example with your explanation.

127. Describe how to multiply conjugates.

128. Describe what it means to rationalize a denominator. Use both $\dfrac{1}{\sqrt{5}}$ and $\dfrac{1}{5+\sqrt{5}}$ in your explanation.

129. When a radical expression has its denominator rationalized, we change the denominator so that it no longer contains any radicals. Doesn't this change the value of the radical expression? Explain.

130. Square the real number $\dfrac{2}{\sqrt{3}}$. Observe that the radical is eliminated from the denominator. Explain whether this process is equivalent to rationalizing the denominator.

Technology Exercises

In Exercises 131–134, determine if each operation is performed correctly by graphing the function on each side of the equation with your graphing utility. Use the given viewing rectangle. The graphs should be the same. If they are not, correct the right side of the equation and then use your graphing utility to verify the correction.

131. $(\sqrt{x}-1)(\sqrt{x}-1)=x+1$
[0, 5, 1] by [−1, 2, 1]

132. $(\sqrt{x}+2)(\sqrt{x}-2)=x^2-4$ for $x \geq 0$
[0, 10, 1] by [−10, 10, 1]

133. $(\sqrt{x}+1)^2=x+1$
[0, 8, 1] by [0, 15, 1]

134. $\dfrac{3}{\sqrt{x+3}-\sqrt{x}}=\sqrt{x+3}+\sqrt{x}$
[0, 8, 1] by [0, 6, 1]

Critical Thinking Exercises

Make Sense? *In Exercises 135–138, determine whether each statement makes sense or does not make sense, and explain your reasoning.*

135. I use the same ideas to multiply $(\sqrt{2}+5)(\sqrt{2}+4)$ that I did to find the binomial product $(x+5)(x+4)$.

136. I used a special-product formula and simplified as follows: $(\sqrt{2}+\sqrt{5})^2=2+5=7$.

137. In some cases when I multiply a square root expression and its conjugate, the simplified product contains a radical.

138. I use the fact that 1 is the multiplicative identity to both rationalize denominators and rewrite rational expressions with a common denominator.

In Exercises 139–142, determine whether each statement is true or false. If the statement is false, make the necessary change(s) to produce a true statement.

139. $\dfrac{\sqrt{3}+7}{\sqrt{3}-2}=-\dfrac{7}{2}$

140. $\dfrac{4}{\sqrt{x+y}}=\dfrac{4\sqrt{x-y}}{x-y}$

141. $\dfrac{4\sqrt{x}}{\sqrt{x}-y}=\dfrac{4x+4y\sqrt{x}}{x-y^2}$

142. $(\sqrt{x}-7)^2=x-49$

143. Solve:
$$7[(2x-5)-(x+1)]=(\sqrt{7}+2)(\sqrt{7}-2).$$

144. Simplify: $(\sqrt{2+\sqrt{3}}+\sqrt{2-\sqrt{3}})^2$.

145. Rationalize the denominator: $\dfrac{1}{\sqrt{2}+\sqrt{3}+\sqrt{4}}$.

Review Exercises

146. Add: $\dfrac{2}{x-2}+\dfrac{3}{x^2-4}$.
(Section 6.2, Example 6)

147. Solve: $\dfrac{2}{x-2}+\dfrac{3}{x^2-4}=0$.
(Section 6.6, Example 5)

148. If $f(x)=x^4-3x^2-2x+5$, use synthetic division and the Remainder Theorem to find $f(-2)$. (Section 6.5, Example 2)

Preview Exercises

Exercises 149–151 will help you prepare for the material covered in the next section.

149. Multiply: $(\sqrt{x+4}+1)^2$.

150. Solve: $4x^2-16x+16=4(x+4)$.

151. Solve: $26-11x=16-8x+x^2$.

SECTION

7.6 Radical Equations 

What am I supposed to learn?

After studying this section, you should be able to:

1. Solve radical equations.
2. Use models that are radical functions to solve problems.

One of the most dramatic developments in the U.S. work force has been the increase in the number of women. In 1960, approximately 38% of women belonged to the labor force. By 2000, that number had increased to over 60%. With higher levels of education than ever before, most women now choose careers in the labor force rather than homemaking.

The function

$$f(x)=3.5\sqrt{x}+38$$

models the percentage of U.S. women in the labor force, $f(x)$, x years after 1960. How can we predict the year when, say, 70% of women will participate in the U.S. work force? Substitute 70 for $f(x)$ in $f(x)=3.5\sqrt{x}+38$ and solve for x:

$$70=3.5\sqrt{x}+38.$$

The resulting equation contains a variable in the radicand and is called a *radical equation*. A **radical equation** is an equation in which the variable occurs in a square root, cube root, or any higher root. Some examples of radical equations are

$$\sqrt{2x+3}=5,\quad \sqrt{3x+1}-\sqrt{x+4}=1,\quad \text{and}\quad \sqrt[3]{3x-1}+4=0.$$

Variables occur in radicands.

In this section, you will learn how to solve radical equations. Solving such equations will enable you to solve new kinds of problems using radical functions.

1 Solve radical equations.

Solving Radical Equations

Consider the following radical equation:

$$\sqrt{x}=9.$$

We solve the equation by squaring both sides:

Squaring both sides eliminates the square root.

$$(\sqrt{x})^2=9^2$$
$$x=81.$$

The proposed solution, 81, can be checked in the original equation, $\sqrt{x}=9$. Because $\sqrt{81}=9$, the solution is 81 and the solution set is $\{81\}$.

In general, we solve radical equations with square roots by squaring both sides of the equation. We solve radical equations with nth roots by raising both sides of the equation to the nth power. Unfortunately, if n is even, all the solutions of the equation raised to the even power may not be solutions of the original equation. Consider, for example, the equation

$$x=4.$$

If we square both sides, we obtain

$$x^2=16.$$

$x^2-16=0$ Subtract 16 from both sides and write the quadratic equation in standard form.

$(x+4)(x-4)=0$ Factor.

$x+4=0$ or $x-4=0$ Set each factor equal to 0.

$x=-4$ $\quad$ $x=4$ Solve the resulting equations.

The equation $x^2=16$ has two solutions, -4 and 4. By contrast, only 4 is a solution of the original equation, $x=4$. For this reason, **when raising both sides of an equation to an even power, always check proposed solutions in the original equation**.

Here is a general method for solving radical equations with nth roots:

Solving Radical Equations Containing nth Roots

1. If necessary, arrange terms so that one radical is isolated on one side of the equation.
2. Raise both sides of the equation to the nth power to eliminate the nth root.
3. Solve the resulting equation. If this equation still contains radicals, repeat steps 1 and 2.
4. Check all proposed solutions in the original equation.

EXAMPLE 1 Solving a Radical Equation

Solve: $\sqrt{2x + 3} = 5$.

Solution

Step 1. Isolate a radical on one side. The radical, $\sqrt{2x + 3}$, is already isolated on the left side of the equation, so we can skip this step.

Step 2. Raise both sides to the *n*th power. Because n, the index, is 2, we square both sides.

$$\sqrt{2x + 3} = 5 \quad \text{This is the given equation.}$$

$$(\sqrt{2x + 3})^2 = 5^2 \quad \text{Square both sides to eliminate the radical.}$$

$$2x + 3 = 25 \quad \text{Simplify.}$$

Step 3. Solve the resulting equation.

$$2x + 3 = 25 \quad \text{The resulting equation is a linear equation.}$$

$$2x = 22 \quad \text{Subtract 3 from both sides.}$$

$$x = 11 \quad \text{Divide both sides by 2.}$$

Step 4. Check the proposed solution in the original equation. Because both sides were raised to an even power, this check is essential.

Check 11:

$$\sqrt{2x + 3} = 5$$

$$\sqrt{2 \cdot 11 + 3} \stackrel{?}{=} 5$$

$$\sqrt{25} \stackrel{?}{=} 5$$

$$5 = 5, \quad \text{true}$$

The solution is 11 and the solution set is $\{11\}$. ■

✓ CHECK POINT 1 Solve: $\sqrt{3x + 4} = 8$.

EXAMPLE 2 Solving a Radical Equation

Solve: $\sqrt{x - 3} + 6 = 5$.

Solution

Step 1. Isolate a radical on one side. The radical, $\sqrt{x - 3}$, can be isolated by subtracting 6 from both sides. We obtain

$$\sqrt{x - 3} = -1.$$

A principal square root cannot be negative. This equation has no solution. Let's continue the solution procedure to see what happens.

Step 2. Raise both sides to the *n*th power. Because n, the index, is 2, we square both sides.

$$(\sqrt{x - 3})^2 = (-1)^2$$

$$x - 3 = 1 \quad \text{Simplify.}$$

Step 3. Solve the resulting equation.

$$x - 3 = 1 \quad \text{The resulting equation is a linear equation.}$$

$$x = 4 \quad \text{Add 3 to both sides.}$$

Step 4. Check the proposed solution in the original equation.

Check 4:

$$\sqrt{x-3}+6=5$$
$$\sqrt{4-3}+6 \stackrel{?}{=} 5$$
$$\sqrt{1}+6 \stackrel{?}{=} 5$$
$$1+6 \stackrel{?}{=} 5$$
$$7=5, \quad \text{false}$$

This false statement indicates that 4 is not a solution. Thus, the equation has no solution. The solution set is $\varnothing$, the empty set. ■

Example 2 illustrates that extra solutions may be introduced when you raise both sides of a radical equation to an even power. Such solutions, which are not solutions of the given equation, are called **extraneous solutions**. Thus, 4 is an extraneous solution of $\sqrt{x-3}+6=5$.

✓ CHECK POINT 2 Solve: $\sqrt{x-1}+7=2$.

EXAMPLE 3 Solving a Radical Equation

Solve:

$$x+\sqrt{26-11x}=4.$$

Great Question!

Can I square the right side of $\sqrt{26-11x}=4-x$ by first squaring 4 and then squaring x?

No. Be sure to square *both sides* of an equation. Do *not* square each term.

Correct:

$$\left(\sqrt{26-11x}\right)^2=(4-x)^2$$

Incorrect!

$$\left(\sqrt{26-11x}\right)^2=4^2-x^2$$

Solution

Step 1. Isolate a radical on one side. We isolate the radical, $\sqrt{26-11x}$, by subtracting x from both sides.

$$x+\sqrt{26-11x}=4 \quad \text{This is the given equation.}$$
$$\sqrt{26-11x}=4-x \quad \text{Subtract } x \text{ from both sides.}$$

Step 2. Square both sides.

$$\left(\sqrt{26-11x}\right)^2=(4-x)^2$$
$$26-11x=16-8x+x^2$$

Simplify. Use the special-product formula $(A-B)^2=A^2-2AB+B^2$ to square $4-x$.

Step 3. Solve the resulting equation. Because of the x^2-term, the resulting equation is a quadratic equation. We need to write this quadratic equation in standard form. We can obtain zero on the left side by subtracting 26 and adding $11x$ on both sides.

$$26-26-11x+11x=16-26-8x+11x+x^2$$
$$0=x^2+3x-10 \quad \text{Simplify.}$$
$$0=(x+5)(x-2) \quad \text{Factor.}$$
$$x+5=0 \quad \text{or} \quad x-2=0 \quad \text{Set each factor equal to zero.}$$
$$x=-5 \qquad\qquad x=2 \quad \text{Solve for } x.$$

Step 4. Check the proposed solutions in the original equation.

Check −5:

$$x+\sqrt{26-11x}=4$$
$$-5+\sqrt{26-11(-5)} \stackrel{?}{=} 4$$
$$-5+\sqrt{81} \stackrel{?}{=} 4$$
$$-5+9 \stackrel{?}{=} 4$$
$$4=4, \quad \text{true}$$

Check 2:

$$x+\sqrt{26-11x}=4$$
$$2+\sqrt{26-11\cdot 2} \stackrel{?}{=} 4$$
$$2+\sqrt{4} \stackrel{?}{=} 4$$
$$2+2 \stackrel{?}{=} 4$$
$$4=4, \quad \text{true}$$

The solutions are −5 and 2, and the solution set is $\{-5, 2\}$. ■

Using Technology

Graphic Connections

You can use a graphing utility to provide a graphic check that $\{-5, 2\}$ is the solution set of $x + \sqrt{26 - 11x} = 4$.

Use the given equation

$$x + \sqrt{26 - 11x} = 4.$$

Enter $y_1 = x + \sqrt{26 - 11x}$ in the $\boxed{y=}$ screen.

Enter $y_2 = 4$ in the $\boxed{y=}$ screen.

Display graphs for y_1 and y_2. The solutions are the x-coordinates of the intersection points. These x-coordinates are -5 and 2. This verifies $\{-5, 2\}$ as the solution set of $x + \sqrt{26 - 11x} = 4$.

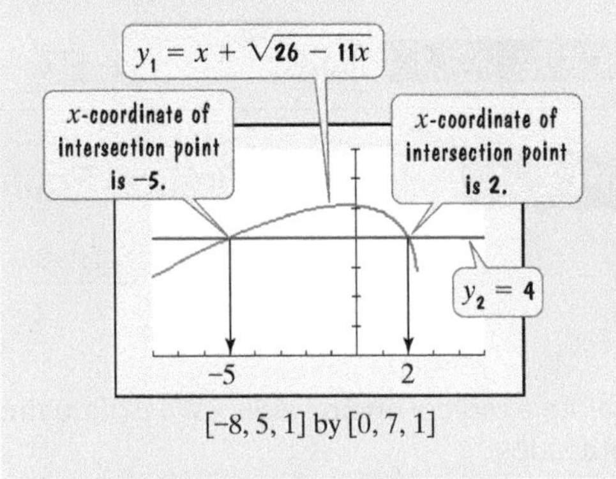

[−8, 5, 1] by [0, 7, 1]

Use the equivalent equation

$$x + \sqrt{26 - 11x} - 4 = 0.$$

Enter $y_1 = x + \sqrt{26 - 11x} - 4$ in the $\boxed{y=}$ screen.

Display the graph for y_1. The solutions are the x-intercepts. The x-intercepts are -5 and 2. This verifies $\{-5, 2\}$ as the solution set of $x + \sqrt{26 - 11x} - 4 = 0$.

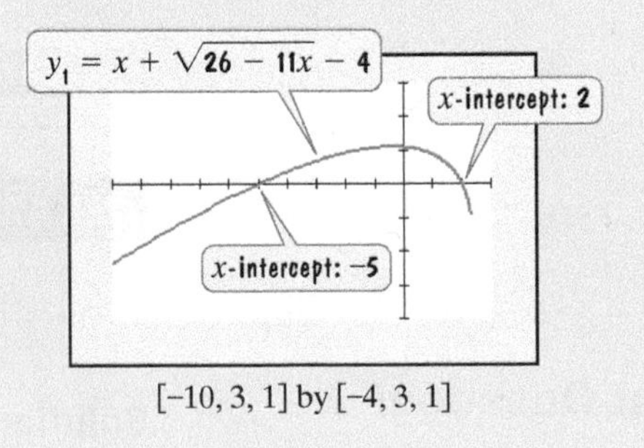

[−10, 3, 1] by [−4, 3, 1]

✓ **CHECK POINT 3** Solve: $\sqrt{6x + 7} - x = 2$.

The solution of a radical equation with two or more square root expressions involves isolating a radical, squaring both sides, and then repeating this process. Let's consider an equation containing two square root expressions.

EXAMPLE 4 Solving an Equation That Has Two Radicals

Solve: $\sqrt{3x + 1} - \sqrt{x + 4} = 1$.

Solution

Step 1. Isolate a radical on one side. We can isolate the radical $\sqrt{3x + 1}$ by adding $\sqrt{x + 4}$ to both sides. We obtain

$$\sqrt{3x + 1} = \sqrt{x + 4} + 1.$$

Step 2. Square both sides.

$$\left(\sqrt{3x + 1}\right)^2 = \left(\sqrt{x + 4} + 1\right)^2$$

Squaring the expression on the right side of the equation can be a bit tricky. We have to use the formula

$$(A + B)^2 = A^2 + 2AB + B^2.$$

Focusing on just the right side, here is how the squaring is done:

$$(A + B)^2 = A^2 + 2 \cdot A \cdot B + B^2$$

$$(\sqrt{x + 4} + 1)^2 = (\sqrt{x + 4})^2 + 2 \cdot \sqrt{x + 4} \cdot 1 + 1^2 = x + 4 + 2\sqrt{x + 4} + 1.$$

Now let's return to squaring both sides.

$(\sqrt{3x+1})^2 = (\sqrt{x+4}+1)^2$	Square both sides of the equation with an isolated radical.
$3x + 1 = x + 4 + 2\sqrt{x+4} + 1$	$(\sqrt{3x+1})^2 = 3x + 1$; square the right side using the formula for $(A + B)^2$.
$3x + 1 = x + 5 + 2\sqrt{x+4}$	Combine numerical terms on the right side: $4 + 1 = 5$.

Can you see that the resulting equation still contains a radical, namely $\sqrt{x+4}$? Thus, we need to repeat the first two steps.

Repeat Step 1. Isolate a radical on one side. We isolate $2\sqrt{x+4}$, the radical term, by subtracting $x + 5$ from both sides. We obtain

$3x + 1 = x + 5 + 2\sqrt{x+4}$	This is the equation from our last step.
$2x - 4 = 2\sqrt{x+4}.$	Subtract x and subtract 5 from both sides.

Although we can simplify the equation by dividing both sides by 2, this sort of simplification is not always helpful. Thus, we will work with the equation in this form.

Repeat Step 2. Square both sides.

Be careful in squaring both sides. Use $(A - B)^2 = A^2 - 2AB + B^2$ to square the left side. Use $(AB)^2 = A^2B^2$ to square the right side.

$(2x - 4)^2 = (2\sqrt{x+4})^2$	Square both sides.
$4x^2 - 16x + 16 = 4(x + 4)$	Square both 2 and $\sqrt{x+4}$ on the right side.

Step 3. Solve the resulting equation. We solve this quadratic equation by writing it in standard form.

$4x^2 - 16x + 16 = 4x + 16$	Use the distributive property.
$4x^2 - 20x = 0$	Subtract $4x + 16$ from both sides.
$4x(x - 5) = 0$	Factor.
$4x = 0$ or $x - 5 = 0$	Set each factor equal to zero.
$x = 0 \qquad x = 5$	Solve for x.

Step 4. Check the proposed solutions in the original equation.

Check 0:	Check 5:
$\sqrt{3x+1} - \sqrt{x+4} = 1$	$\sqrt{3x+1} - \sqrt{x+4} = 1$
$\sqrt{3 \cdot 0 + 1} - \sqrt{0+4} \stackrel{?}{=} 1$	$\sqrt{3 \cdot 5 + 1} - \sqrt{5+4} \stackrel{?}{=} 1$
$\sqrt{1} - \sqrt{4} \stackrel{?}{=} 1$	$\sqrt{16} - \sqrt{9} \stackrel{?}{=} 1$
$1 - 2 \stackrel{?}{=} 1$	$4 - 3 \stackrel{?}{=} 1$
$-1 = 1,$ false	$1 = 1,$ true

The check indicates that 0 is not a solution. It is an extraneous solution brought about by squaring each side of the equation. The only solution is 5 and the solution set is {5}. ■

✓ CHECK POINT 4 Solve: $\sqrt{x+5} - \sqrt{x-3} = 2.$

EXAMPLE 5 Solving a Radical Equation

Solve: $(3x - 1)^{\frac{1}{3}} + 4 = 0$.

Solution Although we can rewrite the equation in radical form

$$\sqrt[3]{3x - 1} + 4 = 0,$$

it is not necessary to do so. Because the equation involves a cube root, we isolate the radical term—that is, the term with the rational exponent—and cube both sides.

$$(3x - 1)^{\frac{1}{3}} + 4 = 0$$ This is the given equation.

$$(3x - 1)^{\frac{1}{3}} = -4$$ Subtract 4 from both sides and isolate the term with the rational exponent.

$$\left[(3x - 1)^{\frac{1}{3}}\right]^3 = (-4)^3$$ Cube both sides.

$$3x - 1 = -64$$ Multiply exponents on the left side and simplify.

$$3x = -63$$ Add 1 to both sides.

$$x = -21$$ Divide both sides by 3.

Because both sides were raised to an odd power, it is not essential to check the proposed solution, -21. However, checking is always a good idea. Do so now and verify that -21 is the solution and the solution set is $\{-21\}$. ■

Example 5 illustrates that a radical equation with rational exponents can be solved by

1. isolating the expression with the rational exponent, and
2. raising both sides of the equation to a power that is the reciprocal of the rational exponent.

Keep in mind that it is essential to check proposed solutions when both sides have been raised to even powers. Thus, equations with rational exponents such as $\frac{1}{2}$ and $\frac{1}{4}$ must be checked.

✓ CHECK POINT 5 Solve: $(2x - 3)^{\frac{1}{3}} + 3 = 0$.

2 Use models that are radical functions to solve problems.

Applications of Radical Equations

Radical equations can be solved to answer questions about variables contained in radical functions.

EXAMPLE 6 Women in the Labor Force

The bar graph in **Figure 7.9** shows the percentage of U.S. women in the labor force from 1960 through 2010. The function $f(x) = 3.5\sqrt{x} + 38$ models the percentage of U.S. women in the labor force, $f(x)$, x years after 1960. According to the model, when will 70% of U.S. women participate in the work force?

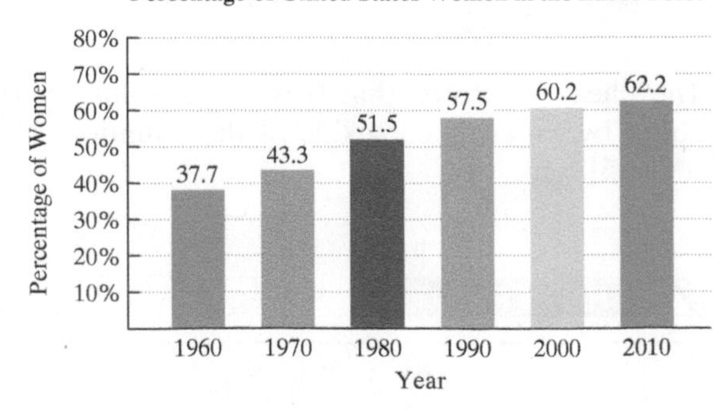

Figure 7.9
Source: U.S. Department of Labor

Solution To find when 70% of U.S. women will be in the work force, substitute 70 for $f(x)$ in the given function. Then solve for x, the number of years after 1960.

$$f(x) = 3.5\sqrt{x} + 38 \quad \text{This is the given function.}$$

$$70 = 3.5\sqrt{x} + 38 \quad \text{Substitute 70 for } f(x).$$

$$32 = 3.5\sqrt{x} \quad \text{Subtract 38 from both sides.}$$

$$\frac{32}{3.5} = \sqrt{x} \quad \text{Divide both sides by 3.5.}$$

$$\left(\frac{32}{3.5}\right)^2 = (\sqrt{x})^2 \quad \text{Square both sides.}$$

$$84 \approx x \quad \text{Use a calculator.}$$

The model indicates that 70% of U.S. women will be in the labor force approximately 84 years after 1960. Because $1960 + 84 = 2044$, this is projected to occur in 2044. ■

✓ **CHECK POINT 6** Use the function in Example 6 to project when 73% of U.S. women will participate in the work force.

CONCEPT AND VOCABULARY CHECK

Fill in each blank so that the resulting statement is true.

1. An equation in which the variable occurs in a square root, cube root, or any higher root is called a/an ___________ equation.
2. Solutions of a squared equation that are not solutions of the original equation are called ___________ solutions.
3. Consider the equation

   $$\sqrt{2x + 1} = x - 7.$$

   Squaring the left side and simplifying results in ______________. Squaring the right side and simplifying results in ______________.
4. Consider the equation

   $$\sqrt{x + 2} = 3 - \sqrt{x - 1}.$$

   Squaring the left side and simplifying results in ________________. Squaring the right side and simplifying results in ________________.
5. Consider the equation

   $$(2x + 3)^{\frac{1}{3}} = 2.$$

   Cubing the left side and simplifying results in ___________. Cubing the right side and simplifying results in ___________.
6. True or false: 4 is a solution of $\sqrt{5x + 16} = x + 2$. ______________
7. True or false: -3 is a solution of $\sqrt{5x + 16} = x + 2$. ______________

7.6 EXERCISE SET

MyMathLab®

Practice Exercises

In Exercises 1–38, solve each radical equation.

1. $\sqrt{3x - 2} = 4$
2. $\sqrt{5x - 1} = 8$
3. $\sqrt{5x - 4} - 9 = 0$
4. $\sqrt{3x - 2} - 5 = 0$
5. $\sqrt{3x + 7} + 10 = 4$
6. $\sqrt{2x + 5} + 11 = 6$
7. $x = \sqrt{7x + 8}$
8. $x = \sqrt{6x + 7}$
9. $\sqrt{5x + 1} = x + 1$
10. $\sqrt{2x + 1} = x - 7$
11. $x = \sqrt{2x - 2} + 1$
12. $x = \sqrt{3x + 7} - 3$
13. $x - 2\sqrt{x} - 3 = 3$
14. $3x - \sqrt{3x + 7} = -5$
15. $\sqrt{2x - 5} = \sqrt{x + 4}$
16. $\sqrt{6x + 2} = \sqrt{5x + 3}$
17. $\sqrt[3]{2x + 11} = 3$
18. $\sqrt[3]{6x - 3} = 3$
19. $\sqrt[3]{2x - 6} - 4 = 0$

20. $\sqrt[3]{4x-3}-5=0$
21. $\sqrt{x-7}=7-\sqrt{x}$
22. $\sqrt{x-8}=\sqrt{x}-2$
23. $\sqrt{x+2}+\sqrt{x-1}=3$
24. $\sqrt{x-4}+\sqrt{x+4}=4$
25. $2\sqrt{4x+1}-9=x-5$
26. $2\sqrt{x-3}+4=x+1$
27. $(2x+3)^{\frac{1}{3}}+4=6$
28. $(3x-6)^{\frac{1}{3}}+5=8$
29. $(3x+1)^{\frac{1}{4}}+7=9$
30. $(2x+3)^{\frac{1}{4}}+7=10$
31. $(x+2)^{\frac{1}{2}}+8=4$
32. $(x-3)^{\frac{1}{2}}+8=6$
33. $\sqrt{2x-3}-\sqrt{x-2}=1$
34. $\sqrt{x+2}+\sqrt{3x+7}=1$
35. $3x^{\frac{1}{3}}=(x^2+17x)^{\frac{1}{3}}$
36. $2(x-1)^{\frac{1}{3}}=(x^2+2x)^{\frac{1}{3}}$
37. $(x+8)^{\frac{1}{4}}=(2x)^{\frac{1}{4}}$
38. $(x-2)^{\frac{1}{4}}=(3x-8)^{\frac{1}{4}}$

Practice PLUS

39. If $f(x)=x+\sqrt{x+5}$, find all values of x for which $f(x)=7$.
40. If $f(x)=x-\sqrt{x-2}$, find all values of x for which $f(x)=4$.
41. If $f(x)=(5x+16)^{\frac{1}{3}}$ and $g(x)=(x-12)^{\frac{1}{3}}$, find all values of x for which $f(x)=g(x)$.
42. If $f(x)=(9x+2)^{\frac{1}{4}}$ and $g(x)=(5x+18)^{\frac{1}{4}}$, find all values of x for which $f(x)=g(x)$.

In Exercises 43–46, solve each formula for the specified variable.

43. $r=\sqrt{\frac{3V}{\pi h}}$ for V
44. $r=\sqrt{\frac{A}{4\pi}}$ for A
45. $t=2\pi\sqrt{\frac{l}{32}}$ for l
46. $v=\sqrt{\frac{FR}{m}}$ for m
47. If 5 times a number is decreased by 4, the principal square root of this difference is 2 less than the number. Find the number(s).
48. If a number is decreased by 3, the principal square root of this difference is 5 less than the number. Find the number(s).

In Exercises 49–50, find the x-intercept(s) of the graph of each function without graphing the function.

49. $f(x)=\sqrt{x+16}-\sqrt{x}-2$
50. $f(x)=\sqrt{2x-3}-\sqrt{2x}+1$

Application Exercises

A basketball player's hang time is the time spent in the air when shooting a basket. The formula

$$t=\frac{\sqrt{d}}{2}$$

models hang time, t, in seconds, in terms of the vertical distance of a player's jump, d, in feet. Use this formula to solve Exercises 51–52.

51. When Michael Wilson of the Harlem Globetrotters slam-dunked a basketball, his hang time for the shot was approximately 1.16 seconds. What was the vertical distance of his jump, rounded to the nearest tenth of a foot?
52. If hang time for a shot by a professional basketball player is 0.85 second, what is the vertical distance of the jump, rounded to the nearest tenth of a foot?

Use the graph of the formula for hang time to solve Exercises 53–54.

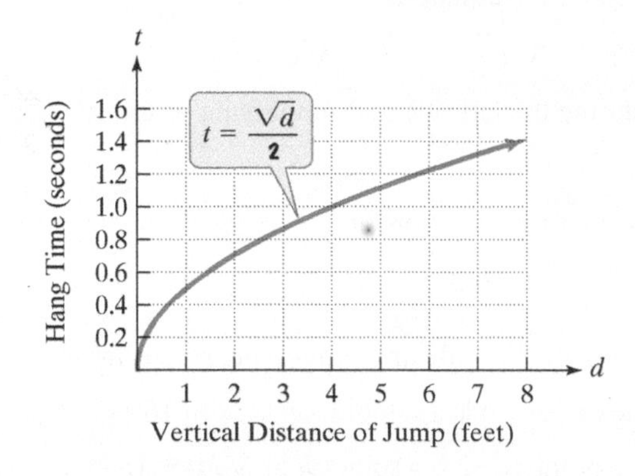

53. How is your answer to Exercise 51 shown on the graph?
54. How is your answer to Exercise 52 shown on the graph?

The graph at the top of the next page shows the less income people have, the more likely they are to report that their health is fair or poor. The function

$$f(x)=-4.4\sqrt{x}+38$$

models the percentage of Americans reporting fair or poor health, f(x), in terms of annual income, x, in thousands of dollars. Use this function to solve Exercises 55–56.

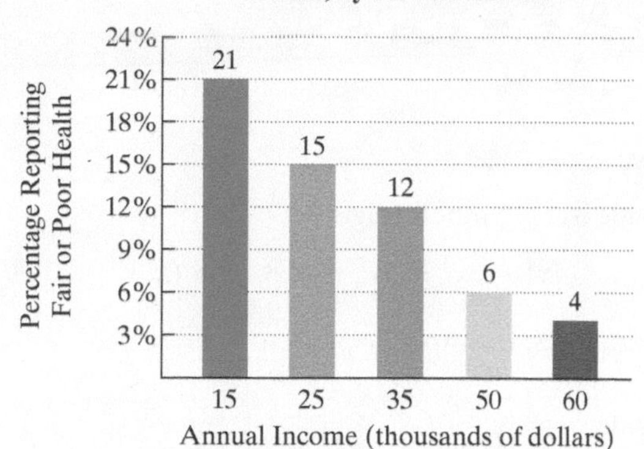

Source: William Kornblum and Joseph Julian, *Social Problems, Twelfth Edition*, Prentice Hall, 2007.

55. a. Find and interpret $f(25)$. Does this underestimate or overestimate the percent displayed by the graph? By how much?

b. According to the model, what annual income corresponds to 14% reporting fair or poor health? Round to the nearest thousand dollars.

56. a. Find and interpret $f(60)$. Round to one decimal place. Does this underestimate or overestimate the percent displayed by the graph? By how much?

b. According to the model, what annual income corresponds to 24% reporting fair or poor health? Round to the nearest thousand dollars.

The function

$$f(x) = 29x^{\frac{1}{3}}$$

models the number of plant species, $f(x)$, on the islands of the Galápagos in terms of the area, x, in square miles, of a particular island. Use the function to solve Exercises 57–58.

57. What is the area of a Galápagos island that has 87 species of plants?

58. What is the area of a Galápagos island that has 58 species of plants?

For each planet in our solar system, its year is the time it takes the planet to revolve once around the sun. The function

$$f(x) = 0.2x^{\frac{3}{2}}$$

models the number of Earth days in a planet's year, $f(x)$, where x is the average distance of the planet from the sun, in millions of kilometers. Use the function to solve Exercises 59–60.

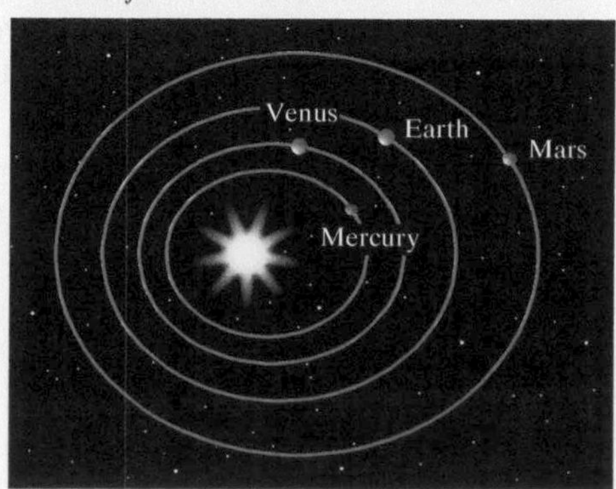

59. We, of course, have 365 Earth days in our year. What is the average distance of Earth from the sun? Use a calculator and round to the nearest million kilometers.

60. There are approximately 88 Earth days in the year of the planet Mercury. What is the average distance of Mercury from the sun? Use a calculator and round to the nearest million kilometers.

Explaining the Concepts

61. What is a radical equation?

62. In solving $\sqrt{2x - 1} + 2 = x$, why is it a good idea to isolate the radical term? What if we don't do this and simply square each side? Describe what happens.

63. What is an extraneous solution to a radical equation?

64. Explain why $\sqrt{x} = -1$ has no solution.

65. Explain how to solve a radical equation with rational exponents.

66. In Example 6 of the section, we used a square root function that modeled an increase in the percentage of U.S. women in the labor force, although the rate of increase in this percentage was leveling off. Describe an event that might occur in the future that could result in an ever-increasing rate in the percentage of women in the labor force. Would a square root function be appropriate for modeling this trend? Explain your answer.

67. The graph for Exercises 55–56 shows that the less income people have, the more likely they are to report fair or poor health. What explanations can you offer for this trend?

Technology Exercises

In Exercises 68–72, use a graphing utility to solve each radical equation. Graph each side of the equation in the given viewing rectangle. The equation's solution set is given by the x-coordinate(s) of the point(s) of intersection. Check by substitution.

68. $\sqrt{2x + 2} = \sqrt{3x - 5}$
$[-1, 10, 1]$ by $[-1, 5, 1]$

69. $\sqrt{x} + 3 = 5$
$[-1, 6, 1]$ by $[-1, 6, 1]$

70. $\sqrt{x^2 + 3} = x + 1$
$[-1, 6, 1]$ by $[-1, 6, 1]$

71. $4\sqrt{x} = x + 3$
$[-1, 10, 1]$ by $[-1, 14, 1]$

72. $\sqrt{x} + 4 = 2$
$[-2, 18, 1]$ by $[0, 10, 1]$

Critical Thinking Exercises

Make Sense? *In Exercises 73–76, determine whether each statement makes sense or does not make sense, and explain your reasoning.*

73. When checking a radical equation's proposed solution, I can substitute into the original equation or any equation that is part of the solution process.

74. After squaring both sides of a radical equation, the only solution that I obtained was extraneous, so $\varnothing$ must be the solution set of the original equation.

75. When I raise both sides of an equation to any power, there's always the possibility of extraneous solutions.

76. Now that I know how to solve radical equations, I can use models that are radical functions to determine the value of the independent variable when a function value is known.

In Exercises 77–80, determine whether each statement is true or false. If the statement is false, make the necessary change(s) to produce a true statement.

77. The first step in solving $\sqrt{x+6} = x + 2$ is to square both sides, obtaining $x + 6 = x^2 + 4$.

78. The equations $\sqrt{x+4} = -5$ and $x + 4 = 25$ have the same solution set.

79. The equation $-\sqrt{x} = 9$ has no solution.

80. The equation $\sqrt{x^2 + 9x + 3} = -x$ has no solution because a principal square root is always nonnegative.

81. Find the length of the three sides of the right triangle shown in the figure.

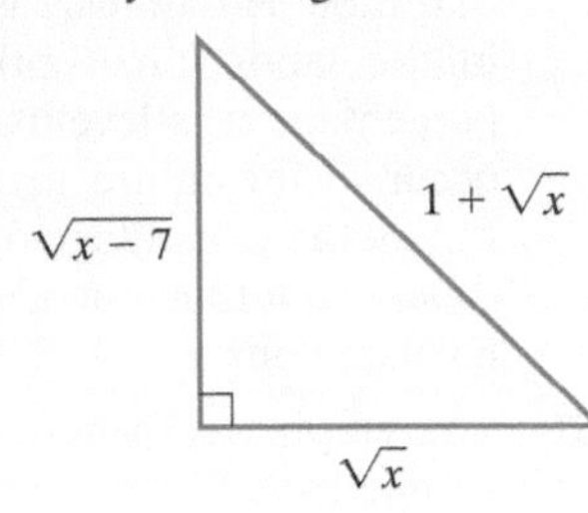

In Exercises 82–84, solve each equation.

82. $\sqrt[3]{x\sqrt{x}} = 9$

83. $\sqrt{\sqrt{x} + \sqrt{x+9}} = 3$

84. $(x - 4)^{\frac{2}{3}} = 25$

Review Exercises

85. Divide using synthetic division:

$$(4x^4 - 3x^3 + 2x^2 - x - 1) \div (x + 3).$$

(Section 6.5, Example 1)

86. Divide:

$$\frac{3x^2 - 12}{x^2 + 2x - 8} \div \frac{6x + 18}{x + 4}.$$

(Section 6.1, Example 7)

87. Factor: $y^2 - 6y + 9 - 25x^2$.
(Section 5.5, Example 6)

Preview Exercises

Exercises 88–90 will help you prepare for the material covered in the next section.

88. Simplify: $(-5 + 7x) - (-11 - 6x)$.

89. Multiply: $(7 - 3x)(-2 - 5x)$.

90. Rationalize the denominator: $\dfrac{7 + 4\sqrt{2}}{2 - 5\sqrt{2}}$.

SECTION

7.7 Complex Numbers 

What am I supposed to learn?

After studying this section, you should be able to:

1. Express square roots of negative numbers in terms of i.
2. Add and subtract complex numbers.
3. Multiply complex numbers.
4. Divide complex numbers.
5. Simplify powers of i.

Who is this kid warning us about our eyeballs turning black if we attempt to find the square root of −9? Don't believe what you hear on the street. Although square roots of negative numbers are not real numbers, they do play a significant role in algebra. In this section, we move beyond the real numbers and discuss square roots with negative radicands.

1 Express square roots of negative numbers in terms of i.

The Imaginary Unit i

In Chapter 8, we will study equations whose solutions involve the square roots of negative numbers. Because the square of a real number is never negative, there is no real number x such that $x^2 = -1$. To provide a setting in which such equations have solutions, mathematicians invented an expanded system of numbers, the complex numbers. The *imaginary number i*, defined to be a solution of the equation $x^2 = -1$, is the basis of this new set.

The Imaginary Unit i

The **imaginary unit i** is defined as

$$i = \sqrt{-1}, \quad \text{where} \quad i^2 = -1.$$

Using the imaginary unit i, we can express the square root of any negative number as a real multiple of i. For example,

$$\sqrt{-25} = \sqrt{25(-1)} = \sqrt{25}\sqrt{-1} = 5i.$$

We can check that $\sqrt{-25} = 5i$ by squaring $5i$ and obtaining -25.

$$(5i)^2 = 5^2i^2 = 25(-1) = -25$$

The Square Root of a Negative Number

If b is a positive real number, then

$$\sqrt{-b} = \sqrt{b(-1)} = \sqrt{b}\sqrt{-1} = \sqrt{b}i \quad \text{or} \quad i\sqrt{b}.$$

EXAMPLE 1 Expressing Square Roots of Negative Numbers as Multiples of i

Write as a multiple of i:

a. $\sqrt{-9}$ **b.** $\sqrt{-3}$ **c.** $\sqrt{-80}$.

Solution

a. $\sqrt{-9} = \sqrt{9(-1)} = \sqrt{9}\sqrt{-1} = 3i$

b. $\sqrt{-3} = \sqrt{3(-1)} = \sqrt{3}\sqrt{-1} = \sqrt{3}i$ — Be sure not to write i under the radical.

c. $\sqrt{-80} = \sqrt{80(-1)} = \sqrt{80}\sqrt{-1} = \sqrt{16 \cdot 5}\sqrt{-1} = 4\sqrt{5}i$

In order to avoid writing i under a radical, let's agree to write i before any radical. Consequently, we express the multiple of i in part (b) as $i\sqrt{3}$ and the multiple of i in part (c) as $4i\sqrt{5}$. ■

Great Question!

Now that we've introduced square roots of negative numbers, can I still use the rule $\sqrt{ab} = \sqrt{a}\sqrt{b}$?

We allow the use of the product rule $\sqrt{ab} = \sqrt{a}\sqrt{b}$ when a is positive and b is -1. However, you cannot use $\sqrt{ab} = \sqrt{a}\sqrt{b}$ when both a and b are negative.

✓ **CHECK POINT 1** Write as a multiple of i:

a. $\sqrt{-64}$ **b.** $\sqrt{-11}$ **c.** $\sqrt{-48}$.

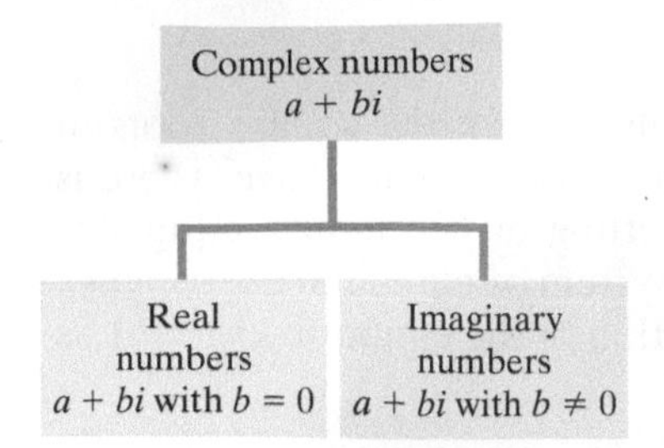

Figure 7.10
The complex number system

A new system of numbers, called *complex numbers*, is based on adding multiples of i, such as $5i$, to the real numbers.

Complex Numbers and Imaginary Numbers

The set of all numbers in the form

$$a + bi,$$

with real numbers a and b, and i, the imaginary unit, is called the set of **complex numbers.** The real number a is called the **real part**, and the real number b is called the **imaginary part** of the complex number $a + bi$. If $b \neq 0$, then the complex number is called an **imaginary number (Figure 7.10).**

Here are some examples of complex numbers. Each number can be written in the form $a + bi$.

$-4 + 6i$ — a, the real part, is -4. b, the imaginary part, is 6.

$2i = 0 + 2i$ — a, the real part, is 0. b, the imaginary part, is 2.

$3 = 3 + 0i$ — a, the real part, is 3. b, the imaginary part, is 0.

Can you see that b, the imaginary part, is not zero in the first two complex numbers? Because $b \neq 0$, these complex numbers are imaginary numbers. By contrast, the imaginary part of the complex number on the right is zero. This complex number is not an imaginary number. The number 3, or $3 + 0i$, is a real number.

2 Add and subtract complex numbers.

Adding and Subtracting Complex Numbers

The form of a complex number $a + bi$ is like the binomial $a + bx$. Consequently, we can add, subtract, and multiply complex numbers using the same methods we used for binomials, remembering that $i^2 = -1$.

Adding and Subtracting Complex Numbers

1. $(a + bi) + (c + di) = (a + c) + (b + d)i$
   In words, this says that you add complex numbers by adding their real parts, adding their imaginary parts, and expressing the sum as a complex number.
2. $(a + bi) - (c + di) = (a - c) + (b - d)i$
   In words, this says that you subtract complex numbers by subtracting their real parts, subtracting their imaginary parts, and expressing the difference as a complex number.

EXAMPLE 2 Adding and Subtracting Complex Numbers

Perform the indicated operations, writing the result in the form $a + bi$:

a. $(5 - 11i) + (7 + 4i)$ **b.** $(-5 + 7i) - (-11 - 6i)$.

Solution

a. $(5 - 11i) + (7 + 4i)$

$= 5 - 11i + 7 + 4i$ Remove the parentheses.

$= 5 + 7 - 11i + 4i$ Group real and imaginary terms.

$= (5 + 7) + (-11 + 4)i$ Add real parts and add imaginary parts.

$= 12 - 7i$ Simplify.

Great Question!

Are operations with complex numbers similar to operations with polynomials?

Yes. The following examples, using the same integers as in Example 2, show how operations with complex numbers are just like operations with polynomials.

a. $(5 - 11x) + (7 + 4x)$
$= 12 - 7x$

b. $(-5 + 7x) - (-11 - 6x)$
$= -5 + 7x + 11 + 6x$
$= 6 + 13x$

b. $(-5 + 7i) - (-11 - 6i)$

$= -5 + 7i + 11 + 6i$ — Remove the parentheses. Change signs of real and imaginary parts in the complex number being subtracted.

$= -5 + 11 + 7i + 6i$ — Group real and imaginary terms.

$= (-5 + 11) + (7 + 6)i$ — Add real parts and add imaginary parts.

$= 6 + 13i$ — Simplify. ■

CHECK POINT 2 Add or subtract as indicated:

a. $(5 - 2i) + (3 + 3i)$

b. $(2 + 6i) - (12 - 4i)$.

---

3 Multiply complex numbers.

Multiplying Complex Numbers

Multiplication of complex numbers is performed the same way as multiplication of polynomials, using the distributive property and the FOIL method. After completing the multiplication, we replace any occurrences of i^2 with -1. This idea is illustrated in the next example.

EXAMPLE 3 Multiplying Complex Numbers

Find the products:

a. $4i(3 - 5i)$ **b.** $(7 - 3i)(-2 - 5i)$.

Solution

a. $4i(3 - 5i)$

$= 4i \cdot 3 - 4i \cdot 5i$ — Distribute $4i$ throughout the parentheses.

$= 12i - 20i^2$ — Multiply.

$= 12i - 20(-1)$ — Replace i^2 with -1.

$= 20 + 12i$ — Simplify to $12i + 20$ and write in $a + bi$ form.

b. $(7 - 3i)(-2 - 5i)$

F O I L

$= -14 - 35i + 6i + 15i^2$ — Use the FOIL method.

$= -14 - 35i + 6i + 15(-1)$ — $i^2 = -1$

$= -14 - 15 - 35i + 6i$ — Group real and imaginary terms.

$= -29 - 29i$ — Combine real and imaginary terms. ■

CHECK POINT 3 Find the products:

a. $7i(2 - 9i)$ **b.** $(5 + 4i)(6 - 7i)$.

Consider the multiplication problem

$$5i \cdot 2i = 10i^2 = 10(-1) = -10.$$

The problem $5i \cdot 2i$ can also be given in terms of square roots of negative numbers:

$$\sqrt{-25} \cdot \sqrt{-4}.$$

Because the product rule for radicals only applies to real numbers, multiplying radicands is incorrect. **When performing operations with square roots of negative numbers, begin by expressing all square roots in terms of *i*.** Then perform the indicated operation.

CORRECT:

$$\sqrt{-25} \cdot \sqrt{-4} = \sqrt{25}\sqrt{-1} \cdot \sqrt{4}\sqrt{-1}$$
$$= 5i \cdot 2i$$
$$= 10i^2 = 10(-1) = -10$$

INCORRECT:

$$\sqrt{-25} \cdot \sqrt{-4} = \sqrt{(-25)(-4)}$$
$$= \sqrt{100}$$
$$= 10$$

EXAMPLE 4 Multiplying Square Roots of Negative Numbers

Multiply: $\sqrt{-3} \cdot \sqrt{-5}$.

Solution

$$\sqrt{-3} \cdot \sqrt{-5} = \sqrt{3}\sqrt{-1} \cdot \sqrt{5}\sqrt{-1}$$
$$= i\sqrt{3} \cdot i\sqrt{5}$$ Express square roots in terms of i.
$$= i^2\sqrt{15}$$ $\sqrt{3} \cdot \sqrt{5} = \sqrt{15}$ and $i \cdot i = i^2$.
$$= (-1)\sqrt{15}$$ $i^2 = -1$
$$= -\sqrt{15}$$ ■

CHECK POINT 4 Multiply: $\sqrt{-5} \cdot \sqrt{-7}$.

---

4 Divide complex numbers.

Conjugates and Division

It is possible to multiply imaginary numbers and obtain a real number. Here is an example:

F O I L

$$(4 + 7i)(4 - 7i) = 16 - 28i + 28i - 49i^2$$
$$= 16 - 49i^2 = 16 - 49(-1) = 65.$$

Replace i^2 with -1.

You can also perform $(4 + 7i)(4 - 7i)$ using the formula

$$(A + B)(A - B) = A^2 - B^2.$$

A real number is obtained even faster:

$$(4 + 7i)(4 - 7i) = 4^2 - (7i)^2 = 16 - 49i^2 = 16 - 49(-1) = 65.$$

The **conjugate** of the complex number $a + bi$ is $a - bi$. The **conjugate** of the complex number $a - bi$ is $a + bi$. The multiplication problem that we just performed involved conjugates. The multiplication of conjugates always results in a real number:

$$(a + bi)(a - bi) = a^2 - (bi)^2 = a^2 - b^2i^2 = a^2 - b^2(-1) = a^2 + b^2.$$

The product eliminates i.

Conjugates are used to divide complex numbers. The goal of the division procedure is to obtain a real number in the denominator. This real number becomes the denominator of a and b in the quotient $a + bi$. By multiplying the numerator and the denominator of the division by the conjugate of the denominator, you will obtain this real number in the denominator. Here are two examples of such divisions:

- $\frac{7+4i}{2-5i}$ — The conjugate of the denominator is $2+5i$.
- $\frac{5i-4}{3i}$ or $\frac{5i-4}{0+3i}$. — The conjugate of the denominator is $0-3i$, or $-3i$.

The procedure for dividing complex numbers, illustrated in Examples 5 and 6, should remind you of rationalizing denominators.

EXAMPLE 5 Using Conjugates to Divide Complex Numbers

Divide and simplify to the form $a + bi$:

$$\frac{7+4i}{2-5i}.$$

Solution The conjugate of the denominator is $2 + 5i$. Multiplication of both the numerator and the denominator by $2 + 5i$ will eliminate i from the denominator while maintaining the value of the expression.

$$\frac{7+4i}{2-5i} = \frac{7+4i}{2-5i}\cdot\frac{2+5i}{2+5i}$$ Multiply by 1.

F O I L

$$= \frac{14+35i+8i+20i^2}{2^2-(5i)^2}$$ Use FOIL in the numerator and $(A-B)(A+B) = A^2 - B^2$ in the denominator.

$$= \frac{14+43i+20i^2}{4-25i^2}$$ Simplify.

$$= \frac{14+43i+20(-1)}{4-25(-1)}$$ $i^2 = -1$

$$= \frac{14+43i-20}{4+25}$$ Perform the multiplications involving -1.

$$= \frac{-6+43i}{29}$$ Combine real terms in the numerator and denominator.

$$= -\frac{6}{29} + \frac{43}{29}i$$ Express the answer in the form $a + bi$. ■

CHECK POINT 5 Divide and simplify to the form $a + bi$:

$$\frac{6+2i}{4-3i}.$$

EXAMPLE 6 Using Conjugates to Divide Complex Numbers

Divide and simplify to the form $a + bi$:

$$\frac{5i-4}{3i}.$$

Solution The denominator of $\frac{5i-4}{3i}$ is $3i$, or $0+3i$. The conjugate of the denominator is $0-3i$. Multiplication of both the numerator and the denominator by $-3i$ will eliminate i from the denominator while maintaining the value of the expression.

$$\frac{5i-4}{3i} = \frac{5i-4}{3i} \cdot \frac{-3i}{-3i} \quad \text{Multiply by 1.}$$

$$= \frac{-15i^2+12i}{-9i^2} \quad \text{Multiply. Use the distributive property in the numerator.}$$

$$= \frac{-15(-1)+12i}{-9(-1)} \quad i^2 = -1$$

$$= \frac{15+12i}{9} \quad \text{Perform the multiplications involving } -1.$$

$$= \frac{15}{9} + \frac{12}{9}i \quad \text{Express the division in the form } a+bi.$$

$$= \frac{5}{3} + \frac{4}{3}i \quad \text{Simplify real and imaginary parts.} \blacksquare$$

✓ CHECK POINT 6 Divide and simplify to the form $a + bi$:

$$\frac{3-2i}{4i}.$$

---

5 Simplify powers of i.

Powers of *i*

Using the fact that $i^2 = -1$, any integral power of i greater than or equal to 2 can be simplified to either $-i$, i, -1, or 1. Here are some examples:

$$i^3 = i^2 \cdot i = (-1)i = -i$$
$$i^4 = (i^2)^2 = (-1)^2 = 1$$
$$i^5 = i^4 \cdot i = (i^2)^2 \cdot i = (-1)^2 \cdot i = i$$
$$i^6 = (i^2)^3 = (-1)^3 = -1$$

Here is a procedure for simplifying powers of i:

Simplifying Powers of *i*

1. Express the given power of i in terms of i^2.
2. Replace i^2 with -1 and simplify. Use the fact that -1 to an even power is 1 and -1 to an odd power is -1.

EXAMPLE 7 Simplifying Powers of *i*

Simplify:

a. i^{12} **b.** i^{39} **c.** i^{50}.

Solution

a. $i^{12} = (i^2)^6 = (-1)^6 = 1$

b. $i^{39} = i^{38}i = (i^2)^{19}i = (-1)^{19}i = (-1)i = -i$

c. $i^{50} = (i^2)^{25} = (-1)^{25} = -1$ ■

✓ CHECK POINT 7 Simplify:

a. i^{16} **b.** i^{25} **c.** i^{35}.

Blitzer Bonus

The Patterns of Chaos

One of the new frontiers of mathematics suggests that there is an underlying order in things that appear to be random, such as the hiss and crackle of background noises as you tune a radio. Irregularities in the heartbeat, some of them severe enough to cause a heart attack, or irregularities in our sleeping patterns, such as insomnia, are examples of chaotic behavior. Chaos in the mathematical sense does not mean a complete lack of form or arrangement. In mathematics, chaos is used to describe something that appears to be random, but actually contains underlying patterns that are far more intricate than previously assumed. The patterns of chaos appear in images like the one on the right and the one in the chapter opener, called the Mandelbrot set. Magnified portions of this image yield repetitions of the original structure, as well as new and unexpected patterns. The Mandelbrot set transforms the hidden structure of chaotic events into a source of wonder and inspiration.

The Mandelbrot set is made possible by opening up graphing to include complex numbers in the form $a + bi$. Each complex number is plotted like an ordered pair in a coordinate system consisting of a real axis and an imaginary axis. Plot certain complex numbers in this system, add color to the magnified boundary of the graph, and the patterns of chaos begin to appear.

29-Fold M-Set Seahorse. © 2011 Richard F. Voss

Achieving Success

Two Ways to Stay Sharp

- **Concentrate on one task at a time.** Do not multitask. Doing several things at once can cause confusion and can take longer to complete the tasks than tackling them sequentially.
- **Get enough sleep.** Fatigue impedes the ability to learn and do complex tasks. (One can only imagine what occurs if you're fatigued and trying to perform complex tasks with complex numbers!)

CONCEPT AND VOCABULARY CHECK

Fill in each blank so that the resulting statement is true.

1. The imaginary unit i is defined as $i =$ ____________, where $i^2 =$ ____________.
2. $\sqrt{-16} = \sqrt{16(-1)} = \sqrt{16}\sqrt{-1} =$ ____________
3. The set of all numbers in the form $a + bi$ is called the set of ____________ numbers. If $b \neq 0$, then the number is also called a/an ____________ number. If $b = 0$, then the number is also called a/an ____________ number.
4. $-9i + 3i =$ ____________
5. $10i - (-4i) =$ ____________
6. Consider the following multiplication problem:
   $$(3 + 2i)(6 - 5i).$$
   Using the FOIL method, the product of the first terms is ____________, the product of the outside terms is ____________, and the product of the inside terms is ____________. The product of the last terms in terms of i^2 is ____________, which simplifies to ____________.
7. The conjugate of $2 - 9i$ is ____________.
8. The division
   $$\frac{7 + 4i}{2 - 5i}$$
   is performed by multiplying the numerator and denominator by ____________.
9. The division
   $$\frac{3 - 2i}{4i}$$
   is performed by multiplying the numerator and denominator by ____________.
10. $i^{16} = (i^2)^8 = (\_)^8 = \_$
11. $i^{35} = i^{34}i = (i^2)^{17}i = (\_)^{17}i = (\_)i = \_$

7.7 EXERCISE SET

Practice Exercises

In Exercises 1–16, express each number in terms of i and simplify, if possible.

1. $\sqrt{-100}$
2. $\sqrt{-49}$
3. $\sqrt{-23}$
4. $\sqrt{-21}$
5. $\sqrt{-18}$
6. $\sqrt{-125}$
7. $\sqrt{-63}$
8. $\sqrt{-28}$
9. $-\sqrt{-108}$
10. $-\sqrt{-300}$
11. $5 + \sqrt{-36}$
12. $7 + \sqrt{-4}$
13. $15 + \sqrt{-3}$
14. $20 + \sqrt{-5}$
15. $-2 - \sqrt{-18}$
16. $-3 - \sqrt{-27}$

In Exercises 17–32, add or subtract as indicated. Write the result in the form a + bi.

17. $(3 + 2i) + (5 + i)$
18. $(6 + 5i) + (4 + 3i)$
19. $(7 + 2i) + (1 - 4i)$
20. $(-2 + 6i) + (4 - i)$
21. $(10 + 7i) - (5 + 4i)$
22. $(11 + 8i) - (2 + 5i)$
23. $(9 - 4i) - (10 + 3i)$
24. $(8 - 5i) - (6 + 2i)$
25. $(3 + 2i) - (5 - 7i)$
26. $(-7 + 5i) - (9 - 11i)$
27. $(-5 + 4i) - (-13 - 11i)$
28. $(-9 + 2i) - (-17 - 6i)$
29. $8i - (14 - 9i)$
30. $15i - (12 - 11i)$
31. $(2 + i\sqrt{3}) + (7 + 4i\sqrt{3})$
32. $(4 + i\sqrt{5}) + (8 + 6i\sqrt{5})$

In Exercises 33–62, find each product. Write imaginary results in the form a + bi.

33. $2i(5 + 3i)$
34. $5i(4 + 7i)$
35. $3i(7i - 5)$
36. $8i(4i - 3)$
37. $-7i(2 - 5i)$
38. $-6i(3 - 5i)$
39. $(3 + i)(4 + 5i)$
40. $(4 + i)(5 + 6i)$
41. $(7 - 5i)(2 - 3i)$
42. $(8 - 4i)(3 - 2i)$
43. $(6 - 3i)(-2 + 5i)$
44. $(7 - 2i)(-3 + 6i)$
45. $(3 + 5i)(3 - 5i)$
46. $(2 + 7i)(2 - 7i)$
47. $(-5 + 3i)(-5 - 3i)$
48. $(-4 + 2i)(-4 - 2i)$
49. $(3 - i\sqrt{2})(3 + i\sqrt{2})$
50. $(5 - i\sqrt{3})(5 + i\sqrt{3})$
51. $(2 + 3i)^2$
52. $(3 + 2i)^2$
53. $(5 - 2i)^2$
54. $(5 - 3i)^2$
55. $\sqrt{-7} \cdot \sqrt{-2}$
56. $\sqrt{-7} \cdot \sqrt{-3}$
57. $\sqrt{-9} \cdot \sqrt{-4}$
58. $\sqrt{-16} \cdot \sqrt{-4}$
59. $\sqrt{-7} \cdot \sqrt{-25}$
60. $\sqrt{-3} \cdot \sqrt{-36}$
61. $\sqrt{-8} \cdot \sqrt{-3}$
62. $\sqrt{-9} \cdot \sqrt{-5}$

In Exercises 63–84, divide and simplify to the form a + bi.

63. $\frac{2}{3 + i}$
64. $\frac{3}{4 + i}$
65. $\frac{2i}{1 + i}$
66. $\frac{5i}{2 + i}$
67. $\frac{7}{4 - 3i}$
68. $\frac{9}{1 - 2i}$
69. $\frac{6i}{3 - 2i}$
70. $\frac{5i}{2 - 3i}$
71. $\frac{1 + i}{1 - i}$
72. $\frac{1 - i}{1 + i}$
73. $\frac{2 - 3i}{3 + i}$
74. $\frac{2 + 3i}{3 - i}$
75. $\frac{5 - 2i}{3 + 2i}$
76. $\frac{6 - 3i}{4 + 2i}$
77. $\frac{4 + 5i}{3 - 7i}$
78. $\frac{5 - i}{3 - 2i}$
79. $\frac{7}{3i}$
80. $\frac{5}{7i}$
81. $\frac{8 - 5i}{2i}$
82. $\frac{3 + 4i}{5i}$
83. $\frac{4 + 7i}{-3i}$
84. $\frac{5 + i}{-4i}$

In Exercises 85–100, simplify each expression.

85. i^{10}
86. i^{14}
87. i^{11}
88. i^{15}
89. i^{22}
90. i^{46}
91. i^{200}
92. i^{400}
93. i^{17}
94. i^{21}
95. $(-i)^4$
96. $(-i)^6$
97. $(-i)^9$
98. $(-i)^{13}$
99. $i^{24} + i^2$
100. $i^{28} + i^{30}$

Practice PLUS

In Exercises 101–108, perform the indicated operation(s) and write the result in the form a + bi.

101. $(2 - 3i)(1 - i) - (3 - i)(3 + i)$
102. $(8 + 9i)(2 - i) - (1 - i)(1 + i)$
103. $(2 + i)^2 - (3 - i)^2$
104. $(4 - i)^2 - (1 + 2i)^2$
105. $5\sqrt{-16} + 3\sqrt{-81}$
106. $5\sqrt{-8} + 3\sqrt{-18}$
107. $\frac{i^4 + i^{12}}{i^8 - i^7}$
108. $\frac{i^8 + i^{40}}{i^4 + i^3}$

109. Let $f(x) = x^2 - 2x + 2$. Find $f(1 + i)$.

110. Let $f(x) = x^2 - 2x + 5$. Find $f(1 - 2i)$.

In Exercises 111–114, simplify each evaluation to the form $a + bi$.

111. Let $f(x) = x - 3i$ and $g(x) = 4x + 2i$. Find $(fg)(-1)$.

112. Let $f(x) = 12x - i$ and $g(x) = 6x + 3i$. Find $(fg)\left(-\frac{1}{3}\right)$.

113. Let $f(x) = \frac{x^2 + 19}{2 - x}$. Find $f(3i)$.

114. Let $f(x) = \frac{x^2 + 11}{3 - x}$. Find $f(4i)$.

Application Exercises

Complex numbers are used in electronics to describe the current in an electric circuit. Ohm's law relates the current in a circuit, I, in amperes, the voltage of the circuit, E, in volts, and the resistance of the circuit, R, in ohms, by the formula $E = IR$. Use this formula to solve Exercises 115–116.

115. Find E, the voltage of a circuit, if $I = (4 - 5i)$ amperes and $R = (3 + 7i)$ ohms.

116. Find E, the voltage of a circuit, if $I = (2 - 3i)$ amperes and $R = (3 + 5i)$ ohms.

117. The mathematician Girolamo Cardano is credited with the first use (in 1545) of negative square roots in solving the now-famous problem, "Find two numbers whose sum is 10 and whose product is 40." Show that the complex numbers $5 + i\sqrt{15}$ and $5 - i\sqrt{15}$ satisfy the conditions of the problem. (Cardano did not use the symbolism $i\sqrt{15}$ or even $\sqrt{-15}$. He wrote R.m 15 for $\sqrt{-15}$, meaning "radix minus 15." He regarded the numbers 5 + R.m 15 and 5 − R.m 15 as "fictitious" or "ghost numbers," and considered the problem "manifestly impossible." But in a mathematically adventurous spirit, he exclaimed, "Nevertheless, we will operate.")

Explaining the Concepts

118. What is the imaginary unit i?

119. Explain how to write $\sqrt{-64}$ as a multiple of i.

120. What is a complex number? Explain when a complex number is a real number and when it is an imaginary number. Provide examples with your explanation.

121. Explain how to add complex numbers. Give an example.

122. Explain how to subtract complex numbers. Give an example.

123. Explain how to find the product of $2i$ and $5 + 3i$.

124. Explain how to find the product of $3 + 2i$ and $5 + 3i$.

125. Explain how to find the product of $3 + 2i$ and $3 - 2i$.

126. Explain how to find the product of $\sqrt{-1}$ and $\sqrt{-4}$. Describe a common error in the multiplication that needs to be avoided.

127. What is the conjugate of $2 + 3i$? What happens when you multiply this complex number by its conjugate?

128. Explain how to divide complex numbers. Provide an example with your explanation.

129. Explain each of the three jokes in the cartoon on page 562.

130. A stand-up comedian uses algebra in some jokes, including one about a telephone recording that announces "You have just reached an imaginary number. Please multiply by i and dial again." Explain the joke.

Explain the error in Exercises 131–132.

131. $\sqrt{-9} + \sqrt{-16} = \sqrt{-25} = i\sqrt{25} = 5i$

132. $(\sqrt{-9})^2 = \sqrt{-9} \cdot \sqrt{-9} = \sqrt{81} = 9$

Critical Thinking Exercises

Make Sense? *In Exercises 133–136, determine whether each statement makes sense or does not make sense, and explain your reasoning.*

133. The humor in this cartoon is based on the fact that "rational" and "real" have different meanings in mathematics and in everyday speech.

134. The word *imaginary* in imaginary numbers tells me that these numbers are undefined.

135. By writing the imaginary number $5i$, I can immediately see that 5 is the constant and i is the variable.

136. When I add or subtract complex numbers, I am basically combining like terms.

In Exercises 137–140, determine whether each statement is true or false. If the statement is false, make the necessary change(s) to produce a true statement.

137. Some irrational numbers are not complex numbers.

138. $(3 + 7i)(3 - 7i)$ is an imaginary number.

139. $\frac{7 + 3i}{5 + 3i} = \frac{7}{5}$

140. In the complex number system, $x^2 + y^2$ (the sum of two squares) can be factored as $(x + yi)(x - yi)$.

In Exercises 141–143, perform the indicated operations and write the result in the form $a + bi$.

141. $\dfrac{4}{(2 + i)(3 - i)}$

142. $\dfrac{1 + i}{1 + 2i} + \dfrac{1 - i}{1 - 2i}$

143. $\dfrac{8}{1 + \dfrac{2}{i}}$

Review Exercises

144. Simplify:

$$\frac{\frac{x}{y^2} + \frac{1}{y}}{\frac{y}{x^2} + \frac{1}{x}}.$$

(Section 6.3, Example 1)

145. Solve for x: $\dfrac{1}{x} + \dfrac{1}{y} = \dfrac{1}{z}$.

(Section 6.7, Example 1)

146. Solve:

$$2x - \frac{x - 3}{8} = \frac{1}{2} + \frac{x + 5}{2}.$$

(Section 1.4, Example 4)

Preview Exercises

Exercises 147–149 will help you prepare for the material covered in the first section of the next chapter.

147. Solve by factoring: $2x^2 + 7x - 4 = 0$.

148. Solve by factoring: $x^2 = 9$.

149. Use substitution to determine if $-\sqrt{6}$ is a solution of the quadratic equation $3x^2 = 18$.

Chapter 7 Summary

Definitions and Concepts	Examples
Section 7.1 Radical Expressions and Functions	
If $b^2 = a$, then b is a square root of a. The principal square root of a, designated $\sqrt{a}$, is the nonnegative number satisfying $b^2 = a$. The negative square root of a is written $-\sqrt{a}$. A square root of a negative number is not a real number. A radical function in x is a function defined by an expression containing a root of x. The domain of a square root function is the set of real numbers for which the radicand is nonnegative.	Let $f(x) = \sqrt{6 - 2x}$. $f(-15) = \sqrt{6 - 2(-15)} = \sqrt{6 + 30} = \sqrt{36} = 6$ $f(5) = \sqrt{6 - 2 \cdot 5} = \sqrt{6 - 10} = \sqrt{-4}$, not a real number Domain of f: Set the radicand greater than or equal to zero. $6 - 2x \geq 0$ $-2x \geq -6$ $x \leq 3$ Domain of $f = (-\infty, 3]$
The cube root of a real number a is written $\sqrt[3]{a}$. $\sqrt[3]{a} = b$ means that $b^3 = a$. The nth root of a real number a is written $\sqrt[n]{a}$. The number n is the index. Every real number has one root when n is odd. The odd nth root of a, $\sqrt[n]{a}$, is the number b for which $b^n = a$. Every positive real number has two real roots when n is even. An even root of a negative number is not a real number. • If n is even, then $\sqrt[n]{a^n} = \lvert a \rvert$. • If n is odd, then $\sqrt[n]{a^n} = a$.	• $\sqrt[3]{-8} = -2$ because $(-2)^3 = -8$. • $\sqrt[4]{-16}$ is not a real number. • $\sqrt{x^2 - 14x + 49} = \sqrt{(x - 7)^2} = \lvert x - 7 \rvert$ • $\sqrt[3]{125(x + 6)^3} = 5(x + 6)$

Definitions and Concepts	Examples

Section 7.2 Rational Exponents

Definitions and Concepts	Examples
• $a^{\frac{1}{n}} = \sqrt[n]{a}$ • $a^{\frac{m}{n}} = (\sqrt[n]{a})^m$ or $\sqrt[n]{a^m}$ • $a^{-\frac{m}{n}} = \dfrac{1}{a^{\frac{m}{n}}}$	• $121^{\frac{1}{2}} = \sqrt{121} = 11$ • $64^{\frac{1}{3}} = \sqrt[3]{64} = 4$ • $27^{\frac{5}{3}} = (\sqrt[3]{27})^5 = 3^5 = 3 \cdot 3 \cdot 3 \cdot 3 \cdot 3 = 243$ • $16^{-\frac{3}{4}} = \dfrac{1}{16^{\frac{3}{4}}} = \dfrac{1}{(\sqrt[4]{16})^3} = \dfrac{1}{2^3} = \dfrac{1}{8}$ • $(\sqrt[3]{7xy})^4 = (7xy)^{\frac{4}{3}}$
Properties of integer exponents are true for rational exponents. An expression with rational exponents is simplified when no parentheses appear, no powers are raised to powers, each base occurs once, and no negative or zero exponents appear.	Simplify: $\left(8x^{\frac{1}{3}}y^{-\frac{1}{2}}\right)^{\frac{1}{3}}$. $= 8^{\frac{1}{3}}\left(x^{\frac{1}{3}}\right)^{\frac{1}{3}}\left(y^{-\frac{1}{2}}\right)^{\frac{1}{3}}$ $= 2x^{\frac{1}{9}}y^{-\frac{1}{6}} = \dfrac{2x^{\frac{1}{9}}}{y^{\frac{1}{6}}}$
Some radical expressions can be simplifed using rational exponents. Rewrite the expression using rational exponents, simplify, and rewrite in radical notation if rational exponents still appear.	• $\sqrt[9]{x^3} = x^{\frac{3}{9}} = x^{\frac{1}{3}} = \sqrt[3]{x}$ • $\sqrt[5]{x^2} \cdot \sqrt[4]{x} = x^{\frac{2}{5}} \cdot x^{\frac{1}{4}} = x^{\frac{2}{5}+\frac{1}{4}}$ $= x^{\frac{8}{20}+\frac{5}{20}} = x^{\frac{13}{20}} = \sqrt[20]{x^{13}}$

Section 7.3 Multiplying and Simplifying Radical Expressions

Definitions and Concepts	Examples
The product rule for radicals can be used to multiply radicals: $\sqrt[n]{a} \cdot \sqrt[n]{b} = \sqrt[n]{ab}.$	$\sqrt[3]{7x} \cdot \sqrt[3]{10y^2} = \sqrt[3]{7x \cdot 10y^2} = \sqrt[3]{70xy^2}$
The product rule for radicals can be used to simplify radicals: $\sqrt[n]{ab} = \sqrt[n]{a} \cdot \sqrt[n]{b}.$ A radical expression with index n is simplified when its radicand has no factors that are perfect nth powers. To simplify, write the radicand as the product of two factors, one of which is the greatest perfect nth power. Then use the product rule to take the nth root of each factor. If all variables in a radicand are positive, then $\sqrt[n]{a^n} = a.$ Some radicals can be simplified after multiplication is performed.	• Simplify: $\sqrt[3]{54x^7y^{11}}$. $= \sqrt[3]{27 \cdot 2 \cdot x^6 \cdot x \cdot y^9 \cdot y^2}$ $= \sqrt[3]{(27x^6y^9)(2xy^2)}$ $= \sqrt[3]{27x^6y^9} \cdot \sqrt[3]{2xy^2} = 3x^2y^3\sqrt[3]{2xy^2}$ • Assuming positive variables, multiply and simplify: $\sqrt[4]{4x^2y} \cdot \sqrt[4]{4xy^3}$. $= \sqrt[4]{4x^2y \cdot 4xy^3} = \sqrt[4]{16x^3y^4}$ $= \sqrt[4]{16y^4} \cdot \sqrt[4]{x^3} = 2y\sqrt[4]{x^3}$

Section 7.4 Adding, Subtracting, and Dividing Radical Expressions

Definitions and Concepts	Examples
Like radicals have the same indices and radicands. Like radicals can be added or subtracted using the distributive property. In some cases, radicals can be combined once they have been simplified.	$4\sqrt{18} - 6\sqrt{50}$ $= 4\sqrt{9 \cdot 2} - 6\sqrt{25 \cdot 2} = 4 \cdot 3\sqrt{2} - 6 \cdot 5\sqrt{2}$ $= 12\sqrt{2} - 30\sqrt{2} = -18\sqrt{2}$

Definitions and Concepts	Examples

Section 7.4 Adding, Subtracting, and Dividing Radical Expressions (continued)

Definitions and Concepts	Examples
The quotient rule for radicals can be used to simplify radicals: $\sqrt[n]{\frac{a}{b}} = \frac{\sqrt[n]{a}}{\sqrt[n]{b}}$.	$\sqrt[3]{-\frac{8}{x^{12}}} = \frac{\sqrt[3]{-8}}{\sqrt[3]{x^{12}}} = -\frac{2}{x^4}$ $\sqrt[3]{x^{12}} = (x^{12})^{\frac{1}{3}} = x^4$
The quotient rule for radicals can be used to divide radicals with the same indices: $\frac{\sqrt[n]{a}}{\sqrt[n]{b}} = \sqrt[n]{\frac{a}{b}}$. Some radicals can be simplified after the division is performed.	Assuming a positive variable, divide and simplify: $\frac{\sqrt[4]{64x^5}}{\sqrt[4]{2x^{-2}}} = \sqrt[4]{32x^{5-(-2)}} = \sqrt[4]{32x^7}$ $= \sqrt[4]{16 \cdot 2 \cdot x^4 \cdot x^3} = \sqrt[4]{16x^4} \cdot \sqrt[4]{2x^3}$ $= 2x\sqrt[4]{2x^3}$.

Section 7.5 Multiplying with More Than One Term and Rationalizing Denominators

Definitions and Concepts	Examples
Radical expressions with more than one term are multiplied in much the same way that polynomials with more than one term are multiplied.	• $\sqrt{5}(2\sqrt{6} - \sqrt{3}) = 2\sqrt{30} - \sqrt{15}$ • $(4\sqrt{3} - 2\sqrt{2})(\sqrt{3} + \sqrt{2})$ F O I L $= 4\sqrt{3} \cdot \sqrt{3} + 4\sqrt{3} \cdot \sqrt{2} - 2\sqrt{2} \cdot \sqrt{3} - 2\sqrt{2} \cdot \sqrt{2}$ $= 4 \cdot 3 + 4\sqrt{6} - 2\sqrt{6} - 2 \cdot 2$ $= 12 + 4\sqrt{6} - 2\sqrt{6} - 4 = 8 + 2\sqrt{6}$
Radical expressions that involve the sum and difference of the same two terms are called conjugates. Use $(A + B)(A - B) = A^2 - B^2$ to multiply conjugates.	$(8 + 2\sqrt{5})(8 - 2\sqrt{5})$ $= 8^2 - (2\sqrt{5})^2$ $= 64 - 4 \cdot 5$ $= 64 - 20 = 44$
The process of rewriting a radical expression as an equivalent expression without any radicals in the denominator is called rationalizing the denominator. When the denominator contains a single radical with an *n*th root, multiply the numerator and the denominator by a radical of index *n* that produces a perfect *n*th power in the denominator's radicand.	Rationalize the denominator: $\frac{7}{\sqrt[3]{2x}}$. $= \frac{7}{\sqrt[3]{2x}} \cdot \frac{\sqrt[3]{4x^2}}{\sqrt[3]{4x^2}} = \frac{7\sqrt[3]{4x^2}}{\sqrt[3]{8x^3}} = \frac{7\sqrt[3]{4x^2}}{2x}$
If the denominator contains two terms, rationalize the denominator by multiplying the numerator and the denominator by the conjugate of the denominator.	$\frac{13}{5 - \sqrt{3}} = \frac{13}{5 - \sqrt{3}} \cdot \frac{5 + \sqrt{3}}{5 + \sqrt{3}}$ $= \frac{13(5 + \sqrt{3})}{5^2 - (\sqrt{3})^2}$ $= \frac{13(5 + \sqrt{3})}{25 - 3} = \frac{13(5 + \sqrt{3})}{22}$

Definitions and Concepts	Examples

Section 7.6 Radical Equations

A radical equation is an equation in which the variable occurs in a radicand.

Solving Radical Equations Containing *n*th Roots

1. Isolate one radical on one side of the equation.
2. Raise both sides to the nth power.
3. Solve the resulting equation.
4. Check proposed solutions in the original equation. Solutions of an equation to an even power that is radical-free, but not the original equation, are called extraneous solutions.

Solve: $\sqrt{6x + 13} - 2x = 1$.

$\sqrt{6x + 13} = 2x + 1$ — Isolate the radical.

$(\sqrt{6x + 13})^2 = (2x + 1)^2$ — Square both sides.

$6x + 13 = 4x^2 + 4x + 1$ — $(A + B)^2 = A^2 + 2AB + B^2$

$0 = 4x^2 - 2x - 12$ — Subtract $6x + 13$ from both sides.

$0 = 2(2x^2 - x - 6)$ — Factor out the GCF.

$0 = 2(2x + 3)(x - 2)$ — Factor completely.

$2x + 3 = 0$ or $x - 2 = 0$ — Set variable factors equal to zero.

$2x = -3$, $x = -\frac{3}{2}$; $x = 2$ — Solve for x.

Check both proposed solutions. 2 checks, but $-\frac{3}{2}$ is extraneous.

The solution is 2 and the solution set is {2}.

Section 7.7 Complex Numbers

The imaginary unit i is defined as

$$i = \sqrt{-1}, \quad \text{where} \quad i^2 = -1.$$

The set of numbers in the form $a + bi$ is called the set of complex numbers; a is the real part and b is the imaginary part. If $b = 0$, the complex number is a real number. If $b \neq 0$, the complex number is an imaginary number.

- $\sqrt{-81} = \sqrt{81(-1)} = \sqrt{81}\sqrt{-1} = 9i$
- $\sqrt{-75} = \sqrt{75(-1)} = \sqrt{25 \cdot 3}\sqrt{-1} = 5i\sqrt{3}$

To add or subtract complex numbers, add or subtract their real parts and add or subtract their imaginary parts.

$(2 - 4i) - (7 - 10i)$
$= 2 - 4i - 7 + 10i$
$= (2 - 7) + (-4 + 10)i = -5 + 6i$

To multiply complex numbers, multiply as if they were polynomials. After completing the multiplication, replace i^2 with -1. When performing operations with square roots of negative numbers, begin by expressing all square roots in terms of i. Then multiply.

F O I L

- $(2 - 3i)(4 + 5i) = 8 + 10i - 12i - 15i^2$
  $= 8 + 10i - 12i - 15(-1)$
  $= 23 - 2i$
- $\sqrt{-36} \cdot \sqrt{-100} = \sqrt{36(-1)} \cdot \sqrt{100(-1)}$
  $= 6i \cdot 10i = 60i^2 = 60(-1) = -60$

The complex numbers $a + bi$ and $a - bi$ are conjugates. Conjugates can be multiplied using the formula

$$(A + B)(A - B) = A^2 - B^2.$$

The multiplication of conjugates results in a real number.

$(3 + 5i)(3 - 5i) = 3^2 - (5i)^2$
$= 9 - 25i^2$
$= 9 - 25(-1) = 34$

Definitions and Concepts	Examples

Section 7.7 Complex Numbers (continued)

Definitions and Concepts	Examples
To divide complex numbers, multiply the numerator and the denominator by the conjugate of the denominator in order to obtain a real number in the denominator. This real number becomes the denominator of a and b in the quotient $a + bi$.	$\frac{5+2i}{4-i} = \frac{5+2i}{4-i}\cdot\frac{4+i}{4+i} = \frac{20+5i+8i+2i^2}{16-i^2} = \frac{20+13i+2(-1)}{16-(-1)} = \frac{20+13i-2}{16+1} = \frac{18+13i}{17} = \frac{18}{17}+\frac{13}{17}i$
To simplify powers of i, rewrite the expression in terms of i^2. Then replace i^2 with -1 and simplify.	Simplify: i^{27}. $i^{27} = i^{26}\cdot i = (i^2)^{13}i = (-1)^{13}i = (-1)i = -i$

CHAPTER 7 REVIEW EXERCISES

7.1 *In Exercises 1–5, find the indicated root, or state that the expression is not a real number.*

1. $\sqrt{81}$
2. $-\sqrt{\frac{1}{100}}$
3. $\sqrt[3]{-27}$
4. $\sqrt[4]{-16}$
5. $\sqrt[5]{-32}$

In Exercises 6–7, find the indicated function values for each function. If necessary, round to two decimal places. If the function value is not a real number and does not exist, so state.

6. $f(x) = \sqrt{2x-5};\ f(15), f(4), f\left(\frac{5}{2}\right), f(1)$
7. $g(x) = \sqrt[3]{4x-8};\ g(4), g(0), g(-14)$

In Exercises 8–9, find the domain of each square root function.

8. $f(x) = \sqrt{x-2}$
9. $g(x) = \sqrt{100-4x}$

In Exercises 10–15, simplify each expression. Assume that each variable can represent any real number, so include absolute value bars where necessary.

10. $\sqrt{25x^2}$
11. $\sqrt{(x+14)^2}$
12. $\sqrt{x^2-8x+16}$
13. $\sqrt[3]{64x^3}$
14. $\sqrt[4]{16x^4}$
15. $\sqrt[5]{-32(x+7)^5}$

7.2 *In Exercises 16–18, use radical notation to rewrite each expression. Simplify, if possible.*

16. $(5xy)^{\frac{1}{3}}$
17. $16^{\frac{3}{2}}$
18. $32^{\frac{4}{5}}$

In Exercises 19–20, rewrite each expression with rational exponents.

19. $\sqrt{7x}$
20. $\left(\sqrt[3]{19xy}\right)^5$

In Exercises 21–22, rewrite each expression with a positive rational exponent. Simplify, if possible.

21. $8^{-\frac{2}{3}}$
22. $3x(ab)^{-\frac{4}{5}}$

In Exercises 23–26, use properties of rational exponents to simplify each expression.

23. $x^{\frac{1}{3}}\cdot x^{\frac{1}{4}}$
24. $\frac{5^{\frac{1}{2}}}{5^{\frac{1}{3}}}$
25. $(8x^6y^3)^{\frac{1}{3}}$
26. $\left(x^{-\frac{2}{3}}y^{\frac{1}{4}}\right)^{\frac{1}{2}}$

In Exercises 27–31, use rational exponents to simplify each expression. If rational exponents appear after simplifying, write the answer in radical notation.

27. $\sqrt[3]{x^9y^{12}}$
28. $\sqrt[9]{x^3y^9}$
29. $\sqrt{x}\cdot\sqrt[3]{x}$
30. $\frac{\sqrt[3]{x^2}}{\sqrt[4]{x^2}}$
31. $\sqrt[5]{\sqrt[3]{x}}$

32. The function $f(x) = 350x^{\frac{2}{3}}$ models the expenditures, $f(x)$, in millions of dollars, for the U.S. National Park Service x years after 1985. According to this model, what were expenditures in 2012?

7.3 *In Exercises 33–35, use the product rule to multiply.*

33. $\sqrt{3x}\cdot\sqrt{7y}$
34. $\sqrt[5]{7x^2}\cdot\sqrt[5]{11x}$
35. $\sqrt[6]{x-5}\cdot\sqrt[6]{(x-5)^4}$
36. If $f(x) = \sqrt{7x^2-14x+7}$, express the function, f, in simplified form. Assume that x can be any real number.

In Exercises 37–39, simplify by factoring. Assume that all variables in a radicand represent positive real numbers.

37. $\sqrt{20x^3}$

38. $\sqrt[3]{54x^8y^6}$

39. $\sqrt[4]{32x^3y^{11}z^5}$

In Exercises 40–43, multiply and simplify, if possible. Assume that all variables in a radicand represent positive real numbers.

40. $\sqrt{6x^3}\cdot\sqrt{4x^2}$

41. $\sqrt[3]{4x^2y}\cdot\sqrt[3]{4xy^4}$

42. $\sqrt[5]{2x^4y^3z^4}\cdot\sqrt[5]{8xy^6z^7}$

43. $\sqrt{x+1}\cdot\sqrt{x-1}$

7.4

Assume that all variables represent positive real numbers. In Exercises 44–47, add or subtract as indicated.

44. $6\sqrt[3]{3}+2\sqrt[3]{3}$

45. $5\sqrt{18}-3\sqrt{8}$

46. $\sqrt[3]{27x^4}+\sqrt[3]{xy^6}$

47. $2\sqrt[3]{6}-5\sqrt[3]{48}$

In Exercises 48–50, simplify using the quotient rule.

48. $\sqrt[3]{\dfrac{16}{125}}$

49. $\sqrt{\dfrac{x^3}{100y^4}}$

50. $\sqrt[4]{\dfrac{3y^5}{16x^{20}}}$

In Exercises 51–54, divide and, if possible, simplify.

51. $\dfrac{\sqrt{48}}{\sqrt{2}}$

52. $\dfrac{\sqrt[3]{32}}{\sqrt[3]{2}}$

53. $\dfrac{\sqrt[4]{64x^7}}{\sqrt[4]{2x^2}}$

54. $\dfrac{\sqrt{200x^3y^2}}{\sqrt{2x^{-2}y}}$

7.5

Assume that all variables represent positive real numbers.

In Exercises 55–62, multiply as indicated. If possible, simplify any radical expressions that appear in the product.

55. $\sqrt{3}(2\sqrt{6}+4\sqrt{15})$

56. $\sqrt[3]{5}(\sqrt[3]{50}-\sqrt[3]{2})$

57. $(\sqrt{7}-3\sqrt{5})(\sqrt{7}+6\sqrt{5})$

58. $(\sqrt{x}-\sqrt{11})(\sqrt{y}-\sqrt{11})$

59. $(\sqrt{5}+\sqrt{8})^2$

60. $(2\sqrt{3}-\sqrt{10})^2$

61. $(\sqrt{7}+\sqrt{13})(\sqrt{7}-\sqrt{13})$

62. $(7-3\sqrt{5})(7+3\sqrt{5})$

In Exercises 63–75, rationalize each denominator. Simplify, if possible.

63. $\dfrac{4}{\sqrt{6}}$

64. $\sqrt{\dfrac{2}{7}}$

65. $\dfrac{12}{\sqrt[3]{9}}$

66. $\sqrt{\dfrac{2x}{5y}}$

67. $\dfrac{14}{\sqrt[3]{2x^2}}$

68. $\sqrt[4]{\dfrac{7}{3x}}$

69. $\dfrac{5}{\sqrt[5]{32x^4y}}$

70. $\dfrac{6}{\sqrt{3}-1}$

71. $\dfrac{\sqrt{7}}{\sqrt{5}+\sqrt{3}}$

72. $\dfrac{10}{2\sqrt{5}-3\sqrt{2}}$

73. $\dfrac{\sqrt{x}+5}{\sqrt{x}-3}$

74. $\dfrac{\sqrt{7}+\sqrt{3}}{\sqrt{7}-\sqrt{3}}$

75. $\dfrac{2\sqrt{3}+\sqrt{6}}{2\sqrt{6}+\sqrt{3}}$

In Exercises 76–79, rationalize each numerator. Simplify, if possible.

76. $\sqrt{\dfrac{2}{7}}$

77. $\dfrac{\sqrt[3]{3x}}{\sqrt[3]{y}}$

78. $\dfrac{\sqrt{7}}{\sqrt{5}+\sqrt{3}}$

79. $\dfrac{\sqrt{7}+\sqrt{3}}{\sqrt{7}-\sqrt{3}}$

7.6

In Exercises 80–84, solve each radical equation.

80. $\sqrt{2x+4}=6$

81. $\sqrt{x-5}+9=4$

82. $\sqrt{2x-3}+x=3$

83. $\sqrt{x-4}+\sqrt{x+1}=5$

84. $(x^2+6x)^{\frac{1}{3}}+2=0$

85. The bar graph shows the percentage of U.S. college freshmen who described their health as "above average" for six selected years.

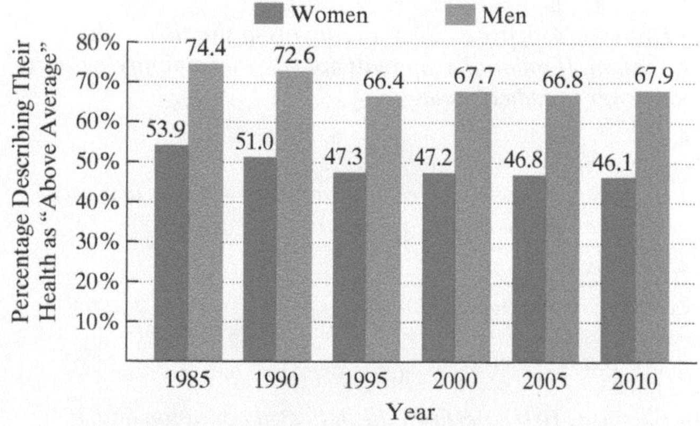

Source: John Macionis, *Sociology, Fourteenth Edition*, Pearson, 2012.

The function

$$f(x) = -1.6\sqrt{x}+54$$

models the percentage of freshmen women who described their health as above average, $f(x)$, x years after 1985.

a. Find and interpret $f(20)$. Round to the nearest tenth of a percent. How does this rounded value compare with the percentage of women displayed by the graph?

b. According to the model, in which year will 44.4% of freshmen women describe their health as above average?

86. Out of a group of 50,000 births, the number of people, $f(x)$, surviving to age x is modeled by the function

$$f(x) = 5000\sqrt{100 - x}.$$

To what age will 20,000 people in the group survive?

7.7 *In Exercises 87–89, express each number in terms of i and simplify, if possible.*

87. $\sqrt{-81}$

88. $\sqrt{-63}$

89. $-\sqrt{-8}$

In Exercises 90–99, perform the indicated operation. Write the result in the form a + bi.

90. $(7 + 12i) + (5 - 10i)$

91. $(8 - 3i) - (17 - 7i)$

92. $4i(3i - 2)$

93. $(7 - 5i)(2 + 3i)$

94. $(3 - 4i)^2$

95. $(7 + 8i)(7 - 8i)$

96. $\sqrt{-8} \cdot \sqrt{-3}$

97. $\dfrac{6}{5 + i}$

98. $\dfrac{3 + 4i}{4 - 2i}$

99. $\dfrac{5 + i}{3i}$

In Exercises 100–101, simplify each expression.

100. i^{16}

101. i^{23}

CHAPTER 7 TEST

Step-by-step test solutions are found on the Chapter Test Prep Videos available in MyMathLab® or on YouTube® (search "BlitzerInterAlg7e" and click on "Channels").

1. Let $f(x) = \sqrt{8 - 2x}$.

a. Find $f(-14)$.

b. Find the domain of f.

2. Evaluate: $27^{-\frac{4}{3}}$.

3. Simplify: $\left(25x^{-\frac{1}{2}}y^{\frac{1}{4}}\right)^{\frac{1}{2}}$.

In Exercises 4–5, use rational exponents to simplify each expression. If rational exponents appear after simplifying, write the answer in radical notation.

4. $\sqrt[8]{x^4}$

5. $\sqrt[4]{x} \cdot \sqrt[5]{x}$

In Exercises 6–9, simplify each expression. Assume that each variable can represent any real number.

6. $\sqrt{75x^2}$

7. $\sqrt{x^2 - 10x + 25}$

8. $\sqrt[3]{16x^4y^8}$

9. $\sqrt[5]{-\dfrac{32}{x^{10}}}$

In Exercises 10–17, perform the indicated operation and, if possible, simplify. Assume that all variables represent positive real numbers.

10. $\sqrt[3]{5x^2} \cdot \sqrt[3]{10y}$

11. $\sqrt[4]{8x^3y} \cdot \sqrt[4]{4xy^2}$

12. $3\sqrt{18} - 4\sqrt{32}$

13. $\sqrt[3]{8x^4} + \sqrt[3]{xy^6}$

14. $\dfrac{\sqrt[3]{16x^8}}{\sqrt[3]{2x^4}}$

15. $\sqrt{3}(4\sqrt{6} - \sqrt{5})$

16. $(5\sqrt{6} - 2\sqrt{2})(\sqrt{6} + \sqrt{2})$

17. $(7 - \sqrt{3})^2$

In Exercises 18–20, rationalize each denominator. Simplify, if possible. Assume all variables represent positive real numbers.

18. $\sqrt{\dfrac{5}{x}}$

19. $\dfrac{5}{\sqrt[3]{5x^2}}$

20. $\dfrac{\sqrt{2} - \sqrt{3}}{\sqrt{2} + \sqrt{3}}$

In Exercises 21–23, solve each radical equation.

21. $3 + \sqrt{2x - 3} = x$

22. $\sqrt{x + 9} - \sqrt{x - 7} = 2$

23. $(11x + 6)^{\frac{1}{3}} + 3 = 0$

24. The function

$$f(x) = 2.9\sqrt{x} + 20.1$$

models the average height, $f(x)$, in inches, of boys who are x months of age, $0 \leq x \leq 60$. Find the age at which the average height of boys is 40.4 inches.

25. Express in terms of i and simplify: $\sqrt{-75}$.

In Exercises 26–29, perform the indicated operation. Write the result in the form a + bi.

26. $(5 - 3i) - (6 - 9i)$

27. $(3 - 4i)(2 + 5i)$

28. $\sqrt{-9} \cdot \sqrt{-4}$

29. $\dfrac{3 + i}{1 - 2i}$

30. Simplify: i^{35}.

CUMULATIVE REVIEW EXERCISES (CHAPTERS 1–7)

In Exercises 1–5, solve each equation, inequality, or system.

1. $\begin{cases} 2x - y + z = -5 \\ x - 2y - 3z = 6 \\ x + y - 2z = 1 \end{cases}$
2. $3x^2 - 11x = 4$
3. $2(x + 4) < 5x + 3(x + 2)$
4. $\dfrac{1}{x+2} + \dfrac{15}{x^2 - 4} = \dfrac{5}{x-2}$
5. $\sqrt{x+2} - \sqrt{x+1} = 1$
6. Graph the solution set of the system:

   $\begin{cases} x + 2y < 2 \\ 2y - x > 4. \end{cases}$

In Exercises 7–15, perform the indicated operations.

7. $\dfrac{8x^2}{3x^2 - 12} \div \dfrac{40}{x - 2}$
8. $\dfrac{x + \frac{1}{y}}{y + \frac{1}{x}}$
9. $(2x - 3)(4x^2 - 5x - 2)$
10. $\dfrac{7x}{x^2 - 2x - 15} - \dfrac{2}{x - 5}$
11. $7(8 - 10)^3 - 7 + 3 \div (-3)$
12. $\sqrt{80x} - 5\sqrt{20x} + 2\sqrt{45x}$
13. $\dfrac{\sqrt{3} - 2}{2\sqrt{3} + 5}$
14. $(2x^3 - 3x^2 + 3x - 4) \div (x - 2)$
15. $(2\sqrt{3} + 5\sqrt{2})(\sqrt{3} - 4\sqrt{2})$

In Exercises 16–17, factor completely.

16. $24x^2 + 10x - 4$
17. $16x^4 - 1$
18. The amount of light provided by a light bulb varies inversely as the square of the distance from the bulb. The illumination provided is 120 lumens at a distance of 10 feet. How many lumens are provided at a distance of 15 feet?
19. You invested $6000 in two accounts paying 7% and 9% annual interest. At the end of the year, the total interest from these investments was $510. How much was invested at each rate?
20. Although there are 2332 students enrolled in the college, this is 12% fewer students than there were enrolled last year. How many students were enrolled last year?

CHAPTER 8

Quadratic Equations and Functions

We are surrounded by evidence that the world is profoundly mathematical. After turning a somersault, a diver's path can be modeled by a quadratic function, $f(x) = ax^2 + bx + c$, as can the path of a football tossed from quarterback to receiver or the path of a flipped coin. Even if you throw an object directly upward, although its path is straight and vertical, its changing height over time is described by a quadratic function. And tailgaters beware: whether you're driving a car, a motorcycle, or a truck on dry or wet roads, an array of quadratic functions that model your required stopping distances at various speeds is available to help you become a safer driver.

Here's where you'll find these applications:

- The quadratic functions surrounding our long history of throwing things appear throughout the chapter, including Example 6 in Section 8.3 and Example 6 in Section 8.5.
- Tailgaters should pay close attention to the Section 8.5 opener, Exercises 73–74 and 88–89 in Exercise Set 8.5, and Exercises 30–31 in the Chapter Review Exercises.

SECTION

8.1 The Square Root Property and Completing the Square

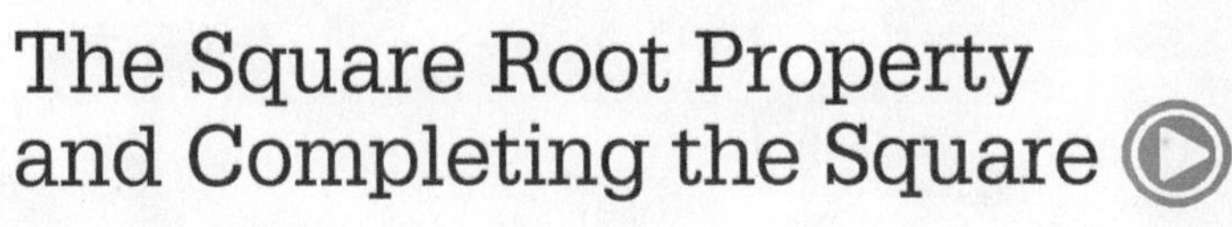

What am I supposed to learn?

After studying this section, you should be able to:

1. Solve quadratic equations using the square root property.
2. Complete the square of a binomial.
3. Solve quadratic equations by completing the square.
4. Solve problems using the square root property.

I'm very well acquainted, too, with matters mathematical,
I understand equations, both simple and quadratical.
About binomial theorem I'm teeming with a lot of news,
With many cheerful facts about the square of the hypotenuse.

—Gilbert and Sullivan,
The Pirates of Penzance

Equations quadratical? Cheerful news about the square of the hypotenuse? You've come to the right place. In this section, you'll enhance your understanding of quadratic equations with two new solution methods, called the *square root method* and *completing the square*. Using these techniques, we return (cheerfully, of course) to the Pythagorean Theorem and the square of the hypotenuse.

Great Question!

Haven't we already discussed quadratic equations and quadratic functions? Other than not calling them equations and functions quadratical, what am I expected to know about these topics in order to achieve success in this chapter?

Here is a summary of what you should already know about quadratic equations and quadratic functions.

1. A **quadratic equation** in x can be written in the standard form
$$ax^2 + bx + c = 0, \quad a \neq 0.$$
2. Some quadratic equations can be solved by factoring.
Solve: $2x^2 + 7x - 4 = 0.$
$$(2x - 1)(x + 4) = 0$$
$$2x - 1 = 0 \quad \text{or} \quad x + 4 = 0$$
$$2x = 1 \qquad\qquad x = -4$$
$$x = \tfrac{1}{2}$$
The solutions are -4 and $\frac{1}{2}$, and the solution set is $\left\{-4, \frac{1}{2}\right\}$.
3. A polynomial function of the form
$$f(x) = ax^2 + bx + c, \quad a \neq 0$$
is a **quadratic function**. Graphs of quadratic functions are shaped like bowls or inverted bowls, with the same behavior at each end.
4. The real solutions of $ax^2 + bx + c = 0$ correspond to the x-intercepts for the graph of the quadratic function $f(x) = ax^2 + bx + c$. For example, the solutions of the equation $2x^2 + 7x - 4 = 0$ are -4 and $\frac{1}{2}$. **Figure 8.1** shows that the solutions appear as x-intercepts on the graph of the quadratic function $f(x) = 2x^2 + 7x - 4$.

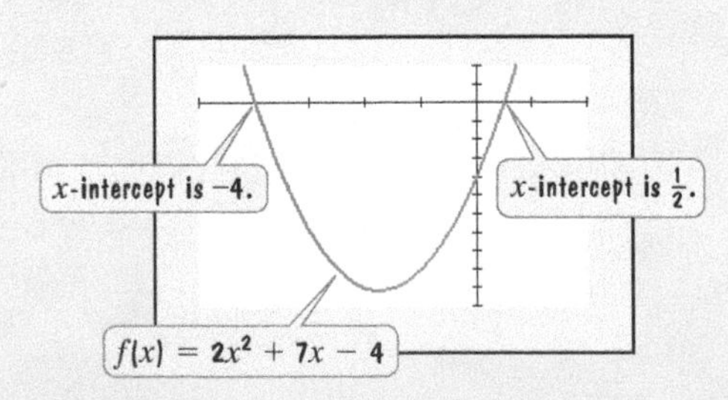

Figure 8.1

Now that we've summarized what we know, let's look at where we go. How do we solve a quadratic equation, $ax^2 + bx + c = 0$, if the trinomial $ax^2 + bx + c$ cannot be factored? Methods other than factoring are needed. In this section and the next, we look at other ways of solving quadratic equations.

1 Solve quadratic equations using the square root property.

The Square Root Property

Let's begin with a relatively simple quadratic equation:

$$x^2 = 9.$$

The value of x must be a number whose square is 9. There are two numbers whose square is 9:

$$x = \sqrt{9} = 3 \quad \text{or} \quad x = -\sqrt{9} = -3.$$

Thus, the solutions of $x^2 = 9$ are 3 and -3. This is an example of the **square root property**.

Discover for Yourself

Solve $x^2 = 9$, or

$$x^2 - 9 = 0,$$

by factoring. What is the advantage of using the square root property?

The Square Root Property

If u is an algebraic expression and d is a nonzero real number, then $u^2 = d$ is equivalent to $u = \sqrt{d}$ or $u = -\sqrt{d}$:

$$\text{If } u^2 = d, \quad \text{then} \quad u = \sqrt{d} \text{ or } u = -\sqrt{d}.$$

Equivalently,

$$\text{If} \quad u^2 = d, \quad \text{then} \quad u = \pm\sqrt{d}.$$

Notice that $u = \pm\sqrt{d}$ is a shorthand notation to indicate that $u = \sqrt{d}$ or $u = -\sqrt{d}$. Although we usually read $u = \pm\sqrt{d}$ as "u equals plus or minus the square root of d," we actually mean that u is the positive square root of d or the negative square root of d.

EXAMPLE 1 Solving a Quadratic Equation by the Square Root Property

Solve: $3x^2 = 18$.

Solution To apply the square root property, we need a squared expression by itself on one side of the equation.

$$3x^2 = 18$$

We want x^2 by itself.

We can get x^2 by itself if we divide both sides by 3.

$3x^2 = 18$	This is the original equation.
$\dfrac{3x^2}{3} = \dfrac{18}{3}$	Divide both sides by 3.
$x^2 = 6$	Simplify.
$x = \sqrt{6}$ or $x = -\sqrt{6}$	Apply the square root property.

Now let's check these proposed solutions in the original equation. Because the equation has an x^2-term and no x-term, we can check both values, $\pm\sqrt{6}$, at once.

Check $\sqrt{6}$ and $-\sqrt{6}$:

$$3x^2 = 18 \quad \text{This is the original equation.}$$
$$3(\pm\sqrt{6})^2 \stackrel{?}{=} 18 \quad \text{Substitute the proposed solutions.}$$
$$3 \cdot 6 \stackrel{?}{=} 18 \quad (\pm\sqrt{6})^2 = 6$$
$$18 = 18, \quad \text{true}$$

The solutions are $-\sqrt{6}$ and $\sqrt{6}$. The solution set is $\{-\sqrt{6}, \sqrt{6}\}$ or $\{\pm\sqrt{6}\}$. ■

✓ CHECK POINT 1 Solve: $4x^2 = 28$.

In this section, we will express irrational solutions in simplified radical form, rationalizing denominators when possible.

EXAMPLE 2 Solving a Quadratic Equation by the Square Root Property

Solve: $2x^2 - 7 = 0$.

Solution To solve by the square root property, we isolate the squared expression on one side of the equation.

$$2x^2 - 7 = 0$$

We want x^2 by itself.

$$2x^2 - 7 = 0 \quad \text{This is the original equation.}$$
$$2x^2 = 7 \quad \text{Add 7 to both sides.}$$
$$x^2 = \frac{7}{2} \quad \text{Divide both sides by 2.}$$
$$x = \sqrt{\frac{7}{2}} \text{ or } x = -\sqrt{\frac{7}{2}} \quad \text{Apply the square root property.}$$

Because the proposed solutions are opposites, we can rationalize both denominators at once:

$$\pm\sqrt{\frac{7}{2}} = \pm\frac{\sqrt{7}}{\sqrt{2}} \cdot \frac{\sqrt{2}}{\sqrt{2}} = \pm\frac{\sqrt{14}}{2}.$$

Substitute these values into the original equation and verify that the solutions are $-\frac{\sqrt{14}}{2}$ and $\frac{\sqrt{14}}{2}$. The solution set is $\left\{-\frac{\sqrt{14}}{2}, \frac{\sqrt{14}}{2}\right\}$ or $\left\{\pm\frac{\sqrt{14}}{2}\right\}$. ■

✓ CHECK POINT 2 Solve: $3x^2 - 11 = 0$.

Some quadratic equations have solutions that are imaginary numbers.

EXAMPLE 3 Solving a Quadratic Equation by the Square Root Property

Solve: $9x^2 + 25 = 0$.

Solution We begin by isolating the squared expression on one side of the equation.

$$9x^2 + 25 = 0$$

We need to isolate x^2.

$$9x^2 + 25 = 0 \quad \text{This is the original equation.}$$
$$9x^2 = -25 \quad \text{Subtract 25 from both sides.}$$
$$x^2 = -\frac{25}{9} \quad \text{Divide both sides by 9.}$$
$$x = \sqrt{-\frac{25}{9}} \quad \text{or} \quad x = -\sqrt{-\frac{25}{9}} \quad \text{Apply the square root property.}$$
$$x = \sqrt{\frac{25}{9}}\sqrt{-1} \qquad x = -\sqrt{\frac{25}{9}}\sqrt{-1}$$
$$x = \frac{5}{3}i \qquad x = -\frac{5}{3}i \qquad \sqrt{-1} = i$$

Using Technology

Graphic Connections

The graph of

$$f(x) = 9x^2 + 25$$

has no x-intercepts. This shows that

$$9x^2 + 25 = 0$$

has no real solutions. Example 3 algebraically establishes that the solutions are imaginary numbers.

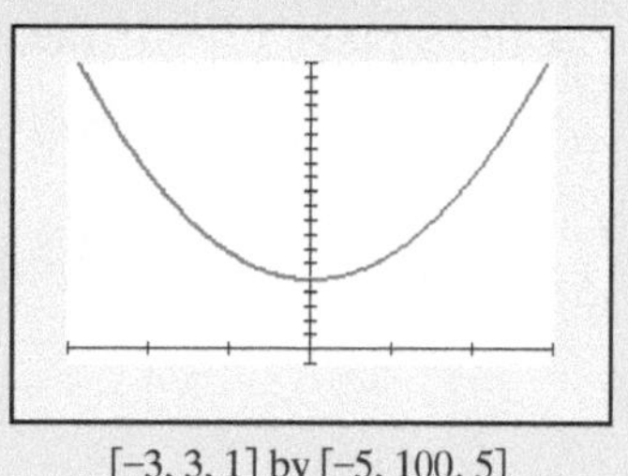

[−3, 3, 1] by [−5, 100, 5]

Because the equation has an x^2-term and no x-term, we can check both proposed solutions, $\pm\frac{5}{3}i$, at once.

Check $\frac{5}{3}i$ and $-\frac{5}{3}i$:

$$9x^2 + 25 = 0 \quad \text{This is the original equation.}$$
$$9\left(\pm\frac{5}{3}i\right)^2 + 25 \stackrel{?}{=} 0 \quad \text{Substitute the proposed solutions.}$$
$$9\left(\frac{25}{9}i^2\right) + 25 \stackrel{?}{=} 0 \quad \left(\pm\frac{5}{3}i\right)^2 = \left(\pm\frac{5}{3}\right)^2 i^2 = \frac{25}{9}i^2$$
$$25i^2 + 25 \stackrel{?}{=} 0 \quad 9\cdot\frac{25}{9} = 25$$

$i^2 = -1$

$$25(-1) + 25 \stackrel{?}{=} 0 \quad \text{Replace } i^2 \text{ with } -1.$$
$$0 = 0, \quad \text{true}$$

The solutions are $-\frac{5}{3}i$ and $\frac{5}{3}i$. The solution set is $\left\{-\frac{5}{3}i, \frac{5}{3}i\right\}$ or $\left\{\pm\frac{5}{3}i\right\}$. ■

✓ CHECK POINT 3 Solve: $4x^2 + 9 = 0$.

Can we solve an equation such as $(x-1)^2 = 5$ using the square root property? Yes. The equation is in the form $u^2 = d$, where u^2, the squared expression, is by itself on the left side.

$$(x-1)^2 \quad = \quad 5$$

This is u^2 in $u^2 = d$ with $u = x - 1$.

This is d in $u^2 = d$ with $d = 5$.

Discover for Yourself

Try solving

$$(x-1)^2 = 5$$

by writing the equation in standard form and factoring. What problem do you encounter?

EXAMPLE 4 Solving a Quadratic Equation by the Square Root Property

Solve by the square root property: $(x-1)^2 = 5$.

Solution

$$(x-1)^2 = 5 \quad \text{This is the original equation.}$$
$$x - 1 = \sqrt{5} \quad \text{or} \quad x - 1 = -\sqrt{5} \quad \text{Apply the square root property.}$$
$$x = 1 + \sqrt{5} \qquad x = 1 - \sqrt{5} \quad \text{Add 1 to both sides in each equation.}$$

Check $1 + \sqrt{5}$:

$$(x-1)^2 = 5$$
$$(1+\sqrt{5}-1)^2 \stackrel{?}{=} 5$$
$$(\sqrt{5})^2 \stackrel{?}{=} 5$$
$$5 = 5, \quad \text{true}$$

Check $1 - \sqrt{5}$:

$$(x-1)^2 = 5$$
$$(1-\sqrt{5}-1)^2 \stackrel{?}{=} 5$$
$$(-\sqrt{5})^2 \stackrel{?}{=} 5$$
$$5 = 5, \quad \text{true}$$

The solutions are $1 \pm \sqrt{5}$, and the solution set is $\{1+\sqrt{5}, 1-\sqrt{5}\}$ or $\{1 \pm \sqrt{5}\}$. ■

✓ CHECK POINT 4 Solve: $(x-3)^2 = 10$.

2 Complete the square of a binomial.

Completing the Square

We return to the question that opened this section: How do we solve a quadratic equation, $ax^2 + bx + c = 0$, if the trinomial $ax^2 + bx + c$ cannot be factored? We can convert the equation into an equivalent equation that can be solved using the square root property. This is accomplished by **completing the square**.

Completing the Square

If $x^2 + bx$ is a binomial, then by adding $\left(\frac{b}{2}\right)^2$, which is the square of half the coefficient of x, a perfect square trinomial will result.

The coefficient of x^2 must be 1 to complete the square.

$$x^2 + bx + \left(\frac{b}{2}\right)^2 = \left(x + \frac{b}{2}\right)^2.$$

EXAMPLE 5 Completing the Square

What term should be added to each binomial so that it becomes a perfect square trinomial? Write and factor the trinomial.

a. $x^2 + 8x$ **b.** $x^2 - 7x$ **c.** $x^2 + \frac{3}{5}x$

Solution To complete the square, we must add a term to each binomial. The term that should be added is the square of half the coefficient of x.

$x^2 + 8x$: Add $\left(\frac{8}{2}\right)^2 = 4^2$. Add 16 to complete the square.

$x^2 - 7x$: Add $\left(\frac{-7}{2}\right)^2$, or $\frac{49}{4}$, to complete the square.

$x^2 + \frac{3}{5}x$: Add $\left(\frac{1}{2}\cdot\frac{3}{5}\right)^2 = \left(\frac{3}{10}\right)^2$. Add $\frac{9}{100}$ to complete the square.

a. The coefficient of the x-term in $x^2 + 8x$ is 8. Half of 8 is 4, and $4^2 = 16$. Add 16. The result is a perfect square trinomial.

$$x^2 + 8x + 16 = (x+4)^2$$

b. The coefficient of the x-term in $x^2 - 7x$ is -7. Half of -7 is $-\frac{7}{2}$, and $\left(-\frac{7}{2}\right)^2 = \frac{49}{4}$. Add $\frac{49}{4}$. The result is a perfect square trinomial.

$$x^2 - 7x + \frac{49}{4} = \left(x - \frac{7}{2}\right)^2$$

c. The coefficient of the x-term in $x^2 + \frac{3}{5}x$ is $\frac{3}{5}$. Half of $\frac{3}{5}$ is $\frac{1}{2} \cdot \frac{3}{5}$, or $\frac{3}{10}$, and $\left(\frac{3}{10}\right)^2 = \frac{9}{100}$. Add $\frac{9}{100}$. The result is a perfect square trinomial.

$$x^2 + \frac{3}{5}x + \frac{9}{100} = \left(x + \frac{3}{10}\right)^2 \quad ■$$

Great Question!

I'm not accustomed to factoring perfect square trinomials in which fractions are involved. Is there a rule or an observation that can make the factoring easier?

Yes. The constant in the factorization is always half the coefficient of x.

$$x^2 - 7x + \frac{49}{4} = \left(x - \frac{7}{2}\right)^2 \qquad x^2 + \frac{3}{5}x + \frac{9}{100} = \left(x + \frac{3}{10}\right)^2$$

Half the coefficient of x, -7, is $-\frac{7}{2}$.

Half the coefficient of x, $\frac{3}{5}$, is $\frac{3}{10}$.

✓ CHECK POINT 5 What term should be added to each binomial so that it becomes a perfect square trinomial? Write and factor the trinomial.

a. $x^2 + 10x$

b. $x^2 - 3x$

c. $x^2 + \frac{3}{4}x$

---

3 Solve quadratic equations by completing the square.

Solving Quadratic Equations by Completing the Square

We can solve *any* quadratic equation by completing the square. If the coefficient of the x^2-term is 1, we add the square of half the coefficient of x to both sides of the equation. **When you add a constant term to one side of the equation to complete the square, be certain to add the same constant to the other side of the equation.** These ideas are illustrated in Example 6.

Great Question!

When I solve a quadratic equation by completing the square, doesn't this result in a new equation? How do I know that the solutions of this new equation are the same as those of the given equation?

When you complete the square for the binomial expression $x^2 + bx$, you obtain a different polynomial. When you solve a quadratic equation by completing the square, you obtain an equation with the same solution set because the constant needed to complete the square is added to *both sides*.

EXAMPLE 6 Solving a Quadratic Equation by Completing the Square

Solve by completing the square: $x^2 - 6x + 4 = 0$.

Solution We begin by subtracting 4 from both sides. This is done to isolate the binomial $x^2 - 6x$ so that we can complete the square.

$x^2 - 6x + 4 = 0$ This is the original equation.

$x^2 - 6x = -4$ Subtract 4 from both sides.

Next, we work with $x^2 - 6x = -4$ and complete the square. Find half the coefficient of the x-term and square it. The coefficient of the x-term is -6. Half of -6 is -3 and $(-3)^2 = 9$. Thus, we add 9 to both sides of the equation.

$x^2 - 6x + 9 = -4 + 9$ Add 9 to both sides to complete the square.

$(x - 3)^2 = 5$ Factor and simplify.

$x - 3 = \sqrt{5}$ or $x - 3 = -\sqrt{5}$ Apply the square root property.

$x = 3 + \sqrt{5}$ $\quad x = 3 - \sqrt{5}$ Add 3 to both sides in each equation.

The solutions are $3 \pm \sqrt{5}$, and the solution set is $\{3 + \sqrt{5}, 3 - \sqrt{5}\}$ or $\{3 \pm \sqrt{5}\}$. ■

If you solve a quadratic equation by completing the square and the solutions are rational numbers, the equation can also be solved by factoring. By contrast, quadratic equations with irrational solutions cannot be solved by factoring. However, all quadratic equations can be solved by completing the square.

✓ **CHECK POINT 6** Solve by completing the square: $x^2 + 4x - 1 = 0.$

We have seen that the leading coefficient must be 1 in order to complete the square. If the coefficient of the x^2-term in a quadratic equation is not 1, you must divide each side of the equation by this coefficient before completing the square. For example, to solve $9x^2 - 6x - 4 = 0$ by completing the square, first divide every term by 9:

$$\frac{9x^2}{9} - \frac{6x}{9} - \frac{4}{9} = \frac{0}{9}$$

$$x^2 - \frac{6}{9}x - \frac{4}{9} = 0$$

$$x^2 - \frac{2}{3}x - \frac{4}{9} = 0.$$

Now that the coefficient of the x^2-term is 1, we can solve by completing the square.

EXAMPLE 7 Solving a Quadratic Equation by Completing the Square

Solve by completing the square: $9x^2 - 6x - 4 = 0.$

Solution

$$9x^2 - 6x - 4 = 0$$ This is the original equation.

$$x^2 - \frac{2}{3}x - \frac{4}{9} = 0$$ Divide both sides by 9.

$$x^2 - \frac{2}{3}x = \frac{4}{9}$$ Add $\frac{4}{9}$ to both sides to isolate the binomial.

$$x^2 - \frac{2}{3}x + \frac{1}{9} = \frac{4}{9} + \frac{1}{9}$$ Complete the square: Half of $-\frac{2}{3}$ is $-\frac{2}{6}$, or $-\frac{1}{3}$, and $\left(-\frac{1}{3}\right)^2 = \frac{1}{9}$.

$$\left(x - \frac{1}{3}\right)^2 = \frac{5}{9}$$ Factor and simplify.

$$x - \frac{1}{3} = \sqrt{\frac{5}{9}} \quad \text{or} \quad x - \frac{1}{3} = -\sqrt{\frac{5}{9}}$$ Apply the square root property.

$$x - \frac{1}{3} = \frac{\sqrt{5}}{3} \qquad x - \frac{1}{3} = -\frac{\sqrt{5}}{3}$$ $\sqrt{\frac{5}{9}} = \frac{\sqrt{5}}{\sqrt{9}} = \frac{\sqrt{5}}{3}$

$$x = \frac{1}{3} + \frac{\sqrt{5}}{3} \qquad x = \frac{1}{3} - \frac{\sqrt{5}}{3}$$ Add $\frac{1}{3}$ to both sides and solve for x.

$$x = \frac{1 + \sqrt{5}}{3} \qquad x = \frac{1 - \sqrt{5}}{3}$$ Express solutions with a common denominator.

The solutions are $\frac{1 \pm \sqrt{5}}{3}$ and the solution set is $\left\{\frac{1 \pm \sqrt{5}}{3}\right\}$. ■

Using Technology

Graphic Connections

Obtain a decimal approximation for each solution of

$$9x^2 - 6x - 4 = 0,$$

the equation in Example 7.

$$\frac{1 + \sqrt{5}}{3} \approx 1.1$$

$$\frac{1 - \sqrt{5}}{3} \approx -0.4$$

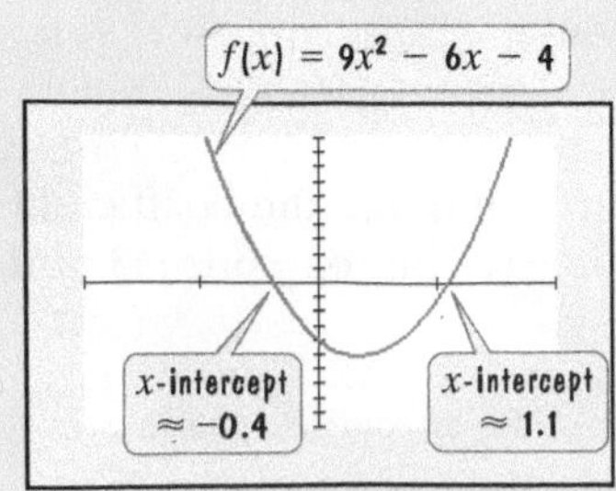

[−2, 2, 1] by [−10, 10, 1]

The x-intercepts of $f(x) = 9x^2 - 6x - 4$ verify the solutions.

✓ **CHECK POINT 7** Solve by completing the square: $2x^2 + 3x - 4 = 0.$

EXAMPLE 8 Solving a Quadratic Equation by Completing the Square

Solve by completing the square: $2x^2 - x + 6 = 0$.

Solution

$$2x^2 - x + 6 = 0$$ This is the original equation.

$$x^2 - \frac{1}{2}x + 3 = 0$$ Divide both sides by 2.

$$x^2 - \frac{1}{2}x = -3$$ Subtract 3 from both sides to isolate the binomial.

$$x^2 - \frac{1}{2}x + \frac{1}{16} = -3 + \frac{1}{16}$$ Complete the square: Half of $-\frac{1}{2}$ is $-\frac{1}{4}$ and $\left(-\frac{1}{4}\right)^2 = \frac{1}{16}$.

$$\left(x - \frac{1}{4}\right)^2 = \frac{-47}{16}$$ Factor and simplify: $-3 + \frac{1}{16} = \frac{-48}{16} + \frac{1}{16} = \frac{-47}{16}$.

$$x - \frac{1}{4} = \pm\sqrt{\frac{-47}{16}}$$ Apply the square root property.

$$x - \frac{1}{4} = \pm i\frac{\sqrt{47}}{4}$$ $\pm\sqrt{\frac{-47}{16}} = \pm\sqrt{\frac{47(-1)}{16}} = \pm i\sqrt{\frac{47}{16}} = \pm i\frac{\sqrt{47}}{\sqrt{16}} = \pm i\frac{\sqrt{47}}{4}$

$$x = \frac{1}{4} \pm i\frac{\sqrt{47}}{4}$$ Add $\frac{1}{4}$ to both sides and solve for x.

The solutions are $\frac{1}{4} \pm i\frac{\sqrt{47}}{4}$ and the solution set is $\left\{\frac{1}{4} \pm i\frac{\sqrt{47}}{4}\right\}$. ■

☑ CHECK POINT 8 Solve by completing the square: $3x^2 - 9x + 8 = 0$.

4 Solve problems using the square root property.

Applications

We all want a wonderful life with fulfilling work, good health, and loving relationships. And let's be honest: Financial security wouldn't hurt! Achieving this goal depends on understanding how money in a savings account grows in remarkable ways as a result of *compound interest.* **Compound interest** is interest computed on your original investment as well as on any accumulated interest. For example, suppose you deposit \$1000, the principal, in a savings account at a rate of 5%. **Table 8.1** shows how the investment grows if the interest earned is automatically added on to the principal.

Table 8.1 Compound Interest on \$1000

Year	Starting Balance	Interest Earned: $I = Pr$	New Balance
1	\$1000	\$1000 × 0.05 = \$50	\$1050
2	\$1050	\$1050 × 0.05 = \$52.50	\$1102.50
3	\$1102.50	\$1102.50 × 0.05 ≈ \$55.13	\$1157.63

A faster way to determine the amount, A, in an account subject to compound interest is to use the following formula:

A Formula for Compound Interest

Suppose that an amount of money, P, is invested at interest rate r, compounded annually. In t years, the amount, A, or balance, in the account is given by the formula

$$A = P(1 + r)^t.$$

Some compound interest problems can be solved using quadratic equations.

EXAMPLE 9 Solving a Compound Interest Problem

You invested \$1000 in an account whose interest is compounded annually. After 2 years, the amount, or balance, in the account is \$1210. Find the annual interest rate.

Solution We are given that

P (the amount invested) = \$1000
t (the time of the investment) = 2 years
A (the amount, or balance, in the account) = \$1210.

We are asked to find the annual interest rate, r. We substitute the three given values into the compound interest formula and solve for r.

$$A = P(1 + r)^t$$ Use the compound interest formula.

$$1210 = 1000(1 + r)^2$$ Substitute the given values.

$$\frac{1210}{1000} = (1 + r)^2$$ Divide both sides by 1000.

$$\frac{121}{100} = (1 + r)^2$$ Simplify the fraction.

$$1 + r = \sqrt{\frac{121}{100}} \quad \text{or} \quad 1 + r = -\sqrt{\frac{121}{100}}$$ Apply the square root property.

$$1 + r = \frac{11}{10} \qquad\qquad 1 + r = -\frac{11}{10}$$ $\sqrt{\frac{121}{100}} = \frac{\sqrt{121}}{\sqrt{100}} = \frac{11}{10}$

$$r = \frac{11}{10} - 1 \qquad\qquad r = -\frac{11}{10} - 1$$ Subtract 1 from both sides and solve for r.

$$r = \frac{1}{10} \qquad\qquad r = -\frac{21}{10}$$ $\frac{11}{10} - 1 = \frac{11}{10} - \frac{10}{10} = \frac{1}{10}$ and $-\frac{11}{10} - 1 = -\frac{11}{10} - \frac{10}{10} = -\frac{21}{10}$.

Because the interest rate cannot be negative, we reject $-\frac{21}{10}$. Thus, the annual interest rate is $\frac{1}{10} = 0.10 = 10\%$.

We can check this answer using the formula $A = P(1 + r)^t$. If \$1000 is invested for 2 years at 10% interest, compounded annually, the balance in the account is

$$A = \$1000(1 + 0.10)^2 = \$1000(1.10)^2 = \$1210.$$

Because this is precisely the amount given by the problem's conditions, the annual interest rate is, indeed, 10% compounded annually. ■

✓ **CHECK POINT 9** You invested \$3000 in an account whose interest is compounded annually. After 2 years, the amount, or balance, in the account is \$4320. Find the annual interest rate.

In Chapter 5, we solved problems using the Pythagorean Theorem. Recall that in a right triangle, the side opposite the 90° angle is the hypotenuse and the other sides are legs. The Pythagorean Theorem states that the sum of the squares of the lengths of the legs equals the square of the length of the hypotenuse. Some problems that involve the Pythagorean Theorem can be solved using the square root property.

EXAMPLE 10 Using the Pythagorean Theorem and the Square Root Property

a. A wheelchair ramp with a length of 122 inches has a horizontal distance of 120 inches. What is the ramp's vertical distance?

b. Construction laws are very specific when it comes to access ramps for the disabled. Every vertical rise of 1 inch requires a horizontal run of 12 inches. Does this ramp satisfy the requirement?

Solution

a. **Figure 8.2** shows the right triangle that is formed by the ramp, the wall, and the ground. We can find x, the ramp's vertical distance, using the Pythagorean Theorem.

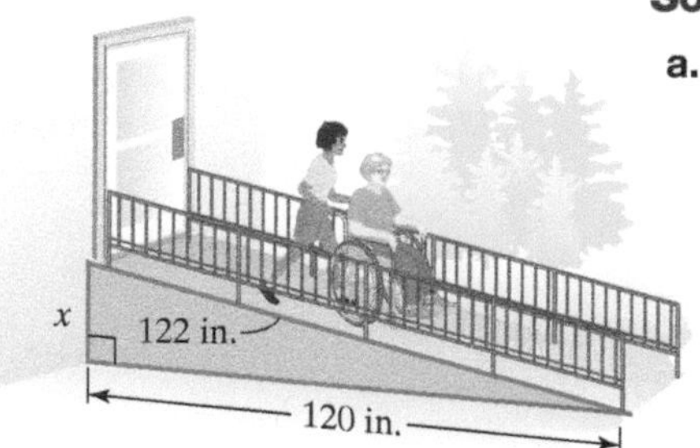

Figure 8.2

(leg)²	plus	(leg)²	equals	(hypotenuse)²
x^2	$+$	120^2	$=$	122^2

We solve this equation using the square root property.

$x^2 + 120^2 = 122^2$ — This is the equation resulting from the Pythagorean Theorem.

$x^2 + 14{,}400 = 14{,}884$ — Square 120 and 122.

$x^2 = 484$ — Isolate x^2 by subtracting 14,400 from both sides.

$x = \sqrt{484}$ or $x = -\sqrt{484}$ — Apply the square root property.

$x = 22$ $\quad x = -22$

Because x represents the ramp's vertical distance, we reject the negative value. Thus, the ramp's vertical distance is 22 inches.

b. Every vertical rise of 1 inch requires a horizontal run of 12 inches. Because the ramp has a vertical distance of 22 inches, it requires a horizontal distance of 22(12) inches, or 264 inches. The horizontal distance is only 120 inches, so this ramp does not satisfy construction laws for access ramps for the disabled. ■

✓ **CHECK POINT 10** A 50-foot supporting wire is to be attached to an antenna. The wire is anchored 20 feet from the base of the antenna. How high up the antenna is the wire attached? Express the answer in simplified radical form. Then find a decimal approximation to the nearest tenth of a foot.

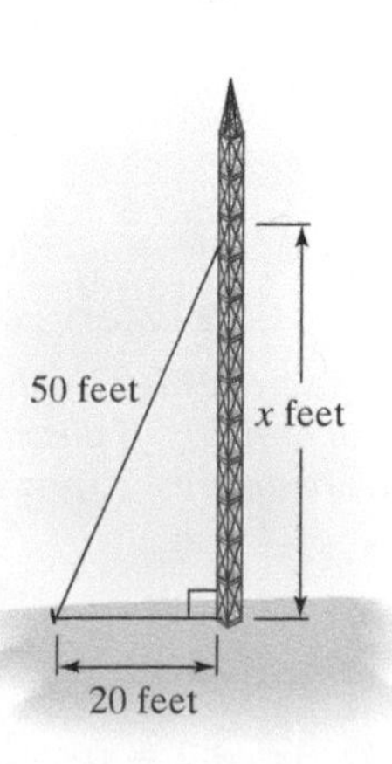

Achieving Success

Many intermediate algebra courses cover only selected sections from this chapter and the book's remaining chapters. Regardless of the content requirements for your course, **it's never too early to start thinking about a final exam.** Here are some strategies to help you prepare for your final:

- Review your back exams. Be sure you understand any error that you made. Seek help with any concepts that are still unclear.
- Ask your professor if there are additional materials to help students review for the final. This includes review sheets and final exams from previous semesters.
- Attend any review sessions conducted by your professor or by the math department.
- Use the strategy first introduced on page 162: Imagine that your professor will permit two 3 by 5 index cards of notes on the final. Organize and create such a two-card summary for the most vital information in the course, including all important formulas. Refer to the chapter summaries in the textbook to prepare your personalized summary.
- For further review, work the relevant exercises in the Cumulative Review at the end of this chapter. The 42 exercises in the Cumulative Review for Chapters 1 through 8 cover most of the objectives in the book's first eight chapters.
- Write your own final exam with detailed solutions for each item. You can use test questions from back exams in mixed order, worked examples from the textbook's chapter summaries, exercises in the Cumulative Reviews, and problems from course handouts. Use your test as a practice final exam.

CONCEPT AND VOCABULARY CHECK

Fill in each blank so that the resulting statement is true.

1. The square root property states that if $u^2 = d$, then $u =$ ________.
2. If $x^2 = 7$, then $x =$ ________.
3. If $x^2 = \frac{11}{2}$, then $x =$ ________. Rationalizing denominators, we obtain $x =$ ________.
4. If $x^2 = -9$, then $x =$ ________.
5. To complete the square on $x^2 + 10x$, add ________.
6. To complete the square on $x^2 - 3x$, add ________.
7. To complete the square on $x^2 - \frac{4}{5}x$, add ________.
8. To solve $x^2 + 6x = 7$ by completing the square, add ________ to both sides of the equation.
9. To solve $x^2 - \frac{2}{3}x = \frac{4}{9}$ by completing the square, add ________ to both sides of the equation.

8.1 EXERCISE SET

MyMathLab®

Practice Exercises

In Exercises 1–22, solve each equation by the square root property. If possible, simplify radicals or rationalize denominators. Express imaginary solutions in the form $a + bi$.

1. $3x^2 = 75$
2. $5x^2 = 20$
3. $7x^2 = 42$
4. $8x^2 = 40$
5. $16x^2 = 25$
6. $4x^2 = 49$
7. $3x^2 - 2 = 0$
8. $3x^2 - 5 = 0$
9. $25x^2 + 16 = 0$
10. $4x^2 + 49 = 0$
11. $(x + 7)^2 = 9$
12. $(x + 3)^2 = 64$
13. $(x - 3)^2 = 5$
14. $(x - 4)^2 = 3$
15. $2(x + 2)^2 = 16$
16. $3(x + 2)^2 = 36$
17. $(x - 5)^2 = -9$
18. $(x - 5)^2 = -4$

19. $\left(x+\frac{3}{4}\right)^2=\frac{11}{16}$

20. $\left(x+\frac{2}{5}\right)^2=\frac{7}{25}$

21. $x^2-6x+9=36$

22. $x^2-6x+9=49$

In Exercises 23–34, determine the constant that should be added to the binomial so that it becomes a perfect square trinomial. Then write and factor the trinomial.

23. x^2+2x

24. x^2+4x

25. x^2-14x

26. x^2-10x

27. x^2+7x

28. x^2+9x

29. $x^2-\frac{1}{2}x$

30. $x^2-\frac{1}{3}x$

31. $x^2+\frac{4}{3}x$

32. $x^2+\frac{4}{5}x$

33. $x^2-\frac{9}{4}x$

34. $x^2-\frac{9}{5}x$

In Exercises 35–58, solve each quadratic equation by completing the square.

35. $x^2+4x=32$

36. $x^2+6x=7$

37. $x^2+6x=-2$

38. $x^2+2x=5$

39. $x^2-8x+1=0$

40. $x^2+8x-5=0$

41. $x^2+2x+2=0$

42. $x^2-4x+8=0$

43. $x^2+3x-1=0$

44. $x^2-3x-5=0$

45. $x^2+\frac{4}{7}x+\frac{3}{49}=0$

46. $x^2+\frac{6}{5}x+\frac{8}{25}=0$

47. $x^2+x-1=0$

48. $x^2-7x+3=0$

49. $2x^2+3x-5=0$

50. $2x^2+5x-3=0$

51. $3x^2+6x+1=0$

52. $3x^2-6x+2=0$

53. $3x^2-8x+1=0$

54. $2x^2+3x-4=0$

55. $8x^2-4x+1=0$

56. $9x^2-6x+5=0$

57. $2x^2-5x+7=0$

58. $4x^2-2x+5=0$

59. If $g(x)=\left(x-\frac{2}{5}\right)^2$, find all values of x for which $g(x)=\frac{9}{25}$.

60. If $g(x)=\left(x+\frac{1}{3}\right)^2$, find all values of x for which $g(x)=\frac{4}{9}$.

61. If $h(x)=5(x+2)^2$, find all values of x for which $h(x)=-125$.

62. If $h(x)=3(x-4)^2$, find all values of x for which $h(x)=-12$.

Practice PLUS

63. Three times the square of the difference between a number and 2 is −12. Find the number(s).

64. Three times the square of the difference between a number and 9 is −27. Find the number(s).

In Exercises 65–68, solve the formula for the specified variable. Because each variable is nonnegative, list only the principal square root. If possible, simplify radicals or rationalize denominators.

65. $h=\frac{v^2}{2g}$ for v

66. $s=\frac{kwd^2}{l}$ for d

67. $A=P(1+r)^2$ for r

68. $C=\frac{kP_1P_2}{d^2}$ for d

In Exercises 69–72, solve each quadratic equation by completing the square.

69. $\frac{x^2}{3}+\frac{x}{9}-\frac{1}{6}=0$

70. $\frac{x^2}{2}-\frac{x}{6}-\frac{3}{4}=0$

71. $x^2-bx=2b^2$

72. $x^2-bx=6b^2$

73. The ancient Greeks used a geometric method for completing the square in which they literally transformed a figure into a square.

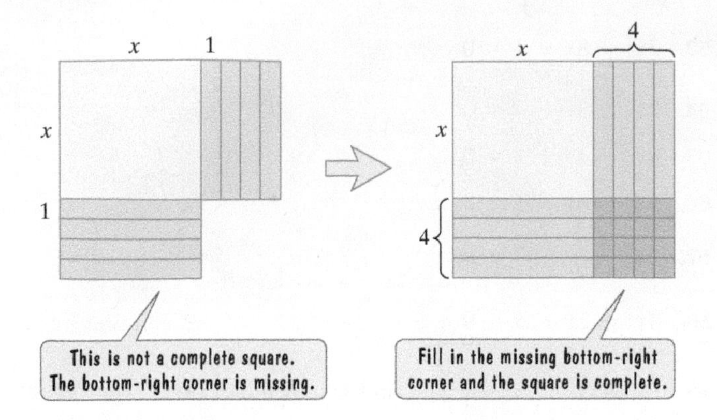

a. Write a binomial in x that represents the combined area of the small square and the eight rectangular stripes that make up the incomplete square on the left.

b. What is the area of the region in the bottom-right corner that literally completes the square?

c. Write a trinomial in x that represents the combined area of the small square, the eight rectangular stripes, and the bottom-right corner that make up the complete square on the right.

d. Use the length of each side of the complete square on the right to express its area as the square of a binomial.

74. An **isosceles right triangle** has legs that are the same length and acute angles each measuring 45°.

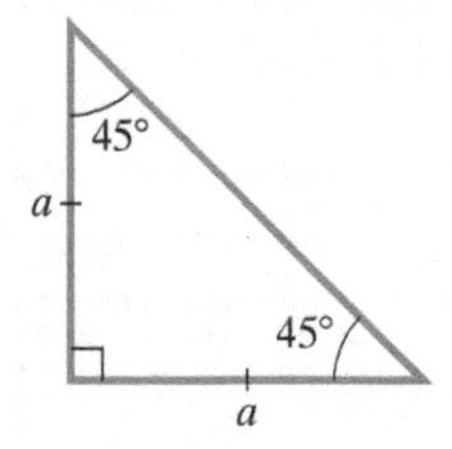

a. Write an expression in terms of a that represents the length of the hypotenuse.

b. Use your result from part (a) to write a sentence that describes the length of the hypotenuse of an isosceles right triangle in terms of the length of a leg.

Application Exercises

In Exercises 75–78, use the compound interest formula

$$A = P(1 + r)^t$$

to find the annual interest rate, r.

75. In 2 years, an investment of \$2000 grows to \$2880.

76. In 2 years, an investment of \$2000 grows to \$2420.

77. In 2 years, an investment of \$1280 grows to \$1445.

78. In 2 years, an investment of \$80,000 grows to \$101,250.

The bar graph shows the number of billionaires worldwide for six selected years from 1987 through 2012.

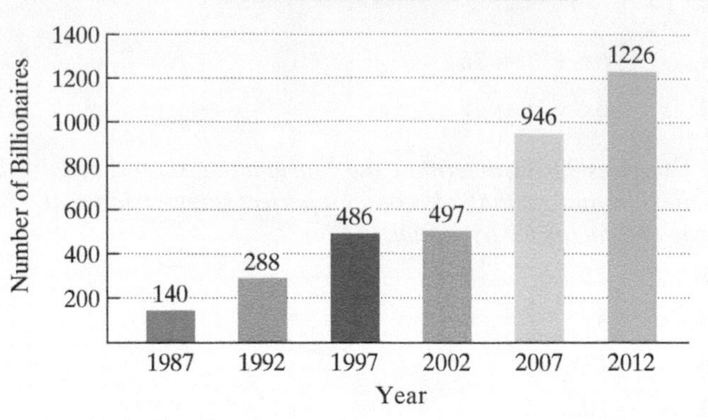

Source: Forbes

The data can be modeled by the function

$$f(x) = 2x^2 + 140,$$

where $f(x)$ represents the number of billionaires worldwide x years after 1987. Use this information to solve Exercises 79–80.

79. a. According to the model, how many billionaires were there in 2007? Does this underestimate or overestimate the number displayed by the graph? By how much?

b. According to the model, in which year will there be 1940 billionaires worldwide?

80. a. According to the model, how many billionaires were there in 2012? Does this underestimate or overestimate the number displayed by the graph? By how much?

b. According to the model, in which year will there be 2062 billionaires worldwide?

The function $s(t) = 16t^2$ models the distance, $s(t)$, in feet, that an object falls in t seconds. Use this function and the square root property to solve Exercises 81–82. Express answers in simplified radical form. Then use your calculator to find a decimal approximation to the nearest tenth of a second.

81. A sky diver jumps from an airplane and falls for 4800 feet before opening a parachute. For how many seconds was the diver in a free fall?

82. A sky diver jumps from an airplane and falls for 3200 feet before opening a parachute. For how many seconds was the diver in a free fall?

Use the Pythagorean Theorem and the square root property to solve Exercises 83–88. Express answers in simplified radical form. Then find a decimal approximation to the nearest tenth.

83. A rectangular park is 6 miles long and 3 miles wide. How long is a pedestrian route that runs diagonally across the park?

84. A rectangular park is 4 miles long and 2 miles wide. How long is a pedestrian route that runs diagonally across the park?

85. The base of a 30-foot ladder is 10 feet from the building. If the ladder reaches the flat roof, how tall is the building?

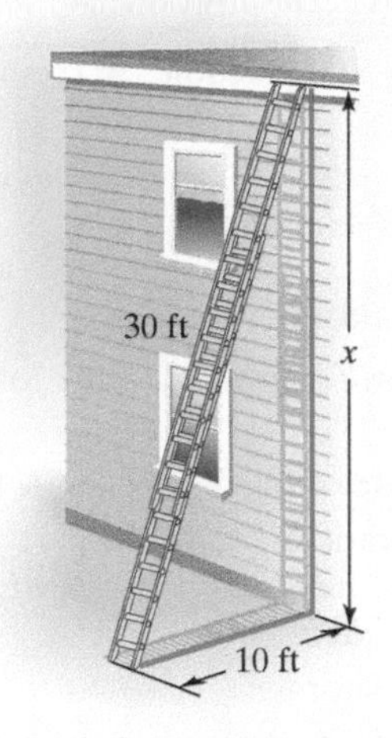

86. The doorway into a room is 4 feet wide and 8 feet high. What is the diameter of the largest circular tabletop that can be taken through this doorway diagonally?

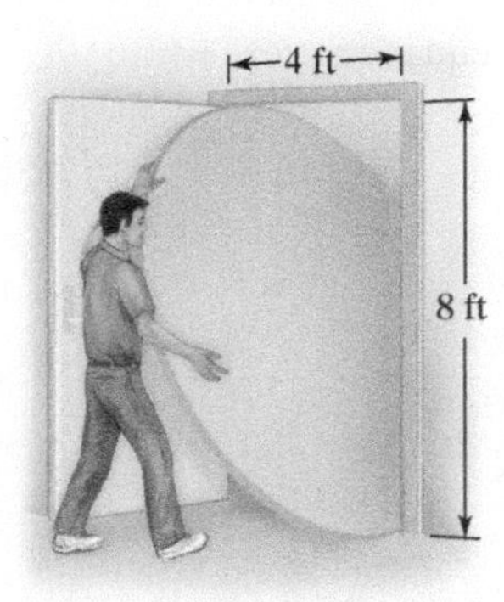

87. A supporting wire is to be attached to the top of a 50-foot antenna. If the wire must be anchored 50 feet from the base of the antenna, what length of wire is required?

88. A supporting wire is to be attached to the top of a 70-foot antenna. If the wire must be anchored 70 feet from the base of the antenna, what length of wire is required?

89. A square flower bed is to be enlarged by adding 2 meters on each side. If the larger square has an area of 196 square meters, what is the length of a side of the original square?

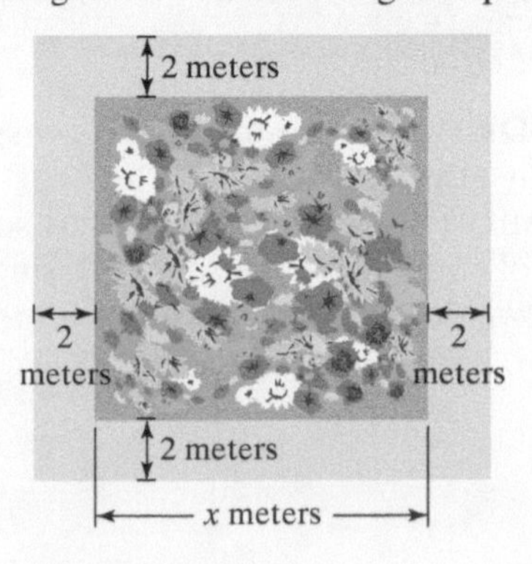

90. A square flower bed is to be enlarged by adding 4 feet on each side. If the larger square has an area of 225 square feet, what is the length of a side of the original square?

Explaining the Concepts

91. What is the square root property?

92. Explain how to solve $(x - 1)^2 = 16$ using the square root property.

93. Explain how to complete the square for a binomial. Use $x^2 + 6x$ to illustrate your explanation.

94. Explain how to solve $x^2 + 6x + 8 = 0$ by completing the square.

95. What is compound interest?

96. In your own words, describe the compound interest formula

$$A = P(1 + r)^t.$$

Technology Exercises

97. Use a graphing utility to solve $4 - (x + 1)^2 = 0$. Graph $y = 4 - (x + 1)^2$ in a $[-5, 5, 1]$ by $[-5, 5, 1]$ viewing rectangle. The equation's solutions are the graph's x-intercepts. Check by substitution in the given equation.

98. Use a graphing utility to solve $(x - 1)^2 - 9 = 0$. Graph $y = (x - 1)^2 - 9$ in a $[-5, 5, 1]$ by $[-9, 3, 1]$ viewing rectangle. The equation's solutions are the graph's x-intercepts. Check by substitution in the given equation.

99. Use a graphing utility and x-intercepts to verify any of the real solutions that you obtained for five of the quadratic equations in Exercises 35–54.

Critical Thinking Exercises

Make Sense? *In Exercises 100–103, determine whether each statement makes sense or does not make sense, and explain your reasoning.*

100. When the coefficient of the x-term in a quadratic equation is negative and I'm solving by completing the square, I add a negative constant to each side of the equation.

101. When I complete the square for the binomial $x^2 + bx$, I obtain a different polynomial, but when I solve a quadratic equation by completing the square, I obtain an equation with the same solution set.

102. When I use the square root property to determine the length of a right triangle's side, I don't even bother to list the negative square root.

103. When I solved $4x^2 + 10x = 0$ by completing the square, I added 25 to both sides of the equation.

In Exercises 104–107, determine whether each statement is true or false. If the statement is false, make the necessary change(s) to produce a true statement.

104. The graph of $y = (x - 2)^2 + 3$ cannot have x-intercepts.

105. The equation $(x - 5)^2 = 12$ is equivalent to $x - 5 = 2\sqrt{3}$.

106. In completing the square for $2x^2 - 6x = 5$, we should add 9 to both sides.

107. Although not every quadratic equation can be solved by completing the square, they can all be solved by factoring.

108. Solve for y: $\dfrac{x^2}{a^2} + \dfrac{y^2}{b^2} = 1$.

109. Solve by completing the square:

$$x^2 + x + c = 0.$$

110. Solve by completing the square:

$$x^2 + bx + c = 0.$$

111. Solve: $x^4 - 8x^2 + 15 = 0$.

Review Exercises

112. Simplify: $4x - 2 - 3[4 - 2(3 - x)]$. (Section 1.2, Example 14)

113. Factor: $1 - 8x^3$. (Section 5.5, Example 9)

114. Divide: $(x^4 - 5x^3 + 2x^2 - 6) \div (x - 3)$. (Section 6.5, Example 1)

Preview Exercises

Exercises 115–117 will help you prepare for the material covered in the next section.

115. a. Solve by factoring: $8x^2 + 2x - 1 = 0$.

b. The quadratic equation in part (a) is in the standard form $ax^2 + bx + c = 0$. Compute $b^2 - 4ac$. Is $b^2 - 4ac$ a perfect square?

116. a. Solve by factoring: $9x^2 - 6x + 1 = 0$.

b. The quadratic equation in part (a) is in the standard form $ax^2 + bx + c = 0$. Compute $b^2 - 4ac$.

117. a. Clear fractions in the following equation and write in the form $ax^2 + bx + c = 0$:

$$3 + \frac{4}{x} = -\frac{2}{x^2}.$$

b. For the equation you wrote in part (a), compute $b^2 - 4ac$.

SECTION

8.2 The Quadratic Formula

What am I supposed to learn?

After studying this section, you should be able to:

1. Solve quadratic equations using the quadratic formula.
2. Use the discriminant to determine the number and type of solutions.
3. Determine the most efficient method to use when solving a quadratic equation.
4. Write quadratic equations from solutions.
5. Use the quadratic formula to solve problems.

Until fairly recently, many doctors believed that your blood pressure was theirs to know and yours to worry about. Today, however, people are encouraged to find out their blood pressure. That pumped-up cuff that squeezes against your upper arm measures blood pressure in millimeters (mm) of mercury (Hg). Blood pressure is given in two numbers: systolic pressure over diastolic pressure, such as 120 over 80. Systolic pressure is the pressure of blood against the artery walls when the heart contracts. Diastolic pressure is the pressure of blood against the artery walls when the heart is at rest.

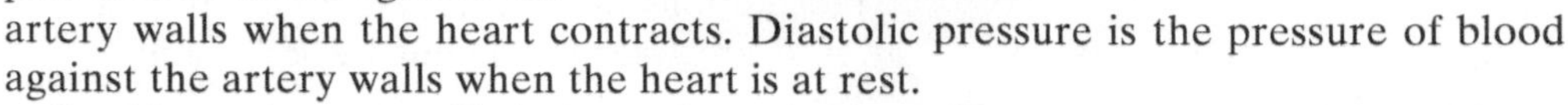

In this section, we will derive a formula that will enable you to solve quadratic equations more quickly than the method of completing the square. Using this formula, we will work with functions that model changing systolic pressure for men and women with age.

1 Solve quadratic equations using the quadratic formula.

Solving Quadratic Equations Using the Quadratic Formula

We can use the method of completing the square to derive a formula that can be used to solve all quadratic equations. The derivation given below also shows a particular quadratic equation, $3x^2 - 2x - 4 = 0$, to specifically illustrate each of the steps.

Deriving the Quadratic Formula

Standard Form of a Quadratic Equation	Comment	A Specific Example
$ax^2 + bx + c = 0, a > 0$	This is the given equation.	$3x^2 - 2x - 4 = 0$
$x^2 + \frac{b}{a}x + \frac{c}{a} = 0$	Divide both sides by the coefficient of x^2.	$x^2 - \frac{2}{3}x - \frac{4}{3} = 0$
$x^2 + \frac{b}{a}x = -\frac{c}{a}$	Isolate the binomial by adding $-\frac{c}{a}$ on both sides.	$x^2 - \frac{2}{3}x = \frac{4}{3}$
$x^2 + \frac{b}{a}x + \left(\frac{b}{2a}\right)^2 = -\frac{c}{a} + \left(\frac{b}{2a}\right)^2$ (half)^2 $x^2 + \frac{b}{a}x + \frac{b^2}{4a^2} = -\frac{c}{a} + \frac{b^2}{4a^2}$	Complete the square. Add the square of half the coefficient of x to both sides.	$x^2 - \frac{2}{3}x + \left(-\frac{1}{3}\right)^2 = \frac{4}{3} + \left(-\frac{1}{3}\right)^2$ (half)^2 $x^2 - \frac{2}{3}x + \frac{1}{9} = \frac{4}{3} + \frac{1}{9}$
$\left(x + \frac{b}{2a}\right)^2 = -\frac{c}{a}\cdot\frac{4a}{4a} + \frac{b^2}{4a^2}$	Factor on the left side and obtain a common denominator on the right side.	$\left(x - \frac{1}{3}\right)^2 = \frac{4}{3}\cdot\frac{3}{3} + \frac{1}{9}$
$\left(x + \frac{b}{2a}\right)^2 = \frac{-4ac + b^2}{4a^2}$ $\left(x + \frac{b}{2a}\right)^2 = \frac{b^2 - 4ac}{4a^2}$	Add fractions on the right side.	$\left(x - \frac{1}{3}\right)^2 = \frac{12 + 1}{9}$ $\left(x - \frac{1}{3}\right)^2 = \frac{13}{9}$
$x + \frac{b}{2a} = \pm\sqrt{\frac{b^2 - 4ac}{4a^2}}$	Apply the square root property.	$x - \frac{1}{3} = \pm\sqrt{\frac{13}{9}}$
$x + \frac{b}{2a} = \pm\frac{\sqrt{b^2 - 4ac}}{2a}$	Take the square root of the quotient, simplifying the denominator.	$x - \frac{1}{3} = \pm\frac{\sqrt{13}}{3}$
$x = \frac{-b}{2a} \pm \frac{\sqrt{b^2 - 4ac}}{2a}$	Solve for x by subtracting $\frac{b}{2a}$ from both sides.	$x = \frac{1}{3} \pm \frac{\sqrt{13}}{3}$
$x = \frac{-b \pm \sqrt{b^2 - 4ac}}{2a}$	Combine fractions on the right side.	$x = \frac{1 \pm \sqrt{13}}{3}$

The formula shown at the bottom of the left column is called the *quadratic formula.* A similar proof shows that the same formula can be used to solve quadratic equations if a, the coefficient of the x^2-term, is negative.

The Quadratic Formula

The solutions of a quadratic equation in standard form $ax^2 + bx + c = 0$, with $a \neq 0$, are given by the **quadratic formula**:

$$x = \frac{-b \pm \sqrt{b^2 - 4ac}}{2a}.$$

x equals negative b plus or minus the square root of $b^2 - 4ac$, all divided by $2a$.

To use the quadratic formula, write the quadratic equation in standard form if necessary. Then determine the numerical values for a (the coefficient of the x^2-term), b (the coefficient of the x-term), and c (the constant term). Substitute the values of a, b, and c into the quadratic formula and evaluate the expression. The $\pm$ sign indicates that there are two (not necessarily distinct) solutions of the equation.

EXAMPLE 1 Solving a Quadratic Equation Using the Quadratic Formula

Solve using the quadratic formula: $8x^2 + 2x - 1 = 0$.

Solution The given equation is in standard form. Begin by identifying the values for a, b, and c.

$$8x^2 + 2x - 1 = 0$$

$a = 8$ $b = 2$ $c = -1$

Substituting these values into the quadratic formula and simplifying gives the equation's solutions.

$$x = \frac{-b \pm \sqrt{b^2 - 4ac}}{2a}$$ Use the quadratic formula.

$$x = \frac{-2 \pm \sqrt{2^2 - 4(8)(-1)}}{2(8)}$$ Substitute the values for a, b, and c: $a = 8$, $b = 2$, and $c = -1$.

$$= \frac{-2 \pm \sqrt{4 - (-32)}}{16}$$ $2^2 - 4(8)(-1) = 4 - (-32)$

$$= \frac{-2 \pm \sqrt{36}}{16}$$ $4 - (-32) = 4 + 32 = 36$

$$= \frac{-2 \pm 6}{16}$$ $\sqrt{36} = 6$

Using Technology

Graphic Connections

The graph of the quadratic function

$$y = 8x^2 + 2x - 1$$

has x-intercepts at $-\frac{1}{2}$ and $\frac{1}{4}$. This verifies that $\{-\frac{1}{2}, \frac{1}{4}\}$ is the solution set of the quadratic equation

$$8x^2 + 2x - 1 = 0.$$

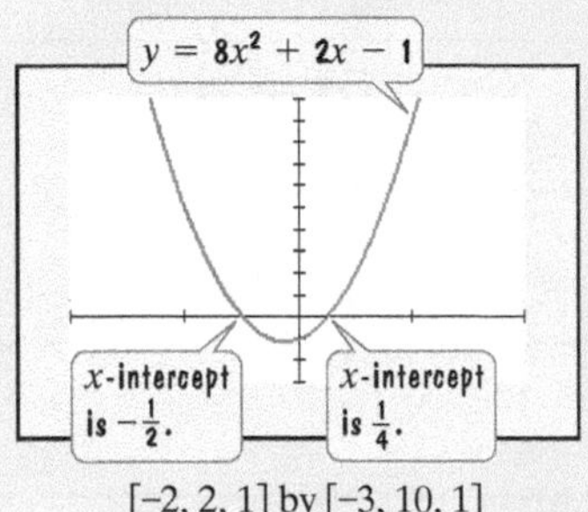

[-2, 2, 1] by [-3, 10, 1]

Now we will evaluate this expression in two different ways to obtain the two solutions. On the left, we will *add* 6 to -2. On the right, we will *subtract* 6 from -2.

$$x = \frac{-2 + 6}{16} \quad \text{or} \quad x = \frac{-2 - 6}{16}$$

$$= \frac{4}{16} = \frac{1}{4} \qquad = \frac{-8}{16} = -\frac{1}{2}$$

The solutions are $-\frac{1}{2}$ and $\frac{1}{4}$, and the solution set is $\{-\frac{1}{2}, \frac{1}{4}\}$. ■

In Example 1, the solutions of $8x^2 + 2x - 1 = 0$ are rational numbers. This means that the equation can also be solved by factoring. The reason that the solutions are rational numbers is that $b^2 - 4ac$, the radicand in the quadratic formula, is 36, which is a perfect square. If a, b, and c are rational numbers, all quadratic equations for which $b^2 - 4ac$ is a perfect square have rational solutions.

✓ **CHECK POINT 1** Solve using the quadratic formula: $2x^2 + 9x - 5 = 0$.

EXAMPLE 2 Solving a Quadratic Equation Using the Quadratic Formula

Solve using the quadratic formula:

$$2x^2 = 4x + 1.$$

Solution The quadratic equation must be in standard form to identify the values for a, b, and c. To move all terms to one side and obtain zero on the right, we subtract $4x + 1$ from both sides. Then we can identify the values for a, b, and c.

$$2x^2 = 4x + 1 \quad \text{This is the given equation.}$$

$$2x^2 - 4x - 1 = 0 \quad \text{Subtract } 4x + 1 \text{ from both sides.}$$

$a = 2$ $\quad b = -4$ $\quad c = -1$

Substituting these values into the quadratic formula and simplifying gives the equation's solutions.

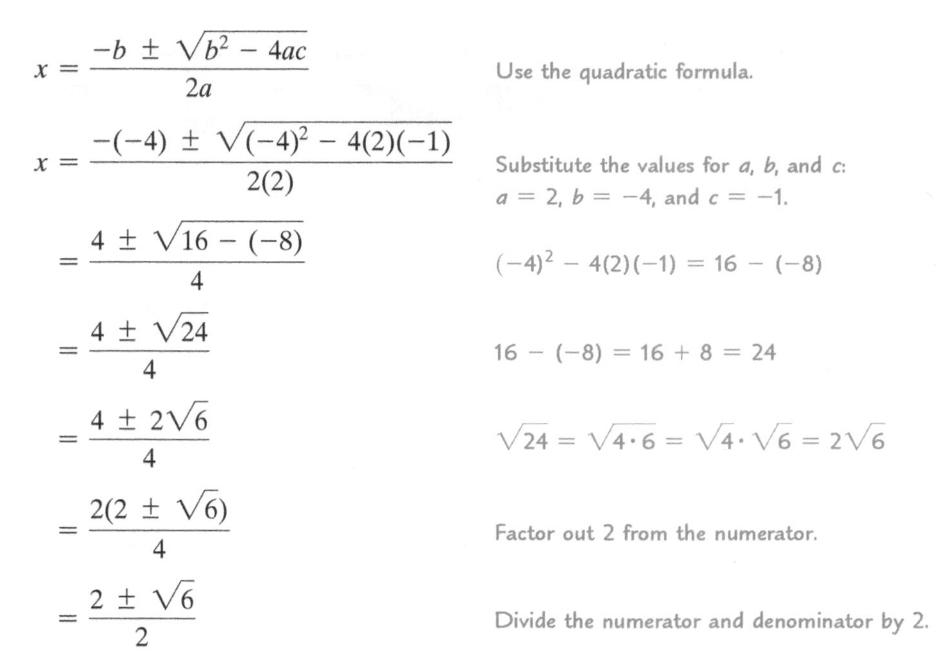

$$x = \frac{-b \pm \sqrt{b^2 - 4ac}}{2a} \quad \text{Use the quadratic formula.}$$

$$x = \frac{-(-4) \pm \sqrt{(-4)^2 - 4(2)(-1)}}{2(2)} \quad \text{Substitute the values for } a, b, \text{ and } c\text{: } a = 2, b = -4, \text{ and } c = -1.$$

$$= \frac{4 \pm \sqrt{16 - (-8)}}{4} \quad (-4)^2 - 4(2)(-1) = 16 - (-8)$$

$$= \frac{4 \pm \sqrt{24}}{4} \quad 16 - (-8) = 16 + 8 = 24$$

$$= \frac{4 \pm 2\sqrt{6}}{4} \quad \sqrt{24} = \sqrt{4 \cdot 6} = \sqrt{4} \cdot \sqrt{6} = 2\sqrt{6}$$

$$= \frac{2(2 \pm \sqrt{6})}{4} \quad \text{Factor out 2 from the numerator.}$$

$$= \frac{2 \pm \sqrt{6}}{2} \quad \text{Divide the numerator and denominator by 2.}$$

The solutions are $\frac{2 \pm \sqrt{6}}{2}$, and the solution set is $\left\{\frac{2 + \sqrt{6}}{2}, \frac{2 - \sqrt{6}}{2}\right\}$ or $\left\{\frac{2 \pm \sqrt{6}}{2}\right\}$. ■

Using Technology

You can use a graphing utility to verify that the solutions of $2x^2 - 4x - 1 = 0$ are $\frac{2 \pm \sqrt{6}}{2}$. Begin by entering $y_1 = 2x^2 - 4x - 1$ in the [Y=] screen. Then evaluate this function at each of the proposed solutions.

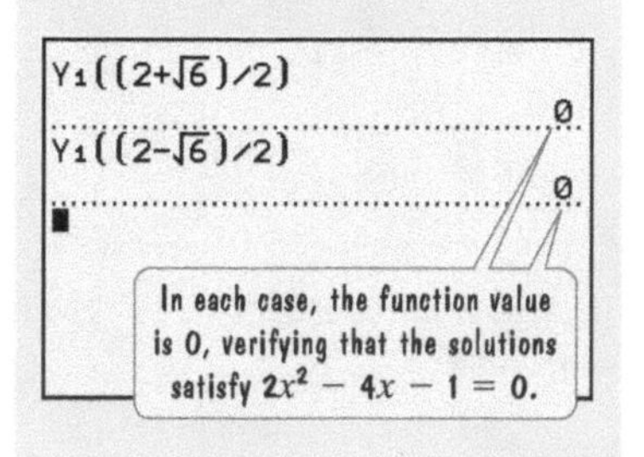
Y1((2+√6)/2)
0
Y1((2-√6)/2)
0
In each case, the function value is 0, verifying that the solutions satisfy 2x² − 4x − 1 = 0.

In Example 2, the solutions of $2x^2 = 4x + 1$ are irrational numbers. This means that the equation cannot be solved by factoring. The reason that the solutions are irrational numbers is that $b^2 - 4ac$, the radicand in the quadratic formula, is 24, which is not a perfect square. Notice, too, that the solutions, $\frac{2 + \sqrt{6}}{2}$ and $\frac{2 - \sqrt{6}}{2}$, are conjugates.

Great Question!

The simplification of the irrational solutions in Example 2 was kind of tricky. Any suggestions to guide the process?

Many students use the quadratic formula correctly until the last step, where they make an error in simplifying the solutions. Be sure to factor the numerator before dividing the numerator and the denominator by the greatest common factor.

$$\frac{4 \pm 2\sqrt{6}}{4} = \frac{2(2 \pm \sqrt{6})}{4} = \frac{\overset{1}{\cancel{2}}(2 \pm \sqrt{6})}{\underset{2}{\cancel{4}}} = \frac{2 \pm \sqrt{6}}{2}$$

You cannot divide just one term in the numerator and the denominator by their greatest common factor.

Incorrect!

$$\frac{\overset{1}{\cancel{4}} \pm 2\sqrt{6}}{\underset{1}{\cancel{4}}} = 1 \pm 2\sqrt{6} \qquad \frac{4 \pm \overset{1}{\cancel{2}}\sqrt{6}}{\underset{2}{\cancel{4}}} = \frac{4 \pm \sqrt{6}}{2}$$

Can all irrational solutions of quadratic equations be simplified? No. The following solutions cannot be simplified:

$$\frac{5 \pm 2\sqrt{7}}{2} \qquad \frac{-4 \pm 3\sqrt{7}}{2}.$$

Other than 1, terms in each numerator have no common factor.

CHECK POINT 2 Solve using the quadratic formula: $2x^2 = 6x - 1$.

EXAMPLE 3 Solving a Quadratic Equation Using the Quadratic Formula

Solve using the quadratic formula:

$$3x^2 + 2 = -4x.$$

Solution Begin by writing the quadratic equation in standard form.

$$3x^2 + 2 = -4x \quad \text{This is the given equation.}$$

$$3x^2 + 4x + 2 = 0 \quad \text{Add } 4x \text{ to both sides.}$$

$a = 3$ $b = 4$ $c = 2$

Substituting these values into the quadratic formula and simplifying gives the equation's solutions.

$$x = \frac{-b \pm \sqrt{b^2 - 4ac}}{2a} \quad \text{Use the quadratic formula.}$$

$$x = \frac{-4 \pm \sqrt{4^2 - 4 \cdot 3 \cdot 2}}{2 \cdot 3} \quad \text{Substitute the values for } a, b, \text{ and } c\text{: } a = 3, b = 4, \text{ and } c = 2.$$

$$= \frac{-4 \pm \sqrt{16 - 24}}{6} \quad \text{Multiply under the radical.}$$

$$= \frac{-4 \pm \sqrt{-8}}{6} \quad \text{Subtract under the radical.}$$

$$= \frac{-4 \pm 2i\sqrt{2}}{6} \quad \sqrt{-8} = \sqrt{8(-1)} = \sqrt{8}\sqrt{-1} = i\sqrt{8} = i\sqrt{4 \cdot 2} = 2i\sqrt{2}$$

$$= \frac{2(-2 \pm i\sqrt{2})}{6} \quad \text{Factor out 2 from the numerator.}$$

$$= \frac{-2 \pm i\sqrt{2}}{3} \quad \text{Divide the numerator and denominator by 2.}$$

$$= -\frac{2}{3} \pm i\frac{\sqrt{2}}{3} \quad \text{Express in the form } a + bi, \text{ writing } i \text{ before the square root.}$$

The solutions are $-\frac{2}{3} \pm i\frac{\sqrt{2}}{3}$, and the solution set is $\left\{-\frac{2}{3} + i\frac{\sqrt{2}}{3}, -\frac{2}{3} - i\frac{\sqrt{2}}{3}\right\}$ or $\left\{-\frac{2}{3} \pm i\frac{\sqrt{2}}{3}\right\}$. ■

Using Technology

Graphic Connections

The graph of the quadratic function

$$y = 3x^2 + 4x + 2$$

has no x-intercepts. This verifies that the equation in Example 3

$$3x^2 + 2 = -4x, \quad \text{or}$$

$$3x^2 + 4x + 2 = 0$$

has imaginary solutions.

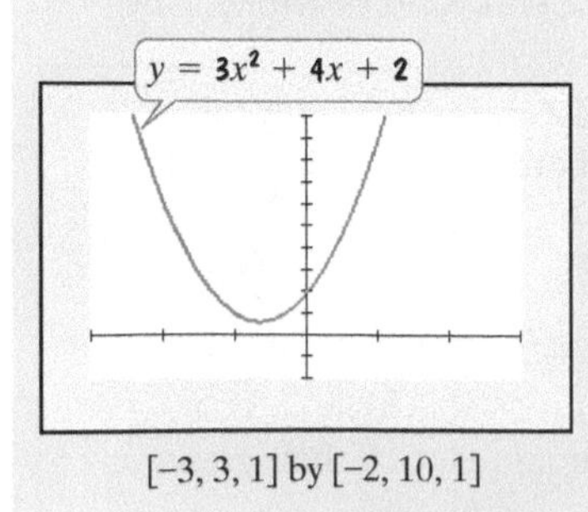

[−3, 3, 1] by [−2, 10, 1]

In Example 3, the solutions of $3x^2 + 2 = -4x$ are imaginary numbers. This means that the equation cannot be solved using factoring. The reason that the solutions are imaginary numbers is that $b^2 - 4ac$, the radicand in the quadratic formula, is -8, which is negative. Notice, too, that the solutions are complex conjugates.

Great Question!

Should I check irrational and imaginary solutions by substitution in the given quadratic equation?

No. Checking irrational and imaginary solutions can be time-consuming. The solutions given by the quadratic formula are always correct, unless you have made a careless error. Checking for computational errors or errors in simplification is sufficient.

✓ CHECK POINT 3 Solve using the quadratic formula: $3x^2 + 5 = -6x$.

Some rational equations can be solved using the quadratic formula. For example, consider the equation

$$3 + \frac{4}{x} = -\frac{2}{x^2}.$$

The denominators are x and x^2. The least common denominator is x^2. We clear fractions by multiplying both sides of the equation by x^2. Notice that x cannot equal zero.

$$x^2\left(3 + \frac{4}{x}\right) = x^2\left(-\frac{2}{x^2}\right),\ x \neq 0$$

$$3x^2 + \frac{4}{x}\cdot x^2 = x^2\left(-\frac{2}{x^2}\right) \quad \text{Use the distributive property.}$$

$$3x^2 + 4x = -2 \quad \text{Simplify.}$$

By adding 2 to both sides of $3x^2 + 4x = -2$, we obtain the standard form of the quadratic equation:

$$3x^2 + 4x + 2 = 0.$$

This is the equation that we solved in Example 3. The two imaginary solutions are not part of the restriction that $x \neq 0$.

2 Use the discriminant to determine the number and type of solutions.

The Discriminant

The quantity $b^2 - 4ac$, which appears under the radical sign in the quadratic formula, is called the **discriminant**. **Table 8.2** shows how the discriminant of the quadratic equation $ax^2 + bx + c = 0$ determines the number and type of solutions.

Table 8.2 The Discriminant and the Kinds of Solutions to $ax^2 + bx + c = 0$

Discriminant $b^2 - 4ac$	Kinds of Solutions to $ax^2 + bx + c = 0$	Graph of $y = ax^2 + bx + c$
$b^2 - 4ac > 0$	**Two unequal real** solutions: If a, b, and c are rational numbers and the discriminant is a perfect square, the solutions are rational. If the discriminant is not a perfect square, the solutions are irrational conjugates.	Two x-intercepts
$b^2 - 4ac = 0$	**One solution (a repeated solution) that is a real number:** If a, b, and c are rational numbers, the repeated solution is also a rational number.	One x-intercept
$b^2 - 4ac < 0$	**No real solution; two imaginary solutions:** The solutions are complex conjugates.	No x-intercepts

EXAMPLE 4 Using the Discriminant

For each equation, compute the discriminant. Then determine the number and type of solutions:

a. $3x^2 + 4x - 5 = 0$ **b.** $9x^2 - 6x + 1 = 0$ **c.** $3x^2 - 8x + 7 = 0.$

Great Question!

Is the square root sign part of the discriminant?

No. The discriminant is $b^2 - 4ac$. It is not $\sqrt{b^2 - 4ac}$, so do not give the discriminant as a radical.

Solution Begin by identifying the values for a, b, and c in each equation. Then compute $b^2 - 4ac$, the discriminant.

a. $3x^2 + 4x - 5 = 0$

$a = 3$ $b = 4$ $c = -5$

Substitute and compute the discriminant:

$$b^2 - 4ac = 4^2 - 4 \cdot 3(-5) = 16 - (-60) = 16 + 60 = 76.$$

The discriminant, 76, is a positive number that is not a perfect square. Thus, there are two real irrational solutions. (These solutions are conjugates of each other.)

b. $9x^2 - 6x + 1 = 0$

$a = 9$ $b = -6$ $c = 1$

Substitute and compute the discriminant:

$$b^2 - 4ac = (-6)^2 - 4 \cdot 9 \cdot 1 = 36 - 36 = 0.$$

The discriminant, 0, shows that there is only one real solution. This real solution is a rational number.

c. $3x^2 - 8x + 7 = 0$

$a = 3$ $b = -8$ $c = 7$

$$b^2 - 4ac = (-8)^2 - 4 \cdot 3 \cdot 7 = 64 - 84 = -20$$

The negative discriminant, -20, shows that there are two imaginary solutions. (These solutions are complex conjugates of each other.) ■

✓ CHECK POINT 4 For each equation, compute the discriminant. Then determine the number and type of solutions:

a. $x^2 + 6x + 9 = 0$

b. $2x^2 - 7x - 4 = 0$

c. $3x^2 - 2x + 4 = 0.$

---

3 Determine the most efficient method to use when solving a quadratic equation.

Determining Which Method to Use

All quadratic equations can be solved by the quadratic formula. However, if an equation is in the form $u^2 = d$, such as $x^2 = 5$ or $(2x + 3)^2 = 8$, it is faster to use the square root property, taking the square root of both sides. If the equation is not in the form $u^2 = d$, write the quadratic equation in standard form ($ax^2 + bx + c = 0$). Try to solve the equation by factoring. If $ax^2 + bx + c$ cannot be factored, then solve the quadratic equation by the quadratic formula.

Because we used the method of completing the square to derive the quadratic formula, we no longer need it for solving quadratic equations. However, we will use completing the square in Chapter 10 to help graph certain kinds of equations.

Table 8.3 summarizes our observations about which technique to use when solving a quadratic equation.

Table 8.3 Determining the Most Efficient Technique to Use When Solving a Quadratic Equation

Description and Form of the Quadratic Equation	Most Efficient Solution Method	Example
$ax^2 + bx + c = 0$ and $ax^2 + bx + c$ can be factored easily.	Factor and use the zero-product principle.	$3x^2 + 5x - 2 = 0$ $(3x - 1)(x + 2) = 0$ $3x - 1 = 0$ or $x + 2 = 0$ $x = \frac{1}{3}$ $\quad$ $x = -2$
$ax^2 + c = 0$ The quadratic equation has no x-term. $(b = 0)$	Solve for x^2 and apply the square root property.	$4x^2 - 7 = 0$ $4x^2 = 7$ $x^2 = \frac{7}{4}$ $x = \pm\frac{\sqrt{7}}{2}$
$u^2 = d$; u is a first-degree polynomial.	Use the square root property.	$(x + 4)^2 = 5$ $x + 4 = \pm\sqrt{5}$ $x = -4 \pm\sqrt{5}$
$ax^2 + bx + c = 0$ and $ax^2 + bx + c$ cannot be factored or the factoring is too difficult.	Use the quadratic formula: $x = \frac{-b \pm \sqrt{b^2 - 4ac}}{2a}$.	$x^2 - 2x - 6 = 0$ $a = 1$ $\quad$ $b = -2$ $\quad$ $c = -6$ $x = \frac{-(-2) \pm \sqrt{(-2)^2 - 4(1)(-6)}}{2(1)}$ $= \frac{2 \pm \sqrt{4 + 24}}{2}$ $= \frac{2 \pm \sqrt{28}}{2} = \frac{2 \pm \sqrt{4}\sqrt{7}}{2}$ $= \frac{2 \pm 2\sqrt{7}}{2} = \frac{2(1 \pm \sqrt{7})}{2}$ $= 1 \pm \sqrt{7}$

4 Write quadratic equations from solutions.

Writing Quadratic Equations from Solutions

Using the zero-product principle, the equation $(x - 3)(x + 5) = 0$ has two solutions, 3 and -5. By applying the zero-product principle in reverse, we can find a quadratic equation that has two given numbers as its solutions.

The Zero-Product Principle in Reverse

If $A = 0$ or $B = 0$, then $AB = 0$.

EXAMPLE 5 Writing Equations from Solutions

Write a quadratic equation with the given solution set:

a. $\left\{-\frac{5}{3}, \frac{1}{2}\right\}$ **b.** $\{-2\sqrt{3}, 2\sqrt{3}\}$ **c.** $\{-5i, 5i\}$.

Solution

a. Because the solution set is $\left\{-\frac{5}{3}, \frac{1}{2}\right\}$, then

$$x = -\frac{5}{3} \quad \text{or} \quad x = \frac{1}{2}.$$

$$x + \frac{5}{3} = 0 \quad \text{or} \quad x - \frac{1}{2} = 0$$ Obtain zero on one side of each equation.

$$3x + 5 = 0 \quad \text{or} \quad 2x - 1 = 0$$ Clear fractions, multiplying by 3 and 2, respectively.

$$(3x + 5)(2x - 1) = 0$$ Use the zero-product principle in reverse: If $A = 0$ or $B = 0$, then $AB = 0$.

$$6x^2 - 3x + 10x - 5 = 0$$ Use the FOIL method to multiply.

$$6x^2 + 7x - 5 = 0$$ Combine like terms.

Thus, one equation is $6x^2 + 7x - 5 = 0$. Many other quadratic equations have $\{-\frac{5}{3}, \frac{1}{2}\}$ for their solution sets. These equations can be obtained by multiplying both sides of $6x^2 + 7x - 5 = 0$ by any nonzero real number.

b. Because the solution set is $\{-2\sqrt{3}, 2\sqrt{3}\}$, then

$$x = -2\sqrt{3} \quad \text{or} \quad x = 2\sqrt{3}.$$

$$x + 2\sqrt{3} = 0 \quad \text{or} \quad x - 2\sqrt{3} = 0$$ Obtain zero on one side of each equation.

$$(x + 2\sqrt{3})(x - 2\sqrt{3}) = 0$$ Use the zero-product principle in reverse: If $A = 0$ or $B = 0$, then $AB = 0$.

$$x^2 - (2\sqrt{3})^2 = 0$$ Multiply conjugates using $(A + B)(A - B) = A^2 - B^2$.

$$x^2 - 12 = 0$$ $(2\sqrt{3})^2 = 2^2(\sqrt{3})^2 = 4 \cdot 3 = 12$

Thus, one equation is $x^2 - 12 = 0$.

c. Because the solution set is $\{-5i, 5i\}$, then

$$x = -5i \quad \text{or} \quad x = 5i.$$

$$x + 5i = 0 \quad \text{or} \quad x - 5i = 0$$ Obtain zero on one side of each equation.

$$(x + 5i)(x - 5i) = 0$$ Use the zero-product principle in reverse: If $A = 0$ or $B = 0$, then $AB = 0$.

$$x^2 - (5i)^2 = 0$$ Multiply conjugates using $(A + B)(A - B) = A^2 - B^2$.

$$x^2 - 25i^2 = 0$$ $(5i)^2 = 5^2 i^2 = 25i^2$

$$x^2 - 25(-1) = 0$$ $i^2 = -1$

$$x^2 + 25 = 0$$ This is the required equation. ■

CHECK POINT 5 Write a quadratic equation with the given solution set:

a. $\{-\frac{3}{5}, \frac{1}{4}\}$

b. $\{-5\sqrt{2}, 5\sqrt{2}\}$

c. $\{-7i, 7i\}$.

---

5 Use the quadratic formula to solve problems.

Applications

Quadratic equations can be solved to answer questions about variables contained in quadratic functions.

EXAMPLE 6 Blood Pressure and Age

The graphs in **Figure 8.3** illustrate that a person's normal systolic blood pressure, measured in millimeters of mercury (mm Hg), depends on his or her age. The function

$$P(A) = 0.006A^2 - 0.02A + 120$$

models a man's normal systolic pressure, $P(A)$, at age A.

a. Find the age, to the nearest year, of a man whose normal systolic blood pressure is 125 mm Hg.

b. Use the graphs in **Figure 8.3** to describe the differences between the normal systolic blood pressures of men and women as they age.

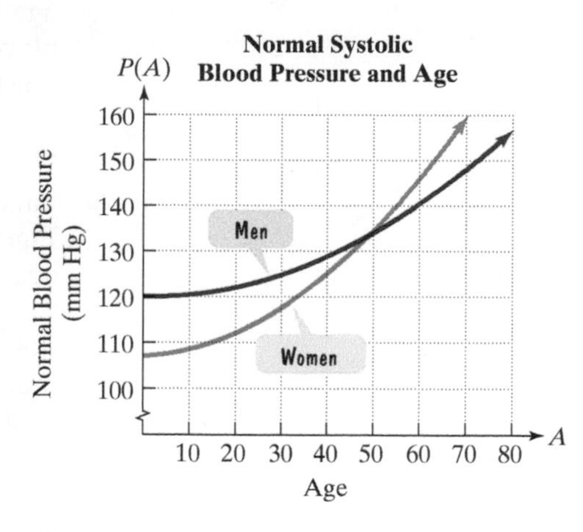

Figure 8.3

Solution

a. We are interested in the age of a man with a normal systolic blood pressure of 125 millimeters of mercury. Thus, we substitute 125 for $P(A)$ in the given function for men. Then we solve for A, the man's age.

$$P(A) = 0.006A^2 - 0.02A + 120$$ This is the given function for men.

$$125 = 0.006A^2 - 0.02A + 120$$ Substitute 125 for $P(A)$.

$$0 = 0.006A^2 - 0.02A - 5$$ Subtract 125 from both sides and write the quadratic equation in standard form.

$a = 0.006$ $b = -0.02$ $c = -5$

Because the trinomial on the right side of the equation is prime, we solve using the quadratic formula.

Notice that the variable is A, rather than the usual x.

$$A = \frac{-b \pm \sqrt{b^2 - 4ac}}{2a}$$ Use the quadratic formula.

$$= \frac{-(-0.02) \pm \sqrt{(-0.02)^2 - 4(0.006)(-5)}}{2(0.006)}$$ Substitute the values for a, b, and c: $a = 0.006$, $b = -0.02$, and $c = -5$.

$$= \frac{0.02 \pm \sqrt{0.1204}}{0.012}$$ Use a calculator to simplify the expression under the square root.

$$\approx \frac{0.02 \pm 0.347}{0.012}$$ Use a calculator: $\sqrt{0.1204} \approx 0.347$.

$$A \approx \frac{0.02 + 0.347}{0.012} \quad \text{or} \quad A \approx \frac{0.02 - 0.347}{0.012}$$

$$A \approx 31 \qquad\qquad A \approx -27$$ Use a calculator and round to the nearest integer.

Reject this solution. Age cannot be negative.

Using Technology

On most calculators, here is how to approximate

$$\frac{0.02 + \sqrt{0.1204}}{0.012}.$$

Many Scientific Calculators

[( ] .02 [+] .1204 [√] [ )]

[÷] .012 [=]

Many Graphing Calculators

[( ] .02 [+] [√] .1204 [ )]

[÷] .012 [ENTER]

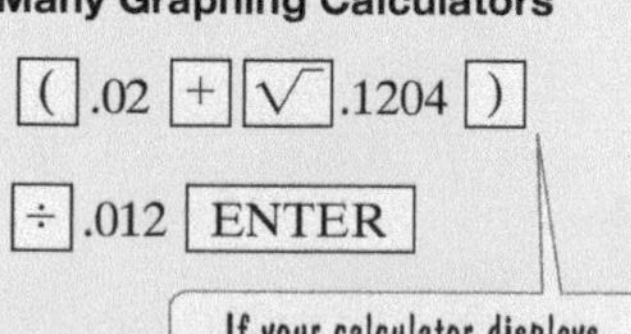

If your calculator displays an open parenthesis after √, you'll need to enter another closed parenthesis here.

The positive solution, $A \approx 31$, indicates that 31 is the approximate age of a man whose normal systolic blood pressure is 125 mm Hg. This is illustrated by the black lines with the arrows on the red graph representing men in **Figure 8.4**.

b. Take a second look at the graphs in **Figure 8.4**. Before approximately age 50, the blue graph representing women's normal systolic blood pressure lies below the red graph representing men's normal systolic blood pressure. Thus, up to age 50, women's normal systolic blood pressure is lower than men's, although it is increasing at a faster rate. After age 50, women's normal systolic blood pressure is higher than men's. ■

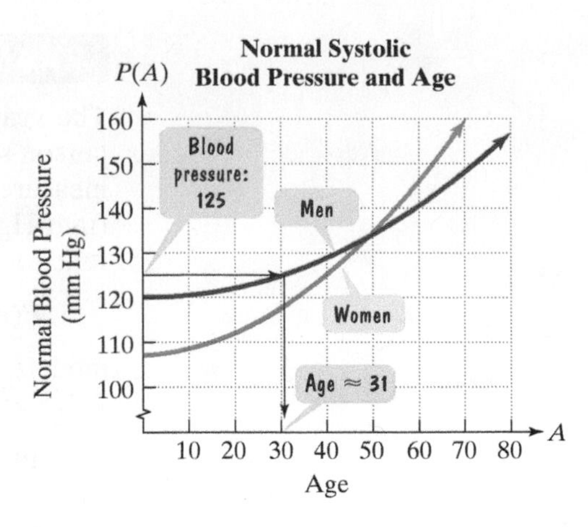

Figure 8.4

✓ **CHECK POINT 6** The function $P(A) = 0.01A^2 + 0.05A + 107$ models a woman's normal systolic blood pressure, $P(A)$, at age A. Use this function to find the age, to the nearest year, of a woman whose normal systolic blood pressure is 115 mm Hg. Use the blue graph in **Figure 8.4** to verify your solution.

Achieving Success

A test-taking tip: Go for partial credit. Always show your work. If worse comes to worst, write something down, anything, even if it's a formula that you think might solve a problem or a possible idea or procedure for solving the problem. Here's an example:

Test Question: Solve by the quadratic formula:

$$2x^2 = -4x + 5.$$

Student Solution:

$2x^2 = -4x + 5$

$2x^2 + 4x - 5 = 0$

$a = 2 \quad b = 4 \quad c = -5$

Egad! I forgot the formula.

$x = \frac{-b \pm \sqrt{?}}{2a} = \frac{-4 \pm \sqrt{?}}{2 \cdot 2}$

I know I need to find ?.
That will give me the solutions.
Perhaps I can simplify after that.
???

CONCEPT AND VOCABULARY CHECK

Fill in each blank so that the resulting statement is true.

1. The solutions of a quadratic equation in standard form $ax^2 + bx + c = 0$, $a \neq 0$, are given by the quadratic formula $x =$ ______.
2. In order to solve $2x^2 + 9x - 5 = 0$ by the quadratic formula, we use $a =$ ______, $b =$ ______, and $c =$ ______.
3. In order to solve $x^2 = 4x + 1$ by the quadratic formula, we use $a =$ ______, $b =$ ______, and $c =$ ______.
4. $x = \dfrac{-(-4) \pm \sqrt{(-4)^2 - 4(1)(2)}}{2(1)}$ simplifies to $x =$ ______.
5. $x = \dfrac{-4 \pm \sqrt{4^2 - 4 \cdot 2 \cdot 5}}{2 \cdot 2}$ simplifies to $x =$ ______.
6. The discriminant of $ax^2 + bx + c = 0$ is defined by ______.
7. If the discriminant of $ax^2 + bx + c = 0$ is negative, the quadratic equation has ______ real solutions.
8. If the discriminant of $ax^2 + bx + c = 0$ is positive, the quadratic equation has ______ real solutions.
9. The most efficient technique for solving $(2x + 7)^2 = 25$ is by using ______.
10. The most efficient technique for solving $x^2 + 5x - 10 = 0$ is by using ______.
11. The most efficient technique for solving $x^2 + 8x + 15 = 0$ is by using ______.
12. True or false: An equation with the solution set $\{2, 5\}$ is $(x + 2)(x + 5) = 0$. ______

8.2 EXERCISE SET

MyMathLab®

Practice Exercises

In Exercises 1–18, solve each equation using the quadratic formula. Simplify solutions, if possible.

1. $x^2 + 8x + 12 = 0$
2. $x^2 + 8x + 15 = 0$
3. $2x^2 - 7x = -5$
4. $5x^2 + 8x = -3$
5. $x^2 + 3x - 20 = 0$
6. $x^2 + 5x - 10 = 0$
7. $3x^2 - 7x = 3$
8. $4x^2 + 3x = 2$
9. $6x^2 = 2x + 1$
10. $2x^2 = -4x + 5$
11. $4x^2 - 3x = -6$
12. $9x^2 + x = -2$
13. $x^2 - 4x + 8 = 0$
14. $x^2 + 6x + 13 = 0$
15. $3x^2 = 8x - 7$
16. $3x^2 = 4x - 6$
17. $2x(x - 2) = x + 12$
18. $2x(x + 4) = 3x - 3$

In Exercises 19–30, compute the discriminant. Then determine the number and type of solutions for the given equation.

19. $x^2 + 8x + 3 = 0$
20. $x^2 + 7x + 4 = 0$
21. $x^2 + 6x + 8 = 0$
22. $x^2 + 2x - 3 = 0$
23. $2x^2 + x + 3 = 0$
24. $2x^2 - 4x + 3 = 0$
25. $2x^2 + 6x = 0$
26. $3x^2 - 5x = 0$
27. $5x^2 + 3 = 0$
28. $5x^2 + 4 = 0$
29. $9x^2 = 12x - 4$
30. $4x^2 = 20x - 25$

In Exercises 31–50, solve each equation by the method of your choice. Simplify solutions, if possible.

31. $3x^2 - 4x = 4$

32. $2x^2 - x = 1$

33. $x^2 - 2x = 1$

34. $2x^2 + 3x = 1$

35. $3x^2 = x - 9$

36. $2x^2 = -6x - 7$

37. $(2x - 5)(x + 1) = 2$

38. $(2x + 3)(x + 4) = 1$

39. $(3x - 4)^2 = 16$

40. $(2x + 7)^2 = 25$

41. $\frac{x^2}{2} + 2x + \frac{2}{3} = 0$

42. $\frac{x^2}{3} - x - \frac{1}{6} = 0$

43. $(3x - 2)^2 = 10$

44. $(4x - 1)^2 = 15$

45. $\frac{1}{x} + \frac{1}{x + 2} = \frac{1}{3}$

46. $\frac{1}{x} + \frac{1}{x + 3} = \frac{1}{4}$

47. $(2x - 6)(x + 2) = 5(x - 1) - 12$

48. $7x(x - 2) = 3 - 2(x + 4)$

49. $x^2 + 10 = 2(2x - 1)$

50. $x(x + 6) = -12$

In Exercises 51–64, write a quadratic equation in standard form with the given solution set.

51. $\{-3, 5\}$

52. $\{-2, 6\}$

53. $\left\{-\frac{2}{3}, \frac{1}{4}\right\}$

54. $\left\{-\frac{5}{6}, \frac{1}{3}\right\}$

55. $\{-\sqrt{2}, \sqrt{2}\}$

56. $\{-\sqrt{3}, \sqrt{3}\}$

57. $\{-2\sqrt{5}, 2\sqrt{5}\}$

58. $\{-3\sqrt{5}, 3\sqrt{5}\}$

59. $\{-6i, 6i\}$

60. $\{-8i, 8i\}$

61. $\{1 + i, 1 - i\}$

62. $\{2 + i, 2 - i\}$

63. $\{1 + \sqrt{2}, 1 - \sqrt{2}\}$

64. $\{1 + \sqrt{3}, 1 - \sqrt{3}\}$

Practice PLUS

Exercises 65–68 describe quadratic equations. Match each description with the graph of the corresponding quadratic function. Each graph is shown in a $[-10, 10, 1]$ by $[-10, 10, 1]$ viewing rectangle.

65. A quadratic equation whose solution set contains imaginary numbers

66. A quadratic equation whose discriminant is 0

67. A quadratic equation whose solution set is $\{3 \pm \sqrt{2}\}$

68. A quadratic equation whose solution set contains integers

a.

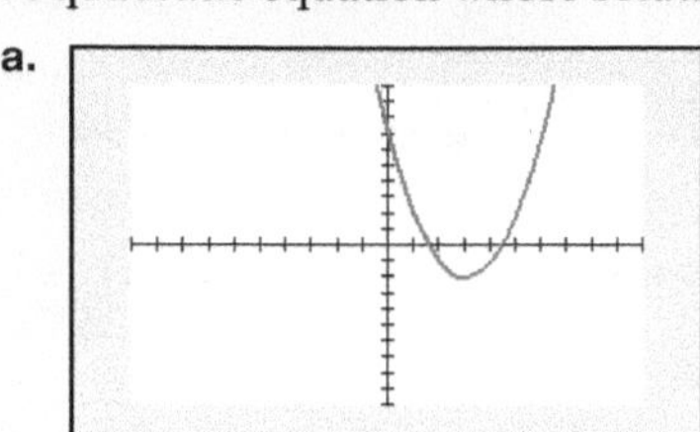

b.

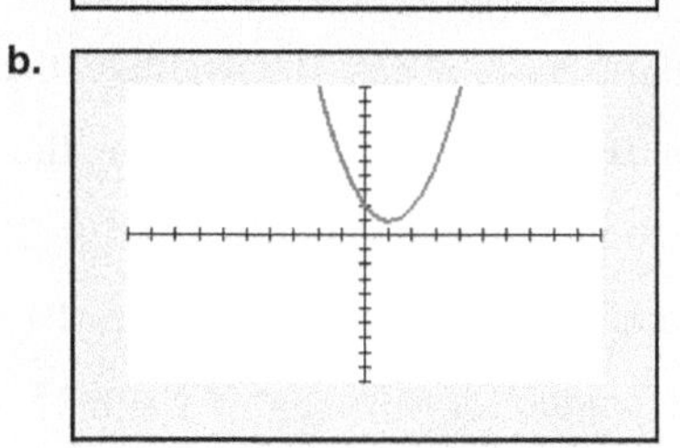

c.

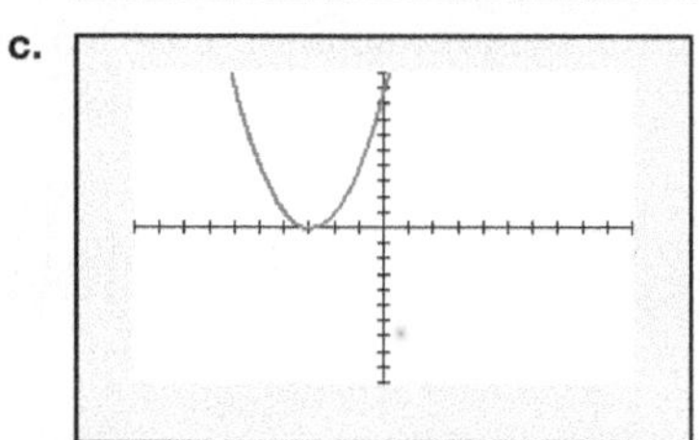

d.

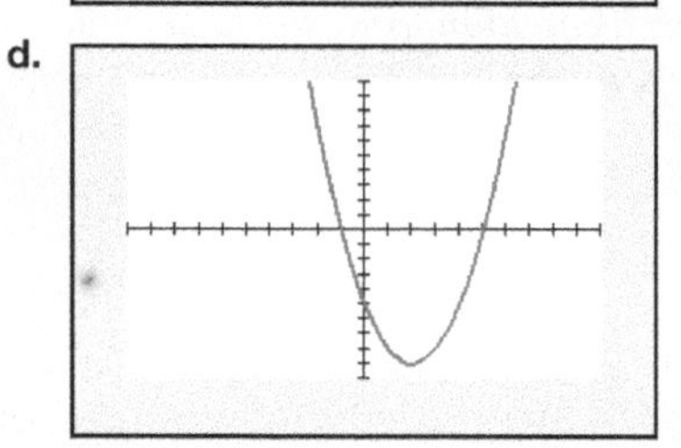

69. When the sum of 6 and twice a positive number is subtracted from the square of the number, 0 results. Find the number.

70. When the sum of 1 and twice a negative number is subtracted from twice the square of the number, 0 results. Find the number.

In Exercises 71–76, solve each equation by the method of your choice.

71. $\frac{1}{x^2 - 3x + 2} = \frac{1}{x + 2} + \frac{5}{x^2 - 4}$

72. $\frac{x - 1}{x - 2} + \frac{x}{x - 3} = \frac{1}{x^2 - 5x + 6}$

73. $\sqrt{2}x^2 + 3x - 2\sqrt{2} = 0$

74. $\sqrt{3}x^2 + 6x + 7\sqrt{3} = 0$

75. $|x^2 + 2x| = 3$

76. $|x^2 + 3x| = 2$

Application Exercises

A driver's age has something to do with his or her chance of getting into a fatal car crash. The bar graph shows the number of fatal vehicle crashes per 100 million miles driven for drivers of various age groups. For example, 25-year-old drivers are involved in 4.1 fatal crashes per 100 million miles driven. Thus, when a group of 25-year-old Americans have driven a total of 100 million miles, approximately 4 have been in accidents in which someone died.

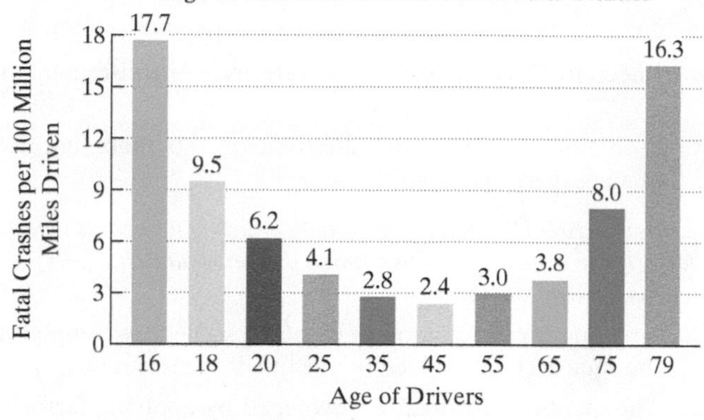

Source: Insurance Institute for Highway Safety

The number of fatal vehicle crashes per 100 million miles, $f(x)$, for drivers of age x can be modeled by the quadratic function

$$f(x) = 0.013x^2 - 1.19x + 28.24.$$

Use the function to solve Exercises 77–78.

77. What age groups are expected to be involved in 3 fatal crashes per 100 million miles driven? How well does the function model the trend in the actual data shown in the bar graph?

78. What age groups are expected to be involved in 10 fatal crashes per 100 million miles driven? How well does the function model the trend in the actual data shown in the bar graph?

Throwing events in track and field include the shot put, the discus throw, the hammer throw, and the javelin throw. The distance that an athlete can achieve depends on the initial velocity of the object thrown and the angle above the horizontal at which the object leaves the hand.

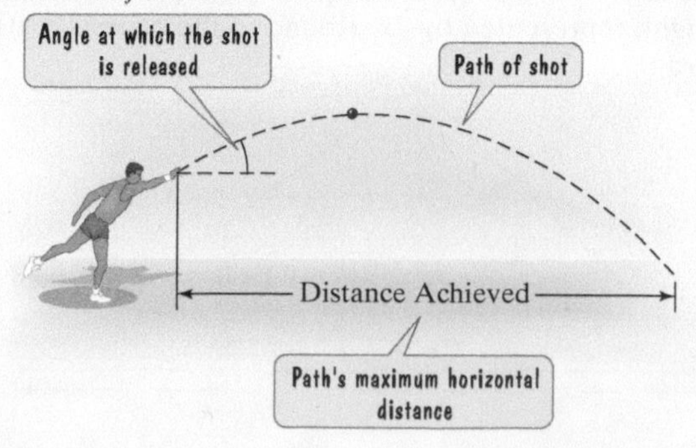

In Exercises 79–80, an athlete whose event is the shot put releases the shot with the same initial velocity, but at different angles.

79. When the shot is released at an angle of 35°, its path can be modeled by the function

$$f(x) = -0.01x^2 + 0.7x + 6.1,$$

in which x is the shot's horizontal distance, in feet, and $f(x)$ is its height, in feet. This function is shown by one of the graphs, (a) or (b), in the figure. Use the function to determine the shot's maximum distance. Use a calculator and round to the nearest tenth of a foot. Which graph, (a) or (b), shows the shot's path?

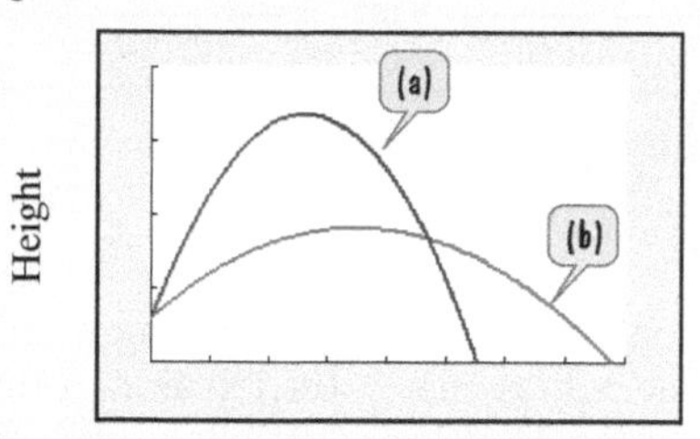

[0, 80, 10] by [0, 40, 10]

80. When the shot is released at an angle of 65°, its path can be modeled by the function

$$f(x) = -0.04x^2 + 2.1x + 6.1,$$

in which x is the shot's horizontal distance, in feet, and $f(x)$ is its height, in feet. This function is shown by one of the graphs, (a) or (b), in the figure above. Use the function to determine the shot's maximum distance. Use a calculator and round to the nearest tenth of a foot. Which graph, (a) or (b), shows the shot's path?

81. The length of a rectangle is 4 meters longer than the width. If the area is 8 square meters, find the rectangle's dimensions. Round to the nearest tenth of a meter.

82. The length of a rectangle exceeds twice its width by 3 inches. If the area is 10 square inches, find the rectangle's dimensions. Round to the nearest tenth of an inch.

83. The longer leg of a right triangle exceeds the shorter leg by 1 inch, and the hypotenuse exceeds the longer leg by 7 inches. Find the lengths of the legs. Round to the nearest tenth of a inch.

84. The hypotenuse of a right triangle is 6 feet long. One leg is 2 feet shorter than the other. Find the lengths of the legs. Round to the nearest tenth of a foot.

85. A rain gutter is made from sheets of aluminum that are 20 inches wide. As shown in the figure, the edges are turned up to form right angles. Determine the depth of the gutter that will allow a cross-sectional area of 13 square inches. Show that there are two different solutions to the problem. Round to the nearest tenth of an inch.

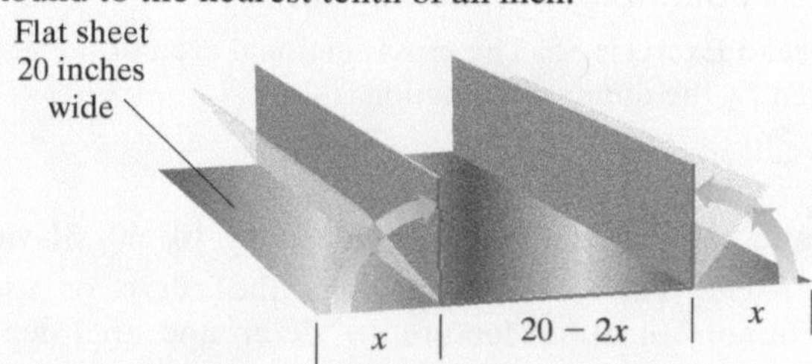

86. A piece of wire is 8 inches long. The wire is cut into two pieces and then each piece is bent into a square. Find the length of each piece if the sum of the areas of these squares is to be 2 square inches.

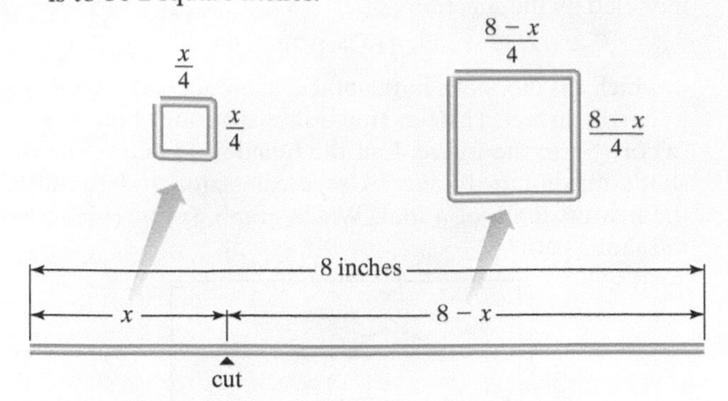

87. Working together, two people can mow a large lawn in 4 hours. One person can do the job alone 1 hour faster than the other person. How long does it take each person working alone to mow the lawn? Round to the nearest tenth of an hour.

88. A pool has an inlet pipe to fill it and an outlet pipe to empty it. It takes 2 hours longer to empty the pool than it does to fill it. The inlet pipe is turned on to fill the pool, but the outlet pipe is accidentally left open. Despite this, the pool fills in 8 hours. How long does it take the outlet pipe to empty the pool? Round to the nearest tenth of an hour.

Explaining the Concepts

89. What is the quadratic formula and why is it useful?

90. Without going into specific details for every step, describe how the quadratic formula is derived.

91. Explain how to solve $x^2 + 6x + 8 = 0$ using the quadratic formula.

92. If a quadratic equation has imaginary solutions, how is this shown on the graph of the corresponding quadratic function?

93. What is the discriminant and what information does it provide about a quadratic equation?

94. If you are given a quadratic equation, how do you determine which method to use to solve it?

95. Explain how to write a quadratic equation from its solution set. Give an example with your explanation.

Technology Exercises

96. Use a graphing utility to graph the quadratic function related to any five of the quadratic equations in Exercises 19–30. How does each graph illustrate what you determined algebraically using the discriminant?

97. Reread Exercise 85. The cross-sectional area of the gutter is given by the quadratic function

$$f(x) = x(20 - 2x).$$

Graph the function in a [0, 10, 1] by [0, 60, 5] viewing rectangle. Then TRACE along the curve or use the maximum function feature to determine the depth of the gutter that will maximize its cross-sectional area and allow the greatest amount of water to flow. What is the maximum area? Does the situation described in Exercise 85 take full advantage of the sheets of aluminum?

Critical Thinking Exercises

Make Sense? *In Exercises 98–101, determine whether each statement makes sense or does not make sense, and explain your reasoning.*

98. Because I want to solve $25x^2 - 169 = 0$ fairly quickly, I'll use the quadratic formula.

99. I simplified $\frac{3 + 2\sqrt{3}}{2}$ to $3 + \sqrt{3}$ because 2 is a factor of $2\sqrt{3}$.

100. I need to find a square root to determine the discriminant.

101. I obtained -17 for the discriminant, so there are two imaginary irrational solutions.

In Exercises 102–105, determine whether each statement is true or false. If the statement is false, make the necessary change(s) to produce a true statement.

102. Any quadratic equation that can be solved by completing the square can be solved by the quadratic formula.

103. The quadratic formula is developed by applying factoring and the zero-product principle to the quadratic equation $ax^2 + bx + c = 0$.

104. In using the quadratic formula to solve the quadratic equation $5x^2 = 2x - 7$, we have $a = 5, b = 2$, and $c = -7$.

105. The quadratic formula can be used to solve the equation $x^2 - 9 = 0$.

106. Solve for t: $s = -16t^2 + v_0 t$.

107. A rectangular swimming pool is 12 meters long and 8 meters wide. A tile border of uniform width is to be built around the pool using 120 square meters of tile. The tile is from a discontinued stock (so no additional materials are available) and all 120 square meters are to be used. How wide should the border be? Round to the nearest tenth of a meter. If zoning laws require at least a 2-meter-wide border around the pool, can this be done with the available tile?

108. The area of the shaded green region outside the rectangle and inside the triangle is 10 square yards. Find the triangle's height, represented by $2x$. Round to the nearest tenth of a yard.

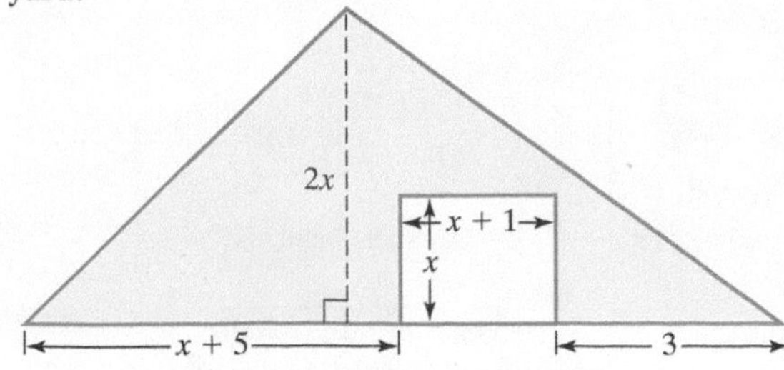

Review Exercises

109. Solve: $|5x + 2| = |4 - 3x|$. (Section 4.3, Example 3)

110. Solve: $\sqrt{2x - 5} - \sqrt{x - 3} = 1$.
(Section 7.6, Example 4)

111. Rationalize the denominator: $\frac{5}{\sqrt{3} + x}$. (Section 7.5, Example 5)

Preview Exercises

Exercises 112–114 will help you prepare for the material covered in the next section.

112. Use point plotting to graph $f(x) = x^2$ and $g(x) = x^2 + 2$ in the same rectangular coordinate system.

113. Use point plotting to graph $f(x) = x^2$ and $g(x) = (x + 2)^2$ in the same rectangular coordinate system.

114. Find the x-intercepts for the graph of $f(x) = -2(x - 3)^2 + 8$.

SECTION

8.3 Quadratic Functions and Their Graphs 

What am I supposed to learn?

After studying this section, you should be able to:

1. Recognize characteristics of parabolas.
2. Graph parabolas in the form $f(x) = a(x - h)^2 + k$.
3. Graph parabolas in the form $f(x) = ax^2 + bx + c$.
4. Determine a quadratic function's minimum or maximum value.
5. Solve problems involving a quadratic function's minimum or maximum value.

We have a long history of throwing things. Before 400 B.C., the Greeks competed in games that included discus throwing. In the seventeenth century, English soldiers organized cannonball-throwing competitions. In 1827, a Yale University student, disappointed over failing an exam, took out his frustrations at the passing of a collection plate in chapel. Seizing the monetary tray, he flung it in the direction of a large open space on campus. Yale students see this act of frustration as the origin of the Frisbee.

In this section, we study quadratic functions and their graphs. By graphing functions that model the paths of the things we throw, you will be able to determine both the maximum height these objects attain and the distance these objects travel.

Graphs of Quadratic Functions

1 Recognize characteristics of parabolas.

The graph of any quadratic function

$$f(x) = ax^2 + bx + c, \quad a \neq 0,$$

is called a **parabola**. Parabolas are shaped like bowls or inverted bowls, as shown in **Figure 8.5** at the top of the next page. If the coefficient of x^2 (the value of a in $ax^2 + bx + c$) is positive, the parabola opens upward. If the coefficient of x^2 is negative, the parabola opens downward. The **vertex** (or turning point) of the parabola is the lowest point on the graph when it opens upward and the highest point on the graph when it opens downward.

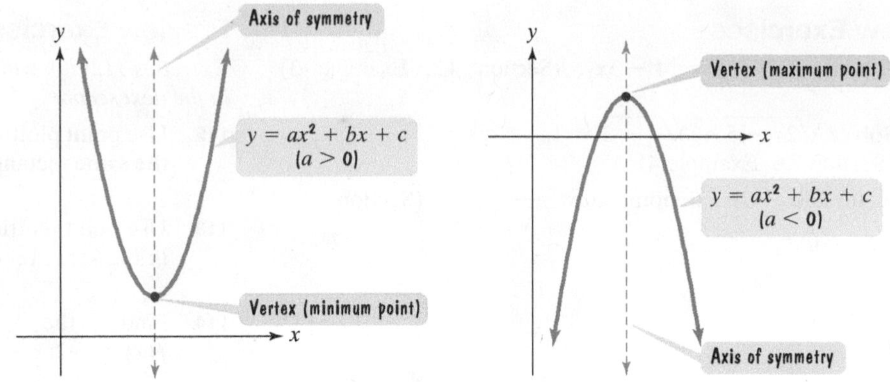

$a > 0$: Parabola opens upward. $a < 0$: Parabola opens downward.

Figure 8.5 Characteristics of graphs of quadratic functions

The two halves of a parabola are mirror images of each other. A "mirror line" through the vertex, called the **axis of symmetry**, divides the figure in half. If a parabola is folded along its axis of symmetry, the two halves match exactly.

2 Graph parabolas in the form $f(x) = a(x - h)^2 + k$.

Graphing Quadratic Functions in the Form $f(x) = a(x - h)^2 + k$

One way to obtain the graph of a quadratic function is to use point plotting. Let's begin by graphing the functions $f(x) = x^2$, $g(x) = 2x^2$, and $h(x) = \frac{1}{2}x^2$ in the same rectangular coordinate system. Select integers for x, starting with -3 and ending with 3. A partial table of coordinates for each function is shown below. The three parabolas are shown in **Figure 8.6**.

x	$f(x) = x^2$	(x, y) or $(x, f(x))$
-3	$f(-3) = (-3)^2 = 9$	$(-3, 9)$
-2	$f(-2) = (-2)^2 = 4$	$(-2, 4)$
-1	$f(-1) = (-1)^2 = 1$	$(-1, 1)$
0	$f(0) = 0^2 = 0$	$(0, 0)$
1	$f(1) = 1^2 = 1$	$(1, 1)$
2	$f(2) = 2^2 = 4$	$(2, 4)$
3	$f(3) = 3^2 = 9$	$(3, 9)$

x	$g(x) = 2x^2$	(x, y) or $(x, g(x))$
-3	$g(-3) = 2(-3)^2 = 18$	$(-3, 18)$
-2	$g(-2) = 2(-2)^2 = 8$	$(-2, 8)$
-1	$g(-1) = 2(-1)^2 = 2$	$(-1, 2)$
0	$g(0) = 2 \cdot 0^2 = 0$	$(0, 0)$
1	$g(1) = 2 \cdot 1^2 = 2$	$(1, 2)$
2	$g(2) = 2 \cdot 2^2 = 8$	$(2, 8)$
3	$g(3) = 2 \cdot 3^2 = 18$	$(3, 18)$

x	$h(x) = \frac{1}{2}x^2$	(x, y) or $(x, h(x))$
-3	$h(-3) = \frac{1}{2}(-3)^2 = \frac{9}{2}$	$\left(-3, \frac{9}{2}\right)$
-2	$h(-2) = \frac{1}{2}(-2)^2 = 2$	$(-2, 2)$
-1	$h(-1) = \frac{1}{2}(-1)^2 = \frac{1}{2}$	$\left(-1, \frac{1}{2}\right)$
0	$h(0) = \frac{1}{2} \cdot 0^2 = 0$	$(0, 0)$
1	$h(1) = \frac{1}{2} \cdot 1^2 = \frac{1}{2}$	$\left(1, \frac{1}{2}\right)$
2	$h(2) = \frac{1}{2} \cdot 2^2 = 2$	$(2, 2)$
3	$h(3) = \frac{1}{2} \cdot 3^2 = \frac{9}{2}$	$\left(3, \frac{9}{2}\right)$

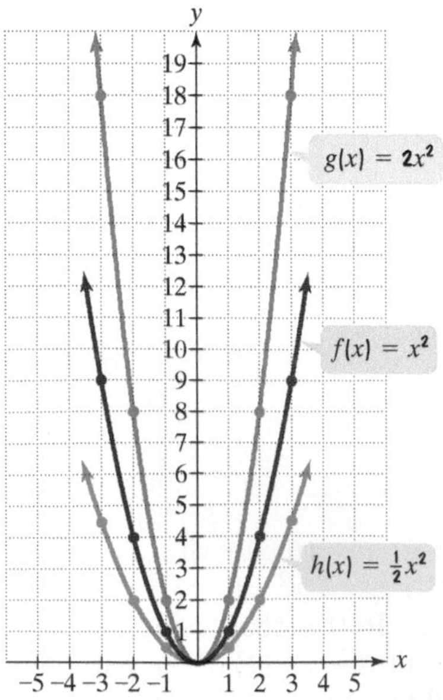

Figure 8.6

Can you see that the graphs of f, g, and h in **Figure 8.6** all have the same vertex, $(0, 0)$? They also have the same axis of symmetry, the y-axis, or $x = 0$. This is true for all graphs of the form $f(x) = ax^2$. However, the blue graph of $g(x) = 2x^2$ is a narrower parabola than the red graph of $f(x) = x^2$. By contrast, the green graph of $h(x) = \frac{1}{2}x^2$ is a flatter parabola than the red graph of $f(x) = x^2$.

Is there a more efficient method than point plotting to obtain the graph of a quadratic function? The answer is yes. The method is based on comparing graphs of the form $g(x) = a(x - h)^2 + k$ to those of the form $f(x) = ax^2$.

In **Figure 8.7(a)**, the graph of $f(x) = ax^2$ for $a > 0$ is shown in black. The parabola's vertex is $(0, 0)$ and it opens upward. In **Figure 8.7(b)**, the graph of $f(x) = ax^2$ for $a < 0$ is shown in black. The parabola's vertex is $(0, 0)$ and it opens downward.

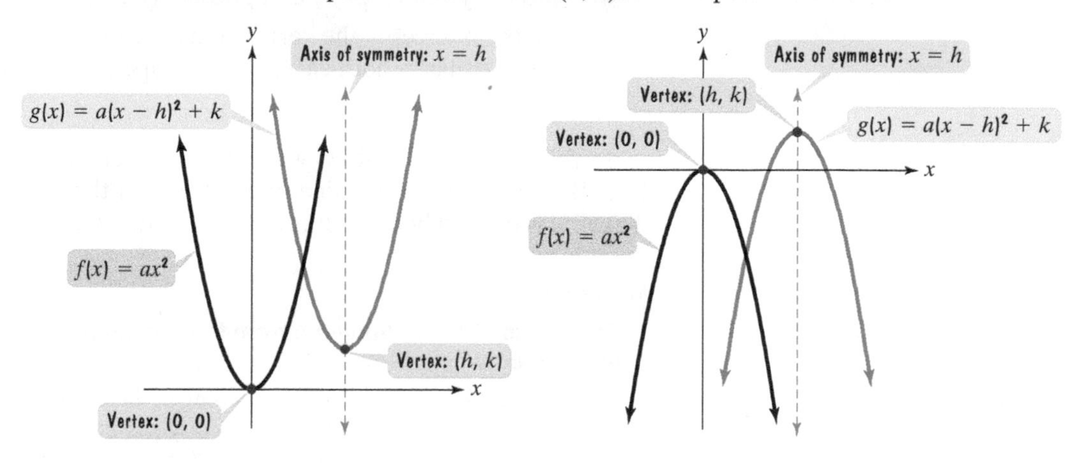

(a) $a > 0$: Parabola opens upward. **(b)** $a < 0$: Parabola opens downward.

Figure 8.7 Moving, or shifting, the graph of $f(x) = ax^2$

Figures 8.7(a) and **8.7(b)** also show the graph of $g(x) = a(x - h)^2 + k$ in blue. Compare these graphs to those of $f(x) = ax^2$. Observe that h determines a horizontal move, or shift, and k determines a vertical move, or shift, of the graph of $f(x) = ax^2$:

$$g(x) = a(x - h)^2 + k.$$

If $h > 0$, the graph of $f(x) = ax^2$ is shifted h units to the right.

If $k > 0$, the graph of $y = a(x - h)^2$ is shifted k units up.

Consequently, the vertex $(0, 0)$ on the black graph of $f(x) = ax^2$ moves to the point (h, k) on the blue graph of $g(x) = a(x - h)^2 + k$. The axis of symmetry is the vertical line whose equation is $x = h$.

The form of the expression for g is convenient because it immediately identifies the vertex of the parabola as (h, k).

Quadratic Functions in the Form $f(x) = a(x - h)^2 + k$

The graph of

$$f(x) = a(x - h)^2 + k, \qquad a \neq 0$$

is a parabola whose vertex is the point (h, k). The parabola is symmetric with respect to the line $x = h$. If $a > 0$, the parabola opens upward; if $a < 0$, the parabola opens downward.

The sign of a in $f(x) = a(x - h)^2 + k$ determines whether the parabola opens upward or downward. Furthermore, if $|a|$ is small, the parabola opens more flatly than if $|a|$ is large. On the next page is a general procedure for graphing parabolas whose equations are in this form.

Graphing Quadratic Functions with Equations in the Form $f(x) = a(x - h)^2 + k$

To graph $f(x) = a(x - h)^2 + k$,

1. Determine whether the parabola opens upward or downward. If $a > 0$, it opens upward. If $a < 0$, it opens downward.
2. Determine the vertex of the parabola. The vertex is (h, k).
3. Find any x-intercepts by solving $f(x) = 0$. The equation's real solutions are the x-intercepts.
4. Find the y-intercept by computing $f(0)$.
5. Plot the intercepts, the vertex, and additional points as necessary. Connect these points with a smooth curve that is shaped like a bowl or an inverted bowl.

In the graphs that follow, we will show each axis of symmetry as a dashed vertical line. Because this vertical line passes through the vertex, (h, k), its equation is $x = h$. The line is dashed because it is not part of the parabola.

Great Question!

I'm confused about finding h from the equation $f(x) = a(x - h)^2 + k$. Can you help me out?

It's easy to make a sign error when finding h, the x-coordinate of the vertex. In

$$f(x) = a(x - h)^2 + k,$$

h is the number that follows the subtraction sign.

- $f(x) = -2(x - 3)^2 + 8$

  The number after the subtraction is 3: $h = 3$.

- $f(x) = (x + 3)^2 + 1$
  $= (x - (-3))^2 + 1$

  The number after the subtraction is -3: $h = -3$.

EXAMPLE 1 Graphing a Quadratic Function in the Form $f(x) = a(x - h)^2 + k$

Graph the quadratic function $f(x) = -2(x - 3)^2 + 8$.

Solution We can graph this function by following the steps in the preceding box. We begin by identifying values for a, h, and k.

$$f(x) = a(x - h)^2 + k$$

$a = -2$ $h = 3$ $k = 8$

$$f(x) = -2(x - 3)^2 + 8$$

Step 1. Determine how the parabola opens. Note that a, the coefficient of x^2, is -2. Thus, $a < 0$; this negative value tells us that the parabola opens downward.

Step 2. Find the vertex. The vertex of the parabola is at (h, k). Because $h = 3$ and $k = 8$, the parabola has its vertex at $(3, 8)$.

Step 3. Find the x-intercepts by solving $f(x) = 0$. Replace $f(x)$ with 0 in $f(x) = -2(x - 3)^2 + 8$.

$$0 = -2(x - 3)^2 + 8$$ Find x-intercepts, setting $f(x)$ equal to 0.

$$2(x - 3)^2 = 8$$ Solve for x. Add $2(x - 3)^2$ to both sides of the equation.

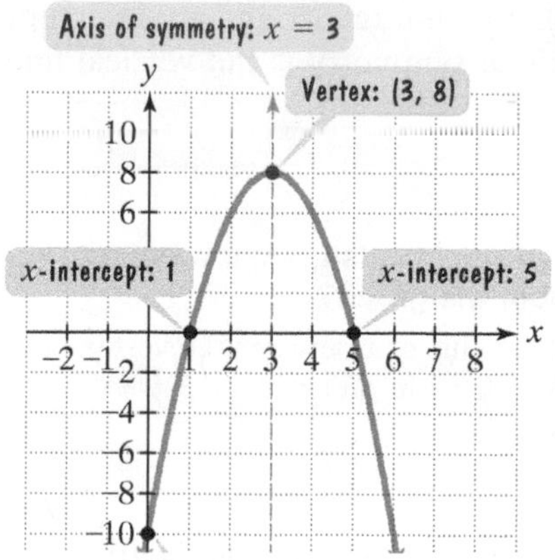

Figure 8.8 The graph of $f(x) = -2(x - 3)^2 + 8$

$$(x - 3)^2 = 4 \quad \text{Divide both sides by 2.}$$

$$x - 3 = \sqrt{4} \quad \text{or} \quad x - 3 = -\sqrt{4} \quad \text{Apply the square root property.}$$

$$x - 3 = 2 \qquad x - 3 = -2 \qquad \sqrt{4} = 2$$

$$x = 5 \qquad x = 1 \qquad \text{Add 3 to both sides in each equation.}$$

The x-intercepts are 5 and 1. The parabola passes through $(5, 0)$ and $(1, 0)$.

Step 4. Find the y-intercept by computing $f(0)$. Replace x with 0 in $f(x) = -2(x - 3)^2 + 8$.

$$f(0) = -2(0 - 3)^2 + 8 = -2(-3)^2 + 8 = -2(9) + 8 = -10$$

The y-intercept is -10. The parabola passes through $(0, -10)$.

Step 5. Graph the parabola. With a vertex at $(3, 8)$, x-intercepts at 5 and 1, and a y-intercept at -10, the graph of f is shown in **Figure 8.8**. The axis of symmetry is the vertical line whose equation is $x = 3$. ■

CHECK POINT 1 Graph the quadratic function $f(x) = -(x - 1)^2 + 4$.

EXAMPLE 2 Graphing a Quadratic Function in the Form $f(x) = a(x - h)^2 + k$

Graph the quadratic function $f(x) = (x + 3)^2 + 1$.

Solution We begin by finding values for a, h, and k.

$$f(x) = a(x - h)^2 + k \quad \text{Form of quadratic function}$$

$$f(x) = (x + 3)^2 + 1 \quad \text{Given function}$$

$$f(x) = 1(x - (-3))^2 + 1$$

$a = 1$ $h = -3$ $k = 1$

Step 1. Determine how the parabola opens. Note that a, the coefficient of x^2, is 1. Thus, $a > 0$; this positive value tells us that the parabola opens upward.

Step 2. Find the vertex. The vertex of the parabola is at (h, k). Because $h = -3$ and $k = 1$, the parabola has its vertex at $(-3, 1)$. See **Figure 8.9**.

Step 3. Find the x-intercepts by solving $f(x) = 0$. Replace $f(x)$ with 0 in $f(x) = (x + 3)^2 + 1$. Because the vertex is at $(-3, 1)$, which lies above the x-axis, and the parabola opens upward, it appears that this parabola has no x-intercepts. We can verify this observation algebraically.

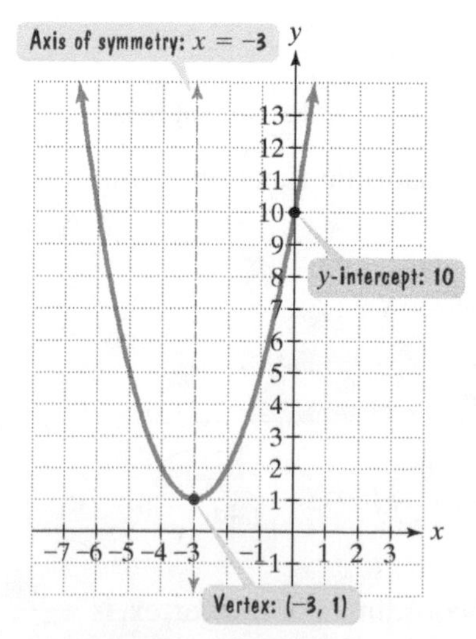

Figure 8.9 The graph of $f(x) = (x + 3)^2 + 1$

$$0 = (x + 3)^2 + 1 \quad \text{Find possible } x\text{-intercepts, setting } f(x) \text{ equal to 0.}$$

$$-1 = (x + 3)^2 \quad \text{Solve for } x. \text{ Subtract 1 from both sides.}$$

$$x + 3 = \sqrt{-1} \quad \text{or} \quad x + 3 = -\sqrt{-1} \quad \text{Apply the square root property.}$$

$$x + 3 = i \qquad x + 3 = -i \qquad \sqrt{-1} = i$$

$$x = -3 + i \qquad x = -3 - i \qquad \text{The solutions are } -3 \pm i.$$

Because this equation has no real solutions, the parabola has no x-intercepts.

Step 4. Find the y-intercept by computing $f(0)$. Replace x with 0 in $f(x) = (x + 3)^2 + 1$.

$$f(0) = (0 + 3)^2 + 1 = 3^2 + 1 = 9 + 1 = 10$$

The y-intercept is 10. The parabola passes through $(0, 10)$.

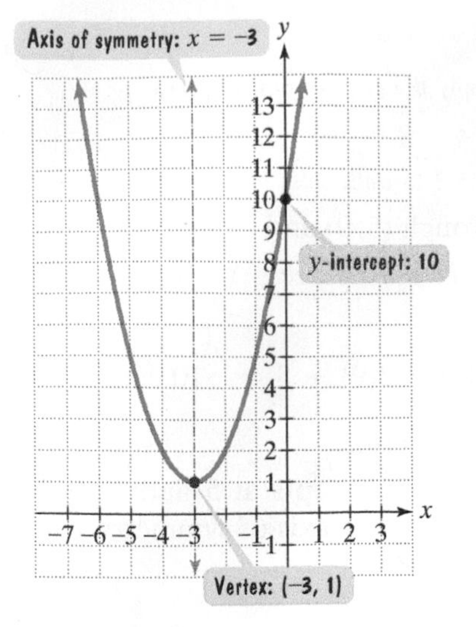

Figure 8.9 (repeated)

Step 5. Graph the parabola. With a vertex at $(-3, 1)$, no x-intercepts, and a y-intercept at 10, the graph of f is shown in **Figure 8.9**. The axis of symmetry is the vertical line whose equation is $x = -3$. ■

Great Question!

You graphed the parabola in Figure 8.9 using only two points. Is there a way I can find some additional points to feel more secure about the graph?

Yes. You can find an additional point or two by choosing a value of x between the vertex and the y-intercept. For example, let $x = -1$. Replace x with -1 in $f(x) = (x + 3)^2 + 1$.

$$f(-1) = (-1 + 3)^2 + 1 = 2^2 + 1 = 4 + 1 = 5$$

The parabola passes through $(-1, 5)$.

More points? You can use the axis of symmetry and mirror both $(-1, 5)$ and $(0, 10)$. Mirroring $(-1, 5)$ gives the point $(-5, 5)$. Mirroring $(0, 10)$ gives the point $(-6, 10)$. Take a moment to identify these points in **Figure 8.9**.

☑ CHECK POINT 2 Graph the quadratic function $f(x) = (x - 2)^2 + 1$.

---

3 Graph parabolas in the form $f(x) = ax^2 + bx + c$.

Graphing Quadratic Functions in the Form $f(x) = ax^2 + bx + c$

Quadratic functions are frequently expressed in the form $f(x) = ax^2 + bx + c$. How can we identify the vertex of a parabola whose equation is in this form? Completing the square provides the answer to this question.

$$f(x) = ax^2 + bx + c$$

$$= a\left(x^2 + \frac{b}{a}x\right) + c \qquad \text{Factor out } a \text{ from } ax^2 + bx.$$

$$= a\left(x^2 + \frac{b}{a}x + \frac{b^2}{4a^2}\right) + c - a\left(\frac{b^2}{4a^2}\right)$$

Complete the square by adding the square of half the coefficient of x.

By completing the square, we added $a \cdot \frac{b^2}{4a^2}$. To avoid changing the function's equation, we must subtract this term.

$$= a\left(x + \frac{b}{2a}\right)^2 + c - \frac{b^2}{4a}$$

Write the trinomial as the square of a binomial and simplify the constant term.

Now let's compare the form of this equation with a quadratic function in the form $f(x) = a(x - h)^2 + k$.

The form we know how to graph: $f(x) = a(x - h)^2 + k$

$h = -\frac{b}{2a}$ $\qquad k = c - \frac{b^2}{4a}$

Equation under discussion: $f(x) = a\left(x - \left(-\frac{b}{2a}\right)\right)^2 + c - \frac{b^2}{4a}$

The important part of this observation is that h, the x-coordinate of the vertex, is $-\frac{b}{2a}$. The y-coordinate can be found by evaluating the function at $-\frac{b}{2a}$.

The Vertex of a Parabola Whose Equation Is $f(x) = ax^2 + bx + c$

Consider the parabola defined by the quadratic function $f(x) = ax^2 + bx + c$. The parabola's vertex is $\left(-\frac{b}{2a}, f\left(-\frac{b}{2a}\right)\right)$. The x-coordinate is $-\frac{b}{2a}$. The y-coordinate is found by substituting the x-coordinate into the parabola's equation and evaluating the function at this value of x.

EXAMPLE 3 Finding a Parabola's Vertex

Find the vertex for the parabola whose equation is $f(x) = 3x^2 + 12x + 8$.

Solution We know that the x-coordinate of the vertex is $x = -\frac{b}{2a}$. Let's identify the numbers a, b, and c in the given equation, which is in the form $f(x) = ax^2 + bx + c$.

$$f(x) = 3x^2 + 12x + 8$$

$a = 3$ $b = 12$ $c = 8$

Substitute the values of a and b into the equation for the x-coordinate:

$$x = -\frac{b}{2a} = -\frac{12}{2 \cdot 3} = -\frac{12}{6} = -2.$$

The x-coordinate of the vertex is -2. We substitute -2 for x into the equation of the function, $f(x) = 3x^2 + 12x + 8$, to find the y-coordinate:

$$f(-2) = 3(-2)^2 + 12(-2) + 8 = 3(4) + 12(-2) + 8 = 12 - 24 + 8 = -4.$$

The vertex is $(-2, -4)$. ■

CHECK POINT 3 Find the vertex for the parabola whose equation is $f(x) = 2x^2 + 8x - 1$.

We can apply our five-step procedure and graph parabolas in the form $f(x) = ax^2 + bx + c$.

Graphing Quadratic Functions with Equations in the Form $f(x) = ax^2 + bx + c$

To graph $f(x) = ax^2 + bx + c$,

1. Determine whether the parabola opens upward or downward. If $a > 0$, it opens upward. If $a < 0$, it opens downward.
2. Determine the vertex of the parabola. The vertex is $\left(-\frac{b}{2a}, f\left(-\frac{b}{2a}\right)\right)$.
3. Find any x-intercepts by solving $f(x) = 0$. The real solutions of $ax^2 + bx + c = 0$ are the x-intercepts.
4. Find the y-intercept by computing $f(0)$. Because $f(0) = c$ (the constant term in the function's equation), the y-intercept is c and the parabola passes through $(0, c)$.
5. Plot the intercepts, the vertex, and additional points as necessary. Connect these points with a smooth curve.

EXAMPLE 4 Graphing a Quadratic Function in the Form $f(x) = ax^2 + bx + c$

Graph the quadratic function $f(x) = -x^2 - 2x + 1$. Use the graph to identify the function's domain and its range.

Solution

Step 1. Determine how the parabola opens. Note that a, the coefficient of x^2, is -1. Thus, $a < 0$; this negative value tells us that the parabola opens downward.

Step 2. Find the vertex. We know that the x-coordinate of the vertex is $x = -\frac{b}{2a}$. We identify a, b, and c in $f(x) = ax^2 + bx + c$.

$$f(x) = -x^2 - 2x + 1$$

$a = -1 \quad b = -2 \quad c = 1$

Substitute the values of a and b into the equation for the x-coordinate:

$$x = -\frac{b}{2a} = -\frac{-2}{2(-1)} = -\left(\frac{-2}{-2}\right) = -1.$$

The x-coordinate of the vertex is -1. We substitute -1 for x into the equation of the function, $f(x) = -x^2 - 2x + 1$, to find the y-coordinate:

$$f(-1) = -(-1)^2 - 2(-1) + 1 = -1 + 2 + 1 = 2.$$

The vertex is $(-1, 2)$.

Step 3. Find the x-intercepts by solving $f(x) = 0$. Replace $f(x)$ with 0 in $f(x) = -x^2 - 2x + 1$. We obtain $0 = -x^2 - 2x + 1$. This equation cannot be solved by factoring. We will use the quadratic formula to solve it.

$$-x^2 - 2x + 1 = 0$$

$a = -1 \quad b = -2 \quad c = 1$

$$x = \frac{-b \pm \sqrt{b^2 - 4ac}}{2a} = \frac{-(-2) \pm \sqrt{(-2)^2 - 4(-1)(1)}}{2(-1)} = \frac{2 \pm \sqrt{4 - (-4)}}{-2}$$

To locate the x-intercepts, we need decimal approximations. Thus, there is no need to simplify the radical form of the solutions.

$$x = \frac{2 + \sqrt{8}}{-2} \approx -2.4 \quad \text{or} \quad x = \frac{2 - \sqrt{8}}{-2} \approx 0.4$$

The x-intercepts are approximately -2.4 and 0.4. The parabola passes through $(-2.4, 0)$ and $(0.4, 0)$.

Step 4. Find the y-intercept by computing $f(0)$. Replace x with 0 in $f(x) = -x^2 - 2x + 1$.

$$f(0) = -0^2 - 2 \cdot 0 + 1 = 1$$

The y-intercept is 1, which is the constant term in the function's equation. The parabola passes through $(0, 1)$.

Step 5. Graph the parabola. With a vertex at $(-1, 2)$, x-intercepts at -2.4 and 0.4, and a y-intercept at 1, the graph of f is shown in **Figure 8.10(a)**. The axis of symmetry is the vertical line whose equation is $x = -1$.

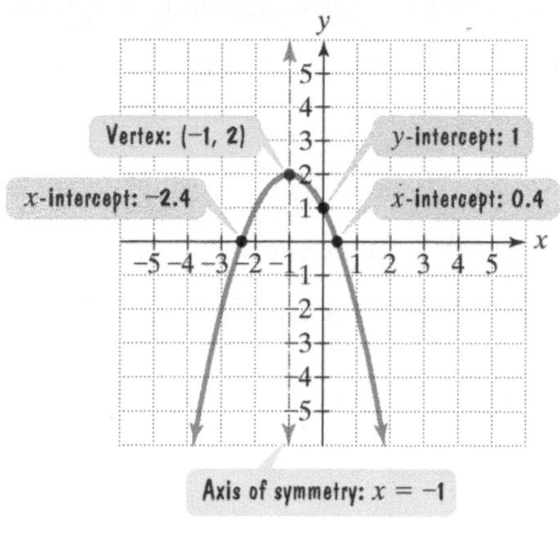

Figure 8.10(a) The graph of $f(x) = -x^2 - 2x + 1$

Domain: Inputs on x-axis include all real numbers.
Range: Outputs on y-axis fall at or below 2.

Figure 8.10(b) Determining the domain and range of $f(x) = -x^2 - 2x + 1$

Now we are ready to determine the domain and range of $f(x) = -x^2 - 2x + 1$. We can use the parabola, shown again in **Figure 8.10(b)**, to do so. To find the domain, look for all the inputs on the x-axis that correspond to points on the graph. As the graph widens and continues to fall at both ends, can you see that these inputs include all real numbers?

$$\text{Domain of } f \text{ is } (-\infty, \infty).$$

To find the range, look for all the outputs on the y-axis that correspond to points on the graph. **Figure 8.10(b)** shows that the parabola's vertex, $(-1, 2)$, is the highest point on the graph. Because the y-coordinate of the vertex is 2, outputs on the y-axis fall at or below 2.

$$\text{Range of } f \text{ is } (-\infty, 2]. \quad \blacksquare$$

Great Question!

Are there rules to find domains and ranges of quadratic functions?

Yes. The domain of any quadratic function includes all real numbers. If the vertex is the graph's highest point, the range includes all real numbers at or below the y-coordinate of the vertex. If the vertex is the graph's lowest point, the range includes all real numbers at or above the y-coordinate of the vertex.

✓ **CHECK POINT 4** Graph the quadratic function $f(x) = -x^2 + 4x + 1$. Use the graph to identify the function's domain and its range.

Great Question!

I feel overwhelmed by the amount of information required to graph just one quadratic function. Is there a way I can organize the information and gain a better understanding of the graphing procedure?

You're right: The skills needed to graph a quadratic function combine information from many of this book's chapters. Try organizing the items you need to graph quadratic functions in a table, something like this:

Graphing $f(x) = a(x - h)^2 + k$ or $f(x) = ax^2 + bx + c$

1. Opens upward if $a > 0$. Opens downward if $a < 0$.	2. Find the vertex. (h, k) or $\left(-\frac{b}{2a}, f\left(-\frac{b}{2a}\right)\right)$
3. Find x-intercepts. Solve $f(x) = 0$. Find real solutions.	4. Find the y-intercept. Find $f(0)$.

4 Determine a quadratic function's minimum or maximum value.

Minimum and Maximum Values of Quadratic Functions

Consider the quadratic function $f(x) = ax^2 + bx + c$. If $a > 0$, the parabola opens upward and the vertex is its lowest point. If $a < 0$, the parabola opens downward and the vertex is its highest point. The x-coordinate of the vertex is $-\frac{b}{2a}$. Thus, we can find the minimum or maximum value of f by evaluating the quadratic function at $x = -\frac{b}{2a}$.

Minimum and Maximum: Quadratic Functions

Consider the quadratic function $f(x) = ax^2 + bx + c$.

1. If $a > 0$, then f has a minimum that occurs at $x = -\frac{b}{2a}$. This minimum value is $f\left(-\frac{b}{2a}\right)$.
2. If $a < 0$, then f has a maximum that occurs at $x = -\frac{b}{2a}$. This maximum value is $f\left(-\frac{b}{2a}\right)$.

In each case, the value of x gives the location of the minimum or maximum value. The value of y, or $f\left(-\frac{b}{2a}\right)$, gives that minimum or maximum value.

EXAMPLE 5 Obtaining Information about a Quadratic Function from Its Equation

Consider the quadratic function $f(x) = -3x^2 + 6x - 13$.

a. Determine, without graphing, whether the function has a minimum value or a maximum value.

b. Find the minimum or maximum value and determine where it occurs.

c. Identify the function's domain and its range.

Solution We begin by identifying a, b, and c in the function's equation:

$$f(x) = -3x^2 + 6x - 13.$$

$a = -3$ $b = 6$ $c = -13$

a. Because $a < 0$, the function has a maximum value.

b. The maximum value occurs at

$$x = -\frac{b}{2a} = -\frac{6}{2(-3)} = -\frac{6}{-6} = -(-1) = 1.$$

The maximum value occurs at $x = 1$ and the maximum value of $f(x) = -3x^2 + 6x - 13$ is

$$f(1) = -3 \cdot 1^2 + 6 \cdot 1 - 13 = -3 + 6 - 13 = -10.$$

We see that the maximum is -10 at $x = 1$.

c. Like all quadratic functions, the domain is $(-\infty, \infty)$. Because the function's maximum value is -10, the range includes all real numbers at or below -10. The range is $(-\infty, -10]$. ■

We can use the graph of $f(x) = -3x^2 + 6x - 13$ to visualize the results of Example 5. **Figure 8.11** shows the graph in a $[-6, 6, 1]$ by $[-50, 20, 10]$ viewing rectangle. The maximum function feature verifies that the function's maximum is -10 at $x = 1$. Notice that x gives the location of the maximum and y gives the maximum value. Notice, too, that the maximum value is -10 and not the ordered pair $(1, -10)$.

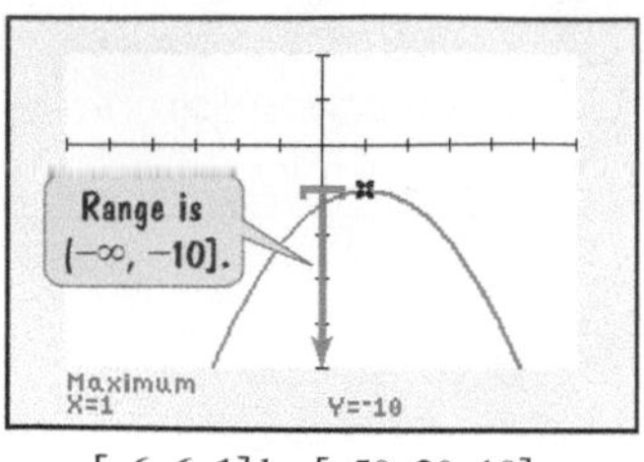

[−6, 6, 1] by [−50, 20, 10]

Figure 8.11

✓ CHECK POINT 5 Repeat parts (a) through (c) of Example 5 using the quadratic function $f(x) = 4x^2 - 16x + 1000$.

---

5 Solve problems involving a quadratic function's minimum or maximum value.

Applications of Quadratic Functions

Many applied problems involve finding the maximum or minimum value of a quadratic function, as well as where this value occurs.

EXAMPLE 6 Parabolic Paths of a Shot Put

An athlete whose event is the shot put releases the shot with the same initial velocity, but at different angles. **Figure 8.12** shows the parabolic paths for shots released at angles of 35° and 65°.

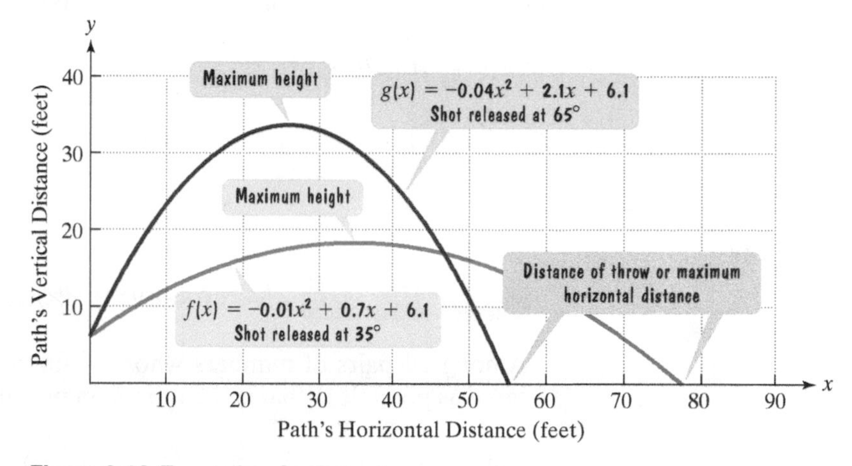

Figure 8.12 Two paths of a shot put

When the shot is released at an angle of 35°, its path can be modeled by the function

$$f(x) = -0.01x^2 + 0.7x + 6.1,$$

in which x is the shot's horizontal distance, in feet, and $f(x)$ is its height, in feet. What is the maximum height of this shot's path?

Solution The quadratic function is in the form $f(x) = ax^2 + bx + c$, with $a = -0.01$ and $b = 0.7$. Because $a < 0$, the function has a maximum that occurs at $x = -\frac{b}{2a}$.

$$x = -\frac{b}{2a} = -\frac{0.7}{2(-0.01)} = -(-35) = 35$$

This means that the shot's maximum height occurs when its horizontal distance is 35 feet. Can you see how this is shown by the blue graph of f in **Figure 8.12**? The maximum height of this path is

$$f(35) = -0.01(35)^2 + 0.7(35) + 6.1 = 18.35$$

or 18.35 feet. ■

CHECK POINT 6 Use function g, whose equation and graph are shown in **Figure 8.12** on the previous page, to find the maximum height, to the nearest tenth of a foot, when the shot is released at an angle of 65°.

Quadratic functions can also be modeled from verbal conditions. Once we have obtained a quadratic function, we can then use the x-coordinate of the vertex to determine its maximum or minimum value. Here is a step-by-step strategy for solving these kinds of problems:

Strategy for Solving Problems Involving Maximizing or Minimizing Quadratic Functions

1. Read the problem carefully and decide which quantity is to be maximized or minimized.
2. Use the conditions of the problem to express the quantity as a function in one variable.
3. Rewrite the function in the form $f(x) = ax^2 + bx + c$.
4. Calculate $-\frac{b}{2a}$. If $a > 0$, f has a minimum at $x = -\frac{b}{2a}$. This minimum value is $f\left(-\frac{b}{2a}\right)$. If $a < 0$, f has a maximum at $x = -\frac{b}{2a}$. This maximum value is $f\left(-\frac{b}{2a}\right)$.
5. Answer the question posed in the problem.

EXAMPLE 7 Minimizing a Product

Among all pairs of numbers whose difference is 10, find a pair whose product is as small as possible. What is the minimum product?

Solution

Step 1. Decide what must be maximized or minimized. We must minimize the product of two numbers. Calling the numbers x and y, and calling the product P, we must minimize

$$P = xy.$$

Step 2. Express this quantity as a function in one variable. In the formula $P = xy$, P is expressed in terms of two variables, x and y. However, because the difference of the numbers is 10, we can write

$$x - y = 10.$$

We can solve this equation for y in terms of x (or vice versa), substitute the result into $P = xy$, and obtain P as a function of one variable.

$$-y = -x + 10$$ Subtract x from both sides of $x - y = 10$.

$$y = x - 10$$ Multiply both sides of the equation by -1 and solve for y.

Now we substitute $x - 10$ for y in $P = xy$.

$$P = xy = x(x - 10).$$

Because P is now a function of x, we can write

$$P(x) = x(x - 10).$$

Step 3. Write the function in the form $f(x) = ax^2 + bx + c$. We apply the distributive property to obtain

$$P(x) = x(x - 10) = x^2 - 10x.$$

$a = 1$ $b = -10$

Numeric Connections

The TABLE feature of a graphing utility can be used to verify our work in Example 7.

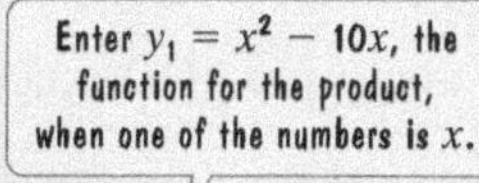

Enter $y_1 = x^2 - 10x$, the function for the product, when one of the numbers is x.

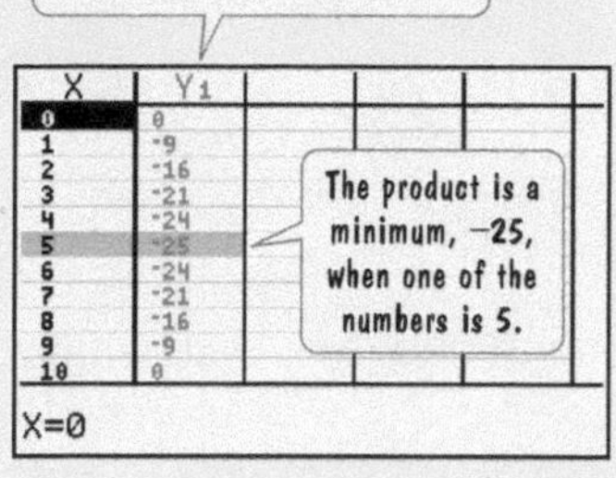

Step 4. Calculate $-\frac{b}{2a}$. If $a > 0$, the function has a minimum at this value. The voice balloons show that $a = 1$ and $b = -10$.

$$x = -\frac{b}{2a} = -\frac{-10}{2(1)} = -(-5) = 5$$

This means that the product, P, of two numbers whose difference is 10 is a minimum when one of the numbers, x, is 5.

Step 5. Answer the question posed by the problem. The problem asks for the two numbers and the minimum product. We found that one of the numbers, x, is 5. Now we must find the second number, y.

$$y = x - 10 = 5 - 10 = -5.$$

The number pair whose difference is 10 and whose product is as small as possible is $5, -5$. The minimum product is $5(-5)$, or -25. ■

✓ CHECK POINT 7 Among all pairs of numbers whose difference is 8, find a pair whose product is as small as possible. What is the minimum product?

You have 100 yards of fencing to enclose a rectangular region. Find the dimensions of the rectangle that maximize the enclosed area. What is the maximum area?

Solution

Step 1. Decide what must be maximized or minimized. We must maximize area. What we do not know are the rectangle's dimensions, x and y.

Step 2. Express this quantity as a function in one variable. Because we must maximize area, we have $A = xy$. We need to transform this into a function in which A is represented by one variable. Because you have 100 yards of fencing, the perimeter of the rectangle is 100 yards. This means that

$$2x + 2y = 100.$$

We can solve this equation for y in terms of x, substitute the result into $A = xy$, and obtain A as a function in one variable. We begin by solving for y.

$$2y = 100 - 2x \quad \text{Subtract } 2x \text{ from both sides.}$$

$$y = \frac{100 - 2x}{2} \quad \text{Divide both sides by 2.}$$

$$y = 50 - x \quad \text{Divide each term in the numerator by 2.}$$

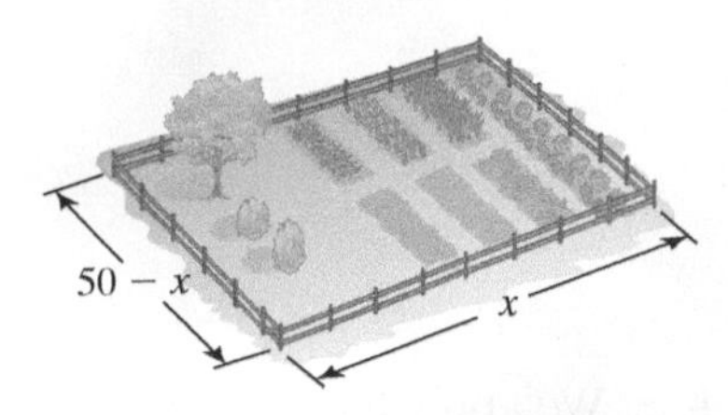

Figure 8.13 What value of x will maximize the rectangle's area?

Now we substitute $50 - x$ for y in $A = xy$.

$$A = xy = x(50 - x)$$

The rectangle and its dimensions are illustrated in **Figure 8.13**. Because A is now a function of x, we can write

$$A(x) = x(50 - x).$$

This function models the area, $A(x)$, of any rectangle whose perimeter is 100 yards in terms of one of its dimensions, x.

Step 3. Write the function in the form $f(x) = ax^2 + bx + c$. We apply the distributive property to obtain

$$A(x) = x(50 - x) = 50x - x^2 = -x^2 + 50x.$$

$a = -1$ $b = 50$

Using Technology

Graphic Connections

The graph of the area function

$$A(x) = x(50 - x)$$

was obtained with a graphing utility using a [0, 50, 2] by [0, 700, 25] viewing rectangle. The maximum function feature verifies that a maximum area of 625 square yards occurs when one of the dimensions is 25 yards.

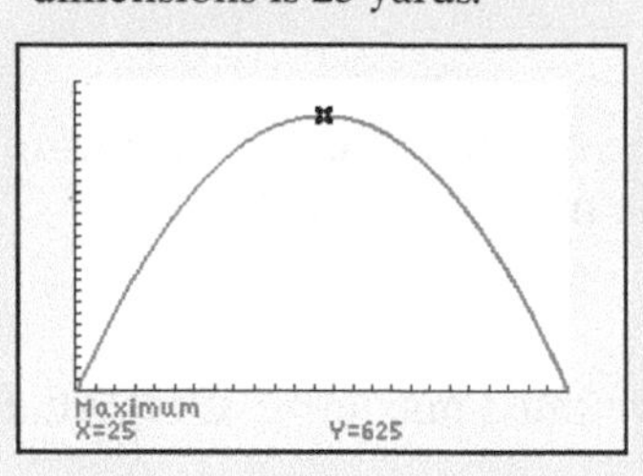

Step 4. Calculate $-\frac{b}{2a}$. If $a < 0$, the function has a maximum at this value. The voice balloons show that $a = -1$ and $b = 50$.

$$x = -\frac{b}{2a} = -\frac{50}{2(-1)} = 25$$

This means that the area, $A(x)$, of a rectangle with perimeter 100 yards is a maximum when one of the rectangle's dimensions, x, is 25 yards.

Step 5. Answer the question posed by the problem. We found that $x = 25$. **Figure 8.13** shows that the rectangle's other dimension is $50 - x = 50 - 25 = 25$. The dimensions of the rectangle that maximize the enclosed area are 25 yards by 25 yards. The rectangle that gives the maximum area is actually a square with an area of 25 yards · 25 yards, or 625 square yards. ■

✓ **CHECK POINT 8** You have 120 feet of fencing to enclose a rectangular region. Find the dimensions of the rectangle that maximize the enclosed area. What is the maximum area?

Achieving Success

Address your stress.

Stress levels can help or hinder performance. The parabola in **Figure 8.14** serves as a model that shows people under both low stress and high stress perform worse than their moderate-stress counterparts.

Moderate stress, high performance

Performance

Low stress, low performance

High stress, low performance

Level of Stress

Figure 8.14

Source: Herbert Benson, *Your Maximum Mind*, Random House, 1987.

CONCEPT AND VOCABULARY CHECK

Fill in each blank so that the resulting statement is true.

1. The graph of any quadratic function, $f(x) = ax^2 + bx + c, a \neq 0$, is called a/an ______________. If $a > 0$, the graph opens ______________. If $a < 0$, the graph opens ______________.
2. The vertex of a parabola is the ______________ point if $a > 0$.
3. The vertex of a parabola is the ______________ point if $a < 0$.
4. The vertex of the graph of $f(x) = a(x - h)^2 + k, a \neq 0$, is the point ______________.
5. The x-coordinate of the vertex of the graph of $f(x) = ax^2 + bx + c, a \neq 0$, is ______________. The y-coordinate of the vertex is found by substituting ______________ into the function's equation and evaluating the function at this value of x.
6. If $f(x) = a(x - h)^2 + k$ or $f(x) = ax^2 + bx + c, a \neq 0$, any x-intercepts are found by solving $f(x) =$ ______________. The equation's real ______________ are the x-intercepts.
7. If $f(x) = a(x - h)^2 + k$ or $f(x) = ax^2 + bx + c, a \neq 0$, the y-intercept is found by computing ______________.

8.3 EXERCISE SET

Practice Exercises

In Exercises 1–4, the graph of a quadratic function is given. Write the function's equation, selecting from the following options:

$$f(x) = (x + 1)^2 - 1, g(x) = (x + 1)^2 + 1,$$
$$h(x) = (x - 1)^2 + 1, j(x) = (x - 1)^2 - 1.$$

1.

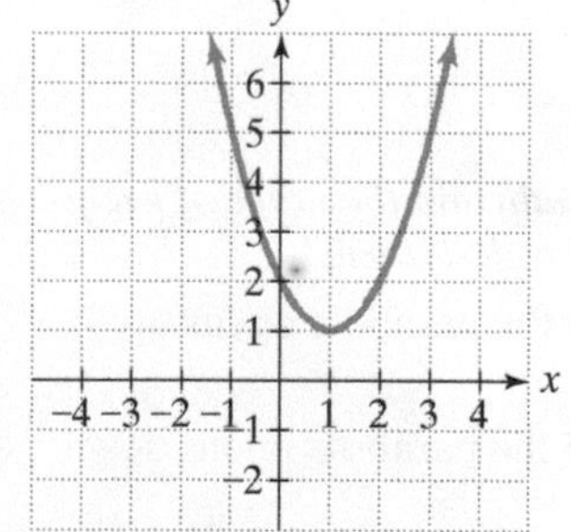

2.

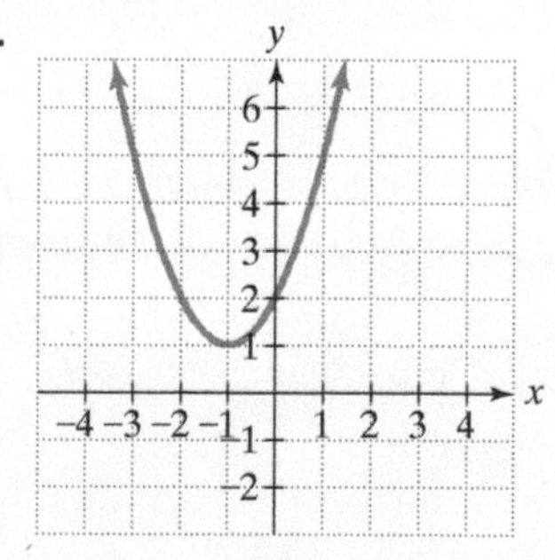

3.

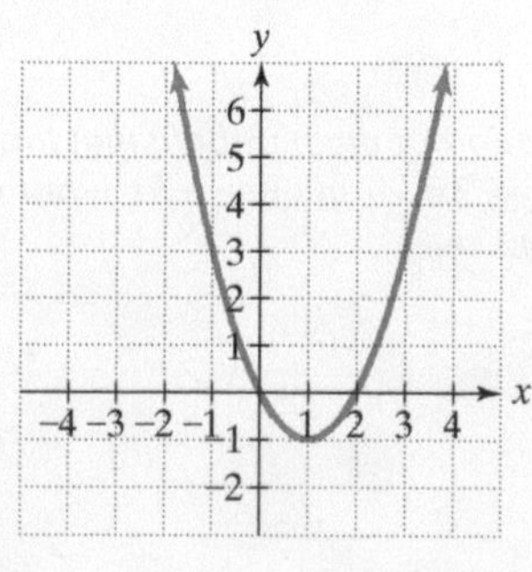

4.

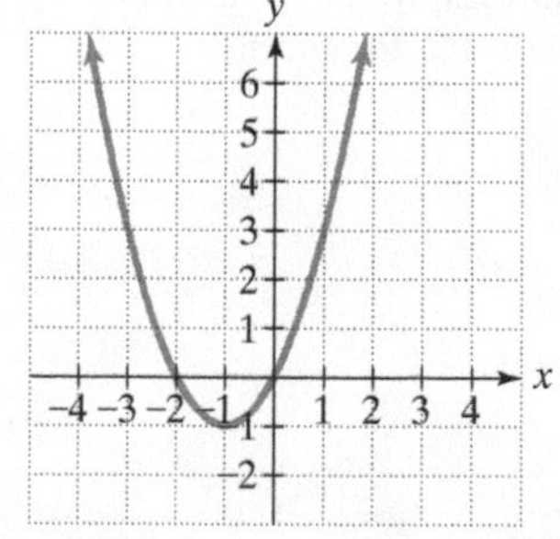

In Exercises 5–8, the graph of a quadratic function is given. Write the function's equation, selecting from the following options:

$$f(x) = x^2 + 2x + 1, g(x) = x^2 - 2x + 1,$$
$$h(x) = x^2 - 1, j(x) = -x^2 - 1.$$

5.

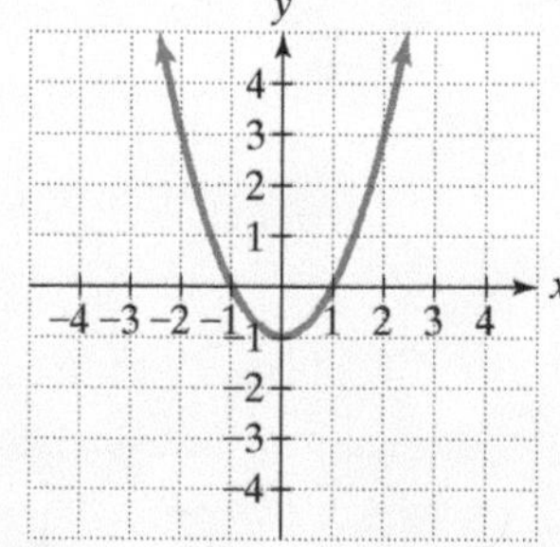

6.

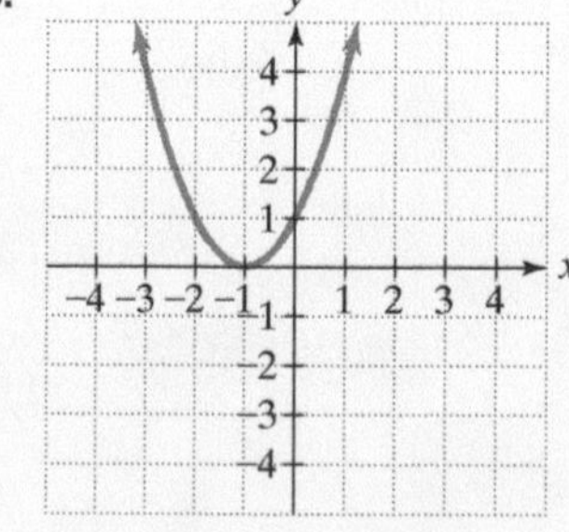

7.

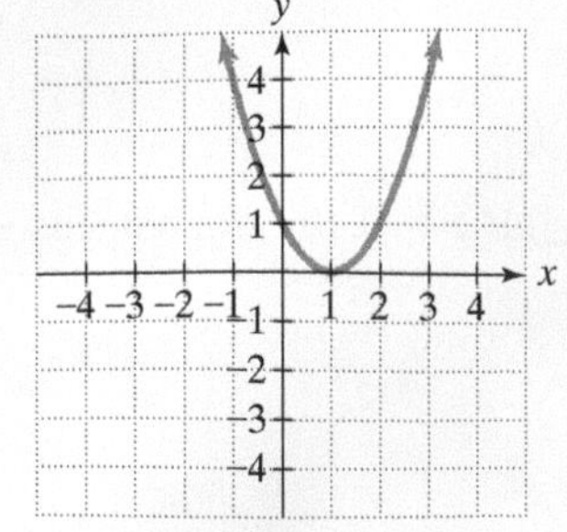

8.

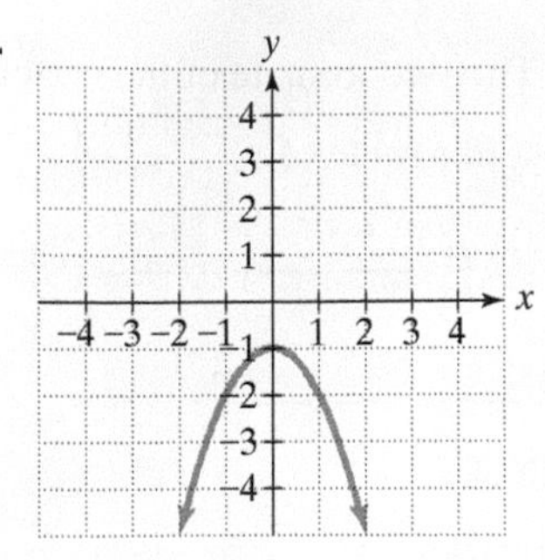

In Exercises 9–16, find the coordinates of the vertex for the parabola defined by the given quadratic function.

9. $f(x) = 2(x - 3)^2 + 1$

10. $f(x) = -3(x - 2)^2 + 12$

11. $f(x) = -2(x + 1)^2 + 5$

12. $f(x) = -2(x + 4)^2 - 8$

13. $f(x) = 2x^2 - 8x + 3$

14. $f(x) = 3x^2 - 12x + 1$

15. $f(x) = -x^2 - 2x + 8$

16. $f(x) = -2x^2 + 8x - 1$

In Exercises 17–38, use the vertex and intercepts to sketch the graph of each quadratic function. Use the graph to identify the function's range.

17. $f(x) = (x - 4)^2 - 1$

18. $f(x) = (x - 1)^2 - 2$

19. $f(x) = (x - 1)^2 + 2$

20. $f(x) = (x - 3)^2 + 2$

21. $y - 1 = (x - 3)^2$

22. $y - 3 = (x - 1)^2$

23. $f(x) = 2(x + 2)^2 - 1$

24. $f(x) = \frac{5}{4} - \left(x - \frac{1}{2}\right)^2$

25. $f(x) = 4 - (x - 1)^2$

26. $f(x) = 1 - (x - 3)^2$

27. $f(x) = x^2 - 2x - 3$

28. $f(x) = x^2 - 2x - 15$

29. $f(x) = x^2 + 3x - 10$

30. $f(x) = 2x^2 - 7x - 4$

31. $f(x) = 2x - x^2 + 3$

32. $f(x) = 5 - 4x - x^2$

33. $f(x) = x^2 + 6x + 3$

34. $f(x) = x^2 + 4x - 1$

35. $f(x) = 2x^2 + 4x - 3$

36. $f(x) = 3x^2 - 2x - 4$

37. $f(x) = 2x - x^2 - 2$

38. $f(x) = 6 - 4x + x^2$

In Exercises 39–44, an equation of a quadratic function is given.

a. *Determine, without graphing, whether the function has a minimum value or a maximum value.*

b. *Find the minimum or maximum value and determine where it occurs.*

c. *Identify the function's domain and its range.*

39. $f(x) = 3x^2 - 12x - 1$

40. $f(x) = 2x^2 - 8x - 3$

41. $f(x) = -4x^2 + 8x - 3$

42. $f(x) = -2x^2 - 12x + 3$

43. $f(x) = 5x^2 - 5x$

44. $f(x) = 6x^2 - 6x$

Practice PLUS

In Exercises 45–48, give the domain and the range of each quadratic function whose graph is described.

45. The vertex is $(-1, -2)$ and the parabola opens up.

46. The vertex is $(-3, -4)$ and the parabola opens down.

47. Maximum $= -6$ at $x = 10$

48. Minimum $= 18$ at $x = -6$

In Exercises 49–52, write an equation of the parabola that has the same shape as the graph of $f(x) = 2x^2$, but with the given point as the vertex.

49. $(5, 3)$

50. $(7, 4)$

51. $(-10, -5)$

52. $(-8, -6)$

In Exercises 53–56, write an equation of the parabola that has the same shape as the graph of $f(x) = 3x^2$ or $g(x) = -3x^2$, but with the given maximum or minimum.

53. Maximum $= 4$ at $x = -2$

54. Maximum $= -7$ at $x = 5$

55. Minimum $= 0$ at $x = 11$

56. Minimum $= 0$ at $x = 9$

Application Exercises

57. A person standing close to the edge on the top of a 160 foot building throws a baseball vertically upward. The quadratic function

$$s(t) = -16t^2 + 64t + 160$$

models the ball's height above the ground, $s(t)$, in feet, t seconds after it was thrown.

a. After how many seconds does the ball reach its maximum height? What is the maximum height?

b. How many seconds does it take until the ball finally hits the ground? Round to the nearest tenth of a second.

c. Find $s(0)$ and describe what this means.

d. Use your results from parts (a) through (c) to graph the quadratic function. Begin the graph with $t = 0$ and end with the value of t for which the ball hits the ground.

58. A person standing close to the edge on the top of a 200-foot building throws a baseball vertically upward. The quadratic function

$$s(t) = -16t^2 + 64t + 200$$

models the ball's height above the ground, $s(t)$, in feet, t seconds after it was thrown.

a. After how many seconds does the ball reach its maximum height? What is the maximum height?

b. How many seconds does it take until the ball finally hits the ground? Round to the nearest tenth of a second.

c. Find $s(0)$ and describe what this means.

d. Use your results from parts (a) through (c) to graph the quadratic function. Begin the graph with $t = 0$ and end with the value of t for which the ball hits the ground.

59. Among all pairs of numbers whose sum is 16, find a pair whose product is as large as possible. What is the maximum product?

60. Among all pairs of numbers whose sum is 20, find a pair whose product is as large as possible. What is the maximum product?

61. Among all pairs of numbers whose difference is 16, find a pair whose product is as small as possible. What is the minimum product?

62. Among all pairs of numbers whose difference is 24, find a pair whose product is as small as possible. What is the minimum product?

63. You have 600 feet of fencing to enclose a rectangular plot that borders on a river. If you do not fence the side along the river, find the length and width of the plot that will maximize the area. What is the largest area that can be enclosed?

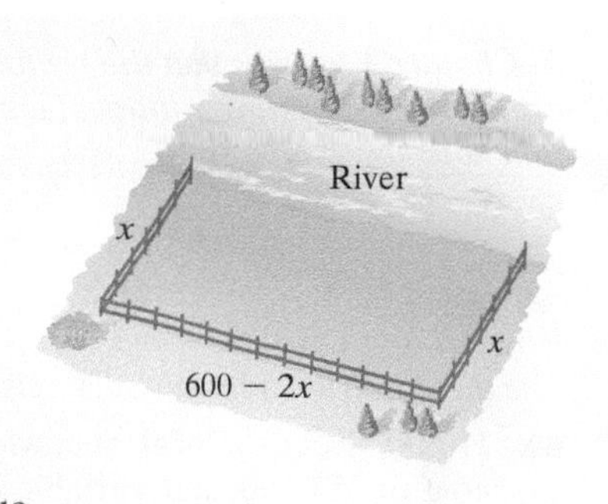

64. You have 200 feet of fencing to enclose a rectangular plot that borders on a river. If you do not fence the side along the river, find the length and width of the plot that will maximize the area. What is the largest area that can be enclosed?

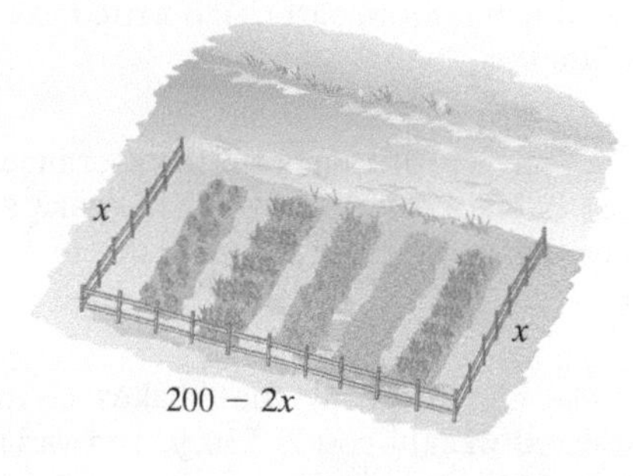

65. You have 50 yards of fencing to enclose a rectangular region. Find the dimensions of the rectangle that maximize the enclosed area. What is the maximum area?

66. You have 80 yards of fencing to enclose a rectangular region. Find the dimensions of the rectangle that maximize the enclosed area. What is the maximum area?

67. A rain gutter is made from sheets of aluminum that are 20 inches wide by turning up the edges to form right angles. Determine the depth of the gutter that will maximize its cross-sectional area and allow the greatest amount of water to flow. What is the maximum cross-sectional area?

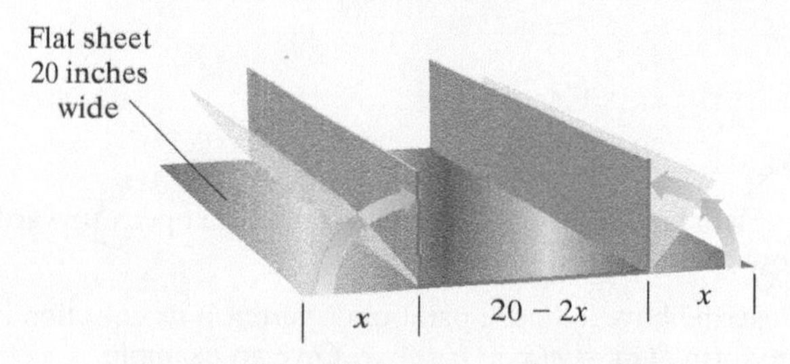

68. A rain gutter is made from sheets of aluminum that are 12 inches wide by turning up the edges to form right angles. Determine the depth of the gutter that will maximize its cross-sectional area and allow the greatest amount of water to flow. What is the maximum cross-sectional area?

In Chapter 3, we saw that the profit, $P(x)$, generated after producing and selling x units of a product is given by the function

$$P(x) = R(x) - C(x),$$

where R and C are the revenue and cost functions, respectively. Use these functions to solve Exercises 69–70.

69. Hunky Beef, a local sandwich store, has a fixed weekly cost of \$525.00, and variable costs for making a roast beef sandwich are \$0.55.

a. Let x represent the number of roast beef sandwiches made and sold each week. Write the weekly cost function, C, for Hunky Beef.

b. The function $R(x) = -0.001x^2 + 3x$ describes the money that Hunky Beef takes in each week from the sale of x roast beef sandwiches. Use this revenue function and the cost function from part (a) to write the store's weekly profit function, P.

c. Use the store's profit function to determine the number of roast beef sandwiches it should make and sell each week to maximize profit. What is the maximum weekly profit?

70. Virtual Fido is a company that makes electronic virtual pets. The fixed weekly cost is \$3000, and variable costs for each pet are \$20.

a. Let x represent the number of virtual pets made and sold each week. Write the weekly cost function, C, for Virtual Fido.

b. The function $R(x) = -x^2 + 1000x$ describes the money that Virtual Fido takes in each week from the sale of x virtual pets. Use this revenue function and the cost function from part (a) to write the weekly profit function, P.

c. Use the profit function to determine the number of virtual pets that should be made and sold each week to maximize profit. What is the maximum weekly profit?

Explaining the Concepts

71. What is a parabola? Describe its shape.

72. Explain how to decide whether a parabola opens upward or downward.

73. Describe how to find a parabola's vertex if its equation is in the form $f(x) = a(x - h)^2 + k$. Give an example.

74. Describe how to find a parabola's vertex if its equation is in the form $f(x) = ax^2 + bx + c$. Use $f(x) = x^2 - 6x + 8$ as an example.

75. A parabola that opens upward has its vertex at (1, 2). Describe as much as you can about the parabola based on this information. Include in your discussion the number of x-intercepts (if any) for the parabola.

Technology Exercises

76. Use a graphing utility to verify any five of your hand-drawn graphs in Exercises 17–38.

77. a. Use a graphing utility to graph $y = 2x^2 - 82x + 720$ in a standard viewing rectangle. What do you observe?

b. Find the coordinates of the vertex for the given quadratic function.

c. The answer to part (b) is $(20.5, -120.5)$. Because the leading coefficient, 2, of the given function is positive, the vertex is a minimum point on the graph. Use this fact to help find a viewing rectangle that will give a relatively complete picture of the parabola. With an axis of symmetry at $x = 20.5$, the setting for x should extend past this, so try Xmin = 0 and Xmax = 30. The setting for y should include (and probably go below) the y-coordinate of the graph's minimum point, so try Ymin = −130. Experiment with Ymax until your utility shows the parabola's major features.

d. In general, explain how knowing the coordinates of a parabola's vertex can help determine a reasonable viewing rectangle on a graphing utility for obtaining a complete picture of the parabola.

In Exercises 78–81, find the vertex for each parabola. Then determine a reasonable viewing rectangle on your graphing utility and use it to graph the parabola.

78. $y = -0.25x^2 + 40x$

79. $y = -4x^2 + 20x + 160$

80. $y = 5x^2 + 40x + 600$

81. $y = 0.01x^2 + 0.6x + 100$

Critical Thinking Exercises

Make Sense? *In Exercises 82–85, determine whether each statement makes sense or does not make sense, and explain your reasoning.*

82. Parabolas that open up appear to form smiles ($a > 0$), while parabolas that open down frown ($a < 0$).

83. I must have made an error when graphing this parabola because its axis of symmetry is the y-axis.

84. I like to think of a parabola's vertex as the point where it intersects its axis of symmetry.

85. I threw a baseball vertically upward and its path was a parabola.

In Exercises 86–89, determine whether each statement is true or false. If the statement is false, make the necessary change(s) to produce a true statement.

86. No quadratic functions have a range of $(-\infty, \infty)$.

87. The vertex of the parabola described by $f(x) = 2(x - 5)^2 - 1$ is at (5, 1).

88. The graph of $f(x) = -2(x + 4)^2 - 8$ has one y-intercept and two x-intercepts.

89. The maximum value of y for the quadratic function $f(x) = -x^2 + x + 1$ is 1.

In Exercises 90–91, find the axis of symmetry for each parabola whose equation is given. Use the axis of symmetry to find a second point on the parabola whose y-coordinate is the same as the given point.

90. $f(x) = 3(x + 2)^2 - 5;\quad (-1, -2)$

91. $f(x) = (x - 3)^2 + 2; (6, 11)$

In Exercises 92–93, write the equation of each parabola in $f(x) = a(x - h)^2 + k$ form.

92. Vertex: $(-3, -4)$; The graph passes through the point $(1, 4)$.

93. Vertex: $(-3, -1)$; The graph passes through the point $(-2, -3)$.

94. A rancher has 1000 feet of fencing to construct six corrals, as shown in the figure. Find the dimensions that maximize the enclosed area. What is the maximum area?

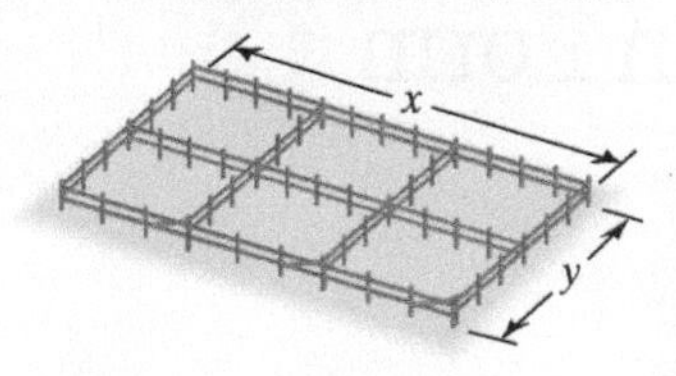

95. The annual yield per lemon tree is fairly constant at 320 pounds when the number of trees per acre is 50 or fewer. For each additional tree over 50, the annual yield per tree for all trees on the acre decreases by 4 pounds due to overcrowding. Find the number of trees that should be planted on an acre to produce the maximum yield. How many pounds is the maximum yield?

Review Exercises

96. Solve: $\frac{2}{x+5} + \frac{1}{x-5} = \frac{16}{x^2 - 25}$. (Section 6.6, Example 5)

97. Simplify: $\dfrac{1 + \frac{2}{x}}{1 - \frac{4}{x^2}}$. (Section 6.3, Example 1)

98. Solve using determinants (Cramer's Rule):

$$\begin{cases} 2x + 3y = 6 \\ x - 4y = 14. \end{cases}$$

(Section 3.5, Example 2)

Preview Exercises

Exercises 99–101 will help you prepare for the material covered in the next section.

In Exercises 99–100, solve each quadratic equation for u.

99. $u^2 - 8u - 9 = 0$

100. $2u^2 - u - 10 = 0$

101. If $u = x^{\frac{1}{3}}$, rewrite $5x^{\frac{2}{3}} + 11x^{\frac{1}{3}} + 2 = 0$ as a quadratic equation in u. [*Hint:* $x^{\frac{2}{3}} = \left(x^{\frac{1}{3}}\right)^2$.]

MID-CHAPTER CHECK POINT Section 8.1–Section 8.3

What You Know: We saw that not all quadratic equations can be solved by factoring. We learned three new methods for solving these equations: the square root property, completing the square, and the quadratic formula. We saw that the discriminant of $ax^2 + bx + c = 0$, namely $b^2 - 4ac$, determines the number and type of the equation's solutions. We graphed quadratic functions using vertices, intercepts, and additional points, as necessary. We learned that the vertex of $f(x) = a(x - h)^2 + k$ is (h, k) and the vertex of $f(x) = ax^2 + bx + c$ is $\left(-\frac{b}{2a}, f\left(-\frac{b}{2a}\right)\right)$. We used the vertex to solve problems that involved minimizing or maximizing quadratic functions.

In Exercises 1–13, solve each equation by the method of your choice. Simplify solutions, if possible.

1. $(3x - 5)^2 = 36$

2. $5x^2 - 2x = 7$

3. $3x^2 - 6x - 2 = 0$

4. $x^2 + 6x = -2$

5. $5x^2 + 1 = 37$

6. $x^2 - 5x + 8 = 0$

7. $2x^2 + 26 = 0$

8. $(2x + 3)(x + 2) = 10$

9. $(x + 3)^2 = 24$

10. $\frac{1}{x^2} - \frac{4}{x} + 1 = 0$

11. $x(2x - 3) = -4$

12. $\frac{x^2}{3} + \frac{x}{2} = \frac{2}{3}$

13. $\frac{2x}{x^2 + 6x + 8} = \frac{x}{x+4} - \frac{2}{x+2}$

14. Solve by completing the square: $x^2 + 10x - 3 = 0$.

In Exercises 15–18, graph the given quadratic function. Give each function's domain and range.

15. $f(x) = (x - 3)^2 - 4$

16. $g(x) = 5 - (x + 2)^2$

17. $h(x) = -x^2 - 4x + 5$

18. $f(x) = 3x^2 - 6x + 1$

In Exercises 19–20, without solving the equation, determine the number and type of solutions.

19. $2x^2 + 5x + 4 = 0$

20. $10x(x + 4) = 15x - 15$

In Exercises 21–22, write a quadratic equation in standard form with the given solution set.

21. $\left\{-\frac{1}{2}, \frac{3}{4}\right\}$

22. $\{-2\sqrt{3}, 2\sqrt{3}\}$

23. A company manufactures and sells bath cabinets. The function

$$P(x) = -x^2 + 150x - 4425$$

models the company's daily profit, $P(x)$, when x cabinets are manufactured and sold per day. How many cabinets should be manufactured and sold per day to maximize the company's profit? What is the maximum daily profit?

24. Among all pairs of numbers whose sum is -18, find a pair whose product is as large as possible. What is the maximum product?

25. The base of a triangle measures 40 inches minus twice the measure of its height. For what measure of the height does the triangle have a maximum area? What is the maximum area?

SECTION

8.4 Equations Quadratic in Form 

What am I supposed to learn?

After studying this section, you should be able to:

1 Solve equations that are quadratic in form.

"My husband asked me if we have any cheese puffs. Like he can't go and lift the couch cushion up himself."
—Roseanne Barr

How important is it for you to have a clean house? The percentage of people who find this to be quite important varies by age. In the Exercise Set (Exercises 47–48), you will work with a function that models this phenomenon. Your work will be based on equations that are not quadratic, but that can be written as quadratic equations using an appropriate substitution. Here are some examples:

Given Equation	Substitution	New Equation
$x^4 - 10x^2 + 9 = 0$ or $(x^2)^2 - 10x^2 + 9 = 0$	$u = x^2$	$u^2 - 10u + 9 = 0$
$5x^{\frac{2}{3}} + 11x^{\frac{1}{3}} + 2 = 0$ or $5\left(x^{\frac{1}{3}}\right)^2 + 11x^{\frac{1}{3}} + 2 = 0$	$u = x^{\frac{1}{3}}$	$5u^2 + 11u + 2 = 0$

1 Solve equations that are quadratic in form.

An equation that is **quadratic in form** is one that can be expressed as a quadratic equation using an appropriate substitution. Both of the preceding given equations are quadratic in form.

In an equation that is quadratic in form, the variable factor in one term is the square of the variable factor in the other variable term. The third term is a constant. By letting u equal the variable factor that reappears squared, a quadratic equation in u will result. Now it's easy. Solve this quadratic equation for u. Finally, use your substitution to find the values for the variable in the given equation. Example 1 shows how this is done.

EXAMPLE 1 Solving an Equation Quadratic in Form

Solve: $x^4 - 8x^2 - 9 = 0$.

Solution Can you see that the variable factor in one term is the square of the variable factor in the other variable term?

$$x^4 - 8x^2 - 9 = 0$$

x^4 is the square of x^2: $(x^2)^2 = x^4$.

We will let u equal the variable factor that reappears squared. Thus,

$$\text{let } u = x^2.$$

Now we write the given equation as a quadratic equation in u and solve for u.

$x^4 - 8x^2 - 9 = 0$	This is the given equation.
$(x^2)^2 - 8x^2 - 9 = 0$	The given equation contains x^2 and x^2 squared.
$u^2 - 8u - 9 = 0$	Let $u = x^2$. Replace x^2 with u.
$(u - 9)(u + 1) = 0$	Factor.
$u - 9 = 0$ or $u + 1 = 0$	Apply the zero-product principle.
$u = 9$ $\quad$ $u = -1$	Solve for u.

We're not done! Why not? We were asked to solve for x and we have values for u. We use the original substitution, $u = x^2$, to solve for x. Replace u with x^2 in each equation shown, namely $u = 9$ and $u = -1$.

$$x^2 = 9 \qquad x^2 = -1$$
$$x = \pm\sqrt{9} \qquad x = \pm\sqrt{-1} \qquad \text{Apply the square root property.}$$
$$x = \pm 3 \qquad x = \pm i$$

Substitute these values into the given equation and verify that the solutions are $-3, 3, -i$, and i. The solution set is $\{-3, 3, -i, i\}$. The graph in the Using Technology box shows that only the real solutions, -3 and 3, appear as x-intercepts. ■

Using Technology

Graphic Connections

The graph of

$$y = x^4 - 8x^2 - 9$$

has x-intercepts at -3 and 3. This verifies that the real solutions of

$$x^4 - 8x^2 - 9 = 0$$

are -3 and 3. The imaginary solutions, $-i$ and i, are not shown as intercepts.

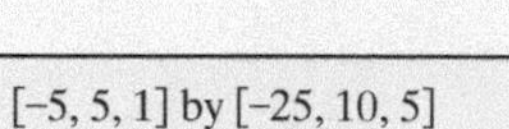

[-5, 5, 1] by [-25, 10, 5]

CHECK POINT 1 Solve: $x^4 - 5x^2 + 6 = 0$.

If checking proposed solutions is not overly cumbersome, you should do so either algebraically or with a graphing utility. The Using Technology box shows a check of the two real solutions in Example 1. Are there situations when solving equations quadratic in form where a check is essential? Yes. **If at any point in the solution process both sides of an equation are raised to an even power, a check is required.** Extraneous solutions that are not solutions of the given equation may have been introduced.

EXAMPLE 2 Solving an Equation Quadratic in Form

Solve: $2x - \sqrt{x} - 10 = 0$.

Solution To identify exponents on the terms, let's rewrite $\sqrt{x}$ as $x^{\frac{1}{2}}$. The equation can be expressed as

$$2x^1 - x^{\frac{1}{2}} - 10 = 0.$$

By expressing the equation as $2x^1 - x^{\frac{1}{2}} - 10 = 0$, we can see that the variable factor in one term is the square of the variable factor in the other variable term.

$$2x^1 - x^{\frac{1}{2}} - 10 = 0$$

x^1 is the square of $x^{\frac{1}{2}}$: $\left(x^{\frac{1}{2}}\right)^2 = x^1$.

We will let u equal the variable factor that reappears squared. Thus,

$$\text{let } u = x^{\frac{1}{2}}.$$

Now we write the given equation as a quadratic equation in u and solve for u.

$2x - \sqrt{x} - 10 = 0$	This is the given equation.
$2x^1 - x^{\frac{1}{2}} - 10 = 0$	This is the given equation in exponential form.
$2(x^{\frac{1}{2}})^2 - x^{\frac{1}{2}} - 10 = 0$	The equation contains $x^{\frac{1}{2}}$ and $x^{\frac{1}{2}}$ squared.
$2u^2 - u - 10 = 0$	Let $u = x^{\frac{1}{2}}$. Replace $x^{\frac{1}{2}}$ with u.
$(2u - 5)(u + 2) = 0$	Factor.
$2u - 5 = 0$ or $u + 2 = 0$	Set each factor equal to zero.
$u = \frac{5}{2}$ $\quad$ $u = -2$	Solve for u.

Use the original substitution, $u = x^{\frac{1}{2}}$, to solve for x. Replace u with $x^{\frac{1}{2}}$ in each of the preceding equations.

$x^{\frac{1}{2}} = \frac{5}{2}$ or $x^{\frac{1}{2}} = -2$	Replace u with $x^{\frac{1}{2}}$.
$\left(x^{\frac{1}{2}}\right)^2 = \left(\frac{5}{2}\right)^2$ $\quad$ $\left(x^{\frac{1}{2}}\right)^2 = (-2)^2$	Solve for x by squaring both sides of each equation.
$x = \frac{25}{4}$ $\quad$ $x = 4$	Square $\frac{5}{2}$ and -2.

Both sides are raised to even powers. We must check.

It is essential to check both proposed solutions in the original equation.

Check $\frac{25}{4}$:

$$2x - \sqrt{x} - 10 = 0$$
$$2 \cdot \frac{25}{4} - \sqrt{\frac{25}{4}} - 10 \stackrel{?}{=} 0$$
$$\frac{25}{2} - \frac{5}{2} - 10 \stackrel{?}{=} 0$$
$$\frac{20}{2} - 10 \stackrel{?}{=} 0$$
$$0 = 0, \text{ true}$$

Check 4:

$$2x - \sqrt{x} - 10 = 0$$
$$2 \cdot 4 - \sqrt{4} - 10 \stackrel{?}{=} 0$$
$$8 - 2 - 10 \stackrel{?}{=} 0$$
$$6 - 10 \stackrel{?}{=} 0$$
$$-4 = 0, \text{ false}$$

The check indicates that 4 is not a solution. It is an extraneous solution brought about by squaring each side of one of the equations. The only solution is $\frac{25}{4}$ and the solution set is $\left\{\frac{25}{4}\right\}$. ■

☑ **CHECK POINT 2** Solve: $x - 2\sqrt{x} - 8 = 0$.

The equations in Examples 1 and 2 can be solved by methods other than using substitutions.

$$x^4 - 8x^2 - 9 = 0 \qquad\qquad 2x - \sqrt{x} - 10 = 0$$

This equation can be solved directly by factoring: $(x^2 - 9)(x^2 + 1) = 0$.

This equation can be solved by isolating the radical term: $2x - 10 = \sqrt{x}$. Then square both sides.

In the examples that follow, solving the equations by methods other than first introducing a substitution becomes increasingly difficult.

EXAMPLE 3 Solving an Equation Quadratic in Form

Solve: $(x^2 - 5)^2 + 3(x^2 - 5) - 10 = 0$.

Solution This equation contains $x^2 - 5$ and $x^2 - 5$ squared. We

$$\text{let } u = x^2 - 5.$$

$(x^2 - 5)^2 + 3(x^2 - 5) - 10 = 0$	This is the given equation.
$u^2 + 3u - 10 = 0$	Let $u = x^2 - 5$.
$(u + 5)(u - 2) = 0$	Factor.
$u + 5 = 0$ or $u - 2 = 0$	Set each factor equal to zero.
$u = -5$ $\quad$ $u = 2$	Solve for u.

Discover for Yourself

Solve Example 3 by first simplifying the given equation's left side. Then factor out x^2 and solve the resulting equation. Do you get the same solutions? Which method, substitution or first simplifying, is faster?

Use the original substitution, $u = x^2 - 5$, to solve for x. Replace u with $x^2 - 5$ in each of the preceding equations.

$x^2 - 5 = -5$ or $x^2 - 5 = 2$	Replace u with $x^2 - 5$.
$x^2 = 0$ $\quad$ $x^2 = 7$	Solve for x by isolating x^2.
$x = 0$ $\quad$ $x = \pm\sqrt{7}$	Apply the square root property.

Although we did not raise both sides of an equation to an even power, checking the three proposed solutions in the original equation is a good idea. Do this now and verify that the solutions are $-\sqrt{7}$, 0, and $\sqrt{7}$, and the solution set is $\{-\sqrt{7}, 0, \sqrt{7}\}$. ■

CHECK POINT 3 Solve: $(x^2 - 4)^2 + (x^2 - 4) - 6 = 0$.

EXAMPLE 4 Solving an Equation Quadratic in Form

Solve: $10x^{-2} + 7x^{-1} + 1 = 0$.

Solution The variable factor in one term is the square of the variable factor in the other variable term.

$$10x^{-2} + 7x^{-1} + 1 = 0$$

x^{-2} is the square of x^{-1}: $(x^{-1})^2 = x^{-2}$.

We will let u equal the variable factor that reappears squared. Thus,

$$\text{let } u = x^{-1}.$$

Now we write the given equation as a quadratic equation in u and solve for u.

$$10x^{-2} + 7x^{-1} + 1 = 0$$ This is the given equation.

$$10(x^{-1})^2 + 7x^{-1} + 1 = 0$$ The equation contains x^{-1} and x^{-1} squared.

$$10u^2 + 7u + 1 = 0$$ Let $u = x^{-1}$.

$$(5u + 1)(2u + 1) = 0$$ Factor.

$$5u + 1 = 0 \quad \text{or} \quad 2u + 1 = 0$$ Set each factor equal to zero.

$$5u = -1 \qquad 2u = -1$$ Solve each equation for u.

$$u = -\frac{1}{5} \qquad u = -\frac{1}{2}$$

Use the original substitution, $u = x^{-1}$, to solve for x. Replace u with x^{-1} in each of the preceding equations.

$$x^{-1} = -\frac{1}{5} \quad \text{or} \quad x^{-1} = -\frac{1}{2}$$ Replace u with x^{-1}.

$$(x^{-1})^{-1} = \left(-\frac{1}{5}\right)^{-1} \qquad (x^{-1})^{-1} = \left(-\frac{1}{2}\right)^{-1}$$ Solve for x by raising both sides of each equation to the -1 power.

$$x = -5 \qquad x = -2$$

$\left(-\frac{1}{5}\right)^{-1} = \frac{1}{-\frac{1}{5}} = -5$ $\quad$ $\left(-\frac{1}{2}\right)^{-1} = \frac{1}{-\frac{1}{2}} = -2$

We did not raise both sides of an equation to an even power. A check will show that both -5 and -2 are solutions of the original equation. The solution set is $\{-5, -2\}$. ■

CHECK POINT 4 Solve: $2x^{-2} + x^{-1} - 1 = 0$.

EXAMPLE 5 Solving an Equation Quadratic in Form

Solve: $5x^{\frac{2}{3}} + 11x^{\frac{1}{3}} + 2 = 0$.

Solution The variable factor in one term is the square of the variable factor in the other variable term.

$$5x^{\frac{2}{3}} + 11x^{\frac{1}{3}} + 2 = 0$$

$x^{\frac{2}{3}}$ is the square of $x^{\frac{1}{3}}$: $\left(x^{\frac{1}{3}}\right)^2 = x^{\frac{2}{3}}$.

We will let u equal the variable factor that reappears squared. Thus,

$$\text{let } u = x^{\frac{1}{3}}.$$

Now we write the given equation as a quadratic equation in u and solve for u.

$$5x^{\frac{2}{3}} + 11x^{\frac{1}{3}} + 2 = 0$$ This is the given equation.

$$5\left(x^{\frac{1}{3}}\right)^2 + 11\left(x^{\frac{1}{3}}\right) + 2 = 0$$ The given equation contains $x^{\frac{1}{3}}$ and $x^{\frac{1}{3}}$ squared.

$$5u^2 + 11u + 2 = 0$$ Let $u = x^{\frac{1}{3}}$.

$$(5u + 1)(u + 2) = 0$$ Factor.

$$5u + 1 = 0 \quad \text{or} \quad u + 2 = 0$$ Set each factor equal to 0.

$$u = -\tfrac{1}{5} \qquad u = -2$$ Solve for u.

Use the original substitution, $u = x^{\frac{1}{3}}$, to solve for x. Replace u with $x^{\frac{1}{3}}$ in each of the preceding equations.

$$x^{\frac{1}{3}} = -\frac{1}{5} \quad \text{or} \quad x^{\frac{1}{3}} = -2$$ Replace u with $x^{\frac{1}{3}}$ in $u = -\frac{1}{5}$ and $u = -2$.

$$\left(x^{\frac{1}{3}}\right)^3 = \left(-\frac{1}{5}\right)^3 \qquad \left(x^{\frac{1}{3}}\right)^3 = (-2)^3$$ Solve for x by cubing both sides of each equation.

$$x = -\frac{1}{125} \qquad x = -8$$

We did not raise both sides of an equation to an even power. A check will show that both -8 and $-\frac{1}{125}$ are solutions of the original equation. The solution set is $\left\{-8, -\frac{1}{125}\right\}$. ■

✓ **CHECK POINT 5** Solve: $3x^{\frac{2}{3}} - 11x^{\frac{1}{3}} - 4 = 0$.

CONCEPT AND VOCABULARY CHECK

Fill in each blank so that the resulting statement is true.

1. We solve $x^4 - 13x^2 + 36 = 0$ by letting $u =$ ______________.
   We then rewrite the equation in terms of u as ______________.
2. We solve $x - 2\sqrt{x} - 8 = 0$ by letting $u =$ ______________.
   We then rewrite the equation in terms of u as ______________.
3. We solve $(x + 3)^2 + 7(x + 3) - 18 = 0$ by letting $u =$ ______________.
   We then rewrite the equation in terms of u as ______________.
4. We solve $2x^{-2} - 7x^{-1} + 3 = 0$ by letting $u =$ ______________.
   We then rewrite the equation in terms of u as ______________.
5. We solve $x^{\frac{2}{3}} + 2x^{\frac{1}{3}} - 3 = 0$ by letting $u =$ ______________.
   We then rewrite the equation in terms of u as ______________.

8.4 EXERCISE SET

MyMathLab®

Practice Exercises

In Exercises 1–32, solve each equation by making an appropriate substitution. If at any point in the solution process both sides of an equation are raised to an even power, a check is required.

1. $x^4 - 5x^2 + 4 = 0$
2. $x^4 - 13x^2 + 36 = 0$
3. $x^4 - 11x^2 + 18 = 0$
4. $x^4 - 9x^2 + 20 = 0$
5. $x^4 + 2x^2 = 8$
6. $x^4 + 4x^2 = 5$
7. $x + \sqrt{x} - 2 = 0$
8. $x + \sqrt{x} - 6 = 0$
9. $x - 4x^{\frac{1}{2}} - 21 = 0$
10. $x - 6x^{\frac{1}{2}} + 8 = 0$
11. $x - 13\sqrt{x} + 40 = 0$
12. $2x - 7\sqrt{x} - 30 = 0$
13. $(x - 5)^2 - 4(x - 5) - 21 = 0$
14. $(x + 3)^2 + 7(x + 3) - 18 = 0$
15. $(x^2 - 1)^2 - (x^2 - 1) = 2$
16. $(x^2 - 2)^2 - (x^2 - 2) = 6$
17. $(x^2 + 3x)^2 - 8(x^2 + 3x) - 20 = 0$
18. $(x^2 - 2x)^2 - 11(x^2 - 2x) + 24 = 0$
19. $x^{-2} - x^{-1} - 20 = 0$
20. $x^{-2} - x^{-1} - 6 = 0$
21. $2x^{-2} - 7x^{-1} + 3 = 0$
22. $20x^{-2} + 9x^{-1} + 1 = 0$
23. $x^{-2} - 4x^{-1} = 3$
24. $x^{-2} - 6x^{-1} = -4$
25. $x^{\frac{2}{3}} - x^{\frac{1}{3}} - 6 = 0$
26. $x^{\frac{2}{3}} + 2x^{\frac{1}{3}} - 3 = 0$

27. $x^{\frac{2}{5}} + x^{\frac{1}{5}} - 6 = 0$

28. $x^{\frac{2}{5}} + x^{\frac{1}{5}} - 2 = 0$

29. $2x^{\frac{1}{2}} - x^{\frac{1}{4}} = 1$

30. $2x^{\frac{1}{2}} - 5x^{\frac{1}{4}} = 3$

31. $\left(x - \frac{8}{x}\right)^2 + 5\left(x - \frac{8}{x}\right) - 14 = 0$

32. $\left(x - \frac{10}{x}\right)^2 + 6\left(x - \frac{10}{x}\right) - 27 = 0$

In Exercises 33–38, find the x-intercepts of the given function, f. Then use the x-intercepts to match each function with its graph. [The graphs are labeled (a) through (f).]

33. $f(x) = x^4 - 5x^2 + 4$

34. $f(x) = x^4 - 13x^2 + 36$

35. $f(x) = x^{\frac{1}{3}} + 2x^{\frac{1}{6}} - 3$

36. $f(x) = x^{-2} - x^{-1} - 6$

37. $f(x) = (x + 2)^2 - 9(x + 2) + 20$

38. $f(x) = 2(x + 2)^2 + 5(x + 2) - 3$

a.

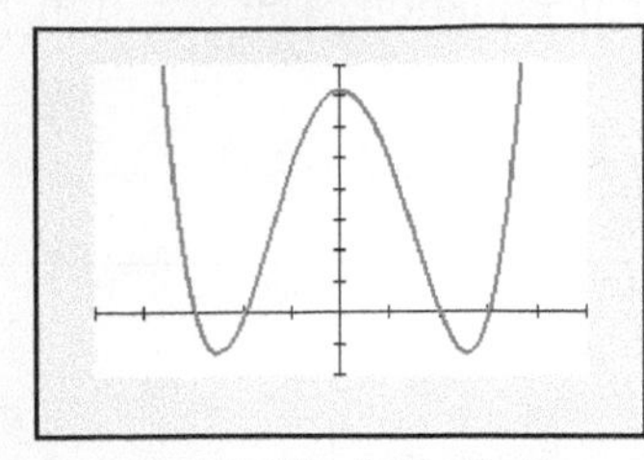

[−5, 5, 1] by [−10, 40, 5]

b.

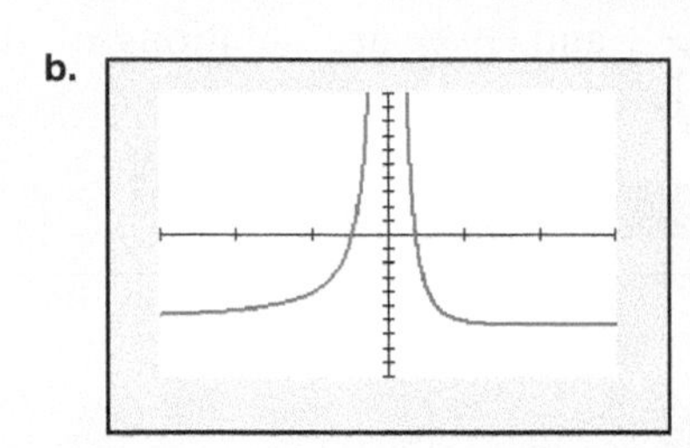

[−3, 3, 1] by [−10, 10, 1]

c.

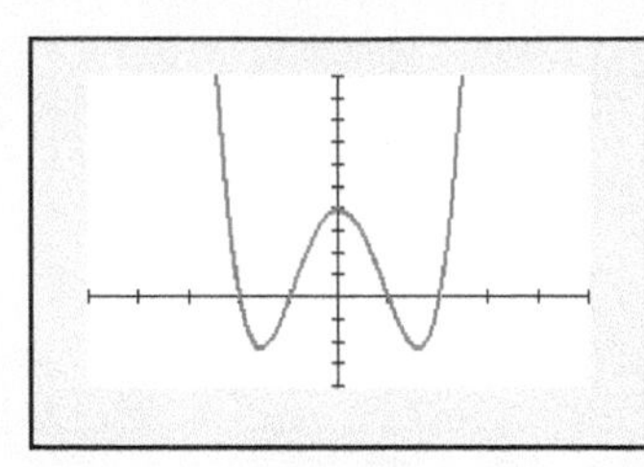

[−5, 5, 1] by [−4, 10, 1]

d.

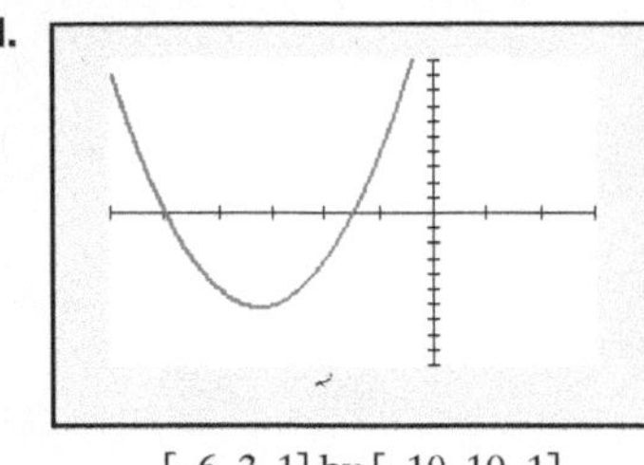

[−6, 3, 1] by [−10, 10, 1]

e.

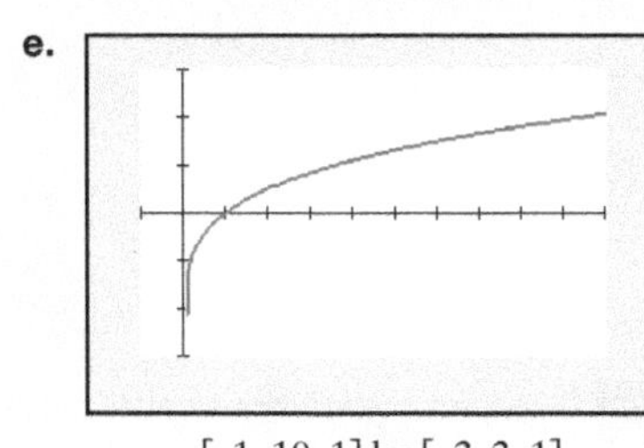

[−1, 10, 1] by [−3, 3, 1]

f.

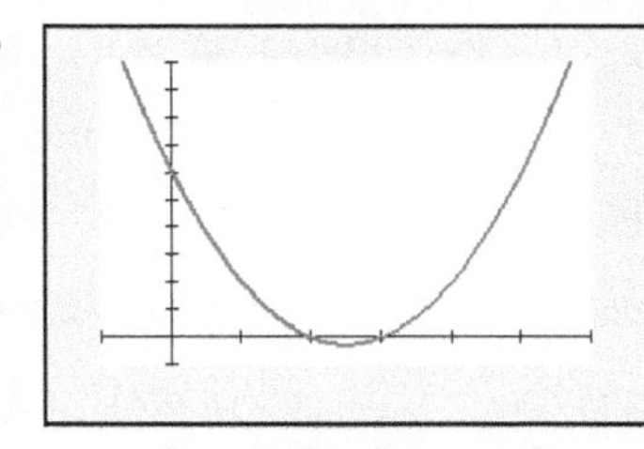

[−1, 6, 1] by [−1, 10, 1]

Practice PLUS

39. Let $f(x) = (x^2 + 3x - 2)^2 - 10(x^2 + 3x - 2)$. Find all x such that $f(x) = -16$.

40. Let $f(x) = (x^2 + 2x - 2)^2 - 7(x^2 + 2x - 2)$. Find all x such that $f(x) = -6$.

41. Let $f(x) = 3\left(\frac{1}{x} + 1\right)^2 + 5\left(\frac{1}{x} + 1\right)$. Find all x such that $f(x) = 2$.

42. Let $f(x) = 2x^{\frac{2}{3}} + 3x^{\frac{1}{3}}$. Find all x such that $f(x) = 2$.

43. Let $f(x) = \frac{x}{x - 4}$ and $g(x) = 13\sqrt{\frac{x}{x - 4}} - 36$. Find all x such that $f(x) = g(x)$.

44. Let $f(x) = \frac{x}{x - 2} + 10$ and $g(x) = -11\sqrt{\frac{x}{x - 2}}$. Find all x such that $f(x) = g(x)$.

45. Let $f(x) = 3(x - 4)^{-2}$ and $g(x) = 16(x - 4)^{-1}$. Find all x such that $f(x)$ exceeds $g(x)$ by 12.

46. Let $f(x) = 6\left(\frac{2x}{x - 3}\right)^2$ and $g(x) = 5\left(\frac{2x}{x - 3}\right)$. Find all x such that $f(x)$ exceeds $g(x)$ by 6.

Application Exercises

How important is it for you to have a clean house? The bar graph indicates that the percentage of people who find this to be quite important varies by age. The percentage, P(x), who find having a clean house very important can be modeled by the function

$$P(x) = 0.04(x + 40)^2 - 3(x + 40) + 104,$$

where x is the number of years a person's age is above or below 40. Thus, x is positive for people over 40 and negative for people under 40. Use the function to solve Exercises 47–48.

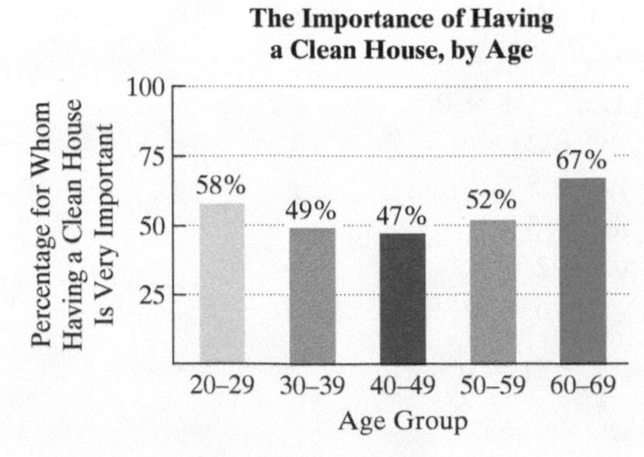

Source: Soap and Detergent Association

47. According to the model, at which ages do 60% of us feel that having a clean house is very important? Substitute 60 for $P(x)$ and solve the quadratic in form equation. How well does the function model the data shown in the bar graph?

48. According to the model, at which ages do 50% of us feel that having a clean house is very important? Substitute 50 for $P(x)$ and solve the quadratic-in-form equation. How well does the function model the data shown in the bar graph?

Explaining the Concepts

49. Explain how to recognize an equation that is quadratic in form. Provide two original examples with your explanation.

50. Describe two methods for solving this equation:

$$x - 5\sqrt{x} + 4 = 0.$$

Technology Exercises

51. Use a graphing utility to verify the solutions of any five equations in Exercises 1–32 that you solved algebraically. The real solutions should appear as x-intercepts on the graph of the function related to the given equation.

Use a graphing utility to find the real solutions of the equations in Exercises 52–59. Check by direct substitution.

52. $x^6 - 7x^3 - 8 = 0$

53. $3(x - 2)^{-2} - 4(x - 2)^{-1} + 1 = 0$

54. $x^4 - 10x^2 + 9 = 0$

55. $2x + 6\sqrt{x} = 8$

56. $2(x + 1)^2 = 5(x + 1) + 3$

57. $(x^2 - 3x)^2 + 2(x^2 - 3x) - 24 = 0$

58. $x^{\frac{1}{2}} + 4x^{\frac{1}{4}} = 5$

59. $x^{\frac{2}{3}} - 3x^{\frac{1}{3}} + 2 = 0$

Critical Thinking Exercises

Make Sense? *In Exercises 60–63, determine whether each statement makes sense or does not make sense, and explain your reasoning.*

60. When I solve an equation that is quadratic in form, it's important to write down the substitution that I am making.

61. Although I've rewritten an equation that is quadratic in form as $au^2 + bu + c = 0$ and solved for u, I'm not finished.

62. Checking is always a good idea, but it's never necessary when solving an equation that is quadratic in form.

63. The equation $5x^{\frac{2}{3}} + 11x^{\frac{1}{3}} + 2 = 0$ is quadratic in form, but when I reverse the variable terms and obtain $11x^{\frac{1}{3}} + 5x^{\frac{2}{3}} + 2 = 0$, the resulting equation is no longer quadratic in form.

In Exercises 64–67, determine whether each statement is true or false. If the statement is false, make the necessary change(s) to produce a true statement.

64. If an equation is quadratic in form, there is only one method that can be used to obtain its solution set.

65. An equation with three terms that is quadratic in form has a variable factor in one term that is the square of the variable factor in another term.

66. Because x^6 is the square of x^3, the equation $x^6 - 5x^3 + 6x = 0$ is quadratic in form.

67. To solve $x - 9\sqrt{x} + 14 = 0$, we let $\sqrt{u} = x$.

In Exercises 68–70, use a substitution to solve each equation.

68. $x^4 - 5x^2 - 2 = 0$

69. $5x^6 + x^3 = 18$

70. $\sqrt{\frac{x+4}{x-1}} + \sqrt{\frac{x-1}{x+4}} = \frac{5}{2}\left(\text{Let } u = \sqrt{\frac{x+4}{x-1}}.\right)$

Review Exercises

71. Simplify:

$$\frac{2x^2}{10x^3 - 2x^2}.$$

(Section 6.1, Example 4)

72. Divide: $\frac{2+i}{1-i}$. (Section 7.7, Example 5)

73. Solve using matrices:

$$\begin{cases} 2x + y = 6 \\ x - 2y = 8. \end{cases}$$

(Section 3.4, Example 2)

Preview Exercises

Exercises 74–76 will help you prepare for the material covered in the next section.

74. Solve: $2x^2 + x = 15$.

75. Solve: $x^3 + x^2 = 4x + 4$.

76. Simplify: $\frac{x+1}{x+3} - 2$.

SECTION

8.5 Polynomial and Rational Inequalities

What am I supposed to learn?

After studying this section, you should be able to:

1. Solve polynomial inequalities.
2. Solve rational inequalities.
3. Solve problems modeled by polynomial or rational inequalities.

Tailgaters beware: If your car is going 35 miles per hour on dry pavement, your required stopping distance is 160 feet, or the width of a football field. At 65 miles per hour, the distance required is 410 feet, or approximately the length of one and one-tenth football fields. **Figure 8.15** shows stopping distances for cars at various speeds on dry roads and on wet roads.

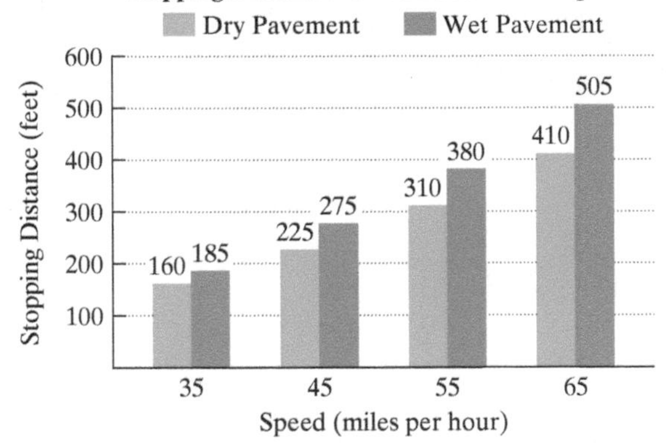

Figure 8.15

Source: National Highway Traffic Safety Administration

Using Technology

We used the statistical menu of a graphing utility and the quadratic regression program to obtain the quadratic function that models stopping distance on dry pavement. After entering the appropriate data from **Figure 8.15**, namely

(35, 160), (45, 225), (55, 310), (65, 410),

we obtained the results shown in the screen.

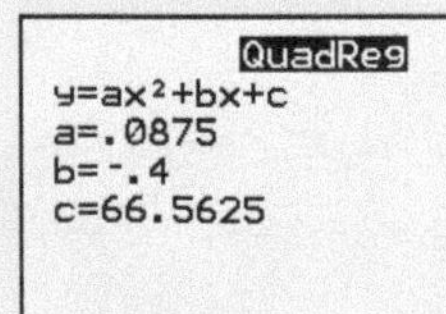

A car's required stopping distance, $f(x)$, in feet, on dry pavement traveling at x miles per hour can be modeled by the quadratic function

$$f(x) = 0.0875x^2 - 0.4x + 66.6.$$

How can we use this function to determine speeds on dry pavement requiring stopping distances that exceed the length of one and one-half football fields, or 540 feet? We must solve the inequality

$$0.0875x^2 - 0.4x + 66.6 > 540.$$

Required stopping distance — exceeds — 540 feet.

We begin by subtracting 540 from both sides. This will give us zero on the right:

$$0.0875x^2 - 0.4x + 66.6 - 540 > 540 - 540$$
$$0.0875x^2 - 0.4x - 473.4 > 0.$$

The form of this inequality is $ax^2 + bx + c > 0$. Such a quadratic inequality is called a *polynomial inequality*.

Definition of a Polynomial Inequality

A polynomial inequality is any inequality that can be put in one of the forms

$$f(x) < 0, \quad f(x) > 0, \quad f(x) \le 0, \quad \text{or} \quad f(x) \ge 0,$$

where f is a polynomial function.

In this section, we establish the basic techniques for solving polynomial inequalities. We will also use these techniques to solve inequalities involving rational functions.

1 Solve polynomial inequalities.

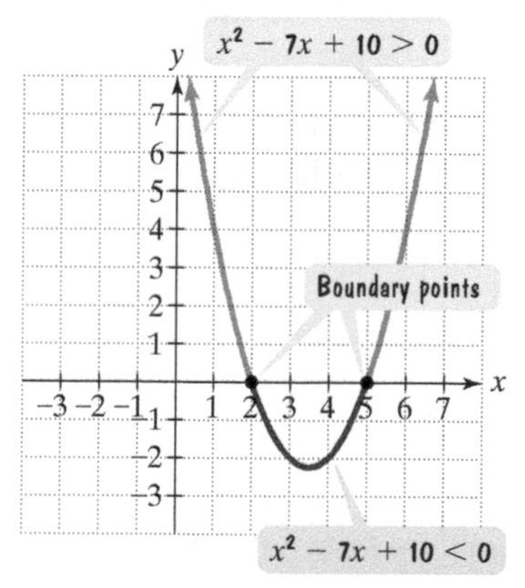

Figure 8.16

Solving Polynomial Inequalities

Graphs can help us visualize the solutions of polynomial inequalities. For example, the graph of $f(x) = x^2 - 7x + 10$ is shown in **Figure 8.16**. The x-intercepts, 2 and 5, are **boundary points** between where the graph lies above the x-axis, shown in blue, and where the graph lies below the x-axis, shown in red.

Locating the x-intercepts of a polynomial function, f, is an important step in finding the solution set for polynomial inequalities in the form $f(x) < 0$ or $f(x) > 0$.

We use the x-intercepts of f as boundary points that divide the real number line into intervals. On each interval, the graph of f is either above the x-axis $[f(x) > 0]$ or below the x-axis $[f(x) < 0]$. For this reason, x-intercepts play a fundamental role in solving polynomial inequalities. The x-intercepts are found by solving the equation $f(x) = 0$.

Procedure for Solving Polynomial Inequalities

1. Express the inequality in the form

$$f(x) < 0 \quad \text{or} \quad f(x) > 0,$$

   where f is a polynomial function.
2. Solve the equation $f(x) = 0$. The real solutions are the **boundary points**.
3. Locate these boundary points on a number line, thereby dividing the number line into intervals.
4. Choose one representative number, called a **test value**, within each interval and evaluate f at that number.
   a. If the value of f is positive, then $f(x) > 0$ for all numbers, x, in the interval.
   b. If the value of f is negative, then $f(x) < 0$ for all numbers, x, in the interval.
5. Write the solution set, selecting the interval or intervals that satisfy the given inequality.

This procedure is valid if $<$ is replaced by $\le$ or $>$ is replaced by $\ge$. However, if the inequality involves $\le$ or $\ge$, include the boundary points [the solutions of $f(x) = 0$] in the solution set.

EXAMPLE 1 Solving a Polynomial Inequality

Solve and graph the solution set on a real number line: $2x^2 + x > 15$.

Solution

Step 1. Express the inequality in the form $f(x) < 0$ or $f(x) > 0$. We begin by rewriting the inequality so that 0 is on the right side.

Using Technology

Graphic Connections

The solution set for

$$2x^2 + x > 15$$

or, equivalently,

$$2x^2 + x - 15 > 0$$

can be verified with a graphing utility. The graph of $f(x) = 2x^2 + x - 15$ was obtained using a $[-10, 10, 1]$ by $[-16, 6, 1]$ viewing rectangle. The graph lies above the x-axis, representing $>$, for all x in $(-\infty, -3)$ or $\left(\frac{5}{2}, \infty\right)$.

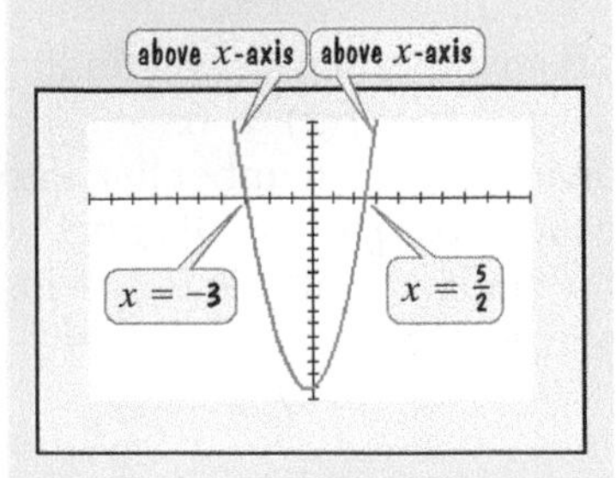

$$2x^2 + x > 15 \quad \text{This is the given inequality.}$$
$$2x^2 + x - 15 > 15 - 15 \quad \text{Subtract 15 from both sides.}$$
$$2x^2 + x - 15 > 0 \quad \text{Simplify.}$$

This inequality is equivalent to the one we wish to solve. It is in the form $f(x) > 0$, where $f(x) = 2x^2 + x - 15$.

Step 2. Solve the equation $f(x) = 0$. We find the x-intercepts of $f(x) = 2x^2 + x - 15$ by solving the equation $2x^2 + x - 15 = 0$.

$$2x^2 + x - 15 = 0 \quad \text{This polynomial equation is a quadratic equation.}$$
$$(2x - 5)(x + 3) = 0 \quad \text{Factor.}$$
$$2x - 5 = 0 \quad \text{or} \quad x + 3 = 0 \quad \text{Set each factor equal to 0.}$$
$$x = \tfrac{5}{2} \qquad x = -3 \quad \text{Solve for } x.$$

The x-intercepts of f are -3 and $\frac{5}{2}$. We will use these x-intercepts as boundary points on a number line.

Step 3. Locate the boundary points on a number line and separate the line into intervals. The number line with the boundary points is shown as follows:

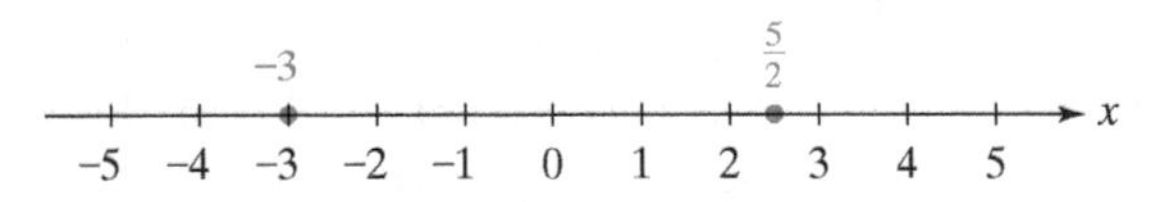

The boundary points divide the number line into three intervals:

$$(-\infty, -3) \quad \left(-3, \tfrac{5}{2}\right) \quad \left(\tfrac{5}{2}, \infty\right).$$

Step 4. Choose one test value within each interval and evaluate f at that number.

Interval	Test Value	Substitute into $f(x) = 2x^2 + x - 15$	Conclusion
$(-\infty, -3)$	-4	$f(-4) = 2(-4)^2 + (-4) - 15$ $= 13$, positive	$f(x) > 0$ for all x in $(-\infty, -3)$.
$\left(-3, \frac{5}{2}\right)$	0	$f(0) = 2 \cdot 0^2 + 0 - 15$ $= -15$, negative	$f(x) < 0$ for all x in $\left(-3, \frac{5}{2}\right)$.
$\left(\frac{5}{2}, \infty\right)$	3	$f(3) = 2 \cdot 3^2 + 3 - 15$ $= 6$, positive	$f(x) > 0$ for all x in $\left(\frac{5}{2}, \infty\right)$.

Step 5. Write the solution set, selecting the interval or intervals that satisfy the given inequality. We are interested in solving $f(x) > 0$, where $f(x) = 2x^2 + x - 15$. Based on our work in step 4, we see that $f(x) > 0$ for all x in $(-\infty, -3)$ or $\left(\frac{5}{2}, \infty\right)$. Thus, the solution set of the given inequality, $2x^2 + x > 15$, or, equivalently, $2x^2 + x - 15 > 0$, is

$$(-\infty, -3) \cup \left(\tfrac{5}{2}, \infty\right).$$

The graph of the solution set on a number line is shown as follows:

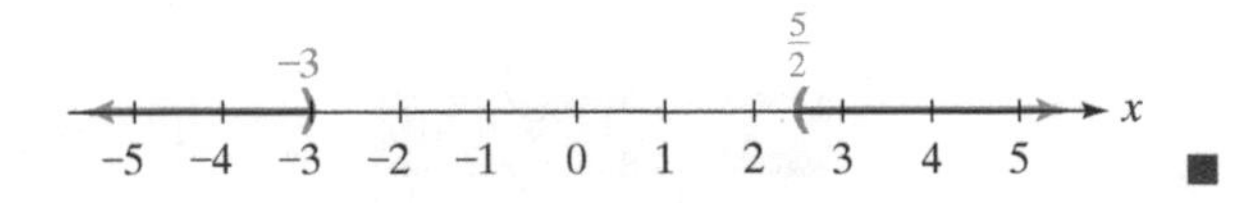

■

✓ **CHECK POINT 1** Solve and graph the solution set on a real number line: $x^2 - x > 20$.

EXAMPLE 2 Solving a Polynomial Inequality

Solve and graph the solution set on a real number line: $4x^2 \le 1 - 2x$.

Solution

Step 1. Express the inequality in the form $f(x) \le 0$ or $f(x) \ge 0$. We begin by rewriting the inequality so that 0 is on the right side.

$$4x^2 \le 1 - 2x \quad \text{This is the given inequality.}$$

$$4x^2 + 2x - 1 \le 1 - 2x + 2x - 1 \quad \text{Add 2x and subtract 1 on both sides.}$$

$$4x^2 + 2x - 1 \le 0 \quad \text{Simplify.}$$

This inequality is equivalent to the one we wish to solve. It is in the form $f(x) \le 0$, where $f(x) = 4x^2 + 2x - 1$.

Step 2. Solve the equation $f(x) = 0$. We will find the x-intercepts of $f(x) = 4x^2 + 2x - 1$ by solving the equation $4x^2 + 2x - 1 = 0$. This equation cannot be solved by factoring. We will use the quadratic formula to solve it.

$$4x^2 + 2x - 1 = 0$$

$a = 4$ $b = 2$ $c = -1$

$$x = \frac{-b \pm \sqrt{b^2 - 4ac}}{2a} = \frac{-2 \pm \sqrt{2^2 - 4 \cdot 4(-1)}}{2 \cdot 4} = \frac{-2 \pm \sqrt{4 - (-16)}}{8}$$

$$= \frac{-2 \pm \sqrt{20}}{8} = \frac{-2 \pm \sqrt{4}\sqrt{5}}{8} = \frac{-2 \pm 2\sqrt{5}}{8} = \frac{2(-1 \pm \sqrt{5})}{8} = \frac{-1 \pm \sqrt{5}}{4}$$

$$x = \frac{-1 + \sqrt{5}}{4} \approx 0.3 \qquad x = \frac{-1 - \sqrt{5}}{4} \approx -0.8$$

The x-intercepts of f are $\frac{-1 + \sqrt{5}}{4}$ (approximately 0.3) and $\frac{-1 - \sqrt{5}}{4}$ (approximately −0.8). We will use these x-intercepts as boundary points on a number line.

Step 3. Locate the boundary points on a number line and separate the line into intervals. The number line with the boundary points is shown as follows:

$\frac{-1 - \sqrt{5}}{4} \approx -0.8$ $\frac{-1 + \sqrt{5}}{4} \approx 0.3$

−5 −4 −3 −2 −1 0 1 2 3 4 5 x

The boundary points divide the number line into three intervals:

$$\left(-\infty, \frac{-1 - \sqrt{5}}{4}\right) \left(\frac{-1 - \sqrt{5}}{4}, \frac{-1 + \sqrt{5}}{4}\right) \left(\frac{-1 + \sqrt{5}}{4}, \infty\right).$$

Step 4. Choose one test value within each interval and evaluate f at that number.

Interval	Test Value	Substitute into $f(x) = 4x^2 + 2x - 1$	Conclusion
$\left(-\infty, \frac{-1 - \sqrt{5}}{4}\right)$	−1	$f(-1) = 4(-1)^2 + 2(-1) - 1$ $= 1$, positive	$f(x) > 0$ for all x in $\left(-\infty, \frac{-1 - \sqrt{5}}{4}\right)$.
$\left(\frac{-1 - \sqrt{5}}{4}, \frac{-1 + \sqrt{5}}{4}\right)$	0	$f(0) = 4 \cdot 0^2 + 2 \cdot 0 - 1$ $= -1$, negative	$f(x) < 0$ for all x in $\left(\frac{-1 - \sqrt{5}}{4}, \frac{-1 + \sqrt{5}}{4}\right)$.
$\left(\frac{-1 + \sqrt{5}}{4}, \infty\right)$	1	$f(1) = 4 \cdot 1^2 + 2 \cdot 1 - 1$ $= 5$, positive	$f(x) > 0$ for all x in $\left(\frac{-1 + \sqrt{5}}{4}, \infty\right)$.

Using Technology

Graphic Connections

The solution set for

$$4x^2 \le 1 - 2x$$

or, equivalently,

$$4x^2 + 2x - 1 \le 0$$

can be verified with a graphing utility. The graph of $f(x) = 4x^2 + 2x - 1$ was obtained using a $[-2, 2, 1]$ by $[-10, 10, 1]$ viewing rectangle. The graph lies on or below the x-axis, representing $\le$, for all x in

$$\left[\frac{-1-\sqrt{5}}{4}, \frac{-1+\sqrt{5}}{4}\right] \approx [-0.8, 0.3].$$

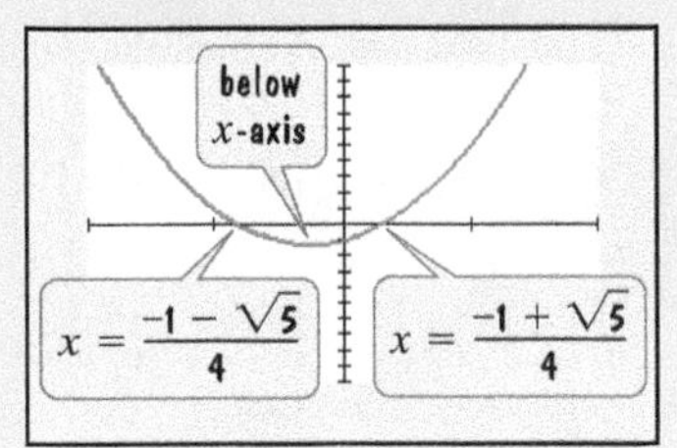

[-2, 2, 1] by [-10, 10, 1]

Step 5. Write the solution set, selecting the interval or intervals that satisfy the given inequality. We are interested in solving $f(x) \le 0$, where $f(x) = 4x^2 + 2x - 1$. Based on our work in step 4, we see that $f(x) < 0$ for all x in $\left(\frac{-1-\sqrt{5}}{4}, \frac{-1+\sqrt{5}}{4}\right)$. However, because the inequality involves $\le$ (less than or *equal to*), we must also include the solutions of $4x^2 + 2x - 1 = 0$, namely $\frac{-1-\sqrt{5}}{4}$ and $\frac{-1+\sqrt{5}}{4}$, in the solution set. Thus, the solution set of the given inequality $4x^2 \le 1 - 2x$, or, equivalently, $4x^2 + 2x - 1 \le 0$, is

$$\left[\frac{-1-\sqrt{5}}{4}, \frac{-1+\sqrt{5}}{4}\right].$$

The graph of the solution set on a number line is shown as follows:

−5 −4 −3 −2 −1 0 1 2 3 4 5 ■

✓ CHECK POINT 2 Solve and graph the solution set on a real number line: $2x^2 \le -6x - 1$.

EXAMPLE 3 Solving a Polynomial Inequality

Solve and graph the solution set on a real number line: $x^3 + x^2 \le 4x + 4$.

Solution

Step 1. Express the inequality in the form $f(x) \le 0$ or $f(x) \ge 0$. We begin by rewriting the inequality so that 0 is on the right side.

$x^3 + x^2 \le 4x + 4$	This is the given inequality.
$x^3 + x^2 - 4x - 4 \le 4x + 4 - 4x - 4$	Subtract $4x + 4$ from both sides.
$x^3 + x^2 - 4x - 4 \le 0$	Simplify.

This inequality is equivalent to the one we wish to solve. It is in the form $f(x) \le 0$, where $f(x) = x^3 + x^2 - 4x - 4$.

Step 2. Solve the equation $f(x) = 0$. We find the x-intercepts of $f(x) = x^3 + x^2 - 4x - 4$ by solving the equation $x^3 + x^2 - 4x - 4 = 0$.

$x^3 + x^2 - 4x - 4 = 0$	This polynomial equation is of degree 3.
$x^2(x + 1) - 4(x + 1) = 0$	Factor x^2 from the first two terms and -4 from the last two terms.
$(x + 1)(x^2 - 4) = 0$	A common factor of $x + 1$ is factored from the expression.
$(x + 1)(x + 2)(x - 2) = 0$	Factor completely.
$x + 1 = 0$ or $x + 2 = 0$ or $x - 2 = 0$	Set each factor equal to 0.
$x = -1$ $\quad x = -2$ $\quad x = 2$	Solve for x.

The x-intercepts of f are -2, -1, and 2. We will use these x-intercepts as boundary points on a number line.

Step 3. Locate the boundary points on a number line and separate the line into intervals. The number line with the boundary points is shown as follows:

−2 −1 2

−5 −4 −3 −2 −1 0 1 2 3 4 5 x

The boundary points divide the number line into four intervals:

$$(-\infty, -2) \quad (-2, -1) \quad (-1, 2) \quad (2, \infty).$$

Step 4. Choose one test value within each interval and evaluate f at that number.

Interval	Test Value	Substitute into $f(x) = x^3 + x^2 - 4x - 4$	Conclusion
$(-\infty, -2)$	-3	$f(-3) = (-3)^3 + (-3)^2 - 4(-3) - 4$ $= -10$, negative	$f(x) < 0$ for all x in $(-\infty, -2)$.
$(-2, -1)$	-1.5	$f(-1.5) = (-1.5)^3 + (-1.5)^2 - 4(-1.5) - 4$ $= 0.875$, positive	$f(x) > 0$ for all x in $(-2, -1)$.
$(-1, 2)$	0	$f(0) = 0^3 + 0^2 - 4 \cdot 0 - 4$ $= -4$, negative	$f(x) < 0$ for all x in $(-1, 2)$.
$(2, \infty)$	3	$f(3) = 3^3 + 3^2 - 4 \cdot 3 - 4$ $= 20$, positive	$f(x) > 0$ for all x in $(2, \infty)$.

Using Technology

Graphic Connections

The solution set for

$$x^3 + x^2 \le 4x + 4$$

or, equivalently,

$$x^3 + x^2 - 4x - 4 \le 0$$

can be verified with a graphing utility. The graph of $f(x) = x^3 + x^2 - 4x - 4$ lies on or below the x-axis, representing $\le$, for all x in $(-\infty, -2]$ or $[-1, 2]$.

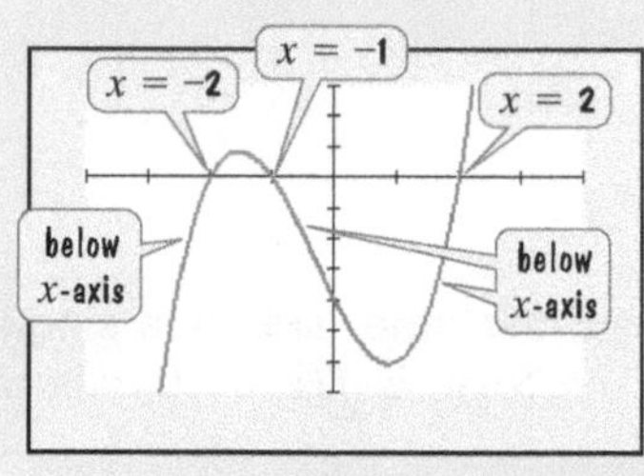

[−4, 4, 1] by [−7, 3, 1]

Step 5. Write the solution set, selecting the interval or intervals that satisfy the given inequality. We are interested in solving $f(x) \le 0$, where $f(x) = x^3 + x^2 - 4x - 4$. Based on our work in step 4, we see that $f(x) < 0$ for all x in $(-\infty, -2)$ or $(-1, 2)$. However, because the inequality involves $\le$ (less than or equal to), we must also include the solutions of $x^3 + x^2 - 4x - 4 = 0$, namely -2, -1, and 2, in the solution set. Thus, the solution set of the given inequality, $x^3 + x^2 \le 4x + 4$, or, equivalently, $x^3 + x^2 - 4x - 4 \le 0$, is

$$(-\infty, -2] \cup [-1, 2].$$

The graph of the solution set on a number line is shown as follows:

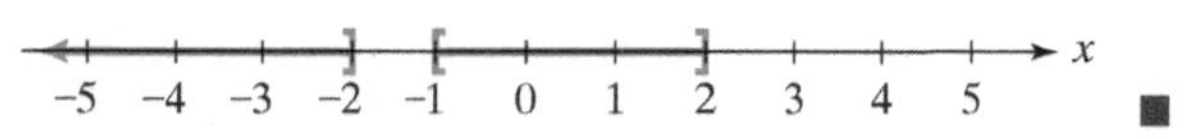

■

CHECK POINT 3 Solve and graph the solution set on a real number line: $x^3 + 3x^2 \le x + 3$.

2 Solve rational inequalities.

Solving Rational Inequalities

A **rational inequality** is any inequality that can be put in one of the forms

$$f(x) < 0, \quad f(x) > 0, \quad f(x) \le 0, \quad \text{or} \quad f(x) \ge 0,$$

where f is a rational function. An example of a rational inequality is

$$\frac{3x + 3}{2x + 4} > 0.$$

This inequality is in the form $f(x) > 0$, where f is the rational function given by

$$f(x) = \frac{3x + 3}{2x + 4}.$$

The graph of f is shown in **Figure 8.17**.

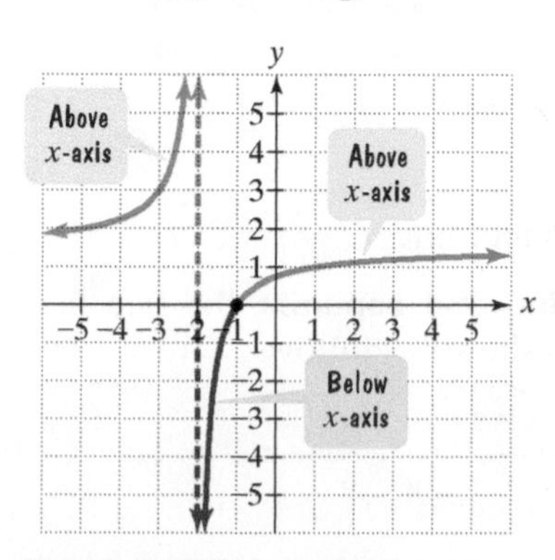

Figure 8.17 The graph of $f(x) = \dfrac{3x + 3}{2x + 4}$

We can find the x-intercept of f by setting the numerator equal to 0:

$$3x + 3 = 0$$
$$3x = -3$$
$$x = -1.$$

f has an x-intercept at -1 and passes through $(-1, 0)$.

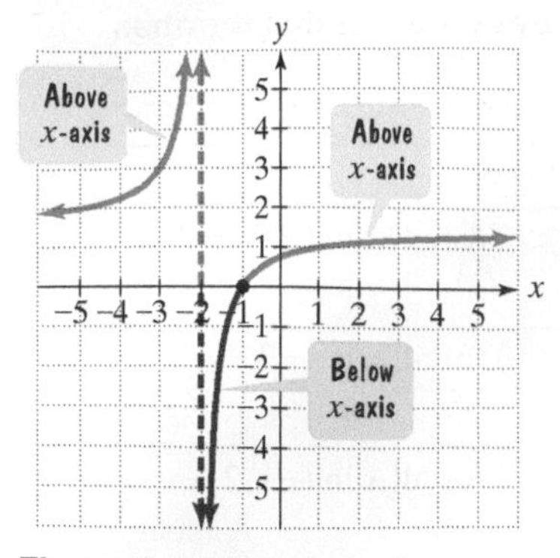

Figure 8.17 The graph of $f(x) = \frac{3x + 3}{2x + 4}$ (repeated)

We can determine where f is undefined by setting the denominator of $f(x) = \frac{3x + 3}{2x + 4}$ equal to 0:

$$2x + 4 = 0$$
$$2x = -4$$
$$x = -2.$$

f is undefined at -2. **Figure 8.17** shows that the function's vertical asymptote is $x = -2$.

By setting both the numerator and the denominator of f equal to 0, we obtained the solutions -2 and -1. These numbers separate the x-axis into three intervals: $(-\infty, -2)$, $(-2, -1)$, and $(-1, \infty)$. On each interval, the graph of f is either above the x-axis $[f(x) > 0]$ or below the x-axis $[f(x) < 0]$.

Examine the graph in **Figure 8.17** carefully. Can you see that it is above the x-axis for all x in $(-\infty, -2)$ or $(-1, \infty)$, shown in blue? Thus, the solution set of $\frac{3x + 3}{2x + 4} > 0$ is $(-\infty, -2) \cup (-1, \infty)$. By contrast, the graph of f lies below the x-axis for all x in $(-2, -1)$, shown in red. Thus, the solution set of $\frac{3x + 3}{2x + 4} < 0$ is $(-2, -1)$.

The first step in solving a rational inequality is to bring all terms to one side, obtaining zero on the other side. Then express the rational function on the nonzero side as a single quotient. The second step is to set the numerator and the denominator of f equal to zero. The solutions of these equations serve as boundary points that separate the real number line into intervals. At this point, the procedure is the same as the one we used for solving polynomial inequalities.

EXAMPLE 4 Solving a Rational Inequality

Solve and graph the solution set: $\frac{x + 3}{x - 7} < 0$.

Solution

Step 1. Express the inequality so that one side is zero and the other side is a single quotient. The given inequality is already in this form. The form is $f(x) < 0$, where $f(x) = \frac{x + 3}{x - 7}$.

Step 2. Set the numerator and the denominator of f equal to zero. The real solutions are the boundary points.

$$x + 3 = 0 \qquad x - 7 = 0$$ Set the numerator and denominator equal to 0. These are the values that make the quotient zero or undefined.

$$x = -3 \qquad x = 7$$ Solve for x.

We will use these solutions as boundary points on a number line.

Step 3. Locate the boundary points on a number line and separate the line into intervals. The number line with the boundary points is shown as follows:

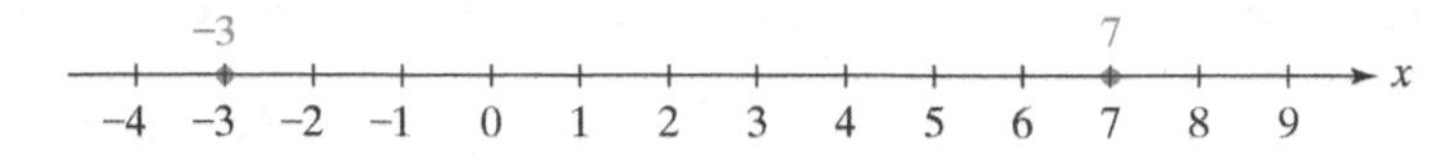

The boundary points divide the number line into three intervals:

$$(-\infty, -3) \quad (-3, 7) \quad (7, \infty).$$

Great Question!

Can I solve

$$\frac{x + 3}{x - 7} < 0$$

by first multiplying both sides by $x - 7$ to clear fractions?

No. The problem is that $x - 7$ contains a variable and can be positive or negative, depending on the value of x. Thus, we do not know whether or not to change the sense of the inequality.

Step 4. Choose one test value within each interval and evaluate *f* at that number.

Interval	Test Value	Substitute into $f(x) = \frac{x+3}{x-7}$	Conclusion
$(-\infty, -3)$	-4	$f(-4) = \frac{-4+3}{-4-7} = \frac{-1}{-11} = \frac{1}{11}$, positive	$f(x) > 0$ for all x in $(-\infty, -3)$.
$(-3, 7)$	0	$f(0) = \frac{0+3}{0-7} = -\frac{3}{7}$, negative	$f(x) < 0$ for all x in $(-3, 7)$.
$(7, \infty)$	8	$f(8) = \frac{8+3}{8-7} = 11$, positive	$f(x) > 0$ for all x in $(7, \infty)$.

Step 5. Write the solution set, selecting the interval or intervals that satisfy the given inequality. We are interested in solving $f(x) < 0$, where $f(x) = \frac{x+3}{x-7}$. Based on our work in step 4, we see that $f(x) < 0$ for all x in $(-3, 7)$.

Because $f(x) < 0$ for all x in $(-3, 7)$, the solution set of the given inequality, $\frac{x+3}{x-7} < 0$, is $(-3, 7)$.

The graph of the solution set on a number line is shown as follows:

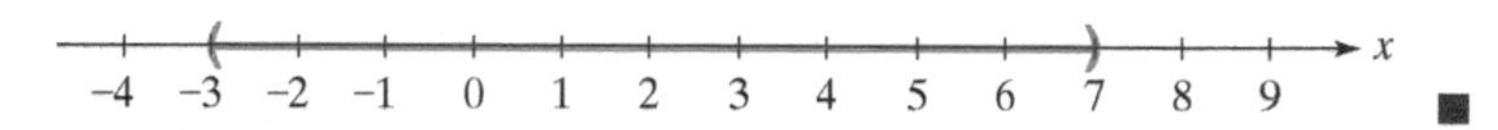

■

CHECK POINT 4 Solve and graph the solution set: $\frac{x-5}{x+2} < 0$.

EXAMPLE 5 Solving a Rational Inequality

Solve and graph the solution set: $\frac{x+1}{x+3} \geq 2$.

Solution

Step 1. Express the inequality so that one side is zero and the other side is a single quotient. We subtract 2 from both sides to obtain zero on the right.

$$\frac{x+1}{x+3} \geq 2 \quad \text{This is the given inequality.}$$

$$\frac{x+1}{x+3} - 2 \geq 0 \quad \text{Subtract 2 from both sides, obtaining 0 on the right.}$$

$$\frac{x+1}{x+3} - \frac{2(x+3)}{x+3} \geq 0 \quad \text{The least common denominator is } x+3. \text{ Express 2 in terms of this denominator.}$$

$$\frac{x+1-2(x+3)}{x+3} \geq 0 \quad \text{Subtract rational expressions.}$$

$$\frac{x+1-2x-6}{x+3} \geq 0 \quad \text{Apply the distributive property.}$$

$$\frac{-x-5}{x+3} \geq 0 \quad \text{Simplify.}$$

This inequality is equivalent to the one we wish to solve. It is in the form $f(x) \geq 0$, where $f(x) = \frac{-x-5}{x+3}$.

Great Question!

Can I begin solving

$$\frac{x+1}{x+3} \geq 2$$

by multiplying both sides by $x + 3$?

No. We do not know if $x + 3$ is positive or negative. Thus, we do not know whether or not to change the sense of the inequality.

Step 2. Set the numerator and the denominator of f equal to zero. We need to solve $f(x) \geq 0$, where $f(x) = \frac{-x-5}{x+3}$. We use $f(x) = \frac{-x-5}{x+3}$. The real solutions obtained by setting the numerator and the denominator equal to zero are the boundary points.

$$-x - 5 = 0 \qquad x + 3 = 0$$ Set the numerator and denominator equal to 0. These are the values that make the quotient zero or undefined.

$$x = -5 \qquad x = -3$$ Solve for x.

We will use these solutions as boundary points on a number line.

Step 3. Locate the boundary points on a number line and separate the line into intervals. The number line with the boundary points is shown as follows:

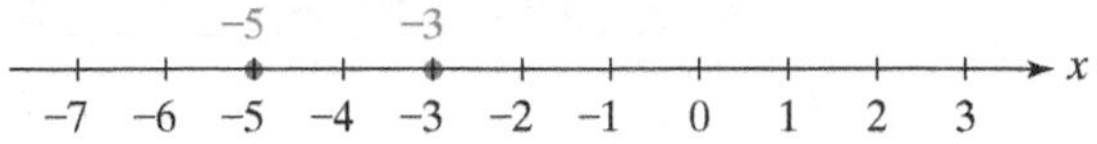

The boundary points divide the number line into three intervals:

$$(-\infty, -5) \quad (-5, -3) \quad (-3, \infty).$$

Step 4. Choose one test value within each interval and evaluate f at that number.

Interval	Test Value	Substitute into $f(x) = \frac{-x-5}{x+3}$	Conclusion
$(-\infty, -5)$	-6	$f(-6) = \frac{-(-6)-5}{-6+3}$ $= -\frac{1}{3}$, negative	$f(x) < 0$ for all x in $(-\infty, -5)$.
$(-5, -3)$	-4	$f(-4) = \frac{-(-4)-5}{-4+3}$ $= 1$, positive	$f(x) > 0$ for all x in $(-5, -3)$.
$(-3, \infty)$	0	$f(0) = \frac{-0-5}{0+3}$ $= -\frac{5}{3}$, negative	$f(x) < 0$ for all x in $(-3, \infty)$.

Step 5. Write the solution set, selecting the interval or intervals that satisfy the given inequality. We are interested in solving $f(x) \geq 0$, where $f(x) = \frac{-x-5}{x+3}$. Based on our work in step 4, we see that $f(x) > 0$ for all x in $(-5, -3)$. However, because the inequality involves $\geq$ (greater than or equal to), we must also include the solution of $f(x) = 0$, namely the value that we obtained when we set the numerator of f equal to zero. Thus, we must include -5 in the solution set. The solution set of the given inequality is $[-5, -3)$.

The graph of the solution set on a number line is shown as follows:

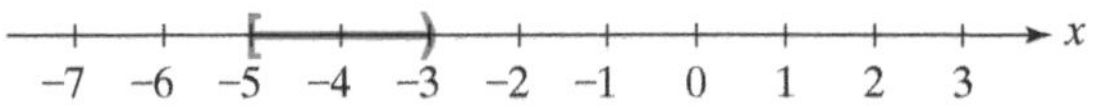

■

Great Question!

Which boundary points must I always exclude from the solution set of a rational inequality?

Never include values that cause a rational function's denominator to equal zero. Division by zero is undefined.

Using Technology

Graphic Connections

The solution set for

$$\frac{x+1}{x+3} \geq 2$$

or, equivalently,

$$\frac{-x-5}{x+3} \geq 0$$

can be verified with a graphing utility. The graph of $f(x) = \frac{-x-5}{x+3}$, shown in blue, lies on or above the x-axis, representing $\geq$, for all x in $[-5, -3)$. We also graphed the vertical asymptote, $x = -3$, to remind us that f is undefined for $x = -3$.

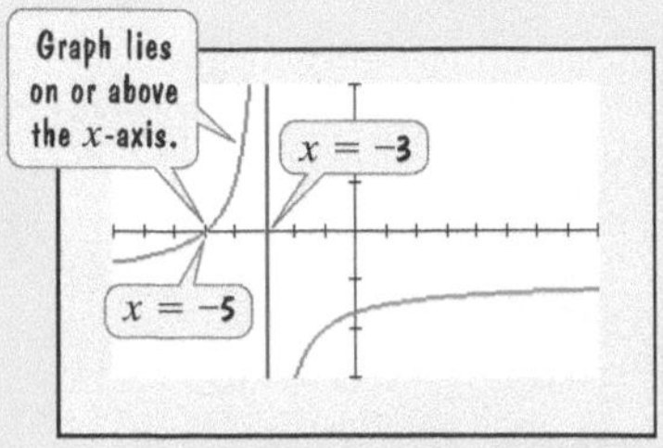

[-8, 8, 1] by [-3, 3, 1]

✓ CHECK POINT 5 Solve and graph the solution set: $\frac{2x}{x+1} \geq 1.$

3 Solve problems modeled by polynomial or rational inequalities.

Applications

Polynomial inequalities can be solved to answer questions about variables contained in polynomial functions.

EXAMPLE 6 Modeling the Position of a Free-Falling Object

A ball is thrown vertically upward from the top of the Leaning Tower of Pisa (190 feet high) with an initial velocity of 96 feet per second (**Figure 8.18**). The function

$$s(t) = -16t^2 + 96t + 190$$

models the ball's height above the ground, $s(t)$, in feet, t seconds after it was thrown. During which time period will the ball's height exceed that of the tower?

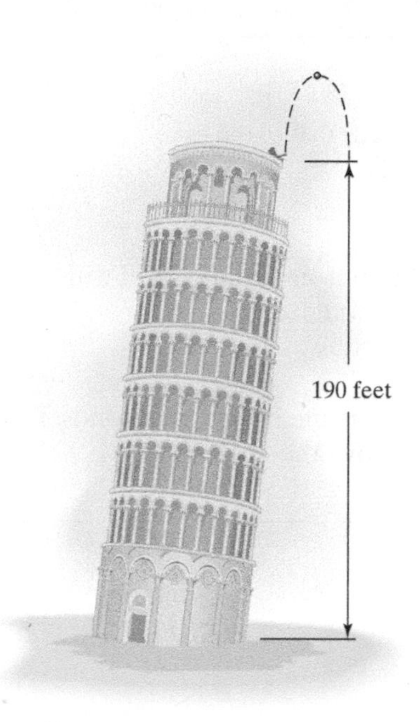

Figure 8.18 Throwing a ball from the top of the Leaning Tower of Pisa

Solution Using the problem's question and the given model for the ball's height, $s(t) = -16t^2 + 96t + 190$, we obtain a polynomial inequality.

$$-16t^2 + 96t + 190 > 190$$

When will the ball's height | exceed that | of the tower?

$-16t^2 + 96t + 190 > 190$ This is the inequality that models the problem's question.

$-16t^2 + 96t > 0$ Subtract 190 from both sides. This inequality is in the form $f(t) > 0$, where $f(t) = -16t^2 + 96t$.

$-16t^2 + 96t = 0$ Solve the equation $f(t) = 0$.

$-16t(t - 6) = 0$ Factor.

$-16t = 0$ or $t - 6 = 0$ Set each factor equal to 0.

$t = 0$ $\quad t = 6$ Solve for t. The boundary points are 0 and 6.

−2 −1 0 1 2 3 4 5 6 7 8 t

Locate these values on a number line.

The intervals are $(-\infty, 0)$, $(0, 6)$, and $(6, \infty)$. For our purposes, the mathematical model is useful only from $t = 0$ until the ball hits the ground. (By setting $-16t^2 + 96t + 190$ equal to zero, we find $t \approx 7.57$; the ball hits the ground after approximately 7.57 seconds.) Thus, we use $(0, 6)$ and $(6, 7.57)$ for our intervals.

Interval	Test Value	Substitute into $f(t) = -16t^2 + 96t$	Conclusion
$(0, 6)$	1	$f(1) = -16 \cdot 1^2 + 96 \cdot 1$ $= 80$, positive	$f(t) > 0$ for all t in $(0, 6)$.
$(6, 7.57)$	7	$f(7) = -16 \cdot 7^2 + 96 \cdot 7$ $= -112$, negative	$f(t) < 0$ for all t in $(6, 7.57)$.

We are interested in solving $f(t) > 0$, where $f(t) = -16t^2 + 96t$. We see that $f(t) > 0$ for all t in $(0, 6)$. This means that the ball's height exceeds that of the tower between 0 and 6 seconds. ■

Using Technology

Graphic Connections

The graphs of

$$y_1 = -16x^2 + 96x + 190$$

and

$$y_2 = 190$$

are shown in a

$$[0, 8, 1] \text{ by } [0, 360, 36]$$

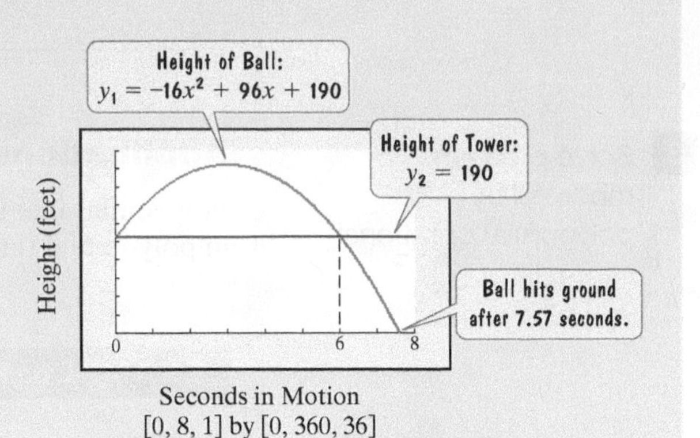

viewing rectangle. The graphs show that the ball's height exceeds that of the tower between 0 and 6 seconds.

☑ **CHECK POINT 6** An object is propelled straight up from ground level with an initial velocity of 80 feet per second. Its height at time t is modeled by

$$s(t) = -16t^2 + 80t,$$

where the height, $s(t)$, is measured in feet and the time, t, is measured in seconds. In which time interval will the object be more than 64 feet above the ground?

CONCEPT AND VOCABULARY CHECK

Fill in each blank so that the resulting statement is true.

1. We solve the polynomial inequality $x^2 + 8x + 15 > 0$ by first solving the equation ______________. The real solutions of this equation, -5 and -3, shown on the number line, are called ______________ points.

$-5 \quad -3$

$-7 \quad -6 \quad -5 \quad -4 \quad -3 \quad -2 \quad -1 \quad 0 \quad 1 \quad 2 \quad 3 \quad x$

2. The points at -5 and -3 shown above divide the number line into three intervals:

______________, ______________, ______________.

3. True or false: A test value for the leftmost interval on the number line shown above could be -10. ______________

4. True or false: A test value for the rightmost interval on the number line shown above could be 0. ______________

5. Consider the rational inequality

$$\frac{x-1}{x+2} \geq 0.$$

Setting the numerator and the denominator of $\frac{x-1}{x+2}$ equal to zero, we obtain $x = 1$ and $x = -2$. These values are shown as points on the number line. Also shown is information about three test values.

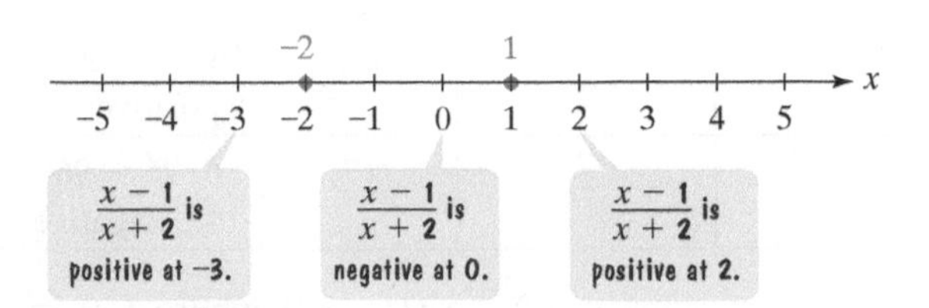

Based on the information shown above, the solution set of $\frac{x-1}{x+2} \geq 0$ is ______________.

8.5 EXERCISE SET

MyMathLab®

Practice Exercises

Solve each polynomial inequality in Exercises 1–40 and graph the solution set on a real number line.

1. $(x - 4)(x + 2) > 0$
2. $(x + 3)(x - 5) > 0$
3. $(x - 7)(x + 3) \le 0$
4. $(x + 1)(x - 7) \le 0$
5. $x^2 - 5x + 4 > 0$
6. $x^2 - 4x + 3 < 0$
7. $x^2 + 5x + 4 > 0$
8. $x^2 + x - 6 > 0$
9. $x^2 - 6x + 8 \le 0$
10. $x^2 - 2x - 3 \ge 0$
11. $3x^2 + 10x - 8 \le 0$
12. $9x^2 + 3x - 2 \ge 0$
13. $2x^2 + x < 15$
14. $6x^2 + x > 1$
15. $4x^2 + 7x < -3$
16. $3x^2 + 16x < -5$
17. $x^2 - 4x \ge 0$
18. $x^2 + 2x < 0$
19. $2x^2 + 3x > 0$
20. $3x^2 - 5x \le 0$
21. $-x^2 + x \ge 0$
22. $-x^2 + 2x \ge 0$
23. $x^2 \le 4x - 2$
24. $x^2 \le 2x + 2$
25. $3x^2 > 4x + 2$
26. $3x^2 > 10x - 5$
27. $2x^2 - 5x \ge 1$
28. $5x^2 + 8x \ge 11$
29. $x^2 - 6x + 9 < 0$
30. $4x^2 - 4x + 1 \ge 0$
31. $(x - 1)(x - 2)(x - 3) \ge 0$
32. $(x + 1)(x + 2)(x + 3) \ge 0$
33. $x^3 + 2x^2 - x - 2 \ge 0$
34. $x^3 + 2x^2 - 4x - 8 \ge 0$
35. $x^3 - 3x^2 - 9x + 27 < 0$
36. $x^3 + 7x^2 - x - 7 < 0$
37. $x^3 + x^2 + 4x + 4 > 0$
38. $x^3 - x^2 + 9x - 9 > 0$
39. $x^3 \ge 9x^2$
40. $x^3 \le 4x^2$

Solve each rational inequality in Exercises 41–56 and graph the solution set on a real number line.

41. $\frac{x - 4}{x + 3} > 0$
42. $\frac{x + 5}{x - 2} > 0$
43. $\frac{x + 3}{x + 4} < 0$
44. $\frac{x + 5}{x + 2} < 0$
45. $\frac{-x + 2}{x - 4} \ge 0$
46. $\frac{-x - 3}{x + 2} \le 0$
47. $\frac{4 - 2x}{3x + 4} \le 0$
48. $\frac{3x + 5}{6 - 2x} \ge 0$
49. $\frac{x}{x - 3} > 0$
50. $\frac{x + 4}{x} > 0$
51. $\frac{x + 1}{x + 3} < 2$
52. $\frac{x}{x - 1} > 2$
53. $\frac{x + 4}{2x - 1} \le 3$
54. $\frac{1}{x - 3} < 1$
55. $\frac{x - 2}{x + 2} \le 2$
56. $\frac{x}{x + 2} \ge 2$

In Exercises 57–60, use the given functions to find all values of x that satisfy the required inequality.

57. $f(x) = 2x^2, g(x) = 5x - 2; f(x) \ge g(x)$
58. $f(x) = 4x^2, g(x) = 9x - 2; f(x) < g(x)$
59. $f(x) = \frac{2x}{x + 1}, g(x) = 1; f(x) < g(x)$
60. $f(x) = \frac{x}{2x - 1}, g(x) = 1; f(x) \ge g(x)$

Practice PLUS

Solve each inequality in Exercises 61–66 and graph the solution set on a real number line.

61. $|x^2 + 2x - 36| > 12$
62. $|x^2 + 6x + 1| > 8$
63. $\frac{3}{x + 3} > \frac{3}{x - 2}$

64. $\dfrac{1}{x+1} > \dfrac{2}{x-1}$

65. $\dfrac{x^2 - x - 2}{x^2 - 4x + 3} > 0$

66. $\dfrac{x^2 - 3x + 2}{x^2 - 2x - 3} > 0$

In Exercises 67–68, use the graph of the polynomial function to solve each inequality.

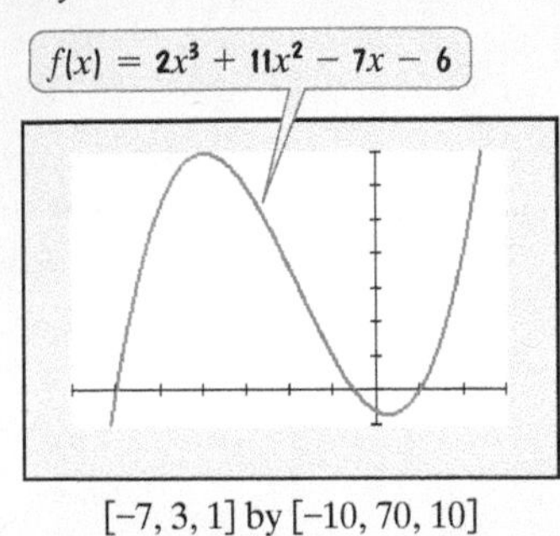

[−7, 3, 1] by [−10, 70, 10]

67. $2x^3 + 11x^2 \geq 7x + 6$

68. $2x^3 + 11x^2 < 7x + 6$

In Exercises 69–70, use the graph of the rational function to solve each inequality.

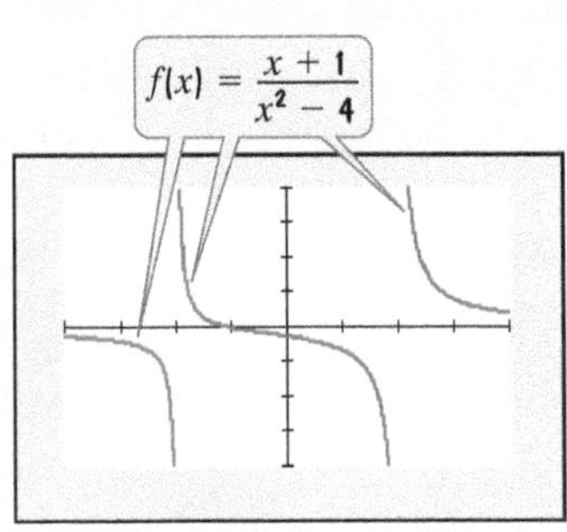

[−4, 4, 1] by [−4, 4, 1]

69. $\dfrac{1}{4(x+2)} \leq -\dfrac{3}{4(x-2)}$

70. $\dfrac{1}{4(x+2)} > -\dfrac{3}{4(x-2)}$

Application Exercises

71. You throw a ball straight up from a rooftop 160 feet high with an initial speed of 48 feet per second. The function

$$s(t) = -16t^2 + 48t + 160$$

models the ball's height above the ground, $s(t)$, in feet, t seconds after it was thrown. During which time period will the ball's height exceed that of the rooftop?

72. Divers in Acapulco, Mexico, dive headfirst from the top of a cliff 87 feet above the Pacific Ocean. The function

$$s(t) = -16t^2 + 8t + 87$$

models a diver's height above the ocean, $s(t)$, in feet, t seconds after leaping. During which time period will the diver's height exceed that of the cliff?

The functions

$$f(x) = 0.0875x^2 - 0.4x + 66.6 \quad \text{(Dry pavement)}$$

and

$$g(x) = 0.0875x^2 + 1.9x + 11.6 \quad \text{(Wet pavement)}$$

model a car's stopping distance, $f(x)$ or $g(x)$, in feet, traveling at x miles per hour. Function f models stopping distance on dry pavement and function g models stopping distance on wet pavement. The graphs of these functions are shown for speeds of 30 miles per hour and greater. Notice that the figure does not specify which graph is the model for dry roads and which is the model for wet roads. Use this information to solve Exercises 73–74.

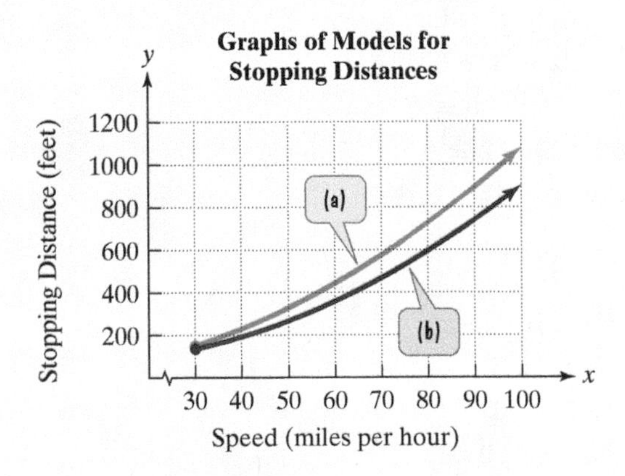

73. **a.** Use the given functions to find the stopping distance on dry pavement and the stopping distance on wet pavement for a car traveling at 35 miles per hour. Round to the nearest foot.

b. Based on your answers to part (a), which rectangular coordinate graph shows stopping distances on dry pavement and which shows stopping distances on wet pavement?

c. How well do your answers to part (a) model the actual stopping distances shown in **Figure 8.15** on page 638?

d. Determine speeds on dry pavement requiring stopping distances that exceed the length of one and one-half football fields, or 540 feet. Round to the nearest mile per hour. How is this shown on the appropriate graph of the models?

74. **a.** Use the given functions to find the stopping distance on dry pavement and the stopping distance on wet pavement for a car traveling at 55 miles per hour. Round to the nearest foot.

b. Based on your answers to part (a), which rectangular coordinate graph shows stopping distances on dry pavement and which shows stopping distances on wet pavement?

c. How well do your answers to part (a) model the actual stopping distances shown in **Figure 8.15** on page 638?

d. Determine speeds on wet pavement requiring stopping distances that exceed the length of one and one-half football fields, or 540 feet. Round to the nearest mile per hour. How is this shown on the appropriate graph of the models on page 650?

A company manufactures wheelchairs. The average cost function, $\overline{C}$, of producing x wheelchairs per month is given by

$$\overline{C}(x) = \frac{500{,}000 + 400x}{x}.$$

The graph of the rational function is shown. Use the function to solve Exercises 75–76.

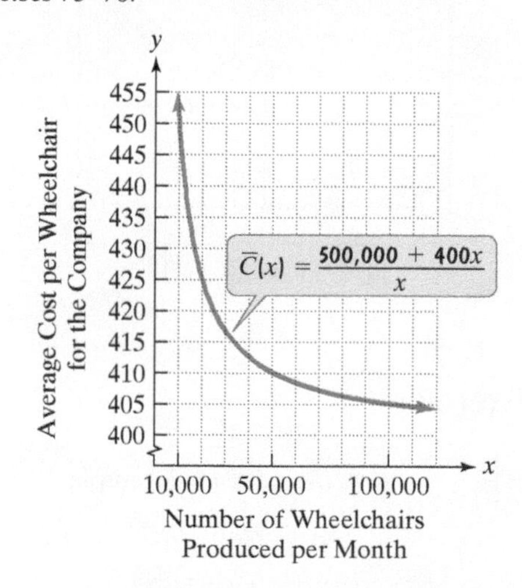

75. Describe the company's production level so that the average cost of producing each wheelchair does not exceed \$425. Use a rational inequality to solve the problem. Then explain how your solution is shown on the graph.

76. Describe the company's production level so that the average cost of producing each wheelchair does not exceed \$410. Use a rational inequality to solve the problem. Then explain how your solution is shown on the graph.

77. The perimeter of a rectangle is 50 feet. Describe the possible length of a side if the area of the rectangle is not to exceed 114 square feet.

78. The perimeter of a rectangle is 180 feet. Describe the possible lengths of a side if the area of the rectangle is not to exceed 800 square feet.

Explaining the Concepts

79. What is a polynomial inequality?

80. What is a rational inequality?

81. Describe similarities and differences between the solutions of

$$(x - 2)(x + 5) \geq 0 \quad \text{and} \quad \frac{x - 2}{x + 5} \geq 0.$$

Technology Exercises

Solve each inequality in Exercises 82–87 using a graphing utility.

82. $x^2 + 3x - 10 > 0$

83. $2x^2 + 5x - 3 \leq 0$

84. $\frac{x - 4}{x - 1} \leq 0$

85. $\frac{x + 2}{x - 3} \leq 2$

86. $\frac{1}{x + 1} \leq \frac{2}{x + 4}$

87. $x^3 + 2x^2 - 5x - 6 > 0$

The graph shows stopping distances for trucks at various speeds on dry roads and on wet roads. Use this information to solve Exercises 88–89.

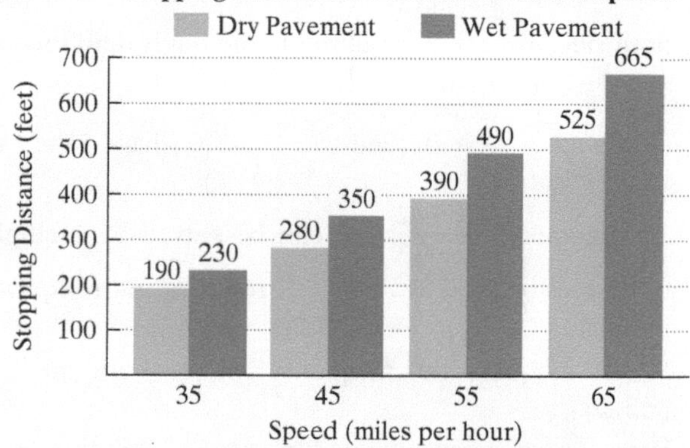

Source: National Highway Traffic Safety Administration

88. a. Use the statistical menu of your graphing utility and the quadratic regression program to obtain the quadratic function that models a truck's stopping distance, $f(x)$, in feet, on dry pavement traveling at x miles per hour. Round the constant term to one decimal place.

b. Use the function from part (a) to determine speeds on dry pavement requiring stopping distances that exceed 455 feet. Round to the nearest mile per hour.

89. a. Use the statistical menu of your graphing utility and the quadratic regression program to obtain the quadratic function that models a truck's stopping distance, $f(x)$, in feet, on wet pavement traveling at x miles per hour. Round the constant term to one decimal place.

b. Use the function from part (a) to determine speeds on wet pavement requiring stopping distances that exceed 446 feet.

Critical Thinking Exercises

Make Sense? *In Exercises 90–93, determine whether each statement makes sense or does not make sense, and explain your reasoning.*

90. When solving $f(x) > 0$, where f is a polynomial function, I only pay attention to the sign of f at each test value and not the actual function value.

91. I'm solving a polynomial inequality that has a value for which the polynomial function is undefined.

92. Because it takes me longer to come to a stop on a wet road than on a dry road, graph (a) for Exercises 73–74 is the model for stopping distances on wet pavement and graph (b) is the model for stopping distances on dry pavement.

93. I began the solution of the rational inequality $\frac{x+1}{x+3} \geq 2$ by setting both $x + 1$ and $x + 3$ equal to zero.

In Exercises 94–97, determine whether each statement is true or false. If the statement is false, make the necessary change(s) to produce a true statement.

94. The solution set of $x^2 > 25$ is $(5, \infty)$.

95. The inequality $\frac{x-2}{x+3} < 2$ can be solved by multiplying both sides by $x + 3$, resulting in the equivalent inequality $x - 2 < 2(x + 3)$.

96. $(x + 3)(x - 1) \geq 0$ and $\frac{x+3}{x-1} \geq 0$ have the same solution set.

97. The inequality $\frac{x-2}{x+3} < 2$ can be solved by multiplying both sides by $(x + 3)^2$, $x \neq -3$, resulting in the equivalent inequality $(x - 2)(x + 3) < 2(x + 3)^2$.

98. Write a quadratic inequality whose solution set is $[-3, 5]$.

99. Write a rational inequality whose solution set is $(-\infty, -4) \cup [3, \infty)$.

In Exercises 100–103, use inspection to describe each inequality's solution set. Do not solve any of the inequalities.

100. $(x - 2)^2 > 0$

101. $(x - 2)^2 \leq 0$

102. $(x - 2)^2 < -1$

103. $\frac{1}{(x-2)^2} > 0$

104. The graphing calculator screen shows the graph of $y = 4x^2 - 8x + 7$.

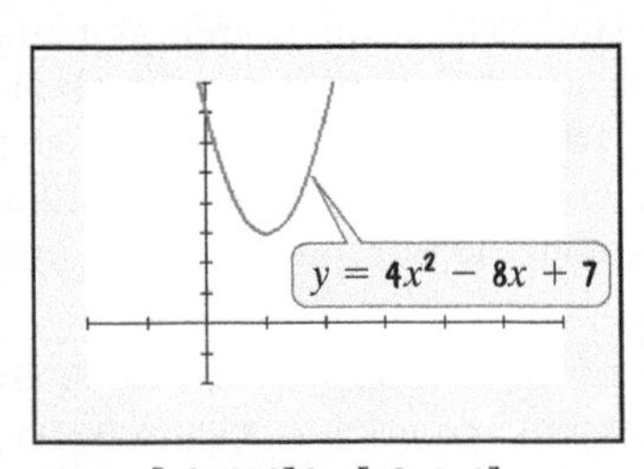

[−2, 6, 1] by [−2, 8, 1]

a. Use the graph to describe the solution set for $4x^2 - 8x + 7 > 0$.

b. Use the graph to describe the solution set for $4x^2 - 8x + 7 < 0$.

c. Use an algebraic approach to verify each of your descriptions in parts (a) and (b).

105. The graphing calculator screen shows the graph of $y = \sqrt{27 - 3x^2}$. Write and solve a quadratic inequality that explains why the graph only appears for $-3 \leq x \leq 3$.

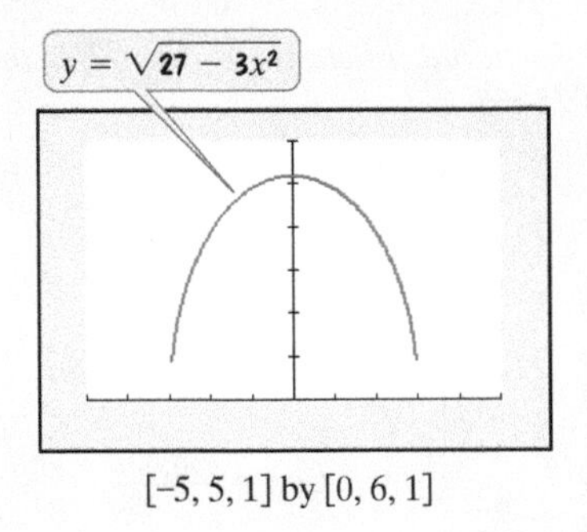

[−5, 5, 1] by [0, 6, 1]

Review Exercises

106. Solve: $\left|\frac{x-5}{3}\right| < 8$. (Section 4.3, Example 4)

107. Divide:

$$\frac{2x+6}{x^2+8x+16} \div \frac{x^2-9}{x^2+3x-4}.$$

(Section 6.1, Example 7)

108. Factor completely: $x^4 - 16y^4$. (Section 5.5, Example 3)

Preview Exercises

Exercises 109–111 will help you prepare for the material covered in the first section of the next chapter. In each exercise, use point plotting to graph the function. Begin by setting up a table of coordinates, selecting integers from −3 to 3, inclusive, for x.

109. $f(x) = 2^x$

110. $f(x) = 2^{-x}$

111. $f(x) = 2^x + 1$

Achieving Success

Avoid coursus interruptus.

Now that you've almost completed intermediate algebra, don't interrupt the sequence until you have completed all your required math classes. You'll have better results if you take your math courses without a break. If you start, stop, and start again, it's easy to forget what you've learned and lose your momentum.

Chapter 8 Summary

Definitions and Concepts	Examples

Section 8.1 The Square Root Property and Completing the Square

The Square Root Property

If u is an algebraic expression and d is a real number, then

$$\text{If } u^2 = d, \quad \text{then} \quad u = \sqrt{d} \quad \text{or} \quad u = -\sqrt{d}.$$

Equivalently,

$$\text{If } u^2 = d, \quad \text{then} \quad u = \pm\sqrt{d}.$$

Solve: $(x - 6)^2 = 50.$

$$x - 6 = \pm\sqrt{50}$$
$$x - 6 = \pm\sqrt{25 \cdot 2}$$
$$x - 6 = \pm 5\sqrt{2}$$
$$x = 6 \pm 5\sqrt{2}$$

The solutions are $6 \pm 5\sqrt{2}$ and the solution set is $\{6 \pm 5\sqrt{2}\}$.

Completing the Square

If $x^2 + bx$ is a binomial, then by adding $\left(\frac{b}{2}\right)^2$, the square of half the coefficient of x, a perfect square trinomial will result. That is,

$$x^2 + bx + \left(\frac{b}{2}\right)^2 = \left(x + \frac{b}{2}\right)^2.$$

Complete the square:

$$x^2 + \frac{2}{7}x.$$

Half of $\frac{2}{7}$ is $\frac{1}{2} \cdot \frac{2}{7} = \frac{1}{7}$ and $\left(\frac{1}{7}\right)^2 = \frac{1}{49}$.

$$x^2 + \frac{2}{7}x + \frac{1}{49} = \left(x + \frac{1}{7}\right)^2$$

Solving Quadratic Equations by Completing the Square

1. If the coefficient of x^2 is not 1, divide both sides by this coefficient.
2. Isolate variable terms on one side.
3. Complete the square by adding the square of half the coefficient of x to both sides.
4. Factor the perfect square trinomial.
5. Solve by applying the square root property.

Solve by completing the square:

$$2x^2 + 16x - 6 = 0.$$

$$\frac{2x^2}{2} + \frac{16x}{2} - \frac{6}{2} = \frac{0}{2} \quad \text{Divide by 2.}$$
$$x^2 + 8x - 3 = 0 \quad \text{Simplify.}$$
$$x^2 + 8x = 3 \quad \text{Add 3.}$$

The coefficient of x is 8. Half of 8 is 4 and $4^2 = 16$. Add 16 to both sides.

$$x^2 + 8x + 16 = 3 + 16$$
$$(x + 4)^2 = 19$$
$$x + 4 = \pm\sqrt{19}$$
$$x = -4 \pm \sqrt{19}$$

Section 8.2 The Quadratic Formula

The solutions of a quadratic equation in standard form

$$ax^2 + bx + c = 0, \quad a \neq 0,$$

are given by the quadratic formula

$$x = \frac{-b \pm \sqrt{b^2 - 4ac}}{2a}.$$

Solve using the quadratic formula:

$$2x^2 = 6x - 3.$$

First write the equation in standard form by subtracting $6x$ and adding 3 on both sides.

$$2x^2 - 6x + 3 = 0$$

$a = 2$ $b = -6$ $c = 3$

$$x = \frac{-(-6) \pm \sqrt{(-6)^2 - 4 \cdot 2 \cdot 3}}{2 \cdot 2} = \frac{6 \pm \sqrt{36 - 24}}{4}$$
$$= \frac{6 \pm \sqrt{12}}{4} = \frac{6 \pm \sqrt{4 \cdot 3}}{4} = \frac{6 \pm 2\sqrt{3}}{4}$$
$$= \frac{2(3 \pm \sqrt{3})}{2 \cdot 2} = \frac{3 \pm \sqrt{3}}{2}$$

Definitions and Concepts	Examples

Section 8.2 The Quadratic Formula (continued)

The Discriminant

The discriminant, $b^2 - 4ac$, of the quadratic equation $ax^2 + bx + c = 0$ determines the number and type of solutions.

Discriminant	Solutions
Positive perfect square, with a, b, and c rational numbers	2 real rational solutions
Positive and not a perfect square	2 real irrational solutions
Zero, with a, b, and c rational numbers	1 real rational solution
Negative	2 imaginary solutions

$$2x^2 - 7x - 4 = 0$$

$a = 2$ $b = -7$ $c = -4$

$$b^2 - 4ac = (-7)^2 - 4(2)(-4)$$
$$= 49 - (-32) = 49 + 32 = 81$$

Positive perfect square

The equation has two real rational solutions.

Writing Quadratic Equations from Solutions

The zero-product principle in reverse makes it possible to write a quadratic equation from solutions:

$$\text{If } A = 0 \quad \text{or} \quad B = 0, \quad \text{then} \quad AB = 0.$$

Write a quadratic equation with the solution set $\{-2\sqrt{3}, 2\sqrt{3}\}$.

$$x = -2\sqrt{3} \qquad x = 2\sqrt{3}$$
$$x + 2\sqrt{3} = 0 \quad \text{or} \quad x - 2\sqrt{3} = 0$$
$$(x + 2\sqrt{3})(x - 2\sqrt{3}) = 0$$
$$x^2 - (2\sqrt{3})^2 = 0$$
$$x^2 - 12 = 0$$

Section 8.3 Quadratic Functions and Their Graphs

The graph of the quadratic function

$$f(x) = a(x - h)^2 + k, \quad a \neq 0,$$

is called a parabola. The vertex, or turning point, is (h, k). The graph opens upward if a is positive and downward if a is negative. The axis of symmetry is a vertical line passing through the vertex. The graph can be obtained using the vertex, x-intercepts, if any, (set $f(x)$ equal to zero and solve), and the y-intercept (set $x = 0$).

Graph: $f(x) = -(x + 3)^2 + 1$.

$$f(x) = -1(x - (-3))^2 + 1$$

$a = -1$ $h = -3$ $k = 1$

- Vertex $(h, k) = (-3, 1)$
- Opens downward because $a < 0$
- x-intercepts: Set $f(x) = 0$.

$$0 = -(x + 3)^2 + 1$$
$$(x + 3)^2 = 1$$
$$x + 3 = \pm\sqrt{1}$$
$$x + 3 = 1 \quad \text{or} \quad x + 3 = -1$$
$$x = -2 \qquad x = -4$$

- y-intercept: Set $x = 0$.

$$f(0) = -(0 + 3)^2 + 1 = -9 + 1 = -8$$

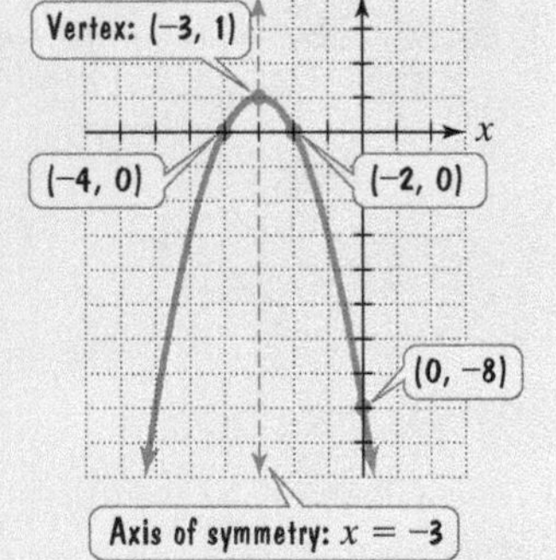

Definitions and Concepts	Examples

Section 8.3 Quadratic Functions and Their Graphs (continued)

A parabola whose equation is in the form

$$f(x) = ax^2 + bx + c, \quad a \neq 0,$$

has its vertex at

$$\left(-\frac{b}{2a}, f\left(-\frac{b}{2a}\right)\right).$$

The parabola is graphed as described in the left column at the bottom of the previous page. The only difference is how we determine the vertex. If $a > 0$, then f has a minimum that occurs at $x = -\frac{b}{2a}$. This minimum value is $f\left(-\frac{b}{2a}\right)$. If $a < 0$, then f has a maximum that occurs at $x = -\frac{b}{2a}$. This maximum value is $f\left(-\frac{b}{2a}\right)$.

Graph:

$$f(x) = x^2 - 6x + 5.$$

$a = 1$ $b = -6$ $c = 5$

- Vertex:

$$x = -\frac{b}{2a} = -\frac{-6}{2 \cdot 1} = 3$$

$$f(3) = 3^2 - 6 \cdot 3 + 5 = -4$$

Vertex is at $(3, -4)$.

- Opens upward because $a > 0$.
- x-intercepts: Set $f(x) = 0$.

$$x^2 - 6x + 5 = 0$$
$$(x - 1)(x - 5) = 0$$
$$x = 1 \quad \text{or} \quad x = 5$$

- y-intercept: Set $x = 0$.

$$f(0) = 0^2 - 6 \cdot 0 + 5 = 5$$

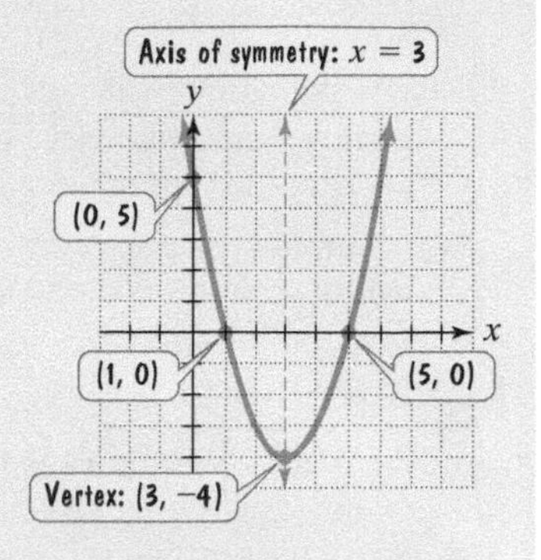

Section 8.4 Equations Quadratic in Form

An equation that is quadratic in form is one that can be expressed as a quadratic equation using an appropriate substitution. In these equations, the variable factor in one term is the square of the variable factor in the other variable term. Let $u =$ the variable factor that reappears squared. If at any point in the solution process both sides of an equation are raised to an even power, a check is required.

Solve:

$$x^{\frac{2}{3}} - 3x^{\frac{1}{3}} + 2 = 0.$$
$$\left(x^{\frac{1}{3}}\right)^2 - 3x^{\frac{1}{3}} + 2 = 0$$

Let $u = x^{\frac{1}{3}}$.

$$u^2 - 3u + 2 = 0$$
$$(u - 1)(u - 2) = 0$$
$$u - 1 = 0 \quad \text{or} \quad u - 2 = 0$$
$$u = 1 \qquad\qquad u = 2$$
$$x^{\frac{1}{3}} = 1 \qquad\qquad x^{\frac{1}{3}} = 2$$
$$\left(x^{\frac{1}{3}}\right)^3 = 1^3 \qquad\qquad \left(x^{\frac{1}{3}}\right)^3 = 2^3$$
$$x = 1 \qquad\qquad x = 8$$

The solutions are 1 and 8, and the solution set is {1, 8}.

Definitions and Concepts	Examples

Section 8.5 Polynomial and Rational Inequalities

Solving Polynomial Inequalities

1. Express the inequality in the form

$$f(x) < 0 \quad \text{or} \quad f(x) > 0,$$

where f is a polynomial function.

2. Solve the equation $f(x) = 0$. The real solutions are the boundary points.
3. Locate these boundary points on a number line, thereby dividing the number line into intervals.
4. Choose one representative number, called a test value, within each interval and evaluate f at that number.
   a. If the value of f is positive, then $f(x) > 0$ for all x in the interval.
   b. If the value of f is negative, then $f(x) < 0$ for all x in the interval.
5. Write the solution set, selecting the interval or intervals that satisfy the given inequality.

This procedure is valid if $<$ is replaced by $\leq$ and $>$ is replaced by $\geq$. In these cases, include the boundary points in the solution set.

Solve: $2x^2 + x - 6 > 0$.

The form of the inequality is $f(x) > 0$ with $f(x) = 2x^2 + x - 6$. Solve $f(x) = 0$.

$$2x^2 + x - 6 = 0$$
$$(2x - 3)(x + 2) = 0$$
$$2x - 3 = 0 \quad \text{or} \quad x + 2 = 0$$
$$x = \frac{3}{2} \qquad\qquad x = -2$$

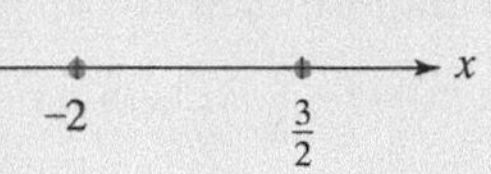

Use -3, 0, and 2 as test values.

$$f(-3) = 2(-3)^2 + (-3) - 6 = 9, \text{ positive}$$
$f(x) > 0$ for all x in $(-\infty, -2)$.
$$f(0) = 2 \cdot 0^2 + 0 - 6 = -6, \text{ negative}$$
$f(x) < 0$ for all x in $\left(-2, \frac{3}{2}\right)$.
$$f(2) = 2 \cdot 2^2 + 2 - 6 = 4, \text{ positive}$$
$f(x) > 0$ for all x in $\left(\frac{3}{2}, \infty\right)$.

The solution set is $(-\infty, -2) \cup \left(\frac{3}{2}, \infty\right)$.

Solving Rational Inequalities

1. Express the inequality in the form

$$f(x) < 0 \quad \text{or} \quad f(x) > 0,$$

where f is a rational function written as a single quotient.

2. Set the numerator and the denominator of f equal to zero. The real solutions are the boundary points.
3. Locate these boundary points on a number line, thereby dividing the number line into intervals.
4. Choose one representative number, called a test value, within each interval and evaluate f at that number.
   a. If the value of f is positive, then $f(x) > 0$ for all x in the interval.
   b. If the value of f is negative, then $f(x) < 0$ for all x in the interval.
5. Write the solution set, selecting the interval or intervals that satisfy the given inequality.

This procedure is valid if $<$ is replaced by $\leq$ and $>$ is replaced by $\geq$. In these cases, include any values that make the numerator of f zero. Always exclude any values that make the denominator zero.

Solve: $\dfrac{x}{x + 4} \geq 2$.

$$\frac{x}{x + 4} - \frac{2(x + 4)}{x + 4} \geq 0$$
$$\frac{-x - 8}{x + 4} \geq 0$$

The form of the inequality is $f(x) \geq 0$ with $f(x) = \dfrac{-x - 8}{x + 4}$.

Set the numerator and the denominator equal to zero.

$$-x - 8 = 0 \qquad x + 4 = 0$$
$$-8 = x \qquad\qquad x = -4$$

x
-8 -4

Use -9, -7, and -3 as test values.

$$f(-9) = \frac{-(-9) - 8}{-9 + 4} = \frac{1}{-5}, \text{ negative}$$
$f(x) < 0$ for all x in $(-\infty, -8)$.
$$f(-7) = \frac{-(-7) - 8}{-7 + 4} = \frac{-1}{-3} = \frac{1}{3}, \text{ positive}$$
$f(x) > 0$ for all x in $(-8, -4)$.
$$f(-3) = \frac{-(-3) - 8}{-3 + 4} = \frac{-5}{1} = -5, \text{ negative}$$
$f(x) < 0$ for all x in $(-4, \infty)$.

Because of $\geq$, include -8, the value that makes the numerator zero, in the solution set.
The solution set is $[-8, -4)$.

CHAPTER 8 REVIEW EXERCISES

8.1 *In Exercises 1–5, solve each equation by the square root property. If possible, simplify radicals or rationalize denominators. Express imaginary solutions in the form $a + bi$.*

1. $2x^2 - 3 = 125$

2. $3x^2 - 150 = 0$

3. $3x^2 - 2 = 0$

4. $(x - 4)^2 = 18$

5. $(x + 7)^2 = -36$

In Exercises 6–7, determine the constant that should be added to the binomial so that it becomes a perfect square trinomial. Then write and factor the trinomial.

6. $x^2 + 20x$

7. $x^2 - 3x$

In Exercises 8–10, solve each quadratic equation by completing the square.

8. $x^2 - 12x + 27 = 0$

9. $x^2 - 7x - 1 = 0$

10. $2x^2 + 3x - 4 = 0$

11. In 2 years, an investment of \$2500 grows to \$2916. Use the compound interest formula

$$A = P(1 + r)^t$$

to find the annual interest rate, r.

12. The function $W(t) = 3t^2$ models the weight of a human fetus, $W(t)$, in grams, after t weeks, where $0 \le t \le 39$. After how many weeks does the fetus weigh 588 grams?

13. A building casts a shadow that is double the length of the building's height. If the distance from the end of the shadow to the top of the building is 300 meters, how high is the building? Express the answer in simplified radical form. Then find a decimal approximation to the nearest tenth of a meter.

8.2 *In Exercises 14–16, solve each equation using the quadratic formula. Simplify solutions, if possible.*

14. $x^2 = 2x + 4$

15. $x^2 - 2x + 19 = 0$

16. $2x^2 = 3 - 4x$

In Exercises 17–19, without solving the given quadratic equation, determine the number and type of solutions.

17. $x^2 - 4x + 13 = 0$

18. $9x^2 = 2 - 3x$

19. $2x^2 + 4x = 3$

In Exercises 20–26, solve each equation by the method of your choice. Simplify solutions, if possible.

20. $3x^2 - 10x - 8 = 0$

21. $(2x - 3)(x + 2) = x^2 - 2x + 4$

22. $5x^2 - x - 1 = 0$

23. $x^2 - 16 = 0$

24. $(x - 3)^2 - 8 = 0$

25. $3x^2 - x + 2 = 0$

26. $\dfrac{5}{x + 1} + \dfrac{x - 1}{4} = 2$

In Exercises 27–29, write a quadratic equation in standard form with the given solution set.

27. $\left\{-\frac{1}{3}, \frac{3}{5}\right\}$

28. $\{-9i, 9i\}$

29. $\{-4\sqrt{3}, 4\sqrt{3}\}$

30. The graph shows stopping distances for motorcycles at various speeds on dry roads and on wet roads.

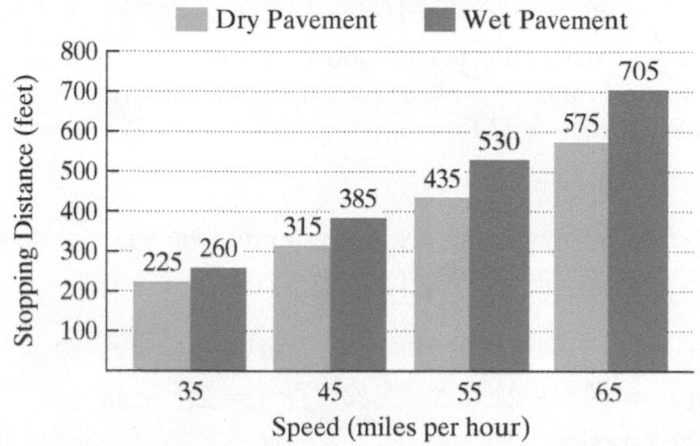

Source: National Highway Traffic Safety Administration

The functions

$$f(x) = 0.125x^2 - 0.8x + 99$$

Dry pavement

and

Wet pavement

$$g(x) = 0.125x^2 + 2.3x + 27$$

model a motorcycle's stopping distance, $f(x)$ or $g(x)$, in feet traveling at x miles per hour. Function f models stopping distance on dry pavement and function g models stopping distance on wet pavement.

a. Use function g to find the stopping distance on wet pavement for a motorcycle traveling at 35 miles per hour. Round to the nearest foot. Does your rounded answer overestimate or underestimate the stopping distance shown by the graph? By how many feet?

b. Use function f to determine a motorcycle's speed requiring a stopping distance on dry pavement of 267 feet.

31. The graphs of the functions in Exercise 30 are shown for speeds of 30 miles per hour and greater.

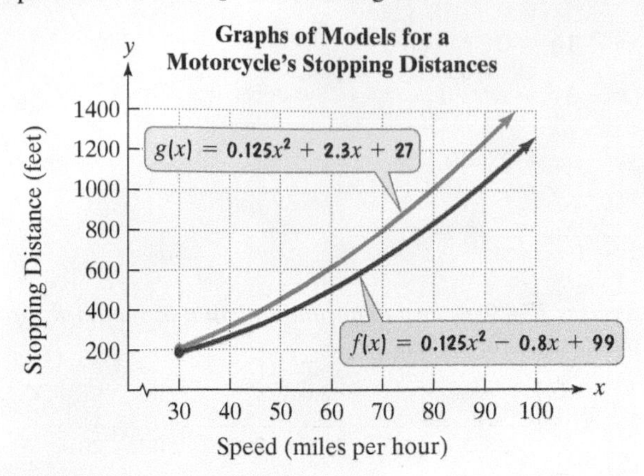

a. How is your answer to Exercise 30(a) shown on the graph of g?

b. How is your answer to Exercise 30(b) shown on the graph of f?

32. A baseball is hit by a batter. The function

$$s(t) = -16t^2 + 140t + 3$$

models the ball's height above the ground, $s(t)$, in feet, t seconds after it is hit. How long will it take for the ball to strike the ground? Round to the nearest tenth of a second.

8.3 *In Exercises 33–36, use the vertex and intercepts to sketch the graph of each quadratic function.*

33. $f(x) = -(x + 1)^2 + 4$ **34.** $f(x) = (x + 4)^2 - 2$

35. $f(x) = -x^2 + 2x + 3$ **36.** $f(x) = 2x^2 - 4x - 6$

37. The function

$$f(x) = -0.02x^2 + x + 1$$

models the yearly growth of a young redwood tree, $f(x)$, in inches, with x inches of rainfall per year. How many inches of rainfall per year result in maximum tree growth? What is the maximum yearly growth?

38. A model rocket is launched upward from a platform 40 feet above the ground. The quadratic function

$$s(t) = -16t^2 + 400t + 40$$

models the rocket's height above the ground, $s(t)$, in feet, t seconds after it was launched. After how many seconds does the rocket reach its maximum height? What is the maximum height?

39. The function

$$f(x) = 104.5x^2 - 1501.5x + 6016$$

models the death rate per year per 100,000 males, $f(x)$, for U.S. men who average x hours of sleep each night. How many hours of sleep, to the nearest tenth of an hour, corresponds to the minimum death rate? What is this minimum death rate, to the nearest whole number?

40. A field bordering a straight stream is to be enclosed. The side bordering the stream is not to be fenced. If 1000 yards of fencing material is to be used, what are the dimensions of the largest rectangular field that can be fenced? What is the maximum area?

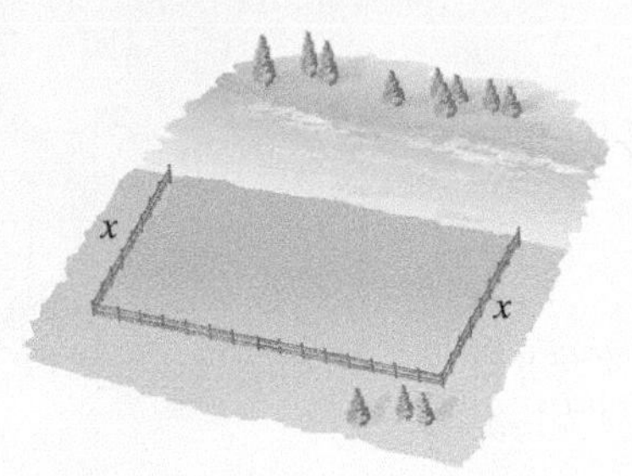

41. Among all pairs of numbers whose difference is 14, find a pair whose product is as small as possible. What is the minimum product?

8.4 *In Exercises 42–47, solve each equation by making an appropriate substitution. When necessary, check proposed solutions.*

42. $x^4 - 6x^2 + 8 = 0$

43. $x + 7\sqrt{x} - 8 = 0$

44. $(x^2 + 2x)^2 - 14(x^2 + 2x) = 15$

45. $x^{-2} + x^{-1} - 56 = 0$

46. $x^{\frac{2}{3}} - x^{\frac{1}{3}} - 12 = 0$

47. $x^{\frac{1}{2}} + 3x^{\frac{1}{4}} - 10 = 0$

8.5 *In Exercises 48–52, solve each inequality and graph the solution set on a real number line.*

48. $2x^2 + 5x - 3 < 0$

49. $2x^2 + 9x + 4 \geq 0$

50. $x^3 + 2x^2 > 3x$

51. $\frac{x - 6}{x + 2} > 0$ **52.** $\frac{x + 3}{x - 4} \leq 5$

53. A model rocket is launched from ground level. The function

$$s(t) = -16t^2 + 48t$$

models the rocket's height above the ground, $s(t)$, in feet, t seconds after it was launched. During which time period will the rocket's height exceed 32 feet?

54. The function

$$H(x) = \frac{15}{8}x^2 - 30x + 200$$

models heart rate, $H(x)$, in beats per minute, x minutes after a strenuous workout.

a. What is the heart rate immediately following the workout?

b. According to the model, during which intervals of time after the strenuous workout does the heart rate exceed 110 beats per minute? For which of these intervals has model breakdown occurred? Which interval provides a more realistic answer? How did you determine this?

CHAPTER 8 TEST

Step-by-step test solutions are found on the Chapter Test Prep Videos available in MyMathLab® or on YouTube (search "BlitzerInterAlg7e" and click on "Channels").

Express solutions to all equations in simplified form. Rationalize denominators, if possible.

In Exercises 1–2, solve each equation by the square root property.

1. $2x^2 - 5 = 0$

2. $(x - 3)^2 = 20$

In Exercises 3–4, determine the constant that should be added to the binomial so that it becomes a perfect square trinomial. Then write and factor the trinomial.

3. $x^2 - 16x$

4. $x^2 + \frac{2}{5}x$

5. Solve by completing the square: $x^2 - 6x + 7 = 0$.

6. Use the measurements determined by the surveyor to find the width of the pond. Express the answer in simplified radical form.

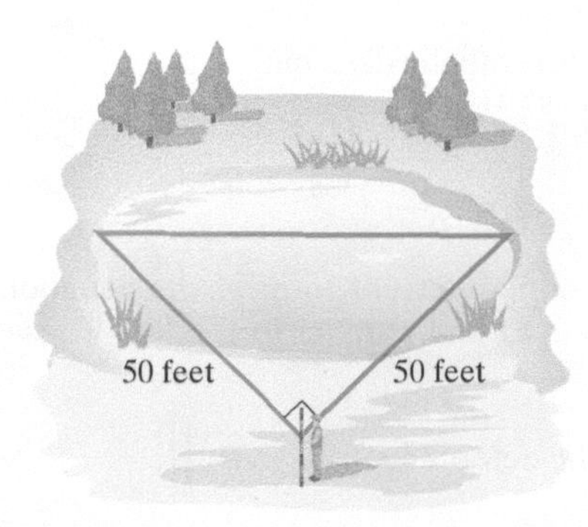

In Exercises 7–8, without solving the given quadratic equation, determine the number and type of solutions.

7. $3x^2 + 4x - 2 = 0$

8. $x^2 = 4x - 8$

In Exercises 9–12, solve each equation by the method of your choice.

9. $2x^2 + 9x = 5$

10. $x^2 + 8x + 5 = 0$

11. $(x + 2)^2 + 25 = 0$

12. $2x^2 - 6x + 5 = 0$

In Exercises 13–14, write a quadratic equation in standard form with the given solution set.

13. $\{-3, 7\}$

14. $\{-10i, 10i\}$

15. As gas prices surge, more Americans are cycling as a way to save money, stay fit, or both. The bar graph in the next column shows the number of bicycle-friendly U.S. communities, as designated by the League of American Bicyclists, for selected years from 2003 through 2014.

Number of U.S. Communities Designated "Bicycle Friendly" by the League of American Bicyclists

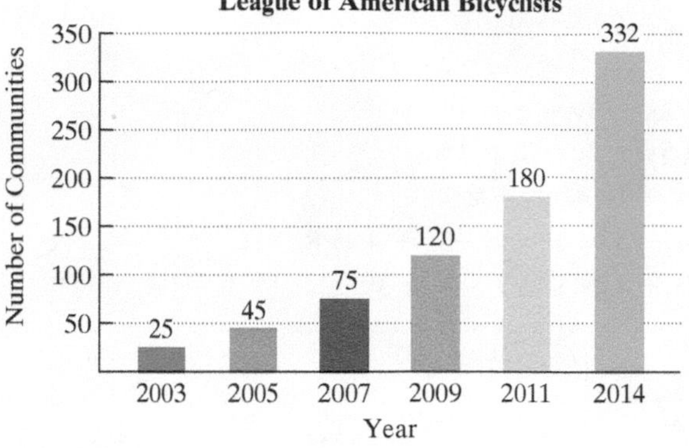

Source: League of American Bicyclists

The function

$$f(x) = 2.4x^2 + 0.7x + 29$$

models the number of bicycle-friendly communities, $f(x)$, x years after 2003.

a. Use the function to find the number of bicycle-friendly communities in 2014. Round to the nearest whole number. Does this rounded value underestimate or overestimate the number shown by the graph? By how much?

b. Use the function to determine the year in which 1206 U.S. communities will be bicycle friendly.

In Exercises 16–17, use the vertex and intercepts to sketch the graph of each quadratic function.

16. $f(x) = (x + 1)^2 + 4$ **17.** $f(x) = x^2 - 2x - 3$

A baseball player hits a pop fly into the air. The function

$$s(t) = -16t^2 + 64t + 5$$

models the ball's height above the ground, $s(t)$, in feet, t seconds after it is hit. Use the function to solve Exercises 18–19.

18. When does the baseball reach its maximum height? What is that height?

19. After how many seconds does the baseball hit the ground? Round to the nearest tenth of a second.

20. The function $f(x) = -x^2 + 46x - 360$ models the daily profit, $f(x)$, in hundreds of dollars, for a company that manufactures x computers daily. How many computers should be manufactured each day to maximize profit? What is the maximum daily profit?

In Exercises 21–23, solve each equation by making an appropriate substitution. When necessary, check proposed solutions.

21. $(2x - 5)^2 + 4(2x - 5) + 3 = 0$

22. $x^4 - 13x^2 + 36 = 0$

23. $x^{\frac{2}{3}} - 9x^{\frac{1}{3}} + 8 = 0$

In Exercises 24–25, solve each inequality and graph the solution set on a real number line.

24. $x^2 - x - 12 < 0$ **25.** $\frac{2x + 1}{x - 3} \leq 3$

CUMULATIVE REVIEW EXERCISES (CHAPTERS 1–8)

In Exercises 1–13, solve each equation, inequality, or system.

1. $9(x - 1) = 1 + 3(x - 2)$
2. $\begin{cases} 3x + 4y = -7 \\ x - 2y = -9 \end{cases}$
3. $\begin{cases} x - y + 3z = -9 \\ 2x + 3y - z = 16 \\ 5x + 2y - z = 15 \end{cases}$
4. $7x + 18 \le 9x - 2$
5. $4x - 3 < 13$ and $-3x - 4 \ge 8$
6. $2x + 4 > 8$ or $x - 7 \ge 3$
7. $|2x - 1| < 5$
8. $\left|\frac{2}{3}x - 4\right| = 2$
9. $\frac{4}{x - 3} - \frac{6}{x + 3} = \frac{24}{x^2 - 9}$
10. $\sqrt{x + 4} - \sqrt{x - 3} = 1$
11. $2x^2 = 5 - 4x$
12. $x^{\frac{2}{3}} - 5x^{\frac{1}{3}} + 6 = 0$
13. $2x^2 + x - 6 \le 0$

In Exercises 14–17, graph each function, equation, or inequality in a rectangular coordinate system.

14. $x - 3y = 6$
15. $f(x) = \frac{1}{2}x - 1$
16. $3x - 2y > -6$
17. $f(x) = -2(x - 3)^2 + 2$

In Exercises 18–28, perform the indicated operations, and simplify, if possible.

18. $4[2x - 6(x - y)]$
19. $(-5x^3y^2)(4x^4y^{-6})$
20. $(8x^2 - 9xy - 11y^2) - (7x^2 - 4xy + 5y^2)$
21. $(3x - 1)(2x + 5)$
22. $(3x^2 - 4y)^2$
23. $\frac{3x}{x + 5} - \frac{2}{x^2 + 7x + 10}$
24. $\dfrac{1 - \frac{9}{x^2}}{1 + \frac{3}{x}}$
25. $\frac{x^2 - 6x + 8}{3x + 9} \div \frac{x^2 - 4}{x + 3}$
26. $\sqrt{5xy} \cdot \sqrt{10x^2y}$
27. $4\sqrt{72} - 3\sqrt{50}$
28. $(5 + 3i)(7 - 3i)$

In Exercises 29–31, factor completely.

29. $81x^4 - 1$
30. $24x^3 - 22x^2 + 4x$
31. $x^3 + 27y^3$

In Exercises 32–34, let $f(x) = x^2 + 3x - 15$ and $g(x) = x - 2$. Find each indicated expression.

32. $(f - g)(x)$ and $(f - g)(5)$
33. $\left(\frac{f}{g}\right)(x)$ and the domain of $\frac{f}{g}$
34. $\frac{f(a + h) - f(a)}{h}$
35. Divide using synthetic division: $(3x^3 - x^2 + 4x + 8) \div (x + 2)$.
36. Solve for R: $I = \frac{R}{R + r}$.
37. Write the slope-intercept form of the equation of the line through $(-2, 5)$ and parallel to the line whose equation is $3x + y = 9$.
38. Evaluate the determinant: $\begin{vmatrix} -2 & -4 \\ 5 & 7 \end{vmatrix}$.
39. The price of a computer is reduced by 30% to \$434. What was the original price?
40. The area of a rectangle is 52 square yards. The length of the rectangle is 1 yard longer than 3 times its width. Find the rectangle's dimensions.
41. You invested \$4000 in two stocks paying 12% and 14% annual interest. At the end of the year, the total interest from these investments was \$508. How much was invested at each rate?
42. The current, I, in amperes, flowing in an electrical circuit varies inversely as the resistance, R, in ohms, in the circuit. When the resistance of an electric percolator is 22 ohms, it draws 5 amperes of current. How much current is needed when the resistance is 10 ohms?

CHAPTER 9

Exponential and Logarithmic Functions

Can I put aside $25,000 when I'm 20 and wind up sitting on half a million dollars by my early fifties? Will population growth lead to a future without comfort or individual choice? Why did I feel I was walking too slowly on my visit to New York City? Are Californians at greater risk from drunk drivers than from earthquakes? What is the difference between earthquakes measuring 6 and 7 on the Richter scale? Can I live forever? And what can possibly be causing merchants at our local shopping mall to grin from ear to ear as they watch the browsers?

The functions that you will be learning about in this chapter will provide you with the mathematics for answering these questions. You will see how these remarkable functions enable us to predict the future and rediscover the past.

Here's where you'll find these applications:

- You'll be sitting on $500,000 in Example 8 of Section 9.5.
- World population growth: Section 9.6, Examples 4 and 5
- Population and walking speed: Section 9.6, Check Point 3
- Alcohol and risk of a car accident: Section 9.5, Example 7
- Earthquake intensity: Section 9.3, Example 9
- Immortality: Blitzer Bonus, page 668

We open the chapter with those grinning merchants and the sound of ka-ching!

SECTION

9.1 Exponential Functions

What am I supposed to learn?

After studying this section, you should be able to:

1. Evaluate exponential functions.
2. Graph exponential functions.
3. Evaluate functions with base *e*.
4. Use compound interest formulas.

Just browsing? Take your time. Researchers know, to the dollar, the average amount the typical consumer spends per minute at the shopping mall. And the longer you stay, the more you spend. So if you say you're just browsing, that's just fine with the mall merchants. Browsing is time and, as shown in **Figure 9.1**, time is money.

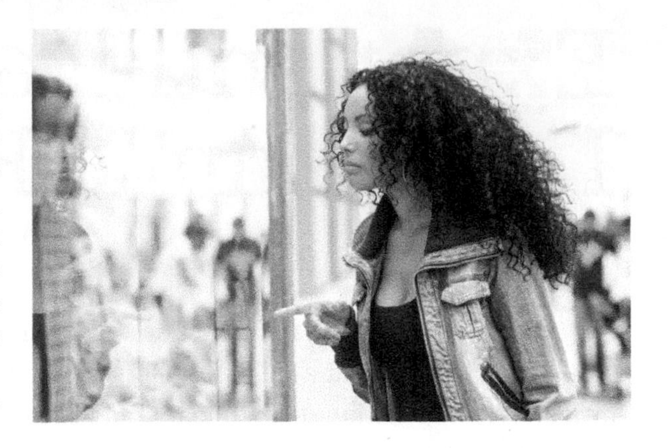

The data in **Figure 9.1** can be modeled by the function

$$f(x) = 42.2(1.56)^x,$$

where $f(x)$ is the average amount spent, in dollars, at a shopping mall after x hours. Can you see how this function is different from polynomial functions? The variable x is in the exponent. Functions whose equations contain a variable in the exponent are called **exponential functions.** Many real-life situations, including population growth, growth of epidemics, radioactive decay, and other changes that involve rapid increase or decrease, can be described using exponential functions.

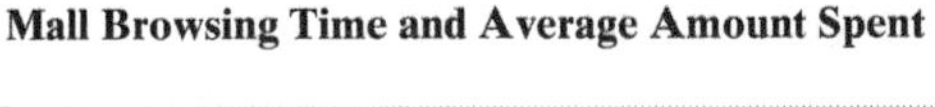

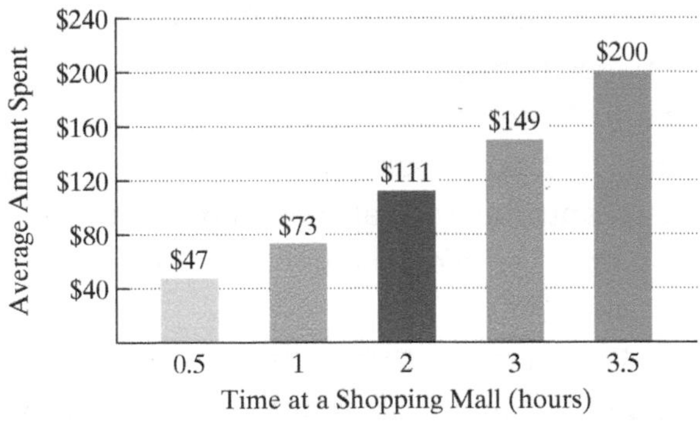

Figure 9.1

Source: International Council of Shopping Centers Research, 2006

Definition of an Exponential Function

The **exponential function *f* with base *b*** is defined by

$$f(x) = b^x \quad \text{or} \quad y = b^x,$$

where b is a positive constant other than 1 ($b > 0$ and $b \neq 1$) and x is any real number.

Here are some examples of exponential functions:

$f(x) = 2^x$ — Base is 2.

$g(x) = 10^x$ — Base is 10.

$h(x) = 3^{x+1}$ — Base is 3.

$j(x) = \left(\frac{1}{2}\right)^{x-1}$ — Base is $\frac{1}{2}$.

Each of these functions has a constant base and a variable exponent.

By contrast, the following functions are not exponential functions:

$F(x) = x^2$ — Variable is the base and not the exponent.

$G(x) = 1^x$ — The base of an exponential function must be a positive constant other than 1.

$H(x) = (-1)^x$ — The base of an exponential function must be positive.

$J(x) = x^x$. — Variable is both the base and the exponent.

Why is $G(x) = 1^x$ not classified as an exponential function? The number 1 raised to any power is 1. Thus, the function G can be written as $G(x) = 1$, which is a constant function.

Why is $H(x) = (-1)^x$ not an exponential function? The base of an exponential function must be positive to avoid having to exclude many values of x from the domain that result in nonreal numbers in the range:

$H(x) = (-1)^x$ — Not an exponential function

$H\left(\frac{1}{2}\right) = (-1)^{\frac{1}{2}} = \sqrt{-1} = i.$ — All values of x resulting in even roots of negative numbers produce nonreal numbers.

1 Evaluate exponential functions.

You will need a calculator to evaluate exponential expressions. Most scientific calculators have a y^x key. Graphing calculators have a ∧ key. To evaluate expressions of the form b^x, enter the base b, press y^x or ∧, enter the exponent x, and finally press = or ENTER.

EXAMPLE 1 Evaluating an Exponential Function

The exponential function $f(x) = 42.2(1.56)^x$ models the average amount spent, $f(x)$, in dollars, at a shopping mall after x hours. What is the average amount spent, to the nearest dollar, after four hours?

Solution Because we are interested in the amount spent after four hours, substitute 4 for x and evaluate the function.

$$f(x) = 42.2(1.56)^x \quad \text{This is the given function.}$$

$$f(4) = 42.2(1.56)^4 \quad \text{Substitute 4 for } x.$$

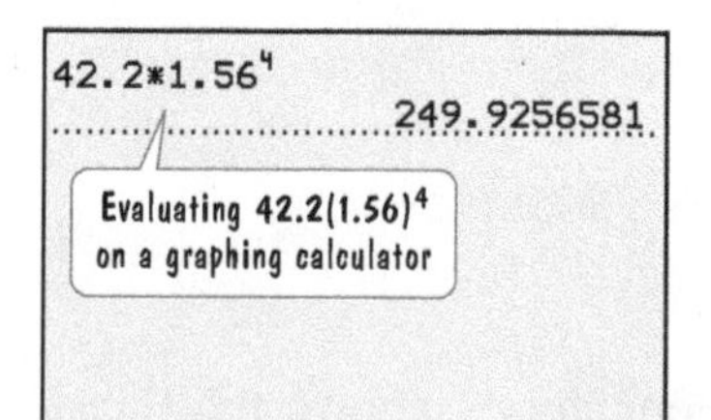

Use a scientific or graphing calculator to evaluate $f(4)$. Press the following keys on your calculator to do this:

Scientific calculator: 42.2 × 1.56 y^x 4 =

Graphing calculator: 42.2 × 1.56 ∧ 4 ENTER.

The display should be approximately 249.92566.

$$f(4) = 42.2(1.56)^4 \approx 249.92566 \approx 250$$

Thus, the average amount spent after four hours at a mall is approximately \$250. ■

✓ CHECK POINT 1 Use the exponential function in Example 1 to find the average amount spent, to the nearest dollar, after three hours at a shopping mall. Does this rounded function value underestimate or overestimate the amount shown in **Figure 9.1**? By how much?

2 Graph exponential functions.

Graphing Exponential Functions

We are familiar with expressions involving b^x where x is a rational number. For example,

$$b^{1.7} = b^{\frac{17}{10}} = \sqrt[10]{b^{17}} \text{ and } b^{1.73} = b^{\frac{173}{100}} = \sqrt[100]{b^{173}}.$$

However, note that the definition of $f(x) = b^x$ includes all real numbers for the domain x. You may wonder what b^x means when x is an irrational number, such as $b^{\sqrt{3}}$ or b^{π}. Using closer and closer approximations for $\sqrt{3}$ ($\sqrt{3} \approx 1.73205$), we can think of $b^{\sqrt{3}}$ as the value that has the successively closer approximations

$$b^{1.7}, b^{1.73}, b^{1.732}, b^{1.73205}, \ldots.$$

In this way, we can graph the exponential function with no holes, or points of discontinuity, at the irrational domain values.

EXAMPLE 2 Graphing an Exponential Function

Graph: $f(x) = 2^x$.

Solution We begin by setting up a table of coordinates.

We selected integers from −3 to 3, inclusive, to include three negative numbers, 0, and three positive numbers. We also wanted to keep the resulting computations for y relatively simple.

x	$f(x) = 2^x$
−3	$f(-3) = 2^{-3} = \frac{1}{8}$
−2	$f(-2) = 2^{-2} = \frac{1}{4}$
−1	$f(-1) = 2^{-1} = \frac{1}{2}$
0	$f(0) = 2^0 = 1$
1	$f(1) = 2^1 = 2$
2	$f(2) = 2^2 = 4$
3	$f(3) = 2^3 = 8$

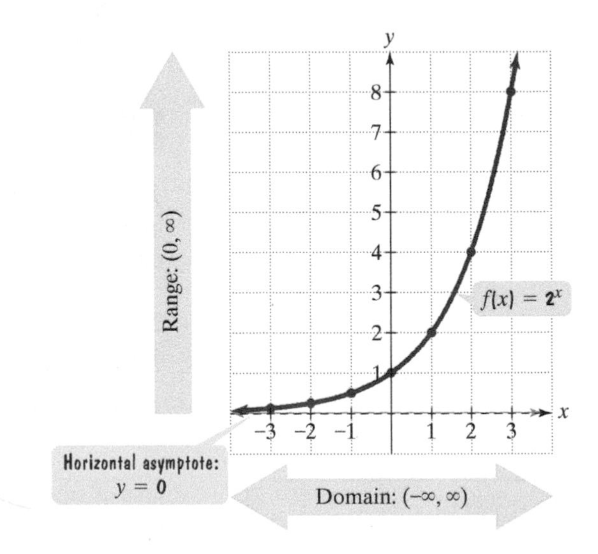

Figure 9.2 The graph of $f(x) = 2^x$

We plot these points, connecting them with a continuous curve. **Figure 9.2** shows the graph of $f(x) = 2^x$. Observe that the graph approaches, but never touches, the negative portion of the x-axis. Thus, the x-axis, or $y = 0$, is a horizontal asymptote. The range is the set of all positive real numbers. Although we used integers for x in our table of coordinates, you can use a calculator to find additional points. For example, $f(0.3) = 2^{0.3} \approx 1.231$ and $f(0.95) = 2^{0.95} \approx 1.932$. The points $(0.3, 1.231)$ and $(0.95, 1.932)$ approximately fit the graph. ■

CHECK POINT 2 Graph: $f(x) = 3^x$.

EXAMPLE 3 Graphing an Exponential Function

Graph: $g(x) = \left(\frac{1}{2}\right)^x$.

Solution We begin by setting up a table of coordinates. We compute the function values by noting that

$$g(x) = \left(\frac{1}{2}\right)^x = (2^{-1})^x = 2^{-x}.$$

x	$g(x) = \left(\frac{1}{2}\right)^x$ or 2^{-x}
-3	$g(-3) = 2^{-(-3)} = 2^3 = 8$
-2	$g(-2) = 2^{-(-2)} = 2^2 = 4$
-1	$g(-1) = 2^{-(-1)} = 2^1 = 2$
0	$g(0) = 2^{-0} = 1$
1	$g(1) = 2^{-1} = \frac{1}{2^1} = \frac{1}{2}$
2	$g(2) = 2^{-2} = \frac{1}{2^2} = \frac{1}{4}$
3	$g(3) = 2^{-3} = \frac{1}{2^3} = \frac{1}{8}$

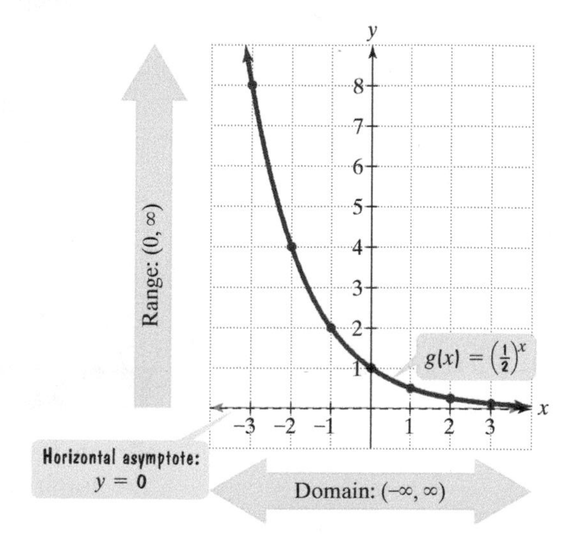

Figure 9.3 The graph of $g(x) = \left(\frac{1}{2}\right)^x$

We plot these points, connecting them with a continuous curve. **Figure 9.3** shows the graph of $g(x) = \left(\frac{1}{2}\right)^x$. This time the graph approaches, but never touches, the *positive* portion of the x-axis. Once again, the x-axis, or $y = 0$, is a horizontal asymptote. The range consists of all positive real numbers. ■

Do you notice a relationship between the graphs of $f(x) = 2^x$ and $g(x) = \left(\frac{1}{2}\right)^x$ in **Figures 9.2** and **9.3**? The graph of $g(x) = \left(\frac{1}{2}\right)^x$ is a mirror image, or reflection, of the graph of $f(x) = 2^x$ about the y-axis.

✓ **CHECK POINT 3** Graph: $f(x) = \left(\frac{1}{3}\right)^x$. Note that $f(x) = \left(\frac{1}{3}\right)^x = (3^{-1})^x = 3^{-x}$.

Four exponential functions have been graphed in **Figure 9.4**. Compare the black and green graphs, where $b > 1$, to those in blue and red, where $b < 1$. When $b > 1$, the value of y increases as the value of x increases. When $b < 1$, the value of y decreases as the value of x increases. Notice that all four graphs pass through (0, 1). These graphs illustrate general characteristics of exponential functions, listed in the box on the next page.

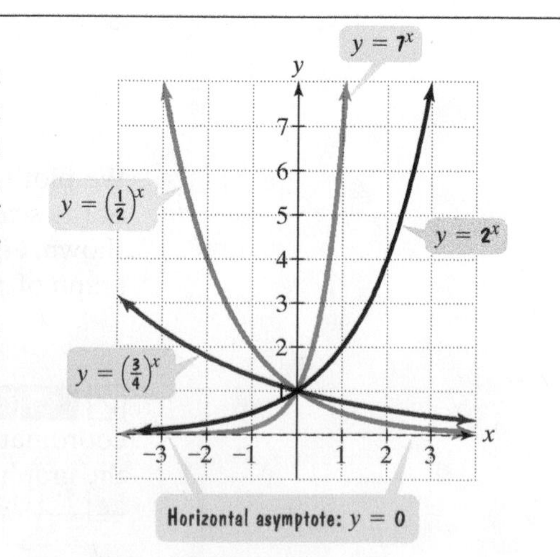

Figure 9.4 Graphs of four exponential functions

Characteristics of Exponential Functions of the Form $f(x) = b^x$

1. The domain of $f(x) = b^x$ consists of all real numbers: $(-\infty, \infty)$. The range of $f(x) = b^x$ consists of all positive real numbers: $(0, \infty)$.
2. The graphs of all exponential functions of the form $f(x) = b^x$ pass through the point $(0, 1)$ because $f(0) = b^0 = 1$ $(b \neq 0)$. The y-intercept is 1.
3. If $b > 1, f(x) = b^x$ has a graph that goes up to the right and is an increasing function. The greater the value of b, the steeper the increase.
4. If $0 < b < 1, f(x) = b^x$ has a graph that goes down to the right and is a decreasing function. The smaller the value of b, the steeper the decrease.
5. The graph of $f(x) = b^x$ approaches, but does not touch, the x-axis. The x-axis, or $y = 0$, is a horizontal asymptote.

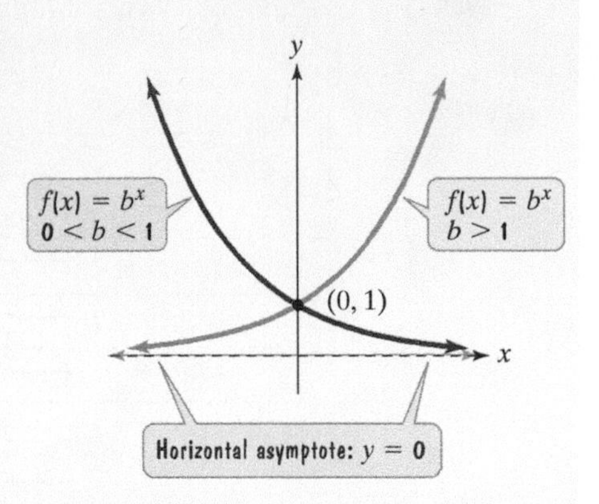

EXAMPLE 4 Graphing Exponential Functions

Graph $f(x) = 3^x$ and $g(x) = 3^{x+1}$ in the same rectangular coordinate system. How is the graph of g related to the graph of f?

Solution We begin by setting up a table showing some of the coordinates for f and g, selecting integers from -2 to 2 for x. Notice that $x + 1$ is the exponent for $g(x) = 3^{x+1}$.

x	$f(x) = 3^x$	$g(x) = 3^{x+1}$
-2	$f(-2) = 3^{-2} = \frac{1}{9}$	$g(-2) = 3^{-2+1} = 3^{-1} = \frac{1}{3}$
-1	$f(-1) = 3^{-1} = \frac{1}{3}$	$g(-1) = 3^{-1+1} = 3^0 = 1$
0	$f(0) = 3^0 = 1$	$g(0) = 3^{0+1} = 3^1 = 3$
1	$f(1) = 3^1 = 3$	$g(1) = 3^{1+1} = 3^2 = 9$
2	$f(2) = 3^2 = 9$	$g(2) = 3^{2+1} = 3^3 = 27$

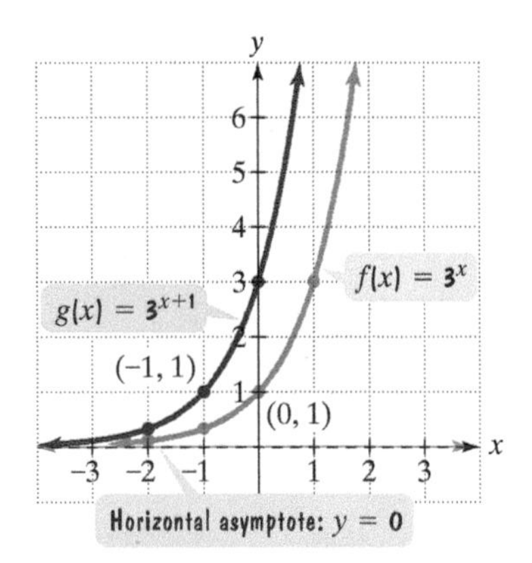

Figure 9.5

We plot the points for each function and connect them with a smooth curve. Because of the scale on the y-axis, the points on each function corresponding to $x = 2$ are not shown. **Figure 9.5** shows the graphs of $f(x) = 3^x$ and $g(x) = 3^{x+1}$. The graph of g is the graph of f shifted one unit to the left. ■

✓ CHECK POINT 4 Graph $f(x) = 3^x$ and $g(x) = 3^{x-1}$ in the same rectangular coordinate system. Select integers from -2 to 2 for x. How is the graph of g related to the graph of f?

EXAMPLE 5 Graphing Exponential Functions

Graph $f(x) = 2^x$ and $g(x) = 2^x - 3$ in the same rectangular coordinate system. How is the graph of g related to the graph of f?

Solution We begin by setting up a table showing some of the coordinates for f and g, selecting integers from -2 to 2 for x.

x	$f(x) = 2^x$	$g(x) = 2^x - 3$
-2	$f(-2) = 2^{-2} = \frac{1}{4}$	$g(-2) = 2^{-2} - 3 = \frac{1}{4} - 3 = -2\frac{3}{4}$
-1	$f(-1) = 2^{-1} = \frac{1}{2}$	$g(-1) = 2^{-1} - 3 = \frac{1}{2} - 3 = -2\frac{1}{2}$
0	$f(0) = 2^0 = 1$	$g(0) = 2^0 - 3 = 1 - 3 = -2$
1	$f(1) = 2^1 = 2$	$g(1) = 2^1 - 3 = 2 - 3 = -1$
2	$f(2) = 2^2 = 4$	$g(2) = 2^2 - 3 = 4 - 3 = 1$

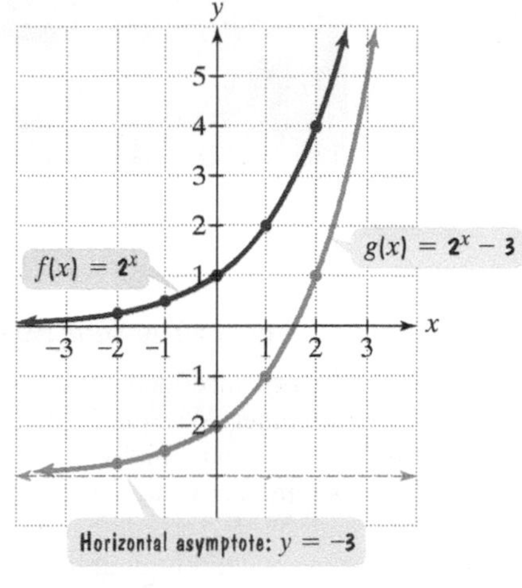

Figure 9.6

We plot the points for each function and connect them with a smooth curve. **Figure 9.6** shows the graphs of $f(x) = 2^x$ and $g(x) = 2^x - 3$. The graph of g is the graph of f shifted down three units. As a result, $y = -3$ is the horizontal asymptote for g. ■

✓ CHECK POINT 5 Graph $f(x) = 2^x$ and $g(x) = 2^x + 3$ in the same rectangular coordinate system. Select integers from -2 to 2 for x. How is the graph of g related to the graph of f?

3 Evaluate functions with base e.

The Natural Base e

An irrational number, symbolized by the letter e, appears as the base in many applied exponential functions. The number e is defined as the value that $\left(1 + \frac{1}{n}\right)^n$ approaches as n gets larger and larger. **Table 9.1** shows values of $\left(1 + \frac{1}{n}\right)^n$ for increasingly large values of n. As n increases, the approximate value of e to nine decimal places is

$$e \approx 2.718281827.$$

The irrational number e, approximately 2.72, is called the **natural base**. The function $f(x) = e^x$ is called the **natural exponential function.**

Table 9.1

n	$\left(1 + \frac{1}{n}\right)^n$
1	2
2	2.25
5	2.48832
10	2.59374246
100	2.704813829
1000	2.716923932
10,000	2.718145927
100,000	2.718268237
1,000,000	2.718280469
1,000,000,000	2.718281827

As n takes on increasingly large values, the expression $\left(1 + \frac{1}{n}\right)^n$ approaches e.

Using Technology

Graphic Connections

As n increases, the graph of $y = \left(1 + \frac{1}{n}\right)^n$ approaches the graph of $y = e$.

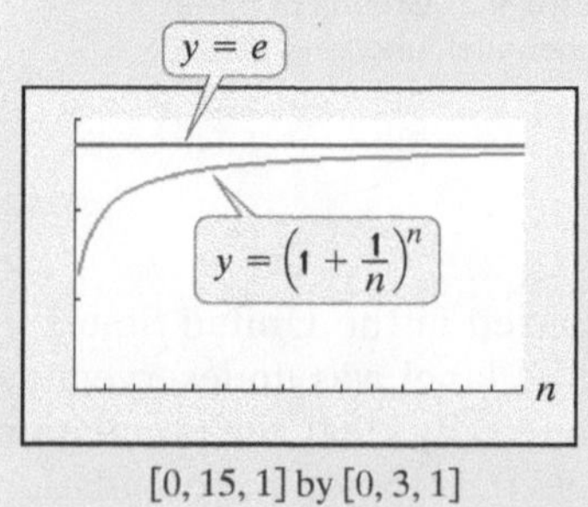

[0, 15, 1] by [0, 3, 1]

Blitzer Bonus

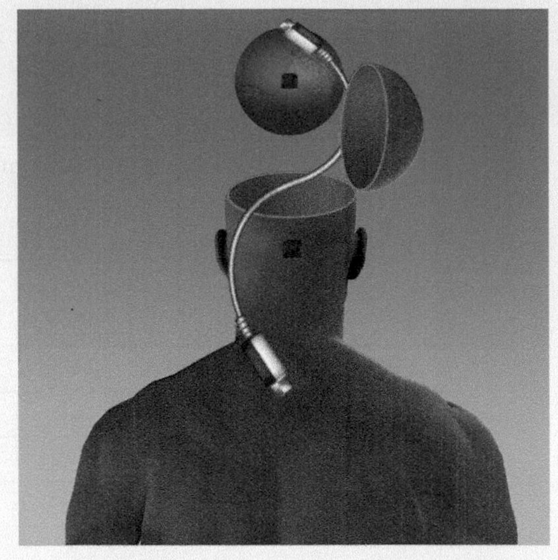

Exponential Growth: The Year Humans Become Immortal

In 2011, *Jeopardy*! aired a three-night match between a personable computer named Watson and the show's two most successful players. The winner: Watson. In the time it took each human contestant to respond to one trivia question, Watson was able to scan the content of one million books. It was also trained to understand the puns and twists of phrases unique to *Jeopardy*! clues.

Watson's remarkable accomplishments can be thought of as a single data point on an exponential curve that models growth in computing power. According to inventor, author, and computer scientist Ray Kurzweil (1948–), computer technology is progressing exponentially, doubling in power each year. What does this mean in terms of the accelerating pace of the graph of $y = 2^x$ that starts slowly and then rockets skyward toward infinity? According to Kurzweil, by 2023, a supercomputer will surpass the brainpower of a human. As progress accelerates exponentially and every hour brings a century's worth of scientific breakthroughs, by 2045, computers will surpass the brainpower equivalent to that of all human brains combined. Here's where it gets exponentially weird: In that year (says Kurzweil), we will be able to scan our consciousness into computers and enter a virtual existence, or swap our bodies for immortal robots. Indefinite life extension will become a reality and people will die only if they choose to.

Use a scientific or graphing calculator with an $\boxed{e^x}$ key to evaluate e to various powers. For example, to find e^2, press the following keys on most calculators:

Scientific calculator: $2\ \boxed{e^x}$

Graphing calculator: $\boxed{e^x}\ 2\ \boxed{\text{ENTER}}$.

The display should be approximately 7.389.

$$e^2 \approx 7.389$$

The number e lies between 2 and 3. Because $2^2 = 4$ and $3^2 = 9$, it makes sense that e^2, approximately 7.389, lies between 4 and 9.

Because $2 < e < 3$, the graph of $y = e^x$ is between the graphs of $y = 2^x$ and $y = 3^x$, shown in **Figure 9.7**.

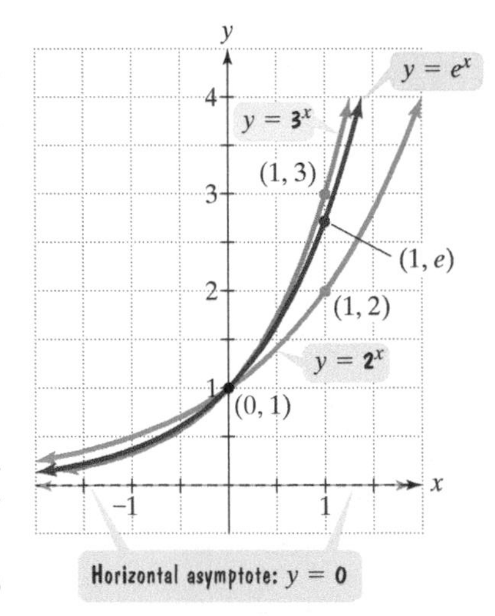

Figure 9.7 Graphs of three exponential functions

EXAMPLE 6 Gray Wolf Population

Insatiable killer. That's the reputation the gray wolf acquired in the United States in the nineteenth and early twentieth centuries. Although the label was undeserved, an estimated two million wolves were shot, trapped, or poisoned. By 1960, the population was reduced to 800 wolves. **Figure 9.8** on the next page shows the rebounding population in two recovery areas after the gray wolf was declared an endangered species and received federal protection.

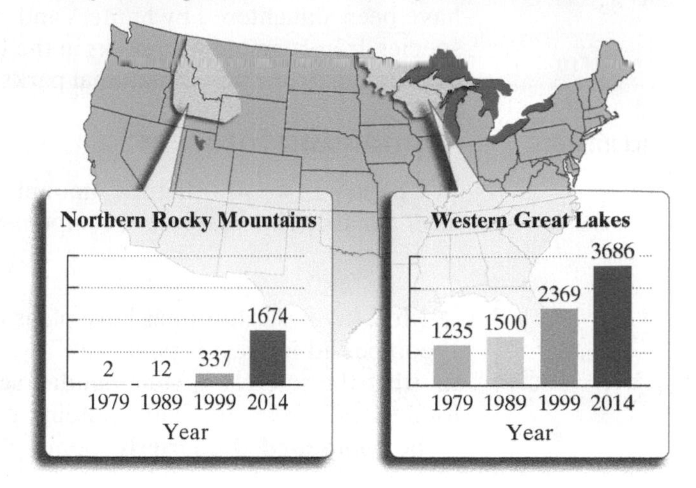

Figure 9.8

Source: U.S. Fish and Wildlife Service

The exponential function

$$f(x) = 1145e^{0.0325x}$$

models the gray wolf population of the Western Great Lakes, $f(x)$, x years after 1978.

a. According to the model, what was the gray wolf population, rounded to the nearest whole number, of the Western Great Lakes in 2014?

b. Does the model underestimate or overestimate the gray wolf population of the Western Great Lakes in 2014? By how much?

Solution

a. Because 2014 is 36 years after 1978 ($2014 - 1978 = 36$), we substitute 36 for x in the given function.

$$f(x) = 1145e^{0.0325x} \quad \text{This is the given function.}$$

$$f(36) = 1145e^{0.0325(36)} \quad \text{Substitute 36 fox } x.$$

Perform this computation on your calculator.

Scientific calculator: 1145 [×] [(] .0325 [×] 36 [)] e^x [=]

Graphing calculator: 1145 [×] [e^x] [(] .0325 [×] 36 [)] [ENTER]

This parenthesis is given on some calculators.

The display should be approximately 3689.181571. Thus,

$$f(36) = 1145e^{0.0325(36)} \approx 3689.$$

According to the model, the gray wolf population of the Western Great Lakes in 2014 was approximately 3689 wolves.

b. **Figure 9.8** shows that the gray wolf population of the Western Great Lakes in 2014 was 3686 wolves. The model in part (a) indicates a population of 3689 wolves. Thus, the model overestimates the population by 3689 – 3686, or by 3 wolves. ■

✓ **CHECK POINT 6** Use $f(x) = 1145e^{0.0325x}$, the model described in Example 6, to project the gray wolf population of the Western Great Lakes, rounded to the nearest whole number, in 2017.

Using exponential functions and projections like those in Example 6 and Check Point 6, in 2011, the U.S. Fish and Wildlife Service removed the Northern Rocky

Mountain gray wolf from the endangered species list. Since then, hundreds of wolves have been slaughtered by hunters and trappers. Without protection as an endangered species, free-roaming wolf packs in the United States will be lucky to survive anywhere outside the protection of national parks.

4 Use compound interest formulas.

Compound Interest

In Chapter 8, we saw that the amount of money, A, that a principal, P, will be worth after t years at interest rate r, compounded annually, is given by the formula

$$A = P(1 + r)^t.$$

Most savings institutions have plans in which interest is paid more than once a year. If compound interest is paid twice a year, the compounding period is six months. We say that the interest is **compounded semiannually**. When compound interest is paid four times a year, the compounding period is three months and the interest is said to be **compounded quarterly**. Some plans allow for monthly compounding or daily compounding.

In general, when compound interest is paid n times a year, we say that there are ***n* compounding periods per year**. The formula $A = P(1 + r)^t$ can be adjusted to take into account the number of compounding periods in a year. If there are n compounding periods per year, the formula becomes

$$A = P\left(1 + \frac{r}{n}\right)^{nt}.$$

Some banks use **continuous compounding**, where the number of compounding periods increases infinitely (compounding interest every trillionth of a second, every quadrillionth of a second, etc.). As n, the number of compounding periods in a year, increases without bound, the expression $\left(1 + \frac{1}{n}\right)^n$ approaches e. As a result, the formula for continuous compounding is $A = Pe^{rt}$. Although continuous compounding sounds terrific, it yields only a fraction of a percent more interest over a year than daily compounding.

Formulas for Compound Interest

After t years, the balance, A, in an account with principal P and annual interest rate r (in decimal form) is given by the following formulas:

1. For n compounding periods per year: $A = P\left(1 + \frac{r}{n}\right)^{nt}$
2. For continuous compounding: $A = Pe^{rt}$.

EXAMPLE 7 Choosing Between Investments

You decide to invest \$8000 for 6 years and you have a choice between two accounts. The first pays 7% per year, compounded monthly. The second pays 6.85% per year, compounded continuously. Which is the better investment?

Solution The better investment is the one with the greater balance in the account after 6 years. Let's begin with the account with monthly compounding. We use the compound interest model with $P = 8000$, $r = 7\% = 0.07$, $n = 12$ (monthly compounding means 12 compounding periods per year), and $t = 6$.

$$A = P\left(1 + \frac{r}{n}\right)^{nt} = 8000\left(1 + \frac{0.07}{12}\right)^{12 \cdot 6} \approx 12{,}160.84$$

The balance in this account after 6 years is \$12,160.84.

For the second investment option, we use the model for continuous compounding with $P = 8000$, $r = 6.85\% = 0.0685$, and $t = 6$.

$$A = Pe^{rt} = 8000e^{0.0685(6)} \approx 12{,}066.60$$

The balance in this account after 6 years is \$12,066.60, slightly less than the previous amount. Thus, the better investment is the 7% monthly compounding option. ■

✓ **CHECK POINT 7** A sum of \$10,000 is invested at an annual rate of 8%. Find the balance in the account after 5 years subject to **a.** quarterly compounding and **b.** continuous compounding.

CONCEPT AND VOCABULARY CHECK

Fill in each blank so that the resulting statement is true.

1. The exponential function f with base b is defined by $f(x) =$ ______________, $b > 0$ and $b \neq 1$. Using interval notation, the domain of this function is ______________ and the range is ______________.
2. The graph of the exponential function f with base b approaches, but does not touch, the ______________-axis. This axis, whose equation is ______________, is a/an ______________ asymptote.
3. The value that $\left(1 + \frac{1}{n}\right)^n$ approaches as n gets larger and larger is the irrational number ______________, called the ______________ base. This irrational number is approximately equal to ______________.
4. Consider the compound interest formula
$$A = P\left(1 + \frac{r}{n}\right)^{nt}.$$
This formula gives the balance, ______________, in an account with principal ______________ and annual interest rate ______________, in decimal form, subject to compound interest paid ______________ times per year.
5. If compound interest is paid twice a year, we say that the interest is compounded ______________. If compound interest is paid four times a year, we say that the interest is compounded ______________. If the number of compounding periods increases infinitely, we call this ______________ compounding.

9.1 EXERCISE SET

MyMathLab®

Practice Exercises

In Exercises 1–10, approximate each number using a calculator. Round your answer to three decimal places.

1. $2^{3.4}$
2. $3^{2.4}$
3. $3^{\sqrt{5}}$
4. $5^{\sqrt{3}}$
5. $4^{-1.5}$
6. $6^{-1.2}$
7. $e^{2.3}$
8. $e^{3.4}$
9. $e^{-0.95}$
10. $e^{-0.75}$

In Exercises 11–16, set up a table of coordinates for each function. Select integers from −2 to 2, inclusive, for x. Then use the table of coordinates to match the function with its graph. [The graphs are on the next page and labeled (a) through (f).]

11. $f(x) = 3^x$
12. $f(x) = 3^{x-1}$
13. $f(x) = 3^x - 1$
14. $f(x) = -3^x$

15. $f(x) = 3^{-x}$

16. $f(x) = -3^{-x}$

a.

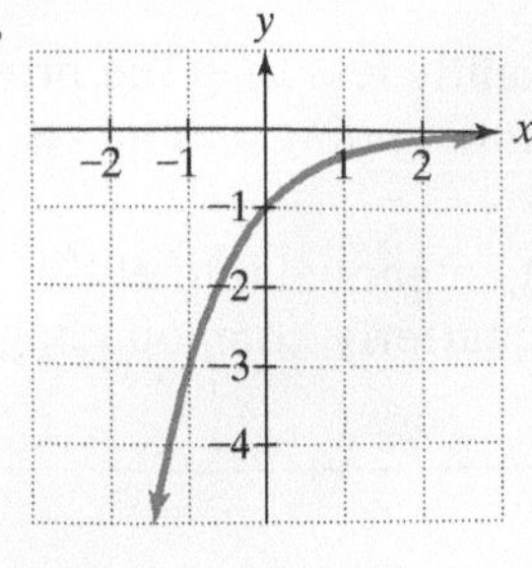

b.

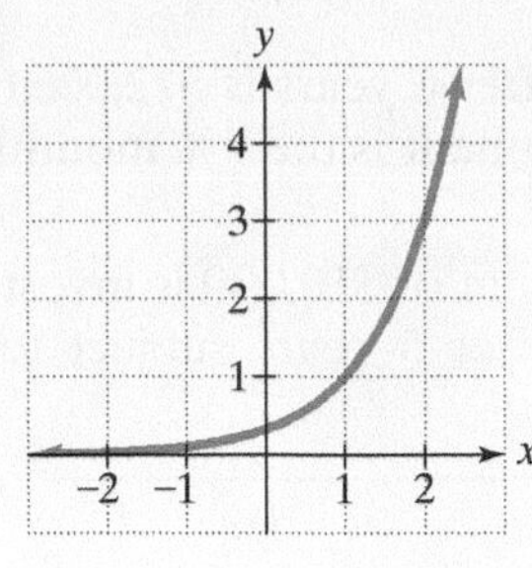

c.

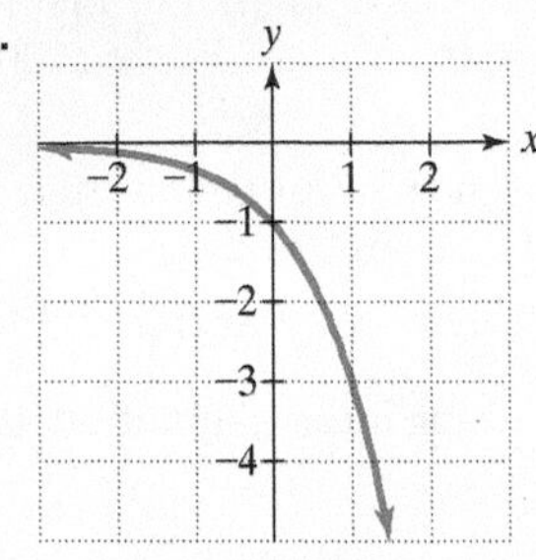

d.

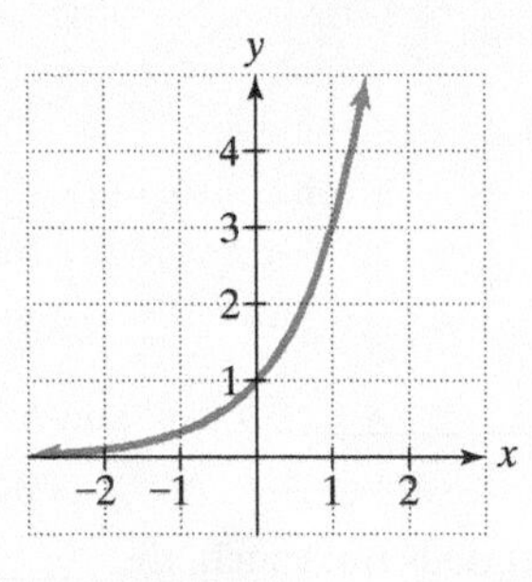

e.

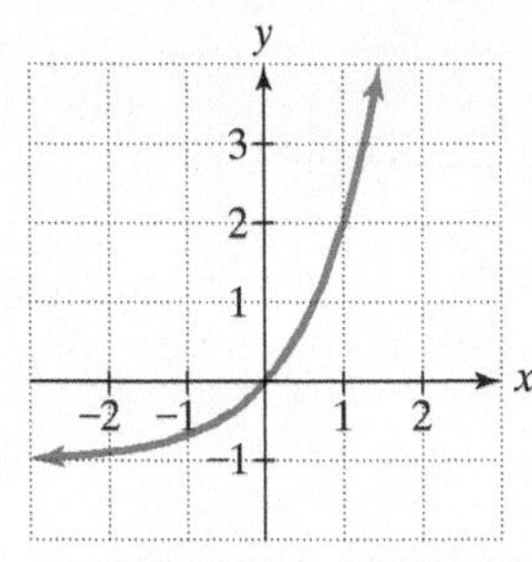

f.

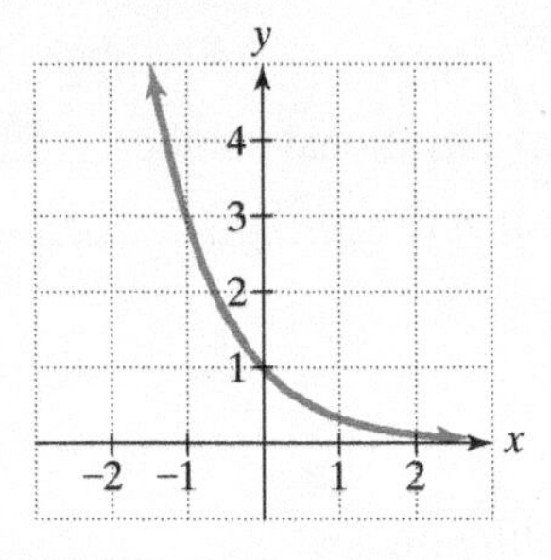

In Exercises 17–24, graph each function by making a table of coordinates. If applicable, use a graphing utility to confirm your hand-drawn graph.

17. $f(x) = 4^x$

18. $f(x) = 5^x$

19. $g(x) = \left(\frac{3}{2}\right)^x$

20. $g(x) = \left(\frac{4}{3}\right)^x$

21. $h(x) = \left(\frac{1}{2}\right)^x$

22. $h(x) = \left(\frac{1}{3}\right)^x$

23. $f(x) = (0.6)^x$

24. $f(x) = (0.8)^x$

In Exercises 25–38, graph functions f and g in the same rectangular coordinate system. Select integers from −2 to 2, inclusive, for x. Then describe how the graph of g is related to the graph of f. If applicable, use a graphing utility to confirm your hand-drawn graphs.

25. $f(x) = 2^x$ and $g(x) = 2^{x+1}$

26. $f(x) = 2^x$ and $g(x) = 2^{x+2}$

27. $f(x) = 2^x$ and $g(x) = 2^{x-2}$

28. $f(x) = 2^x$ and $g(x) = 2^{x-1}$

29. $f(x) = 2^x$ and $g(x) = 2^x + 1$

30. $f(x) = 2^x$ and $g(x) = 2^x + 2$

31. $f(x) = 2^x$ and $g(x) = 2^x - 2$

32. $f(x) = 2^x$ and $g(x) = 2^x - 1$

33. $f(x) = 3^x$ and $g(x) = -3^x$

34. $f(x) = 3^x$ and $g(x) = 3^{-x}$

35. $f(x) = 2^x$ and $g(x) = 2^{x+1} - 1$

36. $f(x) = 2^x$ and $g(x) = 2^{x+1} - 2$

37. $f(x) = 3^x$ and $g(x) = \frac{1}{3} \cdot 3^x$

38. $f(x) = 3^x$ and $g(x) = 3 \cdot 3^x$

Use the compound interest formulas, $A = P\left(1 + \frac{r}{n}\right)^{nt}$ and $A = Pe^{rt}$, to solve Exercises 39–42. Round answers to the nearest cent.

39. Find the accumulated value of an investment of $10,000 for 5 years at an interest rate of 5.5% if the money is **a.** compounded semiannually; **b.** compounded monthly; **c.** compounded continuously.

40. Find the accumulated value of an investment of $5000 for 10 years at an interest rate of 6.5% if the money is **a.** compounded semiannually; **b.** compounded monthly; **c.** compounded continuously.

41. Suppose that you have $12,000 to invest. Which investment yields the greater return over 3 years: 7% compounded monthly or 6.85% compounded continuously?

42. Suppose that you have $6000 to invest. Which investment yields the greater return over 4 years: 8.25% compounded quarterly or 8.3% compounded semiannually?

Practice PLUS

In Exercises 43–48, use each exponential function's graph to determine the function's domain and range.

43.

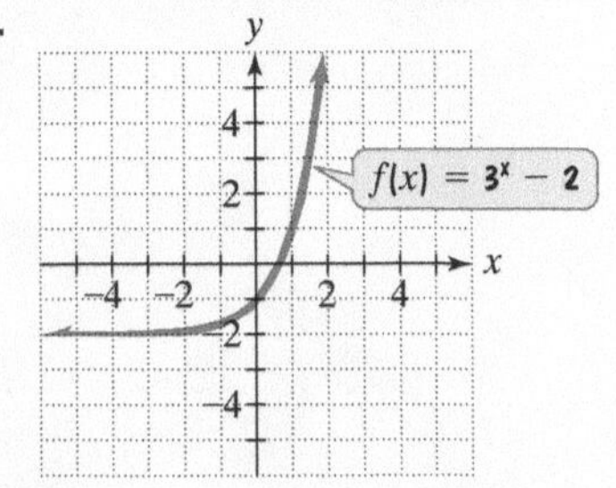

44.

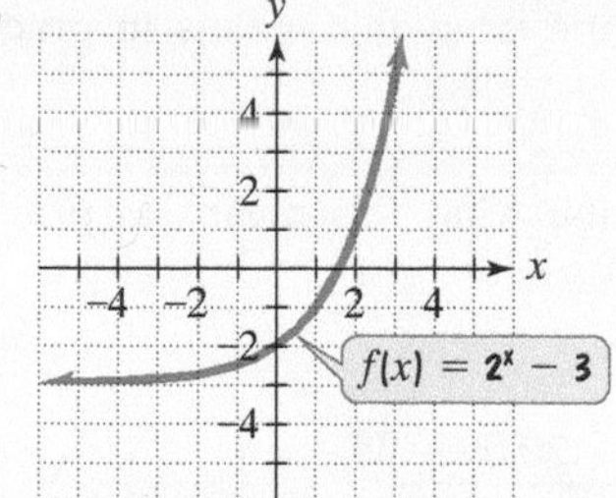

45.

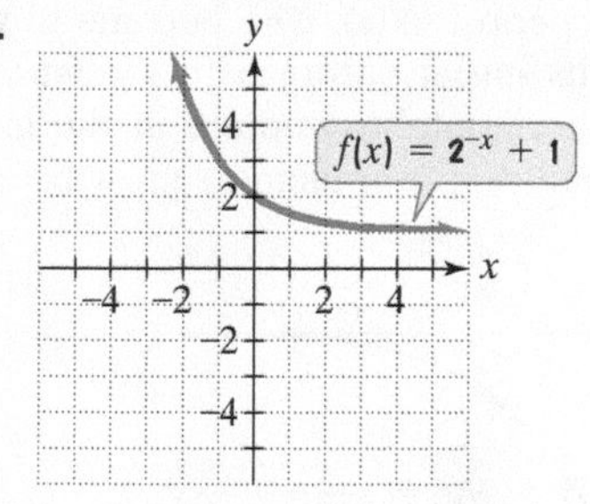

46.

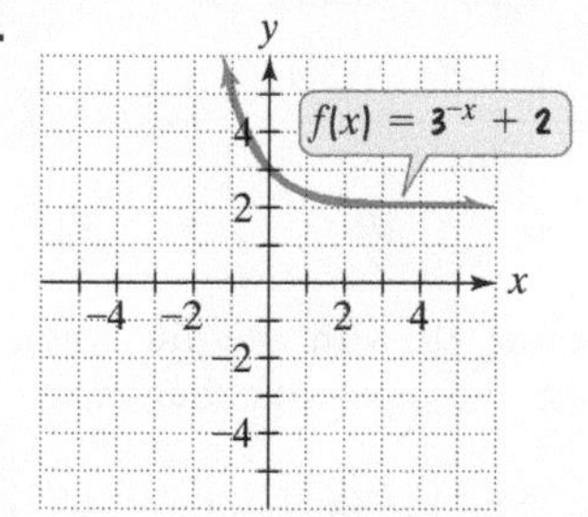

47.

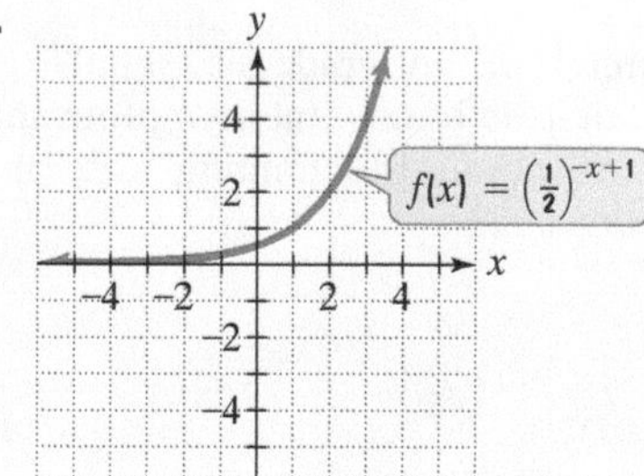

48.

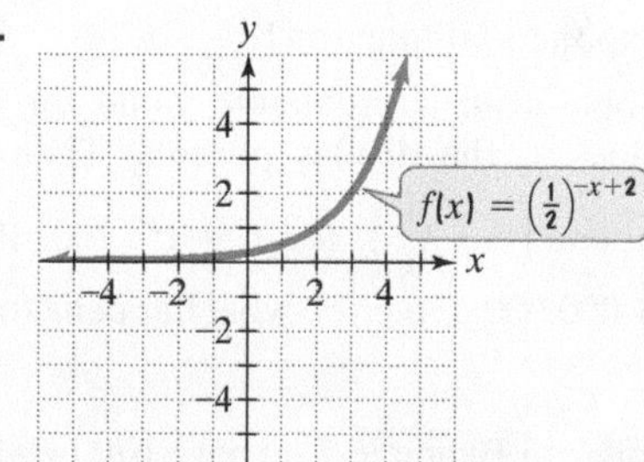

In Exercises 49–50, graph f and g in the same rectangular coordinate system. Then find the point of intersection of the two graphs.

49. $f(x) = 2^x, g(x) = 2^{-x}$

50. $f(x) = 2^{x+1}, g(x) = 2^{-x+1}$

51. Graph $y = 2^x$ and $x = 2^y$ in the same rectangular coordinate system.

52. Graph $y = 3^x$ and $x = 3^y$ in the same rectangular coordinate system.

Application Exercises

Use a calculator with a $\boxed{y^x}$ *key or a* $\boxed{\wedge}$ *key to solve Exercises 53–56.*

53. India is currently one of the world's fastest-growing countries. By 2040, the population of India will be larger than the population of China; by 2050, nearly one-third of the world's population will live in these two countries alone. The exponential function $f(x) = 574(1.026)^x$ models the population of India, $f(x)$, in millions, x years after 1974.

a. Substitute 0 for x and, without using a calculator, find India's population in 1974.

b. Substitute 27 for x and use your calculator to find India's population, to the nearest million, in the year 2001 as modeled by this function.

c. Find India's population, to the nearest million, in the year 2028 as predicted by this function.

d. Find India's population, to the nearest million, in the year 2055 as predicted by this function.

e. What appears to be happening to India's population every 27 years?

54. The 1986 explosion at the Chernobyl nuclear power plant in the former Soviet Union sent about 1000 kilograms of radioactive cesium-137 into the atmosphere. The function $f(x) = 1000(0.5)^{\frac{x}{30}}$ describes the amount, $f(x)$, in kilograms, of cesium-137 remaining in Chernobyl x years after 1986. If even 100 kilograms of cesium-137 remain in Chernobyl's atmosphere, the area is considered unsafe for human habitation. Find $f(80)$ and determine if Chernobyl will be safe for human habitation by 2066.

The formula $S = C(1 + r)^t$ models inflation, where $C =$ the value today, $r =$ the annual inflation rate, and $S =$ the inflated value t years from now. Use this formula to solve Exercises 55–56. Round answers to the nearest dollar.

55. If the inflation rate is 6%, how much will a house now worth \$465,000 be worth in 10 years?

56. If the inflation rate is 3%, how much will a house now worth \$510,000 be worth in 5 years?

Use a calculator with an $\boxed{e^x}$ *key to solve Exercises 57–63.*

The bar graph shows the percentage of U.S. high school seniors who applied to more than three colleges for selected years from 1980 through 2013.

Percentage of U.S. High School Seniors Applying to More Than Three Colleges

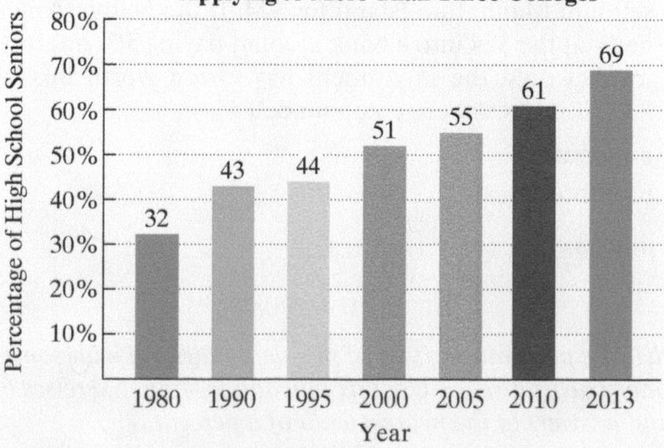

Source: The Higher Education Research Institute

The data in the bar graph at the bottom of the previous page can be modeled by

$$f(x) = x + 31 \quad \text{and} \quad g(x) = 32.7e^{0.0217x},$$

in which f(x) and g(x) represent the percentage of high school seniors who applied to more than three colleges x years after 1980. Use these functions to solve Exercises 57–58. Where necessary, round answers to the nearest percent.

57. **a.** According to the linear model, what percentage of high school seniors applied to more than three colleges in 2013?

b. According to the exponential model, what percentage of high school seniors applied to more than three colleges in 2013?

c. Which function is a better model for the data shown by the bar graph in 2013?

58. **a.** According to the linear model, what percentage of high school seniors applied to more than three colleges in 2010?

b. According to the exponential model, what percentage of high school seniors applied to more than three colleges in 2010?

c. Which function is a better model for the data shown by the bar graph in 2010?

59. In college, we study large volumes of information—information that, unfortunately, we do not often retain for very long. The function

$$f(x) = 80e^{-0.5x} + 20$$

describes the percentage of information, $f(x)$, that a particular person remembers x weeks after learning the information.

a. Substitute 0 for x and, without using a calculator, find the percentage of information remembered at the moment it is first learned.

b. Substitute 1 for x and find the percentage of information that is remembered after 1 week.

c. Find the percentage of information that is remembered after 4 weeks.

d. Find the percentage of information that is remembered after one year (52 weeks).

60. In 1626, Peter Minuit persuaded the Wappinger Indians to sell him Manhattan Island for \$24. If the Native Americans had put the \$24 into a bank account paying 5% interest, how much would the investment have been worth in the year 2005 if interest were compounded

a. monthly?

b. continuously?

The function

$$f(x) = \frac{90}{1 + 270e^{-0.122x}}$$

models the percentage, f(x), of people x years old with some coronary heart disease. Use this function to solve Exercises 61–62. Round answers to the nearest tenth of a percent.

61. Evaluate $f(30)$ and describe what this means in practical terms.

62. Evaluate $f(70)$ and describe what this means in practical terms.

63. The function

$$N(t) = \frac{30{,}000}{1 + 20e^{-1.5t}}$$

describes the number of people, $N(t)$, who become ill with influenza t weeks after its initial outbreak in a town with 30,000 inhabitants. The horizontal asymptote in the graph indicates that there is a limit to the epidemic's growth.

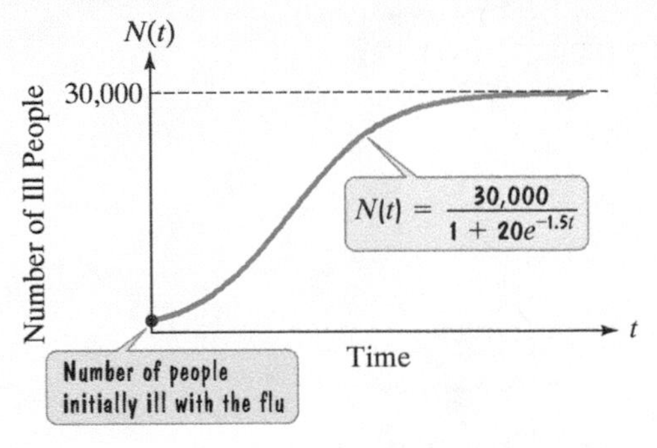

a. How many people became ill with the flu when the epidemic began? (When the epidemic began, $t = 0$.)

b. How many people were ill by the end of the third week?

c. Why can't the spread of an epidemic simply grow indefinitely? What does the horizontal asymptote shown in the graph indicate about the limiting size of the population that becomes ill?

Explaining the Concepts

64. What is an exponential function?

65. What is the natural exponential function?

66. Use a calculator to obtain an approximate value for e to as many decimal places as the display permits. Then use the calculator to evaluate $\left(1 + \frac{1}{x}\right)^x$ for $x = 10, 100, 1000,$ 10,000, 100,000, and 1,000,000. Describe what happens to the expression as x increases.

67. Write an example similar to Example 7 on page 670 in which continuous compounding at a slightly lower yearly interest rate is a better investment than compounding n times per year.

68. Describe how you could use the graph of $f(x) = 2^x$ to obtain a decimal approximation for $\sqrt{2}$.

Technology Exercises

69. You have \$10,000 to invest. One bank pays 5% interest compounded quarterly and the other pays 4.5% interest compounded monthly.

a. Use the formula for compound interest to write a function for the balance in each account at any time t in years.

b. Use a graphing utility to graph both functions in an appropriate viewing rectangle. Based on the graphs, which bank offers the better return on your money?

70. a. Graph $y = e^x$ and $y = 1 + x + \frac{x^2}{2}$ in the same viewing rectangle.

b. Graph $y = e^x$ and $y = 1 + x + \frac{x^2}{2} + \frac{x^3}{6}$ in the same viewing rectangle.

c. Graph $y = e^x$ and $y = 1 + x + \frac{x^2}{2} + \frac{x^3}{6} + \frac{x^4}{24}$ in the same viewing rectangle.

d. Describe what you observe in parts (a)–(c). Try generalizing this observation.

Critical Thinking Exercises

Make Sense? *In Exercises 71–74, determine whether each statement makes sense or does not make sense, and explain your reasoning.*

71. My graph of $f(x) = 3 \cdot 2^x$ shows that the horizontal asymptote for f is $x = 3$.

72. I'm using a photocopier to reduce an image over and over by 50%, so the exponential function $f(x) = \left(\frac{1}{2}\right)^x$ models the new image size, where x is the number of reductions.

73. Taxing thoughts: I'm looking at data that show the number of pages in the publication that explains the U.S. tax code for selected years from 1945 through 2013. A linear function appears to be a better choice than an exponential function for modeling the number of pages in the tax code during this period.

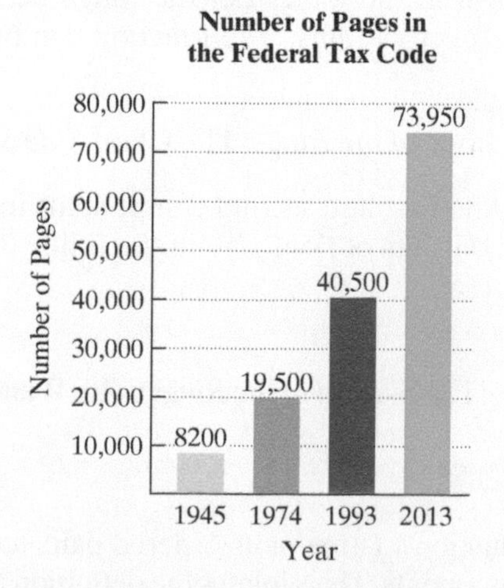

Source: CCH Inc.

74. I use the natural base e when determining how much money I'd have in a bank account that earns compound interest subject to continuous compounding.

In Exercises 75–78, determine whether each statement is true or false. If the statement is false, make the necessary change(s) to produce a true statement.

75. As the number of compounding periods increases on a fixed investment, the amount of money in the account over a fixed interval of time will increase without bound.

76. The functions $f(x) = 3^{-x}$ and $g(x) = -3^x$ have the same graph.

77. If $f(x) = 2^x$, then $f(a + b) = f(a) + f(b)$.

78. The functions $f(x) = \left(\frac{1}{3}\right)^x$ and $g(x) = 3^{-x}$ have the same graph.

79. The graphs labeled (a)–(d) in the figure represent $y = 3^x$, $y = 5^x$, $y = \left(\frac{1}{3}\right)^x$, and $y = \left(\frac{1}{5}\right)^x$, but not necessarily in that order. Which is which? Describe the process that enables you to make this decision.

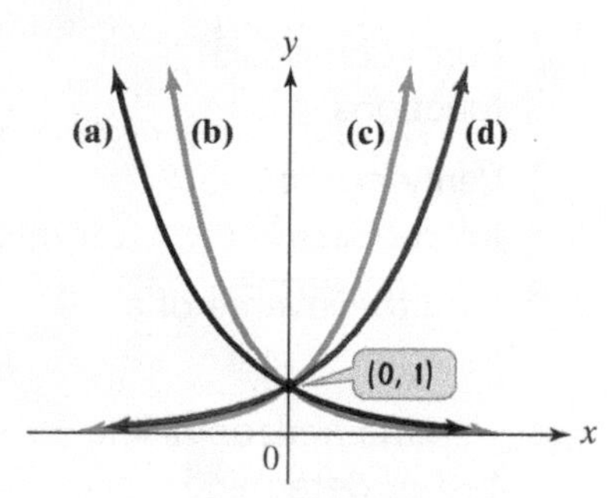

80. The hyperbolic cosine and hyperbolic sine functions are defined by

$$\cosh x = \frac{e^x + e^{-x}}{2} \quad \text{and} \quad \sinh x = \frac{e^x - e^{-x}}{2}.$$

Prove that $(\cosh x)^2 - (\sinh x)^2 = 1$.

Review Exercises

81. Solve for b: $D = \frac{ab}{a + b}$.

(Section 6.7, Example 1)

82. Evaluate: $\begin{vmatrix} 3 & -2 \\ 7 & -5 \end{vmatrix}$.

(Section 3.5, Example 1)

83. Solve: $x(x - 3) = 10$.

(Section 5.7, Example 2)

Preview Exercises

Exercises 84–86 will help you prepare for the material covered in the next section.

84. Let $f(x) = 3x - 4$ and $g(x) = x^2 + 6$.

a. Find $f(5)$.

b. Find $g(f(5))$.

85. Simplify: $3\left(\frac{x - 2}{3}\right) + 2$.

86. Solve for y: $x = 7y - 5$.

SECTION

9.2 Composite and Inverse Functions

What am I supposed to learn?

After studying this section, you should be able to:

1. Form composite functions.
2. Verify inverse functions.
3. Find the inverse of a function.
4. Use the horizontal line test to determine if a function has an inverse function.
5. Use the graph of a one-to-one function to graph its inverse function.

Based on Shakespeare's *Romeo and Juliet*, the film *West Side Story* swept the 1961 Academy Awards with ten Oscars. The top four movies to win the most Oscars are shown in **Table 9.2**.

Table 9.2 Films Winning the Most Oscars

Movie	Year	Number of Academy Awards
Ben-Hur	1960	11
Titanic	1998	11
The Lord of the Rings: The Return of the King	2003	11
West Side Story	1961	10

Source: Russell Ash, *The Top 10 of Everything, 2011*

We can use the information in **Table 9.2** to define a function. Let the domain of the function be the set of four movies shown in the table. Let the range be the number of Academy Awards for each of the respective films. The function can be written as follows:

f: {(*Ben-Hur*, 11), (*Titanic*, 11), (*The Lord of the Rings*, 11), (*West Side Story*, 10)}.

Now let's "undo" f by interchanging the first and second components in each of the ordered pairs. Switching the inputs and outputs of f, we obtain the following relation:

Same first component

Undoing f: {(11, *Ben-Hur*), (11, *Titanic*), (11, *The Lord of the Rings*), (10, *West Side Story*)}.

Different second components

Can you see that this relation is not a function? Three of its ordered pairs have the same first component and different second components. This violates the definition of a function.

If a function f is a set of ordered pairs, (x, y), then the changes produced by f can be "undone" by reversing the components of all the ordered pairs. The resulting relation, (y, x), may or may not be a function. In the next section, we will reverse the components of the ordered pairs of exponential functions. To understand what occurs when we interchange components, we turn to the topics of *composite* and *inverse functions*.

1 Form composite functions.

Composite Functions

In Chapter 2, we saw that functions could be combined using addition, subtraction, multiplication, and division. Now let's consider another way of combining two functions. To help understand this new combination, suppose that your local computer store is having a sale. The models that are on sale cost either \$300 less than the regular price or 85% of the regular price. If x represents the computer's regular price, the discounts can be described with the following functions:

$$f(x) = x - 300 \qquad g(x) = 0.85x.$$

The computer is on sale for \$300 less than its regular price.

The computer is on sale for 85% of its regular price.

At the store, you bargain with the salesperson. Eventually, she makes an offer you can't refuse. The sale price will be 85% of the regular price followed by a \$300 reduction:

$$0.85x - 300.$$

85% of the regular price

followed by a \$300 reduction

In terms of functions f and g, this offer can be obtained by taking the output of $g(x) = 0.85x$, namely $0.85x$, and using it as the input of f:

$$f(x) = x - 300$$

Replace x with $0.85x$, the output of $g(x) = 0.85x$.

$$f(0.85x) = 0.85x - 300.$$

Because $0.85x$ is $g(x)$, we can write this last equation as

$$f(g(x)) = 0.85x - 300.$$

We read this equation as "f of g of x is equal to $0.85x - 300$." We call $f(g(x))$ the **composition of the function f with g**, or a **composite function**. This composite function is written $f \circ g$. Thus,

$$(f \circ g)(x) = f(g(x)) = 0.85x - 300.$$

This can be read "f of g of x" or "f composed with g of x."

Like all functions, we can evaluate $f \circ g$ for a specified value of x in the function's domain. For example, here's how to find the value of the composite function describing the offer you cannot refuse at 1400:

$$(f \circ g)(x) = 0.85x - 300$$

Replace x with 1400.

$$(f \circ g)(1400) = 0.85(1400) - 300 = 1190 - 300 = 890.$$

Because $(f \circ g)(1400) = 890$, this means that a computer that regularly sells for \$1400 is on sale for \$890 subject to both discounts. We can use a partial table of coordinates for each of the discount functions, g and f, to numerically verify this result.

Computer's regular price | 85% of the regular price

x	$g(x) = 0.85x$
1200	1020
1300	1105
1400	1190

85% of the regular price | \$300 reduction

x	$f(x) = x - 300$
1020	720
1105	805
1190	890

Using these tables, we can find $(f \circ g)(1400)$:

$$(f \circ g)(1400) = f(g(1400)) = f(1190) = 890.$$

The table for g shows that $g(1400) = 1190$.

The table for f shows that $f(1190) = 890$.

This verifies that a computer that regularly sells for \$1400 is on sale for \$890 subject to both discounts.

Before you run out to buy a computer, let's generalize our discussion of the computer's double discount and define the composition of any two functions.

The Composition of Functions

The **composition of the function f with g** is denoted by $f \circ g$ and is defined by the equation

$$(f \circ g)(x) = f(g(x)).$$

The **domain of the composite function $f \circ g$** is the set of all x such that

1. x is in the domain of g and
2. $g(x)$ is in the domain of f.

EXAMPLE 1 Forming Composite Functions

Given $f(x) = 3x - 4$ and $g(x) = x^2 + 6$, find each of the following composite functions:

a. $(f \circ g)(x)$ **b.** $(g \circ f)(x)$.

Solution

a. We begin with $(f \circ g)(x)$, the composition of f with g. Because $(f \circ g)(x)$ means $f(g(x))$, we must replace each occurrence of x in the equation for f with $g(x)$.

$$f(x) = 3x - 4$$ This is the given equation for f.

Replace x with $g(x)$.

$$(f \circ g)(x) = f(g(x)) = 3g(x) - 4$$

$$= 3(x^2 + 6) - 4$$ Because $g(x) = x^2 + 6$, replace $g(x)$ with $x^2 + 6$.

$$= 3x^2 + 18 - 4$$ Use the distributive property.

$$= 3x^2 + 14$$ Simplify.

Thus, $(f \circ g)(x) = 3x^2 + 14$.

b. Next, we find $(g \circ f)(x)$, the composition of g with f. Because $(g \circ f)(x)$, means $g(f(x))$, we must replace each occurrence of x in the equation for g with $f(x)$.

$$g(x) = x^2 + 6$$ This is the given equation for g.

Replace x with $f(x)$.

$$(g \circ f)(x) = g(f(x)) = (f(x))^2 + 6$$

$$= (3x - 4)^2 + 6$$ Because $f(x) = 3x - 4$, replace $f(x)$ with $3x - 4$.

$$= 9x^2 - 24x + 16 + 6$$ Use $(A - B)^2 = A^2 - 2AB + B^2$ to square $3x - 4$.

$$= 9x^2 - 24x + 22$$ Simplify.

Thus, $(g \circ f)(x) = 9x^2 - 24x + 22$.

Notice that $f \circ g$ is not the same function as $g \circ f$. ■

✓ CHECK POINT 1 Given $f(x) = 5x + 6$ and $g(x) = x^2 - 1$, find each of the following composite functions:

a. $(f \circ g)(x)$ **b.** $(g \circ f)(x)$.

Inverse Functions

Here are two functions that describe situations related to the price of a computer, x:

$$f(x) = x - 300 \qquad g(x) = x + 300.$$

Function f subtracts \$300 from the computer's price and function g adds \$300 to the computer's price. Let's see what $f(g(x))$ does. Put $g(x)$ into f:

$$f(x) = x - 300$$ This is the given equation for f.

Replace x with $g(x)$.

$$f(g(x)) = g(x) - 300$$

$$= x + 300 - 300$$ Because $g(x) = x + 300$, replace $g(x)$ with $x + 300$.

$$= x.$$ This is the computer's original price.

By putting $g(x)$ into f and finding $f(g(x))$, we see that the computer's price, x, went through two changes: the first, an increase; the second, a decrease:

$$x + 300 - 300.$$

The final price of the computer, x, is identical to its starting price, x.

In general, if the changes made to x by function g are undone by the changes made by function f, then

$$f(g(x)) = x.$$

Assume, also, that this "undoing" takes place in the other direction:

$$g(f(x)) = x.$$

Under these conditions, we say that each function is the *inverse function* of the other. The fact that g is the inverse of f is expressed by renaming g as f^{-1}, read "f-inverse." For example, the inverse functions

$$f(x) = x - 300 \qquad g(x) = x + 300$$

are usually named as follows:

$$f(x) = x - 300 \qquad f^{-1}(x) = x + 300.$$

We can use partial tables of coordinates for f and f^{-1} to gain numerical insight into the relationship between a function and its inverse function.

Computer's regular price | $300 reduction

x	$f(x) = x - 300$
1200	900
1300	1000
1400	1100

Ordered pairs for f: (1200, 900), (1300, 1000), (1400, 1100)

Price with $300 reduction | $300 price increase

x	$f^{-1}(x) = x + 300$
900	1200
1000	1300
1100	1400

Ordered pairs for f^{-1}: (900, 1200), (1000, 1300), (1100, 1400)

The tables illustrate that if a function f is the set of ordered pairs (x, y), then the inverse, f^{-1}, is the set of ordered pairs (y, x). Using these tables, we can see how one function's changes to x are undone by the other function:

$$(f^{-1} \circ f)(1300) = f^{-1}(f(1300)) = f^{-1}(1000) = 1300.$$

The table for f shows that $f(1300) = 1000$. The table for f^{-1} shows that $f^{-1}(1000) = 1300$.

The final price of the computer, $1300, is identical to its starting price, $1300.

With these ideas in mind, we present the formal definition of the inverse of a function:

Great Question!

Is the -1 in f^{-1} an exponent?

The notation f^{-1} represents the inverse function of f. The -1 is not an exponent. The notation f^{-1} does *not* mean $\frac{1}{f}$:

$$f^{-1} \neq \frac{1}{f}.$$

Definition of the Inverse of a Function

Let f and g be two functions such that

$$f(g(x)) = x \qquad \text{for every } x \text{ in the domain of } g$$

and

$$g(f(x)) = x \qquad \text{for every } x \text{ in the domain of } f.$$

The function g is the **inverse of the function** f, and is denoted by f^{-1} (read "f-inverse"). Thus, $f(f^{-1}(x)) = x$ and $f^{-1}(f(x)) = x$. The domain of f is equal to the range of f^{-1}, and vice versa.

2 Verify inverse functions.

EXAMPLE 2 Verifying Inverse Functions

Show that each function is the inverse of the other:

$$f(x) = 5x \qquad \text{and} \qquad g(x) = \frac{x}{5}.$$

Solution To show that f and g are inverses of each other, we must show that $f(g(x)) = x$ and $g(f(x)) = x$. We begin with $f(g(x))$.

$$f(x) = 5x$$ This is the given equation for f.

Replace x with $g(x)$.

$$f(g(x)) = 5g(x) = 5\left(\frac{x}{5}\right) = x$$ Because $g(x) = \frac{x}{5}$, replace $g(x)$ with $\frac{x}{5}$. Then simplify.

Next, we find $g(f(x))$.

$$g(x) = \frac{x}{5}$$ This is the given equation for g.

Replace x with $f(x)$.

$$g(f(x)) = \frac{f(x)}{5} = \frac{5x}{5} = x$$ Because $f(x) = 5x$, replace $f(x)$ with $5x$. Then simplify.

Because g is the inverse of f (and vice versa), we can use inverse notation and write

$$f(x) = 5x \quad \text{and} \quad f^{-1}(x) = \frac{x}{5}.$$

Notice how f^{-1} undoes the change produced by f: f changes x by multiplying by 5 and f^{-1} undoes this change by dividing by 5. ■

✓ CHECK POINT 2 Show that each function is the inverse of the other:

$$f(x) = 7x \quad \text{and} \quad g(x) = \frac{x}{7}.$$

EXAMPLE 3 Verifying Inverse Functions

Show that each function is the inverse of the other:

$$f(x) = 3x + 2 \quad \text{and} \quad g(x) = \frac{x-2}{3}.$$

Solution To show that f and g are inverses of each other, we must show that $f(g(x)) = x$ and $g(f(x)) = x$. We begin with $f(g(x))$.

$$f(x) = 3x + 2$$

Replace x with $g(x)$.

$$f(g(x)) = 3g(x) + 2 = 3\left(\frac{x-2}{3}\right) + 2 = (x-2) + 2 = x$$

$g(x) = \frac{x-2}{3}$

Next, we find $g(f(x))$.

$$g(x) = \frac{x-2}{3}$$

Replace x with $f(x)$.

$$g(f(x)) = \frac{f(x)-2}{3} = \frac{(3x+2)-2}{3} = \frac{3x}{3} = x$$

$f(x) = 3x + 2$

Because g is the inverse of f (and vice versa), we can use inverse notation and write

$$f(x) = 3x + 2 \quad \text{and} \quad f^{-1}(x) = \frac{x-2}{3}.$$

Notice how f^{-1} undoes the changes produced by f: f changes x by *multiplying* by 3 and *adding* 2, and f^{-1} undoes this by *subtracting* 2 and *dividing* by 3. This "undoing" process is illustrated in **Figure 9.9**. ■

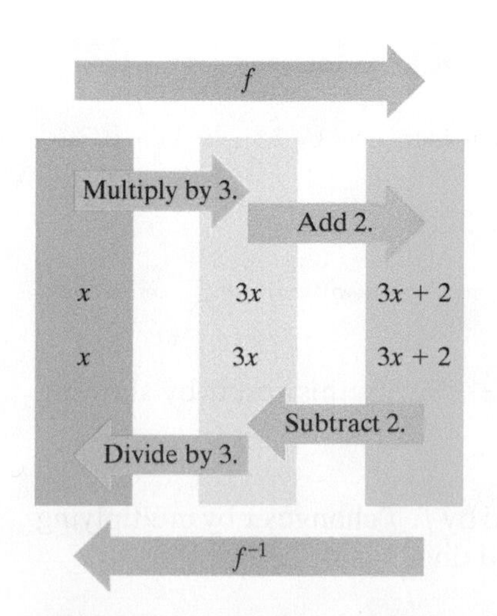

Figure 9.9 f^{-1} undoes the changes produced by f.

CHECK POINT 3 Show that each function is the inverse of the other:

$$f(x) = 4x - 7 \quad \text{and} \quad g(x) = \frac{x+7}{4}.$$

---

3 Find the inverse of a function.

Finding the Inverse of a Function

The definition of the inverse of a function tells us that the domain of f is equal to the range of f^{-1}, and vice versa. This means that if the function f is the set of ordered pairs (x, y), then the inverse of f is the set of ordered pairs (y, x). If a function is defined by an equation, we can obtain the equation for f^{-1}, the inverse of f, by interchanging the role of x and y in the equation for the function f.

Finding the Inverse of a Function

The equation for the inverse of a function f can be found as follows:

1. Replace $f(x)$ with y in the equation for $f(x)$.
2. Interchange x and y.
3. Solve for y. If this equation does not define y as a function of x, the function f does not have an inverse function and this procedure ends. If this equation does define y as a function of x, the function f has an inverse function.
4. If f has an inverse function, replace y in step 3 with $f^{-1}(x)$. We can verify our result by showing that $f(f^{-1}(x)) = x$ and $f^{-1}(f(x)) = x$.

The procedure for finding a function's inverse uses a *switch-and-solve* strategy. Switch x and y, then solve for y.

EXAMPLE 4 Finding the Inverse of a Function

Find the inverse of $f(x) = 7x - 5$.

Solution

Step 1. Replace $f(x)$ with y:

$$y = 7x - 5.$$

Step 2. Interchange x and y:

$$x = 7y - 5. \quad \text{This is the inverse function.}$$

Step 3. Solve for y:

$$x + 5 = 7y \quad \text{Add 5 to both sides.}$$

$$\frac{x+5}{7} = y. \quad \text{Divide both sides by 7.}$$

Step 4. Replace y with $f^{-1}(x)$:

$$f^{-1}(x) = \frac{x+5}{7}. \quad \text{The equation is written with } f^{-1} \text{ on the left.}$$

Thus, the inverse of $f(x) = 7x - 5$ is $f^{-1}(x) = \frac{x+5}{7}$. (Verify this result by showing that $f(f^{-1}(x)) = x$ and $f^{-1}(f(x)) = x$.)

The inverse function, f^{-1}, undoes the changes produced by f. f changes x by multiplying by 7 and subtracting 5. f^{-1} undoes this by adding 5 and dividing by 7. ■

Find the inverse of $f(x) = 2x + 7$.

EXAMPLE 5 Finding the Inverse of a Function

Find the inverse of $f(x) = x^3 + 1$.

Solution

Step 1. Replace $f(x)$ with y: $y = x^3 + 1$.

Step 2. Interchange x and y: $x = y^3 + 1$.

Step 3. Solve for y. We need to solve $x = y^3 + 1$ for y.

Our goal is to isolate y. Because $\sqrt[3]{y^3} = y$, we will take the cube root of both sides of the equation.

$x - 1 = y^3$ Subtract 1 from both sides of $x = y^3 + 1$.

$\sqrt[3]{x-1} = \sqrt[3]{y^3}$ Take the cube root on both sides.

$\sqrt[3]{x-1} = y$ Simplify.

Step 4. Replace y with $f^{-1}(x)$: $f^{-1}(x) = \sqrt[3]{x-1}$.

Thus, the inverse of $f(x) = x^3 + 1$ is $f^{-1}(x) = \sqrt[3]{x-1}$. ■

CHECK POINT 5 Find the inverse of $f(x) = 4x^3 - 1$.

4 Use the horizontal line test to determine if a function has an inverse function.

The Horizontal Line Test and One-to-One Functions

Let's see what happens if we try to find the inverse of the quadratic function $f(x) = x^2$.

Step 1. Replace $f(x)$ with y: $y = x^2$.

Step 2. Interchange x and y: $x = y^2$.

Step 3. Solve for y: We apply the square root property to solve $y^2 = x$ for y. We obtain

$$y = \pm\sqrt{x}.$$

The $\pm$ in this last equation shows that for certain values of x (all positive real numbers), there are two values of y. Because this equation does not represent y as a function of x, the quadratic function $f(x) = x^2$ does not have an inverse function.

We can use a few of the solutions of $y = x^2$ to illustrate numerically that this function does not have an inverse:

Four solutions of $y = x^2$. $(-2, 4), (-1, 1), (1, 1), (2, 4)$,

Interchange x and y in each ordered pair. $(4, -2), (1, -1), (1, 1), (4, 2)$.

The input 1 is associated with two outputs, -1 and 1.

The input 4 is associated with two outputs, -2 and 2.

A function provides exactly one output for each input. Thus, the ordered pairs in the bottom row do not define a function.

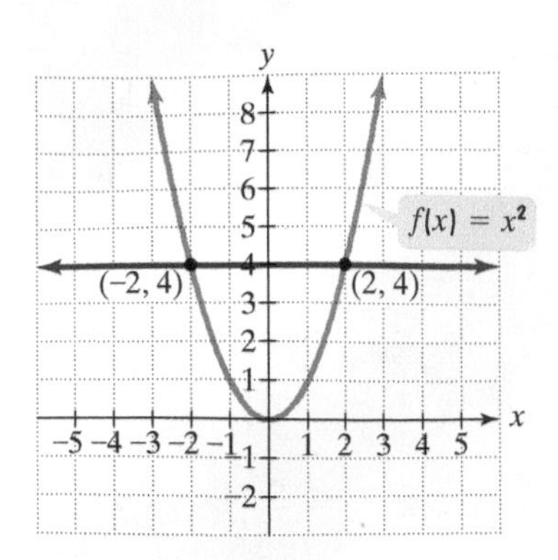

Figure 9.10 The horizontal line intersects the graph twice.

Can we look at the graph of a function and tell if it represents a function with an inverse? Yes. The graph of the quadratic function $f(x) = x^2$ is shown in **Figure 9.10**. Four units above the x-axis, a horizontal line is drawn. This line intersects the graph at two of its points, $(-2, 4)$ and $(2, 4)$. Inverse functions have ordered pairs with the coordinates reversed. We just saw what happened when we interchanged x and y. We obtained $(4, -2)$ and $(4, 2)$, and these ordered pairs do not define a function.

If any horizontal line, such as the one in **Figure 9.10**, intersects a graph at two or more points, the set of these points will not define a function when their coordinates are reversed. This suggests the **horizontal line test** for inverse functions.

The Horizontal Line Test for Inverse Functions

A function f has an inverse that is a function, f^{-1}, if there is no horizontal line that intersects the graph of the function f at more than one point.

EXAMPLE 6 Applying the Horizontal Line Test

Which of the following graphs represent functions that have inverse functions?

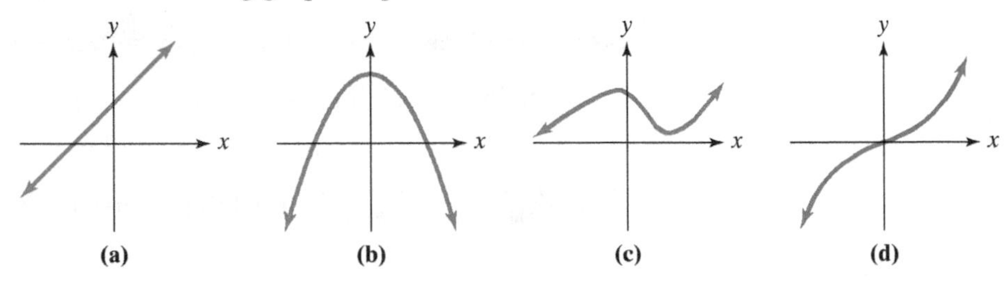

Discover for Yourself

How might you restrict the domain of $f(x) = x^2$, graphed in **Figure 9.10**, so that the remaining portion of the graph passes the horizontal line test?

Solution Notice that horizontal lines can be drawn in graphs **(b)** and **(c)** that intersect the graphs more than once. These graphs do not pass the horizontal line test. These are not the graphs of functions with inverse functions. By contrast, no horizontal line can be drawn in graphs **(a)** and **(d)** that intersects the graphs more than once. These graphs pass the horizontal line test. Thus, the graphs in parts **(a)** and **(d)** represent functions that have inverse functions.

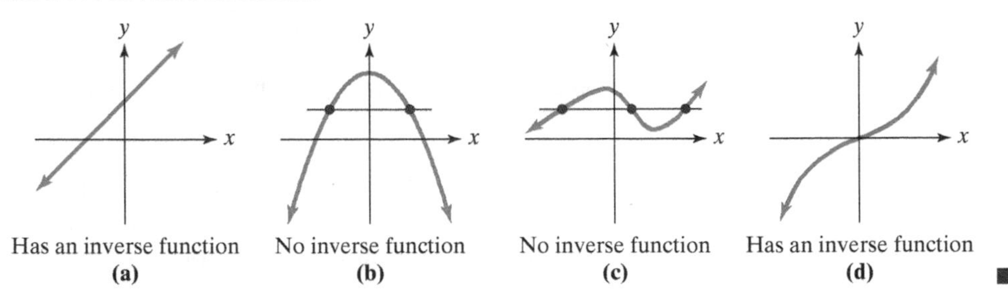

■

CHECK POINT 6 Which of the following graphs represent functions that have inverse functions?

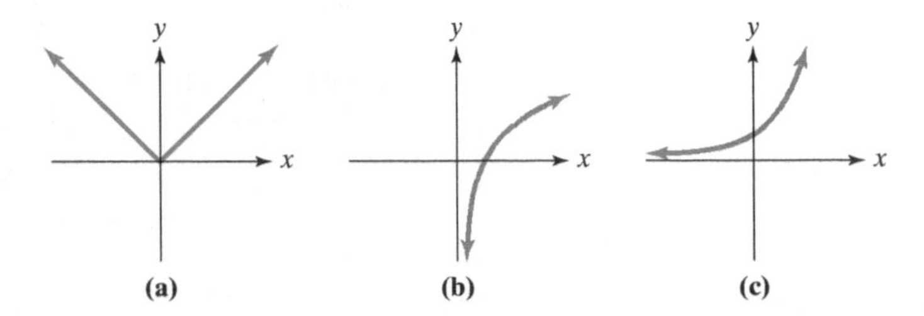

A function passes the horizontal line test when no two different ordered pairs have the same second component. This means that if $x_1 \neq x_2$, then $f(x_1) \neq f(x_2)$. Such a function is called a **one-to-one function.** Thus, **a one-to-one function is a function in which no two different ordered pairs have the same second component. Only**

one-to-one functions have inverse functions. Any function that passes the horizontal line test is a one-to-one function. Any one-to-one function has a graph that passes the horizontal line test.

5 Use the graph of a one-to-one function to graph its inverse function.

Graphs of f and f^{-1}

There is a relationship between the graph of a one-to-one function, f, and its inverse, f^{-1}. Because inverse functions have ordered pairs with the coordinates reversed, if the point (a, b) is on the graph of f, then the point (b, a) is on the graph of f^{-1}. The points (a, b) and (b, a) are symmetric with respect to the line $y = x$. Thus, **the graph of f^{-1} is a reflection of the graph of f about the line $y = x$.** This is illustrated in **Figure 9.11**.

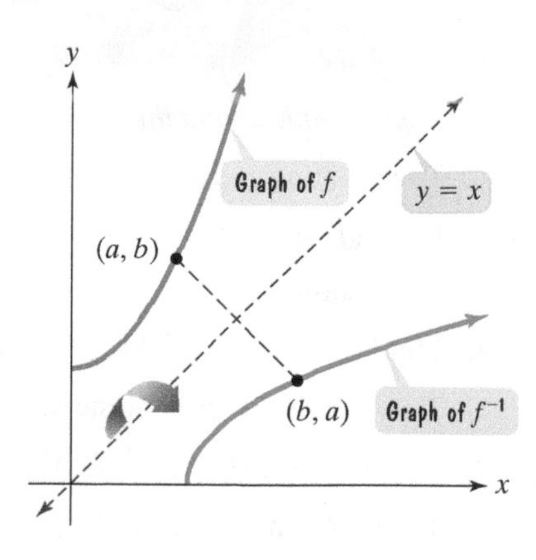

Figure 9.11 The graph of f^{-1} is a reflection of the graph of f about $y = x$.

EXAMPLE 7 Graphing the Inverse Function

Use the graph of f in **Figure 9.12** to draw the graph of its inverse function.

Solution We begin by noting that no horizontal line intersects the graph of f, shown again in blue in **Figure 9.13**, at more than one point, so f does have an inverse function. Because the points $(-3, -2)$, $(-1, 0)$, and $(4, 2)$ are on the graph of f, the graph of the inverse function, f^{-1}, has points with these ordered pairs reversed. Thus, $(-2, -3)$, $(0, -1)$, and $(2, 4)$ are on the graph of f^{-1}. We can use these points to graph f^{-1}. The graph of f^{-1} is shown in green in **Figure 9.13**. Note that the green graph of f^{-1} is the reflection of the blue graph of f about the line $y = x$.

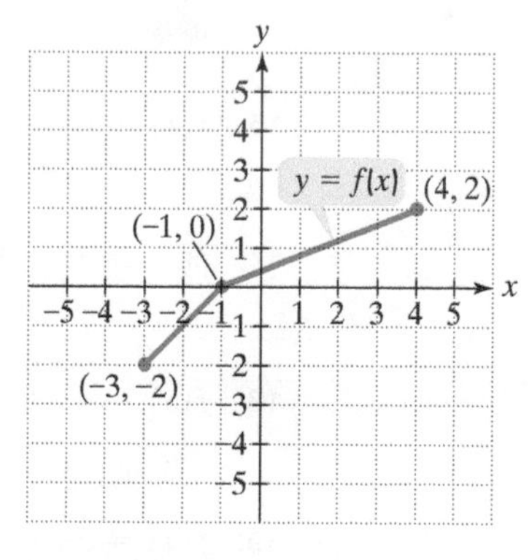

Figure 9.12

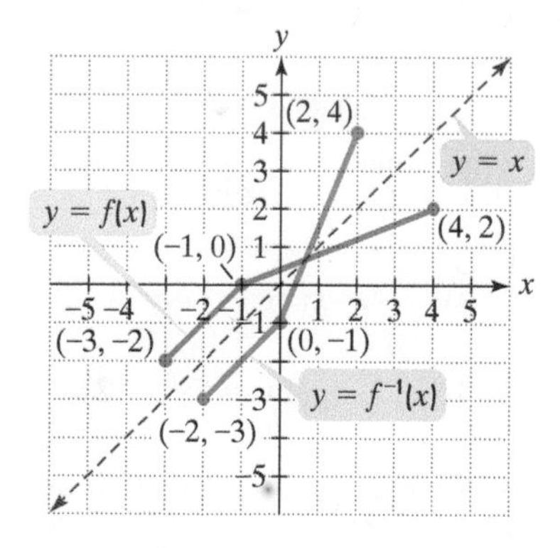

Figure 9.13 The graphs of f and f^{-1} ■

✓ CHECK POINT 7 The graph of function f consists of two line segments, one segment from $(-2, -2)$ to $(-1, 0)$ and a second segment from $(-1, 0)$ to $(1, 2)$. Graph f and use the graph to draw the graph of its inverse function.

Achieving Success

Warm up your brain before starting the assigned homework. Researchers say the mind can be strengthened, just like your muscles, with regular training and rigorous practice. Think of the book's Exercise Sets as brain calisthenics. If you're feeling a bit sluggish before any of your mental workouts, try this warmup:

In the list below say the color the word is printed in, not the word itself. Once you can do this in 15 seconds without an error, the warmup is over and it's time to move on to the assigned exercises.

Blue Yellow Red Green Yellow Green Blue Red Yellow Red

CONCEPT AND VOCABULARY CHECK

Fill in each blank so that the resulting statement is true.

1. The notation $f \circ g$, called the ____________ of the function f with g, is defined by $(f \circ g)(x) =$ ____________.
2. I find $(f \circ g)(x)$ by replacing each occurrence of x in the equation for ____________ with ____________.
3. The notation $g \circ f$, called the ____________ of the function g with f, is defined by $(g \circ f)(x) =$ ____________.
4. I find $(g \circ f)(x)$ by replacing each occurrence of x in the equation for ____________ with ____________.
5. True or false: $f \circ g$ is the same function as $g \circ f$. ____________
6. True or false: $f(g(x)) = f(x) \cdot g(x)$ ____________
7. The notation f^{-1} means the ____________ of the function f.
8. If the function g is the inverse of the function f, then $f(g(x)) =$ ____________ and $g(f(x)) =$ ____________.
9. A function f has an inverse that is a function if there is no ____________ line that intersects the graph of f at more than one point. Such a function is called a/an ____________ function.
10. The graph of f^{-1} is a reflection of the graph of f about the line whose equation is ____________.

9.2 EXERCISE SET

MyMathLab®

Practice Exercises

In Exercises 1–14, find

a. $(f \circ g)(x)$; **b.** $(g \circ f)(x)$; **c.** $(f \circ g)(2)$.

1. $f(x) = 2x, \quad g(x) = x + 7$
2. $f(x) = 3x, \quad g(x) = x - 5$
3. $f(x) = x + 4, \quad g(x) = 2x + 1$
4. $f(x) = 5x + 2, \quad g(x) = 3x - 4$
5. $f(x) = 4x - 3, \quad g(x) = 5x^2 - 2$
6. $f(x) = 7x + 1, \quad g(x) = 2x^2 - 9$
7. $f(x) = x^2 + 2, \quad g(x) = x^2 - 2$
8. $f(x) = x^2 + 1, \quad g(x) = x^2 - 3$
9. $f(x) = \sqrt{x}, \quad g(x) = x - 1$
10. $f(x) = \sqrt{x}, \quad g(x) = x + 2$
11. $f(x) = 2x - 3, \quad g(x) = \dfrac{x + 3}{2}$
12. $f(x) = 6x - 3, \quad g(x) = \dfrac{x + 3}{6}$
13. $f(x) = \dfrac{1}{x}, \quad g(x) = \dfrac{1}{x}$
14. $f(x) = \dfrac{1}{x}, \quad g(x) = \dfrac{2}{x}$

In Exercises 15–24, find $f(g(x))$ and $g(f(x))$ and determine whether each pair of functions f and g are inverses of each other.

15. $f(x) = 4x$ and $g(x) = \dfrac{x}{4}$
16. $f(x) = 6x$ and $g(x) = \dfrac{x}{6}$
17. $f(x) = 3x + 8$ and $g(x) = \dfrac{x - 8}{3}$
18. $f(x) = 4x + 9$ and $g(x) = \dfrac{x - 9}{4}$
19. $f(x) = 5x - 9$ and $g(x) = \dfrac{x + 5}{9}$
20. $f(x) = 3x - 7$ and $g(x) = \dfrac{x + 3}{7}$
21. $f(x) = \dfrac{3}{x - 4}$ and $g(x) = \dfrac{3}{x} + 4$
22. $f(x) = \dfrac{2}{x - 5}$ and $g(x) = \dfrac{2}{x} + 5$
23. $f(x) = -x$ and $g(x) = -x$
24. $f(x) = \sqrt[3]{x - 4}$ and $g(x) = x^3 + 4$

The functions in Exercises 25–44 are all one-to-one. For each function,

a. *Find an equation for $f^{-1}(x)$, the inverse function.*

b. *Verify that your equation is correct by showing that $f(f^{-1}(x)) = x$ and $f^{-1}(f(x)) = x$.*

25. $f(x) = x + 3$

26. $f(x) = x + 5$

27. $f(x) = 2x$

28. $f(x) = 4x$

29. $f(x) = 2x + 3$

30. $f(x) = 3x - 1$

31. $f(x) = x^3 + 2$

32. $f(x) = x^3 - 1$

33. $f(x) = (x + 2)^3$

34. $f(x) = (x - 1)^3$

35. $f(x) = \dfrac{1}{x}$

36. $f(x) = \dfrac{2}{x}$

37. $f(x) = \sqrt{x}$

38. $f(x) = \sqrt[3]{x}$

39. $f(x) = x^2 + 1$, for $x \geq 0$

40. $f(x) = x^2 - 1$, for $x \geq 0$

41. $f(x) = \dfrac{2x + 1}{x - 3}$

42. $f(x) = \dfrac{2x - 3}{x + 1}$

43. $f(x) = \sqrt[3]{x - 4} + 3$

44. $f(x) = x^{\frac{3}{5}}$

Which graphs in Exercises 45–50 represent functions that have inverse functions?

45.

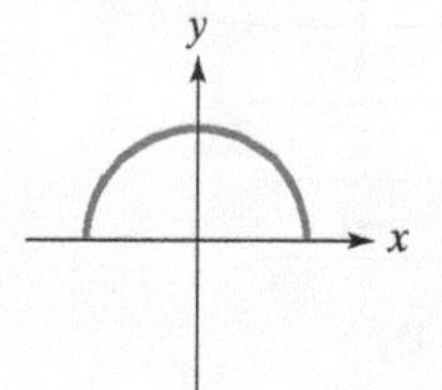

46.

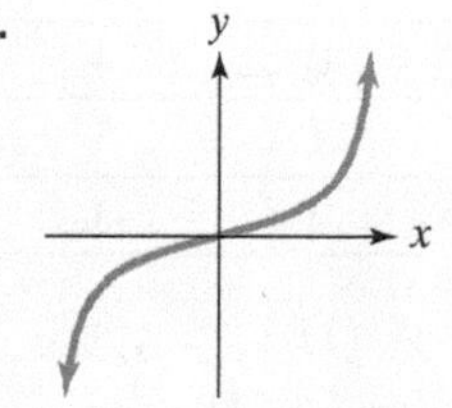

47.

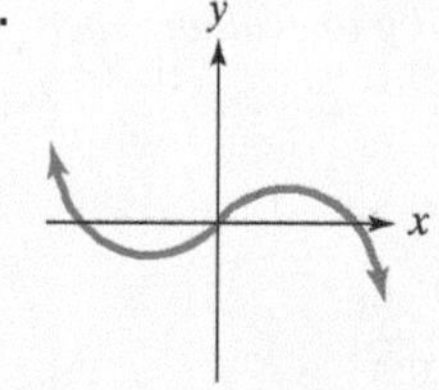

48.

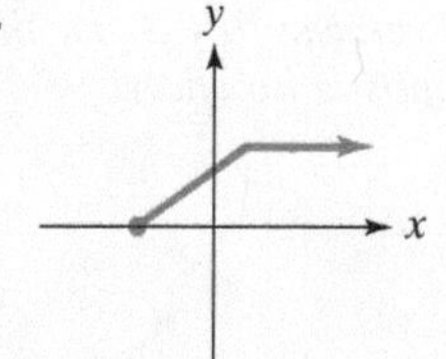

49.

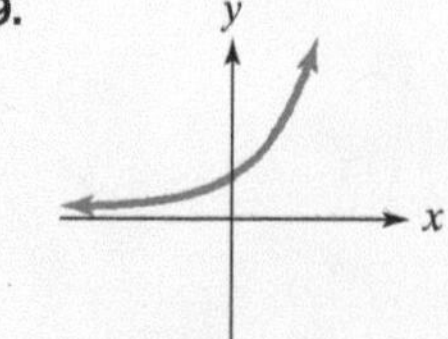

50.

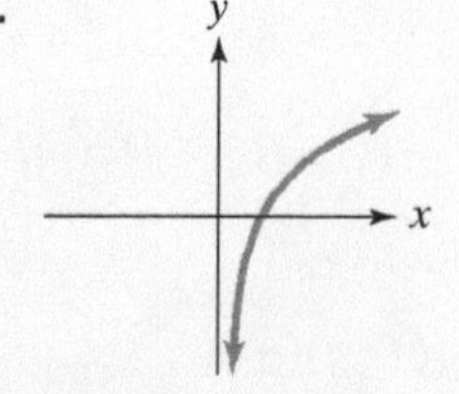

In Exercises 51–54, use the graph of f to draw the graph of its inverse function.

51.

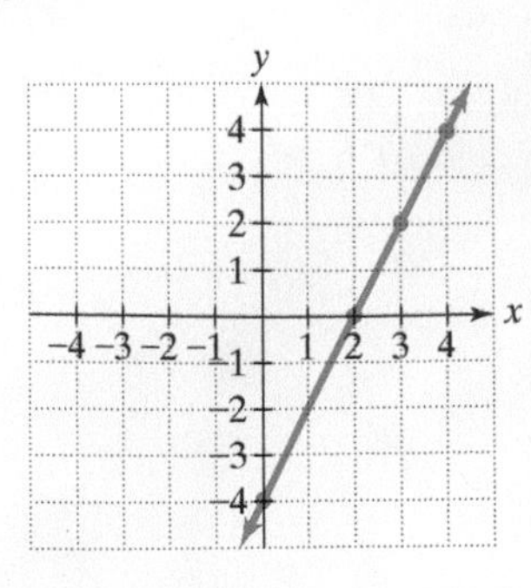

52.

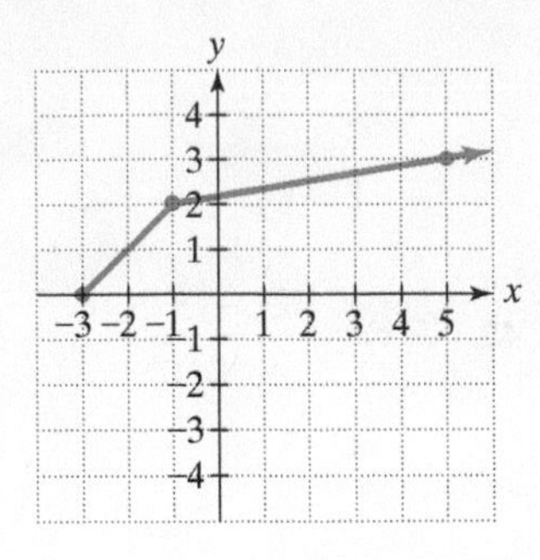

53.

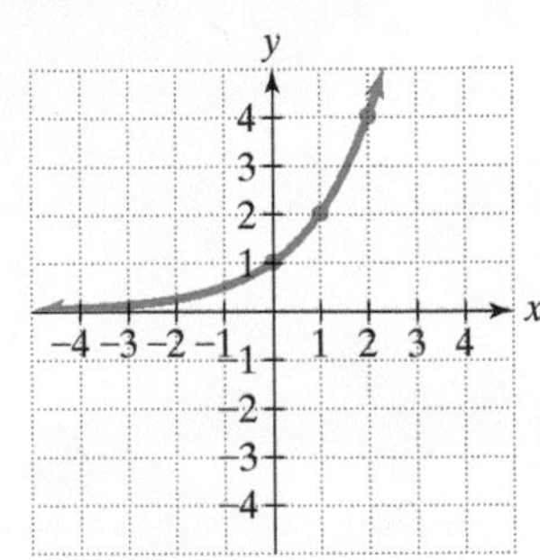

54.

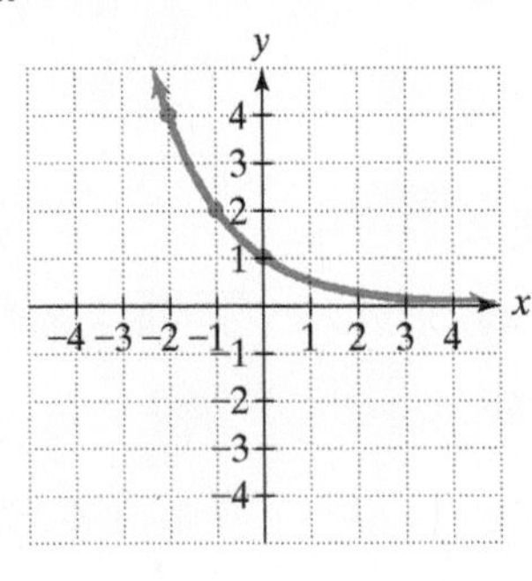

Practice PLUS

In Exercises 55–60, f and g are defined by the following tables. Use the tables to evaluate each composite function.

x	$f(x)$
−1	1
0	4
1	5
2	−1

x	$g(x)$
−1	0
1	1
4	2
10	−1

55. $f(g(1))$

56. $f(g(4))$

57. $(g \circ f)(-1)$

58. $(g \circ f)(0)$

59. $f^{-1}(g(10))$

60. $f^{-1}(g(1))$

In Exercises 61–64, use the graphs of f and g to evaluate each composite function.

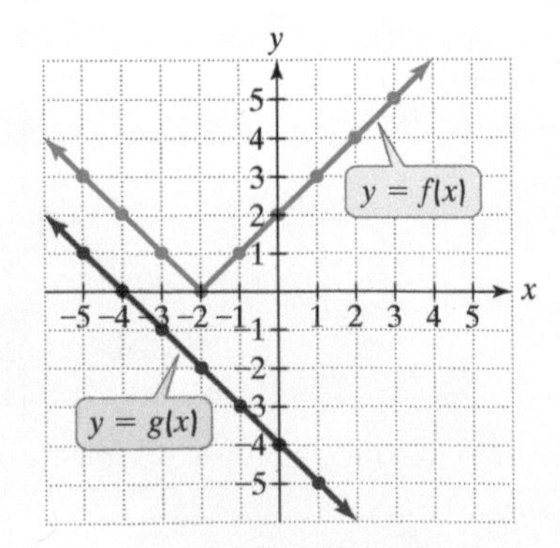

61. $(f \circ g)(-1)$

62. $(f \circ g)(1)$

63. $(g \circ f)(0)$

64. $(g \circ f)(-1)$

In Exercises 65–70, let

$$f(x) = 2x - 5$$
$$g(x) = 4x - 1$$
$$h(x) = x^2 + x + 2.$$

Evaluate the indicated function without finding an equation for the function.

65. $(f \circ g)(0)$

66. $(g \circ f)(0)$

67. $f^{-1}(1)$

68. $g^{-1}(7)$

69. $g(f[h(1)])$

70. $f(g[h(1)])$

Application Exercises

71. The regular price of a computer is x dollars. Let $f(x) = x - 400$ and $g(x) = 0.75x$.

a. Describe what the functions f and g model in terms of the price of the computer.

b. Find $(f \circ g)(x)$ and describe what this models in terms of the price of the computer.

c. Repeat part (b) for $(g \circ f)(x)$.

d. Which composite function models the greater discount on the computer, $f \circ g$ or $g \circ f$? Explain.

e. Find f^{-1} and describe what this models in terms of the price of the computer.

72. The regular price of a pair of jeans is x dollars. Let $f(x) = x - 5$ and $g(x) = 0.6x$.

a. Describe what functions f and g model in terms of the price of the jeans.

b. Find $(f \circ g)(x)$ and describe what this models in terms of the price of the jeans.

c. Repeat part (b) for $(g \circ f)(x)$.

d. Which composite function models the greater discount on the jeans, $f \circ g$ or $g \circ f$? Explain.

e. Find f^{-1} and describe what this models in terms of the price of the jeans.

Way to Go *Holland was the first country to establish an official bicycle policy. It currently has over 12,000 miles of paths and lanes exclusively for bicycles. The graph at the top of the next page shows the percentage of travel by bike and by car in Holland, as well as in four other selected countries. Use the information in the graph to solve Exercises 73–74.*

Modes of Travel in Selected Countries

by bike by car

Percentage of Travel: 90%, 75%, 60%, 45%, 30%, 15%

U.S.: 1%, 84%
U.K.: 8%, 62%
Italy: 5%, 42%
France: 5%, 45%
Holland: 30%, 45%

Country

Source: EUROSTAT

73. a. Consider a function, f, whose domain is the set of the five countries shown in the graph. Let the range be the percentage of travel by bike in each of the respective countries. Write function f as a set of ordered pairs.

b. Write the relation that is the inverse of f as a set of ordered pairs. Is this relation a function? Explain your answer.

74. a. Consider a function, f, whose domain is the set of the five countries shown in the graph. Let the range be the percentage of travel by car in each of the respective countries. Write function f as a set of ordered pairs.

b. Write the relation that is the inverse of f as a set of ordered pairs. Is this relation a function? Explain your answer.

75. The graph represents the probability that two people in the same room share a birthday as a function of the number of people in the room. Call the function f.

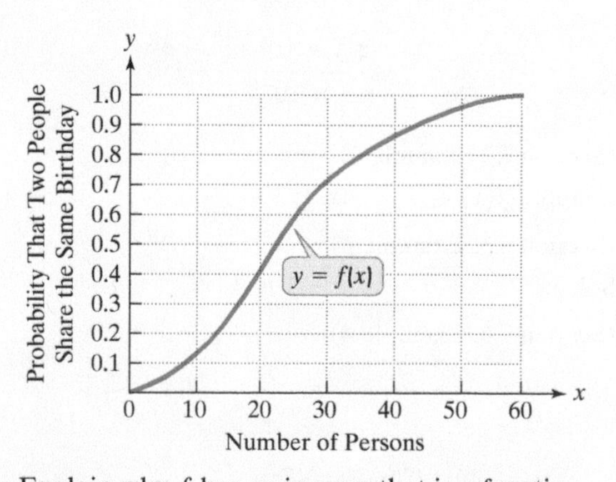

a. Explain why f has an inverse that is a function.

b. Describe in practical terms the meanings of $f^{-1}(0.25)$, $f^{-1}(0.5)$, and $f^{-1}(0.7)$.

76. A study of 900 working women in Texas showed that their feelings changed throughout the day. As the graph indicates, the women felt better as time passed, except for a blip at lunchtime.

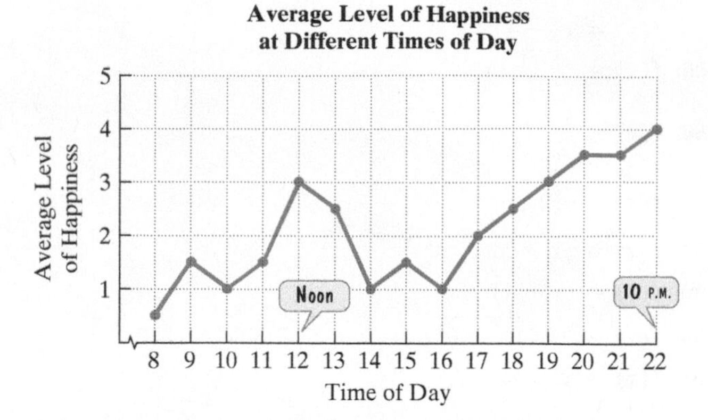

Source: D. Kahneman et al., "A Survey Method for Characterizing Daily Life Experience." *Science*, Vol. 306, No. 5702, Dec. 3, 2004, pp. 1776–1780.

a. Does the graph have an inverse that is a function? Explain your answer.

b. Identify two or more times of day when the average happiness level is 3. Express your answers as ordered pairs.

c. Do the ordered pairs in part (b) indicate that the graph represents a one-to-one function? Explain your answer.

77. The formula

$$y = f(x) = \frac{9}{5}x + 32$$

is used to convert from x degrees Celsius to y degrees Fahrenheit. The formula

$$y = g(x) = \frac{5}{9}(x - 32)$$

is used to convert from x degrees Fahrenheit to y degrees Celsius. Show that f and g are inverse functions.

Explaining the Concepts

78. Describe a procedure for finding $(f \circ g)(x)$.

79. Explain how to determine if two functions are inverses of each other.

80. Describe how to find the inverse of a one-to-one function.

81. What is the horizontal line test and what does it indicate?

82. Describe how to use the graph of a one-to-one function to draw the graph of its inverse function.

83. How can a graphing utility be used to visually determine if two functions are inverses of each other?

Technology Exercises

In Exercises 84–91, use a graphing utility to graph each function. Use the graph to determine whether the function has an inverse that is a function (that is, whether the function is one-to-one).

84. $f(x) = x^2 - 1$

85. $f(x) = \sqrt[3]{2 - x}$

86. $f(x) = \dfrac{x^3}{2}$

87. $f(x) = \dfrac{x^4}{4}$

88. $f(x) = |x - 2|$

89. $f(x) = (x - 1)^3$

90. $f(x) = -\sqrt{16 - x^2}$

91. $f(x) = x^3 + x + 1$

In Exercises 92–94, use a graphing utility to graph f and g in the same viewing rectangle. In addition, graph the line y = x and visually determine if f and g are inverses.

92. $f(x) = 4x + 4, \quad g(x) = 0.25x - 1$

93. $f(x) = \dfrac{1}{x} + 2, \quad g(x) = \dfrac{1}{x - 2}$

94. $f(x) = \sqrt[3]{x} - 2, \quad g(x) = (x + 2)^3$

Critical Thinking Exercises

Make Sense? *In Exercises 95–98, determine whether each statement makes sense or does not make sense, and explain your reasoning.*

95. This diagram illustrates that $f(g(x)) = x^2 + 4$.

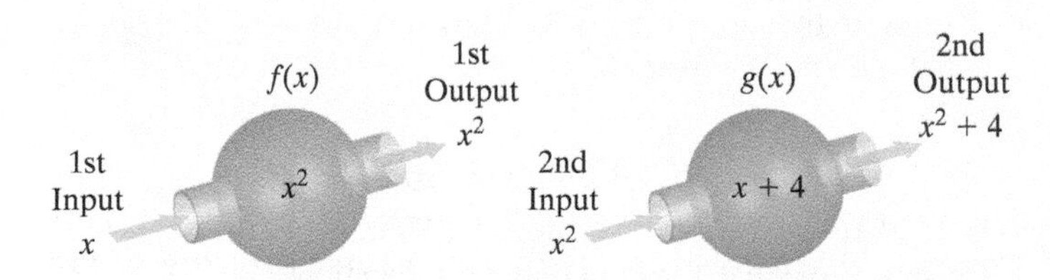

96. I must have made a mistake in finding the composite functions $f \circ g$ and $g \circ f$, because I notice that $f \circ g$ is the same function as $g \circ f$.

97. Regardless of what exponential function I graph, its shape indicates that it always has an inverse function.

98. I'm working with the linear function $f(x) = 3x + 5$ and I do not need to find f^{-1} in order to determine the value of $(f \circ f^{-1})(17)$.

In Exercises 99–102, determine whether each statement is true or false. If the statement is false, make the necessary change(s) to produce a true statement.

99. The inverse of $\{(1, 4), (2, 7)\}$ is $\{(2, 7), (1, 4)\}$.

100. The function $f(x) = 5$ is one-to-one.

101. If $f(x) = \sqrt{x}$ and $g(x) = 2x - 1$, then $(f \circ g)(5) = g(2)$.

102. If $f(x) = 3x$, then $f^{-1}(x) = \dfrac{1}{3x}$.

103. If $h(x) = \sqrt{3x^2 + 5}$, find functions f and g so that $h(x) = (f \circ g)(x)$.

104. If $f(x) = 3x$ and $g(x) = x + 5$, find $(f \circ g)^{-1}(x)$ and $(g^{-1} \circ f^{-1})(x)$.

105. Show that

$$f(x) = \frac{3x - 2}{5x - 3}$$

is its own inverse.

106. Consider the two functions defined by $f(x) = m_1x + b_1$ and $g(x) = m_2x + b_2$. Prove that the slope of the composite function of f with g is equal to the product of the slopes of the two functions.

Review Exercises

107. Divide and write the quotient in scientific notation:

$$\frac{4.3 \times 10^5}{8.6 \times 10^{-4}}.$$

(Section 1.7, Example 4)

108. Graph: $f(x) = x^2 - 4x + 3$.

(Section 8.3, Example 4)

109. Solve: $\sqrt{x + 4} - \sqrt{x - 1} = 1$.

(Section 7.6, Example 4)

Preview Exercises

Exercises 110–112 will help you prepare for the material covered in the next section.

110. What problem do you encounter when using the switch-and-solve strategy to find the inverse of $f(x) = 2^x$?

111. 25 to what power gives 5? $(25^? = 5)$

112. Solve: $(x - 3)^2 > 0$.

SECTION

9.3

Logarithmic Functions

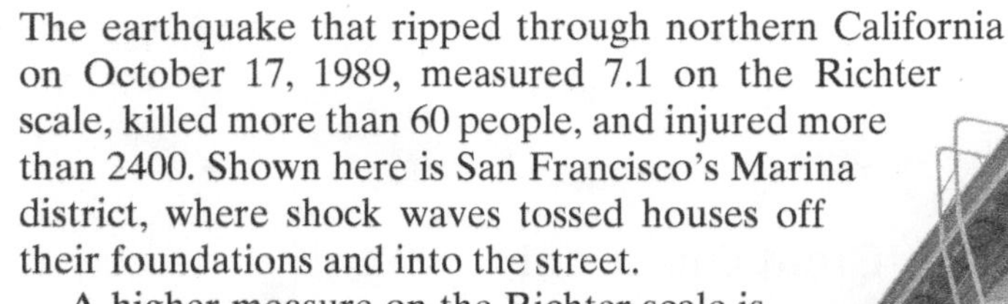

What am I supposed to learn?

After studying this section, you should be able to:

1. Change from logarithmic to exponential form.
2. Change from exponential to logarithmic form.
3. Evaluate logarithms.
4. Use basic logarithmic properties.
5. Graph logarithmic functions.
6. Find the domain of a logarithmic function.
7. Use common logarithms.
8. Use natural logarithms.

The earthquake that ripped through northern California on October 17, 1989, measured 7.1 on the Richter scale, killed more than 60 people, and injured more than 2400. Shown here is San Francisco's Marina district, where shock waves tossed houses off their foundations and into the street.

A higher measure on the Richter scale is more devastating than it seems because for each increase in one unit on the scale, there is a tenfold increase in the intensity of an earthquake. In this section, our focus is on the inverse of the exponential function, called the logarithmic function. The logarithmic function will help you to understand diverse phenomena, including earthquake intensity, human memory, and the pace of life in large cities.

The Definition of Logarithmic Functions

No horizontal line can be drawn that intersects the graph of an exponential function at more than one point. This means that the exponential function is one-to-one and has an inverse. Let's use our switch-and-solve strategy to find the inverse.

All exponential functions have inverse functions.

$$f(x) = b^x$$

Step 1. Replace $f(x)$ with y: $y = b^x$.

Step 2. Interchange x and y: $x = b^y$.

Step 3. Solve for y: ?

The question mark indicates that we do not have a method for solving $b^y = x$ for y. To isolate the exponent y, a new notation, called *logarithmic notation*, is needed. This notation gives us a way to name the inverse of $f(x) = b^x$. **The inverse function of the exponential function with base b is called the *logarithmic function with base b*.**

Definition of the Logarithmic Function

For $x > 0$ and $b > 0, b \neq 1$,

$$y = \log_b x \text{ is equivalent to } b^y = x.$$

The function $f(x) = \log_b x$ is the **logarithmic function with base b.**

The equations

$$y = \log_b x \quad \text{and} \quad b^y = x$$

are different ways of expressing the same thing. The first equation is in **logarithmic form** and the second equivalent equation is in **exponential form**.

Notice that a **logarithm, y, is an exponent. Logarithmic form allows us to isolate this exponent.** You should learn the location of the base and exponent in each form.

Location of Base and Exponent in Exponential and Logarithmic Forms

Logarithmic Form: $y = \log_b x$ (Exponent: y; Base: b)   Exponential Form: $b^y = x$ (Exponent: y; Base: b)

Great Question!

Much of what you've discussed so far involves changing from logarithmic form to the more familiar exponential form. Is there a pattern I can use to help me remember how to do this?

Yes. To change from logarithmic form to exponential form, use this pattern:

$$y = \log_b x \quad \text{means} \quad b^y = x.$$

1 Change from logarithmic to exponential form.

EXAMPLE 1 Changing from Logarithmic to Exponential Form

Write each equation in its equivalent exponential form:

a. $2 = \log_5 x$ **b.** $3 = \log_b 64$ **c.** $\log_3 7 = y$.

Solution We use the fact that $y = \log_b x$ means $b^y = x$.

a. $2 = \log_5 x$ means $5^2 = x$. (Logarithms are exponents.)

b. $3 = \log_b 64$ means $b^3 = 64$. (Logarithms are exponents.)

c. $\log_3 7 = y$ or $y = \log_3 7$ means $3^y = 7$. ■

CHECK POINT 1 Write each equation in its equivalent exponential form:

a. $3 = \log_7 x$ **b.** $2 = \log_b 25$ **c.** $\log_4 26 = y$.

2 Change from exponential to logarithmic form.

EXAMPLE 2 Changing from Exponential to Logarithmic Form

Write each equation in its equivalent logarithmic form:

a. $12^2 = x$ **b.** $b^3 = 8$ **c.** $e^y = 9$.

Solution We use the fact that $b^y = x$ means $y = \log_b x$. In logarithmic form, the exponent is isolated on one side of the equal sign.

a. $12^2 = x$ means $2 = \log_{12} x$. (Exponents are logarithms.)

b. $b^3 = 8$ means $3 = \log_b 8$. (Exponents are logarithms.)

c. $e^y = 9$ means $y = \log_e 9$. ■

CHECK POINT 2 Write each equation in its equivalent logarithmic form:

a. $2^5 = x$ **b.** $b^3 = 27$ **c.** $e^y = 33$.

3 Evaluate logarithms.

Remembering that logarithms are exponents makes it possible to evaluate some logarithms by inspection. The logarithm of x with base b, $\log_b x$, is the exponent to which b must be raised to get x. For example, suppose we want to evaluate $\log_2 32$. We ask, 2 to what power gives 32? Because $2^5 = 32$, we can conclude that $\log_2 32 = 5$.

EXAMPLE 3 Evaluating Logarithms

Evaluate:

a. $\log_2 16$ **b.** $\log_3 9$ **c.** $\log_{25} 5$.

Solution

Logarithmic Expression	Question Needed for Evaluation	Logarithmic Expression Evaluated
a. $\log_2 16$	2 to what power gives 16?	$\log_2 16 = 4$ because $2^4 = 16$.
b. $\log_3 9$	3 to what power gives 9?	$\log_3 9 = 2$ because $3^2 = 9$.
c. $\log_{25} 5$	25 to what power gives 5?	$\log_{25} 5 = \frac{1}{2}$ because $25^{\frac{1}{2}} = \sqrt{25} = 5$.

■

✓ **CHECK POINT 3** Evaluate:

a. $\log_{10} 100$ **b.** $\log_3 3$ **c.** $\log_{36} 6$.

4 Use basic logarithmic properties.

Basic Logarithmic Properties

Because logarithms are exponents, they have properties that can be verified using the properties of exponents.

Basic Logarithmic Properties Involving 1

1. $\log_b b = 1$ because 1 is the exponent to which b must be raised to obtain b. ($b^1 = b$)
2. $\log_b 1 = 0$ because 0 is the exponent to which b must be raised to obtain 1. ($b^0 = 1$)

EXAMPLE 4 Using Properties of Logarithms

Evaluate:

a. $\log_7 7$ **b.** $\log_5 1$.

Solution

a. Because $\log_b b = 1$, we conclude $\log_7 7 = 1$.

b. Because $\log_b 1 = 0$, we conclude $\log_5 1 = 0$. ■

✓ **CHECK POINT 4** Evaluate:

a. $\log_9 9$ **b.** $\log_8 1$.

Now that we are familiar with logarithmic notation, let's resume and finish the switch-and-solve strategy for finding the inverse of $f(x) = b^x$.

Step 1. Replace $f(x)$ with y: $y = b^x$.

Step 2. Interchange x and y: $x = b^y$.

Step 3. Solve for y: $y = \log_b x$.

Step 4. Replace y with $f^{-1}(x)$: $f^{-1}(x) = \log_b x$.

The completed switch-and-solve strategy illustrates that if $f(x) = b^x$, then $f^{-1}(x) = \log_b x$. The inverse of an exponential function is the logarithmic function with the same base.

In Section 9.2, we saw how inverse functions "undo" one another. In particular,

$$f(f^{-1}(x)) = x \quad \text{and} \quad f^{-1}(f(x)) = x.$$

Applying these relationships to exponential and logarithmic functions, we obtain the following **inverse properties of logarithms:**

Inverse Properties of Logarithms

For $b > 0$ and $b \neq 1$,

$\log_b b^x = x$ The logarithm with base b of b raised to a power equals that power.

$b^{\log_b x} = x$ b raised to the logarithm with base b of a number equals that number.

EXAMPLE 5 Using Inverse Properties of Logarithms

Evaluate:

a. $\log_4 4^5$ **b.** $6^{\log_6 9}$.

Solution

a. Because $\log_b b^x = x$, we conclude $\log_4 4^5 = 5$.

b. Because $b^{\log_b x} = x$, we conclude $6^{\log_6 9} = 9$. ■

CHECK POINT 5 Evaluate:

a. $\log_7 7^8$ **b.** $3^{\log_3 17}$.

5 Graph logarithmic functions.

Graphs of Logarithmic Functions

How do we graph logarithmic functions? We use the fact that a logarithmic function is the inverse of an exponential function. This means that the logarithmic function reverses the coordinates of the exponential function. It also means that the graph of the logarithmic function is a reflection of the graph of the exponential function about the line $y = x$.

EXAMPLE 6 Graphs of Exponential and Logarithmic Functions

Graph $f(x) = 2^x$ and $g(x) = \log_2 x$ in the same rectangular coordinate system.

Solution We first set up a table of coordinates for $f(x) = 2^x$. Reversing these coordinates gives the coordinates for the inverse function $g(x) = \log_2 x$.

x	-2	-1	0	1	2	3
$f(x) = 2^x$	$\frac{1}{4}$	$\frac{1}{2}$	1	2	4	8

Reverse coordinates.

x	$\frac{1}{4}$	$\frac{1}{2}$	1	2	4	8
$g(x) = \log_2 x$	-2	-1	0	1	2	3

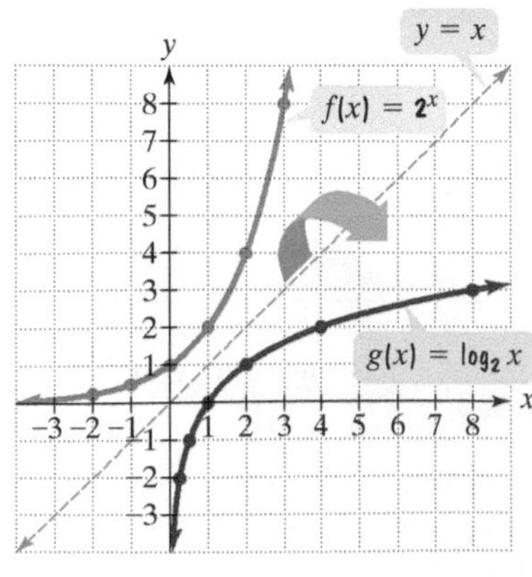

Figure 9.14 The graphs of $f(x) = 2^x$ and its inverse function

We now plot the ordered pairs from each table, connecting them with smooth curves. **Figure 9.14** shows the graphs of $f(x) = 2^x$ and its inverse function $g(x) = \log_2 x$. The graph of the inverse can also be drawn by reflecting the graph of $f(x) = 2^x$ about the line $y = x$. ■

Great Question!

You found the coordinates of $g(x) = \log_2 x$ by reversing the coordinates of $f(x) = 2^x$. Do I have to do it that way?

Not necessarily. You can obtain a partial table of coordinates for $g(x) = \log_2 x$ without having to obtain and reverse coordinates for $f(x) = 2^x$. Because $g(x) = \log_2 x$ means $2^{g(x)} = x$, we begin with values for $g(x)$ and compute corresponding values for x:

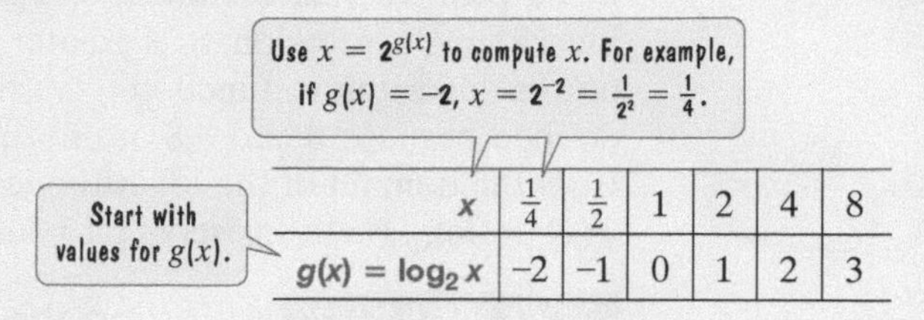

x	$\frac{1}{4}$	$\frac{1}{2}$	1	2	4	8
$g(x) = \log_2 x$	-2	-1	0	1	2	3

CHECK POINT 6 Graph $f(x) = 3^x$ and $g(x) = \log_3 x$ in the same rectangular coordinate system.

Figure 9.15 illustrates the relationship between the graph of an exponential function, shown in blue, and its inverse, a logarithmic function, shown in red, for bases greater than 1 and for bases between 0 and 1. Also shown and labeled are the exponential function's horizontal asymptote ($y = 0$) and the logarithmic function's vertical asymptote ($x = 0$).

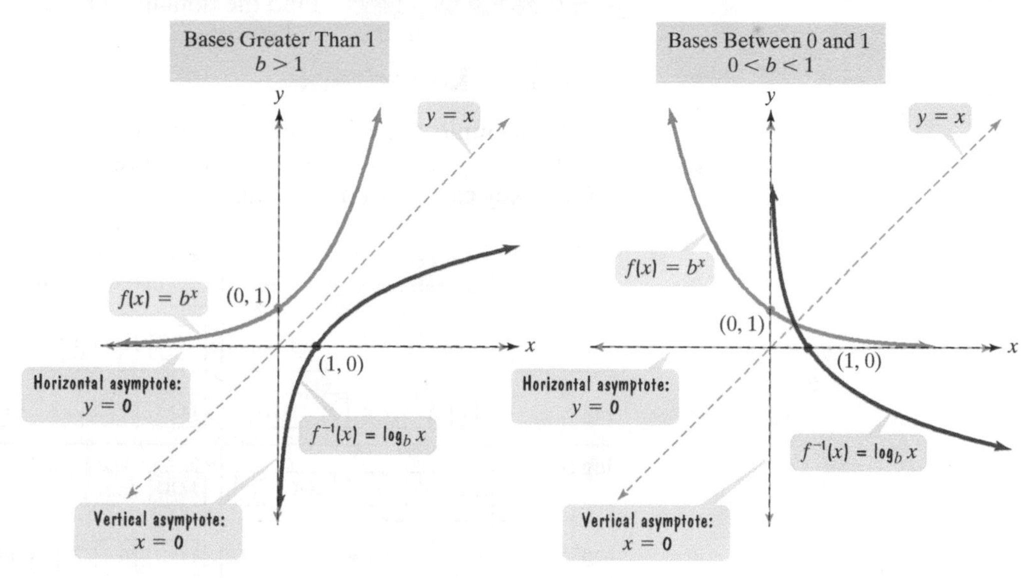

Figure 9.15 Graphs of exponential and logarithmic functions

The red graphs in **Figure 9.15** illustrate the following general characteristics of logarithmic functions:

Characteristics of Logarithmic Functions of the Form $f(x) = \log_b x$

1. The domain of $f(x) = \log_b x$ consists of all positive real numbers: $(0, \infty)$. The range of $f(x) = \log_b x$ consists of all real numbers: $(-\infty, \infty)$.
2. The graphs of all logarithmic functions of the form $f(x) = \log_b x$ pass through the point (1,0) because $f(1) = \log_b 1 = 0$. The x-intercept is 1. There is no y-intercept.
3. If $b > 1$, $f(x) = \log_b x$ has a graph that goes up to the right and is an increasing function.
4. If $0 < b < 1$, $f(x) = \log_b x$ has a graph that goes down to the right and is a decreasing function.
5. The graph of $f(x) = \log_b x$ approaches, but does not touch, the y-axis. The y-axis, or $x = 0$, is a vertical asymptote.

6 Find the domain of a logarithmic function.

The Domain of a Logarithmic Function

In Section 9.1, we learned that the domain of an exponential function of the form $f(x) = b^x$ includes all real numbers and its range is the set of positive real numbers. Because the logarithmic function reverses the domain and the range of the exponential function, the **domain of a logarithmic function of the form $f(x) = \log_b x$ is the set of all positive real numbers**. Thus, $\log_2 8$ is defined because the value of x in the logarithmic expression, 8, is greater than zero and therefore is included in the domain of the logarithmic function $f(x) = \log_2 x$. However, $\log_2 0$ and $\log_2(-8)$ are not defined because 0 and -8 are not positive real numbers and therefore are excluded from the domain of the logarithmic function $f(x) = \log_2 x$. In general, **the domain of $f(x) = \log_b g(x)$ consists of all x for which $g(x) > 0$.**

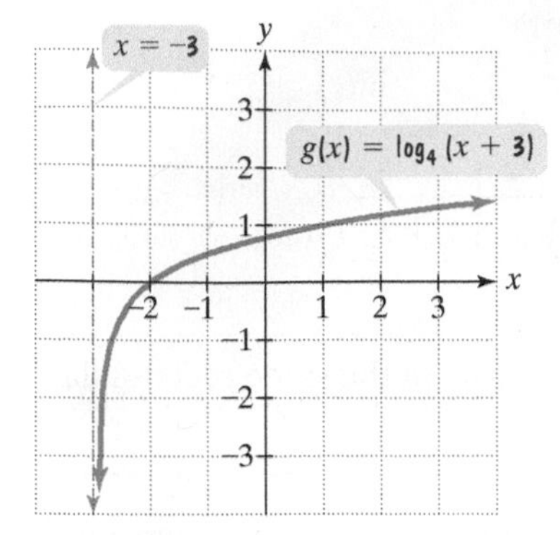

Figure 9.16 The domain of $g(x) = \log_4(x + 3)$ is $(-3, \infty)$.

EXAMPLE 7 Finding the Domain of a Logarithmic Function

Find the domain of $g(x) = \log_4(x + 3)$.

Solution The domain of g consists of all x for which $x + 3 > 0$. Solving this inequality for x, we obtain $x > -3$. Thus, the domain of g is $(-3, \infty)$. This is illustrated in **Figure 9.16**. The vertical asymptote is $x = -3$ and all points on the graph of g have x-coordinates that are greater than -3. ■

✓ CHECK POINT 7 Find the domain of $h(x) = \log_4(x - 5)$.

7 Use common logarithms.

Common Logarithms

The logarithmic function with base 10 is called the **common logarithmic function**. The function $f(x) = \log_{10} x$ is usually expressed as $f(x) = \log x$. A calculator with a [LOG] key can be used to evaluate common logarithms. Here are some examples.

Logarithm	Most Scientific Calculator Keystrokes	Most Graphing Calculator Keystrokes	Display (or Approximate Display)
log 1000	1000 [LOG]	[LOG] 1000 [ENTER]	3
$\log \frac{5}{2}$	[(] 5 [÷] 2 [)] [LOG]	[LOG] [(] 5 [÷] 2 [)] [ENTER]	0.39794
$\frac{\log 5}{\log 2}$	5 [LOG] [÷] 2 [LOG] [=]	[LOG] 5 [÷] [LOG] 2 [ENTER]	2.32193
log(−3)	3 [+/−] [LOG]	[LOG] [(−)] 3 [ENTER]	[ERROR]

Some graphing calculators display an open parenthesis when the [LOG] key is pressed. In this case, remember to close the set of parentheses after entering the function's domain value: [LOG] 5 [)] [÷] [LOG] 2 [)] [ENTER].

The error message or [NONREAL ANSWERS] message given by many calculators for log(−3) is a reminder that the domain of the common logarithmic function, $f(x) = \log x$, is the set of positive real numbers. In general, the domain of $f(x) = \log g(x)$ consists of all x for which $g(x) > 0$.

Many real-life phenomena start with rapid growth and then the growth begins to level off. This type of behavior can be modeled by logarithmic functions.

EXAMPLE 8 Modeling Heights of Children

The percentage of adult height attained by a boy who is x years old can be modeled by

$$f(x) = 29 + 48.8 \log(x + 1),$$

where x represents the boy's age and $f(x)$ represents the percentage of his adult height. Approximately what percentage of his adult height has a boy attained at age eight?

Solution We substitute the boy's age, 8, for x and evaluate the function at 8.

$$f(x) = 29 + 48.8\log(x + 1) \quad \text{This is the given function.}$$

$$f(8) = 29 + 48.8\log(8 + 1) \quad \text{Substitute 8 for } x.$$

$$= 29 + 48.8\log 9$$

Graphing calculator keystrokes: 29 [+] 48.8 [LOG] 9 [ENTER]

$$\approx 76$$

Thus, an 8-year-old boy has attained approximately 76% of his adult height. ■

☑ **CHECK POINT 8** Use the function in Example 8 to answer this question: Approximately what percentage of his adult height has a boy attained at age ten?

The basic properties of logarithms that were listed earlier in this section can be applied to common logarithms.

Properties of Common Logarithms

General Properties		Common Logarithms
1. $\log_b 1 = 0$		**1.** $\log 1 = 0$
2. $\log_b b = 1$		**2.** $\log 10 = 1$
3. $\log_b b^x = x$	Inverse properties	**3.** $\log 10^x = x$
4. $b^{\log_b x} = x$		**4.** $10^{\log x} = x$

The property $\log 10^x = x$ can be used to evaluate common logarithms involving powers of 10. For example,

$$\log 100 = \log 10^2 = 2, \quad \log 1000 = \log 10^3 = 3, \quad \text{and} \quad \log 10^{7.1} = 7.1.$$

EXAMPLE 9 Earthquake Intensity

The magnitude, R, on the Richter scale of an earthquake of intensity I is given by

$$R = \log\frac{I}{I_0},$$

where I_0 is the intensity of a barely felt zero-level earthquake. The earthquake that destroyed San Francisco in 1906 was $10^{8.3}$ times as intense as a zero-level earthquake. What was its magnitude on the Richter scale?

Solution Because the earthquake was $10^{8.3}$ times as intense as a zero-level earthquake, the intensity, I, is $10^{8.3}I_0$.

$$R = \log\frac{I}{I_0} \quad \text{This is the formula for magnitude on the Richter scale.}$$

$$R = \log\frac{10^{8.3}I_0}{I_0} \quad \text{Substitute } 10^{8.3}I_0 \text{ for } I.$$

$$= \log 10^{8.3} \quad \text{Simplify.}$$

$$= 8.3 \quad \text{Use the property } \log 10^x = x.$$

San Francisco's 1906 earthquake registered 8.3 on the Richter scale. ■

☑ **CHECK POINT 9** Use the formula in Example 9 to solve this problem. If an earthquake is 10,000 times as intense as a zero-level quake ($I = 10{,}000I_0$), what is its magnitude on the Richter scale?

---

8 Use natural logarithms.

Natural Logarithms

The logarithmic function with base e is called the **natural logarithmic function**. The function $f(x) = \log_e x$ is usually expressed as $f(x) = \ln x$, read "el en of x." A calculator with an LN key can be used to evaluate natural logarithms. Keystrokes are identical to those shown for common logarithmic evaluations on page 696.

Like the domain of all logarithmic functions, the domain of the natural logarithmic function $f(x) = \ln x$ is the set of all positive real numbers. Thus, the domain of $f(x) = \ln g(x)$ consists of all x for which $g(x) > 0$.

[−10, 10, 1] by [−10, 10, 1]

Figure 9.17 The domain of $f(x) = \ln(3 - x)$ is $(-\infty, 3)$.

EXAMPLE 10 Finding Domains of Natural Logarithmic Functions

Find the domain of each function:

a. $f(x) = \ln(3 - x)$ **b.** $g(x) = \ln(x - 3)^2$.

Solution

a. The domain of $f(x) = \ln(3 - x)$ consists of all x for which $3 - x > 0$. Solving this inequality for x, we obtain $x < 3$. Thus, the domain of f is $(-\infty, 3)$. This is verified by the graph in **Figure 9.17**.

b. The domain of $g(x) = \ln(x - 3)^2$ consists of all x for which $(x - 3)^2 > 0$. It follows that the domain of g is all real numbers except 3. Thus, the domain of g is $(-\infty, 3) \cup (3, \infty)$. This is shown by the graph in **Figure 9.18**. To make it more obvious that 3 is excluded from the domain, we used a DOT format. ■

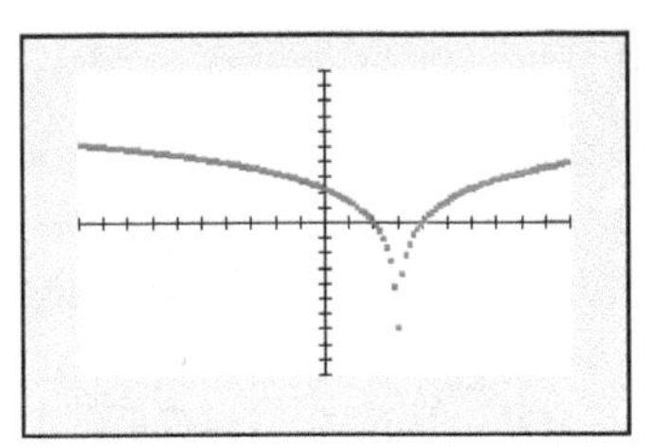

[−10, 10, 1] by [−10, 10, 1]

Figure 9.18 The value 3 is excluded from the domain of $g(x) = \ln(x - 3)^2$.

✓ CHECK POINT 10 Find the domain of each function:

a. $f(x) = \ln(4 - x)$

b. $g(x) = \ln x^2$.

The basic properties of logarithms that were listed earlier in this section can be applied to natural logarithms.

Properties of Natural Logarithms

General Properties		Natural Logarithms
1. $\log_b 1 = 0$		**1.** $\ln 1 = 0$
2. $\log_b b = 1$		**2.** $\ln e = 1$
3. $\log_b b^x = x$	Inverse properties	**3.** $\ln e^x = x$
4. $b^{\log_b x} = x$		**4.** $e^{\ln x} = x$

Examine the inverse properties, $\ln e^x = x$ and $e^{\ln x} = x$. Can you see how ln and e "undo" one another? For example,

$$\ln e^2 = 2, \quad \ln e^{7x^2} = 7x^2, \quad e^{\ln 2} = 2, \quad \text{and} \quad e^{\ln 7x^2} = 7x^2.$$

EXAMPLE 11 Dangerous Heat: Temperature in an Enclosed Vehicle

When the outside air temperature is anywhere from 72° to 96° Fahrenheit, the temperature in an enclosed vehicle climbs by 43° in the first hour. The bar graph in **Figure 9.19** shows the temperature increase throughout the hour. The function

$$f(x) = 13.4 \ln x - 11.6$$

models the temperature increase, $f(x)$, in degrees Fahrenheit, after x minutes. Use the function to find the temperature increase, to the nearest degree, after 50 minutes. How well does the function model the actual increase shown in **Figure 9.19**?

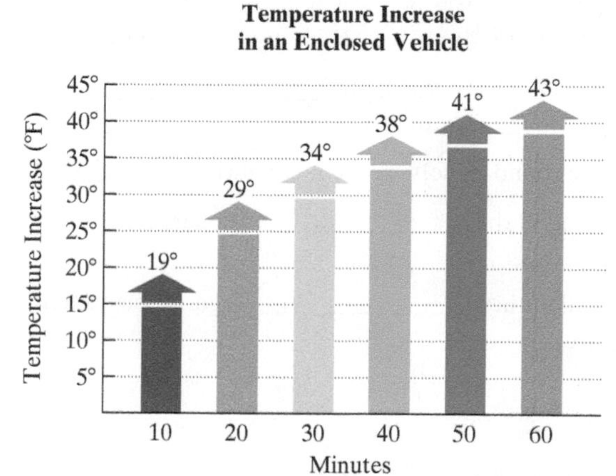

Figure 9.19

Source: Lynn I. Gibbs and David W. Lawrence, "Heat Exposure in an Enclosed Automobile." *Journal of the Lousiana State Medical Society*, Volume 147(12), 1995

Solution We find the temperature increase after 50 minutes by substituting 50 for x and evaluating the function at 50.

$$f(x) = 13.4 \ln x - 11.6$$ This is the given function.

$$f(50) = 13.4 \ln 50 - 11.6$$ Substitute 50 for x.

$$\approx 41$$ Graphing calculator keystrokes: 13.4 [ln] 50 [−] 11.6 [ENTER]. On some calculators, a parenthesis is needed after 50.

According to the function, the temperature will increase by approximately 41° after 50 minutes. Because the increase shown in **Figure 9.19** is 41°, the function models the actual increase extremely well. ■

✓ CHECK POINT 11 Use the function in Example 11 to find the temperature increase, to the nearest degree, after 30 minutes. How well does the function model the actual increase shown in **Figure 9.19**?

Achieving Success

We have seen that **the best way to achieve success in math is through practice**. Keeping up with your homework, preparing for tests, asking questions of your professor, reading your textbook, and attending all classes will help you learn the material and boost your confidence. Use language in a proactive way that reflects a sense of responsibility for your own successes and failures.

Reactive Language	Proactive Language
I'll try.	I'll do it.
That's just the way I am.	I can do better than that.
There's not a thing I can do.	I have options for improvement.
I have to.	I choose to.
I can't.	I can find a way.

CONCEPT AND VOCABULARY CHECK

Fill in each blank so that the resulting statement is true.

1. $y = \log_b x$ is equivalent to the exponential form ________________, $x > 0, b > 0, b \neq 1$.
2. The function $f(x) = \log_b x$ is the ________________ function with base ________________.
3. $\log_b b =$ ________________
4. $\log_b 1 =$ ________________
5. $\log_b b^x =$ ________________
6. $b^{\log_b x} =$ ________________
7. Using interval notation, the domain of $f(x) = \log_b x$ is ________________ and the range is ________________.
8. The graph of $f(x) = \log_b x$ approaches, but does not touch, the ________________-axis. This axis, whose equation is ________________, is a/an ________________ asymptote.
9. The domain of $g(x) = \log_2(5 - x)$ can be found by solving the inequality ________________.
10. The logarithmic function with base 10 is called the ________________ logarithmic function. The function $f(x) = \log_{10} x$ is usually expressed as $f(x) =$ ________________.
11. The logarithmic function with base e is called the ________________ logarithmic function. The function $f(x) = \log_e x$ is usually expressed as $f(x) =$ ________________.

9.3 EXERCISE SET

MyMathLab®

Practice Exercises

In Exercises 1–8, write each equation in its equivalent exponential form.

1. $4 = \log_2 16$
2. $6 = \log_2 64$
3. $2 = \log_3 x$
4. $2 = \log_9 x$
5. $5 = \log_b 32$
6. $3 = \log_b 27$
7. $\log_6 216 = y$
8. $\log_5 125 = y$

In Exercises 9–20, write each equation in its equivalent logarithmic form.

9. $2^3 = 8$
10. $5^4 = 625$
11. $2^{-4} = \frac{1}{16}$
12. $5^{-3} = \frac{1}{125}$
13. $\sqrt[3]{8} = 2$
14. $\sqrt[3]{64} = 4$
15. $13^2 = x$
16. $15^2 = x$
17. $b^3 = 1000$
18. $b^3 = 343$
19. $7^y = 200$
20. $8^y = 300$

In Exercises 21–42, evaluate each expression without using a calculator.

21. $\log_4 16$
22. $\log_7 49$
23. $\log_2 64$
24. $\log_3 27$
25. $\log_5 \frac{1}{5}$
26. $\log_6 \frac{1}{6}$
27. $\log_2 \frac{1}{8}$
28. $\log_3 \frac{1}{9}$
29. $\log_7 \sqrt{7}$
30. $\log_6 \sqrt{6}$
31. $\log_2 \frac{1}{\sqrt{2}}$
32. $\log_3 \frac{1}{\sqrt{3}}$
33. $\log_{64} 8$
34. $\log_{81} 9$
35. $\log_5 5$
36. $\log_{11} 11$
37. $\log_4 1$
38. $\log_6 1$
39. $\log_5 5^7$
40. $\log_4 4^6$
41. $8^{\log_8 19}$
42. $7^{\log_7 23}$
43. Graph $f(x) = 4^x$ and $g(x) = \log_4 x$ in the same rectangular coordinate system.
44. Graph $f(x) = 5^x$ and $g(x) = \log_5 x$ in the same rectangular coordinate system.
45. Graph $f(x) = \left(\frac{1}{2}\right)^x$ and $g(x) = \log_{\frac{1}{2}} x$ in the same rectangular coordinate system.
46. Graph $f(x) = \left(\frac{1}{4}\right)^x$ and $g(x) = \log_{\frac{1}{4}} x$ in the same rectangular coordinate system.

In Exercises 47–52, find the domain of each logarithmic function.

47. $f(x) = \log_5(x + 4)$
48. $f(x) = \log_5(x + 6)$
49. $f(x) = \log(2 - x)$
50. $f(x) = \log(7 - x)$
51. $f(x) = \ln(x - 2)^2$
52. $f(x) = \ln(x - 7)^2$

In Exercises 53–66, evaluate each expression without using a calculator.

53. $\log 100$
54. $\log 1000$
55. $\log 10^7$
56. $\log 10^8$
57. $10^{\log 33}$
58. $10^{\log 53}$

59. $\ln 1$ **60.** $\ln e$

61. $\ln e^6$ **62.** $\ln e^7$

63. $\ln \frac{1}{e^6}$ **64.** $\ln \frac{1}{e^7}$

65. $e^{\ln 125}$ **66.** $e^{\ln 300}$

In Exercises 67–72, simplify each expression.

67. $\ln e^{9x}$

68. $\ln e^{13x}$

69. $e^{\ln 5x^2}$

70. $e^{\ln 7x^2}$

71. $10^{\log \sqrt{x}}$

72. $10^{\log \sqrt[3]{x}}$

Practice PLUS

In Exercises 73–76, write each equation in its equivalent exponential form. Then solve for x.

73. $\log_3(x - 1) = 2$

74. $\log_5(x + 4) = 2$

75. $\log_4 x = -3$

76. $\log_{64} x = \frac{2}{3}$

In Exercises 77–80, evaluate each expression without using a calculator.

77. $\log_3(\log_7 7)$

78. $\log_5(\log_2 32)$

79. $\log_2(\log_3 81)$

80. $\log(\ln e)$

In Exercises 81–86, match each function with its graph. The graphs are labeled (a) through (f), and each graph is displayed in a [−5, 5, 1] by [−5, 5, 1] viewing rectangle.

81. $f(x) = \ln(x + 2)$

82. $f(x) = \ln(x - 2)$

83. $f(x) = \ln x + 2$

84. $f(x) = \ln x - 2$

85. $f(x) = \ln(1 - x)$

86. $f(x) = \ln(2 - x)$

a.

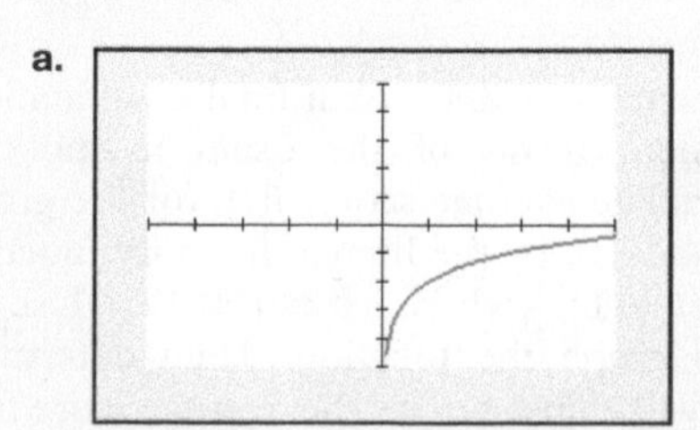

b.

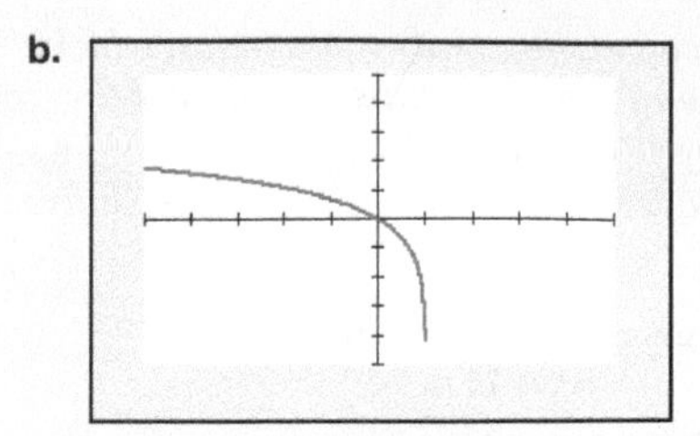

c.

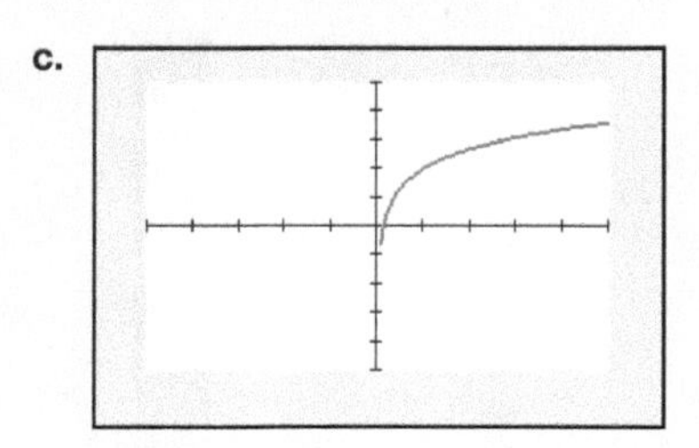

d.

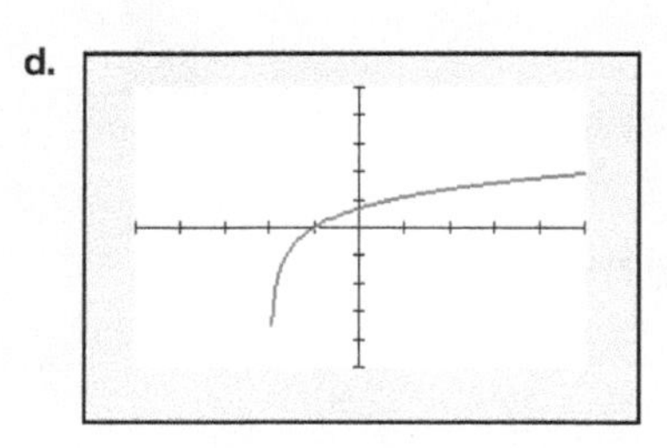

e.

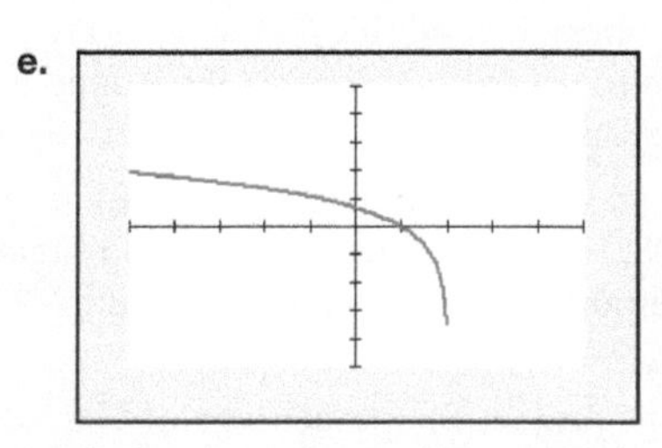

f.

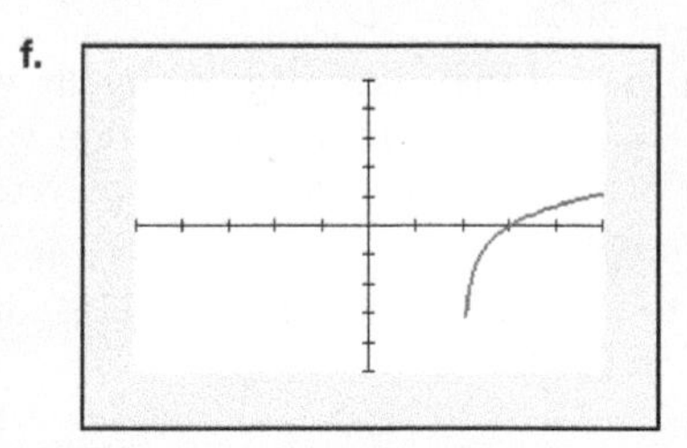

Application Exercises

The percentage of adult height attained by a girl who is x years old can be modeled by

$$f(x) = 62 + 35\log(x - 4),$$

where x represents the girl's age (from 5 to 15) and f(x) represents the percentage of her adult height. Use the formula to solve Exercises 87–88. Round answers to the nearest tenth of a percent.

87. Approximately what percentage of her adult height has a girl attained at age 13?

88. Approximately what percentage of her adult height has a girl attained at age ten?

Obesity is determined by calculating body mass index, which takes into account height and weight. Adults are obese if they have a body mass index of 30 or greater. The bar graph shows the percentage of U.S. adults, ages 20 to 74, who were obese in five selected years.

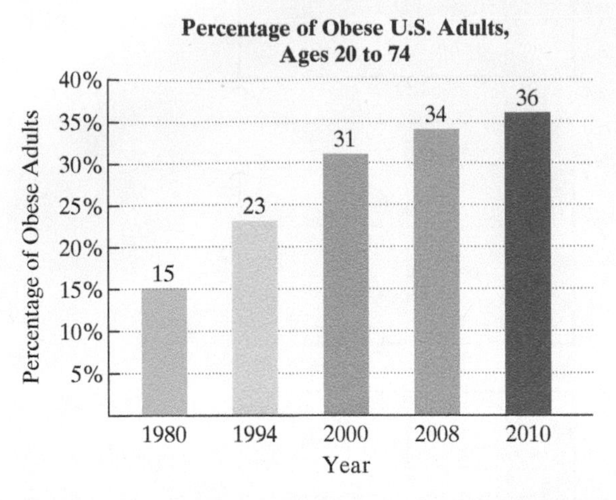

Source: Centers for Disease Control and Prevention

The function

$$f(x) = 6.3 \ln x + 12.8$$

models the percentage of obese U.S. adults, $f(x)$, x years after 1979. Use this information to solve Exercises 89–90. Round all answers to the nearest percent.

89. **a.** Use the function to find the percentage of adults who were obese in 2008. How well does this rounded function value estimate the percent displayed by the graph?

b. Use the function to project the percentage of U.S. adults who will be obese in 2025.

90. **a.** Use the function to find the percentage of adults who were obese in 2000. Does this rounded function value underestimate or overesimate the percent displayed by the graph? By how much?

b. Use the function to project the percentage of U.S. adults who will be obese in 2030.

The loudness level of a sound, D, in decibels, is given by the formula

$$D = 10 \log(10^{12} I),$$

where I is the intensity of the sound, in watts per meter^2. Decibel levels range from 0, a barely audible sound, to 160, a sound resulting in a ruptured eardrum. Use the formula to solve Exercises 91–92.

91. The sound of a blue whale can be heard 500 miles away, reaching an intensity of 6.3×10^6 watts per meter^2. Determine the decibel level of this sound. At close range, can the sound of a blue whale rupture the human eardrum?

92. What is the decibel level of a normal conversation, 3.2×10^{-6} watt per meter^2?

93. Students in a psychology class took a final examination. As part of an experiment to see how much of the course content they remembered over time, they took equivalent forms of the exam in monthly intervals thereafter. The average score for the group, $f(t)$, after t months was modeled by the function

$$f(t) = 88 - 15 \ln(t + 1), \qquad 0 \le t \le 12.$$

a. What was the average score on the original exam?

b. What was the average score, to the nearest tenth, after 2 months? 4 months? 6 months? 8 months? 10 months? one year?

c. Sketch the graph of f (either by hand or with a graphing utility). Describe what the graph indicates in terms of the material retained by the students.

Explaining the Concepts

94. Describe the relationship between an equation in logarithmic form and an equivalent equation in exponential form.

95. What question can be asked to help evaluate $\log_3 81$?

96. Explain why the logarithm of 1 with base b is 0.

97. Describe the following property using words: $\log_b b^x = x$.

98. Explain how to use the graph of $f(x) = 2^x$ to obtain the graph of $g(x) = \log_2 x$.

99. Explain how to find the domain of a logarithmic function.

100. Logarithmic models are well suited to phenomena in which growth is initially rapid, but then begins to level off. Describe something that is changing over time that can be modeled using a logarithmic function.

101. Suppose that a girl is 4 feet 6 inches at age 10. Explain how to use the function in Exercises 87–88 to determine how tall she can expect to be as an adult.

Technology Exercises

In Exercises 102–105, graph f and g in the same viewing rectangle. Then describe the relationship of the graph of g to the graph of f.

102. $f(x) = \ln x$, $g(x) = \ln(x + 3)$

103. $f(x) = \ln x$, $g(x) = \ln x + 3$

104. $f(x) = \log x$, $g(x) = -\log x$

105. $f(x) = \log x$, $g(x) = \log(x - 2) + 1$

106. Students in a mathematics class took a final examination. They took equivalent forms of the exam in monthly intervals thereafter. The average score, $f(t)$, for the group after t months is modeled by the human memory function $f(t) = 75 - 10 \log(t + 1)$, where $0 \le t \le 12$. Use a graphing utility to graph the function. Then determine how many months will elapse before the average score falls below 65.

107. In parts (a)–(c), graph f and g in the same viewing rectangle.

a. $f(x) = \ln(3x)$, $g(x) = \ln 3 + \ln x$

b. $f(x) = \log(5x^2)$, $g(x) = \log 5 + \log x^2$

c. $f(x) = \ln(2x^3)$, $g(x) = \ln 2 + \ln x^3$

d. Describe what you observe in parts (a)–(c). Generalize this observation by writing an equivalent expression for $\log_b(MN)$, where $M > 0$ and $N > 0$.

e. Complete this statement: The logarithm of a product is equal to ____________________

108. Graph each of the following functions in the same viewing rectangle and then place the functions in order from the one that increases most slowly to the one that increases most rapidly.

$$y = x, y = \sqrt{x}, y = e^x, y = \ln x, y = x^x, y = x^2$$

Critical Thinking Exercises

Make Sense? *In Exercises 109–112, determine whether each statement makes sense or does not make sense, and explain your reasoning.*

109. I estimate that $\log_8 16$ lies between 1 and 2 because $8^1 = 8$ and $8^2 = 64$.

110. When graphing a logarithmic function, I like to show the graph of its horizontal asymptote.

111. I can evaluate some common logarithms without having to use a calculator.

112. An earthquake of magnitude 8 on the Richter scale is twice as intense as an earthquake of magnitude 4.

In Exercises 113–116, determine whether each statement is true or false. If the statement is false, make the necessary change(s) to produce a true statement.

113. $\dfrac{\log_2 8}{\log_2 4} = \dfrac{8}{4}$

114. $\log(-100) = -2$

115. The domain of $f(x) = \log_2 x$ is $(-\infty, \infty)$.

116. $\log_b x$ is the exponent to which b must be raised to obtain x.

117. Without using a calculator, find the exact value of

$$\frac{\log_3 81 - \log_\pi 1}{\log_{2\sqrt{2}} 8 - \log 0.001}$$

118. Without using a calculator, find the exact value of $\log_4[\log_3(\log_2 8)]$.

119. Without using a calculator, determine which is the greater number: $\log_4 60$ or $\log_3 40$.

Review Exercises

120. Solve the system:

$$\begin{cases} 2x = 11 - 5y \\ 3x - 2y = -12. \end{cases}$$

(Section 3.1, Example 5)

121. Factor completely:

$$6x^2 - 8xy + 2y^2.$$

(Section 5.4, Example 9)

122. Solve: $x + 3 \le -4$ or $2 - 7x \le 16$.

(Section 4.2, Example 6)

Preview Exercises

Exercises 123–125 will help you prepare for the material covered in the next section. In each exercise, evaluate the indicated logarithmic expressions without using a calculator.

123. a. Evaluate: $\log_2 32$.

b. Evaluate: $\log_2 8 + \log_2 4$.

c. What can you conclude about $\log_2 32$, or $\log_2(8 \cdot 4)$?

124. a. Evaluate: $\log_2 16$.

b. Evaluate: $\log_2 32 - \log_2 2$.

c. What can you conclude about

$$\log_2 16, \text{ or } \log_2\left(\frac{32}{2}\right)?$$

125. a. Evaluate: $\log_3 81$.

b. Evaluate: $2 \log_3 9$.

c. What can you conclude about

$$\log_3 81, \text{ or } \log_3 9^2?$$

SECTION

9.4 Properties of Logarithms

What am I supposed to learn?

After studying this section, you should be able to:

1. Use the product rule.
2. Use the quotient rule.
3. Use the power rule.
4. Expand logarithmic expressions.
5. Condense logarithmic expressions.
6. Use the change-of-base property.

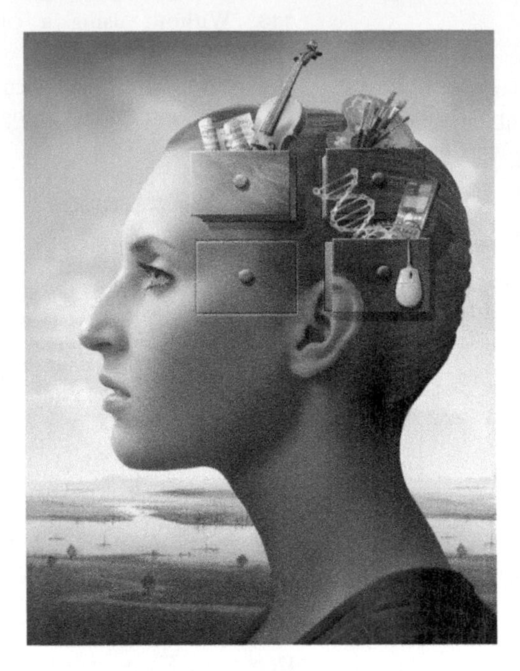

We all learn new things in different ways. In this section, we consider important properties of logarithms. What would be the most effective way for you to learn these properties? Would it be helpful to use your graphing utility and discover one of these properties for yourself? To do so, work Exercise 107 in Exercise Set 9.3 before continuing. Would it be helpful to evaluate certain logarithmic expressions that suggest three of the properties? If this is the case, work Preview Exercises 123–125 in Exercise Set 9.3 before continuing. Would the properties become more meaningful if you could see exactly where they come from? If so, you will find details of the proofs of many of these properties in the appendix. The remainder of our work in this chapter will be based on the properties of logarithms that you learn in this section.

1 Use the product rule.

The Product Rule

Properties of exponents correspond to properties of logarithms. For example, when we multiply exponential expressions with the same base, we add exponents:

$$b^m \cdot b^n = b^{m+n}.$$

This property of exponents, coupled with an awareness that a logarithm is an exponent, suggests the following property, called the **product rule**:

Discover for Yourself

We know that $\log 100{,}000 = 5$. Show that you get the same result by writing 100,000 as $1000 \cdot 100$ and then using the product rule. Then verify the product rule by using other numbers whose logarithms are easy to find.

The Product Rule

Let b, M, and N be positive real numbers with $b \neq 1$.

$$\log_b(MN) = \log_b M + \log_b N$$

The logarithm of a product is the sum of the logarithms.

When we use the product rule to write a single logarithm as the sum of two logarithms, we say that we are **expanding a logarithmic expression**. For example, we can use the product rule to expand $\ln(7x)$:

$$\ln(7x) = \ln 7 + \ln x.$$

The logarithm of a product | is | the sum of the logarithms.

EXAMPLE 1 Using the Product Rule

Use the product rule to expand each logarithmic expression:

a. $\log_4(7 \cdot 5)$ **b.** $\log(10x)$.

Solution

a. $\log_4(7 \cdot 5) = \log_4 7 + \log_4 5$ — The logarithm of a product is the sum of the logarithms.

b. $\log(10x) = \log 10 + \log x$ — The logarithm of a product is the sum of the logarithms. These are common logarithms with base 10 understood.

$= 1 + \log x$ — Because $\log_b b = 1$, then $\log 10 = 1$. ■

✓ **CHECK POINT 1** Use the product rule to expand each logarithmic expression:

a. $\log_6(7 \cdot 11)$ **b.** $\log(100x)$.

2 Use the quotient rule.

The Quotient Rule

When we divide exponential expressions with the same base, we subtract exponents:

$$\frac{b^m}{b^n} = b^{m-n}.$$

This property suggests the following property of logarithms, called the **quotient rule:**

Discover for Yourself

We know that $\log_2 16 = 4$. Show that you get the same result by writing 16 as $\frac{32}{2}$ and then using the quotient rule. Then verify the quotient rule using other numbers whose logarithms are easy to find.

The Quotient Rule

Let b, M, and N be positive real numbers with $b \neq 1$.

$$\log_b\left(\frac{M}{N}\right) = \log_b M - \log_b N$$

The logarithm of a quotient is the difference of the logarithms.

When we use the quotient rule to write a single logarithm as the difference of two logarithms, we say that we are **expanding a logarithmic expression**. For example, we can use the quotient rule to expand $\log \frac{x}{2}$:

$$\log\left(\frac{x}{2}\right) = \log x - \log 2.$$

The logarithm of a quotient | is | the difference of the logarithms.

EXAMPLE 2 Using the Quotient Rule

Use the quotient rule to expand each logarithmic expression:

a. $\log_7\left(\frac{19}{x}\right)$ **b.** $\ln\left(\frac{e^3}{7}\right)$.

Solution

a. $\log_7\left(\frac{19}{x}\right) = \log_7 19 - \log_7 x$ — The logarithm of a quotient is the difference of the logarithms.

b. $\ln\left(\frac{e^3}{7}\right) = \ln e^3 - \ln 7$ — The logarithm of a quotient is the difference of the logarithms. These are natural logarithms with base e understood.

$= 3 - \ln 7$ — Because $\ln e^x = x$, then $\ln e^3 = 3$. ■

✓ **CHECK POINT 2** Use the quotient rule to expand each logarithmic expression:

a. $\log_8\left(\frac{23}{x}\right)$ **b.** $\ln\left(\frac{e^5}{11}\right)$.

3 Use the power rule.

The Power Rule

When an exponential expression is raised to a power, we multiply exponents:

$$(b^m)^n = b^{mn}.$$

This property suggests the following property of logarithms, called the **power rule**:

The Power Rule

Let b and M be positive real numbers with $b \neq 1$, and let p be any real number.

$$\log_b M^p = p \log_b M$$

The logarithm of a number with an exponent is the product of the exponent and the logarithm of that number.

When we use the power rule to "pull the exponent to the front," we say that we are **expanding a logarithmic expression**. For example, we can use the power rule to expand $\ln x^2$:

$$\ln x^2 = 2 \ln x.$$

The logarithm of a number with an exponent | is | the product of the exponent and the logarithm of that number.

Figure 9.20 shows the graphs of $y = \ln x^2$ and $y = 2 \ln x$ in $[-5, 5, 1]$ by $[-5, 5, 1]$ viewing rectangles. Are $\ln x^2$ and $2 \ln x$ the same? The graphs illustrate that $y = \ln x^2$ and $y = 2 \ln x$ have different domains. The graphs are only the same if $x > 0$. Thus, we should write

$$\ln x^2 = 2 \ln x \quad \text{for} \quad x > 0.$$

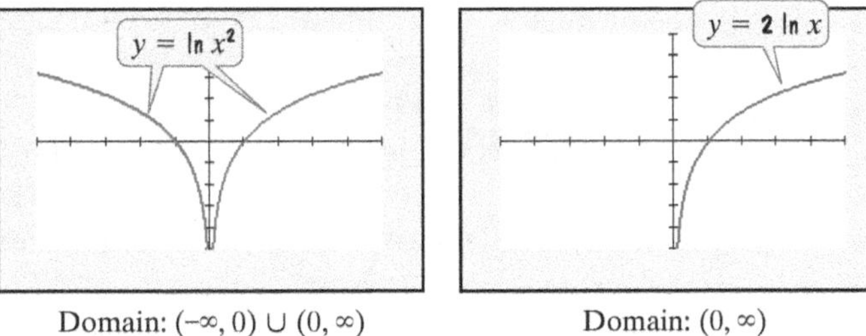

Domain: $(-\infty, 0) \cup (0, \infty)$ Domain: $(0, \infty)$

Figure 9.20
$\ln x^2$ and $2 \ln x$ have different domains.

When expanding a logarithmic expression, you might want to determine whether the rewriting has changed the domain of the expression. For the rest of this section, assume that all variables and variable expressions represent positive numbers.

EXAMPLE 3 Using the Power Rule

Use the power rule to expand each logarithmic expression:

a. $\log_5 7^4$ **b.** $\ln \sqrt{x}$ **c.** $\log(4x)^5$.

Solution

a. $\log_5 7^4 = 4\log_5 7$ — The logarithm of a number with an exponent is the exponent times the logarithm of that number.

b. $\ln\sqrt{x} = \ln x^{\frac{1}{2}}$ — Rewrite the radical using a rational exponent.

$= \frac{1}{2}\ln x$ — Use the power rule to bring the exponent to the front.

c. $\log(4x)^5 = 5\log(4x)$ — We immediately apply the power rule because the entire variable expression, 4x, is raised to the 5th power. ■

✓ CHECK POINT 3 Use the power rule to expand each logarithmic expression:
a. $\log_6 8^9$ **b.** $\ln\sqrt[3]{x}$ **c.** $\log(x+4)^2$.

4 Expand logarithmic expressions.

Expanding Logarithmic Expressions

It is sometimes necessary to use more than one property of logarithms when you expand a logarithmic expression. Properties for expanding logarithmic expressions are as follows:

Properties for Expanding Logarithmic Expressions

For $M > 0$ and $N > 0$:

1. $\log_b(MN) = \log_b M + \log_b N$ — Product rule
2. $\log_b\left(\frac{M}{N}\right) = \log_b M - \log_b N$ — Quotient rule
3. $\log_b M^p = p\log_b M$ — Power rule

Great Question!

Are there some common blunders that I can avoid when using properties of logarithms?

The graphs show

$y_1 = \ln(x+3)$ and $y_2 = \ln x + \ln 3$.

The graphs are not the same:

$$\ln(x+3) \neq \ln x + \ln 3.$$

In general,

$$\log_b(M+N) \neq \log_b M + \log_b N.$$

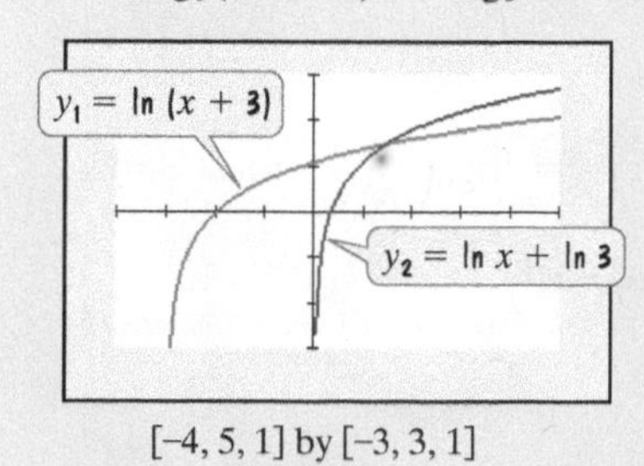

[-4, 5, 1] by [-3, 3, 1]

Try to avoid the following errors:

Incorrect!

$\log_b(M+N) = \log_b M + \log_b N$

$\log_b(M-N) = \log_b M - \log_b N$

$\log_b(M \cdot N) = \log_b M \cdot \log_b N$

$\log_b\left(\frac{M}{N}\right) = \frac{\log_b M}{\log_b N}$

$\frac{\log_b M}{\log_b N} = \log_b M - \log_b N$

$\log_b(MN^p) = p\log_b(MN)$

EXAMPLE 4 Expanding Logarithmic Expressions

Use logarithmic properties to expand each expression as much as possible:

a. $\log_b(x^2\sqrt{y})$ **b.** $\log_6\left(\frac{\sqrt[3]{x}}{36y^4}\right)$.

Solution We will have to use two or more of the properties for expanding logarithms in each part of this example.

a. $\log_b(x^2\sqrt{y}) = \log_b(x^2 y^{\frac{1}{2}})$ Use exponential notation.

$= \log_b x^2 + \log_b y^{\frac{1}{2}}$ Use the product rule.

$= 2\log_b x + \frac{1}{2}\log_b y$ Use the power rule.

b. $\log_6\left(\frac{\sqrt[3]{x}}{36y^4}\right) = \log_6\left(\frac{x^{\frac{1}{3}}}{36y^4}\right)$ Use exponential notation.

$= \log_6 x^{\frac{1}{3}} - \log_6(36y^4)$ Use the quotient rule.

$= \log_6 x^{\frac{1}{3}} - (\log_6 36 + \log_6 y^4)$ Use the product rule on $\log_6(36y^4)$.

$= \frac{1}{3}\log_6 x - (\log_6 36 + 4\log_6 y)$ Use the power rule.

$= \frac{1}{3}\log_6 x - \log_6 36 - 4\log_6 y$ Apply the distributive property.

$= \frac{1}{3}\log_6 x - 2 - 4\log_6 y$ $\log_6 36 = 2$ because 2 is the power to which we must raise 6 to get 36. $(6^2 = 36)$ ■

CHECK POINT 4 Use logarithmic properties to expand each expression as much as possible:

a. $\log_b(x^4\sqrt[3]{y})$

b. $\log_5\left(\frac{\sqrt{x}}{25y^3}\right)$.

5 Condense logarithmic expressions.

Condensing Logarithmic Expressions

To **condense a logarithmic expression**, we write the sum or difference of two or more logarithmic expressions as a single logarithmic expression. We use the properties of logarithms to do so:

Great Question!

Are the properties listed on the right the same as those in the box on page 707?

Yes. The only difference is that we've reversed the sides in each property from the previous box.

Properties for Condensing Logarithmic Expressions

For $M > 0$ and $N > 0$:

1. $\log_b M + \log_b N = \log_b(MN)$ Product rule
2. $\log_b M - \log_b N = \log_b\left(\frac{M}{N}\right)$ Quotient rule
3. $p\log_b M = \log_b M^p$ Power rule

EXAMPLE 5 Condensing Logarithmic Expressions

Write as a single logarithm:

a. $\log_4 2 + \log_4 32$

b. $\log(4x - 3) - \log x$.

Solution

a. $\log_4 2 + \log_4 32 = \log_4(2 \cdot 32)$ — Use the product rule.

$= \log_4 64$ — We now have a single logarithm. However, we can simplify.

$= 3$ — $\log_4 64 = 3$ because $4^3 = 64$.

b. $\log(4x - 3) - \log x = \log\left(\frac{4x - 3}{x}\right)$ — Use the quotient rule. ■

CHECK POINT 5 Write as a single logarithm:

a. $\log 25 + \log 4$ **b.** $\log(7x + 6) - \log x$.

Coefficients of logarithms must be 1 before you can condense them using the product and quotient rules. For example, to condense

$$2 \ln x + \ln(x + 1),$$

the coefficient of the first term must be 1. We use the power rule to rewrite the coefficient as an exponent:

1. Use the power rule to make the number in front an exponent.

$$2 \ln x + \ln(x + 1) = \ln x^2 + \ln(x + 1) = \ln[x^2(x + 1)].$$

2. Use the product rule. The sum of logarithms with coefficients of 1 is the logarithm of the product.

EXAMPLE 6 Condensing Logarithmic Expressions

Write as a single logarithm:

a. $\frac{1}{2}\log x + 4\log(x - 1)$ **b.** $3\ln(x + 7) - \ln x$

c. $4\log_b x - 2\log_b 6 - \frac{1}{2}\log_b y$.

Solution

a. $\frac{1}{2}\log x + 4\log(x - 1)$

$= \log x^{\frac{1}{2}} + \log(x - 1)^4$ — Use the power rule so that all coefficients are 1.

$= \log\left[x^{\frac{1}{2}}(x - 1)^4\right]$ — Use the product rule. The condensed form can be expressed as $\log\left[\sqrt{x}\,(x - 1)^4\right]$.

b. $3\ln(x + 7) - \ln x$

$= \ln(x + 7)^3 - \ln x$ — Use the power rule so that all coefficients are 1.

$= \ln\left[\frac{(x + 7)^3}{x}\right]$ — Use the quotient rule.

c. $4\log_b x - 2\log_b 6 - \frac{1}{2}\log_b y$

$= \log_b x^4 - \log_b 6^2 - \log_b y^{\frac{1}{2}}$ — Use the power rule so that all coefficients are 1.

$= \log_b x^4 - \left(\log_b 36 + \log_b y^{\frac{1}{2}}\right)$ — Rewrite as a single subtraction.

$= \log_b x^4 - \log_b\left(36y^{\frac{1}{2}}\right)$ — Use the product rule.

$= \log_b\left(\frac{x^4}{36y^{\frac{1}{2}}}\right)$ or $\log_b\left(\frac{x^4}{36\sqrt{y}}\right)$ — Use the quotient rule. ■

✓ CHECK POINT 6 Write as a single logarithm:

a. $2 \ln x + \frac{1}{3}\ln(x + 5)$ **b.** $2 \log(x - 3) - \log x$

c. $\frac{1}{4}\log_b x - 2 \log_b 5 - 10 \log_b y$.

6 Use the change-of-base property.

The Change-of-Base Property

We have seen that calculators give the values of both common logarithms (base 10) and natural logarithms (base e). To find a logarithm with any other base, we can use the following change-of-base property:

The Change-of-Base Property

For any logarithmic bases a and b, and any positive number M,

$$\log_b M = \frac{\log_a M}{\log_a b}.$$

The logarithm of M with base b is equal to the logarithm of M with any new base divided by the logarithm of b with that new base.

In the change-of-base property, base b is the base of the original logarithm. Base a is a new base that we introduce. Thus, the change-of-base property allows us to change from base b to *any* new base a, as long as the newly introduced base is a positive number not equal to 1.

The change-of-base property is used to write a logarithm in terms of quantities that can be evaluated with a calculator. Because calculators contain keys for common (base 10) and natural (base e) logarithms, we will frequently introduce base 10 or base e.

Change-of-Base Property	Introducing Common Logarithms	Introducing Natural Logarithms
$\log_b M = \frac{\log_a M}{\log_a b}$ a is the new introduced base.	$\log_b M = \frac{\log_{10} M}{\log_{10} b}$ 10 is the new introduced base.	$\log_b M = \frac{\log_e M}{\log_e b}$ e is the new introduced base.

Using the notations for common logarithms and natural logarithms, we have the following results:

The Change-of-Base Property: Introducing Common and Natural Logarithms

Introducing Common Logarithms

$$\log_b M = \frac{\log M}{\log b}$$

Introducing Natural Logarithms

$$\log_b M = \frac{\ln M}{\ln b}$$

EXAMPLE 7 Changing Base to Common Logarithms

Use common logarithms to evaluate $\log_5 140$.

Solution Because

$$\log_b M = \frac{\log M}{\log b},$$

$$\log_5 140 = \frac{\log 140}{\log 5}$$

$$\approx 3.07.$$

Use a calculator: 140 [LOG] [÷] 5 [LOG] [=] or [LOG] 140 [÷] [LOG] 5 [ENTER]. On some calculators, parentheses are needed after 140 and 5.

This means that $\log_5 140 \approx 3.07$. ■

Discover for Yourself

Find a reasonable estimate of $\log_5 140$ to the nearest whole number: 5 to what power is 140? Compare your estimate to the value obtained in Example 7.

✓ **CHECK POINT 7** Use common logarithms to evaluate $\log_7 2506$.

EXAMPLE 8 Changing Base to Natural Logarithms

Use natural logarithms to evaluate $\log_5 140$.

Solution Because

$$\log_b M = \frac{\ln M}{\ln b},$$

$$\log_5 140 = \frac{\ln 140}{\ln 5}$$

$$\approx 3.07.$$

Use a calculator: 140 [LN] [÷] 5 [LN] [=] or [LN] 140 [÷] [LN] 5 [ENTER]. On some calculators, parentheses are needed after 140 and 5.

We have again shown that $\log_5 140 \approx 3.07$. ■

✓ **CHECK POINT 8** Use natural logarithms to evaluate $\log_7 2506$.

Using Technology

We can use the change-of-base property to graph logarithmic functions with bases other than 10 or e on a graphing utility. For example, **Figure 9.21** shows the graphs of

$$y = \log_2 x \quad \text{and} \quad y = \log_{20} x$$

in a [0, 10, 1] by [−3, 3, 1] viewing rectangle. Because $\log_2 x = \frac{\ln x}{\ln 2}$ and $\log_{20} x = \frac{\ln x}{\ln 20}$, the functions are entered as

$y_1 =$ [LN] x [÷] [LN] 2

and $y_2 =$ [LN] x [÷] [LN] 20.

On some calculators, parentheses are needed after x, 2, and 20.

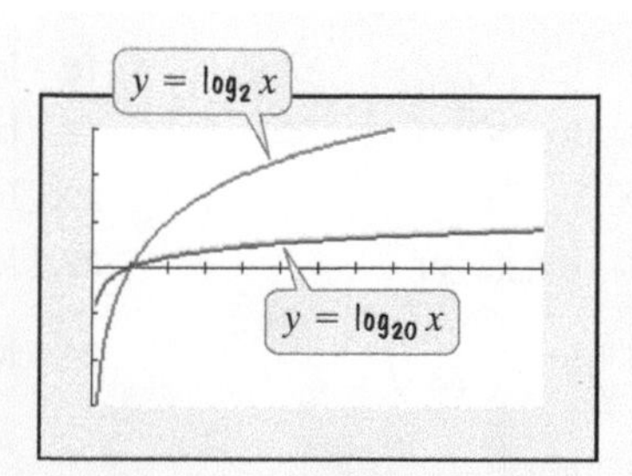

Figure 9.21 Using the change-of-base property to graph logarithmic functions

CONCEPT AND VOCABULARY CHECK

Fill in each blank so that the resulting statement is true.

1. The product rule for logarithms states that $\log_b(MN) =$ ______________. The logarithm of a product is the ______________ of the logarithms.
2. The quotient rule for logarithms states that $\log_b\left(\frac{M}{N}\right) =$ ______________. The logarithm of a quotient is the ______________ of the logarithms.
3. The power rule for logarithms states that $\log_b M^p =$ ______________. The logarithm of a number with an exponent is the ______________ of the exponent and the logarithm of that number.
4. The change-of-base property for logarithms allows us to write logarithms with base b in terms of a new base a. Introducing base a, the property states that

$$\log_b M = \frac{\_\_\_\_}{\_\_\_\_}.$$

9.4 EXERCISE SET

MyMathLab®

Practice Exercises

In all exercises, assume that all variables and variable expressions represent positive numbers.

In Exercises 1–36, use properties of logarithms to expand each logarithmic expression as much as possible. Where possible, evaluate logarithmic expressions without using a calculator.

1. $\log_5(7 \cdot 3)$
2. $\log_8(13 \cdot 7)$
3. $\log_7(7x)$
4. $\log_9(9x)$
5. $\log(1000x)$
6. $\log(10{,}000x)$
7. $\log_7\left(\frac{7}{x}\right)$
8. $\log_9\left(\frac{9}{x}\right)$
9. $\log\left(\frac{x}{100}\right)$
10. $\log\left(\frac{x}{1000}\right)$
11. $\log_4\left(\frac{64}{y}\right)$
12. $\log_5\left(\frac{125}{y}\right)$
13. $\ln\left(\frac{e^2}{5}\right)$
14. $\ln\left(\frac{e^4}{8}\right)$
15. $\log_b x^3$
16. $\log_b x^7$
17. $\log N^{-6}$
18. $\log M^{-8}$
19. $\ln \sqrt[5]{x}$
20. $\ln \sqrt[7]{x}$
21. $\log_b(x^2y)$
22. $\log_b(xy^3)$
23. $\log_4\left(\frac{\sqrt{x}}{64}\right)$
24. $\log_5\left(\frac{\sqrt{x}}{25}\right)$
25. $\log_6\left(\frac{36}{\sqrt{x+1}}\right)$
26. $\log_8\left(\frac{64}{\sqrt{x+1}}\right)$
27. $\log_b\left(\frac{x^2y}{z^2}\right)$
28. $\log_b\left(\frac{x^3y}{z^2}\right)$
29. $\log \sqrt{100x}$
30. $\ln \sqrt{ex}$
31. $\log \sqrt[3]{\frac{x}{y}}$
32. $\log \sqrt[5]{\frac{x}{y}}$
33. $\log_b\left(\frac{\sqrt{x}y^3}{z^3}\right)$
34. $\log_b\left(\frac{\sqrt[3]{x}y^4}{z^5}\right)$
35. $\log_5 \sqrt[3]{\frac{x^2y}{25}}$
36. $\log_2 \sqrt[5]{\frac{xy^4}{16}}$

In Exercises 37–60, use properties of logarithms to condense each logarithmic expression. Write the expression as a single logarithm whose coefficient is 1. Where possible, evaluate logarithmic expressions.

37. $\log 5 + \log 2$
38. $\log 250 + \log 4$
39. $\ln x + \ln 7$
40. $\ln x + \ln 3$
41. $\log_2 96 - \log_2 3$
42. $\log_3 405 - \log_3 5$
43. $\log(2x + 5) - \log x$
44. $\log(3x + 7) - \log x$
45. $\log x + 3 \log y$
46. $\log x + 7 \log y$
47. $\frac{1}{2}\ln x + \ln y$

48. $\frac{1}{3}\ln x + \ln y$

49. $2 \log_b x + 3 \log_b y$

50. $5 \log_b x + 6 \log_b y$

51. $5 \ln x - 2 \ln y$

52. $7 \ln x - 3 \ln y$

53. $3 \ln x - \frac{1}{3}\ln y$

54. $2 \ln x - \frac{1}{2}\ln y$

55. $4 \ln(x + 6) - 3 \ln x$

56. $8 \ln(x + 9) - 4 \ln x$

57. $3 \ln x + 5 \ln y - 6 \ln z$

58. $4 \ln x + 7 \ln y - 3 \ln z$

59. $\frac{1}{2}(\log_5 x + \log_5 y) - 2 \log_5(x + 1)$

60. $\frac{1}{3}(\log_4 x - \log_4 y) + 2 \log_4(x + 1)$

In Exercises 61–68, use common logarithms or natural logarithms and a calculator to evaluate to four decimal places.

61. $\log_5 13$

62. $\log_6 17$

63. $\log_{14} 87.5$

64. $\log_{16} 57.2$

65. $\log_{0.1} 17$

66. $\log_{0.3} 19$

67. $\log_\pi 63$

68. $\log_\pi 400$

Practice PLUS

In Exercises 69–74, let $\log_b 2 = A$ and $\log_b 3 = C$. Write each expression in terms of A and C.

69. $\log_b \frac{3}{2}$

70. $\log_b 6$

71. $\log_b 8$

72. $\log_b 81$

73. $\log_b \sqrt{\frac{2}{27}}$

74. $\log_b \sqrt{\frac{3}{16}}$

In Exercises 75–88, determine whether each equation is true or false. Where possible, show work to support your conclusion. If the statement is false, make the necessary change(s) to produce a true statement.

75. $\ln e = 0$

76. $\ln 0 = e$

77. $\log_4(2x^3) = 3 \log_4(2x)$

78. $\ln(8x^3) = 3 \ln(2x)$

79. $x \log 10^x = x^2$

80. $\ln(x + 1) = \ln x + \ln 1$

81. $\ln(5x) + \ln 1 = \ln(5x)$

82. $\ln x + \ln(2x) = \ln(3x)$

83. $\log(x + 3) - \log(2x) = \frac{\log(x + 3)}{\log(2x)}$

84. $\frac{\log(x + 2)}{\log(x - 1)} = \log(x + 2) - \log(x - 1)$

85. $\log_6\left(\frac{x - 1}{x^2 + 4}\right) = \log_6(x - 1) - \log_6(x^2 + 4)$

86. $\log_6[4(x + 1)] = \log_6 4 + \log_6(x + 1)$

87. $\log_3 7 = \frac{1}{\log_7 3}$

88. $e^x = \frac{1}{\ln x}$

In Exercises 89–92,

a. *Evaluate the expression in part (a) without using a calculator.*

b. *Use your result from part (a) to write the expression in part (b) as a single logarithm whose coefficient is 1.*

89. a. $\log_3 9$

b. $\log_3 x + 4 \log_3 y - 2$

90. a. $\log_2 16$

b. $\log_2 x + 5 \log_2 y - 4$

91. a. $\log_{25} 5$

b. $\log_{25} x + \log_{25}(x^2 - 1) - \log_{25}(x + 1) - \frac{1}{2}$

92. a. $\log_{36} 6$

b. $\log_{36} x + \log_{36}(x^2 - 4) - \log_{36}(x + 2) - \frac{1}{2}$

Application Exercises

93. The loudness level of a sound can be expressed by comparing the sound's intensity to the intensity of a sound barely audible to the human ear. The formula

$$D = 10(\log I - \log I_0)$$

describes the loudness level of a sound, D, in decibels, where I is the intensity of the sound, in watts per meter^2, and I_0 is the intensity of a sound barely audible to the human ear.

a. Express the formula so that the expression in parentheses is written as a single logarithm.

b. Use the form of the formula from part (a) to answer this question. If a sound has an intensity 100 times the intensity of a softer sound, how much larger on the decibel scale is the loudness level of the more intense sound?

94. The formula

$$t = \frac{1}{c}[\ln A - \ln(A - N)]$$

describes the time, t, in weeks, that it takes to achieve mastery of a portion of a task, where A is the maximum learning possible, N is the portion of the learning that is to be achieved, and c is a constant used to measure an individual's learning style.

a. Express the formula so that the expression in brackets is written as a single logarithm.

b. The formula is also used to determine how long it will take chimpanzees and apes to master a task. For example, a typical chimpanzee learning sign language can master a maximum of 65 signs. Use the form of the formula from part (a) to answer this question. How many weeks will it take a chimpanzee to master 30 signs if c for that chimp is 0.03?

Explaining the Concepts

95. Describe the product rule for logarithms and give an example.

96. Describe the quotient rule for logarithms and give an example.

97. Describe the power rule for logarithms and give an example.

98. Without showing the details, explain how to condense $\ln x - 2\ln(x+1)$.

99. Describe the change-of-base property and give an example.

100. Explain how to use your calculator to find $\log_{14} 283$.

101. You overhear a student talking about a property of logarithms in which division becomes subtraction. Explain what the student means by this.

102. Find ln 2 using a calculator. Then calculate each of the following: $1 - \frac{1}{2}$; $1 - \frac{1}{2} + \frac{1}{3}$; $1 - \frac{1}{2} + \frac{1}{3} - \frac{1}{4}$; $1 - \frac{1}{2} + \frac{1}{3} - \frac{1}{4} + \frac{1}{5}$;.... Describe what you observe.

Technology Exercises

103. a. Use a graphing utility (and the change-of-base property) to graph $y = \log_3 x$.

b. Graph $y = 2 + \log_3 x$, $y = \log_3(x+2)$, and $y = -\log_3 x$ in the same viewing rectangle as $y = \log_3 x$. Then describe the change or changes that need to be made to the graph of $y = \log_3 x$ to obtain each of these three graphs.

104. Graph $y = \log x$, $y = \log(10x)$, and $y = \log(0.1x)$ in the same viewing rectangle. Describe the relationship among the three graphs. What logarithmic property accounts for this relationship?

105. Use a graphing utility and the change-of-base property to graph $y = \log_3 x$, $y = \log_{25} x$, and $y = \log_{100} x$ in the same viewing rectangle.

a. Which graph is on the top in the interval $(0, 1)$? Which is on the bottom?

b. Which graph is on the top in the interval $(1, \infty)$? Which is on the bottom?

c. Generalize by writing a statement about which graph is on top, which is on the bottom, and in which intervals, using $y = \log_b x$ where $b > 1$.

Disprove each statement in Exercises 106–110 by

a. *letting y equal a positive constant of your choice, and*

b. *using a graphing utility to graph the function on each side of the equal sign. The two functions should have different graphs, showing that the equation is not true in general.*

106. $\log(x + y) = \log x + \log y$

107. $\log\frac{x}{y} = \frac{\log x}{\log y}$

108. $\ln(x - y) = \ln x - \ln y$

109. $\ln(xy) = (\ln x)(\ln y)$

110. $\frac{\ln x}{\ln y} = \ln x - \ln y$

Critical Thinking Exercises

Make Sense? *In Exercises 111–114, determine whether each statement makes sense or does not make sense, and explain your reasoning.*

111. Because I cannot simplify the expression $b^m + b^n$ by adding exponents, there is no property for the logarithm of a sum.

112. Because logarithms are exponents, the product, quotient, and power rules remind me of properties for operations with exponents.

113. I can use any positive number other than 1 in the change-of-base property, but the only practical bases are 10 and e because my calculator gives logarithms for these two bases.

114. I expanded $\log_4\sqrt{\frac{x}{y}}$ by writing the radical using a rational exponent and then applying the quotient rule, obtaining $\frac{1}{2}\log_4 x - \log_4 y$.

In Exercises 115–118, determine whether each statement is true or false. If the statement is false, make the necessary change(s) to produce a true statement.

115. $\ln\sqrt{2} = \frac{\ln 2}{2}$

116. $\frac{\log_7 49}{\log_7 7} = \log_7 49 - \log_7 7$

117. $\log_b(x^3 + y^3) = 3\log_b x + 3\log_b y$

118. $\log_b(xy)^5 = (\log_b x + \log_b y)^5$

119. Use the change-of-base property to prove that

$$\log e = \frac{1}{\ln 10}.$$

120. If $\log 3 = A$ and $\log 7 = B$, find $\log_7 9$ in terms of A and B.

121. Write as a single term that does not contain a logarithm:

$$e^{\ln 8x^5 - \ln 2x^2}.$$

Review Exercises

122. Graph: $5x - 2y > 10$. (Section 4.4, Example 1)

123. Solve: $x - 2(3x - 2) > 2x - 3$. (Section 4.1, Example 2)

124. Divide and simplify: $\frac{\sqrt[3]{40x^2y^6}}{\sqrt[3]{5xy}}$. (Section 7.4, Example 5)

Preview Exercises

Exercises 125–127 will help you prepare for the material covered in the next section.

125. Simplify: $16^{\frac{3}{2}}$.

126. Evaluate $3\ln(2x)$ if $x = \frac{e^4}{2}$.

127. Solve: $\frac{x+2}{4x+3} = \frac{1}{x}$.

MID-CHAPTER CHECK POINT Section 9.1–Section 9.4

What You Know: We evaluated and graphed exponential functions $[f(x) = b^x, b > 0 \text{ and } b \neq 1]$, including the natural exponential function $[f(x) = e^x, e \approx 2.718]$. We studied composite and inverse functions, noting that a function has an inverse that is a function if there is no horizontal line that intersects the function's graph more than once. The exponential function passes this horizontal line test and we called the inverse of the exponential function with base b the logarithmic function with base b. We learned that $y = \log_b x$ is equivalent to $b^y = x$. We evaluated and graphed logarithmic functions, including the common logarithmic function $[f(x) = \log_{10} x \text{ or } f(x) = \log x]$ and the natural logarithmic function $[f(x) = \log_e x \text{ or } f(x) = \ln x]$. Finally, we used properties of logarithms to expand and condense logarithmic expressions.

In Exercises 1–3, find $(f \circ g)(x)$ and $(g \circ f)(x)$. Are f and g inverses of each other?

1. $f(x) = 3x + 2, g(x) = 4x - 5$

2. $f(x) = \sqrt[3]{7x + 5}, g(x) = \frac{x^3 - 5}{7}$

3. $f(x) = \log_5 x, g(x) = 5^x$

4. Let $f(x) = \frac{x - 1}{x}$ and $g(x) = \sqrt{x + 3}$.

a. Find $(f \circ g)(6)$.
b. Find $(g \circ f)(-1)$.
c. Find $(f \circ f)(5)$.
d. Find $(g \circ g)(-2)$.

In Exercises 5–7, find f^{-1}.

5. $f(x) = \frac{2x + 5}{4}$

6. $f(x) = 10x^3 - 7$

7. $f = \{(2, 5), (10, -7), (11, -10)\}$

In Exercises 8–10, which graphs represent functions? Among these graphs, which have inverse functions?

8.

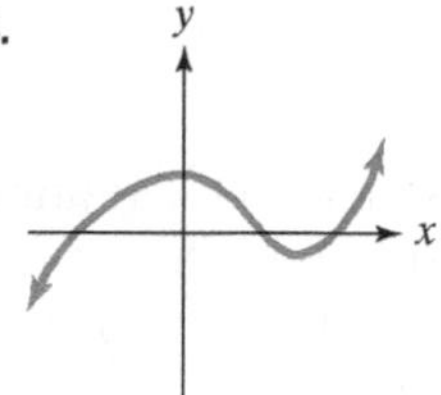

9.

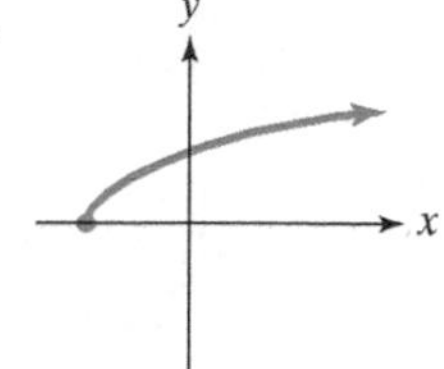

10.

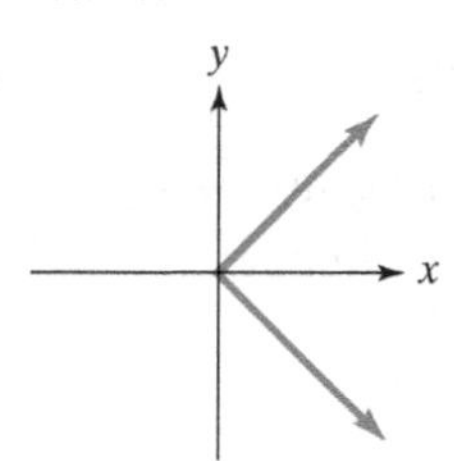

In Exercises 11–14, graph the given function. Give each function's domain and range.

11. $f(x) = 2^x - 3$

12. $f(x) = \left(\frac{1}{3}\right)^x$

13. $f(x) = \log_2 x$

14. $f(x) = \log_2 x + 1$

In Exercises 15–18, find the domain of each function.

15. $f(x) = \log_3(x + 6)$

16. $g(x) = \log_3 x + 6$

17. $h(x) = \log_3(x + 6)^2$

18. $f(x) = 3^{x+6}$

In Exercises 19–29, evaluate each expression without using a calculator. If evaluation is not possible, state the reason.

19. $\log_2 8 + \log_5 25$

20. $\log_3 \frac{1}{9}$

21. $\log_{100} 10$

22. $\log \sqrt[3]{10}$

23. $\log_2(\log_3 81)$

24. $\log_3\left(\log_2 \frac{1}{8}\right)$

25. $6^{\log_6 5}$

26. $\ln e^{\sqrt{7}}$

27. $10^{\log 13}$

28. $\log_{100} 0.1$

29. $\log_\pi \pi^{\sqrt{\pi}}$

In Exercises 30–31, expand and evaluate numerical terms.

30. $\log\left(\dfrac{\sqrt{xy}}{1000}\right)$

31. $\ln(e^{19}x^{20})$

In Exercises 32–34, write each expression as a single logarithm.

32. $8\log_7 x - \dfrac{1}{3}\log_7 y$

33. $7\log_5 x + 2\log_5 x$

34. $\dfrac{1}{2}\ln x - 3\ln y - \ln(z-2)$

35. Use the formulas

$$A = P\left(1+\frac{r}{n}\right)^{nt} \quad \text{and} \quad A = Pe^{rt}$$

to solve this exercise. You plan to invest \$8000 for 3 years at an annual rate of 8%. How much more is the return if the interest is compounded continuously than monthly? Round to the nearest dollar.

SECTION

9.5 Exponential and Logarithmic Equations

What am I supposed to learn?

After studying this section, you should be able to:

1. Use like bases to solve exponential equations.
2. Use logarithms to solve exponential equations.
3. Use exponential form to solve logarithmic equations.
4. Use the one-to-one property of logarithms to solve logarithmic equations.
5. Solve applied problems involving exponential and logarithmic equations.

At age 20, you inherit \$30,000. You'd like to put aside \$25,000 and eventually have over half a million dollars for early retirement. Is this possible? In this section, you will see how techniques for solving equations with variable exponents provide an answer to this question.

Exponential Equations

1 Use like bases to solve exponential equations.

An **exponential equation** is an equation containing a variable in an exponent. Examples of exponential equations include

$$2^{3x-8} = 16, \quad 4^x = 15, \quad \text{and} \quad 40e^{0.6x} = 240.$$

Some exponential equations can be solved by expressing each side of the equation as a power of the same base. All exponential functions are one-to-one—that is, no two different ordered pairs have the same second component. Thus, if b is a positive number other than 1 and $b^M = b^N$, then $M = N$.

Solving Exponential Equations by Expressing Each Side as a Power of the Same Base

$$\text{If } b^M = b^N, \text{ then } M = N.$$

Express each side as a power of the same base.

Set the exponents equal to each other.

1. Rewrite the equation in the form $b^M = b^N$.
2. Set $M = N$.
3. Solve for the variable.

EXAMPLE 1 Solving Exponential Equations

Using Technology

Graphic Connections

The graphs of

$$y_1 = 2^{3x-8}$$

$$\text{and} \quad y_2 = 16$$

have an intersection point whose x-coordinate is 4. This verifies that {4} is the solution set of $2^{3x-8} = 16$.

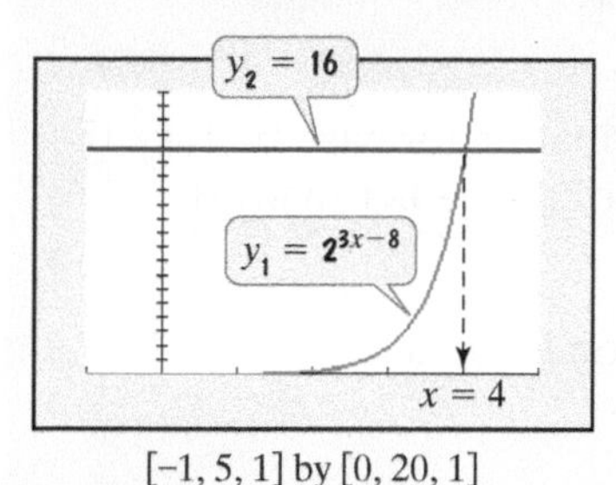

[−1, 5, 1] by [0, 20, 1]

Solve: **a.** $2^{3x-8} = 16$ **b.** $16^x = 64$.

Solution In each equation, express both sides as a power of the same base. Then set the exponents equal to each other.

a. Because 16 is 2^4, we express each side of $2^{3x-8} = 16$ in terms of base 2.

$$2^{3x-8} = 16 \quad \text{This is the given equation.}$$

$$2^{3x-8} = 2^4 \quad \text{Write each side as a power of the same base.}$$

$$3x - 8 = 4 \quad \text{If } b^M = b^N,\ b > 0 \text{ and } b \neq 1, \text{ then } M = N.$$

$$3x = 12 \quad \text{Add 8 to both sides.}$$

$$x = 4 \quad \text{Divide both sides by 3.}$$

Check 4:

$$2^{3x-8} = 16$$

$$2^{3\cdot4-8} \stackrel{?}{=} 16$$

$$2^4 \stackrel{?}{=} 16$$

$$16 = 16, \quad \text{true}$$

The solution is 4 and the solution set is {4}.

b. Because $16 = 4^2$ and $64 = 4^3$, we express each side of $16^x = 64$ in terms of base 4.

$$16^x = 64 \quad \text{This is the given equation.}$$

$$(4^2)^x = 4^3 \quad \text{Write each side as a power of the same base.}$$

$$4^{2x} = 4^3 \quad \text{When an exponential expression is raised to a power, multiply exponents.}$$

$$2x = 3 \quad \text{If two powers of the same base are equal, then the exponents are equal.}$$

$$x = \frac{3}{2} \quad \text{Divide both sides by 2.}$$

Check $\frac{3}{2}$:

$$16^x = 64$$

$$16^{\frac{3}{2}} \stackrel{?}{=} 64$$

$$(\sqrt{16})^3 \stackrel{?}{=} 64 \quad b^{\frac{m}{n}} = (\sqrt[n]{b})^m$$

$$4^3 \stackrel{?}{=} 64$$

$$64 = 64, \quad \text{true}$$

The solution is $\frac{3}{2}$ and the solution set is $\left\{\frac{3}{2}\right\}$. ■

Discover for Yourself

The equation $16^x = 64$ can also be solved by writing each side in terms of base 2. Do this. Which solution method do you prefer?

✓ **CHECK POINT 1** Solve:

a. $5^{3x-6} = 125$ **b.** $4^x = 32$.

---

2 Use logarithms to solve exponential equations.

Most exponential equations cannot be rewritten so that each side has the same base. Here are two examples:

$$4^x = 15 \qquad 10^x = 120{,}000.$$

We cannot rewrite both sides in terms of base 2 or base 4.

We cannot rewrite both sides in terms of base 10.

Logarithms are extremely useful in solving equations such as $4^x = 15$ and $10^x = 120{,}000$. The solution begins with isolating the exponential expression. Notice that the exponential expression is already isolated in both $4^x = 15$ and $10^x = 120{,}000$. Then we take the logarithm on both sides. Why can we do this? All logarithmic relations are functions. Thus, if M and N are positive real numbers and $M = N$, then $\log_b M = \log_b N$.

The base that is used when taking the logarithm on both sides of an equation can be any base at all. If the exponential equation involves base 10, as in $10^x = 120{,}000$, we'll take the common logarithm on both sides. If the exponential equation involves any other base, as in $4^x = 15$, we'll take the natural logarithm on both sides.

Using Logarithms to Solve Exponential Equations

1. Isolate the exponential expression.
2. Take the common logarithm on both sides of the equation for base 10. Take the natural logarithm on both sides of the equation for bases other than 10.
3. Simplify using one of the following properties:

$$\ln b^x = x \ln b \quad \text{or} \quad \ln e^x = x \quad \text{or} \quad \log 10^x = x.$$

4. Solve for the variable.

EXAMPLE 2 Solving Exponential Equations

Solve: **a.** $4^x = 15$ **b.** $10^x = 120{,}000$.

Solution We will use the natural logarithmic function to solve $4^x = 15$ and the common logarithmic function to solve $10^x = 120{,}000$.

a. Because the exponential expression, 4^x, is already isolated on the left side of $4^x = 15$, we begin by taking the natural logarithm on both sides of the equation.

$4^x = 15$	This is the given equation.
$\ln 4^x = \ln 15$	Take the natural logarithm on both sides.
$x \ln 4 = \ln 15$	Use the power rule and bring the variable exponent to the front: $\ln b^x = x \ln b$.
$x = \dfrac{\ln 15}{\ln 4}$	Solve for x by dividing both sides by ln 4.

We now have an exact value for x. We use the exact value for x in the equation's solution set. Thus, the equation's solution is $\dfrac{\ln 15}{\ln 4}$ and the solution set is $\left\{\dfrac{\ln 15}{\ln 4}\right\}$. We can obtain a decimal approximation by using a calculator: $x \approx 1.95$. Because $4^2 = 16$, it seems reasonable that the solution to $4^x = 15$ is approximately 1.95.

b. Because the exponential expression, 10^x, is already isolated on the left side of $10^x = 120{,}000$, we begin by taking the common logarithm on both sides of the equation.

$10^x = 120{,}000$	This is the given equation.
$\log 10^x = \log 120{,}000$	Take the common logarithm on both sides.
$x = \log 120{,}000$	Use the inverse property $\log 10^x = x$ on the left.

The equation's solution is log 120,000 and the solution set is {log 120,000}. We can obtain a decimal approximation by using a calculator: $x \approx 5.08$. Because $10^5 = 100{,}000$, it seems reasonable that the solution to $10^x = 120{,}000$ is approximately 5.08. ■

Discover for Yourself

Keep in mind that the base used when taking the logarithm on both sides of an equation can be any base at all. Solve $4^x = 15$ by taking the common logarithm on both sides. Solve again, this time taking the logarithm with base 4 on both sides. Use the change-of-base property to show that the solutions are the same as the one obtained in Example 2(a).

☑ CHECK POINT 2 Solve:

a. $5^x = 134$ **b.** $10^x = 8000$.

Find each solution set and then use a calculator to obtain a decimal approximation to two decimal places for the solution.

EXAMPLE 3 Solving an Exponential Equation

Solve: $40e^{0.6x} - 3 = 237$.

Solution We begin by adding 3 to both sides and dividing both sides by 40 to isolate the exponential expression, $e^{0.6x}$. Then we take the natural logarithm on both sides of the equation.

$$40e^{0.6x} - 3 = 237 \quad \text{This is the given equation.}$$
$$40e^{0.6x} = 240 \quad \text{Add 3 to both sides.}$$
$$e^{0.6x} = 6 \quad \text{Isolate the exponential factor by dividing both sides by 40.}$$
$$\ln e^{0.6x} = \ln 6 \quad \text{Take the natural logarithm on both sides.}$$
$$0.6x = \ln 6 \quad \text{Use the inverse property } \ln e^x = x \text{ on the left.}$$
$$x = \frac{\ln 6}{0.6} \approx 2.99 \quad \text{Divide both sides by 0.6 and solve for } x.$$

Thus, the solution of the equation is $\frac{\ln 6}{0.6} \approx 2.99$. Try checking this approximate solution in the original equation to verify that $\left\{\frac{\ln 6}{0.6}\right\}$ is the solution set. ■

☑ CHECK POINT 3 Solve: $7e^{2x} - 5 = 58$. Find the solution set and then use a calculator to obtain a decimal approximation to two decimal places for the solution.

---

3 Use exponential form to solve logarithmic equations.

Logarithmic Equations

A **logarithmic equation** is an equation containing a variable in a logarithmic expression. Examples of logarithmic equations include

$$\log_4(x + 3) = 2 \quad \text{and} \quad \ln(x + 2) - \ln(4x + 3) = \ln\left(\frac{1}{x}\right).$$

Some logarithmic equations can be expressed in the form $\log_b M = c$. We can solve such equations by rewriting them in exponential form.

Using Exponential Form to Solve Logarithmic Equations

1. Express the equation in the form $\log_b M = c$.
2. Use the definition of a logarithm to rewrite the equation in exponential form:

$$\log_b M = c \quad \text{means} \quad b^c = M.$$

Logarithms are exponents.

3. Solve for the variable.
4. Check proposed solutions in the original equation. Include in the solution set only values for which $M > 0$.

EXAMPLE 4 Solving Logarithmic Equations

Solve: **a.** $\log_4(x + 3) = 2$ **b.** $3 \ln(2x) = 12$.

Solution The form $\log_b M = c$ involves a single logarithm whose coefficient is 1 on one side and a constant on the other side. Equation (a) is already in this form. We will need to divide both sides of equation (b) by 3 to obtain this form.

a.

$\log_4(x + 3) = 2$ This is the given equation.

$4^2 = x + 3$ Rewrite in exponential form: $\log_b M = c$ means $b^c = M$.

$16 = x + 3$ Square 4.

$13 = x$ Subtract 3 from both sides.

Check 13:

$\log_4(x + 3) = 2$ This is the given logarithmic equation.

$\log_4(13 + 3) \stackrel{?}{=} 2$ Substitute 13 for x.

$\log_4 16 \stackrel{?}{=} 2$

$2 = 2$, true $\log_4 16 = 2$ because $4^2 = 16$.

This true statement indicates that the solution is 13 and the solution set is $\{13\}$.

b.

$3 \ln(2x) = 12$ This is the given equation.

$\ln(2x) = 4$ Divide both sides by 3.

$\log_e(2x) = 4$ Rewrite the natural logarithm showing base e. This step is optional.

$e^4 = 2x$ Rewrite in exponential form: $\log_b M = c$ means $b^c = M$.

$\frac{e^4}{2} = x$ Divide both sides by 2.

Check $\frac{e^4}{2}$:

$3 \ln(2x) = 12$ This is the given logarithmic equation.

$3 \ln\left[2\left(\frac{e^4}{2}\right)\right] \stackrel{?}{=} 12$ Substitute $\frac{e^4}{2}$ for x.

$3 \ln e^4 \stackrel{?}{=} 12$ Simplify: $\frac{\cancel{2}}{1} \cdot \frac{e^4}{\cancel{2}} = e^4$.

$3 \cdot 4 \stackrel{?}{=} 12$ Because $\ln e^x = x$, we conclude $\ln e^4 = 4$.

$12 = 12$, true

This true statement indicates that the solution is $\frac{e^4}{2}$ and the solution set is $\left\{\frac{e^4}{2}\right\}$. ■

Using Technology

Graphic Connections

The graphs of

$y_1 = \log_4(x + 3)$ and $y_2 = 2$

have an intersection point whose x-coordinate is 13. This verifies that $\{13\}$ is the solution set for $\log_4(x + 3) = 2$.

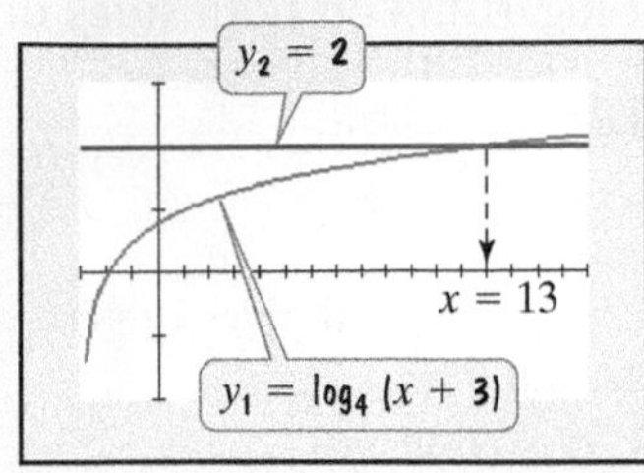

[−3, 17, 1] by [−2, 3, 1]

✓ CHECK POINT 4 Solve:

a. $\log_2(x - 4) = 3$ **b.** $4 \ln(3x) = 8$.

Logarithmic expressions are defined only for logarithms of positive real numbers. **Always check proposed solutions of a logarithmic equation in the original equation. Exclude from the solution set any proposed solution that produces the logarithm of a negative number or the logarithm of 0.**

Great Question!

Can a negative number belong to the solution set of a logarithmic equation?

Yes. Here's an example.

$$\log_2(x+20) = 3 \quad \text{Solve this equation.}$$
$$2^3 = x + 20 \quad \text{Rewrite in exponential form.}$$
$$8 = x + 20 \quad \text{Cube 2.}$$
$$-12 = x \quad \text{Subtract 20 from both sides.}$$

Check −12:

$$\log_2(-12+20) \stackrel{?}{=} 3 \quad \text{Substitute } -12 \text{ for } x.$$
$$\log_2 8 \stackrel{?}{=} 3$$
$$3 = 3, \text{ true} \quad \log_2 8 = 3 \text{ because } 2^3 = 8.$$

The solution set is $\{-12\}$. Although -12 is negative, it does not produce the logarithm of a negative number in $\log_2(x+20) = 3$, the given equation. Note that the domain of the expression $\log_2(x+20)$ is $(-20, \infty)$, which includes negative numbers such as -12.

To rewrite the logarithmic equation $\log_b M = c$ in the equivalent exponential form $b^c = M$, we need a single logarithm whose coefficient is one. It is sometimes necessary to use properties of logarithms to condense logarithms into a single logarithm. In the next example, we use the product rule for logarithms to obtain a single logarithmic expression on the left side.

EXAMPLE 5 Solving a Logarithmic Equation

Solve: $\log_2 x + \log_2(x-7) = 3.$

Solution

$$\log_2 x + \log_2(x-7) = 3 \quad \text{This is the given equation.}$$
$$\log_2[x(x-7)] = 3 \quad \text{Use the product rule to obtain a single logarithm: } \log_b M + \log_b N = \log_b(MN).$$
$$2^3 = x(x-7) \quad \text{Rewrite in exponential form.}$$
$$8 = x^2 - 7x \quad \text{Evaluate } 2^3 \text{ on the left and apply the distributive property on the right.}$$
$$0 = x^2 - 7x - 8 \quad \text{Set the equation equal to 0.}$$
$$0 = (x-8)(x+1) \quad \text{Factor.}$$
$$x - 8 = 0 \quad \text{or} \quad x + 1 = 0 \quad \text{Set each factor equal to 0.}$$
$$x = 8 \qquad\qquad x = -1 \quad \text{Solve for } x.$$

Check 8:

$$\log_2 x + \log_2(x-7) = 3$$
$$\log_2 8 + \log_2(8-7) \stackrel{?}{=} 3$$
$$\log_2 8 + \log_2 1 \stackrel{?}{=} 3$$
$$3 + 0 \stackrel{?}{=} 3$$
$$3 = 3, \text{ true}$$

Check −1:

$$\log_2 x + \log_2(x-7) = 3$$
$$\log_2(-1) + \log_2(-1-7) \stackrel{?}{=} 3$$

The number −1 does not check. It produces logarithms of negative numbers. Neither −1 nor −8 are in the domain of a logarithmic function.

The solution is 8 and the solution set is $\{8\}$. ■

CHECK POINT 5 Solve: $\log x + \log(x-3) = 1.$

4 Use the one-to-one property of logarithms to solve logarithmic equations.

Some logarithmic equations can be expressed in the form $\log_b M = \log_b N$. Because all logarithmic functions are one-to-one, we can conclude that $M = N$.

Using the One-to-One Property of Logarithms to Solve Logarithmic Equations

1. Express the equation in the form $\log_b M = \log_b N$. This form involves a single logarithm whose coefficient is 1 on each side of the equation.
2. Use the one-to-one property to rewrite the equation without logarithms: If $\log_b M = \log_b N$, then $M = N$.
3. Solve for the variable.
4. Check proposed solutions in the original equation. Include in the solution set only values for which $M > 0$ and $N > 0$.

EXAMPLE 6 Solving a Logarithmic Equation

Solve: $\ln(x + 2) - \ln(4x + 3) = \ln\left(\frac{1}{x}\right)$.

Solution In order to apply the one-to-one property of logarithms, we need a single logarithm whose coefficient is 1 on each side of the equation. The right side is already in this form. We can obtain a single logarithm on the left side by applying the quotient rule.

$$\ln(x + 2) - \ln(4x + 3) = \ln\left(\frac{1}{x}\right)$$ This is the given equation.

$$\ln\left(\frac{x + 2}{4x + 3}\right) = \ln\left(\frac{1}{x}\right)$$ Use the quotient rule to obtain a single logarithm on the left side: $\log_b M - \log_b N = \log_b\left(\frac{M}{N}\right)$.

$$\frac{x + 2}{4x + 3} = \frac{1}{x}$$ Use the one-to-one property: If $\log_b M = \log_b N$, then $M = N$.

$$x(4x + 3)\left(\frac{x + 2}{4x + 3}\right) = x(4x + 3)\left(\frac{1}{x}\right)$$ Multiply both sides by $x(4x + 3)$, the LCD.

$$x(x + 2) = 4x + 3$$ Simplify.

$$x^2 + 2x = 4x + 3$$ Apply the distributive property.

$$x^2 - 2x - 3 = 0$$ Subtract $4x + 3$ from both sides and set the equation equal to 0.

$$(x - 3)(x + 1) = 0$$ Factor.

$$x - 3 = 0 \quad \text{or} \quad x + 1 = 0$$ Set each factor equal to 0.

$$x = 3 \qquad\qquad x = -1$$ Solve for x.

Substituting 3 for x into the original equation produces the true statement $\ln\left(\frac{1}{3}\right) = \ln\left(\frac{1}{3}\right)$. However, substituting -1 produces logarithms of negative numbers. Thus, -1 is not a solution. The solution is 3 and the solution set is $\{3\}$. ■

Using Technology

Numeric Connections

A graphing utility's [TABLE] feature can be used to verify that $\{3\}$ is the solution set of

$$\ln(x + 2) - \ln(4x + 3) = \ln\left(\frac{1}{x}\right).$$

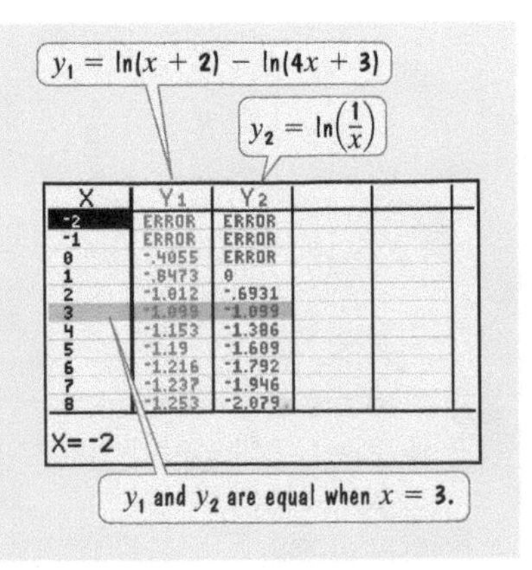

X	Y1	Y2
-2	ERROR	ERROR
-1	ERROR	ERROR
0	-.4055	ERROR
1	-.8473	0
2	-1.012	-.6931
3	-1.099	-1.099
4	-1.153	-1.386
5	-1.19	-1.609
6	-1.216	-1.792
7	-1.237	-1.946
8	-1.253	-2.079

X= -2

✓ CHECK POINT 6 Solve: $\ln(x - 3) = \ln(7x - 23) - \ln(x + 1)$.

5 Solve applied problems involving exponential and logarithmic equations.

Applications

Our first applied example provides a mathematical perspective on the old slogan "Alcohol and driving don't mix." In California, where 38% of fatal traffic crashes involve drunk drivers, it is illegal to drive with a blood alcohol concentration of 0.08 or higher. At these levels, drivers may be arrested and charged with driving under the influence.

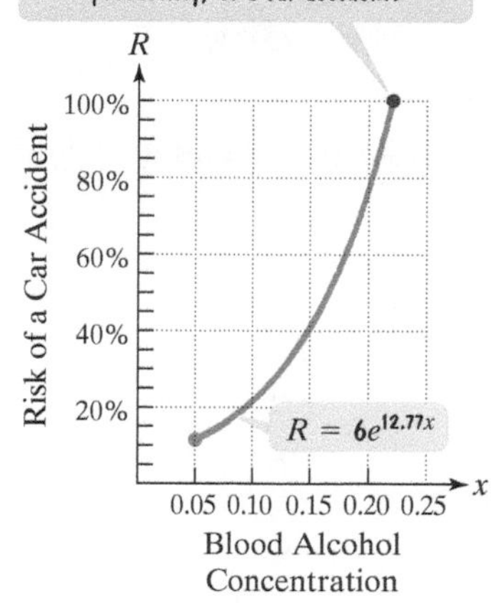

Figure 9.22

EXAMPLE 7 Alcohol and Risk of a Car Accident

Medical research indicates that the risk of having a car accident increases exponentially as the concentration of alcohol in the blood increases. The risk is modeled by

$$R = 6e^{12.77x},$$

where x is the blood alcohol concentration and R, given as a percent, is the risk of having a car accident. What blood alcohol concentration corresponds to a 17% risk of a car accident? How is this shown on the graph of R in **Figure 9.22**?

Solution For a risk of 17%, we let $R = 17$ in the equation and solve for x, the blood alcohol concentration.

$$R = 6e^{12.77x}$$ This is the given equation.

$$6e^{12.77x} = 17$$ Substitute 17 for R and (optional) reverse the two sides of the equation.

$$e^{12.77x} = \frac{17}{6}$$ Isolate the exponential factor by dividing both sides by 6.

$$\ln e^{12.77x} = \ln\left(\frac{17}{6}\right)$$ Take the natural logarithm on both sides.

$$12.77x = \ln\left(\frac{17}{6}\right)$$ Use the inverse property $\ln e^x = x$ on the left side.

$$x = \frac{\ln\left(\frac{17}{6}\right)}{12.77} \approx 0.08$$ Divide both sides by 12.77.

For a blood alcohol concentration of 0.08, the risk of a car accident is 17%. This is shown on the graph of R in **Figure 9.22** by the point (0.08, 17) that lies on the blue curve. Take a moment to locate this point on the curve. In many states, it is illegal to drive with a blood alcohol concentration of 0.08. ■

✓ CHECK POINT 7 Use the formula in Example 7 to solve this problem. What blood alcohol concentration corresponds to a 7% risk of a car accident? (In many states, drivers under the age of 21 can lose their licenses for driving at this level.)

Suppose that you inherit \$30,000 at age 20. Is it possible to invest \$25,000 and have over half a million dollars for early retirement? Our next example illustrates the power of compound interest.

Blitzer Bonus

Playing Doubles: Interest Rates and Doubling Time

One way to calculate what your savings will be worth at some point in the future is to consider doubling time. The following table shows how long it takes for your money to double at different annual interest rates subject to continuous compounding.

Annual Interest Rate	Years to Double
5%	13.9 years
7%	9.9 years
9%	7.7 years
11%	6.3 years

Of course, the first problem is collecting some money to invest. The second problem is finding a reasonably safe investment with a return of 9% or more.

EXAMPLE 8 Revisiting the Formula for Compound Interest

The formula

$$A = P\left(1 + \frac{r}{n}\right)^{nt}$$

describes the accumulated value, A, of a sum of money, P, the principal, after t years at annual percentage rate r (in decimal form) compounded n times a year. How long will it take \$25,000 to grow to \$500,000 at 9% annual interest compounded monthly?

Solution

$$A = P\left(1 + \frac{r}{n}\right)^{nt}$$ This is the given formula.

$$500{,}000 = 25{,}000\left(1 + \frac{0.09}{12}\right)^{12t}$$ A(the desired accumulated value) $= 500{,}000$, P(the principal) $= 25{,}000$, r(the interest rate) $= 9\% = 0.09$, and $n = 12$ (monthly compounding).

Our goal is to solve the equation for t. Let's reverse the two sides of the equation and then simplify within parentheses.

$$25{,}000\left(1 + \frac{0.09}{12}\right)^{12t} = 500{,}000$$ Reverse the two sides of the previous equation.

$$25{,}000(1 + 0.0075)^{12t} = 500{,}000$$ Divide within parentheses: $\frac{0.09}{12} = 0.0075$.

$$25{,}000(1.0075)^{12t} = 500{,}000$$ Add within parentheses.

$$(1.0075)^{12t} = 20$$ Divide both sides by 25,000.

$$\ln(1.0075)^{12t} = \ln 20$$ Take the natural logarithm on both sides.

$$12t\ln(1.0075) = \ln 20$$ Use the power rule to bring the exponent to the front: $\ln b^x = x \ln b$.

$$t = \frac{\ln 20}{12 \ln 1.0075}$$ Solve for t, dividing both sides by $12 \ln 1.0075$.

$$\approx 33.4$$ Use a calculator.

After approximately 33.4 years, the \$25,000 will grow to an accumulated value of \$500,000. If you set aside the money at age 20, you can begin enjoying a life of leisure at about age 53. ■

CHECK POINT 8 How long, to the nearest tenth of a year, will it take \$1000 to grow to \$3600 at 8% annual interest compounded quarterly?

EXAMPLE 9 Revisiting the Model for Heights of Children

We have seen that the percentage of adult height attained by a boy who is x years old can be modeled by

$$f(x) = 29 + 48.8\log(x + 1),$$

where x represents the boy's age (from 5 to 15) and $f(x)$ represents the percentage of his adult height. At what age, rounded to the nearest year, has a boy attained 85% of his adult height?

Solution To find at what age a boy has attained 85% of his adult height, we substitute 85 for $f(x)$ and solve for x, the boy's age.

$$f(x) = 29 + 48.8\log(x + 1) \qquad \text{This is the given function.}$$
$$85 = 29 + 48.8\log(x + 1) \qquad \text{Substitute 85 for } f(x).$$

Our goal is to isolate $\log(x + 1)$ and then rewrite the equation in exponential form.

$$56 = 48.8\log(x + 1) \qquad \text{Subtract 29 from both sides.}$$
$$\frac{56}{48.8} = \log(x + 1) \qquad \text{Divide both sides by 48.8.}$$
$$\frac{56}{48.8} = \log_{10}(x + 1) \qquad \text{Rewrite the common logarithm showing base 10. This step is optional.}$$
$$10^{\frac{56}{48.8}} = x + 1 \qquad \text{Rewrite in exponential form.}$$
$$10^{\frac{56}{48.8}} - 1 = x \qquad \text{Subtract 1 from both sides.}$$
$$13 \approx x$$

Use a calculator:
10 [∧] [(] 56 [÷] 48.8 [)] [−] 1 [ENTER].
Some calculators require that you use the right arrow key to exit the exponent before you subtract 1.

At approximately age 13, a boy has attained 85% of his adult height. This is shown on the graph of the model in **Figure 9.23** by the point (13, 85).

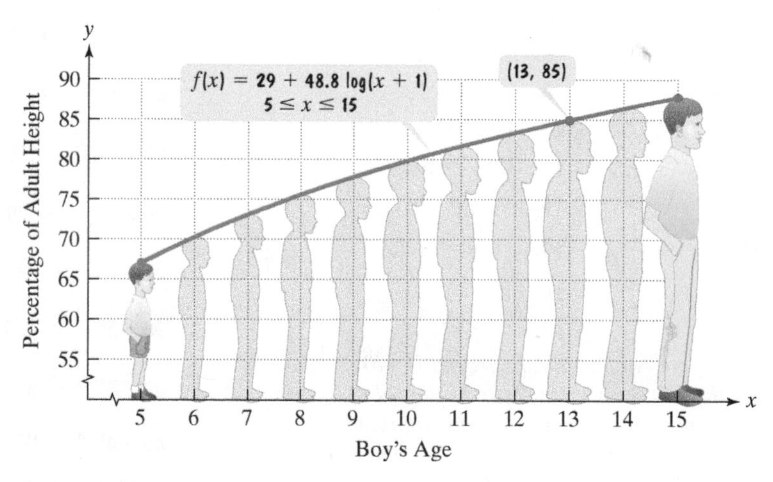

Figure 9.23 Graph of a model for the percentage of adult height attained by a boy ■

✓ **CHECK POINT 9** The percentage of adult height attained by a girl who is x years old can be modeled by

$$f(x) = 62 + 35\log(x - 4),$$

where x represents the girl's age (from 5 to 15) and $f(x)$ represents the percentage of her adult height. At what age has a girl attained 97% of her adult height?

CONCEPT AND VOCABULARY CHECK

Fill in each blank so that the resulting statement is true.

1. If $b^M = b^N$, then ______________.
2. If $2^{4x-1} = 2^7$, then ______________ = 7.

3. If $x \ln 9 = \ln 20$, then $x =$ ______________.

4. If $e^{0.6x} = 6$, then $0.6x =$ ______________.

5. If $\log_5(x + 1) = 3$, then ______________ $= x + 1$.

6. If $\log_3 x + \log_3(x + 1) = 2$, then $\log_3$ ______________ $= 2$.

7. If $\ln\left(\frac{7x - 23}{x + 1}\right) = \ln(x - 3)$, then ______________ $= x - 3$.

8. True or false: $x^4 = 15$ is an exponential equation. ______________

9. True or false: $4^x = 15$ is an exponential equation. ______________

10. True or false: -3 is a solution of $\log_5 9 = 2 \log_5 x$. ______________

9.5 EXERCISE SET

MyMathLab®

Practice Exercises

Solve each exponential equation in Exercises 1–18 by expressing each side as a power of the same base and then equating exponents.

1. $2^x = 64$
2. $3^x = 81$
3. $5^x = 125$
4. $5^x = 625$
5. $2^{2x-1} = 32$
6. $3^{2x+1} = 27$
7. $4^{2x-1} = 64$
8. $5^{3x-1} = 125$
9. $32^x = 8$
10. $4^x = 32$
11. $9^x = 27$
12. $125^x = 625$
13. $3^{1-x} = \frac{1}{27}$
14. $5^{2-x} = \frac{1}{125}$
15. $6^{\frac{x-3}{4}} = \sqrt{6}$
16. $7^{\frac{x-2}{6}} = \sqrt{7}$
17. $4^x = \frac{1}{\sqrt{2}}$
18. $9^x = \frac{1}{\sqrt[3]{3}}$

Solve each exponential equation in Exercises 19–40 by taking the logarithm on both sides. Express the solution set in terms of logarithms. Then use a calculator to obtain a decimal approximation, correct to two decimal places, for the solution.

19. $e^x = 5.7$
20. $e^x = 0.83$
21. $10^x = 3.91$
22. $10^x = 8.07$
23. $5^x = 17$
24. $19^x = 143$
25. $5e^x = 25$
26. $9e^x = 99$
27. $3e^{5x} = 1977$
28. $4e^{7x} = 10{,}273$
29. $e^{0.7x} = 13$
30. $e^{0.08x} = 4$
31. $1250e^{0.055x} = 3750$
32. $1250e^{0.065x} = 6250$
33. $30 - (1.4)^x = 0$
34. $135 - (4.7)^x = 0$
35. $e^{1-5x} = 793$
36. $e^{1-8x} = 7957$
37. $7^{x+2} = 410$
38. $5^{x-3} = 137$
39. $2^{x+1} = 5^x$
40. $4^{x+1} = 9^x$

Solve each logarithmic equation in Exercises 41–90. Be sure to reject any value of x that is not in the domain of the original logarithmic expressions. Give the exact answer. Then, where necessary, use a calculator to obtain a decimal approximation, correct to two decimal places, for the solution.

41. $\log_3 x = 4$
42. $\log_5 x = 3$
43. $\log_2 x = -4$
44. $\log_2 x = -5$
45. $\log_9 x = \frac{1}{2}$
46. $\log_{25} x = \frac{1}{2}$
47. $\log x = 2$
48. $\log x = 3$

49. $\log_4(x+5) = 3$

50. $\log_5(x-7) = 2$

51. $\log_3(x-4) = -3$

52. $\log_7(x+2) = -2$

53. $\log_4(3x+2) = 3$

54. $\log_2(4x+1) = 5$

55. $\ln x = 2$

56. $\ln x = 3$

57. $\ln x = -3$

58. $\ln x = -4$

59. $5\ln(2x) = 20$

60. $6\ln(2x) = 30$

61. $6 + 2\ln x = 5$

62. $7 + 3\ln x = 6$

63. $\ln\sqrt{x+3} = 1$

64. $\ln\sqrt{x+4} = 1$

65. $\log_5 x + \log_5(4x-1) = 1$

66. $\log_6(x+5) + \log_6 x = 2$

67. $\log_3(x-5) + \log_3(x+3) = 2$

68. $\log_2(x-1) + \log_2(x+1) = 3$

69. $\log_2(x+2) - \log_2(x-5) = 3$

70. $\log_4(x+2) - \log_4(x-1) = 1$

71. $\log(3x-5) - \log(5x) = 2$

72. $\log(2x-1) - \log x = 2$

73. $\ln(x+1) - \ln x = 1$

74. $\ln(x+2) - \ln x = 2$

75. $\log_3(x+4) = \log_3 7$

76. $\log_2(x-5) = \log_2 4$

77. $\log(x+4) = \log x + \log 4$

78. $\log(5x+1) = \log(2x+3) + \log 2$

79. $\log(3x-3) = \log(x+1) + \log 4$

80. $\log(2x-1) = \log(x+3) + \log 3$

81. $2\log x = \log 25$

82. $3\log x = \log 125$

83. $\log(x+4) - \log 2 = \log(5x+1)$

84. $\log(x+7) - \log 3 = \log(7x+1)$

85. $2\log x - \log 7 = \log 112$

86. $\log(x-2) + \log 5 = \log 100$

87. $\log x + \log(x+3) = \log 10$

88. $\log(x+3) + \log(x-2) = \log 14$

89. $\ln(x-4) + \ln(x+1) = \ln(x-8)$

90. $\log_2(x-1) - \log_2(x+3) = \log_2\left(\frac{1}{x}\right)$

Practice PLUS

In Exercises 91–98, solve each equation.

91. $5^{2x} \cdot 5^{4x} = 125$

92. $3^{x+2} \cdot 3^x = 81$

93. $3^{x^2} = 45$

94. $5^{x^2} = 50$

95. $\log_2(x-6) + \log_2(x-4) - \log_2 x = 2$

96. $\log_2(x-3) + \log_2 x - \log_2(x+2) = 2$

97. $5^{x^2-12} = 25^{2x}$

98. $3^{x^2-12} = 9^{2x}$

Application Exercises

99. The formula $A = 37.3e^{0.0095t}$ models the population of California, A, in millions, t years after 2010.

a. What was the population of California in 2010?

b. When will the population of California reach 40 million?

100. The formula $A = 25.1e^{0.0187t}$ models the population of Texas, A, in millions, t years after 2010.

a. What was the population of Texas in 2010?

b. When will the population of Texas reach 28 million?

The function $f(x) = 20(0.975)^x$ models the percentage of surface sunlight, $f(x)$, that reaches a depth of x feet beneath the surface of the ocean. The figure shows the graph of this function. Use this information to solve Exercises 101–102.

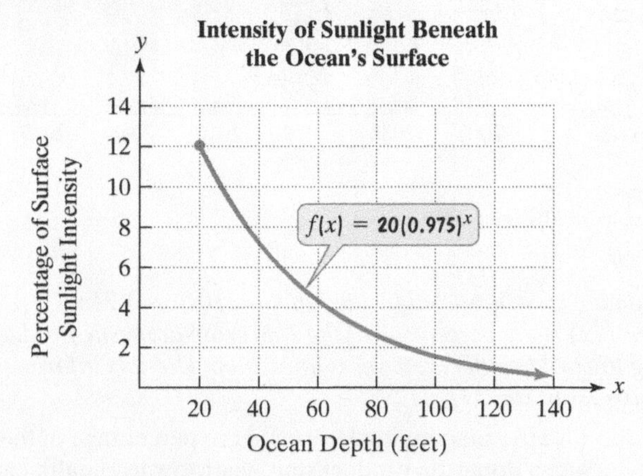

101. Use the function to determine at what depth, to the nearest foot, there is 1% of surface sunlight. How is this shown on the graph of f?

102. Use the function to determine at what depth, to the nearest foot, there is 3% of surface sunlight. How is this shown on the graph of f?

In Exercises 103–106, complete the table for a savings account subject to n compounding periods per year $\left[A = P\left(1 + \frac{r}{n}\right)^{nt}\right]$. *Round answers to one decimal place.*

	Amount Invested	Number of Compounding Periods	Annual Interest Rate	Accumulated Amount	Time t in Years
103.	\$12,500	4	5.75%	\$20,000	
104.	\$7250	12	6.5%	\$15,000	
105.	\$1000	360		\$1400	2
106.	\$5000	360		\$9000	4

In Exercises 107–110, complete the table for a savings account subject to continuous compounding ($A = Pe^{rt}$). *Round answers to one decimal place.*

	Amount Invested	Annual Interest Rate	Accumulated Amount	Time t in Years
107.	\$8000	8%	Double the amount invested	
108.	\$8000		\$12,000	2
109.	\$2350		Triple the amount invested	7
110.	\$17,425	4.25%	\$25,000	

By 2019, nearly \$1 out of every \$5 spent in the U.S. economy is projected to go for health care. The bar graph shows the percentage of the U.S. gross domestic product (GDP) going toward health care from 2007 through 2014, with a projection for 2019.

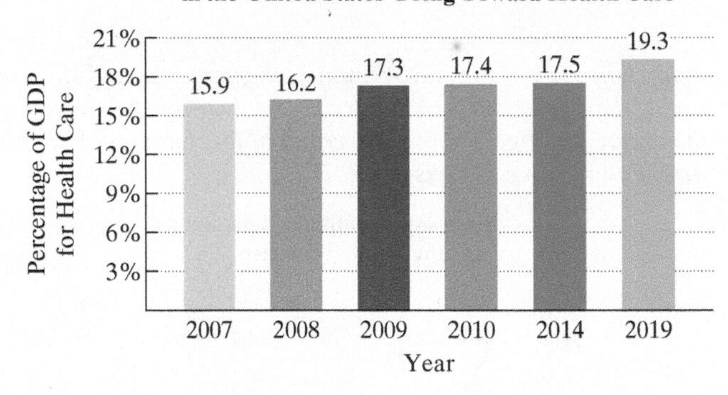

Source: Health Affairs (healthaffairs.org)

The data can be modeled by the function $f(x) = 1.2\ln x + 15.7$, *where* $f(x)$ *is the percentage of the U.S. gross domestic product going toward health care x years after 2006. Use this information to solve Exercises 111–112.*

111. a. Use the function to determine the percentage of the U.S. gross domestic product that went toward health care in 2009. Round to the nearest tenth of a percent. Does this underestimate or overestimate the percent displayed by the graph? By how much?

b. According to the model, when will 18.5% of the U.S. gross domestic product go toward health care? Round to the nearest year.

112. a. Use the function to determine the percentage of the U.S. gross domestic product that went toward health care in 2008. Round to the nearest tenth of a percent. Does this underestimate or overestimate the percent displayed by the graph? By how much?

b. According to the model, when will 18.6% of the U.S. gross domestic product go toward health care? Round to the nearest year.

The function $P(x) = 95 - 30\log_2 x$ *models the percentage,* $P(x)$, *of students who could recall the important features of a classroom lecture as a function of time, where x represents the number of days that have elapsed since the lecture was given. The figure shows the graph of the function. Use this information to solve Exercises 113–114. Round answers to one decimal place.*

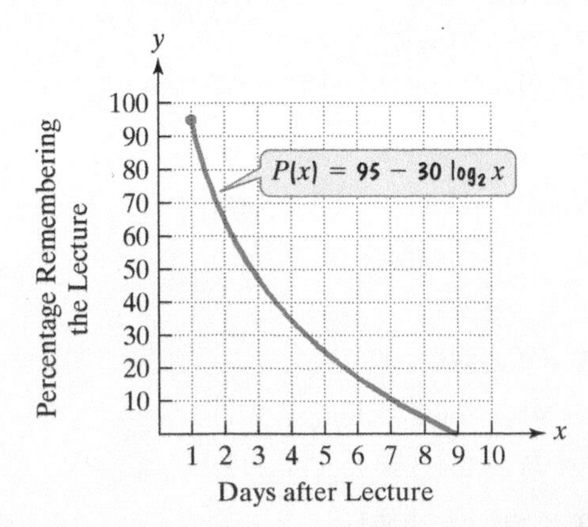

113. After how many days do only half the students recall the important features of the classroom lecture? (Let $P(x) = 50$ and solve for x.) Locate the point on the graph that conveys this information.

114. Refer to the graph and function on the previous page. After how many days have all students forgotten the important features of the classroom lecture? (Let $P(x) = 0$ and solve for x.) Locate the point on the graph that conveys this information.

The pH scale is used to measure the acidity or alkalinity of a solution. The scale ranges from 0 to 14. A neutral solution, such as pure water, has a pH of 7. An acid solution has a pH less than 7 and an alkaline solution has a pH greater than 7. The lower the pH below 7, the more acidic is the solution. Each whole-number decrease in pH represents a tenfold increase in acidity.

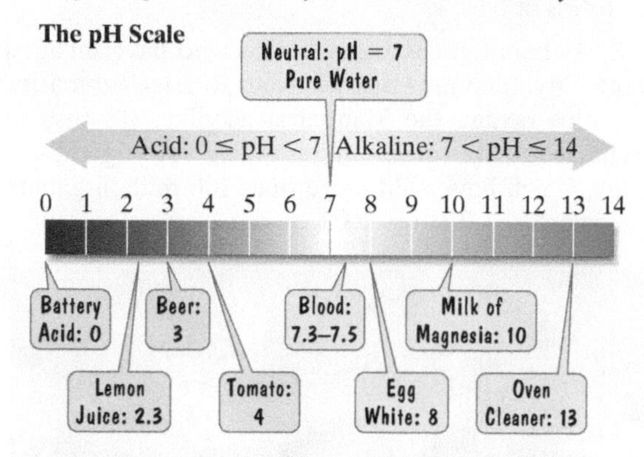

The pH of a solution is given by

$$\text{pH} = -\log x,$$

where x represents the concentration of the hydrogen ions in the solution, in moles per liter. Use the formula to solve Exercises 115–116. Express answers as powers of 10.

115. **a.** Normal, unpolluted rain has a pH of about 5.6. What is the hydrogen ion concentration?

b. An environmental concern involves the destructive effects of acid rain. The most acidic rainfall ever had a pH of 2.4. What was the hydrogen ion concentration?

c. How many times greater is the hydrogen ion concentration of the acidic rainfall in part (b) than the normal rainfall in part (a)?

116. **a.** The figure indicates that lemon juice has a pH of 2.3. What is the hydrogen ion concentration?

b. Stomach acid has a pH that ranges from 1 to 3. What is the hydrogen ion concentration of the most acidic stomach?

c. How many times greater is the hydrogen ion concentration of the acidic stomach in part (b) than the lemon juice in part (a)?

Explaining the Concepts

117. What is an exponential equation?

118. Explain how to solve an exponential equation when both sides can be written as a power of the same base.

119. Explain how to solve an exponential equation when both sides cannot be written as a power of the same base. Use $3^x = 140$ in your explanation.

120. What is a logarithmic equation?

121. Explain the differences between solving $\log_3(x - 1) = 4$ and $\log_3(x - 1) = \log_3 4$.

122. In many states, a 17% risk of a car accident with a blood alcohol concentration of 0.08 is the lowest level for charging a motorist with driving under the influence. Do you agree with the 17% risk as a cutoff percentage, or do you feel that the percentage should be lower or higher? Explain your answer. What blood alcohol concentration corresponds to what you believe is an appropriate percentage?

Technology Exercises

In Exercises 123–130, use your graphing utility to graph each side of the equation in the same viewing rectangle. Then use the x-coordinate of the intersection point to find the equation's solution set. Verify this value by direct substitution into the equation.

123. $2^{x+1} = 8$

124. $3^{x+1} = 9$

125. $\log_3(4x - 7) = 2$

126. $\log_3(3x - 2) = 2$

127. $\log(x + 3) + \log x = 1$

128. $\log(x - 15) + \log x = 2$

129. $3^x = 2x + 3$

130. $5^x = 3x + 4$

Hurricanes are one of nature's most destructive forces. These low-pressure areas often have diameters of over 500 miles. The function $f(x) = 0.48 \ln(x + 1) + 27$ models the barometric air pressure, $f(x)$, in inches of mercury, at a distance of x miles from the eye of a hurricane. Use this function to solve Exercises 131–132.

131. Graph the function in a [0, 500, 50] by [27, 30, 1] viewing rectangle. What does the shape of the graph indicate about barometric air pressure as the distance from the eye increases?

132. Use an equation to answer this question: How far from the eye of a hurricane is the barometric air pressure 29 inches of mercury? Use the TRACE and ZOOM features or the intersect command of your graphing utility to verify your answer.

133. The function $P(t) = 145e^{-0.092t}$ models a runner's pulse, $P(t)$, in beats per minute, t minutes after a race, where $0 \le t \le 15$. Graph the function using a graphing utility. TRACE along the graph and determine after how many minutes the runner's pulse will be 70 beats per minute. Round to the nearest tenth of a minute. Verify your observation algebraically.

134. The function $W(t) = 2600(1 - 0.51e^{-0.075t})^3$ models the weight, $W(t)$, in kilograms, of a female African elephant at age t years. (1 kilogram ≈ 2.2 pounds) Use a graphing utility to graph the function. Then TRACE along the curve to estimate the age of an adult female elephant weighing 1800 kilograms.

Critical Thinking Exercises

Make Sense? *In Exercises 135–138, determine whether each statement makes sense or does not make sense, and explain your reasoning.*

135. Because the equations $2^x = 15$ and $2^x = 16$ are similar, I solved them using the same method.

136. Because the equations

$$\log(3x + 1) = 5 \text{ and } \log(3x + 1) = \log 5$$

are similar, I solved them using the same method.

137. I can solve $4^x = 15$ by writing the equation in logarithmic form.

138. It's important for me to check that the proposed solution of an equation with logarithms gives only logarithms of positive numbers in the original equation.

In Exercises 139–142, determine whether each statement is true or false. If the statement is false, make the necessary change(s) to produce a true statement.

139. If $\log(x + 3) = 2$, then $e^2 = x + 3$.

140. If $\log(7x + 3) - \log(2x + 5) = 4$, then the equation in exponential form is $10^4 = (7x + 3) - (2x + 5)$.

141. If $x = \frac{1}{k}\ln y$, then $y = e^{kx}$.

142. Examples of exponential equations include $10^x = 5.71$, $e^x = 0.72$, and $x^{10} = 5.71$.

143. If \$4000 is deposited into an account paying 3% interest compounded annually and at the same time \$2000 is deposited into an account paying 5% interest compounded annually, after how long will the two accounts have the same balance? Round to the nearest year.

Solve each equation in Exercises 144–146. Check each proposed solution by direct substitution or with a graphing utility.

144. $(\ln x)^2 = \ln x^2$

145. $(\log x)(2 \log x + 1) = 6$

146. $\ln(\ln x) = 0$

Review Exercises

147. Solve: $\sqrt{2x - 1} - \sqrt{x - 1} = 1$.
(Section 7.6, Example 4)

148. Solve: $\frac{3}{x + 1} - \frac{5}{x} = \frac{19}{x^2 + x}$.
(Section 6.6, Example 5)

149. Simplify: $(-2x^3y^{-2})^{-4}$.
(Section 1.6, Example 7)

Preview Exercises

Exercises 150–152 will help you prepare for the material covered in the next section.

150. The formula $A = 10e^{-0.003t}$ models the population of Hungary, A, in millions, t years after 2006.

a. Find Hungary's population, in millions, for 2006, 2007, 2008, and 2009. Round to two decimal places.

b. Is Hungary's population increasing or decreasing?

151. a. Simplify: $e^{\ln 3}$.

b. Use your simplification from part (a) to rewrite 3^x in terms of base e.

152. U.S. soldiers fight Russian troops who have invaded New York City. Incoming missiles from Russian submarines and warships ravage the Manhattan skyline. It's just another scenario for the multi-billion-dollar video games *Call of Duty*, which have sold more than 100 million games since the franchise's birth in 2003.

The table shows the annual retail sales for *Call of Duty* video games from 2004 through 2010. Create a scatter plot for the data. Based on the shape of the scatter plot, would a logarithmic function, an exponential function, or a linear function be the best choice for modeling the data?

Annual Retail Sales for *Call of Duty* Games

Year	Retail Sales (millions of dollars)
2004	56
2005	101
2006	196
2007	352
2008	436
2009	778
2010	980

Source: The NPD Group

SECTION

9.6 Exponential Growth and Decay; Modeling Data

What am I supposed to learn?

After studying this section, you should be able to:

1. Model exponential growth and decay.
2. Choose an appropriate model for data.
3. Express an exponential model in base e.

On October 31, 2011, the world marked a major milestone: According to the United Nations, the number of people on Earth reached 7 billion—and counting. Since the dawn of humankind some 50,000 years ago, an estimated total of 108 billon people have lived on our planet, which means that about 6.5% of all humans ever born are alive today. That's a lot of bodies to feed, clothe, and shelter. Scientists, politicians, economists, and demographers have long disagreed when it comes to making predictions about the effects of the world's growing population. Debates about entities that are growing exponentially can be approached mathematically: We can create functions that model data and use these functions to make predictions. In this section, we will show you how this is done.

1 Model exponential growth and decay.

Exponential Growth and Decay

One of algebra's many applications is to predict the behavior of variables. This can be done with *exponential growth and decay models*. With exponential growth or decay, quantities grow or decay at a rate directly proportional to their size. Populations that are growing exponentially grow extremely rapidly as they get larger because there are more adults to have offspring. For example, the **growth rate** for world population is approximately 1.2%, or 0.012. This means that each year world population is 1.2% more than what it was in the previous year. In 2010, world population was 6.9 billion. Thus, we compute the world population in 2011 as follows:

$$6.9 \text{ billion} + 1.2\% \text{ of } 6.9 \text{ billion} = 6.9 + (0.012)(6.9) = 6.9828.$$

This computation indicates that 6.9828, or approximately 7.0, billion people populated the world in 2011. The 0.0828 billion represents an increase of 82.8 million people from 2010 to 2011, the equivalent of the population of Germany. Using 1.2% as the annual growth rate, world population for 2012 is found in a similar manner:

$$6.9828 + 1.2\% \text{ of } 6.9828 = 6.9828 + (0.012)(6.9828) \approx 7.067.$$

This computation indicates that approximately 7.1 billion people populated the world in 2012.

The explosive growth of world population may remind you of the growth of money in an account subject to compound interest. Just as the growth rate for world population is multiplied by the population plus any increase in the population, a compound interest rate is multiplied by your original investment plus any accumulated interest. The balance in an account subject to continuous compounding and world population are special cases of *exponential growth models*.

Great Question!

Why does the formula for exponential growth look familiar?

You have seen the formula for exponential growth before, but with different letters. It is the formula for compound interest with continuous compounding.

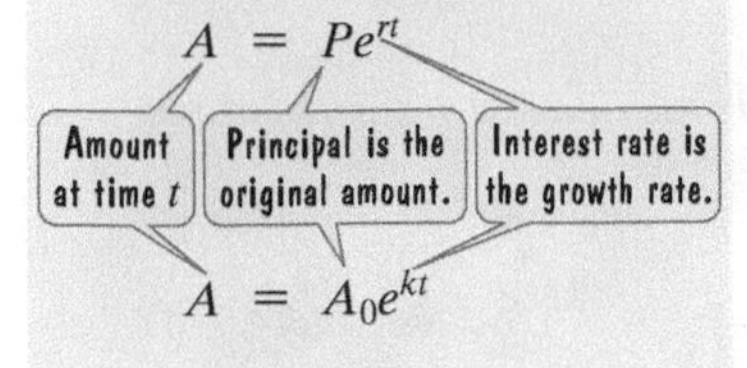

Exponential Growth and Decay Models

The mathematical model for **exponential growth** or **decay** is given by

$$f(t) = A_0e^{kt} \quad \text{or} \quad A = A_0e^{kt}.$$

- **If $k > 0$, the function models the amount, or size, of a *growing* entity.** A_0 is the original amount, or size, of the growing entity at time $t = 0$, A is the amount at time t, and k is a constant representing the growth rate.
- **If $k < 0$, the function models the amount, or size, of a *decaying* entity.** A_0 is the original amount, or size, of the decaying entity at time $t = 0$, A is the amount at time t, and k is a constant representing the decay rate.

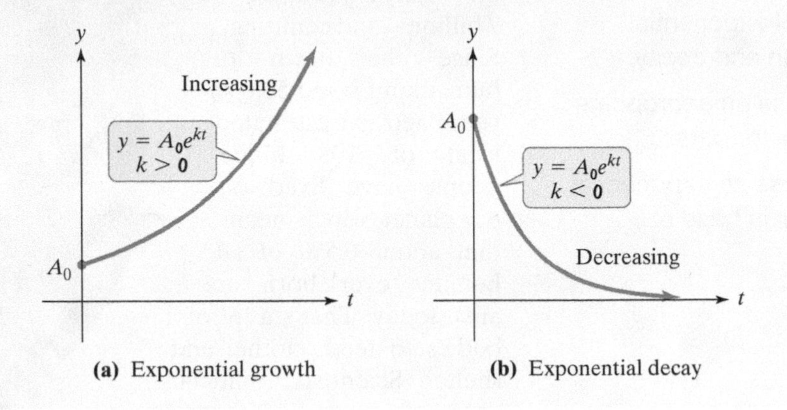

(a) Exponential growth **(b)** Exponential decay

Sometimes we need to use given data to determine k, the rate of growth or decay. After we compute the value of k, we can use the formula $A = A_0e^{kt}$ to make predictions. This idea is illustrated in our first two examples.

EXAMPLE 1 Modeling the Growth of the U.S. Population

The graph in **Figure 9.24** shows the U.S. population, in millions, for five selected years from 1970 through 2010. In 1970, the U.S. population was 203.3 million. By 2010, it had grown to 308.7 million.

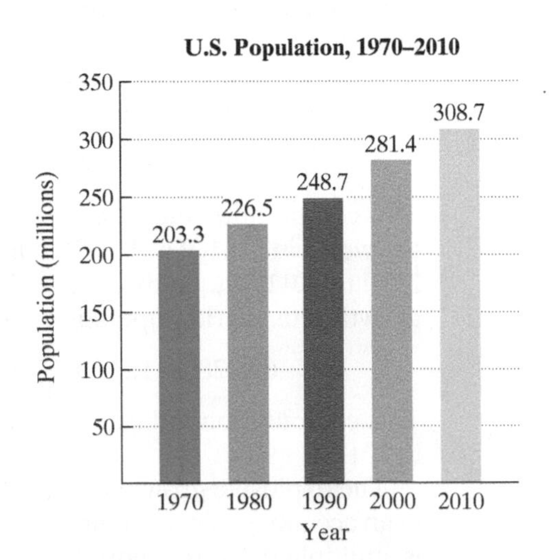

Figure 9.24

Source: U.S. Bureau of the Census

a. Find an exponential growth function that models the data for 1970 through 2010.

b. By which year will the U.S. population reach 335 million?

Solution

a. We use the exponential growth model

$$A = A_0 e^{kt}$$

in which t is the number of years after 1970. This means that 1970 corresponds to $t = 0$. At that time the U.S. population was 203.3 million, so we substitute 203.3 for A_0 in the growth model:

$$A = 203.3e^{kt}.$$

We are given that 308.7 million is the population in 2010. Because 2010 is 40 years after 1970, when $t = 40$ the value of A is 308.7. Substituting these numbers into the growth model will enable us to find k, the growth rate. We know that $k > 0$ because the problem involves growth.

$$A = 203.3e^{kt}$$ Use the growth model with $A_0 = 203.3$.

$$308.7 = 203.3e^{k \cdot 40}$$ When $t = 40$, $A = 308.7$. Substitute these numbers into the model.

$$e^{40k} = \frac{308.7}{203.3}$$ Isolate the exponential factor by dividing both sides by 203.3. We also reversed the sides.

$$\ln e^{40k} = \ln\left(\frac{308.7}{203.3}\right)$$ Take the natural logarithm on both sides.

$$40k = \ln\left(\frac{308.7}{203.3}\right)$$ Simplify the left side using $\ln e^x = x$.

$$k = \frac{\ln\left(\frac{308.7}{203.3}\right)}{40} \approx 0.01$$ Divide both sides by 40 and solve for k. Then use a calculator.

The value of k, approximately 0.01, indicates a growth rate of about 1%. We substitute 0.01 for k in the growth model, $A = 203.3e^{kt}$, to obtain an exponential growth function for the U.S. population. It is

$$A = 203.3e^{0.01t},$$

where t is measured in years after 1970.

b. To find the year in which the U.S. population will reach 335 million, substitute 335 for A in the model from part (a) and solve for t.

$$A = 203.3e^{0.01t}$$ This is the model from part (a).

$$335 = 203.3e^{0.01t}$$ Substitute 335 for A.

$$e^{0.01t} = \frac{335}{203.3}$$ Divide both sides by 203.3. We also reversed the sides.

$$\ln e^{0.01t} = \ln\left(\frac{335}{203.3}\right)$$ Take the natural logarithm on both sides.

$$0.01t = \ln\left(\frac{335}{203.3}\right)$$ Simplify on the left using $\ln e^x = x$.

$$t = \frac{\ln\left(\frac{335}{203.3}\right)}{0.01} \approx 50$$ Divide both sides by 0.01 and solve for t. Then use a calculator.

Because t represents the number of years after 1970, the model indicates that the U.S. population will reach 335 million by $1970 + 50$, or in the year 2020. ■

In Example 1, we used only two data values, the population for 1970 and the population for 2010, to develop a model for U.S. population growth from 1970 through 2010. By not using data for any other years, have we created a model that inaccurately describes both the existing data and future population projections given by the U.S. Census Bureau? Something else to think about: Is an exponential model the best choice for describing U.S. population growth, or might a linear model provide a better description? We return to these issues in Exercises 55–59 in the Exercise Set.

✓ CHECK POINT 1 In 2000, the population of Africa was 807 million and by 2011 it had grown to 1052 million.

a. Use the exponential growth model $A = A_0e^{kt}$, in which t is the number of years after 2000, to find the exponential growth function that models the data.

b. By which year will Africa's population reach 2000 million, or two billion?

Our next example involves exponential decay and its use in determining the age of fossils and artifacts. The method is based on considering the percentage of carbon-14 remaining in the fossil or artifact. Carbon-14 decays exponentially with a *half-life* of approximately 5715 years. The **half-life** of a substance is the time required for half of a given sample to disintegrate. Thus, after 5715 years a given amount of carbon-14 will have decayed to half the original amount. Carbon dating is useful for artifacts or fossils up to 80,000 years old. Older objects do not have enough carbon-14 left to determine age accurately.

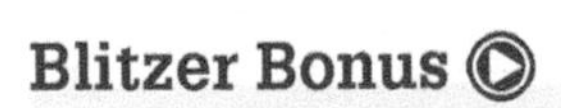

Blitzer Bonus

Carbon Dating and Artistic Development

The artistic community was electrified by the discovery in 1995 of spectacular cave paintings in a limestone cavern in France. Carbon dating of the charcoal from the site showed that the images, created by artists of remarkable talent, were 30,000 years old, making them the oldest cave paintings ever found. The artists seemed to have used the cavern's natural contours to heighten a sense of perspective. The quality of the painting suggests that the art of early humans did not mature steadily from primitive to sophisticated in any simple linear fashion.

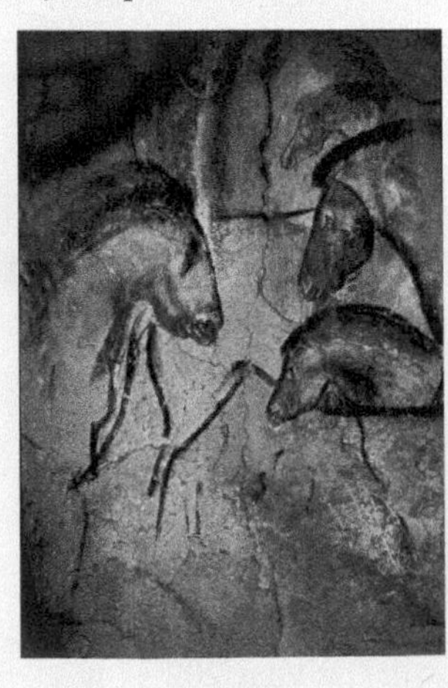

EXAMPLE 2 Carbon-14 Dating: The Dead Sea Scrolls

a. Use the fact that after 5715 years a given amount of carbon-14 will have decayed to half the original amount to find the exponential decay model for carbon-14.

b. In 1947, earthenware jars containing what are known as the Dead Sea Scrolls were found by an Arab Bedouin herdsman. Analysis indicated that the scroll wrappings contained 76% of their original carbon-14. Estimate the age of the Dead Sea Scrolls.

Solution

a. We begin with the exponential decay model $A = A_0e^{kt}$. We know that $k < 0$ because the problem involves the decay of carbon-14. After 5715 years ($t = 5715$), the amount of carbon-14 present, A, is half the original amount, A_0. Thus, we can substitute $\frac{A_0}{2}$ for A in the exponential decay model. This will enable us to find k, the decay rate.

$$A = A_0e^{kt}$$
Begin with the exponential decay model.

$$\frac{A_0}{2} = A_0e^{k \cdot 5715}$$
After 5715 years ($t = 5715$), $A = \frac{A_0}{2}$ (because the amount present, A, is half the original amount, A_0).

$$\frac{1}{2} = e^{5715k}$$
Divide both sides of the equation by A_0.

$$\ln\left(\frac{1}{2}\right) = \ln e^{5715k}$$
Take the natural logarithm on both sides.

$$\ln\left(\frac{1}{2}\right) = 5715k$$
Simplify the right side using $\ln e^x = x$.

$$k = \frac{\ln\left(\frac{1}{2}\right)}{5715} \approx -0.000121$$
Divide both sides by 5715 and solve for k.

Substituting for k in the decay model, $A = A_0e^{kt}$, the model for carbon-14 is

$$A = A_0e^{-0.000121t}.$$

b. In 1947, the Dead Sea Scrolls contained 76% of their original carbon-14. To find their age in 1947, substitute $0.76A_0$ for A in the model from part (a) and solve for t.

$$A = A_0 e^{-0.000121t}$$ This is the decay model for carbon-14.

$$0.76A_0 = A_0 e^{-0.000121t}$$ A, the amount present, is 76% of the original amount, so $A = 0.76A_0$.

$$0.76 = e^{-0.000121t}$$ Divide both sides of the equation by A_0.

$$\ln 0.76 = \ln e^{-0.000121t}$$ Take the natural logarithm on both sides.

$$\ln 0.76 = -0.000121t$$ Simplify the right side using $\ln e^x = x$.

$$t = \frac{\ln 0.76}{-0.000121} \approx 2268$$ Divide both sides by -0.000121 and solve for t.

The Dead Sea Scrolls are approximately 2268 years old plus the number of years between 1947 and the current year. ■

✓ CHECK POINT 2 Strontium-90 is a waste product from nuclear reactors. As a consequence of fallout from atmospheric nuclear tests, we all have a measurable amount of strontium-90 in our bones.

a. Use the fact that after 28 years a given amount of strontium-90 will have decayed to half the original amount to find the exponential decay model for strontium-90.

b. Suppose that a nuclear accident occurs and releases 60 grams of strontium-90 into the atmosphere. How long will it take for strontium-90 to decay to a level of 10 grams?

2 Choose an appropriate model for data.

Modeling Data

Throughout this chapter, we have been working with models that were given. However, we can create functions that model data by observing patterns in scatter plots. **Figure 9.25** shows scatter plots for data that are exponential, logarithmic, and linear.

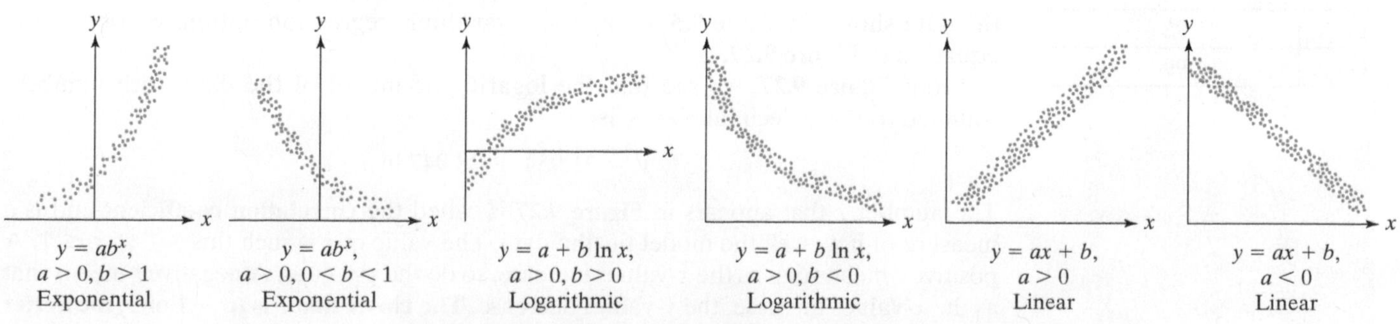

Figure 9.25 Scatter plots for exponential, logarithmic, and linear models

EXAMPLE 3 Choosing a Model for Data

The data in **Table 9.3** indicate that between the ages of 1 and 11, the human brain does not grow linearly, or steadily. A scatter plot for the data is shown in **Figure 9.26**. What type of function would be a good choice for modeling the data?

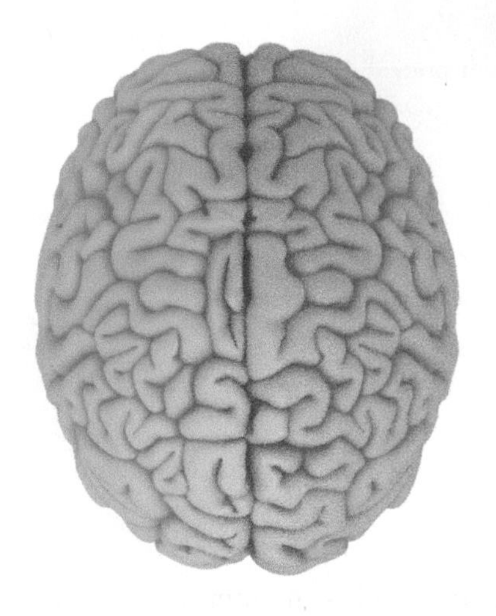

Table 9.3 Growth of the Human Brain

Age	Percentage of Adult Size Brain
1	30%
2	50%
4	78%
6	88%
8	92%
10	95%
11	99%

Source: Gerrig and Zimbardo, *Psychology and Life*, 18th Edition, Allyn and Bacon, 2008.

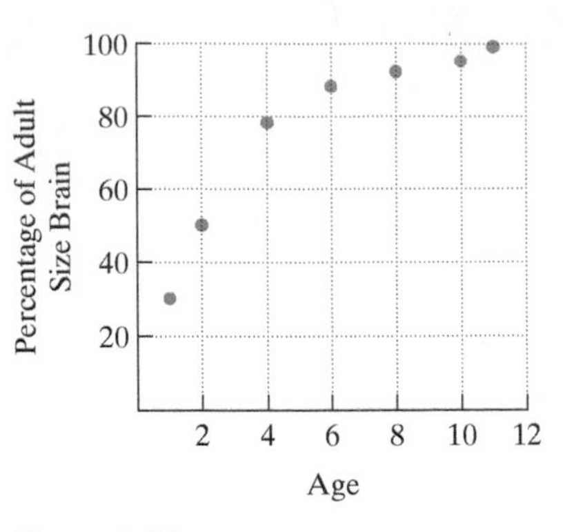

Figure 9.26

Solution Because the data in the scatter plot increase rapidly at first and then begin to level off a bit, the shape suggests that a logarithmic function is a good choice for modeling the data. ■

✓ CHECK POINT 3 **Table 9.4** shows the populations of various cities, in thousands, and the average walking speed, in feet per second, of a person living in the city. Create a scatter plot for the data. Based on the scatter plot, what type of function would be a good choice for modeling the data?

Table 9.4 Population and Walking Speed

Population (thousands)	Walking Speed (feet per second)
5.5	0.6
14	1.0
71	1.6
138	1.9
342	2.2

Source: Mark H. Bornstein and Helen G. Bornstein, "The Pace of Life." *Nature*, 259, Feb. 19, 1976, pp. 557–559

Table 9.5

x, Age	y, Percentage of Adult Size Brain
1	30
2	50
4	78
6	88
8	92
10	95
11	99

```
LnReg
y=a+blnx
a=31.95404756
b=28.94733911
r²=.9806647799
r=.9902852013
```

Figure 9.27 A logarithmic model for the data in **Table 9.5**

How can we obtain a logarithmic function that models the data for the growth of the human brain? A graphing utility can be used to obtain a logarithmic model of the form $y = a + b \ln x$. **Because the domain of the logarithmic function is the set of positive numbers, zero must not be a value for x.** This is not a problem for the data giving the percentage of an adult size brain because the data begin at age 1. We will assign x to represent age and y to represent the percentage of an adult size brain. This gives us the data shown in **Table 9.5**. Using the logarithmic regression option, we obtain the equation in **Figure 9.27**.

From **Figure 9.27**, we see that the logarithmic model of the data, with numbers rounded to three decimal places, is

$$y = 31.954 + 28.947 \ln x.$$

The number r that appears in **Figure 9.27** is called the **correlation coefficient** and is a measure of how well the model fits the data. The value of r is such that $-1 \le r \le 1$. A positive r means that as the x-values increase, so do the y-values. A negative r means that as the x-values increase, the y-values decrease. **The closer that r is to -1 or 1, the better the model fits the data.** Because r is approximately 0.99, the model fits the data very well.

EXAMPLE 4 Choosing a Model for Data

Figure 9.28(a) shows world population, in billions, for seven selected years from 1950 through 2010. A scatter plot is shown in **Figure 9.28(b)**. Suggest two types of functions that would be good choices for modeling the data.

World Population, 1950–2010

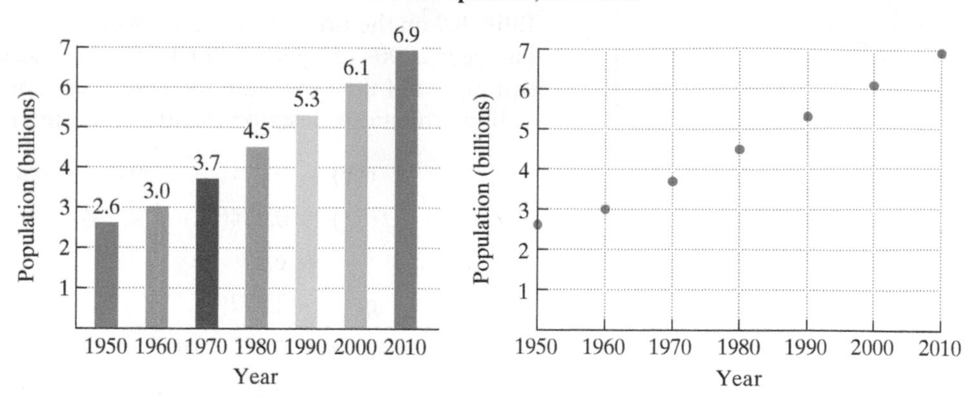

Figure 9.28(a)

Source: U.S. Census Bureau, International Database

Figure 9.28(b)

Solution Because the data in the scatter plot appear to increase more and more rapidly, the shape suggests that an exponential model might be a good choice. Furthermore, we can probably draw a line that passes through or near the seven points. Thus, a linear function would also be a good choice for modeling the data. ■

Table 9.6
Percentage of U.S. Men Who Are Married or Who Have Been Married, by Age

Age	Percent
18	2
20	7
25	36
30	61
35	75

Source: National Center for Health Statistics

✓ **CHECK POINT 4** **Table 9.6** shows the percentage of U.S. men who are married or who have been married, by age. Create a scatter plot for the data. Based on the scatter plot, what type of function would be a good choice for modeling the data?

EXAMPLE 5 Comparing Linear and Exponential Models

The data for world population are shown in **Table 9.7**. Using a graphing utility's linear regression feature and exponential regression feature, we enter the data and obtain the models shown in **Figure 9.29**.

Although the domain of $y = ab^x$ is the set of all real numbers, some graphing utilities only accept positive values for x. That's why we assigned x to represent the number of years after 1949.

Table 9.7

x, Number of Years after 1949	y, World Population (billions)
1 (1950)	2.6
11 (1960)	3.0
21 (1970)	3.7
31 (1980)	4.5
41 (1990)	5.3
51 (2000)	6.1
61 (2010)	6.9

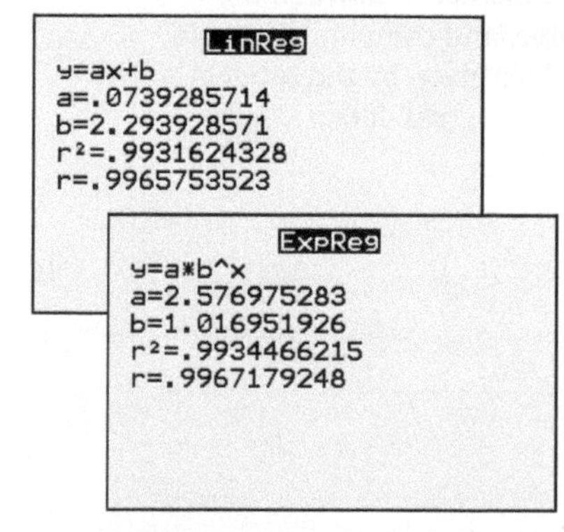

Figure 9.29 A linear model and an exponential model for the data in **Table 9.7**

Because r, the correlation coefficient, is close to 1 in each screen in **Figure 9.29**, the models fit the data very well.

a. Use **Figure 9.29** to express each model in function notation, with numbers rounded to three decimal places.

b. How well do the functions model world population in 2000?

c. By one projection, world population is expected to reach 8 billion in the year 2026. Which function serves as a better model for this prediction?

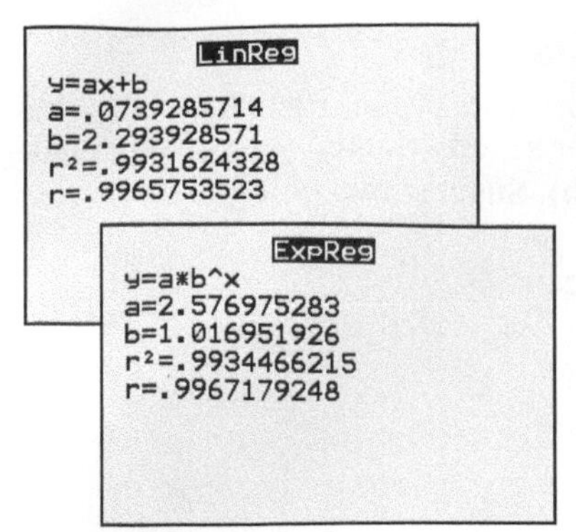

Figure 9.29 (repeated)

Solution

a. Using **Figure 9.29** and rounding to three decimal places, the functions

$$f(x) = 0.074x + 2.294 \quad \text{and} \quad g(x) = 2.577(1.017)^x$$

model world population, in billions, x years after 1949. We named the linear function f and the exponential function g, although any letters can be used.

b. **Table 9.7** on the previous page shows that world population in 2000 was 6.1 billion. The year 2000 is 51 years after 1949. Thus, we substitute 51 for x in each function's equation and then evaluate the resulting expressions with a calculator to see how well the functions describe world population in 2000.

$f(x) = 0.074x + 2.294$ This is the linear model.

$f(51) = 0.074(51) + 2.294$ Substitute 51 for x.

≈ 6.1 Use a calculator.

$g(x) = 2.577(1.017)^x$ This is the exponential model.

$g(51) = 2.577(1.017)^{51}$ Substitute 51 for x.

≈ 6.1 Use a calculator:
2.577 [×] 1.017 [y^x] 51 [=]
or 2.577 [×] 1.017 [∧] 51 [ENTER].

Because 6.1 billion was the actual world population in 2000, both functions model world population in 2000 extremely well.

c. Let's see which model comes closer to projecting a world population of 8 billion in the year 2026. Because 2026 is 77 years after 1949 ($2026 - 1949 = 77$), we substitute 77 for x in each function's equation.

$f(x) = 0.074x + 2.294$ This is the linear model.

$f(77) = 0.074(77) + 2.294$ Substitute 77 for x.

≈ 8.0 Use a calculator.

$g(x) = 2.577(1.017)^x$ This is the exponential model.

$g(77) = 2.577(1.017)^{77}$ Substitute 77 for x.

≈ 9.4 Use a calculator:
2.577 [×] 1.017 [y^x] 77 [=]
or 2.577 [×] 1.017 [∧] 77 [ENTER].

The linear function $f(x) = 0.074x + 2.294$ serves as a better model for a projected world population of 8 billion by 2026. ■

Great Question!

How can I use a graphing utility to see how well my models describe the data?

Once you have obtained one or more models for the data, you can use a graphing utility's [TABLE] feature to numerically see how well each model describes the data. Enter the models as y_1, y_2, and so on. Create a table, scroll through the table, and compare the table values given by the models to the actual data.

CHECK POINT 5 Use the models in Example 5(a) to solve this problem.

a. World population in 1970 was 3.7 billion. Which function serves as a better model for this year?

b. By one projection, world population is expected to reach 9.3 billion by 2050. Which function serves as a better model for this projection?

When using a graphing utility to model data, begin with a scatter plot, drawn either by hand or with the graphing utility, to obtain a general picture for the shape of the data. It might be difficult to determine which model best fits the data—linear, logarithmic, exponential, quadratic, or something else. If necessary, use your graphing utility to fit several models to the data. The best model is the one that yields the value r, the correlation coefficient, closest to 1 or -1. Finding a proper fit for data can be almost as much art as it is mathematics. In this era of technology, the process of creating models that best fit data is one that involves more decision making than computation.

3 Express an exponential model in base e.

Expressing $y = ab^x$ in Base e

Graphing utilities display exponential models in the form $y = ab^x$. However, our discussion of exponential growth involved base e. Because of the inverse property $b = e^{\ln b}$, we can rewrite any model in the form $y = ab^x$ in terms of base e.

Expressing an Exponential Model in Base e

$$y = ab^x \quad \text{is equivalent to} \quad y = ae^{(\ln b)\cdot x}.$$

Blitzer Bonus

Global Population Increase

By the time you finish reading Example 6 and working Check Point 6, more than 1000 people will have been added to our planet. By this time tomorrow, world population will have increased by more than 220,000.

EXAMPLE 6 Rewriting the Model for World Population in Base e

We have seen that the function

$$g(x) = 2.577(1.017)^x$$

models world population, $g(x)$, in billions, x years after 1949. Rewrite the model in terms of base e.

Solution We use the two equivalent equations shown in the voice balloons to rewrite the model in terms of base e.

$y = ab^x$ $\qquad$ $y = ae^{(\ln b)\cdot x}$

$$g(x) = 2.577(1.017)^x \quad \text{is equivalent to} \quad g(x) = 2.577e^{(\ln 1.017)x}.$$

Using $\ln 1.017 \approx 0.017$, the exponential growth model for world population, $g(x)$, in billions, x years after 1949 is

$$g(x) = 2.577e^{0.017x}. \quad \blacksquare$$

In Example 6, we can replace $g(x)$ with A and x with t so that the model has the same letters as those in the exponential growth model $A = A_0e^{kt}$.

$A = A_0 \, e^{kt}$ This is the exponential growth model.

$A = 2.577e^{0.017t}$ This is the model for world population.

The value of k, 0.017, indicates a growth rate of 1.7%. Although this is an excellent model for the data, we must be careful about making projections about world population using this growth function. Why? World population growth rate is now 1.2%, not 1.7%, so our model will overestimate future populations.

✓ CHECK POINT 6 Rewrite $y = 4(7.8)^x$ in terms of base e. Express the answer in terms of a natural logarithm and then round to three decimal places.

CONCEPT AND VOCABULARY CHECK

Fill in each blank so that the resulting statement is true.

1. Consider the model for exponential growth or decay given by

$$A = A_0e^{kt}.$$

If k ____________, the function models the amount, or size, of a growing entity. If k ____________, the function models the amount, or size, of a decaying entity.

2. In the model for exponential growth or decay, the amount, or size, at $t = 0$ is represented by ____________. The amount, or size, at time t is represented by ____________.

For each of the following scatter plots, determine whether an exponential function, a logarithmic function, or a linear function is the best choice for modeling the data.

3.

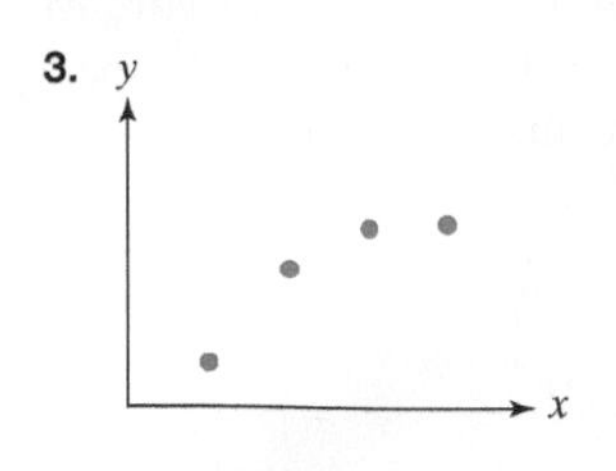

4.

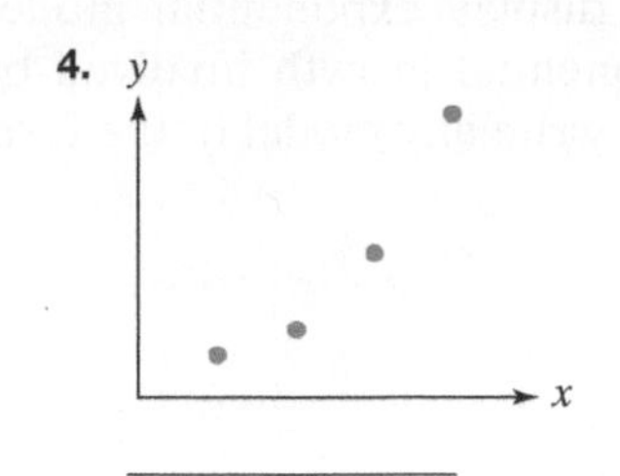

5.

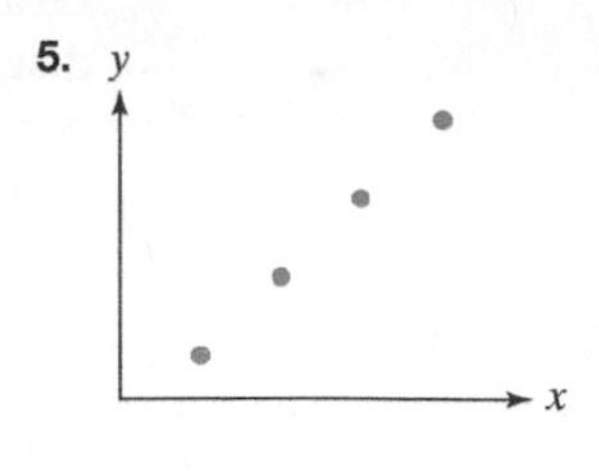

6. $y = 3(5)^x$ can be written in terms of base e as $y = 3e^{(\quad)x}$.

9.6 EXERCISE SET MyMathLab®

Practice Exercises and Application Exercises

The exponential models describe the population of the indicated country, A, in millions, t years after 2010. Use these models to solve Exercises 1–6.

India $A = 1173.1e^{0.008t}$

Iraq $A = 31.5e^{0.019t}$

Japan $A = 127.3e^{-0.006t}$

Russia $A = 141.9e^{-0.005t}$

1. What was the population of Japan in 2010?
2. What was the population of Iraq in 2010?
3. Which country has the greatest growth rate? By what percentage is the population of that country increasing each year?
4. Which countries have a decreasing population? By what percentage is the population of these countries decreasing each year?
5. When will India's population be 1377 million?
6. When will India's population be 1491 million?

About the size of New Jersey, Israel has seen its population soar to more than 6 million since it was established. The graphs show that by 2050, Palestinians in the West Bank, Gaza Strip, and East Jerusalem will outnumber Israelis. Exercises 7–8 involve the projected growth of these two populations.

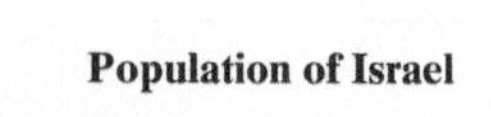

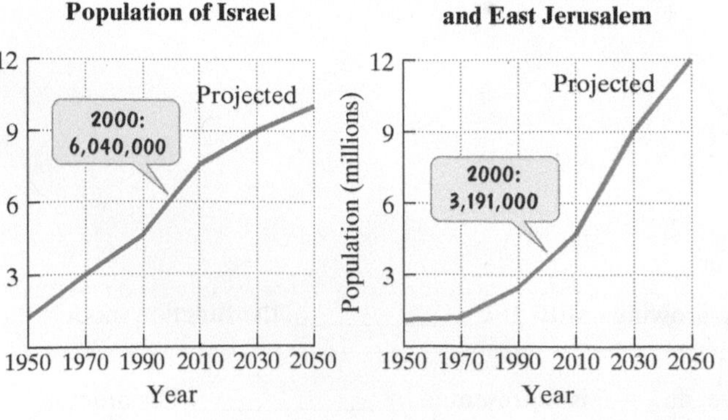

Source: Newsweek

7. a. In 2000, the population of Israel was approximately 6.04 million and by 2050 it is projected to grow to 10 million. Use the exponential growth model $A = A_0e^{kt}$, in which t is the number of years after 2000, to find an exponential growth function that models the data.

   b. In which year will Israel's population be 9 million?

8. a. In 2000, the population of the Palestinians in the West Bank, Gaza Strip, and East Jerusalem was approximately 3.2 million and by 2050 it is projected to grow to 12 million. Use the exponential growth model $A = A_0e^{kt}$, in which t is the number of years after 2000, to find an exponential growth function that models the data.

   b. In which year will the Palestinian population be 9 million?

In Exercises 9–14, complete the table. Round projected populations to one decimal place and values of k to four decimal places.

Country	2010 Population (millions)	Projected 2050 Population (millions)	Projected Growth Rate, k
9. Philippines	99.9		0.0095
10. Pakistan	184.4		0.0149
11. Colombia	44.2	62.9	
12. Madagascar	21.3	42.7	
13. Germany	82.3	70.5	
14. Bulgaria	7.1	5.4	

Source: International Programs Center, U.S. Census Bureau

An artifact originally had 16 grams of carbon-14 present. The decay model $A = 16e^{-0.000121t}$ describes the amount of carbon-14 present after t years. Use this model to solve Exercises 15–16.

15. How many grams of carbon-14 will be present in 5715 years?
16. How many grams of carbon-14 will be present in 11,430 years?

17. The half-life of the radioactive element krypton-91 is 10 seconds. If 16 grams of krypton-91 are initially present, how many grams are present after 10 seconds? 20 seconds? 30 seconds? 40 seconds? 50 seconds?

18. The half-life of the radioactive element plutonium-239 is 25,000 years. If 16 grams of plutonium-239 are initially present, how many grams are present after 25,000 years? 50,000 years? 75,000 years? 100,000 years? 125,000 years?

Use the exponential decay model for carbon-14, $A = A_0e^{-0.000121t}$, to solve Exercises 19–20.

19. Prehistoric cave paintings were discovered in a cave in France. The paint contained 15% of the original carbon-14. Estimate the age of the paintings.

20. Skeletons were found at a construction site in San Francisco in 1989. The skeletons contained 88% of the expected amount of carbon-14 found in a living person. In 1989, how old were the skeletons?

In Exercises 21–26, complete the table. Round half-lives to one decimal place and values of k to six decimal places.

Radioactive Substance	Half-Life	Decay Rate, k
21. Tritium		5.5% per year $= -0.055$
22. Krypton-85		6.3% per year $= -0.063$
23. Radium-226	1620 years	
24. Uranium-238	4560 years	
25. Arsenic-74	17.5 days	
26. Calcium-47	113 hours	

27. The August 1978 issue of *National Geographic* described the 1964 find of bones of a newly discovered dinosaur weighing 170 pounds, measuring 9 feet, with a 6-inch claw on one toe of each hind foot. The age of the dinosaur was estimated using potassium-40 dating of rocks surrounding the bones.

a. Potassium-40 decays exponentially with a half-life of approximately 1.31 billion years. Use the fact that after 1.31 billion years a given amount of potassium-40 will have decayed to half the original amount to show that the decay model for potassium-40 is given by $A = A_0e^{-0.52912t}$, where t is in billions of years.

b. Analysis of the rocks surrounding the dinosaur bones indicated that 94.5% of the original amount of potassium-40 was still present. Let $A = 0.945A_0$ in the model in part (a) and estimate the age of the bones of the dinosaur.

Use the exponential decay model, $A = A_0e^{kt}$, to solve Exercises 28–31. Round answers to one decimal place.

28. The half-life of thorium-229 is 7340 years. How long will it take for a sample of this substance to decay to 20% of its original amount?

29. The half-life of lead is 22 years. How long will it take for a sample of this substance to decay to 80% of its original amount?

30. The half-life of aspirin in your bloodstream is 12 hours. How long will it take for the aspirin to decay to 70% of the original dosage?

31. Xanax is a tranquilizer used in the short-term relief of symptoms of anxiety. Its half-life in the bloodstream is 36 hours. How long will it take for Xanax to decay to 90% of the original dosage?

32. A bird species in danger of extinction has a population that is decreasing exponentially ($A = A_0e^{kt}$). Five years ago the population was at 1400 and today only 1000 of the birds are alive. Once the population drops below 100, the situation will be irreversible. When will this happen?

33. Use the exponential growth model, $A = A_0e^{kt}$, to show that the time it takes a population to double (to grow from A_0 to $2A_0$) is given by $t = \frac{\ln 2}{k}$.

34. Use the exponential growth model, $A = A_0e^{kt}$, to show that the time it takes a population to triple (to grow from A_0 to $3A_0$) is given by $t = \frac{\ln 3}{k}$.

Use the formula $t = \frac{\ln 2}{k}$ that gives the time for a population with a growth rate k to double to solve Exercises 35–36. Express each answer to the nearest whole year.

35. The growth model $A = 4.1e^{0.01t}$ describes New Zealand's population, A, in millions, t years after 2010.

a. What is New Zealand's growth rate?

b. How long will it take New Zealand to double its population?

36. The growth model $A = 107.4e^{0.012t}$ describes Mexico's population, A, in millions, t years after 2010.

a. What is Mexico's growth rate?

b. How long will it take Mexico to double its population?

Exercises 37–42 present data in the form of tables. For each data set shown by the table,

a. *Create a scatter plot for the data.*
b. *Use the scatter plot to determine whether an exponential function, a logarithmic function, or a linear function is the best choice for modeling the data. (If applicable, in Exercise 61 you will use your graphing utility to obtain these functions.)*

37. Percent of Miscarriages, by Age

Woman's Age	Percent of Miscarriages
22	9%
27	10%
32	13%
37	20%
42	38%
47	52%

Source: Time Magazine

38. Savings Needed for Health-Care Expenses During Retirement

Age at Death	Savings Needed
80	$219,000
85	$307,000
90	$409,000
95	$524,000
100	$656,000

Source: Employee Benefit Research Institute

39. Intensity and Loudness Level of Various Sounds

Intensity (watts per meter^2)	Loudness Level (decibels)
0.1 (loud thunder)	110
1 (rock concert, 2 yd from speakers)	120
10 (jackhammer)	130
100 (jet takeoff, 40 yd away)	140

40. Temperature Increase in an Enclosed Vehicle

Minutes	Temperature Increase (°F)
10	19°
20	29°
30	34°
40	38°
50	41°
60	43°

41. Hamachiphobia

Generation	Percentage Who Won't Try Sushi	Percentage Who Don't Approve of Marriage Equality
Millennials	42	36
Gen X	52	49
Boomers	60	59
Silent/Greatest Generation	72	66

Source: Pew Research Center

42. Teenage Drug Use

	Percentage Who Have Used	
Country	Marijuana	Other Illegal Drugs
Czech Republic	22	4
Denmark	17	3
England	40	21
Finland	5	1
Ireland	37	16
Italy	19	8
Northern Ireland	23	14
Norway	6	3
Portugal	7	3
Scotland	53	31
United States	34	24

Source: De Veaux et al., *Intro Stats*, Pearson, 2009.

In Exercises 43–46, rewrite the equation in terms of base e. Express the answer in terms of a natural logarithm and then round to three decimal places.

43. $y = 100(4.6)^x$

44. $y = 1000(7.3)^x$

45. $y = 2.5(0.7)^x$

46. $y = 4.5(0.6)^x$

Explaining the Concepts

47. Nigeria has a growth rate of 0.025 or 2.5%. Describe what this means.

48. How can you tell if an exponential model describes exponential growth or exponential decay?

49. Suppose that a population that is growing exponentially increases from 800,000 people in 1997 to 1,000,000 people in 2000. Without showing the details, describe how to obtain an exponential growth function that models the data.

50. What is the half-life of a substance?

51. Describe the shape of a scatter plot that suggests modeling the data with an exponential function.

52. You take up weightlifting and record the maximum number of pounds you can lift at the end of each week. You start off with rapid growth in terms of the weight you can lift from week to week, but then the growth begins to level off. Describe how to obtain a function that models the number of pounds you can lift at the end of each week. How can you use this function to predict what might happen if you continue the sport?

53. Would you prefer that your salary be modeled exponentially or logarithmically? Explain your answer.

54. One problem with all exponential growth models is that nothing can grow exponentially forever. (Or can it? See the Blitzer Bonus on page 668.) Describe factors that might limit the size of a population.

Technology Exercises

In Example 1 on page 732, we used two data points and an exponential function to model the population of the United States from 1970 through 2010. The data are shown again in the table. Use all five data points to solve Exercises 55–59.

x, Number of Years after 1969	y, U.S. Population (millions)
1 (1970)	203.3
11 (1980)	226.5
21 (1990)	248.7
31 (2000)	281.4
41 (2010)	308.7

55. a. Use your graphing utility's exponential regression option to obtain a model of the form $y = ab^x$ that fits the data. How well does the correlation coefficient, r, indicate that the model fits the data?

b. Rewrite the model in terms of base e. By what percentage is the population of the United States increasing each year?

56. Use your graphing utility's logarithmic regression option to obtain a model of the form $y = a + b \ln x$ that fits the data. How well does the correlation coefficient, r, indicate that the model fits the data?

57. Use your graphing utility's linear regression option to obtain a model of the form $y = ax + b$ that fits the data. How well does the correlation coefficient, r, indicate that the model fits the data?

58. Use your graphing utility's power regression option to obtain a model of the form $y = ax^b$ that fits the data. How well does the correlation coefficient, r, indicate that the model fits the data?

59. Use the values of r in Exercises 55–58 to select the two models of best fit. Use each of these models to predict by which year the U.S. population will reach 335 million. How do these answers compare to the year we found in Example 1, namely 2020? If you obtained different years, how do you account for this difference?

60. The figure shows the number of people in the United States age 65 and over, with projected figures for the year 2020 and beyond.

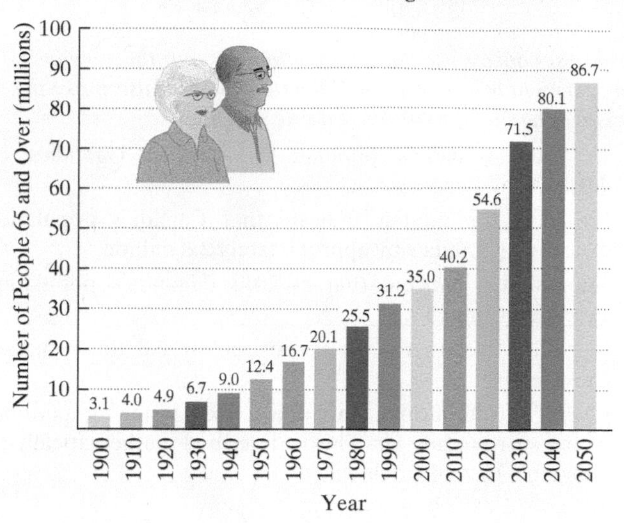

Source: U.S. Bureau of the Census

a. Let x represent the number of years after 1899 and let y represent the U.S. population age 65 and over, in millions. Use your graphing utility to find the model that best fits the data in the bar graph.

b. Rewrite the model in terms of base e. By what percentage is the 65 and over population increasing each year?

61. In Exercises 37–42, you determined the best choice for the kind of function that modeled the data in the table. For each of the exercises that you worked, use a graphing utility to find the actual function that best fits the data. Then use the model to make a reasonable prediction for a value that exceeds those shown in the table's first column.

Critical Thinking Exercises

Make Sense? *In Exercises 62–65, determine whether each statement makes sense or does not make sense, and explain your reasoning.*

62. I used an exponential model with a positive growth rate to describe the depreciation in my car's value over four years.

63. After 100 years, a population whose growth rate is 3% will have three times as many people as a population whose growth rate is 1%.

64. Because carbon-14 decays exponentially, carbon dating can determine the ages of ancient fossils.

65. When I used an exponential function to model Russia's declining population, the growth rate k was negative.

The exponential growth models describe the population of the indicated country, A, in millions, t years after 2006.

Canada — $A = 33.1e^{0.009t}$

Uganda — $A = 28.2e^{0.034t}$

In Exercises 66–69, use this information to determine whether each statement is true or false. If the statement is false, make the necessary change(s) to produce a true statement.

66. In 2006, Canada's population exceeded Uganda's by 4.9 million.

67. By 2009, the models indicate that Canada's population exceeded Uganda's by approximately 2.8 million.

68. The models indicate that in 2013, Uganda's population exceeded Canada's.

69. Uganda's growth rate is approximately 3.8 times that of Canada's.

70. Over a period of time, a hot object cools to the temperature of the surrounding air. This is described mathematically by Newton's Law of Cooling:

$$T = C + (T_0 - C)e^{-kt},$$

where t is the time it takes for an object to cool from temperature T_0 to temperature T, C is the surrounding air temperature, and k is a positive constant that is associated with the cooling object. A cake removed from the oven has a temperature of 210°F and is left to cool in a room that has a temperature of 70°F. After 30 minutes, the temperature of the cake is 140°F. What is the temperature of the cake after 40 minutes?

Review Exercises

71. Divide:

$$\frac{x^2 - 9}{2x^2 + 7x + 3} \div \frac{x^2 - 3x}{2x^2 + 11x + 5}.$$

(Section 6.1, Example 7)

72. Solve: $x^{\frac{2}{3}} + 2x^{\frac{1}{3}} - 3 = 0.$

(Section 8.4, Example 5)

73. Simplify: $6\sqrt{2} - 2\sqrt{50} + 3\sqrt{98}.$

(Section 7.4, Example 2)

Preview Exercises

Exercises 74–76 will help you prepare for the material covered in the first section of the next chapter.

In Exercises 74–75, let $(x_1, y_1) = (7, 2)$ and $(x_2, y_2) = (1, -1)$.

74. Find $\sqrt{(x_2 - x_1)^2 + (y_2 - y_1)^2}$. Express the answer in simplified radical form.

75. Find the point represented by $\left(\frac{x_1 + x_2}{2}, \frac{y_1 + y_2}{2}\right)$.

76. Use a rectangular coordinate system to graph the circle with center (1, −1) and radius 1.

Chapter 9 Summary

Definitions and Concepts	Examples

Section 9.1 Exponential Functions

The exponential function with base b is defined by $f(x) = b^x$, where $b > 0$ and $b \neq 1$. The graph contains the point (0, 1). When $b > 1$, the graph rises from left to right. When $0 < b < 1$, the graph falls from left to right. The x-axis is a horizontal asymptote. The domain is $(-\infty, \infty)$; the range is $(0, \infty)$. The natural exponential function is $f(x) = e^x$, where $e \approx 2.71828$.

Graph $f(x) = 2^x$ and $g(x) = 2^{x-1}$.

x	$f(x) = 2^x$	$g(x) = 2^{x-1}$
−2	$2^{-2} = \frac{1}{4}$	$2^{-3} = \frac{1}{8}$
−1	$2^{-1} = \frac{1}{2}$	$2^{-2} = \frac{1}{4}$
0	$2^0 = 1$	$2^{-1} = \frac{1}{2}$
1	$2^1 = 2$	$2^0 = 1$
2	$2^2 = 4$	$2^1 = 2$

The graph of g is the graph of f shifted one unit to the right.

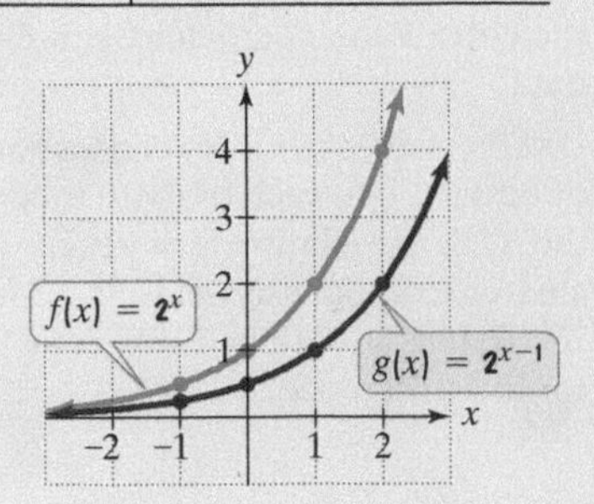

Definitions and Concepts	Examples

Section 9.1 Exponential Functions (continued)

Formulas for Compound Interest
After t years, the balance, A, in an account with principal P and annual interest rate r is given by the following formulas:

1. For n compounding periods per year: $A = P\left(1 + \frac{r}{n}\right)^{nt}$
2. For continuous compounding: $A = Pe^{rt}$

Select the better investment for $4000 over 6 years:

- 6% compounded semiannually

$$A = P\left(1 + \frac{r}{n}\right)^{nt}$$
$$= 4000\left(1 + \frac{0.06}{2}\right)^{2\cdot 6} \approx \$5703$$

- 5.9% compounded continuously

$$A = Pe^{rt} = 4000e^{0.059(6)} \approx \$5699$$

The first investment is better.

Section 9.2 Composite and Inverse Functions

Composite Functions
The composite function $f \circ g$ is defined by

$$(f \circ g)(x) = f(g(x)).$$

The composite function $g \circ f$ is defined by

$$(g \circ f)(x) = g(f(x)).$$

Let $f(x) = x^2 + x$ and $g(x) = 2x + 1$.

- $(f \circ g)(x) = f(g(x)) = (g(x))^2 + g(x)$

Replace x with $g(x)$.

$$= (2x + 1)^2 + (2x + 1) = 4x^2 + 4x + 1 + 2x + 1$$
$$= 4x^2 + 6x + 2$$

- $(g \circ f)(x) = g(f(x)) = 2f(x) + 1$

Replace x with $f(x)$.

$$= 2(x^2 + x) + 1 = 2x^2 + 2x + 1$$

Inverse Functions
If $f(g(x)) = x$ and $g(f(x)) = x$, function g is the inverse of function f, denoted f^{-1} and read "f inverse." The procedure for finding a function's inverse uses a switch-and-solve strategy. Switch x and y, then solve for y.

If $f(x) = 2x - 5$, find $f^{-1}(x)$.

$y = 2x - 5$ Replace $f(x)$ with y.

$x = 2y - 5$ Exchange x and y.

$x + 5 = 2y$ Solve for y.

$\frac{x + 5}{2} = y$

$f^{-1}(x) = \frac{x + 5}{2}$ Replace y with $f^{-1}(x)$.

Definitions and Concepts	Examples

Section 9.2 Composite and Inverse Functions (continued)

The Horizontal Line Test for Inverse Functions
A function, f, has an inverse that is a function, f^{-1}, if there is no horizontal line that intersects the graph of f at more than one point. A one-to-one function is one in which no two different ordered pairs have the same second component. Only one-to-one functions have inverse functions. If the point (a, b) is on the graph of f, then the point (b, a) is on the graph of f^{-1}. The graph of f^{-1} is a reflection of the graph of f about the line $y = x$.

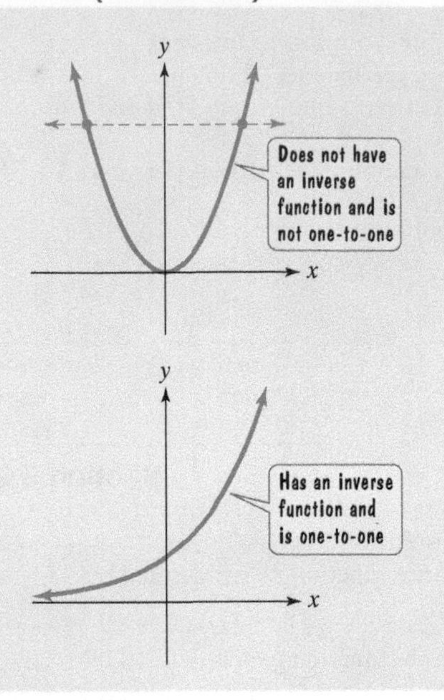

Section 9.3 Logarithmic Functions

Definition of the logarithmic function: For $x > 0$ and $b > 0$, $b \neq 1$, $y = \log_b x$ is equivalent to $b^y = x$. The function $f(x) = \log_b x$ is the logarithmic function with base b. This function is the inverse function of the exponential function with base b.

- Write $\log_2 32 = 5$ in exponential form.

  $2^5 = 32$ $\quad y = \log_b x$ means $b^y = x$.

- Write $\sqrt{49} = 7$, or $49^{\frac{1}{2}} = 7$, in logarithmic form.

  $\frac{1}{2} = \log_{49} 7$ $\quad b^y = x$ means $y = \log_b x$.

The graph of $f(x) = \log_b x$ can be obtained from $f(x) = b^x$ by reversing coordinates. The graph of $f(x) = \log_b x$ contains the point $(1, 0)$. If $b > 1$, the graph rises from left to right. If $0 < b < 1$, the graph falls from left to right. The y-axis is a vertical asymptote. The domain is $(0, \infty)$; the range is $(-\infty, \infty)$. $f(x) = \log x$ means $f(x) = \log_{10} x$ and is the common logarithmic function. $f(x) = \ln x$ means $f(x) = \log_e x$ and is the natural logarithmic function. The domain of $f(x) = \log_b g(x)$ consists of all x for which $g(x) > 0$.

- Find the domain: $f(x) = \log_6(4 - x)$.

  $4 - x > 0$

  $4 > x \quad (\text{or } x < 4)$

  The domain is $(-\infty, 4)$.

Basic Logarithmic Properties

Base b ($b > 0$, $b \neq 1$)	Base 10 (Common Logarithms)	Base e (Natural Logarithms)
$\log_b 1 = 0$	$\log 1 = 0$	$\ln 1 = 0$
$\log_b b = 1$	$\log 10 = 1$	$\ln e = 1$
$\log_b b^x = x$	$\log 10^x = x$	$\ln e^x = x$
$b^{\log_b x} = x$	$10^{\log x} = x$	$e^{\ln x} = x$

- $\log_8 1 = 0$ because $\log_b 1 = 0$.
- $\log_4 4 = 1$ because $\log_b b = 1$.
- $\ln e^{8x} = 8x$ because $\ln e^x = x$.
- $e^{\ln \sqrt[3]{x}} = \sqrt[3]{x}$ because $e^{\ln x} = x$.
- $\log_t t^{25} = 25$ because $\log_b b^x = x$.

Definitions and Concepts	Examples

Section 9.4 Properties of Logarithms

Properties of Logarithms

For $M > 0$ and $N > 0$:

1. *The Product Rule*: $\log_b(MN) = \log_b M + \log_b N$
2. *The Quotient Rule*: $\log_b\left(\frac{M}{N}\right) = \log_b M - \log_b N$
3. *The Power Rule*: $\log_b M^p = p \log_b M$
4. *The Change-of Base Property*:

The General Property	Introducing Common Logarithms	Introducing Natural Logarithms
$\log_b M = \frac{\log_a M}{\log_a b}$	$\log_b M = \frac{\log M}{\log b}$	$\log_b M = \frac{\ln M}{\ln b}$

- Expand: $\log_3(81x^7)$.

$$= \log_3 81 + \log_3 x^7$$
$$= 4 + 7\log_3 x$$

- Write as a single logarithm: $7\ln x - 4\ln y$.

$$= \ln x^7 - \ln y^4 = \ln\left(\frac{x^7}{y^4}\right)$$

- Evaluate: $\log_6 92$.

$$\log_6 92 = \frac{\ln 92}{\ln 6} \approx 2.5237$$

Section 9.5 Exponential and Logarithmic Equations

An exponential equation is an equation containing a variable in an exponent. Some exponential equations can be solved by expressing both sides as a power of the same base. Then set the exponents equal to each other:

If $b^M = b^N$, then $M = N$.

Solve: $4^{2x-1} = 64$.

$$4^{2x-1} = 4^3$$
$$2x - 1 = 3$$
$$2x = 4$$
$$x = 2$$

The solution is 2 and the solution set is {2}.

If both sides of an exponential equation cannot be expressed as a power of the same base, isolate the exponential expression. Take the natural logarithm on both sides for bases other than 10 and take the common logarithm on both sides for base 10. Simplify using

$$\ln b^x = x \ln b \quad \text{or} \quad \ln e^x = x \quad \text{or} \quad \log 10^x = x.$$

Solve: $7^x = 103$.

$$\ln 7^x = \ln 103$$
$$x \ln 7 = \ln 103$$
$$x = \frac{\ln 103}{\ln 7}$$

The solution is $\frac{\ln 103}{\ln 7}$ and the solution set is $\left\{\frac{\ln 103}{\ln 7}\right\}$.

A logarithmic equation is an equation containing a variable in a logarithmic expression. Logarithmic equations in the form $\log_b x = c$ can be solved by rewriting in exponential form as $b^c = x$. When checking logarithmic equations, reject proposed solutions that produce the logarithm of a negative number or the logarithm of zero in the original equation.

Solve: $\log_2(3x - 1) = 5$.

$$2^5 = 3x - 1$$
$$32 = 3x - 1$$
$$33 = 3x$$
$$11 = x$$

The solution is 11 and the solution set is {11}.

Solve: $3 \ln 2x = 15$.

$$\ln 2x = 5$$
$$\log_e 2x = 5$$
$$e^5 = 2x$$
$$\frac{e^5}{2} = x$$

Exponential form

The solution is $\frac{e^5}{2}$ and the solution set is $\left\{\frac{e^5}{2}\right\}$.

Definitions and Concepts	Examples

Section 9.5 Exponential and Logarithmic Equations (continued)

Logarithmic equations in the form $\log_b M = \log_b N$, where $M > 0$ and $N > 0$, can be solved using the one-to-one property of logarithms:

If $\log_b M = \log_b N$, then $M = N$.

Solve: $\log(2x - 1) = \log(4x - 3) - \log x$.

$$\log(2x - 1) = \log\left(\frac{4x - 3}{x}\right)$$
$$2x - 1 = \frac{4x - 3}{x}$$
$$x(2x - 1) = 4x - 3$$
$$2x^2 - x = 4x - 3$$
$$2x^2 - 5x + 3 = 0$$
$$(2x - 3)(x - 1) = 0$$
$$2x - 3 = 0 \quad \text{or} \quad x - 1 = 0$$
$$x = \frac{3}{2} \qquad\qquad x = 1$$

Neither number produces the logarithm of 0 or logarithms of negative numbers in the original equation. The solutions are 1 and $\frac{3}{2}$, and the solution set is $\left\{1, \frac{3}{2}\right\}$.

Section 9.6 Exponential Growth and Decay; Modeling Data

Exponential growth and decay models are given by $A = A_0 e^{kt}$ in which t represents time, A_0 is the amount present at $t = 0$, and A is the amount present at time t. If $k > 0$, the model describes growth and k is the growth rate. If $k < 0$, the model describes decay and k is the decay rate. Scatter plots for exponential and logarithmic models are shown in **Figure 9.25** on page 735. When using a graphing utility to model data, the closer that the correlation coefficient r is to -1 or 1, the better the model fits the data.

The 1970 population of the Tokyo, Japan, urban area was 16.5 million: in 2000, it was 26.4 million. Write an exponential growth function that describes the population, in millions, t years after 1970. Begin with $A = A_0 e^{kt}$.

$A = 16.5e^{kt}$ — In 1970 ($t = 0$), the population was 16.5 million.

$26.4 = 16.5e^{k \cdot 30}$ — When $t = 30$ (in 2000), $A = 26.4$.

$e^{30k} = \frac{26.4}{16.5}$ — Isolate the exponential factor.

$\ln e^{30k} = \ln\left(\frac{26.4}{16.5}\right)$ — Take the natural logarithm on both sides.

$$30k = \ln\left(\frac{26.4}{16.5}\right) \quad \text{and} \quad k = \frac{\ln\left(\frac{26.4}{16.5}\right)}{30} \approx 0.016$$

The growth function is $A = 16.5e^{0.016t}$.

Growth rate is 0.016 or 1.6%.

Expressing an Exponential Model in Base e

$y = ab^x$ is equivalent to $y = ae^{(\ln b)x}$.

Rewrite in terms of base e: $y = 24(7.2)^x$.

$$y = 24e^{(\ln 7.2)x} \approx 24e^{1.974x}$$

CHAPTER 9 REVIEW EXERCISES

9.1 *In Exercises 1–4, set up a table of coordinates for each function. Select integers from −2 to 2, inclusive, for x. Then use the table of coordinates to match the function with its graph. [The graphs are labeled (a) through (d).]*

1. $f(x) = 4^x$

2. $f(x) = 4^{-x}$

3. $f(x) = -4^{-x}$

4. $f(x) = -4^{-x} + 3$

a.

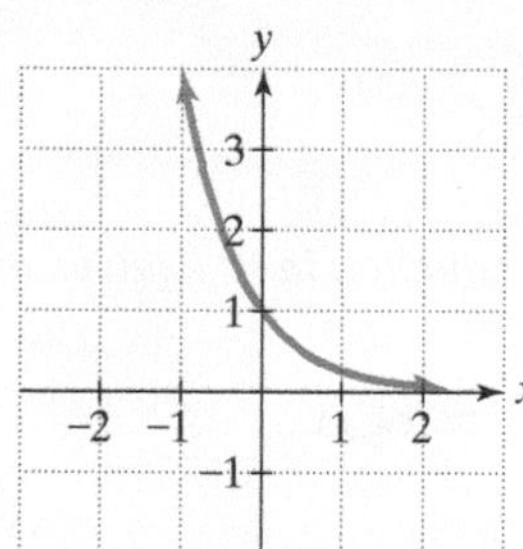

b.

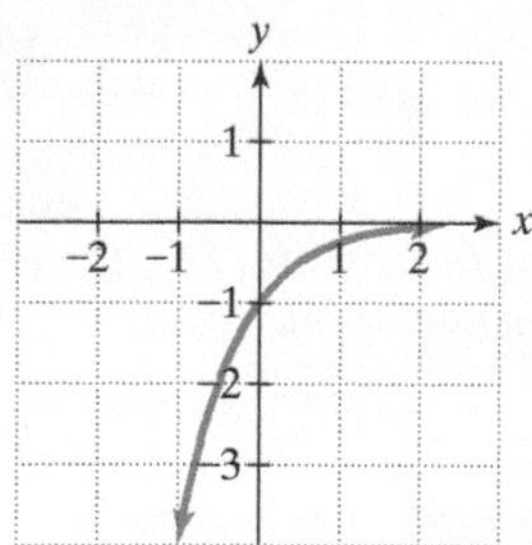

c.

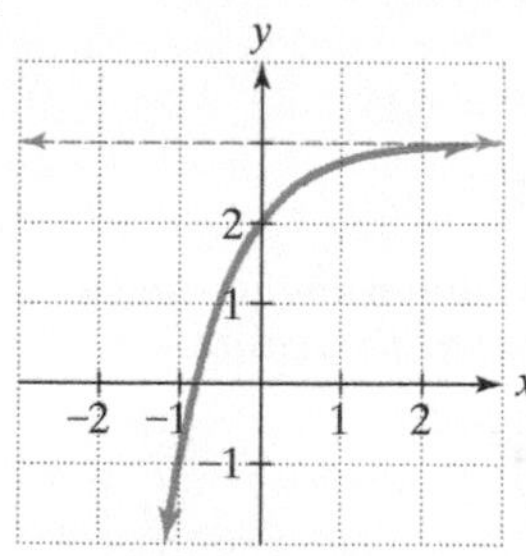

d.

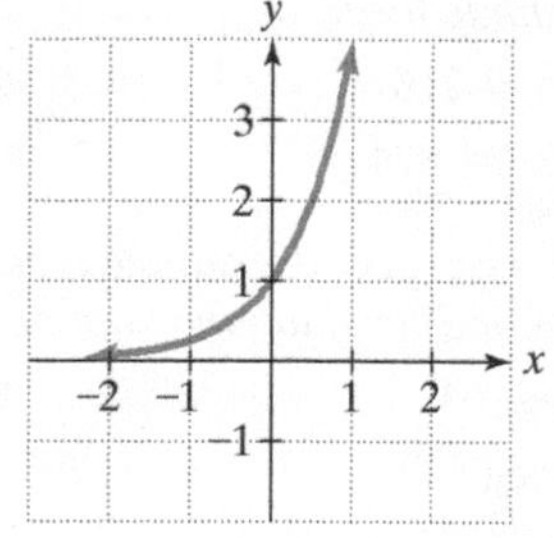

In Exercises 5–8, graph functions f and g in the same rectangular coordinate system. Select integers from −2 to 2, inclusive, for x. Then describe how the graph of g is related to the graph of f. If applicable, use a graphing utility to confirm your hand-drawn graphs.

5. $f(x) = 2^x$ and $g(x) = 2^{x-1}$

6. $f(x) = 2^x$ and $g(x) = \left(\frac{1}{2}\right)^x$

7. $f(x) = 3^x$ and $g(x) = 3^x - 1$

8. $f(x) = 3^x$ and $g(x) = -3^x$

Use the compound interest formulas

$$A = P\left(1 + \frac{r}{n}\right)^{nt} \quad \text{and} \quad A = Pe^{rt}$$

to solve Exercises 9–10.

9. Suppose that you have \$5000 to invest. Which investment yields the greater return over 5 years: 5.5% compounded semiannually or 5.25% compounded monthly?

10. Suppose that you have \$14,000 to invest. Which investment yields the greater return over 10 years: 7% compounded monthly or 6.85% compounded continuously?

11. A cup of coffee is taken out of a microwave oven and placed in a room. The temperature, T, in degrees Fahrenheit, of the coffee after t minutes is modeled by the function $T = 70 + 130e^{-0.04855t}$. The graph of the function is shown in the figure.

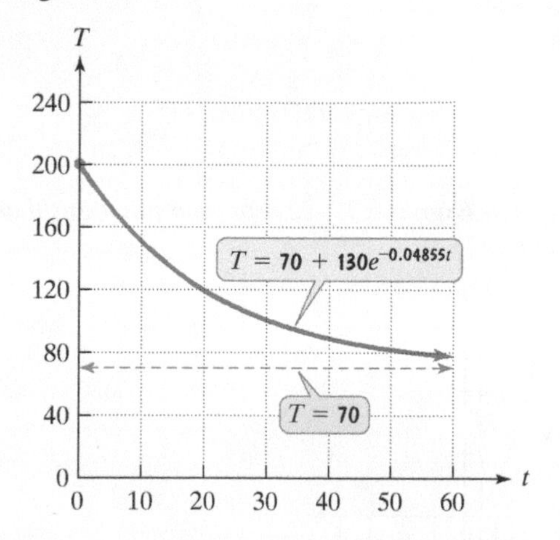

Use the graph to answer each of the following questions.

a. What was the temperature of the coffee when it was first taken out of the microwave?

b. What is a reasonable estimate of the temperature of the coffee after 20 minutes? Use your calculator to verify this estimate.

c. What is the limit of the temperature to which the coffee will cool? What does this tell you about the temperature of the room?

9.2 *In Exercises 12–13, find* **a.** $(f \circ g)(x)$; **b.** $(g \circ f)(x)$; **c.** $(f \circ g)(3)$.

12. $f(x) = x^2 + 3,\ g(x) = 4x - 1$

13. $f(x) = \sqrt{x},\ g(x) = x + 1$

In Exercises 14–15, find f(g(x)) and g(f(x)) and determine whether each pair of functions f and g are inverses of each other.

14. $f(x) = \frac{3}{5}x + \frac{1}{2}$ and $g(x) = \frac{5}{3}x - 2$

15. $f(x) = 2 - 5x$ and $g(x) = \frac{2 - x}{5}$

The functions in Exercises 16–18 are all one-to-one. For each function,

a. *Find an equation of $f^{-1}(x)$, the inverse function.*

b. *Verify that your equation is correct by showing that $f(f^{-1}(x)) = x$ and $f^{-1}(f(x)) = x$.*

16. $f(x) = 4x - 3$

17. $f(x) = \sqrt{x + 2}$

18. $f(x) = 8x^3 + 1$

Which graphs in Exercises 19–22 represent functions that have inverse functions?

19.

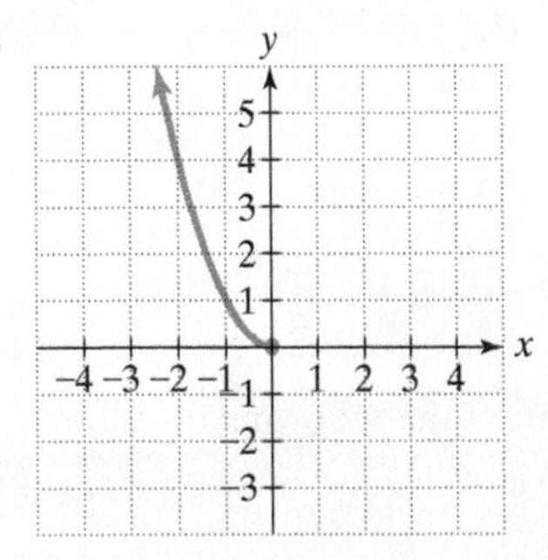

20.

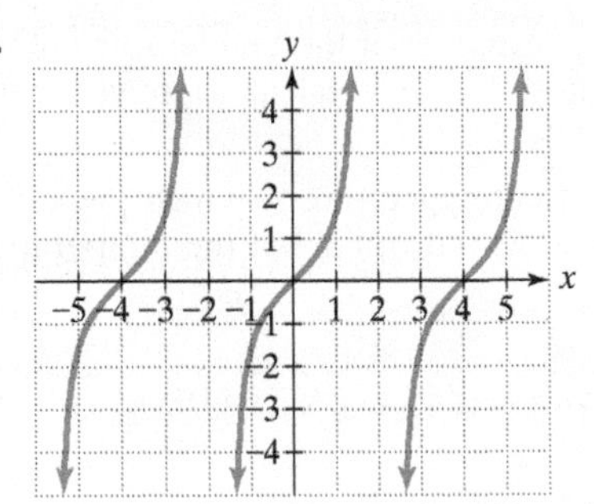

21.

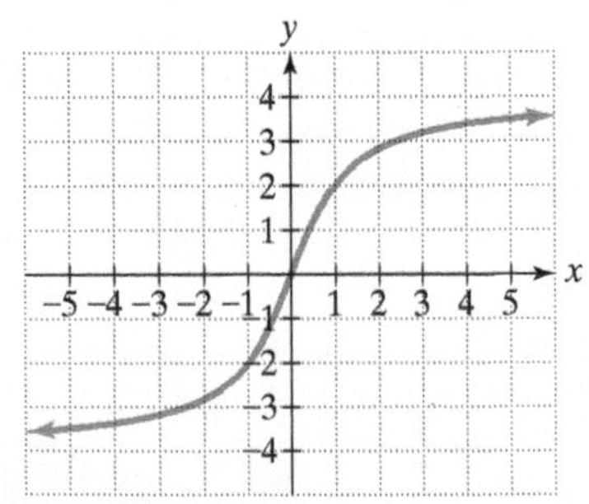

22.

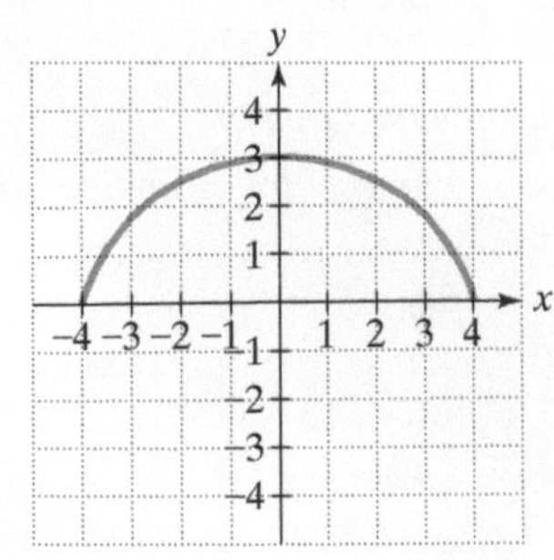

23. Use the graph of f in the figure shown to draw the graph of its inverse function.

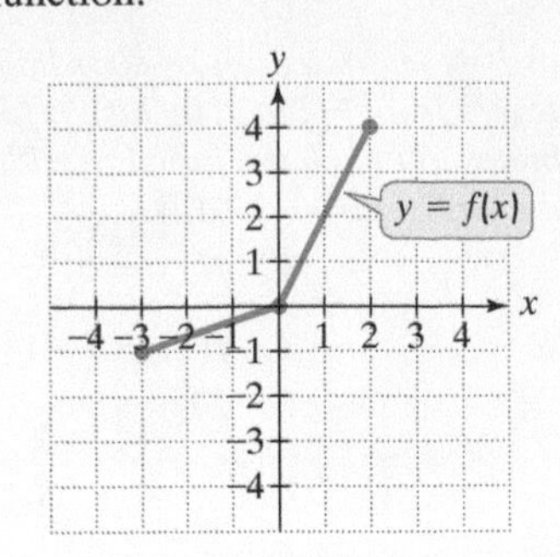

9.3 *In Exercises 24–26, write each equation in its equivalent exponential form.*

24. $\frac{1}{2} = \log_{49} 7$

25. $3 = \log_4 x$

26. $\log_3 81 = y$

In Exercises 27–29, write each equation in its equivalent logarithmic form.

27. $6^3 = 216$

28. $b^4 = 625$

29. $13^y = 874$

In Exercises 30–40, evaluate each expression without using a calculator. If evaluation is not possible, state the reason.

30. $\log_4 64$

31. $\log_5 \frac{1}{25}$

32. $\log_3(-9)$

33. $\log_{16} 4$

34. $\log_{17} 17$

35. $\log_3 3^8$

36. $\ln e^5$

37. $\log_3 \frac{1}{\sqrt{3}}$

38. $\ln \frac{1}{e^2}$

39. $\log \frac{1}{1000}$

40. $\log_3(\log_8 8)$

41. Graph $f(x) = 2^x$ and $g(x) = \log_2 x$ in the same rectangular coordinate system. Use the graphs to determine each function's domain and range.

42. Graph $f(x) = \left(\frac{1}{3}\right)^x$ and $g(x) = \log_{\frac{1}{3}} x$ in the same rectangular coordinate system. Use the graphs to determine each function's domain and range.

In Exercises 43–45, find the domain of each logarithmic function.

43. $f(x) = \log_8(x + 5)$

44. $f(x) = \log(3 - x)$

45. $f(x) = \ln(x - 1)^2$

In Exercises 46–48, simplify each expression.

46. $\ln e^{6x}$

47. $e^{\ln\sqrt{x}}$

48. $10^{\log 4x^2}$

49. On the Richter scale, the magnitude, R, of an earthquake of intensity I is given by $R = \log\frac{I}{I_0}$, where I_0 is the intensity of a barely felt zero-level earthquake. If the intensity of an earthquake is $1000I_0$, what is its magnitude on the Richter scale?

50. Students in a psychology class took a final examination. As part of an experiment to see how much of the course content they remembered over time, they took equivalent forms of the exam in monthly intervals thereafter. The average score, $f(t)$, for the group after t months is modeled by the function $f(t) = 76 - 18\log(t + 1)$, where $0 \le t \le 12$.

a. What was the average score when the exam was first given?

b. What was the average score, to the nearest tenth, after 2 months? 4 months? 6 months? 8 months? one year?

c. Use the results from parts (a) and (b) to graph f. Describe what the shape of the graph indicates in terms of the material retained by the students.

51. The formula

$$t = \frac{1}{c}\ln\left(\frac{A}{A - N}\right)$$

describes the time, t, in weeks, that it takes to achieve mastery of a portion of a task. In the formula, A represents maximum learning possible, N is the portion of the learning that is to be achieved, and c is a constant used to measure an individual's learning style. A 50-year-old man decides to start running as a way to maintain good health. He feels that the maximum rate he could ever hope to achieve is 12 miles per hour. How many weeks will it take before the man can run 5 miles per hour if $c = 0.06$ for this person?

9.4 *In Exercises 52–55, use properties of logarithms to expand each logarithmic expression as much as possible. Where possible, evaluate logarithmic expressions without using a calculator. Assume that all variables represent positive numbers.*

52. $\log_6(36x^3)$

53. $\log_4\left(\frac{\sqrt{x}}{64}\right)$

54. $\log_2\left(\frac{xy^2}{64}\right)$

55. $\ln\sqrt[3]{\frac{x}{e}}$

In Exercises 56–59, use properties of logarithms to condense each logarithmic expression. Write the expression as a single logarithm whose coefficient is 1.

56. $\log_b 7 + \log_b 3$

57. $\log 3 - 3\log x$

58. $3\ln x + 4\ln y$

59. $\frac{1}{2}\ln x - \ln y$

In Exercises 60–61, use common logarithms or natural logarithms and a calculator to evaluate to four decimal places.

60. $\log_6 72{,}348$

61. $\log_4 0.863$

In Exercises 62–65, determine whether each equation is true or false. Where possible, show work to support your conclusion. If the statement is false, make the necessary change(s) to produce a true statement.

62. $(\ln x)(\ln 1) = 0$

63. $\log(x + 9) - \log(x + 1) = \frac{\log(x + 9)}{\log(x + 1)}$

64. $(\log_2 x)^4 = 4\log_2 x$

65. $\ln e^x = x\ln e$

9.5 *In Exercises 66–71, solve each exponential equation. Where necessary, express the solution set in terms of natural logarithms and use a calculator to obtain a decimal approximation, correct to two decimal places, for the solution.*

66. $2^{4x-2} = 64$

67. $125^x = 25$

68. $9^x = \frac{1}{27}$

69. $8^x = 12{,}143$

70. $9e^{5x} = 1269$

71. $30e^{0.045x} = 90$

In Exercises 72–81, solve each logarithmic equation.

72. $\log_5 x = -3$

73. $\log x = 2$

74. $\log_4(3x - 5) = 3$

75. $\ln x = -1$

76. $3 + 4\ln(2x) = 15$

77. $\log_2(x + 3) + \log_2(x - 3) = 4$

78. $\log_3(x - 1) - \log_3(x + 2) = 2$

79. $\log_4(3x - 5) = \log_4 3$

80. $\ln(x + 4) - \ln(x + 1) = \ln x$

81. $\log_6(2x + 1) = \log_6(x - 3) + \log_6(x + 5)$

82. The function $P(x) = 14.7e^{-0.21x}$ models the average atmospheric pressure, $P(x)$, in pounds per square inch, at an altitude of x miles above sea level. The atmospheric pressure at the peak of Mt. Everest, the world's highest mountain, is 4.6 pounds per square inch. How many miles above sea level, to the nearest tenth of a mile, is the peak of Mt. Everest?

83. **Newest Dinosaur: The PC-osaurus?** For the period from 2009 through 2012, worldwide PC sales stayed relatively flat, while tablet sales skyrocketed. The bar graph shows worldwide PC and tablet sales, in millions, from 2009 through 2012.

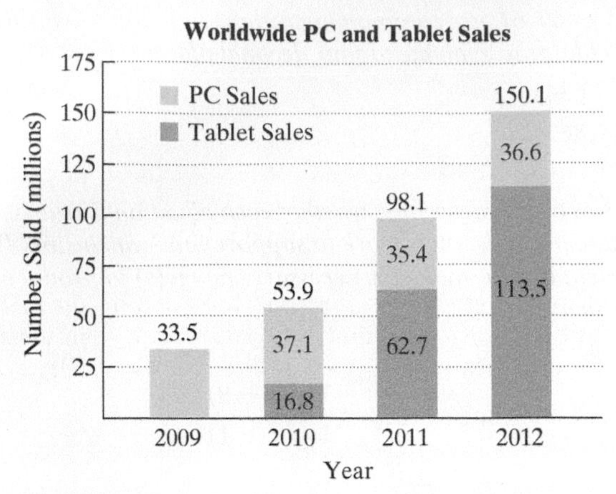

Source: Canalys

The function

$$f(t) = 33.4(1.66)^t$$

models worldwide PC and tablet sales combined, $f(t)$, in millions, t years after 2009. When does this model project that 421 million PC and tablets were sold? Round to the nearest year. Based on the relatively flat PC sales over the four years shown by the graph, estimate the number of tablet sales for that year.

84. **60 + Years after "Brown v. Board"** In 1954, the Supreme Court ended legal segregation in U.S. public schools. The bar graph shows the percentage of black and white U.S. adults ages 25 or older who completed high school since the end of segregation.

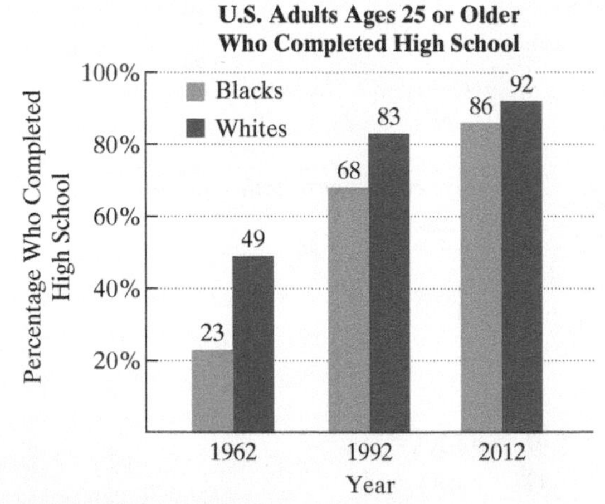

Source: Pew Research Center

The functions

$$B(x) = 16 \ln x + 23 \text{ and } W(x) = 11 \ln x + 49$$

model the percentage of black adults, $B(x)$, and white adults, $W(x)$, who had completed high school x years after 1961.

a. Use the appropriate model to determine the percentage of white adults who had completed high school in 2012. Round to the nearest percent. How well does this rounded value model the percents displayed by the graph?

b. Use the appropriate model to determine when 90% of black adults will have completed high school. Round to the nearest year.

85. Use the compound interest formula

$$A = P\left(1 + \frac{r}{n}\right)^{nt}$$

to solve this problem. How long, to the nearest tenth of a year, will it take \$12,500 to grow to \$20,000 at 6.5% annual interest compounded quarterly?

Use the compound interest formula

$$A = Pe^{rt}$$

to solve Exercises 86–87.

86. How long, to the nearest tenth of a year, will it take \$50,000 to triple in value at 7.5% annual interest compounded continuously?

87. What interest rate is required for an investment subject to continuous compounding to triple in 5 years?

9.6

88. According to the U.S. Bureau of the Census, in 2000 there were 35.3 million residents of Hispanic origin living in the United States. By 2010, the number had increased to 50.5 million. The exponential growth function $A = 35.3e^{kt}$ describes the U.S. Hispanic population, A, in millions, t years after 2000.

a. Find k, correct to three decimal places.

b. Use the resulting model to project the Hispanic resident population in 2015.

c. In which year will the Hispanic resident population reach 70 million?

89. Use the exponential decay model, $A = A_0e^{kt}$, to solve this exercise. The half-life of polonium-210 is 140 days. How long will it take for a sample of this substance to decay to 20% of its original amount?

Exercises 90–92 present data in the form of tables. For each data set shown by the table,

a. *Create a scatter plot for the data.*

b. *Use the scatter plot to determine whether an exponential function, a logarithmic function, or a linear function is the best choice for modeling the data.*

90. Number of Seriously Mentally Ill Adults in the United States

Year	Number of Seriously Mentally Ill Adults
2006	9.0
2008	9.2
2010	9.4
2012	9.6

Source: U.S. Census Bureau

91. Percentage of Moderate Alcohol Users in the United States (Not Binge or Heavy Drinkers)

Age	Percentage of Moderate Drinkers
20	15%
25	24%
30	28%
35	32%
40	34%
45	35%

Source: Substance Abuse and Mental Health Services Administration

92. U.S. Electric Car Sales

Year	Number of Cars Sold
2010	5000
2011	19,000
2012	53,000
2013	98,000

Source: Electric Drive Transportation Association

In Exercises 93–94, rewrite the equation in terms of base e. Express the answer in terms of a natural logarithm and then round to three decimal places.

93. $y = 73(2.6)^x$

94. $y = 6.5(0.43)^x$

95. The figure shows world population projections through the year 2150. The data are from the United Nations Family Planning Program and are based on optimistic or pessimistic expectations for successful control of human population growth. Suppose that you are interested in modeling these data using exponential, logarithmic, linear, and quadratic functions. Which function would you use to model each of the projections? Explain your choices. For the choice corresponding to a quadratic model, would your formula involve one with a positive or negative leading coefficient? Explain.

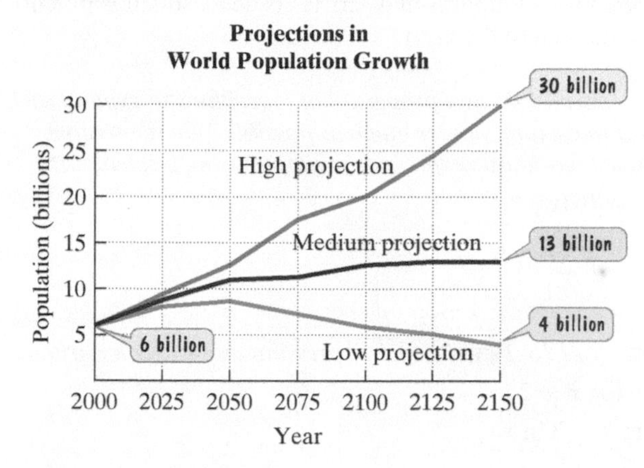

CHAPTER 9 TEST

Step-by-step test solutions are found on the Chapter Test Prep Videos available in MyMathLab® or on YouTube (search "BlitzerInterAlg7e" and click on "Channels").

1. Graph $f(x) = 2^x$ and $g(x) = 2^{x+1}$ in the same rectangular coordinate system.

2. Use $A = P\left(1 + \frac{r}{n}\right)^{nt}$ and $A = Pe^{rt}$ to solve this problem. Suppose you have \$3000 to invest. Which investment yields the greater return over 10 years: 6.5% compounded semiannually or 6% compounded continuously? How much more (to the nearest dollar) is yielded by the better investment?

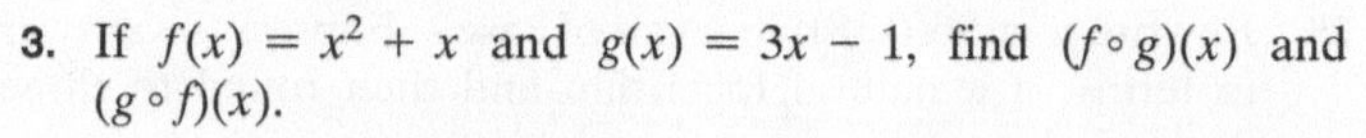

3. If $f(x) = x^2 + x$ and $g(x) = 3x - 1$, find $(f \circ g)(x)$ and $(g \circ f)(x)$.

4. If $f(x) = 5x - 7$, find $f^{-1}(x)$.

5. A function f models the amount given to charity as a function of income. The graph of f is shown in the figure.

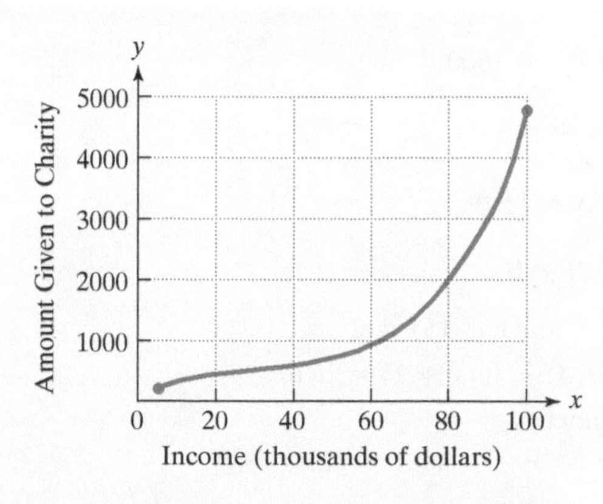

a. Explain why f has an inverse that is a function.

b. Find $f(80)$.

c. Describe in practical terms the meaning of $f^{-1}(2000)$.

6. Write in exponential form: $\log_5 125 = 3$.
7. Write in logarithmic form: $\sqrt{36} = 6$.
8. Graph $f(x) = 3^x$ and $g(x) = \log_3 x$ in the same rectangular coordinate system. Use the graphs to determine each function's domain and range.

In Exercises 9–11, simplify each expression.

9. $\ln e^{5x}$ 10. $\log_b b$ 11. $\log_6 1$

12. Find the domain: $f(x) = \log_5 (x - 7)$.
13. On the decibel scale, the loudness of a sound, in decibels, is given by $D = 10 \log \frac{I}{I_0}$, where I is the intensity of the sound, in watts per meter^2, and I_0 is the intensity of a sound barely audible to the human ear. If the intensity of a sound is $10^{12} I_0$, what is its loudness in decibels? (Such a sound is potentially damaging to the ear.)

In Exercises 14–15, use properties of logarithms to expand each logarithmic expression as much as possible. Where possible, evaluate logarithmic expressions without using a calculator.

14. $\log_4(64x^5)$
15. $\log_3\left(\frac{\sqrt[3]{x}}{81}\right)$

In Exercises 16–17, write each expression as a single logarithm.

16. $6 \log x + 2 \log y$
17. $\ln 7 - 3 \ln x$
18. Use a calculator to evaluate $\log_{15} 71$ to four decimal places.

In Exercises 19–26, solve each equation.

19. $3^{x-2} = 81$
20. $5^x = 1.4$
21. $400e^{0.005x} = 1600$
22. $\log_{25} x = \frac{1}{2}$
23. $\log_6(4x - 1) = 3$
24. $2 \ln(3x) = 8$
25. $\log x + \log(x + 15) = 2$
26. $\ln(x - 4) - \ln(x + 1) = \ln 6$
27. The function

$$A = 82.3e^{-0.004t}$$

models the population of Germany, A, in millions, t years after 2010.

a. What was the population of Germany in 2010?

b. Is the population of Germany increasing or decreasing? Explain.

c. In which year will the population of Germany be 79.1 million?

Use the formulas

$$A = P\left(1 + \frac{r}{n}\right)^{nt} \quad \text{and} \quad A = Pe^{rt}$$

to solve Exercises 28–29.

28. How long, to the nearest tenth of a year, will it take \$4000 to grow to \$8000 at 5% annual interest compounded quarterly?
29. What interest rate is required for an investment subject to continuous compounding to double in 10 years?
30. The 2010 population of Asia was 4121 million; in 2050, it is projected to be 5231 million. Write an exponential growth function that describes the population of Asia, in millions, t years after 2010.
31. Use the exponential decay model for carbon-14, $A = A_0e^{-0.000121t}$, to solve this exercise. Bones of a prehistoric man were discovered and contained 5% of the original amount of carbon-14. How long ago did the man die?

In Exercises 32–35, determine whether the values in each table belong to an exponential function, a logarithmic function, a linear function, or a quadratic function.

32.

x	y
0	3
1	1
2	−1
3	−3
4	−5

33.

x	y
$\frac{1}{3}$	−1
1	0
3	1
9	2
27	3

34.

x	y
0	1
1	5
2	25
3	125
4	625

35.

x	y
0	12
1	3
2	0
3	3
4	12

36. Rewrite $y = 96(0.38)^x$ in terms of base e. Express the answer in terms of a natural logarithm and then round to three decimal places.

CUMULATIVE REVIEW EXERCISES (CHAPTERS 1–9)

In Exercises 1–7, solve each equation, inequality, or system.

1. $8 - (4x - 5) = x - 7$

2. $\begin{cases} 5x + 4y = 22 \\ 3x - 8y = -18 \end{cases}$

3. $\begin{cases} -3x + 2y + 4z = 6 \\ 7x - y + 3z = 23 \\ 2x + 3y + z = 7 \end{cases}$

4. $|x - 1| > 3$

5. $\sqrt{x + 4} - \sqrt{x - 4} = 2$

6. $x - 4 \geq 0$ and $-3x \leq -6$

7. $2x^2 = 3x - 2$

In Exercises 8–12, graph each function, equation, or inequality in a rectangular coordinate system.

8. $3x = 15 + 5y$

9. $2x - 3y > 6$

10. $f(x) = -\frac{1}{2}x + 1$

11. $f(x) = x^2 + 6x + 8$

12. $f(x) = (x - 3)^2 - 4$

13. Evaluate:

$$\begin{vmatrix} 3 & 1 & 0 \\ 0 & 5 & -6 \\ -2 & -1 & 0 \end{vmatrix}.$$

14. Solve for c: $A = \frac{cd}{c + d}$.

In Exercises 15–17, let $f(x) = x^2 + 3x - 15$ and $g(x) = x - 2$. Find each indicated expression.

15. $f(g(x))$

16. $g(f(x))$

17. $g(a + h) - g(a)$

18. If $f(x) = 7x - 3$, find $f^{-1}(x)$.

In Exercises 19–20, find the domain of each function.

19. $f(x) = \frac{x - 2}{x^2 - 3x + 2}$

20. $f(x) = \ln(2x - 8)$

21. Write the equation of the linear function whose graph contains the point $(-2, 4)$ and is perpendicular to the line whose equation is $2x + y = 10$.

In Exercises 22–26, perform the indicated operations and simplify, if possible.

22. $\frac{-5x^3y^7}{15x^4y^{-2}}$

23. $(4x^2 - 5y)^2$

24. $(5x^3 - 24x^2 + 9) \div (5x + 1)$

25. $\frac{\sqrt[3]{32xy^{10}}}{\sqrt[3]{2xy^2}}$

26. $\frac{x + 2}{x^2 - 6x + 8} + \frac{3x - 8}{x^2 - 5x + 6}$

In Exercises 27–28, factor completely.

27. $x^4 - 4x^3 + 8x - 32$

28. $2x^2 + 12xy + 18y^2$

29. Write as a single logarithm whose coefficient is 1:

$$2 \ln x - \frac{1}{2} \ln y.$$

30. The length of a rectangular carpet is 4 feet greater than twice its width. If the area is 48 square feet, find the carpet's length and width.

31. Working alone, you can mow the lawn in 2 hours and your sister can do it in 3 hours. How long will it take you to do the job if you work together?

32. Your motorboat can travel 15 miles per hour in still water. Traveling with the river's current, the boat can cover 20 miles in the same time it takes to go 10 miles against the current. Find the rate of the current.

33. Use the formula for continuous compounding, $A = Pe^{rt}$, to solve this problem. What interest rate is required for an investment of \$6000 subject to continuous compounding to grow to \$18,000 in 10 years?

Answers to Selected Exercises

CHAPTER 1

Section 1.1 Check Point Exercises

1. a. $8x + 5$ **b.** $\frac{x}{7} - 2x$ **2.** 21.8; At age 10, the average neurotic level is 21.8. **3.** 608 **4. a.** \$27,660 **b.** overestimates by \$978
5. a. true **b.** true **6. a.** -8 is less than -2.; true **b.** 7 is greater than -3.; true **c.** -1 is less than or equal to -4.; false
d. 5 is greater than or equal to 5.; true **e.** 2 is greater than or equal to -14.; true
7. a. $\{x|-2 \le x < 5\}$ **b.** $\{x|1 \le x \le 3.5\}$ **c.** $\{x|x < -1\}$

Concept and Vocabulary Check

1. variable **2.** expression **3.** b to the nth power; base; exponent **4.** formula; modeling; models **5.** natural **6.** whole **7.** integers
8. rational **9.** irrational **10.** rational; irrational **11.** left **12.** 2; 5; 2; 5 **13.** greater than **14.** less than or equal to

Exercise Set 1.1

1. $x + 5$ **3.** $x - 4$ **5.** $4x$ **7.** $2x + 10$ **9.** $6 - \frac{1}{2}x$ **11.** $\frac{4}{x} - 2$ **13.** $\frac{3}{5 - x}$ **15.** 57 **17.** 10 **19.** $1\frac{1}{9}$ or $\frac{10}{9}$ **21.** 10 **23.** 44
25. 46 **27.** $\{1, 2, 3, 4\}$ **29.** $\{-7, -6, -5, -4\}$ **31.** $\{8, 9, 10, \dots\}$ **33.** $\{1, 3, 5, 7, 9\}$ **35.** true **37.** true **39.** false **41.** true
43. false **45.** true **47.** false **49.** -6 is less than -2.; true **51.** 5 is greater than -7.; true **53.** 0 is less than -4.; false
55. -4 is less than or equal to 1.; true **57.** -2 is less than or equal to -6.; false **59.** -2 is less than or equal to -2.; true
61. -2 is greater than or equal to -2.; true **63.** 2 is less than or equal to $-\frac{1}{2}$.; false
65. $\{x|1 < x \le 6\}$ **67.** $\{x|-5 \le x < 2\}$ **69.** $\{x|-3 \le x \le 1\}$
71. $\{x|x > 2\}$ **73.** $\{x|x \ge -3\}$ **75.** $\{x|x < 3\}$
77. $\{x|x < 5.5\}$ **79.** true **81.** false **83.** true **85.** false **87.** false **89.** 4.2 **91.** 0.4
93. 62%; underestimates by 1% **95.** 10°C **97.** 60 ft **117.** does not make sense
119. makes sense **121.** false **123.** true **125.** $2 \cdot 4 + 20 = 28$ and $\frac{1}{7}$ of 28 is 4, but $2(4 + 20) = 48$ and $\frac{1}{7}$ of 48 is about 6.86. You can't purchase 6.86 birds; the correct translation is $2 \cdot 4 + 20$. **127.** $8 + 2 \cdot (4 - 3) = 10$ **129.** -5 and 5
130. 8 **131.** 34; 34

Section 1.2 Check Point Exercises

1. a. 6 **b.** 4.5 **c.** 0 **2. a.** -28 **b.** 0.7 **c.** $-\frac{1}{10}$ **3. a.** 8 **b.** $-\frac{1}{3}$ **4. a.** -3 **b.** 10.5 **c.** $-\frac{3}{5}$
5. a. 25 **b.** -25 **c.** -64 **d.** $\frac{81}{625}$ **6. a.** -8 **b.** $\frac{8}{15}$ **7.** 74 **8.** -4 **9.** addition: $9 + 4x$; multiplication: $x \cdot 4 + 9$
10. a. $(6 + 12) + x = 18 + x$ **b.** $(-7 \cdot 4)x = -28x$ **11.** $-28x - 8$ **12.** $14x + 15x^2$ or $15x^2 + 14x$ **13.** $12x - 40$ **14.** $42 - 4x$

Concept and Vocabulary Check

1. negative number **2.** 0 **3.** positive number **4.** positive number **5.** positive number **6.** negative number **7.** positive number
8. divide **9.** subtract **10.** absolute value; 0; a **11.** a; $-a$ **12.** 0; inverse; 0; identity **13.** $b + a$ **14.** $(ab)c$ **15.** $ab + ac$
16. simplified

Exercise Set 1.2

1. 7 **3.** 4 **5.** 7.6 **7.** $\frac{\pi}{2}$ **9.** $\sqrt{2}$ **11.** $-\frac{2}{5}$ **13.** -11 **15.** -4 **17.** -4.5 **19.** $\frac{2}{15}$ **21.** $-\frac{35}{36}$ **23.** -8.2 **25.** -12.4
27. 0 **29.** -11 **31.** 5 **33.** 0 **35.** -12 **37.** 18 **39.** -15 **41.** $-\frac{1}{4}$ **43.** 5.5 **45.** $\sqrt{2}$ **47.** -90 **49.** 33 **51.** $-\frac{15}{13}$
53. 0 **55.** -8 **57.** 48 **59.** 100 **61.** -100 **63.** -8 **65.** 1 **67.** -1 **69.** $\frac{1}{8}$ **71.** -3 **73.** 45 **75.** 0 **77.** undefined

79. $\frac{9}{14}$ **81.** -15 **83.** -2 **85.** -24 **87.** 45 **89.** $\frac{1}{121}$ **91.** 14 **93.** $-\frac{8}{3}$ **95.** $-\frac{1}{2}$ **97.** 31 **99.** 37 **101.** addition: $10 + 4x$; multiplication: $x \cdot 4 + 10$ **103.** addition: $-5 + 7x$; multiplication: $x \cdot 7 - 5$ **105.** $(4 + 6) + x = 10 + x$ **107.** $(-7 \cdot 3)x = -21x$ **109.** $\left(-\frac{1}{3} \cdot -3\right)y = y$ **111.** $6x + 15$ **113.** $-14x - 21$ **115.** $-3x + 6$ **117.** $12x$ **119.** $5x^2$ **121.** $10x + 12x^2$ **123.** $18x - 40$ **125.** $8y - 12$ **127.** $16y - 25$ **129.** $12x^2 + 11$ **131.** $x - (x + 4)$; -4 **133.** $6(-5x)$; $-30x$ **135.** $5x - 2x$; $3x$ **137.** $8x - (3x + 6)$; $5x - 6$ **139.** -8 **141.** 50 **143.** by 7 **145.** -3 **147.** \$116.8 billion; underestimates by \$0.2 billion **149. a.** $1200 - 0.07x$ **b.** \$780 **169.** makes sense **171.** does not make sense **173.** false **175.** false **177.** $(8 - 2) \cdot 3 - 4 = 14$ **179.** -7 **180.** $\frac{10}{x} - 4x$ **181.** 42 **182.** 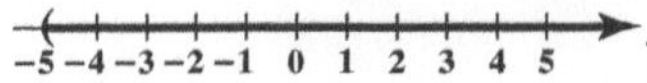

183. $-5; 0; 3; 4; 3; 0; -5$ **184.** $-8; -3; 0; 1; 0; -3; -8$ **185.** $3; 2; 1; 0; 1; 2; 3$

Section 1.3 Check Point Exercises

1.
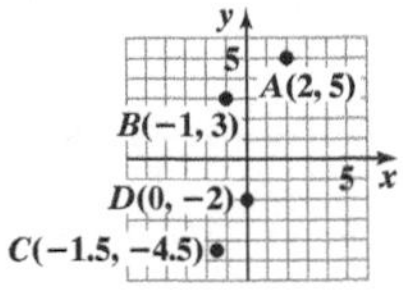

2.
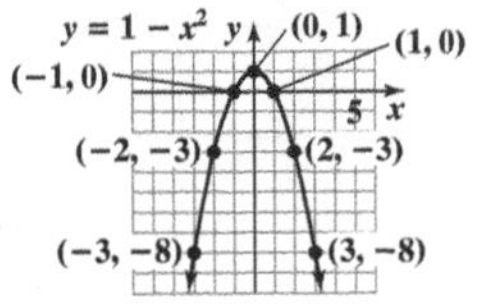

3.
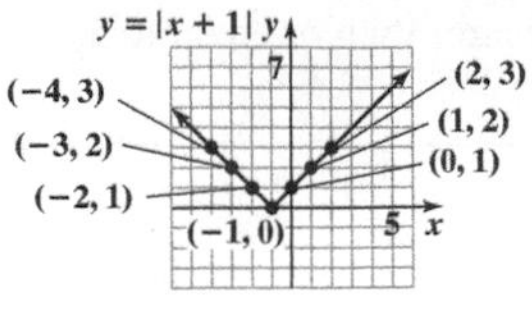

4. a. 0 to 3 hr **b.** 3 to 13 hr **c.** 0.05 mg per 100 ml; after 3 hr **d.** None of the drug is left in the body. **5.** minimum x-value: -100; maximum x-value: 100; distance between tick marks on x-axis: 50; minimum y-value: -100; maximum y-value: 100; distance between tick marks on y-axis: 10
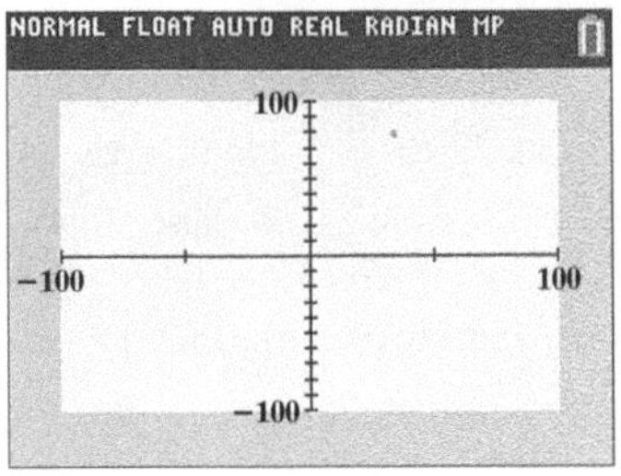

Concept and Vocabulary Check

1. x-axis **2.** y-axis **3.** origin **4.** quadrants; four **5.** x-coordinate; y-coordinate **6.** solution; satisfies

Exercise Set 1.3

1–9.
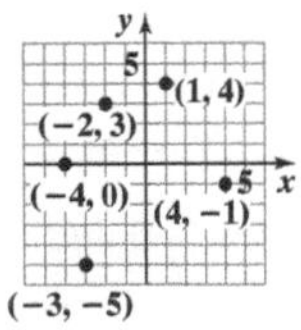

11.
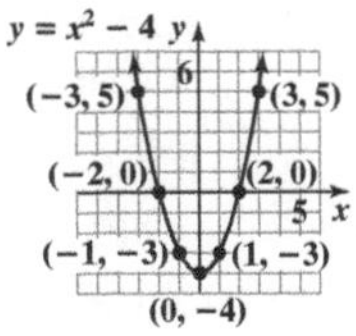

13.
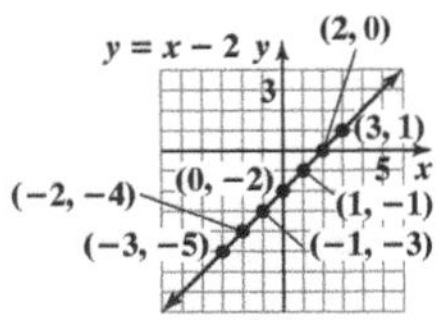

15.
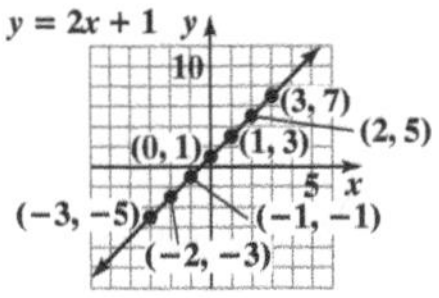

17.
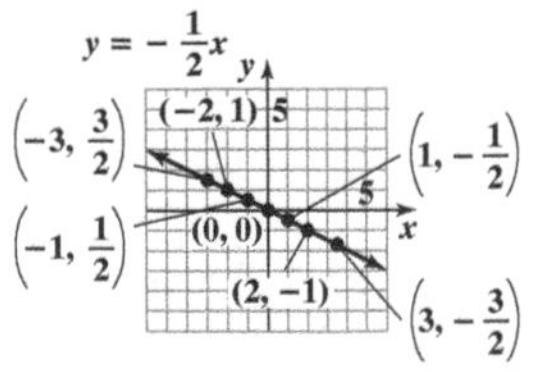

19.
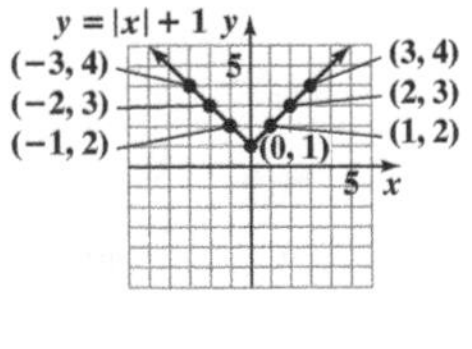

21.
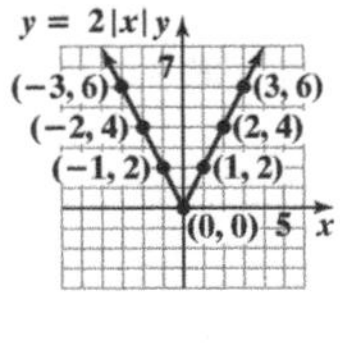

23.
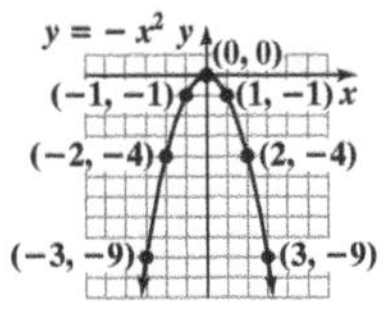

25.
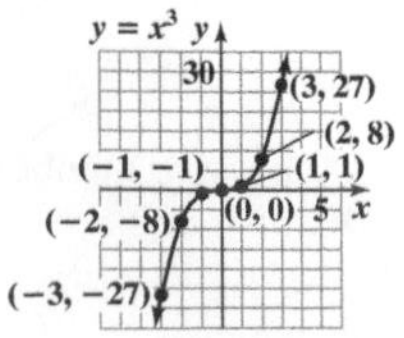

27. c **29.** b **31.** c **33.** no **35.** $(2, 0)$ **37.** $(-2, 4)$ and $(1, 1)$ **39.** $y = 2x + 4$
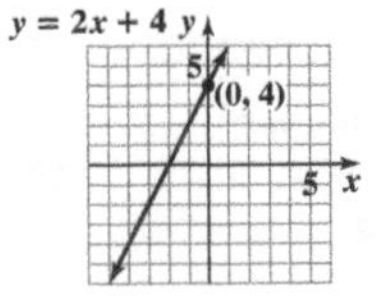

41. $y = 3 - x^2$
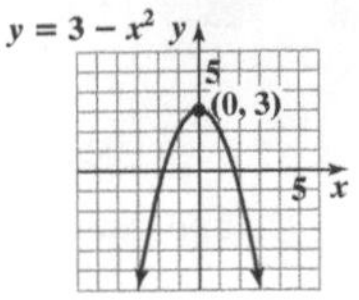

43.
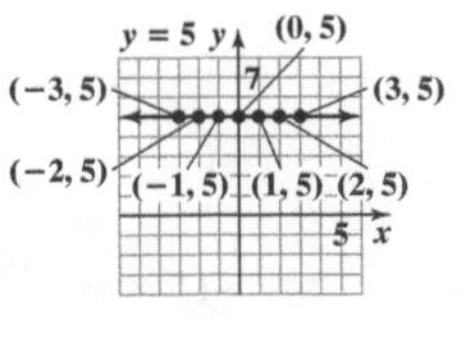

45.
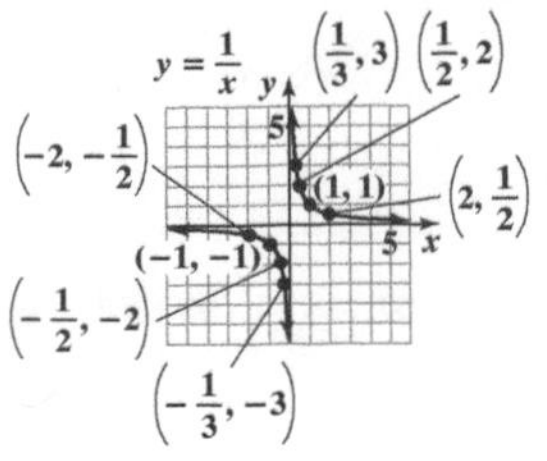

47. 2013; 84% **49.** 2009 and 2010; 71% **51.** 2008 and 2009; 3% **53.** 8 yr old; 1 awakening **55.** about 1.9 awakenings **57.** a **59.** b **61.** b **63.** c **73.** makes sense **75.** makes sense **77.** false **79.** true **81.** $15 **83.** 14.3 **84.** 3 **85.** $-14x - 25$ **86.** true **87.** $7 - 3x$ **88.** $15x + 5$

Section 1.4 Check Point Exercises

1. 6 or {6} **2.** -1 or $\{-1\}$ **3.** 6 or {6} **4.** 1 or {1} **5.** no solution or $\varnothing$; inconsistent equation **6.** $\{x|x$ is a real number$\}$ or $(-\infty, \infty)$ or $\mathbb{R}$; identity **7.** the school year ending 2021

Concept and Vocabulary Check

1. linear **2.** equivalent **3.** $b + c$ **4.** bc **5.** apply the distributive property **6.** least common denominator; 12 **7.** inconsistent; $\varnothing$ **8.** identity; $(-\infty, \infty)$

Exercise Set 1.4

1. 3 or {3} **3.** 11 or {11} **5.** 11 or {11} **7.** 7 or {7} **9.** 13 or {13} **11.** 2 or {2} **13.** -4 or $\{-4\}$ **15.** 9 or {9} **17.** -5 or $\{-5\}$ **19.** 6 or {6} **21.** 19 or {19} **23.** $\frac{5}{2}$ or $\left\{\frac{5}{2}\right\}$ **25.** 12 or {12} **27.** 24 or {24} **29.** -15 or $\{-15\}$ **31.** 5 or {5} **33.** 13 or {13} **35.** -12 or $\{-12\}$ **37.** $\frac{46}{5}$ or $\left\{\frac{46}{5}\right\}$ **39.** $\{x|x$ is a real number$\}$ or $(-\infty, \infty)$ or $\mathbb{R}$; identity **41.** no solution or $\varnothing$; inconsistent equation **43.** 0 or {0}; conditional equation **45.** -10 or $\{-10\}$; conditional equation **47.** no solution or $\varnothing$; inconsistent equation **49.** 0 or {0}; conditional equation **51.** $3(x - 4) = 3(2 - 2x)$; 2 or {2} **53.** $-3(x - 3) = 5(2 - x)$; 0.5 or {0.5} **55.** 2 **57.** -7 **59.** -2 or $\{-2\}$ **61.** no solution or $\varnothing$ **63.** 10 or {10} **65.** -2 or $\{-2\}$ **67. a.** model 1: $31,159; model 2: $31,736; Model 1 underestimates the cost by $542, and model 2 overestimates the cost by $35. **b.** the school year ending 2019 **69. a.** $22,000 **b.** $21,809; reasonably well **c.** $21,726; reasonably well **71.** model 2; overestimates by $26 **73.** 2020

87. 5 or {5}

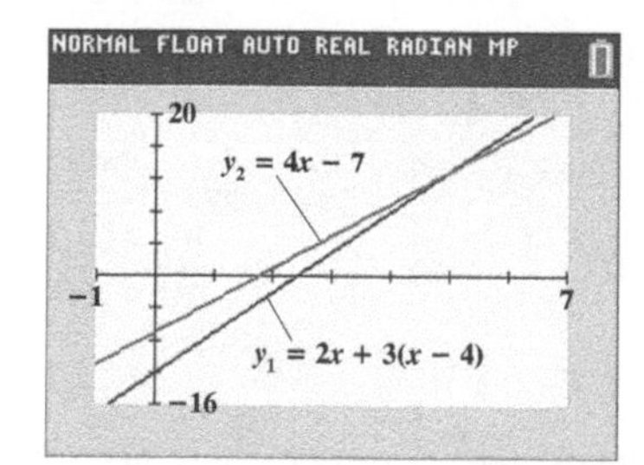

89. -5 or $\{-5\}$

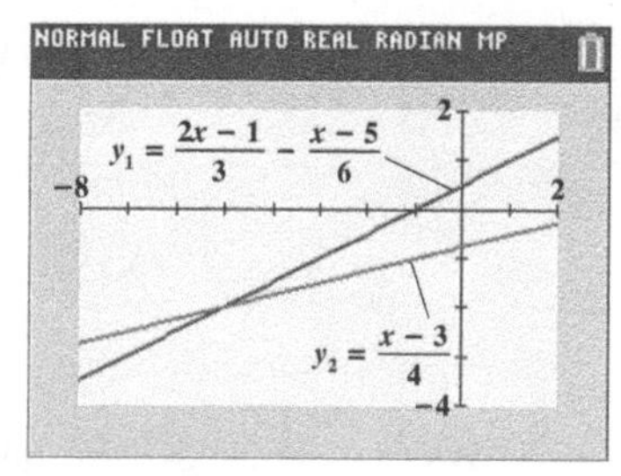

91. makes sense **93.** does not make sense **95.** false **97.** false **101.** 2 **102.** $\frac{3}{10}$ **103.** -60 **104.**

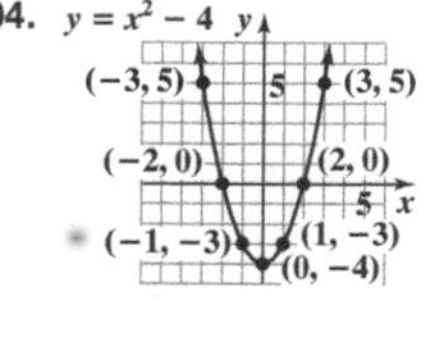

105. a. $3x - 4 = 32$ **b.** 12 **106.** $x + 44$ **107.** $20,000 - 2500x$

Mid-Chapter Check Point Exercises

1. $3x + 10$ **2.** 6 or {6} **3.** -15 **4.** -7 or $\{-7\}$ **5.** 3 **6.** $13x - 23$ **7.** 0 or {0} **8.** $-7x - 34$ **9.** 7 **10.** no solution or $\varnothing$; inconsistent equation **11.** 3 or {3} **12.** -4 **13.** $\{x|x$ is a real number$\}$ or $(-\infty, \infty)$ or $\mathbb{R}$; identity **14.** 2 **15.** $\{x|-2 \le x < 0\}$

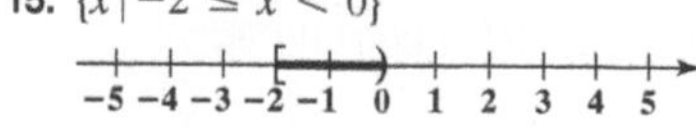

16. $\{x|x \le 0\}$

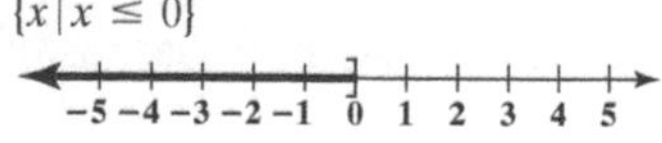

17.

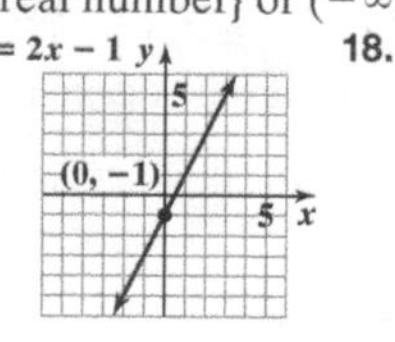

18.

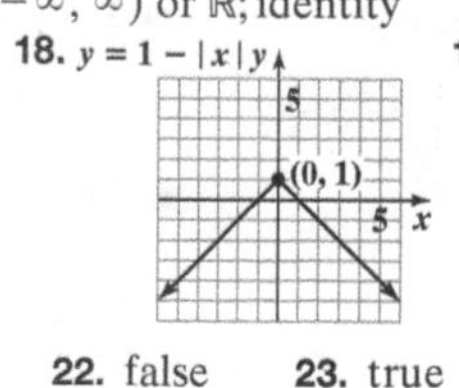

19.

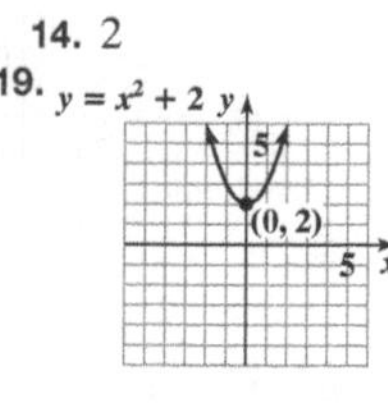

20. true **21.** false **22.** false **23.** true

Section 1.5 Check Point Exercises

1. associate's degree: $33 thousand; bachelor's degree: $47 thousand; master's degree: $59 thousand **2.** by 67 years after 1969, or in 2036 **3.** 20 bridge crossings per month **4.** $1200 **5.** 50 ft by 94 ft **6.** $w = \frac{P - 2l}{2}$ **7.** $h = \frac{V}{lw}$ **8.** $W = 106 + 6H$ or $W = 6H + 106$ **9.** $C = \frac{P}{1 + M}$

Concept and Vocabulary Check

1. $x + 658.6$ **2.** $31 + 2.4x$ **3.** $4 + 0.15x$ **4.** $x - 0.15x$ or $0.85x$ **5.** isolated on one side **6.** distributive

Exercise Set 1.5

1. 6 **3.** 25 **5.** 120 **7.** 320 **9.** 19 and 45 **11.** 2 **13.** 8 **15.** all real numbers **17.** goldfish: 43 years; horse: 64 years; human: 122 years **19.** 94°, 47°, 39° **21.** 59°, 60°, 61° **23.** by 2015 **25.** by 2042 **27. a.** births: 384,000; deaths: 156,000 **b.** 83 million **c.** approximately 4 years **29.** 30 times **31.** 20 times **33. a.** 2017; 22,300 students **b.** $y_1 = 13,300 + 1000x$; $y_2 = 26,800 - 500x$ **35.** $420 **37.** $150 **39.** $467.20 **41.** 50 yd by 100 yd **43.** 36 ft by 78 ft **45.** 2 in. **47.** 11 min **49.** $l = \frac{A}{w}$ **51.** $b = \frac{2A}{h}$ **53.** $P = \frac{I}{rt}$ **55.** $p = \frac{T - D}{m}$ **57.** $a = \frac{2A}{h} - b$ or $a = \frac{2A - hb}{h}$ **59.** $h = \frac{3V}{\pi r^2}$ **61.** $m = \frac{y - y_1}{x - x_1}$ **63.** $d_1 = Vt + d_2$ **65.** $x = \frac{C - By}{A}$ **67.** $v = \frac{2s - at^2}{2t}$ **69.** $n = \frac{L - a}{d} + 1$ or $n = \frac{L - a + d}{d}$ **71.** $l = \frac{A - 2wh}{2w + 2h}$ **73.** $I = \frac{E}{R + r}$

81. 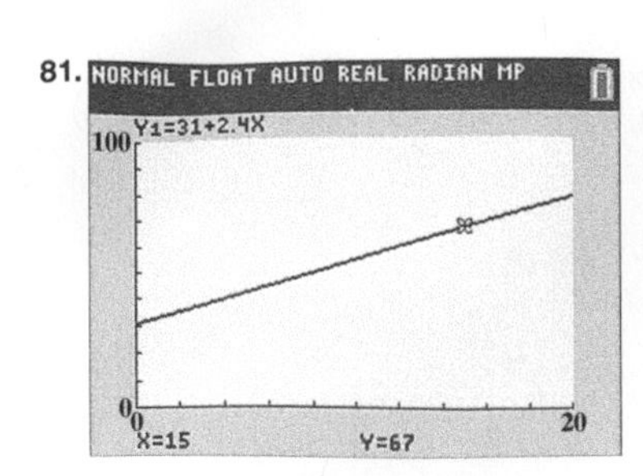

83. does not make sense **85.** makes sense **87.** false **89.** false **91.** \$200 **93.** Mrs. Ricardo received \$4000; boy received \$8000; girl received \$2000. **95.** $C = \dfrac{LV - SN}{L - N}$ **96.** $\{x \mid -4 < x \le 0\}$

−5 −4 −3 −2 −1 0 1 2 3 4 5

97. $\frac{7}{3}$ **98.** -8 or $\{-8\}$ **99. a.** b^7 **b.** b^{10} **c.** Add the exponents. **100. a.** b^4 **b.** b^6 **c.** Subtract the exponents. **101.** -8

Section 1.6 Check Point Exercises

1. a. b^{11} **b.** $40x^5y^{10}$ **2. a.** $(-3)^3$ or -27 **b.** $9x^{11}y^3$ **3. a.** 1 **b.** 1 **c.** −1 **d.** 10 **e.** 1 **4. a.** $\frac{1}{25}$ **b.** $-\frac{1}{27}$ **c.** 16 **d.** $\frac{3y^4}{x^6}$ **5. a.** $\frac{4^3}{7^2} = \frac{64}{49}$ **b.** $\frac{x^2}{5}$ **6. a.** x^{15} **b.** $\frac{1}{y^{14}}$ **c.** b^{12} **7. a.** $16x^4$ **b.** $-27y^6$ **c.** $\frac{y^2}{16x^{10}}$ **8. a.** $\frac{x^{15}}{64}$ **b.** $\frac{16}{x^{12}y^8}$ **c.** $x^{15}y^{20}$ **9. a.** $-12y^9$ **b.** $\frac{4y^{14}}{x^6}$ **c.** $\frac{64}{x^9y^{15}}$

Concept and Vocabulary Check

1. b^{m+n}; add **2.** b^{m-n}; subtract **3.** 1 **4.** $\frac{1}{b^n}$ **5.** false **6.** b^n **7.** true

Exercise Set 1.6

1. b^{11} **3.** x^4 **5.** 32 **7.** $6x^6$ **9.** $20y^{12}$ **11.** $100x^{10}y^{12}$ **13.** $21x^5yz^4$ **15.** b^9 **17.** $5x^5$ **19.** x^5y^5 **21.** $10xy^3$ **23.** $-8a^{11}b^8c^4$ **25.** 1 **27.** 1 **29.** −1 **31.** 13 **33.** 1 **35.** $\frac{1}{3^2} = \frac{1}{9}$ **37.** $\frac{1}{(-5)^2} = \frac{1}{25}$ **39.** $-\frac{1}{5^2} = -\frac{1}{25}$ **41.** $\frac{x^2}{y^3}$ **43.** $\frac{8y^3}{x^7}$ **45.** $5^3 = 125$ **47.** $(-3)^4 = 81$ **49.** $\frac{y^5}{x^2}$ **51.** $\frac{b^7c^3}{a^4}$ **53.** x^{60} **55.** $\frac{1}{b^{12}}$ **57.** 7^{20} **59.** $64x^3$ **61.** $9x^{14}$ **63.** $8x^3y^6$ **65.** $9x^4y^{10}$ **67.** $-\frac{x^6}{27}$ **69.** $\frac{y^8}{25x^6}$ **71.** $\frac{x^{20}}{16y^{16}z^8}$ **73.** $\frac{16}{x^4}$ **75.** $\frac{x^6}{25}$ **77.** $\frac{81x^4}{y^4}$ **79.** $\frac{x^{24}}{y^{12}}$ **81.** x^9y^{12} **83.** a^8b^{12} **85.** $\frac{1}{x^6}$ **87.** $-\frac{4}{x}$ **89.** $\frac{2}{x^7}$ **91.** $\frac{10a^2}{b^5}$ **93.** $\frac{1}{x^9}$ **95.** $-\frac{6}{a^2}$ **97.** $\frac{3y^4z^3}{x^7}$ **99.** $3x^{10}$ **101.** $\frac{1}{x^{10}}$ **103.** $-\frac{5y^8}{x^6}$ **105.** $\frac{8a^9c^{12}}{b}$ **107.** x^{16} **109.** $-\frac{27b^{15}}{a^{18}}$ **111.** 1 **113.** $\frac{81x^{20}}{y^{32}}$ **115.** $\frac{1}{100a^4b^{12}c^8}$ **117.** $10x^2y^4$ **119.** $\frac{8}{3xy^{10}}$ **121.** $\frac{y^5}{8x^{14}}$ **123.** $\frac{1}{128x^7y^{16}}$ **125. a.** 1000 aphids **b.** 16,000 aphids **c.** 125 aphids **127. a.** one person **b.** 10 people **129. a.** (0, 1) **b.** (4, 10) **131.** d **133.** 0.55 astronomical unit **135.** 1.8 astronomical units **147.** makes sense **149.** does not make sense **151.** false **153.** false **155.** false **157.** true **159.** x^{9n} **161.** $x^{3n}y^{6n+3}$

162. $y = 2x - 1$ y; 10; (2, 3); (3, 5); (1, 1); 5; x; (0, −1); (−1, −3); (−2, −5); (−3, −7)

163. $y = \dfrac{C - Ax}{B}$ **164.** 40 m by 75 m **165.** It moves the decimal point three places to the right. **166.** It moves the decimal point two places to the left. **167. a.** 10^5; 100,000 **b.** 10^6; 1,000,000

Section 1.7 Check Point Exercises

1. a. −2,600,000,000 **b.** 0.000003017 **2. a.** 5.21×10^9 **b.** -6.893×10^{-8} **3.** 1.8×10^7 **4. a.** 3.55×10^{-1} **b.** 4×10^8 **5.** \$7756 **6.** 3.1×10^7 mi or 31 million mi

Concept and Vocabulary Check

1. a number greater than or equal to 1 and less than 10; integer **2.** true **3.** false

Exercise Set 1.7

1. 380 **3.** 0.0006 **5.** −7,160,000 **7.** 1.4 **9.** 0.79 **11.** −0.00415 **13.** −60,000,100,000 **15.** 3.2×10^4 **17.** 6.38×10^{17} **19.** -3.17×10^2 **21.** -5.716×10^3 **23.** 2.7×10^{-3} **25.** -5.04×10^{-9} **27.** 7×10^{-3} **29.** 3.14159×10^0 **31.** 6.3×10^7 **33.** 6.4×10^4 **35.** 1.22×10^{-11} **37.** 2.67×10^{13} **39.** 2.1×10^3 **41.** 4×10^5 **43.** 2×10^{-8} **45.** 5×10^3 **47.** 4×10^{15} **49.** 9×10^{-3} **51.** 6×10^{13} or $\{6 \times 10^{13}\}$ **53.** -6.2×10^3 or $\{-6.2 \times 10^3\}$ **55.** 1.63×10^{19} or $\{1.63 \times 10^{19}\}$ **57.** -3.6×10^5 or $\{-3.6 \times 10^5\}$ **59.** $\$7.81 \times 10^{10}$ **61.** $\$1.07 \times 10^{10}$ **63.** approximately 67 hot dogs per person **65.** $2.5 \times 10^2 = 250$ chickens **67. a.** $\$1.0813 \times 10^4$; \$10,813 **b.** \$901 **69.** Medicare; \$3242 **71.** 1.06×10^{-18} g **73.** 3.1536×10^7 **81.** does not make sense **83.** makes sense **85.** false **87.** false **89.** true **91.** 1.25×10^{-15} **94.** $85x - 26$ **95.** 4 or {4} **96.** $\frac{y^6}{64x^8}$ **97.** set 1 **98.** −170 **99.** $5a + 5h + 7$

Review Exercises

1. $2x - 10$ **2.** $6x + 4$ **3.** $\frac{9}{x} + \frac{1}{2}x$ **4.** 34 **5.** 60 **6.** 15 **7.** {1, 2} **8.** {−3, −2, −1, 0, 1} **9.** false **10.** true **11.** true **12.** −5 is less than 2.; true **13.** −7 is greater than or equal to −3.; false **14.** −7 is less than or equal to −7.; true **15.** overestimates by 3.2 million

16. $\{x \mid -2 < x \leq 3\}$ **17.** $\{x \mid -1.5 \leq x \leq 2\}$ **18.** $\{x \mid x > -1\}$

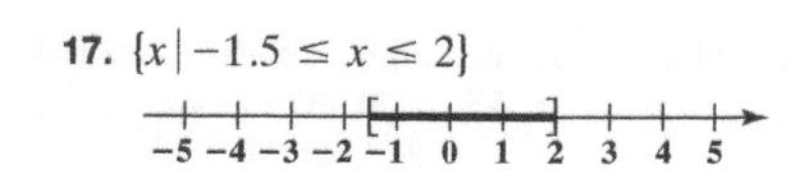

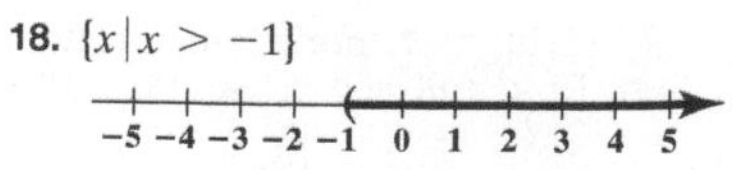

19. 9.7 **20.** 5.003 **21.** 0 **22.** −7.6 **23.** −4.4 **24.** 13 **25.** 60 **26.** $-\frac{1}{10}$ **27.** $-\frac{3}{35}$ **28.** −240 **29.** 16 **30.** −32 **31.** $-\frac{5}{12}$ **32.** 7 **33.** −9.1 **34.** 7 **35.** 9 **36.** −2 **37.** −18 **38.** 55 **39.** 1 **40.** −4 **41.** −13 **42.** $17x - 15$ **43.** $9x^2 + x$ **44.** $5y - 17$ **45.** $10x$ **46.** $-3x - 8$

47–49.

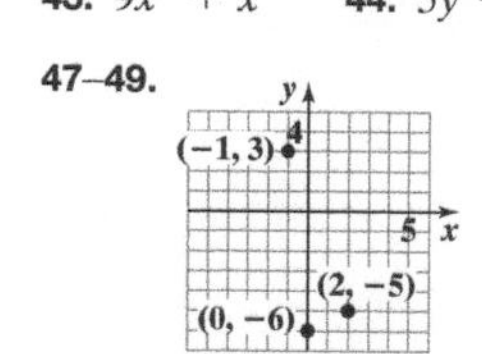

50.

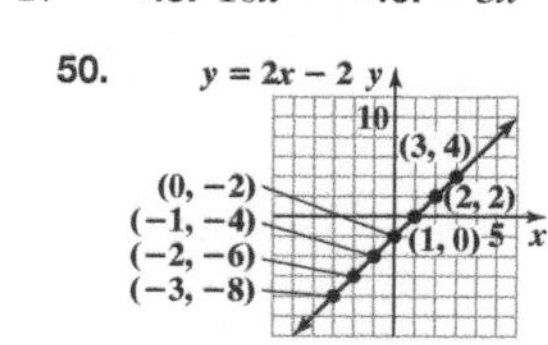

51.

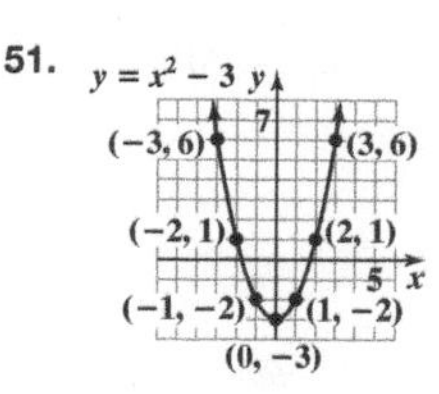

52.

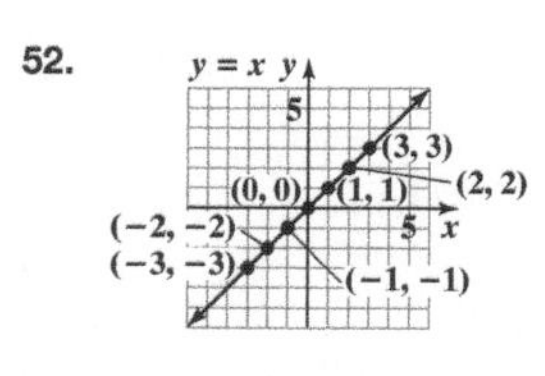

53.

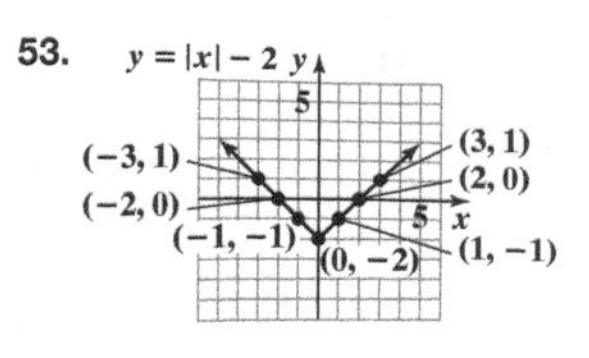

54. minimum x-value: −20; maximum x-value: 40; distance between tick marks on x-axis: 10; minimum y-value: −5; maximum y-value: 5; distance between tick marks on y-axis: 1

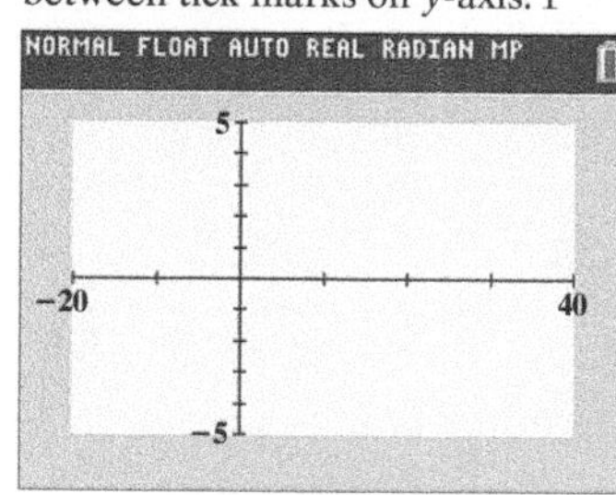

55. 20% **56.** 85 years old **58.** c **59.** 6 or {6} **60.** −10 or {−10} **61.** 5 or {5} **62.** −13 or {−13} **63.** −3 or {−3} **64.** −1 or {−1} **65.** 2 or {2} **66.** 2 or {2} **67.** $\frac{72}{11}$ or $\left\{\frac{72}{11}\right\}$ **68.** −12 or {−12} **69.** $\frac{77}{15}$ or $\left\{\frac{77}{15}\right\}$ **70.** no solution or ∅; inconsistent equation **71.** $\{x \mid x$ is a real number$\}$ or $(-\infty, \infty)$ or ℝ; identity **72.** 0 or {0}; conditional equation **73.** $\frac{3}{2}$ or $\left\{\frac{3}{2}\right\}$; conditional equation

74. no solution or ∅; inconsistent equation **75. a.** underestimates by 1% **b.** 2020 **76.** engineering: \$76 thousand; accounting: \$63 thousand; marketing: \$57 thousand **77.** 25°, 35°, 120° **78. a.** 2018 **b.** \$1117 billion **c.** They are shown by the intersection of the graphs at approximately (2018, 1117). **79.** 5 GB **80.** \$60 **81.** \$10,000 in sales **82.** 44 yd by 126 yd **83. a.** $14{,}100 + 1500x = 41{,}700 - 800x$ **b.** 2027; 32,100 **84.** $h = \frac{3V}{B}$ **85.** $x = \frac{y - y_1}{m} + x_1$ or $x = \frac{y - y_1 + mx_1}{m}$ **86.** $R = \frac{E}{I} - r$ or $R = \frac{E - Ir}{I}$ **87.** $F = \frac{9C + 160}{5}$ or $F = \frac{9}{5}C + 32$ **88.** $g = \frac{s - vt}{t^2}$ **89.** $g = \frac{T}{r + vt}$ **90.** $15x^{13}$ **91.** $\frac{x^2}{y^5}$ **92.** $\frac{x^4y^7}{9}$ **93.** $\frac{1}{x^{18}}$ **94.** $49x^6y^2$ **95.** $-\frac{8}{y^7}$ **96.** $-\frac{12}{x^7}$ **97.** $3x^{10}$ **98.** $-\frac{a^8}{2b^5}$ **99.** $-24x^7y^4$ **100.** $\frac{3}{4}$ **101.** $\frac{y^{12}}{125x^6}$ **102.** $-\frac{6x^9}{y^5}$ **103.** $\frac{9x^8y^{14}}{25}$ **104.** $-\frac{x^{21}}{8y^{27}}$ **105.** 7,160,000 **106.** 0.000107 **107.** -4.1×10^{13} **108.** 8.09×10^{-3} **109.** 1.26×10^8 **110.** 2.5×10^1 **111.** 2.88×10^{13}

Chapter Test

1. $4x - 5$ **2.** 170 **3.** {−4, −3, −2, −1} **4.** true **5.** −3 is greater than −1.; false **6.** $\{x \mid -3 \leq x < 2\}$ **7.** $\{x \mid x \leq -1\}$

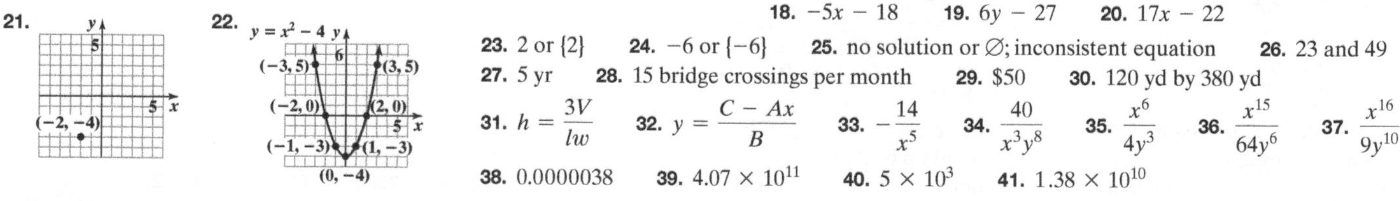

8. underestimates by \$13 **9.** 17.9 **10.** −7.6 **11.** $\frac{1}{4}$ **12.** −60 **13.** $\frac{1}{8}$ **14.** 3.1 **15.** −3 **16.** 6 **17.** −4 **18.** $-5x - 18$ **19.** $6y - 27$ **20.** $17x - 22$

21. (−2, −4) **22.** $y = x^2 - 4$: (−3, 5), (3, 5), (−2, 0), (2, 0), (−1, −3), (1, −3), (0, −4)

23. 2 or {2} **24.** −6 or {−6} **25.** no solution or ∅; inconsistent equation **26.** 23 and 49 **27.** 5 yr **28.** 15 bridge crossings per month **29.** \$50 **30.** 120 yd by 380 yd **31.** $h = \frac{3V}{lw}$ **32.** $y = \frac{C - Ax}{B}$ **33.** $-\frac{14}{x^5}$ **34.** $\frac{40}{x^3y^8}$ **35.** $\frac{x^6}{4y^3}$ **36.** $\frac{x^{15}}{64y^6}$ **37.** $\frac{x^{16}}{9y^{10}}$ **38.** 0.0000038 **39.** 4.07×10^{11} **40.** 5×10^3 **41.** 1.38×10^{10}

CHAPTER 2

Section 2.1 Check Point Exercises

1. domain: {0, 10, 20, 30, 40}; range: {9.1, 6.7, 10.7, 13.2, 21.2} **2. a.** not a function **b.** function **3. a.** 29 **b.** 65 **c.** 46 **d.** $6a + 6h + 9$ **4. a.** Every element in the domain corresponds to exactly one element in the range. **b.** domain: {0, 1, 2, 3, 4}; range: {3, 0, 1, 2} **c.** 0 **d.** 2 **e.** $x = 0$ and $x = 4$

Concept and Vocabulary Check

1. relation; domain; range **2.** function **3.** f; x **4.** r; −2

Exercise Set 2.1

1. function; domain: {1, 3, 5}; range: {2, 4, 5} **3.** not a function; domain: {3, 4}; range: {4, 5} **5.** function; domain: {−3, −2, −1, 0};

range: $\{-3, -2, -1, 0\}$ **7.** not a function; domain: $\{1\}$; range: $\{4, 5, 6\}$ **9. a.** 1 **b.** 6 **c.** -7 **d.** $2a + 1$ **e.** $a + 3$ **11. a.** -2 **b.** -17 **c.** 0 **d.** $12b - 2$ **e.** $3b + 10$ **13. a.** 5 **b.** 8 **c.** 53 **d.** 32 **e.** $48b^2 + 5$ **15. a.** -1 **b.** 26 **c.** 19 **d.** $2b^2 + 3b - 1$ **e.** $50a^2 + 15a - 1$ **17. a.** 7 **b.** -7 **c.** 13 **d.** 12 **19. a.** $\frac{3}{4}$ **b.** -3 **c.** $\frac{11}{8}$ **d.** $\frac{13}{9}$ **e.** $\frac{2a + 2h - 3}{a + h - 4}$ **f.** Denominator would be zero. **21. a.** 6 **b.** 12 **c.** 0 **23. a.** 2 **b.** 1 **c.** -1 and 1 **25.** -2; 10 **27.** -38 **29.** $-2x^3 - 2x$ **31. a.** -1 **b.** 7 **c.** 19 **d.** 112 **33. a.** {(Iceland, 9.7), (Finland, 9.6), (New Zealand, 9.6), (Denmark, 9.5)} **b.** Yes; each country corresponds to exactly one corruption rating. **c.** {(9.7, Iceland), (9.6, Finland), (9.6, New Zealand), (9.5, Denmark)} **d.** No; 9.6 in the domain corresponds to two countries in the range, Finland and New Zealand. **39.** makes sense **41.** makes sense **43.** false **45.** true **47.** true **49.** 3 **51.** $f(2) = 6; f(3) = 9; f(4) = 12$; no **52.** 0 **53.** $\frac{y^{10}}{9x^4}$ **54.** $\{-15\}$

55.

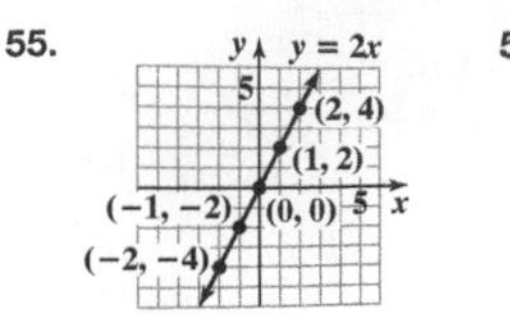

56.

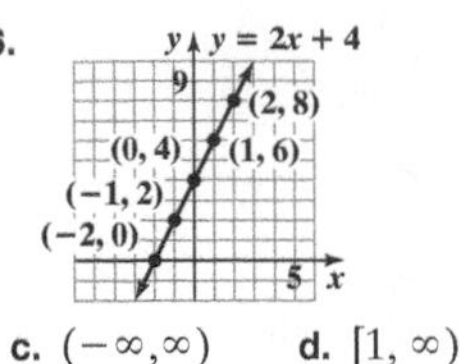

57. a. 3 **b.** -3 and 3 **c.** $(-\infty, \infty)$ **d.** $[1, \infty)$

Section 2.2 Check Point Exercises

1. ; The graph of g is the graph of f shifted down by 3 units.

2. a. function **b.** function **c.** not a function **3. a.** 400 **b.** 9 **c.** approximately 425 **4. a.** domain: $[-2, 1]$; range $[0, 3]$ **b.** domain: $(-2, 1]$; range: $[-1, 2)$ **c.** domain: $[-3, 0)$; range: $\{-3, -2, -1\}$

Concept and Vocabulary Check

1. ordered pairs **2.** more than once; function **3.** $[1, 3)$; domain **4.** $[1, \infty)$; range

Exercise Set 2.2

1.

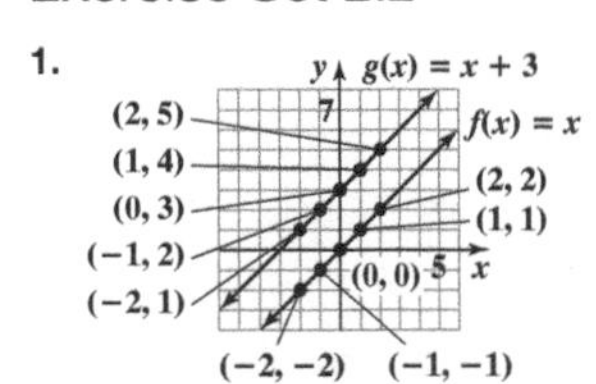

The graph of g is the graph of f shifted up by 3 units.

3. The graph of g is the graph of f shifted down by 1 unit.

5. The graph of g is the graph of f shifted up by 1 unit.

7.

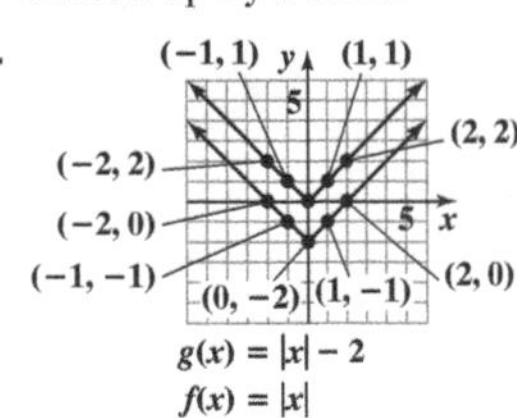

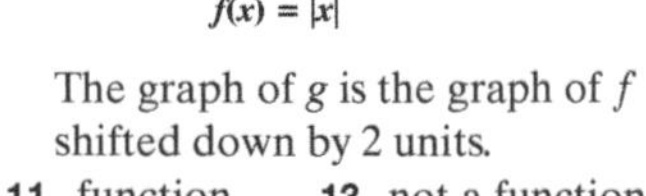

The graph of g is the graph of f shifted down by 2 units.

9. The graph of g is the graph of f shifted up by 2 units.

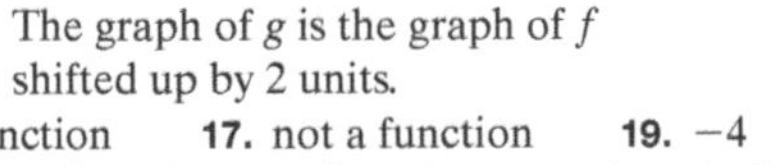
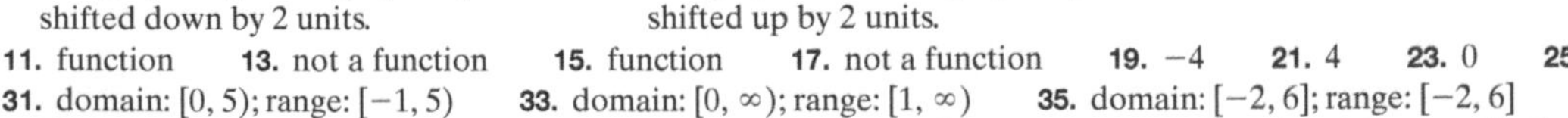
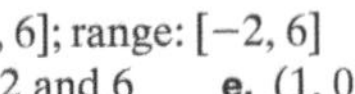

11. function **13.** not a function **15.** function **17.** not a function **19.** -4 **21.** 4 **23.** 0 **25.** 2 **27.** 2 **29.** -2 **31.** domain: $[0, 5)$; range: $[-1, 5)$ **33.** domain: $[0, \infty)$; range: $[1, \infty)$ **35.** domain: $[-2, 6]$; range: $[-2, 6]$ **37.** domain: $(-\infty, \infty)$; range: $(-\infty, 2]$ **39.** domain: $\{-5, -2, 0, 1, 3\}$; range: $\{2\}$ **41. a.** $(-\infty, \infty)$ **b.** $[-4, \infty)$ **c.** 4 **d.** 2 and 6 **e.** (1, 0), (7, 0) **f.** (0, 4) **g.** (1, 7) **h.** positive **43. a.** 81; In 2010, the wage gap was 81%; (30, 81) **b.** underestimates by 2% **45.** 440; For 20-year-old drivers, there are 440 accidents per 50 million miles driven.; (20, 440) **47.** $x = 45$; $y = 190$; The minimum number of accidents is 190 per 50 million miles driven and is attributed to 45-year-old drivers. **49.** 0.91; It costs \$0.91 to mail a 3-ounce first-class letter. **51.** \$0.70

57.

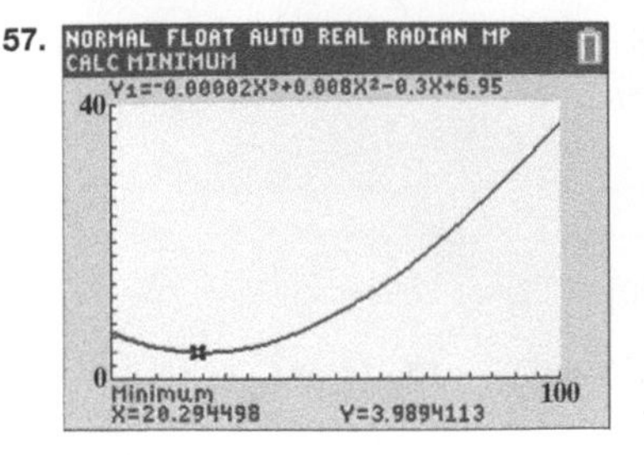

The number of annual visits decreases until about age 20 and then increases; (20.3, 4.0): The minimum is at about age 20 when there are about 4 annual physician visits. **59.** makes sense **61.** does not make sense **63.** true **65.** false **67.** false **69.** -3 **70.** yes **71.** $\frac{3}{2}$ or $\left\{\frac{3}{2}\right\}$ **72.** 76 yd by 236 yd **73.** Division by 0 is undefined. **74.** 19 **75.** $-2x^2 + 42x + 1582$

Section 2.3 Check Point Exercises

1. a. $(-\infty, \infty)$ **b.** $(-\infty, -5)$ or $(-5, \infty)$ **2. a.** $3x^2 + 6x + 6$ **b.** 78 **3. a.** $\frac{5}{x} - \frac{7}{x-8}$ **b.** $(-\infty, 0)$ or $(0, 8)$ or $(8, \infty)$ **4. a.** 23 **b.** $x^2 - 3x - 3; 1$ **c.** $\frac{x^2 - 2x - 7}{x + 3}; \frac{7}{2}$ **d.** -24 **5. a.** $(B + D)(x) = -3.2x^2 + 56x + 6406$ **b.** 6545.2 thousand **c.** overestimates by 7.2 thousand

Concept and Vocabulary Check

1. zero **2.** negative **3.** $f(x) + g(x)$ **4.** $f(x) - g(x)$ **5.** $f(x) \cdot g(x)$ **6.** $\frac{f(x)}{g(x)}; g(x)$ **7.** $(-\infty, \infty)$ **8.** $(2, \infty)$ **9.** $(0, 3); (3, \infty)$

Exercise Set 2.3

1. $(-\infty, \infty)$ **3.** $(-\infty, -4)$ or $(-4, \infty)$ **5.** $(-\infty, 3)$ or $(3, \infty)$ **7.** $(-\infty, 5)$ or $(5, \infty)$ **9.** $(-\infty, -7)$ or $(-7, 9)$ or $(9, \infty)$ **11. a.** $5x - 5$ **b.** 20 **13. a.** $3x^2 + x - 5$ **b.** 75 **15. a.** $2x^2 - 2$ **b.** 48 **17.** $(f + g)(x) = 3x - 3$; $(f - g)(x) = 7x + 3$; $(fg)(x) = -10x^2 - 15x$; $\left(\frac{f}{g}\right)(x) = \frac{5x}{-2x - 3}$ **19.** $(-\infty, \infty)$ **21.** $(-\infty, 5)$ or $(5, \infty)$ **23.** $(-\infty, 0)$ or $(0, 5)$ or $(5, \infty)$ **25.** $(-\infty, -3)$ or $(-3, 2)$ or $(2, \infty)$ **27.** $(-\infty, 2)$ or $(2, \infty)$ **29.** $(-\infty, \infty)$ **31.** $x^2 + 3x + 2; 20$ **33.** 0 **35.** $x^2 + 5x - 2; 48$ **37.** -8 **39.** -16 **41.** -135 **43.** $\frac{x^2 + 4x}{2 - x}; 5$ **45.** -1 **47.** $(-\infty, \infty)$ **49.** $(-\infty, 2)$ or $(2, \infty)$ **51.** 5 **53.** -1 **55.** $[-4, 3]$ **57.** **59.** -4 **61.** -4 **63. a.** $(M + F)(x) = 2.9x + 236$ **b.** 308.5 million **c.** underestimates by 0.5 million **65. a.** $\left(\frac{M}{F}\right)(x) = \frac{1.5x + 115}{1.4x + 121}$ **b.** 0.968 **c.** overestimates by approximately 0.003

71.

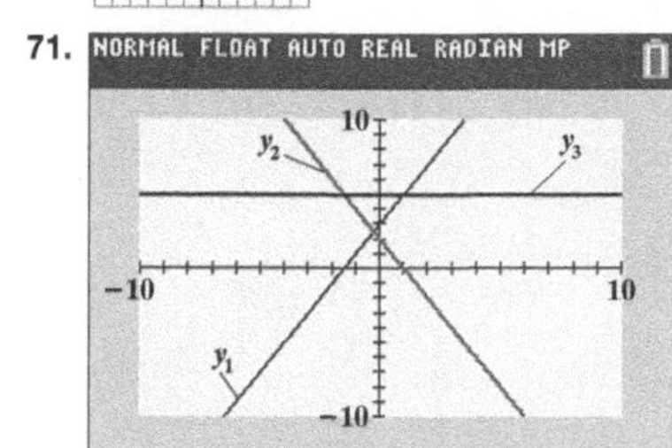

73.

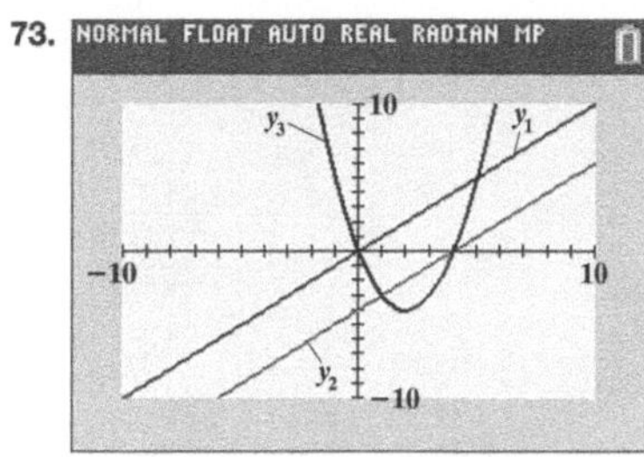

75. No y-value is displayed; y_3 is undefined at $x = 0$. **77.** makes sense **79.** makes sense **81.** true **83.** false **84.** $b = \frac{R - 3a}{3}$ or $b = \frac{R}{3} - a$ **85.** 7 or $\{7\}$ **86.** $6b + 8$ **87. a.** $\frac{3}{2}$ **b.** -2

88. a. **b.** $(-2, 0)$ **c.** $(0, 4)$ **89.** $y = -\frac{5}{3}x - 4$

Mid-Chapter Check Point Exercises

1. not a function; domain: $\{1, 2\}$; range $\{-6, 4, 6\}$ **2.** function; domain: $\{0, 2, 3\}$; range: $\{1, 4\}$ **3.** function; domain: $[-2, 2)$; range: $[0, 3]$ **4.** not a function; domain: $(-3, 4]$; range: $[-1, 2]$ **5.** not a function; domain: $\{-2, -1, 0, 1, 2\}$; range: $\{-2, -1, 1, 3\}$ **6.** function; domain: $(-\infty, 1]$; range: $[-1, \infty)$ **7.** No vertical line intersects the graph of f more than once. **8.** 3 **9.** -2 **10.** -6 and 2 **11.** $(-\infty, \infty)$ **12.** $(-\infty, 4]$ **13.** $(-\infty, \infty)$ **14.** $(-\infty, -2)$ or $(-2, 2)$ or $(2, \infty)$ **15.** 23 **16.** 23 **17.** $a^2 - 5a - 3$ **18.** $x^2 - 5x + 3; 17$ **19.** $x^2 - x + 13; 33$ **20.** -36 **21.** $\frac{x^2 - 3x + 8}{-2x - 5}; 12$ **22.** $\left(-\infty, -\frac{5}{2}\right)$ or $\left(-\frac{5}{2}, \infty\right)$

Section 2.4 Check Point Exercises

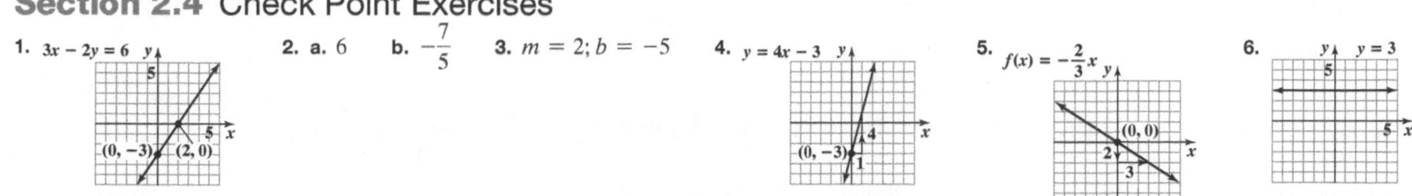

2. a. 6 **b.** $-\frac{7}{5}$ **3.** $m = 2; b = -5$

7.

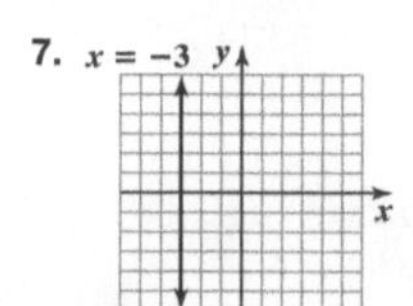

8. -0.05; From 2005 to 2010, waste production for each American decreased by 0.05 pound per day each year. The rate of change per is -0.05 pound per day per year. **9.** 0.01 mg per 100 ml per hr **10. a.** $C(x) = 1.33x + 310$ **b.** 443 parts per million

Concept and Vocabulary Check

1. scatter plot; regression **2.** standard **3.** x-intercept; zero **4.** y-intercept; zero **5.** $\frac{y_2 - y_1}{x_2 - x_1}$ **6.** positive **7.** negative **8.** zero **9.** undefined **10.** $y = mx + b$; slope; y-intercept **11.** $(0, 3)$; 2; 5 **12.** horizontal **13.** vertical **14.** y; x

Exercise Set 2.4

1. 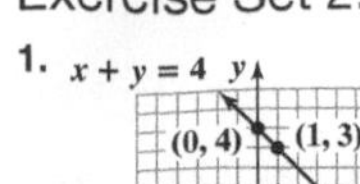 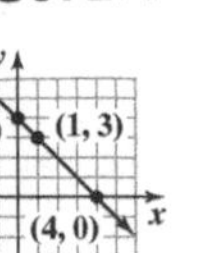

3. x + 3y = 6 (0, 2) (3, 1) (6, 0)

5. 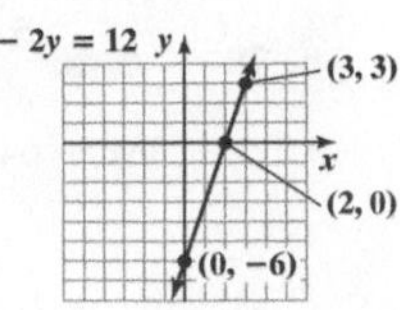

7. 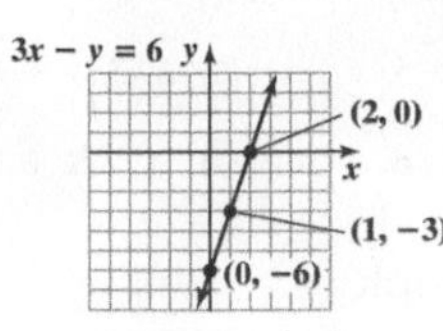

9. 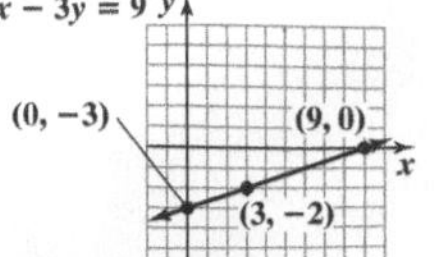

11. 2x = 3y + 6 (0, −2) (6, 2) (3, 0)

13. 6x − 3y = 15 (5/2, 0) (0, −5)

15. $m = 4$; rises **17.** $m = \frac{1}{3}$; rises

19. $m = 0$; horizontal **21.** $m = -\frac{4}{3}$; falls **23.** $m = \frac{5}{2}$; rises **25.** undefined slope; vertical **27.** L_1: $\frac{2}{3}$; L_2: -2; L_3: $-\frac{1}{2}$

29. $m = 2; b = 1;$
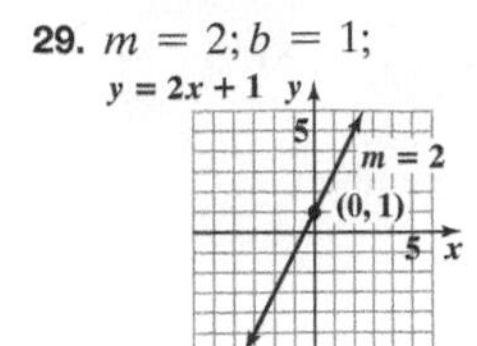

31. $m = -2; b = 1;$
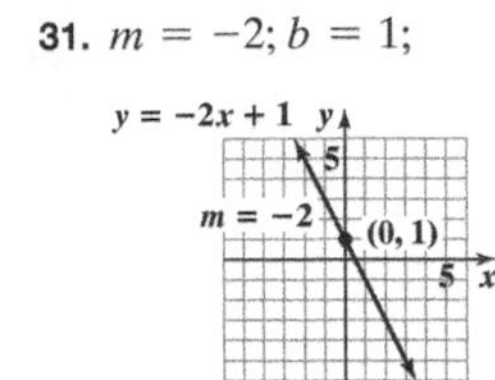

33. $m = \frac{3}{4}; b = -2;$
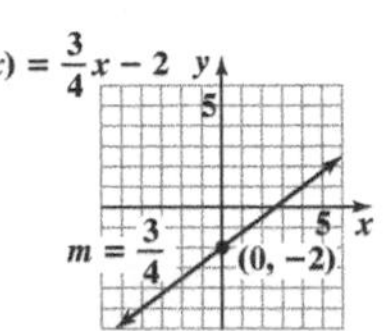

35. $m = -\frac{3}{5}; b = 7;$ f(x) = −3/5 x + 7 (0, 7) m = −3/5

37. $m = -\frac{1}{2}; b = 0;$
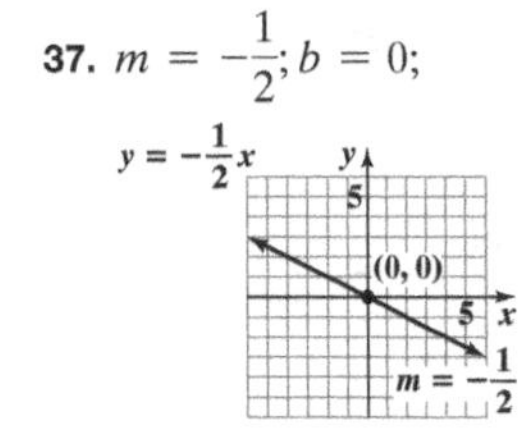

39. $m = 0; b = -\frac{1}{2};$
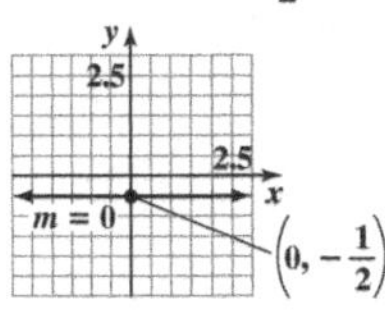

41. a. $y = -2x$
b. $m = -2; b = 0$
c. 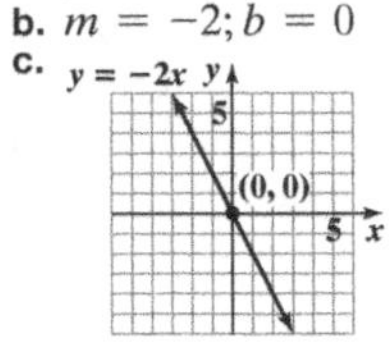

43. a. $y = \frac{4}{5}x$
b. $m = \frac{4}{5}; b = 0$
c. 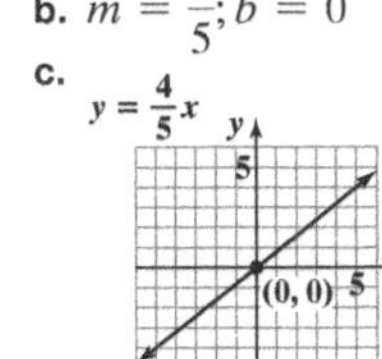

45. a. $y = -3x + 2$
b. $m = -3; b = 2$
c. 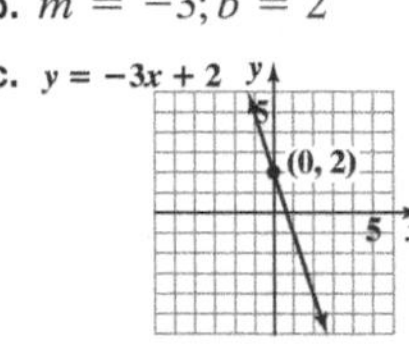

47. a. $y = -\frac{5}{3}x + 5$
b. $m = -\frac{5}{3}; b = 5$
c. 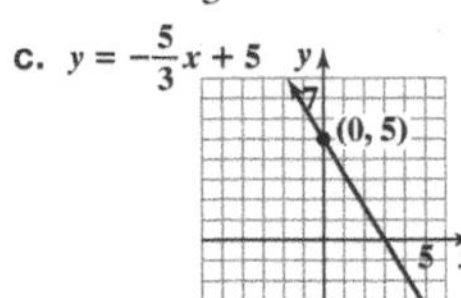

49. y = 3 **51.** f(x) = −2

53. 3y = 18 **55.** 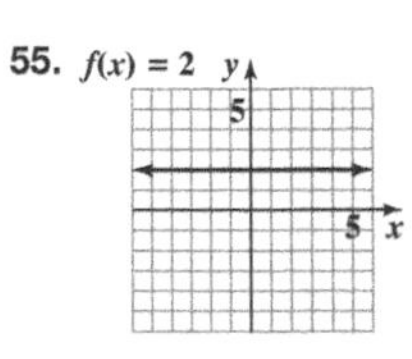

57. x = 5 **59.** 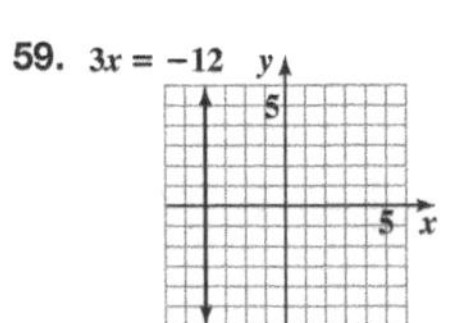

61. x = 0 **63.** $m = -\frac{a}{b}$; falls **65.** undefined slope; vertical **67.** $m = -\frac{A}{B}; b = \frac{C}{B}$ **69.** -2

71. 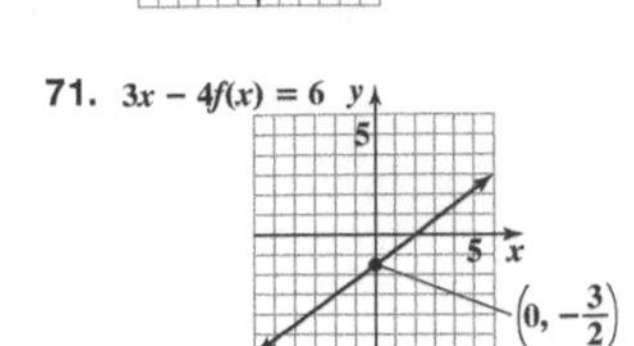

73. 5 **75.** m_1, m_3, m_2, m_4 **77.** $m = 55.7$; Smartphone sales are increasing by 55.7 million each year.
79. $m = -0.52$; The percentage of U.S. adults who smoke cigarettes is decreasing by 0.52% each year.
81. a. 30% **b.** 50% **c.** $m = 4$; average increase of 4% of marriages ending in divorce per year
83. a. 254; If no women in a country are literate, the mortality rate of children under 5 is 254 per thousand.
b. -2.4; For each 1% of adult females who are literate, the mortality rate decreases by 2.4 per thousand.
c. $f(x) = -2.4x + 254$ **d.** 134 per thousand **85.** $P(x) = 0.24x + 29$

105. $m = 2$;

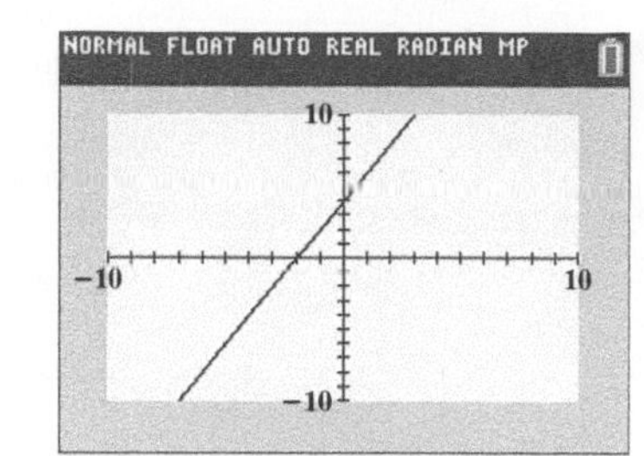

107. $m = -\frac{1}{2}$;

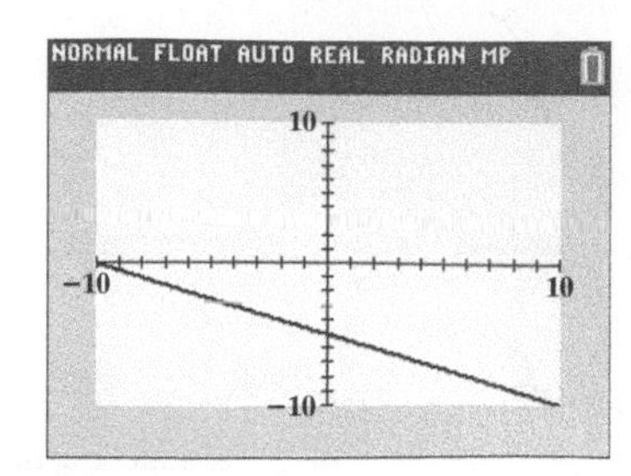

109. does not make sense **111.** does not make sense **113.** false **115.** true **117.** coefficient of x: -6; coefficient of y: 3
119. a. $mx_1 + mx_2 + b$ **b.** $mx_1 + mx_2 + 2b$ **c.** no **120.** $16x^4y^6$ **121.** 3.2×10^{-3} **122.** $3x + 17$ **123.** $y = 7x + 33$
124. $y = -\frac{7}{3}x - \frac{2}{3}$ **125. a.** $y = -\frac{1}{4}x + 2; -\frac{1}{4}$ **b.** 4

Section 2.5 Check Point Exercises

1. $y + 3 = -2(x - 4); y = -2x + 5$ or $f(x) = -2x + 5$ **2. a.** $y + 3 = -2(x - 6)$ or $y - 5 = -2(x - 2)$ **b.** $y = -2x + 9$ or $f(x) = -2x + 9$
3. Answers will vary due to rounding.; $f(x) = 0.17x + 72.9$ or $f(x) = 0.17x + 73$; 83.1 yr or 83.2 yr
4. $y - 5 = 3(x + 2); y = 3x + 11$ or $f(x) = 3x + 11$ **5. a.** $m = 3$ **b.** $y + 6 = 3(x + 2); y = 3x$ or $f(x) = 3x$

Concept and Vocabulary Check

1. $y - y_1 = m(x - x_1)$ **2.** equal/the same **3.** -1 **4.** $-\frac{1}{5}$ **5.** $\frac{5}{3}$ **6.** $-4; -4$ **7.** $\frac{1}{2}; -2$

Exercise Set 2.5

1. $y - 5 = 3(x - 2); f(x) = 3x - 1$ **3.** $y - 6 = 5(x + 2); f(x) = 5x + 16$ **5.** $y + 2 = -4(x + 3); f(x) = -4x - 14$
7. $y - 0 = -5(x + 2); f(x) = -5x - 10$ **9.** $y + \frac{1}{2} = -1(x + 2); f(x) = -x - \frac{5}{2}$ **11.** $y - 0 = \frac{1}{4}(x - 0); f(x) = \frac{1}{4}x$
13. $y + 4 = -\frac{2}{3}(x - 6); f(x) = -\frac{2}{3}x$ **15.** $y - 3 = 1(x - 6)$ or $y - 2 = 1(x - 5); f(x) = x - 3$
17. $y - 0 = 2(x + 2)$ or $y - 4 = 2(x - 0); f(x) = 2x + 4$ **19.** $y - 13 = -2(x + 6)$ or $y - 5 = -2(x + 2); f(x) = -2x + 1$
21. $y - 9 = -\frac{11}{3}(x - 1)$ or $y + 2 = -\frac{11}{3}(x - 4); f(x) = -\frac{11}{3}x + \frac{38}{3}$ **23.** $y + 5 = 0(x + 2)$ or $y + 5 = 0(x - 3); f(x) = -5$
25. $y - 8 = 2(x - 7)$ or $y - 0 = 2(x - 3); f(x) = 2x - 6$ **27.** $y - 0 = \frac{1}{2}(x - 2)$ or $y + 1 = \frac{1}{2}(x - 0); f(x) = \frac{1}{2}x - 1$
29. a. 5 **b.** $-\frac{1}{5}$ **31. a.** -7 **b.** $\frac{1}{7}$ **33. a.** $\frac{1}{2}$ **b.** -2 **35. a.** $-\frac{2}{5}$ **b.** $\frac{5}{2}$ **37. a.** -4 **b.** $\frac{1}{4}$ **39. a.** $-\frac{1}{2}$ **b.** 2
41. a. $\frac{2}{3}$ **b.** $-\frac{3}{2}$ **43. a.** undefined **b.** 0 **45.** $y - 2 = 2(x - 4); y = 2x - 6$ or $f(x) = 2x - 6$
47. $y - 4 = -\frac{1}{2}(x - 2); y = -\frac{1}{2}x + 5$ or $f(x) = -\frac{1}{2}x + 5$ **49.** $y + 10 = -4(x + 8); y = -4x - 42$ or $f(x) = -4x - 42$
51. $y + 3 = -5(x - 2); y = -5x + 7$ or $f(x) = -5x + 7$ **53.** $y - 2 = \frac{2}{3}(x + 2); y = \frac{2}{3}x + \frac{10}{3}$ or $f(x) = \frac{2}{3}x + \frac{10}{3}$
55. $y + 7 = -2(x - 4); y = -2x + 1$ or $f(x) = -2x + 1$ **57.** $f(x) = 5$ **59.** $f(x) = -\frac{1}{2}x + 1$ **61.** $f(x) = -\frac{2}{3}x - 2$
63. $f(x) = 4x - 5$ **65.** $-\frac{A}{B}$ **67. a.** $y - 38.9 = 0.89(x - 20)$ or $y - 47.8 = 0.89(x - 30)$ **b.** $f(x) = 0.89x + 21.1$ **c.** 56.7%
69. a. & b.

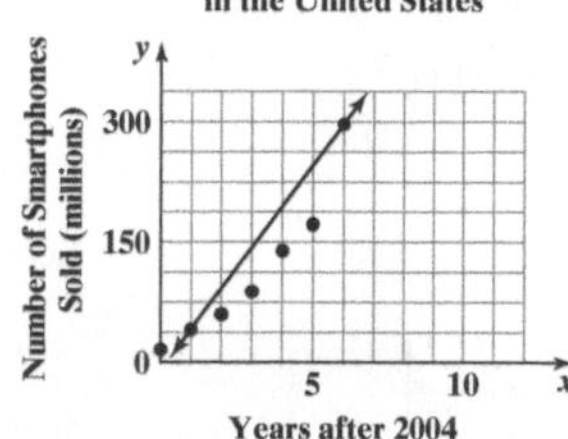

b. $y - 40.8 = 51.16(x - 1)$ or $y - 296.6 = 51.16(x - 6)$: $f(x) = 51.16x - 10.36$ **c.** 552.4 million

71. a. $m \approx 43.1$; The cost of Social Security is projected to increase at a rate of approximately \$43.1 billion per year. **b.** $m \approx 51.4$; The cost of Medicare is projected to increase at a rate of approximately \$51.4 billion per year. **c.** no; The cost of Medicare is projected to increase at a faster rate than the cost of Social Security.

81. a. NORMAL FLOAT AUTO REAL RADIAN MP

L1	L2	L3	L4	L5
0	15.8	------	------	------
1	40.8			
2	60.1			
3	88.6			
4	139.3			
5	172.4			
6	296.6			
------	------			

L1(1)=0

b. & d.

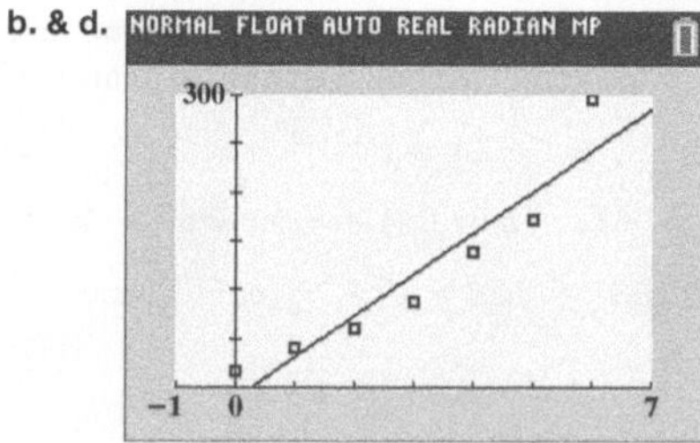

c.

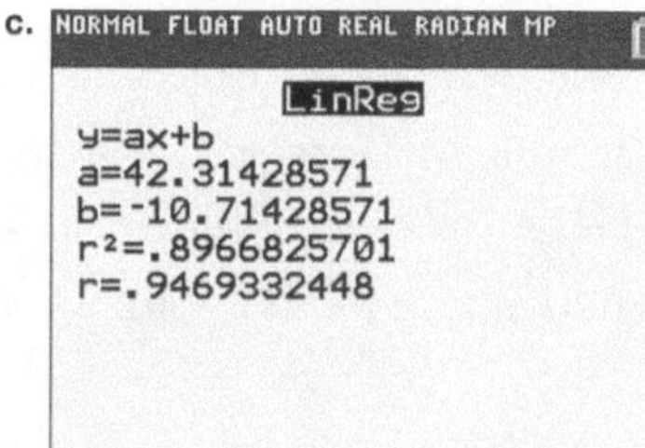

83. makes sense **85.** makes sense **87.** true **89.** true **91.** −4 **93.** (−40, 74) and (97, −200) **95.** 33 **96.** −56 **97.** 40°, 60°, and 80° **98. a.** yes **b.** yes **99.** (3, −4);

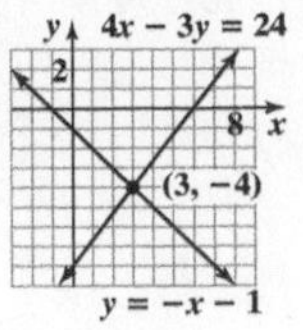

100. 1 or {1}

Review Exercises

1. function; domain: {3, 4, 5}; range: {10} **2.** function; domain: {1, 2, 3, 4}; range: {12, 100, π, −6} **3.** not a function; domain: {13, 15}; range: {14, 16, 17} **4. a.** −5 **b.** 16 **c.** −75 **d.** 14a − 5 **e.** 7a + 9 **5. a.** 2 **b.** 52 **c.** 70 **d.** $3b^2 - 5b + 2$ **e.** $48a^2 - 20a + 2$

6.

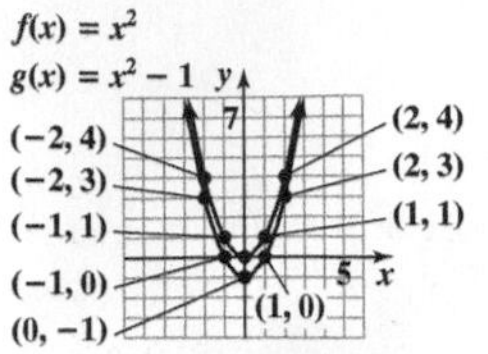

The graph of g is the graph of f shifted down by 1 unit.

7.

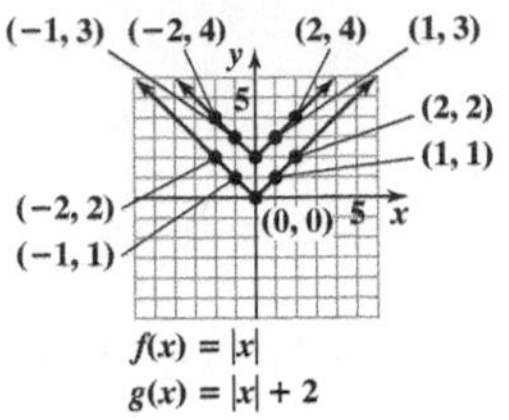

The graph of g is the graph of f shifted up by 2 units.

8. not a function **9.** function **10.** function **11.** not a function **12.** not a function **13.** function **14.** −3 **15.** −2 **16.** 3 **17.** [−3, 5) **18.** [−5, 0]
19. a. For each time, there is only one height.
b. 0; The eagle was on the ground after 15 seconds.
c. 45 m
d. 7 and 22; After 7 seconds and after 22 seconds, the eagle's height is 20 meters.
e. Answers will vary.
20. (−∞, ∞) **21.** (−∞, −8) or (−8, ∞) **22.** (−∞, 5) or (5, ∞) **23. a.** 6x − 4 **b.** 14 **24. a.** $5x^2 + 1$ **b.** 46 **25.** (−∞, 4) or (4, ∞) **26.** (−∞, −6) or (−6, −1) or (−1, ∞) **27.** $x^2 - x - 5$; 1 **28.** 1 **29.** $x^2 - 3x + 5$; 3 **30.** 9 **31.** −120 **32.** $\frac{x^2 - 2x}{x - 5}$; −8 **33.** (−∞, ∞) **34.** (−∞, 5) or (5, ∞)

35.

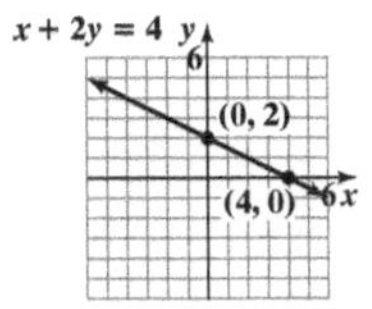

36.

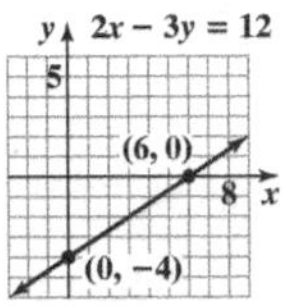

37.

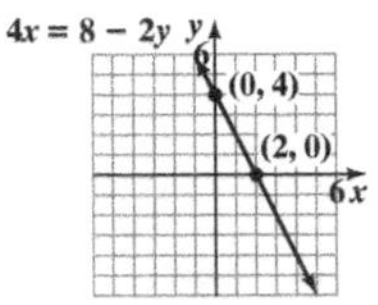

38. 2; rises **39.** $-\frac{2}{3}$; falls **40.** undefined; vertical **41.** 0; horizontal

42. $m = 2$; $b = -1$;

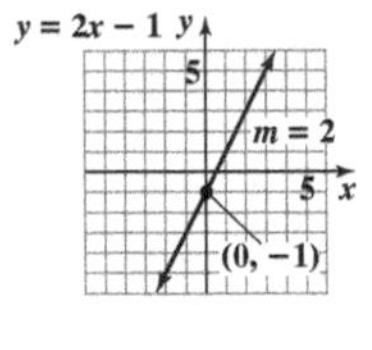

43. $m = -\frac{1}{2}$; $b = 4$;

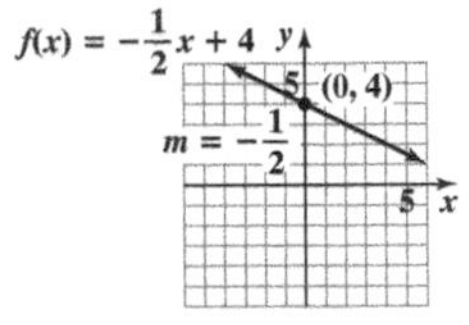

44. $m = \frac{2}{3}$; $b = 0$;

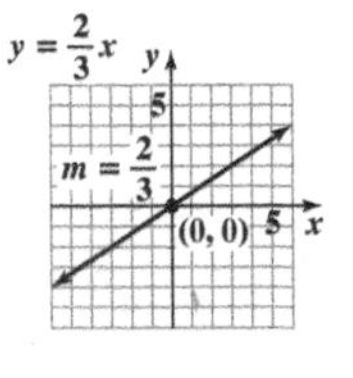

45. $y = -2x + 4$; $m = -2$; $b = 4$ **46.** $y = -\frac{5}{3}x$; $m = -\frac{5}{3}$; $b = 0$ **47.** $y = -\frac{5}{3}x + 2$; $m = -\frac{5}{3}$; $b = 2$

48.

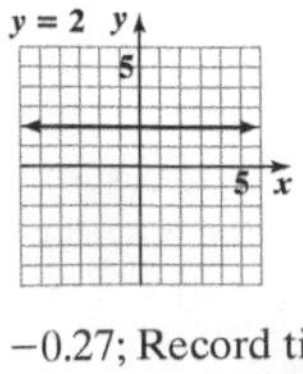

49.

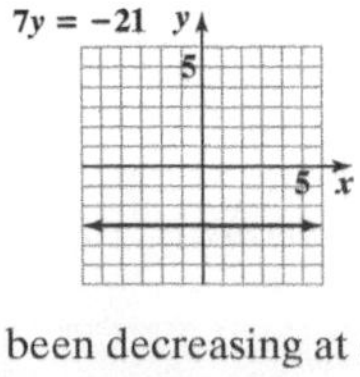

50.

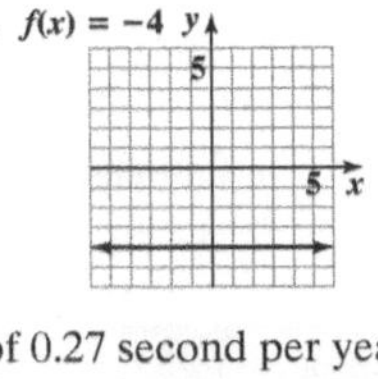

51.

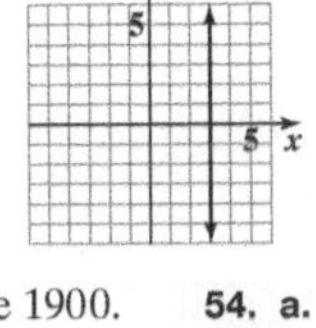

52.

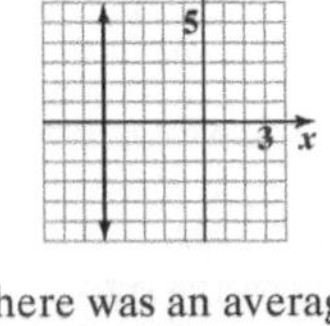

53. −0.27; Record time has been decreasing at a rate of 0.27 second per year since 1900. **54. a.** 137; There was an average increase of approximately 137 discharges per year. **b.** −130; There was an average decrease of approximately 130 discharges per year.
55. a. $F = \frac{9}{5}C + 32$ **b.** 86° **56.** $y - 2 = -6(x + 3)$; $y = -6x - 16$ or $f(x) = -6x - 16$ **57.** $y - 6 = 2(x - 1)$ or $y - 2 = 2(x + 1)$; $y = 2x + 4$ or $f(x) = 2x + 4$ **58.** $y + 7 = -3(x - 4)$; $y = -3x + 5$ or $f(x) = -3x + 5$ **59.** $y - 6 = -3(x + 2)$; $y = -3x$ or $f(x) = -3x$
60. a. $y - 28.2 = 0.2(x - 2)$ or $y - 28.6 = 0.2(x - 4)$ **b.** $f(x) = 0.2x + 27.8$ **c.** 30.2 years

Chapter Test

1. function; domain: {1, 3, 5, 6}; range: {2, 4, 6} **2.** not a function; domain: {2, 4, 6}; range: {1, 3, 5, 6} **3.** $3a + 10$ **4.** 28 **5.** $g(x) = x^2 + 1$; $f(x) = x^2 - 1$; (−2, 5), (−2, 3), (−1, 2), (−1, 0), (0, −1), (0, 1), (2, 5), (2, 3), (1, 2), (1, 0)

The graph of g is the graph of f shifted up by 2 units.

6. function **7.** not a function **8.** −3 **9.** −2 and 3 **10.** $(-\infty, \infty)$ **11.** $(-\infty, 3]$ **12.** $(-\infty, 10)$ or $(10, \infty)$ **13.** $x^2 + 5x + 2; 26$ **14.** $x^2 + 3x - 2; -4$ **15.** −15 **16.** $\frac{x^2 + 4x}{x + 2}; 3$ **17.** $(-\infty, -2)$ or $(-2, \infty)$

18. 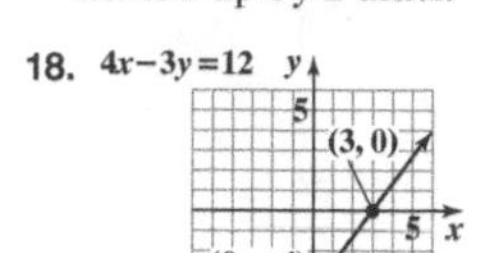

19. 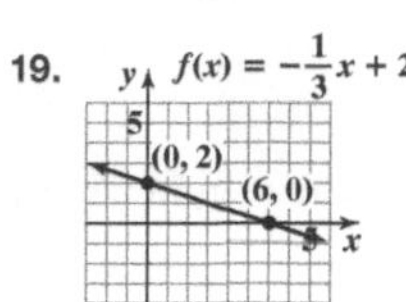

20. 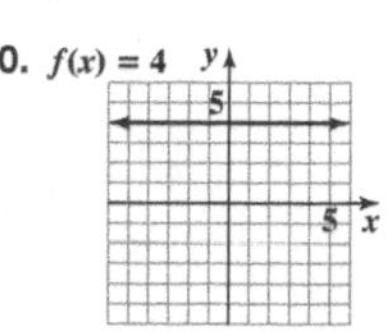

21. $-\frac{1}{2}$; falls **22.** undefined; vertical **23.** 176; In 2005, the number of Super Bowl viewers was 176 million. **24.** 3.6; The number of Super Bowl viewers is increasing at a rate of 3.6 million per year. **25.** $y + 3 = 1(x + 1)$ or $y - 2 = 1(x - 4)$; $y = x - 2$ or $f(x) = x - 2$ **26.** $y - 3 = 2(x + 2)$; $y = 2x + 7$ or $f(x) = 2x + 7$ **27.** $y + 4 = -\frac{1}{2}(x - 6)$; $y = -\frac{1}{2}x - 1$ or $f(x) = -\frac{1}{2}x - 1$ **28. a.** $y - 0.053 = 0.017(x - 3)$ or $y - 0.121 = 0.017(x - 7)$ **b.** $f(x) = 0.017x + 0.002$ **c.** 0.138

Cumulative Review Exercises

1. {0, 1, 2, 3} **2.** false **3.** 7 **4.** 15 **5.** $7 + 3x$ or $3x + 7$ **6.** −4 or {−4} **7.** $\{x \mid x \text{ is a real number}\}$ or $(-\infty, \infty)$ or $\mathbb{R}$; identity **8.** −6 or {−6} **9.** \$2250 **10.** $t = \frac{A - p}{pr}$ **11.** $\frac{y^{10}}{9x^8}$ **12.** $\frac{9x^{10}}{y^{12}}$ **13.** 2.1×10^{-5} **14.** function; domain: {1, 2, 3, 4, 6}; range: {5} **16.** $(-\infty, 15)$ or $(15, \infty)$ **17.** $2x^2 + x + 5; 6$

15. **18.** **19.** 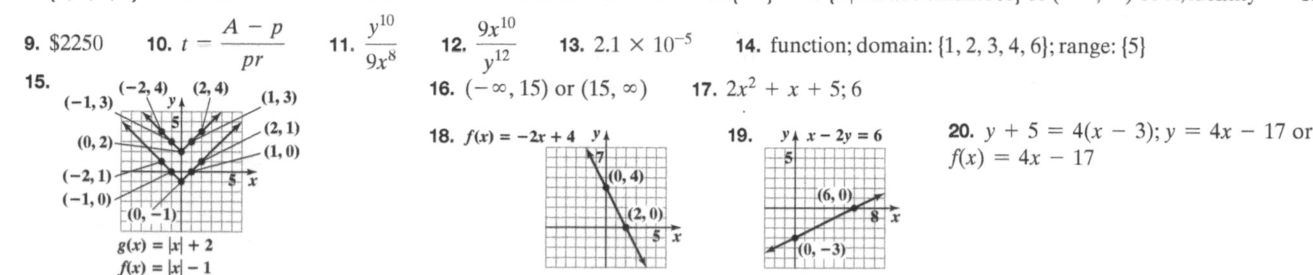

The graph of g is the graph of f shifted up by 3 units.

20. $y + 5 = 4(x - 3)$; $y = 4x - 17$ or $f(x) = 4x - 17$

CHAPTER 3

Section 3.1 Check Point Exercises

1. a. not a solution **b.** solution **2.** (1, 4) or {(1, 4)} **3.** (6, 11) or {(6, 11)} **4.** (−2, 5) or {(−2, 5)} **5.** $\left(-\frac{1}{2}, 2\right)$ or $\left\{\left(-\frac{1}{2}, 2\right)\right\}$ **6.** (2, −1) or {(2, −1)} **7.** $\left(\frac{37}{7}, \frac{19}{7}\right)$ or $\left\{\left(\frac{37}{7}, \frac{19}{7}\right)\right\}$ **8.** no solution or ∅ **9.** $\{(x, y) \mid x = 4y - 8\}$ or $\{(x, y) \mid 5x - 20y = -40\}$

Concept and Vocabulary Check

1. satisfies both equations in the system **2.** the intersection point **3.** $\left\{\left(\frac{1}{3}, -2\right)\right\}$ **4.** −2 **5.** −3 **6.** ∅; inconsistent; parallel **7.** $\{(x, y) \mid x = 3y + 2\}$ or $\{(x, y) \mid 5x - 15y = 10\}$; dependent; are identical or coincide

Exercise Set 3.1

1. solution **3.** not a solution **5.** solution **7.** {(3, 1)} **9.** $\left\{\left(\frac{1}{2}, 3\right)\right\}$ **11.** {(4, 3)} **13.** $\{(x, y) \mid 2x + 3y = 6\}$ or $\{(x, y) \mid 4x = -6y + 12\}$ **15.** {(1, 0)} **17.** ∅ **19.** {(1, 2)} **21.** {(3, 1)} **23.** ∅ **25.** {(2, 4)} **27.** {(3, 1)} **29.** {(2, 1)} **31.** {(2, −3)} **33.** ∅ **35.** {(3, −2)} **37.** ∅ **39.** {(−5, −1)} **41.** $\left\{(x, y) \mid y = \frac{2}{5}x - 2\right\}$ or $\{(x, y) \mid 2x - 5y = 10\}$ **43.** {(5, 2)} **45.** {(2, −3)} **47.** {(−1, 1)} **49.** {(−2, −7)} **51.** {(7, 2)} **53.** {(4, −1)} **55.** $\{(x, y) \mid 2x + 6y = 8\}$ or $\{(x, y) \mid 3x + 9y = 12\}$ **57.** $\left\{\left(\frac{29}{22}, -\frac{5}{11}\right)\right\}$ **59.** {(−2, −1)} **61.** {(1, −3)} **63.** {(1, −3)} **65.** {(4, 3)} **67.** $\{(x, y) \mid x = 3y - 1\}$ or $\{(x, y) \mid 2x - 6y = -2\}$ **69.** ∅ **71.** {(5, 1)} **73.** $\left\{\left(\frac{32}{7}, -\frac{20}{7}\right)\right\}$ **75.** {(−5, 7)} **77.** ∅ **79.** $\{(x, y) \mid x + 2y - 3 = 0\}$ or $\{(x, y) \mid 12 = 8y + 4x\}$ **81.** {(0, 0)} **83.** {(6, −1)} **85.** $\left\{\left(\frac{1}{a}, 3\right)\right\}$ **87.** $m = -4, b = 3$ **89.** $y = x - 4$; $y = -\frac{1}{3}x + 4$ **91. a.** 2039; 37% **b.** by the intersection point (69, 37)

93. a. $y = 0.04x + 5.48$ **b.** $y = 0.17x + 1.84$ **c.** 2028; 6.6%; Medicare **95. a.** $y = -0.54x + 38$ **b.** $y = -0.79x + 40$ **c.** 1993; 33.68% **97. a.** 150 sold; 300 supplied **b.** \$100; 250 **109.** makes sense **111.** makes sense **113.** false **115.** false **117.** $a = 3, b = 2$ **119.** $\left\{\left(\frac{b_2c_1 - b_1c_2}{a_1b_2 - a_2b_1}, \frac{a_1c_2 - a_2c_1}{a_1b_2 - a_2b_1}\right)\right\}$ **120.** $\left\{\frac{10}{9}\right\}$ **121.** $-128x^{19}y^8$ **122.** 11 **123.** $0.15x + 0.07y$ **124.** 15 mL **125.** $80x$

Section 3.2 Check Point Exercises

1. hamburger and fries: 1240; fettuccine Alfredo: 1500 **2.** \$3150 at 9%; \$1850 at 11% **3.** 12% solution: 100 oz; 20% solution: 60 oz **4.** boat: 35 mph; current: 7 mph **5. a.** $C(x) = 300{,}000 + 30x$ **b.** $R(x) = 80x$ **c.** (6000, 480,000); The company will break even when it produces and sells 6000 pairs of shoes. At this level, both revenue and cost are \$480,000. **6.** $P(x) = 50x - 300{,}000$

Concept and Vocabulary Check

1. $1180x + 125y$ **2.** $0.12x + 0.09y$ **3.** $0.09x + 0.6y$ **4.** $x + y; x - y$ **5.** $4(x + y)$ **6.** revenue; profit **7.** break-even point

Exercise Set 3.2

1. 3 and 4 **3.** first number: 2; second number: 5 **5. a.** 1500 units; \$48,000 **b.** $P(x) = 17x - 25{,}500$ **7. a.** 500 units; \$122,500 **b.** $P(x) = 140x - 70{,}000$ **9.** multiple times per day: 24%; once per day: 17% **11.** 22 computers and 14 hard drives **13.** \$2000 at 6% and \$5000 at 8% **15.** first fund: \$8000; second fund: \$6000 **17.** \$17,000 at 12%; \$3000 at a 5% loss **19.** California: 100 gal; French: 100 gal **21.** 18-karat gold: 96 g; 12-karat gold: 204 g **23.** cheaper candy: 30 lb; more expensive candy: 45 lb **25.** 8 nickels and 7 dimes **27.** plane: 130 mph; wind: 30 mph **29.** crew: 6 km/hr; current: 2 km/hr **31.** in still water: 4.5 mph; current: 1.5 mph **33.** 86 and 74 **35.** 80°, 50°, 50° **37.** 70 ft by 40 ft **39.** two-seat tables: 6; four-seat tables: 11 **41.** 500 radios **43.** –6000; When the company produces and sells 200 radios, the loss is \$6000. **45. a.** $P(x) = 20x - 10{,}000$ **b.** \$190,000 **47. a.** $C(x) = 18{,}000 + 20x$ **b.** $R(x) = 80x$ **c.** (300, 24,000); When 300 canoes are produced and sold, both revenue and cost are \$24,000. **49. a.** $C(x) = 30{,}000 + 2500x$ **b.** $R(x) = 3125x$ **c.** (48, 150,000); For 48 sold-out performances, both cost and revenue are \$150,000.

59. (6, 300);
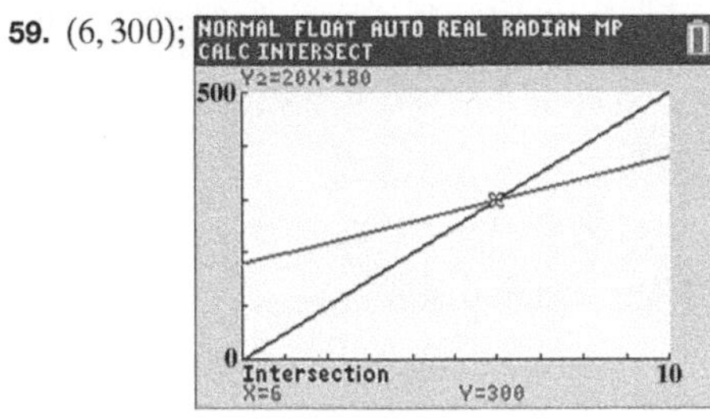

63. does not make sense **65.** makes sense **67.** yes, 8 hexagons and 4 squares **69.** 95 **71.** $y - 5 = -2(x + 2)$ or $y - 13 = -2(x + 6)$; $y = -2x + 1$ or $f(x) = -2x + 1$ **72.** $y - 0 = 1(x + 3)$; $y = x + 3$ or $f(x) = x + 3$ **73.** $(-\infty, 3)$ or $(3, \infty)$ **74.** yes **75.** $11x + 4y = -3$ **76.** $16a + 4b + c = 1682$

Section 3.3 Check Point Exercises

1. $(-1) - 2(-4) + 3(5) = 22$; $2(-1) - 3(-4) - 5 = 5$; $3(-1) + (-4) - 5(5) = -32$ **2.** $(1, 4, -3)$ or $\{(1, 4, -3)\}$ **3.** $(4, 5, 3)$ or $\{(4, 5, 3)\}$ **4.** $y = 3x^2 - 12x + 13$ or $f(x) = 3x^2 - 12x + 13$

Concept and Vocabulary Check

1. triple; all **2.** −2; −4 **3.** z; add Equations 1 and 3 **4.** quadratic **5.** curve fitting

Exercise Set 3.3

1. not a solution **3.** solution **5.** $\{(2, 3, 3)\}$ **7.** $\{(2, -1, 1)\}$ **9.** $\{(1, 2, 3)\}$ **11.** $\{(3, 1, 5)\}$ **13.** $\{(1, 0, -3)\}$ **15.** $\{(1, -5, -6)\}$ **17.** no solution or $\varnothing$ **19.** infinitely many solutions; dependent equations **21.** $\left\{\left(\frac{1}{2}, \frac{1}{3}, -1\right)\right\}$ **23.** $y = 2x^2 - x + 3$ **25.** $y = 2x^2 + x - 5$ **27.** 7, 4, and 5 **29.** $\{(4, 8, 6)\}$ **31.** $y = -\frac{3}{4}x^2 + 6x - 11$ **33.** $\left\{\left(\frac{8}{a}, -\frac{3}{b}, -\frac{5}{c}\right)\right\}$ **35. a.** (0, 5), (50, 31), (100, 15) **b.** $\begin{cases} 0a + 0b + c = 5 \\ 2500a + 50b + c = 31 \\ 10{,}000a + 100b + c = 15 \end{cases}$ **37. a.** $y = -16x^2 + 40x + 200$ **b.** 0; After 5 seconds, the ball hits the ground. **39.** chemical engineering: 22 hours; mathematics: 16 hours; psychology: 14 hours **41.** \$1200 at 8%; \$2000 at 10%; \$3500 at 12% **43.** 200 \$8 tickets; 150 \$10 tickets; 50 \$12 tickets **45.** 4 oz of food A; 0.5 oz of food B; 1 oz of food C **55.** does not make sense **57.** makes sense **59.** false **61.** false **63.** 13 triangles, 21 rectangles, and 6 pentagons

65. $f(x) = -\frac{3}{4}x + 3$
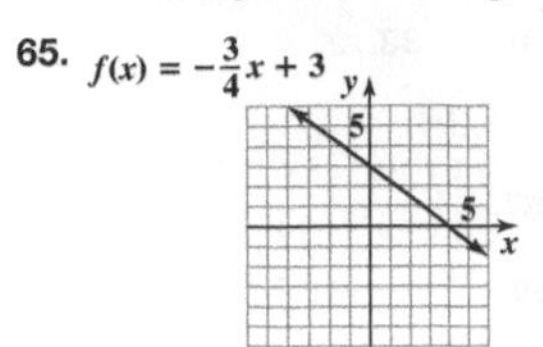

66. $-2x + y = 6$

67. $f(x) = -5$
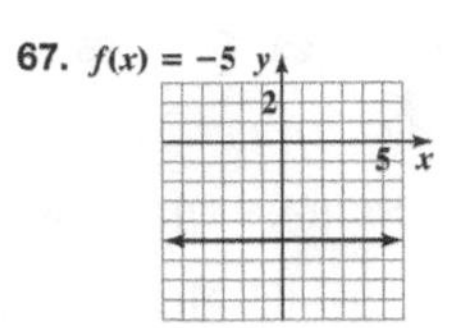

68. $\{(-3, 1)\}$; The value for y is given and the value for x can be found by back-substitution. **69.** $\{(6, 3, 5)\}$; The value for z is given and the values of the other variables can be found by back-substitution. **70.** $\begin{bmatrix} 1 & 2 & -1 \\ 0 & -11 & -11 \end{bmatrix}$

Mid-Chapter Check Point Exercises

1. $\{(-1, 2)\}$ **2.** $\{(1, -2)\}$ **3.** $\{(6, 10)\}$ **4.** $\{(x, y)|y = 4x - 5\}$ or $\{(x, y)|8x - 2y = 10\}$ **5.** $\left\{\left(\frac{11}{19}, \frac{7}{19}\right)\right\}$ **6.** $\varnothing$ **7.** $\{(-1, 2, -2)\}$
8. $\{(4, -2, 3)\}$ **9.** $\{(3, 2)\}$ **10.** $\{(-2, 1)\}$ **11. a.** $C(x) = 400{,}000 + 20x$ **b.** $R(x) = 100x$ **c.** $P(x) = 80x - 400{,}000$
d. (5000, 500,000); The company will break even when it produces and sells 5000 PDAs. At this level, both revenue and cost are \$500,000.
12. 6 roses and 14 carnations **13.** \$6300 at 5% and \$8700 at 6% **14.** 13% nitrogen: 20 gal; 18% nitrogen: 30 gal
15. rowing rate in still water: 3 mph; current: 1.5 mph **16.** \$4500 at 2% and \$3500 at 5% **17.** $y = -x^2 + 2x + 3$
18. 8 nickels, 6 dimes, 12 quarters

Section 3.4 Check Point Exercises

1. a. $\left[\begin{array}{ccc|c} 1 & 6 & -3 & 7 \\ 4 & 12 & -20 & 8 \\ -3 & -2 & 1 & -9 \end{array}\right]$ **b.** $\left[\begin{array}{ccc|c} 1 & 3 & -5 & 2 \\ 1 & 6 & -3 & 7 \\ -3 & -2 & 1 & -9 \end{array}\right]$ **c.** $\left[\begin{array}{ccc|c} 4 & 12 & -20 & 8 \\ 1 & 6 & -3 & 7 \\ 0 & 16 & -8 & 12 \end{array}\right]$ **2.** $(-1, 2)$ or $\{(-1, 2)\}$ **3.** $(5, 2, 3)$ or $\{(5, 2, 3)\}$

Concept and Vocabulary Check

1. matrix; elements
2. $\left[\begin{array}{cc|c} 3 & -2 & -6 \\ 4 & 5 & -8 \end{array}\right]$ **3.** $\left[\begin{array}{ccc|c} 2 & 1 & 4 & -4 \\ 3 & 0 & 1 & 1 \\ 4 & 3 & 1 & 8 \end{array}\right]$ **4.** first; $\frac{1}{2}$ **5.** 3; second; –2; third **6.** false **7.** true

Exercise Set 3.4

1. $\left[\begin{array}{cc|c} 1 & -\frac{3}{2} & 5 \\ 2 & 2 & 5 \end{array}\right]$ **3.** $\left[\begin{array}{cc|c} 1 & -\frac{4}{3} & 2 \\ 3 & 5 & -2 \end{array}\right]$ **5.** $\left[\begin{array}{cc|c} 1 & -3 & 5 \\ 0 & 12 & -6 \end{array}\right]$ **7.** $\left[\begin{array}{cc|c} 1 & -\frac{3}{2} & \frac{7}{2} \\ 0 & \frac{17}{2} & -\frac{17}{2} \end{array}\right]$ **9.** $\left[\begin{array}{ccc|c} 1 & -3 & 2 & 5 \\ 1 & 5 & -5 & 0 \\ 3 & 0 & 4 & 7 \end{array}\right]$ **11.** $\left[\begin{array}{ccc|c} 1 & -3 & 2 & 0 \\ 0 & 10 & -7 & 7 \\ 2 & -2 & 1 & 3 \end{array}\right]$
13. $\left[\begin{array}{ccc|c} 1 & 1 & -1 & 6 \\ 0 & -3 & 3 & -15 \\ 0 & -4 & 2 & -14 \end{array}\right]$ **15.** $\{(4, 2)\}$ **17.** $\{(3, -3)\}$ **19.** $\{(2, -5)\}$ **21.** no solution or $\varnothing$ **23.** infinitely many solutions; dependent equations **25.** $\{(1, -1, 2)\}$ **27.** $\{(3, -1, -1)\}$ **29.** $\{(1, 2, -1)\}$ **31.** $\{(1, 2, 3)\}$ **33.** no solution or $\varnothing$
35. infinitely many solutions; dependent equations **37.** $\{(-1, 2, -2)\}$ **39.** $\begin{cases} w - x + y + z = 3 \\ x - 2y - z = 0 \\ y + 6z = 17 \\ z = 3 \end{cases}$; $\{(2, 1, -1, 3)\}$ **41.** $\left[\begin{array}{cccc|c} 1 & -1 & 1 & 1 & 3 \\ 0 & 1 & -2 & -1 & 0 \\ 0 & 2 & 1 & 2 & 5 \\ 0 & 6 & -3 & -1 & -9 \end{array}\right]$
43. $\{(1, 2, 3, -2)\}$ **45. a.** $s(t) = -16t^2 + 56t$ **b.** 0; The ball hits the ground 3.5 seconds after it is thrown.; (3.5, 0)
47. yes: 34%; no: 61%; not sure: 5% **59.** makes sense **61.** does not make sense **63.** false **65.** false **67.** $-6a + 13$ **68.** 15
69. $-\frac{x^{11}}{3y^{36}}$ **70.** 2 **71.** 6 **72.** –31

Section 3.5 Check Point Exercises

1. a. -4 **b.** -17 **2.** $(4, -2)$ or $\{(4, -2)\}$ **3.** 80 **4.** $(2, -3, 4)$ or $\{(2, -3, 4)\}$

Concept and Vocabulary Check

1. $5 \cdot 3 - 2 \cdot 4 = 15 - 8 = 7$; determinant; 7 **2.** $x = \frac{\begin{vmatrix} 8 & 1 \\ -2 & -1 \end{vmatrix}}{\begin{vmatrix} 1 & 1 \\ 1 & -1 \end{vmatrix}}$; $y = \frac{\begin{vmatrix} 1 & 8 \\ 1 & -2 \end{vmatrix}}{\begin{vmatrix} 1 & 1 \\ 1 & -1 \end{vmatrix}}$ **3.** $3\begin{vmatrix} 3 & 1 \\ 1 & 1 \end{vmatrix} - 4\begin{vmatrix} 2 & 1 \\ 1 & 1 \end{vmatrix} + 5\begin{vmatrix} 2 & 1 \\ 3 & 1 \end{vmatrix}$ **4.** $\frac{\begin{vmatrix} 3 & -8 & 4 \\ 2 & 11 & -2 \\ 1 & 4 & -2 \end{vmatrix}}{\begin{vmatrix} 3 & 1 & 4 \\ 2 & 3 & -2 \\ 1 & -3 & -2 \end{vmatrix}}$

Exercise Set 3.5

1. 1 **3.** –29 **5.** 0 **7.** 33 **9.** $-\frac{7}{16}$ **11.** $\{(5, 2)\}$ **13.** $\{(2, -3)\}$ **15.** $\{(3, -1)\}$ **17.** $\{(4, 0)\}$ **19.** $\{(4, 2)\}$ **21.** $\{(7, 4)\}$
23. inconsistent; no solution or $\varnothing$ **25.** dependent equations; infinitely many solutions **27.** 72 **29.** –75 **31.** 0 **33.** $\{(-5, -2, 7)\}$
35. $\{(2, -3, 4)\}$ **37.** $\{(3, -1, 2)\}$ **39.** $\{(2, 3, 1)\}$ **41.** –42 **43.** $\begin{cases} 2x - 4y = 8 \\ 3x + 5y = -10 \end{cases}$ **45.** $\{-11\}$ **47.** $\{4\}$ **49.** 28 sq units **51.** yes
53. $\begin{vmatrix} x & y & 1 \\ 3 & -5 & 1 \\ -2 & 6 & 1 \end{vmatrix} = 0$; $y = -\frac{11}{5}x + \frac{8}{5}$ **65.** does not make sense **67.** does not make sense **69.** true **71.** false
73. The value is multiplied by –1. **76.** $(-\infty, \infty)$ **77.** $y = \frac{2x + 7}{3}$ **78.** $\{0\}$ **79.** $\{14\}$ **80.** $\{-3\}$ **81.** $(-\infty, \infty)$

Review Exercises

1. not a solution **2.** solution **3.** $\{(2, 3)\}$ **4.** $\{(x, y)|3x - 2y = 6\}$ or $\{(x, y)|6x - 4y = 12\}$ **5.** $\{(-5, -6)\}$ **6.** $\varnothing$ **7.** $\{(3, 4)\}$
8. $\{(23, -43)\}$ **9.** $\{(-4, 2)\}$ **10.** $\left\{\left(3, \frac{1}{2}\right)\right\}$ **11.** $\{(x, y)|y = 4 - x\}$ or $\{(x, y)|3x + 3y = 12\}$ **12.** $\left\{\left(3, \frac{8}{3}\right)\right\}$ **13.** $\varnothing$

14. TV: \$350; stereo: \$370 **15.** \$2500 at 4%; \$6500 at 7% **16.** 10 mL of 34%; 90 mL of 4% **17.** plane: 630 mph; wind 90 mph **18.** 12 ft by 5ft **19.** loss of \$4500 **20.** (500, 42,500); When 500 calculators are produced and sold, both cost and revenue are \$42,500. **21.** $P(x) = 45x - 22{,}500$ **22. a.** $C(x) = 60{,}000 + 200x$ **b.** $R(x) = 450x$ **c.** (240, 108,000); When 240 desks are produced and sold, both cost and revenue are \$108,000. **23.** no **24.** $\{(0, 1, 2)\}$ **25.** $\{(2, 1, -1)\}$ **26.** infinitely many solutions; dependent equations **27.** $y = 3x^2 - 4x + 5$ **28.** war: 124 million; famine: 111 million; tobacco: 71 million **29.** $\left[\begin{array}{cc|c} 1 & -8 & 3 \\ 0 & 1 & -2 \end{array}\right]$ **30.** $\left[\begin{array}{cc|c} 1 & -3 & 1 \\ 0 & 7 & -7 \end{array}\right]$ **31.** $\left[\begin{array}{ccc|c} 1 & -1 & \frac{1}{2} & -\frac{1}{2} \\ 1 & 2 & -1 & 2 \\ 6 & 4 & 3 & 5 \end{array}\right]$ **32.** $\left[\begin{array}{ccc|c} 1 & 2 & 2 & 2 \\ 0 & 1 & -1 & 2 \\ 0 & 0 & 9 & -9 \end{array}\right]$ **33.** $\{(-5, 3)\}$ **34.** no solution or $\varnothing$ **35.** $\{(1, 3, -4)\}$ **36.** $\{(-2, -1, 0)\}$ **37.** 17 **38.** 4 **39.** -86 **40.** -236 **41.** $\left\{\left(\frac{7}{4}, -\frac{25}{8}\right)\right\}$ **42.** $\{(2, -7)\}$ **43.** $\{(23, -12, 3)\}$ **44.** $\{(-3, 2, 1)\}$ **45.** $y = \frac{5}{8}x^2 - 50x + 1150$; 30-year-old drivers are involved in 212.5 accidents daily and 50-year-old drivers are involved in 212.5 accidents daily.

Chapter Test

1. $\{(2, 4)\}$ **2.** $\{(6, -5)\}$ **3.** $\{(1, -3)\}$ **4.** $\{(x, y) \mid 4x = 2y + 6\}$ or $\{(x, y) \mid y = 2x - 3\}$ **5.** one-bedroom: 15 units; two-bedroom: 35 units **6.** \$2000 at 6% and \$7000 at 7% **7.** 6% solution: 12 oz; 9% solution: 24 oz **8.** boat: 14 mph; current: 2 mph **9.** $C(x) = 360{,}000 + 850x$ **10.** $R(x) = 1150x$ **11.** (1200, 1,380,000); When 1200 computers are produced and sold, both cost and revenue are \$1,380,000. **12.** $P(x) = 85x - 350{,}000$ **13.** $\{(1, 3, 2)\}$ **14.** $\left[\begin{array}{ccc|c} 1 & 0 & -4 & 5 \\ 0 & -1 & 26 & -20 \\ 2 & -1 & 4 & -3 \end{array}\right]$ **15.** $\{(4, -2)\}$ **16.** $\{(-1, 2, 2)\}$ **17.** 17 **18.** -10 **19.** $\{(-1, -6)\}$ **20.** $\{(4, -3, 3)\}$

Cumulative Review Exercises

1. 1 **2.** $15x - 7$ **3.** $\{-6\}$ **4.** $\{-15\}$ **5.** no solution or $\varnothing$ **6.** \$2000 **7.** $-\frac{x^8}{4y^{30}}$ **8.** $-4a - 3$ **9.** $(-\infty, -3)$ or $(-3, \infty)$ **10.** $x^2 - 3x - 1$; -1

11. $f(x) = -\frac{2}{3}x + 2$

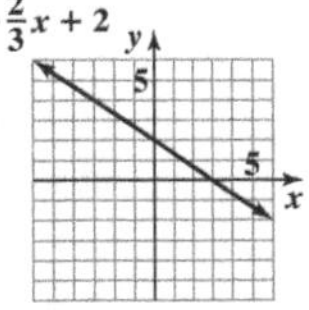

12. $2x - y = 6$

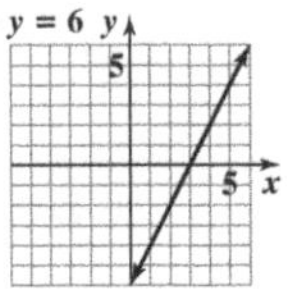

13. $y - 4 = -3(x - 2)$ or $y + 2 = -3(x - 4)$; $y = -3x + 10$ or $f(x) = -3x + 10$ **14.** $y - 0 = -3(x + 1)$; $y = -3x - 3$ or $f(x) = -3x - 3$ **15.** $\left\{\left(7, \frac{1}{3}\right)\right\}$ **16.** $\{(3, 2, 4)\}$ **17.** pad: \$0.80; pen: \$0.20 **18.** 23 **19.** $\{(3, -2, 1)\}$ **20.** $\{(-3, 2)\}$

CHAPTER 4

Section 4.1 Check Point Exercises

1. $(-5, \infty)$ **2.** $(-\infty, 4)$ 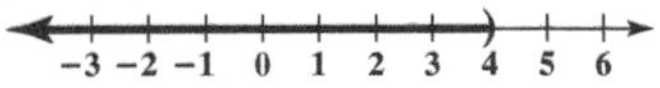 **3.** $[13, \infty)$ **4. a.** $(-\infty, \infty)$ **b.** $\varnothing$ **5.** more than 720 miles

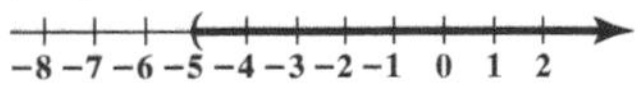

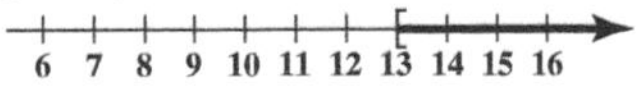

Concept and Vocabulary Check

1. $< b + c$ **2.** $< bc$ **3.** $> bc$ **4.** adding 4; dividing; -3; direction; $>$; $<$ **5.** $\varnothing$ **6.** $(-\infty, \infty)$ **7.** $x \geq 7$ **8.** $x \leq 7$ **9.** $x \leq 7$ **10.** $x \geq 7$

Exercise Set 4.1

1. $(-\infty, 3)$ **3.** $[7, \infty)$ **5.** $(-\infty, -4]$ 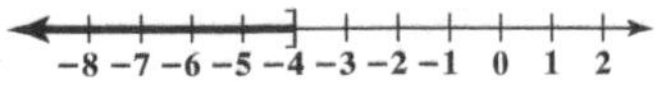 **7.** $\left(-\infty, -\frac{2}{5}\right]$ 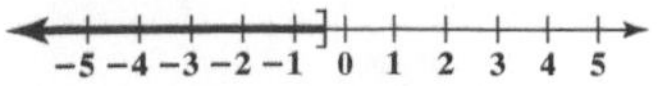 **9.** $[0, \infty)$ **11.** $(-\infty, 1)$ **13.** $[6, \infty)$ **15.** $[-6, \infty)$ **17.** $(-\infty, -6)$ **19.** $[13, \infty)$ **21.** $(-\infty, \infty)$ **23.** $(-\infty, \infty)$ **25.** $\varnothing$ **27.** $(-6, \infty)$ **29.** $[-1, \infty)$

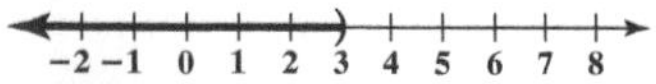

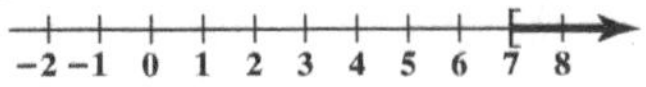

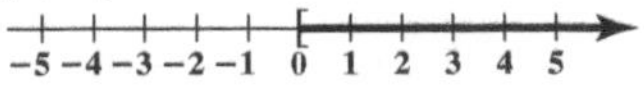

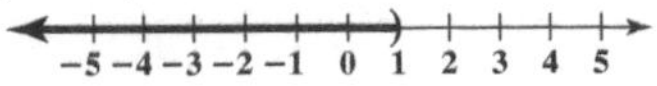

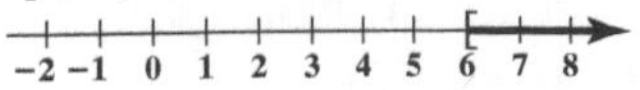

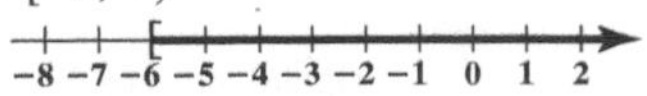

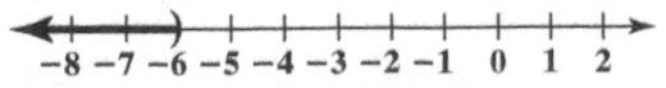

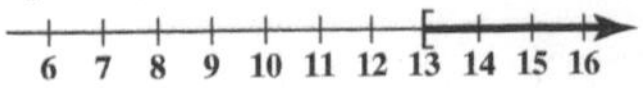

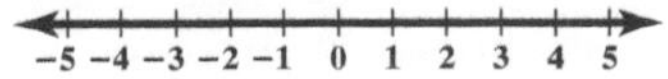

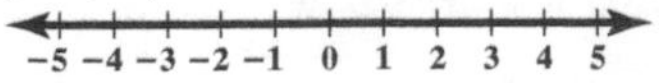

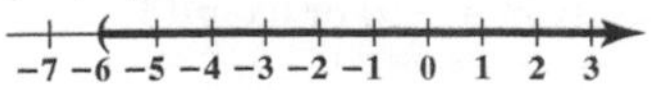

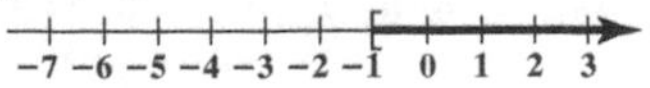

31. $(-\infty, -2)$ **33.** $(-\infty, 5)$ **35.** $\left[-\frac{4}{7}, \infty\right)$ **37.** $[6, \infty)$ **39.** $(-\infty, 2)$

41. $x < \frac{c - b}{a}$ **43.** $(-\infty, -3]$ **45.** $(-1.4, \infty)$ **47.** $(0, 4)$ **49.** intimacy ≥ passion or passion ≤ intimacy

51. commitment > passion or passion < commitment **53.** 9; after 3 years **55. a.** It's exact. **b.** $x > 40$; years after 2020

57. a. −8; Per-capita consumption of carbonated soda is decreasing by eight 12-ounce servings per year on average. **b.** years after 2019

59. more than 100 miles per day **61.** greater than $32,000 **63.** more than 6250 DVDs **65.** 40 bags or fewer

75. $(-\infty, -3)$ **77.** ∅

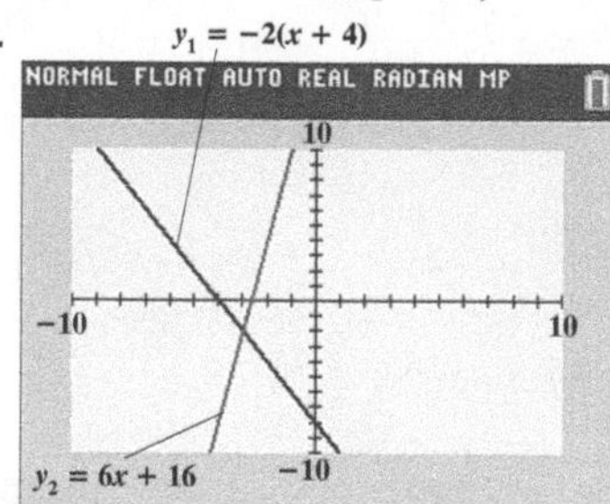

79. a. plan A: $4 + 0.10x$; plan B: $2 + 0.15x$

b.

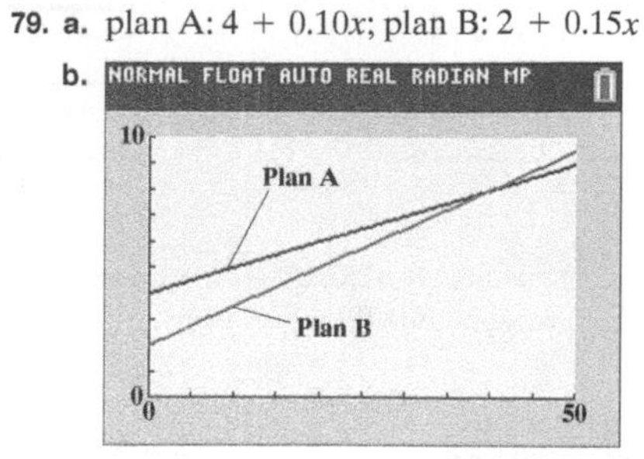

c–d. more than 40 checks per month

81. makes sense **83.** makes sense **85.** false **87.** true **89.** Since $x > y$, $y - x < 0$. Thus, when both sides were multiplied by $y - x$, the sense of the inequality should have been changed. **90.** 29 **91.** $\{(-1, -1, 2)\}$ **92.** $\frac{x^9}{8y^{15}}$ **93. a.** $\{3, 4\}$ **b.** $\{1, 2, 3, 4, 5, 6, 7\}$ **94. a.** $(-\infty, 8)$ **b.** $(-\infty, 5)$ **c.** any number less than 5 **d.** any number in $[5, 8)$ **95. a.** $[1, \infty)$ **b.** $[3, \infty)$ **c.** any number greater than or equal to 3 **d.** any number in $[1, 3)$

Section 4.2 Check Point Exercises

1. $\{3, 7\}$ **2.** $(-\infty, 1)$ **3.** ∅ **4.** $[-1, 4)$; **5.** $\{3, 4, 5, 6, 7, 8, 9\}$ **6.** $(-\infty, 1] \cup (3, \infty)$ **7.** $(-\infty, \infty)$

Concept and Vocabulary Check

1. intersection; $A \cap B$ **2.** union; $A \cup B$ **3.** $(-\infty, 9)$ **4.** $(-\infty, 12)$ **5.** middle

Exercise Set 4.2

1. $\{2, 4\}$ **3.** ∅ **5.** ∅

7. $(6, \infty)$

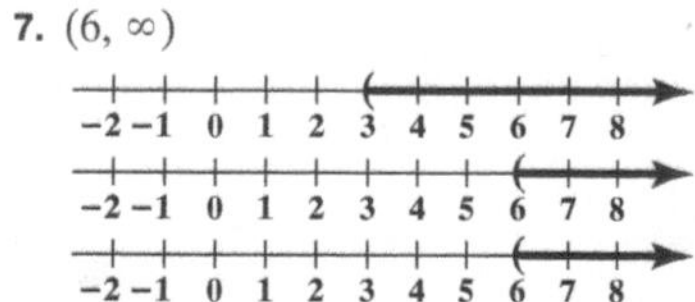

9. $(-\infty, 1]$

11. $[-1, 2)$

13. ∅

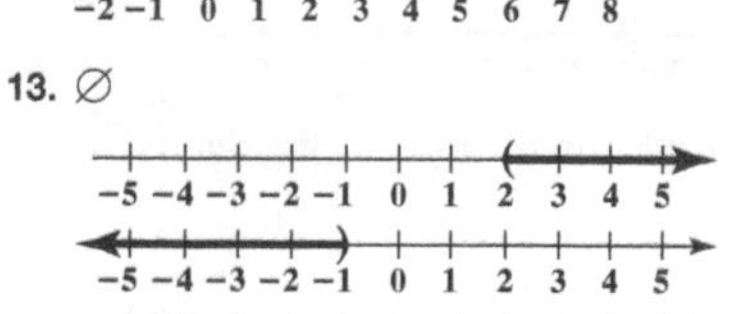

15. $(-6, -4)$

17. $(-3, 6]$

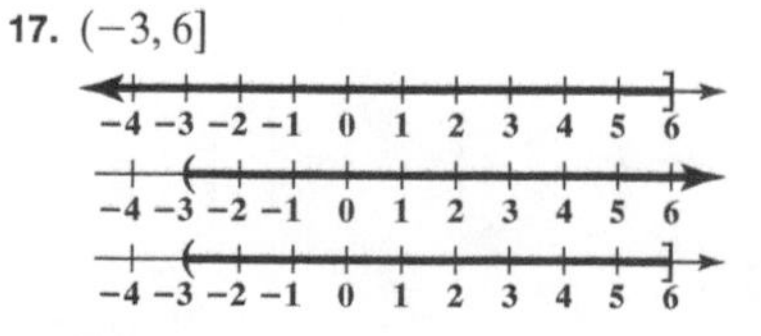

19. $(2, 5)$

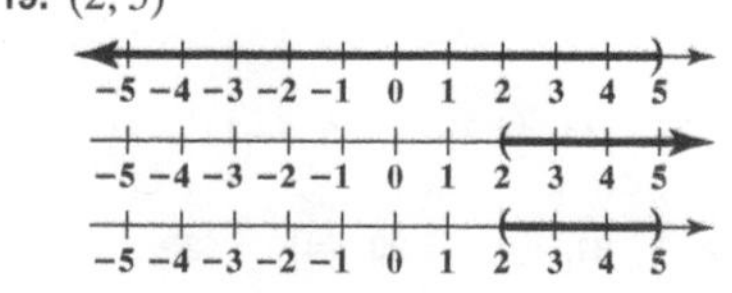

21. ∅

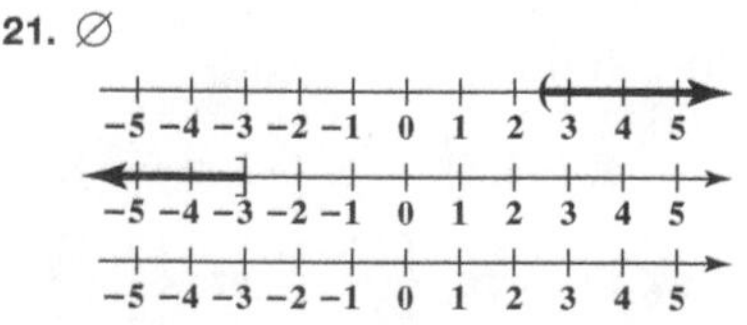

23. $[0, 2)$

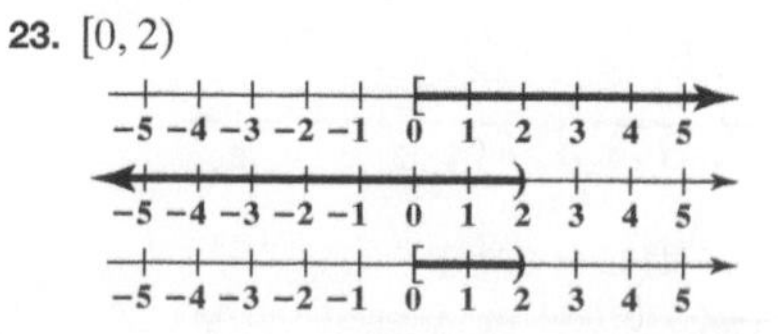

25. $(3, 5)$

29. $(-5, -2]$

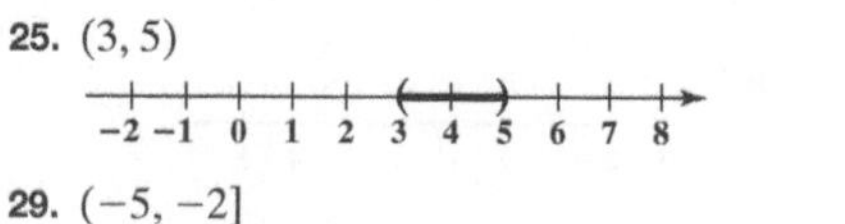

27. $[-1, 3)$

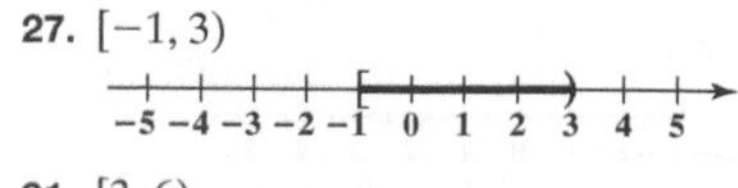

31. $[3, 6)$

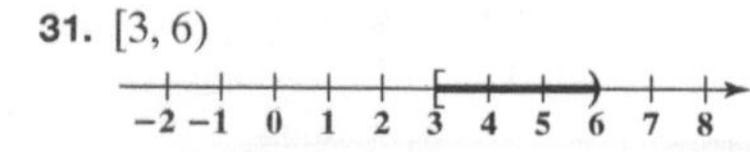

33. $\{1, 2, 3, 4, 5\}$ **35.** $\{1, 2, 3, 4, 5, 6, 7, 8, 10\}$ **37.** $\{a, e, i, o, u\}$

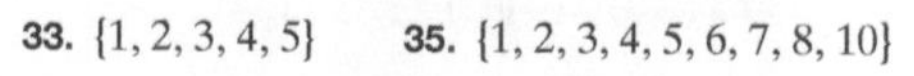

39. $(3, \infty)$

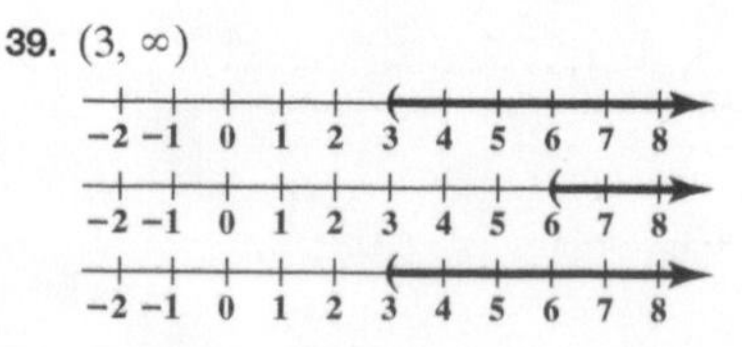

41. $(-\infty, 5]$

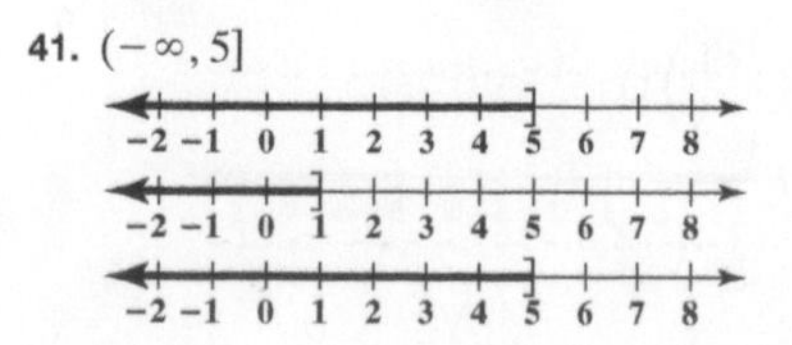

43. $(-\infty, \infty)$

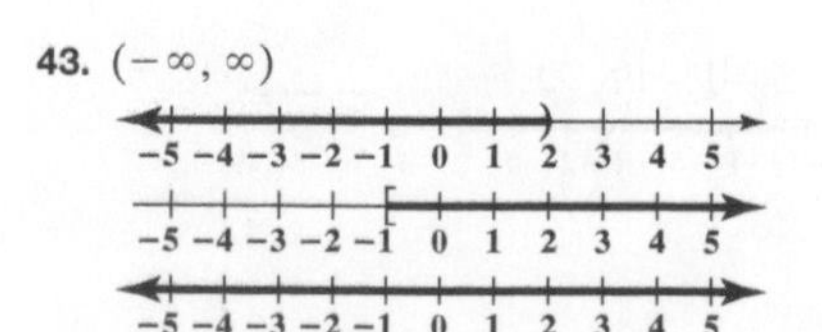

45. $(-\infty, -1) \cup [2, \infty)$

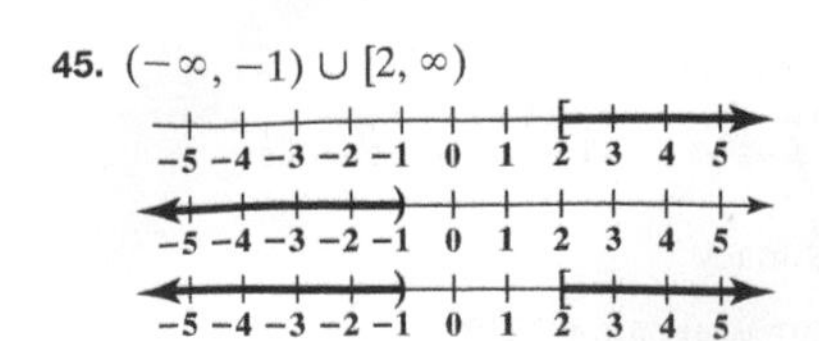

47. $(-\infty, -3) \cup (4, \infty)$

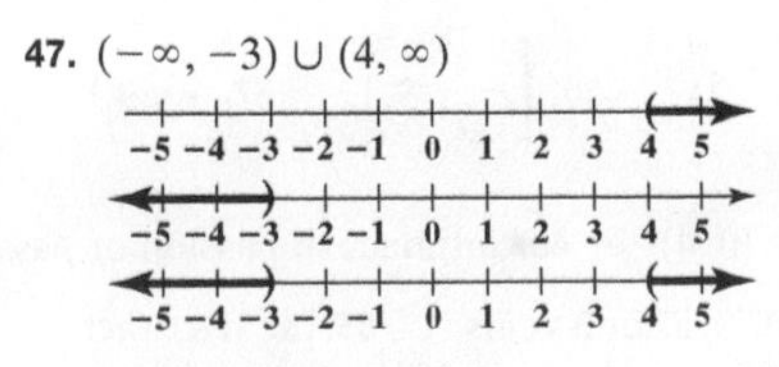

49. $(-\infty, 1] \cup [3, \infty)$

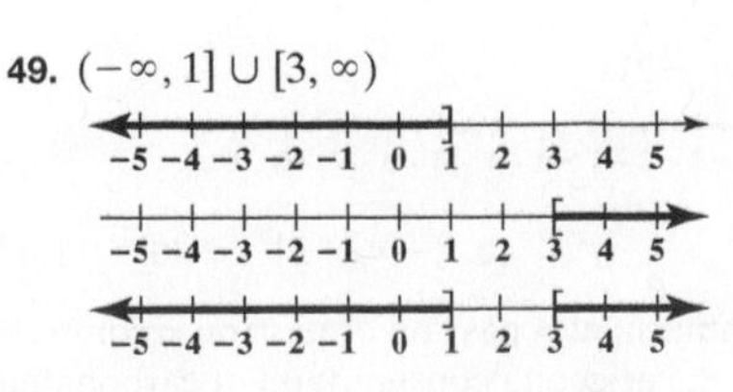

51. $(-\infty, \infty)$

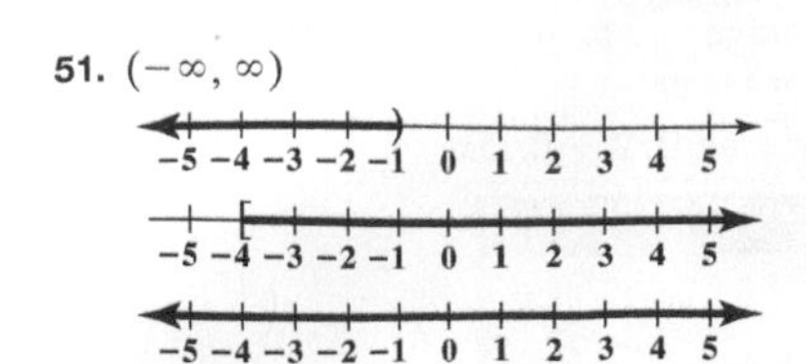

53. $(-\infty, 2)$

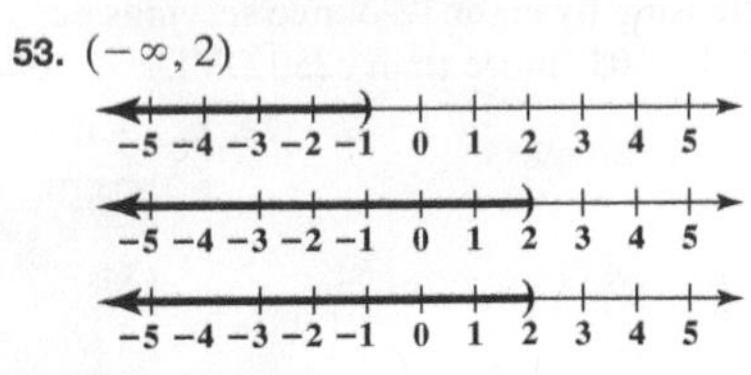

55. $(4, \infty)$ **57.** $(-\infty, 0) \cup (6, \infty)$ **59.** $\frac{b-c}{a} < x < \frac{b+c}{a}$ **61.** $[-1, 3]$ **63.** $(-1, 3)$ **65.** $[-1, 2)$ **67.** $\{-3, -2, -1\}$

69. a. years after 2016 **b.** years after 2020 **c.** years after 2020 **d.** years after 2016 **71.** [5°C, 10°C] **73.** [76, 126); If the highest grade is 100, then [76, 100]. **75.** more than 3 and less than 15 crossings per 3-month period

83. $(-2, 6)$;

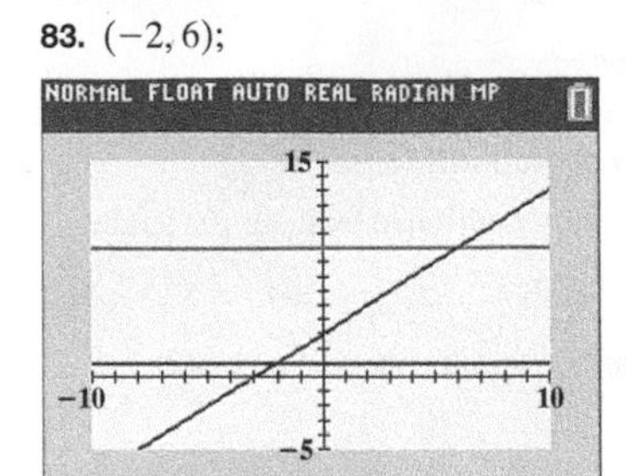

85. $\left[2, \frac{5}{2}\right]$;

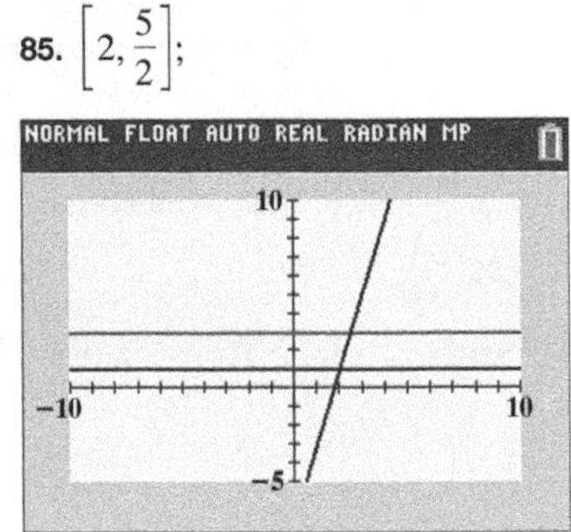

87. Exercise 83:

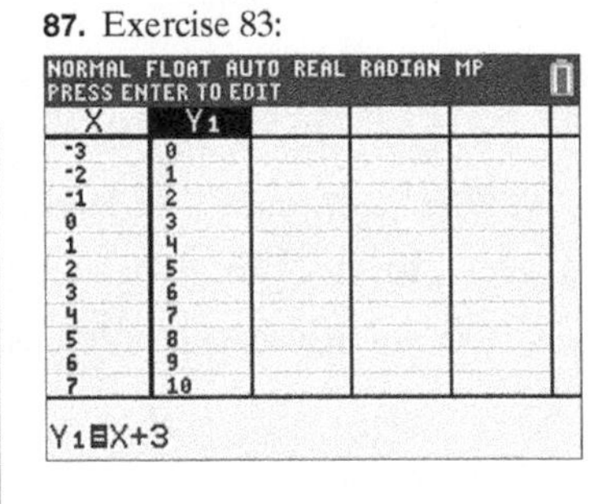

Exercise 85:

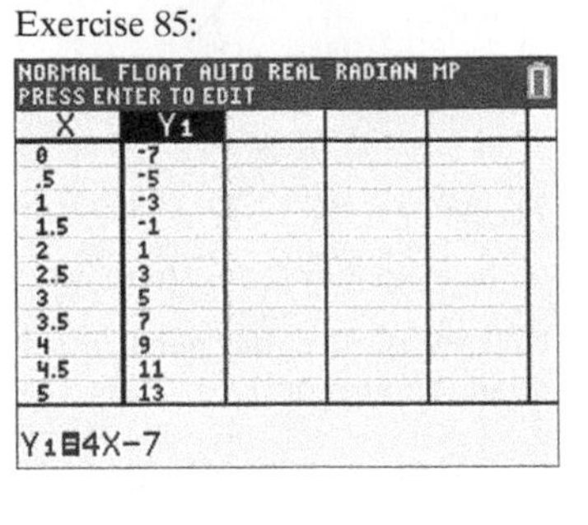

89. makes sense **91.** makes sense **93.** false **95.** false **97.** $(-\infty, 4]$ **99.** $[-1, 4]$ **101.** least: 4 nickels; greatest: 7 nickels **102.** $-x^2 + 5x - 9; -15$ **103.** $f(x) = -\frac{1}{2}x + 4$ **104.** $17 - 2x$ **105.** $-\frac{1}{2}$ and 1 **106.** -1 and 3 **107. a.** -5 satisfies the inequality. **b.** no

Section 4.3 Check Point Exercises

1. -2 and 3, or $\{-2, 3\}$ **2.** $-\frac{13}{3}$ and 5, or $\left\{-\frac{13}{3}, 5\right\}$ **3.** $\frac{4}{3}$ and 10, or $\left\{\frac{4}{3}, 10\right\}$

4. $(-3, 7)$

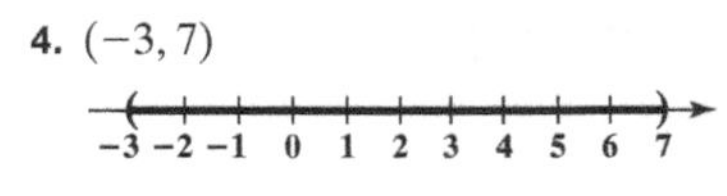

5. $\left[-\frac{11}{5}, 3\right]$;

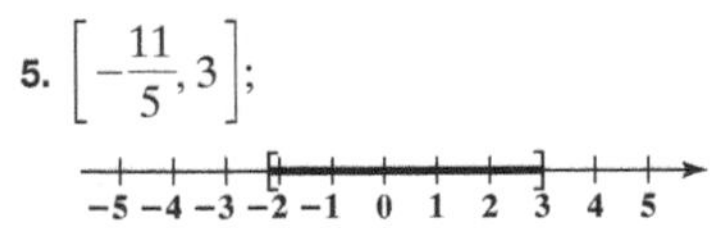

6. $(-\infty, 1] \cup [4, \infty)$

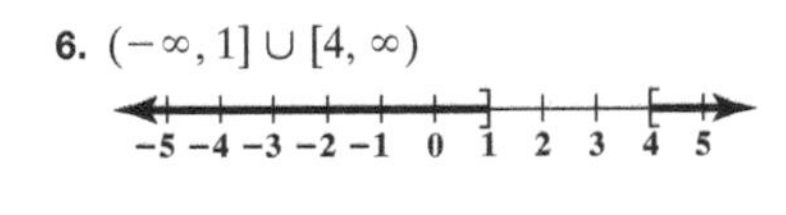

7. [37.8, 44.2]; The percentage of U.S. adults in the population who dread going to the dentist is between a low of 37.8% and a high of 44.2%.

Concept and Vocabulary Check

1. c; $-c$ **2.** v; $-v$ **3.** $-c$; c **4.** $-c$; c **5.** $<$ **6.** $<$ **7.** C **8.** E **9.** A **10.** B **11.** D **12.** F

Exercise Set 4.3

1. $\{-8, 8\}$ **3.** $\{-5, 9\}$ **5.** $\{-3, 4\}$ **7.** $\{-1, 2\}$ **9.** $\varnothing$ **11.** $\{-3\}$ **13.** $\{-11, -1\}$ **15.** $\{-3, 4\}$ **17.** $\left\{-\frac{13}{3}, 5\right\}$ **19.** $\left\{-\frac{2}{5}, \frac{2}{5}\right\}$ **21.** $\varnothing$ **23.** $\varnothing$ **25.** $\left\{\frac{1}{2}\right\}$ **27.** $\left\{\frac{3}{4}, 5\right\}$ **29.** $\left\{\frac{5}{3}, 3\right\}$ **31.** $\{0\}$ **33.** $\{4\}$ **35.** $\{4\}$ **37.** $\{-1, 15\}$

39. $(-3, 3)$ **41.** $(1, 3)$ **43.** $[-3, -1]$

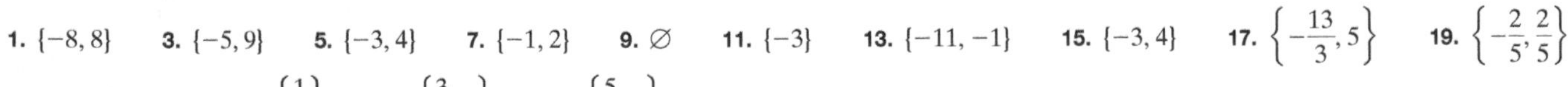

45. $(-1, 7)$ **47.** $(-\infty, -3) \cup (3, \infty)$ **49.** $(-\infty, -4) \cup (-2, \infty)$

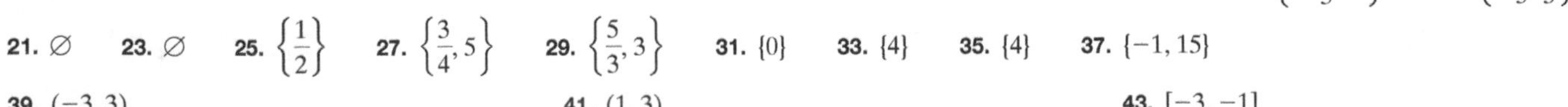

51. $(-\infty, 2] \cup [6, \infty)$ **53.** $\left(-\infty, \frac{1}{3}\right) \cup (5, \infty)$ **55.** $[-5, 3]$

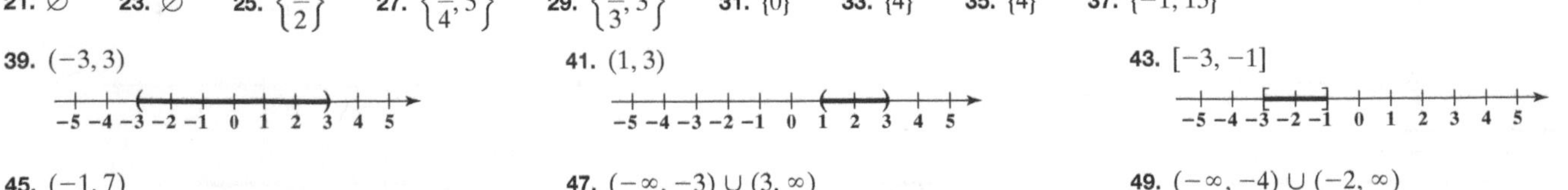

57. $(-6, 0)$

59. $(-\infty, -5] \cup [3, \infty)$

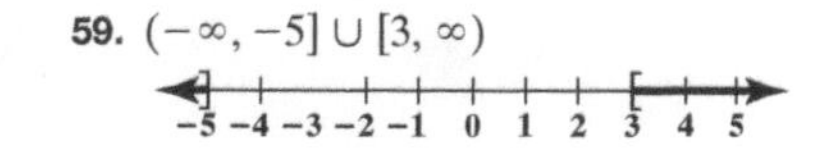

61. $(-\infty, -3) \cup (12, \infty)$

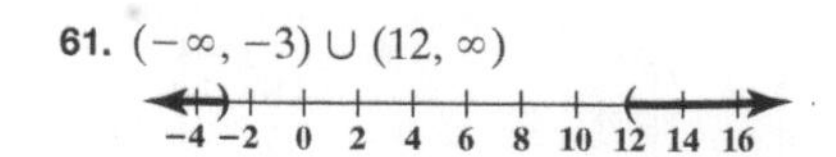

63. $\varnothing$

65. $(-\infty, \infty)$

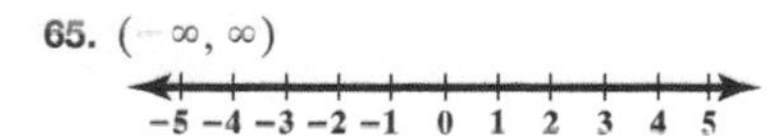

67. $[-9, 5]$

69. $(-\infty, 1) \cup (2, \infty)$

71. $(-\infty, -3) \cup (5, \infty)$

73. $(-\infty, -1] \cup [2, \infty)$

75. $-\frac{3}{2}$ and 4 **77.** -7 and -2 **79.** $\left[-\frac{7}{3}, 1\right]$ **81.** $(-\infty, -1) \cup (4, \infty)$ **83.** $\left(-\infty, -\frac{1}{3}\right] \cup [3, \infty)$ **85.** $\left(\frac{-c-b}{a}, \frac{c-b}{a}\right)$ **87.** $\{3, 5\}$
89. $[-2, 1]$ **91.** $[18, 24]$; The percentage of interviewers in the population turned off by the job applicant being arrogant is between a low of 18% and a high of 24%. **93.** $[50, 64]$; The monthly average temperature for San Francisco, CA, is between a low of 50°F and a high of 64°F. **95.** $[8.59, 8.61]$; A machine part that is supposed to be 8.6 cm is acceptable between a low of 8.59 and a high of 8.61 cm.
97. If the number of outcomes that result in heads is 41 or less or 59 or more, then the coin is unfair.

105. $\{-6, 4\}$;

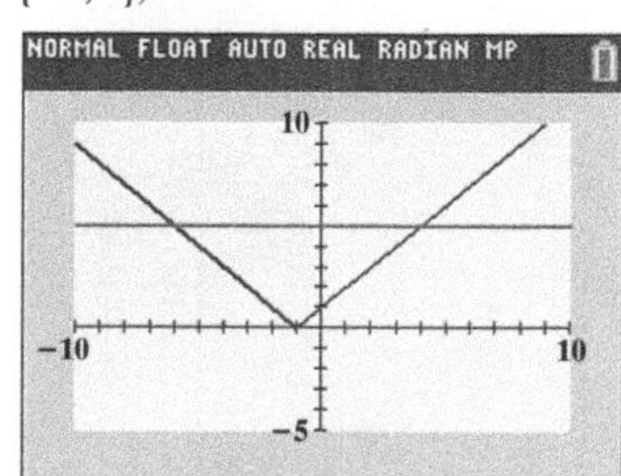

107. $\{2, 3\}$;

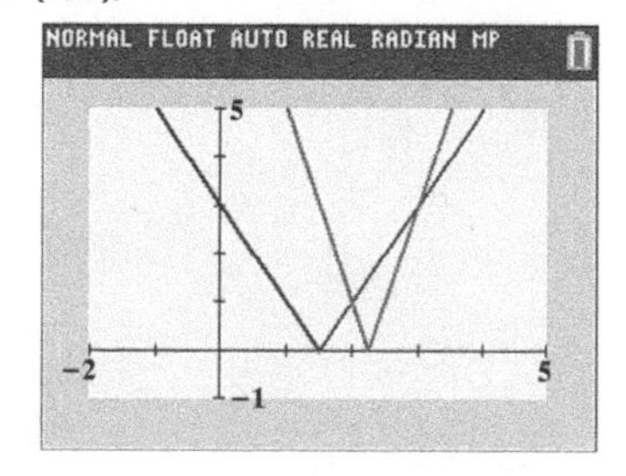

109. $(-2, 3)$;

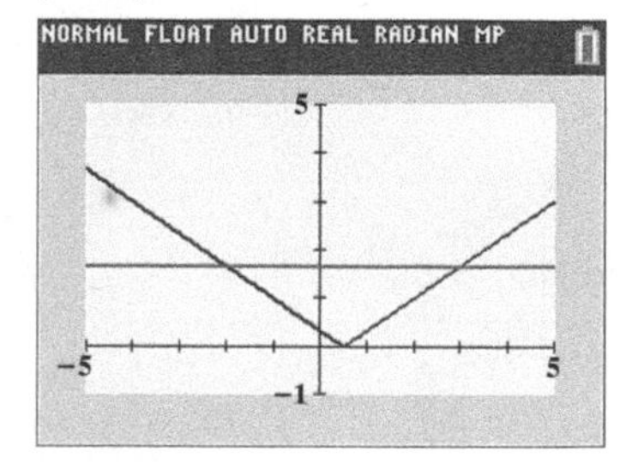

111. $(-\infty, -3) \cup (4, \infty)$;

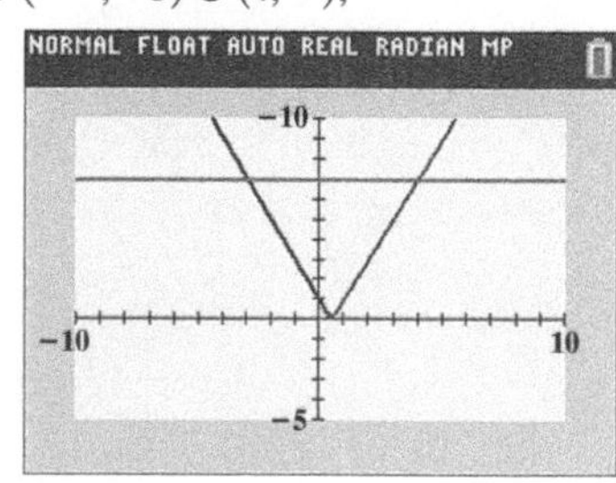

113. $(-\infty, \infty)$;

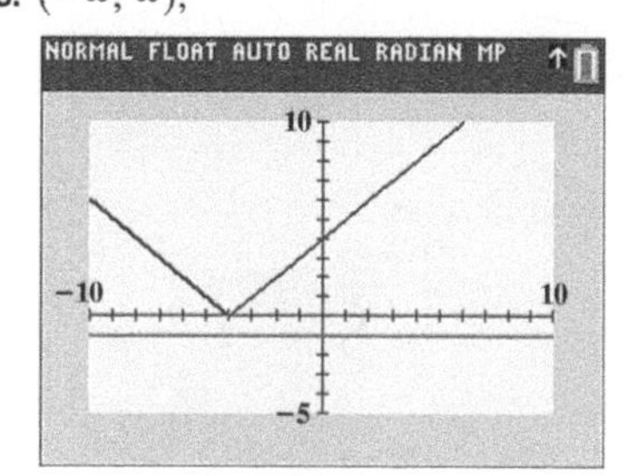

115. does not make sense
117. does not make sense **119.** false
121. true **123. a.** $|x - 4| < 3$
b. $|x - 4| \geq 3$ **125.** $\{1\}$

126. $3x - 5y = 15$

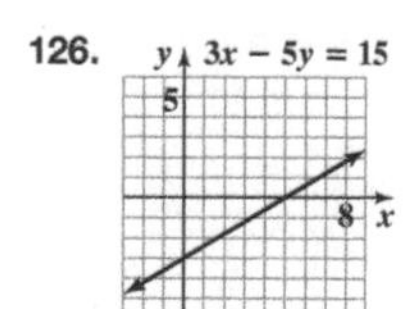

127. $f(x) = -\frac{2}{3}x$

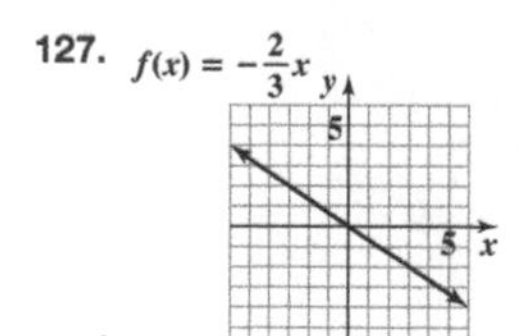

128. $f(x) = -2$

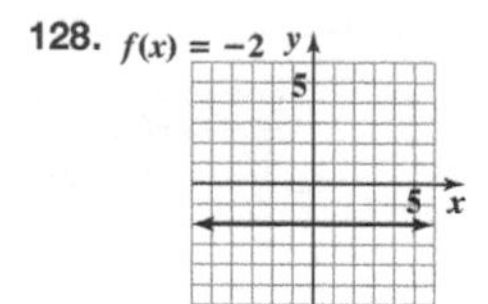

Mid-Chapter Check Point Exercises

1. $(-\infty, -4]$ **2.** $[3, 5)$ **3.** $\left\{\frac{1}{2}, 3\right\}$ **4.** $(-\infty, -1)$ **5.** $(-\infty, -9) \cup (-5, \infty)$ **6.** $\left[-\frac{2}{3}, 2\right]$ **7.** $\left\{\frac{1}{2}, \frac{13}{4}\right\}$ **8.** $(-4, -2]$ **9.** $(-\infty, \infty)$
10. $(-\infty, -3]$ **11.** $(-\infty, -3)$ **12.** $(-\infty, -1) \cup \left(-\frac{1}{5}, \infty\right)$ **13.** $(-\infty, -10] \cup [2, \infty)$ **14.** $\varnothing$ **15.** $\left(-\infty, -\frac{5}{3}\right)$ **16.** $(4, \infty)$ **17.** $\{-2, 7\}$
18. $\varnothing$ **19.** no more than 80 miles per day **20.** $[49\%, 99\%)$ **21.** at least \$120,000 **22.** at least 750,000 discs each month

Section 4.4 Check Point Exercises

1. $4x - 2y \geq 8$

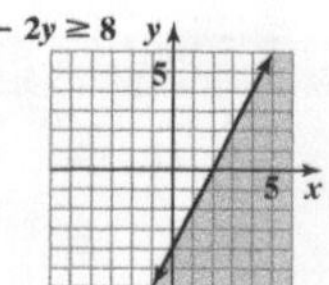

2. $y > -\frac{3}{4}x$

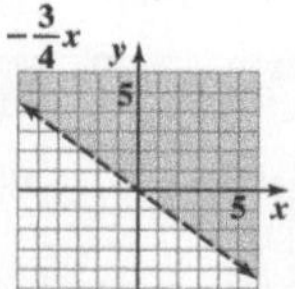

3. a. $y > 1$

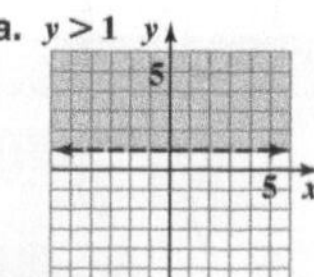

b. $x \leq -2$

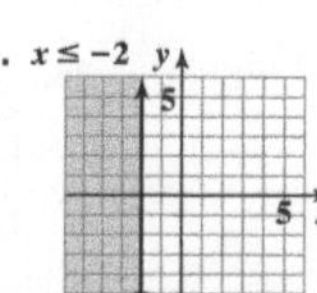

4. $B = (60, 20)$; Using $T = 60$ and $P = 20$, each of the three inequalities for grasslands is true: $60 \geq 35$, true; $5(60) - 7(20) \geq 70$, true; $3(60) - 35(20) \leq -140$, true.

5. $x - 3y < 6$; $2x + 3y \geq -6$ **6.** $x + y < 2$; $-2 \leq x < 1$; $y > -3$

Concept and Vocabulary Check

1. solution; x; y; $5 > 1$ **2.** graph **3.** half-plane **4.** false **5.** true **6.** false **7.** $x - y < 1$; $2x + 3y \geq 12$ **8.** false

Exercise Set 4.4

1. $x + y \geq 3$

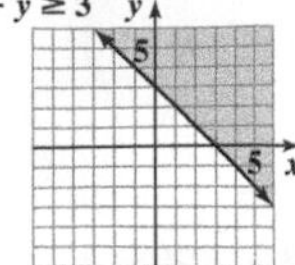

3. $x - y < 5$

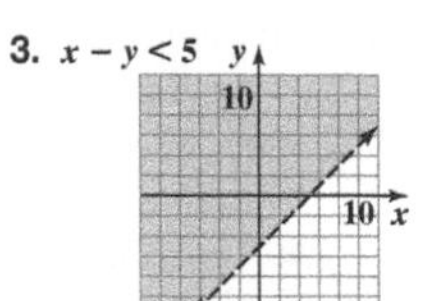

5. $x + 2y > 4$

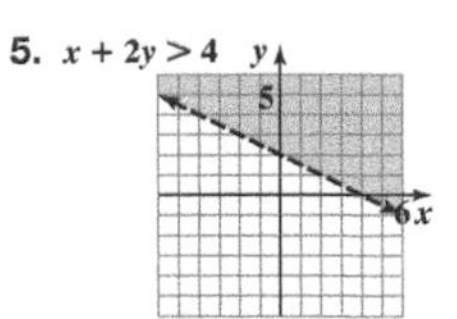

7. $3x - y \leq 6$

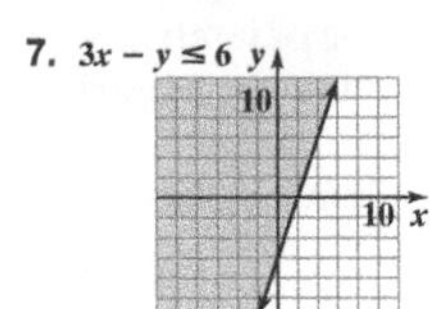

9. $\frac{x}{2} + \frac{y}{3} < 1$

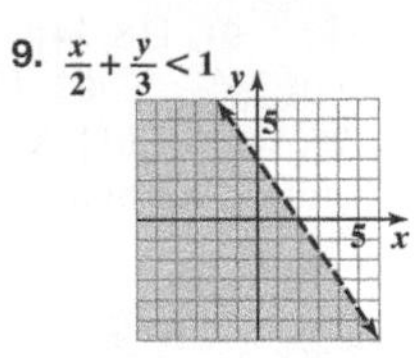

11. $y > \frac{1}{3}x$

13. $y \leq 3x + 2$

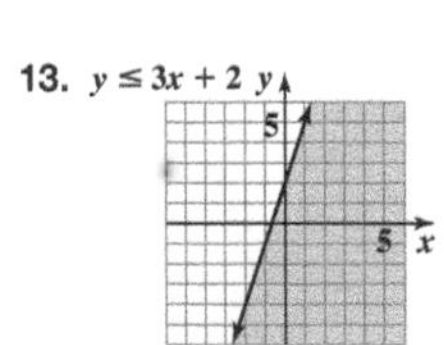

15. $y < -\frac{1}{4}x$

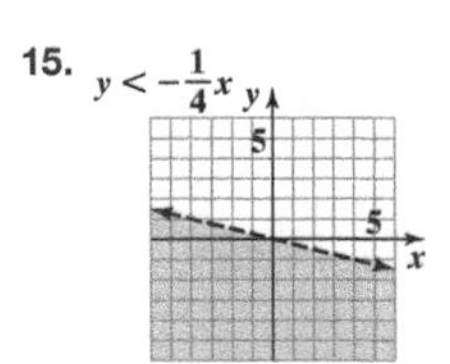

17. $x \leq 2$

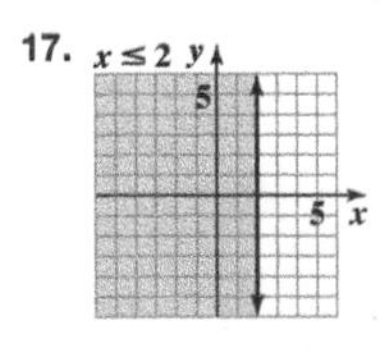

19. $y > -4$

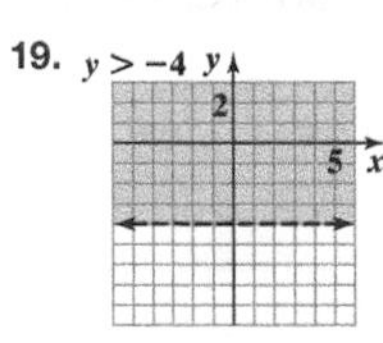

21. $y \geq 0$

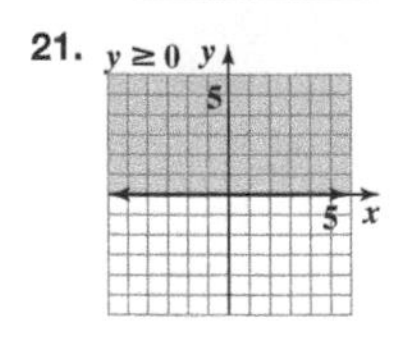

23. $3x + 6y \leq 6$; $2x + y \leq 8$

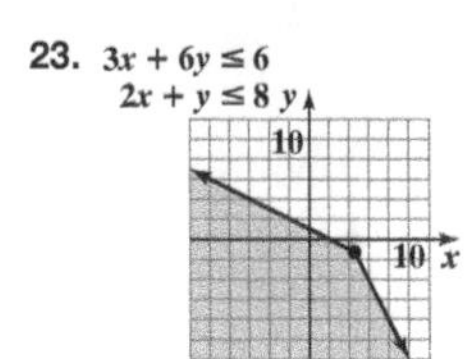

25. $2x - 5y \leq 10$; $3x - 2y > 6$

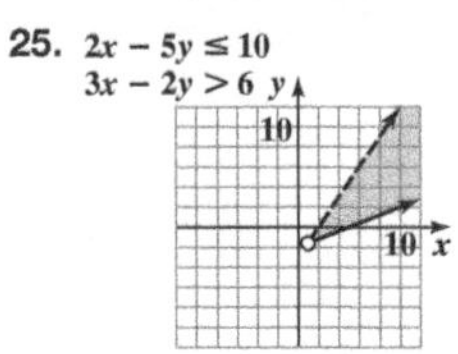

27. $y > 2x - 3$; $y < -x + 6$

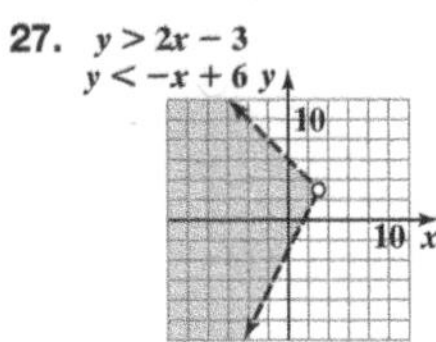

29. $x + 2y \leq 4$; $y \geq x - 3$

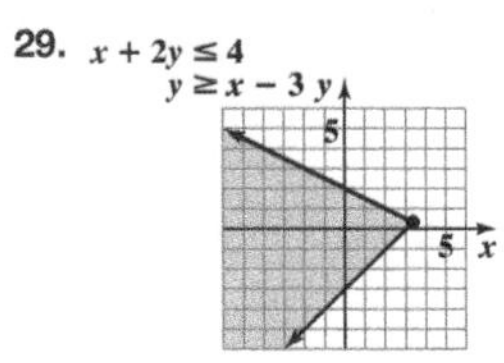

31. $x \leq 2$; $y \geq -1$

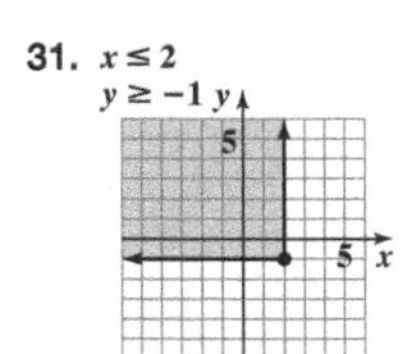

33. $-2 \leq x < 5$

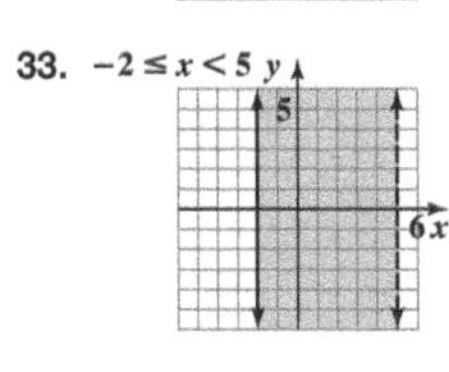

35. $x - y \leq 1$; $x \geq 2$

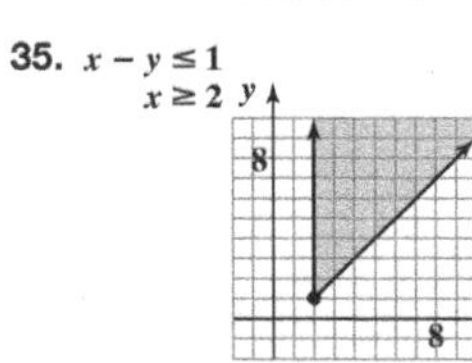

37. $\varnothing$

39. $x + y > 4$; $x + y > -1$

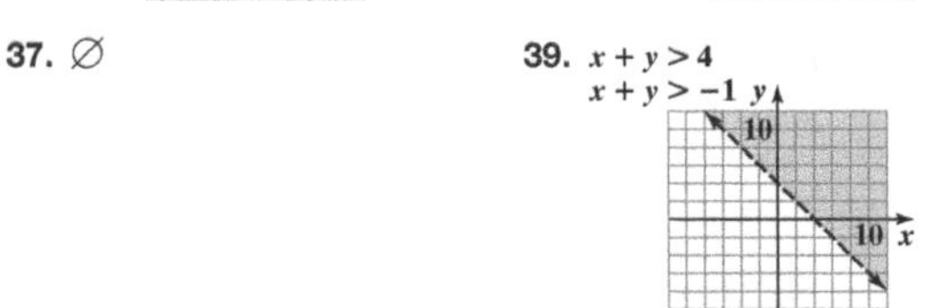

41. $x - y \leq 2$; $x \geq -2$; $y \leq 3$

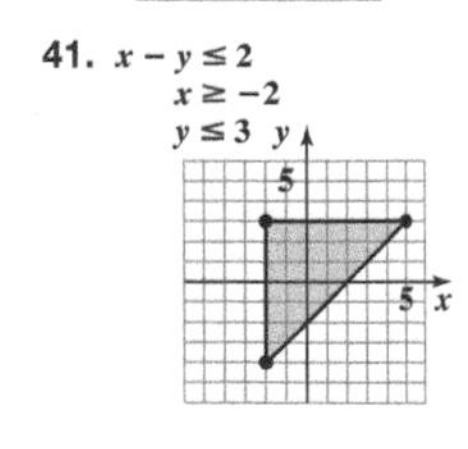

43. $x \geq 0$; $y \geq 0$; $2x + 5y \leq 10$; $3x + 4y \leq 12$

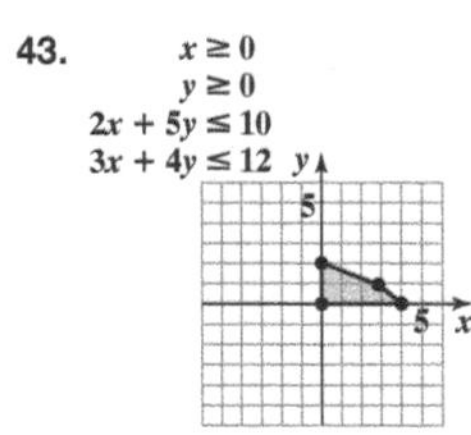

45. $3x + y \leq 6$; $2x - y \leq -1$; $x \geq -2$; $y \leq 4$

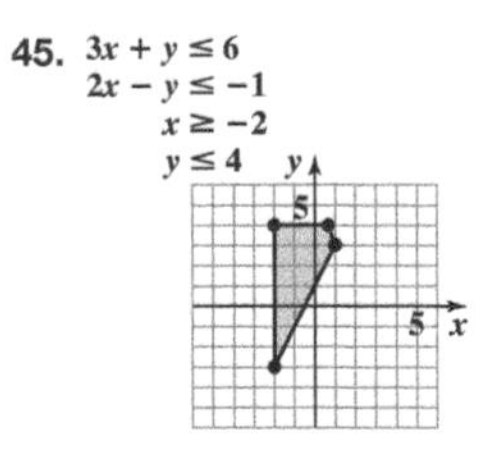

47. $y \geq -2x + 4$

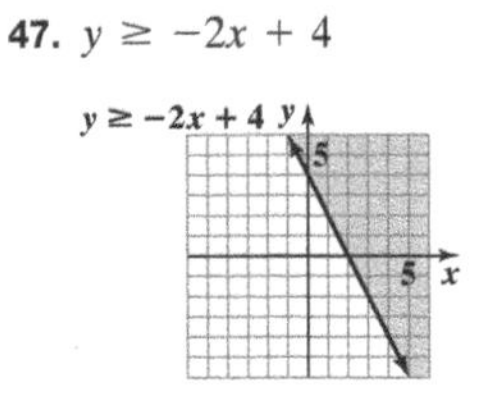

49. $\begin{cases} x + y \leq 4 \\ 3x + y \leq 6 \end{cases}$

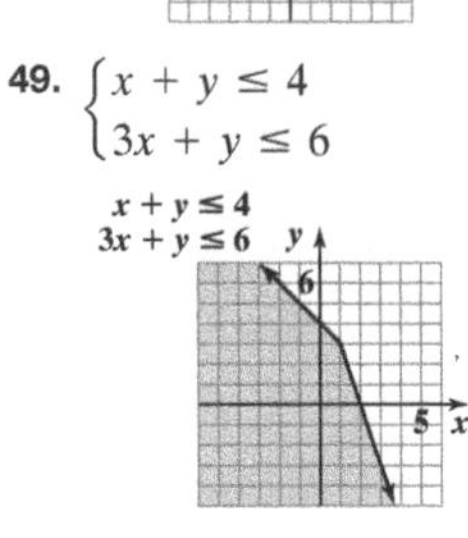

51. $-2 \leq x \leq 2$; $-3 \leq y \leq 3$

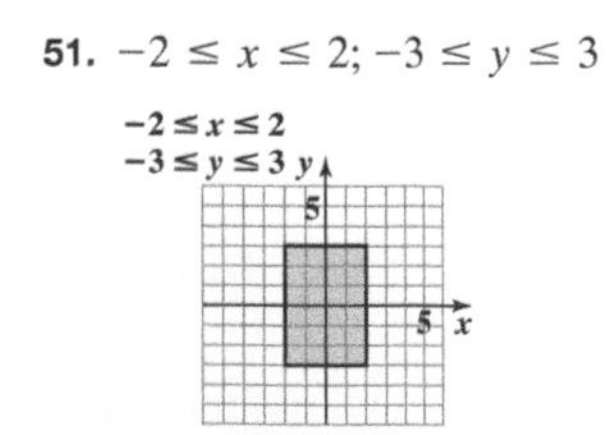

53. $y > \frac{3}{2}x - 2$ or $y < 4$

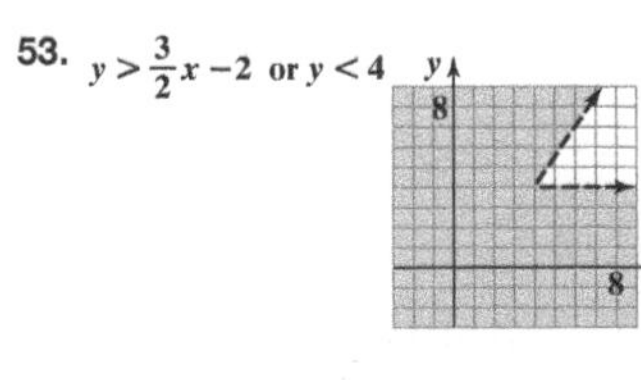

55. no solution **57.** infinitely many solutions **59. a.** $A = (20, 150)$; A 20-year-old with a heart rate of 150 beats per minute is within the target range. **b.** $10 \leq 20 \leq 70$, true; $150 \geq 0.7(220 - 20)$, true; $150 \leq 0.8(220 - 20)$, true **61.** $10 \leq a \leq 70$; $H \geq 0.6(220 - a)$; $H \leq 0.7(220 - a)$

63. a. $\begin{cases} y \geq 0 \\ x + y \geq 5 \\ x \geq 1 \\ 200x + 100y \leq 700 \end{cases}$

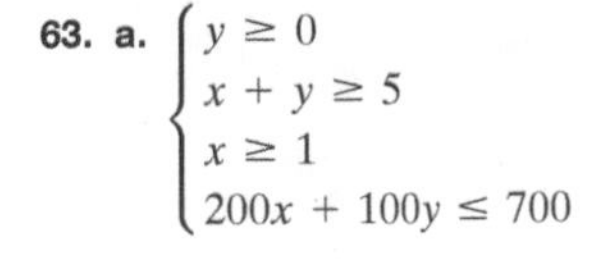

b. $y \geq 0$; $x + y \geq 5$; $x \geq 1$; $200x + 100y \leq 700$

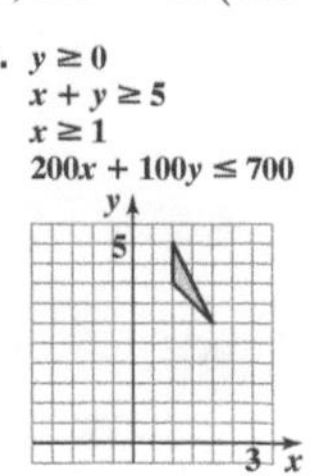

c. 2 nights

75.

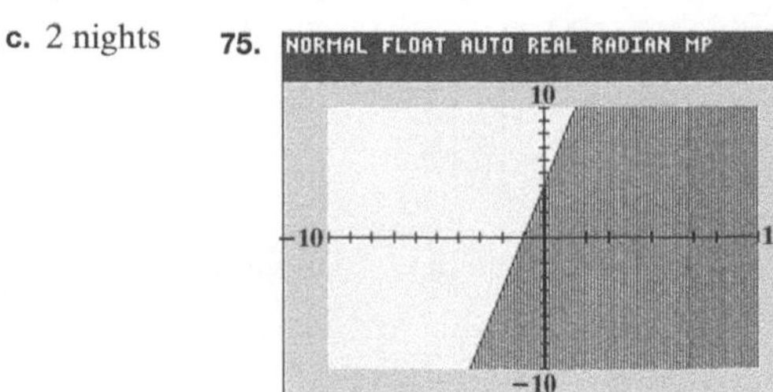

77.

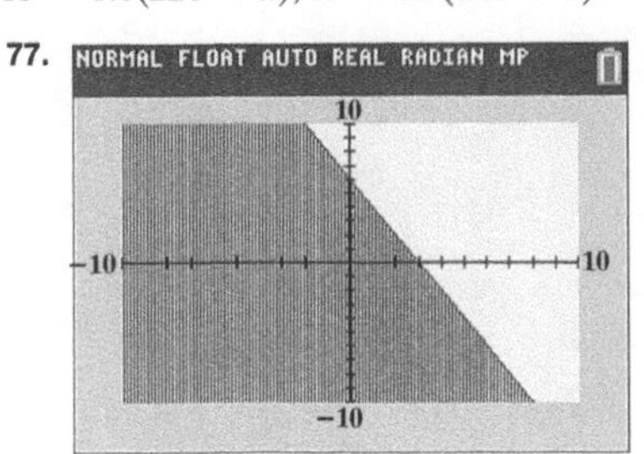

83. makes sense **85.** makes sense **87.** false **89.** true **91.** $\begin{cases} x \geq -2 \\ y > -1 \end{cases}$ **93.** $y \geq 2x - 2$

95. $y \geq nx + b$
$y \leq mx + b$

96. $\{(3, 1)\}$ **97.** $\{(2, 4)\}$ **98.** 165 **99. a.**

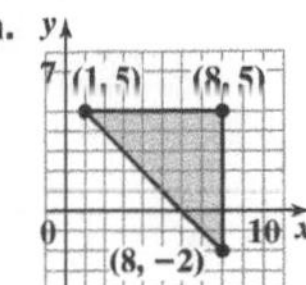

b. $(1, 5), (8, 5), (8, -2)$

c. at $(1, 5)$: 13; at $(8, 5)$: 34; at $(8, -2)$: 20

100. a.

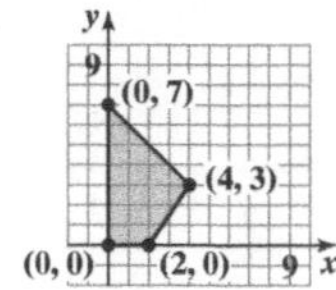

b. $(0, 0), (2, 0), (4, 3), (0, 7)$ **c.** at $(0, 0)$: 0; at $(2, 0)$: 4; at $(4, 3)$: 23; at $(0, 7)$: 35 **101.** $20x + 10y \leq 80{,}000$

Section 4.5 Check Point Exercises

1. $z = 25x + 55y$ **2.** $x + y \leq 80$ **3.** $30 \leq x \leq 80, 10 \leq y \leq 30$; $z = 25x + 55y$; $\begin{cases} x + y \leq 80 \\ 30 \leq x \leq 80 \\ 10 \leq y \leq 30 \end{cases}$

4. 50 bookshelves and 30 desks; \$2900 **5.** 30

Concept and Vocabulary Check

1. linear programming **2.** objective **3.** constraints; corner

Exercise Set 4.5

1. $(1, 2)$: 17; $(2, 10)$: 70; $(7, 5)$: 65; $(8, 3)$: 58; maximum: 70; minimum: 17 **3.** $(0, 0)$: 0; $(0, 8)$: 400; $(4, 9)$: 610; $(8, 0)$: 320; maximum: 610; minimum: 0

5. a.

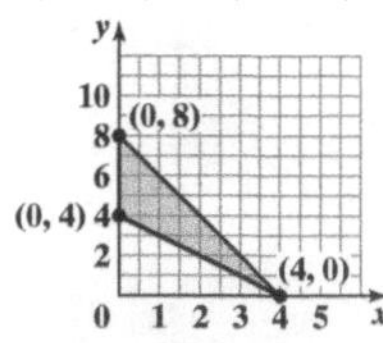

b. $(0, 4)$: 8; $(0, 8)$: 16; $(4, 0)$: 12
c. maximum: 16; at $(0, 8)$

7. a.

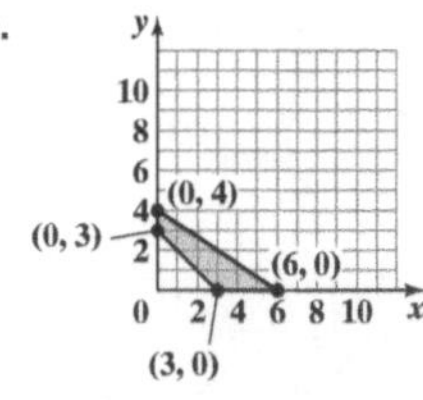

b. $(0, 3)$: 3; $(0, 4)$: 4; $(6, 0)$: 24; $(3, 0)$: 12
c. maximum: 24; at $(6, 0)$

9. a.

b. $(1, 2)$: -1; $(1, 4)$: -5; $(5, 8)$: -1; $(5, 2)$: 11
c. maximum: 11; at $(5, 2)$

11. a.

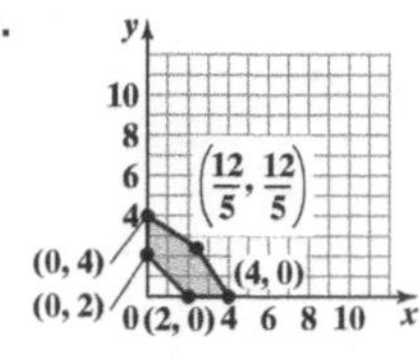

b. $(0, 2)$: 4; $(0, 4)$: 8; $\left(\frac{12}{5}, \frac{12}{5}\right)$: $\frac{72}{5} = 14.4$; $(4, 0)$: 16; $(2, 0)$: 8
c. maximum: 16; at $(4, 0)$

13. a.

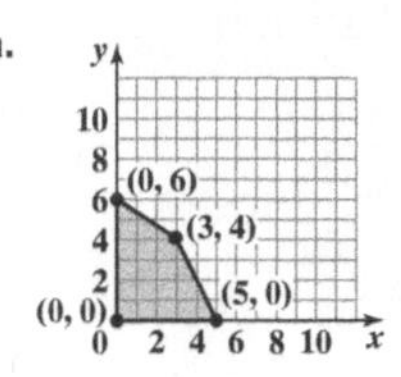

b. $(0, 0)$: 0; $(0, 6)$: 72; $(3, 4)$: 78; $(5, 0)$: 50
c. maximum: 78; at $(3, 4)$

15. a. $z = 125x + 200y$
b. $\begin{cases} x \leq 450 \\ y \leq 200 \\ 600x + 900y \leq 360{,}000 \end{cases}$
c.
d. 0; 40,000; 77,500; 76,250; 56,250
e. 300; 200; 77,500

17. 40 of model A and 0 of model B **19.** 300 boxes of food and 200 boxes of clothing **21.** 100 parents and 50 students **23.** 10 Boeing 727s and 42 Falcon 20s **29.** does not make sense **31.** makes sense **33.** \$5000 in stocks and \$5000 in bonds **35.** $54x^7y^{15}$ **36.** $L = \frac{12P + W}{2}$ **37.** 10 **38.** $4x^3 + 9x^2 - 13x - 3$ **39.** $5x^3 - 2x^2 + 12x - 15$ **40. a.** g **b.** f

Review Exercises

1. $[-2, \infty)$ **2.** $\left[\frac{3}{5}, \infty\right)$ **3.** $\left(-\infty, -\frac{21}{2}\right)$

4. $(-3, \infty)$

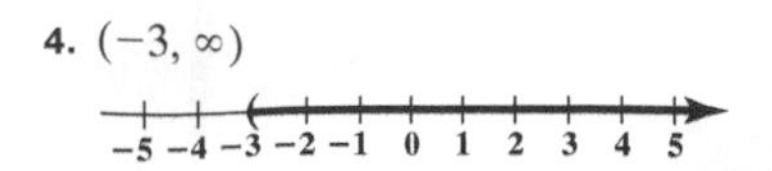

5. $(-\infty, -2]$

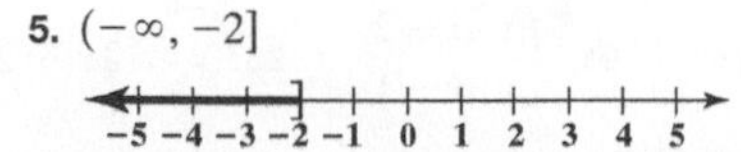

6. $\varnothing$

7. more than 50 checks **8.** more than \$13,500 in sales **9.** $\{a, c\}$ **10.** $\{a\}$ **11.** $\{a, b, c, d, e\}$ **12.** $\{a, b, c, d, f, g\}$

13. $(-\infty, 3]$

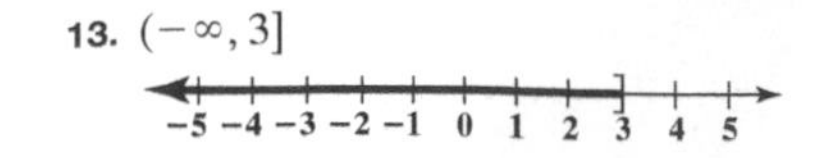

14. $(-\infty, 6)$

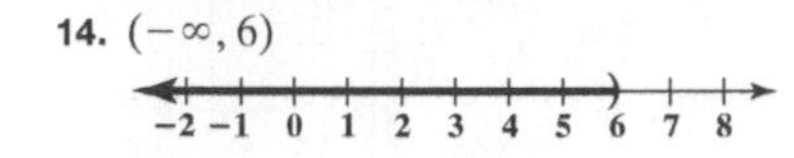

15. $(6, 8)$

16. $(-\infty, 1]$

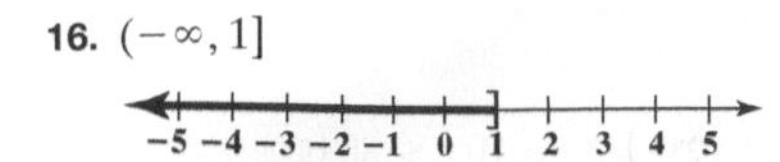

17. $\varnothing$

18. $(-\infty, 1) \cup (2, \infty)$

19. $(-\infty, -4] \cup (2, \infty)$

20. $(-\infty, -2)$

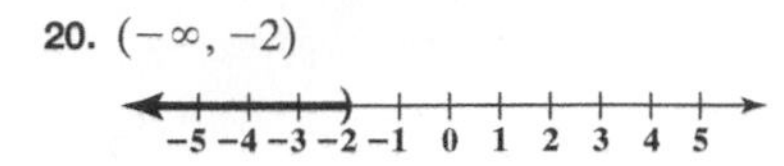

21. $(-\infty, \infty)$

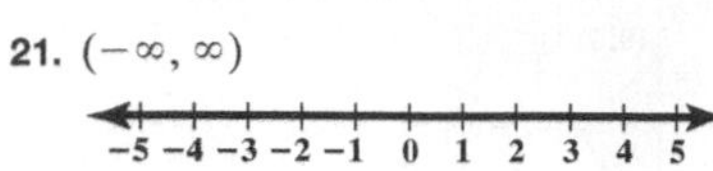

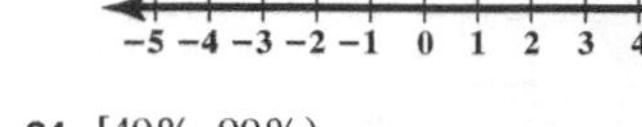

22. $(-5, 2]$

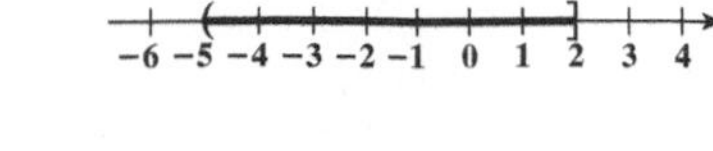

23. $\left[-\frac{3}{4}, 1\right]$

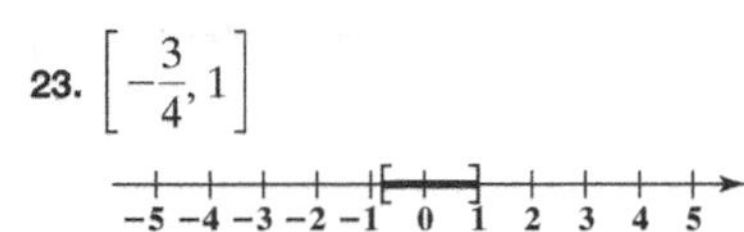

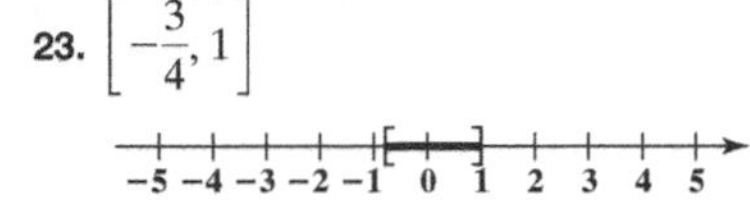

24. $[49\%, 99\%)$

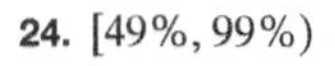

25. $\{-4, 3\}$ **26.** $\varnothing$ **27.** $\left\{-\frac{11}{2}, \frac{23}{2}\right\}$ **28.** $\left\{-4, -\frac{6}{11}\right\}$

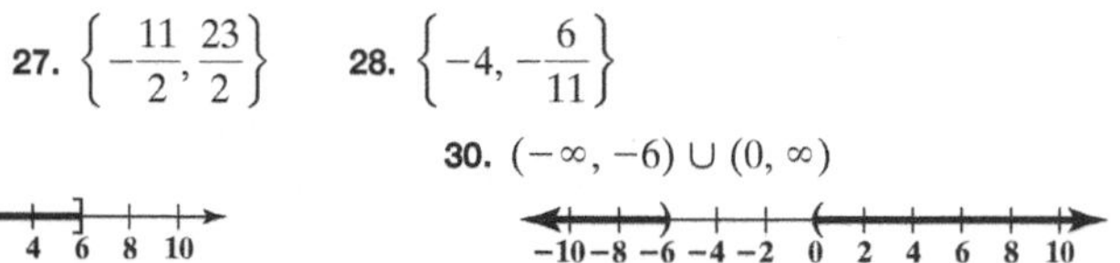

29. $[-9, 6]$

30. $(-\infty, -6) \cup (0, \infty)$

31. $(-3, -2)$

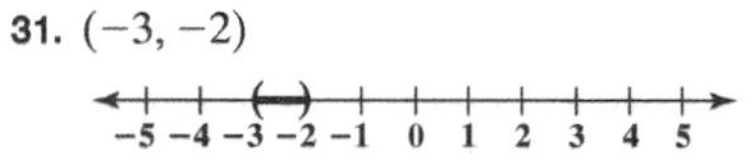

32. $(-\infty, -5] \cup [1, \infty)$;

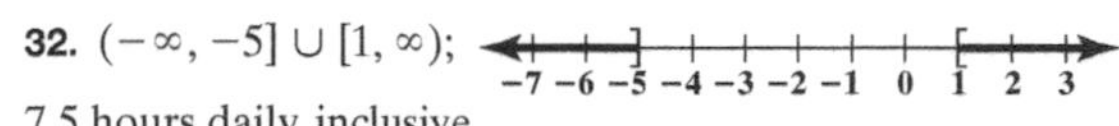

33. $\varnothing$ **34.** Approximately 90% of the population sleeps between 5.5 hours and 7.5 hours daily, inclusive.

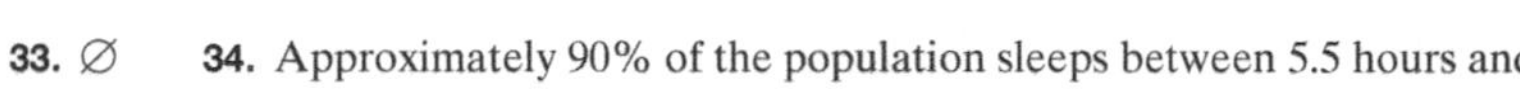

35. $3x - 4y > 12$

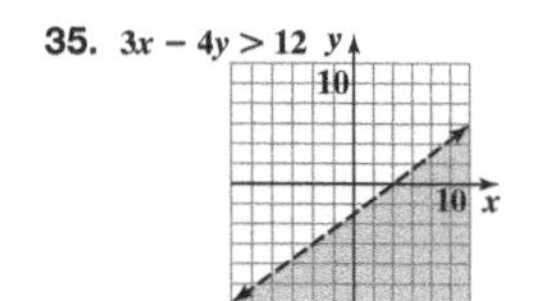

36. $x - 3y \le 6$

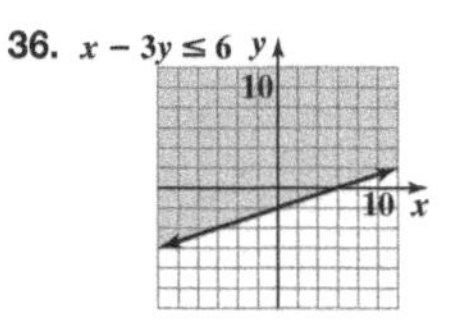

37. $y \le -\frac{1}{2}x + 2$

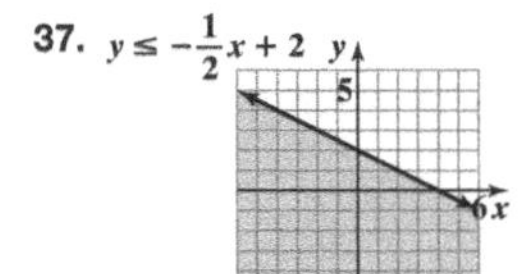

38. $y > \frac{3}{5}x$

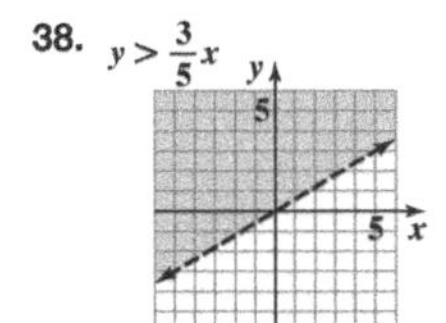

39. $x \le 2$

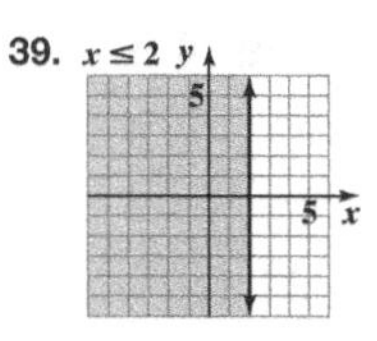

40. $y > -3$

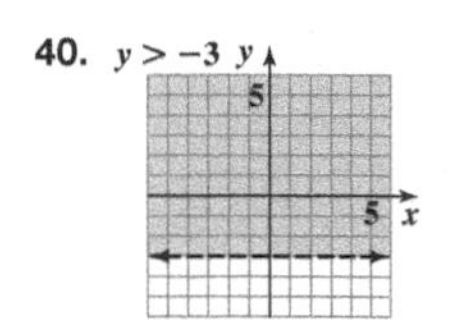

41. $2x - y \le 4$, $x + y \ge 5$

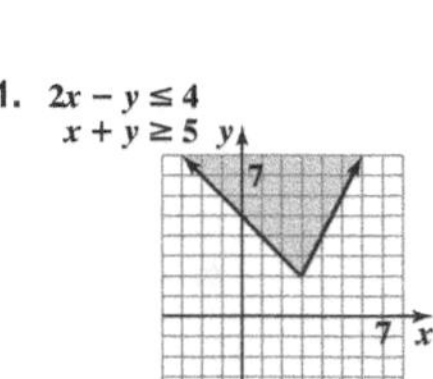

42. $y < -x + 4$, $y > x - 4$

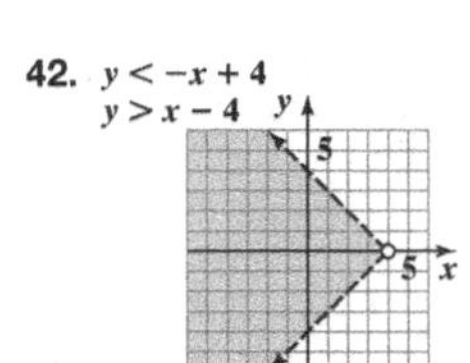

43. $-3 \le x < 5$

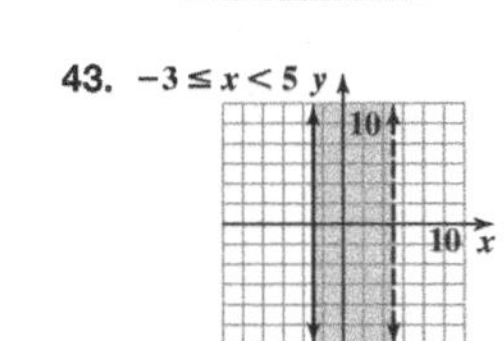

44. $-2 < y \le 6$

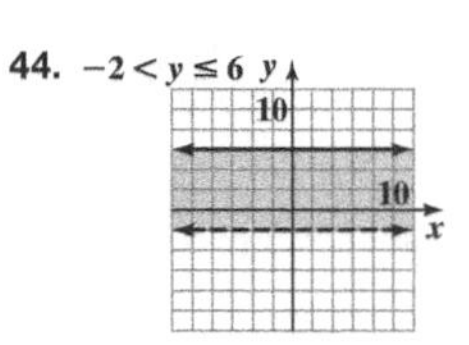

45. $x \ge 3$, $y \le 0$

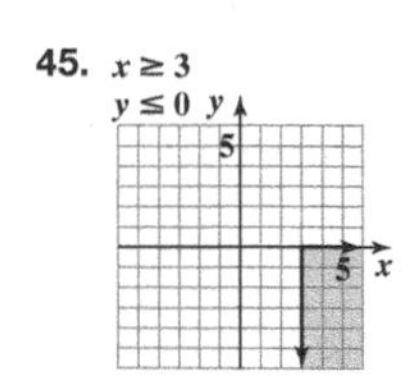

46. $2x - y > -4$, $x \ge 0$

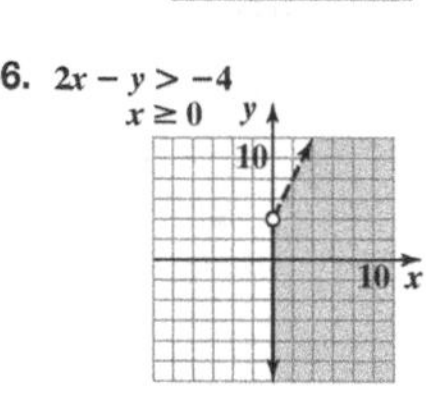

47. $x + y \le 6$, $y \ge 2x - 3$

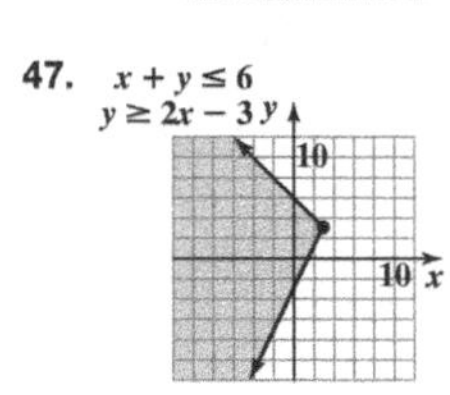

48. $3x + 2y \ge 4$, $x - y \le 3$, $x \ge 0, y \ge 0$

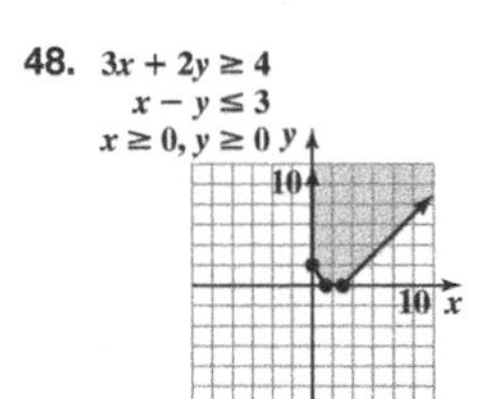

49. $\varnothing$

50. $\left(\frac{1}{2}, \frac{1}{2}\right)$: $\frac{5}{2}$; (2, 2): 10; (4, 0): 8; (1, 0): 2; maximum: 10; minimum: 2

51.

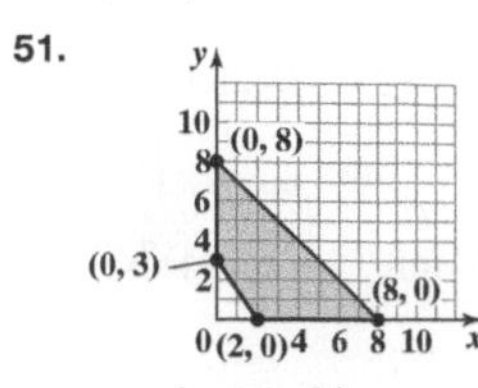

maximum: 24

52.

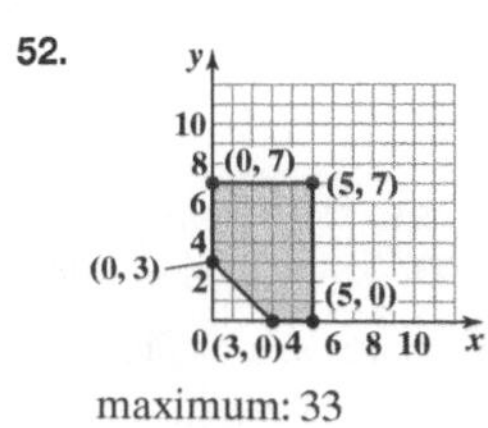

maximum: 33

53.

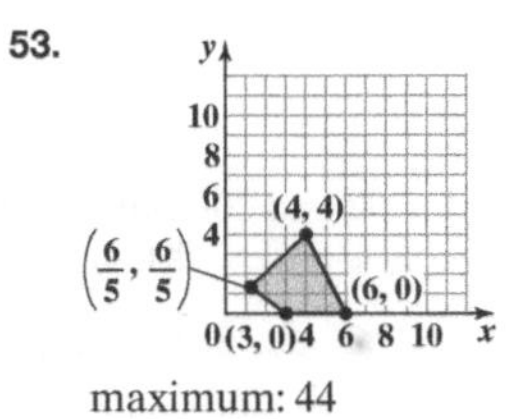

maximum: 44

54. a. $z = 500x + 350y$ **b.** $\begin{cases} x + y \le 200 \\ x \ge 10 \\ y \ge 80 \end{cases}$ **c.**

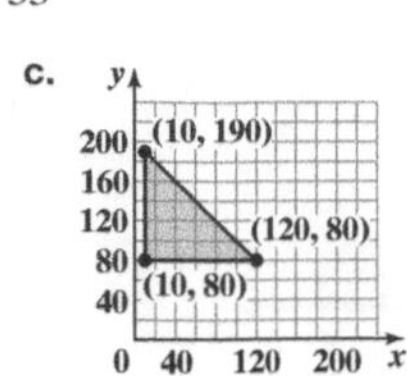

d. (10, 80): 33,000; (10, 190): 71,500; (120, 80): 88,000

e. 120; 80; 88,000

55. 480 of model A and 240 of model B

Chapter Test

1. $(-\infty, 12]$ **2.** $\left[\frac{21}{8}, \infty\right)$ **3.** more than 200 minutes **4.** $\{4, 6\}$ **5.** $\{2, 4, 6, 8, 10, 12, 14\}$ **6.** $(-2, -1)$ **7.** $[-2, \infty)$ **8.** $(-\infty, 4)$ **9.** $(-\infty, -4] \cup (2, \infty)$ **10.** $\left[-7, \frac{13}{2}\right)$ **11.** $\left\{-2, \frac{4}{5}\right\}$ **12.** $\left\{-\frac{8}{5}, 7\right\}$ **13.** $(-3, 4)$ **14.** $(-\infty, -1] \cup [4, \infty)$ **15.** $(-\infty, 90.6) \cup (106.6, \infty)$; Hypothermia: Body temperature below 90.6°F; Hyperthermia: Body temperature above 106.6°F

16. $3x - 2y < 6$

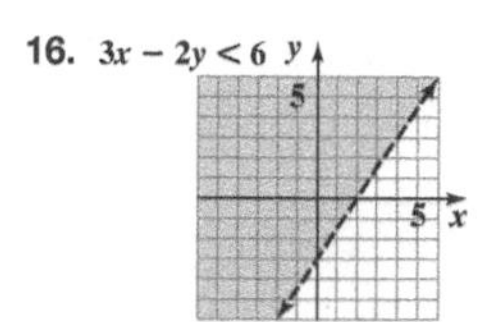

17. $y \geq \frac{1}{2}x - 1$ **18.** $y \leq -1$ **19.** $x + y \geq 2$, $x - y \geq 4$ **20.** $3x + y \leq 9$, $2x + 3y \geq 6$, $x \geq 0$, $y \geq 0$

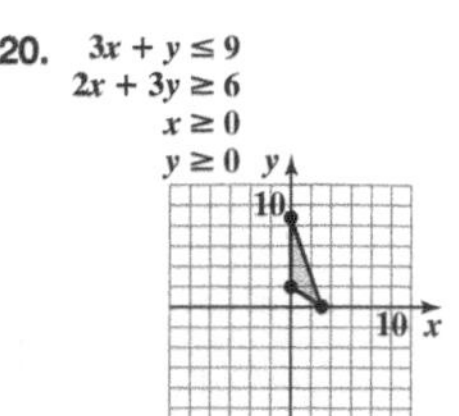

21. $-2 < x \leq 4$ **22.** maximum: 26 **23.** 50 regular and 100 deluxe; $35,000

Cumulative Review Exercises

1. $\{-1\}$ **2.** $\{8\}$ **3.** $-\frac{2y^7}{3x^5}$ **4.** 22; $4a^2 - 6a + 4$ **5.** $2x^2 + x + 2$; 12 **6.** $f(x) = -\frac{1}{2}x + 4$

7. $f(x) = 2x + 1$

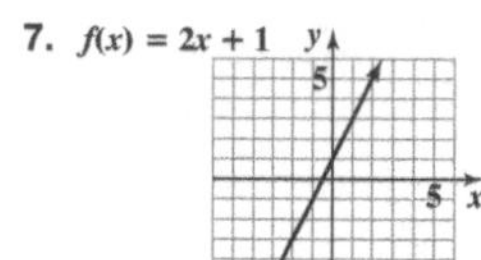

8. $y > 2x$

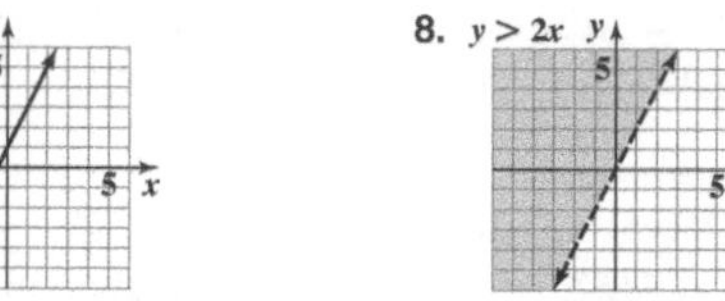

9. $2x - y \geq 6$

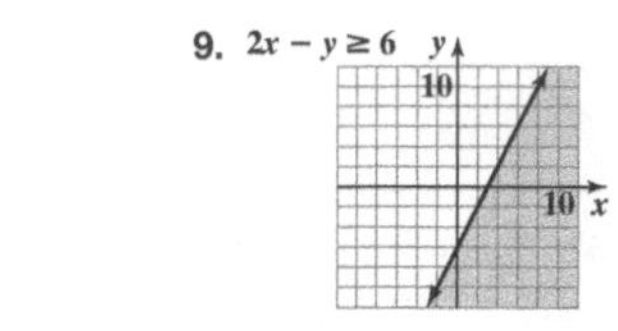

10. $f(x) = -1$

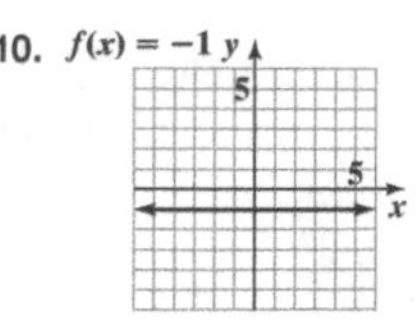

11. $\{(-4, 2, -1)\}$ **12.** $\{(-1, 2)\}$ **13.** -17 **14.** 46 rooms with kitchen facilities and 14 without kitchen facilities **15.** a. and b. are functions. **16.** $[-7, \infty)$ **17.** $(-\infty, -6)$ **18.** $(-\infty, 3] \cup [5, \infty)$

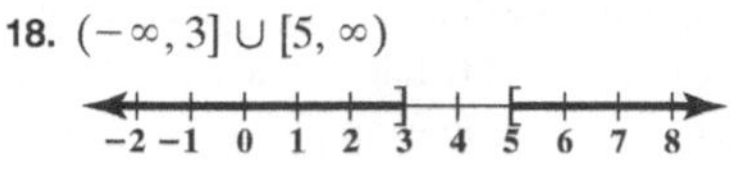

19. $[-10, 7]$ **20.** $\left(-\infty, \frac{1}{3}\right) \cup (5, \infty)$

CHAPTER 5

Section 5.1 Check Point Exercises

1.

Term	Coefficient	Degree
$8x^4y^5$	8	9
$-7x^3y^2$	−7	5
$-x^2y$	−1	3
$-5x$	−5	1
11	11	0

The degree of the polynomial is 9, the leading term is $8x^4y^5$, and the leading coefficient is 8.

2. 16 **3.** The graph rises to the left and to the right. **4.** This would not be appropriate over long time periods. Since the graph falls to the right, at some point the ratio would be negative, which is not possible. **5.** The graph does not show the end behavior of the function. The graph should fall to the left. **6.** $-3x^3 + 10x^2 - 10$ **7.** $9xy^3 + 3xy^2 - 15y - 9$ **8.** $10x^3 - 2x^2 + 8x - 10$ **9.** $13x^2y^5 + 2xy^3 - 10$

Concept and Vocabulary Check

1. whole **2.** standard **3.** monomial **4.** binomial **5.** trinomial **6.** n **7.** $n + m$ **8.** greatest; leading; leading **9.** true **10.** false **11.** end; leading **12.** falls; rises **13.** rises; falls **14.** rises; rises **15.** falls; falls **16.** true **17.** true **18.** like **19.** $-3x^3$ **20.** $-3x^3y$ **21.** $2x^5$ **22.** $10x^5y^2$ **23.** $12xy^2 - 12y^2$

Exercise Set 5.1

1.

Term	Coefficient	Degree
$-x^4$	-1	4
x^2	1	2

The degree of the polynomial is 4, the leading term is $-x^4$, and the leading coefficient is -1.

3.

Term	Coefficient	Degree
$5x^3$	5	3
$7x^2$	7	2
$-x$	-1	1
9	9	0

The degree of the polynomial is 3, the leading term is $5x^3$, and the leading coefficient is 5.

5.

Term	Coefficient	Degree
$3x^2$	3	2
$-7x^4$	-7	4
$-x$	-1	1
6	6	0

The degree of the polynomial is 4, the leading term is $-7x^4$, and the leading coefficient is -7.

7.

Term	Coefficient	Degree
x^3y^2	1	5
$-5x^2y^7$	-5	9
$6y^2$	6	2
-3	-3	0

The degree of the polynomial is 9, the leading term is $-5x^2y^7$, and the leading coefficient is -5.

9.

Term	Coefficient	Degree
x^5	1	5
$3x^2y^4$	3	6
$7xy$	7	2
$9x$	9	1
-2	-2	0

The degree of the polynomial is 6, the leading term is $3x^2y^4$, and the leading coefficient is 3.

11. 0 **13.** 12 **15.** 56 **17.** -29 **19.** -1
21–23. Graph #23 is not that of a polynomial function.
25. falls to the left and falls to the right; graph (b)
27. rises to the left and rises to the right; graph (a)
29. $11x^3 + 7x^2 - 12x - 4$ **31.** $-\frac{2}{5}x^4 + x^3 + \frac{3}{8}x^2$
33. $9x^2y - 6xy$ **35.** $2x^2y + 15xy + 15$
37. $-9x^4y^2 - 6x^2y^2 - 5x^2y + 2xy$ **39.** $5x^{2n} - 2x^n - 6$
41. $12x^3 + 4x^2 + 12x - 14$ **43.** $22y^5 + 9y^4 + 7y^3 - 13y^2 + 3y - 5$
45. $-5x^3 + 8xy - 9y^2$ **47.** $x^4y^2 + 8x^3y + y - 6x$
49. $y^{2n} + 2y^n - 3$ **51.** $8a^2b^4 + 3ab^2 + 8ab$ **53.** $5x^3 + 3x^2y - xy^2 - 4y^3$ **55.** $5x^4 - x^3 + 5x^2 - 5x + 2$ **57.** $-10x^2y^2 + 4x^2 + 3$ **59.** $-4x^3 - x^2 + 4x + 8; 7$ **61.** $-8x^2 - 2x - 1; -29$ **63.** $-9x^3 - x^2 + 13x + 20$ **65. a.** 3167; The world tiger population in 2010 was approximately 3167.; (40, 3167) **b.** underestimates by 33 **67.** rises to the right; no; The model indicates an increasing world tiger population but the tiger population will actually decrease without conservation efforts. **69.** falls to the right; The number of viral particles eventually decreases as the days increase. **71.** falls to the right; Eventually the elk population will be extinct. **85.** Answers will vary; an example is $f(x) = -x^2 - x + 1$. **87.** Answers will vary; an example is $f(x) = x^3 + x + 1$.

89.
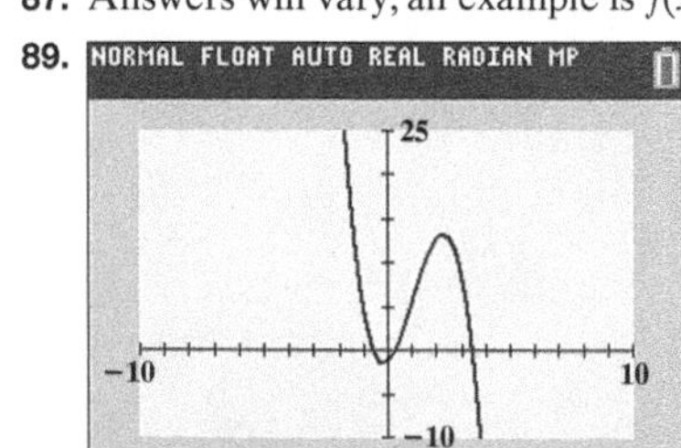

91.
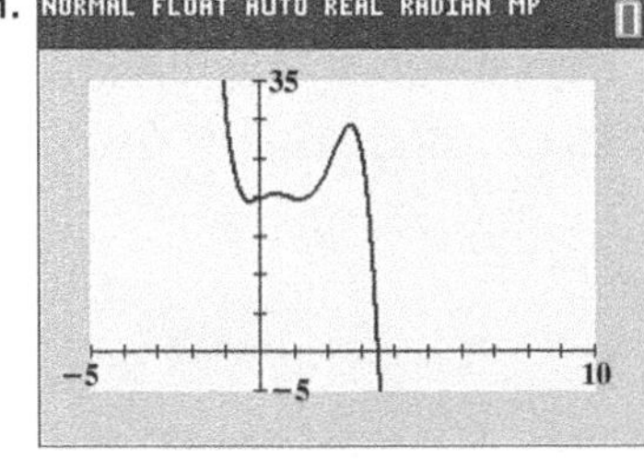

93.
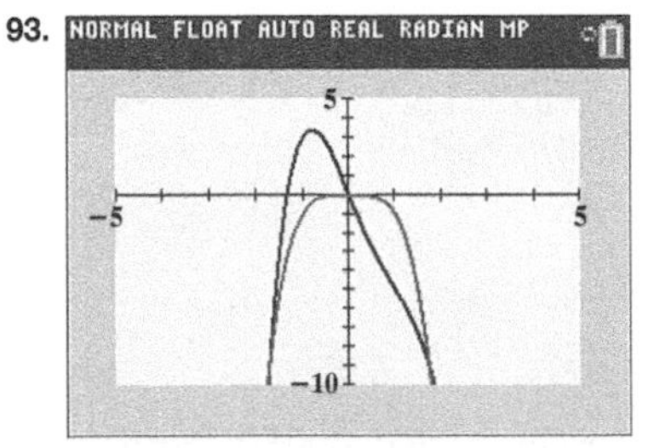

95. does not make sense **97.** makes sense **99.** false **101.** true **103.** $10y^{3n} - 6y^{2n} + 5$ **105.** $\left\{\frac{2}{3}\right\}$

106. $2x - 3y < -6$ **107.** $y - 5 = 3(x + 2)$; $y = 3x + 11$ or $f(x) = 3x + 11$ **108.** $10x^7y^9$ **109.** $16x^8 + 6x^5$ **110.** $3x^3 + 19x^2 + 43x + 35$

Section 5.2 Check Point Exercises

1. a. $-18x^7y^{11}$ **b.** $30x^{10}y^6z^8$ **2. a.** $12x^9 - 18x^6 + 24x^4$ **b.** $10x^5y^9 - 8x^7y^7 - 10x^4y^3$ **3.** $6x^3 - 2x^2 - x + 2$ **4.** $12x^2y^6 - 8x^2y^4 + 6xy^5 + 2y^2$ **5. a.** $x^2 + 8x + 15$ **b.** $14x^2 + xy - 4y^2$ **c.** $4x^6 - 12x^4 - 5x^3 + 15x$ **6. a.** $x^2 + 16x + 64$ **b.** $16x^2 + 40xy + 25y^2$ **7. a.** $x^2 - 10x + 25$ **b.** $4x^2 - 24xy^4 + 36y^8$ **8. a.** $x^2 - 9$ **b.** $25x^2 - 49y^2$ **c.** $25a^2b^4 - 16a^2$ **9. a.** $9x^2 + 12x + 4 - 25y^2$ **b.** $4x^2 + 4xy + y^2 + 12x + 6y + 9$ **10. a.** $x^2 - 10x + 21$ **b.** 5 **11. a.** $a^2 + a - 2$ **b.** $2ah + h^2 - 5h$

Concept and Vocabulary Check

1. add **2.** distributive; $4x^5 - 8x^2 + 6$; $7x^3$ **3.** $5x$; 3; like **4.** $3x^2$; $5x$; $21x$; 35 **5.** $A^2 + 2AB + B^2$; squared; product of the terms; squared **6.** $A^2 - 2AB + B^2$; minus; product of the terms; plus **7.** $A^2 - B^2$; minus **8.** x; $a + h$

Exercise Set 5.2

1. $15x^6$ **3.** $15x^3y^{11}$ **5.** $-6x^2y^9z^9$ **7.** $2x^{3n}y^{n-2}$ **9.** $12x^3 + 8x^2$ **11.** $2y^3 - 10y^2$ **13.** $10x^8 - 20x^5 + 45x^3$ **15.** $28x^2y + 12xy^2$ **17.** $18a^3b^5 + 15a^2b^3$ **19.** $-12x^6y^3 + 28x^3y^4 - 24x^2y$ **21.** $-12x^{3n} + 20x^{2n} - 2x^{n+1}$ **23.** $x^3 - x^2 - x - 15$ **25.** $x^3 - 1$ **27.** $a^3 - b^3$ **29.** $x^4 + 5x^3 + x^2 - 11x + 4$ **31.** $x^3 - 4x^2y + 4xy^2 - y^3$ **33.** $x^3y^3 + 8$ **35.** $x^2 + 11x + 28$ **37.** $y^2 - y - 30$ **39.** $10x^2 + 11x + 3$ **41.** $6y^2 - 11y + 4$ **43.** $15x^2 - 22x + 8$ **45.** $2x^2 + xy - 21y^2$ **47.** $14x^2y^2 - 19xy - 3$ **49.** $x^3 - 4x^2 - 5x + 20$ **51.** $8x^5 - 40x^3 + 3x^2 - 15$ **53.** $3x^{2n} + 5x^ny^n - 2y^{2n}$ **55.** $x^2 + 6x + 9$ **57.** $y^2 - 10y + 25$ **59.** $4x^2 + 4xy + y^2$ **61.** $25x^2 - 30xy + 9y^2$ **63.** $4x^4 + 12x^2y + 9y^2$ **65.** $16x^2y^4 - 8x^2y^3 + x^2y^2$ **67.** $a^{2n} + 8a^nb^n + 16b^{2n}$ **69.** $x^2 - 16$ **71.** $25x^2 - 9$ **73.** $16x^2 - 49y^2$ **75.** $y^6 - 4$ **77.** $1 - y^{10}$ **79.** $49x^2y^4 - 100y^2$ **81.** $25a^{2n} - 49$ **83.** $4x^2 + 12x + 9 - 16y^2$ **85.** $x^2 + 2xy + y^2 - 9$ **87.** $25x^2 + 70xy + 49y^2 - 4$ **89.** $25y^2 - 4x^2 - 12x - 9$ **91.** $x^2 + 2xy + y^2 + 2x + 2y + 1$ **93.** $x^4 - 1$ **95. a.** $x^2 + 4x - 12$ **b.** -15 **c.** -12 **97. a.** $x^3 - 27$ **b.** -35 **c.** -27 **99. a.** $a^2 + a + 5$ **b.** $2ah + h^2 - 3h$ **101. a.** $3a^2 + 14a + 15$ **b.** $6ah + 3h^2 + 2h$ **103.** $48xy$ **105.** $-9x^2 + 3x + 9$ **107.** $16x^4 - 625$ **109.** $x^3 - 3x^2 + 3x - 1$ **111.** $(2x - 7)^2 = 4x^2 - 28x + 49$ **113. a.** $x^2 + 6x + 4x + 24$ **b.** $(x + 6)(x + 4) = x^2 + 10x + 24$ **115. a.** $x^2 + 12x + 27$ **b.** $x^2 + 6x + 5$ **c.** $6x + 22$ **117. a.** $4x^2 - 36x + 80$ **b.** $4x^3 - 36x^2 + 80x$ **119. a.** $V(x) = -2x^3 + 10x^2 + 300x$ **b.** rises to the left and falls to the right **c.** no; Because the graph falls to the right, volume will eventually be negative, which is not possible. **d.** 2000; Carry-on luggage with a depth of 10 inches has a volume of 2000 cubic inches. **e.** (10, 2000) **f.** (0, 15), although answers may vary.

131.

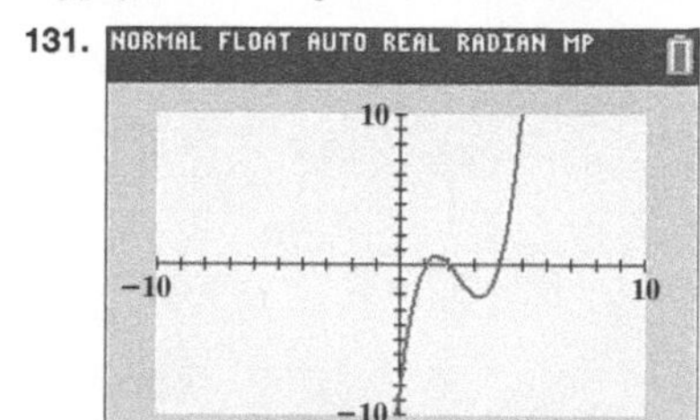

conclusion: $y_1 = y_2$

133.

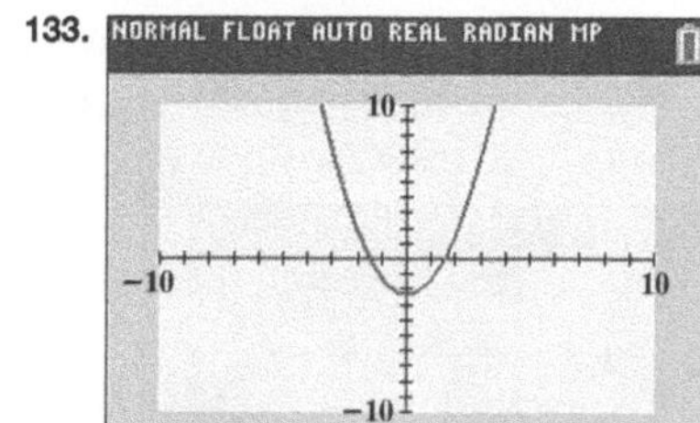

conclusion: $y_1 = y_2$

135. makes sense **137.** makes sense, although answers may vary **139.** false **141.** false **143.** $x^2 + 2x$ **145.** $2x^3 + 12x^2 + 12x + 10$ **147.** 9 and 11 **148.** $\left(-\infty, -\frac{14}{3}\right] \cup [2, \infty)$ **149.** $[-3, \infty)$ **150.** 8.034×10^9 **151. a.** $3x^2$ **b.** $6x^2y^2$ **152.** $x^3 - 5x^2 + 3x - 15$ **153.** $3x^3 - 2xy + 12x - 8y$

Section 5.3 Check Point Exercises

1. $10x(2x + 3)$ **2. a.** $3x^2(3x^2 + 7)$ **b.** $5x^3y^2(3 - 5xy)$ **c.** $4x^2y^3(4x^2y^2 - 2xy + 1)$ **3.** $-2x(x^2 - 5x + 3)$ **4. a.** $(x - 4)(3 + 7a)$ **b.** $(a + b)(7x - 1)$ **5.** $(x - 4)(x^2 + 5)$ **6.** $(x + 5)(4x - 3y)$

Concept and Vocabulary Check

1. factoring **2.** greatest common factor; smallest/least **3.** false **4.** $-2x$ **5.** false

Exercise Set 5.3

1. $2x(5x + 2)$ **3.** $y(y - 4)$ **5.** $x^2(x + 5)$ **7.** $4x^2(3x^2 - 2)$ **9.** $2x^2(16x^2 + x + 4)$ **11.** $2xy(2xy^2 + 3)$ **13.** $10xy^2(3xy - 1)$ **15.** $2x(6y - 3z + 2w)$ **17.** $3x^2y^4(5xy^2 - 3x^2 + 4y)$ **19.** $5x^2y^4z^2(5xy^2 - 3x^2z^2 + 5yz)$ **21.** $5x^n(3x^n - 5)$ **23.** $-4(x - 3)$ **25.** $-8(x + 6)$ **27.** $-2(x^2 - 3x + 7)$ **29.** $-5(y^2 - 8x)$ **31.** $-4x(x^2 - 8x + 5)$ **33.** $-1(x^2 + 7x - 5)$ **35.** $(x + 3)(4 + a)$ **37.** $(y - 6)(x - 7)$ **39.** $(x + y)(3x - 1)$ **41.** $(3x - 1)(4x^2 + 1)$ **43.** $(x + 3)(2x + 1)$ **45.** $(x + 3)(x + 5)$ **47.** $(x + 7)(x - 4)$ **49.** $(x - 3)(x^2 + 4)$ **51.** $(y - 6)(x + 2)$ **53.** $(y + 1)(x - 7)$ **55.** $(5x - 6y)(2x + 7y)$ **57.** $(4x - 1)(x^2 - 3)$ **59.** $(x - a)(x - b)$ **61.** $(x - 3)(x^2 + 4)$ **63.** $(a - b)(y - x)$ **65.** $(a + 2b)(y^2 - 3x)$ **67.** $(x^n + 1)(y^n + 3)$ **69.** $(a + 1)(b - c)$ **71.** $(x^3 - 5)(1 + 4y)$ **73.** $y^6(3x - 1)^4(6xy - 2y - 7)$ **75.** $(x^2 + 5x - 2)(a + b)$ **77.** $(x + y + z)(a - b + c)$ **79. a.** 16; The ball is 16 feet above the ground after 2 seconds. **b.** 0; The ball is on the ground after 2.5 seconds. **c.** $-8t(2t - 5)$; $f(t) = -8t(2t - 5)$ **d.** 16; 0; yes; no; Answers will vary. **81. a.** $(x - 0.4x)(1 - 0.4) = (0.6x)(0.6) = 0.36x$ **b.** no; 36% **83.** $A = P + Pr + (P + Pr)r = P(1 + r) + Pr(1 + r) = (1 + r)(P + Pr) = P(1 + r)(1 + r) = P(1 + r)^2$ **85.** $A = r(\pi r + 2l)$

93.

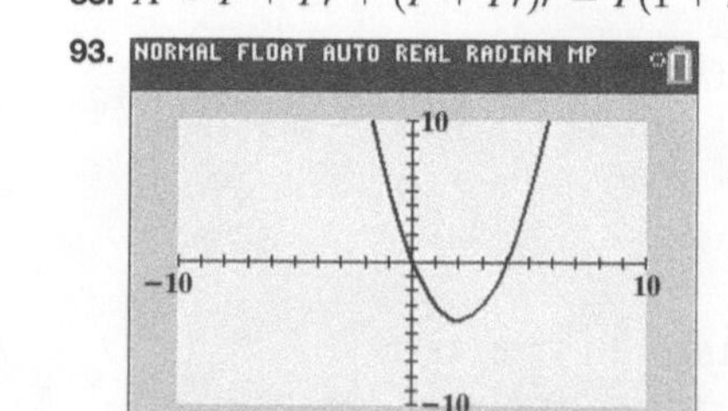

Graphs coincide.; factored correctly

95.

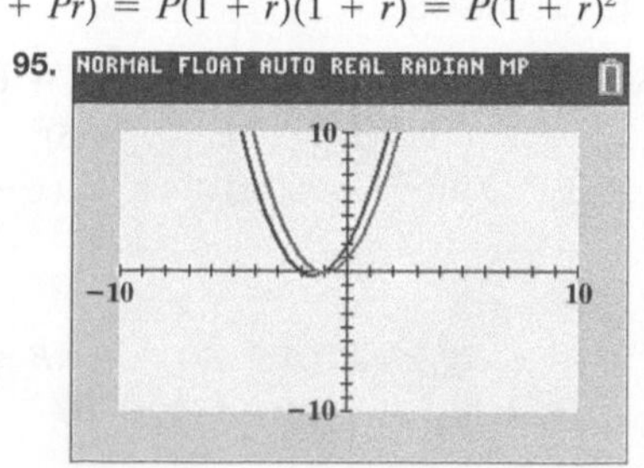

Graphs do not coincide.; $x^2 + 2x + x + 2 = (x + 2)(x + 1)$

97. does not make sense **99.** makes sense **101.** false **103.** true **105.** $x^{2n}(x^{2n} + 1 + x^n)$ **107.** $4y^{2n}(2y^4 + 4y^3 - 3)$
109. Answers will vary; an example is $6x^2 - 4x + 9x - 6$. **110.** $\left\{\left(\frac{20}{11}, -\frac{14}{11}\right)\right\}$ **111. a.** function **b.** not a function
112. length: 8 ft; width: 3 ft **113.** 4 **114.** 2 **115.** 7

Section 5.4 Check Point Exercises

1. $(x + 4)(x + 2)$ or $(x + 2)(x + 4)$ **2.** $(x - 5)(x - 4)$ **3.** $(y + 22)(y - 3)$ **4.** $(x - 3y)(x - 2y)$ **5.** $3x(x - 7)(x + 2)$
6. $(x^3 - 5)(x^3 - 2)$ **7.** $(3x - 14)(x - 2)$ **8.** $x^4(3x - 1)(2x + 7)$ **9.** $(2x - y)(x - 3y)$ **10.** $(3y^2 - 2)(y^2 + 4)$ **11.** $(2x - 5)(4x - 1)$

Concept and Vocabulary Check

1. completely **2.** greatest common factor **3.** $+5$ **4.** -4 **5.** $+16$ **6.** $-3y$ **7.** $2x$; $2x$; -18 **8.** -11 **9.** $2x + 9$ **10.** $-6y$

Exercise Set 5.4

1. $(x + 3)(x + 2)$ **3.** $(x + 6)(x + 2)$ **5.** $(x + 5)(x + 4)$ **7.** $(y + 8)(y + 2)$ **9.** $(x - 3)(x - 5)$ **11.** $(y - 2)(y - 10)$
13. $(a + 7)(a - 2)$ **15.** $(x + 6)(x - 5)$ **17.** $(x + 4)(x - 7)$ **19.** $(y + 4)(y - 9)$ **21.** prime **23.** $(x - 2y)(x - 7y)$
25. $(x + 5y)(x - 6y)$ **27.** prime **29.** $(a - 10b)(a - 8b)$ **31.** $3(x + 3)(x - 2)$ **33.** $2x(x - 3)(x - 4)$ **35.** $3y(y - 2)(y - 3)$
37. $2x^2(x + 3)(x + 16)$ **39.** $(x^3 + 2)(x^3 - 3)$ **41.** $(x^2 - 6)(x^2 + 1)$ **43.** $(x + 6)(x + 2)$ **45.** $(3x + 5)(x + 1)$ **47.** $(x + 11)(5x + 1)$
49. $(y + 8)(3y - 2)$ **51.** $(y + 2)(4y + 1)$ **53.** $(2x + 3)(5x + 2)$ **55.** $(4x - 3)(2x - 3)$ **57.** $(y - 3)(6y - 5)$ **59.** prime
61. $(x + y)(3x + y)$ **63.** $(2x + y)(3x - 5y)$ **65.** $(5x - 2y)(3x - 5y)$ **67.** $(3a - 7b)(a + 2b)$ **69.** $5x(3x - 2)(x - 1)$
71. $2x^2(3x + 2)(4x - 1)$ **73.** $y^3(5y + 1)(3y - 1)$ **75.** $3(8x + 9y)(x - y)$ **77.** $2b(a + 3)(3a - 10)$ **79.** $2y(2x - y)(3x - 7y)$
81. $13x^3y(y + 4)(y - 1)$ **83.** $(2x^2 - 3)(x^2 + 1)$ **85.** $(2x^3 + 5)(x^3 + 3)$ **87.** $(2y^5 + 1)(y^5 + 3)$ **89.** $(5x + 12)(x + 2)$
91. $(2x - 13)(x - 2)$ **93.** $(x - 0.3)(x - 0.2)$ **95.** $\left(x + \frac{3}{7}\right)\left(x - \frac{1}{7}\right)$ **97.** $(ax - b)(cx + d)$ **99.** $-x^3y^2(4x - 3y)(x - y)$
101. $f(x) = 3x - 13$ and $g(x) = x - 3$, or vice versa **103.** $2x + 1$ by $x + 3$ **105. a.** 32; The diver is 32 feet above the water after 1 second.
b. 0; The diver hits the water after 2 seconds. **c.** $-16(t - 2)(t + 1)$; $f(t) = -16(t - 2)(t + 1)$ **d.** 32; 0
107. a. $x^2 + x + x + x + 1 + 1 = x^2 + 3x + 2$ **b.** $(x + 2)(x + 1)$ **c.** Answers will vary.

117.

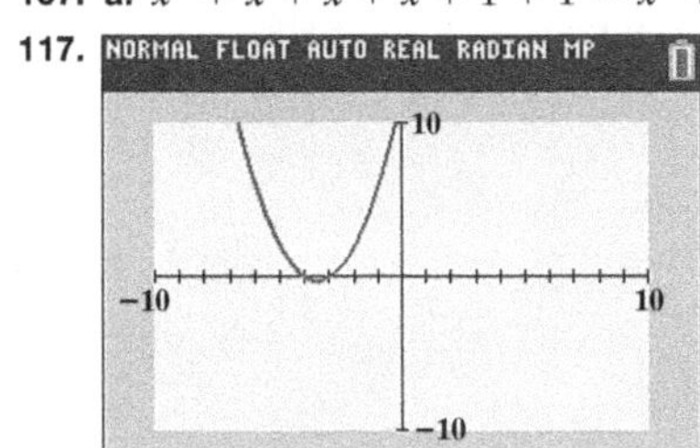

Graphs coincide.; factored correctly

119.

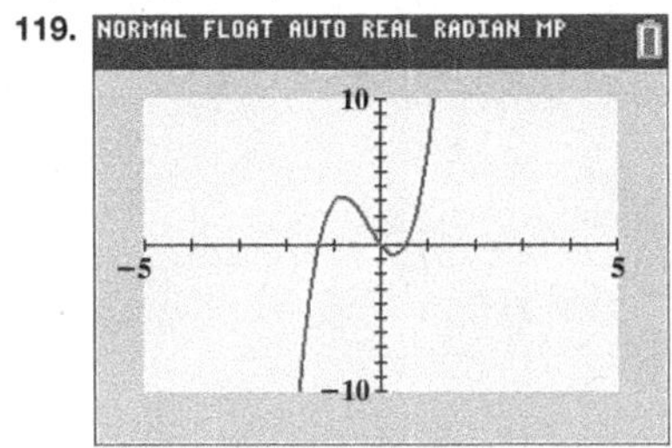

Graphs coincide.; factored correctly

123. makes sense **125.** makes sense **127.** true **129.** false **131.** $-16, -8, 8, 16$ **133.** $(4x^n - 5)(x^n - 1)$ **135.** $b^2(b^n - 2)(b^n + 5)$
137. $d^n(2d - 3)(d - 1)$ **138.** $(5, \infty)$ **139.** $\{(2, -1, 3)\}$ **140.** $(x + 2)(4x^2 - 5)$ **141.** $(x + 7)(x + 7)$ or $(x + 7)^2$
142. $(x - 4)(x - 4)$ or $(x - 4)^2$ **143.** $(x + 5)(x - 5)$

Mid-Chapter Check Point Exercises

1. $-x^3 + 4x^2 + 6x + 17$ **2.** $-2x^7y^3z^5$ **3.** $30x^5y^3 - 35x^3y^2 - 2x^2y$ **4.** $3x^3 + 4x^2 - 39x + 40$ **5.** $2x^4 - x^3 - 8x^2 + 11x - 4$
6. $-x^2 - 5x + 5$ **7.** $-4x^3y - 6x^2y - 7y - 5$ **8.** $8x^2 + 18x - 5$ **9.** $10x^2y^2 - 11xy - 6$ **10.** $9x^2 - 4y^2$ **11.** $6x^3y - 9xy^2 + 2x^2 - 3y$
12. $49x^6y^2 - 25x^2$ **13.** $6xh + 3h^2 - 2h$ **14.** $x^4 - 6x^2 + 9$ **15.** $x^5 + 2x^3 + 2x^2 - 15x - 6$ **16.** $4x^2 + 20xy + 25y^2$
17. $x^2 + 12x + 36 - 9y^2$ **18.** $x^2 + 2xy + y^2 + 10x + 10y + 25$ **19.** $(x - 8)(x + 3)$ **20.** $(5x + 2)(3y + 1)$
21. $(5x - 2)(x + 2)$ **22.** $5(7x^2 + 2x - 10)$ **23.** $9(x - 2)(x + 1)$ **24.** $5x^2y(2xy - 4y + 7)$ **25.** $(3x + 1)(6x + 5)$
26. $(4x - 3y)(3x - 4)$ **27.** $(3x - 4)(3x - 1)$ **28.** $(3x^3 + 5)(x^3 + 2)$ **29.** $x(5x - 2)(5x + 7)$ **30.** $(x^3 - 3)(2x - y)$

Section 5.5 Check Point Exercises

1. a. $(4x + 5)(4x - 5)$ **b.** $(10y^3 + 3x^2)(10y^3 - 3x^2)$ **2.** $6y(1 + xy^3)(1 - xy^3)$ **3.** $(4x^2 + 9)(2x + 3)(2x - 3)$
4. $(x + 7)(x + 2)(x - 2)$ **5. a.** $(x + 3)^2$ **b.** $(4x + 5y)^2$ **c.** $(2y^2 - 5)^2$ **6.** $(x + 5 + y)(x + 5 - y)$ **7.** $(a + b - 2)(a - b + 2)$
8. a. $(x + 3)(x^2 - 3x + 9)$ **b.** $(x^2 + 10y)(x^4 - 10x^2y + 100y^2)$ **9. a.** $(x - 2)(x^2 + 2x + 4)$ **b.** $(1 - 3xy)(1 + 3xy + 9x^2y^2)$

Concept and Vocabulary Check

1. $(A - B)(A + B)$ **2.** $(A + B)^2$ **3.** $(A - B)^2$ **4.** $(A + B)(A^2 - AB + B^2)$ **5.** $(A - B)(A^2 + AB + B^2)$ **6.** $4x$; $4x$
7. $(b + 3)$; $(b + 3)$ **8.** -7 **9.** $4x$ **10.** $+3$; $-3x$ **11.** -10; $+100$ **12.** false **13.** true **14.** false **15.** true **16.** false

Exercise Set 5.5

1. $(x + 2)(x - 2)$ **3.** $(3x + 5)(3x - 5)$ **5.** $(3 + 5y)(3 - 5y)$ **7.** $(6x + 7y)(6x - 7y)$ **9.** $(xy + 1)(xy - 1)$
11. $(3x^2 + 5y^3)(3x^2 - 5y^3)$ **13.** $(x^7 + y^2)(x^7 - y^2)$ **15.** $(x - 3 + y)(x - 3 - y)$ **17.** $(a + b - 2)(a - b + 2)$
19. $(x^n + 5)(x^n - 5)$ **21.** $(1 + a^n)(1 - a^n)$ **23.** $2x(x + 2)(x - 2)$ **25.** $2(5 + y)(5 - y)$ **27.** $8(x + y)(x - y)$
29. $2xy(x + 3)(x - 3)$ **31.** $a(ab + 7c)(ab - 7c)$ **33.** $5y(1 + xy^3)(1 - xy^3)$ **35.** $8(x^2 + y^2)$ **37.** prime
39. $(x^2 + 4)(x + 2)(x - 2)$ **41.** $(9x^2 + 1)(3x + 1)(3x - 1)$ **43.** $2x(x^2 + y^2)(x + y)(x - y)$ **45.** $(x + 3)(x + 2)(x - 2)$
47. $(x - 7)(x + 1)(x - 1)$ **49.** $(x + 2)^2$ **51.** $(x - 5)^2$ **53.** $(x^2 - 2)^2$ **55.** $(3y + 1)^2$ **57.** $(8y - 1)^2$ **59.** $(x - 6y)^2$
61. prime **63.** $(3x + 8y)^2$ **65.** $(x - 3 + y)(x - 3 - y)$ **67.** $(x + 10 + x^2)(x + 10 - x^2)$ **69.** $(3x - 5 + 6y)(3x - 5 - 6y)$
71. $(x^2 + x + 1)(x^2 - x - 1)$ **73.** $(z + x - 2y)(z - x + 2y)$ **75.** $(x + 4)(x^2 - 4x + 16)$ **77.** $(x - 3)(x^2 + 3x + 9)$
79. $(2y + 1)(4y^2 - 2y + 1)$ **81.** $(5x - 2)(25x^2 + 10x + 4)$ **83.** $(xy + 3)(x^2y^2 - 3xy + 9)$ **85.** $x(4 - x)(16 + 4x + x^2)$
87. $(x^2 + 3y)(x^4 - 3x^2y + 9y^2)$ **89.** $(5x^2 - 4y^2)(25x^4 + 20x^2y^2 + 16y^4)$ **91.** $(x + 1)(x^2 - x + 1)(x^6 - x^3 + 1)$
93. $(x - 2y)(x^2 - xy + y^2)$ **95.** $(0.2x + 0.3)^2$ or $\frac{1}{100}(2x + 3)^2$ **97.** $x\left(2x - \frac{1}{2}\right)\left(4x^2 + x + \frac{1}{4}\right)$
99. $(x - 1)(x^2 + x + 1)(x - 2)(x^2 + 2x + 4)$ **101.** $(x^4 + 1)(x^2 + 4)(x + 2)(x - 2)$ **103.** $(x + 1)(x - 1)(x - 2)(x^2 + 2x + 4)$
105. a. $(A + B)^2$ **b.** $A^2; AB; AB; B^2$ **c.** $A^2 + 2AB + B^2$ **d.** $A^2 + 2AB + B^2 = (A + B)^2$; factoring a perfect square trinomial
107. $25x^2 - 9 = (5x + 3)(5x - 3)$ **109.** $49x^2 - 36 = (7x + 6)(7x - 6)$ **111.** $3a^3 - 3ab^2 = 3a(a + b)(a - b)$

117. 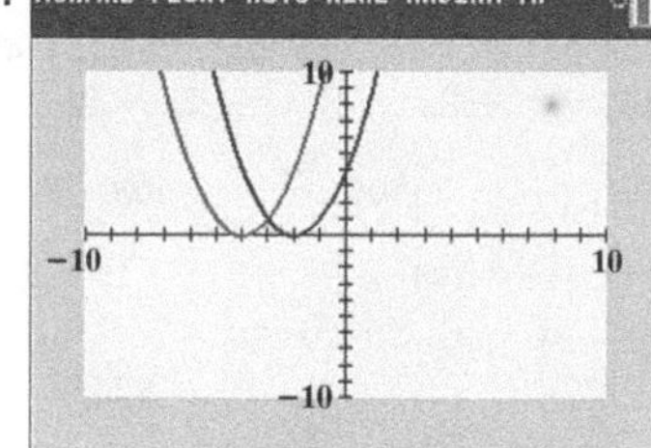

; Graphs do not conincide.; $x^2 + 4x + 4 = (x + 2)^2$

119. 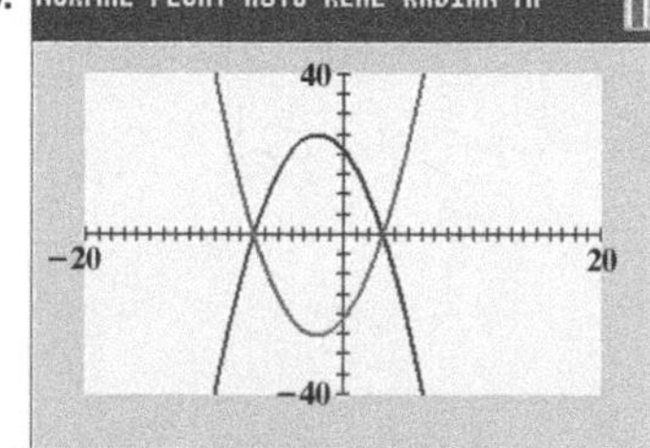

; Graphs do not coincide; $25 - (x^2 + 4x + 4) = (7 + x)(3 - x)$

121. 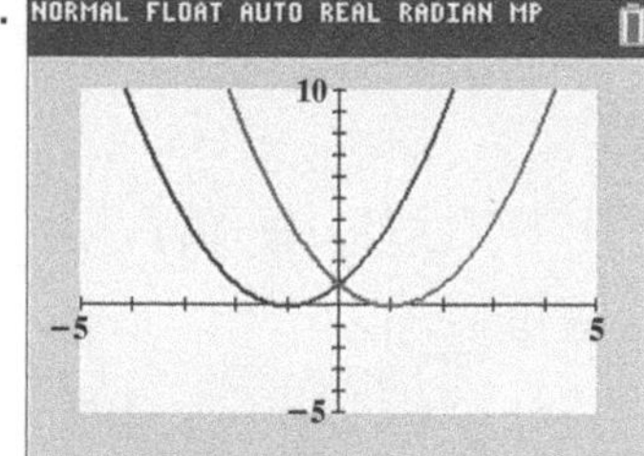

; Graphs do not coincide.; $(x - 3)^2 + 8(x - 3) + 16 = (x + 1)^2$

123. 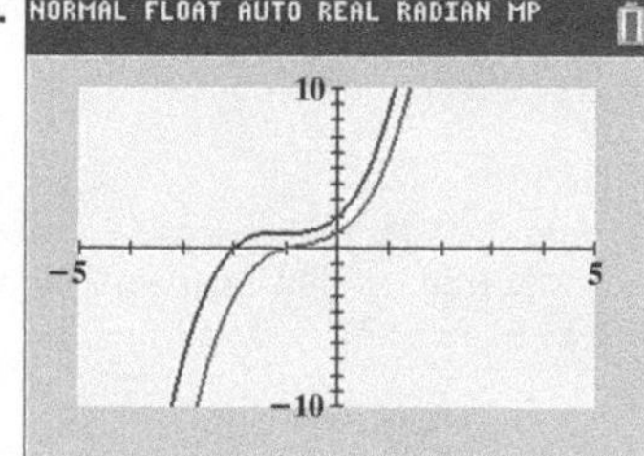

; Graphs do not coincide.; $(x + 1)^3 + 1 = (x + 2)(x^2 + x + 1)$

125. makes sense **127.** does not make sense **129.** false **131.** false **133.** $(y + x)(y^2 - xy + x^2 + 1)$ **135.** $(x^n + y^{4n})(x^{2n} - x^ny^{4n} + y^{8n})$
137. $x^6 - y^6 = (x^3 + y^3)(x^3 - y^3) = (x + y)(x^2 - xy + y^2)(x - y)(x^2 + xy + y^2)$;
$x^6 - y^6 = (x^2 - y^2)(x^4 + x^2y^2 + y^4) = (x + y)(x - y)(x^4 + x^2y^2 + y^4); x^4 + x^2y^2 + y^4 = (x^2 - xy + y^2)(x^2 + xy + y^2)$
139. 1 **140.** $[5, 11]$ **141.** $\{(-2, 1)\}$ **142.** $(x + 7)(3x - y)$ **143.** $2x(x + 2)^2$ **144.** $5x(x - y)(x - 7y)$
145. $(x + y)(3b + 4)(3b - 4)$

Section 5.6 Check Point Exercises

1. $3x(x - 5)^2$ **2.** $3y(x + 2)(x - 6)$ **3.** $(x + y)(4a + 5)(4a - 5)$ **4.** $(x + 10 + 6a)(x + 10 - 6a)$ **5.** $x(x + 2)(x^2 - 2x + 4)(x^6 - 8x^3 + 64)$

Concept and Vocabulary Check

1. b **2.** e **3.** h **4.** c **5.** d **6.** f **7.** a **8.** g

Exercise Set 5.6

1. $x(x + 4)(x - 4)$ **3.** $3(x + 3)^2$ **5.** $3(3x - 1)(9x^2 + 3x + 1)$ **7.** $(x + 4)(x - 4)(y - 2)$ **9.** $2b(2a + 5)(a - 3)$
11. $(y + 2)(y - 2)(a - 4)$ **13.** $11x(x^2 + y)(x^2 - y)$ **15.** $4x(x^2 + 4)(x + 2)(x - 2)$ **17.** $(x - 4)(x + 3)(x - 3)$
19. $2x^2(x + 3)(x^2 - 3x + 9)$ **21.** $3y(x^2 + 4y^2)(x + 2y)(x - 2y)$ **23.** $3x(2x + 3y)^2$ **25.** $(x - 6 + 7y)(x - 6 - 7y)$ **27.** prime
29. $12xy(x + y)(x - y)$ **31.** $6b(x^2 + y^2)$ **33.** $(x + y)(x - y)(x^2 + xy + y^2)$ **35.** $(x + 6 + 2a)(x + 6 - 2a)$ **37.** $(x^3 - 2)(x^3 + 7)$
39. $(2x - 7)(2 + x^2)$ **41.** $2(3x - 2y)(9x^2 + 6xy + 4y^2)$ **43.** $(x + 5 + y)(x + 5 - y)$ **45.** $(x^4 + y^4)(x^2 + y^2)(x + y)(x - y)$
47. $xy(x + 4y)(x - 4y)$ **49.** $x(1 + 2x)(1 - 2x + 4x^2)$ **51.** $2(4y + 1)(2y - 1)$ **53.** $y(14y^2 + 7y - 10)$ **55.** $3(3x + 2y)^2$
57. $3x(4x^2 + y^2)$ **59.** $x^3y^3(xy - 1)(x^2y^2 + xy + 1)$ **61.** $2(x + 5)(x - 5)$ **63.** $(x - y)(a + 2)(a - 2)$
65. $(c + d - 1)(c + d + 1)[(c + d)^2 + 1]$ **67.** $(p + q)^2(p - q)$ **69.** $(x + 2y)(x - 2y)(x + y)(x - y)$
71. $(x + y + 3)^2$ **73.** $(x - y)^2(x - y + 2)(x - y - 2)$ **75.** $(2x - y^2)(x - 3y^2)$ **77.** $(x - y)(x^2 + xy + y^2 - 1)$
79. $(xy + 1)(x^2y^2 - xy + 1)(x - 2)(x^2 + 2x + 4)$ **81.** **a.** $x(x + y) - y(x + y)$ **b.** $(x + y)(x - y)$
83. **a.** $xy + xy + xy + 3x(x) = 3xy + 3x^2$ **b.** $3x(y + x)$ **85.** **a.** $8x^2 - 2\pi x^2$ **b.** $2x^2(4 - \pi)$

89.

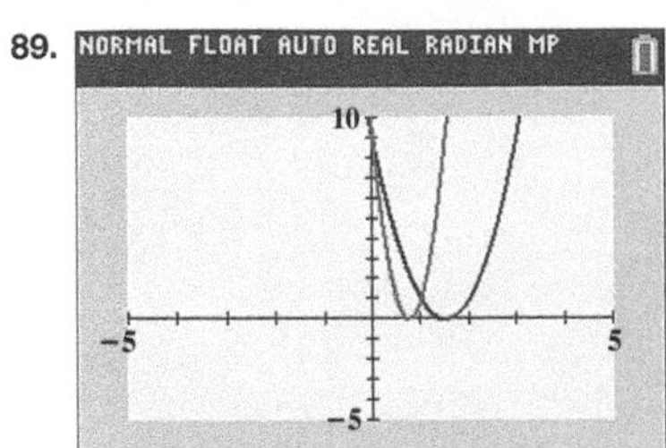

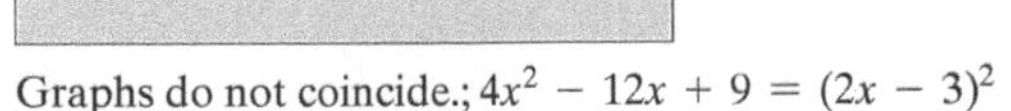
Graphs do not coincide.; $4x^2 - 12x + 9 = (2x - 3)^2$

91.

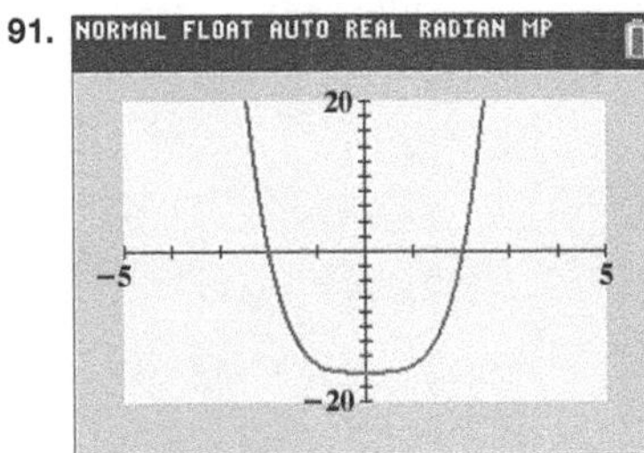

Graphs coincide.; factored correctly

95. makes sense
97. does not make sense **99.** true
101. false **103.** $x^n(3x - 1)(x - 4)$
105. $(x^2 + 2x + 2)(x^2 - 2x + 2)$
107. $\{-1\}$ **108.** $\dfrac{8x^9}{y^3}$ **109.** 52
110. 0 **111.** 0
112. $(x - 5)(x + 3) = 0$

Section 5.7 Check Point Exercises

1. $-\frac{1}{2}$ and 5, or $\left\{-\frac{1}{2}, 5\right\}$ **2.** **a.** 0 and $\frac{2}{3}$, or $\left\{0, \frac{2}{3}\right\}$ **b.** 5 or $\{5\}$ **c.** -4 and 3, or $\{-4, 3\}$ **3.** $-2, -\frac{3}{2}$, and 2, or $\left\{-2, -\frac{3}{2}, 2\right\}$
4. after 3 seconds; (3, 336) **5.** 2 ft **6.** 5 ft

Concept and Vocabulary Check

1. quadratic **2.** $A = 0$ or $B = 0$ **3.** x-intercepts **4.** subtracting $20x$ **5.** subtracting $8x$; adding 12
6. polynomial; 0; descending; highest/greatest **7.** right; hypotenuse; legs **8.** right; legs; the square of the length of the hypotenuse

Exercise Set 5.7

1. $\{-4, 3\}$ **3.** $\{-7, 1\}$ **5.** $\left\{-4, \frac{2}{3}\right\}$ **7.** $\left\{\frac{3}{5}, 1\right\}$ **9.** $\left\{-2, \frac{1}{3}\right\}$ **11.** $\{0, 8\}$ **13.** $\left\{0, \frac{5}{3}\right\}$ **15.** $\{-2\}$ **17.** $\{7\}$ **19.** $\left\{\frac{5}{3}\right\}$
21. $\{-5, 5\}$ **23.** $\left\{-\frac{10}{3}, \frac{10}{3}\right\}$ **25.** $\{-3, 6\}$ **27.** $\{-3, -2\}$ **29.** $\{4\}$ **31.** $\{-2, 8\}$ **33.** $\left\{-1, \frac{2}{5}\right\}$ **35.** $\{-6, -3\}$
37. $\{-5, -4, 5\}$ **39.** $\{-5, 1, 5\}$ **41.** $\{-4, 0, 4\}$ **43.** $\{0, 2\}$ **45.** $\{-5, -3, 0\}$ **47.** 2 and 4; d **49.** -4 and -2; c **51.** $\{-7, -1, 6\}$
53. $\left\{-\frac{4}{3}, 0, \frac{4}{5}, 2\right\}$ **55.** -4 and 8 **57.** $-2, -\frac{1}{2}$, and 2 **59.** -7 and 4 **61.** 2 and 3 **63.** -9 and 5 **65.** 1 second; 0.25, 0.5, 0.75, 1, 1.25
67. 7 **69.** (7, 21) **71.** length: 9 ft; width: 6 ft **73.** 5 in. **75.** 5 m **77.** **a.** $4x^2 + 44x$ **b.** 3 ft **79.** length: 10 in.; width: 10 in.
81. length: 12 ft; width: 5 ft **83.** $30\frac{1}{8}$ ft

97.

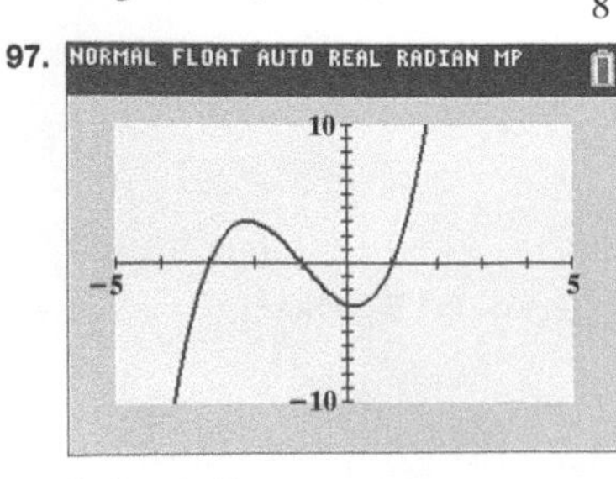

$\{-3, -1, 1\}$

99.

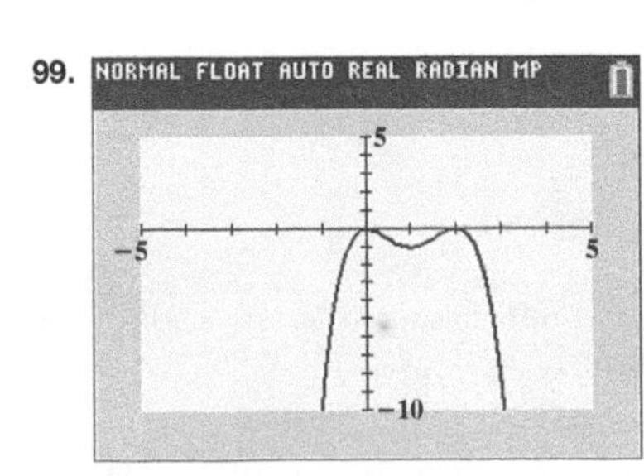

$\{0, 2\}$

101. makes sense **103.** makes sense
105. false **107.** true **108.** Answers will vary; an example is $x^2 - 4x - 21 = 0$.
111. $\left\{-2, \frac{10}{3}\right\}$ **112.** 12
113. \$1700 at 5% and \$1300 at 8%
114. 30 **115.** $(-\infty, 2) \cup (2, \infty)$
116. $\dfrac{(x - 9)(x + 2)}{(2x - 1)(x + 2)}; \dfrac{x - 9}{2x - 1}$

Review Exercises

1.

Term	Coefficient	Degree
$-5x^3$	-5	3
$7x^2$	7	2
$-x$	-1	1
2	2	0

The degree of the polynomial is 3, the leading term is $-5x^3$, and the leading coefficient is -5.

2.

Term	Coefficient	Degree
$8x^4y^2$	8	6
$-7xy^6$	-7	7
$-x^3y$	-1	4

The degree of the polynomial is 7, the leading term is $-7xy^6$, and the leading coefficient is -7.

3. -31 **4. a.** 284; There were 284,000 record daily high temperatures in the United States in the 2000s. **b.** underestimates by 6000 **5.** rises to the left and falls to the right; c **6.** rises to the left and rises to the right; b **7.** falls to the left and rises to the right; a **8.** falls to the left and falls to the right; d **9. a.** rises to the right **b.** no; The model indicates increasing deforestation despite a declining rate at which the forest is being cut down. **c.** falls to the right **d.** no; The model indicates that the amount of forest cleared, in square kilometers, will eventually be negative, which is not possible. **10.** $x^3 - 6x^2 - x - 9$ **11.** $12x^3y - 2x^2y - 14y - 17$ **12.** $15x^3 - 10x^2 + 11x - 8$ **13.** $-2x^3y^2 - 4x^3y - 8$ **14.** $3x^3 + 5x^2y - xy^2 - 8y^3$ **15.** $-12x^6y^2z^7$ **16.** $2x^8 - 24x^5 - 12x^3$ **17.** $21x^5y^4 - 35x^2y^3 - 7xy^2$ **18.** $6x^3 + 29x^2 + 27x - 20$ **19.** $x^4 + 4x^3 + 4x^2 - x - 2$ **20.** $12x^2 - 23x + 5$ **21.** $15x^2y^2 + 2xy - 8$ **22.** $9x^2 + 42xy + 49y^2$ **23.** $x^4 - 10x^2y + 25y^2$ **24.** $4x^2 - 49y^2$ **25.** $9x^2y^4 - 16x^2$ **26.** $x^2 + 6x + 9 - 25y^2$ **27.** $x^2 + 2xy + y^2 + 8x + 8y + 16$ **28.** $2x^2 - x - 15$; 21 **29. a.** $a^2 - 9a + 10$ **b.** $2ah + h^2 - 7h$ **30.** $8x^2(2x + 3)$ **31.** $2x(1 - 18x)$ **32.** $7xy(3xy - 2y + 1)$ **33.** $9x^2y(2xy - 3)$ **34.** $-4(3x^2 - 2x + 12)$ **35.** $-1(x^2 + 11x - 14)$ **36.** $(x - 1)(x^2 - 2)$ **37.** $(y - 3)(x - 5)$ **38.** $(x - 3y)(5a + 2b)$ **39.** $(x + 5)(x + 3)$ **40.** $(x + 20)(x - 4)$ **41.** $(x + 17y)(x - y)$ **42.** $3x(x - 1)(x - 11)$ **43.** $(x + 7)(3x + 1)$ **44.** $(3x - 2)(2x - 3)$ **45.** $(5x + 4y)(x - 2y)$ **46.** $x(3x + 4)(2x - 1)$ **47.** $(x + 3)(2x + 5)$ **48.** $(x^3 + 6)(x^3 - 5)$ **49.** $(x^2 + 3)(x^2 - 13)$ **50.** $(x + 11)(x + 9)$ **51.** $(5x^3 + 2)(x^3 + 3)$ **52.** $(2x + 5)(2x - 5)$ **53.** $(1 + 9xy)(1 - 9xy)$ **54.** $(x^4 + y^3)(x^4 - y^3)$ **55.** $(x - 1 + y)(x - 1 - y)$ **56.** $(x + 8)^2$ **57.** $(3x - 1)^2$ **58.** $(5x + 2y)^2$ **59.** prime **60.** $(5x - 4y)^2$ **61.** $(x + 9 + y)(x + 9 - y)$ **62.** $(z + 5x - 1)(z - 5x + 1)$ **63.** $(4x + 3)(16x^2 - 12x + 9)$ **64.** $(5x - 2)(25x^2 + 10x + 4)$ **65.** $(xy + 1)(x^2y^2 - xy + 1)$ **66.** $3x(5x + 1)$ **67.** $3x^2(2x + 1)(2x - 1)$ **68.** $4x(5x^3 - 6x^2 + 7x - 3)$ **69.** $x(x - 2)(x - 13)$ **70.** $-2y^2(y - 3)(y - 9)$ **71.** $(3x - 5)^2$ **72.** $5(x + 3)(x - 3)$ **73.** $(2x - 1)(x + 3)(x - 3)$ **74.** $(3x - y)(2x - 7y)$ **75.** $2y(y + 3)^2$ **76.** $(x + 3 + 2a)(x + 3 - 2a)$ **77.** $(2x - 3)(4x^2 + 6x + 9)$ **78.** $x(x^2 + 1)(x + 1)(x - 1)$ **79.** $(x^2 - 3)^2$ **80.** prime **81.** $4(a + 2)(a^2 - 2a + 4)$ **82.** $(x^2 + 9)(x + 3)(x - 3)$ **83.** $(a + 3b)(x - y)$ **84.** $(3x - 5y)(9x^2 + 15xy + 25y^2)$ **85.** $2xy(x + 3)(5x - 4)$ **86.** $(2x^3 + 5)(3x^3 - 1)$ **87.** $(x + 5)(2 + xy)$ **88.** $(y + 2)(y + 5)(y - 5)$ **89.** $(a^4 + 1)(a^2 + 1)(a + 1)(a - 1)$ **90.** $(x - 4)(3 + y)(3 - y)$ **91. a.** $2xy + 2y^2$ **b.** $2y(x + y)$ **92. a.** $x^2 - 4y^2$ **b.** $(x + 2y)(x - 2y)$ **93.** $\{-5, -1\}$ **94.** $\left\{\frac{1}{3}, 7\right\}$ **95.** $\{-8, 7\}$ **96.** $\{0, 4\}$ **97.** $\{-5, -3, 3\}$ **98.** 9 seconds **99. a.** 20 miles per hour **b.** (20, 40) **c.** As a car's speed increases, its stopping distance gets longer at increasingly greater rates. **100.** length: 9 ft; width: 6 ft **101.** 2 in. **102.** 50 yd, 120 yd, and 130 yd

Chapter Test

1. degree: 3; leading coefficient: -6 **2.** degree: 9; leading coefficient: 7 **3.** 6; 4 **4.** falls to the left and falls to the right **5.** falls to the left and rises to the right **6.** $7x^3y - 18x^2y - y - 9$ **7.** $11x^2 - 13x - 6$ **8.** $35x^7y^3$ **9.** $x^3 - 4x^2y + 2xy^2 + y^3$ **10.** $21x^2 - 20xy - 9y^2$ **11.** $4x^2 - 25y^2$ **12.** $16y^2 - 56y + 49$ **13.** $x^2 + 4x + 4 - 9y^2$ **14.** $3x^2 + x - 10$; 60 **15.** $2ah + h^2 - 5h$ **16.** $x^2(14x - 15)$ **17.** $(9y + 5)(9y - 5)$ **18.** $(x + 3)(x + 5)(x - 5)$ **19.** $(5x - 3)^2$ **20.** $(x + 5 + 3y)(x + 5 - 3y)$ **21.** prime **22.** $(y + 2)(y - 18)$ **23.** $(2x + 5)(7x + 3)$ **24.** $5(x - 1)(x^2 + x + 1)$ **25.** $3(2x + y)(2x - y)$ **26.** $2(3x - 1)(2x - 5)$ **27.** $3(x^2 + 1)(x + 1)(x - 1)$ **28.** $(x^4 + y^4)(x^2 + y^2)(x + y)(x - y)$ **29.** $4x^2y(3y^3 + 2xy - 9)$ **30.** $(x^3 + 2)(x^3 - 14)$ **31.** $(x^2 - 6)(x^2 + 4)$ **32.** $3y(4x - 1)(x - 2)$ **33.** $y(y - 3)(y^2 + 2)$ **34.** $\left\{-\frac{1}{3}, 2\right\}$ **35.** $\left\{-1, \frac{6}{5}\right\}$ **36.** $\left\{0, \frac{1}{3}\right\}$ **37.** $\{-1, 1, 4\}$ **38.** 7 seconds **39.** length: 5 yd; width: 3 yd **40.** 12 units, 9 units, and 15 units

Cumulative Review Exercises

1. $\{4\}$ **2.** $\left\{\left(-4, \frac{1}{2}\right)\right\}$ **3.** $\{(2, 1, -1)\}$ **4.** (2, 3) **5.** $(-\infty, -2] \cup [7, \infty)$ **6.** $\left\{1, \frac{5}{2}\right\}$ **7.** $\{-5, 0, 2\}$ **8.** $x = \frac{b}{c - a}$ **9.** $f(x) = 2x + 1$ **10.** winner: 1480 votes; loser: 1320 votes

11. $f(x) = -\frac{1}{3}x + 1$ **12.** $4x - 5y < 20$ **13.** $y \le -1$ **14.** $-\frac{y^{10}}{2x^6}$ **15.** 7.06×10^{-5} **16.** $9x^4 - 6x^2y + y^2$ **17.** $9x^4 - y^2$ **18.** $(x + 3)(x - 3)^2$ **19.** $x^2(x^2 + 1)(x + 1)(x - 1)$ **20.** $14x^3y^2(1 - 2x)$

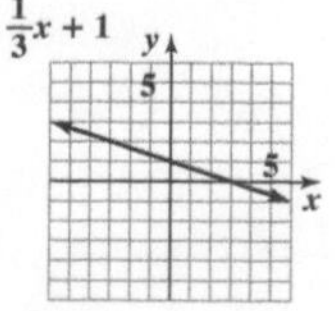

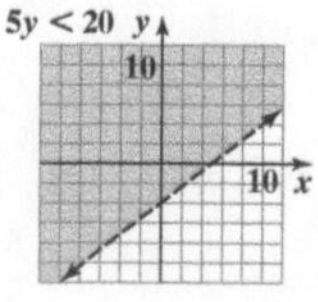

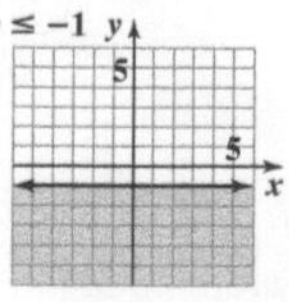

CHAPTER 6

Section 6.1 Check Point Exercises

1. a. 80; The cost to remove 40% of the lake's pollutants is \$80 thousand.; (40, 80) **b.** 180; The cost to remove 60% of the lake's pollutants is \$180 thousand.; (60, 180) **2.** $(-\infty, -3) \cup \left(-3, \frac{1}{2}\right) \cup \left(\frac{1}{2}, \infty\right)$ **3.** $x + 5$ **4. a.** $\frac{x-5}{3x-1}$ **b.** $\frac{x+4y}{3x(x+y)}$ **5.** $\frac{x+3}{x-4}$ **6.** $\frac{-4(x+2)}{3x(3x-2)}$ or $-\frac{4(x+2)}{3x(3x-2)}$ **7. a.** $9(3x+7)$ **b.** $\frac{x+3}{5}$

Concept and Vocabulary Check

1. polynomial; polynomial **2.** zero **3.** asymptote **4.** asymptote **5.** factoring; common factors **6.** $x + 5$ **7.** false **8.** -1 **9.** numerators; denominators **10.** multiplicative inverse/reciprocal; $\frac{S}{R}$; $\frac{PS}{QR}$ **11.** $\frac{x^2}{70}$ **12.** $\frac{10}{7}$

Exercise Set 6.1

1. -5; -3; 2 **3.** 0; does not exist; $-\frac{21}{2}$ **5.** $-\frac{7}{2}$; -5; $\frac{11}{5}$ **7.** $(-\infty, 5) \cup (5, \infty)$ **9.** $(-\infty, -3) \cup (-3, 1) \cup (1, \infty)$ **11.** $(-\infty, -5) \cup (-5, \infty)$ **13.** $(-\infty, 3) \cup (3, 5) \cup (5, \infty)$ **15.** $\left(-\infty, -\frac{4}{3}\right) \cup \left(-\frac{4}{3}, 2\right) \cup (2, \infty)$ **17.** 4 **19.** domain: $(-\infty, -2) \cup (-2, 2) \cup (2, \infty)$; range: $(-\infty, 0] \cup (3, \infty)$ **21.** As x decreases, the function values are approaching 3.; $y = 3$ **23.** There is no point on the graph with x-coordinate -2. **25.** The graph is not continuous. Furthermore, it neither rises nor falls without bound to either the left or the right. **27.** $x + 2$ **29.** $\frac{1}{x-3}$ **31.** $\frac{4}{x}$ **33.** $\frac{4}{y+5}$ **35.** $-\frac{1}{3x+5}$ **37.** $\frac{y+7}{y-7}$ **39.** $\frac{x+9}{x-1}$ **41.** cannot be simplified **43.** $-\frac{x+3}{x+4}$ **45.** $\frac{x+5y}{3x-y}$ **47.** $\frac{x^2+2x+4}{x+2}$ **49.** $x^2 - 3$ **51.** $\frac{3}{2}$ **53.** $\frac{x+7}{x}$ **55.** $\frac{x+3}{x-2}$ **57.** $\frac{x+4}{x+2}$ **59.** $-\frac{2}{y(y+3)}$ **61.** $\frac{y^2+2y+4}{2y}$ **63.** $\frac{x^2+x+1}{x-2}$ **65.** 4 **67.** $\frac{4(x+y)}{3(x-y)}$ **69.** 1 **71.** 1 **73.** $\frac{9}{28}$ **75.** $\frac{7}{10}$ **77.** $\frac{x}{3}$ **79.** $\frac{y-5}{2}$ **81.** $\frac{(x^2+4)(x-4)}{x-1}$ **83.** $\frac{y-7}{y-5}$ **85.** $(x-1)(2x-1)$ **87.** $\frac{1}{x-2y}$ **89.** $3x^2$ **91.** $\frac{x+1}{x(x-1)}$ **93.** $\frac{-(x-y)}{(x+y)(b+c)}$ **95.** $\frac{a^2+1}{b(a+2)}$ **97.** $-\frac{a-b}{4c^2}$ *or* $\frac{b-a}{4c^2}$ **99.** 7 **101.** $2a + h - 5$ **103.** $\left(\frac{f}{g}\right)(x) = -x - 2$; $(-\infty, -2) \cup \left(-2, \frac{1}{2}\right) \cup \left(\frac{1}{2}, \infty\right)$ **105.** 195; The cost to inoculate 60% of the population is \$195 million.; (60, 195) **107.** 100; This indicates that we cannot inoculate 100% of the population. **109.** after 6 minutes; about 4.8 **111.** 6.5; Over time the pH level rises back to normal. **113.** 90; Lung cancer has an incidence ratio of 10 between the ages of 55 and 64. 90% of deaths from lung cancer in this group are smoking related.; (10, 90) **115.** $y = 100$; As the incidence ratio increases, the percentage of smoking-related deaths is approaching 100%.

129.

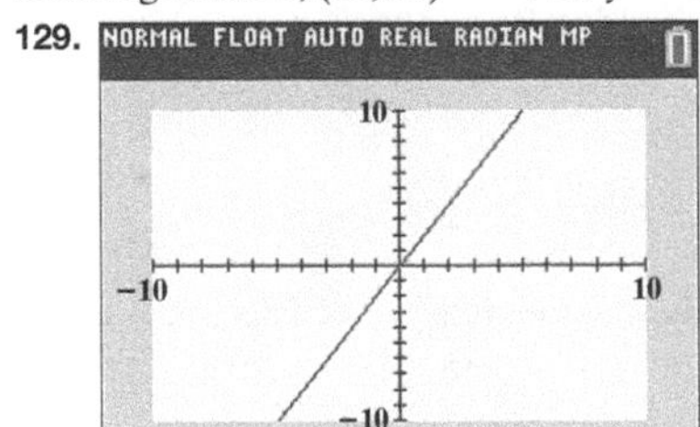

Graphs coincide.; multiplied correctly

131. NORMAL FLOAT AUTO REAL RADIAN MP

Graphs do not coincide; right side should be $x + 3$.

135. makes sense **137.** does not make sense **139.** false **141.** false

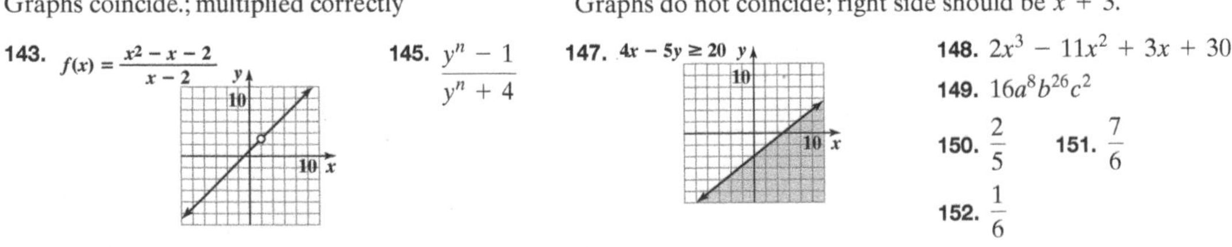

143. $f(x) = \frac{x^2 - x - 2}{x - 2}$ **145.** $\frac{y^n - 1}{y^n + 4}$ **147.** $4x - 5y \geq 20$

148. $2x^3 - 11x^2 + 3x + 30$ **149.** $16a^8b^{26}c^2$ **150.** $\frac{2}{5}$ **151.** $\frac{7}{6}$ **152.** $\frac{1}{6}$

Section 6.2 Check Point Exercises

1. $\frac{x-5}{x+3}$ **2.** $\frac{1}{x-y}$ **3.** $18x^2$ **4.** $5x(x+3)(x+3)$ or $5x(x+3)^2$ **5.** $\frac{21+4x}{18x^2}$ or $\frac{4x+21}{18x^2}$ **6.** $\frac{2x^2-2x+8}{(x-4)(x+4)}$ or $\frac{2(x^2-x+4)}{(x-4)(x+4)}$ **7.** $\frac{x^2-3x+5}{(x-3)(x+1)(x-2)}$ **8.** $\frac{3}{y+2}$ **9.** $\frac{3x-5y}{x-3y}$

Concept and Vocabulary Check

1. $\frac{P+Q}{R}$; numerators; common denominator **2.** $\frac{P-Q}{R}$; numerators; common denominator **3.** $\frac{x-5+y}{3}$ **4.** $\frac{-1}{-1}$ **5.** factor denominators **6.** $x + 3$ and $x - 2$; $x + 3$ and $x + 1$; $(x+3)(x-2)(x+1)$ **7.** $2x$ **8.** $3y + 4$ **9.** -1

Exercise Set 6.2

1. $\frac{2}{3x}$ **3.** $\frac{10x+3}{x-5}$ **5.** $\frac{2x-1}{x+3}$ **7.** $\frac{y-3}{y+3}$ **9.** $\frac{x+1}{4x-3}$ **11.** $\frac{x-3}{x-1}$ **13.** $\frac{4y+3}{2y-1}$ **15.** $\frac{x^2+xy+y^2}{x+y}$ **17.** $175x^2$ **19.** $(x-5)(x+5)$ **21.** $y(y+10)(y-10)$ **23.** $(x+4)(x-4)(x-4)$ **25.** $(y-5)(y-6)(y+1)$ **27.** $(y+2)(2y+3)(y-2)(2y+1)$ **29.** $\frac{3+50x}{5x^2}$ **31.** $\frac{7x-2}{(x-2)(x+1)}$ **33.** $\frac{3x-4}{(x+2)(x-3)}$ **35.** $\frac{2x^2-2x+61}{(x+5)(x-6)}$ **37.** $\frac{20-x}{(x-5)(x+5)}$ **39.** $\frac{13}{(y-3)(y-2)}$ **41.** $\frac{1}{x+3}$ **43.** $\frac{6x^2+14x+10}{(x+1)(x+4)(x+3)}$ **45.** $-\frac{x^2-3x-13}{(x+4)(x-2)(x+1)}$ **47.** $\frac{4x-11}{x-3}$ **49.** $\frac{2y-14}{(y-4)(y+4)}$ **51.** $\frac{x^2+2x-14}{3(x+2)(x-2)}$ **53.** $\frac{16}{x-4}$ **55.** $\frac{12x+7y}{(x-y)(x+y)}$ **57.** $\frac{x^2-18x-30}{(5x+6)(x-2)}$ **59.** $\frac{3x^2+16xy-13y^2}{(x+5y)(x-5y)(x-4y)}$ **61.** $\frac{8x^3-32x^2+23x+6}{(x+2)(x-2)(x-2)(x-1)}$ **63.** $\frac{16a^2-12ab-18b^2}{(3a+4b)(3a-4b)(2a-b)}$ **65.** $\frac{m+6}{(m-1)(m+2)(2m-1)}$ **67.** $\frac{x^2+5x+8}{(x+2)(x+1)}$ **69.** 2 **71.** $-\frac{1}{x(x+h)}$ **73.** $\frac{2d}{a^2+ab+b^2}$ **75.** $(f-g)(x)=\frac{x+2}{x+3}$; $(-\infty,-5)\cup(-5,-3)\cup(-3,\infty)$ **77.** 11; If you average 0 miles per hour over the speed limits, the total driving time is about 11 hours.; (0, 11) **79.** $\frac{720x+48{,}050}{(x+70)(x+65)}$; 11 **81.** 12 mph; Answers will vary. **83. a.** 307; There are 307 arrests for every 100,000 drivers who are 20 years old.; (20, 307) **b.** $f(x)=\frac{-5x^3+27{,}680x-388{,}150}{x^2+9}$ **c.** 25; 356 **85.** $\frac{2x(2x+15)}{(x+7)(x+8)}$ **91.** Answers will vary.; $\frac{b+a}{ab}$ **93.** makes sense **95.** does not make sense **97.** false **99.** false **101.** $\frac{1}{x^{2n}-1}$ **103.** $\frac{x-y+1}{(x-y)(x-y)}$ **104.** $\frac{y^{10}}{9x^4}$ **105.** $\left[-\frac{13}{3},5\right]$ **106.** $2x(5x+3)(5x-3)$ **107.** xy^2+y^3 **108.** $-h$ **109.** $\frac{x+1}{x-3}$

Section 6.3 Check Point Exercises

1. $\frac{y}{x+y}$ **2.** $-\frac{1}{x(x+7)}$ **3.** $\frac{2x}{x^2+1}$ **4.** $\frac{x+2}{x-5}$

Concept and Vocabulary Check

1. complex; complex **2.** $\frac{7x+5}{5x+x^2}$ **3.** $\frac{1}{x+3}$; $\frac{1}{x}$; x; $x+3$; -3; $-\frac{1}{x(x+3)}$

Exercise Set 6.3

1. $\frac{4x+2}{x-3}$ **3.** $\frac{x^2+9}{(x-3)(x+3)}$ **5.** $\frac{y+x}{y-x}$ **7.** $\frac{4-x}{5x-3}$ **9.** $\frac{1}{x-3}$ **11.** $-\frac{1}{x(x+5)}$ **13.** $-x$ **15.** $-\frac{x+1}{x-1}$ **17.** $\frac{(x+y)(x+y)}{xy}$ **19.** $\frac{4x}{x^2+4}$ **21.** $\frac{2y^3+5x^2}{y^3(5-3x^2)}$ **23.** $-\frac{12}{5}$ **25.** $\frac{3ab(b+a)}{(2b+3a)(2b-3a)}$ **27.** $-\frac{x+10}{5x-2}$ **29.** $\frac{2y}{3y+7}$ **31.** $\frac{2b+a}{b-2a}$ **33.** $\frac{4(7x+5)}{69(x+5)}$ **35.** $\frac{3x-2y}{13y+5x}$ **37.** $\frac{m(m+3)}{(m-2)(m+1)}$ **39.** $\frac{3a(a-3)}{(a+4)(3a-5)}$ **41.** $-\frac{4}{(x+2)(x-2)}$ **43.** 6 **45.** $x(x+1)$ **47.** $\frac{x+4}{x+2}$ **49.** $-\frac{3}{a(a+h)}$ **51. a.** $\frac{Pi(1+i)^n}{(1+i)^n-1}$ **b.** \$527 per month **53.** $R=\frac{R_1R_2R_3}{R_2R_3+R_1R_3+R_1R_2}$; about 2.18 ohms

59.

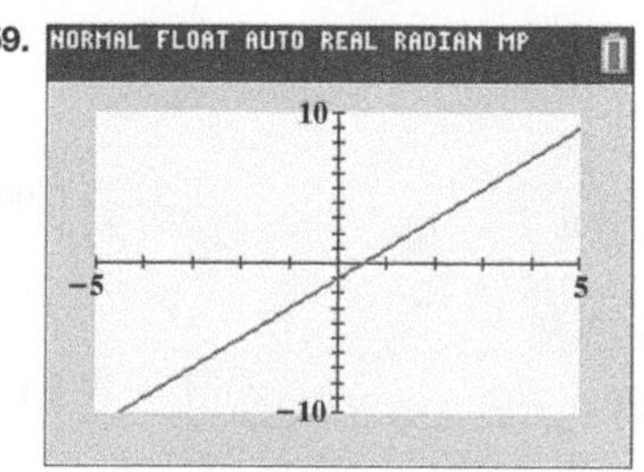

Graphs coincide.; simplified correctly

61.

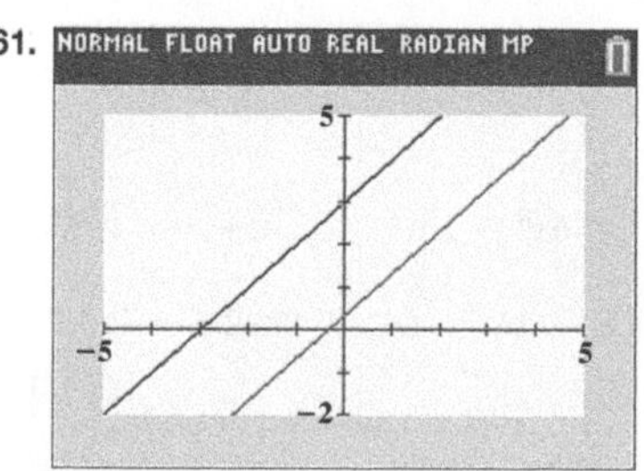

Graphs do not coincide.; $\frac{\frac{1}{x}+\frac{1}{3}}{\frac{1}{3x}}=x+3$

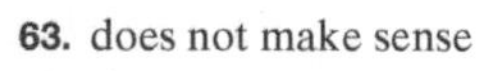

63. does not make sense **65.** does not make sense **67.** $\frac{1}{(x+h+1)(x+1)}$ **69.** $\frac{a+1}{a+2}$ **71.** $\{3,9\}$ **72.** $16x^4-8x^2y+y^2$ **73.** $(1,5)$ **74.** $2xy^3$ **75.** $35+\frac{2}{21}$ **76.** $7x$

Section 6.4 Check Point Exercises

1. $4x^2-8x+\frac{1}{2}+\frac{1}{x}$ **2.** $3x^2y^2-xy+\frac{2}{y}$ **3.** $3x-8$ **4.** $2x^2-x+3-\frac{6}{2x-1}$ **5.** $2x^2+7x+14+\frac{21x-10}{x^2-2x}$

Concept and Vocabulary Check

1. $16x^3-32x^2+2x+4$; $4x$ **2.** $2x^3+0x^2+6x-4$ **3.** $6x^3$; $3x$; $2x^2$; $7x^2$ **4.** $2x^2$; $5x-2$; $10x^3-4x^2$; $10x^3+6x^2$ **5.** $6x^2-10x$; $6x^2+8x$; $18x$; -4; $18x-4$ **6.** 9; $3x-5$; 9; $3x-5+\frac{9}{2x+1}$ **7.** divisor; quotient; remainder; dividend

Exercise Set 6.4

1. $5x^4 - 3x^2 + 2$ **3.** $6x^2 + 2x - 3 - \frac{2}{x}$ **5.** $7x - \frac{7}{4} - \frac{4}{x}$ **7.** $-5x^3 + 10x^2 - \frac{3}{5}x + 8$ **9.** $2a^2b - a - 3b$ **11.** $6x - \frac{3}{y} - \frac{2}{xy^2}$ **13.** $x + 3$ **15.** $x^2 + x - 2$ **17.** $x - 2 + \frac{2}{x - 5}$ **19.** $x + 5 - \frac{10}{2x + 3}$ **21.** $x^2 + 2x + 3 + \frac{1}{x + 1}$ **23.** $2y^2 + 3y - 1$ **25.** $3x^2 - 3x + 1 + \frac{2}{3x + 2}$ **27.** $2x^2 + 4x + 5 + \frac{9}{2x - 4}$ **29.** $2y^2 + y - 2 - \frac{2}{2y - 1}$ **31.** $2y^3 + 3y^2 - 4y + 1$ **33.** $4x^2 + 3x - 8 + \frac{18}{x^2 + 3}$ **35.** $5x^2 + x + 3 + \frac{x + 7}{3x^2 - 1}$ **37.** $\left(\frac{f}{g}\right)(x) = 2x^2 - 9x + 10$ **39.** $\left(\frac{f}{g}\right)(x) = x^3 - x^2 + x - 2$ **41.** $x^3 - x^2y + xy^2 - y^3 + \frac{2y^4}{x + y}$ **43.** $3x^2 + 2x - 1$ **45.** $4x - 7 + \frac{4x + 8}{x^2 + x + 1}$ **47.** $x^3 + x^2 - x - 3 + \frac{-x + 5}{x^2 - x + 2}$ **49.** $4x^2 + 5xy - y^2$ **51.** $\left(\frac{f - g}{h}\right)(x) = 2x^2 - 5x + 8; \left(-\infty, -\frac{1}{4}\right) \cup \left(-\frac{1}{4}, \infty\right)$ **53.** $x = 3a^2 + 4a - 2$
55. 70; At the tax rate of 30%, the government tax revenue is \$70 tens of billions, or \$700 billion.; (30, 70)
57. $f(x) = 80 + \frac{800}{x - 110}$; 70; yes; Answers will vary.

65.
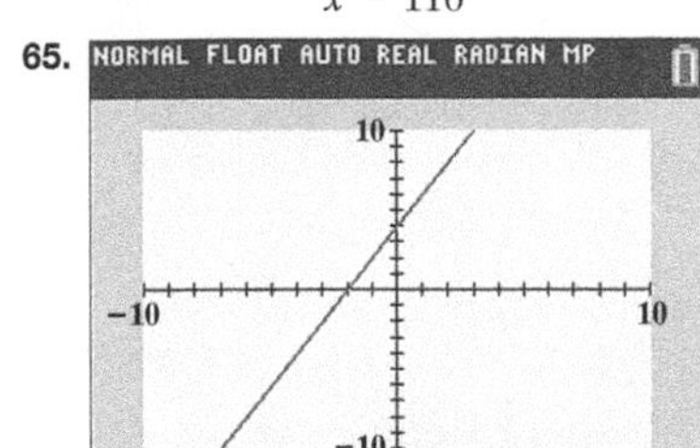

Graphs coincide.; division correct

67.
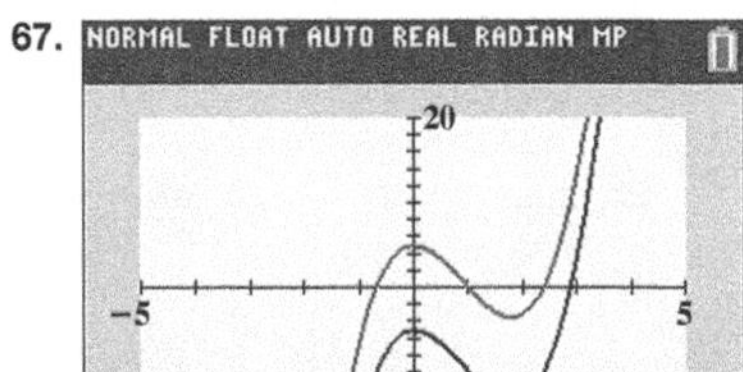

Graphs do not coincide; right side should be $3x^3 - 8x^2 - 5$.

69. makes sense **71.** makes sense **73.** false **75.** false **77.** $x^{2n} + x^n + 3 + \frac{3}{x^n - 5}$ **79.** $x - 2$ **81.** $\left(-\infty, -\frac{1}{2}\right) \cup \left(\frac{7}{2}, \infty\right)$ **82.** 4.061×10^7 **83.** $22x + 12$ **84. a.** $5x^2 - 10x + 26 - \frac{44}{x + 2}$ **b.** -10; 26; -44; The numbers are the coefficients of the quotient and the remainder. **85. a.** $3x^2 - 7x + 9 - \frac{10}{x + 1}$ **b.** -7; 9; -10; The numbers are the coefficients of the quotient and the remainder. **86.** $2x^2 + 3x - 2$; $(x - 3)(2x - 1)(x + 2)$

Mid-Chapter Check Point Exercises

1. $\frac{x + 2}{x + 6}$ **2.** $\frac{3(x - 2)}{x - 1}$ **3.** $\frac{x^2 + 3x + 9}{x - 1}$ **4.** $\frac{5x - 3}{x - 2}$ **5.** $\frac{x^2 + 6x - 4}{x^2 + 6x + 1}$ **6.** $2x^3 - 5x^2 - 3x + 6$ **7.** $x + y$ **8.** $4x^6y^2 - 2x^4y + \frac{3}{7}y$ **9.** $\frac{x^2 - 14x - 16}{(x + 6)(x - 2)}$ **10.** $\frac{10}{(x + 2)(x - 2)}$ **11.** $\frac{2(x - 3)}{x - 1}$ **12.** $\frac{x - 5}{x - 7}$ **13.** $2x^2 - x - 3 + \frac{x + 1}{3x^2 - 1}$ **14.** $\frac{5x + 2}{3x - 1}$ **15.** $\frac{5x}{(x - 6)(x - 1)(x + 4)}$ **16.** $\frac{7x + 4}{4(x + 1)}$ **17.** $x + 2$ **18.** $16x^2 - 8x + 4 - \frac{2}{2x + 1}$ **19.** $\frac{3x + 2}{(x + 2)(x - 1)}$ **20.** Domain: $(-\infty, -7) \cup (-7, 2) \cup (2, \infty)$; $f(x) = \frac{5}{x + 7}$

Section 6.5 Check Point Exercises

1. $x^2 - 2x - 3$ **2.** -105 **3.** The remainder is zero.; -1, $-\frac{1}{3}$, and $\frac{2}{5}$, or $\left\{-1, -\frac{1}{3}, \frac{2}{5}\right\}$

Concept and Vocabulary Check

1. 4; 1; 5; -7; 1 **2.** -5; 4; 0, -8; -2 **3.** true **4.** $f(c)$

Exercise Set 6.5

1. $2x + 5$ **3.** $3x - 8 + \frac{20}{x + 5}$ **5.** $4x^2 + x + 4 + \frac{3}{x - 1}$ **7.** $6x^4 + 12x^3 + 22x^2 + 48x + 93 + \frac{187}{x - 2}$ **9.** $x^3 - 10x^2 + 51x - 260 + \frac{1300}{5 + x}$ **11.** $3x^2 + 3x - 3$ **13.** $x^4 + x^3 + 2x^2 + 2x + 2$ **15.** $x^3 + 4x^2 + 16x + 64$ **17.** $2x^4 - 7x^3 + 15x^2 - 31x + 64 - \frac{129}{x + 2}$ **19.** -25 **21.** -133 **23.** 240 **25.** 1 **27.** The remainder is 0.; $\{-1, 2, 3\}$ **29.** The remainder is 0.; $\left\{-\frac{1}{2}, 1, 2\right\}$ **31.** The remainder is 0.; $\left\{-5, \frac{1}{3}, \frac{1}{2}\right\}$ **33.** 2; The remainder is zero.; -3, -1, and 2, or $\{-3, -1, 2\}$ **35.** 1; The remainder is zero.; $\frac{1}{3}, \frac{1}{2}$, and 1, or $\left\{\frac{1}{3}, \frac{1}{2}, 1\right\}$ **37.** $4x^3 - 9x^2 + 7x - 6$ **39.** $0.5x^2 - 0.4x + 0.3$ **41. a.** The remainder is 0. **b.** 3 mm **49.** makes sense **51.** makes sense **53.** The remainder is 0.; $\{-2, -1, 2, 5\}$ **54.** $(-10, \infty)$ **55.** $\{-5, -1\}$ **56.** $\{(5, 3)\}$ **57.** $6x$ **58.** $3x$ **59.** $(x + 3)(x - 3)$

Section 6.6 Check Point Exercises

1. 6 or $\{6\}$ **2.** 4 or $\{4\}$ **3.** no solution or $\varnothing$ **4.** 4 and 6, or $\{4, 6\}$ **5.** 1 and 7, or $\{1, 7\}$ **6.** 50%

Concept and Vocabulary Check

1. least common denominator (LCD) **2.** 0 **3.** $2x$ **4.** $(x + 5)(x + 1)$ **5.** $x \neq 2$; $x \neq -4$ **6.** $5(x + 3) + 3(x + 4) = 12x + 9$ **7.** true

Exercise Set 6.6

1. {1} **3.** {3} **5.** {4} **7.** {−4} **9.** $\left\{\frac{9}{7}\right\}$ **11.** no solution or ∅ **13.** {2} **15.** {2} **17.** {5} **19.** {−7, −1} **21.** {−6, 3} **23.** {−1} **25.** no solution or ∅ **27.** {2} **29.** no solution or ∅ **31.** $\left\{-\frac{4}{3}\right\}$ **33.** $\left\{-6, \frac{1}{2}\right\}$ **35.** 4, 10 **37.** 3, $\frac{3}{4}$ **39.** $\frac{x-2}{x(x+1)}$ **41.** {2} **43.** {4} **45.** $\frac{-2(x-4)}{(x-2)(x^2+2x+4)}$ **47.** 0 **49.** 1, 7 **51.** 60% **53.** 10 days; (10, 8) **55.** $y = 5$; On average, the students remembered 5 words over an extended period of time. **57.** 11 learning trials; (11, 0.95) **59.** As the number of learning trials increases, the proportion of correct responses increases.; Initially, the proportion of correct responses increases rapidly, but slows down as time increases. **61.** 125 liters

71. 
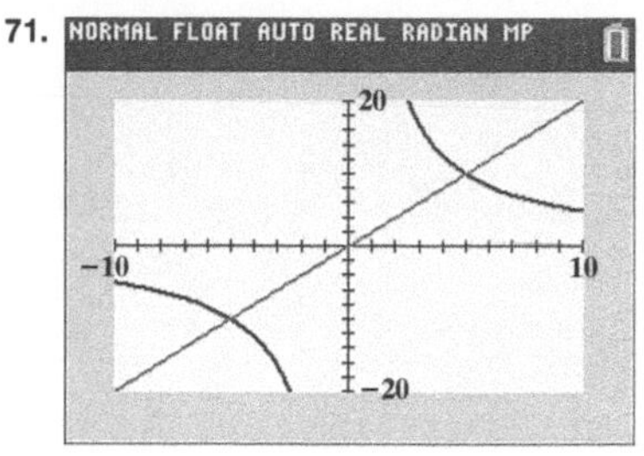

; {−5, 5}

73. 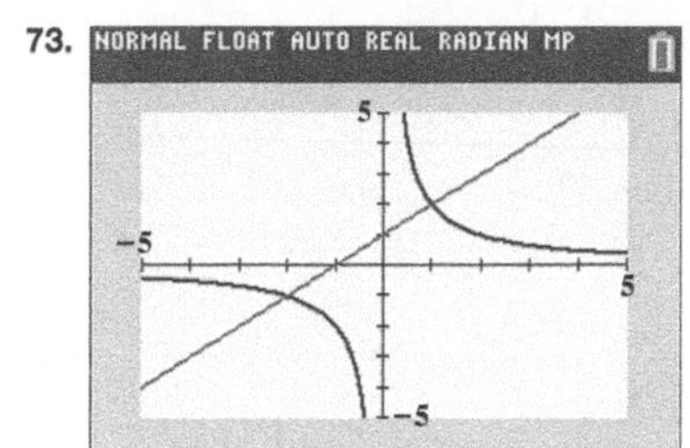

; {−2, 1}

75. does not make sense **77.** makes sense **79.** false **81.** false **83.** no solution or ∅

87. $x + 2y \geq 2$
$x - y \geq -4$

88. {15} **89.** $F = \frac{9C + 160}{5}$ **90.** $p = \frac{qf}{q - f}$ **91.** {−20, 30} **92.** 40 miles per hour

Section 6.7 Check Point Exercises

1. $x = \frac{b - 2a}{a}$ **2.** $x = \frac{yz}{y - z}$ **3. a.** $C(x) = 500{,}000 + 400x$ **b.** $\overline{C}(x) = \frac{500{,}000 + 400x}{x}$ **c.** 10,000 wheelchairs **4.** 12 mph **5.** 6 months **6.** experienced carpenter: 8 hr; apprentice: 24 hr

Concept and Vocabulary Check

1. xyz **2.** fixed; variable **3.** number of units produced **4.** distance traveled; rate of travel **5.** 1 **6.** $\frac{x}{19}$

Exercise Set 6.7

1. $P_1 = \frac{P_2V_2}{V_1}$ **3.** $f = \frac{pq}{q + p}$ **5.** $r = \frac{A - P}{P}$ **7.** $m_1 = \frac{Fd^2}{Gm_2}$ **9.** $x = \overline{x} + zs$ **11.** $R = \frac{E - Ir}{I}$ **13.** $f_1 = \frac{ff_2}{f_2 - f}$ **15.** 50,000 wheelchairs **17.** $y = 400$; Average cost is nearing \$400 as production level increases. **19. a.** $C(x) = 100{,}000 + 100x$ **b.** $\overline{C}(x) = \frac{100{,}000 + 100x}{x}$ **c.** 500 bikes **21.** 5 mph **23.** The time increases as the running rate is close to zero miles per hour. **25.** car: 50 mph; bus: 30 mph **27.** 6 mph **29.** 3 mph **31.** 500 mph **33.** 4.2 ft per sec **35.** 12 miles **37.** 18 min; yes **39.** 4 hr **41.** 12 hr **43.** 10 hr **45.** 10 hr **47.** 10 min **49.** 3 **51.** $\frac{1}{3}$ and 3 **53.** 10 **55.** $x = \frac{ab}{a + b}$ **67.** makes sense **69.** makes sense **71.** false **73.** true **75.** $f = \frac{p - s}{ps - s}$ **77.** 8 hr **78.** $(x + 2 + 3y)(x + 2 - 3y)$ **79.** {(5, −3)} **80.** {(3, −1, 2)} **81. a.** 16 **b.** $y = 16x^2$ **c.** 400 **82. a.** 96 **b.** $y = \frac{96}{x}$ **c.** 32 **83.** 8

Section 6.8 Check Point Exercises

1. 66 gal **2.** about 556 ft **3.** 512 cycles per second **4.** 24 min **5.** 96π cubic feet

Concept and Vocabulary Check

1. $y = kx$; constant of variation **2.** $y = kx^n$ **3.** $y = \frac{k}{x}$ **4.** $y = \frac{kx}{z}$ **5.** $y = kxz$ **6.** directly; inversely **7.** jointly; inversely

Exercise Set 6.8

1. 156 **3.** 30 **5.** $\frac{5}{6}$ **7.** 240 **9.** 50 **11.** $x = kyz$; $y = \frac{x}{kz}$ **13.** $x = \frac{kz^3}{y}$; $y = \frac{kz^3}{x}$ **15.** $x = \frac{kyz}{\sqrt{w}}$; $y = \frac{x\sqrt{w}}{kz}$

17. $x = kz(y + w)$; $y = \frac{x - kzw}{kz}$ **19.** $x = \frac{kz}{y - w}$; $y = \frac{xw + kz}{x}$ **21.** 5.4 ft **23.** 80 in. **25.** about 607 lb **27.** 32°
29. a. $L = \frac{1890}{R}$ **b.** an approximate model **c.** 70 yr **31. a.** 90 beats per minute **b.** 95 beats per minute **c.** by the point (63, 30)
33. 90 milliroentgens per hour **35.** This person has a BMI of 24.4 and is not overweight. **37.** 1800 Btu **39.** $\frac{1}{4}$ of what it was originally
41. a. $C = \frac{kP_1P_2}{d^2}$ **b.** $k \approx 0.02$; $C = \frac{0.02P_1P_2}{d^2}$ **c.** approximately 39,813 daily phone calls

43. a.

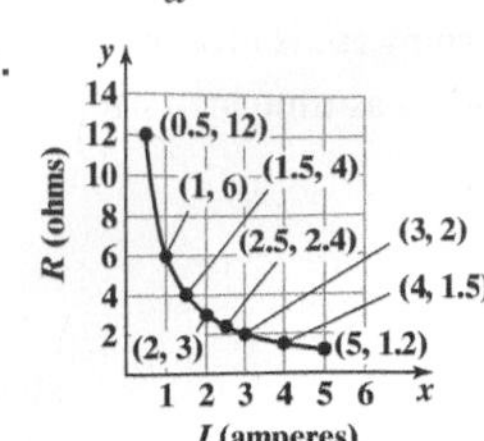

b. Current varies inversely as resistance. **c.** $R = \frac{6}{I}$

49. z varies directly as the square root of x and inversely as the square of y. **53.** does not make sense **55.** makes sense **57.** The wind pressure is 4 times more destructive. **59.** Distance is increased by $\sqrt{50}$, or about 7.07, for the space telescope. **60.** −2 **61.** $(x + 3)(x - 3)(y - 3)$ **62.** 4 **63.** 3 **64.** 6 **65.** domain: $[-4, \infty)$; range: $[0, \infty)$

Review Exercises

1. a. $\frac{7}{4}$ **b.** $\frac{3}{4}$ **c.** does not exist **d.** 0 **2.** $(-\infty, -4) \cup (-4, 3) \cup (3, \infty)$ **3.** $(-\infty, -2) \cup (-2, 1) \cup (1, \infty)$ **4.** $\frac{x^2 - 7}{3x}$ **5.** $\frac{x - 1}{x - 7}$
6. $\frac{3x + 2}{x - 5}$ **7.** cannot be simplified **8.** $\frac{x^2 + 2x + 4}{x + 2}$ **9.** $\frac{5(x + 1)}{3}$ **10.** $\frac{2x - 1}{x + 1}$ **11.** $\frac{x - 7}{x^2}$ **12.** $\frac{1}{3(x + 3)}$ **13.** $\frac{1}{2}$
14. $\frac{(y + 4)(y + 2)}{(y - 6)(y^2 + 4y + 16)}$ **15.** $\frac{3x^2}{2x - 5}$ **16. a.** 50 deer **b.** 150 deer **c.** $y = 225$; The population will approach 225 deer over time.
17. 4 **18.** $\frac{1}{x + 3}$ **19.** $3x + 2$ **20.** $36x^3$ **21.** $(x + 7)(x - 5)(x + 2)$ **22.** $\frac{3x - 5}{x(x - 5)}$ **23.** $\frac{5x - 2}{(x - 3)(x + 2)(x - 2)}$
24. $\frac{2x - 8}{(x - 3)(x - 5)}$ or $\frac{2(x - 4)}{(x - 3)(x - 5)}$ **25.** $\frac{4x}{(3x + 4)(3x - 4)}$ **26.** $\frac{y - 3}{(y + 1)(y + 3)}$ **27.** $\frac{2x^2 - 9}{(x + 3)(x - 3)}$ **28.** $3(x + y)$ or $3x + 3y$
29. $\frac{3}{8}$ **30.** $\frac{x}{x - 5}$ **31.** $\frac{3x + 8}{3x + 10}$ **32.** $\frac{4(x - 2)}{2x + 5}$ **33.** $\frac{3x^2 + 9x}{x^2 + 8x - 33}$ **34.** $\frac{1 + x}{1 - x}$ **35.** $3x - 6 + \frac{2}{x} - \frac{2}{5x^2}$ **36.** $6x^3y + 2xy - 10x$
37. $3x - 7 + \frac{26}{2x + 3}$ **38.** $2x^2 - 4x + 1 - \frac{10}{5x - 3}$ **39.** $x^5 + 5x^4 + 8x^3 + 16x^2 + 33x + 63 + \frac{128}{x - 2}$ **40.** $2x^2 + 3x - 1$
41. $4x^2 - 7x + 5 - \frac{4}{x + 1}$ **42.** $3x^3 + 6x^2 + 10x + 10$ **43.** $x^3 - 4x^2 + 16x - 64 + \frac{272}{x + 4}$ **44.** 3 **45.** 4 **46.** solution
47. not a solution **48.** The remainder is 0.; $\left\{-1, \frac{1}{3}, \frac{1}{2}\right\}$ **49.** {6} **50.** {52} **51.** {7} **52.** $\left\{-\frac{9}{2}\right\}$ **53.** $\left\{-\frac{1}{2}, 3\right\}$ **54.** {−23, 2}
55. {−3, 2} **56.** 80% **57.** $C = R - nP$ **58.** $T_1 = \frac{P_1V_1T_2}{P_2V_2}$ **59.** $P = \frac{A}{rT + 1}$ **60.** $R = \frac{R_1R_2}{R_2 + R_1}$ **61.** $n = \frac{IR}{E - Ir}$
62. a. $C(x) = 50{,}000 + 25x$ **b.** $\overline{C}(x) = \frac{50{,}000 + 25x}{x}$ **c.** 5000 graphing calculators **63.** 12 mph **64.** 9 mph **65.** 2 hr; no
66. faster crew: 36 hr; slower crew: 45 hr **67.** 240 min or 4 hr **68.** \$4935 **69.** 1600 ft **70.** 440 vibrations per second **71.** 112 decibels
72. 16 hr **73.** 800 cubic feet

Chapter Test

1. $(-\infty, 2) \cup (2, 5) \cup (5, \infty)$; $\frac{x}{x - 5}$ **2.** $\frac{x}{x - 4}$ **3.** $\frac{(x + 3)(x - 1)}{x + 1}$ **4.** $\frac{x - 2}{x + 5}$ **5.** $\frac{3x - 24}{x - 3}$ **6.** $\frac{1}{3x + 5}$ **7.** $\frac{x^2 + 2x + 15}{(x + 3)(x - 3)}$
8. $\frac{3x - 4}{(x + 2)(x - 3)}$ **9.** $\frac{5x^2 - 7x + 4}{(x + 2)(x - 1)(x - 2)}$ **10.** $\frac{x + 3}{x + 5}$ **11.** $\frac{x - 2}{x - 10}$ **12.** $\frac{x - 2}{8}$ **13.** $1 - x$ **14.** $3x^2y^2 + 4y^2 - \frac{5}{2}y$
15. $3x^2 - 3x + 1 + \frac{2}{3x + 2}$ **16.** $3x^2 + 2x + 3 + \frac{9 - 6x}{x^2 - 1}$ **17.** $3x^3 - 4x^2 + 7$ **18.** 12 **19.** solution **20.** {−7, 3} **21.** {−1}
22. 3 years **23.** $a = \frac{Rs}{s - R}$ or $a = -\frac{Rs}{R - s}$ **24. a.** $C(x) = 300{,}000 + 10x$ **b.** $\overline{C}(x) = \frac{300{,}000 + 10x}{x}$ **c.** 20,000 players **25.** 12 hr
26. 4 mph **27.** 45 foot-candles

Cumulative Review Exercises

1. $(-\infty, -6)$ **2.** $\left\{-\frac{1}{2}, 4\right\}$ **3.** {(−2, 0, 4)} **4.** $\left[-2, \frac{14}{3}\right]$ **5.** {12} **6.** $s = \frac{2R - Iw}{2I}$ **7.** {(3, 2)} **8.** $f(x) = -3x - 2$

9. $y = |x| + 2$ **10.** $y \geq 2x - 1$, $x \geq 1$

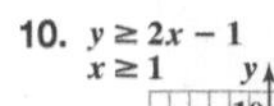

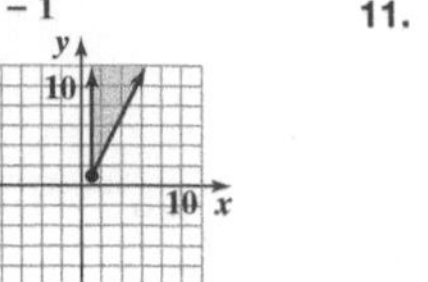

11. $2x - y < 4$

12. $x^2 + 4x + 4 - 9y^2$ **13.** $\frac{x(x + 1)}{2x + 5}$ **14.** $\frac{x + 25}{2(x - 4)(x - 5)}$
15. $3x + 4 + \frac{2}{x + 2}$ **16.** $(y - 6)(x + 2)$
17. $2xy(2x + 3)(6x - 5)$ **18.** 4 sec **19.** basic cable service: \$25; each movie channel: \$10 **20.** 1 ft

CHAPTER 7

Section 7.1 Check Point Exercises

1. a. 8 **b.** -7 **c.** $\frac{4}{5}$ **d.** 0.09 **e.** 5 **f.** 7 **2. a.** 4 **b.** $-\sqrt{24} \approx -4.90$ **3.** $[3, \infty)$ **4.** approximately 15.7 minutes **5. a.** 7 **b.** $|x + 8|$ **c.** $|7x^5|$ or $7|x^5|$ **d.** $|x - 3|$ **6. a.** 3 **b.** -2 **7.** $-3x$ **8. a.** 2 **b.** -2 **c.** not a real number **d.** -1 **9. a.** $|x + 6|$ **b.** $3x - 2$ **c.** 8

Concept and Vocabulary Check

1. principal **2.** 8^2 **3.** $[0, \infty)$ **4.** $5x - 20 \geq 0$ **5.** $|a|$ **6.** 10^3 **7.** $(-5)^3$ **8.** a **9.** $(-\infty, \infty)$ **10.** nth; index **11.** $|a|$; a **12.** true **13.** false **14.** true **15.** false

Exercise Set 7.1

1. 6 **3.** -6 **5.** not a real number **7.** $\frac{1}{5}$ **9.** $-\frac{3}{4}$ **11.** 0.9 **13.** -0.2 **15.** 3 **17.** 1 **19.** not a real number **21.** 4; 1; 0; not a real number **23.** -5; $-\sqrt{5} \approx -2.24$; -1; not a real number **25.** 4; 2; 1; 6 **27.** $[3, \infty)$; c **29.** $[-5, \infty)$; d **31.** $(-\infty, 3]$; e **33.** 5 **35.** 4 **37.** $|x - 1|$ **39.** $|6x^2|$ or $6x^2$ **41.** $-|10x^3|$ or $-10|x^3|$ **43.** $|x + 6|$ **45.** $-|x - 4|$ **47.** 3 **49.** -3 **51.** $\frac{1}{5}$ **53.** $-\frac{3}{10}$ **55.** 3; 2; -1; -4 **57.** -2; 0; 2 **59.** 1 **61.** 2 **63.** -2 **65.** not a real number **67.** -1 **69.** not a real number **71.** -4 **73.** 2 **75.** -2 **77.** x **79.** $|y|$ **81.** $-2x$ **83.** -5 **85.** 5 **87.** $|x + 3|$ **89.** $-2(x - 1)$

91. $f(x) = \sqrt{x} + 3$; points (0, 3), (1, 4), (4, 5), (9, 6)

domain: $[0, \infty)$; range: $[3, \infty)$

93. $f(x) = \sqrt{x - 3}$; points (3, 0), (4, 1), (7, 2), (12, 3)

domain: $[0, \infty)$; range: $[3, \infty)$

95. $(-\infty, 15)$ **97.** $[1, 3)$ **99.** 3 **101. a.** 40.2 in.; underestimates by 0.6 in. **b.** 0.9 in. per month **c.** 0.2 in. per month; This is a much smaller rate of change.; The graph is not as steep between 50 and 60 as it is between 0 and 10. **103.** 70 mph; The officer should not believe the motorist.; Answers will vary.

115. 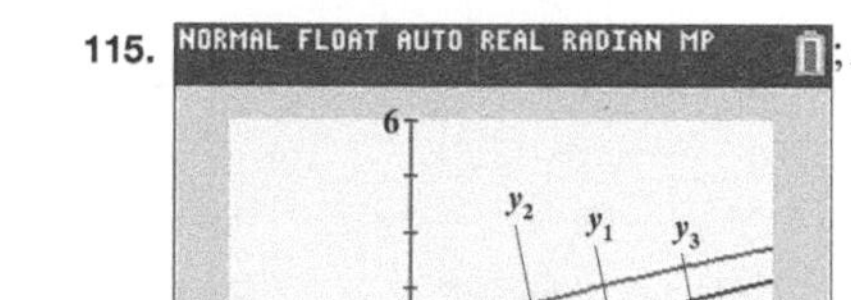

; Answers will vary

117. 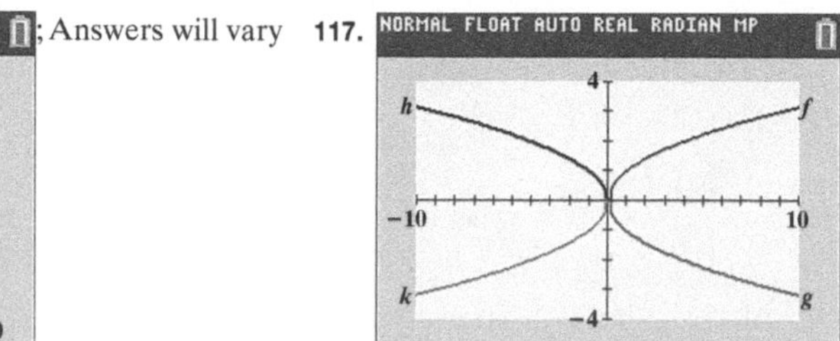

domain of f: $[0, \infty)$; range of f: $[0, \infty)$
domain of g: $[0, \infty)$; range of g: $(-\infty, 0]$
domain of h: $(-\infty, 0]$; range of h: $[0, \infty)$
domain of k: $(-\infty, 0]$; range of k: $(-\infty, 0]$

119. does not make sense **121.** makes sense **123.** false **125.** false **127.** Answers will vary; an example is $f(x) = \sqrt{15 - 3x}$. **129.** $|(2x + 3)^5|$ **131.** The graph of h is the graph of f shifted left 3 units.

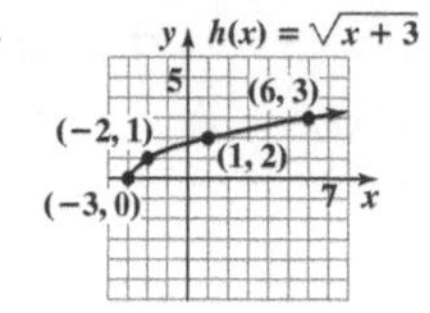

132. $7x + 30$ **133.** $\frac{x^8}{9y^6}$ **134.** $\left(-\infty, -\frac{7}{3}\right) \cup (5, \infty)$ **135.** $\frac{2^7}{x}$ or $\frac{128}{x}$ **136.** $\frac{2}{x^3}$ **137.** $\frac{y^{12}}{x^8}$

Section 7.2 Check Point Exercises

1. a. $\sqrt{25} = 5$ **b.** $\sqrt[3]{-8} = -2$ **c.** $\sqrt[4]{5xy^2}$ **2. a.** $(5xy)^{1/4}$ **b.** $\left(\frac{a^3b}{2}\right)^{1/5}$ **3. a.** $(\sqrt[3]{8})^4 = 16$ **b.** $(\sqrt{25})^3 = 125$ **c.** $-(\sqrt[4]{81})^3 = -27$ **4. a.** $6^{4/3}$ **b.** $(2xy)^{7/5}$ **5. a.** $\frac{1}{100^{1/2}} = \frac{1}{10}$ **b.** $\frac{1}{8^{1/3}} = \frac{1}{2}$ **c.** $\frac{1}{32^{3/5}} = \frac{1}{8}$ **d.** $\frac{1}{(3xy)^{5/9}}$ **6. a.** $7^{5/6}$ **b.** $\frac{5}{x}$ **c.** $9.1^{3/10}$ **d.** $\frac{y^{1/12}}{x^{1/5}}$ **7. a.** $\sqrt{x}$ **b.** $2a^4$ **c.** $\sqrt[4]{x^2y}$ **d.** $\sqrt[6]{x}$ **e.** $\sqrt[6]{x}$

Concept and Vocabulary Check

1. $\sqrt{36} = 6$ **2.** $\sqrt[3]{8} = 2$ **3.** $\sqrt[n]{a}$ **4.** $(\sqrt[4]{16})^3 = (2)^3 = 8$ **5.** $(\sqrt[n]{a})^m$ or $\sqrt[n]{a^m}$ **6.** $\frac{5}{3}$ **7.** $\frac{1}{16^{3/2}} = \frac{1}{(\sqrt{16})^3} = \frac{1}{(4)^3} = \frac{1}{64}$

Exercise Set 7.2

1. $\sqrt{49} = 7$ **3.** $\sqrt[3]{-27} = -3$ **5.** $-\sqrt[4]{16} = -2$ **7.** $\sqrt[3]{xy}$ **9.** $\sqrt[5]{2xy^3}$ **11.** $(\sqrt{81})^3 = 729$ **13.** $(\sqrt[3]{125})^2 = 25$

15. $(\sqrt[5]{-32})^3 = -8$ **17.** $(\sqrt[3]{27})^2 + (\sqrt[4]{16})^3 = 17$ **19.** $\sqrt[7]{(xy)^4}$ **21.** $7^{1/2}$ **23.** $5^{1/3}$ **25.** $(11x)^{1/5}$ **27.** $x^{3/2}$ **29.** $x^{3/5}$ **31.** $(x^2y)^{1/5}$ **33.** $(19xy)^{3/2}$ **35.** $(7xy^2)^{5/6}$ **37.** $2xy^{2/3}$ **39.** $\frac{1}{49^{1/2}} = \frac{1}{7}$ **41.** $\frac{1}{27^{1/3}} = \frac{1}{3}$ **43.** $\frac{1}{16^{3/4}} = \frac{1}{8}$ **45.** $\frac{1}{8^{2/3}} = \frac{1}{4}$ **47.** $\left(\frac{27}{8}\right)^{1/3} = \frac{3}{2}$ **49.** $\frac{1}{(-64)^{2/3}} = \frac{1}{16}$ **51.** $\frac{1}{(2xy)^{7/10}}$ **53.** $\frac{5x}{z^{1/3}}$ **55.** 3 **57.** 4 **59.** $x^{5/6}$ **61.** $x^{3/5}$ **63.** $\frac{1}{x^{5/12}}$ **65.** 25 **67.** $\frac{1}{y^{1/6}}$ **69.** $32x$ **71.** $5x^2y^3$ **73.** $\frac{x^{1/4}}{y^{3/10}}$ **75.** 3 **77.** $27y^{2/3}$ **79.** $\sqrt[4]{x}$ **81.** $2a^2$ **83.** x^2y^3 **85.** x^6y^6 **87.** $\sqrt[5]{3y}$ **89.** $\sqrt[3]{4a^2}$ **91.** $\sqrt[3]{x^2y}$ **93.** $\sqrt[6]{2^5}$ or $6\sqrt{32}$ **95.** $\sqrt[10]{x^9}$ **97.** $\sqrt[12]{a^{10}b^7}$ **99.** $\sqrt[20]{x}$ **101.** $\sqrt{y}$ **103.** $\sqrt[8]{x}$ **105.** $\sqrt[4]{x^2y}$ **107.** $\sqrt[12]{2x}$ **109.** x^9y^{15} **111.** $\sqrt[4]{a^3b^3}$ **113.** $x^{2/3} - x$ **115.** $x + 2x^{1/2} - 15$ **117.** $2x^{1/2}(3 + x)$ **119.** $15x^{1/3}(1 - 4x^{2/3})$ **121.** $\frac{x^2}{7y^{3/2}}$ **123.** $\frac{x^3}{y^2}$ **125.** 58 species of plants **127.** about 1872 calories per day **129. a.** $C = 35.74 + 0.6215t - 35.74v^{4/25} + 0.4275tv^{4/25}$ **b.** 8°F **131. a.** $C(v) = 35.74 - 35.74v^{4/25}$ **b.** $C(25) \approx -24$; When the air temperature is 0°F and the wind speed is 25 miles per hour, the windchill temperature is −24°. **c.** the point (25, −24) **133. a.** $L + 1.25S^{1/2} - 9.8D^{1/3} \leq 16.296$ **b.** eligible **145.** simplified correctly **147.** Right side should be $x^{1/2}$. **149.** does not make sense **151.** does not make sense **153.** false **155.** true **157.** $\frac{1}{4}$ of the cake **159.** $[3, \infty)$ **160.** $y = -2x + 11$ or $f(x) = -2x + 11$ **161.** $y \leq -\frac{3}{2}x + 3$

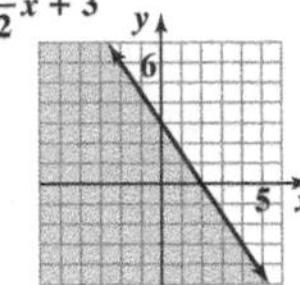

162. $\{(3, 4)\}$ **163. a.** 8 **b.** 8 **c.** $\sqrt{16} \cdot \sqrt{4} = \sqrt{16 \cdot 4}$ **164. a.** 17.32 **b.** 17.32 **c.** $\sqrt{300} = 10\sqrt{3}$ **165. a.** x^7 **b.** y^4

Section 7.3 Check Point Exercises

1. a. $\sqrt{55}$ **b.** $\sqrt{x^2 - 16}$ **c.** $\sqrt[3]{60}$ **d.** $\sqrt[7]{12x^4}$ **2. a.** $4\sqrt{5}$ **b.** $2\sqrt[3]{5}$ **c.** $2\sqrt[4]{2}$ **d.** $10|x|\sqrt{2y}$ **3.** $f(x) = \sqrt{3}|x - 2|$ **4.** $x^4y^5z\sqrt{xyz}$ **5.** $2x^3y^4\sqrt[3]{5xy^2}$ **6.** $2x^2z\sqrt[5]{x^2y^2z^3}$ **7. a.** $2\sqrt{3}$ **b.** $100\sqrt[3]{4}$ **c.** $2x^2y\sqrt[4]{2}$

Concept and Vocabulary Check

1. $\sqrt[n]{ab}$ **2.** 77 **3.** $\sqrt[3]{8} \cdot \sqrt[3]{5} = 2\sqrt[3]{5}$ **4.** $\sqrt{5}|x + 1|$ **5.** x^5; x^4; x^5

Exercise Set 7.3

1. $\sqrt{15}$ **3.** $\sqrt[3]{18}$ **5.** $\sqrt[4]{33}$ **7.** $\sqrt{33xy}$ **9.** $\sqrt[5]{24x^4}$ **11.** $\sqrt{x^2 - 9}$ **13.** $\sqrt[6]{(x - 4)^5}$ **15.** $\sqrt{x}$ **17.** $\sqrt[4]{\frac{3x}{7y}}$ **19.** $\sqrt[7]{77x^5y^3}$ **21.** $5\sqrt{2}$ **23.** $3\sqrt{5}$ **25.** $5\sqrt{3x}$ **27.** $2\sqrt[3]{2}$ **29.** $3x$ **31.** $-2y\sqrt[3]{2x^2}$ **33.** $6|x + 2|$ **35.** $2(x + 2)\sqrt[3]{4}$ **37.** $|x - 1|\sqrt{3}$ **39.** $x^3\sqrt{x}$ **41.** $x^4y^4\sqrt{y}$ **43.** $4x\sqrt{3x}$ **45.** $y^2\sqrt[3]{y^2}$ **47.** $x^4y\sqrt[3]{x^2z}$ **49.** $3x^2y^2\sqrt[3]{3x^2}$ **51.** $(x + y)\sqrt[3]{(x + y)^2}$ **53.** $y^3\sqrt[5]{y^2}$ **55.** $2xy^3\sqrt[5]{2xy^2}$ **57.** $2x^2\sqrt[4]{5x^2}$ **59.** $(x - 3)^2\sqrt[4]{(x - 3)^2}$ or $(x - 3)^2\sqrt{x - 3}$ **61.** $2\sqrt{6}$ **63.** $5\sqrt{2xy}$ **65.** $6x$ **67.** $10xy\sqrt{2y}$ **69.** $60\sqrt{2}$ **71.** $2\sqrt[3]{6}$ **73.** $2x^2\sqrt{10x}$ **75.** $5xy^4\sqrt[3]{x^2y^2}$ **77.** $2xyz\sqrt[4]{x^2z^3}$ **79.** $2xy^2z\sqrt[5]{2y^3z}$ **81.** $(x - y)^2\sqrt[3]{(x - y)^2}$ **83.** $-6x^3y^3\sqrt[3]{2yz^2}$ **85.** $-6y^2\sqrt[5]{2x^3y}$ **87.** $-6x^3y^3\sqrt{2}$ **89.** $-12x^3y^4\sqrt[4]{x^3}$ **91.** $6\sqrt{3}$ miles; 10.4 miles **93.** $8\sqrt{3}$ ft per sec; 14 ft per sec **95. a.** $\frac{7.644}{2\sqrt[4]{2}} = \frac{3.822}{\sqrt[4]{2}}$ **b.** 3.21 liters of blood per minute per square meter; (32, 3.21)

103. Graphs are the same; simplification is correct.

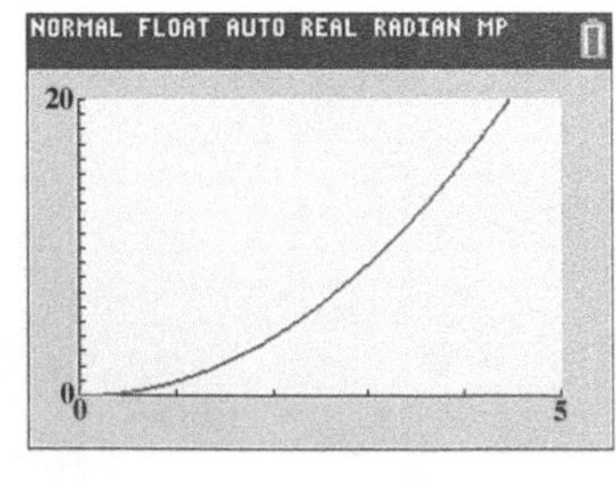

105. Graphs are not the same.; $\sqrt{3x^2 - 6x + 3} = |x - 1|\sqrt{3}$

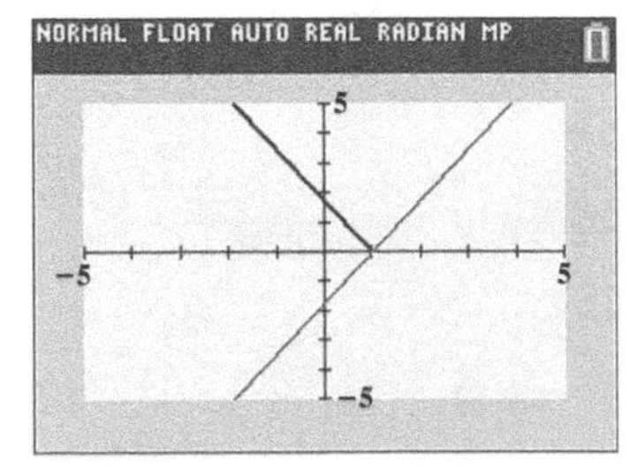

107. makes sense **109.** makes sense **111.** false **113.** false **115.** Its square root is multiplied by $\sqrt{3}$. **117.** $g(x) = \sqrt[3]{4x^2}$ **119.** [5, 11] **120.** $\{(2, -4)\}$ **121.** $(4x - 3)(16x^2 + 12x + 9)$ **122. a.** $31x$ **b.** $31\sqrt{2}$ **123. a.** $-8x$ **b.** $-8\sqrt[3]{2}$ **124.** $\frac{y\sqrt[4]{7y}}{x^3}$

Section 7.4 Check Point Exercises

1. a. $10\sqrt{13}$ **b.** $(21 - 6x)\sqrt[3]{7}$ **c.** $5\sqrt[4]{3x} + 2\sqrt[3]{3x}$ **2. a.** $21\sqrt{5}$ **b.** $-12\sqrt{3x}$ **c.** cannot be simplified **3. a.** $-9\sqrt[3]{3}$ **b.** $(5 + 3xy)\sqrt[3]{x^2y}$ **4. a.** $\frac{2\sqrt[3]{3}}{5}$ **b.** $\frac{3x\sqrt{x}}{y^5}$ **c.** $\frac{2y^2\sqrt[3]{y}}{x^4}$ **5. a.** $2x^2\sqrt{5}$ **b.** $\frac{5\sqrt{xy}}{2}$ **c.** $2x^2y$

Concept and Vocabulary Check

1. $(5+8)\sqrt{3} = 13\sqrt{3}$ **2.** $\sqrt{9\cdot 3} - \sqrt{4\cdot 3} = 3\sqrt{3} - 2\sqrt{3} = \sqrt{3}$ **3.** $\sqrt[3]{27\cdot 2} + \sqrt[3]{8\cdot 2} = 3\sqrt[3]{2} + 2\sqrt[3]{2} = 5\sqrt[3]{2}$
4. $\frac{\sqrt[n]{a}}{\sqrt[n]{b}}$ **5.** $\frac{\sqrt[3]{8}}{\sqrt[3]{27}} = \frac{2}{3}$ **6.** $\sqrt{\frac{72x^3}{2x}} = \sqrt{36x^2} = 6x$

Exercise Set 7.4

1. $11\sqrt{5}$ **3.** $7\sqrt[3]{6}$ **5.** $2\sqrt[5]{2}$ **7.** $\sqrt{13} + 2\sqrt{5}$ **9.** $7\sqrt{5} + 2\sqrt[3]{x}$ **11.** $4\sqrt{3}$ **13.** $19\sqrt{3}$ **15.** $6\sqrt{2x}$ **17.** $13\sqrt[3]{2}$
19. $(9x+1)\sqrt{5x}$ **21.** $7y\sqrt[3]{2x}$ **23.** $(3x-2)\sqrt[3]{2x}$ **25.** $4\sqrt{x-2}$ **27.** $5x\sqrt[3]{xy^2}$ **29.** $\frac{\sqrt{11}}{2}$ **31.** $\frac{\sqrt[3]{19}}{3}$ **33.** $\frac{x}{6y^4}$ **35.** $\frac{2x\sqrt{2x}}{5y^3}$
37. $\frac{x\sqrt[3]{x}}{2y}$ **39.** $\frac{x^2\sqrt[3]{50x^2}}{3y^4}$ **41.** $\frac{y\sqrt[4]{9y^2}}{x^2}$ **43.** $\frac{2x^2\sqrt[5]{2x^3}}{y^4}$ **45.** $2\sqrt{2}$ **47.** 2 **49.** $3x$ **51.** x^2y **53.** $2x^2\sqrt{5}$ **55.** $4a^5b^5$
57. $3\sqrt{xy}$ **59.** $2xy$ **61.** $2x^2y^2\sqrt[4]{y^2}$ or $2x^2y^2\sqrt{y}$ **63.** $\sqrt[3]{x+3}$ **65.** $\sqrt[3]{a^2 - ab + b^2}$ **67.** $\frac{43\sqrt{2}}{35}$ **69.** $-11xy\sqrt{2x}$ **71.** $25x\sqrt{2x}$
73. $7x\sqrt{3xy}$ **75.** $3x\sqrt[3]{5xy}$ **77.** $\left(\frac{f}{g}\right)(x) = 4x\sqrt{x}$; domain: $(0,\infty)$ **79.** $\left(\frac{f}{g}\right)(x) = 2x\sqrt[3]{2x}$; domain: $(-\infty,0)\cup(0,\infty)$
81. $P = 18\sqrt{5}$ ft; $A = 100$ sq ft **83.** $12\sqrt{5}$ m **85. a.** $5\sqrt{10}$; the projected increase in the number of Americans ages 65–84, in millions, from 2020 to 2050 **b.** 15.8; underestimates by 2.7 million
95. Graphs are not the same.; $\sqrt{16x} - \sqrt{9x} = \sqrt{x}$ **97.** makes sense **99.** does not make sense **101.** false **103.** false **105.** $2\sqrt{2}$
107. $x^2y^3a^5b\sqrt{b}$ **108.** $\{0\}$ **109.** $(x-2y)(x-6y)$ **110.** $\frac{3x^2+8x+6}{(x+3)^2(x+2)}$
111. a. $7x + 35$ **b.** $x\sqrt{7} + \sqrt{35}$ **112. a.** $6x^2 + 33x + 15$ **b.** $27 + 33\sqrt{2}$
113. $\frac{5\sqrt[5]{8x^2y^4}}{x}$

Mid-Chapter Check Point Exercises

1. 13 **2.** $2x^2y^3\sqrt{2xy}$ **3.** $5\sqrt[3]{4x^2}$ **4.** $12x\sqrt[3]{2x}$ **5.** 1 **6.** $4xy^{1/12}$ **7.** $-\sqrt{3}$ **8.** $\frac{5x\sqrt{5x}}{y^2}$ **9.** $\sqrt[4]{x^3}$ **10.** $3x\sqrt[3]{2x^2}$ **11.** $2\sqrt[3]{10}$
12. $\frac{x^2}{y^4}$ **13.** $x^{7/12}$ **14.** $x\sqrt[3]{y^2}$ **15.** $(x-2)\sqrt[3]{(x-2)^2}$ **16.** $2x^2y^4\sqrt[4]{2x^3y}$ **17.** $14\sqrt[3]{2}$ **18.** $x\sqrt[3]{x^2y^2}$ **19.** $\frac{1}{25}$ **20.** $\sqrt[6]{32}$
21. $\frac{4x}{y^2}$ **22.** $8x^2y^3\sqrt{y}$ **23.** $-10x^3y\sqrt{6y}$ **24.** $(-\infty, 6]$ **25.** $(-\infty,\infty)$

Section 7.5 Check Point Exercises

1. a. $x\sqrt{6} + 2\sqrt{15}$ **b.** $y - \sqrt[3]{7y}$ **c.** $36 - 18\sqrt{10}$ **2. a.** $11 + 2\sqrt{30}$ **b.** 1 **c.** $a - 7$ **3. a.** $\frac{\sqrt{21}}{7}$ **b.** $\frac{\sqrt[3]{6}}{3}$
4. a. $\frac{\sqrt{14xy}}{7y}$ **b.** $\frac{\sqrt[3]{3xy^2}}{3y}$ **c.** $\frac{3\sqrt[5]{4x^3y}}{y}$ **5.** $12\sqrt{3} - 18$ **6.** $\frac{3\sqrt{5} + 3\sqrt{2} + \sqrt{35} + \sqrt{14}}{3}$ **7.** $\frac{1}{\sqrt{x+3} + \sqrt{x}}$

Concept and Vocabulary Check

1. $350; -42\sqrt{10}; 30\sqrt{10}; -36$ **2.** $(\sqrt{10})^2 - (\sqrt{5})^2 = 10 - 5 = 5$ **3.** rationalizing the denominator **4.** $\sqrt{5}$ **5.** $\sqrt[3]{9}$
6. $7\sqrt{2} - 5$ **7.** $3\sqrt{6} + \sqrt{5}$

Exercise Set 7.5

1. $x\sqrt{2} + \sqrt{14}$ **3.** $7\sqrt{6} - 6$ **5.** $12\sqrt{2} - 6$ **7.** $\sqrt[3]{12} + 4\sqrt[3]{10}$ **9.** $2x\sqrt[3]{2} - \sqrt[3]{x^2}$ **11.** $32 + 11\sqrt{2}$ **13.** $34 - 15\sqrt{5}$
15. $117 - 36\sqrt{7}$ **17.** $\sqrt{6} + \sqrt{10} + \sqrt{21} + \sqrt{35}$ **19.** $\sqrt{6} - \sqrt{10} - \sqrt{21} + \sqrt{35}$ **21.** $-48 + 7\sqrt{6}$ **23.** $8 + 2\sqrt{15}$
25. $3x - 2\sqrt{3xy} + y$ **27.** -44 **29.** -71 **31.** 6 **33.** $6 - 5\sqrt{x} + x$ **35.** $\sqrt[3]{x^2} + \sqrt[3]{x} - 20$ **37.** $2x^2 + x\sqrt[3]{y^2} - y\sqrt[3]{y}$ **39.** $\frac{\sqrt{10}}{5}$
41. $\frac{\sqrt{11x}}{x}$ **43.** $\frac{3\sqrt{3y}}{y}$ **45.** $\frac{\sqrt[3]{4}}{2}$ **47.** $3\sqrt[3]{2}$ **49.** $\frac{\sqrt[3]{18}}{3}$ **51.** $\frac{4\sqrt[3]{x^2}}{x}$ **53.** $\frac{\sqrt[3]{2y}}{y}$ **55.** $\frac{7\sqrt[3]{4x}}{2x}$ **57.** $\frac{\sqrt[3]{2x^2y}}{xy}$ **59.** $\frac{3\sqrt[4]{x^3}}{x}$
61. $\frac{3\sqrt[5]{4x^2}}{x}$ **63.** $x\sqrt[4]{8x^3y}$ **65.** $\frac{3\sqrt{3y}}{xy}$ **67.** $-\frac{5a^2\sqrt{3ab}}{b^2}$ **69.** $\frac{\sqrt{2mn}}{2m}$ **71.** $\frac{3\sqrt[4]{x^3y}}{x^2y}$ **73.** $-\frac{6\sqrt[5]{xy}}{x^2y^3}$ **75.** $8\sqrt{5} - 16$
77. $\frac{13\sqrt{11} + 39}{2}$ **79.** $3\sqrt{5} - 3\sqrt{3}$ **81.** $\frac{a + \sqrt{ab}}{a - b}$ **83.** $25\sqrt{2} + 15\sqrt{5}$ **85.** $4 + \sqrt{15}$ **87.** $\frac{x - 2\sqrt{x} - 3}{x - 9}$ **89.** $\frac{3\sqrt{6} + 4}{2}$
91. $\frac{4\sqrt{xy} + 4x + y}{y - 4x}$ **93.** $\frac{3}{\sqrt{6}}$ **95.** $\frac{2x}{\sqrt[3]{2x^2y}}$ **97.** $\frac{x - 9}{x - 3\sqrt{x}}$ **99.** $\frac{a - b}{a - 2\sqrt{ab} + b}$ **101.** $\frac{1}{\sqrt{x+5} + \sqrt{x}}$ **103.** $\frac{1}{(x+y)(\sqrt{x} - \sqrt{y})}$
105. $\frac{3\sqrt{2}}{2}$ **107.** $-2\sqrt[3]{25}$ **109.** $\frac{7\sqrt{6}}{6}$ **111.** $7\sqrt{3} - 7\sqrt{2}$ **113.** 0 **115.** 6 **117.** $\frac{\sqrt{5} + 1}{2}$; 1.62 to 1 **119.** $P = 8\sqrt{2}$ in.; $A = 7$ sq in.
121. $2\sqrt{6}$ in.

131. 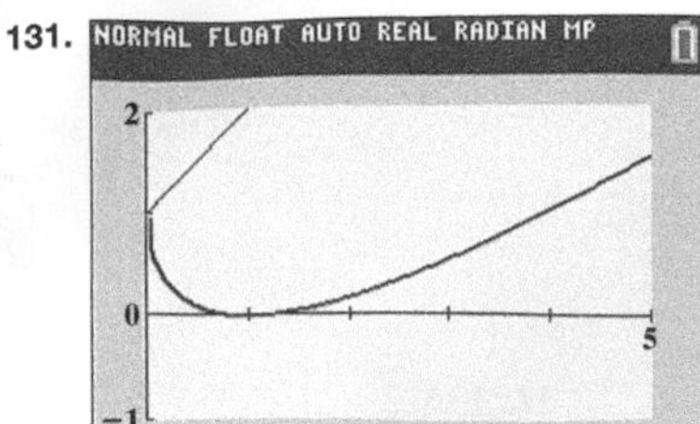

; Graphs are not the same.; $(\sqrt{x}-1)(\sqrt{x}-1)=x-2\sqrt{x}+1$

133. 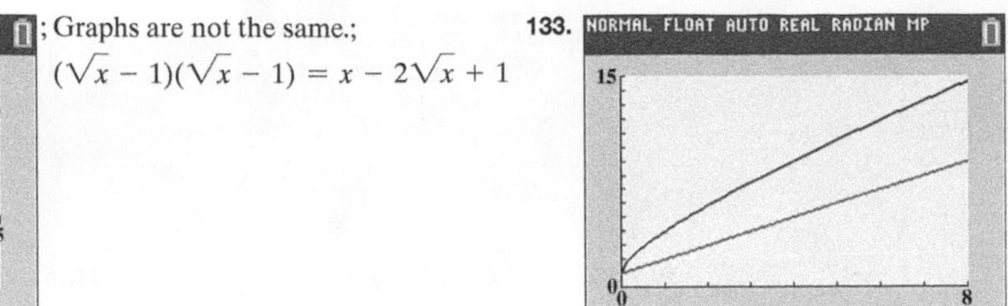

; Graphs are not the same.; $(\sqrt{x}+1)^2=x+2\sqrt{x}+1$

135. makes sense **137.** does not make sense **139.** false **141.** true **143.** $\left\{\frac{45}{7}\right\}$ **145.** $\frac{5\sqrt{2}+3\sqrt{3}-4\sqrt{6}+2}{23}$ **146.** $\frac{2x+7}{x^2-4}$ **147.** $\left\{-\frac{7}{2}\right\}$ **148.** 13 **149.** $x+5+2\sqrt{x+4}$ **150.** $\{0,5\}$ **151.** $\{-5,2\}$

Section 7.6 Check Point Exercises

1. 20 or $\{20\}$ **2.** no solution or $\varnothing$ **3.** -1 and 3, or $\{-1,3\}$ **4.** 4 or $\{4\}$ **5.** -12 or $\{-12\}$ **6.** 2060

Concept and Vocabulary Check

1. radical **2.** extraneous **3.** $2x+1$; $x^2-14x+49$ **4.** $x+2$; $x+8-6\sqrt{x-1}$ **5.** $2x+3$; 8 **6.** true **7.** false

Exercise Set 7.6

1. $\{6\}$ **3.** $\{17\}$ **5.** no solution or $\varnothing$ **7.** $\{8\}$ **9.** $\{0,3\}$ **11.** $\{1,3\}$ **13.** $\{3,7\}$ **15.** $\{9\}$ **17.** $\{8\}$ **19.** $\{35\}$ **21.** $\{16\}$ **23.** $\{2\}$ **25.** $\{2,6\}$ **27.** $\left\{\frac{5}{2}\right\}$ **29.** $\{5\}$ **31.** no solution or $\varnothing$ **33.** $\{2,6\}$ **35.** $\{0,10\}$ **37.** $\{8\}$ **39.** 4 **41.** -7 **43.** $V=\frac{\pi r^2 h}{3}$ or $V=\frac{1}{3}\pi r^2 h$ **45.** $l=\frac{8t^2}{\pi^2}$ **47.** 8 **49.** 9 **51.** 4 ft **53.** by the point (5.4, 1.16) **55. a.** 16; 16% of Americans earning \$25 thousand annually report fair or poor health.; overestimates by 1% **b.** approximately \$30 thousand **57.** 27 sq mi **59.** 149 million km

69. 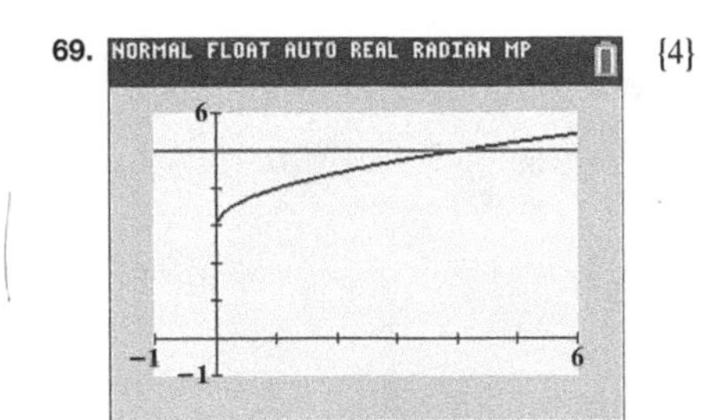

$\{4\}$

71. 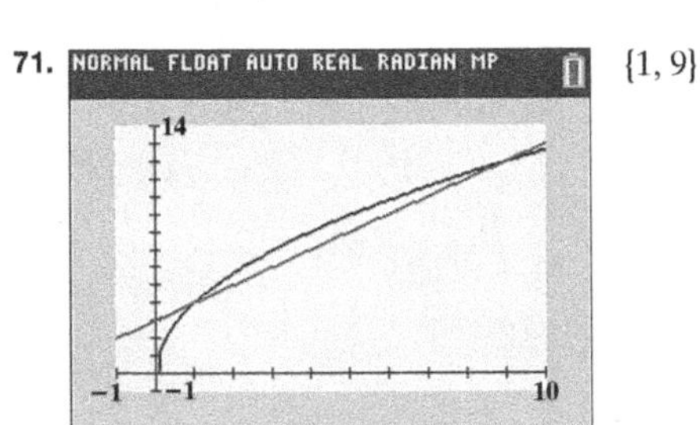

$\{1,9\}$

73. does not make sense **75.** does not make sense **77.** false **79.** true **81.** $\sqrt{x}-7=3$; $\sqrt{x}=4$; $1+\sqrt{x}=5$ **83.** $\{16\}$ **85.** $4x^3-15x^2+47x-142+\frac{425}{x+3}$ **86.** $\frac{x+2}{2(x+3)}$ **87.** $(y-3+5x)(y-3-5x)$ **88.** $6+13x$ **89.** $15x^2-29x-14$ **90.** $\frac{54+43\sqrt{2}}{-46}$

Section 7.7 Check Point Exercises

1. a. $8i$ **b.** $i\sqrt{11}$ **c.** $4i\sqrt{3}$ **2. a.** $8+i$ **b.** $-10+10i$ **3. a.** $63+14i$ **b.** $58-11i$ **4.** $-\sqrt{35}$ **5.** $\frac{18}{25}+\frac{26}{25}i$ **6.** $-\frac{1}{2}-\frac{3}{4}i$ **7. a.** 1 **b.** i **c.** $-i$

Concept and Vocabulary Check

1. $\sqrt{-1}$; -1 **2.** $4i$ **3.** complex; imaginary; real **4.** $-6i$ **5.** $14i$ **6.** 18; $-15i$; $12i$; $-10i^2$; 10 **7.** $2+9i$ **8.** $2+5i$ **9.** $-4i$ **10.** -1; 1 **11.** -1; -1; $-i$

Exercise Set 7.7

1. $10i$ **3.** $i\sqrt{23}$ **5.** $3i\sqrt{2}$ **7.** $3i\sqrt{7}$ **9.** $-6i\sqrt{3}$ **11.** $5+6i$ **13.** $15+i\sqrt{3}$ **15.** $-2-3i\sqrt{2}$ **17.** $8+3i$ **19.** $8-2i$ **21.** $5+3i$ **23.** $-1-7i$ **25.** $-2+9i$ **27.** $8+15i$ **29.** $-14+17i$ **31.** $9+5i\sqrt{3}$ **33.** $-6+10i$ **35.** $-21-15i$ **37.** $-35-14i$ **39.** $7+19i$ **41.** $-1-31i$ **43.** $3+36i$ **45.** 34 **47.** 34 **49.** 11 **51.** $-5+12i$ **53.** $21-20i$ **55.** $-\sqrt{14}$ **57.** -6 **59.** $-5\sqrt{7}$ **61.** $-2\sqrt{6}$ **63.** $\frac{3}{5}-\frac{1}{5}i$ **65.** $1+i$ **67.** $\frac{28}{25}+\frac{21}{25}i$ **69.** $-\frac{12}{13}+\frac{18}{13}i$ **71.** $0+i$ or i **73.** $\frac{3}{10}-\frac{11}{10}i$ **75.** $\frac{11}{13}-\frac{16}{13}i$ **77.** $-\frac{23}{58}+\frac{43}{58}i$ **79.** $0-\frac{7}{3}i$ or $-\frac{7}{3}i$ **81.** $-\frac{5}{2}-4i$ **83.** $-\frac{7}{3}+\frac{4}{3}i$ **85.** -1 **87.** $-i$ **89.** -1 **91.** 1 **93.** i **95.** 1 **97.** $-i$ **99.** 0 **101.** $-11-5i$ **103.** $-5+10i$ **105.** $0+47i$ or $47i$ **107.** $1-i$ **109.** 0 **111.** $10+10i$ **113.** $\frac{20}{13}+\frac{30}{13}i$ **115.** $(47+13i)$ volts **117.** $(5+i\sqrt{15})+(5-i\sqrt{15})=10$; $(5+i\sqrt{15})(5-i\sqrt{15})=25-15i^2=25+15=40$

131. $\sqrt{-9} + \sqrt{-16} = 3i + 4i = 7i$ **133.** makes sense **135.** does not make sense **137.** false **139.** false **141.** $\frac{14}{25} - \frac{2}{25}i$ **143.** $\frac{8}{5} + \frac{16}{5}i$ **144.** $\frac{x^2}{y^2}$ **145.** $x = \frac{yz}{y - z}$ **146.** $\left\{\frac{21}{11}\right\}$ **147.** $\left\{-4, \frac{1}{2}\right\}$ **148.** $\{-3, 3\}$ **149.** $-\sqrt{6}$ is a solution.

Review Exercises

1. 9 **2.** $-\frac{1}{10}$ **3.** -3 **4.** not a real number **5.** -2 **6.** 5; 1.73; 0; not a real number **7.** 2; -2; -4 **8.** $[2, \infty)$ **9.** $(-\infty, 25]$ **10.** $5|x|$ **11.** $|x + 14|$ **12.** $|x - 4|$ **13.** $4x$ **14.** $2|x|$ **15.** $-2(x + 7)$ **16.** $\sqrt[3]{5xy}$ **17.** $(\sqrt{16})^3 = 64$ **18.** $(\sqrt[5]{32})^4 = 16$ **19.** $(7x)^{1/2}$ **20.** $(19xy)^{5/3}$ **21.** $\frac{1}{8^{2/3}} = \frac{1}{4}$ **22.** $\frac{3x}{a^{4/5}b^{4/5}} = \frac{3x}{\sqrt[5]{a^4b^4}}$ **23.** $x^{7/12}$ **24.** $5^{1/6}$ **25.** $2x^2y$ **26.** $\frac{y^{1/8}}{x^{1/3}}$ **27.** x^3y^4 **28.** $y\sqrt[3]{x}$ **29.** $\sqrt[6]{x^5}$ **30.** $\sqrt[6]{x}$ **31.** $\sqrt[15]{x}$ **32.** $3150 million **33.** $\sqrt{21xy}$ **34.** $\sqrt[5]{77x^3}$ **35.** $\sqrt[6]{(x - 5)^5}$ **36.** $f(x) = \sqrt{7}|x - 1|$ **37.** $2x\sqrt{5x}$ **38.** $3x^2y^2\sqrt[3]{2x^2}$ **39.** $2y^2z\sqrt[4]{2x^3y^3z}$ **40.** $2x^2\sqrt{6x}$ **41.** $2xy\sqrt[3]{2y^2}$ **42.** $xyz^2\sqrt[5]{16y^4z}$ **43.** $\sqrt{x^2 - 1}$ **44.** $8\sqrt[3]{3}$ **45.** $9\sqrt{2}$ **46.** $(3x + y^2)\sqrt[3]{x}$ **47.** $-8\sqrt[3]{6}$ **48.** $\frac{2}{5}\sqrt[3]{2}$ **49.** $\frac{x\sqrt{x}}{10y^2}$ **50.** $\frac{y\sqrt[4]{3y}}{2x^5}$ **51.** $2\sqrt{6}$ **52.** $2\sqrt[3]{2}$ **53.** $2x\sqrt[4]{2x}$ **54.** $10x^2\sqrt{xy}$ **55.** $6\sqrt{2} + 12\sqrt{5}$ **56.** $5\sqrt[3]{2} - \sqrt[3]{10}$ **57.** $-83 + 3\sqrt{35}$ **58.** $\sqrt{xy} - \sqrt{11x} - \sqrt{11y} + 11$ **59.** $13 + 4\sqrt{10}$ **60.** $22 - 4\sqrt{30}$ **61.** -6 **62.** 4 **63.** $\frac{2\sqrt{6}}{3}$ **64.** $\frac{\sqrt{14}}{7}$ **65.** $4\sqrt[3]{3}$ **66.** $\frac{\sqrt{10xy}}{5y}$ **67.** $\frac{7\sqrt[3]{4x}}{x}$ **68.** $\frac{\sqrt[4]{189x^3}}{3x}$ **69.** $\frac{5\sqrt[5]{xy^4}}{2xy}$ **70.** $3\sqrt{3} + 3$ **71.** $\frac{\sqrt{35} - \sqrt{21}}{2}$ **72.** $10\sqrt{5} + 15\sqrt{2}$ **73.** $\frac{x + 8\sqrt{x} + 15}{x - 9}$ **74.** $\frac{5 + \sqrt{21}}{2}$ **75.** $\frac{3\sqrt{2} + 2}{7}$ **76.** $\frac{2}{\sqrt{14}}$ **77.** $\frac{3x}{\sqrt[3]{9x^2y}}$ **78.** $\frac{7}{\sqrt{35} + \sqrt{21}}$ **79.** $\frac{2}{5 - \sqrt{21}}$ **80.** $\{16\}$ **81.** no solution or $\varnothing$ **82.** $\{2\}$ **83.** $\{8\}$ **84.** $\{-4, -2\}$ **85. a.** $f(20) \approx 46.8$; In 2005 (20 years after 1985), approximately 46.8% of freshmen women described their health as above average.; The rounded value is the same as the value displayed by the graph. **b.** 36 years after 1985, or in 2021 **86.** 84 years old **87.** $9i$ **88.** $3i\sqrt{7}$ **89.** $-2i\sqrt{2}$ **90.** $12 + 2i$ **91.** $-9 + 4i$ **92.** $-12 - 8i$ **93.** $29 + 11i$ **94.** $-7 - 24i$ **95.** $113 + 0i$ or 113 **96.** $-2\sqrt{6} + 0i$ or $-2\sqrt{6}$ **97.** $\frac{15}{13} - \frac{3}{13}i$ **98.** $\frac{1}{5} + \frac{11}{10}i$ **99.** $\frac{1}{3} - \frac{5}{3}i$ **100.** 1 **101.** $-i$

Chapter Test

1. a. 6 **b.** $(-\infty, 4]$ **2.** $\frac{1}{81}$ **3.** $\frac{5y^{1/8}}{x^{1/4}}$ **4.** $\sqrt{x}$ **5.** $\sqrt[20]{x^9}$ **6.** $5|x|\sqrt{3}$ **7.** $|x - 5|$ **8.** $2xy^2\sqrt[3]{xy^2}$ **9.** $-\frac{2}{x^2}$ **10.** $\sqrt[3]{50x^2y}$ **11.** $2x\sqrt[4]{2y^3}$ **12.** $-7\sqrt{2}$ **13.** $(2x + y^2)\sqrt[3]{x}$ **14.** $2x\sqrt[3]{x}$ **15.** $12\sqrt{2} - \sqrt{15}$ **16.** $26 + 6\sqrt{3}$ **17.** $52 - 14\sqrt{3}$ **18.** $\frac{\sqrt{5x}}{x}$ **19.** $\frac{\sqrt[3]{25x}}{x}$ **20.** $-5 + 2\sqrt{6}$ **21.** $\{6\}$ **22.** $\{16\}$ **23.** $\{-3\}$ **24.** 49 months **25.** $5i\sqrt{3}$ **26.** $-1 + 6i$ **27.** $26 + 7i$ **28.** $-6 + 0i$ or -6 **29.** $\frac{1}{5} + \frac{7}{5}i$ **30.** $-i$

Cumulative Review Exercises

1. $\{(-2, -1, -2)\}$ **2.** $\left\{-\frac{1}{3}, 4\right\}$ **3.** $\left(\frac{1}{3}, \infty\right)$ **4.** $\left\{\frac{3}{4}\right\}$ **5.** $\{-1\}$

6. $x + 2y < 2$
$2y - x > 4$

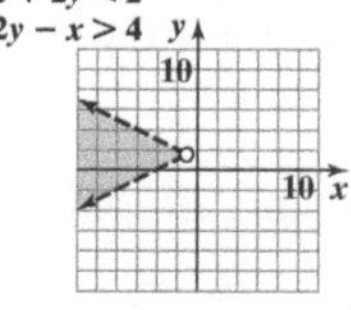

7. $\frac{x^2}{15(x + 2)}$ **8.** $\frac{x}{y}$ **9.** $8x^3 - 22x^2 + 11x + 6$ **10.** $\frac{5x - 6}{(x - 5)(x + 3)}$ **11.** -64 **12.** 0 **13.** $-\frac{16 - 9\sqrt{3}}{13}$

14. $2x^2 + x + 5 + \frac{6}{x - 2}$ **15.** $-34 - 3\sqrt{6}$ **16.** $2(3x + 2)(4x - 1)$ **17.** $(4x^2 + 1)(2x + 1)(2x - 1)$ **18.** about 53 lumens **19.** $1500 at 7% and $4500 at 9% **20.** 2650 students

CHAPTER 8

Section 8.1 Check Point Exercises

1. $\pm\sqrt{7}$ or $\{\pm\sqrt{7}\}$ **2.** $\pm\frac{\sqrt{33}}{3}$ or $\left\{\pm\frac{\sqrt{33}}{3}\right\}$ **3.** $\pm\frac{3}{2}i$ or $\left\{\pm\frac{3}{2}i\right\}$ **4.** $3 \pm \sqrt{10}$ or $\{3 \pm \sqrt{10}\}$ **5. a.** 25; $x^2 + 10x + 25 = (x + 5)^2$ **b.** $\frac{9}{4}$; $x^2 - 3x + \frac{9}{4} = \left(x - \frac{3}{2}\right)^2$ **c.** $\frac{9}{64}$; $x^2 + \frac{3}{4}x + \frac{9}{64} = \left(x + \frac{3}{8}\right)^2$ **6.** $-2 \pm \sqrt{5}$ or $\{-2 \pm \sqrt{5}\}$ **7.** $\frac{-3 \pm \sqrt{41}}{4}$ or $\left\{\frac{-3 \pm \sqrt{41}}{4}\right\}$ **8.** $\frac{3}{2} \pm i\frac{\sqrt{15}}{6}$ or $\left\{\frac{3}{2} \pm i\frac{\sqrt{15}}{6}\right\}$ **9.** 20% **10.** $10\sqrt{21}$ ft; 45.8 ft

Concept and Vocabulary Check

1. $\pm\sqrt{d}$ **2.** $\pm\sqrt{7}$ **3.** $\pm\sqrt{\frac{11}{2}}$; $\pm\frac{\sqrt{22}}{2}$ **4.** $\pm 3i$ **5.** 25 **6.** $\frac{9}{4}$ **7.** $\frac{4}{25}$ **8.** 9 **9.** $\frac{1}{9}$

Exercise Set 8.1

1. $\{\pm 5\}$ **3.** $\{\pm\sqrt{6}\}$ **5.** $\left\{\pm\frac{5}{4}\right\}$ **7.** $\left\{\pm\frac{\sqrt{6}}{3}\right\}$ **9.** $\left\{\pm\frac{4}{5}i\right\}$ **11.** $\{-10, -4\}$ **13.** $\{3 \pm \sqrt{5}\}$ **15.** $\{-2 \pm 2\sqrt{2}\}$ **17.** $\{5 \pm 3i\}$ **19.** $\left\{\frac{-3 \pm \sqrt{11}}{4}\right\}$ **21.** $\{-3, 9\}$ **23.** $1; x^2 + 2x + 1 = (x+1)^2$ **25.** $49; x^2 - 14x + 49 = (x-7)^2$ **27.** $\frac{49}{4}; x^2 + 7x + \frac{49}{4} = \left(x + \frac{7}{2}\right)^2$ **29.** $\frac{1}{16}; x^2 - \frac{1}{2}x + \frac{1}{16} = \left(x - \frac{1}{4}\right)^2$ **31.** $\frac{4}{9}; x^2 + \frac{4}{3}x + \frac{4}{9} = \left(x + \frac{2}{3}\right)^2$ **33.** $\frac{81}{64}; x^2 - \frac{9}{4}x + \frac{81}{64} = \left(x - \frac{9}{8}\right)^2$ **35.** $\{-8, 4\}$ **37.** $\{-3 \pm \sqrt{7}\}$ **39.** $\{4 \pm \sqrt{15}\}$ **41.** $\{-1 \pm i\}$ **43.** $\left\{\frac{-3 \pm \sqrt{13}}{2}\right\}$ **45.** $\left\{-\frac{3}{7}, -\frac{1}{7}\right\}$ **47.** $\left\{\frac{-1 \pm \sqrt{5}}{2}\right\}$ **49.** $\left\{-\frac{5}{2}, 1\right\}$ **51.** $\left\{\frac{-3 \pm \sqrt{6}}{3}\right\}$ **53.** $\left\{\frac{4 \pm \sqrt{13}}{3}\right\}$ **55.** $\left\{\frac{1}{4} \pm \frac{1}{4}i\right\}$ **57.** $\left\{\frac{5}{4} \pm i\frac{\sqrt{31}}{4}\right\}$ **59.** $-\frac{1}{5}, 1$ **61.** $-2 \pm 5i$ **63.** $2 \pm 2i$ **65.** $v = \sqrt{2gh}$ **67.** $r = \frac{\sqrt{AP}}{P} - 1$ **69.** $\left\{\frac{-1 \pm \sqrt{19}}{6}\right\}$ **71.** $\{-b, 2b\}$ **73. a.** $x^2 + 8x$ **b.** 16 **c.** $x^2 + 8x + 16$ **d.** $(x+4)^2$ **75.** 20% **77.** 6.25% **79. a.** 940 billionaires; underestimates by 6 billionaires **b.** 30 years after 1987, or 2017 **81.** $10\sqrt{3}$ sec; 17.3 sec **83.** $3\sqrt{5}$ mi; 6.7 mi **85.** $20\sqrt{2}$ ft; 28.3 ft **87.** $50\sqrt{2}$ ft; 70.7 ft **89.** 10 m **97.** $\{-3, 1\}$ **101.** makes sense **103.** does not make sense **105.** false **107.** false **109.** $\left\{\frac{-1 \pm \sqrt{1 - 4c}}{2}\right\}$ **111.** $\{\pm\sqrt{5}, \pm\sqrt{3}\}$ **112.** $4 - 2x$ **113.** $(1 - 2x)(1 + 2x + 4x^2)$ **114.** $x^3 - 2x^2 - 4x - 12 - \frac{42}{x-3}$ **115. a.** $\left\{-\frac{1}{2}, \frac{1}{4}\right\}$ **b.** 36; yes **116. a.** $\left\{\frac{1}{3}\right\}$ **b.** 0 **117. a.** $3x^2 + 4x + 2 = 0$ **b.** -8

Section 8.2 Check Point Exercises

1. -5 and $\frac{1}{2}$, or $\left\{-5, \frac{1}{2}\right\}$ **2.** $\frac{3 \pm \sqrt{7}}{2}$ or $\left\{\frac{3 \pm \sqrt{7}}{2}\right\}$ **3.** $-1 \pm i\frac{\sqrt{6}}{3}$ or $\left\{-1 \pm i\frac{\sqrt{6}}{3}\right\}$ **4. a.** 0; one real rational solution **b.** 81; two real rational solutions **c.** -44; two imaginary solutions that are complex conjugates **5. a.** $20x^2 + 7x - 3 = 0$ **b.** $x^2 - 50 = 0$ **c.** $x^2 + 49 = 0$ **6.** 26 years old; The point (26, 115) lies approximately on the blue graph.

Concept and Vocabulary Check

1. $\frac{-b \pm \sqrt{b^2 - 4ac}}{2a}$ **2.** 2; 9; -5 **3.** 1; -4; -1 **4.** $2 \pm \sqrt{2}$ **5.** $-1 \pm i\frac{\sqrt{6}}{2}$ **6.** $b^2 - 4ac$ **7.** no **8.** two **9.** the square root property **10.** the quadratic formula **11.** factoring and the zero-product principle **12.** false

Exercise Set 8.2

1. $\{-6, -2\}$ **3.** $\left\{1, \frac{5}{2}\right\}$ **5.** $\left\{\frac{-3 \pm \sqrt{89}}{2}\right\}$ **7.** $\left\{\frac{7 \pm \sqrt{85}}{6}\right\}$ **9.** $\left\{\frac{1 \pm \sqrt{7}}{6}\right\}$ **11.** $\left\{\frac{3}{8} \pm i\frac{\sqrt{87}}{8}\right\}$ **13.** $\{2 \pm 2i\}$ **15.** $\left\{\frac{4}{3} \pm i\frac{\sqrt{5}}{3}\right\}$ **17.** $\left\{-\frac{3}{2}, 4\right\}$ **19.** 52; two real irrational solutions **21.** 4; two real rational solutions **23.** -23; two imaginary solutions **25.** 36; two real rational solutions **27.** -60; two imaginary solutions **29.** 0; one (repeated) real rational solution **31.** $\left\{-\frac{2}{3}, 2\right\}$ **33.** $\{1 \pm \sqrt{2}\}$ **35.** $\left\{\frac{1}{6} \pm i\frac{\sqrt{107}}{6}\right\}$ **37.** $\left\{\frac{3 \pm \sqrt{65}}{4}\right\}$ **39.** $\left\{0, \frac{8}{3}\right\}$ **41.** $\left\{\frac{-6 \pm 2\sqrt{6}}{3}\right\}$ **43.** $\left\{\frac{2 \pm \sqrt{10}}{3}\right\}$ **45.** $\{2 \pm \sqrt{10}\}$ **47.** $\left\{1, \frac{5}{2}\right\}$ **49.** $\{2 \pm 2i\sqrt{2}\}$ **51.** $x^2 - 2x - 15 = 0$ **53.** $12x^2 + 5x - 2 = 0$ **55.** $x^2 - 2 = 0$ **57.** $x^2 - 20 = 0$ **59.** $x^2 + 36 = 0$ **61.** $x^2 - 2x + 2 = 0$ **63.** $x^2 - 2x - 1 = 0$ **65.** b **67.** a **69.** $1 + \sqrt{7}$ **71.** $\left\{\frac{-1 \pm \sqrt{21}}{2}\right\}$ **73.** $\left\{-2\sqrt{2}, \frac{\sqrt{2}}{2}\right\}$ **75.** $\{-3, 1, -1 \pm i\sqrt{2}\}$ **77.** 33-year-olds and 58-year-olds; The function models the actual data well. **79.** 77.8 ft; (b) **81.** 5.5 m by 1.5 m **83.** 17.6 in. and 18.6 in. **85.** 9.3 in. and 0.7 in. **87.** 7.5 hr and 8.5 hr

97. NORMAL FLOAT AUTO REAL RADIAN MP CALC MAXIMUM $Y_1=X(20-2X)$ 60 0 10 Maximum X=5 Y=50 ; depth: 5 in.; maximum area: 50 sq in. **99.** does not make sense **101.** does not make sense **103.** false **105.** true **107.** 2.4 m; yes **109.** $\left\{-3, \frac{1}{4}\right\}$ **106.** $\{3, 7\}$

111. $\frac{5\sqrt{3} - 5x}{3 - x^2}$ **112.** **113.** **114.** 1 and 5

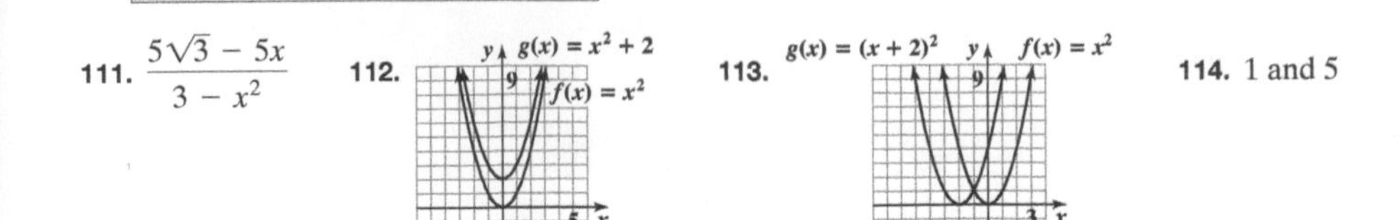

Section 8.3 Check Point Exercises

1.

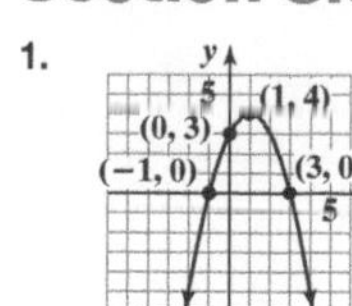

2.

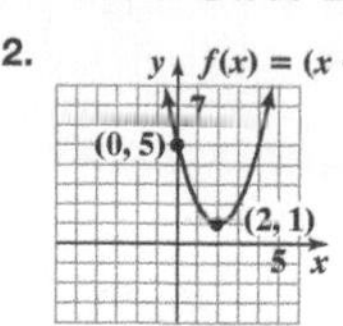

3. $(-2, -9)$ **4.** $f(x) = -x^2 + 4x + 1$; (2, 5); (0, 1); $(2+\sqrt{5}, 0)$; $(2-\sqrt{5}, 0)$ domain: $(-\infty, \infty)$; range: $(-\infty, 5]$

5. a. minimum **b.** Minimum is 984 at $x = 2$. **c.** domain: $(-\infty, \infty)$; range: $[984, \infty)$ **6.** 33.7 ft **7.** 4, −4; −16 **8.** 30 ft by 30 ft; 900 sq ft

Concept and Vocabulary Check

1. parabola; upward; downward **2.** lowest/minimum **3.** highest/maximum **4.** (h, k) **5.** $-\frac{b}{2a}; -\frac{b}{2a}$ **6.** 0; solutions **7.** $f(0)$

Exercise Set 8.3

1. $h(x) = (x-1)^2 + 1$ **3.** $j(x) = (x-1)^2 - 1$ **5.** $h(x) = x^2 - 1$ **7.** $g(x) = x^2 - 2x + 1$ **9.** $(3, 1)$ **11.** $(-1, 5)$
13. $(2, -5)$ **15.** $(-1, 9)$

17.

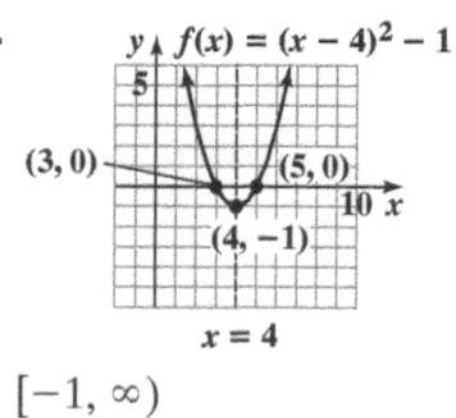

$[-1, \infty)$

19.

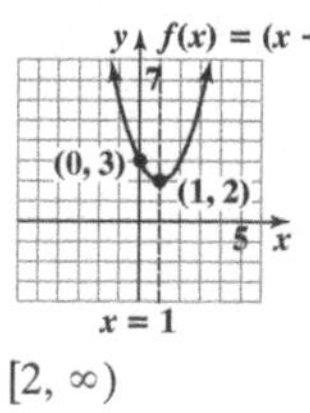

$[2, \infty)$

21.

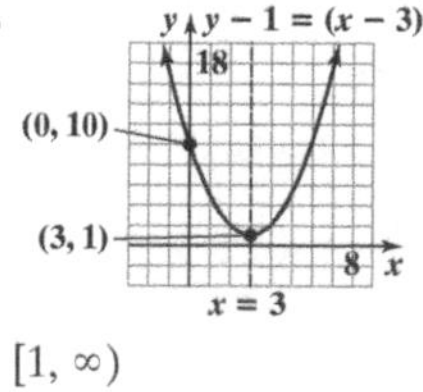

$[1, \infty)$

23.

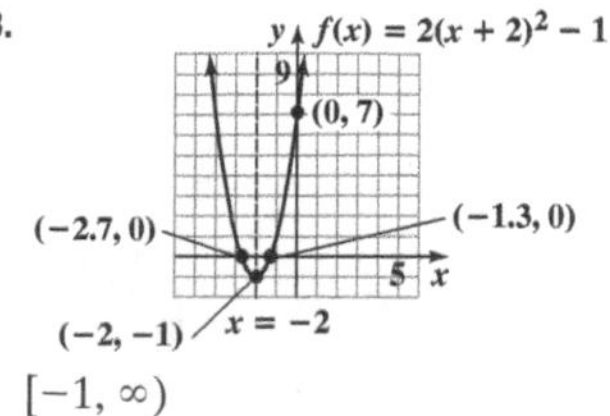

$[-1, \infty)$

25.

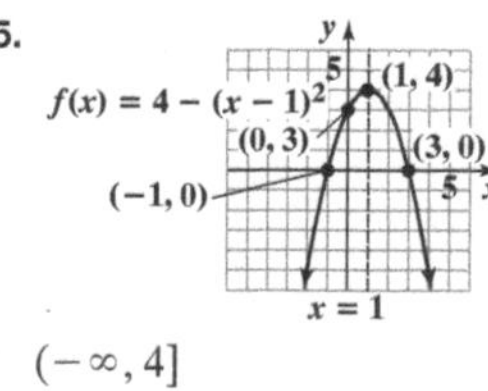

$(-\infty, 4]$

27.

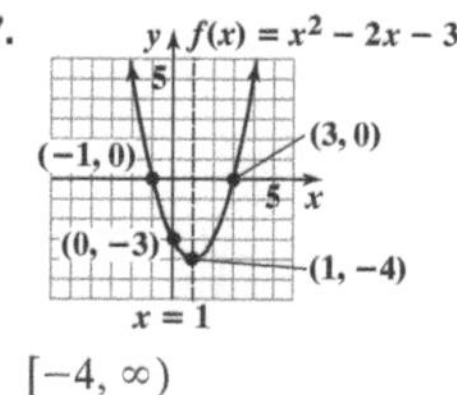

$[-4, \infty)$

29.

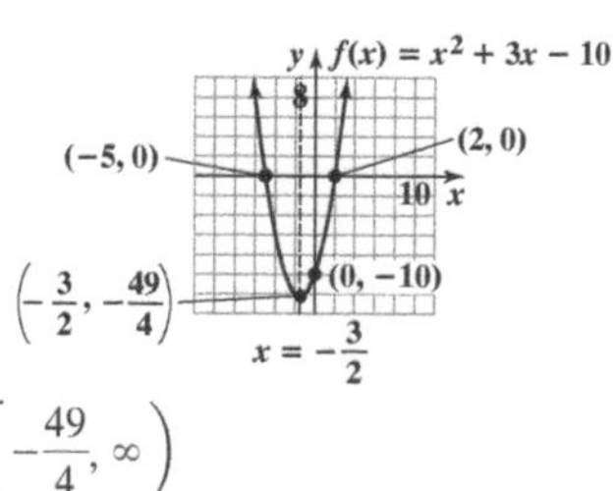

$\left[-\frac{49}{4}, \infty\right)$

31.

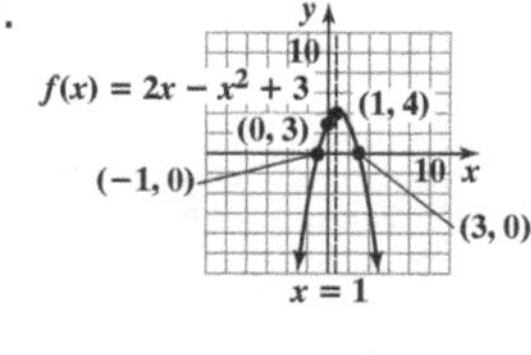

$(-\infty, 4]$

33.

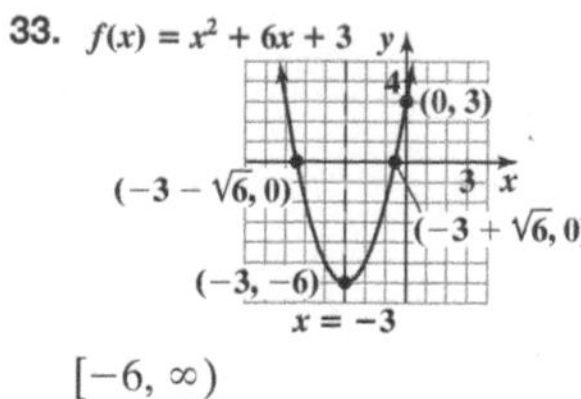

$[-6, \infty)$

35.

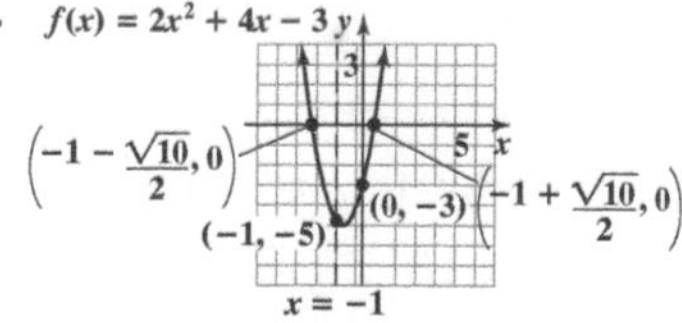

$[-5, \infty)$

37.

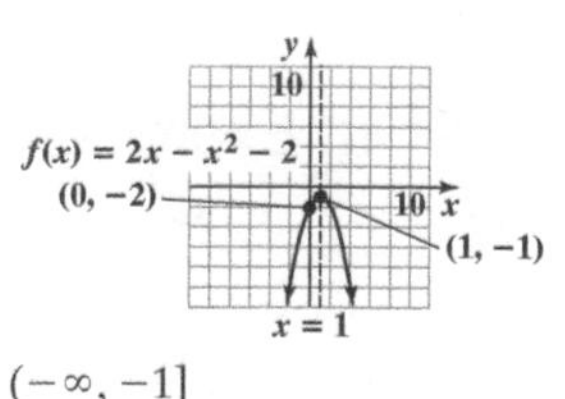

$(-\infty, -1]$

39. a. minimum **b.** Minimum is −13 at $x = 2$. **c.** domain: $(-\infty, \infty)$; range: $[-13, \infty)$ **41. a.** maximum **b.** Maximum is 1 at $x = 1$.
c. domain: $(-\infty, \infty)$; range: $(-\infty, 1]$ **43. a.** minimum **b.** Minimum is $-\frac{5}{4}$ at $x = \frac{1}{2}$. **c.** domain: $(-\infty, \infty)$; range: $\left[-\frac{5}{4}, \infty\right)$
45. domain: $(-\infty, \infty)$; range: $[-2, \infty)$ **47.** domain: $(-\infty, \infty)$; range: $(-\infty, -6]$ **49.** $f(x) = 2(x-5)^2 + 3$ **51.** $f(x) = 2(x+10)^2 - 5$
53. $f(x) = -3(x+2)^2 + 4$ **55.** $f(x) = 3(x-11)^2$ **57. a.** 2 sec; 224 ft **b.** 5.7 sec **c.** 160; 160 feet is the height of the building.

d. $s(t) = -16t^2 + 64t + 160$; (2, 224); (0, 160); (5.7, 0)

59. 8 and 8; 64 **61.** 8, −8; −64 **63.** length: 300 ft; width: 150 ft; maximum area: 45,000 sq ft
65. 12.5 yd by 12.5 yd; 156.25 sq yd **67.** 5 in.; 50 sq in. **69. a.** $C(x) = 525 + 0.55x$
b. $P(x) = -0.001x^2 + 2.45x - 525$
c. 1225 sandwiches; \$975.63

77. a.

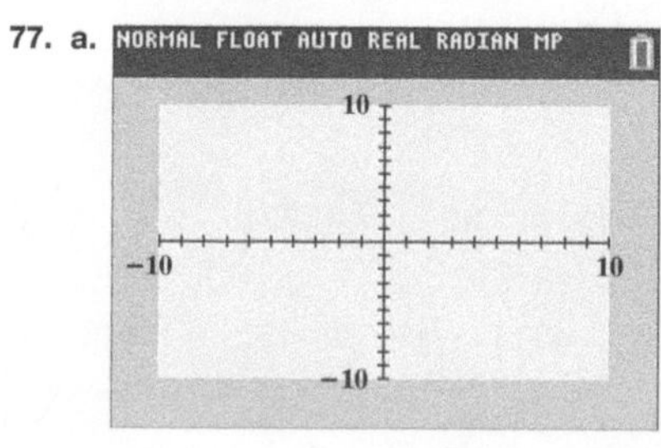

b. $(20.5, -120.5)$ **c.**

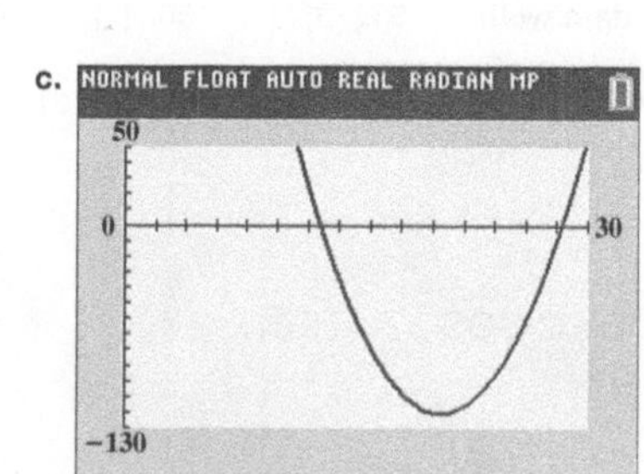

d. Answers will vary.

79. (2.5, 185)

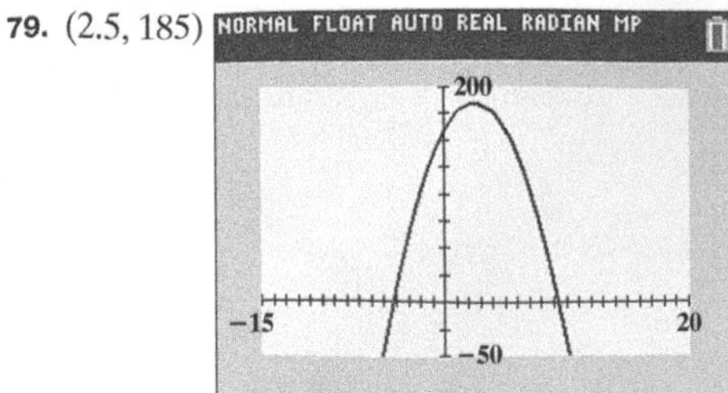

81. (−30, 91)

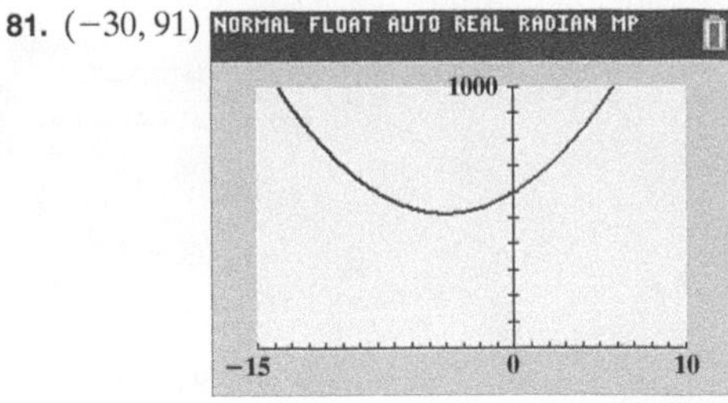

83. does not make sense **85.** does not make sense **87.** false **89.** false **91.** $x = 3$; (0, 11) **93.** $f(x) = -2(x + 3)^2 - 1$

95. 65 trees; 16,900 pounds **96.** {7} **97.** $\frac{x}{x - 2}$ **98.** {(6, −2)} **99.** {−1, 9} **100.** $\left\{-2, \frac{5}{2}\right\}$ **101.** $5u^2 + 11u + 2 = 0$

Mid-Chapter Check Point Exercises

1. $\left\{-\frac{1}{3}, \frac{11}{3}\right\}$ **2.** $\left\{-1, \frac{7}{5}\right\}$ **3.** $\left\{\frac{3 \pm \sqrt{15}}{3}\right\}$ **4.** $\{-3 \pm \sqrt{7}\}$ **5.** $\left\{\pm\frac{6\sqrt{5}}{5}\right\}$ **6.** $\left\{\frac{5}{2} \pm i\frac{\sqrt{7}}{2}\right\}$ **7.** $\{\pm i\sqrt{13}\}$

8. $\left\{-4, \frac{1}{2}\right\}$ **9.** $\{-3 \pm 2\sqrt{6}\}$ **10.** $\{2 \pm \sqrt{3}\}$ **11.** $\left\{\frac{3}{4} \pm i\frac{\sqrt{23}}{4}\right\}$ **12.** $\left\{\frac{-3 \pm \sqrt{41}}{4}\right\}$ **13.** {4} **14.** $\{-5 \pm 2\sqrt{7}\}$

15. $f(x) = (x - 3)^2 - 4$

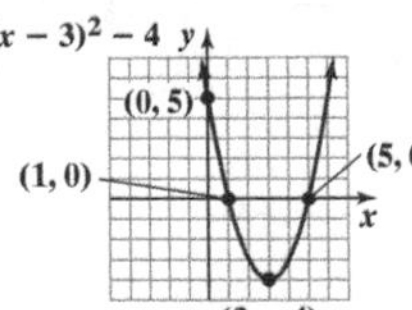

domain: $(-\infty, \infty)$; range: $[-4, \infty)$

16. $g(x) = 5 - (x + 2)^2$

(−2, 5)
(0, 1)
$(-2-\sqrt{5}, 0)$
$(-2+\sqrt{5}, 0)$

domain: $(-\infty, \infty)$; range: $(-\infty, 5]$

17. $h(x) = -x^2 - 4x + 5$

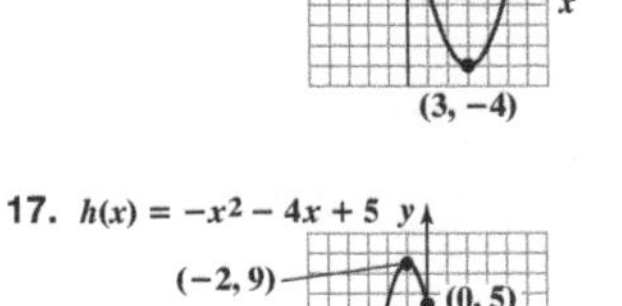

domain: $(-\infty, \infty)$; range: $(-\infty, 9]$

18.

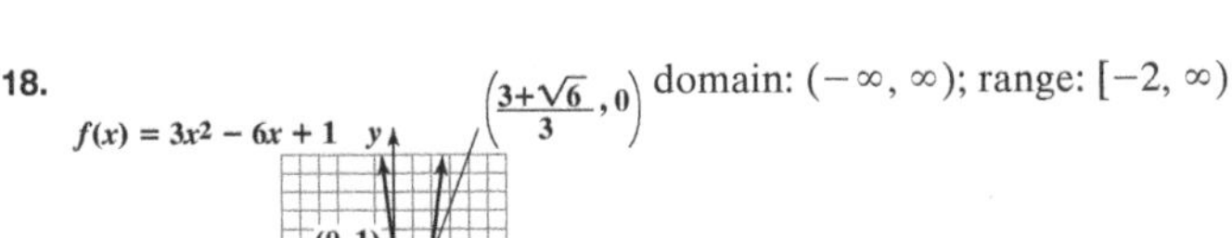

domain: $(-\infty, \infty)$; range: $[-2, \infty)$

19. two imaginary solutions **20.** two real rational solutions **21.** $8x^2 - 2x - 3 = 0$ **22.** $x^2 - 12 = 0$ **23.** 75 cabinets per day; $1200
24. −9, −9; 81 **25.** 10 in.; 100 sq in.

Section 8.4 Check Point Exercises

1. $-\sqrt{3}, -\sqrt{2}, \sqrt{2}$, and $\sqrt{3}$ or $\{\pm\sqrt{2}, \pm\sqrt{3}\}$ **2.** 16 or {16} **3.** $-\sqrt{6}$, −1, 1, and $\sqrt{6}$, or $\{-\sqrt{6}, -1, 1, \sqrt{6}\}$ **4.** −1 and 2, or {−1, 2}

5. $-\frac{1}{27}$ and 64, or $\left\{-\frac{1}{27}, 64\right\}$

Concept and Vocabulary Check

1. x^2; $u^2 - 13u + 36 = 0$ **2.** $x^{1/2}$ or $\sqrt{x}$; $u^2 - 2u - 8 = 0$ **3.** $x + 3$; $u^2 + 7u - 18 = 0$ **4.** x^{-1}; $2u^2 - 7u + 3 = 0$
5. $x^{1/3}$; $u^2 + 2u - 3 = 0$

Exercise Set 8.4

1. {−2, 2, −1, 1} **3.** $\{-3, 3, -\sqrt{2}, \sqrt{2}\}$ **5.** $\{-2i, 2i, -\sqrt{2}, \sqrt{2}\}$ **7.** {1} **9.** {49} **11.** {25, 64} **13.** {2, 12} **15.** $\{-\sqrt{3}, 0, \sqrt{3}\}$

17. {−5, −2, −1, 2} **19.** $\left\{-\frac{1}{4}, \frac{1}{5}\right\}$ **21.** $\left\{\frac{1}{3}, 2\right\}$ **23.** $\left\{\frac{-2 \pm \sqrt{7}}{3}\right\}$ **25.** {−8, 27} **27.** {−243, 32} **29.** {1} **31.** {−8, −2, 1, 4}

33. −2, 2, −1, and 1; c **35.** 1; e **37.** 2 and 3; f **39.** −5, −4, 1, and 2 **41.** $-\frac{3}{2}$ and $-\frac{1}{3}$ **43.** $\frac{64}{15}$ and $\frac{81}{20}$ **45.** $\frac{5}{2}$ and $\frac{25}{6}$
47. ages 20 and 55; The function models the data well. **53.** {3, 5} **55.** {1} **57.** {−1, 4} **59.** {1, 8} **61.** makes sense

63. does not make sense **65.** true **67.** false **69.** $\left\{\sqrt[3]{-2}, \frac{\sqrt[3]{225}}{5}\right\}$ **71.** $\frac{1}{5x - 1}$ **72.** $\frac{1}{2} + \frac{3}{2}i$ **73.** {(4, −2)} **74.** $\left\{-3, \frac{5}{2}\right\}$

75. {−2, −1, 2} **76.** $\frac{-x - 5}{x + 3}$

Section 8.5 Check Point Exercises

1. $(-\infty, -4) \cup (5, \infty)$

−8 −6 −4 −2 0 2 4 6 8 10 12

2. $\left[\frac{-3 - \sqrt{7}}{2}, \frac{-3 + \sqrt{7}}{2}\right]$

−5 −4 −3 −2 −1 0 1 2 3 4 5

3. $(-\infty, -3] \cup [-1, 1]$

−5 −4 −3 −2 −1 0 1 2 3 4 5

4. $(-2, 5)$

5. $(-\infty, -1) \cup [1, \infty)$

6. between 1 and 4 seconds, excluding $t = 1$ and $t = 4$

Concept and Vocabulary Check

1. $x^2 + 8x + 15 = 0$; boundary **2.** $(-\infty, -5), (-5, -3), (-3, \infty)$ **3.** true **4.** true **5.** $(-\infty, -2) \cup [1, \infty)$

Exercise Set 8.5

1. $(-\infty, -2) \cup (4, \infty)$

3. $[-3, 7]$

5. $(-\infty, 1) \cup (4, \infty)$

7. $(-\infty, -4) \cup (-1, \infty)$

9. $[2, 4]$

11. $\left[-4, \frac{2}{3}\right]$

13. $\left(-3, \frac{5}{2}\right)$

15. $\left(-1, -\frac{3}{4}\right)$

17. $(-\infty, 0] \cup [4, \infty)$

19. $\left(-\infty, -\frac{3}{2}\right) \cup (0, \infty)$

21. $[0, 1]$

23. $[2 - \sqrt{2}, 2 + \sqrt{2}]$

25. $\left(-\infty, \frac{2 - \sqrt{10}}{3}\right) \cup \left(\frac{2 + \sqrt{10}}{3}, \infty\right)$;

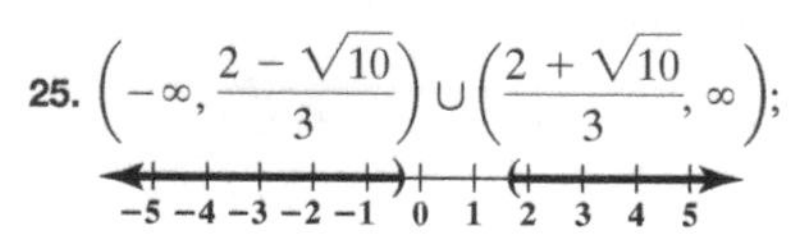

27. $\left(-\infty, \frac{5 - \sqrt{33}}{4}\right] \cup \left[\frac{5 + \sqrt{33}}{4}, \infty\right)$;

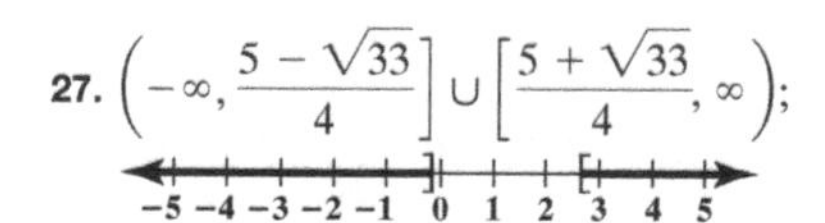

29. no solution or $\varnothing$

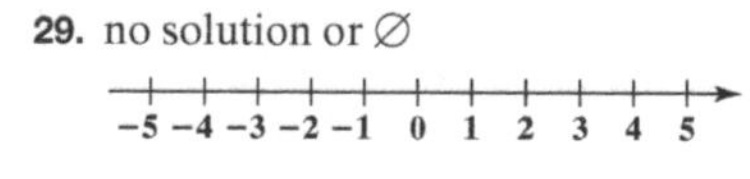

31. $[1, 2] \cup [3, \infty)$

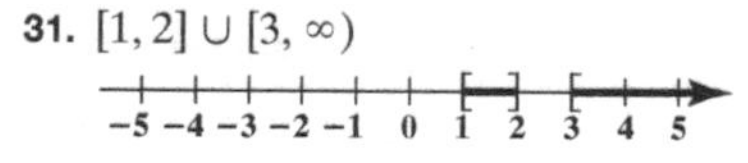

33. $[-2, -1] \cup [1, \infty)$

35. $(-\infty, -3)$

37. $(-1, \infty)$

39. $\{0\} \cup [9, \infty)$

41. $(-\infty, -3) \cup (4, \infty)$

43. $(-4, -3)$

45. $[2, 4)$

47. $\left(-\infty, -\frac{4}{3}\right) \cup [2, \infty)$

49. $(-\infty, 0) \cup (3, \infty)$

51. $(-\infty, -5) \cup (-3, \infty)$

53. $\left(-\infty, \frac{1}{2}\right) \cup \left[\frac{7}{5}, \infty\right)$

55. $(-\infty, -6] \cup (-2, \infty)$

57. $\left(-\infty, \frac{1}{2}\right] \cup [2, \infty)$

59. $(-1, 1)$

61. $(-\infty, -8) \cup (-6, 4) \cup (6, \infty)$

63. $(-3, 2)$

65. $(-\infty, -1) \cup (1, 2) \cup (3, \infty)$

67. $\left[-6, -\frac{1}{2}\right] \cup [1, \infty)$ **69.** $(-\infty, -2) \cup [-1, 2)$ **71.** between 0 and 3 seconds, excluding $t = 0$ and $t = 3$ **73. a.** dry: 160 ft; wet: 185 ft **b.** dry pavement: graph (b); wet pavement: graph (a) **c.** extremely well; Function values and data are identical. **d.** speeds exceeding 76 miles per hour; points on graph (b) to the right of (76, 540) **75.** The company's production level must be at least 20,000 wheelchairs per month. For values of x greater than or equal to 20,000, the graph lies on or below the line $y = 425$. **77.** The length of the shorter side cannot exceed 6 ft. **83.** $\left[-3, \frac{1}{2}\right]$ **85.** $(-\infty, 3) \cup [8, \infty)$ **87.** $(-3, -1) \cup (2, \infty)$ **89. a.** $f(x) = 0.1375x^2 + 0.7x + 37.8$ **b.** speeds exceeding 52 mph **91.** does not make sense **93.** does not make sense **95.** false **97.** true **99.** Answers will vary.; example: $\frac{x - 3}{x + 4} \geq 0$ **101.** $\{2\}$ **103.** $(-\infty, 2) \cup (2, \infty)$ **105.** $27 - 3x^2 \geq 0$; $[-3, 3]$ **106.** $(-19, 29)$ **107.** $\frac{2(x - 1)}{(x + 4)(x - 3)}$ **108.** $(x^2 + 4y^2)(x + 2y)(x - 2y)$

109.

110.

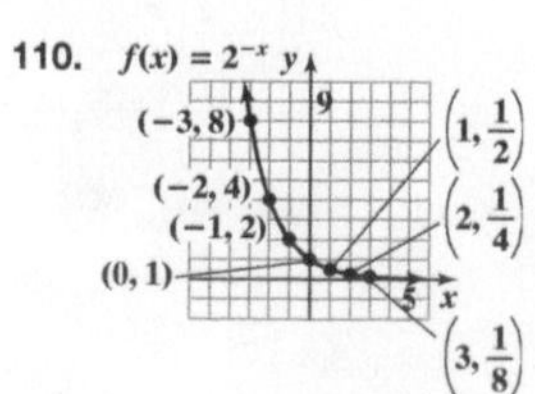

111.

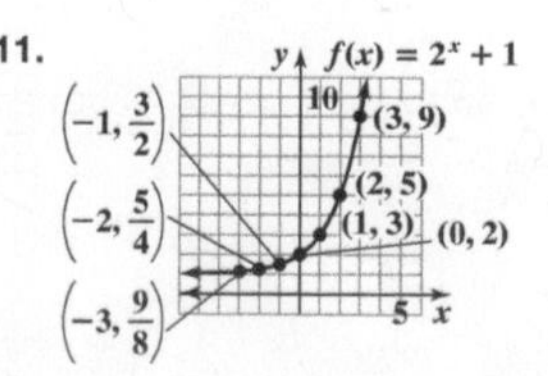

Review Exercises

1. $\{\pm 8\}$ **2.** $\{\pm 5\sqrt{2}\}$ **3.** $\left\{\pm \frac{\sqrt{6}}{3}\right\}$ **4.** $\{4 \pm 3\sqrt{2}\}$ **5.** $\{-7 \pm 6i\}$ **6.** $100; x^2 + 20x + 100 = (x + 10)^2$

7. $\frac{9}{4}; x^2 - 3x + \frac{9}{4} = \left(x - \frac{3}{2}\right)^2$ **8.** $\{3, 9\}$ **9.** $\left\{\frac{7 \pm \sqrt{53}}{2}\right\}$ **10.** $\left\{\frac{-3 \pm \sqrt{41}}{4}\right\}$ **11.** 8% **12.** 14 weeks **13.** $60\sqrt{5}$ m; 134.2 m

14. $\{1 \pm \sqrt{5}\}$ **15.** $\{1 \pm 3i\sqrt{2}\}$ **16.** $\left\{\frac{-2 \pm \sqrt{10}}{2}\right\}$ **17.** two imaginary solutions **18.** two real rational solutions

19. two real irrational solutions **20.** $\left\{-\frac{2}{3}, 4\right\}$ **21.** $\{-5, 2\}$ **22.** $\left\{\frac{1 \pm \sqrt{21}}{10}\right\}$ **23.** $\{-4, 4\}$ **24.** $\{3 \pm 2\sqrt{2}\}$ **25.** $\left\{\frac{1}{6} \pm i\frac{\sqrt{23}}{6}\right\}$

26. $\{4 \pm \sqrt{5}\}$ **27.** $15x^2 - 4x - 3 = 0$ **28.** $x^2 + 81 = 0$ **29.** $x^2 - 48 = 0$ **30. a.** 261 ft; overestimates by 1 ft **b.** 40 mph
31. a. by the point (35, 261) **b.** by the point (40, 267) **8.** 8.8 sec

33.

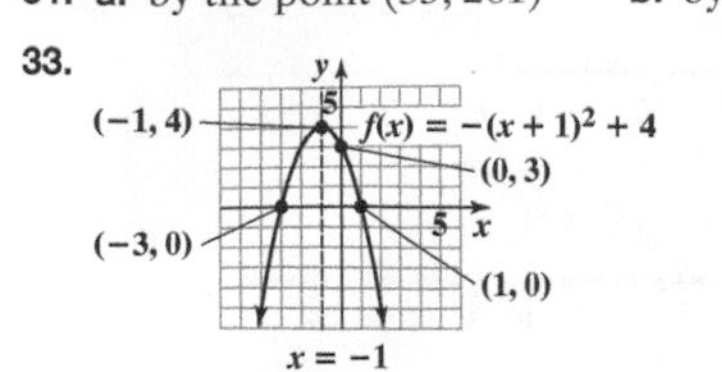

34.

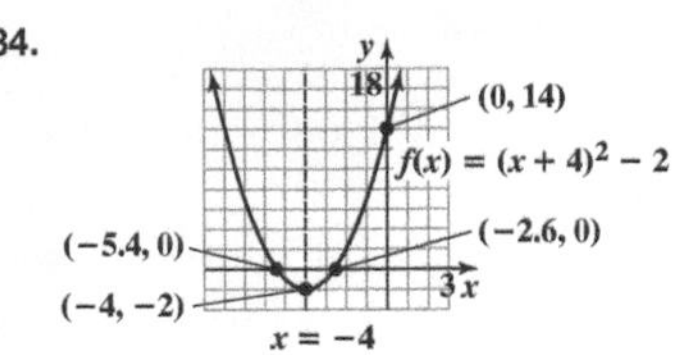

35.

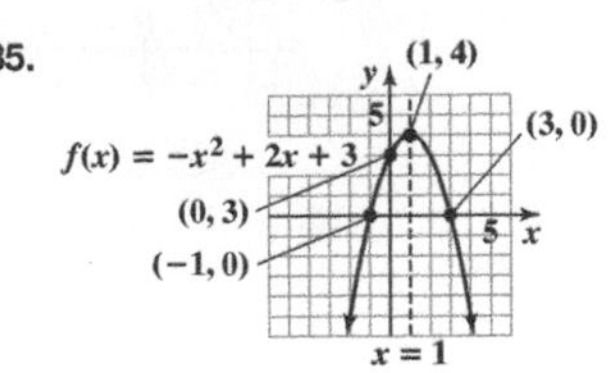

36.

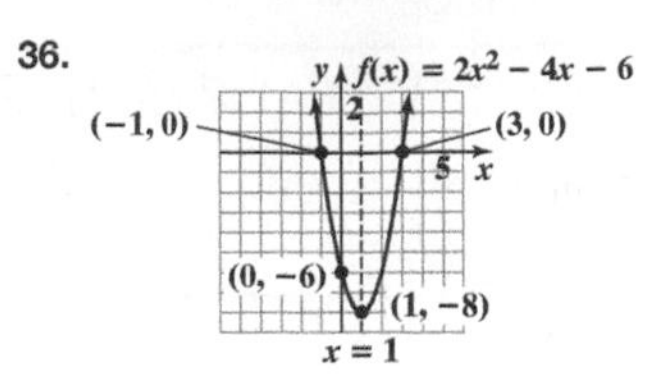

37. 25 in. of rainfall per year; 13.5 in. of growth **38.** 12.5 sec; 2540 feet **39.** 7.2 h; 622 per 100,000 males **40.** 250 yd by 500 yd; 125,000 sq yard

41. −7 and 7; −49 **42.** $\{-\sqrt{2}, \sqrt{2}, -2, 2\}$ **43.** $\{1\}$ **44.** $\{-5, -1, 3\}$ **45.** $\left\{-\frac{1}{8}, \frac{1}{7}\right\}$ **46.** $\{-27, 64\}$ **47.** $\{16\}$

48. $\left(-3, \frac{1}{2}\right)$;

49. $(-\infty, -4] \cup \left[-\frac{1}{2}, \infty\right)$;

50. $(-3, 0) \cup (1, \infty)$;

51. $(-\infty, -2) \cup (6, \infty)$;

52. $(-\infty, 4) \cup \left[\frac{23}{4}, \infty\right)$;

53. between 1 and 2 seconds, excluding $t = 1$ and $t = 2$
54. a. 200 beats per minute
b. between 0 and 4 minutes and more than 12 minutes after the workout; between 0 and 4 minutes; Answers will vary.

Chapter Test

1. $\left\{\pm \frac{\sqrt{10}}{2}\right\}$ **2.** $\{3 \pm 2\sqrt{5}\}$ **3.** $64; x^2 - 16x + 64 = (x - 8)^2$ **4.** $\frac{1}{25}; x^2 + \frac{2}{5}x + \frac{1}{25} = \left(x + \frac{1}{5}\right)^2$ **5.** $\{3 \pm \sqrt{2}\}$ **6.** $50\sqrt{2}$ ft

7. two real irrational solutions **8.** two imaginary solutions **9.** $\left\{-5, \frac{1}{2}\right\}$ **10.** $\{-4 \pm \sqrt{11}\}$ **11.** $\{-2 \pm 5i\}$ **12.** $\left\{\frac{3}{2} \pm \frac{1}{2}i\right\}$

13. $x^2 - 4x - 21 = 0$ **14.** $x^2 + 100 = 0$ **15. a.** $327.1 \approx 327$; underestimates by 5 **b.** 22 years after 2003, or 2025

16.
f(x) = (x + 1)² + 4
(−1, 4) (0, 5)
x = −1

17.

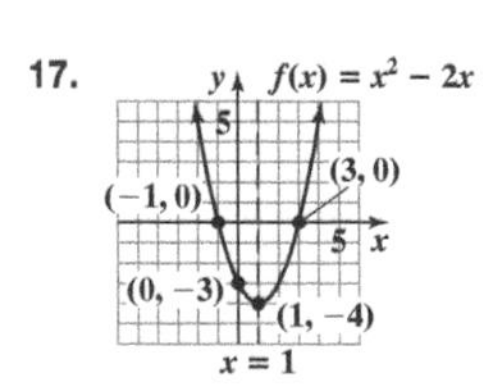

18. after 2 sec; 69 ft **19.** 4.1 sec **20.** 23 computers; $169 hundreds or $16,900
21. $\{1, 2\}$ **22.** $\{-3, 3, -2, 2\}$ **23.** $\{1, 512\}$
24. $(-3, 4)$;
25. $(-\infty, 3) \cup [10, \infty)$;

Cumulative Review Exercises

1. $\left\{\frac{2}{3}\right\}$ **2.** $\{(-5, 2)\}$ **3.** $\{(1, 4, -2)\}$ **4.** $[10, \infty)$ **5.** $(-\infty, -4]$ **6.** $(2, \infty)$ **7.** $(-2, 3)$ **8.** $\{3, 9\}$ **9.** no solution or $\varnothing$ **10.** $\{12\}$

11. $\left\{\frac{-2 \pm \sqrt{14}}{2}\right\}$ **12.** $\{8, 27\}$ **13.** $\left[-2, \frac{3}{2}\right]$

14.

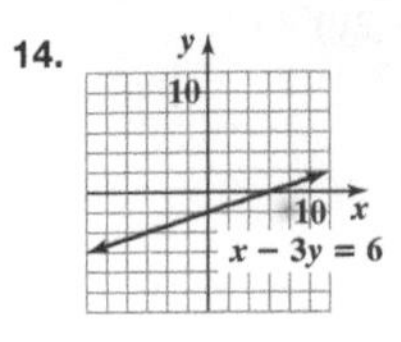

15.

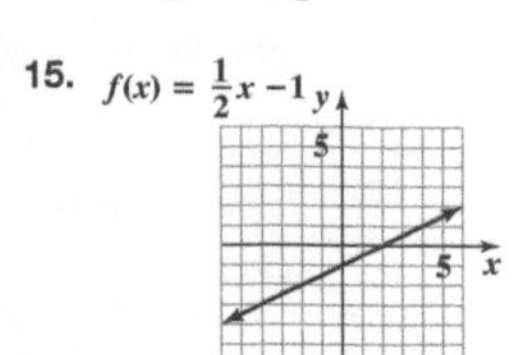

16.

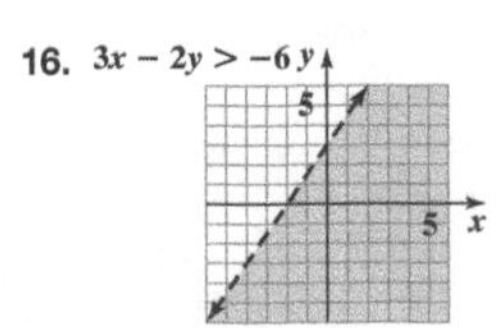

17.

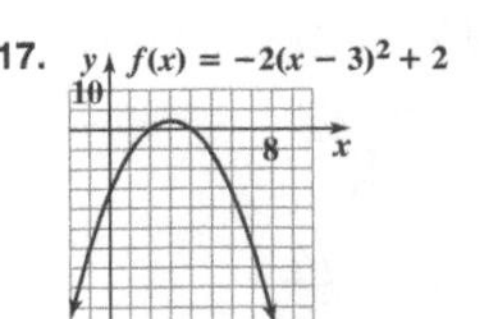

18. $-16x + 24y$ **19.** $-\frac{20x^7}{y^4}$ **20.** $x^2 - 5xy - 16y^2$ **21.** $6x^2 + 13x - 5$ **22.** $9x^4 - 24x^2y + 16y^2$ **23.** $\frac{3x^2 + 6x - 2}{(x + 5)(x + 2)}$ **24.** $\frac{x - 3}{x}$
25. $\frac{x - 4}{3x + 6}$ **26.** $5xy\sqrt{2x}$ **27.** $9\sqrt{2}$ **28.** $44 + 6i$ **29.** $(9x^2 + 1)(3x + 1)(3x - 1)$ **30.** $2x(4x - 1)(3x - 2)$
31. $(x + 3y)(x^2 - 3xy + 9y^2)$ **32.** $x^2 + 2x - 13; 22$ **33.** $x + 5; (-\infty, 2) \cup (2, \infty)$ **34.** $2a + h + 3$
35. $3x^2 - 7x + 18 - \frac{28}{x + 2}$ **36.** $R = -\frac{Ir}{I - 1}$ or $R = \frac{Ir}{1 - I}$ **37.** $y = -3x - 1$ or $f(x) = -3x - 1$ **38.** 6 **39.** \$620
40. 13 yd by 4 yd **41.** \$2600 at 12% and \$1400 at 14% **42.** 11 amps

CHAPTER 9

Section 9.1 Check Point Exercises

1. approximately \$160; overestimates by \$11

2.

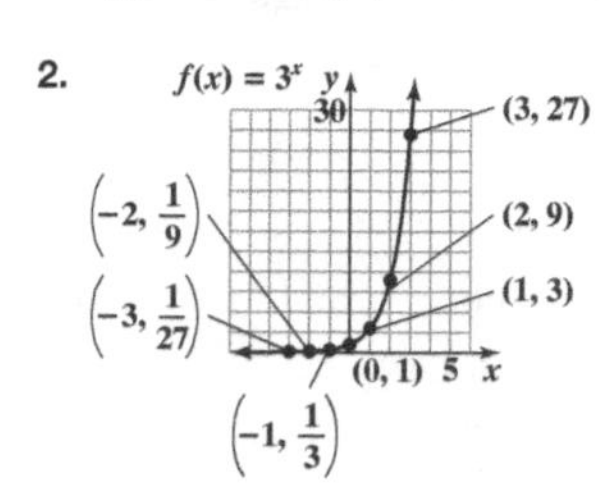

3.

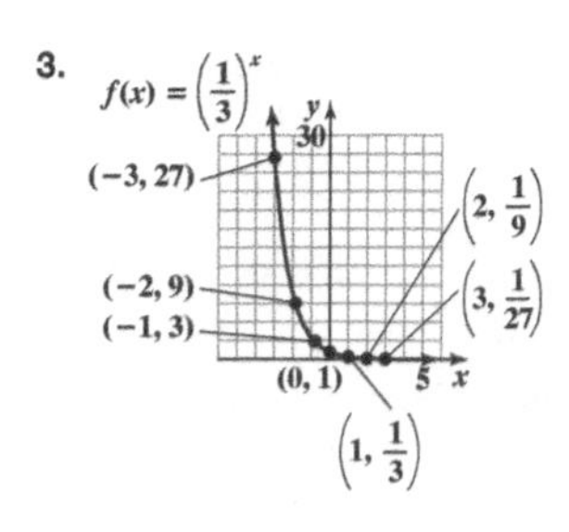

4.

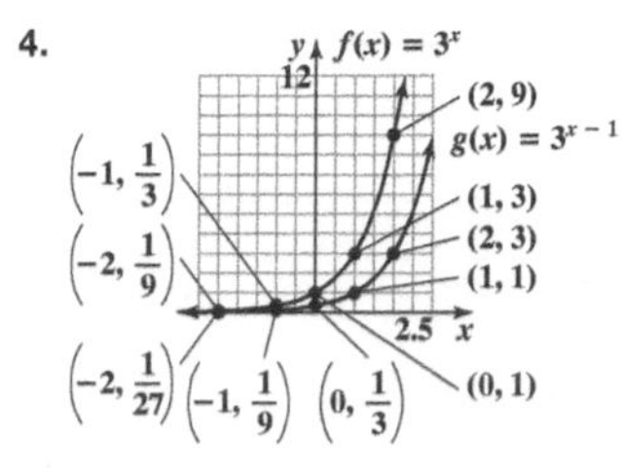

The graph of g is the graph of f shifted 1 unit to the right.

5.

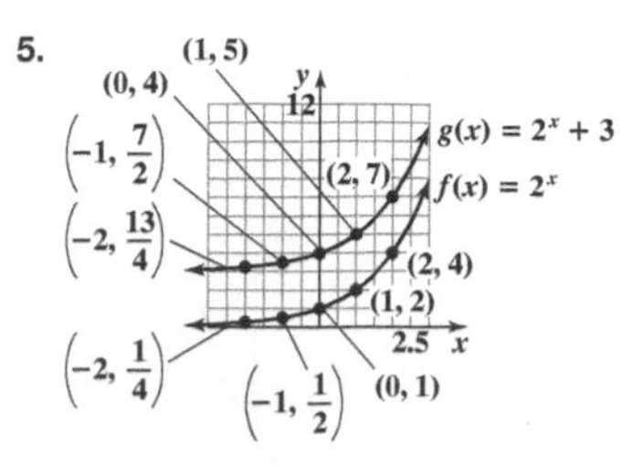

The graph of g is the graph of f shifted up 3 units.

6. approximately 4067 **7. a.** \$14,859.47 **b.** \$14,918.25

Concept and Vocabulary Check

1. b^x; $(-\infty, \infty)$; $(0, \infty)$ **2.** x; $y = 0$; horizontal **3.** e; natural; 2.72 **4.** A; P; r; n **5.** semiannually; quarterly; continuous

Exercise Set 9.1

1. 10.556 **3.** 11.665 **5.** 0.125 **7.** 9.974 **9.** 0.387

11.

x	$f(x)$
−2	$\frac{1}{9}$
−1	$\frac{1}{3}$
0	1
1	3
2	9

; d

13.

x	$f(x)$
−2	$-\frac{8}{9}$
−1	$-\frac{2}{3}$
0	0
1	2
2	8

; e

15.

x	$f(x)$
−2	9
−1	3
0	1
1	$\frac{1}{3}$
2	$\frac{1}{9}$

; f

17.

x	$f(x)$
−2	$\frac{1}{16}$
−1	3
0	1
1	4
2	16

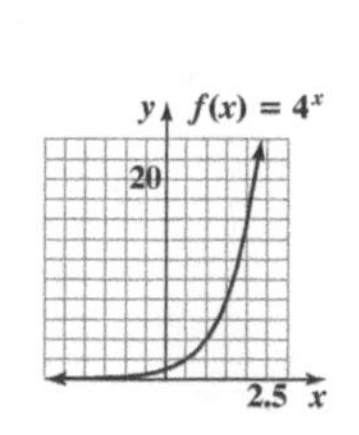

19.

x	$g(x)$
−2	$\frac{4}{9}$
−1	$\frac{2}{3}$
0	1
1	$\frac{3}{2}$
2	$\frac{9}{4}$

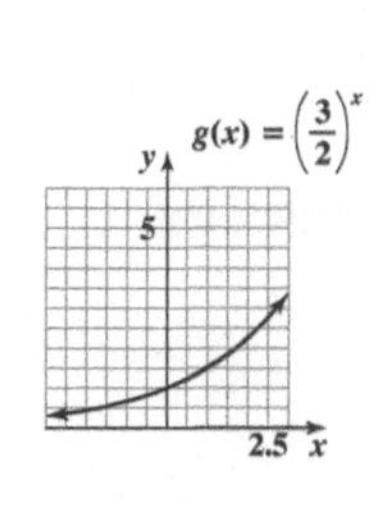

21.

x	$h(x)$
−2	4
−1	2
0	1
1	$\frac{1}{2}$
2	$\frac{1}{4}$

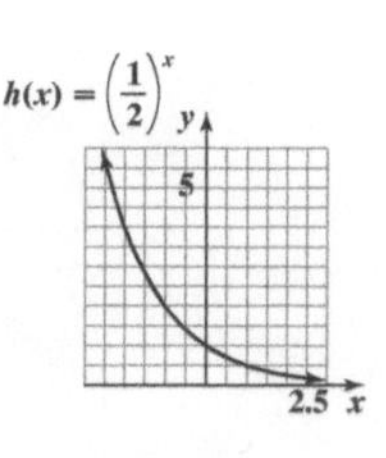

23.

x	$f(x)$
−2	2.78
−1	1.67
0	1
1	0.6
2	0.36

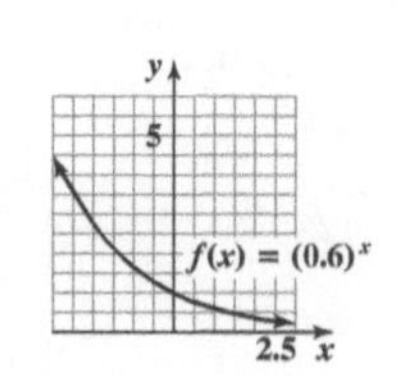

25.

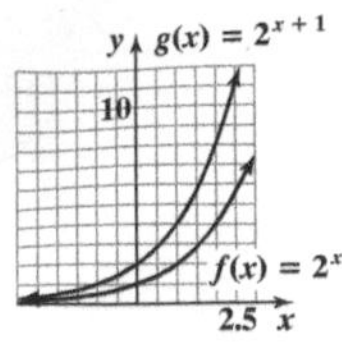

The graph of g is the graph of f shifted 1 unit to the left.

27.

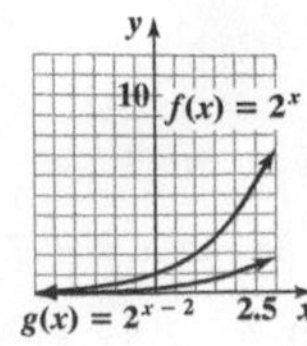

The graph of g is the graph of f shifted 2 units to the right.

29.

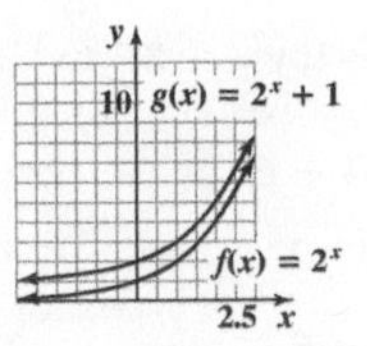

The graph of g is the graph of f shifted up 1 unit.

31.

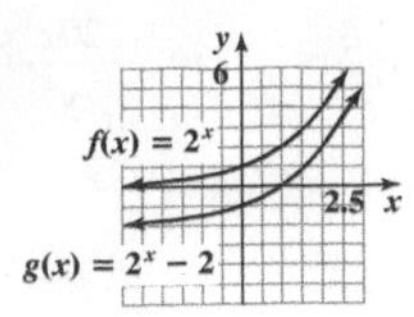

The graph of g is the graph of f shifted down 2 units.

33.

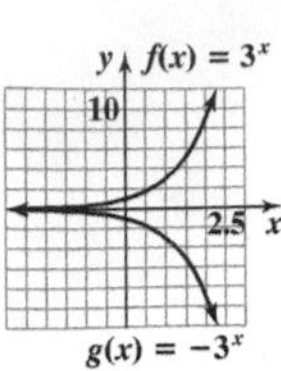

The graph of g is a reflection of the graph of f across the x-axis.

35.

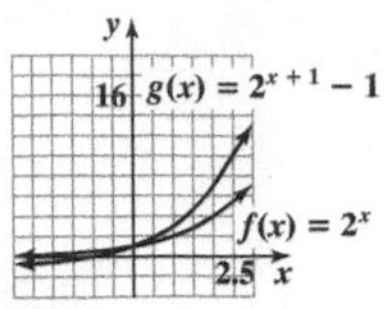

The graph of g is the graph of f shifted 1 unit to the left and 1 unit down.

37.

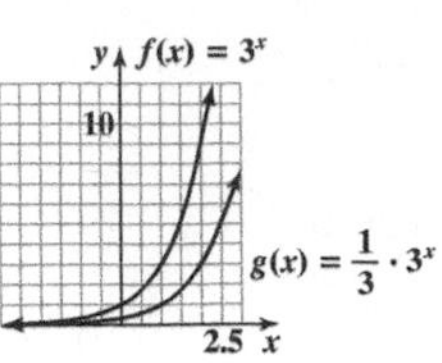

The graph of g is the graph of f stretched vertically by a factor of $\frac{1}{3}$.

39. a. \$13,116.51 **b.** \$13,157.04 **c.** \$13,165.31 **41.** 7% compounded monthly **43.** domain: $(-\infty, \infty)$; range: $(-2, \infty)$ **45.** domain: $(-\infty, \infty)$; range: $(1, \infty)$ **47.** domain: $(-\infty, \infty)$; range: $(0, \infty)$

49. (0, 1)

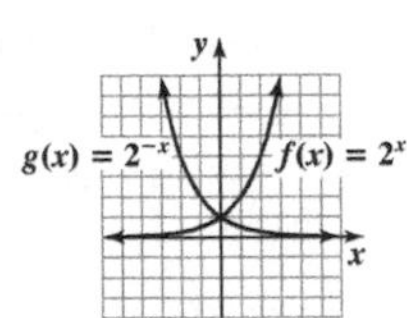

51.

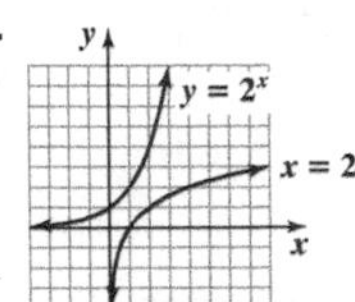

53. a. 574 million **b.** 1148 million **c.** 2295 million **d.** 4590 million **e.** It appears to double. **55.** \$832,744 **57. a.** 64% **b.** 67% **c.** the exponential model **59. a.** 100% **b.** about 68.5% **c.** about 30.8% **d.** about 20% **61.** 11.3; About 11.3% of 30-year-olds have some coronary heart disease.

63. a. about 1429 people **b.** about 24,546 people **c.** The number of ill people cannot exceed the population.; The asymptote indicates that the number of ill people will not exceed 30,000, the population of the town.

69. a. $f(t) = 10{,}000\left(1 + \frac{0.05}{4}\right)^{4t}$; $f(t) = 10{,}000\left(1 + \frac{0.045}{12}\right)^{12t}$

b.

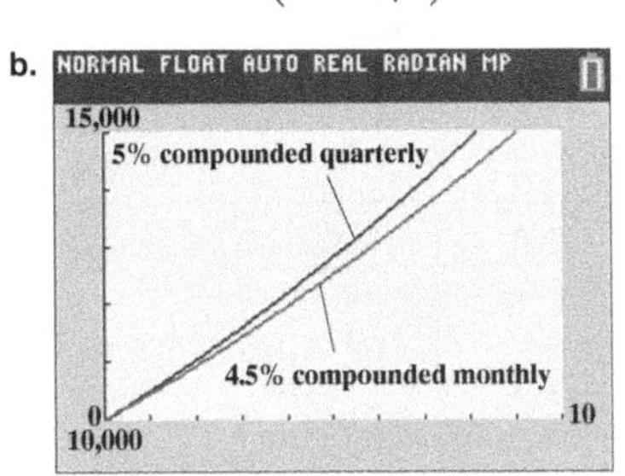

the bank that pays 5% interest compounded quarterly

71. does not make sense **73.** does not make sense **75.** false **77.** false **79. a.** $y = \left(\frac{1}{3}\right)^x$ **b.** $y = \left(\frac{1}{5}\right)^x$ **c.** $y = 5^x$ **d.** $y = 3^x$

81. $b = \frac{Da}{a - D}$ **82.** -1 **83.** $\{-2, 5\}$ **84. a.** 11 **b.** 127 **85.** x **86.** $y = \frac{x + 5}{7}$

Section 9.2 Check Point Exercises

1. a. $5x^2 + 1$ **b.** $25x^2 + 60x + 35$ **2.** $f(g(x)) = 7\left(\frac{x}{7}\right) = x$; $g(f(x)) = \frac{7x}{7} = x$

3. $f(g(x)) = 4\left(\frac{x + 7}{4}\right) - 7 = (x + 7) - 7 = x$; $g(f(x)) = \frac{(4x - 7) + 7}{4} = \frac{4x}{4} = x$ **4.** $f^{-1}(x) = \frac{x - 7}{2}$ **5.** $f^{-1}(x) = \sqrt[3]{\frac{x + 1}{4}}$

6. (b) and (c) **7.**

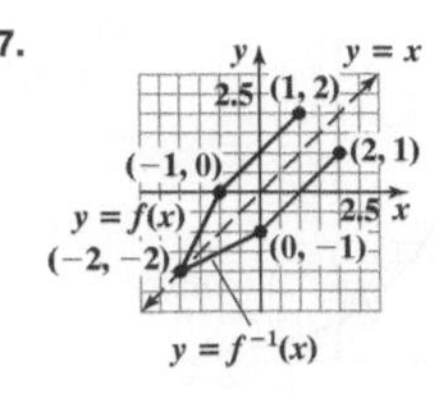

Concept and Vocabulary Check

1. composition; $f(g(x))$ **2.** f; $g(x)$ **3.** composition; $g(f(x))$ **4.** g; $f(x)$ **5.** false **6.** false **7.** inverse **8.** x; x **9.** horizontal; one-to-one **10.** $y = x$

Exercise Set 9.2

1. a. $2x + 14$ **b.** $2x + 7$ **c.** 18 **3. a.** $2x + 5$ **b.** $2x + 9$ **c.** 9 **5. a.** $20x^2 - 11$ **b.** $80x^2 - 120x + 43$ **c.** 69 **7. a.** $x^4 - 4x^2 + 6$ **b.** $x^4 + 4x^2 + 2$ **c.** 6 **9. a.** $\sqrt{x - 1}$ **b.** $\sqrt{x} - 1$ **c.** 1 **11. a.** x **b.** x **c.** 2 **13. a.** x **b.** x **c.** 2 **15.** $f(g(x)) = x$; $g(f(x)) = x$; inverses **17.** $f(g(x)) = x$; $g(f(x)) = x$; inverses **19.** $f(g(x)) = \frac{5x - 56}{9}$; $g(f(x)) = \frac{5x - 4}{9}$; not inverses **21.** $f(g(x)) = x$; $g(f(x)) = x$; inverses **23.** $f(g(x)) = x$; $g(f(x)) = x$; inverses **25. a.** $f^{-1}(x) = x - 3$ **b.** $f(f^{-1}(x)) = (x - 3) + 3 = x$ and $f^{-1}(f(x)) = (x + 3) - 3 = x$ **27. a.** $f^{-1}(x) = \frac{x}{2}$ **b.** $f(f^{-1}(x)) = 2\left(\frac{x}{2}\right) = x$ and $f^{-1}(f(x)) = \frac{2x}{2} = x$ **29. a.** $f^{-1}(x) = \frac{x - 3}{2}$ **b.** $f(f^{-1}(x)) = 2\left(\frac{x - 3}{2}\right) + 3 = x$ and $f^{-1}(f(x)) = \frac{(2x + 3) - 3}{2} = x$ **31. a.** $f^{-1}(x) = \sqrt[3]{x - 2}$ **b.** $f(f^{-1}(x)) = (\sqrt[3]{x - 2})^3 + 2 = x$ and $f^{-1}(f(x)) = \sqrt[3]{(x^3 + 2) - 2} = x$ **33. a.** $f^{-1}(x) = \sqrt[3]{x} - 2$ **b.** $f(f^{-1}(x)) = ((\sqrt[3]{x} - 2) + 2)^3 = x$ and $f^{-1}(f(x)) = \sqrt[3]{(x + 2)^3} - 2 = x$ **35. a.** $f^{-1}(x) = \frac{1}{x}$ **b.** $f(f^{-1}(x)) = \frac{1}{\frac{1}{x}} = x$ and $f^{-1}(f(x)) = \frac{1}{\frac{1}{x}} = x$ **37. a.** $f^{-1}(x) = x^2, x \geq 0$ **b.** $f(f^{-1}(x)) = \sqrt{x^2} = x$ and $f^{-1}(f(x)) = (\sqrt{x})^2 = x$ **39. a.** $f^{-1}(x) = \sqrt{x - 1}$ **b.** $f(f^{-1}(x)) = (\sqrt{x - 1})^2 + 1 = x$ and $f^{-1}(f(x)) = \sqrt{(x^2 + 1) - 1} = x$ **41. a.** $f^{-1}(x) = \frac{3x + 1}{x - 2}$ **b.** $f(f^{-1}(x)) = \frac{2\left(\frac{3x + 1}{x - 2}\right) + 1}{\left(\frac{3x + 1}{x - 2}\right) - 3} = x$ and $f^{-1}(f(x)) = \frac{3\left(\frac{2x + 1}{x - 3}\right) + 1}{\left(\frac{2x + 1}{x - 3}\right) - 2} = x$ **43. a.** $f^{-1}(x) = (x - 3)^3 + 4$ **b.** $f(f^{-1}(x)) = \sqrt[3]{((x - 3)^3 + 4) - 4} + 3 = x$ and $f^{-1}(f(x)) = ((\sqrt[3]{x - 4} + 3) - 3)^3 + 4 = x$ **45.** no inverse **47.** no inverse **49.** inverse function

51.

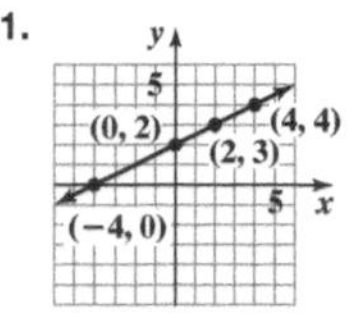

53.

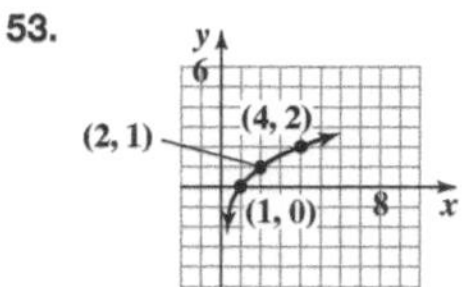

55. 5 **57.** 1 **59.** 2 **61.** 1 **63.** −6 **65.** −7 **67.** 3 **69.** 11

71. a. f represents the price after a \$400 discount, and g represents the price after a 25% discount (75% of the regular price). **b.** $0.75x - 400$; $f \circ g$ represents an additional \$400 discount on a price that has already been reduced by 25%. **c.** $0.75(x - 400) = 0.75x - 300$; $g \circ f$ represents an additional 25% discount on a price that has already been reduced \$400. **d.** $f \circ g$; $0.75x - 400 < 0.75x - 300$, so $f \circ g$ represents the lower price after the two discounts. **e.** $f^{-1}(x) = x + 400$; f^{-1} represents the regular price, since the value of x here is the price after a \$400 discount. **73. a.** f: {(U.S., 1%), (U.K., 8%), (Italy, 5%), (France, 5%), (Holland, 30%)} **b.** f^{-1}: {(1%, U.S.), (8%, U.K.), (5%, Italy), (5%, France), (30%, Holland)}; No; The input 5% is associated with two outputs, Italy and France. **75. a.** No horizontal line intersects the graph of f in more than one point. **b.** $f^{-1}(0.25)$, or approximately 15, represents the number of people who would have to be in the room so that the probability of two sharing a birthday would be 0.25; $f^{-1}(0.5)$, or approximately 23, represents the number of people so that the probability would be 0.5; $f^{-1}(0.7)$, or approximately 30, represents the number of people so that the probability would be 0.7. **77.** $f(g(x)) = \frac{9}{5}\left[\frac{5}{9}(x - 32)\right] + 32 = x$ and

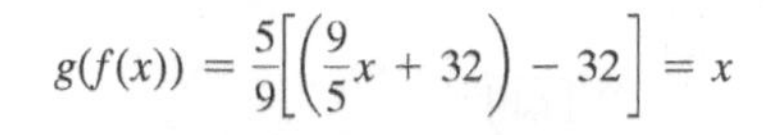

$g(f(x)) = \frac{5}{9}\left[\left(\frac{9}{5}x + 32\right) - 32\right] = x$

85. ; inverse function

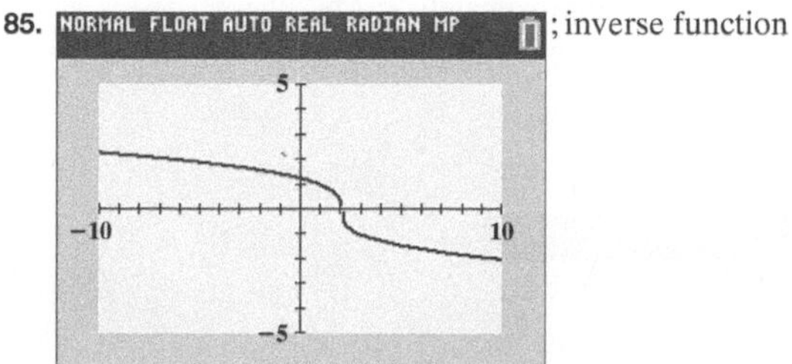
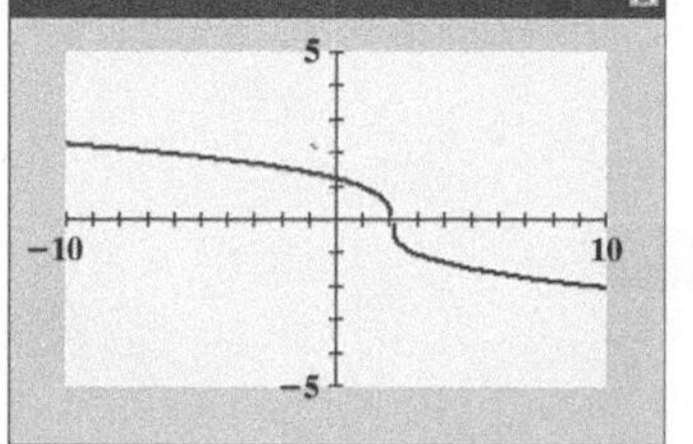

87. ; no inverse function

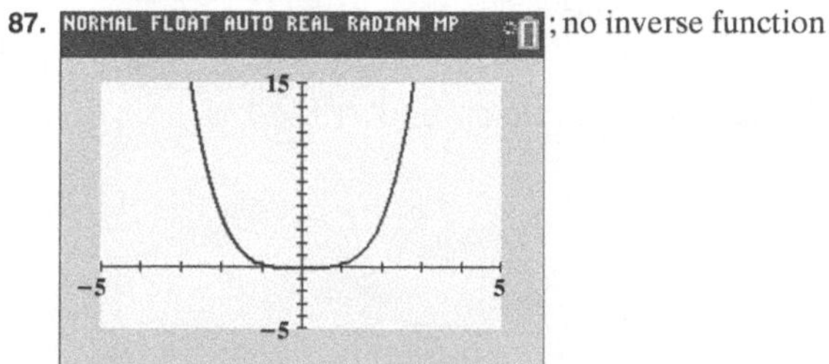

89. ; inverse function

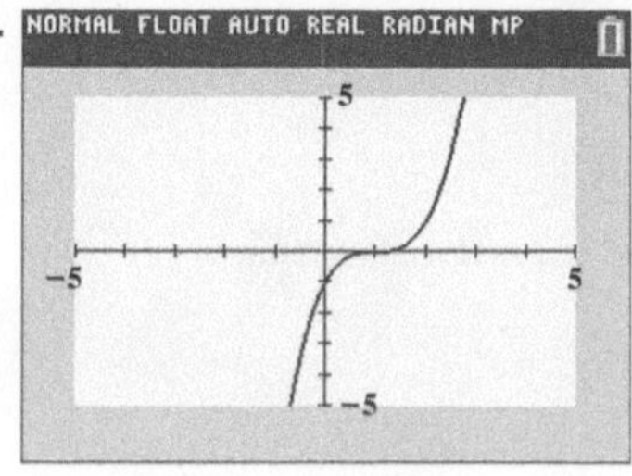

91. ; inverse function

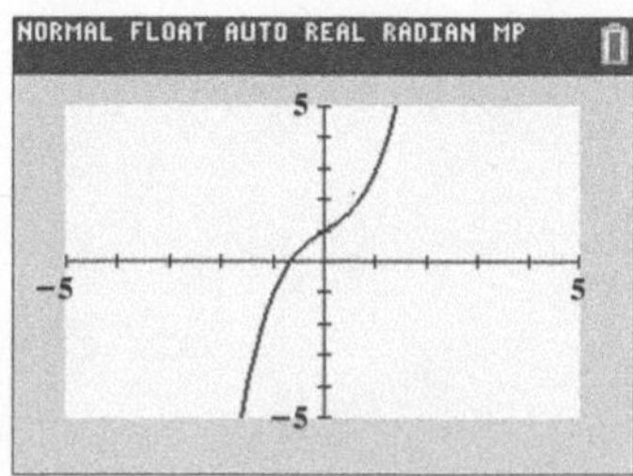

93. 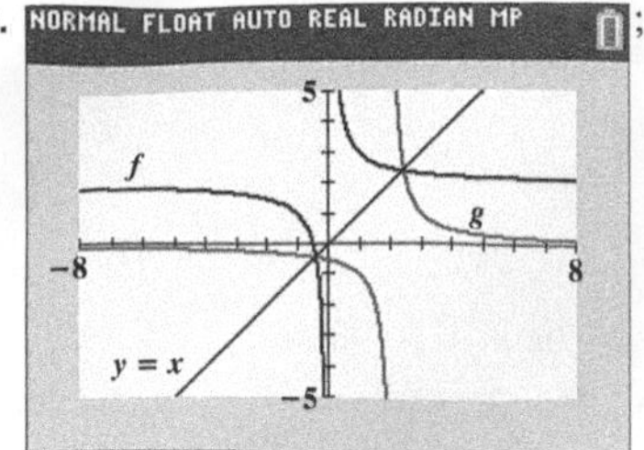

; inverses **95.** does not make sense **97.** makes sense **99.** false **101.** true

103. Answers will vary; Examples are $f(x) = \sqrt{x+5}$ and $g(x) = 3x^2$.

105. $f(f(x)) = \dfrac{3\left(\dfrac{3x-2}{5x-3}\right) - 2}{5\left(\dfrac{3x-2}{5x-3}\right) - 3} = \dfrac{3(3x-2) - 2(5x-3)}{5(3x-2) - 3(5x-3)} = \dfrac{9x - 6 - 10x + 6}{15x - 10 - 15x + 9} = \dfrac{-x}{-1} = x$

107. 5×10^8 **108.** 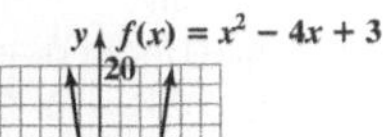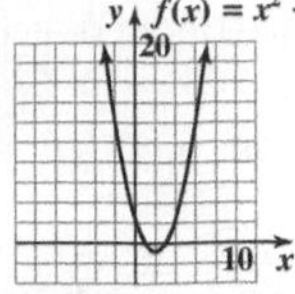

109. $\{5\}$

110. There is no method for solving $x = 2^y$ for y. **111.** $\frac{1}{2}$ **112.** $(-\infty, 3) \cup (3, \infty)$

Section 9.3 Check Point Exercises

1. a. $7^3 = x$ **b.** $b^2 = 25$ **c.** $4^y = 26$ **2. a.** $5 = \log_2 x$ **b.** $3 = \log_b 27$ **c.** $y = \log_e 33$ **3. a.** 2 **b.** 1 **c.** $\frac{1}{2}$ **4. a.** 1 **b.** 0

5. a. 8 **b.** 17 **6.** 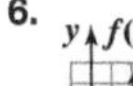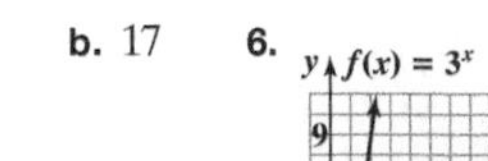

7. $(5, \infty)$ **8.** approximately 80% **9.** 4 **10. a.** $(-\infty, 4)$ **b.** $(-\infty, 0) \cup (0, \infty)$ **11.** 34°; extremely well

Concept and Vocabulary Check

1. $b^y = x$ **2.** logarithmic; b **3.** 1 **4.** 0 **5.** x **6.** x **7.** $(0, \infty)$; $(-\infty, \infty)$ **8.** y; $x = 0$; vertical **9.** $5 - x > 0$ **10.** common; $\log x$ **11.** natural; $\ln x$

Exercise Set 9.3

1. $2^4 = 16$ **3.** $3^2 = x$ **5.** $b^5 = 32$ **7.** $6^y = 216$ **9.** $\log_2 8 = 3$ **11.** $\log_2 \frac{1}{16} = -4$ **13.** $\log_8 2 = \frac{1}{3}$ **15.** $\log_{13} x = 2$

17. $\log_b 1000 = 3$ **19.** $\log_7 200 = y$ **21.** 2 **23.** 6 **25.** −1 **27.** −3 **29.** $\frac{1}{2}$ **31.** $-\frac{1}{2}$ **33.** $\frac{1}{2}$ **35.** 1 **37.** 0 **39.** 7

41. 19 **43.** 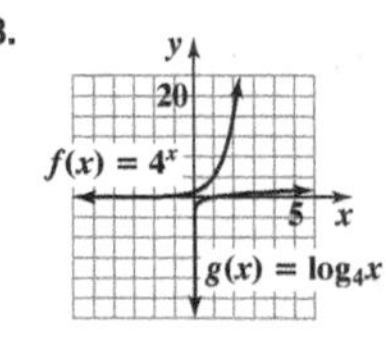

45. 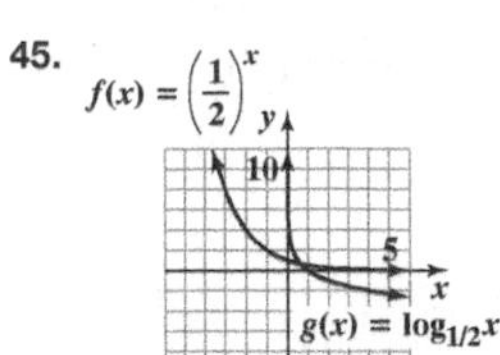

47. $(-4, \infty)$ **49.** $(-\infty, 2)$ **51.** $(-\infty, 2) \cup (2, \infty)$ **53.** 2 **55.** 7 **57.** 33 **59.** 0 **61.** 6 **63.** −6 **65.** 125 **67.** $9x$ **69.** $5x^2$ **71.** $\sqrt{x}$ **73.** $3^2 = x - 1$; $\{10\}$ **75.** $4^{-3} = x$; $\left\{\frac{1}{64}\right\}$ **77.** 0 **79.** 2 **81.** d **83.** c **85.** b **87.** approximately 95.4% **89. a.** 34%; It's the same. **b.** 37%

91. approximately 188 decibels; yes

93. a. 88
b. 71.5; 63.9; 58.8; 55.0; 52.0; 49.5
c. 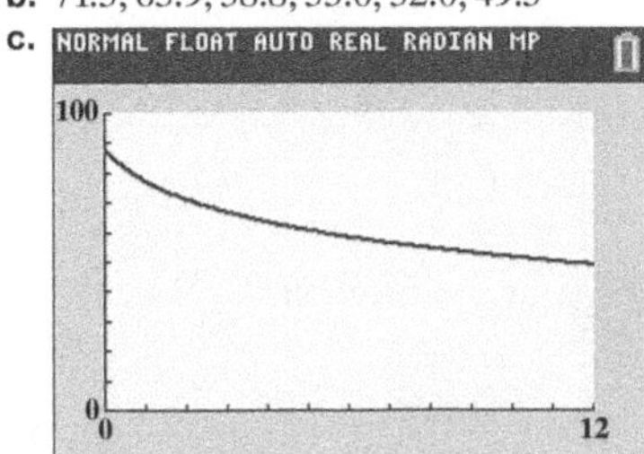

The students remembered less of the material over time.

103. 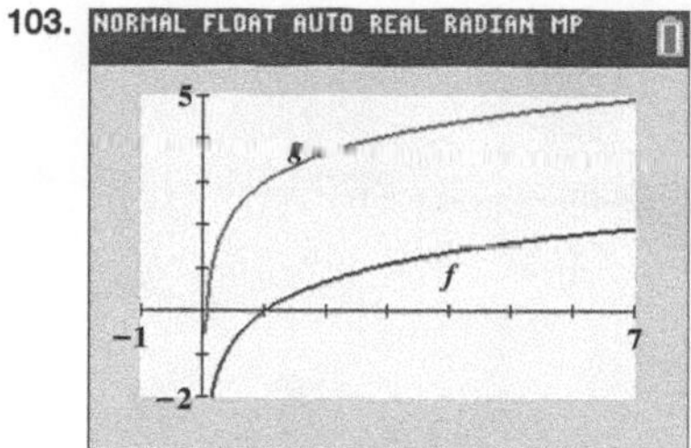

The graph of g is the graph of f shifted up 3 units.

105. 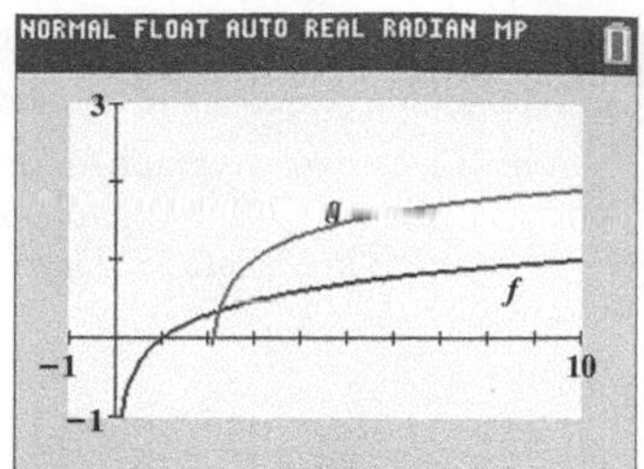

The graph of g is the graph of f shifted 2 units to the right and 1 unit up.

107. a. 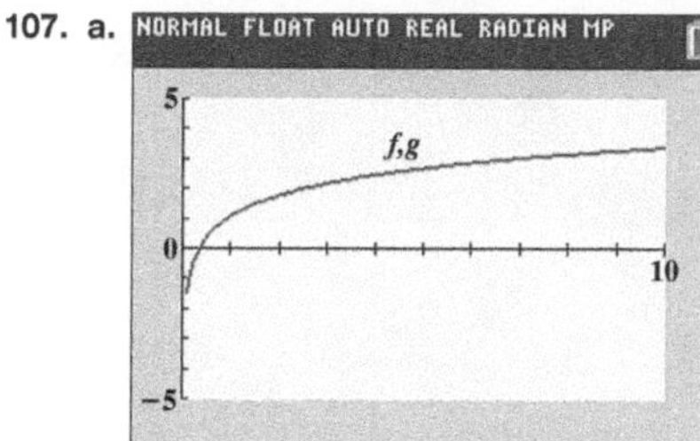 **b.** 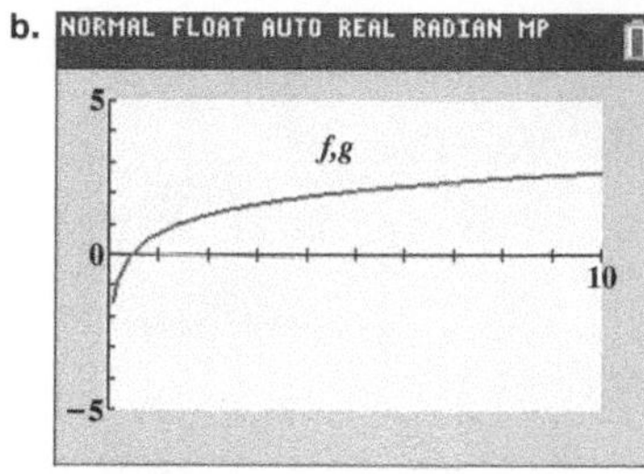 **c.**

d. In each case, the graphs of f and g are the same.; $\log_b(MN) = \log_b M + \log_b N$
e. the sum of the logarithms of the factors

109. makes sense **111.** makes sense **113.** false **115.** false **117.** $\frac{4}{5}$ **119.** $\log_3 40$ **120.** $\{(-2, 3)\}$ **121.** $2(3x - y)(x - y)$
122. $(-\infty, -7] \cup [-2, \infty)$ **123. a.** 5 **b.** 5 **c.** $\log_2(8 \cdot 4) = \log_2 8 + \log_2 4$ **124. a.** 4 **b.** 4 **c.** $\log_2\left(\frac{32}{2}\right) = \log_2 32 - \log_2 2$
125. a. 4 **b.** 4 **c.** $\log_3 9^2 = 2\log_3 9$

Section 9.4 Check Point Exercises

1. a. $\log_6 7 + \log_6 11$ **b.** $2 + \log x$ **2. a.** $\log_8 23 - \log_8 x$ **b.** $5 - \ln 11$ **3. a.** $9 \log_6 8$ **b.** $\frac{1}{3}\ln x$ **c.** $2 \log(x + 4)$
4. a. $4 \log_b x + \frac{1}{3}\log_b y$ **b.** $\frac{1}{2}\log_5 x - 2 - 3 \log_5 y$ **5. a.** $\log 100 = 2$ **b.** $\log\left(\frac{7x + 6}{x}\right)$ **6. a.** $\ln(x^2\sqrt[3]{x + 5})$ **b.** $\log\left[\frac{(x - 3)^2}{x}\right]$
c. $\log_b\left(\frac{\sqrt[4]{x}}{25y^{10}}\right)$ **7.** $\frac{\log 2506}{\log 7} \approx 4.02$ **8.** $\frac{\ln 2506}{\ln 7} \approx 4.02$

Concept and Vocabulary Check

1. $\log_b M + \log_b N$; sum **2.** $\log_b M - \log_b N$; difference **3.** $p \log_b M$; product **4.** $\frac{\log_a M}{\log_a b}$

Exercise Set 9.4

1. $\log_5 7 + \log_5 3$ **3.** $1 + \log_7 x$ **5.** $3 + \log x$ **7.** $1 - \log_7 x$ **9.** $\log x - 2$ **11.** $3 - \log_4 y$ **13.** $2 - \ln 5$ **15.** $3 \log_b x$
17. $-6 \log N$ **19.** $\frac{1}{5}\ln x$ **21.** $2 \log_b x + \log_b y$ **23.** $\frac{1}{2}\log_4 x - 3$ **25.** $2 - \frac{1}{2}\log_6(x + 1)$ **27.** $2 \log_b x + \log_b y - 2 \log_b z$
29. $1 + \frac{1}{2}\log x$ **31.** $\frac{1}{3}\log x - \frac{1}{3}\log y$ **33.** $\frac{1}{2}\log_b x + 3 \log_b y - 3 \log_b z$ **35.** $\frac{2}{3}\log_5 x + \frac{1}{3}\log_5 y - \frac{2}{3}$ **37.** $\log 10 = 1$ **39.** $\ln(7x)$
41. $\log_2 32 = 5$ **43.** $\log\left(\frac{2x + 5}{x}\right)$ **45.** $\log(xy^3)$ **47.** $\ln(y\sqrt{x})$ **49.** $\log_b(x^2y^3)$ **51.** $\ln\left(\frac{x^5}{y^2}\right)$ **53.** $\ln\left(\frac{x^3}{\sqrt[3]{y}}\right)$ **55.** $\ln\left[\frac{(x + 6)^4}{x^3}\right]$
57. $\ln\left(\frac{x^3y^5}{z^6}\right)$ **59.** $\log_5\left[\frac{\sqrt{xy}}{(x + 1)^2}\right]$ **61.** 1.5937 **63.** 1.6944 **65.** −1.2304 **67.** 3.6193 **69.** $C - A$ **71.** $3A$ **73.** $\frac{1}{2}A - \frac{3}{2}C$
75. false; $\ln e = 1$ **77.** false; $\log_4(2x)^3 = 3 \log_4(2x)$ **79.** true **81.** true **83.** false; $\log(x + 3) - \log(2x) = \log\left(\frac{x + 3}{2x}\right)$ **85.** true
87. true **89. a.** 2 **b.** $\log_3\left(\frac{xy^4}{9}\right)$ **91. a.** $\frac{1}{2}$ **b.** $\log_{25}\left[\frac{x(x^2 - 1)}{5(x + 1)}\right] = \log_{25}\left[\frac{x(x - 1)}{5}\right]$ **93. a.** $D = 10 \log\left(\frac{I}{I_0}\right)$ **b.** 20 decibels

103. a. & b.

$y = \log_3 x$ is shifted up 2 units to obtain $y = 2 + \log_3 x$,
$y = \log_3 x$ is shifted to the left 2 units to obtain $y = \log_3(x + 2)$, and
$y = \log_3 x$ is reflected across the x-axis to obtain $y = -\log_3 x$.

105. 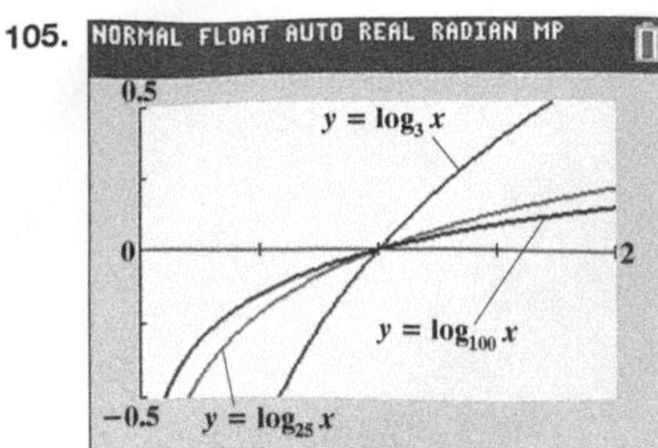

a. $y = \log_{100} x$ is on the top and $y = \log_3 x$ is on the bottom.
b. $y = \log_3 x$ is on the top and $y = \log_{100} x$ is on the bottom.
c. If $y = \log_b x$ is graphed for two different values of b, the graph of the one with the larger base will be on top in the interval $(0, 1)$ and the one with the smaller base will be on top in the interval $(1, \infty)$.

111. makes sense **113.** makes sense **115.** true **117.** false **119.** $\log e = \dfrac{\ln e}{\ln 10} = \dfrac{1}{\ln 10}$ **121.** $4x^3$

122. $5x - 2y > 10$ **123.** $(-\infty, 1)$ **124.** $2y\sqrt[3]{xy^2}$ **125.** 64 **126.** 12 **127.** $\{-1, 3\}$

Mid-Chapter Check Point Exercises

1. $(f \circ g)(x) = 12x - 13$; $(g \circ f)(x) = 12x + 3$; no **2.** $(f \circ g)(x) = x$; $(g \circ f)(x) = x$; yes **3.** $(f \circ g)(x) = x$; $(g \circ f)(x) = x$; yes

4. a. $\frac{2}{3}$ **b.** $\sqrt{5}$ **c.** $-\frac{1}{4}$ **d.** 2 **5.** $f^{-1}(x) = \dfrac{4x - 5}{2}$ **6.** $f^{-1}(x) = \sqrt[3]{\dfrac{x + 7}{10}}$ **7.** $f^{-1} = \{(5, 2), (-7, 10), (-10, 11)\}$

8. function; no inverse function **9.** function; inverse function **10.** not a function

11. 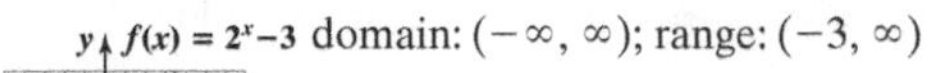
$f(x) = 2^x - 3$ domain: $(-\infty, \infty)$; range: $(-3, \infty)$
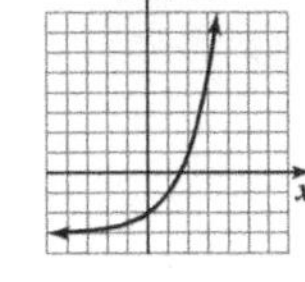

12. $f(x) = \left(\frac{1}{3}\right)^x$ domain: $(-\infty, \infty)$; range: $(0, \infty)$
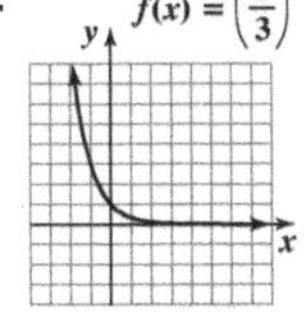

13. $f(x) = \log_2 x$ domain: $(0, \infty)$; range: $(-\infty, \infty)$
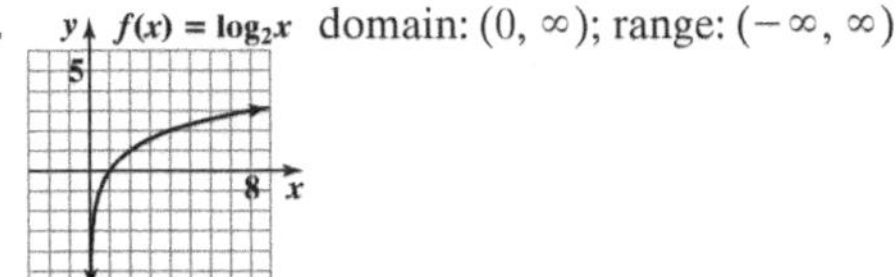

14. $f(x) = \log_2 x + 1$ domain: $(0, \infty)$; range: $(-\infty, \infty)$
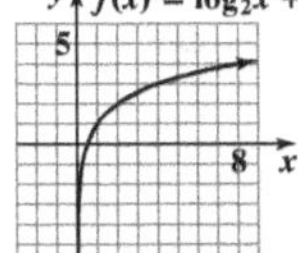

15. $(-6, \infty)$ **16.** $(0, \infty)$ **17.** $(-\infty, -6) \cup (-6, \infty)$ **18.** $(-\infty, \infty)$ **19.** 5 **20.** −2 **21.** $\frac{1}{2}$ **22.** $\frac{1}{3}$ **23.** 2

24. Evaluation is not possible; $\log_2 \frac{1}{8} = -3$ and $\log_3(-3)$ are undefined. **25.** 5 **26.** $\sqrt{7}$ **27.** 13 **28.** $-\frac{1}{2}$ **29.** $\sqrt{\pi}$

30. $\frac{1}{2}\log x + \frac{1}{2}\log y - 3$ **31.** $19 + 20 \ln x$ **32.** $\log_7\left(\dfrac{x^8}{\sqrt[3]{y}}\right)$ **33.** $\log_5 x^9$ **34.** $\ln\left[\dfrac{\sqrt{x}}{y^3(z - 2)}\right]$ **35.** \$8

Section 9.5 Check Point Exercises

1. a. 3 or $\{3\}$ **b.** $\frac{5}{2}$ or $\left\{\frac{5}{2}\right\}$ **2. a.** $\dfrac{\ln 134}{\ln 5} \approx 3.04$ or $\left\{\dfrac{\ln 134}{\ln 5} \approx 3.04\right\}$ **b.** $\log 8000 \approx 3.90$ or $\{\log 8000 \approx 3.90\}$

3. $\dfrac{\ln 9}{2} = \ln 3 \approx 1.10$ or $\{\ln 3 \approx 1.10\}$ **4. a.** 12 or $\{12\}$ **b.** $\dfrac{e^2}{3}$ or $\left\{\dfrac{e^2}{3}\right\}$ **5.** 5 or $\{5\}$ **6.** 4 and 5, or $\{4, 5\}$

7. blood alcohol concentration of 0.01 **8.** 16.2 years **9.** 14

Concept and Vocabulary Check

1. $M = N$ **2.** $4x - 1$ **3.** $\dfrac{\ln 20}{\ln 9}$ **4.** ln 6 **5.** 5^3 **6.** $x^2 + x$ **7.** $\dfrac{7x - 23}{x + 1}$ **8.** false **9.** true **10.** false

Exercise Set 9.5

1. $\{6\}$ **3.** $\{3\}$ **5.** $\{3\}$ **7.** $\{2\}$ **9.** $\left\{\frac{3}{5}\right\}$ **11.** $\left\{\frac{3}{2}\right\}$ **13.** $\{4\}$ **15.** $\{5\}$ **17.** $\left\{-\frac{1}{4}\right\}$ **19.** $\{\ln 5.7 \approx 1.74\}$

21. $\{\log 3.91 \approx 0.59\}$ **23.** $\left\{\dfrac{\ln 17}{\ln 5} \approx 1.76\right\}$ **25.** $\{\ln 5 \approx 1.61\}$ **27.** $\left\{\dfrac{\ln 659}{5} \approx 1.30\right\}$ **29.** $\left\{\dfrac{\ln 13}{0.7} \approx 3.66\right\}$ **31.** $\left\{\dfrac{\ln 3}{0.055} \approx 19.97\right\}$

33. $\left\{\dfrac{\ln 30}{\ln 1.4} \approx 10.11\right\}$ **35.** $\left\{\dfrac{1 - \ln 793}{5} \approx -1.14\right\}$ **37.** $\left\{\dfrac{\ln 410}{\ln 7} - 2 \approx 1.09\right\}$ **39.** $\left\{\dfrac{\ln 2}{\ln 5 - \ln 2} \approx 0.76\right\}$ **41.** $\{81\}$ **43.** $\left\{\frac{1}{16}\right\}$

45. {3} **47.** {100} **49.** {59} **51.** $\left\{\frac{109}{27}\right\}$ **53.** $\left\{\frac{62}{3}\right\}$ **55.** $\{e^2 \approx 7.39\}$ **57.** $\{e^{-3} \approx 0.05\}$ **59.** $\left\{\frac{e^4}{2} \approx 27.30\right\}$ **61.** $\{e^{-1/2} \approx 0.61\}$ **63.** $\{e^2 - 3 \approx 4.39\}$ **65.** $\left\{\frac{5}{4}\right\}$ **67.** {6} **69.** {6} **71.** no solution or ∅ **73.** $\left\{\frac{1}{e-1} \approx 0.58\right\}$ **75.** {3} **77.** $\left\{\frac{4}{3}\right\}$ **79.** no solution or ∅ **81.** {5} **83.** $\left\{\frac{2}{9}\right\}$ **85.** {28} **87.** {2} **89.** no solution or ∅ **91.** $\left\{\frac{1}{2}\right\}$ **93.** $\left\{\pm\sqrt{\frac{\ln 45}{\ln 3}} \approx \pm 1.86\right\}$ **95.** {12} **97.** {−2, 6} **99. a.** 37.3 million **b.** 2017 **101.** 118 ft; by the point (118, 1) **103.** 8.2 **105.** 16.8% **107.** 8.7 **109.** 15.7% **111. a.** 17.0%; underestimates by 0.3% **b.** 2016 **113.** about 2.8 days; (2.8, 50) **115. a.** $10^{-5.6}$ mole per liter **b.** $10^{-2.4}$ mole per liter **c.** $10^{3.2}$ times greater **123.** {2} **125.** {4} **127.** {2} **129.** {−1.39, 1.69}

131.

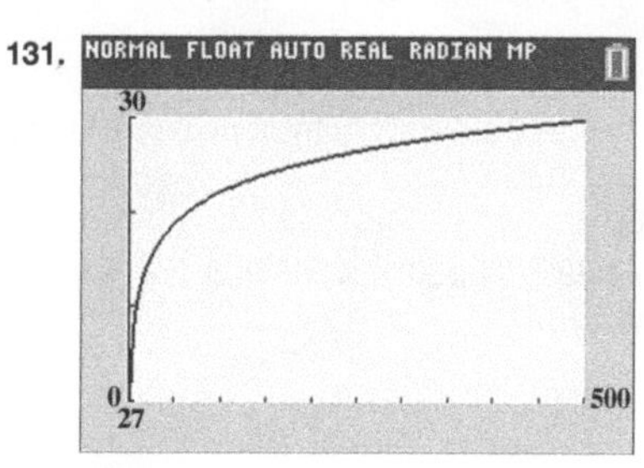

The barometric air pressure increases as the distance from the eye increases.

133.

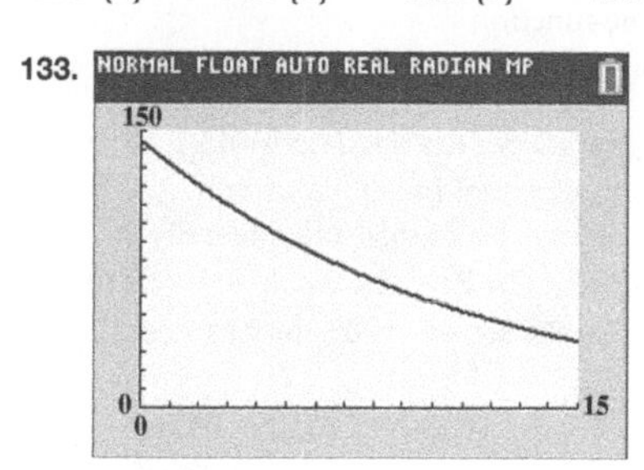

The runner's pulse will be 70 beats per minute after about 7.9 minutes.

135. does not make sense **137.** makes sense **139.** false **141.** true **143.** about 36 yr **145.** $\{10^{-2}, 10^{3/2}\}$ **147.** {1, 5} **148.** {−12} **149.** $\frac{y^8}{16x^{12}}$ **150. a.** 10 million; 9.97 million; 9.94 million; 9.91 million **b.** decreasing **151. a.** 3 **b.** $e^{(\ln 3)x}$

152.

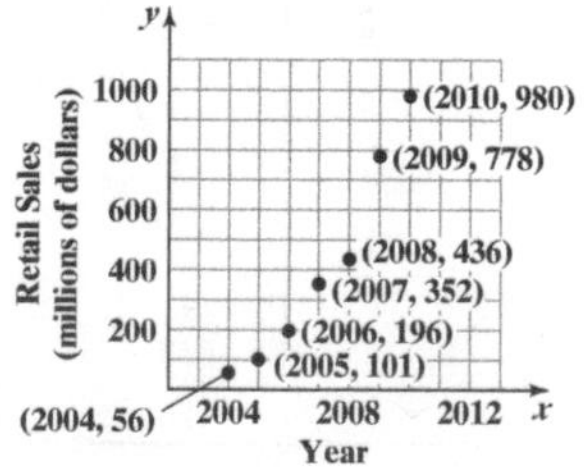

exponential function

Section 9.6 Check Point Exercises

1. a. $A = 807e^{0.024t}$ **b.** 2038 **2. a.** $A = A_0e^{-0.0248t}$ **b.** about 72.2 years

3. logarithmic function **4.** exponential function, although answers may vary

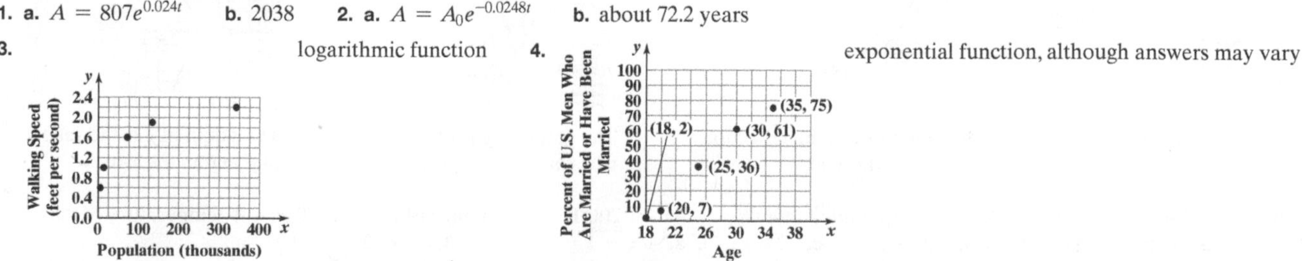

5. a. the exponential function g **b.** the linear function f **6.** $y = 4e^{(\ln 7.8)x}$; $y = 4e^{2.054x}$

Concept and Vocabulary Check

1. > 0; < 0 **2.** A_0; A **3.** logarithmic **4.** exponential **5.** linear **6.** ln 5

Exercise Set 9.6

1. 127.3 million **3.** Iraq; 1.9% **5.** 2030 **7. a.** $A = 6.04e^{0.01t}$ **b.** 2040 **9.** 146.1 **11.** 0.0088 **13.** −0.0039 **15.** approximately 8 grams **17.** 8 grams after 10 seconds; 4 grams after 20 seconds; 2 grams after 30 seconds; 1 gram after 40 seconds; 0.5 gram after 50 seconds **19.** approximately 15,679 years old **21.** 12.6 yr **23.** 0.0428% per yr = −0.000428 **25.** 3.9608% per day = −0.039608 **27. a.** $\frac{1}{2} = e^{1.31k}$ yields $k = \frac{\ln\left(\frac{1}{2}\right)}{1.31} \approx -0.52912$. **b.** about 0.1069 billion or 106,900,000 years old **29.** 7.1 yr **31.** 5.5 hr **33.** $2A_0 = A_0e^{kt}$; $2 = e^{kt}$; $\ln 2 = \ln e^{kt}$; $\ln 2 = kt$; $\frac{\ln 2}{k} = t$ **35. a.** 1% **b.** about 69 years

37.

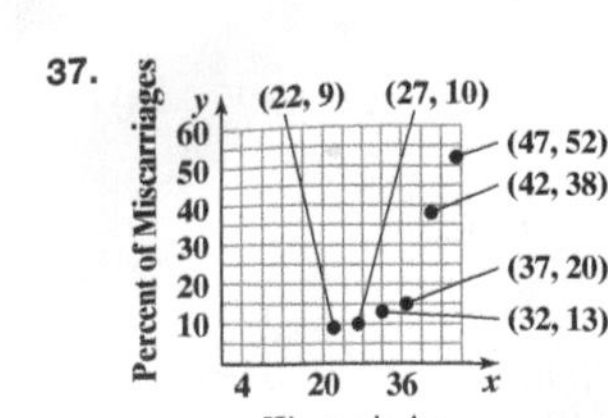

exponential function

39.

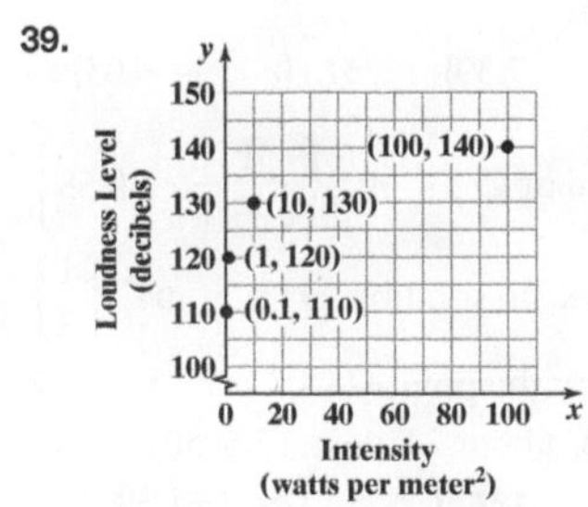

logarithmic function

41.

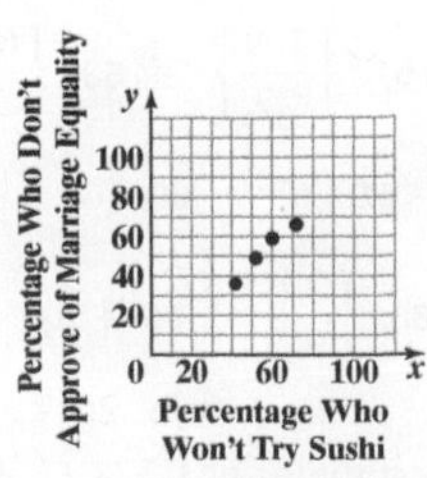

logarithmic function

43. $y = 100e^{(\ln 4.6)x}$; $y = 100e^{1.526x}$ **45.** $y = 2.5e^{(\ln 0.7)x}$; $y = 2.5e^{-0.357x}$

55. a. $y = 201.2(1.011)^x$; $r \approx 0.999$; Since r is close to 1, the model fits the data well. **b.** $y = 201.2e^{\ln(1.011)x}$; $y = 201.2e^{0.0109x}$; by approximately 1%

57. $y = 2.657x + 197.923$; $r \approx 0.997$; Since r is close to 1, the model fits the data well.

59. $y = 201.2(1.011)^x$; $y = 2.657x + 197.923$; using exponential, by 2016; using linear, by 2021; Answers will vary.

61. Models will vary. Examples are given. Predictions will vary. For Exercise 37: $y = 1.402(1.078)^x$; For Exercise 39: $y = 120 + 4.343 \ln x$; For Exercise 41: $y = -175.582 + 56.808 \ln x$ **63.** does not make sense **65.** makes sense **67.** true **69.** true

71. $\dfrac{x+5}{x}$ **72.** $\{-27, 1\}$ **73.** $17\sqrt{2}$ **74.** $3\sqrt{5}$ **75.** $\left(4, \dfrac{1}{2}\right)$ **76.**

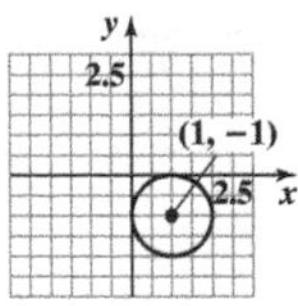

Review Exercises

1.

x	$f(x)$
−2	$\frac{1}{16}$
−1	$\frac{1}{4}$
0	1
1	4
2	16

; d

2.

x	$f(x)$
−2	16
−1	4
0	1
1	$\frac{1}{4}$
2	$\frac{1}{16}$

; a

3.

x	$f(x)$
−2	−16
−1	−4
0	−1
1	$-\frac{1}{4}$
2	$-\frac{1}{16}$

; b

4.

x	$f(x)$
−2	−13
−1	−1
0	2
1	$\frac{11}{4}$
2	$\frac{47}{16}$

; c

5.

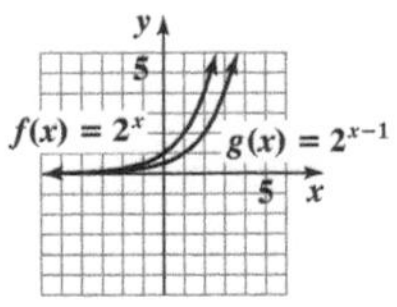

The graph of g is the graph of f shifted to the right 1 unit.

6. The graph of g is the reflection of f across the y-axis.

7.

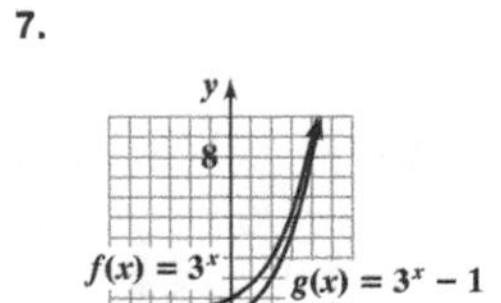

The graph of g is the graph of f shifted down 1 unit.

8.

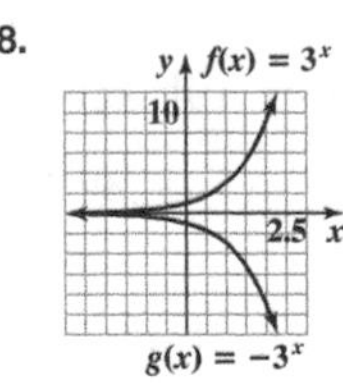

The graph of g is the reflection of f across the x-axis.

9. 5.5% compounded semiannually **10.** 7% compounded monthly **11. a.** 200°F **b.** about 119°F **c.** 70°F; The temperature of the room is 70°F. **12. a.** $16x^2 - 8x + 4$ **b.** $4x^2 + 11$ **c.** 124 **13. a.** $\sqrt{x+1}$ **b.** $\sqrt{x} + 1$ **c.** 2

14. $f(g(x)) = x - \frac{7}{10}$; $g(f(x)) = x - \frac{7}{6}$; not inverses **15.** $f(g(x)) = x$; $g(f(x)) = x$; inverses

16. a. $f^{-1}(x) = \dfrac{x+3}{4}$ **b.** $f(f^{-1}(x)) = 4\left(\dfrac{x+3}{4}\right) - 3 = x$ and $f^{-1}(f(x)) = \dfrac{(4x-3)+3}{4} = x$

17. a. $f^{-1}(x) = x^2 - 2, x \ge 0$ **b.** $f(f^{-1}(x)) = \sqrt{(x^2-2)+2} = x$ and $f^{-1}(f(x)) = (\sqrt{x+2})^2 - 2 = x$

18. a. $f^{-1}(x) = \dfrac{\sqrt[3]{x-1}}{2}$ **b.** $f(f^{-1}(x)) = 8\left(\dfrac{\sqrt[3]{x-1}}{2}\right)^3 + 1 = x$ and $f^{-1}(f(x)) = \dfrac{\sqrt[3]{(8x^3+1)-1}}{2} = x$

19. inverse function **20.** no inverse function **21.** inverse function **22.** no inverse function

23. **24.** $49^{1/2} = 7$ **25.** $4^3 = x$ **26.** $3^y = 81$ **27.** $\log_6 216 = 3$ **28.** $\log_b 625 = 4$ **29.** $\log_{13} 874 = y$ **30.** 3 **31.** −2 **32.** −9 is not in the domain of $y = \log_3 x$. **33.** $\frac{1}{2}$ **34.** 1 **35.** 8 **36.** 5 **37.** $-\frac{1}{2}$ **38.** −2 **39.** −3 **40.** 0

41. domain of f: $(-\infty, \infty)$; range of f: $(0, \infty)$; domain of g: $(0, \infty)$; range of g: $(-\infty, \infty)$

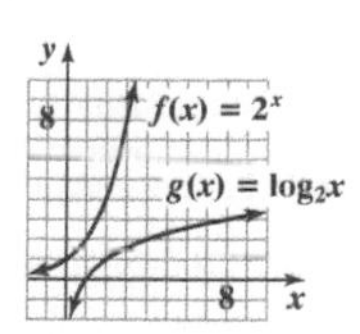

42. domain of f: $(-\infty, \infty)$; range of f: $(0, \infty)$; domain of g: $(0, \infty)$; range of g: $(-\infty, \infty)$

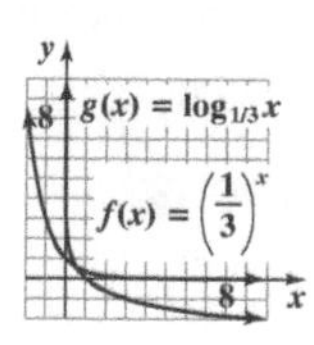

43. $(-5, \infty)$ **44.** $(-\infty, 3)$ **45.** $(-\infty, 1) \cup (1, \infty)$ **46.** $6x$ **47.** $\sqrt{x}$ **48.** $4x^2$ **49.** 3

50. a. 76

b. 67.4 after 2 months; 63.4 after 4 months; 60.8 after 6 months; 58.8 after 8 months; 55.9 after one year

c.

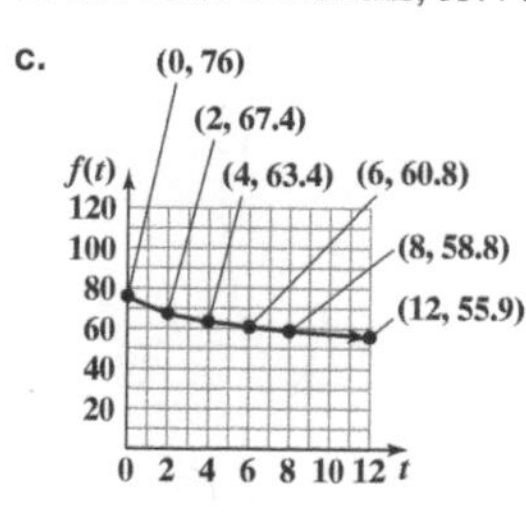

As time increases, the amount of material retained by the students decreases.

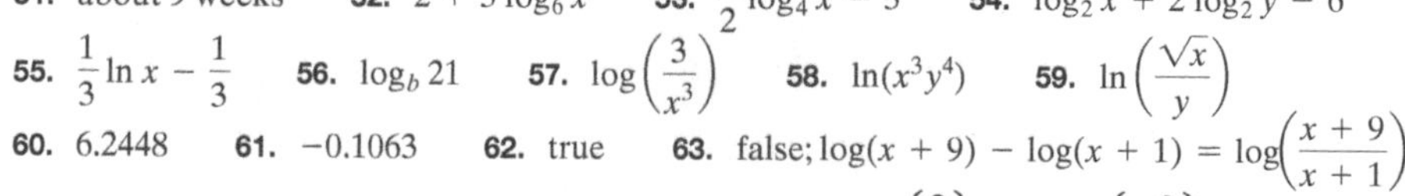

51. about 9 weeks **52.** $2 + 3\log_6 x$ **53.** $\frac{1}{2}\log_4 x - 3$ **54.** $\log_2 x + 2\log_2 y - 6$ **55.** $\frac{1}{3}\ln x - \frac{1}{3}$ **56.** $\log_b 21$ **57.** $\log\left(\frac{3}{x^3}\right)$ **58.** $\ln(x^3y^4)$ **59.** $\ln\left(\frac{\sqrt{x}}{y}\right)$ **60.** 6.2448 **61.** -0.1063 **62.** true **63.** false; $\log(x+9) - \log(x+1) = \log\left(\frac{x+9}{x+1}\right)$ **64.** false; $\log_2 x^4 = 4\log_2 x$ **65.** true **66.** $\{2\}$ **67.** $\left\{\frac{2}{3}\right\}$ **68.** $\left\{-\frac{3}{2}\right\}$ **69.** $\left\{\frac{\ln 12{,}143}{\ln 8} \approx 4.52\right\}$ **70.** $\left\{\frac{\ln 141}{5} \approx 0.99\right\}$ **71.** $\left\{\frac{\ln 3}{0.045} \approx 24.41\right\}$ **72.** $\left\{\frac{1}{125}\right\}$

73. $\{100\}$ **74.** $\{23\}$ **75.** $\left\{\frac{1}{e}\right\}$ **76.** $\left\{\frac{e^3}{2}\right\}$ **77.** $\{5\}$ **78.** no solution or $\varnothing$ **79.** $\left\{\frac{8}{3}\right\}$ **80.** $\{2\}$ **81.** $\{4\}$ **82.** 5.5 mi

83. 5 years after 2009, or 2014; 385 million **84. a.** 92%; It's the same. **b.** 66 years after 1961, or 2027 **85.** 7.3 years **86.** 14.6 years

87. about 22% **88. a.** 0.036 **b.** 60.6 million **c.** 2019 **89.** 325 days

90. Mentally Ill Adults in United States; linear function

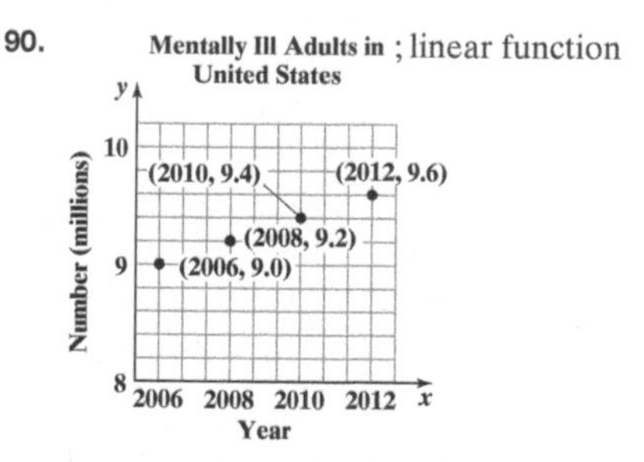

91. Moderate Alcohol Users in the United States; logarithmic function

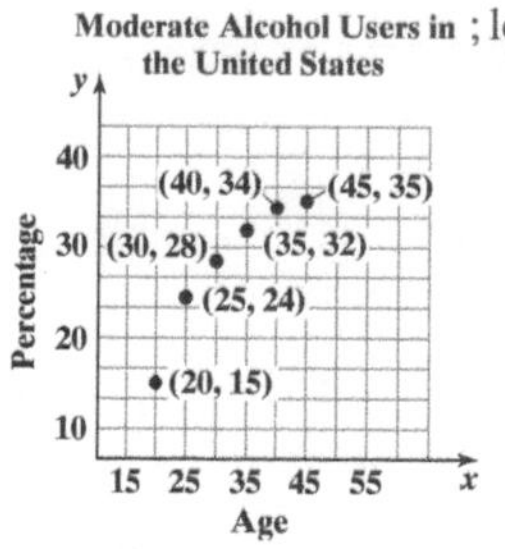

Percentage: 40, 30, 20, 10; points (40, 34), (45, 35), (30, 28), (35, 32), (25, 24), (20, 15); Age: 15 25 35 45 55 x

92. U.S. Electric Car Sales; exponential function

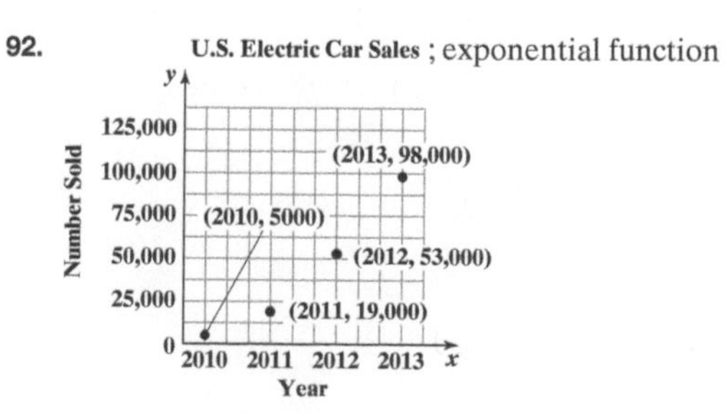

93. $y = 73e^{(\ln 2.6)x}$; $y = 73e^{0.956x}$ **94.** $y = 6.5e^{(\ln 0.43)x}$; $y = 6.5e^{-0.844x}$ **95.** Answers will vary.

Chapter Test

1. $g(x) = 2^{x+1}$; $f(x) = 2^x$ **2.** 6.5% compounded semiannually; \$221 **3.** $(f \circ g)(x) = 9x^2 - 3x$; $(g \circ f)(x) = 3x^2 + 3x - 1$ **4.** $f^{-1}(x) = \frac{x+7}{5}$ **5. a.** No horizontal line intersects the graph of f in more than one point. **b.** 2000 **c.** $f^{-1}(2000)$ represents the income, \$80 thousand, of a family that gives \$2000 to charity. **6.** $5^3 = 125$ **7.** $\log_{36} 6 = \frac{1}{2}$

8. $f(x) = 3^x$; $g(x) = \log_3 x$; domain of f: $(-\infty, \infty)$; range of f: $(0, \infty)$; domain of g: $(0, \infty)$; range of g: $(-\infty, \infty)$

9. $5x$ **10.** 1 **11.** 0 **12.** $(7, \infty)$ **13.** 120 decibels **14.** $3 + 5\log_4 x$ **15.** $\frac{1}{3}\log_3 x - 4$ **16.** $\log(x^6y^2)$ **17.** $\ln\left(\frac{7}{x^3}\right)$ **18.** 1.5741 **19.** $\{6\}$ **20.** $\left\{\frac{\ln 1.4}{\ln 5}\right\}$ **21.** $\left\{\frac{\ln 4}{0.005}\right\}$ **22.** $\{5\}$ **23.** $\left\{\frac{217}{4}\right\}$ **24.** $\left\{\frac{e^4}{3}\right\}$ **25.** $\{5\}$ **26.** no solution or $\varnothing$ **27. a.** 82.3 million **b.** decreasing; The growth rate, -0.004, is negative. **c.** 2020 **28.** 13.9 years **29.** about 6.9% **30.** $A = 4121e^{0.006t}$ **31.** about 24,758 years ago **32.** linear **33.** logarithmic **34.** exponential **35.** quadratic **36.** $y = 96e^{(\ln 0.38)x}$; $y = 96e^{-0.968x}$

Cumulative Review Exercises

1. $\{4\}$ **2.** $\{(2, 3)\}$ **3.** $\{(2, 0, 3)\}$ **4.** $(-\infty, -2) \cup (4, \infty)$ **5.** $\{5\}$ **6.** $[4, \infty)$ **7.** $\left\{\frac{3}{4} \pm i\frac{\sqrt{7}}{4}\right\}$

8. $3x = 15 + 5y$

9.

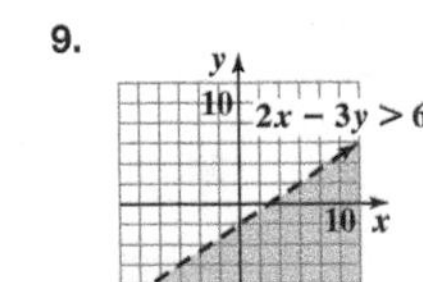

10. $f(x) = -\frac{1}{2}x + 1$

11.

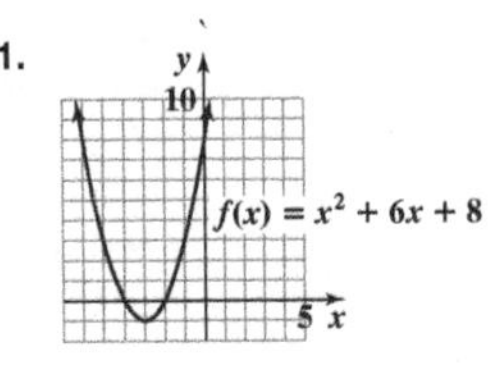

12.

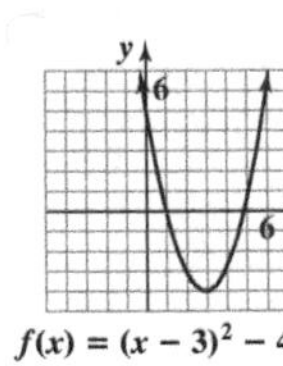

13. -6 **14.** $c = \frac{Ad}{d - A}$ **15.** $x^2 - x - 17$ **16.** $x^2 + 3x - 17$ **17.** h **18.** $f^{-1}(x) = \frac{x+3}{7}$ **19.** $(-\infty, 1) \cup (1, 2) \cup (2, \infty)$ **20.** $(4, \infty)$ **21.** $y = \frac{1}{2}x + 5$ or $f(x) = \frac{1}{2}x + 5$ **22.** $-\frac{y^9}{3x}$ **23.** $16x^4 - 40x^2y + 25y^2$ **24.** $x^2 - 5x + 1 + \frac{8}{5x + 1}$ **25.** $2y^2\sqrt[3]{2y^2}$ **26.** $\frac{4x - 13}{(x - 3)(x - 4)}$ **27.** $(x - 4)(x + 2)(x^2 - 2x + 4)$ **28.** $2(x + 3y)^2$ **29.** $\ln\left(\frac{x^2}{\sqrt{y}}\right)$ **30.** length: 12 ft; width: 4 ft **31.** $\frac{6}{5}$ hr or 1 hr and 12 min **32.** 5 mph **33.** approximately 11%

Applications Index

Subject Index

Credits

Photo Credits

Front Matter

Discpicture/Shutterstock, **p. iii**(t) Jose Luis Pelaez Inc/ Blend Images/Alamy, **p. iii**(b) Shizuo Kambayashi/AP Images, **p. iv** Mega Pixel/Shutterstock, **p. v** 103tnn/ Fotolia, **p. vi** Bob Blitzer, **p. xvii**

Chapter 1

Discpicture/Shutterstock, **p. 1**(t) Jose Luis Pelaez Inc/ Blend Images/Alamy, **p. 1**(b) ASK-Fotografie/Fotolia, **p. 2**(r) Raisa Kanareva/Fotolia, **p. 2**(l) Vectorskills/ Fotolia, **p. 15** Ron Chapple/Dreamstime LLC, **p. 30** Iofoto/Shutterstock, **p. 40**(t) Stephen Coburn/ Shutterstock, **p. 40**(b) Pictorial Press Ltd/Alamy, **p. 54**(l) Pictorial Press Ltd/Alamy, **p. 54**(r) Blitzer, Robert F., **p. 56** Scott Camazine/Alamy, **p. 65** Susumu Nishinaga/Science Source, **p. 70**(l) Dhoxax/ Shutterstock, **p. 70**(r) Slanted Roof Studio/Alamy, **p. 82** Alexey Boldin/Shutterstock, **p. 98**

Chapter 2

Jan Martin Will/Shutterstock, **p. 103** Nancy Kaszerman/Zuma Press, Inc./Alamy, **p. 104**(a) Peter Brooker/Rex Features/Presselect/Alamy, **p. 104**(b) Creative Collection Tolbert Photo/Alamy, **p. 104**(c) Dee Cercone/Everett Collection Inc/Alamy, **p. 104**(d) Hyperstar/Alamy, **p. 104**(e) NIBSC/Science Source, **p. 114** Günay Mutlu/E+/Getty Images, **p. 125** LWA/Sharie Kennedy/Blend Images/Alamy, **p. 136** CP, Frank Gunn/AP Images, **p. 155**(l) Kippa/ ANP/Newscom, **p. 155**(c) Chris Pizzello/AP Images, **p. 155**(r) French Government Tourist Office, **p. 167**

Chapter 3

Shizuo Kambayashi/AP Images, **p. 177** cypher0x/123RF, **p. 178** Hurst Photo/Shutterstock, **p. 194** Jules Selmes/ Pearson Education, Inc., **p. 196**(l) FomaA/Fotolia, **p. 196**(r) Laura Rauch/AP Images, **p. 201** Universal Uclick, **p. 205** Silver-john/Fotolia, **p. 208** TANNEN MAURY/epa european pressphoto agency b.v./Alamy, **p. 221** SSPL/The Image Works, **p. 232**

Chapter 4

Jimmy Chin/National Geographic Magazines/Getty Images, **p. 253** Julián Maldonado/Fotolia, **p. 254** Gary Fabiano/Sipa/Newscom, **p. 266**(tl) UPI UPI Photo Service/Newscom, **p. 266**(tc) Pamela Price/Zumapress/ Newscom, **p. 266**(tr) Corbis/SuperStock, **p. 275** Jim Goldstein/Danita Delimont Photography/Newscom, **p. 287**(l) limitedqstock/Alamy, **p. 287**(r) AP Images, **p. 298** US Navy Photo/Alamy, **p. 299**

Chapter 5

Fuse/Getty Images, **p. 313** Zuma Press, Inc./Alamy, **p. 314** Fancy Collection/SuperStock, **p. 336** Monkey Business/Fotolia, **p. 341** Stock Connection/SuperStock, **p. 350** StudioSmart/Shutterstock, **p. 364** Rido/Fotolia, **p. 374** Warren Photographic Ltd., **p. 381**

Chapter 6

M. Spencer Green/AP Images, **p. 403** Claudia Otte/ Shutterstock, **p. 404** Steve Skjold/Alamy, **p. 418** Mauritius/SuperStock, **p. 430** ginasanders/ 123RF, **p. 438** Nature's Images/Science Source, **p. 448** Songquan Deng/Shutterstock, **p. 455** Charlie Neuman/ U-T San Diego/Zuma Press, Inc/Alamy, **p. 466** Holger Burmeister/Alamy, **p. 478** Gary718/Shutterstock, **p. 479**(tc) Anna Azimi/Shutterstock, **p. 479**(tr) Jana Lumley/Fotolia, **p. 480** Exactostock-1527/ SuperStock, **p. 482** Ty Allison/Photographer's Choice/ Getty Images, **p. 486**(t) David Madison/Getty Images, **p. 486**(b) Ullstein Bild/The Image Works, **p. 488**

Chapter 7

Ian Evans/Alamy, **p. 501** Kike Calvo/V&W/The Image Works, **p. 502** Ryan M. Bolton/Shutterstock, **p. 515** George Tooker "Mirror II," 1963. Egg tempera on gesso panel, 20 × 20 in. 1968.4. Gift of R.H. Donnelley Erdman (PA 1956). Addison Gallery of American Art, Phillips Academy, Andover, Massachusetts. © The Estate of George Tooker. Courtesy of DC Moore Gallery, New York. Photo courtesy of Art Resource. **p. 525** Paul Wootton/Science Source, **p. 533** Universal Uclick, **p. 542** John G. Ross/Science Source, **p. 551** Bikeriderlondon/Shutterstock, **p. 552** Justin Kase Conder/Icon SMI 745/Justin Kase Conder/Icon SMI/Newscom, **p. 560** Roz Chast/The New Yorker Collection/The Cartoon Bank, **p. 562** Richard F. Voss, **p. 569** Gordon Caulkins/ Cartoon Stock, **p. 571**

Chapter 8

Photo courtesy of Professor Andrew Davidhazy/ Rochester Institute of Technology (RIT), **p. 581** Alamy, **p. 582** Kazoka303030/Fotolia, **p. 596** Simon Price/Alamy, **p. 611** Plpchirawong/Fotolia, **p. 630** Copyright © 2011 by Warren Miller/ The New Yorker Collection/The Cartoon Bank, **p. 638**

Chapter 9

Mega Pixel/Shutterstock, **p. 661** Javiindy/Fotolia, **p. 662** Seth Wenig/AP Images, **p. 668**(tl) Rolffimages/ Dreamstime, **p. 668**(tr) Tim Davis/Corbis, **p. 668**(b) Archives du 7eme Art/Photos 12/Alamy, **p. 676**(r) Everett Collection, **p. 676**(l) John Trotter/Sacramento Bee/Newscom, **p. 691** Wieslaw Smetek/Science Source, **p. 704** Mitchel Gray/Superstock, **p. 716** Jamaway/ Alamy, **p. 730** Franck Boston/Shutterstock, **p. 731** Courtesy of the French Ministry of Culture and Communication, **p. 734** Science Photo Library/ SuperStock, **p. 736**

Chapter 10

Stocktrek Images/Getty Images, **p. 757** Millard H. Sharp/Science Source, **p. 758** Kevin Fleming/Corbis, **p. 768** Ronald Royer/Science Source, **p. 776** Pixtal Images/Photolibrary New York/Getty Images, **p. 780** NASA, **p. 790** NASA, **p. 795** Richard E. Prince "The Cone of Apollonius" (detail), fiberglass, steel, paint, graphite, 51 × 18 × 14 in. Collection: Vancouver Art Gallery, Vancouver, Canada. Photo courtesy of Equinox Gallery, Vancouver, Canada., **p. 796** Kolestamas/Fotolia, **p. 802**

Chapter 11

103tnn/Fotolia, **p. 821** Sergey Galushko/iStock/Getty Images, **p. 822** Xixinxing/Fotolia, **p. 832** Ariel Skelley/ Blend Images/Alamy, **p. 842** Jokerpro/Shutterstock, **p. 845** Courtesy U.S. Bureau of Engraving and Printing., **p. 848**(a) Courtesy U.S. Bureau of Engraving and Printing., **p. 848**(b) Courtesy U.S. Bureau of Engraving and Printing., **p. 848**(c) Courtesy U.S. Bureau of Engraving and Printing., **p. 848**(d) Courtesy U.S. Bureau of Engraving and Printing., **p. 848**(e) Dr. Rudolph Schild/Science Source, **p. 859** Anna Omelchenko/Shutterstock, **p. 870**

Text Credits

Chapter 1

Screenshots from Texas Instruments. Courtesy of Texas Instruments, **p. 37**

Chapter 2

Screenshots from Texas Instruments. Courtesy of Texas Instruments, **p. 108**

Chapter 3

Quote by Henny Youngman, **p. 178**
Quote by Jeff Foxworthy, **p. 178**

Chapter 4

Quote by Cha Sa-Soon from Newsweek, **p. 261**

Chapter 7

Quote by Alan Bass, **p. 510**

Chapter 8

Quote from With Amusement for All: A History of American Popular Culture Since 1830 by LeRoy Ashby. Published by University Press of Kentucky, © 2006, **p. 630**

Definitions, Rules, and Formulas

The Real Numbers

Natural Numbers: $\{1, 2, 3, \ldots\}$

Whole Numbers: $\{0, 1, 2, 3, \ldots\}$

Integers: $\{\ldots, -3, -2, -1, 0, 1, 2, 3, \ldots\}$

Rational Numbers: $\{\frac{a}{b} \mid a$ and b are integers, $b \neq 0\}$

Irrational Numbers: $\{x \mid x$ is real and not rational$\}$

Basic Rules of Algebra

Commutative: $a + b = b + a$; $ab = ba$

Associative: $(a + b) + c = a + (b + c)$; $(ab)c = a(bc)$

Distributive: $a(b + c) = ab + ac$; $a(b - c) = ab - ac$

Identity: $a + 0 = a$; $a \cdot 1 = a$

Inverse: $a + (-a) = 0$; $a \cdot \frac{1}{a} = 1 (a \neq 0)$

Multiplication Properties: $(-1)a = -a$;
$(-1)(-a) = a$; $a \cdot 0 = 0$; $(-a)(b) = (a)(-b) = -ab$;
$(-a)(-b) = ab$

Set-Builder Notation, Interval Notation, and Graphs

$(a, b) = \{x \mid a < x < b\}$

$[a, b) = \{x \mid a \leq x < b\}$

$(a, b] = \{x \mid a < x \leq b\}$

$[a, b] = \{x \mid a \leq x \leq b\}$

$(-\infty, b) = \{x \mid x < b\}$

$(-\infty, b] = \{x \mid x \leq b\}$

$(a, \infty) = \{x \mid x > a\}$

$[a, \infty) = \{x \mid x \geq a\}$

$(-\infty, \infty) = \{x \mid x$ is a real number$\} = \{x \mid x \in R\}$

Slope Formula

$$\text{slope } (m) = \frac{\text{Change in } y}{\text{Change in } x} = \frac{y_2 - y_1}{x_2 - x_1}, \quad (x_1 \neq x_2)$$

Equations of Lines

1. Slope-intercept form: $y = mx + b$
   m is the line's slope and b is its y-intercept.
2. Standard form: $Ax + By = C$
3. Point-slope form: $y - y_1 = m(x - x_1)$
   m is the line's slope and (x_1, y_1) is a fixed point on the line.
4. Horizontal line parallel to the x-axis: $y = b$
5. Vertical line parallel to the y-axis: $x = a$

Systems of Equations

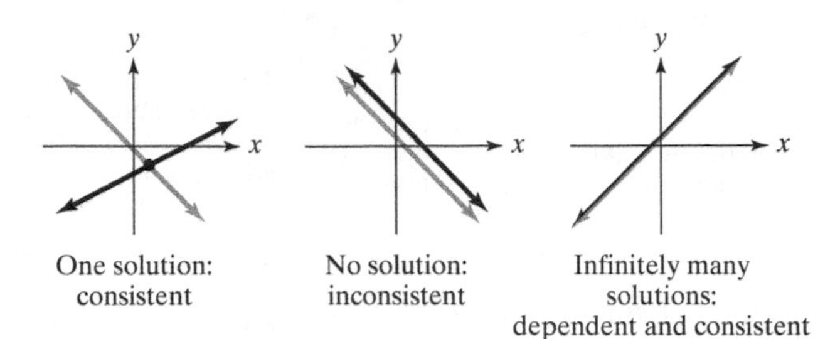

One solution: consistent | No solution: inconsistent | Infinitely many solutions: dependent and consistent

A system of linear equations may be solved: (a) graphically, (b) by the substitution method, (c) by the addition or elimination method, (d) by matrices, or (e) by determinants.

$$\begin{vmatrix} a_1 b_1 \\ a_2 b_2 \end{vmatrix} = a_1 b_2 - a_2 b_1$$

Cramer's Rule:

Given a system of a equations of the form

$$\begin{aligned} a_1 x + b_1 y &= c_1 \\ a_2 x + b_2 y &= c_2 \end{aligned}, \quad \text{then } x = \frac{\begin{vmatrix} c_1 b_1 \\ c_2 b_2 \end{vmatrix}}{\begin{vmatrix} a_1 b_1 \\ a_2 b_2 \end{vmatrix}} \text{ and } y = \frac{\begin{vmatrix} a_1 c_1 \\ a_2 c_2 \end{vmatrix}}{\begin{vmatrix} a_1 b_1 \\ a_2 b_2 \end{vmatrix}}.$$

Absolute Value

1. $|x| = \begin{cases} x & \text{if } x \geq 0 \\ -x & \text{if } x < 0 \end{cases}$
2. If $|x| = c$, then $x = c$ or $x = -c$. $(c > 0)$
3. If $|x| < c$, then $-c < x < c$. $(c > 0)$
4. If $|x| > c$, then $x < -c$ or $x > c$. $(c > 0)$

Special Factorizations

1. Difference of two squares:
   $$A^2 - B^2 = (A + B)(A - B)$$
2. Perfect square trinomials:
   $$A^2 + 2AB + B^2 = (A + B)^2$$
   $$A^2 - 2AB + B^2 = (A - B)^2$$

3. Sum of two cubes:

$$A^3 + B^3 = (A + B)(A^2 - AB + B^2)$$

4. Difference of two cubes:

$$A^3 - B^3 = (A - B)(A^2 + AB + B^2)$$

Variation

English Statement	*Equation*
y varies directly as x.	$y = kx$
y varies directly as x^n.	$y = kx^n$
y varies inversely as x.	$y = \frac{k}{x}$
y varies inversely as x^n.	$y = \frac{k}{x^n}$
y varies jointly as x and z.	$y = kxz$

Exponents

Definitions of Rational Exponents

1. $a^{\frac{1}{n}} = \sqrt[n]{a}$ 2. $a^{\frac{m}{n}} = (\sqrt[n]{a})^m$ or $\sqrt[n]{a^m}$

3. $a^{-\frac{m}{n}} = \frac{1}{a^{\frac{m}{n}}}$

Properties of Rational Exponents

If m and n are rational exponents, and a and b are real numbers for which the following expressions are defined, then

1. $b^m \cdot b^n = b^{m+n}$ 2. $\frac{b^m}{b^n} = b^{m-n}$

3. $(b^m)^n = b^{mn}$ 4. $(ab)^n = a^n b^n$

5. $\left(\frac{a}{b}\right)^n = \frac{a^n}{b^n}$

Radicals

1. If n is even, then $\sqrt[n]{a^n} = |a|$.
2. If n is odd, then $\sqrt[n]{a^n} = a$.
3. The product rule: $\sqrt[n]{a} \cdot \sqrt[n]{b} = \sqrt[n]{ab}$
4. The quotient rule: $\frac{\sqrt[n]{a}}{\sqrt[n]{b}} = \sqrt[n]{\frac{a}{b}}$

Complex Numbers

1. The imaginary unit i is defined as

$$i = \sqrt{-1}, \quad \text{where} \quad i^2 = -1.$$

The set of numbers in the form $a + bi$ is called the set of complex numbers. If $b = 0$, the complex number is a real number. If $b \neq 0$ the complex number is an imaginary number.

2. The complex numbers $a + bi$ and $a - bi$ are conjugates. Conjugates can be multiplied using the formula

$$(A + B)(A \quad D) = A^2 - B^2.$$

The multiplication of conjugates results in a real number.

3. To simplify powers of i, rewrite the expression in terms of i^2. Then replace i^2 with -1 and simplify.

Quadratic Equations and Functions

1. The solutions of a quadratic equation in standard form

$$ax^2 + bx + c = 0, \quad a \neq 0,$$

are given by the quadratic formula

$$x = \frac{-b \pm \sqrt{b^2 - 4ac}}{2a}.$$

2. The discriminant, $b^2 - 4ac$, of the quadratic equation $ax^2 + bx + c = 0$ determines the number and type of solutions.

Discriminant	Solutions
Positive perfect square with a, b, and c rational numbers	2 rational solutions
Positive and not a perfect square	2 irrational solutions
Zero, with a, b, and c rational numbers	1 rational solution
Negative	2 imaginary solutions

3. The graph of the quadratic function

$$f(x) = a(x - h)^2 + k, \quad a \neq 0,$$

is called a parabola. The vertex, or turning point, is (h, k). The graph opens upward if a is positive and downward if a negative. The axis of symmetry is a vertical line passing through the vertex. The graph can be obtained using the vertex, x-intercepts, if any, [set $f(x)$ equal to zero], and the y-intercept (set $x = 0$).

4. A parabola whose equation is in the form

$$f(x) = ax^2 + bx + c, \quad a \neq 0,$$

has its vertex at

$$\left(-\frac{b}{2a}, f\left(-\frac{b}{2a}\right)\right).$$

If $a > 0$, then f has a minimum that occurs at $x = -\frac{b}{2a}$. If $a < 0$, then f has a maximum that occurs at $x = -\frac{b}{2a}$.

Definitions, Rules, and Formulas (continued)

Exponential and Logarithmic Functions

1. Exponential Function: $f(x) = b^x, b > 0, b \neq 1$
   Graphs:

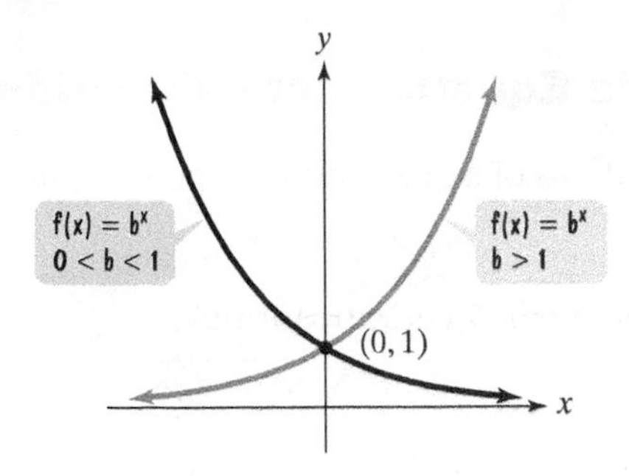

2. Logarithmic Function: $f(x) = \log_b x, b > 0, b \neq 1$
   $y = \log_b x$ is equivalent to $x = b^y$.

   Graphs:

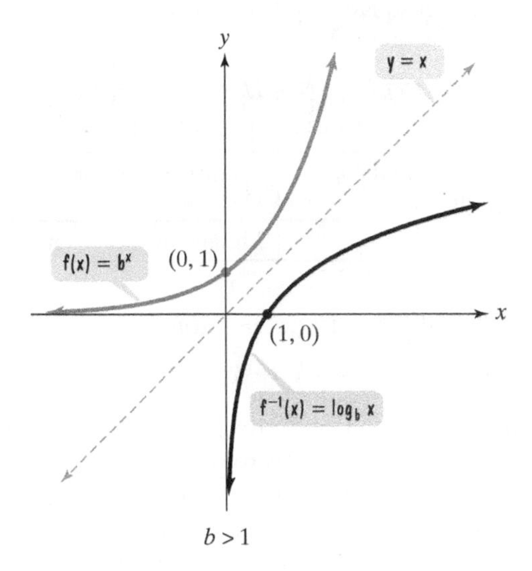

3. Properties of Logarithms
   a. $\log_b(MN) = \log_b M + \log_b N$
   b. $\log_b\left(\frac{M}{N}\right) = \log_b M - \log_b N$
   c. $\log_b M^p = p\log_b M$
   d. $\log_b M = \frac{\log_a M}{\log_a b} = \frac{\ln M}{\ln b} = \frac{\log M}{\log b}$
   e. $\log_b b^x = x; \log 10^x = x; \ln e^x = x$
   f. $b^{\log_b x} = x; 10^{\log x} = x; e^{\ln x} = x$

Distance and Midpoint Formulas

1. The distance from (x_1, y_1) to (x_2, y_2) is
$$\sqrt{(x_2 - x_1)^2 + (y_2 - y_1)^2}.$$

2. The midpoint of the line segment with endpoints (x_1, y_1) and (x_2, y_2) is
$$\left(\frac{x_1 + x_2}{2}, \frac{y_1 + y_2}{2}\right).$$

Conic Sections Circle

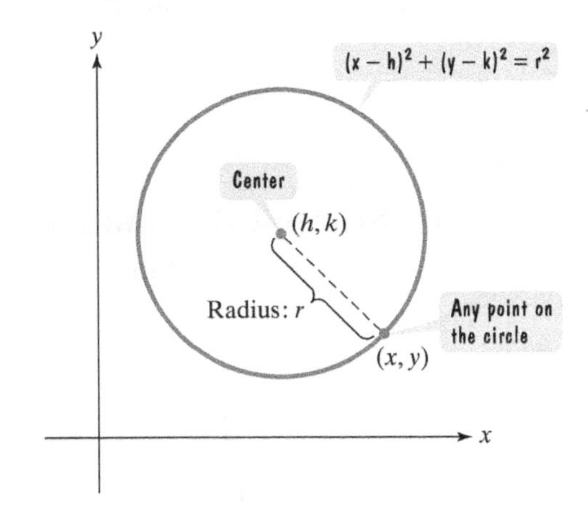

Ellipse

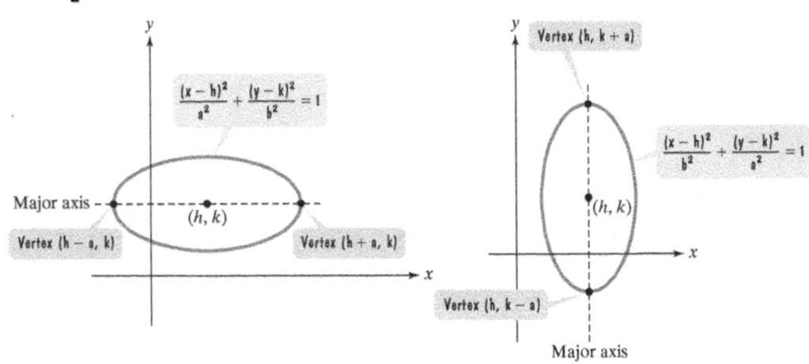

Hyperbola

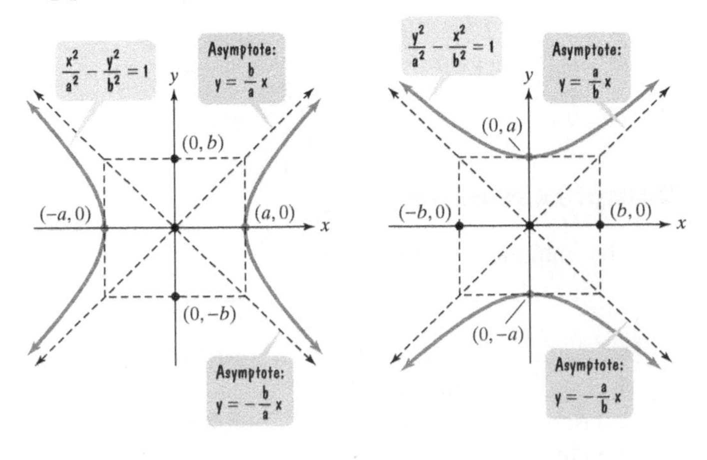

Parabolas Opening to the Right or to the Left

The graphs of

$$x = a(y - k)^2 + h \quad \text{and} \quad x = ay^2 + by + c$$

are parabolas opening to the right or to the left.

1. If $a > 0$, the graph opens to the right. If $a < 0$, the graph opens to the left
2. The vertex of $x = a(y - k)^2 + h$ is (h, k).
3. The y-coordinate of the vertex of $x = ay^2 + by + c$ is $y = -\frac{b}{2a}$.

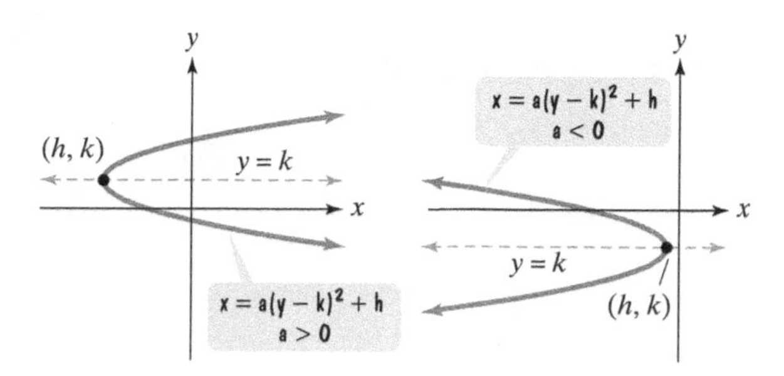

Sequences

1. Infinite Sequence: $\{a_n\} = a_1, a_2, a_3, \ldots, a_n, \ldots$
2. Summation Notation:

$$\sum_{i=1}^{n} a_i = a_1 + a_2 + a_3 + \cdots + a_n$$

3. nth Term of an Arithmetic Sequence:

$$a_n = a_1 + (n - 1)d$$

4. Sum of First n Terms of an Arithmetic Sequence:

$$S_n = \frac{n}{2}(a_1 + a_n)$$

5. nth Term of a Geometric Sequence: $a_n = a_1 r^{n-1}$
6. Sum of First n Terms of a Geometric Sequence:

$$S_n = \frac{a_1(1 - r^n)}{1 - r} \quad (r \neq 1)$$

7. Sum of an Infinite Geometric Series with $|r| < 1$:

$$S = \frac{a_1}{1 - r}$$

The Binomial Theorem

1. $n! = n(n - 1)(n - 2) \cdots 3 \cdot 2 \cdot 1; 0! = 1$
2. $\binom{n}{r} = \frac{n!}{r!(n - r)!}$
3. Binomial Theorem: For any positive integer n,

$$(a + b)^n = \binom{n}{0}a^n + \binom{n}{1}a^{n-1}b + \binom{n}{2}a^{n-2}b^2 + \binom{n}{3}a^{n-3}b^3 + \cdots + \binom{n}{n}b^n.$$

4. ***Finding a Particular Term in a Binomial Expansion***
The $(r + 1)$st term of the expansion of $(a + b)^n$ is

$$\binom{n}{r}a^{n-r}b^r.$$